Organic Electronics

Organic Electronics: Foundations to Applications

Stephen R. Forrest

Departments of Electrical Engineering and Computer Science, Physics, and Materials Science and Engineering, University of Michigan, Ann Arbor

OXFORD
UNIVERSITY PRESS

Organic Electronics: Foundations to Applications. Stephen R. Forrest. © Stephen R. Forrest 2020.
Published in 2020 by Oxford University Press. DOI: 10.1093/oso/9780198529729.001.0001

OXFORD
UNIVERSITY PRESS

Great Clarendon Street, Oxford, OX2 6DP,
United Kingdom

Oxford University Press is a department of the University of Oxford.
It furthers the University's objective of excellence in research, scholarship,
and education by publishing worldwide. Oxford is a registered trade mark of
Oxford University Press in the UK and in certain other countries

First Edition published in 2020

Impression: 1

Published in the United States of America by Oxford University Press
198 Madison Avenue, New York, NY 10016, United States of America

British Library Cataloguing in Publication Data
Data available

Library of Congress Control Number: 2019949835

ISBN 978–0–19–852972–9

Printed and bound in Great Britain by
Bell and Bain Ltd, Glasgow

To Rosamund, who makes everything possible

Acknowledgments

There are so many people to thank for making this book possible. First and foremost, I thank my many gifted students who have worked in my laboratory over the last 35 years. They have taught me almost everything I know about organic electronics (and many other things as well). It is fair to say that everything that is right about this book I owe to them, and everything that is wrong is because of my own mistakes.

I have also have had the pleasure of working with many collaborators who have also been my teachers. Professor Mark Thompson at the University of Southern California has collaborated with me on countless projects, and over the years we have become close friends. He is an amazingly creative chemist who has had the patience to teach me much about chemistry. In fact, I often joke that he has taught me just enough chemistry to make me dangerous.

Professor Vladimir Agranovich provided both the inspiration and motivation for me to take on this massive project. He has in so many ways been a role model and mentor to me, as well as being a giant of exciton physics. I would like to thank the people who have believed in me and supported my work that has led to such an exciting career in organic electronics. In particular I thank the management of Universal Display Corporation (Steve Abramson, Dr. Julie Brown, Sid Rosenblatt, and of course Sherwin Seligsohn) who have long supported my sometimes apparently aimless wanderings through the wilderness of organic electronics. Also, I thank Dr. Charles Lee at the U. S. Air Force Office of Scientific Research for his personal guidance and support over so many years, believing in my judgement and celebrating good results. I owe much of my success and happiness to these and many other supporters of my students and myself.

I owe a special thanks to my editor, Sonke Adlung, at Oxford University Press. Ever cheerful and encouraging, he has patiently tolerated my delays for more years than either of us would like to be reminded of. I thank Joe Matthews, also at Oxford, for his assistance throughout the often complex production process of this volume.

I also owe a debt of gratitude to the Technion, The Israel Institute of Technology, for providing me with a Lady Davis Visiting Fellowship in 2014. It was there that I began the serious effort of writing this book. Without those months at the Technion (and the many weekend walks in the Galilee with my friends Profs. Moti Segev, Gadi Eisenstein, David Gershoni, and many others) it is unclear that this would have ever happened in the first place. I thank the University of Michigan, particularly the Departments of Electrical Engineering and Computer Science, and Physics, who gave me both the space and time to complete this monumental task. Special thanks go to my assistant of many years, Ms. Eva Ruff, who took care of my group when I was otherwise preoccupied by writing, and also for proofreading and doing so much more to help make this project a success. I thank Jonathon (Jack) Smith for helping with organizing the multitude of figures that comprise this text.

Finally, I want to thank my wonderful family, in particular, my mother and father, Esther and Jerry Forrest. They encouraged me to get the education that they never had, and particularly my mother who would never hesitate to let me know when I wasn't living up to expectations. And then, of course, there is Rosamund, my wife, best friend, and self-proclaimed "physics widow." What would I ever do without her love, persistence, support, and that inevitable question as I walked in the door each night, "How many pages did you write today?" Maybe we can all get back to what used to pass as normal, now that this project is finished.

Preface

Organic electronics is the discipline at the core of the enormous, global organic light emitting device (OLED) display industry. Yet it is so much more. OLEDs also appear poised to be used as efficient lighting sources, organic photovoltaic solar cells may soon become a major source for converting solar into electrical energy, and electronic circuits based on organic thin film transistors may find their way into medical and a wide range of other sensing, memory and logic applications. What makes organic electronics such a compelling driver for these emerging applications is the unique features that they offer differ so remarkably from those of conventional semiconductors. Principally, organics serve as a platform for very low cost and high performance optoelectronic and electronic devices that can cover large areas (such as required for lighting, displays and solar cells), are lightweight when deposited on thin plastic substrates, and can be both flexible and conformable to fit onto irregularly shaped surfaces such as foldable smart phones. Due to the scale of the industries addressed by organic electronics, there are possibly 5000–10,000 scientists, engineers and technicians currently working to advance this field. Even so, very few texts are available that comprehensively cover the subject.

Organic electronics is an inherently interdisciplinary field, engaging experts in chemistry, in materials, in the physics of electronic and optical properties of disordered semiconductors, and in the engineering of practical, very high performance devices. To have a working comprehension of such a broad field, and ultimately to be an effective contributor to it, one must have knowledge of several disciplines that stretch well beyond the boundaries of traditional subjects taught in classrooms today. This book is structured to expose the student and the practitioner to the many aspects that define organic electronics, and provide an understanding that can allow the reader to be both knowledgeable and creative in extending our understanding of this exciting and explosively growing field.

This book is a product of both one and two semester courses that I have taught at both Princeton University and the University of Michigan over a period of more than two decades. During that period, the field has undergone dramatic changes, and my course material has changed accordingly. During this same period, many aspects of our fundamental knowledge of the field have also evolved to its current highly sophisticated, albeit still incomplete, level of development. Thus, while the field continues to move forward, it is essential that a text be available to capture the main ideas, and applications that define the subject today. It is these circumstances that have motivated me to write a book that presents the major elements of the field of organic electronics, and at the same time can serve as a resource for students, scientists and engineers alike. The target audience of the book, therefore, is students at all levels of graduate studies, highly motivated senior undergraduates, as well as practicing engineers and scientists.

The book is divided into two major sections. Part I, Foundations, lays down the fundamental principles of the field of organic electronics. It is divided into three principal subject areas: the basic properties and structures that define the broad class of organic semiconductors, followed by in-depth discussions of their optical and electronic properties. For Part I, it is assumed that the reader has an elementary knowledge of quantum mechanics, and electricity and magnetism. The reader is *not* expected to have a background knowledge of organic chemistry. The basic terminology of this field is provided in Chapter 1 to acquaint the reader with the vocabulary and a few important simple concepts that will be used throughout the book.

The foundations built in Part I are then extensively used in Part II, Applications, whose entire focus is on practical realizations of organic electronic devices. For this section, the reader is expected to have knowledge of the basic principles of semiconductor devices. Part II begins with a discussion of organic thin film deposition and patterning. I consider this to be the "how to" section of the book, providing what I hope is valuable information about how devices and structures are made in modern laboratories and manufacturing facilities. This is followed by chapters on the three most important device classes addressed by organic electronics, that is, organic light emitters, detectors, and thin film transistors. These chapters are the longest and most comprehensive of the entire book. The last chapter (Chapter 9) describes several devices and phenomena that are not covered in the previous chapters, since they lie somewhat outside of the current mainstream of the field. Nevertheless, the topics of light emitting electrochemical cells, strongly coupled optical phenomena, thermoelectricity, memories and limited dimensional systems are all central to the field, and indeed may also eventually lead to substantial future applications unique to organic semiconductors.

Given the enormity of the field of organic electronics and its inherently interdisciplinary nature, this volume is correspondingly large. It is primarily organized as a teaching text meant for a two semester course. The first part of the book, Part I, is used in the first semester focused on teaching the fundamentals of organic semiconductors. The second semester is focused on applications (Part II). However, this entire subject matter can be covered at less depth in a single semester by condensing the most important aspects of foundations and applications into a 15 week course with approximately 45 hours of classroom lectures.

Topics are introduced in the beginning of each major section by laying the basic foundations to be discussed. Hence, a general introduction of a topic is followed by theoretical considerations. This lays the groundwork for discussing a few illustrative examples. This organization mimics the format of most easy-to-follow scholarly papers: introduction, theory, experimental results and discussion, and conclusions. The examples provided are never meant to be comprehensive—that type of treatment should be left to review articles. Rather, the descriptions of phenomenological observations or device demonstrations are chosen to illustrate or reinforce points made in the early sections of the topic or chapter.

It should be noted that in discussing a particular topic, I have intentionally avoided including "record results" to the extent possible. Rather, the foci of the chapters are on examples that are particularly innovative, or whose description can provide lessons about how things work. The intentional avoidance of record results is that in such a fast moving field, today's record rapidly becomes "yesterday's newspaper." For this volume to remain relevant well past its publication date, it must resist the temptation (no matter how large) to present only hero results or device demonstrations. Our aim is to provide a practical perspective by maintaining focus on concepts and results that have a realistic possibility for finding their way into future applications.

Each of the chapters contains numerous references to the literature, with the expectation that this book will serve both as a classroom text, and as a source book that directs the reader to important work beyond that treated within the volume. At the end of every chapter is a section called *Further reading*. As its title implies, these sections provide a short, curated list of books, review papers, and other references that the reader may wish to access to deepen his or her comprehension of a particular topic. Finally, every chapter also contains a section of approximately 10 problems. These problems are meant to challenge the readers' understanding of selected topics contained within the chapter. Students in my graduate-level course are set a single problem every week to reinforce the material that they have learned during the previous lectures. Hence, many of the problems tend to be highly conceptual, somewhat lengthy, and challenging. In a few cases, the reader will need to refer to the literature to obtain a materials property or fill in a concept that is not completely provided in the problem or the book. Like any research problem, the solutions may not always be unique, or they may require the use of a computer. Once solved, they should provide a deeper understanding, and perhaps lead to new insights that are not fully addressed in the material presented within the chapters themselves.

Using this book in class

I have taught both one and two semester courses using notes that are the basis for this book. Most recently I have taught these courses at the University of Michigan that has a 13 week semester, amounting to approximately 40 class hours. This is shorter than many schools whose semesters typically last from 15 to 17 weeks. Below is a table that provides example readings, topics and teaching cadences for a 15 week semester, assuming 3 class-hours/week. Sample curricula for both one and two semester courses are shown. There are three possibilities for 1 semester courses: the first column shows a comprehensive but necessarily brief coverage of the entire field. Covering such a breadth of material in one semester makes this more of a survey course that necessarily compresses or omits much important material. Yet another format is to turn the two semester course into either a 1 semester course on fundamentals, or one that focusses only on devices (semesters 1 and 2 of the 2 semester course on the right hand column, respectively). The examples can be modified in many different ways to suit the interests of the class or the needs of the instructor. In all cases, the subject is laid out like the book: it starts by introducing the subject (Chapter 1), laying the foundational principles of organic semiconductors (Chapters 2–5), and then launching into discussions covering the important device operating principles, designs, and applications (Chapters 6–8). Note that for convenience in teaching a two semester course, Chapter 5 is moved into foundations. Several of the detailed sub-sections on materials, and mathematical developments should be omitted. Yet the student can always use the text to dig deeper. The two semester course is much more comprehensive, although student schedules do not always permit such extended concentration on this subject. I encourage the instructor and the students to experiment with topics and formats. If you find a better alternative to this structure, drop me a line.

Companion lecture notes are available for downloading at www.oup.co.uk/companion/organicelectronics2020 or https://organic-electronics.engin.umich.edu. These are once again meant to reinforce the material within the book and provide an aid to teaching by prospective instructors. Please feel free to download the notes, and use and distribute them as you wish. I only ask that you acknowledge their source.

Suggested study guide for one and two semester courses on Organic Electronics (15 week semester)

Week	Topic (1 semester)	Chapter/ Sections reading	Week (2 sem.)	Topic (2 semester)	Chapter/ Sections reading
1	Introduction to OE: Overview, history	1.1–1.3	1–1	Introduction to OE: Overview, history, language	1.1–1.3
2	Common language, crystal structure & binding	1.4, 2.1–2.4	1–2	Common language, crystal structure & binding	1.4, 2.1–2.4
3	Optical Prop. 1: Born-Oppenheimer & Franck-Condon, Fermi's golden rule, transitions	3.1, 3.2, 3.5	1–3	Calculating structure, epitaxy, self-assembly	2.5 – 2.8
4	Optical Prop. 2: Excitons, Spin, Energy transfer	3.6, 3.7.1, 3.7.4, 3.8	1–4	Optical Prop. 1: Born-Oppenheimer and Franck-Condon, LCAO, Fermi's golden rule, transitions	3.1–3.5.2
5	Optical Prop. 3: Exciton diffusion and recombination; Electronic Properties 1: Energy bands, electron transport	3.9–3.10; 4.1–4.3.1, 4.3.2.1, 4.3.2.2	1–5	Optical Prop. 2: Understanding spectra, dimers, excimers, exciplexes	3.5.3–3.6.5
6	Electronic Prop. 2: Conduction, mobility, doping, HJs	4.4–4.7	1–6	Optical Prop. 3: Excitons, CT states, spin, energy transfer	3.6.6–3.8.2
7	Materials growth & purification, device patterning, packaging	5	1–7	Optical Prop. 4: energy transfer, exciton diffusion, recomb. & annihilation; Electron Prop. 1, Energy bands	3.8.2–3.10; 4.1
8	Light emitters 1: Basics, efficiency, fluorescence, phosphorescence, TADF	6.1–6.3.4, 6.4	1–8	Electronic Prop. 2: Energy bands, hopping, conduction, mobility	4.2–4.4
9	Light emitters 2: Rolloff, White OLEDs, outcoupling	6.5–6.6.1	1–9	Electronic Prop. 3: Mobility, doping, metal contacts	4.4–4.6.2
10	Light emitters 3: Outcoupling, reliability	6.6.2–6.7	1–10	Electronic Prop. 4: Contacts, HJs	4.6.3–4.7.2
11	Light detectors 1: Basics	7.1–7.3.2	1–11	Electronic Prop. 5: O-O and O-i HJs	4.7.2–4.8
12	Light detectors 2: Efficiency, architect., materials, transparency	7.3.3–7.4.3	1–12	Purity and crystal growth	5.1–5.4.2.3
13	Light detectors 3: Multijunc. OPV, reliability, modules; Transistors 1: Basics	7.5, 7.8, 7.9; 8.1–8.3.2	1–13	Thin film dep, processing, patterning	5.4.2.4–5.6 (except 5.6.4)
14	Transistors 2: Architectures, morphology, reliability, apps.	8.3.2–8.4, 8.9	1–14	Nanopatterning, R2R, packaging	5.6.4, 5.7–5.9
15	Semester Review		1–15	Semester Review	

(*Continued*)

Week	Topic (1 semester)	Chapter reading	Week (2 sem.)	Topic	Chapter reading
1			2–1	Review Semester 1. Light emitters 1: Basics, Displays	6.1, 6.4
2			2–2	Light emitters 2: OLED basics, efficiency, emission processes, materials	6.1–6.3.3
3			2–3	Light emitters 3: TADF, annihilation, White OLEDs	6.3.4–6.3.5,6.5.1–6.5.4
4			2–4	Light emitters 4: WOLEDs, outcoupling	6.5.4–6.6
5			2–5	Light emitters 5: Reliability, lasers	6.7.4 – 6.8
6			2–6	Detectors 1: Basics, photoconductivity, photodiodes	7.1–7.2
7			2–7	Detectors 2: PD apps, solar cell basics, efficiency, architecture	7.2.2.4–7.4.1
8			2–8	Detectors 3: Morphology, materials, transparency	7.4
9			2–9	Detectors 4: Multijunction OPV, singlet fission, light trapping, reliability	7.5–7.7
10			2–10	Detectors 5: Modules; Transistors 1: Basics	7.9–7.10; 8.1–8.3.3
11			2–11	Transistors 2: Ambipolar, circuits, architectures, phototransistors, morphology, patterning	8.4–8.7
12			2–12	Transistors 3: SAMs, reliability, apps.	8.6.1, 8.7–8.9
13			2–13	Other topics	Selected from Ch. 9 or other sources
14			2–14	Other topics	Selected from Ch. 9 or other sources
	Review		2–15	Review of Semester 2 and the entire course	

A few organizational notes

As mentioned above, the best use of this book will be made by those who have a suitable background in a few key subjects. Thus, a basic familiarity with quantum mechanics, electricity and magnetism, and semiconductor devices is highly recommended, although holes in your understanding of some topics will be filled by reading this book whose intent is to be instructional to a broadly interdisciplinary readership. Also, it is not important to have a background in organic chemistry, although that is helpful. This book is written by a physicist/electrical engineer who has worked for 35 years in the field without a formal background in that subject, although I have picked up a fair amount of it along the way. A few key chemistry definitions and concepts needed to proceed with this volume are provided in Chapter 1. Additional background is provided as it is encountered in the text to enable the understanding of a particular topic.

The chapters are replete with commonly used acronyms (e.g. VTE for vacuum thermal evaporation). Following conventions found in the scientific literature, these acronyms are generally defined by their first use in each chapter. However, a reader who is skipping ahead to research a specialized point may find it challenging to locate the definition of a particular acronym. Hence, many common acronyms are collected for convenience in Appendix A. The chapters describe numerous example experiments, data and devices, each using a plethora of materials and layers. Identifying each compound by its chemical nomenclature based on the International Union of Pure and Applied Chemistry (IUPAC) standards is quite disruptive to the narrative flow. For this reason, in all but a few cases, the compounds are identified by their acronyms (e.g. PTCDA for 3,4,9,10-perylenetetracarboxylic dianhydride), with their IUPAC nomenclatures provided in Appendix B.

Each chapter begins and ends with a summary of the important points being addressed. Particularly useful equations are highlighted in boxes, and the first use of an important technical word or phrase (e.g. exciton) is identified by *italics*, along with a definition of that term.

We often use the very convenient Dirac bracket notation when discussing quantum mechanical wavefunctions and integrals. That is, wavefunctions are expressed using the ket-vector:

$$\psi_A = |A\rangle.$$

and its complex conjugate is given by the bra-vector:

$$\psi_A^* = \langle A|.$$

Similarly, integrals are replaced by the *bracket*:

$$\int \psi_A^* \Omega \psi_B d^3 r = \langle A|\Omega|B\rangle,$$

where Ω is an operator. Finally, the expectation value of an operator is given by:

$$\langle \Omega \rangle = \frac{\int \psi^* \Omega \psi d^3 r}{\int \psi^* \psi d^3 r}.$$

Stephen Forrest
Ann Arbor, Michigan
Rochester, Vermont
2019

Contents

PART I

Foundations

CHAPTER 1

Introduction to organic electronics

"There is no object so soft but it makes a hub for the wheeled universe."
Walt Whitman, American Poet, *Song to Myself*

What makes organic semiconductors such a fascinating topic for study? Beyond the materials and physics vistas that open when we study such materials, we also should ask the question, are they useful? Since you are reading this book, you most likely already know that they are enormously useful. After all, their largest application is in organic light emitting diode (OLED) displays, today a multibillion dollar industry that exhibits massive growth each year. In this chapter, we briefly introduce the topic of organic semiconductors, and by extension their applications in organic electronics, by providing an overview of the scientific and technological scope of the field. We show that intensive study over the last 70 years has created extraordinary insights into the physics of disordered materials, while also launching global industries that employ thousands of scientists and engineers across multiple disciplines that stretch from physics and chemistry, to materials science, and electrical and mechanical engineering. Indeed, organic electronics is the quintessential interdisciplinary subject. To be conversant and innovative in this field, one must understand multiple technical languages and be able to collaborate effectively across the boundaries of traditionally "siloed" fields of study.

This chapter (and book) is meant to motivate the reader to dig deeper into this literally limitless subject, and along the way perhaps explode a few myths that have swirled around the field almost from its beginnings. The chapter is meant to provide a common language that will help the reader bridge the gulf that separates the disciplines, and often creates a barrier to understanding and communication. Hence, at the outset, we define a few essential terms that will aid in understanding the context and terminology that is used throughout the book.

The materials have unique properties that are determined by their fundamental excited state, the exciton. The excitonic nature of organic semiconductors, that is distinct from the band-like character of inorganics, is a direct consequence of the weak intermolecular bonding forces that form the materials into their solid state. Given the frailty of those bonds, the materials tend to be disordered, leading to unusual properties beyond their tendency to form strongly bound excitons. We therefore devote a section to a comparison of the distinct properties of organic and inorganic semiconductors. Indeed, their differences lead to their satisfying very different applications when exploited in optoelectronic devices such as displays, transistors and solar cells.

Following this comparison, we provide a brief history of the field. It has been an interesting and colorful history indeed, extending from the discovery of organic semiconductors, to the awarding of a Nobel Prize, to the present day where organic electronics is the foundation of global industries. And to cap it off, organic electronics was launched into the twenty-first century by being the focal point of the largest fraud in the history of physics. Indeed, such a fraud could only have been perpetrated on an emerging field for materials

Organic Electronics: Foundations to Applications. Stephen R. Forrest. © Stephen R. Forrest 2020.
Published in 2020 by Oxford University Press. DOI: 10.1093/oso/9780198529729.001.0001

whose properties, even today, defy description by comprehensive theories that are nowhere near as complete and precise as that found for inorganic materials.

We next digress to briefly provide some useful terms that will help the reader navigate the more in-depth discussions found in subsequent chapters. There, we will provide important example molecular structures and common terminologies that can serve as a reference for the rest of the book. More specialized terminology will be provided within the context of the topic where it is first introduced. A section is included that confronts some pervasive myths that have been created about this field, for example, organics don't live long enough to be useful, or solution processing is less costly than vapor deposition, etc. These myths need to be confronted, however briefly, with more in-depth consideration of their origins, and what separates these impressions from the truth, found later in the book.

We finish our introduction by describing what makes organic electronics so appealing a topic for engineers and scientists worldwide. That is, we address the question, "What is organic electronics good for?" Is it a solution looking for a problem, or does it enable important optoelectronic device applications niches that are inaccessible to other, conventional semiconductors? We will lay out the rationale for organic electronic devices that have motivated the technical community, and large industries, to devote so much effort and resources over many years toward their development and application. We hope that by the end of this chapter, you too will be motivated to learn more about this exciting and promising field that is now making a difference to so many lives around the globe.

1.1 What is an organic semiconductor?

Organic compounds are formally defined as those that comprise at least one carbon-hydrogen bond. However, this term is often more loosely defined as simply referring to any carbon-rich compound. Thus, solely carbon-containing compounds such as the fullerenes (e.g. C_{60}), or carbon nanotubes are sometimes incorrectly referred to as organic when used in devices. Nevertheless, these two types of materials find extensive use in photodetectors and solar cells. But strictly speaking, they are not organic compounds. Molecules without C–H bonds are known as *inorganic compounds*. In the solid state, organics are a member of a broad class of soft materials that are primarily bonded by weak van der Waals forces, although there may be small admixtures of other forces such as Coulomb attraction due to charge transfer between molecular units, hydrogen bonds, etc. A result of weak bonding forces in soft materials is a relatively high degree of disorder that is a combination of dynamic processes (i.e. thermal motion between neighboring molecules as well as intramolecular vibrations), as well as static disorder from crystalline defects. As we will find in Chapters 2–4, disorder is a significant factor that leads organic solids to have optical and electronic properties that are distinct from those of more strongly bonded solids, such as metals and semiconductors.

The preponderance of organic solids are electrical insulators, that is, they have a large energy gap that substantially limits the number of thermally excited charges into the conduction level at room temperature, and hence their conductivity is very small. Also, the wide energy gap results in transparency in the visible. On the other end of the conductivity spectrum are metallic compounds that are based on charge transfer salts. The materials that have found the most use in organic electronics are those with intermediate energy gaps, ranging between 0.5 and 3 eV. Like all conventional semiconductors, their conductivity can be modified over many orders of magnitude by introducing either an electron-accepting or electron-donating compound to provide the host semiconductor with either p-type or n-type conductive character, respectively. Also, their intermediate energy gaps result in optical absorption or emission across the visible, and into the near infrared (NIR) regions of the spectrum. This feature makes organic semiconductors ideal candidates for optical emitters and detectors (including solar cells) that operate within these spectral zones.

The properties of organic compounds can be tuned over a nearly limitless range by modifying their composition; even the smallest modification, or chemical functionalization can lead to large changes in the optical or electronic properties of the molecule. And when that molecule is part of a solid, the bulk material itself can possess both physical and optoelectronic properties that are distinct and helpful for understanding their fundamental properties, or are useful in applications such as displays, lighting, or solar cells.

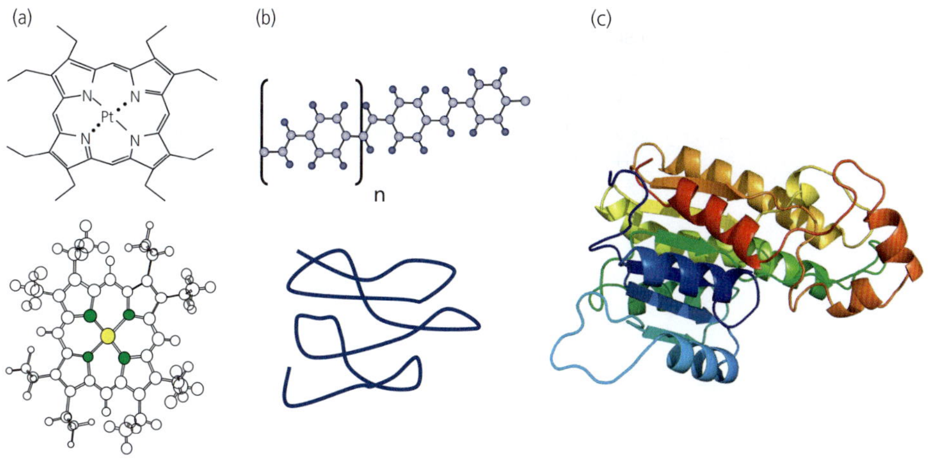

Figure 1.1 Three types of organic molecules. (a) Molecular materials, represented by a monomer of platinum octaethylporphyrin. The upper diagram is the molecular structural formula, and the lower is a "ball and stick" representation of the same molecule. The metal (Pt) atom is shown in yellow, the small circles are H atoms, the larger open circles are C and N (the latter shown in green each bonded to the central Pt). (b) Polymeric materials consisting of an indefinite number (*n*) of monomer units of *p*-phenylenevinylene. The lower diagram represents a long chain comprised of hundreds or thousands or these molecular links. (c) A protein molecule of biological origin (Wikipedia, 2019).

Organic materials are broadly classified into three different types: small molecules, polymers and biological molecules, with examples provided in Fig. 1.1. *Small molecules* are organics with a well-defined molecular weight. The smallest molecular structure consists of a single unit, called a *monomer*. Here, *-mer* comes from the Greek, *meros*, or part. The example shown in Fig. 1.1a is platinum octaethylporphyrin ($C_{36}H_{44}N_4Pt$), which is an intensely emissive red phosphorescent compound. The inclusion of a metal Pt atom in the molecule makes this a member of the class of metalorganic compounds. Its formal International Union of Pure and Applied Chemistry (IUPAC) nomenclature is platinum(II) 2,3,7,8,12, 13,17,18-octaethyl-21*H*,23*H*-porphyrin. Compounds are referred to in this volume by their common acronyms, in this case PtOEP. Their full IUPAC designations are found in Appendix B.

Two or more monomers can be connected into a molecule that consists of a well-defined number or repeat units called dimers (two units), trimers (three units), tetramers (four units), and so on. These are all members of the class of materials comprising a plurality of monomers, called *oligomers* (from Greek *oligoi* meaning "few"). Each monomer is linked by a connecting unit, which may be no more than a C–C bond shared between monomers, or it may consist of a moiety comprising multiple bonds and atoms. Like monomers, oligomers have well defined molecular weights, that is, an oligomer is composed of an exact number of atoms. A schematic representation of a series of oligomers is shown in Fig. 1.2.

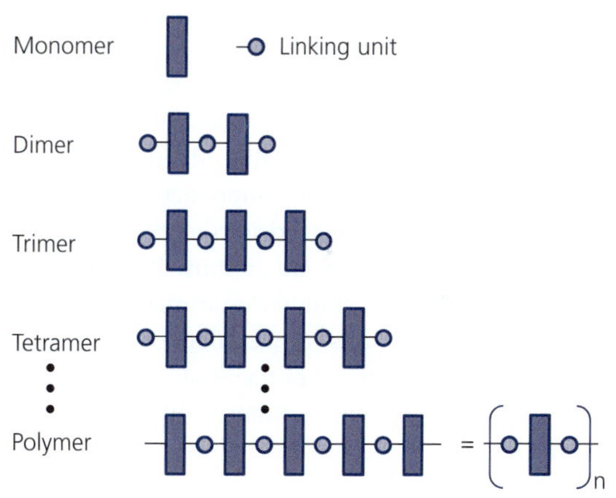

Figure 1.2 Hierarchical assembly of molecules from monomers to polymers. The linking unit may have a stable terminating group indicated by the circles. Also, the polymer can be represented by a single monomer unit repeated *n* times as shown in lower right.

A solid solution containing a population of oligomers may be *monodisperse*, where each member of the population is identical. For example, the solution comprises only trimers. However, solutions that contain populations of oligomers of different molecular weights are *polydisperse*.

Another type of molecular material with monomers arranged in branches originating from a central core is a *dendrimer*. This species of molecule gets its name from its resemblance to dendrites, or branched projections from a single core such as found in hoar frost, ferns, neurons, and other "hub-and-spoke"

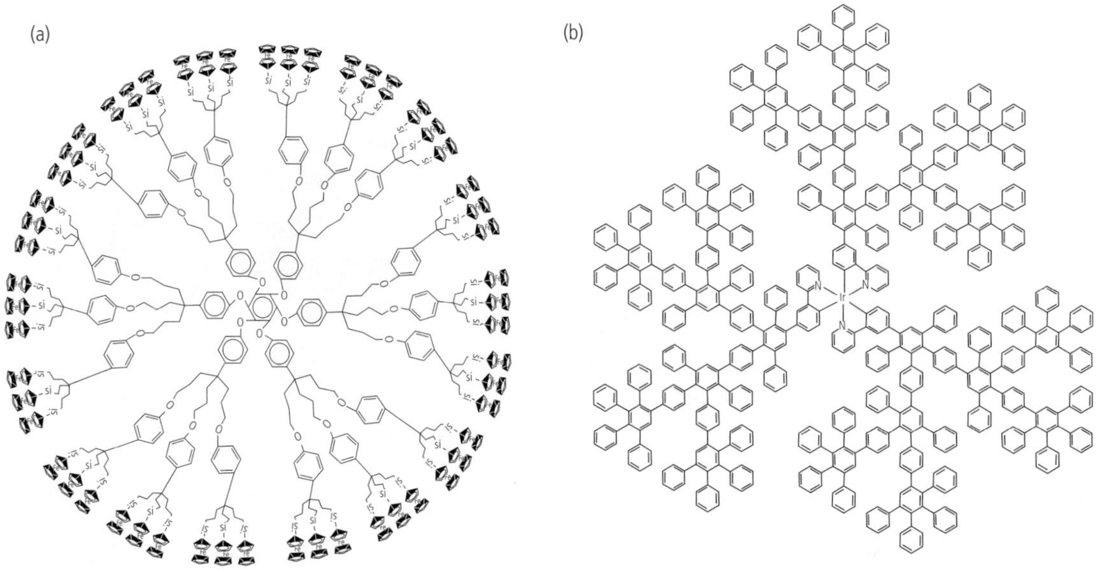

Figure 1.3 Two example dendrimer molecules. (a) 54 ferrocene (so named due to the 54 ferrocene groups on its periphery), and (b) a fourth generation, luminescent metalorganic dendrimer with an Ir atom at its core.

structures. Two example dendrimers are shown in Fig. 1.3. The number of units along a single branch gives the "generation" of the dendrimer. That is, a molecule comprising only a core unit is a monomer. Formally, this is a first generation dendrimer. More interestingly, a second generation dendrimer comprises a core that is branched into a second rank of monomers, and if these are in turn branched yet again, it is a third generation dendrimer, and so on. Since a dendrimer has a well-defined molecular weight, it too is a member of the broad class of small molecular weight materials, although high generation dendrimers may have very large molecular weights, indeed. Dendrimers have found their primary applications in medicine, providing possible routes for directed delivery of drugs to specific sites (Tomalia et al., 2012). Unfortunately, these large, dandelion-shaped molecules do not form strong intermolecular couplings in the solid state. This inhibits both charge and energy transfer between molecules, thereby limiting their use as electronic materials, although they can be intensely luminescent.

Molecular compounds can be synthesized in the laboratory, or they can have natural biological origins. The first artificially synthesized molecule was urea. Credited to Friedrich Wöhler in 1828, this discovery ushered in the field of organic chemistry. A common example of a small molecule that can be synthesized or can be found in nature is anthracene, whose molecular structural formula is shown in Fig. 1.4. Anthracene, which is found in anthracite

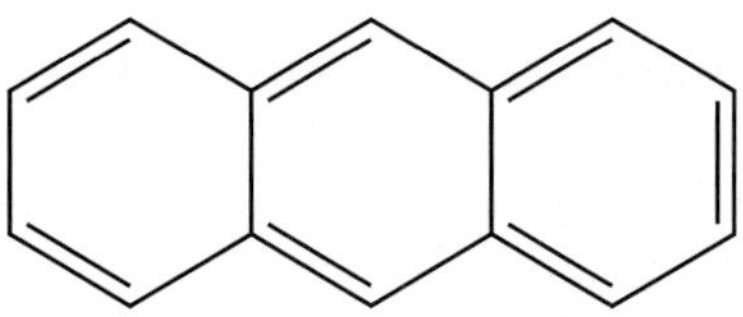

Figure 1.4 The molecular structural formula of anthracene ($C_{14}H_{10}$). Each outer vertex is bonded to an H atom (not shown). Double and single C bonds are indicated.

(known as hard coal) and even in interstellar space (Iglesias-Groth et al., 2010), was discovered to have semiconducting properties over 60 years ago (Karl, 1989). Indeed, it was among the first organic materials to be developed for its optoelectronic properties, emitting dim blue light when electrically excited (Pope et al., 1963). Anthracene comprises three fused phenyl rings with alternating single and double carbon bonds. These single and double bonds are in resonance, that is, the bond arrangement can alternate between carbon atoms, finding equivalent positions shifted 60° from those shown in the figure without a gain or loss in energy. Thus, the electrons in the hybrid p-orbitals of the carbon atoms in the rings are delocalized, that is, they are shared by all the other carbons covalently bonded within the molecular plane. This electronic delocalization forms what is known as a *conjugated electron system*. A cyclic, planar, closed ring molecule with resonant bonds such as anthracene is an *aromatic compound*. Aromatic materials are of considerable interest in organic electronics due to their stability, their ease in forming films and crystals by thermal evaporation in vacuum onto a substrate, and their very strong absorption bands from the visible and into the NIR. They are known as aromatics since many have strong odors. A prime example of this is the simplest, single carbon ring molecule, benzene (C_6H_6), a principal ingredient of fossil fuels. Due to their intense absorption spectra, aromatics have long been used as dyes in textiles and paints, and as inks used in photography, printing, etc. Hence, many compounds are readily available in large quantities, and most importantly, they can be inexpensive, leading to their widespread application as dye stuffs.

In contrast to small molecules, a *polymer* is a chain of monomers or oligomers of indefinite length (see Figs. 1.1b and 1.2). Polymers are the basis for plastics, whose combination of light weight, stability, strength and flexibility has led to their ubiquitous adoption as materials used in construction, packaging, windows, etc. The first synthetic polymer was "Parkesine" based on cellulose introduced by Alexander Parkes in 1856, and the first fully synthetic polymer was Bakelite (Leo Baekland, in 1907) that found applications as an electronic insulator. The vast majority of polymers used today are insulators, and hence are optically transparent. Their stability has both benefits as well as significant deficiencies. Stable polymers resist degradation, which is beneficial in many applications where long-term durability is essential. On the other hand, their stability has generated an immense amount of refuse contributing to global pollution that humankind appears at a loss of how to manage.

The use of polymers as active electronic materials is a relatively new development, extending only over the last few decades. In 1963, Bolto and coworkers discovered conduction in the polymer, polypyrolle (Bolto et al., 1963). This was followed in 1977 when the polymer, polyacetylene, was doped with an electron accepting molecule to vary its conductivity over several orders of magnitude (Shirakawa et al., 1977). This showed that the electronic properties of polymers can be varied by introducing extrinsic impurities, analogous to impurity doping that has made inorganic semiconductors useful as electronic materials. Today, polymer semiconductors are widely investigated for their potential applications in virtually all organic electronic devices.

In the solid state, both small molecules and polymers pack together to form amorphous, polycrystalline or single crystalline structures, depending on the physical and electronic properties of the molecules, and the processes used to form the solid. The morphologies of small molecule and polymer solids can vary significantly due to their compact vs. extended and folded structures, respectively. Since the separate molecular units are not chemically bonded to each other, organic materials tend to be soft and fragile, although they can also be very flexible and even ductile.

Small molecules are deposited both from the vapor phase, or from solution when they are functionalized to have high solubility. In contrast, high molecular weight polymers are only deposited from solution since their mass prevents volatilization by heating below their decomposition temperatures. This results in significant differences in film stability and purity between these classes of compounds. Indeed, their different film-forming properties has led to the widespread adoption of small molecules rather than polymers in electronic devices. This topic will be considered further in Part II, on applications.

The last class of organic materials comprises highly complex molecules of biological origin (Fig. 1.1c). Exemplary molecules include deoxyribonucleic acid (DNA), proteins, and photosynthetic complexes. With only a few exceptions (e.g. DNA), this class of molecules has not been investigated for its electronic properties since the molecules are highly complex and cannot be synthesized in the laboratory, and they typically require an aqueous solution to maintain their function, shape or bulk morphology. Nevertheless, protein complexes can be solidified into well-defined crystal structures, such that detailed information on their molecular structures has been obtained. This understanding is central to our developing an understanding of the relationship between molecular structure and function in biological systems. And there are some examples where biological systems have provided important models for electronic systems as well, such as photosynthesis, neural networks, etc. A significant differentiating feature between biological and artificial electronic systems is that the former have the capacity for self-repair. Cellular damage that occurs during use, or exposure to environmental agents can sometimes be corrected by the biological machinery encoded in the genome. Artificially engineered organic or inorganic electronic devices have no such ability; when damage occurs it permanently impairs the operation of the device or system. Electronic systems manage this problem only poorly through the use of device redundancy in some applications where long-term operation is critical (e.g. in remote space-borne and undersea systems). Given their complexity, today we have only the most primitive understanding of the chemical and electronic properties of biological molecules, and hence they have yet to play a significant role in the field of organic electronics.

1.2 The differences between organic and inorganic semiconductors

Before we can fully appreciate the rationale for using organic semiconductors in practical applications, we must first understand the key differences between this materials species and conventional, inorganic semiconductors. In this section, we address the question, "What are the characteristics of organic electronic materials that distinguish them from their inorganic analogs, and what is the impact of their inherent limitations and advantages on the performance of devices in which they are employed?"

A comparison of properties of organic and inorganic semiconductors is provided in Table 1.1. The single most distinguishing feature of organic semiconductors is their absence of intermolecular *chemical* bonds. Organic solids are primarily bonded by purely electrostatic *van der Waals bonds* due to induced dipole-induced dipole interactions discussed in the following chapter. These bonds can also have an admixture of Coulombic attraction due to charge transfer between neighboring molecules, or hydrogen bonds, but normally these contributions are small compared to those from van der Waals forces. In any event, there are no chemical bonds between the molecules. This is in stark contrast to inorganic materials that consist of chemically bonded atomic species (Ashcroft and Mermin, 1976). Chemical bonds can be covalent due to sharing of electrons between atoms within the semiconductor (as in Si and Ge), or primarily ionic due to Coulombic attraction of anion/cation atomic pairs (CdTe, ZnSe, etc.). In some semiconductor alloys, bonding is via a mixture of both covalent and ionic bonds (e.g. InP, GaAs, GaN, TlP).

While *intra*molecular covalent forces are responsible for binding atoms within the compound, the lack of *inter*molecular chemical bonds results in starkly different electronic and mechanical properties in organic and inorganic materials, summarized in Table 1.1. This distinction is so profound that it governs all the physical, electronic and optical properties of these materials. The overriding importance of the bond type is emphasized by highlighting the entry in the table.

The strong chemical bonds in inorganic solids have a dissociation energy of ~1–5 eV, which is at least an order of magnitude higher than for organics. Thus, organic materials are elastic and flexible, whereas inorganics are stiff and brittle. The low intermolecular binding energy of organics requires a comparatively low temperature to dissociate the molecules from the solid (~100–500°C). However, most inorganic semiconductors thermally decompose at double to triple the highest temperatures tolerated by organic solids,

Table 1.1 Comparison of typical properties of inorganic and organic semiconductors

Property	Organics	Inorganics
Bond type	**van der Waals**	**Covalent/ionic**
Intermolecular bond Energy	0.1 eV	1–5 eV
Energy band width	0.01–0.5 eV	1–4 eV
Charge transport	Polaron hopping	Band transport
Charge mobility	≤ 1 cm^2/V s	~1000 cm^2/V s
Conduction levels	Frontier orbitals	Bands
Dielectric constant (Refractive index)	2–3 (1.5–2)	10–15 (3–5)
Peak absorption (Oscillator strength)	10^5–10^6 cm^{-1} (High)	10^3–10^4 cm^{-1} (Low)
Exciton species	Frenkel	Wannier–Mott
Exciton binding energy	~500–1000 meV	~5–10 meV
Exciton radius	~10 Å	~100 Å
Hardness	Soft	Hard
Flexibility	Flexible	Brittle

although there are compounds whose bond dissociation energies (BDE) are near the higher end of the range experienced by organics (e.g. HgTe). Not surprisingly, decomposition temperatures of organic compounds are often considerably higher than their evaporation temperatures, since the intramolecular chemical bond strengths are similar to those of inorganic materials. The low solid dissociation temperature of organics allows for film deposition at or near room temperature, leading to a low "investment" of energy required for device fabrication. On the other hand, low decomposition temperatures can be a source of thermal instability that shortens device lifetime, or may prohibit its use at even modestly elevated temperatures. At sufficiently high intermolecular bond energies, molecular decomposition occurs at lower temperatures than needed for evaporation. In this case, dissolving molecules in a solvent and then casting or spraying them onto a substrate is a more appropriate means of deposition. Solution processing is almost exclusively used for large molecular weight species such as polymers and dendrimers, as well as for many fragile small molecules (see Chapter 2).

Covalent bonds result in delocalization of electrons among all atoms arranged periodically in a crystal. Thus, charges move relatively unimpeded over large distances, and respond collectively to an applied electric field. This leads to a high charge mobility (100–10^4 cm^2/V-s at room temperature in homogeneous and alloy semiconductors), and wide energy bands. Lattice periodicity results in the opening of a *band gap* of energy, E_G, between the conduction and valence energies. Charge delocalization creates a large *density of electronic states* (DOS) spread across a bandwidth (*BW*) of several eV, as shown in Fig. 1.5. In contrast, organic semiconductors have an *energy gap* between narrow, frontier molecular electronic orbitals. The *highest occupied molecular orbital* (HOMO) is comprised of the most energetic electrons bound to the molecular π-system; they are the highest molecular valence states. Similarly, electrons in the *lowest unoccupied molecular orbital* (LUMO) comprise the least energetic conduction states. Rather than forming a continuous DOS, the deeper orbitals in organic crystals are relatively independent of each other due to weak electronic interactions between buried electronic states in neighboring molecules. The states deeper within the occupied and unoccupied orbitals are denoted HOMO-*i* and LUMO+*i* states, where *i* is the index of an orbital relative to the frontier orbital (where i = 0). Note too, that the DOS for inorganics begins at zero at a

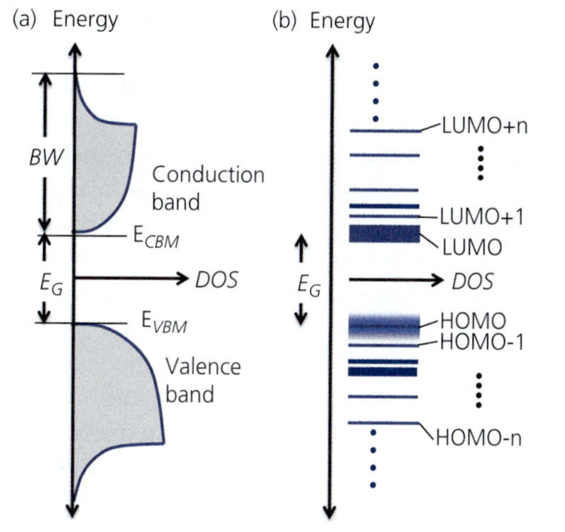

Figure 1.5 Characteristic energy level schemes of (a) an inorganic semiconductor with band gap energy, E_G, and (b) an organic semiconductor with an energy gap of approximately the same magnitude. Here, DOS is the density of electronic states, HOMO is the highest occupied molecular orbital, LUMO is the lowest unoccupied MO, and *BW* is the width of the corresponding band. The "fuzzy" depiction of the HOMO represents the case of a tightly packed organic crystal that has developed incipient, or narrow band-like character. E_{VBM} and E_{CBM} are the energies of the valence band maximum and conduction band minimum, respectively.

band edge, whereas for organics it is relatively constant and high within each narrow orbital. The density of states of each orbital is approximately equal to the molecular number density.

A situation analogous to organics obtains in engineered inorganic semiconductor structures such as quantum wells, wires and dots. In those cases, the bands disaggregate into a series of narrow energy levels characteristic of the size of the independent, nanometer-scale quantum confined regions. Indeed, quantum dots (QDs) play a role in inorganic semiconductors that is similar to the nearly independent molecules comprising a molecular crystal. This is particularly true of molecules that are doped at low concentration into a matrix comprising a different compound. For example, luminous dye molecules (called *lumophores*) within a wide energy gap matrix act independently of each other, exhibiting the same properties as the isolated molecule. Hence, the dopant can be thought of as a QD within a wider energy gap semiconductor, or insulator.

We note that the organic semiconductor energy *BW* is never 0 eV, as is characteristic of completely isolated atoms and molecules. If this were the case, there could be no carrier momentum within the level, and hence charge transport between molecules would be impossible. In fact, for the most tightly packed organic crystals, relatively large energy band widths of ~500 meV are observed. These levels are said to form *incipient bands* that give such organics many properties similar to those found in inorganic semiconductors. An excellent example of this class of molecules is PTCDA with a small intermolecular van der Waals spacing of only 3.21 Å. As a result of its band structure, PTCDA has been the subject of investigation for over 30 years, with properties that bridge the chasm between organic and inorganic materials (Forrest, 1997). This highlights a fascinating characteristic of organics: they have properties shared by both insulators and conventional semiconductors, and by the individual molecules and their bulk condensed state.

The distinction between band gap energy and energy gap is an important one that is often confused in referring to the energetic properties of organics. The existence of a band gap implies that there is an energy band with a continuum of electronic states that begins at the valence band maximum energy (E_{VBM}) and extending to much lower energies, and from the conduction band minimum energy (E_{CBM}), extending to much higher energies. The narrow bands characteristic of organics, therefore, make it more appropriate to simply refer to the energy separation of the HOMO and LUMO as the energy gap.

Band-like transport leads to a relatively high charge mobility. The bulk charge mobility of inorganic semiconductors is 10^3–10^6 times higher than organic semiconductors, whose charge transport is due to incoherent *hopping* from one molecule to the next. Hopping refers to the process where momentum of a charge is randomized at each molecular site along its transport path. The transport, therefore, is incoherent: the history of where the charge has been has little or no influence on the next step that it takes as it drifts from molecule to molecule in an electric field. The hopping process can be activated either by temperature or an applied electric field. These are, respectively, known as thermally or tunneling-assisted hopping.

The optical characteristics of the solids, too, are dramatically modified by the nature of their cohesive forces. Chemical bonding leads to a high polarizability of a substance due to the collective response of a delocalized sea of electrons to an external electric field. This leads to relatively high dielectric constants and the corresponding refractive indexes compared to van der Waals bonded solids.

An incident photon can be absorbed by a single molecule in the solid. The excited state can subsequently transfer from a molecule to its neighbor, thus resulting in the diffusive energy transport over substantial distances, sometimes as large as 50–100 nm. A mobile molecular excited state is known as an *exciton*. Since transitions from the ground to the excited state have a very high oscillator strength, the excitonic properties of organic materials dominates their optical response (Silinsh, 1980, Pope and Swenberg, 1982). For this reason, organic materials are often called *excitonic materials*, and the study of their optical properties is known as *excitonics*. In contrast, delocalized valence electrons and their distribution in energy over a broad DOS lead to a much lower oscillator strength in inorganic semiconductors. These transitions are thus between bands vs. discrete orbitals.

An archetype molecule can be represented by a two-level energetic system shown graphically in Fig. 1.6. The arrows represent electrons, one with spin up and the other with spin down in the ground state, S_0. The spin down electron is elevated in energy via photon absorption into the first excited state, S_1. As we show in Chapter 3, S_0 and S_1 connote quantum mechanical spin singlet states. The absent electron in S_0 is a hole that is Coulombically bound to the unpaired, excited electron. The pair can hop from molecule to molecule, transporting energy contained in the excited state across the solid. The bound pair is the exciton, illustrated by the shaded oval in Fig. 1.6.

When the electron and hole are localized on a single molecule, the species is called a *Frenkel exciton*. Its radius is on the order of the dimensions of the molecule, as depicted in Fig. 1.7a. Due to the low dielectric constant of the material, its binding energy ranges from 0.5 to 1 eV, and hence it is stable at room temperature. This stability allows it to diffuse long distances before the excited electron relaxes back to the ground state, thus eliminating the exciton. The

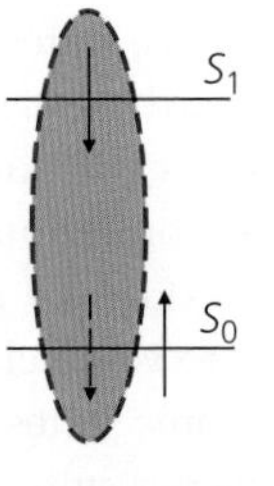

Figure 1.6 A molecular excited state. The arrows represent electrons with spins pointing up or down, the dashed arrow is the absence of an electron in the ground state, and the dashed oval is an exciton, or bound electron–hole pair. The singlet ground state is labeled S_0, and the excited state is S_1.

molecular relaxation can be either radiative or non-radiative via defect states within the material.

During transport, the excited electron can momentarily transfer to a neighboring molecule, creating an anion-cation pair of neighbors. The delocalization of the charge over a pair, or between only a few neighboring molecules is known as a *charge transfer* (CT) *exciton*, or charge transfer state, as shown in Fig. 1.7b. The binding energy of the CT state is substantially smaller (~0.1–0.5 eV) than that of a Frenkel exciton due the reduced Coulomb attraction of the spatially separated charges. Finally, the high dielectric constant of inorganic semiconductors leads to effective field screening between the photogenerated electron–hole pair, resulting in a very loosely bound (5–10 meV), and consequently a large *Wannier–Mott* (W-M) *exciton* that extends over many atomic sites (see Fig. 1.7c). This excitation is weakly absorbing, and is only stable at low temperatures. Structures such as quantum wells, wires or dots can be engineered to confine the electron and hole within a nanometer scale energetically bounded region, giving the W-M state more stable, Frenkel-like character. Hence, the optical properties of such artificial structures can also be dominated by excitons compared to the native, bulk semiconductor. In the absence of quantum confinement, the absorption of a photon in inorganic materials primarily results in the direct band-to-band generation of free charge, bypassing an exciton intermediate. These materials are, therefore, called "band-like," with absorption features that are broad, extending from the band gap to much higher energies.

The very large oscillator strengths of primary transitions in organics make them highly absorptive over relatively narrow spectral bands (typically ~100 nm full width at half maximum), that are broadened by thermal motions (phonons) between atoms comprising the molecule. The absorption coefficients of organic semiconductors are ~10^5–10^6 cm^{-1}, which is on the order of that of atomic systems. This is another outcome of the similarity between molecular and two-level atomic systems, whose densities of states are approximately equal to the number density of molecules in the solid.

The narrow and intense absorption spectra typical of molecular thin films are shown in Fig. 1.8 for several members of a class of non-fullerene acceptor molecules used in organic photovoltaic (OPV) cells. The excitonic absorption of each molecule is rigidochromically shifted from a peak wavelength of 650 nm to 850 nm, depending on the particular molecular structure. A *rigidochromic spectral shift* refers to the apparent translation of the spectra without significant line broadening or changes in shape from molecule to molecule. The double peaks of each exciton line are due to different intramolecular vibrational modes of the molecules; in this case they arise from the C–C bond stretch with energy of ~1500 cm^{-1} (Chapter 3). Several interesting properties of organics are readily apparent from these spectra. For example, the peak absorption wavelength blue shifts as the molecular backbone consisting of sulfur-containing (thiophene) rings is lengthened. Attaching Cl to the end-capping groups also increases the peak absorption wavelength. This series of molecules and spectra dramatically illustrates the versatility of function obtained by subtle modifications in the molecular design.

The large oscillator strengths of organics also lead to their highly efficient optical emission resulting from radiative recombination of excitons; the process opposite to exciton generation via absorption. In Fig. 1.9, we show several Ir-based organometallic compounds used in high efficiency, electrophosphorescent OLEDs, or PHOLEDs (Thompson et al., 2007). *Organometallic* refers to compounds that contain at least one bond between an organic *ligand*, or molecular grouping, that has coordinating bonds to a metal atom. The compounds in Fig. 1.9 are thus known as *coordination complexes*. The various compounds each

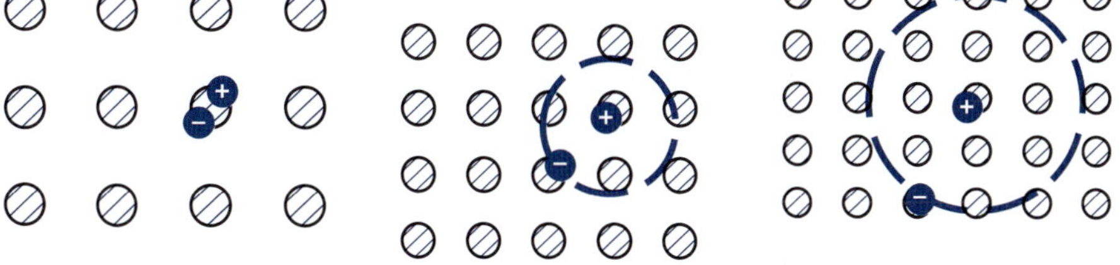

Figure 1.7 (a) Frenkel, (b) charge transfer (CT), and (c) Wannier–Mott excitons. The molecules or atoms are the hatched circles on square lattice sites, and the electron and hole are denoted by "−" and "+," respectively. The dashed circles are the approximate spatial limits of the corresponding excitons.

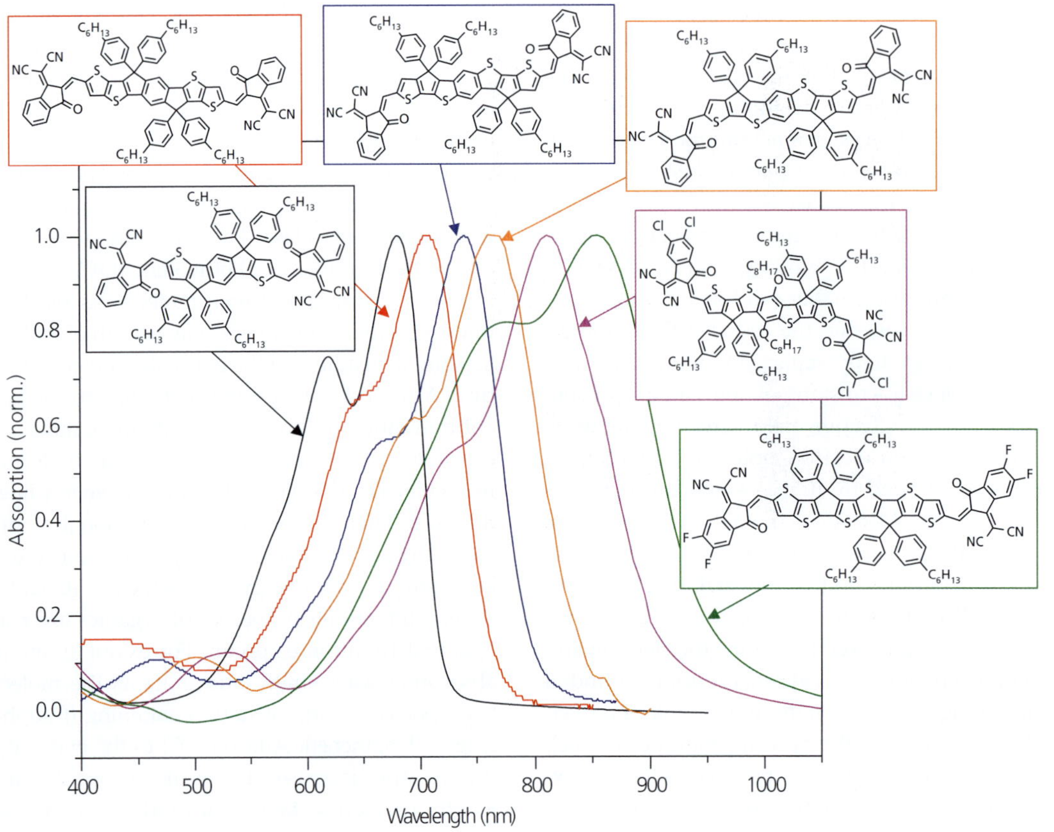

Figure 1.8 Absorption spectra of thin films comprising the several molecules shown in the insets. These molecules are employed in efficient organic photovoltaics, and are members of a class of non-fullerene acceptors.

emit at a different wavelength, with their characteristic narrow emission spectra shown in the figure. As in the compounds in Fig. 1.8, the spectra are almost rigidochromically shifted from the blue to the NIR, depending on the exciton energy of the ligand. The color coordinates of the various PHOLEDs are located on the Commission Internationale d'Eclairage (CIE) chromaticity chart. This collection of molecules, whose emission ranges from cyan to deep red, is another example of the power of ligand functionalization—that is, by making subtle changes to their composition, the emission wavelengths of this set of molecules can span the visible spectrum.

In contrast to the spectra in Fig. 1.8, inorganic semiconductors tend to be broadly but weakly absorbing, with absorption coefficients one to two orders of magnitude smaller than organics. The oscillator strengths in inorganic materials are lower since the interaction with a photon is spread over a large number of delocalized electrons. Many common semiconductors such as Si, Ge, and GaAs have a black appearance due to their very broad bandwidths that stretch from their band edge in the NIR and across the visible, although wider band gap semiconductors such as GaN are transparent since they only absorb from the blue into the ultraviolet (UV).

We noted above that the weak van der Waals bonds give molecular solids and polymers their soft mechanical character, opening up vast new opportunities for applications that are inaccessible to conventional, inorganic semiconductors. For example, organic solids can be flexible, whereas inorganic semiconductors are often very hard, brittle and subject to fracture. While organics are soft, and hence easily damaged by handling, they are also bendable, can conform to any number of oddly shaped surfaces, and can even be folded without cracking.

In this context, one of the drawbacks of weak intermolecular bonds is that the materials are prone to developing structural defects such as dislocations, and the incorporation of interstitial impurities during growth. Hence, it is challenging to grow large single crystals or thin films with low defect densities, as is commonly achieved in inorganic semiconductors. As

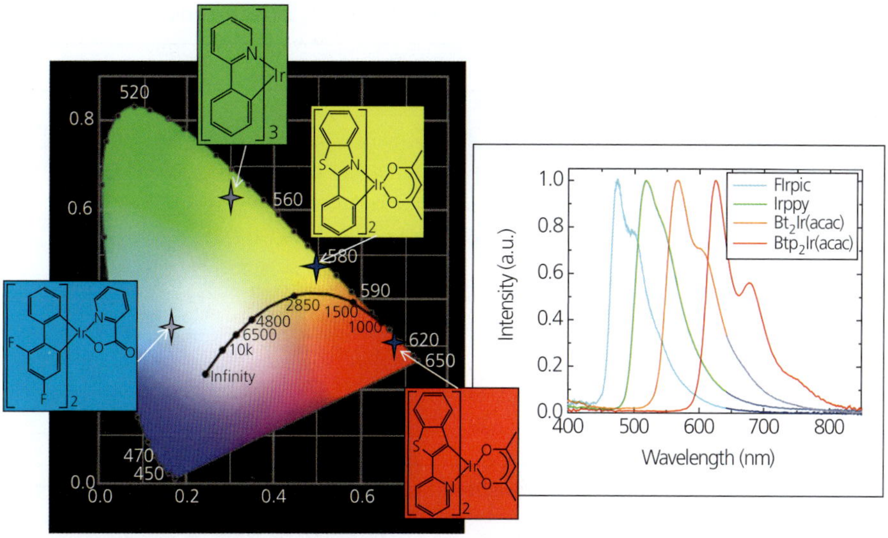

Figure 1.9 Several Ir-based phosphorescent compounds used in OLED displays along with their color coordinates on the chromaticity chart (stars on graph at left), and the emission spectra of the molecules when used in phosphorescent OLEDs (PHOLEDs). Small changes in molecular structure result in large changes in emission properties. The subscripts for each compound indicate the number of identical ligands enclosed in the brackets that are coordinated with the central Ir atom. These ligands are distributed around the three-dimensional space surrounding the Ir atom. Courtesy, M. E. Thompson, 2019.

an example, 450 mm diameter single crystalline Si boules have been produced, whereas single crystals of organics are generally no larger than 10 mm on a side. The lack of chemical bonds also allows for ingress of foreign atoms and molecules, causing rapid oxidation, hydration or attack by solvents. Finally, the relatively weak intramolecular bonds can be easily dissociated by photophysical (e.g. exposure to high energy radiation such as UV light), chemical, photochemical (e.g. oxidation that is accelerated by simultaneous exposure to oxygen and light), or thermal means. For example, the C=C bond energy, one of the strongest in an organic molecule, is 602 kJ/mol, whereas for Si–Si it is 222 kJ/mol. However, in a Si crystal, each Si atom is bonded to four neighbors, resulting in a total bonding energy of nearly 900 kJ/mol. The relatively weak bonds in organic molecules are vulnerable to bond rupture and other routes to decomposition. The film crystalline morphology is similarly vulnerable to environmentally-induced changes. For this reason, organic electronic devices must be protected from exposure to atmosphere and high energy radiation if their long-term operational stability is to be assured.

An interesting feature of organics is the pronounced asymmetry of their properties along different crystalline directions. For example, intermolecular van der Waals bonds provide weak adhesion between polymer chains, although along the chains, strong chemical bonds give the material properties that are more band-like. Similarly, small molecules are often asymmetric, with very different stacking habits parallel and perpendicular to a particular molecular axis. Hence, very large differences in refractive index, mobility, conductivity, and absorption (i.e. birefringence) are found along different crystalline directions. Even their stress tolerance has a well-defined directionality for ordered polymer or molecular solids (Forrest, 1997).

A qualitative summary of some of the attributes of organics compared to inorganic semiconductors is given in Table 1.2. This table provides a partial answer to the question, "What are organic semiconductors good for?" A positive attribute is noted by a "+," a very positive attribute by multiple "++," whereas negative and neutral attributes are noted by "−" and "0," respectively. Heading the list of differentially positive attributes for organics is their ability to be deposited on very large area substrates at low cost. Hence, organic semiconductors are ideally suited to fill the niche of "big devices" such as displays, lighting and solar cells whose performance and attractiveness scale with size. Inorganics fill the niche of the "very small." Their enablement of enormously complex circuits containing billions of transistors per square centimeter lies at the foundation of modern electronic technology. This is a domain that is completely unsuitable for organic electronic circuits whose smallest components (e.g. an OLED pixel in a high definition display) are at least three orders of

Table 1.2 Qualitative comparison of attributes of organic and inorganic semiconductors when applied to electronic devices

Attribute	Inorganics[a]	Organics
Large area	−−−	+++
Cost	−−	++
Complexity	++	0
Green processing	−−	+
Easy to pattern	+	0
Tunable properties	0	++
Optical absorption	−/+	++
Optical emission	−/++	++
Conductivity	+	−−
Device reliability	++	−

[a] "−" = negative, "+" = positive, and "0" = neutral attribute of a material. Repetition of a symbol indicates a higher emphasis on this attribute, either positive or negative.

magnitude larger than transistors in commercial circuits with dimensions of 5–10 nm. For this same reason, the "complexity" of organic electronic circuits is far lower than that of inorganics. In Chapter 8 we will see that organic transistor circuits comprising several thousand devices have been demonstrated, yet this pales by comparison to today's computer central processing units (CPUs) with 5 billion transistors, or memories with 10–100 times that many devices. Of course, complexity is only a virtue in applications where it is needed. Hence, if we adhere to the principle that we only use technologies in applications that derive benefit from their use (i.e. we do not want to provide a solution in search of a problem), then large area devices such as displays and solar cells do not obviously improve simply by making more complex, and smaller pixels or device structures.

Large area devices are also ubiquitous devices. Solar cells, displays, lighting, and medical electronic devices (the latter provides a potential niche for organic transistor circuits), are found virtually in every corner of the globe, in almost every household and on every electronic appliance. In producing, using and ultimately disposing of these devices we must be concerned with their environmental impact. As noted above, the low bond energies of organics leads to low processing temperatures in their deposition and film formation processes. Hence, they can be produced at an energy expense far below that needed for strongly bonded inorganic substances such as Si or GaAs, the two most common inorganic electronic materials. Furthermore, the materials themselves are relatively non-toxic compared to, for example, perovskites and CdTe used in solar cells due to their heavy metal content, unstable bonds, etc. Hence, at the end of their useful life, organic devices can be disposed of, and even recycled with a relatively small negative impact on the environment. Of course, plastics provide an enormous and growing environmental threat due to their lack of biodegradability. Generally, these types of plastics are not electronic materials. However, they are used as substrates for organic electronics, and in packaging of conventional electronic devices. This environmental threat requires immediate attention, and represents a challenge that extends far beyond the boundaries of electronic technology alone.

Circuit fabrication requires patterning that extends from the macro (i.e. >1 m^2 substrates) to the nanometer scales. As we will discuss in Chapter 5, device patterning is primarily based on photolithography. This requires the deposition of a thin, photosensitive polymer resist onto the circuit surface, followed by selective optical exposure and removal of the resist in areas where it is not needed. The resist itself is deposited, developed and stripped using a combination of solution and dry processing steps. A typical electronic circuit may require 50–100 applications, exposures and removals of resist during a fabrication sequence to create the numerous device layers and patterns required by the circuit. Unfortunately, organic materials can be damaged by solution-based photoresist processing. Also, there is often little "materials contrast" between an organic polymer resist and the organic active layer. This makes it difficult to selectively remove a resist from an underlying electronic layer based on chemical differences. For these reasons, patterning of organic devices creates challenges that are absent in conventional electronic device fabrication. Patterning techniques such as direct printing, stamp transfer, laser patterning, etc. have been developed to accommodate the material vulnerabilities of organics. These techniques, however, are not nearly as well developed nor as versatile as is photolithography. This has proven to be an impediment to making organic circuits whose complexity is not anywhere near that of inorganic semiconductors. Of course, we have already noted that large area electronics generally do not need the same level of precision in features and complexities required for nanometer scale transistor circuits.

The entry in Table 1.2 entitled "tunable properties" refers to the facility with which a material or structural composition can be changed to meet a specific requirement of the target application. The vast flexibility in properties easily accessed through the synthesis of organic compounds precisely designed to meet an application need is a truly unique advantage of organic electronics; a field founded on the extraordinary versatility of organic chemistry. The solids themselves are soft, allowing for their layering into a variety of structures, albeit creating films that have a large degree of structural disorder. The situation is considerably different for inorganic semiconductors. Their stiff lattices require that a layer of the desired composition has a lattice structure that is matched in symmetry and size to that of the substrate on which it is grown. Hence, inorganic alloy semiconductors are constrained to lattice-matched, or to only minimally strained systems, such as GaAs/AlGaAs, GaN/InGaN, or InP/InGaAs(P). Engineered, quantum confined structures expand the range of properties and materials used, but ultimately these constraints are far more limiting than those hindering organic materials.

We have already mentioned that optical absorption and emission are exceptional strengths of organic materials due to their excitonic nature. But what about inorganics? The most pervasive of all semiconductor materials is, of course, Si. While this material is ideal for electronic applications, it is a poor optical material due to its indirect band gap. Direct absorption, particularly at energies near the energy gap of $E_G = 1.1$ eV, is absent. Thus, optical detectors must have thick absorption regions to yield a high external quantum efficiency, that is, the ratio of photogenerated electron–hole pairs collected at the contacts to the number of photons incident on the photodiode. We will find in Chapter 7 that this forces a compromise between quantum efficiency and detector bandwidth that can be surpassed by organic materials. Nevertheless, if bandwidth is not an issue, the perfection achieved in Si purity and device performance can result in near unity quantum efficiency in solar cells and detectors. The indirect band gap of Si prohibits its use as a light emitting diode or laser.

Direct band gap materials such as GaAs, GaN, and InP have excellent optical properties, although once again the range of wavelengths accessed with ease is not as great as for organics. Yet, fiber optic communication is based on extremely high efficiency light emission from InP-based lasers, and high bandwidth/high quantum efficiency detectors using this same materials system. Similarly, efficient and long-lived lighting based on InGaN technology is widespread. While inorganic semiconductors have inherent materials limitations that are absent in organics, time and again inorganics have been tailored to meet a diversity of optical device applications at the highest device performance.

Low charge mobility results from electronic states localized on individual molecules, with very limited overlap with adjacent molecules in the solid. As noted above, the conduction is via incoherent hopping from molecule to molecule. In a device, this property manifests itself as a very high resistivity. And high resistance, in turn, limits the current that can be transported in transistors, light emitters, and so on. Additional consequences of resistive materials are a low bandwidth response to optical or electrical signals, as well as Joule heating that can reduce the device lifetime or increase energy consumption. These deficiencies are not nearly as pronounced in inorganic devices due to their band transport nature, leading to charge mobilities orders of magnitude larger than for organics. Note, however, that the low bandwidth and current carrying capacity of organics can be partially mitigated by keeping molecular layers thin. The very high absorption coefficient of organics requires layers that are sometimes less than 10% of the thickness used in inorganic photodiodes with similarly high quantum efficiencies. Further, the emission efficiency in PHOLEDs routinely approaches 100% from layers that are <5 nm thick. Hence, while high resistance is undoubtedly a drawback, device and materials designs have been advanced that minimize the disadvantages that this property would otherwise engender.

The last entry in Table 1.2 brings up the topic of reliability. The weak chemical bonds in many compounds lead to their rapid decomposition under heat, atmosphere and/or light. Hence, organic materials have the reputation of being unstable. It is logical to expect that the devices employing such compounds will have an unacceptably short operational lifetime. While it is generally true that organic devices can be highly unstable and thus impractical for use, it is also true that by using high purity, state-of-the-art materials in an appropriate package that prevents contamination from the environment, very long lifetimes can be achieved. In Chapter 6 we will show that OLEDs can have lifetimes extending to millions of hours, and OPV cells can possibly operate for tens to thousands of years. While reliability is an over-riding concern for all organic devices, there are many examples where this problem has been solved or reduced to provide electronic appliances that are as reliable as their inorganic semiconductor analogs. Ultimately, the standard for acceptable reliability is defined by the target application. A thermally unstable material cannot be employed in a device whose application demands use at elevated temperatures. Yet that same device may fill a niche where high temperatures are never experienced. In Part II, the issue of device reliability, and how to quantify it is addressed for OLEDs, OPVs and organic thin film transistors (OTFTs). We will find in each case that acceptable reliabilities have been achieved, although organic devices are rarely if ever as robust as those based on Si or alloy semiconductors employed in high bandwidth, high temperature applications.

1.3 70 years of advances in organic electronics

One cannot understand where one stands today if one is unaware of the advances made in the past. To fully appreciate the subject of organic electronics it is, therefore, worthwhile to briefly review its history, and some notable discoveries made along the way that led to its current state of advance. But the question is, where to start? Not with the synthesis of urea by Friedrich Wöhler in 1828. While that discovery arguably marked the foundation of the field of organic chemistry, a discipline that underpins every advance made in organic electronics, urea is not an electronic material, nor was it ever identified as such.

With this criterion in mind, I somewhat arbitrarily choose the foundation of the field of organic electronics by the discovery of semiconducting properties of violanthrone by Akamatu and Inokuchi (1950). Indeed, Hiroo Inokuchi is often regarded as the "father of organic electronics." Akamatu and Inokuchi measured the resistivity of this molecular solid, which was found to decrease with increasing temperature. The thermal activation of the resistivity has a characteristic energy, $\Delta\epsilon$, that these investigators correctly identified as twice the energy gap of the material. This advance, more than any that had preceded it, clearly assigned semiconducting properties to an organic material, launching the field of organic electronics.

A timeline showing selected milestones in the subsequent development of the field is shown in Fig. 1.10. This timeline is admittedly brief, and only shows a few such milestones. I apologize in advance for all the important (even momentous) discoveries that have been omitted. Specific references made in the figure are provided in Table 1.3, listing seminal papers that have forever changed this exciting, and still evolving interdisciplinary field of science and technology. In the remainder of this section, we will briefly note the

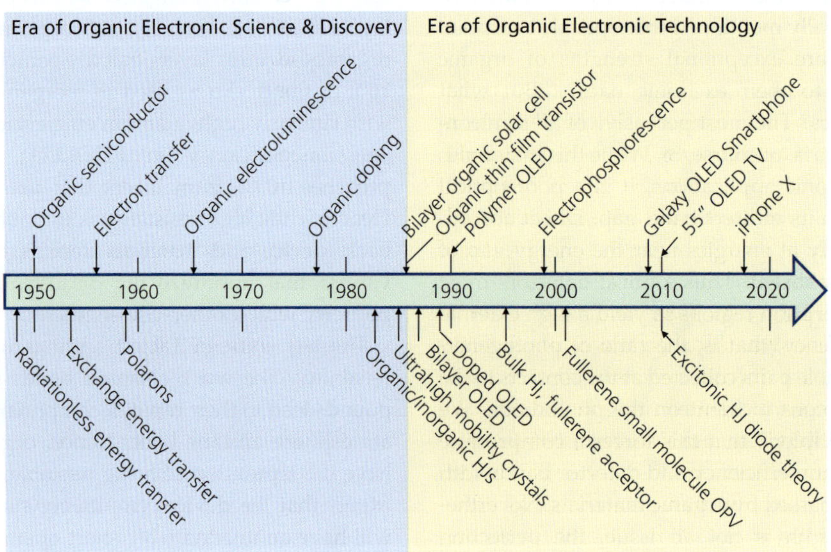

Figure 1.10 Historical timeline of organic electronics, highlighting a few significant milestones met along the way.

Table 1.3 Several discoveries and innovations made during the 70 year history of organic electronics

Discovery/advance	Date	Type[a]	Reference
Radiationless energy transfer	1948	F	(Förster, 1948)
Organic semiconductors	1950	F	(Akamatu and Inokuchi, 1950)
Exchange energy transfer	1953	F	(Dexter, 1953)
Theory of electron transfer	1956	F	(Marcus, 1956)
Polaron theory	1959	F	(Holstein, 1959a)
Organic electroluminescence	1965	F	(Helfrich and Schneider, 1965)
High conductivity doped polymers	1977	F	(Shirakawa et al., 1977)
Organic/inorganic junctions	1982	F	(Forrest et al., 1982)
Mobility of ultrapurified organics	1985	F	(Warta et al., 1985)
Organic solar cell	1986	T	(Tang, 1986)
Polymer TFT	1986	T	(Tsumura et al., 1986)
Bilayer OLED	1987	T	(Tang and VanSlyke, 1987)
Doped OLED	1989	T	(Tang et al., 1989)
Polymer OLED	1990	T	(Burroughes et al., 1990)
Bulk heterojunction	1995	T	(Halls et al., 1995; Yu et al., 1995)
Fullerene acceptors	1995	T	(Yu et al., 1995)
Electrophosphorescence	1998	T	(Baldo et al., 1998)
C_{60} acceptor	2001	T	(Peumans and Forrest, 2001)
i7500 Galaxy phone AMOLED display	2009	T	
Diode theory of organic junctions	2010	F	(Giebink et al., 2010b)
55″ OLED TV	2012	T	
iPhoneX AMOLED display	2018	T	

[a] Type refers to either fundamental (F) or technological (T).

specifics of these milestones, leaving more detailed descriptions for the reader to discover through the papers themselves, and in the remainder of this book where much of the important content of the papers can be found.

An interesting feature to note in Table 1.3, is that among the various discoveries and innovations, there is a dramatic shift between a focus on fundamental discoveries (denoted by "F" in the "Type" column) and those that are primarily technological (denoted by "T"). This change abruptly occurred with the demonstration of the first efficient, all-organic bilayer rectifying solar cell by C. W. Tang, at Kodak, in 1986 (Tang, 1986). This was soon followed by the announcement of a 1% quantum efficiency, low voltage (~10 V) bilayer organic light emitting device, later to be named an OLED (Tang and VanSlyke, 1987). In the midst of these announcements came that of the first organic polymer thin film transistor based on polythiophene (Tsumura et al., 1986). After these innovations, the organic electronics community turned much of its attention from past pursuits aimed at understanding the fundamental properties of organics, toward developing these materials for use in what was apparently becoming a practical device technology. This is not to say that all important work was purely fundamental prior to these seminal discoveries, nor that afterwards only technological innovations were sought. Far from it. Indeed, such a broad categorization of any scientific or engineering work into fundamental or applied is admittedly arbitrary. Yet there is no denying that once an organic rectifier, a light emitter and a transistor were demonstrated to have potentially practical performance characteristics, that the attention of the scientific community swung hard toward the pursuit of even higher performance that would eventually fulfill the long-held hope that low cost and high performance devices based on soft, organic electronic materials would catalyze new global industries, realizing applications that were otherwise inaccessible to conventional semiconductors.

Given this unique turn of events, the historical timeline in Fig. 1.10 has been, again somewhat arbitrarily, divided into two periods. The first is the "Era of Organic Electronic Science and Discovery," and the second is the "Era of Organic Electronic Technology," neatly divided in 1986 with the invention of the bilayer organic solar cell.

We begin our timeline two years before the landmark work of Akamatu and Inokuchi, with the theory of electrostatic energy transfer elucidated by Förster (1948, 1959). While the theory was derived with the purpose of describing energy transfer in photosynthetic complexes, it is now widely applied to describe energy transfer and exciton diffusion in organic electronic solids in general. The transfer process is known as *fluorescent resonant energy transfer*, or FRET (also known as *Förster resonant energy transfer*). While FRET occurs over distances of a few nanometers via dipole-dipole coupling between molecules, a tunneling-like transfer dominates at shorter distances. This is known as *exchange energy transfer*, or *Dexter energy transfer*, named after the person who first derived its properties in a landmark paper in 1953 (Dexter, 1953).

The process of charge transfer from molecule to molecule that underlies hopping conduction was developed by Rudolf Marcus, who won the Nobel Prize in Chemistry in 1992 for his seminal contributions. Known as *Marcus electron transfer*, it describes the general process of charge transfer in organic electronic materials, although it was first developed to understand charge transfer in chemical reactions (Marcus, 1956, 1957). This deceptively simple theory predicts unusual behaviors of electron transfer in molecular systems such as the so-called Marcus inverted region that governs charge transfer from the donor to acceptor molecules in organic solar cells. A natural extension of Marcus theory is the concept of the small *polaron* as the charge carrier in organic solids. An electron or hole within an organic significantly polarizes the surrounding medium, creating a shallow self-trapping site. Thus, the charge finds itself heavily "dressed" by the molecules in its immediate vicinity, taking on a large effective mass. This is a very different situation than exists in inorganic semiconductors, where a high density of surrounding, mobile charges screen the effects of a free charge, significantly limiting its polarization of the medium. The concept of the polaron, therefore, is central to our understanding of hopping transport that leads to low charge mobility in organic materials. This concept was quantitatively described by Holstein in a pair of papers in 1959 (Holstein, 1959a, 1959b).

Demonstrations of organic electroluminescence in anthracene (Pope et al., 1963, Helfrich and Schneider, 1965) introduced a new phenomenon to the organic electronics community. Injecting charge at high voltages (100 V to >1000 V) from aqueous electrodes into anthracene crystals produced faint blue emission. This was distinct, however, from work done 20 years later by Tang and co-workers, who employed very thin film bilayer devices that operated at very low voltages, and that emitted bright green light at much higher efficiencies.

Shirakawa, Heeger, and MacDiarmid were the first to electronically dope an organic polymer to vastly alter its conductive properties (Chiang et al., 1977, Shirakawa et al., 1977). This team doped the insulating polymer, *trans*-polyacetylene, with AsF_5 to increase its conductivity from 10^{-5} S/cm to > 100 S/cm. Doping is essential in fabricating semiconductor junctions by converting intrinsic semiconductors to either a p- or n-type conductor. Likewise, the ability to dope an organic semiconductor opened the possibility that their conductive properties, too, would usher in new generations of organic electronic devices. This team was recognized for its contribution by receiving the Nobel Prize in Chemistry in 2000.

The rectifier is the most basic, active (i.e. nonlinear) electronic component. Considerable research into the mid-1980s was therefore directed at demonstrating such a device using organics. Possibly the first such successful device was the demonstration of highly asymmetric current characteristics in the forward vs. reverse biased directions obtained by layering an organic semiconductor onto the surface of a Si wafer (Forrest et al., 1982). Today, organic-inorganic junctions are widely investigated for their use in QD solar cells, nonlinear optical devices, and as platforms for multiexciton generation in solar cells (Dexter, 1979). Interestingly, this organic–inorganic rectifier was the first device to employ the molecule, PTCDA, as an electronic material. As we will see in subsequent chapters, PTCDA is a remarkable material given its excellent film-forming capabilities, large and anisotropic charge mobility, and closely coupled electronic states from which an incipient band structure emerges,. Indeed, it has become one of only a handful of model materials broadly studied to understand the properties of organic semiconductors.

It had long been thought that organic materials have very low charge mobilities. Thus, the work of Warta, Stehle, and Karl made a profound impact on our understanding of organics with their work on ultrapurified naphthalene and perylene (Warta et al., 1985). They found that by zone refining, mobilities of purified crystals of several hundred cm^2/V s can be obtained at low temperatures. Mobilities this high clearly suggest band transport, a property that was thought to be the sole domain of inorganic semiconductors. But that preconception changed due to these experiments. To this day, this work stands alone as having reported the highest mobilities for organic systems, largely due to the careful removal of impurities from the nearly perfect molecular crystals.

The work of Warta et al. marks the end of the Era of Organic Electronic Science and Discovery, since it was followed a year later by the landmark papers from Kodak on bilayer junction solar cells and OLEDs. One additional important, later contribution from the Kodak group was the demonstration of doping of the emission region of an OLED with a fluorescent lumophore (Tang et al., 1989). It was discovered that the efficiency (and ultimately the lifetime) of OLEDs can be increased by including only a few percent of the luminescent molecule within a conducting, wider energy gap matrix, whose combination forms the emission layer. In a doped device, excitons are formed on the matrix material by electrical injection, and are then transferred with near unity efficiency via FRET to the lumophore. The lumophore itself radiates with higher efficiency than if it were used as a neat layer in the OLED due to the elimination of *concentration quenching*: the process whereby an excited state on a molecule is non-radiatively quenched due to vibronic coupling to neighboring molecules. The excited state is readily quenched by contact between similar molecules that dissipates by coupling to inter- and intramolecular vibrational modes. Isolation of the individual fluorescent molecules traps the excited state on the luminescent species, reducing its ability to dissipate energy. Hence, the demonstration of doping was an important step toward realizing efficient OLEDs.

The work of Tang and VanSlyke focused on vacuum deposited small molecule devices. However, an unfulfilled promise of organic electronics was that solution deposition can significantly simplify the fabrication of device into a one-step, one beaker procedure. The demonstration of the first solution-processed polymer OLED was a step toward realizing this dream (Burroughes et al., 1990). The active material used in this single layer device was poly(phenylenevinylene) (PPV). Its yellow emission exhibited only 0.05% external quantum efficiency, although it operated at reasonably low voltage and demonstrated how a simple, solution-processed device might eventually become a practical light source.

Polymers were soon shown to have uses in solar energy harvesting with the invention of the bulk heterojunction (BHJ) solar cell.(Halls et al., 1995, Yu et al., 1995) The BHJ comprises a bicontinuous, or interpenetrating network of donor (D) and acceptor (A) materials that creates a large surface area for exciton dissociation at the D–A junction. By blending these materials excitons dissociate into free charges very near to their point of origin (i.e. the point of photon absorption), thereby decreasing losses from exciton diffusion that are experienced in bilayer junctions. The BHJ is now found in many high efficiency OPVs. An additional advance by Yu et al. was the introduction of a solution-processed fullerene acceptor, $PC_{61}BM$, into the device active region (Yu et al., 1995). The rapid charge transfer from a donor to the fullerene with its high thin film electron mobility has since led to its widespread use in OPVs.

Fluorescent OLEDs are limited through spin statistics to only 25% internal quantum efficiency. The internal efficiency is the ratio of the number of photons emitted to the number of electron–hole pairs that undergo radiative recombination within the device. The injection of electrons produces a random population of spins, resulting in the formation of singlet and triplet excited states at a statistical ratio of 1:3 within the emission layer of the OLED. Unfortunately, only singlet states are quantum mechanically permitted to radiatively recombine on fluorescent molecules, thus wasting the energy used to generate the 75% triplet population (see Chapters 3 and 6). In 1998, a solution to this problem was offered. Employing metalorganic molecules comprising a heavy metal atom such as Pt or Ir couples the spin on the luminescent molecular ligand with the orbital angular momentum of the heavy metal atom. Using a heavy metal with a large electronic orbital momentum ensures strong mixing of the singlet and triplet states (Baldo et al., 1998, 1999). This so-called *spin–orbit coupling* allows the singlet state to rapidly cross to the triplet state manifold on the ligand, and for the triplet to become radiative. This enabled what is now known as *electrophosphorescence*. When heavy metal atoms are used in electrophosphorescent OLEDs, or PHOLEDs, the internal quantum efficiency increases to 100%. This innovation led to intense interest in PHOLEDs for use in high efficiency displays and lighting, giving birth to today's multibillion dollar OLED display industry.

A similar design used in PHOLEDs was then incorporated into a double heterojunction small molecule OPV, employing the first vacuum-deposited fullerene, C_{60}, in an organic solar cell (Peumans and

Forrest, 2001). This work, combined with that of Yu on solution-processed polymer photodiodes, was followed by a series of innovations on fullerene acceptor-based solar cells that has led to consistent improvement in efficiency persisting for over two decades.

No history of organic electronics would be complete without mentioning the singular events surrounding the largest fraud in physics, perpetrated by J. H. Schoen at AT&T Bell Laboratories, that extended from 2000 to 2001. The understanding of the properties of disordered soft materials is sufficiently incomplete that it allowed for a breathtaking abuse of scientific norms that were reported in a series of astonishing papers that regularly appeared in *Science* and *Nature* magazines, and a few other scientific journals. These papers appeared to literally rewrite all of our understanding of condensed matter physics. Over a period of approximately 2 years, startling discoveries were reported, ranging from organic lasers to extremely high performance monolayer OTFTs employing ultrahigh mobility materials. Almost all major phenomena known to condensed matter physics including the quantum Hall effect, the formation of Wigner crystals, and superconductivity were reported to have been observed in organic semiconductor crystals in a series of publications until, inevitably, it was discovered to be a fraud. Numerous papers and books have been written about this period, and its causes (Beasley, 2002, Reich, 2009). Had it not been for the emerging industrial success of OLEDs, it is plausible that the entire field of organic electronics could have been set back, or even irreparably damaged by the scale of this abuse. Also, too many verifiable and exciting results were emerging from other labs, admittedly not nearly as apparently profound as those from Bell Labs, that helped to prevent the damage of scientific trust from having too great an impact. Yet as is typical of all aspects of organic electronics, there is much to be learned from this experience by the larger scientific community. It opened up questions about the responsibilities, both technical and ethical, of collaborators when one of their own is serially, and somewhat openly abusing scientific norms. What is the role of the scientific supervisor and of the larger scientific community for calling out, and halting this abuse? What is the role of the journals to ensure that standards of careful scientific review are followed? These will remain open, but nevertheless essential questions that the Schoen affair has placed uncomfortably before the scientific establishment. In spite of the unprecedented scale of the fakery, the field of organic electronics was not mortally wounded, although many individual victims suffered from pursuing false leads and results created over the 2 years that Schoen was most active.

In spite of this setback, advances in organic electronic science and applications have been sustained to the present day. A pivotal event in its success was the launch in 2009 of the first major organic electronic product: Samsung's Galaxy smartphone series employing active matrix OLED (AMOLED) displays. Today, more than 2 billion such phones have been sold, along with OLED tablets, phablets, and so on, leading to a rapidly expanding US$20 billion market. Next, LG Electronics introduced 55″, 65″, and 77″ OLED TVs in 2012. Now these TVs are available in high definition versions, and work is progressing toward commercializing "roll-up" OLED TVs that can be stowed in a compact box. Most recently, Apple has entered the OLED display world by the introduction of the iPhoneX. Beyond OLEDs, major products are soon to emerge in lighting and organic solar cells. The future is certainly bright for organic electronics, following its long history of discovery and innovation.

One other fundamental advance worth noting that applies particularly to organic solar cells, was the derivation in 2010 of an ideal diode equation for excitonic heterojunctions. In 1949, William Shockley derived the first ideal diode equation for inorganic semiconductor p-n junctions (Shockley, 1949). As noted above, the rectifier is the most basic active element of an electronic circuit. The Shockley equation created the foundation on which semiconductor electronics has been built. However, we have seen that the fundamental physics of charge generation and transport is fundamentally different for excitonic semiconductors. Giebink and co-workers undertook the task of deriving a similar theory for organic junctions starting with the generation of an exciton in systems governed by the nearly isolated electronic orbitals of molecular systems (Giebink et al., 2010a, 2010b). This resulted in a theoretical framework similar to that of Shockley, but faithful to the underlying physics peculiar to molecular solids. This framework is helpful in understanding many important aspects of organic semiconductors, particularly in the context of photogeneration such as in predicting the thermodynamic limits to organic solar cell efficiency (Giebink et al., 2011). This work is an example of a fundamental discovery that has emerged well within the "Era of Organic Electronic Technology."

The foregoing provides but a few of the many examples of significant advances that have led to the current high state of sophistication in our understanding of the fundamental phenomena of soft organic materials. And this understanding has led to their exploitation in devices that are now enjoying widespread commercial success. It bears repeating that this historical account omits many important discoveries and innovations, as it is only meant to give a thumbnail sketch of a rich and vibrant history of this fascinating subject. Likewise, separating the timeline into two "eras" is in itself wholly artificial, yet it helps to clarify the arc of progress within the field, and it highlights some of the most transformational events within that history. Ultimately, this "division" has also informed the structure of this volume, dividing it between Part I: Foundations, and Part II: Applications.

1.4 Aligning on language: a few useful definitions

This is not a chemistry text. It is written almost entirely from a physics/engineering perspective that is reflective of the training of its author. But organic electronics is a deeply interdisciplinary subject, where chemists, physicists, engineers and materials scientists must all convene to effect the progress essential for new discoveries and innovations. It is nevertheless hoped that this text will be accessible to readers from all of these communities, independent of background in fields that lie beyond the scope of one's personal training. For this objective to be met, we must establish a common language. The goal of this section is to provide a handful of chemical terms that can serve as a useful reference when they are encountered in later chapters. No effort is made to be comprehensive, as there are abundant resources one can draw upon to learn more about organic chemistry. Here, we provide an introduction, with more specialized terminology defined in the text where it first appears.

Conjugation: Organic electronic compounds comprise conjugated electron systems, that is, those systems that share p-orbitals between neighboring atoms that merge to form a delocalized orbital both below and above the molecular plane. The delocalized electronic systems are known as π-orbitals, or π-bonds. They increase molecular stability by ensuring structural planarity. Conjugation results from resonant, alternating single and double C–C bonds that delocalize electrons along the molecular backbone.

Aromaticity: Aromatic compounds form a class of conjugated molecules that are highly stable, have excellent film forming characteristics when deposited in vacuum, are generally insoluble, and are commonly found in organic electronic devices. An aromatic compound is one that has an uninterrupted cloud of π-electrons that are located above and below a planar core. The π-orbital must contain an odd number of pairs of π-electrons. This is referred to as *Hückel's rule*, where the π orbital of an aromatic must contain a total of $4n+2$ π-electrons, where n is an integer. A subclass of aromatics are the linear polyacenes of benzene (a single C ring with $n = 1$), naphthalene ($n = 2$), anthracene (3 fused rings) and so on, with examples shown in Fig. 1.11. These constitute a homologous series of fused ring, planar molecules containing only C and H atoms. For example, tetracene (compound **4**) has nine pairs of π electrons (count the double bonds in the figure), or 18 such electrons, giving $n = 4$. This series of compounds is found throughout organic electronics, often used as archetypes to reveal physical or chemical attributes of a specific material of interest.

Linear polyacenes are members of a larger class of *polycyclic aromatic hydrocarbons* (PAHs). A few important examples of these planar, fused ring aromatics are shown in Fig. 1.12. PAHs can be found in nature, particularly in fossil fuel deposits. They are also products from burning organic matter, and are found in abundance in interstellar space, having been forged in the interiors of stars. Likewise, they are easily synthesized in the laboratory. Their stability and rich hues make them useful in clothing dyes, paints, and as ink

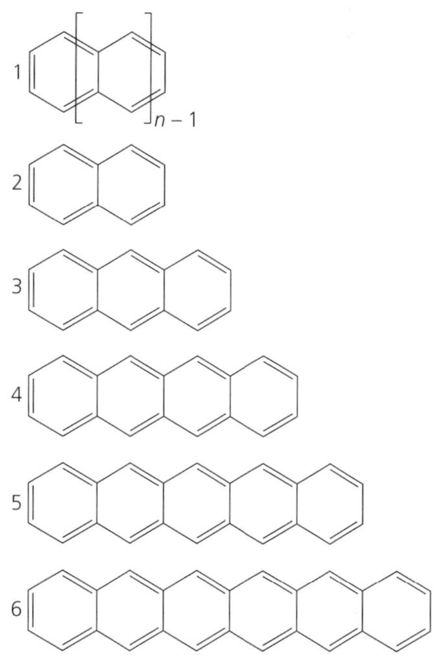

Figure 1.11 The six lowest order linear polyacenes. **1** = general formula for the linear polyacenes. $n = 1$ corresponds to benzene. **2** = naphthalene, **3** = anthracene, **4** = tetracene, **5** = pentacene, **6** = hexithene.

jet printing pigments. Like their linear analogs, PAHs are insoluble, but are sufficiently stable to be thermally evaporated (most PAHs sublime in vacuum on heating, although a few compounds first go through a liquid phase).

Functionalization: If the conjugated molecules and their analogs described above were the end of the story, the list of organic compounds used in organic electronics would indeed be short. However, the versatility of organics rests on their ability to be functionalized to gain attributes not present in the parent molecule. Functionalization is the substitution of a chemical group (a substituent) on a bond or bonds of a parent molecule to alter its characteristics. For example, the molecule benzene in Fig. 1.13 is functionalized by replacing one of its peripheral H atoms with an R-group. An R-functionalization implies the attachment of a non-aromatic, non-conjugated hydrocarbon. While it more loosely can include other atoms besides C and H, it formally applies to a series of C–H molecules known as *alkanes*, with the formula, $R = C_nH_{2n+1}$. A series of lowest order alkanes is shown in Fig. 1.14. The alkanes are also products of fossil fuels, and are

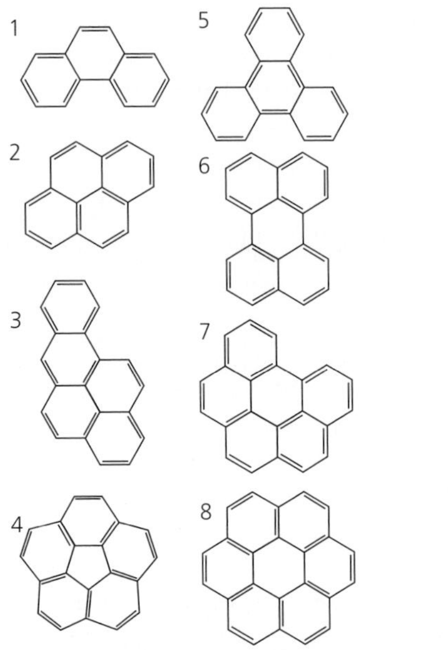

Figure 1.12 A series of polycyclic aromatic hydrocarbons. **1** = phenanthrene, **2** = pyrene, **3** = benzopyrene, **4** = corannulene, **5** = triphenylene, **6** = perylene, **7** = benzoperylene, **8** = coronene.

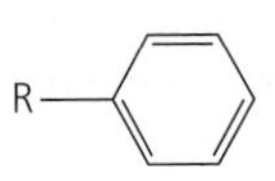

Figure 1.13 Benzene functionalized by substitution of an H atom with an R-group.

almost all found in the liquid or gas state at room temperature. Alkanes are most often attached to PAHs and other molecules to make them soluble in organic solvents.

Another functionalizing substituent is the Ar- or *aryl group*. This is a group derived from an aromatic ring. Most commonly, Ar- refers to a phenyl (i.e. benzo) group, but it can also refer to naphthyl derived from naphthalene, or xylene, pyrene, etc. Hence, an Ar group is one that derives from an aromatic hydrocarbon parent.

The last functionalizing group to be considered comprises heterocyclic substituent molecules. These are aromatic rings where one (monocycle) or more (polycycle) carbon atoms are replaced by other atomic species, such as O, S, N, etc. These heterocyclic additions can introduce substantial changes in molecular character, and hence are pervasively used to modify molecules employed in organic electronics. Figure 1.15 provides a reference of many common heterocyclic substituents found in organic semiconductors.

Ligand: In inorganic chemistry, a ligand is a functional group that is attached to a metal atom (e.g. Al, Cu, Ir, Pt, etc.) to form a *coordination complex*. Metalorganic compounds are ubiquitous in organic electronics. For example, PHOLEDs employ such compounds to induce spin-orbit coupling via metal-ligand charge transfer between a heavy metal atom and the light emitting ligand. Also, the metal complex, Alq_3, was used in the first bilayer OLED due to its intense green fluorescence when electrically excited.(Tang and VanSlyke, 1987) One or more ligands can coordinate with the metal atom, depending on the valences of the atom. The metal-ligand bond can range from the purely ionic to covalent. In Fig. 1.16 we show the molecular formula of $Ir(ppy)_3$ comprising three identical ligands attached to an Ir core. This is known as a *tridentate* molecule (having 3 ligands, or "teeth") that is *homoleptic* (all ligands are identical). In contrast, $Ir(ppy)_2acac$ is a *bidentate, heteroleptic* molecule. Both are phosphors that emit in the green, although $Ir(ppy)_2acac$ is by far the most efficient of the pair, having been used in PHOLEDs with 100% internal quantum efficiency (Adachi et al., 2001). The acetylacetate (acac) ligand, is a non-emissive *ancillary ligand* that is attached to satisfy the bonds of the Ir atom. The general formulation of these molecules is IrL_3 for the homoleptic compounds, and IrL_2X, for heteroleptic molecules with one ancillary (X) ligand.

Porphyrins and phthalocyanines: These two classes of compounds have optical and electronic properties

Name	Formula	Structure	State (at room temp.)
Methane	CH_4	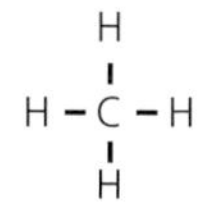	Gas
Ethane	C_2H_6	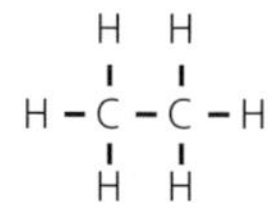	Gas
Propane	C_3H_8	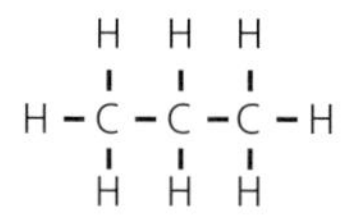	Gas
Butane	C_4H_{10}	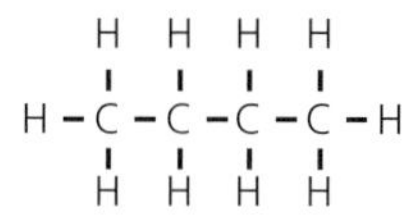	Gas
Pentane	C_5H_{12}	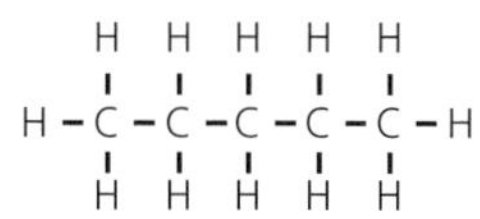	Liquid
Hexane	C_6H_{14}	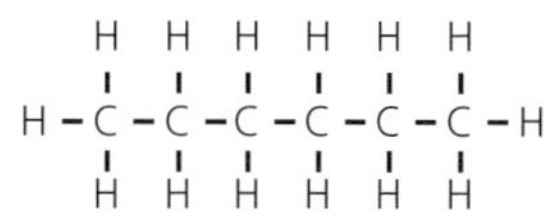	Liquid

Figure 1.14 Lowest order alkane molecules, from $n = 1$ (methane) to $n = 6$ (hexane).

that are commonly found in devices ranging from light emitters, to optical detectors, and transistors. A porphyrin is a heterocyclic compound comprising four pyrrole subgroups. The heterocycle consists of 26 π-electrons, but only 18 of those electrons form a continuous, conjugated path around the periphery of the molecule, as shown for PtOEP in Fig. 1.17. The parent porphyrin, comprising two H atoms coordinated with the pyrrole groups is known as porphine. When the H atoms are replaced by a metal coordinated to the pyrroles, the molecule is known as a porphyrin. These molecules are characterized by rich colors and crystal packing arrangements that can lead to high charge mobilities along the resulting discotic stacks. *Discotic* refers to the planar, disc-like molecules that can stack like poker chips to form continuous conducting pathways due to overlapping π-electron systems on adjacent molecules. Porphyrins are found in biological systems. Examples are Mg porphyrin which is a light absorbing molecule in photosynthetic complexes, and Fe porphyrin (so-called heme) found in red blood cells as a part of the hemoglobin protein. In organic electronics, heavy metal porphyrin complexes, particularly those containing Pt, are used as highly efficient phosphor dopants in organic electronics. Indeed, PtOEP was used in the first PHOLED, efficiently emitting in the deep red (Baldo et al., 1998). Similarly, tetraphenyl porphyrin (TPP), which lacks a heavy metal atom, is an efficient fluorescent molecule also emitting in the deep red (Shen et al., 1997). An attractive feature of both the porphyrins and phthalocyanines is that their optical properties are strongly affected by the choice of coordinating metal. Hence, minor chemical adjustments can be made to result in metal porphyrin phosphors that narrowly emit from the NIR into the blue spectral regions.

The phthalocyanines (Pcs) are similar to porphyrins in that they form a class of highly stable, easily deposited compounds that are deeply absorbing across the visible, depending on the transition metal used to coordinate between the four, peripheral isoindole groups. Like the porphyrins, 18 π-electrons make up an extensive, conjugated macrocycle. The parent, or

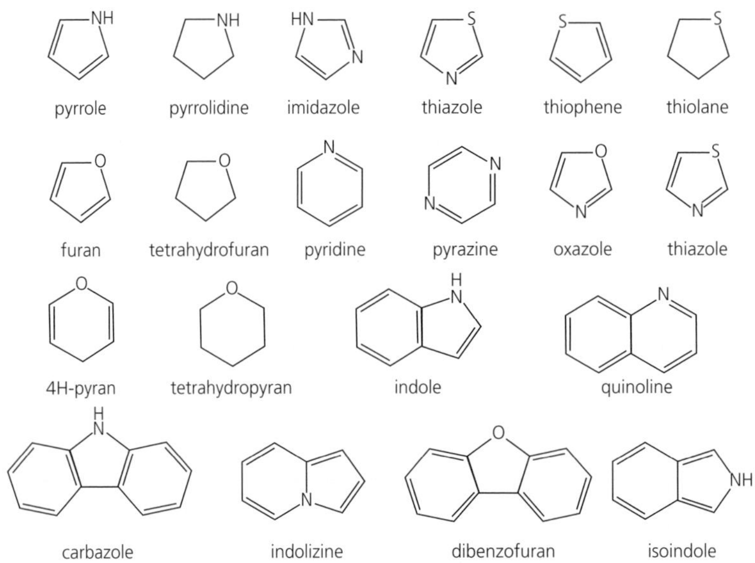

Figure 1.15 Common heterocycles found in organic electronic molecules.

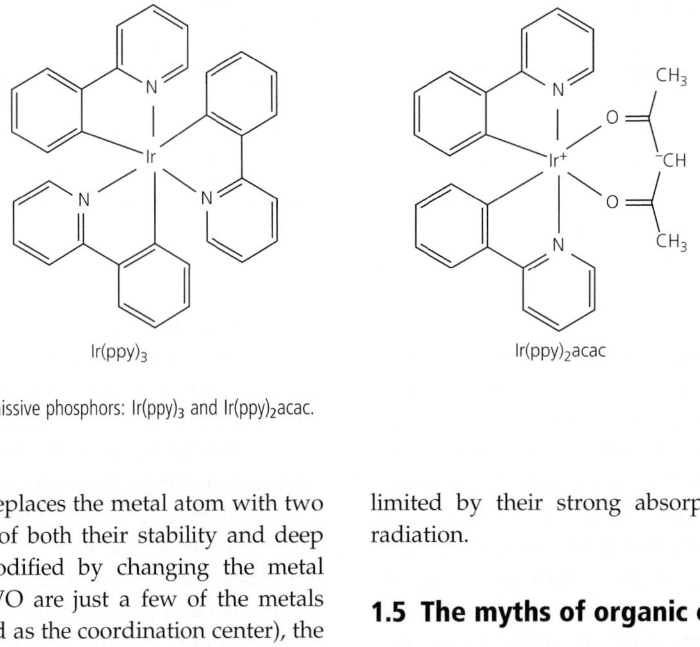

Figure 1.16 Two green emissive phosphors: $Ir(ppy)_3$ and $Ir(ppy)_2acac$.

free-base Pc, H_2Pc, replaces the metal atom with two H atoms. By virtue of both their stability and deep coloration easily modified by changing the metal atom (Cu, Zn, Pb, VO are just a few of the metals and compounds used as the coordination center), the phthalocyanines are used in dyes and even in hydrodynamic therapy. CuPc shown in Fig. 1.17 absorbs in the red, and hence appears to be blue-green to the eye. Indeed, CuPc was used as a donor molecule in the first bilayer OPV, exhibiting a power conversion efficiency of approximately 1% under simulated solar illumination (Tang, 1986). Their excellent film-forming characteristics have also led to uses in transistors. Much less frequently, metal Pc (MPc) molecules have been incorporated into OLEDs as contact layers, although this application has been limited by their strong absorption of the emitted radiation.

1.5 The myths of organic electronics

Since the earliest years of its development, myths have swirled around unfounded suppositions about the physical properties, and capabilities of organic electronics. After all, organics could not possibly be useful in practical electronic devices since they are inherently unstable, right? We know that living organisms that are comprised of organic molecules eventually die, they ordinarily cannot be subjected to high temperatures, and they cannot be exposed to bright sunlight for any length of time without degrading. All of these preconceptions,

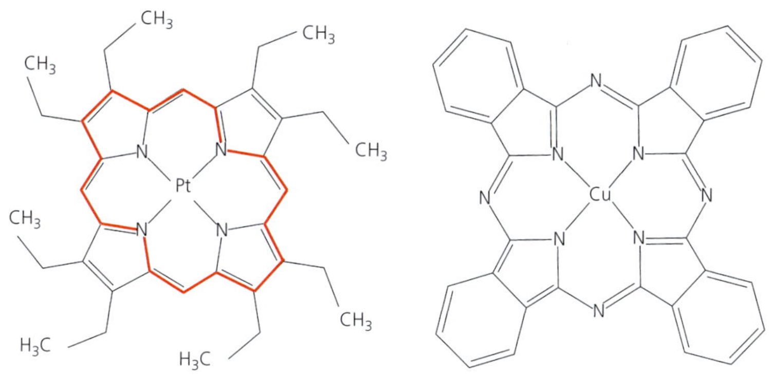

Figure 1.17 Molecular structural formulae of the porphyrin, PtOEP, and the phthalocyanine, CuPc. The red line in the PtOEP molecule traces the 18 π-electron cycle.

and many more have slowed the investment in, and ultimately the progress of organic electronics science and technology.

But long experience has told us that many of these prejudices are not founded in fact. After all, some OLEDs have demonstrated lifetimes of $>10^6$ h, and OPVs have been shown to slowly degrade over periods as long as decades. These findings have led to the development of global industries. Indeed, many organic electronic devices are fit for countless unexpected, but wholly practical applications, many of which that are yet to be developed. My opinion, with caveats, is that organics can "do anything" that ordinary semiconductors can do, within the appropriate application space.

Within this context, it is helpful to enumerate some pervasive myths, and to explain the current state of our understanding that contradicts common preconceptions. A partial list is provided in Table 1.4, along with a brief explanation. The chapters in which these myths are more fully addressed are found in the last column on the right.

Many myths involve supposed costs of organic electronic devices and processes. While these are difficult to estimate without data supplied by manufacturers, the costs of semiconductors are rarely dominated by the cost of equipment used in their production. For example, a semiconductor production facility (a "fab") typically costs over US$5 billion to build and fully equip, yet the cost of semiconductor chips represents only a minimal fraction of the application in which they are employed (e.g. in a laptop computer). The costs of high precision equipment, and an expensive fabrication facility that maintains extreme levels of purity in a dust-free environment are amortized over the volume of chips produced. That is, the most important determinant in cost is not the capital expense, but rather by the throughput and yield of the end product.

There are many other myths (and facts) than those listed in the table. It is an interesting challenge to think about our notions of the strengths and weaknesses of organic materials and devices. This listing should get one started down the path of separating fact from fiction that will increase our appreciation of the advantages and weaknesses that characterize this rich and useful field of study, and its applications. And with this knowledge, we are prepared to meet the challenge of exploding yet another myth.

1.6 Summing up

In this chapter, we have provided the definition of an organic semiconductor, and how its properties differ in fundamental ways from those of conventional, inorganic semiconductors. We have given a thumbnail sketch of the 70 year history of this most exciting scientific field, and through that discussion have seen why it motivates so many scientists to uncover the mysteries of disordered, soft materials. The ability to modify materials characteristics offered via organic chemical synthesis promises unprecedented opportunities for scientists and technologists to engineer properties "on demand" that satisfy the needs of a particular application unserved by other electronic technologies. Indeed, organic electronics is unique in its versatility and ability to change the way we interact with, and sense the world around us.

Yet, we have not directly answered the question that we posed at the beginning of this chapter: "What is organic electronics good for?" In fact, we have already answered this question to some extent. After all,

Table 1.4 Facts and myths in organic electronics

Myth	Fact	Chapters
Surface morphology can be used to reveal the crystallinity of an organic thin film	While rough surfaces may indicate a bulk crystalline structure, or a smooth surface an amorphous one, often the opposite is the case. X-ray and other structural probes are the only way to tell the true film morphology.	2, 5
Charge mobilities in organics can approach those of crystalline inorganics	Hopping dominates in organics, and without exception at room temperature, dynamic disorder limits bulk mobilities to < 5 cm^2/V s.	4
Organic thin film transistors (OTFTs) can lead to reliable mobility measurements.	Yes and no, it depends on what you mean by mobility. OTFTs can give a value for the mobility at the dielectric interface, but not the bulk. The most reliable mobilities are obtained from OTFT frequency response.	4, 8
Organic devices will always be unreliable	Red PHOLEDs have lifetimes in excess of 10^6 h. Longer lifetimes are possible, depending on the device. OPVs, too are showing lifetimes well in excess of 25 years.	6–8
Organic devices are inexpensive since the fabrication processes and materials do not need the high purity of inorganics	There is no example of a practical devices, either organic or inorganic, whose performance is not improved by using the purest materials and most controlled processes possible.	5–8
Ir and Pt complexes used in PHOLEDs need to be replaced by materials containing atoms less expensive than Ir or Pt.	The cost of materials rarely determines the cost of electronic devices. However, the cost of synthesizing materials and fabricating devices do impact device cost. Also, the amounts of these precious metals used in PHOLEDs are negligible.	6
Solution processing is less costly than deposition in vacuum, hence making solution processes preferable.	In any manufacturing environment, product throughput is the most important factor determining cost, not the expense of the machines employed. Further, whether solution or vapor phase processing is used, it must be done in an ultraclean environment with highly controllable and therefore expensive equipment, and there is always likely to be at least one vacuum deposited layer (e.g. metal contacts) on a completed device.	5
Single layer polymer devices should be less expensive than multilayer, complicated structures used for small molecule devices.	All photonic devices, and most electronic devices require multiple layers to achieve high performance and long lifetime.	6, 7
The performance and cost of organic transistors is insufficient to make them competitive with emerging metal-oxide transistors that can also be solution processed.	For many applications, this is undoubtedly true. However, organic transistors with their form factors, design flexibility, etc. have opportunities in large area, conformable electronics, medical devices, stretchable electronics, etc. that are difficult or impossible to address with other materials systems.	8
LED lighting is so efficient and low cost that OLEDs will not compete.	The conformability, transparency, fixture design capabilities, attractiveness, and high efficiency make OLEDs extremely well suited to many spatial lighting applications.	6
QD LCDs have a color gamut as large as OLEDs. They will eventually dominate the TV and monitor industries.	QD LCDs are quite attractive, but they still suffer from almost all other disadvantages of LCDs that are not drawbacks of OLEDs: slow response, viewing angle and temperature dependence, etc.	6
OPV efficiencies will always be too low to become important for solar generation	OPV efficiency has gone through numerous "steps" over the years. Today at >15%, it is poised to make another step toward the predicted single cell thermodynamic limit of 20–25%. This, combined with lightweight, semitransparency, etc. make it promising for many unique solar energy harvesting applications.	7
An electrically pumped organic laser will be achieved by finding materials with sufficient conductivity and robustness to create population inversion.	At high currents, the losses of triplets confined to the lasing medium overwhelm the gain from electrical pumping. Hence, achieving *direct* electrical pumping of the gain medium is unlikely.	6
Plastic substrates are inexpensive.	The multilayer organic/inorganic stacks needed to provide a barrier to moisture and oxygen penetration through plastic substrates for long device lifetime are costly to manufacture. Glass, and especially metal foils are much less expensive. The latter is the least costly if lack of transparency is acceptable.	5-8

OLEDs have generated a massive mobile and TV display industry that is poised to unseat the long time, but in many ways inferior incumbent liquid crystal display technology. And given their very high efficiency and superb color rendition, OLED lighting is emerging as a new application opportunity for organics. OPVs, too, appear to be heading toward commercialization, and perhaps later, a new medical sensing technology based on flexible OTFTs will develop. We do not know for sure what the future holds, but we do know that the enormous strides made in understanding organic semiconductors, driven in large part by the intense interest in creating and exploiting the new market opportunities they have created, is opening vistas in physics, chemistry and materials science that daily increase our understanding of the nature of disordered materials.

In this context, we now define those applications where the distinct characteristics of organic electronics provide a clear advantage over other electronic technologies.

Large area: With their ability to be deposited over very large (square meter and larger), reasonably smooth substrates without regard to the constraints of matching its lattice to that of the substrate, makes organics ideally suited for applications requiring large devices. These include displays, lighting and solar cells where large scale (sometimes the larger the better) is a virtue.

Flexibility, conformability, foldability: Organic semiconductors are soft materials, suggesting that they can be twisted, stretched and folded without creating large stress within the thin films comprising the devices. Combined with their ability to be deposited on almost any smooth surface allows, them to serve applications where flexibility (e.g. in a roll-up OPV), conformability (e.g. on a display used on a curved instrument panel or a shaped lighting fixture), or even where the device must have a crease, such as in a fold-up smart phone, are necessary. There are many challenges to be met to achieve this vision, since the substrate and the contacts must also allow for the necessary distortions during use without breakage or degradation.

Environmental compatibility: The use of generally non-toxic materials within the application and in its manufacture is important when we consider the life-cycle of a product, from chemical synthesis, to fabrication, to its use, and finally to disposal when it is no longer needed. Since organics are destined for use as large area, ubiquitous consumer products, attention to environmental impact becomes especially important. The low energy investment required for their near-room-temperature fabrication is also an advantage compared to conventional semiconductors such as Si whose process temperatures are ~1500°C.

Low cost: Here it can be argued that organics have a significant advantage over other semiconductor technologies. The materials are often dye compounds that can be manufactured in bulk. We will see in subsequent chapters that thin film deposition on flexible substrates opens the door to roll-to-roll (R2R) manufacture of organic devices as conceptualized in Fig. 1.18. This requires a significant paradigm shift away from producing electronics by the batch as we do today, to generating product by the kilometer tomorrow! High throughput, R2R manufacturing however, must be accompanied by high manufacturing yield. This is a significant challenge for displays that require micrometer-scale pixel patterning. However, significant strides have been made in both demonstrating R2R manufacturing, as well as nanoimprint patterning that are well suited to continuous web production. The combination of these manufacturing functions is also illustrated in Fig. 1.18. As suggested in Section 1.4, the R2R environment must be contaminant-free, that is, organics are as susceptible to degradation due to impurities, dust, and other extrinsic agents as is every other electronics technology. Indeed, there are few, if any examples of an electronic device technology in widespread use that has not gone to great lengths to use the highest purity materials and manufacturing processes. The experience of the OLED industry in constantly improving device and materials purity to improve performance is a prime example of the importance of maintaining the highest standards from materials purification, all the way through to device fabrication and packaging. As noted previously, however, the cost of maintaining materials purity and processing integrity is more than offset by increased throughput offered via high speed, continuous R2R production.

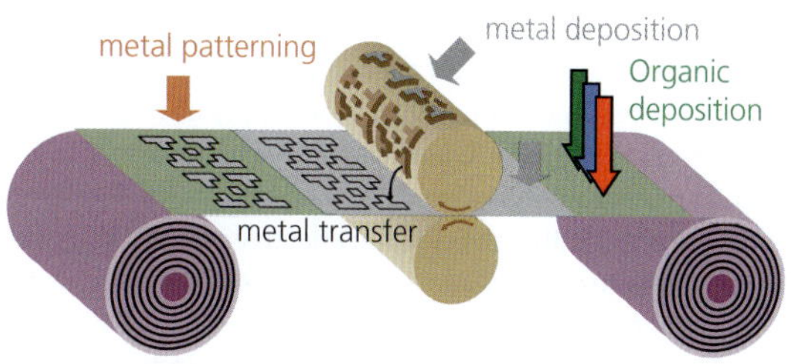

Figure 1.18 Roll-to-roll concept for the deposition and patterning of organic devices consisting of organic thin films and printed metal contacts (Forrest, 2004).

First published in Forrest, S. R. 2004. Nature (London), 428, 911.

In aggregate, organic materials can lead to dramatic transformations in how we manufacture, and use semiconductor devices—allowing for low temperature solution or vapor-phase-compatible R2R, processing and deposition on flexible substrates comprised of plastic, ultrathin glass, or metal foils. One vision of the future enabled by this unique combination of properties is that custom circuits, prepared to meet the needs of specialized applications, can be produced "on demand" on the desktop using an organic electronic printer connected to a laptop computer. This concept might be particularly attractive to a physician who needs to create a custom monitoring device for personalized medical applications. Indeed, this customization paradigm is already making inroads in the manufacture of mechanical structures using 3D printers. Surely, printing of organic optoelectronic circuits cannot be far behind!

Further reading

1. P. Yurkanis Bruice, 2017. *Organic Chemistry*, 8th ed. Upper Saddle River, NJ: Prentice Hall.
2. S. R. Forrest, 2004. The path to ubiquitous and low cost organic electronic appliances on plastic. *Nature*, 428, 911.
3. M. Geoghegan and G. Hadziioannou, 2013. *Polymer Electronics*. Oxford: Oxford University Press.
4. A. Köhler and H. Bässler, 2015. *Electronic Processes in Organic Semiconductors: An Introduction*. Weinheim, Germany: Wiley VCH.
5. E. Reich, 2009. *Plastic Fantastic: How the Biggest Fraud in Physics Shook the Scientific World*. New York: Palgrave MacMillan.

Problems

1. Table 1.4 provides a list of "myths" that have pervaded the field of organic electronics over many years.
   a. Select a myth, and briefly describe why, or why not it is a myth. You should find a paper or papers to back up your arguments: do not just state opinion but provide support backed up by evidence that your conclusion is correct. Fully reference your sources (authors, date, title, volume, page numbers).
   b. Find an important and pervasive myth that is not on the list. The same standards of proof apply as in part (a).
2. An organic semiconductor has a dielectric constant of 2.5, and an inorganic has a dielectric constant of 12. Assuming for the organic that the electron effective mass is $m_e^* = 1m_0$ and the hole effective mass is $m_h^* = 0.5m_0$, where m_0 is the electron mass:
   a. Calculate the exciton radius, a_0, and binding energy, E_B, assuming the Bohr model of the atom is an appropriate model for an exciton.
   b. Redo part (a) for an inorganic semiconductor with $m_e^* = 0.1m_0$.
   c. A CT exciton in the organic material is found to have $E_B = 0.4$ eV. Calculate its radius. Assuming a molecular dimension of 0.5 nm separated by 0.35 nm, what is the maximum number of molecules over which this CT state can spread?
3. This problem requires that you find some information about materials from the literature. For the materials Si, TiO_2, CdSe, anthracene, and assuming the Bohr model of an atom, (put your answers to this question in tabular or graphical form):
   a. Calculate their exciton binding energies and exciton radii.
   b. Which would you consider to be Wannier–Mott and which would be Frenkel excitons?
   c. What is a typical intermolecular bond energy in each case?
4. Akamatu and Inokuchi were the first to observe semiconductor-like behavior in an organic by measuring the dependence of conductivity on temperature.
   a. Starting with the basic theory of Fermi statistics in semiconductors, derive expressions that describe the conductivity (σ) as a function of temperature, T, from high temperature to near 0 K. Assume that the semiconductor is n-type with band gap E_G, effective densities of states N_C, N_V for the conduction and valence band edges, respectively, and a donor concentration $N_D << N_C, N_V$ at energy E_D below the conduction band edge. To keep things simple, we assume that the T-independent mobility of electrons $\mu_e >> \mu_h$ (the mobility of holes), and that the effective masses are m_e, m_h, respectively.
   b. Plot the results in (a) for Si and Ge with $N_D = 10^{15}$ cm^{-3} and $E_D = 20$ meV. The plot should be of the form, log(σ) vs. $1000/T$.
   c. From Akamatu and Inokuchi's resistivity data (in units of Ω cm) in Fig. P1.4, determine the energy gap of

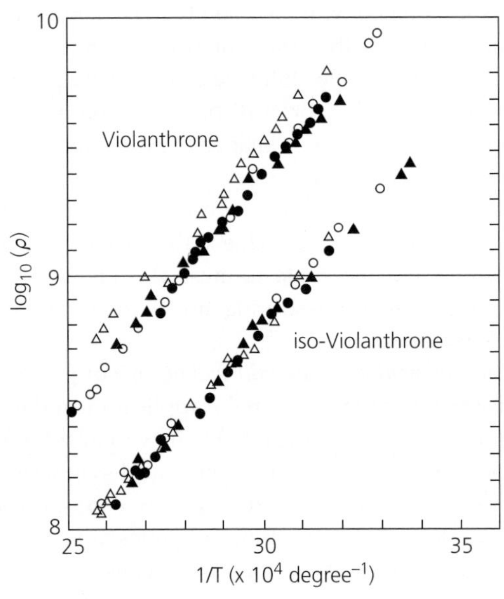

Figure P1.4 Resistivity data from Akamatu and Inokuchi (1950).

violanthrone, as well as the lower limit to the donor energy, E_D. Assume $m_e^* = m_0$, the rest mass of the electron, $\mu_e = 0.001\ \mathrm{cm^2/V\ s}$, and that there is one electronic state per molecule. The value of E_D is rather insensitive to these choices, so you can make intelligent estimates.

5. Several molecules used in organic electronics are shown in the accompanying Fig. P1.5.
   a. Which of the compounds are aromatics? How many paired electrons exist in the π-orbitals?
   b. Which of the molecules can potentially emit light efficiently?
   c. Can you identify a homoleptic compound? A heteroleptic compound?
   d. Write the chemical formulae for each of the compounds in the figure.

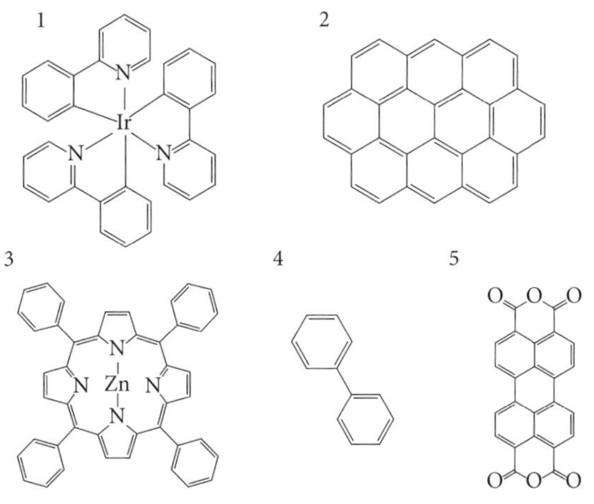

Figure P1.5 Molecules used in organic electronics.

References

Adachi, C., Baldo, M. A., Thompson, M. E. & Forrest, S. R. 2001. Nearly 100% internal phosphorescence efficiency in an organic light emitting device. *J. Appl. Phys.*, 90, 5048.

Akamatu, H. & Inokuchi, H. 1950. On the electrical conductivity of violanthrone, iso-violanthrone, and pyranthrone. *J. Chem. Phys.*, 18, 810.

Ashcroft, N. W. & Mermin, N. D. 1976. *Solid State Physics*. New York: Holt, Rinehart and Winston.

Baldo, M. A., Lemansky, S., Burrows, P. E., Thompson, M. E. & Forrest, S. R. 1999. Very high efficiency green organic light emitting devices based on electrophosphorescence. *Appl. Phys. Lett.*, 76, 4.

Baldo, M. A., O'Brien, D. F., You, Y., Shoustikov, A., Thompson, M. E. & Forrest, S. R. 1998. High efficiency phosphorescent emission from organic electroluminescent devices. *Nature*, 395, 151.

Beasley, M. R. 2002. Report of the Investigation Committee on the possibility of scientific misconduct in the work of Hendrik Schon and coauthors. Available at: http://www.lucent.com/news events/pdf/researchreview.pdf.

Bolto, B. A., McNeill, R. & Weiss, D. E. 1963. Electronic conduction in polymers. III. Electronic properties of polypyrrole. *Aust. J. Chem.*, 16, 1090.

Burroughes, J. H., Bradley, D. D. C., Brown, A. R., Marks, R. N., Mackay, K., Friend, R. H., Burns, P. L. & Holmes, A. B. 1990. Light-emitting diodes based on conjugated polymers. *Nature*, 347, 539.

Chiang, C. K., C. R. Fincher, J., Park, Y. W., Heeger, A. J., Shirakawa, H., Louis, E. J., Gau, S. C. & MacDiarmid, A. G. 1977. Electrical conductivity in doped polyacetylene. *Phys. Rev. Lett.*, 39, 1098.

Dexter, D. L. 1953. A theory of sensitized luminescence in solids. *J. Chem. Phys.*, 21, 836.

Dexter, D. L. 1979. Two ideas on energy transfer phenomena: ion-pair effects involving the OH stretching mode, and sensitization of photovoltaic cells. *J. Lumin.*, 18–19, 779.

Forrest, S. R. 1997. Ultrathin organic films grown by organic molecular beam deposition and related techniques. *Chem. Rev.*, 97, 1793.

Forrest, S. R. 2004. The path to ubiquitous and low cost organic electronic appliances on plastic. *Nature*, 428, 911.

Forrest, S. R., Kaplan, M. L., Schmidt, P. H., Feldmann, W. L. & Yanowski, E. 1982. Organic-on-inorganic semiconductor contact barrier devices. *Appl. Phys. Lett.*, 41, 90.

Förster, T. 1948. Zwischenmolekulare Energiewanderung und Fluoreszenz. *Ann. Phys.*, 55, 2.

Förster, T. 1959. 10th Spiers Memorial Lecture: transfer mechanisms of electronic excitations. *Disc. Faraday Soc.*, 27, 7.

Giebink, N. C., Lassiter, B. E., Wiederrecht, G. P., Wasielewski, M. R. & Forrest, S. R. 2010a. The ideal diode equation for organic heterojunctions. II. The role of polaron pair recombination. *Phys. Rev. B*, 82, 155306.

Giebink, N. C., Wiederrecht, G. P., Wasielewski, M. R. & Forrest, S. R. 2010b. The ideal diode equation for organic heterojunctions. I. Derivation and application. *Phys. Rev. B*, 82, 155305.

Giebink, N. C., Wiederrecht, G. P., Wasielewski, M. R. & Forrest, S. R. 2011. The thermodynamic efficiency limit of excitonic solar cells. *Phys. Rev. B*, 83, 195326.

Halls, J. J. M., Walsh, C. A., Greenham, N. C., Marseglia, E. A., Friend, R. H., Moratti, S. C. & Holmes, A. B. 1995. Efficient photodiodes from interpenetrating polymer networks. *Nature*, 376, 498.

Helfrich, W. & Schneider, W. G. 1965. Recombination radiation in anthracene crystals. *Phys. Rev. Lett.*, 14, 229.

Holstein, T. 1959a. Studies of polaron motion. Part I. The molecular-crystal model. *Ann. Phys.*, 8, 325.

Holstein, T. 1959b. Studies of polaron motion. Part II. The 'small' polaron. *Ann. Phys.*, 8, 343.

Iglesias-Groth, S., Manchado, A., Rebolo, R., González Hernández, J. I., García-Hernández, D. A. & Lambert, D. L. 2010. A search for interstellar anthracene towards the Perseus anomalous microwave emission region. *Mon. Not. Roy. Astron. Soc.*, 407, 2157.

Karl, N. 1989. Studies of organic semiconductors for 40 years. III. *Mol. Cryst. Liq. Cryst.*, 171, 31.

Marcus, R. A. 1956. On the theory of oxidation-reduction reactions involving electron transfer. I. *J. Chem. Phys.*, 24, 966.

Marcus, R. A. 1957. On the theory of oxidation-reduction reactions involving electron transfer. II. Applications to data on the rates of isotopic exchange reactions. *J. Chem. Phys.*, 26, 867.

Peumans, P. & Forrest, S. R. 2001. Very high efficiency double heterostructure copper phthalocyanine/C_{60} photovoltaic cells. *Appl. Phys. Lett.*, 79, 126.

Pope, M. & Swenberg, C. E. 1982. *Electronic Processes in Organic Crystals*. Oxford: Clarendon Press.

Pope, M., Kallmann, H. & Magnante, P. 1963. Electroluminescence in organic crystals. *J. Chem. Phys.*, 38, 2042.

Reich, E. S. 2009. *Plastic Fantastic: How the Biggest Fraud in Physics Shook the Scientific World*. New York: Palgrave MacMillan.

Shen, Z., Burrows, P. E., Bulovic, V., Forrest, S. R. & Thompson, M. E. 1997. Three color, tunable organic light emitting device. *Science*, 276, 2009.

Shirakawa, H., Louis, E. J., MacDiarmid, A. G., Chiang, C. K. & Heeger, A. J. 1977. Synthesis of electrically conducting organic polymers: halogen derivatives of polyacetylene, $(CH)_x$. *Chem. Commun.*, 578.

Shockley, W. 1949. The theory of p-n junctions in semiconductors and p-n junction transistors. *Bell Syst. Tech. J.*, 28, 435.

Silinsh, E. A. 1980. *Organic Molecular Crystals*. Berlin: Springer-Verlag.

Tang, C. W. 1986. Two-layer organic photovoltaic cell. *Appl. Phys. Lett.*, 48, 183.

Tang, C. W. & VanSlyke, S. A. 1987. Organic electroluminescent devices. *Appl. Phys. Lett.*, 51, 913.

Tang, C. W., VanSlyke, S. A. & Chen, C. H. 1989. Electroluminescence of doped organic thin films. *J. Appl. Phys.*, 65, 3610.

Thompson, M. E., Djurovich, P. E., Barlow, S. & Marder, S. 2007. Organometallic complexes for optoelectronic applications. *Comprehensive Organometallic Chem. III*, 12, 101.

Tomalia, D. A., Christensen, J. B. & Boas, U. 2012. *Dendrimers, Dendrons, and Dendritic Polymers*. Cambridge: Cambridge University Press.

Tsumura, A., Koezuka, H. & Ando, T. 1986. Macromolecular electronic device: field-effect transistor with a polythiophene thin film. *Appl. Phys. Lett.*, 49, 1210.

Warta, W., Stehle, R. & Karl, N. 1985. Ultrapure, high mobility organic photoconductors. *Appl. Phys. A*, 36, 163.

Wikipedia. 2019. *DHRS7B*. Available at: https://en.wikipedia.org/wiki/DHRS7B [accessed May 2, 2019].

Yu, G., Gao, J., Hummelen, J., Wudl, F. & Heeger, A. J. 1995. Polymer photovoltaic cells: enhanced efficiencies via a network of internal donor–acceptor heterojunctions. *Science*, 270, 1789.

CHAPTER 2

Bulk and thin film organic crystal structures

"Beware of over-confidence; especially in matters of structure."
Cass Gilbert, American Architect

Crystal structure ultimately determines the properties of a solid. Its binding forces, symmetry, and electrical and optical properties all arise from the structure of the material. In this chapter we begin our discussion of the basic concepts of crystalline structures by defining the various types and symmetries that a crystal lattice might take. Various definitions and conventions used to describe crystal structure, including the concept of the unit cell, symmetry space groups, identifying and indexing crystal planes, and lattice types, will be discussed with particular focus on those features most pertinent to organic materials. Next we consider the fundamental forces that bind materials into solids. These cohesive forces include *ionic*, *covalent*, *metallic*, and most importantly for organic materials, *van der Waals* and *hydrogen bonds*. Our treatment of these bonding forces is based on seeking the minimum total energy of the crystal—an approach that is conceptually straightforward but in practice is very difficult to implement on a macroscopic scale.

Almost all optoelectronic devices are comprised of thin films rather than bulk crystalline materials. This is particularly true for organic semiconductors. The films are typically grown on a supporting substrate that may consist of a structure very similar to the layers themselves, or they may have a completely different structure, ranging from amorphous, to nanocrystalline, to single crystalline morphologies. In every case, the growth mode and resulting film morphology is determined by both the energy between film and substrate, and the thermodynamic conditions employed during layer growth. We will discuss factors relating to the lattice potential energy that result in a particular film structure or morphology. We find that the modes of growth are determined by the thin film structure and its relationship to the substrate that results from the particular binding forces at play. Here, we distinguish between the principal forms of layer templating with the substrate, which include *epitaxy*, *van der Waals epitaxy*, and *quasiepitaxy*. Indeed, since molecular materials are primarily bonded via flexible (i.e. soft) van der Waals forces, the resulting structures are a result of the many degrees of freedom that soft bonding allows. Moving beyond simple layered structures, self-assembly provides a driving force for many organic and indeed inorganic structures that are of both scientific and practical interest. Indeed, self-assembly is an example of a multiscale process whereby individual units at one length scale (e.g. atoms or molecules) are functionalized to assemble into organized units of a much larger scale (from nanometers in the case of thin films, to microns in biological systems, to even the centimeter scales and beyond). Both molecular and polymeric organic material structures are strongly influenced by forces that lead to larger assemblies that produce a broad and rich range of structures.

While this chapter considers the fundamental properties of molecules and assemblies that lead to micro- or mesoscopic order, it does not provide guidance on how to achieve a desired structure. Further discussions on growth kinetics leading to target morphologies can be found in Chapter 5, and the following chapters in Part II when it relates to the performance of a particular device.

Organic Electronics: Foundations to Applications. Stephen R. Forrest. © Stephen R. Forrest 2020.
Published in 2020 by Oxford University Press. DOI: 10.1093/oso/9780198529729.001.0001

2.1 Molecular materials: definitions

As discussed in Chapter 1, organic materials are broadly categorized into the large class of "soft materials." That is, the materials have a low bulk modulus (i.e. they are compressible and elastic) compared to covalently or ionically bonded materials. It is the lack of chemical bonds between the individual cohesive units (e.g. the molecules comprising the solids) that lends them their numerous physical, optical and electronic properties. Generally, the term molecular material refers to that class of materials whereby the individual cohesive units are not chemically bonded, but rather cohere due to the presence of purely electrostatic van der Waals and/or hydrogen bonds. An alternative definition of a molecular material coined by Kitaigorodsky (Robertson, 1958, Kitaigorodsky, 1973) is a material where the atoms within a molecular unit in the crystal are more closely spaced than the atoms residing on neighboring molecules. For example, the nearest point of contact between molecules in a solid ranges from 3.1 Å to 4.5 Å. However, the C—C bond distance ranges between 1.2 Å and 1.5 Å, whereas a H—C or H—N bond length is ~1 Å. Hence by this metric, in the solid state, small molecules and polymers fall within the general class of molecular materials.

Almost all molecular materials are organic. There are notable exceptions, however. These include non-carbon containing molecules such as O_2, N_2, and solid phases of the noble gases such as Ar, Ne, Xe, etc. In addition, the fullerenes (e.g. C_{60}, C_{70}, C_{84}, and so on) and collections of carbon nanotubes are inorganic materials that also fall within the class of molecular materials.

In this chapter, we are primarily concerned with organic small molecule materials and the forces that lead to their peculiar physical properties. We begin by discussing the crystal lattice structure, crystal symmetry and the packing of molecules in the solid state. This provides a framework for understanding the specific properties of organic solids. We note that this discussion covers all classes of structure, but concentrates primarily on those found in molecular materials. Since equilibrium structures are those found at the lowest lattice free energy, we need to consider crystalline cohesion; examining the major bonding forces including ionic, covalent, metallic, van der Waals and hydrogen bonds.

2.2 Lattices and crystal structure

A *Bravais lattice* is an infinite array of points that define the symmetry of the crystal. A three-dimensional (3D) vector that translates one point to another, identical point in space, defines the lattice itself. That is, we introduce the *lattice translation vector* as

$$\mathbf{R} = l\mathbf{a} + m\mathbf{b} + n\mathbf{c}, \tag{2.1}$$

where l, m, and n are integers corresponding to the number of lattice translations, and **a**, **b**, and **c** are *primitive* translation vectors that, taken together, define the smallest *unit cell* within the lattice that can be translated to any other identical position via **R**. That is, we consider an arbitrary vector, **r**, and then any translation to a point **r** + **R** will be at a position equivalent to **r** within the lattice. The magnitudes of the translation vectors, **a**, **b**, and **c** are called *lattice constants* or *lattice parameters*. Figure 2.1 gives examples of **R** and the primitive translation vectors of a two-dimensional (2D) square lattice comprising two atoms (symbolized by open and closed circles). We see that the choice of primitive vectors is not unique. Indeed, the only requirement is that their choice is established by Eq. 2.1, in that every point in the lattice can be accessed by integer multiples of the displacement vectors, **a** and **b**.

In an actual material, the atoms or molecules lie within the unit cell, defined by the primitive translation vectors. To be a Bravais lattice, the unit cell must be space-filling; that is, when replicated and translated by vector **R**, it must completely fill the crystal volume with no gaps or spaces between adjacent cells. A notable exception to this rule is the quasicrystal with fivefold symmetry. Such a structure is not

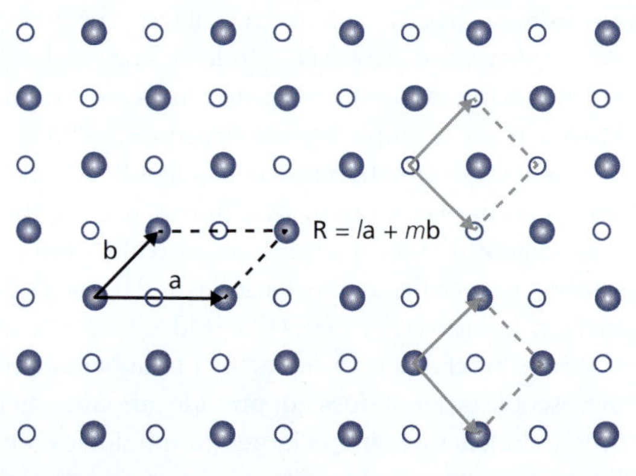

Figure 2.1 An example two-dimensional square lattice with a basis of two atoms (one solid, the other hollow). The primitive displacement vectors are **a**, **b**, and the translation vector **R** takes each lattice point into another equivalent point for all integers, l, m. The unit cell is the area bounded by the parallelogram, and is space filling. Other possible unit cells are shown in gray.

space filling, yet is found in some insulators (Shechtman et al., 1984). When one or more atoms or molecules are located within each unit cell, the Bravais lattice then defines the crystal structure. The number of atoms or molecules comprising the cell is known as the *basis*. For example, NaCl has a basis of 2, consisting of 1 Na^+ and 1 Cl^- ion within each primitive unit cell. Figure 2.1 is an example of a square lattice with a basis of two.

In three dimensions, there are 14 possible Bravais lattices in seven lattice systems shown in Table 2.1. Molecular crystals form close packed arrangements that almost always are in the cubic (face and body centered), orthorhombic, monoclinic, or triclinic unit lattice systems, with a basis of one, two or four molecules. The most common lattices found in organic materials are highlighted by the shaded box in the table. As we will find in Section 2.4, close packing arrangements result in the lowest energy, equilibrium molecular lattice.

The number of neighboring units (i.e. atoms or molecules) in a lattice contacting a molecule is known as the *coordination number*. For a simple cubic lattice the coordination number is 6, increasing to 8 for body centered cubic (bcc), and 12 for face centered cubic (fcc). Interestingly, the diamond structure of Si and Ge is also cubic consisting of two interpenetrating fcc lattices, with one atom at the vertex of one fcc lattice positioned at the origin, (0,0,0), and its nearest neighbor anchoring the second fcc lattice at (¼,¼,¼)**a**, where $a=|\mathbf{a}|$ is the lattice constant. The coordination number of the diamond structure is 4 and the basis is 2. Note that the zincblende structure characteristic of many alloy semiconductors such as GaAs and InP is also a diamond structure, but where the nearest neighbor atoms are different (e.g. Ga and As). The volume of the unit cell is

Table 2.1 Classification of Bravais lattices

Lattice system	Bravais lattice
Triclinic	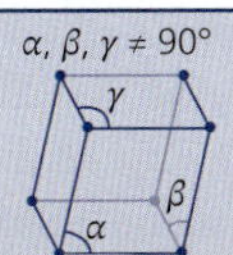
Monoclinic	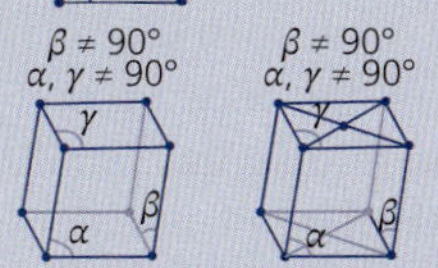
Orthorhombic	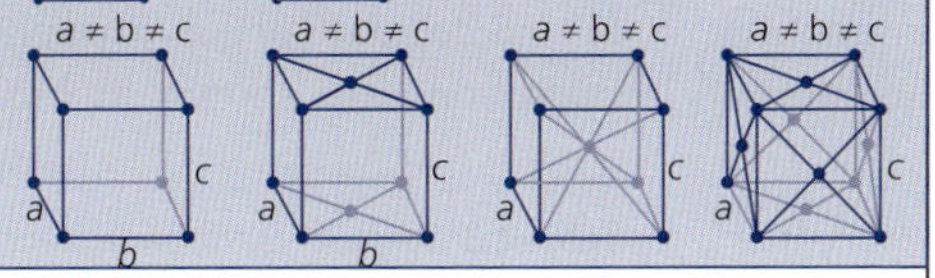
Tetragonal	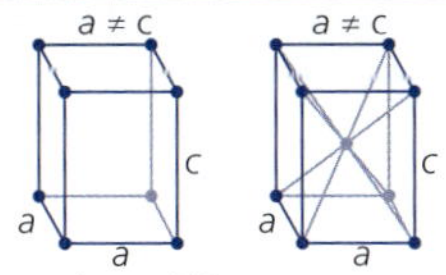
Rhombohedral	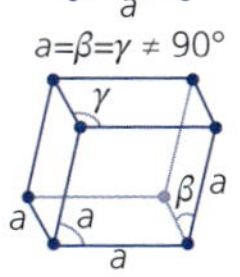
Hexagonal	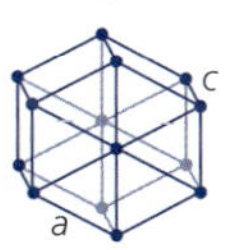
Cubic	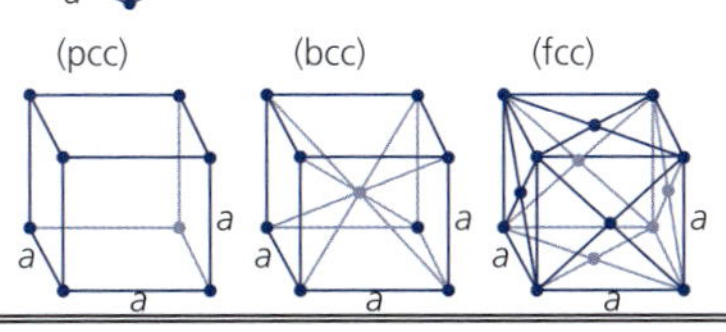

$$V_{Cell} = \mathbf{a} \bullet (\mathbf{b} \times \mathbf{c}). \tag{2.2}$$

It is clear that a cubic structure, where all sides are mutually orthogonal, has $a = b = c$ and $V_{cell} = a^3$.

To minimize lattice energy, van der Waals solids almost always form either monoclinic or triclinic structures. Indeed, the total free energy that must be minimized in an equilibrium structure where the closest packing is achieved is $E_{tot} = E_{lat} + E_{vib}$. Here, E_{lat} and E_{vib} are the lattice and vibrational energies, respectively. The vibrational energy comes from internal degrees of freedom both within and between the molecules or atoms. There are no known exceptions where the equilibrium structure is not close-packed, whereby each molecule is separated by the van der Waals radii (i.e. the distance at which the repulsive nuclear core potential energy equals that of the attractive binding energy, see Section 2.4) of the nearest atoms from adjacent molecules. We define the *packing coefficient* or *packing fraction* of a lattice as the fraction of space that is filled (normalized to a unit volume), taking into consideration the minimum intermolecular contact distances (i.e. the van der Waals radii), the bond lengths between atoms comprising the molecules, and the valence angles. Then, the lowest packing fraction of a molecular crystal in its lowest energy phase is 0.65–0.77 (Kitaigorodsky, 1973). The packing fractions of several different common lattice structures are provided in Table 2.2 if the molecules are approximated by spheres. We see that molecular materials fall within the ranges predicted for the packing of spheres, even though the shapes of and symmetries molecules can differ significantly from this simple approximation.

The symmetry of a crystal structure is defined by its *space group*, which gives the number and type of translational, rotational or mirror operations that a lattice can undergo that will bring it back onto itself; that is, for the lattice to find an orientation equivalent to its original. One space group is the *point group* where the symmetry operation is around a fixed point that once more leaves the lattice invariant.

Table 2.2 Packing fractions for several cubic lattices

Lattice structure	Packing fraction
Simple (point centered) cubic (pcc)	$\pi/6 = 0.52$
Body centered cubic (bcc)	$\pi\sqrt{3}/8 = 0.68$
Face centered cubic (fcc)	$\pi\sqrt{2}/6 = 0.74$
Hexagonal close packed (hcp)	$\pi/3\sqrt{2} = 0.74$
Diamond	$\sqrt{3}\pi/16 = 0.34$

There are 230 symmetry or space groups for a lattice with a basis, but only a small subset of these groups are found in molecular crystals due to the limited number of crystal habits that such close-packed, asymmetric materials can assume. In addition, there are 32 point groups. It is beyond the scope of this discussion to enumerate all of the various space and point groups, so we will confine ourselves to considering the basic symmetry operations, and then listing those groups in which molecular crystals are found.

Table 2.3 lists the symmetry operations and their notations, from which the nomenclature for all groups can be decoded. In identifying a space group for a given lattice, it is essential to find the highest possible symmetry. The space groups for molecular crystals are the primitive groups for close packing of molecules without symmetry: P1, $P2_1$, $P2_1/c$, Pca, Pna, $P2_12_12_1$; and with a center of symmetry, the space groups are $P\bar{1}$, $P2_1/c$, C2/c, Pbca. These notations are defined as follows: P = Primitive, C = centered on the **a** or **b** face of the lattice. Then, for example, $P2_1/c$ is a primitive lattice, with twofold screw axis along **b** for a monoclinic cell, followed by a glide along the **c** axis (here, the subscript following the 2 indicates the number of *half* translations along an axis). A glide plane is a reflection about a plane followed by a translation along the same plane. An n-glide is along half a diagonal of the face of the plane, and a glide is along a body diagonal. Similarly, Pca is a primitive lattice with glide planes along **c** and **a**. Examples of 2D molecular plane groups and their symmetry operations are shown in Fig. 2.2. The coordination

Table 2.3 Symmetry operations and notations

Symmetry operation	Element	Notation
Point symmetries		
Inversion	Point	$\bar{\mathbf{1}}$
Rotation	Line	**n**
Rotation + inversion	Line	$\bar{\mathbf{n}}$
Mirror	Plane	**m**
Identity		**1**
Translational symmetries		
Screw axis	Rotation + translation	$\mathbf{n_r}$
Glide plane	Reflection + translation	Parallel to vectors **a**, **b**, **c**; Face diagonal (**n**) Body diagonal (**d**)

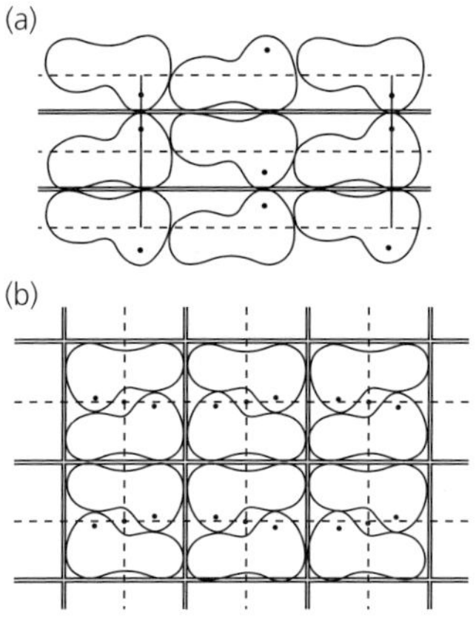

Figure 2.2 Molecular plane groups for an arbitrary molecular shape, with a spatial reference shown by the point in each molecule. Layers with (a) *cm* and (b) *cmm* symmetries (Kitaigorodsky, 1973).

number of the structure in Fig. 2.2a is 4, and the space group is *cm*, that is, there is one mirror symmetry operation possible about the horizontal planes that take the molecular lattice sites back onto themselves when folded about the planes. In Fig. 2.2b, the coordination number is also 4, with two mirror symmetry operations about the perpendicular and horizontal planes denoted *cmm*.

2.3 The reciprocal lattice, Miller indexes, and the Brillouin zone

Crystals are periodic structures. As with all periodic systems, it is convenient to analyze not the periodic data set describing the system, but rather its Fourier transform where its periodicity can be more easily managed analytically. In the case of crystals, the Fourier transform is useful for analyzing X-ray diffraction data that reveal the structure, the electron density surrounding the atomic or molecular sites on equally spaced locations within the lattice, and so on. To simplify the analysis of the physical properties of periodic (lattice) structures, we construct the *reciprocal lattice*, which is the Fourier space equivalent of the real space lattice. Thus, the reciprocal lattice vector, **G**, defines a plane wave at position, **r**, given by $e^{i\mathbf{G}\cdot\mathbf{r}}$, which is equivalent under any lattice translation, **R**, in Eq. 2.1 such that

$$e^{i\mathbf{G}\cdot\mathbf{r}} = e^{i\mathbf{G}\cdot(\mathbf{R}+\mathbf{r})}. \tag{2.3}$$

In this case, we have the condition for **G** where

$$\mathbf{G}\cdot\mathbf{R} = 2\pi \tag{2.4}$$

such that $e^{i\mathbf{G}\cdot(\mathbf{R}+\mathbf{r})} = e^{i2\pi}e^{i\mathbf{G}\cdot\mathbf{r}} = e^{i\mathbf{G}\cdot\mathbf{r}}$, thus satisfying condition in Eq. 2.3. Note that the reciprocal lattice, as defined by **G**, has an identical symmetry to the physical lattice defined by **R**.

It is then straightforward to show that the primitive reciprocal lattice vectors are defined in terms of the real space lattice constants via

$$\bar{\mathbf{a}} = 2\pi\frac{\mathbf{b}\times\mathbf{c}}{V_{Cell}}, \tag{2.5}$$

$$\bar{\mathbf{b}} = 2\pi\frac{\mathbf{c}\times\mathbf{a}}{V_{Cell}}, \tag{2.6}$$

$$\bar{\mathbf{c}} = 2\pi\frac{\mathbf{a}\times\mathbf{b}}{V_{Cell}}. \tag{2.7}$$

Now, the unit cell volume in reciprocal space is: $\bar{V}_{Cell} = \mathbf{a}\cdot\mathbf{b}\times\mathbf{c} = (2\pi)^3/V_{Cell}$.

The primitive unit cell in the reciprocal lattice that is centered on a lattice point, and that is constructed by bisecting all distances between the lattice point and its nearest neighbors, forms a *Wigner–Seitz primitive cell* (Ashcroft and Mermin, 1976). This cell encloses the volume of the first *Brillouin zone* in reciprocal space, which is to say the primitive volume in momentum- or *k*-space. For example, for a cubic lattice with lattice constant *a*, the first Brillouin zone has sides of length $k_x = k_y = k_z = 2\pi/a$. Throughout the rest of this text, we will use the brief discussion in this and previous sections, including the construct of the reciprocal lattice, to describe and understand the physical properties of crystals that are comprised of periodic arrays of molecules.

Miller indexes are used to identify crystal planes and directions within the solid. This plane identification is a critical component to identifying crystal structure based on X-ray diffraction data. The axes in a crystal are defined by its symmetry. Hence, for a cubic lattice, the **a**, **b**, and **c** axes are orthogonal forming a Cartesian system. However, in a monoclinic lattice, the angles between **a** and **c**, and **b** and **c** are $\alpha = \beta = 90^\circ$, whereas the angle between **a** and **b** is $\gamma \neq 90^\circ$, thus defining a non-Cartesian, skewed coordinate system.

The following discussion focuses on the simple case of the cubic lattice with lattice constant, *a*, shown in Fig. 2.3, although the indexing system is easily adapted to any Bravais lattice type. A vector normal to its surface defines the plane that intersects one or more axes. Accordingly, the reciprocal lattice vectors $\bar{\mathbf{a}}, \bar{\mathbf{b}}, \bar{\mathbf{c}}$ in Eqs. 2.5 to 2.7, are normal to the real lattice vectors, **a**, **b**, **c**. Hence, the vectors in reciprocal space define the planes in the real lattice. For example, in Fig. 2.3 (left to right) we show three planes whose normal vectors are defined by the coordinates (a,0,0), (a,0,a),

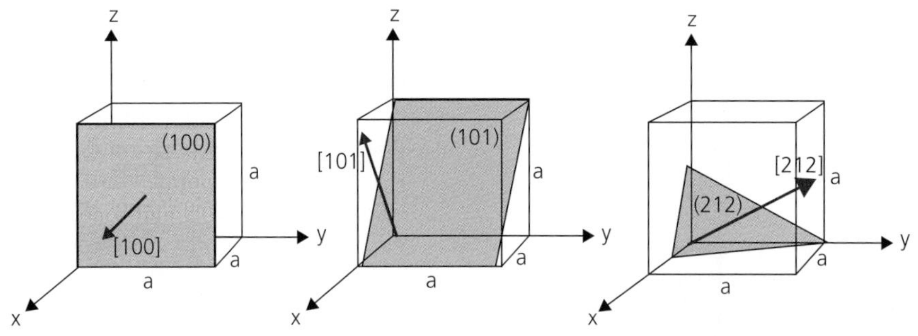

Figure 2.3 Miller indexes of three planes (in parentheses) and their normal direction vectors (in square brackets) in a cubic lattice.

and (½a,a,½a). To identify the plane, we use the convention of Miller indices, h, k, l, whereby its normal vector given by $\mathbf{G} = h\bar{\mathrm{a}} + k\bar{\mathrm{b}} + l\bar{\mathrm{c}}$ defines the plane. Here, h, k, l are the lowest integers that are the inverse of the intercepts between the plane and the axes. Thus, the planes in Fig. 2.2 are defined by the indices (100), (101) and (212). By convention, the commas are omitted. To avoid confusion with negative intercepts, the minus sign is written above the index; for example, (−111) is written as $(\bar{1}11)$. Similarly, the coordinates of a vector direction are written between square brackets; thus [111] is the real space vector given by $\mathbf{R} = a(\widehat{\mathbf{x}} + \widehat{\mathbf{y}} + \widehat{\mathbf{z}})$. The vector index values are referenced to the origin.

Oftentimes, planes are equivalent due to symmetry of the specific Bravais lattice. For example, in a cubic lattice, the planes of (110), (101), (011), $(\bar{1}10)$, $(0\bar{1}1)$, etc. are all equivalent. In this case, the set of equivalent planes is denoted {110}. Similarly, equivalent vector directions are enclosed in angled brackets as <110>.

X-ray diffraction data are used to determine the separation of crystal planes. If we define the normal to the plane identified by the coordinates $h, k. l$, then the spacing between planes with these Miller indices is called the *d-spacing*, denoted by d_{hkl}. Once again for the example of the cubic lattice, we have $d_{100} = a$, $d_{101} = \sqrt{2}a$, and so on.

2.4 Crystal energy and cohesion

Crystals are held together by cohesive forces between their constituent atoms and/or molecules. The *lattice potential*, or *lattice energy*, is the total ground state energy contained in the bonds that comprise the crystal structure. That is, when energy comparable to the lattice energy is supplied to the crystal, it will fragment into its smallest constituent parts. The lattice energy, therefore, provides a measure of the thermal energy (applied, for example, during evaporation or sublimation) needed for the material to undergo a phase change from the solid state.

The lattice energy arises from the competition between the short-range repulsion between atomic or molecular cores that keeps the solid from collapsing, and the longer-range attraction of the electronic systems of the atoms or molecules. In addition, the *zero-point energy* at $T = 0\mathrm{K}$ is a quantum mechanical phenomenon whose origins are in the uncertainty principle with no classical analog. The uncertainty in the atomic (or molecular) momentum in the lattice, Δp, is related to its positional uncertainty, Δx, via: $\Delta p \Delta x > \hbar$, where $\hbar$ is Planck's constant divided by 2π. For a molecule of mass M, the zero point energy is

$$E_{00} \simeq \frac{\hbar^2}{M\Delta x^2}. \tag{2.8}$$

This energy, which is due to the zero point molecular vibrational motion, increases for molecules with smaller masses. We can estimate the magnitude of E_{00} for a "typical" organic molecule used in organic electronics. In this case, a molecule consisting of approximately 20 carbon atoms and with a length of $\Delta x \sim 1$ nm will have $E_{00} \sim 10^{-5}$ kcal/mol. As this is several orders of magnitude lower than even the weakest cohesive forces binding most crystals stable at room temperature, it can be neglected.

As temperature is increased, the nuclear cores gain thermal motion, and the lattice potential must be amended to include entropic contributions via a calculation of the Helmholtz free energy. This becomes important when considering the kinetics of crystal growth where the structure is a sensitive function of the thermodynamic conditions used in the formation of the solid itself. However, these considerations are beyond the scope of this discussion, and the interested reader is encouraged to consult the literature for detailed treatments (see *Further reading*).

In Chapter 1 we saw that many if not all of the unique properties of organic semiconductors come from the fact that there are no chemical bonds between adjacent molecules, although such bonds exist between atoms within a given molecule. This is the case for both molecular as well as polymeric materials. By chemical bond, we imply that the electrons within the valence shells of atoms forming the crystal are shared, that is, they are *delocalized* between neighboring ionic cores. The cohesive forces that bind molecular materials are purely electrostatic without the sharing of electron systems, which are therefore highly localized to the molecules themselves. These bonds are typically via van der Waals forces in combination with hydrogen bonds. Chemical bonds include ionic, covalent, and metallic. In this section, we will discuss all bond types, as they play a central role in determining the properties of the materials. Conveniently, van der Waals bonds lend themselves to simple analytical expressions that allow for calculating many of the most important properties of molecular crystals, although longer-range ionic forces can also be calculated with reasonable accuracy in some simple cases. We begin our discussion with ionic bonds, then consider metallic and covalent bonds. Following that, we will discuss the physical origins of the van der Waals force, and apply it to the calculation of several molecular solids to determine its accuracy in predicting crystal structure. The discussion will conclude with a consideration of hydrogen bonds that are also important in the cohesion of organic solids.

Ignoring the zero point energy and entropic contributions, the energy between two atoms or molecules separated by distance $|\mathbf{r}_i - \mathbf{r}_j|$ within a crystal is given by

$$U_{tot}\left(|\mathbf{r}_i - \mathbf{r}_j|\right) = U_{repulse}\left(|\mathbf{r}_i - \mathbf{r}_j|\right) - U_{attract}\left(|\mathbf{r}_i - \mathbf{r}_j|\right). \tag{2.9}$$

In equilibrium U_{tot} is negative, that is, $U_{attract} > U_{repulse}$. Now the total crystal potential is obtained by summing Eq. 2.9 over all N atoms in the solid, while ensuring that each bond is counted only once (thus the factor of $N/2$), to give

$$U_{crystal} = \frac{1}{2}\sum_{i \neq j}^{N} U_{tot}\left(|\mathbf{r}_i - \mathbf{r}_j|\right). \tag{2.10}$$

The equilibrium (i.e. lowest energy) configuration of the crystal is determined by setting the first spatial derivative of $U_{crystal}$ to zero. There are too many degrees of freedom to use this energy minimization approach to calculate actual crystal structures of any meaningful scale (i.e. with N larger than only a few tens of lattice sites). Indeed, determining whether or not the energy minimum found by a given calculation method is a global minimum (corresponding to the true equilibrium crystal structure) or is only a local minimum (corresponding to a metastable structure) is also difficult to determine. In the remainder of this chapter, we will present the basic physical concepts by which the lattice potential can be determined, and in some cases provide examples whereby small molecular crystal structures are predicted with reasonable accuracy.

2.4.1 Ionic bonds

Ionic bonds exist in substances comprised of an assembly of an equal number of negatively and positively charged ions that are held together by long-range Coulomb forces that decrease as $1/r^2$, where r is the distance between ions. Clearly, because the solid consists of ions of both charges, the force is both attractive (between opposite charges) and repulsive (between like charges). However, for the solid as a whole, the net force must be attractive. The energy between two singly charged ions in vacuum, one at $\mathbf{r}_i$ and the other at $\mathbf{r}_j$ is simply

$$U_{ij} = \pm\frac{q^2}{4\pi\varepsilon_0|\mathbf{r}_i - \mathbf{r}_j|}, \tag{2.11}$$

where q is the electronic charge and ε_0 is the permittivity of free space. Here, the "+" is for like charges, and the "−" for opposite charges. For a solid, these pairwise ionic interactions must be summed over all N ions, which leads to a net attractive energy of

$$U_{attract} = \frac{q^2}{4\pi\varepsilon_0}\sum_{i,j}^{N}\left(\frac{1}{\mathbf{R}_{ij}} - \frac{1}{|\mathbf{R}_{ij} - \mathbf{a}|}\right), \tag{2.12}$$

where similarly charged ions are on lattice sites $\mathbf{R}_{ij} = \mathbf{R}_i - \mathbf{R}_j$, and oppositely charged ions are displaced by $\mathbf{a}$ from the lattice site. That is, while the sum is over both positively and negatively charged ions, the largest contribution to the energy is from nearest neighbors of the opposite sign.

Ionic bonds are found in both inorganic and organic materials whenever there is significant charge delocalization over a particular ion pair. The most well-known example of such a material is the insulator, NaCl, or table salt. Inorganic semiconductors consisting of alloys of elements found in columns III and V of the periodic chart (e.g. GaAs and InP) are bound by covalent forces with a slight ionic content, with even greater ionic content found in compounds

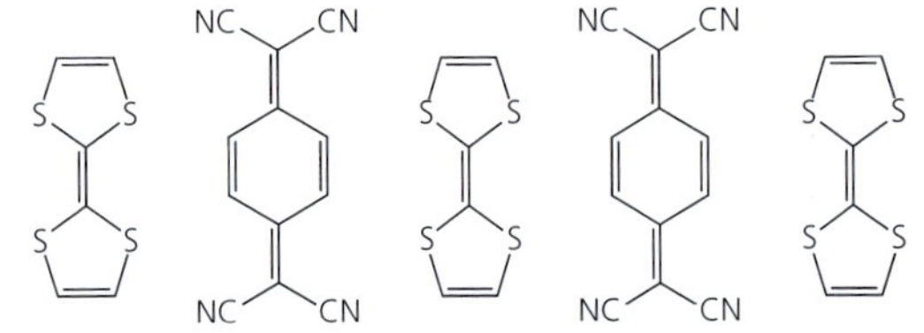

Figure 2.4 Chemical structural formulae of molecules of the charge transfer complex, TTF-TCNQ. From the left, the sulfur-containing TTF serves as the electron rich anion which extracts its electron density from the neighboring TCNQ cation.

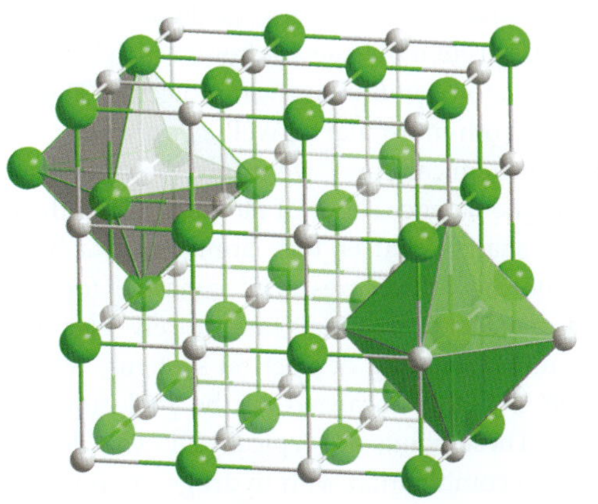

Figure 2.5 NaCl fcc crystal structure. Here, Na^+ is represented by small gray spheres, and Cl^- by the larger green spheres. The shaded regions enclose the volume containing nearest neighbors to the central ion. (chemistry.stackexchange.com)

more widely separated in the periodic chart (e.g. the II–VI compounds such as CdTe and CdS, and the I–VII compounds such as LiF with even larger ionic content). Organic salts are also partially bound by ionic forces via electron transfer from the cation to the anion. Examples of charge transfer salts include DAST or TTF-TCNQ. This latter molecular pair is shown in Fig. 2.4. Such organic salts are also known as *charge transfer complexes*.

The series in Eq. 2.12 can be difficult to evaluate in practice, since it is slowly convergent, conditionally convergent (i.e. convergence is only achieved dependent on the order in which the sum is taken), or even non-converging. However, for cubic lattices with a nearest neighbor distance, r, we can replace the series by its summation, α, such that

$$U_{attract}(r) = -\frac{\alpha q^2}{4\pi\varepsilon_0}\frac{1}{r}. \tag{2.13}$$

Here, α is a sum over all lattice sites, known as the *Madelung constant*, that depends on the crystal structure. Note that the dielectric constant of the material is not included in Eq. 2.13, since the expression is for the energy required to assemble the crystal from atoms in vacuum.

Consider the NaCl structure in Fig. 2.5 which forms an fcc lattice, coordination number 6, basis Z = 2, and with nearest neighbors of Na^+ and Cl^- spaced at a distance of $r = a/2$. In this case, there are 6 nearest neighbors, 12 second nearest neighbors at $r = a\sqrt{2}$, 8 neighbors at $r = a\sqrt{3}$, and so on, such that

$$\alpha_{fcc} = \left(6 - \frac{12}{\sqrt{2}} + \frac{8}{\sqrt{3}} - \frac{6}{2} + \ldots\right) = 1.7476. \tag{2.14}$$

In Table 2.4 we show Madelung constants calculated for several cubic lattice examples.

Extending these calculations to lattices beyond the cubic examples above is complex. Much work has been done to simplify the calculation of the Madelung constant for a range of geometries, with one convenient approach given by Nijboer and deWette (1957).

Table 2.4 Madelung constants for three cubic lattices

Lattice	Example	Coord. no.	Basis (*Z*)	α
fcc	NaCl	6	2	1.7476
pcc	CsCl	8	2	1.7627
Zincblende (fcc)	CdS	4	2	1.6381

The attractive potential in Eq. 2.13 becomes infinite as the atomic cores approach each other (i.e. in the limit of $r_{12} \to 0$). Thus, the equilibrium condition of $U_{crystal}(r) < 0$ includes the core repulsion potential that prevents the crystal from collapse. The physical origin of the core repulsion is due to the combination of the nuclear repulsion and the Pauli exclusion principle, which asserts that a quantum state can be occupied by only a single particle. The repulsion can be thought of as a "hard core" whose potential becomes abruptly positive at the equilibrium separation, r_0. More realistically, it can be modeled as a high, positive power in r, that is, $U_{repulse}(r) \sim 1/r^l$, where l is typically (and somewhat arbitrarily chosen) between 8 and 12. Thus, $U_{repulse}(r)$ is a very short range but strongly positive function of distance. In our discussion of van der Waals bonding, we will show that $l = 12$ is chosen as a convenient repulsive core potential power law, although the crystal energy is relatively insensitive to this choice. Often, an exponential form of the repulsive interaction, known as the Born–Mayer potential, is used to more accurately approximate its spatial dependence (Abrahamson, 1969). In this case,

$$U_{repulse}(r) = \beta \exp[-\gamma r], \tag{2.15}$$

where β and γ are empirical constants. Hence, the energy per ion is

$$U_{tot}(r) = \beta \exp[-\gamma r] - \frac{\alpha q^2}{4\pi\varepsilon_0}\frac{1}{r}. \tag{2.16}$$

Alternatively, assuming a power law repulsive potential:

$$U_{tot}(r) = \frac{\sigma}{r^l} - \frac{\alpha q^2}{4\pi\varepsilon_0}\frac{1}{r}, \tag{2.17}$$

where σ is also an empirical constant. While Eq. 2.16 may more accurately predict the equilibrium spacing of ions, Eq. 2.17 tends to be more mathematically manageable. For example, the equilibrium interionic separation is found by setting

$$\left.\frac{\partial U_{tot}}{\partial r}\right|_{r=r_0} = 0, \tag{2.18}$$

whence

$$r = \left[\frac{l\sigma}{\alpha q^2}4\pi\varepsilon_0\right]^{1/(l-1)} \approx \left[\frac{l\sigma}{\alpha q^2}4\pi\varepsilon_0\right]^{1/l}, \tag{2.19}$$

corresponding to a minimum energy of

$$U_{tot}(r_0) = \left(\frac{\alpha q^2}{4\pi\varepsilon_0 l}\right)\frac{1}{r_0}(1-l). \tag{2.20}$$

For large l, the total energy is negative and is simply equal to the value of the Coulomb attraction in Eq. 2.13 at $r = r_0$, which corresponds to the equilibrium atomic separation. From Eq. 2.10, the total crystal energy is then obtained by summing the contributions in Eq. 2.20 from all ions over the N lattice sites. Their pairwise interactions are accounted for by the Madelung constant, α.

Equation 2.17 is plotted in Fig. 2.6. The hard-core repulsion is a very rapidly increasing function for $r < r_0$, whereas the $1/r$ dependence of attraction leads to a very slow asymptotic approach to $U_{tot} = 0$ at large r. This slow decay makes the calculation of crystal structure problematic due to the long range of Coulombic interactions.

The *elastic constant*, k, or restoring force for molecules displaced from their equilibrium separation provides a quantitative measure of the elasticity of the crystal. Now $\mathcal{F} = -k\Delta r$ is the restoring force on a simple harmonic oscillator in a parabolic potential displaced from r_0 by a distance, Δr. Then the displacement energy is simply given by $\Delta U = -\frac{1}{2}k\Delta r^2$. The parabolic potential is a reasonable approximation for the shape of the Coulomb potential near r_0. For uniaxial strain exerted on a cubic lattice, k is obtained from the second derivative of $U(r)$ with respect to r at equilibrium:

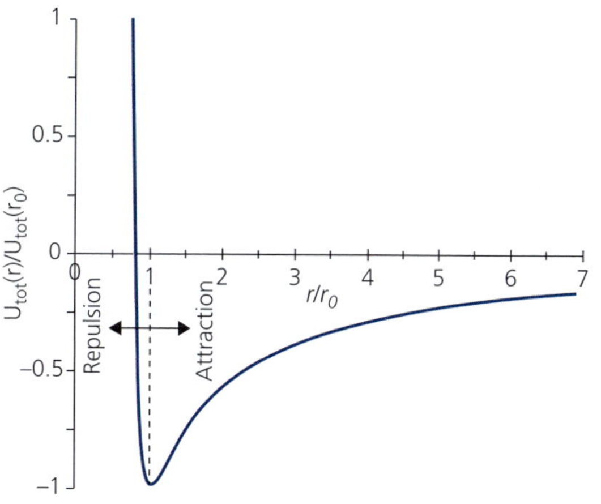

Figure 2.6 Calculation of energy vs. distance from an ionic core assuming Coulomb interactions between neighboring ions.

$$k = \left.\frac{\partial^2 U(r)}{\partial r^2}\right|_{r=r_0} = \left(\frac{\alpha q^2}{4\pi\varepsilon_0}\right)\frac{1}{r_0^3}(l-1). \tag{2.21}$$

Uniaxial pressure, or pressure exerted along a single crystal axis, will compress the lattice constant along that axis. However, minimum energy is generally maintained by conserving the unit cell volume. As a result, the lattice expands along the remaining crystal directions. In a cubic lattice, pressure along the (100) direction will tend to increase the lattice constants along (010) and (001). This is known as *tetragonal distortion*. The distortion is quantified using *Poisson's ratio*, defined as the fraction of linear expansion along one direction by the fraction of compression along a different direction. We express this ratio for small strains as

$$v = -\frac{d\varepsilon_t}{d\varepsilon_a}, \tag{2.22}$$

where ε_t is the strain transverse to the (axial) strain, ε_a. Here, $\varepsilon_t < 0$ and $\varepsilon_a > 0$ for axial tension, with the signs reversed when the solid is compressed. Note that in some structures, compression (or tension) along one axis can induce compression (or expansion) along one or more of the other axes. In this case, the signs of the inequalities are the same, such that $v < 0$. Such structures are said to have a negative Poisson's ratio.

Similar to the elastic constant, the *bulk modulus* quantifies the compressibility or stiffness of a solid, and hence its measurement can provide information about both cohesive forces and symmetry of a crystal. The bulk modulus, B, is defined as the change in volume, V, due to a change pressure, P, at a constant temperature, T. That is,

$$B = -V\left(\frac{\partial P}{\partial V}\right)_T = V\frac{\partial^2 U_{tot0}}{\partial V^2}, \qquad (2.23)$$

where the last term is obtained at $T = 0$ at pressure $P = -\partial U_{tot0}/\partial V$. Then, solving for B at $r = r_0$, and assuming uniaxial compression, we get

$$B = \left(\frac{\alpha q^2}{4\pi\varepsilon_0}\right)\frac{1}{r_0^4}(l-1). \qquad (2.24)$$

Hence, measuring the bulk modulus and the density for a given crystal structure, we obtain a value for the core potential exponent, l, which as noted above, typically ranges from 8 to 12.

2.4.2 Covalent and metallic bonds

The structure of individual molecules is determined by the sharing of valence charges of separate atoms that are covalently bonded. This situation is visualized for the H_2 molecule in Fig. 2.7. The two, shared 1s orbital electrons of each H atom have their maximum density in the region of nearest atomic contact. Inorganic covalent materials are exemplified by Si and Ge, each with four valence electrons that are shared with their nearest neighbors in a diamond structure. As noted in Section 2.4.1, Group III-V compound semiconductors such as InP and GaAs are also primarily bonded by covalent forces, with a small ionic bonding content arising from charge exchange between nearest atomic neighbors.

Similarly, the ionic cores in metals are bonded by the valence electron "gas" that is delocalized throughout the solid. Metallic bonds are similar to covalent bonds although there is an even greater degree of electron delocalization in the former. The shared electron distribution leads to solids that are highly stable at room temperature, requiring very high energies (>100 kcal/mol) to decompose. In addition, the delocalized electron systems result in material hardness and in some metals, ductility and high conductivity.

Calculating covalent bond energies and the resulting equilibrium crystal structure cannot be accomplished by determining individual atom–atom (or molecule–molecule) bond energies since the valence electrons are delocalized; a situation unlike that for

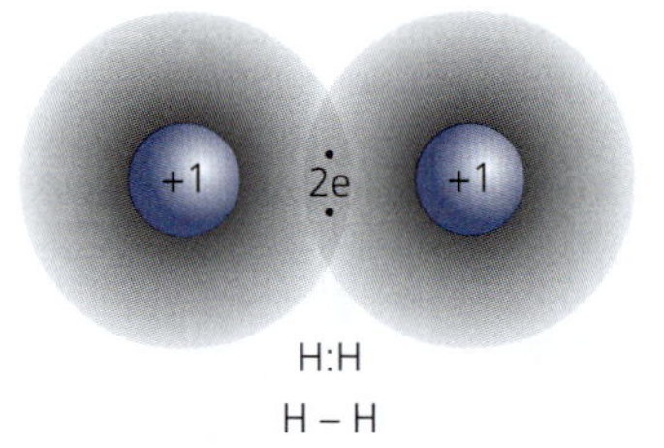

Figure 2.7 Illustration of covalent bonding of the H_2 molecule. Here the two valence electrons denoted by • are shared by the two cationic (proton) cores, with the maximum electron density located in the region between the ions. The localized electrons form the molecular bond denoted by −.

ionic, and van der Waals bonds considered below. Hence a full quantum mechanical treatment is required to determine the electron distribution, and ultimately the bond energy of the solid. One such treatment is called *density functional theory* (DFT), which is useful for calculating electron distributions in relatively small molecular ensembles, and will be considered in Section 3.3 (Giustino, 2014).

An example calculation of the electronic bond configurations in the anthracene molecule ($C_{14}H_{10}$) is shown in Fig. 2.8a and b. Anthracene consists of three fused benzene rings where the covalent carbon bonds forming the structure distribute the highest energy bound electrons in a cloud both above and below the molecular plane. In this case, the electrons are in the atomic p-orbitals of the molecule, and hence have their greatest density approximately 1 Å from the plane of the atomic nuclei. The collection of p-orbitals forms the π-electron system that ultimately binds the molecules into a crystal by intermolecular van der Waals attraction, as discussed in Section 2.3.3.

Before describing the full DFT analysis use in calculating electron distributions in complex molecules which is the topic of Sections 3.3 and 3.4, it is instructive to start here by considering the simplest case of the hydrogen molecular ion, H_2^+ as a means for understanding the basic nature of covalent bonds. The H_2^+ molecules comprise two protons sharing a single electron. Given that this is a three-particle system, approximations must be made to allow for an analytical solution. Recognizing that the electron distribution can adjust to perturbations on a time scale much faster than that of the far more massive nuclei, we assume that the nuclei are stationary immediately following such a perturbation. This central assumption for calculating molecular ground states is known as the *Born–Oppenheimer approximation* (cf. Section 3.2).

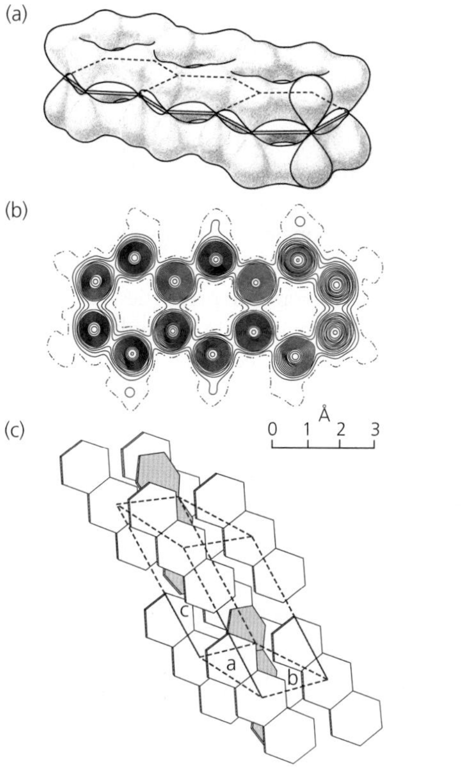

Figure 2.8 (a) An anthracene molecule viewed from the edge of the molecular plane. The electronic p-orbitals extend both above and below the molecular plane to form a conjugated π-electron system (Pope, 1966). (b) Calculated electronic density of states in anthracene. The positions of the 14 C atomic cores are clearly apparent. The denser contour line packing corresponds to a higher electron density (Robertson, 1958). (c) The equilibrium packing of anthracene molecules in a monoclinic lattice showing the lattice translation vectors **a**, **b**, and **c**, along with the unit cell (frame). Adapted from Pope (1966).

While this approximation is reasonable for ground states, it can lead to significant inaccuracies when calculating molecular excited state energies.

To calculate the ground state energy of the H_2^+ ion, we start by writing the Hamiltonian for both the electronic and nuclear energies for a multinuclear molecule:

$$H = -\frac{\hbar^2}{2m}\nabla_{\mathbf{r}}^2 - \sum_i \frac{\hbar^2}{2M_i}\nabla_{\mathbf{R}_i}^2 + V(\mathbf{r}, \mathbf{R}_i), \qquad (2.25)$$

where $\hbar$ is Planck's constant divided by 2π, m is the electron mass, M_i is the mass of the ith nucleus, and $V(\mathbf{r}, \mathbf{R}_i)$ is the potential for an electron at $\mathbf{r}$ and the nuclei at $\mathbf{R}_i$. The square of the gradients, $\nabla_{\mathbf{r}}^2$ and $\nabla_{\mathbf{R}}^2$ are relative to the electronic and nuclear coordinates, respectively. The total energy is given by a solution to Schrödinger's equation:

$$H\Psi(\mathbf{r}, \mathbf{R}) = E\Psi(\mathbf{r}, \mathbf{R}), \qquad (2.26)$$

where $\Psi(\mathbf{r}, \mathbf{R})$ is the total molecular wavefunction that depends on the positions of both the electron and protons. For the case of H_2^+, $i = 1, 2$ and Eq. 2.26 reduces to

$$\begin{aligned} H\Psi(\mathbf{r}, \mathbf{R}_1, \mathbf{R}_2) &= E\Psi(\mathbf{r}, \mathbf{R}_1, \mathbf{R}_2) \\ &= E\psi(\mathbf{r}, \mathbf{R}_1, \mathbf{R}_2)\Phi(\mathbf{R}_1, \mathbf{R}_2), \end{aligned} \qquad (2.27)$$

where the second term on the right assumes that the wavefunction (and hence the solution) can be separated into an electronic part, $\psi(\mathbf{r}, \mathbf{R}_1, \mathbf{R}_2)$, that depends on both the electron and nuclear positions, and a purely nuclear component, $\Phi(\mathbf{R}_1, \mathbf{R}_2)$. Finally, the potential energy of the system is

$$V(\mathbf{r}, \mathbf{R}_1, \mathbf{R}_2) = -\frac{q^2}{4\pi\varepsilon_0}\left(\frac{1}{|\mathbf{r}-\mathbf{R}_1|} + \frac{1}{|\mathbf{r}-\mathbf{R}_2|} - \frac{1}{|\mathbf{R}_1-\mathbf{R}_2|}\right), \qquad (2.28)$$

where the first two terms are the attractive potentials of the electron to the two nuclei, and the third term is the repulsive interaction of the protons.

Now, inserting Eq. 2.25 into Eq. 2.27 gives

$$\begin{aligned} H\psi(\mathbf{r})\Phi(\mathbf{R}_1, \mathbf{R}_2) &= -\frac{\hbar^2}{2m}\Phi(\mathbf{R}_1, \mathbf{R}_2)\nabla_{\mathbf{r}}^2\psi(\mathbf{r}, \mathbf{R}_1, \mathbf{R}_2) \\ &\quad -\sum_i \frac{\hbar^2}{2M_i}\left[\left(2\nabla_{\mathbf{R}_i}\psi(\mathbf{r}, \mathbf{R}_1, \mathbf{R}_2)\nabla_{\mathbf{R}_i}\Phi(\mathbf{R}_1, \mathbf{R}_2)\right.\right. \\ &\quad +\Phi(\mathbf{R}_1, \mathbf{R}_2)\nabla_{\mathbf{R}_i}^2\psi(\mathbf{r}, \mathbf{R}_1, \mathbf{R}_2)) \\ &\quad \left.\left.+\psi(\mathbf{r}, \mathbf{R}_1, \mathbf{R}_2)\nabla_{\mathbf{R}_i}^2\Phi(\mathbf{R}_1, \mathbf{R}_2\right)\right] \\ &\quad +V(\mathbf{r}, \mathbf{R}_1, \mathbf{R}_2)\psi(\mathbf{r}, \mathbf{R}_1, \mathbf{R}_2)\Phi(\mathbf{R}_1, \mathbf{R}_2) \end{aligned} \qquad (2.29)$$

Now, the Born–Oppenheimer approximation implies that the first two terms in the square brackets are insignificant compared to the other terms due to the large nuclear mass, M. Then our problem is simplified to the following two eigenvalue equations, one for the nuclear and the other for the electronic wavefunctions:

$$\begin{aligned} E\Phi(\mathbf{R}_1, \mathbf{R}_2) &= -\sum_i \frac{\hbar^2}{2M_i}\nabla_{\mathbf{R}_i}^2\Phi(\mathbf{R}_1, \mathbf{R}_2) \\ &\quad +E^*(\mathbf{R}_1, \mathbf{R}_2)\Phi(\mathbf{R}_1, \mathbf{R}_2) \end{aligned} \qquad (2.30)$$

and

$$\begin{aligned} E^*(\mathbf{R}_1, \mathbf{R}_2)\psi(\mathbf{r}, \mathbf{R}_1, \mathbf{R}_2) &= -\frac{\hbar^2}{2m}\nabla_{\mathbf{r}}^2\psi(\mathbf{r}, \mathbf{R}_1, \mathbf{R}_2) \\ &\quad + V(\mathbf{r}, \mathbf{R}_1, \mathbf{R}_2)\psi(\mathbf{r}, \mathbf{R}_1, \mathbf{R}_2). \end{aligned} \qquad (2.31)$$

The solution to Eq. 2.31 using the potential in Eq. 2.28 provides the locus of electron energies corresponding to different nuclear separations $|\mathbf{R}_1 - \mathbf{R}_2|$ in the H_2^+ molecule, and whose minimum provides the equilibrium molecular radius.

Even for the three-body problem considered, however, the solution is complex. Nevertheless, analytical solutions can be found using elliptical coordinates (Atkins and Friedman, 1997). The solutions show that the relative internuclear distance results in a positive, repulsive term as the nuclear separation $|\mathbf{R}_1 - \mathbf{R}_2| \to 0$, and a negative, or attractive interaction that vanishes at large nuclear separations, as illustrated in Fig. 2.9a. Only the potential energies of the two lowest energies, labeled 1σ and 2σ, are shown. The 1σ bonding state has a potential minimum at an internuclear separation of $R = |\mathbf{R}_1 - \mathbf{R}_2| \approx 2a_0$, where a_0 is the Bohr radius of an H atom. Furthermore, as shown in Fig. 2.9b, the electron concentration is highest in the region between the atoms, forming a stable covalent bond between the nuclei. On the other hand, the electron density of the 2σ antibonding state is considerably less in the internuclear region, being more highly localized at the nuclear cores themselves. The 2σ state increases for all $|\mathbf{R}_1 - \mathbf{R}_2|$, and hence the ion is unconditionally unstable.

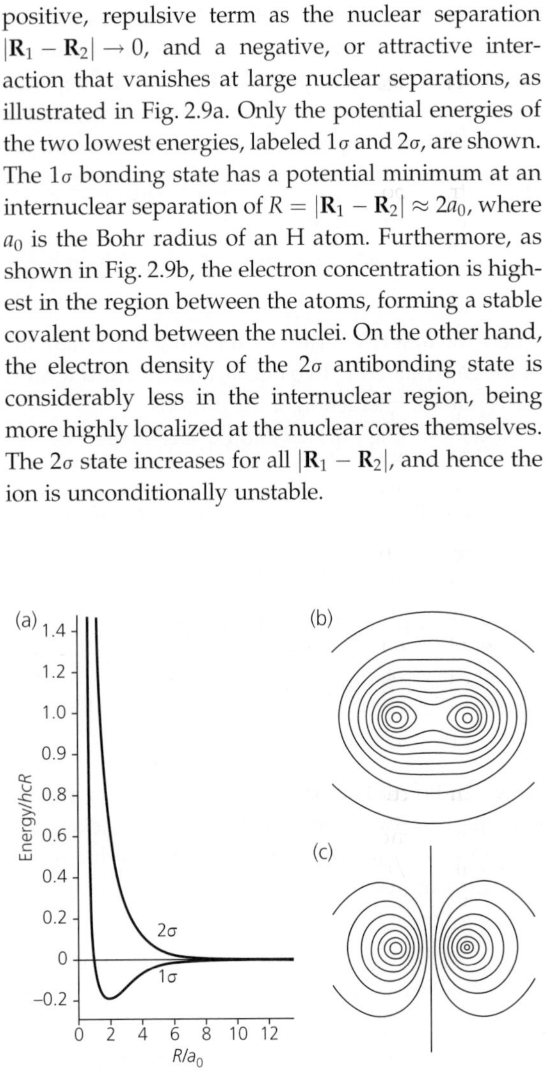

Figure 2.9 (a) The two lowest potential energies of a H_2^+ molecular ion corresponding to the bonding (1σ) and antibonding (2σ) states vs. the internuclear separation, R/a_0. The electron density contour plots for the states are shown in (b) and (c), respectively. The 1σ state has a higher concentration of electron density between the nuclear cores than the 2σ state shown by the connecting contours in the former in contrast to the localized contours in the latter. Here, $a_o = 0.5$ Å is the Bohr radius. The energies are in units of $hcR_\infty = 13.6$eV (the Rydberg energy), which is the ionization energy of an H atom, with $h =$ Planck's constant and $c =$ speed of light. From Atkins and Friedman (1997).

Moving beyond the simplest case of a H_2^+ molecular ion is extremely complex, and does not lead to exact, analytical solutions. Hence, several approaches have been developed to determine the lowest energy bond configurations, most notably the method of *linear combination of atomic orbitals* (LCAO). The molecular structures resulting from intramolecular covalent bonds are responsible for many of the morphological, electronic and optical properties of organic solids. The dependence of optical properties on the molecular configuration in the ground and excited states can be accurately predicted by LCAO and other methods, and is the subject of Chapter 3.

2.4.3 Dipolar interactions and van der Waals bonds

The most prevalent cohesive force binding molecular crystals and polymeric films is the van der Waals force. This attractive force is due to induced dipole–induced dipole interactions between nonpolar molecules. However, in many cases the molecules themselves have a fixed dipole moment in the ground state, in which case the strength of the bond is determined by the magnitudes of the individual dipoles.

In induced dipole interactions, the valance electrons are spatially distributed over a large proportion of the molecular surface. The covalent carbon bonds within the molecule result in the distribution of the highest energy bound electrons both above and below the molecular plane. The electrons in anthracene, for example, occupy the π-orbital of the molecule, with the greatest density approximately 1 Å above and below the atomic nuclear plane, as shown in Fig. 2.8a.

A dipole or higher order multipole is formed by an instantaneous fluctuation in the equilibrium electron distribution of neighboring molecules. Regions that experience this instantaneous increase in electron density are balanced by regions where the density is depleted, thereby forming an electrostatic dipole. This dipole then induces an oppositely directed instantaneous dipole in a neighboring molecule. According to the Born–Oppenheimer approximation, the electronic charge redistribution occurs on a time scale too short for the nuclei to respond, and hence we consider the molecular structure to be stationary.

The two oppositely directed dipoles Coulombically attract, thereby forming the van der Waals bond as illustrated Fig. 2.10. Since the dipoles are perturbations in the electronic distributions, they continue to fluctuate on a time scale on the order of the

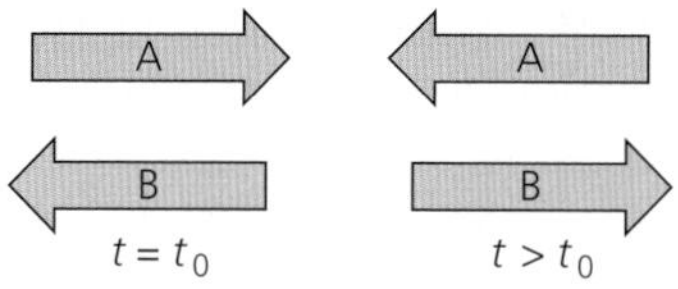

Figure 2.10 Illustration of the fluctuating induced dipole–induced dipole van der Waals attractive interaction. The arrow points in the direction of the dipole (from a negative to a positive charge at the arrow point). "A" and "B" refer to two adjacent molecules. The fluctuation is implied by the time (t) dependence of the relative dipole orientations.

interatomic vibrational time (or phonon lifetime) of τ_{ph}~10^{-14} to 10^{-12} s, continually inducing the opposite response in the neighboring molecule, thus forming a stable, electrostatic bond. Now the highest energy phonon in anthracene is due to the C-C stretch vibrational mode, with energy E_{ph} ~ 0.12 eV, leading via the uncertainty principle to $\tau_{ph} = \hbar/E_{ph}$ ~30 fs. The quantum mechanical origin of the fluctuations suggest that they are present even at $T = 0$ K as a result of the zero-point energy of the molecule.

A similar attraction can exist between two molecules with fixed dipole (or higher order multipole) moments, or a molecule with a fixed moment and one with an induced moment. In all cases, the force is attractive, with the molecular distance ultimately reaching its equilibrium value depending on the magnitude of the ground state dipole moments, the relative molecular orientations, and the repulsive force of the atomic cores.

To determine the molecular potential due to the van der Waals interaction, we first consider the electric field due to charge multipoles. Consider a molecule with a charge density $\rho(\mathbf{r})$ determined from the covalent bonds between atoms comprising the compound, such as that in Fig. 2.8a and b, where $\mathbf{r}$ is the distance to an arbitrary origin. Then, the total (monopole) charge, is given by

$$q = \int \rho(\mathbf{r}) d^3\mathbf{r}, \qquad (2.32)$$

where the integral is over all space. The higher moments of charge are then

$$\mathbf{p} = \int \mathbf{r}\rho(\mathbf{r}) d^3\mathbf{r}, \quad \text{(dipole)} \qquad (2.33)$$

$$Q_{ij} = \int (3r_i r_j - r^2\delta_{ij})\rho(\mathbf{r}) d^3\mathbf{r}, \quad \text{(quadrupole)} \qquad (2.34)$$

with increasingly complex expressions for higher order multipoles (Jackson, 1998). Here, r_i and r_j are relative coordinates within the charge distribution, and δ_{ij} is the Kronecker delta. Now, the expansion of

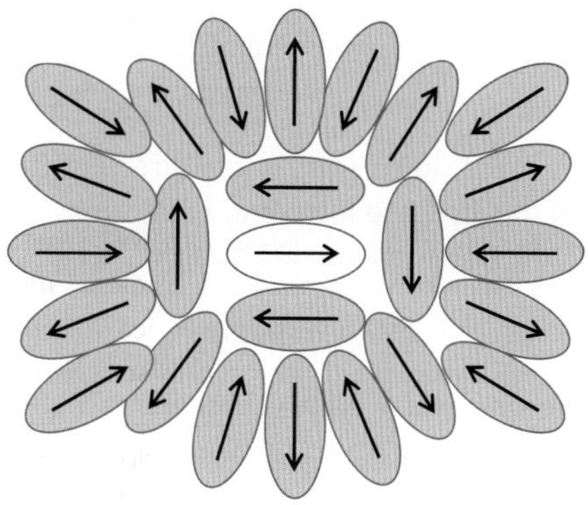

Figure 2.11 Polarization of the surrounding medium by the fixed dipole in the center. The tendency to polarize gives rise to the dielectric properties of the medium.

the electrical potential in a medium with relative dielectric constant, ε_r, can be written

$$\Phi(\mathbf{r}) = \frac{1}{4\pi\varepsilon_0\varepsilon_r}\left[\frac{q}{r} + \frac{\mathbf{p}\cdot\mathbf{r}}{r^3} + \frac{1}{2}\sum_{i,j} Q_{ij}\frac{r_i r_j}{r^5} + \ldots\right], \qquad (2.35)$$

with higher order multipoles decreasing more rapidly with increasing r. When we consider two dipoles due to adjacent molecules, $\varepsilon_r \approx 1$. However, when placed in a medium containing other polarizable molecules, their collective response is a physical reorientation of the induced dipoles. This screens the local electric field at the original molecule, which by superposition alters the net electric field, as illustrated in Fig. 2.11. For a solution of mobile molecules, this is known as the *solvation effect*, or in a solid, the *solid-state solvation effect*, which is treated analytically in Section 3.6.7.

2.4.3.1 Interactions between fixed dipoles

Since the potential of the nth charge moment is proportional to $1/r^{n+1}$, it is generally not useful to consider multipole orders higher than four due to their rapid decrease with distance. Indeed, in many practical cases, only the zeroth and first order moments are required. For the remainder of this treatment, we will only consider the effects of charges and their first moment. For neutral molecules, q=0, leaving only the first, or ground state dipole moment contributing significantly beyond the first nearest neighboring molecule in a crystal. Hence, we write

$$\Phi(\mathbf{r}) \cong \frac{\mathbf{p}\cdot\mathbf{r}}{4\pi\varepsilon_0\varepsilon_r r^3} \qquad (2.36)$$

for the dipole potential of a neutral molecule. Poisson's equation gives the electric field: $\mathbf{F}(\mathbf{r}) = -\Delta\Phi(\mathbf{r})$. Then, Eq. 2.36 yields (Jackson, 1998)

$$\mathbf{F}(\mathbf{r}) = \frac{3\widehat{\mathbf{n}}(\mathbf{p}\cdot\widehat{\mathbf{n}}) - \mathbf{p}}{4\pi\varepsilon_0\varepsilon_r r^3}, \tag{2.37}$$

where $\widehat{\mathbf{n}}$ is the unit vector in the direction from the center of the dipole to the point, $\mathbf{r}$.

Now, the dipole energy is given by

$$U(\mathbf{r}) = -\mathbf{p}\cdot\mathbf{F}(\mathbf{r}). \tag{2.38}$$

Then the interaction energy between two fixed dipoles, $\mathbf{p_1}$ and $\mathbf{p_2}$ separated by distance r_{12} oriented in the direction $\widehat{\mathbf{n}}_{12}$ (i.e. $\mathbf{r}_{12} = r_{12}\widehat{\mathbf{n}}_{12}$; see Fig. 2.12) is

$$U(r_{12}) = \frac{\mathbf{p}_1\cdot\mathbf{p}_2 - 3(\widehat{\mathbf{n}}_{12}\cdot\mathbf{p}_1)(\widehat{\mathbf{n}}_{12}\cdot\mathbf{p}_2)}{4\pi\varepsilon_0\varepsilon_r r_{12}^3}. \tag{2.39}$$

While calculating the dipole interaction energy in a solid can be complicated due to its dependence on relative orientation, it is clear that the energy decreases rapidly as $1/r^3$ compared to $1/r$ for a point charge, q. Hence, the dipole–dipole interaction is short range; vanishing rapidly beyond nearest neighbor interactions in a lattice.

Assuming that two in-plane dipoles are anti-parallel (corresponding to the lowest energy configuration), with r_{12} at angle θ relative to the dipole axes, Eq. 2.39 can be written

$$U_{12}(r_{12}) = -\frac{p_1p_2}{4\pi\varepsilon_0\varepsilon_r r^3}\left(1 - 3cos^2\theta\right) = -\frac{p_1p_2}{4\pi\varepsilon_0\varepsilon_r r^3}f(\theta), \tag{2.40}$$

where we have replaced the dipole vectors with their scalar magnitudes, p_1 and p_2.

To calculate two randomly oriented dipoles (at angles θ_1 and θ_2 relative to $\widehat{\mathbf{n}}_{12}$, and rotated by azimuthal angle, ϕ relative to each other), we must use the full expression for the fixed dipole interaction, Eq. 2.39. As noted above, the lowest energy of the system is when $\mathbf{p}_1$ and $\mathbf{p}_2$ are anti-parallel (and hence $U(r_{12}) < 0$), and is highest when they are parallel. Hence, the most probable orientation in an ensemble of dipoles is the one that achieves the lowest energy, requiring that we weight each orientation by a Boltzmann factor, $exp(-U(r_{12})/k_BT)$. Here, k_B is Boltzmann's constant.

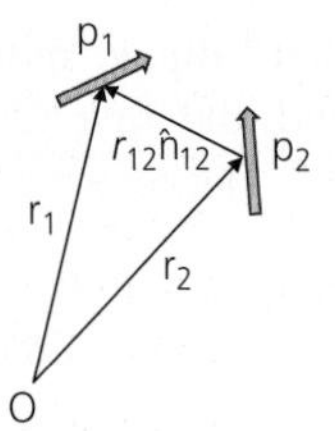

Figure 2.12 Two dipoles and positions, $\mathbf{r}_1$, $\mathbf{r}_2$, relative to the observation point, O.

The *potential distribution theorem* then gives a spatially averaged value for the energy according to (Beck et al., 2006)

$$e^{-U(\mathbf{r}_{12})/k_BT} = \frac{\int e^{-U(r_{12},\theta_1,\theta_2,\phi)/k_BT}d\Omega_{12}}{\int d\Omega_{12}}, \tag{2.41}$$

where $d\Omega_{12} = \sin\theta_1\sin\theta_2 d\theta_1 d\theta_2 d\phi$. Then the denominator is equal to 8π, and the fixed dipole energy can be solved when $U(\mathbf{r}_{12}) < k_BT$. Expanding the exponential gives

$$\langle e^{-U(\mathbf{r}_{12})/k_BT}\rangle = \left\langle 1 - \frac{U(\mathbf{r}_{12})}{k_BT} + \frac{1}{2}\left(\frac{U(\mathbf{r}_{12})}{k_BT}\right)^2 - \ldots\right\rangle. \tag{2.42}$$

Writing the spatial average in terms of the expansion in $U(r_{12})$ we obtain

$$\begin{aligned} e^{-U(r_{12})/k_BT} &= 1 - \frac{U(r_{12})}{k_BT} + \frac{1}{2}\left(\frac{U(r_{12})}{k_BT}\right)^2 - \ldots \\ &= \left\langle 1 - \frac{U(\mathbf{r}_{12})}{k_BT} + \frac{1}{2}\left(\frac{U(\mathbf{r}_{12})}{k_BT}\right)^2 - \ldots\right\rangle. \end{aligned} \tag{2.43}$$

Solving this to first order, we get

$$\frac{U(r_{12})}{k_BT} = \left\langle \frac{U(\mathbf{r}_{12})}{k_BT} + \frac{1}{2}\left(\frac{U(\mathbf{r}_{12})}{k_BT}\right)^2 - \ldots\right\rangle. \tag{2.44}$$

The final step is to integrate over all angles using the relative dipole orientation function $f(\theta,\phi) = 2\cos\theta_1\cos\theta_2 - \sin\theta_1\sin\theta_2\cos\phi$, where $-2 \le f(\theta,\phi) \le 2$. Using the expression for the dipole energy in Eq. 2.40, this finally gives the *Keesom interaction* between fixed dipoles:

$$U(r_{12}) = -\frac{2p_1^2p_2^2}{3(4\pi\varepsilon_0\varepsilon_r)^2 k_BT}\frac{1}{r_{12}^6} = -\frac{A_{DD}}{r_{12}^6}. \tag{2.45}$$

This expression has the rapidly decaying $1/r^6$ dependence characteristic of all dipole interactions. Furthermore, by superposition, the potential is additive to other molecular attractive energies such as dipole–induced dipole and induced dipole–induced dipole interactions to be discussed below.

2.4.3.2 Dipole–induced dipole interactions

Since a fixed dipole induces a dipole in a neighboring molecule, the pair will orient in the lowest energy, anti-parallel, position. The geometry of the system is illustrated in Fig. 2.13. The polarization of the electron

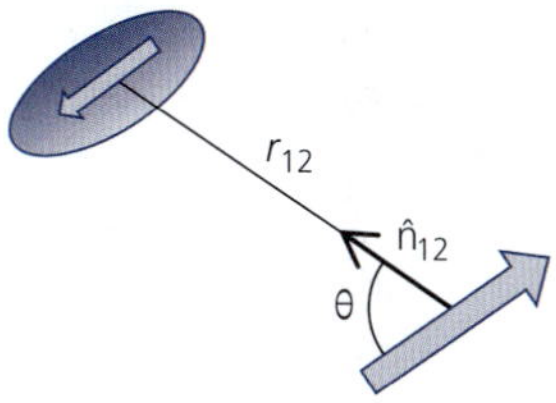

Figure 2.13 Fixed dipole (large arrow) inducing a dipole (small arrow) in the charge distribution of an adjacent molecule.

system of the molecule leads, to first order, to an induced dipole moment whose magnitude is proportional to the electric field:

$$\mathbf{p}_{ind}(\mathbf{r}) = \alpha \mathbf{F}(\mathbf{r}), \tag{2.46}$$

where α is the molecular polarizability. Generally, α is a tensor ($\overleftrightarrow{\alpha}$) although in this treatment we take it as a scalar for simplicity. The molecules have a linear polarizability that depends on the extent of the conjugated π-electron system and the molecular structure. That is, planar, conjugated molecules such as anthracene, tetracene, and pentacene (consisting of 3, 4, and 5 fused carbon rings, respectively) can have a very large α, whereas molecules that are smaller or have higher symmetry (e.g. Alq_3 used in electroluminescent devices) have a considerably smaller polarizability.

We can write Eq. 2.46 for a pair of point charges separated by distance d, using the scalar relationship $p_{ind} = qd = \alpha F$. Since $F = q/4\pi\varepsilon_0 d^2$ in vacuum, we have $\alpha = 4\pi\varepsilon_0 d^3$; that is the molecular polarizability increases as the third power of the charge separation. The units of p are Debye (D), where 1 D = 3.336 × 10^{-30} C m. Hence, a molecule that supports a full electron charge across its length of 1 nm has p = 4.8 D.

Using Eq. 2.37, we calculate the field at distance r_{12} from the dipole for the system in Fig. 2.13:

$$\mathbf{F}(r_{12}) = \frac{p_1}{4\pi\varepsilon_0\varepsilon_r}\frac{1}{r_{12}^3}\left(2cos\theta\,\widehat{\mathbf{n}}_{12} + sin\theta\,\widehat{\boldsymbol{\theta}}\right), \tag{2.47}$$

which has a magnitude of

$$F(r_{12}) = \frac{p_1}{4\pi\varepsilon_0\varepsilon_r}\frac{1}{r_{12}^3}\left(1 + 3cos^2\theta\right)^{1/2}. \tag{2.48}$$

Induction leads to a displacement of the charge in adjacent molecules, exerting a force, $\mathcal{F}$, due to electric field, **F**. Hence, the free energy of the system is given by

$$U(r_{12}) = -\int \mathcal{F}(r_{12})\bullet d^3\mathbf{r} = -q\int \mathbf{F}(r_{12})\bullet d^3\mathbf{r}. \tag{2.49}$$

Averaging over all angles, θ, we are left with the radial component of the electric field:

$$F_r(r_{12}) = \frac{2p_1}{4\pi\varepsilon_0\varepsilon_r}\frac{1}{r_{12}^3}. \tag{2.50}$$

Substituting $p_{ind} = p_1$, we thus obtain for the induced field:

$$F_{ind}(r_{12}) = -\frac{2\alpha}{4\pi\varepsilon_0\varepsilon_r}\frac{F_r}{r_{12}^3} = -\frac{2\alpha q}{(4\pi\varepsilon_0\varepsilon_r)^2}\frac{1}{r_{12}^5}. \tag{2.51}$$

Finally, integrating over r we arrive at the energy of an induced dipole for a polarizable molecule of

$$U(r_{12}) = -\frac{\alpha q^2}{2(4\pi\varepsilon_0\varepsilon_r)^2}\frac{1}{r_{12}^4} = \frac{1}{2}\alpha F_{ind}^2. \tag{2.52}$$

The factor of ½ on the right hand side comes from the energy expended in inducing the dipole in the neighboring molecule. Using this relationship along with Eq. 2.48, we obtain the energy for the fixed dipole-induced dipole interaction:

$$U(r_{12}) = -\frac{\alpha p_1^2}{2(4\pi\varepsilon_0\varepsilon_r)^2}\frac{1}{r_{12}^6}\left(1 + 3cos^2\theta\right). \tag{2.53}$$

Finally, taking the spatial average of $\langle cos^2\theta\rangle = 1/3$, we arrive at the dipole-induced dipole interaction energy that once more is a function of $1/r^6$:

$$U(r_{12}) = -\frac{\alpha p_1^2}{(4\pi\varepsilon_0\varepsilon_r)^2}\frac{1}{r_{12}^6} = -\frac{A_{D-ind}}{r_{12}^6}. \tag{2.54}$$

2.4.3.3 Induced dipole–induced dipole (London) interactions

The exact derivation of induced dipole-induced dipole interactions (also known as London or dispersion interactions) is considerably more difficult than the cases of fixed dipoles since its origin is inherently quantum mechanical. Its first derivation by London was based on perturbation theory (London, 1930), where the instantaneous formation of a dipole or higher order multipole in the delocalized electron distribution of one molecule generates an electric field that induces a dipole pointing in the opposite direction in a neighboring molecule. This instantaneous multipole arises from perturbations of the electrons that can occur at T = 0 K due to zero point fluctuations, or at higher temperatures due to thermally excited molecular vibrations, or phonons.

The fluctuating charge distribution in a molecule induces net attractive responses in all of the surrounding molecules. Due to the short range of the dipole interaction—we have seen that dipolar interactions decrease as $1/r^6$—the charge distributions of only the nearest neighbors oscillate in phase to the molecule at the origin. Since the zero point energy cannot be dissipated, the induction process is non-radiative. Clearly this electrostatic attraction between many neighbors depends on the lattice type and basis, and becomes increasingly complex for molecular structures that are non-planar or otherwise lack inversion symmetry.

Induced multipole forces exist between all molecules and atoms, independent of other binding forces that may exist. However, due to their relatively short range, these bonds are often considerably weaker than, and are thus dominated by covalent or ionic forces when they coexist in a solid. Complications arise where there are many neighboring molecules, all responding to the initial fluctuation and hence creating an instantaneous response as illustrated in Fig. 2.11.

Induced dipole effects are extremely important in determining the optical, electronic and structural properties of organic solids. That is, they comprise the dominant binding forces between neutral, non-polar molecules, and they also participate in energy and electron transfer between molecules. Importantly, since they are far weaker than ionic or covalent bonds, the sublimation temperatures of molecular solids (~100°C to 500°C) tend to be far lower than for materials bonded by Coulombic attraction (~600° C to 1500°C).

London has shown that the quantum *dispersion interaction* can be described using second order perturbation theory (London, 1930, 1937):

$$U(r_{12}) = -\frac{A_{disp}}{r_{12}^6}, \quad (2.55)$$

where

$$A_{disp} = \frac{2}{3}\sum_{f1}{}'\sum_{f2}{}' \frac{|\langle i_1|p_1|f_1\rangle|^2|\langle i_2|p_2|f_2\rangle|^2}{(E_{f1}+E_{f2})-(E_{i1}+E_{f2})}. \quad (2.56)$$

Here, the initial and final states of molecules 1, 2 are $|i_1\rangle, |i_2\rangle$ and $|f_1\rangle, |f_2\rangle$, respectively, with corresponding energies E_{i1}, E_{i2} and E_{f1}, E_{f2}. The sums are over all final states, excluding those states where $|i\rangle = |f\rangle$.

Deriving and then calculating Eq. 2.56 is complicated and adds little insight to our understanding of the dispersion interaction. A simple approximation based on the polarizability of a pair of coupled dipoles shown in Fig. 2.14 can be illustrative of induction effects. For this calculation, consider the zero point motion of the electron distribution around a neutral atom or molecule as a simple harmonic oscillator. The dipoles have lengths, Δr_1 and Δr_2, which are much smaller than their separation distance, r_{12}. The individual dipoles are parallel since they each exist in response to anti-parallel fluctuations in the charge of their neighbor. The total state Hamiltonian of the system is then written

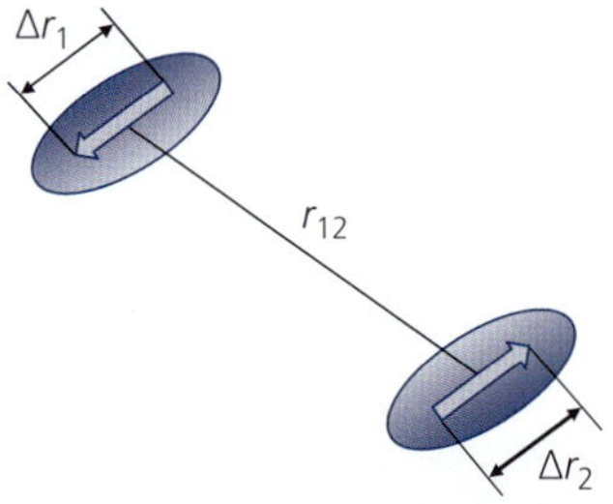

Figure 2.14 Configuration of two induced dipoles. The electron distributions around molecules 1 and 2 are illustrated by ellipses, and the instantaneous dipole moments are shown by arrows.

$$H = -\frac{\hbar}{2m_0}(\nabla_1^2+\nabla_2^2) + \frac{1}{2}k_0(\Delta r_1^2+\Delta r_2^2) + V(r_{12},\Delta r_1,\Delta r_2), \quad (2.57)$$

where the first two terms are for the two independent oscillators, and the third term is the interaction potential.

The restoring force is related to the normal mode vibrational frequency of the electron density, ω_0, via $k_0 = m_0\omega_0^2$, and m_0 is the electron mass. The intermolecular coupling is then given by the electrostatic potentials between the various charges:

$$V(r_{12},\Delta r_1,\Delta r_2) = \frac{q^2}{4\pi\varepsilon_0\varepsilon_r}\left(\frac{1}{r_{12}} + \frac{1}{r_{12}+\Delta r_1-\Delta r_2} - \frac{1}{r_{12}+\Delta r_1} - \frac{1}{r_{12}-\Delta r_2}\right), \quad (2.58)$$

where the signs for the dipole displacements arise from the oppositely directed dipole vectors, $\hat{\mathbf{p}}_1 = -\hat{\mathbf{p}}_2$. Now, in the far-field approximation, we have $\Delta r_1, \Delta r_2 \ll r_{12}$, in which case expansion of Eq. 2.58 to second order yields

$$V(r_{12},\Delta r_1,\Delta r_2) = -\frac{q^2\Delta r_1\Delta r_2}{4\pi\varepsilon_0\varepsilon_r r_{12}^3}. \quad (2.59)$$

Schrodinger's equation is solved by substituting the symmetric and antisymmetric momentum and position vectors into Eq. 2.58 and 2.59:

$$\Delta r_{\pm} = \frac{1}{\sqrt{2}}(\Delta r_1 \pm \Delta r_2)$$
$$\text{and } \nabla^2_{\pm} = \frac{1}{\sqrt{2}}(\nabla^2_1 \pm \nabla^2_2). \quad (2.60)$$

Then Eq. 2.57 becomes:.

$$H = -\frac{\hbar}{2m_0}(\nabla^2_+ + \nabla^2_-) + \frac{1}{2}\left(\left[k_0 - \frac{2q^2}{4\pi\varepsilon_0\varepsilon_r r^3_{12}}\right]\Delta r^2_+ + \left[k_0 + \frac{2q^2}{4\pi\varepsilon_0\varepsilon_r r^3_{12}}\right]\Delta r^2_-\right). \quad (2.61)$$

By comparing Eq. 2.57 and 2.61, we find that the normal mode dipole oscillation frequency, ω_0 is now broken into two symmetric harmonics:

$$\omega_{\pm} = \sqrt{\frac{k_0 \pm 2q^2/r^3_{12}}{m}} \simeq \omega_0\left(1 \pm \left(\frac{q^2}{4\pi\varepsilon_0\varepsilon_r k_0 r^3_{12}}\right) - \frac{1}{2}\left(\frac{q^2}{4\pi\varepsilon_0\varepsilon_r k_0 r^3_{12}}\right)^2 \pm \ldots\right). \quad (2.62)$$

The lowest energy of the normal mode of a quantum oscillator is $\frac{1}{2}\hbar\omega_0$. Thus, the induced dipole interactions shift this zero-point energy by the amount:

$$\Delta U(r_{12}) = \frac{1}{2}\hbar(\omega_+ + \omega_-) = -\frac{1}{2}\hbar\omega_0\left(\frac{q^2}{4\pi\varepsilon_0\varepsilon_r k_0 r^3_{12}}\right)^2. \quad (2.63)$$

Equating the Lorentz force ($\mathcal{F} = q\mathbf{F}$) to the restoring force of a harmonic oscillator, we find $k_0 = q^2/\alpha$. Then, we finally arrive at the dispersion interaction energy for the induced dipole–induced dipole energy:

$$\Delta U(r_{12}) = -\frac{1}{2}\hbar\omega_0\left(\frac{\alpha}{4\pi\varepsilon_0\varepsilon_r}\right)^2\frac{1}{r^6_{12}} = -\frac{A_{disp}}{r^6_{12}}. \quad (2.64)$$

Since the magnitude of the energy decrease is proportional to $\hbar$, this is a purely quantum mechanical effect, whereby the zero-point energy is lowered by the dipole-dipole interaction.

Another, intuitive approach suggested by London (1937) is to take $\hbar\omega_0$ as the ionization energy of the atom (e.g. He, Ar, Ne, etc.). Generalizing this to a molecule, the electron that occupies the highest occupied molecular orbital (HOMO) can be removed with energy equal to its *ionization potential* (*IP*). Given that the induction between two molecules follows a relationship similar to that used to derive Eq. 2.45, and recognizing that induced dipole energies can be superposed, the following simplified approximate relationship for Eqs. 2.54 is:

$$U(r_{12}) \approx -\frac{3}{2}\frac{\alpha_1\alpha_2}{(4\pi\varepsilon_0\varepsilon_r)^2}\frac{IP_1 IP_2}{IP_1 + IP_2}\left(\frac{1}{r^6_{12}}\right). \quad (2.65)$$

By inspection, $\hbar\omega_0 \approx \frac{3}{2}IP$, where we assume identical molecules as in Eq. 2.64 such that $\alpha_1 = \alpha_2$ and $IP = IP_1 = IP_2$. In fact these expressions are identical since Eq. 2.64 is solved for a one dimensional coordinate system, whereas Eq. 2.65 more realistically assumes each molecule has three degrees of freedom.

Up to this point, we have only considered the attractive binding forces from fixed electronic multipoles due to molecules that contain electron donating or withdrawing groups, or are spontaneously induced by zero-point and/or thermal fluctuations of the electronic distribution within a molecule. The total potential due to attractive dipole moments that includes nuclear core repulsion is thus given by

$$U(r_{12}) = \frac{A_{core}}{r^{12}_{12}} - \frac{A_{disp}}{r^6_{12}}. \quad (2.66)$$

Equation 2.66 is known as the *Lennard-Jones 6-12 potential*. As in the case of ionic bonds (Eq. 2.17), the repulsive power law of $1/r^{12}$ is chosen for convenience. The very large power reflects the steep and short range of core-core repulsion potentials. An alternative form is known as the *Buckingham potential* (Buckingham, 1938):

$$U(r_{12}) = \alpha_{12}exp(-\beta_{12}r_{12}) - \frac{A_{disp}}{r^6_{12}}. \quad (2.67)$$

In the following discussion, we will use the 6-12 potential in its common, simplified form:

$$U(r_{12}) = 4\varepsilon_{12}\left[\left(\frac{\sigma_{12}}{r_{12}}\right)^{12} - \left(\frac{\sigma_{12}}{r_{12}}\right)^6\right], \quad (2.68)$$

where $\varepsilon_{12} = \left(A^2_{disp}/4A_{core}\right)$ and $\sigma_{12} = \left(A_{core}/A_{disp}\right)^{1/6}$ are empirical constants obtained from pairwise interactions of atoms and/or molecules in the gas phase (Vargasa et al., 2001). That is, they are obtained by measuring deviations of atomic species from the ideal gas law, $PV = Nk_BT$, where P is the pressure, V is the volume, and N is the number of atoms or molecules. The deviations from this expression as a function of gas density are quantified by the *virial coefficients* which are related to many-particle interactions in the gas. These coefficients then provide the equilibrium

Table 2.5 Self-consistent Lennard-Jones 6-12 parameters between several atomic species commonly found in organic molecules (Scott and Scheraga, 1965, Abe et al., 1966)

Atoms *i–j*	$r_{0,ij}$ Å	ε_{ij} kcal mol^{-1}	Atoms *i-j*	$r_{0,ij}$ Å	ε_{ij} kcal mol^{-1}
C-C	4.00	0.150	O-S	3.60	0.200
C-N	3.75	0.155	O-H	2.60	0.063
C-O	3.60	0.173	S-C	4.00	0.173
C-S	4.00	0.173	S-N	3.75	0.179
C-H	3.00	0.055	S-O	3.60	0.200
N-C	3.75	0.155	S-S	4.00	0.200
N-N	3.50	0.160	S-H	3.00	0.063
N-O	3.35	0.179	H-C	3.00	0.055
N-S	3.75	0.179	H-N	2.75	0.057
N-H	2.75	0.057	H-O	2.60	0.063
O-C	3.60	0.173	H-S	3.00	0.063
O-N	3.35	0.179	H-H	2.00	0.020
O-O	3.20	0.200			

interatomic distances and energies used to calculate their interaction energies in Eq. 2.68.

Now the equilibrium separation, $r_{0,12}$, of the molecules is found using Eq. 2.18 which gives $r_{0,12} = \sqrt[6]{2}\sigma_{12} = 1.12\sigma_{12}$, and $U(r_0) = -\varepsilon_{12}$. Here, $r_{0,12}$ is known as the *van der Waals radius*, which is the distance of closest approach of adjacent molecules. A list of Lennard-Jones coefficients for several common interatomic interactions is provided in Table 2.5. The Lennard-Jones potential and the definition of the equilibrium values for U and r are shown in Fig. 2.15. The short-range nature of the van der Waals energy (solid line) is compared with that of the longer-range Coulomb energy (dashed line).

The elastic constant, k, for a van der Waals solid is (cf. Eq. 2.21):

$$k = \left.\frac{\partial^2 U(r)}{\partial r^2}\right|_{r=r_0} = \frac{72\,\varepsilon}{\sqrt[3]{2}\sigma^2} = 57.16\frac{\varepsilon}{\sigma^2}. \tag{2.69}$$

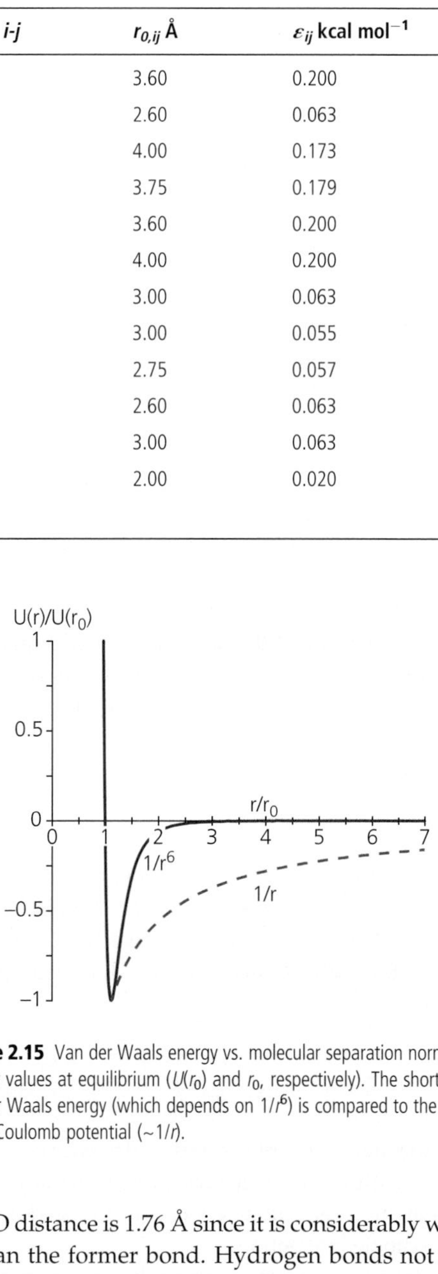

Figure 2.15 Van der Waals energy vs. molecular separation normalized to their values at equilibrium ($U(r_0)$ and r_0, respectively). The short range van der Waals energy (which depends on $1/r^6$) is compared to the longer range Coulomb potential ($\sim 1/r$).

2.4.4 Hydrogen bonds

Hydrogen bonds exist between molecules when a hydrogen atom is covalently bonded to an electronegative atom. The classic case of an H-bonded substance is water where neighboring molecules align due to interactions between an H atom on one dipolar H_2O molecule with the O atom on an adjacent molecule, as illustrated in Fig. 2.16. The hydrogen bond is represented by the dashed line: H· · ·O – H. Here, the covalent O–H bond distance is 1 Å, whereas the H· · ·O distance is 1.76 Å since it is considerably weaker than the former bond. Hydrogen bonds not only exist between H and O, but also are found with N, F, and Cl. The bond strength depends on the electronegativity of the atom bonded to the H atom. The H bond results from Coulombic attraction between an H atom bound to one molecule and the nearby molecule with which it coordinates. It arises from electrostatic dipole–dipole interactions whose bond strengths can be considerably larger and more directional than van der Waals bonds. Hydrogen bonds owe their properties to a limited but nevertheless significant covalent content between the electron on

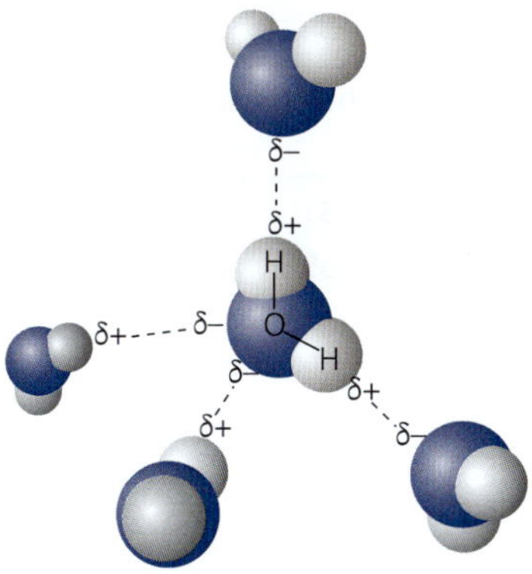

Figure 2.16 Five H_2O molecules coordinated through H bonds denoted by dashed lines. The oxygen atoms (large spheres) have 8 valence electrons. The increase (decrease) in charge density is denoted by (+).

the H atom shared with the electronegative atom on the adjacent molecule. This leaves the H atom with a slight positive charge that increases the electronegativity of the adjacent molecule (Jeffrey, 1997).

The H-bond strength and directionality leads to the high boiling points of polar liquids as H_2O, and can play a significant role in the bonding of molecular crystals, and between polymers where H atoms are suitably positioned on each monomer along the molecular backbone. Importantly, H-bonding is primarily responsible for the double helix structure of DNA, providing the attractive force between base pairs on the two strands that comprise the molecule.

The directionality and strength of the bond that leads to the complex structures of DNA, proteins and other molecular species is qualitatively illustrated in the plot of $U(\theta)$ in Fig. 2.17, where θ is the angle between the two bonded molecules (inset). The H-bond is more directional than the van der Waals interaction, but is far less directional than are covalent bonds. The bond reaches a minimum energy at $\theta = 0$. This forces it to be as linear as steric constraints between the molecules allow. Otherwise, in the example of H_2O, the O–O repulsion would dominate over the H··O attraction, considerably reducing surface tension and all of the other unique properties of this substance.

The larger directionality of the H bond compared with that of the van der Waals bond suggests that its influence attenuates less rapidly than $1/r^6$. Indeed, the bond can be modeled by Coulomb attraction from an extended (molecular) charge distribution of dipole moment, p, at angle, θ, and distance, r, from a point charge. This yields an attractive energy (Israelachvili, 2011):

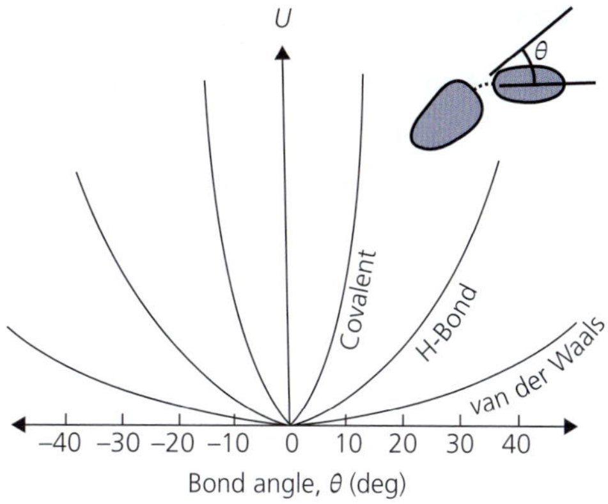

Figure 2.17 The directionality of the hydrogen bond lies between the highly directional covalent bond and the non-directional induced dipolar van der Waals bond, sharing characteristics of both. Inset: The angle between molecules bonded by H interactions (Israelachvili, 2011).

$$U(r, \theta) = -\delta^+ \frac{qp\cos\theta}{r^2}. \qquad (2.70)$$

Here, δ^+ is the fraction of a charge transferred from the H atom to the neighboring molecule. Note that while Eq. 2.70 is approximate, it provides the essential details of directionality and relatively slowly decreasing energy with distance of the H bond. As in the case of other bonding mechanisms, the hard nuclear core repulsion becomes dominant as the distance between molecules approaches their van der Waals radii.

2.4.5 Comparison of bond strengths

Ionic, covalent, van der Waals, and hydrogen bonds are all found to play a role in determining the structure of organic materials, including molecular crystalline solids and polymers. The strength of the bonds determines the hardness (elastic strength), as well as melting and boiling points of the solid. A sampling of materials bonded by these different forces is provided in Table 2.6, along with some of their physical characteristics. We note that in general, the higher the bond energy, the higher the sublimation or melting point, and the density of the solid (i.e. higher bond strengths lead to tighter packing within the crystal). A seeming exception to this rule in Table 2.6 is the van der Waals bond. However, examination of the energies in Table 2.5 shows that van der Waals bonds are roughly 0.2 kcal/mol *per bond between atomic species*. This is in contrast to the molecular compounds in Table 2.6 that consist of 20 or more atoms. To obtain the total molecular bond strength, we show in Section 2.5 that energies from each atom within the molecule must

Table 2.6 Example materials, their bond types and physical properties

Bond type	Material	Melting point (°C)	Boiling point (°C)	Sublimation point[a] (°C)	Density[b] (g/cm^3) STP	Bond energy[a] (kJ/mol)
Covalent	Si	1414	3265		2.33	222
	Ge	938	2833		5.35	188
	GaAs	1238			5.32	210
Ionic	NaCl	801	1413		2.17	787
	LiF	845	1676		2.64	1046
van der Waals	Anthracene	216	340		1.28	129
	Alq_3			350		162
	PTCDA			550	1.70	240
	Pentacene	>300		372	1.3	166
Hydrogen	Water	0	100		1	21
	Ammonia	−77	−33		0.82 (*s*)	46
	Ethanol	−114	78		0.79 (*l*)	38

[a] Approximate values are given for sublimation points and several bond energies.
[b] STP = standard temperature and pressure, *s* = solid, *l* = liquid.

be summed to result in the total molecular bond energies. Then, these are only slightly less than covalent bond energies between individual Si or Ge atoms. As a rule of thumb, we find that van der Waals bond energies are ~k_BT at room temperature, compared with $10k_BT$ for H bonds, $100k_BT$ for covalent bonds, and perhaps 5–10 times higher still for ionic bonds.

2.5 Equilibrium crystal structures

Using the interaction energy between two induced dipoles, in principle it is straightforward to determine the energy of an ensemble of molecules to determine the minimum energy that determines the equilibrium crystal structure. In practice, Eq. 2.68 is solved to determine simple, pairwise interatomic interactions. It can, however, also be extended to molecular binding as inferred from our discussion in Section 2.4.3, although the determination of the Lennard-Jones coefficients, ε and σ, is more complex and is unavailable for most molecules. Recall that these coefficients are determined from deviations at high density from the ideal gas law for atomic species. Such data rarely exist for molecules, and hence a different approach is required.

Let us assume that the interaction energies are $U(R_{ij})$ for individual molecules, i, j separated by a distance $R_{ij} = |\mathbf{R}_i - \mathbf{R}_j|$ are known. Then we can write the total crystal energy as the sum of the individual molecule–molecule binding energies, which are expected to have a form similar to Eq. 2.68. The total crystal energy is given by the sum over all molecule–molecule interactions in a manner identical to that used between atoms, viz.

$$U_{crystal} = \frac{1}{2}\sum_{i \neq j} U(R_{ij}). \tag{2.71}$$

The minimum free energy of a solid is then obtained by taking the first derivative of $U_{crystal}$ relative to position, R_{ij} and then setting this derivative to zero to find the equilibrium intermolecular separation, R_0.

In this discussion we have implicitly assumed that the minimum energy is achieved in a crystal structure, rather than in an amorphous collection of molecules. However, often conditions do not exist to kinetically drive the molecules into the energy structure minimum. Also, some solids comprising heterogeneous mixtures of molecular isomers (e.g. both *cis* and *trans* isomers), or more than a single compound, may not have a well-defined structure with a distinct minimum energy. In these instances, *metastable* crystalline phases corresponding to a *local* energy minimum, or even amorphous solids can result, depending on the conditions used to achieve crystal growth. It is important to emphasize, however, that growth kinetics notwithstanding, the minimum energy structure of a homogeneous material is crystalline whereby neighboring molecules achieve the closest possible equilibrium packing habit, as discussed in Section 2.2.

While in principle the determination of the minimum energy structure is straightforward, in practice the process generally consumes extraordinary amounts of computer resources. The absence of data for molecular van der Waals coefficients requires that the energy calculation includes all pairwise interactions between *atoms* comprising the interacting molecules. Then, their energy contributions are determined using Eq. 2.68 with the help of the coefficients in Table 2.5, summed using Eq. 2.71, and then minimized. This calculation must be done for each molecule with respect to all others in the ensemble, and for all possible orientations for each molecule relative to all others in the neighborhood that extends until the contributions from more distant molecules becomes negligible. To accomplish this for almost any crystal structure, the number of degrees of freedom rapidly leads to an intractably large computational challenge. However, there are several simplifications that can be safely applied to many molecular solids that can provide a reasonably accurate estimation of the lattice energy and structure.

The process of reducing the molecule to its atomic constituents, and then summing over all the pairwise interactions between atoms to calculate their individual bond energies is known as the *atom–atom potential method*. In this case, the energy between molecules i, j is

$$U(R_{ij}) = \frac{1}{2}\sum_{m,n} U(r_{mn}), \tag{2.72}$$

where m, n are atoms in different molecules, i, j, whose potential is given by Eq. 2.68. The validity of this approach is based on several assumptions: (i) The calculation applies to equilibrium at $T = 0$ K, and hence it cannot predict dynamic growth processes, nor does it consider vibrational contributions or phase changes that may occur at $T > 0$ K. Thus, the model is static. (ii) The van der Waals forces between atoms on neighboring molecules are isotropic, and are not significantly perturbed by the molecular structure in which they reside. (iii) The molecules are rigid, that is, there are no internal degrees of freedom where they might flex or twist when placed in proximity to another molecule. In effect, the ground state configuration of the isolated molecule is unaffected by its environment. This constraint is quite limiting since there are many molecules that have extended, linear structures that do indeed conform to the forces exerted by their neighbors. (iv) There is no significant contribution to the intermolecular energy arising from Coulombic forces, hydrogen bonds, higher order multipoles, etc. Hence, intermolecular charge transfer is ignored. In principle, these forces can be added although their long-range nature extends the neighborhood of molecules that must be included, and hence can greatly increase calculational intensity. On the other hand, the calculation can be simplified by eliminating configurations prohibited by crystal symmetry or unlikely molecular orientations. For example, we can start the structural calculations by assuming either a monoclinic or triclinic lattice with the adjacent molecules stacked in either a lamellar or herringbone configuration, which captures a large majority of organic structures (see Table 2.1). Previous calculations for planar, nonpolar molecules such as anthracene and coronene have shown that the assumptions used in the atom–atom potential calculations generally lead to accurate, first-order estimations of equilibrium structures (Kitaigorodsky, 1973).

The accuracy of the method has been tested by calculating the bulk crystal structures of NTCDA and PTCDA, and then comparing these results with crystallographic data (Forrest, 1997). Perspective views of the measured bulk PTCDA and NTCDA unit cells are provided in Fig. 2.18a and b, respectively (Forrest et al., 1987). Here, PTCDA and anthracene (Fig. 2.8) form lamellar structures, whereas NTCDA is in one of many possible herringbone arrangements. Table 2.7 provides the details of the lattice structures of both materials. In this table we see there are two possible structures of PTCDA (called the α and β phases), and only one known *isomorph* of NTCDA. PTCDA is *polymorphic*; that is, it can take on several different stable or metastable forms, each one known as an isomorph. Polymorphism is a common attribute of molecular solids, with each isomorph achieved via a different kinetic route. Also, the unusually close intermolecular spacing of only 3.21 Å in α-PTCDA leads to many of its remarkable physical as well as electronic properties, that will be discussed in the following chapters.

As noted above, tackling the full structure is a complex 3D problem involving numerous degrees of freedom between the several molecules in the cells. Hence, only limited aspects of the 3D structure are calculated to demonstrate the application and accuracy of the atom–atom potential method. In Fig. 2.19, the energy surfaces of two PTCDA molecules and two NTCDA molecules are stacked one above the other (forming physical dimers). The surfaces are obtained by calculating the total energies (Eq. 2.72) while translating the molecules along the glide plane normal to the stacking axis using the atomic van der Waals coefficients in Table 2.5.

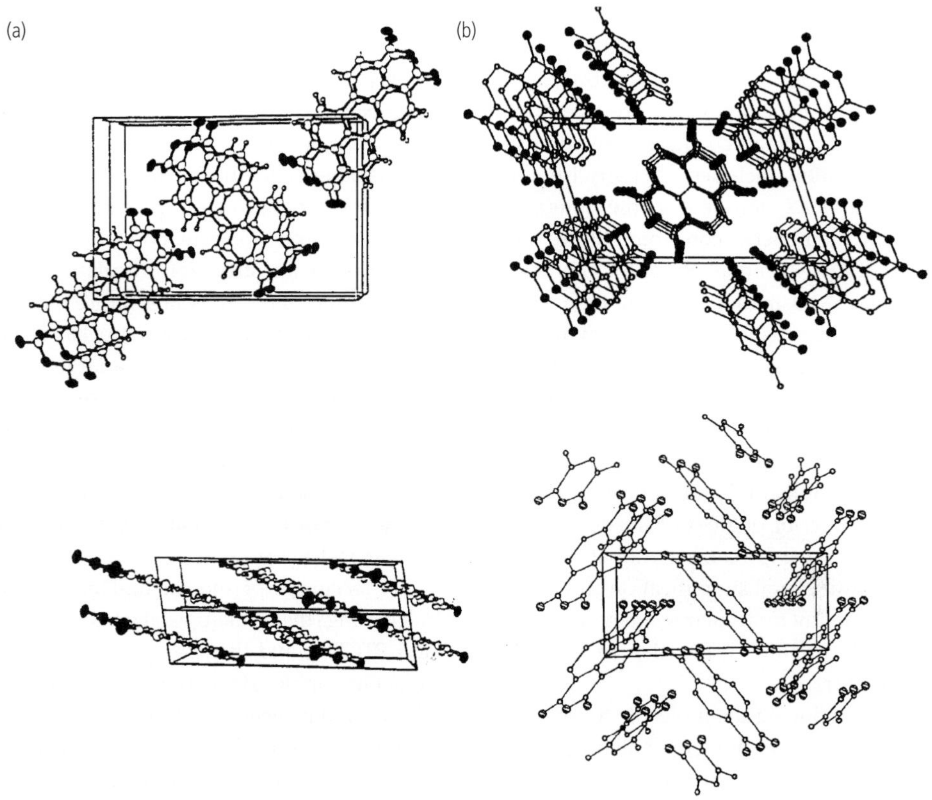

Figure 2.18 Views of the crystal structures of (a) α-PTCDA and (b) NTCDA. PTCDA is a closely packed, monoclinic lamellar structure, with a minimum distance of 3.21 Å between stacked molecules. NTCDA forms a more open herringbone monoclinic lattice. The unit cells are shown by the frames. Carbon atoms are shown as open circles, O as closed circles, and H are the small circles at each outer vertex at the carbon rings. Adapted from Forrest (1997).

From these calculations, the equilibrium distances (i.e. the van der Waals radius) in the stacking direction are between 3.20 Å and 3.26 Å for these planar molecules. It has been assumed that the molecular planes of adjacent molecules are parallel, which is not the case for the herringbone structure of NTCDA. The binding energy of the PTCDA dimer is approximately −1.3 eV, corresponding to 24 kcal/mol (see Table 2.7). This value is consistent with the sublimation energies of many aromatic molecules similar to PTCDA and NTCDA (Cox and Pilcher, 1970). Indeed, the sublimation temperatures or evaporation temperatures that range between 80°C and 500°C for most organic crystals (compared to 700°C to 2000°C characteristic of covalently bonded semiconductors) are a direct consequence of the bond strengths. Stronger bonds result in higher boiling or evaporation temperatures. Thus, van der Waals bonds are typically in the range of 1–10 meV/atom (Kitaigorodsky, 1973), leading to a binding energy/molecule $\geq$ 1 eV, whereas for covalent or ionic bonding, the energies range from 100 meV to ~5 eV/atom (Ashcroft and Mermin, 1976).

There are two minima located at $\pm$1.1 Å from the center of the PTCDA molecules forming the dimer, and $\pm$1.0 Å for NTCDA. The energy minima are displaced from the molecular centers of mass, and are located along the long molecular axis. The equilibrium offset is experimentally observed in the PTCDA bulk structure, resulting in the molecular lamella stacking along the (102) axis, forming a slip-stacked structure. That is, there is an 11° tilt of the molecular plane within the unit cell, which has been widely observed for PTCDA deposited on such insulating substrates as glass (Forrest et al., 1984b) and polymers (Taylor et al., 1995, 1997). Interestingly, this 11° tilt is not observed for PTCDA on conducting substrates such as Au or GaAs (Hirose et al., 1996), suggesting that more complex charge exchange interactions not included in the calculations may exist for these latter situations.

Already we have seen that the computational intensity of even these simple dimer structures involves a considerable investment of computer time. As a simplification, the molecule can, in some cases be substituted by a proxy structure for the individual atoms. The proxy has a similar spatial symmetry of the actual molecule. For example, PTCDA and NTCDA have been replaced by ellipsoidal "atoms" whose effective van der Waals coefficients approximate the total intermolecular interactions (Forrest and Zhang, 1994). In the ellipsoidal approximation, all the atom–atom potentials between two planar molecules (e.g. PTCDA and NTCDA) are reduced to ellipsoidal potentials, one for each molecule, but where the Buckingham coefficients, α_{12}, β_{12}, and A_{disp} in Eq. 2.67 are replaced by phenomenological constants that depend on angle and distance to the nearest neighbor molecule.

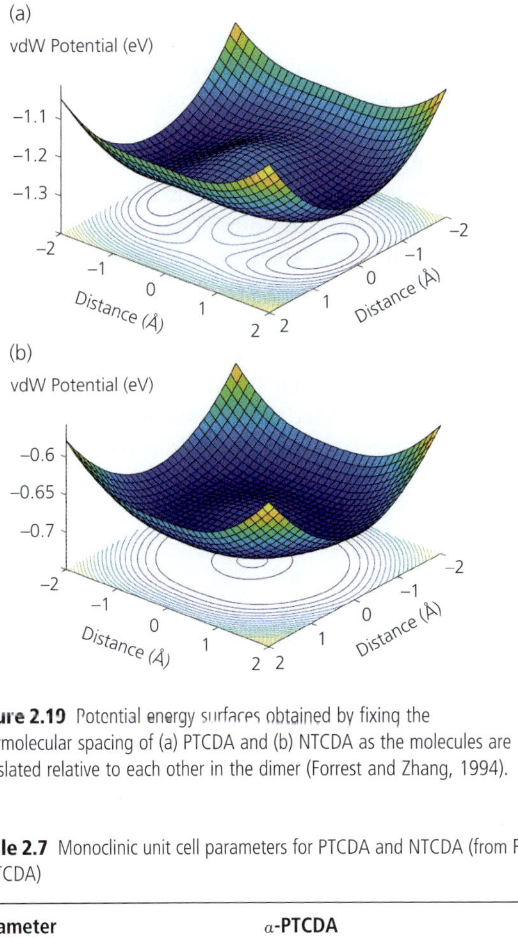

Figure 2.19 Potential energy surfaces obtained by fixing the intermolecular spacing of (a) PTCDA and (b) NTCDA as the molecules are translated relative to each other in the dimer (Forrest and Zhang, 1994).

To account for more complex molecular shapes, the molecular structure can be broken into a limited set of interaction sites representing a cluster of atoms in a molecule, and spatially arranged to approximate the shape of the actual molecule. The Lennard-Jones potential is then used, along with "effective" atom–atom interaction coefficients derived for each site to calculate the lowest energy crystal structures. For example, the 54 atoms comprising the CuPc molecule are replaced by 13 spherically symmetric sites, thereby eliminating 95% of the pairwise calculations required to determine the interaction between two like molecules (Liu et al., 1996). The number and distribution of sites chosen is determined by a compromise between the requirement to accurately represent molecular geometry and symmetry, and the need to include as much of the interaction physics as possible without making the calculation impractically unwieldy.

Unfortunately, none of these potential approximations can be employed without some *a priori* experimental verification of their accuracy. Yet, obtaining such data (via, for example, examining the structure of the actual solid under study) can ultimately minimize or even eliminate the need for the calculation in the first place.

It is inevitable, therefore, that approximations must be employed to make structural calculations tractable, such as limiting the interactions to short range van der Waals forces, eliminating unlikely structures from the outset, or by replacing all the atom–atom potentials with a few effective potentials. Even with such approximations, the predicted structures may be difficult or even impossible to reach by conventional growth methods. Given these significant limitations,

Table 2.7 Monoclinic unit cell parameters for PTCDA and NTCDA (from Forrest and Zhang, 1994 for α-PTCDA and NTCDA, and Mobus et al., 1992 for β-PTCDA)

Parameter	α-PTCDA	β-PTCDA	NTCDA
Formula	$C_{24}O_6H_8$	$C_{24}O_6H_8$	$C_{14}O_6H_4$
Space group	$P2_1/c(C_{2h}^5)$	$P2_1/c(C_{2h}^5)$	$P2_1/c(C_{2h}^5)$
a (Å)	3.72	3.78	7.89
b (Å)	11.96	19.30	5.33
c (Å)	17.34	10.77	12.74
α (°)	90	90	90
β (°)	98.8	83.6	109.04
Z (basis)	2	2	2

structures are often approximated using *molecular dynamics* (MD) methods, whereby a limited set of interactions for an ensemble of molecules are applied at a temperature, $T > 0$ K. The non-zero temperature provides kinetic energy to the ensemble. The calculation of the structure is then carried out for a time sufficient for the molecules to reach an equilibrium, or near-equilibrium configuration. This is equivalent to structural *annealing*, whereby the individual molecules are jostled around to find an energetically favorable structure that may actually be reached under realistic laboratory conditions.

MD simulations, for example, have been used to understand the structure of an interface between films of C_{60} and BP4mPy and its evolution over time after its growth by *organic vapor phase deposition* (OVPD, see Section 5.4.2) (Baldo et al., 1998, Song and Forrest, 2016). The molecular structural formulas of C_{60} and BP4mPy are shown in Schemes **1** and **2**, respectively.

Scheme 1 C_{60}

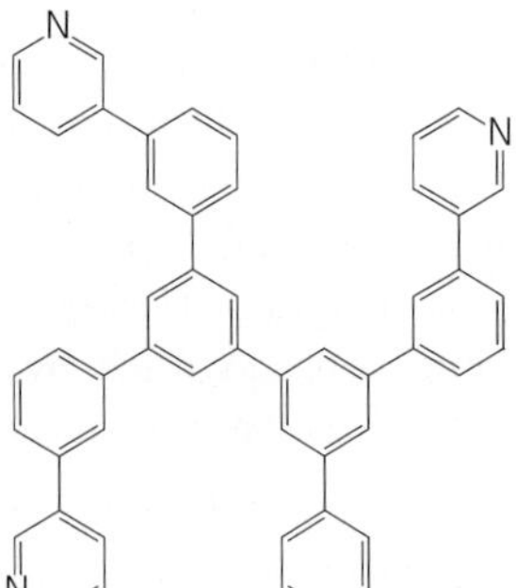

Scheme 2 BP4mPy

In OVPD, organic molecules are volatilized by a hot inert carrier gas that transports them to a cold substrate where they are adsorbed (Shtein et al., 2001). The carrier gas imparts kinetic energy to adsorbate molecules to promote diffusion along the surface until the molecules find suitably low energy sites where they stabilize. In MD calculations, the annealing energy, however, is provided by the ambient temperature. For OVPD-based growth, the dependence of kinetic energy on pressure can be substituted by an equivalent small change in substrate temperature according to the ideal gas law. That is, if the temperature at pressure P_1 is T_1, then at a different pressure, P, the incremental kinetic energy is given by the canonical relationship:

$$\begin{aligned} \delta E_{kin}(T,P) &= \frac{k_B(T_{eff}-T_1)}{k_BT_1}\delta E_{kin,0}(T_1,P_1) \\ +\delta E_{kin,0}(T_1,P_1) &= \left[\frac{k_BT_1(P/P_1-1)}{k_BT_1}+1\right]\delta E_{kin,0}(T_1,P_1), \end{aligned} \tag{2.73}$$

where T_{eff} is the equivalent substrate temperature that leads to the increased energy. Note that T_{eff} is not the actual temperature; it is the temperature needed to supply the same kinetic energy to the molecules as the change in pressure in the actual growth environment. The increase in carrier gas pressure allows molecules to find lower energy sites during the annealing process due to their larger kinetic energy. Equation 2.73 depends on the activation energy barrier, $\delta E_{kin,0}$, that must be overcome for a molecule to have sufficient surface mobility to seek a low energy site, rather than fixed at its point of initial incidence on the surface. For the system considered, nanocrystalline films formed at $P_1 = 0.17$ torr, whereas in vacuum only amorphous films were grown. From these data, it was inferred that $T_1 = 350$ K, from which the activation energy for crystallite formation was $\delta E_{kin,0} = 4.5$meV (Song and Forrest, 2016).

Molecular dynamic simulation results were obtained for an ensemble of 32 C_{60} and 12 BP4mPy molecules that were initially segregated into amorphous regions bounded by an abrupt, planar interface. This is clearly a far more challenging problem than seeking the lowest energy of a homogeneous pair of flat, polyaromatic hydrocarbons. In this case, we are examining an inherently disordered structure comprised of 44 molecules with vastly different volumes, shapes and atomic distributions. Whereas C_{60} has spherical symmetry, BP4mPy is approximately planar, although its single C–C bonds between phenyl groups allows for numerous twisting and bending degrees of freedom, all of which are considered in the simulations. Molecular dynamic simulations based on a 6-12 Lennard-Jones potential, and that included simulated annealing for times up to 400 ps were performed using commercial software (Bovia Corp, 2016). Convergence toward structural equilibrium in <150 ps is found for $P_2 = 0.28$ torr (with an equivalent temperature of $T_{eff} = 576$ K; see Fig. 2.20).

The simulation proceeds by allowing the ensemble of molecules to reach different equilibrium arrangements at each background pressure. At P_1 (T_{eff} = 350 K), Fig. 2.21a shows that there is minor reorganization within the C_{60} and BP4mPy regions with a slight intermixing at the interface. At P_2 (corresponding to δE_{kin} = 7.4meV) in Fig. 2.21b, C_{60} molecules form into their characteristic, tight packed fcc lattice that limits interdiffusion of BP4mPy. As the growth pressure increases to P_3 = 0.49 torr (corresponding to δE_{kin} = 13.0meV in Fig. 2.21c), the C_{60} region becomes more disordered, allowing BP4mPy molecules to interdiffuse across the interface. The molecules become completely blended in an amorphous mixture at the highest pressure, P_4 = 0.82 torr (δE_{kin} = 21.7meV in Fig. 2.21d). The corresponding van der Waals energies of each configuration are −0.833 eV/molecule at P_1, −0.872 eV/molecule at P_2, −0.935 eV/molecule at P_3, and −1.01 eV/molecule at P_4. This implies that the ensemble energy decreases as the BP4mPy:C_{60} blend becomes more intermixed. Increasing pressure (and hence kinetic energy) drives the morphology to an increasingly stabilized structure.

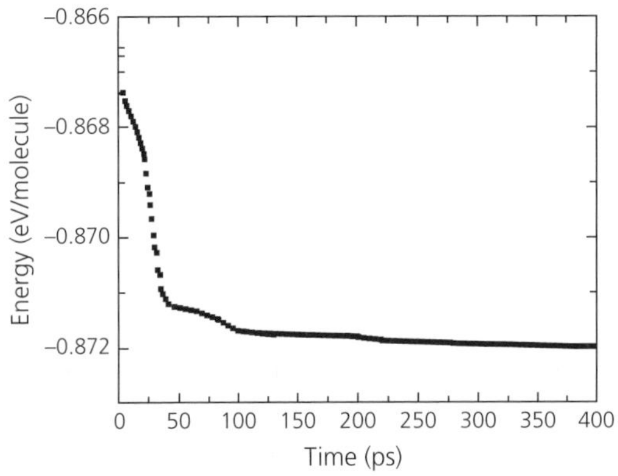

Figure 2.20 Potential energy of the simulated molecular configuration annealed at T = 576 K (corresponding to P_2 = 0.28 torr) as a function of simulation time. The equilibrium structure is achieved after approximately 150 ps (Song and Forrest, 2016).

2.6 Molecular layer structures

2.6.1 Epitaxial growth modes

For practical applications, the optoelectronically active material is layered onto a supporting substrate. Unlike inorganic semiconductors, it is impractical or exceptionally difficult to grow molecular crystals of a size useful for studying and exploiting their properties in the bulk. Hence, almost all electronic devices

(a) (b) (c) (d)

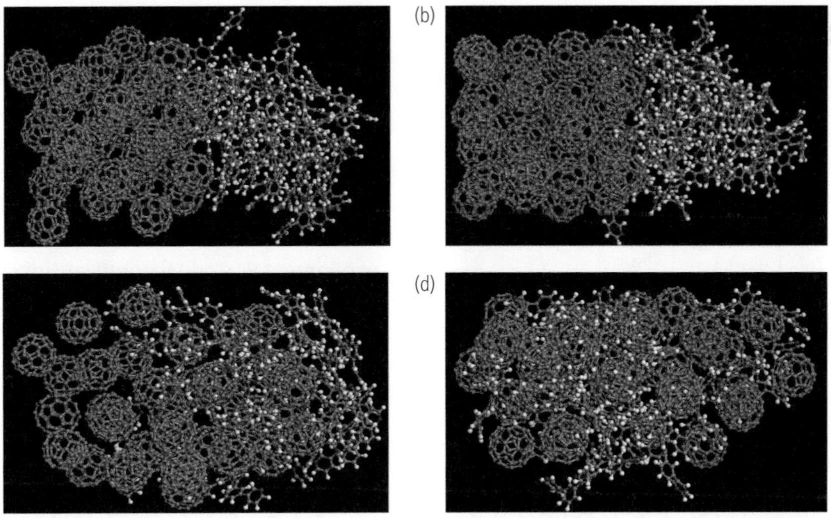

Figure 2.21 Molecular dynamic simulation results of an ensemble of 12 BP4mPy and 32 C_{60} molecules after simulated annealing with effective molecular kinetic energies of (a) δE_{kin} = 4.5 meV, (b) 7.4 meV, (c) 13 meV, (d) 21.7 meV corresponding to OVPD growth pressures of 0.17 torr, 0.28 torr, 0.49 torr, 0.82 torr, respectively. The initial condition corresponds to (a) with C_{60} to the left and BP4mPy to the right of the interface in the center of the diagram. Note the ordered fcc C_{60} lattice in (b), and the almost complete intermixing of the equilibrium structure achieved at the highest pressure of 0.82 torr in (d) (Song and Forrest, 2016).

employ thin films of active materials grown onto a supporting substrate. Epitaxial growth, or *epitaxy*, is achieved when the lattice structure of the film matches that of the substrate, that is, when the lattice of the adlayer atoms or molecules is identical to that of the underlying substrate. When this condition is met, the substrate acts as a template on which to grow the layer. For materials bonded by strong and long range covalent and/or ionic forces (e.g. Si, GaAs, InP, etc.), lattice-matching between the substrate and epitaxial layer is essential in achieving a thin film free of dislocations or other defects that are generated to relieve lattice strain.

Strain is defined as the degree of lattice mismatch between two contacting layers. The lattice mismatch for a cubic structure is given by

$$\Delta a = |a_S - a_E|, \quad (2.74)$$

where a_S is the lattice constant of the substrate and a_E is that of the epitaxial layer. For example, due to the strong interatomic bonding of inorganic semiconductors, a close lattice-match (typically with strains of $f = \Delta a/a_S < 10^{-3}$) is required when growing films thicker than a critical value beyond which a very high density of strain-energy relieving misfit dislocations are generated (Matthews and Blakeslee, 1974). While such mismatched epitaxy can occasionally be useful for device applications, the short carrier lifetimes and high junction leakage currents that accompany high defect densities are often unacceptable. This severely restricts the choice of materials available to the device engineer.

The *Matthews–Blakeslee thickness limit* is an estimate of the thickness at which a high density of dislocations is generated in an epitaxial layer. The critical Matthews–Blakeslee thickness is

$$t_{crit} = \frac{\|\mathbf{a}\|(1-\nu/4)}{4\sqrt{2}\pi f(1+\nu)}\left(log\left[\frac{\sqrt{2}t_{crit}}{\|\mathbf{a}\|}\right]+1\right). \quad (2.75)$$

This is proportional to $1/f$, that is, t_{crit} decreases inversely with the degree of lattice mismatch between the epitaxial layer and the substrate. Also, it scales linearly with $\|\mathbf{a}\|$, the magnitude of the *Burgers vector* of the dislocation, which depends on both the magnitude and direction of the lattice distortion due to a lattice dislocation. The relationship between the Burgers vector and a dislocation is shown in Fig. 2.22. It points in the direction, and has the magnitude of the gap in a clockwise path originating and ending at the dislocation.

Ultimately, the tolerable strain is determined by how much distortion the lattice can withstand before a dislocation is formed. Distortion into the out-of-plane direction (i.e. normal to the substrate) relieves strain by attempting to maintain a constant unit cell volume, and hence maintaining its minimum lattice energy. In cubic lattices, this is known as *tetragonal distortion*, and is determined by the magnitude of the Poisson's ratio (Eq. 2.22). The effect of tetragonal distortion on a square lattice is illustrated in Fig. 2.23.

The substantially weaker cohesive forces exerted by van der Waals bonds between the adlayer and the substrate leads to a significantly reduced Poisson's ratio ($\nu \sim 0.01$) compared to that for inorganic materials ($\nu \sim 0.3$–0.5). The magnitude of the lattice distortion is correspondingly reduced, thereby resulting in an increase in the critical thickness. For strains of $\sim 10^{-3}$ that are tolerable in covalently bonded materials, the critical thickness is ~100 nm, whereas this value can be at least an order of magnitude larger for organics.

While the properties of organic molecular crystals have been studied for nearly 70 years, only recently has long range structural ordering of layers been achieved on a wide range of substrates (Forrest et al., 1984a, Hara et al., 1989, So et al., 1990, Umbach et al., 1998, Lunt et al., 2011). There has been both experimental and theoretical investigation of the

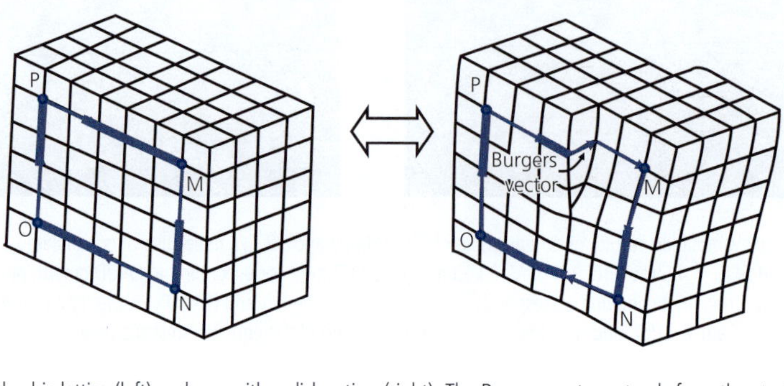

Figure 2.22 An undistorted cubic lattice (left) and one with a dislocation (right). The Burgers vector extends from the start to the end of the clockwise path of MNOP.

conditions that lead to these ordered molecular thin films (Forrest and Zhang, 1994, Hillier and Ward, 1996, Yang and Forrest, 2008).

In the preceding discussion, we have emphasized the effects that the substrate lattice can have on the quality of subsequently grown layers, and the degree to which mismatch-induced strain can have on the quality of the epitaxy. To categorize the various types of structures, it is interesting to note that the etymology of the word *epitaxy* is a combination of the Greek *epi* for "upon", plus *taxis* for "arrangement" or "in the form of." Several such overlayer arrangements are shown in Fig. 2.24, where the deposited layer is shown by the grid with black circles, and the substrate with open circles. A conventional epitaxial layer is both *commensurate* with the underlying lattice (i.e. both lattices precisely fit), and is aligned to it. An epitaxial layer therefore has a lattice with an identical symmetry as the substrate, as well as a lattice constant that is an integer multiple (usually no larger than 2) of the substrate. This motif results in an unstrained layer since there are no periodic mismatches between the upper and lower lattices. Thus, analogous to Fig. 2.24a, the lattice sites in the upper layer lie directly on top of those in the lower layer. In this example, the epitaxial layer is tilted such that its lattice points match every second substrate lattice point in both the x and y directions. This is known as a 2 x 2 surface reconstruction.

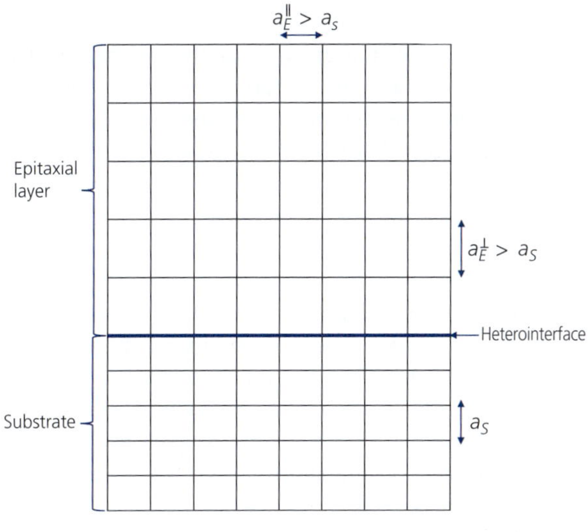

Figure 2.23 An epitaxial layer whose lattice constant, a_E, is slightly larger than that of the substrate, a_S. The strain energy is relieved in the layer by tetragonal distortion to maintain the original unit cell volume. In this case, the lattice constant perpendicular to the substrate plane, $a_E^{\perp}$, is enlarged relative to that parallel to the substrate, where $a_E^{\parallel} = a_S$.

If such a registry is not possible, the *coincident* layer illustrated in Fig. 2.24b may be achieved. The particular configuration shown is known as *point-on-line coincidence*. Notice that the epitaxial lattice is registered along each column of lattice sites on the substrate. This registration minimizes misfit strain, although it is not entirely eliminated. Coincident structures find "magic angles" of alignment to produce the minimum energy structure. Coincidence occurs for lattices with similar, but not identical symmetries (in the diagram a rectangular layer is registered with a square substrate lattice), and the lattice constants are related by a simple, rational fraction or integer multiple of each other. For example, if the substrate has a square lattice with lattice constant a, the overlayer can find a coincident registration if it is a rectangular lattice with lattice constants $a, \frac{3}{2}a$.

Finally, the most common situation found in organics is the nearly total lack of periodicity between the upper and lower layers. This results in an incommensurate registration between the lattices. *Incommensurate lattices* have both different symmetries and lattice constants. Since there is no perfect fit, the lattice constants differ by irrational numbers.

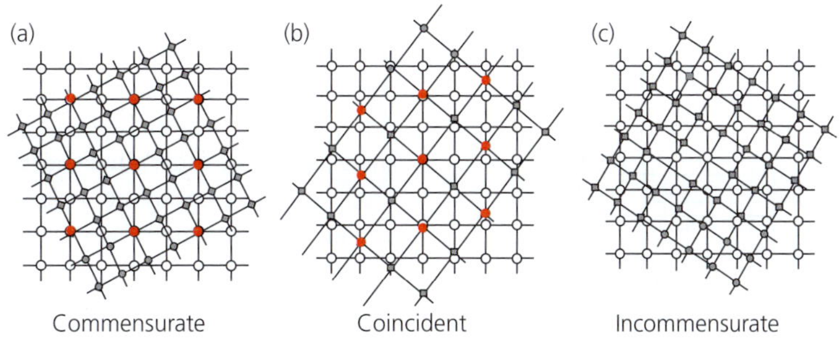

Figure 2.24 Three possible layer schemes of a misfit lattice on a substrate. (a) Commensurate, (b) coincident, and (c) incommensurate. The square substrate lattice is shown by open circles and the overlayer structure is shown by filled circles. Coincidence in (a) and (b) is highlighted by red points.

Such a layer will always be strained, although as shown below, it can nevertheless find a preferred alignment with the substrate that can produce long-range order. Indeed, given the highly elastic nature van der Waals solids, incommensurability does not necessarily result in structural defects, as it would surely do in covalent or ionically bonded materials.

To distinguish these possible relationships between the layer and the substrate, van der Waals film growth has been identified by various terms including epitaxy (Hara et al., 1989), quasiepitaxy (So and Forrest, 1991) and epitaxy (Koma, 1992), depending on the degree of strain or morphology of the resulting film. Here, we will clarify this terminology as well as identify the primary characteristics of molecular films grown on a variety of substrates.

In conventional epitaxy, the molecules are chemisorbed onto the substrate surface through the relatively high energy covalent bond, whereas in van der Waals epitaxy adhesion to the substrate is through purely electrostatic bonds (Burrows et al., 1992). Inevitable mismatches between the substrate and film lattices result in strained layer growth that promotes three different surface morphologies illustrated in Fig. 2.25. In order of decreasing strain between the adsorbate and the substrate, the growth mode is (Smith, 1995): (i) layer-by-layer, or *Frank–van der Merwe growth* (Fig. 2.25a), (ii) wetting layer-plus-island, or *Stranski–Krastanov growth* (Fig. 2.25b), and (iii) island, or *Volmer–Weber growth* (Fig. 2.25c).

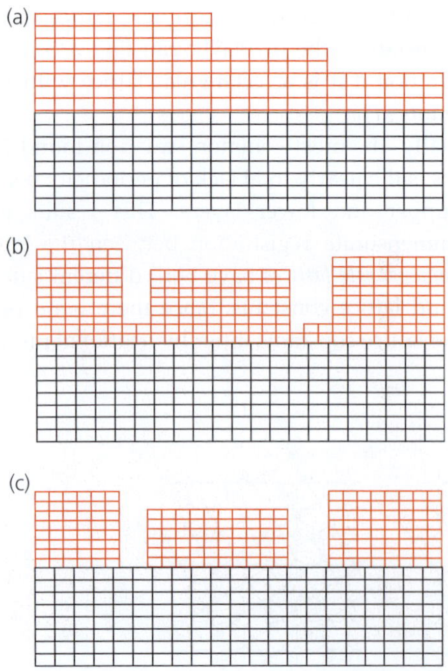

Figure 2.25 Growth modes of epitaxial layers. (a) Layer-by-layer (Frank–van der Merwe), (b) layer-plus-island (Stranski–Krastanov), and (c) island (Volmer–Weber) growth. The substrate lattice is in black, and the epitaxy is in red.

In the absence of significant strain, epitaxial growth occurs layer-by-layer, with each layer lattice matched to the one beneath. In this case, the Matthews–Blakeslee critical thickness is substantially larger than the total layer thickness. The surface morphology is either flat or terraced. If the strain is large (i.e. the epitaxial lattice is significantly mismatched to, or incommensurate with the substrate lattice), only a few molecular layers ("wetting layers") can grow before t_{crit} is exceeded. The strain in subsequent layers exceeds the elastic limit of the material, and hence the strain energy is released by the formation of defects. The defects separate strain-relaxed islands that nucleate on the wetting layers. Finally, for the most highly strained layers, the growth commences by isolated island nucleation on the substrate surface. The size of the islands is determined by the degree of strain; the mean island diameter decreases with increasing lattice mismatch. As the layer thickness is increased, the islands grow in both height and width in a layer-by-layer mode, eventually merging into a highly disrupted and often discontinuous, polycrystalline layer with grain boundaries separating each island. The process of island nucleation and increasing size with thickness is known as *Ostwald ripening* (Ratke and Voorhees, 2002). The rate and scale of island growth are determined by the wetting properties of the nucleation sites on the substrate (i.e. the mismatch in surface energies between substrate and adsorbate), and the available thermal energy that promotes molecular migration across the surface to the growing islands.

2.6.2 van der Waals epitaxy and quasiepitaxy

Perhaps the most "classical" example of a van der Waals bonded solid is graphite. The carbon atoms within a layer are covalently bonded into graphene sheets, yet the forces bonding the sheets together to form a three dimensional solid are by the weaker van der Waals interaction. Like mica, graphite is a layered material whose individual sheets can easily be separated. This gives graphite a significant asymmetry in tensile strength along these orthogonal axes.

As noted in the previous section, it is difficult to find commensurate molecular crystal/substrate combinations leading to unstrained van der Waals epitaxy. Thus, Stranski–Krastanov growth tends to be the most frequently observed mode, albeit with a wetting layer that is considerably thicker than found in inorganic semiconductor materials combinations. Somewhat compensating the lack of a matching substrate is the large number of degrees of freedom that the adlayer molecules can assume due to their large size and complex structure. For example, with appropriate substrate surface preparation, the molecules can assume various configurations that compensate for strain, such as standing on edge, or lying flat on the substrate plane (Shtein et al., 2002, Lassiter et al., 2010, Yook et al., 2011). As a result, all of the modes in Fig. 2.25 have been realized in crystalline organic layers. The mode that develops depends on the kinetic conditions present during growth (i.e. substrate lattice mismatch to the equilibrium film structure, bonding energy between molecules and between molecules and substrate, rate of film growth, substrate temperature, background pressure in the case of vapor phase deposition, etc.).

Examples of different growth modes of organic films on various substrates are given in Figs. 2.26–2.28. Figure 2.26 shows the surface of a PTCDA film grown on highly oriented pyrolitic graphite (HOPG). The growth is by thermal evaporation in ultrahigh vacuum. This is an example of layer-by-layer, Frank-van der Merwe growth, with a monolayer at the bottom left, and a terrace leading to an additional layer, forming a bilayer, at the upper right. The crystalline molecular planes are clearly visible and oriented relative to the graphite substrate. Also, a few defects appear in the monolayer toward the top center of the image. These are apparently holes in the film where PTCDA molecules are absent, exposing the underlying HOPG substrate.

Figure 2.27 shows Stranski–Krastonov, layer-plus-island growth of pentacene on an amorphous polyimide substrate. The polyimide does not have the structure to provide a growth template. Thus, the first layer of thermally deposited pentacene molecules in high vacuum form randomly distributed nucleation sites on the substrate. As subsequent molecules arrive, they migrate to those sites leading to the formation of nanocrystalline islands. The islands continue to grow until they contact a neighboring island, eventually covering the entire substrate area. Further growth only increases the island thicknesses, but the deep fissures between the islands remain. These fissures (the darkest regions in the image) are grain boundaries that possibly extend all the way to the wetting layer. The crystallinity of the islands themselves is inferred from the fern-like, dendritic appearance that radiates from a central location corresponding to the original nucleation site.

Finally, Fig. 2.28 shows an isolated island oligoethylene-bridged ferrocene grown by thermal deposition in ultrahigh vacuum on the surface of Ag (110). This is an example of Volmer–Weber, or island growth. The island has a mesa-like morphology, that is, it is a multilayer structure that rises above the substrate. Growth within the island itself is layer-by-layer, since the island is sufficiently small such that dislocations and grain boundaries are absent. The terraced features characteristic of layer-by-layer growth are clearly apparent across the island surface.

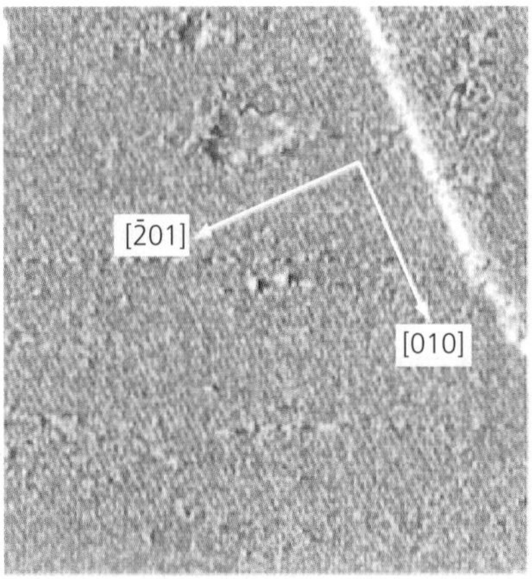

Figure 2.26 Scanning tunneling microscope image of layer-by-layer growth of PTCDA on graphite. Note the molecular planes and the step edge of the second layer. Also indicated are the crystal plane directions of PTCDA. The image size is approximately (150 nm)^2 (Kendrick et al., 1996).

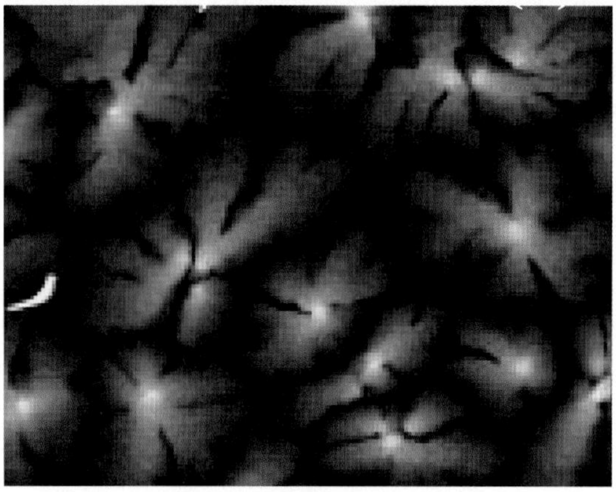

Figure 2.27 Stranski–Krastonov growth of pentacene on polyimide. Note the deep fissures and rough morphology formed by the islands. Approximate dimension of the image is 4×3 μm (Yang et al., 2005).

Figure 2.28 (a) Image of an island of oligoethylene-bridged ferrocene grown in ultrahigh vacuum on Ag (110). The terraces on the island show layer-by-layer growth. (b) Faceting of the island denoted by E1, E2, E3 correspond to the (0,1), (1,1), and (1,0) directions in the $(\bar{2}11)$ plane, respectively. (c) Thickness profile of the island taken along a line between the short arrows. It is roughly 2 μm at its widest point (Zhong et al., 2008).

As in the other growth processes, nucleation on the substrate is random. Once an island is nucleated, subsequently deposited molecules migrate toward those nucleation sites, growing the size and height of the islands. Insufficient material has been deposited for the separate islands to grow together in what would undoubtedly be a very uneven surface with many grain boundaries. The absence of a nearby island in this image, in spite of its multilayer thickness of 40 nm and edge faceting, suggests that the binding energy of the molecules to themselves (i.e. to the island) is significantly larger than the binding energy to the Ag surface. Thus, molecules can migrate relatively large distances until they find an island on which to attach. The substrate temperature of 300 K is sufficiently high to promote this long-distance molecular migration. The energy available to molecular surface motion increases with substrate temperature. Thus, the island thickness is understandably temperature dependent, ranging from 38 nm at 250 K to 51 nm at 350 K.

Quasiepitaxy (QE) has been used to describe, and understand how ordered films can be grown on incommensurate substrates. That is, quasiepitaxial ordering appears to be a unique feature of highly elastic van der Waals solids, where large crystal domains (sometimes extending across entire substrates) can be grown, even though there is no clear integer relationship between the dimensions of the unit cells of the adlayer (called *unit meshes*) and substrate (Fenter et al., 1995, Sellam et al., 2001, Kilian et al., 2006).

The epitaxial relationship between lattices is expressed using a 2D vector:

$$\begin{pmatrix} \mathbf{a}' \\ \mathbf{b}' \end{pmatrix} = \begin{pmatrix} a_{11} & a_{12} \\ b_{21} & b_{22} \end{pmatrix} \begin{pmatrix} \mathbf{a} \\ \mathbf{b} \end{pmatrix} = \mathbf{T}_{\mathrm{a,b}} \begin{pmatrix} \mathbf{a} \\ \mathbf{b} \end{pmatrix}, \qquad (2.76)$$

where $\mathbf{T_{a,b}}$ is the spatial transform matrix. The matrix elements, a_{ij}, b_{ij} $(i, j = 1, 2)$ are rational numbers that transform the substrate lattice unit mesh with primitive vector, (**a**, **b**), into the adlayer lattice vector, (**a′**, **b′**). Even in lattice-mismatched QE growth, however, there remains an orientational relationship between the film and substrate that results in a defined azimuthal orientation between the lattices. Residual stress can develop in the film as it "attempts" to conform to the substrate lattice; this is known as strained QE. Strained QE films can be distorted from their bulk lattice structure, although there is no significant relaxation of this distortion with film thickness. It has been suggested that strain relaxation in QE films results from local periodic variations in molecular density of the adsorbate molecules along the substrate surface. That is, the unit cells are distorted, similar to tetragonal distortion in the vertical growth direction, with the compression of cells in one region followed by expansion farther along the surface. This results in an oscillating energy strain along the surface that is insufficient to generate dislocations (Zhang and Forrest, 1993). Periodic longitudinal lattice distortions, known as *mass density waves*, have been proposed as a strain-limiting mechanism in Ar adsorbed on graphite (McTague and Novaco, 1979).

The ability of van der Waals-bonded molecular crystals to accommodate strain over large distances is due to their small elastic constants. In contrast to van der Waals epitaxy that results in relaxed and often highly disordered systems, QE is often achieved under non-equilibrium growth conditions, that is, on cold substrates or at high growth rates. Since strain relief can occur without inducing defects, Frank–van der Merwe as well as Stranski–Krastanov QE growth modes have been observed.

The primary requirement for QE is that there exists a range over which a surface molecule can be translated relative to the substrate without a significant change in energy. If the total potential between molecules within a layer is U_{intra}, and between molecules in different layers is U_{inter} then this condition met when the relative magnitudes of the inter- and intralayer compressibilities follow

$$\frac{\partial^2 U_{intra}(r)}{\partial r^2} \gg \frac{\partial^2 U_{inter}(r)}{\partial r^2}. \qquad (2.77)$$

This is the governing condition for QE growth. In this admittedly "static" model, the condition for QE is relatively independent of the absolute magnitudes of the energies themselves. As in many cases involving large planar molecules, $U_{inter} > U_{intra}$ even though the inequality of the elastic constants in Eq. 2.77 is reversed.

The QE condition highlights an inherent difference with "orientational epitaxy" of inert gas atoms bonded to surfaces whose periodicities are similar (Itoh et al., 1988). In QE growth, large molecules are often bound to an inorganic substrate with a considerably smaller lattice constant than the adsorbed layer. Hence, the molecule can "sample" many substrate atoms to find its minimum energy position, which is not the case for adsorbed inert gas atoms.

The condition for QE can be understood from the plot of elastic constants for a pair PTCDA molecules shown in Fig. 2.29, where the elastic constant within a layer (proportional to the second derivative, U_{intra}'', as in Eq. 2.77) is clearly larger than the elastic shear constant (proportional to U_{inter}'') near the equilibrium molecular separation corresponding to a displacement of $\Delta r = 0$. Indeed, attempting to compress the molecules in the plane even further (at $\Delta r \to 1$ Å) results in a dramatic increase in lattice stiffness for separations less than the van der Waals radius.

Thus, the QE condition implies that the surface mesh of the overlayer is only slightly distorted when deposited onto a substrate. This small lateral compressibility allows for minor but important adjustments of the layer orientation relative to the substrate. The potential surface of the molecule-substrate interaction has a relatively broad minimum (leading to a small U_{inter}''), allowing for the orientational conformation between lattices without a large expense of strain energy. Rafts of large molecules thus have a range of energy-equivalent positions, and therefore can form (under appropriate growth conditions) an incommensurate overlayer with only a small interfacial strain energy. These assumptions suggest that a QE thin film is a rigid overlayer that is largely undistorted when contacting the incommensurate

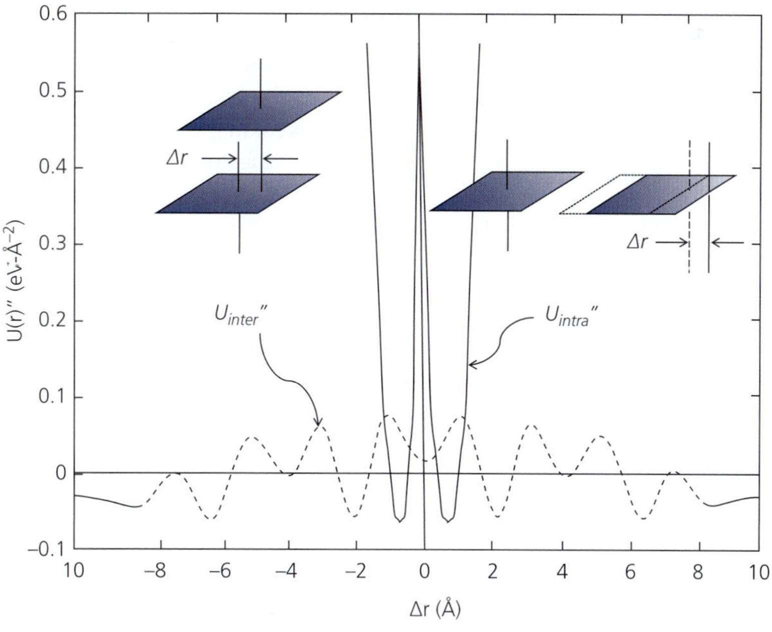

Figure 2.29 Calculated elastic constants of two PTCDA molecules that are positioned perpendicular to the molecular plane (with a second spatial derivative of U_{inter}''), and in-plane (U_{intra}'') as a function of relative molecular displacements, Δr defined in the insets. For calculating U_{inter}'', the "test molecule" is located at an equilibrium molecular distance of $r_0 = 3.2$Å over an infinite, 1D lattice of substrate molecules. The calculation of U_{intra}'' corresponds to that of two adjacent, co-planar molecules. The equilibrium molecular distance within the plane is calculated to be $r_0 \sim 0.7$ Å. Adapted from Forrest and Burrows (1997).

substrate lattice. Since the overlayer has a very low spatial periodicity as compared with the substrate, strain may be relieved through modifications of the internal degrees of freedom within the larger, adsorbed molecular lattice. The elastic distortion of the film does not appear to relax even in the thickest films studied. The details of QE structures are considerably more complex than predicted by the model discussed here, although their general features can be predicted by the somewhat simplified assumptions used to make the calculations both intuitively useful and tractable.

In many respects, the conditions leading to QE are similar to those responsible for self-assembly of materials, a topic we discuss further in Section 2.7 and in Part II. Self-assembly is the process by which the forces between atoms or molecules are sufficiently high that an equilibrium structure, whether it is a colloidal particle, an island grown on a substrate, or even a complete and macroscopically thick layer, is achieved largely independent of its energetic environment, although such assembly is often driven by forces between the layer and the substrate that can serve as a growth template. Hence, crystalline QE layers are essentially self-assembled films grown in their native equilibrium crystal habit on a substrate whose own structure may influence their relative orientation or the degree of strain within the adlayer. The internal organization of the film (i.e. its assembly into its equilibrium structure) arises from strong internal molecular forces that are greater than the adhesive forces to the substrate itself.

2.6.3 Modeling and growth of ordered layers

In previous sections we introduced methods for finding equilibrium structures of bulk molecular crystals. We now turn to methods that can predict the structures of epitaxial layers whose lattices might have significant misfits with the substrate. In particular, we need to develop an understanding of the factors that determine the relative axial orientation of incommensurate lattices. The QE condition explains how strain is accommodated by the asymmetric stiffness between and within a layer, but it does not provide a predictive model for the film structure itself. Several models have been advanced to understand epitaxial growth of covalent/ionic bonded epitaxial layers that can provide a foundation for understanding van der Waals bonded systems (Reiss, 1968, van der Merwe, 1982). The growth dynamics, and indeed the various growth modes achieved are the same for these apparently different systems, although the energies (and hence the temperatures) involved in growth are a function of the different bond strengths. Epitaxial growth is typically modeled using a harmonic potential for the atomic cores, with a period equal to the atomic spacing of the substrate (van der Merwe, 1982). This is different for incommensurate lattices whose potential between overlayer and substrate is anharmonic. Hence, the analytical solutions attained for epitaxial systems must be modified using often impractically computationally intensive methods. One approach is to employ the atom–atom potential calculation simplified using one or more of the following strategies:

1. If we consider a set of n atoms in the epitaxial layer and m atoms in the substrate, the total number of atomic pairwise interactions is $\frac{1}{2}nm$. To make these calculations more efficient, we recognize that van der Waals interactions decrease as $1/r^6$. Hence, only nearest and next-nearest neighbor interactions need to be considered.
2. Extending the model to include molecules where long-range (but weak) intra- and interlayer Coulomb binding plays a role (e.g. where charge transfer between molecules does not occur) can significantly complicate the problem, and to our knowledge has not been treated in detail. However, the calculational methods employed for van der Waals solids can, in principle, be extended to include other, long-range bonding forces. When such interactions are dominant, conventional epitaxy results, and a harmonic potential can be employed. In this case, *intralayer* interactions between molecules need not be included, and the equilibrium configuration of the first monolayer is accurately determined by the single molecule/substrate binding interactions (Hashimoto et al., 1995).
3. Since the number of degrees of freedom within and between the molecules is large, an accurate and predictive calculation must include minimization of total energy with respect to all degrees of freedom. Where possible, therefore, it is essential to limit the degrees of freedom. For example, molecules can be restricted to lie within the substrate plane, and the molecules and substrate are both assumed to be rigid. This is known as the *rigid lattice approximation*. Whether or not it is an accurate representation of reality depends on the details of both the molecular and substrate structures.

Whatever assumptions are made, the calculations must include sufficient physics to lead to an accurate, equilibrium lattice configuration. Example simplifications are to assume that there is an approximate lattice match, or to recognize and exploit symmetries common to both the substrate and the adlayer. Then there are a very limited number of preferred alignments that lead to a minimum energy configuration.

An example of this is to find directions where there is an apparent "fit" between substrate and adlayer. It would then seem logical that this orientation should lead to a minimum energy configuration. For example, for a given situation there may be an angle that results in point-on-line coincidence (Hoshino et al., 1994, 1995), where the lattice points of the adlayer are aligned along a primary (i.e. low Miller index) direction of the substrate, as illustrated in Fig. 2.24b. Coincidence models have also been proposed whereby a large epitaxial "super cell" consisting of many unit cells is found that achieves approximate coincidence along a particular substrate direction (Hillier and Ward, 1996). While these simple coincidence models are appealing, they simply seek "geometric" fits between substrate and supercell. This approach may provide intuitive insights, although it is not based on a rigorous energy minimization that seeks the equilibrium crystal structure. Indeed, point-on-line and super cell approaches can result in several equivalent orientations of the adlayer with the substrate, but they do not distinguish between those with higher or lower energy.

In contrast to coincidence models, energy minimizations employing simplified molecular structures have also been attempted. That is, the full, atom–atom potential summation between two complex molecules can be replaced by an ellipsoidal, site specific, or other potential approximations discussed in Section 2.5. These simplifications reduce the massive computational challenge required in modeling layers consisting of a sufficient number of molecules to approximate actual physical systems, although they have the shortcomings of being applicable to only specific molecular types or substrate-layer morphologies. Hence, they lack a clearly predictive quality of molecular combinations that lead to a particular structure.

In the context of this discussion, the validity of the atom–atom potential method has been directly compared to coincidence models to understand the long range, and perfect QE ordering of multilayer stacks of NTCDA and the neutral charge transfer complex, DB-TCNQ grown by OVPD on KBr substrates (Lunt et al., 2011). Figure 2.30 shows the molecular structural formulae for NTCDA, DB-TCNQ, and KBr. The multilayers were grown on crystalline substrates via OVPD (see Chapter 5). Sustained ordering of NTCDA/DB-TCNQ pairs initiated on a KBr substrate was maintained for several periods, with a clear quasiepitaxial relationship between adjacent, 5 nm thick layers in the stacks.

Figure 2.31 illustrates the multilayer stack along with structural models of the molecules within each layer. The structure was monitored *in situ* during growth using reflection high energy electron diffraction (RHEED) whereby a beam of electrons incident at 1–2° from the surface plane is scattered by the surface mesh of the growing layer. The shallow e-beam incidence angle results only in surface scattering since the electron beam penetration into the film is only ~10 Å. A flat surface results in diffraction streaks whose spacing gives the Bragg spacings of d_{xy} from the (x, y) surface mesh. The pattern for a single crystal appears only when the Bragg condition of

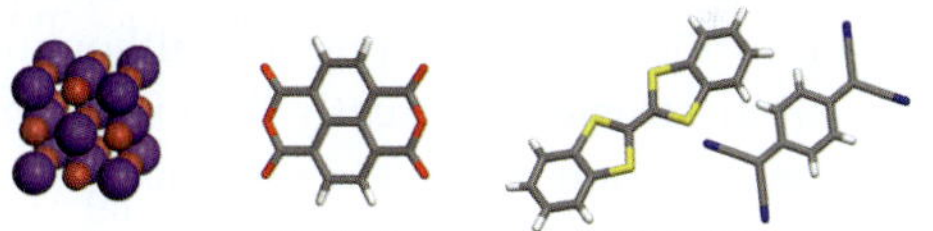

Figure 2.30 Molecular structures of the molecules used in the multilayer stacks in Fig. 2.31. The KBr substrate is an ionic crystal, NTCDA is a neutral molecule and DB-TCNQ is a charge transfer salt. K = red, Br = purple spheres. In NTCDA, C = gray, O = red and H = white. In DB-TCNQ the color code is the same as for NTCDA with S = yellow (in DB) and N = blue.

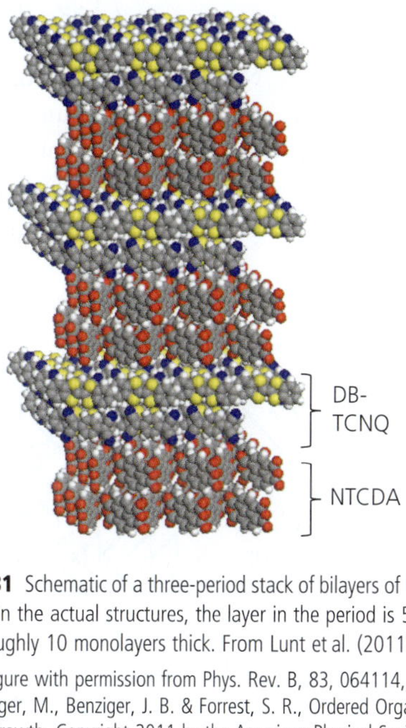

Figure 2.31 Schematic of a three-period stack of bilayers of NTCDA and DB-TCNQ. In the actual structures, the layer in the period is 5 nm thick, which is roughly 10 monolayers thick. From Lunt et al. (2011).

$$m\lambda = 2d_{xy}sin\theta \tag{2.78}$$

is met. Here, λ is the electron wavelength ($= hc/E_{RHEED}$ where E_{RHEED} is the electron beam energy), m = 0, 1, 2…is the diffraction order, and θ is the angle between diffraction streaks.

The growth of the multilayer stack commenced with the NTCDA deposited onto KBr. The diffraction streaks were continuous, indicating a flat surface, and their intensities varied along different azimuthal angles corresponding to different crystal directions in the NTCDA lattice. This indicated single crystalline, layer-by-layer growth across the entire 2 × 2 cm^2 substrate covered by the glancing incidence beam (Lunt et al., 2011). The bulk lattice of NTCDA(100) has unit mesh dimensions of a_1 = 0.531 nm, a_2 = 1.257 nm, and β = 90°, which is slightly different than the reconstructed values measured for the first layer. The in-plane NTCDA lattice constants were not found to vary for thicknesses up to 100 nm, suggesting that although the lattice is reconstructed differently than in the bulk, the maximum strain of $\Delta a/a = 6.4\%$ accumulated even in the thick layers is insufficient to generate dislocations. Note that this amount of strain is more than an order of magnitude larger than can be accommodated in covalently bound crystals without inducing a high density of misfit dislocations, attesting to the elasticity inherent in the van der Waals bond.

The epitaxial relationship between the KBr and NTCDA lattices is $\mathbf{T}_{\mathbf{KBr,NTCDA}} = \begin{pmatrix} 1.99 & 0.009 \\ 0.01 & 0.753 \end{pmatrix}$ (cf. Eq. 2.76), where the NTCDA molecules contact the KBr substrate in its characteristic herringbone structure. Hence, an approximately coincident structure is observed within the error of the measured surface mesh, as shown in Fig. 2.32.

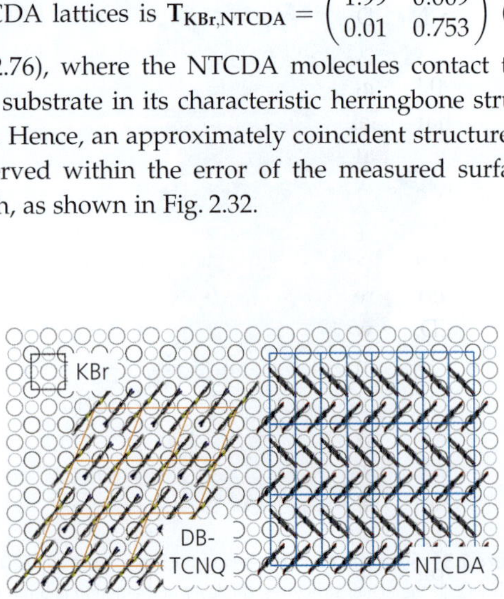

Figure 2.32 Measured orientations of the surface unit meshes of DB-TCNQ and NTCDA on KBr (atoms shown as circles) (Lunt et al., 2011).

The lattice surface mesh of DB-TCNQ(001) on KBr is a_1 = 0.922 nm, a_2 = 1.064 nm, and β = 67.66°, which is similar to its bulk dimensions. Remarkably, although DB and TCNQ molecules were deposited from separate sources in the OVPD system, a 1:1 concentration of these two constituents is achieved in the layer. This is an example of *congruent growth* where the layer has the same stoichiometric composition as the bulk. This implies that excess DB or TCNQ is not incorporated into the lattice and hence is removed by the carrier gas in the growth chamber. Similar congruent growth is observed in binary inorganic compounds such as GaAs and InP reinforcing the conclusion that the growth of organics is similar in almost all ways to that of inorganic compounds once the different binding energies are taken into account.

Finally, the transform matrix between the DB-TCNQ and NTCDA lattices is $\mathbf{T}_{\mathbf{NTCDA,TCNQ}} = \begin{pmatrix} 0.70 & 0.02 \\ 0.32 & 1.96 \end{pmatrix}$. The absence of simple fractions shows that there is no apparent epitaxial relationship (corresponding to exact coincidence or commensurality) between these two molecular structures. The existence of a 5 nm thick, strained single crystal DB-TCNQ layer covering the NTCDA layer with a well-defined orientation even though the lattices are incommensurate provides a compelling example of strained, QE growth. The crystalline alignment, single crystallinity of all layers, and surface flatness are maintained by alternating 5 nm thick NTCDA followed by 5 nm thick DB-TCNQ layers for up to at least five periods (Lunt et al., 2011).

The geometric lattice potential (Hillier and Ward, 1996) that follows (in one dimension) the form $U(r)/U_0 = \cos(2\pi r/a)$, was compared to atom–atom potential calculations as a function of angle between lattice directions. In the geometric approach, (U/U_0) vs. relative lattice angles of the overlayer and substrate can describe the degree of lattice commensurality. For example, if $U(\theta)/U_0 = 1.0$ for all relative lattice orientations, θ, the lattices are incommensurate and there is no preferred direction. However if $U(\theta)/U_0 = 0.0$ at a symmetry equivalent orientations, the lattices are fully commensurate.

The atom–atom potential calculations have to account for charge transfer between the anion and cation pair of DB and TCNQ, respectively, which makes the calculations considerably more complicated. Nevertheless, by only including van der Waals forces in the simulations confirmed that the intermolecular distances between DB and TCNQ, and between DB-TCNQ pairs of 3.29 Å and 3.27 Å, respectively, were similar to their experimental values, indicating that charge transfer contributions were small and could be ignored.

Structural simulations using the geometric lattice and the full atomistic van der Waals potential energy between the substrate and the adlayer as a function of substrate–adlayer azimuthal rotation angle, θ, are compared in Fig. 2.33a. The geometric potential tends toward coincidence (with a minimum $U(\theta)/U_0 \sim 0.6$) at the symmetry equivalent angles of 0°, 90°, and 180° for NTCDA lattices comprising 3 x 3 unit cells on KBr. For larger overlayer cell meshes (15 × 15 cells) the coincidence becomes less pronounced, with only a shallow minimum of $U(\theta)/U_0 \sim 0.95$ at $\theta = 0°$. In contrast,

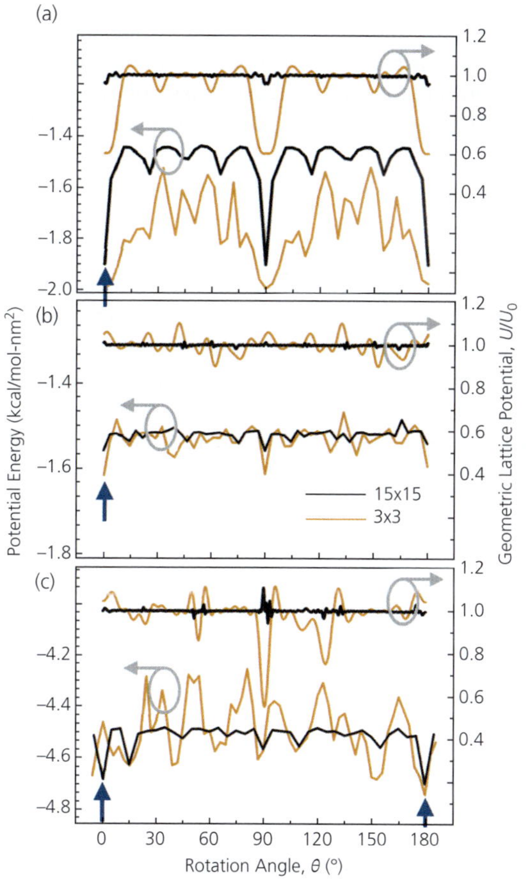

Figure 2.33 Lattice potential energy (via the atom–atom potential method and via the coincident geometric method) for different lattice combinations. (a) NTCDA on KBr, (b) DB-TCNQ on KBr, and (c) DB-TCNQ on NTCDA. The orange line is for a 3 × 3, whereas the black line is for a 15 × 15 surface lattice. Arrows indicate measured rotation angles, θ (Lunt et al., 2011).

the atom–atom energy calculations indicate an energy minimum (baseline minus bottom of potential well) of approximately −0.3 kcal/mol nm^2 at $\theta = 0°$ that grows deeper and narrower with increasing cell size. The depth of the energy well for a 3 × 3 surface mesh is 75meV, or ~$3k_BT$ at room temperature where growth proceeds. This suggests the route taken to achieve ordered growth. As the NTCDA nucleates, the first monolayer samples the energy landscape to find a preferred geometric alignment. As the grain grows, the alignment becomes fixed since the energy cost for rotation increases.

The preferred alignments predicted by both models are consistent with electron diffraction data. A significant difference between the approaches, however, is that the atom–atom potential calculation predicts that the total binding energy increases with the area of the film, whereas the coincidence model suggests that the energy decreases under this circumstance. This latter result is both unintuitive and unphysical.

For DB-TCNQ grown on KBr (see Fig. 2.33b), the geometric model fails to predict the existence of any preferred angle, that is, it suggests an incommensurate lattice at every rotation, $U(\theta)/U_0 \sim 1$. On the other hand, the potential calculations exhibit minima at 0° and 90° consistent with the experimental data. As the lattice size increases, the potential well depth decreases while the minima remain at the same angles. The depth of the well is ~0.1 kcal/mol nm^2, which is smaller than for NTCDA on KBr, indicating a weaker driving force for alignment. For DB-TCNQ deposited on NTCDA in Fig. 2.23c, the geometric model predicts one coincident ($U(\theta)/U_0 \sim 0.6$) alignment at $\theta = 90.5°$ (a_1(DB-TCNQ)∥a_2(NTCDA)) for small cells that disappear for larger surface meshes. In contrast, the atom–atom potential calculations indicate several energy minima for small domain sizes, with the deepest energy minimum at 180°. For the larger 15 × 15 unit cell domains, the preferred alignments are at $\theta = 0°$ and 180° (a_1(DB-TCNQ)∥ a_1(NTCDA)) with a well depth similar to NTCDA of ~0.2 kcal/mol nm^2. The 0° and 180° orientations are experimentally observed.

The energy-based calculations predict that preferred alignments can be observed for entirely incommensurate organic lattices, which are absent in purely geometric (i.e. coincidence or periodic potential) approximations. This highlights the importance of molecular structure over lattice geometry in determining the lowest energy layer-to-substrate orientations.

2.6.4 Dependence of strain on molecular size and shape

Meeting the QE condition of $U_{inter}'' \ll U_{intra}''$ depends on many different factors, including details of the molecular size and shape, the symmetry match between substrate and adlayer, substrate lattice structure and surface morphology, etc. Molecular size, for example, strongly influences the shear forces between molecules. For example, the energy surfaces calculated for a (small) benzene (C_6H_6) and a much larger planar coronene ($C_{24}H_{12}$) molecule on a PTCDA substrate lattice are shown in Fig. 2.34. Unsurprisingly, the energy minimum of the larger molecule is broad and flat as compared to that of benzene. Hence, the intermolecular shear stress (U_{inter}'') at the coronene/PTCDA interface is therefore much smaller than that of benzene, which should enhance the probability for long range QE ordering of this materials combination.

Molecular symmetry also plays a role in determining strain between layers. Figure 2.35 plots the minimum energies of adhesion on PTCDA for the linear polyacene series that includes benzene, naphthalene, anthracene, tetracene and pentacene, with 1, 2, 3, 4, and 5 fused rings, respectively. The shear stress decreases monotonically for the three smallest molecules, and then it increases with a further increase in molecular length. Since the perylene core of PTCDA extends to only three fused rings, longer molecules (e.g. tetracene and pentacene) extend beyond the central molecule in the PTCDA surface cell over the carboxyl end groups, thus increasing the resistance to shear.

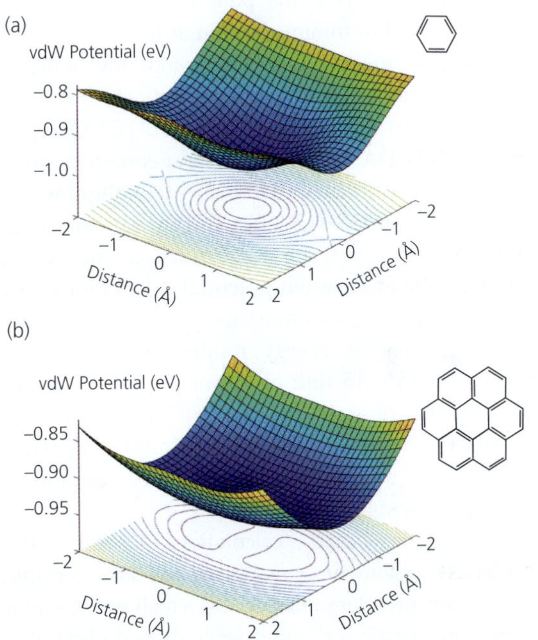

Figure 2.34 Potential energy surfaces calculated for (a) benzene and (b) coronene molecules positioned 3.2 Å above a PTCDA molecule. Notice the broad, flat minimum for the larger coronene, indicating a lower shear stress compared to benzene. Molecular formulas for the admolecules are shown. Adapted from Forrest and Zhang (1994).

From the foregoing, it is apparent that the shear stress normalized to the equilibrium energies between layers decreases monotonically with increasing molecular size, or number of carbon rings. Hence, we expect that the ability to grow ordered QE layers increases as we progress from smaller (benzene) to larger, "rounder" molecules (coronene). As in all epitaxial growths, QE is favored when both the molecular shape and size of the substrate and overlayer molecules are approximately matched.

We once more emphasize that while the energy calculations predict the favored, or minimum energy configuration of the grown layers, the degree and extent of layer ordering that is actually achieved depends critically on the non-equilibrium thermodynamic conditions under which growth occurs. Static models do not predict these conditions. However, we find that certain low-energy thresholds between theoretically predicted structural isomorphs imply that growth at low substrate temperatures is favored for achieving uniform crystalline order, as has often been experimentally observed. Nevertheless, molecular dynamic models that consider the kinetic energy available during deposition are required to find the preferred crystal structures. In Part II we will describe how such dynamic simulations can be

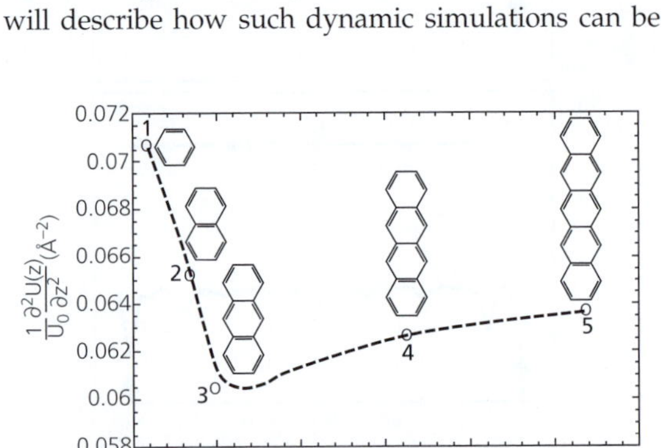

Figure 2.35 Normalized shear stress vs. "molecular shape" (as defined as the moment of inertia of a molecule rotated about an axis centered in the molecule and perpendicular to the molecular plane). The series of linear polyacenes are (**1**) benzene, (**2**) naphthalene, (**3**) anthracene, (**4**) tetracene, and (**5**) pentacene. Adapted from Forrest and Zhang (1994).

useful in describing and predicting the charge and energy transport characteristics of a range of molecular structures that are employed in high performance organic devices.

2.7 Self-assembly

Self-assembly is the process by which the structure of a solid is guided by the cohesive energy and symmetry of its components. Self-assembly can spontaneously occur at several length scales, from the atomic or molecular, to the nano-, all the way to the macroscale. Indeed, self-assembly of solids can span all of these scales, resulting in what is often termed *multiscale assembly* (Whitesides and Boncheva, 2002). Self-assembly occurs ubiquitously in nature, driving the structure from cells to entire organisms. A classic example of self-assembly is the growth of hoarfrost. Water molecules nucleate into crystallites on dust or other imperfections on a cold window pane. As further molecules are adsorbed, the crystal grows into fern-like, self-assembled dendritic structures. Similarly, QE ordering is partially driven by self-assembly of the adsorbed film: the structure of the film is ultimately determined by strong intermolecular cohesion within a layer, with lesser adhesive forces binding the layer to the substrate. Self-assembly has been observed almost independent of the method used in structural formation, and has been achieved by growth in the vapor phase as well as in solution.

Molecular materials are particularly suited to layer formation by self-assembly due to weak intermolecular forces that allows for adjustment of molecules to form near-perfect structures even at room temperature. In Part II we will describe numerous high performance device structures that rely on self-assembled mono and multilayers to achieve desired film morphologies. Self-assembly has been found particularly useful in solar cells, transistors and in limited-dimensional devices (Chapter 7–9).

Self-assembly relies on molecules or molecular sub-units that interact with each other with sufficient strength to form a stable structure. As we have seen in Section 2.4, the subunits must interact with an appropriate balance of attraction and repulsion. If the molecules adhere with excessively high binding forces, a glassy structure results. However, if the binding is on the order of the thermal energy, then the molecules can adjust to find the most energetically favorable configuration, which as we have seen, is often a crystal structure with order that extends over considerable distances. In this sense, the assembly process must be reversible, where errors can be rapidly corrected before the structural defects become embedded in an imperfect assembly. This emphasizes the importance of suppling external energy (most probably, thermal) to allow for structural annealing and mass transport. The formation of self-assembled structures with different degrees of order, depending on the balance of attractive and repulsive forces between units, is illustrated in Fig. 2.36.

Self-assembly plays an important role in the structural organization of organic thin films and crystals, whether they are based on monomers, oligomers, dendrimers, or polymers. The most elemental and commonly used form of self-assembly in organic electronics is where molecules are designed and synthesized that have functional groups, steric properties or employ other strategies that force the assembly of molecules into large-scale structures. The adhesive forces between constituents are rarely covalent due to the strength of the bond, but rather are assembled via electrostatic Coulomb, van der Waals, or hydrogen bonds.

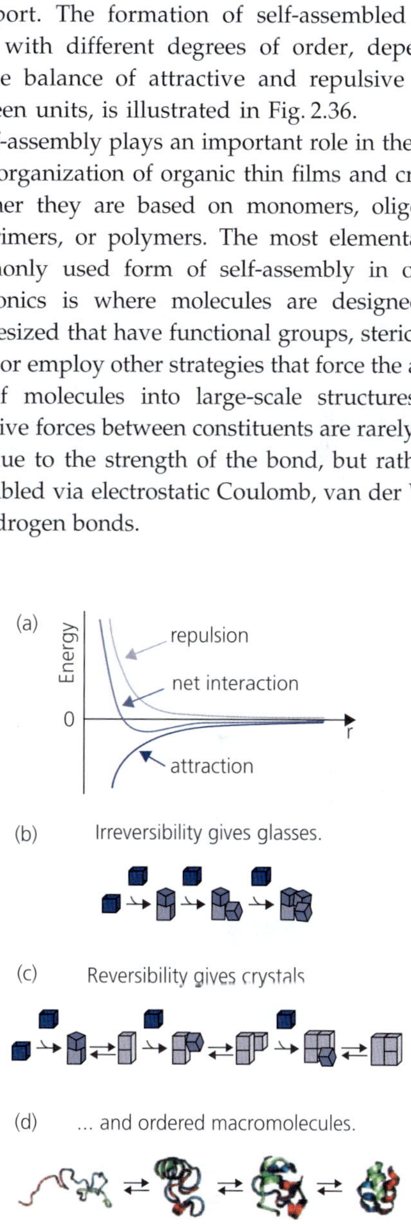

Figure 2.36 (a) A balance of repulsive and attractive forces between structural elements is required for self-assembly. For example (b) shows that when the attractive forces are large, a disordered glass is formed, whereas (c) when thermal energy is available for weakly bonded elements, they can adjust to find the lowest energy configuration to form a crystal. (d) Likewise, large macromolecules such as proteins or cells can assemble with an appropriate balance of repulsive and attractive interactions (Whitesides and Boncheva, 2002).

A particularly useful electronic application of self-assembly is substrate space filling via deposition of *self-assembled monolayers* (SAMs). Alkanes with sulfur end groups (known as alkane thiols) can attach to Au substrates. Organosilicon groups can also attach to appropriately prepared surfaces (Ullman, 1996). A functionalized, space-filling layer pre-assembled on a substrate can serve as a template for the ordered growth of a subsequently deposited active organic molecular species (see Fig. 2.37). This strategy has been used to initiate ordered growth of pentacene channel layers on pre-deposited octadecyltrichlorosilane (OTS) SAMs to achieve high mobility organic thin film transistors (Section 8.6.1) (Shtein et al., 2002). Numerous strategies for using SAMs to create a function or to prevent metallic particles from sticking to each other on contact have been employed in the design of materials and structures to suit a particular application need.

Beyond molecular assemblies, structures several nanometers in size have been demonstrated to exhibit interesting optoelectronic as well as mechanical properties. For example, composites consisting of an organic matrix where semiconductor or metallic quantum dots, nanorods, tripods, tetrapods and other more complex structures are embedded have applications to organic light emission, photodetection, and solar energy harvesting (Murray et al., 1993, Coe et al., 2002, Gur et al., 2007). An example of a population of CdS tetrapod nanoparticles grown from solution from CdSe seeds in Fig. 2.38 illustrates the control that can be obtained in creating self-assembled structures driven by the chemical composition of the individual constituents comprising the whole. Indeed, using self-assembled structures in colloidal solids has become an area of intense investigation from both a practical as well as a fundamental materials science perspective. Their application to organic photodetectors, memories and light emitting devices is found in Part II of this volume.

Beyond their optoelectronic properties, self-assembled structures can serve as nano-scaffolds on which larger, ordered mesoscopic structures can be built. For example, highly complex assemblies based on folded proteins are responsible for such varied structures as molecular motors, photosynthetic complexes and artificial materials with a wide range of mechanical properties. An example of a natural self-assembled structure is *the light harvesting complex*, LH2, shown in Fig. 2.39. This is a nine-fold symmetric system of two concentric cylinders supported by protein coils that optimize the photogeneration of charge in the bacteria, *Rhodopsuedomonas acidophila*. The total volume of the assembly is 4287 nm^3. Indeed, cells found in all organisms, from the most primitive to the most complex are, in themselves examples of complex, mesoscopic self-assembled units with volumes 10^6 times larger than LH2. These units feature many diverse components, each providing a specific function to the living organism

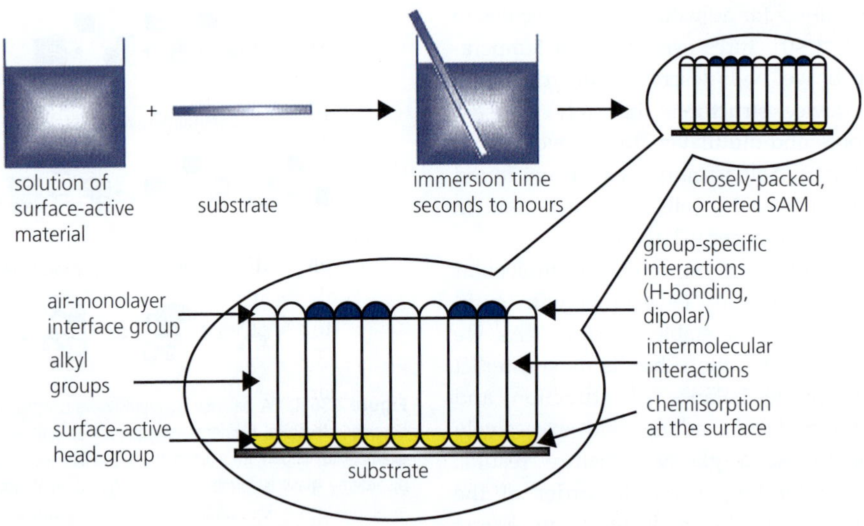

Figure 2.37 Illustration of the formation of a self-assembled monolayer onto a substrate. In this case, an alkyl chain is functionalized on one end to allow for chemical bonding to the substrate atoms, whereas the opposite end is functionalized to accommodate a different molecule that is later deposited, or that prevents the continued deposition of more than a single monolayer (Ullman, 1996).

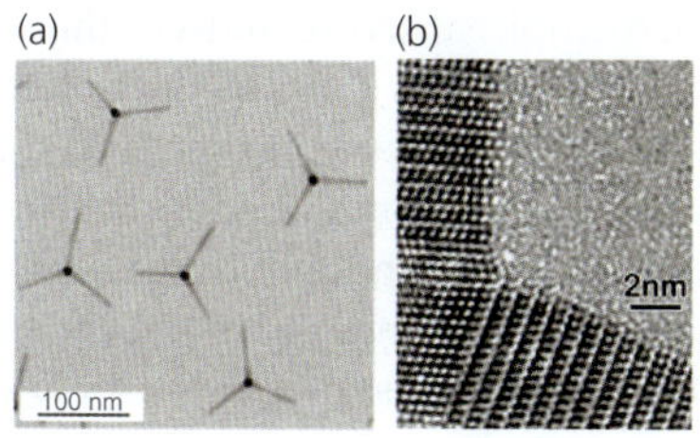

Figure 2.38 (a) Tetrapods consisting of a CdSe seed with three, radiating CdS arms. (b) Transmission electron microscope image showing the zincblende structure of the seed (upper right) and the wurtzite structure of two arms (Nelson et al., 2007).

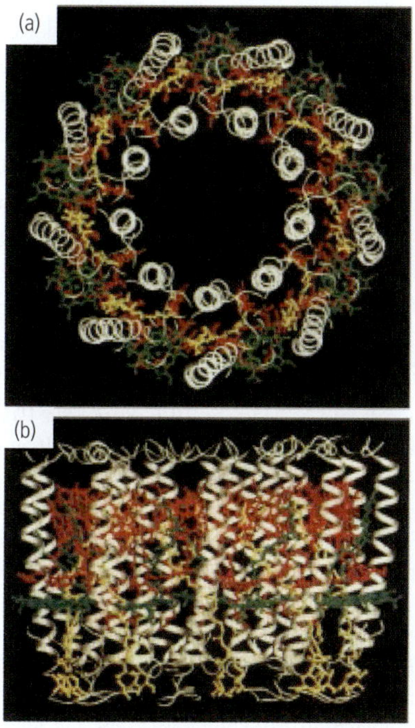

Figure 2.39 Two views of the photosynthetic light harvesting complex, LH2 showing the incredible complexity and diversity of this mesoscopic biological, self-assembled system. The white coils are the protein scaffolds, with other colors representing smaller molecular units active in the charge generating process. (a) Top view of the nine-segment complex, (b) side view. The dimensions of the unit cell are $a = b = 120.3$ Å, $c = 296.2$ Å (McDermott et al., 1995).

as a whole. Considerable research has been done over the years to mimic (largely without success) various aspects of biological systems using self-assembly.

The process of self-organization of a macrostructure based on the function and composition of its microscopic components has been taken beyond the molecular, nano-, and mesoscopic scales discussed above, to realize large-scale mechanical systems. For example small multifaceted objects and even robots can be programmed to self-replicate and then organize into a larger system (Gross and Dorigo, 2008). While this type of self-assembly is beyond the scope of this book, it is nevertheless illustrative of the power and versatility of multiscale self-assembly, where building blocks at the nanoscale create sub-assemblies at the microscale. These, in turn assemble into larger functional systems at the millimeter and centimeter scales. This is, in fact known as a *biomimetic* approach, whereby cells build to substructures that assemble into organs, and ultimately into a fully functional living organism. Self-assembly similarly provides a route to creating a very diverse range of functions in organic optoelectronics. It owes its usefulness and consequent nearly ubiquitous adaptation in molecular devices to the comparatively weak van der Waals bonds that allows for facile adjustment of structure and properties during the assembly process.

2.8 Summing up

In this chapter we have considered the important role that structure plays in determining the properties of organic (and all other) materials. We found that materials can be bonded by a variety of forces whose strengths range over many orders of magnitude. All of these forces are electrostatic, but their strengths depend on whether the electrons are shared between atoms as in the strongest covalent, metallic or ionic bonds, or whether the attractive forces are based on relatively weak dipolar or higher multipole forces, as in van der Waals and hydrogen bonds.

Based on our understanding of the cohesive forces, we showed that molecular materials are bonded primarily by van der Waals attraction. The equilibrium crystal structures are found, to first order, using atom–atom potential calculations, although this static model does not have the capability for predicting the thermodynamic conditions that are needed to achieve a given structure. More comprehensive (and necessarily complex) MD simulations are required to achieve this goal. Ultimately, the materials scientist is challenged to use a combination of intuition backed by deep knowledge and experience to grow high quality molecular layers useful for both scientific as well as technological applications. We will discuss the methods used to achieve such crystals and films in Chapter 5.

A particular attribute of weak van der Waals forces is the ability to achieve long range order in layers grown onto substrates where there is substantial strain between the contacting lattices. Even incommensurate layers with strains approaching 10% can be grown without incurring a high density of defects. Such a large strain accommodation is unheard of in covalent and ionically bonded inorganic semiconductors. Thus, van der Waals solids avail themselves to applications that are inaccessible to conventional semiconductors, enabling one of the most attractive features of organic electronics: the ability to achieve very high thin film device performance without regard to lattice matching constraints with the substrate.

A qualitative summary of the growth modes of materials with different interlayer binding energies is shown in Fig. 2.40. Covalent and ionically bonded structure can accommodate very small strains (typically $< 10^{-3}$). Use of strained layer epitaxy, whereby alternating layers with strains of opposite signs can increase the strain tolerance, allows for relaxation of the Matthews–Blakeslee thickness limit. However, strain balance requires the use of layers of different compositions which may not be desirable for all applications. In contrast, van der Waals structures have bonding strengths at least an order of magnitude smaller than inorganic semiconductors, allowing them to accommodate a significantly larger strain energy. The growth modes for this class of materials are variously termed van der Waals epitaxy, strained van der Waals epitaxy, and quasiepitaxy. The illustration attempts to distinguish these different modes, although the terminology is often used interchangeably in the literature. Finally, for all solids, excessive strain results in relaxed polycrystalline or amorphous structure. We will find in subsequent chapters that the vast majority of organic devices exploit this class of film structure.

We also found that self-assembly of molecules can be used to drive structural organization of a higher order, macroscopic structure. Indeed, the process of quasiepitaxy has many features common to self-assembled systems. Quasiepitaxial layers are those whose lattices are incommensurate, or have large lattice mismatches with the substrate lattice, yet they can exhibit long-range order and a well-defined orientation with the substrate. This structural organization is attributed to the pronounced asymmetry of bonding forces within the layer, and between the layer and the substrate. Self-assembly is found to play a significant role in the organization of molecular and polymeric solids, and it is the governing principle that drives the formation of mesoscopic biological assemblies such as proteins, and functional biological complexes. It is responsible for such fundamental processes as photosynthesis and even the functioning of cells and entire organisms. Indeed, self-assembly appears to be one of the most fundamental principles governing the form and function of all biological systems. As a result, biomimicry provides a foundation on which many concepts in organic electronics are built.

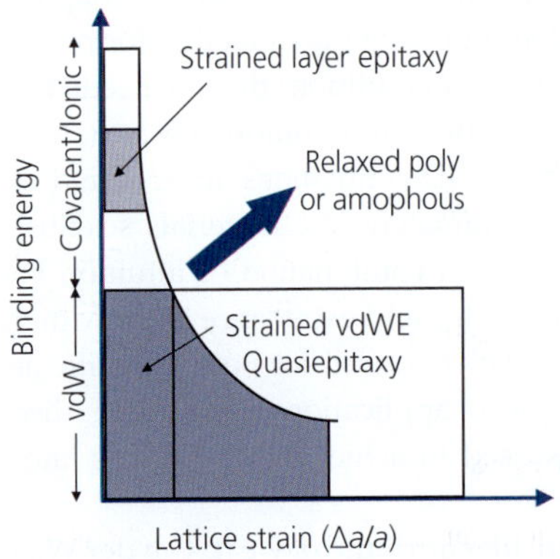

Figure 2.40 A qualitative plot of the binding energy (which spans several decades) vs. lattice strain, and the resulting possible growth modes over this range of parameters.

Further reading

1. N. Ashcroft and N. Mermin, 1976. *Solid State Physics*. New York: Holt, Rinehart and Winston.
2. P. W. Atkins and R. S. Friedman, 1997. *Molecular Quantum Mechanics*, 3rd ed. New York: Oxford University Press.
3. S. R. Forrest, 1997. Ultrathin organic films grown by organic molecular beam deposition and related techniques. *Chem. Rev.*, 97, 1793.
4. F. Giustino, 2014. *Materials Modelling using Density Functional Theory: Properties and Predictions*. New York: Oxford University Press.
5. J. N. Israelachvili, 2011. *Intermolecular and Surface Forces*, 3rd ed. Waltham, MA: Academic Press.
6. J. D. Jackson, 1998. *Classical Electrodynamics*, 3rd ed. New York: John Wiley and Sons.
7. K. Jacobs, 2008. Fundamentals of equilibrium thermodynamics of crystal growth. In: *Crystal Growth Technology: From Fundamentals and Simulation to Large-scale Production* (ed. H. J. Scheel and P. Capper). Weinheim, Germany: Wiley-VCH Verlag.
8. A. I. Kitaigorodsky, 1973. *Molecular Crystals and Molecules*. New York: Academic Press.
9. G. M. Whitesides and M. Boncheva, 2002. Beyond molecules: self-assembly of mesoscopic and macroscopic components. *Proc. Natl. Acad. Sci.*, 99, 4769.

Problems

1. Crystal symmetry groups.
   (a) What is the symmetry group of the diamond lattice?
   (b) What is its coordination number?

(c) What is the symmetry group of the zincblende lattice?
(d) Calculate the d_{444} spacing for a zincblende structure with lattice constant, a.
(e) Answer (a) and (c) for graphite.

2. For the crystal structures in Table 2.2,
(a) Derive the packing fractions.
(b) What are the coordination numbers of the lattices considered?
(c) The wurtzite structure is a lattice with hexagonal symmetry and a basis of $Z = 2$, as shown in Fig. P2.2. Do parts (a) and (b) for a wurtzite structure.

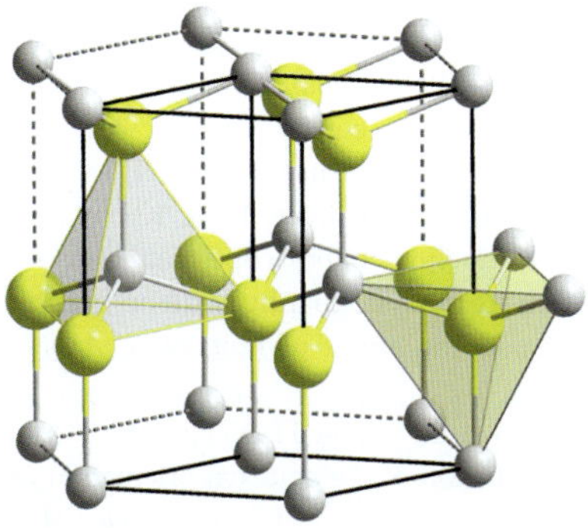

Figure P2.2 Wurtzite structure consisting of a basis of two atoms (gray and yellow) (M. Khaksar, et al., 2015).

3. Assume that a charge transfer molecular pair form a hexagonal close-packed structure with dimensions a and $c = 2a$. Assume that one of the pair of molecules is positioned in the center of the unit cell, that is, the basis is $Z = 2$, and that $\delta = 0.6q$ is transferred between the anion to the cation.
(a) Calculate the first five terms of the Madelung constant for this structure.
(b) Assuming that only ionic forces are important, calculate the total crystal energy using the result in part (a).
(c) Calculate the bulk modulus of the crystal assuming uniaxial compression along the c-axis.

4. (a) Fill in the missing steps leading to the derivations of the expressions for the fixed dipole-fixed dipole (Keesom) and fixed dipole-induced dipole interaction energies in Eqs. 2.45 and 2.54, respectively.
(b) The quantum mechanical treatment leading to the dispersion energy in Eq. 2.64 is based on a one dimensional interaction between dipoles. Extend this to three dimensions and compare with the London equation in Eq. 2.65.

5. The "self-energy" of a molecule in a medium is the energy equal to the sum of all of its interactions with its neighbors (see Fig. 2.11).
(a) By integrating the attractive energy over all space, show that this self-energy of an isolated molecule approximated as a hard sphere of radius, R, in a medium of density, ρ, consisting of induced dipoles with an attractive pair potential $u(r) = -A/r^6$ at $r > 2R$ is $U_{self} = -\pi A\rho/6R^3$.
(b) If a spherical molecule is placed into a close-packed medium, it will have 12 nearest neighbors. Show that by placing the molecule into its own medium, the total energy change is $U_{self} = 6u(2R)$.
(c) From (a), show that $U_{self} = -4u(2R)$.
(d) What is the source of the difference between (b) and (c)?

6. Assume that two benzene molecules (C_6H_6) form a face-to-face dimer. Using the atom–atom potential method, calculate:
(a) the equilibrium distance between the molecules forming the dimer;
(b) the equilibrium energy;
(c) the elastic constant;
(d) the bulk modulus.
(e) Estimate the boiling point of benzene from this result.

7. Atomic lattice potentials can be approximated by a periodic potential of the form (in one dimension):

$$U(r) = U_0 cos(2\pi r/a)$$

where U_0 is the strength of the interatomic potential, r is the displacement, and a is the substrate (1D) lattice constant.
(a) Write the generalized 2D geometric potential.
(b) Defining the transformation matrix, **T**, which depends on the relative azimuthal relationship between the adsorbed layer and the substrate lattice, show that:

$$T = \begin{pmatrix} b_1 sin(\alpha - \theta)/a_1 sin(\alpha) & b_1 sin(\theta)/a_2 sin(\alpha) \\ b_2 sin(\alpha - \theta - \beta)/a_1 sin(\alpha) & b_2 sin(\theta + \beta)/a_2 sin(\alpha) \end{pmatrix}$$

where i, j are the lattice sites of the adsorbed layers, and $a_{1,2}, \alpha$ and $b_{1,2}, \beta$ are the primitive vectors and angles of the substrate and adlayer unit meshes, respectively, and θ is the angle between principal lattice directions in the two lattices.
(c) Show that $U(\theta)/U_0 = 0$ at all symmetry equivalent angles corresponds to the case of perfect commensurality, whereas complete inccommensurality corresponds to $U/U_0 = 1$ at all angles.

8. (a) Calculate the zero point energies of Ar and anthracene.
(b) How does this compare to the van der Waals energy and the thermal energy (at room temperature) of these two species?

9. (a) Solve Eq. 2.30 and 2.31 for the H_2^+ ion and show that the result for the two lowest energy levels is given in Fig. 2.9. *Hint*: This solution requires a transformation to elliptical coordinates.
(b) What is the expectation value of the Hamiltonian, H? How does this compare with literature values for the energies of H_2^+?

References

Abe, A., Jernigan, R. L. & Flory, P. J. 1966. Conformational energies of n-alkanes and the random configuration of higher homologs including polymethylene. *J. Am. Chem. Soc.*, 88, 631.

Abrahamson, A. A. 1969. Born–Mayer type interatomic potential for neutral ground-state atoms with Z = 2 to Z = 105. *Phys. Rev.*, 178, 76.

Ashcroft, N. W. & Mermin, N. D. 1976. *Solid State Physics.* New York: Holt, Rinehart and Winston.

Atkins, P. W. & Friedman, R. S. 1997. *Molecular Quantum Mechanics.* New York: Oxford University Press.

Baldo, M., Deutsch, M., Burrows, P., Gossenberger, H., Gerstenberg, M., Ban, V. & Forrest, S. 1998. Organic vapor phase deposition. *Adv. Mater.*, 10, 1505.

Beck, T. L., Paulaitis, M. E. & Pratt, L. R. 2006. *The Potential Distribution Theorem and Models of Molecular Solutions.* Cambridge: Cambridge University Press.

Bovia Corp. 2016. *Materials Studio v8.0, Forcite Module.* San Diego, CA: Bovia Corp.

Buckingham, R. A. 1938. The classical equation of state of gaseous helium, neon and argon. *Proc. Roy. Soc. Ser. A*, 168, 264.

Burrows, P. E., Zhang, Y., Haskal, E. I. & Forrest, S. R. 1992. Observation and modelling of quasi-epitaxial growth of a crystalline organic thin film. *Appl. Phys. Lett.*, 61, 2417.

Coe, S., Woo, W.-K., Bawendi, M. & Bulovic, V. 2002. Electroluminescence from single monolayers of nanocrystals in molecular organic devices. *Nature*, 420, 800.

Cox, J. D. & Pilcher, G. 1970. *Thermochemistry of Organic and Organometallic Compounds.* New York: Academic Press.

Fenter, P., Burrows, P. E., Eisenberger, P. & Forrest, S. R. 1995. Layer-by-layer quasi-epitaxial growth of a crystalline organic thin film. *J. Cryst. Growth*, 152, 65.

Forrest, S. R. 1997. Ultrathin organic films grown by organic molecular beam deposition and related techniques. *Chem. Rev.*, 97, 1793.

Forrest, S. R. & Burrows, P. E. 1997. Growth modes of organic semiconductor thin films using organic molecular beam deposition: epitaxy, van der Waals epitaxy, and quasi-epitaxy. *Supramol. Sci.*, 4, 127.

Forrest, S. R. & Zhang, Y. 1994. Ultra-high vacuum quasi-epitaxial growth of model van der Waals thin films. I. Theory. *Phys. Rev. B*, 49, 11297.

Forrest, S. R., Kaplan, M. L. & Schmidt, P. H. 1984a. Organic-on-inorganic semiconductor contact barrier diodes. I. Theory with applications to organic thin films and prototype devices. *J. Appl. Phys.*, 55, 1492.

Forrest, S. R., Kaplan, M. L. & Schmidt, P. H. 1984b. Organic-on-inorganic semiconductor contact barrier diodes. II. Dependence on organic film and metal contact properties. *J. Appl. Phys.*, 56, 543.

Forrest, S. R., Kaplan, M. L. & Schmidt, P. H. 1987. Organic thin films for semiconductor wafer diagnostics. *Annu. Rev. Mater. Sci.*, 17, 189.

Giustino, F. 2014. *Materials Modelling using Density Functional Theory: Properties and Predictions.* New York: Oxford University Press.

Gross, R. & Dorigo, M. 2008. Self-assembly at the macroscopic scale. *Proc. IEEE*, 96, 1490.

Gur, I., Fromer, N. A., Chen, C.-P., Kanaras, A. G. & Alivisatos, A. P. 2007. Hybrid solar cells with prescribed nanoscale morphologies based on hyperbranched semiconductor nanocrystals. *Nano Lett.*, 7, 409.

Hara, M., Sasabe, H., Yamada, A. & Garito, A. F. 1989. Epitaxial growth of organic thin films by organic molecular beam epitaxy. *Jpn. J. Appl. Phys.*, 28, L306.

Hashimoto, S., Ogawa, T., Asaka, N., Isoda, S. & Kobayashi, T. 1995. Pseudomorphic vanadyl-phthaocyanine and its stable orientation on KBr. *J. Cryst. Growth*, 146, 649.

Hillier, A. C. & Ward, M. D. 1996. Epitaxial interactions between molecular overlayers and ordered substrates. *Phys. Rev. B*, 54, 14037.

Hirose, Y., Aristov, V., Soukiassian, P., Bulovic, V., Forrest, S. R. & Kahn, A. 1996. Chemistry and electronic properties of metal/organic semiconductor interfaces: Al, Ti, In, Sn, Ag and Au on PTCDA. *Phys. Rev. B*, 54, 13748.

Hoshino, A., Isoda, S., Kurata, H. & Kobayashi, T. 1994. Scanning tunneling microscope contrast of perylene 3,4,9,10 tetracarboxylic-dianhydride on graphite and its application to the study of epitaxy. *J. Appl. Phys.*, 76, 4113.

Hoshino, A., Isoda, S., Kurata, H., Kobaysahi, T. & Yamashita, Y. 1995. Prediction of the epitaxial orientation of ultrathin organic films on graphite. *Jpn. J. Appl. Phys.*, 34, 3858.

Israelachvili, J. N. 2011. *Intermolecular and Surface Forces.* Waltham, MA: Academic Press.

Itoh, C., Miyazaki, T., Aizawa, K., Aoki, H. & Okazaki, M. 1988. Monte Carlo study of lattice-mismatched epitaxy. *J. Phys. C, Solid State Phys.*, 21, 4527.

Jackson, J. D. 1998. *Classical Electrodynamics.* New York: John Wiley and Sons.

Jeffrey, G. A. 1997. *An Introduction to Hydrogen Bonding.* New York: Oxford University Press.

Kendrick, C., Kahn, A. & Forrest, S. R. 1996. STM study of the organic semiconductor PTCDA on highly-oriented pyrolytic graphite. *Appl. Surf. Sci.*, 104/105, 586.

Khaksar, M., Amini, M., Boghaei, D. M. 2015. Efficient and green oxidative degradation of methylene blue using Mn-doped ZnO nanoparticles ($Zn_{1-x}Mn_xO$). J. Experimental Nanosci., doi:10.1080/17458080.2014.998300

Kilian, L., Umbach, E. & Sokolowski, M. 2006. A refined structural analysis of the PTCDA monolayer on the reconstructed Au(111) surface: 'rigid or distorted carpet?'. *Surf. Sci.*, 600, 2633.

Kitaigorodsky, A. I. 1973. *Molecular Crystals and Molecules.* New York: Academic Press.

Koma, A. 1992. New epitaxial growth method for modulated structures using van der Waals interactions. *Surf. Sci.*, 267, 29.

Lassiter, B. E., Lunt, R. R., Renshaw, C. K. & Forrest, S. R. 2010. Structural templating of multiple polycrystalline layers in organic photovoltaic cells. *Opt. Express*, 18, A444.

Liu, D.-J., Selinger, R. L. B. & Weeks, J. D. 1996. Representing molecular shape and interactions: a reduced intermolecular potential for copper phthalocyanine. *J. Chem. Phys.*, 105, 4751.

London, F. 1930. Zur Theorie und Systematik der Molekularkrafte. *Z. Phys.*, 63, 245.

London, F. 1937. The general theory of molecular forces. *Trans. Faraday Soc.*, 33, 8.

Lunt, R. R., Sun, K., Kröger, M., Benziger, J. B. & Forrest, S. R. 2011. Ordered organic–organic multilayer growth. *Phys. Rev. B*, 83, 064114.

Matthews, J. W. & Blakeslee, A. E. 1974. Defects in epitaxial multilayers. I. Misfit dislocations. *J. Cryst. Growth*, 27, 118.

McDermott, G., Prince, S. M., Freer, A. A., Hawthornthaite-Lawless, A. M., Papiz, M. Z., Cogdell, R. J. & Isaacs, N. W. 1995. Crystal structure of an integral membrane light-harvesting complex from photosynthetic bacteria. *Nature*, 374, 517.

McTague, J. P. & Novaco, A. D. 1979. Substrate-induced strain and orientational ordering in adsorbed monolayers. *Phys. Rev. B*, 19, 5299.

Mobus, M., Karl, N. & Kobayashi, T. 1992. Structure of perylene-tetracarboxylic-dianhydride thin films on alkali halide crystal substrates. *J. Cryst. Growth*, 116, 495.

Murray, C. B., Norris, D. J. & Bawendi, M. G. 1993. Synthesis and characterization of nearly monodisperse CdE (E = sulfur, selenium, tellurium) semiconductor nanocrystallites. *J. Am. Chem. Soc.*, 115, 8706.

Nelson, J. H., Shevchenko, E. V., Aloni, S., Sadtler, B. & Alivisatos, A. P. 2007. Seeded growth of highly luminescent CdSe/CdS nanoheterostructures with rod and tetrapod morphologies. *Nano Lett.*, 7, 2951.

Nijboer, B. R. A. & deWette, F. W. 1957. The calculation of lattice sums. *Physica*, 23, 309.

Pope, M. 1966. Electric currents in organic crystals. *Sci. Am.*, 216, 86.

Ratke, L. & Voorhees, P. W. 2002. *Growth and Coarsening: Ostwald Ripening in Materials Processing*, Berlin, Springer-Verlag.

Reiss, H. 1968. Rotation and translation of islands in the growth of heteroepitaxial films. *J. Appl. Phys.*, 39, 5045.

Robertson, J. M. 1958. Structure of naphthalene and anthracene. *Rev. Modern Phys.*, 30, 155.

Scott, R. A. & Scheraga, H. A. 1965. Method for calculating internal rotation barriers. *J. Chem. Phys.*, 42, 2209.

Sellam, F., Schmitz-Hubsch, T., Toerker, M., Mannsfeld, S., Proehl, H., Fritz, T., Leo, K., Simpson, C. & Mullen, K. 2001. LEED and STM investigations of organic-organic heterostructures grown by molecular beam epitaxy. *Surf. Sci.*, 478, 113.

Shechtman, D., Blech, I., Gratias, D. & Cahn, J. W. 1984. Metallic phase with long-range orientational order and no translational symmetry. *Phys. Rev. Lett.*, 53, 1951.

Shtein, M., Gossenberger, H. F., Benziger, J. B. & Forrest, S. R. 2001. Material transport regimes and mechanisms for growth of molecular organic thin films using low-pressure organic vapor phase deposition. *J. Appl. Phys*, 89, 1470.

Shtein, M., Mapel, J., Benziger, J. B. & Forrest, S. R. 2002. Effects of film morphology and gate dielectric surface preparation on the electrical characteristics of organic vapor phase deposited pentacene thin-film transistors. *Appl. Phys. Lett.*, 81, 268.

Smith, D. L. 1995. *Thin Film Deposition: Principles and Practice*. New York: McGraw-Hill.

So, F. F. & Forrest, S. R. 1991. Evidence for exciton confinement in crystalline organic multiple quantum wells. *Phys. Rev. Lett.*, 66, 2649.

So, F. F., Forrest, S. R., Shi, Y. Q. & Steier, W. H. 1990. Quasi-epitaxial growth of organic multiple quantum well structures by organic molecular beam deposition. *Appl. Phys. Lett.*, 56, 674.

Song, B. & Forrest, S. R. 2016. Nanoscale control of morphology in fullerene-based electron-conducting buffers via organic vapor phase deposition. *Nano Lett.*, 16, 3905.

Taylor, R. B., Burrows, P. E. & Forrest, S. R. 1997. An integrated, crystalline organic waveguide-coupled InGaAs photodetector. *IEEE Photonics Technol. Lett.*, 9, 365.

Taylor, R. B., Shen, Z. & Forrest, S. R. 1995. Polarization insensitive crystalline organic waveguides. *Top. Mtg. Org. Thin Films Photonics Appl. Tech. Dig.*, 38, Paper MB5.

Ullman, A. 1996. Formation and structure of self-assembled monolayers. *Chem. Rev.*, 96, 1533.

Umbach, E., Glockler, K. & Sokolowski, M. 1998. Surface 'architecture' with large ordered molecules: Interface order and epitaxy. *Surf. Sci.*, 402–404, 20.

van der Merwe, J. H. 1982. Analytical selection of ideal epitaxial configurations and some speculations on the occurrence of epitaxy. *Philos. Mag. A*, 45, 127.

Vargasa, P., Muñoza, E. & Rodriguez, L. 2001. Second virial coefficient for the Lennard-Jones potential. *Physica A*, 290, 92.

Whitesides, G. M. & Boncheva, M. 2002. Beyond molecules: self-assembly of mesoscopic and macroscopic components. *Proc. Natl. Acad. Sci.*, 99, 4769.

Yang, F. & Forrest, S. R. 2008. Photocurrent generation in nanostructured organic solar cells. *ACS Nano*, 2, 1022.

Yang, S. Y., Shin, K. & Park, C. E. 2005. The effect of gate-dielectric surface energy on pentacene morphology and organic field effect transistor characteristics. *Adv. Funct. Mater.*, 15, 1806.

Yook, K. S., Chin, B. D., Lee, J. Y., Lassiter, B. E. & Forrest, S. R. 2011. Vertical orientation of copper phthalocyanine in organic solar cells using a small molecular weight organic templating layers. *Appl. Phys. Lett.*, 99, 043308.

Zhang, Y. & Forrest, S. R. 1993. Mechanisms of quasi-epitaxial ordering at organic molecular thin film interfaces. *Phys. Rev. Lett.*, 71, 2765.

Zhong, D. Y., Hirtz, M., Wang, W. C., Dou, R. F., Chi, L. F. & Fuchs, H. 2008. Kinetics of island formation in organic film growth. *Phys. Rev. B*, 77, 113404.

CHAPTER 3

Optical properties of organic semiconductors

"Light brings us news of the Universe."
Sir William Bragg, *The Universe of Light*, 1933

The optical properties of organic semiconductors have their origins in the generation of a fundamental excitation known as the *exciton*. Excitons are mobile molecular excited states that consist of a tightly bound electron and hole pair. The large binding energies of molecular excitons are stable due to the low dielectric constants characteristic of organics which, in turn, is a consequence of the absence of chemical bonds between molecules. The excitonic nature of organic semiconductors leads to their exceptionally rich, complex and ultimately useful optical properties. They can be extremely strong light absorbers whose absorption coefficients can be comparable to individual atoms, and they also often can reach 100% luminescent quantum yields. The understanding of organic semiconductors, therefore, begins by understanding their optical properties. It is the purpose of this chapter to introduce the reader to the many interesting optical phenomena of organics, beginning with a discussion of the origins of light absorption and emission that arise from the energy level structure of individual molecules. A fascinating aspect of organics is that the optical properties of individual molecules are replicated, often with little modification, by their collective properties in the solid state. That is, collective phenomena do not entirely overwhelm and mask the properties of the molecules themselves. Indeed, molecular materials provide an ideal system whereby we can study the transitions between isolated species and their behavior in ordered and disordered solids. We will find that properties emerging from interactions of just a few molecules, such as excimer and exciplex emission, provide insights into molecular characteristics that can be exploited in practical device applications.

Unlike conventional semiconductors where bonds lead to degeneracy of the electronic states, the spin multiplicity of excited states in organic materials, that is, whether the exciton is a singlet, triplet or a higher order multiplet state, plays a defining role in its characteristics. We, therefore, make an effort to clarify the origins and properties of both singlets and triplets, and the electronic transitions that lead to their absorption and emission spectra. Within this discussion, we will find that other transitions between vibronic (i.e. phonon) states and non-radiative transfer between spin states within an electronic manifold also influence in the optical characteristics of organic materials.

Next, we discuss the quantum mechanical origin of energy transfer between molecules. Transfer can occur by *electron exchange* (also known as *Dexter transfer*) from one molecule to the next, or by longer-range electrostatic interactions between molecules due to *Förster* or *radiative* processes. Exciton transfer is intimately related to exciton–exciton interactions such as annihilation, fusion and fission. Long range transfer that consists of a series of individual steps is called *energy migration* (or *exciton diffusion*). We include in this treatment a discussion of the optical (i.e. excitonic) properties of very thin multilayer, and even "multiple

Organic Electronics: Foundations to Applications. Stephen R. Forrest. © Stephen R. Forrest 2020.
Published in 2020 by Oxford University Press. DOI: 10.1093/oso/9780198529729.001.0001

quantum wells" consisting of alternating films of different composition, each only a few monolayers thick. Excited states in such thin layers provide information about their spatial extent.

Our treatment is necessarily based on several quantum mechanical principles. We introduce each phenomenon starting with the wavefunction of electrons comprising molecular bonds, and extending this to the understanding collective phenomena in the solid state. The interested reader should be able to follow all of the discussion, although some of the more involved derivations are left to the problems at the end of the chapter. For those only seeking to use important results in analyzing their data, relevant expressions are highlighted by shading throughout the chapter, as is the custom throughout this book.

To paraphrase Sir William Bragg, we will see that the optical properties of organics do indeed bring us news of their universe!

3.1 The electronic structure of molecules

In the following three sections, we describe increasingly accurate approaches to calculate the electronic structure of complex molecular species. The principal objective of the calculations is to derive the energy levels from which emerge many rich and diverse optical properties. Indeed, the molecular spectra are exploited in all photonic applications ranging from light emission to detection. The molecular energetics are essential to computing molecular conformations (e.g. rotational and vibrational modes), and even transition energies leading to intermolecular reactions. Understanding their fundamental origins ultimately enables rational molecular design based on sound physical principles. For this reason, a great deal of effort has been dedicated over many decades to accurately calculate the energetics of molecular and polymeric systems. The discussion of energy fundamentals in Sections 3.2 and 3.3 will be followed in Section 3.4 by a brief introduction to computational approaches used to approximate electronic energies of individual molecules, extended to include to those of ensembles of interacting molecules in the solid state.

3.2 The Born–Oppenheimer approximation and the Franck–Condon principle

In Section 2.4.2, we calculated the electron distribution, and its role in forming covalent bonds in the simplest of all molecular species, the H_2^+ cation. This molecule is, in fact, the most complex situation that can be solved analytically. As more atoms and valence electrons are added, the solutions become increasingly approximate and unreliable, and require considerable computational resources to solve. We start by describing the physical principles leading to estimations of the energies of multi-electron molecular systems.

The molecular wavefunction has three components: electronic, nuclear, spatial and electron spin. These components can be decomposed into the product of the individual wavefunctions. The total wavefunction for a molecule with N electrons is given by

$$\Psi_t(\{\mathbf{r}_i\},\{\mathbf{R}_I\};\{S_i\}) = \Phi(\{\mathbf{r}_i\},\{\mathbf{R}_I\})\sigma(\{S_i\}), \quad (3.1)$$

where $\mathbf{r}_i$ represents the position of the ith electron, and the brackets, $\{\}$, represent all N position vectors, $\mathbf{r}_1$, $\mathbf{r}_2, \ldots \mathbf{r}_N$, $\{\mathbf{R}_I\}$ are the nuclear coordinates, $\{S_i\}$ are the electron spins, Φ is the spatial wavefunction, and σ is the electronic spin wavefunction. Nuclear spins are ignored since coupling to the electron spin gives rise to the *hyperfine interaction* that does not play a significant role in determining the electronic properties of the molecule.

The *Pauli Exclusion Principle* demands that the entire wavefunction be antisymmetric under the exchange of two electrons. For electrons at $\mathbf{r}_1$ and $\mathbf{r}_2$ with total spin quantum number, S, and its projection along the z-axis, m_S, the total electronic wavefunctions are given by $\psi(\mathbf{r}_1,\mathbf{r}_2;S,m_S)$. Thus, if the spatial wavefunction is symmetric, then the spin function must be antisymmetric, and vice versa. For the $S=0$ state, $m_S=0$. This results in a single, totally antisymmetric function:

$$\psi(\mathbf{r}_1,\mathbf{r}_2;0,0) = \frac{1}{\sqrt{2}}(\phi_a(\mathbf{r}_1)\phi_b(\mathbf{r}_2)+\phi_a(\mathbf{r}_2)\phi_b(\mathbf{r}_1)) \times (\alpha_1\beta_2-\alpha_2\beta_1). \quad (3.2)$$

Here, $\phi_i(\mathbf{r}_k)$ is the spatial wavefunction of an electron in atomic or molecular orbital i at position $\mathbf{r}_k$, and α corresponds to a spin with $m_S=+\frac{1}{2}$ (also denoted as $\alpha = |\uparrow\rangle$) and β corresponds to $m_S=-\frac{1}{2}$ (i.e. $\beta=|\downarrow\rangle$) for electrons 1 and 2. The factor of ½ normalizes the wavefunctions.

Similarly, the antisymmetric spatial wavefunction, S=1 results in three possible spin states corresponding to $m_S = 1$, 0, and −1:

$$\psi(\mathbf{r}_1,\mathbf{r}_2;1,1) = \frac{1}{\sqrt{2}}(\phi_a(\mathbf{r}_1)\phi_b(\mathbf{r}_2)-\phi_a(\mathbf{r}_2)\phi_b(\mathbf{r}_1))\alpha_1\alpha_2, \quad (3.3a)$$

$$\psi(\mathbf{r}_1, \mathbf{r}_2; 1, 0) = \frac{1}{\sqrt{2}}(\phi_a(\mathbf{r}_1)\phi_b(\mathbf{r}_2) - \phi_a(\mathbf{r}_2)\phi_b(\mathbf{r}_1)) \times (\alpha_1\beta_2 + \alpha_2\beta_1), \quad (3.3b)$$

and

$$\psi(\mathbf{r}_1, \mathbf{r}_2; 1, -1) = \frac{1}{\sqrt{2}}(\phi_a(\mathbf{r}_1)\phi_b(\mathbf{r}_2) - \phi_a(\mathbf{r}_2)\phi_b(\mathbf{r}_1))\beta_1\beta_2. \quad (3.3c)$$

For convenience, we replace the combined spatial and spin functions with a *spinorbital function*, $\Phi^\sigma(\mathbf{r}; S, m_S) = \phi(\mathbf{r})\sigma(S, m_S)$. For example, we write $\Phi^\uparrow(\mathbf{r}_1) = \phi(\mathbf{r}_1)\alpha_1$. The total wavefunction asymmetry can then be expressed using a determinant for the two particle system:

$$\psi(\mathbf{r}_1, \mathbf{r}_2) = \frac{1}{\sqrt{2}}\det\begin{vmatrix} \Phi^\uparrow(\mathbf{r}_1) & \Phi^\downarrow(\mathbf{r}_1) \\ \Phi^\uparrow(\mathbf{r}_2) & \Phi^\downarrow(\mathbf{r}_2) \end{vmatrix}. \quad (3.4)$$

Extending this to an N-electron system we can write the asymmetric wavefunction in terms of an N-dimensional *Slater determinant*:

$$\psi(\{\mathbf{r}_i\}) = \frac{1}{\sqrt{N!}}\det\begin{vmatrix} \Phi^a(\mathbf{r}_1) & \Phi^b(\mathbf{r}_1) & .. & .. & \Phi^z(\mathbf{r}_1) \\ \Phi^a(\mathbf{r}_2) & \Phi^b(\mathbf{r}_2) & .. & .. & \Phi^z(\mathbf{r}_2) \\ .. & .. & .. & .. & .. \\ .. & .. & .. & .. & .. \\ \Phi^a(\mathbf{r}_N) & \Phi^b(\mathbf{r}_N) & .. & .. & \Phi^z(\mathbf{r}_N) \end{vmatrix}. \quad (3.5)$$

The Pauli Exclusion Principle is clearly expressed by this formulation since if any two rows are the same, which corresponds to two electrons in identical states, then the determinant, and hence $\psi(\{\mathbf{r}_i\})$, vanishes.

The lone antisymmetric spin state in Fig. 3.1 is called the *singlet state* (corresponding to $S = 0$, $m_S = 0$), and the three symmetric combinations of spin under exchange in Eq. 3.3 comprise the *triplet state*. The two electron spin quantum states are visualized by their vector representations in Fig. 3.1. Here, S is the vector sum of the two separate spins whose projections along the z-axis give their quantum numbers m_s. The two *coupled* spins for a given state precess in phase. For the singlet state, the individual spins are at 180° relative to each other while they precess about the z-axis to give $S = 0$, which is equal to $m_S = ½ - ½ = 0$. Interchange of the spins results in a π-phase change, resulting in its antisymmetric property. In contrast, the three triplet states require no phase change on spin exchange, and hence they are spin symmetric. Nevertheless, the spin precession must be coupled to result in S=1 in all three cases. Note that for $m_S = 0$, the total spin vector rotates within the x–y plane, thereby lacking a projection along z, as required.

We will return to the spin wavefunction and symmetry, which play vital roles in the optical properties of molecules, in Section 3.5. Since spin and spatial wavefunctions are separable, we will first treat the spatial wavefunction to determine the electronic distribution across the molecule.

The time independent Schrödinger's equation for the total system of energy, E_T, consisting of both nuclei and electrons is

$$H_T\Phi(\{\mathbf{r}_i\}, \{\mathbf{R}_I\}) = E_T\Phi(\{\mathbf{r}_i\}, \{\mathbf{R}_I\}), \quad (3.6)$$

where the Hamiltonian is:

$$H_T = -\frac{\hbar^2}{2m_e}\sum_i^N \nabla^2_{\mathbf{r}_i} - \frac{\hbar^2}{2}\sum_I^M \frac{1}{m_{NI}}\nabla^2_{\mathbf{R}_I} + \frac{q^2}{4\pi\varepsilon_0} \times \left(\sum_{i>j}^N \frac{1}{|\mathbf{r}_i - \mathbf{r}_j|} - \sum_{i,I}^{N,M} \frac{Z_I}{|\mathbf{r}_i - \mathbf{R}_I|} + \sum_{I>J}^M \frac{Z_I Z_J}{|\mathbf{R}_I - \mathbf{R}_J|}\right) \quad (3.7)$$

(cf. Eqs. 2.25 and 2.28). Here, i and I are the indices of the ith electron and Ith nucleus of atomic number Z_I,

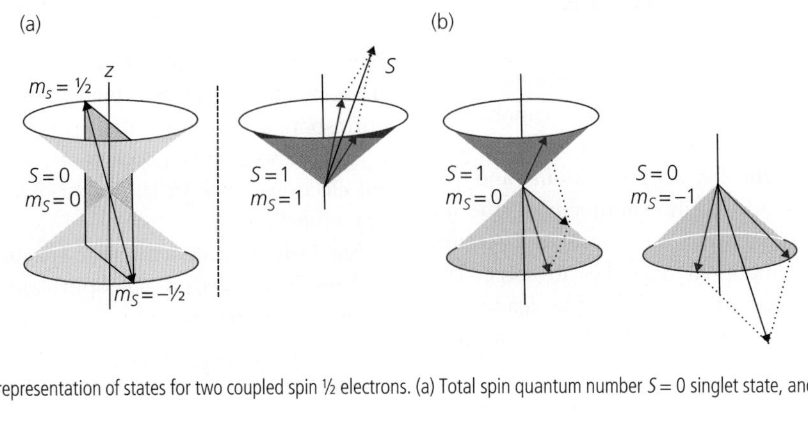

Figure 3.1 Vector representation of states for two coupled spin ½ electrons. (a) Total spin quantum number $S = 0$ singlet state, and (b) three $S = 1$ triplet states.

and m_e (m_N) is the electronic (nuclear) mass. The sums are over all electrons and nuclei in the molecule. Now, since $m_N >> m_e$, we can neglect the kinetic energy of the nuclei. Solving this problem for N electrons and M nuclei has no analytical solutions for $N > 1$, $M > 2$. To simplify the calculation, we apply the *Born–Oppenheimer approximation* where we assume that the electronic motion can be separated from that of the nuclei. Then, since the last term in Eq. 3.7 is independent of electron position, it can be eliminated to simplify an already complex problem. The nuclei thus contribute a constant additive energy term. Due to their orthogonality, we can then separate the nuclear, $\phi_N(\{\mathbf{R}_I\})$, and electronic, $\phi_e(\{\mathbf{r}_i\},\{\mathbf{R}_I\})$, wavefunctions accordingly:

$$\Phi(\{\mathbf{r}_i\},\{\mathbf{R}_I\}) = \phi_e(\{\mathbf{r}_i\},\{\mathbf{R}_I\})\phi_N(\{\mathbf{R}_I\}). \tag{3.8}$$

The electronic contribution to the energy is thus

$$\begin{aligned} H_e\phi_e(\{\mathbf{r}_i\};\{\mathbf{R}_I\}) &= \left[-\frac{\hbar^2}{2m_e}\sum_i^N \nabla^2_{\mathbf{r}_i} + \frac{q^2}{4\pi\varepsilon_0}\right. \\ &\quad \left.\times\left(\sum_{i>j}^N \frac{1}{|\mathbf{r}_i - \mathbf{r}_j|} - \sum_{i,I}^{N,M} \frac{Z_I}{|\mathbf{r}_i - \mathbf{R}_I|}\right)\right]\phi_e(\{\mathbf{r}_i\};\{\mathbf{R}_I\}) \\ &= E_e\phi_e(\{\mathbf{r}_i\};\{\mathbf{R}_I\}) \end{aligned} \tag{3.9}$$

where E_e is the total electronic energy.

The problem is further simplified if we write the electronic wavefunction, $\phi_{e,i}(\mathbf{r}_i)$, as the product of one-electron wavefunctions for the M atoms in the molecule, that is,

$$\phi_e^0(\{\mathbf{r}_i\}) = \prod_{i=1}^N \phi_{e,i}^0(\mathbf{r}_i). \tag{3.10}$$

Here, the "0" superscript indicates that this is a first order approximation where the N electrons comprising the atomic systems of the M molecules are non-interacting. Once these electrons are shared within the molecule, the approximation of Eq. 3.10 must be adjusted to account for their interactions. The problem has now been simplified from the intractable situation of a single system consisting of N interacting electrons, to that of N single-electron systems perturbed by their interactions.

Even with this simplification, the solution to Eq. 3.9 cannot be found in a closed form, analytical approach. It must be solved for the entire electron system of all the atoms in the molecule, and minimized for each possible nuclear configuration. For this purpose, it is convenient to introduce the normal coordinates, Q_k, which represent a particular (kth) configuration of all the nuclei. This configuration has a corresponding electron energy, $E_e(Q_k)$, and wavefunction, $\phi_e(\{\mathbf{r}_i\},Q_k)$. The equilibrium electron energy is thus obtained by minimizing E_e for each Q_k, and set of electronic coordinates, $\{\mathbf{r}_i\}$. That is, each set of nuclear coordinate configurations has an associated minimum electronic energy. However, this may only be a "local" minimum; the true or "global" minimum energy of both nuclear and electronic coordinates that defines the most stable molecular configuration can in fact be difficult to find for a complex molecule that can assume many equivalent configurations. This situation is shown graphically in Fig. 3.2 for a molecule with six carbon and six hydrogen atoms that is well known to have a global minimum configuration in the form of the cyclic benzene molecule.

The various configurations that a compound can assume as it approaches its global minimum are called *geometric isomers*. When the same molecule can have different topologies (i.e. DNA where both helices and knots are possible), the differently shaped structures are referred to as *topological isomers*. It is also possible that two isomeric forms have identical energies: the classical example being the ammonia molecule that consists of a pyramidal structure formed by three equally spaced H atoms forming the base, and with the N atom at the apex. The symmetry of this configuration demands that the energy be equal whether the N atom is above or below the plane of H atoms, as shown in Fig. 3.3. The energy surface thus consists of two minima, where the N atom tunnels between the two. The isomers, NH_3 and H_3N, are mirror symmetric, and are known as *inversion isomers*.

The general approach to finding a solution is to determine the energy surface corresponding to a

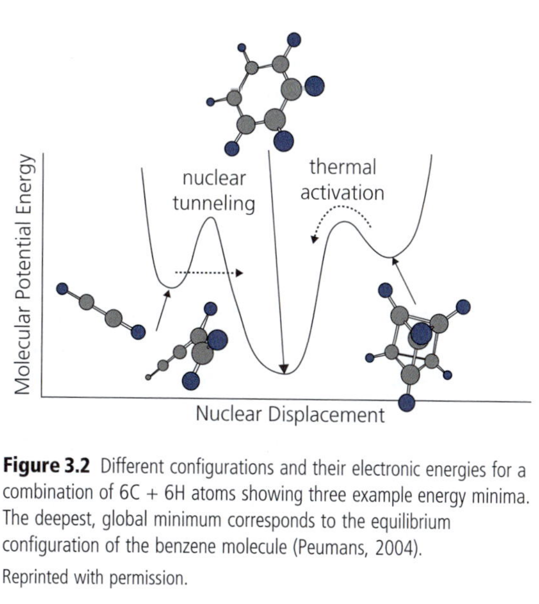

Figure 3.2 Different configurations and their electronic energies for a combination of 6C + 6H atoms showing three example energy minima. The deepest, global minimum corresponds to the equilibrium configuration of the benzene molecule (Peumans, 2004). Reprinted with permission.

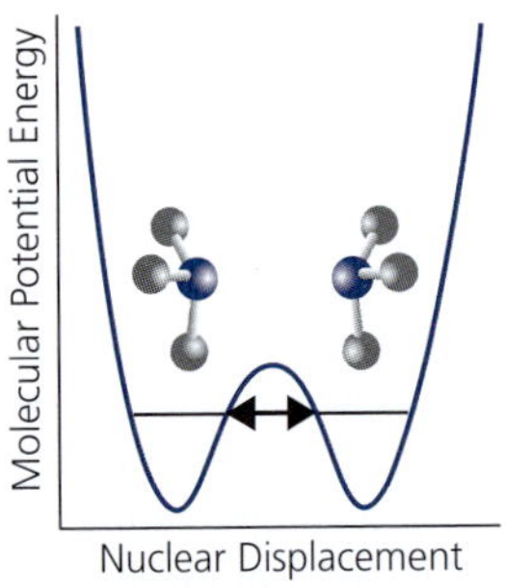

Figure 3.3 Molecular potential energy vs. N atom displacement for the two inversion isomers of ammonia (N is shown in blue, H in brown). The ground state energy level of the atom is shown by the horizontal line, indicating identical energies for both isomers. The dashed arrow indicates isoenergetic tunneling between the coupled states that approximately form two harmonic oscillator potentials.

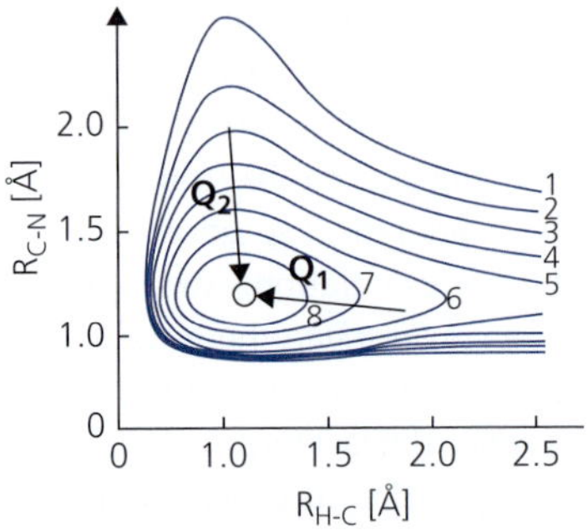

Figure 3.4 Ground state energy landscape for the HCN molecule whereby the bond lengths of H–C and C–N are varied. The contours correspond to energies of **1** $= 8 \times 10^4$ cm^{-1}, **2** $= 7 \times 10^4$ cm^{-1}, **3** $= 6 \times 10^4$ cm^{-1} ... **8** $= 1 \times 10^4$ cm^{-1}. Two possible nuclear configurations, Q_1 and Q_2 are also shown to indicate how the minimum is found (Peumans, 2004).

Reprinted with permission.

particular Q_k, and then incrementally changing the nuclear coordinates to seek yet a lower value of E_e. This application of the Born–Oppenheimer approximation continues until the lowest value of E_e is found (corresponding to E_e^0). Without a guarantee that the absolute global minimum has been found, it is often necessary that a new starting configuration for the molecule be chosen and the procedure repeated once again.

The energy landscape of a simple linear molecule, HCN, serves to illustrate this approach in Fig. 3.4. Two different nuclear configurational "trajectories," Q_1 and Q_2, arrive at the same equilibrium energy. The shape of these curves follows that for a covalently bonded solid similar to that in Fig. 2.9. For example, the electronic potential surface along Q_1 assumes a low attractive slope as the H-C separation distance is increased from equilibrium at $R_{H\text{-}C} = 1.1$ Å, but becomes steep and repulsive as the distance is reduced from that same point (indicated by "bunching" of the contours for small $R_{H\text{-}C}$ and $R_{C\text{-}N}$).

The lowest energy attained by calculating the energy surface for each nuclear configuration corresponds to the ground state of the molecule. This problem is computationally challenging due to the nearly unlimited possible configurations that even the simplest molecule can assume. For a molecule consisting of M nuclei, there are $3M - 6$ degrees of freedom, where three translational and three rotational dimensions are eliminated. The energy is invariant under translation, and, in general, rotational mode energies are small compared to those associated with nuclear displacements and are ignored. Even the simple benzene molecule with 12 nuclei presents a daunting calculational challenge with its 30 independent nuclear configurations. To simplify the problem, efficient calculations of molecular orbital energies start by only considering known stable isomers of a molecule, thereby confining the calculations to a realistic set of configurations.

We now write the nuclear coordinates in terms of the relative coordinate, Q, that is, $\phi_e(r,Q) = \phi(\{\mathbf{r}_i\}, \{\mathbf{R}_I\})$. Then, Eq. 3.7 becomes

$$H_T = T_e(r) + T_N(Q) + V_e(r) + V_{eN}(r,Q) + V_N(Q), \tag{3.11}$$

where T and V are the kinetic and potential energy operators, respectively. It is further convenient to write integrals in terms of bracket vectors to arrive at the full solution to Eq. 3.6 as follows:

$$E_T = \langle \phi_e \phi_N | T_e(r) + T_N(Q) + V_e(r) + V_{eN}(r,Q) + V_N(Q) | \phi_N \phi_e \rangle \tag{3.12a}$$

or

$$E_T = \langle \phi_N | T_N(Q) + V_N(Q) + E_e(Q) | \phi_N \rangle, \tag{3.12b}$$

where

$$\langle \phi_N | E_e(Q) | \phi_N \rangle = \langle \phi_e \phi_N | T_e(r) + V_e(r) + V_{eN}(r,Q) | \phi_N \phi_e \rangle \tag{3.13}$$

is the electronic energy corresponding to the nuclear configuration, Q.

Since we have already selected the configuration near equilibrium, the total system energy can be expressed as a small deviation from its minimum at $E_e(0) + V_N(0)$, giving:

$$E_T = E_e(0) + V_N(0) + \Delta E_e(Q) + \Delta V_N(Q). \tag{3.14}$$

As discussed above, different molecular isomers result in different energy surfaces, each defined by a separate equation with the form of Eq. 3.14. If the energy surfaces of two isomers intersect, they are degenerate at the value of Q where the intersection appears. In most cases, the degeneracy occurs at large deviations from equilibrium, in which case there is a significant activation energy barrier separating the isomers, as represented by the tunneling barriers in Figs. 3.2 and 3.3. Transformations between such energetically isolated isomers are difficult, making them independent and chemically distinct. Hence, one does not influence the energetic environment of the other.

As in Figs. 2.9 and 3.3, the Coulomb attraction between neighboring nuclei forms an approximately parabolic surface near their equilibrium separation. Thus, the nuclear vibration about equilibrium results in a new set of quantum states defined by $E_N = \Delta E_e(Q) + \Delta V_N(Q)$ in Eq. 3.14. These vibrational modes, or *vibronics*, form discrete levels. Since the potential well near its minimum is approximately parabolic, the total vibronic energy in level j follows the well-known solutions for a simple harmonic oscillator (SHO):

$$E_{N,j} = \hbar \sum_{l=1}^{3M-6} \omega_l(E_{e,j}) \left[n_l(E_{e,j}) + \frac{1}{2} \right]. \tag{3.15}$$

Here, $n_l = 0, 1, 2, \ldots$ is the quantum number of the nth vibronic state of the lth normal nuclear mode in a molecule consisting of M atoms. Each electronic state, itself, is defined by $E_{e,j}$ (where $j = 0, 1, 2 \ldots$) and has its own associated set of vibronic sub-levels. Since the nuclear masses are much larger than the electron mass, we infer that $E_N << E_e$; that is, the vibronic states are much closer spaced than the electronic states, as illustrated in Fig. 3.5.

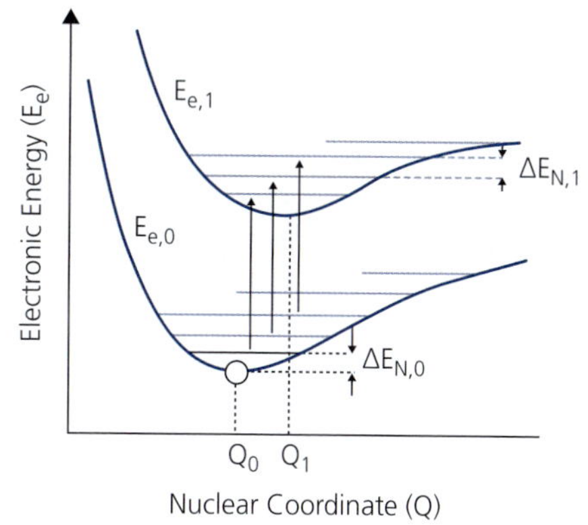

Figure 3.5 Energy minima for the two lowest electronic states ($E_{e,0}$ and $E_{e,1}$) of a polyatomic molecule, showing the vibronic modes within each level with spacing $\Delta E_N << E_{e,1} - E_{e,0}$ due to the much larger nuclear *vs.* electronic mass. The minimum energy configurations corresponding to the nuclear coordinates Q_0 and Q_1 are shown. The three vertical arrows represent transitions between similar vibronic states within the two corresponding electronic states.

The actual interatomic potential including nuclear repulsion is not well approximated by the infinite parabolic potential of a SHO with its equal spacing of an unlimited number of vibronic modes in Eq. 3.15. The anharmonicity of the potential is more accurately represented by the *Morse potential* that was first applied to approximate the potential of a diatomic molecule. This is given by

$$V(r) = V_0(1 - \exp[-a(r - r_0)])^2, \tag{3.16}$$

where r_0 is the equilibrium interatomic spacing, and $V_0 = V(\infty) - V(r_0)$ is the well depth. Since the reference energy is arbitrary, it is convenient to set $V(\infty) = 0$, in which case $V_0 < 0$. Finally, $a = \sqrt{k_0/2V_0}$, where k_0 is the force constant in the limit of $r \to r_0$ (see Eq. 2.21). A comparison of the Morse potential to a SHO in an infinite potential well is shown in Fig. 3.6.

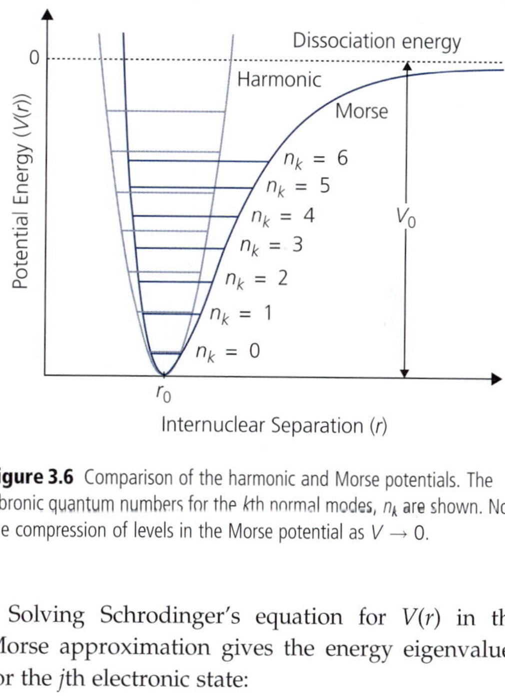

Figure 3.6 Comparison of the harmonic and Morse potentials. The vibronic quantum numbers for the kth normal modes, n_k are shown. Note the compression of levels in the Morse potential as $V \to 0$.

Solving Schrodinger's equation for $V(r)$ in the Morse approximation gives the energy eigenvalues for the jth electronic state:

$$E_{N,j} = \hbar \sum_{l=1}^{3M-6} \omega_l(E_{e,j}) \left[\left\{ n_l(E_{e,j}) + \frac{1}{2} \right\} - \left\{ \frac{1}{4V_0} \hbar\omega_l(E_{e,j}) \left(n_l(E_{e,j}) + \frac{1}{2} \right)^2 \right\} \right], \tag{3.17}$$

where the lth normal mode frequency is

$$\omega_l = \frac{a}{2\pi} \sqrt{\frac{2V_0}{m_e}}. \tag{3.18}$$

Note that the energy eigenvalues are the same as those predicted in Eq. 3.15 for the SHO, modified by a second order term in $\hbar\omega_l$ that becomes increasingly important with increasing n_l. This difference arises since the Morse potential widens as $V \to 0$, resulting in compression of the spacing between the modes, which contrasts with the equally spaced vibronics of the SHO. The energy between two adjacent vibronics in the Morse potential approximation is

$$\Delta E_l = E(n_{l+1}) - E(n_l) = \hbar\omega_l\left(1 - \frac{\hbar\omega_l(n_l+1)}{2V_0}\right). \tag{3.19}$$

This expression is valid until we reach the highest bound state at $n_{l,max}$ near $V = 0$, corresponding to

$$n_{l,max} = \frac{2V_0 - \hbar\omega_l}{\hbar\omega_l}. \tag{3.20}$$

While the Morse potential more accurately represents the anharmonicity of the interatomic potential, it remains insufficient to accurately predict the spectral characteristics of most multi-atom systems. To accommodate this inaccuracy, Eq. 3.16 has been further adjusted using the Morse long-range potential that improves agreement with measured spectra (LeRoy et al., 2009).

The approximations in Eqs. 3.15 and 3.17 are only valid if the electronic states have an energy minimum. We will find in Section 3.5 that this is not always the case. For example, an *excimer* is a state shared between two neighboring molecules that is bound only when excited. Since there is no bound ground state, the vibronic progression is absent.

The equilibrium coordinate Q_1 in Fig. 3.5 is depicted at a different location than Q_0. This indicates that when the electronic state changes, the nuclear configuration of the molecule must also adjust. Often, excited states have a larger separation between nuclei than in the ground state since the size of electronic orbitals generally increases with quantum number (i.e. d-orbitals tend to be larger than p-orbitals, which in turn are larger than s-orbitals). Because of the larger mass of the nuclei compared to that of the electron, the reconfiguration of the nuclear positions takes place on a time scale much longer than that of the electrons, which is known as the *Franck–Condon principle*. Like the Born–Oppenheimer approximation, application of this principle greatly simplifies calculations and the interpretation of absorption and emission spectra since it implies that the molecular structure relaxes into its new equilibrium state *after* the electronic transition has occurred.

3.3 Linear combination of atomic orbitals, hybridization, and the aufbau principle

In the previous section we described a model to understand the absorption and emission spectra of molecules consisting of M nuclei and N electrons based on the quantum mechanical nature of the individual electronic states. While that treatment was based on first principles, it is not obvious how to extend it to an understanding of more complex molecules, where the electronic system is distributed over many atoms. In particular, as we build up the electrons from the deepest atomic orbitals to those at the highest (least tightly bound) valence states, the electrons find themselves sharing valences among several nearby and symmetrically equivalent atoms. These shared valence states are called *hybrid orbitals*. The most energetic bound state and the lowest energy excited state that were treated in Section 3.2 are known as the *frontier molecular orbital energies* between which the dominant optical transitions occur. Calculation of these energies forms the basis of molecular orbital theory.

The frontier energies are established using the *aufbau principle* (or "building up" principle). This is based on the Pauli Exclusion Principle whereby no two electrons can simultaneously occupy the same state. Since there are two possible spin states ($|\alpha\rangle$ and $|\beta\rangle$) for fermions, each molecular orbital can be occupied by two electrons, resulting in the five possible configurations illustrated in Fig. 3.7. In the ground state, the electrons fill each orbital until the highest occupied molecular orbital (HOMO) energy is reached. The LUMO is the lowest unoccupied molecular orbital energy level. State filling up to the LUMO is characteristic of closed shell systems that consist of an even number, N, of electrons that fill the $N/2$ lowest energy orbitals. In open shell systems, a single unpaired electron exists in the HOMO (and perhaps one or more HOMO $- i$ levels, with $i = 1, 2, 3\ldots$). A singlet spin state consists of one electron promoted from the HOMO to the LUMO. Following excited state nuclear positional relaxation, the state can emit light as the electron transitions back into the HOMO, or ground state. Electrons promoted from deeper occupied molecular orbitals (HOMO $- i$) to higher unoccupied electronic manifolds (LUMO $+ i$) can also be observed under the appropriate excitation conditions.

The difference in energy between the HOMO and LUMO is called the *energy gap*, that is,

$$E_G = E_{LUMO} - E_{HOMO}. \tag{3.21}$$

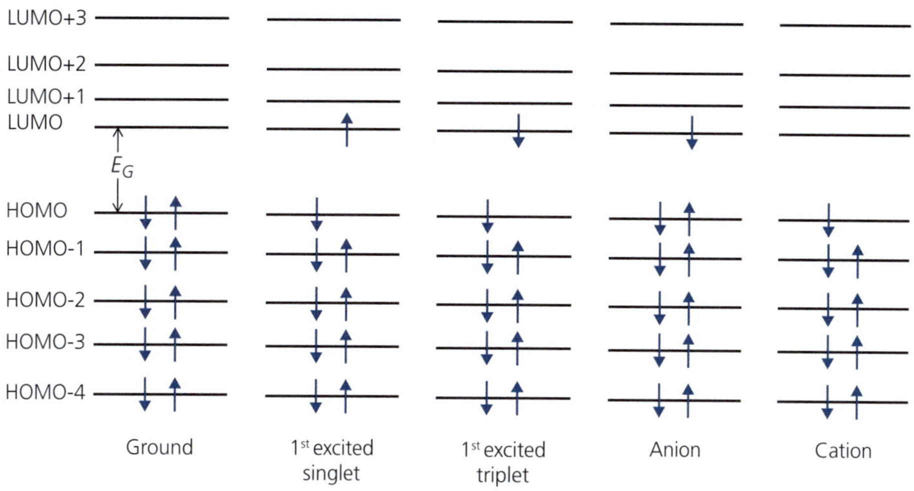

Figure 3.7 Several possible electronic state configurations of the frontier orbitals of a molecule. E_G is the energy gap.

If the excited electron in the LUMO has the same spin as the electron remaining in the HOMO, the state is a spin-symmetric triplet state, $|\beta_1\beta_2\rangle$. A molecule with one additional electron is a negatively charged anion, whereas a positively charged cationic molecule is missing an electron. In the solid state, *anions* and *cations* are also known as negatively and positively charged *polarons*, respectively. When these species move from molecule to molecule, they result in the conductivity of organic materials, as discussed in Chapter 4. If a charge is transferred from one molecule to its neighbor, the resulting state is a cation-anion pair, also known as a *charge transfer state* or polaron pair.

In Section 3.2 the Schrödinger equation comprised of both the electronic and nuclear potentials was introduced. We now extend this formalism to calculate the full molecular orbital structure. A common starting point is using the *linear combination of atomic orbitals* (LCAO) method. Here, the molecule is assumed to be comprised of atoms whose orbitals are only slightly perturbed from their individual atomic states. This allows us to choose a basis set of electronic orbitals that have already been calculated for the individual atomic constituents, and then to correct them based on small perturbations that occur when they are combined in the molecule itself.

The ith orbital of a molecule consisting of L atoms obtained from a sum over all M atomic orbitals is expressed via:

$$\psi_i(\mathbf{r}_i) = \sum_{j=1}^{M}\sum_{k=1}^{L} c_{ijk}\phi_{jk}(\mathbf{r}_i - \mathbf{R}_j). \qquad (3.22)$$

Here, the individual orbital wavefunctions, ϕ_{ij}, are defined for an electron at position $\mathbf{r}_i$ relative to an atom whose nucleus is at position $\mathbf{R}_j$. The coefficients c_{ijk} weight the functions according to their contributions to ψ_i. This is precisely the method used in Chapter 2 to calculate the solutions to the H_2^+ molecular ion. Unfortunately, even a calculation of the atomic orbitals using the Slater determinant in Eq. 3.5 can be daunting.

To make the calculations tractable, it is convenient to choose approximate basis set functions that lead to relatively rapid solutions that arrive at the wavefunctions, $\psi_i(\mathbf{r}_i)$. One example basis set is comprised of Gaussian orbitals, which in Cartesian coordinates take the form (Leach, 1996):

$$\phi(r) = Ax^a y^b z^c \exp(-\alpha r^2), \qquad (3.23)$$

where α is the rate of decay of the wavefunction with distance that is proportional to the atomic number, Z, and the exponents a, b, and c express the orbital symmetry (i.e. s- or p-orbital) such that $a + b + c$ equals 0, 1, or 2. Also, A is a normalizing constant. Alternatively, the Slater basis set has a form similar to that of the atomic hydrogen state with quantum numbers n, l, m:

$$\begin{aligned}\phi(r,\theta,\varphi) &= R_{nl}(r)Y_{lm}(\theta,\varphi)\\ &= A(2\alpha)^{n+\frac{1}{2}}r^{n-1}\exp(-\alpha r)Y_{lm}(\theta,\varphi).\end{aligned} \qquad (3.24)$$

Here, $R_{nl}(r)$ is the radial wavefunction component, and $Y_{lm}(\theta, \varphi)$ are the spherical harmonics describing the angular distribution of the wavefunction. Hence, the 1s state is simply $\phi(r) = 2A\alpha^{3/2}\exp(-\alpha r)$. A plot of the 1s Gaussian and Slater functions is shown in Fig. 3.8. A single Gaussian function does not accurately reproduce the tail at large r, nor does it provide an adequate probability amplitude near the origin.

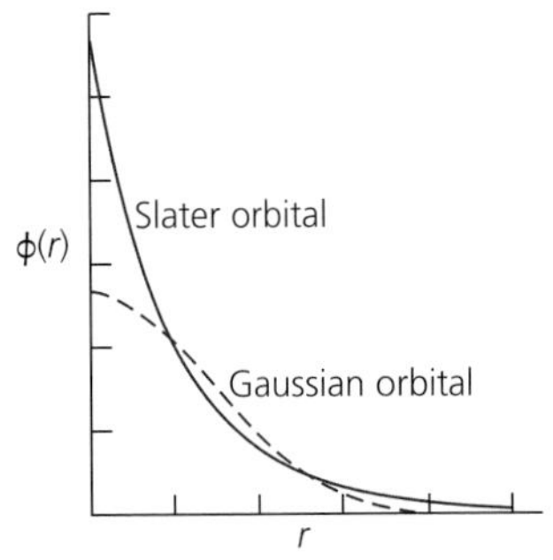

Figure 3.8 Calculated 1s states based on Gaussian or Slater wavefunctions.

These problems can be partially fixed by summing over several higher order Gaussian functions. There are many other basis sets and useful variants that are even more accurate than those in Eqs. 3.23 and 3.24 (see *Further reading* and Leach, 1996, for example).

We now simplify Eq. 3.22 by invoking the Born–Oppenheimer approximation, whereby the nuclear coordinates (and interactions between nuclei) can be ignored. Then:

$$\psi_i(\mathbf{r}_i) = \sum_{r=1}^{M} c_{ir}\phi_r'(\mathbf{r}_i), \tag{3.25}$$

where c_{ir} are the coefficients with respect to the electronic coordinates $\mathbf{r}_i$, and $\phi_r'(\mathbf{r}_i)$ comprise the new electronic orbital basis set. From time-dependent perturbation theory, the solution to Schrödinger's equation using Eq. 3.25 becomes (Atkins and de Paula, 2006):

$$\sum_{r=1}^{M} c_r\left(\langle\phi_s'|H_{int}|\phi_r'\rangle - E_i\langle\phi_s'|\phi_r'\rangle\right) = 0, \tag{3.26}$$

where H_{int} is the Coulombic interaction Hamiltonian in Eq. 3.9, and $S_{rs} = \langle\phi_s'|\phi_r'\rangle$ is the *overlap integral* between orbitals r and s. Its magnitude depends on the relative proximity of the atomic orbitals: as the separation between two atoms increases, S_{rs} rapidly decreases. Hence, only nearest neighbor interactions are generally considered. The first term in Eq. 3.26 can be rewritten as H_{rs}. The diagonal matrix elements are $H_{rr} = H_{ss}$, known as the Coulomb integral that provides the kinetic and Coulomb potential energies due to attraction between all nuclei and electrons within a particular orbital. The off-diagonal term $H_{rs} = H_{sr}$ $(r \neq s)$ is the resonance or bond integral that gives the electron energy in regions between atoms where the electron distributions overlap. As in the case of S_{rs}, H_{rs} decreases as atoms become increasingly separated, and hence only nearest neighbor interactions need be considered.

The solution to the eigenvalue Eq. 3.46 for the ith electronic state is found by requiring the determinant to vanish, viz.

$$\det|\mathbf{H} - E_i\mathbf{S}| = 0, \tag{3.27}$$

where $\mathbf{H}$ and $\mathbf{S}$ represent the matrices consisting of elements H_{rs} and S_{rs}, respectively. The solutions to Eq. 3.27 are computationally intensive for even the simplest molecules (Atkins and Friedman, 1997). Indeed, finding a suitable basis set $|\phi_r'\rangle$ is in itself a complicated undertaking and is usually approached via the *Hartree–Fock approximation* (Callaway, 1974, Konishi and Paffuti, 2009) that builds up the atom starting with hydrogen. The interested reader is referred to the extensive literature on this subject for further detail.

It is instructive to return to the simple problem of the H_2^+ molecule that we encountered in Section 2.4.2, now solving the problem using the LCAO method. This will allow for us to build the theory for more complex and larger molecules. Assuming the separate H atoms are in their 1s states, we can explicitly obtain the matrix elements as follows (cf. Eq. 3.7):

$$H_{rr} = \langle\phi_r'|H|\phi_r'\rangle = E_{1s} - \frac{q^2}{4\pi\varepsilon_0}\langle\phi_r'|\frac{1}{|\mathbf{r}_r - \mathbf{R}_s|}|\phi_r'\rangle + \frac{q^2}{4\pi\varepsilon_0|\mathbf{R}_r - \mathbf{R}_s|}. \tag{3.28}$$

Here, E_{1s} corresponds to the interaction of an electron with its own proton, which is simply the sum of the Coulombic attraction and kinetic energy of a 1s H atom. The second two terms are interactions of the electron with the other proton, and the repulsion of the two H nuclei. The second term in Eq. 3.28 is solved using the 1s H ground state wavefunction, which requires the use of elliptical coordinates (see Problem 3.2). We can rewrite Eq. 3.28 as:

$$H_{rr} = H_{ss} = E_{1s} - J + \frac{q^2}{4\pi\varepsilon_0|\mathbf{R}_r - \mathbf{R}_s|}. \tag{3.29}$$

where J is the *Coulomb integral*. The off-diagonal terms are:

$$H_{rs} = H_{sr} = \langle\phi_r'|H|\phi_s'\rangle = \left[E_{1s} + \frac{q^2}{4\pi\varepsilon_0|\mathbf{R}_r - \mathbf{R}_s|}\right]S_{rs} - \frac{q^2}{4\pi\varepsilon_0}\left\langle\phi_r'\left|\frac{1}{|\mathbf{r}_r - \mathbf{R}_s|}\right|\phi_s'\right\rangle \tag{3.30}$$

or

$$H_{rs} = H_{sr} = \left[E_{1s} + \frac{q^2}{4\pi\varepsilon_0|\mathbf{R}_r - \mathbf{R}_s|}\right]S_{rs} - K, \tag{3.31}$$

where K originates from interactions of the overlap of the electronic wavefunctions in the states r and s. With these matrix elements in hand, we can solve the secular equation using the determinant, Eq. 3.27 to obtain the new eigenenergies for the H_2^+ molecular ion:

$$E = \left[E_{1s} + \frac{q^2}{4\pi\varepsilon_0|\mathbf{R}_r - \mathbf{R}_s|}\right] \pm \frac{J \mp K}{1 \mp S_{rs}}. \tag{3.32}$$

Thus, the original 1s states are now split into two new orbitals: the lower orbital is the 1σ (corresponding to the linear combination $\psi = \frac{1}{\sqrt{2}}(\phi_r' + \phi_s')$, where $c_r = c_s = 1$ in Eq. 3.25) *bonding orbital* with energy E_+, and the upper 2σ orbital (with $\psi = \frac{1}{\sqrt{2}}(\phi_r' - \phi_s')$ and $c_r = 1 = -c_s$) is the *antibonding orbital* with energy E_-, as shown in Fig. 3.9. In Fig. 3.9, we show that the degeneracy between 1σ and 2σ at $|\mathbf{R}_r - \mathbf{R}_s| \to \infty$ is lifted once this distance becomes comparable to the Bohr radius, a_0. At ~$2a_0$, E_+ reaches a minimum equal to the energy of the equilibrium bound state, and then both E_+ and E_- become large and positive as the atomic spacing is further reduced due to electron repulsion. Also, from Fig. 3.9 we see that the antibonding 2σ orbital has a higher energy than the 1σ state due to repulsion of the electron densities away from the region between the nuclei, whereas the electron density for the 1σ state is higher between the protons, resulting in a lower energy at all nuclear separation distances (c.f. Fig. 2.9). Hence, the names antibonding and bonding for these two different situations.

Moving to more complex organic molecules necessarily requires further approximations to allow for a reasonably tractable, yet predictive solution. The most important case in organic electronics is that of a conjugated electron system, which is a characteristic of both semiconducting polymer and small molecular weight compounds. Conjugated molecules allow for significant simplification by taking full advantage of molecular symmetry. We consider two atoms in a molecule to be calculationally equivalent if under a symmetry operation they can be transformed into each other. For example, two atoms in their s-states are equivalent since their wavefunctions are spherically symmetric. Hence, any spatial transformation leaves such molecules unchanged. In contrast, higher order orbitals (e.g. p or d) have a more restricted set of symmetry operations that are invariant under exchange. Linear combinations of atomic orbitals of adjacent but symmetrically equivalent atoms therefore results in an orbital whose symmetry is shared with that of the constituent atoms; these linear combinations are known as *hybrid atomic orbitals* (HAOs) that determine the physical, chemical and electronic properties of the molecules that they comprise.

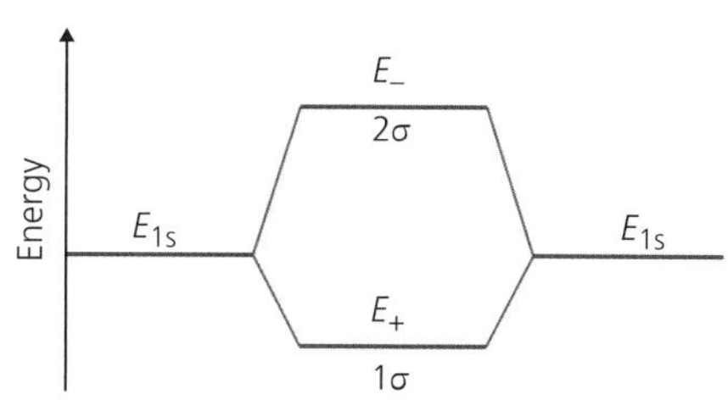

Figure 3.9 Bonding (1σ) and antibonding (2σ) orbital energies of the H_2^+ molecular ion compared to H atom ground state energies, E_{1s}.

An example of hybridization is the heteronuclear ethyne molecule, with the structure HC ≡ CH. In this case, each spherically symmetric C 2s and $2p_z$ orbital forms two hybrid sp-hybridized orbitals. The $2p_z$-orbital is not spherically symmetric since it changes sign when reflected at the x–y plane, as shown in Fig. 3.10. The $2p_z$-orbital can have two equivalent phase configurations, whereas the s-orbital has only one (see the two left hand illustrations). When added, there is larger electron content where the phases are additive compared to when they are of opposite sign. When linearly superposed, the resulting sp_z π-orbital (right hand illustration) is symmetrically disposed about the x–y plane and oriented along the $\hat{\mathbf{z}}$ axis.

The hybrid orbitals from Eq. 3.25 are therefore comprised of linear combinations of the s and p_z states:

$$|\psi_1^H\rangle = c_{1s}|\phi_s\rangle + c_{1p}|\phi_{pz}\rangle \tag{3.33a}$$

and

$$|\psi_2^H\rangle = c_{2s}|\phi_s\rangle + c_{2p}|\phi_{pz}\rangle, \tag{3.33b}$$

where $|\psi^H\rangle$ are the hybridized wavefunctions. These hybrids lie along the axis between the nuclei, and have the same symmetry as the p_z-orbitals of which they are comprised. Orthonormality of the individual AOs in Eq. 3.33 leads to the following conditions:

$$\langle\psi_1^H|\psi_2^H\rangle = c_{1s}c_{2s} + c_{1p}c_{2p} = 0 \tag{3.34a}$$

and

$$\langle\psi_{1,2}^H|\psi_{1,2}^H\rangle = c_{1s}^2 + c_{1p}^2 = c_{2s}^2 + c_{2p}^2 = 1. \tag{3.34b}$$

Furthermore, from symmetry: $|\phi_s(z)\rangle = |\phi_s(-z)\rangle$ and $|\phi_{pz}(z)\rangle = -|\phi_{pz}(-z)\rangle$, from which we obtain $c_{1s} = c_{2s}$ and $c_{1p} = -c_{2p}$. It then follows that the sp_z-orbitals are given by

$$|\psi_1^H\rangle = \frac{1}{\sqrt{2}}\left(|\phi_s\rangle + |\phi_{pz}\rangle\right) \tag{3.35a}$$

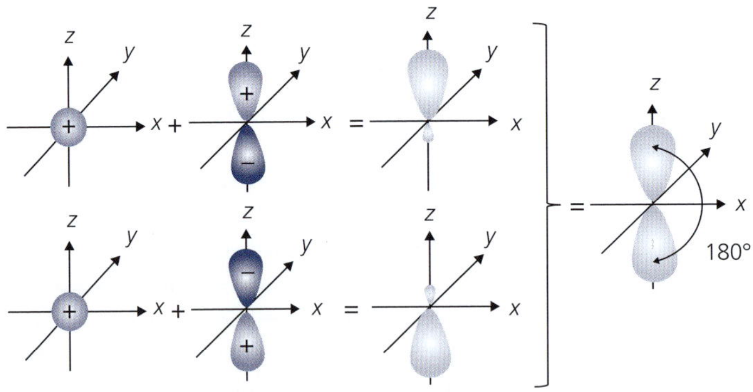

Figure 3.10 Illustration of the hybridization of the s and p_z orbitals to form the symmetric sp-hybrid orbital on the right. The signs indicate the relative phases of the wavefunctions across the *x–y* plane.

and

$$|\psi_2^H\rangle = \frac{1}{\sqrt{2}}\left(|\phi_s\rangle - |\phi_{pz}\rangle\right). \quad (3.35b)$$

These two hybrids are illustrated by the larger and smaller p-like orbitals in Fig. 3.10 whose orientations are along the internuclear (C ≡ C) axis.

The next level of hybridization is formed by a superposition of an s-orbital and two p-orbitals (p_x, p_y), forming three sp^2-hybrid orbitals. An example molecule with this hybridization is ethene (also called ethylene), $H_2C{=}CH_2$. The molecule is planar with relative bond angles between the HCH and HCC atomic groups of 120°. The procedures for determining the hybrid wavefunctions is similar to that used for sp-hybridization, leading to:

$$|\psi_1^H\rangle = \frac{1}{\sqrt{3}}|\phi_s\rangle + \frac{2}{\sqrt{6}}|\phi_{px}\rangle, \quad (3.36a)$$

$$|\psi_2^H\rangle = \frac{1}{\sqrt{3}}|\phi_s\rangle - \frac{1}{\sqrt{6}}|\phi_{px}\rangle + \frac{1}{\sqrt{2}}|\phi_{py}\rangle, \quad (3.36b)$$

$$|\psi_3^H\rangle = \frac{1}{\sqrt{3}}|\phi_s\rangle - \frac{1}{\sqrt{6}}|\phi_{px}\rangle - \frac{1}{\sqrt{2}}|\phi_{py}\rangle. \quad (3.36c)$$

These three hybrids are formed from the two, 2p electrons of the C atom. All of the H atoms are symmetrically equivalent. The orbitals lie in the *x–y* (or $\sigma^{(3)}$) plane between the two C nuclei, as shown in Fig. 3.11. The third electron in the $2p_z$ orbital is excluded from the hybridization and lies perpendicular to the plane along the $\hat{\mathbf{z}}$-axis. This electron pairs with that of the second carbon atom to form a cylindrical π-orbital in the $\sigma^{(2)}$ (*x-z*) plane, holding the molecule rigid within its planar structure. As we will see below, the electrons occupying the π-orbitals contribute significantly toward the optical and electronic properties of conjugated molecules.

The final case is where the s-orbital electron hybridizes with all three p-electrons to form sp^3 hybrid orbitals, characteristic of methane, CH_4. The four tetragonally oriented orbitals are linear combinations of the four valence 2s orbitals with the three 2p-orbitals of the C atom, viz.

$$|\psi_1^H\rangle = \frac{1}{2}|\phi_s\rangle + \frac{1}{2}|\phi_{px}\rangle + \frac{1}{2}|\phi_{py}\rangle + \frac{1}{2}|\phi_{pz}\rangle, \quad (3.37a)$$

$$|\psi_2^H\rangle = \frac{1}{2}|\phi_s\rangle - \frac{1}{2}|\phi_{px}\rangle - \frac{1}{2}|\phi_{py}\rangle + \frac{1}{2}|\phi_{pz}\rangle, \quad (3.37b)$$

$$|\psi_3^H\rangle = \frac{1}{2}|\phi_s\rangle - \frac{1}{2}|\phi_{px}\rangle + \frac{1}{2}|\phi_{py}\rangle - \frac{1}{2}|\phi_{pz}\rangle, \quad (3.37c)$$

$$|\psi_4^H\rangle = \frac{1}{2}|\phi_s\rangle + \frac{1}{2}|\phi_{px}\rangle - \frac{1}{2}|\phi_{py}\rangle - \frac{1}{2}|\phi_{pz}\rangle. \quad (3.37d)$$

The C atom is at the vertex of the tetragon with the electron density shared in σ-bonds with the four H atoms, as shown in Fig. 3.12.

Given these symmetry arguments for arriving at the hybrid orbitals, we now can manage the general case of conjugated molecular systems that consist of alternating single and double bonds between adjacent atoms. This is referred to as *π-orbital conjugation*, whereby electrons within the hybridized π-system are distributed between atoms along the molecular plane. The calculations of the eigenfunctions and energies of such systems are enabled by using the simplifications of the *Hückel approximation* that takes full account of the molecular and bond symmetries between nearest neighbors as previously. The Hückel approximation assumes the following:

1. The different σ- and π-orbital symmetries allow for separation of their respective wavefunctions, ψ_σ and ψ_π, such that the total wavefunction is $\psi_T = \psi_\sigma\psi_\pi$.

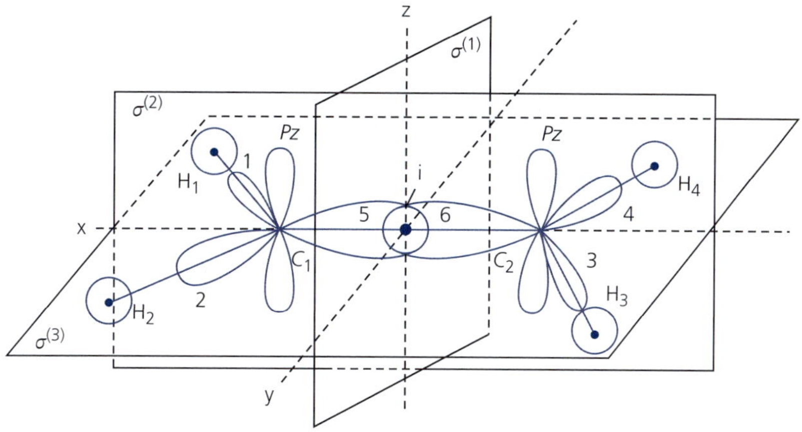

Figure 3.11 Orbital arrangements for ethene. The four H s-orbitals, four sp^2 C orbitals (labeled 1-4) and two sp^2 orbitals between the C atoms (5,6) are indicated. Also, the unhybridized p_z-orbitals lie in the $\sigma^{(2)}$ plane, forming the molecular π-bond. The several symmetry planes of the molecule are indicated.

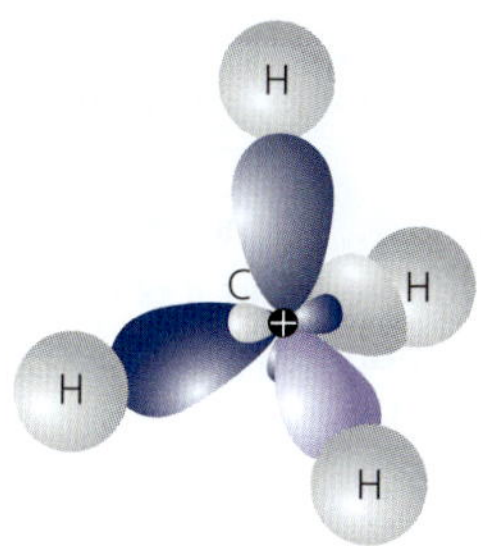

Figure 3.12 Hybrid sp^3 orbitals of methane.

2. Nearest neighbor interactions are set equal to a constant. That is, $H_{rr} = \alpha$, and $H_{rs} = \beta$ for $|r - s| = 1$, otherwise $H_{rs} = 0$. Both α and β are negative (bonding) energies. Here, α is simply an energy offset, and β is the energy due to the hybridization of the orbitals themselves.
3. The overlap integrals $S_{rs} = \langle \phi_s' | \phi_r' \rangle = \delta_{rs}$, where the Kronecker delta function has values of $\delta_{rs} = 1$ when $r = s$; and $\delta_{rs} = 0$ otherwise. This is the most problematic of the assumptions since the overlap from adjacent orbitals can be substantial; ≈ 0.2 or larger in some cases.

With this approximation, the secular determinant in Eq. 3.27 now becomes:

$$det|(\alpha - E)\mathbf{I} + \beta\mathbf{B}| = 0. \tag{3.38}$$

Here, $\mathbf{I} = \delta_{rs}$ is the unity matrix, and $\mathbf{B} = \delta_{r,r\pm 1}$.

Benzene (C_6H_6) provides a simple example of the utility of the Hückel approximation in estimating the energies of a conjugated system. Here, each C atom is sp^2 hybridized with its neighbors, with its third p_z-orbital lying perpendicular to the molecular plane. As in the case of ethene, the p_z-orbitals from adjacent C atoms overlap to form an extended π-system that maintains molecular planarity. For benzene, therefore, Eq. 3.38 becomes:

$$det\begin{vmatrix} \alpha - E & \beta & 0 & 0 & 0 & \beta \\ \beta & \alpha - E & \beta & 0 & 0 & 0 \\ 0 & \beta & \alpha - E & \beta & 0 & 0 \\ 0 & 0 & \beta & \alpha - E & \beta & 0 \\ 0 & 0 & 0 & \beta & \alpha - E & \beta \\ \beta & 0 & 0 & 0 & \beta & \alpha - E \end{vmatrix} = 0. \tag{3.39}$$

The solutions to this equation give the following four energies of the six π-orbitals:

$$\begin{aligned} E(a_{2u}) &= \alpha + 2\beta; \\ E(e_{1g}) &= \alpha + \beta; \\ E(e_{2u}) &= \alpha - \beta; \\ E(b_{2g}) &= \alpha - 2\beta. \end{aligned} \tag{3.40}$$

Since both α and β are negative, the first two energy terms are the lowest bonding π-orbitals, and the second two are π^*-antibonding orbitals illustrated in Fig. 3.13. Note that there are two degenerate orbitals (labeled e_1 and e_2). Also, since α is a common additive energy offset, it is ignored.

The various symmetry states in Eq. 3.40 are identified by their spectroscopic nomenclatures based on group theory. The subscript g refers to states that have even symmetry, or *gerade* in German, under spatial inversion across the molecular plane. Even orbital symmetry (or *parity*) refers to the situation where the wavefunction is unchanged due to this symmetry operation, that is, $\psi(\mathbf{r}) = \psi(-\mathbf{r})$. On the other hand, u is *ungerade*, or odd symmetry under such an

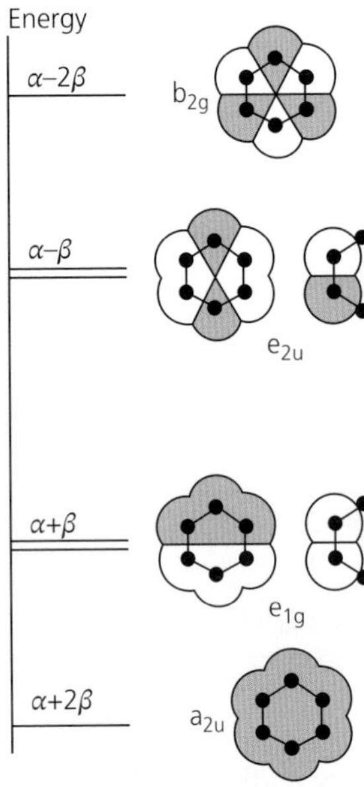

Figure 3.13 The π-orbitals of benzene and their relative energies. Light and dark shaded areas correspond to regions of opposite wavefunction phase. Note that both the e_1 (bonding) and e_2 (antibonding) orbitals are doubly degenerate (Atkins, 1983).

operation. The other notations, *a*, *e*, and *b* refer to different point group symmetries whose details can be found elsewhere (see Chapter 2 as well as many texts on molecular spectroscopy, e.g. Harris and Bertolucci, 1989). Since there are six valence electrons, and two spin states per energy level, the ground state of the benzene molecule has the electronic configuration of $a_{2u}^2 e_{1g}^4$, which is called the $^1A_{1g}$ state in the spectroscopic notation of Platt (1949). Thus, the a_{2u} state corresponds to the HOMO-1 state, whereas the frontier orbital, or HOMO, is the e_{1g} state with its four electrons. Similarly, the opposite (unoccupied) frontier orbital, or LUMO, corresponds to the e_{2u} state, and b_{2g} is the LUMO+1 orbital. In benzene, the spectroscopic nomenclature for the first excited state is ^1B_u (Platt, 1949).

The total energy of six unhybridized electrons is $6(\alpha + \beta)$ However, according to our LCAO calculation, the ground state energy of benzene is $2(\alpha + 2\beta) + 4(\alpha + \beta) = 6\alpha + 8\beta$. Hence, the ground state benzene molecule has an energy 2β lower than six individual C atoms. This difference is the *stabilization* or *delocalization* energy, and leads to the ground state benzene ring configuration. As we move to higher energies in the LUMO, the molecule becomes less stable since the orbitals are now antibonding.

An admittedly oversimplified but intuitive prediction of the π-system orbital energies of aromatic hydrocarbons can be found using *perimeter-free electron orbital* (PFEO) theory (Platt, 1949). The PFEO method can be applied to the *catacondensed molecules* comprising N fused rings. For polyaromatic hydrocarbons with the formula $C_{4n+2}H_{2n+4}$, each ring contributes $4n + 2$ π-electrons. The basic premise is that π-electrons in the rings are completely decoupled from the inner σ-electrons, and are responsible for establishing the electronic ground and excited states of the molecule (Hückel approximation #1). Furthermore, the π-electrons contributed by each C atom in catacondensed molecules belong to no more than two rings and lie on the molecular periphery. The electron is free to move along the perimeter of length $L = 2\pi r = (4n + 2)a_{CC}$, which for the purposes of this approximation is distorted into a circle whose radius is r. Then L is equal to the number of C–C bonds of length, a_{CC}. Hence, the π-electron wavefunctions have periodic boundary conditions with the form:

$$\psi_l(\theta) = \frac{1}{\sqrt{2\pi}} \exp(il\theta), \tag{3.41}$$

where $l = 0, \pm1, \pm2, \ldots$ is the orbital angular momentum quantum number, and θ is the azimuthal angle. The appropriate boundary condition is $\psi_l(0) = \psi_l(2\pi)$, as shown in Fig. 3.14.

The wavefunction can expressed using the LCAO formulation by summing over the basis set of p-atomic orbitals on N atoms (Moffitt, 1954):

$$\psi_l(\theta) = \frac{1}{\sqrt{2\pi N}} \sum_{k=0}^{N-1} \exp(ik\theta_l)\phi_k(\theta). \tag{3.42}$$

where $\theta_l = 2\pi l/N$ Solving Schrödinger's equation for the wavefunctions in Eq. 3.41 gives the PFEO eigenvalues:

$$E_l = \alpha + \frac{\hbar^2}{2mr^2} l^2. \tag{3.43}$$

There are two degenerate counter propagating wavefunctions (e.g. $l = \pm1$) resulting in four electrons occupying each level, except for the $l = 0$ state which is occupied by only two electrons following the aufbau principle. The energy gap is between the highest filled $2n + 1$ orbital (the HOMO) and the orbital at $2n+2$ corresponding to the LUMO. In PFEO nomenclature, the highest filled level is called the "*f* state," with letters ascending (descending) through the alphabet with increasing (decreasing) l. This is consistent with the picture for benzene in Fig. 3.13 where the six π-electrons fill levels up to the HOMO energy corresponding to the e_{1g} orbitals. Unlike the more precise solution obtained via the Hückel approximation, however, the difference between energy levels

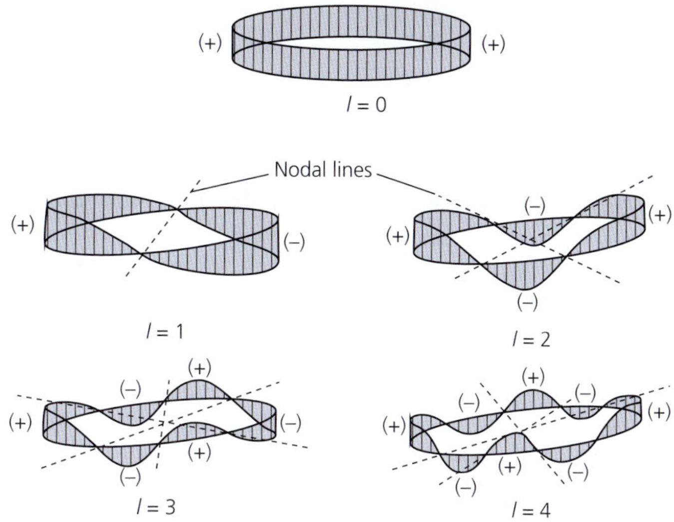

Figure 3.14 Wavefunctions, $\psi(\theta)$, for the lowest five orbitals with angular momentum quantum numbers, l. The number of nodal lines is equal to l. The relative wavefunction phases (+, −) above and below the molecular plane are shown. Adapted from Pope and Swenberg (1982).

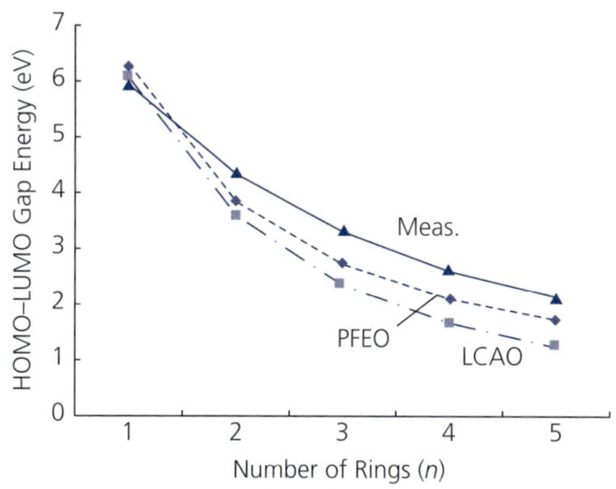

Figure 3.15 The HOMO–LUMO gap energy calculated using a modified Gaussian basis set vs. the number of fused rings (n) in the first five linear polyacenes (benzene to pentacene). The differences between experiment, PFEO and LCAO theories increases with n. For the LCAO calculation, $\beta - 2.85$ eV and $S_{rs} - 0.25$.

increase in separation with l^2. Using the more accurate basis of Eq. 3.42, it can be shown that the energies follow:

$$E_l = \alpha + 2\beta\cos\left(\theta_l\right), \qquad (3.44)$$

which leads to "compression" of the energy levels with the largest energy gap between states f and g (corresponding to the first empty shell, or LUMO energy), similar to predictions using LCAO theory.

A comparison of the calculations by both the PFEO and LCAO approaches along with the measured long wavelength absorption arising from equal HOMO–LUMO transitions is shown in Fig. 3.15 for linear polyacenes with an increasing number of rings, starting with benzene ($n = 1$), then naphthalene ($n = 2$), anthracene ($n = 3$), tetracene ($n = 4$), and pentacene ($n = 5$). Clearly, the approximations used lead to increasing inaccuracy with molecular size. Except in the case of benzene where both models are reasonably consistent with experiment, the PFEO and LCAO methods lead to an underestimation of the gap energy for the somewhat arbitrary choice of $S_{rs} = 0.25$ in Fig. 3.15 (Scherr, 1953). For $S_{rs} = 0$, the LCAO model significantly underestimates the energy for all of the polyacenes shown.

3.4 Improving accuracy: numerical models

To improve these admittedly coarse approximations that become increasingly inaccurate with molecular complexity, several of the approximations used in the LCAO method must be abandoned. A complete discussion of the more sophisticated approaches outlined here are found in the extensive literature on this subject (see *Further reading*).

There are two principal methods for improving computational accuracy. One is the *ab initio* approach based on first principles, as its name implies. These approaches all start with the Hartree–Fock equations of the individual atomic electron energies. It is then assumed that when placed in a molecule, each electron moves in a potential that is the average spherical superposition of all the other, $n - 1$ valence electrons in the molecule. The *semi-empirical approach*, on the other hand, starts with the Hückel approximation, and then lifts assumptions that lead to the greatest inaccuracies by employing a few empirical constants for a particular molecular system or class of structures. As already discussed, assuming $S_{rs} = 0$ leads to significant inaccuracies. Hence, the first step is to include the overlap integrals, S_{rs} ($|r - s| = 1$) between nearest neighbors. We first describe the most popular *ab initio* approach known as *density functional theory* (DFT). Next, we consider the suite of common semi-empirical methods starting with the Pariser–Parr–Pople (PPP) method, and then the complete neglect of differential overlap (CNDO), the intermediate neglect of differential overlap (INDO), and a few variations on these approaches (MINDO and ZINDO). There are many approximation methods and an equally large number of software packages that are available (many of them open source) that can be employed to calculate the orbital energetics of a wide variety of molecular systems, albeit with varying precision.

3.4.1 *Ab initio* approaches

Ab initio approaches start with the assumption that each electron in an orbital is affected by the mean potential due to the $n - 1$ other electrons in the molecule. Thus, we decompose the n-electron problem into n one-electron problems, where the electron is in an orbital of the ith hydrogenic atom with nuclear charge $Z_i q$. The one-electron spin–orbital wavefunction is then $\phi^0_{e,i}(\mathbf{r}_i, \mathbf{R}_I)$ (see Eq. 3.8), and the total wavefunction for the system is given by the product of single electron wavefunctions, Eq. 3.10. Now the energy for each electron is found using

$$H^0\phi^0_{e,i}(\mathbf{r}_i, \mathbf{R}_I) = E^0\phi^0_{e,i}(\mathbf{r}_i, \mathbf{R}_I) \tag{3.45}$$

and

$$H^0 = \sum_{i=1}^{n} h_i + \frac{1}{2}\sum_{i\neq j}\frac{q^2}{4\pi\varepsilon_0|\mathbf{r}_i - \mathbf{r}_j|}, \tag{3.46}$$

where h_i are the one-electron hydrogenic core Hamiltonians. It can be shown that Eq. 3.45 leads to n Hartree–Fock equations for the ith electron following:

$$f_i\phi^0_{e,i}(\mathbf{r}_i, \mathbf{R}_I) = \varepsilon_i\phi^0_{e,i}(\mathbf{r}_i, \mathbf{R}_I) \tag{3.47}$$

with

$$f_i = h_i + \sum_{j=1}^{j\neq i}\left(J_j(i) - K_i(i)\right). \tag{3.48}$$

Here f_i is called the Fock operator, and

$$J_j = \frac{q^2}{4\pi\varepsilon_0}\langle\phi^0_j\left|\frac{1}{|\mathbf{r}_i - \mathbf{r}_j|}\right|\phi^0_j\rangle \tag{3.49a}$$

and

$$K_j = \frac{q^2}{4\pi\varepsilon_0}\langle\phi^0_j\left|\frac{1}{|\mathbf{r}_i - \mathbf{r}_j|}\right|\phi^0_i\rangle, \tag{3.49b}$$

where J_j is the *Coulomb integral* and is the energy due to repulsion of electrons i and j. Similarly, K_j is the *exchange integral* that gives the energy expense when the two electrons are interchanged.

This set of equations is solved for each of n electrons with a trial set of n spinorbital functions to solve for the energies, ε_i. Reasonable trial basis sets are the Gaussian or Slater functions in Eqs. 3.23 and 3.24, although other functional forms are often used. This, of course, generates a new set of spinorbitals, and the process is iterated until the changes in ε_i and $\phi^0_{e,i}(\mathbf{r}_i, \mathbf{R}_I)$ between successive iterations are acceptably small. This iterative procedure continually readjusts the electron distribution such that it is consistent with the functional form of the wavefunctions, and hence is called the *self-consistent field* (SCF) *approximation*.

The most important approximation used in the SCF approach is that an electron only sees the average potential of all other electrons. Yet, as the single additional electron is added to the system, it instantaneously changes the original distribution, which changes the potential environment for the added electron. In effect, the distribution of $n - 1$ electrons is *polarized* by the extra charge. This is known as *electron correlation*, and is treated using further approximations that account for this so-called *configuration interaction* between electrons (see Atkins and Friedman, 1997 for a review of this method).

DFT provides a second approach to the Hartree–Fock method by replacing the electron distribution with a charge distribution, $\rho(\mathbf{r})$, comprising the entire system of n-electrons. The energy due to this distribution is $E(\rho)$. The initial wavefunction with $3n$ degrees of freedom ($\{\mathbf{r}_i\}$) is reduced to a single density, with only three degrees of freedom ($\mathbf{r}$) (Kohn and Hohenberg, 1964) The challenge is to determine the charge density functional that can be expressed in terms of the electronic probability density due to n single electron spatial orbitals, viz.

$$\rho(\mathbf{r}) = \sum_{i=1}^{n}|\phi_{e,i}(\mathbf{r}_i)|^2. \tag{3.50}$$

Thus, the energy is a function of the density, following:

$$E(\rho) = -\frac{\hbar^2}{2m_e}\sum_{i=1}^{n}\int \phi_{e,i}^*(\mathbf{r})\nabla^2\phi_{e,i}^*(\mathbf{r})d^3r$$
$$-\sum_{I=1}^{M}\int \frac{Z_I q^2}{4\pi\varepsilon_0|\mathbf{r}-\mathbf{R}_I|}\rho(\mathbf{r})d^3r$$
$$+\frac{1}{2}\sum_{i=1}^{i\neq j}\int\int \frac{\rho(\mathbf{r}_i)\rho(\mathbf{r}_j)}{4\pi\varepsilon_0|\mathbf{r}_i-\mathbf{r}_j|}d^3r_i d^3r_j$$
$$+E_{XC}(\rho), \tag{3.51}$$

where $E_{XC}(\rho)$ is the exchange-correlation energy. This energy is quantum mechanical in origin, and is comprised of both electron exchange and correlation effects on the distribution. The calculation of $E_{XC}(\rho)$ is approximate, and can lead to the largest departures from accuracy since it is determined by placing an electron in the charge distribution using the unperturbed orbital basis set, $\phi_{e,i}(\mathbf{r})$ in Eq. 3.50. Then the one-electron energies are minimized via the variational principle by choosing an initial $\rho(\mathbf{r})$. Once again we proceed with iterative solutions solved to achieve self-consistency between the orbitals and energies thus derived. Calculation of $E_{XC}(\rho)$ is simplified by considering its value for a single electron, $\varepsilon_{XC}(\rho)$, and then integrating over the entire distribution:

$$E_{XC} = \int \rho(\mathbf{r})\varepsilon_{XC}(\rho(\mathbf{r}))d^3r, \tag{3.52}$$

in what is known as the *local density approximation*.

Figure 3.16 shows calculated electron densities in the HOMO and LUMO of the pentacene molecule. The HOMO (Fig. 3.16a) shows four, and the LUMO (Fig. 3.16b) six nodal planes perpendicular to the long molecular axis. In addition, the HOMO has a nodal plane along the long molecular axis. The phases of the electron probability amplitudes invert across each of the nodal planes.

In the presence of time varying electric or magnetic fields such as encountered when light is incident on a molecule, time dependence must be included using *time dependent DFT*, or TD-DFT. The Hamiltonian now includes a time dependent external potential, $V_{ext}(t)$, as follows:

$$H_T(t) = T_e + V_e(t) + v_{ext}(t), \tag{3.53}$$

which is solved using the time dependent Schrödinger equation:

$$H_T(t)\Phi(t) = i\hbar\frac{\partial}{\partial t}\Phi(t), \tag{3.54}$$

where the total molecular orbital wavefunction is decomposed into its *time dependent* basis set of n single electron orbitals to give a density:

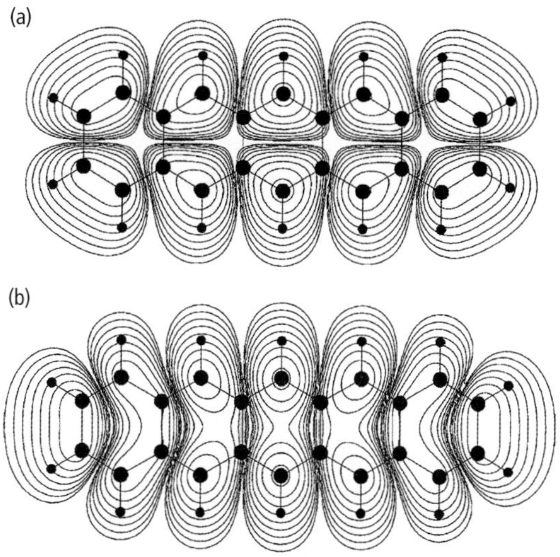

Figure 3.16 (a) HOMO and (b) LUMO π-electron densities along the pentacene molecular plane. The contour lines decrease logarithmically from the carbon atom cores (large circles). Hydrogen atoms shown as small circles. Relative phases of densities are not shown (de Wijs et al., 2003).
Reprinted from Synthetic Metals, 139, de Wijs, G. A., Mattheus, C. C., de Groot, R. A. & Palstra, T. T. M. Anisotropy of the mobility of pentacene from frustration. 109-114, Copyright (2003)

$$\rho(\mathbf{r},t) = \sum_{i=1}^{n}\left|\phi_{e,i}(\mathbf{r}_i,t)\right|^2. \tag{3.55}$$

Here we have included all of the Coulomb terms (electron–electron, electron–nuclei, etc.) as well as the exchange-correlation terms in the potential, $V_e(t)$. Note that $V_{ext}(t)$ is replaced with $v_{ext}(t)$ in Eq. 3.53 to indicate that this formulation is simplified when we determine the electron distribution under small external field conditions that ensure linear response of the system.

In TD-DFT, we iterate the solution to achieve an accurate estimation of energy in both space and time. That is, the charge distribution responds to the external field as well as to a charge added to the system. This latter problem is solved using Slater determinants as in DFT, and then the time response is calculated by solving Eq. 3.54.

3.4.2 Semi-empirical approaches

In contrast to *ab initio* methods that utilize an atomic orbital basis founded on the Hartree–Fock equations, semi-empirical approaches take a computational short cut by making use of experimentally derived properties of molecules when it is appropriate, and when the data are available. In particular, semi-empirical approaches focus only on the π-electron system, greatly reducing computational intensity. This information is combined with the LCAO method to arrive at an improved estimate (compared to using the Hückel approximation) of the molecular orbitals. Given the significant complexity of larger molecules, semi-empirical calculations can provide a sometimes orders-of-magnitude more efficient path to finding the energies and wavefunctions than purely *ab initio* techniques. However, with the continual improvement of computing power, semi-empirical approaches are losing popularity since DFT calculations are now becoming possible for even the most complex systems. Nevertheless, it is instructive to review some of the principles underlying semi-empirical calculations since they provide insight into the fundamental physics leading to observed molecular properties.

The procedure begins identically to *ab initio* approaches, with the exception that only the n_π π-electrons are considered. Then we develop a Hamiltonian analogous to Eq. 3.46, but only include outer electrons thus:

$$H^\pi = \sum_{i=1}^{n} h_i^\pi + \frac{1}{2}\sum_{i \neq j} \frac{q^2}{4\pi\varepsilon_0 |\mathbf{r}_i - \mathbf{r}_j|}, \tag{3.56}$$

where the core Hamiltonian for the ith π-electron is given by

$$h_i^\pi = -\frac{\hbar^2}{2m_e}\nabla_i^2 + V_i^\pi. \tag{3.57}$$

Here, V_i^π is the potential due to all the nuclei comprising the molecule, as well as all of the σ-bonds between the nuclei. Then the total π-orbital wavefunction $\Phi^\pi(\mathbf{r})$ can be written as the product of the individual molecular orbitals:

$$\Phi^\pi(\mathbf{r}) = \prod_{i=1}^{n_\pi} \phi_{e,i}(\mathbf{r}), \tag{3.58}$$

where each molecular orbital is given by the sum over atomic orbitals (φ_k) following Eq. 3.22. The individual orbitals are found via a solution to:

$$h_i^\pi \phi_{e,i}(\mathbf{r}) = \varepsilon_i^\pi \phi_{e,i}(\mathbf{r}). \tag{3.59}$$

Then, the total molecular energy is simply equal to the sum of all the orbital energies as:

$$E^\pi = \sum_{i=1}^{n_\pi} n_i \varepsilon_i^\pi, \tag{3.60}$$

where n_i is the total number of electrons in the ith π-orbital. Hence:

$$\varepsilon_i = \frac{\sum_{k=1}^{L}\sum_{l=1}^{L} c_k^* c_l \langle \varphi_k | h_i^\pi | \varphi_l \rangle}{\sum_{k=1}^{L}\sum_{l=1}^{L} c_k^* c_l \langle \varphi_k | \varphi_l \rangle} = \frac{\sum_{k=1}^{L}\sum_{l=1}^{L} c_k^* c_l H_{i,kl}^\pi}{\sum_{k=1}^{L}\sum_{l=1}^{L} c_k^* c_l S_{i,kl}^\pi}, \tag{3.61}$$

where $H_{i,kl}^\pi$ and $S_{i,kl}^\pi$ are the Coulomb and overlap integrals, analogous to those used in the *ab initio* formalism, and L is the number of atomic orbitals in the basis. Equation 3.61 must be minimized using variational theory for each orbital to obtain the total energy in Eq. 3.60.

The solutions to this system of equations are found by solving the secular determinant

$$\det|\mathbf{H}^\pi - E\mathbf{S}^\pi| = 0 \tag{3.62}$$

using the usual Hückel approximations. When this semi-empirical approach was first introduced, the electron–electron repulsive interaction terms in Eq. 3.56 were ignored. This *Hückel molecular orbital theory* (HMO) was relatively easy to solve for conjugated systems, but its rather coarse approximation introduces significant errors. This has led to more robust and extended HMO methods that attempt to correct the damage done by this simplification.

The *Pariser-Parr-Pople* (PPP) method was introduced by Pariser and Parr (Pariser and Parr, 1953), and separately by Pople (Pople, 1953). The PPP method and its extensions attempt to eliminate some of the most troublesome approximations of HMO theory, particularly recovering electron-electron repulsion. The PPP method thus provides the foundation for many popular semi-empirical approaches in use today. The HMO approach assumes that the single π-electron takes into account the screening of all other electrons in their orbitals. Then, by adding in the repulsive term in Eq. 3.56, the interelectronic interactions will end up being over-counted (by a factor of two).

In the PPP method, we start with the π-system molecular orbital ($\Phi^\pi(\mathbf{r})$) as the product of π spinorbitals ($\psi_i^\pi(\mathbf{r})$) following Eq. 3.10. The atomic orbitals are given by the sum in Eq. 3.22 weighted by the coefficients, c_{ir}. Then the spinorbitals are found by minimizing the sum of secular equations:

$$\sum_{r=1}^{N} c_{ir}[F_{rs} - \varepsilon_i S_{rs}] = 0. \tag{3.63}$$

From the previous treatment, it follows that

$$F_{rs} = H_{rs}^\pi + \sum_{t=1}^{N}\sum_{u=1}^{N} P_{tu}\left[\langle rt|su\rangle - \frac{1}{2}\langle rt|us\rangle\right], \tag{3.64}$$

where H_{rs}^π includes the full core Hamiltonian in Eq. 3.46 along with the electron-electron repulsion terms. Here, we have used the shorthand notation

$$\langle rt|su\rangle = \frac{q^2}{4\pi\varepsilon_0}\langle \phi_r(\mathbf{r}_1)\phi_t(\mathbf{r}_2)| \frac{1}{|\mathbf{r}_1 - \mathbf{r}_2|} |\phi_s(\mathbf{r}_1)_u(\mathbf{r}_2)\rangle \tag{3.65}$$

to denote two-electron Coulomb integrals over the atomic orbitals. Finally, we define

$$P_{tu} = 2\sum_{w=1}^{N} c_{wt}^{*} c_{wu}, \tag{3.66}$$

which leads us to the total electronic energy (Pople, 1953):

$$E^{\pi} = \frac{1}{2}\sum_{r=1}^{N}\sum_{s=1}^{N} P_{rs}\left[H_{rs}^{\pi} + F_{rs}\right]. \tag{3.67}$$

The differences between HMO and the PPP method are contained in the two electron terms, $\langle rt \mid su\rangle$, otherwise known as the *differential overlap integrals*. As in the case of all such integrals, they decrease rapidly as the electron separation increases, thus allowing us to make some approximations. In PPP theory, we take:

$$\langle rt|su\rangle = \delta_{rs}\delta_{tu}\langle rr|tt\rangle, \tag{3.68}$$

which corresponds to the case of zero differential overlap. This is the very simplest and approximate condition that nevertheless is a considerable improvement over HMO that neglects electron-electron interactions entirely by setting all such integrals to zero.

The approach is inherently semi-empirical since electrons in neighboring atoms that share a bond replace the off-diagonal core Hamiltonian values with an empirical parameter, $h_{ij}^{\pi} = \beta_{ij}$, whereas in *ab initio* methods this term is calculated, or perhaps even worse, it is ignored.

The CNDO method takes all electrons (not just the π-system) into account, and then follows an identical procedure to the PPP approach, including only integrals of the form of Eq. 3.68. The next level of precision is achieved by including the exchange integrals: $\langle rs|rs\rangle = \langle rs|sr\rangle$. This procedure is known as the INDO method. As the approximations are lifted one after another in an effort to achieve ever greater accuracy, their computational complexities increase accordingly. These methods go by the names of modified INDO (MINDO), Zerner's INDO (ZINDO) (Zerner, 1991) that includes more atomic species than INDO, and so on. Numerous software packages exist to calculate orbitals by these more exact means, but their further discussion is beyond the scope of this text.

A comparison of the electronic distributions in the HOMO and LUMO manifolds calculated for anthracene by both the *ab initio* LCAO Hückel method using a modified Gaussian basis set, and by the semi-empirical Austin model 1 (AM1 which is a MINDO-based method) is shown in Fig. 3.17. Close examination of these distributions indicates subtle but nevertheless important differences that are produced using the different approaches.

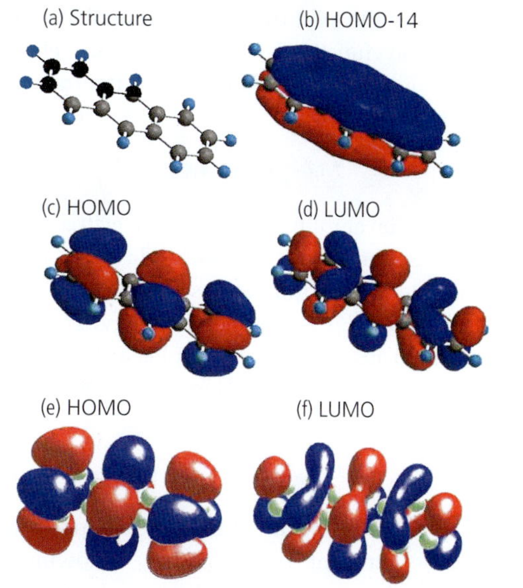

Figure 3.17 (a) Structure of anthracene ($C_{14}H_{10}$). AM1 based calculation of the π-electron systems in anthracene for the (b) HOMO-14 level showing complete delocalization of the electrons between the p_z atomic orbitals, (c) HOMO and (d) LUMO orbitals. Similar calculations based on the LCAO model for the (e) HOMO and (f) LUMO orbitals. The red and blue regions indicate opposite phases of the spatial wavefunctions that invert across each molecular plane (cf. Fig. 3.16) (Peumans, 2004). Reprinted with permission.

3.5 Transitions between states

The symmetry and composition of electron wavefunctions directly impacts the spectra and conducting properties of organics. The most important ingredients for making the leap from molecular orbitals to the optoelectronic properties of a compound is by understanding the transition rates between energy levels and states, and their quantum mechanical selection rules. Those properties are the subject of this section.

3.5.1 Fermi's Golden Rule

Now that we are armed with expressions for the electronic and vibronic wavefunctions, $|\phi_e\rangle$ and $|\phi_N\rangle$, we are prepared to calculate the allowed transitions between the ground and excited electronic states to determine the absorption and emission spectra of a molecule. The transition rate between initial state, i, and final state, f, is given by *Fermi's Golden Rule*:

$$k_{if} = \frac{2\pi}{\hbar}|\langle\psi_f|H_{int}|\psi_i\rangle|^2\rho(E_{if}). \tag{3.69}$$

Here, $\rho(E_{if})$ is the joint density of states of the initial and final wavefunctions, ψ_i and ψ_f, respectively, and H_{int} is the interaction Hamiltonian for the transition. The term

$$M_{if} = \langle \psi_f | H_{int} | \psi_i \rangle \tag{3.70}$$

is the transition amplitude, or *transition matrix element*, and M_{if}^2 is the transition probability for $i \rightarrow f$. For discrete initial and final states, $\rho(E_{if}) = \delta(E_i - E_f \mp E_{ph})$, where E_{ph} is the photon energy. The minus sign refers to absorption, and the plus to emission. To first order, the interaction Hamiltonian is approximated by the electric dipole transition:

$$H_{int} = -\boldsymbol{\mu}_{\mathbf{r},\mathbf{R}} \cdot \mathbf{F}, \tag{3.71}$$

where $\mathbf{F}$ is the optical electric field and $\boldsymbol{\mu}_{\mathbf{r},\mathbf{R}}$ is the dipole moment operator that, following the Born–Oppenheimer approximation, can be separated into its nuclear and electronic components as follows:

$$\boldsymbol{\mu}_{\mathbf{r},\mathbf{R}} = \boldsymbol{\mu}_e + \boldsymbol{\mu}_N = -q\left[\sum_k \mathbf{r}_k - \sum_K Z_K \mathbf{R}_K(Q)\right]. \tag{3.72}$$

Here the nuclear positions are given relative to their normal coordinates, Q, and the sum is over k valence electrons and K nuclei. Hence, the dipole matrix element is:

$$\mu_{if} = \langle \phi_{e,f}(\mathbf{r},Q)\phi_{N,f}(Q) | \mu_{\mathbf{r},\mathbf{R}} | \phi_{e,i}(\mathbf{r},Q)\phi_{N,i}(Q)\rangle. \tag{3.73}$$

The final and initial states of the wavefunctions are orthonormal, for example,

$$\langle \phi_{N,f}(Q) | \phi_{N,i}(Q)\rangle = \int \phi^*_{N,f}(Q)\phi_{N,i}(Q)dQ = 0 \tag{3.74}$$

where $\langle \phi_{N,i}(Q) | \phi_{N,i}(Q)\rangle = 1$. Insertion of Eq. 3.72 into Eq. 3.73 yields the integrals:

$$\begin{aligned}\mu_{if} = &-q\left[\sum_k \int \phi^*_{e,f}(\mathbf{r},Q)\mathbf{r}_k\phi_{e,i}(\mathbf{r},Q)d^3r \times \left(\phi^*_{N,f}(Q)\phi_{N,i}(Q)dQ\right)\right] \\ &+q\left[\sum_K Z_K \int \phi^*_{e,f}(\mathbf{r},Q)\phi_{e,i}(\mathbf{r},Q)d^3r \times \left(\phi^*_{N,f}(Q)\mathbf{R}_K\phi_{N,i}(Q)dQ\right)\right].\end{aligned} \tag{3.75}$$

The second integral vanishes due to orthogonality of the different electronic wavefunctions in the initial and final electronic states, $\phi_{e,i}(\mathbf{r})$ and $\phi_{e,f}(\mathbf{r})$, respectively. In contrast, to first order during the transition, the nuclei are fixed (the Franck-Condon principle), and hence only the first integral survives. Thus we arrive at:

$$\mu_{if} = -q\left[\int \phi^*_{N,f}(Q)\phi_{N,i}(Q)dQ \sum_k \phi^*_{e,f}(\mathbf{r},Q)\mathbf{r}_k\phi_{e,i}(\mathbf{r},Q)d^3r\right]. \tag{3.76}$$

3.5.2 Quantum mechanical selection rules

From the treatment thus far, we can determine which transitions are allowed, and which are forbidden, thereby developing a set of quantum mechanical selection rules. From Eq. 3.69 and 3.70, we see that a transition is allowed if the matrix element, M_{if}, is non-zero. Since the electronic, nuclear and spin wavefunctions are separable according to the Born–Oppenheimer approximation, and since μ_{if} operates only on the electronic wavefunction (Eq. 3.76), the matrix element leading to an allowed transition in the electric dipole approximation is given by:

$$\begin{aligned}M_{if} = q\langle \phi_{e,f}(\mathbf{r}_f) | \mathbf{r}\cdot\mathbf{F} | \phi_{e,i}(\mathbf{r}_i)\rangle \\ \langle \phi_{N,f}(Q_f) | \phi_{N,i}(Q_i)\rangle\langle \sigma_f(S_f) | \sigma_i(S_i)\rangle \neq 0.\end{aligned} \tag{3.77}$$

The first two terms are given by the integral in Eq. 3.76. The spatial term enforces a change in parity between $\phi_{e,i}(\mathbf{r}_i)$ and $\phi_{e,f}(\mathbf{r}_f)$. That is, since the dipole operator has odd parity defined as $\boldsymbol{\mu}_\mathbf{r}(\mathbf{r}) = -\boldsymbol{\mu}_\mathbf{r}(-\mathbf{r})$, and since the integral is over all space, then the integral is non-zero only when $\phi_{e,i}(\mathbf{r}_i)$ and $\phi_{e,f}(\mathbf{r}_f)$ are of opposite parity. Generally, the ground state of a molecule with inversion symmetry has even parity (e.g. the A_{1g} state Section 3.3). Hence allowed transitions are to or from odd parity excited states (e.g. B_{2u}). This results in the selection rule that *electric dipole transitions are allowed only between electronic states of opposite spatial symmetry*. The higher the symmetry of the molecule, the more likely it is to find transitions whereby the parity between the initial and final states is ostensibly the same, hence allowing for forbidden transitions. Finally, the spatial integral will also be very small (and hence $M_{if} \approx 0$) if there is little or no overlap between the electron orbitals in $\phi_{e,i}(\mathbf{r}_i)$ and $\phi_{e,f}(\mathbf{r}_f)$.

The integrals in Eq. 3.76 and 3.77 are also non-zero only if the transitions are between the same nuclear (i.e. vibronic) sub-state in the final and initial electronic states, provided that there is sufficient overlap between the vibronic wavefunctions $\phi_{N,i}(Q_i)$ and $\phi_{N,f}(Q_f)$. This second selection rule implies that *transitions are only allowed between vibronics with the same quantum number, n_l,* in Eq. 3.15. This is illustrated in Fig. 3.5 by the vertical arrows that represent allowed absorption transitions between the same vibronic levels in two different electronic states. This assumption is partially justified by the Franck–Condon principle whereby the relaxation of the electronic states follows the transition, implying that Q is constant.

In the common circumstance that the vibronic quantum number is different in i and f, that is, $\Delta n_l \neq 0$, then orthogonality of the wavefunctions demands that $\mu_{if} = 0$ (cf. Eq. 3.76). However, if there is a substantial shift in the molecular coordinates between E_i and E_f, this symmetry-based selection rule is broken. This is apparent when we expand Eq. 3.76 about equilibrium (Sponer and Teller, 1941, Gribov and Orville-Thomas, 1988):

$$\begin{aligned}\mu_{if} &= -q\int \phi^*_{N,f}(Q)\phi_{N,i}(Q)dQ\sum_k \phi^*_{e,f}(r,Q)\mathbf{r}_k\phi_{e,i}(r,Q)d^3r \\ &-q\sum_{k,l}\int \phi^*_{N,f}(Q)Q_k\phi_{N,i}(Q)dQ\int \left.\frac{\partial\left(\phi^*_{e,f}\phi_{e,i}\right)}{\partial Q_k}\right|_0 \mathbf{r}_l d^3r \\ &-\frac{q}{2}\sum_{k,l,m}\int \phi^*_{N,f}(Q)Q_kQ_m\phi_{N,i}(Q)dQ \\ &\times\int \left.\frac{\partial^2\left(\phi^*_{e,f}\phi_{e,i}\right)}{\partial Q_k\partial Q_m}\right|_0 \mathbf{r}_l d^3r + \ldots.\end{aligned} \tag{3.78}$$

This is known as the *Herzberg–Teller expansion*. The first integral is the overlap between vibronics in the electronic manifolds and is known as the *Franck–Condon integral*. The intensity of the transition is thus related to the square of the integral, whose value is called the *Franck–Condon factor* (FC_{if}) for the transition. This is conveniently written as:

$$FC_{if} = |\langle\phi_{N,f}(Q_f)|\phi_{N,i}(Q_i)\rangle|^2. \tag{3.79}$$

As noted above, if there is no change in Q from the initial to the final state, then only the first term in the expansion around the potential minimum survives, and only transitions corresponding to $\Delta n_l = 0$ are allowed.

It is quite rare for there to be no nuclear positional adjustments between two different electronic states. This situation corresponds to a simple vertical translation of the electronic potential. Often, an electronic transition results in a significant nuclear reconfiguration of $\Delta Q = Q_f - Q_i$, corresponding to a horizontal as well as vertical translation of the potentials in Fig. 3.5. This causes symmetry breaking, that is, the vibronic wavefunctions in the two electronic manifolds are no longer orthogonal since they become *mixed* with the electronic states. In this case, higher order terms in the Herzberg–Teller expansion survive. Transitions between vibronic states with $\Delta n_l \neq 0$ are now allowed with a probability determined by time independent perturbation theory. In other words, the transition results in the *simultaneous* excitation of the vibrational as well as electronic modes. The second term in the Herzberg–Teller approximation corresponds to a transition between states when $\Delta n_l \neq 0$. The product of $|\phi_{N,f}(Q_f)\rangle$ and $|\phi_{N,i}(Q_i)\rangle$ has odd spatial symmetry which, when combined with the odd spatial coordinate, Q, results in an integrand with even symmetry. Hence, the integral is non-vanishing, whereas the first and third terms vanish.

Similarly, for the next higher order transition where two vibronics on different electronic states are excited, the third integral is non-vanishing, and so on. Each of these possible transitions is at a higher energy of

$$\Delta E_l = |E_f - E_i| + \Delta n(\hbar\omega_l) \tag{3.80}$$

in both absorption and emission. As in Eq. 3.15, ω_l is the frequency of the lth normal vibrational mode that is excited in the transition. Thus, the transition energy differs from the energy of the electronic transition by the vibronic energies with quantum numbers that differ by n.

The final selection rule requires that the spin of the initial and final states is conserved. Since the symmetry of the $S = 0$ and $S = 1$ states are odd and even, respectively, the spin integral $\langle\sigma_f(S_f)|\sigma_i(S_i)\rangle$ *is non-zero only for transitions between states of the same spin multiplicity*, that is, when $S_i \to S_f$ or $T_i \to T_f$. However, an important exception to this rule occurs in the presence of spin–orbit or spin–spin coupling, a topic that is the focus of Section 3.7.

A summary of these three selection rules, and exceptions that lead to the observation of otherwise quantum mechanically forbidden transitions, is provided in Table 3.1. We note that transitions between states of opposite parity can arise from two photon transitions, that is, absorption or emission that involves more than a single photon. Also, higher order multipole transitions can also lead to violations of the parity rule.

3.5.3 Understanding molecular spectra

The *vibronic progression* of transitions in both absorption and emission are illustrated in Fig. 3.18. Here, the ground (S_0) and first excited (S_1) singlet states are shifted by a small but noticeable configuration shift, ΔQ. Since the electron population (N) of the vibronic states follows a Boltzmann distribution, that is,

Table 3.1 Summary of selection rules for electronic, nuclear and spin transitions for electric dipole interactions

Transition	Selection rule	Matrix element	Exception
Between electronic states	Parity of $\phi_{e,f}$ and $\phi_{e,i}$ must be different (e.g. even $\rightarrow$ odd)	$\langle\phi_{e,f}(\mathbf{r})\vert\mathbf{r}\vert\phi_{e,i}(\mathbf{r})\rangle$	Low symmetry molecules, two photon transitions, higher order multipoles
Between vibronic states in different electronic manifolds	Vibronic quantum number $n_i - n_f = 0$	$\langle\phi_{N,f}(Q)\vert\phi_{N,i}(Q)\rangle$	$\Delta Q_{if}\neq 0$: nuclear reconfiguration between $\phi_{e,f}(\mathbf{r},Q)$ and $\phi_{e,i}(\mathbf{r},Q)$
Between spin states	$S_i \rightarrow S_f$; $T_i \rightarrow T_f$	$\langle\sigma_f(S_f)\vert\sigma_i(S_i)\rangle$	Spin–orbit coupling
			Spin–spin coupling

$$N(n_l) = N(0)\exp(-n_l\hbar\omega_l/k_BT), \qquad (3.81)$$

then N decreases with increasing n_l, where $N(0)$ corresponds to the electron density of the lowest vibronic ($n = 0$). Equation 3.81 implies that the most intense absorption and emission spectral features are due to excitations from the lowest vibronic level in the initial electronic state, S_0. This is known as *Kasha's rule*, and it results in approximately mirror-symmetric replicas of the absorption and emission spectra, separated by an energy shift known as the *Franck–Condon shift* or *Stokes shift*, ΔFC. The relative intensity of each transition depends on the overlap of the wavefunctions as determined by the Franck–Condon integral, Eq. 3.79. Thus, larger nuclear coordinate shifts result in a reduction in intensity in the lowest order (i.e. 0–0) transition, which is strongest for $\Delta Q \approx 0$. The 0–0 transition, itself is known as the zero phonon transition. The quantized vibrational levels arise from the many intramolecular phonon modes of polyatomic molecules.

Recall that the Franck-Condon principle states that molecules relax to their equilibrium configurations following an electronic transition. Hence, *absorption occurs from the ground state into the unrelaxed* (and, therefore, higher energy) *excited state*. This rapid relaxation occurs within the time it takes the nuclei to reconfigure to their new equilibrium positions, that is, on the order of a phonon lifetime of 10^{-12} to 10^{-14} s. In contrast, *emission occurs from the relaxed excited state* (known as the Frank–Condon state) back to the ground state. The Franck-Condon shift corresponds to this relaxation energy. The singlet exciton energy (denoted in Fig. 3.18 as S_1) corresponds to the energy gap, E_G, reduced by the relaxation energy that is lost due to the Coulomb binding energy (E_B), which is the difference in energy between the excited electron and the hole left in the ground state.

In Fig. 3.18 we use spectroscopic notation of $S_1 \rightarrow S_0$ to denote emission, and $S_1 \leftarrow S_0$ for absorption. In this convention, the highest energy state is always on the left, and the arrow points in the direction of the transition. This convention is not always rigorously followed, where the transition between the initial and final states, i and f is simply written $i \rightarrow f$ along with an identification of the highest energy state.

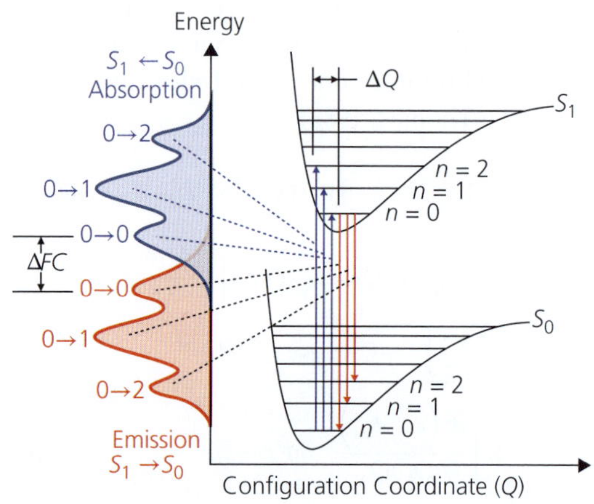

Figure 3.18 Ground (S_0) and first excited (S_1) state potentials *vs.* the normal mode, or configuration coordinate, Q. Each vibronic level is labeled by its corresponding quantum number ($n = 0, 1, 2, 3\ldots$). Emission and absorption spectra corresponding to the transitions are indicated on the left by the ascending (absorption) and descending (emission) arrows. The label for each peak indicates the vibronic quantum number of the initial and final states (i.e. in emission, 0–2 corresponds to a transition denoted as $S_1 \rightarrow S_0$ from $n = 0$ to 2). The Franck–Condon, or Stokes shift is labeled ΔFC, and the configuration coordinate shift between electronic manifolds is ΔQ.

An illustration of molecular reconfiguration leading to $|\Delta Q| > 0$ is illustrated in Fig. 3.19 for FIPAC dissolved in a solution of MeTHF. There is a dramatic configuration shift between two isomers when the temperature is changed between cryogenic and room temperature. The ground state and the high energy S* state have a saddle like conformation. Addition of thermal energy (ΔH) induces an isomeric transformation into the lower energy planar P* conformation. The conversion from P* to S* is reversed by cooling.

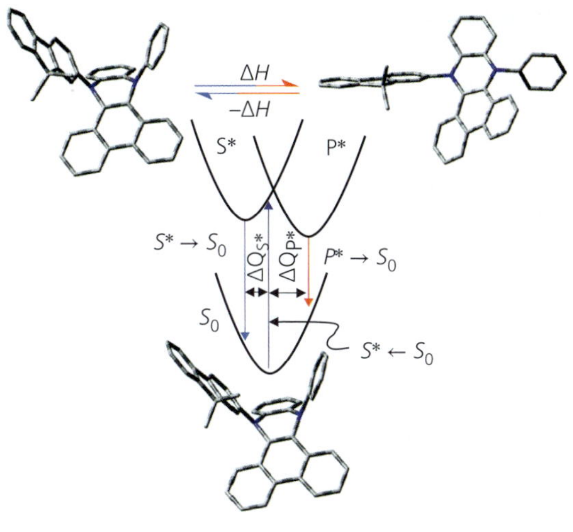

Figure 3.19 Conversion between two structural isomers (S* and P*) of FIPAC by heating (ΔH) or cooling ($-\Delta H$). The configuration shifts (ΔQ) and transitions to the ground state (S_0) are indicated. C atoms are shown in grey, N in blue. Adapted from Chen et al. (2015).

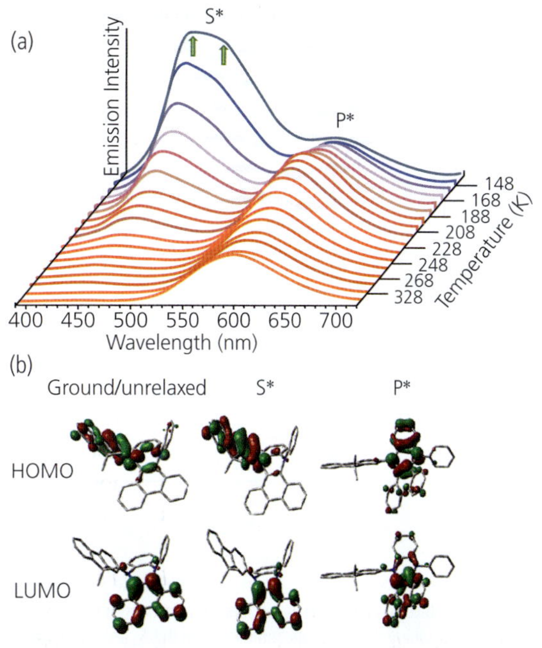

Figure 3.20 (a) Evolution of the emission spectra of FIPAC vs. temperature. Arrows show two vibronics (from left to right they correspond to 0-0 and 0-1) due to $S^* \rightarrow S_0$ transitions. (b) DFT calculations of the HOMO and LUMO states of the unrelaxed states, as well as the S* and P* isomers. After Chen et al. (2015).

The configuration change is accompanied by a large change in nuclear coordinates that results in concomitant changes in the emission spectrum, as shown in Fig. 3.20a. The energy from the ground to the unrelaxed LUMO ($S^* \leftarrow S_0$) is 3.43 eV. At room temperature, the molecule relaxes into S*, but the small activation energy between S* and P* results in transfer of the excited state to P* followed by a radiative transition at 595 nm (2.08 eV). At low temperature, the excitation is trapped in S*, shifting the emission to 435 nm (2.85 eV). Note that the emission at $T < 200$ K shows two clearly resolved peaks corresponding to the 0–0 and 0–1 transitions. That is, the smaller, negative configuration shift of S* compared with the positive shift for P* increases the overlap with the ground state, allowing for transitions to higher order ground state vibronics.

Figure 3.20b shows the electron density distributions calculated by TD-DFT using the B3LYP/6-31G* functional. Opposite phases of the probability density are indicated by the red and green regions. Unlike nonpolar polyacenes such as PTCDA and pentacene whose HOMO and LUMO electron densities are evenly distributed across the molecular plane, the twist of FIPAC creates separate regions occupied by the frontier orbitals. The HOMO is localized on the fluorene group, apart from the LUMO on the triphenyl. Furthermore, the HOMO is largely unperturbed by transformation between the isomers. The spatial separation of the orbitals results in larger charge transfer from the HOMO to LUMO in the unrelaxed and S* conformations compared to that in the P* state.

Returning to Fig. 3.18, the 0–1 transitions have the highest intensity. However, for molecules whose electronic systems are strongly coupled across a rigid molecular plane (e.g. in polyacenes such as perylene or anthracene), the 0-0 transition is often the most intense, and decreases for transitions between higher order vibronic modes. In Section 3.6.1 we will find that the strength of a transition is proportional to the magnitude of the overlap between vibronic wavefunctions in a transition between two electronic states, as expressed by FC_{if}. Figure 3.21 shows both emission and absorption transitions between S_1 and S_2 that are offset by a molecular reorganization of ΔQ. Superimposed on each vibronic state is the corresponding probability amplitude of the wavefunction in the SHO approximation. The maximum overlap for the $S_1 \leftarrow S_0$ absorption is between the $n = 0$ vibronic in the ground state and $n = 2$ in the first excited state. Due to the Boltzmann factor in Eq. 3.81, the ground state population is concentrated in the S_0, $n = 0$ state; a realization of Kasha's rule. Similarly, the S_1, $n = 0$ state also has the largest electron population after relaxation due to phonon scattering. Hence, emission in this case is most likely to the S_0, $n = 2$ vibronic which has the largest overlap with the S_1, $n = 0$ state (corresponding to a 0–2 transition). Both absorption

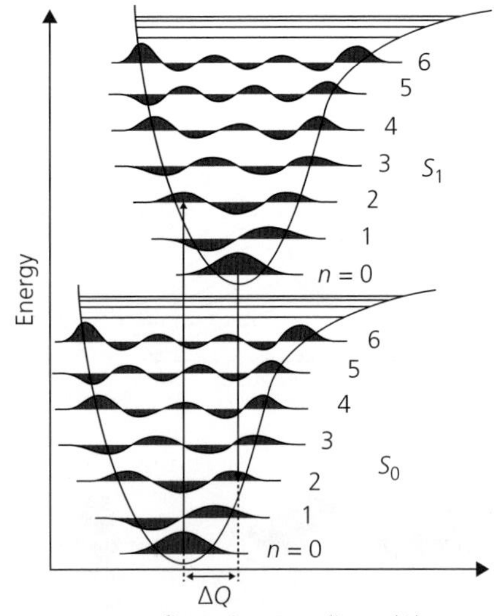

Figure 3.21 Transitions between S_0 and S_1 showing the probability amplitudes of the vibronic states in each electronic manifold. Transitions are most likely where the wavefunction overlap in the S_0 and S_1 manifolds is highest. In this case absorption is most likely from the filled $n = 0$ vibronic in S_0 to the n = 2 vibronic in S_1, and emission from $n = 0$ in S_1 to $n = 2$ in S_0, accounting for a configuration change of ΔQ.

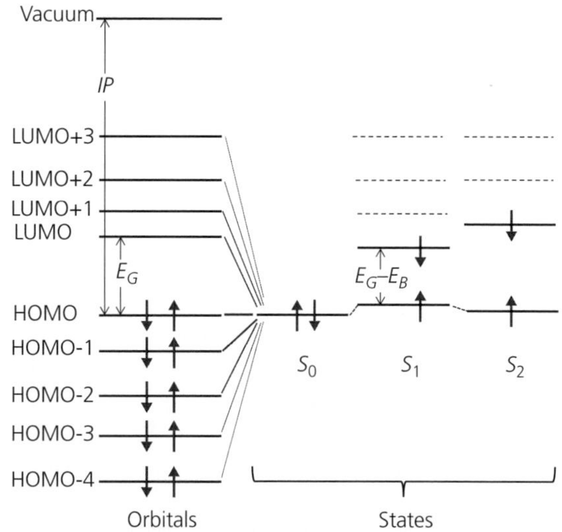

Figure 3.22 Reconciling the differences between the electron orbital picture and that of electronic states. A state is comprised of a linear combination of orbitals, with the nearest orbital having the greatest influence. In this case, we show that the ground, S_0 state is a superposition of the HOMOs, with lesser influence from the empty LUMO levels. The S_1 and higher excited states are primarily determined by the LUMOs. The diagonal lines represent the relative contributions of the orbital energies to S_0, whose strengths are suggested by the boldness of the lines. While orbital energies are referenced to vacuum, states are not. S_1 is comprised of one half filled HOMO orbital and one half filled LUMO. S_2 is similar to S_1 except the Coulombically shifted LUMO+1 is now half filled. The shifts of the state energies relative to the orbitals (dashed lines in S_1 and S_2) are indicated. Here, E_G = energy gap, E_B = binding energy between the excited electron and the hole in the ground state, and IP = ionization potential energy.

and emission transitions are vertical since the photon carries very small angular momentum compared to the electron. Usually, the Frank–Condon factor decreases with increasing difference in the vibronic quantum numbers, n, between the initial and final states, resulting in a lower transition probability, and hence a concomitant decrease in spectral intensity. However, this is by no means always the case for molecules that are less rigid (and hence undergo a larger ΔQ), or of higher molecular weight. The intensity may be *borrowed* from one vibronic by another (Orlandi and Siebrand, 1973), creating a progression of vibronic overtones on the electronic absorption or emission spectra whose intensities are due simply to the value of the corresponding Franck–Condon integrals in Eq. 3.79.

At this point, we need to clearly distinguish between electronic orbitals as represented in Fig. 3.7, and electronic states in Fig. 3.18. Orbitals refer to energies of electrons and holes referenced to vacuum, as shown in Fig. 3.22. That is, the HOMO and LUMO energies are all negatively disposed relative to the vacuum level since these represent the energies of charges bound to a molecule. The energy needed to remove an electron from the HOMO is the *ionization potential*, *IP* (see Section 3.6). In contrast, *states* refer to Coulombically bound electrons in the relaxed HOMO and LUMO levels that, according to the LCAO model, are comprised of a linear superposition of all the other relaxed orbitals in the molecule. State energies are referenced relative to each other (i.e. S_0 to S_1 in Fig. 3.18), and hence it does not make sense to position the state diagram relative to the vacuum level. As a consequence, states and orbitals cannot be accurately represented on the same diagram, although when referring to device energy level diagrams it is often convenient to do so, as we will see in Part II.

As above, the state, or exciton energy is given by

$$E_X = E_G - E_B. \tag{3.82}$$

In the solid, both free charges (i.e. polarons) and excitons are free to move from molecule to molecule.

Perturbation theory implies that the influence of orbitals to a particular state decreases with the energy difference on that state. Hence, the principal contribution to the ground, or S_0 state is due to the HOMO level, with diminishing contributions from states HOMO $- i$ as i is increased from 1 to 2 to 3, etc. Similarly, the first excited state S_1 consisting of one electron in the relaxed HOMO and a second in the relaxed LUMO has a major contribution from the LUMO, with decreasing influence from LUMO + i as i increases.

A distinction between charges occupying energy levels and excitons is that while charges have spins equal to ½, states are comprised of two or more spins whose combination results in the state spin multiplicity (e.g. singlets, triplets, quintets, etc.). These multiplicities are denoted by S, T, and Q, respectively.

Figure 3.23 shows examples of the emission and absorption spectra of three molecules of varying stiffness and conjugation lengths. We notice that the absorption energy is higher than the fluorescence emission energy, with a crossing of the spectra at the *adiabatic transition energy*, E_{00}. This is often taken as equal to the energy gap, E_G, between the HOMO and LUMO energies. The absorption spectra show well-resolved vibronic progressions at low temperature that are considerably broadened at room temperature, particularly for the molecules that have multiple rotational or vibrational degrees of freedom (Fig. 3.23b and c). One reason that the vibronics in the fluorescence spectra are poorly resolved at high temperatures is that their inter-ring bonds are shortened in the excited state, resulting in a closer overlap between higher frequency vibronics and rotational modes. At low temperatures, many of these modes are "frozen out," which results in spectral line narrowing.

The vibronic progression is often dominated by the highest energy phonon mode, which in the polyacenes and related compounds is due to the C–C stretch. An example of vibronic progressions in the mirror-symmetric absorption and fluorescence spectral at low temperature is shown for DBP in Fig. 3.24.

The ability to resolve individual vibronic features in the fluorescence (or photoluminescence) and absorption spectra increases with decreasing temperature. In addition, there is a *hypsochromic* (blue) shift in E_{00} with increasing temperature, and the intensity of the 0–1 PL peak increases relative to the 0–0 (highest energy) emission peak at room temperature. The nearly featureless PL spectrum at $T = 300$ K is due to thermal broadening of the many intramolecular vibrational, rotational and intermolecular librational features that allows for greater intensity borrowing between the substates, consistent with Eq. 3.81. Also, the room temperature PL spectral intensity is approximately 5% of that at low temperature since non-radiative exciton decay competes with radiative decay with increasing temperature. The energetic separation between vibronics, $\Delta E = E_{0-n} - E_{0-(n+1)} \approx 0.1$ eV, corresponds to the C–C stretch mode. The absorption feature at wavelengths < 350 nm is due to $S_2 \leftarrow S_0$ transitions.

The relationship between unrelaxed excited and ground states is shown in Fig. 3.25. Here, ΔE_{eq} is the lowest equilibrium energy within a particular manifold (corresponding to the energy of the lowest, $n_l = 0$ vibronic) relative to the electronic energy state minimum. We have the following relations for the unrelaxed, vertical electronic transition energies:

$$E_{vert}(\mathrm{Abs}) = E_{00} + \Delta E_{eq}(\mathrm{Abs}) \tag{3.83a}$$

and

$$E_{vert}(\mathrm{PL}) = E_{00} + \Delta E_{eq}(\mathrm{PL}) \tag{3.83b}$$

with $\Delta E(PL) \approx \Delta E(Abs)$, where PL and Abs denote emission and absorption transitions, respectively

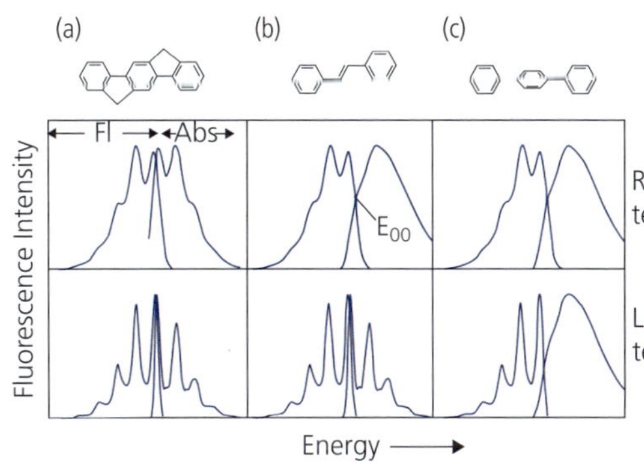

Figure 3.23 Fluorescence (Fl) and absorption (Abs) spectra of three molecules with different degrees of conjugation and stiffness at two different temperatures. (a) Ladder-type oligophenylene, (b) diphenylvinylene, (c) triphenylene. E_{00} is the adiabatic transition energy. From Gierschner et al. (2007).

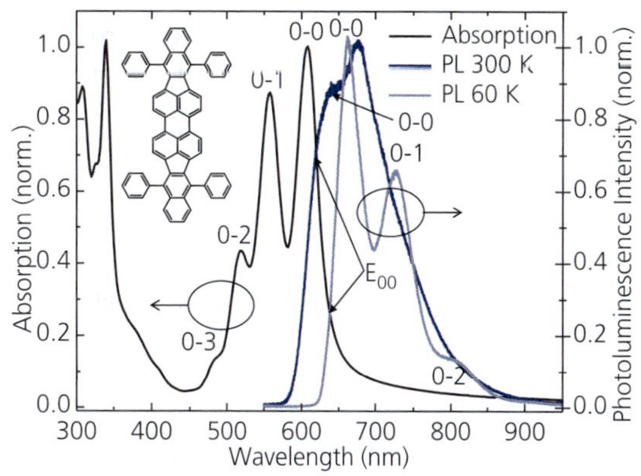

Figure 3.24 Absorption at 300 K, and fluorescence (or photoluminescence, PL) at 60 K and 300 K) spectra of DBP shown in the inset. The labeling of each peak provides the vibronic quantum numbers for a particular transition. The intensity distribution, energies and line broadening of the vibronics are strongly temperature dependent.

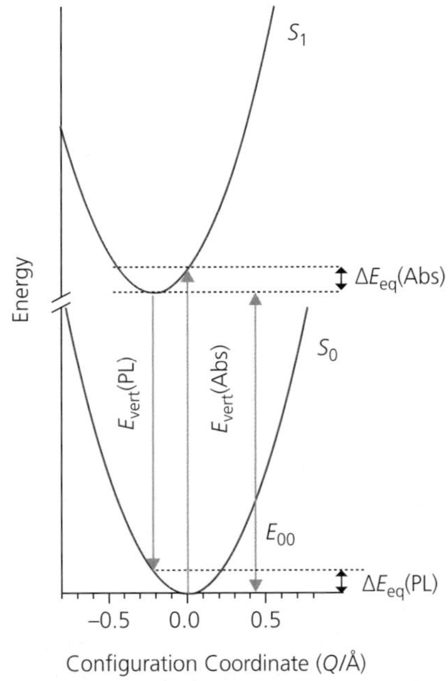

Figure 3.25 Relationship between energies of the unrelaxed ground (S_0) and excited (S_1) molecular states (Gierschner et al., 2007).

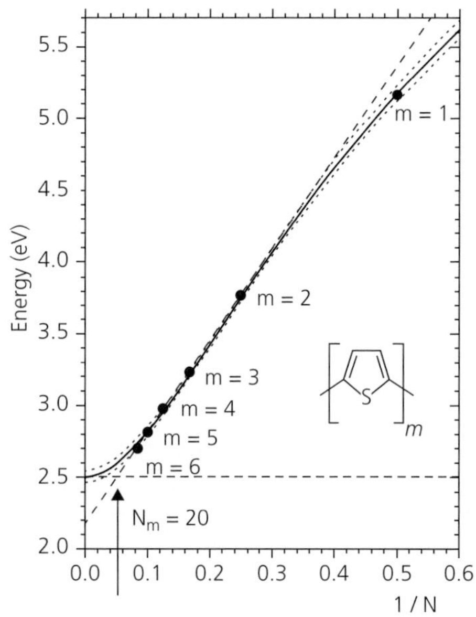

Figure 3.26 Lowest absorption energy of a series of thiophene oligomers as a function of the number of monomer segments, m. The extrapolated energy (diagonal dashed line) for $m \to \infty$ is lower than the actual lowest energy shown by the horizontal dashed line that corresponds to the number of double bonds in a chain ($N_m = 20$ corresponding to m = 10). The solid line is a fit to a simple harmonic oscillator model (see text), with dotted lines at the 95% confidence level (Gierschner et al., 2007).

That is, the lowest (0–0) transition is approximately equal to the energy difference between vibronics at the bottom of the S_0 and S_1 energy surfaces and the vertical transition energy.

While this analysis has been developed for monomeric species, with some important modifications it can also accurately provide the emission and absorption spectra of polymers consisting of a chain of individual monomers of indeterminate number. The lowest energy for a single monomer in the chain is E_{00}. However, as the chain lengthens to more than a single unit, the conjugation length also increases, resulting in accessing even lower energies of the more spatial extended states. The conjugation length corresponds to the extent of the π-electron system that is supported by a chain of alternating single and double C–C bonds. Thus, the long wavelength (low energy) absorption cut-off energy, $E_0 < E_{00}$, now corresponds to excitation of the longest continuous electronic conjugation length of the chain. This length is set by the mean spacing of defects in the chain structure that ultimately limit the spatial extent of the excited state. Defects can arise from intrinsic, thermodynamically-induced irregularities in the chain structure (e.g. due to rotation between adjacent units), or they may be of extrinsic origin (e.g. due to a kink, impurity, contact with another molecule, or other inhomogeneity along the chain).

The dependence of absorption energy on the number of monomer units comprising a thiophene oligomer is shown in Fig. 3.26, which gives the equilibrium (lowest) energy as a function of the total number of double bonds (N) in the oligomer. Each thiophene monomer unit ($m = 1$) contains two double bonds, corresponding to $N = 2$. In this case, $E_{00} = E_0 = 5.2$ eV. As m increases, the relationship between E_0 and $1/N$ is nearly linear, until $m \geq 10$ ($N \geq 20$), at which point the equilibrium energy reaches an asymptote of $E_0 = 2.5$ eV (horizontal dashed line).

The dependence of E_0 on N can be estimated by assuming that each monomer is a SHO consisting of double bonds that are coupled by a force constant, k_0. Then the energy of the normal modes of the ith unit in the polymer is given by $\varepsilon = \hbar\sqrt{k_0/m_r^*}$, where m_r^* is the reduced mass:

$$\frac{1}{m_r^*} = \frac{1}{m_e^*} + \frac{1}{m_h^*}, \tag{3.84}$$

and m_e^* (m_h^*) is the effective mass of the electron (hole). For a chain of N double bonds, the equilibrium energy is then given by

$$E_0 = \varepsilon\sqrt{1 - 2\frac{k_N}{k_0}\cos\left(\frac{\pi}{N+1}\right)}, \tag{3.85}$$

where $k_N \sim 0.4k_0 > 0$ is the force constant of the total of N oscillatory modes, which can be extracted from the energy obtained from an infinitely long ($N >> 20$) chain (Kuhn, 1948). A fit of the harmonic oscillator model to the data for the thiophene oligomers in Fig. 3.26 is indicated by the solid line. The fit is reasonably predictive for shorter chain lengths, and importantly, the cosine functional behavior results in a minimum energy of 2.5 eV for a polymer of infinite length, suggesting that the limited conjugation length is indeed an intrinsic property of these molecular species.

We conclude that the maximum conjugation length in polythiophenes is limited to only 10 or fewer molecular units. This is, in general, true of most highly pure polymers: thermodynamic or structural defects limit the effective conjugation length of the molecule to only ten or so individual monomeric units. This results in a lowest energy transition that is considerably higher than the minimum $E_{00} = 2.2$ eV for polythiophene obtained by linear extrapolation of the data in Fig. 3.26.

The electron density distribution along the polymer backbone under both ideal (i.e. completely straight and rigid) as well as more realistic chain configurations where defects, folds or torsional stresses can impact the extent of the electronic system, is illustrated for polyoligoenes in Fig. 3.27. The mean electron–hole spatial separation along the polymer backbone, $\langle r_{eh}\rangle = |i-j|$, calculated using DFT is plotted vs. the length of the oligoene, L, in units of the number of carbon bonds. The electron is located on polymer site i and the hole on site j. In the ideal case, the effective conjugation length (equal to $\langle r_{eh}\rangle$) is $5.1L$, which is reached when $L > 40$. That is, the asymptotic conjugation length is achieved only when the ends of the chain are well removed from the excitation such that edge interactions no longer influence the electron distribution of the excited state. The size of the excited state is further reduced when the chain is twisted, as shown when the torsional deviation is less than $\theta_{max} = 9°$ and $21°$, resulting in $\langle r_{eh}\rangle = 4.9L$ and $4.4L$, respectively.

The electron probability distributions ($P_{i,j}$) in the lower insets of Fig. 3.27 show the first three excited states. The fact that the probability of finding the electron and hole confined in a narrow strip spread out only slightly along the diagonal indicates that the effective separation between them is small, even for very long and perfect chains. This calculation also shows the nodes of the electron probability density that suggests a particle confined along a one dimensional (quantum wire) potential defined by the oligomer backbone.

From elementary quantum mechanics, the confined wavefunctions follow:

$$\phi_{e,l} = \sqrt{\frac{2}{L}}sin\left(\frac{l\pi x}{L}\right) \tag{3.86}$$

with energies

$$E_m(L) = \frac{(l\pi\hbar)^2}{2m_r^* L^2}, \tag{3.87}$$

where l is an integer. To first order, the $1/L^2$ dependence of the energy on the number of polymer units, such as in poly-p-phenylenevinylene, is observed until the number of carbon bonds approaches the limit of $L = 1$. Then, the lowest absorption energy edge approaches the adiabatic transition energy, E_{00}, of the monomer.

The electronic density is even more spatially confined for polymers containing torsional defects, as seen in the upper insets in Fig. 3.27, with greater confinement occurring for a larger maximum torsional angle, θ_{max}. Confinement is shown by the electronic distributions in the first two excited states of a thiophene oligomer containing a kink in Fig. 3.28. The kink decreases the effective conjugation length with the higher excited state more confined than the lower.

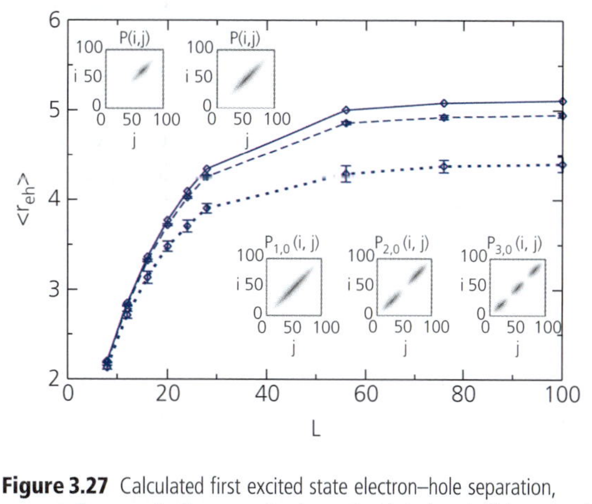

Figure 3.27 Calculated first excited state electron–hole separation, $<r_{eh}>$, along a polyoligoene molecule vs. the chain length (L) in units of C–C bonds. A defect free chain (solid line), is compared to one with a maximum torsion angle along the chain of $\theta_{max} = 9°$ (dashed line) and $\theta_{max} = 21°$ (dotted line). *Lower insets*: Electron probability density along j and hole density along i, for (*left to right*) the first, second, and third excited states of a defect-free oligomer. *Upper insets*: (*left*) A chain with $\theta_{max} = 21°$, and (*right*) $\theta_{max} = 9°$ (Rissler, 2004).

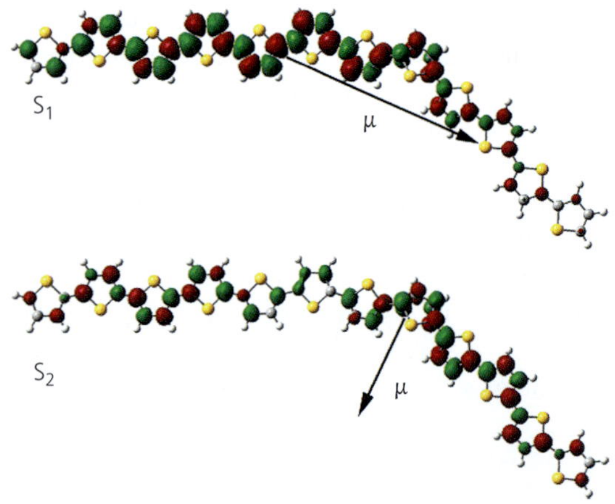

Figure 3.28 Electron densities for the S_1 and S_2 states in a horizontally kinked oligomer, dodecithiophene. Also shown are the molecular transition dipole moment vectors, μ (Beenken and Pullerits, 2004).

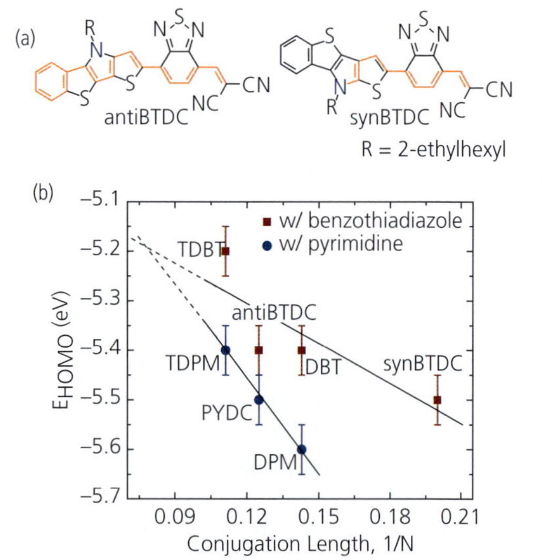

Figure 3.29 (a) Molecular formulae of the d-a-a′ type molecules antiBTDC and synBTDC. The orange line traces the electron conjugation path along the molecular backbone. (b) HOMO energy vs. the number of conjugated bonds along the d-a-a′ backbone. The two sets of data correspond to molecules bridged by either a benzothiadiazole or pyrimidine group (Che et al., 2016).

The kink itself does not perturb the distribution sufficiently to prevent electronic leakage beyond the position of the defect.

Similar dependences on conjugation length (i.e. the extent of the electron system) have been observed in a class of small molecules that are used in high efficiency organic photovoltaics. The design of these molecules follows a donor–acceptor–acceptor′ motif, where there is a single electron donating (i.e. electron deficient) moiety (the d-group), followed by two electron accepting (i.e. electron rich) groups labeled a and a′. Two isomers of the archetype d-a-a′ molecule, BTDC are shown in Fig. 3.29a, with the electron conjugation path traced in orange. The synBTDC isomer has a shorter conjugation length than antiBTDC due to *cross conjugation*. That is, the transposition of the two thiophene groups in the d-moiety in synBTDC breaks the conjugation path. The HOMO energy (E_{HOMO}) corresponding to the ionization potential of a series of heterotetracene-based d-a-a′ donors containing a benzothiadiazole or pyrimidine central electron-withdrawing block that bridges the a′ unit of dicyanovinylene, and an electron-donating dithieno[3,2-b:2′,3′-d]pyrrole (DTP) group is plotted vs. the inverse of the conjugation length, $1/N$, in Fig. 3.29b. Here, N = 5 to 9 is defined as the number of double bonds along the shortest path connecting the terminal carbon atoms of the backbone. Since the LUMO energies of this molecular series are all approximately equal, the energy gap scales approximately with E_{HOMO} (Gierschner et al., 2007)

The group of molecules with benzothiadiazole is shown by squares, while those with pyrimidine are shown by circles. For both molecular groups a linear relationship is observed between E_{HOMO} and $1/N$: the longer conjugation (smaller $1/N$) leads to more extended π-electron delocalization, resulting in red-shifted absorption (and hence a reduced E_G) and a shallower HOMO. With reduced aromaticity, the molecules with pyrimidine generally exhibit a larger energy gap and deeper HOMO than their benzothiadiazole counterparts with the same N (PYDC-antiBTDC; DPM-DBT; TDPM-TDBT), and their HOMO levels are more sensitive to the change of molecular conjugation as indicated by the larger fitted slope. The crossing point of the two fits at $N = 13$ indicates an upper limit of $E_{HOMO} = -5.2$ eV. Similar to the polymers discussed above, extended small molecules are subject to kinetically induced disruptions (e.g. twist or kinks) in the backbone, limiting the spatial extent of the orbitals in excited state. Nevertheless, further tuning of E_{HOMO} may be possible for $N > 9$.

From the foregoing, we see that the Born–Oppenheimer and Franck–Condon approximations are essential for interpreting electronic transitions in both small molecules and polymers. To simplify our analysis, we have intentionally neglected all but the most dominant vibronic modes corresponding to the high energy C–C bond stretch. The remaining degrees of freedom of a molecule appear as features corresponding to lower energy vibronic and rotational intramolecular phonon modes (10–100 meV, typical), as well as by very low energy intermolecular libratory modes (1–10 meV). These finer spectral features are typically masked by *inhomogeneous thermal line broadening* that resolve into separate absorption or emission lines only at very low temperature. The low and room temperature absorption and emission spectra of perylene in Fig. 3.30a and b provide a dramatic illustration of inhomogeneous broadening. As temperature is decreased, the various modes are distinguishable as narrow, homogenously broadened features.

Possible transitions in singly charged molecules are shown in Fig. 3.31. Charged molecules can be either *cationic* (positive) or *anionic* (negative), with transitions occurring from *singly occupied molecular orbitals* (SOMOs). To observe these transitions, the molecule must be prepared in a charged state by charge injection, or by doping with a second electron donating or accepting species. This corresponds to *oxidation* or *reduction*, respectively (cf. Section 4.4.1). That is, if a dopant is introduced whose LUMO is lower than the HOMO of the subject molecular species, an electron can transfer from its HOMO to that of the dopant, thereby oxidizing the molecule. Conversely, reduction occurs when the HOMO energy of the dopant is sufficiently high that it donates an electron to the LUMO of the host molecule. By a similar mechanism, ionic species such as the alkali metals, Na or K, serve as strong reductants, or the halogens Cl or F can serve as oxidants when doped into an organic material.

Charged PAHs have been a subject of considerable interest for understanding their optical and conducting properties (Shida and Iwata, 1973). In fact, charged PAH spectra found in interstellar space have been identified by their spectra (Halasinski et al., 2000). Low temperature (4.2 K) near infrared (NIR) spectra from both cations and anions of pentacene in solid Ne are shown in Fig. 3.32a. The radical molecular species are formed by exchanging charge with Na and K atoms that are co-deposited with pentacene in the inert gas matrixes. The cationic and anionic transitions are labeled by (+) and (−), respectively.

The lowest excited state neutral pentacene transitions are centered around 525 nm. Irradiating the pentacene/Na mixtures with a Na lamp ionizes the alkali metal which donates an electron to form

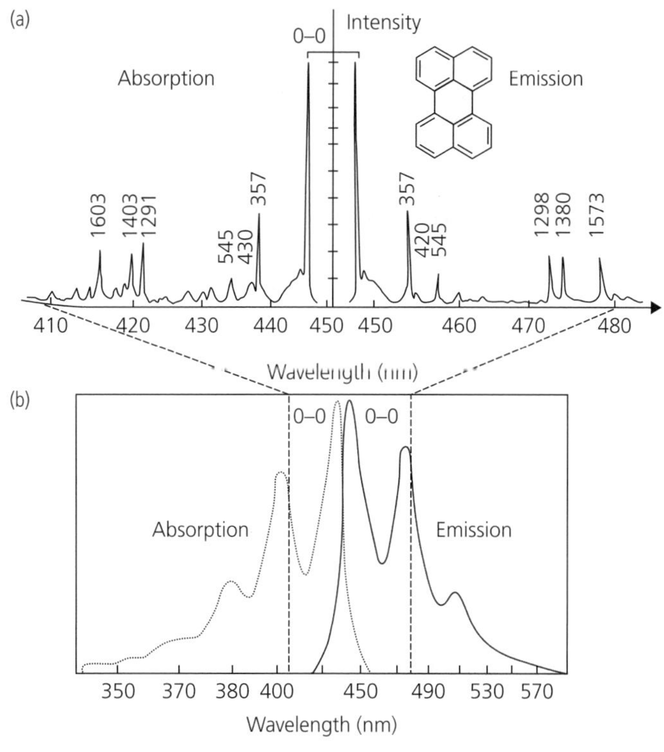

Figure 3.30 (a) Mirror symmetric, low temperature ($T = 4.2$ K) absorption and emission spectra of perylene in an *n*-hexane solution. The narrow peaks are due to discrete phonon modes. The energies (in wavenumbers) of each mode are indicated (Gribov and Orville-Thomas, 1988). Inset: Molecular structural formula of perylene. (b) The modes merge to form inhomogeneously broadened spectral features at room temperature. The dominant 0–0 transitions are indicated.

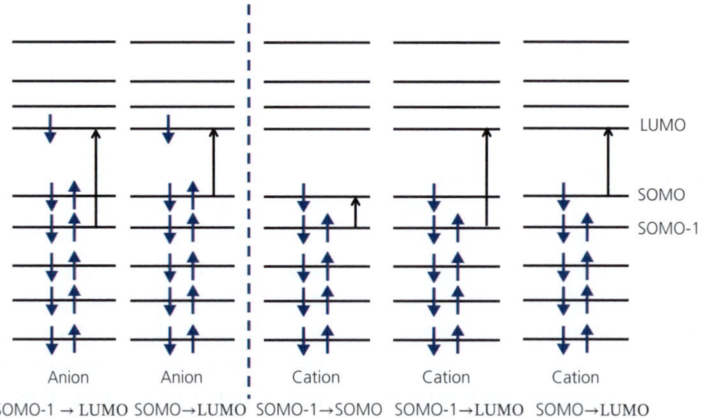

Figure 3.31 Several example transitions from singly occupied molecular orbitals (SOMOs) from anionic and cationic molecules.

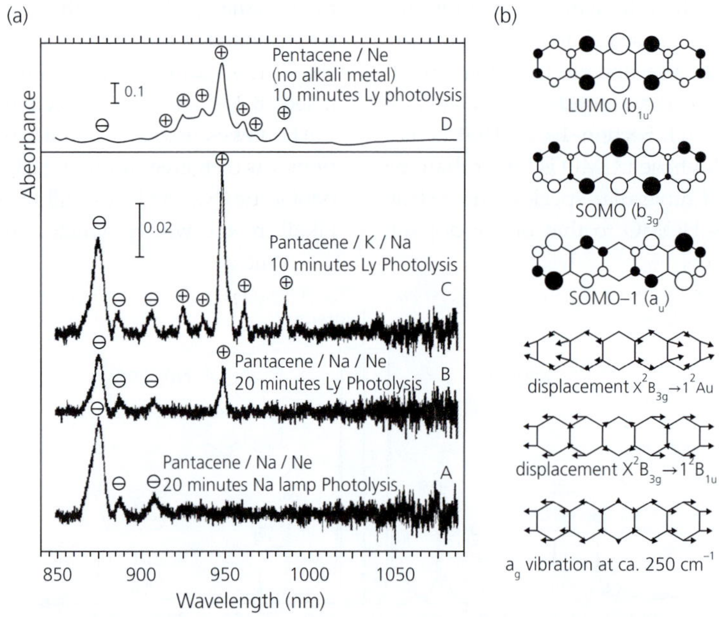

Figure 3.32 (a) Absorption spectra of pentacene anions (−) and cations (+) in solid Ne. The pentacene radicals are formed by photolysis in the presence of ionized Na or K. Light sources are used in catalysis are Na or high energy Lyman-α illumination. (b) TD-DFT calculations of the LUMO and SOMO levels of pentacene (*top*), and several high energy stretching modes (bottom three diagrams). The state symmetries are noted below each diagram (Halasinski et al., 2000).

pentacene anions (Fig. 3.32a, spectrum A). Irradiation by the 10.2 eV Lyman-α line from a hydrogen lamp is sufficient to directly ionize the pentacene, giving rise to a series of long wavelength features apparent in spectra B and C. The direct ionization of pentacene is confirmed by Lyman-α illumination of pentacene in the absence of Na or K where both cation and anion features are apparent.

Comparing the measured spectra to simulations (Halasinski et al., 2000), the feature at 979.3 nm in Fig. 3.32a is assigned to the cationic LUMO $\leftarrow$ SOMO ($1^2B_{1u} \leftarrow X^2B_{3g}$ in spectroscopic notation) transition, whereas the transition at 943.9 nm is assigned to the SOMO $\leftarrow$ SOMO $-$ 1 ($1^2A_u \leftarrow X^2B_{3g}$) transition.

Simulation results are illustrated by the electron density distributions of the LUMO, SOMO, and SOMO-1 orbitals in Fig. 3.27b. The opposite phase of the probability amplitudes are indicated by the filled and open circles, and the magnitude of the wavefunction is represented by their diameters. The anionic satellite features are separated from the parent excitations by 250 nm^{-1}. These correspond to the progression of higher order a_g vibronic stretching modes, illustrated in the bottom panel of Fig. 3.32b. The vibronics manifest themselves as symmetric stretching modes of the transitions in the SOMO-1 and SOMO states shown immediately above. Similarly, the anionic modes at 904.0 nm and 872.6 nm are identified with SOMO $\leftarrow$ SOMO $-$ 1 and LUMO $\leftarrow$ SOMO transitions, respectively, with the secondary peak at 883.8 nm due to the a_g vibronic.

3.6 Excitons

Thus far, we have been concerned only with isolated molecules. However, when the molecules are condensed into solids, the excited states can diffuse from molecule to molecule. These mobile excited states are referred to as excitons, whose properties dominate the optical characteristics of organic electronic materials. An exciton is a quasiparticle that can transport energy (i.e. the molecular excited state) between molecules. Since it carries no charge, its motion is not influenced by external electric fields. Given their central role in transporting energy in organic materials, the remainder of this chapter will be concerned with excitonic properties in dielectric media, their spin characteristics, and exciton diffusion.

The relationship between excited states, spins, vibronic levels and the transition processes between the states is illustrated by the *Jablonski diagram* (Jablonski, 1933) in Fig. 3.33. The ground state is nearly always a singlet, and hence is denoted S_0. The most common source of emission is known as *fluorescence* (subscript, F). It arises from spin-conserving transitions from the singlet manifolds to the ground state, that is, via $S_1 \rightarrow S_0$, $S_2 \rightarrow S_0$, as well as $T_2 \rightarrow T_1$, for example. The process of *phosphorescence* (subscript, Ph) is spin forbidden since it originates via a transition from the spin-symmetric triplet ($S = 1$) manifold, typically the T_1 state, to S_0, violating the conservation of spin angular momentum (Table 3.1). In the absence of other decay processes, weak phosphorescence can be observed due to its very small transition rate, k_{Ph}.

Note that the energy of T_1 is lower than S_1, as is the case of all triplet and singlet states within the same energy manifold (i.e. $T_2 < S_2$, $T_3 < S_3$, etc.). This is understood in terms of the spinorbital wavefunctions in Eqs. 3.2 and 3.3. For triplet states, the spatial part of the wavefunction is antisymmetric. Thus, as the two electrons converge (i.e. $\mathbf{r}_1 \rightarrow \mathbf{r}_2$), the spatial wavefunction gives $\phi_a(\mathbf{r}_1)\phi_b(\mathbf{r}_2) - \phi_a(\mathbf{r}_2)\phi_b(\mathbf{r}_1) \rightarrow 0$. On average, therefore, the electrons are located far apart, reducing Coulomb repulsion, compared to singlet states where $\phi_a(\mathbf{r}_1)\phi_b(\mathbf{r}_2) + \phi_a(\mathbf{r}_2)\phi_b(\mathbf{r}_1) \rightarrow$ large as $\mathbf{r}_1 \rightarrow \mathbf{r}_2$. Due to

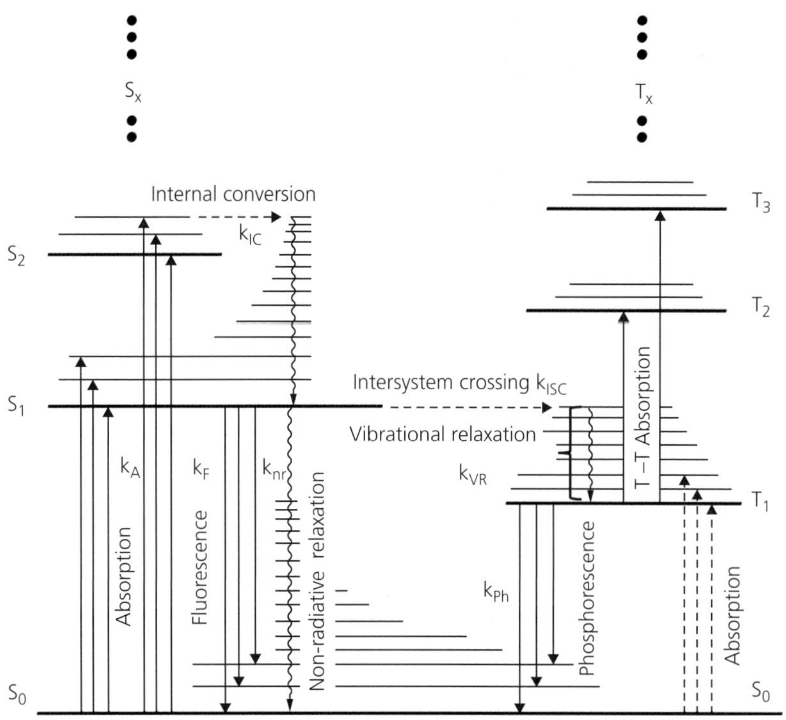

Figure 3.33 Jablonski diagram for a molecular system showing the lowest energy singlet (S) and triplet (T) manifolds, along with vibronic levels in each manifold and the several possible transition processes. Here, k are the rates for each process, with each subscript referring to a specific transition. Adapted from McGlynn et al. (1969).

the reduced Coulomb repulsion, therefore, triplets lie at lower energies (typically between 0.5 eV and 1.0 eV) than singlets. Energies differ by the magnitude of the exchange integral (see Eq. 3.49).

While $T \to S$ transitions are forbidden, transitions within the triplet manifold, viz. $T_i \to T_f$ have $\Delta S = 0$, and therefore are allowed. Transitions between triplet and singlet manifolds result from *intersystem crossing* (ISC), and can be exothermic, endothermic or isoenergetic. Clearly, this type of transition requires an increase or decrease in spin angular momentum, and hence is not simply mediated by a single photon. A spin conserving transition between electronic states is known as *internal conversion* (IC). *Vibronic relaxation* (VR) results in the decay of an excitation from a higher to a lower vibronic within a particular electronic state manifold. Vibronic relaxation occurs via the emission of one or more phonons (i.e. vibrational modes) within the molecule. Alternatively, absorption of a phonon can result in the promotion from a lower to higher vibronic state. Finally, *non-radiative relaxation* (NR) occurs from either the singlet or triplet state to the ground state, S_0. In the Jablonski diagram, the transitions are characterized by their respective rate constants, k, using the appropriate identifying subscript.

While the ground state is almost always a singlet, under unusual conditions it can in fact be a triplet state. If the ground state is filled but is doubly degenerate (or nearly degenerate), two electrons of the same spin can be shared between the degenerate states without violating the Pauli Exclusion Principle. The property of "near degeneracy" allows for equal populations of both states at or below room temperature. While possible, triplet ground states exist only in rare instances, and will not be considered further.

3.6.1 Transition probabilities, oscillator strength, and the energy gap law

The Jablonski diagram delineates the various available transition pathways. To determine the strength of a particular transition (and hence its prominence in the molecular spectrum) we must calculate its transition probability. For this, we need the *Einstein coefficients* for *spontaneous absorption, spontaneous emission* and *stimulated emission*, see Fig. 3.34. Now the Einstein coefficient for spontaneous absorption, B_{lu}, can be written in terms of the transition probability from the lower (l) to the upper (u) state (Hilborn, 1982):

$$B_{ln,um} = \frac{q^2\pi}{3\varepsilon_0\hbar^2}|\langle\phi_{ln}||\mathbf{r}_n - \mathbf{r}_m||\phi_{um}\rangle|^2 = \frac{\pi}{3\varepsilon_0\hbar^2}|\mu_{ln,um}|^2, \tag{3.88}$$

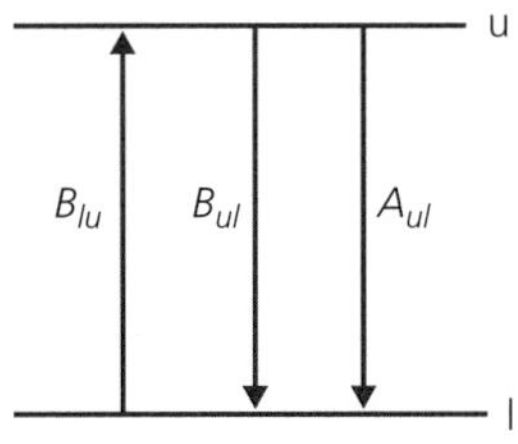

Figure 3.34 Einstein coefficients for spontaneous absorption (B_{lu}), spontaneous emission (A_{ul}), and stimulated emission (B_{ul}) for a two-level system whose upper (u) and lower (l) levels are indicated.

where, as in Eq. 3.76, the wavefunctions in the electric dipole approximation include both electronic and nuclear contributions. The latter are labeled n in the lower, and m in the upper electronic state. Here, we have assumed that the degeneracy, g, of final (u) and initial states (l) is the same. If this is not the case, Eq. 3.88 is multiplied by their ratio, g_u/g_l.

Now, Einstein showed that $B_{ln,um}$ is related to the coefficient for absorption, $A_{um,ln}$ via (Yariv, 1989):

$$A_{um,ln} = \frac{\hbar}{\pi^2}\left(\frac{\omega n_r}{c}\right)^3 B_{ln,um}, \tag{3.89}$$

where n_r is the refractive index of the medium. (Note that care needs to be taken when writing these coefficients in terms of angular frequency, ω, frequency, ν, or wavenumber, $\tilde{\nu} = \nu/c$. In Eq. 3.88 and 3.89, $B_{ln,um}$ is written in terms of the angular frequency.) The total radiative lifetime (τ_R) is equal to the sum of $A_{u0,ln}$ over all vibronic levels in the initial and final electronic states. We also apply Kasha's rule by assuming all transitions occur from the lowest ($m = 0$) vibronic in the excited state, while it can access a large number of final states, n. Then,

$$\tau_R^{-1} = \sum_m A_{u0,ln}. \tag{3.90}$$

Comparing Eqs. 3.89 and 3.90, we find that the radiative lifetime can also be obtained from the absorption spectrum of the molecule. For this, we integrate over the populations in the initial and final states:

$$\tau_R^{-1} = \frac{8\pi^2 n_r{}^3}{3\varepsilon_0\hbar c^3}\sum_m \nu_{u0,ln}^3|\mu_{u0,ln}|^2, \tag{3.91}$$

where we have replaced ω with frequency, ν, for convenience in analyzing the measured spectra. To complete our treatment, we weight each contribution by a Boltzmann factor to account for the probability of finding an electron in the initial state.

The time averaged incident power on a molecule is called the *irradiance*, I_{inc}. From electromagnetic theory, the irradiance due to an electric field of amplitude, F, is given by:

$$I_{inc} = \frac{1}{2} c\varepsilon_0 F^2 = \int i(\omega) d\omega, \tag{3.92}$$

where $i(\omega)$ is the incremental irradiance between frequencies ω and $\omega + d\omega$. Since $B_{ln,um}$ is the probability for absorbing a photon, and taking $\rho(\omega)$ as the energy density of the incident photon field, then the incremental power per unit area absorbed within the length, dx, by N_l atoms populating the lower level is

$$dI_a = \hbar\omega B_{ln,um} N_i \rho(\omega)\gamma(\omega) d\omega dx, \tag{3.93}$$

where $\gamma(\omega)$ is the normalized lineshape function defined by $\int \gamma(\omega) d\omega = 1$. Thus, within a wavelength interval, $d\omega$, we rewrite Eq. 3.93 in terms of $i(\omega)$:

$$\begin{aligned} \frac{1}{i(\omega)} \frac{di}{dx} &= -\hbar\omega n_r n_l B_{ln,um} \gamma(\omega)/c \\ &= -n_r n_l \sigma(\omega) = -\alpha(\omega), \end{aligned} \tag{3.94}$$

where n_l is the number density of molecules in the lower level, and $\alpha(\omega)$ is the absorption coefficient. We thus define the normalized absorption cross section in terms of $B_{ln,um}$ from Eq. 3.94:

$$\sigma(\omega) = \hbar\omega B_{ln,um} \gamma(\omega)/c. \tag{3.95}$$

Finally, we can express the Einstein B-coefficient using the absorption cross section:

$$B_{ln,um} = \frac{c}{\hbar n_r} \int \frac{\sigma(\omega) d\omega}{\omega}, \tag{3.96}$$

where the lineshape is now included in $\sigma(\omega) = \sigma_0 \gamma(\omega)$. For experimental convenience, we can replace the cross-section with the light frequency-dependent *decadic molar extinction coefficient* (in units of l cm^{-1} M^{-1}) as a function of frequency, $\varepsilon(\nu)$, in which case we write

$$B_{in,fm} = \frac{2303\ c}{N_A n_r \hbar} \int \frac{\varepsilon(\nu) d\nu}{\upsilon}, \tag{3.97}$$

where N_A is Avogadro's number. The factor, 2303, arises since the irradiance, $I(0)$ incident on a sample with absorption coefficient, $\alpha(\nu)$, will decrease to $I(d)$ after travelling a distance, x (cf. Eq. 3.94):

$$I(d) = I(0) e^{-\alpha(\nu)x}. \tag{3.98}$$

This is known as the *Beer–Lambert law*. Here, $\alpha(\nu) = ln10\varepsilon(\nu)C = 2.303\varepsilon(\nu)C$, and C is the concentration of the molecule in solution (units of [M/l]).

The radiative rate is given by the *Strickler–Berg expression* (Strickler and Berg, 1962):

$$\tau_R^{-1} = 8\pi n_r^3 \hbar \langle \tilde{\nu}^{-3} \rangle^{-1} B_{ln,um} = \frac{9212 n_r^2 c}{N_A} \langle \tilde{\nu}^{-3} \rangle^{-1} \int [\varepsilon(\tilde{\nu})\tilde{\nu}] d\tilde{\nu}, \tag{3.99}$$

where the average transition frequency, $\langle \tilde{\nu}^{-3} \rangle$ is obtained by integration over the measured spectral intensity, via

$$\langle \tilde{\nu}^{-3} \rangle = \frac{\int \tilde{\nu}^{-3} I(\tilde{\nu}) d\tilde{\nu}}{\int I(\tilde{\nu}) d\tilde{\nu}}. \tag{3.100}$$

The *oscillator strength* is the ratio of the emission or absorption rate of the molecule to that of a classical oscillator. This is defined by

$$\begin{aligned} f_{in,fm} &= \frac{4\pi m_e c}{3\hbar} \langle \tilde{\nu}_{in,fm} \rangle |\langle \phi_{in} || \mathbf{r}_i - \mathbf{r}_f || \phi_{fm} \rangle|^2 \\ &= \frac{4\pi m_e c}{3q^2\hbar} \langle \tilde{\nu}_{in,fm} \rangle |\boldsymbol{\mu}_{in,fm}|^2, \end{aligned} \tag{3.101}$$

where we generalize by substituting i and f for u and l. Comparing this with Eq. 3.88 and 3.96, the oscillator strength can be written in terms of the cross-section and absorption coefficients:

$$\begin{aligned} f_{in,fm} &= \frac{2m_e \varepsilon_0 c}{\pi q^2 n_r} \int \sigma(\omega) d\omega = \frac{2303 m_e c^2}{N_A \pi q^2 n_r} \int \varepsilon(\tilde{\nu}) d\tilde{\nu} \\ &= \frac{4.39 \times 10^{-9}}{n_r} \int \varepsilon(\tilde{\nu}) d\tilde{\nu}, \end{aligned} \tag{3.102}$$

where we have removed the frequency, ω, from the integral in Eq. 3.96 by assuming that it is approximately constant over the relatively narrow energy band subtended by the initial and final states. It is often more important to compare the relative probabilities for transitions, rather than to calculate their absolute values. For this, we normalize the oscillator strength to an easily measured quantity—the intensity of the 0–0 transition ($f_{i0,f0}$). Since, the electric dipole transition probability is the same for all vibronics contained within the same electronic manifold, Eq. 3.101, then we only need to consider the ratio of Franck–Condon factors. That is, from Eq. 3.79, $|\mu_{in,fm}|^2/|\mu_{i0,f0}|^2 = |\langle \phi_{N,in} | \phi_{N,fm} \rangle|^2 = FC_{n,m}$. This yields

$$\bar{f}_{in,fm} = \frac{\tilde{\nu}_{in,fm}}{\tilde{\nu}_{i0,f0}} FC_{n,m}, \tag{3.103}$$

where $\bar{f}_{in,fm} = f_{in,fm}/f_{i0,f0}$. For absorptive transitions $f \leftarrow i$, the initial state is i, 0, that can access the vibronic levels, m, in state f. Thus, the relative intensity of the peaks in the vibronic progression depends on the overlap between the wavefunction of the lowest vibronic in the initial state with the vibronic wavefunctions in the final state.

We can now calculate the relative intensities of the several peaks within the vibronic progression, which are the solutions for a SHO. There are two possibilities that are generally considered to enable estimation of the relative intensities of the vibronic states. The first is the *displaced oscillator* (Fig. 3.35a), whereby the electronic potential in state f is identical to that in i, but is rigidly displaced by a change in the nuclear positions by ΔQ. The energy separation between vibronic eigenstates remains unchanged by this displacement, which implies that the normal frequencies, ω_l in Eq. 3.15 are also unchanged. The second is the case of a *distorted oscillator* (Fig. 3.35b), where the electronic potential shape changes from i to f. In this case, the normal modes are changed: in the figure the frequencies are increased for the distorted oscillator excited state potential (Siebrand, 1967).

To treat the general problem of the transition $i \rightarrow f$, we return to Fermi's Golden Rule, separating out only the nuclear part of the wavefunctions, viz.

$$k_{if} = \frac{2\pi}{\hbar}\rho(E)|\langle\phi_{f,N}|H_N|\phi_{i,N}\rangle|^2, \tag{3.104}$$

where H_N is the nuclear Hamiltonian. This can be approximated by the first term in a Taylor series expansion in the nuclear displacement, $(Q_n - Q^0)$ (Siebrand, 1967, Engleman and Jortner, 1970):

$$H_N = \sum_{n=1}^{N}\frac{\partial\tilde{J}}{\partial Q_n}(Q_n - Q^0) \tag{3.105}$$

with

$$\tilde{J} = \langle\phi_{f,e}|J_N|\phi_{i,e}\rangle, \tag{3.106}$$

where the sum is over all N normal modes, J_N is the nuclear kinetic energy operator, $\partial/\partial Q_n$, Q_n is the nuclear coordinate of the nth normal mode in the final state, and Q^0 is the equilibrium coordinate in the initial state in which the molecule is "prepared" prior to the transition. It is assumed that there is only a single, dominant pth mode in the initial state.

Since the vibronic modes are coupled, we need to sum over all possible final vibronic states,

$$H_N = \tilde{J}|FC_{if}(E)|^{1/2} = \tilde{J}\prod_{n\neq p=1}^{N}|\langle\phi^f_{N,n}(\omega_n)|\phi^i_N(\omega_p)\rangle|. \tag{3.107}$$

Equation 3.107 includes the product of Franck–Condon factors for transitions between the dominant initial, and the other n vibronics. Also, the initial and final states of the nuclear wavefunctions are denoted by superscripts for clarity. It is assumed that the system is initially in mode ω_i, followed by a transition to one of n modes with frequencies ω_p. Then the Franck–Condon factor for the transition is found by summing over all possible permutations, P:

$$FC_{if} = \sum_E FC_{if}(E) = \sum_P P\left[\prod_{n\neq p}|\langle\phi^f_{N,n}(\omega_n)|\phi^i_N(\omega_p)\rangle|^2\right]. \tag{3.108}$$

The lineshape function is then obtained by multiplying the Franck–Condon factors over the modal occupancy, which is determined by a Boltzmann

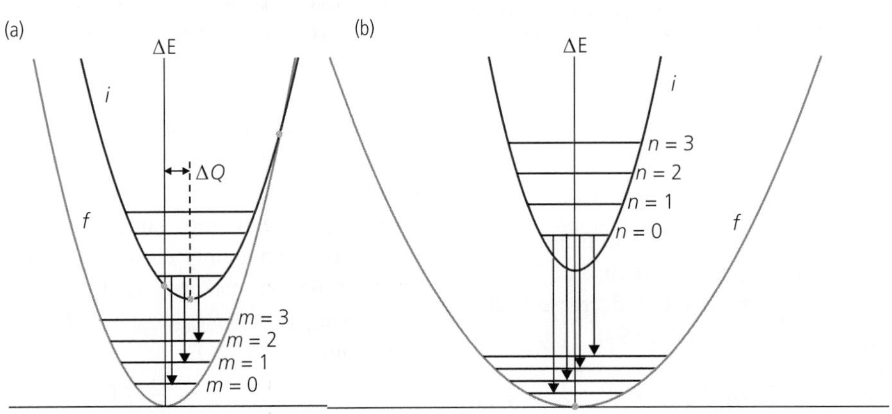

Figure 3.35 (a) Displaced oscillator and (b) distorted oscillator potential energy surfaces. The displaced oscillator undergoes a coordinate shift of ΔQ, whereas the distorted oscillator has $\Delta Q = 0$. Example transitions from the $n = 0$ initial state to various vibronics, m, in the final state are indicated.

factor, Eq. 3.81. Thus the line shape is (Engleman and Jortner, 1970)

$$F(E) = \left[\sum_{i,f}^{N} FC_{ij}\exp(-E_i/k_BT)\delta(E_f - E_i - E)\right] \Big/ \sum_{i}^{N} exp(-E_i/k_BT). \quad (3.109)$$

For the displaced, but undistorted oscillator, we assume a Poisson distribution for the oscillator strength across the m-vibronic levels in the final state. Then, it can be shown (see Problem 3.7) that Eq. 3.108 yields $FC_{n,m} = Z^m e^{-Z}/m!$, such that (Siebrand, 1967):

$$\bar{f}_{in,fm} = \frac{\tilde{\nu}_{in,fm}}{\tilde{\nu}_{i0,f0}}\frac{Z^m}{m!}e^{-Z}, \quad (3.110)$$

where Z is known as the *Huang–Rhys parameter*. The restoring force of the harmonic potential is k, and hence the energy due to a displacement, $\Delta Q = |Q_i - Q_f|$, of that state is $\frac{1}{2}k\Delta Q^2$. Since Q represents the configuration of all nuclear coordinates, it then follows that ΔQ corresponds to

$$\Delta Q = \sqrt{\sum_{j=1}^{M} \Delta \mathbf{R}_j^2}, \quad (3.111)$$

where $\Delta\mathbf{R}_j$ is the displacement of the jth nucleus in the molecule containing M atoms from its equilibrium position, $\mathbf{R}_j$. Thus, in Eq. 3.110 we have

$$Z = \frac{1}{2}\frac{k\Delta Q^2}{\hbar\omega_{i0,fm}} = \frac{1}{2}\frac{M\omega_{i0,fm}}{\hbar}\Delta Q^2 \quad (3.112)$$

for each final state, f, m. While this treatment assumes a Poisson distribution, this is often not the case, although its use leads to considerable simplicity and often acceptable fits to data.

Equation 3.110 is also used to describe emission. Figure 3.36a shows the low temperature fluorescence spectra of very thin films of PTCDA sandwiched between layers of the higher band gap NTCDA (Haskal et al., 1995). The layer pairs, grown by thermal evaporation in ultrahigh vacuum, are stacked such that the luminescence from one pair adds to that of another to form easily detectable PTCDA emission at T = 4.2 K. No NTCDA emission is observed between 1.3 and 2.1 eV shown in the figure. Solutions to Eq. 3.110 as functions of ΔQ^2 are shown in Fig. 3.36b for the four highest energy features in the spectrum

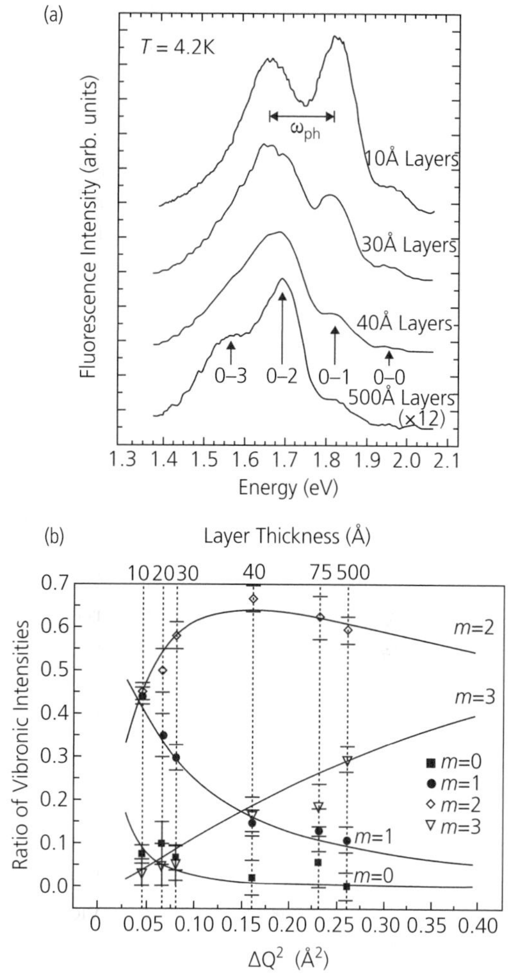

Figure 3.36 (a) Fluorescence spectrum at T = 4.2 K of PTCDA in PTCDA/NTCDA multilayer stacks, labeling the $S_1 \rightarrow S_0$ series of vibronic transitions. The individual layer thicknesses in the stacks are noted which are the same for both NTCDA and PTCDA. (b) Fits of the relative peak intensities in (a) for each layer thickness to the Poisson distribution in Eq. 3.110 are shown as solid lines. The values of n correspond to the vibronic level in the final state, S_0 (Haskal et al., 1995).

(corresponding to the transitions from the S_1, n = 0 state, to S_0 in descending order of energies from $m = 0$ to $m = 1$, 2 and $m = 3$). Each layer thickness yields only one value of ΔQ^2 indicated by the vertical, dashed lines. The data for each layer thickness intersect the values of Eq. 3.110, from which we infer there is only a single, displaced electronic state manifold. Also note from Fig. 3.36b that ΔQ^2 increases from 0.04 Å^2 to 0.25 Å^2 as the layer thickness increases from 10 Å to 500 Å, which is attributed to the reduced energetic confinement of the excited states with increasing layer thickness (Haskal et al., 1995).

For distorted oscillators, the Franck–Condon factors follow (Siebrand, 1967):

$$FC_{if} = \left(1 - \xi^2\right)^{1/2} \xi^m \times \frac{1 \cdot 3 \cdot 5 \cdots (m-1)}{2 \cdot 4 \cdot 6 \cdots m}; \qquad m \text{ even}, \quad (3.113a)$$

$$FC_{if} = 0; \qquad m \text{ odd}. \quad (3.113b)$$

Here, $\xi = \left(\omega_{i0} - \omega_{fm}\right) / \left(\omega_{i0} + \omega_{fm}\right) = \Delta\omega / \left(\omega_{i0} + \omega_{fm}\right)$ for each final mode of frequency, m, where $\hbar\Delta\omega$ is the equilibrium energy difference for the undistorted oscillator, and $\hbar\omega_{fm}$ is the energy of the relaxed state. Simplified, and somewhat more intuitive expressions for the distribution of spectral intensities that takes into consideration the population of states in the distorted vs. undistorted molecular potentials is given for absorption by (Gierschner et al., 2002):

$$G(\Delta\omega) = \exp(-\hbar\Delta\omega/(D-1)k_B T) \quad (3.114)$$

and for emission:

$$G(\Delta\omega) = \exp(-\hbar\Delta\omega D/(D-1)k_B T), \quad (3.115)$$

where $D = \omega_{i0}/\omega_{fm}$.

A distorted oscillator is used to describe the changes in the vibronic spectra of NTCDA under hydrostatic pressure (Jayaraman et al., 1985). The peaks in Fig. 3.37 are the Raman spectra of the molecular vibronic modes. They exhibit a rigid blue shift, as well as a decrease in peak spacing with increasing pressure. The *rigidochromic shift* (i.e. a peak shift without broadening) corresponds to the behavior of a displaced oscillator, while the changes in peak separation result from intramolecular distortions. For example, at atmospheric pressure (1 bar), there are peaks at 83, 75, and 50 cm^{-1}, with frequency differences of 8 and 25 cm^{-1}, respectively. As pressure increases, the peak energies monotonically decrease and split into doublets in some cases. Thus, at 18.5 kbar, the peaks are found at 121, 98, and 69 cm^{-1}, with separations of 23 and 27 cm^{-1}. Apparently, the distortion is most pronounced for the higher frequency modes with the lowest vibronic quantum numbers. Furthermore, there is a considerable increase in intensity of the highest energy mode at the expense of the lower energy modes, indicating that the vibronics are in resonance.

A further indication of potential distortion due to molecular configuration changes is the breakdown of mirror symmetry with temperature, as in Fig. 3.23b for diphenylvinylene. This distortion is due to thermally activated torsional motion of the phenyl groups, which can also result in an increase in the Stokes shift between absorption and emission. An even more extreme example of potential distortion is found for a series of methyl-substituted quarterthiophenes (4T) in Fig. 3.38.

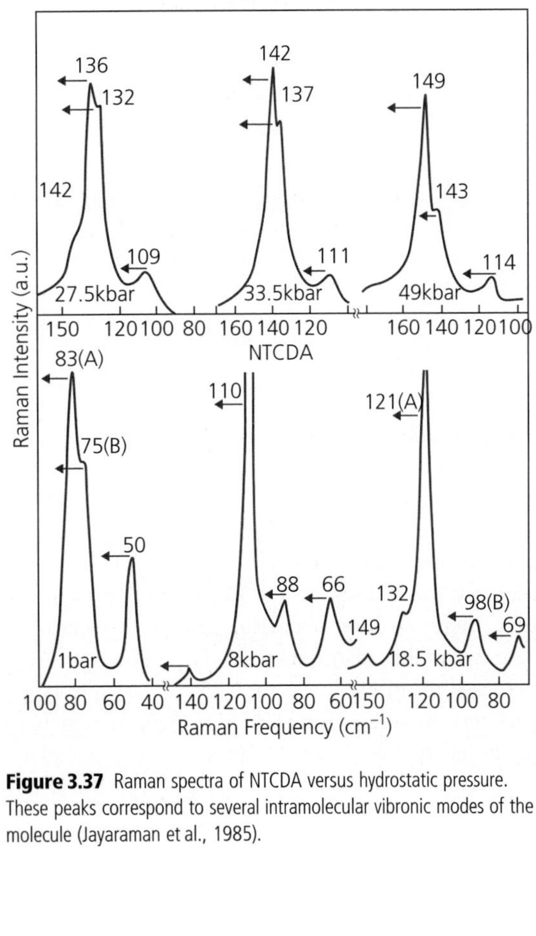

Figure 3.37 Raman spectra of NTCDA versus hydrostatic pressure. These peaks correspond to several intramolecular vibronic modes of the molecule (Jayaraman et al., 1985).

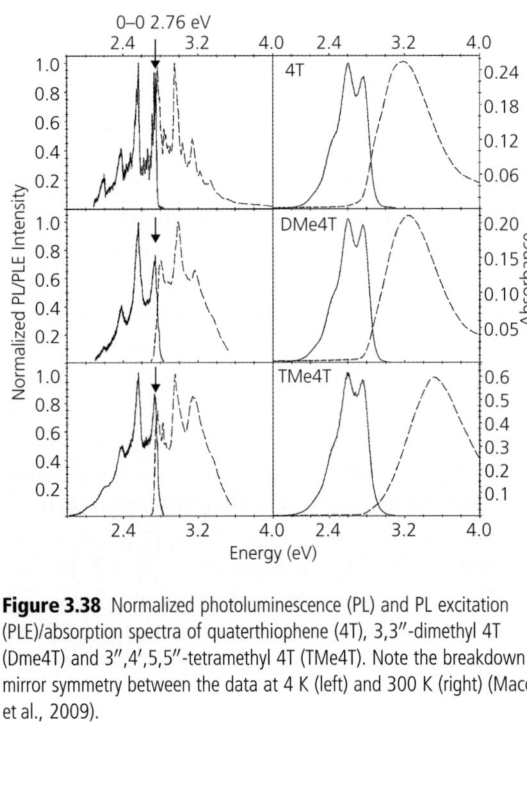

Figure 3.38 Normalized photoluminescence (PL) and PL excitation (PLE)/absorption spectra of quaterthiophene (4T), 3,3″-dimethyl 4T (Dme4T) and 3″,4′,5,5″-tetramethyl 4T (TMe4T). Note the breakdown of mirror symmetry between the data at 4 K (left) and 300 K (right) (Macchi et al., 2009).

The intensities provided by this analysis are only approximate, avoiding common departures from idealized situations, such as anharmonicity of the equilibrium potential, non-Boltzmann excited state distribution statistics, coupling between states, and the presence of multiple overlapping (i.e. degenerate) vibronics. It is nevertheless instructive since it demonstrates the relationship between transition probabilities, Franck–Condon factors, and the oscillator strengths.

Non-radiative transitions between vibronic states (in this case from the pth mode in the initial state to one of the N normal modes in the final state) also have an exponential dependence of the transition rates on the energy gap, $E_G = E_i - E_f$. Assuming once again that there is a single, dominant mode of frequency, ω_p, the non-radiative rate follows (Engleman and Jortner, 1970):

$$k_{if} = \frac{C^2\sqrt{2\pi}}{\hbar\sqrt{\hbar\omega_p E_G}}\exp(-\gamma E_G/\hbar\omega_p) = A\exp(-\gamma E_G/\hbar\omega_p), \tag{3.116}$$

where $C \sim \sum_k^N J_k \langle \phi_{N,ik} | \frac{\partial}{\partial Q_n} | \phi_{N,fk} \rangle$ is the kinetic energy of the nuclei. This expression is valid in the weak coupling limit, that is, when the relative displacement of the states, ΔQ, is small, which is the primary assumption in using the expansion in Eq. 3.105. Also,

$$\gamma = \log\left(\frac{E_G}{\Omega E_p}\right) - 1, \tag{3.117}$$

where Ω is the number of modes contributing the maximum energy, $E_p = \frac{1}{2}\hbar\omega_p\Delta_p^2$, which is *half the Stokes shift*. It is assumed that there are Ω *equivalent* modes of energy $\hbar\omega_p$ in the final state. Here, $\Delta_p = \sqrt{M\omega_p/\hbar}(Q_p - Q^0)$ is the reduced (unitless) displacement between the excited and equilibrium modes with coordinates Q_p and Q^0, respectively, and M is the molecular mass. We note that Eq. 3.116 is dependent on E_G (i.e. as E_G increases, k_{if} exponentially decreases). Also, γ is a function of energy gap. Its weak logarithmic dependence does not strongly influence k_{if}.

Equation 3.116 is known as the *energy gap law*, which states that the larger the energy gap, the lower the probability for non-radiative recombination. That is, as the energy gap of a molecular species decreases, competition between radiative vs. non-radiative decay of the excited state increases.

As we will find in Chapter 6, organics are far more likely to have a high radiative efficiency as their energy gap increases, that is, while it is common to observe very high red, green and blue emission efficiencies, it is less common to observe emission from molecules with energy gaps at near and mid-infrared wavelengths. The energy gap law is followed in both polymers and small molecules, and is widely observed for singlet, triplet, and charge transfer transitions (Siebrand and Williams, 1967, Caspar et al., 1982, Bixon et al., 1994, Chynwat and Frank, 1995, Wilson et al., 2001). An example of the energy gap law is shown in Fig. 3.39 for triplet transitions in a series of aromatic hydrocarbons, and their deuterated isotopes. (Siebrand and Williams, 1967). The dominant vibrational energy is $E_p = 4000\ \text{cm}^{-1}$ due to the C–H bond stretch. In the deuterated isotopes, the stiffer C–D bond has a higher $E_p = 5500\ \text{cm}^{-1}$, also leading to a higher slope vs. energy for k_{if} according to Eq. 3.116.

The difference in behavior for the two different isotopes of the aromatic hydrocarbons can be used to determine the energy of the dominant vibronics. That is, by replacing a H with a D atom within the molecule, the force constant $k = M\omega^2$ undergoes a corresponding change. To identify the dominant excitation, we take the derivative of FC_{if} in Eq. 3.108 (Siebrand and Williams, 1967):

$$\left.\frac{\partial \log FC_{if,H}(E)}{\partial E}\right|_{E'} = \left(\sqrt{\frac{M_H}{M_L}}\right)\left.\frac{\partial \log FC_{if,L}(E)}{\partial E}\right|_{E'}. \tag{3.118}$$

Here, $FC_{if,H}$ refers to the Stokes shift of the heavier isotope with mass M_H, compared to the lighter with M_L. The ratio of the masses of the heavier to the lighter isotopes determines the slope of the line in Fig. 3.39. This dependence of the shift in oscillator strength with isotopic mass is known as the *isotope rule*.

The non-radiative rates in Fig. 3.39 are for $S_1 \to S_0$ transitions. The generality of the energy gap law is demonstrated by triplet transitions, $T_1 \to S_0$ of a homologous series of *thermally assisted delayed fluorescent* (TADF) metal–organic Cu complexes whose molecular structural formulae are shown in **Scheme 1**. Varying the ligand composition tunes the energy gap of the complexes from the red with a HOMO-LUMO gap energy of 2.04 eV (compound 6), to the blue, with $E_G = 3.54$ eV. Correspondingly, emission from the T_1 state ranges from 704 nm to 432 nm when the phosphor is dispersed at very low concentration (1%) in a wide energy gap polystyrene matrix.

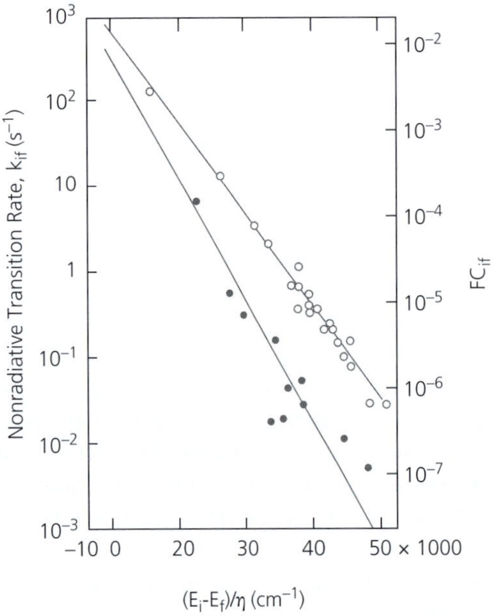

Figure 3.39 Non-radiative triplet transition rates measured for a series of aromatic hydrocarbons (open circles) and their deuterated isotopes (filled circles). The slopes of the solid lines fit to the energy gap law are in the ratio of 4000/5500 = 0.73 which is compared to the predicted value of 0.55 for these isotopes. Here, FC_{if} is the Frank–Condon factor, and η is the relative number of H or D atoms in the molecule (Siebrand and Williams, 1967).

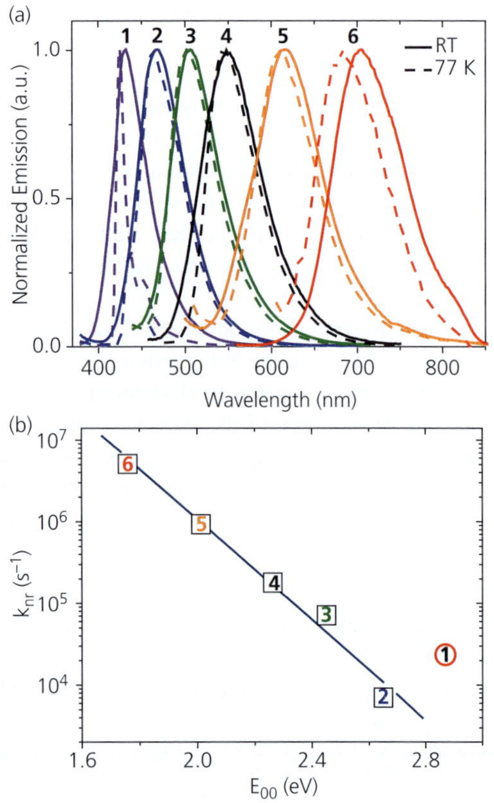

Figure 3.40 (a) The emission spectra of complexes **1–6** (**Scheme 1**) in polystyrene at room temperature and 77 K. (b) Non-radiative recombination rate vs. E_{00} (Shi et al., 2019).

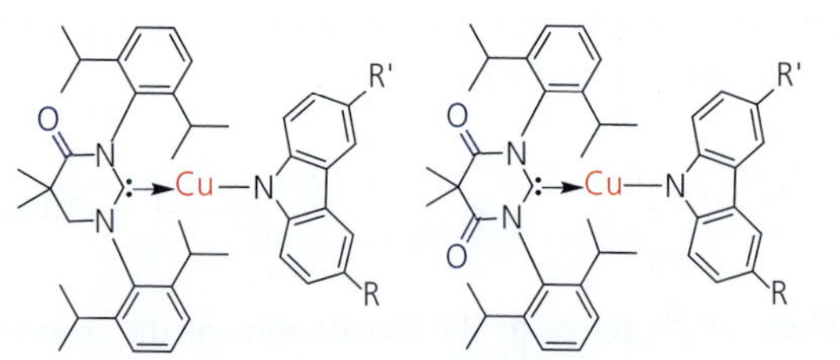

Carbene = MAC*	Carbene = DAC*
R, R' = CN (**1**)	R, R' = CN (**4**)
R = CN, R' = H (**2**)	R = CN, R' = H (**5**)
R, R' = H (**3**)	R, R' = H (**6**)

Scheme 1 Cu-based TADF compounds.

The non-radiative transition rate is related to the photoluminescence quantum yield, Φ_{PL}, via:

$$\Phi_{PL} = \frac{k_r}{k_r + k_{nr}}, \tag{3.119}$$

where k_r is the radiative rate. Hence, the non-radiative rate is measured from the values of Φ_{PL}, taking $k_{nr} \to 0$ at low temperatures, where k_r is measured from the luminescent decay transient. Using this analysis, k_{nr} is plotted vs. E_{00} in Fig. 3.40b. Consistent with the energy gap law, the non-radiative rate decreases with increasing E_{00}, except for compound **1** which falls considerably off of the trend. This is attributed to a different composition for the orbitals on **1** vs. the other compounds, which is also reflected in its comparitively lower Φ_{PL} (Shi et al., 2019).

3.6.2 Dimers

Understanding exciton transport in solids begins by consideration of the interaction of an isolated pair of identical molecules, known as a *dimer*, which is the smallest possible molecular aggregate. If the pair forms a bond, the system is known as a *chemical dimer*, whereas if they interact by van der Waals or other electrostatic attraction, the pair forms a *physical dimer*. In the following, we consider only physical dimers. Aggregation of molecules into dimers results in energy level splitting and the development of band structure for the excited states. Level splitting red shifts the broadened electronic states, resulting in differences between the spectra of individual molecules and the solid.

The wavefunction of one molecule at $\mathbf{r}_1$ of a dimer pair is $\psi_1(\mathbf{r}_1)$, and the other at $\mathbf{r}_2$ is $\psi_2(\mathbf{r}_2)$. If their coupling is small, then the *Heitler–London formalism* for the ground state dimer wavefunction gives: $\psi_0(\mathbf{r}_1, \mathbf{r}_2) = \psi_1(\mathbf{r}_1)\psi_2(\mathbf{r}_2)$, each with ground state energies $E_1 = E_2 = E_0$. The total ground state energy must also include the intermolecular interaction energy, E_I. In the simple case of van der Waals bonding, then $H_{int} = -\frac{A_{disp}}{r_{12}^6}$, where A_{disp} is the dispersion coefficient and $r_{12} = |\mathbf{r}_1 - \mathbf{r}_2|$ (c.f. Eq. 2.66). Then:

$$E_I = \langle \psi_1(\mathbf{r}_1)\psi_2(\mathbf{r}_2)|H_{\text{int}}|\psi_1(\mathbf{r}_1)\psi_2(\mathbf{r}_2)\rangle, \qquad (3.120)$$

leading to a ground state energy of $E_G = 2E_0 + E_I$. Since $E_I < 0$, the bound ground state energies of the two individual molecules is less than that of the dimer that is, therefore, the more stable species. If the molecules have fixed dipole moments, H_{int} is modified according to the analysis in Section 2.4.3.

If one molecule is excited, time dependent perturbation theory predicts that the excited state oscillates between the molecules forming the dimer. Hence, the system is described by a linear superposition of the two possible excited states, or two orthogonal normal modes of the system:

$$\psi'_\pm(\mathbf{r}_1, \mathbf{r}_2) = \frac{1}{\sqrt{2}}\left(\psi'_1(\mathbf{r}_1)\psi_2(\mathbf{r}_2) \pm \psi_1(\mathbf{r}_1)\psi'_2(\mathbf{r}_2)\right), \quad (3.121)$$

where $\psi'(\mathbf{r})$ denotes an excited state. These modes are similar to two SHOs coupled by a spring: one (+) corresponds to both oscillators moving coherently in phase, while the other (−) is for the oscillators moving 180° out of phase.

The total energy is the sum of the energies of the excited state of one molecule, $E'_1 = E'_2 = E'$, the ground state of the other molecule, E_0, and interaction terms from the wavefunctions in Eq. 3.121. This yields the Coulomb energy:

$$E'_I = \langle \psi'_1(\mathbf{r}_1)\psi_2(\mathbf{r}_2)|H_{\text{int}}|\psi'_1(\mathbf{r}_1)\psi_2(\mathbf{r}_2)\rangle \qquad (3.122)$$

plus the resonance energy:

$$E_\pm = \pm\langle \psi'_1(\mathbf{r}_1)\psi_2(\mathbf{r}_2)|H_{\text{int}}|\psi_1(\mathbf{r}_1)\psi'_2(\mathbf{r}_2)\rangle \qquad (3.123)$$

giving the total energy:

$$E'_T = E_0 + E' + E'_I + E_\pm. \qquad (3.124)$$

Recall that $E'_I < 0$ since H_{int} itself is negative, resulting in the dimer energy diagram in Fig. 3.41. Letting $E_+ = \beta = E_-$, then the total energy splitting is $\Delta E = 2E_\pm = 2\beta$.

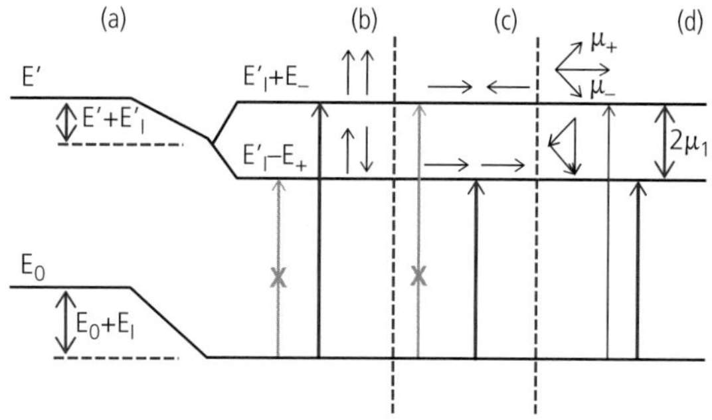

Figure 3.41 Energy level splitting for a physical dimer. Energy of the (a) ground (E_0) and first excited (E') state of the isolated monomers, (b) ground and first excited state of the dimer showing the dominant absorption transition into the upper and (c) lower doublet levels. (d) Transition energies for dimer dipoles that are neither parallel nor antiparallel (arrows). The doublet splitting energy is $2\mu_1 = 2\mu_2$.

The transition dipole moment is obtained using the Hamiltonian in Eq. 3.71 along with the dipole operator Eq. 3.72. Thus

$$\mu_\pm = -\frac{q}{\sqrt{2}}\langle \psi'_1(\mathbf{r}_1)\psi_2(\mathbf{r}_2) \pm \psi_1(\mathbf{r}_1)\psi'_2(\mathbf{r}_2)|\mathbf{r}|\psi_1(\mathbf{r}_1)\psi_2(\mathbf{r}_2)\rangle$$
$$= \frac{1}{\sqrt{2}}(\mu_1 \pm \mu_2), \qquad (3.125)$$

where μ_1 and μ_2 are the individual transition moments of the respective monomers between their ground and excited states (e.g. $\mu_1 = -q\langle \psi'_1(\mathbf{r}_1)\psi_2(\mathbf{r}_2)|\mathbf{r}|\psi_1(\mathbf{r}_1)\psi_2(\mathbf{r}_2)\rangle$). Since $\mu_1 = \mu_2$ for identical molecules, transitions into the antisymmetric state ψ_- have a vanishing transition dipole, $\mu_- = \mu_1 - \mu_2 = 0$, see the lower level, Fig. 3.41b. Figures 3.41b and c show situations where the dipoles are translationally invariant. Figure 3.41b shows the molecular dipoles aligned in a side-by-side arrangement. The lowest energy is achieved when the dipoles are antiparallel, that is $\mu_- = 0$. Then only transitions to the upper antisymmetric state at $E' + E'_I + E_+$ are allowed. When the dipoles are arranged end-on-end (Fig. 3.41c), the vector sum of the transition dipole is zero in the upper level, allowing only transitions to the lower state. Figure 3.41d shows a situation where neither of the parallel arrangements exist (such as in a herringbone crystal structure). Thus, transitions are possible into both the upper and lower levels. The energy separation of the levels is equal to $2\mu_1 = 2\mu_2$.

An example of the red shift in the dimer absorption spectrum compared with that of a monomer is shown in Fig. 3.42 for the molecule 4-*n*-propyl-5-ethyl farnesyl bacteriochlorophyll-c in dichloromethane (Olson

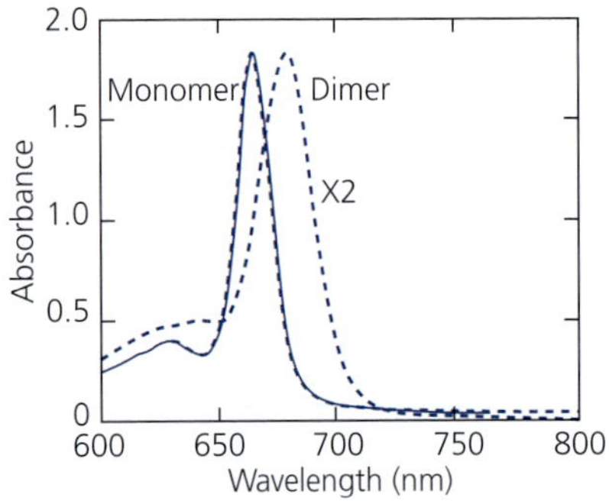

Figure 3.42 Monomer and dimer absorption spectra for bacteriochlorophyll in dichloromethane (Olson and Cox, 1991).

and Cox, 1991). The peak shows a rigidochromic shift from a wavelength of $\lambda = 665$ nm to 679 nm as the solution is concentrated, thereby increasing the probability for molecular aggregation into dimers, trimers, etc. In the concentrated solution, the dimers coexist with the monomers, and hence the dimer peak was extracted by subtracting the monomer contribution obtained in a dilute solution. Two dimer peaks, one at 670 nm and the second at ~690 nm are superposed to result in the peak at 679 nm, although the contribution from the lower energy peak is significantly less than the shorter wavelength peak since it corresponds to transitions into a forbidden dipole state. Since both peaks are present, this corresponds to the situation in Fig. 3.41d.

3.6.3 Excimers and exciplexes

Physical dimers support an excitation that resonantly oscillates between two electrostatically bound molecules. When the spacing of the molecules is close, the pair may not necessarily be bound in the ground state. When one of the pair is excited, the resonance interaction, β, becomes large, thus stabilizing the pair in a closely spaced, bound configuration. A dimer that has no bound ground state but is bound in the excited state is called an *excimer*, which is a contraction of the words "excited dimer." If the pair of excited molecules are of different composition, the state is an *exciplex*.

The difference between dimers and excimers is illustrated in Fig. 3.43, which shows the dimer splitting into E_+ and E_-, along with possible transitions to the ground (S_0) state. In the dimer, S_0 is a bound state, and hence the vibronic progression is observable in the luminescence spectrum shown at the right. In this diagram, the excited state has a smaller displacement than the ground state (i.e. the molecules move toward each other when one of the pair is excited). This is not

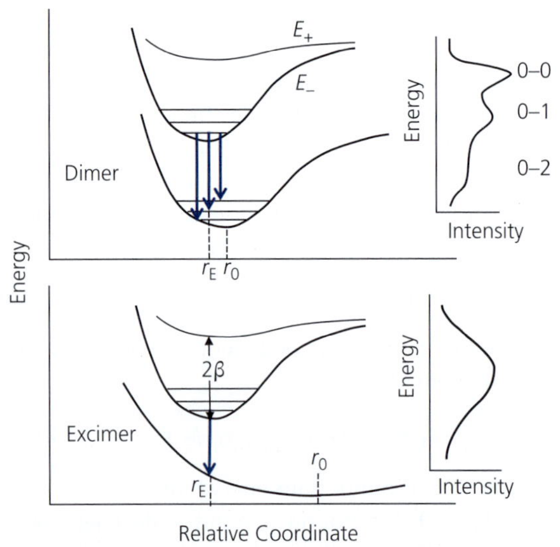

Figure 3.43 Configurational energy diagrams distinguishing a dimer and an excimer. To the right are characteristic luminescence spectra due to excited-to-ground state transitions. A dimer shows a clear vibronic progression, whereas excimer luminescence is broad and featureless. Luminescent transitions are indicated by the vertical arrows, r_E and r_0 are the equilibrium excited and ground state molecular coordinates. Note the lack of a bound ground state for the excimer.

the case for all dimers since the displacement depends on the relative contributions from the Coulomb and resonance integrals. For the excimer, the coordinate change is considerably larger and always negative, that is, the individual molecules transition from an unbound to a stable and more closely spaced bound pair after excitation. The lack of a ground state replaces the resolved vibronic features of the monomer with a broad and featureless spectrum. The calculation of excimer energies is identical to that used for dimers, although the relative magnitudes of β and E_I' are different.

Excimers are typically observed in solution: as an excited molecule collides with a ground state molecule, they form a bound pair that shares the excitation. This pair exists until a photon is emitted, or until a random event results in non-radiative return to the ground state, at which point the molecules once more disassociate. Evidence for excimers is the dependence of a broad luminescence feature on molecular concentration in solution. As the solution is increasingly diluted, the probability for excimer formation decreases proportionately.

In the solid state, random molecular encounters are absent. However, in some crystals, molecules can closely stack in a facial arrangement, forming physical dimers. When one of the molecules is excited, the lattice can distort to decrease the stacking distance, forming a lower energy excimer state. By polarizing the surrounding lattice, the free exciton becomes *self-trapped* on the excimer, where it remains until it recombines in the ground state (Song and Williams, 1993). Thus, the unexcited crystal is said to be in a *pre-excimeric* configuration. In solid amorphous blends of a high concentration of luminescent molecules in a high energy gap (i.e. transparent) matrix, the chromophores may find themselves in closely spaced pre-excimeric states. Then, both excimer and monomer luminescence are simultaneously observed due to the coexistence of aggregates and isolated molecules. As the chromophore concentration increases, the excimer and monomer luminescence proportionately increase and decrease, respectively.

An additional difference between monomer and excimer emission is that the broad excimer spectrum does not narrow nor vary significantly at very low temperatures (Birks and Kazzaz, 1968). To excite the molecular pair, the monomer transition energy (corresponding to $S_1 \leftarrow S_0$) must overcome the excimer binding energy at point B, by either thermionic emission or tunneling to point C in Fig. 3.44. Excimers lower the system energy by reducing both the Coulomb and resonance energies, ultimately resulting in emission from the excimer, E^*. The multistep dynamical process leading to excimer emission is described by

$$S_0 + h\nu_M \rightarrow S_1,$$

$$S_1 \rightarrow E^* + h\nu_{ph}, \qquad (3.126)$$

$$E^* \rightarrow S_0 + h\nu_{E^*}.$$

Here, the energies $h\nu_M$, $h\nu_{ph}$, and $h\nu_{E^*}$ correspond to the monomer excitation energy, the phonon energy leading to internal conversion to E^*, and the excimer emission energy, respectively. As the exciton moves from a free to a self-trapped state, additional thermal (phonon) energy is required to surmount the barrier, B. As temperature is decreased, the probability of excimer vs. monomer emission is also decreased.

Planar molecules are particularly susceptible to excimer generation since they tend to form tight, facial π–π stacks that result in the sharing of excitations between molecular neighbors. An archetype excimeric PAH is pyrene, whose structure is shown in Fig. 3.45a. In the ground state, the equilibrium intermolecular spacing is $r_0 = 3.53$ Å (Robertson and White, 1947), decreasing to $r_E = 3.34$ Å for the self-trapped excimer (Birks, 1975). The changes in dimension as well as the energies of the transitions are indicated in the energy scheme in Fig. 3.45b. Excimer

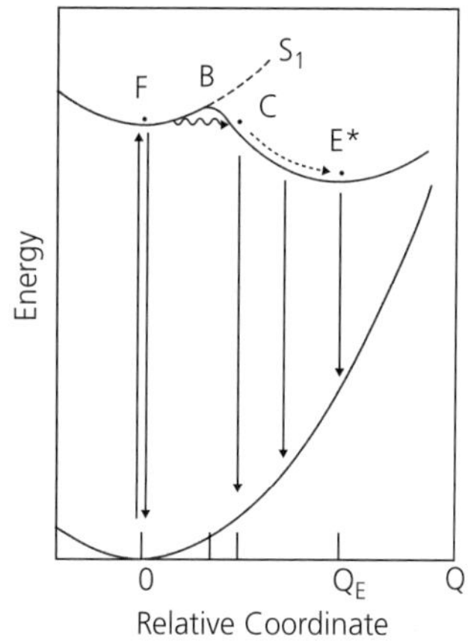

Figure 3.44 Energy transfer from a free exciton (F) to a self-trapped, or excimer state (E^*). Possible transitions of the free, excimer and intermediate states are shown as the energy surface is traversed between initial and final states. Excimers are formed by overcoming of the barrier B by either thermionic emission, or tunneling to C. After Nasu (1987).

emission from a dilute solution of pyrene in is shown in Fig. 3.46 (Birks and Christophorou, 1963). In the most dilute solutions (10^{-4}–10^{-3} M), the monomer spectrum coexists with the broad excimer feature, illustrating the differences between these two excited states. The monomer spectrum has noticeable structure, whereas the excimer spectrum is broad and featureless. Note the very strong dependence of the ratio of excimer-to-monomer fluorescence intensity with increasing pyrene concentration.

An example of excimer phosphorescence in a solid mixture is shown in Fig. 3.47 for the triplet emitting platinum complex, FPt1, diluted in CBP (D'Andrade et al., 2002). The energy gap of CBP is 3.1 eV and that of FPt1 is 2.6 eV. The data are collected by using short wavelength illumination to selectively excite only the $S_1 \leftarrow S_0$ transition in CBP, which then transfers its excited state by a Förster process to the FPt1 triplet state (see Section 3.8.1). The triplet then emits in the 450–550 nm wavelength range, following $T_1 \rightarrow S_0$. At very low concentrations, the planar FPt1 molecules are isolated in the solid CBP matrix, and primarily light blue monomer emission is observed, as characterized by the vibronic progression in the spectrum from Film 2. The long wavelength emission tail at > 550 nm is due to excimer emission from FPt1 aggregates present even at low concentrations. At the higher phosphor concentration (7 wt%) of Film 3, excimer emission is dominant.

An exciplex is analogous to an excimer except that the pair of molecules are of different composition. Exciplex, therefore, refers to an "excited molecular

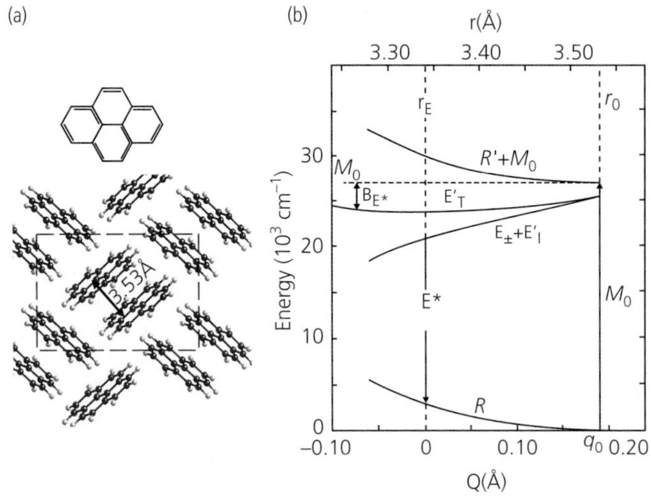

Figure 3.45 (a) Molecular structural formula of pyrene ($C_{15}H_{10}$) and its equilibrium crystal structure, showing four molecules per unit cell and an equilibrium interplanar spacing of 3.53 Å. A projection of the *a–b* plane of the unit cell along the *c*-axis is shown by the dashed line. (b) Energy level diagram of the excimer (E'_T) and monomer (M_0) energies, along with the attractive interaction term ($E_\pm + E'_I$), and the repulsive monomer energy in the excited (R') and ground (R) states. B_{E*} is the excimer binding energy. Both the absolute (r) and relative coordinates (Q) are shown (Birks, 1975).

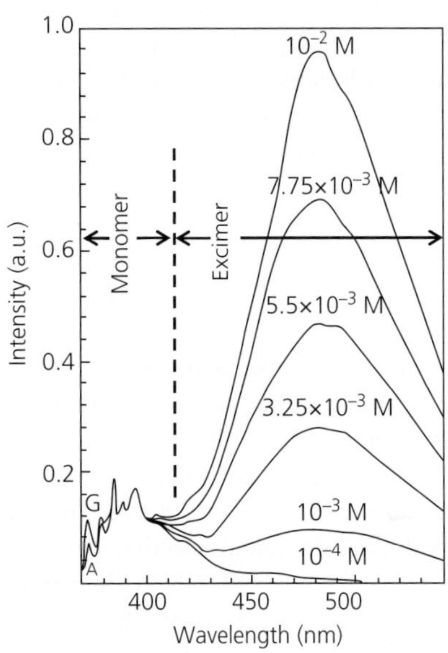

Figure 3.46 Fluorescence spectra of pyrene in cyclohexane at various pyrene concentrations. The approximate boundary between the monomer and excimer spectra is shown. The intensities are normalized to that of the monomer fluorescence (Birks and Christophorou, 1963).

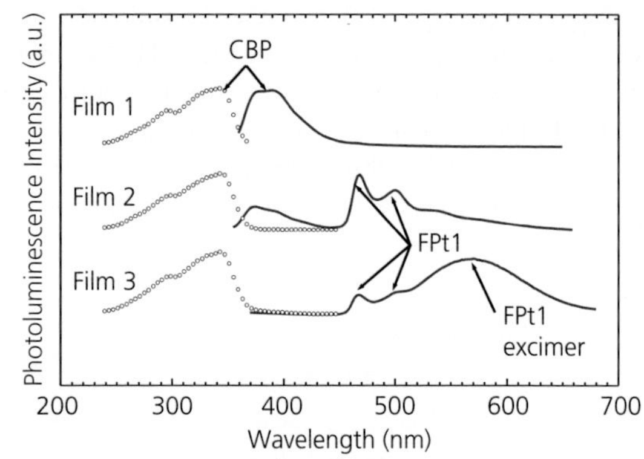

Figure 3.47 Photoluminescence (solid lines) and excitation spectra (circles) of a neat CBP (Film 1), <1% (by weight) FPt1 in CBP (Film 2) and 7% FPt1 in CBP (Film 3) (D'Andrade et al., 2002).

complex." Due to the obvious asymmetry of the system, the charge density is not shared equally between molecules; in effect there is a limited degree of charge transfer from one to the other. Thus, one molecule serves as an anionic electron acceptor (A), and the other as the cationic donor (D), depending on the relative positions of the HOMO and LUMO levels of the pair (see Fig. 3.48). Only partial charge transfer

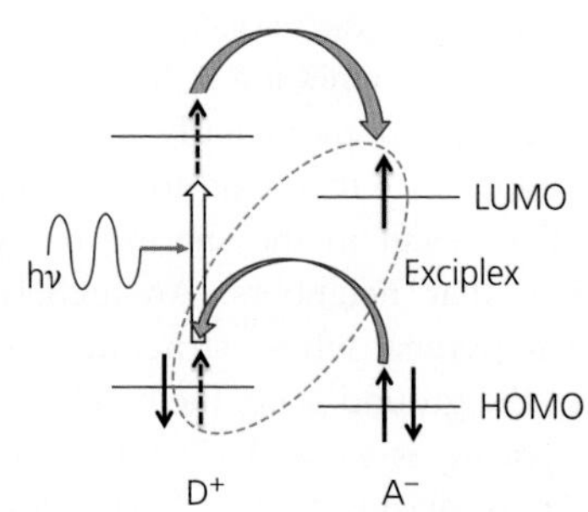

Figure 3.48 Energetics of an exciplex. Individual electron spins are shown, with transfer of charge from the donor (D) to the acceptor (A) molecule forming the exciplex (dashed ellipse).

occurs that comprises a fractional charge ($\delta^{\pm}$) redistribution of $\delta^+ = \delta^- < q$ across the exciplex. We can express the charge transfer reaction as follows:

$$D^* + A^0 \rightarrow (D^+A^-), \tag{3.127}$$

where D^* is the excited donor, A^0 is the ground state acceptor, and the term in parentheses represents the bound exciplex state.

Charge transfer can occur in both the ground and excited states, again depending on their energy level offsets. When there is ground state charge exchange, the molecular pair is called a *charge transfer complex*, discussed in Section 3.6.6.

The exciplex wavefunction is described as a linear combination of individual molecular wavefunctions. This differs from the treatment of a dimer (and excimer) in Eq. 3.121 since the two wavefunctions are now associated with *different* molecules. The ground and excited state wavefunctions are thus:

$$\psi_T^0 = \alpha^0\psi_1\psi_2 + \gamma^0\psi_1^+\psi_2^- + \delta^0\psi_1^-\psi_2^+, \tag{3.128}$$

$$\psi_T' = \alpha\psi_1'\psi_2 + \beta\psi_1\psi_2' + \gamma\psi_1^+\psi_2^- + \delta\psi_1^-\psi_2^+, \tag{3.129}$$

where the coefficients, α, β, γ, δ, etc., can be either positive or negative, and their magnitudes give the strengths of their contributions to the total exciplex wavefunction. The subscripts refer to molecules 1, 2 forming the complex, superscript 0 indicates the ground state, $\psi_{1,2}'$ is the wavefunction for the corresponding individual molecular excited state, and $\psi^{\pm}$ indicates the net charge on the molecule. Note that the charge can transfer in either direction, depending how large an energy step exists between frontier energy levels in the two molecules.

There are many examples of exciplexes in mixtures of aromatic hydrocarbons such as derivatives of benzene, naphthalene, anthracene, etc. due to their planar (and hence close packing) structures (Birks, 1970). An example of exciplex fluorescence for a series of acceptors in a PVK donor matrix is shown in Fig. 3.49. All acceptors have closely matched HOMO-LUMO energy gaps of 2.70 eV for CN-MBE, to 2.84 eV for CN-TFPA. However, their HOMO energies, measured by ultraviolet photoelectron spectroscopy, (Appendix C) are offset due to a differing amount of electron withdrawing character afforded by the cyano (–CN) and trifluoromethyl (–CF_3) groups comprising each molecule. The molecules are also mixed into the very wide energy gap PMMA matrix that does not interact with the various acceptors. These mixtures luminesce only from the acceptor molecules themselves without

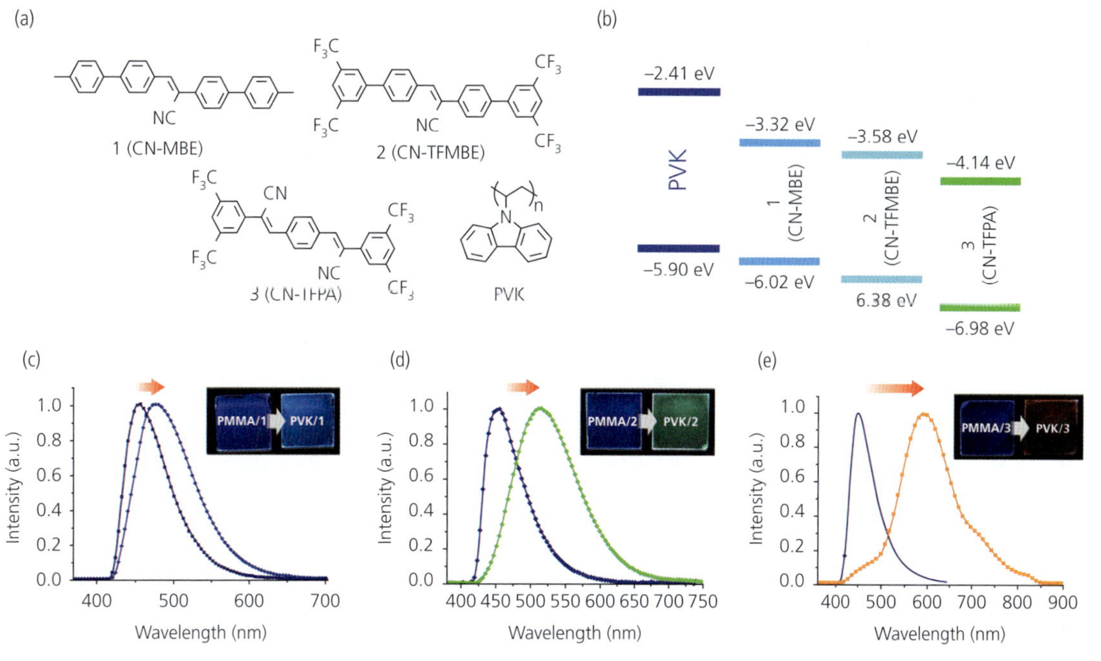

Figure 3.49 (a) Molecular structural formulae for a series of three acceptor molecules and the electron donor, PVK. (b) Frontier orbitals of the compounds in (a) indicating their HOMO (lower) and LUMO (upper) energies. The LUMO energy is estimated by subtracting the low energy cut-off of the optical absorption tail of the compound (i.e. the *optical energy gap*) from its ionization energy. (c)–(e) The spectra of compounds **1**, **2**, and **3** in PMMA (purple line) and PVK (blue, green and orange lines). The red shift for the mixtures in PVK is due to exciplex emission, whereas the unshifted spectra for the PMMA-based mixtures are due to the individual acceptor molecules themselves (Kim et al., 2014).

charge or energy transfer. Thus, the red shift in the spectra is presumed due only to luminescence from exciplexes formed at the PVK/acceptor interface.

To understand the red shifts in the exciplex spectra in Fig. 3.49, the transition energy $(D^+A^-) \rightarrow S_0$ is (Giebink et al., 2010):

$$\Delta E = |E_{A,LUMO} - E_{D,HOMO}| - \frac{q^2}{4\pi\varepsilon_0\varepsilon_r a_0}, \quad (3.130)$$

where $E_{A,LUMO}$ is the LUMO energy of the acceptor, $E_{D,HOMO}$ is the HOMO energy of the donor, ε_r is the relative dielectric constant of the medium, and a_0 is the radius of the charge transfer state. The last term expresses the Coulomb binding energy (E_B) of the exciplex. For convenience, we assume that a full unit charge is transferred from D to A. If we assume that $a_0 \sim 1$ nm, or approximately the size of a molecule in the pair, and $\varepsilon_r \sim 3$ typical of organic materials, then $E_B \sim 0.2$ eV. Hence, for the blends in Fig. 3.49, we expect the red shift and the peak energies to follow the trends implied by the energy offsets in Fig. 3.49b. The PVK/**1** mixture should have the highest energy transition of approximately $E_G = E_{A,LUMO} - E_{D,HOMO} = 5.90 - 3.32 - 0.2$ eV $= 2.38$ eV, with the greatest red shift corresponding to PVK/**3**, with $E_G = 1.56$ eV.

While the energy trends qualitatively match expectations, the magnitude of the shift diverges as the LUMO energy is decreased from PVK/**1** to PVK/**3**, as summarized in Table 3.2. Given the approximate means for determining $E_{A,LUMO}$ by adding the optical energy gap (corresponding to S_1, not the conduction level) to $E_{D,LUMO}$, it is not surprising that the divergence is large. Furthermore, it is incorrect to assume that E_B is independent of the details of molecular structure and morphology. Nevertheless, given the consistency of the trends with variations in frontier orbital energies, we conclude that emission from this system of D–A pairs is due to exciplex formation.

Table 3.2 Calculated and measured energies for exciplex systems in Fig. 3.49 (Kim et al., 2014)

Blend	$E_{D,LUMO}$	$E_{A,HOMO}$	E_{AD}	ΔE (meas.)
PVK/**1**	−5.90	−3.32	2.58	2.59
PVK/**2**	−5.90	−3.58	2.32	2.40
PVK/**3**	−5.90	−4.14	1.76	2.10

All energies in eV.

3.6.4 Excited states in crystals

Davydov extended considerations of identical molecules to a crystal unit cell consisting of an arbitrary number of translationally *inequivalent* molecules. It was found that the spectrum splits into branches whose number is equal to the number of inequivalent molecules. Generally a crystal consisting of N unit cells, with a basis of $m \geq 1$ has the following Hamiltonian (Sibley et al., 1965):

$$H = \sum_i^N \sum_\mu^m H_{i\mu} + \sum_{i\mu < kv}^{N,m} V_{i\mu,kv}, \quad (3.131)$$

where $H_{i\mu}$ is the Hamiltonian for the individual molecule in ith unit cell at site μ, and $V_{i\mu,k\nu}$ is the interaction between molecules at i, μ and k, ν. The ground state wavefunction is

$$\psi_{i\mu}^0 = A \prod_{i\mu}^{N,m} \phi_{i\mu}, \quad (3.132)$$

where A is an operator that antisymmetrizes the wavefunction under electron exchange between individual ground state molecular wavefunctions, $\phi_{i\mu}$. When a single molecule at site k, ν is excited, the "one-site" exciton wavefunction becomes

$$\psi'_{kv} = A\phi'_{kv} \prod_{i\mu \neq kv}^{N,m} \phi_{i\mu}. \quad (3.133)$$

Since the excited state (i.e. exciton) can move between all molecules in the crystal, we use Bloch's theorem to describe the delocalized wavefunction to yield

$$\Psi_\mu = \frac{1}{\sqrt{N}} \sum_i^N \psi'_{i\mu} \exp(i\mathbf{k} \cdot \mathbf{R}_{i\mu}). \quad (3.134)$$

That is, we assume that the exciton can be described by a periodic plane wave with wavevector **k**, that can propagate among equivalent positions in the lattice, $\mathbf{R}_{i\mu}$.

Continuing to follow the procedure used to determine the energy of a dimer, we write the excited state energy for the crystal as

$$E(\mathbf{k}) = E_0 + E' + E'_I + E_\pm(\mathbf{k}), \quad (3.135)$$

where E'_I and $E_\pm(\mathbf{k})$ are the molecular interaction energies analogous to their definitions in Eqs. 3.122 and 3.123. Thus, $E(\mathbf{k})$ defines the dispersion relationship between energy and exciton momentum, **k**. Using Eq. 3.132 to 3.134 with the Hamiltonian, Eq. 3.131, we have:

$$E'_I = \sum_{k\nu} \left(\langle \psi'_{i\mu}\psi'_{i\mu} | V_{i\mu,k\nu} | \psi^0_{k\nu}\psi^0_{k\nu} \rangle - \langle \psi^0_{i\mu}\psi^0_{i\mu} | V_{i\mu,k\nu} | \psi^0_{k\nu}\psi^0_{k\nu} \rangle \right), \quad (3.136)$$

where the second term eliminates contributions from molecular pairs in the ground state. This is known as the *self-polarization energy* of the molecule in the crystal, see Section 3.6.7. The interaction energy, $V_{i\mu,k\nu} < 0$, and thus shifts the center of the band of exciton states from its initial value of E' as in Fig. 3.50a. Similarly:

$$\begin{aligned} E_{\pm}(\mathbf{k}) &= \sum_{i,k}^{N}\sum_{\mu}^{m}\langle\phi'_{i\mu}\phi_{k\mu}|V_{i\mu,k\mu}|\phi_{i\mu}\phi'_{k\mu}\rangle\exp\left(i\mathbf{k}\cdot\Delta\mathbf{R}_{i\mu,k\mu}\right) \\ &\pm\sum_{i,k}^{N}\sum_{\nu\neq\mu}^{m}\langle\phi'_{i\mu}\phi_{k\nu}|V_{i\mu,k\nu}|\phi_{i\mu}\phi'_{k\nu}\rangle\exp\left(i\mathbf{k}\cdot\Delta\mathbf{R}_{i\mu,k\nu}\right) \\ &= E_{\mu\mu}\pm E_{\mu\nu} \end{aligned} \quad (3.137)$$

where the first term corresponds to sites that are translationally equivalent within the unit cells (i.e. the expression is found by diagonalizing the matrices in Eqs. 3.132 and 3.133), and the second to translationally inequivalent sites. Then, $\Delta\mathbf{R}_{i\mu,k\nu} = \mathbf{R}_{i\mu} - \mathbf{R}_{k\nu}$ is the lattice translation vector.

This situation is considerably simplified when only nearest neighbor interactions are considered, which is consistent with the short range of van der Waals bonds in molecular crystals. Then the wavefunctions in Eq. 3.134 are

$$\Psi_{\mu} = \frac{1}{\sqrt{N}}\sum_{i}^{N}\psi'_{i\mu}\exp\left(i\mathbf{k}\cdot\Delta\mathbf{R}_{i\mu}\right), \quad (3.138)$$

where $\Delta\mathbf{R}_{i\mu}$ is now restricted to distances between nearest neighbors. For a one-dimensional (1D) crystal of length N unit cells (which serves as a reasonable approximation to transport along a molecular stacking axis), or for a cubic lattice of lattice constant a, then $\exp\left(i\mathbf{k}\cdot\Delta\mathbf{R}_{i\mu}\right) = \exp(ika)$, with N values for k, viz. $k = 0, \pm\frac{2\pi}{Na}, \ldots, \pm\frac{\pi}{a}$. Then it is straightforward to show that

$$E(\mathbf{k}) = E_0 + E' + E'_I + 2E_{\mu\mu}\cos(\mathbf{k}\cdot\Delta\mathbf{R}). \quad (3.139)$$

The energies of the two dimer states are now split into $2N$ different, but nearly degenerate states whose wavefunctions are given in Eq. 3.133. These states form the *Davydov band* whose bandwidth is $\Delta E = 4E_{\mu\mu}$. The exciton states that occupy this band are known as *Frenkel excitons*.

Note that Eq. 3.137 is analogous to the solution obtained for a dimer, Eq. 3.123. However, the energy splitting between the bands in the crystal is (from Eq. 3.139) equal to $\Delta E = 4E_{\mu\mu}$, which is twice that for the dimer. The difference arises since a molecule in the 1D crystal stack has inversion symmetry, and hence each molecule has two nearest neighbors—one above and another below each site.

When more than one translationally inequivalent molecule forms the crystal basis, the approximation of Eq. 3.139 is no longer valid since $E_{\mu\nu}\neq E_{\mu\mu}$ for $\mu\neq\nu$. Referring to Eq. 3.136, there is an additional splitting of the exciton energy. For example, anthracene consists of a basis of two, translationally inequivalent molecules (see Fig. 2.8c), thereby splitting the spectrum into two absorption bands sensitive to light polarized parallel and perpendicular to the **b** crystal symmetry axis. This polarization-dependent spectral shift is known as *Davydov splitting* with energy ΔE_D (Davydov, 1962). Band splitting and example spectra showing the Davydov energy for the anthracene triplet state are shown in Fig. 3.50b. In the three-dimensional crystal, each molecule has four nearest neighbors that share the excitation, such that $\Delta E_D \approx 8E_{\mu\nu}$.

From a solution to the eigenvalue equation using the Hamiltonian in Eq. 3.131 along with the excited state wavefunctions in Eq. 3.133, one arrives at the Frenkel exciton band energy dispersion, $E(\mathbf{k})$, first calculated for anthracene by Davydov and Sheka (1965). Figure 3.51 shows the lowest two Frenkel

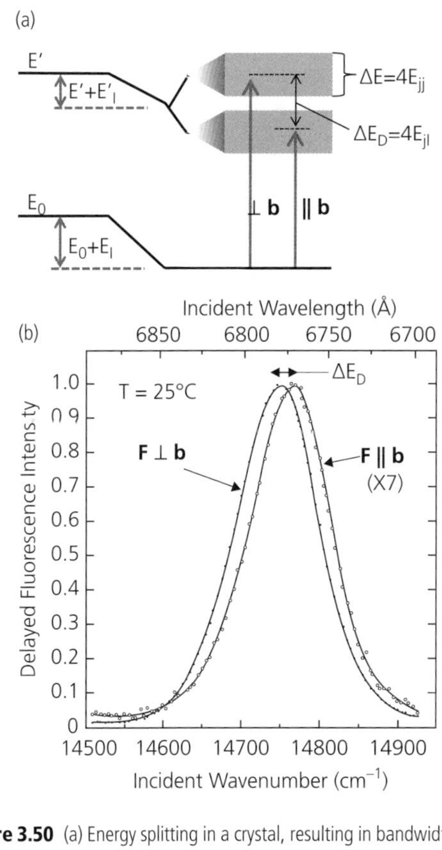

Figure 3.50 (a) Energy splitting in a crystal, resulting in bandwidth, ΔE, and a Davydov splitting energy, ΔE_D. The transitions for anthracene with the incident field (**F**)⊥**b** and ||**b** crystal axes are shown. (b) Spectra due to delayed fluorescence originating from the triplet state of anthracene (Avakian et al., 1968).

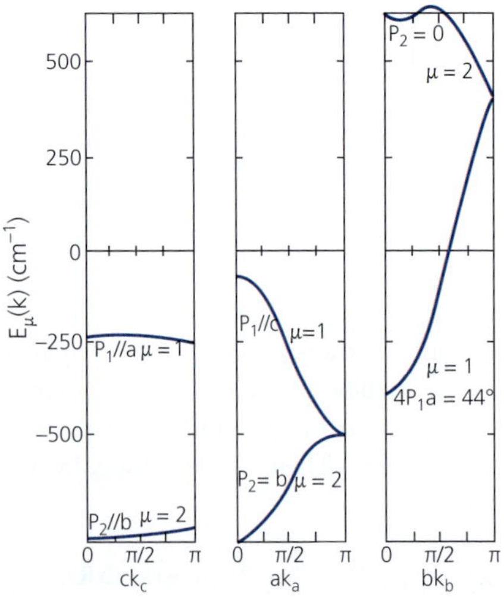

Figure 3.51 Dispersion relationships, $E_\mu(\mathbf{k})$, of the lowest two exciton bands in anthracene in the first Brillouin zone along the (a) **c**, (b) **a**, and (c) **b** crystal axes. The light polarization directions P_μ for the different bands (μ) are indicated (Davydov and Sheka, 1965).

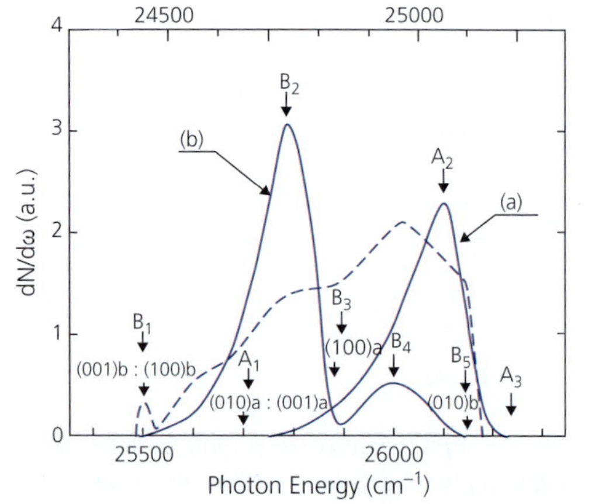

Figure 3.52 Thermoabsorption spectra give the density of states ($dN/d\omega$) of the first exciton bands of anthracene along the (a) **a** and (b) **b** directions. The low energy, peak and high energy edges along these directions are denoted by the letters with subscripts 1, 2, 3, respectively. Features B_4 and B_5 along **b** are also noted. The top abscissa gives the measured energy, and the lower axis is the calculated energy. The scales are offset by 1050 cm^{-1} to align the calculation to the measured fundamental 0–0 transition. The dashed curve is the calculated density of states (Matsui et al., 1973).

bands in the first Brillouin zone along the three symmetry directions (the **a**, **b** and **c** axes), using the folded zone representation. The $S_1 \leftarrow S_0$ transition energy is $\tilde{\nu} = 27{,}570\,\text{cm}^{-1} = 3.42\,\text{eV}$, corresponding to a wavelength of 275.8 nm.

Optical transitions occur near to the $\mathbf{k} = 0$ point since the photon carries only a small momentum compared with the crystal (i.e. the transition is restricted to momentum changes of $\Delta\mathbf{k} \approx 0$). The electron effective mass tensor is due to the band curvature, and is defined by

$$\frac{1}{m^*_{i,j}} = \left[\frac{1}{\hbar^2}\frac{\partial^2 E(\mathbf{k})}{\partial \mathbf{k}^2}\right]_{i,j} = \frac{1}{\hbar^2}\frac{\partial^2 E(\mathbf{k})}{\partial k_i \partial k_j}. \tag{3.140}$$

The bandwidths in molecular solids (i.e. the energy difference between the bottom of a band, typically at $k = 0$ for direct bandgap systems, and the top of the band often found at the edge of the Brillouin zone) are small ($\leq$300 cm^{-1}) compared to inorganic semiconductors with bandwidths ~1 eV (equivalent to ~8065 cm^{-1}). The largest bandwidth in anthracene is along the **b**-axis which, among the three symmetry axes, has the smallest intermolecular spacing of 6.04 Å (Robertson, 1958). The bandwidth along this axis is approximately 0.09 eV (725 cm^{-1}), which is consistent with measurements of the singlet exciton band density of states using thermoabsorption spectroscopy, with results shown in Fig. 3.52 (Matsui et al., 1973). The Davydov splitting between the peaks, $A_2 - B_2$, is 210 cm^{-1}, which is consistent with calculated values. Note that this is greater than 10 cm^{-1} for the triplet state in Fig. 3.50 due to the considerably larger oscillator strength of the singlet transition.

The anthracene bandwidths are smaller than bandwidths typical of inorganic semiconductors, the latter therefore having correspondingly smaller effective masses. Very tightly packed organic molecular solids can have broader bandwidths than that of anthracene. For example, PTCDA (with ΔE ~ 0.1 eV, see Fig. 3.53) and rubrene support excited states that bridge more than just nearest neighbors, and hence share characteristics common to both organic and inorganic materials. These intermediate systems and their excitonic properties are discussed further in the following section.

This chapter has taken us from the consideration of the energetics of individual molecules, to physical dimers, and finally to a crystal with a basis consisting of more than one translationally inequivalent molecule. The crystal exhibits a band structure for Frenkel excitons following the usual Bloch formalism used for describing electronic states in infinite crystal lattices. The transition from an isolated molecule to an infinite solid provides a picture that is useful in understanding molecular materials that apparently share many features of both the solid and isolated molecules of which it is comprised. This is in striking contrast to covalent or ionic materials whose spectroscopic properties seem to bear no resemblance to that of the individual atoms that make up the lattice. The differences in these systems, of course, is a result of the very weak electrostatic bonds of van der Waals solids that only slightly perturb the electronic systems of their molecular constituents, whereas strong chemical bonds result in the sharing of, and strongly influence the valence electrons in inorganic semiconductors. Analogously, the covalent bonding within the molecule itself results in energetics that bear only a faint resemblance to the constituent atoms.

The transition from molecule to solid is shown graphically in Fig. 3.54. This picture is a modified version of the *Lyons model* (Silinsh, 1980), which provides a bridge between the Jablonski diagram (Fig. 3.33) for individual molecules and the properties of the solid. Figure 3.54 shows the various transition energies observed for the molecule and the crystal referenced to the vacuum level. The *electron affinity* (*EA*) is the energy required to remove an electron from the lowest conduction state in the solid to the vacuum level. The *ionization energy* (often stated as the *ionization potential*, *IP*) is the energy difference between the ground state of the molecule, or HOMO, to the vacuum. The *polarization energy* of the solid for both electrons and holes (P_e, P_h, respectively) lowers their *frontier orbital* (i.e. HOMO and LUMO) energies from values in the isolated molecules, as discussed in Sections 3.6.6 and 3.6.7.

The changes in frontier orbital energies and their segregation into bands are accessed by different polarizations of the incident radiation. Their comparative spectra are shown in Fig. 3.55 for isolated anthracene molecules in a dilute ethanol solution. The solid state spectra are red-shifted and broadened relative to the solution spectra due to interactions between molecules (cf. Eq. 3.139). Similar phenomenological features for PTCDA are discussed below.

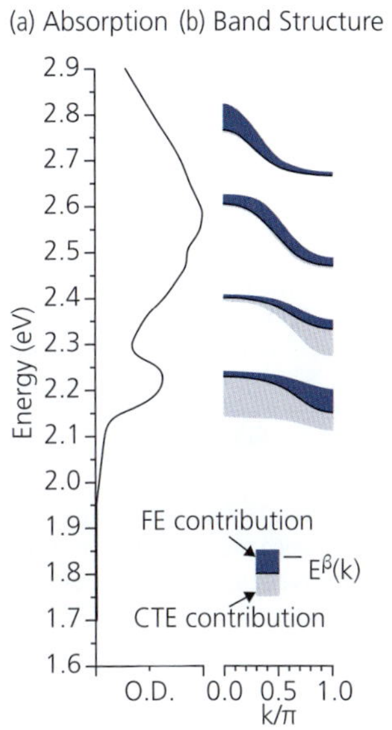

Figure 3.53 (a) The absorption spectrum and (b) calculated exciton band structure for PTCDA. Here, FE is the Frenkel and CTE is the charge transfer exciton contributions to the band structure. See also Section 3.6.6 (Hoffmann et al., 2000).

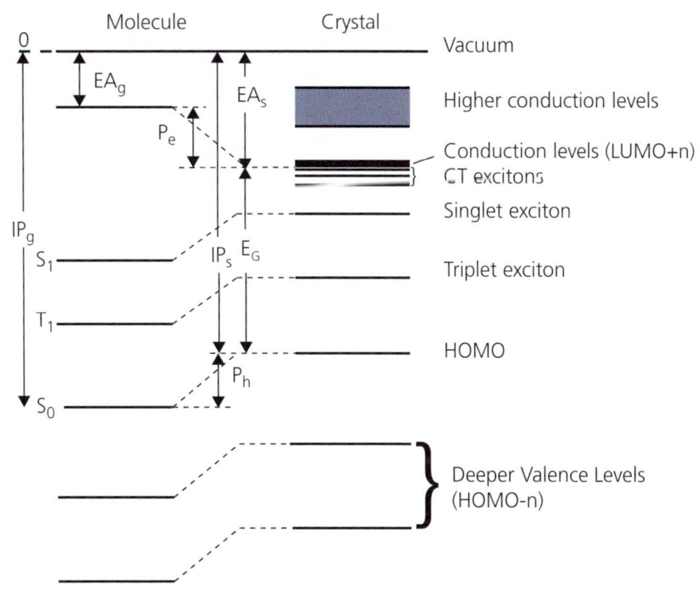

Figure 3.54 The correspondence between the energetics of isolated molecules and their crystals. The vacuum level energy is $E = 0$. Other energies shown are: EA = electron affinity, IP = ionization energy (subscript s = solid, g = gas), P_e (P_h) = polarization energy of the electron (hole), and E_G = energy gap.

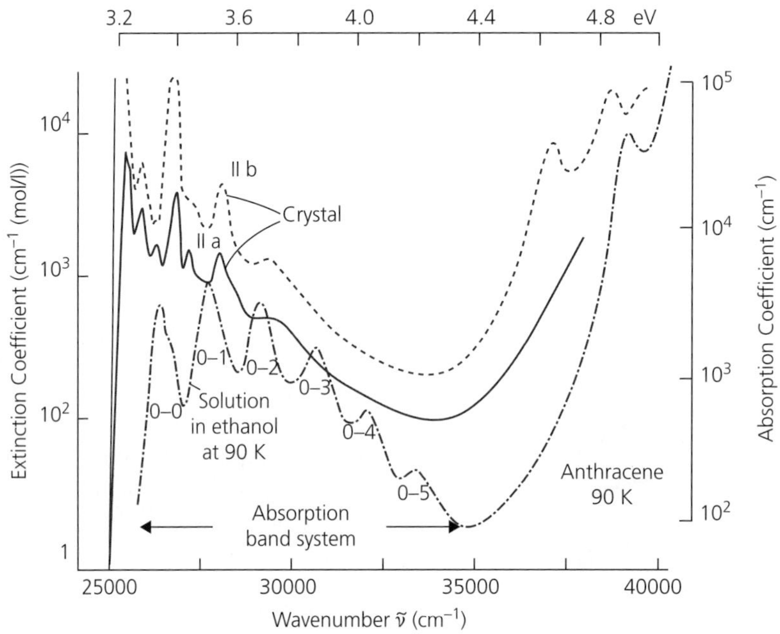

Figure 3.55 Solution and solid state spectra of anthracene in ethanol. The solid state spectra are taken along two crystal axes, **a** and **b**. Note the red shift and broadening of the crystal spectra indicate the formation of energy bands. The vibronic progression is clearly visible in the solution spectrum; each transition is identified (Wolf, 1959).

3.6.5 Classification of excitons

If we approximate the exciton as a hydrogenic state, that is, the hole and electron are Coulombically bound, then from the Bohr model we obtain a binding energy of

$$E_B = -\frac{q^2}{8\pi a_0(\varepsilon_r\varepsilon_0)}\frac{1}{n^2} \tag{3.141}$$

with Bohr radius

$$a_0 = \frac{4\pi(\varepsilon_0\varepsilon_r)\hbar^2}{m_r^* q^2}. \tag{3.142}$$

Here, n is the principal quantum number of the excited state and m_r^* is the reduced effective mass of the exciton (Eq. 3.84). From this simplified treatment, we see that $E_B \propto 1/\varepsilon_r^2$ and $a_0 \propto \varepsilon_r$. The exciton binding energies and radii for different semiconductors are listed in Table 3.3.

There are several differences between inorganic and organic semiconductors. The most notable is that the dielectric constants for the organic materials are significantly smaller, and the effective masses are larger than for inorganic semiconductors. As noted previously, these differences stem from the purely electrostatic bonds of organics that give rise to a smaller dielectric constant, and thus a concomitant increase in exciton binding energy and decrease in radius compared to chemically bonded inorganic materials. Figure 3.56 classifies the three types of excited states that characterize these different materials.

Table 3.3 Comparison of exciton radii and binding energies for a variety of organic and inorganic semiconductors

Medium	ε_r	m_r^*/m_e	a_0 (Å)	E_B (eV)
Vacuum (H atom)	1	1	0.5	13.6
Anthracene[a]	2.4–4.1	1	1.2–2.1	0.8–2.4
PTCDA[b]	2.0–4.4	0.14–1	2.2–7.1	0.5–0.7
ZnS[c]	5.2	0.4	5.2	0.2
Si[c]	11.9	0.14	43	0.014
GaAs[c]	13.1	0.07	94	0.006

[a] From Cummins and Dunmur (1974).
[b] From Forrest (1997) and van der Meer (2014).
[c] From Sze (1981).

(i) *Frenkel excitons*: Molecular excitons are typically confined to a single molecule in the solid (Fig. 3.56a). These are known as Frenkel excitons. The exciton energy is influenced by polarization of the surrounding medium. Further, the Frenkel state can readily diffuse within the material by hopping from a molecule to its neighbors. Their very high binding energies (~ 0.5–1 eV) make

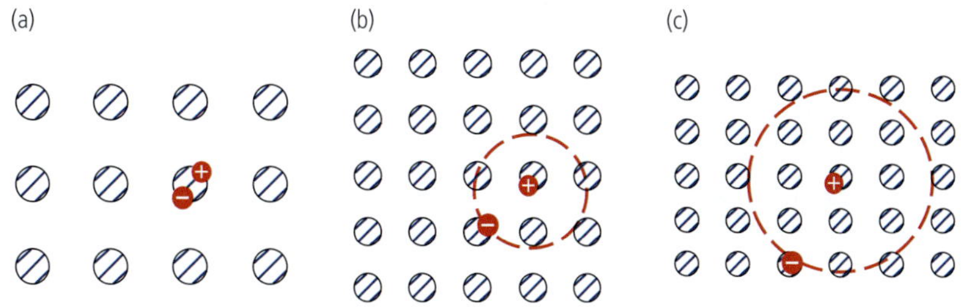

Figure 3.56 Three types of excitons illustrated by a positive and negative charge separated by an effective Bohr radius (dashed lines) in a lattice consisting of atoms or molecules (shaded circles): (a) Frenkel; (b) charge transfer; and (c) Wannier–Mott excitons.

these states stable at room temperature and above. Hence, Frenkel excitons dominate the optical properties of organics.

(ii) *Charge transfer excitons*: During transport, the electron and hole may briefly reside on adjacent molecules. These molecules can be similar, or two dissimilar materials can form a *heterojunction* that is spanned by the Coulombically bound electron and hole. As inferred from the data in Fig. 3.36, the excited state may extend across several nearby molecules when their stacking distance is particularly small. For example, the intermolecular spacing along the (102) axis of PTCDA is only 3.21 Å (Forrest, 1997). PTCDA forms extended stacks that support a charge transfer (CT) state. These intermediate species are illustrated in Fig. 3.56b. Organic materials with prominent charge transfer characteristics can simultaneously exhibit Frenkel-like features in their optical properties. Such materials primarily behave as conventional excitonic materials, yet they share some important characteristics with chemically bonded inorganic semiconductors.

(iii) *Wannier–Mott excitons*: Chemically bonded inorganic materials have high dielectric constants, leading to very low exciton binding energies (~5–50 meV) that are unstable at room temperature. These excited states, with concomitantly large radii that span many tens of lattice constants, are known as Wannier–Mott (W-M) excitons. Given their instability and low oscillator strengths, materials that support W-M excitons have optical and transport properties that are almost entirely independent of the exciton. Rather, inorganic materials exhibit optical and electronic properties that are dominated by band-to-band absorption and emission, and band transport phenomena. Engineered inorganic materials such as quantum wells, wires and dots comprising a very thin, low band gap semiconductor sandwiched between two larger energy gap semiconductors can artificially reduce the size (and dimensionality) of the exciton while increasing its binding energy. In this way, engineered inorganic materials can be made to approximate Frenkel-like behavior.

We notice from Table 3.3 that organic materials exhibit pronounced anisotropies in their electronic properties: typically the coupling between molecules along the π-stacking direction is far stronger than in the other directions where there is comparatively little orbital overlap. This is in contrast to inorganic semiconductors that tend to have a higher level of lattice symmetry (e.g. cubic or hexagonal), leading to more isotropic properties.

The absorption spectra along different crystal axes reflect the large anisotropies in dielectric properties. Indeed, for PTCDA, the difference in the real part of the refractive index (known as the *birefringence*) is particularly large. Thus, $\Delta n_r = 0.75$–1.0 (depending on wavelength), with the larger index in the plane of the molecule and the smaller along the stacking direction. In contrast, nearly isotropic dielectric properties are observed in the molecular plane due to the herringbone structure along this direction (see Fig. 2.18a). Figure 3.57 illustrates the directional dependence of the birefringence in both the real (n) and imaginary (k) parts of the refractive index for a single crystal of PTCDA both parallel and perpendicular to the (102) stacking direction (Zang et al., 1991, Alonso et al., 2002).

Anisotropy is also found in the molecular conductivities, as will be discussed in Chapter 4. Both Si and GaAs, for example, are found in lattices with cubic symmetry, whose properties are the same properties along the three orthogonal lattice directions.

Note that ZnS is the most ionic semiconductor in Table 3.3. It has a particularly large binding energy and small exciton radius for an inorganic material.

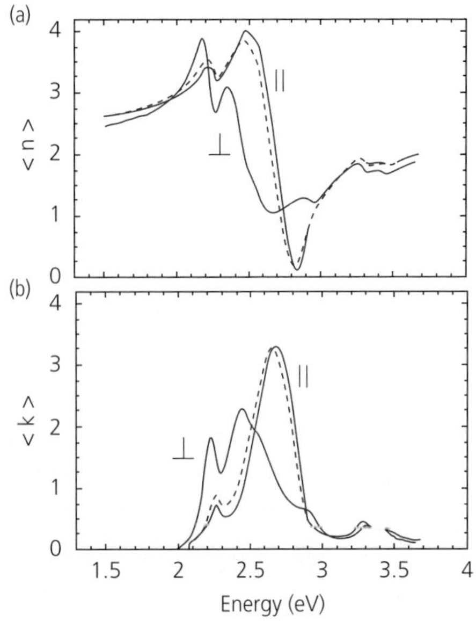

Figure 3.57 Real (n) and imaginary (k) parts of the refractive index of a single crystal of PTCDA. Here, || refers to light incident parallel to the crystal **b**-axis (i.e. the in-plane direction) and ⊥ is along the stacking axis (perpendicular to the **b**-axis). Slight differences in the || spectra are observed for measurements at 180° relative to each other. From Alonso et al. (2002).

Reprinted from Org. Electron., 3, 23 Alonso, M. I., Garriga, M., Karl, N., Osso, J. O. & Schreiber, F., Anisotropic optical properties of single crystalline PTCDA studied by spectroscopic ellipsometry. Copyright 2002 with permission from Elsevier.

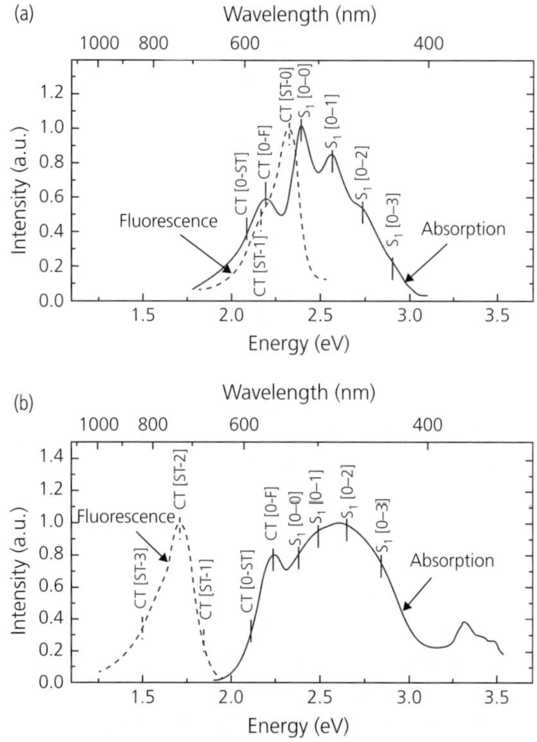

Figure 3.58 (a) Solution and (b) 100 nm thick film absorption and emission (fluorescence) spectra of PTCDA. The dilute solution consists of 2 μM PTCDA dissolved in DMSO (Bulovic et al., 1996).

Ionic materials can exhibit stable, charge transfer-like properties due to their electron distributions localized among neighboring atomic cores in the lattice.

3.6.6 Charge transfer states

Typically, the CT state features are strongest at the low energy limb of the Frenkel exciton absorption and emission spectra in tightly stacked molecular solids. Furthermore, the large dielectric anisotropies characteristic of organics can result in the *mixing* of both Frenkel and CT properties across the entire spectral feature. That is, the CT state can extend along the π-stacking direction where the molecular orbitals significantly overlap, whereas perpendicular to the stacking direction the overlap is small, leading to nearly pure Frenkel-like behavior. The coexistence of these distinct excitonic states is particularly pronounced in the solution and thin film solid-state absorption and emission spectra of PTCDA shown in Fig. 3.58. In these spectra, S_1 is the lowest Frenkel state. Hence, in spectroscopic notation, the several transitions correspond to $S_1 \leftrightarrow S_0$ and $CT \leftrightarrow S_0$. Notice that there is a significant shift (~0.6 eV) of the fluorescence spectrum in the solid state compared to that in solution. This is attributed to emission from the relaxed CT state that lies below the Frenkel exciton, but is absent in solutions where the density of aggregates (and hence the CT state that depends on the cooperative excitation of several neighboring molecules) is low (see Section 3.6.2).

These spectral features are understood from the energy diagram in Fig. 3.59. Two CT states are identified: CT^F is a free charge transfer state, whereas CT^{ST} corresponds to a self-trapped state due to local lattice polarization by the extended CT state (Song and Williams, 1993). No direct $S_1 \leftrightarrow S_0$ emission is observed since internal conversion transitions of $S_1 \rightarrow CT^F$ and $S_1 \rightarrow CT^{ST}$ are extremely rapid (~10 ns vs. 10 fs, the latter corresponding to the intramolecular phonon lifetime). In solids, energy stabilization in the vicinity of a lattice defect, impurity or free surface gives rise to self-trapping where the dielectric environment is locally distorted to immobilize the excited state (Craig et al., 1977). This is the proposed origin of the self-trapped CT exciton in PTCDA, whose energy is red-shifted by ~0.15 eV from that of the free, or mobile CT^F exciton.

The solution spectrum shows clear evidence for a Frenkel-like vibronic progression, which is obscured in the thin film spectrum due to line broadening from interactions with the intermolecular vibronic modes. While the solution is very dilute, there is, nevertheless, a sufficient concentration of aggregates to allow for easily detectable absorption at energies below the Frenkel manifold, that is, between 1.7 eV and 2.2 eV. However, the strength of CT vs. Frenkel absorption is less than in the solid state.

The CT states can be unambiguously identified by their absorption strength as a function of molecular concentration in the solvent. That is, since the CT states result from aggregation, we expect the absorption intensity to decrease superlinearly with dilution, compared to the linear concentration dependence of monomer states. In Fig. 3.60 we show the PTCDA absorption spectra as a function of concentration in a DMSO solution. The 2.25 eV peak intensity (i.e. its oscillator strength, f, relative to the $S_1(0\text{–}0)$ peak at a concentration of $[C_{PTCDA}] = 2\ \mu M$) falls off superlinearly with concentration, and indeed completely vanishes at $[C_{PTCDA}] < 0.2\ \mu M$, whereas the decrease in the higher energy Frenkel peak intensities between 2.3 and 2.8 eV are apparent at even the lowest molarity. Since the absorption intensity depends on the aggregate density, it must decrease with dilution until the solution becomes unsaturated, at which point neither aggregation nor their associated CT states exist. On the other hand, the monomer intensity should scale approximately linearly with $[C_{PTCDA}]$. However, the spectra in Fig. 3.60 show that the monomer concentration follows $[C_M] \sim [C_{PTCDA}]^{0.64}$ whereas at lower energies there is an aggregate concentration dependence of $[C_A] \sim [C_{PTCDA}]^{1.75}$. The sublinear dependence of the monomer intensity suggests that residual aggregates are present in even the most dilute solutions, and that S_1 and CT are mixed across the entire PTCDA absorption spectrum.

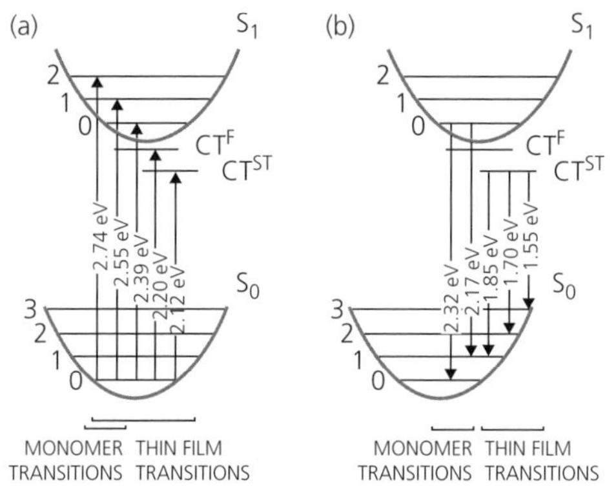

Figure 3.59 Frenkel (monomer) and thin film CT transitions identified from the PTCDA spectra in Fig. 3.58 (Bulovic et al., 1996).

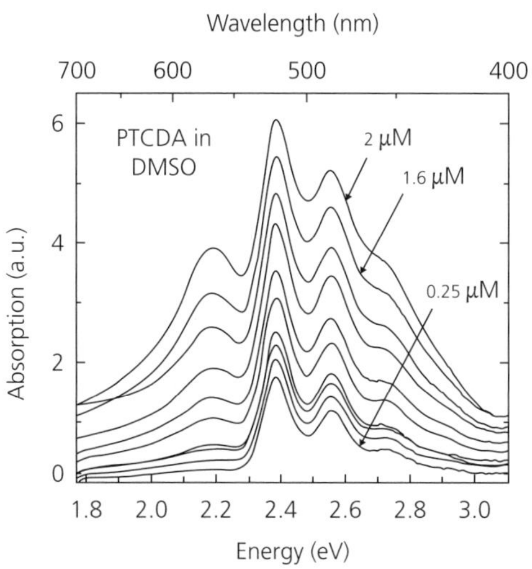

Figure 3.60 Spectrum of a dilute solution of PTCDA in DMSO as a function of concentration (Bulovic et al., 1996).

Charge transfer and Frenkel exciton mixing has been theoretically predicted for PTCDA, as shown in Figs. 3.53 and 3.61 (Hoffmann and Soos, 2002). The model considers all observed excited states, that are represented by the vertical bars in Fig. 3.61b and c. It accurately reproduces the positions of the measured absorption peaks in the spectrum in Fig. 3.61a that

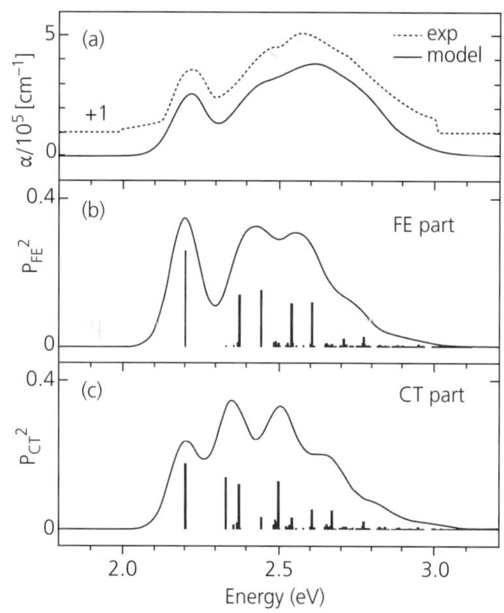

Figure 3.61 (a) Calculated and measured solid-state spectra of PTCDA. (b) Transition spectral weights due to Frenkel (FE) and (c) charge transfer (CT) excitons in the absorption spectrum. P_{FE}^2 and P_{CT}^2 are the corresponding transition moments. Vertical bars in (b) and (c) correspond to the eigenstate positions and oscillator strengths contributing to the spectra before including line broadening. From Hoffmann and Soos (2002).

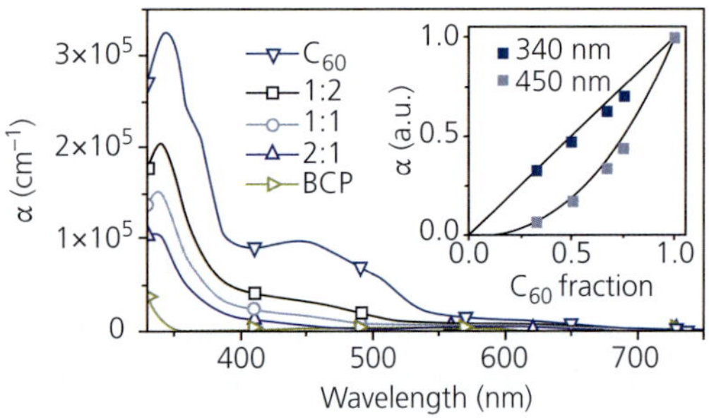

Figure 3.62 Absorption spectra of C_{60} mixed at varying dilutions in a wide energy gap bathocuproine (BCP) host. The ratios shown are X:Y (by volume) of BCP:C_{60}. Inset: Normalized absorption peak intensities at the two wavelengths indicated, with the peak at 450 nm due to CT state absorption, and at 340 nm to Frenkel exciton absorption (Bartynski et al., 2013).

includes line broadening due to vibronic coupling, excited state lifetime, and other extrinsic sources. The closely coupled molecular orbitals within the PTCDA stacks result in an admixture of both CT and Frenkel states across the observed spectra.

A second example of a strong spectral contribution from CT states is C_{60}, whose spherical symmetry allows for a tightly spaced hexagonal close-packed lattice. The low energy tail of the C_{60} absorption spectrum is dominated by a CT state centered at a wavelength of $\lambda \sim 450$ nm (Kazaoui et al., 1998). In Fig. 3.62, we show the absorption spectra of films of C_{60} diluted in a wide energy gap matrix of bathocuproine (BCP) (Bartynski et al., 2013). As in the case of the solution spectra of PTCDA, the CT peak intensity falls off rapidly with decreasing C_{60} concentration (as $[C_{60}]^{2.7}$), whereas the monomer peak at $\lambda = 340$ nm follows a linear dependence on $[C_{60}]$, as expected from Beer's law. DFT calculations show that the CT and Frenkel excitons are not entirely independent, which is similar to the situation with PTCDA. Indeed, the entire C_{60} spectrum contains contributions from both localized and extended molecular states (Kazaoui et al., 1998).

3.6.7 Solvatochromism and dielectric effects

We have seen that the dielectric environment can play a significant role on the excited state energies. This is especially true for molecules whose excited states and extended π-electron systems interact with neighboring molecules. The spectral shifts that result are the origin of the phenomenon broadly referred to as *solvatochromism*, which is most pronounced for dipolar molecules in dipolar solvents where intermolecular interactions are strongest (Reichardt, 2003). The

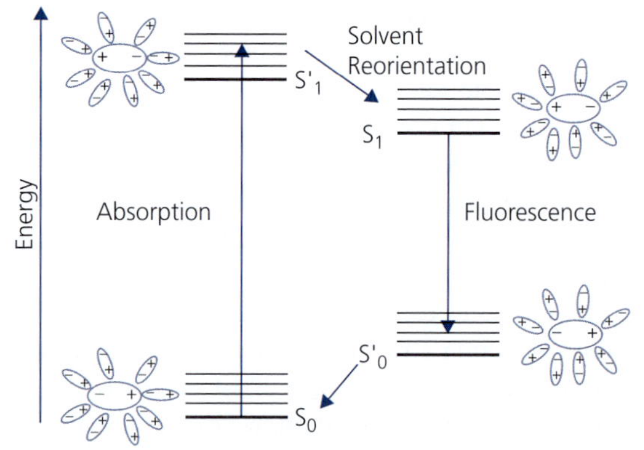

Figure 3.63 The origin and spectral consequences of solvatochromism. Shown adjacent to each state is a dipolar fluorescent molecule (large ovals) surrounded by solvent molecules (small ovals) whose dipoles orient to respond to that of the solute in both the ground and excited states. The lines above the lowest state energies represent the vibronic states (Reichardt, 2003).

solvent can be in either the liquid or solid phase. While the treatment of solvatochromism has been widely studied in liquid solution, we are ultimately concerned with energy shifts due to self-polarization (or self-trapping as discussed in Section 3.6.6) from solvation in the solid state due to its relevance to electronic devices.

Addition of a dipole into a dielectric polarizes the surrounding medium that either lowers the energy of the molecule, thereby causing a red shift in its emission that is added to the Franck–Condon energy, or less frequently it raises the dipole energy inducing a blue shift. In Fig. 3.63 we show the dipolar orientation of solvent molecules surrounding a polar solute in both the ground and first excited states. This reorientation lowers the total system energy, thereby stabilizing the molecular states. The red shifted emission is called *positive solvatochromism*, and the shift to lower energy is known as a *bathochromic shift*. A blue, or *hypsochromic shift*, is known as *negative solvatochromism*.

Three different situations can occur in solution (and analogously, in solids). (i) Both the solvent and the solute are nonpolar, in which case the solvatochromatic energy stabilization is due to weak van der Waals or London forces. Their very short range ($\sim 1/r^6$) only contributes minor shifts in the absorption and emission spectra of the solute molecule. (ii) One of the species (solvent or solute) has a fixed ground state dipole moment. Then the forces between solvent and solute are due to longer range fixed dipole-induced dipole forces ($\sim 1/r^3$). And finally, (iii) a dipolar solute is in a polar solvent. This is the situation in Fig. 3.63 where the solvent molecules form a "cage" around the solute that screens the electric field, stabilizing both solvent and solute. If the solute dipole moment in the excited state increases relative to the ground state, the excited state is more highly stabilized in the solvent resulting in a bathochromic shift of the absorption spectrum. Alternatively, if the excited dipole moment is reduced, a hypsochromic shift ensues.

The more polar the solvent, the larger the spectral shift, as shown in Fig. 3.64 for the dipolar, red fluorophore DCM2 in different solvents. A fluorescence emission shift of ~0.5 eV is observed between dilution in the nonpolar benzene to self-solvation in its highly polar thin film (with a dipole moment of $\mu = 11.3$ D). From these data, it is apparent that solvent shifts can occur both in liquids and solids, and have a common

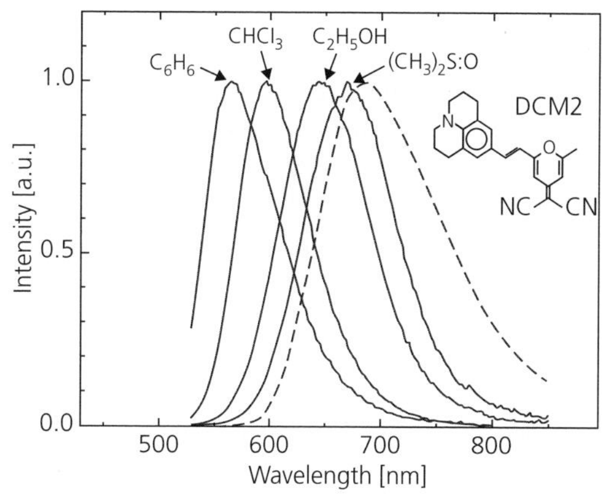

Figure 3.64 Fluorescence spectrum of DCM2 (shown in inset) dissolved in increasingly polar solvents, from the nonpolar benzene (C_6H_6 with a dipole moment of $\mu = 0$ D) to 1.15 D for $CHCl_3$ (chloroform), 1.69 D for C_2H_5OH (ethanol), to 3.9 D for $(CH_3)_2S{:}O$ (DMSO). Fluorescence from a neat thin film of DCM2 is also shown (dashed line, $\mu = 11.3$ D) (Bulovic et al., 1998).

Reprinted from Chem. Phys. Lett., 287, 455 Bulovic, V., Shoustikov, A., Baldo, M. A., Bose, E., Kozlov, V. G., Thompson, M. E. & Forrest, S. R., Bright, saturated red-to-yellow organic light-emitting devices based on polarization-induced spectral shifts copyright 1998 with permission from Elsevier.

origin—stabilization of the excited or ground state molecules within the dielectric environment. Of course, the solvent dipoles can fully reorient in liquids, whereas only induced dipoles can reorient in the solid state.

Molecular reorientation due to lattice polarization occurs slowly compared to the time scale required for absorption, and hence the Franck–Condon principle can be applied to analyze solvent effects. On the other hand, the time between absorption and emission is generally sufficiently long to allow for dielectric relaxation due to electron-phonon interactions within the molecule itself. Phonon lifetimes result in spectral shifts on times scales ranging from 10^{-13} to 10^{-11} s.

We model solvation effects by considering the *reaction field* created by the dipoles of the solvent responding to that of the solute, forming the solvent cage in Fig. 3.65a and b. If the solute molecule is removed from the center of the cage, we are left with a reaction field, $\mathbf{F}_R$ resulting from the solvent molecular alignment (Fig. 3.65c). This field opposes the solute dipole field, consequently enhancing both the permanent and induced dipoles, leading to a spectral energy shift. To determine the magnitude of the shifts, we must calculate $\mathbf{F}_R$.

Onsager calculated the reaction field using the *continuum approximation*, whereby the dipole is placed in a homogeneous nonpolar dielectric medium (Onsager, 1936). The total electric dipole moment in an electric field, $\mathbf{F}_0$, is the sum of the permanent and induced dipole moments ($\alpha\mathbf{F}$):

$$\mathbf{p}_T = p_M\hat{\mathbf{r}} + \alpha\mathbf{F}_0, \tag{3.143}$$

where $\mathbf{p}_T$ is the dipole moment of the solute, p_M is that of the solute in vacuum, α is the polarizability of the medium, and $\hat{\mathbf{r}}$ is a unit vector along the direction of

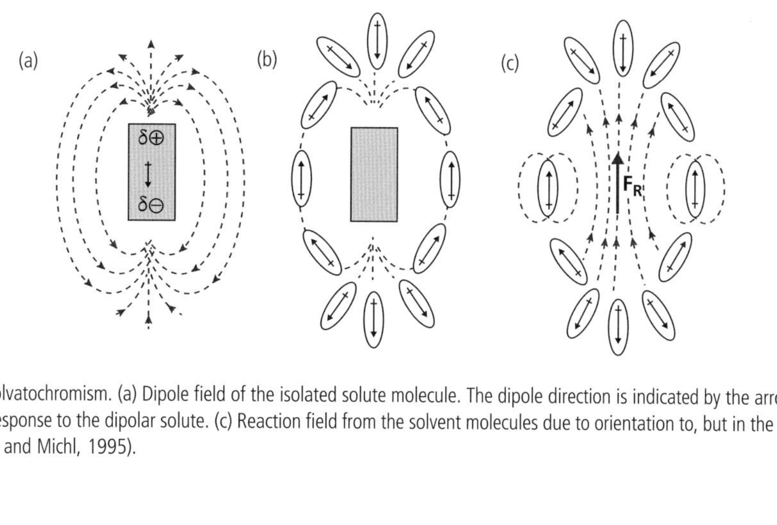

Figure 3.65 Origin of solvatochromism. (a) Dipole field of the isolated solute molecule. The dipole direction is indicated by the arrow. (b) Orientation of the solvent molecules in response to the dipolar solute. (c) Reaction field from the solvent molecules due to orientation to, but in the absence of the solute molecule itself (Klessinger and Michl, 1995).

the dipole moment. From Debye's expression for the dielectric constant (ε_r), the reaction field is

$$\mathbf{F}_\mathrm{R} = -\frac{2(\varepsilon_r - 1)}{2\varepsilon_r + 1}\frac{\mathbf{p}_T}{a^3}, \quad (3.144)$$

where the molecule is assumed to be spherical, with radius a. For an induced dipole (see Section 2.4.3), the energy is $E = -\frac{1}{2}\mathbf{p}\cdot\mathbf{F}$, giving a solvatochromatic energy shift of

$$\Delta E_{SOL} = -\frac{1}{2}\frac{2(\varepsilon_r - 1)}{2\varepsilon_r + 1}\frac{p_T^2}{a^3}. \quad (3.145)$$

This treatment ignores changes in $\mathbf{p}$ from the ground to the excited states, and it assumes that the dipole in solution is free to align with $\mathbf{F}_\mathrm{R}$.

For polar solvents, we must also add to the energy contributions due to reorientation of the dipoles in both the solvent and solute. When these are included we obtain (Karelson and Zerner, 1992)

$$\Delta E_{SOL} = -\frac{1}{2a}\left[\frac{2(\varepsilon_r - 1)}{2\varepsilon_r + 1}(\mathbf{p}_G - \mathbf{p}_E)\cdot\mathbf{p}_G + \frac{2(n_r^2 - 1)}{2n_r^2 + 1}(\mathbf{p}_G - \mathbf{p}_E)\cdot\mathbf{p}_E\right], \quad (3.146)$$

where $\mathbf{p}_\mathrm{G}$ ($\mathbf{p}_\mathrm{E}$) is the ground (excited) state dipole moment of the solute molecule. Also, n_r is the index of refraction of the solvent ($\varepsilon_r = n_r^2$ at optical frequencies, $\omega \to \infty$). This expression is reasonably accurate for predicting bathochromic and hypsochromic spectral shifts, depending on whether p_G is larger or less than p_E, respectively.

Solvatochromism in liquids using the continuum approximation has been rigorously analyzed by Liptay based on the Helmholtz free energy of the system, A_S, with results that can also be applied with some modification to the solid state (Liptay, 1974). The free energy is

$$A_\mathrm{S} = A_{sol}^0 + VN\sum_k w_{k\lambda}\{W_k^0 + W_{lk\lambda} + W_{Ck} + W_{Dk} + W_{Pk} - \mathbf{p}_k\cdot\mathbf{F}_{Rk} - \frac{1}{2}\alpha_k F_{Rk}^2\}. \quad (3.147)$$

Here, k refers to the vibrational and λ to the librational states of the solute molecule, V is the volume, and N is the number density. Librations are relative motions between molecules, excited by intermolecular phonons that are at a considerably lower energy (by more than 10 times) than the most energetic intramolecular phonon modes. This expression is instructive, since it decomposes the total energy of the solvent–solute system into each of its contributions. Thus, w_{kl} is the probability for the solute molecule to be in state (k, λ), A_{sol}^0 and W_k^0 are the energies of the solvent and isolated solute molecules, respectively, $W_{lk\lambda}$ is the change of molecular librational energy due to solvent-solute interactions, W_{Ck} is the work required to form the cage in Fig. 3.65b, W_{Dk} is the dispersion interaction energy between solute and solvent, and W_{pk} is the polarization energy due to the dipole in solution. The last two terms are the work required to place the dipole, $\mathbf{p}_k$, in solution, and the energy arising from the polarization of the surrounding solvent due to that dipole (see Eq. 3.143). The equilibrium energies of the solvent and solute do not contribute to the spectral shift, leaving

$$\Delta E_{SOL} = A_\mathrm{S} - A_{sol}^0 - VN\sum_k w_{k\lambda}\{W_k^0 + W_{lk\lambda}\}. \quad (3.148)$$

Each of the terms in Eq. 3.147 is model-dependent, and hence we will not explore their derivation further here. The reader is referred to the extensive literature for more details (see, for example, Liptay, 1974, Reichardt, 2003).

Given the extended and directional nature of CT states that involves at least two neighboring molecules, we need to consider the limits of the continuum approximation. That is, at what point does a system transition from one consisting of discrete molecules in a perhaps highly anisotropic crystal, to that of a homogeneous dielectric medium? This situation has been generally treated in anthracene (Bounds and Siebrand, 1980, Bounds et al., 1981), and subsequently for the archetype molecular solids of PTCDA and NTCDA (Shen and Forrest, 1997). While NTCDA is approximately isotropic, PTCDA is a highly anisotropic molecular crystal, hence providing an interesting contrast between these structures.

The dipole moment of the electron and hole comprising an exciton results in a polarization energy shift, $p_E(\mathbf{r})$, that is particularly pronounced for highly interacting electronic systems such as in closely stacked PTCDA. To calculate $p_E(\mathbf{r})$, consider a molecule at position, $\mathbf{r}(i,\mu)$, in a lattice with a basis whose unit cell volume is v. Here, $i = \{i_a, i_b, i_c\}$ is the lattice index, and $\mu = \{1, 2, 3...\}$ is the index of a particular basis molecule within the unit cell. In the presence of an external electric field, $\mathbf{F}_o$, the total field at site (i, μ) is

$$\mathbf{F}_{tot}(i,\mu) = \mathbf{F}_o(i,\mu) + F_{ind}(i,\mu), \quad (3.149)$$

where $\mathbf{F}_{ind}(i,\mu)$ is the sum of the fields due to the dipole at the same site (i.e. the self-field, $\mathbf{F}_{self}(i,\mu)$), and that due to the surrounding dipoles. Thus, we have

$$\mathbf{F}_{ind}(i,\mu) = \mathbf{F}_{self}(i,\mu) + \sum_{k,\nu\neq i,\mu}^{N,m} \overleftrightarrow{t}(i,\mu;k,\nu)\mathbf{p}_{ind}(k,\nu), \tag{3.150}$$

where, as in Section 3.6.4, N is the number of unit cells and m is the number of molecules in the basis. Also, $\overleftrightarrow{t}$ is the dipole tensor (Jackson, 1998):

$$\overleftrightarrow{t}(i,\mu;k,\nu) = \lim_{r\rightarrow r(k,\nu)} \nabla^2 \left| \frac{1}{\mathbf{r}(i,\mu) - \mathbf{r}} \right|. \tag{3.151}$$

This is the equivalent of the reaction field in solution. Assuming a spherical dipole for homogenous and isotropic media (which is different from the case of anisotropic molecular crystals), we can approximate $\mathbf{F}_{self}(i,\mu) = -\mathbf{p}_{ind}(i,\mu)/3\nu\varepsilon_o$ (Jackson, 1998). Then the induced dipole moment in the solid is (cf. Eq. 3.143)

$$\mathbf{p}_{ind}(i,\mu) = \overleftrightarrow{\alpha}(\mu)\mathbf{F}(i,\mu), \tag{3.152}$$

where $\overleftrightarrow{\alpha}(\mu)$ is the effective polarizability tensor of the molecule at μ, and $\mathbf{F}_{loc}(i,\mu)$ is the local (average) field at site (i,μ) excluding the induced field due to the molecule itself. Hence,

$$\mathbf{F}_{loc}(i,\mu) = \mathbf{F}_{tot}(i,\mu) - \mathbf{F}_{self}(i,\mu). \tag{3.153}$$

Finally, we arrive at the induced dipole moment at (i,μ) due to a reaction from dipoles at (k,ν) in an external field:

$$\mathbf{p}_{ind}(i,\mu) = \overleftrightarrow{\alpha}(\mu)\left[\mathbf{F}_0(i,\mu) + \sum_{k,\nu\neq i,\mu}^{N,m} \overleftrightarrow{t}(i,\mu;k,\nu)\mathbf{p}_{ind}(k,\nu)\right] \tag{3.154}$$

The total crystal polarization energy is found by summing over all individual dipoles to obtain (Shen and Forrest, 1997)

$$p_E(k,\nu) = -\frac{1}{2}\sum_{i,\mu}^{N,m}{}' \mathbf{p}_{ind}(i,\mu)\cdot\mathbf{F}_0(i,\mu). \tag{3.155}$$

Thus, the total electron–hole interaction energy for the exciton is the sum of the unscreened Coulomb attraction and the crystal polarization energy resulting from the local lattice response from the charge pair, via

$$E_{int} = -\frac{q^2}{4\pi\varepsilon_0|\mathbf{r}(i,\mu) - \mathbf{r}(k,\nu)|} + p_E(k,\nu). \tag{3.156}$$

The relative dielectric constant of the medium arises from the crystal polarization given by the second term in Eq. 3.156.

The energy, E_{int}, is calculated for NTCDA and PTCDA using known dielectric tensor elements, with the results shown in Fig. 3.66 (Shen and Forrest, 1997). These results are compared with the "screened" Coulomb potential,

$$E_{int} = -\frac{q^2}{4\pi\varepsilon_0 \overleftrightarrow{\varepsilon} r} \tag{3.157}$$

where $\overleftrightarrow{\varepsilon}$ is the *macroscopic* dielectric tensor for the crystal. Since diagonal elements of the NTCDA dielectric tensor are approximately equal (characteristic of a nearly isotropic herringbone structure, see Fig. 2.18b), the electron–hole interaction energy is comparable to that obtained with a single, spatially averaged dielectric constant, ε_{avg}, as shown by the solid line in Fig. 3.66 (top), along with the calculation using the actual dielectric constants from Eq. 3.157 along the **a**, **b**, and **c** crystal directions. The separate points correspond to interaction energies due to CT states where the electron is located on discrete molecular lattice sites at different distances from the hole to which it is Coulombically

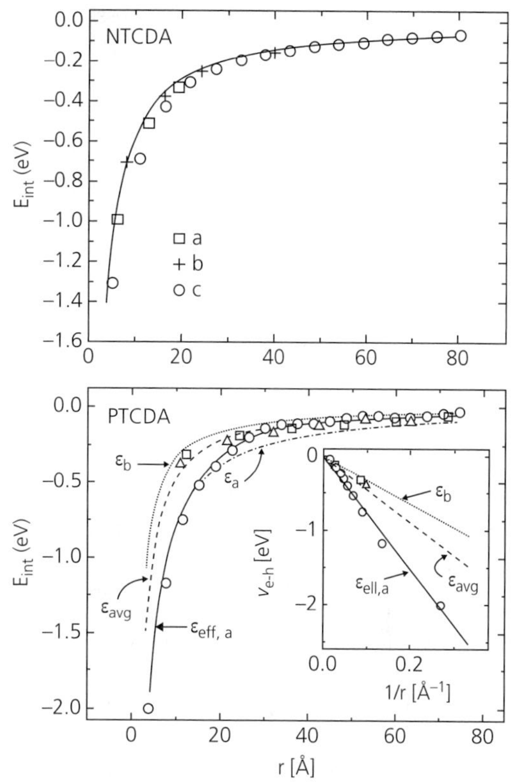

Figure 3.66 Calculated interaction energies for two molecular crystals. NTCDA is a herringbone structure with nearly isotropic dielectric properties, evidenced by the similarities in interaction energies of the CT state (E_{int}) along the **a**, **b**, and **c** lattice directions. The solid line corresponds to an isotropic crystal with a single average dielectric constant, ε_{avg}. PTCDA is a planar stacking crystal with large dielectric anisotropies. The dashed line corresponds to the spatial average, and the lower solid line shows a phenomenological fit to an "effective" isotropic dielectric constant, $\varepsilon_{eff,a}$. Inset: Data replotted vs. $1/r$ (Shen and Forrest, 1997).

bound. Clearly, $E_{int} = -\frac{q^2}{4\pi\varepsilon_0\varepsilon_{avg}r}$, characteristic of a nearly spherical Wannier–Mott-like state provides a good fit to the calculated values along all crystalline directions, and even for excitons that sample only nearest neighbors (i.e. for $r \sim 4$ Å). Indeed, we can consider the interaction energy to follow the Bohr progression of $E_B \sim 1/n^2$ (Eq. 3.141) where $n \geq 1$ is the principal quantum number of a nearly spherical CT state whose electron and hole are separated by increasing, but discrete distances since each is located on a distinct molecule at site i (Bounds and Siebrand, 1980).

For anisotropic PTCDA, however, E_{int} depends sensitively on the crystal direction for $r < 20$ Å. The dashed curve in Fig. 3.66 is a plot of $E_{int} = -\frac{q^2}{4\pi\varepsilon_0\varepsilon_{avg}r}$, where the dielectric constants along the three axes are averaged following $\varepsilon_{avg} = 3/(1/\varepsilon_\perp + 2/\varepsilon_\parallel) = 3.2$ (Shen and Forrest, 1997). Here, $\varepsilon_\perp$ is the dielectric constant along the (102) π-stacking axis, and $\varepsilon_\parallel$ is along the more loosely packed, in-plane directions. The "isotropic approximation" results in reasonable fits to E_{int} only for large r (>25 Å), suggesting that the potential becomes spatially averaged at electron–hole separations of several molecular lattice constants ($\geq$5) from the origin. However, for small r (<15 Å), E_{int} noticeably deviates from the spatial average. This deviation is more apparent in the inset of Fig. 3.66, where E_{int} is plotted vs. $1/r$ to accentuate variances with the isotropic approximation as $r \to 0$. For small r, the mean dielectric constant can be approximated using $E_{int} = -\frac{q^2}{4\pi\varepsilon_0\varepsilon_i r}$, where i refers to the three crystal lattice directions.

The foregoing results are, in some respects, similar to the case of conventional semiconductors where the crystal lattice is regarded as a macroscopic polarizable medium with spatially averaged dielectric constants, as long as the electron–hole distance is large compared to the lattice constant. Fundamentally, an electron in the LUMO as well as a hole in the HOMO in molecular crystals are accompanied by significant electronic polarization of the surrounding medium. This suggests that CT excitons with radii of only a very few lattice constants sample a sufficiently large volume to respond as if the medium were macroscopic, even though at that scale individual molecules exist in an environment that is considerably more discrete than found in inorganic materials.

An example of solvatochromism in organic light emitting diodes (OLEDs) whose emissive region comprises solid binary mixtures (called the *solid state solvation effect*, SSSE) of the fluorophores DCM2 and Alq_3

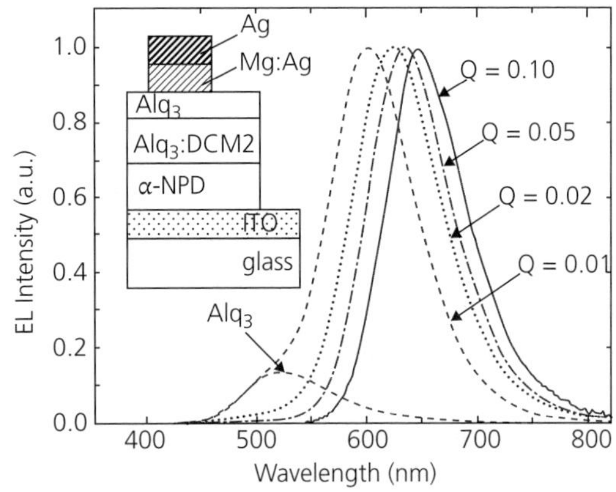

Figure 3.67 Solid state solvation effect observed in the electroluminescence spectrum of an OLED with different concentrations (Q) of the red emitting fluorophore, DCM2, in Alq_3. Here, ITO is indium tin oxide, and the hole transport layer is α-NPD. As the DCM2 concentration increases from 0.01 to 0.10, the mixture becomes increasingly dipolar, resulting in a pronounced red shift of the emission peak from 575 nm to 650 nm. The neat Alq_3 spectrum is shown for reference. (Bulovic et al., 1998)

is shown in Fig. 3.67. A rigid bathochromic shift of the emission peak from $\lambda = 575$ nm to 650 nm (corresponding to an energy shift of 0.26 eV) is observed as the concentration is increased from 1% to 10% DCM2. The DCM2 molecule has a dipole moment of 11 D, compared with its host, Alq_3, at 5.5 D. Hence, the average dipole moment of the solid solution increases with DCM2 concentration, resulting in the shift in the OLED emission spectrum. Bathochromic shifts are often attributed to excimer formation (Tang et al., 1989). However, the shift is not accompanied by spectral broadening. Furthermore, similar spectral changes are observed for DCM2 in dipolar solutions in Fig. 3.64. This is evidence that the shifts are due to solvation effects in the solid, binary mixture (Bulovic et al., 1998). We will show in Section 6.3.2 that the SSSE is an effective means for controlling the color emission from OLEDs.

The above analysis suggests that the solvatochromatic energy shift linearly increases with the dipole field (cf. Eqs. 3.155 and 3.156). This, in turn, should increase with the dipole density. The change in relative dielectric constant, $\Delta\varepsilon_r$, is expected to depend on the change in dipole (i.e. fluorophore) density, ΔN, due to the applied pressure, P. This approach eliminates red shifts due to aggregation or other extrinsic effects that may accompany increasing concentrations of the dipolar fluorophore in a host matrix.

Relating the density to the uniaxial pressure, it can be shown that (Chang et al., 2015)

$$\frac{\Delta\varepsilon_r}{\varepsilon_r - 1} = \frac{\Delta N}{N_0} \approx \frac{1-2\nu}{Y}P = aP, \tag{3.158}$$

where N_0 is the equilibrium dipole density, ν is Poisson's ratio (Chapter 2), and α is a constant. Also, Y is *Young's modulus*, which is the ratio of the stress to the strain, that is, it relates the elongation or compression of a solid to the amount of pressure applied. Thus,

$$Y = \frac{\mathcal{F}a_0}{A\Delta a}, \tag{3.159}$$

where $\mathcal{F}$ is the force applied, Δa is the strain (i.e. its change from its equilibrium lattice constant, a_0), and A is the sample area.

The pressure dependence of solvatochromism in DCM2 doped at 2 vol.% in Alq_3 was investigated using a micrometer head that was advanced to apply pressure to the doped film while measuring the peak emission wavelength, spectral width and radiative lifetime of DCM2. As shown in Fig. 3.68, the peak red shifts with the change in pressure from 0 to 0.4 GPa (and hence the DCM2 density). Recall from our discussion of dimers in Section 3.6.2 that aggregation results in a broadened and featureless emission spectrum, and it can reduce the PL quantum yield due to strong coupling to intermolecular vibrational modes. Since neither the peak width nor its radiative lifetime show significant changes with pressure, we conclude that molecular aggregation effects are absent, and the spectral shifts are due entirely to increased dipole coupling over the range of pressures investigated.

The dashed line in Fig. 3.68 is a fit to the data following Eq. 3.158, with a pressure coefficient of $\alpha = 1.1$ GPa^{-1}. This is at least a factor of 2–4 higher than predicted using measured values of ν and Y. The discrepancy is attributed to simplifications leading to Eq. 3.158, as well as to errors in estimating the pressure in the experiments (Chang et al., 2015). Nevertheless, the pressure-dependent peak shifts illustrate the importance of dipole density, and hence the local dipole-induced electric field in influencing the energetics of organic semiconductors.

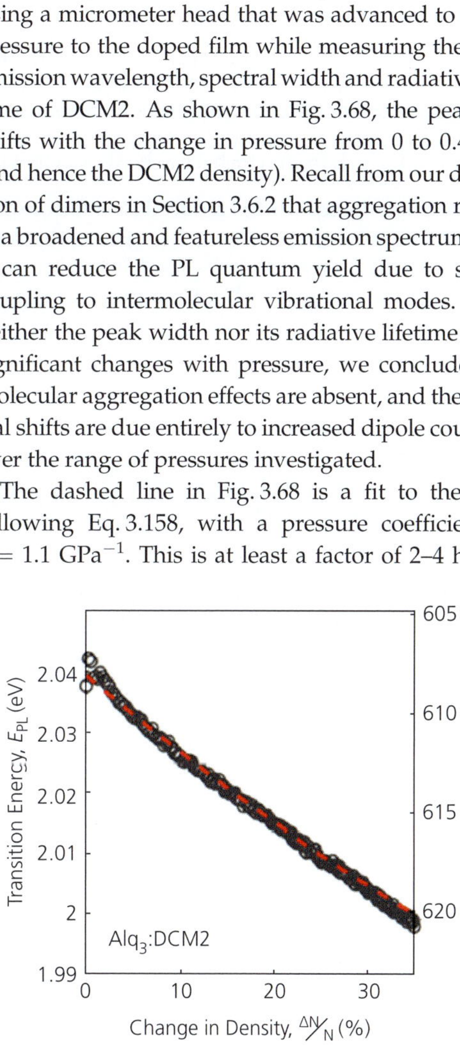

Figure 3.68 DCM2 peak energy vs. the change in molecular density, $\Delta N/N$, which is linearly proportional to the pressure applied. Data points are overlaid with a phenomenological model (dashed line) described in the text (Chang et al., 2015).

Reprinted with permission from ACS Nano, 9, 4412-4418, Chang, W., Akselrod, G. M. & Bulović, V., Solid-State Solvation and Enhanced Exciton Diffusion in Doped Organic Thin Films under Mechanical Pressure. Copyright 2015 American Chemical Society.

3.7 Spin

In Section 3.1 we noted that excitonic states can have either an odd or even symmetry under spin exchange. This results in either a singlet (with spin total quantum number $S = 0$) or a triplet ($S = 1$) quantum state. The triplet and singlet wavefunctions are provided in Eqs. 3.2 and 3.3, respectively. As discussed, the Pauli Exclusion Principle requires that the total spinorbital wavefunction, $\Phi^{\sigma}(\mathbf{r})$, must be asymmetric under electron exchange. Since the singlet state is antisymmetric, its spatial (orbital) wavefunction is symmetric (Eq. 3.2), and for the symmetric triplet states, the spatial wavefunction is antisymmetric (Eq. 3.3). Now, the dominant optical transition between states is due to electric dipole interactions whose Hamiltonian contains the dipole operator $\boldsymbol{\mu}_e = -q\mathbf{r}$. Hence, the expectation value of electronic transitions is determined by the integral:

$$\langle \psi_f(\mathbf{r}_1, \mathbf{r}_2; S, m_S)|H_{int}|\psi_i(\mathbf{r}_1, \mathbf{r}_2; S, m_S)\rangle$$
$$\backsim \langle \psi_f(\mathbf{r}_1, \mathbf{r}_2; \mathrm{S}, \mathrm{m_S})|\mathbf{r}|\psi_i(\mathbf{r}_1, \mathbf{r}_2; \mathrm{S}, \mathrm{m_S})\rangle. \tag{3.160}$$

Since the operator symmetry is odd, allowed transitions can only occur between states whose spatial symmetry is different (this is known as the *Laporte selection rule*, Table 3.1). That is, there must be a change in parity between the initial and final states. However, since the dipole operator carries no spin-related terms (it operates only on the spatial wavefunction), the initial spin states are conserved in dipole transitions. Since the ground state is typically a singlet (S_0), this implies that the dominant transition from the ground state is to a singlet excited state (e.g. $S_1 \leftarrow S_0$). Transitions between triplet states (e.g. $T_3 \leftarrow T_1$) are also dipole allowed.

Other factors can lead to otherwise spin-forbidden transitions. Most important are the disallowed triplet-singlet transitions that lead to phosphorescence in organic heavy metal complexes. According to Fig. 3.33, $T \leftrightarrow S$ transitions can arise from ISC between spin states of different symmetry. Once such a transition occurs in the excited state, rapid and radiative fluorescent de-excitation to the ground state is observed. The opposite transition, if it occurs on a time scale shorter than the fluorescence lifetime (~1–10 ns), can result in phosphorescent emission.

3.7.1 Spin–orbit coupling

Violation of the spin selection rule can occur when the triplet and singlet states are mixed. An effective source of mixing is *spin–orbit coupling*, whereby the spin of the electron is perturbed by interacting with its orbital angular momentum. Spin–orbit coupling can be understood by calculating the field generated by a charge moving in an orbital. The *Biot–Savart law* gives the magnetic field in the reference frame of a charge of velocity, **v**, in an electric field, **F**, as follows:

$$\mathbf{B} = -\frac{\mathbf{v}\times\mathbf{F}}{c^2}. \tag{3.161}$$

Since the field of a point charge is isotropic, we can write **F** in terms of the radial potential, V:

$$\mathbf{F}(\mathbf{r}) = -\hat{\mathbf{r}}\frac{dV(r)}{dr}, \tag{3.162}$$

from which we obtain:

$$\mathbf{B} = \mathbf{v}\times\frac{\hat{\mathbf{r}}}{c^2}\frac{dV(r)}{dr}. \tag{3.163}$$

Now, the angular momentum of the charge of mass, m_e, is

$$\mathbf{l} = \mathbf{r}\times m_e\mathbf{v}, \tag{3.164}$$

which leads to

$$\mathbf{B} = \frac{1}{m_e r c^2}\frac{dV(r)}{dr}\mathbf{l}. \tag{3.165}$$

Hence, an electron in an orbital of radius, r, produces a magnetic field whose strength is proportional to the gradient of the electrostatic potential. This field perturbs the spin whose magnetic moment is given by

$$\mathbf{m}_s = -g_s\mu_B\frac{\mathbf{s}}{\hbar} = -g_s\frac{q}{2m_e}\mathbf{s}, \tag{3.166}$$

where $\mu_B = q\hbar/2m_e = 9.27\times10^{-24}$ J/T is the *Bohr magneton* and $g_S \approx 2$ is the electron spin *g-factor*. Also, it is helpful to define the *gyromagnetic ratio*, $\gamma_e = -\mu_B/\hbar = 1.76\times10^{11}\,\mathrm{s^{-1}\,T^{-1}}$ (also known as the *magnetogyric ratio*). The interaction energy of a magnetic field and a magnetic dipole is

$$E_{int} = -\mathbf{m}\cdot\mathbf{B}. \tag{3.167}$$

Combining Eqs. 3.163–3.165, we obtain the Hamiltonian for the spin–orbit interaction in the reference frame of the electron:

$$H_{SO} = -g_S\frac{q}{2m_e^2c^2}\frac{1}{r}\frac{dV(r)}{dr}\mathbf{l}\cdot\mathbf{s}. \tag{3.168}$$

Thus far, we have considered the field in the rest frame of the electron. However, Llewellyn Thomas recognized that in the rest frame of the stationary nucleus, relativistic effects need to be included. That is, the rate of precession of the spin in the trajectory of the electron of velocity, v, around the nucleus needs to be corrected by the Lorentz factor, $\gamma = (1 - v^2/c^2)^{-1/2}$ giving the *Thomas precession frequency*:

$$\Omega_T = \omega(\gamma - 1), \tag{3.169}$$

where $\omega = v/r$. To first order, the Lorentz factor results in $\Omega_T \to \omega\left(\frac{1}{2}\frac{v^2}{c^2}\right)$, from which it can be shown that the magnetic field in Eq. 3.161 is reduced by a factor of ½. That is, in the rest frame of the nucleus (Kroemer, 2004):

$$\mathbf{B} \to -\frac{\mathbf{v}\times\mathbf{F}}{2c^2}. \tag{3.170}$$

This introduces the well-known Thomas relativistic correction to the spin–orbit coupling leading to the following SO Hamiltonian:

$$H_{SO} = -\frac{q}{2m_e^2c^2}\frac{1}{r}\frac{dV(r)}{dr}\mathbf{l}\cdot\mathbf{s} = \xi(r)\mathbf{l}\cdot\mathbf{s}, \tag{3.171}$$

where we have used $g_S = 2$.

Our final task is to evaluate the spin–orbit coupling factor, $\xi(r)$. For a hyrdogenic atom with atomic number Z, $V(r) = -Zq/4\pi\varepsilon_0 r$, in which case Eq. 3.171 gives

$$\xi(r) = -\frac{Zq^2}{8\pi\varepsilon_0 m_e^2c^2}\frac{1}{r^3}. \tag{3.172}$$

The spin–orbit interaction energy is found by calculating the expectation value of H_{SO} (given by $\langle H_{SO}\rangle$) for the hydrogenic wavefunction: $|nlm_l\rangle = R_{nl}(r)Y_{lm_l}(\theta,\phi)$, where R_{nl}(r) is the radial part of the wavefunction, and $Y_{lm_l}(\theta,\phi)$ is the angular part given by the spherical harmonic functions for angular momentum quantum numbers l, m_l. Since H_{SO} is only a function of r, we simply need to calculate the expectation value $\langle 1/r^3\rangle$ via

$$\left\langle\frac{1}{r^3}\right\rangle = \langle nl|\frac{1}{r^3}|nl\rangle = \int_0^\infty \frac{1}{r}R_{nl}^2 dr = \frac{(Z/a_0)^3}{n^3 l(l+\frac{1}{2})(l+1)}, \tag{3.173}$$

where a_0 is the Bohr radius (Eq. 3.142). Hence, the spin–orbit interaction energy is:

$$E_{SO} = \langle H_{SO} \rangle = \frac{Z^4 q^2}{8\pi\varepsilon_0 a_0^3 m_e^2 c^2} \frac{1}{n^3 l(l+\frac{1}{2})(l+1)} \langle \mathbf{l} \cdot \mathbf{s} \rangle, \tag{3.174}$$

which in more compact form is expressed as

$$E_{SO} = \frac{2\pi \alpha^2 R_\infty Z^4 c}{n^3 l(l+\frac{1}{2})(l+1)\hbar} \langle \mathbf{l} \cdot \mathbf{s} \rangle, \tag{3.175}$$

where $\alpha = q^2/4\pi\varepsilon_0 \hbar c \approx 1/137$ is the *fine structure constant* and the *Rydberg constant* is $R_\infty = m_e q^4/64\pi^3 \varepsilon_0^2 \hbar^3 c \approx 1.097 \times 10^7\ \mathrm{m}^{-1}$.

The interaction scales as Z^4, hence we expect that spin–orbit interactions are far stronger in molecules containing an atom with a high atomic number. Thus, emission from Ir ($Z = 77$) and Pt ($Z = 78$) and other heavy metal-based organic complexes is entirely from their triplet states, even though the singlet state might initially be excited. That is, spin–orbit coupling in heavy-metal containing molecules promotes rapid non-radiative ISC from the first singlet to the triplet excited state, followed by triplet emission as the excited state relaxes back into the singlet ground state. This process leads to 100% internal quantum efficiency in phosphorescent OLEDs (Baldo et al., 1998, Adachi et al., 2001).

Finally, we must evaluate $\langle \mathbf{l} \cdot \mathbf{s} \rangle$ to obtain the spin–orbit interaction energy. Since the total angular momentum operator of a composite system with contributions from both orbital and spin momenta is

$$\mathbf{j} = \mathbf{l} + \mathbf{s}, \tag{3.176}$$

it follows that

$$j^2 = |\mathbf{l} + \mathbf{s}|^2 = l^2 + s^2 + 2\mathbf{l} \cdot \mathbf{s}. \tag{3.177}$$

Now, the orbital angular momentum operator gives the relationships $l^2|nlm\rangle = \hbar^2 l(l+1)|nlm\rangle$, with analogous expressions for j^2 and s^2, from which we get

$$E_{SO} = \frac{\alpha^2 R_\infty Z^4 hc[j(j+1) - l(l+1) - s(s+1)]}{2n^3 l(l+\frac{1}{2})(l+1)}. \tag{3.178}$$

As the principal quantum number (n) increases, spin–orbit coupling decreases, whereas it increases with l, as expected. That is, an s-state has $l = 0$ and $j = s$, from which it follows that $E_{SO} = 0$.

While Eqs. 3.175 and 3.178 provide the energy of the spin–orbit interaction, they do not explain how the multiplicity of the spin state changes due to this interaction. That is, we have yet to show that $S \leftrightarrow T$ transitions are allowed due to symmetry breaking via singlet-triplet mixing.

The $S \rightarrow T$ transition is non-radiative, implying that it is an isoenergetic transition from the relaxed singlet (i.e. from the $\nu = 0$ vibronic) to the triplet manifold where $\nu > 0$. Then, ISC is followed by rapid internal conversion to the $\nu = 0$ triplet state. To quantitatively treat ISC, we follow the analysis of McClure based on group theory (McClure, 1949). Using the Hamiltonian Eq. 3.171 for a system of $N > 1$ electrons in a central field:

$$H_{SO} = \sum_i^N \xi_i(r) \mathbf{l}_i \cdot \mathbf{s}_i. \tag{3.179}$$

Under exchange of two electrons, i and j ($i \neq j$), Eq. 3.179 becomes

$$H_{SO} = \sum_i^N \sum_j^N \left[\xi_i(r) \mathbf{l}_i \cdot \mathbf{s}_i + \xi_j(r) \mathbf{l}_j \cdot \mathbf{s}_j \right]. \tag{3.180}$$

This can be expanded to

$$H_{SO} = \frac{1}{2N} \sum_i^N \sum_j^N \left[\left(\xi_i(r)\mathbf{l}_i + \xi_j(r)\mathbf{l}_j \right) \cdot (\mathbf{s}_i + \mathbf{s}_j) + \left(\xi_i(r)\mathbf{l}_i - \xi_j(r)\mathbf{l}_j \right) \cdot (\mathbf{s}_i - \mathbf{s}_j) \right]. \tag{3.181}$$

The terms to the left correspond to states of even symmetry under exchange, and on the right to odd symmetry. The even symmetry term preserves the symmetry of the state on which it operates: if either its orbital or spin wavefunction is symmetric or antisymmetric under exchange, its symmetry will remain unchanged when operated on by this term. However, the second (odd) term results in a symmetry change: if the initial wavefunction has even spin symmetry (a triplet state), it will be transformed into an odd symmetry singlet. Furthermore, the orbital angular momentum undergoes a symmetry change due to the second term. While the first term preserves the symmetry of the initial and final states, it leads to multiplet splitting. That is, it lifts the degeneracy of the three $m_S = 0, \pm 1$ substates of the $S = 1$ manifold. This separation of the spin state energies in the absence of an external magnetic field is known as the zero-field splitting energy. In contrast, the second term leads directly to S–T mixing without a significant change in energy. The strength of mixing determines the rate of ISC, k_{ISC} (Fig. 3.33). In fact, it can be shown that the z-axis operators of Eq. 3.181 mix singlet and triplet states with $m_s = 0$, and the x and y operators mix S and T states with a change in quantum numbers of $\Delta m_S = \pm 1$ (McGlynn et al., 1969).

Now that we have shown that H_{SO} mixes the S and T states via simple symmetry arguments, we can calculate the radiative phosphorescence emission rate from the triplet state, k_{Ph}, using perturbation theory. From the Jablonski diagram (Fig. 3.33) we find several levels in both the singlet and triplet manifolds. For example, the triplet state is comprised of $T_1, T_2, \ldots T_N$, and similarly for singlets. Not shown are higher order spin states such as the $S = 2$ quintet manifold, where each quintet consists of five degenerate spin states corresponding to $m_S = 0, \pm 1, \pm 2$, and so on. Since we are only interested in the fundamental $S_0 \leftrightarrow T_1$ transition, the effects of quintets are ignored. Now, the triplet spinorbital perturbed by an admixture of singlet content is, to first order:

$$|T_1'^{m_S}\rangle = |T_1^{m_S}\rangle + \sum_i^N |S_i\rangle \frac{\langle S_i|H_{SO}|T_1^{m_S}\rangle}{E(T_1) - E(S_i)} + \sum_i^N \sum_{k=-1}^{1} |T_i^k\rangle \frac{\langle T_i^k|H_{SO}|T_1^{m_S}\rangle}{E(T_1) - E(T_i)}. \quad (3.182)$$

Here, $|T_1'^{m}\rangle$ is the perturbed spinorbital for the first triplet state (T_1') whose spin quantum number projection along the z-axis is denoted by m_S. We also note that all the triplets within a particular manifold are degenerate (i.e. the energies of the $m_S = 0, \pm 1$ states are equal). It follows that there are three $|T_1'^{m_S}\rangle$ that comprise the first triplet state manifold; these three substates are reflected by their contributions in the second summation over $k = -1, 0, 1$ in Eq. 3.182. Thus, the sums are over all N singlet and triplet states, although since the energy denominator increases for higher order excited states, in practice only the states nearest in energy are required.

Similarly, the perturbed S-state has only a single expression corresponding to $m_S = 0$:

$$|S_0'\rangle = |S_0\rangle + \sum_i^N \sum_{k=-1}^{1} |T_i^k\rangle \frac{\langle T_i^k|H_{SO}|S_0\rangle}{E(S_0) - E(T_i)}. \quad (3.183)$$

Since the energy gap between S_0 and all the triplet states is large, we can approximate Eq. 3.183 by: $|S_0'\rangle \approx |S_0\rangle$. Furthermore, it is often assumed that only one S-state contributes significantly to the $S_0 \leftrightarrow T_1$ transitions. State mixing leading to ISC is illustrated in Fig. 3.69. Denoting the dominant perturbing singlet as S_D, the second term on the right in Eq. 3.183 can be removed from the summation. Similarly, contributions to $|T_1'^{m_S}\rangle$ from other separate triplet states are small. Thus:

$$|T_1'^{m_S}\rangle \approx |T_1^{m_S}\rangle + |S_D\rangle \frac{\langle S_D|H_{SO}|T_1^{m_S}\rangle}{E(T_1) - E(S_D)}. \quad (3.184)$$

Generally, S_D can be replaced by S_1 since this approximation leads to the smallest energy denominator due to the larger separation between $E(T_1)$ and $E(S_i), i \neq 1$.

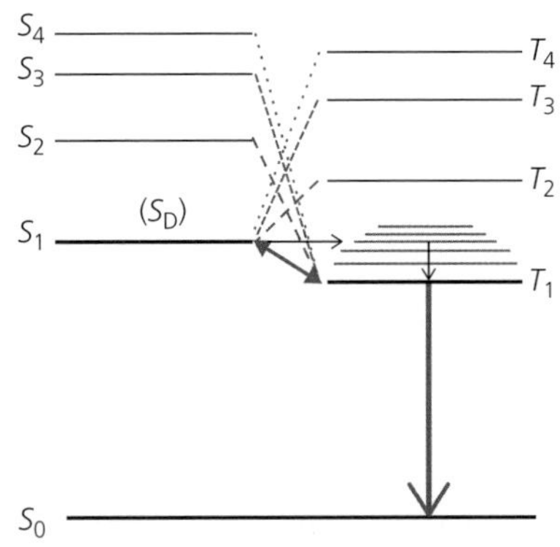

Figure 3.69 Representation of the admixtures of S_1 and T_1 leading to ISC and phosphorescence. The stronger interactions are shown as bolder dashed lines. Also, the dominant mixed state, $S_D = S_1$, and ISC from S_1 to T_1 followed by internal conversion is shown.

From perturbation theory, the energy, E'^{m_S}, of $|T_1'^{m_S}\rangle$ is stabilized by spin–orbit coupling from its unperturbed value, $E(T_1)$ as follows:

$$E'^{m_S} = E(T_1) + \sum_i^N \frac{|\langle S_i|H_{SO}|T_1^{m_S}\rangle|^2}{E(T_1) - E(S_i)} + \sum_i^N \sum_{k=-1}^{1} \frac{|\langle T_i^k|H_{SO}|T_1^{m_S}\rangle|^2}{E(T_1) - E(T_i)}, \quad (3.185)$$

where we assume that the energy splitting between m_S substates is negligible in the absence of spin–orbit coupling. Since the denominator is negative (see Fig. 3.69), the energy of the perturbed state is lower (hence stabilized) compared to the original state, T_1. As in Eq. 3.184, we note that generally only a single state exists that minimizes the denominator in this expression, leading to the simplified expression:

$$E'^{m_S} = E(T_1) + \frac{|\langle S_D|H_{SO}|T_1^{m_S}\rangle|^2}{E(T_1) - E(S_D)} \quad (3.186)$$

To calculate k_{Ph} we turn to Fermi's Golden Rule, Eq. 3.69. For electric dipole transitions from T_1 to S_0 (denoted $T_1 \leftarrow S_0$), we obtain a triplet decay rate between spin–orbit mixed states of:

$$k_{Ph} = \frac{8\pi\omega^3}{3hc^3} \sum_{k=-1}^{1} |\langle S_0|q\mathbf{r}|T_1'^k\rangle|^2 \quad (3.187)$$

In this expression, we have used the photon density of states for incident illumination at frequency, ω:

$$\rho(E_{fi}) = \frac{2\omega^3}{3\pi c^3}. \quad (3.188)$$

Substituting Eq. 3.184 into Eq. 3.187:

$$k_{Ph} = \frac{8\pi\omega^3}{3hc^3} \sum_{k=-1}^{1} \left| \frac{\langle S_D|H_{SO}|T_1'^k\rangle}{E(T_1) - E(S_D)} \right|^2 |\langle S_0|q\mathbf{r}|S_D\rangle|^2. \quad (3.189)$$

The same spin–orbit coupling process that leads to phosphorescent emission (e.g. $T_1 \to S_0$) also leads to ISC ($S_1 \leftarrow T_1$). We note that a calculation of k_{ISC} introduces additional complications since the transfer between manifolds must take non-radiative as well as radiative contributions into consideration. Still, application of Fermi's Golden Rule that accounts for these additional terms remains valid for both small molecules and polymers (Henry and Siebrand, 1971, Beljonne et al., 2001, Tatchen et al., 2007).

Including non-radiative contributions to k_{ISC} requires the full Born–Oppenheimer wavefunction that comprises both electronic as well as nuclear (i.e. vibronic) coordinates (see Eq. 3.8). Then the initial and final wavefunctions are: $|S_1(\mathbf{r}_i, \mathbf{R}_i)\rangle = \phi_e(\mathbf{r}_i, \mathbf{R}_i)\,\phi_N(\mathbf{R}_i)$ and $|T_f(\{\mathbf{r}_i\}, \{\mathbf{R}_i\})\rangle = \phi_e(\{\mathbf{r}_f\}, \{\mathbf{R}_f\})\phi_N(\mathbf{R}_f)$, respectively, from which it follows:

$$k_{ISC} = \frac{2\pi}{\hbar}\sum_{\{f\}}\left|\langle S_1(\mathbf{r}_i, \mathbf{R}_i)\middle|H_{SO}\middle|T_f(\{\mathbf{r}_f\}, \{\mathbf{R}_f\})\rangle\right|^2 \times\delta\left(E^0_{S_1} - E^0_{T_f}\right), \tag{3.190}$$

where the unperturbed singlet and triplet energies are provided in the Kronecker delta-function and $\{f\}$ is the set of all final states. We can reduce the nuclear coordinate terms to wavefunctions labeled by their vibronic quantum numbers, viz. $|\nu_k\rangle$, such that $\langle S_1(\mathbf{r}_i, \mathbf{R}_i)|H_{SO}|T_f(\{\mathbf{r}_f\}, \{\mathbf{R}_f\})\rangle = \langle S_1(\nu_i)|H_{SO}|T_f\{\nu_f\}\rangle$. We now expand each term in Eq. 3.190 about the equilibrium nuclear coordinate, $Q_0 = 0$, to obtain (Tatchen et al., 2007)

$$\langle S_1(\nu_i)|H_{SO}|T_f(\nu_f)\rangle = \langle S_1(\mathbf{r}_i)|H_{SO}|T_f(\mathbf{r}_f)\rangle|_{Q_0}\langle\nu_i|\nu_f\rangle + \sum_k\left(\frac{\partial}{\partial Q_k}\langle S_1(\mathbf{r}_i)|H_{SO}|T_f(\mathbf{r}_f)\rangle\right)_{Q_0}\times\langle\nu_i|Q_k|\nu_f\rangle + O(|Q^2|). \tag{3.191}$$

This is the Herzberg–Teller expansion of Eq. 3.78. Note that $FC_{if} = \langle\nu_i|\nu_f\rangle$ is the Franck–Condon factor of Eq. 3.79. The second term in the expansion is known as the *vibronic spin–orbit coupling factor*. Inserting the sum over all final vibronic and electronic states, $|T_f(\{\mathbf{r}_f\}, \{\mathbf{R}_f\})\rangle$ in Eq. 3.190 thus leads to the rate for ISC. This is clearly a much more involved calculation than that used to obtain k_{Ph} due to the vibronic couplings, although we see that the spin–orbit interaction is the same and is responsible for both processes, albeit at considerably different rates.

3.7.2 Spin–spin coupling

The magnetic field that arises from the spin of an electron interacts with the magnetic moment of a second spin, resulting in a spin–spin interaction energy of $E_{SS} = -\mathbf{m}\cdot\mathbf{B}$ (see Eq. 3.167). For singlet states, the two spins are 180° out of phase, and hence $E_{SS} = 0$. For triplets, the spins are in phase leading to an energy splitting even when the external field, $\mathbf{B} = 0$. This *zero-field splitting* (ZFS) of the triplet state is resolved in the fine structure of the molecular spectral features. Other interactions between the electron and nuclear spins (giving rise to *hyperfine interactions*), or between the nuclei themselves (nuclear *Overhauser effects*) are comparatively small and hence do not play a significant role in the spectra of molecular electronic materials.

Analogous to the electric dipole interactions in Eq. 2.39, magnetic dipole field due to a spin of angular momentum, $\mathbf{s}_1$, is given by (Jackson, 1998)

$$\mathbf{B}(\mathbf{r}) = -\left(\frac{gq\mu_B}{8\pi m_e r^3}\right)\left(\mathbf{s}_1 - \frac{3\mathbf{r}(\mathbf{r}\cdot\mathbf{s})}{r^2}\right) = \left(\frac{g\mu_B\gamma_e}{4\pi r^3}\right)\left(\mathbf{s}_1 - \frac{3\mathbf{r}(\mathbf{r}\cdot\mathbf{s})}{r^2}\right). \tag{3.192}$$

It follows that the Hamiltonian for the interaction of two spins summed over all electrons in the molecule is

$$H_{SS} = \left(\frac{g^2\mu_B\gamma_e^2}{4\pi}\right)\sum_{i>j}^{N}\left(\frac{\mathbf{s}_i\cdot\mathbf{s}_j}{r_{ij}^3} - \frac{3(\mathbf{r}_{ij}\cdot\mathbf{s}_i)(\mathbf{r}_{ij}\cdot\mathbf{s}_j)}{r_{ij}^5}\right), \tag{3.193}$$

where $\mathbf{r}_{ij} = \mathbf{r}_i - \mathbf{r}_j$ is the relative coordinate of spins $\mathbf{s}_i$ and $\mathbf{s}_j$. The energy splitting between states is calculated by diagonalizing the matrix consisting of the three elements $\langle S, m_S|H_{S-S}|S, m_S\rangle$ (corresponding to $S = 1$ and $m_S = 0, \pm1$). The ZFS energy terms for interacting electrons, $i\neq j$ are thus (McGlynn et al., 1969)

$$D_{ij} - \left(\frac{3}{4}\right)\frac{\mu_0 g^2\gamma_e^2}{4\pi}\langle\phi_e(i,j)|\frac{1}{r_{ij}^3} - \frac{3z_{ij}^2}{r_{ij}^5}|\phi_e(i,j)\rangle \tag{3.194a}$$

and

$$E_{ij} = \left(\frac{3}{4}\right)\frac{\mu_0 g^2\gamma_e^2}{4\pi}\langle\phi_e(i,j)|\frac{3\left\{y_{ij}^2 - x_{ij}^2\right\}}{r_{ij}^5}|\phi_e(i,j)\rangle, \tag{3.194b}$$

where $\phi_e(i, j)$ is the spatial part of the wavefunction. The energies for the three terms are found by solving the secular determinant:

$$\begin{vmatrix} \frac{1}{3}D_{ij} - E_{SS} & 0 & E_{ij} \\ 0 & -\frac{2}{3}D_{ij} - E_{SS} & 0 \\ E_{ij} & 0 & \frac{1}{3}D_{ij} - E_{SS} \end{vmatrix} = 0, \tag{3.195}$$

yielding the roots

$$E_{SS}(0) = -\frac{2}{3}D_{ij}, \tag{3.196a}$$

$$E_{SS}(\pm) = -\frac{1}{3}D_{ij} \mp E_{ij}. \tag{3.196b}$$

Here, we have used the notation $E_{SS}(0)$ corresponding to the $m_s = 0$ state $|0\rangle$, and energies $E_{SS}(\pm)$ to the symmetric (+) and antisymmetric (-) mixing of the $|1\rangle = |\alpha_1\alpha_2\rangle$ and $|-1\rangle = |\beta_1\beta_2\rangle$ states (cf. Eqs. 3.3a and 3.3c), as illustrated in Fig. 3.70a. Indeed, the extent of ZFS and the relative sizes of E_{ij} and D_{ij} is a direct consequence of the overlap of electronic orbitals within a molecule; as the overlap increases so does the splitting (see Eq. 3.194).

Absent spin–spin interactions, the three triplet states are degenerate and there is no ZFS. A first order approximation applies to $E_{ij} = 0$, where interactions between states $|1\rangle$ and $|-1\rangle$ vanish. In this limit, state $|0\rangle$ has the lowest energy $(-2D_{ij}/3)$ since its spin dipoles are antiparallel and hence are attractive, whereas $|1\rangle$ and $|-1\rangle$ have their dipoles pointing in the same direction, thus raising the total system energy, as shown in Fig. 3.70b. When the full interaction is turned on, there is an additional splitting due to mixing of the otherwise degenerate $|1\rangle$ and $|-1\rangle$ states (Fig. 3.70c). Applying an external field lifts this degeneracy even further via the *Zeeman effect*, giving rise to well-resolved triplet energies.

3.7.3 Metal–ligand and ligand-centered interactions in organic–transition metal complexes

An important class of organic electronic compounds that exhibit high efficiency emission via the triplet state are the organic-transition metal complexes based on heavy metal atoms such as Ir(III), Os(III), and Ru(II), Pt(II), Pd(II), etc. In this section, we discuss two major classes of these transition metal complexes, the quasi-octahedral chelating complexes, and the dihedral, distorted square planar complexes. Here, *chelate* refers to a compound containing an organic ligand that is attached to a central metal atom at two or more points. While so-called *metal–ligand charge transfer* (MLCT) in the former systems leads to strong spin–orbit coupling, and consequently very fast and efficient $T_1 \rightarrow S_0$ transitions, similar transitions in the latter compounds are due to the somewhat weaker *ligand-centered* (LC) couplings.

Spin–orbit interactions leading to efficient phosphorescence in transition metal complexes are classified into two categories illustrated in Fig. 3.71 with example molecules that support these processes. Type I molecules are based on ligand-centered triplet (^{3}LC) states. Although Type I triplet emitting molecules can

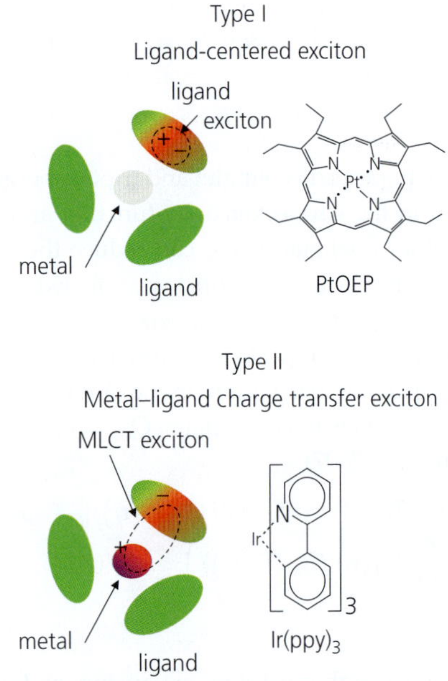

Figure 3.71 Two types of triplet excitons formed on metal complexes, with example molecules supporting these excited states.

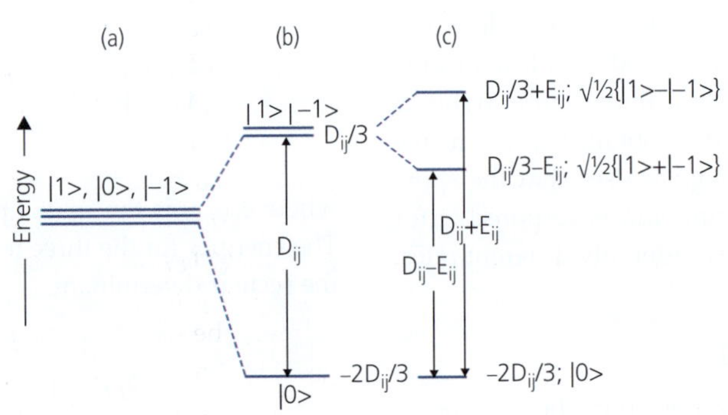

Figure 3.70 Zero-field splitting of the triplet state for (a) no spin–spin interactions, (b) spin–spin interactions due only to diagonal terms, and (c) a full solution to the secular determinant in Eq. 3.195. From McGlynn et al. (1969).

have very high quantum yield as is the case for PtOEP, the radiative decay rate of $k_{Ph} \sim 10^5$ s^{-1} competes with parasitic non-radiative down-conversion processes that occur at approximately the same rates.

In Type II phosphors, the electron density in the ligand is shared with the heavy metal atom forming a triplet ^{3}MLCT state. The sharing of the electron density in complexes such as Ir(ppy)$_3$ leads to a large H_{SO}, and hence rapid ISC that results in highly efficient phosphorescent emission (Lamansky et al., 2001).

The symmetries of the chelating M(III) and square planar M(II) complexes (M = metal) are illustrated in Fig. 3.72. The M(III) complexes such as the Ir(C^N)$_3$ tris-bidentate phosphors have quasi- or distorted octahedral symmetry (O_h) of their *d*-orbitals, and depending on the atoms A, B can form a facial (*fac*) or meridonal (*mer*) isomer. For example, the C^N cyclometalating ligand in Ir(ppy)$_3$, has A = C and B = N. In the *fac*-isomer, the N-atoms lie in a plane. However, by exchanging N and C on one of the three ligands, the three N atoms now lie on two different, intersecting planes in the *mer*-isomer. The octahedral symmetry is apparent from the four A, B atoms around the equator, capped by A and B at opposite molecular azimuths. In contrast, the distorted square molecule has a fourfold dihedral symmetry (D_{4h}) characteristic of the planar Pt(II) and Pd(II) complexes.

The pseudo-octahedral symmetry of metalorganic the *fac*- and *mer*-isomers is illustrated in more detail in Fig. 3.73. The overlap of the metal and ligand wavefunctions is distinct for the isomers. The *fac*-isomer

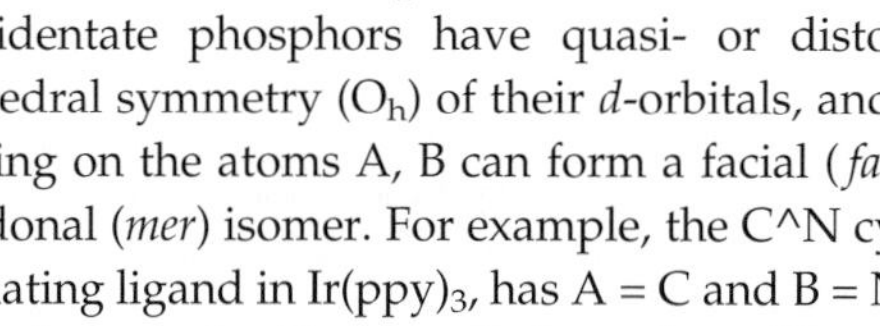

Figure 3.72 (a) Structure of quasi-octahedral, O_h, and (b) square planar dihedral, D_{4h}, transition metal–organic complexes. Here A and B represent atoms that coordinate with the transition metal atom M. A, B may be the same or different atoms such as C, N, O, etc.

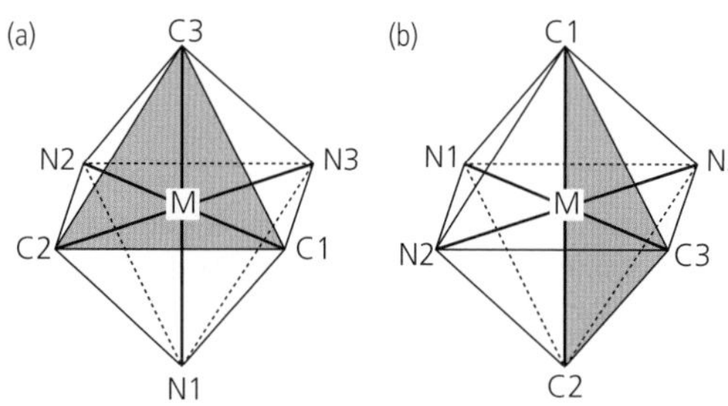

Figure 3.73 (a) *Fac* and (b) *mer* isomer crystal structural symmetries for pseudo-octahedral M(III)(C^N)$_3$ complexes.

has the three identical ligands occupying the face of the octahedron. There are three symmetry-equivalent positions for the ligands that transform into each other by rotations of $2\pi/3$, resulting in C_3 symmetry. The ligand plane of the *mer*-isomer intersects the metal atom, resulting in significant overlap of the metal and ligand orbitals. It lacks the rotational symmetry of the *fac*-complex, having a C_1 symmetry. Overlap of the metal and organic orbitals result in a broad, red-shifted spectrum which contrasts with the vibronically distinct, blue-shifted emission features characteristic of *fac*-isomers.

To understand the relationship between spin–orbit coupling and emission efficiency in transition metal complexes, and moreover the relationship between this coupling and the magnitude of the ZFS, we begin by considering a simplified energy landscape of the M(III) chelates. Figure 3.74a shows that the *d*-orbitals of an undistorted O_h compound are split into the three low energy, bonding (σ) d_{xy}, d_{yz}, and d_{xz}, t_{2g} symmetry states (labeled π_M), and two higher lying antibonding e_g, σ^* orbitals labeled $d_{x^2-y^2}$ and d_{z^2}. The electronic spatial distribution of the orbitals are illustrated in Fig. 3.75. The low lying t_{2g} orbitals are triply degenerate, whereas the e_g orbitals are twofold degenerate. They both have even spatial symmetry under inversion, that is, they are both *gerade* states. Hence, electric dipole transitions between them ($t_{2g} \leftrightarrow e_g$) are symmetry forbidden by the Laporte selection rule. The difference in energy, $10Dq$, between the t_{2g} and e_g substates results since the latter states are located closer to the ligand. Consequently, there is increased Coulomb repulsion that raises their relative energies. Here, Dq is known as the *ligand field parameter*. The excited antibonding ligand state, π_L^*, is also shown. In contrast, the $\pi_M \leftrightarrow \pi_L^*$ and $\pi_L \leftrightarrow \pi_L^*$ correspond to ^{3}MLCT and ^{3}LC transitions, respectively, and are spin allowed.

The degeneracy of the *d*-orbitals is lifted in the distorted octahedral transition metal chelates, resulting in a small energy splitting (~10–100 cm^{-1}) of the t_{2g} states into the relabeled substates, d_1, d_2, and d_3. Similarly, $d_{x^2-y^2}$ and d_{z^2} are split in the distorted molecule.

In Fig. 3.74b we show that even larger ZFS of the *d*-orbitals occur by distortion of square-planar dihedral D_{4h} molecules; a characteristic feature of Ir(I), Pt(II), and Pd(II). Initially, the lowest *d*-orbitals correspond to the degenerate d_{xz} and d_{yz} in the e_g state, d_{z^2} in the a_{1g}, and d_{xy} in b_{2g}. Finally, $d_{x^2-y^2}$ lies in the upper b_{1g} orbital. The distortion then results in lifting the degeneracy in e_g, giving rise to four lowest lying *d*-orbitals, $d_1, \ldots d_4$. Note that the splittings are far larger (~1000–4000 cm^{-1}) than in the octahedral compounds, resulting in reduced mixing between d_1 and

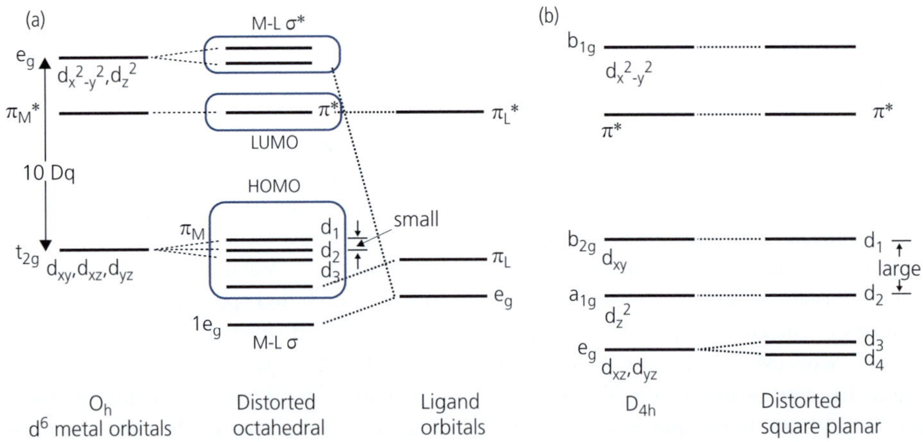

Figure 3.74 Zero-field splitting (ZFS) of the *d*-orbitals in (a) distorted octahedral and (b) distorted dihedral transition metal complexes. The distorted octahedral molecule has an energetic landscape that combines the metal plus the ligand energetics. The bonding and antibonding orbitals are σ and σ^*, respectively, and the π^* orbital is a linear combination of similar states in the metal (π_M*) and the ligand (π_L*). The ZFS is small compared to the distorted square planar ZFS between the d_1 and d_2 orbitals (Rausch et al., 2010).

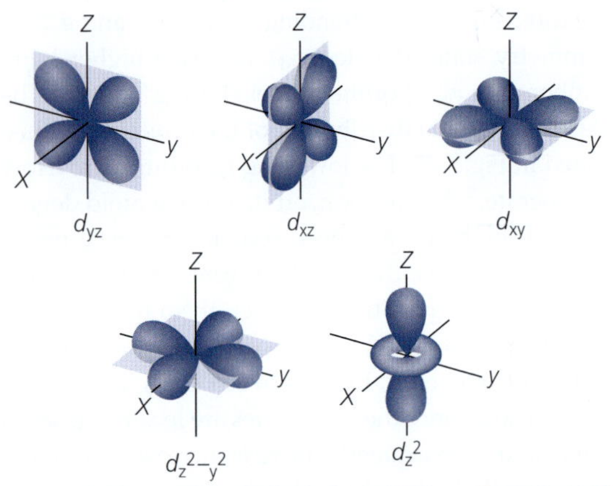

Figure 3.75 Electron distributions in the 5d orbitals. Note the different rotational positions of d_{xy} and $d_{x^2-y^2}$.

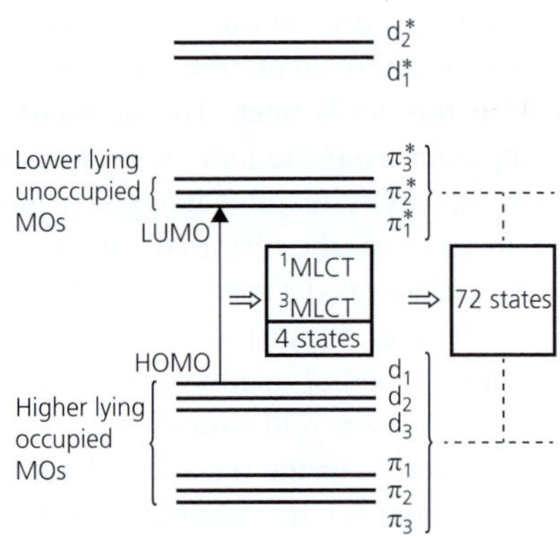

Figure 3.76 Highest molecular orbitals π_1, π_2, π_3 corresponding to the three ligands, and the three lowest metal d-orbitals, d_1, d_2, d_3 that support the $LUMO \leftarrow HOMO$ MLCT transitions to the three lowest unoccupied molecular orbitals, π_1^*, π_2^*, π_3^*. Higher lying e_g orbitals are also shown (Yersin et al., 2011).

the other *d*-orbitals. This, ultimately, reduces the ZFS, giving a weaker ^{3}LC character to the transitions compared with ^{3}MLCT.

To understand spin–orbit interactions in the quasi-octahedral transition metal complexes, we turn to a simplified set of diagrams that single out the energy levels responsible for the lowest energy transitions in Fig. 3.76 (Yersin et al., 2011). The primary $LUMO \leftarrow HOMO$ MLCT transitions occur from occupied metal *d*- and occupied molecular orbitals to one of three unoccupied ligand π^* substates localized on each of the three equivalent ligands. Each of these transitions share oscillator strength with 3 triplets or one singlet state. Hence there are a total of six initial $d + \pi$-orbitals, times three π^* states, times four possible spin combinations between the initial and final state, equaling 72 total states that can intermix via spin–orbit interactions. We neglect higher lying *d*-orbitals and even lower lying (non-frontier) molecular orbitals. While these too may mix with the lowest lying states, it was already noted that those states lying closest to the one supporting the transition result in the largest perturbation of the final state (Eq. 3.184).

Since the spin–orbit Hamiltonian in Section 3.7.1 is a single electron operator, the *Slater–Condon selection rule* requires that the final and initial states differ by only one spinorbital (Atkins and de Paula, 2006). Furthermore, the operator imposes the selection rule that the final and initial states can differ by a change of spin quantum number of $\Delta m_S = 0, \pm 1$. One further simplifying assumption is that the exponential decay of the wavefunction with distance allows us to consider interactions with only a single ligand, and hence a single π-orbital. With these conditions, we can limit our consideration of *LUMO* $\rightarrow$ *HOMO* transitions affected by spin–orbit interactions to those shown in Fig. 3.77. Both singlet and triplet LC and MLCT transitions to π^* are shown from the upper two d or π orbitals. Admixtures from d_3 and orbitals not shown are also possible, although they become increasingly unlikely as the energy denominator in Eq. 3.184 increases due to the increased energy difference with the radiative state.

Since the closed shell configuration of the complexes is $d_1^2 d_2^2 \pi^2$, we will consider, for example, only the $m_S = +1$, triplet substate denoted $|^3(d_1\pi^*)_{+1}\rangle$, shown in the second row of Fig. 3.77 (Yersin et al., 2011). The notation used is $|^S(state)_{m_S}\rangle$. First, we can ignore coupling between $|^3(d_1\pi^*)_{+1}\rangle$ and the ligand-centered substates, $|^{1,3}(\pi\pi^*)\rangle$ since these two- and three-centered integrals are small due to the spatial separation of orbitals on different ligands. Furthermore, a d-orbital cannot couple to itself according to the Slater–Condon selection rule. Hence, we are left with the couplings of $|^3(d_1\pi^*)_{+1}\rangle$ to the d_2 triplet and singlet substates, that is, $|^3(d_2\pi^*)_{+1}\rangle$, $|^3(d_2\pi^*)_0\rangle$ and $|^1(d_2\pi^*)\rangle$. Note that the $m_S = 0$ subscript is ignored in writing $|^1(d_2\pi^*)\rangle$ since it is the only possibility for a singlet state, and hence is redundant.

Returning to Section 3.7.1, using perturbation theory, and Eq. 3.182, we write the modified eigenstate as:

$$|^3(d_1\pi*)'_{+1}\rangle = |^3(d_1\pi*)_{+1}\rangle + |^1(d_2\pi*)\rangle \times \frac{\langle^1(d_2\pi*)|H_{SO}|^3(d_1\pi*)_{+1}\rangle}{E\left(^3(d_1\pi*)_{+1}\right) - E\left(^1(d_2\pi*)\right)} + |^3(d_2\pi*)_0\rangle \frac{\langle^3(d_2\pi*)_0|H_{SO}|^3(d_1\pi*)_{+1}\rangle}{E\left(^3(d_1\pi*)_{+1}\right) - E\left(^3(d_2\pi*)_0\right)} \quad (3.197)$$

or, taking only contributions from the nearest singlet state,

$$|^3(d_1\pi^*)'_{+1}\rangle \approx |^3(d_1\pi^*)_{+1}\rangle + |^1(d_2\pi^*)\rangle \times \frac{\langle^1(d_2\pi^*)|H_{SO}|^3(d_1\pi^*)_{+1}\rangle}{E\left(^3(d_1\pi^*)_{+1}\right) - E\left(^1(d_2\pi^*)\right)}. \quad (3.198)$$

Ignoring the second triplet term in Eq. 3.197 is justified since its transition dipole oscillator

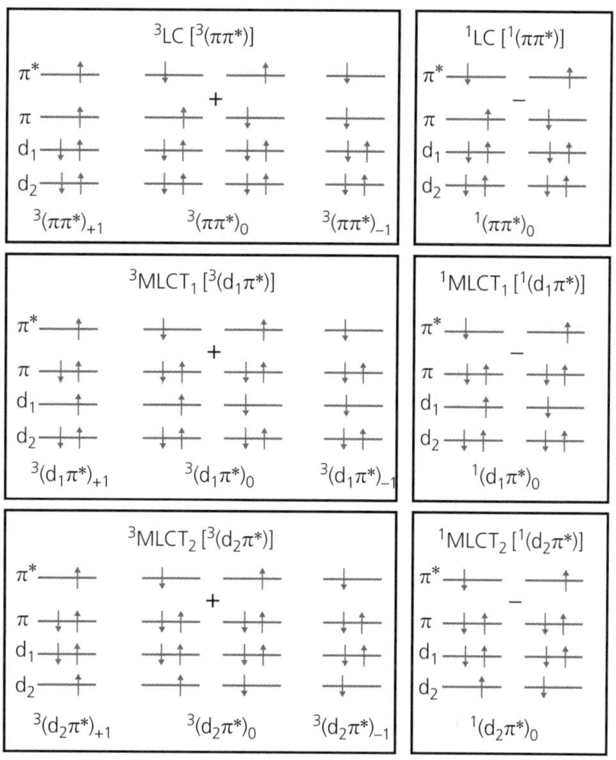

Figure 3.77 Lowest energy single electron ligand-centered (LC) and metal ligand charge transfer (MLCT) states in a pseudo-octahedral transition metal-organic system (Yersin et al., 2011).

strength to the ground state, S_0, is negligible. This latter expression is equivalent to the approximation of Eq. 3.184 that assumes the spin–orbit coupling is dominated by a lone singlet state, $|^1(d_1\pi^*)\rangle$. Following this same treatment, we derive the orbital stabilization energy due to spin–orbit interactions from Eq. 3.186:

$$E'\left({}^3(d_1\pi^*)_{+1}\right) = E\left({}^3(d_1\pi^*)_{+1}\right) + \frac{|\langle{}^1(d_2\pi^*)|H_{SO}|^3(d_1\pi^*)_{+1}\rangle|^2}{E\left({}^3(d_1\pi^*)_{+1}\right) - E\left({}^1(d_2\pi^*)\right)}, \qquad (3.199)$$

where once again, we note that the triplet lies at a lower energy than the singlet, such that $E'\left({}^3(d_1\pi^*)_{+1}\right) < E\left({}^3(d_1\pi^*)_{+1}\right)$. This treatment has been simplified to ignore interactions with other d- and π-orbitals.

As noted, the magnitudes of the couplings are expected to decrease rapidly as their separations in energy and distance increase. Hence, the approximations employed are adequate to describe the observed effects without unnecessarily complicating the discussion. Finally, we obtain the triplet or phosphorescent radiative rate from $|^3(d_2\pi^*)_{+1}\rangle$ in Eq. 3.189:

$$k_{Ph} = \frac{8\pi\omega^3}{3hc^3}\left|\frac{\langle{}^1(d_2\pi*)|H_{SO}|^3(d_1\pi^*)'_{+1}\rangle}{E\left({}^3(d_1\pi*)_{+1}\right) - E\left({}^1(d_2\pi*)\right)}\right|^2 \times|\langle S_0|q\mathbf{r}|^1(d_2\pi*)\rangle|^2. \qquad (3.200)$$

Clearly, a smaller energy denominator, and hence small exchange energy splitting of T_1 and S_1 in Eq. 3.200 leads to an increased radiative triplet rate. Furthermore, k_{Ph} also depends on the strength of the singlet transition dipole moment, as seen by the last term in this expression. And most importantly, in the absence of spin–orbit coupling the first term in Eq. 3.200 vanishes, and the triplet state is nonradiative. Note that Eq. 3.199 considers contributions from only a *single* substate interaction. Of course, there are additional states that intermix, which only increase ISC and the radiative rate, k_{Ph}.

The last step in our analysis is to consider transitions from the lowest ligand centered triplet, $|^3(\pi\pi^*)\rangle$, shown in the top row of Fig. 3.76. We have already seen that two- and three-centered integrals are small compared to single-centered integrals that characterize MLCT transitions. However, MLCT interactions are small in dihedral molecules such as the square planar PtOEP, due to the large energy gap between d-orbitals in these complexes (see Fig. 3.74). Treating the LC transitions is considerably more complex than MLCT interactions in distorted octahedral complexes since it involves both MLCT and LC couplings. We can understand this via the following arguments: Following the selection rules outlined above, $|^3(\pi\pi^*)\rangle$ cannot mix with itself; hence it necessarily involves two and three center integrals that are weaker than the one-center integrals employed to describe MLCT. To estimate the ZFS from LC states, we once again focus on a single substate: $|^3(\pi\pi^*)_{+1}\rangle$. This can couple to the MLCT$_1$ and MLCT$_2$ states $|^3(d_1\pi^*)_{+1}\rangle$ and $|^3(d_2\pi^*)_{+1}\rangle$, respectively (see Fig. 3.77) (Yersin et al., 2011). The mixing with $|^3(d_1\pi^*)_{+1}\rangle$ is due to *electronic configuration interactions* (CI), whose selection rules are $\Delta S = \Delta m_S = 0$. As discussed in Section 3.4.1, CI are used in the calculation of changes in energies of n-electron systems that arise from reorganization of the π-orbitals with the addition of a single electron.

Hence, CI lead to a perturbation from each of the three ^{3}MLCT states of the same magnitude:

$$|^3(\pi\pi^*)'_{+1}\rangle \approx |^3(\pi\pi^*)_{+1}\rangle + |^3(d_1\pi^*)_{+1}\rangle \frac{\langle{}^3(d_1\pi^*)_{+1}|H_{CI}|^3(\pi\pi^*)_{+1}\rangle}{E\left({}^3(\pi\pi^*)_{+1}\right) - E\left({}^3(d_1\pi^*)\right)}. \qquad (3.201)$$

where H_{CI} is the configuration Hamiltonian. Now the admixed $|^3(d_1\pi^*)_{+1}\rangle$ can, in turn, interact with the singlet $|^1(d_1\pi^*)\rangle$ as we have seen in the foregoing discussion of MLCT. Thus, LC interactions involve two processes: the first due to CI with the ligand electron system, and the second via interaction of the CI-perturbed π-system with the d-orbitals via MLCT. Given that this requires a hybridization of two orbital systems, its intensity (and hence resulting ZFS) is weaker. For this reason, ^{3}LC transitions in distorted square planar systems are generally weaker, and their transition rates slower than ^{3}MLCT transitions in M(III) transition metal complexes.

The importance of ZFS on the luminescent properties of triplet emitting materials should be apparent from Eqs. 3.199 and 3.200. The calculated dependence of the average lifetime, $\tau_{av} \approx k_{Ph}^{-1}$ on ZFS energy, ΔE, is shown in Fig. 3.78. The energies range from >200 cm^{-1} (25 meV) for Os- and Ir-based phosphors with strong ^{3}MLCT transitions, to <1 cm^{-1} for ^{3}LC-dominated Pd, Re, and Rh complexes. This energy range corresponds to three orders of magnitude change in τ_{av}, from ~400 ns for octahedral, to milliseconds for square planar complexes. Indeed, a key observation from these data are that the square planar molecules are dominated by ^{3}LC behavior with longer lifetimes, whereas the octahedral compounds exhibit rapid decay due to the large ^{3}MLCT content of the spin–orbit coupling.

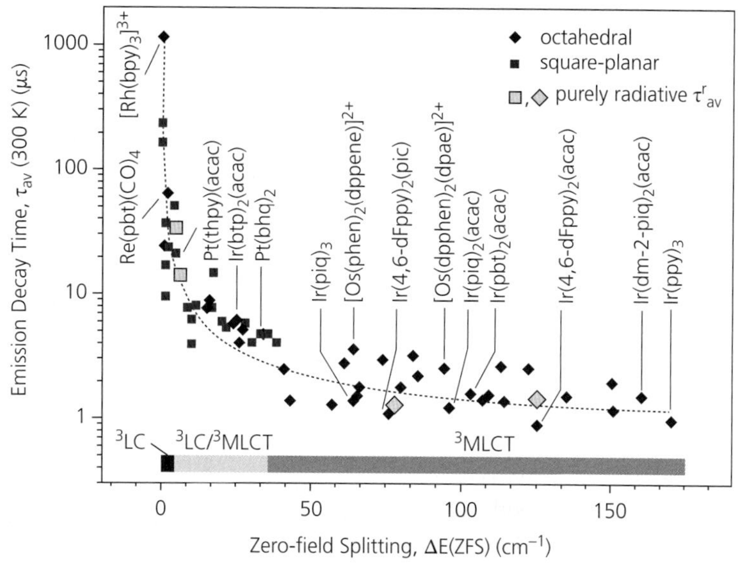

Figure 3.78 Measured triplet emission decay time vs. calculated ZFS for a number of transition metal organic compounds with differing ^{3}MLCT and ^{3}LC excited state content. The larger ZFS energy results in a decrease in triplet radiative lifetime, which in turn generally leads to a higher luminescent efficiency due to more favorable competition of radiative with non-radiative recombination (Yersin et al., 2011).

The emission efficiency generally is higher for materials with a smaller radiative lifetimes. This is due to two principal reasons. The first is that the PLQY depends on the competition between radiative and non-radiative recombination according to Eq. 3.119. Thus, $\Phi_{PL} \rightarrow 1$ for $k_r = k_{Ph} >> k_{nr}$. The second is that the density of triplets increases with increasing lifetime under constant illumination. Thus, the probability for triplet–triplet annihilation, which is a mechanism for non-radiative loss, increases with τ_{av}, thus resulting in a decrease in Φ_{PL}, particularly at high intensities (see Section 3.10). For these reasons, the highest emission efficiencies are from phosphors based on either Ir or Pt with strongly ^{3}MLCT character and ZFS energies typically > 10 cm^{-1}.

It is noteworthy that the increase of ZFS > 100 cm^{-1} does not result in a further decrease in τ_{av}. We have shown that there is a clear correlation between the magnitude of the spin–orbit coupling and the ZFS energy, which in turn results in a higher k_{Ph} according to Eq. 3.200. Recall, however, that *total* ZFS results from the sum over all T_1 substates in Eq. 3.185. We have seen that spin–orbit coupling lifts the degeneracy of these three states that are separated by energies ΔE_{ij}, where *i,j* denote the two states that are mixed. Hence, the emission rate is determined by the Boltzmann factors for each substate. The average emission rate, $\bar{k}_{Ph}$, is the sum of several rates, each weighted by its activation energy following (Yersin et al., 2011):

$$\bar{k}_{Ph} = \frac{k_1 + k_2\exp(-\Delta E_{12}/k_BT) + k_3\exp(-\Delta E_{13}/k_BT)}{1 + \exp(-\Delta E_{12}/k_BT) + \exp(-\Delta E_{13}/k_BT)}, \tag{3.202}$$

where we have labeled the participating substates $i = 1, 2, 3$. Hence, the rate is ultimately limited by that substate that has the greatest energy separation from the lowest energy level $i = 1$ (corresponding to $m_S = 0$, see Fig. 3.70), giving rise to the asymptotic dependence of τ_{av} on ZFS in Fig. 3.78.

Measurement of ZFS energies and the corresponding transition rates is primarily done via high resolution spectroscopy (Yersin, 2004). However, the temperature dependence of the PL transients using Eq. 3.202 presents an alternative method for obtaining the energies, transition rates and Φ_{PL}. While this expression has several variables, it is simplified by measuring the rates at low and high temperatures. Thus, in the limit, $k_BT >> \Delta E_{12}, \Delta E_{13}$, we have $\bar{k}_{Ph} = k_1$, and in the high temperature limit, $\bar{k}_{Ph} = (k_1 + k_2 + k_3)/3$.

This approach has been used to characterize a series of phosphorescent pyrazolate-bridged cyclometalated platinum(II) complexes with the general formula: C^NPt(μ-*pz*)$_2$PtC^N shown in **Scheme 2,** where C^N represents the cyclometalating ligand (Qi et al., 2008). The pyrazolate (*pz*) ligand controls the degree of metal–metal interaction by adjusting the angle between opposing ligands, and thus the emission spectra.

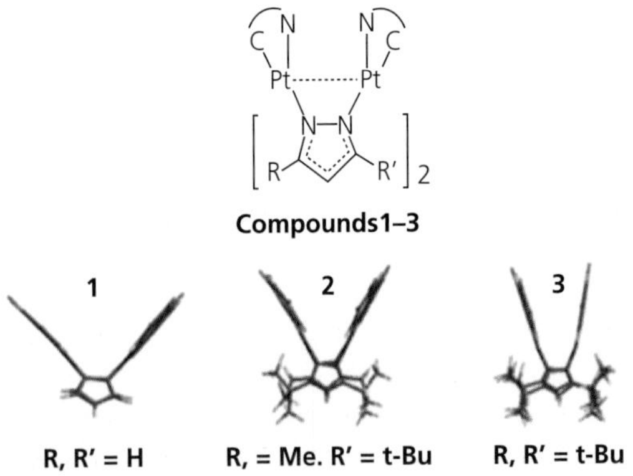

Scheme 2 Binuclear C^NPt (μ-pz)$_2$PtC^N complexes.

Compounds **1**, **2**, and **3** have Pt · · · Pt spacings of 3.19 Å, 3.05 Å, and 2.83 Å, respectively (Ma et al., 2005). Their phosphorescence spectra in the blue, green and red, are shown in Fig. 3.79a. The bridging ligands rigidly link the two metal centers on opposing pyridyl rings. For compound **1**, the well resolved vibronic progression in the emission spectrum indicates only weak interaction between the two metal centers. In contrast, broad and featureless emission spectra of compounds **2** and **3** are observed due to a mixed ^{3}LC/^{3}MLCT state, and are assigned to strong triplet metal-metal ligand charge transfer (^{3}MMLCT) enhanced by the proximity of the emissive ligands (Lai et al., 2002). Note that the phosphorescent efficiencies of OLEDs at low current whose active region employs these compounds doped into a wide energy gap UGH2 matrix are 3.8%, 6.6% and 6.6% for compounds **1**, **2**, and **3**, respectively. The electroluminescent efficiencies are functions of Φ_{PL} (among other factors arising from the OLED current injection and emission properties, see Chapter 6). Hence, the differences in OLED efficiencies suggest that the blue emitting compound **1** has the lowest Φ_{PL} of the three phosphors.

Transient PL data in Fig. 3.79b were obtained by exciting a film consisting of compound **3** doped at 8 wt% in an UGH2 matrix using a N_2 laser with a wavelength of λ = 337 nm. A fit to the data using Eq. 3.202 is shown by the line. The resulting transition rates, k_1, k_2, and k_3 obtained from the fits to these and similar data for compounds **1** and **2** are provided in Table 3.4. The large ZFS of compounds **2** and **3** is apparent from values of ΔE_{13} and ΔE_{12} that are 4–8 times higher than that from compound **1**. This leads to correspondingly larger decay rates for the latter compounds, which is consistent with differences in their quantum yields. The large ZFS energies of these

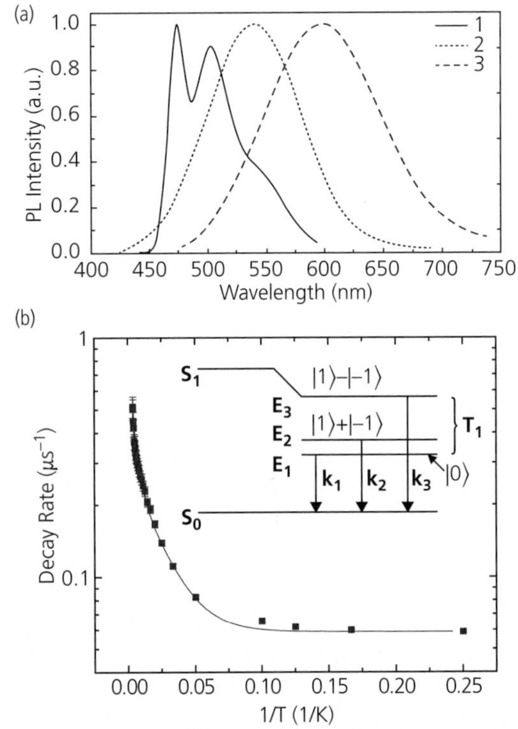

Figure 3.79 (a) Room temperature photoluminescence (PL) spectra of compounds **1**, **2**, and **3**. (b) Measured PL decay rate (data points) vs. $1/T$ for 8 wt% compound **3** doped into UGH2. The line is a fit to Eq. 3.202 in text. Inset: Energy level scheme of the triplet manifold indicating the ZFS energies and corresponding decay rates. Also indicated are the substate wavefunctions where the number in the ket vectors correspond to the m_S values (Qi et al., 2008).

Table 3.4 Zero-field splitting energies and transition rates for the binuclear Pt complexes (Qi et al., 2008)

Compound	ΔE_{13} (cm^{-1})	ΔE_{12} (cm^{-1})	k_1 (μs^{-1})	k_2 (μs^{-1})	k_3 (μs^{-1})
1	28±3	<4	0.12	0.11	0.25
2	142±9	31±5	0.07	0.42	0.81
3	113±10	33±6	0.06	0.32	1.12

binuclear complexes are more characteristic of octahedral than dihedral compounds, as inferred from Fig. 3.78, providing support for the strong ^{3}MMLCT coupling of these complexes.

The metal participation in the triplet state, which is expected to influence the magnitude of the ZFS, is estimated from the simulated electron orbital distributions in Fig. 3.80. No pronounced change in triplet spin density takes place at the pyridyl ring, whereas the spin density surfaces at the Pt–Pt center for compounds **2** and **3** are double that of **1**. According to calculation, the complexes exhibit significant overlap

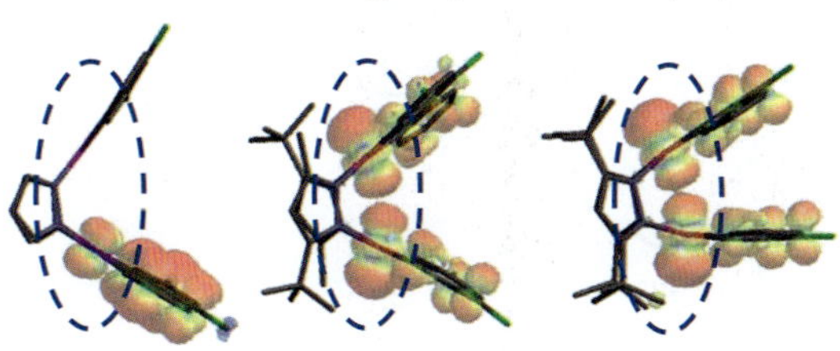

Figure 3.80 DFT simulations of the triplet spin density surfaces for compounds **1**, **2** and **3**. The spin density at the Pt · · · Pt centers is double for the latter two compounds compared to that of compound **1**. The dashed oval approximately locates the region of Pt · · · Pt interactions between the ligands. After Qi et al. (2008).

of the *d*-orbitals of the individual metal atoms. The strong interactions lift the triplet energy degeneracy, leading to the pronounced ZFS and high PL efficiencies.

The relationship between ZFS for ^{3}MLCT vs. ^{3}LC dominated transitions is also apparent from the comparison of (N^C)Pt(acac) and (N^C)$_2$Ir(acac) analogs in Table 3.5. Here, acac is the ancillary, acetylacetonate ligand. The effects of strong ^{3}MLCT character in the Ir-complexes compared with the ^{3}LC dominance of the square planar Pt complexes are striking. The red shift in λ_{peak} is higher for ^{3}MLCT vs. ^{3}LC transitions. From Eq. 3.199, the energy stabilization is larger in the former than in the latter case. Furthermore, the ZFS energy shift is considerably larger for Ir complexes, where $ZFS \simeq$ 20–100 cm^{-1} compared with typically ten times smaller shifts for the Pt analogs. All of these characteristics are consistent with the much larger triplet emission rate (again, often by a factor of ten) for the octahedral vs. the dihedral complexes.

Finally, we note that only subtle differences in structure of the same molecule can also alter its excitonic character from ^{3}LC to ^{3}MLCT with dramatic consequences to the photophysical characteristics of a molecular species. The facial (*fac*-) and meridonal (*mer*-) isomers of Ir(pmp)$_3$, respectively shown in Fig. 3.81 provide a striking illustration of the effects of the charge transfer character of the triplet excited state (Lee et al., 2016). Based on DFT calculations, the HOMO and LUMO of the *fac*-isomer are formed on the same ligand due to its C_3 symmetry, resulting in considerable localization of the triplet state. In contrast, the HOMO of *mer*-Ir(pmp)$_3$ is formed on two oppositely-disposed ligands, and the LUMO is on the third. This leads to more delocalized excited state that efficiently couples to the metal atomic orbitals.

Table 3.5 Peak emission wavelength (λ_{peak}), zero-field splitting energy (ZFS), and radiative lifetimes (k_{Ph}) of analogous Pt(II) and Ir(III) transition metal complexes. From Yersin et al. (2011)

Compound[a]	λ_{peak} (nm)	ZFS (cm^{-1})	k_{Ph} ($\times 10^4$ s^{-1})
Ir(4,6-dFppy)$_2$(acac)	484	109	64
Pt(4,6-dFppy)(acac)	467	8	8.0
Ir(thpy)$_2$(acac)	567	33	12
Pt(thpy)(acac)	558	4	2.4
Ir(s1-thpy)$_2$(acac)	586	16	9.8
Pt(s1-thpy)(acac)	577	<1	2.1
Ir(bt)$_2$(acac)	557	100	14
Pt(bt)(acac)	539	10	6.0

[a] (4,6-dFppy)$_2$(acac) = bis (4′,6′-difluorophenyl) pyridindate, acac = acetylacetonate; thpy = 2-(2-thienyl)pyridinatoN,C^3; s1-thpy = 5,2-bis (2-thienyl)pyridinate; bt = bis(2-phenylbenzo-thiozolato-N,C2′)

Several photophysical characteristics of this pair of isomers confirm that the *fac*- isomer phosphorescence is predominantly of ^{3}LC (Type I phosphor) character, whereas emission from *mer*-Ir(pmp)$_3$ is from the ^{3}MLCT (Type II) state. Due to the lower E_{SO} of ^{3}LC states, the photoluminescence of *fac*-Ir(pmp)$_3$ is considerably blue-shifted from that of the *mer*-isomer (see Fig. 3.82a). Their room temperature transient decay rates in Fig. 3.82b are equal to within a factor of two, while at low temperature the decay rate of *fac*-Ir(pmp)$_3$ is ~10 times smaller than for *mer*-Ir(pmp)$_3$, ($k_{Ph} = 6.4 \times 10^5$ s^{-1} vs. 10×10^5 s^{-1} at 295 K; 1.1×10^5 s^{-1} vs. 1.0×10^6 s^{-1} at 77 K for *fac* and *mer*-Ir(pmp)$_3$, respectively) as is characteristic of Type I vs. Type II phosphors.

A further indication of the more polar nature expected for the ^{3}MLCT state is found in the relative solvatochromatic shifts in a solvents of different polarity in Fig. 3.83. The larger excited state dipole moment of *mer*-Ir(pmp)$_3$ is a consequence of the spatially extended ^{3}MLCT state compared to that of the more compact ^{3}LC state localized onto a single ligand. This leads to a significantly larger bathochromic shift for *mer*-Ir(pmp)$_3$ as the polarity of the three solvents is increased (see Section 3.6.7). The dipole moments of the solvents used are 0 D, 0.375 D, and 1.6D for PMMA, toluene, and DCM, respectively. We summarize these differences between Type I and II Ir(pmp)$_3$ phosphors in Table 3.6.

3.7.4 Relative phosphorescent and fluorescent rates and efficiencies

Phosphorescence is formally defined as emission observed after a "perceptible" time has elapsed from the end of the excitation. A more reasonable definition, however, is emission that is controlled by triplets. The

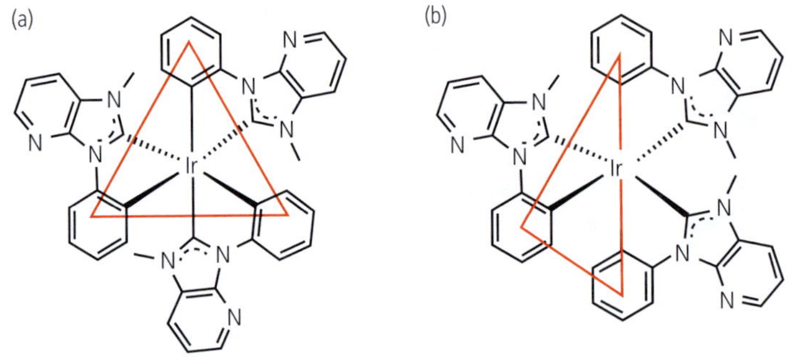

Figure 3.81 Structural formulae for the (a) *fac*- and (b) *mer*-isomers of Ir(pmp)$_3$. These carbenes have two C atoms coordinated from each ligand to the Ir atom, forming a Ir(C^C)$_3$ complex.

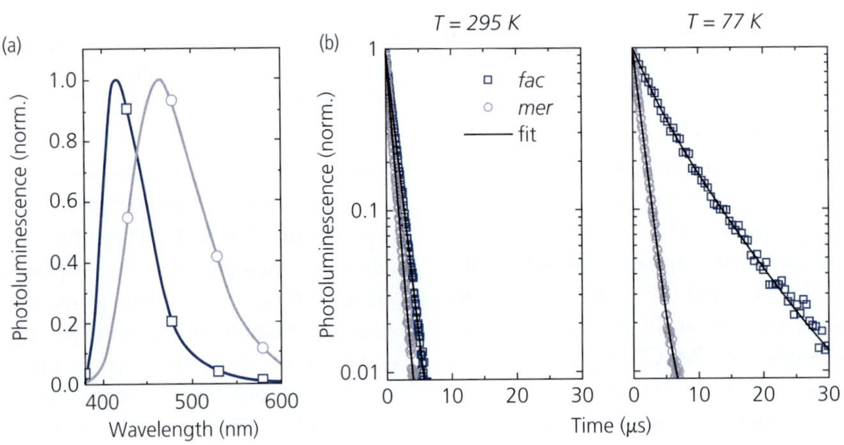

Figure 3.82 (a) Room temperature PL spectra of *fac*- (squares) and *mer*-Ir(pmp)$_3$ (circles) doped into a wide energy gap (and hence transparent) host, TSPO1 and excited at $\lambda = 325$ nm. (b) PL decay transients of *fac*- and *mer*-Ir(pmp)$_3$ in 2-methyltetrahydrofuran at $T = 295$ and 77 K, with the fits based on mono- or multi-exponential decays (Lee et al., 2016).

First published in Lee, J., et al., 2016 Nature Mater., 14, 92

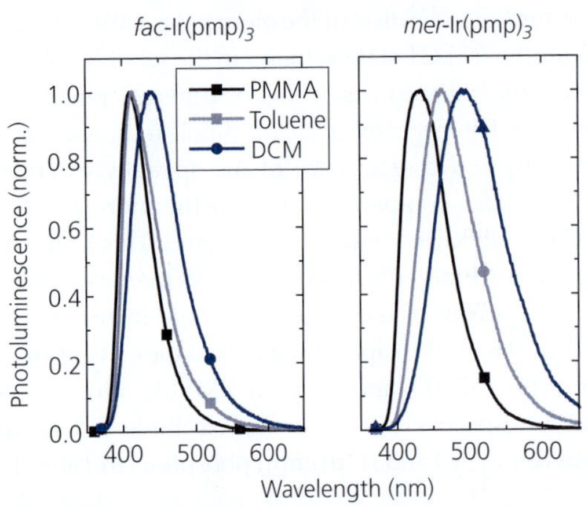

Figure 3.83 Solvatochromism of the two Ir(pmp)$_3$ isomers dissolved in solvents of increasing polarity, from PMMA (0 D) to DCM (1.6 D) (Lee et al., 2016).

First published in Lee, J., et al., 2016 Nature Mater., 14, 92

Table 3.6 Comparison of the properties of Type I (*fac*), and Type II *mer*-Ir (pmp)$_3$

Property	Type I (^{3}LC)	Type II (^{3}MLCT)
Isomer	*fac*-Ir(pmp)$_3$	*mer*-Ir(pmp)$_3$
Excited state dipole	small (localized excited state)	large (extended excited state)
Red shift in DCM (1.6 D)	0.19 eV	0.33 eV
Emission energy	high (3.1 eV, deep blue)	moderate (2.7 eV, light blue)
Emission rate (295 K)	6.4×10^5 s^{-1}	10×10^5 s^{-1}
Emission rate (77 K)	1.1×10^5 s^{-1}	1.0×10^6 s^{-1}

relatively low phosphorescent emission rate depends primarily on two factors: the strength of electric dipole transitions between the dominant singlet excited state to S_0, and the atomic number of the heavy atom or electronic system that governs the magnitude of the spin–orbit interaction, H_{SO}. Phosphorescent emission depends on the ISC rate (k_{ISC}) of $S_1 \to T_1$, followed by radiative decay (rate k_{Ph}) from $T_1 \to S_0$. Efficient phosphorescence requires that these rates are larger than competing processes, particularly non-radiative decay pathways via defects or impurities, such that $k_{nr} << k_{ISC}, k_{Ph}$. Spin–orbit coupling also leads to rapid ISC from S_D to T_1 which suppresses fluorescence from singlets (i.e. the fluorescence lifetime is $\tau_F \sim$ 1–10 ns compared with the phosphorescence lifetime for heavy metal complexes of $\tau_{Ph} \sim$ 1–10 μs). The hierarchy of transition rates that govern radiative and non-radiative transitions in organics are shown in Fig. 3.84. Note that k_{ISC} covers a broad range, approaching the rate of internal conversion (k_{IC}) for heavy metal complexes where H_{SO} is large, that is, $k_{ISC} > 10^{10}\,\mathrm{s}^{-1} > k_F$. In contrast, $k_{ISC} < 10^5\,\mathrm{s}^{-1}$ for low molecular weight fluorescent materials (e.g. naphthalene) where $H_{SO} \to 0$ (Beljonne et al., 2001, Forster, 2006).

We can understand the relative importance of the processes in Fig. 3.84 by comparing expressions for their *photoluminescence quantum yields*, which is defined as the ratio of the number of photons emitted into 4π solid angle, to the total number of exciting photons that are absorbed in the sample. The fluorescent quantum yield is (cf. Eq. 3.119)

$$\Phi_F = \frac{k_F}{k_F + k_{nrS} + k_{ISC}} \tag{3.203}$$

with an analogous expression for phosphorescence:

$$\Phi_P = \frac{\Phi_{ISC} k_{Ph}}{k_{Ph} + k_{nrT}}, \tag{3.204}$$

where k_{nrS} (k_{nrT}) is the non-radiative decay rate from the singlet (triplet) state, and Φ_{ISC} is the ISC quantum yield:

$$\Phi_{ISC} = \frac{k_{ISC}}{k_{ISC} + k_F + k_{nrS}} \tag{3.205}$$

That is, a condition for strong phosphorescent emission is that $k_{ISC} >> k_F$, such that $\Phi_{ISC} \to 1$ and $\Phi_F \to 0$. It follows that for a small non-radiative rate from the triplet state (which is often the case for heavy-metal-based triplet emitting molecules), then we arrive at *Kasha's intersystem crossing ratio*:

$$\frac{\Phi_{Ph}}{\Phi_F} = \frac{k_{ISC}}{k_F}. \tag{3.206}$$

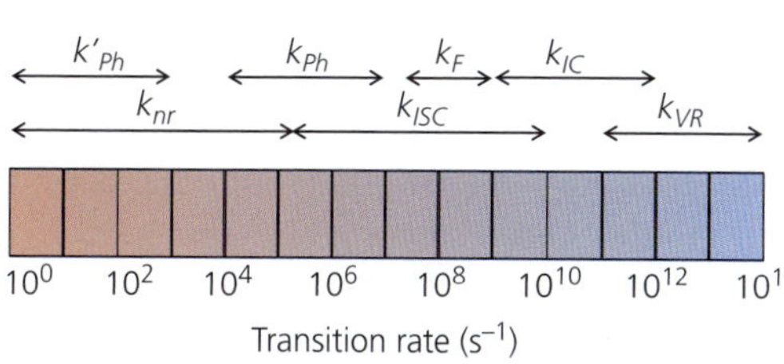

Figure 3.84 Characteristic transition rates in organic molecules. Here, k'_{Ph} is the phosphorescence transition rate in the absence of heavy-metal-induced spin–orbit coupling, k_{Ph} is the phosphorescence rate for heavy metal complexes. Other rates include fluorescence (k_F), ISC for heavy metal complexes (k_{ISC}), vibrational relaxation (k_{VR}), internal conversion (k_{IC}), and non-radiative relaxation (k_{nr}). See Fig. 3.33.

While we have concentrated our discussion on Frenkel excitons, charge transfer excitons also have either a singlet or triplet spin multiplicity, and the same is true for excimers and exciplexes. The notations used to denote the multiplicities of CT states are ^{1}CT and ^{3}CT for singlets and triplets, respectively.

3.7.5 Delayed fluorescence

Rapid ISC occurs when the singlet and triplet energies are approximately in resonance. Based on first order perturbation theory, k_{ISC} becomes large when $E(T_1) - E(S_D) \to 0$ in the energy denominator in Eq. 3.189. We caution that in this situation, first order perturbation theory is no longer strictly valid. Nevertheless, in near resonance, a triplet is *endothermically* promoted into the singlet manifold, followed by fluorescent emission. Since the direct emission from the triplet is electric dipole forbidden, the fluorescence emission rate is limited by the thermally activated transfer rate of the long-lived triplets, as illustrated in the Jablonski diagram in Fig. 3.85. This phenomenon is known as E-type delayed fluorescence, first observed in the molecule, eosin (Parker and Hatchard, 1961). More commonly, the process is termed *thermally activated delayed fluorescence* (TADF). TADF is one of two processes leading to delayed fluorescence; the other is due to lossy *triplet–triplet annihilation* (TTA) involving the participation of two triplets (see Section 3.10). Formation of a singlet via TTA is sometimes referred to as P-type delayed fluorescence due to its first observation in pyrene (Parker, 1963).

While spin–orbit coupling is required to mix the spin states in compounds undergoing TADF, the molecules do not require a heavy metal atom since the near-

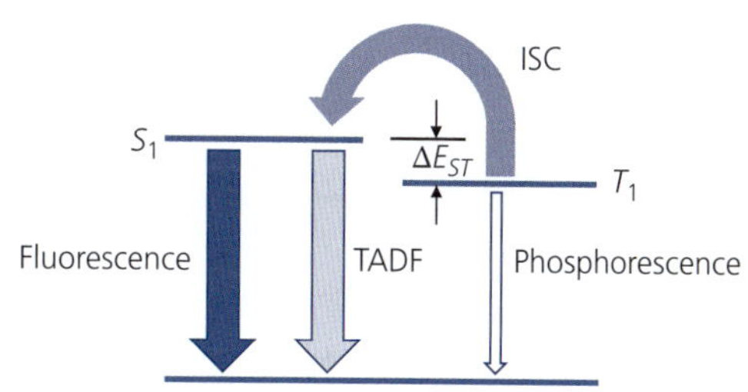

Figure 3.85 The TADF process due to endothermic conversion from T_1 to S_1 separated by a small activation energy ΔE_{ST}.

resonance between S_1 and T_1 reduces the reliance of the transition on the magnitude of E_{SO}. However, in the absence of strong spin–orbit coupling, $k_{Ph} << k_F$. Hence, while back transfer from S_1 to T_1 is also rapid and exothermic, as long as the non-radiative vibronic relaxation rate is $k_{nr} < k_{Ph}$, the excited state will be repeatedly thermionically excited into S_1 until a fluorescent $S_1 \rightarrow S_0$ transition occurs. The emission cycle of $T_1 \rightarrow S_1 \rightarrow T_1$ persists for the duration of

$$\frac{1}{\tau_{TADF}} \approx k_{nr} + k_{Ph}, \quad (3.207)$$

which is approximately equal to the natural phosphorescent lifetime.

The kinetic scheme leading to S–T mixing and ultimately to TADF is illustrated in Fig. 3.86. Here, the ISC rate, $k_{ISC,T}$, from $T_1 \rightarrow S_1$ is thermally activated, following:

$$k_{ISC,T} = k_{ISC,T}^{\infty} \exp(-\Delta E_{ST}/k_B T), \quad (3.208)$$

where ΔE_{ST} is the *singlet–triplet exchange energy*, and $k_{ISC,T}^{\infty}$ is the value of $k_{ISC,T}$ as $T \rightarrow \infty$. Since the opposite rate, $k_{ISC,S}$, is exothermic, it is much larger than $k_{ISC,T}$. A closer energy match of S_1 to T_1, and a longer triplet lifetime ($\tau_{Ph} = 1/k_{Ph}$) leads to more efficient TADF, which in some circumstances can approach 100%. The probability that higher order vibronics in T_1 are occupied at higher temperatures follows the Boltzmann distribution in Eq. 3.208. Hence, as temperature increases, the transfer between T_1 and S_1 becomes increasingly more efficient.

The average number of cycles, $\bar{n}$, between S_1 and T_1 prior to emission via $T_1 \rightarrow S_0$ (Fig. 3.86) depends on the relative magnitudes of the radiative and non-radiative rates, as well as on $k_{ISC,T}$. Defining the quantum yields for singlet and triplet formation as:

$$\Phi_S = \frac{k_{ISC,T}}{k_{ISC,T} + k_{Ph} + k_{nrT}} \quad (3.209)$$

and

$$\Phi_T = \frac{k_{ISC,S}}{k_{ISC,S} + k_F + k_{nrS}}, \quad (3.210)$$

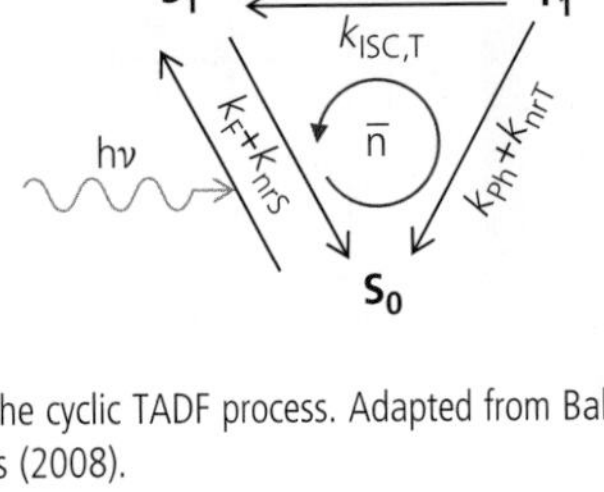

Figure 3.86 The cyclic TADF process. Adapted from Baleizao and Berberan-Santos (2008).

respectively, then it can be shown that the number of cycles is

$$\bar{n} = \frac{\Phi_T \Phi_S}{1 - \Phi_T \Phi_S}. \quad (3.211)$$

The kinetics of TADF have been studied in the fullerenes, C_{60} and C_{70}. Astonishingly, for C_{60} $\bar{n} \approx 100$ at 150°C (Baleizao and Berberan-Santos, 2008). However, due to the activation energy barrier between S_1 and T_1, the contribution of TADF to the emission intensity falls off rapidly as temperature is reduced.

More recently TADF has been employed in very high efficiency OLEDs (Uoyama et al., 2012), a topic discussed further in Section 6.3.4. For this purpose, fluorescent molecules with a minimal ΔE_{ST} are required for high efficiency emission. As noted in Section 3.7.1, the exchange energy is typically from 0.5 eV to 1.0 eV, in which case, $\Delta E_{ST} >> k_B T$ and hence the reverse transfer rate, $k_{ISC,T}$ is small. The key to reducing the exchange energy, K_j in Eq. 3.49b, is to design molecules that minimize the overlap between orbitals in S_1 and S_0, while keeping the molecular structure sufficiently rigid to inhibit rapid $T_1 \rightarrow S_0$ non-radiative decay.

Metal-free TADF compounds with high efficiency emission across the visible spectrum are exemplified by the archetype green fluorescent compound 4CzIPN shown in Fig. 3.87a (Uoyama et al., 2012). Time dependent DFT calculations show in Fig. 3.87b that the lowest unoccupied natural transition orbital (LUNTO) is centered on the dicyanobenzene group at the molecular core, with little overlap with the highest occupied NTO (HONTO) that is delocalized over the four carbazoyl groups (see Fig. 3.87c). In these calculations, the NTO is obtained from a weighted sum over all empty HOMO and filled LUMO states to obtain an estimate of the most likely $S_1 \rightarrow S_0$ transition probability density distribution (Martin, 2003). The limited orbital overlap between the excited and ground states results in a small $\Delta E_{ST} = 83$ meV, thereby introducing a substantial oscillator strength to the intersystem $S_1 \leftarrow T_1$ transition. These molecules also exhibit a relatively small Stokes shift, suggesting minimal reconfiguration during transition that is enforced by steric effects introduced by the cyano groups. Taken along with suppressed k_{nr}, a quantum yield of $\Phi_S = 94\%$ is observed.

A feature of TADF emission is the absence of vibronic overtones typically observed in monomeric $S_1 \rightarrow S_0$ spectra. Indeed, TADF spectra are comparatively broad and featureless, as shown in Fig. 3.88. This results from the lack of overlap of the HOMO and LUMO on the same molecule, creating a situation similar to that found for excimer transitions. In that

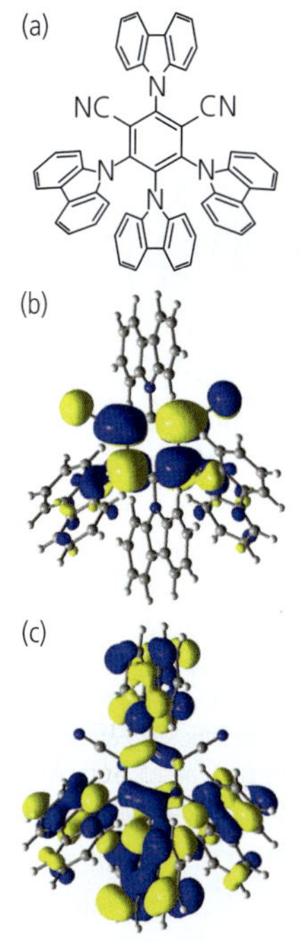

Figure 3.87 The compound, 4CzIPN exhibiting high efficiency green TADF emission. (a) Molecular structural formula, (b) calculated lowest unoccupied, and (c) highest occupied natural transition orbitals (Uoyama et al., 2012).

Reprinted by permission from Springer Nature, Highly efficient organic light-emitting diodes from delayed fluorescence, Uoyama, H., Goushi, K., Shizu, K., Nomura, H. & Adachi, C. Copyright 2012.

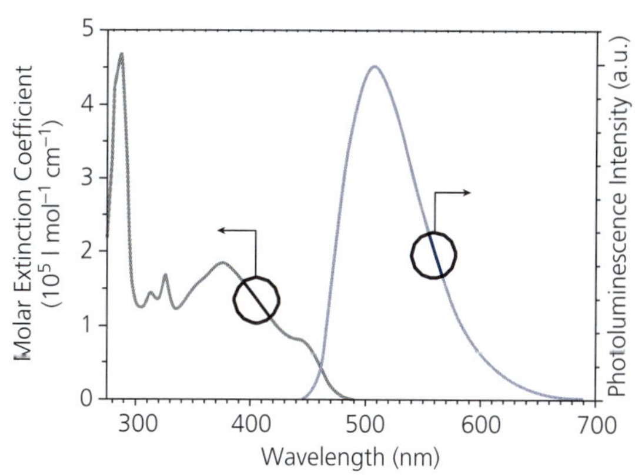

Figure 3.88 Absorption and photoluminescence spectra of 4CzIPN in Fig. 3.87. Note the broad, featureless emission spectrum of the molecule characteristic of TADF (Uoyama et al., 2012).

case, the absence of a bound ground state is responsible for the featureless shape of the emission spectra.

Thermally activated delayed fluorescence is also commonly observed in metal–organic complexes such as the Cu carbenes in **Scheme 1**, Section 3.6.1. Indeed, TADF in Cu complexes have been extensively studied due to their high emission efficiencies that are useful in OLEDs (Wallesch et al., 2014, Shi et al., 2017,

LUMO HOMO

3

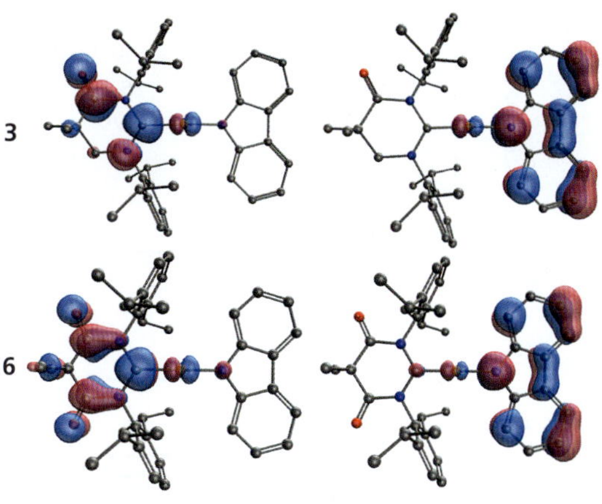

6

Figure 3.89 Frontier orbitals. LUMO (left) and HOMO (right) of compounds **3** (top) and **6** (bottom), **Scheme 1** (Shi et al., 2019).

Reprinted with permission from J. Am. Chem. Soc., Shi, S., Jung, M. C., Coburn, C., A.Tadle, D. Sylvinson, Djurovich, P. I., Forrest, S. R. & Thompson, M. E., Highly Efficient Photo and Electroluminescence from Two Coordinate Cu(I) Complexes Featuring Non conventional N heterocyclic Carbenes. Copyright 2019 American Chemical Society.

Yersin et al., 2017). The frontier molecular orbitals of Compounds **3** and **6** in **Scheme 1** calculated by DFT and TD-DFT and shown in Fig. 3.89 provide insight into the effectiveness of this molecular motif in promoting high efficiency delayed fluorescence (Shi et al., 2019). The Cu linkage between the carbazoyl (to the right) and the carbene (MAC or DAC) separates the electron densities of the HOMOs and LUMOs such that there is little or no overlap between them. Transitions thus require intramolecular charge transfer across a small ΔE_{ST}, as required for efficient TADF at room temperature. The Cu-complexes show featureless emission (see Fig. 3.40) and absorption spectra that are due to excimer-like charge transfer between the excited and ground state orbitals.

The photophysics of the Cu–carbene complexes have been investigated to understand the mechanisms and energetics leading to highly efficient TADF. The measured transient PL decay vs. temperature of compound **3** is shown in Fig. 3.90a (data points). The lifetime changes rapidly from a plateau of > 200 μs at low temperature, to approximately 1 μs as temperature is increased above 175 K. The data are fit by assuming that the lower triplet levels are populated according to Boltzmann statistics. Then, following Eq. 3.202, we can write an analogous expression for the TADF PL lifetime:

$$\bar{k}_{TADF} = \frac{2k_{12} + k_3\exp(-\Delta E_{13}/k_BT) + k_F\exp(-\Delta E_{S_1,1}/k_BT)}{2 + \exp(-\Delta E_{13}/k_BT) + \exp(-\Delta E_{S_1,1}/k_BT)}, \tag{3.212}$$

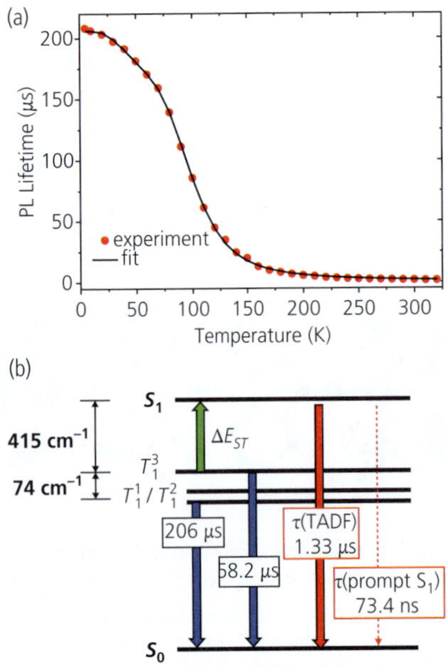

Figure 3.90 (a) Photoluminescent transient decay vs. temperature of Cu–carbene compound **3** in Fig. 3.89. Measurements shown as circles, fit to theory in text as lines. (b) Energy level scheme inferred from the fit in (a). After Shi et al. (2019).

where the factors of 2 are due to the twofold degeneracy of states T_1^1 and T_1^2. The PL rate is determined by the rates at which S_1 is populated via excitations from the three triplet substates, 1, 2, and 3. This must compete against direct fluorescence from S_1 at rate $k_F = 1.2 \times 10^7\ \mathrm{s}^{-1}$ which is weighted by the factor, $\exp\left(-\Delta E_{S_1,1}/k_BT\right)$. A fit to the data in Fig. 3.90a (solid line) using this expression yields the energies in Fig. 3.90b. A substantial ZFS energy of 74 cm^{-1} and an exchange energy splitting of 415 cm^{-1} (51.5 meV) leads to a high room temperature photoluminescent quantum yield of 90% (Shi et al., 2019).

3.8 Energy transfer

The mobile character of the exciton arises from dispersion, $E(\mathbf{k})$, in its energy band structure discussed in Section 3.6. Hence, the effective mass of the exciton is finite, resulting in kinetic energy, $\hbar^2k^2/2m_r^*$, where $\hbar k$ is the momentum. That is, if the bands are dispersionless, as is the case for isolated atoms and molecules, then $\partial^2E/\partial k^2 \to 0$, in which case $m^* \to \infty$ which prohibits exciton motion. Its ability to move freely about the solid implies that exciton migration occurs via energy transfer from molecule to molecule.

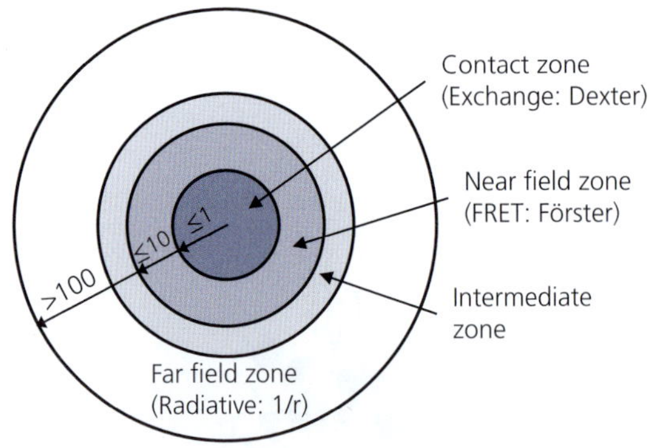

Figure 3.91 Interaction zones and approximate distances (in nm) over which they occur (shown as numbers) for $D^* \to A^0$ intermolecular energy transfer transitions. The contact zone is where D* and A^0 are in physical contact, radiationless dipole–dipole transfer occurs in the near field zone where Förster theory is valid, the intermediate zone is a transition region between radiationless and radiative transfer, and in the far field zone, emission by D* is absorbed by A^0. In this zone, interactions scale as $1/r$, where r is the intermolecular distance. After Förster (1959) and van der Meer (2014).

Energy transfer is governed by three principal processes, depending on the distance over which it occurs. These are: (i) long range *radiative transfer* (also known as *trivial energy transfer*), (ii) near field *Förster* (or *fluorescent*) *resonant energy transfer* (or FRET), and (iii) *exchange transfer* (or *Dexter transfer*) between molecules that are in direct physical contact. The interaction zones and characteristic interaction distances are shown in Fig. 3.91. In the transfer process, one molecule is a donor in its excited state, D*, which delivers its energy to an acceptor in its ground state, A^0 (terminology is the same as used for excimers and exciplexes in Section 3.6.3). Sometimes, the donor is referred to as the *sensitizer molecule*, and the acceptor is called the activator, but here we will use the former and more common terminology. Furthermore, the donor and acceptor can be molecules of similar or different compositions.

Following the Second Law of Thermodynamics, the transfer from D* to A^0 is always accompanied by energy loss. That is, if the exciton energy of D* is $E_D{}^*$ and of A is E_{A^0}, then efficient transfer requires $E_D{}^* > E_{A^0}$. This corresponds to exothermic transfer. The energy loss arises from relaxation of the molecule following excitation, for example, via internal conversion or lattice distortions that result in a Stokes shift. Furthermore, in the subsequent transfer step, the acceptor molecule becomes the donor, that once more undergoes the same relaxation process, lowering the energy even further from that of the original excitation. Nevertheless, if these losses are not excessive, energy can be extracted from the system (e.g. in the form of heat via phonon absorption) to enable endothermic energy transfer even when $E_D* < E_{A^0}$.

The primary objective of this section is to determine the energy transfer rates, k_{ET}, for each process. These are calculated using Fermi's Golden Rule (Eq. 3.69), where the interaction is Coulombic: that is the transition multipole expansion can be used for H_{int}, with different terms in the expansion dominating at different distances. The full quantum mechanical treatment used to describe these several process in a unified way is useful but complicated, for which the reader is referred to Andrews (1989). Here, we focus on the three primary regions of interest, omitting a discussion of transfer within the intermediate zone. It is convenient to begin our discussion with FRET in the near field region, and then move to exchange interactions, finishing up with radiative transfer. In addition to providing a derivation based on quantum mechanics, where necessary we will provide a semiclassical description that may be more useful in the practical understanding of energy transfer.

3.8.1 Near field radiationless energy transfer: FRET

In Chapter 2 we showed how induced dipole-induced dipole interactions provide the adhesive force for van der Waals solids. This same interaction is responsible for radiationless coupling of the excitation of a donor molecule to a nearby, unexcited acceptor. The first treatment of this widely observed phenomenon is attributed to J. Perrin (Perrin, 1927) and F. Perrin (Perrin, 1932). While their work introduced the concept of energy exchange between coupled oscillators, errors in their treatments led to a considerable overestimation of the distances over which transfer occurs: ~20–100 nm compared to actual distances of only ~5 nm. These inconsistencies were corrected in a landmark paper by T. Förster that presented the first comprehensive and accurate analysis of radiationless energy transfer in the near field (Förster, 1948).

In FRET, an excited singlet donor ($^1D^*$) is coupled to an unexcited acceptor ($^1A^0$) via dipole–dipole interactions. After a time $\tau_{ET} = 1/k_{ET}$ the excitation is transferred according to

$$^1D^* + {}^1A^0 \rightarrow {}^1D^0 + {}^1A^*, \qquad (3.213)$$

a process illustrated in Fig. 3.92. That is, the excited singlet donor transfers its energy to a nearby (~1–10 nm) ground state acceptor. This process requires that the D and A transitions be quantum

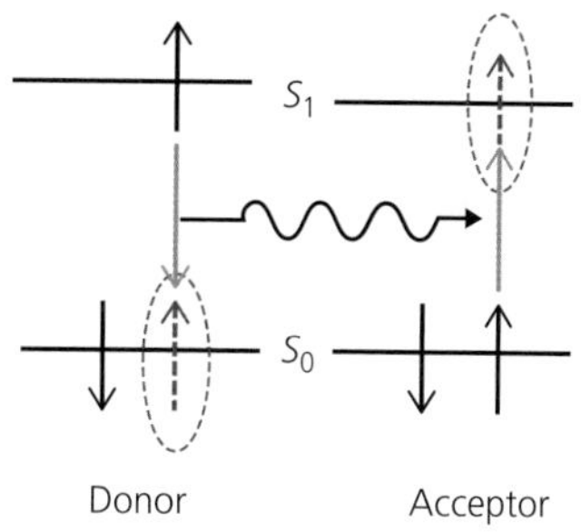

Figure 3.92 Förster resonant energy transfer from a singlet excited state donor to a ground state acceptor. The small solid and dashed arrows are the electron and hole spins, respectively, and the large arrows show the transitions.

mechanically allowed, and therefore the initial and final spin states of the D and A must be identical. However, transfer to an acceptor between triplet manifolds is also allowed via

$$^1D^* + {}^3A_m \rightarrow {}^1D^0 + {}^3A_n, \qquad (3.214)$$

where $1 \leq m < n$ are the indexes of the initial and final acceptor triplet states (e.g. $T_2 \leftarrow T_1$ or $T_3 \leftarrow T_1$). This transition requires that ^3A_m is already populated, that is, it is already in an excited state. Finally, we note that when the donor molecule is capable of efficient triplet emission (e.g. it contains a heavy metal atom), then process Eq. 3.213 is modified as follows:

$$^3D^* + {}^1A^0 \rightarrow {}^1D^0 + {}^1A^*. \qquad (3.215)$$

This process, known as *phosphor sensitized fluorescence*, was predicted by Förster (1959) and has been demonstrated as a route to high efficiency fluorescent OLEDs (Baldo et al., 2000) (see Chapter 6.3).

The transfer process is governed by electrostatic multipole interactions between the neutral D* and A molecules (Eq. 2.39 *ff.*), viz.

$$H_{int} = \frac{q^2}{4\pi\varepsilon_0\varepsilon_r R^3}\left\{\mathbf{r}_D \cdot \mathbf{r}_A - \frac{3(\mathbf{r}_D \cdot \mathbf{R})(\mathbf{r}_A \cdot \mathbf{R})}{R^2}\right\} + O\left(\frac{1}{R^4}\right) + \ldots \qquad (3.216)$$

Here, R is the distance between D and A, and $\mathbf{r}_D$ and $\mathbf{r}_A$ are the positions of dipoles of moment $\mu = -q\mathbf{r}$. The term in brackets is due to dipole–dipole interactions. The higher order terms represent dipole-quadrupole and higher multipole interactions, which due to their very short range, can be safely ignored. The vector relationships between these quantities are illustrated in Fig. 3.93. It follows from Eq. 3.69 that the energy transfer rate is:

$$k_{ET}(E) = \frac{2\pi}{\hbar}\left|\langle\Phi_f|H_{int}|\Phi_i\rangle\right|^2\rho(E) = \frac{2\pi}{\hbar}\left|M_{fi}\right|^2\rho(E) \qquad (3.217)$$

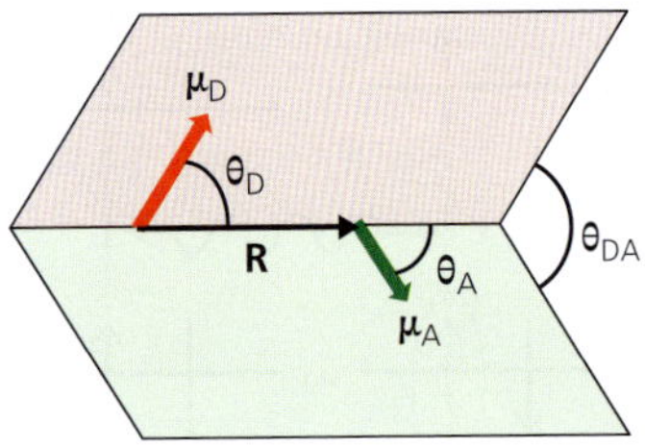

Figure 3.93 Vector relationships between the donor and acceptor electric dipole moments, $\mathbf{\mu}_D$ and $\mathbf{\mu}_A$, respectively, and their separation, **R**.

where $\Phi_{i,f}$ are the wavefunctions, including both electronic and nuclear components in the initial (i) and final (f) states.

In the very weak coupling limit of Förster transfer, there is no unique correspondence between the vibronic levels of the initial and final states. That is, due to numerous degrees of freedom, the vibronic levels are inhomogeneously broadened, resulting in a distribution of transition densities across the excited donor and ground state acceptor electronic manifolds. This must be considered when we determine $\rho(E)$. A graphical representation of this situation is shown in Fig. 3.94.

To determine k_{ET}, we first calculate the matrix element in Eq. 3.216. The wavefunctions explicitly include both the nuclear (ϕ_N) and electronic parts (ϕ_e), such that

$$\Phi_i\left(E_{D^*}, E_{A^0}\right) = \phi_e^{D^*}\phi_e^{A^0}\phi_N^{D^*}\left(E_{D^*}\right)\phi_N^{A^0}(E_{A^0}) \quad (3.218)$$

and

$$\Phi_f\left(E_{D^0}, E_{A^*}\right) = \phi_e^{D^0}\phi_e^{A^*}\phi_N^{D^0}\left(E_{D^*} - E\right)\phi_N^{A^*}(E_{A^0} + E), \quad (3.219)$$

where the energy transferred from D* to A^0 is E, and the other energies correspond to the molecular states as denoted in their subscripts. Then,

$$\begin{aligned} |M_{fi}|^2 = |\langle\Phi_f|H_{\text{int}}|\Phi_i\rangle|^2 &= \left|\left\langle\phi_e^{D^0}\phi_e^{A^*}\middle|H_{\text{int}}\middle|\phi_e^{D^*}\phi_e^{A^0}\right\rangle\right|^2 \\ &\times\left|\langle\phi_N^{D^0}\left(E_{D^*} - E\right)\middle|\phi_N^{D^*}\left(E_{D^*}\right)\right\rangle\right|^2 \\ &\times\left|\left\langle\phi_N^{A^*}(E_{A^0} + E)\middle|\phi_N^{A^0}(E_{A^0})\right\rangle\right|^2 \\ &= |M_{e;fi}|^2 FC_D\left(E_{D^*}; E_{D^*} - E\right) \times FC_A(E_{A^0}; E_{A^0} + E), \end{aligned} \quad (3.220)$$

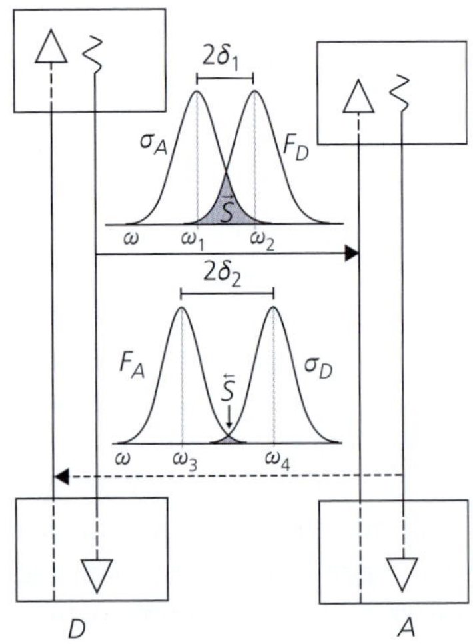

Figure 3.94 Coupling between the broad donor (D) and acceptor (A) bands involved in FRET. Here, $F_{A,D}(\omega)$ are the emission spectra, and $\sigma_{A,D}(\omega)$ the absorption spectra of the (acceptor, donor) molecules. Their respective Stokes shifts are $2\delta_1$ and $2\delta_2$. S is the overlap between the two (Andrews and Rodriguez, 2007).

where the overlap integrals of the ground and excited state vibronic levels are given by their respective Franck-Condon factors. Also, we have used $M_{e;fi}$ for the Coulomb integral resulting from the dipole–dipole interaction in Eq. 3.216.

The next step is to calculate the Coulomb integral. For this, we refer to the vector relationships of $\mathbf{r}_A$ and $\mathbf{r}_B$ in Fig. 3.93, to rewrite the second term in Eq. 3.216 as follows:

$$q^2\left\{\mathbf{r}_D \cdot \mathbf{r}_A - \frac{3(\mathbf{r}_D \cdot \mathbf{R})(\mathbf{r}_A \cdot \mathbf{R})}{R^2}\right\}^2 = \kappa_F^2 \mu_D^2 \mu_A^2. \quad (3.221)$$

Here, κ_F is the *orientation factor* that relates to the coupling between dipoles:

$$\kappa_F^2 = (cos\theta_{DA} - 3cos\theta_D cos\theta_A)^2. \quad (3.222)$$

This quantity has a value between 0 and 4, depending on the relative angles between dipoles. It is apparent that if the dipoles are orthogonal to each other, no energy transfer can occur. In a medium in which the dipoles are uncorrelated and hence randomly oriented, the spatial average is $\kappa_F^2 = \frac{2}{3}$. Thus, the efficiency of FRET can be used to determine the degree of orientational correlation between donor and acceptor. Specifically, the transfer efficiency in a solid can be measured as a function polarization of the incident radiation that is used to induce the transition $D^* \leftarrow D^0$ (Andrews and Bradshaw, 1999).

Using Eqs. 3.216 and 3.220, we obtain

$$|M_{fi}|^2 = \frac{\kappa_F^2 \mu_D^2 \mu_A^2}{(4\pi\varepsilon_0)^2 n_r^4 R_{DA}^6} FC(E_{D^*}; E_{D^*} - \hbar\omega) \times FC(E_{A^0}; E_{A^0} + \hbar\omega), \quad (3.223)$$

where R_{DA} is the distance between dipoles, and the real part of the index of refraction is related to the permittivity via $n_r = \sqrt{\varepsilon_r}$. Finally, integrating over the density of initial and final states within the electronic manifolds of D* and A^0, we obtain the total, energy-independent rate of:

$$k_{ET} = \frac{\kappa^2 \mu_D^2 \mu_A^2}{(2\pi)^3 (4\pi\varepsilon_0)^2 n_r^4 \hbar^2 R_{DA}^6} \times \int g^*_D(E_{D^*}) g_A^0(E_{A^0}) FC_D(E_{D^*}; E_{D^*} - \hbar\omega) \times FC_A(E_{A^0}; E_{A^0} + \hbar\omega) d\omega. \quad (3.224)$$

In this expression, we have introduced the Boltzmann factors, g^*_D and g_A^0, that account for the distribution of excitations over the several vibronic levels within the particular electronic manifold (i.e. D* or A^0). The dipole overlap integrals are conveniently expressed via

$$M_D(\omega) = \mu_D^2 \int g^*_D(E_{D^*}) FC_D d(E_{D^*}) \quad (3.225a)$$

and

$$L_A(\omega) = \mu_D^2 \int g_A^0(E_{A^0}) FC_A d(E_{A^0}). \quad (3.225b)$$

The integral Eq. 3.224 is proportional to the overlap of the normalized emission spectrum of the donor ($f_D(\omega)$) with the molar absorption spectrum of the acceptor. Thus, in terms of the un-normalized donor fluorescence spectrum, $F_D(\nu)$:

$$f_D(\nu) = \frac{F_D(v)}{\int F_D(v) dv} = \frac{F_D(\nu)}{\Phi_D}, \quad (3.226)$$

where Φ_D is the total fluorescence quantum yield of the donor. To obtain the exact expressions for these terms, Förster has shown (Förster, 1951) from the Einstein coefficients for spontaneous emission, A_{fi}, and stimulated absorption, B_{if} that:

$$M_D(\nu) = \frac{3}{4} \frac{\varepsilon_0 \hbar c^3 \Phi_D f_D(\nu)}{\pi n_r \tau_D \nu^3} \quad (3.227)$$

and

$$L_A(\nu) = \frac{6\varepsilon_0 \alpha_A(\nu) n_r \hbar c}{N_A \nu} \frac{M_{wA}}{\rho_{MA}}. \quad (3.228)$$

Here, M_{wA} is the molar molecular mass, ρ_{MA} is the mass density of the acceptor in the mixture, and τ_D is the total radiative lifetime (including non-radiative transitions) of the donor in the absence of the acceptor molecule. Then combining Eqs. 3.224, 3.225, 3.227, and 3.228 we arrive at the so-called Förster integral:

$$k_{ET} = \frac{9c^4}{128\pi^5} \frac{\kappa^2 \Phi_D}{N_A n_r^4 \tau_D R_{DA}^6} \frac{M_{wA}}{\rho_{MA}} \int \frac{f_D(\nu) \alpha_A(\nu)}{\nu^4} d\nu. \quad (3.229)$$

Often, it is convenient to use the imaginary part of the complex refractive index,

$$\tilde{n} = n_r + ik, \quad (3.230)$$

instead of the absorption coefficient, in which case we substitute $\alpha_A(\nu) = 4\pi k_A(\nu)/\lambda$ in Eq. 3.229. (Here, k is the extinction coefficient that should not be mistaken for a rate, Boltzmann's constant or any number of other confusing assignments used for this variable!) Alternatively, Eq. 3.229 can be restated in terms of the decadic molar extinction coefficient (cf. Eq. 3.97):

$$k_{ET} = \frac{9(2303)c^4}{128\pi^5} \frac{\kappa^2 \Phi_D}{n_r^4 \tau_D N_A R_{DA}^6} \int \frac{f_D(\nu) \varepsilon_A(\nu)}{\nu^4} d\nu. \quad (3.231)$$

Finally, we can write this expression in terms of the acceptor absorption cross section, $\sigma_A = \frac{4\pi k_A}{\lambda} \frac{M_{wA}}{N_A \rho_{MA}}$ to obtain

$$k_{ET} = \frac{9c^4}{8\pi} \frac{\kappa^2 \Phi_D}{n_r^4 \tau_D R_{DA}^6} \int \frac{f_D(\omega) \sigma_A(\omega)}{\omega^4} d\omega. \quad (3.232)$$

Equation 3.229 (and its other versions) has several notable features. First, it is independent of $\hbar$, even though our treatment has been based on the quantum mechanical interaction of the donor and acceptor molecules! Indeed, the premise of Förster's original treatment of this phenomenon was that it is completely classical (Förster, 1948), although at a later date he came to an identical formulation based on quantum mechanics. A review and comparison of these various approaches has been summarized by B. Wieb van der Meer (2014). A further important feature of FRET is its dependence on $1/R_{DA}^6$ as is characteristic of very short range induced dipole-induced dipole interactions. Finally, the transfer rate follows ~Φ_D/τ_D, that is, it scales proportionately with the ratio of quantum yield to fluorescence lifetime of the donor. As the quantum yield increases and lifetime decreases, the transfer becomes increasingly efficient. This is a consequence of the fact that the coupling involves the exchange of quanta, and hence the efficiency of this exchange must be proportional to the radiative quantum yield. The inverse proportionality with lifetime implies that other competing processes can reduce the transfer efficiency. If τ_D is sufficiently short, FRET will dominate.

Finally, and most importantly, the integral is proportional to the degree of *overlap* between the donor fluorescence and acceptor absorption spectra. This, in effect, expresses the extent that the energy levels of these states are in resonance. The Förster integrals can be simplified if we rearrange Eq. 3.229 as follows:

$$k_{ET} = \frac{1}{\tau_D}\left(\frac{R_0}{R_{DA}}\right)^6, \qquad (3.233)$$

where

$$R_0^6 = \frac{9c^4}{128\pi^5}\frac{\kappa^2\Phi_D}{N_A n_r^4}\frac{M_{wA}}{\rho_{MA}}\int\frac{f_D(\nu)\alpha_A(\nu)}{\nu^4}d\nu = A_\nu J_\nu \qquad (3.234)$$

is the *Förster radius* or the *transfer distance*, which is proportional to the spectral overlap integral, J_ν. Thus, when $R_{DA} = R_0$, then $k_{ET} = 1/\tau_D$. In this circumstance, the transfer time is equal to the radiative lifetime, reducing the transfer efficiency by 50%:

$$\eta_{ET} = \frac{\tau_D}{\tau_{ET} + \tau_D} \rightarrow \frac{1}{2}. \qquad (3.235)$$

Here, $1/\tau_D = 1/\tau_F + 1/\tau_{NR}$ where τ_F is the natural fluorescence lifetime of the donor, and τ_{NR} is the time for non-radiative decay via parasitic loss channels. The Förster radius is thus strongly dependent on the spectral overlap integral.

An example of overlap, and the resulting calculated values of R_0 using Eq. 3.234 are the emission and absorption spectra shown in Fig. 3.95 for a layer of a DPT film in contact with rubrene (Rbn), as well as Rbn in contact with DBP. The spectral overlap region is shaded. Clearly, J_ν for Rbn–DBP is much larger than for the DPT–Rbn system, yet the Förster radius differs by less than a factor of two ($R_0 = 2.5$ nm vs. 4.7 nm). This seeming contradiction points to the fact that $R_0 \sim J_\nu^{1/6}$, and hence is very insensitive to the degree of overlap and the other physical and optical properties of the molecular system. Most effective molecular species that exhibit efficient FRET have Förster radii between 3 nm and 5 nm (van der Meer et al., 1994).

It is worth discussing the effects of the orientation factor, κ_F^2, that carries a wealth of information about the morphology of the D–A blend, since the degree of coupling between molecules depends on their relative angles, θ_{DA}, θ_D, and θ_A (Fig. 3.93). Recovery of this information is complicated by the many other parameters in the prefactor to the overlap integral.

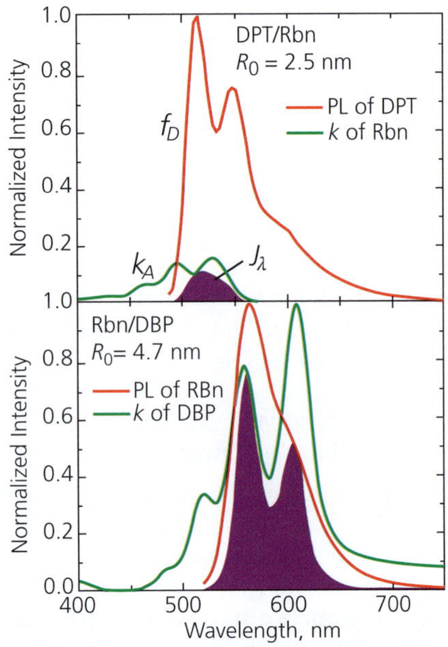

Figure 3.95 Normalized photoluminescence spectrum (corresponding to the fluorescence spectrum f_D) and extinction coefficient (k) for the (a) DPT/Rbn and (b) Rbn/DBP D–A pairs. The overlap between spectra are shaded, along with the calculated Förster radii (R_0 shown in legends) (Griffith and Forrest, 2014).

However, the coupling of polarized light to the donor molecule is proportional to $cos^2\theta$, where θ is the angle between the incident light and the dipole moment, $\boldsymbol{\mu}_D$. Hence, polarized light selectively excites only those molecules aligned to the incident polarization. If energy transfer occurs on a time scale less than the time to randomize the excited donors, then information about the polarization state is passed on to the acceptor population with an efficiency proportional to κ_F^2. We then can determine if there is a preferred orientation between D and A by measuring the polarization of light emitted by the acceptor. The degree polarization of the emitted light is

$$P = \frac{I_\parallel - I_\perp}{I_\parallel + I_\perp}, \qquad (3.236)$$

where $I_\parallel$ and $I_\perp$ are the polarized emission intensities parallel and perpendicular to the excitation polarization. If all absorbing dipoles are aligned parallel to the incident polarization, then $P = 0.5$. Further, for a sample where all emitting dipoles are perfectly aligned, the 3D rotational average should have a ratio of $I_\parallel : I_\perp = 3 : 1$. Differences from this ratio provide information about the degree of misalignment, or of the rate of randomization of the dipoles on time scales comparable to k_{ET} (Andrews and Bradshaw, 1999). Furthermore, the time dependence of this ratio can be used to study the dynamics of molecular orientation and motion within the solid or liquid that hosts the D and A molecules.

As noted above, FRET plays a significant role in the diffusion of excitons, and hence has been studied and exploited in OLEDs, OPVs, lasers, and many other organic devices. These phenomena will be discussed in the context of those devices in Part II. Also, we would be remiss in omitting the essential role that FRET plays in living organisms, most notably in the process of photosynthesis. The literature abounds with detailed studies of FRET in proteins, cells, nucleic acids, nanoparticles and so on (see, for example, Medintz and Hildebrandt, 2014). For example, the light harvesting complexes in purple bacteria comprise ring structures that contain optical absorbing porphyrins. The photogenerated exciton is transferred from porphyrin to porphyrin, eventually finding its way to a reaction center (RC) in the middle of the ring that transforms the excitation into free charge. Hence, the rings serve as optical antennas that funnel excitons to the charge separation center. An example system comprised of two light harvesting complexes, LH1 and LH2 that serve as coupled antennae in *Rhodobacter sphaeroides* is shown in Fig. 3.96 (Sundström et al., 1999).

The complex has been optimized to assure that the path from light absorption to charge generation is nearly 100% efficient. Its properties include: (i) optimization of the spectral characteristics of the donor–acceptor bacteriochlorophyll (BChl) pair B800 and B850 (with absorption between wavelengths of 800 nm to 850 nm, respectively) to assure resonant coupling between molecular units. (ii) Precise positioning of molecules involved in the transfer to less than a Förster radius. The spacing is maintained by a peptide scaffold that also forms two concentric rings to accurately positions the BChl molecules. The separation between B800 and B850 is estimated to be only 1.5 nm to 1.9 nm (see Fig. 3.96a). (iii) Extremely fast and nearly lossless transfer between molecular units around the nine-membered (nonomer) LH2 ring structure, and between LH2 and LH1 and the RC at the central axis of the antenna assembly. Indeed, dipole–dipole coupling energies of ~15 cm^{-1} between B800 and B850 ensure almost no energy loss during each transfer step.

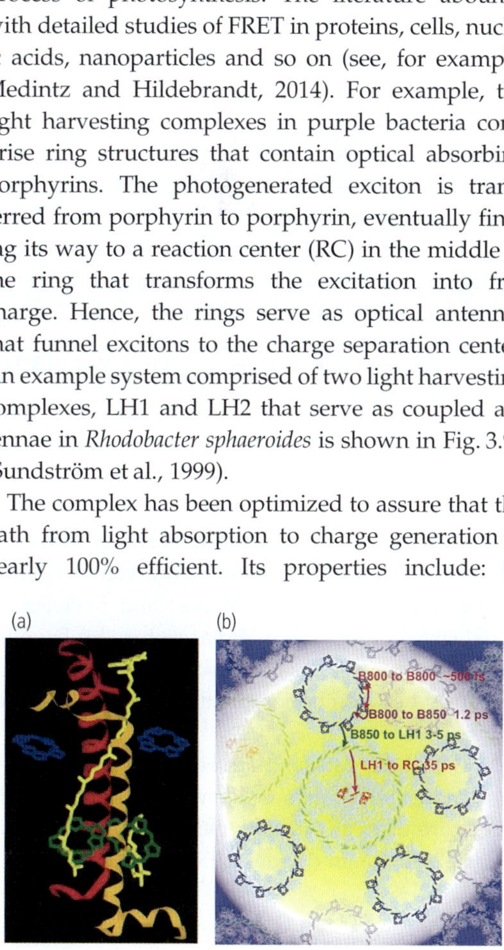

Figure 3.96 The photosynthetic light harvesting complex found in *Rhodobacter sphaeroides.* (a) A detail of one unit of the nine-sided light harvesting complex, LH2, shown as ring structures in (b). The red and orange ribbons are the *a* and *b* polypeptide helices, respectively, that form a scaffold to support the light absorbing B800 (blue) and B850 (green) bacteriochlorophylls (BChl). Carotenoids are shown in yellow. (b) The BChl are precisely positioned to maximize FRET from the LH2 antennae to the LH1 ring centered on the reaction complex (RC) where the excitons are separated into free charge. The transfer times (τ_{ET}) between moieties are indicated (Sundström et al., 1999).

In Fig. 3.96b, we find that the rates of transfer around the rings, and between rings and RC range from 500 fs to 35 ps. There are no other processes in this photosynthetic unit that can compete on these time scales, resulting in near unity quantum yield. Finally, (iv) the carotenoid molecules are extremely tightly coupled (<1 nm spacing) with the pigments, particularly B850 in Fig. 3.96a. The carotenoids (Crt) perform two functions, one in energy transfer and the other in stabilizing the pigment molecules by scavenging damaging triplet states and reactive singlet oxygen. The Crt, S_2 state is very short lived, and hence it efficiently transfers energy to the Q-band absorption of B850, thus broadening the spectral sensitivity of the antenna system. Due to the very close intermolecular spacings, transfer occurs in only ~200 fs, thus competing favorably with parasitic non-radiative processes.

Fluorescent resonant energy transfer plays a central role in this complex antenna network that has been optimized by evolution to ensure rapid and highly efficient energy harvesting in plants and microbial organisms. The foundations to our understanding of energy transfer laid by Förster nearly 70 years ago has opened many new fields in the biological and physical sciences.

3.8.2 Energy transfer in the contact zone: Dexter transfer by electron exchange

When the excited donor is in physical contact with an acceptor molecule, the electron from the donor can be directly transferred to the acceptor simultaneously with the transfer of an electron between their ground states. The exchange can occur between states of either the same or different spin

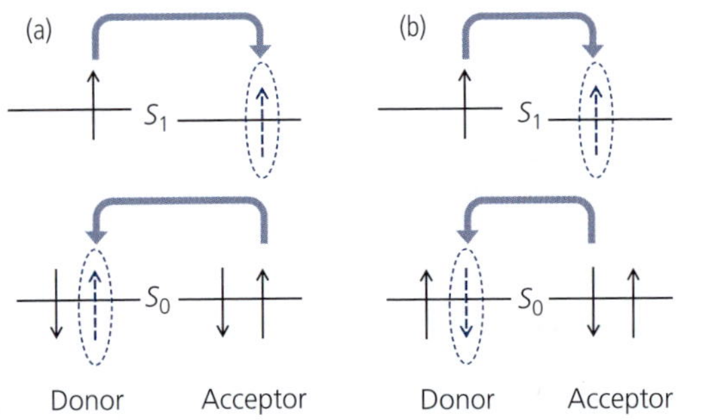

Figure 3.97 (a) Singlet–singlet and (b) triplet–triplet electron exchange energy transfer in the contact zone.

multiplicities, as long as the excited and ground states do not change multiplicity during transfer. In Fig. 3.97 we illustrate that charge exchange, or Dexter transfer, occurs via

$$^1D^* + {}^1A \rightarrow {}^1D + {}^1A^*_, \tag{3.237a}$$

$$^3D^* + {}^1A \rightarrow {}^1D + {}^3A^*_, \tag{3.237b}$$

with obvious variations where the triplet and singlet states are interchanged. Reactions in Eq. 3.237a and b are analogous to those in Fig. 3.97a and b, respectively.

Exchange transfer was first considered by Dexter who extended the work of Förster to include both the exchange and Coulomb integrals in the analysis (Dexter, 1953). We can derive the transfer rate by starting with the initial (i) and final (f) state antisymmetrized donor–acceptor spinorbital wavefunctions as follows:

$$\Psi_i = \frac{1}{\sqrt{2}}\left[\Psi_{D^*}(\mathbf{r}_1, \mathbf{Q}_1; S_{1D^*})\Psi_{A^0}(\mathbf{r}_2, \mathbf{Q}_2; S_{2A^0}) - \Psi_{D^*}(\mathbf{r}_2, \mathbf{Q}_2; S_{2D^*})\Psi_{A^0}(\mathbf{r}_1, \mathbf{Q}_1; S_{1A^0})\right] \tag{3.238}$$

and

$$\Psi_f = \frac{1}{\sqrt{2}}\left[\Psi_{D^0}(\mathbf{r}_1, \mathbf{Q}_1; S_{1D^0})\Psi_{A^*}(\mathbf{r}_2, \mathbf{Q}_2; S_{2A^*}) - \Psi_{D^0}(\mathbf{r}_2, \mathbf{Q}_2; S_{2D^0})\Psi_{A^*}(\mathbf{r}_1, \mathbf{Q}_1; S_{1A^*})\right]. \tag{3.239}$$

The coupling matrix element, M_{fi}, Eq. 3.220, is then:

$$\begin{aligned} M_{fi} = \frac{1}{2}\Big[&\langle\Phi_{D^0}(\mathbf{r}_1, \mathbf{Q}_1)\Phi_{A^*}(\mathbf{r}_2, \mathbf{Q}_2) \\ &\times|H_{int}|\Phi_{D^*}(\mathbf{r}_1, \mathbf{Q}_1)\Phi_{A^0}(\mathbf{r}_2, \mathbf{Q}_2)\rangle \\ &\times\langle\sigma_{D^0}(S_{1D^0})\sigma_{A^*}\left(S_{2A^*}\right)|\sigma_{D^*}\left(S_{1D^*}\right)\sigma_{A^0}(S_{2A^0})\rangle \\ &-\langle\Phi_{D^0}(\mathbf{r}_1, \mathbf{Q}_1)\Phi_{A^*}(\mathbf{r}_2, \mathbf{Q}_2)|H_{int}|\Phi_{D^*}(\mathbf{r}_2, \mathbf{Q}_2)\Phi_{A^0}(\mathbf{r}_1, \mathbf{Q}_1)\rangle \\ &\times\langle\sigma_{D^0}(S_{2D^0})\sigma_{A^*}\left(S_{1A^*}\right)|\sigma_{D^*}\left(S_{1D^*}\right)\sigma_{A^0}(S_{2A^0})\rangle\Big] \end{aligned} \tag{3.240}$$

The first term is the Coulomb integral that couples the excited and ground states of the donor and acceptor pair. It is the same integral that leads to Förster transfer in the previous section. Since the spin eigenfunctions, σ, are orthogonal, the spin integral in the first term vanishes unless $S_{1D^0} = S_{1D^*}$ and $S_{2A^0} = S_{2A^*}$; that is the spin multiplicity on a given site must be conserved in a Coulomb-mediated interaction between the initial and final states. Thus, there can be no spin-flip between the initial and final states of either the donor or acceptor. Indeed, this is a condition of Förster transfer, as discussed in Section 3.8.1, except for situations where there is strong mixing between states of different multiplicities (e.g. in triplet-emitting metalorganic complexes).

The second term in Eq. 3.240 is the exchange integral: its magnitude depends on the spatial overlap of the wavefunctions, Φ_D and Φ_A between sites $\mathbf{r}_1$ and $\mathbf{r}_2$. This integral is only significant when the donor and acceptor molecules are in close contact. The interaction is purely Coulombic, that is,

$$H_{int} = -q^2/4\pi\varepsilon_0\varepsilon_r|\mathbf{r}_1 - \mathbf{r}_2| = -q^2/4\pi\varepsilon_0\varepsilon_r R_{DA}. \tag{3.241}$$

In contrast to near-field interactions, the spin integral in the second term in Eq. 3.240 places no prohibition on spin-flip during transfer. That is, the integral is non-vanishing when $S_{1A^*} = S_{1D^*}$ and $S_{2D^0} = S_{2A^0}$, where the spins are exchanged between molecular sites. Then, the spin of the excited acceptor is equal to the spin of the excited donor, and likewise the initial states of the donor and acceptor also have the same spin. These spins must therefore transfer simultaneously between donor and acceptor, as illustrated in Fig. 3.97.

Now that the transfer matrix element is defined, we once more use Fermi's Golden Rule along with the procedure used to calculate the Förster integral (e.g. Eq. 3.232) to obtain (Dexter, 1953)

$$k_{ET} = \frac{2\pi}{\hbar}\Gamma^2\int f_D(\omega)\sigma_A(\omega)d\omega. \tag{3.242}$$

(Note: consistent with spectroscopic notation in Eq. 3.232, σ_A is the absorption cross section of the acceptor, as opposed to its spin wavefunctions.) Here, Γ^2 is the overlap integral of the second term in Eq. 3.240 by summing $|M_{fi}|^2$ over all final and initial states, weighted by the appropriate Boltzmann factors that account for the state occupancies. Also, the $1/\omega^4$ dependence in the Förster integral is absent in the denominator of Eq. 3.242 as a result of the $1/R_{DA}$ Coulomb interaction in the latter, in contrast to dipole–dipole coupling with its $1/R_{DA}^3$ dependence in FRET. In the simplest approximation, the wavefunctions of the initial and final states are hydrogenic, such that:

$$\Gamma^2 \propto \frac{q^2}{(4\pi\varepsilon_0)^2 n_r R_{DA}^2} \exp(-2R_{DA}/L), \quad (3.243)$$

where L is the effective Bohr radius (or more accurately the distance of closest molecular spacing as determined by van der Waals forces, see Section 2.3.3) of the contacting molecules. The exponential is proportional to the probability for an electron on the donor to tunnel to a neighboring acceptor. Combining Eq. 3.242 and 3.243 results in the approximate expression for exchange transfer in the contact zone:

$$k_{ET} = K\frac{q^2}{8\pi\varepsilon_0^2 \hbar n_r R_{DA}^2}\left\{\int f_D(\omega)\sigma_A(\omega)d\omega\right\} \times \exp(-2R_{DA}/L) = \frac{K' J_\omega}{R_{DA}^2}\exp(-2R_{DA}/L). \quad (3.244)$$

Here, K and K' are numerical constants that account for the degeneracy of the donor and acceptor molecules, the quantum yield of the donor, and other quantum mechanical effects, and J_ω is the spectral overlap integral given by the term in brackets. In contrast to the Förster process, Dexter transfer is a purely quantum mechanical phenomenon, which is clear from the appearance of $\hbar$ in Eq. 3.244. Similar to FRET, we see that exchange transfer also depends on energetic resonance between the donor and acceptor states expressed by the non-vanishing overlap (J_ω) between their emission and absorption spectra, respectively. We further note that there is no dependence on the orientation factor, κ^2. In the spherical wavefunction approximation, there can be no preferred direction of transfer. In realistic molecular systems, however, the transfer occurs preferentially along directions of closest intermolecular contact. Hence, we expect that the exchange transfer will be dependent on the molecular shape, steric influences, crystal structure (and hence anisotropy), direction, and film morphology. Since the orbital overlap between molecules can be small, depending on whether the material is crystalline, polycrystalline or amorphous, and since the rate decreases exponentially with distance, we expect the maximum transfer distance to extend to only one or a few lattice constants from the donor molecule in almost all circumstances.

The rate of exchange is comparable to the tunneling time, which for transfer between nearest neighbors is $k_{ET} \sim 10^{10}$–10^{11} s^{-1}. We will show in Section 3.9 that both FRET and exchange interactions play a crucial role in exciton diffusion in molecular solids. Whereas FRET is the prevalent means for energy migration in fluorescent (singlet) systems, exchange processes dominate for triplets. There are exceptions to this situation when highly emissive phosphorescent molecules are involved.

3.8.3 Radiative energy transfer

Radiative, or trivial, energy transfer is a long range process whereby the excited donor radiates a photon of energy $\hbar\omega$ that is subsequently absorbed by the acceptor after travelling a distance, R_{DA}, through the medium (see Fig. 3.98). Radiative transfer occurs at distances >50 Å where neither exchange nor FRET are active. In this limit, the donor and acceptor are decoupled; hence there are no spin selection rules that govern the interaction although it is most likely to occur between a radiative donor (which can be a fluorescent or heavy metal containing phosphorescent molecule) and the singlet state of the acceptor. Note that $T_n \leftarrow T_m$ ($n > m \geq 1$) transitions are also possible in radiative transfer, although the acceptor must already be prepared in its excited triplet state, T_m, by, for example, strong optical pumping.

The two-step radiative transfer process follows

$$D^* \to D^0 + \hbar\omega \quad (3.245a)$$

and

$$A^0 + \hbar\omega \to A^*. \quad (3.245b)$$

The isotropic Coulomb interaction between donor and acceptor is once again found via the same method as used for FRET, giving the transfer rate (Andrews and Juzeliūnas, 1992):

$$k_{ET} = \frac{9\kappa_{rad}^2 \Phi_D}{8\pi n_r \tau_D R_{DA}^2}\int f_D(\omega)\sigma_A(\omega)d\omega, \quad (3.246)$$

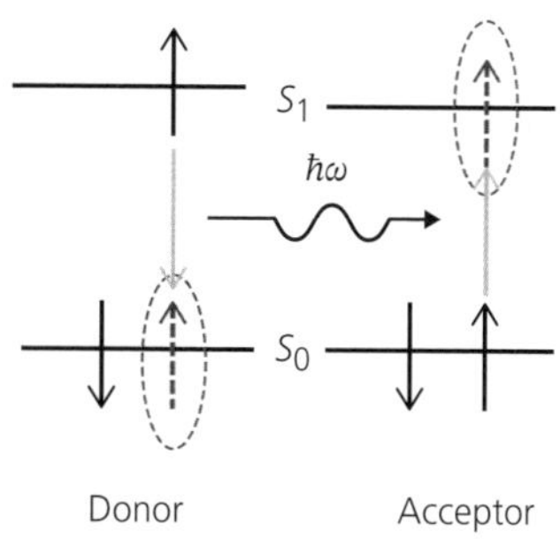

Figure 3.98 Radiative energy transfer involving emission of a photon by the donor and absorption by the acceptor in the far field zone.

where

$$\kappa_{rad}^2 = (\cos\theta_{DA} - \cos\theta_D\cos\theta_A)^2 \tag{3.247}$$

is the orientation factor between donor and acceptor dipoles that accounts for the spherically symmetric radiation field of the Coulomb potential.

Equation 3.246 assumes that a particular donor–acceptor pair at distance R_{DA} is involved in the transfer. This transfer rate scales as $1/R_{DA}^2$, but the number of acceptor molecules contained in a volume centered on the donor increases as R_{DA}^3, which implies that the transfer rate is a linearly increasing function of the size of the sample. This, of course is not the case since the probability of finding a photon at distance R is decreased by the probability that it has been absorbed by acceptors in the intervening volume with radius $<R$. We resolve this problem by weighting the cross section by the absorption factor $\exp(-\alpha_A(\omega)R_{DA})$, using $\alpha_A(\omega) = N_A\rho_{MA}\sigma_A(\omega)/M_{wA} = \sigma_A(\omega)\rho_{NA}$ where ρ_{NA} is the number density of acceptors. Thus Eq. 3.246 is corrected to yield

$$k_{ET} = \frac{9\kappa_{rad}^2\Phi_D}{8\pi n_r\tau_D R_{DA}^2}\int f_D(\omega)\sigma_A(\omega)\exp(-\sigma_A(\omega)\rho_{NA}R_{DA})d\omega. \tag{3.248}$$

The transfer efficiency (η_{rad}) is obtained by calculating the total number of photons emitted to those absorbed in the sample:

$$\eta_{rad} = \frac{1}{\Phi_D}\int f_D(\omega)[1 - \exp(-\sigma_A(\omega)\rho_{NA}d)]d\omega, \tag{3.249}$$

which scales with the inverse exponential of the total sample thickness, d.

Trivial energy transfer typically occurs over distances $1/\alpha_A$ ~10–50 nm. This lossy exothermic process implies only a few transfers typically occur before the photon escapes from the sample, and hence it does not significantly contribute to exciton migration.

The rate for radiative energy transfer is determined by the emission rate, that is, $k_{ET} \sim k_r$ which is typically 10^8–10^9 s^{-1} for fluorescence (cf. Fig. 3.84), although k_r for metalorganic triplet emitters are in the range 10^7–10^5 s^{-1}. Hence, its presence is detected by measuring the increase in radiative lifetime in a sample compared to the natural lifetime of the emitting molecule, and the dependence of this lifetime on sample size.

To summarize, we note that the total energy transfer rate is the sum of all active processes. Therefore, the total rate is given by

$$k_{ET,tot} = k_{ET,exch} + k_{ET,FRET} + k_{ET,rad}. \tag{3.250}$$

The functional dependencies of each of these transfer rates are summarized in Table 3.7.

3.9 Exciton diffusion

Multiple, sequential exciton transfers between individual molecules prior to recombination results in the migration of energy across macroscopic distances within the solid. The transfer from D^* to A^0 by one of the several processes in Section 3.8 is the shortest step taken in what is ultimately a random walk of the exciton, resulting in transport of energy over a characteristic distance, L_D, known as the *exciton diffusion length*. After a *diffusion time*, τ_D, the exciton recombines. We define the *diffusivity* (or *diffusion constant*) of the material as

$$D = \frac{L_D^2}{\tau_D}. \tag{3.251}$$

Whether or not diffusion over extended distances is efficient or even possible ultimately depends on the distance over which transfer must occur between donor and acceptor molecules. In homogeneous thin films, the donor is surrounded by acceptor molecules, and provided that the transfer process is not significantly exothermic (i.e. there is not a large energy loss in the $D^* \to A^0$ process), transfer occurs by simple diffusion for a time, $\tau_D = \tau_r$, the natural recombination lifetime of the exciton. However, if the donor and acceptor molecules are mixed into a matrix consisting of *host* molecules that have an energy gap larger than that of the donor (or *guest*), then the diffusion length is shorter than predicted in Eq. 3.251 if the separation

Table 3.7 Energy transfer processes and their properties

Process	Transfer rate	Distance dependence	Zone	Characteristic transfer distance
Exchange (Dexter)	$k_{ET,exch}$	$\frac{1}{R_{DA}^2}\exp(-2R_{DA}/L)$	Contact	<1 nm
FRET	$k_{ET,FRET}$	$\frac{1}{R_{DA}^6}$	Near	<10 nm
Radiative	$k_{ET,rad}$	$\frac{1}{R_{DA}^2}\exp(-\alpha_A R_{DA})$	Far	>100 nm

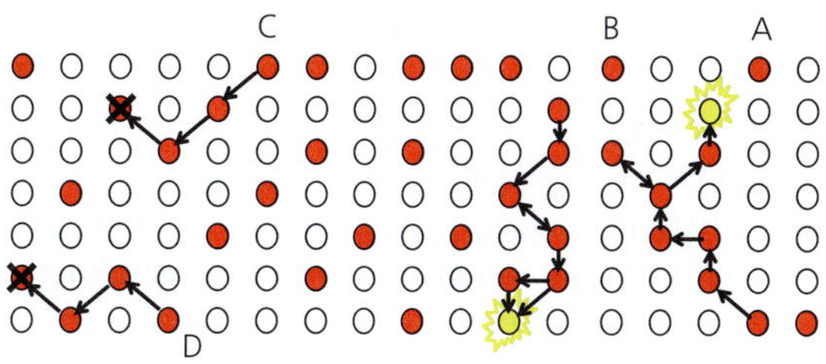

Figure 3.99 Diffusion of an excited state via sequential transfer from donor to acceptor molecules (filled circles) in a wide energy gap host matrix (open circles). The wide energy gap of the matrix prevents energy transfer from the host. Paths A and B are examples of percolation, resulting in radiative recombination (starbursts) of the guest after a diffusion length, L_D, is reached. Paths C and D are too short for the exciton radiate before it non-radiatively recombines (denoted by **X**).

between guest molecules exceeds a critical distance beyond which the exciton has a greater probability for recombination than transfer. In this case, $\tau_D < \tau_r$ for a guest concentration $[G]$ that is less than a critical value, $[G_c]$ that allows for an uninterrupted percolating path for diffusion, illustrated in Fig. 3.99. The dependence of diffusion length on $[G]$ is known as *dynamic percolation* (Monberg and Kopelman, 1978).

The abrupt onset of percolation for triplets governed by Dexter exchange is determined using the exchange integral, K. Guest molecules that are separated, on average by m host molecules, gives (Klafter and Jortner, 1977)

$$K = \beta(\beta/\Delta)^m, \tag{3.252}$$

where β is the exchange integral between nearest neighbors, and Δ is the energy difference between the guest and host excited states. The process whereby the excitation moves from D^* to A^0 over a bridge of m intervening host molecules is known as *superexchange* (Lambert et al., 2002). When the distance between D and A is small, a narrow impurity band develops over which transport (of excitons or charges) can occur, and the exciton is thereby delocalized. Alternatively, the exciton can be trapped within a cluster. In this case, no band exists, and the solid is transformed from a conductor to an insulator. This transition is known as *Anderson localization* (Anderson, 1958), and is observed for a range of physical phenomena including spin waves, acoustic waves, and conductor-to-insulator transitions in disordered media.

For inhomogeneously broadened spectral lines arising from off-diagonal disorder (i.e. the interaction occurs between two spatially separated molecules), the uncertainty principle implies that the hopping time is $\tau_{hop} = \hbar/K \leq \tau_r$, where the latter inequality is required for dynamic percolation to occur. Then,

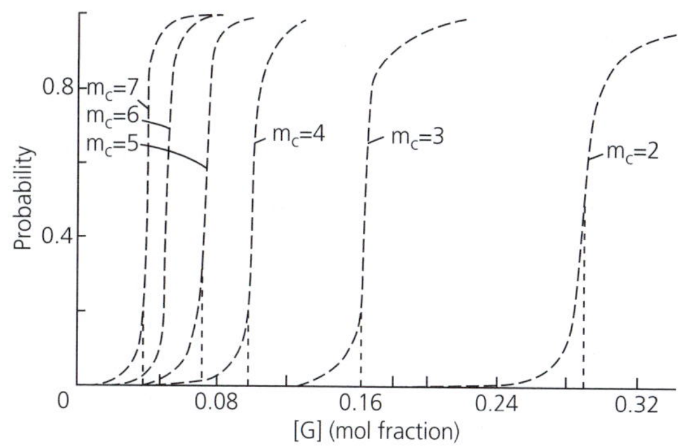

Figure 3.100 Percolation probability for a square lattice of $C_{10}H_8$ (guest) molecules doped in $C_{10}D_8$ (host) as a function of guest concentration, $[G]$. The concentration at which the probability that the guest molecule is found in an "infinite cluster," that is, where the exciton diffusion length is not limited by its lifetime, corresponds to the intersection of the straight dashed lines with the $[G]$ axis. From Monberg and Kopelman (1978).

from Eq. 3.252, the number of steps leading to percolation, is

$$m_c = ln(\beta\tau_r/\hbar)/ln(\Delta/\beta). \tag{3.253}$$

The relationship between m_c and $[G_c]$ depends on the dimensionality of the solid and its lattice structure.

Interestingly, m_c for triplets is typically larger than for singlets. For example, for the system consisting of guest $C_{10}H_8$ molecules in a $C_{10}D_8$ host, the singlet lifetime is $\tau_r \approx 128$ ns vs. 2.6 s for triplets. For this system, $\beta \approx$ 1–10 cm^{-1}, and $\Delta \approx 30$–100 cm^{-1}, leading to $K \approx 10^{-2}$–10^{-4} cm^{-1} (Klafter and Jortner, 1977). Thus, for singlets, $m_c \approx 2$, whereas for triplets, $m_c \approx 6$. That is, owing to the significantly longer lifetime of triplets, the number of sites over which superexchange can occur is correspondingly larger, resulting in a concomitant reduction in $[G_c]$. The probability for the onset of percolation for a square lattice is shown in Fig. 3.100.

Diffusion is governed by *Fick's law*. Thus, the time-dependent exciton population density of $n(\mathbf{r},\lambda)$, at position $\mathbf{r}$ due to absorption at wavelength λ is given by the diffusion equation:

$$D\nabla^2 n(\mathbf{r},\lambda,t) - \frac{n(\mathbf{r},\lambda,t)}{\tau_D} + G(\mathbf{r},\lambda,t) = \frac{\partial n(\mathbf{r},\lambda,t)}{\partial t}, \tag{3.254}$$

where $G(\mathbf{r}, \lambda, t)$ is the generation rate of excitons in the medium at time, t. In steady state, the right hand side of Eq. 3.254 is zero.

In Fig. 3.101 we show the solution to Eq. 3.254 for light propagating in the $+x$ direction with intensity I_0

and incident on the sample surface at $x = 0$. From the Beer–Lampert law, $G(x, \lambda, t) = I_0(\lambda, t)\exp[-\alpha(\lambda)x]$ where $\alpha(\lambda)$ is the absorption coefficient. The calculations are for two different values of α, and for the boundary conditions of $n(x = 0) = 0$, or $\frac{\partial n(0)}{\partial x} = 0$, corresponding to complete, or zero quenching (known as blocking), respectively, of excitons at the surface. For blocking conditions, the exciton density is at a maximum at the light incident surface. Far from the surface, the exciton density simply follows the exponential decay of the generation rate profile. The diffusion length can be accurately measured in luminescent samples possessing one of these two boundary conditions by determining their luminescence intensity as a function of sample thickness. For example, if the excitons are quenched at the surface of incidence, the fluorescence intensity increases with film thickness since an increasing fraction of photons are absorbed deeper in the sample. By fitting the intensity from a series of samples of different thicknesses to the profiles calculated using Eq. 3.254, one can determine L_D. Alternatively, this same procedure can be applied to a single sample by measuring the luminescent intensity as a function of incident wavelength. Since absorption is wavelength dependent, this allows one to vary the *optical path length* within the sample for each wavelength chosen, from which a fit to the diffusion equation provides L_D.

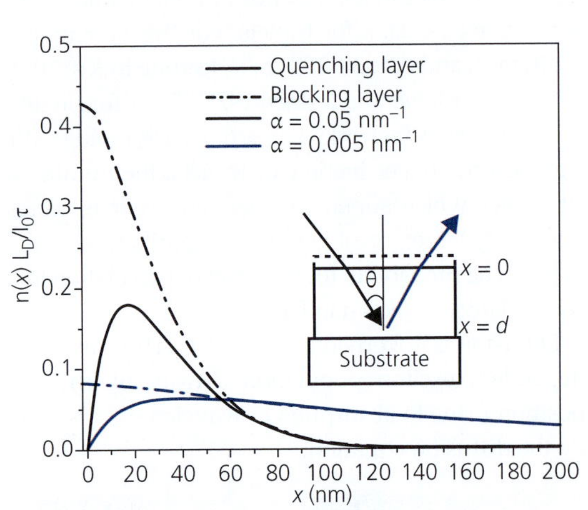

Figure 3.101 Exciton density from the surface at $x = 0$ as a function of boundary conditions normalized to the incident intensity. The sample thickness is $d >> L_D = 15$ nm. The exciton density is given for two different values of absorption coefficient, α. Here, I_0 is the incident light intensity and τ is the exciton lifetime. Inset: Geometry employed in measuring L_D (Lunt et al., 2009).

Reprinted from Lunt, R. R., Giebink, N. C., Belak, A. A., Benziger, J. B. & Forrest, S. R. 2009. Exciton Diffusion Lengths of Organic Semiconductor Thin Films Measured by Spectrally Resolved Photoluminescence Quenching. J. Appl. Phys., 105, 053711 with the permission of AIP Publishing

A particularly accurate method for measuring L_D is *spectrally resolved photoluminescence quenching* (SR-PLQ), where the photoluminescence intensity as a function of wavelength is measured using a capping layer that either quenches or blocks excitons at the material surface (Lunt et al., 2009). The experimental configuration for the SR-PLQ measurement is shown in Fig. 3.101 (inset). A wavelength dependent exciton distribution, $n(x)$, in a thin film is generated using a monochromatic light source incident at $x = 0$ through either a transparent exciton blocking or capping layer. For sufficiently thick films, $n \to 0$ as $x \to d$, the film thickness. The exciton distribution is then described by Eq. 3.254 modified to take into account the angle of the incident illumination, θ:

$$\frac{L_D{}^2}{\tau_r}\frac{\partial^2 n(x)}{\partial x^2} - \frac{n(x)}{\tau_r} + \frac{I_0\alpha(\lambda)}{cos\theta}\exp\left(-\frac{\alpha(\lambda)x}{cos\theta}\right) = 0. \tag{3.255}$$

By comparing the luminescence intensity of the sample with a blocking (PL_B) to that with a quenching layer (PL_Q), the normalized quenching ratio, η, is obtained using:

$$\eta(\alpha) = \frac{PL_B(\alpha)}{PL_Q(\alpha)} = \frac{\int_0^\infty n_B(x,\alpha)dx}{\int_0^\infty n_Q(x,\alpha)dx}. \tag{3.256}$$

Here, n_Q is the exciton density in the film with the quenching layer, n_B is the density in the film with the blocking layer,. Solving for $n(x)$ in Eqs. 3.255 and 3.256 we obtain

$$\eta(\alpha) = \frac{\alpha(\lambda)L_D}{cos\theta} + 1 = \alpha'(\lambda)L_D + 1. \tag{3.257}$$

where $\alpha'(\lambda)$ is the angle-corrected value of the absorption coefficient. Thus, the slope of a plot of PL_B/PL_Q versus $\alpha'(\lambda)$ yields the diffusion length in situations where there is negligible energy transfer between the film and the quenching layer. The analysis becomes somewhat more complicated when the films are thin, there is energy transfer between the film and the quenching layer, or there is overlap between the spectra of the sample and the blocking or quenching layers. While these situations can and should be avoided, they can nevertheless be accounted for by optical modelling of the sample structures, and including independent measurements of transfer radii, etc. (Lunt et al., 2009, Bergemann and Forrest, 2011).

Example results from SR-PLQ analysis of three archetype fluorophores used in organic photovoltaics and OLEDs are shown in Fig. 3.102. Excitation spectra

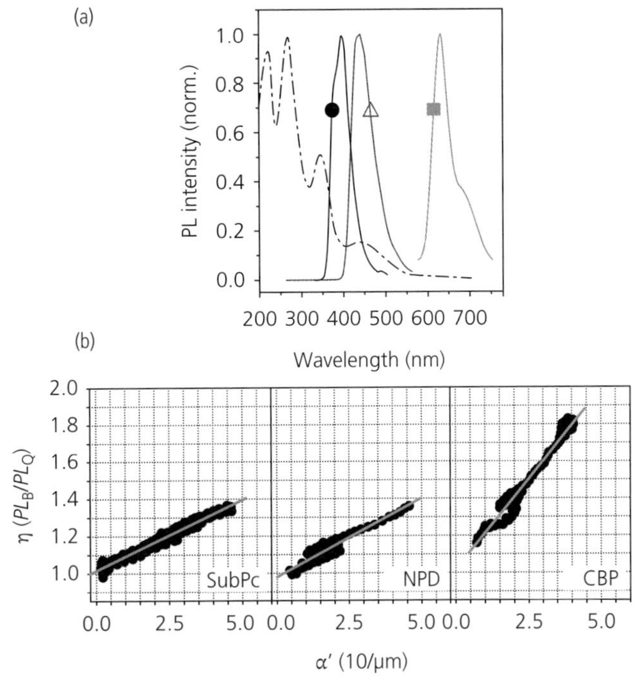

Figure 3.102 (a) Normalized photoluminescence spectra for CBP (filled circle), NPD (triangle), and SubPc (filled square). The normalized C_{60} absorption spectrum (dotted black line) highlights the overlap with CBP and NPD emission that must be considered in the diffusion length analysis. (b) Quenching ratio versus absorption coefficient for the materials in (a). The quenching data were fit using Eq. 3.257 (lines). For NPD and CBP, effects of Förster energy transfer are included (Lunt et al., 2009).

Reprinted from Lunt, R. R., Giebink, N. C., Belak, A. A., Benziger, J. B. & Forrest, S. R. 2009. Exciton Diffusion Lengths of Organic Semiconductor Thin Films Measured by Spectrally Resolved Photoluminescence Quenching. J. Appl. Phys., 105, 053711 with the permission of AIP Publishing

were collected at a fixed wavelength located near a peak in the emission are shown in Fig. 3.102a. Quenching and blocking layers were selected for each material based on the alignments of the HOMO and LUMO energies to the material under test, as well as their absorption characteristics to minimize energy transfer-enhanced quenching. For these samples, C_{60} was used for exciton quenching, and either BCP or NTCDA (in the case of CBP) for blocking.

The linear relationship between η and α' is shown in Fig. 3.102b, suggesting the presence of only a single, resolvable diffusion length (and hence a single excitonic species) that is independent of exciting wavelength. Note that this method is limited to materials with strong oscillator strengths in both absorption and emission, and hence is insensitive to non-radiative triplet or weakly emissive charge transfer states.

Table 3.8 provides diffusion lengths and diffusion coefficients obtained for several archetype organic electronic materials by SR-PLQ (Lunt et al., 2009). From these data we make two observations. In most cases, the diffusion length in crystals is longer than in amorphous solids. This is not surprising, since excitons are easily trapped and quenched at grain boundaries or disordered regions where energy transfer to neighboring molecular sites is impeded. The second, perhaps more surprising finding is that L_D for triplets in PtOEP is nearly equal to that of singlet excitons in fluorescent materials, even though the triplet lifetime is $\tau_D \sim 20$ μs compared to the more than three orders of magnitude smaller singlet lifetime. While the ratio of the triplet to singlet lifetime is not always this large, the triplet diffusion constant is nevertheless several orders of magnitude smaller than for singlets. This results from the lower energy of the triplet, resulting in self-trapping at molecular sites, thereby impeding transfer (by exchange interactions) to neighboring molecules.

Since diffusion over large distances occurs by a series of random transfer steps from donor to acceptor, the relationship between the acceptor capture rate and the diffusion constant can be obtained using the diffusion equation, Eq. 3.254. The analysis assumes that the probability for capture is unity when the exciton arrives within a distance, R_c, from the acceptor. Here, R_c is the *exciton capture radius*.

In the absence of other losses (e.g. recombination at defects or impurities), Eq. 3.254 simplifies to

$$D\nabla^2 n(\mathbf{r},t) = \frac{\partial n(\mathbf{r},t)}{\partial t}. \tag{3.258}$$

Table 3.8 Measured diffusion lengths (L_D) for singlet (S) and triplet (T) excitons of crystalline (C) and amorphous (A) films measured by spectrally resolved photoluminescence quenching

Material[a]	Exciton	Morphology[b]	Quenching/blocking layer	L_D (nm)	D (10^{-4}cm^2/s)
NPD	S	A	C_{60}/BCP	5.1	0.7
CBP	S	A	C_{60} (or NTCDA)/Free[c]	16.8	40
SubPc	S	A	C_{60}/bare	8.0	≥6.4
PTCDA	S	C, 55 nm (flat)	C_{60} (or NPD)/NTCDA	10.4	3.4
DIP	S	C, >150 nm (up)	C_{60}/Free[c]	16.5	15
DIP	S	C, 30 nm (flat)	C_{60}/Free[c]	21.8	26
C_{60}	S	A	NPD/BPhen	34	20
C_{70}	S	A	NPD/BPhen	10	
PtOEP	T—Mon.	C, >150 nm (up)	C_{60}/BCP	18.0	0.041
PtOEP	T—Dim.	C, >150 nm (up)	C_{60}/BCP	13.1	0.00061

[a] All data are from Lunt et al. (2009) except for C_{60} and C_{70} from Bergemann et al. (2015).
[b] Up/flat refers to whether the molecular plane is perpendicular/parallel to the substrate.
[c] Free = no layer.

The unity capture efficiency condition leads to the boundary condition: $n(r,t)|_{r=R_c} = 0$. Taking $n(\mathbf{r},0) = n_0$, in spherical coordinates, the solution to Eq. 3.258 is (Smoluchowski, 1915, Chandrasekhar, 1943)

$$n(r,t) = n_0\left[1 - \frac{R_c}{r} + \frac{2R_c}{r\sqrt{\pi}}\int_0^{x_c}\exp(-x^2)dx\right], \quad (3.259)$$

where $x_c = (r - R_c)/2\sqrt{Dt}$. Then the capture rate is

$$k_c = \frac{dn_0}{dt} = 4\pi DR_c^2\left(\frac{\partial n(r,t)}{\partial r}\right)\bigg|_{r=R_c}, \quad (3.260)$$

which finally leads to

$$k_c = 4\pi DR_c\left(1 + \frac{R_c}{\sqrt{\pi Dt}}\right) \approx 4\pi DR_c. \quad (3.261)$$

The second term in parentheses is rarely considered since the capture radius is $R_c \leq 1$ nm (corresponding to the Förster radius or Dexter hopping distance) and $D \sim 10^{-2}$ cm^2/s (Lunt et al., 2009), with transfer times $t = 1/k_{ET,\ tot} \leq 1$ ns (cf. Eq. 3.250). Hence, the second term is $\lesssim 0.01$, and can usually be neglected.

Assuming that the donor sits at the corner of a cube of side a, then for a hop in time, τ_H, to any neighboring site, we obtain

$$D = \frac{L_D^2}{\tau} = \frac{a^2}{\tau_H}\Theta, \quad (3.262)$$

where

$$\Theta^{-1} = \left[\frac{6}{\sqrt{1}} + \frac{12}{\sqrt{2}} + \frac{8}{\sqrt{3}} + \ldots\right]. \quad (3.263)$$

Here, hops to the 6 nearest neighbors, 12 second nearest neighbors, and so on are summed.

If the hopping process is via FRET, then $k_{FRET} \sim 1/\tau_H$, and assuming that nearest neighbor transfers are the most likely due to the $1/R^6$ dependence, then

$$\frac{\tau}{\tau_H} = \left(\frac{R_{DA}}{a}\right)^6. \quad (3.264)$$

Using this and Eq. 3.262, we obtain a relationship between the diffusion length and the Förster radius:

$$L_{D,FRET} = \frac{1}{\sqrt{6}}\frac{R_{DA}^3}{a^2}. \quad (3.265)$$

If the transfer is via exchange interactions (see Eq. 3.244), similar arguments yield:

$$L_{D,exch} = \sqrt{\frac{\tau K' J_\lambda}{6}}\exp(-a/L), \quad (3.266)$$

where we recall that L is the effective orbital radius of the initial and final electronic states. Comparison of Eqs. 3.265 and 3.266 show a clear distinction between the dependence of the diffusion length on concentration, C (and hence the mean separation between donor and acceptor molecules, a). In FRET, the diffusion length scales as $C^{2/3}$, whereas for exchange, the dependence is exponential.

In addition to its dependence on the efficiency of energy transfer between molecules, exciton diffusion also depends on crystalline morphology. As the number of grain boundaries increases, the potential for trapping within the grains increases. The dependence of L_D on crystallinity has been shown for a series of PTCDA films grown on KBr substrates, where the substrate temperature during growth was varied to achieve morphologies ranging from amorphous (with a mean grain size of $\bar{\Sigma} = 0\pm5$ nm at $T_{sub} = 77$ K using vacuum thermal evaporation) to nearly single crystalline (with $\bar{\Sigma} = 440$ nm at $T_{sub} = 440$ K using organic vapor phase deposition) (Lunt et al., 2010). Micrographs of three representative films with differing PTCDA grain sizes are shown in Fig. 3.103.

Following the treatment that leads to Eq. 3.265, a fit in Fig. 3.104 to the dependence of L_D determined by SR-PLQ on $\bar{\Sigma}$ clearly shows that energy transfer is via FRET. We also find that the fluorescence yield in Fig. 3.104a increases with L_D as expected, since excitons are non-radiatively quenched at the grain boundaries. Indeed, the quenching is extremely efficient: excitons are 10^4 times more likely to be quenched at grain boundaries in PTCDA than from other structural irregularities such as point defects or vacancies.

Interestingly, the maximum diffusion length of 25 nm is achieved only for films whose grain size is $\bar{\Sigma} = 400$ nm, or nearly 20 times larger. This results since excitons are optically generated across the grain under uniform illumination across macroscopic sample areas. Excitons generated near a quenching boundary have a greater probability of diffusing to that boundary, and hence their mean diffusion length will be less than excitons generated in the grain center. Boundary effects become negligible only when the grain diameter is much larger than L_D. The solid line in Fig. 3.104b is a fit to the Förster transfer theory of Eq. 3.265, taking the increase in quantum yield into account for this analysis (Lunt et al., 2010). A transfer radius of $R_{DA} = 2.8\pm0.2$ nm is inferred from these data.

3.10 Exciton recombination and annihilation

Thus far we have considered the diffusion of a single exciton. However, it is never the case that only one excited state exists in a solid. Since the states can

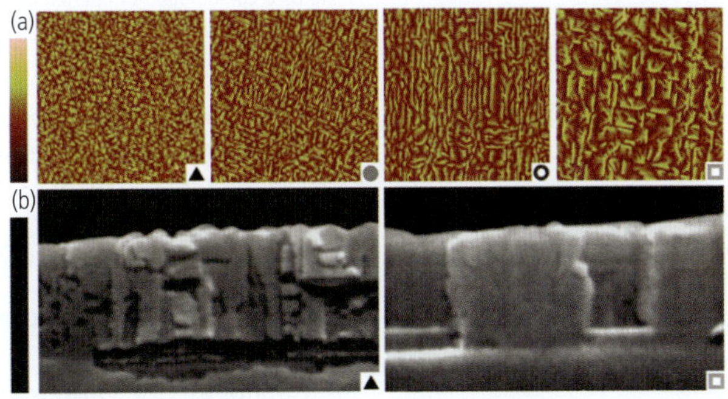

Figure 3.103 (a) Atomic force microscope (AFM) images of 350 nm thick PTCDA films grown under various conditions to control crystal domain diameters. The image widths and heights are 5 μm and 50 nm, respectively. (b) Cross-sectional scanning electron microscope (SEM) images of the PTCDA films (left and right SEM images correspond to the left- and right-most AFM images, respectively). The scale bar on the left corresponds to 500 nm. The symbols correspond to different substrate temperatures used during growth: 440 K (□), 430 K (○), 420 K (●), and 295 K (▲) (Lunt et al., 2010).

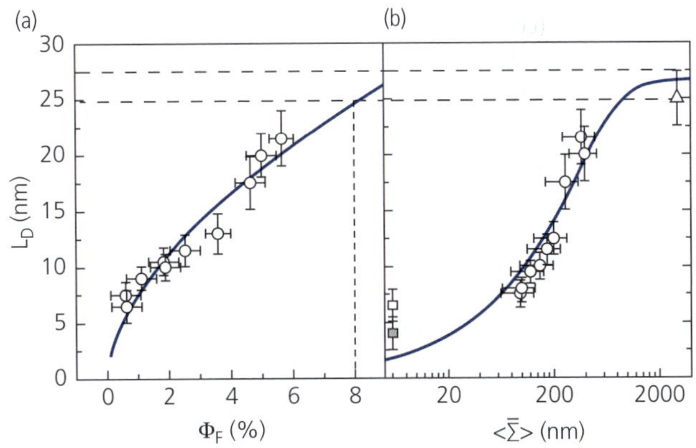

Figure 3.104 Measured exciton diffusion lengths (L_D) versus (a) fluorescence quantum yield (Φ_F) and (b) mean crystal diameter ($\bar{\Sigma}$). Hollow circles correspond to measurements for crystalline films, hollow squares for amorphous films. The hollow triangle gives the estimated diffusion length from the measured quantum yield limit in (a) for PTCDA single crystals. The single crystalline limit is indicated by the region bounded by the dashed lines (Lunt et al., 2010).

migrate, as the excitation density increases, their opportunities for encountering each other or a charge correspondingly increase. These exciton–exciton and exciton–charge (i.e. polaron) encounters often lead to annihilation of one or more of the excitons. That is, the energy of an exciton on one molecule is delivered to an exciton on its neighbor. This is an alternative energy transfer mechanism that results in briefly promoting the neighboring exciton to a high energy state from which it quickly relaxes, while returning the original molecule to its ground state.

This type of interaction is known as exciton annihilation and can occur by one of several routes: singlet–triplet (STA), singlet–polaron (SPA), singlet–singlet (SSA), triplet–triplet annihilation that produces singlets and triplets (TTA-S and TTA-T, respectively), and triplet–polaron annihilation (TPA). Their shared characteristic is that one excited state is the product of a collision between two initial states. Several of the most important reactions and their associated rates (shown above the arrows) are listed here:

$$S + T \overset{k_{STA}}{\rightarrow} T^n + S_0 \overset{k_{Tn}}{\rightarrow} T + S_0 \qquad \text{(STA)} \qquad (3.267a)$$

$$S + P \overset{k_{SPA}}{\rightarrow} P^n + S_0 \overset{k_{Pn}}{\rightarrow} P + S_0 \qquad \text{(SPA)} \qquad (3.267b)$$

$$S + S \overset{k_{SSA}}{\rightarrow} S^n + S_0 \overset{k_{Sn}}{\rightarrow} S + S_0 \qquad \text{(SSA)} \qquad (3.267c)$$

$$T + T \overset{k_{TTA-S}}{\rightarrow} S^n + S_0 \overset{k_{Sn}}{\rightarrow} S + S_0 \qquad \text{(TTA} - \text{S)} \qquad (3.267d)$$

$$T + T \overset{k_{TTA-T}}{\rightarrow} T^n + S_0 \overset{k_{Tn}}{\rightarrow} T + S_0 \qquad \text{(TTA} - \text{T)} \qquad (3.267e)$$

$$T + P \overset{k_{TPA}}{\rightarrow} P^n + S_0 \overset{k_{Pn}}{\rightarrow} P + S_0 \qquad \text{(TPA)} \qquad (3.267f)$$

The energetics of these several properties are provided in the Jablonski diagrams in Fig. 3.105. Note that TTA can result in either a singlet (Eq. 3.267d) or a triplet (Eq. 3.267e). Here, P denotes a polaron of either charge polarity, and the superscript, n, denotes an excited state higher than the lowest, $n = 1$ electronic manifold. Also, TTA-S constitutes *triplet fusion*. This process occurs when two T_1 triplets, each with half of the energy of the S_1 state, collide to form a singlet. Similarly, a singlet may split into two triplets, each with half of the singlet energy in the complementary process of *singlet fission* (Smith and Michl, 2010). Both fusion and fission are potentially useful in photogeneration of charge in organic photovoltaics and emission in OLEDs. This subject will be discussed further in Chapters 6 and 7.

Since annihilation events involve the encounter of *two* particles, the capture radius is double that for a

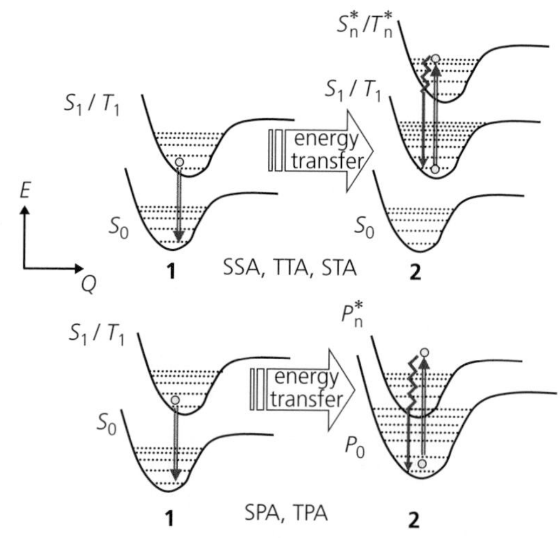

Figure 3.105 Various annihilation processes whereby an exciton in molecule **1** encounters an exciton (top) or polaron (bottom) in molecule **2**. This subsequently de-excites **1** (downward arrow) while promoting **2** into a higher, spin-symmetry-allowed upper state (upward arrow to S_n, T_n, or P_n). This excitation rapidly relaxes back to its first excited state (S_1, T_1, or P_0) by a combination of internal conversion and vibronic relaxation.

single excited state encountering a trap. For example, the relationship between the rate of TTA and the diffusion constant becomes (cf. Eq. 3.261) (Chandrasekhar, 1943)

$$k_{TTA} = 8\pi D R_c, \qquad (3.268)$$

with analogous expressions for the other bimolecular processes.

Cataloging the various transition processes is helpful in writing the time-dependent dynamical expressions for the exciton populations that undergo bimolecular reactions. For example, in an optically and/or electrically pumped film of thickness d, and current density j, the triplet T and singlet S population densities are expressed by a pair of coupled equations:

$$\frac{dS}{dt} = G_S - k_S S - \frac{1}{2} k_{SSA} S^2 - k_{STA} ST - k_{SPA} SP + k_{TTA-S} T^2, \qquad (3.269)$$

$$\frac{dT}{dt} = G_T - k_T T - 2k_{TTA-S} T^2 - \frac{1}{2} k_{TTA-T} T^2 - k_{TPA} TP, \qquad (3.270)$$

where all variables are functions of (r, λ, j, t), as appropriate. If generation occurs by current injection, then $G_{T,S}(j) = \chi_{T,S}\eta_j j/qd$, where $\chi_{T,S}$ is the triplet or singlet generation ratio due to encounters of charges with uncorrelated spins. Also, η_J is the electron-to-exciton conversion ratio, that is, it is the quantum efficiency for exciton generation. The factor of two in Eq. 3.270 results from the loss of two triplets to produce one singlet. Similarly the factors of ½ in both expressions arise since two excitons collide to result in a single exciton of the same spin multiplicity.

It has been shown that the singlet/triplet generation ratio via current injection is determined by random statistics. Thus, unless samples are specifically prepared to have a particular singlet/triplet ratio via, for example, the application of a magnetic field, then χ_S=¼ for singlets and χ_T=¾ for triplets (Baldo et al., 1999). Generally, under optical pumping, $G_T = 0$ due to the spin forbidden $T_1 \leftarrow S_0$ transition.

Equations 3.269 and 3.270 represent simple enumerations of exciton sources and sinks indicated by either a + or – sign, respectively, that accompanies each term. Note that the annihilation terms have a square dependence on density (e.g. SSA has a S^2, TTA has a T^2, and SPA has a SP dependence, respectively) since the reaction requires the encounter of two excitons. As expected, therefore, these bimolecular interactions dominate as the excitation density (either via j or I_0) increases.

An example where several of the processes in Eq. 3.267 are active is in fluorescent OLEDs. For exciton generation via electrical pumping, we have $G = j/qd$, where d is the thickness of the optically pumped emission layer. Triplet density dynamics in this case are determined by the processes of triplet non-radiative decay, TTA, and TPA, as described in Eqs. 3.267d–f. Assuming *Langevin recombination* which is due to the random collisions between polarons to form an exciton (see Section 4.5.4), we obtain

$$\frac{dP}{dt} = \frac{j(t)}{qd} - k_{rec}P^2, \qquad (3.271)$$

where k_{rec} is the Langevin recombination rate given by (Langevin, 1903)

$$k_{rec} = \frac{q}{\varepsilon_r \varepsilon_0}(\mu_h + \mu_e). \qquad (3.272)$$

Here, μ_h and μ_e are the mobilities of holes and electrons in the layer, respectively. Then, from Eqs. 3.270 and 3.272, the triplet concentration is given by

$$\frac{dT}{dt} = \frac{3}{4}k_{rec}P^2 - k_T T - \frac{5}{4}k_{TTA}T^2 - k_{TPA}TP. \qquad (3.273)$$

Here, 3/4 and 5/4 arise from the spin statistics of triplet generation and TTA, respectively.

The singlet density dynamics are a function of the singlet decay rate as well as STA. Furthermore, in the

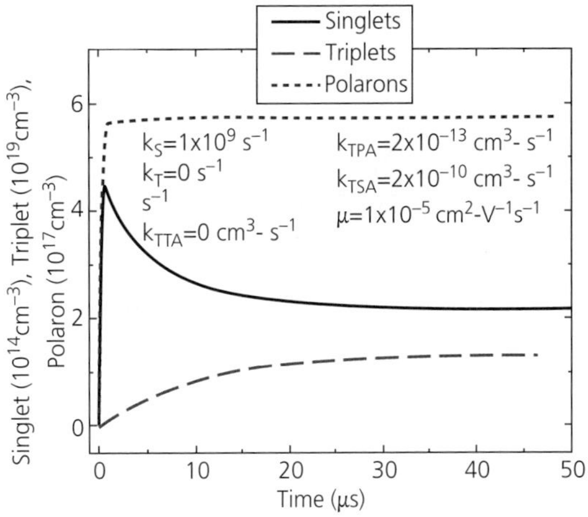

Figure 3.106 Simulated density dynamics at a current density $j = 0.8$ A/cm^2 in a film of thickness $d = 25$ nm, using the parameters listed (Zhang et al., 2010).

presence of STA, the corresponding rate equation for singlets follows:

$$\frac{dS}{dt} = \frac{1}{4}k_{rec}P^2 - k_S S + \frac{1}{4}k_{TTA}T^2 - k_{STA}ST. \qquad (3.274)$$

The simulated singlet, triplet, and polaron densities under a constant current density of $j = 0.8$ A/cm^2 using Eqs. 3.271, 3.273, and 3.274 are shown in Fig. 3.106 using parameters typical of many doped fluorescent systems (Zhang et al., 2010). Initially, the singlet density is large since it only depends on the natural singlet decay rate of k_S. With time, the triplet density increases, leading to an increase in STA that results in a decrease in S that is proportional to the increase in T. Since the light output in fluorescent OLEDs is proportional to the singlet density, that is, the luminance intensity follows $L(t) = k_S S$, STA leads to a decrease in luminance to a steady state value as the relative concentrations of singlets and triplets approaches equilibrium.

The theory of singlet and triplet dynamics accurately models the transient decay of luminescent intensity in many fluorescent and phosphorescent systems. For example, the luminescence transients in DCM:Alq_3-based OLEDs are shown as a function of current pulse amplitude in Fig. 3.107. The rise time (<1 μs) is significantly shorter than the decay time (~20 μs); the former determined by the transit time of electrons and holes into the light emitting layer, while the latter is determined by the triplet decay rate. Fits to the data according to the transient analysis for STA are indicated by solid lines. In this model, only the electron mobility is important since it is considerably higher than holes in the Alq_3 host (Tang et al., 1989, Shi and

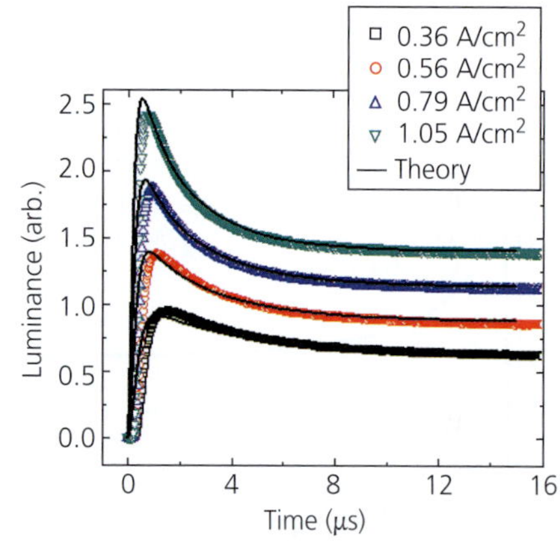

Figure 3.107 Transient electroluminescence of 3% DCM:Alq_3 OLEDs in response to 100 μs current pulses of different amplitudes. The steady state current densities are indicated. Solid lines are the fits to the STA model, with $k_{STA} = 1.9 \times 10^{-10}$ cm^3/s (Zhang et al., 2010).

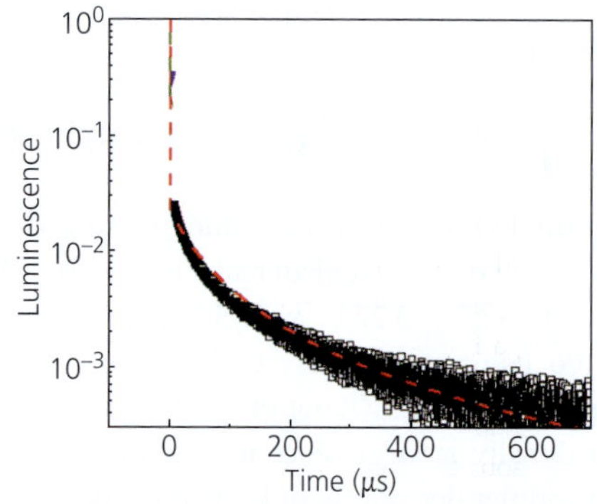

Figure 3.108 Delayed fluorescence from DCM:Alq_3 based OLEDs after the turn-off of the current pulse at $t = 0$ s. The steady state luminescence is normalized at the pulse onset. The fit (dashed line) gives a TTA rate of $k_{TTA} = (2.2 \pm 0.2) \times 10^{-15}$cm^3/s (Zhang et al., 2010).

Tang, 2002). Furthermore, non-radiative triplet decay and TTA are neglected since these processes are small compared to TPA during the device turn-on. The predicted $L(t)$ slightly deviates from the experiment during the initial 1 μs due to neglect of the charge transit time in Eq. 3.271. Nevertheless, it is clear that the simple theory of Eqs. 3.271–3.274 provides an accurate description of the singlet dynamics in electrically pumped organic thin films.

Since TTA results in the generation of singlets, transient solutions to Eq. 3.274 predict that singlet emission should persist long after the singlet natural lifetime, $1/k_S$, due to their formation of singlets following the annihilation of very long lived triplets. The delayed, persistent fluorescence (analogous to TADF in Section 3.7.5) following the DCM:Alq_3 device turn-off is observed in Fig. 3.108. A fit (dashed line) to the clearly non-exponential transient due to the T^2 dependence of TTA gives $k_{TT} = 2.2 \times 10^{-15}$ cm^3/s (Zhang et al., 2010).

Annihilation processes are particularly important in films where triplet emission dominates over singlets, such as in phosphorescent OLEDs (Baldo and Forrest, 2000). The decay rate of triplets (k_T) in metal-organic compounds such as Ir(ppy)$_3$ is typically three orders smaller than k_S, indicating that even at modest triplet densities, annihilation depends on the triplet concentration as ~T^2 for TTA, or TP for TPA. Hence, optimizing the optical properties of triplet emitting devices requires a complete understanding of the dynamics of the various annihilation, energy transfer and transport processes.

3.11 Summing up

The most distinctive properties of organic semiconductors that separates them from the wide diversity of conventional inorganic semiconductors, are their optical characteristics. We have seen that these characteristics are due to the low dielectric constants and lack of chemical bonds between molecules within organic solids. This gives rise to excited states where the electron and hole form an exciton whose binding energy is ~0.5 eV which is considerably larger than the thermal energy of 26 meV at room temperature. The stable excitonic states transport energy over relatively long distances, in marked contrast to conventional semiconductors whose much larger dielectric constants result in unstable excitons with binding energies more than an order of magnitude smaller. As a result, excitons dominate the optical properties of organics, that are, consequently, referred to as excitonic materials. In this chapter, we have developed the theory of excitonic materials, starting from individual bonds between atoms forming a molecule, to the properties of the entire molecule, and concluding with the properties of organic solids comprised of large ensembles of molecules or molecular mixtures.

Starting with the eigenstates of the individual orbitals, we have found that the electronic and vibronic (i.e. phonon) structure of molecules can be conveniently described by the linear combination of electronic wavefunctions of the individual atoms of which they are comprised. Once placed in close proximity, the molecular energy levels split to form bands, with the band structure dispersion implying that the excited states are mobile. The mobility is enabled via electrostatic and quantum mechanical coupling between adjacent molecules due to exchange interactions, to distances of a micron or more via radiative transfer. Further, the Pauli Exclusion Principle gives rise to energy level splitting, revealing a plethora of optical phenomena within solid mixtures. The spin multiplicities (e.g. singlet vs. triplet) of the excited states also play a defining role in organic materials, another clear departure from inorganic semiconductors whose band-like nature tends to obscure these effects.

Ultimately, we observe this rich variety of optical characteristics via the spectroscopic properties of the individual molecules as well as in their collective properties in the solid state. These processes are conveniently quantified in terms of transition rates that are determined using one simple framework of time dependent perturbation theory: Fermi's Golden Rule. Of course we recognize that this framework can be accurately applied to phenomena that only slightly perturb the unexcited system, which is not always the case in excitonic systems.

Unlike conventional semiconductors, the properties of individual molecules are apparent even for molecules bound together in the solid state. This property provides insights as to how the properties of individual units of a solid can lead to the characteristics of the whole. Many of the classic works and advances in our understanding of organics have been developed to describe their fundamental optical processes, including energy transfer and migration. This has formed the foundational knowledge, and application of organics to optoelectronics. It has revealed the properties of diverse systems found in electronics, optics, biology, and so on.

Beyond just providing a laboratory in which one can test the fundamental theories of solid state physics and physical chemistry, the unique optical properties of organics have been extensively exploited in high performance devices such as OLEDs, OPVs, and transistors. Before we can move on to consider these practical applications, we must first lay the foundations for understanding the electrical properties of organics; the subject of the following chapter. Indeed, the transport of energy via exciton diffusion is due among other things, to electron exchange between adjacent molecules. This is similar to the processes of electron and hole transport in the solid.

Further reading

1. P. W. Atkins and R. S. Friedman, 1997. *Molecular Quantum Mechanics*, 3rd ed. New York: Oxford University Press.
2. J. B. Birks, 1970. *Photophysics of Aromatic Molecules*. London: Wiley-Interscience.
3. L. Dexter, 1953. A theory of sensitized luminescence in solids. *J. Chem. Phys.*, 21, 836.
4. T. Förster, 1965. Delocalized excitation and excitation transfer. In: O. Sinanoglu (ed.) *Modern Quantum Chemistry: Istanbul Lectures*, Vol. 3, Sec. IIIB, Ch. 1. New York: Academic Press.
5. M. Klessinger and J. Michl, 1995. *Excited States and Photochemistry of Organic Molecules*. New York: VCH Publishers.
6. A. Köhler and H. Bässler, 2015. *Electronic Processes in Organic Semiconductors: An Introduction*. Weinheim, Germany: Wiley VCH Verlag.
7. A. R. Leach, 1996. *Molecular Modeling: Principles and Applications*. Singapore: Addison Wesley Longman.
8. S. P. McGlynn, T. Azumi, and M. Kinoshita, 1969. *Molecular Spectroscopy of the Triplet State*. Englewood Cliffs, NJ: Prentice-Hall.
9. M. Pope and C. E. Swenberg, 1982. *Electronic Processes in Organic Crystals*. Oxford: Clarendon Press.
10. M. Schworer and H. V. Wolf, 2007. *Organic Molecular Solids*. Weinheim, Germany: Wiley-VCH.
11. N. J. Turro, 1991. *Modern Molecular Photochemistry*. Sausalito, CA: University Science Books.

Problems

1. If each of the individual spatial and spin wavefunctions in Eqs. 3.2 and 3.3 are normalized, show that the combined wavefunctions are also normalized.
2. Assuming a H_2^+ molecule with both hydrogen atoms in the 1s states, calculate
   (a) the Coulomb integral,
   (b) the resonance integral,
   (c) the energy eigenvalues of the molecular ground state.
3. For the Morse potential in Eq. 3.16, show that
   (a) the difference between adjacent energy levels is given by Eq. 3.19,
   (b) the quantum number of the highest bound state is given by Eq. 3.20.
4. Butadiene is the simple organic molecule C_4H_6 shown in the figure below.

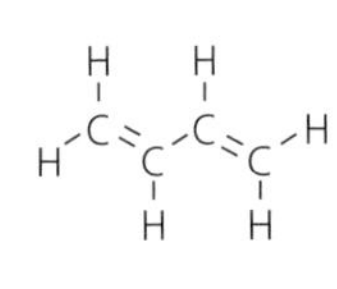

   (a) Construct the 4 × 4 secular determinant and solve for the energy eigenvalues of the π-orbitals of this molecule. The energies can be written in terms of the on- and off-diagonal matrix elements, a, b, respectively.
   (b) Find the four molecular orbitals corresponding to these energies (which is the same as finding the coefficients, c_{ni}, in $|n\pi\rangle = \sum_{i=1}^{4} c_{ni}\phi_i$, where ϕ_i is the orbital of the ith carbon atom, and $n = 1, 2, 3, 4$). You may use a computer to calculate the c values.
   (c) Repeat (a) for cyclobutadiene shown below.

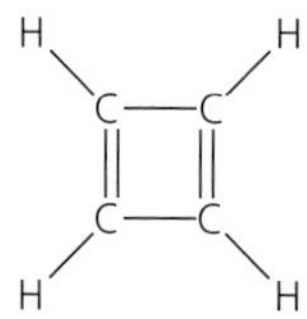

5. One approximation to calculating orbitals that is reasonably accurate for fused ring systems is the perimeter free electron orbital model. One shortcoming of this model is that the energies diverge as the square of the orbital angular momentum quantum number, l (see Eq. 3.43). A modification of the theory is to use the basis set (Eq. 3.42): $\psi_l(\theta) = \frac{1}{\sqrt{2\pi N}} \sum_{k=0}^{N-1} \exp(ik\theta_l)\phi_k(\theta)$.
   (a) Using this basis set, show that the energy eigenvalues are given by: $E_l = \alpha + 2\beta cos(2\pi l/M)$ where M is the number of atoms, α is an energy offset and β corresponds to the off-diagonal energy terms.
   (b) Plot E_l along with that of a simple harmonic oscillator, the solution to the simple PFEO theory, and a Morse potential. Here, assume that l is the vibronic quantum number. Which do you think gives the best approximation to actual molecular systems?
6. An alternative and quite simple model of polyenes is the free electron orbital model. In this model, we can take a simple linear molecule such as β-carotene (Scheme P3.6) as a one dimensional quantum mechanical "potential box" consisting of a chain of N conjugated carbon atoms, with bond length R_0, as forming a box of length $L = (N-1)R_0$. β-Carotene is a molecule that gives carrots their color.

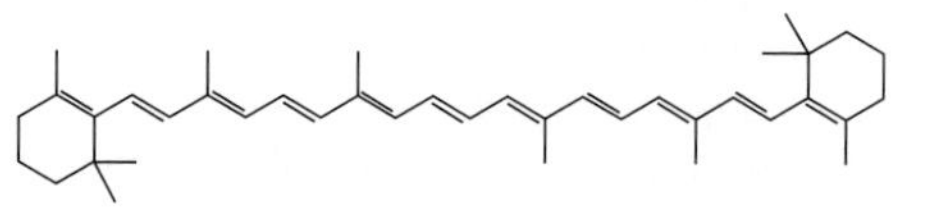

Scheme P3.6

   (a) Calculate the length of the box.
   (b) Find the wavefunctions and their energies for an electron along the chain.
   (c) Suppose that the electrons enter the states in pairs. Thus, the lowest ½N states are occupied. Estimate the wavelength of the lowest energy transition.
   (d) Calculate the expected color of carrots. Is this consistent with your idea of what a carrot should look like?
7. In Section 3.6.1 we showed that the Franck–Condon factor could be written as $FC_{n,m} = Z^m e^{-Z}/m!$, where Z is the Huang–Rhys factor. We want to derive this expression in this problem. We start by assuming that the nuclear positions for 0–0 transitions are expressed by: $|0\rangle = \left(\frac{\sigma}{\pi}\right)^{1/4} \exp\left(-\sigma(R-R_0)^2/2\right)$.
   (a) Using this wavefunction, show that the nuclear overlap between the ground and excited state is given by: $Z_{0\text{-}0} = \langle 0'|0\rangle = \exp\left(-\sigma(R_0-R_e)^2/4\right)$ where the excited state equilibrium position is R_e.
   (b) Similarly, if the first vibronic is given by $|1'\rangle = \sqrt{2\alpha}(R-R_E)|0\rangle$, calculate $Z_{0\text{-}1}$.
   (c) Continuing along this vein, assuming only the $|0\rangle$ state is occupied, show that the series of higher orders of Z results in $FC_{n,m} = Z^m e^{-Z}/m!$
8. Using the relativistic correction that leads to Thomas precession:
   (a) Show that the magnetic field for a moving electron in an electric field **F** at velocity **v** in the laboratory reference frame is given by $\mathbf{B} = -\frac{\mathbf{v}\times\mathbf{F}}{2c^2}$, which leads to the spin–orbit Hamiltonian.
   (b) $H_{SO} = -\frac{q}{2m_e^2c^2}\frac{1}{r}\frac{dV}{dr}\mathbf{l}\cdot\mathbf{s}$, where **l** and **s** are the orbital and spin angular momentum operators, respectively.
   (c) Using this Hamiltonian in the central field of an electron, show for hydrogenic states that the spin–orbit coupling energy is given by: $E_{SO} = \frac{Z^4q^2}{8\pi\varepsilon_0 a_0^3 m_e^2 c^2}\frac{1}{n^3 l(l+\frac{1}{2})(l+1)}\langle \mathbf{l}\cdot\mathbf{s}\rangle$, where Z is the atomic number of the atom, a_0 is the Bohr radius, n is the principal, and l the orbital angular momentum quantum numbers.
   (d) Using the result in (c), plot E_{SO} vs. Z for the following atoms commonly found in organic compounds: C, Al, Si, Cu, Ag, Os, Ir, Pt, Pb.
9. The Slater–Condon selection rules introduced in the context of spin–orbit coupling state that two n-electron spinorbitals can differ by only a single spinorbital when they are coupled via a one-electron Hamiltonian, H_{12}. Consider two wavefunctions, Φ_1, Φ_2 that differ by only a single spinorbital, $\phi_m \leftrightarrow \phi_p$. That is, $\Phi_{1,2} = \sqrt{n!}\,|\phi_1 \ldots \phi_{m,p}\phi_i \ldots \phi_n|$. From this, derive the Slater–Condon selection rule:

$$\langle\Phi_1|H_1|\Phi_2\rangle = \langle\phi_m(1)|H_{12}|\phi_p(1)\rangle + \sum_i \{\langle\phi_m(1)\phi_p(1)|H_{12}|\phi_i(2)\phi_i(2)\rangle - \langle\phi_m(1)\phi_i(1)|H_{12}|\phi_i(2)\phi_p(2)\rangle\}$$

   Here, (1) and (2) are the spinorbital coordinates of electrons, 1, 2.
10. Show that the average number of cycles between the S_1 and T_1 states leading to thermally assisted delayed fluorescence is given by: $\bar{n} = \frac{\Phi_T\Phi_S}{1-\Phi_T\Phi_S}$, where Φ_T and Φ_S are the quantum yields for triplet and singlet emission, respectively.
11. Derive the Förster overlap integrals $M_D(\nu) = \frac{3}{4}\frac{\varepsilon_0\hbar c^3\Phi_D f_D(\nu)}{\pi n_r \tau_D \nu^3}$ and $L_A(\nu) = \frac{6\varepsilon_0\alpha_A(\nu)n_r\hbar c}{N_{A\nu}}\frac{M_{wA}}{\rho_{MA}}$ in Eqs. 3.225a and b using the Einstein A and B coefficients for emission and absorption.
12. Figure P3.12 shows the room temperature absorption spectra of a molecule in solution. In one case the solution is quite dilute, in the other it is more concentrated.
   (a) Which of the two spectra (labeled A and B) corresponds to the less and which to the more dilute solution? Explain your reasoning based on the shapes and broadness of the two spectra.

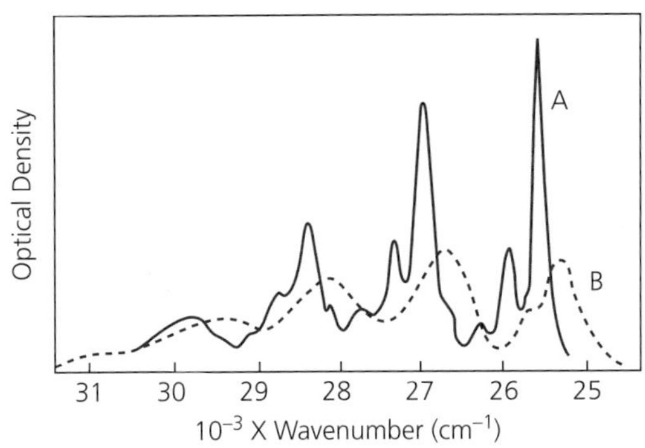

Figure P3.12 Spectra for a molecule in two solutions.

(b) Relabel the horizontal axis in terms of energy units. Label each of the peaks using spectroscopic notation in terms of the transition between vibronic levels.

(c) Calculate the parabolic energy state that accurately describes spectrum A. That is, specify the spring constant and the energy gap from the ground to excited states.

(d) What physical vibrational mode (e.g. a C–H bending mode) do you think these energies correspond to? Is there more than one such mode present?

13. Figure P3.13 shows the room temperature absorption spectrum of the planar aromatic hydrocarbon molecule shown. Also shown is the room temperature emission (i.e. photoluminescence) spectrum of the molecule.

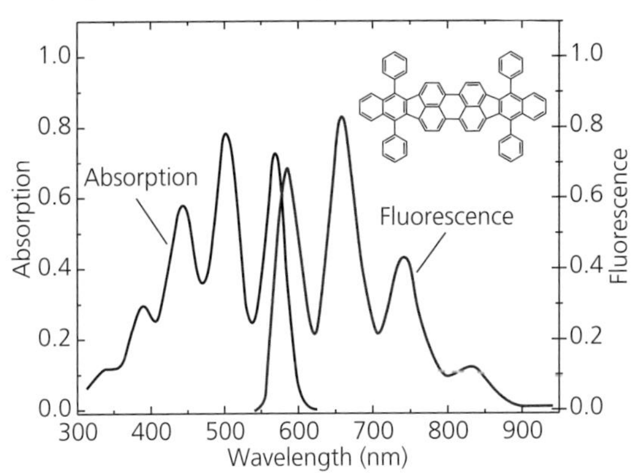

Figure P3.13 Spectra of the molecule in the inset.

(a) What is the chemical formula of the molecule?

(b) Label all of the peaks in the spectrum using standard spectroscopic notation to identify both electronic and vibronic transitions. Also, identify the zero-crossing energy, E_{00}. Note that the absorption peaks at <400 nm are in a different electronic manifold than at >450 nm. What is the spin multiplicity of the states, and why?

(c) What is the binding energy of the lowest exciton state assuming $\varepsilon_r = 3.2$?

(d) Plot the vibronic energy vs. quantum number n_1 for the lowest electronic state (subscript 0) as taken from the absorption spectrum. On this plot, fit the progression to a simple harmonic oscillator and a Morse potential: $V(r) = V_0(1 - \exp[-a(r - r_0)])^2$ Which fits better?

(e) What is the maximum bound quantum state energy for the Morse potential that provides the best fit to the spectra? What is the force constant?

14. In Fig. P3.14, we show the fluorescence and absorption spectra of two different molecular species (data for molecule **A** in blue, for **B**, in grey). The normalized absorption spectra have maximum absorptions of $\alpha_{\mathbf{A}} = 5 \times 10^5$ cm^{-1} and $\alpha_{\mathbf{B}} = 2 \times 10^5$ cm^{-1}. Furthermore, the mass density of **A** is 1.5 gm/cm^3 and index of refraction of 1.8 in the visible, and for **B** these values are 2.1 gm/cm^3 and 2.0. Both have quantum yields approaching 100%.

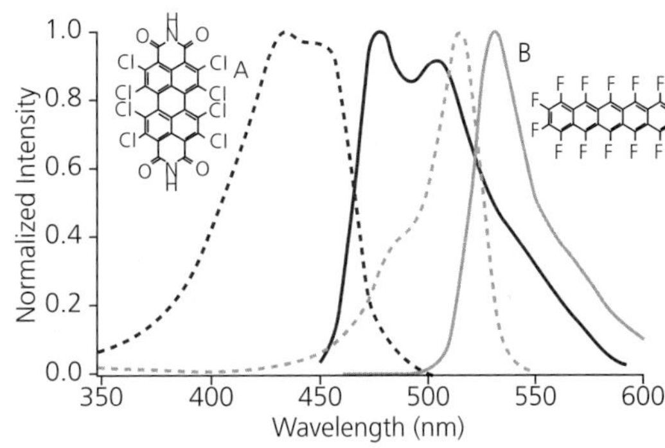

Figure P3.14 Spectra and molecules of a FRET transfer system.

(a) If these molecules are blended, which will act as the most effective donor, and which as an acceptor?

(b) Calculate the Förster transfer radius, R_{DA}, for this donor/acceptor system.

(c) Given that the natural lifetime of **A** is 3 ns, and that of **B** is 6 ns, plot the FRET transfer rate, k_{ET} vs. doping concentration of **B** in **A**.

(d) What is the diffusion coefficient and diffusion length of each of these molecules when they are undoped in a thin film.

15. Show that a solution to the exciton diffusion equation gives an exciton capture rate of $k_c = 4\pi D R_c \left(1 + \frac{R_c}{\sqrt{\pi D t}}\right)$.

16. Assume that a thin squaraine film of thickness d is bounded by two, 3 eV energy gap layers to form the structure in Fig. P3.16a. Light is incident normal to the plane of the film from the left and has intensity, I_0. The recombination lifetime in the film is $\tau_D = 6$ ns, and the absorbance spectrum is given by Fig. P.16b.

(a) For $d = 100$ nm, calculate the steady state exciton density distribution across the entire structure for incident wavelengths of $\lambda = 400, 550, 675$ nm.

(b) The incident illumination at $\lambda = 675$ nm is switched off at $t = 0$. Plot the time evolution of the exciton density distribution across the layer for the next 50 ns.

(c) For the conditions in (b), if the quantum yield of the material is $\Phi_{PL} = 60\%$, calculate the time evolution of the luminescence intensity emitted in the forward direction (i.e. toward the right) if the index of refraction of the outermost layer is $n_r = 1.5$, and that of the material under investigation is 1.7.

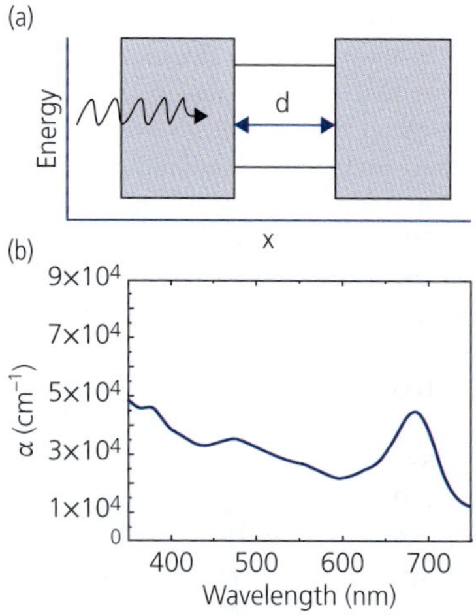

Figure P3.16 (a) Energy diagram for Problem 16. (b) Absorption spectrum of the material between the barriers of thickness *d*.

17. In a fluorescent OLED, excitons are generated by injecting current, j into an emission region of thickness, d. Then, the singlets (density S) and triplets (density T) can undergo many different paths to recombination. In this problem we will only consider triplet–triplet annihilation and singlet–triplet annihilation, and radiation.
    (a) Write down the various singlet and triplet reactions that can occur in this device and label the appropriate rate constants.
    (b) Write down the coupled rate equations for singlets and triplets. Assume that in the active region of the device that all electrons and holes combine to form excitons (i.e. charge balance is unity), and that singlet/triplet formation follows their statistical ratio.
    (c) Solve the coupled equations in (b) for the singlet population vs. time ($S(t)$).
    (d) Calculate the internal quantum efficiency of the device (defined as the ratio of photons emitted to the number of charges injected) as a function of j from 0.01 to 100 A/cm^2, assuming that 20% of the triplets that annihilate form singlets, for the parameters shown in the table below:

Parameter	Value	Units
Singlet lifetime	5	ns
Triplet lifetime	60	μs
TTA rate	10^{-13}	cm^{-3} s^{-1}
STA rate	10^{-10}	cm^{-3} s^{-1}

    (e) Calculate the electroluminescence turn-on transient (assuming a step function current pulse) over the first 100 μs at $j = 0.01$, 1, and 100 A/cm^2.

References

Adachi, C., Baldo, M. A., Thompson, M. E. & Forrest, S. R. 2001. Nearly 100% internal phosphorescence efficiency in an organic light emitting device. *J. Appl. Phys.*, 90, 5048.

Alonso, M. I., Garriga, M., Karl, N., Osso, J. O. & Schreiber, F. 2002. Anisotropic optical properties of single crystalline PTCDA studied by spectroscopic ellipsometry. *Org. Electron.*, 3, 23.

Anderson, P. W. 1958. Absence of diffusion in certain random lattices. *Phys. Rev.*, 109, 1492.

Andrews, D. L. 1989. A unified theory of radiative and radiationless molecular energy transfer. *Chem. Phys.*, 135, 195.

Andrews, D. L. & Bradshaw, D. S. 1999. Resonance energy transfer. In: Andrews, D. L. & Demidov, A. A. (eds.) *Resonance Energy Transfer*. Chichester, UK: John Wiley and Sons.

Andrews, D. L. & Juzeliūnas, G. 1992. Intermolecular energy transfer: retardation effects. *J. Chem. Phys.*, 96, 6606.

Andrews, D. L. & Rodriguez, J. 2007. Resonance energy transfer: spectral overlap, efficiency, and direction. *J. Chem. Phys.*, 127, 084509.

Atkins, P. & de Paula, J. 2006. *Physical Chemistry*. New York: W. H. Freeman and Co.

Atkins, P. W. 1983. *Molecular Quantum Mechanics*. Oxford: Oxford University Press.

Atkins, P. W. & Friedman, R. S. 1997. *Molecular Quantum Mechanics*. New York: Oxford University Press.

Avakian, P., Ern, V., Merrifield, R. E. & Suna, A. 1968. Spectroscopic approach to triplet exciton dynamics in anthracene. *Phys. Rev.*, 165, 974.

Baldo, M. A. & Forrest, S. R. 2000. Transient analysis of organic electrophosphorescence. I. Transient analysis of triplet energy transfer. *Phys. Rev. B*, 62, 10958.

Baldo, M. A., O'Brien, D. F., Thompson, M. E. & Forrest, S. R. 1999. The excitonic singlet–triplet ratio in a semiconducting organic thin film. *Phys. Rev. B*, 60, 14422.

Baldo, M. A., O'Brien, D. F., You, Y., Shoustikov, A., Thompson, M. E. & Forrest, S. R. 1998. High efficiency phosphorescent emission from organic electroluminescent devices. *Nature*, 395, 151.

Baldo, M. A., Thompson, M. E. & Forrest, S. R. 2000. High-efficiency fluorescent organic light-emitting devices using a phosphorescent sensitizer. *Nature*, 403, 750.

Baleizao, C. & Berberan-Santos, M. N. 2008. Thermally activated delayed fluorescence in fullerenes. *Ann. N.Y. Acad. Sci.*, 1130, 224.

Bartynski, A. N., Trinh, C., Panda, A., Bergemann, K., Lassiter, B. E., Zimmerman, J. D., Forrest, S. R. & Thompson, M. E. 2013. A fullerene-based organic exciton blocking layer with high electron conductivity. *Nano Lett.*, 7, 3315.

Beenken, W. J. & Pullerits, T. 2004. Spectroscopic units in conjugated polymers: a quantum chemically founded concept? *J. Phys. Chem. B*, 108, 6164.

Beljonne, D., Shuai, Z., Pourtois, G. & Bredas, J. L. 2001. Spin–orbit coupling and intersystem crossing in conjugated polymers: a configuration interaction description. *J. Phys. Chem. A*, 105, 3899.

Bergemann, K. J. & Forrest, S. R. 2011. Measurement of exciton diffusion lengths in optically thin organic films. *Appl. Phys. Lett.*, 99, 274.

Bergemann, K. J., Liu, X., Panda, A. & Forrest, S. R. 2015. Singlets lead to photogeneration in C_{60}-based organic heterojunctions. *Phys. Rev. B*, 92, 035408.

Birks, J. B. 1970. *Photophysics of Aromatic Molecules*. London: Wiley-Interscience.

Birks, J. B. 1975. Excimers. *Rep. Prog. Phys.*, 38, 903.

Birks, J. B. & Christophorou, L. G. 1963. Excimer fluorescence spectra of pyrene derivatives. *Spectrochim. Acta*, 19, 401.

Birks, J. B. & Kazzaz, A. A. 1968. Excimer fluorescence. XII. The pyrene crystal interaction potential. *Proc. Roy. Soc. A*, 304, 291.

Bixon, M., Jortner, J., Cortes, J., Heitele, H. & Michel-Beyerie, M. E. 1994. Energy gap law for nonradiative and radiative charge transfer in isolated and in solvated supermolecules. *J. Phys. Chem.*, 98, 7289.

Bounds, P. J. & Siebrand, W. 1980. Charge-transfer excitons in anthracene crystals and their role in optical charge carrier generation. *Chem. Phys. Lett.*, 75, 414.

Bounds, P. J., Petelenz, P. & Siebrand, W. 1981. Charge-transfer excitons in anthracene crystals. A theoretical investigation of their optical absorption and thermal dissociation. *Chem. Phys.*, 63, 303.

Bulovic, V., Burrows, P. E., Forrest, S. R., Cronin, J. A. & Thompson, M. E. 1996. Study of localized and extended excitons in semiconductor 3,4,9,10-perylenetetracarboxylic dianhydride. I. Spectroscopic properties of thin films and solutions. *Chem. Phys.*, 210, 1.

Bulovic, V., Shoustikov, A., Baldo, M. A., Bose, E., Kozlov, V. G., Thompson, M. E. & Forrest, S. R. 1998. Bright, saturated red-to-yellow organic light-emitting devices based on polarization-induced spectral shifts. *Chem. Phys. Lett.*, 287, 455.

Callaway, J. 1974. *Quantum Theory of the Solid State*. New York: Academic Press.

Caspar, J. V., Sullivan, B. P., Kober, E. M. & Meyer, T. J. 1982. Application of the energy gap law to the decay of charge transfer excited states. Solvent effects. *Chem. Phys. Lett.*, 91, 91.

Chandrasekhar, S. 1943. Stochastic problems in physics and astronomy. *Rev. Mod. Phys.*, 15, 1.

Chang, W., Akselrod, G. M. & Bulović, V. 2015. Solid-state solvation and enhanced exciton diffusion in doped organic thin films under mechanical pressure. *ACS Nano*, 9, 4412.

Che, X., Chung, C.-L., Liu, X., Chou, S.-H., Y.-H. Liu, Wong, K.-T. & Forrest, S. R. 2016. Regioisomeric effects of donor–acceptor–acceptor′ small-molecule donors on the open circuit voltage of organic photovoltaics. *Adv. Mater.*, 28, 8248.

Chen, J., Wu, Y., Wang, X., Yu, Z., Tian, H., Yao, J. & Fu, H. 2015. A soluble cryogenic thermometer with high sensitivity based on excited-state configuration transformations. *Phys. Chem. Chem. Phys.*, 17, 27658.

Chynwat, V. & Frank, H. A. 1995. The application of the nergy gap law to the S1 energies and dynamics of carotenoids. *Chem. Phys.*, 194, 237.

Craig, D. P., Dissado, L. A. & Walmsley, S. H. 1977. Exciton trapping and self-trapping in molecular crystals. *Chem. Phys. Lett.*, 46, 191.

Cummins, P. G. & Dunmur, D. A. 1974. The electric permittivity of crystalline anthracene. *J. Phys. D: Appl. Phys.*, 7, 451.

D'Andrade, B. W., Brooks, J., Adamovich, V., Thompson, M. E. & Forrest, S. R. 2002. White light emission using triplet excimers in electrophosphorescent organic light-emitting devices. *Adv. Mater.*, 14, 1032.

Davydov, A. S. 1962. *Theory of Molecular Excitons*. New York: McGraw-Hill.

Davydov, A. S. & Sheka, E. F. 1965. Structure of exciton bands in crystalline anthracene. *Phys. Stat. Sol.*, 11, 877.

de Wijs, G. A., Mattheus, C. C., de Groot, R. A. & Palstra, T. T. M. 2003. Anisotropy of the mobility of pentacene from frustration. *Synth. Met.*, 139, 109.

Dexter, L. 1953. A theory of sensitized luminescence in solids. *J. Chem. Phys.*, 21, 836.

Engleman, R. & Jortner, J. 1970. The energy gap law for radiationless transitions in large molecules. *Mol. Phys.*, 18, 145.

Forrest, S. R. 1997. Ultrathin organic films grown by organic molecular beam deposition and related techniques. *Chem. Rev.*, 97, 1793.

Forster, L. S. 2006. Intersystem crossing in transition metal complexes. *Coord.Chem. Rev.*, 250, 2023.

Förster, T. 1948. Zwischenmolekulare Energiewanderung und Fluoreszenz. *Ann. Phys.*, 55, 2.

Förster, T. 1951. *Fluoreszenz Organischer Verbindunger*. Göttingen: Vandenhoeck & Ruprecht.

Förster, T. 1959. 10th Spiers Memorial Lecture: transfer mechanisms of electronic excitations. *Disc. Faraday Soc.*, 27, 7.

Giebink, N. C., Wiederrecht, G. P., Wasielewski, M. R. & Forrest, S. R. 2010. The ideal diode equation for organic heterojunctions. I. Derivation and application. *Phys. Rev. B*, 82, 155305.

Gierschner, J., Cornil, J. & Egelhaaf, H.-J. 2007. Optical bandgaps of π-conjugated organic materials at the polymer limit: experiment and theory. *Adv. Mater.*, 19, 173.

Gierschner, J., Mack, H.-G., Lüer, L. & Oelkrug, D. 2002. Fluorescence and absorption spectra of oligophenylenevinylenes: vibronic coupling, band shapes, and solvatochromism. *J. Chem. Phys.*, 116, 8596.

Gribov, L. A. & Orville-Thomas, W. J. 1988. *Theory and Methods of Calculation of Molecular Spectra*. Chichester, UK: John Wiley and Sons.

Griffith, O. L. & Forrest, S. R. 2014. Exciton management in organic photovoltaic multi-donor energy cascades. *Nano Lett.*, 14, 2353.

Halasinski, T. M., Hudgins, D. M., Salama, F., Allamandola, L. J. & Bally, T. 2000. Electronic absorption spectra of neutral pentacene ($C_{22}H_{14}$) and its positive and negative ions in Ne, Ar and Kr matrices. *J. Phys. Chem. A*, 104, 7484.

Harris, D. C. & Bertolucci, M. D. 1989. *Symmetry and Spectroscopy: An Introduction to Vibrational and Electronic Spectroscopy*. Mineola, NY: Dover.

Haskal, E. I., Shen, Z., Burrows, P. E. & Forrest, S. R. 1995. Excitons and exciton confinement in crystalline organic thin films grown by organic molecular-beam deposition. *Phys. Rev. B*, 51, 4449.

Henry, B. R. & Siebrand, W. 1971. Spin–orbit coupling in aromatic hydrocarbons. Analysis of nonradiative transitions between singlet and triplet states in benzene and naphthalene. *J. Chem. Phys.*, 54, 1072.

Hilborn, R. C. 1982. Einstein coefficients, cross sections, f values, dipole moments, and all that. *Am. J. Phys.*,50, 982.

Hoffmann, M. & Soos, Z. G. 2002. Optical absorption spectra of the Holstein molecular crystal for weak and intermediate electronic coupling. *Phys. Rev. B*, 66, 024305.

Hoffmann, M., Schmidt, K., Fritz, T., Hasche, T., Agranovich, V. M. & Leo, K. 2000. The lowest energy Frenkel and charge transfer excitons in quasi-one-dimensional structures: application to MePTCDI and PTCDA crystals. *Chem. Phys.*, 258, 73.

Jablonski, A. 1933. Efficiency of anti-Stokes fluorescence in dyes. *Nature*, 131, 839.

Jackson, J. D. 1998. *Classical Electrodynamics*. New York: John Wiley and Sons.

Jayaraman, A., Kaplan, M. L. & Schmidt, P. H. 1985. Effect of pressure on the raman and electronic absorption spectra of naphthalene- and perylenetetracarboxylic dianhydrides. *J. Chem. Phys.*, 82, 1682.

Karelson, M. M. & Zerner, M. C. 1992. Theoretical treatment of solvent effects on electronic spectroscopy. *J. Phys. Chem.*, 96, 6949.

Kazaoui, S., Minami, N., Tanabe, Y., Byrne, H. J., Eilmes, A. & Petelenz, P. 1998. Comprehensive analysis of intermolecular charge-transfer excited states in C_{60} and C_{70} films. *Phys. Rev. B*, 58, 7689.

Kim, J.-H., An, B.-K., Yoon, S.-J., Park, S. K., Kwon, J. E., Lim, C.-K. & Park, S. Y. 2014. Highly fluorescent and color-tunable exciplex emission from poly(*N*-vinylcarbazole) film containing nanostructured supramolecular acceptors. *Adv. Funct. Mater.*, 24, 2746.

Klafter, J. & Jortner, J. 1977. Critical concentration for electronic energy transfer in mixed organic solids. *Chem. Phys. Lett.*, 49, 410.

Klessinger, M. & Michl, J. 1995. *Excited States and Photochemistry of Organic Molecules*. New York: VCH.

Kohn, W. & Hohenberg, P. 1964. Inhomogeneous electron gas. *Phys. Rev.*, 136, B864.

Konishi, K. & Paffuti, G. 2009. *Quantum Mechanics: A New Introduction*. New York: Oxford University Press.

Kroemer, H. 2004. The Thomas precession factor in spin–orbit interaction. *Am. J. Phys.*, 72, 51.

Kuhn, W. 1948. Uber das Absorptionsspektrum der Polyene. *Helv. Chim. Acta*, 31, 1780.

Lai, S.-W., Lam, H.-W., Lu, W., Cheung, K.-K. & Che, C.-M. 2002. Observation of low-energy metal- metal-to-ligand charge transfer absorption and emission: electronic spectroscopy of cyclometalated platinum (II) complexes with isocyanide ligands. *Organometallics*, 21, 226.

Lamansky, S., Djurovich, P., Murphy, D., Abdel-Razzaq, F., Lee, H.-E., Adach, C., P. E. Burrows, Forrest, S. R. & Thompson, M. E. 2001. Highly phosphorescent bis-cyclometalated iridium complexes: synthesis, photophysical characterization, and use in organic light emitting diodes. *J. Am. Chem. Soc.*, 123, 4304.

Lambert, C., Nöll, G. & Schelter, J. 2002. Bridge-mediated hopping or superexchange electron-transfer processes in bis(triarylamine) systems. *Nat. Mater.*, 1, 69.

Langevin, P. 1903. Recombinaison et mobilites des ions dans les gaz. *Ann. Chim. Phys*, 28, 122.

Leach, A. R. 1996. *Molecular Modeling: Principles and Applications*. Harlow, UK: Addison Wesley Longman Ltd.

Lee, J., Chen, H.-F., Batagoda, T., Coburn, C., Djurovich, P. I., Thompson, M. E. & Forrest, S. R. 2016. Deep blue phosphorescent organic light-emitting diodes with very high brightness and efficiency. *Nat. Mater.*, 14, 92.

LeRoy, R. J., Dattani, N. S., Coxon, J. A., Ross, A. J., Crozet, P. & Linton, C. 2009. Accurate analytic potentials for Li_2(X) and Li_2(A) from 2 to 90 Angstroms, and the radiative lifetime of Li(2p). *J. Chem. Phys.*, 131, 204309.

Liptay, W. 1974. Dipole moments and polarizabilities of molecules in excited electronic states. In: Lim, E. C. (ed.) *Excited States*. New York: Academic Press.

Lunt, R. R., Benziger, J. B. & Forrest, S. R. 2010. Relationship between crystalline order and exciton diffusion length in molecular organic semiconductors. *Adv. Mater.*, 22, 1233.

Lunt, R. R., Giebink, N. C., Belak, A. A., Benziger, J. B. & Forrest, S. R. 2009. Exciton diffusion lengths of organic semiconductor thin films measured by spectrally resolved photoluminescence quenching. *J. Appl. Phys.*, 105, 053711.

Ma, B., Djurovich, P. I., Yousufuddin, M., Bau, R. & Thompson, M. E. 2005. Synthetic control of Pt ... Pt separation and photophysics of binuclear Pt complexes. *J. Am. Chem. Soc.*, 127, 28.

Macchi, G., Medina, B. M., Zambianchi, M., Tubino, R., Cornil, J., Barbarella, G. & Gierschner, J. 2009. Spectroscopic signatures for planar equilibrium geometries in methyl-substituted oligiothiophenes. *Phys. Chem. Chem. Phys.*, 11, 984.

Martin, R. L. 2003. Natural transition orbitals. *J. Chem. Phys.*, 118, 4775.

Matsui, A., Tomioka, K., Oeda, Y. & Tomokita, T. 1973. Exciton band structure of crystalline anthracene. *Surf. Sci.*, 37, 849.

McClure, D. S. 1949. Selection rules for singlet-triplet perturbations in polyatomic molecules. *J. Chem. Phys.*, 17, 665.

McGlynn, S. P., Azumi, T. & Kinoshita, M. 1969. *Molecular Spectroscopy of the Triplet State*. Englewood Cliffs, NJ: Prentice-Hall.

Medintz, I. & Hildebrandt, N. (eds.) 2014. *FRET-Förster Energy Transfer: From Theory to Applications*. Weinheim: Wiley-VCH Verlag.

Moffitt, W. 1954. The electronic spectra of cata-condensed hydrocarbons. *J. Chem. Phys.*, 22, 320.

Monberg, E. M. & Kopelman, R. 1978. Critical concentrations for triplet exciton tunneling in binary napthalene crystals: the case for percolation. *Chem. Phys. Lett.*, 58, 492.

Nasu, K. 1987. Tunneling and relaxation from the free state to the self-localized state of the exciton: nonadiabatic multi-steptheory. *J. Lumin.*, 38, 90.

Olson, J. M. & Cox, R. P. 1991. Monomers, dimers and tetramers of 4-*n*-propyl-5-ethyl farnesyl bacteriochlorophull c in dichloromethane and carbon tetrachloride. *Photosynthesis Res.*, 30, 35.

Onsager, L. 1936. Electric moments of molecules in liquids. *J. Am. Chem. Soc.*, 58, 1486.

Orlandi, G. & Siebrand, W. 1973. Theory of vibronic intensity borrowing. Comparison of Herzberg–Teller and Born–Oppenheimer coupling. *J. Chem. Phys.*, 58, 4513.

Pariser, R. & Parr, R. G. 1953. A semi-empirical theory of the electronic spectra and electronic structure of complex unsaturated molecules. II. *J. Chem. Phys.*, 21, 767.

Parker, C. A. 1963. Sensitized p-type delayed fluorescence. *Proc. R. Soc. London, Ser. A*, 276, 125.

Parker, C. A. & Hatchard, C. G. 1961. Triplet–singlet emission in fluid solutions. Phosphorescence of eosin. *Trans. Faraday Soc.*, 57, 1894.

Perrin, F. 1932. Théorie quantique des transferts d'activation entre molecules de meme espéce. Cas des solutions fluorescents. *Ann. Chim. Phys.*, 17, 283.

Perrin, J. 1927. Fluorescence et induction moléculaire par résonance. *C. R. Hebdo. Séances Acad. Sci.*, 184, 1097.

Peumans, P. 2004. *Organic thin-film photodiodes.* Ph.D. thesis, Princeton University.

Platt, J. R. 1949. Classification of spectra of cata-condensed hydrocarbons. *J. Chem. Phys.*, 17, 484.

Pope, M. & Swenberg, C. E. 1982. *Electronic Processes in Organic Crystals.* Oxford: Clarendon Press.

Pople, J. A. 1953. Electron interaction in unsaturated hydrocarbons. *Trans. Faraday Soc.*, 49, 1375.

Qi, X., Djurovich, P. I., Giebink, N. C., Thompson, M. E. & Forrest, S. R. 2008. Evidence for enhanced interactions between Pt centers in binuclear phosphorescent complexes. *Chem. Phys. Lett.*, 458, 323.

Rausch, A. F., Homeier, H. H. H. & Yersin, H. 2010. Organometallic Pt(II) and Ir(III) triplet emitters for OLED applications and the role of spin orbit coupling: a study based on high-resolution optical spectroscopy. In: Lees, A. J. (ed.) *Photophysics of Organometallics.* Berlin: Springer-Verlag.

Reichardt, C. 2003. *Solvents and Solvent Effects in Organic Chemistry.* Weinheim, Germany: Wiley-VCH.

Rissler, J. 2004. Effective conjugation length of π-conjugated systems. *Chem. Phys. Lett.*, 395, 92.

Robertson, J. M. 1958. Structure of naphthalene and anthracene. *Rev. Mod. Phys.*, 30, 155.

Robertson, J. M. & White, J. G. 1947. The crystal structure of pyrene. A quantitative X-ray investigation. *J. Chem. Soc.*, 358.

Scherr, C. W. 1953. Free-electron network model for conjugated systems. II. Numerical calculations. *J. Chem. Phys.*, 21, 1582.

Shen, Z. & Forrest, S. R. 1997. Quantum size effects of charge transfer excitons in nonpolar molecular organic thin films. *Phys. Rev. B*, 55, 10578.

Shi, J. & Tang, C. W. 2002. Anthracene derivatives for stable blue-emitting organic electroluminescence devices. *Appl. Phys. Lett.*, 80, 3201.

Shi, S., Collins, L. R., Mahon, M. F., Djurovich, P. I., Thompson, M. E. & Whittlesey, M. K. 2017. Synthesis and characterization of phosphorescent two-coordinate copper (I) complexes bearing diamidocarbene ligands. *Dalton Trans.*, 46, 745.

Shi, S., Jung, M. C., Coburn, C., A. Tadle, D. Sylvinson, Djurovich, P. I., Forrest, S. R. & Thompson, M. E. 2019. Highly efficient photo and electroluminescence from two coordinate Cu(I) complexes featuring non conventional N heterocyclic carbenes. *J. Am. Chem. Soc.*, 141, 3576.

Shida, T. & Iwata, S. 1973. Electronic spectra of ion radicals and their molecular interpretation. III. Aromatic hydrocarbons. *J. Am. Chem. Soc.*, 95, 3473.

Sibley, R., Jortner, J. & Rice, S. A. 1965. On the singlet-exciton states of crystalline anthracene. *J. Chem. Phys.*, 42, 1515.

Siebrand, W. 1967. Radiationless transitions in polyatomic molecules. I. Calculation of Franck–Condon factors. *J. Chem. Phys.*, 46, 440.

Siebrand, W. & Williams, D. F. 1967. Isotope rule for radiationless transitions with an application to triplet decay in aromatic hydrocarbons. *J. Chem. Phys.*, 46, 403.

Silinsh, E. A. 1980. *Organic Molecular Crystals.* Berlin: Springer-Verlag.

Smith, M. B. & Michl, J. 2010. Singlet fission. *Chem. Rev.*, 110, 6891.

Smoluchowski, M. v. 1915. Über Brownsche Molekularbewegung unter Einwirkung außerer Kräfte und deren Zusammenhang mit der verallgemeinerten Diffusionsgleichung. *Ann. Phys.*, 48, 1103.

Song, K. S. & Williams, R. T. 1993. *Self-Trapped Excitons.* Berlin: Springer-Verlag.

Sponer, H. & Teller, E. 1941. Electronic spectra of polyatomic molecules. *Rev. Mod. Phys.*, 13, 75.

Strickler, S. J. & Berg, R. A. 1962. Relationship between absorption intensity and fluorescence lifetime of molecules. *J. Chem. Phys.*, 37, 814.

Sundström, V., Pullerits, T. & Grondelle, R. v. 1999. Photosynthetic light-harvesting: reconciling dynamics and structure of purple bacterial LH2 reveals function of photosynthetic unit. *J. Phys. Chem. B*, 103, 2327.

Sze, S. M. 1981. *Physics of Semiconductor Devices.* New York: Wiley-Interscience.

Tang, C. W., VanSlyke, S. A. & Chen, C. H. 1989. Electroluminescence of doped organic thin films. *J. Appl. Phys.*, 65, 3610.

Tatchen, J., Gilka, N. & Marian, C. M. 2007. Intersystem crossing driven by vibronic spin–orbit coupling: a case study on psoralen. *Phys. Chem. Chem. Phys.*, 9, 5209.

Uoyama, H., Goushi, K., Shizu, K., Nomura, H. & Adachi, C. 2012. Highly efficient organic light-emitting diodes from delayed fluorescence. *Nature*, 492, 234.

van der Meer, B. W. 2014. Förster theory. In: Medintz, I. & Hildebrandt, N. (eds.) *FRET-Förster Energy Transfer: From Theory to Applications.* Weinheim: Wiley-VCH Verlag.

van der Meer, B. W., G. Coker, I. & Chen, S.-Y. S. 1994. *Resonance Energy Transfer: Theory and Data*. New York: VCH.

Wallesch, M., Volz, D., Zink, D. M., Schepers, U., Nieger, M., Baumann, T. & Bräse, S. 2014. Bright coppertunities: multinuclear CuI complexes with N–P ligands and their applications. *Chem. Eur. J.*, 20, 6578.

Wilson, J. S., Chawdhury, N., Al-Mandhary, M. R. A., Younus, M., Khan, M. S., Raithby, P. l. R., A. Kohler & Friend, R. H. 2001. The energy gap law for triplet states in Pt-containing conjugated polymers and monomers. *J. Am. Chem. Soc.*, 123, 9412.

Wolf, H. C. 1959. The electronic spectra of aromatic molecular crystals. In: Seitz, F. & Turnbull, D. (eds.) *Advances in Solid State Physics*. New York: Academic Press.

Yariv, A. 1989. *Quantum Electronics*. New York: John Wiley & Sons.

Yersin, H. 2004. Organometallic triplet emitters for OLED applications: controlling emission properties by chemical variation. In: *Organic Light-Emitting Materials and Devices VII*. Bellingham, WA: International Society for Optics and Photonics, pp.124–133.

Yersin, H., Czerwieniec, R., Shafikov, M. Z. & Suleymanova, A. F. 2017. TADF material design: photophysical background and case studies focusing on CuI and AgI complexes. *ChemPhysChem*, 18, 3508.

Yersin, H., Rausch, A. F., Czerwieniec, R., Hofbeck, T. & Fischer, T. 2011. The triplet state of organo-transition metal compounds. Triplet harvesting and singlet harvesting for efficient OLEDs. *Coord. Chem. Rev.*, 255, 2622.

Zang, D. Y., So, F. F. & Forrest, S. R. 1991. Giant anisoptropies in the dielectric properties of quasi-epitaxial crystalline organic semiconductor thin films. *Appl. Phys. Lett.*, 59, 823–825.

Zerner, M. 1991. *Reviews in Computational Chemistry*. New York: VCH.

Zhang, Y., Whited, M., Thompson, M. E. & Forrest, S. R. 2010. Singlet–triplet quenching in high intensity fluorescent organic light emitting diodes. *Chem. Phys. Lett.*, 495, 161.

CHAPTER 4

Charge transport in organic semiconductors

"When talking about semiconductors, if you can't draw a band diagram then you don't know what you're talking about."

***Kroemer's Lemma*, Herbert Kroemer, ca. 1990.**

It is a small step from the optical to the electronic. Indeed, we have seen that all of the optical properties in organic materials emerge from elementary transitions of electrons from one energy state to the next. But electrons and holes, stripped of their binding from each other, contribute to the conductivity of solids. Indeed, it was the conduction of electrons in molecular anthracene that first opened the door to the field of organic electronics. In this chapter, we consider the origin of conductivity in organic semiconductors. We will find that it differs in many ways from inorganic semiconductors. The purely electrostatic and relatively weak van der Waals bonds result in electronic energies in the solid state that are quite similar to that of the individual molecules. Hence, the well-developed band structure that characterizes inorganic semiconductors is absent. But without even narrow bands, the electron mass becomes infinite and there can be no conductivity. That, of course, is the opposite of what we observe.

The links between energy levels, bands and conductivity can be understood in terms of tight binding theory. But the lack of significant energy bandwidths leads to low mobility hopping transport of electrons and holes from molecule to molecule, in contrast to high mobilities due to transport of delocalized charge in bands in inorganics. In fact, charge transport can be incoherent, or dispersive where a single mobility is absent in many solids. The physics of charge transfer between molecules is analogous to exciton transfer, and needs to be considered before we can address the source of the low charge mobility.

Ultimately, the distinguishing characteristics of organic conductivity once again result from the lack of intermolecular chemical bonds. This, in turn leads to structural disorder, an intrinsic property leading to the low charge mobility (and short exciton diffusion lengths) in the solid. Even in the presence of disorder, nearly ideal junctions between dissimilar materials can be formed, resulting in strong rectification of charge across these interfaces.

We start our discussion on junctions with metal–organic contacts. When the injection barrier is considerably larger than the thermal energy, a Schottky diode is formed. When it is small, Ohmic contacts ensue. We continue on to develop ideal diode equations for organic (i.e. excitonic) heterojunctions analogous to treatments used to describe conduction across p-n junctions by Shockley in 1949. However, even though we can derive such an ideal diode equation, it does not mean that all (or even any) excitonic junctions follow such a formulation. Non-idealities such as recombination, lattice defects, disorder, and impurities are always present, and ultimately cause a deviation of behavior from that predicted based only on the most fundamental, irreducible physical processes. We will discuss both the ideal and non-ideal case of conductivity of organic heterojunctions; structures that form the basis of all organic photonic devices such as OLEDs and OPVs.

Beyond purely organic heterojunctions, there is a wealth of exciting and important phenomena found in organic/ inorganic semiconductor heterojunctions. Such heterojunctions have long been utilized in such devices as dye sensitized solar cells and colloidal quantum dot emitters and detectors, yet the dynamics of charge transport across such hybrid interfaces has not, until recently, been understood from a first-principles perspective. In the latter part of

Organic Electronics: Foundations to Applications. Stephen R. Forrest.
Published in 2020 by Oxford University Press. DOI: 10.1093/oso/9780198529729.001.0001

this chapter, we will develop the theory of transport across hybrid heterojunctions, and find that it bears striking similarities to both inorganic and fully organic junctions.

We note that photoconductivity forms a bridge between the optical and electronic properties of materials. We leave a full discussion of this topic to Chapter 7, where photodetectors of all varieties are considered.

Earlier, we considered the structural and optical properties of organics. With this chapter on charge transport, we complete our discussion of the fundamental physics of organic semiconductor materials. This leads us directly to their applications that will be the subject of Part II.

4.1 From energy levels to energy bands

Due to the intimate bonding between atoms arranged periodically in a crystal, inorganic semiconductors are characterized by a band gap separating broad and continuous conduction and valence bands, themselves characterized by a large *electronic density of states* (DOS). This is illustrated schematically on the left of Fig. 4.1. The bandwidths are typically comparable to the band gap, E_G, that is, on the order of a few electron volts. The peak densities of states depend on the atomic composition of the lattice, and the lattice symmetry itself. The peak DOS ranges from 10^{17}–10^{20} eV^{-1} cm^{-3}, which is significantly less than the atomic density (~10^{22} cm^{-3}). The comparatively low DOS and broad energy bands arise from sharing the outermost bonds between neighboring atoms.

Organic semiconductors do not have broad bands. In this context, it is more precise to call E_G simply an energy gap rather than a band gap. Then E_G is the difference between the highest occupied molecular orbital (HOMO) for the most energetic bound valence states, and the lowest unoccupied MO (LUMO) for the least energetic conduction states. Thus, the HOMO is analogous, but not identical to the *valence band maximum* (VBM) of an inorganic semiconductor, and similarly the LUMO is analogous to the *conduction band minimum* (CBM).

In organic materials the DOS is no longer continuous with a wide bandwidth (BW). Near the HOMO and LUMO, the energy levels from nearest-neighbor molecules overlap to a limited degree, leading to a narrow band. Such is the case with tightly packed molecules typified by PTCDA, rubrene, and pentacene, where BW energies approaching a few hundred meV are found. Yet far more frequently, the spacing between molecules, or their mutual orientation leads to only a limited degree of orbital overlap, such that BWs of only a few tens of meV are more the rule than the exception. In the former case, an *incipient energy band* is formed, giving the molecular solid some characteristics of a band-like inorganic semiconductor while preserving many of the properties of a more conventional organic semiconductor. As temperature is lowered, however, molecular motion is frozen, resulting in increased time-averaged orbital overlap between neighbors that can give a molecular solid with even a small BW at room temperature, band-like characteristics.

The orbitals deeper than the frontier HOMO and LUMO are denoted by HOMO-*i* and LUMO+*i* in Fig. 4.1 (here, *i* is the integer orbital index relative to the frontier orbital). These remain comparatively independent in the solid, that is, there is even less overlap with similar levels from adjacent molecules since these orbitals are more tightly bound to the molecular core. Hence, their lack of participation in the electronic properties of the solid further limits the total BW compared to an inorganic material.

Since the character of the frontier orbitals is not significantly changed in the solid, there is approximately one state per molecule: an enormous increase

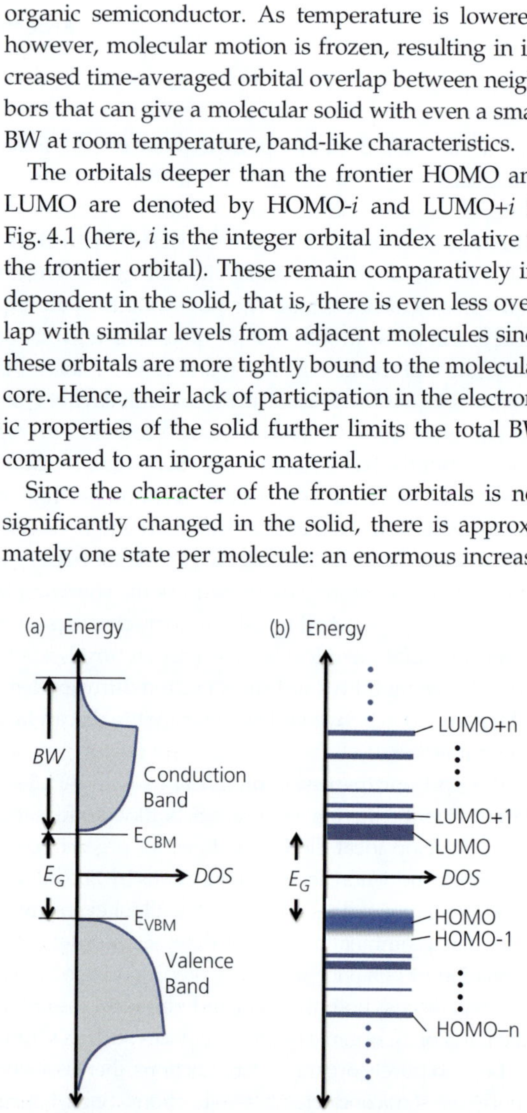

Figure 4.1 Characteristic energy level schemes of (a) an inorganic semiconductor with band gap energy, E_G, and (b) an organic semiconductor with an energy gap of approximately the same magnitude. Here, *DOS* is the density of electronic states, HOMO is the highest occupied molecular orbital, LUMO is the lowest unoccupied MO, and *BW* is the width of the corresponding band. The "fuzzy" depiction of the HOMO suggests that in the case of a tightly packed organic crystal, an incipient, or narrow band-like character emerges.

compared with inorganics. One consequence of this is that the optical absorption coefficient of direct band gap inorganic semiconductors is $\lesssim 10^4$ cm^{-1}, but can be as much as 100 times higher for organics, making even the thinnest films extremely strong absorbers.

The differences in energy level and band structure compared to inorganics play a determining role in their conductive properties. Whereas an electron in an inorganic semiconductor moves relatively effortlessly around the solid within its energy band due to the intimately shared orbitals of adjacent atoms, the reduced intermolecular coupling of molecules results in polarization. The electron therefore self-traps on each molecule as it moves under the influence of an electric field. Given its ability to polarize the surrounding medium, the "free" electron is called a *polaron*, and rather than experiencing band-like transport, it *hops* from molecule to molecule. The immediate consequence is a dramatically reduced mobility: organics have maximum charge mobilities no more than 1–5 cm^2/V s, whereas inorganic semiconductors have mobilities (and hence conductivities) that can exceed this by a factor to 10^3–10^4. We will show that both the field and temperature dependence of the charge mobilities in these two systems can also be vastly different, leading to distinctive conductive character.

We begin our discussion by showing how conduction levels emerge when individual molecules are assembled into a solid. While the lack of chemical bonds between molecules allows them to preserve their individual character to some degree, we have shown in Chapter 3 that molecular assemblies have emergent properties of their own that give rise to energy (i.e. exciton) transport within the solid. These same collective effects also allow for charge transport with a semiconductor-like temperature dependence that was discovered for molecular violanthrone (Akamatu and Inokuchi, 1950).

4.1.1 Tight binding method for calculating energy bands

A convenient means for calculating energy bands is the use of the *tight-binding approximation*. It is a somewhat limited method for calculating the very complex situation of molecules in solids, but it has the advantages of providing an intuitive formalism for appreciating the emergence of band-like character that leads to conduction in organic semiconductors.

In Chapter 2 we found that bringing two H atoms into close proximity to form an H_2^+ molecule results in a splitting of the 1s state into a symmetric bonding, and an antisymmetric antibonding state whose energy difference increases with decreasing atomic separation. This splitting is a result of Coulomb repulsion of the two valence electrons. When N atoms are brought into proximity in a lattice, the result is a splitting into N electronic levels. As is the case for a H_2^+ molecule, the splitting increases as the lattice constant decreases, ultimately resulting in overlap of many adjacent levels to form a continuous band whose width scales with N.

In molecular systems, the situation is in many ways similar to that of atoms but with a significant difference; in covalently bonded solids, an electron is shared between all N atoms whereas in molecular systems the orbital overlap between molecules is comparatively small. Hence, the influence on the orbital energies of individual molecules separated by even a few lattice constants is small. When considering an assembly of N molecules in the solid, therefore, only those molecules within a limited neighborhood need to be considered, and this results in a concomitant reduction in the BW. As for atomic systems, when molecules are brought closer together, the Coulomb interaction between electrons in nearby orbitals increases, broadening the bands even further. Broadening continues until the molecules reach their equilibrium separation distance, R_0, when the attractive van der Waals force is balanced by the repulsive force of the inner shell electrons.

A schematic representation of the formation of bands as the intermolecular spacing decreases is provided in Fig. 4.2. As the spacing decreases, the BW increases until neighboring orbitals (e.g. HOMO and HOMO-1) begin to interact, resulting in further merging and broadening of the bands.

To calculate the perturbation of the electronic energy levels, we employ the tight binding approximation that has been widely applied to describe inorganic materials such as metals, and was first applied to organic semiconductors by LeBlanc (1961),

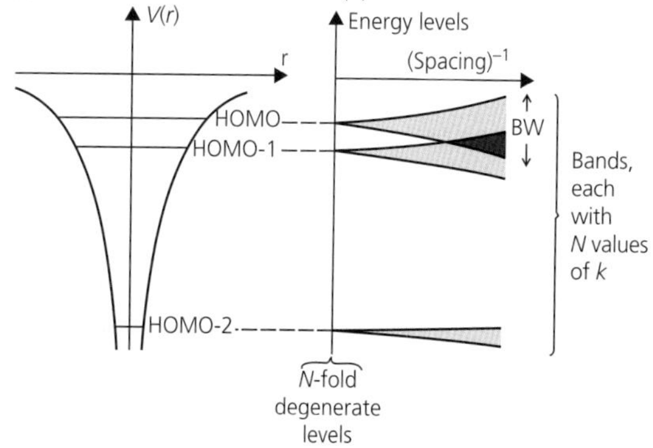

Figure 4.2 Schematic illustration of band formation due to orbital overlap as the molecular spacing is decreased. (a) The orbitals of unperturbed molecules showing the single electron potential, $V(r)$. (b) Bands formed due to overlap of orbitals from N molecules. BW = electronic bandwidth of the HOMO which increases with decreased spacing.

and then extended by Katz et al. (1963). The important features of this method applied to molecular organic semiconductors are now described.

Each orbital level results in a band. For example, if there are six atoms forming the molecule, there are three resulting electronic states (with two spins per state). The bound orbitals are the HOMO, HOMO-1, and HOMO-2. In a crystal of N such molecules therefore, three valence bands emerge, one corresponding to each HOMO level. The conduction bands corresponding to the LUMO, LUMO +1, LUMO+2..., extending to the vacuum level, are also generated.

We calculate the BW and shape by assuming that the interaction energy between electrons on separate molecules is small compared to the orbital energies within an individual molecule. Furthermore, we neglect all interactions between electrons within an orbital; this is known as the *single-electron approximation*. Then, using H_{mol} for the Hamiltonian of a molecular orbital, the total Hamiltonian of the system is

$$H_{tot} = H_{mol} + H_{int}, \tag{4.1}$$

where $H_{int}(\mathbf{r}) = U(\mathbf{r})$ is the interaction at lattice site, $\mathbf{r}$, and has the periodicity of the lattice, that is,

$$U(\mathbf{r} + \mathbf{R}) = U(\mathbf{r}), \tag{4.2}$$

where $\mathbf{R}$ is the lattice translation vector (Section 2.2). The interaction, $H_{int}(\mathbf{r})$, is the perturbation when a molecule is placed into an infinite crystal, and should contain all of the symmetry properties of the lattice. Due to the potential having the same periodicity as the lattice, the *Bloch condition* imposes the same periodicity on the wavefunction, viz.

$$\psi_{\mathbf{k}}(\mathbf{r} + \mathbf{R}) = e^{i\mathbf{k}\cdot\mathbf{R}}\psi_{\mathbf{k}}(\mathbf{r}), \tag{4.3}$$

where $\mathbf{k}$ is the electronic wavevector.

As in Chapter 3, the electronic wavefunction can be solved by the LCAO method, in which case we expand Eq. 4.3 using

$$\psi_{n\mathbf{k}}(\mathbf{r}) = \sum_{i}^{N} c_{\mathbf{k}i}\psi_n(\mathbf{r} - \mathbf{R}_i), \tag{4.4}$$

where the sum is over all N lattice points, and $\psi_n(\mathbf{r} - \mathbf{R}_i)$ is the contribution from the molecule at $\mathbf{R}_i$ and from the electron in its nth orbital. The basis functions, $\psi_n(\mathbf{r})$, are solutions to the individual (isolated) molecular eigenvalue equations, $H_{mol}\psi_n(\mathbf{r}) = E_n\psi_n(\mathbf{r})$. Thus, the lattice wavefunction contains the characteristics of the orbitals of the individual molecules although it must also be consistent with Bloch's theorem for a periodic lattice, as in Eq. 4.3. This requires that the coefficients have the property: $c_{\mathbf{k}i} = N^{-\frac{1}{2}}e^{-i\mathbf{k}\cdot\mathbf{R}_i}$. We can test that this satisfies the condition for periodicity by translating $\psi(\mathbf{r})$ through a lattice vector, viz.

$$\begin{aligned}\psi_{n\mathbf{k}}(\mathbf{r} + \mathbf{R}) &= \frac{1}{\sqrt{N}}\sum_{i}^{N} e^{-i\mathbf{k}\cdot\mathbf{R}_i}\psi_n(\mathbf{r} - \mathbf{R}_i + \mathbf{R}) \\ &= \frac{1}{\sqrt{N}}e^{i\mathbf{k}\cdot\mathbf{R}}\sum_{i}^{N} e^{-i\mathbf{k}\cdot(\mathbf{R}_i - \mathbf{R})}\psi_n(\mathbf{r} - \mathbf{R}_i + \mathbf{R}).\end{aligned} \tag{4.5}$$

But $\tilde{\mathbf{R}}_i = \mathbf{R} - \mathbf{R}_i$ is also a lattice translation vector. Thus,

$$\begin{aligned}\psi_{n\mathbf{k}}(\mathbf{r} + \mathbf{R}) &= \frac{1}{\sqrt{N}}e^{i\mathbf{k}\cdot\mathbf{R}}\sum_{i}^{N} e^{i\mathbf{k}\cdot\tilde{\mathbf{R}}_i}\psi_n(\mathbf{r} - \tilde{\mathbf{R}}_i) \\ &= e^{i\mathbf{k}\cdot\mathbf{R}}\psi_{n\mathbf{k}}(\mathbf{r}),\end{aligned} \tag{4.6}$$

as required by the Bloch boundary condition, Eq. 4.3.

When placed near its neighbors, these functions and the system energy are perturbed. Assuming that the perturbation is small, we write the new wavefunctions as linear combinations of the wavefunctions of the isolated molecules. We allow that the lattice basis can contain more than one molecule, in which case the number of molecules is M. Then we have

$$\phi(\mathbf{r}) = \sum_{j}^{M} a_j\psi_j(\mathbf{r}), \tag{4.7}$$

where, in keeping with only minor perturbations, the coefficients a_j are small. Then we can rewrite the total wavefunction as

$$\begin{aligned}\psi_{\mathbf{k}}(\mathbf{r}) &= \frac{1}{\sqrt{N}}\sum_{i}^{N} e^{i\mathbf{k}\cdot\mathbf{R}_i}\phi(\mathbf{r} - \mathbf{R}_i) \\ &= \frac{1}{\sqrt{N}}\sum_{j}^{N} e^{i\mathbf{k}\cdot\mathbf{R}_i}\sum_{j}^{N} a_j\psi_{j\mathbf{k}}(\mathbf{r} - \mathbf{R}_i).\end{aligned} \tag{4.8}$$

Using the Hamiltonian for the crystal in Eq. 4.1, we then arrive at

$$H_{tot}\psi_{\mathbf{k}}(\mathbf{r}) = (H_{mol} + H_{int}(\mathbf{r}))\psi_{\mathbf{k}}(\mathbf{r}) = E(\mathbf{k})\psi_{\mathbf{k}}(\mathbf{r}). \tag{4.9}$$

The energy eigenvalues are determined by multiplying Eq. 4.9 by the original, molecular wavefunctions, $\psi_m(\mathbf{r})$, which in bracket notation becomes

$$\begin{aligned}\langle\psi_m(\mathbf{r})|H_{tot}|\psi_{\mathbf{k}}(\mathbf{r})\rangle &= \langle\psi_m(\mathbf{r})|(H_{mol} + H_{int}(\mathbf{r}))|\psi_{\mathbf{k}}(\mathbf{r})\rangle \\ &= \langle\psi_m(\mathbf{r})|E(\mathbf{k})|\psi_{\mathbf{k}}(\mathbf{r})\rangle.\end{aligned} \tag{4.10}$$

Now, the eigenvalues of the unperturbed molecular Hamiltonian, H_{mol}, are simply E_m, in which case we can rewrite Eq. 4.10 as

$$(E(\mathbf{k}) - E_m)\langle\psi_m(\mathbf{r})|\psi_{\mathbf{k}}(\mathbf{r})\rangle = \langle\psi_m(\mathbf{r})|H_{int}(\mathbf{r})|\psi_{\mathbf{k}}(\mathbf{r})\rangle. \tag{4.11}$$

Substituting Eq. 4.8 into Eq. 4.11, and from the orthogonality of the wavefunctions, we require that $n = m$. Then Eq. 4.11 becomes

$$
\begin{aligned}
(E(\mathbf{k}) - E_m)a_m &= -(E(\mathbf{k}) - E_m) \\
&\times \sum_{n}^{M}\left[\sum_{i\neq 0}^{N}\int \psi_m^*(\mathbf{r})\psi_n(\mathbf{r}-\mathbf{R}_i)e^{i\mathbf{k}\cdot\mathbf{R}_i}d^3\mathbf{r}\right]a_n \\
&+\sum_{n}^{M}\left[\int \psi_m^*(\mathbf{r})H_{int}(\mathbf{r})\psi_n(\mathbf{r})\right]a_n \\
&+\sum_{n}^{M}\left[\sum_{i\neq 0}^{N}\int \psi_m^*(\mathbf{r})H_{int}(\mathbf{r})\psi_n(\mathbf{r}-\mathbf{R}_i)e^{i\mathbf{k}\cdot\mathbf{R}_i}d^3\mathbf{r}\right]a_n
\end{aligned}
\tag{4.12}
$$

Since we are assuming the collective effects of the crystal are small and the magnitude of the wavefunctions, $\psi_n(\mathbf{r}-\mathbf{R}_i)$ must decrease rapidly with $\mathbf{R}_i$, only interactions between immediate neighbors (i.e. $i \approx 1$) are included. Hence, terms on both sides of Eq. 4.12 must be small, which means that $E(\mathbf{k}) \approx E_m$. This is equivalent to saying that the HOMO energy leads to a broadening of the highest (i.e. valence) band, but it does not influence the energies of the HOMO-1, HOMO-2, etc. levels. Those orbitals themselves are also impacted as the molecules are brought into close proximity but, like the HOMO, are only broadened by other, similar HOMO-*i* levels, forming the corresponding HOMO-*i* bands. While the various terms in Eq. 4.12 are small, they are nevertheless finite. If they were not, then we revert to the case of non-interacting molecules, that is, $E(\mathbf{k}) = E_m$.

The first term on the right corresponds to contributions to the energy from all of the other molecules in the crystal, although as mentioned above, $\psi_n(\mathbf{r}-\mathbf{R}_i)$ falls off rapidly with distance. The second and third terms are due to perturbations of the electronic potential due to intermolecular interactions which, in themselves, are assumed to be small. The smallest term in this expression is the last, which is the product of the interaction potential and the wavefunction at distances of one or more lattice constants from the origin. Of course, the interaction may be large far from the origin when $\psi_n(\mathbf{r}-\mathbf{R}_i)$ is small, or vice versa.

In solving the Eq. 4.12, we must account for the degeneracy of a particular band. That is, if the band has s-like symmetry, then there is a single equation to be solved, resulting in a single energy band, $E(\mathbf{k})$. Likewise, a level with p-symmetry results in three such expressions (with three a_j values) that must be solved with a 3×3 secular determinant (see Section 3.3). If the energies from two bands are sufficiently close, or if they are broadened by interactions from nearby molecules such that they overlap with another energy level, then that energy level must also be included. For example, if the HOMO with s-symmetry interacts with the HOMO-1 with p-symmetry, then the energy $E(\mathbf{k})$ of the new HOMO band is calculated via a 4×4 secular determinant that considers the contribution from all of the interacting orbitals. Finally, spin must be taken into account if the states are not degenerate. In this case, the number of equations used to calculate the energy levels is doubled yet again.

As an example, we consider an s-symmetric HOMO level. Then Eq. 4.12 reduces to:

$$
\begin{aligned}
E(\mathbf{k}) - E_{HOMO} &= -(E(\mathbf{k}) - E_{HOMO})\sum_{i\neq 0}^{N}\alpha(\mathbf{R}_i)e^{i\mathbf{k}\cdot\mathbf{R}_i} \\
&+\beta - \sum_{i\neq 0}^{N}J(\mathbf{R}_i)e^{i\mathbf{k}\cdot\mathbf{R}_i},
\end{aligned}
\tag{4.13}
$$

which leads to

$$
E(\mathbf{k}) - E_{HOMO} = \frac{\beta + \sum_{i\neq 0}^{N}J(\mathbf{R}_i)e^{i\mathbf{k}\cdot\mathbf{R}_i}}{1 + \sum_{i\neq 0}^{N}\alpha(\mathbf{R}_i)e^{i\mathbf{k}\cdot\mathbf{R}_i}}
\tag{4.14}
$$

with

$$
\begin{aligned}
\alpha(\mathbf{R}_i) &= \int \psi_{HOMO}^*(\mathbf{r})\psi_n(\mathbf{r}-\mathbf{R}_i)d^3\mathbf{r} \\
&= \langle\psi_{HOMO}(\mathbf{r})|\psi_n(\mathbf{r}-\mathbf{R}_i)\rangle,
\end{aligned}
\tag{4.15}
$$

$$
\beta = \langle\psi_{HOMO}(\mathbf{r})|H_{int}(\mathbf{r})|\psi_n(\mathbf{r})\rangle,
\tag{4.16}
$$

and

$$
J(\mathbf{R}_i) = -\langle\psi_{HOMO}(\mathbf{r})|H_{int}(\mathbf{r})|\psi_n(\mathbf{r}-\mathbf{R}_i)\rangle.
\tag{4.17}
$$

Hence, we have a band centered on the HOMO, but whose width and dependence on **k** are determined by the right hand side of Eq. 4.14. Clearly, a larger overlap between adjacent molecules, and a stronger interaction between molecules leads to an increase in the overlap integral, $J(\mathbf{R}_i)$. Since the overlap ultimately is small in the tight binding approximation, $\alpha(\mathbf{R}_i) \ll 1$, and hence the properties of the HOMO-band are almost entirely determined by the magnitude and functional form of $J(\mathbf{R}_i)$.

Recall from our treatment of excited states in crystals that the energy is perturbed from that of the monomer by nearest neighbor interactions that are modeled as Bloch wavefunctions of the same form as Eq. 4.3 (c.f. Eq. 3.138). Hence, our solution to Eq. 4.14 should also reduce to the dispersion relation of Eq. 3.139, where (LeBlanc, 1961)

$$E(\mathbf{k}) \sim \Delta E' + 2E_W\cos(\mathbf{k}\cdot\mathbf{R}). \tag{4.18}$$

Here, $\Delta E'$ is the sum of the ground (E_{HOMO}) and excited monomer energies, and $BW = 4E_W$ is the BW resulting from intermolecular interactions.

It is straightforward to show that Eq. 4.14 follows this familiar form by its solution for the simple case of a face centered cubic (fcc) structure with lattice constant, a. The molecule at (0,0,0) has 12 nearest neighbors at positions (±½,±½,0), (±½,0,±½), (0,±½,±½). We consider only the s-states in this example. The inversion symmetry of the cubic lattice implies that the interaction Hamiltonian follows $H_{int}(\mathbf{R}) = H_{int}(-\mathbf{R})$. Similarly, the wavefunctions have even parity, i.e. $\psi(\mathbf{r}) = \psi(-\mathbf{r})$, from which it follows that the coefficients α, β, and J have this same symmetry. Finally, as noted above, $\alpha(\mathbf{R}_i)$ is neglected. This leads to a simplified expression for $E(\mathbf{k})$ in Eq. 4.14 with the sum over only the 12 nearest neighbors:

$$E(\mathbf{k}) - E_{HOMO} = -\beta - \sum_{i=1}^{12} J(\mathbf{R}_i)\cos(\mathbf{k}\cdot\mathbf{R}_i) \tag{4.19}$$

with

$$J(\mathbf{R}_i) = J_{fcc} = -\left\langle \psi_{HOMO}(\mathbf{r})|H_{int}(\mathbf{r})|\psi\left(x-\frac{a}{2}, y-\frac{a}{2}, z\right)\right\rangle. \tag{4.20}$$

The expression for $J(\mathbf{R}_i)$is identical for all 12 nearest neighbors (J_{fcc}); here we have written it explicitly only for the molecule at $(\frac{1}{2}, \frac{1}{2}, 0)$. The sum in Eq. 4.19 for the fcc lattice is, therefore:

$$E(\mathbf{k}) - E_{HOMO} = -\beta - 4J_{fcc}\left\{\cos\left(\frac{1}{2}k_x a\right)\cos\left(\frac{1}{2}k_y a\right) + \cos\left(\frac{1}{2}k_x a\right)\cos\left(\frac{1}{2}k_z a\right) + \cos\left(\frac{1}{2}k_y a\right)\cos\left(\frac{1}{2}k_z a\right)\right\} \tag{4.21}$$

The bands have a cosine shape that is symmetric along all three orthogonal axes, and the BW is determined by J_{fcc} that is a function of both the overlap of nearest neighbor orbitals and the strength of the interaction potential. In fact, H_{int} can take any functional form that decays rapidly from the origin. One such function is shown in Fig. 4.3 and follows

$$U(\mathbf{r}) = -U_0\frac{\sin^2(\pi\mathbf{R}/a)}{(\mathbf{R}/a)^2}. \tag{4.22}$$

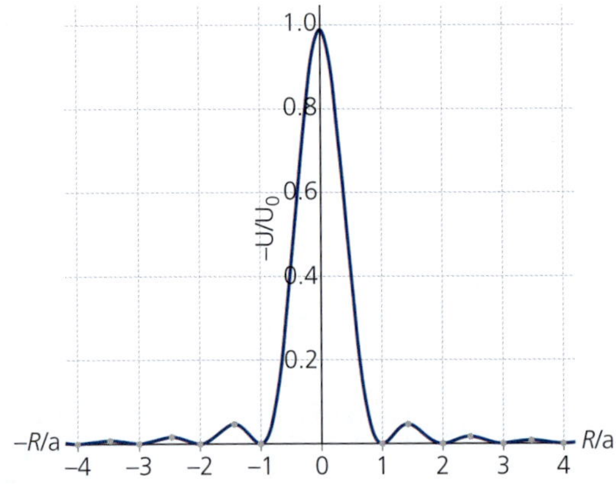

Figure 4.3 Perturbation potential used in tight binding calculation.

One important feature of this potential is that it is large when the wavefunction (and hence the overlap) is small, and vice versa. That is, the potential is maximum midway between the molecular sites (i.e. at $R/a = (2i+1)/2$, $i = 1, 2, \ldots$, except the self energy at $i = 0$), and the wavefunction is small at the origin where the s-state probability density vanishes. This ensures that J_{fcc} remains a small perturbation of the original crystal structure, as required.

In the limit of $ka = 1$ (i.e. for electrons near the center of the Brillouin zone), then $\cos(\frac{1}{2}ka) \approx 1 - \frac{1}{2}(\frac{1}{2}ka)^2$ and Eq. 4.21 becomes

$$E(\mathbf{k}) - E_{HOMO} \approx -\beta - 12J_{fcc} + J_{fcc}(ka)^2. \tag{4.23}$$

Now the slope of the dispersion, $\partial E(\mathbf{k})/\partial\mathbf{k}$, toward the zone edge (e.g. along the (100) direction where $k_x a = 2\pi$) is

$$\partial E(k_x)/\partial k_x = 4J_{fcc}a\sin\left(\frac{1}{2}k_x a\right) \to 0. \tag{4.24}$$

Hence, the band has a cosine shape with width of $BW = 8\ J_{fcc}$ in the (100) direction, and centered on the energy $E_{HOMO} - \beta$. Along other directions, the slope may not vanish at the band edge, and the width can differ from 8 J_{fcc}.

Figure 4.2 implies that the deeper orbitals are not as broadened by proximity with neighboring molecules as are the shallower (frontier) orbitals. This is a consequence of the stronger localization of electrons in the deeper valence states of the molecule, and hence their considerably reduced overlap with similar states in adjacent molecules. We also note that the calculation of conduction (i.e. LUMO) states by the tight binding approximation is far less reliable given the considerable spatial extent of such states compared with bound valence states. Hence, the inclusion of only nearest neighbor interactions becomes dubious as the conduction electron energy increases toward the vacuum level. Finally, the tight binding model as described thus far has only imposed a Coulomb interaction in H_{int}. However, as we have seen in Chapter 3, many other interactions can play a role, including dipole interactions, and spin-orbit interactions that strongly impact the orbitals of molecules containing heavy metals. As an example, the inclusion of first order singlet–triplet coupling requires solution of a 4×4 secular determinant.

A consequence of band structure is that the electron has a velocity given by

$$\mathbf{v}(\mathbf{k}) = \frac{1}{\hbar}\frac{\partial E(\mathbf{k})}{\partial \mathbf{k}}, \tag{4.25}$$

and a corresponding effective mass tensor along crystal directions i, j of

$$\overleftrightarrow{m}_{ij}(\mathbf{k}) = \left[\frac{1}{\hbar^2}\frac{\partial^2 E(\mathbf{k})}{\partial k_i \partial k_j}\right]^{-1} = \left[\frac{1}{\hbar}\frac{\partial v_j(k_i)}{\partial k_i}\right]^{-1}. \tag{4.26}$$

For an isotropic medium, the effective mass reduces to

$$m^* = \left[\frac{1}{\hbar^2}\frac{\partial^2 E(\mathbf{k})}{\partial k^2}\right]^{-1} = \left[\frac{1}{\hbar}\frac{\partial v(\mathbf{k})}{\partial k}\right]^{-1}. \tag{4.27}$$

Holes have negative effective masses, but otherwise are determined by the same expressions as Eqs. 4.25–4.27.

The existence of the bands, no matter how narrow, ultimately allows electrons and holes to move from molecule to molecule within the solid. Since bands do not exist in dilute solutions of non-interacting molecules, we conclude that the conductivity of the material depends entirely on the magnitude of the orbital overlap integral, $J(R)$. This is easily seen by once more taking the example of an fcc lattice near the Brillouin zone center where

$$v(k) = \frac{2J_{fcc}ka^2}{\hbar} \tag{4.28}$$

and

$$m^* = \frac{\hbar^2}{2J_{fcc}a^2}. \tag{4.29}$$

That is, the velocity increases and the effective mass decreases with increasing molecular orbital overlap, J_{fcc}. In contrast, near the (100) zone edge, $v(k_x) = \frac{1}{\hbar}\frac{\partial E(k_x)}{\partial k_x} \to 0$, which leads to the electron effective mass $m* \to \infty$. That is, the electron slows and is reflected at the zone edge.

An intuitive outcome of the tight binding approximation is that the spacing between molecules should decrease as the crystal is compressed, thereby resulting in a broadening of the bands as well as a decrease in the lowest energy transition from the ground to the excited state. Figure 4.4 shows the energies of the long wavelength absorption edges (which are inversely proportional to E_G) of NTCDA and PTCDA as functions of hydrostatic pressure (Jayaraman et al., 1985). The energy gap is reduced with pressure due to a decrease in a. Indeed, as PTCDA is compressed, its appearance changes from red to black. The change in NTCDA absorption at low pressures is non-linear, with a weaker dependence on pressure. Both responses are functions of the compressibility (i.e. the bulk modulus, Eq. 2.23) of these materials. The more open herringbone structure of NTCDA (with an intermolecular spacing of a = 3.506 Å) is more easily compressed than is the very tightly π-stacked PTCDA (a = 3.21 Å). Hence, the changes are more pronounced in NTCDA until the molecular core repulsion becomes dominant. However, the dependence of the energy shift on pressure, which is directly proportional to molecular spacing, is linear over the entire range tested in Fig. 4.4.

Tight binding approximations have been extensively used to calculate energy band dispersions in a number of molecules. Figure 4.5 shows the excess hole and electron three dimensional band structures for pentacene along several crystal symmetry directions. (Here, X = ($\mathbf{a}^*$/2,0,0), M = ($\mathbf{a}^*$/2,$\mathbf{b}^*$/2,0),

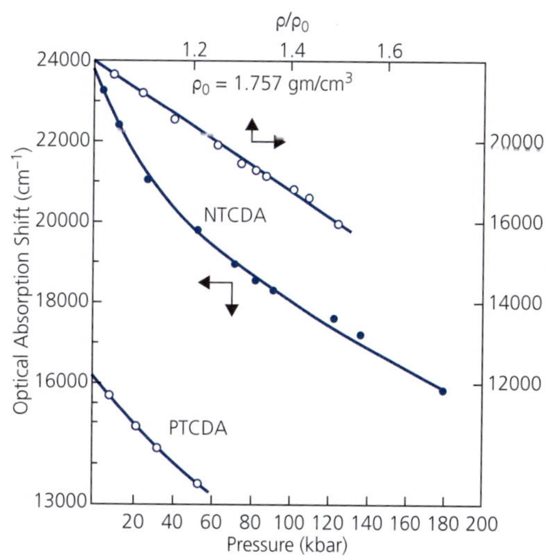

Figure 4.4 Shift in the low energy absorption edge of NTCDA and PTCDA excitonic features vs. hydrostatic pressure. The upper straight line corresponds to the spectral shift as a function of NTCDA density (Jayaraman et al., 1985).

Y = (0,**b***/2,0), Z = (0,0,**c***/2), N = (0,**b***/2,**c***/2), H = (**a***/2,**b***/2,**c***/2), and O = (**a***/2,0,**c***/2). Also, **a***, **b***, and **c*** are the reciprocal lattice vectors, Γ is the center of the Brillouin zone, and the direction from Γ to M is ½(**a*** + **b***).) The bands are calculated using both density functional theory and tight binding with almost identical results (de Wijs et al., 2003).

The bands resulting from the crystal structure and the electron density distributions are shown in Fig. 4.5a. The relative position of the wavefunction phases between adjacent molecules in the crystal (see Fig. 3.16) significantly affects the band structure (i.e. the splitting of the bands within the LUMO and HOMO manifolds) and consequently, the charge mobility. There are two branches in both the HOMO and LUMO bands since two symmetry-inequivalent molecules form the unit cell basis. The BWs are ~600 meV and 700 meV for the HOMO and LUMO, respectively. Note that the calculated band gap is 0.7 eV, which is substantially less than the measured E_G = 2.2 eV (Kang et al., 2006). This difference is sometimes observed in tight binding and DFT calculations (Nabok et al., 2007), although it does not appear to significantly affect other features of the band structure. While the energy band doublets are somewhat distorted due to asymmetries in the transfer integrals, $J(\mathbf{R})$, along different symmetry directions, they nevertheless have approximately the cosine shape in Eq. 4.19. This periodicity is particularly apparent for the upper and lowermost bands in the LUMO and HOMO, respectively.

The effective masses in the **a–b** plane calculated from the band structure are plotted in Fig. 4.5c. The mass varies from approximately $6.5m_0$ along the $\hat{\mathbf{a}}$-axis, to $1.5m_0$ at approximately 15° from the $\hat{\mathbf{b}}$ -axis. It is also found that $m^* = 5.2m_0$ along $\hat{\mathbf{c}}$. Here, m_0 is the free electron mass. The asymmetries in m^* are also due to different magnitudes in the coupling integrals, $J(\mathbf{R})$, along different symmetry axes.

The tight binding method has also been used to understand the charge transfer complex, bis (ethylenedithio)tetrathiafulvalene ($(BEDT\text{-}TTF)_2ClO_4(C_2H_3Cl_3)_{0.5}$) with the highest four bands shown in Fig. 4.6 (Mori et al., 1984). The bands arise from the four molecules per unit cell, presenting a more

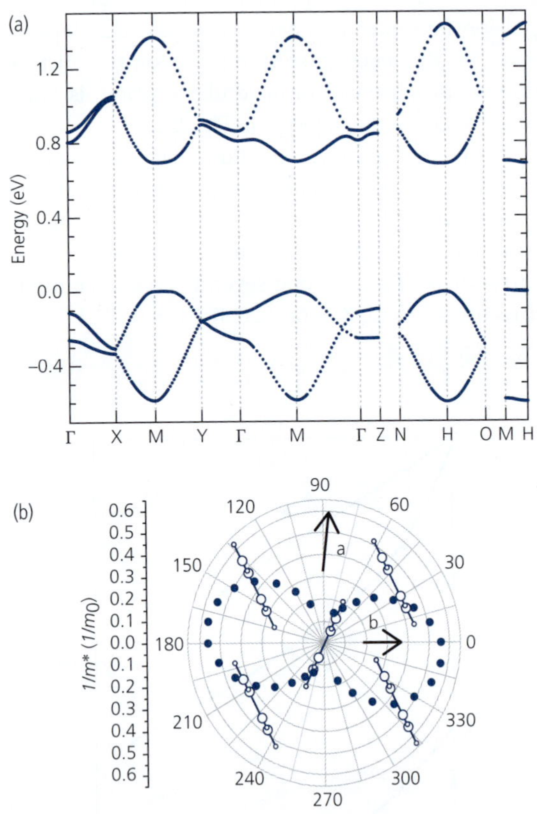

Figure 4.5 (a) Calculated band structure along several crystal symmetry directions. The upper two lines are the LUMO levels, and the lower two are the HOMO levels. There are two symmetry-inequivalent molecules per unit cell, resulting in a split band along each crystallographic direction. (c) The effective mass, m^* relative to the free electron mass, m_0, as a function of direction in the **a–b** plane. The molecular orientations in this plane are shown for reference (de Wijs et al., 2003).

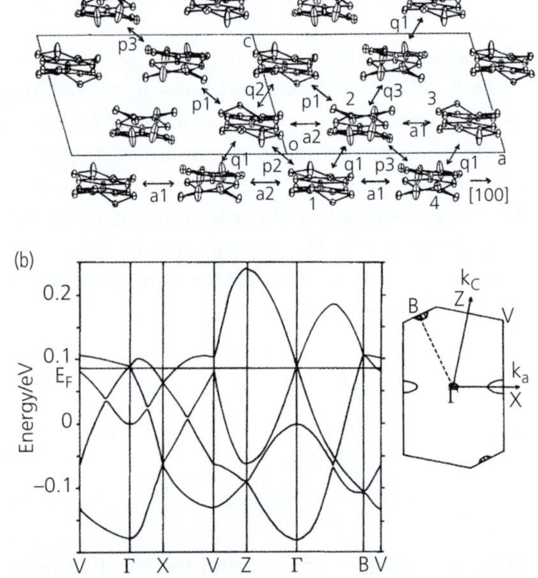

Figure 4.6 (a) Intermolecular interactions (p, q) along different crystal directions for $(BEDT\text{-}TTF)_2ClO_4(C_2H_3Cl_3)_{0.5}$. (b) Four bands arise from the four molecules per unit cell of the molecule along different crystal symmetry directions shown at right. The Γ-point is at the center of the Brillouin zone (Mori et al., 1984).

complex system than polyphenyls such as pentacene. While a small (0.1 eV) band gap around the Fermi energy (E_F) opens up between the third and fourth bands along the V-symmetry direction, band overlap along different crystal directions indicates that this is not a semiconductor, but rather a semimetal with a ¾-filled frontier orbital. Of particular note is the small BW ~200 meV compared to pentacene, resulting from relatively large intermolecular distances of $R \simeq 3.80$ Å.

4.1.2 Experimental dispersion relationships

There are several methods for measuring the frontier orbitals of organic semiconductors, most notably *ultraviolet photoelectron spectroscopy* (UPS, also known simply as *photoelectron spectroscopy*, PES) and *cyclic voltammetry* (CV). These methods are described and compared in Appendix C. Here, we will discuss the use of a modified version of UPS, known as *angle resolved ultraviolet photoelectron spectroscopy* (ARUPS) that can resolve the dispersion relation, $E(\mathbf{k})$. Briefly, UPS is based on the photoelectric effect, whereby a UV light source illuminates the surface of the sample. The energy difference between the HOMO and the vacuum level (known as the ionization potential, IP) is determined by measuring the kinetic energy of electrons ejected from the sample surface for a given photon energy (Hüfner, 2003). The angular dependence of the kinetic energy yields the component of the electronic wavevector perpendicular to the surface, $k_\perp$. The parabolic electronic potential (or *mean inner potential*), V_0, is then determined from ARUPS spectra using

$$k_\perp = (2m/\hbar^2)^{1/2}(E_{kin}\cos^2\theta_e + V_0)^{1/2}, \tag{4.30}$$

where E_{kin} is the kinetic energy determined from the electric field required to retard the motion of the photoemitted electron, and θ_e is its emission angle relative to the surface normal. Tight binding theory provides the band shape perpendicular to the substrate via (Mori et al., 1984)

$$E(k_\perp) = E_{HOMO} - 2J\cos(a_\perp k_\perp), \tag{4.31}$$

where the lattice constant perpendicular to the substrate is $a_\perp$, E_{HOMO} is the energy at the band center at $k_\perp = 0$, and $BW = 4J$. ARUPS has been applied to determine the band structure of BTQBT on graphite with results shown in Fig. 4.7 (Hasegawa et al., 1994). The presence of sulfur atoms in BTQBT results in a close in-plane intermolecular contact distance of only 3.26 Å, with a more relaxed interplanar spacing of 3.45 Å which, due to the molecular tilt relative to the substrate, gives $a_\perp \approx 3.4$ Å. Consistent with the tight binding approximation, the HOMO-centered band has a cosine-like dependence with a BW of almost 0.4 eV (corresponding to $J = 0.092$ eV). This is considerably larger than the next highest HOMO, or NHOMO, centered on HOMO-1 whose BW is ~0.08 eV. The narrower BW is a consequence of the more tightly bound inner shell electrons that make up the deeper HOMO levels. Fits to the ARUPS data give a $V_0 = -12.5$ eV and $E_{HOMO} = 5.4$ eV. Note that the phase of the NHOMO is opposite to that of the HOMO, suggesting that the velocity and effective masses also have opposite signs in the two bands.

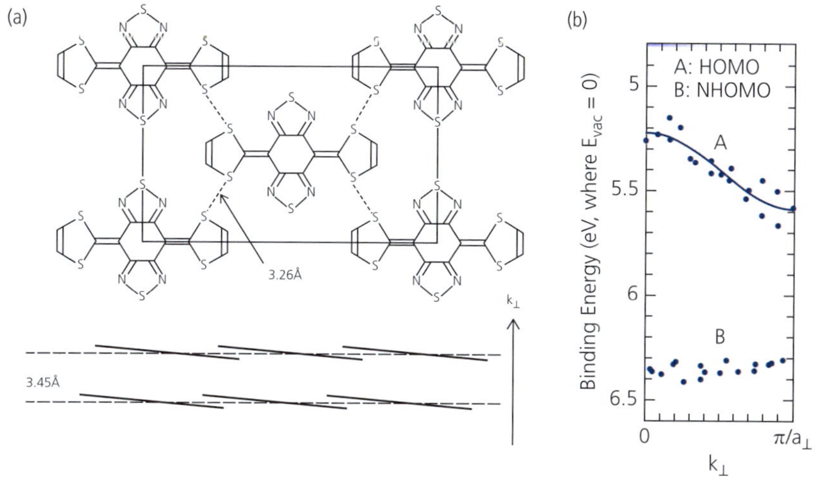

Figure 4.7 (a) Crystal structure and molecular structural formula of BTQBT. (b) Structure of the HOMO and the next highest HOMO (NHOMO) bands measured by ARUPS (data points), and a fit to the tight binding approximation (lines) (Hasegawa et al., 1994).

The BW along $a_\perp$ is substantial for an organic solid, and suggests a high charge mobility and low effective mass. Indeed, from the fits to the tight binding model (Fig. 4.7b, solid line), the hole effective mass at the top of the HOMO band is $m_h^* \simeq 3.1m_0$. Then the charge mobility in the crystal is

$$\mu_h = \frac{q\tau}{m_h^*}, \tag{4.32}$$

where the scattering relaxation time is given by the uncertainty principle to be $\tau > \hbar/k_BT$ for thermally broadened bands. From this, we find that $\mu_h > 6.5$ cm^2/V s (Hasegawa et al., 1994), which is comparable to the measured Hall mobility of 4 cm^2/V s. (Imaeda et al., 1992). There is little question, therefore, that the close intermolecular packing of BTQBT leads to band transport similar to that of inorganic semiconductors.

An even closer π-stacking distance of 3.21 Å for PTCDA suggests an even larger carrier BW. As we have seen in Chapter 3, this close intermolecular spacing results in extended charge transfer states that span several molecules in the stack. In one ARUPS study (Yamane et al., 2003), PTCDA was grown nearly epitaxially onto MoS_2 to produce large crystalline domains. The spacing of $a_\perp \approx 3.8$ Å between PTCDA and MoS_2 is considerably larger than the native intermolecular stacking distance of neat PTCDA crystals. Nevertheless, the fit of the data to a cosine dependence in Fig. 4.8 yields $m_h^* = 5.28m_0$ and $\mu_h > 3.8$ cm^2/V s. These values are expected to differ for bulk crystals with tighter molecular stacking. The data lead us to conclude that, like BTQBT, PTCDA has many band-like properties consistent with inorganic semiconductors.

Similar findings regarding the HOMO energy dispersion have been calculated and measured for high mobility semiconductors such as pentacene (Koch et al., 2006) and rubrene (Yanagisawa et al., 2013). In cases where high mobility is observed, HOMO BWs in the range 150–250 meV are typical, although to date there has not been evidence for a BW higher than 400 meV, as observed for BTQBT. The tight binding approximation clearly provides a reasonable estimate of the energy dispersion of high mobility (and thus high conductivity) organic materials, suggesting that Bloch theory applied to inorganic semiconductors can also be adapted to organic materials with a significant orbital overlap between neighbors.

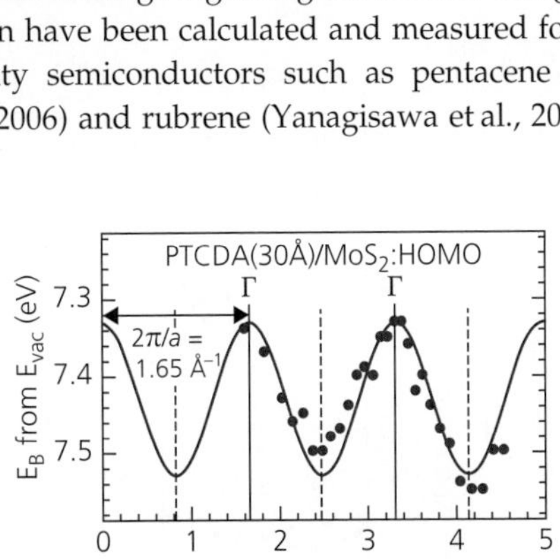

Figure 4.8 Measured HOMO level dispersion of PTCDA deposited on MoS_2 (data points) giving a bandwidth of 200 meV. The solid line is a fit to the cosine-dependent tight binding calculation with an electronic potential of $V_0 = -5.1$ eV and $E_{HOMO} = 7.43$ eV (Yamane et al., 2003).

4.2 Charge transfer

Now that we have established that the band structure of an organic medium is primarily determined by nearest neighbor interactions, we can develop the theory of charge transfer between molecules. The approaches taken to understand the optical properties of organics (and hence exciton transfer and energy migration) are similar to those needed to understand charge migration and conductivity. Before addressing the process of charge transfer, we define two important macroscopic quantities.

Charge mobility ($\overleftrightarrow{\mu}$) is the constant of proportionality between the electric field (**F**) charge velocity (**v**). That is,

$$\mathbf{v} = \overleftrightarrow{\mu}\mathbf{F}. \tag{4.33}$$

It is a tensor whose value depends on crystalline direction. The mobility for electrons (μ_e) and holes (μ_h) is different in virtually all semiconductor materials since they each depend separately on their corresponding intermolecular charge transfer (CT) properties that are sensitive to the crystal and molecular structures. As we will show in the ensuing discussion, the mobility is a function of temperature, T, electric field, and the charge density of states.

Conductivity, $\overleftrightarrow{\sigma}$, is defined via Ohm's law:

$$\mathbf{j} = q(n\mathbf{v}_e + p\mathbf{v}_h) = \overleftrightarrow{\sigma}\mathbf{F}, \tag{4.34}$$

from which we obtain

$$\overleftrightarrow{\sigma} = q(n\overleftrightarrow{\mu}_e + p\overleftrightarrow{\mu}_p). \tag{4.35}$$

Here, **j** is the current density and n and p are the electron and hole densities, respectively.

Given the central importance that μ plays in quantifying conductive properties, calculating and measuring this quantity has been a significant focus of studies of organic electronic materials. The remainder of this section is devoted to understanding its origins, and in Section 4.4.3 we discuss how it is accurately measured.

While charge mobilities >1 cm^2/V s are suggestive of band-like transport, it is far more typical to find organics whose mobilities are < 0.1 cm^2/V s. This is in contrast to mobilities in inorganic solids that are at least 10 times, and even >1000 times larger. The primary difference lies in the strength of the electron–phonon interaction. In inorganics, electron exchange interactions within the bands are dominant. Their large dielectric constants ($\varepsilon_r > 10$) lead to electric field screening. Indeed, the large BWs and weak couplings to phonons lead to a charge *mean free path*, $\lambda \gg a$, where a is the lattice constant. Hence, the *mean free time* between scattering events, τ, is much longer than the time to move between adjacent atoms in the lattice. The uncertainty principle sets a lower limit to the electronic BW of $BW > \hbar/\tau$.

The mobility is related to the diffusion constant, D, via the *Einstein–Smoluchowski relationship*:

$$\frac{D}{\mu} = \frac{k_B T}{q}. \tag{4.36}$$

Since the charge diffusion length, L_D, is given by

$$L_D = \sqrt{D\tau}, \tag{4.37}$$

then band-like properties require that $L_D > a$. In this case, provided that $BW > k_B T$ we have the condition that (LeBlanc, 1961),

$$\mu > \frac{qa^2}{\hbar}\left(\frac{BW}{k_B T}\right). \tag{4.38}$$

To estimate the magnitude of this limiting band-like mobility, we assume that $a \sim 5$ Å, in which case $\mu > 5$ (BW/k_BT) (in units of cm^2/V s). Thus at room temperature where $BW \geq 25$ meV, band transport dominates when $\mu > 5$ cm^2/V s, a high value that is rarely observed in organic materials.

In contrast to inorganic semiconductors, organic semiconductors have a relatively small dielectric constant ($\varepsilon_r \sim 3$), and hence a concomitantly reduced electric field screening. Thus, electron–phonon interactions both within a charged molecule as well as between molecules involved in electron transfer overwhelm the exchange interactions. This leads to charge scattering and localization at each molecular site. In this limit, $\lambda \sim a$ such that transport is *incoherent*, where the charge trajectories are highly randomized. Band transport gives way to hopping transport of a charge that self-traps on each molecular site due to polarization of the local orbital environment. The polarization energy is "carried" with the charge from molecule to molecule, considerably increasing its effective mass. The charge-plus-polarization is called a *polaron*. Since the polarization effect is confined to only nearest or next nearest neighbors (i.e. it is on the order of a lattice constant, a), the polaron is often referred to as a *small polaron* (Holstein, 1959a).

In contrast to the short range interactions of electrons in organics, interactions in ionic compounds are governed by Coulombic effects whose potential decays as $1/r$. Polarization in this case decays slowly with distance, with concomitantly reduced effects on the charge effective mass and other transport properties. This long range lattice distortion is known as a *large polaron*. Since its effects are of relatively minor importance in organic materials, large polarons are not considered further here.

Electron–phonon interactions dominate transport in molecular solids when $\hbar\omega_0 > BW$, where $\hbar\omega_0$ is approximately equal to the highest energy vibration. Thus, mobilities in materials with strong electron–phonon coupling with magnitudes

$$\mu > \frac{qa^2}{\hbar}\left(\frac{\hbar\omega_0}{k_B T}\right) \tag{4.39}$$

suggest the existence of band transport. Typically, $\hbar\omega_0 \sim 100$ meV. Then at room temperature, $\mu > 20$ cm^2/V s is a signature of coherent transport. Such high room temperature mobilities in molecular and polymeric materials are not observed. The transition between hopping and band transport, however, can be observed at far lower mobilities than predicted by the generalities leading to Eqs. 4.38 and 4.39.

Quantitative analysis of intermolecular electron transfer proceeds by solving the Schrödinger equation using the single electron Hamiltonian:

$$H_T = H_e^0 + H_{ph}^0 + H_e^{tr} + H_{e-ph}^{loc} + H_{e-ph}^{non} + H_e^{stat}, \tag{4.40}$$

which is decomposed into the several energetic contributions to the transfer process. The excited electron Hamiltonian for a perfectly ordered lattice is given by (Coropceanu et al., 2007, Bässler and Köhler, 2012)

$$H_e^0 = \sum_n^N E_n^0 a_n^+ a_n, \tag{4.41}$$

where E_n^0 is the electron energy and $a_n^+(a_n)$ is the creation (annihilation) operator for an electron at the nth lattice site in the absence of coupling to surrounding molecules. For an n-particle eigenstate, $|n\rangle$, these operators have the following properties:

$$a^+|n-1\rangle = \sqrt{n}|n\rangle \tag{4.42}$$

and

$$a|n\rangle = \sqrt{n}|n-1\rangle \tag{4.43}$$

such that the number operator

$$\mathcal{N}_i = a_i^\dagger a_i \tag{4.44}$$

counts the number of quantum states. Thus, $\mathcal{N}_i|n\rangle = n|n\rangle$. Similarly, for phonons we have

$$H_{ph}^0 = \sum_{\mathbf{q},j} \hbar\omega_{\mathbf{q},j}(b_{\mathbf{q},j}^{\mathrm{T}} b_{\mathbf{q},j} + \tfrac{1}{2}), \tag{4.45}$$

where $\omega_{\mathbf{q},j}$ is the frequency of the phonon mode with wavevector $\mathbf{q}$ on the jth phonon branch. Also $b_{\mathbf{q},j}^\dagger (b_{\mathbf{q},j})$ is the phonon creation (annihilation) operator. Hence, $H_e = H_e^0 + H_{ph}^0$ gives the energy of an excited molecule that does not interact with the phonon population. In writing Eqs. 4.41 and 4.45 and the expressions for the various terms in H_T that follow, for simplicity only first order (linear) interactions are considered.

Electron transfer from the mth to the nth molecule is expressed by

$$H_e^{tr} = \sum_{m \neq n}^{N} J_{nm} a_n^\dagger a_m, \tag{4.46}$$

where the transfer integral, J_{nm}, gives the electronic interaction, or coupling between molecules m and n:

$$J_{nm} = \langle \psi_n(\mathbf{r} - \mathbf{R}_n) | H_e^{tr} | \psi_m(\mathbf{r} - \mathbf{R}_m) \rangle. \tag{4.47}$$

This is in contrast to the *on-site energy*, with expectation value of ε_{nn} at site n:

$$\varepsilon_{nn} = \langle \psi_n(\mathbf{r} - \mathbf{R}_n) | H_e^{tr} | \psi_n(\mathbf{r} - \mathbf{R}_n) \rangle. \tag{4.48}$$

Interactions with the phonon population are expressed by the electron–phonon coupling terms as follows:

$$H_{e-ph}^{loc} = \frac{1}{\sqrt{N}} \sum_{\mathbf{q},j} \sum_{m}^{N} \hbar\omega_{\mathbf{q},j} (g_m(\mathbf{q},j) b_{\mathbf{q},j} + g_m^*(\mathbf{q},j) b_{-\mathbf{q},j}^\dagger) a_m^\dagger a_m \tag{4.49}$$

and

$$H_{e-ph}^{non} = \frac{1}{\sqrt{N}} \sum_{\mathbf{q},j} \sum_{\substack{m,n \\ m \neq n}}^{N} \hbar\omega_{\mathbf{q},j} \big(g_{nm}(\mathbf{q},j) b_{\mathbf{q},j} + g_{nm}^*(\mathbf{q},j) b_{-\mathbf{q},j}^{+}\big) a_n^\dagger a_m. \tag{4.50}$$

Equation 4.49 is the dynamic disorder term due to electron–phonon coupling within the charged molecule. The strength of the coupling is $g_m(\mathbf{q},j)$, with complex conjugate $g_m^*(\mathbf{q},j)$. Since the interaction only involves intramolecular interactions within the mth molecule, the matrix elements lie only on the diagonal (m, m), and hence Eq. 4.49 is known as *dynamic diagonal disorder*. Similarly, *dynamic off-diagonal disorder* due to intermolecular interactions with the vibronic manifold, is expressed by Eq. 4.50, with an electron–phonon coupling strength of $g_{nm}(\mathbf{q},j)$. These terms are used to calculate the reduction in electron energy arising from intrinsic or polaronic effects (and hence interaction with the phonon population) during transfer. Importantly, when g_m or g_{nm} are large compared to J_{nm}, the transport is dominated by polarons where the mass of the charge is "dressed" by local lattice distortions. In the opposite circumstance when $J_{nm} > g_m, g_{nm}$, the effects of polarons are small, and the transport becomes band-like. This corresponds to the situation in inorganic semiconductors where polaron energies are negligible.

Finally, actual crystal structures always contain a finite number of physical defects, and this is particularly true for weakly bonded organic van der Waals crystals. These imperfections in lattice structure can significantly influence charge transfer between neighboring molecules, and hence the conductivity of the solid. To account for this so-called static disorder, we write

$$H_e^{stat} = \sum_{n}^{N} \delta E_n a_n^+ a_n + \sum_{\substack{m,n \\ m \neq n}}^{N} \delta J_{nm} a_n^+ a_m. \tag{4.51}$$

The terms δE_n and δJ_{nm} are small deviations in the energy and coupling constants due to local variations in the structural morphology. As above, the first term expresses static diagonal disorder and the second is for static off-diagonal disorder. Note that there are no interactions with the phonon population since static disorder is only a property of crystalline morphology. This term is largest in amorphous materials or noncrystalline polymers where there is considerable variation in the molecular environment and configuration from site to site. In sufficiently disordered solids, this term can dominate other effects in Eq. 4.40.

A conceptual illustration of the physical processes and the terms quantifying transfer between molecules in a lattice is shown in Fig. 4.9.

Now that the total Hamiltonian is defined, our task is to calculate the various terms to determine their relative strengths. That is, when $g_m(\mathbf{q},j)$ and $g_{nm}(\mathbf{q},j)$ are large, charge transfer (and hence charge mobility) is dominated by polaronic effects, and the primary mechanism for charge transport is via incoherent hopping between nearest neighbor molecules. Conversely, band-like transport occurs when these coupling constants are small.

In the absence of intermolecular coupling (i.e. $J_{nm} = 0$) and considering only local polaron (i.e. on-diagonal terms), the Hamiltonian simplifies to $H_T = H_e^0 + H_{ph}^0 + H_{e-ph}^{loc}$ which yields the following energy eigenvalues (Coropceanu et al., 2007):

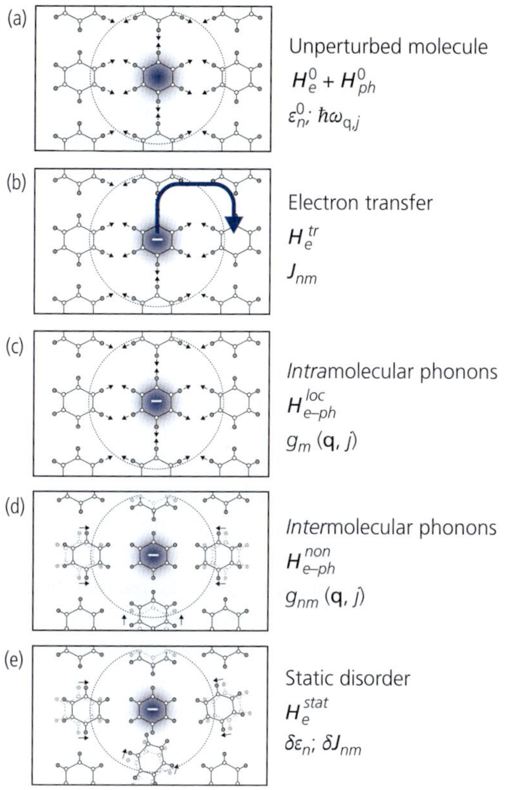

Figure 4.9 Various processes involved in charge transfer. The relevant Hamiltonians and coupling constants are indicated. The unpaired electron (−) undergoes transfer in the direction indicated by the arrow.

$$E_m = E_m^0 + \sum_{\mathbf{q},j} \hbar\omega_{\mathbf{q},j}\left(n_{\mathbf{q}j} + \frac{1}{2}\right) - \frac{1}{N}\sum_{\mathbf{q},j} \hbar\omega_{\mathbf{q},j}|g_m(\mathbf{q},j)|^2. \tag{4.52}$$

The first two terms are the familiar solution to the Schrödinger equation using H_e in Eq. 3.11 *ff.*, whereas the last term results from on-site perturbations of the vibronic modes in response to the excess polaron (Fig. 4.9c). That is, the intramolecular polarization energy which equals the polaron binding energy is

$$E_{pol}^{loc} = -\frac{1}{N}\sum_{\mathbf{q},j} \hbar\omega_{\mathbf{q},j}|g_m(\mathbf{q},j)|^2. \tag{4.53}$$

To determine the coupling strength, we assume that the transfer of an electron results in only a small distortion of the electronic potential. Then we can expand $E_m(Q)$ about the equilibrium nuclear coordinates, $Q(j)$ as defined in Fig. 4.10:

$$E_m(Q) = E_m^0 + \sum_i \frac{\partial E_m}{\partial Q_m(j)}\bigg|_{Q=0} Q_m(j) + \frac{1}{2}\sum_j M_j\omega_j^2 Q_m^2(j) + \ldots. \tag{4.54}$$

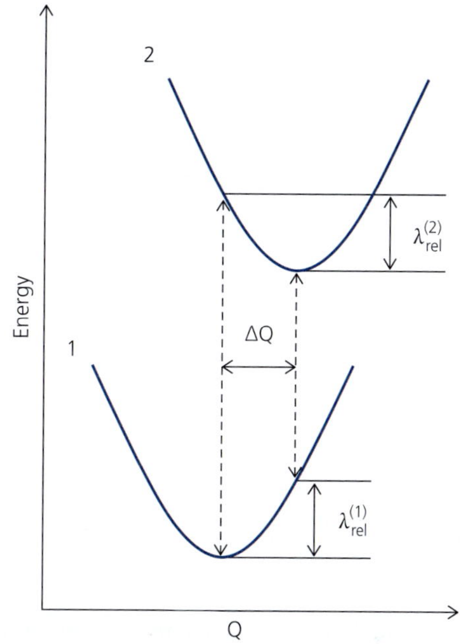

Figure 4.10 Relative coordinate shift (ΔQ) between the initial, neutral state (1), and the final charged molecule state (2) involved in charge transfer. The polarization energies are $\lambda_{rel}^{(1)}$ and $\lambda_{rel}^{(2)}$, respectively.

The second term, which is linear in $Q_m(j)$, is the restoring force near equilibrium. Here, M_j is the molecular mass (Holstein, 1959a). This distortion results in a shift in the nuclear coordinates, $\Delta Q(j)$ that increases the energy by $M_j\omega_j^2\Delta Q(j)$. Assuming that the vibrational modes in the ground and the singly charged excited states are unchanged, and that λ_{rel} is the sum of the energies of the j vibrational modes, $\lambda_{rel} = \sum_J \lambda_j$ then the total polarization energy is

$$E_{pol}^{loc} = \lambda_{rel} = \frac{1}{2}\sum_j M_j\omega_j^2 \Delta Q^2(j) \tag{4.55}$$

(see Fig. 4.10). At equilibrium, the first derivative in Eq. 4.54 is zero. It follows that the shift in coordinates by the transfer of charge is (Austin and Mott, 1969)

$$\Delta Q_m(j) = \left[\frac{1}{2M_j\omega_j^2}\right]\frac{\partial E_m}{\partial Q_m(j)}\bigg|_{Q=0} \tag{4.56}$$

Then from Eqs. 4.53, 4.54, and 4.56, we obtain a relationship for on-site coupling:

$$|g_m(\mathbf{q},j)|^2 = \frac{\lambda_j}{\hbar\omega_{\mathbf{q},j}} = \frac{1}{2M_j\hbar\omega_{\mathbf{q},j}^3}\left|\frac{\partial E_m}{\partial Q_m(j)}\bigg|_{Q=0}\right|^2, \tag{4.57}$$

or

$$|g_m(\mathbf{q},j)|^2 = \frac{2M_j\omega_{\mathbf{q},j}}{\hbar}|\Delta Q_m(j)|^2. \tag{4.58}$$

The relationship between λ_{rel} and ΔQ is shown in the energy diagram of the neutral and ionic molecular states in Fig. 4.10. The coupling constants are proportional to the coordinate displacements, ΔQ, that are due to changes in intramolecular vibrational modes in the presence of an excess charge. This change results in a molecular relaxation of the local polarization of magnitude $\lambda_{rel}^{(1)}$ as the electron is transferred to a neighboring neutral molecule, which in turn relaxes by energy, $\lambda_{rel}^{(2)}$. That is, the charge "carries" the polarization from molecule to molecule. The total reorganization energy upon charge transfer is, therefore,

$$\lambda_{reorg} = \lambda_{rel}^{(1)} + \lambda_{rel}^{(2)}, \tag{4.59}$$

and in the special case that the initial and final states of the molecules are the same, then

$$E_{pol}^{loc} = \lambda_{reorg}/2. \tag{4.60}$$

For rigid molecules such as the polyacenes, the reorganization energies have in fact been found to be approximately equal (E_{pol} ~ 50–100 meV) (Coropceanu et al., 2007). This is not the case for more flexible molecules that can undergo substantial distortion during charge transfer, and where the perturbation approach breaks down.

The reorganization energy was initially introduced by Marcus (1957), who used a classical approach to charge transfer between a donor and acceptor molecule, yielding

$$\lambda_{reorg} = \frac{\Delta q^2}{4\pi\varepsilon_0}\left(\frac{1}{2a_D} + \frac{1}{2a_A} - \frac{1}{R_{DA}}\right)\left(\frac{1}{\varepsilon_\infty} - \frac{1}{\varepsilon_s}\right), \tag{4.61}$$

where a_D (a_A) is the radius of the donor (acceptor) outer sphere electron orbital, R_{DA} is the distance between acceptor and donor molecules, ε_0 is the permittivity of free space, and ε_∞ and ε_s are the optical and static dielectric constants, respectively. The first two terms in Eq. 4.61 are the energies required to charge the donor and acceptor molecules by an amount, Δq, and the third term is the reduction in energy due to the transfer of Δq from D to A. The difference in dielectric constants is a result of the Born–Oppenheimer approximation in that the nuclear degrees of freedom cannot instantaneously adjust during the transfer process. The reorganization energy increases with increasing donor–acceptor separation, as expected.

Calculation of off-diagonal dynamic distortion coupling constants, $g_{nm}(\mathbf{q},j)$ follows an analogous procedure to that used for $g_m(\mathbf{q},j)$. This interaction is governed by the generation of acoustic phonons during transfer. The strength of the intermolecular coupling due to off-diagonal dynamic disorder is often lower (E_{pol}^{non} ~ 10 meV), largely as a result of the lower energy of inter vs. intramolecular phonons in organic crystals (Friedman, 1965). The presence of a polaron can result in substantial displacements of the lattice if it is sufficiently soft, leading to contributions as large as 30% to the total polaron energy (Coropceanu et al., 2007).

While the foregoing discussion has focused on electron transfer, the same arguments are made to describe hole transfer, albeit with different values for the coupling constants. The strength of the coupling determines the transfer rate via Fermi's Golden Rule. Ultimately it is the rate that determines the carrier velocity and hence its mobility. While both electrons and holes apparently follow the same process of charge transfer, it is observed that hole mobilities are often larger than those of electrons. It has been argued that this difference has extrinsic origins; that electrons transported along the LUMO levels are at a higher energy and hence more readily available for oxidation or trapping within the energy gap (Bässler and Köhler, 2012). Intrinsic properties of solids may also lead to a higher mobility for one of the two carrier types: the most important being the orbital overlap between the molecules. In Eq. 4.47 we see that J_{nm} increases proportionately with the orbital overlap integral. Thus, the overlap of the HOMOs and LUMOs between molecules ultimately determines the probability for charge transfer. The charge mobility is, therefore, intimately related to both the spatial distribution of the respective orbitals, and the relative positions (i.e. stacking habit, distance of closest approach defined by the van der Waals radius, steric hindrance, etc.) of adjacent molecules. From tight binding, we have argued that the HOMO overlap leads to the broadest BW. Hence, it is not surprising that holes transported near the top of this band experience the highest mobility, as is often the case.

4.3 Charge transport

Charge transport over macroscopic distances occurs through a sequence of intermolecular charge transfer steps. Mechanisms of charge transport, therefore, determine the macroscopic electronic properties of charge mobility, diffusion and conductivity. We now relate the quantum mechanical couplings in Section 4.2 to these macroscopic quantities.

In Fig. 4.11a we illustrate the process of band transport that develops when $J_{nm} > g_m^2$ and g_{nm}^2, that is, when $H_e^{tr} > H_{e-ph}^{loc}$, and H_{e-ph}^{non} (see Eq. 4.40). This situation arises when the BW is greater than the polaron

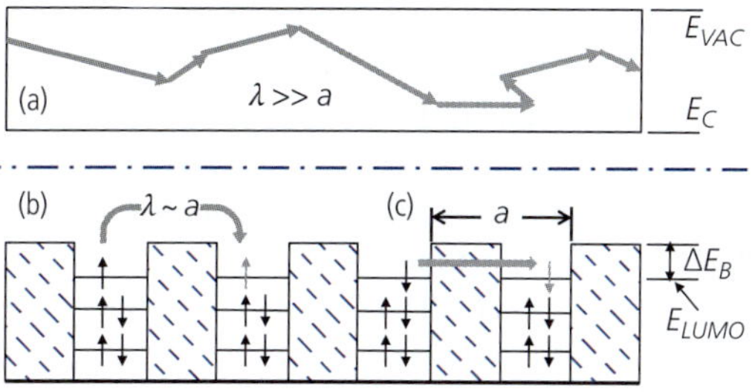

Figure 4.11 Modes of charge conduction in semiconductors. (a) Band transport where the mean free path (λ) of an electron is large and the transport is coherent (i.e. the path of the electron is correlated between collisions). E_C is the energy of the conduction band minimum, and E_{VAC} is the vacuum energy. (b) Hopping transport from molecule to molecule. The orbitals of each molecule are shown. The LUMO energy (E_{LUMO}) is indicated by the topmost level containing an unpaired spin. (c) Tunneling occurs between states of equal energy in adjacent molecules. ΔE_B is energy barrier to hopping or tunneling, and a is the molecular spacing. The energy scales and energy references for (a) are independent of (b) and (c).

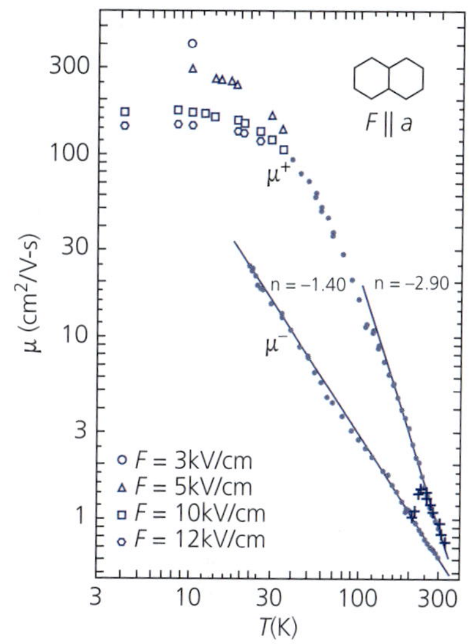

Figure 4.12 Dependence of the electron (μ^-) and hole (μ^+) mobilities on temperature along the **a**-axis of an ultrapurified naphthalene crystal. Also shown is the electric field dependence of the hole mobility at $T < 40$ K. The (+) data between $T = 200$ K and 300 K are for naphthalene, that has not undergone ultrapurification (Warta et al., 1985).

energy in Eq. 4.55, such that the distance between scattering events is much larger than a. The charge motion is then limited by acoustic phonon scattering at the lattice sites. To determine the temperature dependence of the band-like mobility, the root mean square thermal velocity of charges that follow Boltzmann statistics is given by

$$\langle v \rangle = \sqrt{\frac{3k_BT}{m^*}}. \tag{4.62}$$

Now the scattering time for an electron or hole is $\tau^{-1} \sim \rho\sigma_X\langle v \rangle$, where σ_X is the scattering cross section and ρ is the molecular number density in the medium. The temperature dependence of σ_X due to phonon scattering is $\sigma_X \sim T$, resulting in $\tau \sim T^{-3/2}$. It then follows from Eq. 4.32 that temperature dependence of the mobility for band-like conduction is $\mu \sim T^{-3/2}$. More generally, the temperature dependence of band-like mobility is

$$\mu \sim T^{-n}, \tag{4.63}$$

where for most common semiconductors such as Si, Ge, and GaAs, n is in the range from $\frac{3}{2}$ to $\frac{5}{2}$ for both electrons and holes.

Intrinsic band-like behavior has been observed in ultrapurified naphthalene and perylene single crystals (Warta et al., 1985, see Chapter 5.3), where great care was taken to remove impurities and their associated scattering processes that can introduce significant deviations from Eq. 4.63. For example, charged defects can lead to $\sigma_X \sim \langle v \rangle^{-4} \sim T^{-2}$, from which we infer $\mu \sim T^{3/2}$. In Fig. 4.12 we show the dependence of the time of flight (TOF) mobility on T for both electrons and holes for such highly pure naphthalene. The mobility shows a T^{-n} dependence down to $T = 20$ K in Fig. 4.12. This is a clear indication of band-like conduction extending to room temperature. As in the case of perylene, however, n is somewhat larger than commonly observed in inorganic materials, ranging from 2.19 (electrons in perylene) to 2.90 (holes in naphthalene) suggestive of strong interactions with high energy acoustic phonons (Warta et al., 1985). Notably, the electron mobility is 5–10 times less than the hole mobility. Since the single crystals studied in Fig. 4.12 have been thoroughly refined, it is unlikely that this rather significant difference between hole and electron mobilities can simply be ascribed to the presence of impurities. As noted in the previous section, differences in HOMO vs. LUMO overlaps between molecules in a crystal can lead to intrinsically different mobilities due to different magnitudes of the transfer integral, J_{nm}, in Eq. 4.47.

Even though highly pure, the room temperature mobility does not exceed ~1 cm^2/V s, and indeed is comparable to values for less purified materials. The low mobility arises from strong electron–phonon interactions (i.e. small polarons) due to dynamic lattice disorder. This is an intrinsic property of even the highest quality organic materials. Hence, reports of exceptionally high mobilities, often obtained from interface conduction in transistors at room temperature, should be taken with healthy skepticism (cf. Section 4.4.3). Nevertheless, phonon scattering is reduced with temperature, leading to the power-law dependence in Fig. 4.12, with naphthalene achieving hole mobilities of >200 cm^2/V s. This is similar to data for ultrapure perylene in Fig. 5.6 with an electron mobility of 80 cm^2/V s at 30 K.

Another phenomenon observed for ultrapurified naphthalene is the dependence of the hole drift velocity on applied electric field, F, shown in Fig. 4.13. The differential mobility,

$$\mu_{diff} = \frac{\partial v}{\partial F}, \tag{4.64}$$

is given by the slope of the data. At all temperatures, $\mu_{diff} \to 0$ as $F \to$ large, characteristic of velocity saturation due to increased acoustic phonon scattering in the bands. The data in Fig. 4.12 are taken from the

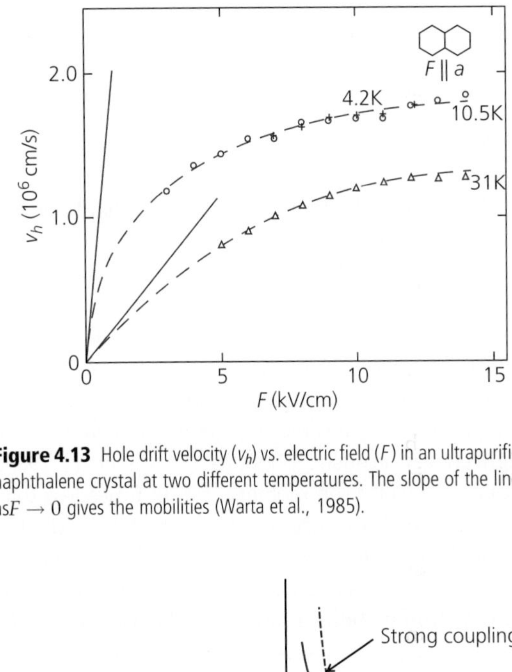

Figure 4.13 Hole drift velocity (v_h) vs. electric field (F) in an ultrapurified naphthalene crystal at two different temperatures. The slope of the lines as$F \to 0$ gives the mobilities (Warta et al., 1985).

linear interpolation of the drift velocities to $F = 0$. Extremely high mobilities (400 cm^2/V s at $F = 3$ kV/cm) are attained at $T = 4.2$ K where scattering is minimal.

Polaron-limited transport is important when $BW < \hbar\omega_0$, k_BT. In this case dynamic disorder leads to $H_{e-ph}^{loc}, H_{e-ph}^{non} > H_e^{tr}$. The polaron is scattered at each molecular site, leading to transport dominated by thermally-assisted hopping at high temperature. As temperature is decreased, tunneling between sites dominates. These two processes are illustrated in Fig. 4.11b and 4.11c, respectively. Tunneling is isoenergetic, that is, it involves the transfer between equal energies. Therefore, it is only weakly temperature dependent arising from thermal phonon broadening of the electron DOS. Like band transport, tunneling leads to *coherent transport* where the mean free path of the carrier is $\lambda \gg a$. In thermally assisted hopping, however, the polaron requires energy to hop over the barriers between adjacent molecules. Since scattering and relaxation occur on time scales less than the hopping rate, the transfer is *incoherent*. That is, there is no "memory" of past events that determines the direction and time of subsequent hops.

The total mobility is given by

$$\mu_T = \mu_{hop} + \mu_{tun}. \tag{4.65}$$

The various mobility regimes vs. temperature have the functional forms illustrated in Fig. 4.14. The dominance of either band or polaron transport is a determined by the strength of the electron–phonon coupling constants, for on-diagonal dynamic disorder, and for

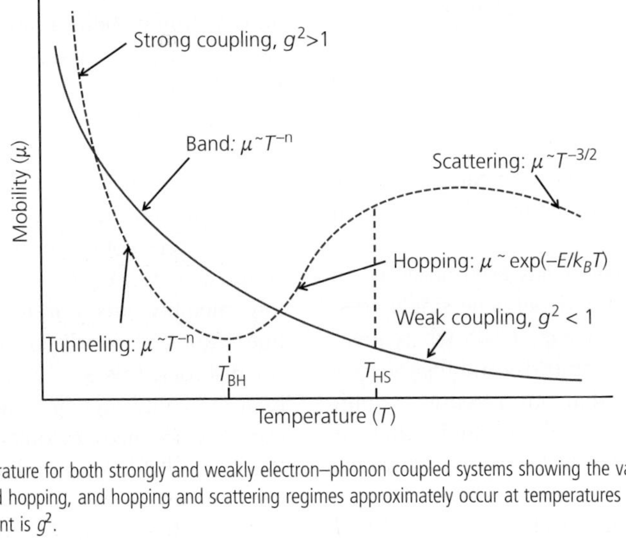

Figure 4.14 Mobility vs. temperature for both strongly and weakly electron–phonon coupled systems showing the various transport regimes. The transitions between the band and hopping, and hopping and scattering regimes approximately occur at temperatures T_{BH} and T_{HS}, respectively. The electron–phonon coupling constant is g^2.

off-diagonal disorder. These couplings exceed the electron transfer integral, J_{nm}, when hopping dominates. Since hopping is thermally assisted, the mobility follows: $\mu_{hop} \sim \exp(-\Delta E/k_B T)$ with an activation energy, ΔE. As temperature is increased, hopping over barriers no longer is the rate-limiting step in the transport process, and residual acoustic phonon scattering dominates, leading to $\mu \sim T^{-3/2}$.

4.3.1 Hopping in the presence of dynamic disorder

Having calculated the transfer energies and integrals in Section 4.2, we now determine the polaron mobilities in the presence of both dynamic and static disorder. First, we will consider the effects of on-diagonal and off-diagonal dynamic disorder where H^{loc}_{e-ph}, $H^{loc}_{e-ph} \gg H^{tr}_e, H^{stat}_e$. This is the small polaron, or polaron transport regime first described by Holstein (1959b). Second, we will consider disorder effects, where $H^{stat}_e \gg H^{loc}_{e-ph}, H^{loc}_{e-ph}, H^{tr}_e$.

Since only nearest neighbor interactions need to be considered, we can describe electron transfer from initial site, p, to the neighboring site, $p \pm 1$. Then for a three dimensional, cubic lattice with lattice constant a, the diffusion constant is given by (Holstein, 1959b)

$$D = \frac{1}{6} k_{ET}(p \to p\pm 1)a^2. \tag{4.66}$$

The factor of 1/6 arises from the choice of hop in any of 6 directions. We assume that molecules at all sites p are identical, and hence positive and negative hops occur with equal probability. For diagonal transitions, the vibrational quantum numbers remain unaltered during transfer, whereas for off-diagonal transitions, the quantum numbers change by ± 1. Then, from Eq 4.36 and 4.66, we can relate the transfer rate k_{ET} to the hopping mobility:

$$\mu = \frac{q}{6k_B T} k_{ET}(p \to p\pm 1)a^2. \tag{4.67}$$

The calculation of k_{ET} follows from the Schrödinger equation employing the Hamiltonian in Eq. 4.40, but excluding disorder, H^{stat}_e. Thus (Holstein, 1959b),

$$i\hbar \frac{\partial \phi_n(Q)}{\partial t} = \left[\sum_m^N \left(-\frac{\hbar^2}{2M} \nabla_m^2 + \frac{1}{2} M\omega_0^2 Q_m^2 + \frac{1}{2} M\omega_0^2 Q_m Q_{m+1} \right) + V_Q(m) Q_n \right] \phi_n(Q) - J_{n,n\pm 1}(\phi_{n-1}(Q) + \phi_{n+1}(Q)), \tag{4.68}$$

where ϕ_n is the wavefunction at site p whose vibrational coordinates are Q_m, and $V_Q(m)$ is the electron–lattice interaction with the mth phonon mode. Also J are the orbital overlap integrals as previously. Here, V_Q is taken to be a small perturbation (Holstein, 1959a), and results in a linear displacement in relative coordinates, Q_m. Solutions to Eq. 4.68 have been derived by Holstein (1959b) for the transfer rate based on time-dependent perturbation theory to yield

$$k_{ET} = \frac{J^2}{\hbar^2 \omega_0} \left(\frac{\pi}{g^2 \mathrm{csch}\left(\frac{\hbar\omega_0}{2k_B T}\right)} \right)^{1/2} \exp\left(-2g^2 \tanh\left(\frac{\hbar\omega_0}{4k_B T} \right) \right), \tag{4.69}$$

taking $g = g_m = g_{nm}$. Here ω_0 is the optical phonon frequency. Thus, using $E_{pol} = \hbar\omega_0 g_m^2$ from Eq. 4.53 in the high temperature limit of $k_B T \gg \hbar\omega_0$ this simplifies to

$$k_{ET} = \frac{J^2}{\hbar} \left[\frac{\pi}{2E_{pol} k_B T} \right]^{1/2} \exp\left(-\frac{E_{pol}}{2k_B T} \right), \tag{4.70}$$

which, from Eq. 4.67, leads to the following thermally activated mobility:

$$\mu_{hop} = \frac{qJ^2 a^2}{6k_B T\hbar} \left[\frac{\pi}{2E_{pol} k_B T} \right]^{1/2} \exp\left(-\frac{E_{pol}}{2k_B T} \right). \tag{4.71}$$

At large T where the thermal energy is higher than the polaron energy, the value of the exponential term is 1, leading to $\mu \sim T^{-3/2}$, characteristic of band transport. Below that temperature, the transport is primarily by diffusion due to hopping to nearest neighbor sites over a potential barrier equal to the polarization energy.

Comparing Eq. 4.68 with the energy expansion in Eq. 4.54, we see that $V_Q(m) = \frac{\partial E_m}{\partial Q_m}|_{Q=0}$. Then, in the special case of the symmetric transfer illustrated in Fig. 4.10 along with Eq. 4.55, we obtain for the polaron binding energy:

$$E^{loc}_{pol} = \lambda_{rel} = \frac{1}{2} \sum_m^N M_m \omega_m^2 \Delta Q_m^2(j) = \sum_m^N V_Q^2(m)/2M_m \omega_m^2. \tag{4.72}$$

As noted above, J is the orbital overlap integral related to the electronic bandwidth via $BW = 4J$ in a simple cubic lattice (Holstein, 1959a). Hence, it follows that a small polaron exists when

$$BW = 4J \ll \lambda_{rel} \approx V_Q^2/2M\omega_0^2, \tag{4.73}$$

and large polarons dominate when the inequality is reversed. This implies that small polarons dominate at low temperatures and for narrow BWs less than the reorganization energy during charge transfer.

An important prediction of the small polaron model is that the BW *narrows* with increasing temperature. This is an outcome of the temperature dependence of the overlap integral (Holstein, 1959b, Coropceanu et al., 2007):

$$J(T) = J\exp\left(-g^2\coth\left(\hbar\omega_0/2k_BT\right)\right) \approx J\exp\left(-g^4\frac{2k_BT}{E_{pol}}\right), \tag{4.74}$$

where, to first order, the approximation on the right is valid for $E_{pol} < k_BT$. For $T \to 0$, then $J(T) = J$. From tight binding theory (Section 4.1.1), the effective mass is $m^* = \hbar^2/2Ja^2$ (c.f. Eq. 4.29). But if the bands are sufficiently narrow in weakly coupled molecular systems, all states are equally populated. Furthermore, since the electron can hop forward or backward in all three dimensions, the relaxation time is given by $\tau = 1/3k_{ET}$. Then the tunneling mobility at low temperature is

$$\mu_{tun} = \frac{qa^2}{3k_BT\hbar^2k_{ET}}J^2(T) \tag{4.75}$$

(cf. Eq. 4.71). The temperature dependence of the mobility separates the hopping from the tunneling regimes at T_{BH} in Fig. 4.14. This transition is shown in detail in Fig. 4.15 for a coupling, $g^2 = 10$ and a BW equal to twice the optical phonon frequency, corresponding to $J = \hbar\omega_0$.

Finally, we note that the hopping mobility is only accurately determined when both on-diagonal and off-diagonal (corresponding to local and non-local) couplings are considered. The Holstein model treats only on-diagonal terms ($g_{nm} = 0$) whereas the Peierls model considers systems where $g_m = 0$, that is, only non-local terms are present. Both of these effects have been incorporated to provide explicit expressions combining both local and non-local couplings while also accounting for crystalline anisotropy (Munn and Silbey, 1985, Hannewald et al., 2004). It is found that the polaron transport BW due solely to local coupling is ~100 meV, but both the shape and the width of the transport bands are significantly impacted by off-diagonal disorder. These couplings add ~10 – 30 meV to the polaron binding energy arising from intermolecular phonon interactions, and hence cannot be ignored. Their inclusion in our analysis, however, does not qualitatively change the conclusions although they impact the magnitude of the observed effects.

In Fig. 4.16 we compare the HOMO and LUMO BWs as functions of temperature for naphthalene using only on-site local (Holstein) interactions, with a combination of local and non-local (Holstein–Peierls) couplings that indicates the influence of intermolecular vibronic interactions (Hannewald et al., 2004). There are several features to note in this comparison. The first is that the BWs are considerably overestimated, and the decrease in BW with T is reduced if off-diagonal disorder is ignored. That is, the LUMO BW from the Holstein model ($g_{nm} = 0$) is 600 meV at $T = 0$ K, decreasing by only ~30 meV as T is increased to 300 K. This compares to $BW = 75$ meV at $T = 0$ K and a decrease of approximately that same magnitude between 0 K and 200 K when incorporating both on- and off-diagonal couplings. It is also striking that the BW of the LUMO level for the Holstein approximation

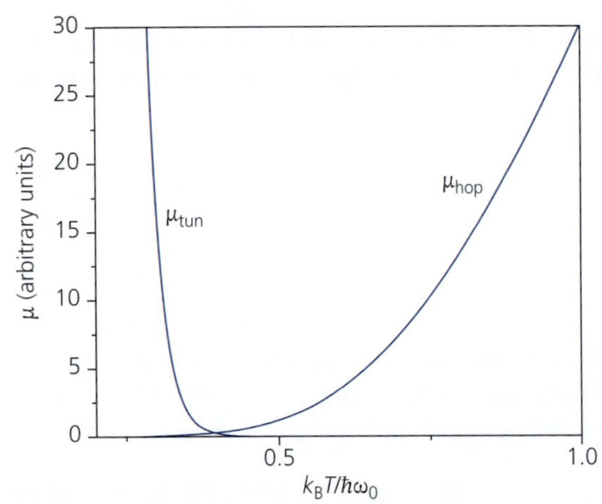

Figure 4.15 Mobility transition between tunneling and hopping as determined from small polaron theory. The bandwidth, $2J$, decreases with increasing temperature, resulting in a decrease in tunneling probability and hence μ_{tun}, along with an increase in thermally activated hopping. For this calculation, $g^2 = 10$ and $J = \hbar\omega_0$ (Coropceanu et al., 2007).

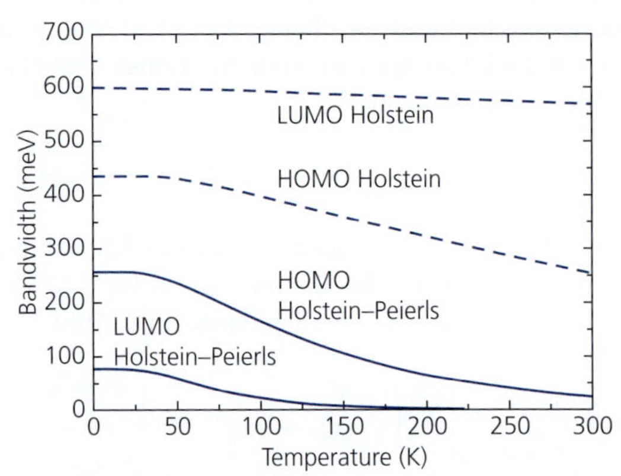

Figure 4.16 Calculations of the LUMO and HOMO bandwidths of naphthalene assuming only local (on-diagonal) couplings corresponding to the Holstein model, along with local plus non-local (off-diagonal) couplings of a combine Holstein–Peierls model (Hannewald et al., 2004).

Reprinted figure with permission from Hannewald, K., Stojanovic, W. M., Schellekens, J. M. T., Bobbert, P. A., Kresse, G. & Hafner, J., Physical Review B, 69, 075211, 2004. Copyright 2004 by the American Physical Society.

is larger than for the HOMO, whereas the opposite obtains in the case of the Holstein–Peierls solution. Indeed, it has been experimentally determined that the hole mobility in naphthalene is larger than its electron mobility (see Fig. 4.12), suggesting that intermolecular polaronic interactions giving rise to the larger HOMO BW (~250 meV at $T = 0$ K) in the Holstein–Peierls model are an essential part of an accurate calculation of the molecular transport properties. When considering both on- and off-diagonal couplings, the LUMO and HOMO BWs are smaller and their dependence on temperature is larger than for cases where only intramolecular couplings are included.

In Fig. 4.17 we show the dependence of BW on temperature for several linear polyacenes. Most noticeable is that the LUMO BW at $T = 0$ K increases monotonically with the number of fused benzene rings, from naphthalene (with 2 rings) where $BW = 20$ meV, to tetracene (4 rings) with $BW = 400$ meV. A similar monotonic dependence is not seen for the HOMO levels, although the ordering is approximately consistent with this "rule." The trend with molecular size follows from the more extended π-system that permits a larger orbital overlap (and hence intermolecular interactions), which results in the increased BW. However, crystal symmetry and anisotropy, molecular shape and lattice structure play important roles in determining the couplings, g and J, and hence these factors can affect the observed trends in unexpected ways, as apparent in Fig. 4.17.

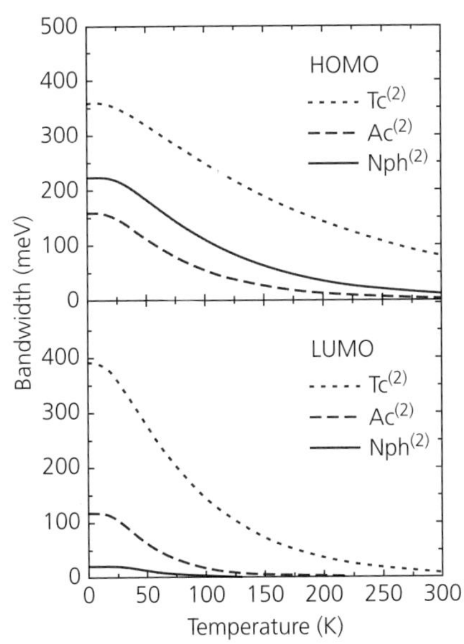

Figure 4.17 Bandwidth of the HOMO and LUMO levels as functions of temperature of three linear polyacenes: Nph = naphthalene, Ac = anthracene, and Tc = tetracene (Hannewald et al., 2004).

An interesting observation is that the high temperature BWs for electrons are considerably smaller than for holes, leading to a concomitant reduction in the mobility of the former. These calculations do not include all effects (particularly disorder) and hence are only approximations to the physical properties of the systems themselves. The asymmetry in BWs may be due to intermolecular interactions of the filled HOMO orbitals that stabilize the orbital overlap (and hence crystal structure) even as temperature is increased. In contrast, the empty LUMO bands are not similarly stabilized, and hence their BW decreases more rapidly as temperature is increased. This leads to the higher mobilities for holes than electrons observed for polyacenes and associated molecules such as PTCDA (Forrest et al., 1984b).

Note that small BW narrowing from 240 meV at $T = 120$ K, to 190 meV at room temperature has been observed in pentacene nanocrystals using ARUPS (Koch et al., 2006). This is within a factor of 2-3 of calculations for the HOMO levels in similar linear polyacenes in Fig. 4.17.

4.3.2 Hopping in the presence of static disorder

Since organic crystals are particularly prone to defects, the effects of static disorder cannot be ignored. Referring to Eq. 4.40, static disorder is important when $H_e^{stat} \gg H_{e-ph}^{loc}, H_{e-ph}^{loc}, H_e^{tr}$. Unlike dynamic disorder, the potential environment of nearest neighbors is not identical, and hence the probabilities for a forward *vs.* a reverse hop are not always equal. There are two models used for site hopping in the presence of disorder; one due to Miller and Abrahams, and the other to Marcus.

4.3.2.1 Miller–Abrahams electron transfer

Miller and Abrahams developed a theory for impurity band conduction in semiconductors that was later successfully adapted to molecular materials. Impurity band conduction occurs by hopping from one dopant atom to the next in a heavily doped (degenerate) semiconductor. Hence, the process is analogous to that which occurs in organics, where hopping between individual molecules leads to scattering at each hopping site. In this formulation, the rate of hopping from site i to site j is given by (Miller and Abrahams, 1960, Ambegaokar et al., 1971, Fishchuk et al., 2002)

$$k_{ET,ij} = \upsilon_0 \exp(-2\gamma_{ij}R_{ij}) \begin{cases} \exp\left(-\dfrac{E_j - E_i}{k_B T}\right) & E_j > E_i \\ 1 & E_j < E_i \end{cases}, \tag{4.76}$$

where E_i and E_j are the energies of the initial and final sites, respectively, separated by distance, R_{ij}, γ_{ij} is the orbital overlap factor, and ν_0 is the "attempt frequency," that is, the rate at which a charge attempts to overcome the energy barrier between nearest neighbor molecules. To reasonable approximation, ν_0 is equal to the highest energy (optical phonon) frequency. The overlap factor is the degree to which the wavefunction on site i ($\phi_j(\mathbf{r}_i)$) overlaps that on site j, that is, $\gamma_{ij} \sim \langle\phi_i(\mathbf{r}_i)|\phi_j(\mathbf{r}_i - \mathbf{R}_j)\rangle$ (cf. Eq. 4.15). The exponential prefactor in Eq. 4.76 thus gives the spatial decay of the wavefunction with distance from the original molecular site, similar to the dependence of exchange transfer for excitons (Section 3.8.2). This points to an important aspect of charge transfer: the probability for transfer decays rapidly with distance since it entails tunneling between adjacent molecules. This results in more rapid transfer, and hence higher charge mobility, for tightly π-stacked molecular crystals where the disorder is minimal. As disorder increases, due for example to the formation of boundaries at the edge of nanocrystalline domains, we expect the mobility to decrease accordingly, similar to observations in a decrease in exciton diffusion length with structural disorder (Lunt et al., 2010 and Chapter 3.9).

Another feature of Miller–Abrahams theory is that the rate from E_i to E_j is thermally activated when $E_j > E_i$, that is, endothermic processes require the absorption of a phonon whose probability is expressed by a Boltzmann factor. On the other hand, exothermic transfer occurs when $E_i > E_j$ with unity probability. In both cases, the rate is modified by the magnitude of orbital overlap between the initial and final electronic states.

4.3.2.2 Marcus electron transfer

An alternative and more physically consistent model for CT reactions in solutions was introduced by Marcus; an accomplishment that earned him the Nobel Prize in 1992 (Marcus, 1956, 1993). This theory is equally applied to charge transfer in solids, and predicts phenomena that are absent from the Miller–Abrahams formulation. In the Marcus picture, electron transfer occurs between the outer sphere (i.e. orbital) of a donor molecule to that of an acceptor illustrated in the free energy (G) vs. reaction coordinate (Q) diagram of Fig. 4.18. It is convenient to express the total system energy in terms of the *Gibbs free energy* which avoids multidimensional considerations of the many hundreds, if not thousands, of electronic, vibronic, and spin coordinates of the reactant and product involved in the transfer.

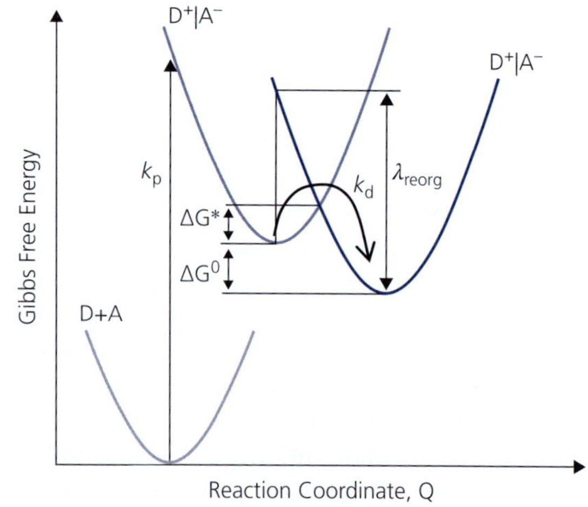

Figure 4.18 Electronic manifolds involved in a charge transfer reaction according to Marcus theory. The diagram shows the initial (D+A), precursor ($D^+|A^-$), and final (D^++A^-) dissociated states.

Indeed, it this insight of Marcus that allows conversion to an N-dimensional space of molecular coordinates for the initial, precursor and dissociated systems to a single energetic representation that makes the problem of charge transfer tractable. A further assumption is that the Franck–Condon principle is operative; that is electron transfer occurs before the molecular nuclei can adjust.

Charge transfer occurs by a process that starts with the *precursor* $D\,|\,A$, moving through an intermediate *successor* $D^+\,|\,A^-$, finally resulting in *dissociation* via the following steps:

$$\mathrm{D} + \mathrm{A} \underset{k_{-p}}{\overset{k_p}{\rightleftharpoons}} \mathrm{D}\Big|\mathrm{A} \underset{k_{-ET}}{\overset{k_{ET}}{\rightleftharpoons}} \mathrm{D}^+\Big|\mathrm{A}^- \overset{k_d}{\rightarrow} \mathrm{D}^+ + \mathrm{A}^-. \qquad (4.77)$$

The rates of each process are indicated, including their reverse, or back-transfer rates. Thus, k_p is the forward precursor formation rate, k_{ET} is the rate of electron transfer to form the successor (or reactant) state, and k_d is the dissociation rate into a charge pair (or product state) that completes the transfer. In Fig. 4.19 we show the case where $k_{ET} >> k_p$, k_d, and hence $D|A \to D^+|A^-$ is not the rate-limiting step. The changes in free energy, ΔG^0 between the $D^+\,|\,A^-$ and the dissociated state, and the barrier between them, ΔG^*, are shown. As previously, molecular reorganization that occurs during transfer results in a reduction of the total energy by λ_{reorg}, see Eqs. 4.55 and 4.59.

There are two possibilities for electron transfer: the first is *non-adiabatic transfer* where the reactant–product (i.e. donor–acceptor) coupling integral, J_{RP}, is small, and the second is *adiabatic electron transfer*

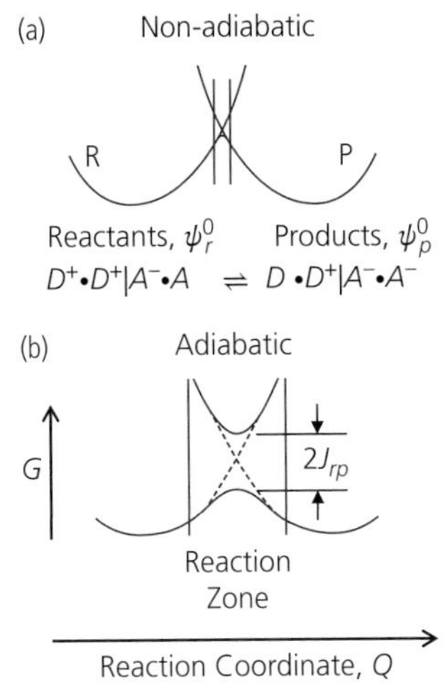

Figure 4.19 (a) Non-adiabatic and (b) adiabatic electron transfer. Reaction zones are delineated by vertical lines. After Newton (1991).

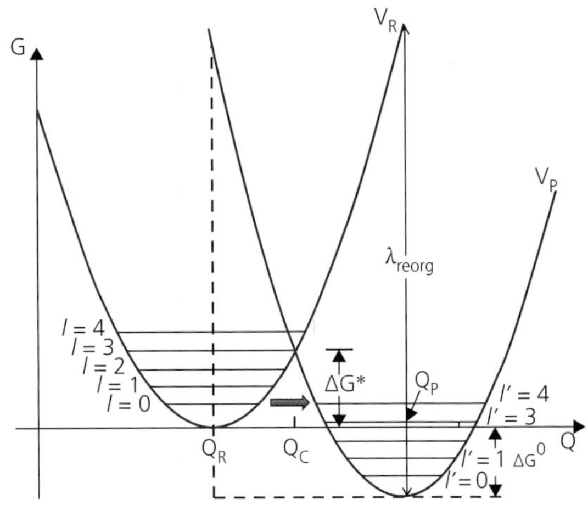

Figure 4.20 Representation of the reactant and product states with potentials $V_R(Q)$ and $V_P(Q)$, respectively. Tunneling (block arrow) from the $l = 0$ to the $l' = 4$ vibronic mode and thermal activation over barrier ΔG^* are two processes leading to charge transfer.

where J_{RP} is large. In analogy to Eq. 4.47, we can express the coupling as

$$J_{RP} = \langle \psi_R^0(\mathbf{r} - \mathbf{R}_R)|H_e|\psi_P^0(\mathbf{r} - \mathbf{R}_P)\rangle, \tag{4.78}$$

where $\psi_{R,P}^0$ are the initial wavefunctions of the reactants and products, respectively, and H_e is the Born–Oppenheimer electronic Hamiltonian for the system. The rate of transfer for weak, or non-adiabatic coupling is limited by the energy barrier height, ΔG^*, that can be surmounted by thermal or optical excitation, or by tunneling following the path $R \rightleftharpoons P$ in Fig. 4.19a. The rate of transfer in this case can be calculated using Fermi's Golden Rule. For strong coupling, the upper and lower branches are split at their intersection by energy $2J_{RP}$, and adiabatic transfer between the reactant and product states occurs as in Fig. 4.19b. Note that the *reaction zone*, corresponding to the region over which the coordinate changes to move from the reactant to the product state, is larger, and the transition more gradual for strongly coupled adiabatic vs. non-adiabatic systems.

To determine the Gibbs free energy barrier, ΔG^*, we refer to Fig. 4.20. Assuming that the potential surfaces for the approximately parabolic reactant and product states are undistorted by the transfer, then at the point of crossing:

$$V_R(Q_C) = V_P(Q_C). \tag{4.79}$$

If the curvature (i.e. the force constant) of these states is κ, then from the crossing point, Q_C:

$$\frac{1}{2}\kappa(Q_C - Q_R)^2 = \Delta G^0 + \frac{1}{2}\kappa(Q_C - Q_P)^2. \tag{4.80}$$

Furthermore, the reorganization energy is given by

$$\lambda_{reorg} = \frac{1}{2}\kappa(Q_P - Q_R)^2 \tag{4.81}$$

and the energy barrier is

$$\Delta G^* = \frac{1}{2}\kappa(Q_C - Q_R)^2. \tag{4.82}$$

Combining the results of Eqs. 4.80–4.82 we arrive at:

$$\Delta G^* = \frac{1}{4\lambda_{reorg}}(\lambda_{reorg} + \Delta G^0)^2. \tag{4.83}$$

In classical Marcus theory, the transfer across this energy barrier is thermally activated, in which case the rate of transfer is given by:

$$k_{ET} = A\exp\left[-\frac{\Delta G^*}{k_BT}\right] = A\exp\left[-\frac{(\lambda_{reorg} + \Delta G^0)^2}{4\lambda_{reorg}k_BT}\right]. \tag{4.84}$$

A striking difference between this expression and Eq. 4.76 for the Miller–Abrahams model is that there is no separate treatment for downward or upward steps; the electron transfer rate is simply due to the change in free energy of the entire system, ΔG^*.

One prediction of Eq. 4.84 is that the transfer rate vanishes in the limit of $T \to 0$. However, this is not found to be the case; the transfer rate is indeed thermally activated at high temperature but achieves a non-zero asymptotic value at low temperature. The lack of complete quenching of the rate of oxidation in *Chromatium* is evident from the data in Fig. 4.21. The persistence of electron transfer at low temperature is due to quantum mechanical tunneling through the

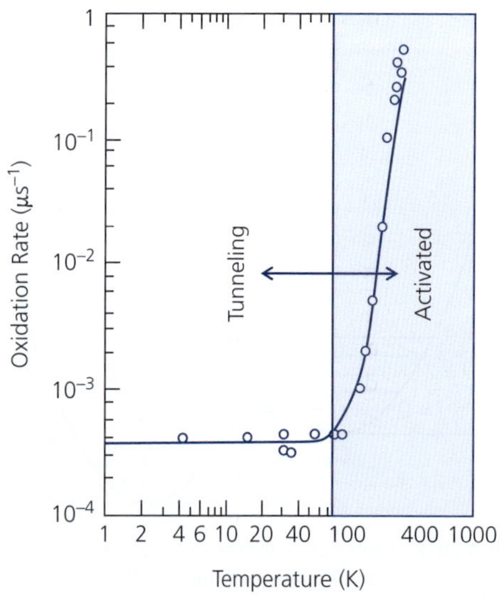

Figure 4.21 Rate of electron transfer from cytochrome to chlorophyll in *Chromatium* molecules as determined by changes in optical density vs. temperature. The rate decreases to a steady state, tunneling-limited value of ~4 × 10^{-4} μs^{-1} at $T < 80$ K. At higher temperatures the rate is thermally activated (Jortner, 1976)

barrier separating the reactant and product states. This process occurs between the vibronic states of the R and P electronic manifolds (Fig. 4.20), and hence the full wavefunction including electronic and nuclear coordinates, $\Phi(\mathbf{r}, Q) = \phi_e(\mathbf{r}, Q)\phi_N(Q)$, must be employed to calculate the transition rate.

We can estimate the boundary between the quantum and classical regimes based on the phonon (i.e. vibronic) modes in the initial and final states that are coupled via tunneling transfer. That is, the process is adequately described by the classical picture if the thermal energy is greater than the energy of the lth vibronic with energy $\hbar\omega_l$ in V_P into which the electron transfers. At $k_BT < \hbar\omega_l$, the treatment fails and quantum mechanical tunneling dominates the transfer reaction.

From Fermi's Golden Rule in the Franck–Condon approximation, the transfer rate is

$$k_{ET} = \frac{2\pi}{\hbar}|\langle\phi_{eR}|H|\phi_{eP}\rangle|^2|\langle\phi_{NR}|\phi_{NP}\rangle|^2\rho(E_P), \quad (4.85)$$

where $\rho(E_P) = \delta(E_R - E_P)$ is the density of product states around $E_P = E_R$. The first term is due to electronic states, and the second is the nuclear Franck–Condon factor weighted by the DOS, that is,

$$FC_\rho = |\langle\phi_{NR}|\phi_{NP}\rangle|^2\rho(E_P). \quad (4.86)$$

Writing the energy expended in mixing the initial and final electronic states as the familiar transfer integral $J_{RP} = \langle\phi_{eR}|H_e|\phi_{eP}\rangle$ that involves only the electronic part of the Hamiltonian, H_e, Eq. 4.85 becomes

$$k_{ET} = \frac{2\pi}{\hbar}J_{RP}^2 FC_\rho. \quad (4.87)$$

The remaining challenge is to evaluate the matrix elements. For this treatment we consider only non-adiabatic (i.e. weakly coupled) transfer. Then the total Hamiltonian of the system is given by (Barbara et al., 1996)

$$\begin{aligned} H = {} & |\phi_{eR}\rangle\langle\phi_{eR}|\left(E_R^0 + \frac{1}{2}\sum_l \kappa_l(Q_l - Q_{R,l})^2\right) \\ & + |\phi_{eP}\rangle\langle\phi_{eP}|\left(E_P^0 + \frac{1}{2}\sum_l \kappa_l(Q_l - Q_{P,l})^2\right) \\ & + (|\phi_{eR}\rangle\langle\phi_{eP}| + |\phi_{eP}\rangle\langle\phi_{eR}|)J_{RP} \end{aligned} \quad (4.88)$$

Here, E_R^0 and E_P^0 are the unperturbed initial reactant and product energies, and κ_l is the force constant of the lth mode. This generalization of the parabolic potential approximation requires only a single value of κ. The summations in Eq. 4.88 are over all vibronic modes. The first term corresponds to the unperturbed reactant state, the second to the product state, and the last term is the interaction that leads to electron transfer due to overlap of the electronic orbitals, ϕ_{eR} and ϕ_{eP}. Indeed, due to the orthogonality of $|\phi_{eR}\rangle$ and $|\phi_{eP}\rangle$, it is clear from Eq. 4.85 that only the term in J_{RP} affects the transfer rate.

The rate has been calculated for non-adiabatic transfer in the low temperature limit for a single coupled mode between two electronic states to yield (Jortner, 1976)

$$k_{ET} = \frac{J_{RP}^2}{\hbar}\left(\frac{\pi}{\lambda_{reorg}k_BT}\right)^{1/2}\sum_{l'}\frac{S^{l'}}{l'!}\exp(-S) \times\exp\left[\frac{-(\lambda_{reorg} + l'\hbar\omega + \Delta G^0)^2}{4\lambda_{reorg}k_BT}\right]. \quad (4.89)$$

In this expression, it is assumed that the transition occurs from the lowest ($l = 0$) vibronic with frequency ω in R, to l' in P. The sum takes into account the transfer from the initial mode into a distribution of final modes whose probability is determined by Boltzmann statistics. Here, S^l is the *Huang–Rhys parameter* (Huang and Rhys, 1950, Jortner, 1976) that gives the electron–phonon coupling strength in the localized molecular states, see Eq. 3.110. The energy released in electron transfer is thus modified such that $\Delta G^0 \to \Delta G^0 + l'\hbar\omega$ in the transfer from $0 \to l'$. Furthermore, the maximum term in the sum is at $-|\Delta G^0| \sim \lambda_{reorg} + l'\hbar\omega$ where there is the closest energy match between initial and final states. When this condition is met, the tunneling probability is high and the thermally activated exponential term to the right of Eq. 4.89 vanishes. Then the temperature dependence arises only weakly from the prefactor. In Fig. 4.20 we show the example of isoenergetic $0 \rightarrow 4'$ tunneling transition.

The summation is simply the Franck–Condon factor, *FC* in Eq. 3.79. For electron transfer, we express S^l in terms of the *intramolecular* reorganization energy, λ_I, due to an excess charge on the electronic manifolds described by ϕ_{eR} to ϕ_{eP}. Thus, if there is a shift in electronic coordinates, ΔQ_e, then the corresponding reorganization energy is $\lambda_I = \frac{1}{2}\kappa\Delta Q_e^2$, yielding for the lth vibronic mode:

$$S^l = \frac{\lambda_I}{\hbar\omega_l} = \frac{\kappa\Delta Q_e^2}{2\hbar\omega_l}. \tag{4.90}$$

When ΔG^0 is small, the electronic manifolds for *R* and *P* are aligned in Fig. 4.20, leading to symmetrical electron transfer. Then transfer occurs between modes $l = 0$ to $l' = 0$, which allows us to simplify Eq. 4.89 as follows:

$$k_{ET} = \frac{J_{RP}^2}{\hbar}\left(\frac{\pi}{\lambda_{reorg}k_BT}\right)^{1/2} \exp\left[\frac{-(\lambda_{reorg} + \Delta G^0)^2}{4\lambda_{reorg}k_BT}\right]. \tag{4.91}$$

This expression is remarkably similar to that obtained for dynamic disorder in Eq. 4.70, but where we have replaced the local polarization energy with $E_{pol}^{loc} = \lambda_{reorg}/2$ (Eq. 4.60), and the system releases free energy ΔG^0 during transfer. Disorder, then, is captured in the transfer integral, J_{RP} where (Coropceanu et al., 2007)

$$J_{RP} = J_{RP0}\exp(-\gamma_{DA}R_{DA}), \tag{4.92}$$

as applied in Miller–Abrahams charge transfer in Eq. 4.76.

A fit to Marcus theory in the adiabatic approximation including tunneling in the low temperature limit (Jortner, 1976) is shown by the solid line in Fig. 4.21. The fit accurately reproduces both the classical, thermally activated rate dependence at high temperature, whereas at low temperature tunneling dominates and the rate becomes constant.

Marcus transfer predicts phenomena that are absent in the Miller-Abrahams treatment where only upward hops are thermally activated. Notably, the rate predicts the existence of the *Marcus inverted region*, that is, the rate increases with $|\Delta G^0|$ until $\Delta G^0 = -\lambda_{reorg}$, at which point it decreases. This can be understood in terms of the reaction coordinate parabolas shown in Fig. 4.22. There, ΔG^0 ranges from the normal region where $\Delta G^0 < \lambda_{reorg}$, to the case of symmetrical transfer $\Delta G^0 = 0$, to the inverted region where $|\Delta G^0| > \lambda_{reorg}$. Importantly, the barrier to transfer ΔG^* is positive except for the case of $|\Delta G^0| = \lambda_{reorg}$, at which point $\Delta G^* = 0$ and the transfer rate achieves its maximum value of $k_{ET} = \frac{J_{RP}^2}{\hbar}\left(\frac{\pi}{\lambda_{reorg}k_BT}\right)^{1/2}$, determined only by the transfer integral, J_{RP}^2, and the reorganization energy.

We can understand the inverted region via the following scheme:

$$\rightarrow D^+ + A^0 \rightleftarrows D^+|A^- \rightleftarrows D^0 + A^- \rightarrow \quad (k_{ET})$$

A donor is either optically or thermally excited from which charge transfer occurs from donor to acceptor across the bridge, or precursor state, $D^+|A^-$. The final state on the right can either proceed to a further charge transfer step (or chemical reaction), or it can back transfer to the initial state at rate k_{ET}. It is the back transfer process that occurs in the inverted region. From Fig. 4.22, we see that the barrier in this region can be lower, and particularly its width is reduced from that experienced in forward transfer, resulting in an increased tunneling probably. Also, as the change in free energy, ΔG^0, during transfer increases, k_{ET} must decrease according to Eq. 4.91.

The first clear evidence for the existence of the inverted region was observed for *intra*molecular

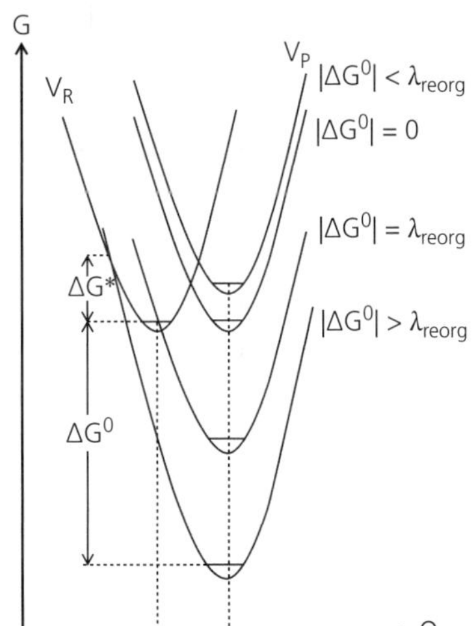

Figure 4.22 Free energy vs. reaction coordinate for several different magnitudes (both positive and negative) of the free energy, ΔG^0, released during charge transfer.

charge transfer for a series of compounds with an A–Sp–B structure, where A is one of eight π-conjugated molecules ranging from 4-biphenylyl to 5,6-dichlorobenzoquinon-2-yl, Sp is a rigid 5α-androstane skeleton spacer, and B is 4-biphenylyl. The molecules were dissolved in MeTHF. Transient optical absorption probed the time for the compound to equilibrate following the CT reaction in Fig. 4.23. The charge transfer rate between B and A across the molecular bridge vs. free energy was determined for each compound, with results in Fig. 4.24. The solid line is a fit to Marcus theory, assuming that the reorganization is equal to the sum of the contributions from solvent reorganization (0.75 eV) and the change in internal vibrational modes during transfer (0.45 eV) (Miller et al., 1984). Since a large fraction of λ_{reorg} is due to the solvent, data were also taken by dissolving the molecules in isooctane, where it was found that the peak of the curve shifts according to predictions of Marcus theory.

An example of *inter*molecular Marcus inversion has been observed between contacting molecules across a series of heterojunctions (HJs) comprising neat films of the donor and acceptor molecules, with results shown in Fig. 4.25 (Rand et al., 2007). The electron transfer rate parameters, λ_{reorg} and J_{RP}^2, and consequently k_{ET} in Eq. 4.91 were determined from the dark current of each donor–acceptor pair listed in Table 4.1. The dark current is measured by sandwiching the HJ between metal electrodes, and then applying a small voltage. The current is due to the difference in the probability of forward transitions from a filled initial state with occupation probability f_i as determined by Fermi-Dirac statistics, to an empty final state with occupation probability $(1 - f_f)$, and reverse transitions from an occupied final to an empty initial state. The rates of forward and reverse transfer are equal, at $k_{ET,if}$. Since $f \to i$ must overcome a barrier due to relaxation after transfer, its rate is weighted by a Boltzmann factor, $\exp(-\Delta E_{if}/k_BT)$. Then, the current density is given by (Nelson et al., 2004)

$$j = q\rho d k_{ET,if}\left(f_i(1-f_f) - \exp(-\Delta E_{if}/k_BT)f_f(1-f_i)\right). \tag{4.93}$$

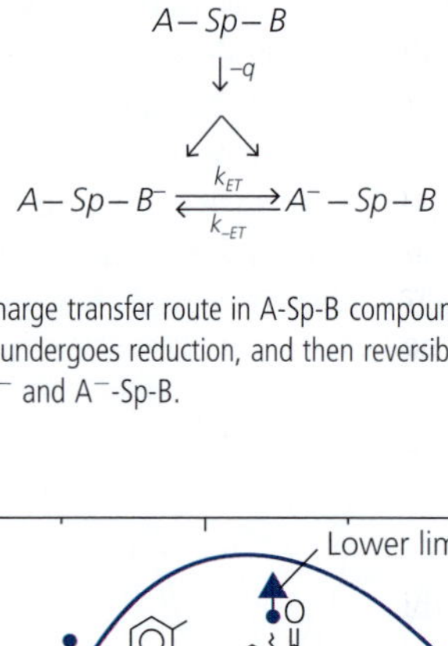

Figure 4.23 Charge transfer route in A-Sp-B compounds. The neutral compound (top) undergoes reduction, and then reversibly transfers between A-Sp-B⁻ and A⁻-Sp-B.

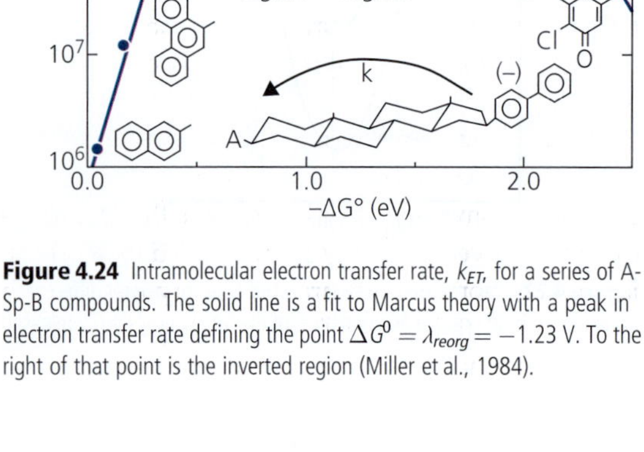

Figure 4.24 Intramolecular electron transfer rate, k_{ET}, for a series of A-Sp-B compounds. The solid line is a fit to Marcus theory with a peak in electron transfer rate defining the point $\Delta G^0 = \lambda_{reorg} = -1.23$ V. To the right of that point is the inverted region (Miller et al., 1984).

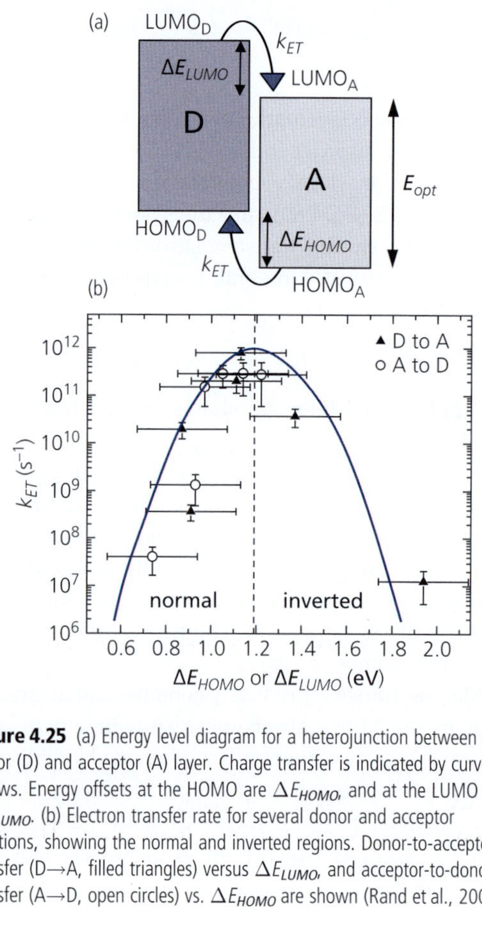

Figure 4.25 (a) Energy level diagram for a heterojunction between a donor (D) and acceptor (A) layer. Charge transfer is indicated by curved arrows. Energy offsets at the HOMO are ΔE_{HOMO}, and at the LUMO are ΔE_{LUMO}. (b) Electron transfer rate for several donor and acceptor junctions, showing the normal and inverted regions. Donor-to-acceptor transfer (D→A, filled triangles) versus ΔE_{LUMO}, and acceptor-to-donor transfer (A→D, open circles) vs. ΔE_{HOMO} are shown (Rand et al., 2007).

Table 4.1 Parameters obtained from fitting the current vs. voltage characteristics for the various donor/acceptor interfaces in Fig. 4.26 (Rand et al., 2007)

Donor/acceptor interface	Donor to acceptor		Acceptor to donor		λ_{cv} (eV)[c]
	k_{ccD} (s^{-1})[a]	λ_{ccD} (eV)[b]	k_{ccA} (s^{-1})[a]	λ_{ccA} (eV)[a]	
Pentacene/C_{60}	8×10^{11}	0.9	3×10^{11}	1.1	0.8
CuPc/C_{60}	2×10^{10}	1.0	2×10^{11}	0.6	0.7
CuPc/PTCBI	4×10^{8}	0.8	1×10^{9}	0.6	0.7
NPD/C_{60}	1×10^{7}	1.3	4×10^{7}	0.4	0.9
CuPc/C_{70}	2×10^{11}	0.7	3×10^{11}	0.7	0.7
Pentacene/C_{70}	4×10^{10}	0.9	3×10^{11}	0.7	0.5

[a] k_{ccD} is the charge transfer rate from $D \to A$, k_{ccA} is for $A \to D$.
[b] λ_{ccD} and λ_{ccA} are the reorganization energies for D and A, respectively.
[c] λ_{cv} is the reorganization energy for transitions from HOMO to LUMO, that is assumed are equal for D and A.

Here d is the layer thickness and ρ is the volume density of molecules. Also, the activation energy, ΔE_{if}, is equal to either the difference in HOMO energies between donor and acceptor (ΔE_{HOMO}) for transitions $A \to D$, or the difference in LUMO energies (ΔE_{LUMO}) for transitions $D \to A$. It is assumed that the change in system free energy is induced by transfer across the HJ, such that $E_{if} = \Delta G^0$, which can be directly measured from the open circuit voltage of the HJ diode under simulated solar illumination (see Section 4.7.2). We will show in Chapter 7 that Marcus transfer in both the inverted and normal regions significantly contributes to voltage losses in organic solar cells.

4.3.2.3 Charge mobility in organic semiconductors with static disorder: the effective medium approximation

Mobilities in almost all small molecule and polymer systems have activation energies of $\Delta E_a = 0.3$–0.6 eV. Since $\Delta E_a = E_{pol}/2$ in the non-adiabatic limit (Eq. 4.71), polarization energies of 0.6–1.2 eV are too large to be due solely to reduced coupling arising from purely dynamic disorder. Furthermore, there is often a transition from dispersive to non-dispersive transport as temperature is decreased. Here, *dispersive transport* refers to mobilities that cannot be defined by a single value (see Section 4.4.3). This is inconsistent with dynamic disorder, where transport is governed by small polarons, and hence is temperature independent.. Thus, the effects of static disorder must be included to provide a complete description of most organic systems.

As in the case of dynamic disorder, the coupling between molecules is short range, requiring consideration of only nearest neighbor interactions. The master equation for the change in occupancy of site i during hopping is

$$\frac{\partial f_i(t)}{\partial t} = \sum_{j \neq i}^{N} \{-k_{ij} f_i(t)[1 - f_j(t)] + k_{ji}(t)[1 - f_i(t)]\} - k_{rec} f_i(t), \tag{4.94}$$

where k_{ij} is the hopping rate from $i \to j$ as determined by either Miller–Abrahams or Marcus theory, and $f_i(t) = \{1 + \exp[(E_j - E_F)/k_B T]\}^{-1}$ from Fermi–Dirac statistics. It is the probability for site j to be occupied at time t, E_F is the Fermi energy, and k_{rec} is the rate of charge recombination, leading to its loss.

This expression is complicated, and hence we make the following simplifications: (i) only low charge densities are considered, in which case terms in f^2 can be ignored, and (ii) recombination is much slower than charge transfer, allowing us to neglect the last term. Thus, the master equation simplifies to

$$\frac{\partial f_i(t)}{\partial t} = \sum_{j \neq i}^{N} \{-k_{ij} f_i(t) + k_{ji} f_j(t)\}. \tag{4.95}$$

From this we can write the current density generated in a layer of thickness, d:

$$j(\mathbf{r}, t) = q \int g(E) dE \sum_{i}^{N} \left[\frac{\partial f_i(\mathbf{r}_i, E_i, t)}{\partial t} - \frac{\partial f_j(\mathbf{r}_j, E_j, t)}{\partial t} \right]. \tag{4.96}$$

The first term in brackets corresponds to transfers, $i \to j$, whereas the second term accounts for back transfer, $i \leftarrow j$. Furthermore, $g(E)$ is the DOS, such that $Ng(E) f(E)$ is the total number of occupied states at energy, E.

The challenge for determining the mobility, therefore, is to calculate the *net* hopping rate $i \to j$, and $g(E)$. It is convenient to assume that the states are distributed over an energy space defined by a Gaussian distribution centered around energy E_0 and with width, σ_{GDM}. (Bässler, 1993) This physically intuitive and mathematically manageable assumption is known as the *Gaussian disorder model* (GDM). The GDM distribution is illustrated in Fig. 4.26, and can be expressed by

$$g(E_i) = \frac{N}{\sqrt{2\pi\sigma_{GDM}^2}} \exp\left(-\frac{(E_i - E_0)^2}{2\sigma_{GDM}^2}\right), \tag{4.97}$$

where N is the density of hopping sites. Since the mean energy can be arbitrarily referenced, it is convenient to set $E_0 = 0$, in which case Eq. 4.97 is referred to as the *occupational density of states* (ODOS). This is the equivalent of the actual DOS reduced to a Gaussian centered about the average energy, E_∞, given by:

$$\langle E_\infty \rangle = \frac{\int_{-\infty}^{\infty} Eg(E)\exp(-E/k_BT)dE}{\int_{-\infty}^{\infty} g(E)\exp(-E/k_BT)dE} = -\frac{\sigma_{GDM}^2}{k_BT}. \tag{4.98}$$

The assumption in this analysis is that the level of injection is low (i.e. the fraction of free charge, n to the number of available sites, N is $\zeta = n/N \ll 1$). Thus, the system is non-degenerate, allowing for a replacement of the Fermi–Dirac distribution with the Boltzmann approximation. Further, the organic medium is only weakly disordered and homogeneous. This implies that a single probability function can be used throughout the film volume. The probability for a hopping site to be available is determined by the Gaussian DOS along with Fermi–Dirac statistics. This is known as the *effective medium approximation* (EMA).

In Fig. 4.26a we show the case of moderate charge injection, where the filled states have a tail of width ~$2k_BT$ centered around the Fermi energy, E_F. Free polarons hop to the empty states into the transport level, which defines the *mobility edge*. Only charges occupying states near E_F can acquire sufficient energy to hop to unoccupied states, and hence contribute to conduction. For low charge densities, the hopping energy required to reach the mobility edge must be correspondingly larger (Fig. 4.26b), and consequently, the conductivity is reduced.

Figure 4.27a shows the mobility edge for the transport DOS. Monte Carlo calculations have been employed to determine the rate at which equilibrium is reached following the injection of a hot (i.e. at $E \gg E_F$) charge into the transport band. Results in Fig. 4.27b are shown for a distribution of width $\sigma_{GDM} = 2k_BT$. The dwell time between each step is given by $1/t_0 = 6v_0 \exp(-2\gamma a)$ (see Eq. 4.76), where a simple cubic lattice with edge length, a, is assumed, and the factor of 6 accounts for all possible nearest neighbor hops. Since the

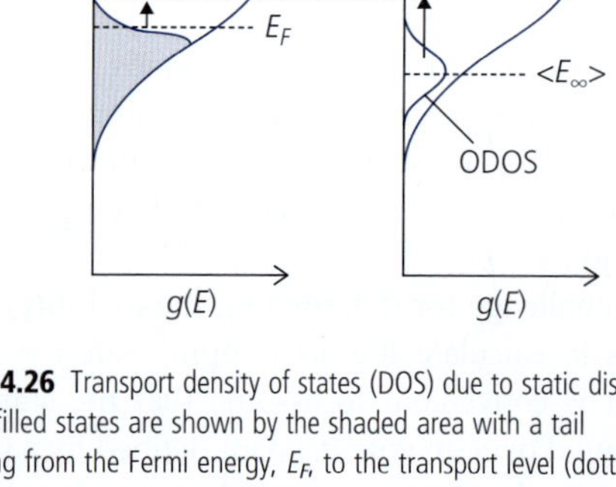

Figure 4.26 Transport density of states (DOS) due to static disorder. (a) The filled states are shown by the shaded area with a tail extending from the Fermi energy, E_F, to the transport level (dotted line). (b) The same as in (a) except at very low current injection. The mean energy of the occupational DOS (ODOS) is $\langle E_\infty \rangle$. After Köhler and Bässler (2015).

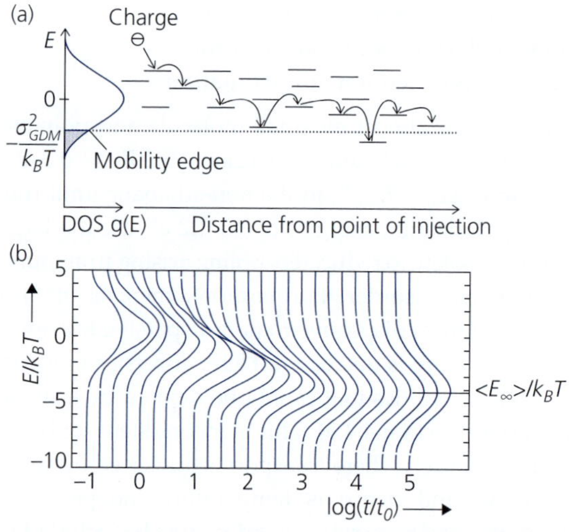

Figure 4.27 (a) Equilibration of a hot charge in both space and energy via a series of intermolecular hops from disordered states whose energy distribution is described by the Gaussian disorder model. The mobility edge, that is, the equilibrium energy of the filled states is shown.
(b) Monte Carlo calculation of the relaxation to the equilibrium energy, $\langle E_\infty \rangle$, of a distribution of charges injected at reference energy $E = 0$ into a density of states of width $\sigma_{GDM} = 2k_BT$. The dwell time between steps is t_0. After Bässler (1993).

time follows a logarithmic scale, it is apparent that it takes many hops and a considerable distance before the charge distribution fully equilibrates.

Note that the GDM represents only one of several possible functions that can be used to represent the transport DOS, $g(E)$ (Tessler et al., 2009). A uniform DOS with $g(E)$ = constant can be employed for its calculational simplicity, although it does not accurately represent the situation in disordered media. Alternatively, the tail in a Gaussian distribution can be approximated by an exponential DOS given by

$$g(E) = \frac{N}{k_B T_T} \exp\left(-\frac{E}{k_B T_T}\right), \tag{4.99}$$

where T_T is the *characteristic trap temperature*. A comparison between a Gaussian and exponential DOS is provided in Fig. 4.28. The exponential function has a longer, more gentle roll-off than the Gaussian. Given that the exponential distribution is readily adaptable to analytical solutions of the master transport equation, it is often used to model trap-limited transport and injection from contacts, as discussed in greater detail in Section 4.4.

With the assumption of a Gaussian density distribution, we can proceed to calculate the mobility in a one-dimensional electric field $\mathbf{F} = \{F, 0, 0\}$. Once more assuming molecules are arranged on a cubic lattice with side, a, then the mobility is given by the difference in the rate of jumps from $i \rightarrow j$ vs. $i \leftarrow j$ (Fishchuk et al., 2003):

$$\mu = a \frac{k_{ij} - k_{ji}}{F}. \tag{4.100}$$

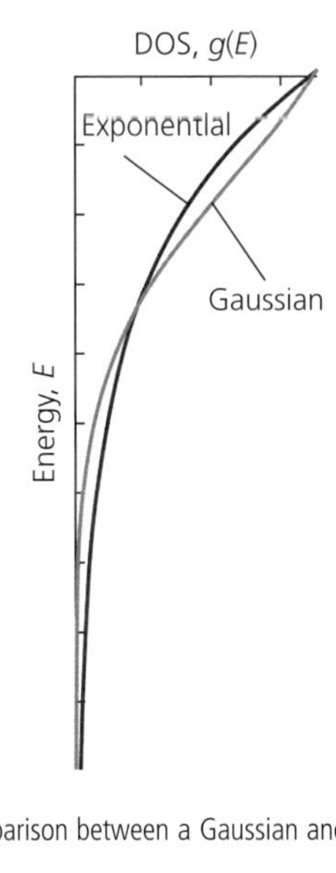

Figure 4.28 Comparison between a Gaussian and exponential DOS.

The rates are given by Miller–Abrahams theory (Eq. 4.76) where the energy difference now includes the driving force provided by the electric field, viz.

$$k_{ij} = \upsilon_0 \exp\left(-2\gamma_{ij} a\right) \times \exp\left[-\frac{|E_j - E_i| + E_j - E_i + q(\mathbf{r}_j - \mathbf{r}_i)\cdot\mathbf{F}}{2k_B T}\right]. \tag{4.101}$$

This must be weighted by the Gaussian DOS in Eq. 4.97 to account for hops from states at or near the mobility edge. Alternatively, the hopping rate can be calculated using Marcus theory (Eq. 4.89):

$$k_{ij} = \frac{J_{RP}^2}{\hbar}\left(\frac{\pi}{\lambda_{reorg} k_B T}\right)^{1/2} \sum_{l'} \frac{S^{l'}}{l'!} \exp(-S) \times \exp\left[\frac{-(\lambda_{reorg} + l'\hbar\omega + \Delta G^0 + q(\mathbf{r}_i - \mathbf{r}_j)\cdot\mathbf{F})^2}{4\lambda_{reorg} k_B T}\right] \tag{4.102}$$

with $\Delta G^0 = E_j - E_i$. We must similarly weight this expression by the DOS.

Solving Eq. 4.100, we arrive at two analytical expressions for the mobility depending on the hopping rate employed. For Miller–Abrahams hopping at "small" electric fields in the range: $k_B T \ll qaF \ll \langle E_\infty \rangle$, that is, where the field-induced hopping is larger than thermally activated hops, yet it is smaller than the mean energy of the DOS, then (Fishchuk et al., 2003)

$$\mu = \frac{\sigma_{GDM}\nu_0}{\sqrt{2\pi} q F^2} \exp(-2\gamma a) \cdot \exp\left[-\frac{1}{2}\left(\frac{\sigma_{GDM}}{k_B T}\right)^2 + \frac{1}{\sqrt{2}}\left[\left(\frac{\sigma_{GDM}}{k_B T}\right)^{3/2} - \left(\frac{\sigma_{GDM}}{k_B T}\right)^{1/2}\right]\sqrt{\frac{qaF}{\sigma_{GDM}}}\right]. \tag{4.103}$$

Similarly, for Marcus theory in the low temperature-low field limit, and assuming that transfer only occurs from $l = 0$ to the lowest vibronic state in the product, $l' = 0$, then

$$\mu = \frac{\sigma_{GDM}\xi}{\sqrt{2\pi} q F^2}\left[\frac{J_{RP}^2}{\hbar}\sqrt{\frac{\pi}{4E_A k_B T}}\right] \cdot \exp\left[-\frac{E_A}{k_B T} - \frac{1}{8\xi^2}\left(\frac{\sigma_{GDM}}{k_B T}\right)^2 + \frac{1}{2\sqrt{2}\xi^2}\left[\left(\frac{\sigma_{GDM}}{k_B T}\right)^{3/2} - \left(\frac{\sigma_{GDM}}{k_B T}\right)^{1/2}\right]\sqrt{\frac{qaF}{\sigma_{GDM}}}\right] \tag{4.104}$$

where $\xi = [1 - \langle E_\infty \rangle / (8E_A)]^{1/2}$ and we have replaced the reorganization energy with the mobility activation energy, that is, $\lambda_{reorg} = E_A$. This allows for a direct comparison with experiment.

There are clear similarities, and some notable differences between the expressions in Eqs. 4.103 and 4.104:

(i) Both Miller–Abrahams and Marcus theories predict a dependence on electric field that follows $\mu \sim \exp(-A\sqrt{F})$. This is a characteristic of the *Frenkel–Poole effect* (Sze, 1981) where the barrier to electron emission from a molecule to its neighbor is lowered by the applied electric field.

(ii) The mobility in both cases decreases exponentially with molecular separation (see Eq. 4.92). Again, this is consistent with charge tunneling between orbitals on nearest neighbors, which increases rapidly as molecular separation, a, decreases.

(iii) Both expressions predict that $\mu \sim \exp\left[-B/(k_BT)^2\right]$ at high temperature, which is strikingly different from conventional thermal activation that follows $\mu \sim \exp[-E_A/k_BT]$. However, at low temperatures there is a deviation from the $1/T^2$ behavior, with notable differences between predictions by Miller-Abrahams vs. Marcus theories, although distinguishing these differences from mobility data may not be possible unless the transport is clearly non-dispersive (see Section 4.4.3).

(iv) The expressions primarily differ by the reorganization energy in the exponential term, E_A, and the transfer integral, J_{RP} that appear in the Marcus formula, Eq. 4.104.

(v) Both expressions are consistent with Monte Carlo calculations of the mobility that result in the following empirical temperature- and electric field-dependent mobility (Bässler, 1993):

$$\mu = \mu_0 \exp\left(-\left(\frac{2\sigma_{GDM}}{3k_BT}\right)^2\right) \begin{cases} \times\exp\left[C\left(\left(\frac{\sigma_{GDM}}{k_BT}\right)^2 - \Sigma^2\right)\sqrt{F}\right] & \Sigma \geq 1.5 \\ \times\exp\left[C\left(\left(\frac{\sigma_{GDM}}{k_BT}\right)^2 - 2.25\right)\sqrt{F}\right] & \Sigma < 1.5 \end{cases} \quad (4.105)$$

For these expressions, it is assumed that the overlap parameter, $2\gamma a$, also has a Gaussian form. This results from application of the *central limit theorem* to account for the varying separation and orientation of molecules in a disordered solid. The variance is given by Σ, and determined by the magnitude of the off-diagonal disorder. Then the ensemble average of the mobility is $\langle\mu\rangle = \mu(\Sigma = 0)\exp(\Sigma^2/2)$.

The mobility dependence on F and T for the polymer, PMPSi measured using the time of flight method (Bässler et al., 1994) has been fit to Eq. 4.104, with results shown in Fig. 4.29. The temperature dependence is reasonably well fit by an activation that follows $\mu \sim \exp[-B/(k_BT)^2]$, although it is often difficult to distinguish between this functional dependence and that due to simple thermal activation of $\mu \sim \exp[-E_A/k_BT]$ (Fishchuk et al., 2003). The high field dependence is consistent with Frenkel–Poole emission. Note, however, that the mobility is not expected to increase indefinitely since scattering

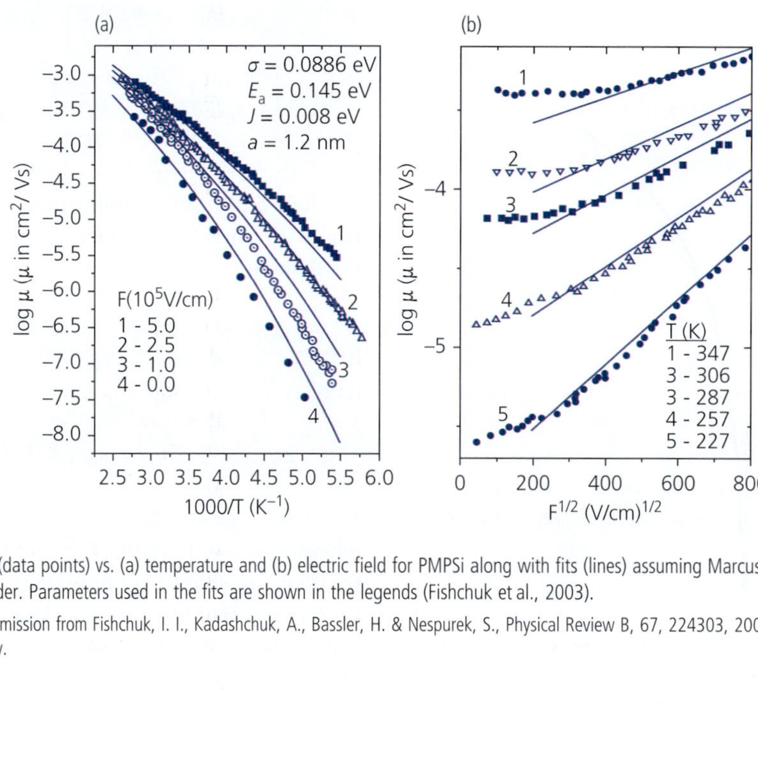

Figure 4.29 Mobility (data points) vs. (a) temperature and (b) electric field for PMPSi along with fits (lines) assuming Marcus electron transfer in the presence of static disorder. Parameters used in the fits are shown in the legends (Fishchuk et al., 2003).

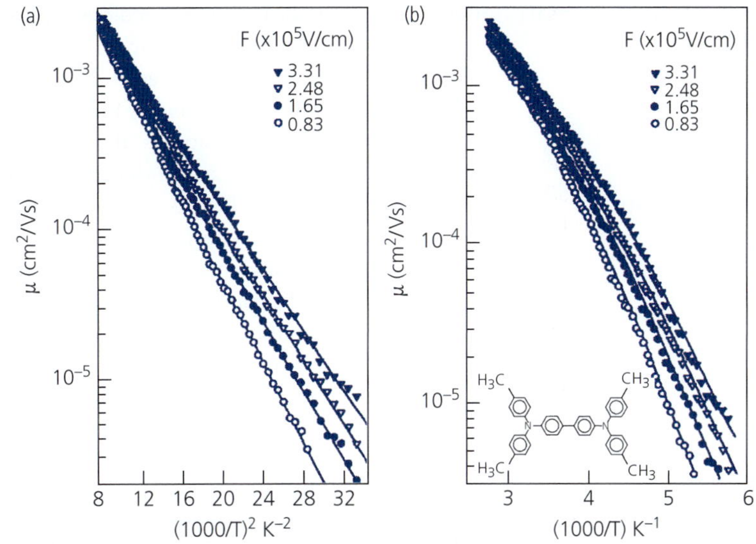

Figure 4.30 Mobility vs. temperature with electric field as parameter for the molecule, TTB. (a) $\mu \sim \exp[-B/(k_BT)^2]$ characteristic of the static disorder model. (b) $\mu \sim \exp[-E_A/k_BT]$ expected for transport limited by small polaron hopping in dynamically disordered systems. Inset: Molecular structural formula of TTB. The lack of curvature in (a) indicates that static disorder limits the measured mobility. Lines in both cases are fits assuming a $1/T^2$ behavior (Heun and Borsenberger, 1995).

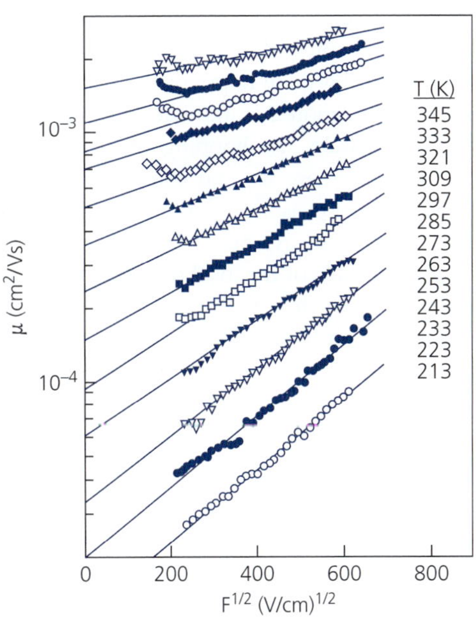

Figure 4.31 Mobility vs. electric field with temperature as parameter for TTB (Heun and Borsenberger, 1995).

becomes important at high electric fields, at which point the mobility saturates, as in Fig. 4.13 for ultra-pure naphthalene crystals.

Results from similar TOF measurements on amorphous films of the small molecule, TTB, are shown in Figs. 4.30 and 4.31. The TTB molecular structural formula is provided in the inset of Fig. 4.30b. The single bond between the two triarylamine end groups allows for torsional modes that lead to dynamic disorder that adds to the temperature dependence of the static disorder in the amorphous solid. Plots of μ vs. an exponential with either a $1/T^2$ (Fig. 4.30a) or a $1/T$ dependence (Fig. 4.30b) assuming Miller–Abrahams electron transfer show fits that are easily distinguished, favoring the $1/T^2$ behavior characteristic of static disorder (Heun and Borsenberger, 1995). In Fig. 4.37, we show the field dependence (with temperature as a parameter), exhibiting the $\mu \sim \exp(-A\sqrt{F})$ dependence anticipated for Frenkel–Poole dominated electron transfer.

4.3.2.4 Beyond the Einstein relation and the EMA

In the previous section, we derived the mobility from the diffusion constant that are proportional, using the Einstein relation in Eq. 4.36. Strictly speaking, this relationship is only valid at or very near equilibrium in an undoped semiconductor where the ratio of the free charges to hopping sites, $\zeta \ll 1$. That is, we assume that the system is non-degenerate, thereby allowing for the replacement of the Fermi–Dirac distribution with the Boltzmann approximation. However, this condition is rarely met in disordered organic semiconductors where the Fermi ODOS overlaps the transport level (E') due to the extended width of the distribution, $\sigma_{GDM} > E' - E_F$. This condition is also violated in high current devices such as OLEDs and thin film transistors. A more general relationship between μ and D is found from the formal expressions for electrons and holes, viz.

$$\mu_n = qD_n \frac{\partial \ln(n)}{\partial E_F} \tag{4.106a}$$

and

$$\mu_p = qD_p \frac{\partial \ln(p)}{\partial E_F}, \tag{4.106b}$$

where the electron (hole) density is n (p) and the Fermi energy, E_F is the chemical potential. Now the hole concentration is given by

$$p(E_F) = \int g(E)f(E, E_F)dE \tag{4.107}$$

with an analogous expression for electrons. Substituting the Fermi–Dirac distribution and the Gaussian DOS into Eqs. 4.107 and 4.106a gives a hole mobility of

$$\mu_p = \frac{qD_p}{k_BT} \frac{\int \exp\left(-\frac{(E-E_0)^2}{2\sigma_{GDM}^2}\right) \frac{\exp((E-E_F)/k_BT)}{[1+\exp((E-E_F)/k_BT)]^2} dE}{\int \exp\left(-\frac{(E-E_0)^2}{2\sigma_{GDM}^2}\right) \frac{1}{[1+\exp((E-E_F)/k_BT)]} dE}. \tag{4.108}$$

A plot of the ratio μ_p/D_p in units of q/k_BT is shown vs. $\zeta = p/N_{HOMO}$ for several different values of the normalized distribution width, $\hat{\sigma} = \sigma_{GDM}/k_BT$ in Fig. 4.32. Here, we assume that the density of hopping sites, N, is equal to the DOS at the HOMO, i.e. N_{HOMO}. For $\hat{\sigma} = 7$ and 10, the distribution has been cut off at -19 k_BT (dotted lines) and $-40\,k_BT$ (dashed lines) to prevent overlapping into the HOMO states. The deviation from the Einstein relation is quite significant as the Gaussian width increases. This is evidence for the degeneracy that arises from the wide ODOS that can overlap well into the frontier orbitals, causing a significant deviation from the mobilities predicted by Eq. 4.36. Furthermore, a narrow DOS leads to adherence to the classical Einstein relationship even at charge densities approaching that of the total density of hopping sites.

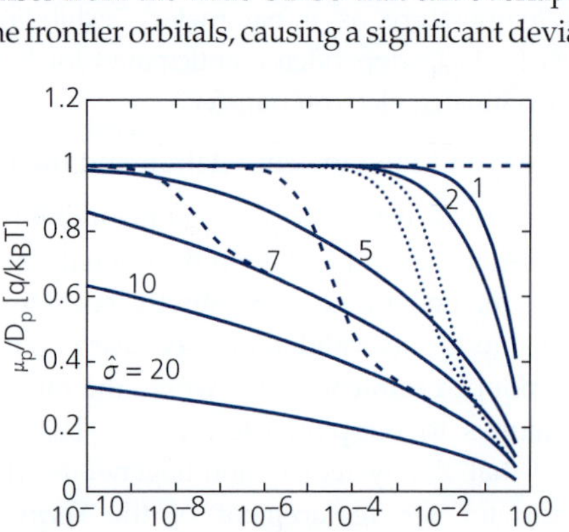

Figure 4.32 Normalized hole mobility vs. the normalized charge density for various Gaussian DOS widths (solid lines). Dashed lines correspond to a cutoff in the DOS of -40 k_BT, and the dotted lines at -19 k_BT for widths of $\hat{\sigma} = 7$ and 10. Here, $\mu_p/D_p = 1$ corresponds to the Einstein relation. After Roichman and Tessler (2002).

The conclusion is that the *rate* of mobility increase with charge density depends on the Gaussian width. That is, as the width is increased, there are higher energy states available for transport as the Fermi energy moves towards the conduction level. In contrast, the states in the tails of a narrow distribution are rapidly filled, resulting in a decrease in the change in mobility. This is illustrated in Fig. 4.33 where we plot the low field mobility vs. ζ for different values of $\hat{\sigma}$.

From the foregoing, we see that the mobility is a function of the charge density, and hence the position of the Fermi energy relative to the conduction levels. Furthermore, the assumption of an effective medium where all charges have the same probability of hopping into a nearby site fails to capture the details of the physics of disordered media, since the available final hopping sites are not evenly distributed about the initial site. Given the very short range nature of the electron wavefunction captured by the exponential, $\nu_0 \exp(-2\gamma a)$ in the Miller–Abrahams formalism, or the overlap integral, J_{RP}, in Marcus theory, the distance from charge to hopping site is critically important, yet is ignored in the effective medium approach. Ultimately, the hopping process is one of percolation; there are favored continuous pathways that lead to lower resistance conduction channels than others. These pathways are ultimately governed by the random details of the local disorder. Conduction is limited by those steps within the favored paths that are most difficult to overcome. That is, the average hop is replaced by the "most difficult but still relevant" hop within a percolating path (Coehoorn et al., 2005). These are the unavoidable steps that,

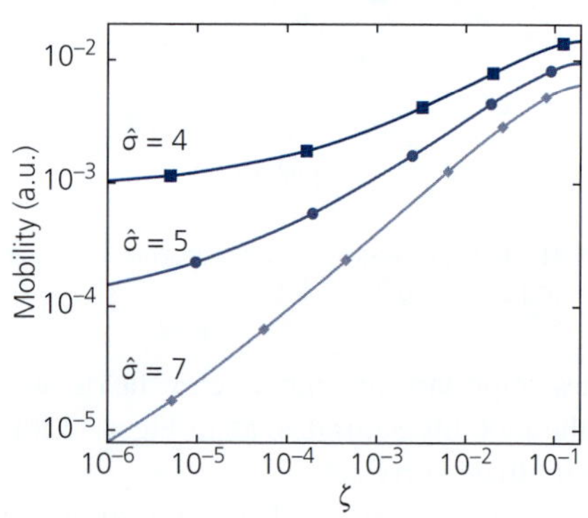

Figure 4.33 Mobility vs. normalized charge density for various Gaussian widths of the DOS. After Roichman et al. (2004).

statistically must be accounted for in determining the macroscopic, measurable quantities of mobility, diffusion constant, and conductivity. And, as noted above, these quantities not only have a complex temperature dependence, but also depend on the details of the distribution assumed for the DOS (e.g. uniform, Gaussian, or exponential), and on the morphology and the steric nature of the molecules comprising the solid. To complicate matters even more, dipolar materials can result in locally organized domains, as well as polaron-dipole interactions that influence the percolation paths available. This so-called *spatially correlated disorder* can create significant departures in the field and temperature dependences of the mobility predicted by the simplified randomness assumed in the GDM (Gartstein and Conwell, 1995, Dunlap et al., 1996, Van Mensfoort et al., 2010).

Several models have been developed to describe the complexities of transport within organic media. These models have been reviewed in the literature (Coehoorn et al., 2005, Tessler et al., 2009); here we will describe and compare several of the most prominent such approaches with the caveat that the results of such models are sensitive to the materials systems considered, the film morphologies, the temperature and injection conditions extant, and the DOS distribution function that is assumed. Furthermore, as in the previous section, most models are developed in the limit of $\zeta \to 0$, and thus must be modified when used to model systems (e.g. thin film transistors) at high current densities.

For most semi-analytical, semi-empirical models, the mobility in the zero charge density limit follows the general form of (Coehoorn et al., 2005)

$$\mu(\zeta \to 0) = \mu_0 \exp\left(-a - \beta\hat{\sigma} - \chi\hat{\sigma}^2\right). \quad (4.109)$$

With the exception of some additional but perhaps minor deviations in the powers of $\hat{\sigma}$ in the argument of the exponent, this is similar to that predicted by the effective medium approach (Eqs. 4.103–4.105). A tabulation of μ_0 and a for several important examples is provided in Table 4.2. In these models, it is assumed that hopping is governed by a Miller-Abrahams process, and that the DOS is defined by Gaussian disorder.

In each of the models with the exception of RT, we provide a factor, B, that depends on the statistics governing the onset of percolation. Also, $\xi^{1/3} = \gamma/N^{1/3}$ is the ratio of the decay length of the wavefunction to the average distance between hopping sites. The basic elements of the models are as follows:

(i) MS is the Movaghar–Schirmacher model based on hopping controlled by the average distance between sites, $\bar{R} \propto 1/N^{1/3}$. This is effectively the EMA, but is corrected by the inclusion of $B = \exp(1)$ to discount the importance of hops to states that are far removed from E_F.

(ii) VM is the model of Vissenberg and Matters. It assumes that the medium is equivalent to a random network of resistors, whose resistance is determined by Miller-Abrahams hopping energetics. Thus, the conductance between sites, i, j is $G_{ij} = (q^2\nu_0/k_BT)\exp(-\hat{\varepsilon}_{ij})$, where $\hat{\varepsilon}_{ij}$ is the reduced energy given by the exponential terms in the expression in Eq. 4.101. The path taken by the hopping charge is determined from percolation theory, whereby there is a critical energy value, $\hat{\varepsilon}_C$ above which the conductance drops to

Table 4.2 Terms used in defining the generalized low-charge-density mobility of Eq. 4.109 for several models assuming Gaussian disorder. Adapted from Coehoorn et al. (2005)

Model	μ_0	a	B	Ref.
MS	$\frac{q\nu_0}{k_BT}\frac{\bar{R}^2}{6}B$	$\left(\frac{6B}{\pi}\xi\right)^{1/3}$	2.7	(Movaghar and Schirmacher, 1981)
VM	$\frac{\sigma_0}{qN}$	$\left(\frac{6B}{\pi}\xi\right)^{1/3}$	2.7	(Vissenberg and Matters, 1998)
Arkhipov[a]	$\frac{q\nu_0\bar{R}^2}{k_BT}$	$\Gamma\left(\frac{4}{3}\right)\left(\frac{6B}{\pi}\xi\right)^{1/3}$	3.933	(Arkhipov et al., 2002)
Martens	$\frac{\sigma_0}{qN}$	$\left(\frac{6B}{\pi}\xi\right)^{1/3}$	1.969	(Martens et al., 2003)
Baranovskii	$\frac{q\nu_0}{k_BT}\frac{R(E_t)^2}{6}\Psi$	$\left(\frac{6B}{\pi}\xi\right)^{1/3}$	1.969	(Baranovskii et al., 2000)
RT	$\frac{q\nu_0}{N^{2/3}k_BT}$	$ln\left(\frac{1}{\pi}\xi^{5/3}\right)$	1	(Roichman et al., 2004)

[a] $\Gamma(\frac{4}{3}) = 0.893$ is the Euler gamma function.

zero. This is the point whereby the "most difficult but still relevant" path is broken, and percolation paths form only isolated clusters. The percolation factor is identical to that of the MS approach.

(iii) Arkhipov invoked a "favored first hop" formulation that asserts that hops occur between neighboring sites where the Miller–Abrahams rate is largest. The available site energies are determined by applying a step function in the Miller–Abrahams hopping energy argument. Then, only those available sites whose average density is less than B participate in the percolation process. As in the EMA, the mobility is assumed to be proportional to D according to the Einstein relation. Unlike VM, the Arkhipov model does not take Fermi–Dirac statics into consideration, and hence a charge may hop into an already occupied site. For this reason, it is only valid at very low charge concentrations.

(iv) Martens specifically requires that hopping occurs from a site at energy E_F to a neighboring site at distance R^*, whose energy is E^*. Hence, the conductivity is given by $\sigma = \sigma_0 \exp(-2\gamma R^* -(E^* - E_F)/k_B T)$, where σ_0 is the conductivity for $R^* = 0$, and $E^* = E_F$. The primary assumption of this model that is absent in VM and MS is that each site is connected to only one other favored site, similar to the model of Arkhipov.

(v) Baranovskii and co-workers developed a model very similar to Martens, but assume the most favored transfer occurs within a sphere whose radius lies in a *range*, $R(E_t)$. The charge can access sites whose energies lie below E_t and the transport level. The parameter Ψ is a function of E_t and the Gaussian width, and is on the order of unity. The Baronovskii model also faithfully reproduces the temperature dependence of Eq. 4.109, containing both the linear and quadratic dependences by setting $E_t = \langle E_\infty \rangle = \beta\hat{\sigma}$ where $\beta = \sigma_{GDM}$ and $\xi = \frac{1}{2}$.

(vi) RT denotes the Roichman–Tessler model which is distinct from the other five models in Table 4.2. Indeed, this is another EMA theory, where percolation does not play a role (i.e. $B = 1$). Thus, the material is considered to be homogeneous, with each carrier randomly selected from the Gaussian DOS. Starting with the conventional definition for the current density, $j = ne\bar{v}$, an average velocity, $\bar{v}$ is calculated for each carrier which is the product of the displacement, dx, parallel to the applied electric field, $F(x)$, and k_{ij} in Eq. 4.101. This has the advantage of including the effects of electric field, and thus μ can be obtained at all fields as in the GDM. It also accounts for site occupancy as determined by Fermi–Dirac statistics, which avoids disallowed hops into occupied final states. The fundamentally different nature of the RT model thus predicts temperature and concentration dependences that are not found in the other, percolation-based analyses that are based on the principle of carriers hopping along "lucky," continuous but largely independent pathways.

One further model is that of Pasveer that is based on solving the master occupancy-based Eq. 4.94 (Pasveer et al., 2005). We will show in the following section that this can lead to satisfactory fits to the mobility as a function of both electric field and charge density over a moderate temperature range in PPV-based polymers. However, it is a largely parametric computational approach and is not based on Eq. 4.109 as are the other percolation models.

In the foregoing, we have not explicitly included the concentration dependence of the mobility. Increased charge concentration results in a change in the Fermi energy, as implied in Figs. 4.32 and 4.33. In effect, the states closer to the center of the DOS have a higher probability of occupancy, resulting in the preferential selection of conduction states in the high energy tails of the DOS. The outcome of state filling is an abrupt increase in mobility with charge density, shown in Fig. 4.34. The onset of the high mobility region is independent of temperature, which is consistent with percolation being linked to a physical property of the material (i.e. the density of sites). The lines are fits to the data using the VM model. Interestingly, it has been found through numerical comparisons that for $\hat{\sigma} > 2$, all the models predict that, at charge densities corresponding to the condition $E_F = \langle E_\infty \rangle$ (Eq. 4.98), the mobility is twice that of its value at $p = 0$ (Coehoorn et al., 2005). That is,

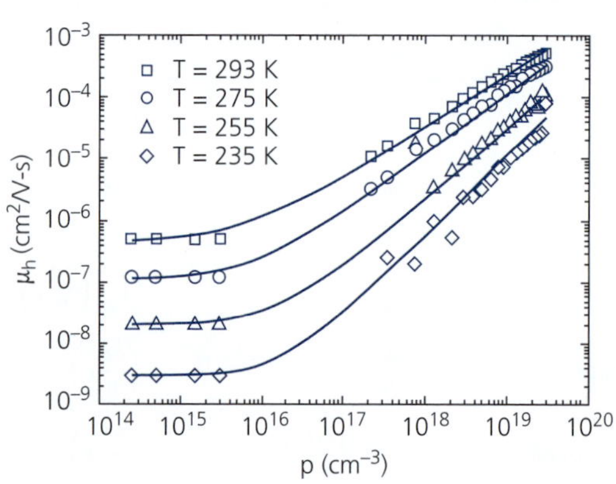

Figure 4.34 Hole mobility vs. hole concentration in a OC1C10-PPV thin film transistor as a function of temperature. Lines are fits to the VM percolation model (Tanase et al., 2004).

$$\frac{\mu(p(E_F = \langle E_\infty \rangle))}{\mu(p = 0)} \simeq 2. \tag{4.110}$$

Furthermore, the carrier concentration at this "cross-over" point is:

$$\zeta(E_F = \langle E_\infty \rangle) = \frac{1}{2}\exp(-\hat{\sigma}/2) \tag{4.111}$$

By measuring the mobility vs. charge concentration, these two expressions combined can yield the width of the DOS. Indeed, for the data in Fig. 4.34 we obtain $\sigma_{GDM} = 23k_BT$, which at 300 K corresponds to 600 meV. This is a significant fraction of the energy gap, justifying our skepticism in using the Einstein relation to deduce the mobility from the diffusion constant.

The several models in Table 4.2 are compared in the mobilities vs. $\hat{\sigma}$ and ζ in Figs. 4.35a and b, respectively. The reduced mobility is given by Eq. 4.109, that is: $\mu_{red} = \exp(-\alpha)$, where α is provided in Table 4.2 for all except the model of Pasveer (Pasveer et al., 2005). The width dependence of the several percolation models is approximately equivalent. The two outliers are the Arkhipov model (upper curve) and that of RT. The higher Arkhipov curve corresponds to the situation where $B = 1$. This result obtains when hops can occur between neighboring sites whether they are within continuous or discontinuous percolating paths. When only continuous paths are included, $B = 3.933$, yielding the lower curve that conforms to results of the other percolation models. Since the RT model assumes the EMA, it generates a considerably larger mobility.

Figure 4.35b compares the predicted dependences of the reduced mobilities vs. charge concentration for several different Gaussian widths normalized to their values at $\hat{\sigma} = 2$. The vertical lines show that the cross-over into the high concentration conduction regime occurs at approximately twice the mobility value at $\zeta \to 0$ for all models. Note that results from RT are not provided since they show a much higher mobility and concentration dependence than the models in the plot. The percolation models predict an increase in μ with charge concentration, with the largest changes occurring for the broadest distributions. However, as the charge density approaches that of the hopping sites (i.e. as $\zeta \to 1$), the available sites become saturated. Fermi–Dirac statistics prohibit hopping into filled sites, thus resulting in a decrease in mobility, as shown for all models except Arkhipov. We noted above that this latter picture does not account for the occupancy of the final states, and hence it continues to increase with ζ.

An empirical relationship that accounts for the features of the several percolation-based models is as follows (Coehoorn et al., 2005):

$$\mu = \frac{q\nu_0}{N^{2/3}k_BT}\Phi\exp\left[-\alpha - \ln\zeta - \beta'\hat{\sigma} - \chi'\hat{\sigma}^2\right], \tag{4.112}$$

where $\beta' = \alpha - E_F/\sigma_{GDM}$ and $\chi' = -d/\alpha$. Here, Φ and d are dimensionless fitting parameters on the order of unity. We note, once again, that such a semi-analytical expression is dependent on many factors that are possibly unique to a given materials system. These include the shape of the DOS function, the sterics of

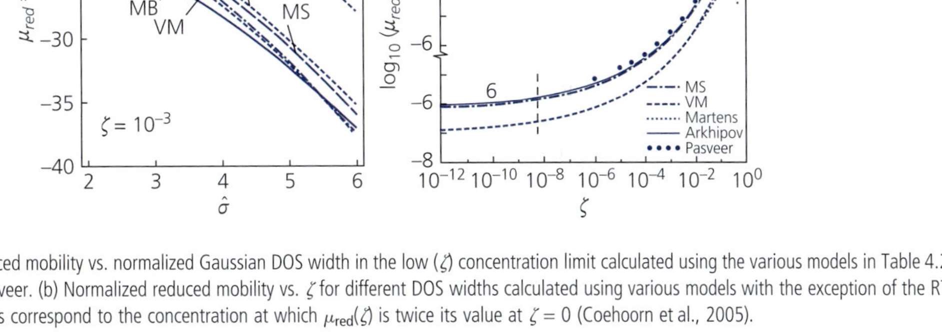

Figure 4.35 (a) Reduced mobility vs. normalized Gaussian DOS width in the low (ζ) concentration limit calculated using the various models in Table 4.2 including results by Pasveer. (b) Normalized reduced mobility vs. ζ for different DOS widths calculated using various models with the exception of the RT model. The vertical lines correspond to the concentration at which $\mu_{red}(\zeta)$ is twice its value at $\zeta = 0$ (Coehoorn et al., 2005).

individual molecules that control the hopping process, morphology of the solid, and so on. However, given the calculational intensity and the uncertainties that accompany the analysis of transport in disordered systems, empirical expressions such as Eq. 4.105 and 4.112 are helpful in analyzing the results of mobility measurements in bulk and thin film organic semiconductors. These measurements and other factors that impact conductivity such as doping are the subjects of the next two sections.

4.4 Conduction in organic thin films

A distinctive feature of semiconductors in device applications is the ability to change their conductivities over several orders of magnitude by doping. Organic electronic materials typically have large energy gaps (1–4 eV), and consequently have very low background intrinsic carrier concentrations. Furthermore, as we have seen in previous sections, the room temperature charge mobility is often <1 cm^2/V s, and the narrow energy BWs of organic semiconductors (typically with a maximum at ~ 500 meV) are considerably smaller than for inorganics (with $BW =$ 1–4 eV). This implies an effective mass $m^* > m_0$. Taking all of these properties together, organic semiconductors have exceptionally low conductivities in their native state. Of course, there are examples of ultrapurified and highly ordered molecular materials that have higher mobilities and large BWs. Nevertheless, these represent exceptions rather than the norm for this materials class.

Our discussion up to this point has concentrated on understanding the mobility of both ordered and disordered organics. However, conductivity—the primary property of an electronic material—is a result of both mobility *and* charge concentration. Both quantities can be complicated functions of temperature, electric field, defect and dopant concentrations, and film morphology. In Sections 4.4.1 and 4.4.2, we describe the conductive properties of both pure and doped organic materials, their dependence on film morphology (i.e. crystalline vs. amorphous), and on defects. Closely associated with dark conductivity is the photoconductivity of organic semiconductors. Indeed, photoconductivity is a signature property of all semiconductors, and it governs the performance of photodetectors and solar cells. However, we will leave the discussion of the photoconductive properties to the chapter on photodetectors, Chapter 7. In Section 4.4.3, we discuss commonly used measurement methods of conductivity and mobility. We have already introduced this topic in our discussions of band structure and hopping, where we reported the results of time of flight measurements of mobility in single crystals and polymers. However, time of flight represents only one of several methods of determining these properties. It is important to fully understand the limitations and the range of validity of the principal methods employed to quantify the electronic properties of organic semiconductors, which is the topic of Section 4.4.3.

4.4.1 Ohmic conduction and doping

Ohm's law predicts a linear relationship between current and voltage, that is, $\mathbf{j} = \overleftrightarrow{\sigma}\mathbf{F}$, where the conductivity tensor is $\overleftrightarrow{\sigma} = q(n\overleftrightarrow{\mu}_e + p\overleftrightarrow{\mu}_e)$ (see Eq. 4.35). A measurement of the conductivity yields the *product* of mobility and the background charge concentration, $n\mu_e + p\mu_p$. For an undoped semiconductor in equilibrium, $n = p = n_i$, where n_i is the intrinsic carrier concentration:

$$n_i^2 = N_{HOMO}N_{LUMO}\exp\left[-\frac{E_G}{k_BT}\right]. \tag{4.113}$$

Here, E_G is the energy gap, and N_{HOMO} and N_{LUMO} are the densities of states at the frontier HOMO and LUMO energies, respectively. In many cases, the mobility of one charge carrier is much larger than its counter charge, allowing us to neglect the smaller charge-mobility products. Due to the large energy gaps of most organic electronic semiconductors, the background carrier concentration is much larger than n_i due to either intentional or adventitious doping.

Intentional doping of inorganic semiconductors is the means by which the conductivity of the material is controlled. This is accomplished by introducing atomic species that have either one more valence electrons than the host lattice (for electron donors), or one less (for acceptors). The *substitutional doping* of As into a Si lattice for n-type doping via the excess and loosely bound valence electron (As has a valence of 5 compared to 4 for Si) and B for p-type doping due to its lack of a fourth valence electron is

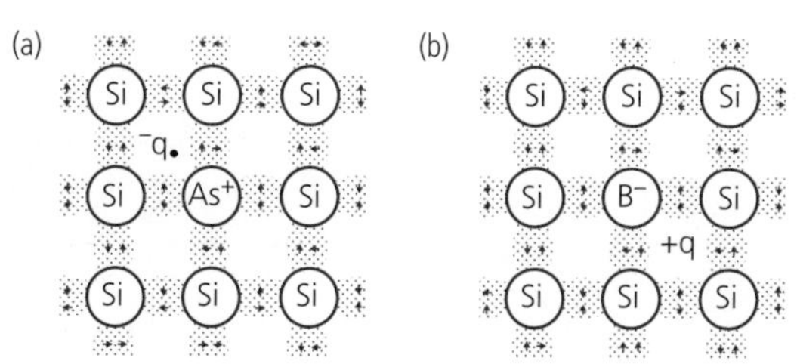

Figure 4.36 Doping by substitutional impurities in a Si lattice: (a) n-type doping via introduction of As atoms, (b) p-type doping via B atoms.

illustrated in Fig. 4.36. Alternatively, dopants can be introduced as interstitial impurities between lattice atoms, although due to the lack of shared bonds with the lattice molecules, the doping efficiency is reduced as there may not be one electron, or hole, released per dopant atom, as there is with substitutional impurities.

The lack of chemical bonds between organic molecules implies that only *interstitial doping* is possible, that is, the dopant atom or molecule is placed between molecules in either crystalline or amorphous solids. Doping to control conductivity in organic semiconductors is, therefore, inefficient, although it can be effective in increasing the conductivity of otherwise insulating solids. While interstitials can introduce additional charge carriers, they can also disrupt the lattice order. This additional static disorder, in turn, reduces the charge mobility, as discussed in Section 4.3. This partially counteracts the effects of increased charge density, and hence may decrease the anticipated gains in conductivity. *Unintentional doping* of organics by oxygen and other contaminants has been observed since the earliest work on organic semiconductors (Eley, 1948, Kearns et al., 1960, Martin et al., 1983), whereas intentional doping has been only recently been effectively used to control the conductivity of organic electronic devices (Chiang et al., 1977, Maitrot et al., 1986, Kido and Matsumoto, 1998, Pfeiffer et al., 1998, Parthasarathy et al., 2001, Lüssem et al., 2013).

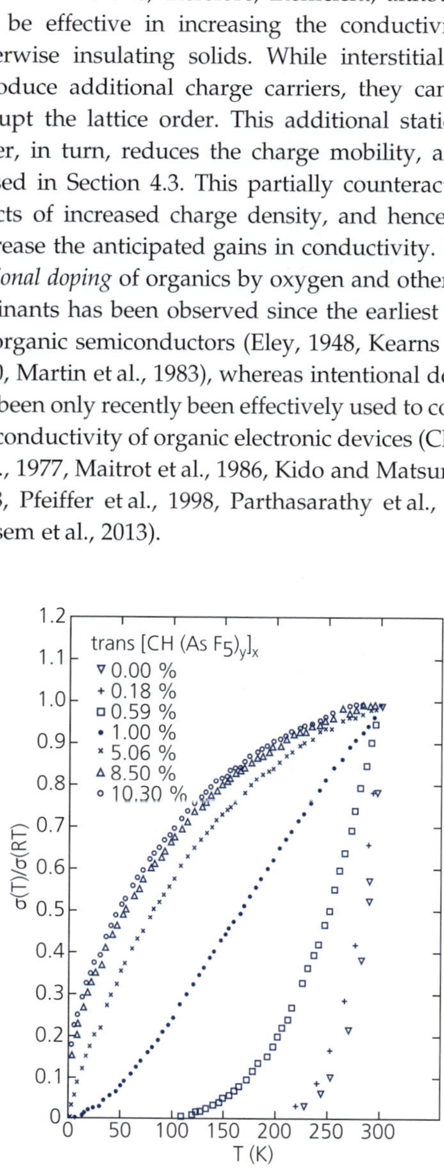

Figure 4.37 Normalized (to room temperature) conductivity of a polyacetylene film vs. temperature doped with different molar ratios of AsF_5 (Chiang et al., 1977).

The first report of doping of an organic semiconductor was the addition of AsF_5 into a 0.1–0.5 mm thick film of *trans*-CH_x (Chiang et al., 1977). The room temperature conductivity of the film changed by >7 orders of magnitude as the *p*-type AsF_5 concentration was varied from 0% (with $\sigma_0 = 10^{-5}$ mS/cm) to 10.3% (with $\sigma = 220$ mS/cm). The dependence of conductivity on temperature for films of several different AsF_5 concentrations is shown in Fig. 4.37. The temperature dependence can be understood in terms of the Fermi–Dirac functions for the occupancy of the dopant levels. That is, for an acceptor energy at E_A, the hole concentration is

$$p(T) = N_A \int g(E) f(E_A, T) dE, \tag{4.114}$$

where the acceptor level occupancy is

$$f(E_A, T) = \frac{1}{1 + \exp((E_A - E_F)/k_B T)}. \tag{4.115}$$

Similar expressions are obtained for electrons and donor levels.

The temperature dependence of the charge concentration is shown in Fig. 4.38. At low temperatures, the exponential function is large, implying that not all acceptor molecules are ionized, that is, $N_A^-/N_A < 1$. This corresponds to the ionization regime where the charge concentration follows a Boltzmann distribution, viz. $p(T) \sim \exp((E_F - E_A)/k_B T)$. Once all the acceptors are ionized, the semiconductor enters the extrinsic region where its conductivity is determined by the free holes. Finally, at high temperature, $n_i = p_i > N_A$, where the intrinsic properties of the semiconductor dominate (corresponding to the *intrinsic regime*). Returning to Fig. 4.37 doped polyacetylene

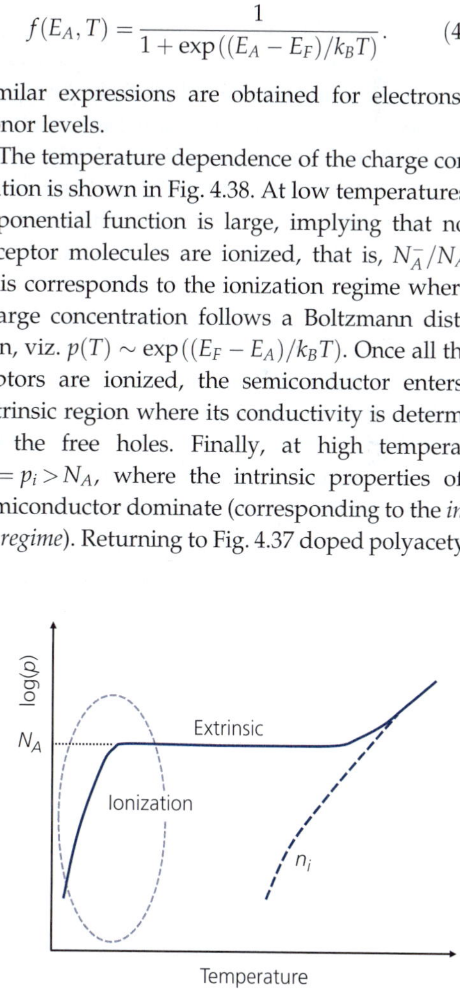

Figure 4.38 Carrier concentration vs. temperature for an p-type semiconductor showing the various charge concentration regimes. Here, N_A = total acceptor concentration, n_i = intrinsic charge density, and the dashed oval corresponds to the data for doped polyacetylene in Fig. 4.37.

moves from the ionization to the intrinsic regime as the dopant concentration is increased.

Atomic dopants can be introduced that either release or capture an electron at an incomplete valence. For example, Li has the valence of $1s^2 2s^1$, and therefore can contribute an excess electron from its 2s shell, thereby serving as a donor in the following reaction: $Li \rightarrow Li^+ + q^-$ (Kido and Matsumoto, 1998, Parthasarathy et al., 2001). Calcium has also been identified as an atomic donor in Alq_3 (Choong et al., 1998) and Cs in Bphen (Reineke et al., 2009). A shortcoming of using metallic dopants, however, is that they are small compared to the molecular host, and therefore they can diffuse away from their initial location, and dope unintended regions of a device. Larger atoms (e.g. Cs vs. Li) have a smaller diffusion constant, and hence are preferred.

Molecular dopants can contribute electrons or holes while being less susceptible to diffusion within the solid due to their larger size. Doping is effectively achieved by choosing molecules with suitable frontier orbital energies relative to the host organic semiconductor, as illustrated in Fig. 4.39. For example, a molecular acceptor molecule should have a very deep LUMO energy relative to vacuum such that it is positioned close to the HOMO of the host (Fig. 4.39a). Subsequently, an electron is transferred from the host HOMO to the acceptor LUMO, leaving behind a semiconductor cation while adding that electron to the acceptor, forming a counter anion. Alternatively, a molecule with a very small electron affinity is easily oxidized via transfer of an electron to the LUMO of the host (Fig. 4.39b). The donor cation is then left with an excess hole, and the host acquires an excess electron.

Several candidate molecular dopants are shown in Fig. 4.40. An example donor, Li quinolate (Liq), releases an electron into the host as an n-type dopant (Liua et al., 2002). Fluorinated tetracyanoquinodimethanes have been identified as effective p-type dopants due to their exceptionally large electron affinities (EA) that increase with the degree of fluorination, making this a highly reductive species.

Figure 4.41 shows the relationship between the conductivity of MeO-TPD with concentration of the p-type dopant, F_4-TCNQ. Interestingly, the conductivity is superlinear with molar dopant ratio, that is, $\sigma \sim p^{1.85}$ up to $MR = 0.24$. Above this concentration, the conductivity saturates at $\sigma \sim 2 \times 10^{-3}$ S/cm; a value that is approximately 10^4 times higher than the neat host semiconductor. Taking a mobility of $\sim 10^{-6}$ cm^2/V s, a conductivity of 10^{-3} S/cm at a $MR \sim 0.1$ suggests that $p \sim 10^{22}$ cm^{-3}. This implies that a maximum of 10% of the dopant molecules provide free holes to the host. This contrasts with inorganic semiconductors where the efficiency of substitutional doping approaches 100%.

The superlinear dependence of conductivity on dopant concentration is due to electric field screening in the low dielectric constant organic semiconductor. That is,

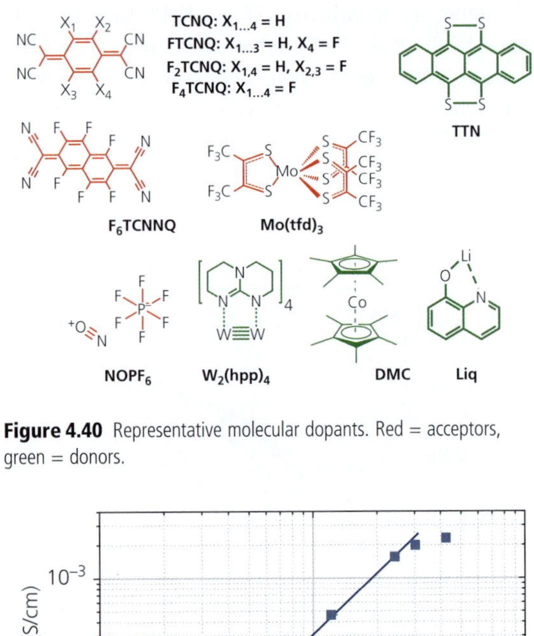

Figure 4.40 Representative molecular dopants. Red = acceptors, green = donors.

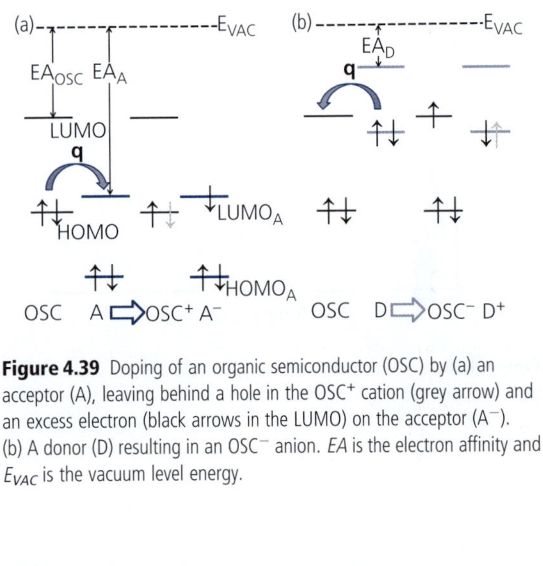

Figure 4.39 Doping of an organic semiconductor (OSC) by (a) an acceptor (A), leaving behind a hole in the OSC^+ cation (grey arrow) and an excess electron (black arrows in the LUMO) on the acceptor (A^-). (b) A donor (D) resulting in an OSC^- anion. EA is the electron affinity and E_{VAC} is the vacuum level energy.

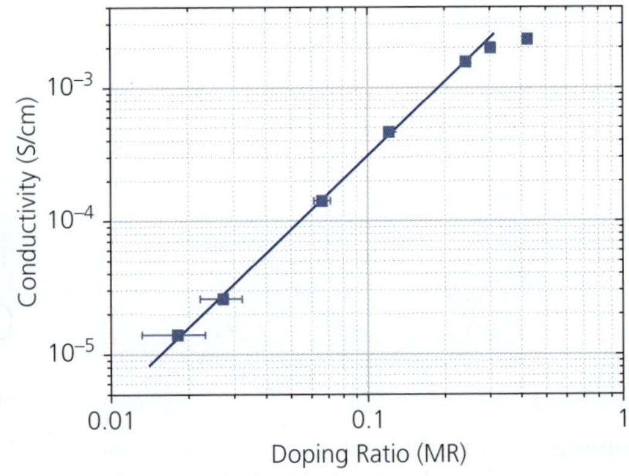

Figure 4.41 Conductivity vs. molar doping ratio (MR) of MeO-TPD with F_4-TCNQ (Olthof et al., 2009).

the film polarizability is increased by electrons bound to the acceptor molecules. The screening field increases with the number of trapped charges given that the free charge density is $p = p_0 + N_A^-$. In the limit where the background carrier concentration, p_0, is much less than the number of charges provided by the dopant, then $p = N_A^- \exp(-\Delta E_a/k_B T)$, where the activation energy, ΔE_a, of the charge is equal to the difference between the electron affinity of the acceptor and the Fermi energy of the host semiconductor. The superlinearity in Fig. 4.41 suggests that ΔE_a is a function of N_A^-, that is, it is not a constant as expected in an inorganic semiconductor. The power law dependence, $\sigma \sim p^{1.85}$ implies that the change in activation energy arises from a decrease in binding energy of the charge to the dopant. The energy thus scales roughly linearly with the inverse distance between dopant molecules, that is, $\Delta E_a = \Delta E_{a0} - \beta (N_A^-)^{1/3}$. Here, ΔE_{a0} is the activation energy of the undoped solid and β is a constant (Gregg et al., 2004).

The effect of dopant polarization is understood using the energy level diagram in Fig. 4.42 (Winkler et al., 2015). We focus our attention on the addition of a cationic molecule (with an excess hole) illustrated in Fig. 4.42a, with identical, symmetrical arguments applying to anions in Fig. 4.42b. When an excess electron is added to a molecule, its HOMO is split into two levels. One is the singly ionized electron affinity, EA^+ which is increased above the HOMO by the reorganization energy, λ_{reorg}. The second is the ionization energy, IE^+, which is increased by the on-site repulsive energy, U (left panel, Fig. 4.42a). The on-site energy is due to Coulomb repulsion of electrons experienced by a doubly ionized molecule, and is calculated using the *Hubbard Hamiltonian*:

$$H = -\sum_{\langle \mathbf{j},\mathbf{i}\rangle,\sigma} t_{\mathbf{ji}} c^\dagger_{\mathbf{j}\sigma} c_{\mathbf{i}\sigma} + U \sum_{\mathbf{j}} \left(\mathcal{N}_{\mathbf{j}\uparrow} - \tfrac{1}{2}\right)\left(\mathcal{N}_{\mathbf{j}\downarrow} - \tfrac{1}{2}\right) - \sum_{\mathbf{j}} \mu_{\mathbf{j}} (\mathcal{N}_{\mathbf{j}\uparrow} + \mathcal{N}_{\mathbf{j}\downarrow}). \quad (4.116)$$

The first term is the kinetic energy resulting in transfer of an electron from site $\mathbf{i}$,σ to site $\mathbf{j}$,σ, and $t_{\mathbf{ji}}$ is the overlap of wavefunctions on these two sites. Here,

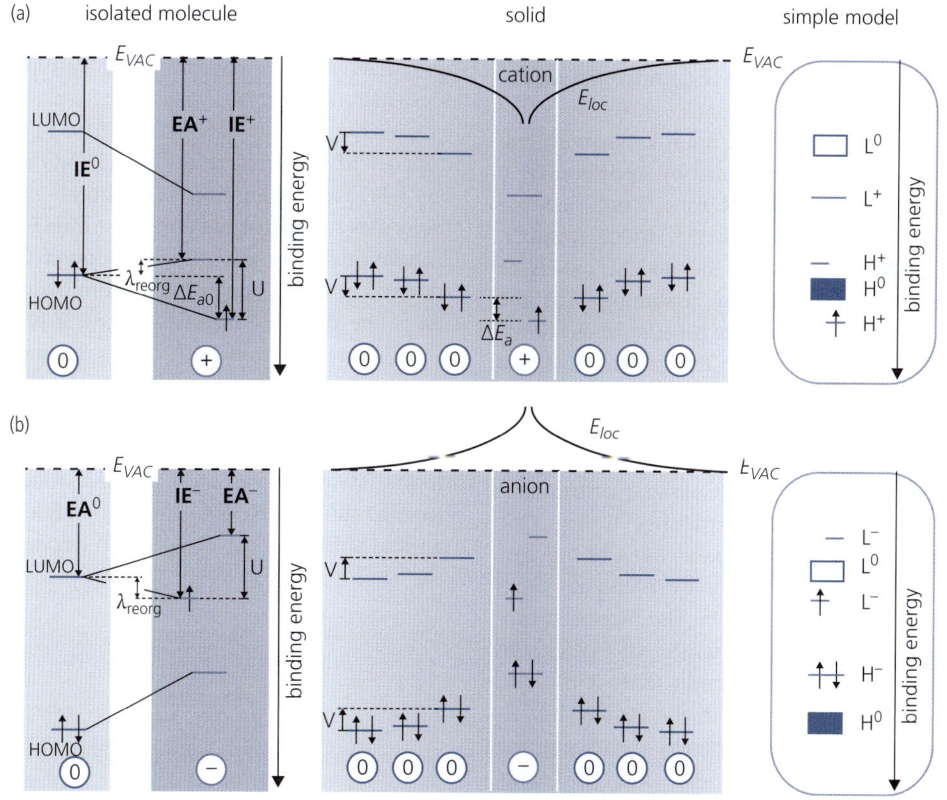

Figure 4.42 Effect on the energy levels of an isolated molecule and a solid in the presence of a (a) cationic dopant and a (b) anionic dopant. Here, E_{VAC} is the vacuum level, E_{loc} is the distorted local potential in the presence of a charged molecule, ΔE_a is the conductivity activation energy, ΔE_{a0} is the activation energy without taking into account polarization of the surrounding medium, λ_{reorg} is the reorganization energy, V is the total energy shift due to the ionized impurity, and U is the on-site repulsive energy for a doubly ionized dopant. EA and IE are the electron affinity and ionization energy, respectively. The diagram at the far right is a simplified representation of the solid where L = LUMO and H = HOMO. The superscripts signify "0" = neutral, "−" = anion, and "+" = cation molecules (Winkler et al., 2015).

$c_{j\sigma}^{\dagger}(c_{j\sigma})$ is the creation (annihilation) operator at lattice site **j** with spin σ, and the sum is over all sites, **j**,**i**,σ, where the brackets $\langle \mathbf{j}, \mathbf{i} \rangle$ in the summation indicates that only nearest neighbor hops are included due to the short range interaction between molecules. As previously, $\mathcal{N}_{j\uparrow}$ is the number operator for site **j**, and in this case for a spin-up electron (subscript ↑). The second term is the on-site repulsive energy for two electrons with strength U, known as the *Hubbard U*. Following the Pauli Exclusion Principle, the electrons have opposite spins. The final term is the chemical potential (μ_j) of site **j**, which determines the state occupancy.

The unperturbed activation energy is obtained from the isolated molecule. From Fig. 4.42a, this is approximately $\Delta E_{a0} = U - \lambda_{reorg}$. PES measurements indicate that $U \approx 1.4$ eV for the acceptor C_{60}, and as in the previous section, $\lambda_{reorg} \sim 0.5$–1 eV for most molecules. When placed in a solid, the long range Coulomb potential of the ionized impurity polarizes the local molecules, resulting in a total HOMO shift by $V \approx 0.5$ eV (Winkler et al., 2015). Then, the activation energy decreases to $\Delta E_a = \Delta E_{a0} - V$. The amount of energy needed to remove a charge from the dopant is thus reduced to that of the first (largest) step required to overcome the local potential (E_{loc}) surrounding the cation.

For an electron polaron (i.e. an anionic state), the energy levels are shifted upward by approximately the same energy as for hole polarons. Thus, the same arguments leading to estimates of ΔE_a and ΔE_{a0} apply. The microscopic diagrams in the middle panels of Fig. 4.42 can be simplified by assuming that the HOMO and LUMO energies are broadened as shown on the right hand panels, with the ionized hole and electron dopant levels residing either above or below their corresponding frontier orbitals.

We can put this discussion in the context of the GDM by assigning a second, Gaussian distribution to the polarization induced splittings in Fig. 4.42. That is, the introduction of dopants releases free carriers that increase the mobility by filling the high energy tail of the Gaussian DOS. The increased mobility, along with the higher density of free charge both conspire to increase the conductivity. However, the ionized dopants introduce an electrostatic retardation of the mobile charges, thereby decreasing the mobility via energetic disorder. Also, increased static disorder due to lattice disruption by the high density of dopants ultimately limits the maximum mobility to less than if disorder were absent. The Coulomb traps formed by the dopants can be visualized by the one-dimensional energy level diagram in Fig. 4.43 for an organic with ionized dopants separated by distance, $d = N_d^{-1/3}$, where N_d is the dopant density (i.e. either the donor or acceptor concentration). The electrostatic attraction of the dopant creates a potential minimum of depth Δ from the maximum at a distance $r_m - r_0$ from the dopant itself. Then it follows that

$$U(r) = -qFr + \frac{q^2}{4\pi\varepsilon_0\varepsilon_r r} + \frac{q^2}{4\pi\varepsilon_0\varepsilon_r |r-d|}, \qquad (4.117)$$

where ε_r is the relative dielectric constant of the material. The energy barrier is at the maximum of U(r), yielding

$$\Delta = \frac{q^2}{4\pi\varepsilon_0\varepsilon_r a} + U(r_m), \qquad (4.118)$$

where a is the distance between the dopant and the nearest hopping site. Then, assuming a Gaussian distribution for both hopping and dopant sites, the total DOS is (Arkhipov et al., 2005a, 2005b)

$$g(E) = \frac{N - N_A^-}{N} g(E) + \frac{N_A^-}{N} g\left(E + \frac{q^2}{4\pi\varepsilon_0\varepsilon_r a} + U(r_m)\right). \qquad (4.119)$$

As previously, N is the density of hopping sites, and for convenience we have assumed a p-type dopant is used such that $p \approx N_A^-$.

A calculation of the HOMO DOS for a P3HT film p-doped with different ratios of F_4TCNQ is shown in Fig. 4.44. The doping ratio is the number of F_4TCNQ molecules per thiophene unit in the P3HT chains. The calculations assume that the undoped Gaussian is centered at E_{HOMO} determined from the surface Kelvin probe potential (see Chapter 8). It is found that E_{HOMO} increases from −4.83 eV for the undoped sample to −4.94 eV for the 1:200 sample. Furthermore, $N = 7 \times 10^{20}$ cm, N_d is taken from the doping ratio, $\sigma_{HOMO} = 78$ meV, and $a = 0.57$ nm. Importantly, the doping states have a DOS that is a function of the doping ratio. The HOMO distribution is also a Gaussian that shifts toward E_{HOMO} with doping ratio. The

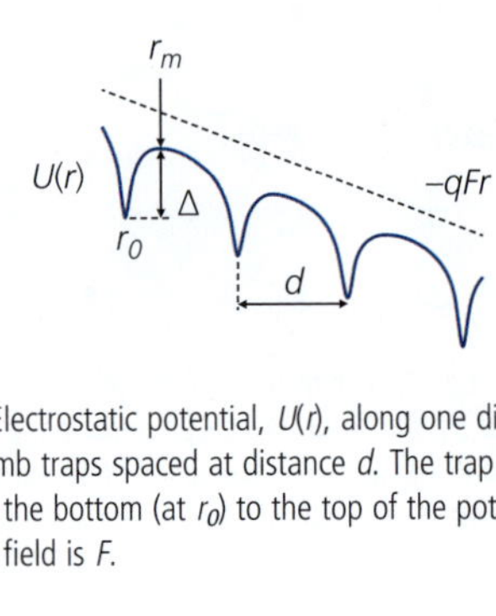

Figure 4.43 Electrostatic potential, $U(r)$, along one dimension for a series of Coulomb traps spaced at distance d. The trap depth is Δ measured from the bottom (at r_0) to the top of the potential (at r_m). The applied electric field is F.

small DOS for the undoped sample is attributed to Al diffusion from the back contact. The width of this distribution is the Hubbard U, ultimately determined by the splitting of the HOMO level due to polarization described in Fig. 4.42a.

Direct evidence for charge transfer from the dopant to the host is obtained using absorption spectroscopy, shown for the F_4TCNQ:P3HT films in Fig. 4.45. The singly ionized (anionic) F_4TCNQ features at 1.43 eV and 1.62 eV are identified from optical measurements on the neat species themselves. The rectangular background between 2 eV and 1 eV lying below the neutral P3HT energy gap at 2 eV is attributed to P3HT hole polarons, as is the rising absorption at <1 eV. The free polaron concentration is then extracted from the correspondence between OD and doping ratio, x.

The measured conductivity vs. x is provided in Fig. 4.46, with a fit to the data (solid line) employing the mobility model extracted using the DOS in Eq. 4.119 (Arkhipov et al., 2005b) and the hole concentration from Fig. 4.45. The slope of the conductivity is sublinear at low doping ratios, becoming linear at moderate ratios, and then superlinear as in Fig. 4.41 at high doping ratios. At low ratios, the dopants compensate the native traps in the P3HT. As doping is increased, the Gaussian tail states are filled, resulting in an increased number of free holes, leading to a rapid rise in hole concentration. Note the decrease in calculated mobility at low to moderate doping concentrations, as shown in the inset of Fig. 4.46. The drop in mobility that leads to the sublinear dependence of the conductivity with x is due to Coulomb traps from the cationic molecular dopants, discussed above. At the highest doping levels, both the mobility and the charge concentration increase, leading to the superlinear dependence of σ on x.

The low doping regime and the facility for dopant charge to compensate native traps induced by structural disorder was studied in the n-type semiconductor system comprising C_{60} doped at very low levels using the donor, $[RuCp^*(mes)]^+$ (Olthof et al., 2012).

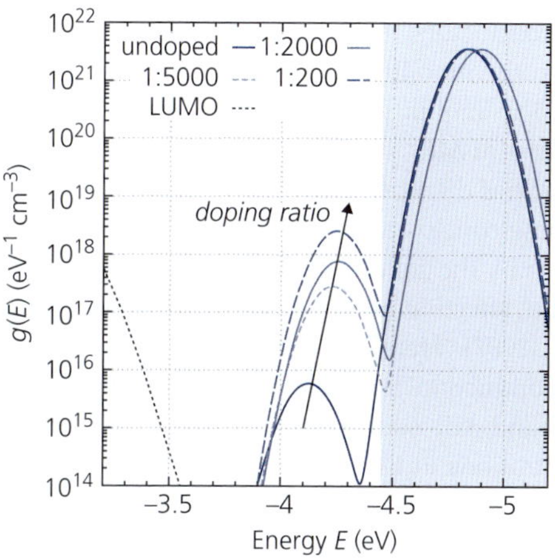

Figure 4.44 Density of HOMO states as a function of F_4TCNQ in P3HT, where the doping ratio (1:x) is equal to the number of dopant molecules per x thiophene units. The LUMO is shown by the dashed lie. The shaded region encompasses the states in the undoped P3HT (Pingel and Neher, 2013).

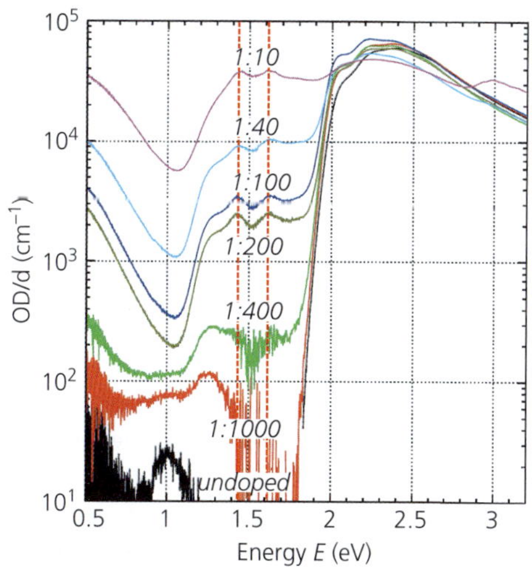

Figure 4.45 Optical density normalized to the film thickness of F_4TCNQ: P3HT (1: x) at various doping ratios. The vertical red dashed lines indicate the positions of singly negatively ionized F_4TCNQ molecules (Pingel and Neher, 2013).

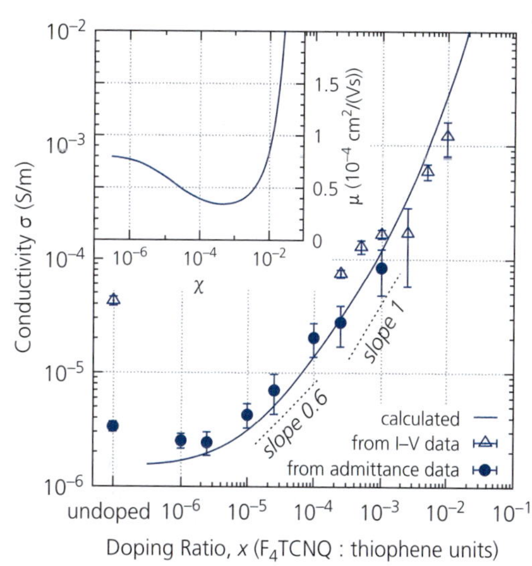

Figure 4.46 Conductivity vs. doping ratio. The solid line fit is to Arkhipov's Gaussian disorder model. Inset: Plot of mobility using the disorder model at low doping ratios (Pingel and Neher, 2013).

The samples were grown in ultrahigh vacuum, and the molar ratio of the dopant was controlled by its relative deposition rate with C_{60}. The free charge concentration (n) was obtained from UPS and IPES measurements of the Fermi energies and frontier orbitals. From this, the mobility was deduced from conductivity measurements, viz. $\mu_n = \sigma/qn$.

The electron mobility vs. dopant *MR* is shown for three different temperatures in Fig. 4.47a. There is a striking change from a region of high slope at $MR < 2 \times 10^{-3}$, to a much lower slope at higher *MR*. This inflection appears to be relatively temperature independent. The transition also corresponds to an abrupt transition from a large dependence of the work function and HOMO energy on *MR* to a much slower dependence as *MR* is increased beyond this apparent threshold value. The explanation for this transition is the compensation of native traps in the disordered C_{60} film at low *MR*. The Fermi energy moves toward the conduction level with doping until those traps are filled. This has the combined effects of neutralizing charged traps while increasing the mean energy of conducting charges. Once compensated, additional dopants simply contribute more electrons. These fill the tail states in the Gaussian DOS, thereby resulting in hopping from higher energy, and hence higher mobility states. This is also supported by the slower approach of the Fermi energy toward the LUMO at the rate of ~60 meV/decade in *MR*, consistent with the theoretical behavior of conventional inorganic semiconductors. However, at these higher doping levels, the mobility no longer shows a significant increase in mobility with doping; indeed it begins to decrease at $MR > 0.05$ due to increased disorder from impurities disrupting the C_{60} lattice.

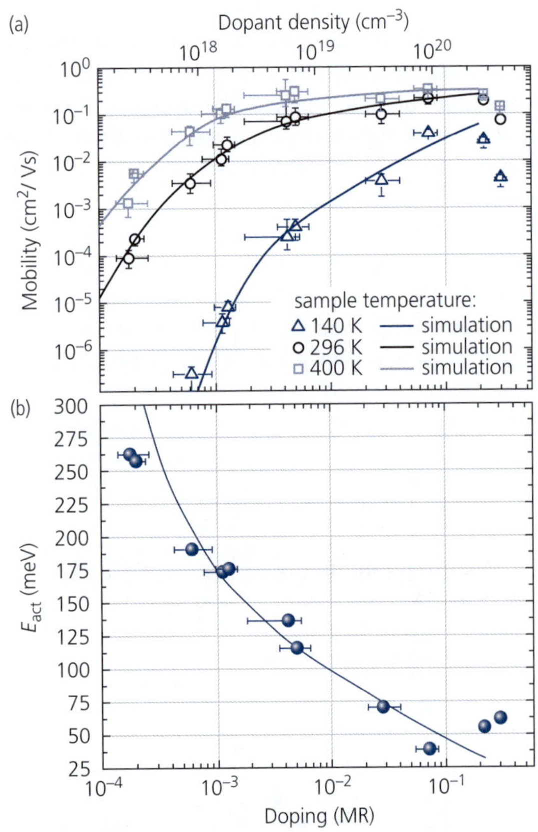

Figure 4.47 (a) Electron mobility in C_{60} as a function of *n*-type doping concentration (*MR* = molar ratio) by the cation $[RuCp^*(mes)]^+$. Data points are measurements, lines are fits based on kinetic Monte Carlo modelling using the GDM effective medium approximation. The data are taken from conductivity vs. temperature along with photoelectron spectroscopy measurements of the Fermi energy to obtain *n*.
(b) Conductivity activation energy vs. *MR*. The line is a fit to the theory in (a) (Olthof et al., 2012).

These conclusions are supported by fits to the data shown by the solid lines in the figure using the GDM in the effective medium approximation (Fishchuk et al., 2007). Here, a Gaussian DOS is assumed for the conduction level states, and an exponential for the band tail state. The Gaussian and exponential distributions fit the data assuming widths of 64 meV and 128 meV, respectively, and the ratio of the total number of exponential to Gaussian states is 0.007 (Olthof et al., 2012). Note that the low temperature mobility does not saturate, even at high *MR*. This results since traps near to the conduction level are not thermalized as temperature is reduced.

A plot of the conductivity activation energy, ΔE_a, is shown in Fig. 4.47b. The line follows $\sigma = \sigma_0 \exp(-\Delta E_a/k_BT)$, where σ_0 is a constant. Since the activation energy is a function of the donor energy relative to E_F, it moves rapidly toward the LUMO energy at low doping, until it saturates at approximately 39 meV at $MR \sim 0.07$. The fit to the EMA model used in Fig. 4.47a is shown by the line in Fig. 4.47b.

In the very high doping regime, the host material can no longer be simply considered to be doped, but rather it is a blend of two organics (host and dopant). Thus, conductivity can occur on either species. In some cases the introduction of a dopant can form an intermediate neutral CT complex rather than releasing a free polaron into the host. Often, an ion pair is never formed, and doping does not result in free charge. In cases when free charge is released, the CT state levels are split above and below the transport energy level. For p-type doping for example, the *EA* of the CT state lies above, and its IP lies below that of the host (Salzmann et al., 2016).

Finally, note that thermally activated diffusion of small dopant molecules can lead to their migration from their intended location (e.g. in a highly doped charge transport layer or in the emission zone of an OLED) over time; a process whose rate is increased at high temperature. While atomic dopants such as Li and Cs are particularly prone to rapid diffusion along the interstices between molecules, even small molecular dopant species can rapidly migrate at even modest temperatures. Similarly, the spatial distribution of binary mixtures of molecules can change over time, and junctions between layers of different compositions can also intermix, changing the electronic properties of the original structure.

Simple, single species diffusion as a function of time, t, in three dimensions is described by *Fick's second law*:

$$\frac{\partial c(r,t)}{\partial t} = D_c \nabla^2 c(r,t), \tag{4.120}$$

where we have assumed that the diffusion constant, D_c, is independent of concentration, $c(r, t)$ of the diffusing species. Furthermore, this expression assumes that D_c is a scalar, independent of direction within the sample. This is not the case for molecular crystals where there is usually substantial anisotropy along the various crystal axes.

Lateral diffusion of F_4TCNQ doped into P3HT has been followed as a function of both time and temperature using fluorescence microscopy to determine the diffusion constant and the mechanisms leading to dopant migration. The sample preparation steps are illustrated in Fig. 4.48a. The 50 nm thick P3HT host film is spun onto the surface of a glass substrate, followed by deposition of a 2.2 nm thick layer of the dopant through a fine Cu shadow mask with square holes, 38 μm on a side on a 66 μm pitch. The dopant is prevented from escaping from its location by depositing an encapsulating layer of 50 nm MoO_3/100 nm Ag. The MoO_3 prevents the Ag layer from diffusing into the sample. Then the entire unit is encapsulated using a glass lid epoxied around its periphery to form a moisture and oxygen-proof seal. The thin F_4TCNQ provides a finite amount of dopant, thus establishing well-defined boundary conditions that simplify diffusion analysis.

Since the dopant quenches the luminescence of the P3HT film, the spatial distribution of F_4TCNQ is monitored over time using fluorescence microscopy. The sample was excited at 488 nm wavelength and imaged through a microscope objective focused onto the P3HT film via the glass substrate. The fluorescence intensity around the periphery of the original F_4TCNQ deposit provides a relative measurement of the amount of dopant that is within that region, providing a spatial-temporal map of the diffusion front. The image on the left shows the initial conditions (at $t = 0$) with the dopant confined to the regions where it was deposited. After 50 min at 348 K, the dopant molecules have diffused far from their point of origin, almost contacting in the 26 μm wide regions that were masked during deposition.

The diffusion profile is found to deviate from predictions made using Eq. 4.120. To provide an improved fit, it was assumed that both the neutral F_4TCNQ and the anionic F_4TCNQ^- acceptor molecule compete in the diffusion process. Since only one molecule can occupy a site at a given time, the neutral F_4TCNQ has to "jump" over the anions, thus resulting in an increased diffusion constant of the neutral compared to the ionic species. At room temperature, fits to the diffusion profiles yield the anion diffusion constant for F_4TCNQ^- of $D_{F_4TCNQ-} = 1.5 \times 10^{-11}$ cm^2/s, and for the neutral species, $D_{F4TCNQ} = 3 \times 10^{-10}$ cm^2/s. Measurement of the diffusion constants at several different temperatures yields activation energies via

$$D_c = D_{c0} \exp(-\Delta E_{ac}/k_B T). \tag{4.121}$$

Measurements of F_4TCNQ in P3HT yield $\Delta E_{ac} = 74.2$ kJ/mol and 55.9 kJ/mol for the neutral and anionic

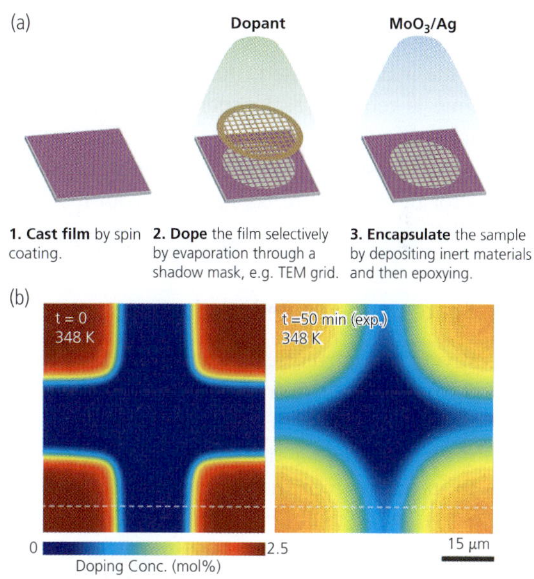

Figure 4.48 (a) Fabrication sequence for creating grid patterns useful in visualizing diffusion in an organic film by a dopant. (b) Relative F_4TCNQ density distribution in P3HT before (left image) and after (right) 50 min diffusion at 348 K. The distribution maps are obtained from analysis of P3HT fluorescence. The color key and scale bars are provided at the bottom of the images (Li et al., 2017).

Adapted with permission from Li, J., Koshnick, C., Diallo, S. O., Ackling, S., Huang, D. M., Jacobs, I. E., Harrelson, T. F., Hong, K., Zhang, G., Beckett, J., Mascal, M. & Moulé, A. J. 2017. Quantitative Measurements of the Temperature-Dependent Microscopic and Macroscopic Dynamics of a Molecular Dopant in a Conjugated Polymer. Macromolecules, 50, 5476-5489. Copyright 2017 American Chemical Society.

F_4TCNQ species, respectively. We note that the activation energies and diffusion constants are not separately measured, but are taken from fits to the diffusion front using a two species, two dimensional diffusion analysis of the concentration profiles (i.e. the fluorescence intensity maps) as functions of time and temperature.

The diffusion constants and activation energies found for F_4TCNQ are similar to those of $PC_{61}BM$ in P3HT which can be fit using only a single species, one-dimensional model as Eq. 4.120 (Treat et al., 2013). The diffusion constants in both cases are sufficiently large that dopants can lead to significant, temperature dependent changes in morphology, and ultimately the optical and electronic properties of materials over time.

We will show in Chapters 6 and 7 that molecular doping can lead to very small internal resistances in OLEDs and OPVs. However, from the foregoing discussion it should be clear that interstitial doping of organics leads to complex and as yet not fully understood dependencies of the number of free charges (and hence conductivity) on such factors as temperature, dopant and host molecular species, dopant concentration, and film morphology. We will find that the stability of organic electronic devices is thermally activated, that is, their operational lifetime is reduced as temperature is increased. Diffusion of species across interfaces, and within the layers themselves is one of the primary mechanisms for device failure, as implied from the foregoing discussion.

4.4.2 Space charge limited conduction

At relatively low currents, the doping and the background free charge concentration in organics allow for Ohmic conduction. However, as the current density is increased, a deficiency in bulk charge leads to injection from the contacts, leading to the *space charge limited* (SCL) *transport* regime. Two situations are considered in this section: SCL current in the absence or presence of traps. While space charge conduction can be supported by injection of a single carrier type (i.e. an electron or hole), double carrier injection is also possible. For simplicity we consider only single carrier injection. The analysis of the current density *vs.* voltage (j–V) characteristics of a semiconductor in the presence of bipolar conduction is quite complex since it requires consideration of charge recombination in the material bulk (Lampert, 1962, Kao and Hwang, 1981). Interested readers can find references that treat this topic in *Further reading* at the end of the chapter.

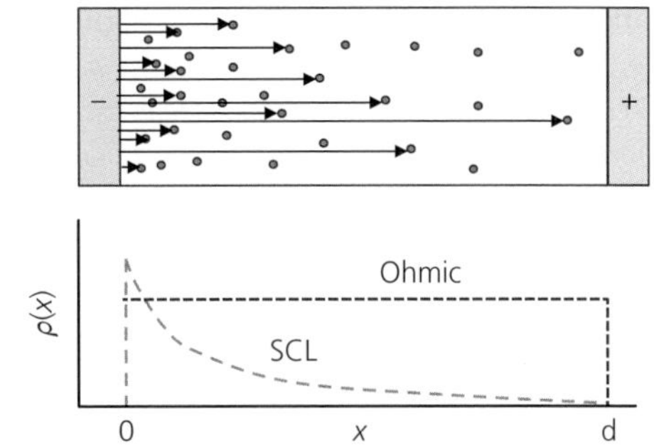

Figure 4.49 (a) Distribution of charge and field lines for a semiconductor dominated by space charge injection from a contact. (b) Charge distribution for Ohmic and SCL currents.

4.4.2.1 Trap-free SCL conduction

The origins of SCL current can be understood from the illustration in Fig. 4.49. The applied voltage is sufficiently high to inject electrons from the cathode. Since field lines terminate on individual charges, the electric field near the cathode is higher than elsewhere in the bulk. In Fig. 4.49b, we illustrate the differences in the field distribution for Ohmic and SCL currents in a sample of thickness, d. Ohmic conduction is characterized by a uniform field, whereas SCL conduction results in a high field near the injecting contact, falling rapidly into the bulk.

In the following derivation, we make several simplifying assumptions, each of which can be removed to consider more general circumstances. (i) Only a single carrier type (holes) is injected, (ii) the hole mobility (μ_p) is independent of field, $\mathbf{F}$, (iii) the free carrier distribution is governed by Boltzmann statistics, (iv) $\mathbf{F}$ is sufficiently small that there is no field emission from the contacts, (v) carrier diffusion from contacts is small and therefore ignored, and (vi) only one-dimensional transport along the $\hat{\mathbf{x}}$ -direction is considered. Then from Gauss' law:

$$\frac{dF(x)}{dx} = \frac{qp(x)}{\varepsilon_r\varepsilon_0} = \frac{q(p_{inj}(x)+p_0)}{\varepsilon_r\varepsilon_0} \approx \frac{qp_{inj}(x)}{\varepsilon_r\varepsilon_0}, \tag{4.122}$$

where the total free hole polaron density, p, consists of the sum of injected hole density, p_{inj}, and the background charge density, p_0. In the SCL regime, $p_{inj} \gg p_0$. Now, the current is given by

$$j(x) = q\mu_p p_{inj}(x)F(x). \tag{4.123}$$

Using Eq. 4.123, we can rewrite Eq. 4.122 as follows:

$$\frac{dF^2(x)}{dx} = 2F(x)\frac{dF(x)}{dx} = \frac{2qp_{inj}(x)F(x)}{\varepsilon_0\varepsilon_r} = \frac{2j(x)}{\varepsilon_0\varepsilon_r\mu_p}. \tag{4.124}$$

Then,

$$F^2(x) = \frac{2j_{SCL}x}{\varepsilon_0\varepsilon_r\mu_p}. \tag{4.125}$$

Current continuity requires that $j(x) = j_{SCL}$, a constant that is independent of position. In this regime, the field is no longer constant across the sample, but according to Eq. 4.125 increases quadratically. Thus $F^2(x) \sim x$ until it reaches a peak value at $x = d$. Also, Eqs. 4.123 and 4.125 give:

$$p_{inj}(x) = \left(\frac{\varepsilon_r\varepsilon_0 j_{SCL}}{2q^2\mu_p}\right)^{1/2}\frac{1}{\sqrt{x}}. \tag{4.126}$$

The charge density is, therefore, also quadratic in current (vs. linear for Ohmic conduction) and is inversely proportional to $\sqrt{x}$. These various relationships are illustrated in Fig. 4.50.

Now, the potential is simply $F(x) = -\frac{dV}{dx}$, and integrating between $x = 0$ where V is the applied voltage, and $x = d$ where $V = 0$, we obtain for SCL current in the absence of traps:

$$j_{SCL} = \frac{9}{8}\mu_p\varepsilon_0\varepsilon_r\frac{V^2}{d^3}. \tag{4.127}$$

This is known as the *Mott–Gurney relationship*, which is analogous to *Child's law* used to describe space charge currents in vacuum tubes. Equation 4.127 is independent of the injected carrier density, p_{inj}, with the mobility and dielectric constant the only materials-dependent properties. It differs from Ohmic conduction due to its quadratic (versus linear) dependence on voltage and inverse cubic (again versus linear) dependence on sample thickness. Hence, the charge mobility in the space charge region can be extracted directly from the slope of j vs. V^2, only requiring knowledge of the film thickness and dielectric constant (the latter can be obtained from the sample capacitance).

Thus, at low voltages, the current is carried by the background charge with density, p_0, and at higher voltages by injected space charge. The point at which these two current densities are equal is the cross-over current, $j_X = j_\Omega = j_{SCL}$. From Eqs. 4.34 and 4.127 we obtain

$$V_X = \frac{8}{9}\frac{qp_0}{\varepsilon_0\varepsilon_r}d^2, \tag{4.128}$$

where V_X is the cross-over voltage. These relationships are shown in Fig. 4.51 for a 200 nm thick film of PTCDA sandwiched between a Ti Ohmic contact and a rectifying Au contact (Forrest et al., 1984a). The asymmetry in current depends on whether the Ti contact is positive (corresponding to forward bias) or negative (reverse bias) relative to the Au. The Au/PTCDA contact has an energy barrier of (0.61 ± 0.05) eV as determined from the temperature dependence of the current in Au/PTCDA/Au devices. Accurate analysis of the *j-V* characteristics requires that the built-in voltage (V_{bi}) of the contacts be subtracted

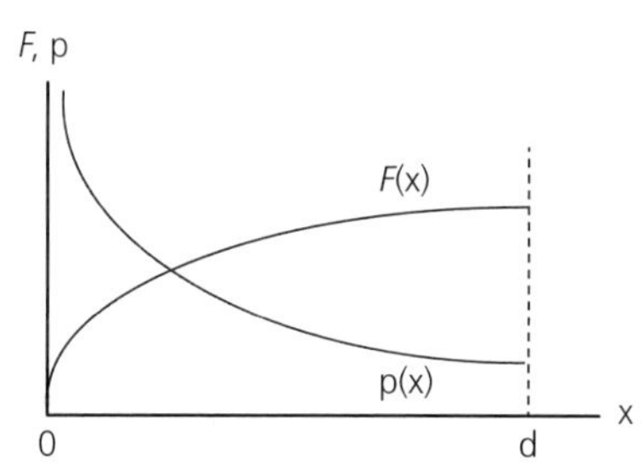

Figure 4.50 Electric field and charge density vs. distance in the SCL current regime.

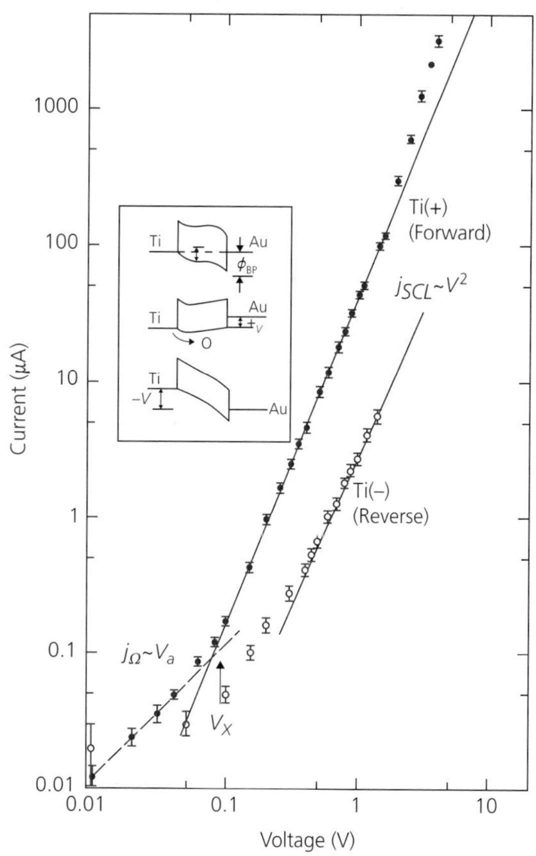

Figure 4.51 Current regimes of a 200 nm thick PTCDA film sandwiched between Ti and Au contacts. The device area is $A = 2.5 \times 10^{-4}$ cm^2. Films with the Ti electrode positive relative to Au are forward biased, indicating a higher built-in voltage at the Au contact. V_X is the cross-over voltage, corresponding to a cross over current of $j_XA = 0.1$ μA. The dashed line is a fit to Ohm's law, and the solid lines are fits to the Mott-Gurney relationship. Inset: Energy level diagrams at different biases (Forrest et al., 1984a).

from the applied voltage. In Eq. 4.127 we should, therefore, replace V with $V - V_{bi}$.

The cross-over from the Ohmic to the SCL regime is at $V_X = 0.1$ V. Fits to the data, shown by the lines in Fig. 4.51, yield a mobility of $\mu_p = 3.5 \times 10^{-3}$ cm^2/V s and a free hole concentration of $p_0 = 5 \times 10^{14}$ cm^{-3}. The fits extend over four decades, suggesting that the assumption of little or no electric field dependence of the mobility is justified up to fields as high as ~4 V/200 nm = 2×10^5 V/cm.

SCL conduction has been observed in semiconducting polymers as well as small molecule organics (Blom et al., 1997b, Tanase et al., 2004, Reid et al., 2008, Nicolai et al., 2010). In Fig. 4.52 we show the j–V characteristics of PFO films of three different thicknesses sandwiched between Ohmic electrodes. The data are consistent with the Mott-Gurney relation, following $(V - V_{bi})^2/d^3$ over the entire range of voltages and currents measured (Nicolai et al., 2010). The hole mobility is relatively constant at $\mu_p = 1.3 \times 10^{-5}$ cm^2/V s, a value that is characteristic of many amorphous semiconducting polymers.

As discussed in Section 4.3.2, static disorder leads to a deviation from a constant mobility with electric field. Indeed, the mobility is generally found to depend not only on F, but also on charge density due to trap filling. It is difficult to separate the contributions of these two variables, since an increase in field also leads to increased charge injection. The increase in current at high voltage in Fig. 4.51 is an indication of changes in mobility at high field. These mobility-dependent effects have been studied for the hole conducting polymer NRS-PPV, sandwiched between ITO and Au contacts (Pasveer et al., 2005). The dotted line in Fig. 4.53 corresponds to the simple V^2 dependence of the Mott-Gurney relationship. The dashed lines are fits to Eq. 4.127 assuming that the mobility depends on hole concentration, and hole transport is via percolation in a Gaussian DOS at the HOMO energy edge. Percolation occurs by charge hopping from conducting polymer cluster to cluster. At a critical density of conducting clusters (which in this case consists of neighboring empty sites into which holes can hop), all clusters are connected and the charge can move from one end of a sample to the other. Within the GDM, the mobility can be approximated by $\mu_p(T, p) \sim \exp(p^\delta)$ where δ is a parameter depending on the width of the site distribution, σ_{GDM} (cf. Table 4.2). The fits are reasonable at the highest temperatures, but the theoretical dependence of j vs. V deviates from a simple charge density dependence at lower temperatures. This difference is attributed to the field dependence of the mobility discussed in Section 4.3.2. The solid lines correct for both the field and charge dependence of the mobility (Pasveer et al., 2005).

We caution that while the fits are accurate over a limited range of temperature and voltage, they employ numerous assumptions and fitting parameters. Hence, it is unclear how generally this fitting procedure can be applied to other materials, and what interdependence one variable has on the

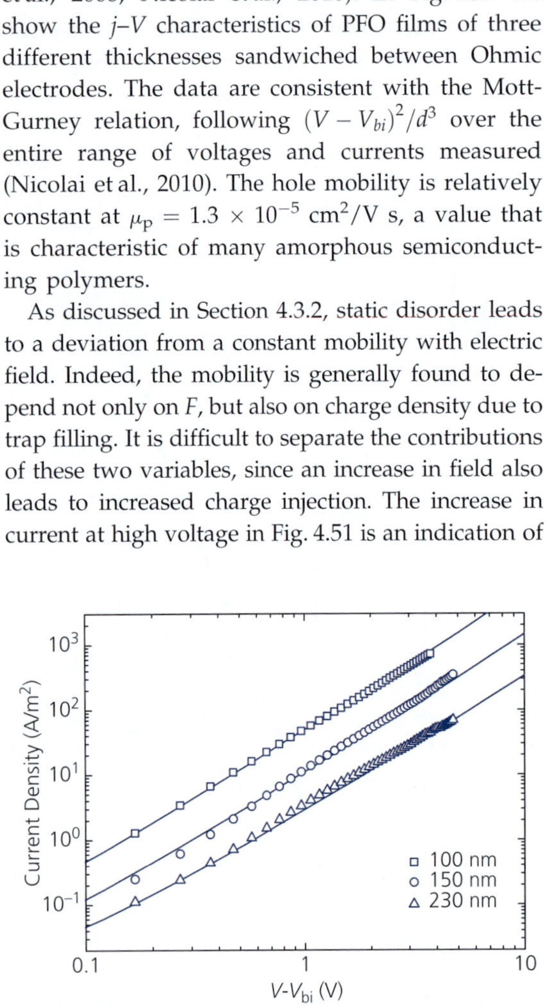

Figure 4.52 Space charge limited current for three different thicknesses of PFO sandwiched between Ohmic electrodes of PEDOT:PSS and MoO_3. Lines are fits to the Mott–Gurney relation, Eq. 4.127, after accounting for the contact built-in voltage, V_{bi} (Nicolai et al., 2010).

Reprinted from Nicolai, H. T., Wetzelaer, G. A. H., Kuik, M., Kronemeijer, A. J., deBoer, B. & Blom, P. W. M. 2010. Space-charge-limited hole current in poly (9,9-dioctylfluorene) diodes. Appl. Phys. Lett., 96, 172107, with the permission of AIP Publishing.

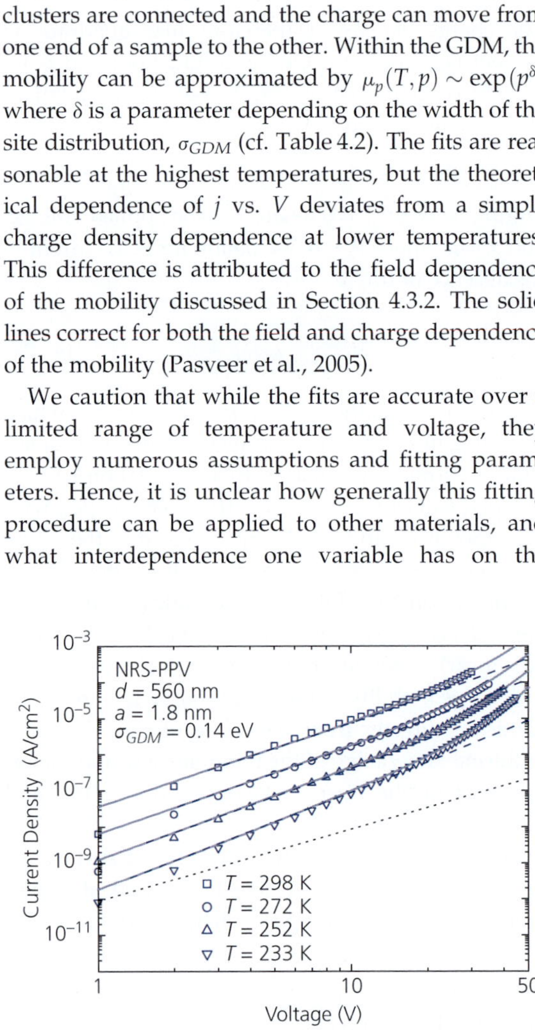

Figure 4.53 Current vs. voltage characteristics for the hole conducting polymer, NRS-PPV, at various temperatures. The solid lines represent fits to the data assuming SCL conduction where mobility is a function of hole concentration and electric field. The dashed lines consider only the contribution of the excess hole concentration. The dotted line is for the constant mobility expression of Eq. 4.127. Here, d is the film thickness, a is the distance between molecules, and σ_{GDM} is the Gaussian distribution half width (Pasveer et al., 2005).

others. This ultimately limits the depth of understanding such a parametric model can provide in revealing the origins of the transport processes in organic semiconductors.

4.4.2.2 SCL conduction in the presence of traps

Disorder in organics leads to deep levels or trap states that exist in the energy gap between the HOMO and LUMO. Hence, it is rare indeed when we can apply the Mott–Gurney relation to these materials systems without considering the effects of traps. Inclusion of traps in our analysis, however, is accompanied by assumptions whose validities are often difficult to verify. We must always bear in mind that the measured current is actually the sum over all charge generating mechanisms, making it challenging or impossible to separate out the individual contributions. The strongest leverage for isolating these mechanisms is to determine the dependence of j on voltage (and hence electric field), and temperature. Even then, ambiguity can arise as we have seen in the analysis of data in Fig. 4.53, necessitating further experiments to access the dependence of current on illumination and frequency to help sort out the origins of the background current. In analyzing the data, therefore, it is useful to use the rule of Occam's razor whereby *"Among competing hypotheses, the one with the fewest assumptions should be selected."*

In this section, we expand our understanding of space charge conduction to include the more general case of conduction in an organic semiconductor in the presence of traps, which may be of either intrinsic (i.e. dynamic and static disorder) or extrinsic (e.g. impurities) origin.

We first consider the simplest, albeit somewhat unrealistic case of a single, discrete defect, or trap state located at energy E_T within the HOMO–LUMO gap. Assuming that the traps are uniformly spatially distributed across the sample, then

$$h_T(E) = H_T\delta(E - E_T) \tag{4.129}$$

is their DOS, H_T is the total trap density, and the Kronecker delta is $\delta(E)$. Then the number of trapped holes (assuming hole-only conduction, with simple substitutions used for electron-only situations) is

$$p_T = \int_{E_{HOMO}}^{E_{LUMO}} h_T(E) f_p(E)dE = \int_{E_{HOMO}}^{E_{LUMO}} \frac{H_T\delta(E - E_T)}{1 + g_p \exp\left[(E_{Fp} - E)/k_B T\right]} dE, \tag{4.130}$$

where

$$f_p(E) = \frac{1}{1 + g_p \exp[(E_{Fp} - E)/k_B T]} \tag{4.131}$$

is the Fermi–Dirac distribution for the number of filled hole states, E_{Fp} is the hole Fermi energy above the HOMO energy E_{HOMO}, and g_p is the degeneracy of the state. Typically $g_p = 2$ due to the two spin states allowed in each level. Evaluating the integral leads to

$$p_T = \frac{H_T}{1 + g_p \dfrac{N_{HOMO}}{p} \exp[(E_{HOMO} - E_T)/k_B T]} \tag{4.132}$$

where

$$p = N_{HOMO} \exp[-(E_{Fp} - E_{HOMO})/k_B T] \tag{4.133}$$

is the total free hole density which is the sum of the equilibrium (p_0) and injected (p_{inj}) densities, and N_{HOMO} is the DOS at the HOMO level. It is convenient to use the free charge factor, $\theta_f = \dfrac{g_p N_{HOMO}}{H_T} \cdot \exp[(E_{HOMO} - E_T)/k_B T]$ such that Eq. 4.132 becomes

$$p_T = \frac{H_T}{1 + H_T \theta_f / p}. \tag{4.134}$$

Now, returning to Eq. 4.122 but including trapped charge, we have

$$\begin{aligned} \frac{dF(x)}{dx} &= \frac{q(p_{inj}(x) + p_0 + p_T)}{\varepsilon_r \varepsilon_0} \\ &\approx \frac{q}{\varepsilon_r \varepsilon_0} \left\{ p_{inj}(x) + \frac{H_T}{1 + H_T \theta_f / p_{inj}(x)} \right\}, \end{aligned} \tag{4.135}$$

where the small background carrier density is ignored in the space charge limit. To solve this analytically, we make the further assumption that the trapped charge density is significantly larger than that of free charges, that is, $H_T\, \theta_f > p_{inj}\,(x)$. Then,

$$\frac{dF(x)}{dx} = \frac{q}{\varepsilon_r \varepsilon_0 \theta_f} p_{inj}(x)(\theta_f + 1). \tag{4.136}$$

Following the same treatment used to calculate the trap-free SCL current, we arrive at the modified Mott–Gurney relation in the presence of a discrete, high density of traps:

$$j_{SCL} = \frac{9}{8} \left[\frac{\theta_f}{(\theta_f + 1)} \right] \mu_p \varepsilon_0 \varepsilon_r \frac{V^2}{d^3}. \tag{4.137}$$

But Eq. 4.135 implies that when $p_T > p_{inj}$, then the ratio of free to trapped charge is $\theta_f \approx p_{inj}/p_T < 1$, yielding (Rose, 1955)

$$j_{SCL} = \frac{9}{8} \theta_f \mu_p \varepsilon_0 \varepsilon_r \frac{V^2}{d^3}. \tag{4.138}$$

Thus, the mobility is decreased to $\mu_{eff} = \theta_f \mu_p < \mu_p$ as the trap density increases. As expected, the reduced mobility also increases the voltage at which Ohmic gives way to SCL conduction. Similar to Eq. 4.128,

$$V_X = \frac{8}{9}\frac{q}{\theta_f \varepsilon_r \varepsilon_0} p_0 d^2. \quad (4.139)$$

A more physically realistic model assumes that an exponential trap distribution continuously extends from the conduction (frontier) orbital level into the energy gap. We use an exponential DOS in place of a nearly equivalent Gaussian (cf. Fig. 4.28) since the former leads to simple analytical solutions. Then, for the hole trap density (again, assuming a uniform distribution of traps throughout the sample bulk):

$$h_T(E) = \frac{H_T}{k_B T_0} \exp[-(E - E_{HOMO})/k_B T_0] \quad (4.140)$$

where T_0 is the *characteristic trap temperature* that sets the position of the distribution relative to the HOMO energy. This trap distribution is illustrated in Fig. 4.54. States between E_{Fp} and E_{HOMO} are filled with holes, and those above E_{Fp} are empty. The maximum trap state density of $H_T/k_B T_0$ is found at E_{HOMO}. An analogous expression can be employed to describe electron traps extending into the energy gap from the LUMO.

For simplicity we take $f_p \approx 1$ at $E > E_{Fp}$, and $f_p \approx 0$ at $E < E_{Fp}$. This gives the number of trapped holes:

$$p_T = \int_{E_{Fp}}^{\infty} h_T(E) dE = \frac{H_T}{k_B T_0} \int_{E_{Fp}}^{\infty} \exp[-(E - E_{HOMO})/k_B T_0] dE$$
$$= H_T \exp\left[-(E_{Fp} - E_{HOMO})/k_B T_0\right] \quad (4.141)$$

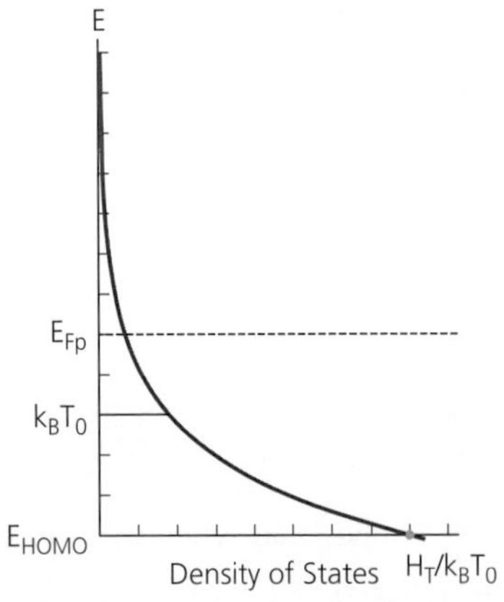

Figure 4.54 Density of trap states from the HOMO level (E_{HOMO}) extending into the energy gap. The maximum density is $H_T/k_B T_0$, and the Fermi level for holes is E_{Fp}. Also, T_0 is the characteristic trap temperature.

We can now write this in terms of the hole density in Eq. 4.133 to obtain

$$p_T = H_T \left(\frac{p}{N_{HOMO}}\right)^{T/T_0} = H_T \left(\frac{p}{N_{HOMO}}\right)^{1/m}, \quad (4.142)$$

where

$$m = T_0/T = E_T/k_B T, \quad (4.143)$$

with $E_T = k_B T_0$ equal to the characteristic trap energy.

Multiplying both sides of Gauss' law, Eq. 4.122, by $[(m+1)/m] F(x)^{1/m}$, we get

$$\left(\frac{m+1}{m}\right) F(x)^{1/m} \frac{dF(x)}{dx} = \frac{dF(x)^{1/m+1}}{dx}$$
$$= \frac{q}{\varepsilon_r \varepsilon_0} \left(\frac{m+1}{m}\right) (pF(x))^{1/m} \left(p^{1-1/m} + H_T N_{HOMO}^{-1/m}\right) \quad (4.144)$$

or

$$F(x)^{1/m+1} = \frac{q}{\varepsilon_r \varepsilon_0} \left(\frac{m+1}{m}\right) \left(\frac{j_{SCL}}{q\mu_p}\right)^{1/m} \times \left(p^{1-1/m} + H_T N_{HOMO}^{-1/m}\right) x$$
$$= \frac{qH_T}{\varepsilon_r \varepsilon_0} \left(\frac{m+1}{m}\right) \left(\frac{j_{SCL}}{q\mu_p N_{HOMO}}\right)^{1/m} \times \left(\frac{p^{1-1/m} N_{HOMO}^{1/m}}{H_T} + 1\right) x \quad (4.145)$$

In the high trap density limit:

$$\frac{p^{1-1/m} N_{HOMO}^{1/m}}{H_T} = p \frac{p^{-1/m} N_{HOMO}^{1/m}}{H_T} = \frac{p}{p_T} \approx 0, \quad (4.146)$$

which leaves the simplified expression for the electric field:

$$F(x) = \left(\frac{qH_T}{\varepsilon_r \varepsilon_0} \frac{m+1}{m}\right)^{\frac{m}{m+1}} \left(\frac{j_{SCL}}{q\mu_p N_{HOMO}}\right)^{\frac{1}{m+1}} x^{m/(m+1)}. \quad (4.147)$$

Integrating this to obtain the voltage, and solving for current, we finally arrive at

$$j_{SCL} = q\mu_p N_{HOMO} \left(\frac{2m+1}{m+1}\right)^{m+1} \left(\frac{m}{m+1} \frac{\varepsilon_r \varepsilon_0}{qH_T}\right)^m \frac{V^{m+1}}{d^{2m+1}}. \quad (4.148)$$

By equating this to Ohm's law, we can again calculate the cross-over voltage:

$$V_X = \frac{qH_T}{\varepsilon_r \varepsilon_0} \frac{m+1}{m} \left(\frac{m+1}{2m+1}\right)^{\frac{m+1}{m}} \left(\frac{p_0}{N_{HOMO}}\right)^{\frac{1}{m}} d^2, \quad (4.149)$$

which, as in the case of SCL current, scales as d^2.

For our treatment of SCL currents, it was assumed that the material is spatially homogeneous (i.e. the traps are uniformly distributed throughout the bulk of the sample), and the mobility is field-independent. In fact, given the several parameters in Eqs. 4.148 and 4.149, it is not possible to unambiguously determine μ_p without *a priori* knowledge of N_{HOMO}, H_T, and m. Thus, mobility should be obtained from an independent time of flight or alternative measurement technique discussed in Section 4.4.3. One further assumption is that there are no injection barriers. This is particularly important since such barriers themselves can also lead to a power-law dependence of j_{SCL} on V. Effects of injection can be separated from SCL-like behavior by measuring the dependence of j_{SCL} on d: SCL currents have a thickness dependence $\sim 1/d^{2m+1}$, whereas injection, in principle, should be layer thickness independent. More often, however, both phenomena exist in the same sample and it is not always possible to distinguish between the limiting sources of the dark current. The topic of injection from contacts is treated in more detail in Section 4.6.

It follows from Eq. 4.148 that the current for the special case of $m = 1$ reverts to the Mott-Gurney relation in the presence of a discrete trap, with $\theta_f = p/p_T \to N_{HOMO}/H_T$. Also, in the limit of a uniform distribution of traps within the HOMO-LUMO gap, $T_0 \to \infty$ in which case $m \to 0$. Then $p_T = H_T$ (Eq. 4.142) and the cross-over voltage becomes $V_X = qH_Td^2/(2\varepsilon_r\varepsilon_0)$.

For completeness, we consider the slightly different case of a Gaussian density of trap states with a distribution of width σ_T and a maximum at energy E_{Tm}. Then we have:

$$h_T(E) = \frac{H_T}{\sqrt{2\pi}\sigma_T}\exp\left[-(E-E_{Tm})^2/2\sigma_T^2\right]. \qquad (4.150)$$

Following the previous approach for a distribution of traps that lies deeper than the Fermi energy, that is, for $E_{Tm} > E_{Fp}$, the density of trapped holes is found by integrating the product $h_T(E)f_p(E)$ over energy:

$$\begin{aligned} p_T &= \frac{H_T}{\sqrt{2\pi}\sigma_T}\int\frac{\exp\left[-(E-E_{Tm})^2/2\sigma_T^2\right]}{1+g_p\exp\left[(E_{Fp}-E)/k_BT\right]}dE \\ &= H_T^G\left(\frac{p}{N_{HOMO}}\right)^{1/m}, \end{aligned} \qquad (4.151)$$

where

$$H_T^G = \left(\frac{H_T}{2g_p}\right)\exp(E_{Tm}/mk_BT) \qquad (4.152)$$

and

$$m = \left(1 + 2\pi\sigma_T^2/16k_B^2T^2\right)^{1/2}. \qquad (4.153)$$

Here, g_p is the degeneracy of the hole trap state. It is then straightforward to show that the current in the presence of a Gaussian density of trap states is given by Eq. 4.135 using the substitution of $H_T \to H_T^G$ (see Prob. 4.6).

Interestingly, if the Gaussian distribution is shallow, then the Fermi–Dirac distribution $f_p(E)$ can be replaced by the Boltzmann approximation, to yield

$$\begin{aligned} p_T &= \int h_T(E)g_p^{-1}\exp\left[(E-E_{Fp})/k_BT\right]dE \\ &= H_Tg_p^{-1}\exp\left[(E_{Tm}-E_{Fp})/k_BT\right]\exp\left[-\left(\sigma_T/\sqrt{2}k_BT\right)^2\right] \\ &= p/\theta_G, \end{aligned} \qquad (4.154)$$

which in the limit of a large trap density is an expression similar to Eq. 4.134 for a single, discrete trap level. For a Gaussian distribution of shallow traps, we can use the j_{SCL} vs. V^2 expression for a discrete trap distribution in Eq. 4.138 by replacing θ_f with θ_G. Thus, a Gaussian distribution cannot be distinguished from an exponential or discrete trap distribution without further independent determination such as via optical excitation or bleaching of trapped charge.

SCL currents for an exponential distribution of traps within the energy gap have been routinely observed in organic semiconductors such as in α-CuPc (Hamann, 1968) and anthracene (Schadt and Williams, 1969). SCL conduction has also been identified in both small molecule and polymer materials that are commonly used in OLEDs. For example, SCL electron transport in the presence of an exponential trap distribution in Mq_3 small molecule materials, where M = Al, Ga or In and q = 8-hydroxyquinoline, has been studied, with the j–V characteristics for Alq_3 at several temperatures shown in Fig. 4.55 (Burrows et al., 1996). The devices comprise a bilayer of the hole conducting TPD and a 55 nm thick the electron conducting Alq_3 layer sandwiched between an ITO anode and Mg–Ag cathode. The 27 nm thick TPD layer has very low resistance and hence its contribution to the total applied voltage drop is ignored. In the inset, the temperature exponent, m is plotted vs. $1/T$ (Eq. 4.143). The slope of the data yields a characteristic trap temperature of $T_0 = 1780$ K, corresponding to an energy of 0.15 eV below the LUMO of Alq_3. Several key parameters are obtained from the fit that characterize the amorphous Alq_3 film. Thus, $H_T = (3.1 \pm 0.1) \times 10^{18}$ cm^{-3}, and the DOS in the LUMO

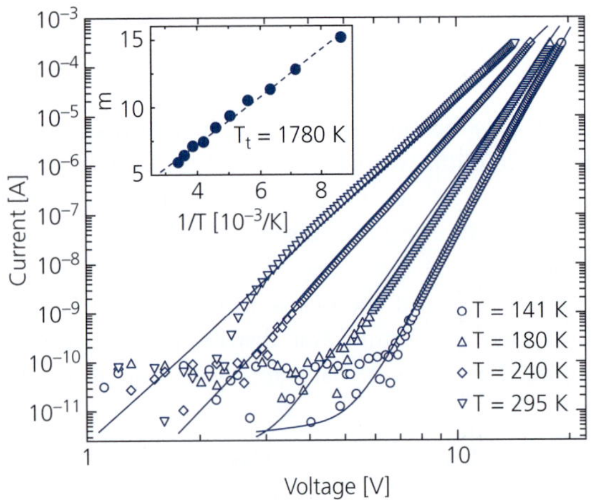

Figure 4.55 Current–voltage characteristics of a TPD/Alq_3 bilayer OLED at four different temperatures. The solid lines are fits to the data of the form $j \sim V^{m+1}$ assuming an exponential trap distribution. The inset shows the linear dependence of m vs. $1/T$, consistent with trap-dominated SCL current. A trap temperature of $T_0 = 1780$ K is obtained from the fit (Burrows et al., 1996).

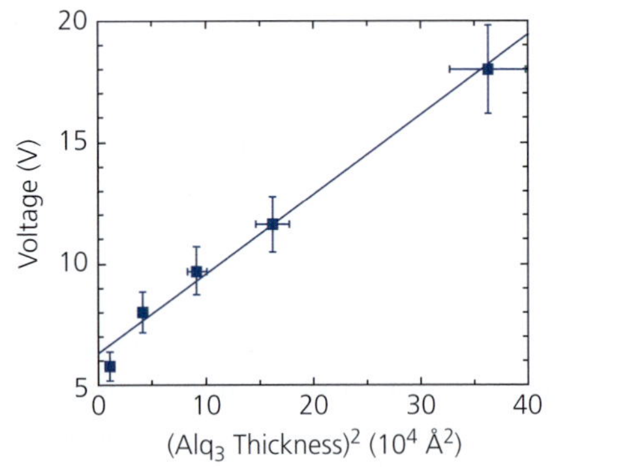

Figure 4.56 Dependence of voltage on the square of the thickness (d) of Alq_3 films at constant current and room temperature (Burrows et al., 1996).

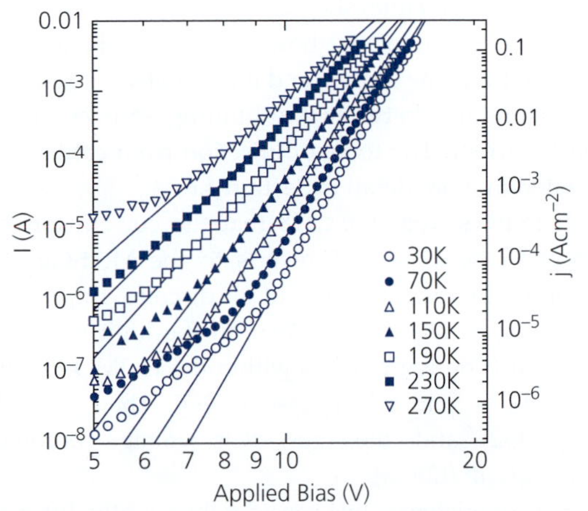

Figure 4.57 Current–voltage characteristics of a 94 nm thick PPV layer sandwiched between ITO and Al Ohmic contacts. The solid lines are fits based on SCL current in the trap limited regime. For these fits, m varies from 6 at 290 K to 17 at 11 K (Campbell et al., 1997).

is $N_{LUMO} = (1.0 \pm 0.5) \times 10^{19}$ cm^{-3}, which are determined using the mobility measured from time of flight data of $\mu_n = (5 \pm 2) \times 10^{-5}$ cm^2/V s. From V_X, the background free electron concentration is $n_0 < 10^{11}$ cm^{-3}. Since H_T is near to N_{LUMO}, it was proposed that H_T is an intrinsic property of the disordered film, that is, H_T is the result of static disorder rather than arising from extrinsic sources such as impurities.

An essential test for the existence of SCL current is its dependence on sample thickness. This is shown for a series of Alq_3 samples in Fig. 4.56. In Eq. 4.148 we see that at a constant current, $V^{m+1} \sim d^{2m+1}$, which for large m, $V \sim d^2$. The fit to this behavior is apparent in the data obtained for d between 10 nm to 60 nm at 20 μA. Due to the pronounced dependence of j_{SCL} on d^{2m+1}, small errors in determining film thickness can lead to large errors in measurements of the thickness dependence of j on V.

Similar analysis of the electroluminescent polymer, PPV, has also shown SCL conduction limited by an exponential distribution of traps (Campbell et al., 1997). The data in Fig. 4.57 are fit (solid lines) to Eq. 4.148 by varying m from 6 at 290 K to 17 at 11 K. The fit of m vs. $1/T$ follows a linear relationship from 300 K to approximately 120 K, with T_0 nearly the same as for Alq_3, although at lower temperatures T_0 begins to fall off rapidly. This dependence of trap energy (given by $E_T = k_B T_0$) arises from assuming that the mobility is independent of both temperature and electric field. We have seen that neither of these approximations is valid over large variations in either parameter. For example, as temperature is lowered, we expect that mobility changes from hopping-limited, consistent with dynamic disorder, to a tunneling-limited regime. Hence, the decrease in trap energy observed for PPV at low temperature is most likely due to the failure of the constant mobility assumption.

The dependence of current on thickness of the PPV layers is shown in Fig. 4.59a. The data are plots of K/d^{2m+1} vs. m obtained from the j–V characteristics which from Eq. 4.148 is written as $j_{SCL} = KV^{m+1}/d^{2m+1}$. The data are for PPV thicknesses of 94 nm and 125 nm, and m is taken from the high voltage range of the j–V characteristics in the range 11 K $< T <$ 290 K. These data conform to the thickness dependence of the trap-limited SCL current in Eq. 4.148. This is supported by replotting these data in Fig. 4.58b after multiplying by d^{2m+1} to remove the

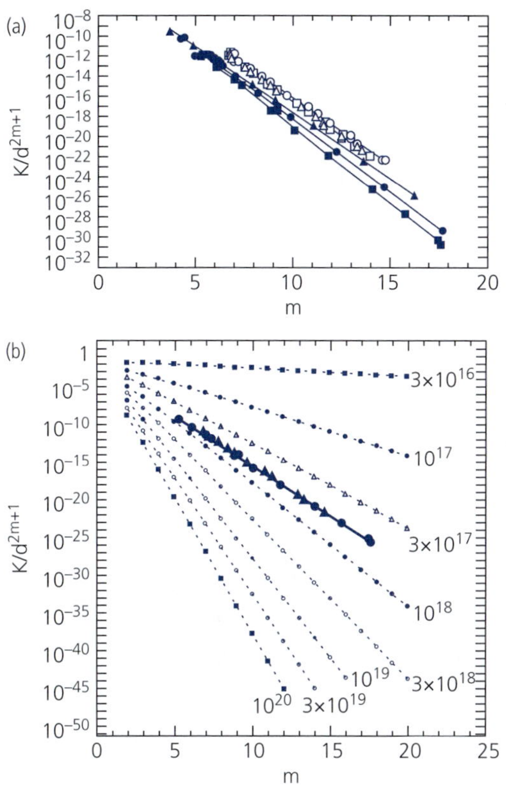

Figure 4.58 (a) Dependence of K/d^{2m+1} obtained from the j–V characteristics, viz. $j_{SCL} = KV^{m+1}/d^{2m+1}$ (cf. Eq. 4.148) vs. m from temperature dependent measurements ranging from 11 K to 290 K. PPV thicknesses are $d = 94$ nm (open symbols), 125 nm (full symbols). (b) Data from (a) replotted by multiplying by d^{2m+1} to remove the thickness dependence. Dashed lines are theoretical fits for various H_T (Campbell et al., 1997).

thickness dependence. All data fit on the same line corresponding to a trap density of approximately $H_T = 5 \times 10^{17}$ cm^{-3}, inferred from their position among the theoretically calculated lines.

Parametric fits to polymer data similar to those in Fig. 4.57 have used a Gaussian vs. an exponential trap distribution (Nicolai et al., 2011). While those fits are also convincing, as noted above they cannot be clearly distinguished from the exponential model although additional parameters (E_{Tm} *and* σ_T) are required. Unless there are compelling physical reasons for choosing the more complex Gaussian trap model, this may indeed be a case where Ockham's razor can justifiably be applied.

4.4.3 Measuring conductivity and mobility

Time of flight of charges in an electric field, and analysis of the j–V characteristics discussed in the foregoing sections are two of several means used for measuring charge mobility and conductivity in organic semiconductors. Keep in mind that different techniques go about the measurement of a quantity using different approaches, and hence there can be differences in the values inferred for μ and σ. For example, *time of flight* (TOF) measurements are based on analysis of transient currents induced by pulsing the sample either with an external voltage or illumination. On the other hand, j–V measurements discussed in the previous sections are obtained in the steady state. In the SCL region, μ is obtained directly from the slope of the characteristics once the dielectric constant (obtained from capacitance-voltage measurements) and film thickness are known. In the Ohmic regime, only the product, $n\mu$, is measured, from which a separate measurement of the charge density is required to obtain μ.

Traps that are almost always present in organics can have different occupancies in the steady or transient states, thus altering the internal electric field experienced by a charge as it moves between contacts. Hence, the mobilities obtained from TOF and SCL currents measure mobilities under very different conditions of trap occupancy, and hence these methods will not necessarily return the same value. To complicate matters even more, several methods measure the charge conductivity along a surface or interface where, again, the presence of traps or structural defects can influence conductive properties. Particularly, thin film transistor characteristics measure currents along the organic/insulator interface to yield a *field effect mobility* (μ_{FE}) that should not be confused with a bulk property of the material.

The inherent differences between measurement methods have led to many ambiguities and contradictory values for mobilities (and conductivities) of organics. Indeed, some materials have exhibited mobilities that differ by several orders of magnitude, depending on the measurement technique used. In this section, we discuss several means for measuring bulk mobilities, pointing out, wherever possible, the limitations and advantages of a particular method. In particular, we will discuss TOF mobility, dark injection space charge limited current, charge extraction by linearly increasing voltage (CELIV), and Hall effect mobility. Field effect mobilities, which are applicable to organic thin film transistors but not necessarily for determining an intrinsic materials property, will be discussed in Chapter 8. For reference, the most common methods for measuring bulk charge mobilities are listed in Table 4.3.

Table 4.3 Methods of mobility measurement

Technique	Sample Geometry	Typical Data
1. TOF (Time-of-flight)	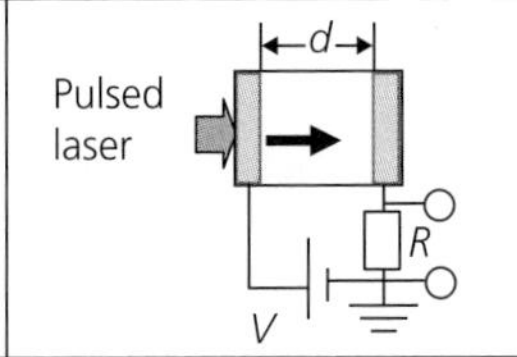	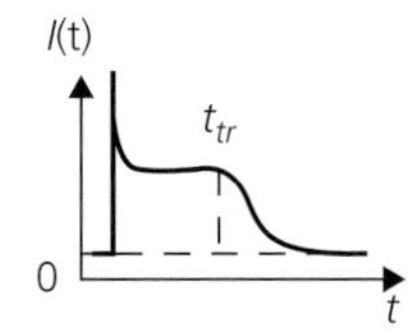
2. DISLC (Dark-injection space-charge-limited current)	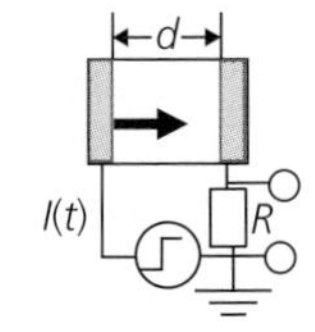	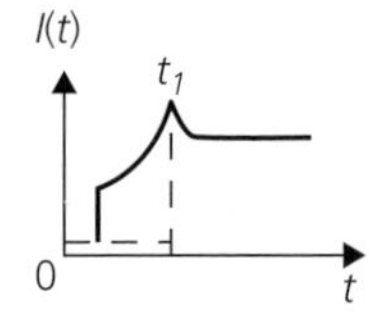
3. *J*-*V* Analysis (current-voltage characteristics)	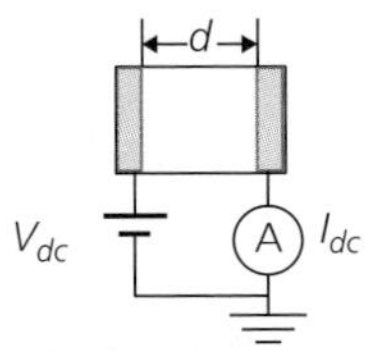	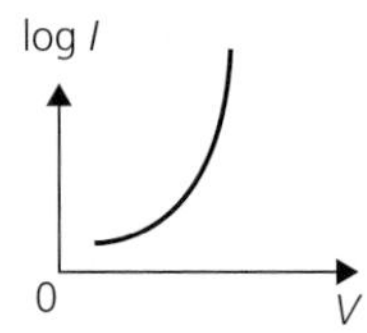
4. CELIV (Charge extraction by linearly increasing voltage)	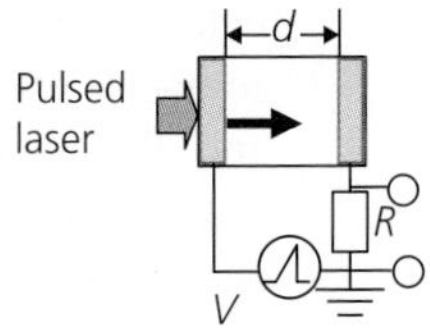	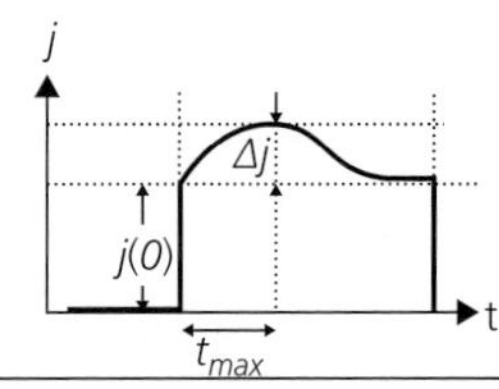
5. Hall Effect	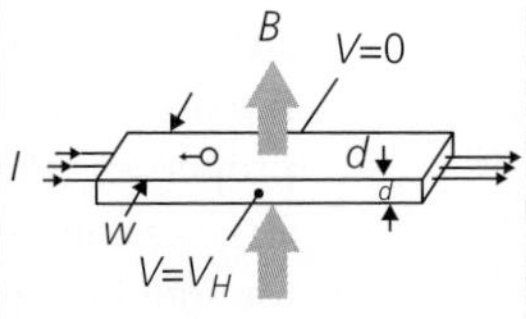	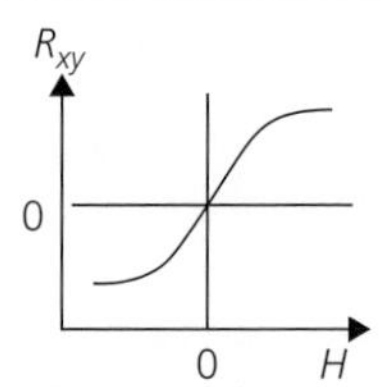
6. OTFT (Organic Thin Film Transistor)	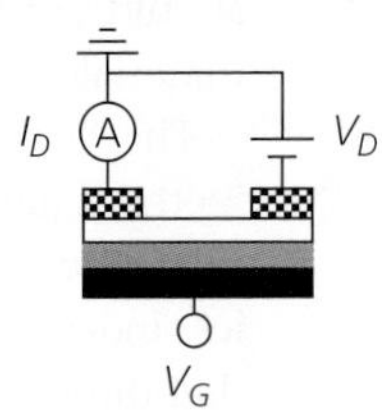	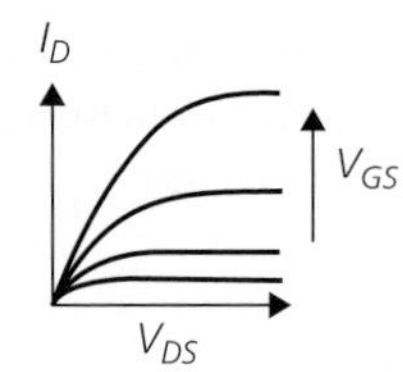

4.4.3.1 Time of flight mobility

TOF measurements have been extensively used to determine the mobilities of both inorganic and inorganic semiconductors. The method introduced by Haynes and Shockley is based on application of a voltage between contacts positioned at the opposite ends of a slab of a semiconductor (Haynes and Shockley, 1949, Shockley et al., 1949). A light pulse whose energy $E = h\nu > E_G$ illuminates the sample of length, L, near one of the electrodes. In an inorganic semiconductor, this generates a free electron–hole pair. Charges of the polarity opposite to that of the nearby electrode are rapidly collected, and the counter charges drift under the applied electric field to the opposite electrode (see Fig. 4.59).

If the field is sufficiently small to avoid space charge injection, the conduction is Ohmic, giving an electron transit time of $t_{tr} = L/\mu_n F = L^2/\mu_n V$. Hence, by measuring the arrival time of the charge pulse at the electrode farthest from the pump location, we extract the electron mobility as a function of the applied voltage, V. This treatment assumes that there is a single well-defined mobility, and that charge trapping and detrapping (which would affect both the width and the shape of the arriving current pulse) are negligible. It further requires that the illumination area be well-defined and whose extent is small relative to L. That is, the point of generation must be well separated from the counter electrode such that the charge packet can develop and be allowed to transit the semiconductor without uncertainty about their point of origin.

Organics seldom have sufficient conductivity to use this lateral thin film geometry. Thus, it is common to employ a vertical, or sandwich structure such as that in Fig. 4.60. The sample of thickness, d, is biased between two electrodes, at least one of which must be Ohmic (i.e. non-blocking). Light is incident on one semi-transparent or transparent (e.g. ITO) electrode, and the current response is measured vs. time. The sample

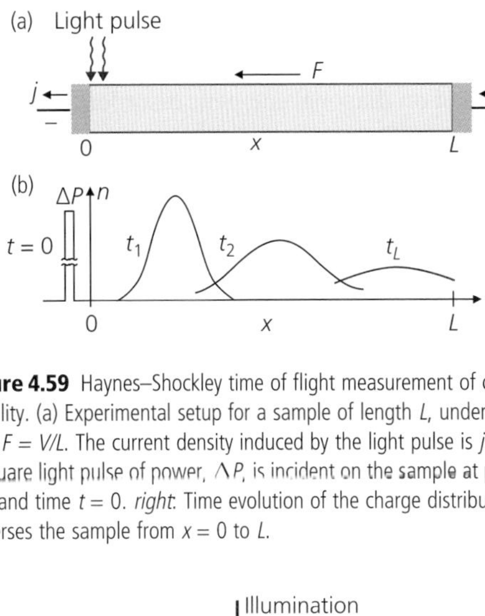

Figure 4.59 Haynes–Shockley time of flight measurement of charge mobility. (a) Experimental setup for a sample of length L, under electric field $F = V/L$. The current density induced by the light pulse is j. (b) *left*: A square light pulse of power, ΔP, is incident on the sample at position $x = 0$ and time $t = 0$. *right*: Time evolution of the charge distribution as it traverses the sample from $x = 0$ to L.

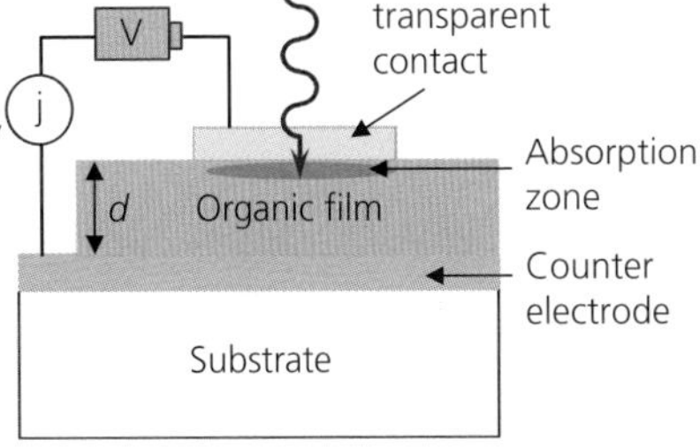

Figure 4.60 Device geometry commonly used in measuring mobility in organic thin films of thickness, d.

must be sufficiently thick to avoid substantial generation far into the film bulk. If this is not the case, then the transit distance will not be the same for all charges, leading to a range of charge arrival times at the opposite electrode. Near-surface generation at the transparent electrode/film interface is accomplished by illumination with a short wavelength pulse that is very strongly absorbed by the organic. To increase exciton dissociation and hence photocurrent, the semi-transparent contact can be photoemitting. Alternatively, an exciton dissociating HJ can be placed immediately beneath the contact (see Section 4.7). In this vertical Haynes–Shockley configuration, the transit time becomes

$$t_{tr} = \frac{d^2}{\mu V}. \qquad (4.155)$$

In Fig. 4.59, we show that the charge packet broadens with time due to diffusion. The one-dimensional electron diffusion equation in the absence of recombination is

$$\frac{\partial n(x,t)}{\partial t} = D_n \frac{\partial^2 n(x,t)}{\partial x^2}, \qquad (4.156)$$

with a similar equation for holes. Assuming that the Einstein relation obtains, then $qD_n = \mu_n k_B T$. The boundary conditions for a sample of width w are $n(0,0) = n_0/w$ and $n(L,0) = n(L,\infty) = 0$. Then Eq. 4.156 has solutions:

$$n(x,t) = \frac{n_0}{2\sqrt{\pi D_n t}} \exp\left(-x^2/4D_n t\right). \qquad (4.157)$$

The time of arrival at L of the full width at $1/e$ of the maximum current is, from Eq. 4.157: $\Delta t = 4\sqrt{D_n t_{tr}^3/L^2}$. The current peak amplitude is proportional to $n_{peak} = n_0/\sqrt{4\pi D_n t_{tr}}$. The peak magnitude, its arrival time, and width give the charge diffusion constant, and hence the mobility.

As noted in Chapter 3, illumination of organics does not result in the direct generation of free charge, but rather in excitons that must be separated into electrons and holes. Charge separation can occur at the metal contact/organic interface, albeit with low efficiency. To increase the charge generation efficiency, it is convenient to insert a thin organic zone near the contact that comprises a donor–acceptor HJ. Charge generation at donor–acceptor junctions found in photovoltaic cells or photodetectors can approach 100% efficiency, as will be discussed in Chapter 7.

Figure 4.61 shows the electron and hole current transients due to optical excitation of an ultrapurified naphthalene crystal (Warta and Karl, 1985). The initial peak is the displacement current due to the capacitance of the organic layer, followed by a plateau due to conduction through the film. Initially the internal field is high, resulting in the injection of a sheet of charge into the sample. The displacement current is then constant as the charge propagates through the sample until it reaches the counter electrode. After this, the charge density decreases, producing the diffusive edge of the transient indicated by the arrows in Fig. 4.61. The mobility is in fact not a scalar, but rather is a tensor that describes the transport properties along different crystalline directions. Hence, when measuring μ on single-crystalline samples such as in Fig. 4.62, care must be taken to apply the field along a known crystalline axis to obtain an accurate value.

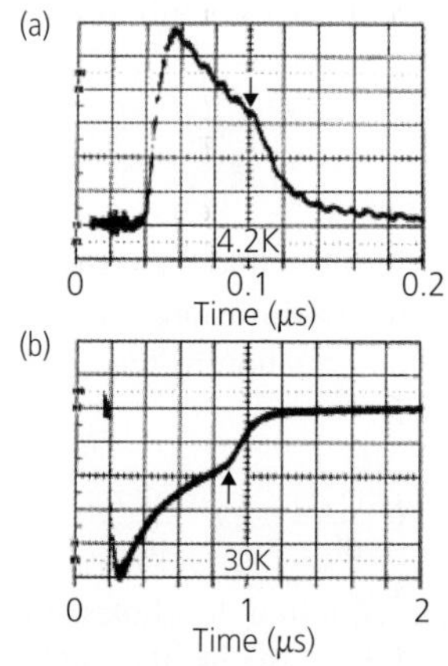

Figure 4.61 (a) Current transient due to hole transport in an ultrapurified naphthalene crystal of length L = 1.01 mm at T = 4.2 K. (b) Current transient due to electron transport at T = 30 K for the sample in (a) but with the electric field direction reversed. An 8 ns optical pulse at a wavelength of 249 nm generated the charges. The vertical scale divisions are 40 μA/cm^2. The arrows mark the time of arrival of the peak current, from which μ_n and μ_p are extracted (Warta and Karl, 1985).

In the presence of energetic disorder with Gaussian width, σ_{GDM}, the arrival times undergo dispersive spreading. That is, the mean charge velocity decreases with time since the initially hot carriers injected into the sample thermalize in a time longer than their transit times. Dispersive transport is due to the arrival of charges with a distribution of mobilities, as opposed to non-dispersive transport where only a single mobility is present. Mobilities measured for dispersive samples depend on the applied voltage, sample length and temperature.

The transition from dispersive to non-dispersive hole transport as a function of temperature has been studied in 5.6 μm long samples of DEH, with results shown in Fig. 4.62 (Borsenberger et al., 1992). The samples were sandwiched between an a-Se photoemitting anode and an Au cathode, and illuminated by 440 nm wavelength, 3 ns duration pulsed radiation. The transport becomes dispersive at T < 243 K for samples whose Gaussian width is $\hat{\sigma} = \sigma_{GDM}/k_BT = 4.4$. Even at higher temperatures, there is some droop in the plateau similar to that seen in Fig. 4.61 due to the presence of on-diagonal energetic disorder. The transition from non-dispersive to dispersive transport is particularly apparent in the double logarithmic plots of the current transients in Fig. 4.62b. The plateau corresponding to non-dispersive transport yields to a continually increasing slope for transients at lower

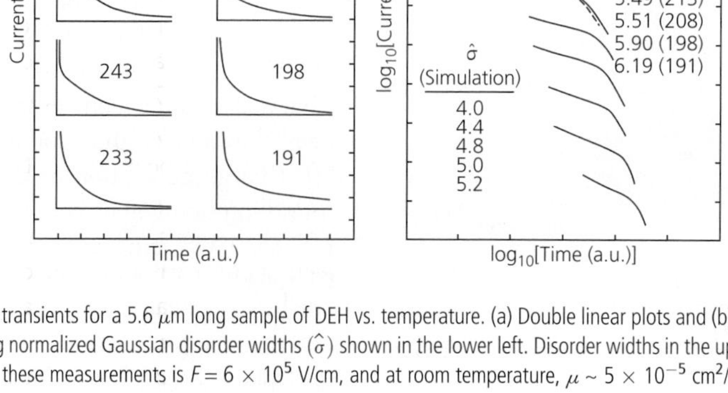

Figure 4.62 TOF hole current transients for a 5.6 μm long sample of DEH vs. temperature. (a) Double linear plots and (b) double logarithmic plots along with fits (dashed lines) assuming normalized Gaussian disorder widths ($\hat{\sigma}$) shown in the lower left. Disorder widths in the upper right are extracted from fits to the data. The electric field in these measurements is $F = 6 \times 10^5$ V/cm, and at room temperature, $\mu \sim 5 \times 10^{-5}$ cm^2/V s (Borsenberger et al., 1992).

temperatures. The long, exponential tail at low temperature is due to slow hopping of holes from the edge of the density of HOMO states that extends deep into the HOMO–LUMO energy gap.

We emphasize that a single value of charge mobility cannot be extracted from samples that exhibit dispersive transport. An important advantage, therefore, is that it can be used to distinguish between non-dispersive and dispersive transport regimes. This differs from measurements based on *j-V* analysis, transistor characteristics, or CELIV that only obtain an effective mobility without revealing details of the transport dynamics.

4.4.3.2 Dark injection space charge limited current

The transient response to a voltage pulse is an alternative means to determine mobility. This process is known as *dark injection space charge limited current* (DISCL). The accuracy and applicability of DISCL is essentially the same as for optically induced TOF measurements, and is equally susceptible to errors arising from charge trapping and uncertainties in the field distribution across the sample. Its inherent simplicity, however, makes this an attractive method for measuring mobilities.

In DISCL, charge of one carrier type is injected from an Ohmic contact on one side of the sample using a short voltage pulse. The resulting charge packet transits the semiconductor under a uniform electric field. The collecting electrode should block injection to avoid simultaneous double carrier injection from both contacts. Measurement of the current response, therefore, provides the charge transit time, from which the mobility is extracted.

The time-dependent current due to an applied electric field in a hole-only device in one dimension is given by

$$j(t) = j_c(x,t) + j_d(x,t) = qp(x,t)\mu_p F(x,t) + \varepsilon_r \varepsilon_0 \frac{\partial F(x,t)}{\partial t}, \tag{4.158}$$

with obvious modifications for electron-only devices. Here, $j_c(x, t)$ is the conduction current and $j_d(x, t)$ is the displacement current. From Poisson's equation,

$$\frac{\partial F(x,t)}{\partial x} = \frac{qp(x,t)}{\varepsilon_r \varepsilon_0} = \frac{Q(x,t)}{\varepsilon_r \varepsilon_0}. \tag{4.159}$$

Equation. 4.158 thus becomes

$$j(t) = \varepsilon_r \varepsilon_0 \left(\frac{\mu_p}{2} \frac{\partial F^2(x,t)}{\partial x} + \frac{\partial F(x,t)}{\partial t} \right). \tag{4.160}$$

This can be solved with the boundary condition that the field at the anode, $F_a(0,t) = 0$, due to the high density of injected holes in the space charge limit. A constant drop in voltage from cathode to anode at the moment the charge sheet is injected into a sample of thickness, d, results in a uniform field: $F(x,0) = V/d$. Then, Eq. 4.160 reduces to

$$j(t) = \left(\frac{\varepsilon_r \varepsilon_0 \mu_p}{2d} \right) F_c^2(t). \tag{4.161}$$

Let t_1 be the time at which the leading edge of charge reaches the cathode. Prior to t_1, the current is due solely to displacement, whence Eq. 4.158 yields

$$\frac{\partial F_c}{\partial t} = \left(\frac{\mu_p}{2d} \right) F_c^2(t) \tag{4.162}$$

In the absence of traps, the charge per area at the anode is $Q_a = \varepsilon_r \varepsilon_0 V/d$. Then integrating Eq. 4.161, we obtain

$$F_c(t) = \frac{Q_a(t)}{\varepsilon_r \varepsilon_0 [1 - t/2t_{tr}]}, \tag{4.163}$$

where t_{tr} is the hole transit time. From Eq. 4.161,

$$j(t) = \frac{\varepsilon_r \varepsilon_0 \mu_p}{2[1 - t/2t_{tr}]^2} \frac{V^2}{d^3} = j_0/[1 - t/2t_{tr}]^2, \tag{4.164}$$

where $j_0 = j(0) = \frac{4}{9} j_{SCL}$. Here, j_{SCL} is the steady state SCL current given by Eq. 4.127. The final step is to determine the arrival time of charge at the cathode. This is done by calculating the integral of the velocity over the time it takes the charge to transit the length of the sample, yielding

$$L = \mu_p \int_0^{t_1} F_c(t) dt = \frac{2\mu_p t_{tr} V}{d} \ln(1 - t_1/t_{tr})^{-1} \tag{4.165}$$

or

$$t_1 = 2t_{tr}(1 - e^{-1/2}) = 0.786 t_{tr}. \tag{4.166}$$

That is, the leading edge of the charge front reaches the cathode approximately 21% earlier than the transit time of the center of the charge distribution.

The current at t_1 relative to its value at $t = 0$ is $j(t_1)/j_0 = e = 2.72$, whereas compared to the current in steady state, $j(t_1)/j_{SCL} = 4e/9 = 1.21$. Hence, the current has a peak at t_1, at which point the sample contains the maximum amount of charge. As time progresses, the charge density decreases by collection at the cathode. The decrease continues until it reaches its steady state value for SCL conduction where the amount injected is equal to that extracted. A measurement of the time, t_1 of the current peak therefore yields t_{tr} from Eq. 4.166, from which we calculate the mobility from Eq. 4.155.

The full time dependent solutions to the drift-diffusion equation in the absence or presence of traps is more complex and beyond the scope of our discussion. For a detailed treatment the reader is referred to the work of Many and Rakavy (1962). The current transient solutions for space charge injection from a voltage step function that turns on at $t = 0$ are shown in Fig. 4.63. The several curves correspond to solutions in the presence or absence of traps, defined by the ratio, $R = t_p/t_{tr}$, where t_p is the mean free time for hole recombination in the presence of traps. That is, as the trap density increases, t_p decreases accordingly. Here, $R = \infty$ corresponds to the case of trap-free SCL current. The peak current is at t_1, and then falls immediately thereafter until steady state is reached. The small dip following the peak arises from the decreased electric field at the anode due to the non-equilibrium oversupply of injected charge. As this excess charge is depleted, the current relaxes to its equilibrium value. A decreased recombination time results in a decrease in the steady state and peak currents. Except for the most extreme cases, t_1 is easily identified at all R, from which μ can be accurately determined.

An example DISCL measurement and fits to the data at several different pulse voltages is shown in Fig. 4.64 for a 300 nm thick film of the first generation hexabutyl triarylamine dendrimer (hexabutyl-MTDATA, Louie et al., 1997) sandwiched between Au contacts. There are several complicating effects that need to be considered to achieve the fits to the data that extend from $t = 0$ to steady state. The best fits, are obtained for $\mu = 6.1 \times 10^{-6}$ cm^2/V s, including the small falling transient at low current at the onset of the transient. This feature results from trap filling. The parameters used in the best fits are a contact barrier height of 0.28 eV, a Gaussian distribution of states at the contact/semiconductor interface of width, 75 meV, a bulk trap density of 7.37×10^{15} cm^{-3}, and charge trapping and detrapping times of 3.39×10^{-12} cm^3/s and 2820 s^{-1}, respectively. Including a Gaussian DOS within the bulk of the semiconductor does not noticeably improve the fits (Szymanski et al., 2012). While the fits require several independent variables, the mobility extracted from the peak is insensitive to the choice of the other parameters, as inferred from the alternative fit to 8.5×10^6 cm^2/V s.

We note that current transient for a Ti/PTCDA/Au device in Fig. 4.52 has been used to obtain an electron mobility of $\mu_n = 0.018$ cm^2/V s. This is approximately five times higher than obtained from analysis of the SCL currents (Forrest et al., 1984a). This difference is to be expected, since the transient response is obtained before deep levels can be filled, reducing Coulomb trapping effects, hence resulting in a higher transient than steady state mobility.

4.4.3.3 Charge extraction by linearly increasing voltage

A variation of the DISCL current method that eliminates displacement current transients is the CELIV technique first introduced to measure mobility in microcrystalline Si (Juška et al., 2000). The method was later applied to polymers and other materials employed in organic solar cells (Lorrmann et al., 2010), and extended to include photoexcitation of charges in addition to electrical injection (Bange et al., 2010).

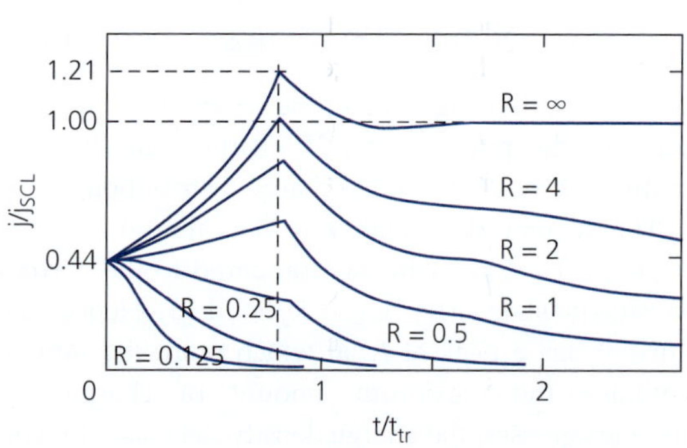

Figure 4.63 Current transient in units of the steady state trap-free SCL current (j_{SCL}) vs. time in units of the charge transit time, t_{tr}. Here, $R = t_p/t_{tr}$ is the ratio of the mean free time for recombination of the mobile hole in the presence of traps to the transit time. From Many and Rakavy (1962).

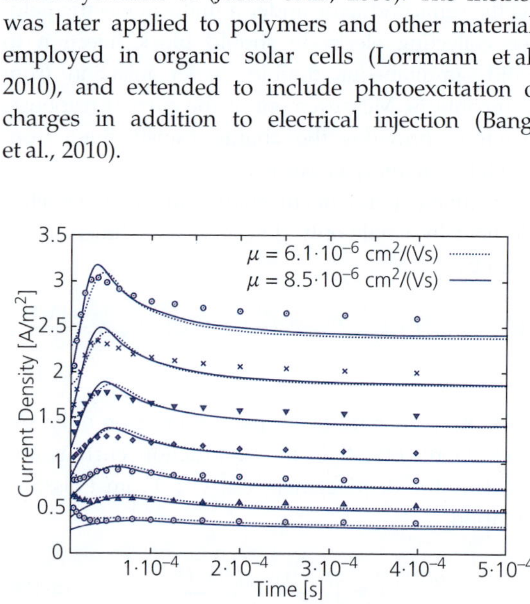

Figure 4.64 Current transients (points) and theoretical fits (lines) for a 300 nm thick triarylamine dendrimer film at several different pulse voltages from 1.0 – 2.5 V. The pulse is turned on at $t = 0$. The fits correspond to two different hole mobilities shown in legend, with the best fit obtained at 6.1×10^{-6} cm^2/V s (Szymanski et al., 2012).

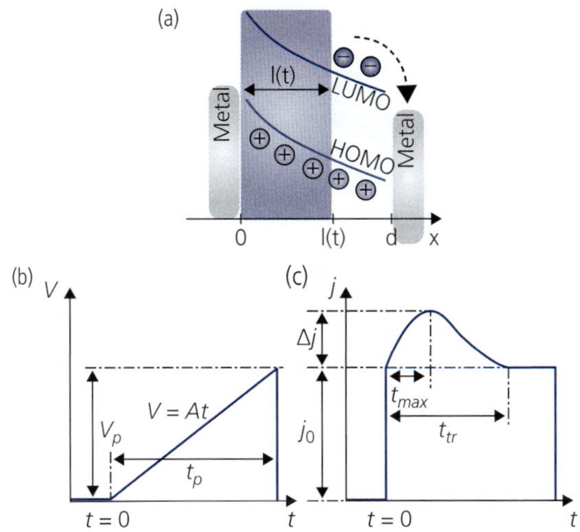

Figure 4.65 (a) Charge distribution and sample geometry used in CELIV. (b) Sawtooth voltage waveform with slope A used to excite the sample. (c) Current response to the linear voltage transient (Lorrmann et al., 2010).

The method is shown schematically in Fig. 4.65. It employs the same vertical sample geometry used in TOF and DISLC, where the semiconductor is sandwiched between two charge-blocking metal electrodes. A linearly increasing voltage, $V = At$ is applied to the sample that sweeps charge from a region of thickness, $l(t)$. This is called the *charge extraction length*, which corresponds to the charge depletion region width. The response is thus a combination of the displacement current, which is a square wave in response to a sawtooth voltage transient plus the charge extracted from the edge of the depletion region. This superposed charge on the square background creates a pulse with a maximum at time t_{max}. From Gauss' law we solve for the charge density, n:

$$F(0,t) - F(d,t) = \frac{qnl(t)}{\varepsilon_r\varepsilon_0}. \tag{4.167}$$

As voltage is ramped, the current response is

$$j(t) = \sigma F(d,t) = qn\frac{dl(t)}{dt}, \tag{4.168}$$

where the total charge per area from Gauss' law is $Q(t) = qnl(t) = \varepsilon_r\varepsilon_0(F(0,t) - F(d,t))$. Now, Poisson's equation gives

$$V(t) = At = -\int_0^d F(x,t)dx = F(d,t)d - \frac{F(d,t) - F(0,t)}{2}l(t), \tag{4.169}$$

where the electric field in an undoped semiconductor decreases linearly within the depletion region in the range $0 < x < l(t)$. Then combining Eqs. 4.167–4.169, we obtain the quadratic Ricatti equation:

$$\frac{dl(t)}{dt} = \frac{\mu At}{d} - \frac{l^2(t)}{2\tau_\sigma d}, \tag{4.170}$$

where the dielectric response time is given by

$$\tau_\sigma = \varepsilon_r\varepsilon_0/\sigma = \varepsilon_r\varepsilon_0/nq\mu. \tag{4.171}$$

The current can be written in terms of the conduction and displacement components as in Eq. 4.158 to give the time-dependent solution:

$$j(t) = \frac{\varepsilon_r\varepsilon_0 A}{d} + \frac{\sigma}{d}\int_{l(t)}^{d} F(d,t)dx = \frac{\varepsilon_r\varepsilon_0 A}{d} + \frac{\sigma F(d,t)}{d}(d - l(t)), \tag{4.172}$$

where the first term is the displacement current ($\varepsilon_r\varepsilon_0\partial F/\partial t$) which is constant for a linearly increasing voltage transient. This eliminates the spike at the onset of the square voltage pulse in DISCL. Then, it follows using Eq. 4.168 and 4.170 that the current transient is described by (Juška et al., 2000)

$$j(t) = \frac{\varepsilon_r\varepsilon_0 A}{d} + \frac{\sigma}{\mu d}(d - l(t))\left(\frac{\mu At}{d} - \frac{l^2(t)}{2\tau_\sigma d}\right) \tag{4.173}$$

From the initial slope at $t = 0$, when $l(0) = 0$, we obtain the film permittivity using $j(0) = \frac{\varepsilon_r\varepsilon_0 A}{d}$, and the conductivity,

$$\sigma = \varepsilon_r\varepsilon_0\frac{1}{j(0)}\frac{dj(t)}{dt}\Big|_{t=0}. \tag{4.174}$$

To extract further information from the transients in Fig. 4.65c, we must make a few approximations. For low conductivity, which is generally (although not always) the case for organics, then τ_σ t_{tr}. Then the electric field is unchanged by the extraction of charge during the transient, such that in Eq. 4.170, the $l^2(t)$ term can be neglected. At $t > t_r$, the displacement current is significantly larger than the conduction current such that $j(t > t_{tr}) = j(0) = \frac{\varepsilon_r\varepsilon_0 A}{d}$ (see Fig. 4.66c). From this assumption, $l(t) \approx \mu At/d$, then at $t \le t_{tr}$:

$$j(t) = \frac{A'}{d}\left[\varepsilon_r\varepsilon_0 + \sigma t\left(1 + \frac{\mu A' t^2}{2d^2}\right)\right]. \tag{4.175}$$

At the peak of the transient, the derivative of Eq. 4.175 gives

$$t_{max} = \frac{t_{tr}}{\sqrt{3}} = d\sqrt{\frac{2}{3\mu A'}}, \tag{4.176}$$

from which we extract the mobility. For the less common case of high conductivity, then $t_{tr} \gg \tau_\sigma$, and

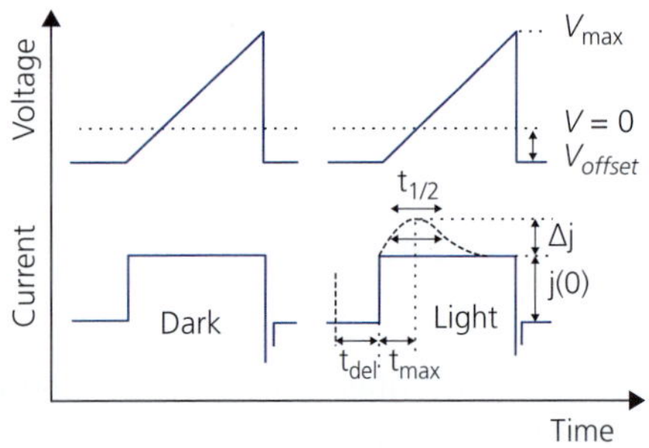

Figure 4.66 (*top*) Sawtooth voltage sequence that swings through $V=0$ in each cycle. The V_{offset} is used to counter the built-in potential from the contacts. (*bottom*) Current response. In the first cycle, the sample is in the dark and a square reactive current response is observed. After a voltage delay of t_{del}, the response is the sum of the extracted dark plus photocurrent.

$$t_{max} = \left(\frac{\tau_\sigma d^2}{\mu A}\right)^{1/3}. \tag{4.177}$$

Finally, in many conductive organic samples, the dielectric relaxation time is comparable to the transit time, in which case the full Ricatti equation must be solved (Lorrmann et al., 2010). This leads to more complicated analysis of CELIV data that makes it less attractive for determining the mobility and other film properties compared to TOF and DISCL. Other complicating factors not considered here are bimolecular recombination (see Section 4.5), space charge effects, and bipolar charge injection.

The carrier lifetime can be determined using a variation of CELIV whereby an optical pulse is temporally offset from the electrical pulse by varying delay times. By measuring the decay rate of the CELIV peak in this so-called *photo-CELIV* method, both the charge recombination time and the mobility are obtained (Mozer et al., 2005, Bange et al., 2010). In photo-CELIV, a reverse bias pulse is applied, followed by a short optical pulse, as shown in Fig. 4.67. Under reverse bias, the only response is the square displacement current transient since no conduction charge is extracted. After a short delay following a nanosecond laser pulse, the voltage ramp is applied. The current following electrical and photoexcitation is the sum of the displacement plus the photocurrent. By varying the delay to the application of the forward bias voltage, the recombination time can be determined, and the mobility is calculated from the position of the current peak using Eq. 4.177. A voltage offset is used to counter the built-in potential from the contacts.

A set of photo-CELIV data for a 1:4 blend of the donor, MDMO-PPV, and the acceptor, $PC_{61}BM$, are shown in Fig. 4.67 (Mozer et al., 2005). In Fig. 4.67a,

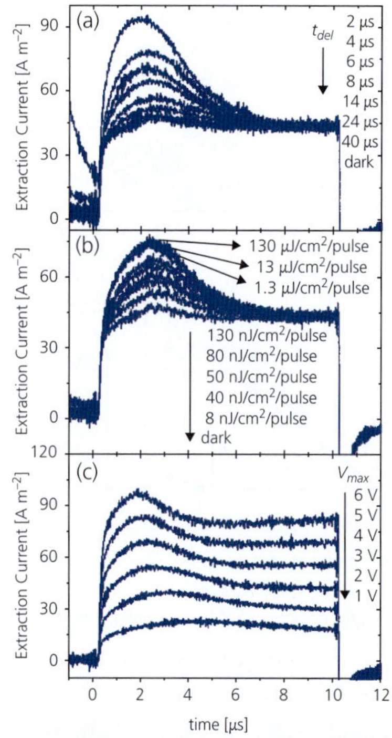

Figure 4.67 Photo-CELIV data for a 1:4 MDMO-PPV:$PC_{61}BM$ sandwiched between a ITO/PEDOT:PSS anode and an Al cathode. (a) The voltage delay time is varied from 2 μs to 40 μs. (b) The light intensity is varied at a constant delay of 5 μs. (c) The extraction voltage V_{max} is applied after a 15 μs delay (Mozer et al., 2005).

the offset voltage is set at -0.75V which is close to the built-in potential, as determined by the near absence of a peak in conduction current in the dark. When the sample is illuminated by a 3 ns, 532 nm wavelength laser pulse, excitons are generated in the blend. The excitons dissociate at the donor–acceptor junction, and are subsequently extracted, giving rise to the peak after ~2 μs. The slight shift to longer times as the voltage delay is increased is due to the field-dependent mobility that decreases as the amount of charge (and hence field) is reduced. In Fig. 4.67b, the delay is fixed but the laser intensity is increased. The current saturates at a beam energy $>1.3\ \mu J/cm^2$ due to bimolecular recombination (see Section 3.10). Finally, Fig. 4.67c shows the dependence on the extraction voltage, V_{max}. The peak position moves to shorter times with V_{max}, also due to the field dependence of the mobility.

The data in Fig. 4.67a are fit by assuming that the electrons and holes recombine via a Langevin process (see Section 4.5.4). In this process, the hole density returns to equilibrium via $\partial p/\partial t = -\gamma p^2$, where γ is the bimolecular recombination coefficient. The fits yield a recombination time of $t_{rec} = 1.7\ \mu$s and $\gamma = 6 \times 10^{-11}\ cm^3/s$. From the peak position, the mobility is approximately $\mu_p = 2 \times 10^{-4}$ cm^2/V s. Clearly, a rich set of materials data, including the dielectric constant, conductivity, mobility and bimolecular recombination time can all be extracted from CELIV and photo-CELIV, albeit after making several assumptions whose validity should be confirmed by other, independent measurements.

4.4.3.4 Hall effect mobility

The final technique we consider, and that is reliable yet rarely applied to organics due to their large contact and internal resistances, is the *Hall effect*. This method has the distinct advantage in that it provides both the charge density and mobility, as well as the carrier type (i.e. hole vs. electron). It employs samples in the lateral geometry, as shown in Fig. 4.68. If the series resistance of the material is high, the Hall current (j_x) is small, and is difficult or even impossible to accurately measure.

The measurement entails placing the sample of length, L, width, w, and thickness d, in a magnetic field, B_z, oriented perpendicular to the sample plane and the direction of the current, j_x (see Fig. 4.68). Due to the Lorentz force, electrons or holes traversing the sample in the $\hat{\mathbf{x}}$-direction are driven toward the edges of the sample in the $\hat{\mathbf{y}}$-direction, creating a voltage drop along $\hat{\mathbf{y}}$ known as the Hall voltage, V_H. Both electrons and holes are driven in the same direction since they have opposite signs, but their current flow is also in opposite directions.

The Lorentz force on an electron is given by

$$\mathcal{F} = q\mathbf{F} + q\mathbf{v} \times \mathrm{B}. \tag{4.178}$$

In the $\hat{\mathbf{y}}$-direction, the net forces on holes and electrons are

$$\mathcal{F}_{py} = qF_y - qv_{px}B_z \tag{4.179a}$$

and

$$\mathcal{F}_{ny} = -qF_y - qv_{nx}B_z, \tag{4.179b}$$

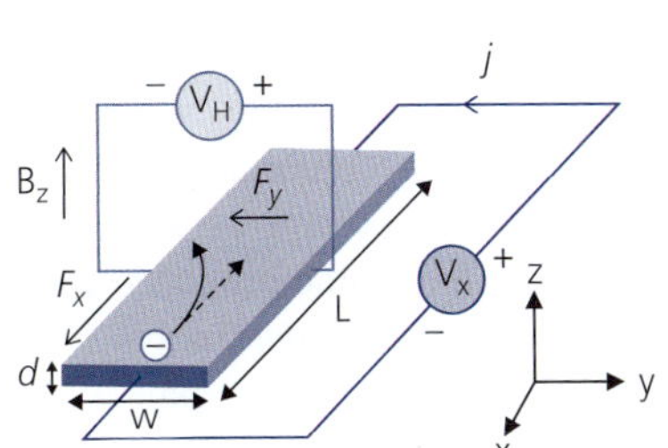

Figure 4.68 Hall effect measurement setup and sample geometry.

respectively. Since the drift velocity for holes is $v_{py} = \mu_p F_y = \mu_p \mathcal{F}_{py}/q$, with an analogous expression for electrons, we can substitute this into Eq. 4.179 to obtain

$$\frac{v_{py}}{\mu_p} = F_y - \mu_p F_x B_z \tag{4.180a}$$

and

$$\frac{v_{ny}}{\mu_n} = F_y + \mu_n F_x B_z. \tag{4.180b}$$

Now the net current in the $\hat{\mathbf{y}}$-direction must be zero:

$$j_y = j_{ey} + j_{hy} = q(nv_{ey} + pv_{hy}) = 0, \tag{4.181}$$

which leads to $nv_{ny} = -pv_{py}$. Finally, for the current in the $\hat{\mathbf{x}}$-direction we write

$$j_x = j_{nx} + j_{px} = (n\mu_n + p\mu_p)qF_x. \tag{4.182}$$

Substituting Eq. 4.182 and the solution of Eqs. 4.181 into Eq. 4.179 then gives

$$(n\mu_n + p\mu_p)^2 qF_y = (p\mu_p^2 - n\mu_n^2)j_x B_z. \tag{4.183}$$

Now the Hall coefficient is defined as

$$R_H = \frac{(p\mu_p^2 - n\mu_n^2)}{q(p\mu_p + n\mu_n)^2} = \frac{F_y}{j_x B_z} = -\frac{V_H}{wj_x B_z}. \tag{4.184}$$

Measurement of the Hall coefficient provides the mobilities and densities of each of the carriers. Also, R_H can be positive or negative, depending on the sign of the dominant charge carrier. The mobility or density of one charge should dominate to obtain clear results from Hall measurements. Thus, for example, if we take $p\mu_p^2 > n\mu_n^2$, then $R_H \approx \frac{1}{qp} > 0$. A measurement of the current, the magnetic field, and sign and magnitude of the Hall voltage therefore gives the hole density. Then, from j_x we can obtain the Hall effect mobility, μ_p.

We caution that the Hall mobility can be different from that measured using drift techniques such as TOF, DISCL, and CELIV, since carrier scattering dynamics in a magnetic field are different than in an electric field. Indeed, when charge transport is via hopping, or there is a high density of traps, the treatment based on the classical Lorentz force is no longer applicable, leading to anomalous results for the Hall coefficient. Finally, while the treatment described is for an isotropic conductive medium, asymmetries in crystal structure make the Hall measurement, and indeed all mobility measurements, sensitive to

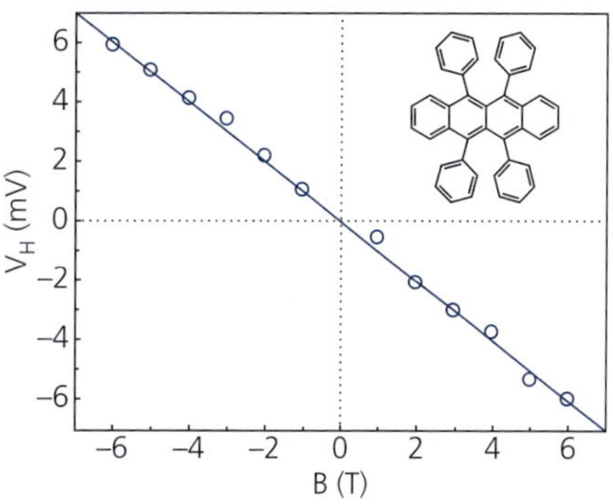

Figure 4.69 Hall voltage vs. magnetic field for a rubrene single crystal in a thin film transistor structure. Inset: Molecular formula of rubrene. After Podzorov et al. (2005).

crystalline direction. Nevertheless, provided that sufficiently large currents can be obtained, the simplicity and accuracy of the Hall measurement makes it a powerful technique for measuring charge concentrations and mobilities.

Due to their high thin film resistance, there are only a handful of reports of Hall measurements of organic semiconductors (Heilmeier and Harrison, 1963, Imaeda et al., 1992, Podzorov et al., 2005). In Fig. 4.69 we show the Hall voltage vs. B for a single crystal of rubrene in an organic thin film field effect transistor (OTFT), where the gate insulator is a 1 μm vacuum gap between the crystal and the gate electrode. As predicted in Eq. 4.184, V_H is proportional to B (for a constant R_H), and it changes sign with the field. The negative slope indicates the majority carriers are holes, consistent with the OTFT measurements. The Hall effect mobility and the hole concentration were both extracted as functions of temperature, and compared with data obtained from OTFT characteristics. The results of these measurements are provided in Fig. 4.70a. The mobility extracted from OTFT data (corresponding to the field effect mobility, μ_{FE}) is due to conduction of charges at the semiconductor/vacuum interface, and hence cannot be directly compared to the bulk mobility. Note that the Hall mobility at low temperature exceeds μ_{FE} by over a factor of two.

The surface hole concentration determined from the OTFT characteristics is

$$p = \frac{C_{GS}}{q}\left(V_{GS} - V_{TH}(T)\right), \tag{4.185}$$

where C_{GS} is the gate-source capacitance, V_{GS} is the gate-source voltage, and $V_{TH}(T)$ is the temperature dependent gate-source threshold voltage (see Chapter 8). The charge concentration obtained from

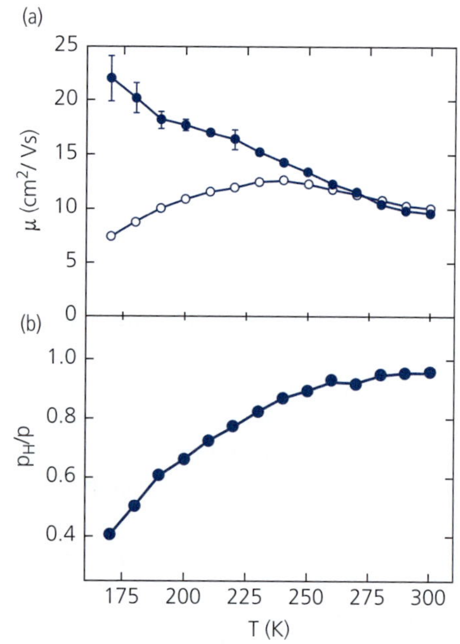

Figure 4.70 (a) Hall mobility (closed circles) vs. that extracted from the conductivity and gate capacitance of an OTFT (open circles). (b) Ratio of the hole concentration measured from the Hall coefficient (p_H) vs. that obtained from the OTFT (p) (Podzorov et al., 2005).

Hall measurements differs from OTFT analysis (Fig. 4.70a), which is attributed to charge trapping that decreases the number of free charges at low temperatures (Podzorov et al., 2005). From the difference between p_H and p, the density of surface traps is found to be ~10^{10} cm^{-2}. This also results in a higher Hall mobility at low temperature than that derived from the conductivity ($\sigma = qp\mu_p$) using Eq. 4.185, also due to trap filling inferred from Fig. 4.70a.

4.5 Charge recombination

In discussing charge transport, we noted that the mobility, and ultimately the current is strongly influenced by traps that can promote charge recombination. Traps can originate from many diverse sources, including static and/or dynamic disorder, impurities, crystalline imperfections, grain boundaries, or a myriad of other defects. In Chapter 3 we considered the various processes leading to exciton recombination, for example, triplet–triplet, singlet–polaron, singlet–triplet annihilation, etc. Exciton recombination is commonly referred to as *geminate recombination*, in that it involves a bound electron–hole pair (i.e. between oppositely charged "twins" with a common origin). In this section we consider the processes leading to recombination of electrons and holes whose origins are from independent sources such as injection from contacts. This is known as *bimolecular recombination*, and can have either intrinsic (e.g. *Langevin* and *Auger recombination*) or extrinsic (e.g. *Shockley–Read–Hall*, or SRH, *recombination* via defects) origins.

To determine the probability for bimolecular recombination, we calculate the recombination rates for electrons or holes, R_n or R_p, respectively. The diffusion equations for the change in electron concentration in the absence of an electric field are given by

$$\frac{\partial n}{\partial t} = \frac{1}{q}\nabla\cdot\mathbf{j}_n - R_n + G_n, \tag{4.186}$$

and for holes by

$$\frac{\partial p}{\partial t} = -\frac{1}{q}\nabla\cdot\mathbf{j}_p - R_p + G_p. \tag{4.187}$$

Here, the total current density is equal to the sum of the hole and electron currents, viz. $\mathbf{j} = \mathbf{j}_n + \mathbf{j}_p$, and G_p (G_n) is the hole (electron) generation rate. As shown below, there are conditions for which these variables are not independent, resulting in coupling of Eqs. 4.186 and 4.187.

From Fick's law of diffusion, the divergence of $\mathbf{j}$ in terms of the diffusion current is:

$$\mathbf{j}_n = qD_n\nabla n \tag{4.188}$$

and

$$\mathbf{j}_p = -qD_p\nabla p, \tag{4.189}$$

where D_n and D_p are the electron and hole diffusion coefficients, respectively. Hence, Eqs. 4.186 and 4.188 now give the familiar charge diffusion equations:

$$\frac{\partial n}{\partial t} = D_n\nabla^2 n - R_n + G_n \tag{4.190}$$

and

$$\frac{\partial p}{\partial t} = D_p\nabla^2 p - R_p + G_p. \tag{4.191}$$

4.5.1 Direct LUMO–HOMO recombination

Recombination is the mechanism that returns a semiconductor to equilibrium after it has been excited by electrical injection, illumination or a thermal transient. The non-equilibrium density of electrons is n_1, or holes, p_1, will therefore relax to their values at thermal equilibrium values, n_0 and p_0, respectively. Then the total electron density is $\delta n = n_1 + n_0$, and Eq. 4.177 becomes

$$\frac{\partial \delta n}{\partial t} = D_e\nabla^2\delta n - R_e + G_e, \tag{4.192}$$

with a similar expression for excess holes (δp) following Eq. 4.191. For direct electron–hole recombination in thermal equilibrium, the net recombination rate is $R_e - G_e$.

Band-to-band recombination in inorganic semiconductors occurs when minority carrier in loses energy to a distribution of majority carriers. Then, in the absence of diffusion currents (i.e. from injection or the presence of a junction), Eq. 4.192 yields the net recombination rate (assuming electrons are the minority carrier):

$$R_{tot} = R_n - G_n = k_n\left(n_p - n_0\right), \tag{4.193}$$

where $k_n = 1/\tau_n$ is the electron recombination rate and τ_n is its lifetime. This expression gives the rate that the electron density in p-type material, n_p, returns to its equilibrium value, n_0. In thermal equilibrium, $R_n = G_e$, and from the *law of mass action* (Sze, 1981), which relates the electron and hole densities to the intrinsic carrier density (see Eq. 4.113),

$$n_0 p_0 = n_i^2. \tag{4.194}$$

Organic semiconductors lack broad energy bands. Hence, the minority carrier (e.g. an electron) must *directly* recombine with its complementary charge (e.g. a hole). This can occur via one-step direct (radiative) recombination by an electron in the LUMO with a hole in the HOMO, as illustrated in Fig. 4.71. Alternatively, a multiphonon cascade is shown in Fig. 4.71b involves recombination from the LUMO to the HOMO via the non-radiative emission of several phonons through a distribution of defect state energies within E_G. We have seen that the phonon energy

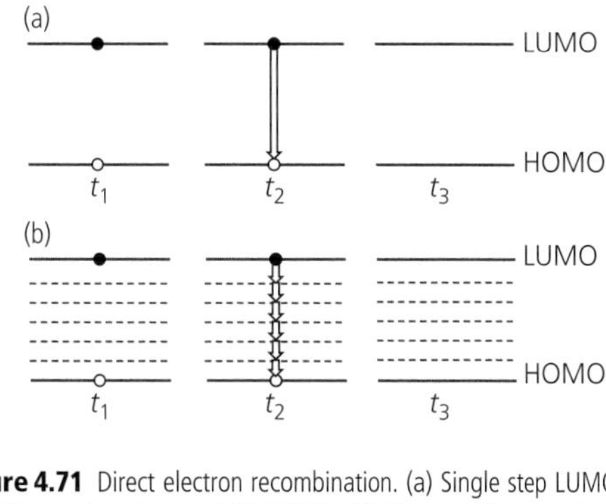

Figure 4.71 Direct electron recombination. (a) Single step LUMO–HOMO recombination, and (b) LUMO–HOMO recombination via a multi-step cascade. The times $t_1 < t_2 < t_3$ for each step are shown.

is approximately 0.15 eV or less in organic molecules. Thus, the cascade can dissipate the full gap energy in only a few steps.

Since direct recombination requires that the electron and hole occupy the same molecule, the rate is proportional to the product of their densities, that is, $R_n = k_{nc}np$ for an electron capture rate, k_{nc}. Note that k_{nc} is a bimolecular recombination constant, and hence differs from k_n. This is reflected in their units ([cm^3/s] vs. [s^{-1}], respectively). For direct recombination, therefore, the net recombination rate is

$$R_n^{dir} = k_{nc}\left(np - n_i^2\right). \qquad (4.195)$$

An identical expression is obtained for holes. As noted in Chapter 3, direct recombination in organics is via an intermediate, bound exciton. Hence, the photon energy must equal that of the relaxed molecular state rather than that of the LUMO-HOMO gap, which would be analogous to band-to-band recombination in inorganics.

4.5.2 Recombination via mid-gap states

In Fig. 4.72 we show two additional sources of recombination that result in a loss of an electron polaron from the LUMO of an anionic molecule back to the HOMO. In Fig. 4.72a we show the process of sequential recombination and generation that occurs via a single defect state (or distribution of states) that lies approximately in the center of the energy gap. For this illustration, we assume that the mid-gap state is a hole trap occupied at $t = t_1$. At t_2, an electron in the LUMO is captured by the trap, that is, the electron and hole recombine, leaving the state unoccupied. At t_3, the trap captures a hole. The hole repopulates the state via transfer from the HOMO, returning the state to its original condition at t_4. The sequence can be repeated indefinitely. Recombination and generation processes are often thermally activated and may arise from optical generation or injection from contacts. This process is governed by Shockley–Read–Hall (SRH) statistics (Hall, 1951, Shockley and Read, 1952).

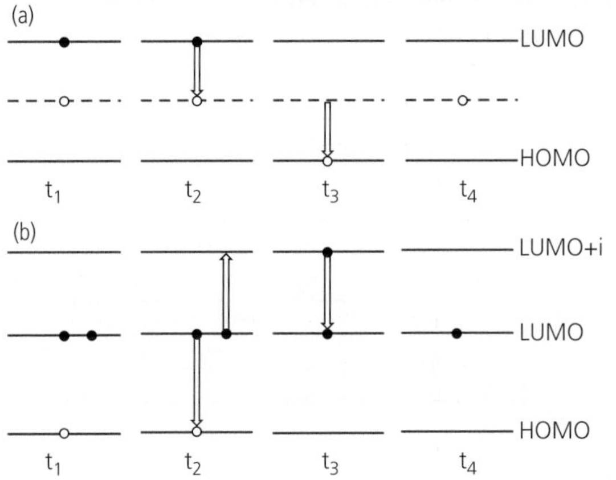

Figure 4.72 Two possible routes to non-radiative electron recombination. (a) Recombination-generation via a mid-energy gap state and (b) Auger recombination. The time sequence for each process, $t_1 < t_2 < t_3 < t_4$ is indicated.

SRH recombination is a special case of the cascade process in Fig. 4.71b, where multiple steps have been replaced by a single step, although the most efficient recombination occurs via traps located at or near the center of the energy gap. In a cascade, the slowest transitions for either electrons or holes create rate-limiting bottlenecks that determine the overall rate (and hence the efficiency) of the recombination event.

Another non-radiative process is Auger recombination in Fig. 4.72b. Two electrons in the LUMO at t_1 undergo a collision at t_2, resulting in the recombination of one electron with a hole in the HOMO, while delivering the excess electron energy to a second electron that is promoted into a higher excited state (i.e. to LUMO + i, $i > 0$). This "hot" electron polaron rapidly thermalizes, returning back to the LUMO at t_3 via non-radiative emission of one or several phonons. Hence, while we started with two excess electrons, Auger recombination is complete when only a single excited electron remains at t_4. Auger recombination can also occur by the promotion of a hole deep into the bound (HOMO - i, $i > 0$) orbitals in a symmetric reaction to that in Fig. 4.72b. This will be discussed further in Section 4.5.3.

We now calculate the rates of generation and recombination via a single mid-gap state in Fig. 4.72a. First, consider the electron capture in the presence of a total trap density:

$$N_T = N_T^0 + N_T^-, \qquad (4.196)$$

where N_T^0 is the number of empty (or neutral) traps and N_T^- is the number of filled (or anionic) traps. Then, for an electron DOS in the LUMO, $g_{LUMO}(E)$, we obtain the incremental capture rate by the trap at energy, E:

$$dk_{nc} = k_{nc}(E)N_T^0 f_e(E)g_{LUMO}(E)dE. \qquad (4.197)$$

Here, $f_e(E)$ is the Fermi–Dirac distribution. Similarly, for hole capture into the same trap:

$$dk_{pc} = k_{pc}(E)N_T^- f_h(E)g_{HOMO}(E)dE, \qquad (4.198)$$

where $g_{HOMO}(E)$ is the HOMO DOS. Since generation results in emission from traps, we can write analogous expressions for these processes, viz.

$$dk_{ng} = k_{ng}(E)N_T^-(1 - f_e(E))g_{LUMO}(E)dE \qquad (4.199)$$

and

$$dk_{pg} = k_{pg}(E)N_T^0(1 - f_h(E))g_{HOMO}(E)dE, \qquad (4.200)$$

where the subscripts ng and pg denote electron and hole generation, respectively, and $1 - f(E)$ is the probability that a state at E will be unoccupied.

From this analysis, we obtain a net incremental recombination rate for electrons of:

$$\begin{aligned} dR_n^{SRH} &= dR_n - dG_n = dk_{nc} - dk_{ng} \\ &= \left[k_{nc}(E)N_T^0 f_e(E) - k_{ng}(E)N_T^-(1 - f_e(E))\right] \\ &\quad \times g_{LUMO}(E)dE. \end{aligned} \tag{4.201}$$

Now we can write the fraction of occupied and unoccupied traps as $f_T = N_T^-/N_T$, and $1 - f_T = N_T^0/N_T$, respectively, where f_T is

$$f_T(E_T) = \frac{1}{1 + \exp((E_T - E_F)/k_BT)}, \tag{4.202}$$

where E_T is the trap energy. In equilibrium, $dk_{ng} = dk_{nc}$. Then from Eqs. 4.197 and 4.199 we find that

$$\frac{k_{ng}(E)}{k_{nc}(E)} = \frac{1 - f_T}{f_T}\frac{f_e(E)}{1 - f_e(E)}. \tag{4.203}$$

Now,

$$\frac{f_e(E)}{1 - f_e(E)} = \exp(-(E - E_F)/k_BT) \tag{4.204}$$

such that Eq. 4.203 becomes

$$\frac{k_{ng}(E)}{k_{nc}(E)} = \exp(-(E_T - E)/k_BT). \tag{4.205}$$

Substituting this back into Eq. 4.201 leads, after some algebra, to

$$\begin{aligned} dR_n^{SRH} &= [1 - \exp(-(E_T - E_F)/k_BT)] \\ &\quad \times (1 - f_T)f_e(E)k_{nc}(E)g_{LUMO}(E)N_T dE. \end{aligned} \tag{4.206}$$

To find the net recombination rate, R_n^{SRH}, Eq. 4.206 is integrated over energies greater than E_{LUMO}. Except at very high n, or low temperatures, $f_e(E)$ can be replaced by the Maxwell–Boltzmann approximation. Also, the capture rate can be expressed in terms of the electron capture cross section by the trap, $\sigma_e(E)$, and the thermal velocity of the electron, v_e, viz.

$$k_{nc}(E) = \sigma_e(E)v_e, \tag{4.207}$$

where $v_e = \sqrt{3k_BT/m^*}$. Then we can write for the product of the electron density, n, and the weighted cross section, $\bar{\sigma}_e$:

$$\begin{aligned} n\bar{\sigma}_e &= \int_{E_{LUMO}}^{\infty} f_e(E)\sigma_e(E)g_{LUMO}(E)dE \\ &\approx \int_{E_{LUMO}}^{\infty} \exp(-(E - E_F)/k_BT)\sigma_e(E)g_{LUMO}(E)dE. \end{aligned} \tag{4.208}$$

Now, the electron density can be expressed in terms of the LUMO DOS, N_{LUMO}, as:

$$n = N_{LUMO}\exp(-(E_{LUMO} - E_F)/k_BT), \tag{4.209}$$

from which we finally obtain for the electron recombination rate:

$$\begin{aligned} R_n^{SRH} &= [n(1 - f_T) - N_{LUMO}\exp(-(E_{LUMO} - E_T)/k_BT)f_T]N_T v_e\bar{\sigma}_e \\ &= [n(1 - f_T) - n_1 f_T]N_T v_e\bar{\sigma}_e, \end{aligned} \tag{4.210}$$

where

$$n_1 = N_{LUMO}\exp(-(E_{LUMO} - E_T)/k_BT). \tag{4.211}$$

By a similar procedure we obtain the hole recombination rate:

$$R_p^{SRH} = [pf_T - p_1(1 - f_T)]N_T v_h\bar{\sigma}_h, \tag{4.212}$$

with

$$p_1 = N_{HOMO}\exp((E_{HOMO} - E_T)/k_BT). \tag{4.213}$$

In steady state, there must be detailed balance, that is, the rate of electron recombination must be equal that of hole recombination: $R^{SRH} = R_n^{SRH} = R_p^{SRH}$. Further, we can use Eqs. 4.210 and 4.212 to obtain f_T in terms of the energy weighted rates, $\bar{k}_{nc} = v_e\bar{\sigma}_e$ and $\bar{k}_{pc} = v_h\bar{\sigma}_h$:

$$f_T = \frac{\bar{k}_{nc}n + \bar{k}_{pc}p_1}{\bar{k}_{nc}(n + n_1) + \bar{k}_{pc}(p + p_1)}. \tag{4.214}$$

It is convenient to write the trap lifetimes in terms of their respective rates as follows: $\tau_n = 1/N_T\bar{k}_{nc} = 1/N_T\bar{\sigma}_e v_e$ and $\tau_p = 1/N_T\bar{k}_{pc} = 1/N_T\bar{\sigma}_h v_h$. Using these relationships, and Eq. 4.214 in Eq. 4.210 or 4.212, we finally arrive at the SRH recombination rate (Hall, 1951, Shockley and Read, 1952):

$$R^{SRH} = \frac{np - n_1p_1}{\tau_e(p + p_1) + \tau_h(n + n_1)}. \tag{4.215}$$

Now, if the trap exists far from the center of the energy gap, for example if it lies near to the LUMO, we have $n_1 \gg p_1$, and vice versa for traps near to the HOMO. In the former case, $n_1 \to n \gg p$, and the trap behaves as an electron donor rather than a recombination center. Also, we would expect that in this situation, $\tau_p \gg \tau_n$, such that the rate limiting step is recombination at the state farther from its respective transport energy level. All of these factors lead to $R^{SRH} \to 0$. Hence, effective recombination centers are mid-energy gap states where $E_T \approx E_G/2$. With these approximations and Eqs. 4.211 and 4.213, we find

$$n_1p_1 = N_{LUMO}N_{HOMO}\exp(-E_G/k_BT) = n_i^2, \tag{4.216}$$

in which case the SRH recombination rate simplifies to

$$R^{SRH} = \frac{np - n_i^2}{\tau_e(p + p_1) + \tau_h(n + n_1)}. \tag{4.217}$$

The recombination rate has an exponential dependence on temperature arising from the charge densities, p_1, n_1, and n_i^2. Indeed, one of the signatures of SRH recombination is the dark current that varies exponentially with T for large trap densities.

Disordered organic semiconductors are characterized by a distribution of traps whose density distribution is $g_T(E_T)$ rather than a single, discrete mid-gap trap. For this situation, we integrate over the trap distribution to obtain the modified expression:

$$R^{SRH} = \int_{E_{HOMO}}^{E_{LUMO}} \frac{np - n_i^2}{\tau_e(p+p_1) + \tau_h(n+n_1)} g_T(E_T) dE_T. \tag{4.218}$$

4.5.3 Auger recombination

Auger recombination involves the recombination of an electron and a hole as in Fig. 4.72b. The energy gained in recombination is transferred to a second electron that is elevated to a higher energy than the LUMO. The hot electron subsequently "cools," that is, it loses its excess energy by returning to the LUMO level, and subsequently relaxes. The initial condition is the existence of two electrons and a hole, and the final condition is only a single electron (with an analogous argument made for Auger recombination of holes). Thus, the probability for Auger recombination requires a three particle interaction proportional to n^2p (or p^2n for holes). Following the procedure used to calculate SRH recombination, we define the electron and hole capture rates as $k_{nc} = C_{nc}n^2p$ and $k_{pc} = C_{pc}np^2$, where C_{nc} and C_{pc} are the Auger recombination capture coefficients for electrons and holes, respectively. This process must conserve both energy and momentum: that is the energy transferred to the secondary electron must be equal to that lost by the initial electron during recombination, and the momentum (**k**) must also be equal in the initial and final states.

Auger recombination is the opposite of the *avalanche multiplication* process, where a high energy (hot) electron accelerated in an electric field suffers an ionizing collision with a molecule, generating a secondary electron–hole pair. In that case, the initial condition is a single electron, and the final state is two electrons plus a hole.

Electron and hole generation rates are defined as $k_{ng} = C_{ng}n$ and $k_{pg} = C_{pg}p$. In thermal equilibrium the emission and capture rates are equal. From the law of mass action, we obtain $C_{ng} = C_{nc}n_i^2$ and $C_{pg} = C_{pc}n_i^2$.

The net Auger recombination rate is given by: $R^{Auger} = (k_{nc} - k_{ng}) + (k_{pc} - k_{pg})$. Hence,

$$R^{Auger} = (C_{nc}n + C_{pc}p)(np - n_i^2). \tag{4.219}$$

Due to its dependence on n^2 and p^2, Auger recombination is important at very high charge density, which is in contrast to R^{SRH} that is present even as $n, p \to 0$.

4.5.4 Langevin recombination

Finally, we consider bimolecular Langevin recombination where a free electron polaron is captured by a hole polaron, followed by non-radiative recombination, as illustrated in Fig. 4.73. When the carriers approach within a *capture radius*, r_c, their Coulomb attraction exceeds the thermal energy, preventing escape. Thus, r_c is the distance at which the Coulomb energy is equal to k_BT:

$$r_c = \frac{q^2}{4\pi\varepsilon_r\varepsilon_0 k_B T}. \tag{4.220}$$

For a relative dielectric constant of $\varepsilon_r = 3$ typical of many organic materials, the capture radius at room temperature is $r_c \approx 15$ nm. In organics, the mean free paths are $\lambda \sim 0.4$ nm, which is on the order of the molecular separation (Helfrich and Schneider, 1965). Thus, $r_c \gg \lambda$. This meets the condition for Langevin recombination where the charges are fully thermalized and hence do not follow ballistic trajectories (Langevin, 1903).

The classical treatment of Langevin recombination assumes the material is isotropic, and that the charge motion is three dimensional. As we showed in Chapter 2, organic materials are rarely isotropic, and the presence of an electric field imposes a direction on the current. Nevertheless, these approximations can lead to simple expressions for the Langevin recombination rate, R^L.

The hole current due to the field, F, from a nearby electron is given by

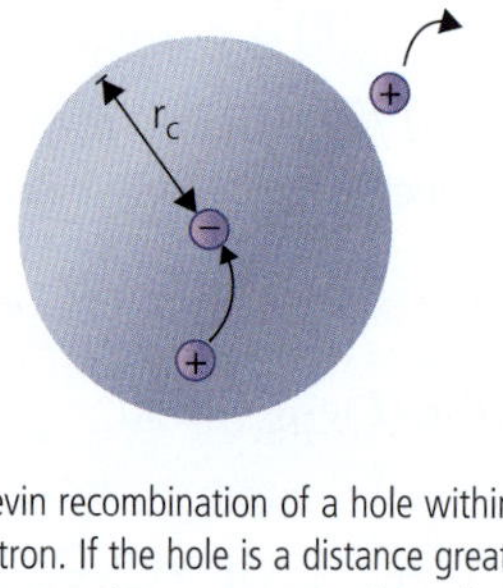

Figure 4.73 Langevin recombination of a hole within the capture radius, r_c, of an electron. If the hole is a distance greater than r_c, there is insufficient Coulomb attraction to overcome thermal energy, and no recombination occurs.

$$\mathbf{j}_h = qp_1 v_{rel} = qp_1(\mu_e + \mu_h)\mathbf{F} = \frac{q^2 p_1(\mu_e + \mu_h)}{4\pi\varepsilon_r\varepsilon_0 r_c^2}\hat{\mathbf{r}}, \tag{4.221}$$

where the relative approach velocity of the electron and hole is proportional to the sum of their individual mobilities: $v_{rel} = (\mu_e + \mu_h)F$. The thermalized hole current flowing toward the electron is along the radial direction, $\hat{\mathbf{r}}$. The sphere of radius r_c has a surface area $4\pi r_c^2$, resulting in a current flux of

$$\phi_h = \frac{q^2 p_1(\mu_e + \mu_h)}{\varepsilon_r\varepsilon_0} = \gamma_L(qp_1), \tag{4.222}$$

where the Langevin recombination rate constant is

$$\gamma_L = \frac{q}{\varepsilon_r\varepsilon_0}(\mu_e + \mu_h) = \frac{q}{\varepsilon_r\varepsilon_0}\mu_T. \tag{4.223}$$

Here, μ_T is sum of the mobilities of electrons and holes. The total rate of recombination is equal to the product of the hole current flux and the electron density, $\phi_n n$, giving

$$R^L = \gamma_L(p_1 n_1) = \gamma_L(pn - n_i^2). \tag{4.224}$$

As for other recombination processes, R^L is important when p and n are large, in which case Eq. 4.224 simplifies to $R^L \approx \gamma_L pn$.

The dominant role of Langevin recombination in both small molecule and polymer systems has been verified by several experiments spanning at least five decades. An investigation of bipolar conduction in anthracene was perhaps the first direct substantiation of this recombination mechanism (Kepler and Coppage, 1966). In that work, anthracene crystals were irradiated with ~0.25 MeV X-rays to create a high density of secondary electrons and holes that were subsequently collected in an external circuit. The total charge generated by each 0.1 μs X-ray pulse gives the non-equilibrium electron and hole densities. In the absence of recombination, the relationship between charge collected and X-ray intensity should be linear. Deviations from linearity therefore provide the number of electrons and holes that recombine prior to collection. That is, since $n = p$ in optically pumped materials, Eq. 4.224 gives $\frac{\partial n}{\partial t} \cong -\gamma_L n^2$.

Figure 4.74 shows the charge collected vs. total x-ray energy deposited (which is proportional to n), at different applied voltages. The straight line corresponds to no recombination (i.e. the charge collected is proportional to the number generated), whereas the curved lines are fits to Eq. 4.224 assuming $\gamma_L = 3 \times 10^{-6}$ cm^3/s. The density of charge

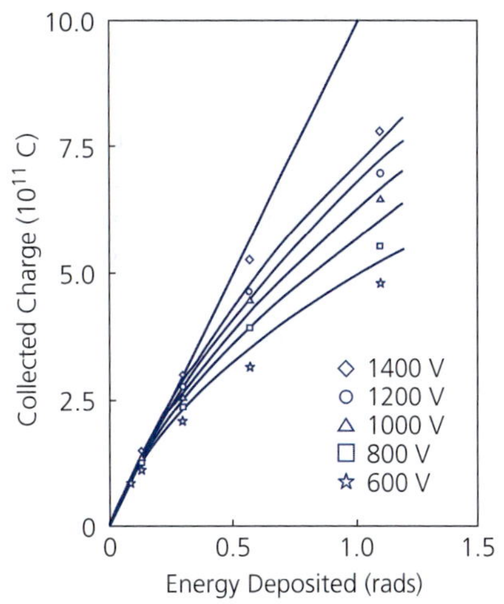

Figure 4.74 Charge collected vs. X-ray energy deposited in a single crystal of anthracene grown by the Bridgeman process (see Section 5.4.1). The data (points) are taken for several different applied voltages. The curved lines are fits to Langevin recombination theory assuming a voltage-independent recombination rate of $\gamma_L = 3 \times 10^{-6}$ cm^3/s. The straight line corresponds to $\gamma_L = 0$ (Kepler and Coppage, 1966).

decreases with increasing voltage as a result of their higher velocities.

Langevin recombination has also been identified by comparing the j–V characteristics of hole-only and double injection devices comprising a PPV active region. Charge injection is controlled by the contact barriers. The ITO/PPV/Au devices permit only hole injection, whereas ITO/PPV/Ca devices inject holes from ITO and electrons from the Ca cathode. From the diffusion equations Eqs. 4.186 and 4.187, we model the one-dimensional j–V characteristics under double injection using (Blom et al., 1997a)

$$j = q\mu_p F(x)p(x) + q\mu_n F(x)n(x), \tag{4.225}$$

$$\frac{\partial F(x)}{\partial x} = \frac{q}{\varepsilon_r\varepsilon_0}[p(x) - n(x) - n_T(x)], \tag{4.226}$$

and

$$\frac{\partial n}{\partial t} = -\frac{\partial p}{\partial t} = \gamma_L p(x)n(x). \tag{4.227}$$

Here, $n_T(x)$ is the trap density at distance x from the Au or Ca cathode given by $n_T(x) = N_T\left(\frac{n(x)}{N_{LUMO}}\right)^{T/T_0}$ (cf. Eq. 4.142). The hole mobility is assumed to be thermally activated, viz.

$$\mu_p(F) = \mu_0 \exp(-\Delta E_a/k_B T)\exp\left[\frac{\beta}{k_B}\left(\frac{1}{T} - \frac{1}{T_1}\right)\sqrt{F}\right], \tag{4.228}$$

with a similar expression for electrons (see Section 4.3.2). It is assumed $\mu_n = \mu_p$. Here, ΔE_a is the activation energy and μ_0, β, and T_1 are constants. In Fig. 4.75a we show fits to a numerical solution to Eqs. 4.225–4.227 for the hole only ($n(x) = 0$), and double injection devices using $\Delta E_a = 0.59$ eV, $\mu_0 = 1.7 \times 10^{-5}$ cm^2/V s, $\beta = 3.13 \times 10^{-4}$ eV(cm/V)$^{1/2}$, $T_1 = 540$ K, $N_T = 5 \times 10^{17}$ cm^{-3}, and $T_0 = 1500$ K. The difference between the two sets of data is attributed to Langevin recombination in the double injection device which is absent when the current is due only to holes in the ITO/PPV/Au device.

From the double injection fits, the Langevin coefficient is found and plotted vs. 1000/T in Fig. 4.75b. At room temperature, $\gamma_L = 2 \times 10^{-12}$ cm^3/s, with thermal activation arising from the electron and hole mobilities. The theoretical fit (solid line) gives the same trends but with a value slightly below those observed. This difference is due to the asymmetry between the electron and hole mobilities, where the "faster" charge reaches a grain boundary within the film, and then "waits" for the slower charge to arrive and recombine (Groves and Greenham, 2008). Although there is a minor quantitative difference in γ_L between theory and experiment, the trends are similar, indicating the importance of Langevin recombination and its dependence on mobility in polymers.

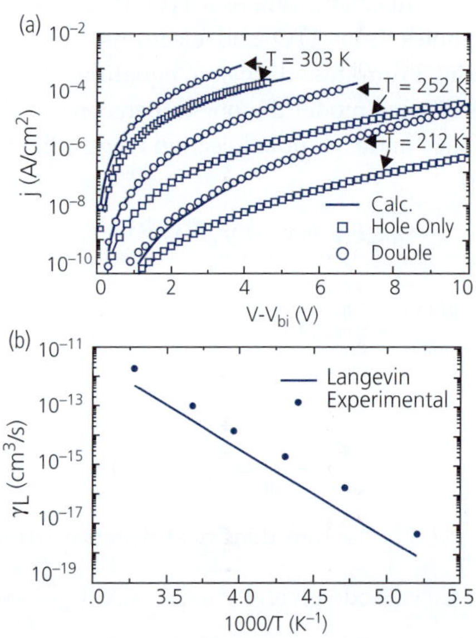

Figure 4.75 (a) Fits (lines) to the measured j–V characteristics (data points) of hole-only and double injection PPV devices based on space charge limited and diffusion theory described in text. (b) Langevin recombination coefficient vs. temperature (T) as obtained from a theoretical fit (line) to the data in (a) (Blom et al., 1997a).

4.5.5 Long range charge diffusion in organic thin films

We have seen that numerous recombination processes can conspire in organics to decrease the charge lifetime, and hence its diffusion length $L_D = \sqrt{D\tau}$. The origins range from intrinsic disorder, to both structural and impurity-related defects within the films or crystals. Yet, if the structures are sufficiently free of defects, the charge carrier diffusion lengths in organics can be as long as those found in the highest quality inorganic semiconductors. Indeed, electron diffusion lengths (L_{De}) exceeding several centimeters have been reported for fullerene channels that have nearly defect-free interfaces with energy barriers that confine the charges within the conducting material (Burlingame et al., 2018).

Devices with the structure shown in Fig. 4.76a were used to quantify electron transport over long distances in a fullerene (C_{60} and C_{70}) channel. The device active region beneath a Ag cathode comprises a DTDCTB:C_{70} (1:1) donor/acceptor junction abutting a neat fullerene electron transport channel. A wide energy gap Bphen layer forms the second channel surface. From the energy level diagram in Fig. 4.76b, the conduction

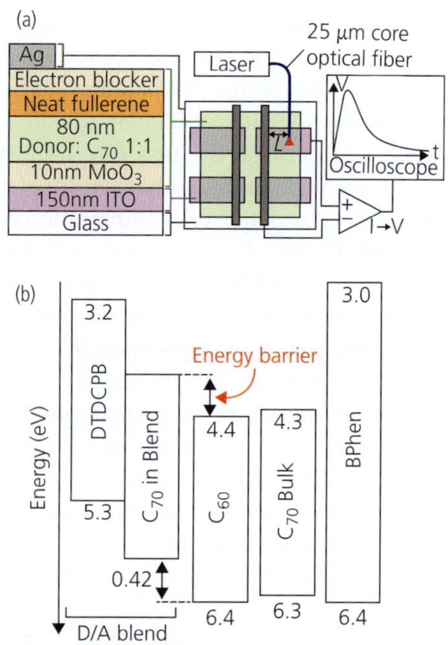

Figure 4.76 (a) Device structure, layout and experimental setup for measuring electron diffusion lengths in a neat fullerene channel. (b) Energy levels of the materials used in the structure in (a) (Burlingame et al., 2018).

(LUMO) level of the channel is well below the barriers presented by the Bphen and mixed HJ.

Electrons are injected into the channel by locally illuminating the active region with a focused optical beam of 500 μs pulses at 405 nm wavelength that is primarily absorbed by the donor. The generated excitons are dissociated at the donor–acceptor junction. The resulting hole is swept out at the anode while the electron enters the channel. By varying the distance, L, of the excitation illumination from the collecting cathode, the time of flight of electrons, as well as the attenuation of the signal due to recombination losses, can be determined.

As L increases from 1 to 10 mm, the amplitude and arrival time of the current in the C_{60} channel varies by nearly two orders of magnitude, although the total integrated charge collected for each transient is the same, independent of distance to the excitation source. This indicates that the electron diffusion length, L_{Dn}, is considerably greater than 1 cm. In contrast, the integrated signal decreased by 50% over a distance of 5 mm in C_{70} channels, suggesting that L_D is small compared with C_{60}.

The temperature dependence of the current transients in C_{60} and C_{70} devices was also obtained, with results shown in Figs. 4.77a and b at $L = 2$ mm and $L = 1$ mm, respectively. The current transients were fit using the electron diffusion model in Eq. 4.186 as shown by solid lines in Figs. 4.77c and d.

The lifetime of electrons in the channel is determined by the rates of trapping at defects and thermal emission into the HJ where recombination can occur. A sufficiently high HJ energy barrier and low defect densities are therefore required to enable transport over macroscopic distances. In devices with a DTDCTB donor, the energy levels of C_{70} undergo a polarization shift due to the large dipole moment (14.5 D) of this asymmetric molecule (Ting et al., 2014, Griffith et al., 2015). This shift forms the required energy barrier at the C_{60} channel/HJ interface (0.42 ± 0.1 eV for DTDCPB) that confines electrons within the channel. Materials presenting smaller barriers result in electron emission from the channel and subsequent recombination with holes in the underlying HJ. The centimeter-scale L_{Dn} observed suggests that in addition to the large E_B, the channel and its

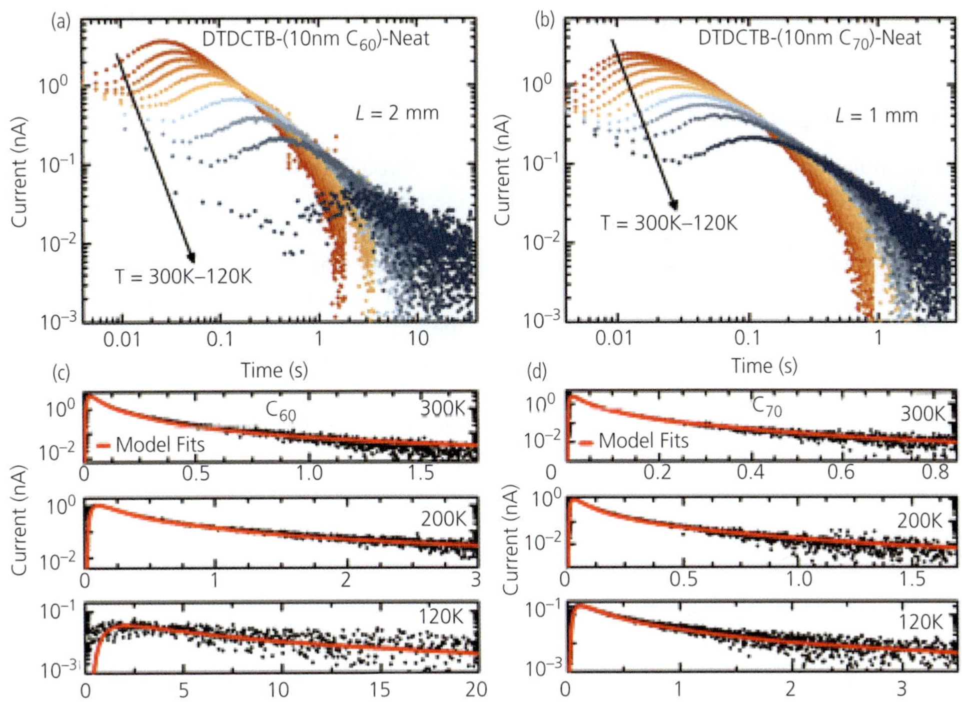

Figure 4.77 Temperature dependent transient photocurrent data at 20 K intervals from 300 K to 120 K in response to 2 ms pulses of 637 nm wavelength illumination on a (a) 10 nm C_{60} channel at $L = 2$ mm and (b) a 10 nm C_{70} channel at $L = 1$ mm. Data (points) and corresponding charge diffusion model fits (lines) at 300 K, 200 K, and 120 K for the (c) C_{60} and (d) C_{70} devices. Lines show best fits to the data, from which diffusivity, D_n, and recombination rate, k_n, are obtained (Burlingame et al., 2018).

interfaces have a remarkably low density of deep electron traps and recombination centers. This is surprising for fullerenes, which, despite their high mobility and diffusivity among molecular solids, form disordered and phase-separated amorphous and crystalline domains (Liu et al., 2016b).

Drift-dominated lateral spreading of charges over long time periods has also been observed in unipolar devices at organic/insulator interfaces due to the lack of recombination (Bürgi et al., 2003). In this case, electron transport is entirely diffusive through the bulk, and is observed despite the presence of optically generated holes.

Replacing the C_{60} channel with C_{70} reduces D_n from 0.67 cm^2/s to 0.16 cm^2/s at room temperature. Using the Einstein relation, the room temperature mobilities of $\mu_n = 26$ cm^2/V s for C_{60} and 6 cm^2/V s for C_{70} are inferred. In general, mobilities in bulk semiconductors are larger than those at interfaces due to interfacial traps, while mobilities and diffusivities in disordered semiconductors typically increase with charge density due to filling of shallow traps and low-energy states in the conduction band tail, as discussed in Section 4.3. The devices in Fig. 4.76 are nearly ideal for achieving high mobility, as charge transport occurs in the bulk of the channel with electron densities $>10^{17}$ cm^{-3}.

4.6 Injection from contacts

The most simple rectifying junction is the semiconductor–metal contact, or *Schottky barrier diode.* Schottky barrier devices were used in the earliest demonstrations of organic solar cells (Simon and André, 1985). In contrast to the Schottky barrier, some metals and process conditions lead to very small or non-existent energy barriers. The barriers are thus easily surmounted by charge under low injection conditions. The resulting Ohmic contacts are formed by doping the surface of the organic with a charge donating impurity, or by damage to the organic during metal contact deposition. We will show in Section 4.6.3, that molecular disorder near the metal–organic interface can also result in efficient charge injection. Before one can develop the contact chemistry and properties suited to a particular application, it is necessary to understand the factors that determine the energy barrier at the organic surface. In the following section we consider many of the most important properties of contacts, and means for their analysis and ultimately, their control.

4.6.1 The ideal Schottky barrier

The metal–semiconductor contact is an essential element in every electronic device. Up until this point, we have considered current limited by the properties of the thin film itself, whether that limitation arises from intrinsic properties such as disorder, or from extrinsic characteristics such as traps, dopants, or impurities. However, in many cases, the conduction is limited by injection from contacts due to an energy barrier at the interface between the electrode and the semiconductor.

The fundamental origin and treatment of charge injection from contacts was developed by Schottky and Mott for the case of a trap-free metal–semiconductor junction (Mott, 1938, Schottky, 1938). The formation of this ideal Schottky barrier diode is illustrated in Fig. 4.78. A metal with a work function, ϕ_m is brought into contact with a semiconductor with electron affinity, EA (Fig. 4.78a). We will assume for this example that the material is n-type, and that $EA < \phi_m$. As the gap, δ, between the metal and semiconductor is reduced, alignment of E_F with $q\phi_m$ results in the formation of a surface dipole that repels electrons from the interface that terminates further transfer of electrons (Fig. 4.78b) once the Fermi energies of the metal and semiconductor are completely aligned (Fig. 4.78c). The energy (or Schottky) barrier is then

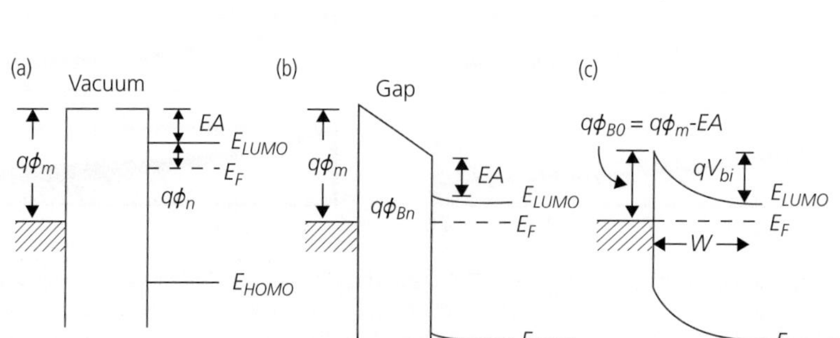

Figure 4.78 Schottky barrier formation defining energies and dimensions used in text.

simply equal to the difference between the work function and electron affinity:

$$q\phi_{B0} = q\phi_m - EA. \tag{4.229}$$

The band bending required to maintain an equilibrium charge distribution extends over the depletion region width, W. The junction built-in potential is then

$$qV_{bi} = q\phi_{B0} - E_F. \tag{4.230}$$

According to Eq. 4.229, the barrier height varies linearly with ϕ_m. A compilation of work functions for elemental metals is shown in Fig. 4.79. In practice, the surface of the semiconductor is never completely free of defect states. Hence, $q\phi_m$ measured at the metal surface (using, for example, UPS) is not the same as the bulk EA of that same metal (Michaelson, 1978). Furthermore, in organics, the deposition of the metal induces defect states due to the loss of kinetic energy of the metal atom as it is adsorbed onto the organic surface. Most metals are evaporated at temperatures >1000°C. The latent heat of condensation as the metal atoms are adsorbed can be sufficient to break molecular bonds, resulting in a defect at the initial point of contact. The resulting traps typically reside within the HOMO–LUMO gap, reducing the energy barrier to below the ideal value calculated using Eq. 4.229 (Ioannidis et al., 1998, Parthasarathy et al., 1998).

Trap formation during electrode deposition is found to be an effective means for forming Ohmic contacts, that are defined by $q\phi_{Bn} \lesssim k_BT$. Here, $q\phi_{Bn}$ is the barrier height including image charge effects (Section 4.6.2). An Ohmic contact is one whose j–V characteristics follow Ohm's law, that is, there is no inflection of the current about $V = 0$. To ensure Ohmic contact formation, a low work function metal (e.g. Mg, Ca, Cs) is preferred for injection into an electron transporting layer, or a high work function metal (e.g. Al, Ag, Pd, Au) for hole injection. We will return to the effects of interface traps and interface dipole layers that result in deviations in barrier height from Eq. 4.229 in subsequent sections.

To calculate the current across a metal–semiconductor barrier, we consider all potential sources illustrated in Fig. 4.80. Process **1** is due to electron thermionic emission over the barrier, **2** is thermionic emission-assisted tunneling, **3** is via recombination of electrons and holes, **4** is electron diffusion, and **5** is minority carrier hole diffusion. Schottky barrier injection is typically unipolar, and hence we can ignore process **5** due to the very low minority carrier concentration.

Most treatments of transport in inorganic semiconductors focus on thermionic emission. To surmount the barrier, the charges must have a

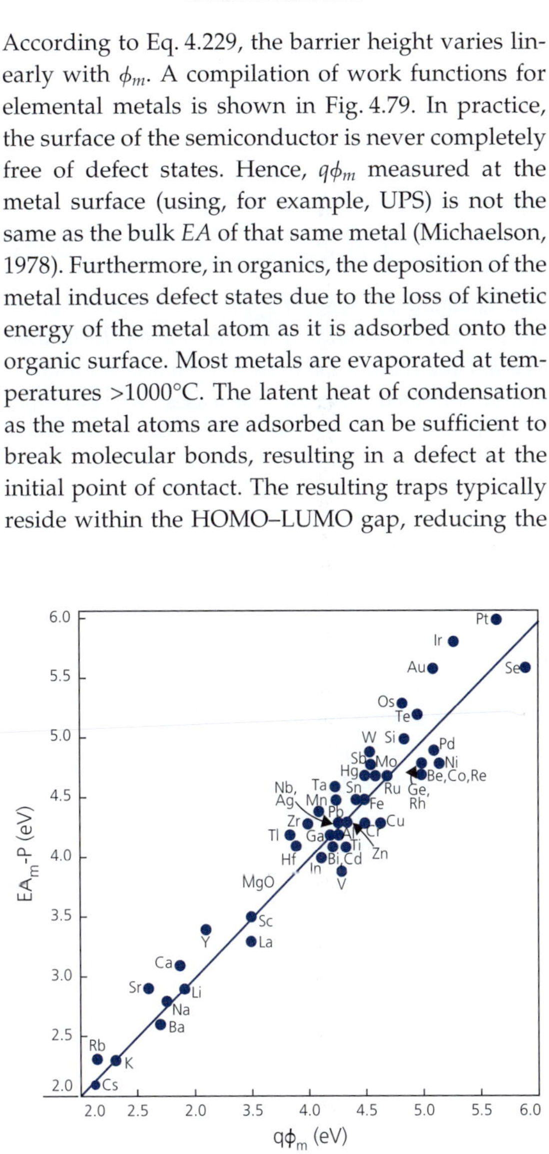

Figure 4.79 Work functions of elemental metals. The vertical axis is the metal electron affinity to within a correction, P, for each periodic grouping of metals (e.g. alkali halides, transition metals, etc.) Here EA_m is a bulk property of the metal and $q\phi_m$ is a surface property. The factor P (which can be as large as 1 eV) corrects for the measured differences between these quantities (Michaelson, 1978).

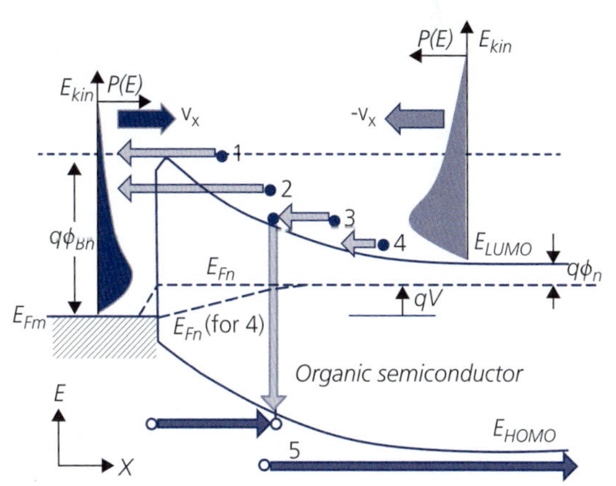

Figure 4.80 Sources of current across a metal/n-type Schottky junction under forward bias, V. The barrier height is $q\phi_{Bn}$. The sources are **1**, thermionic emission, **2**, tunneling, **3**, recombination, **4** majority carrier (electron) diffusion, and **5** minority carrier (hole) diffusion. The electron kinetic energy (E_{kin}) probability distributions ($P(E)$) impinging on the junction from the metal (left) and semiconductor (right) are shown. The upper horizontal dashed line is the energy above which thermionic emission occurs. The directions of electron velocities (v_x) are shown.

minimum kinetic energy, $E_{kin}^{-} = \frac{1}{2}mv_x^2 > qV_{bi}$ for charges traveling with velocity $-v_x$ toward the junction. That is, $E_{kin}^{+} > q\phi_{Bn}$ for charges with velocity $+v_x$. Then, to calculate the total current, we add the contributions from each direction, $j = j(+v_x) + j(-v_x)$, by integrating over a *Maxwell–Boltzmann distribution* of energies (and hence velocities) multiplied by the DOS. From the kinetic theory of gasses, the probability of finding a particle with velocity between v_x and $v_x + dx$ is

$$f(v_x)dv_x = \left(\frac{m^*}{2\pi}\right)^{1/2} \exp\left(-\frac{1}{2}m^*v_x^2/k_BT\right)dv_x, \quad (4.231)$$

where only velocities in one dimension are used to model electrons incident on the interface. The probability distribution function at different thermal energies is plotted in Fig. 4.81. As temperature increases, the distribution broadens and the peak decreases.

The classical picture for inorganic semiconductors depicted in Fig. 4.80 must be modified for organics that lack broad, parabolic conduction and valence bands. As we have shown previously, the DOS of the conduction level (for electrons, the LUMO) is more accurately represented by a Gaussian distribution whose origin lies in the inherent disorder of the solid. The Gaussian width may be sufficiently broad that it overlaps into higher lying conduction levels (e.g. LUMO+1) that are also broadened. Thus, as previously, the total number of conduction electrons incident on the barrier with velocities, $-v_x$, is

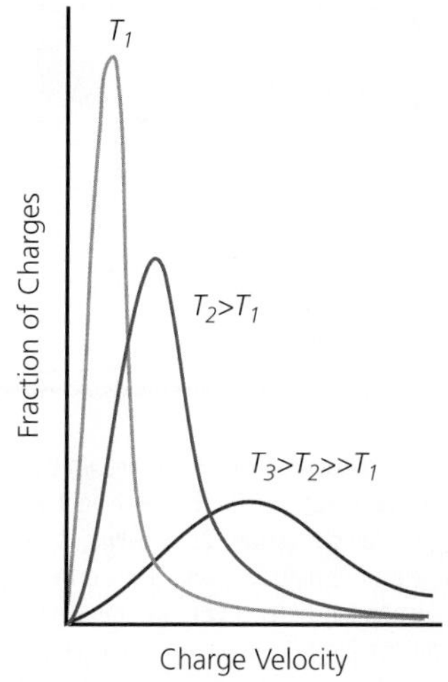

Figure 4.81 Maxwell–Boltzmann distribution at several different temperatures.

$$n(-v_x) = N_{\text{LUMO}} \int g_{LUMO}(E_{kin}) f(E_{kin}) dE_{kin}. \quad (4.232)$$

In forward bias, V measured relative to the metal contact is positive. Then $E_{kin}^{-} > E_{kin}^{+}$ and the current increases exponentially with voltage; that is, $j(-v_x) \sim \exp(qV/k_BT)$. In reverse bias, $V < 0$, such that $j \to j(v_x) \sim \exp(-q\phi_{Bn}/k_BT)$. Thus, it is limited by thermionic emission over the Schottky barrier. Summing these contributions yields the familiar expression for the current across an ideal (i.e. trap-free) metal–inorganic semiconductor junction (Sze, 1981):

$$j = j_0(\exp(qV/k_BT) - 1), \quad (4.233)$$

where for the specific case of thermionic emission of a Maxwell–Boltzmann distribution of electron kinetic energies, the saturation current is

$$j_0 = A^*T^2\exp(-q\phi_{Bn}/k_BT). \quad (4.234)$$

The constant $A^* = 4\pi qm^*k_B^2/h^3$ for a conventional semiconductor is known as the *effective Richardson constant* for thermionic emission. Hence, thermionic emission-limited current decreases exponentially with increasing barrier height and decreasing temperature. The Richardson constant is derived by assuming a parabolic DOS at the semiconductor band edge. It is left to Problem 4.9 to derive the constant for a Gaussian DOS more relevant to organic semiconductors.

We will show in Section 4.7 that, independent of the physical origins of the current (i.e. diffusion vs. thermionic emission, etc.) or the nature of the junction (e.g. metal–semiconductor, p-n junction, or HJ) the ideal diode current always has the form in Eq. 4.233 since the underlying process leading to the dark current is charge recombination. This can occur at contacts, within the semiconductor bulk, or at junctions between dissimilar semiconductors.

In the presence of illumination at wavelength, λ, carriers generated introduce an additional photocurrent:

$$j_{ph}(\lambda) = \eta_{EQE}(\lambda)\frac{q\lambda P_{opt}(\lambda)}{hc}, \quad (4.235)$$

where $P_{opt}(\lambda)$ is the optical power, and $\eta_{EQE}(\lambda)$ is the external quantum efficiency of the diode, equal to the ratio of photogenerated charges detected in an external circuit to the total number of incident photons. If the incident light comprises more than a single wavelength, then j_{ph} is obtained by a suitable integration over λ. The photocurrent flows in the direction of the leakage, j_0, and is therefore subtracted from the expression, Eq. 4.233.

Equation 4.233 is generalized to include the effects of charge trapping (see Section 4.7) and photocurrent to give

$$j = j_0(\exp(qV/nk_BT) - 1) - j_{ph}, \tag{4.236}$$

where n is the *ideality factor*. For an ideal diode without recombination, $n = 1$ as in Eq. 4.233. However, in the presence of recombination, $1 < n < 2$, where $n = 2$ corresponds to recombination via mid-gap states. The dark current characteristics of diodes with $n = 1$ and 2 and under illumination are shown in Fig. 4.82.

Although Eq. 4.234 has frequently been used to model metal/organic junction characteristics (Singh and Prakash, 2012, Liu et al., 2014), it is almost certainly incorrect (Arkhipov et al., 1998). Indeed, it has been found that A^* obtained for conventional semiconductors can be ten orders of magnitude larger than observed in organic thin films (see Problem 4.9) (Campbell et al., 1999, Wolf et al., 1999). We have shown previously that charge transport is primarily dominated by incoherent hopping, where the charge mean free path is on the order of the intermolecular separation. Even in ultrapure or ordered organic semiconductor crystals, coherent band-like transport can only occur over a few lattice constants due to their large effective masses and narrow energy BWs. Hence, "hot" carriers arriving at the barrier from the semiconductor side of the junction never gain sufficient kinetic energy to allow for thermionic emission. It is, therefore, more reasonable to consider majority carrier diffusion (process **4** in Fig. 4.80) as the dominant transport mechanism. The electron current then found from the drift-diffusion equation:

$$\mathbf{j} = qn\overleftrightarrow{\mu}\mathbf{F} + qD\nabla n. \tag{4.237}$$

Considering that the current in one dimension is dominated by electrons, then

$$\begin{aligned} j &= q\left(n(x)\mu_n F(x) + D_n\frac{\partial n(x)}{\partial x}\right) \\ &= qD_n\left(\frac{n(x)}{k_BT}\frac{\partial E_{LUMO}(x)}{\partial x} + \frac{\partial n(x)}{\partial x}\right). \end{aligned} \tag{4.238}$$

Here, we use $F(x) = \frac{1}{q}\frac{\partial E_{LUMO}(x)}{\partial x}$ along with the Einstein relation in Eq. 4.36. Equation 4.238 is solved by multiplying both sides by $\exp(E_{LUMO}(x)/k_BT)$:

$$\begin{aligned} &j\exp(E_{LUMO}(x)/k_BT) \\ &= qD_n\left(\frac{n(x)}{k_BT}\frac{\partial E_{LUMO}(x)}{\partial x} + \frac{\partial n(x)}{\partial x}\right)\exp(E_{LUMO}(x)/k_BT) \\ &= qD_n\frac{\partial}{\partial x}[n(x)\exp(E_{LUMO}(x)/k_BT]. \end{aligned} \tag{4.239}$$

Integrating both sides from $x = 0$ (i.e. the location of the metal/organic contact) to the edge of the depletion region at $x = W$ yields

$$j = \frac{qD_n\{n(x)\exp(E_{LUMO}(x)/k_BT)\}|_0^W}{\int_0^W \exp(E_{LUMO}(x)/k_BT)dx}. \tag{4.240}$$

Using the relationships between various energies in Fig. 4.80 and listed in Table 4.4, it follows that

$$j = \frac{N_{LUMO}\mu_n k_BT(\exp(qV/k_BT))}{\int_0^W \exp(E_{LUMO}(x)/k_BT)dx}. \tag{4.241}$$

Our final task is to determine W, and evaluate the integral in Eq. 4.240. From a solution to Poisson's equation for a one-sided abrupt junction, which is indeed the case for a metal–semiconductor junction, the depletion region width is (Streetman and Banerjee, 2006)

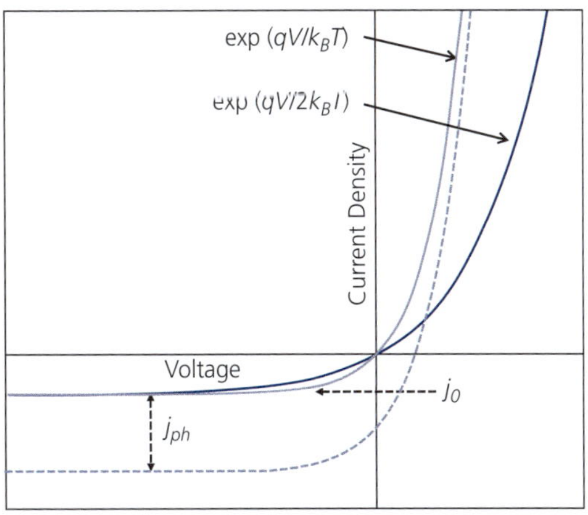

Figure 4.82 Diode characteristics in the dark (solid lines) and under illumination (dashed line) following Eq. 4.236 with ideality factors, $n = 1$ and 2, and photocurrent, j_{ph}.

Table 4.4 Variables used in Fig. 4.80.

x	$n(x)$	$E_{LUMO}(x)$
0	$N_{LUMO}\exp(-q\phi_{Bn}/k_BT)$	$q\phi_{Bn}$
W	$N_{LUMO}\exp(-q\phi_n/k_BT)$	$q(\phi_n + V)$

$$W = \begin{cases} \sqrt{\dfrac{2\varepsilon_r\varepsilon_0}{qN_D}(V_{bi}-V)} & W<d \\ d & \text{otherwise} \end{cases} \tag{4.242}$$

and the maximum field at the contact at $x = 0$ is

$$F_m(0) = \sqrt{\frac{2qN_D}{\varepsilon_r\varepsilon_0}(V_{bi}-V)} = \frac{2(V_{bi}-V)}{W}, \tag{4.243}$$

where N_D is the donor concentration (again, assuming an n-type semiconductor) and $V < 0$ in reverse bias. In many cases there are few if any free carriers in these typically wide energy gap organic semiconductors. Then, $N_D \approx 0$, and $W = d$, the total film thickness at all voltages. Solving Gauss' law we obtain

$$F(x) = F_m - \frac{qN_D}{\varepsilon_r\varepsilon_0}x, \tag{4.244}$$

from which it follows that

$$E_{LUMO}(x) = q\phi_{B0} - \frac{q^2N_D}{\varepsilon_r\varepsilon_0}\left(Wx - \frac{1}{2}x^2\right). \tag{4.245}$$

Substituting this into Eq. 4.241, we arrive at the current density across a metal–organic contact of the general form of Eq. 4.233 with a saturation current of

$$\begin{aligned} j_{0D} &= q\mu_n N_{LUMO}F_m \exp(-q\phi_{B0}/k_BT) \\ &= 2q\mu_n N_{LUMO}\frac{(V_{bi}-V)}{W}\exp(-q\phi_{B0}/k_BT) \end{aligned} \tag{4.246}$$

for $W < d$. When the organic thin film is undoped, and for $W > d$, then $F_m = (V_{bi} - V)/d$.

Note that the diffusion-limited current is similar to that for thermionic emission, although the former is dependent on the applied voltage-mobility product, whereas the latter has a T^2 dependence. Their ratio is given by

$$\frac{j_{0D}}{j_{0TE}} = \frac{2q\mu_n N_{LUMO}(V_{bi}-V)}{WA^*T^2}. \tag{4.247}$$

Hence, by measuring the forward biased current, the distinction between these mechanisms should be resolved. The assumption is that both the mobility and the DOS are temperature and voltage independent. While that is generally true of N_{LUMO} which is proportional to the molecular density, we have shown in Section 4.4 that μ_n is a function of both of these variables. Distinguishing these differences under reverse bias is possible if the only leakage source is the saturation current. However, j_0 tends to be small, and is therefore easily overwhelmed by other sources such as surface current around the periphery of the contact, shunt resistance, or recombination (process **3** in Fig. 4.80).

4.6.2 Barrier lowering and tunneling at metal–organic junctions

As charge diffuses toward the barrier, it can be transported from the semiconductor into the metal (or vice versa) via tunneling if the barrier is sufficiently low or narrow, corresponding process **2** in Fig. 4.80. Due to image-force and barrier lowering in an electric field, tunneling can substantially increase the current beyond that predicted in Eq. 4.233 for either thermionic emission or diffusion.

Image force lowering, known as the *Schottky effect*, is a result of classical electrostatics: a charge in the semiconductor is attracted to its image in the metal. This attraction results in an effective barrier height lowering of $q\Delta\phi$. From Coulomb's law the attractive force of the image charge at a distance x from the interface is

$$\mathcal{F}(x) = -\frac{q^2}{16\pi\varepsilon_r\varepsilon_0 x^2}. \tag{4.248}$$

In the presence of an electric field, an additional force of $-qF$ is added, such that the total potential energy (equal to the integral of $\mathcal{F}$ over x) is

$$q\Delta\phi(x) = -\frac{q^2}{16\pi\varepsilon_r\varepsilon_0 x^2} - qFx. \tag{4.249}$$

Setting the first derivative of Eq. 4.249 to zero, the energy is maximum at

$$x_m = \sqrt{\frac{q}{16\pi\varepsilon_r\varepsilon_0 F_m}}, \tag{4.250}$$

resulting in an energy barrier lowering of

$$q\Delta\phi = \sqrt{\frac{q^3F_m}{4\pi\varepsilon_r\varepsilon_0}}. \tag{4.251}$$

Substituting Eq. 4.243 for F_m then gives

$$\begin{aligned} q\Delta\phi(V) &= q\left[\frac{q^3N_D|\phi_{B0}-\phi_{Bn}|}{8\pi^3(\varepsilon_r\varepsilon_0)^3}\right]^{1/4} \\ &= q\left[\frac{q^3N_D|\phi_{B0}-V_{bi}-V|}{8\pi^3(\varepsilon_r\varepsilon_0)^3}\right]^{1/4}, \end{aligned} \tag{4.252}$$

and for a fully depleted sample of thickness d:

$$q\Delta\phi(V) \approx \left[\frac{q^3|\phi_{B0}-V_{bi}-V|}{2\pi\varepsilon_r\varepsilon_0 d}\right]^{1/2}. \tag{4.253}$$

An illustration of the energy levels in the presence of barrier lowering as a function of applied voltage is shown in Fig. 4.83. The total barrier height is therefore voltage dependent, and is reduced from ϕ_{B0} by

$$q\phi_{Bn} = q\phi_{B0}(V) - q\Delta\phi(V). \tag{4.254}$$

When the sample is forward biased, the barrier lowering is reduced ($q\phi_F < q\phi(0)$), whereas under reverse bias it is increased ($q\phi_F > q\phi(0)$); see Fig. 4.83.

Accounting for image force lowering, the diffusion saturation current for an undoped semiconductor becomes (cf. Eq. 4.246):

$$j_{0D} = q\mu_n N_{LUMO} \frac{V_{bi} - V}{d} \exp[-(q\phi_{B0} - q\Delta\phi(V))/k_B T]. \tag{4.255}$$

In Fig. 4.84 we show evidence for the Schottky effect in 0.73 μm thick dialkoxy-substituted PPV (OPPV) with a low work function Ca cathode (Rikken et al., 1994). Using internal photoemission spectroscopy, the energy barrier was determined by measuring the quantum yield (QY) of electrons excited from the Ca as a function of applied voltage. The QY is the ratio of the charge detected vs. the number of incident photons, and hence is proportional to j_{ph}. Only photons with sufficient energy to surmount the barrier result in photocurrent (block arrows, Fig. 4.84, inset). By varying the bias across the illuminated sample, a plot of QY vs. the difference between the barrier energy and the photon energy should follow (Fowler, 1931)

$$QY = j_{ph} \propto (E_{ph} - q\phi_{Bn})^2. \tag{4.256}$$

This expression implies that when $QY = 0$, then $q\phi_{Bn}(V) = E_{ph}$.

The data in Fig. 4.84 vs. reverse bias show the decrease in barrier height is consistent with Eqs. 4.253 and 4.254. The results of this experiment, including fits to Eqs. 4.252 and 4.253, are shown in Fig. 4.85. At low applied bias, the barrier maximum is far from the interface at $x_m = 30$ nm, implying that this region is nearly completely depleted. Hence, Eq. 4.253 applies and the data lie along a line where $q\phi_{Bn} \sim V^{1/2}$. As the bias is increased to $(V - V_{bi})^{1/2} = 1.2\text{V}^{1/2}$, the maximum moves towards the contact such that $x_m = 6.5$ nm, and the quartic dependence (i.e. $q\phi_{Bn} \sim V^{1/4}$) of

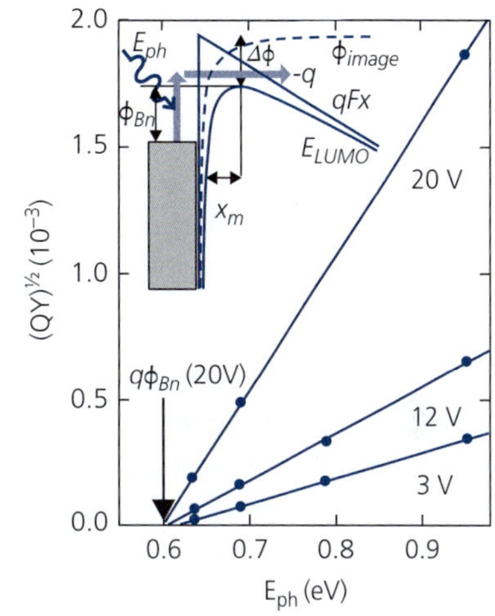

Figure 4.84 Photoemission quantum yield (QY) vs. photon energy (E_{ph}) for a Ca/OPPV junction. The intercepts at $QY = 0$ give the barrier height vs. voltage. The barrier height, ϕ_{Bn}, at 20 V is indicated. Inset: Photoemission experiment, showing the contributions to barrier lowering by the electric field F, and the image potential, ϕ_{image}. After Rikken et al. (1994).

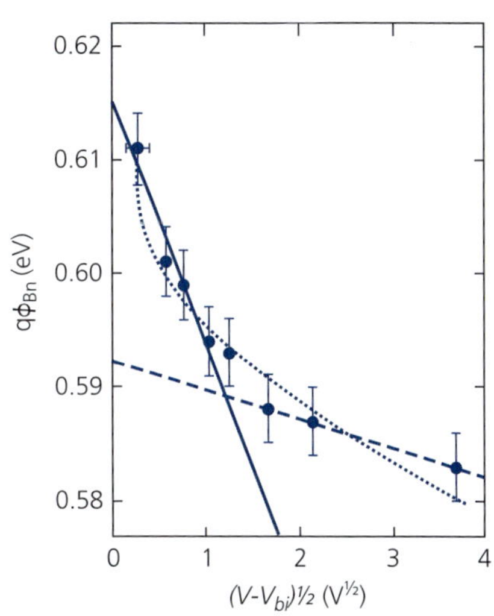

Figure 4.85 Barrier height of a Ca/OPPV Schottky barrier diode vs. reverse bias (data points) measured using data in Fig. 4.84. Straight lines are fits to the high and low voltage data assuming a quadratic dependence. The curved line is a fit to Eq. 4.253 with $N_D = 2 \times 10^{13}$ cm^{-3} and $V_{bi} = 1.4$V (Rikken et al., 1994).

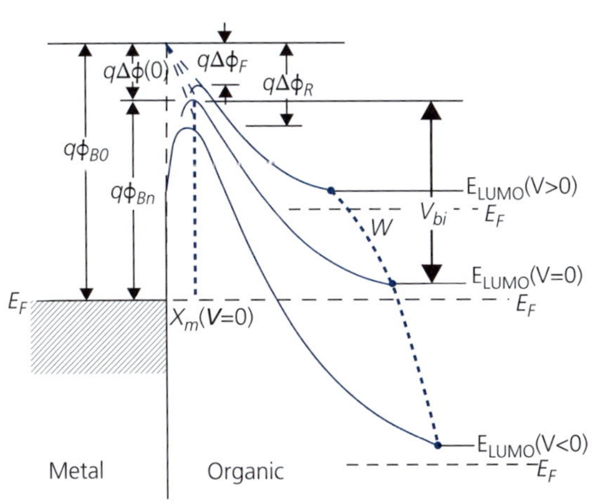

Figure 4.83 A Schottky barrier contact at equilibrium ($V = 0$), forward bias ($V > 0$), and reverse bias ($V < 0$). The equilibrium barrier height is $q\phi_{Bn} = q\phi_{B0} - q\Delta\phi(0)$. The curved dashed line shows the limit of the depletion region under the condition of $W < d$, the sample thickness. Also shown is the position of the barrier maximum at equilibrium, $x_m(V = 0)$.

Eq. 4.252 is observed. A fit to the data is shown by the dotted line in Fig. 4.85.

Since the width of the barrier decreases as energy is increased from the Fermi energy of the metal towards ϕ_{Bn}, it becomes increasingly transparent to charge conduction, giving rise to tunneling (process **2**, Fig. 4.80). In effect, this is equivalent to an additional lowering of the barrier beyond that due to image charges and the external electric field. The tunneling process can be calculated for these approximately triangular barriers using the Fowler–Nordheim model (Fowler and Nordheim, 1928). We briefly describe the approach to this derivation, leaving a detailed treatment to Problem 4.10.

As previously, we assume that the current in the $\hat{\mathbf{x}}$ -direction is equal to the difference of the currents injected from the metal to the semiconductor ($m \rightarrow s$) and vice versa ($s \rightarrow m$):

$$j_{tun} = j_{m\rightarrow s} - j_{s\rightarrow m}. \tag{4.257}$$

In terms of the densities of states, $g(k_x)$, Fermi–Dirac functions, $f(E)$, velocities, barrier transmission coefficients, $T(k_x)$, and the wavevector along $\hat{\mathbf{x}}$, k_x, the incremental current densities are given by

$$dj_{m\rightarrow s} = qT(k_x)v_x g_m(k_x) f_m(E)(1 - f_s(E))dk_x \tag{4.257a}$$

and

$$dj_{s\rightarrow m} = qT(k_x)v_x g_s(k_x) f_s(E)(1 - f_m(E))dk_x. \tag{4.257b}$$

Since $g(\mathbf{k}) = 1/4\,\pi^3$ and

$$g(k_x) = \int_0^\infty\int_0^\infty g(\mathbf{k})dk_y dk_z, \tag{4.259}$$

then, using $E_x = \hbar^2 k_x^2/2m^*$ and $v_x = \frac{1}{\hbar}\frac{\partial E}{\partial k_x}$, Eq. 4.258 becomes

$$dj_{m\rightarrow s} = \frac{q}{4\pi^3\hbar} T(E_x)dE_x \int_0^\infty\int_0^\infty f_m(E)(1 - f_s(E))dk_y dk_z \tag{4.260a}$$

and

$$dj_{s\rightarrow m} = \frac{q}{4\pi^3\hbar} T(E_x)dE_x \int_0^\infty\int_0^\infty f_s(E)(1 - f_m(E))dk_y dk_z. \tag{4.260b}$$

In polar coordinates, we substitute $E_r = \hbar^2(k_y^2 + k_z^2)/2m^*$ into Eq. 4.260. Subtracting the two currents gives

$$j_{tun} = \frac{qm^*}{2\pi^2\hbar^3} \int_{E_{\min}}^{E_{\max}} T(E_x)dE_x \int_0^\infty\int_0^\infty (f_m(E) - f_s(E))dE_r. \tag{4.261}$$

To substantially simplify the solution, we assume that the Fermi functions can be replaced by step functions, that is, $f_m(E) = 1$ for $E < E_{Fm}$ and $f_m(E) = 0$ for $E > E_{Fm}$, with a similar relationship for $f_s(E)$. For an n-type semiconductor, Eq. 4.261 thus becomes

$$j_{tun} = \frac{qm^*}{2\pi^2\hbar^3}\left(\int_{-\infty}^{E_{Fn}} T(E_x)(E_{Fm} - E_{Fn})dE_x + \int_{E_{Fn}}^{E_{Fm}} T(E_x)(E_{Fm} - E_x)dE_x\right). \tag{4.262}$$

Here, we have assumed that $E_{Fm} > E_{Fn}$. Since tunneling is between isoenergetic initial and final states. Since states below E_{Fn} are in the energy gap, the first integral vanishes.

To determine the tunneling probability, $T(E_x)$, we assume an approximately triangular barrier (Fig. 4.86). Using the WKB approximation, we assume that the electrons are plane waves (Merzbacher, 1961). Thus,

$$T(E_x) = \exp\left(-2\int_0^{x_T} k_x dx\right) = \exp\left(-\frac{2}{\hbar}\int_0^{x_T} \sqrt{2m^*(E_{LUMO}(x) - E_x)}dx\right), \tag{4.263}$$

where the factor of 2 comes from the relationship of $T \sim |\psi_e|^2$ (ψ_e is the electron wavefunction), and the integral is between the classical turning points of $x = 0$ and x_T in Fig. 4.86. Also, $E_{LUMO}(x) = q\phi_{Bn} - qFx$, where we have arbitrarily chosen E_{Fm} to be the reference energy. One classical turning point is at $x_T = (q\phi_{Bn} - E_x)/qF$. Substituting these expressions into 4.263 and integrating gives a transmission coefficient:

$$T(E_x) = \exp\left(-\frac{4\sqrt{2m^*}}{3\hbar qF}(q\phi_{Bn} - E_x)^{3/2}\right) \tag{4.264}$$

The argument in this expression approximates the barrier height "experienced" by a tunneling charge, which is less than $q\phi_{Bn}$.

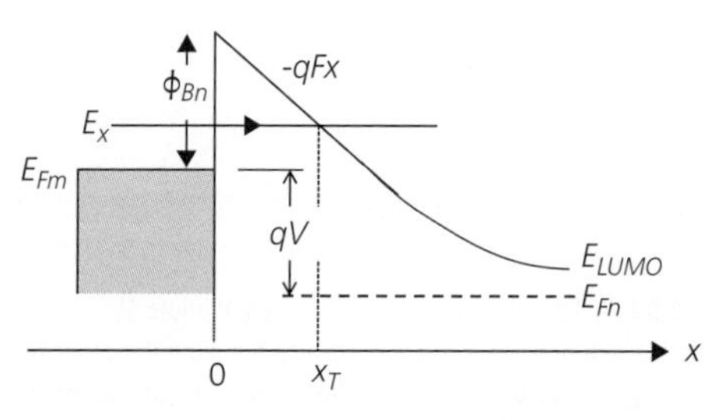

Figure 4.86 Tunneling of an electron of energy E_z through a triangular barrier of slope, $-qF$. The classical turning points are at $x = 0$ and x_T.

The transmission coefficient used with Eq. 4.262 results in an integral that cannot be solved analytically unless we expand the term $(q\phi_{Bn} - E_x)^{3/2} \approx (q\phi_{Bn})^{3/2} + \frac{3}{2}E_x(q\phi_{Bn})^{1/2} + \ldots$ Then we arrive at

$$j_{tun} = \frac{q^2}{(4\pi)^2\hbar}\frac{F^2}{\phi_{Bn}}\exp\left(-\frac{4\sqrt{2m^*}}{3\hbar qF}(q\phi_{Bn})^{3/2}\right). \quad (4.265)$$

This expression is similar to tunneling in a p-n junction, where the Schottky barrier height is replaced by the energy gap, E_G (Sze, 1981). The tunneling current is an exponential function of the barrier height and the effective mass. The tunneling probability increases exponentially with $1/F$, and hence $1/V$, for fully depleted films. Thus, for organics where $m^* \sim m_0$, the tunneling current is generally negligible except for large electric fields or small barrier heights. Unlike currents originating entirely from recombination, tunneling is not thermally activated. In fact the only component with a weak temperature dependence is ϕ_{Bn}, which increases with decreasing temperature due to lattice contraction, as inferred from tight binding in Section 4.1.

Tunneling currents have been numerically calculated based on an analytical theory of charges hopping among localized states. The distribution of tunneling states is assumed to be Gaussian with width $\sigma = 0.08$ eV, and the intermolecular distance is $a = 0.6$ nm. Results of these calculations, shown in Fig. 4.87, illustrate both the weak temperature and strong barrier height dependence predicted by Eq. 4.265, where the temperature dependence primarily arises from the thermally activated hops between an energetic distribution of sites. Fits of the model to the j–V characteristics for several archetype polymers suggest that tunneling can play an important role for organics with contact barriers $\phi_{Bn} \sim 0.5$ eV or larger (Arkhipov et al., 1998).

4.6.3 Metal–organic junctions in the presence of traps

Traps arising either from the properties of the organic itself, from purposely functionalizing the organic surface prior to contact deposition, or from process conditions (whether intended or not) employed in fabricating the contact (Liu et al., 2014) can affect the magnitude of the injection barrier. The role of traps in determining the Schottky barrier height is well known in inorganic semiconductors such as Si, GaAs, and InP, where a high density of defect levels arising from dangling atomic bonds at the surface can "pin" the Fermi level at the defect energy. This results in a barrier height that is independent of applied voltage.

The role of interface defects in determining the barrier height is illustrated in Fig. 4.88. The quantitative relationship between the interfacial layer of thickness, δ, the density of interface charge states, D_{IT}, and $q\phi_{Bn}$, is determined by the net surface charge on the organic, Q_{SO}. Defining the neutral energy or *charge neutrality level*, $q\phi_0$ as the energy above which the states are acceptor-like (i.e. they are neutral when empty and negative when charged), and below which they are donor-like (neutral when charged,

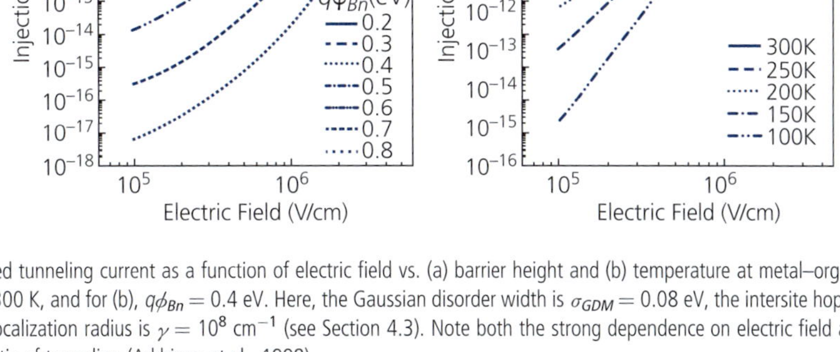

Figure 4.87 Calculated tunneling current as a function of electric field vs. (a) barrier height and (b) temperature at metal–organic semiconductor junctions. For (a), $T = 300$ K, and for (b), $q\phi_{Bn} = 0.4$ eV. Here, the Gaussian disorder width is $\sigma_{GDM} = 0.08$ eV, the intersite hopping distance is $a = 0.6$ nm, and the hopping localization radius is $\gamma = 10^8$ cm^{-1} (see Section 4.3). Note both the strong dependence on electric field and weak temperature dependence characteristic of tunneling (Arkhipov et al., 1998).

and positive when empty), then at equilibrium, we have

$$Q_{S0} = -qD_{IT}(E_G - q\phi_0 - \phi_{Bn}). \tag{4.266}$$

Adding in the space charge within the semiconductor, $Q_{SC} = qN_DW$, the total charge on the metal surface required to achieve neutrality is $Q_M = -(Q_{S0} + Q_{SC})$. The interface dipole energy, Δ, is obtained from Gauss' law and Fig. 4.88,

$$\Delta = \frac{q\delta Q_M}{\varepsilon_r\varepsilon_0} = q\phi_m - EA - q\phi_{Bn}. \tag{4.267}$$

From the preceding discussion and Eq. 4.242 for W, such that $Q_{SC} = -\sqrt{2q\varepsilon_r\varepsilon_0N_D(\phi_{Bn} - \phi_n)}$ where $q\phi_n = E_{Fn}$, we obtain for the energy barrier height in the presence of an interfacial layer:

$$q\phi_{Bn} = -\sqrt{\frac{2q^2\varepsilon_rN_D\delta^2}{\varepsilon_\delta^2\varepsilon_o}(q\phi_{Bn}\text{-}q\phi_n)} + \frac{q^2\delta D_{IT}}{\varepsilon_\delta\varepsilon_o}(E_G - q\phi_{Bn} - q\phi_0) - EA + q\phi_m. \tag{4.268}$$

Here, ε_δ is the dielectric constant of the interface layer.

As the density of surface states, or the width of the interfacial layer increases, so does the barrier height. Indeed, for $D_{IT} \to \infty$, then $q\phi_{Bn} = E_G - q\phi_0$. That is, the barrier height is independent of the metal work function and is pinned at the neutral point, $q\phi_0$. In the opposite limit of δ and $D_{IT} \to 0$, Eq. 4.268 simplifies to $q\phi_{Bn} = q\phi_m - EA$, which is the barrier height for the ideal Schottky junction (cf. Eq. 4.229).

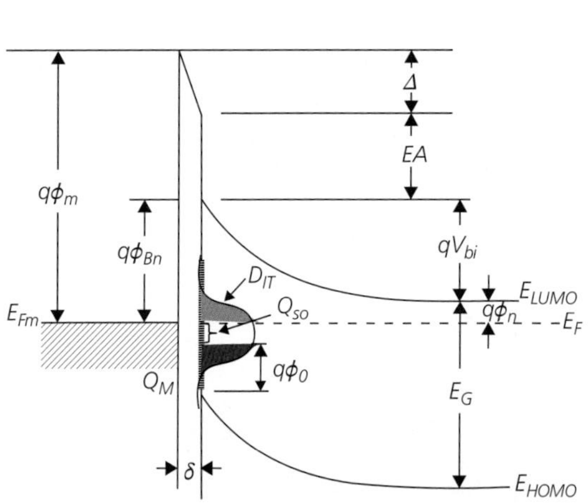

Figure 4.88 A metal–semiconductor junction in the presence of an interface density of trap states, D_{IT}. The light shaded area indicates filled acceptor traps, and the dark shaded area are filled donor traps. Δ = dipole energy across the interface barrier of thickness δ, Q_{SO} is the density of surface charge in the organic, Q_M is the surface charge in the metal, and $q\phi_o$ is the charge neutrality energy.

Clearly, the energy barrier at the metal–organic interface is controlled by the interface charge, giving rise to a charge dipole. For example, an Ohmic contact on the surface of an n-type or preferentially electron-conducting organic layer can be realized by introducing negative space charge (i.e. acceptors) in a thin layer at its surface. This drives the neutral energy toward the LUMO, and $q\phi_{Bn} \to 0$ as required.

Methods for intentional surface modification to lower the metal contact barrier with efficient charge injection properties have included chemical treatment of the electrode, insertion of reactive layers between the contact and the organic (Piromreun et al., 2000), and insertion of barrier-lowering organic monolayers between the metal and the organic. Both cathode and anode treatments of these types have proven successful in creating Ohmic contacts with good injection properties (Shen et al., 2004, Liu et al., 2014). Unintentional extrinsic effects such as cathode-metal interdiffusion (Probst and Haight, 1997, Rajagopal and Kahn, 1998) of damage induced at the organic surface during contact deposition are other sources of barrier pinning.

Interface dipoles can also have intrinsic origins. The j–V characteristics limited by injection over barriers induced by dipole layers can result in characteristics similar to those dominated by traps in the bulk. Dipole layers have been found to significantly shift the metal organic barrier by energies exceeding 1 eV (Hill et al., 1998, Ishii et al., 1999, Hill et al., 2000b, Crispin et al., 2002). Values of Δ measured for several different metal contacts to Alq_3 are provided in Table 4.5. The dipoles are a product of polarization of molecular orbitals in the region adjacent to the electron-rich metal surface. Then, the disorder in the local interfacial dipole field broadens the energy distribution of organic transport states. For a sufficiently small $q\phi_{Bn}$, charges can be injected directly from the metal into the interfacial states, and then subsequently hop to adjacent molecular layers at higher energies until the transport level (i.e. the LUMO) is reached. This process is illustrated in Fig. 4.89a for the Mg:Ag/Alq_3 interface. The density of interface states narrows as its mean value moves toward E_{LUMO} with increasing distance from the metal/organic junction (Baldo and Forrest, 2001).

Within each near-interface layer, the dipoles of individual molecules suffer orientational randomness influenced by dipoles in the surrounding layers. To quantitatively analyze transport across the layers, the dipoles are assumed to have the usual Gaussian angular distribution centered perpendicular to the contact ($\theta = 0$), and with an angular half width of σ_θ

(Fig. 4.89b). It is also assumed that these dipoles have a strength of $p = 30$ D, corresponding to charges separated by ~ 6 Å, which is on the order of the expected metal–organic molecular separation at the interface.

With these parameters, the mean energy of the distribution is calculated to be -0.7 eV below the LUMO (see Fig. 4.89c), consistent with the $\Delta \sim 1$ eV in Table 4.5. With increasing distance from the interface, the center of the distribution approaches the reference $E_{LUMO} = 0$ eV, and its width ($\sigma_B = 0.13$ eV) also approaches that of the bulk. For a dipole angular distribution of $\sigma_\theta = 1.5$ rad, the energetic distribution in the first organic layer is broadened to $\sigma_I = 0.35$ eV.

To describe the effects of the interfacial sites on charge injection and the j-V characteristics, the charge density per unit energy near the metal is determined

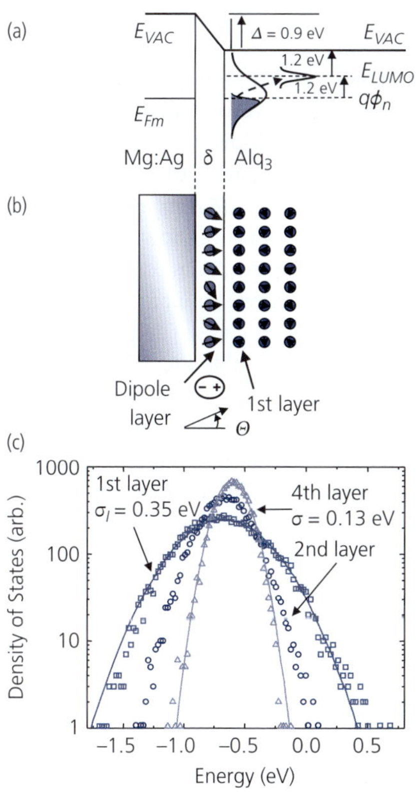

Figure 4.89 (a) Charges are initially injected into the interface between Mg:Ag and Alq_3 where the distribution of energy sites is broadened by randomness in molecular interface dipole orientations. (b) The total interface dipole energy is due to a collection of molecular dipoles oriented in a Gaussian distribution about the perpendicular ($\theta = 0$). (c) Calculation of the electron DOS relative to E_{LUMO}, for interface dipoles $p = 30$ D and a Gaussian angular distribution of width $\sigma_\theta = 1.5$ rad. The resulting energy distribution width is $\sigma_I = 0.35$ eV for the layer immediately adjacent to the interface. With increasing distance into the film, the distribution approaches the bulk value of $\sigma_B = 0.13$ eV, with a mean energy of $E = 0$ eV referenced to the LUMO After Baldo and Forrest (2001).

by Fermi–Dirac statistics, weighted by the energy distribution of dipole states. Thus,

$$n_I(E_I, q\phi_n) = \frac{N_I/\sqrt{2\pi\sigma_I^2}}{1+\exp[(E_I - q\phi_n)/k_BT]} \exp\left[-\frac{1}{2}\left(\frac{E_I}{\sigma_I}\right)^2\right], \tag{4.269}$$

where N_I and E_I are the density and energy of interfacial sites, respectively. Assuming that hopping out of these sites limits the current density, the distribution of transport bulk sites are initially assumed to be unoccupied. Then the current density is given by (Baldo and Forrest, 2001)

$$j(q\phi_n) = \frac{aq}{\sqrt{2\pi\sigma_B^2}} \int_{-\infty}^{\infty}\int_{-\infty}^{\infty} n_I(E_I, q\phi_n) k_{ET} \times (E_B - E_I - aqF) \exp\left[-\frac{1}{2}\left(\frac{E_B}{\sigma_B}\right)^2\right] dE_I dE_B, \tag{4.270}$$

where E_B is the energy of the charge in the bulk, and a is the average intermolecular spacing. The hopping rate is $k_{ET}(E)$, where E is the energy difference between adjacent hopping sites. To simplify calculations, the hopping occurs from one monolayer to the next, until the bulk is reached at a distance of $\delta = 150$ Å. The charge at the interface determines the electric field, F, and voltage $V = \frac{qn_T d\delta}{\varepsilon_r \varepsilon_0}$, where n_T is obtained by integrating Eq. 4.269 over all E_I. As expected for injection-limited transport, the bias voltage is directly proportional to the film thickness, $d \gg \delta$.

The injection model is completed by describing the dependence of the hopping rate, k_{ET}, on electric field and temperature. We assume the small polaron model of intermolecular hopping in disordered systems (Section 4.3.2). Then the low temperature hopping regime follows (Emin, 1975)

Table 4.5 Energies characteristic of several metal/Alq_3 interfaces

Metal contact	Dipole energy (Δ, eV)	Reference
Al/LiF	1.7	(Lee et al., 1998, Ishii et al., 1999)
	1.4	(Schlaf et al., 1998)
Al	1	(Ishii et al., 1999)
Ag	1.1	(Ishii et al., 1999, Hill et al., 2000b)
Au	1.1	(Lee et al., 1998, Rajagopal and Kahn, 1998, Ishii et al., 1999)

$$k_{ET}(E) \propto \exp\left(-\frac{E+|E|}{2k_BT}\right)\frac{(2E_b/\hbar\omega_0)^{|E|/\hbar\omega_0}}{(|E|/\hbar\omega_0)!}, \quad (4.271)$$

and at higher temperatures,

$$k_{ET}(E) \propto (E_b k_B T)^{-1/2} \exp\left(-\frac{E_b}{2k_BT}\left(1+\frac{E}{2E_b}\right)^2\right). \quad (4.272)$$

This latter expression is from Marcus electron transfer over an energy barrier, $E_b = \lambda_{reorg}/2$, and $E = \Delta G^0$, the change in Gibbs free energy (cf. Eq. 4.89).

Thus, there are *two* processes that affect the temperature dependence of injection: energetic disorder that determines the width of the DOS and the likelihood of resonant, temperature-independent hops, and the temperature dependence of the phonon distribution. For a charge carrier that hops from initial energy E_i to final energy E_f, Eqs. 4.270–4.272 give

$$j = CJ_0 \int_{-\infty}^{\infty} \exp\left[\frac{(2E_b + E_f - E_i)^2}{8E_b k_B T}\right] \times \exp\left[-\frac{1}{2}\left(\frac{E_f}{\sigma_B}\right)^2\right] dE_i \propto \exp\left[-\frac{(2E_b - E_i)^2}{8E_b k_B T + 2\sigma_B^2}\right], \quad (4.273)$$

where C is a constant, and J_0 is the intermolecular overlap integral.

A significant complication in distinguishing between bulk- and interface-limited transport arises from the field and temperature dependence of the mobility. Because of the low density of free charge, the injected space charge determines the electric field in the bulk transport limit. The thickness dependence of the current allows for discrimination between injection and bulk limits, assuming an approximately field-independent mobility (Brütting et al., 2001). In injection-limited transport, the voltage at a given current density is proportional to the film thickness, that is, $V \propto d$, whereas for bulk-limited transport, $V^{m+1} = Kd^{2m+1}$ (cf. Eq. 4.148) where $m \geq 1$ and K is a constant. For $m \gg 1$ as is often the case, then $V = Kd^2$.

The predictions of the interface disorder model have been tested using thin-film electron-injection-only amorphous Alq_3 devices. The device structure comprised an injecting cathode, various thicknesses of Alq_3, and a 100 nm-thick 25:1 Mg:Ag anode that prevents hole injection. Cathodes consisted of either Al, 25:1 Mg:Ag, and a 0.5 nm thick LiF layer followed by Al. The Alq_3 thickness dependence for each cathode is given in Fig. 4.90 vs. applied voltage at j = 10 mA/cm^2. The three cathodes show that $V \propto d$,

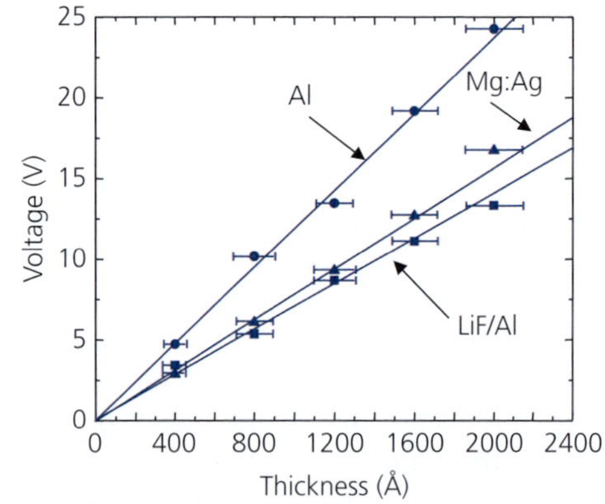

Figure 4.90 Drive voltage at j = 10 mA/cm^2 vs. Alq_3 thickness using LiF/Al, Mg:Ag and Al cathodes. All three cathodes exhibit a linear thickness dependence in the range 200 nm $\geq d >$ 40 nm (Baldo and Forrest, 2001).

Reprinted figure with permission from Baldo, M. A. & Forrest, S. R. 2001. Interface-limited injection in amorphous organic semiconductors. Phys. Rev. B, 64, 085201. Copyright 2001 by American Physical Society.

consistent with injection-limited transport, although similar devices have also exhibited $V \propto d^2$ consistent with SCL transport (Burrows et al., 1996, Campbell et al., 1997, Brütting et al., 2001). At T = 30 K the plot of V versus d at j = 0.1 mA/cm^2 remains linear.

Fits to the j–V–T data are shown in Fig. 4.91a using Eq. 4.273 for the Mg:Ag/Alq_3, consistent with injection limited conduction. Similar fits are obtained for the Al and LiF/Al contacts. The parameters used in the fits are listed in Table 4.6. The temperature dependence of the voltage power law, m, may provide a clue as to the origin of the dark current. Recall for SCL current, we expect $m = T_0/T$ for a characteristic trap temperature, T_0. In contrast, the prediction for injection-limited transport in Eq. 4.273 provides a more complicated temperature behavior. Both functional forms are plotted in Fig. 4.91b. Unfortunately, there is no clear distinction or obvious contradiction with either model that would allow us to claim that the current is limited by either injection or bulk processes.

At very low temperatures (T < 30 K), all the cathodes possess nearly identical j–V characteristics. This unambiguously breaks the link between the cathode work function and its injection efficiency, demonstrating that the initial hop from the metal Fermi level into the organic is not the most energetically costly event in current injection. Since the injection characteristics are determined by the interface energy distribution, the similarity between contacts at T ~ 30 K suggests that the organic interfacial site distributions are similar in all cases.

Of the various models competing to explain the unipolar transport characteristics of amorphous semiconductor materials, we have seen two approaches have proved attractive: starting from the observed shape and temperature dependence of transport, the trapped charge models in Section 4.4 can accurately describe characteristics at $T > 100$ K; whereas starting from a fundamental understanding of charge transport in amorphous materials, disorder models in Section 4.3 have successfully described the field and temperature dependence of charge mobility.

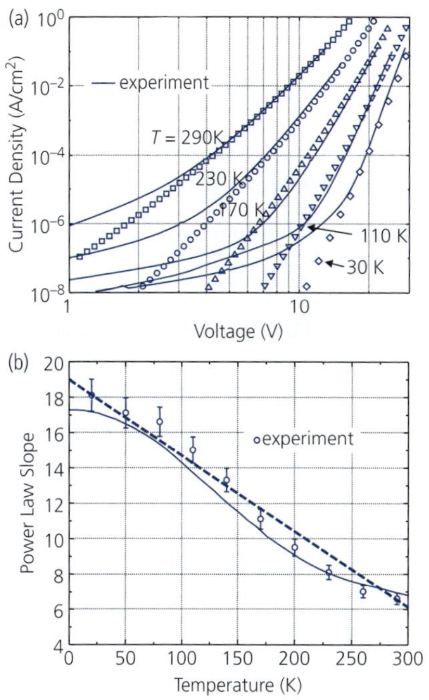

Figure 4.91 (a) Temperature dependence of the j–V characteristics (solid lines) of Mg:Ag/Alq$_3$ devices compared to the interface hopping theory (points). The parameters used in the fits are given in Table 4.6. (b) Temperature dependence of the power law slope, m, where $j \propto V^m$. Circles denote the experimental data between j= 0.1 and 10 mA/cm^2. The solid line is a calculation based on injection limited current. The straight dashed line is a fit to bulk limited SCL transport theory. After Baldo and Forrest (2001).

In some respects, the interfacial model is a synthesis of these approaches. The broad DOS at the interface and the narrower bulk distribution is qualitatively similar to trapped charge limited models, and it can generate the correct power law dependencies. The interfacial model involves limited energetic disorder in the bulk and larger dipole-induced disorder at the interface to explain these distributions. In doing so, it seeks to provide a microscopic model of charge hopping in amorphous semiconductors.

So where does all of this leave us? How do we unambiguously determine whether the current is due to bulk or interface limited conduction? We have seen in this and previous sections that the models often lead to similar results, both providing reasonable fits to the data often over large ranges of temperature and electric field. As noted above, the effects of transport across interfaces often have similar origins to those in the bulk: intermolecular hopping that is well described by small polaron theory. In spite of these ambiguities, careful experimental procedures and analysis can be helpful in eliminating uncertainties in interpretation. Below are a few methods one can use to distinguish between the sources of conduction:

1. The thickness dependence of the voltage at constant current provides perhaps the most reliable means to discriminate between bulk and interface limited transport. Particularly as d is increased, the effects of injection are decreased, and those of the bulk are increased.
2. Samples and contacts that support only single carrier injection eliminate ambiguities that arise from double injection.
3. Ohmic contacts can be assured in samples that exhibit a linear j–V characteristic (i.e. one without inflections or asymmetries independent of the voltage polarity) over a wide range of temperature and applied voltage. Indeed, symmetric j–V

Table 4.6 Parameters the interface dipole model (Baldo and Forrest, 2001)

	Fixed parameters			Fit parameters		
Device structure	**Polaron transition temperature**	**Width of bulk distribution**	**Width of interface distribution**	**Polaron binding energy**	**Interface site density**	**Intermolecular overlap integral**
	$T_P = \dfrac{\hbar\omega_0}{3k_B}$	σ_B (eV)	σ_I (eV)	E_b (eV)	$N_I d_I$ (cm^{-2})	J_0 (meV)
Mg:Ag/Alq$_3$	60	0.13	0.34	0.16	50×10^{12}	6
LiF/Al/Alq$_3$	60	0.13	0.34	0.16	43×10^{12}	10
Al/Alq$_3$	60	0.13	0.40	0.10	30×10^{12}	2

characteristics may be the single most important indicator that Ohmic conduction is dominant.

4. Independent measurement of $\mu(F, T)$ is important for use in the several expressions describing bulk and interface limited injection. The most reliable means to determine this most important quantity is via TOF measurements described in Section 4.4.3.

4.7 Organic semiconductor junctions

When two semiconductors are brought into contact, they form a junction. If prior to contact, the Fermi levels of the two materials are at different energies relative to the vacuum level, they must equalize to achieve equilibrium. This introduces energy level bending and a built-in potential, V_{bi}, due to the differences in chemical potentials of the materials. When the materials have the same composition, but their background free carrier concentrations differ, the contact and subsequent energy level bending results in an energy barrier. This is known as a *homojunction*. More commonly, the contacting organic materials are different, resulting in a *heterojunction* (HJ) whose built-in potential is the sum of the difference in chemical potentials and energy level (e.g. HOMO or LUMO) offsets of the two materials. When the majority carrier type is the same on both sides of the junction (i.e. both n-type or p-type), it is known as an *isotype HJ*, whereas if they are different the contact is called an *anisotype HJ*.

In the previous section we discussed the special case of a metal/organic HJ. In this section we consider the fundamental governing principles of both organic homo- and heterojunctions. Indeed, HJs are found in nearly all organic optoelectronic devices, sometimes in electronic devices such as OTFTs.

Homojunctions are the simplest junctions that can be formed between organic materials, and are discussed along with the fundamental theory of p-n junction operation in Section 4.7.1. The concepts of inorganic semiconductors have often been borrowed to describe the *j–V–T* characteristics of organic HJs. However, the principles governing charge conduction in materials systems as different as organic and inorganic semiconductors must reflect those striking dissimilarities. A theory developed specifically for the transport across organic junctions is the subject of Section 4.7.2. The theory has also been adapted to the increasingly important class of hybrid HJs consisting of contacts between organic and inorganic semiconductors, a topic left for Section 4.7.3.

One final note: while our purpose here is to derive "ideal diode equations" describing charge transport across p-n junctions, excitonic junctions and *hybrid organic–inorganic HJs* (or OI-HJs), it is very rare indeed that a diode exhibits purely ideal characteristics. That is, just because you can derive an ideal diode equation, does not mean you have an ideal diode. Indeed, since Shockley developed the first theory of the ideal diode (for Ge) in 1949, the literature has overflowed with examples of devices that do not follow the predicted fundamental behavior. While it might be puzzling, then, that we spend effort deriving such formalisms, it is essential to remember that understanding the fundamental nature of the ideal p-n junction serves as a trustworthy and extremely powerful foundation on which all of our subsequent understanding of junction diode behavior can be built.

4.7.1 Organic homojunctions

The treatment of inorganic p-n junctions was developed by Shockley in a classic paper that presents the derivation of what is known as the *ideal diode*, or *Shockley equation* (Shockley, 1949). In the following, we give the derivation, allowing us to determine the physics that underlies transport across homojunctions, and ultimately to understand how it fails to adequately describe the properties of excitonic junctions.

The energy band diagram of a p-n junction is shown in Fig. 4.92. The current is derived using the drift-diffusion model (Eq. 4.237) for both electrons and holes. At equilibrium, or close to it (i.e. so-called *quasi-equilibrium*), the current is given by:

$$j = q\left(n(x)\mu_n F(x) + D_n \frac{\partial n(x)}{\partial x}\right) \approx 0 \qquad (4.274a)$$

for electrons and

$$j = q\left(p(x)\mu_p F(x) + D_p \frac{\partial p(x)}{\partial x}\right) \approx 0 \qquad (4.274b)$$

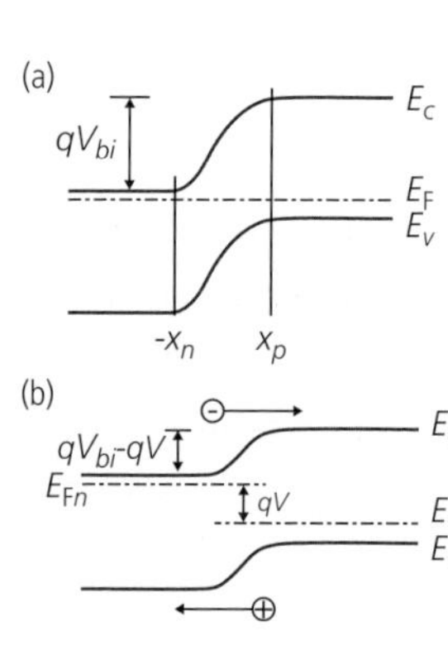

Figure 4.92 Energy bands for a p-n junction diode (a) in equilibrium and (b) under forward bias, *V*.

for holes. The total current is the sum of these two currents, that is, $j = j_n + j_p$. Without loss of generality, we assume that the current is dominated by electrons. Then solving for $F(x)$ and the applied voltage, V,

$$F(x) = -\frac{D_n}{\mu_n}\frac{1}{n}\frac{dn(x)}{dx} = -\frac{k_BT}{q}\frac{1}{n}\frac{dn(x)}{dx}, \quad (4.275)$$

yielding a junction voltage of

$$V_J = V_{bi} - V = -\int_{-x_n}^{x_p} F(x)dx = \frac{k_BT}{q}\ln n(x)\Big|_{n(-x_n)}^{n(x_p)}. \quad (4.276)$$

This results in an expression for the free electron concentration at the two edges of the depletion region, $-x_n$ and x_p:

$$n(-x_n) = n(x_p)\exp(q(V - V_{bi})/k_BT). \quad (4.277)$$

Here, the total depletion region width is $W = x_n + x_p$.

Equation 4.277 can be simply rewritten as $n_n = n_p \exp(q(V - V_{bi})/k_BT)$, where $n_p = n(x_p)$ is the minority electron concentration on the p-side of the junction, and $n_n = n(-x_n)$ is the majority carrier concentration on the n-side. Now,

$$\begin{aligned} n_{n0}p_{p0} &= N_cN_v \exp(-E_G/k_BT)\exp(qV_{bi}/k_BT) \\ &= n_i^2 \exp(qV_{bi}/k_BT), \end{aligned} \quad (4.278)$$

where $qV_{bi} = E_{Fp} - E_{Fn}$ is the difference in the quasi-Fermi levels on the two sides of the junction, and $n_{n0}(p_{p0})$ is the equilibrium electron (hole) concentration on the n-(p-)side of the junction. Using the law of mass action, that is, $n_{p0}p_{p0} = n_{n0}p_{n0} = n_i^2$ (Eq. 4.194), we obtain for the *excess* minority carrier density on each side of the junction, that is, $\Delta p_n = p_{n0} - p_n$ and $\Delta n_p = n_{p0} - n_p$ at voltage, V,

$$\Delta n_p = n_{p0}(\exp(qV/k_BT) - 1) \quad (4.279a)$$

and

$$\Delta p_n = p_{n0}(\exp(qV/k_BT) - 1). \quad (4.279b)$$

These expressions are now used in the diffusion equation at $j_e, j_h \approx 0$ (and therefore in the absence of drift), Eqs. 4.188 and 4.189, to obtain

$$\Delta n_p(x) = n_{p0}(\exp(qV/k_BT) - 1)\exp(-x/L_n), \quad (4.280a)$$

$$\Delta p_n(x) = p_{no}(\exp(qV/k_BT) - 1)\exp(-x/L_p). \quad (4.280b)$$

Here, $L_n^2 = D_n\tau_n$ and $L_p^2 = D_p\tau_p$ are the electron and hole diffusion lengths, respectively. Finally, the electron current is

$$\begin{aligned} j_n(x) &= -qD_n\frac{d\Delta n(x)}{dx} \\ &= \frac{qD_nn_{p0}}{L_n}(\exp(qV/k_BT) - 1)\exp(-x/L_n). \end{aligned} \quad (4.281)$$

But current is constant throughout the sample, such that $j_n(x) = j_n(0)$. Using a similar relationship for j_p, the two components are summed to obtain the total current:

$$j = q\left[\frac{n_i^2}{N_A}\sqrt{\frac{D_n}{\tau_n}} + \frac{n_i^2}{N_D}\sqrt{\frac{D_p}{\tau_p}}\right](\exp(qV/k_BT) - 1) - j_{ph}. \quad (4.282)$$

Here, as in Eq. 4.236, we have added the contribution from the photocurrent, j_{ph}. In this expression, we have once more used the law of mass action, for example, $n_{p0}N_A = n_i^2$ where N_A is the number of ionized acceptors (and hence the number of majority carrier holes at the p-side of the junction) and N_D is the number of ionized donors.

Equation 4.282 is in the same form as Eq. 4.236 for the Schottky barrier diode. We note that the current is determined by the minority carrier diffusion constant and lifetime, as well as the doping concentrations on each side of the junction. This equation is based on current generation processes that are substantially different from those relevant to organic HJs, where exciton generation and the formation of tightly bound and oppositely charged polaron pairs at the heterointerface are the sources of dark and photocurrent. This deficiency will be addressed in the following section.

To complete our analysis, we calculate the ideality factor, n, in the presence of recombination. For SRH recombination in Section 4.5.2 to be effective, the trap state must be near the middle of the energy gap. Then the hole and electron recombination rates are approximately equal, that is, the effective recombination time is $\tau_{eff} = \tau_e \approx \tau_h$. Under forward bias, there is an excess of space charge. Thus, both n and p are much larger than charges released from traps residing at mid-gap, and $n = p \gg n_1 = p_1 \approx n_i$. Since recombination is only efficient when the number of holes and electrons are approximately equal, then Eq. 4.217 is simplified under forward bias:

$$R^{SRH} = \frac{np - n_i^2}{\tau_e(p + p_1) + \tau_h(n + n_1)} \approx \frac{n}{2\tau_{eff}}. \quad (4.283)$$

Recombination is most probable in the depletion region due to the coexistence of a high density of both electrons and holes. Multiplying Eq. 4.283 by the volume and the total number of charges in the depletion region yields the forward-biased recombination current of

$$j_{rec} = \frac{qWn}{2\tau_{eff}} = \frac{qWn_i}{2\tau_{eff}} \exp(qV/2k_BT), \qquad (4.284)$$

where, in the center of the forward-biased space charge region, $n = p = n_i \exp(qV/2k_BT)$.

Under reverse bias, $n_i^2 \gg np$ since there is very little charge in the depletion region. Also, again for mid-gap states, $n_1 = p_1 \approx n_i$ such that

$$j_{rec} = -j_0 = -\frac{qWn_i}{2\tau_{eff}}. \qquad (4.285)$$

Combining Eqs. 4.284 and 4.285 gives the expression for the total recombination current:

$$j_{rec} = \frac{qWn_i}{2\tau_{eff}}(\exp(qV/2k_BT) - 1). \qquad (4.286)$$

This differs from the ideal diode equation based on diffusion since the reverse current (at $V < 0$) is no longer voltage independent. That is, $j_0 \propto W \propto \sqrt{V}$ for uniformly doped semiconductors, with a somewhat different dependence of W on V depending on the doping profile in the junction region. In any case, this voltage dependence results in a small slope in the reverse-biased characteristics that can be useful in distinguishing between current sources. Note that if the sample is completely depleted at $V = 0$, or becomes depleted by an applied voltage, then $W = d$ and recombination currents, too, are voltage independent.

The total current is generally a combination of both recombination and diffusion, and may also be limited by other processes such as series and shunt resistances. To account for contributions from the several non-ideal current sources, the ideality factor, n, is introduced in the exponential argument, where $1 < n < 2$. For $n = 1$, diffusion is the dominant source of current, and as $n \to 2$, recombination becomes increasingly important. It is frequently observed in organic (and inorganic) junctions that $n > 2$. There is no analytical interpretation of such high n-values, which are typically due to a combination of series and/or shunt resistances, space charge effects, etc. Including the photocurrent, we can thus rewrite the Shockley equation as

$$j_{rec} = j_0(\exp(qV/nk_BT) - 1) - j_{ph}, \qquad (4.287)$$

which is identical to Eq. 4.236 for Schottky diodes. The ideality factor is found from the forward biased slope of the j–V characteristics, that is,

$$n = \frac{q}{k_BT} j \frac{\partial V}{\partial j}. \qquad (4.288)$$

The theory thus far presented does not consider the key physical phenomena that govern organics. Hence, it is not surprising that the Shockley equation does not often accurately model the j–V characteristics of organic homojunction p-n diodes (Harada et al., 2005, Tsuji et al., 2009, Cai et al., 2011, Tripathia and Mohapatra, 2012, Kubo et al., 2013). An example of the characteristics of a p-i-n diode (i = intrinsic or undoped region separating the p and n zones) and a Schottky-type metal-i-n (M-i-n) diode is shown in Fig. 4.93. The p-i-n homojunction diode comprises a 30 nm thick ZnPc i-layer sandwiched between a 15 nm thick ZnPc layer doped p-type with F_4-TCNQ, and a third, 40 nm thick n-type ZnPc doped with Ru(terpy)$_2$. The n-layer is contacted with a metal cathode (Al or Au), and the anode was ITO. The M-i-n device eliminates the p-layer and

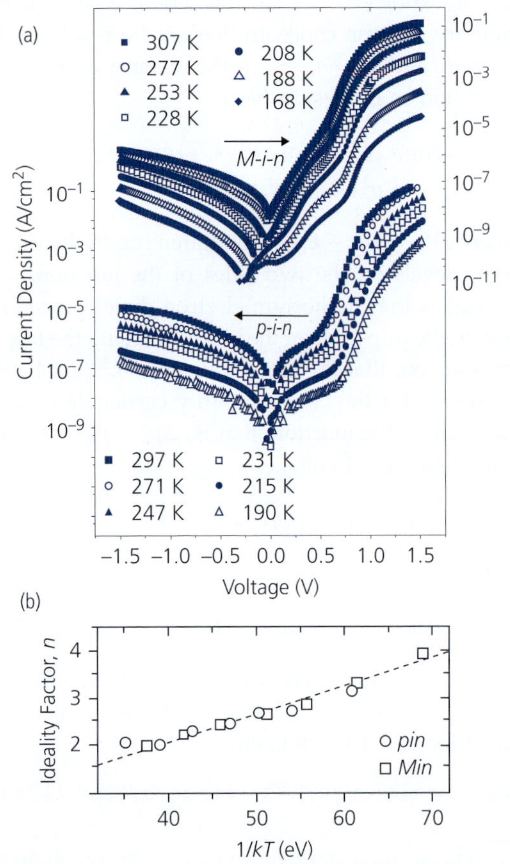

Figure 4.93 (a) Dark current vs. voltage characteristics as a function of temperature of a metal (M)-i-layer-n organic Schottky junction, and an all-organic p-i-n diode. (b) Ideality factor vs. $1/k_BT$ for diodes in (a) (Harada et al., 2005).

uses ITO to form the Schottky barrier to the i-region (Harada et al., 2005).

The forward biased current shows an inflection at 0.5 V, above which an exponential increase with voltage is observed until it rolls off due to series resistance at $V > 1.0$ V. Further, there is no apparent dependence on the *i*-layer thickness in the p-i-n diodes. Thus, space charge effects are ruled out as limiting the current. Under reverse bias, the voltage dependence is much more pronounced than predicted for recombination in Eq. 4.286.

The Shockley equation for both p-i-n and M-i-n devices implies that the ideality factor is temperature independent. According to Eq. 4.288, the forward biased slope of the j–V characteristics for a constant n-factor decreases with increasing temperature. However, the data in Fig. 4.93 do not show this behavior: the slopes are nearly constant over the range 307 K $> T >$ 168 K. Furthermore, the current should increase rapidly with voltage under both forward and reverse bias if it is thermally activated with an energy approaching half the energy gap of ZnPc. Again, this does not appear to be the case.

The discrepancies with the ideal diode theory, that is, the temperature independence of n and the inflection in the forward characteristics at $V < 1$ V are eliminated if we modify the conventional semiconductor equations to account for recombination governed by an effective medium with a Gaussian density of transport states of width, σ (Section 4.3.2). The junction region of the M-i-p device can be understood using the energy level diagram in Fig. 4.94, with only minor adjustments needed to describe a p-i-n homojunction. For purposes of this discussion, the p- and i-regions comprise MTDATA, and the p-type dopant is, once again, F_4TCNQ.

Recombination occurs via tunneling at the junction over a very short distance between filled, nearly degenerate hole states from the p-region into only partially filled states within the i-region. That is, as voltage is increased under forward bias, the density of available, unoccupied states in the i-region increases, thereby resulting in an exponentially increasing current. At $V < V_{th}$ (where V_{th} is the threshold voltage), the current is small due to an overlap of the filled tail of the *i*-region with that of the p-region. This accounts for the low current "shelf" at low voltages in Fig. 4.93a. The current under forward bias is, therefore, equal to the integral of the product of the DOS from the both sides of the junction. Thus, the current is given by

$$j = j_0 \int \exp\left[-(E(V) - E_0)^2/2\sigma^2\right] \exp\left[-E_0^2/2\sigma^2\right] f_p(1 - f_i)dE, \tag{4.289}$$

where $E(V)$ is the energy offset between the peaks of the Gaussian DOS which is a function of applied voltage, and E_0 is the energy of the peak in the p-region which, due to the near degeneracy of the organic, is independent of voltage. Also, f_p and f_i are the occupancies of states on the p- and i-sides of the junction, respectively, and we have assumed that the Gaussian widths are the same on both sides of the junction. This latter assumption may not be strictly valid, given that a large concentration of dopants in the semiconductor may increase the degree of disorder.

Setting $E_0 = qV_0$, and $q\Delta V = E(V) - E_0$, and eliminating all but linear terms in $\Delta V/V_0$ yields (Tripathia and Mohapatra, 2012)

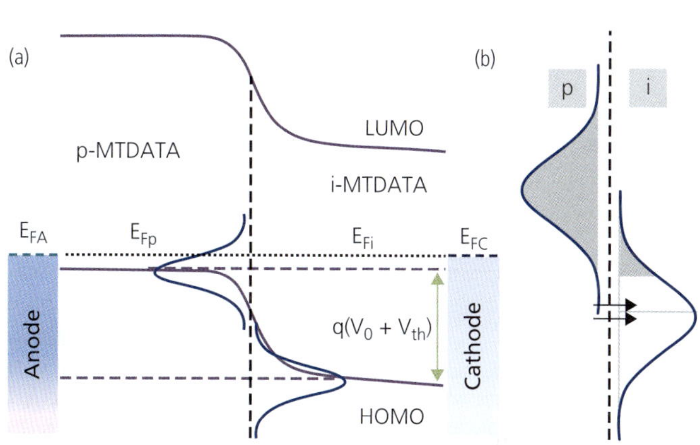

Figure 4.94 (a) Energy level diagram of an M-i-p diode defining the voltages and densities of states in the junction region. (b) A detail of the Gaussian density of states at the junction. Shaded areas denote states filled with holes (Tripathia and Mohapatra, 2012).

$$j = j_0 \exp\left(-\frac{q^2 V_0}{2\sigma^2}\left(\frac{V_0}{2} - \Delta V\right)\right) \quad \text{for} \quad V > V_{th}. \tag{4.290}$$

Here, we assume that $V > V_{th}$ corresponds to the condition that $f_p(1 - f_i) \approx 1$, that is, there are empty states in the i-region that are at the same energy as filled states in the p-region, allowing for tunneling to occur. A key feature of this expression is that the slope of the $\ln j - \Delta V$ characteristics is temperature independent, and is equal to $\zeta = q^2V_0/2\sigma^2$. That is, the slope gives the Gaussian width of the DOS, and is independent of thickness of the *i*-region, as shown in Fig. 4.95a. Note at high voltages, the slope of the *j*-*V* characteristics in both Figs. 4.93 and 4.95 is reduced. The dependence on layer thickness, temperature and voltage in this regime are signatures of space charge limited transport.

The last step in the derivation is to determine V_{th} which is obtained by fitting the exponential region of the *j*–*V* characteristics to Eq. 4.290. Figure 4.95 shows that the threshold voltage increases linearly as temperature is decreased. This can be found from the relationship between voltage and charge density:

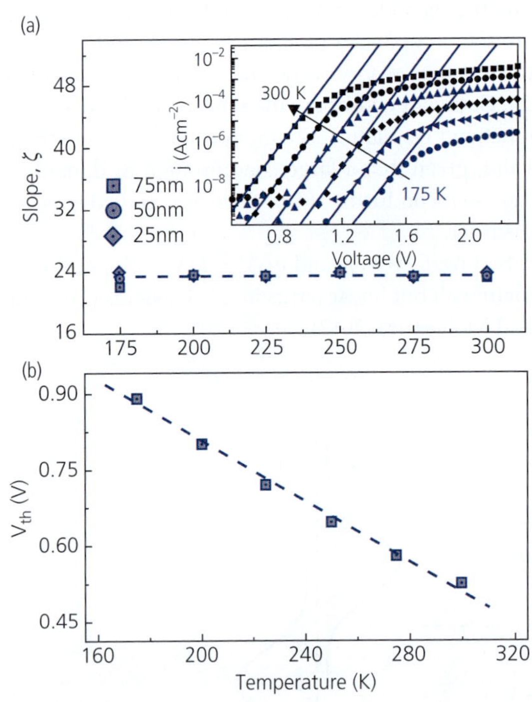

Figure 4.95 (a) Slope of the forward biased *j*–*V* characteristics vs. temperature for different i-layer thicknesses of an Al-i-p diode, where the i- and p-regions are MTDATA. Inset: Forward biased *j*–*V* characteristics vs. temperature. The lines are fits to the data in the exponentially increasing *j* region. (b) V_{th} vs. temperature for the devices in (a) (Tripathia and Mohapatra, 2012).

$$p_j = p_0 \exp(qV_{th}/k_BT), \tag{4.291}$$

where p_j is the hole concentration at the edge of the depleted i-region, and from Boltzmann statistics:

$$p_0 = N_{HOMO} \exp(-(E_{Fi} - E_{HOMO})/k_BT) \tag{4.292}$$

(see Fig. 4.94). Using these two expressions and solving for V_{th} gives

$$qV_{th} = k_BT\log\left(\frac{p_j}{N_{HOMO}}\right) + (E_{Fi} - E_{HOMO}). \tag{4.293}$$

This yields a linear dependence on temperature consistent with the data in Fig. 4.95b. The slope of the data give p_j/N_{HOMO}, and the intercept yields the position of the Fermi energy of the intrinsic region, E_{Fi}. For the p-i MTDATA devices in Fig. 4.95, the value of p_j is vanishingly small within the depletion region. Also, the Fermi energy is only slightly thickness dependent between 25 nm and 75 nm, possibly due to depletion of the entire *i*-region to the cathode. It is found that $E_{Fi} - E_{HOMO} \simeq 1.4$ eV.

As a final note, the tunneling model is supported by the rapid rise of current in the reverse direction (see Fig. 4.93a). This dependence has been fit by *Fowler–Nordheim tunneling*, viz.

$$j = AF^2 \exp\left(-8\pi\sqrt{2m^*}\phi^{3/2}/3qhF\right), \tag{4.294}$$

where ϕ is the tunneling barrier height. Indeed, the reverse characteristics of similar M-i-p devices are reasonably consistent with this dependence over a temperature range of 200–300 K and over six decades of current using $\phi \sim 0.5$–0.7 eV, weakly depending on the i-layer thickness (Kumar et al., 2017).

4.7.2 Excitonic heterojunctions

Nearly every high performance organic electronic device comprises a combination of layers of different composition. This allows each individual material to be optimized to perform a function: electron and hole transport materials are selected for their correspondingly high electron or hole mobilities, light emission layers use materials that are optimized for emission efficiency at a particular wavelength, and materials used for solar cells and photodetectors are selected based on their exciton transport and optical absorption properties. The contact between two dissimilar semiconductors is known as a HJ. A feature of HJs is that an energy barrier forms at the interface (the heterointerface) between the semiconductors. The barrier controls charge and energy transport, as well as exciton dissociation into free charges (in the case of organic photovoltaics). In this section we consider both the origin of the rectifying energy barrier, and its energy and transport properties. Our theoretical treatment will also be briefly adapted to understanding the dynamics of charge and exciton transport in organic p–n homojunctions, clearing up some of the mysteries encountered in their analysis in Section 4.7.1.

4.7.2.1 Excitonic HJ fundamentals

In Fig. 4.96 we show the relevant energies of two contacting semiconductors with electron affinities, EA_1 and EA_2. In the absence of traps or other non-ideal HJ properties, the difference in conduction band energies following their contact is

$$\Delta E_{LUMO} = |EA_1 - EA_2|, \tag{4.295}$$

and the difference in the HOMO energies is equal to the difference in IPs:

$$\begin{aligned}\Delta E_{HOMO} &= |IP_1 - IP_2| = |EA_1 - EA_2 + E_{G1} - E_{G2}| \\ &= \Delta E_G - \Delta E_{LUMO}.\end{aligned} \tag{4.296}$$

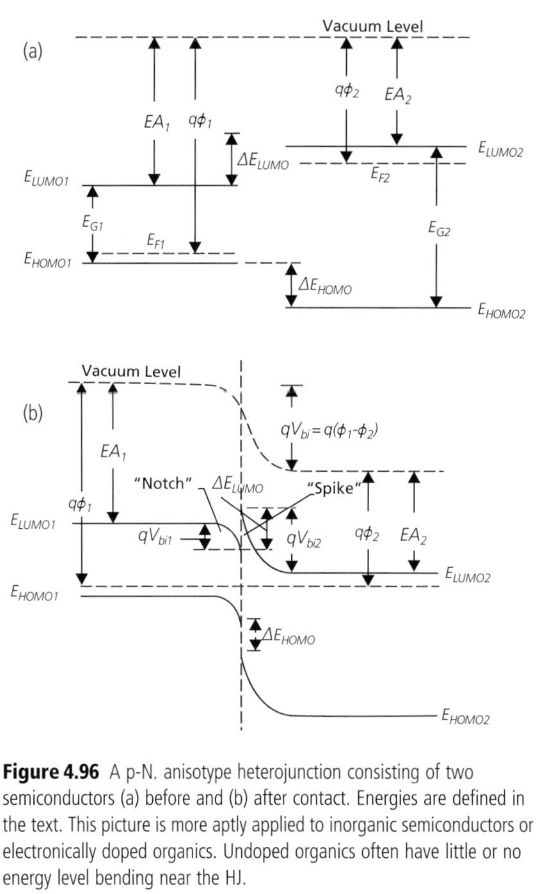

Figure 4.96 A p-N. anisotype heterojunction consisting of two semiconductors (a) before and (b) after contact. Energies are defined in the text. This picture is more aptly applied to inorganic semiconductors or electronically doped organics. Undoped organics often have little or no energy level bending near the HJ.

Equation 4.295 is known as the *Anderson rule* (Anderson, 1960) and for many years was accepted as the means for determining HJ energy band offsets in inorganic semiconductor systems. While the Anderson rule provides a useful conceptual framework for understanding HJ energy level offsets, its predictions are inaccurate for several fundamental reasons. (i) It ignores the importance of charge redistribution and the subsequent formation of dipoles at the interface required to establish equilibrium between the semiconductors. (ii) The electron affinity, and in practice the IPs of the semiconductors are measured using surface sensitive probes such as UPS. These probes can measure the surface energies to within only a few monolayers with reasonable accuracy (typically ~50–100 meV), but the HJ lies deep within the bulk of the material whose Fermi energy may differ considerably from its value at the surface. (iii) UPS and similar surface probes are very sensitive to surface charge, surface lattice reconstruction and other properties unique to a free surface. These may not be characteristic of the bulk, but can significantly impact the energies measured. The effects of surface defects and traps can also strongly distort the surface potential from the bulk. Hence, the Anderson model is not considered to be an accurate method for determining energy level offsets at HJs buried within the bulk. Additional complications arise at organic semiconductor junctions that further influence the measured energy level offsets.

As shown in Fig. 4.96, redistribution of charge at equilibrium due to matching of the Fermi energies of the contacting semiconductors results in depletion (the energy "spike") on one side of the junction, and accumulation of an equal amount of charge on the opposite side (resulting in an energy "notch"). The HJ shown consists of a p-N junction (and hence is an anisotype HJ) with $EA_1 > EA_2$. The semiconductor convention is that the small letter (p) refers to the small band gap, and the capital letter (N) refers to the large band gap constituent. The spike and notch are reversed if the relative magnitudes of the electron affinities are likewise inverted. The total built-in potential is equal to the sum of the built-in potentials on each side of the HJ (i.e. $V_{bi} = V_{bi1} + V_{bi2}$). In organics where there may be very little free charge, and in the absence of interface dipoles, the band bending at the HJ is small, resulting in flat bands that extend up to the interface, with $V_{bi} = 0$.

In Fig. 4.97 we show an isotype, n-N HJ. As for the anisotype junction, a spike and notch form at the heterointerface, and the junction offsets result in rectification, although the total barrier height is reduced to only the energy level offsets of ΔE_{LUMO} and ΔE_{HOMO}.

The energy level alignments in Figs. 4.96 and 4.97 show that the conduction and valence levels of the narrow energy gap semiconductor are nested within the gap of the wide energy gap constituent. This is known as a Type I heterojunction. Different alignments are possible, and indeed are required for different device purposes. Possible alignments are classified in Fig. 4.98. The *Type I HJ*, known as a *nested* or *straddling gap*, is useful for confining charges in luminescent devices such as OLEDs and lasers. The *staggered gap* (*Type II heterojunction*) is essential for the dissociation of excitons into free charge employed in photodetectors and solar cells. And finally, the less common *broken gap* (*Type III heterojunction*) is found in some thin film organic transistor applications. In this offset arrangement, electrons in the semiconductor with the large electron affinity recombine with holes from the contacting semiconductor since the LUMO of the first material is below the HOMO of the second.

The HJs depicted have been constructed to describe inorganic semiconductors with their broad energy bands, small effective masses, and a significant density of background free carriers. Earlier in this chapter, we have found that this picture is not directly applicable to organic semiconductors, characterized instead by very narrow conduction levels for holes and electrons, considerable static and dynamic disorder that results in large effective masses and hence low mobilities, and often very low densities of free charge. Nevertheless, the analogy to inorganic semiconductors is a helpful starting point, as long as these differences are recognized. At this point, we will extend our view to accommodate the important differences that exist between inorganic and organic systems in an effort to adapt the energy diagrams of Figs. 4.96 and 4.97 to the specific physics of organic semiconductors.

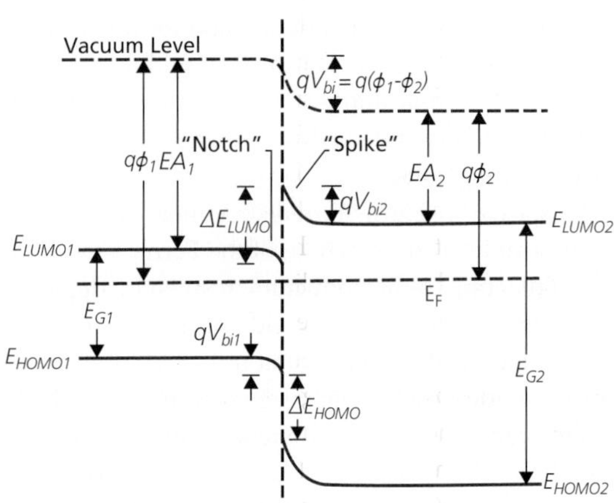

Figure 4.97 An n-N isotype heterojunction.

One distinct property of organics is that measurements of the energetics of the contact using such techniques as UPS and Kelvin probes, reveals an interface energy dipole of $\Delta \sim 0.1$–0.5 eV. This results from charge redistribution across the interface following contact of the two layers (Vázquez et al., 2005, Molodtsova and Knupfer, 2006, Osikowicz et al., 2007). We have seen that dipoles are found at metal–organic contacts (Fig. 4.89 and Table 4.5) which is attributed to static disorder of molecules at the junction (Baldo and Forrest, 2001). Also, the presence of surface traps can lead to the creation of a dipole (Fig. 4.88).

One of the most interesting effects is that *the dipole between organics at a HJ depends on the surface on which it is deposited* (Tang et al., 2006, Zhao et al., 2008). A central assumption of an ideal HJ is that the order of layer deposition should not affect the magnitude of the band offsets. That is, depositing material A on material B should result in the same energy level offsets as B on A (Margaritondo and Perfetti, 1987). This *commutative property* of the HJ holds true for interface-trap-free inorganic HJs, yet has not been found to obtain for organic semiconductors. In Fig. 4.99 we plot the interface dipole energies vs. the difference between the substrate work function

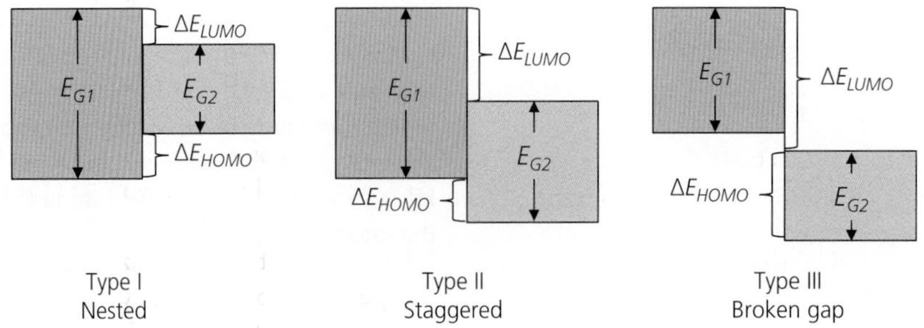

Figure 4.98 Classification of HJ types.

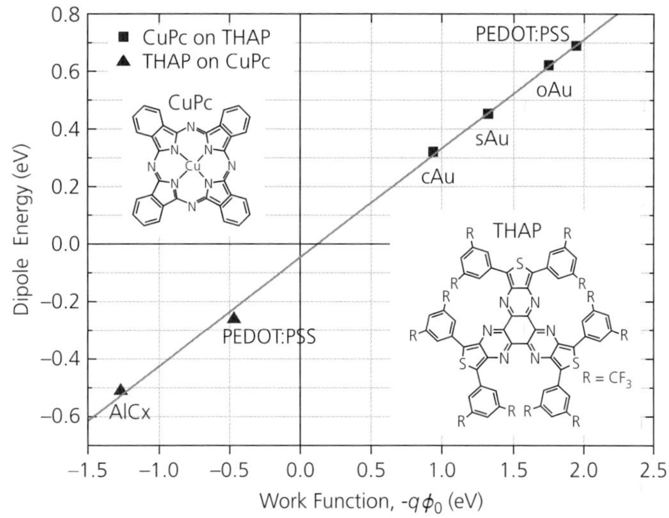

Figure 4.99 Energy dipole at HJs between CuPc and THAP on several different substrates. Molecular structural formulae for the compounds are shown. The letters preceding the Au contact refer to different surface preparations: c = solvent cleaned, s = sputter cleaned in 1 keV Ar^+, and o = treated in UV ozone for 1 h. The charge neutrality energy is $q\phi_0$. After Zhao et al. (2008).

and the charge neutrality level, $q\phi_0$ (see Section 4.6.3) for two organic semiconductors, CuPc and THAP. Starting with the substrate, CuPc is deposited on THAP, and vice versa (Zhao et al., 2008). These data show both the violation of the commutative rule and the strong dependence on substrate preparation. The non-commutative property is exemplified by deposition on PEDOT:PSS, where the dipole energy varies from +0.7 eV for CuPc on THAP, to −0.25 eV for THAP on CuPc.

A comprehensive explanation for the substrate dependence of the dipole energy, and the resulting charge redistribution across organic HJs relies on understanding the role played by disorder (Oehzelt et al., 2015). Figure 4.100a shows an approximate representation of the transport levels at an Alq_3/CuPc HJ. The energetic disorder is approximated by a Gaussian DOS of full width, 2σ, extending both above and below the LUMO or HOMO level, represented by the rectangles in the figure. The individual molecular layers are represented by their own Gaussian, separated by an interlayer distance, Δz. The fractional occupation of the DOS at each molecule is determined by the separation of the DOS from the Fermi energy, E_F, at each position, z. Once the materials are in contact, thermodynamic equilibrium is established, resulting in a flat E_F. The ability to establish local and global equilibria is a condition for current flow without charge build-up within the layers.

Figure 4.100b shows the Alq_3 and CuPc energies prior to contact. Apparently, if nothing is perturbed by the contact, the Fermi energy extends into the LUMO of CuPc with its high density of electrons. The metal contact E_F is taken at the same energy as E_F of the first layer of Alq_3. The total internal electrostatic potential across the film, $V(z)$, is calculated from Poisson's equation. The contact results in band bending, shown in Fig. 4.100c, such that the Fermi energy now lies outside of the LUMO band of the CuPc. Also, pinning the Alq_3 Fermi energy the metal E_F yields an Ohmic contact. This impacts the charge distribution all the way to the HJ located at $z = 10$ nm. The calculated charge distribution is shown in Fig. 4.100d. We see that electrons accumulate on the CuPc side (with charge neutrality established by an equal and oppositely charged hole distribution at the contact), and the Alq_3 layer is completely depleted of charge, resulting in a uniform electric field in this layer. Hence, $V(z)$, and the vacuum level vary linearly across the Alq_3, and then flatten in the CuPc. The maximum electrostatic potential that is achieved at $\frac{\partial V}{\partial z} = 0$ is the total dipole energy.

This layer-by-layer shift in $V(z)$ is observed using Kelvin probe and UPS measurements for numerous semiconductor systems. For example, a series of UPS spectra taken for various levels of coverage of a 10 nm thick layer of Alq_3 pre-deposited on a low work function Mg thin film is shown in Fig. 4.101a (Tang et al., 2006). The shift in the ionization energy is noted by the peak at approximately 2 eV below E_F, which corresponds to the onset of the photoelectron current (Hüfner, 2003). The Alq_3 valence features are resolved

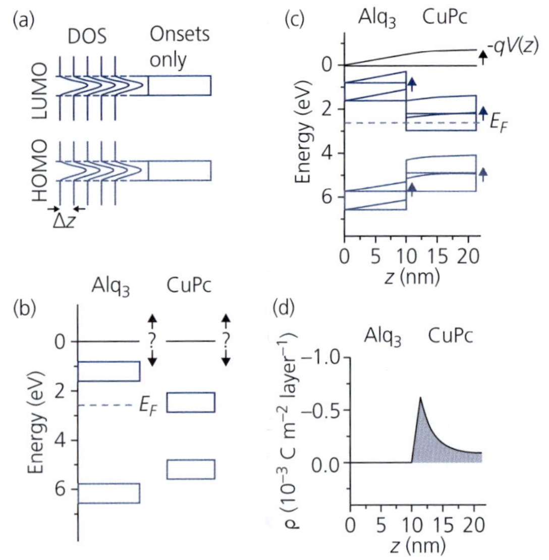

Figure 4.100 Interpretation of the Alq_3/CuPc heterojunction offset dependence on the substrate. (a) Representation of the transport band as the envelope of the disorder-induced Gaussian density of states (DOS). Each monolayer is indicated by its own Gaussian DOS separated by distance, Δz. The Alq_3 rests on a metal layer located at $z \leq 0$. (b) The positions of the LUMO (upper rectangles) and HOMO (lower rectangles) transport and Fermi (E_F) levels prior to, and (c) following HJ formation. Here, $V(z)$ is the internal potential at position z. (d) Electron distribution, $\rho(z)$, after contact (Oehzelt et al., 2015).

at 0 Å coverage, and become increasingly attenuated, and then dominated by CuPc features as its thickness is increased. The IP shifts at the high energy cutoffs of the spectra are accompanied by a similar shift of 0.7 eV in the vacuum level energy as coverage is increased from 0 Å to ~25 Å. At greater layer thicknesses the energy level bending is complete, and thermal equilibrium is established. These data are also shown in Fig. 4.101b. The LUMO energy equals the sum of the IP and the charge transport gap. The latter quantity is obtained from the energy difference between the onset of the CuPc UPS spectrum (which yields the ionization energy) and the inverse photoelectron spectrum (IPES), which gives the lower energy edge of the LUMO (Hill et al., 2000a, Tang et al., 2007). In IPES, low energy electrons (typically ≤ 20 eV) are incident on the organic surface. Photons are emitted as the electrons relax to the LUMO. The resulting photoemission spectrum vs. electron energy is measured, with the lowest energy photons providing the LUMO edge relative to E_F (see Appendix C) (Hill et al., 2000a, Hüfner, 2003, Tang et al., 2007). The LUMO follows the HOMO energy, as required for a material with a defined energy gap.

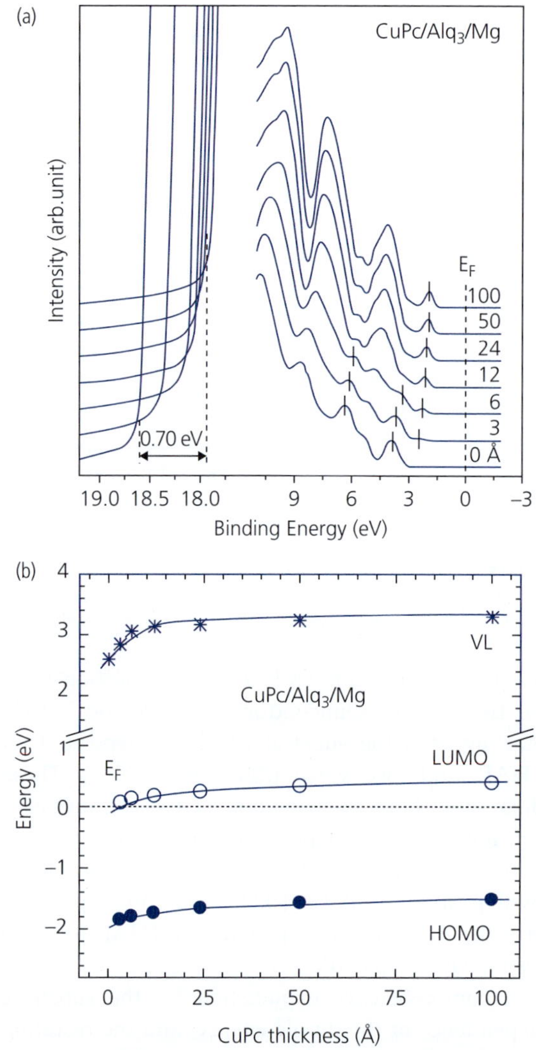

Figure 4.101 Ultraviolet photoelectron He I spectra of CuPc deposited on Alq_3 on a Mg thin film. The CuPc layer thicknesses increasing from bottom to top, are in Å (Tang et al., 2006). (b) The HOMO, LUMO, and vacuum level (VL) as functions of CuPc thickness. Also shown is the Fermi energy, E_F (dashed line) (Tang et al., 2007).

The extent of the band bending in Fig. 4.103 due to charge accumulation on the CuPc side of the HJ (forming the potential "notch" at the heterointerfaces) is consistent with calculations leading to the energy level diagram in Fig. 4.100. Hence, the UPS data lead to a consistent explanation of the origin of the HJ dipole energy level diagrams for Type I, II, and III HJs (Oehzelt et al., 2015). For example, Fig. 4.102 shows a series of Type II staggered HJs for contacts with a range of Fermi energies. In Fig. 4.102a, E_F is positioned near the HOMO of the material with the smallest IP energy in the left-most diagram, with E_F

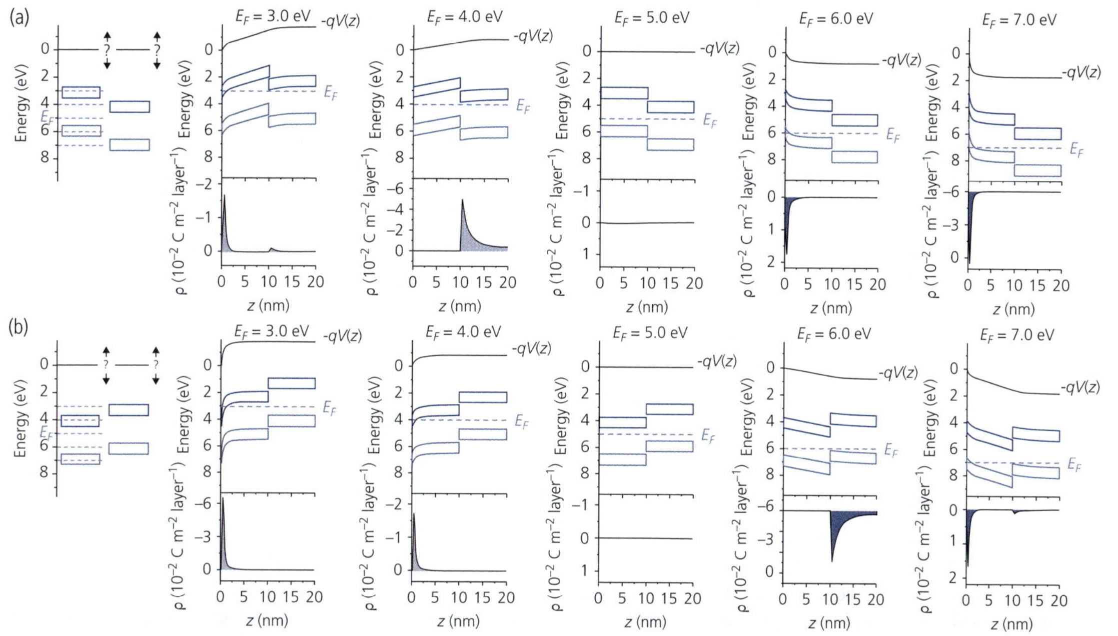

Figure 4.102 Energy level diagrams for Type II staggered organic HJs for various Fermi energy positions relative to the semiconductor frontier orbitals. (a) The material with a small ionization potential contacts the surface, and (b) the situation where the material positions are reversed. The charge distributions $\rho(z)$ are shown for each HJ along with the spatially dependent electrostatic potential. To the left are the relative energies assumed for the individual organics prior to their contact (Oehzelt et al., 2015).

becoming progressively larger, and hence closer to the HOMO from left to right. For $E_F = 3$ eV, there is bending in the energy levels for both materials, resulting in electron accumulation. The Fermi energy is centered within both energy gaps at 5 eV, resulting in flat bands, and consequently there is no electrostatic potential (or dipole) nor charge accumulation within the layers. Finally, as E_F approaches the HOMO, excess holes are accumulated at the contact, and the dipole energy is reversed.

In Fig. 4.102b the layers are reversed. It is clear that the dipole energies in this case are different from in Fig. 4.102a, accounting for the absence of the commutative property for excitonic HJs. The analysis leading to Figs. 4.100 and 4.102 assumes that the interface is free of defect states.

The influence of the substrate on the HJ energy level alignment can be used to create an Ohmic contact where $\phi_{Bn,p} \rightarrow 0$. Indeed, it has been shown that insertion of a very thin organic layer between a high work function metal oxide electrode (e.g. MoO_3, V_2O_5, WO_3) and a thick organic "device" layer can reduce the hole barrier to zero as long as the IE of the interlayer is larger than the IP of the organic (Kotadiya et al., 2018). This strategy apparently is successful in achieving Ohmic contacts independent of the conducting oxide or the organic layer as long as this relative energy criterion is met. The interlayer, itself an organic, must be sufficiently thin to allow for charge transport via tunneling or only a few hopping transitions. As in the HJ energy level alignment, the interposition of the ultrathin layer results in broadening of the surface DOS, allowing for overlap between the electronic states in the contact with the relatively narrow conduction states of the organic. This overlap promotes injection while also possibly pinning the organic layer Fermi energy at the metal E_F.

4.7.2.2 Ideal diode equation for current transport in a trap-free excitonic HJ

The j–V characteristics of organic semiconductor HJs are often similar to those of inorganic p-n junctions. As a consequence, theoretical treatments based on the generalized Shockley equation derived for inorganic devices have been extended to model the operation of organic solar cells (Dyakonov, 2004, Rand et al., 2007, Potscavage et al., 2008, Giebink and Forrest, 2009, Li et al., 2009). While this phenomenological approach often yields a reasonably accurate fit to the data, the application of inorganic semiconductor models obscures the inherently different physics of transport in organic semiconductors. In particular, the Shockley equation in Section 4.7.1 applies to inorganic p-n junctions with well-developed energy band structures with a sea of delocalized free charge carriers. We have seen throughout this chapter, however, that organic semiconductors are characterized by hopping transport and tightly bound, localized exciton states that require significant energy to dissociate into free charges.

In this section, we follow the derivation of Giebink, et al. for an ideal diode equation adapted to describe fully organic (i.e. excitonic) Type II HJs (Giebink et al., 2010a, 2010b). Note that the dynamics of Types I and III HJs are different from those of Type II junctions, and hence are beyond the scope of this particular analysis. Type II junctions used in organic photovoltaics must consider bipolar injection, that is, the staggered energy level line-up allows holes and electrons injected from their opposite electrodes to traverse the heterointerface. Hence, the Type II symmetry allows for rectification similar to that experienced by carriers in a conventional p-n junction. In contrast, Type I, nested junctions support unipolar injection since the hole and electron barriers are positioned to prevent the free flow of charge of both polarities across the interface. As a result, the current in these HJs is typically Ohmic or space charge limited, and is modeled in Section 4.4. Finally, the current in Type III HJs is limited by electron–hole recombination at the broken energy gap in Fig. 4.98, and hence conduction occurs by transfer from the LUMO to HOMO levels across the HJ. These will be discussed further in the context of OPV blocking and injection layers in Chapter 7.

Polaron pair generation, recombination, and dissociation at a Type II HJ provide a distinct departure from considerations of charge diffusion that lead to the Shockley equation. Among the differences between organic and inorganic semiconductor junctions, a more physically justifiable representation predicts the correct temperature dependence of the dark current, the intensity and temperature dependence of the open circuit voltage (V_{oc}) and short-circuit current (j_{sc}), and the maximum V_{oc} attainable under one sun illumination for a given small molecule or polymer HJ material pair. Additionally, the temperature-dependent diode ideality factor in Section 4.7.1 arises from recombination via disorder-induced traps at the heterointerface. In the following section we specifically treat the case of traps with an exponential energetic distribution extending into the energy gap.

This treatment is ideal, insofar in that it assumes that current is governed solely by generation and recombination at the HJ, and that both processes proceed through the *polaron pair* (PP) intermediate state (Giebink et al., 2010a, Vandewal et al., 2014). That is, we ignore all losses that occur prior to the exciton arriving within the HJ "interaction volume" of width a_0, which gives the polaron pair separation distance across the heterointerface. These extrinsic losses include incomplete optical absorption in the donor or acceptor layers, geminate and bipolar recombination, thermal generation and thermalization. As in the Shockley diode equation, these non-idealities can be added later to describe the characteristics of less than ideal devices.

We define a polaron pair as the lowest energy CT state at the HJ, where the electron on the acceptor side is Coulombically bound to a hole on the donor side of the heterointerface. The configuration of the Type II excitonic HJ is shown in Fig. 4.103a. Current outside of the active region is unipolar, with pure hole (j_h) and electron (j_e) currents flowing in the donor and acceptor bulks, respectively. The interfacial energy gap, ΔE_{HL}, in Fig. 4.103a is the difference between the donor HOMO and the acceptor LUMO that includes shifts due to an interface dipole. The hole and electron injection barriers at the anode and cathode are $q\phi_a$ and

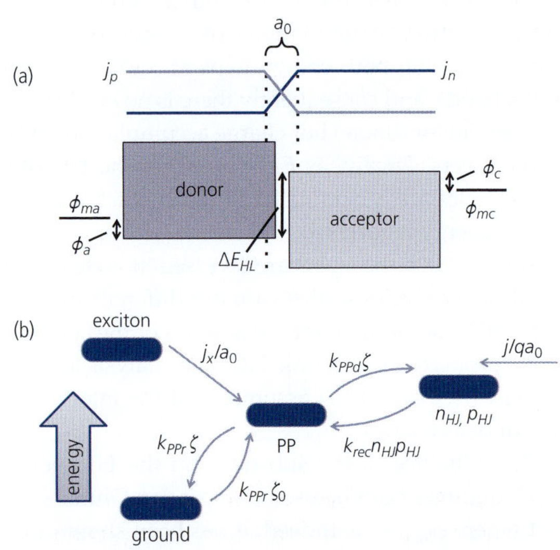

Figure 4.103 (a) Energy level diagram of a Type II excitonic (donor–acceptor) heterojunction defining the hole and electron currents (j_p, j_n, respectively) and the "interaction volume," a_0, within which the polaron pair (PP) is formed. ΔE_{HL} is the HOMO–LUMO energy difference. (b) State energies involved in PP production, showing the various transition rates (k) involved in the generation of charge, and exciton currents (j, j_X, respectively). Terms are defined in the text (Giebink et al., 2010b).

$q\phi_c$, respectively. These energies include interface dipoles that might exist. The built-in potential is given by the corresponding difference in contact work functions: $qV_{bi} = q\phi_{ma} - q\phi_{mc}$.

Figure 4.103b shows the processes that occur within the HJ volume. The steady state recombination of polaron pairs is described via two coupled equations:

$$\frac{d\zeta}{dt} = \frac{j_X}{a_0} - k_{PPr}(\zeta - \zeta_0) - k_{PPd}\zeta + k_{rec}n_{HJ}p_{HJ} = 0, \tag{4.297}$$

and for free charges:

$$\frac{dp_{HJ}}{dt} = k_{PPd}\zeta - k_{rec}n_{HJ}p_{HJ} + \frac{j}{qa_0} = 0, \tag{4.298}$$

where ζ is the PP density, j_X is the exciton current density diffusing to the interface, and n_{HJ} and p_{HJ} are the free electron and hole densities at opposite sides of the HJ, respectively.

Time-dependent solutions that elucidate the dynamical nature of these processes can be obtained by evaluating the full transient solutions to Eqs. 4.297 and 4.298. These solutions provide the time evolution of both the polaron and hole populations, respectively (Giebink et al., 2010a).

Polaron pairs recombine to the ground state at rate k_{PPr}, which is also linked to the thermal equilibrium PP population, ζ_0, determined by detailed balance (Kirchartz and Rau, 2008). Polaron pairs dissociate at rate k_{PPd}, which is a function of temperature and the electric field at the interface according to the *Onsager–Braun model* (see below). Finally, free carriers bimolecularly recombine to form PPs with rate constant, k_{rec}, approximated by its bulk Langevin value, $q\mu_{tot}/\varepsilon_r\varepsilon_0$.

Solving Eq. 4.297 for the PP density and substituting the result into Eq. 4.298 gives

$$j = qa_0k_{rec}\left(\frac{k_{PPr}}{k_{PPd}+k_{PPr}}\right)\left(n_{HJ}p_{HJ} - \frac{k_{PPd}}{k_{PPd0}}n_{HJ0}p_{HJ0}\right) - qj_X\left(\frac{k_{PPd}}{k_{PPd}+k_{PPr}}\right). \tag{4.299}$$

The subscript "0" indicates the value at thermal equilibrium. Similar to the Shockley equation, we assume quasi-equilibrium. Hence, the carrier densities at the interface (n_{HJ}, p_{HJ}) and contacts (n_C, p_C) are related via

$$n_{HJ} = n_C \exp[q\delta_A(V - V_{bi})/k_BT] \tag{4.300a}$$

and

$$p_{HJ} = p_C \exp[q\delta_D(V - V_{bi})/k_BT], \tag{4.300b}$$

where δ_D, and δ_A are the fractions of the potential dropped across the donor (D) and acceptor (A) layers, respectively. Thus, $\delta_D + \delta_A = 1$. These relations are strictly valid only when $j = 0$, but are a good approximation at low current when both diffusion and drift are small.

Substituting Eqs. 4.300 into Eq. 4.299 yields

$$j = qa_0k_{rec}N_{HOMO}N_{LUMO}(1 - \eta_{PPd})\exp(-\Delta E_{HL}/k_BT) \times\left[\exp(qV/k_BT) - \frac{k_{PPd}}{k_{PPd0}}\right] - q\eta_{PPd}j_X, \tag{4.301}$$

where $\eta_{PPd} = k_{PPd}/(k_{PPd} + k_{PPr})$ is the PP dissociation probability (Braun, 1984, Mihailetchi et al., 2004), and we have assumed that the density of charge at the cathode can be written in terms of the LUMO DOS, N_{LUMO}, such that $n_c = N_{LUMO}\exp(-q\phi_c/k_BT)$. At the anode, $p_a = N_{HOMO}\exp(-q\phi_a/k_BT)$. Here, effects that are appreciable at high electric field and low temperature such as tunneling have been neglected. Also, we define the HOMO–LUMO offset energy (cf. Fig. 4.103a):

$$\Delta E_{HL} = q\phi_a + q\phi_c + qV_{bi}. \tag{4.302}$$

Hence, Eq. 4.301 is the ideal diode equation for an excitonic Type II HJ in the absence of traps, and can be written in the form of Eq. 4.236:

$$j = j_0\left(\exp(qV/k_BT) - \frac{k_{PPd}}{k_{PPd0}}\right) - j_{ph}, \tag{4.303}$$

where the photocurrent is given by

$$j_{ph} = q\eta_{PPd}j_X. \tag{4.304}$$

That is, j_{ph} is simply the product of the exciton current, its dissociation efficiency at the HJ, and q. Importantly, Eq. 4.303 differs from the Shockley equation in that it includes the rate of polaron pair dissociation, k_{PPd}, which is a function of both electric field and temperature. An applied field can polarize the PP, inducing pair ionization at high reverse bias. At $V = 0$, the system is in equilibrium such that $k_{PPd}(V = 0) = k_{PPd0}$, in which case the term in parentheses vanishes.

The field dependence of PP dissociation is obtained from the theory of Onsager for ion pairs (Onsager, 1938), and extended to symmetric excitonic states by Braun (1984). This is a classical model based on Langevin recombination dynamics: if the charge pair is at a capture (or thermalization) radius $r_{th} < q^2/(4\pi\varepsilon_r\varepsilon_0k_BT)$, the charges recombine, otherwise they escape their mutual Coulomb attraction and remain free charges. This radius is ~2–10 nm at room temperature. In an electric field, the charge pair is polarized, resulting in a higher probability for dissociation. Thus, there is a competition between field-induced ionization, and thermalization leading to geminate recombination.

Figure 4.104a shows the probability for dissociation as a function of PP radius, a_0, and at three different electric fields in an isotropic medium whose dielectric constant, $\varepsilon_r = 3.02$, is equal that of anthracene (Batt et al., 1969). For this calculation, it is assumed that the initial pair orientation and distribution is independent of the external electric field. The escape probability is shown in Fig. 4.104b, with the initial PP radii ranging from 0.6 to 1.0 nm. The probability increases with field, until it saturates at $P(F) = 1$ at $F \approx 10^5$ V/cm.

The dissociation rate k_{PPd} is given by (Braun, 1984, Giebink et al., 2010b)

$$k_{PPd} = \frac{3}{4\pi a_0^3} k_{rec} \exp(-E_B/k_B T) \frac{J_1(2\sqrt{-2b})}{\sqrt{-2b}}, \quad (4.305)$$

where E_B is the PP binding energy, $b = q^3 F_{HJ}/(8\pi\varepsilon_r\varepsilon_0 k_B^2 T^2)$, J_1 is the Bessel function of order one, and F_{HJ} is the electric field at the HJ. The validity of Eq. 4.305 for negative fields (i.e. when F_{HJ} reverses direction under forward bias) is unclear, and its numerical evaluation becomes unstable for $F_{HJ} < -10^4$ V/cm.

The theory of Braun has been extended to consider electrons and holes that can only "sample" the half-space occupied by the acceptor and donor layers, respectively. This is a situation found in Type II HJs with ΔE_{LUMO} and ΔE_{HOMO} sufficiently large to prevent electron or hole transfer back over the HJ energy offsets at the HOMO and LUMO edges where they would recombine (Fig. 4.105a) (Peumans and Forrest, 2004). The energy of the bound pair is given by

$$E_{PP} = \frac{q^2}{4\pi\varepsilon_r\varepsilon_0|\mathbf{r}_e - \mathbf{r}_h|} - q\mathbf{F}\cdot(\mathbf{r}_e - \mathbf{r}_h) + E_{LUMO,A}(\mathbf{r}_e) + E_{HOMO,D}(\mathbf{r}_h), \quad (4.306)$$

where $\mathbf{r}_e$ ($\mathbf{r}_h$) is the position of the electron (hole), $E_{LUMO,A}$ is the energy of the acceptor LUMO, and $E_{HOMO,D}$ is the energy of the donor HOMO. Given that the electron must reside on the acceptor and the hole on the donor, the initial condition requires that the PP dipole moment is preferentially aligned perpendicular to the plane of the heterointerface (i.e. the orientation distribution of dipoles has a width, $\sigma_\theta = 0$); a striking departure from the conditions assumed in the Onsager–Braun model. Hence, the orientation of the PP dipole is no longer random relative to the applied field, which is also assumed to be normal to the interface plane.

Results of Monte Carlo simulations of the dissociation (or escape) probability of the electron and hole show a strong dependence on the asymmetry of mobilities of the two carriers on opposite sides of the HJ (see inset, Fig. 4.105b). That is, the volume sampled by the more mobile carrier is far larger than that of the less mobile carrier, resulting in a higher opportunity for escape than if both charges have the same

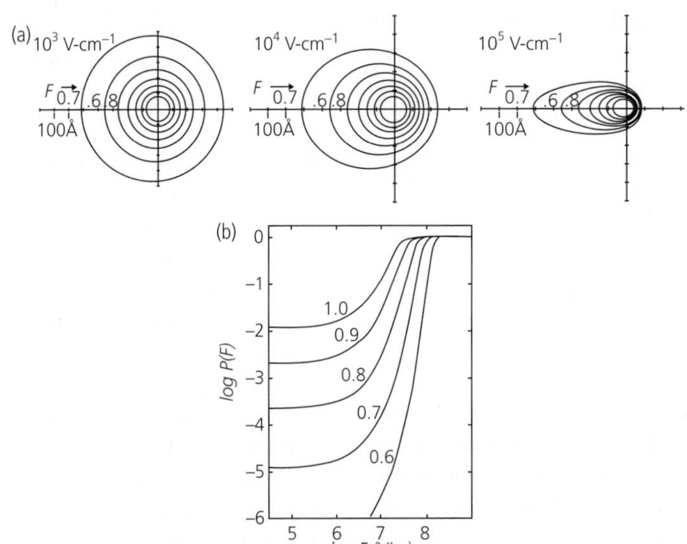

Figure 4.104 (a) Ionization probability at three different electric fields for a polaron pair in anthracene at $T = 300$ K. Each contour is separated by a probability interval of 0.1 (Batt et al., 1969). (b) Ionization probability vs. electric field for an initial starting radius (in nm) shown for each curve. For these calculations, the total Langevin mobility is $\mu_{tot} = 1$ cm^2/V s (Braun, 1984).

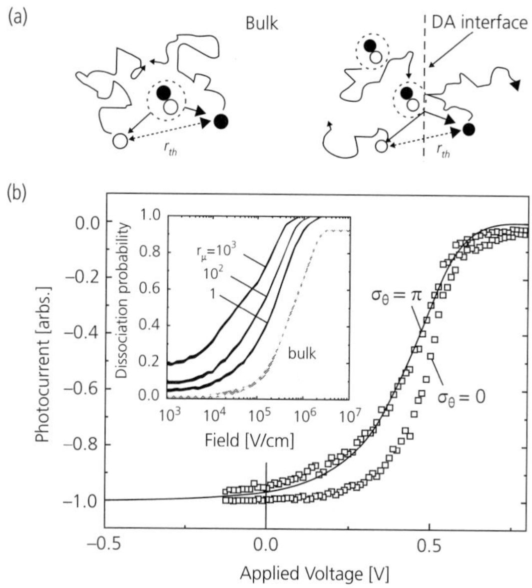

Figure 4.105 (a) The capture process of an electron and hole in the bulk and at a donor–acceptor HJ. Here, r_{th} is the thermalization radius. The dashed circles denote bound electron–hole pairs. (b) Monte Carlo simulations of the photocurrent from a CuPc/C_{60} heterojunction (squares) assuming a thermalization distance for both electrons and holes of r_{th} = 24 Å, an initial polaron pair separation of 8 Å, and T = 300 K. The electrons and holes are either randomly distributed relative to the HJ ($\sigma_\theta = \pi$) or are oriented perpendicular to it ($\sigma_\theta = 0$). Here, σ_θ is the width of the distribution of dipole orientations. The line indicates data. Inset: Simulated dissociation probability at the HJ vs. electric field for various ratios of electron-to-hole mobilities r_μ (assuming the larger mobility is in the numerator) (Peumans and Forrest, 2004).

mobility, and hence are pinned at the interface where the recombination probability increases.

The inset in Fig. 4.105b shows the dissociation probability at a DA HJ as a function of electric field for several mobility ratios, $r_\mu \geq 1$, where the larger mobility is in the numerator. This calculation shows that the dissociation probability substantially increases with electric field, saturating for $r_\mu = 1$ at $F > 10^5$ V/cm. This is higher than the bulk dissociation probability (with randomly oriented dipoles) of ~ 0.9 at the highest fields (Peumans and Forrest, 2004).

This extension of Onsager–Braun theory to a Type II excitonic HJ has been tested by measuring the photocurrent from a 20 nm CuPc/20 nm C_{60} junction as a function of applied voltage, with results shown in Fig. 4.105b. The fit by integrating over the half-space is substantially improved from that obtained assuming a totally random, bulk orientation (squares). The photocurrent saturates at high voltages, consistent with the probabilities calculated in Fig. 4.104a.

The field dependence of k_{PPd} in Eq. 4.305 suggests that the reverse current increases with voltage, contrary to predictions of the Shockley equation where the diffusion current is voltage independent. This voltage dependence is consistent with observation for excitonic HJs (Potscavage et al., 2008, Li et al., 2009). Under forward bias, k_{PPd} is similar to or less than k_{PPd0}. Further, the current density increases exponentially with an ideality factor $n = 1$, and the photocurrent (Eq. 4.304) is directly proportional to the PP dissociation efficiency that diminishes with increasing forward bias. Note that while the formulation of Onsager and Braun extended to ionization at a HJ provides an analytical form for k_{PPd} in Eq. 4.305, it is unlikely that it is valid at fields $\geq 10^4$ V/cm, as will be discussed in Chapter 7 (Renshaw et al., 2012).

The central assumption of the preceding analysis is that the relaxed state polaron pair (or CT exciton) is a precursor to charge generation. That is, according to Fig. 4.103, excitons that migrate to the heterointerface form the PP precursor state that is driven to dissociation via the energy level offsets, ΔE_{LUMO} and ΔE_{HOMO}. If $k_{PPr} \gg k_{PPD}$, the efficiency for dissociation is less than for recombination, and the excitation is lost to thermalization.

In Chapter 3 we discussed the dynamics and photophysics of CT states in DA complexes, or between molecules of similar composition (i.e. excimeric states). In the context of the current analysis, we must consider whether or not the dissociation of the CT or PP state bridging a HJ comprising a contact between otherwise homogeneous donor and acceptor layers is the precursor to charge generation.

Evidence that such states lead to free charge generation at CuPc/C_{60} and SubPc/C_{60} junctions has been reported (Giebink et al., 2010a), and the participation of CT states in a variety of small molecule HJs has been inferred from comprehensive studies of their open circuit voltages (Rand et al., 2007). However, what was lacking in those earlier studies was convincing evidence that the photocurrent predominantly originates from a relaxed CT versus a hot, charge separated (CS) state. These competing scenarios for the generation of charge from an excited donor molecule ($D^* = S_1$ in this example) are illustrated by the Jablonski diagram in Fig. 4.106. Direct generation to D^* is denoted by G_{D^*}. Then D^* can decay to the CT manifold at rate k_{relax}, followed by internal conversion to CT_1. Due to the spatially distributed nature of CT and CS states, there are no well-defined vibronic levels. Once the excitation has reached CT_1, it can fluoresce at rate k_f, or it can gain energy to reach the

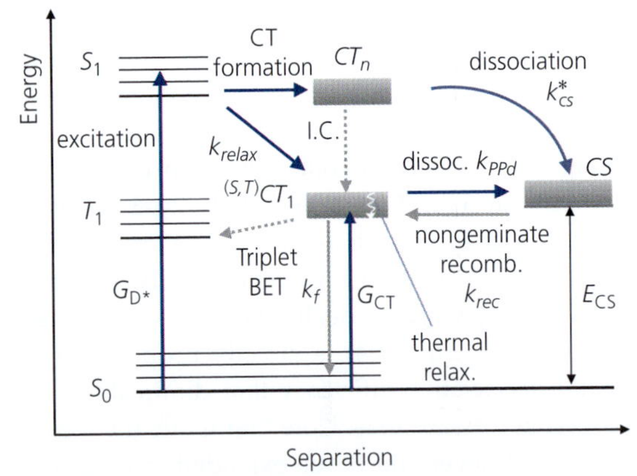

Figure 4.106 Processes leading to charge generation. Here, $S_1 = D^* =$ excited singlet donor state, $CT =$ charge transfer state, $G_{CT} =$ optical energy to excite a polaron to the lowest CT state (CT_1), $CS =$ charge separated state, BET = back electron transfer, IC = internal conversion. The rates are labeled k with the subscripts: *relax* = relaxation, f = fluorescence, *rec* = reverse $CS \rightarrow CT$ transfer, *PPd* = forward $CT \rightarrow CS$ transfer. G indicates direct generation processes.

charge separated state, CS, ultimately generating free charges. This transition occurs at rate k_{PPd}, which competes with direct CS formation from D* at rate, k^*_{CS}, and with back transfer from CS to CT at rate k_{rec}. The CT state can be directly excited at rate G_{CT}, providing a powerful tool for investigating its oscillator strength and energetics. We show in the following that direct transitions of D* to CS are unlikely, and that the most likely route to free charge generation is via CT_1 and subsequently to CS.

Direct transfer from D* to CS at the HJ bears similarities to direct charge generation in the bands of an inorganic semiconductor, where a photon of energy E_{D^*} results in generation of a free electron polaron (Bakulin et al., 2012, Grancini et al., 2013). Its direct excitation requires that the transfer occur without thermalization into CT, which is unlikely due to the very short mean free path of charges (equal approximately to the intermolecular separation), and the high phonon energies that lead to rapid thermalization in organic semiconductors.

The question is not whether CS states can be produced directly from exciton dissociation, but whether these, or CT states are the immediate precursor to photocurrent generation. The CT states can be directly accessed by optical pumping at wavelengths longer than those needed to excite the donor and acceptor states (i.e. $E_{CT} \simeq \Delta E_{HL} < E_{D^*,A^*}$), and then observing the resulting CT state photoluminescence (Loi et al., 2007, Liu et al., 2016b). Alternatively, the HJ can be forward biased, and the quantum efficiency of the long wavelength electroluminescence from the CT

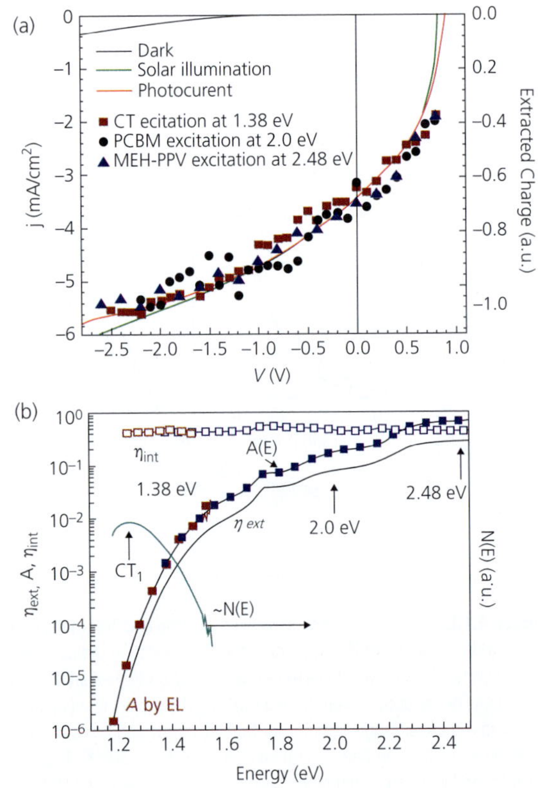

Figure 4.107 (a) Photo and dark current characteristics of a MEH-PPV/PC_{61}BM HJ. The photocurrent (red line) is the difference between response under simulated solar illumination and the dark current (black line). (b) External quantum efficiency (η_{ext}), absorption ($A(E)$), and internal quantum efficiency (η_{int}) of the HJ in (a). $N(E)$ is the CT photon flux spectrum (Vandewal et al., 2014).

state can be used to quantitatively determine the population of these states (Vandewal et al., 2010, Rivan et al., 2014).

Figure 4.107a shows the dark and photocurrent vs. voltage characteristics of a MEH-PPV/PC_{61}BM HJ. Also shown is the photocurrent arising from direct CT excitation at 1.38 eV, and from direct excitation of the D^* (MEH-PPV) state at 2.48 eV, and A^* (PC_{61}BM) at 2.0 eV. The similarities between the field dependence of the current generated by direct excitation of the CT state, D^* and A^*, and the total photocurrent suggests that all photocurrent sources have the same origin. In contrast, no field dependence is expected from an unbound CS state since it does not undergo Onsager-like ionization.

A quantitative comparison of the measured external quantum efficiency (η_{ext}) spectrum for this HJ with calculations assuming 100% of the photocurrent arises via an intermediate CT state is shown in Fig. 4.107b. To determine what fraction of the photogenerated charge arises from CT dissociation, the CT state absorption spectrum, $A(E)$ is needed. In thermal equilibrium, $A(E) \propto N(E)\exp(E/k_BT)/E^2$, where $N(E)$ is the flux of photons emitted at energy, E (Vandewal et al., 2014). The CT emission spectrum shown in Fig. 4.107b has been measured at low injection to assure that near-equilibrium conditions exist. The calculated absorption spectrum is given by the solid line. The ratio between the measured η_{ext} and calculated $A(E)$ then gives the internal quantum efficiency (η_{int}), also shown in Fig. 4.107b (squares). That is, η_{int} is the ratio of charges generated to photons absorbed in the active region, that is, $\eta_{int}(E) = \eta_{ext}(E)/A(E)$. The lack of spectral dependence of the η_{int} spectrum extending to low energies where only the CT state is excited (<1.4 eV) is evidence that the CT, otherwise known as the bound PP state, is the intermediate between exciton and photocurrent generation. This analysis has been applied to both polymer as well as small molecule systems, some with a field-dependent photocurrent and others without, but all leading to the same conclusions regarding the pivotal role played by the CT state (Vandewal et al., 2014).

Another, less model-dependent example of photocurrent generation via CT states is shown in Fig. 4.108 for the small molecular weight HJ consisting of the donor, DBP, and the acceptor, C_{70} (Liu et al., 2016b). Both the electroluminescence and photoluminescence spectra show intense emission from two clearly distinguishable CT states; the lowest energy feature (CT_1) is identified with electrons confined in amorphous phases of C_{70}, and the higher energy (corresponding to lower binding energy) CT_2 state is due to electrons in C_{70} crystalline domains (see Section 7.4.3). Both CT states are observed at energies well below the C_{70} and DBP singlet emission lines. In Fig. 4.108 η_{int} is obtained from the η_{ext} spectrum after directly measuring the absorption spectrum of the blended HJ thin films. Since η_{int} approaches 100% at energies as low as 1.4 eV, this unambiguously implicates the CT state as the precursor to photogeneration, even via absorption of photons at the highest energies measured.

The long wavelength features are associated with the CT state since it is absent in devices consisting solely of C_{70} or DBP. These latter devices exhibit very low η_{ext} due to the lack of a charge-separating interface. Calculations using time dependent density functional theory were employed to investigate the CT spectra and to ensure that the origin of the low energy states are indeed due to charge transfer between the monomeric species. The charge density distribution of CT states for several different C_{70} dimer and trimer configurations with DBP clusters are shown in Fig. 4.109a. These calculations show the electron

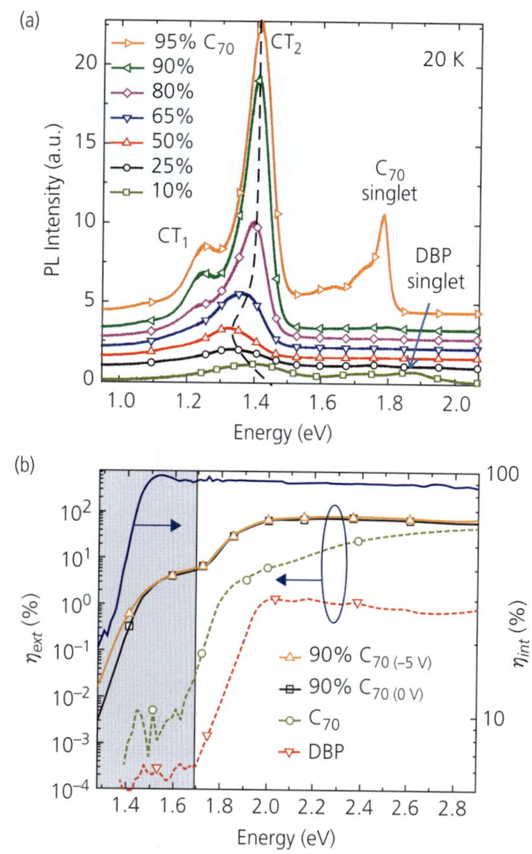

Figure 4.108 (a) Photoluminescence (PL) intensity as a function of energy for HJ blends where the ratio of the acceptor, C_{70} to the donor, DBP is varied. Emission from two CT states is identified well below the A* and D* emission: CT_1 is associated with the electron in amorphous regions of C_{70}, and CT_2 is for electrons in crystalline islands of C_{70}. (b) External (η_{ext}) and internal quantum efficiency (η_{int}) spectra at 0 V and −5 V for 10% DBP/90% C_{70} HJs. Also shown are the η_{ext} spectra for non-HJ devices consisting of only DBP or C_{70}. singlet features (Liu et al., 2016b).

(purple) and hole (yellow) densities comprising the CT states span the C_{70} and DBP molecules, respectively. Including additional C_{70} molecules into the complex to form trimers and tetramers leads to moderate electron delocalization and creates several higher energy CT states. These calculations consider only a few of the many possible DBP-C_{70} molecular configurations, with those provided serving as examples. The energies of the molecular combinations are provided in Fig. 4.109b. The C_{70} monomer has a singlet energy of $E_S = 2.04$ eV, while the DBP monomer has $E_S = 2.18$ eV. Other high order singlet state energies are also shown. The singlet energies of the DBP-C_{70}

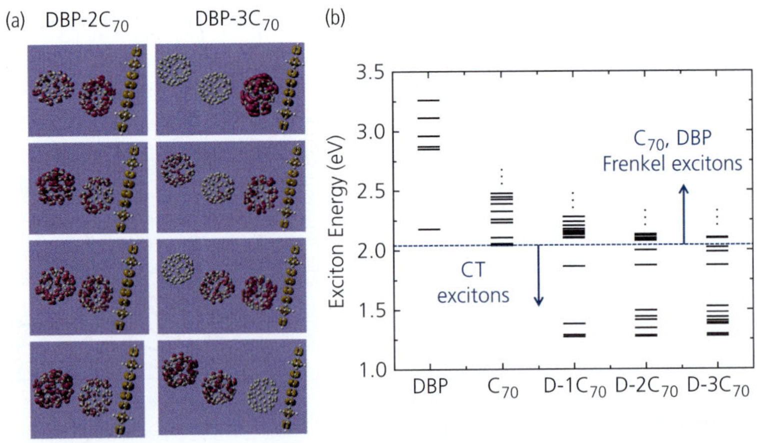

Figure 4.109 (a) Charge density of the first four CT states of the DBP-C_{70} dimer (left) and trimer (right) complexes calculated using TD-DFT. The electron (purple) and the hole (yellow) densities are located on the C_{70} and DBP molecules, respectively. (b) Calculated singlet energies of different monomer, dimer (D-1C_{70}), trimer (D-2C_{70}), and tetramer (D-3C_{70}) combinations of DBP and C_{70}. DBP is abbreviated as D (Liu et al., 2016b).

Adapted with permission from Liu, X., Ding, K., Panda, A. & Forrest, S. R. 2016b. Charge transfer states in dilute donor-acceptor blend organic heterojunctions. ACS Nano, 10, 7619. Copyright 2016 by the American Chemical Society.

physical dimer, trimer (one DBP-two C_{70} molecules) and tetramer (one DBP-three C_{70} molecules) that lie below the DBP and C_{70} monomer states are CT states, consistent with the spectra in Fig. 4.108. The calculated oscillator strengths of dimer CT states are in the range 10^{-4}–10^{-3}, which are 100 to 1000 times lower than for the monomers, but higher than for trimer and tetramer CT states.

There is apparently no voltage dependence of η_{ext} due to CT_2, although there is indication that CT_1 does undergo limited field ionization at $E < 1.24$ eV. Quantitative analysis of the field dependence of the ratio of η_{int} at $E < 1.4$V where current originates, from the more tightly bound CT_1 vs. η_{int} at higher energies, suggests that CT_2 contributes 88% of the total photocurrent, with the remainder due to CT_1 (Liu et al., 2016b).

While these results show a quantitative dependence of the generation of free polarons and the PP precursor states at the HJ, there remains the question of how the $CT \to CS$ transition occurs, which can often require an increase in energy of a few hundred meV. It is likely that this excess energy is supplied by energy level bending at the heterointerfaces that segregates the electron and hole in the PP, reducing the recombination probability as charges migrate away from the HJ (Peumans and Forrest, 2004). Asymmetries in the electron and hole mobilities in their respective layers encourages dissociation, further reducing k_{PPr}, as shown in Fig. 4.105. Morphology also plays an important role, whereby the electron mobility in fullerene blended HJs is increased with increasing domain size, thereby promoting the $CT \to CS$ transition (Bernardo et al., 2014, Liu et al., 2016b).

Returning to Eq. 4.301 and solving for the open-circuit voltage by setting $j = 0$ gives

$$qV_{OC} = \Delta E_{HL} - k_B T \ln\left[\left(\frac{k_{PPr}}{k_{PPd}}\right)\frac{a_0 k_{rec} N_{HOMO} N_{LUMO}}{j_X}\right], \tag{4.307a}$$

or

$$qV_{OC} = \Delta E_{HL} - k_B T \ln\left[\left(\frac{k_{rec} k_{PPr}}{k_{PPd} + k_{PPr}}\right)\frac{q a_0 N_{HOMO} N_{LUMO}}{j_{ph}}\right]. \tag{4.307b}$$

Equation 4.307 predicts that V_{OC} should increase with decreasing temperature, and with the logarithm of j_X (proportional to the illumination intensity) with slope $k_B T/q$. When polaron pairs are strongly coupled to the ground state, k_{PPr} is large and V_{OC} decreases, that is, this analysis predicts a decrease in open circuit voltage with increased PP recombination. At V_{OC}, the electric field, F_{HJ}, is small or even reverses direction (i.e. aiding recombination), such that k_{PPd} takes on its zero-field value, $k_{PPd}(F_{HJ} = 0) \approx k_{PPd} = \frac{3}{4\pi a_0^3} k_{rec} \exp(-E_B/k_B T)$. Since $4\pi a_0^3/3$ is the volume occupied by a polaron pair, this relationship is inserted into Eq. 4.307 to give

$$qV_{OC} = (\Delta E_{HL} - E_B) - k_B T \ln\left[\frac{a_0 k_{PPr} N_{HOMO} N_{LUMO}}{j_X \zeta_{\max}}\right], \tag{4.308}$$

where ζ_{max} is the maximum PP density that is sustained at the interface (i.e. all states are occupied).

At $T = 0$ or at very high illumination intensities, $\zeta_{max} \to N_{HOMO}, N_{LUMO}$. Under these conditions, all available molecular sites are occupied by a polaron or an exciton. Furthermore, the maximum exciton current is $j_X = k_{PPr} a_0 N_{HOMO}$; at higher intensities there are no further HOMO states to excite and the material becomes transparent. Thus, the argument of the logarithm in Eq. 4.308 approaches unity, giving the maximum possible open circuit voltage:

$$qV_{OC}^{max} = \Delta E_{HL} - E_B. \tag{4.309}$$

This expression is a direct result of the foregoing analysis, and is analogous to expressions previously suggested from experimental observation, but never rigorously proven (Rand et al., 2007, Cheyns et al., 2008, Vandewal et al., 2010).

The linear dependence of Eq. 4.309 for a selection of small molecular weight HJs listed in Table 4.7 is shown in Fig. 4.110, which is a plot of V_{OC}^{max} vs. the difference between the ionization energy of the donor (IE_D) and the optical electron affinity of the acceptor (EA_A(opt)). Here, EA_A(opt) is the sum of the energy determined from the long wavelength cut-off of the exciton absorption spectrum, that is, the optical energy gap, and the IP. The maximum open circuit voltage is measured at $T = 300$ K or at low temperature (~77 K). It is striking that the best fit to the data is a straight line with a unity slope and an intercept of approximately zero. To relate these data to Eq. 4.309, we note that Rand et al. (2007) postulated that $EA_A(\text{opt}) \approx E_{LUMO,A} - \frac{1}{2}E_{BA} - \frac{1}{2}E_{BD} \approx E_{LUMO,A} - E_{BD}$. That is, EA_A(opt) is equal to the difference between the $E_{LUMO,A} = EA_A$, and approximately half the polaron pair binding energy due to the electron on the acceptor (E_{BA}), and the other half of the PP energy due to the hole on the donor (E_{BD}). With this reasonable approximation, the abscissa in Fig. 4.110 is $IE_D - EA_A(\text{opt}) = E_{LUMO,A} - E_{HOMO,D} - E_B = \Delta E_{HJ} - E_B$. Thus, the linear fit directly follows from Eq. 4.309.

A similar linear correspondence between V_{OC}^{max} and $E_{HOMO,D}$ of a series of polymer donor molecules combined with $PC_{61}BM$ is shown in Fig. 4.111. In that study, both $E_{HOMO,D}$ and $E_{LUMO,A}$ were measured using cyclic voltammetry (CV, see Appendix C). A linear fit with a unity slope to the data

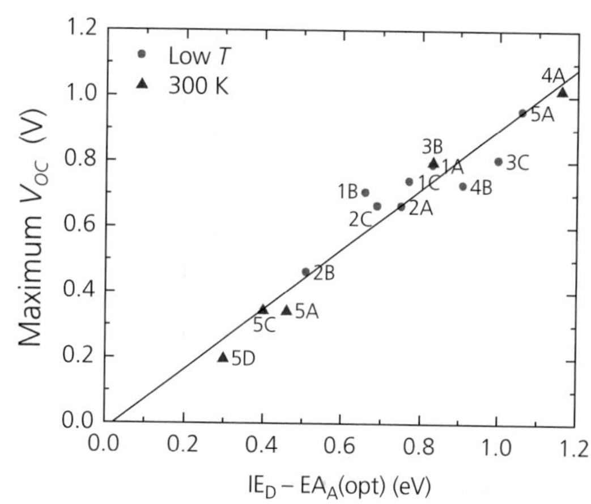

Figure 4.110 Maximum V_{OC} measured at two different temperatures vs. the difference of the ionization energy of the donor (IE_D) and the optically determined electron affinity of the acceptor ($EA_A(opt)$) for a number of Type II HJ donor–acceptor pairs listed in Table 4.7. Adapted from Rand et al. (2007).

Table 4.7 Donor and acceptor materials in Fig. 4.110 and their corresponding ionization energies (IE), electron affinities (EA), and optical energy gaps (E_{opt})

Label[a]	Material	IE (eV)[b]	EA (eV)[c]	E_{opt} (eV)	Reference
1	CuPc	5.2	3.2	1.7	(Kahn et al., 2003)
2	Pentacene	5.1	3	1.8	(Kahn et al., 2003)
3	NPD	5.5	1.7	3.1	(Kahn et al., 2003)
4	SubPc	5.6	-	2.0	(Mutolo et al., 2006)
5	$Ru(acac)_3$	4.9	-	2.1	(Rand et al., 2005)
A	C_{60}	6.2	3.6	1.8	(Benning et al., 1992)
B	C_{70}	6.4	4.3	1.7	(Benning et al., 1992)
C	PTCBI	6.2	3.6	1.7	(Kahn et al., 2003)
D	PTCDA	6.8	4.6	2.2	(Kahn et al., 2003)

[a] Numbers represent donors, letters represent acceptors.
[b] Measured via UPS with an error of ± 0.1 eV.
[c] Measured via IPES with an error of ± 0.5 eV.

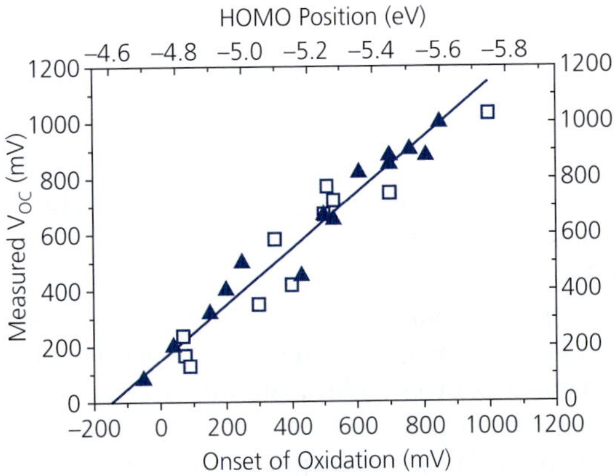

Figure 4.111 Maximum open circuit voltage vs. oxidation potential measured by cyclic voltammetry and the HOMO energies of a series of donor/$PC_{61}BM$ junctions. The straight line is a fit to the data with unity slope and an offset of −0.3 V (Sharber et al., 2006).

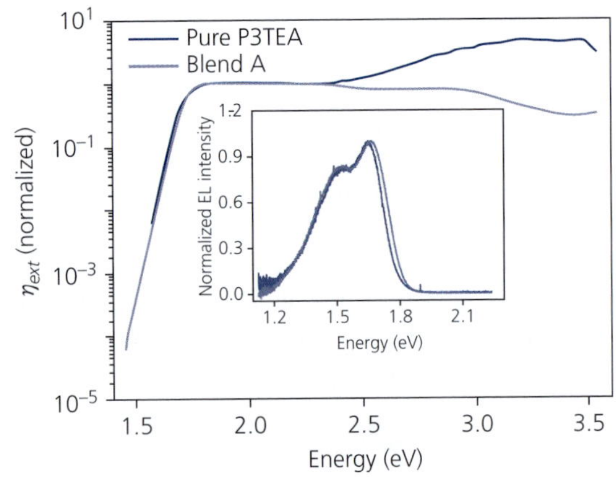

Figure 4.112 Fourier transform photocurrent spectroscopy is used to obtain the η_{ext} spectra of pure P3TEA and a 1:1.5 P3TEA:SF-PDI_2 blend (Blend A). Inset: Electroluminescence spectra of these two devices (Liu et al., 2016a).

following, $qV_{OC}^{max} = \Delta E_{HL} - 0.3$ eV was found. The offset energy of 0.3 eV was attributed to losses from diode non-idealities and internal fields within the device. Yet, in the context of Eq. 4.309, this offset again can be attributed to the PP binding energy.

Finally, we note that Eq. 4.309 is the organic analog to the band-gap limitation of V_{OC}^{max} in inorganic solar cells established by Shockley and Queisser (1961), of which we will have more to say in Chapter 7. Indeed, the PP binding energy, E_B, represents a significant source of energy loss in the photogeneration process. This is readily apparent in Fig. 4.108, where the energies of CT_1 and CT_2 are greater than 0.4 eV below the singlet state energies of the organics comprising the HJ. As we will show in Section 7.4.3, this is the energy needed to drive charge separation. Considerable research has, therefore, been devoted to designing molecules and active region morphologies that minimize the PP binding energy.

An example system where the driving energy is nearly eliminated and yet there is a high $\eta_{ext} = 66\%$ for charge separation, comprises the P3TEA donor combined with the spiro-fluorene bridged bis-perylenediimide acceptor, SF-PDI_2 (see Section 7.4.3, Yan et al., 2013). The normalized η_{ext} for both the pure donor and the P3TEA:SF-PDI_2 blend are shown in Fig. 4.112, and their electroluminescence spectra are shown in the inset. Unlike the η_{ext} spectra for the fullerene-based devices in Fig. 4.108, here we see complete overlap between the spectra of the blend and the pure donor. Furthermore, the CT electroluminescence spectrum is only slightly shifted from that of the pure donor. There is apparently no difference between the energies of the singlet exciton of the donor and the PP state, implying that $E_B \approx 0$. Furthermore, the CT state transfer rate, $k_{relax} = 0.3 \times 10^{12}\ s^{-1}$, is 10 times larger than k_f, which leads to the high external efficiency observed.

The high dissociation efficiency is attributed to several factors. The large volume occupied by SF-PDI_2 results in intramolecular charge delocalization, which greatly reduces the exciton binding energy (Liu et al., 2016a). Thus, the transfer between the exciton and CT state is greatly facilitated, resulting in $k_{relax} \gg k_f$. The reduction in exciton binding energy has led to reduction of the energy losses in OPVs that employ large, non-fullerene acceptor molecules such as SF-PDI_2. The quantitative relationship between molecular size and E_B is treated in greater detail Chapter 7. Details of the morphology of the donor–acceptor blend may also lead to rapid CT dissociation, even for small E_B. Morphology is yet another factor controlled by molecular size and shape. Finally, disorder within the domains of the blends may also reduce the CT binding energies.

4.7.2.3 Current–voltage characteristics in the presence of traps

The DOS centered at the HOMO and LUMO energies is broadened by ground state interactions between the donor and acceptor molecules at the heterointerfaces (Sreearunothai et al., 2006, Groves et al., 2010). Here, we extend our calculation of the j-V characteristics to include an exponential distribution of traps or other sources of disorder at the heterointerface. Their presence changes the kinetics of recombination, and introduces a temperature dependent ideality factor $n > 1$ into the diode equation.

Assuming trapped (n_T) and free (n) electron densities in the acceptor along with the trap distribution of Eq. 4.140, and using similar relationships for the trapped hole density, p_T, in the donor, then Eq. 4.299 becomes

$$j = qa_0\left(\frac{k_{PPr}}{k_{PPd}+k_{PPr}}\right) \times \left\{ \begin{array}{l} k_{rec,n}\left(n_{HJ}p_{HJT} - \frac{k_{PPd}}{k_{PPd0}} n_{HJ0}\, p_{HJT0}\right) \\ +k_{rec,p}\left(p_{HJ}n_{HJT} - \frac{k_{PPd}}{k_{PPd0}} p_{HJ0}\, n_{HJT0}\right) \end{array} \right\} - qj_X\left(\frac{k_{PPd}}{k_{PPd}+k_{PPr}}\right). \quad (4.310)$$

Hence, $k_{rec,n}$ and $k_{rec,p}$ are the rate constants due to recombination at the HJ between a free electron in the acceptor (n_{HJ}) with a trapped hole in the donor (p_{HJT}), and vice versa. Referring to our treatment of trap-free diodes, then,

$$j = qa_0(1-\eta_{PPd}) \times \left[k_{rec,n}N_{LUMO}H_{TD}\exp(-\alpha_D/k_BT)\left(\exp(qV/k_BT) - \frac{k_{PPd}}{k_{PPd0}}\right) + k_{rec,p}N_{HOMO}H_{TA}\exp(-\alpha_A/k_BT)\left(\exp(qV/k_BT) - \frac{k_{PPd}}{k_{PPd0}}\right)\right] - j_{ph}, \quad (4.311)$$

where

$$\alpha_D = \frac{\Delta E_{HL}}{n_D} + \frac{m_D - 1}{m_D}(\delta_D\phi_C - \delta_A\phi_A) \quad (4.312a)$$

and

$$\alpha_A = \frac{\Delta E_{HL}}{n_A} - \frac{m_A - 1}{m_A}(\delta_D\phi_C - \delta_A\phi_A). \quad (4.312b)$$

Here, $m_D = T_{OD}/T$ and $m_A = T_{OA}/T$ are the trap distribution exponents similar to Eq. 4.143 with the characteristic temperatures, T_{OD} and T_{OA}, on the donor and acceptor sides of the HJ, respectively. Also, H_{TD} *and* H_{TA} are the total trap densities on the donor and acceptor sides of the junction, and the respective ideality factors, n_D and n_A are given by

$$n_D = \frac{m_D}{\delta_A(m_D - 1) + 1} \quad (4.313a)$$

and

$$n_A = \frac{m_A}{\delta_D(m_A - 1) + 1}. \quad (4.313b)$$

Note that, in a striking difference with conventional Shockley theory, the ideality factors now have an implicit temperature dependence since $m \sim 1/T$.

More compactly, Eq. 4.311 becomes the ideal diode equation in the presence of traps (Giebink et al., 2010b):

$$j = j_{0D}\left(\exp(qV/n_Dk_BT) - \frac{k_{PPd}}{k_{PPd0}}\right) + j_{0A}\left(\exp(qV/n_Ak_BT) - \frac{k_{PPd}}{k_{PPd0}}\right) - j_{ph}, \quad (4.314)$$

which simplifies to

$$j = j_{0D}(\exp(qV/n_Dk_BT) - 1) + j_{0A}(\exp(qV/n_Ak_BT) - 1) - j_{ph} \quad (4.315)$$

when $k_{PPd} \leq k_{PPd,eq}$ under forward bias. Here J_{OD} and J_{OA} are the dark saturation currents given by the prefactors in Eq. 4.311, and j_{ph} is the photocurrent defined in Eq. 4.304.

Thus, in general, the sources of dark current stem from the recombination of free electrons with trapped holes, and vice versa. Each pathway produces its own temperature-dependent ideality factor (n_D and n_A, respectively) and dark saturation current, both of which depend on the balance of the voltage drop across the two contacting layers, as well as their characteristic trap temperatures.

It is instructive to consider the important case of a perfectly symmetric device, that is, one with identical transport properties and injection barriers for electrons and holes. Then Eq. 4.311 simplifies to

$$j = 2qa_0(1-\eta_{PPd})k_{rec}N_SH_{TS}\exp(-\Delta E_{HL}/n_Sk_BT) \times \left(\exp(qV/n_Sk_BT) - \frac{k_{PPd}}{k_{ppd0}}\right) - q\eta_{PPd}j_X \quad (4.316a)$$

or, in our usual form,

$$j = j_{0S}\left(\exp(qV/n_Sk_BT) - \frac{k_{PPd}}{k_{ppd0}}\right) - q\eta_{PPd}j_X, \quad (4.316b)$$

where the saturation current j_{OS} is obtained via comparison with Eq. 4.316a, and we have allowed $N_S = N_{LUMO} = N_{HOMO}$, and $H_{TS} = H_{TD} = H_{TA}$. The symmetric HJ ideality factor is

$$n_S = \frac{2m_S}{m_S + 1}, \quad (4.317)$$

where m_S has the same definition that relates it to the characteristic trap temperature, viz. $m_S = T_{TS}/T$ and $T_{TS} = T_{TA} = T_{TD}$. Comparing Eq. 4.317 to Eq. 4.313 implies that the voltage is now equally divided between the donor and acceptor sides of the junction. One approximation made is that the HOMO and LUMO densities of states are equal. Although this is reasonable, it would be more precise to allow $N_s = \sqrt{N_{LUMO}N_{HOMO}}$. Then, solving Eq. 4.316 for V_{OC} leads to

$$qV_{OC} = \Delta E_{HL} - n_Sk_BT\ln\left[\left(\frac{k_{PPr}}{k_{PPd}}\right)\frac{a_0k_{rec}N_S^2}{j_X}\right]. \quad (4.318)$$

The maximum open circuit voltage, V_{OC}^{max}, still reduces to $\Delta E_{HL} - E_B$, since in the limit of high light intensity, the assumption of n_T, p_T n,p used for trap limited transport no longer holds. In this regime all traps are filled and recombination is no longer trap-limited such that Eq. 4.318 reverts to Eq. 4.307b.

The symmetric diode can apply to a p–i–n excitonic homojunction if it too is symmetric, that is, the majority carrier concentrations of both the p- and n-regions are equal. Then the voltage is dropped equally across both sides of the junction. However, the barrier to charge recombination in the p-n junction is E_G/n_S, which must be substituted into the expression for saturation current. Furthermore, recombination occurs across the depleted i-region, introducing a quadratic voltage dependence to the saturation current as in Eq. 4.286.

A somewhat different expression for the homojunction current is derived in Section 4.7.1. The primary discrepancy arises since the model presented here is based on recombination across the depletion region, in contrast to tunneling across the junction. Recombination is ultimately a thermally activated process, although the Gaussian or exponential DOS results in a temperature dependent ideality factor that counters the thermal energy in such a way that the product, $n_S(T)k_BT$ is approximately temperature independent, as observed in the devices in Figs. 4.93 and 4.95. Since both models assume a broad density of transport states (one whose alignment allows for tunneling, and the other that factors into the recombination statistics that are reflected in the ideality factor), it is not entirely surprising that both models predict approximately the same functional behavior in the j–V–T characteristics, although the various constants for j_0 may differ due to the differing physical models employed.

Figure 4.113a shows the dark j–V characteristics calculated using Eq. 4.314 over the temperature range from 120 K to 296 K for a device using the parameters typical of organic photovoltaic cells listed in Table 4.8. The series resistance is $R_s = 1\ \Omega\ \mathrm{cm}^2$ (in which case, V in the exponential argument is replaced by $V - jR_s$) and it is assumed that most of the potential is dropped across the donor layer (i.e. $\delta_A = 0.1$, $\delta_D = 0.9$), resulting in different ideality factors, n_A and n_D, plotted in Fig. 4.113b. Both ideality factors increase with decreasing temperature due to the corresponding increases in m_A and m_D. Also, $n\ln(j_0)$ is nonlinear when plotted vs. $1/k_BT$. This contrasts with previous analyses based on the Shockley equation (Rand et al., 2007, Perez et al., 2009), which predicts

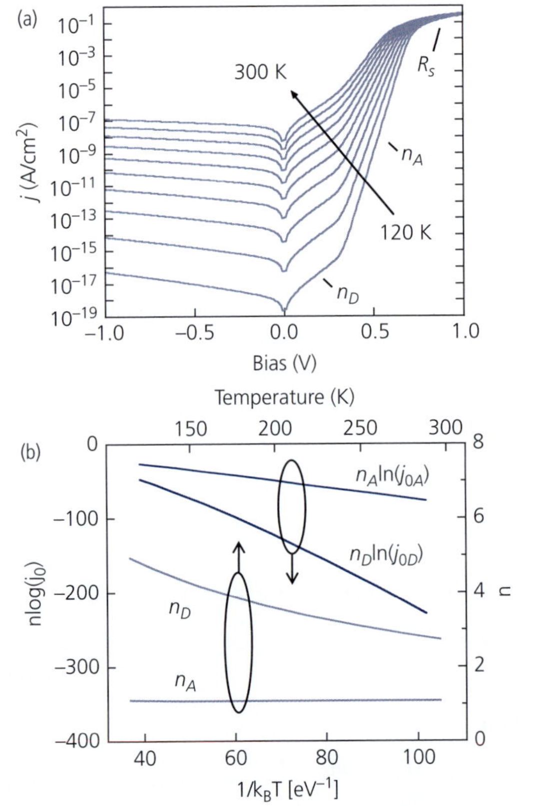

Figure 4.113 Simulated (a) dark j–V characteristics and (b) ideality factors (n_A, n_D) and $n\ln(j_0)$ (where j_0 is the saturation current density normalized to 1 A/cm²) vs. temperature for an excitonic Type II HJ with an exponential trap distribution in both the donor and acceptor layers. Parameters used in the simulations are given in Table 4.8 (Giebink et al., 2010b).

a linear relationship for $n\ln(j_0)$ with a slope of $-\Delta E_{HL}/2$.

In the generalized Shockley equation, increasing dark current with reverse bias is phenomenologically accounted for by recombination and generation in the depletion region, whereas in Fig. 4.113 it is due to the field dependent dissociation of thermally generated PPs. The two slopes often observed in semi-log plots under forward bias (Rand et al., 2007, Potscavage et al., 2008, Li et al., 2009) result from the two ideality factors that reflect recombination with trapped carriers at each side of the HJ.

The asymmetry in voltage dropped across the donor and acceptor layers is due to the sheet of trapped interface charge, which is the primary cause of the difference in ideality factors. At low

Table 4.8 Parameters used in calculating j–V characteristics in Fig. 4.113

Parameter	Value
donor, acceptor thicknesses	40 nm
ΔE_{HL}	1.2 eV
V_{bi}	0.5 V
$T_{0A} = T_{0D}$	1000 K
$H_{TA} = H_{TD}$	10^{18} cm^{-3}
$N_{HOMO} = N_{LUMO}$	10^{21} cm^{-3}
δ_A	0.1
a_0	1.5 nm
k_{PPr}	1 μs^{-1}
ε_r	3
$\mu_n = \mu_p$	10^{-3} cm^2/V s
R_s	1 Ω cm^2

bias, the current is predominantly mediated by PPs formed from the recombination of free holes in the donor with trapped electrons in the acceptor, such that the current is limited by n_D. At higher bias ($0.4 < V_a < 0.7$), the reverse process dominates, and the slope is determined by n_A. As temperature decreases, carriers freeze into the trap states in the tail of the DOS, and recombination becomes further trap-limited, leading to the increase of both ideality factors at low temperature. In forward bias $V > 0.7$ V, series resistance from contacts and the layer bulk limits the current.

The excitonic diode theory has been applied to 20 nm CuPc/40 nm C_{60} and 11 nm SubPc/40 nm C_{60} HJs sandwiched between a 10 nm thick BCP buffer layer beneath an Al cathode, and an ITO anode. The j-V-T characteristics and fits to Eq. 4.314 are shown in Fig. 4.114a and b. The fits are considerably improved compared to those obtained from the Shockley equation in Fig. 4.114c and d, particularly at low temperature. Series resistance (Table 4.8) accounts for the roll-off in j at high forward bias. The inflection in the slopes at $V \sim 0.5$–0.7 V results from the different values of ideality factors, n_A and n_D.

The ideality factors follow the trends predicted in Fig. 4.113b, providing additional validation to the excitonic diode theory (Giebink et al., 2010b). Since the acceptor layers are identical in both the CuPc and SubPc devices, n_A is also identical for both junctions, and varies from 4 at low temperature, to 2 at $T =$ 300 K. In contrast, the n_D for CuPc- and SubPc-based junctions are larger than n_A, and have a more pronounced temperature dependence. Since δ_D and δ_A do not cancel as in the trap-free case, changes in the potential distribution across the D and A layers with bias and/or temperature influences the values of the ideality factors. For example, an increase in the potential dropped across the acceptor with decreasing temperature leads to a correspondingly greater increase in n_A over the same interval. This leads to ambiguities in uniquely determining δ and $n_{A,D}$ from fits to the exponential trap distribution model. Thus, an important feature of the model is the existence of two ideality factors that increase superlinearly with decreasing temperature.

We have seen that the trap-free and trap-limited excitonic diode equations are similar in form to the Shockley equation, but the interpretation of the fitting parameters (ideality factors, saturation current densities) is considerably different, reflecting the different excited states and transport dynamics in organic versus inorganic semiconductors. In inorganic semiconductors, excitation directly results in unbound, free carriers, whereas in organic semiconductors, tightly bound excitons are generated. The exciton has a low probability of dissociating unless charge transfer occurs at a DA HJ to separate the electron and hole to a more loosely bound polaron pair. Thus, in contrast to inorganic p-n junctions where current is due to drift diffusion and/or recombination within the depletion region, the current in organic HJs depends on polaron pair recombination and dissociation that occur within a small volume (only a few monolayers at most) at the HJ interface. In the absence of shunt paths or other junction defects, generation and recombination via interfacial polaron pairs is the sole source of dark current. This assumption is reasonable at low bias since PPs (or their eponymously named CT state) provide the lowest energy recombination pathway in DA HJs with Type II (i.e. staggered) HJ energy level offsets. They are, therefore, the most heavily populated state at small quasi-Fermi level separations at low voltage or illumination intensities. Disorder or other defects complicate the trap-free picture, as the recombination kinetics depend on the particular trap distribution chosen. Regardless of the choice, however, the double exponential form of the j–V relationship given by Eq. 4.314 is a general result, as it stems from the two possible recombination pathways of free electrons with trapped holes at the HJ, and vice versa.

The fundamental treatment of current in this section provides a basis for analyzing the often complex j–V characteristics of excitonic junctions both as a function of time (Giebink et al., 2010a) and in steady

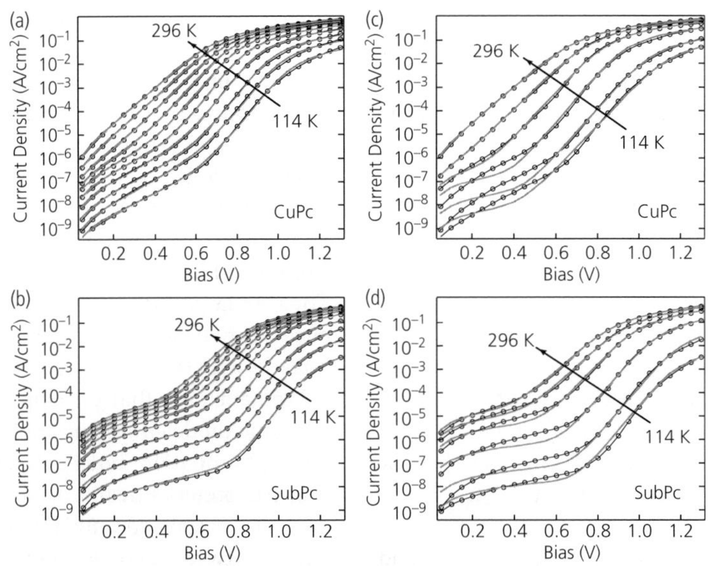

Figure 4.114 Measured (data points) and simulated (lines) forward biased dark *j–V–T* characteristics under forward bias for (a) CuPc/C_{60} and (b) SubPc/C_{60} excitonic HJs. The intermediate temperature data were obtained at 128, 145, 155, 171, 193, 218, 247, and 275 K. Simulations to the excitonic diode equation include an exponential trap distribution employing parameters in Table 4.8. (c) and (d) plot the same data using the conventional Shockley equation generalized to include series resistance (Giebink et al., 2010b).

state, as described by Eqs. 4.297 and 4.298. More importantly, this analysis provides predictions for charge generation when used in solar cells and photodetectors. These predictions have been directly tested and confirmed, leading to improved device performance, as discussed further in Chapter 7.

4.7.3 Organic–inorganic heterojunctions

Intermediate between pure excitonic and inorganic HJs is the combination of these two materials, forming a hybrid organic/inorganic heterojunction (OI-HJ) at their point of contact. The first OI-HJs consisted of a polyacetylene, p-$(CH)_x$, on CdS (Ozaki et al., 1980), and a PTCDA layer deposited onto a Si surface (Forrest et al., 1982). The OI-HJs exhibited remarkable rectifying properties that, prior their discovery, had not been observed in any organic-based device. Since then, many promising technologies have employed OI-HJs as their charge-separating interface, including OI photodiodes (So and Forrest, 1989) and solar cells (Li et al., 2011, Brus et al., 2013), dye-sensitized solar cells (Vlachopoulos et al., 1988, Hagfeldt et al., 2010), and colloidal quantum dot solar cells and light-emitting devices (Coe-Sullivan et al., 2003, Sargent, 2012, Wright and Uddin, 2012).

While there have been many examples of devices that exploit OI-HJs, a comprehensive understanding of their influence on charge generation, recombination and extraction has only recently been developed. Like excitonic HJs, early models have relied on the properties of inorganic HJs to describe their *j–V* characteristics (Forrest et al., 1984a, 1984b). While those models provide a reasonable phenomenological description of the observed characteristics, they fail to provide an accurate physical framework of the dynamics of hybrid state formation and dissociation at the OI-HJ.

More recently, a comprehensive model based on the injection and generation of charge in the inorganic semiconductor and Frenkel excitons in the organic layer that migrate to the interface to form a *hybrid charge transfer exciton* (HCTE) has been introduced (Panda et al., 2014, Renshaw and Forrest, 2014). The HCTE is analogous to a polaron pair state at an excitonic junction, in that the PP and the HCTE are both Coulombically bound charge pairs across a heterointerface. Unlike the polaron pair where both charge species are localized due to disorder, in the HCTE the charge in the inorganic is spread over several lattice sites, while the polaron is localized on one or only a few molecules.

In this section, we derive the *j–V* characteristics of OI-HJs both in the presence and absence of interface traps. The model is developed for the specific example of an n–P anisotype Type II staggered OI-HJ, but is easily generalized to both isotype and Type I junctions (Renshaw and Forrest, 2014). We use the convention where the lower case "n" refers to the majority carrier type of the inorganic semiconductor, while the upper case "P" refers to the organic layer whose Fermi energy is below the center of the energy gap, and hence is closer to the HOMO.

4.7.3.1 Conduction in organic–inorganic heterojunctions

In Fig. 4.115a depicts an ideal n–P OI-HJ in the absence of interface defects or traps, analogous to that for excitonic junctions in Fig. 4.103. Figure 4.115b shows the interface state energy diagram for the processes leading to charge generation. Unlike excitonic junctions, there is substantial band bending on the inorganic side of the HJ. Photocarrier generation can occur via three mechanisms: (i) direct band-to-band absorption of a photon in the doped inorganic semiconductor that results in minority carrier diffusion to the OI-HJ (Forrest et al., 1984a, So and Forrest, 1989, Li et al., 2011); (ii) photon absorption in the organic

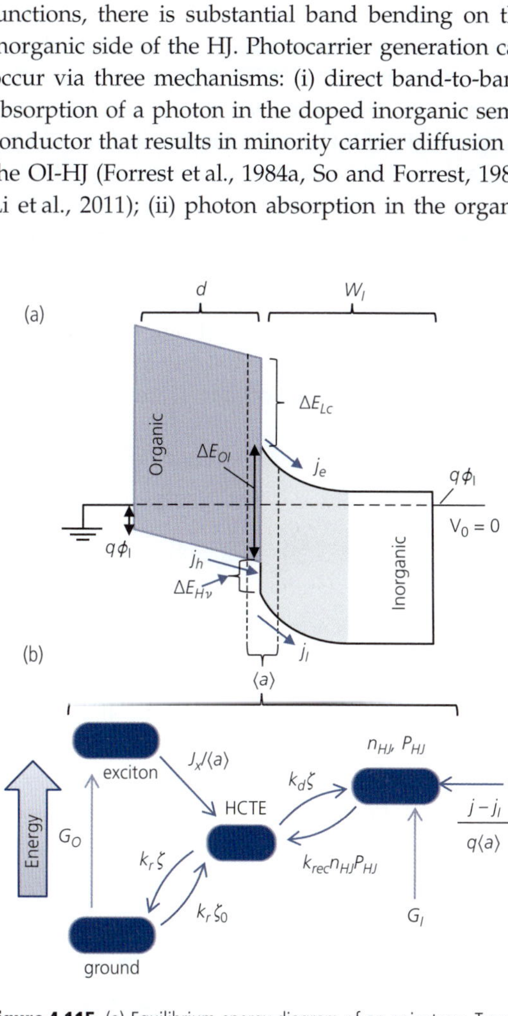

Figure 4.115 (a) Equilibrium energy diagram of an anisotype, Type II n–P OI-HJ. Band bending on the inorganic side of the HJ gives a depletion region width W_I. A uniform field develops across a depleted organic layer of thickness, d. (b) Processes leading to dark and photocurrent generation. The rates (k), contact potentials (ϕ_0, ϕ_I) and polaron, ζ, and charge (n_{HJ}, P_{HJ}) concentrations are defined in text (Renshaw and Forrest, 2014).

leading to exciton generation and diffusion to the junction, with an exciton current of j_X (Panda et al., 2014); or (iii) direct absorption by the HCTE at the OI-HJ (Haeldermans et al., 2008, Vaynzof et al., 2012, Panda et al., 2016).

Bound HCTE states have been experimentally detected at ZnO/organic (Vaynzof et al., 2012, Panda et al., 2016) and CdTe/polymer (Bansal et al., 2013) interfaces. However, the high dielectric constant and delocalized nature of carriers in the inorganic semiconductor leads to an HCTE binding energy (E_B) much smaller than for polaron pairs at excitonic HJs. In some cases $E_B/k_BT \leq 1$ at room temperature, such that the interface exciton may be unstable, or at best metastable (i.e. characterized by a very short lifetime ~ picoseconds) at room temperature.

To estimate E_B, the HCTE can be thought of as a hybrid between a Frenkel and Wannier–Mott exciton bound at the HJ, as illustrated in Fig. 4.116a. The hole is localized to within ~1 nm of the interface; while the electron tunnels into the inorganic forming a delocalized negative charge density extending 5–10 nm over many lattice constants. The charge comprising the exciton rapidly dissociates into an electron on the inorganic semiconductor conduction band that is bound to the hole polaron localized at the OI-HJ, thereby forming the HCTE. The hybrid quasi-particle is then dissociated by either thermal excitation or by the built-in junction field that forces the electron to drift toward the bulk, freeing the hole polaron from the interface. The dynamical properties of the HCTE are also dependent on whether the transfer is between energetically resonant or non-resonant states at the two sides of the interface (Agranovich et al., 2011).

The electron spatial distribution of an untrapped, or free HCTE at the CBP/ZnO interface has been calculated by solving Schrödinger's equation with the following Hamiltonian (Panda et al., 2016):

$$H = -\frac{\hbar^2}{2m_r^*(z)}\nabla^2 - \sum_{i=1}^{k}\frac{\alpha(z)q^2/k}{4\pi\langle\varepsilon_r\rangle\varepsilon_0|\mathbf{r}_e - \mathbf{r}_i|} - \frac{\beta(z)q^2}{16\pi\langle\varepsilon_r\rangle\varepsilon_0 z} + \Delta E_{OI} \tag{4.319}$$

The first term is the kinetic energy of the electron, the second is the Coulomb energy due to the hole screened by the dielectric polarization of CBP, and the third is the image potential energy from the induced charge at the interface. Here, z is the direction perpendicular to the interface, and the reduced effective mass is $m_r^* = (1/m_I + 1/m_O)^{-1}$. Also, m_I (m_O) is the effective mass of the inorganic (organic) layer, $\langle\varepsilon_r\rangle$ is the effective dielectric permittivity given by

$\langle \varepsilon_r \rangle = (a_I\varepsilon_I + a_O\varepsilon_O)/(a_I + a_O)$, where ε_I (ε_O) is the dielectric constant of the inorganic (organic) layer, $\mathbf{r}_e$ is the position of the electron, and ΔE_{OI} is the difference between the ZnO CBM and the CBP LUMO energies. Also, a_I and a_O are the extents of the exciton on the inorganic and organic sides of the HJ, respectively. For simplicity, the hole distribution on the HOMO of CBP is replaced with a summation over k discrete, fractional point charges at position, $\mathbf{r}_i$. For this calculation, it is adequate to let $k = 2$, accounting for two half-charges located at the symmetry points on the two halves of the CBP molecule. The screening factor for the hole is $\alpha(z) = 2\varepsilon_I/(\varepsilon_I + \varepsilon_O)$ in ZnO, and $\alpha(z) = 1$ in CBP, and $\beta(z)$ is the image charge factor with $\beta(z) = \pm(\varepsilon_I - \varepsilon_O)/(\varepsilon_I + \varepsilon_O)$ where the plus sign is for CBP, and the minus for ZnO. Results of this calculation are shown in Fig. 4.116b, illustrating the highly asymmetric charge distribution across the heterointerface.

The Bohr model is used to estimate the radius of the HCTE as (Agranovich et al., 1998)

$$\langle a \rangle = a_I + a_O = \frac{8\pi\langle \varepsilon_r \rangle \hbar^2}{m_r^* q^2}, \tag{4.320}$$

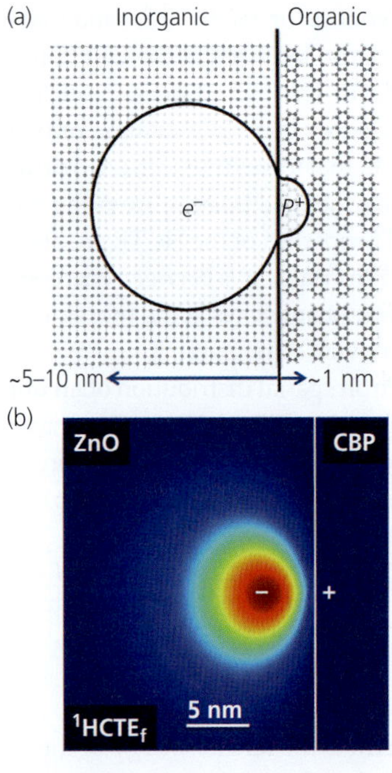

Figure 4.116 (a) Conceptualization of a hybrid charge transfer exciton illustrating its loosely bound, Wannier–Mott character on the inorganic side, and more tightly bound, Frenkel-like nature on the organic side of the HJ (Renshaw and Forrest, 2014). (b) Calculated electron density of the singlet, free ^{1}HCTE$_f$ observed at CBP/ZnO OI HJs. The "+" and "−" indicate the approximate centers of the hole polaron and electron densities. The red color indicates higher electron density (Panda et al., 2016).

This radius approximately defines the "interaction volume" over which recombination or dissociation of the HCTE occurs. It follows that the HCTE binding energy is:

$$E_B = \frac{q^2}{8\pi\varepsilon_0\langle \varepsilon_r \rangle \langle a \rangle}. \tag{4.321}$$

Figure 4.117 shows the exciton binding energy and characteristic exciton radius as functions of the effective dielectric constant for electron effective masses of $m_r^* = 1m_0, 0.2m_0$, and $0.05m_0$, which are consistent with the electron masses for many inorganics (Sze, 1981).

4.7.3.2 Current–voltage characteristics of an ideal OI-HJ

From the preceding discussion, we conclude that at room temperature, the HCTE is a generally unstable precursor to dissociation via the generation of free charges at the OI-HJ (with densities n_{HJ}, P_{HJ} in the inorganic and organic layers, respectively). The free carriers are captured at a rate of $k_{rec}n_{HJ}P_{HJ}$ to form HCTEs with a density of ζ and with a characteristic separation between the electron and polaron, $\langle a \rangle$, across the interface.

Using similar notation and approaches as in Section 4.7.2, the corresponding rate equations for ζ, n_{HJ}, and P_{HJ} as functions of time, t, are given by:

$$\frac{d\zeta}{dt} = -k_r(\zeta - \zeta_0) - k_d\zeta + k_{rec}n_{HJ}P_{HJ} + \frac{j_X}{\langle a \rangle} = 0, \tag{4.322}$$

$$\frac{dn_{HJ}}{dt} = -k_{rec}n_{HJ}P_{HJ} + k_d\zeta + \frac{j_e}{q\langle a \rangle} = 0, \tag{4.323}$$

$$\frac{dP_{HJ}}{dt} = -k_{rec}n_{HJ}P_{HJ} + k_d\zeta + \frac{j_h - j_I}{q\langle a \rangle} = 0, \tag{4.324}$$

where the final equality holds for steady-state conditions. Here, k_r and k_d are the rates that HCTEs

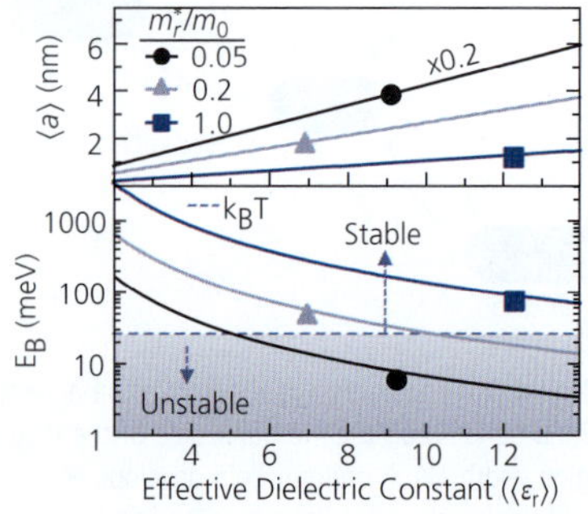

Figure 4.117 Radius, $\langle a \rangle$ and binding energy, E_B of the HCTE as functions of the spatially averaged dielectric constant, $\langle \varepsilon_r \rangle$. Zones of stability at room temperature are indicated for several different reduced effective masses, m_r^*. Adapted from Renshaw and Forrest (2014).

recombine to the ground state and dissociate into free carriers, respectively. The thermal generation rate of an HCTE is $k_r\zeta_0$, where ζ_0 is its equilibrium density given by $\zeta_0 = k_{rec}n_{HJ,0}P_{HJ,0}/k_{d0}$.

Minority carriers photogenerated in the inorganic must be extracted through the organic layer and populate P_{HJ} at a rate determined by the photocurrent density sourced from the inorganic layer, j_I. Also, j_e and j_h are the electron current in the inorganic and the hole current in the organic, respectively. It is assumed that the hole density and mobility are greater than the electron density and mobility in the organic, and hence $j_h = j$. Continuity demands that $\nabla \cdot j = 0$ everywhere in steady-state. At the interface, this implies that $j_e + j_I = j_h = j$. Hence, Eqs. 4.323 and 4.324 are identical, resulting in lossless coupling between n_{HJ} and P_{HJ}.

In steady-state, the rate equations near equilibrium are solved to eliminate ζ, yielding

$$j = q\langle a\rangle k_{rec}(1-\eta_d)\left(n_{HJ}P_{HJ} - \frac{k_d}{k_{d,eq}}n_{HJ,0}P_{HJ,0}\right) - j_{ph}, \tag{4.325}$$

where $j_{ph} = qj_X\eta_d - j_I$ is the total photocurrent, which is the sum of the exciton current generated in the organic and the direct carrier generation current in the inorganic. Also, $j_I < 0$ and $j_X > 0$, resulting in $j_{ph} > 0$, by convention. Here, $\eta_d = k_d/(k_d + k_r)$ is the HCTE dissociation efficiency. Equilibrium corresponds to $V = 0$ in the absence of illumination.

Recall for excitonic junctions we used Onsager–Braun theory (Braun, 1984) to determine k_{PPd} (see Section 4.7.2, Eq. 4.305), where

$$k_d = A_{OB}k_{rec}\exp\left(-\frac{E_B}{k_BT}\right)\frac{J_1[2\sqrt{-2b}]}{\sqrt{-2b}}. \tag{4.326}$$

This model can also be adapted to describe the field dependence of the HCTE. The prefactor, ${A_{OB}}^{-1}$, is the volume of an ion pair of radius, $\langle a\rangle$, that is, $A_{OB} = 3/4\pi\langle a\rangle^3$. In the case where the HCTE is expected to have a partial Wannier–Mott character, we use $A_{OB} = (m^*k_BT/3\pi\hbar^2)^{3/2}$ where m^* is the effective mass of the electron in the inorganic semiconductor.

The bimolecular recombination rate for low mobility solids is diffusion-limited, and therefore follows Langevin recombination statistics where $k_{rec} = q\langle\mu\rangle/\langle\varepsilon_r\rangle$, with an effective mobility of $\langle\mu\rangle = \mu_I + \mu_O$ (cf. Section 4.5.4). In the case of a highly mobile electron in the inorganic semiconductor, $k_{rec} = v_{th}\sigma_X$, where $v_{th} = \sqrt{3k_BT/m^*}$ is the thermal velocity of the electron and $\sigma_X = \pi r_c^2$ is its trap capture cross-section. Electric field screening in the high-dielectric constant inorganic semiconductor allows for neglect of the effects of the electrostatic potential on the capture rate.

If j is sufficiently small so that the electron and hole quasi-Fermi levels (E_{Fn} and E_{Fp}) are flat throughout the respective inorganic and organic layers, then the interface carrier densities are given by

$$n_{HJ} = N_c\exp\left(-\frac{q\phi_I}{k_BT}\right)\exp\left(\frac{qV_I}{k_BT}\right) = n_c\exp\left(\frac{qV_I}{k_BT}\right) \tag{4.327}$$

and

$$P_{HJ} = N_{HOMO}\exp\left(-\frac{q\phi_O}{k_BT}\right)\exp\left(\frac{qV_O}{k_BT}\right) = P_c\exp\left(\frac{qV_O}{k_BT}\right). \tag{4.328}$$

Here, N_c is the effective DOS at the CBM of the inorganic, $q\phi_O$ and $q\phi_I$ are the injection barriers into the organic (from the anode) and inorganic (from the cathode) shown in Fig. 4.115a, and V_O and V_I are the applied voltages across the organic and inorganic layers, respectively. In the case of a thick inorganic layer where the depletion width does not extend to the contacts, then $q\phi_I = E_c - E_{Fn}$ in the undepleted equilibrium region, where E_c is the energy of the CBM. Now, n_c is the electron density at equilibrium as determined by the ionized dopant density (N_D). Also, P_c is the hole density in the organic at the anode contact.

Using these definitions and following procedures analogous to those of Giebink et al. (2010b), we obtain the ideal OI-HJ diode equation by combining Eqs. 4.325, 4.327, and 4.328:

$$j = q\langle a\rangle k_{rec}N_{HOMO}N_c(1-\eta_d)\exp\left(-\frac{\Delta E_{OI}}{k_BT}\right) \times\left(\exp\left(\frac{qV}{k_BT}\right) - \frac{k_d}{k_{d,eq}}\right) - j_{ph}. \tag{4.329}$$

Here $\Delta E_{OI} = qV_{bi} + q\phi_O + q\phi_I$ as defined in Fig. 4.115a, and V_{bi} is the built-in voltage determined by the inorganic Fermi level and the anode work function (modified by energy level shifts, such as those due to interface dipoles). The applied voltage is related to the voltage dropped across each layer using: $V = V_O + V_I + V_{bi}$.

4.7.3.3 Current–voltage characteristics of an OI-HJ in the presence of interface traps

As previously, we consider an organic film with an exponential distribution of trapped interface charge with a total trap density, H_O, a characteristic trap temperature, T_O, and a trap parameter: $m_O = T_O/T$. For convenience, a similar trap profile near the *inorganic* CBM is defined by the parameters H_I, T_I, and m_I. While an exponential might serve as a suitable distribution function for disordered inorganic semiconductors, crystalline inorganics are typically characterized by a discreet trap level where the density of trapped charges is $n_T = H_I\exp(-(E_c - E_{Fn} - E_T)/k_BT)$ at trap energy, E_T.

Assuming transport is dominated by free rather than trapped carriers on either side of the junction, we arrive at

$$j = q\langle a\rangle(1-\eta_d) \times\left[k_{rec,n}N_cH_O\exp\left(-\frac{\alpha_O}{k_BT}\right)\left(\exp\left(\frac{qV}{n_Ok_BT}\right)-\frac{k_d}{k_{d0}}\right) + k_{rec,P}N_{HOMO}H_I\exp\left(-\frac{\alpha_I}{k_BT}\right)\left(\exp\left(\frac{qV}{n_Ik_BT}\right)-\frac{k_d}{k_{d0}}\right)\right] - j_{ph}. \tag{4.330}$$

Analogous to Eqs. 4.312 and 4.313, we have

$$\alpha_O = \frac{\Delta E_{OI}}{n_O} + \frac{m_O-1}{m_O}q(\delta_O\phi_I - \delta_I\phi_O), \tag{4.331a}$$

$$\alpha_I = \frac{\Delta E_{OI}}{n_I} + \frac{m_I-1}{m_I}q(\delta_I\phi_O - \delta_O\phi_I), \tag{4.331b}$$

and

$$n_O = \frac{m_O}{\delta_I(m_O-1)+1}, \tag{4.332a}$$

$$n_I = \frac{m_I}{\delta_O(m_I-1)+1}. \tag{4.332b}$$

The latter two expressions are the ideality factors, n_O and n_I that arise from recombination at traps on the organic and inorganic sides of the HJ, respectively. These factors provide the ratio of the voltage dropped across the corresponding layers of the device. Also, $k_{rec,n}$ and $k_{rec,P}$ are the recombination rates for a free electron with a trapped hole, and a free hole with a trapped electron, respectively. Finally, $\delta_O = V_O/(V - V_{bi})$ and $\delta_I = V_I/(V - V_{bi})$ are the fractions of the applied voltage dropped across the respective organic and inorganic layers. Contrary to the trap-free case, the voltage distribution across the layers affects the current even in the near-equilibrium approximation used here. This is due to the asymmetry of the electronic properties of the organic and inorganic layers. For discrete traps in the inorganic semiconductor, Eqs. 4.331b must be modified such that $\alpha_I = E_T$.

Following the conventions and simplifications in Section 4.7.2 for the trap-free case, we obtain:

$$j = q\langle a\rangle k_{rec}N_{HOMO}N_c\exp\left(-\frac{\Delta E_{OI}}{k_BT}\right)\left(\exp\left(\frac{qV}{k_BT}\right)-1\right) - j_{ph}, \tag{4.333}$$

and for an exponential trap distribution,

$$j = q\langle a\rangle\left[k_{rec,n}N_cH_O\exp\left(-\frac{\alpha_O}{k_BT}\right)\left(\exp\left(\frac{qV}{n_Ok_BT}\right)-1\right) + k_{rec,P}N_{HOMO}H_I\exp\left(-\frac{\alpha_I}{k_BT}\right)\left(\exp\left(\frac{qV}{n_Ik_BT}\right)-1\right)\right] - j_{ph}. \tag{4.334}$$

These expressions are nearly identical to Eqs. 4.329 and 4.330, although they are independent of HCTE dissociation dynamics by letting $k_d/k_{d0} \to 1$. Thus, they are the limiting case where HCTEs dissociate with an efficiency $\eta_d \to 1$. Note that $k_{rec}(1-\eta_d)$ reduces to $A_{OB}k_r$ in the limit that $k_d = k_{rec}/A_{OB} \gg k_r$.

Now, defining the saturation current density:

$$j_0 = q\langle a\rangle k_{rec}N_{HOMO}N_c\exp(-\Delta E_{OI}/k_BT) \tag{4.335}$$

then Eq. 4.333 simplifies to the familiar form of the ideal diode equation, Eq. 4.236. In a similar fashion, an interface with traps yields the current (cf. Eq. 4.314):

$$j = \left[j_{SO}\left(\exp\left(\frac{qV}{n_Ok_BT}\right)-1\right) + j_{SI}\left(\exp\left(\frac{qV}{n_Ik_BT}\right)-1\right)\right] - j_{ph}, \tag{4.336}$$

making the appropriate substitutions for the saturation currents for the organic, j_{SO}, and inorganic, j_{SI}, semiconductors.

The treatment of the ideal diode relationships leading to Eqs. 4.333 and 4.334 assumes that the diode is operating near equilibrium. However, as the current density increases, the asymmetric voltage distribution across the organic and inorganic layers can change dramatically, particularly when there is a high density of interface traps. For example, at high forward voltages, the OI-HJ current is limited by space charge effects in the organic. In this case, the high current regime is reached at the voltage where the space charge current is equal to that of Eq. 4.333. Thus, in the high current regime, the voltage distribution ratio, δ_O/δ_I, must be calculated using a self-consistent solution to the drift-diffusion Eq. 4.237:

$$j = q\mu_{PO}PF - qD_p\nabla P, \tag{4.337}$$

with

$$\nabla \cdot F = \frac{q(P - P_O)}{\varepsilon_O}. \tag{4.338}$$

Here, μ_{PO} is the hole mobility in the organic, P_O is the background free hole polaron concentration, and D_p is the hole diffusivity. This system of equations does not offer an analytical solution, and must be solved numerically. The interested reader should refer to the literature for a more comprehensive treatment (Renshaw and Forrest, 2014).

Equation 4.335 indicates that ΔE_{OI} determines the saturation current. There are several means for measuring interface energetics, although all methods are subject to the effects of traps and Fermi level pinning. As we showed in Section 4.6.1, the space charge region width for a uniformly doped semiconductor that is not fully depleted is given by $W(V) = \sqrt{\frac{2\varepsilon_r\varepsilon_0}{qN_D}(V_{bi} - V)}$. Since the capacitance per area of this one-sided junction is $C(V) = \frac{\varepsilon_r\varepsilon_0}{W(V)}$, then,

$$\frac{1}{C^2(V)} = \frac{2}{qN_D\varepsilon_r\varepsilon_0}(V_{bi} - V). \tag{4.339}$$

Hence, a measurement of $1/C^2$ vs. V has a slope determined by the free carrier concentration, and the intercept yields the built-in potential from which the junction energies (e.g. ΔE_{OI}) can be derived. This technique for barrier measurement can lead to errors in the presence of traps, spatial non-uniformities in N_D over the depletion region width, artifacts arising from fully depleted organic layers (in the case of Schottky diodes or excitonic HJs), and capacitance measurement frequency. At low frequencies the emission of trapped charge can follow the a.c. capacitance measurement voltage (Sharma et al., 2011). While C–V measurements of barrier heights can lead to errors or difficult to interpret results, this is partially offset by their simplicity.

The interface barrier can also be determined from the activation energy of the j–V characteristics, extrapolated to $V = 0$, which yields $j_0(T)$. Plotting the logarithm of the saturation current versus $1/T$ gives a slope of $\Delta E_{OI}/k_B$ for an anisotype n–P or p–N OI-HJ.

For an isotype HJ, the measured barrier is the HJ offset energy, ΔE_{Hv}. This is the offset measured in p-P junctions between the organic HOMO and the inorganic VBM energy, and ΔE_{Lc} is the offset measured at n–N HJs between the organic LUMO and the inorganic CBM energy (see Fig. 4.115). This method has been applied to measuring the offset energy of isotype PTCDA/p-Si ($p_0 = 3 \times 10^{16}\ \text{cm}^{-3}$) junctions, as shown in Fig. 4.118 (So and Forrest, 1988). The slope of the saturation current data gives the barrier potential, $q\phi_{Bp} = 0.56{\pm}0.02$ eV. The valence offset energy is found by adjusting for the equilibrium positions of the Fermi levels to the HOMO and VBM, E_{Fp} and E_{FP}, respectively, such that $\Delta E_{Hv} = q\phi_{Bp} + E_{FP} - E_{FP} = 0.48 \pm 0.02$ ev.

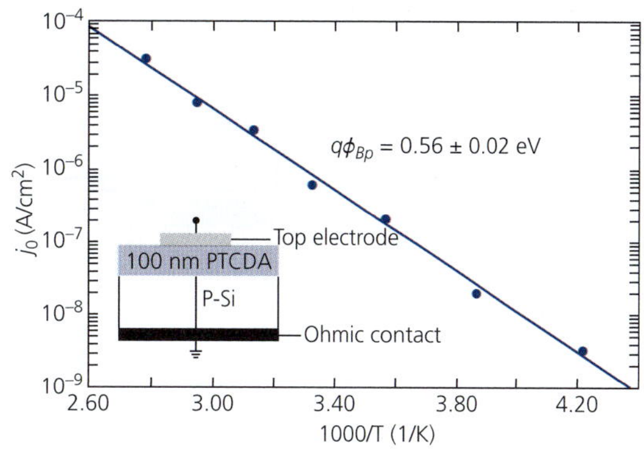

Figure 4.118 Temperature dependence of the saturation current of a PTCDA/p-Si OI-HJ taken from an extrapolation to $V = 0$ where $j \to j_0$ in the forward-biased characteristics. The activation energy obtained from the slope of the line fit to the data is $q\phi_{Bp} = 0.56 \pm 0.02$ eV. Inset: Schematic view of the test device. From So and Forrest (1988).

The energy offset was independently measured using internal photoemission, which was previously shown to be useful for measuring Schottky barrier energies in Section 4.6.2. Photocurrent spectra obtained by illumination of a PTCDA/p-Si OI-HJ with a broad spectral source through the thick Si substrate are shown in Fig. 4.119a. Absorption through the substrate filters out all wavelengths $<1.1\ \mu$m, ensuring only long wavelength light gives rise to the observed photocurrent. The photocurrent has a peak at 0.57 eV, falling off steeply at both lower and higher energies. The spectrum is explained using the inset diagram. At the low energy edge, the photocurrent decreases as the photon energy falls below that of the energy barrier, ΔE_{Hv} (Fig. 4.119a, inset, process "a"). It reaches a peak once the photon energy is larger than the barrier (process "b"), and then decreases once more when the energy exceeds that of the relatively narrow width of the PTCDA HOMO (process "c").

As in Eq. 4.256, internal photoemission data should follow $j_{ph} \propto (E_{ph} - E_{Hv})^2$ (Fowler, 1931, Haase et al., 1987). In Fig. 4.119b, the data on the long wavelength side of the photocurrent spectrum follows the expected quadratic dependence, thus yielding a valence energy offset of $\Delta E_{Hv} = 0.50 \pm 0.01$ eV. This is

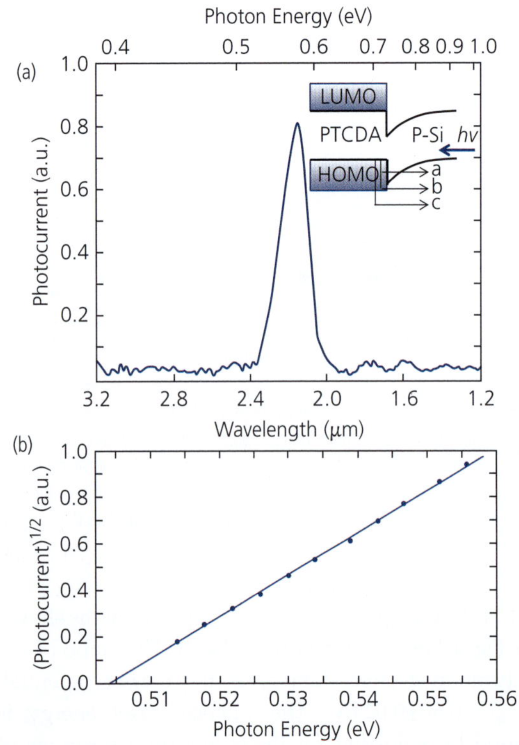

Figure 4.119 (a) Photocurrent spectra of a PTCDA/p-Si OI-HJ at energies much less than the energy gap of PTCDA (2.2 eV) or Si (1.1 eV) due to emission over the discontinuity at the valence energy edge. Inset: Processes responsible for the photocurrent. The energy level alignment assumes that the PTCDA is adventitiously lightly p-type doped. (b) Plot of the $j_{ph}^{1/2}$ vs. photon energy. The intercept with the abscissa gives $\Delta E_{Hv} = 0.50\pm0.01$ eV (So and Forrest, 1988).

consistent with the measurement obtained via the thermal activation of the saturation current in Fig. 4.118. As in transport measurements, it is important to rule out photoemission from unrelated sources such as defects within the organic layer. In this experiment, an In/PTCDA/ITO photoconductor was also illuminated, and no photocurrent was observed, thus ruling out this potential source. However, it remains uncertain from these data whether the energy measured is due to the intrinsic nature of the OI-HJ interface, or whether it is due to Fermi energy pinning at interface traps. In either case, the barrier measured is that which controls the transport, and hence can be used in the analysis of the j–V characteristics of this OI-HJ.

The OI-HJ open circuit voltage is calculated by setting $j = 0$ in the presence of a photocurrent. In the case of a stable HCTE (i.e. $E_B > k_BT$), the derivation of the open circuit voltage is analogous to that of an excitonic junction in Section 4.7.2. That is, for the trap-free case of Eq. 4.333, we obtain

$$qV_{OC} = \Delta E_{OI} - k_BT\ln\left(\frac{k_r}{k_r+k_d}\frac{q\langle a\rangle N_{HOMO}N_c k_{rec}}{j_{ph}}\right). \tag{4.340}$$

The treatment for an unstable exciton ($E_B < k_BT$) is nearly identical, yielding

$$qV_{OC} = \Delta E_{OI} - n_O k_BT\ln\left(\frac{k_r}{k_r+k_d}\frac{q\langle a\rangle N_{HOMO}N_c k_{rec}}{j_{ph}}\right). \tag{4.341}$$

Hence, V_{OC} ultimately depends on the recombination rate, k_r, of the HCTE. The probability for recombination in OI-HJs, however, is significantly lower than in fully excitonic junctions. That is, in OI-HJs, the internal field at the interface rapidly separates charges following migration to the interface (i.e. $k_d \gg k_r$). Thus, the maximum open circuit voltage as $T \to 0$ is

$$qV_{OC}^{max} \approx \Delta E_{OI} - nE_B, \tag{4.342}$$

independent of morphology or other factors that influence V_{OC} in excitonic junctions.

The foregoing analysis has been applied to several different OI-HJ systems with both free (Li et al., 2011, Panda et al., 2014) and trapped (Panda et al., 2016) HCTEs. Evidence for the generation of HCTEs at DBP/TiO_2 and pentacene/InP interfaces is provided by their external quantum efficiency spectra in Fig. 4.120. The devices employed in these experiments are shown in the insets.

The band gap of TiO_2 is 3.3 eV, which exceeds the energy gap of DBP (2.0 eV); hence the η_{ext} spectrum in Fig. 4.120a is due solely to absorption in DBP resulting in singlet excitons. The excitons subsequently migrate to the OI-HJ where they dissociate into an electron in TiO_2 and a hole in DBP via an intermediate HCTE. The liberated charges migrate to their respective electrodes and are collected.

The spectrum precisely replicates that of the DBP absorption spectrum, hence we conclude that the η_{ext} is initiated by singlet exciton formation in that layer. It can be argued that photoconductivity in the DBP can result in a photocurrent without requiring the generation of an HCTE. However, η_{ext} of an Au/DBP/ITO photoconductor was 100 times smaller than that achieved with the OI-HJ. We note that the η_{ext} increases with increasing temperature, with an exponential activation energy of $\Delta E_a = 77 \pm 10$ meV. This is due to the thermally activated exciton transport via hopping of excitons to the heterointerface (Panda et al., 2014).

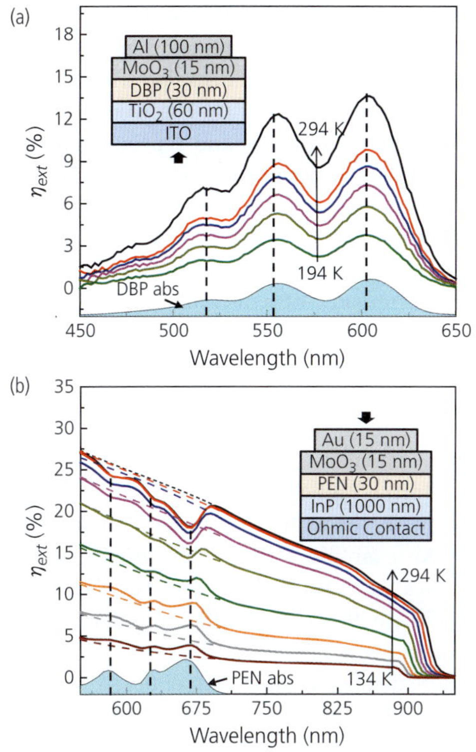

Figure 4.120 (a) External quantum efficiency vs. wavelength and temperature (in 20 K intervals) for a DBP/TiO_2 OI-HJ diode shown in the inset. Also shown is the room temperature absorption spectrum of DBP (DBP abs). Absorption in the DBP is at lower energies than the band gap of TiO_2 (3.3 eV). (b) As in (a) for a pentacene (PEN)/InP OI-HJ shown in the inset. In this device, the absorption spectra of the organic and inorganic semiconductors overlap. Diagonal dashed lines show the extrapolated InP absorption in the absence of pentacene. Note in both cases the absorption spectral features in the diodes align with those of the neat films (Panda et al., 2014).

The results for the moderate band gap InP (1.35 eV)/pentacene (1.8 eV) OI-HJ are shown in Fig. 4.123b. Here, the absorption spectra of the two semiconductors overlap. At high temperatures, absorption by the top pentacene layer results in a decrease in η_{ext}. This suggests that excitons generated in the organic recombine prior to reaching the HJ, and hence the addition of this layer results in a net loss in photocurrent. However, as the temperature is reduced, the feature inverts, and loss converts to gain. The temperature at which the pentacene layer becomes transparent is approximately $T = 200$ K. That is, at sufficiently low temperatures, the excitons can reach the OI-HJ prior to recombination. There, they have an opportunity to form an HCTE that is dissociated into free charge, and thus contributes to j_{ph}. This effect is also thermally activated, with $\Delta E_a = 210\pm40$ meV. The opposite temperature dependence of this sample compared to that of the wide gap TiO_2/DBP junction is attributed to traps at the interface. As temperature is reduced, the interface traps become occupied, and no longer serve as recombination centers for the arriving excitons.

The photocurrent efficiencies for the devices in Fig. 4.120 of ~10–15% are remarkably high for OI-HJs, yet they remain comparatively low compared to excitonic detectors with efficiencies ranging from 30% to 90% (see Chapter 7, and Peumans et al., 2003). The low η_{ext} has frequently been reported in OI-HJ detectors and solar cells (O'Regan and Grätzel, 1991, Olson et al., 2006, Li et al., 2011, Vaynzof et al., 2012). Unlike optimized excitonic OPVs that contain either mixed layers (Uchida et al., 2004) or bulk HJs (Yu et al., 1995) with a very high DA contact surface area and hence a high exciton dissociation efficiency, an OI-HJ has a planar junction. The efficiency is therefore limited by the short diffusion lengths of excitons that must reach the HJ to dissociate into free carriers. The short L_D necessitates the use of a thin organic layer which cannot absorb much of the incident radiation.

Finally, trap states at the OI-HJ can lead to a bound HCTE that has a higher probability for recombination than dissociation. Indeed, such bound states leading to significant loss in η_{ext} have been spectroscopically probed at P3HT or F8TBT on ZnO OI-HJs (Vaynzof et al., 2012). It was found that approximately 50% of the excitons that arrived at the ZnO surface formed an HCTE bound to a surface defect, and were likely to recombine rather than generate free charge. When even a monolayer of a "passivating" acceptor, PCBA, is interposed between either F8TBT or P3HT and the inorganic, the number of bound excitons dropped by half, with a concomitant increase in η_{ext}.

The dynamics of trapped HCTEs have been investigated both theoretically and experimentally in small molecule CBP/ZnO OI-HJs. A free (i.e. unbound to traps) singlet $^1HCTE_f$ contributes to the η_{ext} reaching a maximum efficiency of 6%, which is somewhat lower than observed for DBP/TiO_2. The free exciton, whose electron density distribution is shown in Fig. 4.116, has a binding energy of only 9 meV. Hence, it is very short lived and has a low oscillator strength, resulting in no detectable photoluminescence. However, under electrical injection, strong electroluminescence is observed from a second, lower energy trapped $^1HCTE_T$ state with a relatively high binding energy of 60–430 meV. The electroluminescence from $^1HCTE_T$ is attributed to the state bound to defects on the ZnO surface. Spectroscopic analysis indicates that the occupied defects leading to this trapped, high oscillator strength feature are centered approximately 500 meV below the ZnO CBM, with an occupational

DOS width of $\sigma_T \sim 300$ meV. The energy level scheme describing these various states is shown in Fig. 4.121a (Panda et al., 2016).

The bound state has been modeled by adding the trap potential,

$$V_{trap} = -Sq^2/(4\pi\varepsilon_{rI}\varepsilon_o|\mathbf{r} - \mathbf{r}_{\text{trap}}|), \tag{4.343}$$

to the electron Hamiltonian in Eq. 4.319 that was used in calculating the electron distribution in Fig. 4.116. Here, $\mathbf{r}_{\text{trap}}$ is the position of the localized defect from the electron at $\mathbf{r}$, and S is the trap dielectric screening factor. The electron density as a function of distance from the OI-HJ along the diameters of both $^1\text{HCTE}_\text{f}$ and $^1\text{HCTE}_\text{T}$ are shown in Fig. 4.121b. The trapped state is much smaller and more tightly bound to the surface than the free state. This accounts for its larger oscillator strength (and hence high electroluminescence intensity) than for the free state that exhibits no measurable electroluminescence. It also clearly demonstrates that trapping at interface defects can decrease η_{ext} due to enhanced recombination. Passivation of the interface to reduce inorganic semiconductor surface defects is essential to realizing a high detection efficiency.

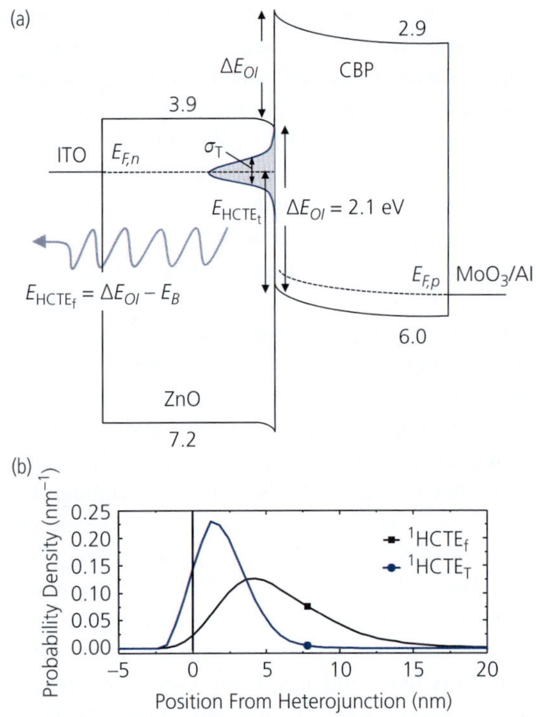

Figure 4.121 (a) Energy level diagram for the CBP/ZnO OI-HJ. The singlet free $^1\text{HCTE}_\text{f}$ has an energy nearly equal to ΔE_{OI} due to a binding energy of only 9 meV. The $^1\text{HCTE}_\text{T}$ state is bound at interface traps filling the distribution shown in green with binding energies ranging from 60 to 430 meV and centered ~ 500 meV below the ZnO conduction band minimum. (b) Probability density of the electron in the trapped and free HCTEs as a function of distance from the OI-HJ (Panda et al., 2016).

Adapted with permission from Panda, A., Ding, K. & Forrest, S. R. 2016. Free and trapped charge transfer excitons at a ZnO/small molecule heterojunction. Phys. Rev. B, 94, 125429. Copyright 2016 by the American Physical Society.

Initial investigations of OI-HJs were encouraged by their very asymmetric and reproducible rectifying characteristics when organics were layered onto a variety of inorganic semiconductors. Particularly the small molecules PTCDA and NTCDA layered onto Si, GaAs and In(GaAs)P show nearly ideal j–V characteristics of a p-n junction made *only* of the inorganic semiconductor (Forrest et al., 1982, 1984a, 1985a, 1985b). This was particularly remarkable since at that time, no organic diodes showing significant rectification had been reported. It was initially and incorrectly assumed that the organic simply formed a Schottky barrier rather than a HJ barrier with the inorganic material (Forrest et al., 1982). An example of the extraordinary rectifying properties of OI-HJs for the two samples in Fig. 4.120 are illustrated by their bipolar j–V characteristics in Fig. 4.122.

A more detailed examination of the j-V characteristics for the PEN/InP junction is shown in Fig. 4.123. Fits to Eq. 4.334 (lines) assume that the junction characteristics are influenced by interface traps, as inferred from its optical properties. The fits replicate all of the details exhibited by the data under both forward and reverse bias. Similarly high quality fits are obtained for DBP/TiO_2 and CBP/ZnO devices (Panda et al., 2014, 2016).

The measured valence band–HOMO offset energy of the PEN/InP OI-HJ inferred from the saturation current in Eq. 4.335 is $\Delta E_{Hv} = 0.24 \pm 0.04$ eV, which agrees well with UPS measurements. Also, the fits to the forward-biased characteristics yield an ideality

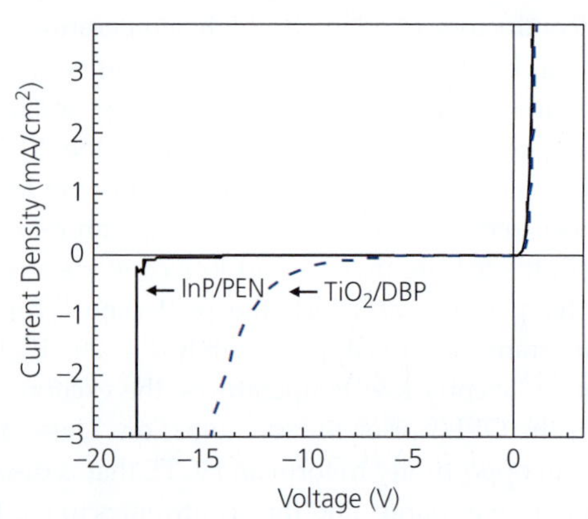

Figure 4.122 Bipolar j–V characteristics of the wide- and moderate-energy gap devices in Fig. 4.120. Note the very large reverse breakdown voltages of ~ −12 V for DBP/TiO_2 and −18 V for PEN/InP, as well as the apparent exponential increase in j with V under forward bias (Panda et al., 2014).

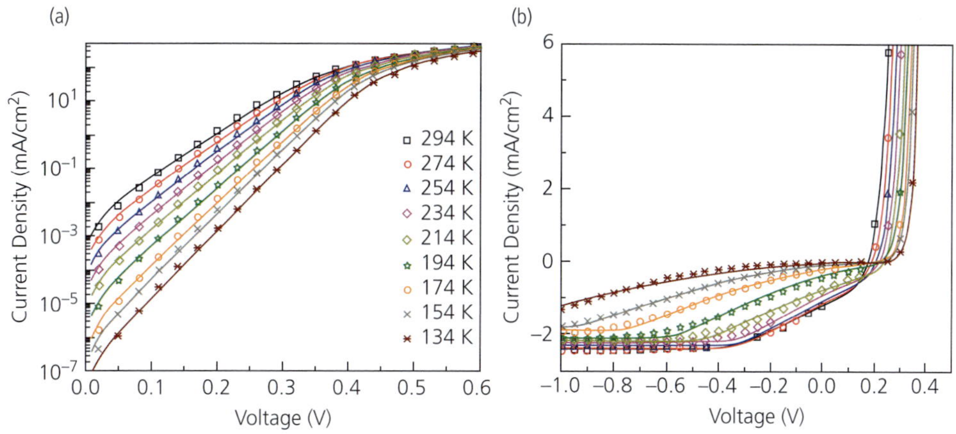

Figure 4.123 (a) Forward-biased and (b) bipolar *j–V* characteristics of the PEN/InP OI-HJ device in Fig. 4.120b. Solid lines are fits to the data (symbols) as a function of temperature (Panda et al., 2014).

factor ranging from $n_O = 1.25$ at room temperature to 1.96 at $T = 134$ K. We have seen in Section 4.7.2 that this temperature dependence for n is indicative of a distribution of traps within the organic energy gap. Furthermore, the thermally activated mobility $\mu(T) = \mu_\infty \exp(-E\mu/\ k_BT)$ with activation energy E_μ= 71 meV and prefactor $\mu_\infty = 10^{-5}$ cm^2/V s, is consistent with independent measurements from SCL transport of the amorphous pentacene films (Panda et al., 2014).

While the fits validate the diode theory developed for OI-HJs, we caution that analyses of all diode equations are inherently parametric. This renders their interpretation based on the fitting parameters subject to error, and also can lead to unsupported speculation. Indeed, *the current measured is effectively an integral of all charge generation, recombination, and leakage mechanisms* (including shunts, contact resistances, surface currents, etc.) present. As such, it is often difficult or even impossible to isolate fundamental from parasitic processes, although we have seen that varying temperature and voltage give particularly powerful insights into the various current sources. Thus, it is essential that as many of the parameters used to fit a set of j–V data be validated via independent measurements to provide confidence in their interpretation. Both the optical and electrical characteristics of the DBP/TiO_2 and PEN/InP samples studied here are consistent with the conclusions that HCTEs do indeed mediate photocurrent generation at these junctions.

4.7.4 Universal ideal diode behavior

We have seen that the ideal diode equations for a variety of junctions, from conventional inorganic p-n junctions, organic p-n junctions, fully organic (or excitonic) HJs, to hybrid OI-HJ devices all apparently follow the same universal form:

$$j = j_0(\exp(qV/nk_BT) - f(V)) - j_{ph}. \tag{4.344}$$

Here, the saturation current density (j_0), the ideality factor (n), the bias-dependent reverse-bias factor ($f(V)$, where $f(0) \to 1$), and the photocurrent (j_{ph}) have different functional forms in the various material systems due to the different physical processes that govern the junction current. The definition of these parameters for each materials system for one-sided junctions (i.e. where current is controlled by only one of the two contacting materials for simplicity of the comparison) are listed in Table 4.9.

The universal form of the diode equation originates with the commonality of the physics at play in all of these systems. The splitting of quasi-Fermi levels and Boltzmann statistics govern the free carrier distribution and result in an exponential current response to an applied voltage. The non-equilibrium condition imposed by the applied bias is sustained by a balance between carrier generation and recombination. The primary distinguishing feature among these junctions is the specific physics governing the generation and recombination processes, which is embodied in the precise form of the saturation current density, j_0.

We note that this commonality of form only exists when the dominant mechanism for current transport is recombination. We have shown that for tunneling dominated transport, the current follows a very different, non-thermally activated behavior that depends exponentially on the energy gap and electric field. Tunneling and its associated process of Frenkel-Poole

Table 4.9 Comparison of diode equations[a]

Equation	j_0	j_{ph}	f(V)[b]	n
Inorganic (diffusion)	$q\frac{D_n n_i^2}{L_n N_A}$	j_l	1	1
Inorganic[c] (generation, recombination)	$\frac{qn_i}{\tau_t}\left(\frac{K_B T}{q}\frac{2\varepsilon}{qWN_D}\right)$	j_l	$\frac{W}{\left(\frac{K_B T}{q}\frac{2\varepsilon}{qWN_D}\right)}$	2
Inorganic Schottky	$\frac{4\pi q m^* k_B^2 T^2}{h^3}$	j_l	1	1–2
Organic Schottky	$q\mu_n N_{LUMO} F_m \cdot \exp(-q\phi_{B0}/k_B T)$	j_0	1	1–2
Organic HJ (Type II)	$qa_0 k_{rec} N_{HOMO} N_{LUMO}(1-\eta_{PPd})\exp\left(-\frac{\Delta E_{HL}}{k_B T}\right)$	$q\eta_{PPd} j_X$	$\frac{k_{PPd}}{k_{PPd0}}$	$\frac{T_A/T}{\delta_D(T_A/T-1)+1}$
Organic homojunction[d]	$A(V)(1-\eta_{PPd})\cdot\exp\left(-\frac{E_G}{nk_B T}\right)$	$q\eta_{PPd} j_X$	$\frac{k_{PPd}}{k_{PPd0}}$	$\frac{2T_S/T}{(T_S/T+1)}$
OI-HJ	$q\langle a\rangle k_{rec} N_{HOMO} N_c(1-\eta_d)\exp\left(-\frac{\Delta E_{OI}}{k_B T}\right)$	$q\eta_d j_0 + j_l$	$\frac{k_d}{k_{d0}}$	$\frac{T_O/T}{\delta_I(T_O/T-1)+1}$

[a] Only one-sided junctions are considered. Definitions of variables given in Sections 4.6 and 4.7. Generalizations of these expressions to more complex situations (e.g. double sided junctions) are found in the literature on inorganic semiconductor diodes.
[b] $k_{ppd} = k_{ppd0}$ and $k_d = k_{d0}$ at $V = 0$. $W =$ depletion width on the low-doped side of the junction.
[c] This considers the simplest case of SRH recombination via a single mid-gap trap level in a symmetric p–n homojunction with $N_C = N_V$, $N_D = N_A$, and $|V| \gg k_B T/q$.
[d] The tunneling based model does not fit the form of Eq. 4.344 which is based only on recombination statistics. $A(V)$ is a voltage dependent constant.

emission over barriers, and impact ionization (Sze, 1981) are mechanisms that can be found under appropriate conditions (Loi et al., 2007) for all of these various species of junctions. Their presence ultimately leads to phenomena that must be taken into account before applying ideal diode theory to a particular junction. Detailed descriptions of these processes can be found in texts listed in *Further reading*.

4.8 Summing up

This chapter concludes Part I, whose purpose has been to lay out the fundamental principles underpinning the field of organic semiconductors. This chapter focused on the transport of charge, which is intimately linked to the physics of molecular excited states and optical properties discussed in Chapter 3. We began by considering the most fundamental of all properties that determines charge transport: the existence of energy bands. Without bands, there can be no charge mobility, and without mobility there is no conductivity. Of course, we learned that the bands in organic semiconductors are, except in a few unusual but important cases, quite narrow (on the order of a few tens of meV) compared with inorganic semiconductors where BW >1 eV is common. Narrow bands and structural disorder play defining roles in the conductive properties of organics, clearly distinguishing them from ordered inorganic semiconductors. Importantly, these features result in charge mobility and diffusion governed by hopping through an energetically broad distribution of transport states.

SCL transport in trap-free or trap-dominated solids is found to be the predominant current conduction mechanism in homogeneous, amorphous thin films. Organics can be doped with donors and acceptors that significantly modify their conductive properties, although the doping efficiency (i.e. the number of free charges per impurity molecule or atom) is considerably less than unity since organic solids are electrostatically rather than chemically bonded.

Charge recombination was found to dominate the charge transport characteristics of all organics. The natural outcome of disorder is a high concentration and energetically broadened density of trap states within the HOMO–LUMO gap that promotes recombination. While the precise functional form of the density of traps may be unknown, it is convenient to use an exponential or Gaussian function since they lend themselves to closed-form analytical expressions. Indeed, the effective medium, Gaussian disorder approximation is found to provide reliable descriptions of the mobility, and hence it is reasonable to assume that this functional form can also describe the distribution of disorder-induced traps.

In the final part of the chapter, we explored transport in systems with junctions. Indeed, all organic electronic devices have contacts, and contacts form junctions. If the contact barrier is large compared to the thermal energy, then the resulting Schottky junction is rectifying, and controls injection into the material. If the barrier is small, then the contact is Ohmic, which is desirable for high current applications such as OLEDs and organic thin film transistors. In all cases, the contact junction properties are determined by disorder resulting in interface dipoles and morphological defects that impact the size and magnitude of the energy barrier.

There are several other types of junctions that play an integral role in transport. Homojunctions consisting of p–n doped regions abutting one another are by far the most fundamental junction in semiconductors, yet are infrequently reported in organics due to the difficulty in efficiently and precisely doping the layers. In contrast, photonic devices contain HJs consisting of two different organic semiconductors in contact with each other. Type I (nested) HJs are found useful for confining charge in the emission zones in OLEDs, whereas Type II (staggered) HJs are used for charge separation in OPVs. Type III (broken junction) HJs can promote charge injection in transistors and photonic devices alike.

We presented semi-classical approaches to describe charge transport across all of these various junction types. The approaches stem from the ideal diode equation of Shockley; an insight that now underpins all modern semiconductor device theory and engineering. Similar considerations lead to descriptions of transport across organic and OI-HJs, providing an understanding of the many rich and varied properties of these important materials systems. The idealized descriptions can be extended in a straightforward manner to consider the many variations that occur in practice. Given the complexities of real junctions consisting of imperfect materials, these so-called ideal diode equations can only go so far in describing actual observations. We continue to caution that although one has at one's fingertips an ideal diode equation, that does not mean that one also has an ideal diode. Such a device, to my knowledge, has never been realized.

While the foundations have been laid, the transport properties of organics represent a topic of immense depth. A truly comprehensive treatment of transport in structures with, or even without a junction, is well beyond the scope of this text. Once again, the interested reader is referred to *Further reading* to continue exploration of this enormously interesting topic.

In spite of our cautions about the effects of materials imperfections, structural and dynamic disorder, and possibly high trap density, the relatively simple theories and analyses described in this an the preceding chapters in Part I provide a remarkably accurate description of the properties of organic semiconductors. While continuously evolving with each new material, measurement and device, this framework has led to advances in optoelectronic applications in recent years that are generating new industries and revolutionary changes in our understanding of what organic semiconductors can achieve. These technological advances are the topic of Part II.

Further reading

1. N. W. Ashcroft and N. D. Mermin, 1976. *Solid State Physics*. New York: Holt, Rinehart and Winston.
2. P. F. Barbara, T. J. Meyer, and M. A. Ratner, 1996. Contemporary issues in electron transfer research. *J. Phys. Chem.*, 100, 13148.
3. F. Capasso and G. Margaritondo (Eds.), 1987. *Heterojunction Band Discontinuities: Physics and Applications*. Amsterdam: North-Holland.
4. V. Coropceanu, J. Cornil, D. A. da Silva Filho, Y. Olivier, R. Silbey, and J.-L. Brédas, 2007. Charge transport in organic semiconductors. *Chem. Rev.*, 107, 926.
5. S. Hüfner, 2003. *Photoelectron Spectroscopy: Principles and Applications*, 3rd ed. Heidelberg: Springer-Verlag.
6. K. C. Kao and W. Hwang, 1981. *Electrical Transport in Solids*. Oxford: Pergamon Press.
7. C. Kittel, 1986. *Introduction to Solid State Physics*, 6th ed. New York: John Wiley and Sons.
8. M. A. Lampert and P. Mark, 1970. *Current Injection in Solids*. New York: Academic Press.
9. M. Pope and C. E. Swenberg, 1982. *Electronic Processes in Organic Crystals*. Oxford: Clarendon Press.
10. A. Rose, 1963. *Concepts in Photoconductivity and Allied Problems*. New York: Interscience.
12. S. M. Sze, 1981. *Physics of Semiconductor Devices*. New York: Wiley-Interscience.

Problems

1. The band structure of organic molecules can be approximated using the tight binding approximation.
   (a) Why is this approximation valid in both polymers and small molecules?
   (b) Assume that the molecules are arranged in a body-centered cubic lattice with lattice constant, a. Give expressions for the bandwidth, velocity, and effective mass of the solid along the (100) and (111) directions. Plot these functions in the first Brillouin zone.
   (c) Write an expression for the effective mass and velocity at the zone center.

2. In tight binding theory, one assumption is that the wavefunction is small when the potential perturbation from nearest neighbors is large. In Fig. 4.3 we illustrated such a potential, described by the following function:

$$U(\mathbf{r}) = -U_0 \frac{\sin^2(\pi\mathbf{R}/a)}{(\mathbf{R}/a)^2}$$

where $\mathbf{R}$ is the lattice translation vector for a simple cubic lattice with side a.
   (a) Calculate the tight binding parameters α, β, and J assuming that the molecular wavefunction, $\phi(r)$ is described by a hydrogenic s-state. Hint: Expand $U(\mathbf{r})$ to second order.
   (b) Using results in (a), calculate BW and the electron effective mass along the (100) direction.
   (c) Plot the dispersion relationship, $E(k_{(100)})$.
3. Assume that the electronic states in an organic film are well described by Gaussian disorder with a width, σ_{GDM}.
   (a) Show that the mean thermal equilibrium energy of the distribution is $\langle E_\infty \rangle = -\frac{\sigma_{GDM}^2}{k_B T}\langle \varepsilon_\infty \rangle = -\frac{\sigma^2}{k_B T}$.
   (b) Assume that a measurement of mobility proceeds by injecting electrons into a thin film slab of length L, thickness, d and then measuring the current at the opposite electrode. Write a suitable expression for mobility in the low temperature, low field regime from time of flight data.
   (c) Based on the drift-diffusion model in the low field, low temperature regime, calculate an approximate expression for, and plot the shape of the current pulse at the opposite electrode.
4. In Fig. P4.4 is shown the reaction coordinates for parabolic potentials for the reactants ($V_R(Q)$) and products ($V_P(Q)$). If ΔG^0 is the Gibbs Free Energy of reaction corresponding to the difference between the uncharged and ionized state,
   (a) Show that the activation energy barrier for charge transfer from the reactant (i.e. donor) to the product (acceptor) molecule is given by:

$$\Delta G^* = \frac{\lambda}{4}\left[1 + \frac{\Delta G^0}{\lambda}\right]^2$$

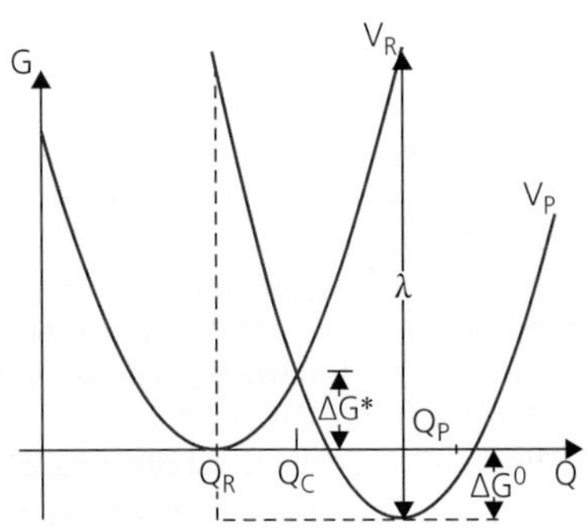

Figure P4.4 Marcus reaction manifolds.

where λ is the reorganization energy for the transfer.
   (b) The diagram is shown for one case of normal Marcus transfer corresponding to $-\Delta G^0 < \lambda$. Replot these parabolas for the cases of $-\Delta G^0 < 0$; $-\Delta G^0 = \lambda$; and for the Marcus inverted region where $-\Delta G^0 > \lambda$.
   (c) If we assume that the rate of electron transfer is given by $k_{ET} = A(\lambda k_B T)^{-1/2}\exp(-\Delta G^*/k_B T)$, plot k_{ET} vs. ΔG^* and find the value of ΔG^0 leading to a maximum transfer rate. Plot k_{ET}/A vs. ΔG^0 for a given value of λ and T from below to above the maximum, clearly labeling the regions of normal and inverted electron transfer.
5. Measuring the energetics of organics and their contacts is key to understanding transport. The most common means for measuring ionization potentials and electron affinities is ultraviolet photoelectron spectroscopy (UPS) and inverse photoelectron spectroscopy (IPES, see Appendix C). The spectra of an organic semiconductor with progressively thicker layers of Au on its surface are shown in Fig. P4.5. To gather the UPS spectra, the He I emission line (21.22 eV) was used as the photon source.
   (a) On the spectra, label all the relevant transitions needed to determine the energetic properties of a semiconductor (the vacuum level, VL or E_{VAC}, the LUMO (E_{LUMO}) and HOMO (E_{HOMO}) energies, the deeper energy levels away from the frontier orbitals that can be observed, the energy gap, E_G, and the Fermi energy, E_F), and discuss how these energies are determined.
   (b) In the data in the figure, the UPS data are shown in varying levels of detail and at different energies. Clearly identify the *all* energies discussed in (a) on the diagrams. The lower black lines in (a) and (b) are the bare organic semiconductor surface, and all other lines correspond to progressively thicker layers of Au.
   (c) From the data in Fig. P4.5, draw the energy level bending as Au thickness increases, and the organic equilibrium energy level diagram for the VL, E_{LUMO}, and E_{HOMO} as functions of Au coverage.
6. The space charge limited current in the presence of a Gaussian trap distribution assumes the trap distribution follows:

$$h_T(E) = \frac{H_T}{\sqrt{2\pi}\sigma_T}\exp\left[-(E - E_{Tm})^2/2\sigma_T^2\right]$$

where H_T is the total density of traps, E_{Tm} is the mean trap energy, and σ_T is the width.
   (a) Show the the trapped charge density is given by: $p_T = H_T^G\left(\frac{p}{N_{HOMO}}\right)^{1/m}$ where $H_T^G = \left(\frac{H_T}{2g_p}\right)\exp(E_{Tm}/mk_B T)$ and $m = \left(1 + 2\pi\sigma_T^2/16k_B^2T^2\right)^{1/2}$. Here, g_p is the degeneracy of the trap states.
   (b) Show that the current voltage relationship is given by:

$$j_{SCL} = q\mu_p N_{HOMO}\left(\frac{2m+1}{m+1}\right)^{m+1}\left(\frac{m}{m+1}\frac{\varepsilon_r\varepsilon_0}{qH_T^G}\right)^m \frac{V^{m+1}}{d^{2m+1}}.$$

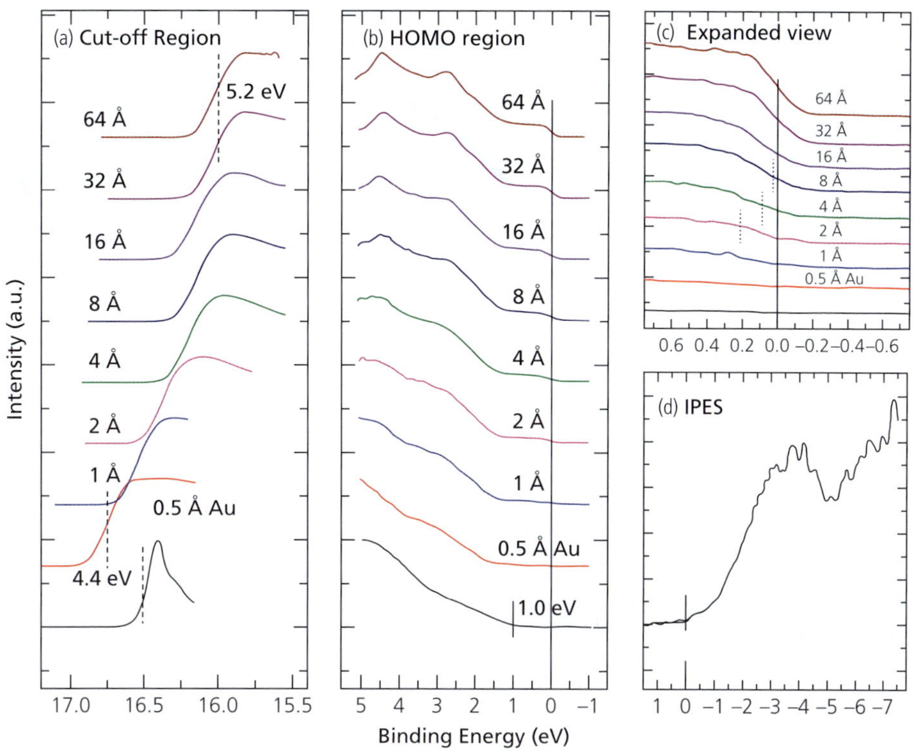

Figure P4.5 (a, b) UPS spectra of an organic in different energy regions; (c) a detail of (b) near the Fermi energy. (d) IPES data for the same material. (Liu et al., 2015)

(c) Calculate the cross-over voltage, V_X, where SCL current equals to the Ohmic current.

(d) Plot the j–V characteristics as a function of temperature and trap density by assuming $m = 5\ E_{TM} = \sigma_T = 10\ k_B T$ and $H_T = 4 \times 10^{19}\ \text{cm}^{-3}$ at $T = 100$ K, 200 K, and 300 K. There is one state per molecule with a density of 5 g/cm^3 and an atomic weight of 550 g/M, the dielectric constant is $\varepsilon_r = 3$, $\mu_p = 0.1\ \text{cm}^2/\text{V s}$, and the sample thickness is $d = 100$ nm.

7. In Fig. P4.7 we show the current–voltage–temperature (I–V–T) characteristics of an organic sample that is 9.3 μm thick, and has an area of 1 cm^2. From these data, and assuming that $m_n^* = m_p^* = m_0$, the mass of the electron, and the density of states, $N_{HOMO} = N_{LUMO}$, determine the following:

   (a) The physical origins of the currents at low and high V.

   (b) Since this is a hole conducting material, what is the hole mobility, μ_p?

   (c) Plot the conductivity (σ) as a function of temperature.

   (d) Calculate the energy gap and N_{HOMO} of this material. Any guesses as to what the material is? (Hint: it is a common, small molecule organic semiconductor.)

   (e) What is the trap density and energy depth assuming an exponential trap distribution?

8. Figure P4.8 shows data taken from the j–V characteristics and by charge extraction by linearly increasing voltage (CELIV) for two regioisomeric donors, A and B. Both CELIV and the j–V characteristics provide hole mobility measurement of two donors, A and **B**. The device

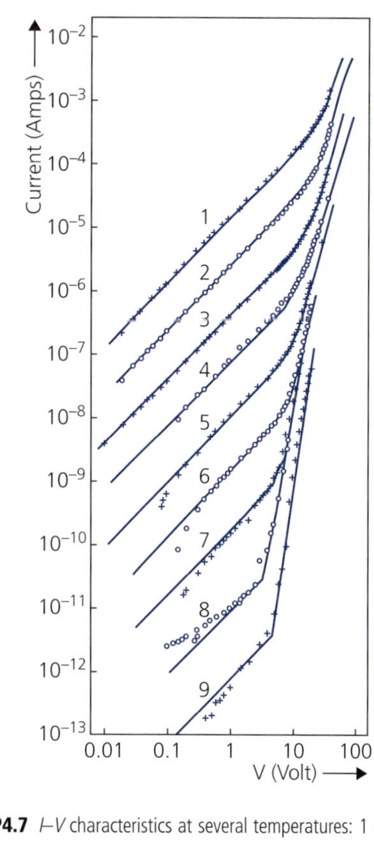

Figure P4.7 I–V characteristics at several temperatures: 1 = 250; 2 = 215; 3 = 180; 4 = 155; 5 = 125; 6 = 100; 7 = 75; 8 = 55; 9 = 17°C

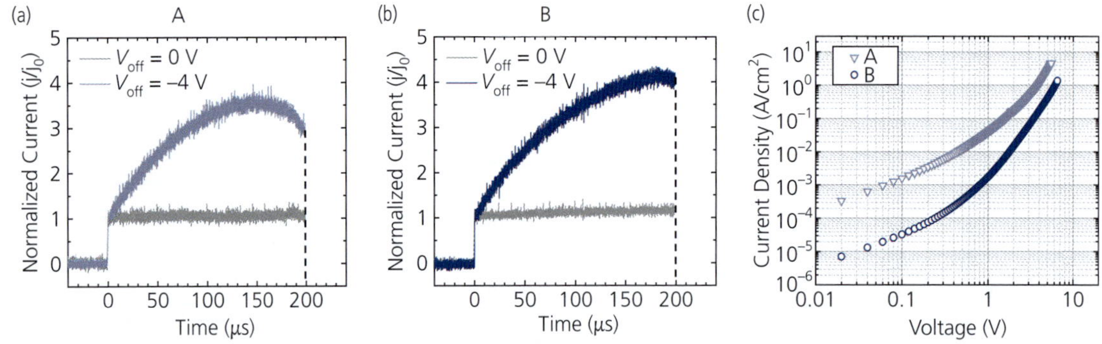

Figure P4.8 CELIV data for (a) sample A and (b) sample B. (c) Current–voltage characteristics of samples A and B.

structure is: ITO/ donor (40 nm)/MgF_2 (15 nm)/Al. The ITO contact is Ohmic and hole injecting. For CELIV, a waveform generator shaped triangular voltage pulse increases by 2 V in 200 μs. The transient current is shown in Fig. P4.8a and b for these two materials at two different voltages that were negatively pulsed at V_{eff} prior to the triangular pulse.

(a) What is the mobility extracted from the current transients shown in Fig. P4.8a and b? Why is there no signal at $V_{eff} = -4$ V?

(b) From the data in Fig. P4.8c calculate the hole mobility, and the background free hole concentration for both donors A and B.

(c) Estimate the errors in both measurements from the data presented. Can you explain why there are differences in mobilities measured by CELIV and the j–V characteristics, and are they significant?

9. The j–V characteristics for Schottky barrier diodes have a saturation current given by $j_0 = A^*T^2\exp(-q\phi_{Bn}/k_BT)$.

(a) Calculate the effective Richardson constant, A^*, by assuming a Gaussian distribution of LUMO states whose center energy is at E_{LUMO}, and whose width is σ.

(b) Calculate the ratio of the constant obtained for A^* to that used for conventional semiconductors. Is there a way to distinguish if the Schottky barrier diode employs an organic or an inorganic semiconductor based on this result?

10. In calculating the tunneling current in Schottky/organic diodes, we made several claims and assumptions. We need to clarify those assumptions.

(a) The total current is equal to the difference of the current from the metal to the semiconductor ($j_{m\to s}$) and from the semiconductor to the metal ($j_{s\to m}$), that is, $j_{tun} = j_{m\to s} - j_{s\to m}$. Show that the tunneling current can be expressed as:

$$j_{tun} = \frac{qm^*}{2\pi^2\hbar^3}\int_{E_{min}}^{E_{max}} T(E_x)dE_x \int_0^\infty\int_0^\infty (f_m(E) - f_s(E))dE_r,$$

where $T(E_x)$ is the tunneling probability of an electron with energy $E_x > E_{LUMO}, E_{metal}$ through the energy barrier along $\hat{\mathbf{x}}$, and the Fermi–Dirac functions are given for the metal and the organic semiconductor by $f_m(E)$ and $f_s(E)$, respectively. Here, E_{LUMO}, E_{metal} are the energies of the organic LUMO and the metal work function, respectively.

(b) Show that by assuming $f_m(E) = 1$ for $E < E_{Fm}$, and $f_m(E) = 0$ for $E > E_{Fm}$, then:

$$j_{tun} = \frac{qm^*}{2\pi^2\hbar^3}\left(\int_{-\infty}^{E_{Fn}} T(E_x)(E_{Fm} - E_{Fn})dE_x.\right.$$
$$\left.+\int_{E_{Fn}}^{E_{Fm}} T(E_x)(E_{Fm} - E_x)dE_x\right).$$

Show that the first integral vanishes when the electron energy is $E_x < E_{Fn}$, the Fermi energy of the n-type organic semiconductor. This is known as the Tsu–Esaki model.

11. An equilateral trapezoidal Schottky barrier with top width, d, and base width, W, is shown in Fig. P4.11.

(a) For an electron of mass m^* in an electric field, F, calculate the tunneling probability using the WKB approximation or other approach.

(b) Calculate the tunneling current for the trapezoidal barrier and plot the current vs. applied voltage, V over 4 decades in j_{tun} for $d/W = 0, 0.5, 1.0$. Compare

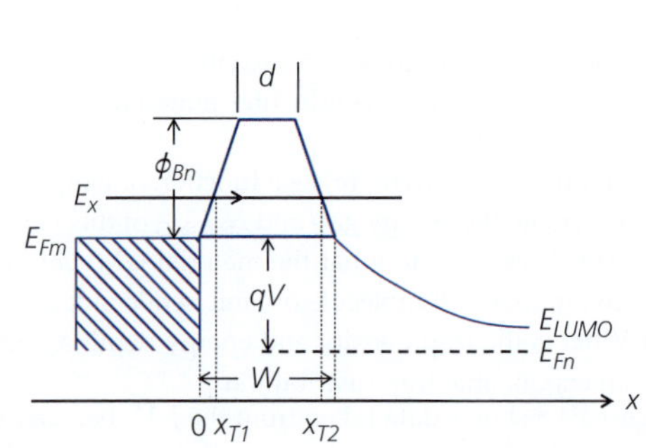

Figure P4.11 Tunneling Schottky barrier energy diagram.

this to tunneling over a triangular barrier of similar barrier height and width, W.

12. The data in Fig. P4.12 are for a homojunction Zn phthalocyanine p–i–n diode and a Schottky metal (M)–i–n diode.
    (a) From the reverse bias $1/C^2$-V data, obtain the donor concentrations and the built-in voltage of both devices.
    (b) Using the built-in voltage from (a), fit the forward bias characteristics of the p–i–n diode to the organic homojunction diode equation 4.290. Assume the device is symmetric (a good assumption in this case), and a normalized Gaussian disorder width of $\hat{\sigma} = 4$.
    (c) Now assuming a recombination model for the p-i-n diode, plot the trap temperature exponent, m vs. $1/T$. What is your physical interpretation of the result?
    (d) From the voltage and temperature dependence of the reverse biased p-i-n, j–V characteristics, can you determine the source of the leakage from the following possibilities: generation-recombination from traps (either discrete or a distribution), parasitic junction shunt resistance, series resistance, tunneling, or other source? Justify your answer via analysis of the data.

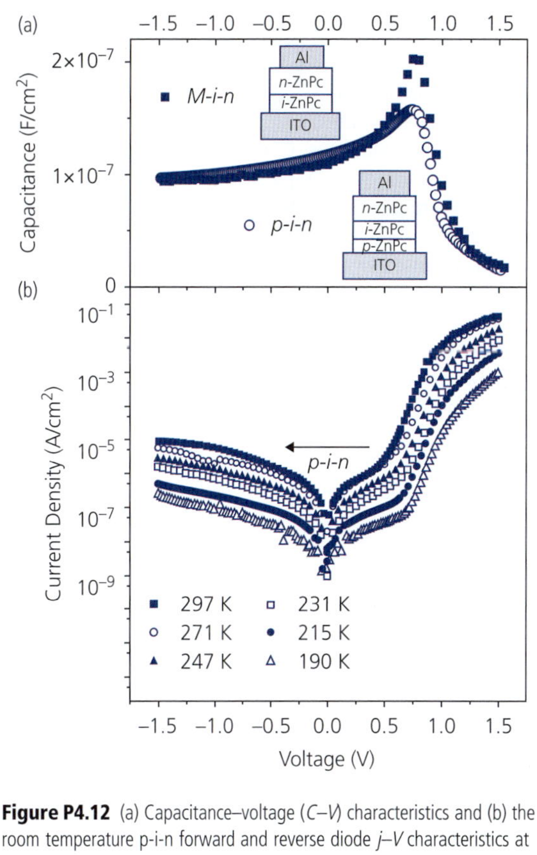

Figure P4.12 (a) Capacitance–voltage (C–V) characteristics and (b) the room temperature p-i-n forward and reverse diode j–V characteristics at several temperatures. Data from Harada et al. (2005).

References

Agranovich, V. M., Basko, D. M., Rossa, G. C. L. & Bassani, F. 1998. Excitons and optical nonlinearities in hybrid organic–inorganic nanostructures. *J. Phys.: Condens. Matter*, 10, 9369.

Agranovich, V. M., Gartstein, Y. N. & Litinskaya, M. 2011. Hybrid resonant organic–inorganic nanostructures for optoelectronic applications. *Chem. Rev.*, 111, 5179.

Akamatu, H. & Inokuchi, H. 1950. On the electrical conductivity of violanthrone, iso-violanthrone, and pyranthrone. *J. Chem. Phys.*, 18, 810.

Ambegaokar, V., Halperin, B. I. & Langer, J. S. 1971. Hopping conductivity in disordered systems. *Phys. Rev. B*, 4, 2612.

Anderson, R. L. 1960. Germanium–gallium arsenide heterojunction. *IBM J. Res. Dev.*, 4, 283.

Arkhipov, V., Emelianova, E., Heremans, P. & Bässler, H. 2005a. Analytic model of carrier mobility in doped disordered organic semiconductors. *Phys. Rev. B*, 72, 235202.

Arkhipov, V., Heremans, P., Emelianova, E., Adriaenssens, G. & Bässler, H. 2002. Weak-field carrier hopping in disordered organic semiconductors: the effects of deep traps and partly filled density-of-states distribution. *J. Phys.: Condens. Matter*, 14, 9899.

Arkhipov, V., Heremans, P., Emelianova, E. & Baessler, H. 2005b. Effect of doping on the density-of-states distribution and carrier hopping in disordered organic semiconductors. *Phys. Rev. B*, 71, 045214.

Arkhipov, V. I., Emelianova, E. V., Tak, Y. H. & Bässler, H. 1998. Charge injection into light-emitting diodes: theory and experiment. *J. Appl. Phys.*, 84, 848.

Austin, I. G. & Mott, N. F. 1969. Polarons in crystalline and non-crystalline materials. *Adv. Phys.*, 18, 41.

Bakulin, A. A., Rao, A., Pavelyev, V. G., Loosdrecht, P. H. M. v., Pshenichnikov, M. S., Niedzialek, D., Cornil, J., Beljonne, D. & Friend, R. H. 2012. The role of driving energy and delocalized states for charge separation in organic semiconductors. *Science*, 335, 1340.

Baldo, M. A. & Forrest, S. R. 2001. Interface-limited injection in amorphous organic semiconductors. *Phys. Rev. B*, 64, 085201.

Bange, S., Schubert, M. & Neher, D. 2010. Charge mobility determination by current extraction under linear increasing voltages: case of nonequilibrium charges and field-dependent mobilities. *Phys. Rev. B*, 81, 035209.

Bansal, N., Reynolds, L. X., MacLachlan, A., Lutz, T., Ashraf, R. S., Zhang, W., Nielsen, C. B., McCulloch, I., Rebois, D. G., Kirchartz, T., Hill, M. S., Molloy, K. C., Nelson, J. & Haque, S. A. 2013. Influence of crystallinity and energetics on charge separation in polymer–inorganic nanocomposite films for solar cells. *Sci. Rep.*, 3, 1531.

Baranovskii, S. D., Cordes, H., Hensel, F. & Leising, G. 2000. Charge-carrier transport in disordered organic solids. *Phys. Rev. B*, 62, 7934.

Barbara, P. F., Meyer, T. J. & Ratner, M. A. 1996. Contemporary issues in electron transfer research. *J. Phys. Chem.*, 100, 13148.

Bässler, H. 1993. Charge transport in disordered organic photoconductors. *Phys. Stat. Sol.*, 175, 15.

Bässler, H. & Köhler, A. 2012. Charge transport in organic semiconductors. *Top. Curr. Chem.*, 312, 1.

Bässler, H., Borsenberger, P. M. & Perry, R. J. 1994. Charge transport in poly(methylphenylsilane): the case for superimposed disorder and polaron effects. *J. Polym. Sci. Part B: Polym. Phys.*, 32, 1677.

Batt, R. H., Braun, C. L. & Hornig, J. F. 1969. Field and temperature dependent recombination in anthracene. *Appl. Opt. Suppl.*, 3, 20.

Benning, P. J., Poirier, D. M., Ohno, T. R., Chen, Y., Jost, M. B., Stepniak, F., Kroll, G. H., Weaver, J. H., Fure, J. & Smalley, R. E. 1992. C-60 and C-70 fullerenes and potassium fullerides. *Phys. Rev. B*, 45, 6899.

Bernardo, B., Cheyns, D., Verreet, B., Schaller, R. D., Rand, B. P. & Giebink, N. C. 2014. Delocalization and dielectric screening of charge transfer states in organic photovoltaic cells. *Nat. Commun.*, 5, 3245.

Blom, P. W. M., Jong, M. J. M. d. & Breedijk, S. 1997a. Temperature dependent electron–hole recombination in polymer light-emitting diodes. *Appl. Phys. Lett.*, 71, 930.

Blom, P. W. M., Jong, M. J. M. d. & vanMunster, M. G. 1997b. Electric-field and temperature dependence of the hole mobility in poly(*p*-phenylene vinylene). *Phys. Rev. B*, 55, R656.

Borsenberger, P. M., Pautmeier, L. T. & Bässler, H. 1992. Nondispersive-to-dispersive charge-transport transition in disordered molecular solids. *Phys. Rev. B*, 46, 12145.

Braun, C. L. 1984. Electric field assisted dissociation of charge transfer states as a mechanism of photocarrier production. *J. Chem. Phys.*, 80, 4157.

Brus, V. V., Zellmeier, M., Zhang, X., Greil, S. M., Töfflinger, A. J., Rappich, J. & Nickel, N. H. 2013. Electrical and photoelectrical properties of P3HT/n-Si hybrid organic–inorganic heterojunction solar cells. *Org. Electron.*, 14, 3109.

Brütting, W., Berleb, S. & Mückl, A. G. 2001. Space-charge limited conduction with a field and temperature dependent mobility in Alq light-emitting devices. *Synth. Met.*, 122, 99.

Bürgi, L., Friend, R. & Sirringhaus, H. 2003. Formation of the accumulation layer in polymer field-effect transistors. *Appl. Phys. Lett.*, 82, 1482.

Burlingame, Q., Coburn, C., Che, X., Panda, A., Qu, Y. & Forrest, S. R. 2018. Centimetre-scale electron diffusion in photoactive organic heterostructures. *Nature*, 554, 77.

Burrows, P. E., Shen, Z., Bulovic, V., McCarty, D. M., Forrest, S. R., Cronin, J. A. & Thompson, M. E. 1996. The relationship between electroluminescence and current transport in organic heterojunction light emitting devices. *J. Appl. Phys.*, 79, 7991.

Cai, C., Su, S.-J., Chiba, T., Sasabe, H., Pu, Y.-J., Nakayama, K. & Kido, J. 2011. High-efficiency red, green and blue phosphorescent homojunction organic light-emitting diodes based on bipolar host materials. *Org. Electron.*, 12, 843.

Campbell, A. J., Bradley, D. D. C., Laubender, J., & Sokolowski, M. 1999. Thermally activated injection limited conduction in single layer *N,N′*-diphenyl-*N,N′*-bis(3-methylphenyl)-1,1′-biphenyl-4,4′-diamine light emitting diodes. *J. Appl. Phys.*, 86, 5004.

Campbell, A. J., Bradley, D. D. C. & Lidzey, D. G. 1997. Space-charge limited conduction with traps in poly(phenylene vinylene) light emitting diodes. *J. Appl. Phys.*, 82, 6326.

Cheyns, D., Poortmans, J., Heremans, P., Deibel, C., Verlaak, S., Rand, B. P. & Genoe, J. 2008. Analytical model for the open-circuit voltage and its associated resistance in organic planar heterojunction solar cells. *Phys. Rev. B*, 77, 165332.

Chiang, C. K., C. R. Fincher, J., Park, Y. W., Heeger, A. J., Shirakawa, H. & Louis, E. J. 1977. Electrical conductivity in doped polyacetylene. *Phys. Rev. Lett.*, 39, 1098.

Choong, V.-E., Mason, M. G., Tang, C. W. & Gao, Y. 1998. Investigation of the interface formation between calcium and tris-(8-hydroxy quinoline) aluminum. *Appl. Phys. Lett.*, 72, 2689.

Coe-Sullivan, S., Woo, W.-K., Steckel, J. S., Bawendi, M. & Bulović, V. 2003. Tuning the performance of hybrid organic/inorganic quantum dot light-emitting devices. *Org. Electron.*, 4, 123.

Coehoorn, R., Pasveer, W. F., Bobbert, P. A. & Michels, M. A. J. 2005. Charge-carrier concentration dependence of the hopping mobility in organic materials with Gaussian disorder. *Phys. Rev. B*, 72, 155206.

Coropceanu, V., Cornil, J., Filho, D. A. d. S., Olivier, Y., Silbey, R. & Brédas, J.-L. 2007. Charge transport in organic semiconductors. *Chem. Rev.*, 107, 926.

Crispin, X., Geskin, V., Crispin, A., Cornil, J., Lazzaron, R., Salaneck, W. R. & Brédas, J. L. 2002. Characterization of the interface dipole at organic/metal interfaces. *J. Am. Chem. Soc.*, 124, 8131.

de Wijs, G. A., Mattheus, C. C., de Groot, R. A. & Palstra, T. T. M. 2003. Anisotropy of the mobility of pentacene from frustration. *Synth. Met.*, 139, 109.

deVault, D. & Chance, B. 1966. Studies of photosynthesis using a pulsed layers. I. Temperature dependence of cytochrome oxidation rate in Chromatium. Evidence for tunneling. *Biophys. J.*, 6, 825.

Dunlap, D. H., Parris, P. E. & Kenkre, V. M. 1996. Charge-dipole model for the universal field dependence of mobilities in molecularly doped polymers. *Phys. Rev. Lett.*, 77, 542.

Dyakonov, V. 2004. Electrical aspects of operation of polymer-fullerene solar cells. *Thin Solid Films*, 451, 493.

Eley, D. D. 1948. Phthalocyanines as semiconductors. *Nature*, 162, 819.

Emin, D. 1975. Phonon-assisted transition rates. I. Optical phonon-assisted hopping in solids. *Adv. Phys.*, 24, 305.

Fishchuk, I. I., Arkhipov, V. I., Kadashchuk, A., Heremans, P. & Bässler, H. 2007. Analytic model of hopping mobility at large charge carrier concentrations in disordered organic semiconductors: polarons versus bare charge carriers. *Phys. Rev. B*, 76, 045210.

Fishchuk, I. I., Hertel, D., Bässler, H. & Kadashchuk, A. K. 2002. Effective-medium theory of hopping charge-carrier transport in weakly disordered organic solids. *Phys. Rev. B*, 65, 125201.

Fishchuk, I. I., Kadashchuk, A., Bassler, H. & Nespurek, S. 2003. Nondispersive polaron transport in disordered organic solids. *Phys. Rev. B*, 67, 224303.

Forrest, S. R., Kaplan, M. L. & Schmidt, P. H. 1984a. Organic-on-inorganic semiconductor contact barrier diodes. I. Theory with applications to organic thin films and prototype devices. *J. Appl. Phys.*, 55, 1492.

Forrest, S. R., Kaplan, M. L. & Schmidt, P. H. 1984b. Organic-on-inorganic semiconductor contact barrier diodes. II. Dependence on organic film and metal contact properties. *J. Appl. Phys.*, 56, 543.

Forrest, S. R., Kaplan, M. L., Schmidt, P. H., Feldmann, W. L. & Yanowski, E. 1982. Organic-on-inorganic semiconductor contact barrier devices. *Appl. Phys. Lett.*, 41, 90.

Forrest, S. R., Kaplan, M. L., Schmidt, P. & Gates, J. V. 1985a. Evaluation of III–V semiconductor wafers using non-destructive organic-on-inorganic contact barriers. *J. Appl. Phys.*, 57, 2982.

Forrest, S. R., Kaplan, M. L., Schmidt, P. H. & J. M. Parsey, J. 1985b. Organic-on-GaAs contact barrier diodes. *J. Appl. Phys.*, 58, 867.

Fowler, R. H. 1931. The analysis of photoelectric sensitivity curves for clean metals at various temperatures. *Phys. Rev.*, 38, 45.

Fowler, R. H. & Nordheim, L. 1928. Electron emission in intense electric fields. *Proc. R. Soc. London, Ser. A*, 119, 173.

Friedman, L. 1965. Electron–phonon interaction in organic molecular crystals. *Phys. Rev.*, 140, A1649.

Gartstein, Y. N. & Conwell, E. 1995. High-field hopping mobility in molecular systems with spatially correlated energetic disorder. *Chem. Phys. Lett.*, 245, 351.

Giebink, N. C. & Forrest, S. R. 2009. Temporal response of optically pumped organic semiconductor lasers and its implication for reaching threshold under electrical excitation. *Phys. Rev. B*, 79, 073302.

Giebink, N. C., Lassiter, B. E., Wiederrecht, G. P., Wasielewski, M. R. & Forrest, S. R. 2010a. The ideal diode equation for organic heterojunctions. II. The role of polaron pair recombination. *Phys. Rev. B*, 82, 155306.

Giebink, N. C., Wiederrecht, G. P., Wasielewski, M. R. & Forrest, S. R. 2010b. The ideal diode equation for organic heterojunctions. I. Derivation and application. *Phys. Rev. B*, 82, 155305.

Grancini, G., Maiuri, M., Fazzi, D., Petrozza, A., Egelhaaf, H.-J., Brida, D., Cerullo, G. & Lanzani, G. 2013. Hot exciton dissociation in polymer solar cells. *Nat. Mater.*, 12, 29.

Gregg, B. A., Chen, S.-G. & Branz, H. M. 2004. On the superlinear increase in conductivity with dopant concentration in excitonic semiconductors. *Appl. Phys. Lett.*, 84, 1707.

Griffith, O. L., Liu, X., Amonoo, J. A., Djurovich, P. I., Thompson, M. E., Green, P. F. & Forrest, S. R. 2015. Charge transport and exciton dissociation in organic solar cells consisting of dipolar donors mixed with C_{70}. *Phys. Rev. B*, 92, 085404.

Groves, C. & Greenham, N. C. 2008. Bimolecular recombination in polymer electronic devices. *Phys. Rev. B*, 78, 155205.

Groves, C., Blakesley, J. C. & Greenham, N. C. 2010. Effect of charge trapping on geminate recombination and polymer solar cell performance. *Nano Lett.*, 10, 1063.

Haase, M. A., Emanuel, M. A., Smith, S. C., Coleman, J. J. & Stillman, G. E. 1987. Band discontinuities in $GaAs/Al_xGa_{1-x}As$ heterojunction photodiodes. *Appl. Phys. Lett.*, 50, 404.

Haeldermans, I., Vandewal, K., Oosterbaan, W. D., Gadisa, A., D'Haen, J., Van Bael, M. K., Manca, J. V. & Mullens, J. 2008. Ground-state charge-transfer complex formation in hybrid poly(3-hexyl thiophene):titanium dioxide solar cells. *Appl. Phys. Lett.*, 93, 223302.

Hagfeldt, A., Boschloo, G., Sun, L., Kloo, L. & Pettersson, H. 2010. Dye-sensitized solar cells. *Chem. Rev.*, 110, 6595.

Hall, R. N. 1951. Germanium rectifier characteristics. *Phys. Rev.*, 83, 228.

Hamann, C. 1968. On the trap distribution in thin films of copper phthalocyanine. *Phys. Stat. Sol.*, 26, 311.

Hannewald, K., Stojanovic, W. M., Schellekens, J. M. T., Bobbert, P. A., Kresse, G. & Hafner, J. 2004. Theory of polaron bandwidth narrowing in organic molecular crystals. *Phys. Rev. B*, 69, 075211.

Harada, K., Werner, A. G., Pfeiffer, M., Bloom, C. J., Elliott, C. M. & Leo, K. 2005. Organic homojunction diodes with a high built-in potential: interpretation of the current-voltage characteristics by a generalized Einstein relation. *Phys. Rev. Lett.*, 94, 036601.

Hasegawa, S., Mori, T., Imaeda, K., Tanaka, S., Yamashita, Y., Inokuchi, H., Fujimoto, H., Seki, K. & Ueno, N. 1994. Intermolecular energy-band dispersion in oriented thin films of bis(1,2,5-thiadiazolo)-p-quinobis(1,3 dithiole) by angle-resolved photoemission. *J. Chem. Phys.*, 100, 6969.

Haynes, J. & Shockley, W. 1949. Investigation of hole injection in transistor action. *Phys. Rev.*, 75, 691.

Heilmeier, G. H. & Harrison, S. E. 1963. Charge transport in copper phthalocyanine single crystals. *Phys. Rev.*, 132, 2010.

Helfrich, W. & Schneider, W. G. 1965. Recombination radiation in anthracene crystals. *Phys. Rev. Lett.*, 14, 229.

Heun, S. & Borsenberger, P. M. 1995. A comparative study of hole transport in vapor-deposited molecular glasses of *N*, *N'*,*N''*,*N''*-tetrakis(4-methylphenyl)-(1,1'-biphenyl)-4,4'-diamine and *N*,*N'*-diphenyl-*N*,*N'*-bis(3-methylphenyl)-(1,1'-biphenyl)-4,4'-diamine. *Chem. Phys.*, 200, 245.

Hill, I. G., Kahn, A., Soos, Z. G. & R.A. Pascal, J. 2000a. Charge-separation energy in films of p-conjugated organic molecules. *Chem. Phys. Lett.*, 327, 181.

Hill, I. G., Mäkinen, A. J. & Kafafi, Z. H. 2000b. Distinguishing between interface dipoles and band bending at metal/tris-(8-hydroxyquinoline) aluminum interfaces. *Appl. Phys. Lett.*, 77, 1825.

Hill, I. G., Rajagopal, A., Kahn, A. & Hu, Y. 1998. Molecular level alignment at organic semiconductor–metal interfaces. *Appl. Phys. Lett.*, 73, 662.

Holstein, T. 1959a. Studies of polaron motion. Part I. The molecular-crystal model. *Ann. Phys.*, 8, 325.

Holstein, T. 1959b. Studies of polaron motion. Part II. The 'small' polaron. *Ann. Phys.*, 8, 343.

Huang, K. & Rhys, A. 1950. Theory of light absorption and non-radiative transitions in F-centres. *Proc. R. Soc. London, Ser. A*, 204, 406.

Hüfner, S. 2003. *Photoelectron Spectroscopy: Principles and Applications*. Heidelberg: Springer-Verlag.

Imaeda, K., Yamashita, Y., Y. Li, Mori, T., Inokuchi, H. & Sanob, M. 1992. Hall-effect observation in the new organic semiconductor bis(1,2,5-thiadiarolo)-*p*-quinobis(1,3-dithiole) (BTQBT). *J. Mater. Chem.*, 2, 115.

Ioannidis, A., Facci, J. S. & Abkowitz, M. A. 1998. Evolution in the charge injection efficiency of evaporated Au contacts on a molecularly doped polymer. *J. Appl. Phys.*, 84, 1439.

Ishii, H., Sugiyama, K., Ito, E. & Seki, K. 1999. Energy level alignment and interfacialelectronic structures at organic/metaland organic/organic interfaces. *Adv. Mater.*, 11, 605.

Jayaraman, A., Kaplan, M. L. & Schmidt, P. H. 1985. Effect of pressure on the Raman and electronic absorption spectra of naphthalene- and perylenetetracarboxylic dianhydrides. *J. Chem. Phys.*, 82, 1682.

Jortner, J. 1976. Temperature dependent activation energy for electron transfer between biological molecules. *J. Chem. Phys.*, 64, 4860.

Juška, G., Arlauskas, K., Viliūnas, M. & Kočka, J. 2000. Extraction current transients: New method for study of charge transport in microcrystalline silicon. *Phys. Rev. Lett.*, 84, 4946.

Kahn, A., Koch, N. & Gao, W. Y. 2003. Electronic structure and electrical properties of interfaces between metals and π-conjugated molecular film. *Polym. Sci. B: Polym. Phys.*, 41, 2529.

Kang, S. J., Yi, Y., Kima, C. Y., Cho, S. W., Noh, M., Jeong, K. & Whang, C. N. 2006. Energy level diagrams of C_{60}/pentacene/Au and pentacene/C_{60}/Au. *Synth. Met.*, 156, 32.

Kao, K. C. & Hwang, W. 1981. *Electrical Transport in Solids*. Oxford: Pergamon Press.

Katz, J. I., Rice, S. A., Choi, S.-I. & Jortner, J. 1963. On the excess electron and hole band structures and carrier mobility in naphthalene, anthracene and several polyphenyls. *J. Chem. Phys.*, 39, 1683.

Kearns, D. R., Tollin, G. & Calvin, M. 1960. Electrical properties of organic solids. II. Effects of added electron acceptor on metal-free phthalocyanine. *J. Chem. Phys.*, 32, 1020.

Kepler, R. G. & Coppage, F. N. 1966. Generation and recombination of holes and electrons in anthracene. *Phys. Rev.*, 151, 610.

Kido, J. & Matsumoto, T. 1998. Bright organic electroluminescent devices having a metal-doped electron-injecting layer. *Appl. Phys. Lett.*, 73, 2866.

Kirchartz, T. & Rau, U. 2008. Detailed balance and reciprocity in solar cells. *Phys. Stat. Sol.*, 205, 2737.

Koch, N., Vollmer, A., Salzmann, I., Nickel, B., Weiss, H. & Rabe, J. P. 2006. Evidence for temperature-dependent electron band dispersion in pentacene. *Phys. Rev. Lett.*, 96, 156803.

Köhler, A. & Bässler, H. 2015. *Electronic Processes in Organic Semiconductors: An Introduction*. Weinheim, Germany: Wiley VCH Verlag GmbH.

Kotadiya, N. B., Lu, H., Mondal, A., Ie, Y., Andrienko, D., Blom, P. W. & Wetzelaer, G.-J. A. 2018. Universal strategy for Ohmic hole injection into organic semiconductors with high ionization energies. *Nat. Materi.*, 17, 329.

Kubo, M., Kaji, T. & Hiramoto, M. 2013. pn-homojunction organic solar cells formed in phase-separated co-deposited films. *Appl. Phys. Lett.*, 103, 263303.

Kumar, S., Tripathi, D. C. & Mohapatra, Y. N. 2017. Organic doped/undoped interface based diode structure: distinct mechanisms underlying forward and reverse bias. *Org. Electron.*, 50, 331.

Lampert, M. A. 1962. Double injection in insulators. *Phys. Rev.*, 125, 126.

Langevin, P. 1903. Recombinaison et mobilites des ions dans les gaz. *Ann. Chim. Phys*, 28, 122.

LeBlanc, O. H., Jr. 1961. Band structure and transport of holes and electrons in anthracene. *J. Chem. Phys.*, 35, 1275.

Lee, S. T., Hou, X. Y., Mason, M. G. & Tang, C. W. 1998. Energy level alignment at Alq/metal interfaces. *Appl. Phys. Lett.*, 72, 1593.

Li, J., Koshnick, C., Diallo, S. O., Ackling, S., Huang, D. M., Jacobs, I. E., Harrelson, T. F., Hong, K., Zhang, G., Beckett, J., Mascal, M. & Moulé, A. J. 2017. Quantitative measurements of the temperature-dependent microscopic and macroscopic dynamics of a molecular dopant in a conjugated polymer. *Macromolecules*, 50, 5476.

Li, N., Lassiter, B. E., Lunt, R. R., Wei, G. & Forrest, S. R. 2009. Open circuit voltage enhancement due to reduced dark current in small molecule photovoltaic cells. *Appl. Phys. Lett.*, 94, 023307.

Li, N., Lee, K., Renshaw, C. K., Xiao, X. & Forrest, S. R. 2011. Improved power conversion efficiency of InP solar cells using organic window layers. *Appl. Phys. Lett.*, 98, 053504.

Liu, C., Xu, Y. & Noh, Y. Y. 2014. Contact engineering in organic field-effect transistors. *Mater. Today*, 18, 79.

Liu, X., Wang, C., Lu, L., Wang, C., Xiao, Z., Bi, C., Huang, J. & Gao, Y. 2015. Electronic structures at the interface between Au and $CH_3NH_3PbI_3$. Phys. Chem. Chem. Phys., 17, 896.

Liu, J., Chen, S., Qian, D., Gautam, B., Yang, G., Zhao, J., Bergqvist, J., Zhang, F., Ma, W. & Ade, H. 2016a. Fast charge separation in a non-fullerene organic solar cell with a small driving force. *Nat. Energy*, 1, 16089.

Liu, X., Ding, K., Panda, A. & Forrest, S. R. 2016b. Charge transfer states in dilute donor–acceptor blend organic heterojunctions. *ACS Nano*, 10, 7619.

Liua, Z., Salataa, O. V. & Maleb, N. 2002. Improved electron injection in organic LED with lithium quinolate/aluminium cathode. *Synth. Met.*, 128, 211.

Loi, M., Toffanin, S., Muccini, M., Forster, M., Scherf, U. & Scarber, M. 2007. Charge transfer excitons in bulk heterojunctions of a polyfluorene copolymer and a fullerene derivative. *Adv. Funct. Mater.*, 17, 2111.

Lorrmann, J., Badada, B. H., Inganäs, O., Dyakonov, V. & Deibel, C. 2010. Charge carrier extraction by linearly increasing voltage: analytic framework and ambipolar transients. *J. Appl. Phys.*, 108, 113705.

Louie, J., Hartwig, J. F. & Fry, A. J. 1997. Discrete high molecular weight triarylamine dendrimers prepared by palladium-catalyzed amination. *J. Am. Chem. Soc.*, 119, 11695.

Lunt, R. R., Benziger, J. B. & Forrest, S. R. 2010. Relationship between crystalline order and exciton diffusion length in molecular organic semiconductors. *Adv. Mater.*, 22, 1233.

Lüssem, B., Riede, M. & Leo, K. 2013. Doping of organic semiconductors. *Phys. Stat. Sol.*, 210, 9.

Maitrot, M., Guillaud, G., Boudjema, B., André, J. J. & Simon, J. 1986. Molecular material-based junctions: formation of a Schottky contact with metallophthalocyanine thin films doped by the cosublimation method. *J. Appl. Phys.*, 60, 2396.

Many, A. & Rakavy, G. 1962. Theory of transient space-charge-limited currents in solids in the presence of trapping. *Phys. Rev.*, 126, 1980.

Marcus, R. A. 1956. On the theory of oxidation-reduction reactions involving electron transfer. I. *J. Chem. Phys.*, 24, 966.

Marcus, R. A. 1957. On the theory of oxidation-reduction reactions involving electron transfer. II. Applications to data on the rates of isotopic exchange reactions. *J. Chem. Phys.*, 26, 867.

Marcus, R. A. 1993. Electron transfer reactions in chemistry. Theory and experiment. *Rev. Mod. Phys.*, 65, 599.

Margaritondo, G. & Perfetti, P. 1987. The problem of heterojunction band discontinuities. In: Capasso, F. & Margaritondo, G. (eds.) *Heterojunction Band Discontinuities: Physics and Device Applications*. Amsterdam: North Holland.

Martens, H., Hulea, I., Romijn, I., Brom, H., Pasveer, W. & Michels, M. 2003. Understanding the doping dependence of the conductivity of conjugated polymers: Dominant role of the increasing density of states and growing delocalization. *Phys. Rev. B*, 67, 121203.

Martin, M., André, J. J. & Simon, J. 1983. Influence of dioxygen on the junction properties of metallophthalocyanine based devices. *J. Appl. Phys.*, 54, 2792.

Merzbacher, E. 1961. *Quantum Mechanics*. New York: John Wiley and Sons.

Michaelson, H. B. 1978. Relation between an etomic electronegativity scale and the work function. *IBM J. Res. Dev.*, 22, 72.

Mihailetchi, V. D., L. J. A, K., Hummelen, J. C. & Blom, P. W. M. 2004. Photocurrent generation in polymer–fullerene bulk heterojunctions. *Phys. Rev. Lett.*, 93, 216601.

Miller, A. & Abrahams, E. 1960. Impurity conduction at low concentrations. *Phys. Rev.*, 120, 745.

Miller, J. R., Calcaterra, L. T. & Closs, G. L. 1984. Intramolecular long distance electron transfer in radical anions. The effects of free energy and solvent on reaction rates. *J. Am. Chem. Soc.*, 106, 3047.

Molodtsova, O. V. & Knupfer, M. 2006. Electronic properties of the organic semiconductor interfaces CuPc/C_{60} and C_{60}/CuPc. *J. Appl. Phys.*, 99, 053704.

Mori, T., Kobayashi, A., Sasaki, Y., Kobayashi, H., Saito, G. & Inokouchi, H. 1984. The intermolecular interaaction of tetrathiafulvalene and bis(ethylenedithio) tetrathaifulvalene in organic metals. Calculation of orbital overlaps and models of energy-band structures. *Bull. Chem. Soc. Japan*, 57, 627.

Mott, N. F. 1938. Note on the contact between a metal and an insulator or semiconductor. *Math. Proc. Cambridge Philos. Soc.*, 34, 568.

Movaghar, B. & Schirmacher, W. 1981. On the theory of hopping conductivity in disordered systems. *J. Phys. C: Solid State Phys.*, 14, 859.

Mozer, A. J., Sariciftci, N. S., Lutsen, L., Vanderzande, D., Österbacka, R., Westerling, M. & Juška, G. 2005. Charge transport and recombination in bulk heterojunction solar cells studied by the photoinduced charge extraction in linearly increasing voltage technique. *Appl. Phys. Lett.*, 86, 112104.

Munn, R. W. & Silbey, R. 1985. Theory of electronic transport in molecular crystals. II. Zeroth order states incorporating nonlocal linear electron–phonon coupling. *J. Chem. Phys.*, 83, 1843.

Mutolo, K. L., Mayo, E. I., Rand, B. P., Forrest, S. R. & Thompson, M. E. 2006. Enhanced open-circuit voltage in subphthalocyanine/C_{60} organic photovoltaic cells. *J. Am. Chem. Soc.*, 128, 8108.

Nabok, D., Puschnig, P., Ambrosch-Draxl, C., Werzer, O., Resel, R. & Smilgies, D.-M. 2007. Crystal and electronic structures of pentacene thin films from grazing-incidence x-ray diffraction and first-principles calculations. *Phys. Rev. B*, 76, 235322.

Nelson, J., Kirkpatrick, J. & Ravirajan, P. 2004. Factors limiting the efficiency of molecular photovoltaic devices. *Phys. Rev. B*, 69, 035337.

Newton, M. D. 1991. Quantum chemical probes of electron-transfer kinetics: the nature of donor–acceptor interactions. *Chem. Rev.*, 91, 767.

Nicolai, H. T., Mandoc, M. M. & Blom, P. W. M. 2011. Electron traps in semiconducting polymers: exponential versus Gaussian trap distributions. *Phys. Rev. B*, 83, 195204.

Nicolai, H. T., Wetzelaer, G. A. H., Kuik, M., Kronemeijer, A. J., deBoer, B. & Blom, P. W. M. 2010. Space-charge-limited hole current in poly(9,9-dioctylfluorene) diodes. *Appl. Phys. Lett.*, 96, 172107.

O'Regan, B. & Grätzel, M. 1991. A low-cost, high-efficiency solar cell based on dye-sensitized colloidal TiO_2 films. *Nature*, 353, 737.

Oehzelt, M., Akaike, K., Koch, N. & Heimel, G. 2015. Energy-level alignment at organic heterointerfaces. *Sci. Adv.*, 1, e1501127.

Olson, D. C., Piris, J., Collins, R. T., Shaheen, S. E. & Ginley, D. S. 2006. Hybrid photovoltaic devices of polymer and ZnO nanofiber composites. *Thin Solid Films*, 496, 26.

Olthof, S., Mehraeen, S., Mohapatra, S. K., Barlow, S., Coropceanu, V., Brédas, J.-L., Marder, S. R. & Kahn, A. 2012. Ultralow doping in organic semiconductors: evidence of trap filling. *Phys. Rev. Lett.*, 109, 176601.

Olthof, S., Tress, W., Meerheim, R., Lüssem, B. & Leo, K. 2009. Photoelectron spectroscopy study of systematically varying doping concentrations in an organic semiconductor layer using a molecular p-dopant. *J. Appl. Phys.*, 106, 103711.

Onsager, L. 1938. Initial recombination of ions. *Phys. Rev.*, 54, 554.

Osikowicz, W., Jong, M. P. d. & Salaneck, W. R. 2007. Formation of the interfacial dipole at organic–organic interfaces: C_{60}/polymer interfaces. *Adv. Mater.*, 19, 4213.

Ozaki, M., Peebles, D., Weinberger, B. R., Heeger, A. J. & MacDiarmid, A. G. 1980. Semiconductor properties of polyacetylene p-$(CH)_x$:n-CdS heterojunctions. *J. Appl. Phys.*, 51, 4252.

Panda, A., Ding, K. & Forrest, S. R. 2016. Free and trapped charge transfer excitons at a ZnO/small molecule heterojunction. *Phys. Rev. B*, 94, 125429.

Panda, A., Renshaw, C. K., Oskooi, A., Lee, K. & Forrest, S. R. 2014. Excited state and charge dynamics of hybrid organic/inorganic heterojunctions. II. Experiment. *Phys. Rev. B*, 90, 045303.

Parthasarathy, G., Burrows, P. E., Khalfin, V., Kozlov, V. G. & Forrest, S. R. 1998. A metal-free cathode for organic semiconductor devices. *Appl. Phys. Lett.*, 72, 2138.

Parthasarathy, G., Shen, C., Kahn, A. & Forrest, S. R. 2001. Lithium doping of semiconducting organic charge transport materials. *J. Appl. Phys.*, 89, 4986.

Pasveer, W. F., Cottar, J., Tanase, C., Coehoorn, R., Bobbert, P. A., Blom, P. W. M. & deLeeuw, D. M. 2005. Unified description of charge-carrier mobilities in disordered semiconductors. *Phys. Rev. Lett.*, 94, 206601.

Perez, M. D., Borek, C., Forrest, S. R. & Thompson, M. E. 2009. Molecular and morphological influences on the open circuit voltages of organic photovoltaic devices. *J. Am. Chem. Soc.*, 131, 9281.

Peumans, P. & Forrest, S. R. 2004. Separation of geminate charge-pairs at donor–acceptor interfaces in disordered solids. *Chem. Phys. Lett.*, 398, 27.

Peumans, P., Yakimov, A. & Forrest, S. R. 2003. Small molecular weight organic thin-film photodetectors and solar cells. *J. Appl. Phys.*, 93, 3693.

Pfeiffer, M., Beyer, A., Fritz, T. & Leo, K. 1998. Controlled doping of phthalocyanine layers by cosublimation with acceptor molecules: a systematic Seebeck and conductivity study. *Appl. Phys. Lett.*, 73, 3202.

Pingel, P. & Neher, D. 2013. Comprehensive picture of p-type doping of P3HT with the molecular acceptor F 4 TCNQ. *Phys. Rev. B*, 87, 115209.

Piromreun, P., Oh, H. S., Shen, Y., G. G. Malliaras, J. Campbell Scott & Brock, P. J. 2000. Role of CsF on electron injection into a conjugated polymer. *Appl. Phys. Lett.*, 77, 2403.

Podzorov, V., Menard, E., Rogers, J. A. & Gershenson, M. E. 2005. Hall efect in the accumulation layers on the surface of organic semiconductors. *Phys. Rev. Lett.*, 95, 226601.

Potscavage, W. J., Yoo, S. & Kippelen, B. 2008. Origin of the open-circuit voltage in multilayer heterojunction organic solar cells. *Appl. Phys. Lett.*, 93, 413.

Probst, M. & Haight, R. 1997. Diffusion of metals into organic films. *Appl. Phys. Lett.*, 70, 1420.

Rajagopal, A. & Kahn, A. 1998. Photoemission spectroscopy investigation of magnesium–Alq_3 interfaces. *J. Appl. Phys.*, 84, 355.

Rand, B. P., Burk, D. P. & Forrest, S. R. 2007. Offset energies at organic semiconductor heterojunctions and their influence on the open-circuit voltage of thin-film solar cells. *Phys. Rev. B*, 75, 115327.

Rand, B. P., Li, J., Xue, J., Holmes, R. J., Thompson, M. E. & Forrest, S. R. 2005. Organic double-heterostructure photovoltaic cells employing thick tris(acetylacetonato)ruthenium (III) exciton-blocking layers. *Adv. Mater.*, 17, 2714.

Reid, O. G., Munechika, K. & Ginger, D. S. 2008. Space charge limited current measurements on conjugated polymer films using conductive atomic force microscopy. *Nano Lett.*, 8, 1602.

Reineke, S., Lindner, F., Schwartz, G., Seidler, N., Walzer, K., Lüssem, B. & Leo, K. 2009. White organic light-emitting diodes with fluorescent tube efficiency. *Nature*, 459, 234.

Renshaw, C. K. & Forrest, S. R. 2014. Excited state and charge dynamics of hybrid organic/inorganic heterojunctions. I. Theory. *Phys. Rev. B*, 90, 045302.

Renshaw, C. K., Zimmerman, J. D., Lassiter, B. E. & Forrest, S. R. 2012. Photoconductivity in donor–acceptor heterojunction organic photovoltaics. *Phys. Rev. B*, 86, 085324.

Rikken, G. L. J. A., Braun, D., Staring, E. G. J. & Demandt, R. 1994. Schottky effect at a metal–polymer interface. *Appl. Phys. Lett.*, 65, 219.

Rivan, N. A., Kuik, M., Love, J. A., Proctor, C. M., Nagao, I., Bazan, G. C. & Nguyen, T.-Q. 2014. Understanding the charge-transfer state and singlet exciton emssion from solution-processed small molecule organic solar cells. *Adv. Mater.*, 26, 7405.

Roichman, Y. & Tessler, N. 2002. Generalized Einstein relation for disordered semiconductors: implications for device performance. *Appl. Phys. Lett.*, 80, 1948.

Roichman, Y., Preezant, Y. & Tessler, N. 2004. Analysis and modeling of organic devices. *Phys. Stat. Sol. (a)*, 201, 1246.

Rose, A. 1955. Space-charge-limited currents in solids. *Phys. Rev.*, 97, 1538.

Salzmann, I., Heimel, G., Oehzelt, M., Winkler, S. & Koch, N. 2016. Molecular electrical doping of organic semiconductors: fundamental mechanisms and emerging dopant design rules. *Accts Chem. Res.*, 49, 370.

Sargent, E. H. 2012. Colloidal quantum dot solar cells. *Nat. Photonics*, 6, 133.

Schadt, M. & Williams, D. F. 1969. Low-temperature hole injection and hole trap distribution in anthracene. *J. Chem. Phys.*, 50, 4364.

Schlaf, R., Parkinson, B. A., Lee, P. A., Nebesny, K. W., Jabbour, G., Kippelen, B., Peyghambarian, N. & Armstrong, N. R. 1998. Photoemission spectroscopy of LiF coated Al and Pt electrodes. *J. Appl. Phys.*, 84, 6729.

Schottky, W. 1938. Halbleitertheorie der Sperrschicht. *Naturwissenschaften*, 26, 843.

Sharber, M. C., Mühlbacher, D., Koppe, M., Denk, P., Waldauf, C., Heeger, A. J. & Brabec, C. J. 2006. Design rules for donors in bulk-heterojunction solar cells: towards 10% energy-conversion efficiency. *Adv. Mater.*, 18, 789.

Sharma, A., Kumar, P., Singh, B., Chaudhuri, S. R. & Ghosh, S. 2011. Capacitance–voltage characteristics of organic Schottky diode with and without deep traps. *Appl. Phys. Lett.*, 99, 023301.

Shen, Y., Hosseini, A. R., Wong, M. H. & Malliaras, G. G. 2004. How to make Ohmic contacts to organic semiconductors. *ChemPhysChem*, 5, 16.

Shockley, W. 1949. The theory of p-n junctions in semiconductors and p-n junction transistors. *Bell Syst. Tech. J.*, 28, 435.

Shockley, W. & Queisser, H. J. 1961. Detailed balance limit of efficiency of p-n junction solar cells. *J. Appl. Phys.*, 32, 510.

Shockley, W. & Read, W. T. 1952. Statistics of the recombinations of holes and electrons. *Phys. Rev.*, 87, 835.

Shockley, W., Pearson, G. L. & Haynes, J. R. 1949. Hole injection in germanium: quantitative studies and filamentary transistors. *Bell Syst. Tech. J.*, 28, 344.

Simon, J. & André, J.-J. 1985. *Molecular Semiconductors: Photoelectrical Properties and Solar Cells*. Berlin: Springer-Verlag.

Singh, A. K. & Prakash, R. 2012. Organic Schottky diode based on conducting polymer–nanoclay composite. *RSC Adv.*, 2, 5277.

So, F. F. & Forrest, S. R. 1988. Measurement of the valence band discontinuities for molecular organic semiconductor/inorganic semiconductor heterojunctions. *Appl. Phys. Lett.*, 52, 1341.

So, F. F. & Forrest, S. R. 1989. Organic-on-inorganic semiconductor photodetector. *IEEE Trans. Electron Dev.*, 36, 66.

Sreearunothai, P., Morteani, A. C., Avilov, I., Cornil, J., Beljonne, D., Friend, R. H., Phillips, R. T., Silva, C. & Herz, L. M. 2006. Influence of copolymer interface orientation on the optical emission of polymeric semiconductor heterojunctions. *Phys. Rev. Lett.*, 96, 117403.

Streetman, B. G. & Banerjee, S. K. 2006. *Solid State Electronic Devices*. Upper Saddle River, NJ: Prentice Hall.

Sze, S. M. 1981. *Physics of Semiconductor Devices*. New York: Wiley-Interscience.

Szymanski, M. Z., Kulszewicz-Bajer, I., Faure-Vincent, J. & Djurado, D. 2012. Comparison of simulations to experiment for a detailed analysis of space-charge-limited transient current measurements in organic semiconductors. *Phys. Rev. B*, 85, 195205.

Tanase, C., Blom, P. W. M. & deLeeuw, D. M. 2004. Origin of the enhanced space-charge-limited current in poly(*p*-phenylene vinylene). *Phys. Rev. B*, 70, 193202.

Tang, J. X., Lau, K. M., Lee, C. S. & Lee, S. T. 2006. Substrate effects on the electronic properties of an organic/organic heterojunction. *Appl. Phys. Lett.*, 88, 232103.

Tang, J. X., Lee, C. S. & Lee, S. T. 2007. Electronic structures of organic/organic heterojunctions: from vacuum level alignment to Fermi level pinning. *J. Appl. Phys.*, 101, 064504.

Tessler, N., Preezant, Y., Rappaport, N. & Roichman, Y. 2009. Charge transport in disordered organic materials and it relevance to thin-film devices: a tutorial review. *Adv. Mater.*, 21, 2741.

Ting, H. C., Chen, Y. H., Lin, L. Y., Chou, S. H., Liu, Y. H., Lin, H. W. & Wong, K. T. 2014. Benzochalcogenodiazole-based donor–acceptor–acceptor molecular donors for organic solar cells. *ChemSusChem*, 7, 457.

Treat, N. D., Mates, T. E., Hawker, C. J., Kramer, E. J. & Chabinyc, M. L. 2013. Temperature dependence of the diffusion coefficient of PCBM in poly(3-hexylthiophene). *Macromolecules*, 46, 1002.

Tripathia, D. C. & Mohapatra, Y. N. 2012. Ideal organic homojunction diode obtained using controlled alignment of localized density of states across doped/undoped interface. *Org. Electron.*, 13, 1680.

Tsuji, H., Mitsui, C., Sato, Y. & Nakamura, E. 2009. Bis(carbazolyl)benzodifuran: a high-mobility ambipolar material for homojunction organic light-emitting diode devices. *Adv. Mater.*, 21, 3776.

Uchida, S., Xue, J., Rand, B. P. & Forrest, S. R. 2004. Organic small molecule solar cells with a homogeneously mixed copper phthalocyanine: C_{60} active layer. *Appl. Phys. Lett.*, 84, 4218.

Van Mensfoort, S., Shabro, V., De Vries, R., Janssen, R. & Coehoorn, R. 2010. Hole transport in the organic small molecule material α-NPD: evidence for the presence of correlated disorder. *J. Appl. Phys.*, 107, 113710.

Vandewal, K., Albrecht, S., Hoke, E. T., Graham, K. R., Widmer, J., Douglas, J. D., Schubert, M., Mateker, W. R., Bloking, J. T., Burkhard, G. F., Sellinger, A., Fréchet, J. M. J., Amassian, A., Riede, M. K., McGehee, M. D., Neher, D. & Salleo, A. 2014. Efficient charge generation by relaxed charge-transfer states at organic interfaces. *Nat. Mater.*, 13, 63.

Vandewal, K., Tvingstedt, K., Gadisa, A., Inganas, O. & Manca, J. V. 2010. Relating the open-circuit voltage to interface molecular properties of donor:acceptor bulk heterojunction solar cells. *Phys. Rev. B*, 81, 125204.

Vaynzof, Y., Bakulin, A. A., Gélinas, S. & Friend, R. H. 2012. Direct observation of photoinduced bound charge-pair states at an organic–inorganic semiconductor surface. *Phys. Rev. Lett.*, 108, 246605.

Vázquez, H., Gao, W., Flores, F. & Kahn, A. 2005. Energy level alignment at organic heterojunctions: role of the charge neutrality level. *Phys. Rev. B*, 71, 041306.

Vissenberg, M. & Matters, M. 1998. Theory of the field-effect mobility in amorphous organic transistors. *Phys. Rev. B*, 57, 12964.

Vlachopoulos, N., Liska, P., Augustynski, J. & Graetzel, M. 1988. Very efficient visible light energy harvesting and conversion by spectral sensitization of high surface area polycrystalline titanium dioxide films. *J. Am. Chem. Soc.*, 110, 1216.

Warta, W. & Karl, N. 1985. Hot holes in naphthalene: high, electric-field-dependent mobilities. *Phys. Rev.*, 32, 1172.

Warta, W., Stehle, R. & Karl, N. 1985. Ultrapure, high mobility organic photoconductors. *Appl. Phys. A*, 36, 163.

Winkler, S., Amsalem, P., Frisch, J., Oehzelt, M., Heimel, G. & Koch, N. 2015. Probing the energy levels in hole-doped molecular semiconductors. *Mater. Horiz.*, 2, 427.

Wolf, U., Arkhipov, V. I. & Bässler, H. 1999. Current injection from a metal to a disordered hopping system. I. Monte Carlo simulation. *Phys. Rev. B*, 59, 7507.

Wright, M. & Uddin, A. 2012. Organic–inorganic hybrid solar cells: a comparative review. *Solar Energy Mater. Solar Cells*, 107, 87.

Yamane, H., Kera, S., Okudaira, K. K., Yoshimura, D., Seki, K. & Ueno, N. 2003. Intermolecular energy-band dispersion in PTCDA multilayers. *Phys. Rev. B*, 68, 033102.

Yan, Q., Zhou, Y., Zheng, Y.-Q., Pei, J. & Zhao, D. 2013. Towards rational design of organic electron acceptors for photovoltaics: a study based on perylenediimide derivatives. *Chem. Sci.*, 4, 4389.

Yanagisawa, S., Morikawa, Y. & Schindlmayr, A. 2013. HOMO band dispersion of crystalline rubrene: effects of self-energy corrections within the GW appproximation. *Phys. Rev. B*, 88, 115438.

Yu, G., Gao, J., Hummelen, J., Wudl, F. & Heeger, A. J. 1995. Polymer photovoltaic cells: enhanced efficiencies via a network of internal donor–acceptor heterojunctions. *Science*, 270, 1789.

Zhao, W., Salomon, E., Zhang, Q., Barlow, S., Marder, S. R. & Kahn, A. 2008. Substrate-dependent electronic structure of an organic heterojunction. *Phys. Rev. B*, 77, 165336.

PART II

Applications

CHAPTER 5

Materials purity, growth, and patterning

"Nature uses only the longest threads to weave her patterns, so that each small piece of her fabric reveals the organization of the entire tapestry."

Richard P. Feynman, Physicist

In Part I, Foundations, we discussed the characteristics of organic semiconductors that determine their structural, optical and electrical properties. We found that almost all differences between organic and inorganic semiconductors arise from their cohesive forces: soft van der Waals bonds dominate in organics, whereas in inorganics, covalent bonds, sometimes with an admixture of ionic bonding, result in comparatively rigid and brittle materials. In Chapter 2, we showed that van der Waals bonds allow for growth of self-assembled crystalline and nanocrystalline structures on virtually any flat substrate. The resulting films, whose morphologies can range from crystalline to highly disordered form the basis for many practical devices, ranging from light emitting diodes to solar cells.

In this first chapter of Part II, Applications, we examine the important role that materials purity plays in determining materials and device properties, and describe several methods used to achieve the highest possible purity. This is followed by a discussion of bulk crystal growth. We note that bulk crystals are not of great interest for device applications, yet they are highly useful for understanding the fundamental optical and electrical characteristics of organic materials. This section relies heavily on the insights gained in Chapter 2 on the bonding forces governing film morphology. The control of both materials purity and film structure can be achieved only if they can be quantitatively evaluated.

Next, we discuss how thin films are deposited, often with monolayer precision, by both vapor and solution phase processes. Integral to achieving the desired film thickness is to also achieve the needed film morphology. This is necessarily varied from single crystalline films, to those with nanocrystalline domains, and to amorphous structures. Further, many devices require precisely controlled blend morphologies comprising two or more constituents. We will discuss many of the growth and post-growth processes used in Sections 5.4 and 5.5. In Section 5.6, we introduce the practical and important topics of patterning of organic films and contacts with nanoscale feature resolution necessary for modern electronic devices. Electronic devices generally require several sequential heterogeneous materials deposition steps that are registered to underlying layers. Patterning of conventional inorganic semiconductor circuits is a highly complex, costly, and sophisticated process that demands the registry of nanometer-scale features from one layer to the next. It is enabled by a photolithographic sequence of addition and removal of materials of widely varying physical and chemical properties. It is not uncommon for a circuit to require 20–50 such precise layering steps in a well-defined sequence to create the end product. Such complexity, however, is well beyond the capabilities of modern organic electronics and may, in fact, not even be desirable. The attraction of organic electronics is its

Organic Electronics: Foundations to Applications. Stephen R. Forrest. © Stephen R. Forrest 2020.
Published in 2020 by Oxford University Press. DOI: 10.1093/oso/9780198529729.001.0001

ability to provide relatively large devices (e.g. displays, lighting, solar cells, biosensors, etc.) at extremely low cost. Hence a large variety of patterning methods have been adapted to the needs specific to organic materials.

This leads to a discussion of manufacturing methods by continuous roll processing, ending with consideration of device packaging to prevent performance degradation due to environmental exposure. One way to view organic electronic device manufacturing is as an exercise in "making small things big." How do we take the quantum properties of molecules, create films and morphologies at the nano- and microscales, then develop individual devices at the micron to millimeter scales, and to finally realize 77″ televisions, large room lighting fixtures, or solar cells that cover entire roof tops? It is these last topics that clearly delineate the "worlds" of organic and inorganic devices. As mentioned in Chapter 1, the ability to print organic devices in batches or even on continuous rolls of lightweight and flexible substrates creates distinct opportunities to generate entirely new classes of applications that conventional semiconductors are unable to address. However, their fragility, and the absence of chemical bonds between molecules also makes organics vulnerable to attack by moisture, oxygen and other environmental contaminants. Hence, a discussion of materials growth and purity would be incomplete without also considering means for protecting the as-grown materials from degradation through environmental exposure.

We intentionally omit a discussion of materials synthesis and chemical structure in this volume. While it is understood that the enormous control of property and structure unique to organic materials arises from the ability to synthesize a staggering variety of complex molecules, we cannot do this field justice by including a necessarily brief overview of the topic. Interested readers will find ample discussion of these topics elsewhere. Here, we will only refer to a small subset of molecules that provide illustrative models for the growth and applications that are currently the most widespread.

5.1 The organic semiconductor difference

In Part I, we discussed the foundational principles and properties of organic semiconductors; it was basically about the "how" in how things work. Part II, then is all about the "what." What do we do with these materials? Can we use their unique physical, electronic and optical properties to create things that can make a difference in our everyday lives, or that can even possibly change the very way we live and interact with the technological world? But before we can address these applications that span the gamut from displays and lighting, to solar cells and many other optical and electronic applications, we first need to understand how to *make* things with organic materials. So, this chapter is all about making things and should give a firm foundation on how to build sophisticated, reliable and high performance devices out of electronic-grade materials deposited in precisely controlled layers.

The ability to synthesize, purify, process, and fabricate high performance organic devices has benefitted from over 60 years of research and technology development. Yet, achieving high materials quality, and the ability to fabricate complex devices sometimes requiring processes that can produce nanometer scale features, is still not nearly as straightforward as it is for conventional semiconductors. The principal difference lies in the relative "softness" of materials: they tend to readily degrade when exposed to many solvents or the ambient, they can be easily damaged by handling, and they are difficult to purify. As in all electronic devices, the starting materials for organic semiconductors must be highly pure to be useful, yet as we will see, this remains a daunting task, and results to date in achieving very high purities have been mixed.

Given the range of materials processing techniques that have been developed, and the extraordinary creativity shown by the engineering communities in their development, a full description of all of the possibilities is impractical. Here, we only provide a sampling of the most important (although in all cases not necessarily the best) methods used. Even so, the discussion in this chapter illustrates many challenges and degrees of freedom in materials deposition and structural control afforded by organic semiconductors. The end product of any electronic application is a tapestry, woven from many threads that together create a whole that is unquestionably greater than the sum of its parts. So, too, does this chapter attempt to weave together the threads that form a high performance organic electronic device.

Before we go further, it is useful to remind ourselves of that which makes organic electronics unique. It is these properties that serve as a guide when we consider the efficacy of a particular growth, patterning or manufacturing method. In Chapter 1, we noted that organic electronics are useful when the application demands a large area, low cost, lightweight, conformable or flexible form factor. Furthermore, the most interesting applications for organics are often those that have a photonic property, such as light emitters or detectors. In these cases, at least one surface of the device must be transparent to the wavelength band in which it operates. This broad application space, and the weak bonds of organic thin films, requires fabrication processes near room temperature. So too, organics are vulnerable to exposure to liquids and other processing agents commonly employed in electronic device processing. Taken together, these constraints limit the types of processes that are suitable for the fabrication of thin film organic devices. Nevertheless, techniques have been adapted to overcome many of the apparent limitations set by the varied substrates and materials employed. Throughout the following discussion, we need to keep these constraints in mind to focus our attention on methods most likely to lead to high performance devices.

5.2 Effects of impurities on materials properties

A fundamental property of all high performance electronic devices that exhibit a long operational lifetime and can be produced at high yield, is that the materials of which they are comprised are of high purity. To access their intrinsic properties, the materials must be as free as possible from extrinsic defects that result from the unintentional incorporation of foreign species. As in real estate where value is claimed to be all about "location, location, location," in electronics it is all about "purity, purity, purity." Unfortunately, the literature on organic materials and devices is replete with studies that report what are assumed to be intrinsic materials characteristics, but are instead features of contaminants adventitiously included during growth or processing. Hence, considerable care must be taken to remove impurities from the source materials if reproducible, stable and comprehensible characteristics are to be achieved.

The different cohesive forces binding inorganic and organic semiconductors lead to different outcomes in their properties in the presence of impurities. As we have seen in Part I, a high concentration of impurities, or defect density leads to significant variations in the conductivities and optical properties compared to pristine material. Furthermore, the operational stability of devices comprised of materials unintentionally doped with impurities can be adversely affected. That is, the various device performance characteristics can change over time, depending on the concentration, reactivity, diffusivity and other properties of the contaminants.

As discussed in Chapter 4, impurities that take the place of a lattice atom in a covalently bonded material can either give up an extra electron to the lattice (if it has one more outer shell electron than the lattice atom itself). Alternatively, if it has one fewer valence electron, it can accept an electron and become negatively charged. In the first case, the impurity is known as an *electron donor*, and in the second it is an *electron acceptor*. These types of impurities are known as *substitutional dopants*. Substitution can result in an increase in the conductivity of covalently bound semiconductors (e.g. Si or Ge) by as much as 10^9 from its intrinsic value at room temperature. As a result, controlled doping is a common strategy for modifying the conductivity in almost all semiconductor devices.

The introduction of extrinsic impurities can create lattice strain: that is the "fit" of an impurity atom into a lattice is never perfect. Hence, the lattice must locally conform to the size and valence electron distribution of the impurity. This creates an additional energy associated with the defect that influences the conductivity of the host material.

Another type of impurity is that which is located at unoccupied lattice sites; these are called *interstitial* dopants. Like substitutional dopants, interstitials can also lead to significant changes in the lattice conductivity even though they do not share a bond with the lattice atoms. Under the right circumstances, alkali atoms and the halogens with their partially filled outer valences can serve as electron donors or acceptors, respectively. Interstitial defects are more difficult to control than substitutional impurities, making the latter more desirable for doping.

A third and final type of defect is the *vacancy*, where an atom is missing from a particular position in the lattice. In Si, this results in four unfulfilled, or dangling bonds from the atoms neighboring the vacant site, thereby potentially donating as many as four electrons to the conduction band of the semiconductor. Unlike impurities which, in principle (but never in practice), can be completely eliminated, the density of lattice vacancies is determined by thermodynamics, and hence can never be avoided. If the enthalpy of formation of a vacancy is ΔH^{vac}, then the density of vacancies in the solid is

$$N_{vac} = N\exp(-\Delta H^{vac}/RT), \tag{5.1}$$

where N is the total number of lattice sites, and R is the universal gas constant. An associated site is the *self-interstitial* where an extra lattice atom locates in a region between filled lattice sites. The probability of formation of a self-interstitial is also given by Eq. 5.1, although the energy of formation is far higher than for a vacancy, since squeezing an extra atom into the lattice is accompanied by significant local lattice distortion.

Unlike inorganic semiconductors, impurities in most organic materials do not share valences within the lattice since there are no bonds to make or break in these electrostatically bonded materials. The defects, therefore, typically occupy interstitial sites where charges remain tightly bound to the impurity itself. Nevertheless, if the impurity species is readily ionized (e.g. alkalis such as Na^+ and Li^+, or halogens Cl^- and F^-), the interstitial defects can also contribute free charge and hence, unintentionally lead to a dramatic change in conductivity. One common example impurity is the oxygen molecule, which readily accepts electrons to form an O_2^- complex with the copper atom in CuPc, increasing the background density of holes (Simon and André, 1985). Also, depending on the position of their frontier orbitals relative to that of the host lattice, larger molecular impurities can also contribute charges to the host. For example, in Section 4.4.1 we showed that F_4TCNQ is an acceptor in such hosts as ZnPc (Maennig et al., 2001). That is, introduction of impurities can result in the reduction or oxidation of the host species, thereby contributing free charge. This strategy has been used to modify the conductivity of a wide range of organic materials.

As noted above, impurities in organic thin films can disrupt the crystalline order in molecular solids. That is, the foreign species can intercalate into the lattice, creating a stacking fault or dislocation. The resulting disorder results in a reduced charge mobility, where the carriers are forced to surmount energy barriers formed at grain boundaries between individual crystallites comprising the solid (Forrest et al., 1984). The size of the impurity species determines the significance of the defect that results: individual atomic or ionic impurities may not be sufficiently large to induce stacking faults due to the large gaps between neighboring molecules, whereas molecular impurities or molecular fractions left over from synthesis or that result from decomposition during processing are more likely to disrupt the lattice.

The presence of impurities also can have a dramatic impact on the luminescent properties of materials. Intentionally doping a conductive host material with luminescent impurity molecules is a well-established means for obtaining high efficiency light emission in OLEDs (Tang et al., 1989, Baldo et al., 1998b). However, unintentional incorporation of molecular species at even trace levels can lead to luminescence quenching due to exciton trapping and non-radiative recombination, parasitic luminescence at undesired wavelengths, and even a reduction in device operational lifetime and stability. A dramatic example that illustrates the sensitivity of the luminescent properties to impurities is illustrated in Fig. 5.1. The emission zones of two OLEDs containing a 20 nm thick CBP host are doped with only a 0.2 Å layer of a highly emissive phosphor. One is doped with the green emitting $Ir(ppy)_3$, and the other with the commercial red-emitting dopant, RD-15 (Universal Display, Inc., Ewing, NJ, USA). Emission from both structures is easily detectable, although that of the lower energy red emitting phosphor is 3–4 times higher than for the green. It is striking that the red dopant concentration constitutes only 10^{-4} of that of the host. The instrumental resolution of this experiment indicates that the red emission is observable at a thickness of 1/500 of a monolayer, or a concentration equivalent to only one part in 10^6.

Impurity quenching of excitons has also been observed in organic photovoltaics. The primary mechanism for photocurrent generation in organics is via the dissociation of optically generated excitons that diffuse to an interface between two dissimilar materials (see Chapter 7). The presence of impurities has been directly linked to a decrease in the exciton diffusion length in CuPc-based organic photovoltaics, thereby preventing their migration to the interface,

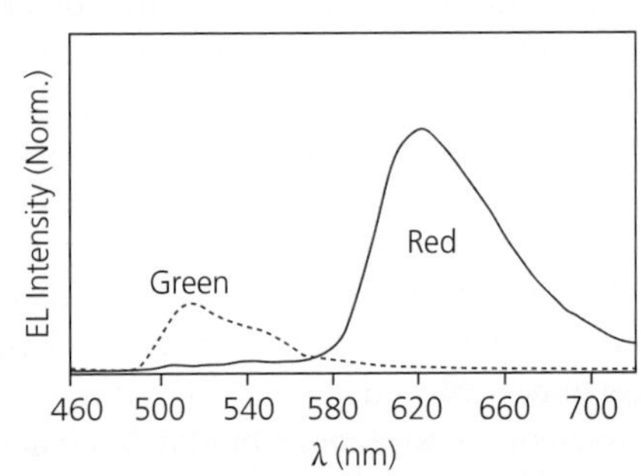

Figure 5.1 Electroluminescence spectra from organic light emitting devices (OLED) whose emission layers contain either a 0.2 Å thick layer of a green or red phosphorescent dopant in a large energy gap host layer. While the green doped device intensity is only ~25% of the red device, the instrumental detection limit of the red emission corresponds to only 1/500 of a monolayer of dopant. After McGraw and Forrest (2013).

and hence reducing their power conversion efficiencies (Salzman et al., 2005).

While there have been relatively few detailed quantitative studies, it is clear that the reliability of almost all electronic devices depends critically on the purity of the materials used in their manufacture. For example, the efficiency of polymer solar cells degrades rapidly over time due to contamination by low molecular weight species, unreacted monomers that comprise the polymer chains, or reaction byproducts that are not completely removed following synthesis (Mateker et al., 2013). These species react with the host polymer to form exciton quenching sites, similar to the effects of extrinsic contaminants in CuPc-based cells. Reaction products that emerge during device operation, such as polymerization of C_{60} in fullerene-based organic solar cells have been implicated in a long term decrease in solar cell performance (Wang et al., 2014). Similar molecular dissociation effects are also known to negatively impact the luminescence efficiency of organic light emitting devices over time via the production of exciton quenching sites (Giebink et al., 2008, Jeong et al., 2019).

Whether the defects are induced during device operation, catalyzed by the presence of extrinsic contaminants, or are an intrinsic materials property, impurities play a major if not dominant role in device performance. While the mechanisms leading to impurity effects may differ between organics and inorganics, their impacts can be equally important. The question remains, however: how pure must a material be for it to be useful? The answer is as ambiguous as the question since it is contingent on the particular application and the materials properties (i.e. both the source and the impurity) themselves. Unfortunately, it is often more difficult to measure the impurity concentration in organics with the same precision achieved for inorganics such as Si and Ge. The best means to determine if the material has achieved the appropriate level of purity is the most practical: the residual contamination following purification should not measurably affect the host material or the devices of which they are comprised, including their electrical, morphological or optical properties from batch-to-batch or over extended periods of use.

In the following section we will discuss means for achieving high materials purity, followed by discussions on how to deposit films in a desired morphology while maintaining the purity of the source. We will also discuss several methods for evaluating both purity and morphology of bulk and thin film materials.

5.3 Materials purification

Virtually all effective means for purification rely on separating the host from its contaminants based on differences in molecular weight or size. Thus, it is easier to achieve highly refined substances when they are *monodisperse*, that is, they have a single, well-defined molecular weight or isomeric content. Monomers and oligomers provide excellent candidates for this type of purification process. In contrast, the indeterminate chain lengths of polymers result in a similarly indeterminate molecular weight and size, making their purification correspondingly more difficult. In the following, we will focus on the most effective and frequently employed means for purification of both classes of molecules. We note that to achieve high purity, often one or more techniques are sequentially employed, each being chosen due to a combination of effectiveness in removing specific impurities, or efficiency in reducing the processing time (and hence, cost). For this discussion, we assume that the materials have gone through the standard regime of purification during synthesis including distillation, precipitation, filtration, etc. The purity of materials used in organic electronics following synthesis is generally specified to be ~90% to 95%, although far lower purities are typically encountered. Moreover, this level of purity is far too low for mass-produced components whose operating characteristics fall within an acceptably narrow range to produce high yield, long lived, and high quality electronic appliances.

5.3.1 Zone refining

The highest purity organic crystals have been achieved via *zone refining*—a situation that also obtains for inorganic semiconductors (Hoesterey and Letson, 1963, Warta et al., 1985). Zone refining is suitable for materials that undergo a liquid phase when heated. Thus, its application has been limited since a large fraction of the useful organic semiconductors go directly from the solid to the gas phase on heating at atmospheric pressure or below (i.e. they *sublime*). Zone refining takes advantage of the difference in the saturated impurity concentration in the source between its liquid and solid phases. The equilibrium concentration of an impurity at temperature, T, is given by its *distribution* (or *segregation*) *coefficient*, k. That is, if c_S is the mole fraction of the impurity in the solid and c_L is that in the liquid, then,

$$k = \frac{c_S}{c_L}. \tag{5.2}$$

By melting a local region of a solid ingot of the source material in a thermally resistant boat (consisting of graphite, quartz, alumina, etc.), and then translating the melted zone along its length, impurities either collect in the melted or the solid zone, depending on whether $k < 1$ or $k > 1$, respectively. That is, when $k < 1$, the mixed impurity/source phase has a lower melting point (at temperature, T_M) than the pure host, and hence the impurities remain in the melted zone as it is slowly, thermally dragged along the length of the ingot, as shown in Fig. 5.2. For an impurity diffusion constant D_I and a melt interface of width w, the speed of zone translation must be kept below $v = D_I/w$ to assure efficient segregation of the impurities from the source. Once the end of the ingot is reached, the zone containing the highest concentration of impurities is removed (Fig. 5.3). The process can be repeated multiple times to continually increase the purity of the source.

This method of isolating impurities can be applied as long as their concentration is low, and that different species do not interact to impact their several distribution coefficients in the host. Furthermore, if the contaminant and host form a *eutectic*, zone refining cannot be employed. Eutectics, however, require a high concentration of both species where zone refining is ineffective.

The impurity concentration in the ingot at zone position, x, along its length L after the nth zone refining cycle is $c_n(x)$. Here, x is an integer index of the zone at that position along the furnace. Here, $x = 0$ indexes the initial zone, and N is the final zone in a furnace N zones long. It follows that the length of each melt zone is $l = L/N$. If the zone is translated a distance x, then the change in concentration in the solid zone is

$$dc_n(x) = k\left[c_{n-1}(x+1) - c_n(x)\right]dx, \quad 0 \le x \le N-1. \tag{5.3}$$

That is, $c_{n-1}(x+1)$ is the concentration in the next solid zone ($x + 1$) that achieved a concentration, c_{n-1} in the previous cycle. Then, at the end of the ingot where $x = L$, we have in the last increment:

$$\frac{dc_n(x)}{dx} = \frac{1-k}{N-x}c_n(x), \quad N-1 \le x \le N. \tag{5.4}$$

After a single pass ($n = 1$), solutions to Eqs. 5.3 and 5.4 are, respectively,

$$c_1(x) = 1 - (1-k)\exp(-kx), \quad 0 \le x \le N-1, \tag{5.5}$$

and

$$c_1(x) = c_1(N-1)\cdot(N-x)^{k-1} \quad N-1 \le x \le N \tag{5.6}$$

Since $N > x$ and $k < 1$ under the most common circumstance where the melting temperature is lowered by the introduction of impurities, we see that $c_1(x > N-1)$ in the end zone (Eq. 5.6) is greater than $c_1(x)$ along the ingot (Eq. 5.5). A further implication is that, as $k \to 1$, then $c_1(N) \to c(x < N)$, that is, purification is ineffective since there is no means to segregate the impurity by melting the host.

As previously, the zone refining process is iterative, with each additional cycle removing more of the remaining impurities. Hence it is interesting to determine the highest purity that can be attained after an unlimited number of zone refining steps. For this

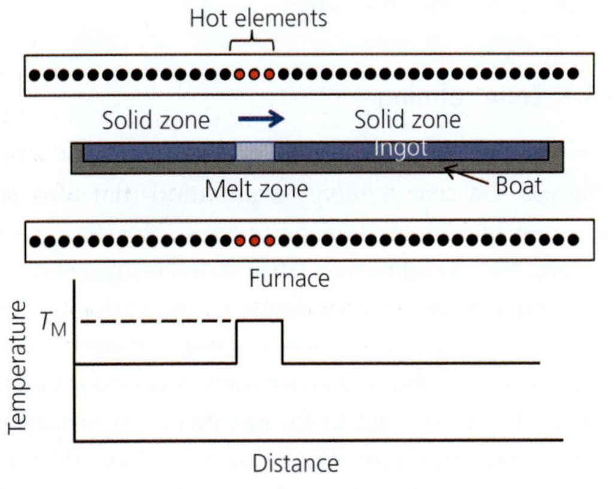

Figure 5.2 Schematic of a zone refining apparatus. The solid source material ingot is contained in a thermally resistant boat inserted into a linear furnace. A selected group of heating elements (red circles) is heated to induce local melting at temperature T_M in the ingot. The hot zone is then slowly translated along the length of the ingot (arrow) until it reaches one end. When $k < 1$, the impurity concentration will be highest at the terminus of the ingot. This section of the ingot is removed to eliminate the segregated impurities. The process is repeated until a suitably high purity is achieved. The temperature profile in the furnace at a given time in the process is shown in the lower graph.

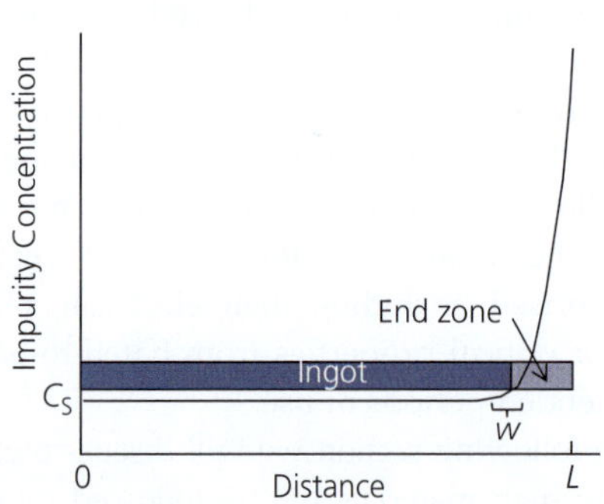

Figure 5.3 Impurity concentration profile of an ingot of length L following zone refining. Here, w is the approximate width of the liquid-solid interface during zone refining, and c_S is the concentration of impurities in the solid host following refinement.

calculation, we assume that $k < 1$, and the impurity concentration is sufficiently low that k is unaffected as the material purity is increased. Complications arise since the concentration c_n, depends on the concentration $c_{n\text{-}1}$ following the procedure for a single zone refining cycle (Helfand and Kornegay, 1966). Derivation of the highest purity obtained (at $n \to \infty$) for a given distribution coefficient can be found elsewhere (see Problem 4.1); here we will simply report the result (Pfann, 1966):

$$c_{n\to\infty}(x) = k_1 \exp(k_2 x), \tag{5.7}$$

where k_1 and k_2 depend implicitly on k, viz.:

$$k = \frac{k_2}{(\exp(k_2) - 1)} \tag{5.8}$$

and

$$k_1 = \frac{Nk_2}{(\exp(Nk_2) - 1)}. \tag{5.9}$$

As $x \to N$, the impurity concentration in the end zone rises exponentially. Since impurities are distributed along the length of the ingot, the ultimate material purity is determined by how much of the source material is discarded. This tradeoff is illustrated in Fig. 5.4 (Karl, 1980). Note that for $k > 1$, the impurity *increases* the melting point, and hence is preferentially incorporated into the solid. Then, the impurity concentration is highest at the beginning of the ingot, in contrast to $k < 1$ where impurities concentrate at the end of the ingot.

Finally, it is useful to determine how many zone passes are required to obtain an *acceptable* purity concentration. In Fig. 5.5, therefore, we show the purity for the case of $k = 0.1$ and the number of zones is $N = 5$, and the number of passes varies from $n = 1$ to ∞ (Karl, 1980). We observe that c_1 is nearly 10 times smaller than c_0, and that $c_9 \approx 10^{-6} c_0$. Clearly, for these conditions there is little reason to proceed beyond nine zone refinement steps, setting a practical limit on the level of purity attainable.

Zone refining can be implemented in both horizontal and vertical furnaces. Typically, the ingot is maintained in an inert atmosphere to prevent contamination of the source material due to environmental exposure at elevated temperatures. For many materials that sublime at atmospheric pressure, the sample can be refined by increasing the pressure such that the source undergoes a solid–liquid transition. To accelerate the refinement rate, many closely spaced zones are used, with several zones simultaneously heated above the melting point, separated by zones that allow the intervening regions to cool and freeze. In this way, multiple zones undergo refinement in a single pass. Very cold regions must be avoided since evaporation of the source and impurities from the heated zones will tend to condense on these cold spots. Then, subsequent passes of the hot zone will re-evaporate the condensate, potentially reintroducing contamination back into previously purified regions of the ingot. Finally, the hot zones must be kept at temperatures well below the decomposition point of the source material.

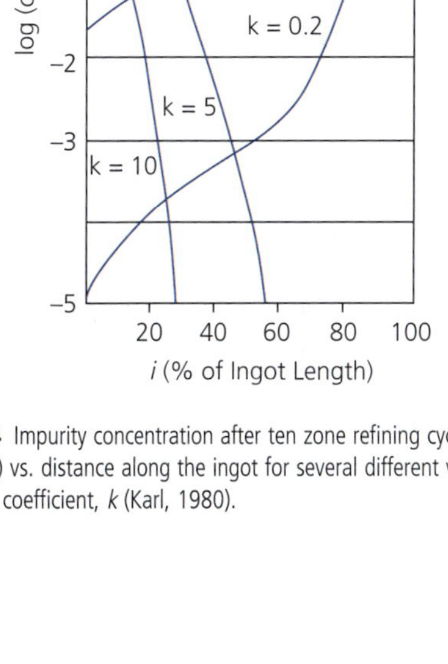

Figure 5.4 Impurity concentration after ten zone refining cycles (for $n = 10$) vs. distance along the ingot for several different values of distribution coefficient, k (Karl, 1980).

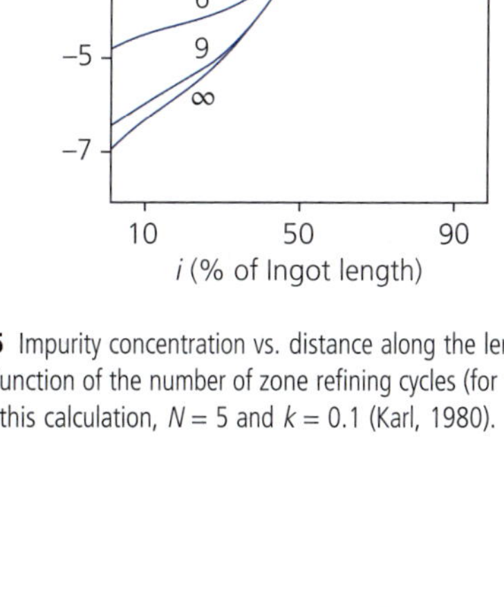

Figure 5.5 Impurity concentration vs. distance along the length of the ingot as a function of the number of zone refining cycles (for $n = 1, 3, 6, 9, \infty$). For this calculation, $N = 5$ and $k = 0.1$ (Karl, 1980).

There are several considerations that must be addressed to achieve the highest possible purities. The most important is the speed of zone movement. As noted, the speed cannot be higher than the rate of diffusion from the melt to the solid phases across the interfacial zone. Since many impurity diffusion coefficients range from $D \sim 10^{-3}$–10^{-6} cm^2/s, and a typical interface width is $w \sim 0.1$ mm in a precision zone refining apparatus, zone translation velocities of $v \sim 10^{-5}$–10^{-7} cm/s are optimal. Thus, a 10 cm long ingot may take as long as 2 years to purify! Furthermore, the rate must be sufficiently slow to avoid accumulation of crystal defects and even bubbles at the melt interface due to thermal expansion, rapid cooling, etc. Such defects can accumulate, or *getter*, impurities, thereby preventing them from following the melted zone. Zone refining is therefore a slow and laborious process requiring an automated furnace consisting of many (10–100) short, and precisely controlled temperature zones. In spite of these shortcomings, zone refining has the potential of achieving the highest materials purity.

Warta and Karl have produced what are arguably the highest purity, electronic grade organic crystals via meticulous zone refining of perylene and naphthalene combined with several pre-purification steps (Warta et al., 1985). In that work, the source material was initially purified by column chromatography and vacuum sublimation, to be discussed below. This was followed by an initial single zone refining step in an apparatus comprised of 100 separate zones that thermally transited the ingot at 1 cm/h. After this, the impurity concentrations of b-methylnaphthalene and thionaphthalene were at 1 ppm. At this concentration, $k \approx 1$, making further purification impossible. Thus, the source was subsequently exposed to molten potassium that reacts with the impurities to form products with significantly lower distribution coefficients. The material exposed to this impurity reaction process was then "ultrapurified" in three additional zone refining steps using an apparatus with 500 zones.

The purified materials were grown into single crystals from the melt via the vertical Bridgman process (see Section 5.4.1). The electronic properties of the crystals were characterized to assess their quality. This process, although time consuming, resulted in low temperature hole mobilities of 400 cm^2/V s at $T <$ 30 K for naphthalene, and 80 cm^2/V s at T = 30 K for perylene; the highest yet reported for an organic solid (Section 4.4.3). The electron mobility in perylene vs. T increases monotonically until T = 30 K, where it peaks at μ_e = 80 cm^2/V s, as shown in Fig. 5.6. It follows a power law dependence of $\mu \sim T^{-n}$ clearly indicative of band-like conductivity (Section 4.3). This is in striking

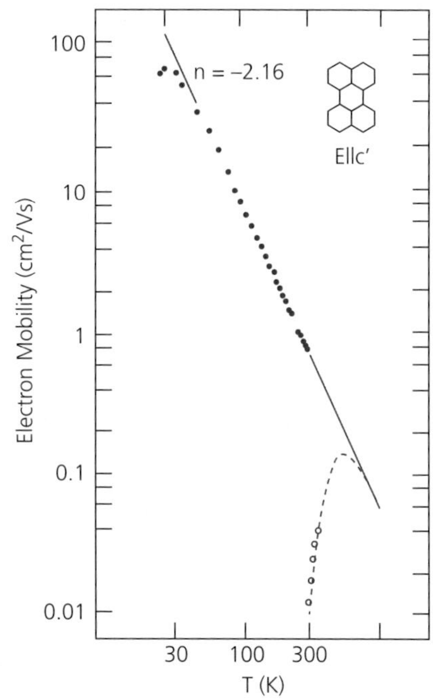

Figure 5.6 Electron mobility vs. temperature for a zone refined perylene crystal along its c′ crystal axis (closed dots). Also shown are data for the same material that was not zone refined (open dots). The mobility has a temperature dependence of $\mu \sim T^{-n}$ (Warta et al., 1985).

contrast to a sample that was not as thoroughly purified, with peak of $\mu_e \sim 0.08$ cm^2/V s at room temperature (Maruyama et al., 1973).

The effect of traps of density N_T in a host of density N_H on the mobility can be derived using Shockley-Reed-Hall recombination statistics to yield (Hoesterey and Letson, 1963, Sze, 1981)

$$\mu(T) = \mu_0(T)[1 + N_T/N_H \exp(\Delta E/k_B T)]^{-1}, \quad (5.10)$$

where μ_0 is the trap-free mobility, $\Delta E = |E_T - E_{FO}|$ is the difference in the trap (E_T) and the frontier orbital (E_{FO}) energies (i.e. the LUMO for electrons, or the HOMO for holes), and k_B is the Boltzmann constant. Fits to the data in Fig. 5.6 using this relation are indicated by the solid and dashed lines. The analysis yields ΔE = 270 meV and $N_T/N_H = 5 \times 10^{-4}$ mol/mol for the ultrapurified crystal, and 17.5 meV and 1.7×10^{-3} mol/mol for the lightly purified film. Hence, the trap density is not only lower in the ultrapurified sample, but it lies deeper in the energy gap. Hence, the traps are frozen out at low temperatures, leading to a significant thermal activation and an increase in μ_e. The source of the traps may be due to impurities or structural defects;

two factors that can be related. That is, interstitial defects can result in stacking faults, leading to trap energies lying deep within the energy gap that determines the temperature dependence of the mobility in Fig. 5.6. In any case, it is clear that zone refining can result in high purity crystals whose properties are largely determined by the *intrinsic* characteristics of the host rather than by extrinsic, impurity-related effects.

5.3.2 Thermal gradient sublimation

Since a significant number of materials used in organic electronic applications sublime rather than melt at atmospheric pressure or below, the most commonly used means for purification is *thermal gradient sublimation* (Gutmann and Lyons, 1981, Forrest, 1997). Thermal gradient sublimation is a variant of zone refining based on different distribution coefficients of the impurity and host during a solid-gas phase transition. A schematic of a gradient sublimation apparatus is shown in Fig. 5.7.

The purification process typically proceeds as follows. Several grams of the unpurified source material are loaded into the end of a long glass ampoule closed at one end, with the open end connected to a vacuum pump. A glass sleeve is placed inside the tube to capture the sublimed material, and a wad of glass wool or other semi-permeable membrane (e.g. an alumina plug) is positioned adjacent to the vacuum port to prevent the solid (powder) source material from blowing into, and contaminating the pump. The tube is evacuated to $< 1 \times 10^{-6}$ torr using a turbomolecular or another dry pump. Conventional organic pump oils used in rotary or diffusion pumps are avoided since they can back-stream into the ampoule and contaminate and/or react with the source material. The end containing the source material is inserted into a tube furnace that consists of several separately controlled zone heaters (typically between 3 and 5) that are adjusted to impose an approximately linear temperature gradient along the tube longitudinal axis. The source material lies in the hottest zone whose temperature should be lower than the source decomposition temperature.

To begin the process, the maximum temperature is adjusted until the pressure within the tube shows a slight increase, indicating that the evaporation of the most volatile impurities has commenced. Once the pressure returns to the base pressure of the cold system (often taking several hours or more), the temperature is incrementally increased across the gradient once again, and the process of waiting for the pressure to return to its base value is repeated. This cycle is iterated until the purified source material is condensed along a band within the sleeve in the glass tube, and the pressure no longer rises significantly when the temperature is incrementally raised one last time. At this point, most volatile impurities are either condensed in the tube cold zones, or have been extracted by the vacuum pump. The less volatile impurities remain in the hottest zones, or at the position where the source material was originally placed at the closed end of the tube.

The sleeve insert is then removed, and the deposit is extracted and used as the source material by placing it in a freshly cleaned ampoule for a subsequent purification cycle carried out in the same way. The purest organic compounds are obtained after multiple cycles, when there is no significant pressure increase after incrementally raising the furnace temperature. After

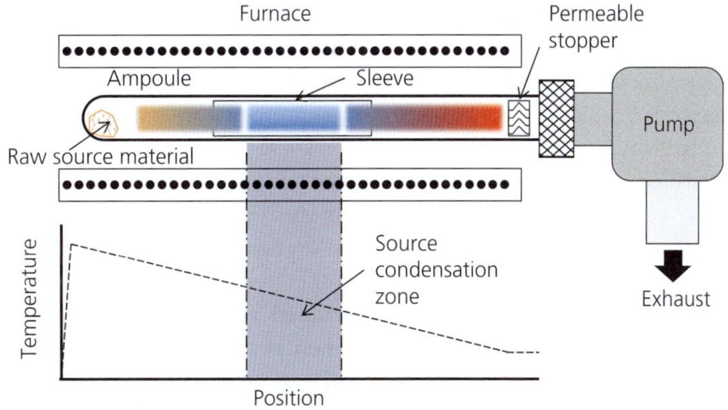

Figure 5.7 Schematic of a vacuum thermal gradient sublimation apparatus. The furnace consists of several (typically 3–5) independently controlled thermal zones to create a temperature gradient. The purified material condensate is captured in a glass or quartz sleeve inserted into the appropriate zone within the furnace. The purified source is shown in blue; other colors along the tube length suggest condensation of impurity-rich fractions.

concluding the process, typically ~50% of the original source material has been removed, and the background pressure of the sublimation system approaches $\sim 10^{-7}$ torr. Often there are losses of the source along with the impurities, making this process somewhat inefficient.

The success of sublimation purification requires that there is a significant difference between the sublimation temperatures of the source and its impurities. This is analogous to requiring a difference in distribution coefficients for zone refining. Furthermore, the sublimation and decomposition temperatures of the host must also be substantially different to avoid degradation of the source. Given the poorly defined width of the source condensation zone (see Fig. 5.7), impurities and host are often intermingled, particularly at the zone edges. This problem can be somewhat reduced although not eliminated by increasing the temperature gradient on each purification cycle. If the process is done quickly, the host condensate will be amorphous. If it proceeds more slowly, single crystalline flakes can deposit within the sleeve. Finally, it is important that the temperature gradient, particularly in the region both in front of, and behind the starting source material, monotonically decrease toward the cold end of the ampoule. This ensures that condensation at a cold, closed end of the tube does not scavenge material that would otherwise be deposited within the capture sleeve.

The differences between single cycle vacuum gradient sublimed and unpurified CuPc materials used in organic photovoltaics are shown in Fig. 5.8. Using a combination of mass spectrometry and Fourier transform infrared (FTIR) spectroscopy, it is apparent that the unpurified source material (labeled as 98% pure by mass, by the manufacturer) has a high concentration of the metal free phthalocyanine (H_2Pc) along with some CuPc fractions that are completely absent in the purified material (Salzman et al., 2005).

Interestingly, film deposition by *vacuum thermal evaporation* (VTE, see Section 5.4.2) can also be used for thermal gradient purification, although this is inefficient. When unpurified or partially purified material is evaporated in vacuum prior to, or even during thin film deposition, the first deposit will have a higher concentration of volatile impurities than in subsequently deposited films since the source is partially purified by the previous deposition cycles. Purification during thermal evaporation results in an increase in the room temperature hole mobilities of a series of CuPc films sequentially deposited in vacuum from the same source, as shown in Fig. 5.9 (Salzman et al., 2005). Between the first and sixth

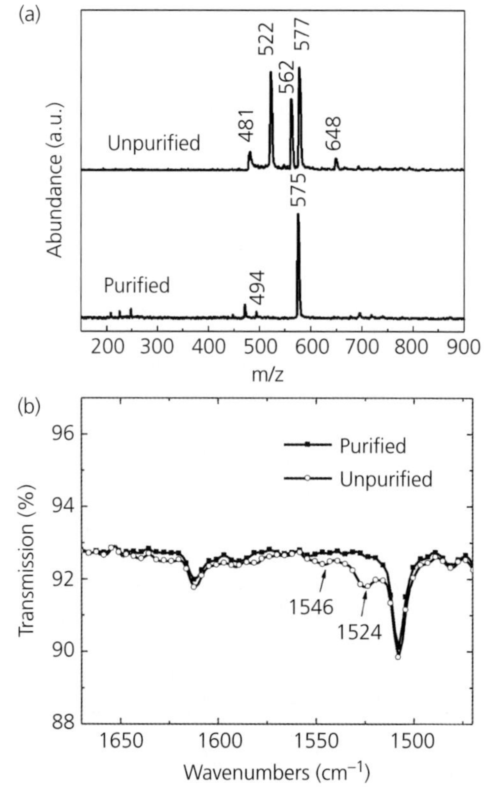

Figure 5.8 (a) Mass (in atomic number, Z) and (b) infrared spectra for CuPc as-purchased, and after a single purification cycle by thermal gradient sublimation of CuPc. The additional peaks in the unpurified samples are associated with H_2Pc (at $m = 522$ amu) and other impurity fractions compared to the purified CuPc, with peaks at $m = 575$ amu in the mass spectra and infrared peaks at ~1510 cm^{-1} and 1615 cm^{-1} (Salzman et al., 2005).

depositions from an originally unpurified batch of CuPc, the mobility increases by nearly three orders of magnitude, from 4×10^{-7} cm^2/V s for growth #1 to 3×10^{-4} cm^2/V s for growth #6. These data indicate the effectiveness of purification via sublimation even under the poorly controlled conditions of *in situ* VTE.

There are numerous variants to the process described above. For example, sublimation can occur in an inert gas environment as opposed to vacuum (Wagner et al., 1982, Laudise et al., 1998). Sublimation often results in purified single crystal flakes, discussed further in Section 5.4.1. Sublimation in an inert gas has been used to achieve both pristine and doped materials with exceptional electrical properties, suggesting the effectiveness of this method (Laudise et al., 1998, Hiramoto and Sakai, 2008). A particular advantage of using an inert carrier gas is the reduction of condensation behind the source in

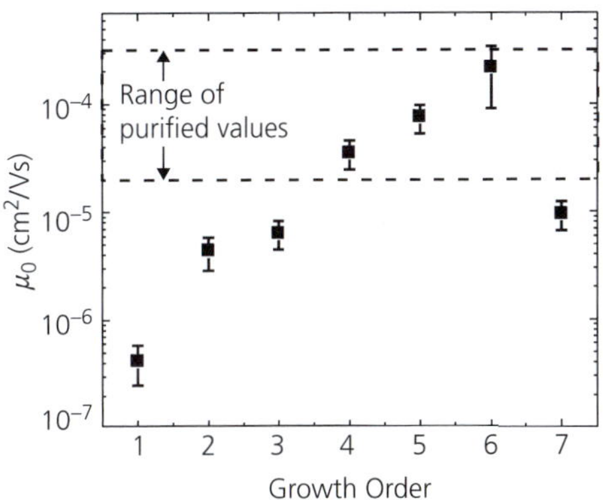

Figure 5.9 Hole mobility (measured via space charge limited current–voltage characteristics, Section 4.4.2) vs. order of thin film growth by vacuum deposition of an unpurified CuPc source. Also shown is the range of mobility obtained for CuPc purified by a conventional single thermal gradient sublimation step (Salzman et al., 2005).

Reprinted from Organic Electron., 6, 242, Salzman, R. F., Xue, J., Rand, B. P., Alexander, A., Thompson, M. E. & Forrest, S. R., The effects of copper phthalocyanine purity on organic solar cell performance. Copyright 2005 with permission from Elsevier.

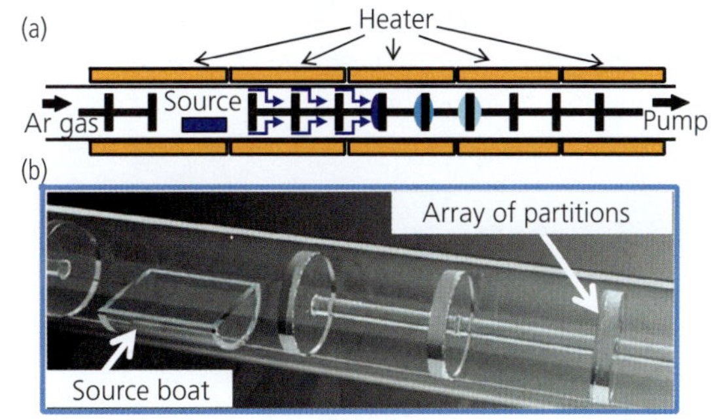

Figure 5.10 Gradient sublimation apparatus using a five-zone furnace with a series of partitions inserted within the thermal gradient. The various material fractions found in the source selectively deposit on baffles located at their particular condensation temperatures. The baffles located in zones where source condensation is favored contain the purified material (Jeon et al., 2008).

Reprinted from Org. Electron., 9, 903, Jeon, H.-G., Inoue, M., Hiramatsu, N., Ichikawa, M. & Taniguchi, Y., A modified sublimation purification system using arrays of partitions, copyright 2008 with permission from Elsevier.

the cold region at the furnace opening by forcing mass transport along the gas flow direction. Using an inert carrier gas can also increase the rate of material transport, hence leading to a shorter time for purification.

Gradient sublimation can also be achieved by completely filling the ampoule with a plug of source material, and then vaporizing a single zone within the plug. Moving the vaporized region toward one end of the plug is analogous to zone refining. It moves those impurities with different distribution coefficients in the vapor vs. the solid to an extreme end of the plug once the zone has been completely transited by the thermal gradient (Weinberg and Rossi, 1960). Care must be taken to avoid decomposing the source, or in translating the hot zone too rapidly such that there is incomplete segregation of the impurities.

To further improve impurity separation, a series of partitions, or baffles, can replace the inner sleeve along the length of the thermal gradient, as shown in Fig. 5.10. The baffles "digitally" separate the impurity fractions by allowing condensation to occur in discretely defined temperature zones along the thermal gradient. Since there is condensation both on the partitions and on the furnace tube walls, a tradeoff must be made between material yield and the level of purity of the refined source material (Jeon et al., 2008).

5.3.3 Solution-based purification methods

There are several methods of purifying materials using solvent-based mass selection, distillation and filtration. These techniques have the advantage of being rapid compared to vapor sublimation and zone refining, and are compatible with both small molecules and polymers. Their disadvantage is that the level of purity achieved is not nearly as high as the upper limits attained using zone refining and thermal gradient sublimation. As noted above, solvent purification is often used as a means for preparing crude source materials for subsequent purification by other more rigorous methods.

Solution based purification has many variants. Here we discuss the three most commonly used in the refinement of organic electronic materials: column chromatography (or simply, chromatography), *size exclusion chromatography* (SEC), and *solvent extraction*. It will become apparent that the principles of these methods are not significantly different than those of sublimation and zone refining: the concept relies on spatially segregating unwanted impurities from the host material by leveraging differences in their ability to be dissolved in a particular solvent. Hence, solvent selectivity (versus temperature selectivity) becomes the determining factor that leads to purity of the source.

Column chromatography was introduced in the early twentieth century by M. S. Tswett who was developing techniques for separating chlorophyll from plants (Gehrke et al., 2001). The process relies on introducing a solution containing the impure source in a column packed with a porous material that selectively captures various solute impurities for different durations. A schematic of the process is shown in Fig. 5.11. The unpurified source is dissolved in a solvent, called the *mobile phase*. This solution is

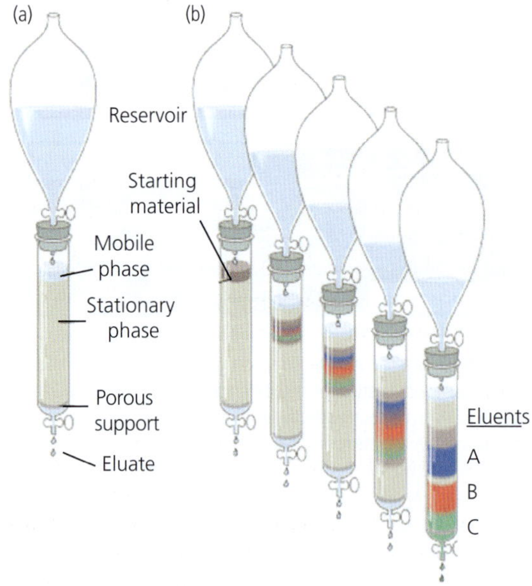

Figure 5.11 Schematic of column chromatography distillation. (a) Initial set up; (b) time progression of eluents as they move along the column, resulting in separate bands (labelled A, B, C) moving at different velocities that are collected at the exit of the column (Ebioe, 2010).

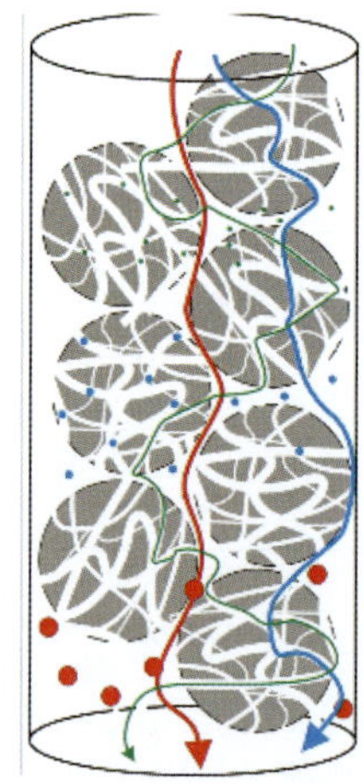

Figure 5.12 Schematic of the size exclusion chromatography separation process. The column is filled with beads of varying pore sizes. Small particles penetrate the smallest pores (green line) while the largest particles pass between the beads. The length of the path taken determines the duration a given size population spends within the column: smaller particles take the longest, and the largest particles transit at the highest average velocity (Li et al., 2013).

placed on top of a column packed with a solid absorbent called the *stationary phase*. Typically, the absorbent is comprised of silica (SiO_2) or alumina (Al_2O_3) beads only a few tens of microns in diameter. The molecular species and their solvents are trapped for differing lengths of time on the particles in the column, depending on their polarizability. Hence, highly "sticky" dipolar fractions, known as *eluents*, will take longer to travel along the column to its exit than will the less polar eluents. By capturing the eluate at different times following their introduction into the column, different materials are collected, thus allowing for separation of the purified source (McGhie, 1974).

When gravity is relied upon to move the solution along the column, the process is known as *gravity column chromatography*. This method tends to be slow, although multiple large diameter columns operating in parallel can be used to significantly expedite the purification of large volumes of the source. Applying pressure with an inert gas can significantly accelerate the process. This is known variously as *flash chromatography*, or *high performance* (or *high pressure*) *liquid chromatography* (HPLC).

The effectiveness of column chromatography depends on the degree to which impurity and source transit times through the column differ. This in turn depends on the adsorption rates on the beads within the column, and the polarity (or non-polarity) of the eluent. Indeed, improved selectivity can be obtained by slowly varying the polarity of the solvent introduced from the reservoir as the process evolves. While this process is useful for both polymers and small molecules, the degree of segregation of source from impurity is never high since non-polar impurities are uniformly dispersed along the column, and the distribution of any given species along the column is a rather broad Gaussian determined by its particular retention time.

Size exclusion chromatography is effective for separating polymers according to their hydrodynamic volume (Moore, 1964). Size exclusion (also known as *gel filtration* or *gel permeation chromatography*) is primarily used to narrow a polydisperse distribution of macromolecules, such as polymer chains of varying mass. It is also capable of separating low molecular weight impurities from the source. Like column chromatography, the process sorts molecules based on differences in their transit times through a column of porous beads, as illustrated in Fig. 5.12. For example, a mixture of polymers of varying molecular weights is introduced into the top of the column. Unlike column chromatography, the molecules themselves do not interact with the beads, that is, there is no trapping or sticking on the surface of the stationary phase that retards their transit. Rather, the size of the molecule determines the path length that is sampled: larger molecules that cannot penetrate into the pores of the spheres transit the column via the interstices between them, whereas smaller molecules can enter the pores within the beads (as well as the interstices) and hence will sample a greater volume of the column before exiting. The largest particles exit first, followed by distributions of successively smaller molecules, thereby effecting their separation. The eluent populations are Gaussian distributions of molecular size, and the pore sizes define the resolution limits of the process. Molecular populations with masses above the exclusion limit are not differentiated by transit times, with a similar situation obtaining for molecules smaller than the smallest pores in the stationary phase. In that case, the molecules are smaller than the permeation limit.

The transit times in SEC are accurately determined using the hydrodynamic volume (V_H) of the molecules. Here, V_H is the volume of the polymer coil in the solvent that is determined by the details of the interactions between the polymer and solvent. The hydrodynamic volume can written in terms of the Einstein viscosity law, viz. (Grubisic et al., 1967)

$$V_H = K\eta M_w. \tag{5.11}$$

Here, η is the viscosity, K is a constant, and M_w is the molecular weight. That is, the viscosity of the material in solution, which determines its transit time through the column, is directly proportional to its hydrodynamic volume. Indeed, this relationship holds for polymers (and even mixtures of polymers and small molecules) independent of their shape, chemical composition or morphology as long as they do not undergo electrostatic or other interactions with the beads in solution, as is typically the case in conventional column chromatography.

The resolution of the process is determined by the differences in particle velocity, which is a function of temperature. Since separation is determined by the elapsed time in the column, care must be taken to introduce the entire population of source and impurity molecules at one time. As with other methods, the best results are obtained when the crude material is first purified by other means before being sorted by SEC. Due to its simplicity and utility in purifying and isolating molecular populations by mass, SEC is frequently employed in the refinement of polymers (Ashraf et al., 2013, Mateker et al., 2013).

Another solution purification technique that is used for both large and small molecules is solvent extraction. Extraction is a means of filtration, or washing of the host materials, and hence is effective when used as a pre-purification step of the crude source material. A convenient implementation uses a Soxhlet distillation apparatus, shown in Fig. 5.13. The crude source material is placed in a semi-porous extraction thimble consisting of alumina, or filter paper. A high boiling point, pure solvent is placed in the lower reservoir. Extraction can proceed by two paths. If the desired product is insoluble, it remains in the thimble, while the impurities wash through to be collected in the reservoir. Alternatively, if the product itself is soluble, it is dissolved and collects in the reservoir.

Solvent extraction has been used in the purification of PTCDA that is insoluble in xylene (with a boiling point of 140°C). Purification then proceeds as follows: The solvent is heated to its boiling point. The vapors travel up the tube on the left in Fig. 5.13 and are cooled in the condenser. The distilled solvent then flows into the upper reservoir where it dissolves the impurities in the source powder. This solvent/impurity solution flows out of the thimble, and into the upper reservoir. With time, the upper reservoir fills to the top of the return tube, at which point the reservoir siphons into the lower reservoir, and the extraction cycle is repeated. In the case of PTCDA (a red powder), the xylene in the lower

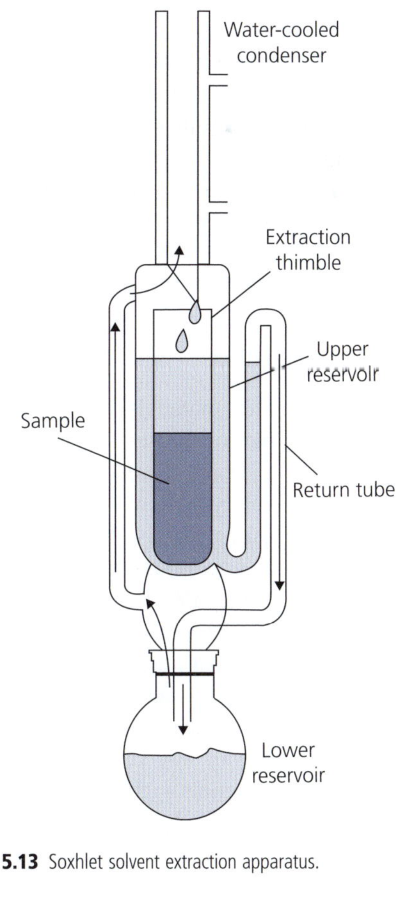

Figure 5.13 Soxhlet solvent extraction apparatus.

Table 5.1 Several high boiling point (BP) solvents useful in solvent extraction

Solvent	BP (°C)	Solvent	BP (°C)
Acetic acid	118.0	Dioxane	101.0
Acetic acid anhydride	139.0	Heptane	98.4
Isobutanol	107.7	*n*-Propanol	97.2
n-Butanol	117.7	Pyridine	115.3
Chlorobenzene	131.7	Toluene	110.6
Dimethyl acetamide	166.1	Water	100.0
Dimethyl formamide	153.0	Xylene	140.0
Dimethyl sulfoxide	189.0		

reservoir becomes saturated with impurities after several cycles, turning from clear to an opaque greenish color. Once the solvent becomes excessively contaminated, it is replaced by fresh liquid. This process is continued until there is no more visible contamination of the solvent after several reflux cycles; a procedure that may take hours or days, depending on the level of contamination of the crude material. At the end of the extraction process, the remaining material is dried and prepared for use, or for further purification.

To remove as many soluble impurities as possible, both polar and non-polar solvents may be sequentially used. Table 5.1 lists several common solvents used in extraction purification. This method can only succeed if there is a difference in solubility between the source and the impurities. Hence, high purity materials can only be obtained when solvent extraction is used in conjunction with other, more rigorous processes such as thermal gradient sublimation or zone refining.

5.3.4 Ultracentrifugation

Centrifugation separates materials having different hydrodynamic volumes or masses suspended in solution by applying a strong, directional force to the suspension by rapid rotary motion. *Ultracentrifugation* is a class of centrifugation where >5000*g* (where *g* is the acceleration due to gravity) is applied to the sample. Another variant known as *density gradient ultracentrifugation* (DGU) (Miller et al., 2014) is employed in separating semiconductor and metallic nanoparticles useful in colloids, and in separating single from multi-walled carbon nanotubes (CNTs). In principle, centrifugation can separate any sufficiently polydisperse population of materials such as polymers, proteins, and other macromolecules.

An illustration of DGU separation is shown in Fig. 5.14. A vial containing a fluid mixture with a

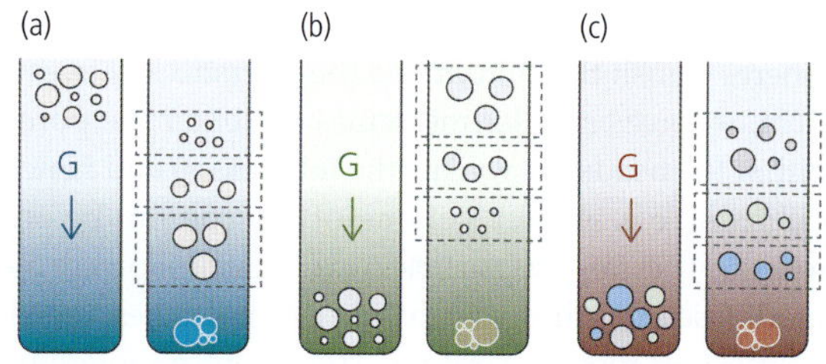

Figure 5.14 Density gradient ultracentrifugation. A vial is filled with a solvent whose density is graded upward from its maximum at the bottom. G is the centrifugal force applied to a polydisperse population that is introduced at the top of the vial. After centrifugation, the volume-disperse population stratifies along the density gradient for particle densities that are: (a) higher, or (b) lower than that of the solvent. (c) Isopycnic DGU for particles of different *densities* (indicated by different color circles) that lie within the density extremes of the solution. The material at the bottom of each vial corresponds to separations of high density impurities (Miller et al., 2014).

density gradient is loaded with the population of particles with a range of hydrodynamic volumes or masses. If the particles are less dense than the liquid, they float to its surface; otherwise they sink (Figs. 5.14a and b, respectively). Ultracentrifugation at ~20,000–80,000 rpm results in stratification of the population into regions of different liquid density. If the population itself consists of particles of different densities, they will separate from (starting at the bottom) the most to the least dense species. High density impurities are trapped at the bottom of the vial. A micropipette is inserted at various distances within the container to extract the particles within a given stratum, thereby completing the separation process.

DGU creates a particle flux with terminal velocity $v\hat{\mathbf{z}}$ along the direction of acceleration, $G\hat{\mathbf{z}}$ of (Miller et al., 2014)

$$v = \frac{2GR^2}{9\eta_s}(\rho_c - \rho_s). \tag{5.12}$$

Here, the solution with density ρ_s and viscosity η_s contains a colloid of density ρ_c and radius $R \sim V_H{}^{1/3}$. From Eq. 5.12, successful separation depends on a large density "contrast" between solution and colloid, and that volumetric differences within a population will result in large spatial separations. For $\rho_c > \rho_s$ (Fig. 5.14a), particles with smaller radii move more slowly than larger particles, and collect near the top of the fluid. For the opposite relationship between ρ_c and ρ_s, the velocity is negative, and the larger particles drift toward the top (Fig. 5.14b), with the smallest particles moving at the highest terminal velocity. *Isopycnic separation* in Fig. 5.14c segregates particles of different densities along the solvent gradient independent of size.

The degree of separation between particles of different radii along the length of the vial can be obtained assuming conservation of mass of an incompressible fluid. Then the radius is:

$$R(z) = \sqrt{\frac{9\eta_s z}{2Gt|\rho_c - \rho_s|}}, \tag{5.13}$$

where z is the position. It is thus important to halt the process at a time, $t < t_M$, where the largest (smallest) particles in the solution with $\rho_c > \rho_s$ ($\rho_c < \rho_s$) have yet to travel the entire length of the vial. Note that since R depends on the hydrodynamic volume of the particle population, the particle shape also determines its position as a function of centrifugation time. For example, elongated nanowires are expected to distribute differently than would nanospheres. Indeed, in mixed particle shape populations, it has been shown that the increased friction of the wires leads to their collecting more toward the top of the solution as compared to spheres (Bai et al., 2010). Using *analytical centrifugation* where optical or other measurement tools are employed during the rotational separation, the particle size distribution, $R(z)$ can be determined in real time, adding a degree of control to the process.

An example of separation of a polydisperse population of Au nanoparticles of varying diameters in a gradient of a mixture of cyclohexane in tetrachloromethane (CCl_4) is shown in Fig. 5.15 (Bai et al., 2010). The solvents are useful in creating the separations, while also eliminating clustering by preventing contact between the nanoparticles which would otherwise adhere to each other due to their attractive van der Waals forces.

The Au nanoparticle population stratification was obtained by centrifugation for 12 min at 25,000 rpm (corresponding to 80,000g). The size distributions before and after ultracentrifugation are shown at the lower right in Fig. 5.15b. The width of each diameter distribution is <1.5 nm, whereas the starting population had a width of ~12 nm.

Semiconducting nanoparticles suspended in polymer matrices have found uses in light emitting devices (Coe et al., 2002) and photodetectors (Konstantatos and Sargent, 2010) (see Chapters 6 and 7). An advantage of DGU is that the polymer matrix can be dissolved in the solvent density gradient along with the particles forming the colloid. Both the polymer and the selected fraction of the distribution can be simultaneously decanted and subsequently used in devices. Figure 5.16 shows the results of DGU of a population of CdSe nanoparticles mixed with polystyrene in a solution gradient used in their separation. The nanoparticles range in radius from 2.5 nm to 5 nm.

DGU has also been used in separating populations of CNTs of varying diameters and conductive properties. Indeed, one of the major roadblocks to employing CNTs for electronic applications is often only a single type of tube is needed for a particular purpose. The efficient separation of tubes according to their chirality (and hence their conductive properties), and the elimination of bundles and clusters has impeded their widespread use in organic electronics applications. Nevertheless, mixtures of single-walled CNTs of different diameters and chiralities along with

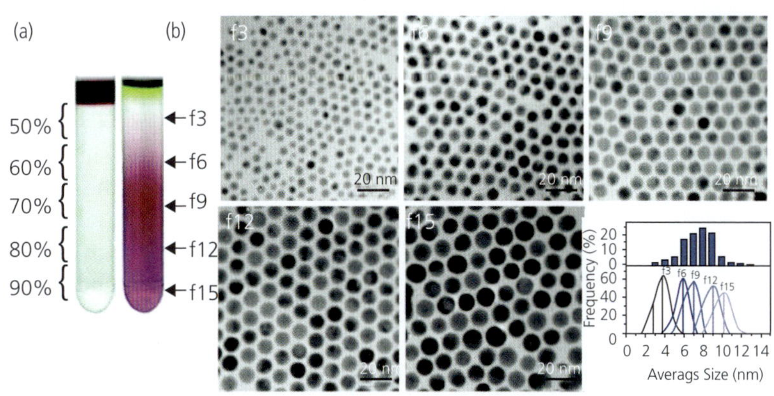

Figure 5.15 (a) A stratified density gradient obtained by varying the mixture of two solvents, and then layering the mixtures in the vial. The positions of the bands of concentration of CCl_4 mixed in cyclohexane are indicated, with the Au colloid on the surface (dark band). The stratified Au nanoparticle gradient after DGU is shown in the image on the right. (b) Images of the Au nanoparticle populations extracted from the corresponding positions in the vial. Scale bars are 20 nm. A histogram of the diameter distribution before DGU is shown in the upper graph at the lower right. The diameter distribution for each population shown after DGU is plotted in the lower graph (Bai et al., 2010).

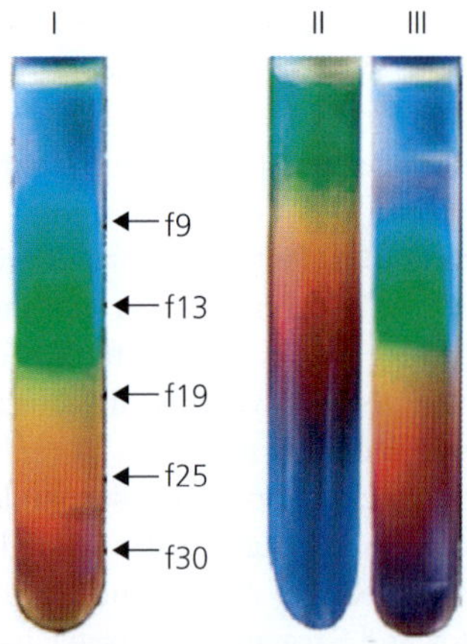

Figure 5.16 Luminescence images from CdSe in polystyrene nanoparticle colloids following DGU in (I) a gradient of CCl_4 and cyclohexane spun at 50,000 rpm for 60 min, (II) as in (I) but containing polystyrene and spun for 60 min, and (III) as in (II) but spun for 110 min. Blue luminescence is from the smallest CdSe spheres, red from the largest (Bai et al., 2010).

Reprinted with permission from Bai, L., Ma, X., Liu, J., Sun, X., Zhao, D. & Evans, D. G., Rapid separation and purification of nanoparticles in organic density gradients. J. Am. Chem. Soc., 132, 2333. Copyright 2010 American Chemical Society

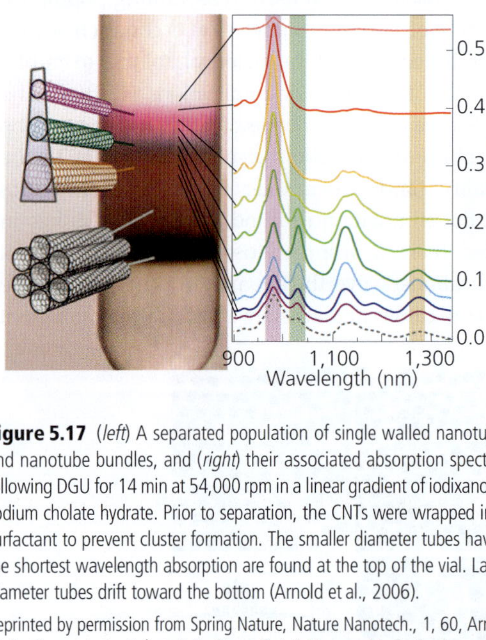

Figure 5.17 (*left*) A separated population of single walled nanotubes and nanotube bundles, and (*right*) their associated absorption spectra following DGU for 14 min at 54,000 rpm in a linear gradient of iodixanol and sodium cholate hydrate. Prior to separation, the CNTs were wrapped in a surfactant to prevent cluster formation. The smaller diameter tubes having the shortest wavelength absorption are found at the top of the vial. Large diameter tubes drift toward the bottom (Arnold et al., 2006).

Reprinted by permission from Spring Nature, Nature Nanotech., 1, 60, Arnold, M. S., Green, A. A., Hulvat, J. F., Stupp, S. I. & Hersam, M. C., Sorting carbon nanotubes by electronic structure using density differentiation. Copyright 2006

bundles wrapped in surfactant molecules, have been effectively separated by DGU as in Fig. 5.17 (Arnold et al., 2006).

5.4 Bulk crystal and thin film growth

In this section we describe several methods to achieve high quality, ordered bulk and thin film organic crystals, as well as amorphous films. The quality of the film ultimately depends not only on its purity, but also on its morphology. Since high purity is prerequisite to good device performance, we will henceforth assume that the materials source is adequately purified. Then, structural order results in improved energy (i.e. exciton) and charge transport, whereas amorphous materials often have higher luminescent efficiencies. Controlled doping during growth to increase the conductivity or modify the luminescent properties of materials is another means for achieving the desired properties. Hence, morphological control is an essential ingredient of all optoelectronic devices. Here, we concentrate on many ways to achieve this in bulk crystals as well as in thin films using vapor or liquid phase growth.

5.4.1 Bulk single crystal growth

Single crystals of covalently bonded materials are generally grown from the melt, and then the crystals are cut into thin slabs, or wafers. The wafers serve as seeds to the active epitaxial layers subsequently grown on their surfaces. Single crystalline layers are achieved via matching of the epitaxial lattice structure to that of the substrate (Section 2.6). This limits the choice of materials that can be compatibly grown without introducing strain defects into the epitaxial layer.

Due to the far higher elasticity of organic materials arising from their electrostatic (van der Waals) adhesive forces, a strict condition of lattice matching is not as important to achieve unstrained growth. Furthermore, given the enormous diversity of organic molecules and their equally diverse crystalline structures, it is often difficult to match the lattice of the epitaxy to that of the bulk crystal. When the compositions of the epitaxy and the substrate are the same, the growth is known as *homoepitaxy*. In contrast, *heteroepitaxy* refers to the situation when the compositions differ. Although the growth of bulk organic crystals has not received substantial attention in recent years, very pure, large ($\sim$1 cm^3 or larger) single crystals are useful for understanding their fundamental electronic, linear and nonlinear optical, and mechanical properties, and their dependence on crystal direction. Single crystal growth also finds uses in laboratory-scale devices that can approach the fundamental limits to performance.

There are two principal methods for growing bulk crystals: *Bridgeman* and *Czochralski growth*. The Bridgeman method starts by inserting a sealed ampoule containing the purified source material into a tube furnace that supports an abrupt, although small (<1°C) temperature step along its length (see Fig. 5.18 for three possible vertical Bridgeman configurations) (Karl, 1980). One region is cooled to just below the solidus temperature of the melt, the other, just above. This sharp temperature gradient can be created by using a multi-zone, or bias-wound furnace (Fig. 5.18a), or by placing a reflecting surface outside of the furnace windings that intensify the heat within its volume, as in Fig. 5.18b. Alternatively, in Fig. 5.18c, we show the ampoule inserted into a container with two immiscible liquids, and surrounded by a dual-zone heater. If the melting temperatures of the materials in solution are below that in the ampoule, they will conduct heat to the ampoule without intermixing (provided the temperature difference is small), thereby sustaining the needed temperature step gradient.

Zone melting is a variant of the Bridgeman technique. It uses a multi-zone furnace similar to that used in zone refining shown in Fig. 5.2. The crystal is grown and simultaneously purified by slowly translating the melted zone along the length of a polycrystalline or amorphous ingot.

In the Bridgeman method, a meniscus forms at the solid–liquid interface. The temperature step is slowly scanned (~1 mm/h) from one end of the ampoule to the other by either changing the position of the ampoule or the reflector. This increases the size of the crystal as the cooler region expands along its length. Great care must be taken to scan the growth front sufficiently slowly to ensure single crystal formation. Figure 5.19 shows an anthracene crystal grown by the Bridgeman method.

While in principle the Bridgeman process is simple, it has several shortcomings. Seeding the crystal growth occurs via spontaneous condensation of the supercooled solution at a nucleation site (e.g. an imperfection on the surface of the ampoule), and thus it is not possible to choose the crystal orientation along which growth proceeds. However, growth along a particular crystalline direction can be induced using a seed crystal inserted at the cold end of the ampoule, and the temperature is adjusted such that the seed does not fully dissolve at the onset of growth.

Due to self-seeding, the crystal can adhere to the ampoule wall, and must be carefully removed by either slight dissolving in solution, breaking the ampoule, or using a very thin knife blade inserted between the crystal and the ampoule wall; all methods that present risks of damage. Using an ampoule material whose coefficient of thermal expansion is smaller than that of the material it contains can also aid in separation on cooling to room temperature. However, thermal stresses introduced at the crystal/ampoule interface can propagate into the crystal, resulting in dislocations. Finally, as the growth front moves, impurities are segregated in the same manner used in zone refining. Given the different distribution coefficients of the impurity and the source material, the Bridgeman process does not allow for doping of the single crystals.

Many of these shortcomings are eliminated in Czochralski growth, commonly used for producing inorganic semiconductor single crystal boules. It was first introduced for organic crystals by the growth of benzyl and benzophenone (Bleay et al., 1978). The growth of a single crystal commences from a seed crystal positioned at the tip of a cooled needle, as shown in Fig. 5.20. The seed is lowered into a crucible heated to just above ($\Delta T \sim$ 0.2–2°C) the melting temperature of the purified source material. It is then slowly withdrawn at speeds of 0.1–1 mm/h from the melt into a

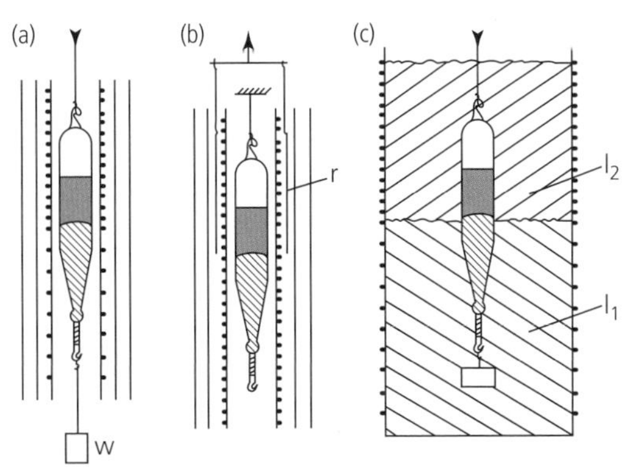

Figure 5.18 Three configurations used in the vertical Bridgeman process. Suspension of an ampoule of material in a (a) bias-wound vertical furnace, (b) furnace with a translating cylindrical reflector; and (c) a reservoir containing two immiscible liquids, l_1 and l_2. Here, w = weight used to draw the ampoule downward. The melt is indicated by the blue region, and the solid is cross-hatched.

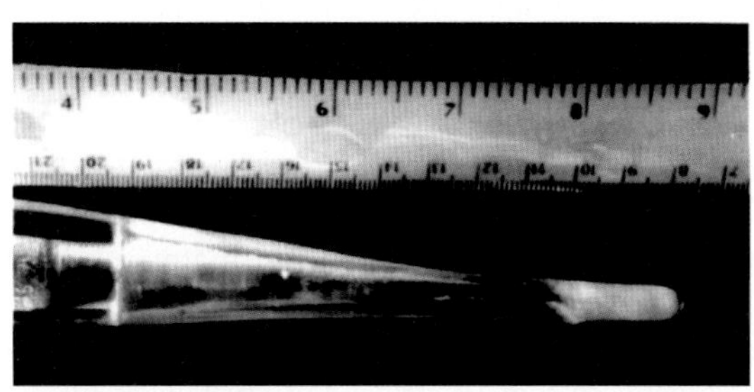

Figure 5.19 An anthracene crystal grown by the Bridgeman process (Arulchakkaravarthi et al., 2002)

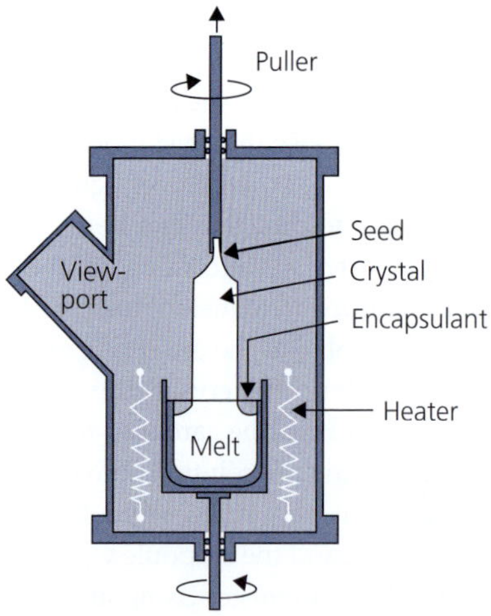

Figure 5.20 Czochralski growth reactor showing its basic components (Singh, 2003).

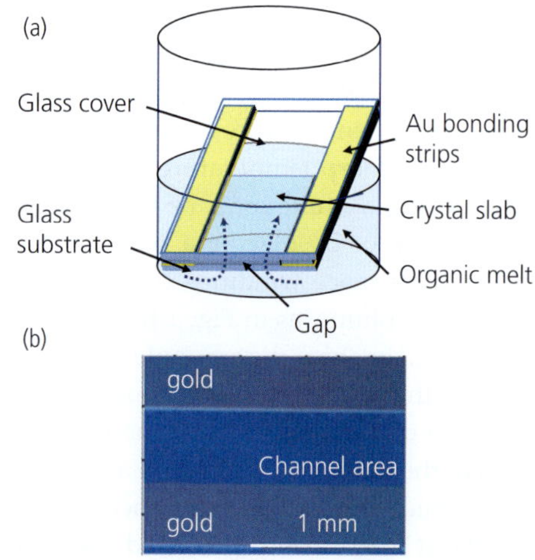

Figure 5.21 (a) Growth of a thin, single crystalline organic slab between two glass sheets cold-weld bonded together by Au stripes. The sum of the thicknesses of the stripes on the substrate and cover determine the crystal thickness, and their separation its width. The molten solution of the source organic (the melt) is drawn by capillary action (dashed arrows) into the gap between the slabs. As the melt is cooled, the slab solidifies, nucleating at the edges of the Au strips. (b) Micrograph of a single crystal of anthracene grown in a channel between two Au bonding strips (Kéna-Cohen et al., 2008).

chamber heated to just below the solidus temperature of the crystal. Slow "pulling" of the seed lifts the crystal from the melt; the speed determines both the perfection of the resulting boule, and its diameter (i.e. smaller diameter boules are grown as the pulling speed is increased). The vessel pressure is maintained by addition of gas into the chamber, thereby preventing decomposition or deposition of material onto the vessel walls. Multicomponent crystals and alloys can be grown under positive pressure to keep both constituents in the melt throughout the growth process.

Czochralski growth can result in crystals of many cubic centimeters in volume. By careful positioning of the seed crystal, growth can be initiated and maintained along any desired crystal axis. Finally, doped and heterogeneous crystals can be grown under the appropriate conditions of pressure and temperature.

Establishing single crystalline growth does not necessarily require the presence of a seed. For example, single crystalline anthracene has been grown in the gap formed between two glass slabs separated by 140 nm thick Au metal spacers, as illustrated in Fig. 5.21. The sandwich is placed in a container partially filled with molten anthracene in a nitrogen atmosphere at 240°C; slightly above its melting point. The liquid anthracene is drawn by capillary action into the channels between the Au strips, and crystallizes by decreasing the temperature at a rate of $\sim$1°C/min. Crystallization is initiated at the Au/liquid interface (Kéna-Cohen et al., 2008). For narrow channels ($<$3 µm wide) and slow cooling ($<$1.5°C/min), the process reproducibly yields larger than centimeter-sized anthracene single crystal slabs. The crystallinity was confirmed by electron and x-ray diffraction that indicated the crystal **a** and **b** crystal axes lay in the glass substrate plane. For most of the channels, the **b**-axis was found to lie along, or very nearly along the channel length.

There are numerous variants for growth of small (microns to millimeters) crystals from the source material dissolved in a solvent such as DCM, chloroform, etc. Several solvent-assisted growth techniques are illustrated in Fig. 5.22. In the solvent evaporation method (Fig. 5.22a), the purified source material is dissolved, and then the solvent is allowed to slowly evaporate from a loosely covered beaker into a controlled atmosphere such as ultrapure nitrogen. At some point, the solution becomes supersaturated, and the organic material spontaneously condenses. Eventually, all the solvent evaporates, leaving behind a crystalline solid that is subsequently removed from the container. Solvent exclusion from the solid can also occur by slow cooling of the solution until it is supersaturated, similarly resulting in a polycrystalline precipitate, as shown in Fig. 5.22b.

Figure 5.22c illustrates the vapor diffusion method. Here, the source material is dissolved into a solvent (solvent **2**) in which it is highly soluble, producing a supersaturated solution. The open

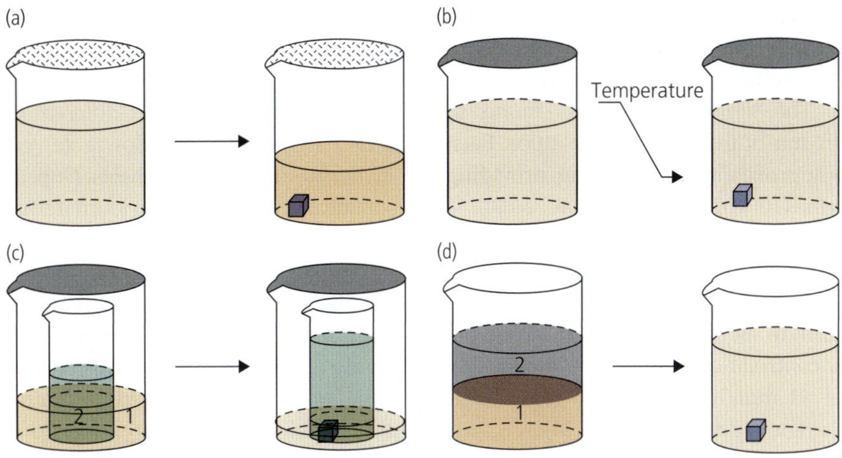

Figure 5.22 Several methods for solvent-assisted single crystal growth. Growth by (a) solvent evaporation, (b) solvent cooling, (c) vapor diffusion, and (d) liquid interdiffusion in a solution containing two solvents, **1** and **2** (Jiang and Kloc, 2013).

Jiang, H. & Kloc, C. 2013. Single-crystal growth of organic semiconductors. MRS bulletin, 38, 28-33 reproduced with permission

vessel containing the solution is placed in a second, high volatility solvent (solvent **1**) in which the source material is less soluble. As solvent **1** evaporates, it enters the inner vessel, leading to supersaturation and eventual precipitation of the source. The rate of crystal growth can be controlled by temperature that varies the relative rates of evaporation of the two solvents. Liquid diffusion, shown in Fig. 5.22d, is similar to vapor diffusion, except that the two solvents, including one containing the source, are added to the same beaker. The two solutions interdiffuse at their interface. Since the solubility of the source is lower in solvent **1**, the interface becomes supersaturated, initiating growth. Growth continues until the two solvents are completely intermixed and the solute is exhausted. We will show in Section 8.7.1 that the liquid interdiffusion process results in high quality single crystal TIPS-pentacene microribbons employed as the channels in thin film transistors (Kim et al., 2007).

Solvent-assisted growth has the advantages of being simple to implement, and it allows for the direct observation of growth as the solvent evaporates. However, it is difficult to control and almost invariably results in polycrystalline growth since there is no well-defined seeding process. The solvent must be completely removed and highly pure to ensure good quality crystal growth with a minimum of point defects, impurity incorporation, high angle grain boundaries, and so on. Also, it is difficult if not impossible in most circumstances to uniformly dope the crystal since impurity incorporation depends on the distribution coefficients of the various constituents. The coefficients are a function of the relative

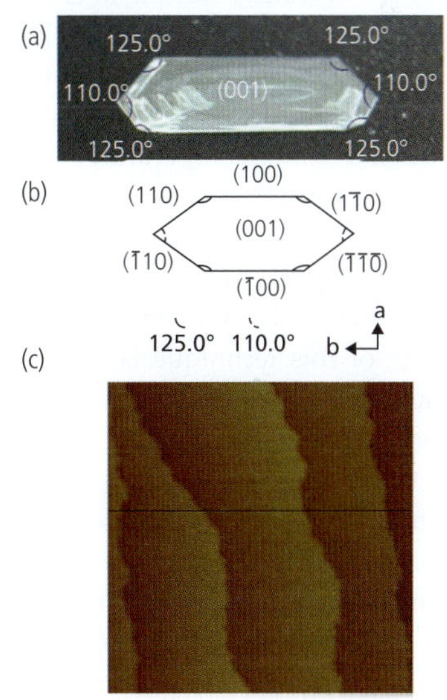

Figure 5.23 (a) Approximately 1 cm long single anthracene crystal grown by slow evaporation from solution. (b) Identification of lattice planes of the crystal in (a). (c) Atomic force microscope image of the (001) surface of the crystal. A terrace structure is observed. Each step in the terrace is one monolayer high. (Zhang et al., 2009)

Reprinted from J. Crystal Growth, Zhang, P., Deng, J., Zeng, X., Liu, Z., Qiu, Y., Zhong, H., Fan, Y., Huang, J., Zhang, J. & Xub, K. Growth mechanism of large-size anthracene single crystals grown by a solution technique, 311, 4708 copyright 2009, with permission from Elsevier.

concentrations of the dopant, host and solvent which varies continuously as the solvent and solute are depleted.

An example of anthracene crystals grown from solution is shown in Fig. 5.23. Several crystals, ranging in size from millimeters to >2 cm, were harvested from a solution that was allowed to evaporate for 4 weeks. Importantly, the growth proceeded layer-by-layer, which results in stepwise terraces on the crystal surface. The terrace height was found to equal d_{001} = 0.92 nm of the bulk crystal lattice. Each terrace length is approximately 1.5 μm long, providing a slight relief to the (100) plane (Zhang et al., 2009).

Growth along a particular crystal axis is achieved by placing a seed within the supersaturated solution with only a single crystal face exposed to the liquid. This imposes growth along the direction facing the liquid. Then, by controlling the solution temperature, growth rates of ∼1 mm/h can result in highly uniform and large single crystals (see Fig. 5.24).

Vapor phase growth, also known as *physical vapor transport* (PVT) growth, is done in vacuum or in the presence of a dilute, ultrapure carrier gas such as H_2, Ar, or N_2 in vessels similar to that used in thermal gradient sublimation (see Fig. 5.7). Crystal nucleation occurs at the vessel walls, resulting in crystallites that are typically only a few millimeters thick by a centimeter in length. Growth proceeds at the material sublimation temperature, which needs to be well below the decomposition temperature. This is in contrast to Bridgeman and Czochralski techniques where growth at the melting point can be very close to its decomposition temperature. Vapor phase growth is unsuitable for materials that cannot be evaporated. An advantage of this technique is that sublimation in a thermal gradient is effective at eliminating impurities, resulting in a high quality crystal without needing to take steps to remove excess solvents or other contaminants found in liquid solutions.

Vapor phase growth results in the development of free, unstrained surfaces, and when not intentionally seeded, several crystals that grow along different axes can be harvested in a single sublimation. As shown in Fig. 5.25, rubrene crystals grow as needles of varying lengths and diameters, whereas other materials (e.g. tetracene) and under different conditions (temperature gradient, gas flow rate, gas composition, etc.) grow as flakes and sheets. Doped and multicomponent (i.e. binary, ternary, quaternary, etc.) crystal mixtures can be grown simply by careful addition, and homogeneous mixing of multiple organic sources into the growth reactor (Karl, 1980). Doping and organic mixtures are achieved by using a carrier gas during growth that accurately dilutes the multiple components to their desired concentrations, similar to the conditions used in *organic vapor phase deposition* (OVPD) of thin organic films (see Section 5.4.2).

High quality, large single crystals are grown in a furnace tube containing a temperature gradient considerably smaller than used in thermal gradient purification. The source material is evaporated either in vacuum or in a slowly flowing inert carrier gas into a slightly cooled (<1°C) zone of the tube. There, nucleation occurs adventitiously on irregularities on the vessel walls. The crystals are allowed to grow slowly (∼0.1 mm/h) in a layer-by-layer mode due to the shallow temperature gradient. Near-equilibrium growth is essential to minimize the formation of dislocations due to strain, incorporation of impurities, and to prevent dendritic growth.

Growth along a particular lattice direction can be achieved by placing a seed at the cooler end of an ampoule positioned within the sublimation zone. The seed temperature is elevated slightly above its sublimation temperature to present a fresh growth surface, and then the temperature is lowered to promote nucleation at the seed surface. Thereafter, growth proceeds along the lattice direction imposed by the seed. Alternatively, the seed can be placed at the end of a small heat sink such as a metal needle inserted into the growth zone. Condensation preferentially occurs at this cold point containing the seed.

Figure 5.24 Single crystals of the nonlinear optical material benzophenone grown from seeded solution along the (110) direction. The boules measure 1–2 cm in diameter (Sankaranarayanana and Ramasamy, 2005).

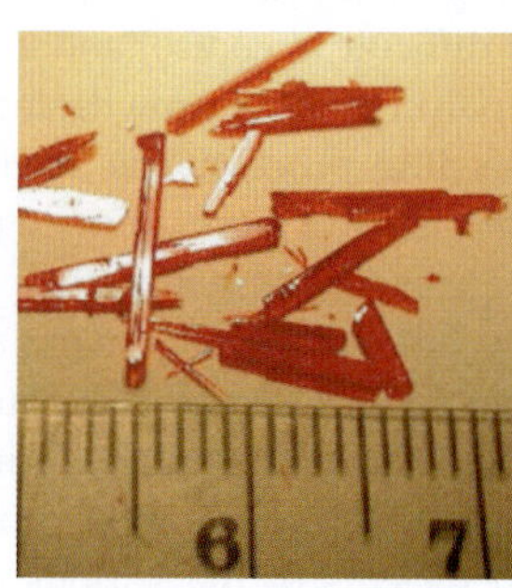

Figure 5.25 Centimeter-long needles of rubrene grown by physical vapor transport (de Boer et al., 2004).

The geometry in Fig. 5.7 is not ideal for the growth of large crystals, which can only controllably occur within a narrow zone where the temperature difference allows for slow, near equilibrium growth. For this purpose, vessels are used that have two parallel surfaces (called plate sublimators); one controlled at the sublimation temperature and the other in close proximity that is at the condensation temperature, perhaps only a fraction of a degree lower than the sublimation plate. Numerous crystals can nucleate over the large surface area for condensation, and can grow to large volumes, as in Fig. 5.26.

Single crystals have been extensively used as the channel materials for organic thin film transistors as a means to achieve very high field effect mobilities, with selected results found in Table 5.2. To fabricate the transistors (with details on structures provided in Chapter 8), the single crystals are mechanically attached to the gate dielectric and the device contacts. The mobilities are generally higher than achieved for similar materials directly deposited from the vapor or solution phase (Reese and Bao, 2007).

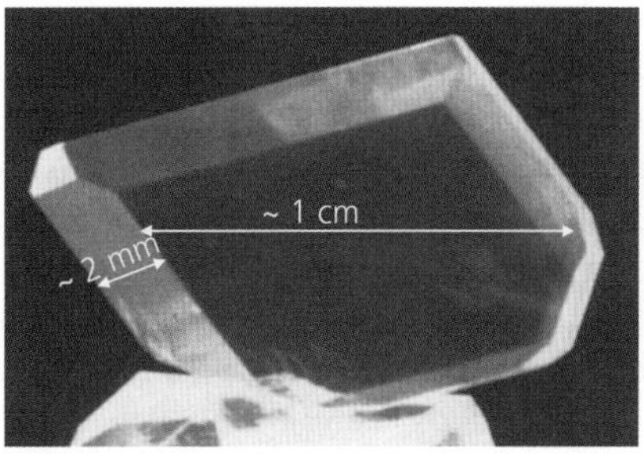

Figure 5.26 A 2 mm thick by 10 mm long pyrene single crystal grown by plate sublimation (Karl, 1980).

While single crystals are useful for understanding fundamental properties, they must be individually positioned between contacts. Further, they must be handled with care to avoid surface damage as well as generation of bulk defects that can be induced by flexing or stretching during device fabrication. An example procedure for fabricating organic transistors using rubrene crystals grown by physical vapor transport is shown in Fig. 5.27. The base layers of this bottom gate/bottom contact configuration (see Chapter 8) form a flexible stamp used to pick up the rubrene single crystal flake. The substrate and gate dielectric material is the flexible elastomer, PDMS. Once the gate, drain and source contacts along with the gate insulator are deposited, the flexible stamp picks up the organic crystal which adheres to it by van der Waals forces. The crystal itself is 0.2–1.0 mm thick along the **c**-axis of rubrene. The dimensions along the basal **a–b** plane are approximately 2–3 mm on a side. The gentle pick-up of the crystal avoids strain and flexing that inevitably degrade the crystal quality.

An interesting feature of this stamp transfer process is that it is both reversible and repeatable. Thus, the crystal can be attached and released numerous times, orienting it along different axes to measure the field effect mobility as a function of crystalline direction. A polar plot of the mobility is shown in Fig. 5.28, which is a compilation of measurements taken by rotating the sample through two entire 360° cycles. The points, obtained from both the linear and saturation transistor operating regimes over both cycles shows the reproducibility, and hence non-destructive nature of stamp transfer. Interestingly the highest mobilities of 4.4 cm^2/V s are obtained along the $\widehat{b}$ -axis, which is four times that along the $\widehat{a}$ -axis. This is consistent with the closer molecular spacing of 7.2 Å along $\widehat{b}$, vs. 14.4 Å along $\widehat{a}$.

Table 5.2 Mobilities measured for OTFTs that employ single crystal channels. From Reese and Bao (2007)

Channel material	Growth process	Charge carrier	Mobility (cm^2/V s)	Structure[a]	Gate dielectric
Pentacene	Vapor	Hole	2.2	TC, TG	Parylene
Tetracene	Vapor	Hole	2.4	BC, BG	PDMS
Anthracene	Vapor	Hole	0.02	BC, BG	SiO_2
CuPc	Vapor	Hole	1	TC, TG	Parylene
FCuPc	Vapor	Electron	0.2	TC, BG	SiO_2
TCNQ	Vapor	Electron	1.6	BC, BG	Air gap
DT-TTF	Solution	Hole	1.4	BC, BG	SiO_2
CH4T	Solution	Hole	0.2	BC, BG	SiO_2
TIPS-Pn	Solution	Hole	0.17	BC, BG	SiO_2
Anthradithiophene	Solution	Hole	1.0	BC, BG	SiO_2

[a] BC = bottom contact; TC = top contact; BG = bottom gate; TG = top gate.

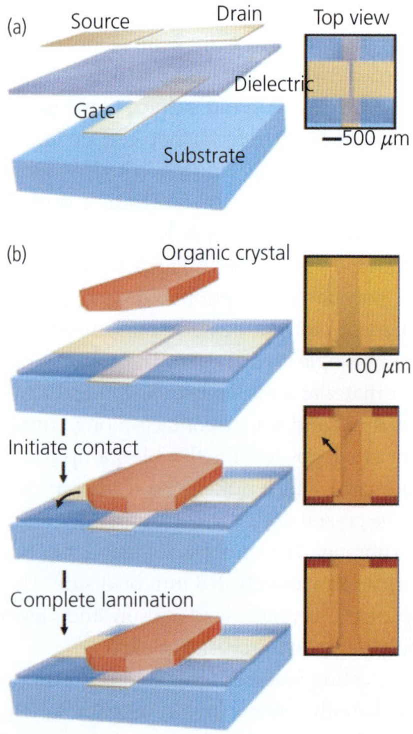

Figure 5.27 Fabrication of a rubrene single crystal field effect transistor. (a) Preparation of the stamp comprising a flexible PDMS substrate, gate electrode, 2–4 mm thick PDMS gate dielectric and the source and drain electrodes. All electrodes are Ti/Au, where the Ti is used for adhesion to the PDMS. These base layers are used as a stamp. (b) The organic crystal is picked up by the stamp along one edge, and then adheres by van der Waals forces to the electrodes. The images at right are of the device before and after organic channel attachment. The arrow in the center image shows the direction of attachment of the crystal after making initial contact along the edge at the lower right corner (Sundar et al., 2004).

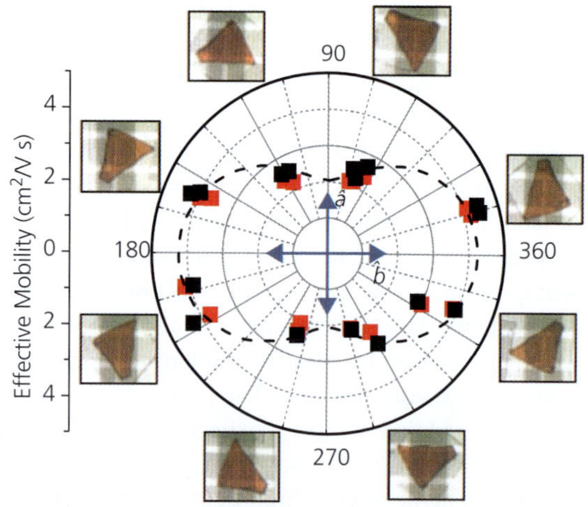

Figure 5.28 Field effect mobility as a function of rubrene crystal orientation relative to the source-drain direction. The $\hat{\mathbf{a}}$ and $\hat{\mathbf{b}}$ crystal axis directions are indicated. Mobilities are obtained from both the linear (black squares) and saturation (red squares) regimes of transistor operation. Also shown are photographs taken at several different orientations (Sundar et al., 2004).

We caution that field effect mobilities obtained from the transistor characteristics are not equivalent to time-of-flight mobility measurements as described in Section 4.4.3. Indeed, the field effect mobilities are extracted from measurements of the conductivity along the surface of the organic at its interface with the gate insulator. Hence, the field effect mobility depends on preparation, contact quality, pressure applied to the organic (Nisikawa et al., 2006), and interface contamination from the environment (Pflaum et al., 2006). Hence, very high room temperature transistor mobilities (>5 $cm^2/V\ s$) occasionally claimed for these values should be viewed with caution, since they may not reflect an intrinsic material property.

5.4.2 Thin film growth

Soft materials enable many different methods for deposition. This ability lies at the heart of their versatility. Self-assembled, epitaxial, and quasiepitaxial structures can develop on a variety of surfaces without the constraints of lattice matching that dominates the structures of "hard" materials such as covalently bonded semiconductors. Given the range of possible deposition methods available to organics, and the variety of structures that can be achieved, in this section we necessarily confine ourselves to a discussion of only the most commonly used methods. Both deposition from vapor (whether in vacuum or in an inert gas) and from solution are described. The interested reader is directed to the literature to find out about other methods, using what follows as an introduction to the topic.

5.4.2.1 Vacuum thermal evaporation

Thermal evaporation in vacuum is by far the most common means for depositing small molecule organic thin films. The configuration used for VTE is illustrated in Fig. 5.29. The organic source material is loaded into a "boat," or crucible, that is resistively heated in a high vacuum chamber to the source evaporation temperature. Materials that either sublime or evaporate from the liquid phase at elevated temperatures are compatible with VTE. By co-evaporating two or more materials from different crucibles, blends and doped films are controllably produced. Sequential deposition of several materials results in the growth of multi-layer structures whose functionalities are determined by composition and layering scheme. The rate of film growth and the total thickness deposited *in real time* are typically measured by a quartz crystal thickness monitor positioned near the substrate. The

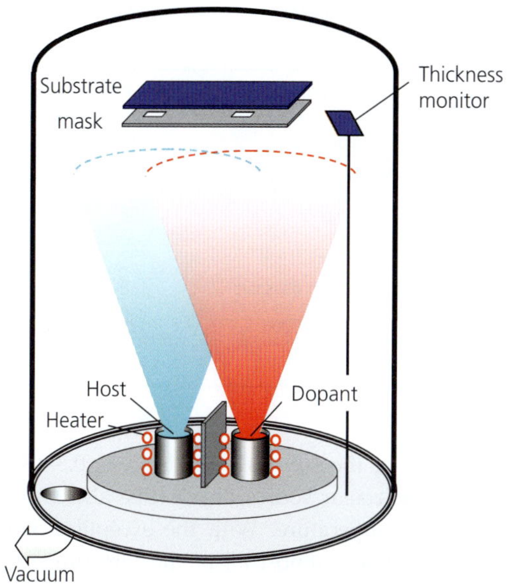

Figure 5.29 Schematic of the basic elements of a vacuum thermal evaporation system.

Table 5.3 Approximate "mother" glass substrate sizes used in display manufacturing

Substrate generation[a]	Dimensions (in mm)[b]
1	300 × 400
2	400 × 500
3	550 × 650
4	680 × 880 or 730 × 920
5	1000 × 1200 or 1100 × 1300
6	1500 × 1800
7	1900 × 2200
8	2200 × 2400
9	2400 × 2800
10	2850 × 3050
11	3200 × 3600

[a] The generation is typically labeled as "Gen," that is, a generation 10 substrate is referred to as Gen 10. As a rule of thumb, the substrate area (in m^2) is roughly equal to its generation.
[b] Actual sizes depend on manufacturer. Thicknesses also vary, but are typically 0.3–0.7 mm.

monitor can also be used in a feedback loop to control the temperatures of the crucibles, thus providing highly controlled film deposition. Alternatively, the film properties can be measured and controlled using optical absorption in either reflection or transmission modes, or by ellipsometry. For these purposes, an optical beam is obliquely incident onto the substrate surface, and the intensity and spectrum of its specular reflection is measured using a photodetector and monochromator placed opposite to the light source at the angle of incidence.

The substrate is positioned at a sufficient distance from the thermal sources to prevent heating, and to ensure uniform thickness and composition of the deposit. On the other hand, the distance must be much less than the ballistic range of the evaporated molecules from the crucible. To achieve high uniformity, the source-to-substrate distance should be at least three to five times the maximum substrate diameter, although this becomes impractical for very large substrates used in manufacturing displays, lighting or solar cells. Table 5.3 lists the approximate substrate sizes use in display manufacturing. The use of linear evaporation sources and other strategies (see below) can relax this distance requirement. To improve uniformity even further, the substrate is rotated at a few revolutions per minute around an axis that is off-center from the crucible orifice. The preferred orientation is shown in Fig. 5.29 where the organic materials are evaporated upward toward the substrate. This avoids dust within the chamber from falling onto the substrate surface, and also prevents the organic source from falling out of the crucible, or boat.

VTE is done in a high vacuum environment, typically ranging from 10^{-6} torr, to ultrahigh vacuum (UHV) approaching 10^{-11} torr. When deposition occurs at UHVs (loosely defined as pressures, $P < 10^{-8}$ torr), the process is known as *organic molecular beam deposition* (OMBD).

The mean free path of the background gas molecules, *mfp*, should be much larger than the chamber size in a vacuum deposition system. From kinetic theory,

$$mfp = \frac{k_B T}{\sqrt{2}\pi d^2 P}, \tag{5.14}$$

where d is the background gas atomic or molecular diameter. The *mfp* is plotted vs. pressure in Fig. 5.30 for common background gases. In most deposition systems, the background gases with the highest partial pressures are H_2 and N_2. At $P < 10^{-6}$ torr, then $mfp > 10$ m, which is considerably larger than chambers used in VTE.

The concentration of contaminants in the high vacuum environment is very small, and hence the purity of the starting organic material is maintained, and even improved by evaporation (Salzman et al., 2005). As in the case of gradient sublimation, vacuum pumps used in VTE should avoid organic oils used in mechanical rotary pumps that rough the chamber from atmospheric pressure, and in high vacuum

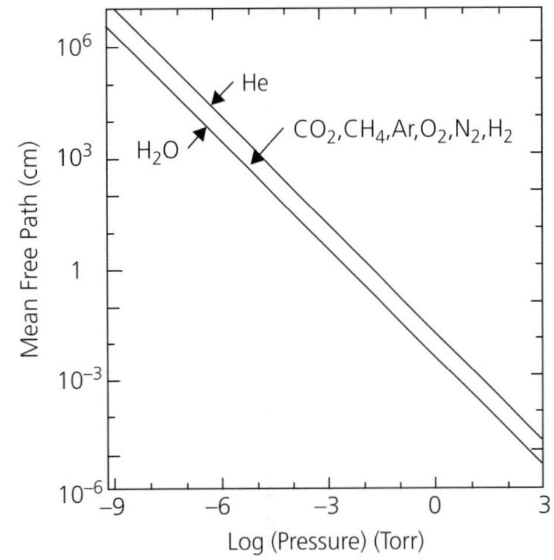

Figure 5.30 Mean free path of several gases as a function of pressure.

Table 5.4 Enthalpies of evaporation of several common organic compounds (Shtein et al., 2001)

Material	Function	ΔH^{vap} (kcal/mol)
DCM2	Fluorophore	110
PtOEP	Phosphor	128
$Ir(ppy)_3$	Phosphor	166
CBP	Electron Transport	143
Alq_3	Electron transport	138
α-NPD	Hole transport	139
CuPc	OPV donor	176

diffusion pumps. The most convenient pump for rapid, high vacuum pumping is the turbo pump, although it is not effective in removing the lightest gasses such as H_2. Capture pumps such as cryopumps are also clean and fast, although care must be taken during regeneration to avoid contact with vapors that are released from many pumping cycles of potentially toxic organic compounds. It is also important to use crucible materials (e.g. Mo, W, ceramics) that do not react with the evaporant, and that the evaporation temperature be kept well below the decomposition temperature of the organic compound.

The evaporation rate is exponentially dependent on temperature, viz.

$$r_{evap} = r_0 \exp(-\Delta H^{vap}/RT), \tag{5.15}$$

where ΔH^{vap} is the enthalpy of evaporation. Hence, the rate can be adjusted between 0.01 Å/s and 100 Å/s with only a small change in crucible temperature. In Table 5.4 we list the evaporation enthalpies of a few example compounds commonly found in organic electronic devices.

Growth rates of 0.1–10 nm/s are commonly used since achieving precise thickness control at higher rates is challenging in depositing the thin layers (~100 nm) used in organic electronic devices. If the rate is too slow, a high concentration of impurities is incorporated into the films in systems with only modest vacuums. In a study of film quality, OLED hole and electron transporting layers of α-NPD and Alq_3, respectively, were grown at rates of from 0.01 to 1000 nm/s at a system base pressure of $<10^{-6}$ torr (Matsushima et al., 2012). The growth rate was varied by positioning the substrate at distances 0.5–18 cm from the molecular source boat, while also varying the source temperature. With the exception of the charge mobilities, almost all of the film properties, including surface roughness, PL lifetimes, and morphology measured by X-ray diffraction, were unaffected over this six order of magnitude variation in deposition rate. The most significant change was a three-fold increase in hole current in α-NPD by a factor of three as the rate increased to 10 nm/s, compared to a decrease in electron current in Alq_3 by a factor of 0.17. This is attributed to changes in bulk mobilities related to morphology within these films. Optical birefringence measurements indicate that the planar α-NPD molecule forms more ordered stacks in the vertical direction at the higher rates, whereas it is likely that the more isotropic Alq_3 simply becomes more porous. Indeed, very high deposition rates are likely to yield a high density of voids not found in layers that are deposited more slowly. That is, the adsorbed molecules have more time to seek equilibrium positions within the film before being buried at energetically unfavorable locations by subsequently arriving molecules.

Although this tradeoff between film quality and deposition rate is weak for α-NPD and Alq_3, it may be more pronounced in other molecular systems. While manufacturing throughput depends on a high rate to complete the film deposition in a minimum amount of time with thickness variations of $< 10\%$, this need must be balanced against the rigorous demands for high performance devices. With these considerations in mind, ~10 nm/s is a practical upper limit to the deposition rate in most circumstances.

In-line organic source boats illustrated in Fig. 5.31 are used for deposition on very large substrates. In this configuration, the substrate is held close to a boat of comparable width. The substrate is translated over the boat opening throughout the layer growth. The

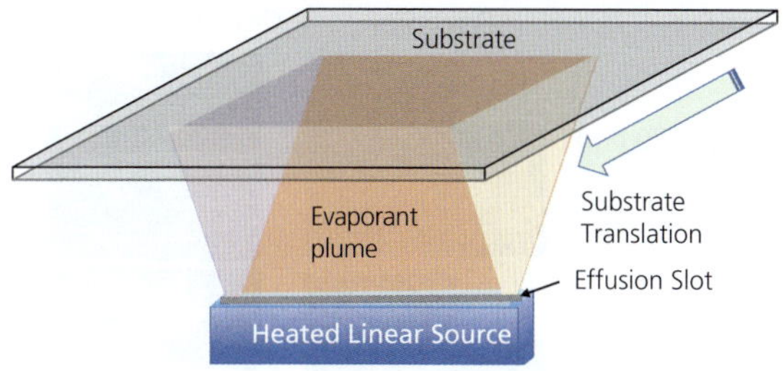

Figure 5.31 Schematic layout of a linear source evaporation system. The source boat is uniformly heated above the source sublimation temperature. The molecules exit the effusion slot, forming a large plume that covers the substrate as it is translated to achieve full coverage.

film thickness is determined by the width of the plume, W, and the velocity of the translating substrate, v, via $d = Wr_{evap}/v$. For multilayer structures, the substrate is translated from source-to-source placed in a series along the length of the vacuum chamber, each containing a different organic species. Doped films are grown by placing two or more linear sources in close proximity, and then tilting them such that the deposition plumes overlap at the substrate plane. The substrate can be oriented as in Fig. 5.31 (bottom-up evaporation), or the entire system can be rotated by 90° to allow for the substrate to stand vertically as it is translated—a relatively compact configuration that is convenient for handling the largest substrates.

In-line sources have proven effective for the manufacture of large OLED displays (typically Gen 8 and larger). They do, however, present challenges beyond those encountered using point sources. Care must be taken to ensure complete overlap of the evaporants when doping thin films to avoid variations in dopant concentrations at the edges of the deposit. Also, the thin film thickness depends on the control of two variables: evaporation *and* substrate translation rate.

Film morphology (i.e. its roughness, crystalline order and orientation, etc.) can be controlled by varying the deposition rate, molecular structure and composition, and substrate temperature from below room temperature to as high as ~100°C. A dramatic demonstration of the effects of deposition rate on morphology is shown by the X-ray diffraction pole figures of PTCDA grown on a glass substrate at room temperature in Fig. 5.32. These pole figures reveal the crystalline texture of the film by plotting the intensity of the intermolecular, (102) Bragg reflection as a function of polar angle with respect to the film normal. Rapid, non-equilibrium growth at a high deposition rate of 50 Å/s results in a surprisingly high degree of order.

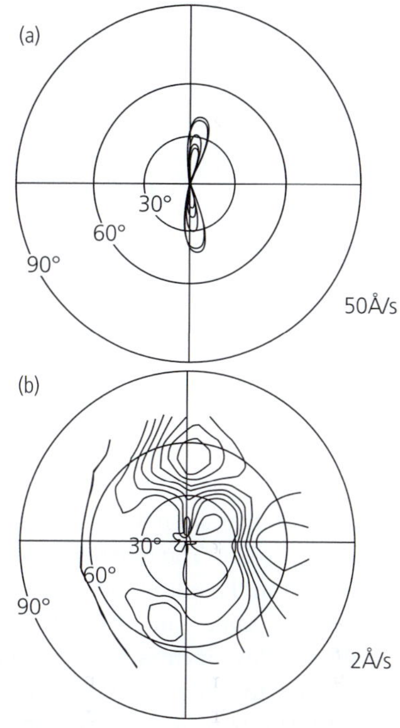

Figure 5.32 X-ray pole figure of the (102) PTCDA diffraction intensity vs. polar angle. A 100 nm thick film is deposited at (a) 50 Å/s and (b) 2 Å/s. Each contour in (a) corresponds to a 10%, and in (b) to a 1% difference in X-ray intensity between adjacent contours (Forrest et al., 1984).

Reprinted from Forrest, S. R., Kaplan, M. L. & Schmidt, P. H. 1984. Organic-on-inorganic semiconductor contact barrier diodes. II. Dependence on Organic Film and Metal Contact Properties. J. Appl. Phys., 56, 543 with permission of AIP Publishing.

The double-lobed x-ray intensity pattern aligned to the 0–180° polar axis, and with peaks at ±11° from the substrate normal in Fig. 5.32a reveal a twinned crystal whose (102) planes are tilted at these same angles (Forrest et al., 1984). In contrast, at a lower deposition rate of 2 Å/s shown in Fig. 5.32b, there is no apparent pattern in the X-ray contours. This suggests that the film is a randomly oriented powder in the substrate plane, that is, it is amorphous.

As can be inferred from the discussion of quasiepitaxy in Section 2.6.2, the kinetics of growth are critical factors in determining morphology. The order at high growth rates is indicative of rapid self-assembly that occurs under non-equilibrium conditions: nucleation begins at only a few locations around the substrate followed by rapid and ordered growth that covers the substrate surface. Pretreatment of the substrate (e.g. by handling, rubbing, etc.) may in fact *template* a directionality to the substrate, whereby there is a preferred alignment direction that results in the lowest crystal energy (Yang et al., 2015). At lower deposition rates, the molecules have an opportunity to move along the substrate surface, creating random

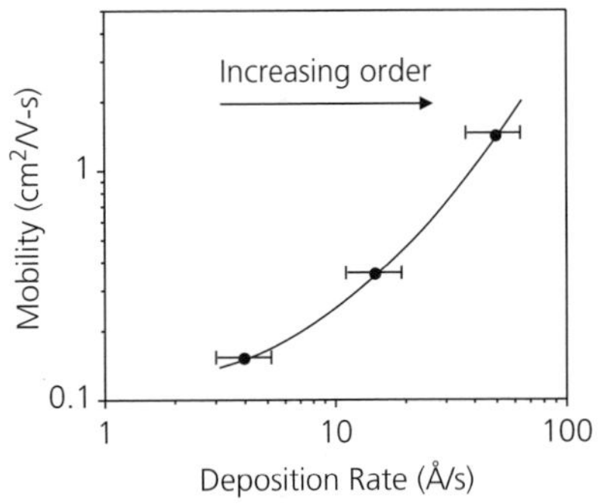

Figure 5.33 Out of plane hole mobility vs. deposition rate. Comparing this result to that in Fig. 5.32, higher rates result in increasing order, and hence a higher charge mobility (Forrest et al., 1984).

nanocrystalline piles that are not oriented relative to each other (see Fig. 5.32b). The rate dependent ordering is also reflected in the out of plane hole mobility measured for these same samples with results in Fig. 5.33. The mobility increases with deposition rate, corresponding to more ordered and extensive crystalline stacks. This is expected since single crystals lack high angle grain boundaries that result in carrier scattering, which would be reflected in a decrease in charge mobility.

The order in the films in Figs. 5.32 and 5.33 arise from a combination of adventitious templating by the substrate, as well as by the strong tendency of PTCDA to assemble into large crystalline "rafts" with molecules lying flat, or near flat on the glass surface. Deposition of very thin layers of PTCDA and other molecules with similar tendencies to self-order can thus be used as templates that drive the order of subsequently deposited layers that have weaker intermolecular interactions. Indeed, such templates can be used to force molecules in the adlayer to lie flat on the template, whereas they would ordinarily stand on edge, or vice versa, depending on the adlayer molecules interactions with the template (Lassiter et al., 2010, Yang et al., 2015). Intentional manipulation of order and orientation of the thin film structure is critical to device performance, resulting in increased emission efficiency in OLEDs, power conversion efficiency of OPVs, and the conductive properties of OTFTs.

While exogenous layers can form a template to control the morphology of subsequently deposited films, the range of control is limited. That is, a structure that is forced by the template eventually relaxes into its equilibrium crystal structure as the thickness of the adlayer is increased. Relaxation, in turn, results in accumulation of disorder. Hence, the layer should only be deposited to a thickness that is within

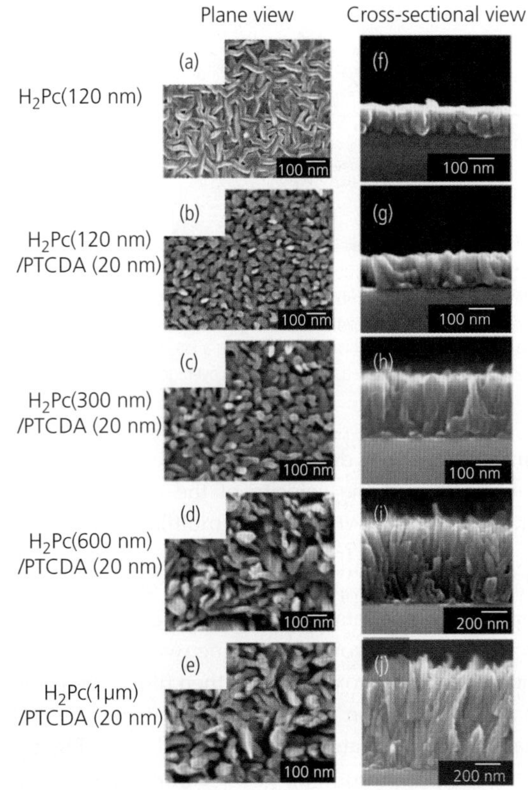

Figure 5.34 Plane and cross-sectional SEM views of H_2Pc films grown on a bare glass substrate (a, f) and at various thicknesses on a 20 nm thick, pre-deposited PTCDA template layer (b–e; g–j) (Sakurai et al., 2005).

the Matthews–Blakeslee strain limit discussed in Chapter 2. Lattice relaxation is exemplified by an H_2Pc film deposited onto a 20 nm thick ordered PTCDA template layer pre-deposited on a glass substrate. Micrographs of the H_2Pc film surface and cross sections for various layer thicknesses are shown in Fig. 5.34. The layer grown directly on glass (Fig. 5.34a and f) shows no preferred order, with crystalline needles distributed in random directions. Cross-sectional views of the films grown on template layers show a preferential orientation of the domains perpendicular to the surface. The crystal facets are also apparent in the corresponding plane views. However, the thickest films (>300 nm, Figs. 5.34d, e, i, and j) show that the domains diverge from a completely vertical aspect. These micrographs, along with X-ray diffraction data, indicate that the H_2Pc molecular layers in the thinner films lie flat on the similarly flat lying PTCDA layer, while as the H_2Pc film thickness is increased, the structure relaxes into its α-phase equilibrium, herringbone crystal habit. This tendency

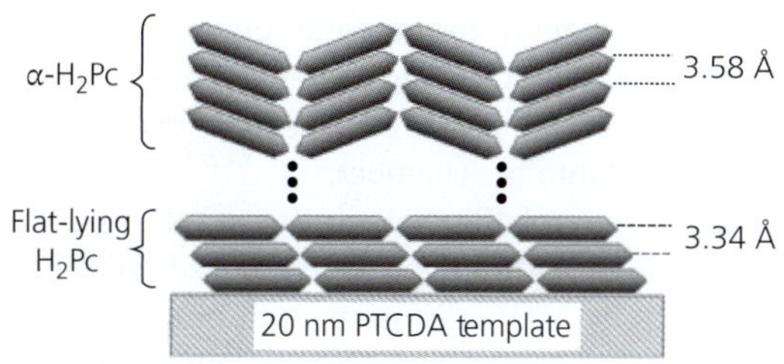

Figure 5.35 Crystal structure evolution of an H_2Pc film on a PTCDA template on glass as the H_2Pc film thickness increases. Molecular plane spacings for the two possible film structures are shown. After Sakurai et al. (2005).

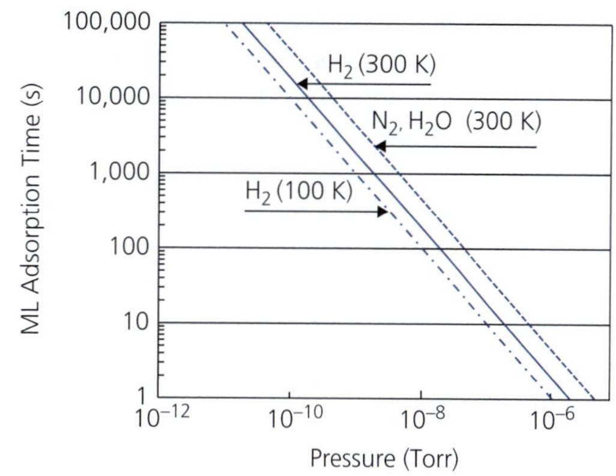

Figure 5.36 Monolayer (ML) adsorption time vs. background chamber pressure, assuming, $\kappa = 1$, $T = 300$ K, and $N_{surf} = 10^{14}$ molecules/cm^2. The H_2 adsorption time at 100 K is also shown.

to assume its equilibrium structure as the film thickness is increased is illustrated in Fig. 5.35.

While the films in Figs. 5.34 were grown in UHV by OMBD, templating appears to be independent of the level of vacuum. Indeed, it has also been observed in films grown via OVPD. It is simply a result of the balance of forces between the substrate and the layers grown on its surface. These forces are modified by either increasing or decreasing the van der Waals attraction between the adlayer and the template, or with the original substrate. Numerous template molecules have been employed, although planar aromatic hydrocarbons such as PTCDA, NTCDA, and larger molecules with similar structures appear to be the most effective at creating order in subsequent layers. This is due to their extended π-systems that form highly attractive surfaces to a range of conjugated molecular structures.

5.4.2.2 Organic molecular beam deposition

The UHV process of OMBD has led to unprecedented advances in the ultraprecise control over both the thickness and morphology of small molecule organic materials (Forrest, 1997). This is primarily a result of the uniformly controlled heating of organic source materials in an ultraclean environment using Knudsen evaporation cells that ensure both controlled material flux as well as *in situ* purification of the source material The UHV environment also leads to nearly complete elimination of the incorporation of impurities during film growth.

To preserve the purity of the starting or source material, the incorporation rate of exogenous contaminants must be substantially less than the film growth rate. From kinetic theory, the time for a monolayer (ML) of background gas atoms with partial pressure, P, to be adsorbed onto a substrate surface is (Weston, 1985)

$$\tau_{ML} = \frac{N_{surf}\sqrt{2\pi m_g k_B T}}{\kappa P}, \tag{5.16}$$

where N_{surf} is the density of surface atoms required to form a complete monolayer, m_g is the mass of the gas molecule, and $\kappa \leq 1$ is the atomic sticking coefficient (i.e. the ratio of atoms that stick to a surface to those incident on it). Equation 5.16 is plotted in Fig. 5.36 assuming that the principal residual gas in a vacuum system is H_2, N_2, or H_2O, where $\kappa = 1$, and $N_{surf} = 10^{14}$ molecules/cm^2. The time required to adsorb a monolayer of H_2 is $\tau_{ML} \sim 20$ s at a background pressure of 10^{-7} torr, and 40 min at 10^{-9} torr. Since N_2 and H_2O are somewhat more massive, their presence can lead to additional contamination, albeit at a somewhat reduced adsorption rate. Some morphologies are optimally produced on cooled substrates, in which case the substrate itself serves as a sync, or pump, for background impurities (Forrest, 1997). The plot in Fig. 5.36 shows that cooling to 100 K reduces the monolayer diffusion time by nearly a factor of two from its value at room temperature. To minimize the incorporation of impurities, it is important to employ the fastest growth rate possible that is compatible with achieving the desired film morphology and thickness control. Practical growth rates in UHV thus lie in the range from 0.1 to 5 ML/s.

Films of Ga phthalocyanine chloride (GaPcCl) grown at a rate of 0.5 Å/min on (001) KBr substrates at a base vacuums 3×10^{-9} torr show a higher degree of crystalline order than films grown at only 5×10^{-6} torr (Tanigaki et al., 1991). Furthermore, scanning tunneling microscope images reveal layer-by-layer growth when PTCDA is deposited on graphite in UHV, whereas clumping into islands is prevalent at lower vacuum (Forrest et al., 1994). It was speculated that impurities and water incorporated during growth at lower vacuum provide sites for crystal nucleation, leading to island growth. Maintaining the source material at elevated temperatures in UHV for long periods, as is done in conventional OMBD systems, reduces moisture, and hence leads to improved source purity.

UHV systems employ stainless steel chambers, as in Fig. 5.37. Typically, growth occurs by evaporation of the source material in a background vacuum ranging from 10^{-9} to 10^{-11} torr using highly purified organic source material heated from a Knudsen cell (Herman and Sitter, 1989). The evaporant is collimated by passing through a series of orifices, after which it is deposited some distance away (~10–20 cm) on a substrate held perpendicular to the beam. The molecular flux is controlled by a combination of oven temperature that is maintained above the sublimation temperature of the source but well below its decomposition temperature, and a mechanical shutter that can switch the beam flux from "on" to "off." Structures consisting of alternating layers of different compounds are grown by sequentially shuttering the beam flux from more than one Knudsen cell (So et al., 1990, So and Forrest, 1991).

Samples enter and exit the OMBD chamber via vacuum load locks, avoiding times needed to pump the system down to UHV. The source materials are stored in the Knudsen cells for extended periods of time at temperatures somewhat below the evaporation temperature in the UHV environment. Since many of the organic molecular materials are hygroscopic, vacuum storage at elevated temperatures leads to thorough drying and outgassing of the sources, thereby achieving material purity over the long term exceeding that of the source materials initially loaded into the chamber.

Due to the volatility of many organic materials, high throughput pumps are employed to achieve UHV. As in VTE, diffusion and other oil-based pumps should be avoided since the oils can contaminate, and even react with the organic compounds. A preferred arrangement is to use a high capacity cryopump during growth, and an ion pump between growths to maintain UHV. The chamber itself is lined with liquid N_2 filled shrouds on which residual gases, and particularly H_2O are condensed (or gettered). This combination can lead to a base pressure of 5×10^{-11} torr.

Varying the substrate temperature is one means for controlling film structure (Vincett et al., 1981, Debe et al., 1988, Mobus et al., 1989, So et al., 1990, Forrest et al., 1994). At low substrate temperatures, the sticking coefficient of impurities and residual gas atoms approaches unity (cf. Fig. 5.36). High deposition rates can minimize impurity incorporation.

The substrate temperature can also be used to control the molecular desorption rate, thereby enabling ML growth of organic molecular materials and heterostructures (Yoshimura et al., 1991). *Molecular layer deposition* (MLD) is similar to *atomic layer epitaxy* (ALE) used for growing inorganic semiconductor multiple quantum well structures, where the growth of each layer is self-terminating due to the

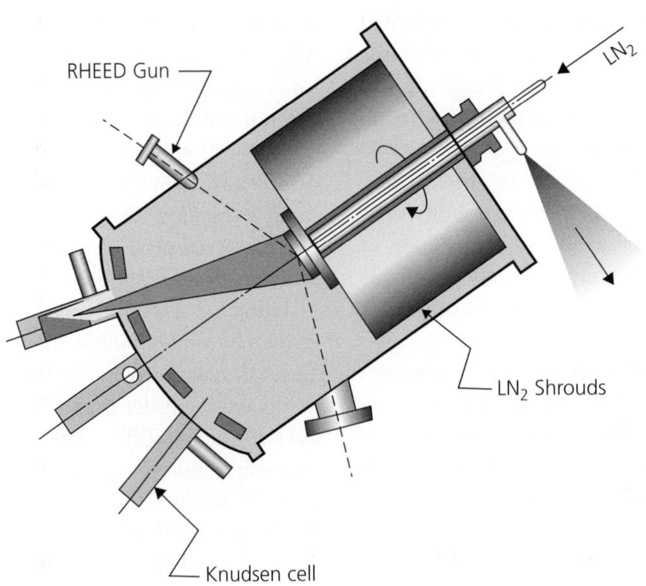

Figure 5.37 OMBD chamber used in the deposition of organic thin films in an ultrahigh vacuum environment (Forrest, 1997).

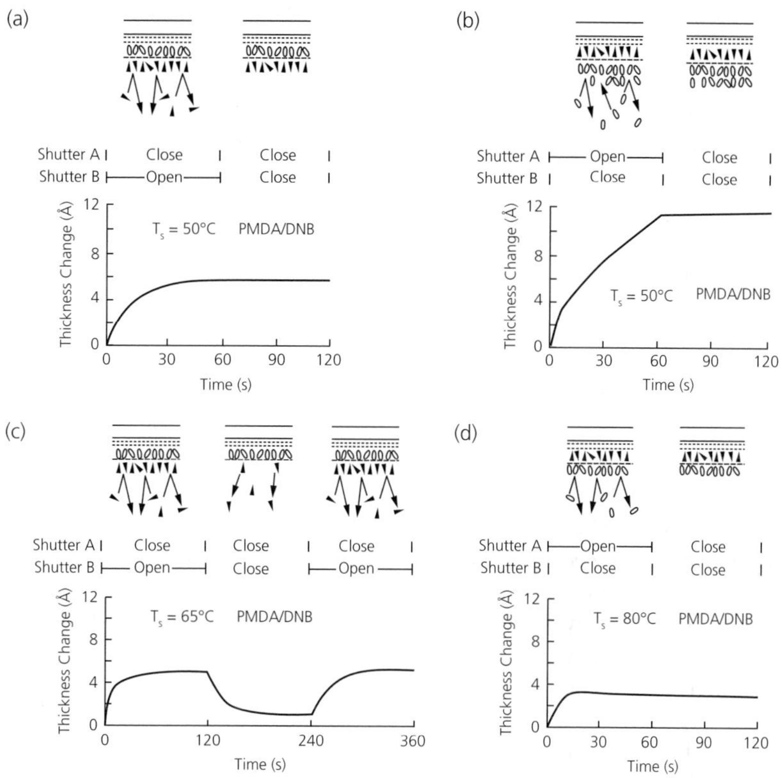

Figure 5.38 Molecular layer deposition of PMDA (open ellipses) and DNB (filled triangles) using temperature to control the layer saturation. Separate shutters A and B are positioned immediately above the evaporation sources for PMDA and DNB, respectively, enabling independent control of their molecular fluxes (Yoshimura et al., 1991).

thermodynamic conditions and the composition of the molecular constituents. In Fig. 5.38 we show the MLD growth of PMDA alternated with DNB. These molecules have considerably different desorption temperatures (T_{desorp}) from the substrate and each other. When the substrate temperature is $T_s > T_{desorp}$, the molecular sticking coefficient is $\kappa \ll 1$, preventing deposition. MLD proceeds by first depositing molecule A whose binding energy ($E_{A\text{-}B} = k_B T_{A\text{-}B}$) to molecule B is higher than that of the molecule to itself (e.g. $E_{B\text{-}B} = k_B T_{B\text{-}B} < E_{A\text{-}B}$). When the substrate temperature is adjusted to $T_{B\text{-}B} < T_s < T_{A\text{-}B}$, only a single monolayer of molecule B is deposited on a layer of molecule A. To deposit a second ML of molecule A, the substrate temperature is changed to $T_{A\text{-}A} < T_s < T_{B\text{-}A}$, and the process is repeated. This combination of steps results, therefore, in the saturated growth of single monolayer films in an alternating layer-by-layer mode by controlling the substrate temperature while introducing a beam of molecules of the desired composition.

The MLD process applied to PMDA and DNB in Fig. 5.38 utilizes the strong bonding between carbonyl-oxy-carbonyl groups of PMDA to the amino group in DNB *vs.* the relatively weak bond energies of these functional moieties to themselves. At T_s = 50°C, the thermal energy is greater than the intermolecular binding energy of DNB ($T_s > T_{B\text{-}B}$), thereby permitting only a single, saturated ML of DNB to deposit on an initial layer of PMDA where $T_s < T_{A\text{-}B}$ (see Fig. 5.38a). However, growth of PMDA on DNB at T_s = 50°C < $T_{A\text{-}A}$ is halted only by closing the shutter positioned between the molecular effusion cell and the substrate, as indicated in Fig. 5.38b. When the substrate temperature is increased to T_s = 65°C, DNB is desorbed (Fig. 5.38c). The alternate process of saturated single monolayer growth of PMDA on DNB is therefore achieved at T_s = 80°C, as shown in Fig. 5.38d. This procedure is slow since the temperature must be stabilized between each layer deposition, although it can be effective for achieving controlled layer-by-layer growth.

One further advantage of UHV growth is that several in-situ film diagnostics can be employed both during and immediately after growth (Herman and Sitter, 1989). In particular, *reflection high energy electron diffraction* (RHEED) provides in-situ and real time information of the thin film crystalline structure

(Haskal et al., 1992, Tada et al., 1992). *Residual gas analysis* (RGA) is used to determine the molecular species as well as any contaminants within the OMBD chamber, and also measures the film-substrate adhesive energy by *thermal desorption spectroscopy*.

Given its combination of ultrahigh purity and *in situ* diagnostics, OMBD has led to significant advances in our understanding of growth dynamics as well as fundamental optoelectronic properties of organic molecular crystals. Many of these same advantages have been effectively used in growing extremely high quality inorganic semiconductors such as GaAs and InP using the analogous process of molecular beam epitaxy.

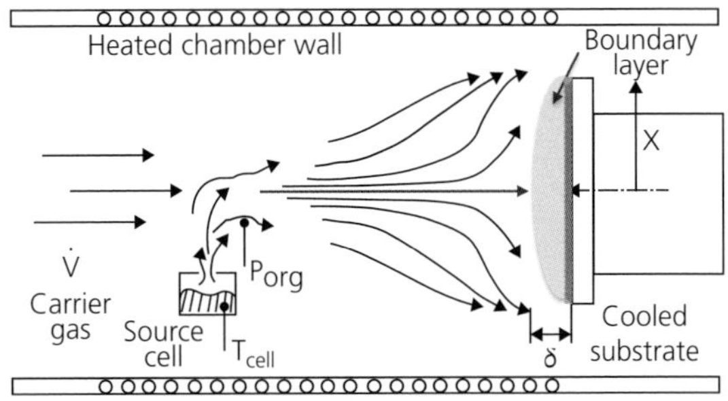

Figure 5.39 Schematic diagram of an OVPD chamber. The source material is located in the inert gas stream, allowing for equilibrium pick-up of the molecules. These are carried in a hot-walled chamber by the carrier gas to a cooled substrate where they are physisorbed. Immediately above the substrate is a diffusive gas boundary layer (shaded region) (Shtein et al., 2001).

5.4.2.3 Organic vapor phase deposition

OVPD differs from VTE since the molecules are evaporated into a hot inert carrier gas (e.g. N_2 or Ar) that transports them to a cold substrate where adsorption occurs (Baldo et al., 1998a, Shtein et al., 2001). The pressure of a carrier gas during growth imparts kinetic energy to the molecules adsorbed onto the substrate, as well as resulting in diffusive transport of the molecules in the gas phase. The additional kinetic energy (beyond that provided by temperature) allows the adsorbate to seek the most energetically favorable attachment sites on the underlying layers, while diffusive transport randomizes the direction of incidence. The combination of these and other attributes discussed below give OVPD the ability to control film morphologies that are not possible using other growth methods.

The pressure within the chamber typically ranges from 0.1 torr to 1 atm, with flow rates having a low *Reynold's number* such that laminar flow is established. The Reynold's number (*Re*) is a dimensionless ratio of the inertial to viscous forces. For a *kinematic viscosity*, v, and a volumetric flow rate of the gas, $\dot{V}$, then,

$$\mathrm{Re} = \frac{\dot{V}D}{vA}. \tag{5.17}$$

Here, D is the chamber diameter and A its cross-sectional area. For "fully developed" laminar flow, $Re < 2300$, whereas turbulent flow ensues at $Re > 4000$. Intermediate Reynold's numbers correspond to the transition flow regime. Typically in OVPD, $Re \sim 1$–10, or well within the laminar flow regime.

The OVPD apparatus is shown schematically in Fig. 5.39. The hot inert carrier gas infuses a boat containing the organic source material, uniformly evaporating it into the chamber where the solute (i.e. the molecules) are further diluted in the inert gas (the solvent) stream. The chamber itself is heated, such that no condensation occurs on its walls. This is in contrast to VTE where the chamber walls are cold. Hence in VTE, material accumulates in areas other than the substrate, ultimately resulting in inefficient materials use, and the necessity for periodic chamber cleaning to remove thick deposits. These eventually flake off the walls, creating dust within the vacuum environment. In OVPD, however, the hot chamber walls are free of deposit, and thus never require cleaning. Furthermore, the infusion of hot gas throughout the source material, which is often a powder with low thermal conductivity, uniformly volitilizes molecules. This results in thermal equilibrium within the source cell whereby the molecules form a saturated vapor whose concentration is easily controlled via adjustment of the cell temperature. In contrast, evaporation in vacuum occurs only at points where the organic is in contact with the heated source cell. The evaporation rate can abruptly change as materials settle into voids developed during sublimation. This makes the deposition rate irregular and more difficult to control.

In vacuum, the molecules travel along ballistic trajectories (i.e. along the line of sight) from the evaporation source to the substrate. This differs from OVPD where the ballistic range of the evaporated molecules is very short. Here, the *ballistic range* is the distance between the evaporation source and where the molecules reach their terminal velocity and, hence, no longer follow straight, line-of-sight paths. Kinetic theory can be used to estimate the ballistic range which limits the dimensions of the source cell and chamber. When molecules are evaporated from the source cell at $T_{source} \sim 200$–$500°C$, they undergo collisions with the residual gas which results in their deceleration and loss of kinetic energy. For a molecule of radius $r \ll mfp$, the position of the molecule of mass M *vs.* time from evaporation in the absence of significant external forces is

$$x(t) = v_0\tau(1 - e^{-t/\tau}), \tag{5.18}$$

where τ is the characteristic time it takes to reach terminal velocity, and v_0 is the initial velocity of the molecule at the source. Assuming all intermolecular collisions are elastic (which is not necessarily the case due to the internal degrees of freedom of motion of the molecules), then (Kubo, 1971)

$$\tau = M/\alpha = \frac{M}{P\pi r^2}\frac{3}{16}\sqrt{\frac{2\pi k_B T_g}{m_g}} \tag{5.19}$$

where α is the drag constant of the molecules used in their equation of motion, and T_g is the temperature in the vacuum chamber. Again, from kinetic theory,

$$v_0 = \sqrt{\frac{8k_B T_{source}}{\pi M}}, \tag{5.20}$$

which gives a ballistic range of a molecule of mass, M:

$$R_B = \frac{3}{4}\frac{k_B}{P\pi r^2}\sqrt{\frac{M}{m_g}T_{source}T_g}. \tag{5.21}$$

As an example, we consider the evaporation of molecule with $M = 300$ amu at a pressure of $P = 0.1$–1 torr, and where the principal background gas is N_2 with $m_g = 14$ amu, giving $R_B \sim 0.4$–4 mm. Hence, in OVPD the molecules remain in the vicinity of the source until they are swept out by the carrier gas, allowing for the source itself to be supersaturated with molecular vapor. Material pick-up by the carrier gas occurs in a controlled, equilibrium environment rather than in the kinetic regime of VTE.

Under equilibrium conditions, the supply of organic species is limited either by the flow velocity of the carrier gas or by the rate of evaporation. Material evaporates from the surface of the organic solid at rate, r_{evap}, given by (Shtein et al., 2001)

$$r_{evap} = k_{evap} \cdot P_{org}^{eq}. \tag{5.22}$$

where the equilibrium vapor pressure of the organic material is P_{org}^{eq}, and k_{evap} is a kinetic constant of proportionality. Vapor re-condenses at a rate, r_{cond}, proportional to the pressure of organic in the source cell, P_{org}, viz.

$$r_{cond} = k_{cond} \cdot P_{org}. \tag{5.23}$$

where k_{cond} is a kinetic constant for condensation. Organic vapors are carried out of the source at a rate, r, that is proportional to the volumetric flow velocity of the carrier gas, and the concentration of organic vapor in the source cell, P_{org}/RT_{cell}:

$$r = \dot{V} \cdot \frac{P_{org}}{RT_{cell}}. \tag{5.24}$$

Here, T_{cell} is the evaporation cell temperature, and R is the ideal gas constant. From this analysis we obtain the partial pressure of the organic evaporant in the carrier gas stream:

$$\frac{P_{org}}{P_{org}^{eq}} = \frac{k_{evap}}{\dot{V}/RT_{cell} + k_{cond}}. \tag{5.25}$$

As noted above, the pressure is related to evaporation rate, which in turn is exponentially dependent on the enthalpy of evaporation of the organic species, viz.

$$P_{org}^{eq} = P_0 \cdot \exp(-\Delta H^{vap}/RT_{cell}), \tag{5.26}$$

where P_0 is an empirical constant specific to each compound.

At high evaporation temperatures and sufficiently low gas flow velocities, the vapor and solid in the source region equilibrate (i.e. $P_{org} \approx P_{org}^{eq}$), and the concentration of organic exiting the source is constant. The resulting flux of organic species in the equilibrium evaporation regime is thus proportional to the vapor pressure and the carrier gas flow rate. At the other extreme of low evaporation temperature and high gas flow, the carrier gas sweeps the organic out of the source region as quickly as it evaporates, forcing the system away from equilibrium. In this case, the concentration of organics in the gas stream is proportional to r_{evap}, and inversely proportional to $\dot{V}$. In the kinetic evaporation regime, therefore, the flux of organic material leaving the source is independent of the carrier gas flow. Figure 5.40 illustrates how the vapor pressure of organics exiting the source varies with temperature and flow rate in both the equilibrium and kinetic evaporation regimes.

In addition to the two modes of evaporation, there are two regimes of deposition—equilibrium- and transport-limited. Similar to evaporation, in equilibrium-limited deposition the substrate temperature is sufficiently high to establish dynamic adsorption/desorption equilibria between the solid and the gas phases. Since the probability of condensation increases with decreasing surface temperature, the substrate should be kept sufficiently cold to ensure a molecular sticking coefficient, $\kappa \approx 1$.

As the carrier gas transports the organic vapors from the source to the substrate, a stagnant *boundary layer* of thickness, δ, develops above the substrate across which organic molecules must diffuse prior to

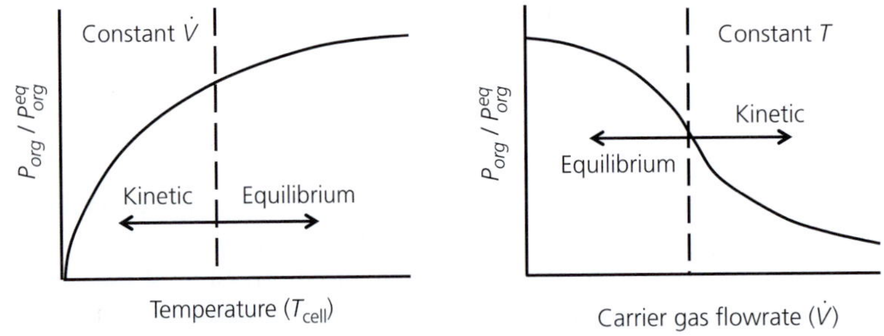

Figure 5.40 Gas transport regimes (kinetic vs. equilibrium) in OVPD and their dependence on thermodynamic conditions within the chamber (Shtein et al., 2001).

Reprinted from Shtein, M., Gossenberger, H. F., Benziger, J. B. & Forrest, S. R. 2001. Material Transport Regimes and Mechanisms for Growth of Molecular Organic Thin Films Using Low-Pressure Organic Vapor Phase Deposition. J. Appl. Phys, 89, 1470 with the permission of AIP Publishing

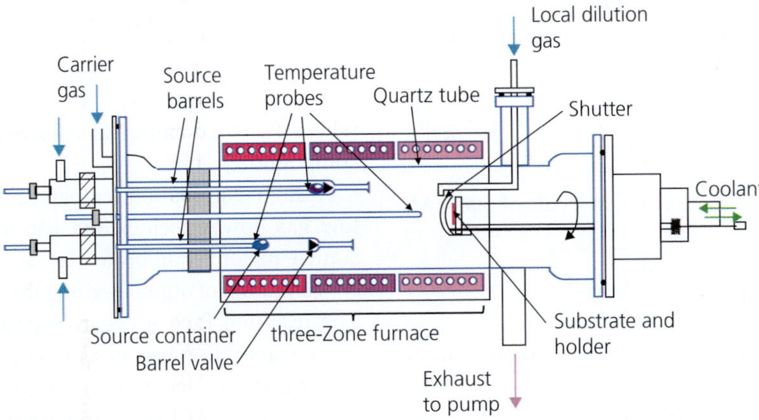

Figure 5.41 Schematic illustration of a laboratory OVPD reactor.

condensation (see Fig. 5.39). Thus, the film deposition rate depends on the relative rates of the convective and diffusive transport within the boundary layer. The molar rate of transport of organic molecules in the convection-limited regime is given by

$$r_{conv} = \frac{P_{org}\dot{V}}{RT}, \tag{5.27}$$

where P_{org} and T are the partial pressure of the organic vapor and the temperature of the carrier gas in the deposition chamber, respectively. The rate of diffusion across the boundary layer is (Shtein et al., 2001)

$$r_{diff} = D_{org} \cdot \frac{P_{org}/RT}{\delta}. \tag{5.28}$$

The net rate of deposition, r_{dep}, across the boundary layer is, then

$$r_{dep} = \frac{P_{org}}{RT} \cdot \frac{\dot{V}}{1 + \dot{V}\delta/D_{org}} \tag{5.29}$$

with

$$\delta = \sqrt{\frac{\nu \cdot x}{\dot{V}}}, \tag{5.30}$$

where x is the radial distance from the axis along the plane of the substrate holder. Equation 5.29 indicates that, at low carrier gas flows the deposition rate is limited by convection, while at the higher gas flows it is limited by diffusion across the stagnant boundary layer, which contracts quadratically with $\dot{V}^{-1}$.

A simple, practical design of a laboratory-scale horizontal OVPD reactor is shown schematically in Fig. 5.41. The furnace tube (quartz) is surrounded by a multiple zone furnace, with the hottest zones toward the gas and source inputs to the left, and the coldest zone near the substrate. This sets up a thermal gradient similar to that used in thermal gradient sublimation that forces diffusion of gases and evaporants toward the cold substrate. The substrate itself is actively cooled to prevent re-evaporation. Each molecular source material is held in a source container within its own tube or "barrel" to prevent cross-contamination. The source barrel is valved at its output to regulate flow of a source material into the main furnace tube. The valve must be designed to withstand the reactor temperatures which can be > 500°C. The source barrels are fed with their own carrier gases, joining an additional dilution gas stream in the main tube. Since the gasses enter from the hot end of the furnace, they are heated before arriving in the source and growth zones. A shutter positioned in front of the substrate can be included to further interrupt growth, although considerable source control is afforded by the barrel valves alone. Thermocouples are placed within the source barrels and at the substrate to monitor temperatures in each zone.

The gas dynamics can be measured *in situ*, and in real time using *laser induced fluorescence* (LIF) and ellipsometry (Rolin et al., 2012, 2014). In LIF, a laser beam locally excites organic species dissolved in the carrier gas while transiting the hot reactor vessel, resulting in fluorescence emission whose intensity and spectrum can indicate the organic concentration and gas temperature, respectively (see Fig. 5.42). At $P < 1$ torr typical of OVPD, the boundary layer extends to a distance of $\delta > 10$ cm from the cooled substrate surface in a chamber with a diameter of 25 cm. The calculated flow pattern in the vicinity of the substrate in a vertical OVPD chamber showing the extent of the boundary layer is illustrated in Fig. 5.43 under typical conditions used in organic film growth.

Most practical organic optoelectronic devices use multi-component or doped thin films. Because gas diffusivity is high, host and dopant molecules interdiffuse in the deposition chamber en route to the substrate, provided that the transit time is longer than the characteristic diffusion time. It has been shown that complete intermixing is achieved when the chamber length is approximately equal to its diameter (Lunt et al., 2009). An alternative for achieving mixing is to use a shower-head gas diffuser that is placed in close proximity to the substrate, well within the boundary layer itself (Schwambera et al., 2002). In both cases, the doping of one material into a second is controlled by differentially regulating the introduction of the various species into a diluting gas stream by adjusting the gas flow rate from each source cell, as

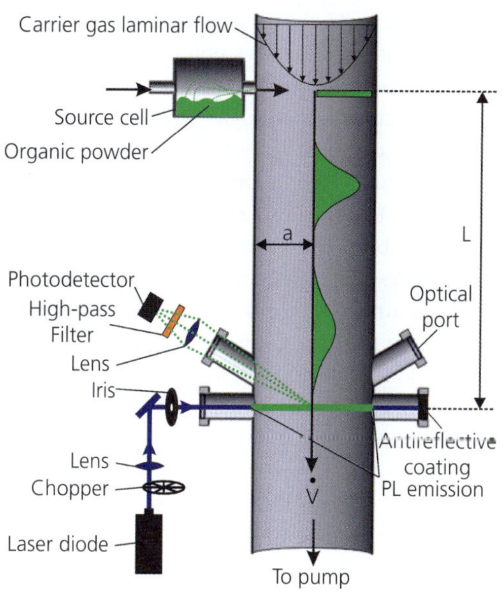

Figure 5.42 Schematic of an OVPD vessel (length, *L*, radius, *a*) with a laser induced fluorescence (LIF) monitor. Molecular concentrations from pulsed material injection along the length of the tube are illustrated by the spreading bell-shaped curves. Here, $\dot{V}$ = molar velocity, and PL = photoluminescence from the molecular species. In LIF, a laser is used to optically pump the molecules whose resulting PL is measured with a photodetector. The intensity of the signal provides the concentration of molecules in the laser beam path in real time, and the temperature is determined by the peak PL wavelength (Rolin et al., 2012).

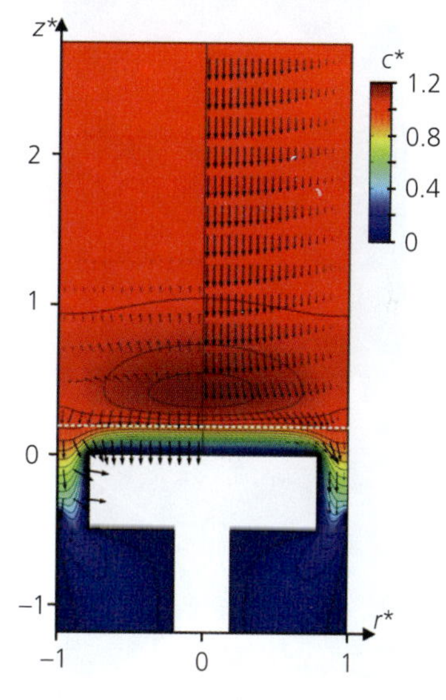

Figure 5.43 Organic concentration c^* in the OVPD chamber calculated for a chamber pressure of 1.5 torr, N_2 carrier gas flow rate of 200 sccm, and an organic concentration of 7.5 μmol/m^3. The position along the chamber axis, z^*, and radially, r^*, are normalized to the chamber radius, and c^* is normalized to the total organic concentration. Arrows in the left and right panels depict, respectively, the relative diffusive and convective fluxes of the organic species. The lengths of the arrows indicate the magnitude of the flux. Thin black contour lines above the substrate holder (the white "T") indicate the extent of the boundary layer, where the diffusive flux is opposite to the direction of convective flux of the organic material (Rolin et al., 2014).

well as the temperature of the carrier gas. By adjusting the gas flow rate, a high degree of stability and control over the relative concentration of species is possible. This situation cannot be easily achieved using only temperature control employed in VTE due to the exponential dependence of evaporation rate on temperature (cf. Eq. 5.15).

The morphologies of α-NPD thin films in Fig. 5.44 illustrate the surface morphologies attained under different thermodynamic conditions used in OVPD. At T_{cell} = 270°C, a carrier gas flow rate of > 10 standard cubic centimeters per minute (sccm), and a deposition rate < 1 nm/s, lead to molecularly smooth films, similar to those grown by VTE. For higher evaporation temperatures and deposition rates, the films roughen due to incorporation of >20 nm diameter spherical clusters. Increasing $\dot{V}$ reduces the deposition rate, thereby resulting in smoother films.

Homogeneous nucleation occurs when organic molecules in the gas phase undergo rapid cooling in proximity to the substrate within the boundary layer. If the boundary layer becomes supersaturated with the organic solute, homogeneous nucleation creates an aerosol of organic clusters within the inert carrier gas. The clusters settle onto the substrate, producing rough surfaces in Fig. 5.44. Larger deposition rates increase the organic concentration near the cold substrate, and hence the number of clusters increases. Cluster growth is reduced by maintaining a sharper temperature gradient near the substrate, or by using higher carrier gas flows where the mass transfer rate to the substrate exceeds the cluster formation rate.

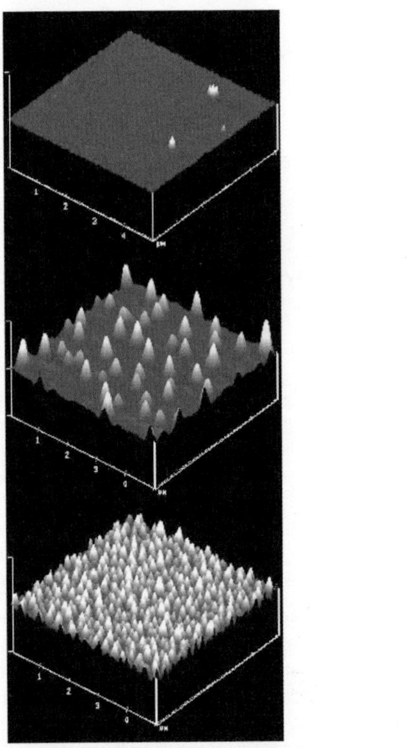

Figure 5.44 Atomic force micrographs of 200 nm thick α-NPD films deposited on silicon by OVPD. Each scan is a 5 μm square, with a vertical range of 50 nm. Films are deposited at rates of (in order, from top to bottom): 0.84, 1.07, and 1.18 nm/s, having rms surface roughnesses of 0.8 nm, 3.6 nm, and 4.5 nm, respectively (Shtein et al., 2001).

Reprinted from Shtein, M., Gossenberger, H. F., Benziger, J. B. & Forrest, S. R. 2001. Material Transport Regimes and Mechanisms for Growth of Molecular Organic Thin Films Using Low-Pressure Organic Vapor Phase Deposition. J. Appl. Phys, 89, 1470 with the permission of AIP Publishing

Considerable control over morphology can be asserted well before gas phase nucleation is observed. Indeed, it is precisely the ability to fine tune morphology that makes OVPD a powerful growth technique in organic electronics. Figure 5.45 provides a matrix of CuPc thin film morphologies accessible to OVPD as the substrate and source cell temperatures are varied at a fixed flow rate. In this matrix. The gas flow rates, deposition times and pressures are held constant for all samples. Except for T_{cell} = 460°C and T_{sub} = 2°C where the high-evaporation rate and rapid cooling near the substrate result in an aerosol of particles, the surface texture shows a monotonic dependence on temperature. The arrows point in the direction of conditions closer to equilibrium. That is, the introduction of molecules into the gas stream exponentially slows with a decrease in cell temperature. Thus, adsorbed molecules have a longer duration to seek low energy (near-equilibrium) surface sites before being buried by subsequently arriving material. Similarly, increasing the substrate temperature adds kinetic energy to the surface molecules, allowing them to move more quickly to sites of lower energies. Films generated near to equilibrium show a needle-like morphology, indicating the growth of crystallites that protrude above the otherwise flat film surface. Amorphous films are produced far from equilibrium.

Figure 5.46 shows the complementary data obtained at fixed temperatures, but at varying carrier gas flow rates. These micrographs are in order of increasing rate (from Fig. 5.46a to d), and hence toward less equilibrium growth conditions. As in Fig. 5.45, near-equilibrium growth results in long and narrow crystallites, whereas at high flow rates, the growth is far from equilibrium, resulting in flat and amorphous film morphologies.

Apparently, the presence of a carrier gas during deposition and the independent control of multiple chamber pressure and temperature conditions can be exploited to create a broad variation in thin film morphology. Thus, the films can approach an equilibrium morphology, depending on the balance between kinetic and the diffusion limited conditions.

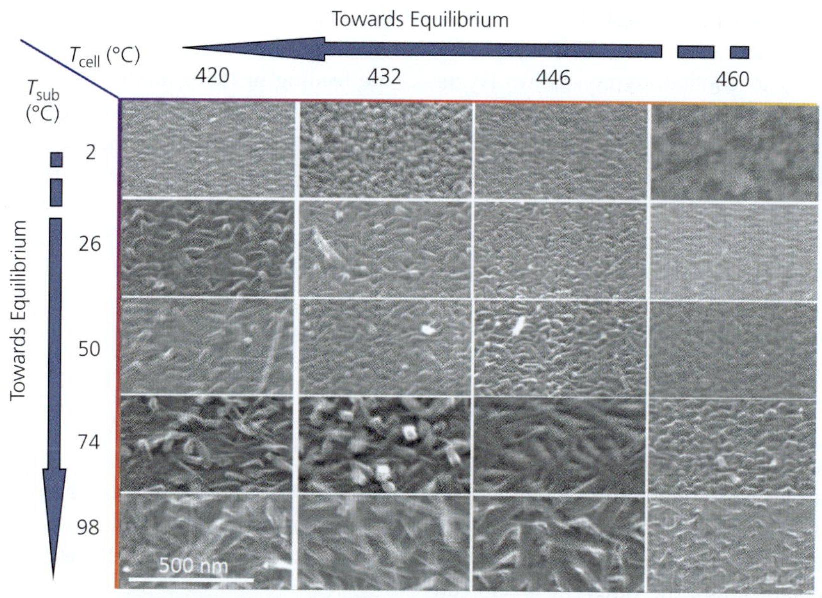

Figure 5.45 Scanning electron microscope (SEM) images of CuPc films grown on ITO vs. substrate and source cell temperatures at a constant The carrier gas flow rate is 100 sccm, and the chamber pressure is 0.45 torr (Yang et al., 2005b).

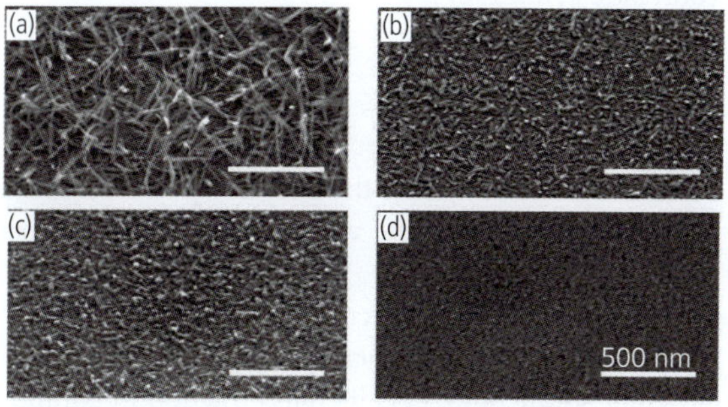

Figure 5.46 SEM images of the surface morphologies of CuPc obtained at flow rates of (a) 100, (b) 125, (c) 150, and (d) 200 sccm at constant source and cell temperatures.

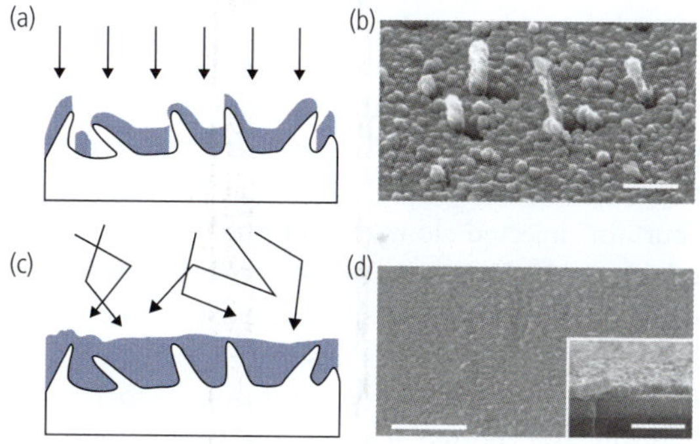

Figure 5.47 Deposition by (a) VTE or the kinetic regime of OVPD results in directional trajectories from source to substrate. If there are irregularities on the substrate surface, directional deposition can result in a coating (shaded regions) that leaves voids and gaps. (b) SEM image of OVPD-grown CuPc nanocrystalline protrusions resulting from strained Stranski–Krastanov growth on a polycrystalline ITO substrate. The CuPc is coated with a layer comprising PTCBI/BCP/Ag grown by VTE. Voids are visible to the right of each crystallite. (c) In the diffusive OVPD growth regime, the adsorbed layer conformally coats the substrate surface, burying the irregularities. (d) SEM image of a CuPc film as in (b), coated by a 50 nm thick layer of PTCBI showing complete substrate coverage. Inset: Cross-section of the film showing the absence of voids. Scale bars (white) in the SEM images are 500 nm (Yang et al., 2005a).

First publication in Nature Materials, 2005.

An example of the different morphological control exerted by OVPD compared to VTE is the growth of CuPc on a microcrystalline ITO surface shown in Fig. 5.47. Local strain imparted by the ITO results in a Stranski–Krastanov growth mode, whereby a thin wetting layer gives way to long crystalline protrusions that rise well above the film surface, similar to the morphology in Fig. 5.46a. The directional deposition of by VTE results in shadowing of the evaporant by the protrusions leaving voids in the film (see Fig. 5.47a). Indeed, this is observed in the SEM image in Fig. 5.47b of a PTCBI/BCP/Ag composite layer deposited onto the CuPc surface. Voids are clearly apparent to the right of each protrusion since the deposition source is offset slightly to the left of the center of the image. Directional deposition is also accessible in the kinetic growth regime of OVPD. However, in the diffusive regime, the molecular paths within the boundary layer are randomized, resulting in complete conformal coating of the surface features (Fig. 5.47c), shown for OVPD grown PTCBI in Fig. 5.47d. There are also no voids in the cross-sectional image in Fig. 5.47d, inset. Such morphological control, ranging from uniform blends to nanocrystallites (Yang et al., 2007), to continuous crystalline layers extending across entire substrate surfaces (Lunt et al., 2007) are made possible by varying conditions used during growth. In Chapters 6–8 we show that this leverage over morphology provides a means for creating thin film structures that result in higher performance OLEDs (Zhou et al., 2005), transistors (Shtein et al., 2002), and solar cells (Yang et al., 2007, Song et al., 2014).

While hot OVPD reactor walls prevent condensation, if two materials with widely different evaporation and decomposition temperatures are co-deposited, the low evaporation/decomposition temperature species can potentially decompose on contact with walls heated to prevent adsorption of the species with the higher evaporation temperature. To prevent decomposition, the different layers can be evaporated in separate chambers held at temperatures appropriate to the particular species being evaporated, with the substrate translated between the chambers as required. When two such materials must be co-doped into a single film, their widely different sublimation and decomposition temperatures may be accommodated by decreasing collisions with the chamber walls by increasing the carrier gas flow rate. Alternatively, a lamellar protective gas "curtain" injected along the periphery of the furnace tube can minimize molecular collisions with the walls. These complexities do not exist in VTE deposition, where each species is separately heated to its individual evaporation temperature and then is ballistically transported through the cold-walled vacuum chamber to the substrate.

5.4.2.4 Film deposition from liquid solution

Casting films from the solution phase is widely used for the deposition of large molecular weight or low decomposition temperature materials that cannot be deposited from the vapor phase. Hence, solution processing has been almost universally adopted for the deposition of polymers and many fragile, or large molecular weight species. The organics must be soluble to be deposited from solution, which may require chemical modification or functionalization. This modification may, in turn, adversely impact the optical and conductive properties of the molecule, since the solubilizing groups may sterically hinder or alter the desired molecular packing and film morphology.

Spin-on and casting are the two most frequently used means for solution deposition. Additional means include spray-on, uniform spreading using a *doctor blade*, *slot dye coating*, and attachment of free standing films by lamination. *Solution casting* entails the feeding of the solubilized molecules onto a substrate that is either stationary, or is translated beneath the solution source (Siemann, 2005). The spin-on method is accomplished by placing a drop of solvent containing the solute onto a substrate that is rapidly rotated. Ultrathin films are produced by the competition between centrifugal force that draws the film toward the substrate edges where the excess is ejected, and evaporation of solvent that results in an increase in viscosity as the film solidifies.

Determining film thickness vs. spin rate is complicated by the interplay between fluid flow in the surface plane, and evaporation. The change of film thickness, h_f, with time, t, of a liquid with initial viscosity, η_0, and density, ρ, on a disc rotating at angular velocity, ω, is (Emslie et al., 1958)

$$\frac{dh_f}{dt} = -\frac{2\rho\omega^2 h_f^3}{3\eta_0}. \tag{5.31}$$

This is counterbalanced by the evaporation rate (Bornside et al., 1987):

$$r_{ev} = \left[\frac{cD_g p_s M_s}{\eta_g^{1/2}\rho RT}\right]\sqrt{\omega}\left(x_{s0} - x_g\right). \tag{5.32}$$

Here, c is a constant, D_g is the diffusivity of the solvent in the gas phase, p_s is the vapor pressure of the pure solvent at temperature T, η_g is the kinematic viscosity of the gas, M_s is the molecular weight of the solvent, x_{s0} is the initial mass fraction of the solvent to the polymer solute, and x_g is its mass fraction in equilibrium with the gas phase. From these two expressions, we obtain the film thickness when the fluid flow ceases (i.e. the wet film becomes immobile and resistant to further changes in thickness) (Hall et al., 1998):

$$h_{wet} = \left[\frac{3\eta_0}{2\rho^2\omega^{3/2}}\frac{cD_g p_s M_s}{\eta_g^{1/2}\rho RT}\left(x_{s0} - x_g\right)\right]^{1/3}. \tag{5.33}$$

Since the final film thickness is determined by evaporation only, we have

$$h_f = h_{wet}(1 - x_{s0}) \propto \left[\frac{\eta_0}{\omega^{3/2}}\right]^{1/3}(1 - x_{s0}). \tag{5.34}$$

That is, the film thickness dependence on rotational velocity follows $1/\sqrt{\omega}$. This dependence is illustrated in Fig. 5.48 for the case of polystyrene of different concentrations (and hence initial viscosities) in a toluene.

The functional dependence on rotational velocity holds for a wide range of film thicknesses, from hundreds of microns to less than 10 nm, with typical rotational velocities from 500 to 5000 rpm. While spin-on deposition is useful for laboratory demonstrations, it is not scalable to substrate sizes that exceed a few tens of centimeters. Also, solvent flow around sharp edge features on a patterned substrate can cause non-uniformities in film thickness that can affect the device properties. Conversely, balancing film viscosity, temperature, and rotational speed can planarize an otherwise irregular surface (Bornside, 1990). Spray-coating or doctor blades that are translated across the substrate surface can produce nanometer thick films on substrates without size limitations.

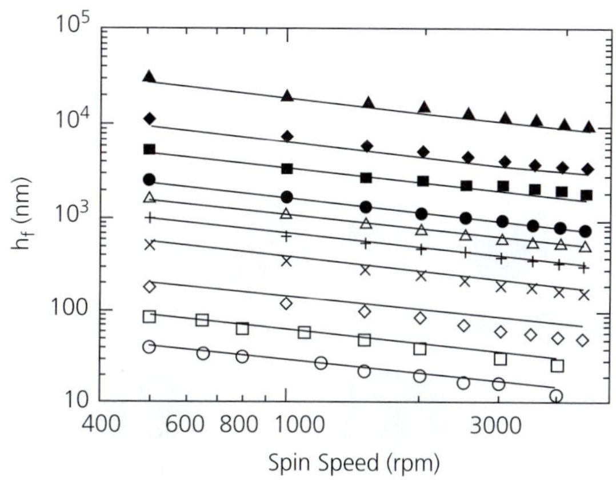

Figure 5.48 Polystyrene (PS) film thickness vs. spin speed for different PS concentrations in a toluene solution. From bottom, the wt% of PS in toluene is 0.5, 1, 2, 4, 6, 8, 10, 15, 20, and 30. The straight lines corresponding to Eq. 5.33, at $T = 303$ K, $m_g = 0$, $\eta_\gamma = 0.1553$ cm^2/s, $c = 0.5474$, $\rho = 0.87$ g/cm^3, $M_s = 92$ g/mol, $p_s = 0.029$ atm, $D_g = 0.086$ cm^2/s, and η_0 is an inverse function of m_{s0} (Hall et al., 1998).

Monolayer deposition control, and even layer-by-layer growth can be achieved by solution casting using the *Langmuir–Blodgett* (L-B) technique (Blodgett and Langmuir, 1937). This method is based on casting a single layer of molecules onto the surface of a shallow trough filled with water. The molecules are functionalized with a *hydrophilic* unit on one end, and a *hydrophobic* unit on the other. By withdrawing a substrate with a hydrophilic surface from the trough, a single layer of molecules attaches to the substrate surface. The surface energy of the molecules on the substrate must be greater than that on the water. As shown in Fig. 5.49a, this method is capable of applying one monolayer at a time until the desired total film thickness is achieved. Furthermore, the functionality and orientation of the molecules can be varied layer-by-layer by choosing whether to submerge or withdraw the substrate from the solution at a particular step. Several molecular orientations that can be achieved by variations in the dipping sequence are shown in Fig. 5.49b. To deposit a continuous, nearly defect-free film, barriers surrounding the trough are moved inward to exert pressure on the floating monolayer of source molecules. After each translation of the substrate into the solution, the

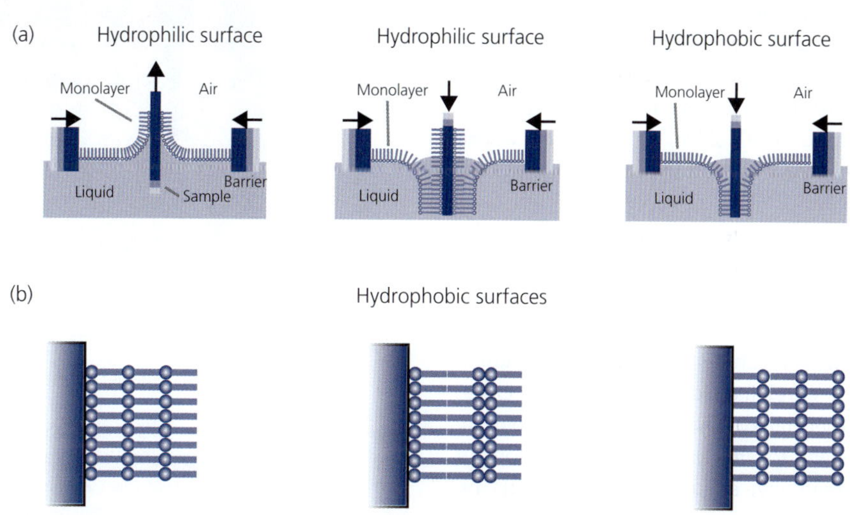

Figure 5.49 Langmuir–Blodgett process for layer-by-layer growth of molecules having a hydrophilic end group (circles) and a hydrophobic tail. (a) Using a substrate with a hydrophilic surface, a monolayer of molecules is attached as the substrate is withdrawn from the solution. On re-submergence, the now hydrophobic surface picks up a second ML of hydrophobic molecules. Alternatively, a hydrophobic surface picks up the ML during submersion. The submersion/withdrawal cycle can be repeated many times to build up thick films one monolayer at a time. After each cycle, barriers at the edge of the tough are moved inward to maintain an unbroken monolayer on the solution surface. (b) Various layer orientations that can be assembled on a hydrophobic surface depending on the direction of motion of the substrate relative to the solution.

barriers are moved to eliminate gaps due to depletion of the molecules now attached to the substrate.

While capable of depositing films many thousands of monolayers thick, this technique has several limitations. The molecules, substrate and solution must be chosen to create a space-filling monolayer on each dip of the substrate that is free of defects such as tears or gaps, and to ensure that the orientation of the molecules in solution is uniform. Also, the surface of the solution must be unperturbed to ensure a defect-free molecular raft. Hence, the process is slow and impractical for coating large surface areas. A different strategy based on selecting charged molecules circumvents some of the shortcomings of the L-B technique. Here, alternating layers of cationic and anionic molecules are picked up using a charged, or neutral dielectric substrate (Iler, 1966). This method has been used to selectively deposit monolayers of large molecules and polymers as illustrated in Fig. 5.50 (Decher, 1997).

A shortcoming of solution deposition is the difficulty in growing multilayer structures where each layer has a different composition. After depositing the first film and removing all of the residual solvent by drying in vacuum or at elevated temperatures, the second layer is then deposited. However, if a similar solvent is used for the second layer as for the first, it can redissolve the first layer, thereby creating defects or non-uniformities in the film. To prevent layer dissolution, the second layer must use an "orthogonal" solvent, that is, a solvent that does not interact with the first deposited film. Examples of two orthogonal classes are hydrophilic and hydrophobic solvents (Gong et al., 2005, Meerholz, 2005). There is a third class of hydrofluoroether-based solvents that can also be introduced to provide further process flexibility (Zakhidov et al., 2008, Lassiter et al., 2013), although these have not been exploited to a significant extent.

Even by the use of three orthogonal solvents, the compositions and complexities of multilayer structures that are needed to achieve high performance electronic devices are severely limited. Hence, other approaches have been sought to meet this challenge. One example is to harden (or cure) the deposited layer by cross-linking using thermal or ultraviolet (UV) radiation, making it resistant to attack by subsequent solvent exposure. Alternatively, a film can be protected by capping it with a continuous, chemically inert layer prior to subsequent solvent-based depositions. This strategy has been employed, for example, in solution-processed tandem organic solar cells, whereby the intervening ZnO layer serves the dual purpose of protecting the first cell deposited on the substrate, as well as providing a transparent charge injection layer separating the two sub-solar cells forming the stack (Gilot et al., 2007). Similarly complex structures use using metal nanoparticle layers combined with a material such as PEDOT:PSS that is water soluble (and hence impermeable to organic solvents used in depositing the top and bottom cells) prevent penetration of the solvent from the upper to the lower cells (see Fig. 5.51) (Hadipour et al., 2006, Che et al., 2018). Although barrier layers can prevent deterioration of active device regions, even small defects or irregularities can result in pin-holes that allow penetration of solvents into the underlying layer. Furthermore, the insertion of such

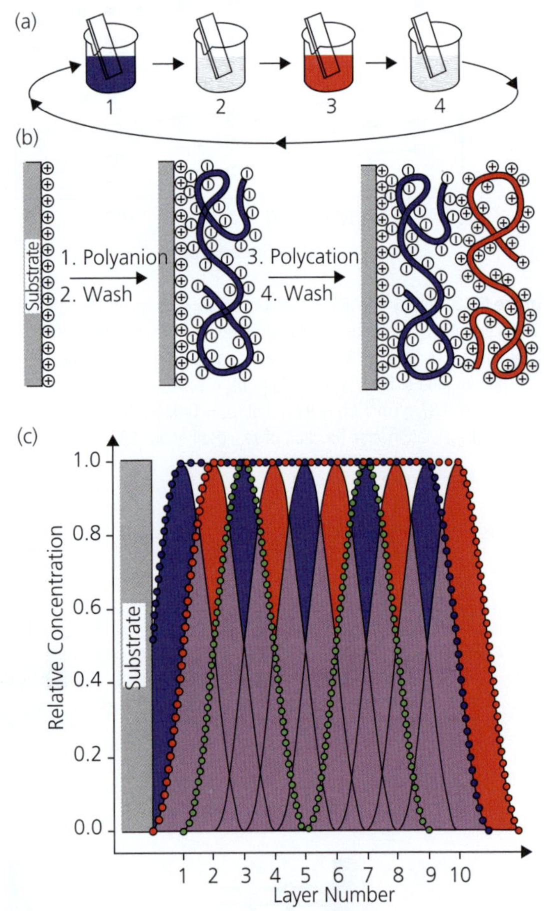

Figure 5.50 Method for layer-by-layer deposition of high molecular weight compounds using electrostatic selection. (a) Each beaker contains a different solution: **1** contains the polyanion, **2** a wash, **3** polycation, and **4** a wash. (b) Moving the functionalized substrate between the beakers alternately deposits a single polyanion and polycation layer, building up the structure as shown. (c) Schematic diagram of the distribution of molecules within each layer, using the same color scheme for anionic (blue) and cationic (red) species as in (a) and (b) (Decher, 1997).

From Decher, 1997. "Fuzzy Nanoassemblies: Toward layered polymeric nanocomposites," Science, 277, 1232. Reprinted with permission from AAAS.

barriers is incompatible with device operation, and hence their use can limit design flexibility.

Finally, solvents can dissolve or physically damage small molecule layers that have been previously vacuum-deposited onto a substrate. When combining solution and vapor-deposited compounds, it is preferred to start by depositing the solution-based structure, followed by vacuum deposition of the small molecule layers.

From the foregoing discussion, we see that solution deposition is generally incompatible for use in devices with more than just a few functional layers. Nevertheless, solution processing remains attractive due to its compatibility with patterning via ink-jet printing, as well as its potential for low cost deposition over large substrate areas. Indeed, when it is followed by vacuum deposition of small molecule based multilayers, complex and functional layer systems and devices can be achieved. While providing constraints in device engineering, many promising methods for avoiding its limitations have been devised, making it an essential process in organic electronics.

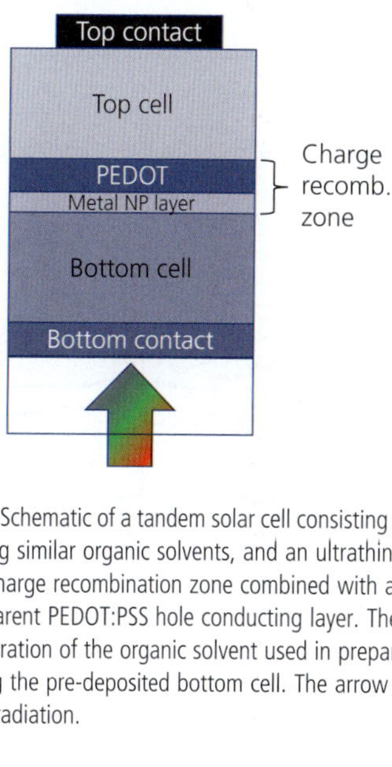

Figure 5.51 Schematic of a tandem solar cell consisting of two sub-cells deposited using similar organic solvents, and an ultrathin metallic nanoparticle charge recombination zone combined with a thicker, water soluble transparent PEDOT:PSS hole conducting layer. The PEDOT prevents penetration of the organic solvent used in preparing the top cell from dissolving the pre-deposited bottom cell. The arrow indicates the incident solar radiation.

5.5 Post-growth control of structure

Often the desired morphology is not achieved during deposition. Indeed, the kinetics of growth often prevent molecules from finding their lowest energy, equilibrium positions within the solid, resulting in microcrystalline or even more frequently, amorphous thin film morphologies. To drive the film away from these unstable, or metastable morphologies to one closer to equilibrium, various post-deposition annealing strategies have been developed for both polymers and small molecules. The principle of annealing is to impart energy into the solid over a duration sufficiently long to allow the molecules to locally rearrange into phase-segregated crystallites in a structure closer to equilibrium. The methods used generally fall into two categories: *thermal annealing* and *solvent vapor annealing* (SVA).

The processes and effects of annealing of a solution-cast film are illustrated in Fig. 5.52. The source solution comprising one or more molecular species dissolved in a solvent is deposited onto a substrate where it assumes a non-equilibrium morphology. Then the energy is added by either heating (thermal annealing) or by exposure to a solvent vapor in a sealed vessel (SVA). The molecules are able to migrate over limited distances within the film, allowing them to reorganize into a structure that is closer to equilibrium. If there is more than one molecular species, they may phase separate into entangled regions. Independent of the number of species, the annealed sample can develop crystalline domains that vary in size from nanometers to micrometers. The crystallites may also be separated by intervening amorphous regions, depending on the annealing conditions used. Similar processes can also be applied to vapor-deposited materials. In the following, we will explore both thermal and solvent annealing in greater detail. Indeed, structural annealing is a powerful method that can lead to remarkable improvements in organic solar cells and transistors, both of which require a degree of crystalline order to achieve high performance.

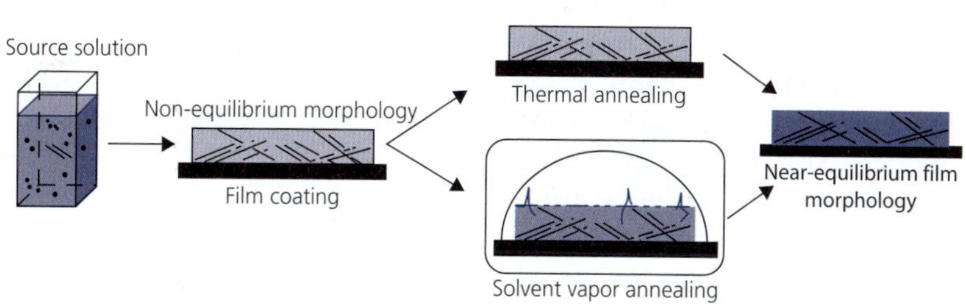

Figure 5.52 A mixture of the material to be deposited and the solvent provides the source solution. The as-deposited material assumes a non-equilibrium morphology (*top*). Using either solvent vapor or thermal annealing, the film morphology reorganizes into a structure closer to equilibrium (*bottom*).

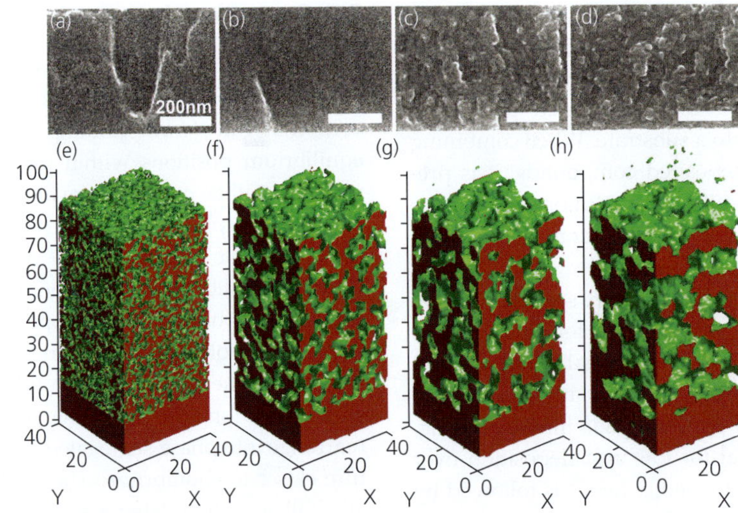

Figure 5.53 Scanning electron micrographs of 4:1 CuPc:PTCBI films annealed under different conditions. (a) As-deposited, and annealed for 15 min at (b) 450 K, (c) 500 K, and (d) 550 K. (e–h) Monte Carlo simulations of the equilibrium configurations of a 1:1 CuPc:PTCBI mixture obtained at the temperatures corresponding to the micrographs immediately above. These configurations are based on the cohesive energy of each molecule, viz. $E_{CO} = E_0 exp(-\Delta H^{vap}/RT)$, where E_0 is a constant (Peumans et al., 2003).

First publication in Springer Nature, 2003.

5.5.1 Thermal annealing

Thermal annealing of mixtures of the small molecules, CuPc and PTCBI is shown in Fig. 5.53. Homogeneous blends of 4:1 CuPc:PTCBI were thermally annealed while the surface was confined to its as-grown flat morphology by a 100 nm thick Ag film (Peumans et al., 2003). Cross-sectional SEM images of films annealed at different temperatures are presented in Fig. 5.53a-d, along with spatial "maps" that simulate the structural evolution during annealing in Fig. 5.53e-h. The as-grown homogeneous mixture in Fig. 5.53a and e forms nanocrystalline domains consisting nearly exclusively of pure regions of each molecular constituent. The calculation assumes an equilibrium configuration is achieved when the sum of the free energy of all contacts between dissimilar molecules is minimized. The internal film structure is modified by annealing even though the film surface is held flat by the Ag cap.

In addition to forming clusters, annealing promotes crystallization within the domains themselves. Figure 5.54 shows X-ray diffraction patterns of the same blends immediately after growth, and following annealing for 15 min at 400 K, 450 K, 500 K, and 550 K. Several peaks emerge after annealing at 450 K and above. These peaks become more intense at higher annealing temperatures, indicating the growth of CuPc crystalline domains within the blends. It is inferred, therefore, that PTCBI is being excluded from the C_{60} domains. No peaks are observed that can be attributed to PTCBI. This suggests that domains containing this species are amorphous, and they may also contain residual C_{60} molecules. It was found that surfaces without an Ag capping film roughen considerably on annealing at these same temperatures and times, and large crystalline needles emerge from the surface at high temperatures.

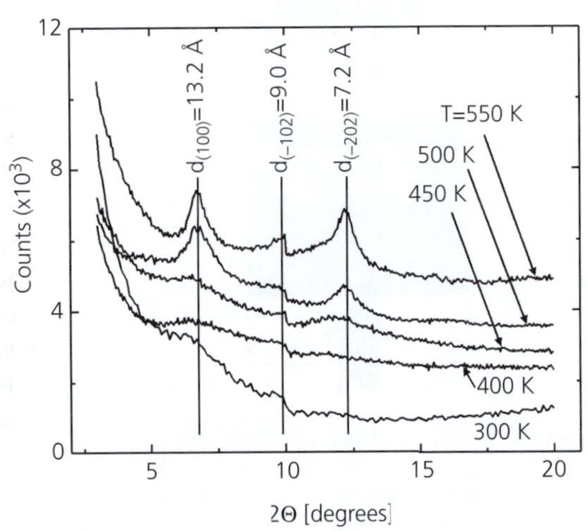

Figure 5.54 X-ray diffraction patterns of 1:4 CuPc:PTCBI films annealed at different temperatures. The lattice *d*-spacings refer to diffraction from the CuPc crystal planes indicated. The domain size of (12±1) nm is calculated from the width of the diffraction peaks, and is consistent with calculations in Fig. 5.53 (Peumans et al., 2003).

First publication in Springer Nature, 2003.

Similar crystallization has been observed after annealing OPV active regions consisting of P3HT mixed with $PC_{61}BM$. Both the individual constituents as well as the blends crystallize on heating to just above the melting temperature of P3HT (T_m = 205°C), or the $PC_{61}BM$ *glass transition temperature* (T_g = 155°C) (Ma et al., 2005, Verploegen et al., 2010). The T_g is the temperature where an amorphous (glassy) solid changes to a molten or elastic-like phase.

Crystallization of a polymer maintained at or above its melting or glass transition temperatures for time t, is described by the *Avrami equation* (Augis and Bennett, 1978):

$$V(t) = V(0)[1 - \exp(-(kt)^n)], \tag{5.35}$$

where k is the reaction rate given by

$$k = k_\infty \exp(-\Delta E/k_B T). \tag{5.36}$$

Here, n is a dimensionless exponent in the range 1–3 that roughly corresponds to the dimensionality of the crystallites of volume $V(t)$ formed during annealing over time, t, ΔE is the activation energy for annealing, and is a constant (Avrami, 1940).

Crystallization starts from a "germ" nucleus that exists in the as-grown film. The nuclei arise from native structural heterogeneities that develop during deposition or synthesis, or they can form around extrinsic impurities or substrate defects. Once heated, crystallites nucleate around these irregularities and grow until the entire film volume is consumed. The dimensionality of the growth depends not only on the temperature (and time), but also on the density of nucleation sites. In the case of *heterogeneous nucleation* in Fig. 5.55, a high density of nucleation sites prevents lateral growth over large distances. This confines the crystallites, which must therefore grow vertically into an ordered orientation.

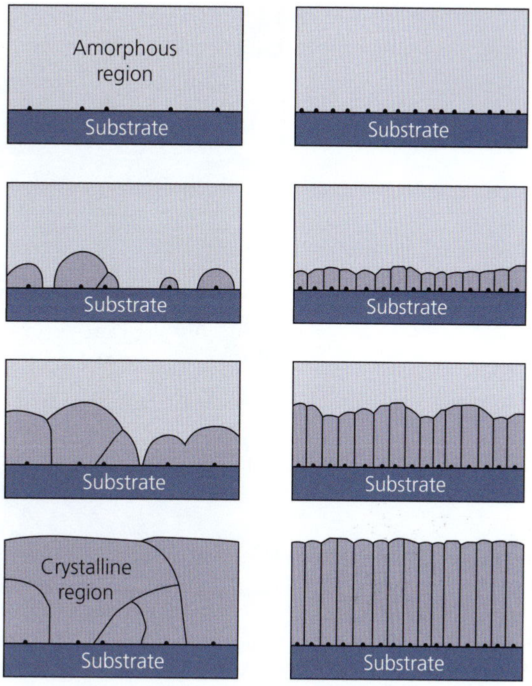

Figure 5.55 Two different modes proposed for crystallization of $PC_{61}BM$ as inferred from X-ray data. (*left*) Nucleation at a low density of germ nuclei, with an Avrami exponent of n = 1.8, and (*right*) at a higher density of sites with n = 1. The higher exponent is indicative of nearly two-dimensional (2D) growth, whereas n = 1 suggests ordered one-dimensional domain growth (Verploegen et al., 2010).

5.5.2 Solvent annealing

Control over crystallization following film growth can also be achieved when the deposit is exposed to a solvent, or its vapor. Indeed, SVA can lead to highly controlled nucleation and growth of crystallites in both polymer and small molecular weight solids. This process is typically used in conjunction with solution deposited films. During the initial film deposition, rapid evaporation of the solvent freezes the film into a morphology far from its most stable state. Once the film is free of its initial solvent, it is exposed for several minutes to the vapors of a second solvent of different composition from the first. The secondary solvent molecules penetrate the film, enabling the molecules to seek a configuration closer to equilibrium.

Exposure to the solvent, DCM, of the active regions of solution-deposited DPASQ capped with VTE-grown C_{60}/PTCBI films provides a dramatic example of the morphological modifications that films can undergo by SVA (Zimmerman et al., 2013). The small DCM molecules induce 2D, inhomogeneous spherulitic growth of crystals that proceeds rapidly along the substrate plane, radiating out from nucleation sites within the DPASQ film. When capped with C_{60} and then subsequently exposed to the solvent vapor, crystallites form in the DPASQ that align to self-assembled nanocrystalline domains that spontaneously form during vacuum deposition of the C_{60} layer. Indeed, C_{60} has a strong tendency to crystallize on almost any surface into close-packed layers.

The crystallization of the underlying amorphous layer whose domains are aligned to the overlying crystal is an example of *inverse quasiepitaxial growth*. As in the case of conventional quasiepitaxy (Section 2.6.2), there is no well-defined commensurate relationship between the first layer and the cap. Inverse

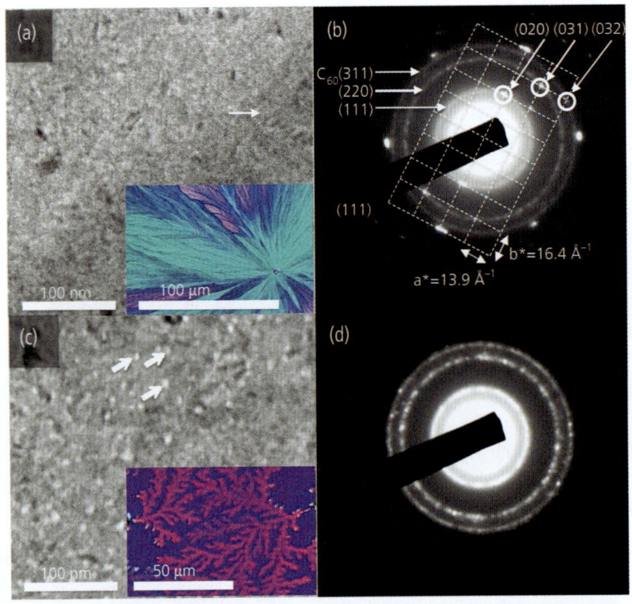

Figure 5.56 (a, c) Transmission electron micrographs and (b, d) selected area electron diffraction (SAED) patterns after solvent vapor annealing (SVA) of (a, b) DPASQ/C_{60}/PTCBI spherulite structures, and (c, d) blended squaraine/C_{60}/PTCBI layers. (a) There is a relatively unstructured surface morphology in the spherulite. (c) Irregularities (arrows) suggest phase separation of the blend. Inset optical micrographs show (a) a spherulite and (b) dendrite formation due to phase separation in the blend. (b) The SAED pattern of DPASQ/C_{60} spherulites show the presence of single-crystal DPASQ and polycrystalline C_{60} with DPASQ lattice registrations at the (020), (031), and (032) C_{60} planes (circled). (d) SAED pattern of the blended film shows decoration of the C_{60} diffraction rings with DPASQ crystallites (Zimmerman et al., 2013).

quasiepitaxy differs since alignment of the underlying layers to the cap occurs following growth of the cap. In conventional quasiepitaxy, alignment is to the substrate. This suggests that there is a strong energy-driven tendency to crystallize even when a confining cap is present—a situation similar to that in Fig. 5.53.

The crystalline texture and morphological properties of an inverse quasi-epitaxial DPASQ/C_{60} structure are elaborated in Fig. 5.56. A bright field transmission electron micrograph as well as an optical micrograph of a spherulitic crystal structure formed during SVA are shown in Fig. 5.56a. The DPASQ aligns to the predeposited C_{60} cap, but is not in perfect registry, as indicated by the DPASQ diffraction spots that lie slightly off of the C_{60} rings in Fig. 5.56b. However, when the asymmetric DPASQ is blended with its symmetric molecular analog, DPSQ, in a 4:6 ratio, the diffraction spots decorate the C_{60} diffraction rings shown in Fig. 5.56d. The micrographic image of the dendritic structure is shown in Fig. 5.56c. This indicates that inverse quasiepitaxial growth does not proceed along the substrate plane, but rather perpendicular to it.

The morphological studies of Fig. 5.56 lead to the growth scenario in Fig. 5.57a. The upper layer (circles) depicts the self-assembled nanocrystalline cap C_{60} layer, and the lower squaraine layer (blue rectangles) shows the direction of fast crystallization (arrows) either in the film plane (thus forming a *spherulite*, Sph.), or into the film bulk (resulting in inverse

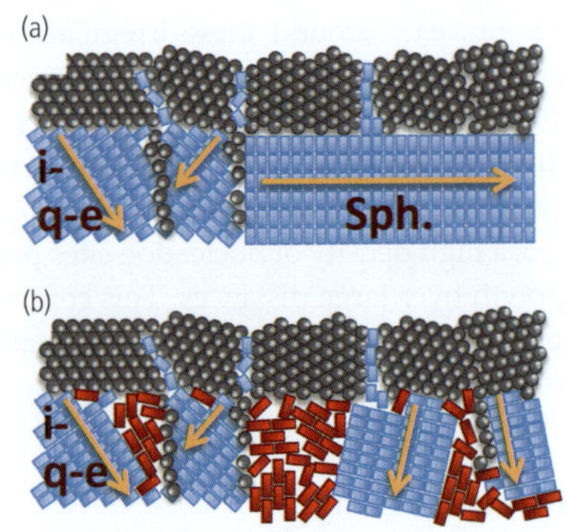

Figure 5.57 Illustration of the crystallization during SVA of (a) DPASQ/C_{60} and (b) a blended DPSQ:DPASQ/C_{60} bilayer. The arrows show the direction of crystalline domain growth, initiated at the interface between the squaraine and C_{60} (Zimmerman et al., 2013).

quasiepitaxial growth, i-q-e). In the blended squaraine in Fig. 5.57b, the squaraine layer phase separates into nanocrystalline DPASQ-rich areas (blue rectangles) that interdiffuse into the C_{60}, and the DPSQ-rich areas (red rectangles).

The impact that molecular order at interfaces has on charge recombination dynamics was discussed in Chapter 4. In general, a high degree of order between molecules increases conductivity and exciton diffusion, but it can also increase charge recombination. A balance between these competing processes has been achieved by SVA of small molecule OPV cells, whereby disorder at the interface is pinned, although order is developed within the bulk of the contacting layers (see Section 7.4.3). This apparently complex morphological tradeoff is obtained by capping the amorphous squaraine film with a vacuum deposited C_{60} layer. Adjacency of molecules at the interface is not disrupted by subsequent exposure to the solvent vapor that reorganizes the bulk layer structure, thereby nucleating the growth of conductive nanocrystalline channels (Zimmerman et al., 2012). Similar strategies for using SVA has also been applied to polymer OPVs with similar success at manipulating the nanomorphologies of donor–acceptor blends (Li et al., 2007). The process is now frequently used to improve performance of devices where crystalline order is advantageous.

Finally, SVA can control crystallinity to allow for high conductivity in the channels of organic field effect transistors. For example, when the solution processed hole conducting organic small molecule, triethylsilylethynyl anthradithiophene, is exposed to 1,2-dichloroethane, the morphology changes from an amorphous morphology attained on spin-coating, to a polycrystalline structure whose grains are as large as 1 μm in the transistor channel. The consequence of crystallization is an increase in the transistor field effect mobility from 0.002 $cm^2/V\ s$ for the as-spun-on sample, to 0.2 $cm^2/V\ s$ after SVA—a 100-fold improvement. The changes observed on SVA do not strongly depend on the polarity or the boiling temperatures of the solvents employed (Dickey et al., 2006). This provides further evidence that the solvent vapors act simply by mechanical intercalation between the larger molecules, allowing them to move to more energetically favorable locations during the process.

5.6 Film patterning

The primary goal in film growth is to cover the substrate surface with a highly uniform and pure layer (or layers) of the desired composition and morphology, using minimal energy and materials. However, almost every conceivable electronic device requires patterning to define its active area. This entails the definition of electrode size and shape, and often the active organic material itself must be patterned to provide a well-defined path for optical, current or thermal transfer. Pattern sizes ranging from tens of nanometers to centimeters are common for optoelectronic components, ranging from quantum-effect devices to large area lighting appliances and solar cells.

An illustration of the challenges faced in patterning of electronic devices is in defining the red (R), green (G), and blue (B) pixels that comprise a full color OLED display. These sub-pixel elements can be as small as 10 μm, depending on the resolution and overall size of the display (larger displays have larger pixels since the viewer is positioned farther from the screen). An example of the sub-pixel arrangement of an OLED display, and a cross section illustrating the patterning challenges is shown in Fig. 5.58. A display requires patterning of millions of very small pixels precisely spaced across very large (several meters) substrates at ~100% yield. Pixel yield is the ratio of functioning devices to the total number fabricated. Furthermore, materials used for one device (or pixel) must not cross over into the region of its neighbor. This cross-contamination inevitably decreases the color purity of the image. Hence, the multiscale fabrication challenges for patterning very small devices over very large substrates are peculiar to the field of organic electronics, which uniquely benefits from scaling spatial dimensions over many orders of magnitude.

5.6.1 Shadow mask patterning

The most common means for patterning organic devices grown from the vapor phase, and for depositing contacts to solution processed thin films, is via the use of shadow masks. The mask is comprised of a thin metal such as Mo. The mask apertures through which

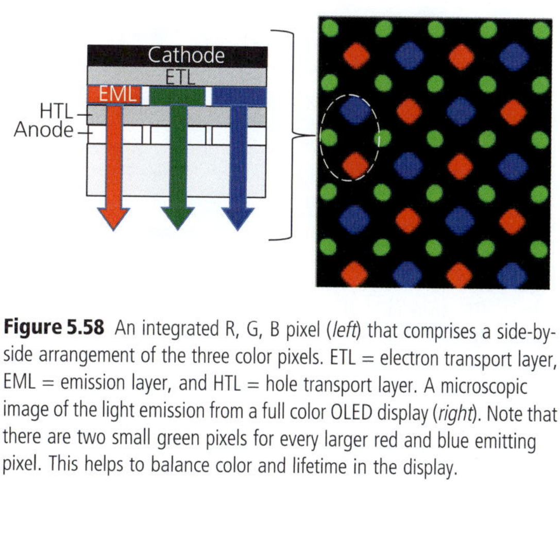

Figure 5.58 An integrated R, G, B pixel (*left*) that comprises a side-by-side arrangement of the three color pixels. ETL = electron transport layer, EML = emission layer, and HTL = hole transport layer. A microscopic image of the light emission from a full color OLED display (*right*). Note that there are two small green pixels for every larger red and blue emitting pixel. This helps to balance color and lifetime in the display.

material is deposited are defined by conventional photolithography. The mask is placed in close proximity, or in contact with the substrate in the direction facing the evaporation sources. In VTE, the evaporant travels ballistically along a line of sight from source to substrate, depositing through the mask apertures. The pattern resolution is determined by the thickness of the mask, t, the width of the aperture, w, the distance from source to substrate, h, the separation between the mask and the substrate, s, and the linear dimension of the source, l (see Fig. 5.59a). Directly behind the aperture, the deposit reaches its full thickness that is monitored in real time with a quartz crystal microbalance. Due to the finite dimensions of s, t and l, some material is deposited in the shadowed area, forming a penumbra of deposit of width (Shtein et al., 2003)

$$r \approx \frac{(s + 2t)l}{2h} \tag{5.37}$$

As the pattern size decreases, spreading of the deposit beneath the mask becomes a larger fraction of w, ultimately limiting the achievable pattern resolution. To improve resolution (i.e. by minimizing r), the mask thickness, t, and s also must be decreased, while l is increased, although this latter dimension is fixed by the size of the vacuum system.

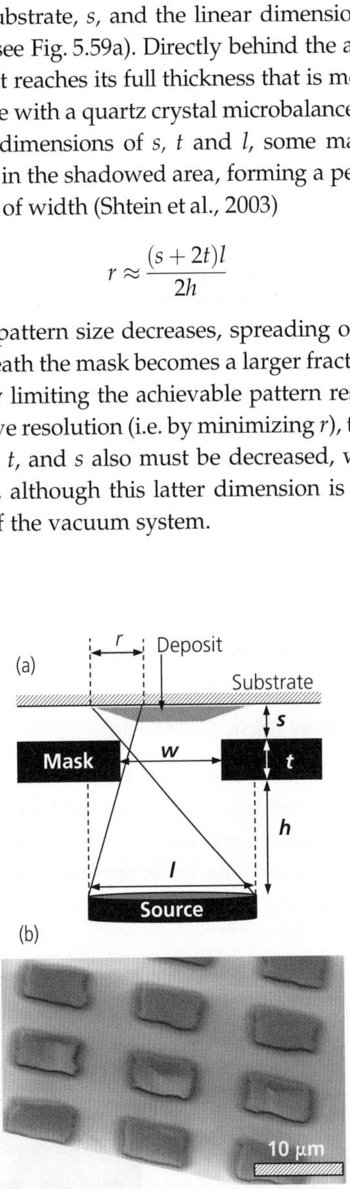

Figure 5.59 (a) Shadow mask configuration used in patterning thin films by VTE, where molecular trajectories from source to substrate are ballistic. (b) Micrograph of a high resolution array of Alq_3 squares deposited through a shadow mask with $w = 7.5$ µm and $s \ll w \sim 0$ µm (Shtein et al., 2003).

Reprinted from Shtein, M., Peumans, P., Benziger, J. B. & Forrest, S. R. 2003. Micropatterning of organic thin films for device applications using organic vapor phase deposition. J. Appl. Phys, 93, 4005 with the permission of AIP Publishing

One limitation of shadow mask patterning is that the aperture width and its distance to an adjacent aperture are limited to approximately the mask thickness. But as the mask gets thinner, its mechanical rigidity decreases, thereby making it more difficult to handle. Also, deposition through a shadow mask results in inefficient materials use since the deposit that falls in areas outside of the apertures is wasted. And the mask must be frequently cleaned to prevent clogging, which is a particularly acute problem for masks with very small apertures.

In spite of these limitations, shadow mask patterning is the dominant process used in production of OLED displays, and is currently used on substrates as large as Gen 8.5 (Table 5.3) to manufacture high resolution displays as large as 77″ diagonal. Very large shadow masks need to be placed beneath the substrate as in Fig. 5.59a to prevent particles from falling onto the substrate surface. The mask must be held in intimate contact with the substrate by magnetic or electrostatic attraction. In the latter case, the substrate and mask are oppositely charged during deposition, and then are separated by reversing the electric field. Figure 5.59b shows an array of small (~8 µm wide) squares of Alq_3 patterned using shadow masks where $s \approx 0$ µm (Shtein et al., 2003). The penumbra around the center pattern is due to the finite mask thickness.

Shadow mask patterning can also be used in OVPD, although it is more challenging to achieve high resolution due to the non-ballistic trajectories of the arriving molecules. Since $s \geq mfp$ (cf. Eq. 5.14), the molecule

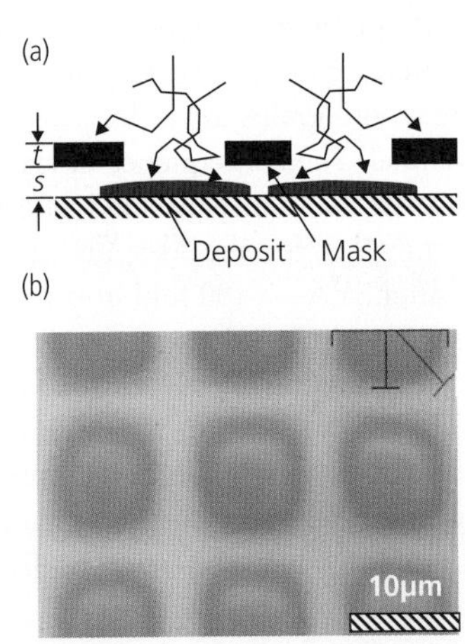

Figure 5.60 (a) Illustration of shadow mask patterning via OVPD. The arrows show the molecular paths as they arrive at the substrate within the boundary layer. (b) Micrograph of an array of Alq_3 squares achieved by shadow mask patterning in OVPD at 0.1 torr N_2 background gas pressure, with $w = 7.5$ µm, $t = 3.5$ µm, and $s < 1$ µm (Shtein et al., 2003).

Reprinted from Shtein, M., Peumans, P., Benziger, J. B. & Forrest, S. R. 2003. Micropatterning of organic thin films for device applications using organic vapor phase deposition. J. Appl. Phys, 93, 4005 with the permission of AIP Publishing undergoes several collisions with the carrier gas molecules within the boundary layer before reaching the substrate, as shown in Fig. 5.60a. This has the effect of spreading the deposit beyond the line-of-site distance, *r*. By clamping the mask to the substrate, the pattern definition can equal that attained by VTE, as shown in Fig. 5.60b. Indeed, it has been calculated that patterns as small as 1–2 μm are possible using shadow masks in the OVPD environment (Shtein et al., 2003).

5.6.2 Photolithographic patterning

Patterning conventional semiconductor devices at the nanometer scale uses photolithography, a photographic process whereby a photosensitive polymer (the photoresist) is spread onto the surface to be patterned. There are many variations to this mature process, including the use of positive or negative resists, multilayer resists, pattern lift-off, and so on. Three principal variants of the photolithographic process are illustrated in in Fig. 5.61, which we only briefly describe here. We refer the interested reader to the extensive literature on this subject for more details (see *Further reading*).

When using a positive photoresist (Fig. 5.61a), the desired pattern is exposed to UV irradiation of the resist through a mask. The UV light breaks the photosensitive bonds in the polymer, making it vulnerable to solvent attack. The sample is dipped into a solvent developer that dissolves the exposed areas, leaving the unexposed regions untouched ("what shows, goes"). The remaining photoresist pattern is the baked (i.e. cured) to render it chemically inert to process chemicals that are used on the underlying substrate to replicate the pattern in the exposed areas. Thus, pattern replication proceeds by dipping the sample into an acid, or exposing it to a reactive gas that selectively attacks the semiconductor, but not the resist. By this means, the open areas of the resist establish the same pattern in the semiconductor, with pattern resolutions down to 10 nm. The final step is the removal of the photoresist by a second chemical exposure, this time using one that does not damage the underlying semiconductor.

Negative photoresists simply reverse the process, that is, the UV exposed areas cross-link the photopolymer whose patterns then remain when developed (Fig. 5.61b). Lift-off is the third process whereby a hard layer (e.g. a metal or a dielectric) coats the semiconductor along with the pre-patterned resist surface (Fig. 5.61c). The resist is dissolved away beneath the coating, leaving it in exposed areas of the semiconductor where the resist was previously removed. Lift-off provides high resolution patterning of metals and insulators that are coated across the entire substrate surface.

A major advantage enjoyed by photolithography is its very high resolution, and the ability to sequentially add or substrate layers in a multi-step patterning campaign, with each layer aligned to the previous pattern to within only a few tens of nanometers. Hence, extraordinarily complex electronic circuits have benefitted from the flexibility and versatility of this highly evolved patterning technology.

Unfortunately, photoresists are seldom employed in the patterning of organic semiconductors (except in preparation of the substrate surface prior to organic film deposition) due to the likelihood of damage from

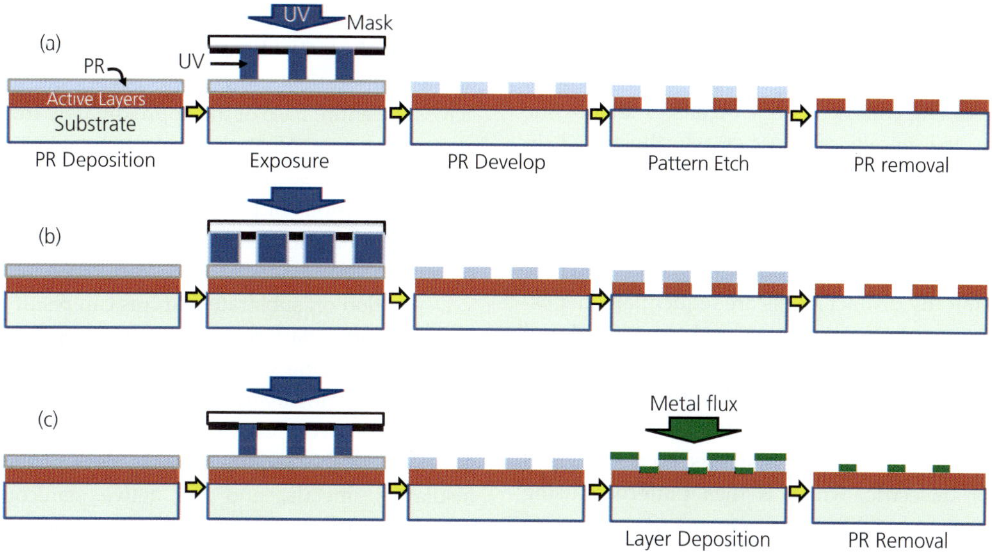

Figure 5.61 Process flow for photolithographic patterning. (a) Positive photoresist (PR), (b) negative resist, and (c) photoresist lift off.

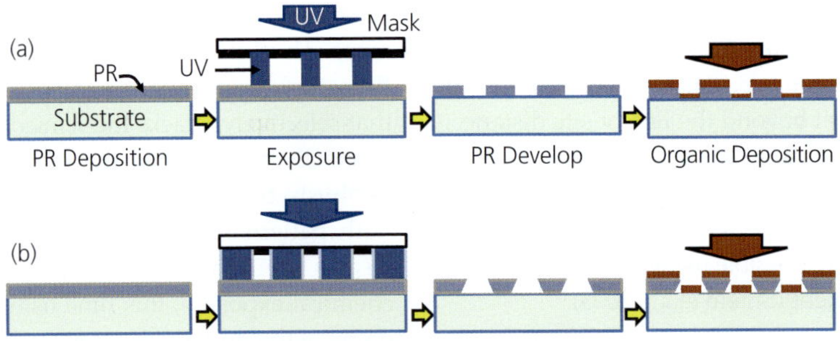

Figure 5.62 Photolithographic patterning to achieve very high resolution vacuum deposited organic structures. (a) Deposition on a photolithographically patterned resist layer, and (b) deposition on an undercut feature forming an *in situ* shadow mask.

the wet chemistry involved in resist deposition, development and removal. Organic semiconductors are often attacked by the same solvents as those used in photolithography. This has led to modifications of conventional photolithography to achieve similarly high resolution patterning of organics.

One method of achieving nanometer-scale features in small molecular weight semiconductors is to prepattern the substrate surface itself using conventional photolithography prior to deposition of the active organic layers. By deposition of the organic on the resulting substrate ridges or troughs, the film replicates the morphology of the underlying topography, as shown in Fig. 5.62a. A problem occurs when the organic is incident from a non-normal direction, as illustrated in Fig. 5.63a. This results in shadowing from one of the sidewalls of the feature, breaking the deposit on only one feature edge, but giving continuous coverage on the opposite sidewall. While this may not be a problem for resistive organics where lateral conductivity along a sidewall is low, patterning a metal contact by this method to define the device area results in shorts between adjacent devices.

To confine deposited patterns at the micron scale while avoiding conductive line breaks or shorts, one can use the *in situ* shadow mask illustrated in Fig. 5.62b (Tang, 2001). A trapezoidal opening is created in either the photoresist itself, or in a pattern etched into the underlying substrate. The trapezoid is formed using a two-level photoresist system, where two chemically distinct resists are sequentially deposited. The top photoresist is exposed, developed and cured. This exposes the underlying resist layer which is uncured or softer than the top layer. Then a second solvent that only dissolves the lower resist creates the undercut. Alternatively, the first resist is coated with a metal or dielectric, which is then patterned using conventional photolithography to create openings to the resist. The metal acts as a mask that is undercut by

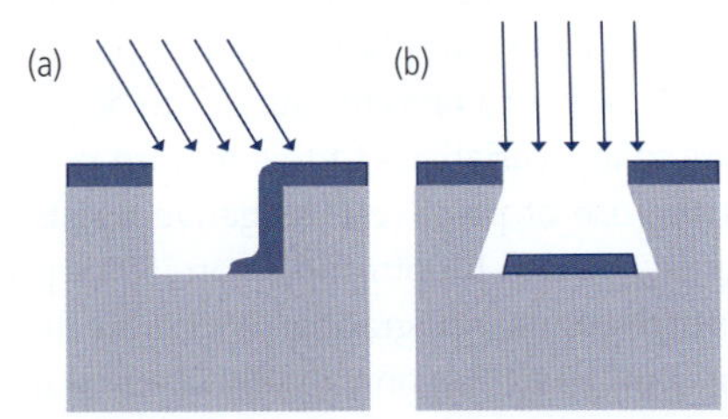

Figure 5.63 (a) Incomplete sidewall coverage of a substrate feature due to off-center deposition of the incident molecules. (b) A reentrant trench in the substrate can produce a self-aligned deposited pattern when the incident molecular flux is normal to the surface. If at an angle, the coverage of the base of the feature can be complete. The coatings are shown in dark blue, the patterned substrate in light blue, and the incident molecular flux by arrows.

the solvent or acid etchant of the underlying layer. Once the undercut is formed, the organic semiconductor is directionally deposited to form the patterns, as shown in Fig. 5.63b.

Patterning contacts on an organic semiconductor using an undercut mask proceeds as follows: The organic is deposited by rotating the substrate relative to the evaporation source. This covers the substrate across the entire base of the approximately trapezoidal resist opening. Next, the contact metal is directionally evaporated such that its edges are confined only to the dimensions of the trapezoidal openings, thereby preventing metal shorts across the organic film pattern.

Deposition on substrate features can result in very high-resolution patterns without exposing the organic to often-damaging wet chemistry required in photolithography. However, the patterning is limited to only one or two levels. Complex electronic systems almost invariably require many levels of patterning of insulators, metals, and the active semiconductor layers themselves. As we will see in the following section, precision patterning of complex organic structures can be achieved by the direct, additive printing of materials, rather than full surface deposition followed by material removal by conventional processes.

5.6.3 Laser induced thermal imaging

Rapid microscale patterning of multiple organic layers over large areas is required for many applications, none more challenging and relevant than the separate deposition of the micron-scale red, green and blue emitting layers required in full color OLED displays. Closely spaced R, G, and B pixels based on both polymers (Lee et al., 2002) and small molecules (Lamansky et al., 2005) can be patterned by *laser induced thermal imaging* (LITI) shown in Fig. 5.64. LITI is based on the local thermal evaporation of the organic material held in intimate contact with the target substrate. The light emitting polymer (LEP) used in the OLED active region is deposited on a transparent sheet with a light-to-heat conversion (LTHC) layer that consists of a material that strongly absorbs at the wavelength of a high power, focused laser beam scanned across its surface. The LTHC transfers the heat thus generated to the LEP, which delaminates from the donor sheet while attaching itself to the substrate. After printing, the sheet with LTHC and active semiconductor layers are replaced. This can be done in a continuous, roll-to-roll procedure for mass production of patterned substrates.

For the process to proceed quickly with uniform adhesion of the active semiconductor to the substrate, the laser power must be high and completely absorbed by the LTHC layer, which also must have a high thermal conductivity. The active organic material must be transparent to the laser radiation to avoid damage. Illumination from 1 W to 10 W Nd:YAG lasers (at 1.06 μm wavelength) is sufficient to deposit material after only a few milliseconds (Lamansky et al., 2005). Effective heat transfer layer materials include graphite or carbon black. An important criterion is that the LEP must have a higher surface energy on the target substrate than on the LTHC layer such that the material is readily transferred. Furthermore, the strength of the film must be sufficiently low to allow for a clean break at the edges of the heated region to achieve high pattern resolution; conditions that limit the range of materials that are suitable for LITI. Rapid RGB pixel patterning has been demonstrated with edge resolutions and positioning accuracies of only a few microns using both small molecules and polymers as target organic semiconductors (Lee et al., 2004).

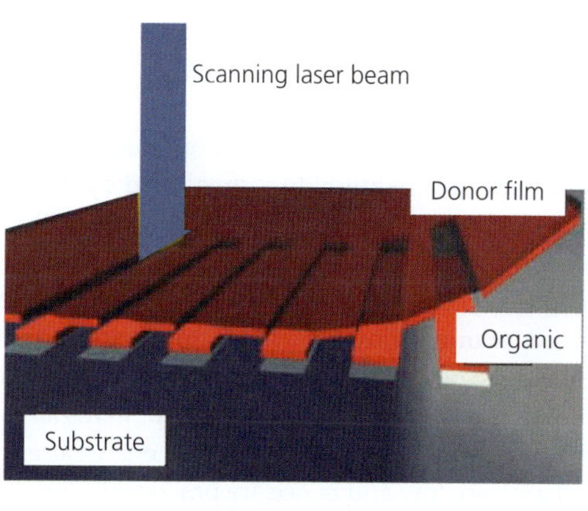

Figure 5.64 Patterning of OLEDs using laser induced thermal imaging. LEP is the light emitting polymer. Adapted from Lee et al. (2002).

5.6.4 Nanoimprinting and stamping

Due to the low intrinsic conductivity of organic semiconductors, the device area is determined by the overlap of the anode and cathode contacts. Several strategies have been developed to rapidly pattern contacts at the micron (or smaller) scale. Patterning the contact that is pre-deposited onto the substrate (e.g. ITO) is done using photolithography prior to the deposition of the organics. However, patterning of the top contact presents a greater challenge since it must avoid processes that degrade the underlying organic semiconductor. Commonly, patterning is achieved using deposition of the metal through a shadow mask whose resolutions are often insufficient for many applications, and is problematic when deposition is over a very large substrate area (see Section 5.6.1).

To avoid these limitations while achieving nanometer-scale resolution, the dry process of transfer printing has been developed. For example, "cold-welding" metal transfer has very high pattern resolution, and is adaptable to a variety of device types, including OLEDs, OPVs and thin film transistors. An illustration of two methods of cold-weld metal transfer—one additive and the other subtractive—is shown in Fig. 5.65 (Kim et al., 2000, Kim and Forrest, 2003).

Cold welding involves applying pressure across an interface between two metallic surfaces. When the pressure is sufficiently high to break through energy barriers due to surface contamination, oxides or debris, the interface barrier collapses, bringing the now pristine metal atoms from each layer into contact. The electron systems are shared across the interface, which effectively disappears across the "perfect"

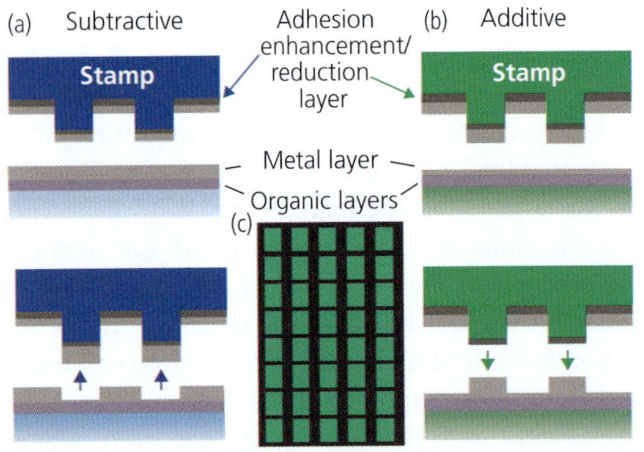

Figure 5.65 Cold-weld metal transfer by (a) the subtractive method whereby metal deposited onto the organic layers is "lifted-off" using a hard stamp, and (b) the additive method that transfers the metal electrode from the stamp to the substrate by cold-welding to a very thin (~5nm) "strike layer" pre-deposited on the substrate. Following transfer, the contacts are isolated by removing the strike layer by plasma etching. (c) An OLED passive matrix array using whose contacts are defined by subtractive patterning.

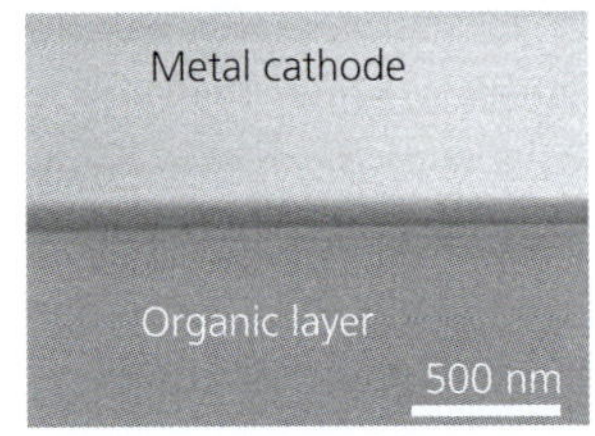

Figure 5.66 Edge of a metal cathode patterned by subtractive cold-weld bonding (Kim et al., 2000).

From Kim, C., Burrows, P. E. & Forrest, S. R. 2000. Micropatterning of Organic Electronic Devices by Cold-Welding. Science, 288, 831. Reprinted with permission from AAAS.

bond (Taylor et al., 1991). The interfacial collapse occurs within only a few tens of picoseconds once a threshold pressure is exceeded.

In the subtractive process in Fig. 5.65a, the cathode metal is deposited on the organic layers, covering the full substrate surface. Next, a hard stamp (e.g. Si) is prepared by photolithographically defining and then etching protrusive features that are shaped into the desired contact pattern. The stamp is subsequently coated with the same metal as used for the cathode, using a layering scheme that ensures that the metal strongly adheres to the stamp surface. The stamp is pressed against the substrate to create a cold-weld bond between the metal on the protruding stamp surface and that on the organics. Sufficient pressure is applied to shear the metal at the sharp stamp edges, thereby separating it from the continuous metal blanket on the organic. The stamp and substrate are then parted, lifting away material that was under pressure. The adhesion of the metal to the organic must be weaker than to the stamp for the separation to be complete. The subtractive method creates a "negative" pattern, similar to the use of a negative photoresist in photolithography.

The subtractive method has been used to demonstrate passive matrix OLED displays (Kim et al., 2000). Given the need to shear the underlying films, very high pressures (> 100 MPa) are required, which are impractical when applied over large substrate areas. Nevertheless, the pattern resolution achieved is very high. Figure 5.66 shows edge definition down to only a few tens of nanometers for a cathode formed on an organic layer surface.

To reduce the pressure of contact formation, the additive method, which is analogous to a positive photoresist process, has been developed (Kim and Forrest, 2003). In Fig. 5.65b, a soft elastomeric stamp comprised of a material such as PDMS replaces the hard stamp used for subtractive patterning. To form the stamp, a solution containing the elastomer in a solvent is poured into a mold and cured. Peeling the hardened stamp form the mold produces protrusions whose highest surfaces are in the shape of the desired material to be transferred. Next, an adhesion reduction layer is deposited onto the stamp. This layer consists of a material such as BCP or pentacene that reduces the adhesion of the layer to be transferred from the stamp to the organic. Hence, the layer is easily and completely released from the stamp when it contacts the target substrate. The substrate preparation itself is also different than for subtractive patterning. Rather than depositing the metal at its final thickness onto the active organics, only a thin (~5 nm) *strike layer* of the metal is deposited. The full thickness of the contact metal is deposited onto the stamp itself. By pressing the stamp against the strike layer, a cold weld bond is formed. When the stamp is detached from the organic thin film, the metal contact is left behind. The thin strike layer metal is removed by exposing the surface to a light dry etch such as an Ar^+ plasma that selectively removes the thin metal without attacking the underlying organic layer.

The pressures required to form a bond by the additive process are significantly lower (~100–200 kPa) than for the hard stamp used in subtractive bonding. This difference is due to the conformal compliance of the soft stamp, allowing it to "flow" around dust particles or other surface irregularities that prevent intimate contact when harder stamp materials are employed (Cao et al., 2005). While the pressures applied are very low and hence are practical for use over large substrate areas, a disadvantage of the use of soft stamps is the considerably reduced pattern resolution that results from distortions of the stamp when even a slight pressure is applied. An edge resolution of ~1–2 μm is common, as shown by the micrograph in Fig. 5.67. To increase the resolution, a stiffer composition of the stamp (e.g. *h*-PDMS) can be employed, although this results in a concomitant increase in pressure required to form the cold-weld bond.

The ability of a material to preferentially adhere to the substrate or the stamp is critically dependent on the surface energies of the various materials. Care must be taken in selecting adhesion-reduction materials to ensure clean and complete release of the transferred material from the stamp. Similarly, in the subtractive process, very strong adhesion between the metal and the stamp ensures that the pick-up of the metal from the organic surface is complete.

It is difficult to overestimate the versatility and number of uses that stamp transfer of material provides. It has been used not only for the application of metals, but also for the transfer of the active organic layers themselves (Kim et al., 2005). Furthermore, it is useful for patterning contacts on distorted or shaped surfaces, such as in transferring photodiode contacts in hemispherical focal plane arrays (HFPA) shown in Fig. 5.68. The array consists of active organic layers of the photodiodes that are sandwiched between two orthogonal arrays of contacts patterned by cold-weld bonding. Each cross-point in the array is independently addressed by the corresponding row and column electrodes. Such an array is useful in forming high definition images using only a single convex lens, similar to the architecture of the human eye (see Section 7.2.2). The soft organic materials allow for deposition on arbitrarily shaped surfaces that serve application needs inaccessible using conventional, brittle semiconductors. Cold-weld bonding allows for patterning on shaped surfaces at very high feature resolutions.

The process for fabricating the HFPA is shown in Fig. 5.69. A stamp is produced with long parallel ridges coated with a thin (~10 nm), semitransparent Au layer. The stamp is placed in a hemispherical mold through which vacuum is applied to draw it into the required shape. The topological transformation of the stamp from its original 2D plane into a three-dimensional (3D) hemisphere introduces considerable strain that does not damage the metal strips since they only weakly adhere to the stamp. Thus, they are allowed to slip along the ridges as shear forces are applied during shaping. It is interesting to note that shear-slip strain relief inspired by this process has also been applied in the fabrication of HFPAs employing inorganic (GaAs) photodiodes (Fan et al., 2019).

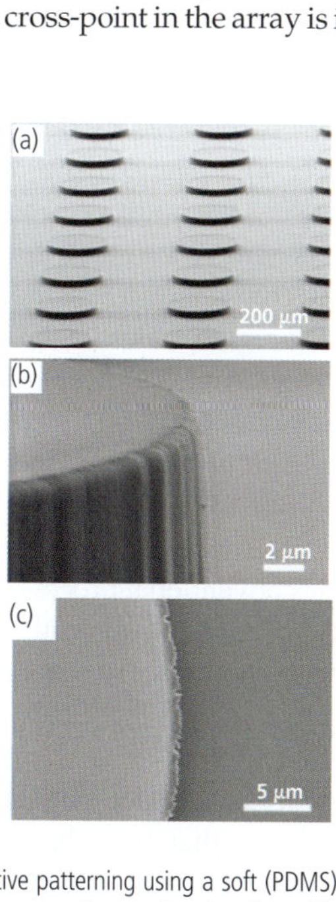

Figure 5.67 Additive patterning using a soft (PDMS) stamp. (a) The stamp consists of an array of protrusive dots formed by injection into a mold. (b) Edge of the stamp exactly replicating depressions in the mold. Notice the rounded, rough edges. (c) Transferred metal contact onto an organic layer surface (Kim and Forrest, 2003).

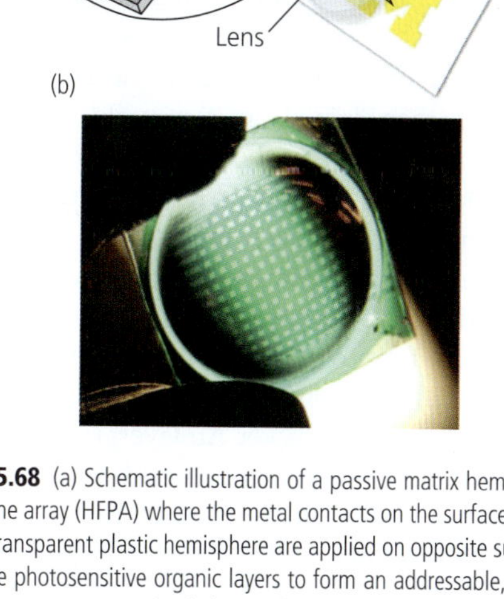

Figure 5.68 (a) Schematic illustration of a passive matrix hemispherical focal plane array (HFPA) where the metal contacts on the surface of a preformed transparent plastic hemisphere are applied on opposite surfaces of the active photosensitive organic layers to form an addressable, passive matrix detector array. A simple lens and the imaged object are also shown. (b) A 1 cm diameter HFPA fabricated by cold-weld contact transfer (Xu et al., 2008).

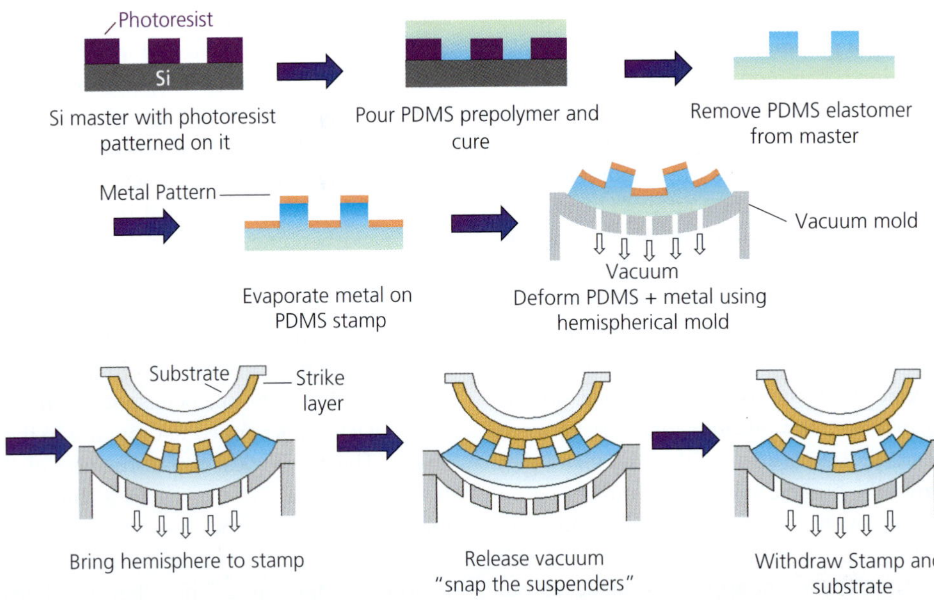

Figure 5.69 Process sequence for fabricating the hemispherical focal plane array in Fig. 5.68. The top two rows show the fabrication of the soft elastomeric stamp consisting of an array or ridges and troughs. It is formed into a hemisphere by applying vacuum from a mold following deposition of the contact metal onto the stamp. The bottom row shows the transfer process of the metal contact lines onto the transparent plastic hemisphere. The hemisphere itself is thermoformed by heating above its softening temperature and drawn by vacuum into the same mold. A thin strike metal is deposited onto its full surface. The thin, semitransparent contact array is transferred from the deformed stamp followed by removal of the strike layer by Ar^+ plasma etching. Not shown is the following full surface deposition of the organic photodetector layers over the metal contacts, followed by a second strike layer. The entire sequence is then repeated, with the second contact array positioned perpendicular to the first, thus completing the passive matrix (Xu et al., 2008).

The transparent, thermoformable, glycol-modified PET (PETg) substrate is shaped using the same mold by heating while applying vacuum. The mold is cooled, freezing the substrate into a hemispherical shape. A thin Au strike layer is deposited along the outer surface of the hemisphere. The substrate and stamp are then joined in the mold (lower left, Fig. 5.69), at which point the vacuum is released. The relaxed elastomeric stamp presses against the strike layer, completing the cold-weld transfer. After a brief exposure to an Ar^+ plasma to remove the strike layer that electrically shorts the contact stripes, the organic active layers forming the photodiode are deposited across the full hemispherical surface. The top contact array, consisting of a thicker Au layer, is then applied using the same strike layer/cold-weld process with the stamp now rotated 90° relative to the first contact array. The resulting device shown in Fig. 5.68b is a passive matrix array of detectors with orthogonal row and column contacts used to address each pixel.

The lower contact array employs a semitransparent Au layer that reduces light coupling into the detectors. While ITO is a more transparent anode material, it is brittle and hence cannot endure the stress of distortion into a spherical shape. To solve this problem, ITO can be deposited *after* the substrate is thermoformed into a hemisphere. This is followed by deposition of a thin metal strike layer, and then cold-welding an array of thick Au lines using the additive process in Fig. 5.70. The strike layer is removed using Ar^+ plasma, leaving a thick array of Au columns on top of the ITO. The ITO is removed from the regions between the Au lines using a suitable selective etchant that does not attack the Au lines that serve as a resist. The final step is removing the Au using a different etchant that does not attack the ITO. Completion of the array by organic deposition and attachment of the top, thick Au array of row contacts follows the conventional process in Fig. 5.69. The transparent ITO contacts, along with the detector active region and orthogonal Au contacts of the completed HFPA, are shown in Fig. 5.70b.

While we have described only one transfer and nanoimprint process, transfer patterning has taken many diverse forms that have been developed for use of fragile materials that are incompatible with conventional lithography. It has proven to be a versatile process that is scalable to large substrates. The interested reader is directed to the substantial literature on this subject for further details (Quake and Scherer, 2000, Rogers and Nuzzo, 2005).

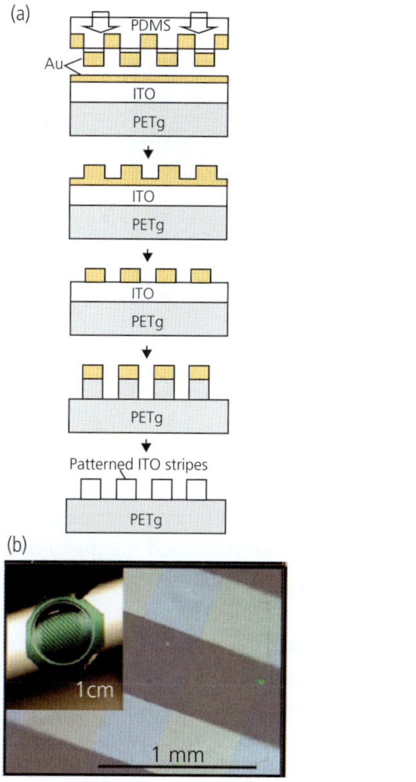

Figure 5.70 (a) Patterning of ITO using a metal mask applied by additive cold-welding. (b) A completed HFPA using an ITO bottom contact array. The ITO array is faintly visible running nearly vertically, and the top Au contacts are the opaque, nearly horizontal lines in the image. Inset: Image of the HFPA (Xu et al., 2009).

Reprinted from Xu, X., Mihnev, M., Taylor, A. & Forrest, S. R. 2009. Organic Photodetector Arrays with Indium Tin Oxide Electrodes Patterned Using Directly Transferred Metal Masks. Appl. Phys. Lett., 94, 043313 with the permission of AIP Publishing.

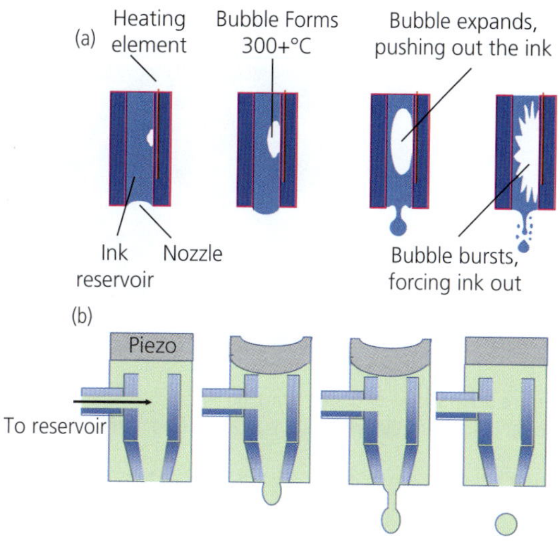

Figure 5.71 Two inkjet print head designs. (a) Thermal and (b) piezoelectric inkjet printing engines.

5.6.5 Inkjet printing

Inkjet printing has become an ubiquitous means for printing, copying and reproducing documents and graphics for home, office, and business. It is a mature, high image resolution technology based on the precise projection of ink droplets of only a few picoliters from a nozzle onto a substrate (typically paper or plastic) held a few millimeters away. Since dye compounds used in ink jet printers are similar to those used for electronic applications, it is reasonable to adapt the technology for the printing of organic devices. Indeed, such an adaptation has been underway, with modifications to conventional inkjet print engines to suit the peculiar properties of electronic polymers and soluble small molecules. While many organics have been optimized as optoelectronic materials, their use in inkjet patterning requires that their fluidic (i.e. *rheological*) properties also be suitable for precise droplet formation and deposition. This added performance criterion is not always compatible with the optoelectronic characteristics of the material, and hence inkjet printing of electronic devices has, to this date, not entered mainstream industrial use. Nevertheless, progress is being made, motivated by and the maturity, flexibility and resolution of the process.

Several inkjet printing engine designs have been developed. Thermal inkjet heads contain a small heater within the nozzle injection system, as shown in Fig. 5.71a. The temperature of the fluid containing a solvent and a suspension of the electronic material is heated, forming a bubble. As the bubble expands, it forces a small amount of fluid out, projecting it toward the substrate. The rheological properties of the suspension must be suitable for controlled droplet formation at temperatures required to create the bubble without degradation of the semiconducting solute. This criterion limits the range of materials that can be adapted for printing, and hence is not generally used in electronics applications.

An alternative approach employs a piezoelectric transducer within the fluid container, Fig. 5.71b. A voltage is applied to the transducer, resulting in its flexure that reduces the container volume, thus expelling a precise volume of fluid. The method is fast (a few milliseconds are required to deflect the piezoelectric diaphragm) and highly precise. It is also nondestructive of the materials in the suspension. Piezoelectric inkjet printing is therefore a preferred means for local deposition of organic semiconductors.

The highest pattern resolution attainable via inkjet printing when the fluid, nozzle and deposition processes are optimized is $\sim$ 50 μm. The ability to

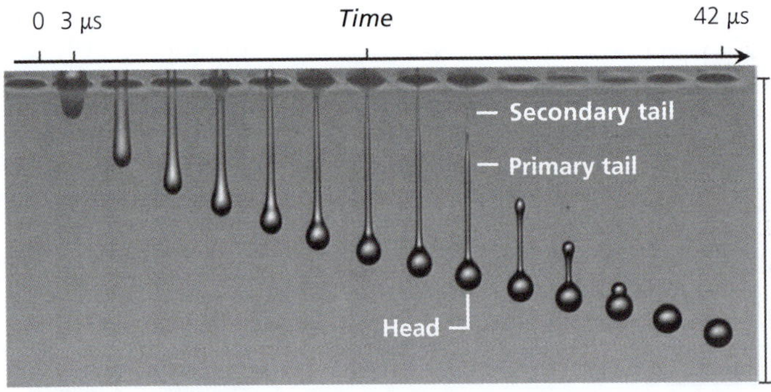

Figure 5.72 Silicone oil droplet formation and ejection from a nozzle visualized using flash photography with a frame spacing of 3 μs. The width of the droplet is 4 μm and the primary tail is ~ 25 μm long (van der Bos et al., 2014).

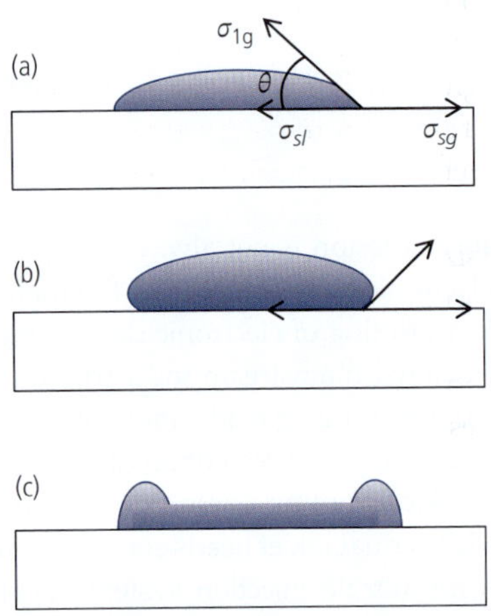

Figure 5.73 Fluid on a (a) wetting and (b) non-wetting substrate surface. (c) Drying pattern of a droplet that exhibits the "coffee stain" effect.

consistently generate droplets of a well-defined volume depends on several properties of the solvent–organic semiconductor solution (or "ink"), including its surface tension, viscosity, temperature, wetting properties on both the nozzle and substrate surfaces, and its tendency to form precipitates or clusters that can clog the very small nozzle orifices. For example, if the surface tension is too high, droplet separation from the nozzle is uneven, and if it is too low the fluid can spray out of the nozzle without forming a droplet of the desired size. Surface tension in the range 20–70 mN/m is optimal to achieve controlled inkjet printing.

Likewise, viscosity plays a significant role in droplet formation. Figure 5.72 shows the droplet ejection process, where a tail forms on the droplet as well as on the fluid retained at the nozzle orifice. The droplet breakup distance, *bu*, depends critically on the viscosity of the fluid. If *bu* is too long due to a high viscosity, the drop volume is larger than required, causing the resulting deposit to exceed the desired dimensions, or resulting in fluid splatter at the substrate. Additionally, a long droplet forms a non-spherical droplet whose tail can break up into multiple segments, thereby creating several additional droplets that trail the primary drop. Typical fluid viscosities that are optimal for high resolution printing are in the range 1–50 cP. An overly viscous fluid leads to poor separation of the droplet tail from the nozzle and nozzle clogging.

Complicating these processes is the formation of threads connecting the droplets due to viscoelasticity of the suspension. When a viscous fluid is passed through a small orifice, the shear stress at the surface of the orifice causes the viscosity to increase. This feature of a non-Newtonian fluid is known as *shear thickening*. A *Newtonian fluid* is one where the viscosity is invariant across its flow field, independent of external stresses (e.g. shear). The opposite effect is known as *shear thinning* whose occurrence depends on the composition and rheological properties of the suspension. Fluid thickening is particularly troublesome since it can completely prevent separation of the droplets, and can clog the very fine nozzle opening.

Pattern formation also depends on the interaction between the fluid and the target substrate surface. The relative surface energy of the substrate and the fluid determines if the fluid flows, or beads up. In Fig. 5.73a, the fluid is in contact with a wettable surface, that is, where the interfacial surface energy between the fluid and the solid, σ_{sl}, is greater than the surface tension of the solid and the ambient, σ_{sg}, in which case the contact angle is $\theta < 90°$. In the opposite case where $\sigma_{sg} > \sigma_{sl}$, the surface is non-wetting and the droplets form beads that attempt to minimize their contact area with the substrate, see Fig. 5.73b. The contact angle depends on these surface energies via the balance of forces expressed by the *Young equation* (Israelachvili, 2011):

$$\sigma_{sg} = \sigma_{sl} + \sigma_{lg}\cos\theta. \tag{5.38}$$

Here, σ_{lg} is the energy difference between liquid and the surrounding atmosphere. This balance of forces gives rise to a range of film morphologies that depend on materials properties, including the composition of the solvent and polymer and their relative concentrations, the substrate morphology (e.g. the crystal face to which the droplet is attached, and roughness), and substrate surface preparation (e.g. its level of hydrophobicity or hydrophilicity). For example, the surface can be functionalized using wetting agents such as OTS or self-assembled monolayers with functional groups that specifically modify the substrate surface energy. Other means include exposure to oxygen or fluorine-based plasmas (see Section 8.6.1).

As the solvent evaporates, the surface energies also determine the final shape of the deposit. An example is the "coffee ring effect" illustrated in Fig. 5.73c, where the polymer piles up near the edge of the deposit, as is the case of the thick dark residue of particles that are left behind when coffee is spilled onto a flat surface (Deegan et al., 1997). This profile results from pinning of the edge of the droplet to its initial position on the substrate, combined with preferential evaporation of the solvent at the periphery of the liquid/solid interface where molecules can escape with little chance for re-adsorption. This contrasts with the center of the deposit where re-adsorption on the large surface is relatively efficient. Hence, a net flow of solvent radiating from the center to the edges develops, carrying the solute (e.g. the polymer) that ultimately piles up along the edge, creating an uneven thickness profile. Edge accumulation can be reduced by exploiting the *Marangoni effect*, whereby a mixture of two solvents with different boiling points creates a reverse flow from the edge to the center of the liquid deposit (Hu and Larson, 2006). In this case, the latent heat of evaporation of the solvents reduces the temperature in the center of the droplet. Since surface tension decreases with increasing temperature, cooling creates a diffusive flow from the (warmer) edges toward the center. When two solvents are used, the higher boiling point constituent moves toward the edges, with the lower boiling point solvent depleted rapidly from the center. This dynamic balance sets up Marangoni flow that prevents large concentrations of the solute from accumulating anywhere within the droplet as the solvent evaporates.

Surface irregularities and rounded profiles are undesirable side effects of the rheological properties of the solvent-solute mixture, and can never be completely eliminated in the inkjet deposition process. Hence, their presence ultimately limits the feature resolution, although many strategies have been employed to minimize their impact (Sirringhaus et al., 2006). One is to print into confined areas defined by polymer "wells" patterned on the substrate surface. The active organic inks are deposited into the appropriate well that confines the material. This method has been developed for printing red, green and blue light emitting layers of OLEDs in full color displays, as shown in Fig. 5.74. Contacts are subsequently deposited to complete the device (Kobayashi et al., 2000).

Although many of these strategies facilitate the printing of small electronic patterns, inkjet techniques have not yet been adopted for widespread application in organic electronics. As noted above, the difficulty in synthesizing a material that leads to high device performance with a stability that is required for most electronic applications, while also optimizing its rheological properties to make it suitable for inkjet deposition, represents a complex materials design challenge that in many cases has been unattainable. Yet the promise of inkjet printing to generate devices with the same ease as document printing continues to motivate considerable research in this area.

5.6.6 Organic vapor jet printing

The method of *organic vapor jet printing* (OVJP) is analogous to inkjet printing except that it is a dry, rather than a solution-based process. OVJP enables the direct, high-resolution, rapid patterning of

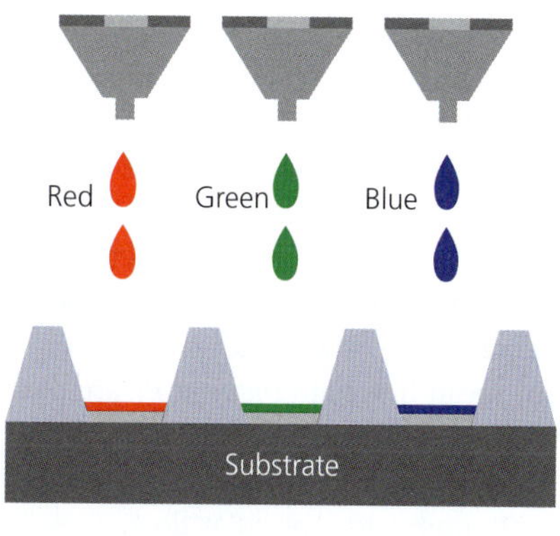

Figure 5.74 Inkjet printing of a full color OLED display into an array of wells used to confine the ink within each pixel.

molecular organic semiconductors on a variety of substrates (Shtein et al., 2004). Like OVPD used in deposition over entire substrate surfaces, OVJP is inherently a local deposition technique whereby a hot inert carrier gas is seeded with organic vapor and then expanded through a microscopic nozzle to form a collimated beam of gas that carries molecules through a vacuum or low pressure environment. The organic molecules are accelerated by the lighter carrier gas, exiting the nozzle at hyperthermal velocities. When the resulting heterogeneous beam impinges on a cold substrate, the carrier gas quickly disperses. However, due their greater mass, the organic molecules retain their axial momentum, resulting in a confined deposit profile.

The deposition profile is controlled by the shape of the gas jet and substrate temperature (which determines the sticking coefficient), avoiding the need for liquid solvents or droplet-confining wells. The pattern itself is "written" by translating the substrate beneath the nozzle array. Consequently, this method broadens the choice of substrate material and shape, while reducing deposition and patterning of the organic layer to a single step. OVJP takes advantage of the high organic-to-carrier gas molecular mass ratio to achieve high resolution patterning. The method has been successfully used for depositing multicolored OLEDs for both lighting (Arnold et al., 2008) and displays (McGraw and Forrest, 2013), and in defining the channels of organic thin film transistors (Shtein et al., 2004).

The OVJP concept and apparatus are schematically illustrated in Fig. 5.75a. The molecular organic sources, A and B, are heated while an inert carrier gas (e.g. N_2) is injected into the source cells, picking up the organic vapor and expanding it through a collimating nozzle. Although the carrier gas flow field rapidly diverges due to the proximity of the substrate to the nozzle outlet, the relatively heavy organic molecules acquire trajectories substantially more collimated than the volatile carrier gas that escapes laterally along the substrate plane. The interplay between diffusive and convective transport at the nozzle orifice that ultimately dictates pattern shape is determined by such factors as nozzle radius, a, nozzle-to-substrate separation, s, and the deposition chamber pressure, P_L. The scaling is usually such that s, the pattern resolution, and the molecular mean free path (mfp) at P_L are of the same magnitude, as indicated in Fig. 5.75b. Hence, downstream from the nozzle, transport is intermediate between continuum and molecular flow.

The gas jet monotonically diverges after exiting the nozzle. When a substrate is placed close to the nozzle, the jet is further dispersed relative to the free-expanding beam due to a stagnation front formed immediately above the surface. This axial enrichment of the jet by the organic molecular species constitutes a key feature of OVJP. Since $s \sim mfp$ at P_L, the organic molecules suffer few collisions within the nozzle-substrate gap. Assuming fully developed flow inside the nozzle, a low (<1% molar) concentration of the organic species and incompressible flow, the organic molecules travel radially outward from their original position in the nozzle by a distance χ (Shtein et al., 2004):

$$\frac{\chi}{a} = \frac{m_g}{M}\frac{s}{mfp} + \sqrt{\frac{1}{3}\frac{\bar{c}s(mfp)}{\bar{u}a^2}}, \tag{5.39}$$

where m_g/M is the carrier gas-to-organic molecular mass ratio, $\bar{c}$ is the mean molecular thermal velocity, and $\bar{u}$ is the mean flow velocity inside the nozzle. The first term in Eq. 5.39 is due to horizontal momentum transfer to the organic molecules from collisions with the diverging carrier gas, while the second term is the scaling of the radial diffusion to the convective transport rate normal to the substrate.

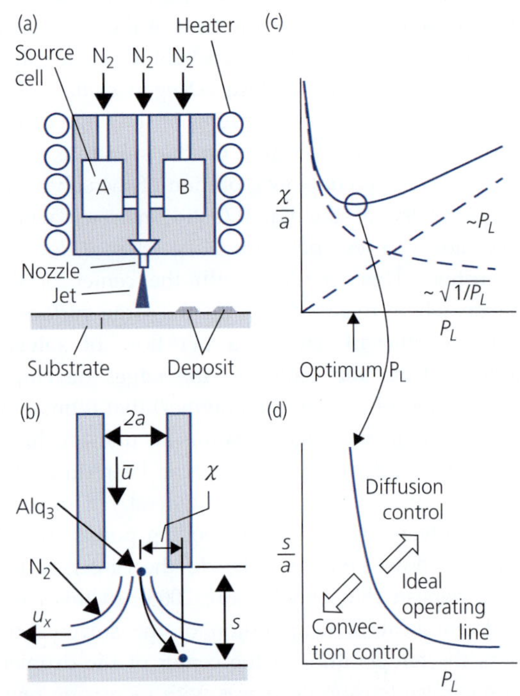

Figure 5.75 (a) Conceptual diagram of an OVJP apparatus. (b) Nozzle and exit gas flow defining the variables described in the text. The example of an Alq_3 molecule swept through the nozzle in a hot N_2 carrier gas is shown. (c) Plot of the normalized molecular beam width, χ/a, vs. the background pressure, P_L, with the optimum pressure resulting in a minimum deposited feature size. (d) Normalized substrate-to-nozzle separation distance, s, as a function of P_L. The various flow regimes encountered are noted (Shtein et al., 2004).

Equation 5.39 shows the relative influence of process conditions on pattern resolution. In particular, given that $mfp = k_B T/2\sigma P_L$, where σ is the cross-sectional area of the molecule (in the hard sphere approximation), the pattern dispersion has a minimum at some value of P_L, as shown in Fig. 5.75c. The maximum resolution is achieved when P_L is in the range 1–50 torr, typical of OVJP. Equation 5.39 also suggests that pattern resolution is enhanced by the use of a lighter carrier gas (e.g. He instead of N_2). Practically, $\bar{u}$ is fixed by the desired deposition rate via the concentration of the organic vapor. Thus, for a given nozzle radius, a, the remaining adjustable parameters are s and P_L.

The operating conditions for maximum pattern resolution are plotted on the process diagram, Fig. 5.75d, where the operating line dictates values of s for any given P_L. For example, to maintain high pattern resolution ($\chi/a \to 1$) even at large separation, s, the downstream pressure, P_L, must be decreased. The region above the operating line represents printing in the diffusion-limited regime, while the region below corresponds to convection-limited transport. Finally, the local dynamic pressure in the region between the nozzle and the substrate generally exceeds P_L, and scales inversely with s.

Details of the flow calculated by *direct simulation Monte Carlo* (DSMC) are shown in Fig. 5.76a and b. The vertical velocity component of the flow field is plotted as a color-map in Fig. 5.76a, with the corresponding trajectories of the carrier gas and the organic molecules (in this case, Alq_3) plotted in Fig. 5.76b. The velocity map shows the acceleration of the flow through the nozzle, reaching ~200 m/s at the nozzle exit. The stagnation front lies immediately above the substrate surface where the dynamic pressure generally exceeds the ambient pressure, P_L. The heavy organic molecule trajectories, however, cross the carrier gas flow lines, resulting in a well-defined deposit.

Directional deposition is demonstrated in the micrograph in Fig. 5.76c for a pentacene line printed on SiO_2 using a 50 μm diameter nozzle at a deposition rate >300 Å/s and s = 35 μm (Shtein et al., 2004). The central line demonstrates that flat deposit profiles are achieved. Furthermore, SEM images reveal that the pentacene crystallites to the left and right of the jet axis tilt in toward the molecular source. Seeding the organic molecules in a fast-flowing carrier stream at hyperthermal velocities determines the incident molecular kinetic energy. This decouples the film crystallization dynamics from surface temperature, leading to highly ordered films even on relatively cold substrates. This effect has important implications for the performance of devices such as polycrystalline channel TFTs (Shtein et al., 2002), and it significantly broadens the choice of substrates to include fragile and/or bio-compatible materials.

For this technology to be practical, the rate of organic deposition must be sufficiently fast to enable high resolution printing on large area substrates in times comparable to that of competing methods such as deposition through a shadow mask. Inkjet printing of documents is, in fact, quite rapid due to parallelism arising from print heads comprising more than 1000 nozzles. Efforts have therefore focused on the scale-up of OVJP print heads from single nozzles, to nozzle arrays using Si fabrication-based photolithographic methods (McGraw et al., 2011, McGraw and Forrest, 2012). The print head is mounted on a feed-through flange mounted on a high vacuum chamber. Nitrogen carrier gas is supplied to the head via a heated metal manifold that connects it to the feedthrough. The manifold in Fig. 5.77a consists of tubes that contain

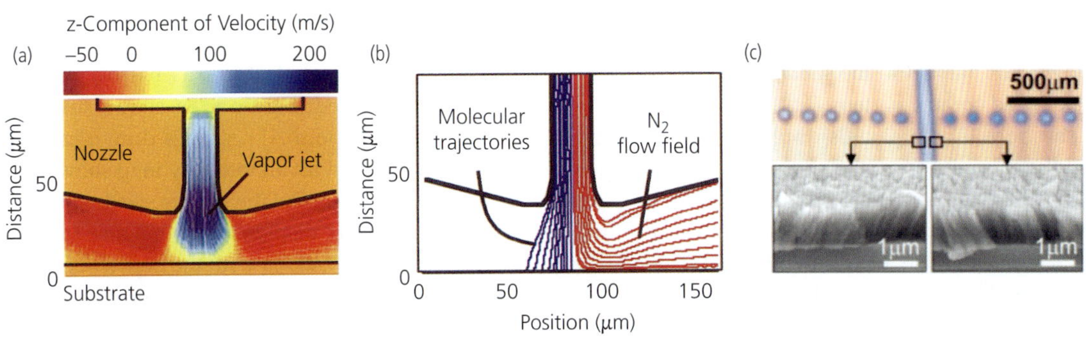

Figure 5.76 (a) Calculated velocity flow field beneath an OVJP nozzle. (b) Trajectories of the molecules and carrier gas illustrating the "focusing" of the molecular deposit beneath the nozzle orifice. (c) Electron micrograph of a pentacene transistor channel grown by OVJP. The upper image is a plane view of the deposit with several dots grown adjacent to the channel stripe. The cross-section view (below) shows how directional flow from the nozzle results in pentacene nanocrystalline growth entrained along the gas streamlines as in (b) (Shtein et al., 2004).

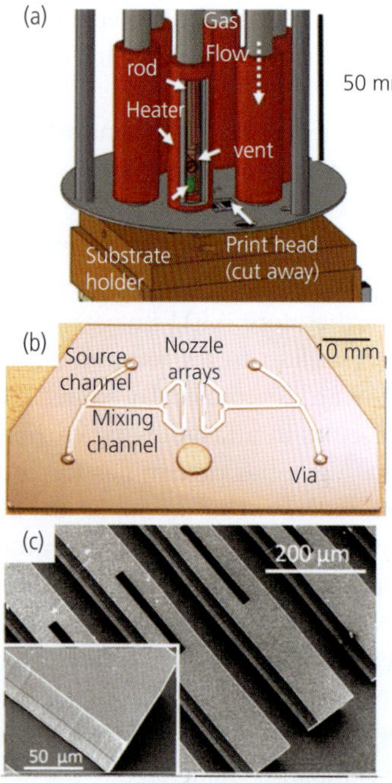

Figure 5.77 (a) Scalable print head and substrate holder assembly (McGraw et al., 2011). (b) Microfluidic plate that directs and mixes organic vapors from the cylindrical material containers in (a) to the nozzle arrays. (c) Electron micrographs of a Si nozzle array consisting of 20 μm wide by 200 μm long nozzle slits. A detailed view of the nozzle and sidewall is shown in inset. The substrate is translated parallel to the long nozzle axis, leading to ~ 20 μm wide deposits (McGraw and Forrest, 2012).

Reprinted from McGraw, G. J. & Forrest, S. R. 2012. Fluid Dynamics and Mass Transport in Organic Vapor Jet Printing. J. Appl. Phys., 111, 043501 with the permission of AIP Publishing

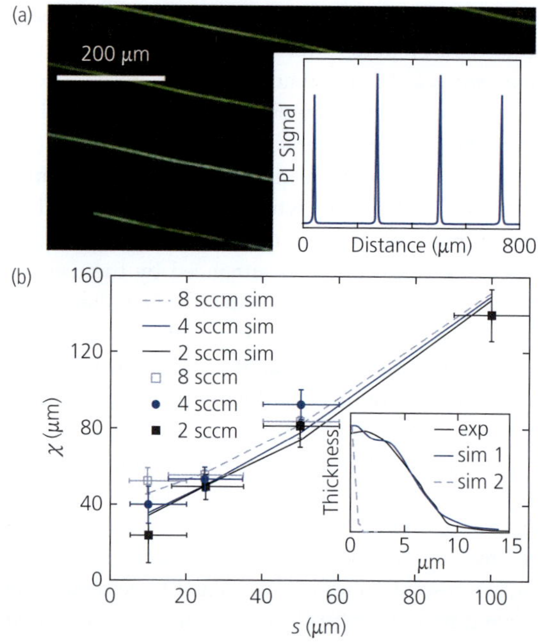

Figure 5.78 (a) Fluorescent emission from an array of Alq_3 stripes printed by OVJP. The stripes were deposited at a N_2 carrier gas flow rate of 2 sccm and a nozzle-to-substrate separation of $s = 10$ μm. Inset: Photoluminescence intensity profiles vs, position along a path perpendicular to the lines. (b) Feature size, χ, as a function of nozzle-to-substrate distance, s, and N_2 flow rate. Simulated deposition feature sizes (lines) match the experimentally observed trends. Inset: Normalized thickness of an Alq_3 line measured using a profilometer (exp). The measured profile agrees with the simulated deposition profile generated for a 10 μm nozzle width, flow rate of 2 sccm, and $s = 5$ μm (sim 1). Simulations suggest that a 0.5 μm nozzle positioned at $s = 0.5$ μm can produce lines as narrow as 1.5 μm wide (sim 2) (McGraw et al., 2011).

Reprinted from McGraw, G. J., Peters, D. L. & Forrest, S. R. 2011. Organic vapor jet printing at micrometer resolution using microfluidic nozzle arrays. Appl. Phys. Lett., 98, 013302 with the permission of AIP Publishing

separate organic vapor sources needed for a particular application. Each tube is independently heated to the sublimation temperature of the material contained within a glass or Pyrex vial within the tube. Hot carrier gas enters the vial, picks up the organic vapor, and carries it through a microfluidic channel plate (Fig. 5.77b) to the print head. A chilled substrate holder sits below the manifold. It is translated in two dimensions parallel to the substrate plane to "draw" the desired organic pattern.

The microfluidic gas distributor in Fig. 5.77b carries the organics from the source vials, mixes them with other compounds in situations where dopants or mixtures are required, and transports the molecules to the nozzle arrays in Fig. 5.77c. The nozzles are formed by etching the silicon-on-insulator (SOI) wafer device layer along the (111) Si plane, extending down from the top wafer surface. Rectangular nozzle orifices are defined by the intersection of the etched zone with the SOI wafer oxide layer, creating nozzles only a few microns wide. Since the process technology is Si-based, large arrays containing >1000 precisely placed and shaped nozzles can be fabricated using this approach.

Using 10 μm wide nozzles located within 10 μm of the substrate, organic features as narrow as 16 μm have been achieved. Figure 5.78a shows fluorescent images obtained from Alq_3 stripes drawn using this nozzle geometry, with an intensity map shown in the inset. The dependence of the feature size for the 10 μm wide nozzle on its distance from the substrate, s, is shown in Fig. 5.78b, along with the deposition profiles shown in the inset. The profile for the nozzle, as well as its dependence on s is fit well by DSMC. Importantly, these simulations suggest that a profile as narrow as 1.5 μm can be achieve with a 0.5 μm wide nozzle positioned at $s = 1.5$ μm (McGraw et al., 2011). As noted above, lines become wider and their edges become less well defined with increasing s. For a 10 μm wide nozzle and $s = 10$ μm, a diffuse edge of material extends approximately 4 μm beyond the line boundary. This degree of "overspray" is within the tolerances required for high definition display patterning.

Based on experimental results, the speed of printing of large organic electronic appliances can be estimated. For example, an array of 3000 nozzles, each with a 20 μm by 800 μm orifice, can simultaneously pattern 30 nm thick red-green-blue emissive layers to produce a full color, high definition 60″ (1 m^2) OLED display in approximately 25 s (McGraw et al., 2011). Multiple vapor streams can be integrated into a single print head to allow for the deposition of a full color pixel triad in a single pass without a loss of printing speed. The 3000 nozzle array would have a width of 810 mm, which can be produced using standard Si fabrication processes.

The most significant challenge to the large scale implementation of OVJP is the control of the nozzle-substrate distance, s, which determines the degree of parasitic material overspray into neighboring regions of the substrate beyond those directly targeted. As discussed, $s < 10$ μm is optimal for printing very small features. Yet this must be maintained over large, rapidly translated substrates where dust, prepatterned topographic features, surface roughness and other surface irregularities can interfere with the print head. Also, the gas exiting the nozzle in close proximity to the substrates imparts a high sideways velocity that can sweep the organic molecules laterally before they can adsorb onto the substrate, as in Fig. 5.76. Several strategies have been developed to counteract lateral spreading, including shaping the nozzle to include relief that reduces the sideways gas acceleration beneath the nozzles, actively removing the exhaust by local application of vacuum (i.e. "skimming"), and placing dielectric barriers on the substrate between printed regions similar to wells used in liquid-based inkjet printing (McGraw and Forrest, 2013).

Confinement of nozzle flow into the desired area can also be accomplished by supplying a *guard flow*, as shown in Fig. 5.79. Here, a concentric nozzle is employed, with the center nozzle conducting the molecular source and its carrier gas, and the outer annular nozzle conducting an additional hot carrier gas (Biswas et al., 2010). The outer guard flow provides hydrodynamic pressure that "focuses" the molecular beam, preventing excessive spreading. This approach may also prevent the entrainment of contaminants into the molecular jet, allowing for deposition of device-quality materials even in atmospheric environments. A heavier gas than the carrier (e.g. Ar) is particularly effective at confinement of the central molecular jet.

Visualization of the effects of guard flows is shown in the simulation based on the Navier–Stokes equations for fluid transport in Fig. 5.80. This analysis provides evidence, as experimentally confirmed, that guard flows both focus the molecular beam while preventing ingress of species from the environment.

Fabricating arrays of concentric nozzles for printing over large areas complicates nozzle design and

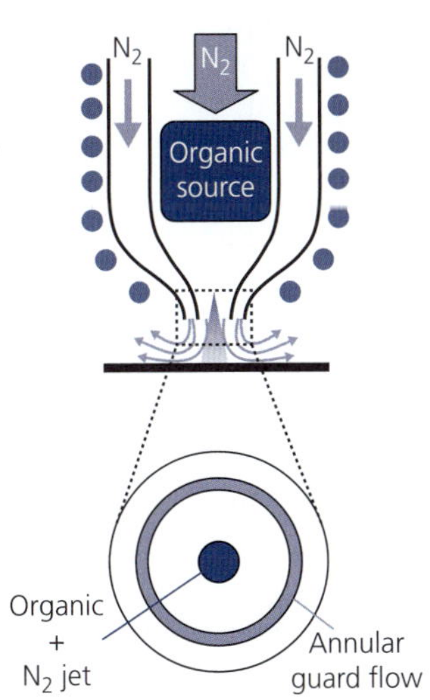

Figure 5.79 Guard flow nozzle showing the central region where the organic source material is carried by a hot N_2 gas to the substrate. The outer annulus carries hot N_2 that serves as the guard flow. The circles surrounding the nozzle assembly represent the heating elements (Biswas et al., 2010).

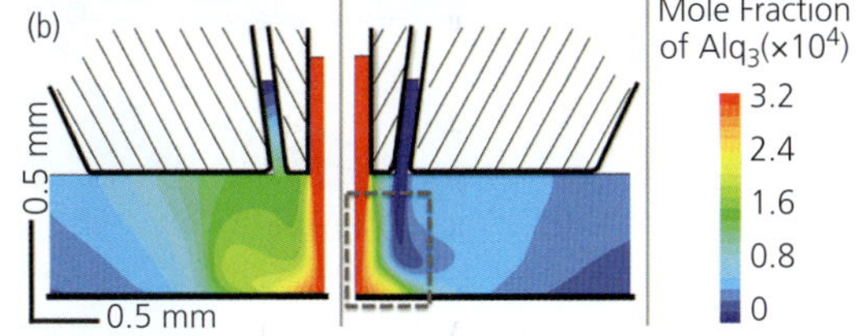

Figure 5.80 Simulation of (a) the velocity field and (b) the concentration of a heavy molecular species (Alq_3) in N_2 gas exiting a nozzle with and without the presence of a N_2 guard flow (Biswas et al., 2010).

fabrication. Further, the increased gas flow from the guard beam added to that of the organic source beam slow the removal of the large gas volume between the nozzle and substrate; a problem that becomes increasingly acute as the number and density of nozzles increases.

OVJP has several advantages over liquid-based inkjet printing. While it is compatible only with molecules that can be evaporated, no molecular modification is required to specifically adapt a species for use in vapor jet printing. The very high feature resolution approaching 1 μm makes it suitable for a wide range of transistor and light emitting device patterning scenarios. Also, the absence of surface tension produces a predictable deposit profile. Since the processing is free of liquid solvents, there is no interaction such as re-dissolution with previously deposited layers. Indeed, OVJP presents no constraints beyond those encountered in VTE or OVPD. Hence, wide flexibility in device design and materials choice is compatible with this dry printing method.

5.6.7 Other patterning techniques

There are numerous forms of rapid printing that have been developed for the graphics and print publication industries. Several of these technologies are also adaptable to patterning organic electronic devices. After all, the inks used in graphics printing are closely related to organic semiconductor materials. Hence, it is logical to assume that similar printing techniques can be useful in both applications. Yet, there are important differences that distinguish these two printing regimes. Organic electronic devices demand precise control of the morphology of the printed region, and often require extremely high pattern resolutions when compared to passive printed media. In both instances, high throughput leads to low cost. Here we discuss several printing technologies that are adaptable to organic semiconductor device applications. Each has its benefits and limitations, but in general none has been developed to the extent that would warrant their use in volume device manufacturing. The various methods are introduced roughly in terms of their maximum throughput.

Screen printing (throughput $\sim$ 2–3 m^2/s) is widely used for printing fabrics, signs, and other large graphical images. Thus, it has been adapted to very high speed and simple, low resolution printing of organic semiconductors and even metallic contacts. Two realizations of screen printers are illustrated in Fig. 5.81. Figure 5.81a shows screen printing to a flat plate, and

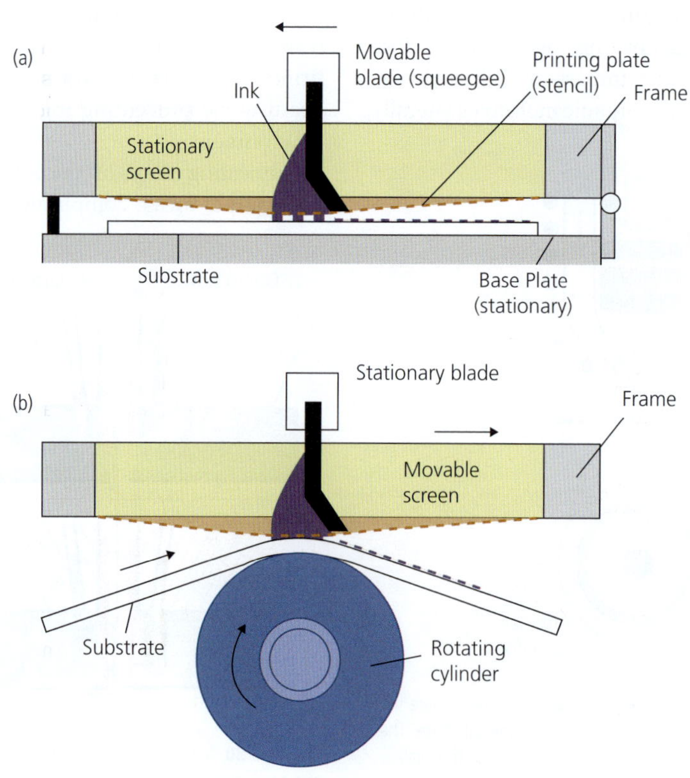

Figure 5.81 A screen printing apparatus for printing on (a) sheets and (b) rolls (Kipphan, 2001).

Fig. 5.81b is from a flat plate to a cylindrical roller. A silk screen with a fine mesh is placed in proximity (1–2 mm, typical) to the substrate on which the material is to be deposited. To create a pattern, a stencil material (typically a plastic or metal film) is embedded into the screen, or is placed onto its surface to mask those areas that are not to be printed. A semi-viscous electronic polymer or metal paste is then loaded onto the film surface and is pushed through the mesh by a rubber blade (often called a *doctor blade* or *squeegee*) that deforms the screen into contact with the substrate. The blade is translated at several centimeters per second across the substrate. The fluid penetrates the un-masked regions of the screen, leaving behind the desired pattern which is dried to remove excess solvent. The solution containing the molecular solute rides in front of the blade and is continuously replenished to avoid gaps in the printed pattern. Screen printing resolution is limited by the screen mesh openings, which are typically ~100 μm, although smaller meshes have been produced. The highest pattern resolution is approximately twice this dimension. Also, the method favors thicker films that are limited by the size and thickness of the mesh. Viscoelastic effects can also affect film uniformity, similar to that encountered in inkjet printing.

Given its limitations, screen printing is not widely used to pattern organic electronics. Nevertheless, given its very low cost and high speed, research laboratories have pursued its use for printing metal contacts, OPV bulk heterojunctions (Shaheen et al., 2001, Brabec and Durrant, 2008), OLEDs (Pardo et al., 2000, Lee et al., 2009), and OTFT active regions (Bao et al., 1997, Rogers et al., 1998).

Offset printing (throughput $\sim$ 5–30 m^2/s) is a form of lithographic printing whereby the ink is applied to the plate, or in Fig. 5.82, the plate cylinder. The pattern to be printed is defined on the plate by having inked regions that are *oleophilic* (and hence hydrophobic), and hydrophilic ink-repellent regions. The plate is prepared by applying inks from the inking unit on a surface that has initially been treated with a film of water from the dampening unit. Dampening the cylinder ensures that no ink will transfer to the regions that should not be printed. The patterns are then transferred from the plate onto the blanket cylinder, which ultimately delivers the pattern to the substrate that rolls over the surface of the final, impression cylinder.

Offset printing is the most common form of printing, resulting in features that are finer than screen printing, typically 10–50 μm. Importantly, for organic electronics, the control of film thickness of from 0.5–1.5 μm is quite high compared to other high volume printing technologies. Also, registration of multiple layer patterns can approach 10 μm, allowing for a modest level of circuit complexity. The advantages of offset printing are somewhat countered by the complexity of the system comprising multiple contacting elements, and controlling the chemistry of the inks and dampening solutions in ways that are compatible with high device performance, yield and

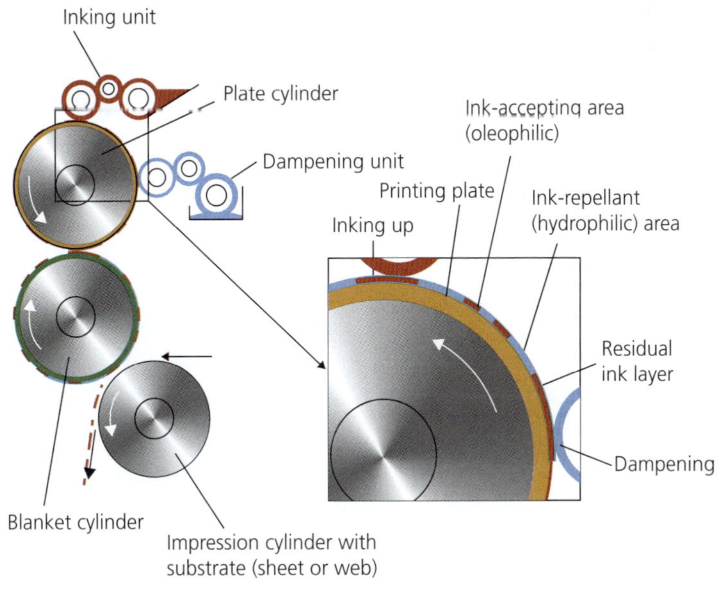

Figure 5.82 Schematic of an offset printing apparatus (Kipphan, 2001).

pattern resolution. Offset printing has been used in producing PEDOT (Zielke et al., 2005) and metal nanoparticle ink-based (Fukuda et al., 2015) transistor source and drain contacts, as well as for transparent OPV electrodes (Kim et al., 2015).

Flexographic printing (throughput $\sim 10\ \mathrm{m^2/s}$) is a variation on *letterpress printing*, and is shown schematically in Fig. 5.83. The image to be printed on the substrate is first embossed on a roller consisting of a soft material such as a hardened rubber or photopolymer that can hold ink on its surface, or absorb the ink. The printing plate itself is held on a hard cylinder or plate to give it structural stability. Prior to transfer to the plate, the ink is first spread over a cylinder containing a high density of ink-retaining wells. This ink-transporting cylinder is known as an *anilox roller*. The ink has a low viscosity compared to that used in offset printing since it must be retained by the roller and the printing plate. The uniformity of the liquid layer is maintained by a doctor blade. The printing plate finally transfers the inks to the substrate to form the desired image.

The combination of low-viscosity fluid with the soft, elastic printing plate, which is distorted under the contact pressure with the substrate, results in a low-resolution image. Flexography bears similarities with so-called pad printing (Kipphan, 2001), as well as the additive soft-stamping process described in Section 5.6.4 that also suffers from lower resolution than hard-stamping methods due to stamp distortion. Thus, feature sizes ~100 μm are typical of flexographic printing, which is at least two orders of magnitude larger than achieved by soft stamp transfer done on a batch, substrate-by-substrate process. Flexography has been used to generate diacrylate gate dielectrics in polymer thin film transistors (Cosnahan et al., 2018). However, given its pattern resolution limitations, it has not been widely exploited in the patterning of organic electronic devices.

Gravure printing (throughput $\sim 50\ \mathrm{m^2/s}$) differs from offset printing in that the image to be transferred is etched or engraved into the plate or cylinder, as shown in Fig. 5.84. The wells thus formed are filled

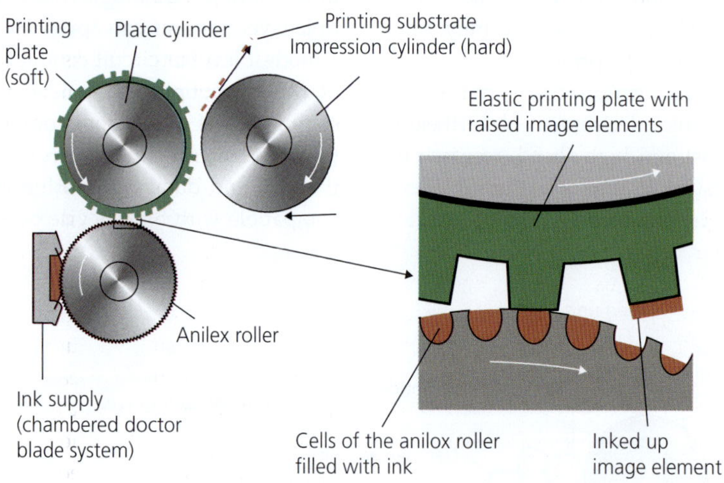

Figure 5.83 Schematic of a flexographic printing apparatus (Kipphan, 2001).

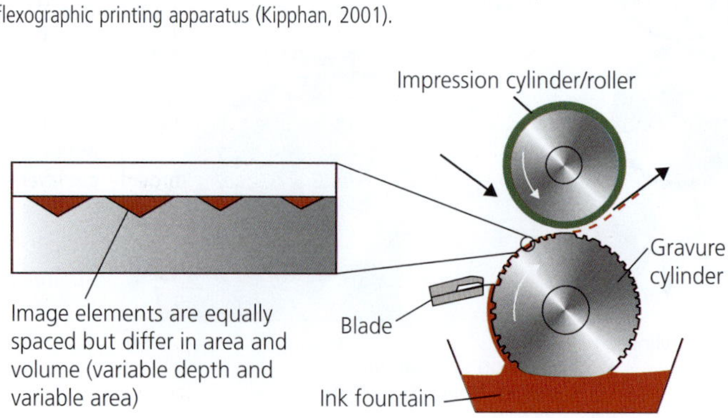

Figure 5.84 Schematic of a gravure printing apparatus (Kipphan, 2001).

with ink, with the excess removed by a doctor blade. The gravure plate transfers the ink to the substrate that is forced against the impression roller. The amount of ink and the resolution of the pattern is determined by the width and depth of the wells. Sidewall control is essential since ink residues can clog the wells or otherwise decrease the pattern resolution achieved. However, the wells themselves are well adapted to retaining low viscosity fluid, whereas in offset printing, the viscosity must be sufficiently high for the oleophilic regions to maintain pattern definition. The technique is capable of printing layer thicknesses < 1 μm.

The engraving or etching process to create the image on the cylinder is costly, and each change in image (corresponding to a particular circuit or device layout) requires the fabrication of a new cylinder. Furthermore, each separate layer (e.g. the R, G, and B pixels in a display) requires a separate cylinder or plate precisely aligned to the others in the production chain. Countering this complexity is the extremely high throughput of gravure printing. Features tend to be large due to the limited precision with which the ink wells can be fabricated, and the low viscosity fluid that makes pattern confinement difficult. Feature sizes of 75 μm or larger are common, making this technique most appropriate for contact deposition or for the active regions in large area devices such as OLED lighting sources (Kopola et al., 2009, Hernandez-Sosa et al., 2013) and solar cells (Kopola et al., 2011, Voigt et al., 2011). Very high resolution patterns (< 10 μm feature sizes) of the polymer gate dielectric in thin film transistors have been achieved using precise engraving of the gravure cylinder. This *microgravure* process also demonstrated linear throughput speeds of nearly 1 m/s (Kang et al., 2012).

We conclude this discussion with a comparison of several printing methods in Table 5.5. While they remain at an immature stage of development for electronics applications, they continue to be of interest due to a primary promise of organic electronics—that of manufacturing large-scale devices and circuits on flexible substrates at extremely low cost. The inherent similarities between inks used in creating graphical images and those used as electronic materials will continue to motivate development of high-speed printing for use in manufacturing of on-demand, organic electronic appliances.

5.7 Roll-to-roll production of organic electronics

A significant driver of low cost electronics is the fabrication of devices in a continuous process, where circuits and complex devices are rapidly produced "by the meter" rather than currently employed, substrate-by-substrate semiconductor batch processes. Increased output is enabled by continuous, *roll-to-roll* (R2R) manufacturing that employs one of the many rapid print patterning methods discussed in previous sections. Most of these patterning processes are already implemented, or can readily be adapted to the R2R environment.

Roll-to-roll, also referred to as *reel-to-reel* or *web* processing, is a general term that implies fabrication of the devices and circuits on a substrate by rapidly translating it from process to process in a continuous, uninterrupted manner. The processes themselves include the addition, subtraction, annealing (either thermally, by illumination, or solvent exposure), and patterning of multiple layers that comprise the electronic or photonic devices. Hence, while process speed is enhanced by continuous web-based manufacturing schemes, it becomes challenging to achieve high tolerance alignment of patterns on successive layers, simply due to the short residence time of the substrate at each process step, as well as the tendency to distort (via stretching, compression, twisting or other mechanical stress) the substrate web as it proceeds through the various manufacturing stages. Hence, web-based processing is only practical where very high alignment

Table 5.5 Comparison of printing methods used for producing graphic images. Adapted from Sirringhaus et al. (2006)

Printing method	Viscosity (MP)	Layer thickness (μm)	Feature size (μm)	Throughput (m^2/s)	Feature registration (μm)
Screen	500–50,000	30–100	20–100	2–3	>25
Offset	40,000–100,000	0.5–1.5	10–50	5–30	>10
Flexography	50–500	0.8–2.5	80	10	<200
Gravure	50–200	0.8–8	75	60	>10
Inkjet	1–30	<0.5	20–50	0.01–0.5	5–20
OVJP	N/A	>0.01	1.5	1–2	5

tolerances (<20 μm) are not required, or where automatic, or self-alignment between layers is built directly into the process flow. Furthermore, since most active organic electronic materials degrade when exposed to air or moisture, some or all of the fabrication processes must occur in highly controlled, contaminant- and dust-free atmospheres—typically in vacuum or in high purity nitrogen. Indeed, while considerable work has focused on solution-based deposition in ambient due to its simplicity, it is unlikely that high performance and reliable devices and processes can be realized under such conditions. In contrast, R2R vacuum coating of extremely high volumes of materials is a standard process employed by the food industry, and has even been successfully used for organic electronic devices (Chaug et al., 2004). Nevertheless, enclosing large processing lines in a contamination-free environment can be difficult and costly, making it convenient to do some of the processing in separate chambers, whereby the web enters and exits via differentially pumped plenums, or where the entire web is included within the vacuum space. The last step in the fabrication sequence is packaging or lamination to protect the electronic devices. This is followed by cutting the web into discrete device or circuit segments. The substrate segmentation is called *singulation*.

Several web processing system layouts are shown schematically in Fig. 5.85. Roll-to-roll implies that the substrate is thin and flexible, typically comprised of a polymer such as PET or PEN, metal foil, very thin and flexible glass sheet, or even paper. While glass is brittle and fragile, if made sufficiently thin (<100 μm) it can be very flexible without being distorted by stretching. Since it is extremely smooth and impervious to atmospheric contaminants, if handled carefully, flexible glass provides a nearly ideal substrate for web processing. The substrate is deployed from a feed reel through rollers to each process stage, and then the finished product is collected by a take-up reel (Fig. 5.85a). Lateral pattern registration is done at the edge of the web, and is adjusted along its length by active control of the tension. This is usually accomplished by rollers placed intermediate between process stations. Their relative positions are actuated in real time to dynamically tighten or loosen the web.

Another means to transport large substrates in a continuous or semi-continuous fashion is sheet-to-sheet, in which the substrates are already cut into manageable sizes, and continuously transported through a series of process stations (Fig. 5.85b). The translation from the feed stack to the output stack is via grippers, or robots that remove the substrate from the previous station, and then place it in the next while also performing the task of registration and alignment required in multilayer fabrication processes. This configuration has been widely adopted by the OLED display industry, where multiple layers are deposited in sequence on large, very thin glass substrates. Sheet-to-sheet manufacturing systems can extend to over 100 m in length to complete the multilayer film deposition process.

In the sheet-on-shuttle process (Fig. 5.85c), the substrates are placed on a continuous carrier or track, each substrate registered in its own fixture within the shuttle. The shuttle itself is translated using a gear mechanism that ensures precise alignment at each process position. Large substrates can be laid either horizontally or held upright. If a process step requires physical contact with the substrate (e.g. stamp patterning of metal contacts), the shuttle must be sufficiently robust to support the substrate as pressure is applied.

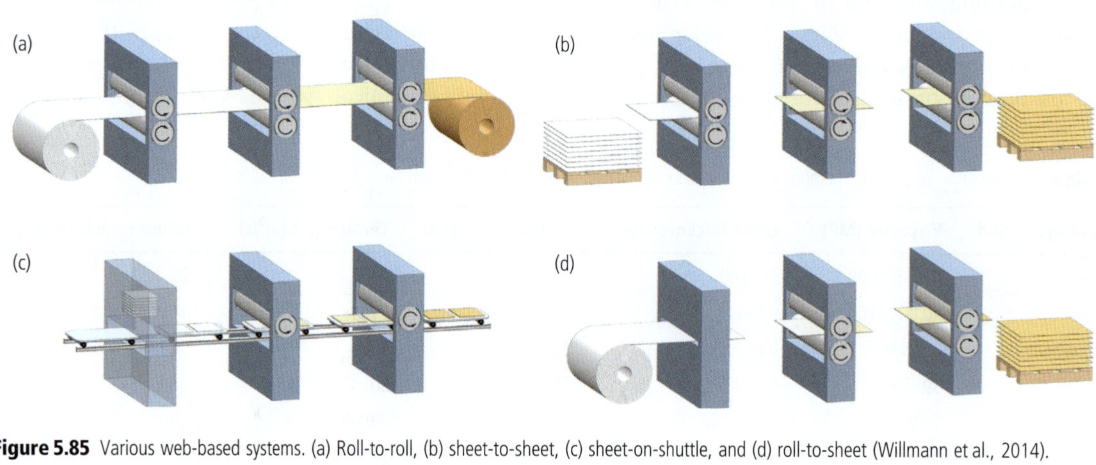

Figure 5.85 Various web-based systems. (a) Roll-to-roll, (b) sheet-to-sheet, (c) sheet-on-shuttle, and (d) roll-to-sheet (Willmann et al., 2014).

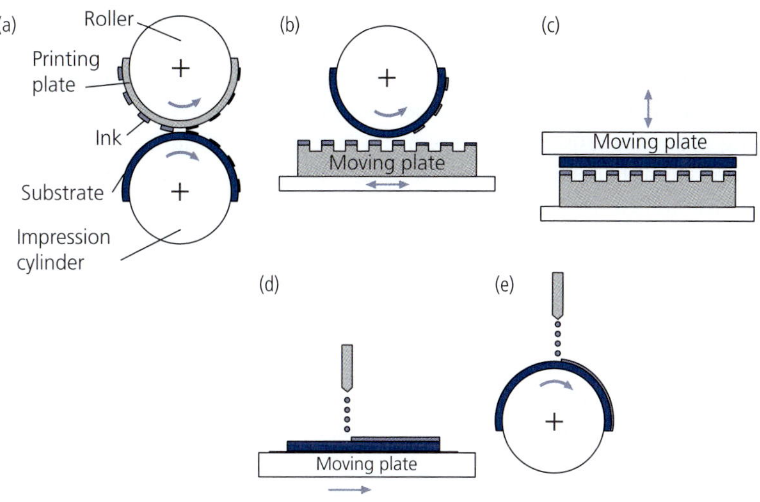

Figure 5.86 Various printing configurations. (a) Contacting rollers; (b) roller-plate; (c) plate-plate; (d) non-contact planar; and (d) non-contact roller printing (Willmann et al., 2014).

Finally, combinations of the substrate handling processes in continuous or semi-continuous web environments is also possible, as shown in Fig. 5.85d. There are many situations where these combinations are essential; for example when some of the processes in circuit fabrication require a controlled environment whereas others do not. In this case, the web roll is removed from one process line and moved to another where sheet-based processing is more convenient. This movement decreases the manufacturing line output, or productivity, requiring the use of parallel systems to complete the slower process steps to ensure that they keep pace with the more rapid processes. That is, the rate of production is determined by the slowest step in the continuous chain. Another consideration is that the process may be continuous (e.g. printing stripes by gravure printing on a roller) or intermittent, in which case the substrate movement pauses while the process occurs (e.g. deposition of two dimensional patterns through a shadow mask). The duration of a particular process step is called the *takt time*, that is, the time that a machine must take to complete the step. The takt time can refer to that needed to complete a single step, multiple steps in a series, or even the entire fabrication process itself.

Figure 5.86 shows several different fabrication configurations that the substrate might encounter during the process sequence. They fall into two categories: contact and non-contact printing. Examples of contact printing described in Section 5.6 include stamping, gravure, or screen printing. Non-contact methods include inkjet and vapor jet printing. Full surface

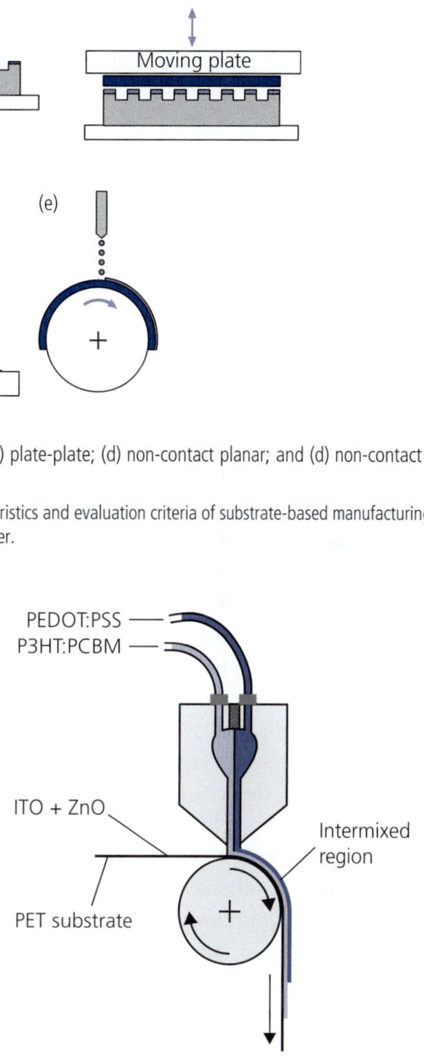

Figure 5.87 Continuous, double slot-die coating of the OPV active region along with the top cathode. When combined with the bottom anode already applied to the plastic substrate, the OPV cell structure is completed in a single process step (Sondergaard et al., 2012).

deposition of films via vapor phase deposition (e.g. OVPD and VTE), or liquid spray-on or spreading with a knife edge (i.e. a doctor blade or squeegee) are also compatible with continuously transported substrates in web fabrication environments. The application of materials can be done in either a flat sheet or continuous, roller-based configuration.

An example of application of a multiple component layer by slot-die coating is shown in Fig. 5.87. Here, the electrode material, PEDOT:PSS, is combined with a continuous flow of a donor–acceptor OPV mixture of P3HT and $PC_{61}BM$, allowing for the simultaneous deposition of two layers with significantly different functions and rheological properties. This is layered onto a PET substrate that supports the transparent anode consisting of the metal oxides, ITO and ZnO. Slot-die coating is similar to blade coating, in that the layers are formed by a sharp edge that ensures a constant thickness determined by the temperature, viscosity, translation velocity, and the surface tension (and hence the meniscus) of the liquid being spread over the substrate surface. As in all liquid-based processing methods, care must be taken in choosing compatible materials such that two or more layers do not interdiffuse or dissolve. In Fig. 5.87, PEDOT:PSS has an aqueous base, whereas P3HT:$PC_{61}BM$ uses an organic (hydrophobic) solvent to minimize interactions between layers. Slot-die coating is potentially a high productivity film deposition procedure, with substrate speeds in the range 0.1–200 m/min.

Some of the stations along the processing line must be devoted to metrology and control. These include *in situ* film thickness measurement, pattern alignment, and particle detection. These properties are monitored in real time, with results fed back to adjust the process flow to prevent it from producing large volumes of useless, non-functional product. All semiconductor manufacturing lines, whether done by batch or web, require such process diagnostics to ensure a high production yield of devices that perform at the required standard.

A further consideration is the method for recharging the deposition stations with fresh material. As substrates are translated through each station, source materials are consumed. To avoid halting the line to refill materials reservoirs, the manufacturing lines allow for continuous feed of materials to the deposition site (e.g. roller or plate) from an external location. Similarly, cleaning of components contained within the system (e.g. shadow masks, screens, and print heads) should also be an integral part of the systems, since routine maintenance ensures maximum productivity, as well as a minimum down time where the system is taken off-line to restore it to optimal operational order. Perhaps the biggest single problem in maintaining a web-based processing line is ensuring that dust and other debris do not accumulate on the substrate as it passes from station to station, or between rollers. Organic devices are typically only a few tens to hundreds of nanometers thick. Hence, even the smallest particles or irregularities on the surface of the web can result in pinholes in the device active regions, creating shorts or paths for the ingress of contaminants. Defect densities of 100–1000 m^{-2} with radii ranging from 0.1 to 3 μm are commonly observed on PET substrates, depending on the coatings applied; a level that may prevent high productivity (Graff et al., 2004). Furthermore, dust can cause substrate damage during stamping or mask attachment where pressure is applied to its surface.

To date, there has been insufficient research on R2R processing to fully assess its usefulness in the manufacture of organic electronic circuits. The equipment is necessarily complex and large, and the consumption of costly source materials is also an impediment to its development. However, once devices, materials and processes that can benefit from such large scale manufacture are defined, it is anticipated that research efforts in this area will grow apace. The most likely applications to benefit from continuous R2R processing are OPVs and low cost organic transistor circuits (Sondergaard et al., 2012). Since the technologies leading to their successful and widespread deployment have not yet been fully developed, it is understandable that their use in large-scale manufacturing of devices has not yet been realized. Nevertheless, one of the major benefits of organic electronics is the promise of very low cost circuits serving a diversity of custom as well as low cost needs. Provided sufficiently high performance can overcome the compromises that are inherent in continuous processing environments, R2R and other web-based fabrication technologies may yet become a mainstay in the manufacture of organic electronics (Forrest, 2004).

5.8 Packaging

The final step in the fabrication of an electronic device or circuit is its packaging. This has several purposes: the package protects the device from degradation due to environmental exposure, it can be sufficiently large and robust to make handling of the electronics contained within it manageable, and it enables connections to an external circuit. The vulnerability of organic materials to degradation when exposed to oxygen (particularly in the presence of light) and moisture makes some form of packaging imperative. Packaging of flexible electronics (requiring flexible packages) presents particular challenges.

For rigid displays used in many smartphones (see Fig. 5.88), the device is deposited onto a glass substrate. It is then sealed in an ultrahigh purity N_2 environment by a metal (or glass) lid to the substrate using a bead of epoxy that is cured by brief exposure to UV light. Alternatively, a cold-weld bond or solder seal is used.

The lid often contains a thin film desiccant such as the hygroscopic compounds BaO or CaO, or a high surface area *zeolite* that is deposited onto its inner

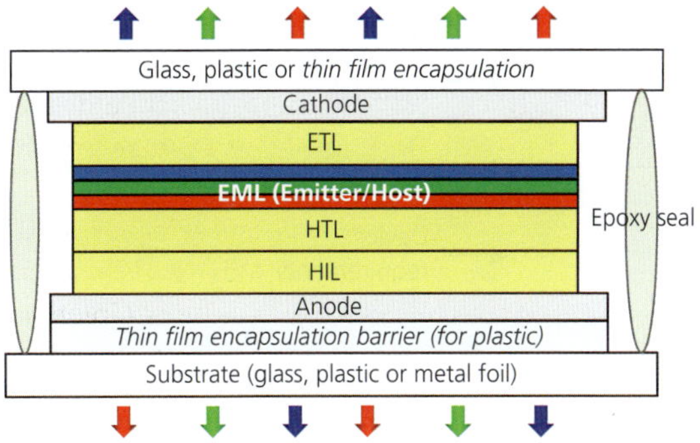

Figure 5.88 Packaged OLED for lighting, showing its various functional layers and two possible emitting surfaces. An epoxy seal is placed around the package periphery, sealing the organic materials within.

surface. The desiccant scavenges the residual moisture that is captured in the package during sealing. Accommodation must also be made to provide electrical access to the device via contacts pre-patterned onto the substrate surface. The contacts must not be shorted nor interrupted by the seal, nor can they introduce leaks into the package interior. A moisture indicator that changes color (e.g. $CoCl_2$) or transparency (Ca) when it is saturated with water can also be deposited inside the lid as a test of package integrity. This latter step is useful in diagnosing the integrity of the sealing strategy employed.

Packaging devices on flexible substrates provides a far greater challenge than on rigid media (Logothetidis, 2008). Metal foils can provide robust, impermeable flexible substrates. However, they must be coated with a planarizing polymer to overcoat surface irregularities absent in plastic or glass foils. On the other hand, plastic substrates are permeable to moisture and oxygen, presenting only a weak barrier to these contaminants that lead to irreversible degradation of device performance. For this reason, multilayer coatings shown in Fig. 5.89 that alternate between a plastic and an inorganic oxide layer have been developed for organic electronic applications. Plasma deposition of oxides such as alumina (Al_2O_3), AlO_xN_y, SiO_xN_y, or SiO_2 are typically interleaved with spin- or vapor-deposited parylene, PET, or other acrylate (Affinito et al., 1996, Dennler et al., 2006). The inorganic and organic layers are typically only a few tens, and a few hundreds of nanometers thick, respectively. If the encapsulated device emits or absorbs light, the laminate must be transparent.

Such laminates have very low moisture permeation rates that are limited by the density and size of defects on the substrate. These defects are responsible for pinholes in the inorganic layers. The layer

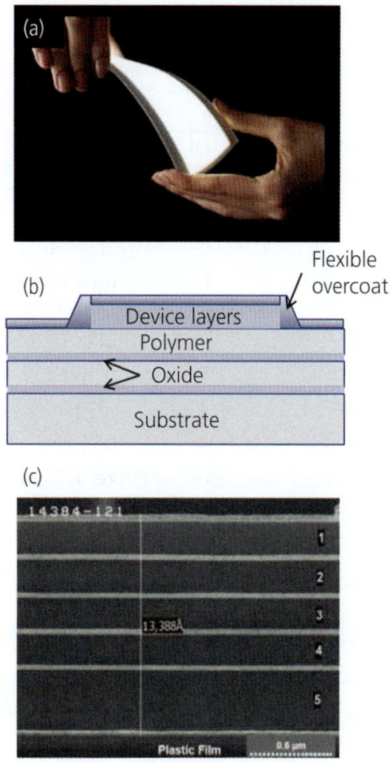

Figure 5.89 (a) Flexible OLED lighting panel demonstrator fabricated by Konica-Minolta: (b) Schematic of a flexible (plastic) package on a substrate comprising two dyads of a polymer/oxide barrier layer, and a flexible encapsulating coating. (c) Transmission electron micrograph of the cross-section of a 5 dyad barrier layer on a plastic substrate. The total barrier layer thickness (vertical line) is 1.34 μm. The layers are produced by Vitex Systems (Lewis, 2006).

permeabilities themselves are no different than for thicker films, which is generally too high to result in device lifetimes greater than a few hundred to a thousand hours. However, when placed in lamellar form such as in Fig. 5.89b and c, the pinholes are nonuniformly distributed at low density in each of the inorganic layers. A water molecule that enters the plastic substrate will meander within the plane of a polymer layer until it finds a pinhole that allows it to cross into the next layer in the laminate. Hence, penetration of moisture into the device volume is retarded simply due to the longer path a water molecule must travel to reach it (Graff et al., 2004). In Table 5.6, we show the *water vapor transmission rate* (WVTR) vs. number of pairs of a 0.34 μm thick PMMA/0.037 μm thick sputter-deposited AlO_x layer on a 177.7 μm thick PET substrate. This organic/inorganic pair is known as a *dyad*.

The permeation rates for several substrates as well as the expectations for various device applications are shown in Fig. 5.90. The display industry standard for an acceptable level of water permeation is

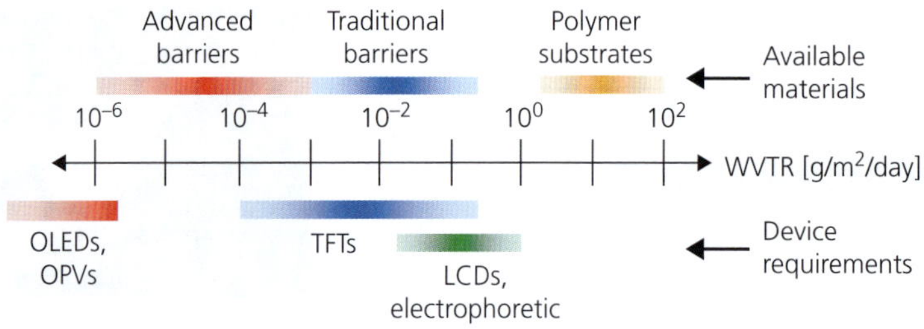

Figure 5.90 Flexible package types (top) and requirements needed to be met by several different device types leading to their acceptable operational lifetimes (bottom) vs. water vapor transmission rate. Advanced barriers are multilayer structures as in Fig. 5.89, traditional barriers are single layer structures. LCDs and electrophoretic displays are not organic electronics-based, and thus are less susceptible to water damage. After Lewis (2006).

Table 5.6 Water vapor transmission rate (WVTR) vs. lamellar substrate structure on PET[a] (Graff et al., 2004).

No. of dyads[b]	WVTR ($g/m^2/day$)
0	4.7
1	0.07
2	<0.005
3	<0.005
4	<0.005
5	$\sim 10^{-6}$

[a] PET substrate thickness = 1.77 μm.
[b] The dyads comprise 0.34 μm PMMA/0.037 μm AlO_x.

Figure 5.91 Schematic of R2R organic electronic encapsulation via lamination between rollers in a continuous web process. The electronic circuits are patterned on the underside of the top layer, and are combined by pressure lamination with a flexible encapsulating "lid" deployed from the lower roller. Both the substrate and lid materials must provide a barrier against atmospheric contaminants. The seal can consist of an adhesive, or a cold-weld bond between metal tracks on the substrate and lid.

WVTR $< 10^{-6}$ $g/m^2/day$, which is met when five or more dyads are employed. Unfortunately, the substrate fabrication process can be expensive, ultimately making the use of metal foil or thin glass substrates preferable.

To complete the flexible electronics packaging, a thin and highly flexible cover is required. While a flexible glass or metal foil lid can be sealed to the flexible substrate similar to the method used in rigid substrate encapsulation, a bilayer of two foils bonded together by an adhesive significantly limits the package flexibility. Alternatively, a laminate consisting of alternating layers of an oxide and a thin polymer can result in a barrier with properties similar to an analogously prepared substrate in Fig. 5.89b. When the top of the device is coated with a multilayer film, there is no need for epoxy or solder sealing the edge of the package. There is also no space within the package between the device and its sealant. Thus, the entire packaging process can be carried out in the same system used to grow the device itself without exposing the sensitive electronic layers to exogenous contamination. Of course, deposition conditions must not result in device degradation, and accommodation must still be made to bring the electrical contacts to a region outside of the device volume.

Finally, packaging and encapsulation on flexible (or rigid) substrates can be made compatible with web processing. A conceptual diagram of a web-based device lamination system included at the end of a R2R fabrication sequence is illustrated in Fig. 5.91. As with all process steps, R2R packaging must be carried out in a contaminant-free and controlled environment to ensure that no impurities are captured within the package volume.

5.9 Summing up

In this chapter, we learned about the "how to" of organic electronics: that is, how to take a material from its crude source to a large scale electronic device created on a continuous web, ultimately leading to potentially low cost and unique optoelectronic applications inaccessible to other electronic technologies. There are many steps from synthesizing the source material to its application.

It starts with refinement of the crude material, to the growth of thin films with precisely controlled morphologies. Indeed, there has yet to be an instance in electronic or photonic technology where purity of the source materials could be sacrificed without making unacceptable compromises in device performance and lifetime. In this regard, organics are no different.

Once the layers are deposited in the desired scheme, the devices must be patterned and ultimately packaged. One of the most captivating attractions of organic electronics is their enablement of large devices such as displays, lighting, solar cells and complex circuits adapted for biological monitoring. Hence, they often must be lightweight, conformal and flexible. While these attributes offer many unique possibilities for applications not served by conventional inorganic semiconductor electronics, they nevertheless present complicated challenges to materials and device engineers. We have seen that organics can be deposited on sharply curved substrates, resulting in detector focal plane arrays that mimic the form and function of the human eye. We have also shown that the similarity of materials used in graphical and document printing, and those used in organic electronics enable similar and potentially very low cost patterning technologies such as inkjet, gravure and offset printing.

While this chapter covered many of the materials preparation, deposition and processing aspects common to all device types, the topics covered represent only a small fraction of all the techniques available. In fact, a unique feature of organic semiconductors is that they have encouraged the development of an extraordinarily wide range of processes from purification to printing. Hopefully, this chapter has provided the foundational knowledge that will be expanded in subsequent chapters when discussing specific details of devices in the remainder of this book.

Covering all aspects of this broad field would itself be the subject of an entire book; hence this chapter is meant only as an introduction. The reader is encouraged to learn more about materials growth and device fabrication processes by referring to the extensive literature devoted to the subject. Several references included in *Further reading* can serve as a starting point for exploration of the materials science and device engineering of organic semiconductors.

Further reading

1. S. Biswas, O. Shalev, and M. Shtein, 2013. Thin-film growth and patterning techniques for small molecular organic compounds used in optoelectronic device applications. *Annu. Rev. Chem. Biomol. Eng.*, 4, 289.
2. S. R. Forrest, 1997. Ultrathin organic films grown by organic molecular beam deposition and related techniques. *Chem. Rev.*, 97, 1793.
3. H. Frey and H. R. Khan (Eds.), 2015. *Handbook of Thin-Film Technology*. Berlin: Springer-Verlag.
4. F. Gutmann and L. E. Lyons, 1981. *Organic Semiconductors*. Malabar, FL: Robert E. Kreiger.
5. N. Karl, 1980. High purity organic molecular crystals. In: *Organic Crystals, Germanates, Semiconductors*. Berlin: Springer.
6. E. H. Klauk, 2006. *Organic Electronics: Materials, Manufacturing and Applications*. Berlin: Wiley-VCH Verlag.
7. M. M. Ling and Z. Bao, Thin film deposition, patterning and printing in organic thin film transistors. *Chem. Mater.*, 16, 4824.
8. C. Mack, 2007. *Fundamental Principles of Optical Lithography: The Science of Microfabrication*. Chichester, UK: John Wiley and Sons.
9. S. M. Sze, 1981. *Physics of Semiconductor Devices*, 2nd ed. New York: Wiley-Interscience.
10. Y. Xia and G. M. Whitesides, 1998. Soft lithography. *Annu. Rev. Mater. Sci.*, 28, 153.

Problems

1. Zone refining is an iterative process. It involves a thermal scan across a material ingot and removing the contaminated section at the end of a scan, and then starting over until the desired purity is reached.
   (a) Starting with Eq. 5.3 for the change in impurity concentration at location x, derive the concentrations after a single pass given by Eqs. 5.5 and 5.6.
   (b) Similarly, in the limit of an infinite number of scans (i.e. $n \to \infty$), show that the impurity concentration is given by Eqs. 5.7–5.9.
2. Ultracentrifugation separates a mixture of particles according to their densities, radii, etc. along a vial containing a fluid of varying density along its length, L.
   (a) Derive Eq. 5.13 for the radius of a fraction of a population at position z along a vial.
   (b) If the initial distribution of radii of nanoparticles of constant density in a colloid is a Gaussian of width σ_i, with a mean radius of $\langle R \rangle$, plot $R(z)$ after centrifugation time, t, for a rotational angular velocity ω using a centrifuge of radius, r. Assume a linear density gradient of the solvent from ρ_{top} at the top, to ρ_{bot} at the bottom of the vial.
   (c) Estimate the maximum time of centrifugation needed to obtain a reasonable separation of the population.
3. Derive Eq. 5.34 for the thickness of a spun-on film.
4. Patterning of displays employs vacuum thermal evaporation (VTE) through a shadow mask. Assume that a shadow mask of thickness, d_M, with a square opening of side a, is used in a VTE deposition system. Furthermore, assume a source-to-substrate distance of D.
   (a) Derive an expression that describes the peripheral shape of the deposit based on all parameters in Fig. 5.59a.

(b) What is the relationship between the pattern width, w, vs. a, D, d_M and the mask-to-substrate distance $d_{M\text{-}S}$ (as measured from the side of the mask facing the substrate). Plot w/a vs. D for $d_M = 20\ \mu m = a/2$.

(c) What shape of mask opening would provide the smallest pattern?

(d) What do you think is the smallest feature size obtainable with a practical VTE deposition apparatus? How does this compare to the highest resolution smart phone pixel size?

5. Estimate the force required to pattern contacts across an entire Gen 8 substrate using
   (a) A hard stamp via the subtractive cold-weld process. Assume that the contacts occupy 80% of the substrate area (a number typical of a high resolution display).
   (b) A soft stamp via the additive process.
   (c) Do the same calculations in (a) and (b) for a R2R system where the pressure is applied between 20 cm diameter rollers, assuming a rigid glass substrate and a contact area equal to 1% of the roller circumference.
6. Calculate the approximate melting temperature of an anthracene dimer using the atom-atom potential method, assuming that the anthracene molecules are facially stacked directly on top of each other. Can this material be evaporated without decomposition (determined by the bond energy of a benzene ring)?
7. Assume that the rate-limiting step in a web process is determined by the organic film printing rate by OVJP of 1 µg/s of Alq_3. The process step consists of writing a continuous, 40 µm wide stripe of material that is 300 nm thick using a single 300 µm long by 20 µm wide nozzle.
   (a) What is the nozzle-to-substrate separation (s) that must be maintained to provide a deposit whose half width is equal to the target half-width?
   (b) What is the web printing rate limitation by this step in meters per second?
   (c) What control mechanism would you use to maintain a constant s to within ±2 µm? Provide details of estimated response time limits of the mechanism.
8. One limitation of OVPD is that in depositing materials with significantly different sublimation and decomposition temperatures, the reactor walls must be maintained at a temperature sufficient to prevent deposition of the less volatile material, while it is kept low enough to prevent decomposition of the lower temperature material. One means to prevent decomposition is to limit the collisions of a molecule travelling along the reactor axis from having too many collisions with the hot walls.
   (a) Derive an approximate expression for the number of collisions (N) a molecule will have from the point of injection along the axis of a reactor tube of length L and radius r. Here, we assume a N_2 carrier gas velocity, v, pressure, P, and reactor temperature, T. A molecule of mass M is entrained (at low Reynold's number, Re) in the carrier gas having reached terminal velocity.
   (b) Assuming L = 1 m long reactor tube with radius r = 10 cm, P = 0.1 torr, T=350°C, and a flow rate of 10 sccm, calculate Re. Is the flow laminar or turbulent?
   (c) Assuming the same conditions as in (b) for a low concentration of Alq_3 molecules injected into the gas stream, calculate N.
   (d) If the reactor temperature is raised to 500°C (approximately 50°C higher than the decomposition temperature of Alq_3), estimate the probability for its decomposition prior to exiting the tube. Can you suggest a means to decrease the probability of decomposition?

References

Affinito, J., Gross, M., Coronado, C., Graff, G., Greenwell, I. & Martin, P. 1996. A new method for fabricating transparent barrier layers. *Thin Solid Films*, 290, 63.

Arnold, M. S., Green, A. A., Hulvat, J. F., Stupp, S. I. & Hersam, M. C. 2006. Sorting carbon nanotubes by electronic structure using density differentiation. *Nat. Nanotechnol.*, 1, 60.

Arnold, M. S., McGraw, G. C., Lunt, R. R. & Forrest, S. R. 2008. Direct vapor jet printing of three color segment organic light emitting devices for white light illumination. *Appl. Phys. Lett.*, 92, 053301.

Arulchakkaravarthi, A., Lakshmanaperumal, C. K., Santhanaraghavan, P., Jayavel, P., Selvan, R., Sivaji, K., Gopalakrishnan, R. & Ramasamy, P. 2002. Investigations on the growth of anthracene and trans-stilbene single crystals using vertical Bridgman technique. *Mater. Sci. Eng. B*, 95, 236.

Ashraf, R. S., Schroeder, R. C., Bronstein, H. A., Huang, Z., Thomas, S., Kline, R. J., Brabec, C. J., Rannou, P., Anthopoulos, T. D., Durrant, J. R. & McCulloch, I. 2013. The influence of polymer purification on photovoltaic device performance of a series of indacenodithiophene donor polymers. *Adv. Mater.*, 25, 2029.

Augis, J. A. & Bennett, J. E. 1978. Calculation of the Avrami parameters for heterogeneous solid state reactions using a modification of the Kissinger method. *J. Thermal Anal.*, 13, 283.

Avrami, M. 1940. Kinetics of phase change. II. *J. Chem. Phys.*, 8, 212.

Bai, L., Ma, X., Liu, J., Sun, X., Zhao, D. & Evans, D. G. 2010. Rapid separation and purification of nanoparticles in organic density gradients. *J. Am. Chem. Soc.*, 132, 2333.

Baldo, M., Deutsch, M., Burrows, P. E., Gossenberger, H., Gerstenberg, M., Ban, V. S. & Forrest, S. R. 1998a. Organic vapor phase deposition. *Adv. Mater.*, 10, 1505.

Baldo, M. A., O'Brien, D. F., You, Y., Shoustikov, A., Thompson, M. E. & Forrest, S. R. 1998b. High efficiency phosphorescent emission from organic electroluminescent devices. *Nature*, 395, 151.

Bao, Z., Feng, Y., Dodabalapur, A., Raju, V. R. & Lovinger, A. J. 1997. High-performance plastic transistors fabricated by printing techniques. *Chem. Mater.*, 9, 1299.

Biswas, S., Pipe, K. P. & Shtein, M. 2010. Solvent-free direct printing of organic semiconductors in atmosphere. *Appl. Phys. Lett.*, 96, 263301.

Bleay, J., Hooper, R. M., Narang, R. S. & Sherwood, J. N. 1978. The growth of single crystals of some organic compounds by the Czochralski technique and the assessment of their perfection. *J. Cryst. Growth*, 43, 589.

Blodgett, K. B. & Langmuir, I. 1937. Built-up films of barium stearate and their optical properties. *Phys. Rev.*, 51, 964.

Bornside, D. E. 1990. Mechanism for the local planarization of microscopically rough surfaces by drying thin films of spin-coated polymer/solvent solutions. *J. Electrochem. Soc.*, 137, 2589.

Bornside, D. E., Macosko, C. W. & Scriven, L. E. 1987. On the modeling of spin coating. *J. Imaging Technol.*, 13, 122.

Brabec, C. J. & Durrant, J. R. 2008. Solution-processed organic solar cells. *MRS Bull.*, 33, 670.

Cao, Y., Kim, C., Forrest, S. R. & Soboyejo, W. 2005. Effects of dust particles and layer properties on organic electronic devices fabricated by stamping. *J. Appl. Phys.*, 98, 033713.

Chaug, Y. S., Haubrich, J. E., Sereda, M. & Liang, R. C. 2004. Roll-to-roll processes for the manufacturing of patterned conductive electrodes on flexible substrates. *MRS Proc.*, 814, I9.6.

Che, X., Li, Y., Qu, Y. & Forrest, S. R. 2018. High fabrication yield organic tandem photovoltaics combining vacuum- and solution-processed subcells with 15% efficiency. *Nat. Energy*, 3, 422.

Coe, S., Woo, W.-K., Bawendi, M. & Bulovic, V. 2002. Electroluminescence from single monolayers of nanocrystals in molecular organic devices. *Nature*, 420, 800.

Cosnahan, T., Watt, A. A. R. & Assender, H. E. 2018. Flexography printing for organic thin film transistors. *Mater Today: Proc.*, 5, 16051.

de Boer, R. W. I., Gershenson, M., Morpurgo, A. & Podzorov, V. 2004. Organic single-crystal field-effect transistors. *Phys. Stat. Sol. (a)*, 201, 1302.

Debe, M. K., Kam, K. K., Liu, J. C. & Poirier, R. J. 1988. Vacuum vapor deposited thin films of a perylene dicarboximide derivative: microstructure versus deposition parameters. *J. Vac. Sci. Technol. A*, 6, 1907.

Decher, G. 1997. Fuzzy nanoassemblies: toward layered polymeric nanocomposites. *Science*, 277, 1232.

Deegan, R. D., Bakajin, O., Dupont, T. F., Huber, G., Nagel, S. R. & Witten, T. A. 1997. Capillary flow as the cause of ring stains from dried liquid drops. *Nature*, 389, 827.

Dennler, G., Lungenschmied, C., Neugebauer, H., Sariciftci, N. S., Latreche, M., Czeremuszkin, G. & Wertheimer, M. R. 2006. A new encapsulation solution for flexible organic solar cells. *Thin Solid Films*, 511, 349.

Dickey, K. C., Anthony, J. E. & Loo, Y.-L. 2006. Improving organic thin film transistor performance through solvent-vapor annealing of solution-processable triethylsilylethynyl anthradithiophene. *Adv. Mater.*, 18, 1721.

Ebioe, S. 2010. Protein purification and analysis. Available at: http://shiyan.ebioe.com/ProteinPurificationAnalysis.htm.

Emslie, A. G., Bonner, F. T. & Peck, L. G. 1958. Flow of a viscous liquid on a rotating disc. *J. Appl. Phys.*, 29, 858.

Fan, D., Lee, B., Coburn, C. & Forrest, S. R. 2019. From 2D to 3D: strain- and elongation-free topological transformations of optoelectronic circuits. *Proc. Natl. Acad. Sci.*, 116, 3968.

Forrest, S. R. 1997. Ultrathin organic films grown by organic molecular beam deposition and related techniques. *Chem. Rev.*, 97, 1793.

Forrest, S. R. 2004. The path to ubiquitous and low cost organic electronic appliances on plastic. *Nature*, 428, 911.

Forrest, S. R., Burrows, P. E., Haskal, E. I. & So, F. F. 1994. Ultra-high vacuum, quasi-epitaxial growth of model van der waals thin films. II. Experiment. *Phys. Rev. B*, 49, 11309.

Forrest, S. R., Kaplan, M. L. & Schmidt, P. H. 1984. Organic-on-inorganic semiconductor contact barrier diodes. II. Dependence on organic film and metal contact properties. *J. Appl. Phys.*, 56, 543.

Fukuda, K., Yoshimura, Y., Okamoto, T., Takeda, Y., Kumaki, D., Katayama, Y. & Tokito, S. 2015. Reverse-offset printing optimized for scalable organic thin-film transistors with submicrometer channel lengths. *Adv. Electron. Mater.*, 1, 1500145.

Gehrke, C. W., Wixom, R. L. & Bayer, E. 2001. *Chromatography: A Century of Discovery 1900–2000*. Amsterdam: Elsevier.

Giebink, N. C., D'Andrade, B. W., Weaver, M. S., Mackenzie, P. B., Brown, J. J., Thompson, M. E. & Forrest, S. R. 2008. Intrinsic luminance loss in phosphorescent small-molecule organic light emitting devices due to bimolecular annihilation reactions. *J. Appl. Phys.*, 103, 044509.

Gilot, J., Wienk, M. M. & Janssen, R. A. J. 2007. Double and triple junction polymer solar cells processed from solution. *Appl. Phys. Lett.*, 90, 143512.

Gong, X., Wang, S., Moses, D., Bazan, G. C. & Heeger, A. J. 2005. Multilayer polymer light-emitting diodes: white-light emission with high efficiency. *Adv. Mater.*, 17, 2053.

Graff, G. L., Williford, R. E. & Burrows, P. E. 2004. Mechanisms of vapor permeation through multilayer barrier films: lag time versus equilibrium permeation. *J. Appl. Phys.*, 96, 1849.

Grubisic, Z., Rempp, P. & Benoit, H. 1967. A universal calibration for gel permeation chromatography. *Polym. Lett.*, 5, 753.

Gutmann, F. & Lyons, L. E. 1981. *Organic Semiconductors, Part A*. Malabar, FL: Robert E. Krieger Publishing.

Hadipour, A., deBoer, B., Wildeman, J., Kooistra, F. B., Hummelen, J. C., Turbiez, M. G. R., Wienk, M. M., Janssen, R. A. J. & Blom, P. W. M. 2006. Solution-processed organic tandem cells. *Adv. Funct. Mater.*, 16, 1897.

Hall, D. B., Underhill, P. & Torkelson, J. M. 1998. Spin coating of thin and ultrathin polymer films. *Polym. Eng. Sci.*, 38, 2039.

Haskal, E. I., So, F. F., Burrows, P. E. & Forrest, S. R. 1992. Evolution of quasi-epitaxial growth of a crystalline organic semiconductor on graphite. *Appl. Phys. Lett.*, 60, 3223.

Helfand, E. & Kornegay, R. L. 1966. Solution of the zone refining equation. *J. Appl. Phys.*, 37, 2484.

Herman, M. A. & Sitter, H. 1989. *Molecular Beam Epitaxy: Fundamentals and Current Status*. Berlin, Springer-Verlag.

Hernandez-Sosa, G., Bornemann, N., Ringle, I., Agari, M., Dörsam, E., Mechau, N. & Lemmer, U. 2013. Rheological and drying considerations for uniformly gravure-printed layers: towards large-area flexible organic light-emitting diodes. *Adv. Funct. Mater.*, 23, 3164.

Hiramoto, M. & Sakai, K. 2008. Efficient organic p-i-n solar cells having very thick codeposited i-layer composed of highly purified organic semiconductors *Proc. SPIE*, 7052, 70520H.

Hoesterey, D. C. & Letson, G. M. 1963. The trapping of photocarriers in anthracene by anthraquinone and naphthacene. *J. Phys. Chem. Solids*, 24, 1609.

Hu, H. & Larson, R. G. 2006. Marangoni effect reverses coffee-ring depositions. *J. Phys. Chem. B*, 110, 7090.

Iler, R. K. 1966. Multilayers of colloidal particles *J. Colloid Interface Sci.*, 21, 569.

Israelachvili, J. N. 2011. *Intermolecular and Surface Forces*. Waltham, MA: Academic Press.

Jeon, H.-G., Inoue, M., Hiramatsu, N., Ichikawa, M. & Taniguchi, Y. 2008. A modified sublimation purification system using arrays of partitions. *Org. Electron.*, 9, 903.

Jeong, C., Coburn, C., Idris, M., Li, Y., Djurovich, P. I., Thompson, M. E. & Forrest, S. R. 2019. Understanding molecular fragmentation in blue phosphorescent organic light-emitting devices. *Org. Electron.*, 64, 15.

Jiang, H. & Kloc, C. 2013. Single-crystal growth of organic semiconductors. *MRS Bull.*, 38, 28.

Kang, H., Kitsomboonloha, R., Jang, J. & Subramanian, V. 2012. High-performance printed transistors realized using femtoliter gravure-printed sub-10 μm metallic nanoparticle patterns and highly uniform polymer dielectric and semiconductor layers. *Adv. Mater.*, 24, 3065.

Karl, N. 1980. High purity organic molecular crystals. In: *Organic Crystals, Germanates, Semiconductors*. Berlin: Springer.

Kéna-Cohen, S., Davanço, M. & Forrest, S. R. 2008. Strong exciton–photon coupling in an organic single crystal microcavity. *Phys. Rev. Lett.*, 101, 116401.

Kim, C. & Forrest, S. R. 2003. Fabrication of organic light-emitting devices by low pressure cold welding. *Adv. Mater.*, 15, 541.

Kim, C., Burrows, P. E. & Forrest, S. R. 2000. Micropatterning of organic electronic devices by cold-welding. *Science*, 288, 831.

Kim, C., Cao, Y., Soboyejo, W. O. & Forrest, S. R. 2005. Patterning of active organic materials by direct transfer for organic electronic devices. *J. Appl. Phys.*, 97, 113512.

Kim, D. H., Lee, D. Y., Lee, H. S., Lee, W. H., Kim, Y. H., Han, J. I. & Cho, K. 2007. High-mobility organic transistors based on single-crystalline microribbons of triisopropylsilylethynyl pentacene via solution-phase self-assembly. *Adv. Mater.*, 19, 678.

Kim, I., Kwak, S.-W., Ju, Y., Park, G.-Y., Lee, T.-M., Jang, Y., Choi, Y.-M. & Kang, D. 2015. Roll-offset printed transparent conducting electrode for organic solar cells. *Thin Solid Films*, 580, 21.

Kipphan, H. (Ed.) 2001. *Handbook of Print Media: Technologies and Production Methods*. Berlin: Springer-Verlag.

Kobayashi, H., Kanbe, S., Seki, S., Kigchi, H., Kimura, M., Yudasaka, I., Miyashita, S., Shimoda, T., Towns, C. R., Burroughes, J. H. & R, H. F. 2000. A novel RGB multicolor light-emitting polymer display. *Synth. Met.*, 111, 125.

Konstantatos, G. & Sargent, E. H. 2010. Nanostructured materials for photon detection. *Nat. Nanotechnol.*, 5, 391.

Kopola, P., Aernouts, T., Sliz, R., Guillerez, S., Ylikunnari, M., Cheyns, D., Välimäki, M., Tuomikoski, M., Hast, J., Jabbour, G., Myllylä, R. & Maaninen, A. 2011. Gravure printed flexible organic photovoltaic modules. *Solar Energy Mater. Solar Cells*, 95, 1344.

Kopola, P., Tuomikoski, M., Suhonen, R. & Maaninen, A. 2009. Gravure printed organic light emitting diodes for lighting applications. *Thin Solid Films*, 517, 5757.

Kubo, R. 1971. *Statistical Mechanics*. Amsterdam: North-Holland.

Lamansky, S., Jr., Le, H., Jones, V., Wold, M. B. & Tolbert, W. A. 2005. Laser-induced thermal imaging of vacuum-coated OLED materials. *Proc. SPIE*, 5937, 593702.

Lassiter, B. E., Lunt, R. R., Renshaw, C. K. & Forrest, S. R. 2010. Structural templating of multiple polycrystalline layers in organic photovoltaic cells. *Opt. Express*, 18, A444.

Lassiter, B. E., Zimmerman, J. D. & Forrest, S. R. 2013. Tandem organic photovoltaics incorporating two solution-processed small molecule donor layers. *J. Appl. Phys.*, 103, 123305.

Laudise, R. A., Kloc, C., Simpkins, P. G. & Siegrist, T. 1998. Physical vapor growth of organic semiconductors. *J. Cryst. Growth*, 187, 449.

Lee, D. H., Choi, J. S., Chae, H., Chung, C. H. & Cho, S. M. 2009. Screen-printed white OLED based on polystyrene as a host polymer. *Curr. Appl. Phys.*, 9, 161.

Lee, S. T., Chin, B. D., Kim, M. H., Kang, T. M., Song, M. W., Lee, J. H., Kim, H. D., Chung, H. K., Wolk, M. B., Bellmann, E., Baetzold, J. P., Lamansky, S., Savvateev, V., Jr., T. R. H., Staral, J. S., Roberts, R. R. & Li, Y. 2004. A novel patterning method for full-color organic light-emitting devices: laser induced thermal imaging (LITI). *SID Symp. Dig. Tech Pap.*, 35, 1008.

Lee, S. T., Lee, J. Y., Kim, M. H., Suh, M. C., Kang, T. M., Choi, Y. J., Park, J. Y., Kwon, J. H., Chung, H. K., Baetzold, J., Bellmann, E., Savvateev, V., Wolk, M. & Webster, S. 2002. A new patterning method for full color polymer light emitting devices: laser induced thermal imaging (LITI). *SID Symp. Dig. Tech. Pap.*, 33, 784.

Lewis, J. 2006. Material challenge for flexible organic devices. *Mater. Today*, 9, 38.

Li, G., Yao, Y., Yang, H., Shrotriya, V., Yang, G. & Yang, Y. 2007. 'Solvent annealing' effect in polymer solar cells based on poly(3-hexylthiophene) and methanofullerenes. *Adv. Funct. Mater.*, 17, 1636.

Li, J., Han, W. & Yu, Y. 2013. Chromatography method. In: Ogawa, T. (ed.) *Protein Engineering: Technology and Application*. London: IntechOpen.

Logothetidis, S. 2008. Flexible organic electronic devices: materials, process and applications. *Mater. Sci. Eng. B*, 152, 96.

Lunt, R. R., Benziger, J. B. & Forrest, S. R. 2007. Growth of an ordered crystalline organic heterojunction. *Adv. Mater.*, 19, 4229.

Lunt, R. R., Lassiter, B. E., Benziger, J. B. & Forrest, S. R. 2009. Organic vapor phase deposition for the growth of large area organic electronic devices. *Appl. Phys. Lett.*, 95, 233305.

Ma, W., Yang, C., Gong, X., Lee, K. & Heeger, A. J. 2005. Thermally stable, efficient polymer solar cells with nanoscale control of the interpenetrating network morphology. *Adv. Funct. Mater.*, 15, 1617.

Maennig, B., Pfeiffer, M., Nollau, A., Zhou, X. & Leo, K. 2001. Controlled p-type doping of polycrystalline and amorphous organic layers: self-consistent description of conductivity and field-effect mobility by microscopic percolation model. *Phys. Rev. B*, 64, 195208.

Maruyama, Y., Kobayashi, T., Inokuchi, H. & Iwashima, S. 1973. Charge-carrier drift mobility in perylene single crystals. *Mol. Cryst. Liq. Cryst.*, 20, 373.

Mateker, W. R., Douglas, J. D., Cabanetos, C., Sachs-Quintana, I. T., Bartelt, J. A., Hoke, E. T., Labban, A. E., Beaujuge, P. M., Frechet, J. M. J. & McGehee, M. D. 2013. Improving the long-term stability of PBDTTPD polymer solar cells through material purification aimed at removing organic impurities. *Energy Environ. Sci.*, 6, 2529.

Matsushima, T., Shiomura, K., Naka, S. & Murata, H. 2012. Optical, morphological, structural, electrical, molecular orientation, and electroluminescence characteristics of organic semiconductor films prepared at various deposition rates. *Thin Solid Films*, 520, 2283.

McGhie, A. R. 1974. An all glass apparatus for 'continuous' chromatography and its application to chrysene. *Mol. Cryst. Liq. Cryst.*, 27, 431.

McGraw, G. J. & Forrest, S. R. 2012. Fluid dynamics and mass transport in organic vapor jet printing. *J. Appl. Phys.*, 111, 043501.

McGraw, G. J. & Forrest, S. R. 2013. Vapor-phase microprinting of multicolor phosphorescent organic light emitting device arrays. *Adv. Mater.*, 25, 1583.

McGraw, G. J., Peters, D. L. & Forrest, S. R. 2011. Organic vapor jet printing at micrometer resolution using microfluidic nozzle arrays. *Appl. Phys. Lett.*, 98, 013302.

Meerholz, K. 2005. Enlightening solutions. *Nature*, 437, 327.

Miller, J. B., Harris, J. M. & Hobbie, E. K. 2014. Purifying colloidal nanoparticles through ultracentrifugation with implications for interfaces and materials. *Langmuir*, 30, 7936.

Mobus, M., Schreck, M. & Karl, N. 1989. Epitaxial growth of thin films of perylenetetracaroxylic dianhydride and coronene on NaCl(001). *Thin Solid Films*, 175, 89.

Moore, J. C. 1964. Gel permeation chromatography. I. A New method for molecular weight distribution of high polymers. *J. Polym. Sci. Part A*, 2, 835.

Nisikawa, T., Moriguchi, N., Anezaki, T., Unno, A., Tachibana, M. & Kojima, K. 2006. Contact pressure dependence of carrier mobility in cleaved tetracene single-crystal field-effect transistors. *Jpn. J. Appl. Phys.*, 45, 5238.

Pardo, D. A., Jabbour, G. E. & Peyghambarian, N. 2000. Application of screen printing in the fabrication of organic light emitting devices. *Adv. Mater.*, 12, 1249.

Peumans, P., Uchida, S. & Forrest, S. R. 2003. Efficient bulk heterojunction photovoltaic cells based on small molecular weight organic thin films. *Nature*, 425, 158.

Pfann, W. G. 1966. *Zone Melting*. New York: John Wiley and Sons.

Pflaum, J., Niemax, J. & Tripathi, A. K. 2006. Chemical and structural effects on the electronic transport in organic single crystals. *Chem. Phys.*, 325, 152.

Quake, S. R. & Scherer, A. 2000. From micro- to nanofabrication with soft materials. *Science*, 290, 1536.

Reese, C. & Bao, Z. 2007. Organic single-crystal field-effect transistors. *Mater. Today*, 10, 20.

Rogers, J. A. & Nuzzo, R. G. 2005. Recent progress in soft lithography. *Mater. Today*, 8, 50.

Rogers, J. A., Bao, Z. & Raju, V. R. 1998. Nonphotolithographic fabrication of organic transistors with micron feature sizes. *Appl. Phys. Lett.*, 72, 2716.

Rolin, C., Song, B. & Forrest, S. R. 2014. Mass transport through the carrier gas boundary layer in organic vapor phase deposition. *Phys. Rev. Appl.*, 1, 034002.

Rolin, C., Vartanian, G. & Forrest, S. R. 2012. Laser induced fluorescence monitoring of the transport of small organic molecules in an organic vapor phase deposition system. *J. Appl. Phys.*, 112, 113502.

Sakurai, T., Kawai, S., Fukasawa, R., Shibata, J. & Akimoto, K. 2005. Influence of 3,4,9,10-perylenetetracarboxylic dianhydride intermediate layer on molecular orientation of phthalocyanine. *Jpn. J. Appl. Phys.*, 44, 1982.

Salzman, R. F., Xue, J., Rand, B. P., Alexander, A., Thompson, M. E. & Forrest, S. R. 2005. The effects of copper phthalocyanine purity on organic solar cell performance. *Org. Electron.*, 6, 242.

Sankaranarayanana, K. & Ramasamy, P. 2005. Unidirectional seeded single crystal growth from solution of benzophenone. *J. Cryst. Growth*, 280, 467.

Schwambera, M., Meyer, N., Leder, S., Reinhold, M., Dauelsberg, M., Strauch, G., Heuken, M., Juergensen, H., Zhou, T., Ngo, T., Brown, J., Shtein, M. & Forrest, S. R. 2002. Modeling and fabrication of organic vapor phase deposition (OVPD) equipment for OLED display manufacturing. *SID Symp. Dig. Tech. Pap*, 33, 894.

Shaheen, S. E., Radspinner, R., Peyghambarian, N. & Jabbour, G. E. 2001. Fabrication of bulk heterojunction plastic solar cells by screen printing. *Appl. Phys. Lett.*, 79, 2996.

Shtein, M., Gossenberger, H. F., Benziger, J. B. & Forrest, S. R. 2001. Material transport regimes and mechanisms for growth of molecular organic thin films using low-pressure organic vapor phase deposition. *J. Appl. Phys*, 89, 1470.

Shtein, M., Mapel, J., Benziger, J. B. & Forrest, S. R. 2002. Effects of film morphology and gate dielectric surface preparation on the electrical characteristics of organic vapor phase deposited pentacene thin-film transistors. *Appl. Phys. Lett.*, 81, 268.

Shtein, M., Peumans, P., Benziger, J. B. & Forrest, S. R. 2003. Micropatterning of organic thin films for device applications using organic vapor phase deposition. *J. Appl. Phys*, 93, 4005.

Shtein, M., Peumans, P., Benziger, J. B. & Forrest, S. R. 2004. Direct mask- and solvent-free printing of molecular organic semiconductors. *Adv. Mater.*, 16, 1615.

Siemann, U. 2005. Solvent cast technology: a versatile tool for thin film production. In: *Scattering Methods and the Properties of Polymer Materials*. Berlin: Springer.

Simon, J. & André, J.-J. 1985. *Molecular Semiconductors: Photoelectrical Properties and Solar Cells*. Berlin: Springer-Verlag.

Singh, J. 2003. *Electronic and Optoelectronic Properties of Semiconductor Structures*. Cambridge: Cambridge University Press.

Sirringhaus, H., Sele, C. W., vonWerne, T. & Ramsdale, C. 2006. Manufacturing of organic transistor circuits by solution-based printing. In: Klauk, H. (ed.) *Organic Electronics, Materials, Manufacturing and Applications*. Berlin: Wiley-VCH Verlag.

So, F. F. & Forrest, S. R. 1991. Evidence for exciton confinement in crystalline organic multiple quantum wells. *Phys. Rev. Lett.*, 66, 2649.

So, F. F., Forrest, S. R., Shi, Y. Q. & Steier, W. H. 1990. Quasi-epitaxial growth of organic multiple quantum well structures by organic molecular beam deposition. *Appl. Phys. Lett.*, 56, 674.

Sondergaard, R. R., Hosel, M. & Krebs, F. C. 2012. Roll-to-roll fabrication of large area functional organic materials. *J. Polym. Sci. Part B: Polym. Phys.*, 51, 16.

Song, B., Rolin, C., Zimmerman, J. D. & Forrest, S. R. 2014. Effect of mixed layer crystallinity on the performance of bulk heterojunction organic photovoltaic cells. *Adv. Mater.*, 26, 2914.

Sundar, V. C., Zaumseil, J., Podzorov, V., Menard, E., Willett, R. L., Someya, T., Gershenson, M. E. & Rogers, J. A. 2004. Elastomeric transistor stamps: reversible probing of charge transport in organic crystals. *Science*, 303, 1644.

Sze, S. M. 1981. *Physics of Semiconductor Devices*. New York: Wiley-Interscience.

Tada, H., Saiki, K. & Koma, A. 1992. Structural analysis of lead phthalocyanine ultrathin films grown on cleaved surfaces of alkali halides by reflection high energy electron diffraction. *Surf. Sci.*, 268, 387.

Tang, C. W. 2001. Brightness on display. *SPIE Newsroom*. SPIE.

Tang, C. W., VanSlyke, S. A. & Chen, C. H. 1989. Electroluminescence of doped organic thin films. *J. Appl. Phys.*, 65, 3610.

Tanigaki, K., Kuroshima, S., Ebbesen, T. W. & Ichihashi, T. 1991. Growth of thin film crystals of monochloro(-meso-tetraphenylporphyrinato) gallium by organic MBE. *J. Cryst. Growth*, 114, 3.

Taylor, P. A., Nelson, J. S. & Dodson, B. W. 1991. Adhesion between atomically flat metallic surfaces. *Phys. Rev. B*, 44, 5834.

van der Bos, A., van der Meulen, M.-J., Driessen, T., van den Berg, M., Reinten, H., Wijshoff, H., Versluis, M. & Lohse, D. 2014. Velocity profile inside piezoacoustic inkjet droplets in flight: comparison between experiment and numerical simulation. *Phys. Rev. Appl.*, 1, 014004.

Verploegen, E., Mondal, R., Bettinger, C. J., Sok, S., Toney, M. F. & Bao, Z. 2010. Effects of thermal annealing upon the morphology of polymer-fullerene blends. *Adv. Funct. Mater.*, 20, 3519.

Vincett, P. S., Popovic, Z. D. & McIntyre, L. 1981. A novel structural singularity in vacuum deposited thin films: the mechanism of critical optimization of thin film properties. *Thin Solid Films*, 82, 357.

Voigt, M. M., Mackenzie, R. C. I., Yau, C. P., Atienzar, P., Dane, J., Keivanidis, P. E., Bradley, D. D. C. & Nelson, J. 2011. Gravure printing for three subsequent solar cell layers of inverted structures on flexible substrates. *Solar Energy Mater. Solar Cells*, 95, 731.

Wagner, H. J., Loutfy, R. O. & Hsiao, C.-K. 1982. Purification and characterization of phthalocyanines. *J. Mater. Sci.*, 17, 2781.

Wang, N., Tong, X., Burlingame, Q., Yu, J. & Forrest, S. R. 2014. Photodegradation of small-molecule organic photovoltaics. *Solar Energy Mater. Solar Cells*, 125, 170.

Warta, W., Stehle, R. & Karl, N. 1985. Ultrapure, high mobility organic photoconductors. *Appl. Phys. A*, 36, 163.

Weinberg, L. R. & Rossi, F. D. 1960. Vapor zone refining. *Rev. Sci. Instrum.*, 31, 206.

Weston, G. F. 1985. *Ultrahigh Vacuum Practice*. London: Butterworths.

Willmann, J., Stocker, D. & Dorsam, E. 2014. Characteristics and evaluation criteria of substrate-based manufacturing. Is roll-to-roll the best solution for printed electronics? *Org. Electron.*, 15, 1631.

Xu, X., Davanco, M., Qi, X. & Forrest, S. R. 2008. Direct transfer patterning on three dimensionally deformed surfaces at micrometer resolutions and its application to hemispherical focal plane detector arrays. *Org. Electron.*, 9, 1122.

Xu, X., Mihnev, M., Taylor, A. & Forrest, S. R. 2009. Organic photodetector arrays with indium tin oxide electrodes patterned using directly transferred metal masks. *Appl. Phys. Lett.*, 94, 043313.

Yang, F., Shtein, M. & Forrest, S. R. 2005a. Controlled growth of a molecular bulk heterojunction photovoltaic cell. *Nat. Mater.*, 4, 39.

Yang, F., Shtein, M. & Forrest, S. R. 2005b. Morphology control and material mixing by high-temperature organic vapor-phase deposition and its application to thin-film solar cells. *J. Appl. Phys.*, 98, 014906.

Yang, F., Sun, K. & Forrest, S. R. 2007. Efficient solar cells using all-organic nanocrystalline networks. *Adv. Mater.*, 19, 4166.

Yang, J., Yan, D. & Jones, T. S. 2015. Molecular template growth and its applications in organic electronics and optoelectronics. *Chem. Rev.*, 115, 5570.

Yoshimura, T., Tatsuura, S. & Sotoyama, W. 1991. Polymer films formed with monolayer growth steps by molecular layer deposition. *Appl. Phys. Lett.*, 59, 482.

Zakhidov, A. A., Lee, J.-K., Fong, H. H., DeFranco, J. A., Chatzichristidi, M., Taylor, P. G., Ober, C. K. & Malliaras, G. G. 2008. Hydrofluorethers as orthogonal solvents for the chemical processing of organic electronic materials. *Adv. Mater.*, 20, 3481.

Zhang, P., Deng, J., Zeng, X., Liu, Z., Qiu, Y., Zhong, H., Fan, Y., Huang, J., Zhang, J. & Xub, K. 2009. Growth mechanism of large-size anthracene single crystals grown by a solution technique. *J. Cryst. Growth*, 311, 4708.

Zhou, T. X., Ngo, T., Brown, J. J., Shtein, M. & Forrest, S. R. 2005. Stable and efficient electrophosphorescent organic light emitting devices grown by organic vapor phase deposition. *Appl. Phys. Lett.*, 86, 021107.

Zielke, D., Hübler, A. C., Hahn, U., Brandt, N., Bartzsch, M., Fügmann, U., Fischer, T., Veres, J. & Ogier, S. 2005. Polymer-based organic field-effect transistor using offset printed source/drain structures. *Appl. Phys. Lett.*, 87, 123508.

Zimmerman, J. D., Lassiter, B. E., Xiao, X., Sun, K., Dolocan, A., Gearba, R., VandenBout, D. A., Stevenson, K. J., Wickramasinghe, P., Thompson, M. E. & Forrest, S. R. 2013. Control of interface order by inverse quasi-epitaxial growth of squaraine/fullerene thin film photovoltaics. *ACS Nano*, 7, 9268.

Zimmerman, J. D., Xiao, X., Renshaw, C. K., Wang, S., Diev, V. V., Thompson, M. E. & Forrest, S. R. 2012. Independent control of bulk and interfacial morphologies of small molecular weight organic heterojunction cells. *Nano Lett.*, 12, 4366.

CHAPTER 6

Organic light emitters

"What we have here is a new form of light!"

Sherwin Seligsohn, visionary co-founder of Universal Display Corporation on his first sight of emission from an OLED

Where would organic electronics be without the organic light emitting device, or OLED? It is hard to say. Certainly, it would remain a topic of fascination and interest to physicists, chemists, and materials scientists. The unique properties of organics are just too tempting to be ignored by the scientific community. This has been the case for at least the last 70 years. But my guess is that only a relatively small group of scientists and some engineers would be engaged in organic electronics scientific discoveries and applications, compared to the enormous cohort that works on this topic today, and whose careers depend on its success. This transformation was brought about by the invention of a relatively high efficiency, green bilayer diode, first named an organic electroluminescent device (OED), but then renamed the more pronounceable and now ubiquitously termed OLED. The success of OLEDs has led to a massive global industry in information displays, and an emerging industry in lighting. Organic light emitting devices (or equivalently, organic light emitting diodes) are rapidly overcoming competition from liquid crystal displays (LCDs) due to their unparalleled attributes, from their simplicity, to brilliant emission colors, and finally to their very high efficiency. And the many attractive features of OLED lighting, such as ultrahigh efficiency and color-tunable light emitting panels that can possibly be produced at exceptionally low cost, give every indication that OLEDs will soon find a major role in lighting our homes and businesses.

OLEDs are electroluminescent sources, in that current injection from the cathode and anode contacts generates an exciton on the emissive molecules within the structure, which then radiate with up to 100% internal quantum efficiency. The chapter begins with a brief discussion of the applications and basic operating principles of the OLED that is treated from an historical perspective. When was electroluminescence first discovered, and when was a device introduced that suggested this might actually be a practical source of light? Next, we discuss basic, modern OLED structures and techniques for quantifying their performance in their two main application spaces: displays and lighting. Indeed, how we think about color, lighting and efficiency are considerably different for these two applications.

The discussions of OLED characteristics rely heavily on all of the fundamental aspects of organic optical and electronic properties described in Chapters 3 and 4. As such, the discussion on OLEDs first presents principles, and then provides example devices and their operation based on those principles. This organization provides for a somewhat iterative treatment: devices introduced early in the chapter to demonstrate an architecture or principle will sometimes reappear later when a deeper look is warranted to understand an underlying physical mechanism of a class of materials or device structures. It is hoped that this revisiting of concepts throughout the chapter will provide the reader with a deeper appreciation of how multiple problems are solved in light emitting devices, sometimes using a single structure or material.

Organic Electronics: Foundations to Applications. Stephen R. Forrest. © Stephen R. Forrest 2020.
Published in 2020 by Oxford University Press. DOI: 10.1093/oso/9780198529729.001.0001

OLEDs fall into two different categories: somewhat low efficiency fluorescent devices and up to 100% efficient phosphorescent devices. Excitation of the fluorophores or phosphors depends on the spins of the injected electrons. We discuss the statistics of singlet and triplet formation, and the resulting efficiency of OLEDs using these chromophores.

Almost all efficient OLEDs use a doped emission zone, where energy transfer from the host, or charge trapping by the dopant, localizes the excited state on the luminescent dopant molecule. It has been found that doped OLEDs, where the chromophore concentration is generally less than 10% (with low doping if the OLED is fluorescent, and higher doping if it is a phosphorescent device), also have longer operational lifetimes than OLEDs with "neat" emission zones containing only the luminescent molecules themselves. Furthermore, as we have seen in Chapter 4, doping the charge conducting regions with electron donating or accepting molecules can reduce series resistance. Hence, conductivity-doped OLEDs tend to have lower operating voltages (leading to higher power efficiencies) and again, longer lifetimes, than devices lacking this type of doping.

The remainder of the discussion on OLEDs treats practical realizations and problems encountered by devices designed specifically for display and lighting applications. We will show examples of the countless clever ways that technologists have found to optimize the emission color, efficiency and lifetime of OLEDs used in a variety of environments, some benign and others quite harsh. Of particular importance is optical outcoupling: when grown on a conventional glass substrate, only ~20% of the emitted light reaches the viewer. An obvious way to improve OLED efficiency, therefore, is to improve outcoupling from the glass substrate. This sounds simple, but in fact it can be very difficult to extract a large fraction of the trapped light with simple, low cost solutions.

We complete our discussion by addressing a central concern of all organic devices: their reliability. Atoms comprising organic compounds are covalently bonded, and hence the stability of the molecules is ultimately limited by the weakest chemical bonds. Finding means to improve molecular robustness and stable device architectures has been a major focus of contemporary OLED engineering. This is a particularly challenging problem to address for high energy (i.e. blue) emitters, where the photon energy can exceed the bond energy of many compounds. To understand the limitations to device reliability, therefore, we must develop an understanding of the physics of failure: what are the fundamental mechanisms that lead to luminance loss in OLEDs, particularly in the blue part of the spectrum? Of course, there are failure mechanisms that are not fundamental but are equally difficult to eliminate, such as contact and layer delamination, impurity incorporation, attack by the ambient, and so on. All of these factors are discussed in the latter sections of this chapter.

Finally, we close with a discussion of organic semiconductor lasers. There remains the open question of whether or not electrically (rather than optically) pumped organic lasers are possible. We discuss the basic principles of organic semiconductor lasers, and the challenges faced in electrical pumping. We will show that its achievement remains a daunting challenge that has yet to be met in spite of numerous attempts at doing so.

Organic light emission is indeed an exciting topic. This chapter serves to provide both the fundamental as well as practical aspects of organic light emitters, with the hope of giving the reader the necessary tools to both understand and innovate in this fast moving field that is of both scientific and commercial interest.

6.1 OLED basics

An organic light emitting device (OLED) is a thin film structure deposited on a supporting substrate that emits light via the process of *electroluminescence* (EL). The emission results from the radiative recombination of electrons and holes injected from anode and cathode contacts located on opposite sides of the organic emissive medium. Hence, an OLED operates by *bipolar injection*. The color of the light is determined by the excitonic energy of the emitting molecule, and hence the device can be designed to radiate across a wide spectral range, from the infrared to the ultraviolet. Emission can occur via either a fluorescent or phosphorescent process. Due to their ultrahigh efficiencies, the range of possible emission colors achieved by modification of the chemical structure of the emissive molecule (i.e. the *lumophore*), and their potentially low cost, OLEDs are finding widespread use in mobile displays such as smart phones, computer tablets and laptops, as well as in wall-mounted and roll-up televisions (Tsujimura, 2012). Their high efficiencies and color tunability have also led to applications in solid-state lighting in the interiors of residences and businesses (Kassakian, 2013).

Practical applications almost universally employ thin film (~100 nm total), multilayer small molecular weight structures. As we have seen in Chapter 5, this choice is driven by the ability to produce highly purified source materials, a key criterion to achieve long device operational lifetimes. Furthermore, the ability to grow multiple layer structures with each layer optimized for a particular purpose (e.g. charge conduction, light emission, etc.) is also an essential ingredient for achieving the highest performance. The ease of deposition without concern for compatibility of the chemistry of a layer with others in the stack makes small molecules most suitable for OLED applications.

In Chapter 5 we showed that small molecular weight thin films are primarily deposited by vacuum thermal evaporation, and patterned into the desired structures (e.g. pixelated displays) using shadow masks. There are other potential technologies that can be employed to grow the OLED structure, including organic vapor phase deposition (OVPD), and various printing technologies such as organic vapor jet and ink jet printing. The OLED can be deposited on a variety of substrates, including glass, plastic or metal foils, or on semiconductors serving as the transistor driving circuits. Since OLEDs comprise amorphous organic thin films, there is no requirement for lattice matching with the substrate. This is a major differentiating characteristic of OLEDs with other LED technologies based on the InGaN or InGaP materials systems, where epitaxial growth is an essential ingredient for high performance. This feature provides extraordinary flexibility in the choice of materials used to form the OLED stack.

The high resolution patterning of the OLED film during deposition permits devices emitting in different spectral ranges to reside in close proximity on the same substrate. Hence, integration of a very high density of red, green and blue pixel triads onto a very large area substrate all deposited in a single growth cycle results in low cost production of displays with millions of pixels. Again, this is different than what can be achieved using LEDs where red, green and blue emitters must all be separately attached onto a substrate; a prohibitively costly process that prevents their use in high information content, large area displays.

We have seen that most organic semiconductors are dyes similar to those used as pigments in fabrics, in inkjet printed documents, or as paints. These synthetic processes often lend themselves to production in large batches at low cost. This feature, however, does not readily translate to electronic materials that are carefully optimized to precisely achieve a desired emission color through chemical modifications that are not always easy or efficient to synthetically access. Further, they need to be tailored for deposition within a narrow range of conditions that are compatible with other materials deposited in complex, multilayer device structures. Ultrahigh purity required for long device lifetime, and to avoid exciton quenching or charge recombination at defects, can involve time consuming and costly synthesis. Hence, while OLEDs have the *potential* for being inexpensive, as of this writing they have not shown a clear cost advantage over the incumbent LCD technology.

The organic materials and the processes used in their synthesis or deposition are environmentally friendly when compared to many other illumination or display technologies. The very low toxicity of most organic thin films, the absence of hazardous metals such as Hg or As found in fluorescent or LED lighting, and the very low energy expense incurred during production due to the low sublimation temperatures of organics, leads to cost advantages and a reduced environmental impact in their production. Since displays and lighting fixtures are very large area devices with widespread consumer applications, minimizing the environmental impact of their production, use and disposal at the end of their life cycle are essential to their broad adoption by the public.

OLEDs possess several distinctive features that open up applications that other illumination technologies cannot access. For example, the OLED can be made fully transparent such that when it is turned off, it can serve as a window on buildings or in head-up displays in automobiles, aircraft, etc. Due to their high efficiency, they can achieve high intensity without significant heating above ambient temperature. Since the device itself can be arbitrarily shaped if placed on a flexible substrate, it can be used as a lighting fixture without requiring an expensive or bulky luminaire used contain and shape the output of conventional lighting sources. Their very low operating voltages (2–10 V) also reduce power consumption. This makes them compatible with low voltage driver circuits in power-constrained mobile applications.

The list of advantages and opportunities afforded by OLEDs is long and growing every day, which helps to explain their popularity with the public. First and foremost, however, OLEDs provide an extraordinarily attractive display technology owing to their brilliant emission colors, very high on/off contrast ratio, ability to be shaped for a particular aesthetic or practical purpose, and even rolled into compact holders and stowed away when not in use. Singlehandedly, OLEDs have provided impetus for driving the robust research and commercialization of many other synergistic organic electronic device technologies to be discussed in subsequent chapters. A few images of OLED displays and lighting fixtures are shown in Figs. 6.1 and 6.2, respectively. While they do seem to represent "a new form of light" suggested by the epigram to this chapter, they are really only a new means for emitting light that opens up an unprecedented array of opportunities for their use.

Figure 6.1 The remarkable world of OLED displays. Clockwise from top left: Samsung 77″ 4K curved TV monitor, LG rollable display, Samsung Galaxy Edge smartphone, Apple iWatches, and transparent OLED showcase windows.

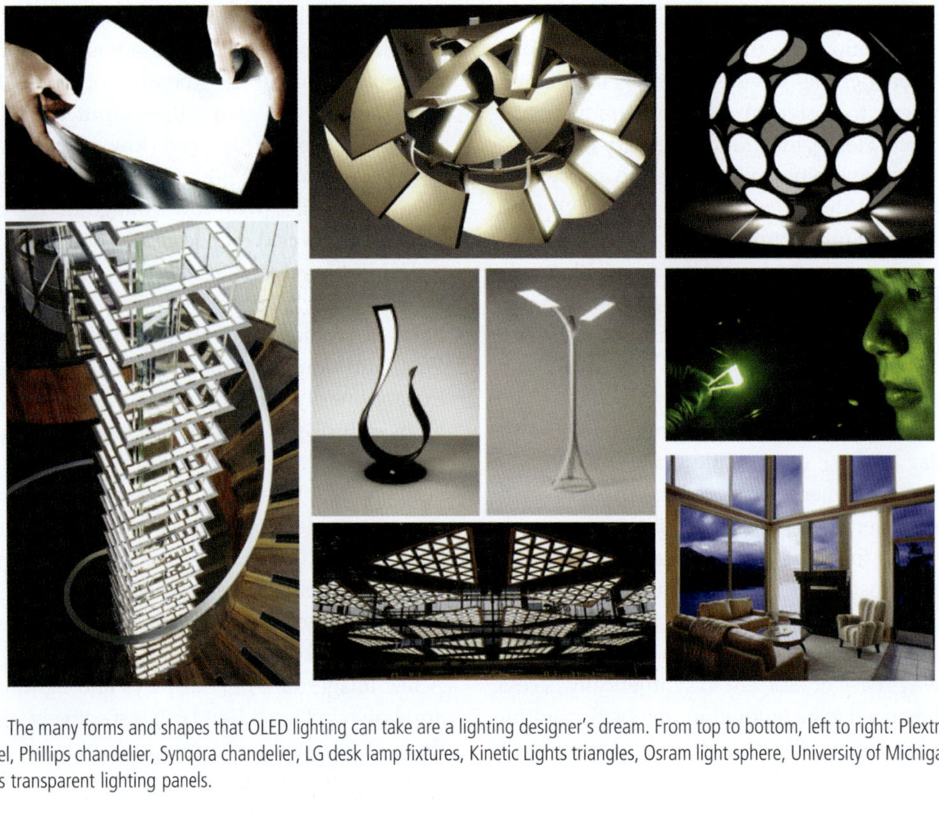

Figure 6.2 The many forms and shapes that OLED lighting can take are a lighting designer's dream. From top to bottom, left to right: Plextronics flexible lighting panel, Phillips chandelier, Synqora chandelier, LG desk lamp fixtures, Kinetic Lights triangles, Osram light sphere, University of Michigan OLED spot light, Orbeos transparent lighting panels.

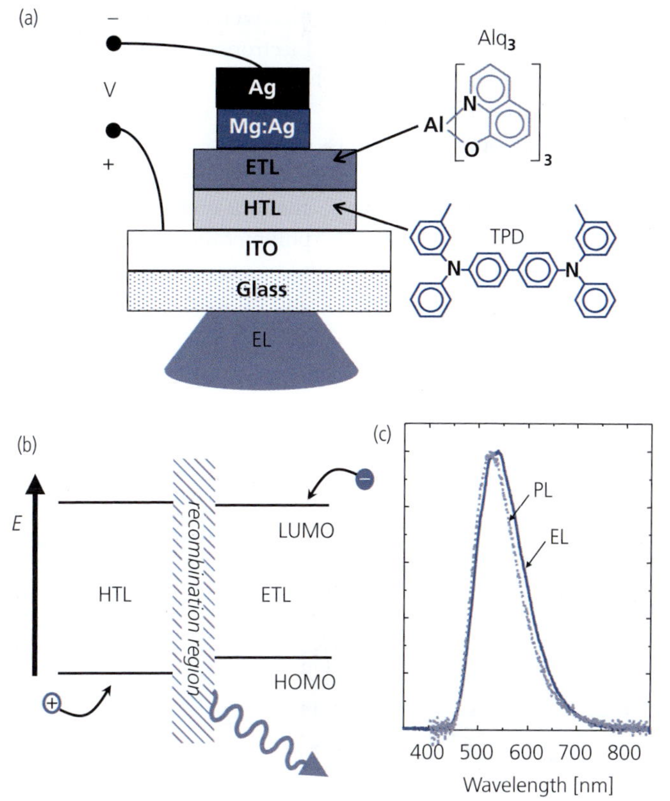

Figure 6.3 Bilayer small molecule OLED. (a) The bilayer structure introduced by Tang and VanSlyke (1987). The molecules shown are used for electron transport and fluorescent emission, Alq_3, and hole transport, TPD. (b) Energy level diagram of the OLED forming a type II heterojunction. The recombination region leading to light emission is at the interface between the electron (ETL) and hole transport (HTL) layers. (c) Electroluminescent (EL) and solution photoluminescent (PL) spectra from the OLED and a film of Alq_3, respectively.

The bilayer, thin film OLED structure similar to that in Fig. 6.3a, was introduced by Tang and VanSlyke (1987). The OLED comprises a type II, staggered heterojunction (HJ) formed between an electron transport layer (ETL) and a hole transport layer (HTL). Both the ETL and HTL thicknesses are only a few tens of nanometers thick, resulting in a total device thickness (not counting the substrate) of ~100 nm. The charge transport layers are sandwiched between a low work function, electron-injecting cathode, and a high work function, transparent anode typically consisting of ITO. The HOMO–LUMO offset energies of the HJ in Fig. 6.3b result in current rectification. Interestingly, the EL spectrum of the OLED is nearly identical to the solution photoluminescence (PL) spectrum of the emissive Alq_3 molecule, with only slight shifts due to dielectric polarization in the solid vs. the solution (Fig. 6.3c). This is in striking contrast with covalently bonded inorganic semiconductors where the PL spectra of the constituent atoms (e.g. Si atoms) bear little resemblance to that of the solid.

OLED light emission is based on the four-step process shown in Fig. 6.4 for a three-layer device consisting of a light emission layer (EML) sandwiched between an ETL and an HTL. (1) Charges are injected from opposing contacts, and (2) subsequently localize on the emissive molecules in the EML. Next, (3) a hole on the HOMO and an electron on the LUMO form a bound exciton. Finally, (4) radiative recombination of the exciton emits light that emerges via the transparent anode, through the substrate, and toward the viewer. The bimolecular recombination region extends only a few nanometers into the ETL. Efficient operation requires the emissive molecule to have a high fluorescence quantum yield. The luminescent molecule determines the emission color. With only minor variations of the ligand structure, efficient emission from the deep blue to the near infrared (NIR) has been achieved, thereby enabling full color displays with large color gamuts, and white light emitting fixtures with a range of color temperatures and pleasing colors.

We will show in subsequent sections that the functions of light emission, and charge transport can be independently optimized in separate layers to deliver higher device performance than the original bilayer design in Fig. 6.3a.

The device in Fig. 6.3 is known as an *electrofluorescent OLED*. We show in Section 6.3 that significantly higher efficiencies can be obtained via phosphorescent emission in *electrophosphorescent OLEDs* (Baldo et al., 1998).

Polymer OLEDs also support bipolar injection. When originally introduced, they consisted of only a single multifunctional layer, as shown in Fig. 6.5 (Burroughes et al., 1990). The device consists of a light emitting polymer, PPV, sandwiched between low and high work function contacts (Al and ITO, respectively). Electrons and holes are both transported by the polymer, and the rectifying characteristics of the device are due to the Schottky barriers at one or both of the contacts, as inferred from the energy level diagram in Fig. 6.5b. In the zone where the overlap between their concentrations is highest, the polarons relax onto a polymer unit to form an exciton that eventually fluoresces via radiative recombination. Electroluminescent and PL spectra for a PPV device are shown in Fig. 6.5c. Similar to small molecule bilayer OLEDs, the two

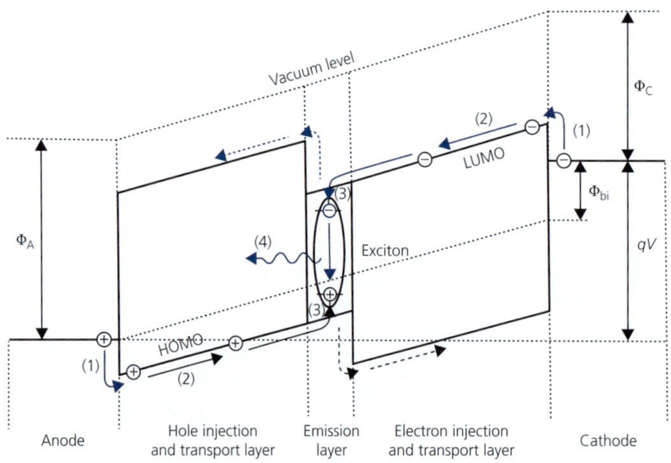

Figure 6.4 Electroluminescence generation process in an OLED (Brütting and Frischeisen, 2012).

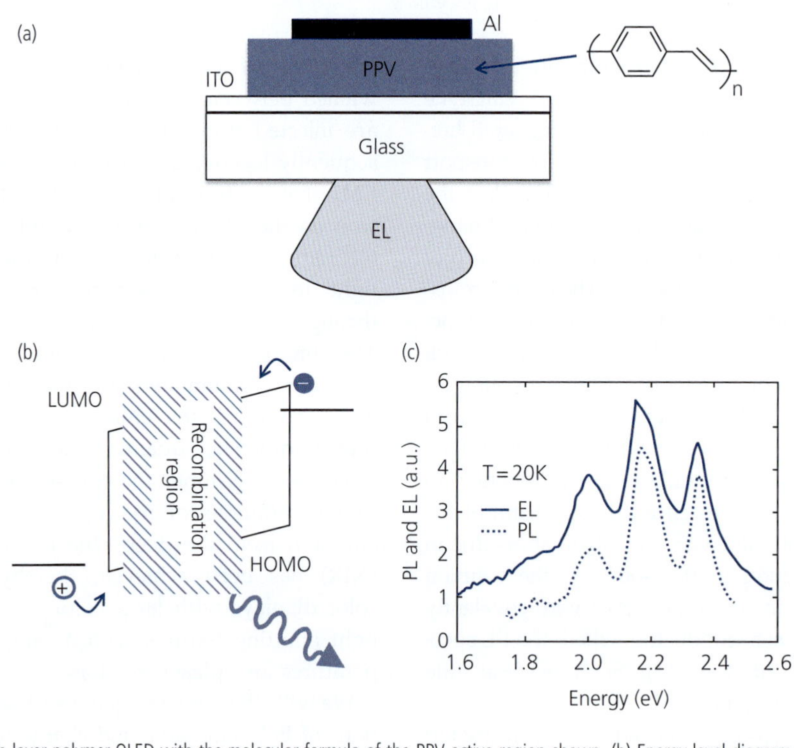

Figure 6.5 (a) Single layer polymer OLED with the molecular formula of the PPV active region shown. (b) Energy level diagram of the polymer OLED. (c) Low temperature photoluminescence and electroluminescence spectra of a PPV film and OLED, respectively. The spectra are obtained at low temperature to increase efficiency from its value at room temperature; (c) is from Swanson et al. (1993).

spectra are nearly identical, indicating that EL is due to exciton recombination following molecular excitation from the injected electron and hole polarons.

Unlike small molecule OLEDs, the polymer device does not have an emission region that confines electrons and holes, resulting in a spatially diffuse emission zone. Thus, the efficiency is low unless doped, multiple layer structures are used to separately optimize charge conduction and light emission. Hence, the seemingly simple single layer design in Fig. 6.5 is deceiving: to achieve very high performance, increased complexity in molecular design is required.

6.2 Design and characterization of electroluminescent devices

In this section we describe the elements of OLED design and operating characteristics. We will describe common approaches to designing OLED structures that go well beyond the original bilayer and single layer structures discussed in the previous section. Then we discuss methods for quantifying and measuring device properties such as quantum efficiency and power efficiency. Accurate measurement and understanding of performance parameters are key to providing valid comparisons between results obtained in different laboratories, and are detailed in Section 6.2.4.

6.2.1 Efficiency

We use two efficiencies to describe the OLED emission characteristics: the *external electroluminescent quantum efficiency*, η_{ext} (also written *EQE*) and the *power efficiency*, η_P (or *PE*). The external quantum efficiency is the ratio of the number of photons emitted from the OLED to the number of electrons injected. It is the product of the efficiencies of several processes that contribute to light emission, as follows:

$$\eta_{ext} = \eta_{int}\eta_{out} = \gamma\chi_{ST}\eta_{PL}\eta_{out}. \tag{6.1}$$

Here, η_{int} is the *internal electroluminescence quantum efficiency*, which is the ratio of the number of photons *generated* to the number of injected electrons. It is expressed as a number ≤ 1 or as a percentage. This efficiency includes photons emitted by the active region of the device but that may not emerge from the OLED structure into the viewing direction due to internal absorption and reflection from optical interfaces within the multilayer stack. Then, η_{out} is the *outcoupling efficiency*, equal to the ratio of photons that are coupled into air modes and hence can be viewed, to the total number of photons emitted. Outcoupling efficiencies of $\eta_{out} \sim 20\%$ are typical of OLEDs emitting through a glass substrate. Methods for increasing η_{out} beyond 20% are the topic of Section 6.6.

Other contributions to η_{ext} include the electroluminescent specific losses, γ, due to such factors as imperfect balance of electrons and holes in the emission zone, recombination via charge traps, etc. Also, η_{PL} is the photoluminescence efficiency or quantum yield of the molecule or molecular mixture in the emission zone. For electrofluorescent devices, η_{PL} is the fluorescent efficiency, and for electrophosphorescent devices, it is the phosphorescent efficiency of the luminescent molecule, or lumophore.

Finally, χ_{ST} is the *spin formation ratio*, that is, it is the ratio of luminescent states generated for each electron–hole pair injected into the EML. Given completely random formation of singlets and triplets due to pairing of uncorrelated spins of injected electrons and holes, spin formation statistics imply that only one in four of such pairings result in the formation of a singlet exciton that emits by the process of fluorescence. In this case, fluorescent devices have $\chi_{ST} = ¼$. On the other hand, triplets are generated with a 75% probability since the three lowest energy triplet (T_1) states can be formed compared to one S_1 singlet. In electrophosphorescent devices, spin–orbit coupling that leads to a radiative triplet state also leads to efficient intersystem crossing from S_1 to T_1 (see Section 3.7). Hence, in electrophosphorescent devices, $\chi_{ST} = 1$. For similar reasons, devices based on *thermally activated delayed fluorescence* can also reach $\chi_{ST} = 1$. The determination of spin statistics in both polymer and small molecule systems will be discussed in Section 6.3.1.

The power efficiency is the ratio of the output optical power to the electrical input power, and has units of [W/W] or [%]. This is calculated using

$$\eta_P = \eta_{ext}\frac{P_{EL}}{jV} = \eta_{ext}\frac{hc}{qV\langle\lambda_{em}\rangle}, \tag{6.2}$$

where j is the OLED injection current density at voltage, V, P_{EL} is the total emitted electroluminescent power density, h is Planck's constant, c is the speed of light in vacuum, q is the electronic charge, and $\langle\lambda_{em}\rangle$ is the mean emission wavelength given by

$$\langle\lambda_{em}\rangle = \frac{\int_0^\infty \lambda\phi_{em}(\lambda)d\lambda_{em}}{\int_0^\infty \phi_{em}(\lambda)d\lambda_{em}}, \tag{6.3}$$

where $\phi_{em}(\lambda)$ is the flux emitted between λ and $\lambda + d\lambda$.

6.2.2 OLED architectures

From Eq. 6.1, we see that the highest efficiencies are obtained for unity charge balance in the emission zone ($\gamma = 1$), and by employing phosphorescent molecules ($\chi_{ST} = 1$) with quantum yields $\phi_p = 1$. Further, OLED efficiencies are greatly improved using schemes that result in efficient outcoupling of light from the device active regions and emitting surfaces. Finally, the highest power efficiency requires devices that operate at very low voltage, that is, they have a minimal series resistance, R_S.

OLED structures that optimize all of these separate efficiency factors can be quite complex. We show for example in Fig. 6.6 many of the layers employed in the highest efficiency top and bottom emitting OLEDs. A particular OLED may not require all of the layers shown; these structures simply define the possible layers that *can* be included. Whereas bottom emitting devices are used in lighting and laboratory demonstrations, top emitters are commonly used in displays where the substrate might be covered with opaque circuit elements used to address the individual pixels. The primary difference between bottom and top emitting OLEDs is that the transparent and opaque contacts are reversed. While this is apparently a simple modification, transparent materials with low work functions necessary for efficient electron injection are not available. There are three remedies for this situation. The first is to invert the stacking order in the top emitting device such that hole transport and injection layers are on the top of the stack adjacent to the anode. A second is to use the stacking order in Fig. 6.6, but employ conditions during the deposition of the transparent contact deposition that introduce damage, and hence deep levels that reduce the barrier to electron injection (see Section 4.6) (Parthasarathy et al., 1998).

A third option is to maintain the stacking order in Fig. 6.6b, but use a thin, semitransparent, low work function metal cathode. The thickness must be considerably less than the penetration depth of the optical electric field, **F**, into the metal. The penetration depth, or *optical skin depth*, δ, is obtained from a solution to Maxwell's equations for an oscillating electric field at frequency, ω:

$$\nabla^2 \mathbf{F} = i\omega\mu_r\mu_0\sigma\mathbf{F}. \tag{6.4}$$

Here, μ_r and μ_0 are the relative and the vacuum permeabilities, respectively, and σ is the conductivity of the metal. If a plane wave is incident in the $\hat{\mathbf{x}}$-direction with initial field amplitude, F_0, this has solutions:

$$F_z = F_{0z}e^{-\tau x}, \tag{6.5}$$

where

$$\tau = (1+i)\sqrt{\omega\mu_r\mu_0\sigma/2} = (1+i)/\delta. \tag{6.6}$$

Thus, Eq. 6.5 becomes

$$F_z = F_{0z}e^{-x/\delta}e^{-ix/\delta}. \tag{6.7}$$

Hence the skin depth is the attenuation length of the field given by

$$\delta = \sqrt{\frac{2}{\omega\mu_r\mu_0\sigma}}. \tag{6.8}$$

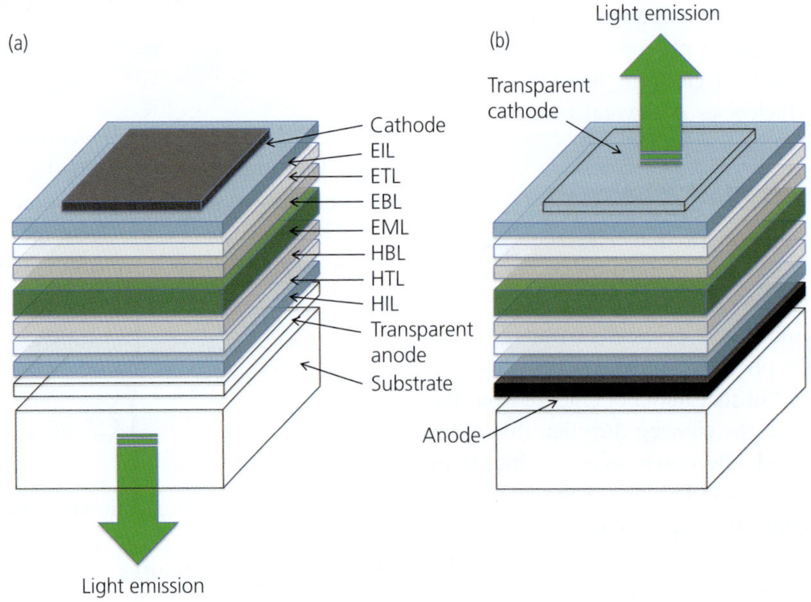

Figure 6.6 Layer structures for (a) bottom emitting and (b) top emitting OLEDs.

In Table 6.1 we show the skin depth in the green spectral region at a wavelength of $\lambda = 550$ nm ($\omega = 3.42 \times 10^{15}$ s^{-1}) for several different metals at room temperature. The skin depth is only a few nanometers, hence in using thin metals there is a trade-off between contact resistance and transparency. Nevertheless, many top emitting OLEDs employ this approach due to its simplicity.

Finally, both the cathode and anode can be made transparent, thus rendering the entire OLED transparent (Bulovic et al., 1996). As we have seen in Figs. 6.1 and 6.2, transparency provides unique opportunities for OLEDs that are finding applications in both lighting and displays.

We can understand the functionality of each of the layers using the corresponding energy level diagram shown in Fig, 6.7, which is more complex than the basic structure in Fig. 6.4. This diagram reflects the energy levels under forward bias of a high performance OLED, where the tilt in the bands that results from applying a negative potential at the cathode for electron injection is not shown. We also neglect the presence of dipole layers at interfaces between different organic layers (see Chapter 4). We now identify and describe the individual purposes of each of the layers.

Table 6.1 Conductivities and skin depths (at $\lambda = 550$ nm) of selected low work function metals

Metal	σ (10^6 S/m)	δ (nm)
Ag	63.0	2.7
Au	45.2	3.2
Al	37.7	3.5
Ca	29.8	3.9
Mg	22.6	4.5

- HIL and EIL: The *hole* and *electron injection layers* at the anode and cathode interfaces, respectively, reduce the charge injection barriers from the adjacent contacts, increasing the charge injection efficiency. Often these layers have their frontier orbitals (i.e. the HOMO of the HIL and the LUMO of the EIL) intermediate between the contact and that of the next layer in the stack (e.g. the HTL or ETL) to "broaden" the energy step between the contact and the subsequent layer. This reduces the internal field opposing charge injection. The injection layer can be doped to increase conductivity, and hence reduce layer internal resistance while also lowering the energy barrier. These layers are typically very thin (<10 nm) since their only purpose is to improve injection from the contacts.
- HTL and ETL: As above, these are the hole and electron transport layers. They preferentially transport charges due to their high mobilities for the particular carrier, and their energy level alignment with the corresponding contact. Often, the HTL and ETL are doped to increase their conductivities. Increased conductivity thus decreases device operating voltage, and hence increases its power efficiency, see Eq. 6.2. They are thicker than the HIL and ETL (typically 10–50 nm) to provide "bulk" to the device that can eliminate shorts across the OLED layer stack.
- HBL and EBL: The *hole* and *electron blocking layers* are used to confine holes and electrons in the EML. From Eq. 6.1, the external efficiency is proportional to the *charge balance factor*. This accounts for electroluminescent specific losses; γ is the ratio of electrons to holes within the emission zone. If the LUMO of the HBL is too low to block electrons, then the emission zone will be electron deficient and γ

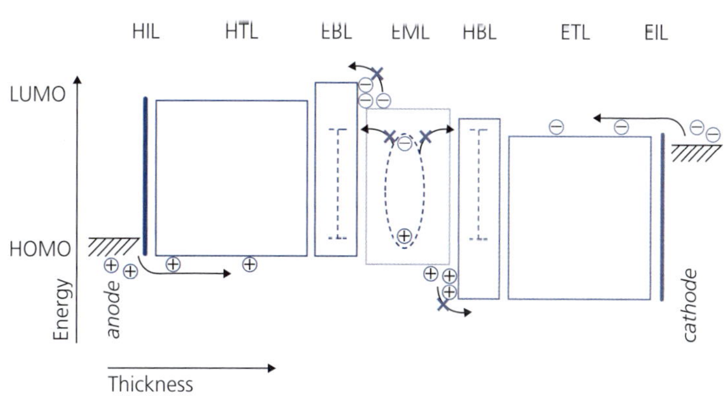

Figure 6.7 Forward biased energy level diagram of a multilayer OLED ignoring the presence of interface dipole layers. Tilting of the levels under an electric field, and dipole layers at heterointerfaces have been omitted. Charge and recombination processes are schematically illustrated. Exciton levels are shown by dashed lines (Reineke et al., 2013).

will decrease. Similarly, if the HOMO of the HTL lies above that of the HOMO of the EML, then holes injected from the anode will traverse the EML without recombination, and thus the electron population will exceed that of holes, once again resulting in a decrease in γ. The purpose of the HBL and EBL is to decrease the probability of escape of these carriers from the emission zone by having a shallow LUMO or deep HOMO respectively, relative to that of the EML. The HBL should be transparent to electron transport to avoid increasing series resistance, and hence its LUMO should align with that of ETL. Similarly, the HOMO of the EBL should align with that of the HTL to avoid energy barriers to hole injection. These layers should also be as thin as possible (5 – 10 nm) to minimize OLED series resistance.

Another essential function of the HBL and EBL is exciton blocking. Indeed, the HBL was first introduced as an exciton blocker in phosphorescent OLEDs (or PHOLEDs) (O'Brien et al., 1999). The long triplet lifetime can often lead to exciton diffusion lengths in excess of that for singlets (Section 3.9). Hence, to prevent triplet migration out of the EML where they are quenched at the cathode metal, an exciton blocking layer whose triplet energy is larger than that of triplets in the EML is employed. The large triplet energy of the blockers (indicated by the vertical dashed lines in the EBL and HBL in Fig. 6.7) discourages energy transfer, thereby confining excitons within the EML where the radiative recombination probability is highest.

- EML: This is the *emission layer*. It comprises a host material that can transport charges of one or both polarities, doped with the lumophore (i.e. a phosphor or fluorophore). A host that transports both electrons and holes is *ambipolar*. If the host transports only one charge effectively due to asymmetric charge mobilities, then the dopant is needed to transport the counter charge. In this instance, the dopant must be at sufficiently high concentrations to provide an adequate conductivity. Asymmetries in charge conductivity result in pile-up of one of the two carriers near an injecting interface. This, in turn, results in a high density of excitons near that interface where exciton–exciton or exciton–polaron annihilation becomes significant as the current density is increased (see Section 6.3.5). To achieve high OLED brightness, the electron and hole distributions within the EML should be as uniform as possible by balancing the mobilities. Furthermore, the EML should be sufficiently thick to avoid high exciton densities that encourage annihilation, yet should be thin enough to allow for low voltage operation. The design of the EML, therefore, requires trade-offs in several materials and electronic properties.

6.2.3 Quantifying OLED performance

We now turn to the topic of understanding and quantifying OLED performance. This requires a knowledge of how we perceive color. Then we quantify the basic quantities of luminescence and efficiency for both displays and lighting; measurements that must be done with precision if we are to compare the performance of materials and devices emitting in different spectral ranges, and produced in different laboratories. The methods discussed in this section follow accepted procedures, with some of the essential details found elsewhere (Forrest et al., 2003).

6.2.3.1 Radiometry and photometry: color perception and efficiency in OLED displays

The color perception and efficiency of displays are intimately linked to the response of the human eye, but also to more readily quantifiable characteristics such as the quantum and power efficiencies. This has led to the development of two systems for quantifying the performance of light emitters. *Radiometric properties* are electromagnetic performance metrics of the OLED that are measured using calibrated photodetectors, spectrometers, and so forth. *Photometric properties* are perceptual quantities, such as the perceived color or brightness of the source. Ultimately both the photometric and radiometric properties are required to fully understand OLED performance.

Radiometric properties are defined based on total power input into the device (i.e. jV) vs. power emitted (see Eq. 6.2). Photometric properties, on the other hand, are based on the responsivity to light power of the human eye. This is known as the *luminosity function*, $\bar{y}(\lambda)$ shown in Fig. 6.8. In fact, two functions are shown, $\bar{y}(\lambda)$, which is the response of the eye in normal daylight conditions, known as the *photopic response*, and $\bar{y}'(\lambda)$ for nighttime vision, called the *scotopic response*.

These luminosity functions are based on the 1931 *Commission Intenationale d'Eclairage* (CIE) standards. As we will see below, the CIE has provided a quantitative correlation between the perception of color due to the response of the three color-sensitive cone cells in the retina, and the spectrum of a particular color source.

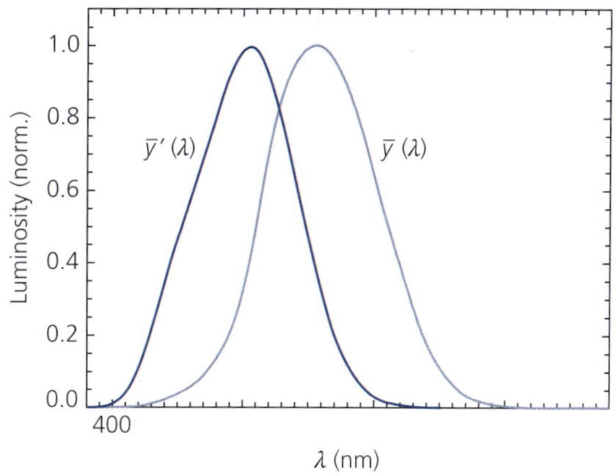

Figure 6.8 1931 CIE luminosity functions for photopic ($\bar{y}(\lambda)$) and scotopic ($\bar{y}'(\lambda)$) vision.

Photopic vision engages three color-sensitive cones with peak sensitivities at wavelengths of $\lambda = 420$ nm (blue), 534 nm (blue-green), and 564 nm (yellow-green). By combining the response of all three cones at light of intensities $>10^{-3}$ cd/m^2, full color sensitivity is achieved. The photometric unit used to quantify the perceived *luminous intensity* emitted per solid angle is the candela (cd). The luminosity function in Fig. 6.8 has a maximum of 683 lm/W at $\lambda = 555$ nm, where the perceived *luminous flux* has units of lumens (lm).

In very dim light ($<10^{-3}$ cd/m^2), the response of the eye is dominated by excitation of the rod cells that transfer the luminosity signal to the cones. In this process, color information is lost, and the peak sensitivity is blue shifted to $\lambda = 500$ nm. Thus, in dim light the image lacks color and is represented by an average luminosity across the visible spectrum. A dramatic demonstration of photopic and scotopic vision is simulated in Fig. 6.9, which is the same image of a winter scene taken in moonlight; the top image shows the grayscale appearance of this scene perceived by an observer with dark-adapted, scotopic vision. Only the relative luminosities of the image elements provide contrast. The actual colors of the scene as recorded by the camera (which does not have the brightness-dependent limitations of the human eye) are shown in the lower image. As expected, the sky is blue due to Rayleigh scattering of the moonlight, the meadow grass is yellow, and the trees are green. This is a low contrast version of the scene perceived during daylight due to our photopic capabilities. For displays and lighting, the intensity ranges from a few tens of cd/m^2 to several thousand cd/m^2, and hence only the photopic response needs to be considered.

We can appreciate the relationship between the photometric properties of luminous flux and the luminous intensity using the diagram in Fig. 6.10. Consider a point source located at point A in Fig. 6.10a. Then the luminous flux (Φ) through incremental area dS_1 that subtends a solid angle, $d\Omega$, is the same as the flux through area dS_2, since it occupies the same solid angle as dS_1. Thus, the luminous intensity is

$$L_\Omega = \frac{d\Phi}{d\Omega}. \tag{6.9}$$

An actual light source is not a point source, but rather can be considered to be comprised of numerous point

Figure 6.9 Photographs of an identical scene that (top) simulates scotopic (luminosity) and (bottom) photopic (color) perception of the human eye. The images were taken on a moonlit night in winter in Yosemite National Park, California.

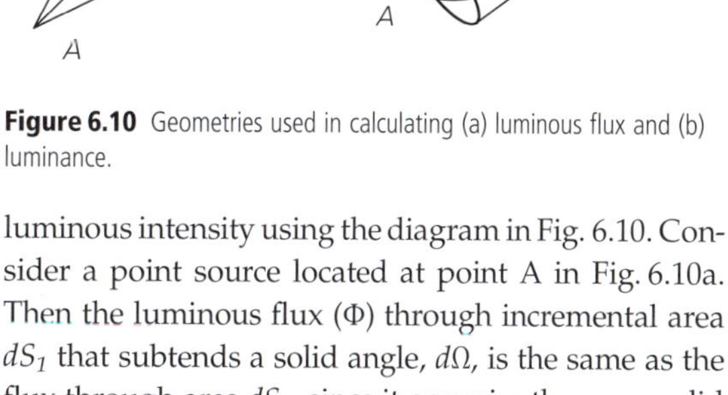

Figure 6.10 Geometries used in calculating (a) luminous flux and (b) luminance.

sources occupying an incremental area at position A in Fig. 6.10b. The luminance is thus the luminous intensity per unit solid angle per area. It depends on the angle (θ) of the source area, dS, relative to the observer at point B. From the diagram we find that the luminance that gives the perceived brightness of the source is

$$L = \frac{dL_\Omega}{dS\cos\theta} = \frac{d\Phi}{dSd\Omega\cos\theta}. \quad (6.10)$$

As noted above, the units of luminance are [cd/m^2] or [nits]; hence [candela] = [lumen/steradian].

The amount of flux that is incident on a secondary surface is called the *illuminance*. Correspondingly, the flux that exits that surface is the *luminous excitance*. This is a quantity that characterizes the brightness of an object, and is useful for quantifying the results of illumination of a volume by a white light source. Illuminance has the units of lux, where 1 lux = 1 lm/m^2. The relationship between luminous flux and illuminance is understood from Fig. 6.11. If the area illuminated is dS, then the amount of light exiting from surface B due to a source of area ΔS at point A is the illuminance (or luminous flux density). From inspection of Fig. 6.11, this is given by

$$E = \frac{d\Phi}{dS} = \frac{L\Delta S\cos\theta\cos\phi}{d^2} \quad (6.11)$$

where d is the distance between the source and the illuminated surface.

Finally, a source that has a constant luminance in all directions has an angular luminance intensity that follows a cosine dependence, that is,

$$L_\Omega(\theta) = L\cos\theta \cdot dSd\Omega. \quad (6.12)$$

This dependence defines a *Lambertian source*, which, for example, can be achieved using a perfectly light diffusing surface. Often an OLED is assumed to be a Lambertian source, although this is generally incorrect since such devices are microcavity emitters, as discussed in Section 6.6.

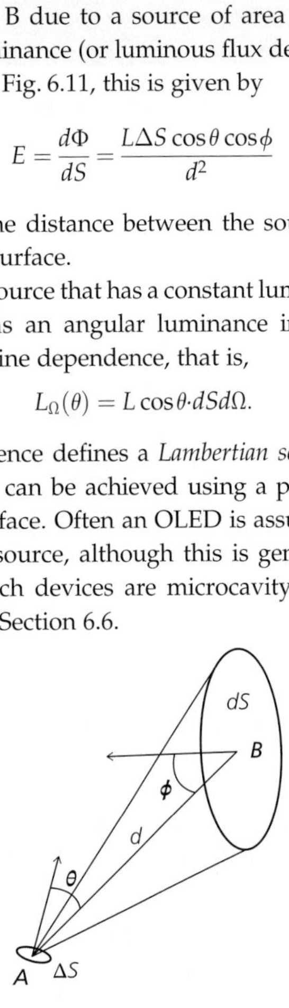

Figure 6.11 Relationship between luminance and illuminance.

The radiometric equivalents (along with their symbols and [units]) of the quantities of luminous flux, luminance, luminous intensity, and illuminance or excitance (if incident or emitted, respectively) are *radiant flux* (Φ_e, [W]), *radiance* (L_e, [W/m^{2}sr]), *radiant intensity* (I_e, [W/sr]), *irradiance* or *radiant excitance* (E_e, M_e, [W/m^2]) if incident or emitted, respectively.

The photometric efficiencies of the OLED are defined as the *luminous efficiency* and the *luminous power efficiency* (also known as the *efficacy*). These are analogous to the radiometric units of external quantum efficiency and power efficiency in Eqs. 6.1 and 6.2, respectively. The luminous efficiency (η_L) is equal to the luminance output divided by the current density, j:

$$\eta_L = \frac{L}{j}, \quad (6.13)$$

and has units of [cd/A]. Likewise, the luminous power efficiency is the luminous power emitted to the total electrical power input into the device:

$$\eta_{LP} = \frac{\Phi}{jV\Delta S} = \frac{E}{jV}. \quad (6.14)$$

A summary of the relationships between photometric and radiometric quantities and their units is provided in Table 6.2.

Photometric units are strongly dependent on the spectral content of the emitted light, and hence small differences in the spectrum of a source can result in significant differences in the luminous and luminous power efficiencies. Hence, for direct comparisons of device performance that are not dependent on the sensitivity of the eye to a particular spectrum, photometric quantities are more reliable and revealing of the device characteristics. Hence, photometric units find their greatest use in quantifying the *appearance* of a light source, rather than in its fundamental emission properties.

6.2.3.2 Chromaticity

The biological origin of the photopic response arises from excitation of three color-sensitive cone cells in the retina. Their individual spectral responses are shown in Fig. 6.12a. The responses extend from the deep red at a wavelength of ~550–650 nm for the long wavelength (L) sensitive cells (see inset, Fig. 6.12b), to a peak at ~530–540 nm for the middle wavelength (M), and 420–440 nm for the short wavelength (S) cones. These responses have been mapped into the *color matching functions*, $\bar{x}(\lambda), \bar{y}(\lambda)$, and $\bar{z}(\lambda)$, based on the 1931 CIE standards in Fig. 6.12b. The color matching functions are useful in quantifying the spectrum of an illumination source to determine the perceived

Table 6.2 Radiometric and photometric units for quantifying properties of illumination sources

Radiometric units				**Photometric units**			
Quantity	**Symbol**	**Expression**	**Unit**	**Quantity**	**Symbol**	**Expression**	**Unit**
Radiant flux	Φ_e		W	Luminance flux	Φ		lm
External quantum efficiency	η_{ext}	$\eta_{int}\eta_{out}$	%	Luminous efficiency	η_L	$\frac{L}{j}$	cd/A
Power efficiency	η_P	$\frac{1}{jV}\frac{d\Phi_e}{dS}=\frac{E_e}{jV}$	%, or W/W	Luminous power efficiency	η_{LP}	$\frac{1}{jV}\frac{d\Phi}{dS}=\frac{E}{jV}$	lm/W
Radiant intensity	I_e	$\frac{d\Phi_e}{d\Omega}$	W/sr	Luminance intensity	L_Ω	$\frac{d\Phi}{d\Omega}$	lm/sr
Radiance	L_e	$\frac{d\Phi_e}{dSd\Omega cos\theta}$	W/sr-m^2	Luminance	L	$\frac{d\Phi}{dSd\Omega cos\theta}$	cd/m^2 = lm/sr-m^2
Irradiance	E_e	$\frac{d\Phi_e}{dS}$	W/m^2	Illuminance	E	$\frac{d\Phi}{dS}$	lm/m^2 = lux
Radiant exitance	M_e	$\frac{d\Phi_e}{dS}$	W/m^2	Luminous exitance	M	$\frac{d\Phi}{dS}$	lm/m^2

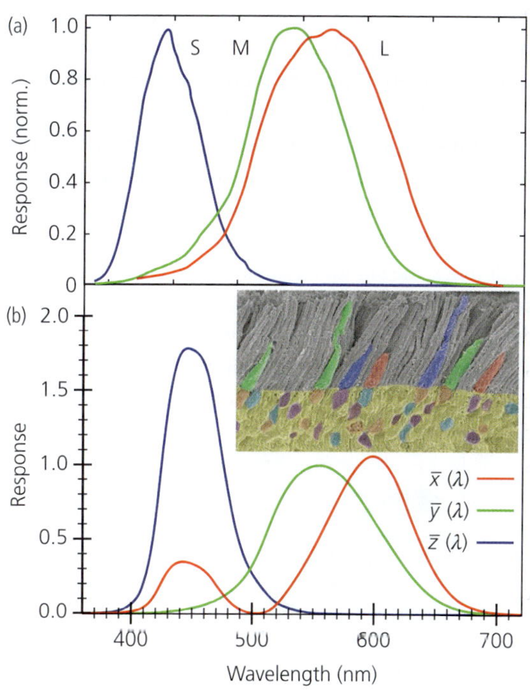

Figure 6.12 (a) Human eye response of the short (S), middle (M), and long (L) wavelength cone cells. (b) 1931 CIE color matching functions, $\bar{x}(\lambda)$, $\bar{y}(\lambda)$, and $\bar{z}(\lambda)$. Inset: Micrograph of color sensing cone cells (colors indicate their respective wavelength sensitivities) and rod cells (gray) that are sensitive only to luminosity.

color response. The photopic luminosity response curve in Fig. 6.8 is the same as the color matching function, $\bar{y}(\lambda)$ in Fig. 6.12b. The sensation, or perception of color is not directly related to the actual source color, which is often a combination of hues. Based on the responses of a population of test subjects who were asked to distinguish between a standard series of sample colors, a *Standard Colorimetric Observer* was used to define the *tristimulus values*, X, Y, and Z, for a source with radiance, $L_e(\lambda)$, as follows:

$$X=\int_{380}^{780} L_e(\lambda)\bar{x}(\lambda)d\lambda, \tag{6.15a}$$

$$Y=\int_{380}^{780} L_e(\lambda)\bar{y}(\lambda)d\lambda, \tag{6.15b}$$

and

$$Z=\int_{380}^{780} L_e(\lambda)\bar{z}(\lambda)d\lambda. \tag{6.15c}$$

The Standard Colorimetric Observer is also known as a 2° Standard Observer since during the tests it was assumed that the color sensitive cone cells were located within a 2° arc defining the fovea, the most sensitive region of the retina. The X, Y, and Z values are used to define the *1931 CIE XYZ color space* that is shown in Fig. 6.13. This color space assigns a coordinate pair, (x, y), to every color in the visible spectrum, which are related to X, Y, and Z by

$$x=\frac{X}{X+Y+Z}, \tag{6.16a}$$

$$y=\frac{Y}{X+Y+Z}, \tag{6.16b}$$

and

$$z=\frac{Z}{X+Y+Z}=1-x-y. \tag{6.16c}$$

Since z depends on x and y, only the latter two coordinates are required to fully define a perceived color. While several other color matching functions and color spaces have been defined, the 1931 CIE XYZ color space is convenient and straightforward for defining the properties of OLEDs and other standard illumination sources, since Y defines the total luminance of the source. The coordinates (x, y, Y) provide both the color and the luminosity, fully defining the properties of the source. This triad of coordinates is called the *1931 CIE xyY color space*. Transformation between xyY and XYZ, and other color spaces is straightforward, and hence we will only use the 1931 CIE xyY standard in the following discussion.

The visible color space shown in Fig. 6.13 represents the total *color gamut* that can be perceived by the Standard Colorimetric Observer. Saturated (pure) colors whose wavelengths (in nanometers) are found along the curved, outer arc of the color space is the *color locus*, and color mixtures are found within the interior of the color space. The inner arc shows the Planckian locus of white points, with each point along the arc corresponding to a *color temperature* listed (in Kelvin) adjacent to the locus. The lines of constant color temperature are shown intersecting the arc. The triangles enclose the space contained within two common color gamuts, one corresponding to super red, green, blue (sRGB) used as a photographic standard, and the other to the National Television Standards Committee (NTSC) used as a video display standard. If the R, G, and B color sources (e.g. pixels in a display) are located at the vertices of the color gamut triangle, then mixtures of these primary colors can produce any color enclosed within that triangle. The color obtained by adjusting the relative luminosities of the sub-pixel triad then results in a total pixel chromaticity defined by coordinates (x, y). Colors outside of the perimeter of the triangle are inaccessible to the light source, and hence cannot be produced. No three color sources (e.g. R, G, and B pixels in a display) can be positioned to cover the entire color space perceived by the eye. Nevertheless, larger color gamuts enclosed by particular sub-pixel triads result in a broader range of accessible display colors.

When designing illumination sources, it is reasonable to ask how sensitive the human eye is in distinguishing between colors with different coordinates in the chromaticity diagram. This question was answered in a painstaking experiment by MacAdam (1942) whereby subjects were asked to match two colors at a fixed luminosity that were viewed through an apparatus that constrained the viewing angle to 2° within the range of foveal vision. The experiment extended over 25,000 color-matching trials across the entire color gamut. This resulted in 25 ellipses drawn to demark the standard deviation in chromaticity that could not be distinguished by an observer. These so-called *MacAdam ellipses*, drawn 10 times their actual

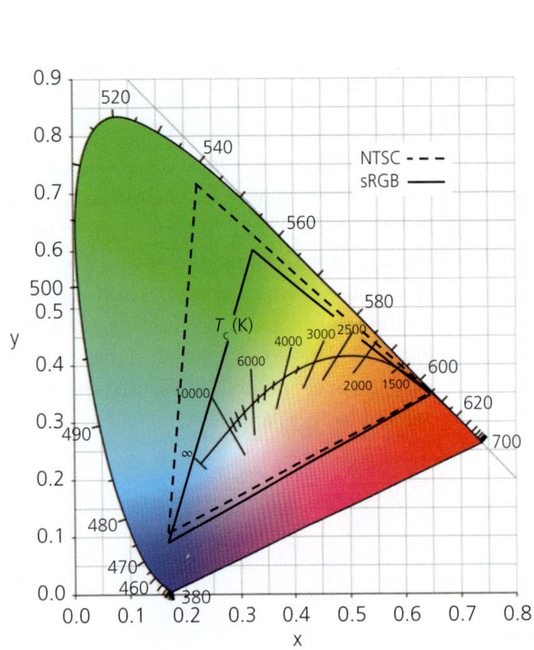

Figure 6.13 1931 CIE xyY color space showing the white locus and the color gamuts for sRGB (solid triangle) and NTSC (dashed triangle) displays.

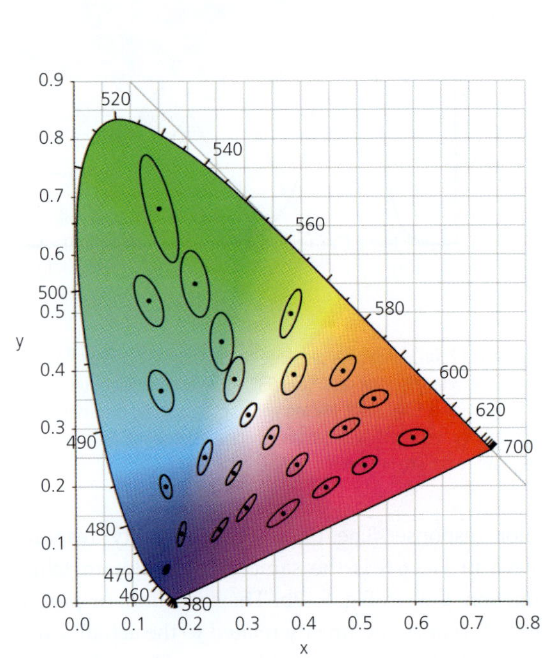

Figure 6.14 1931 CIE chromaticity chart showing the 25 MacAdam ellipses at several color locations. The ellipses are drawn 10 times their actual size.

sizes, are shown in Fig. 6.14. Interestingly, there is no single transformation that can bring these ellipses into equal-sized circles, indicating that the perception of color is a nonlinear function of the spectral content of the source. Furthermore, it is apparent from Fig. 6.14 that color acuity is highest for blue, less so for red, and not nearly so high in the green. This provides an explanation why apparently saturated display colors can be obtained using a gamut that extends nearly to the color locus in red and blue, yet does not extend far toward the limb at $\lambda = 520$ nm (at coordinates of (0.07, 0.83)).

An example color gamut that results from emission from three PHOLEDs emitting in the red, green and blue spectral regions is shown in Fig. 6.15. The phosphors used are the blue emitting $Ir(pmp)_3$, the green emitting $Ir(ppy)_2(acac)$, and the red emitting PQIr. The intensity spectra in Fig. 6.15a must be converted to luminance to determine the color coordinates of each device, denoted by the open circles in Fig. 6.15b. The resulting color gamut is enclosed by the triangle. The gamut is smaller than required by sRGB standards (Fig. 6.13), largely due to the slightly red-shifted and spectrally broad green PHOLED sub-pixel.

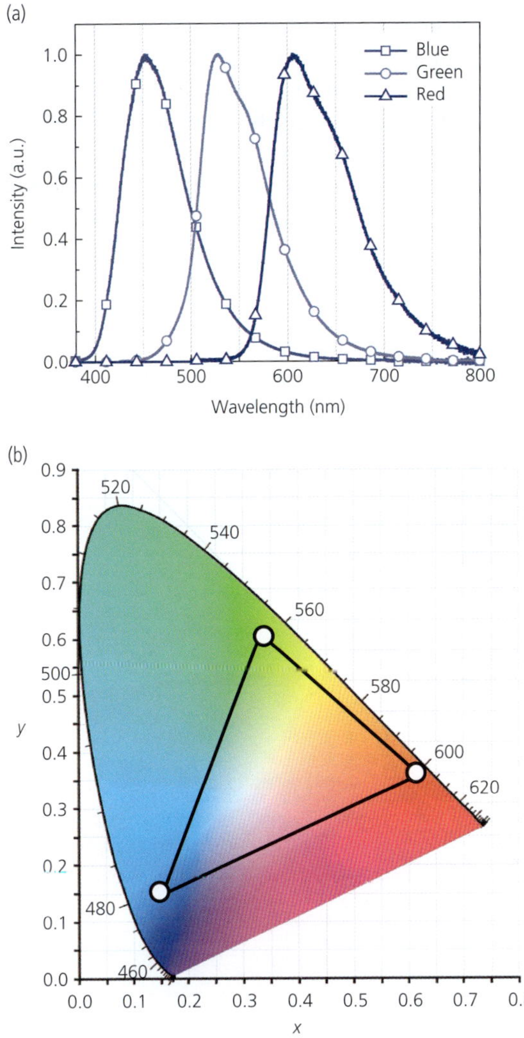

Figure 6.15 (a) Emission spectra from three phosphorescent OLEDs. (b) The color gamut (triangle) due to emission from the three OLEDs in (a). Open circles indicate the color coordinates for each for the three OLEDs.

6.2.3.3 White light

Thus far we have concentrated on quantifying color perception primarily for displays. Another application of OLEDs is in the generation of white light for illumination. The method for characterizing the appearance of white illumination sources is considerably different than that required for information displays, yet ultimately it, too, relies on the nature of human vision and its range of color perception bounded by the chromaticity diagram in Fig. 6.13. The primary difference between a white light source and one that emits a particular hue is the way that objects appear when illuminated by the source. Do the colors look natural as they would under sunlight, or does the source give an unnatural coloration to the objects, yet when directly viewed it nevertheless looks satisfactorily white?

There are two quantities that are used to characterize illumination source color quality: the *color rendering index* (CRI) and the *correlated color temperature* (CCT). The CRI quantifies how close a color source matches a black body radiator at a particular temperature. The rationale for this quantity is that we assume that solar radiation follows that of a Planckian black body at temperature T. Then the spectral radiance, $B(\lambda, T)$, follows Planck's law:

$$B(\lambda, T) = \frac{2hc^2}{\lambda^5} \frac{1}{\exp(hc/\lambda k_B T) - 1}. \tag{6.17}$$

With the exception of spectral loss due to absorption by water and other gases in the atmosphere, the sun is indeed an ideal black body radiator with a surface temperature of 5780 K. However, its apparent color temperature varies throughout the day, shifting to the red in the morning and evening (with an apparently lower temperature of ~5000 K), and becoming bluish-white at mid-day (with a much higher temperature of ~ 6000 K). Perceptually, we consider bluish casts to be "cool white" with color temperatures >5000 K, and redder hues are "warm white" with temperatures between 2700 K and 3000 K (see

Fig. 6.16). Of course, this convention is opposite to the temperature trends predicted by Eq. 6.17.

The CRI is the degree to which the spectrum of the test illumination source differs from that of a black body of a specified temperature. Hence, to quantify a white light source, we need to know its temperature. The Planckian locus indicating the color coordinates of black bodies radiating at a particular temperature is shown by the arc in Fig. 6.13 with lines of constant color temperature as noted. A white light source, other than an incandescent bulb, does not have a perfect black body spectrum. Hence, for white light sources such as OLEDs we define the CCT as the black body temperature that most closely resembles the test source. The 1960 CIE color space shown in Fig. 6.17 is commonly used to determine the CCT of a particular illumination source (i.e. an *illuminant*). The

Figure 6.16 Examples of white LED sources of different color temperatures. From Novalite.com (http://www.nova-lite.com/index.php?route=module/kbm/article&kbm_article_id=7)

color temperatures are shown as well as the isotherms perpendicular to the Planckian locus, whose lengths define the chromaticity range over which meaningful comparison with a black body can be made.

This white color space is often defined in terms of the coordinates (u, v). Via a simple transformation, (u, v) can be converted to the 1931 CIE coordinates (x, y) using

$$u = \frac{4x}{-2x + 12y + 3} \tag{6.18a}$$

and

$$v = \frac{6y}{-2x + 12y + 3}. \tag{6.18b}$$

Deviations from a black body lie above or below the locus at distance Δ_{uv}, shown in Fig. 6.17. Comparison of the illuminant spectrum and a black body is considered meaningful when $\Delta_{uv} \leq 5.4 \times 10^{-3}$. In this case, the CRI obtained for the source will also be accurately defined since its spectrum will approximate that of a black body.

Now the *equal energy point* (E) corresponds to a spectrum that has equal energy at all visible wavelengths. This point is at $(x, y) = (0.33, 0.33)$, or according to Eq. 6.18, $(u, v) = (0.209, 0.315)$.

Once the CCT of the illuminant is determined, the CRI is found by comparing the color coordinates of the reflected light from eight standard *test color samples* (TCS) under the test source, compared to a reference black body at the same color temperature. The TCS in Fig. 6.18 range from light grayish red (TCS01)

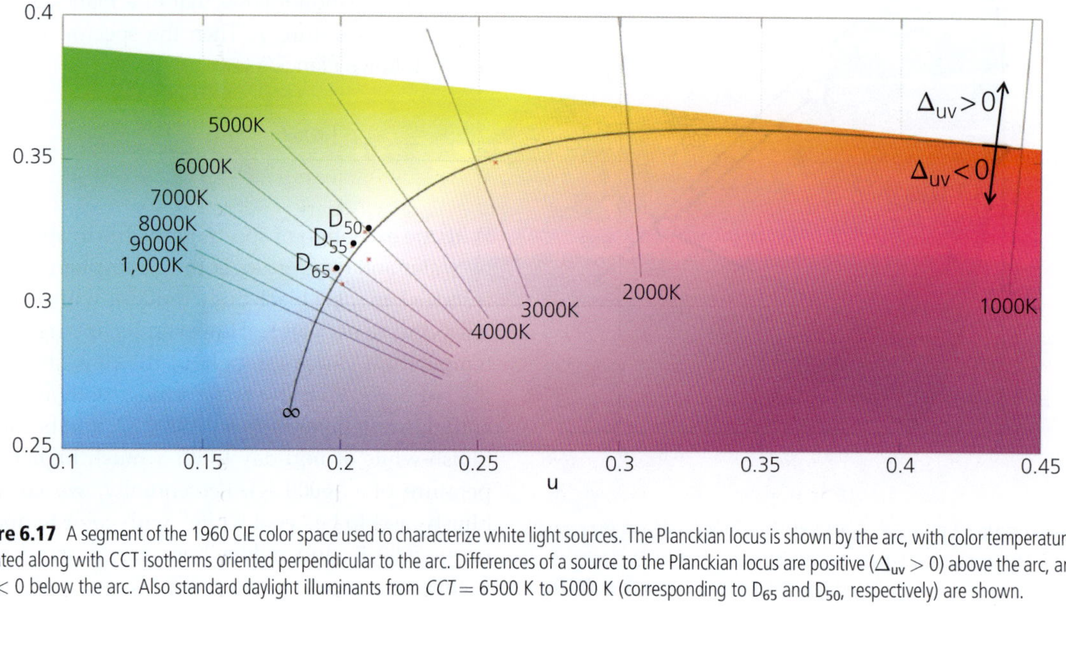

Figure 6.17 A segment of the 1960 CIE color space used to characterize white light sources. The Planckian locus is shown by the arc, with color temperatures indicated along with CCT isotherms oriented perpendicular to the arc. Differences of a source to the Planckian locus are positive ($\Delta_{uv} > 0$) above the arc, and $\Delta_{uv} < 0$ below the arc. Also standard daylight illuminants from $CCT = 6500$ K to 5000 K (corresponding to D_{65} and D_{50}, respectively) are shown.

Figure 6.18 Test color samples (or patches) used for determining the color rendering index.

to light reddish purple (TCS08). The specific CRI, or R_i, is found from the reflectance deviations by calculating their Euclidean distances, ΔE_i, between the test (u_t, v_t), and the reference white point (u_r, v_r) coordinates. A commonly used reference is the D_{65} illuminant in Fig. 6.17. From the tristimulus values we define the 1964 CIE color space:

$$W^* = 23\sqrt[3]{Y} - 17, \tag{6.19a}$$

$$U^* = 13W^*(u_t - u_r), \tag{6.19b}$$

and

$$V^* = 13W^*(v_t - v_r). \tag{6.19c}$$

This differs from the 1960 CIE color space used for determining u, v in that it has information about the luminosity of the source (Y), which is found in the coordinate, W^*. Thus, the 1960 CIE space is used to determine CCT, whereas the 1964 CIE standard (also known as CIE UVW) is useful for determining the CRI.

To find R_i, we determine the differences, $\Delta U_i^*, \Delta V_i^*$, and ΔW_i^* between the CCT of the source and the chromaticity coordinates from the ith TCS to obtain

$$\Delta E_i = \sqrt{(\Delta U_i^*)^2 + (\Delta V_i^*)^2 + (\Delta W_i^*)^2}. \tag{6.20}$$

Then the specific CRI is

$$R_i = 100 - 4.6\Delta E_i \tag{6.21}$$

and the general color rendering index, R_a (also known as simply *CRI*), is the mean of all eight R_i, viz.

$$CRI = \frac{1}{8}\sum_{i=1}^{8} R_i. \tag{6.22}$$

From Eqs. 6.21 and 6.22, we see that *CRI* has a maximum value of 100, a value only achieved by a perfect black body where $\Delta E_i = 0$.

The acceptable *CRI* and CCT of a lighting source depend on the intended application. For general room illumination, $CRI = 90$ or higher is deemed acceptable, whereas for lighting industrial or utility areas, $CRI = 80$–85 are often used. Similarly, the color temperature chosen depends on the application. Room lighting typically employs "soft white" which is warm with a lower color temperature than outdoor lighting, where daylight CCTs of up to 6000 K are often found.

A characteristic of lighting is that there are inevitable differences in the color content of otherwise identical lighting sources that are manufactured at different times or fabrication facilities. To match these different illuminants, we must quantify the perceptual acuity of the observer: that is, we must determine the difference in color coordinates of a lighting source that can be perceived. This question is answered in the same manner that MacAdam quantified perceptual acuity in viewing hues at different locations in the 1931 CIE color space. In Fig. 6.19 we show the MacAdam ellipses drawn along the Planckian locus in the 1931 CIE color space. The ellipses represent three times the standard deviation for perceptual distinction between two lighting sources. The black quadrangles provide so-called "seven-step bins" established by the *American National Standards Institute* (ANSI) in which similar illuminants can be grouped without undue variation between sources in a given bin. However, higher uniformity lighting sources are often placed into bins as small as two and three standard deviations (i.e. two step and three step bins) from the distinguishable limit. Three step bins used to group LED sources are shown in orange.

6.2.4 Measuring OLED performance

The accurate measurement of OLED efficiency is essential for comparing device performances achieved in different laboratories. Also, the procedure for measuring efficiency is different for display and lighting sources. For displays, only photons emitted into the half space in front of the display (i.e. in the viewing direction) are counted, whereas for general lighting, photons emitted in all directions add to the illumination intensity.

If the OLED emission is Lambertian, one simply measures the device performance using a commercial luminance meter. The meter collects light emitted from a fixed diameter spot on the display surface, into a specified (small) solid angle about the

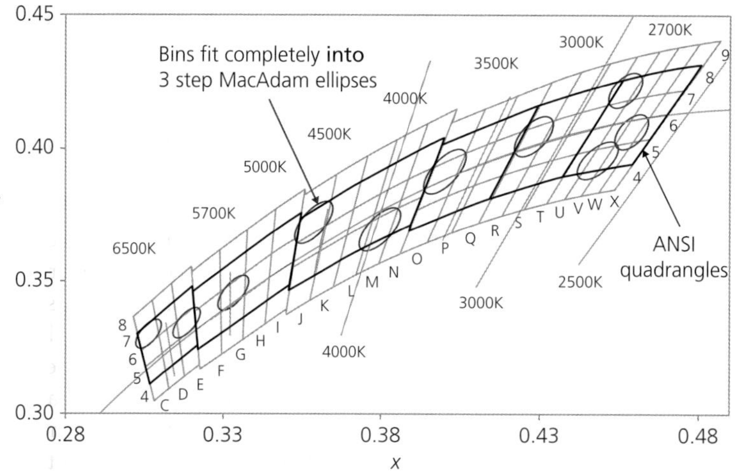

Figure 6.19 MacAdam "ellipses" along the Planckian locus. The ellipses are bounds to areas three times the standard deviation at the perceptual limit to distinguish between white sources, and hence are known as three step ellipses. The black boxes are the ANSI C78.377 standard for variations of an acceptable lighting source at a particular CCT, and the light blue boxes show bins used to group illuminants whose CCT and CRI fit within approximately a three step ellipse.

surface normal. From the intensity, the instrument calculates a value of the surface luminance, L, in cd/m^2, or nits. The meter uses a calibrated photodiode to detect the light, and the conversion of the diode photoresponse into a luminance value is done by the instrument itself, with the implicit assumption that the emission pattern is Lambertian, and that the meter is oriented perpendicular to the emitting surface. The recorded luminance then allows for a calculation of the luminous efficiency η_L (cd/A) using the drive current density, $j_{OLED} = I_{OLED}/A$, where A is the area and I_{OLED} is the OLED current. Given the voltage at which η_L is measured, it is then straightforward to obtain η_P (see Table 6.2). While simple, this method is vulnerable to inaccuracies, and hence should be used only for characterizing the performance of an actual display.

A direct and accurate means for measuring external quantum efficiency employs a calibrated photodetector used in a measurement setup requiring a minimal amount of correction for losses due to lenses, overfilling of the detector active area with light, conversion between lumens to percent, etc. The generally broad spectral emission of OLEDs, however, will introduce errors in measurement if the wavelength dependence of the photodetector is itself ignored. The wavelength dependence of the photodiode *responsivity* between wavelengths λ and $\lambda + d\lambda$ is $R(\lambda) = I_{det}(\lambda)/f(\lambda)P_{OLED}(\lambda)$, where $I_{det}(\lambda)$ is the incremental photocurrent generated in the photodetector by the incident power, $P_{OLED}(\lambda)$, emitted at center wavelength, λ, and $f(\lambda) \leq 1$ is the ratio of light emitted to that coupled into the detector. The external OLED quantum efficiency is, therefore,

$$\eta_{ext} = \frac{q\int \lambda I_{\mathrm{det}}(\lambda)d\lambda}{hcfI_{OLED}\int R(\lambda)d\lambda}. \tag{6.23}$$

Since the OLED may exhibit microcavity effects that result in a non-Lambertian angular dependence of the emission intensity on wavelength (Greenham et al., 1994, Bulovic et al., 1998a, Kim et al., 2000), accurate measurement requires the use of large aperture light collection optics. Often, the detector sensitivity is quoted in terms of its own external quantum efficiency, η_{det}. This is simply related to the detector responsivity via $\eta_{det} = hcR(\lambda)/q\lambda$. Then, Eq. 6.23 can be rewritten as

$$\eta_{ext} = \frac{\int \lambda I_{\mathrm{det}}(\lambda)d\lambda}{fI_{OLED}\int \lambda\eta_{\mathrm{det}}(\lambda)d\lambda}. \tag{6.24}$$

As defined, the quantum efficiency is simply a ratio that, if multiplied by 100%, can be stated as a percentage. For "engineering" display applications, η_{ext} is typically quoted independent of the emission wavelength, as in Eqs. 6.23 and 6.24.

From the foregoing, the optimum measurement geometry has a coupling factor, f, as close to unity as possible. A particularly accurate and convenient configuration for such a measurement is shown in Fig. 6.20a. The primary objective is to detect every

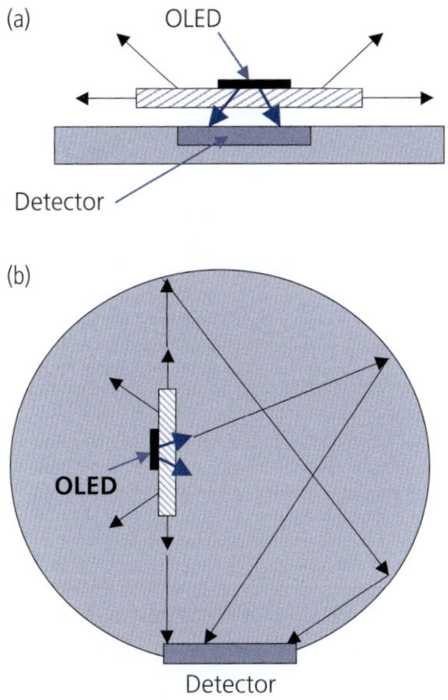

Figure 6.20 Efficiency measurement configurations for an OLED for (a) displays where the OLED is placed directly on the surface of a detector, and (b) lighting where the OLED is inserted into an integrating sphere (Forrest et al., 2003).

photon emitted into the forward direction using a calibrated detector, such that $f \sim 1$ is independent of λ. This can be achieved by using a detector whose area is considerably larger than that of the OLED (a condition known as "underfilling" of the detector active area), and then placing the OLED very close to the detector without any intervening optics (Burrows et al., 1996). It is also necessary to mask trapped photons that may emerge from the substrate edges. Coating the edges with black paint or wax, or by ensuring that the detector is considerably smaller than the OLED substrate can accomplish this objective (Adachi et al., 2001a).

A common method for measuring internal quantum efficiency is to place the OLED into an integrating sphere containing a calibrated detector, and then measuring all photons output from the device (Fig. 6.20b) (Gong et al., 2002). Several sources of inaccuracy need to be considered. (i) The response of the photodiode is generally not uniform over the broad spectral output of an OLED. As in the case of the external efficiency measurement, the spectral response of the photodiode must be calibrated across the entire OLED emission spectrum to accurately obtain the efficiency from the photocurrent. (ii) All significant sources of internal loss in the structure, such as reabsorption in the organic films, total internal reflection (TIR) at the substrate/air interface, propagation and other losses in the transparent anode, surface plasmon polaritons, free carrier absorption, and other effects associated with contacts (see Section 6.6). Due to the uncertainties inherent in these estimates, the internal quantum efficiency can contain significant systematic errors (Bulovic et al., 1998a, Kim et al., 2000). (iii) Losses due to shadowing by the probes and other fixtures (e.g. the baffle used to prevent direct irradiation of the detector) within the sphere, and less than unity reflectivity from the interior surface of the sphere, must be "backed out" of the measurement. (iv) Losses due to reabsorption of rays traversing the sphere by the OLED itself.

The internal efficiency can be several times greater than the external efficiency, since the integrating sphere captures nearly all the photons emitted in the device active region and that are subsequently emitted into all directions, not just in the forward viewing direction (Gong et al., 2002).

Using an integrating sphere is *not* a measurement of the display-relevant (forward emitted) *external* quantum efficiency; the source of some confusion in the literature. To measure the forward-viewing external quantum efficiency in an integrating sphere requires that the edge-emitted light be prevented from reaching the detector. The integrating sphere measurement is useful for assessing the OLEDs for lighting since the source is typically attached to a fixture (also known as a *luminaire*) that directs photons emitted in all directions into the space being illuminated.

Once the external efficiency is measured as a function of I_{OLED} and V, and the device emission spectrum is also determined using calibrated detectors. These measurements allow for calcuation of L, η_L, and η_p. Doing this process in "reverse order" (i.e. deriving the radiometric from the photometric parameters) requires that the output emission pattern, meter calibrations and other geometric factors are accurately known. Deviations from Lambertian emission of most OLEDs (Tokito et al., 1997) can lead to significant errors, and hence direct measurements of the photometric characteristics are the most reliable.

To fully characterize the OLED, therefore, three groups of data are required: the current-voltage characteristics, the external quantum efficiency-current characteristics, and the emission spectrum. An example data set for the PHOLED structure in Fig. 6.21 is shown in Fig. 6.22. The j–V characteristics at $j > 0.01$ mA/cm^2 in Fig. 6.22a are consistent with space charge limited transport (see Section 4.4.2) showing a power-law behavior of $j \sim V^m$. Below 1.5 V, the characteristics are approximately Ohmic. The onset of substantial

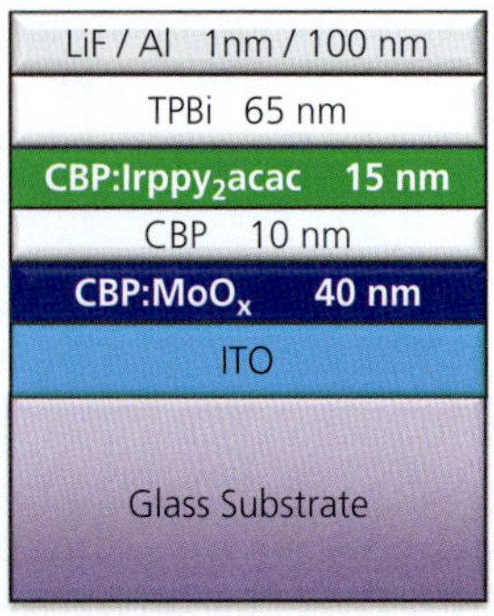

Figure 6.21 Structure of the PHOLED used in Figs. 6.22 and 6.23. Materials are: cathode: LiF/Al; exciton blocking layer: TPBi; emission layer: CBP doped with 10 vol.% Ir(ppy)$_2$acac; hole injection layer: CBP doped with MoO$_x$; anode: ITO.

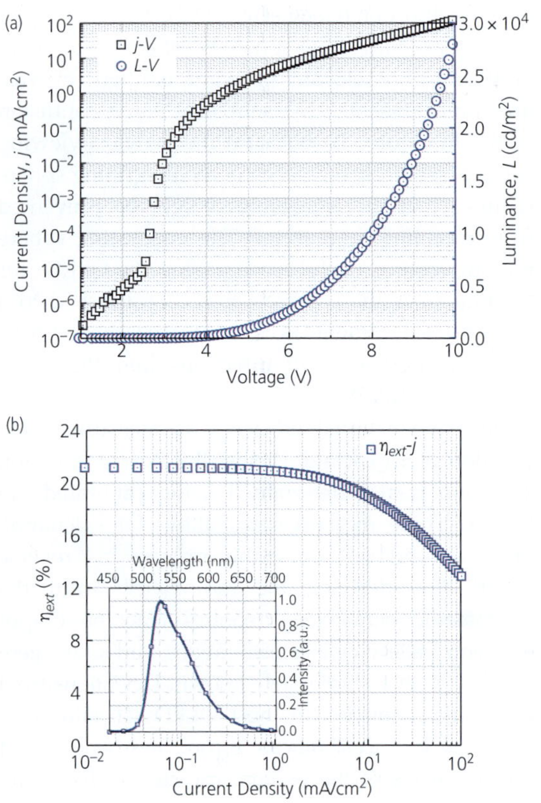

Figure 6.22 (a) Current density–voltage and luminance–voltage characteristics and (b) external quantum efficiency vs. j of the device in Fig. 6.21. The emission spectrum is shown in the inset.

luminance occurs at approximately 0.1 μA/cm^2 at 4V, reaching 3×10^4 cd/m^2 at 10 V.

The external quantum efficiency is provided in Fig. 6.22b. Here, $\eta_{ext} = 21\%$ is approximately constant up to $j = 1$ mA/cm^2, and then begins to decrease

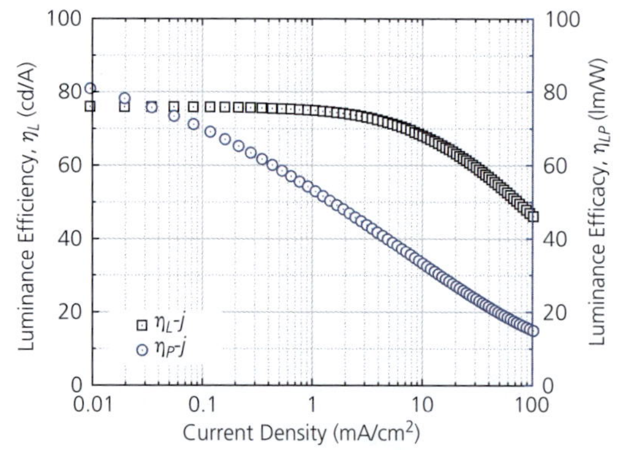

Figure 6.23 Luminance efficiency and efficacy vs. current for the device in Fig. 6.21.

rapidly as current (and brightness) is further increased. We show in Section 6.6 that the external quantum efficiency of 21% corresponds to an internal quantum efficiency, $\eta_{int} \sim 100\%$, which is expected for high performance PHOLEDs. Furthermore, in Section 6.3.5 we show that the roll-off at high current is a result of triplet annihilation. The inset, Fig. 6.22b, provides the last piece of information needed to fully quantify the performance of the OLED—the emission spectrum. This is due to triplet emission from Ir(ppy)$_2$acac with a peak at $\lambda = 525$ nm, with one vibronic overtone observed at $\lambda = 560$ nm.

From the data in Fig. 6.22, we arrive at the luminance efficiency and the luminance power efficiency shown in Fig. 6.23. To obtain η_L, we use both η_{ext} and the spectral luminance in the inset, Fig. 6.22b, as discussed in Section 6.2.3.1. Similar to η_{ext}, the η_L is constant at 75 cd/A until roll-off is reached at $j \sim 10$ mA/cm^2.

Finally, the efficacy is obtained from Eq. 6.14 along with the voltage obtained from the j-V characteristics in Fig. 6.20a. Here, η_{LP} decreases monotonically from 80 lm/W at 10 μA/cm^2, to 20 lm/W at 100 mA/cm^2 due to the monotonic increase in OLED voltage. Efficacy is most useful when comparing white light sources. When combined with the device CRI and CCT, it provides an important quantity in evaluating the properties of the illuminant. It is far less useful for characterizing the actual efficiency performance of monochromatic devices such as the green PHOLED in Fig. 6.21. In this case, the radiometric quantity of power efficiency is more useful.

6.3 Electroluminescent processes

There are two spin-dependent mechanisms for OLED emission: fluorescence due to radiative recombination of electrically generated singlet states, and phosphorescence due to triplet-mediated emission. The efficiency of the device is dependent on which of these two processes are responsible for controlling the emission. In Section 6.2.1 we showed that the maximum OLED internal quantum efficiency is determined by the spin statistical splitting of the injected charges. Random statistics (i.e. the spin of the injected electron and hole are uncorrelated) suggest that the spin formation ratio in Eq. 6.1 is $\chi_{ST} = 25\%$ for singlets and $\chi_{ST} = 75\%$ for triplets due to the multiplicities of these two spin manifolds (Chapter 3). In the following section we describe the spin formation process, and means for measuring χ_{ST}. We show that in almost all materials, random statistics appear to govern the spin formation ratio. This will be followed in subsequent sections by a discussion of the luminescence processes that ensue once the excitons are formed in their various spin manifolds.

6.3.1 Spin statistics

The exciton formation process following the injection of uncorrelated spins from the cathode and anode results in subsequent formation of a precursor electron–hole *polaron pair* (PP). This transfers to an even lower energy *charge transfer* (CT) *state* on adjacent molecules. At this point, the spins are completely uncorrelated and hence must form according to the 1:3 singlet:triplet statistical ratio. The bimolecular pair then scatters onto a single molecule to form the exciton that subsequently recombines.

The energetic landscape traversed from free charges to excitons is shown in Fig. 6.24. The process begins by injection of charge from the anode and cathode, whose collective spin symmetries randomly form either a singlet or triplet polaron pair (denoted S^{+-}, T^{+-}, respectively). Once the polarons approach to within a capture radius where the Coulomb energy is equal to the thermal energy, $r_c(T) = q^2/4\pi\varepsilon_r\varepsilon_0 k_B T$, either a singlet or triplet CT state (^{1}CT, ^{3}CT, respectively) is formed. Here, ε_r is the material dielectric constant and ε_0 is the permittivity of free space. The CT state relaxes to an exciton of radius $r_K(T)$. That is, the polaron-pair collapses from the point at which the two charges are nearly free at r_c (corresponding to the *charge separated state*), to where the oppositely charged polarons locate on nearest neighbors, and the exchange energy is large (at separation, r_K). At this point, the degeneracy is lifted and the spin symmetry no longer changes unless there is strong spin–orbit coupling due, for example, by a heavy metal atom within the molecular complex. Prior to

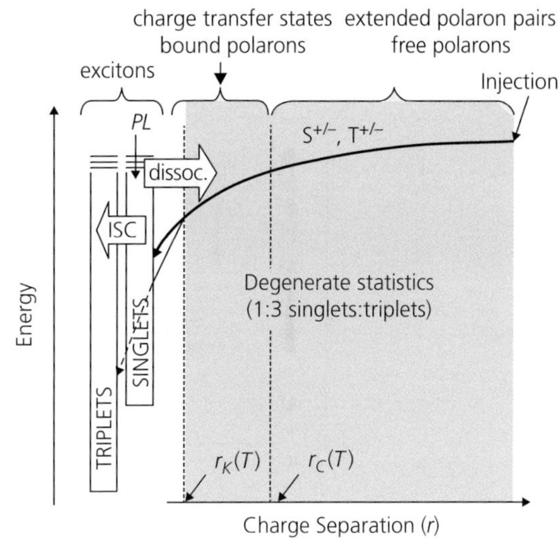

Figure 6.24 Free polaron-to-exciton formation process. The free polarons are injected from contacts, randomly forming a singlet and/or triplet polaron pair ($S^{+/-}$, $T^{+/-}$, respectively). When the polarons are within a capture radius, r_c, a CT state is formed. When the electron and hole of the CT state approach at distance r_K, a bound singlet or triplet exciton forms, depending on the spin symmetry of the original polaron pairs. Here, dissoc. = exciton dissociation, ISC = intersystem crossing (Segal et al., 2003).

reaching radius r_K, there can be mixing at rate k_{SL} between polarons in the charge separated state due to spin-lattice relaxation, or the presence of spin–orbit or hyperfine coupling discussed in Section 3.7 (McClure, 1949, 1952).

Figure 6.25a illustrates the rates governing the generation of excitons from the precursor CT states. Triplet, ^{3}CT, states are statistically generated at three times ($3G$) the probability of singlet, ^{1}CT, states (G). Once formed, the CT states bind into either singlet or triplet excitons at rates k_S or k_T, respectively. However, if the *spin exchange energies* between CT or exciton singlets and triplets (ΔE_{CT} or ΔE_{EX}) are small (i.e. on the order of or less than the thermal energy, k_BT), intersystem crossing between spin manifolds at rates k_{ISCT}, k_{ISC*}, and k_{ISC} can result in exciton spin distributions that differ from the statistical 1:3 ratio. That is, if k_{ISC} is higher than the exciton formation rates, k_S and k_T, then the spin statistical splitting can deviate from 1:3 singlets-to-triplets. This situation occurs for nearly degenerate singlet and triplet polaron-pair (i.e. CT) states.

The CT state formed on a pair of adjacent Alq_3 molecules is shown in Fig. 6.25b. Here, CT formation leads to substantial overlap of the polaron wavefunctions, ensuring efficient exciton formation that occurs via the hop of one of the charges from one molecule to its neighbor. We have seen that both Miller-Abrahams

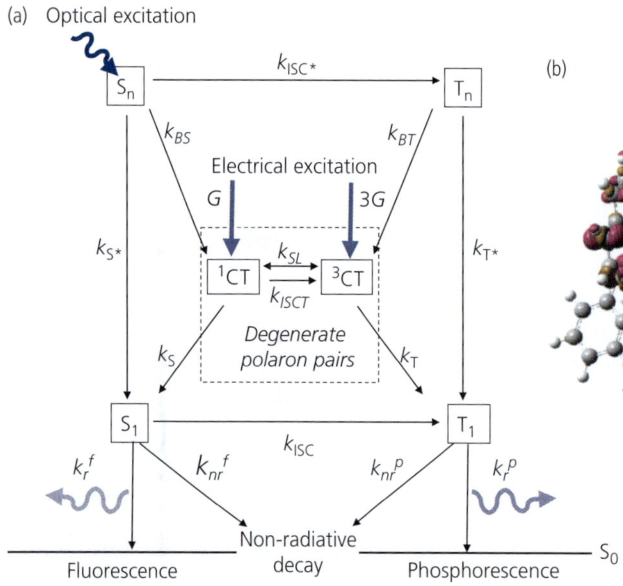

Figure 6.25 (a) Detailed exciton formation process via both optical and electrical excitation. Energies involved in triplet and singlet exciton formation are implied by vertical distances in the diagram. Here, G is the electrical-to-exciton generation probability. Once formed, CT states bind into either singlets (S) or triplet (T) excitons at rates k_S or k_T, respectively. The intersystem crossing between spin manifolds occurs at rates k_{ISCT}, k_{ISC^*}, and k_{ISC} for CT, upper manifold singlet (S*) and triplet (T*) states, and lowest energy (S, T) excitons, respectively. Radiative (k_r^f, k_r^p) and non-radiative (k_{nr}^f, k_{nr}^p) rates from fluorescent (superscript *f*) or phosphorescent (*p*) states are shown. (b) Polaron pair charge distributions on two adjacent Alq_3 molecules. The negative polaron distribution (orange) is primarily on the molecule on the right, and the hole distribution (red) is on the left (Segal et al., 2007).

and Marcus transfer theories (Section 4.3.2) predict that the probability for hopping decays exponentially with intermolecular distance. That is, the hopping rate follows $k_{fi} \sim \exp(-2\gamma a)\exp\left(\frac{E_f - E_i}{k_B T}\right)$ for a final state energy $E_f < E_i$. On the other hand, if $E_i < E_f$, then $k_{fi} = 0$. Here, γ is the hopping parameter and is related to the tunneling distance between the donor and acceptor molecules spaced at distance, a. The hopping rate is thus highly sensitive to the intermolecular spacing, yet the long-range Coulombic attraction between the electron and hole results in a near unity probability for the formation of a bound exciton once the charges are located on neighboring molecules.

The several routes affecting exciton formation statistics are shown schematically in Fig. 6.26. As noted, it is safe to assume that uncorrelated spins of the precursor electron–hole polaron pair combine according to the singlet–triplet statistical ratio of 1:3 as shown in Fig. 6.26a. These subsequently form singlet and triplet CT states (^{1}CT and ^{3}CT) in the same ratio. If the exchange energy splitting is higher than a few $k_B T$ (Kadashchuk et al., 2004), or if the rate of singlet exciton formation, k_S, is greater than for triplets, k_T, it is likely that this ratio will be preserved, and the ultimate ratio of singlet (S^*) to triplet (T^*) excited states will also be 1:3. The resulting emission is thus almost entirely due to singlet recombination ($S^* \rightarrow S_0$) via fluorescence occurring within 1-10 ns of exciton generation.

When there is strong spin–orbit coupling (**L·S**), the spins within the exciton manifold are mixed, resulting in rapid intersystem crossing, k_{ISC}, prior to recombination. As shown in Chapter 3, the spin–orbit interaction strength scales as Z^4, where Z is the atomic number (cf. Eq. 3.175). Hence, incorporation of heavy metal atoms has been shown to be effective in increasing k_{ISC} well above k_S. For these systems, the singlets rapidly transfer to the triplet manifold, and the spin selection rule that ordinarily forbids radiative recombination to the ground state following $T^* \rightarrow S_0$ is allowed, also due to the strong **L·S** coupling. In these instances, the emission is via triplets and is known as phosphorescence. The natural triplet lifetime is typically from 500 ns to 100 μs (Baldo et al., 1998, Baldo et al., 1999a). This situation is depicted in Fig. 6.26b.

In some special cases, the exchange energy of the CT state manifold may be less than, or on the order of the thermal energy, leading to spin mixing prior to exciton formation (see Fig. 6.26c). This leads to a departure from the statistical spin ratio. The relative population of singlets and triplets depends on the ratio of their relative formation rates, k_S/k_T. Typically, $k_S >> k_T$, leading to a disproportionate fraction of

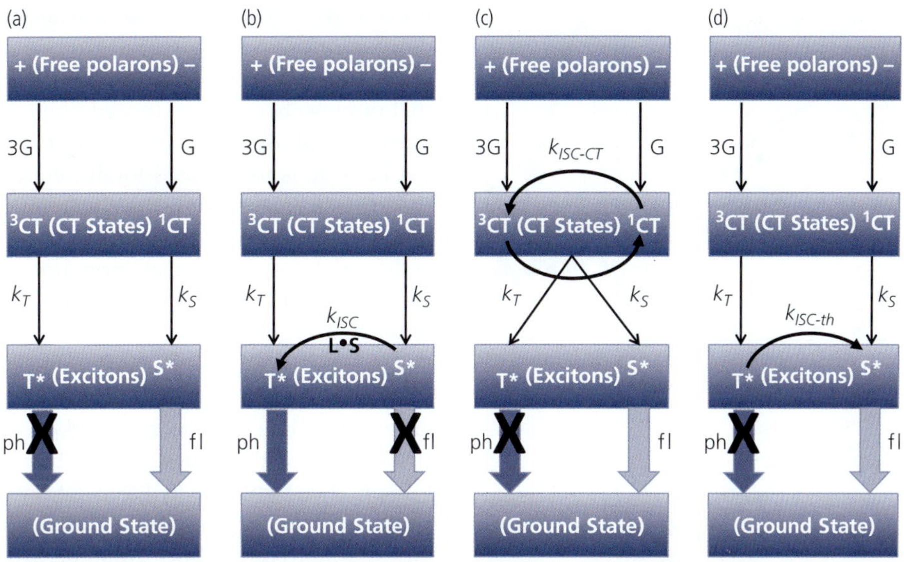

Figure 6.26 Exciton formation processes starting with a free electron–hole polaron pair. The random generation statistics imply that singlet and triplet CT states are generated (G) in a ratio of 1:3. (a) CT states form singlet and triplets without mixing, resulting in a 1:3 excited state singlet (S*) and triplet (T*) exciton ratio. Without mixing, fluorescence (fl) is dominantly observed. (b) Fast intersystem crossing at rate k_{ISC} due to enhanced spin–orbit coupling (**L·S**) results in predominantly phosphorescent (ph) emission. (c) Rapid mixing of spins between CT states at rate $k_{ISC\text{-}CT}$ in the absence of ISC results in fl emission. (d) If the exchange energy splitting between singlet and triplet is small, then thermally activated intersystem crossing at rate $k_{ISC\text{-}th}$ between the T* and S* is possible, resulting in delayed fluorescence. After Reineke and Baldo (2012).

singlets. In the absence of spin–orbit coupling, this process ultimately leads only to fluorescent OLED emission.

Finally, we consider a case analogous to a small CT state exchange energy: that corresponding to a small exchange energy in the final exciton state, shown in Fig. 6.26d. The triplets can rapidly transfer to the singlet manifold via thermal excitation, thus leading to a disproportionately large singlet population, even though the initial formation statistics are still one in four. The singlet undergoes radiative recombination at a timescale of a few nanoseconds, but with the system achieving an internal quantum efficiency greater than 25%. This process is known as *thermally activated delayed fluorescence* (TADF, see Section 3.7.5). It has been exploited in OLEDs to deliver high efficiencies, approaching 100%, and is the topic of Section 6.3.4.

The following question remains: do spin statistics change in the transition from $PP \rightarrow CT \rightarrow S, T$? This question was addressed by electric field polarization of the PP state during formation to determine if the ratio of S to T changes over time (Reufer et al., 2005). A ladder-type polymer, poly(*p*-phenylene) or PhLPPP (see lower inset, Fig. 6.27) slightly contaminated by ~0.1% Pd atoms during synthesis yields both fluorescent and phosphorescent PL. Phosphorescence is made possible by spin–orbit coupling due to the Pd impurities. The separate luminescent processes are resolved in the transient EL spectra of OLEDs fabricated

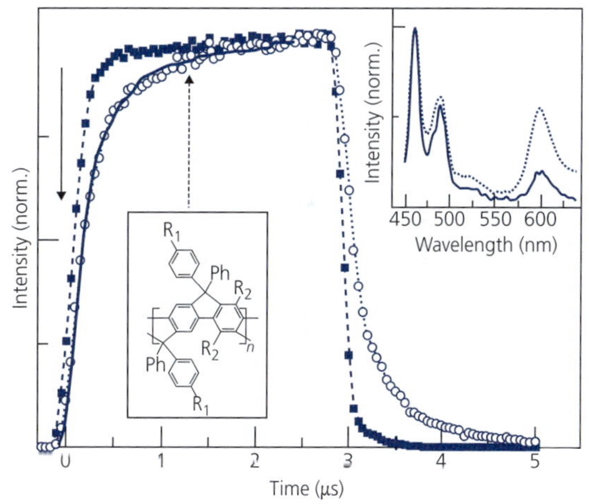

Figure 6.27 Singlet (filled symbols) and triplet (open symbols) EL response of the ladder-type polymer, PhLPPP whose molecular structural formula is shown in the lower inset. The arrows indicate the times that the two spectra in the upper inset are recorded. The S emission has a peak at a wavelength of 480 nm, and T emission is at 600 nm (Reufer et al., 2005).

using this compound. The phosphorescent lifetime puts a lower limit to the triplet formation time of $k_T^{-1} > 100$ns, which is primarily due to triplet diffusion to the widely separated Pd sites. That is, the phosphorescent and fluorescent signals occur on significantly

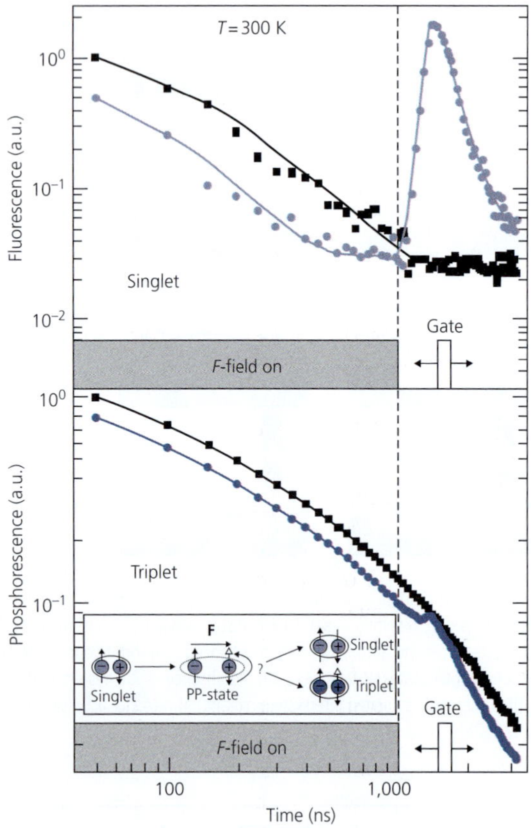

Figure 6.28 Fluorescent (top) and phosphorescent (bottom) emission transients in the absence (black symbols) and presence (color symbols) of an electric field. The electric field separates the exciton state into a weakly interacting polaron pair, and then the field is switched off after 1 μs to determine if the spin state of the singlet has changed. The process of singlet polarization and recombination is illustrated in the inset, bottom panel (Reufer et al., 2005).

different times scales (Fig. 6.27, upper inset). As expected for singlet and triplet splitting, phosphorescence is at lower energy, and hence at a longer wavelength than fluorescence.

To determine the strength of singlet–triplet mixing in the PP state, the samples were illuminated under reverse bias. A strong electric field (>1 MV/cm) physically separates (via polarization) the singlet into a *PP* (inset, Fig. 6.28) where the charges are held for the duration of the voltage pulse. In this state, the orbital overlap is minimal, preventing exchange interactions. After a delay greater than 100 ns (the triplet formation time), the field is turned off, and the *PP* is allowed to once more form an exciton. Essentially, this experiment allows for singlet state production, and then reverses the process by an applied field. Lastly, the field is removed and recombination occurs once again. Once the field is turned off, there is a burst of fluorescence and phosphorescence, as shown in the delayed emission in Fig. 6.28.

The fluorescence signal is 1000 times stronger than that of phosphorescence. Indeed, there is almost no detectable phosphorescence even at 4 K, although the PP state is maintained for 10 times longer than the triplet formation time. It is, therefore, concluded that spin mixing is absent in the PP state and the statistical 1:3 singlet:triplet formation ratio in this conjugated polymer is preserved over the timescale of the F-field delay.

We note that degeneracy of the singlet and triplet CT manifolds might upset the perfect statistical balance, depending on the ratio k_S/k_T as in Fig. 6.26 (Difley et al., 2008). Indeed, it has been shown that CT spin state formation in degenerate systems can be controlled by CT formation at material interfaces in an extrafluorescent OLED, resulting in an apparent increase in the singlet emission efficiency (Segal et al., 2007). While the experiment in Fig. 6.28 suggests that the formation statistics do not change over time, direct and accurate measurement of these statistics has been challenging (Baldo et al., 1999b, Cao et al., 1999, Kim et al., 2000, Wilson et al., 2001, Segal et al., 2003). The primary assumption of most such measurements is that the singlet fraction can be determined from the relationship between the EL and PL efficiencies in Eq. 6.1. Hence, once the EL efficiency in the OLED is measured, it would a appear to be a simple matter of independently determining the PL efficiency, the losses specific to EL emission, and the outcoupling efficiency. From this, it is straightforward to calculate the exciton formation ratio, χ_{ST}. The accurate measurement of these several parameters within an OLED structure, however, can lead to significant inaccuracies. Also, effects such as exciton annihilation where two singlets combine to form one singlet or triplet, or similarly where two triplets can annihilate to form one triplet or singlet, can also be active under high electrical injection conditions. Many of these processes are dependent on factors including device structure, voltage, current and temperature. Hence, measurement of the spin formation statistics has often been the subject of controversy. Nevertheless, its accurate determination is necessary to understand the exciton formation process under electrical injection.

The greatest accuracy is achieved by reducing the number of absolute measurements of the parameters in Eq. 6.1. Then, χ_{ST} can be obtained by measuring the ratio of the EL to PL efficiencies for an OLED in a single experimental configuration (Segal et al., 2003). We now outline an approach that has been used to successfully determine χ_{ST} for both polymeric and small molecular weight materials.

In fluorescent materials, almost 100% of the species generated on optical excitation are singlets (although this too can lead to ~1–5% error, see Section 6.8). Hence, the PL efficiency of a sample provides a reference from which we obtain χ_{ST} from η_{ext}. The PL power emitted by an OLED is given by (Segal et al., 2003)

$$P_{PL} = \frac{hc}{\langle\lambda_{em}\rangle}\eta_{out}\phi_{ph}\frac{k_r}{k_r + k_{nr} + k_Q}, \qquad (6.25)$$

where k_r, k_{nr}, and k_Q are the radiative, non-radiative, and electric field quenching rates, respectively, $\langle\lambda_{em}\rangle$ is given by Eq. 6.3, and ϕ_{ph} is the photon flux *absorbed* by the film. Here, we assume that the PL power is measured in the presence of an electric field across the same OLED structure used to determine its external quantum efficiency, and hence we cannot ignore the current injection-specific losses, γ_L, due, for example, to ionization, quenching at defects, etc. Also, the ratio of rates in Eq. 6.25 is the emission efficiency. Thus, $k_r\eta_{out}\phi_{ph}/(k_r + k_{nr} + k_Q)$ is the photoluminescent flux outcoupled from the device (which depends on the thin film structure) and into the detector, which is a function of the experimental setup.

Under weak reverse bias where $k_Q \ll (k_r + k_{nr})$, an OLED under illumination, generates a photocurrent, j_{ph}. This is proportional to the optical pump intensity, that is,

$$j_{ph} = q\phi_{ph}\frac{k_Q}{k_r + k_{nr} + k_Q} \approx q\phi_{ph}\frac{k_Q}{k_r + k_{nr}}. \qquad (6.26)$$

The difference in PL efficiency in the presence and absence of an electric field gives

$$\begin{aligned}\Delta P_{PL} &= P_{PL}(k_Q) - P_{PL}(k_Q = 0)\\ &\approx -\frac{hc}{\langle\lambda_{em}\rangle}\eta_{out}\eta_{PL}\phi_{ph}\frac{k_Q}{k_r + k_{nr}},\end{aligned} \qquad (6.27)$$

where the PL efficiency in the low field limit is $\eta_{PL} = k_r/(k_r + k_{nr})$. Combining Eqs. 6.26 and 6.27 yields:

$$\eta_{out}\eta_{PL} \approx -\frac{q\langle\lambda_{em}\rangle}{hc}\frac{\Delta P_{PL}}{j_{ph}}. \qquad (6.28)$$

Thus, by measuring the *change* in photoluminescent power efficiency and the photocurrent, we obtain the product of the PL and optical coupling efficiencies. The EL efficiency measured under forward bias is, from Eq. 6.2,

$$\eta_{ext} = \frac{hc}{q\langle\lambda_{em}\rangle}\frac{P_{EL}}{j_{OLED}}. \qquad (6.29)$$

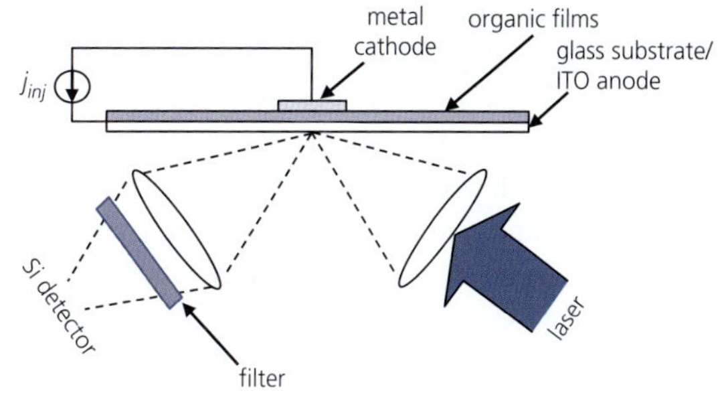

Figure 6.29 Experimental setup used in the measurement of OLED PL and EL using a single optical system for both (Segal et al., 2003).

Then, from Eq. 6.1 we arrive at the spin statistics using

$$\gamma\chi_{ST} = \frac{\eta_{ext}}{\eta_{PL}\eta_{out}} = -\left(\frac{P_{EL}}{\Delta P_{PL}}\right)\frac{j_{ph}}{j_{OLED}}. \qquad (6.30)$$

If the PL and EL optical powers are measured in the same experimental configuration, then the coupling efficiency to the detector remains unchanged for the measurement of both quantities, removing a significant potential source of error. Figure 6.29 shows such an experimental setup whereby the OLED PL is obtained by pumping with a laser source, while its EL is independently obtained by electrical injection. For both experiments, a single optical system comprising a lens and a filter to remove the laser emission is used.

From Eq. 6.30, the only uncertainty remaining is the determination of EL-specific losses, γ. The primary source of loss is imperfect charge balance in the OLED EML. As discussed in Section 6.2.1, employing blocking layers, particularly at the cathode side of the EML, can ensure unity charge balance. Another source of loss is triplet–polaron annihilation (TPA); this can be reduced or nearly eliminated by measuring the EL intensity at low current densities before there is any evidence of roll-off, as in Fig. 6.22 at $j > 2$ mA/cm^2. If TPA is insignificant, η_{ext} vs. j_{OLED} should be constant over the range of currents used.

It is also important to eliminate recombination of counter charges at traps. That is, a charge injected into the EML may become trapped and then non-radiatively recombine with its opposite at a defect. There are several means to eliminate this source of error. Since the trap density should be far lower than the dopant density within the EML, traps would be expected to fill at high currents. That is, the losses via traps will saturate, leading to an increase in η_{ext} with j_{OLED}, although the current may be sufficiently high to introduce annihilation, as above. Furthermore, since traps lie within the energy gap of the guest and/or host in the EML, their contributions to losses should be temperature dependent. Hence, as T is reduced, traps may become more or less active, depending on their energy levels. Models for these various effects have been developed and provide a reasonable approach to limiting potential errors. Employing this analysis, it has been found the singlet fraction in the archetype fluorescent small molecule Alq_3, and the polymer MEH-PPV is, to within experimental error, 25% (Baldo et al., 1999b, Segal et al., 2003).

Another means to measure spin statistics is to use a structure where exciton transfer to a fluorescent dopant from a fluorescent host (the subject of the measurement) occurs in the presence or absence of a phosphorescent "sensitizing" molecule. The fluorescent dopant in the absence of the phosphor emits only via singlet recombination. The phosphor sensitizer doped into the host harvests both singlets and triplets in the host, followed by conversion of singlets to the radiative triplet manifold via ISC. If a fluorescent guest is also doped in the same region as the phosphor, the sensitizer can transfer its excitation to the guest which is now capable of emitting at 100% efficiency from both host singlets and triplets (see Fig. 6.30). The phosphor sensitizer is chosen to ensure rapid Förster transfer to the fluorescent dopant. Hence, measurement of the relative intensity of the fluorophore-only device to that employing both the phosphor and the fluorophore provides the ratio of singlets to the total number of excitons generated in the host, thus providing a direct and uncomplicated measurement of χ_{ST}.

A schematic of a device that exploits the sequential transfer of excitons from host to phosphor to fluorophore, is shown in Fig. 6.30a. This process is known as *phosphor sensitized fluorescence*. It was first proposed by Förster (1959), and is used in OLEDs to generate 100% internal quantum efficiency via the use of fluorescent dyes (Baldo et al., 2000b). This process results in the transfer of all electrically excited excitons (regardless of spin) in a host by a high energy phosphor, to a fluorescent dopant where they can radiatively recombine at the wavelength determined by the fluorophores. In Fig. 6.30, the blue fluorescent CBP host (emitting with a peak wavelength of $\lambda = 400$ nm) and the green $Ir(ppy)_3$ phosphor is combined in the EML with the red fluorescent dye DCM2. DCM2 absorbs in the green and emits at wavelengths between $\lambda = 570$ nm and 650 nm. The ETL is Alq_3, the hole and exciton blocking layer is BCP, and the HTL is TPD.

The transfer process in the phosphor sensitized fluorescent OLED is shown in Fig. 6.30b. Both singlets and triplets are generated on the host, and subsequently transferred to the phosphor whose S and T energies are lower than that of the host. Singlets transfer via the Förster process, whereas triplets transfer via exchange, or Dexter interactions. Hence, the concentration of the phosphor dopant must be sufficiently high to allow for direct contact between the host donor and phosphor acceptor molecules, which typically ranges from 8% to 10%. The phosphor is codoped with the fluorescent "target" molecular species. The transfer from phosphor to the dopant is again via the Förster mechanism. Hence, the fluorophore concentration must be sufficiently high to allow for efficient Förster transfer but low enough ($\leq 1\%$) to discourage direct Dexter transfer of triplets from the host to the fluorophore (Baldo et al., 2000b, D'Andrade et al., 2001).

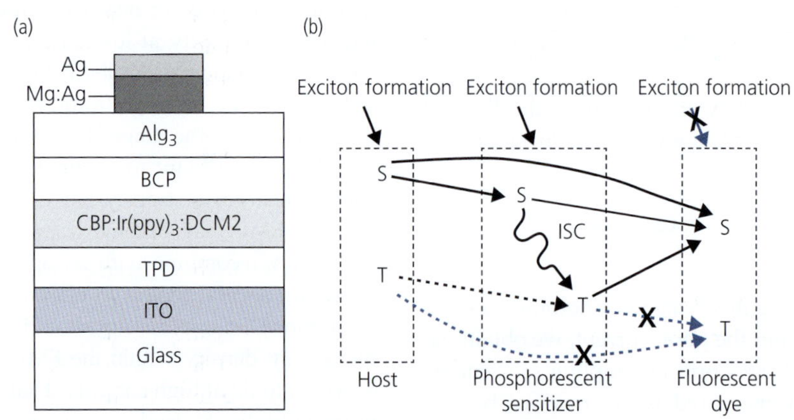

Figure 6.30 Measuring spin formation statistics in CBP. (a) An OLED with an emission zone employing a CBP fluorescent host, doped either with 0.2% of the fluorescent molecule DCM2, or with 8% $Ir(ppy)_3$ + 0.2% DCM2. (b) Exciton formation pathways via electrical injection are shown, as well as energy transfer pathways for both the singlet (S) and triplet (T) manifolds. The relative energies are indicated by vertical displacements of S and T.

To quantify χ_{ST} in the device in Fig. 6.30, we rewrite Eq. 6.1 for the external EL quantum efficiency as

$$\eta_{ext} = \left[\chi_S \eta^f_{PL} \eta_S + \chi_T \eta^p_{PL} \eta_T\right] \gamma \eta_{out}. \tag{6.31}$$

Here, $\chi_T = (1 - \chi_S)$ is the fraction of triplets and χ_S is the singlet fraction. Also, η^f_{PL} and η^p_{PL} are the PL efficiencies of fluorescence and phosphorescence of the acceptor, η_S and η_T are the transfer efficiencies of singlet and triplet excitons from donor to acceptor.

By quantitatively comparing the EL efficiencies (η^f_{ext} and η^p_{ext}) of separate devices employing either the fluorescent or phosphorescent dye, we can use Eq. 6.31 to determine χ_S and χ_T provided that γ and η_{out} are identical in both devices. As noted above, this can be a source of error since it relies on an identical setup for both experiments, and the outcoupling efficiencies of both devices must also be the same. This latter requirement can be particularly challenging since the layer structures of fluorescent and phosphorescent OLEDs are inherently different, leading to differences in the optical field within the OLEDs.

If all singlets are transferred (i.e. $\eta_S = \eta_T \sim 1$), we can obtain χ_S from the ratio of fluorescent and phosphorescent efficiencies to obtain (Baldo et al., 1999b)

$$\chi_S = \frac{\eta_T}{\left(\dfrac{\eta^p_{ext}}{\eta^f_{ext}} \cdot \dfrac{\eta^f_{PL}}{\eta^p_{PL}}\right) - (1 - \eta_T)}. \tag{6.32}$$

Since the ratios reflect relative measurements, the only absolute measurement required is the triplet state transfer efficiency from the host to the phosphor (i.e. η_T).

The efficiency of the phosphor sensitized fluorescent device in Fig. 6.30a is shown in Fig. 6.31. Here the sensitized device, consisting of CBP doped with 8 wt% $Ir(ppy)_3$ and 0.2% DCM2 has $\eta_{ext} = 6\%$ at $j = 0.1\ \mathrm{mA/cm^2}$, decreasing to 3% at $j = 500\ \mathrm{mA/cm^2}$. The roll off is due to *triplet–triplet annihilation* (TTA, see Section 6.3.5). For an analogous fluorescent device doped only with 0.2% DCM2, the efficiency is only 1.3% at $j = 0.1\ \mathrm{mA/cm^2}$ which is characteristic of this particular compound. This decreases to 0.6% at $500\ \mathrm{mA/cm^2}$ (D'Andrade et al., 2001). In the sensitized device, both triplets and singlets are harvested, whereas in the DCM2-only device, singlets are harvested. Hence, the ratio of these two efficiencies gives the singlet fraction of $\chi_S = 1.3/6 = 0.22$, independent of current density, assuming complete energy transfer. The energy transfer efficiency can also be obtained from the relative fluorescence and/or phosphorescence intensities in the emission spectra of the various devices. Should both spectral features coexist in a single device, their relative intensities can be used to infer η_S/η_T. The singlet fraction obtained from phosphor sensitized fluorescence is consistent with the results obtained using the PL/EL experiments applied to MEH-PPV and Alq_3 in Fig. 6.29.

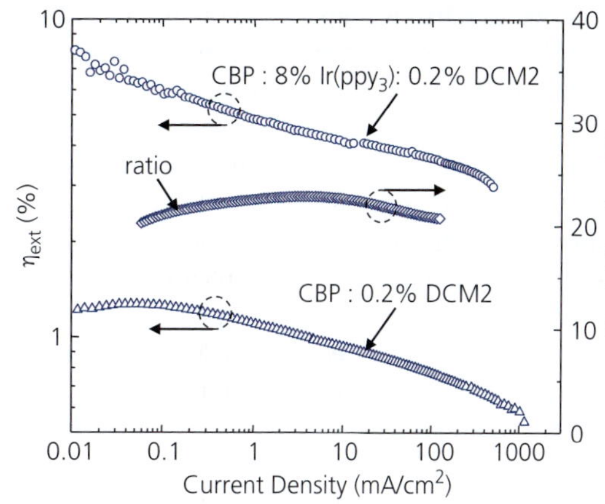

Figure 6.31 External quantum efficiency vs. current density for the fluorescent as well as for the phosphor sensitized fluorescent OLEDs. The ratio of the quantum efficiencies between these two devices is a direct measurement of the singlet fraction (D'Andrade et al., 2001, Segal et al., 2003).

Reprinted figure with permission from Segal, M., Baldo, M. A., Holmes, R. J., Forrest, S. R. & Soos, Z. G. 2003. Excitonic singlet-triplet ratios in molecular and polymeric organic materials. Phys. Rev. B, 68, 075211 Copyright 2003 by the American Physical Society.

Hence, we have seen that fluorescent OLEDs based on small molecules (CBP and Alq_3) and polymers (MEH-PPV) have a singlet–triplet production ratio of 1:3. Indeed, there do not appear to be convincing examples where that ratio is violated in other fluorescent systems, although other measurement approaches have led to a diversity of singlet fractions, ranging from 15% to 60% (Cao et al., 1999, Wilson et al., 2001, Wohlgenannt et al., 2001, Wohlgenannt and Vardeny, 2003). These latter approaches, however, are subject to greater uncertainties than those described above, and hence are not considered further.

6.3.2 Fluorescent OLEDs

The first low voltage OLED was a bilayer fluorescent device based on small molecules (Tang and VanSlyke, 1987), followed three years later by a fluorescent polymer OLED (Burroughes et al., 1990). These technological breakthroughs were preceded by the observation of EL in a film of brilliant acridine orange E (II) (Bernanose and Vouaux, 1953, Bernanose, 1955), and later in a pure anthracene crystal, and one doped at 0.1 mol. % with tetracene (Pope et al., 1963). In the anthracene devices, contacts were made using silver paste, hence this device was capable of unipolar charge injection that was space-charge limited, following $j \sim V^{10}$. At 400 V and at a large current density of 100 μA/cm^2, anthracene luminescence in the undoped sample was observed, whereas tetracene fluorescence was observed in doped anthracene. Due to the very high currents and voltages necessary to observe weak EL from unipolar injection, the phenomenon was attributed to recombination following *impact ionization* of the anthracene molecules. The flood of electrons liberated on impact by high energy injected electrons recombine with holes left behind on the ionized molecules. Luminescence was achieved at even higher voltages (2000 V) applied to the acridine derivatives, and was likely due to similar field ionization effects.

Bimolecular recombination in a 1–5 mm thick anthracene crystal due to injection of electrons and holes from opposite electrolyte cathodes and anodes, respectively, resulted in emission of intense blue fluorescence (Helfrich and Schneider, 1965). The *I*–*V* and luminance voltage characteristics are shown in Fig. 6.32. The light intensity was proportional to bipolar space charge limited current. It was estimated that the blue emission quantum efficiency due to bimolecular recombination was ~100%. The internal resistance of the large anthracene crystals resulted in operating voltages >100 V.

The demonstration of bipolar injection into an anthracene crystal arguably represents a structure similar to a polymer OLED (Helfrich and Schneider, 1965). However, anthracene has a much higher hole than electron mobility, thereby requiring very high voltage operation. Furthermore, OLEDs are thin film devices as opposed to free-standing single crystals. This is an important distinction, since an OLED is interesting primarily because it has the potential for large-scale integration in displays or lighting. The thin films can be applied over large surface area glass, plastic, metal or a range of other substrate materials that provide support as well as mechanical durability. This feature cannot be replicated by bulk crystals that are typically very small (a few cubic millimeters, see Chapter 2) and fragile.

A significant advance was, therefore, the demonstration of the bilayer OLED. The small molecular weight device used an aromatic diamine HTL (which is similar to the high hole mobility TPD molecule, shown in Fig. 6.3a) and an emissive electron transporting Alq_3 layer sandwiched between a low work function Mg:Ag alloy cathode and an ITO anode (Tang and VanSlyke, 1987). Radiative recombination occurs on Alq_3 with a PL quantum yield of ~30% (Garbuzov et al., 1996) emitting intense green fluorescence at a peak wavelength of $\lambda = 550$ nm. The entire device comprised

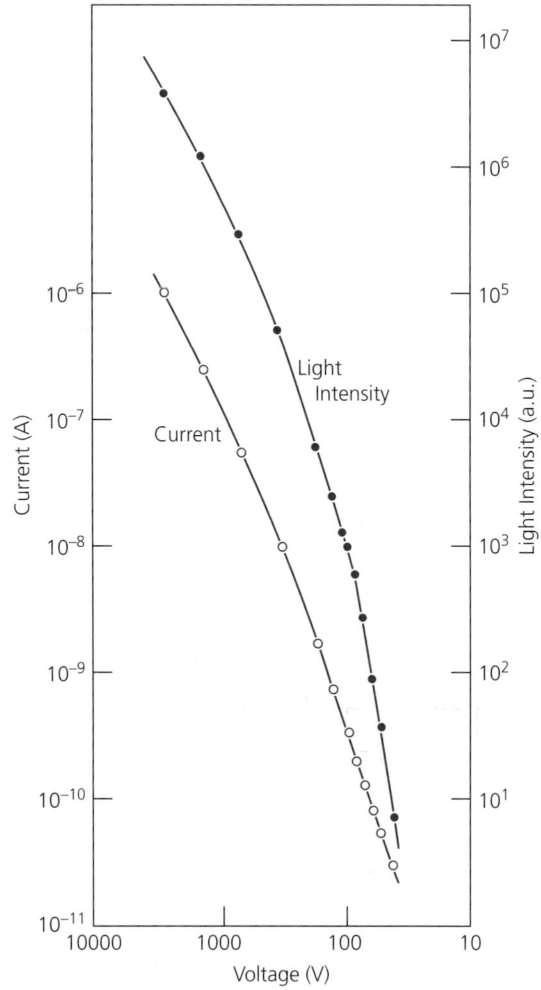

Figure 6.32 Current and luminous intensity vs. voltage due to bimolecular recombination of electrons and holes in an anthracene single crystal (Helfrich and Schneider, 1965).

amorphous organic thin films deposited by vacuum thermal evaporation. In the first reported bilayer OLED, the external quantum efficiency was ~1%. Importantly, the voltage needed to achieve an easily visible emission was only ~ 5V.

6.3.2.1 Exciton diffusion and confinement in multilayer small molecule OLEDs

A disadvantage of the bilayer structure in Fig. 6.3 is that it requires individual layers to serve multiple purposes, for example, both light emission and charge conduction. Further, single layer or bilayer (single HJ) devices lack a region that confines a high density of excitons required for bright light emission. The exciton distribution in the bilayer device, nevertheless, is far more confined than in earlier, single layer structures.

The exciton density distribution, and its effects on the emission properties of OLEDs was first explored in the bilayer and triple layer devices shown in Fig. 6.33 (Adachi et al., 1991). The exciton density distribution was inferred by placing a 1 nm thick layer of the dopant, thiadiazole, that forms a charge transfer complex with the host (NSD) emitting at 630 nm (Fig. 6.33) at different distances from the contacts or HJs, and then measuring the emission from that fluorophore. The thiadiazole thus acts as a *sensor layer* whose emission is distinct from the layer in which it is embedded, and whose intensity is proportional to the local exciton concentration. Figure 6.33a shows a bilayer, single HJ OLED (SH-A) comprised of a TPD HTL and a naphthylstyrylamine derivative (NSD) EML that is preferentially electron conducting. The second device in Fig. 6.33a (SH-B) replaces the HTL with a PBD ETL. Finally, the double heterostructure (DH) comprises a sandwich with the emitting layer placed between an independently optimized ETL and HTL. The cathodes consist of a low work function Mg-Ag contact, and the transparent anode was ITO (Adachi et al., 1990).

The exciton distribution obtained from the relative EL intensities of the thiadiazole and the transport layers in a single layer device in Fig. 6.33b is spread over 75 nm, with the peak nearest to the cathode due to the larger hole mobility of the NSD EML compared to electron mobility of the PBD ETL. The shape of the distribution is exponential due to the dependence of the exciton density ($N(x)$) on its diffusion length (L_D) given by

$$N(x) = N_0 \exp(-x/L_D). \tag{6.33}$$

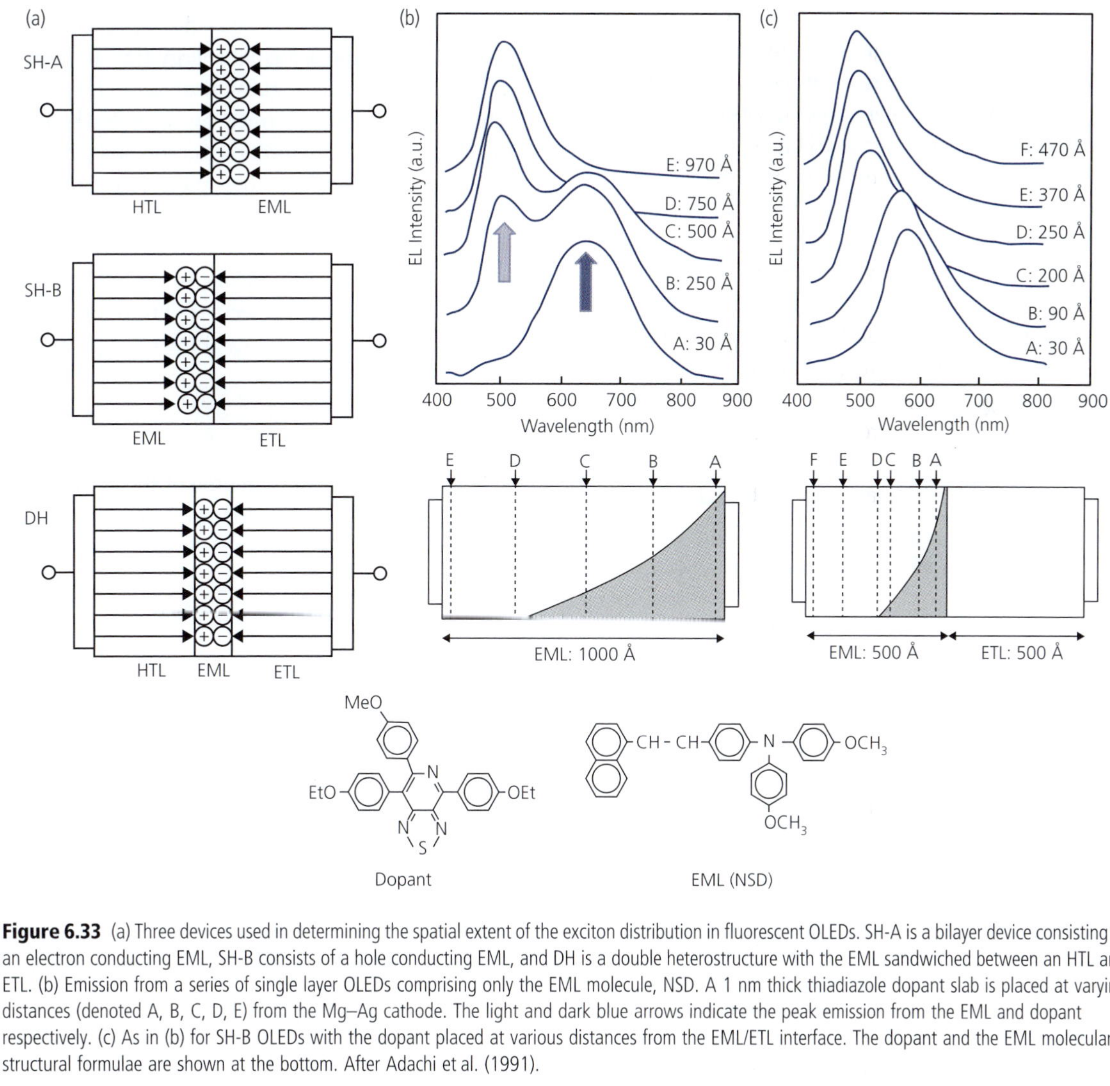

Figure 6.33 (a) Three devices used in determining the spatial extent of the exciton distribution in fluorescent OLEDs. SH-A is a bilayer device consisting of an electron conducting EML, SH-B consists of a hole conducting EML, and DH is a double heterostructure with the EML sandwiched between an HTL and ETL. (b) Emission from a series of single layer OLEDs comprising only the EML molecule, NSD. A 1 nm thick thiadiazole dopant slab is placed at varying distances (denoted A, B, C, D, E) from the Mg–Ag cathode. The light and dark blue arrows indicate the peak emission from the EML and dopant respectively. (c) As in (b) for SH-B OLEDs with the dopant placed at various distances from the EML/ETL interface. The dopant and the EML molecular structural formulae are shown at the bottom. After Adachi et al. (1991).

Here, N_0 is the exciton density at the point of origin. The dopant emission dominates the spectrum at a distance of only 50 nm from the cathode, indicating that excitons generated beyond this point do not significantly contribute to the luminance. This broad distribution also results in a luminance that is 10^4–10^6 times less than for SH-A and SH-B, respectively. The more confined distribution using a hole conducting EML is apparent for SH-B, as shown in Fig. 6.33c. Here, the exciton population extends only as far as 20 nm. The emission is almost entirely from the dopant at longer distances, and the device is far brighter than the single layer device. A similar result is obtained for SH-A, which is the most efficient of the several structures investigated.

These observations suggest that almost all excitons are generated within 5–10 nm of the charge transport/EML or injecting contact/EML interfaces in bilayer and single layer devices, respectively. This conclusion is supported by comparing the luminance of the DH and SH-B devices as functions of EML thickness in Fig. 6.34. Since excitons in the DH structure are confined by the larger energy gaps (and hence energy barriers) of the ETL and HTL layers in the EML, quenching is eliminated. Indeed, the lack of dependence of the luminance on layer thicknesses (in contrast to the strong dependence of the SH-B device) as small as 5 nm suggests that excitons are formed within only a few monolayers of the injecting interfaces. This is apparently the first example where a double heterostructure was used to achieve balanced hole and electron injection (i.e. where $\gamma \to 1$) where injected charges are confined by the energy barriers surrounding the EML (Adachi et al., 1990).

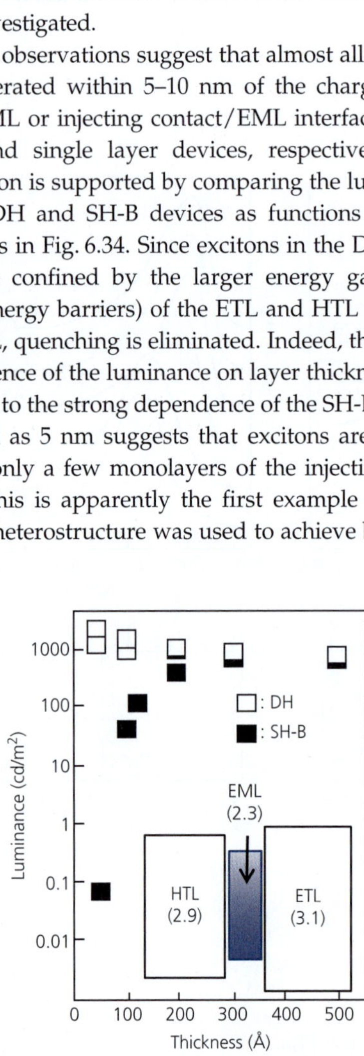

Figure 6.34 Luminance of DH and SH-B devices vs. EML thickness. Inset: Relative energies (in parentheses) of the ETL, HTL and EML taken from the EL spectra of OLEDs. Since the HOMO and LUMO energies are unknown relative to vacuum, a space is left between each material indicating only the magnitude of the optical energy gaps (Adachi et al., 1990).

6.3.2.2 Energy transfer and luminescent layer doping in fluorescent OLEDs

An important advance toward high efficiency and long lifetime was achieved by doping the EML with a fluorescent acceptor guest molecule in a donor host (Tang et al., 1989). This separates charge conduction on the host from emission on the guest, thereby allowing for independent optimization of the materials to carry out their separate functions. The exciton can be formed on the conductive host molecule, in which case complete transfer to the guest is required, or it can form directly on the emissive guest molecule itself. Thus, there are three mechanisms for exciting the guest molecule:

(i) *Fluorescence (or Förster) resonant energy transfer (FRET)*: This is the principal energy transfer mechanism in fluorescent guest–host OLED systems where the exciton is formed by recombination of the injected electron and hole on the host (donor), and then transferred to the guest (acceptor). As discussed in Section 3.8.1, FRET requires that the host molecule have a high radiative efficiency, and hence is active in the transfer of singlet excitons between fluorescent molecules. From Eqs. 3.233 and 3.234, the rate of energy transfer is given by

$$k_{ET} = \frac{1}{\tau_D}\left(\frac{R_0}{R_{DA}}\right)^6, \tag{6.34}$$

where τ_D is the natural lifetime of the donor, R_{DA} is the distance between donor and acceptor molecule, and R_0 is the Förster radius:

$$R_0^6 = A_\lambda \int \lambda^4 f_D(\lambda)\alpha_A(\lambda)d\lambda. \tag{6.35}$$

Here, $f_D(\lambda)$ is the emission spectrum of the donor, $\alpha_A(\lambda)$ is the absorption spectrum of the acceptor molecule, and A_λ is a constant. When $R_0 = R_{DA}$, the transfer efficiency is 50%. As in Chapter 3, $R_0 =$ 3–5 nm is typical for many fluorescent guest–host systems. Since a common molecular dimension is ~1 nm, concentrations of 1–3 wt% of the dopant are sufficient to achieve a very high FRET efficiency.

(ii) *Exchange (or Dexter) energy transfer*: When the host exciton is non-emissive, direct contact between guest and host molecules allows for exchange energy transfer. This is not a useful process for fluorescent molecules since such a high density of guest molecules (~5–10 wt%) leads to *concentration quenching* of the luminescence. This is due to intimate coupling between nearest neighbor molecules that opens up non-radiative recombination channels, such as phonon-assisted relaxation of the excited state. Also, at low doping, there is a reduced opportunity for excitons on adjacent excited molecules to annihilate; a phenomenon that becomes particularly pronounced at high intensities. However, transfer of the host triplet and singlet states to a triplet-emitting molecule such as a transition metal phosphorescent complex can occur by the spin-conserving Dexter mechanism (Baldo et al., 1998). In. Section 6.3.3, we showed that the phosphor dopant concentration must be considerably higher than that used for fluorescent OLEDs to achieve high efficiency. Again, referring to Section 3.8.2, the exchange transfer rate is given by (Eqs. 3.242 and 3.243)

$$k_{ET} = \frac{2\pi}{\hbar}\Gamma^2 \int f_D(\lambda)\sigma_A(\lambda)d\lambda, \tag{6.36}$$

where the orbital overlap between donor and acceptor is given by

$$\Gamma^2 \propto \frac{q^2}{(4\pi\varepsilon_0)^2 n_r R_{DA}^2} \exp(-2R_{DA}/L). \tag{6.37}$$

Here, L is the intermolecular spacing as determined by van der Waals forces of the molecules, and n_r is the refractive index. Hence, the rate of exciton transfer falls off exponentially with distance between guest and host.

(iii) *Direct charge trapping on the dopant:* Direct charge injection onto the emissive guest molecule from the ETL and/or HTL can result in exciton formation that bypasses the host entirely. Similarly, the host molecule can be used to transport one of the charge carriers to the emitting guest molecule, taking advantage of the higher mobility of the host than the guest for that particular charge. Charge trapping is illustrated in Fig. 6.35. Trapping can excite either fluorescent or phosphorescent molecules since the processes are spin-independent. However, the exciton formation statistics are the same for direct charge trapping as for exchange and FRET processes. That is, 25% of excitons formed by trapping will be singlets, with the remainder being triplets. This excitation mechanism is particularly prevalent in blue emitting OLEDs and PHOLEDs, where the choice of high energy hosts that can efficiently exothermically transfer excited states from host to guest is limited (Holmes et al., 2003a).

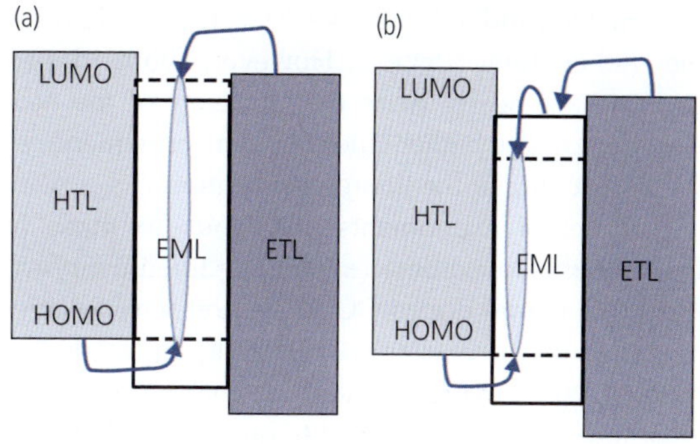

Figure 6.35 Two example direct trapping processes. (a) Electron and hole trapping from the LUMO of the ETL and the HOMO of the HTL onto the LUMO and HOMO levels of the dopant (dashed lines). (b) Trapping from the HOMO of the HTL to that of the dopant, and direct trapping from the host (solid line in the EML) LUMO to that of the guest.

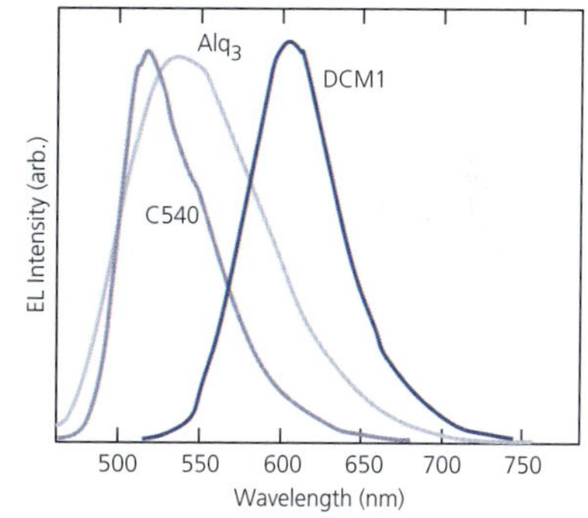

Figure 6.36 Spectra of C540 and DCM1 doped fluorescent OLEDs with an Alq_3 host compared with the emission spectrum of neat Alq_3 (Tang et al., 1989).

Figure 6.36 shows the OLED EL spectra due to FRET in an EML comprising an electron conducting Alq_3 host doped with 1 mol. % of the green fluorophore, coumarin 540, or 0.25 mol. % of the red emitting, DCM1. The OLED emission from the doped devices is compared with that from an undoped Alq_3 EML. The EML is 60 nm thick, and is sandwiched between a Mg:Ag cathode and a 75 nm thick TAPC HTL on an ITO anode. The lack of Alq_3 emission from the DCM1 OLED indicates complete Förster energy transfer, even at only 0.25 mol. % dopant concentration. However, a small shoulder on the long wavelength edge of the coumarin 540 spectrum indicates residual Alq_3 emission. The higher energy of this dopant reduces the overlap between its emission spectrum and the absorption spectrum of the host, thus making FRET somewhat less efficient than for the red-emitting device. However, both doped OLEDs have nearly twice the efficiency of the neat Alq_3 device. The higher efficiency is a consequence of the low doping of the fluorescent centers.

In a set of experiments analogous to those in Fig. 6.33, the doped devices were modified by inserting a thin (4 nm) doped DCM2 exciton sensing layer at various distances from the HTL/Alq_3 interface. The ratio of green emission intensity from Alq_3 to red emission from DCM1 is used to quantitatively determine the exciton diffusion length, L_D, from the interface using (Tang et al., 1989)

$$\eta_{ext}(\lambda) = (1 - \exp(-x/L_D))\eta_{Alq}(\lambda) + \exp(-x/L_D)\eta_{DCM2}(\lambda). \quad (6.38)$$

Here, η_{Alq} and η_{DCM2} are the external quantum efficiencies of the neat Alq_3 and the Alq_3/DCM2 doped OLEDs, respectively. Using this procedure, an exciton diffusion length of 20 nm is obtained for Alq_3. This is somewhat longer than obtained via more direct methods, where $L_D \sim$ 8–15 nm is typical of singlet excitons in several fluorescent materials (Wu et al., 2005, Lunt et al., 2009). The difference may be due to uncertainties in sensing layer thickness, exciton generation at sites near, but not at the HJ, charge trapping by the low energy LUMO or HOMO of the dopant compared to the host, or quenching at electrodes.

Doping of a luminescent molecule into a host can result in spectral shifts arising from the solid state solvation effect (SSSE, see Section 3.6.7). Here, the dipole field from a host molecule interacts with the dipole of the emitting molecular species, changing the local electric field. This results in a change in the relative stability of the ground and excited states, leading to a concomitant change in the exciton energy. Devices based on the dipolar fluorophore, DCM2 (with a dipole moment of $p = 11$ D), exhibited a rigid red shift attributed to solvation of the dipolar DCM2 in the Alq_3 matrix ($p = 5.5$ D) (Bulovic et al., 1998c). Since that early experiment, SSSE has been observed in numerous fluorescent systems (Green et al., 2013, Chang et al., 2015).

Color tuning of OLEDs via solvatochromism is shown in Fig. 6.37. The device in the inset consists of DCM2 mixed into the relatively non-polar TPD HTL ($p = 1.5$ D). A rigid and dramatic bathochromic shift of 0.25 eV in the OLED EL spectra as the relative concentration of DCM2, Q, increases from 0.009 to 0.11 results from an increase in the average dipole moment of the binary mixture (Bulovic et al., 1999).

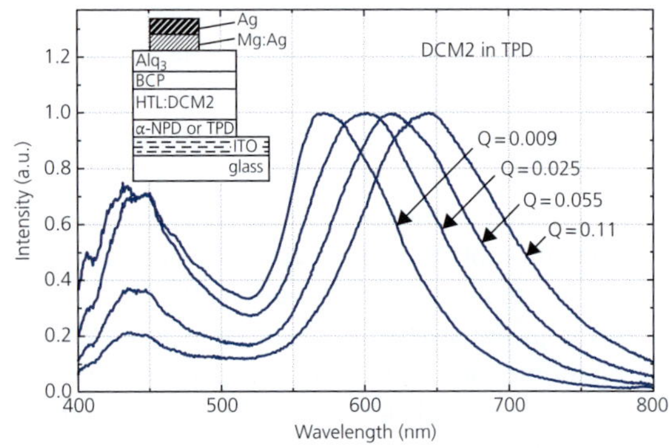

Figure 6.37 Solid state solvation effect observed in the EL spectrum of an OLED with different fractions (*Q*) of the red emitting fluorophore, DCM2 in TPD. The peak at 420 nm is due to TPD emission resulting in incomplete energy transfer to DCM2 (with peak emission at 575 nm) as *Q* is decreased (Bulovic et al., 1999).

Reprinted from Chemical Physics Letters, 308, Bulovic, V., Deshpande, R., Thompson, M. E. & Forrest, S. R., Tuning the color emission of thin film molecular organic light emitting devices by the solid state solvation effect., 317-322, 1999, with permission from Elsevier.

At the smallest concentrations, the mixture is only weakly polar due to the TPD, but as the DCM2 fraction increases, the emitting molecule is in an increasingly dipolar environment due to nearby DCM2 molecules.

Including Alq_3 as a second dipolar dopant into the EML increases the color tuning range. The peak emission wavelength as a function of both DCM2 and Alq_3 fraction (Z) in TPD is shown in Fig. 6.38a. The spectral peak shifts from 570 nm at low DCM2 concentration in TPD, to as long as 650 nm for DCM2 in Alq_3. In all devices, the DCM2 spectral shift was rigidochromic, that is, there was no apparent spectral broadening as the peak position changes. *Rigidochromism* is a distinguishing characteristic of SSSE.

Since Alq_3 emits in the green, and TPD in the blue, a large color gamut extending across a significant fraction of the visible spectrum (including the isochromic white point) is accessible by variations of concentration of the molecular constituents (see Fig. 6.38b). However, the quantum efficiency is also a function of DCM2 and Alq_3 concentrations. The peak $\eta_{ext} = 0.7\%$ at $j = 1$ mA/cm^2 and $Z = 0.2$, but falls to 0.2% at $Z = 0.1$ for a DCM2 concentration of 1.5%. The loss in η_{ext} is due to the reduced Förster energy transfer efficiency from TPD to Alq_3 to DCM2 at low Z as the Förster radius becomes comparable to the molecular spacing. At higher Alq_3 and TPD concentrations than $Z = 0.2$, the reduced efficiency results from increased concentration quenching

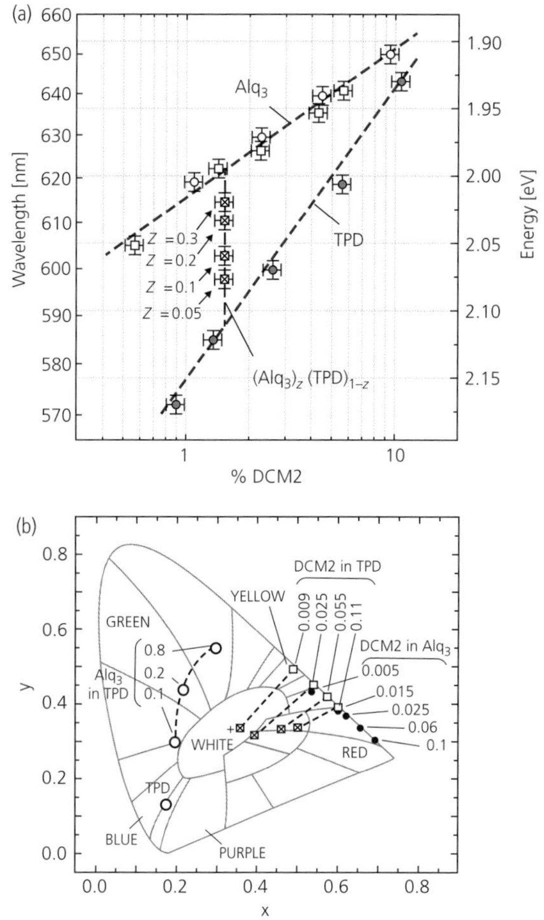

Figure 6.38 (a) Color tuning the OLED EL emission spectrum by varying the relative concentrations of DCM2 and Alq_3 in a TPD hole transport layer. *Z* is the Alq_3 fraction. The TPD is weakly polar. The wavelength can be tuned from 573 nm to 650 nm by the addition of the more polar DCM2 and Alq_3. (b) Colors accessible as a function of the concentrations of the constituents. TPD emits in the blue, Alq_3 in the green and DCM2 in the red, availing a large color gamut to this ternary system. Filled circles = DCM2 in Alq_3, open circles = TPD in Alq_3, crossed squares = DCM2 in TPD. Numbers correspond to the dopant fraction (Bulovic et al., 1999).

Reprinted from Chemical Physics Letters, 308, Bulovic, V., Deshpande, R., Thompson, M. E. & Forrest, S. R., Tuning the color emission of thin film molecular organic light emitting devices by the solid state solvation effect., 317-322, 1999, with permission from Elsevier.

between host molecules. Hence, while SSSE can be useful in tuning the OLED color coordinates, it comes at the cost of reduced efficiency as the concentrations of dopant and hosts are adjusted to reach the desired emission color.

6.3.2.3 Small molecule fluorescent dopants and hosts

The variety of materials used in OLEDs, whether they are for injection, charge transport or emission, is enormous, and its proper treatment would consume a volume all by itself. Due to the advent of exceptionally high efficiency of PHOLEDs compared to their fluorescent analogues, work on fluorophore discovery has recently been almost entirely directed at blue emitting materials combinations whose operational lifetime still exceeds that of blue PHOLEDs. In this section, we will briefly highlight advances in red, green and blue fluorescent dopants and hosts. The purpose is to introduce representative materials and the challenges faced in finding practical materials for use in modern OLEDs. For a more comprehensive discussion, the reader is referred to the many review papers on this topic (Chen, 2004, Wen et al., 2005, Jeong et al., 2015, Jou et al., 2015).

Light emitting dopants must satisfy several conditions to be of interest. (i) They must emit with the desired color coordinates. For displays, the emission colors should be close to the coordinates required either for sRGB or NTSC standards in the red (R), green (G), and blue (B) (see Section 6.2). For lighting, the white isochromic point can be accessed by three (or more) colors (R, G, B), or by only two colors at the end of a line extending from blue-green to yellow. There is little need, except for specialized purposes to develop OLEDs emitting at other color coordinates. (ii) Their luminance intensity must be sufficiently high for the intended application (typically at least 500 cd/m^2 for displays and 3000 cd/m^2 lighting). (iii) They must have high efficiency, and in particular their PL quantum yield must approach unity. (iv) They must be stable, leading to a long OLED operating lifetime.

In Tables 6.3, 6.4, and 6.5, we list example dopant/host combinations employed in red, green, and blue fluorescent OLEDs, respectively. The materials included in these tables do not undergo TADF, a topic that is the subject of Section 6.3.4. However, devices with exceptionally high (>5%) fluorescent external quantum efficiencies may have some degree of delayed fluorescence, or they may preferentially align with their transition dipole moments (TDMs) parallel to the substrate plane, thus increasing the outcoupling efficiency, η_{out}, to greater than ~20% expected for OLEDs on flat glass substrates (Section 6.6.5). The external quantum efficiencies listed are generally < 5%, which is consistent with a maximum singlet spin formation ratio of $\chi_S = 25\%$.

With four exceptions (guest–host systems **5R**, **7R**, **10R**, **11R**, Table 6.3), all red emitting devices employ only Alq_3 as the electron conducting host material. The alternative host is rubrene doped with DBP (Okumoto et al., 2006a) and TPD doped with DCM2 (Bulovic et al., 1999). The latter device emits in the

Table 6.3 Representative red fluorescent guest–host systems

No.	Dopant	Host	Dopant conc.	λ_{max} (nm)	η_{ext} (%)	Lum. eff. (cd/A)	CIE coord.	Comments	Ref.
1R	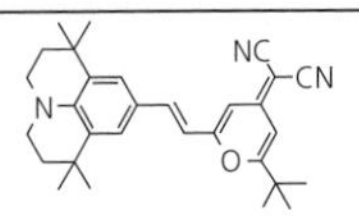 DCJTB	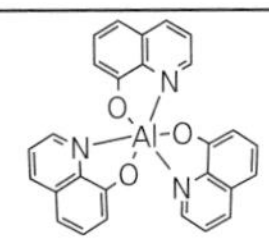 Alq_3	1%	616		2.45	0.60, 0.39		(Lee et al., 2010b)
2R	 Rubrene + DCJTB	Alq_3	2% DCJTB + 5% rubrene	620		3.24	0.64, 0.35	Rubrene assists in energy transfer from host to dopant	(Liu et al., 2003)
3R	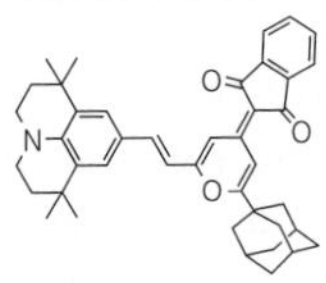 Red 2	Alq_3	0.5%	638		2.31	0.63, 0.36		(Lee et al., 2010b)
4R	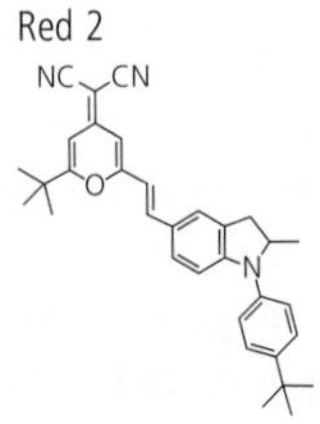 OJ2	Alq_3	1%	614	3.27	6.46	0.62, 0.38	Charge balance established using bathocuproine blocker	(Chang and Chow, 2011)
5R	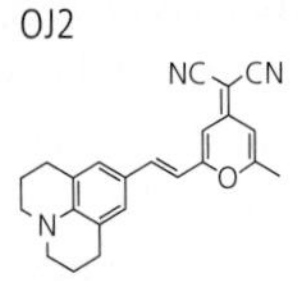 DCM2	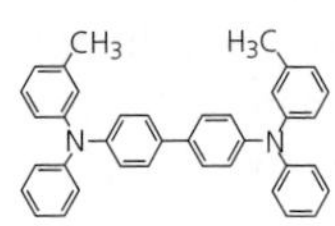 TPD	1.5%	590	0.38		0.52, 0.48	Orange	(Bulovic et al., 1999)
6R	DCM2	Alq_3	1.5%	623	0.38		0.60, 0.39		(Bulovic et al., 1999)
7R	DCM2	20%Alq_3+80% TPD	1.5%	612	0.8		0.23, 0.44	Yellow due to Alq_3 emission	(Bulovic et al., 1999)
8R	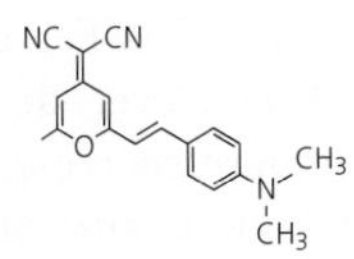 DCM1	Alq_3	0.25%	590	2.3		0.52, 0.48	Orange	(Tang et al., 1989)
9R	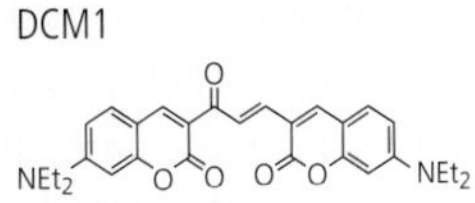 NKX2221	Alq_3	1%	575	1.6		0.46, 0.51	Orange	(Fujikawa, 2002)
10R	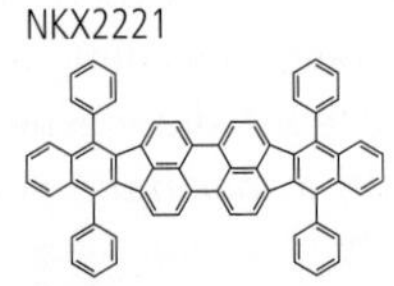 DBP	Rubrene	1%	610	4.7	5.4	0.66, 0.34		(Okumoto et al., 2006a)
11R	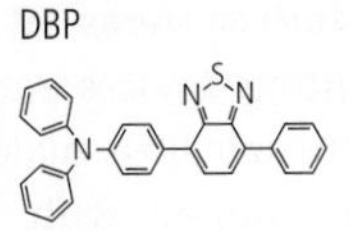 TPA-BZP			588	3.8	8.84	0.55, 0.45	Undoped	(Wang et al., 2016)

Table 6.4 Representative green fluorescent guest–host systems

No.	Dopant	Host	Dopant conc.	λ_{max} (nm)	η_{ext} (%)	Lum. eff. (cd/A)	CIE coord.	Comments	Ref.
1G	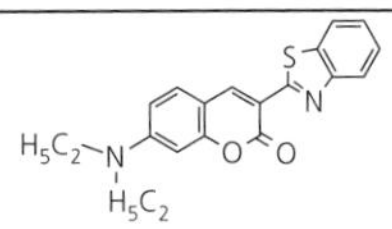 Coumarin 540	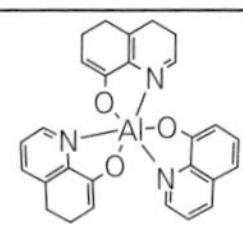 Alq_3	1%	510	2.5				(Tang et al., 1989)
2G	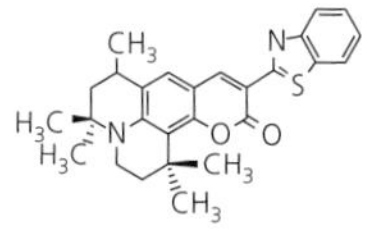 Coumarin 545P	Alq_3	1%	480		11.3	0.31, 0.65		(Lee et al., 2004)
3G	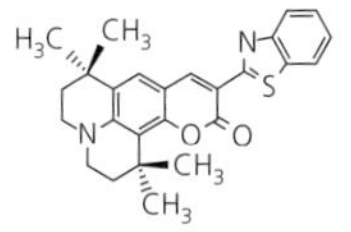 Coumarin 545T	Alq_3	1%	480		10.4	0.30, 0.64		(Lee et al., 2004)
4G	Coumarin 545T	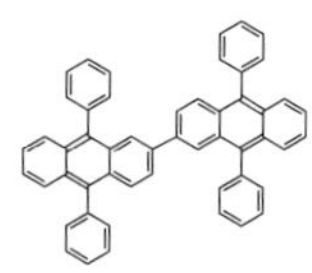 TPBA	0.8%	500	9.3	29.8	0.24, 0.62	High efficiency possibly results from molecular alignment	(Okumoto et al., 2006b)
5G	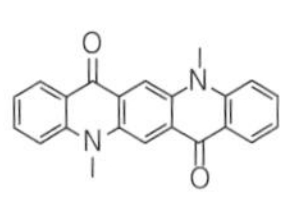 DMQA	Alq_3	0.8	544		7.3	0.39, 0.59		(Shi and Tang, 1997)
6G	Alq_3			560	1.0		0.30, 0.55	Undoped	(Tang and VanSlyke, 1987)
7G	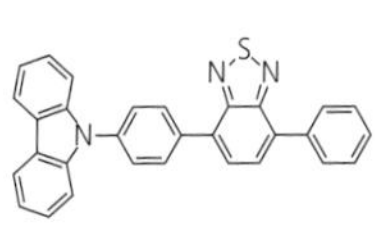 CzP-BZP			538	6.95	24	0.34, 0.60	Undoped	(Wang et al., 2016)

orange due to incomplete energy transfer from the blue fluorescent TPD, thus allowing for dilution of the pure red emission of DCM2 by a small blue peak.

Interestingly, a high efficiency, orange-red device (**11R**) is obtained with a neat thiazole-containing moiety, TPA-BZP (Wang et al., 2016). This is probably enabled by the twist of the molecule of approximately 40–50° at each of the three single C–C bonds along the molecular longitudinal axis that sterically hinders the molecules from close packing in the solid state. The limited packing reduces concentration quenching often found in more planar molecular systems. A similar molecule (CzP-BZP) also shows very high efficiency green emission in an undoped EML (**7G**, Table 6.4). The non-planar motif is used in the design of blue emitting chromophores due to their tendency to exhibit red-shifted exciplex emission when closely packed.

It is notable that the difference in structures between the red and green emitting BZP molecules lies only in a single bond: the carbazole donor of the green emitting CzP-BZP that links the two phenyl groups is absent in TPA-BCP. This bond results in a shift in emission to well into the green. The strong donor character of the latter molecule creates a low energy CT state that is only weakly hybridized with the local, singlet exciton. This is in contrast to the reduced donor character of the carbazole group in CzP-BZP that results in a higher degree of singlet localization, thus lifting the energy of emission into the green. This pair of molecules is an example of how minor molecular structural modifications can result in significant changes in the donor or acceptor character and charge distributions, and consequently the excited state energies.

There are many other molecules that have been investigated for their saturated red emission. Most

Table 6.5 Representative blue fluorescent guest–host systems

No.	Dopant	Host	Dopant conc.	λ_{max} (nm)	η_{ext} (%)	Lum. eff. (cd/A)	CIE coord.	Comments	Ref.
1B	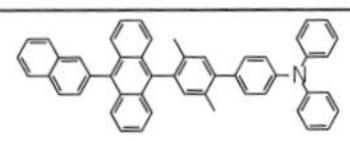 TPAXAN			428	4.62		0.16, 0.05	Undoped	(Kim et al., 2013a)
2B	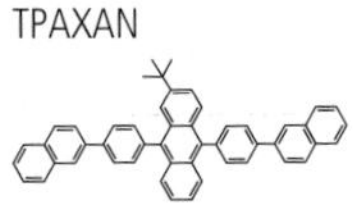 TBDNPA			444	5.17	2.6	0.15, 0.09	Undoped	(Zheng et al., 2010)
3B	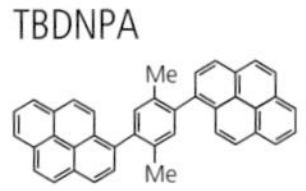 DMPPP			446	5.2	5.2	0.15,0.11	Undoped	(Wu et al., 2008)
4B	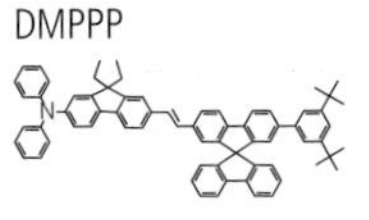	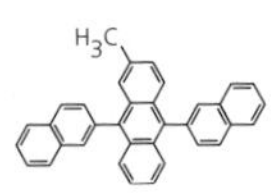 MADN	10%	460, 484	7.72	10.9	0.15, 0.20	Two spectral peaks due to vibronics	(Lee et al., 2011b)
5B	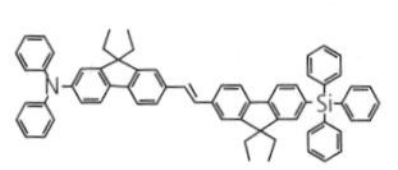	MADN	5%	458, 481	7.35	9.8	0.15, 0.20	Two spectral peaks due to vibronics	(Lee et al., 2010a)
6B	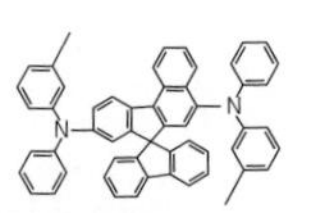 BD-6MDPA	MADN	5%	462	8.16	9.1	0.14, 0.16		(Jeon et al., 2010)
7B	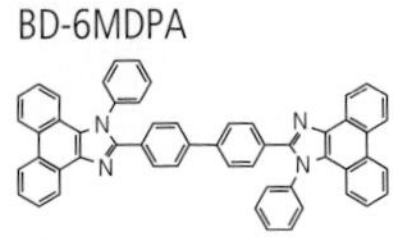 PPIP			462	6.31	7.5	0.15, 0.14	Undoped	(Kuo et al., 2009)
8B	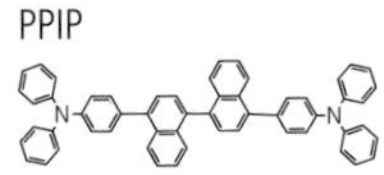 BN1	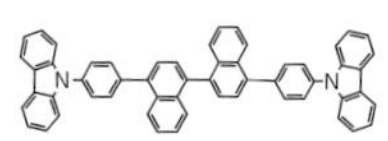 BN2	5%	450	8.6	4.9	0.16, 0.11		(Wei et al., 2010)
9B	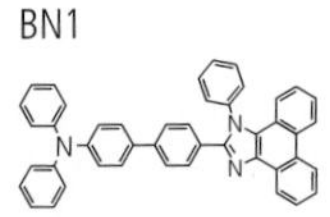 TPA-PPI			434	5.0	5.7	0.15, 0.11	Undoped	(Li et al., 2012)
10B	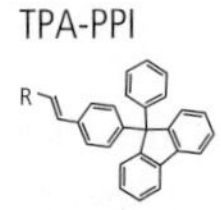 DPFS	MADN	15%	456	5.6	6.0	0.15, 0.13		(Kim et al., 2010)

notable are the metal-free phenyl porphyrin derivatives (Shen et al., 1997). A comprehensive description of many of these compounds is found elsewhere (Chen, 2004). They have not been included in Table 6.3 due to their low efficiencies (typically <1%), and hence are of limited technological significance.

Almost all efficient green fluorescent dopants are based on some variation of either coumarin or quinacridone. As in the case of red devices, the most commonly used host is Alq_3. An exceptionally high efficiency is achieved by doping Coumarin 545T in TPBA (**4G**) (Okumoto et al., 2006b). It is unclear why this particular device has an external quantum efficiency of 9.3% that implies a singlet fraction > 25%. Indeed, assuming $\eta_{out} \sim 20\%$, an external quantum efficiency of ~5% is the maximum value achievable for 100% singlet emission efficiency. Violation of spin statistics is unlikely, given that a similarly doped device employing Alq_3 as the host shows efficiency well below 5%. Two possibilities that would explain the high efficiency achieved with coumarin 545T/TPBA is preferential horizontal alignment of the excited state dipole moment relative to the substrate arising from the planarity of both dopant and host molecules, or additional emission via delayed fluorescence.

Table 6.5 lists several representative deep blue emitting fluorophores. Many approaches to blue emission start by employing a small polycyclic aromatic hydrocarbon with high fluorescence efficiency in the blue, such as anthracene, pyrene, or fluorene. These planar emitting molecules have the advantage of being exceptionally thermally stable. Unfortunately, their planarity leads to a tendency to form extended π-stacks, and hence in the solid state they typically emit from low energy excimers. This results in undesirably broad and low efficiency emission shifted to the green or red. Hence, the design motif leading to efficient blue emission involves the addition of bulky side or end groups that are attached to the luminescent center along a conjugated molecular backbone. These groups twist out-of-plane to prevent close packing.

Examples of this molecular design strategy are the xylene- and dinaphthal-anthracenes (compounds **1B**, **2B**, Table 6.5) that connect bulky end groups via intermediate phenyl moieties. The twist along the backbone of TPAXAN shown in Fig. 6.39, is due to repulsion of electronic orbitals in neighboring molecular groups. A similarly sterically hindered design is employed for the dimethylphenyl dipyrene molecule, DMPP (**3B**). In that case the emitting pyrene moiety appears as end capping groups about a central phenyl ring. An alternative strategy to create steric hindrance is to use a *spiro* architecture such as the spirofluorene and similarly bulky molecules **4B** and **6B**, or to attach a very bulky end cap, such as the Si-based group in **5B**.

Blue luminance in the absence of a planar aromatic molecular core can be achieved by a dipolar molecule that contains both an electron donating (d) and accepting (a) group at opposite ends of its conjugated backbone, such as compound **9B**. This approach shifts the emission from the red to the green for the donor-acceptor species **11R** and **7G**. The challenge of employing d-a structures in the blue is that they form CT states as precursors to the emitting localized exciton (LE) state. The CT state has a low energy and hence may result in unacceptable red shifts in emission. Hence, **9B** consists of a very high energy, violet-emitting PPI donor group with a peak wavelength of $\lambda_{max} = 369$ nm attached to a triphenylamine (TPA) acceptor via a biphenyl moiety that introduces a twist between the TPA and PPI. This results in a partial decoupling of their electronic orbitals, red shifting the emission. OLEDs using **9B** reach 5% external efficiency in the deep blue at $\lambda_{max} = 438$ nm (Li et al., 2012).

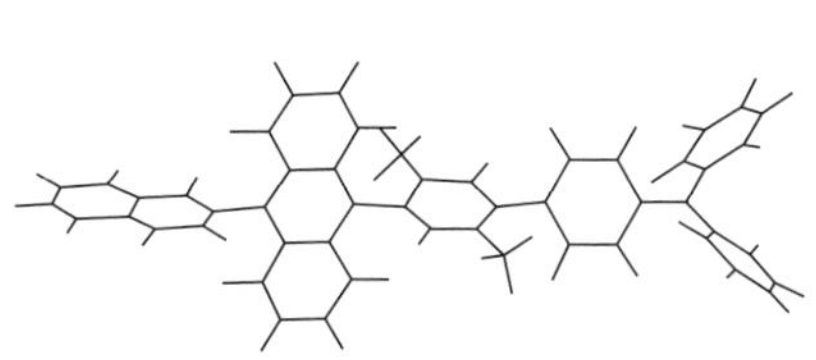

Figure 6.39 Three-dimensional (3D) molecular structure of TPXAN (compound **1B**, Table 6.5) showing a pronounced twist of the groups attached to the C–C bonds along the molecular backbone. The twist is due to electronic repulsion between neighboring functional groups (Kim et al., 2013a).

Although blue emitting molecules are designed to sterically inhibit stacking, several unusually high external efficiencies > 5% (e.g. **4B–8B**) suggest that these species orient with their planar emitting moieties parallel to the substrate plane, that is, the *transition dipole moment* (TDM) is oriented such that emission is principally at angles less than that which results in TIR in the substrate (approximately within a cone with half angle of 20°, see Section 6.6). Alternatively, the high energy of the singlet excited state may lead to additional delayed fluorescence from TTA.

Steric hindrance often eliminates the need for dilution in a wide energy gap host. Indeed, the very high energy of the blue singlet state significantly limits the choice of hosts that do not trap charges or excitons at lower lying energy states. Trapping siphons energy away from the emitting molecule, thus reducing efficiency. Hence, with only a few exceptions (**4B–6B**, **8B**, **10B**), blue fluorescent OLEDs do not employ a host molecule in the EML.

To confine the blue excitons within the EML, many devices include a blocking layer with a higher energy gap than the dopant (or host) in the EML. These layers can be occasionally found in green and red fluorescent OLEDs as well. As discussed throughout this chapter, the OLED comprises many layers ranging from charge transport and injection layers, to the guest–host emitting system, and blocking layers. The materials used for these purposes are highly diverse, with many different solutions applied to OLEDs emitting in different regions of the spectrum. Those interested in exploring this topic are referred to texts in *Further reading* to learn more about the challenges and solutions arrived at for achieving a diversity of high performance OLED structures. In particular, comprehensive treatments of charge transporting, blocking, and injecting materials can be found in O'Brien et al. (1998), Kulkarni et al. (2004), Shirota and Kageyama (2007), Gaspar and Plikarov (2015), Ho et al. (2015), and Jou et al. (2015).

6.3.2.4 Fluorescent polymer OLEDs

Polymer OLEDs were first demonstrated by spin coating a 70 nm thick film of PPV onto the surface of a glass substrate with a pre-deposited ITO anode (Burroughes et al., 1990). While this thin film device had a relatively low operating voltage of ~13 V, the electric field required to produce measurable luminescence was 2×10^6 V/cm. This is three orders of magnitude higher than required in the anthracene single crystal EL device in Fig. 6.32, but only marginally higher than in the Alq_3-based bilayer device in Fig. 6.3. The PPV voltage, therefore, is low only due to the thinness of the PPV film. Due to the limited ability of PPV to conduct electrons whose mobility is considerably less than that of holes (Blom et al., 1996), and its low PL quantum yield of 8%, the external fluorescence quantum efficiency was only ~0.05% for emission in the yellow-green spectral region with a peak at 2.2 eV (~565 nm wavelength).

In previous sections we have shown the benefits of using multilayer structures to achieve high performance. However, the challenges encountered in making devices comprising several solution-processed layers have given rise to strategies that circumvent this limitation. For example, blending a second polymer with improved electron conducting properties into an otherwise hole conducting emissive polymer layer, or using polymers that can support bipolar charge conducting via side chain or main chain functionalization, are two approaches to reducing the operating voltage. Significant early improvements were realized by employing more luminescent polymers with improved charge balance (i.e. where the electron and hole mobilities are approximately equal) compared to PPV-based devices. Improved charge injection is achieved by replacing the Al cathode (which forms an insulating oxide barrier) with Ca or Ba (Cao et al., 1999).

Many improvements have also been realized by employing multilayer structures along with the light emitting polymer blends that have both hole and electron conducting functionality. As discussed previously, charge injection and exciton formation can occur by numerous routes, even in simple two and three layer devices. Mechanisms for charge transport and exciton formation in single layer blend and bilayer polymer OLEDs are summarized in Fig. 6.40. In Fig. 6.40a, a large energy gap host polymer transports both electrons and holes that become trapped by the smaller energy gap, light emitting guest. A disadvantage of this structure is that the host is called upon to transport both charge carriers with nearly equal mobilities to avoid forming the emission zone near one of the contacts. This can lead to exciton quenching at the electrodes, and an emission zone position that is dependent on current and voltage. The shifting position of the emission zone results in a variable optical outcoupling efficiency, depending on the brightness.

Figure 6.40b partially resolves this problem by using the guest lumophore to conduct one of the charge carriers (electrons), with the host conducting the opposite charge (holes). Thus, the guest may be optimized for both its luminescence efficiency and its ability to conduct one of the two charges, while the host provides a high mobility medium for transport of the counter charge. Figure 6.40c has the same problems as Fig. 6.40a since the guest is responsible for both electron and hole conduction and exciton formation.

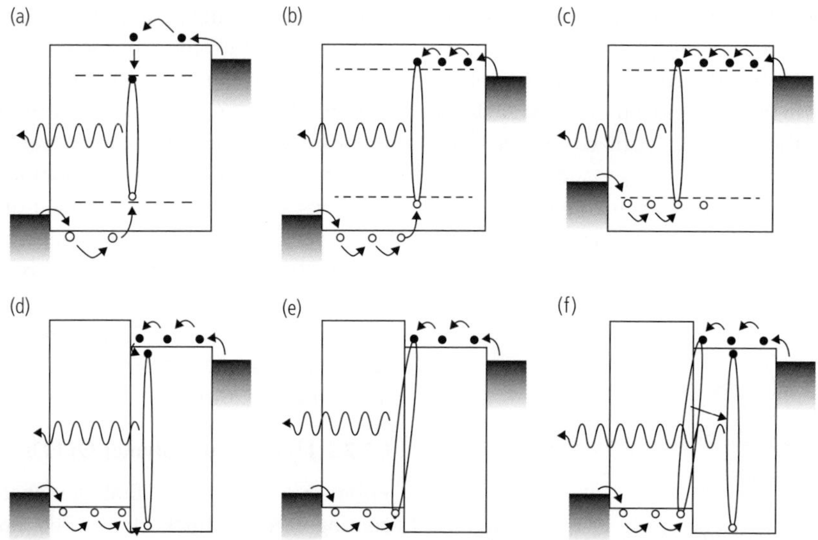

Figure 6.40 Charge transport and exciton formation in single layer blend (a–c) and bilayer type II heterojunction (e–f) OLEDs. The dashed lines in a–zc indicate the energy levels of the light emitting polymer species. The ellipses represent excitons (McNeill and Greenham, 2009).

Doping a luminescent guest within a "passive" host may result in higher luminance due to a reduction in concentration quenching, although a low guest–host blend ratio also decreases the conductivity, thereby increasing operating voltage.

Figures 6.40d–f show bilayer devices where the functionalities of the two materials forming the type II (staggered) HJ are separately optimized. Figure 6.40d represents a device in which excitons are formed in the high electron mobility layer due to leakage of holes from the opposite layer into the ETL/EML interface region. Of course, the opposite situation can also be engineered, where the EML is within the HTL. In Fig. 6.40e, emission is from an exciplex (i.e. a CT state) formed across the interface. This route dominates when the energy level offsets at either the HOMO or LUMO edges of the contacting semiconductors are too high to allow for leakage into the adjacent layer (Fig. 6.40d), thus preventing direct exciton formation. The final case, Fig. 6.40f shows the formation of a precursor CT state that subsequently transfers into one of the polymers to form a localized exciton. As in Fig. 6.40d, this structure relies on sufficiently small energy barriers ($\sim k_B T$) at either the HOMO or LUMO energies to allow for charge hopping across the heterointerface at room temperature.

An efficient, green emitting multilayer polymer OLED comprising separate layers for hole injection, emission and electron injection is shown in Fig. 6.41. This ~130 nm thick device is effectively a combination of strategies shown in Fig. 6.40a and d, where the 40 nm thick PEDOT:PSS/10-15 nm thick TFP layer delivers holes to the 70 nm thick ambipolar conducting and emitting F8BT EML, and electrons are transported to the EML by the ~10 nm thick CPE EIL. The CPE comprises a poly[(9,9-bis(8′-(3″-methyl-1″-imidazolium) octyl)-2,7-fluorene)] brominated polyelectrolyte backbone functionalized by one of the two groups shown (2,7-(9,9-dioctylfluorene) in F8im-Br, and benzo(2,1,3) thiadiazol-4,8-diyl in F8imBT-Br). In spite of the presence of the ionic Br^- and imidazole cation, these devices showed surprisingly fast electroluminescent switching (<10 μs), along with a maximum $\eta_{ext} = 5.1\%$, and luminous efficiency of 10.8 cd/A at 100 cd/m^2 (for F8imBT-Br). The F8BT emits in the green with peak intensity at approximately 550 nm. Taken together, this multilayer structure shows a remarkable improvement in characteristics compared to single layer polymer OLEDs.

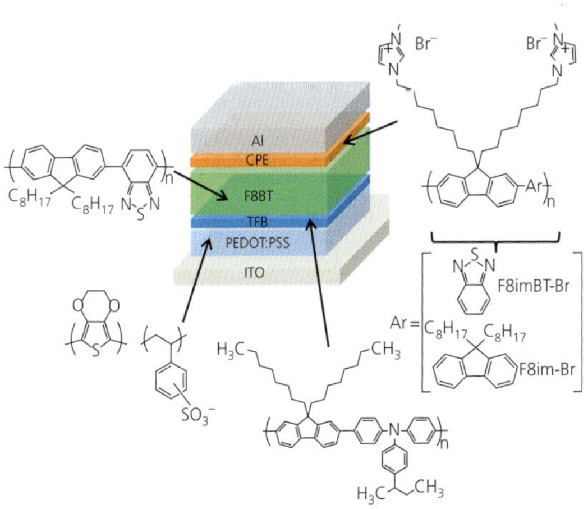

Figure 6.41 Multilayer, green emitting fluorescent polymer OLED showing the structural formulae of polymers used in each layer. After Suh et al. (2015).

The TFB layer in the OLED is used as both an exciton and electron blocking layer. Its energy gap (>3.0 eV) is wider than that of F8BT (~2.3 eV), hence preventing excitons from diffusing to the PEDOT:PSS HIL where they are quenched. Also, electrons injected from the cathode stop at this interlayer (Kim et al., 2005). Due to these benefits, interlayers are now commonly employed in efficient polymer OLEDs.

Interestingly, TFB:F8BT OLEDs form type II HJs, resulting in exciplex emission, as illustrated in Fig. 6.40e. A characteristic of exciplex emission is that it is potentially "lossless," that is, the exciplex is formed without encountering an energy barrier between the free polarons and the excited state (Morteani et al., 2003). This is apparent from the low (< 2V) voltage needed at the onset of current and luminance in Fig. 6.42. The photon emission from F8BT is 2.29 eV (corresponding to $\lambda_{max} = 538$ nm), which is below the threshold for luminance turn-on. The exciplex emits at 2.1 eV (580 nm), with a tail extending into the NIR at 1.8 eV (Fig. 6.42, inset), which is accessible at the low turn-on voltage observed. The "energy barrier free" exciplex emission occurs >10 ns after the onset of the exciton emission; a delay that arises from slow diffusion of the exciton and charges from the F8BT to the HJ.

Photon emission at applied voltages less than the band gap is a well-known phenomenon in inorganic light emitting diodes. This "sub-bandgap emission" is due to charge diffusion to and band-to-band recombination in the quasi-neutral regions adjacent to the p-n junction. Thus, weak emission can be observed with sufficiently sensitive detection equipment at an arbitrarily low voltage. Similarly, it has been shown that OLEDs can emit at well below the energy gap of the emitting material as a result of similar drift-diffusion driven by the built-in potential of the organic HJ (Engmann et al., 2019). A dramatic

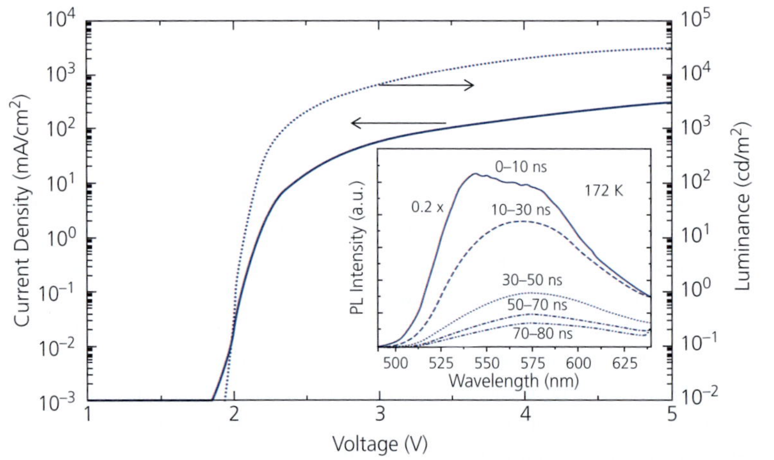

Figure 6.42 Current density and luminance vs. voltage characteristics of a TFB:F8BT OLED. The onset of both luminance and current is at ~1.9 V. Inset: PL spectra of the active layer as a function of time delay from the optical pump pulse. The peak F8BT emission is at 538 nm, and the exciplex peak emerges after a short delay at 580 nm. After Morteani et al. (2003).

demonstration of emission at half the energy gap of rubrene is observed for OLEDs whose energy level diagrams are shown in Fig. 6.43a. The devices comprise a rubrene EML (with optical energy gap of 2.2 eV), and a C_{60} ETL. These are spaced apart by a wide energy gap layer of BCP whose thickness is varied from $x = 0$ (no spacer) to 5 nm, or by a co-evaporated, blended region of rubrene and C_{60} (in a 1:1 ratio) that is either $y = 2$ or 5 nm thick. The BCP spacer is used to prevent formation and recombination of low energy charge transfer states at the rubrene/C_{60} interface, while the blended device should enhance the formation of such states.

The j–V characteristics show a rapid turn-on at <1 V for devices with spacers, and with a forward-biased ideality factor of $n = 1.44$ for $x > 0$ and 1.62 for $x = 0$, suggesting that a mixture of diffusion and recombination determines the current. The neat rubrene device has a turn-on at 2 V and $n = 2$ indicative of pure recombination (see Section 4.5.2). The luminance intensity is observed to turn-on at approximately the same voltage of ~ 1 V, independent of x or y. Remarkably, this is less than half of the energy gap of rubrene. The neat rubrene device, however, shows a luminance turn-on at 2.8 V. Importantly, the emission spectrum is the same for all devices tested, as shown in Fig. 6.43b.

Since rubrene has a triplet energy equal to $\frac{1}{2}E_G$, it is reasonable to expect that the device undergoes singlet fission into two triplet states (Section 3.10). This may account for the very low voltage (approximately 50% of the energy gap) turn-on of the emission. However, the lack of change of spectrum from that characteristic of rubrene $S_1 \rightarrow S_0$ transitions, coupled to the fact that the emission intensity *increases* by a factor of 3 by the interposition of a BCP interlayer, suggesting that singlet fission is not the source of the low voltage luminance turn-on. Furthermore, Auger recombination that involves the promotion of excited states to higher energies via bimolecular interactions, also cannot account for the increased efficiency. Auger processes require that two excitons combine to form a higher energy state, thus *decreasing* the efficiency by at least 50% (see Section 4.5.3). The lack of CT emission and the increase in efficiency achieved when the BCP layer is inserted at the HJ are both indications that the low voltage turn-on is not caused by the emission from these lower energy states. Indeed, the mixed rubrene: C_{60} interface devices should show very strong CT emission, yet they have efficiencies even lower than those of the conventional device with $x = 0$. Hence, we conclude that the sub-energy-gap emission is due to charge drift and diffusion at the HJ; a mechanism that is the same as that attributed to sub-bandgap emission in inorganic LEDs (Engmann et al., 2019).

As discussed in Chapter 5, the ability to form multilayer structures using solution-based processes is somewhat limited since the solvent used in the deposition of one layer can degrade materials previously deposited in the device stack. In the device in Fig. 6.41, this problem is partially mitigated by using layers with orthogonal solvents (i.e. PEDOT:PSS uses an aqueous solvent, followed by TFB in a p-xylene solution). F8BT also uses p-xylene, and CPE is dissolved in 2-methoxyethanol, which is a non-orthogonal solvent that can attack the underlying

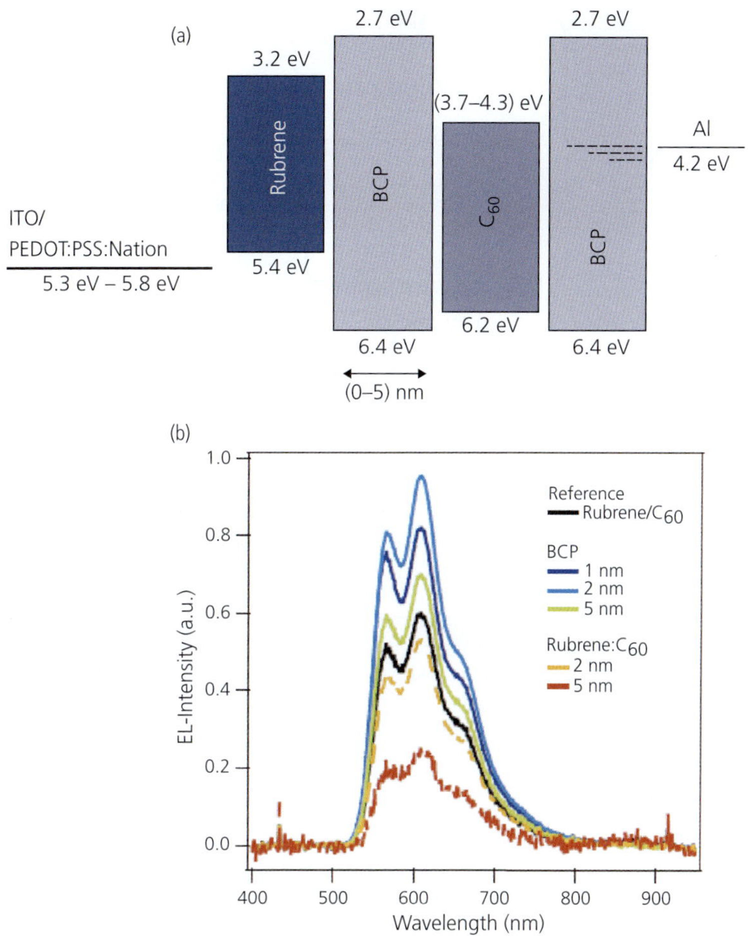

Figure 6.43 (a) Energy level diagram of a rubrene/C_{60} OLED employing a $x = 0$–5 nm thick BCP interface layer and a BCP exciton blocker. (b) Emission spectra for the devices in (a) (Engmann et al., 2019).

layers. To minimize layer interactions, they are hardened by baking prior to application of the subsequent layers. An alternative strategy to fabrication of multilayer structures is to employ polymers that undergo cross-linking under ultraviolet radiation or at elevated temperatures (Sekine et al., 2014). This alters their solubility in solvents used in their deposition. In any case, fabrication of solution processed multilayer structures with reproducible characteristics can present significant challenges in a high volume manufacturing environment (Ho et al., 2015).

6.3.2.5 Materials for fluorescent polymer OLEDs

As in our treatment of small molecule materials, here we discuss only a few example polymer systems and the molecular design strategies leading to high performance devices. There are two primary routes to engineering high efficiency *light emitting polymers* (LEP) summarized in Fig. 6.44. The first is to use a conjugated, emissive polymer that falls into one of several different classes exemplified by PPV, polyfluorene (PF), poly-*p*-phenylene (PPP), polyvinyl carbazole (PVK), and polythiophene (PT). Examples of these conjugated polymers are shown in Fig. 6.45.

Copolymerization is also used to achieve desired properties, as illustrated in Fig. 6.46. Starting with a conjugated backbone moiety such as a fluorine or PPP, fractions of electron and hole transporting units (e.g. benzothiophenes or amines) along with emitting centers such as small molecules or dendrimers are blended with, or copolymerized into the backbone. Thus, the molecular composition can be precisely controlled to achieve the optimized monomer ratios and sequences that combine multiple functionalities into a single molecular structure.

Conjugated polymers have the advantage of a simplified device structure that requires only a single, or at most two different materials (i.e. the emitting and

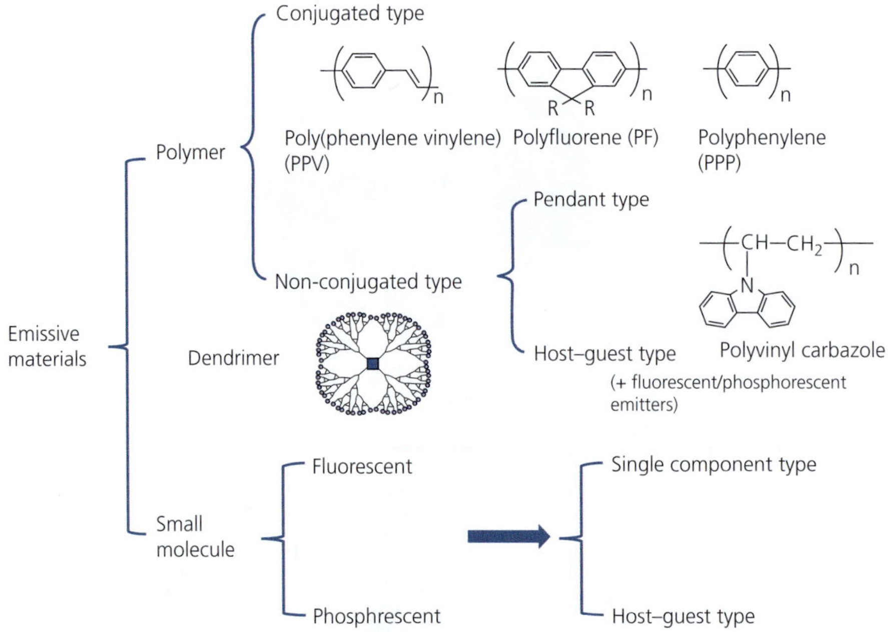

Figure 6.44 Strategies for the design of light emitting polymers used in OLEDs. Adapted from Sekine et al. (2014).

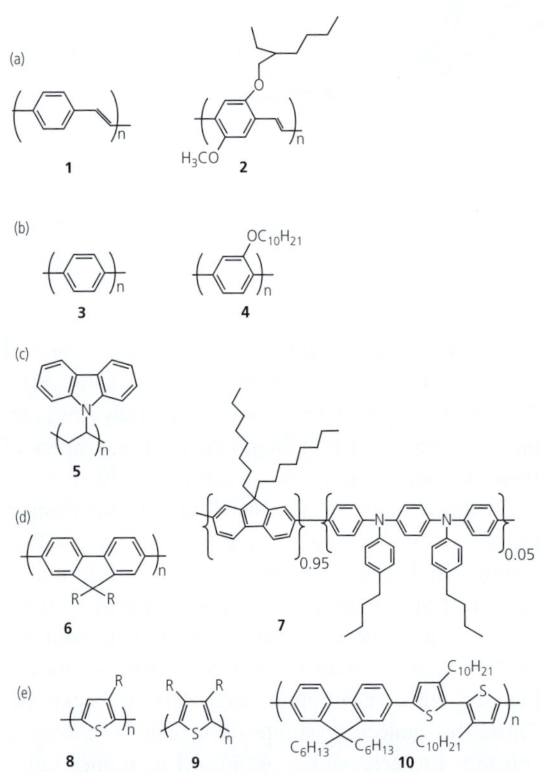

Figure 6.45 Example light emitting polymers and their derivatives that are to the right of the first molecular structural formula shown in the same row. (a) **1** PPV and **2** MEH-PPV; (b) **3** PPP and **4** MEH-PPP; (c) **5** PVK; (d) **6** PF and **7** (King et al., 2011); (e) **8**, **9** PT, with $R =$ alkanes for increased solubility, and **10** (Perepichka et al., 2005).

charge transporting polymer and a blocking interlayer). The emitting chromophores are often linked, or doped within the conjugated polymers backbone. To prevent emission or siphoning off of energy, the backbone units such as PF must be at higher energy than the emissive segments. Furthermore, using conjugated polymers requires that a single molecular species be both a good electron and hole conductor, and often it must also serve as an efficient emitting material. OLEDs following this approach often sacrifice efficiency for simplicity.

A second approach is to use pendant or side chain groups attached to a non-conjugated polymer backbone, or host–guest blends whose constituents are separately optimized to perform the charge transport and/or emissive functions. One example in Fig. 6.45 is PVK, where the emitting carbazole is pendant to a polyvinyl backbone. This separate functionalization is a potentially simpler route to synthesis and optimization.

Up to this point we have discussed strategies for achieving high efficacy fluorescent polymer OLEDs. Clearly, the scope for achieving the desired functionalities via molecular variation is enormous, and has been the subject of considerable research over many years. In Table 6.6 we provide a few illustrative examples of polymer OLED materials. The reader is once more directed to several reviews, and *Further reading* to obtain a comprehensive appreciation for this subject (Kulkarni et al., 2004,

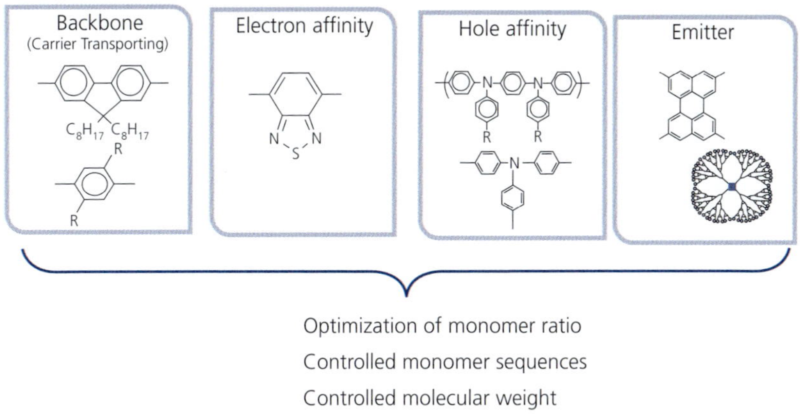

Figure 6.46 Strategy for copolymerization to achieve high conductivity and emitting efficiency in a single polymer (Sekine et al., 2014).

Perepichka et al., 2005, Cheng et al., 2008, Gong, 2013, Ho et al., 2015).

In Table 6.6, we observe that almost all efficient systems (with the exception of **3O** and **9B**) comprise conjugated copolymers with a PVK core. Polyfluorene copolymerized with polyacene emitting groups with increasingly long conjugation lengths (and hence lower singlet energies) are exemplified by compounds **1R**, **4G**, and **7B**. The dependence of wavelength on polyacene molecular weight is shown by the spectra of these compounds in Fig. 6.47. These compounds are, in descending order of molecular weight, copolymers with pentacene, tetracene, and anthracene. The conjugation length of each successively larger polyacene group is increased, shifting the emission from the blue into the red. Hence, the peak wavelengths are λ_{peak} = 625 nm for pentacene, to 525 nm for tetracene and 435 nm for anthracene (Tokito et al., 2001).

In the table we have not listed any thiophene-based compounds as those do not tend to have high efficiency. Furthermore, nearly all devices have a PEDOT:PSS hole injection layer, and **6G** and **9B** employ multiple hole injecting and electron blocking interlayers at the anode side. To avoid dissolution of the deposited layers, the hole transporting 2-NPD in **5G** is thermally cross-linked at 180°C for 30 min prior to the deposition of the copolymer PFBT5. This process ensures resistance to dissolution in solvents used in subsequent layer depositions (Cheng et al., 2008). Most of the highest efficiency devices include a PEDOT:PSS HIL, a hole conducting layer, and an interlayer (the latter two layers often being combined into a single layer) as for the device in Fig. 6.41. For example, **9B** with saturated blue color coordinates of (0.16, 0.09), comprises both a HQTZ HTL and a PVK interlayer, followed by an emissive poly(silafluorene) (PSF) blue EML. All layers were directly spin-coated onto each other, so it is reasonable to assume that a degree of material diffusion and intermixing occurs at the interfaces. Nevertheless, the efficiency of nearly 5% is high for a fluorescent OLED (Xia et al., 2015).

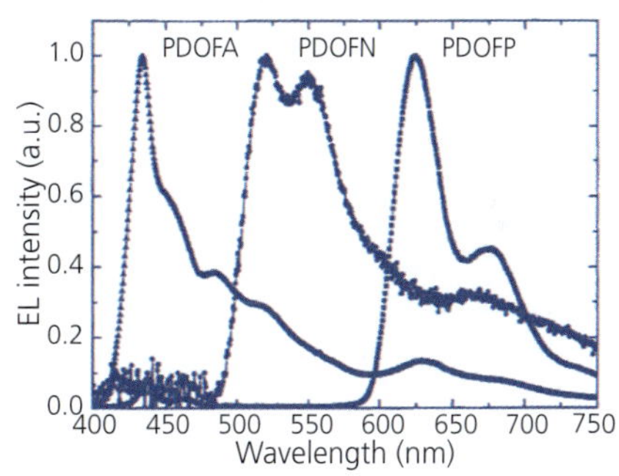

Figure 6.47 Electroluminescence spectra of OLEDs comprising three fluorine-based copolymers in Table 6.6 based on polyacenes of increasing conjugation length from anthracene (PDOFA, compound **7B**), to tetracene (PDOFN, **4G**), to pentacene (PDOFP, **1R**). This copolymer design strategy shifts the emission peak from the deep blue into the deep red (Tokito et al., 2001).

A further notable blue structure is **8B**, comprising the spirofluorene polymer, PF-TPA-OXD, with bipolar transporting groups copolymerized with the PF emitting backbone. The oxadiazole (OXD) group is preferentially electron transporting, whereas the TPA is hole transporting. This is exemplary of the "all-in-one" design strategy illustrated in Fig. 6.46. While this design unquestionably leads to processing simplicity, it is also apparent that performance is sacrificed when employed in an OLED, where only η_{ext} = 1.2% is achieved (Shu et al., 2003).

Table 6.6 Representative polymers used in fluorescent OLEDs

No.	Emitting polymer	Conducting polymer	λ_{max} (nm)	η_{ext} (%)	Lum. eff. (cd/A)	CIE coord.	Ref.
1R	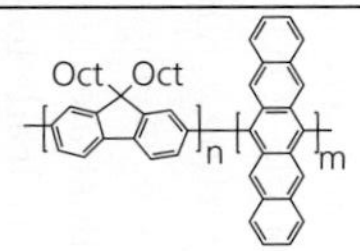 PDOFP[a]		623		0.32	0.64, 0.30	(Tokito et al., 2001)
2R	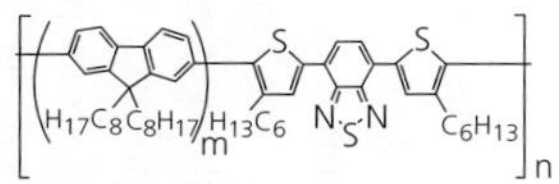 90/10 (PFO-DHBT10)[b]		638	2.54	1.45	0.66, 0.34	(Hou et al., 2004)
3O	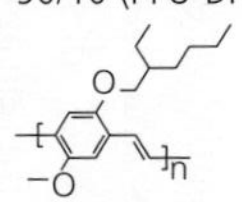 MEH-PPV	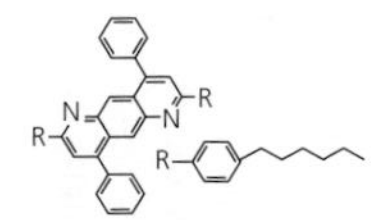 PDA[c]	570	3.1	7.0		(Tonzola et al., 2003)
4G	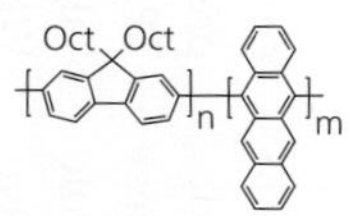 PDOFN[a]		520		0.18	0.37, 0.56	(Tokito et al., 2001)
5G	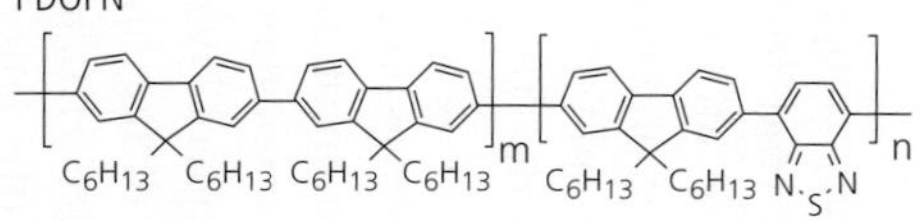 PFBT5[b]	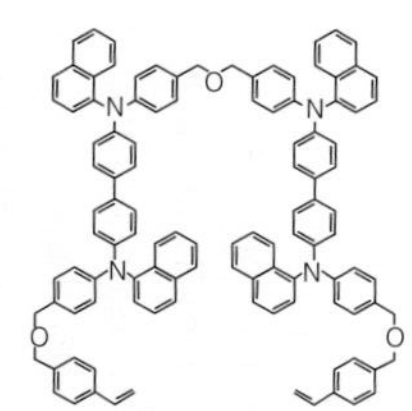 2-NPD[d]	540	3.20	10.8		(Cheng et al., 2008)
6G	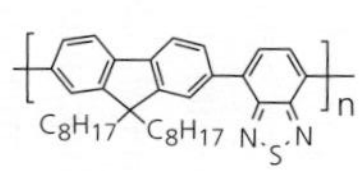 F8BT	F8imBT-Br interlayer	535	5.1	15.5		(Suh et al., 2015)

(Continued)

Table 6.6 Continued

No.	Emitting polymer	Conducting polymer	λ_{max} (nm)	η_{ext} (%)	Lum. eff. (cd/A)	CIE coord.	Ref.
7B	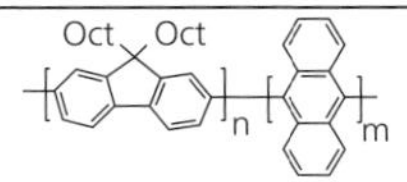 PDOFA[a]		435		0.43	0.21, 0.19	(Tokito et al., 2001)
8B	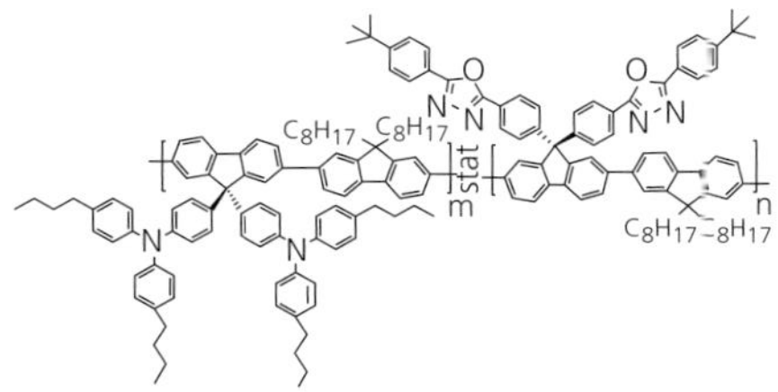 PF-TPA-OXD[e]		425	1.21	1.3	0.19,0.14	(Shu et al., 2003)
9B	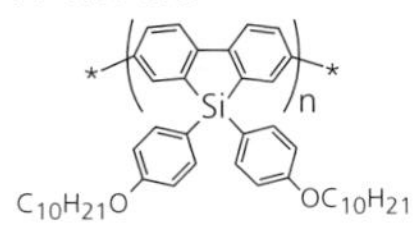 PSF	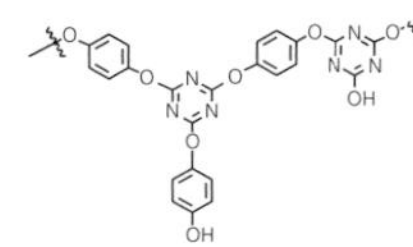 HQTZ[f] + PVK	440	4.93	4.50	0.16,0.09	(Xia et al., 2015)

[a] $m/(n + m) = 0.1$.
[b] $m/(n + m) = 0.9$.
[c] Electron conducting layer.
[d] A pair of NPD molecules linked by a styryl group as a hole transporting compound.
[e] Polyfluorene-1.04 oxydiazole-1.98 triphenylamine statistical copolymer.
[f] Polymerization of hydroquinone and 2,4,6-trichloro-1,3,5-triazine.

Finally, we point out that the MEH-PPV orange-emitting device, **3O**, differs from the other structures in that it employs an ETL near the cathode. This OLED has the structure of ITO anode/PEDOT:PSS HIL/MEH-PPV EML/PDA ETL/Al. The problem of layer dissolution is resolved by employing an aqueous HIL, followed by deposition of MEH-PPV using an organic solvent. This is followed by VTE of the small molecule ETL, PDA. As a result, the layers have sharp interfaces, leading to a reasonably high $\eta_{\text{ext}} = 3.1\%$.

There are numerous other approaches to achieving high efficiency polymer OLEDs. Perhaps the most common is to simply blend in a high efficiency small molecules or dendrimers into the polymer conductive host. This method relies on efficient energy transfer from the polymer to the guest. But an unavoidable conclusion in comparing polymer to small molecule devices is that the potential advantages gained in the simplified layer structures of polymer OLEDs is outweighed by difficulties in device processing, and the consequent loss in device performance arising from interlayer mixing and material incompatibilities. For this reason, small molecule OLEDs have been overwhelmingly adopted for practical applications compared with their polymer analogues.

6.3.3 Phosphorescent OLEDs (PHOLEDs)

A significant disadvantage of the fluorescent OLED is that 75% of the electrons and holes injected from the contacts generate non-radiative triplet states, and hence are wasted. This results in devices with low external efficiencies, typically between only 1% and 5%. The situation changed dramatically with the introduction of electrophosphorescence, that is, the use of molecules containing a heavy metal atom that mixes the singlet and triplet states via spin–orbit coupling (Baldo et al., 1998). Electrophosphorescence is characterized by a spin utilization factor of χ_{ST} =1, and hence a possible η_{int} =100% (Adachi et al., 2001a). PHOLEDs also feature longer radiative emission lifetimes (~500 ns to several microseconds) compared with fluorescent devices (1–10 ns). With the prospect of achieving high efficiency PHOLEDs across the visible spectrum, the display industry underwent a tectonic shift that has resulted in OLED domination of the mobile display market where high efficiency leads to longer battery life. More recently, PHOLEDs are making significant inroads into the television and monitor markets for the same performance-related reasons. Indeed, PHOLEDs are now contenders for use in interior illumination due to their ultrahigh efficiencies, flexible and conformable form factors, and attractive, tunable color coordinates.

Before we consider the technological details, we must first define what is meant by the term, phosphorescence. In Chapter 3 it was found that phosphorescence is conventionally defined as a radiative process that occurs a "noticeable" time after the excitation source has been removed. In the context of organic light emission, it is more accurate to state that *phosphorescence refers to all processes limited by triplet state dynamics*. Direct triplet emission from molecules, whether it originates from slow, spin-forbidden or relatively fast spin-allowed (due to strong spin–orbit coupling) transitions is unambiguously classified as phosphorescence. Triplet-dominated emission also refers to both delayed fluorescence due to TTA, as well as TADF, where the molecules spend typically 99% of the time in a triplet state.

6.3.3.1 Small molecule PHOLED architectures

Electroluminescence from 100% of the molecular excited states using heavy-metal-induced spin–orbit coupling was first demonstrated in a red PHOLED comprising the metalloporphyrin, PtOEP doped into Alq_3 (Baldo et al., 1998). Triplet emission was identified from its high external quantum efficiency (approximately 4% compared to 1% for red fluorescent dopants such as DCM2), but more importantly, due to the slow luminescence decay of 20–30 μs, which is nearly three orders of magnitude longer than for fluorescent emission. A further finding suggesting its triplet origin was the relatively high concentration of the dopant (6%) required to achieve maximum efficiency. As we have seen in Section 6.3.2, fluorescent dopants exchange energy with the host via FRET, which is effective over distances of a few nanometers. However, triplets are most efficiently transferred via Dexter exchange interactions. This process requires the donor and acceptor to be nearly in contact. The doping concentration to achieve high efficiency is, therefore, significantly higher than Förster-mediated transfer. Indeed, it was found that the PtOEP/Alq_3 device had $\eta_{ext} = 2\%$ even at concentrations as high as 20%, where Dexter transfer alone is dominant.

A further development in improving the efficiency of triplet-harvesting devices was the introduction of a wide energy gap exciton/hole blocking layer (EBL, or alternatively, HBL) between the EML and the

ETL. The very long lifetime of triplet excitons is accompanied by a long diffusion length, particularly in a host that does not contain a heavy metal atom to promote radiative recombination of the triplet population. As a result, triplets can migrate out of the EML where they are subsequently quenched at the cathode contact. Insertion of a wide energy gap blocking layer prevents quenching by corralling all triplets within the EML, ultimately allowing them to transfer to the heavy metal phosphor. This architecture, applied to a PtOEP based PHOLED, is shown in Fig. 6.48. The host is CBP, bounded by a α-NPD HTL and a BCP exciton blocking layer, placed between the Alq_3 ETL and the Mg-Ag cathode contact. The blocking layer must have an energy gap higher than that of either of the materials comprising the EML to prevent energy transfer, and if it is on the electron-injecting side of the device, its LUMO should be approximately aligned to that of the ETL to avoid impeding the electron current.

This device achieves a maximum $\eta_{ext} = 5.6\%$, compared to 4% for devices lacking a blocking layer (O'Brien et al., 1999). The blocking efficiency of the structure is apparent from the spectra obtained at several different current densities for devices with and without the EBL as shown in Fig. 6.49. The device lacking a blocking layer shows emission from the Alq_3 ETL, whereas that emission is absent in the blocked device. Thus, the EBL completely eliminates both hole and exciton transfer from the EML into the ETL. Due to their effectiveness in increasing efficiency, blocking layers are now a standard feature of PHOLEDs.

The final major step in bringing PHOLEDs into the mainstream was the introduction of Ir-based heavy metal phosphors whose ligand structure can be modified to emit with ~100% internal quantum efficiency from the deep blue into the deep red. The first such device contained a CBP/$Ir(ppy)_3$ EML that employed a layer structure analogous to that in Fig. 6.48 (Baldo et al., 1999a). There, it was found that the green emission, centered at 520 nm wavelength had a surprisingly short decay time of 500 ns when used in a device, compared with 2 μs in solution. We have seen in Section 3.6 that this short decay time is a feature of the strong spin–orbit coupling of the triplet metal–ligand charge transfer (^{3}MLCT) state of the Ir $(ppy)_3$, and most other pseudo-octahedral Ir-complexes. This is in contrast to the lower interstate mixing (and hence reduced intersystem crossing rate) of ligand centered (^{3}LC) states of square planar Pt complexes, typified by PtOEP. As a consequence, the $Ir(ppy)_3$-based PHOLED external efficiency had a maximum at low current of 8% with a roll-off in efficiency as the current was increased above 100 mA/cm^2.

Shortly after the demonstration of the $Ir(ppy)_3$ OLED, a similar structure using the bidentate, $Ir(ppy)_2acac$ doped at 12 vol.% into the electron transporting TAZ achieved $\eta_{ext} = 19\%$, corresponding to $\eta_{int} = 87\%$. This constituted the first demonstration of an OLED with ostensibly 100% electron-to-photon conversion efficiency (Adachi et al., 2001a). It was proposed that the energy transfer mechanism was by direct hole injection from the HMTPD HTL onto the hole-conducting Ir phosphor, with electrons injected from the ETL onto this same molecule. There was no energy transfer from host to dopant in the optimized $Ir(ppy)_2acac$/TAZ EML.

A convincing indication that this device had unity internal efficiency was obtained by measuring the EL and PL efficiencies as functions of temperature, from

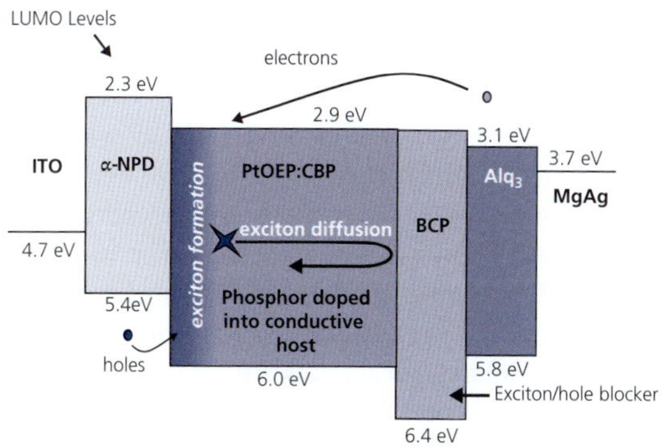

Figure 6.48 Energy level diagram of a PHOLED comprising an exciton/hole blocking layer.

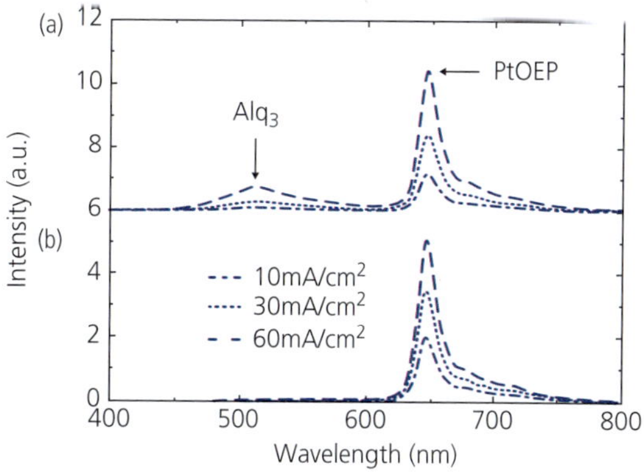

Figure 6.49 PHOLED emission spectra at three different drive currents for devices (a) without and (b) with a BCP blocking layer (O'Brien et al., 1999).

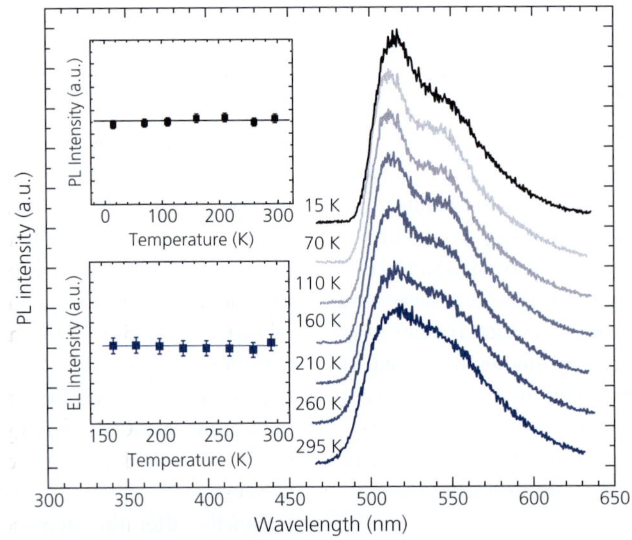

Figure 6.50 Temperature dependent PL spectrum of a 12% Ir(ppy)$_2$acac/TAZ film, with the PL and EL intensity of an OLED using this same emitting region shown in the insets (Adachi et al., 2001a).

300 K to 15 K (Fig. 6.50). The lack of change in efficiency over this wide temperature range indicates that there are no temperature-dependent losses due to non-radiative recombination. As shown in Chapters 3 and 4, recombination via defect states or coupling to disorder is expected to be thermally activated, depending on the position of the defect within the energy gap of the host. The temperature independence of the data in Fig. 6.50 attests to the lack of loss channels. Hence, in the space of little more than two years from the first introduction of electrophosphorescence in PtOEP, $\eta_{int} = 100\%$ was shown to be possible using heavy metal organic complexes with strong spin–orbit coupling.

As we move toward deeper blue emission needed in displays with large color gamuts, the choice of materials becomes limited. The first blue PHOLEDs were based on *endothermic energy transfer* from host to guest. That is, the CBP host triplet energy was higher than that of the cyan-emitting phosphor, FIrpic. The energy level diagram of this system is shown in Fig. 6.51 (Adachi et al., 2001b). Although the CBP triplet energy is lower than that of the dopant by 60 meV, the reverse, or endothermic transfer of triplets to CBP is much faster than the competing process of the disallowed host emission at rate, k_H. As long as the guest emission rate is $k_G \gg k_R + k_H$, the triplet emission can be efficient. Note that endothermic transfer is similar to

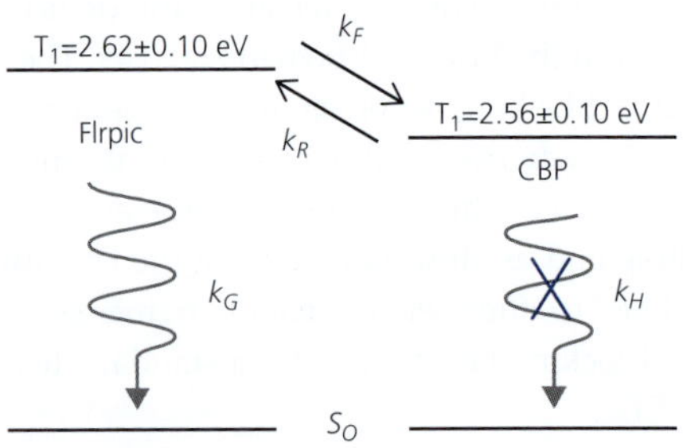

Figure 6.51 Energy level scheme for a FIrpic/CBP guest/host system. The various transfer rates are indicated.

that used in thermally activated delayed fluorescent OLEDs discussed in Section 6.3.4.

Of course, it is preferable to use hosts that allow for exothermic energy transfer to eliminate competition with processes such as k_H that reduce PHOLED efficiency. This was achieved by replacing CBP with mCP with its higher triplet energy of $T_1 = 2.90$ eV. The PL spectra from which the various triplet energies are obtained are shown in Fig. 6.52, inset. The triplet energy of mCP is higher than that of FIrpic, whereas CBP is considerably lower, leading to less efficient energy transfer (Holmes et al., 2003b). The FIrpic:CBP PHOLED had a peak efficiency of 6.1% in contrast to the FIrpic:mCP device that reaches $\eta_{ext} = 7.5\%$.

A signature of endothermic transfer is the pronounced temperature dependence of its emission. In

Fig. 6.52 we show a comparison of the PL intensity between the endo- and exothermic FIrpic-doped thin films. The CBP-based PL shows a peak PL intensity near 200 K, and then a rapid decrease at lower temperatures. These data suggest that at the highest temperatures, the thermally activated non-radiative recombination rate via defect states is higher than k_R, leading to a reduction in emission efficiency. Similarly, as temperature is lowered, defect-related recombination is frozen out and the efficiency increases until at $T < 200$ K, k_R must compete, and the efficiency drops once more with direct host emission, k_H. In contrast, the exothermic mCP system shows relatively temperature independent phosphorescent emission intensity until $T < 50$ K. At the lowest temperatures, host exciton diffusion slows to the extent that the excitons recombine before reaching a phosphor dopant. The PL transient response of the mCP-based film reinforces this picture. That is, as the temperature is lowered to 10K, the FIrpic radiative lifetime is 23 μs, compared to only 1 μs at room temperature since thermally activated exciton diffusion on mCP substantially increases the probability of non-radiative recombination prior to transfer to a FIrpic molecule at low T.

Blue PHOLEDs require both wide energy gap dopants and hosts. The LUMO energies of these materials necessarily lie close to the vacuum level that can lead to potentially damaging molecular oxidation. This limits the choice of effective and stable blocking materials. For example, in Fig. 6.53 we show the energy level diagrams for two blue OLEDs based on the *cyclometallated carbine* ($Ir(C^\wedge C)_3$) compound, ($Ir(pmp)_3$). Recall from Section 3.7.3 that the ^{3}LC-dominated *fac*-isomer emission is blue-shifted from that of the ^{3}MLCT-dominated *mer*-isomer due to the stronger spin–orbit coupling of the latter compound. For both isomers, a high energy gap electron-conducting host, TSPO1 is used for the ETL, CzSi for the HTL, and TPBi and a blend of CzSi at 15% vol. concentration in MoO_3 are used as the EIL and HIL, respectively. The LUMO energies of these compounds are considerably less than 2 eV, and in some cases as small as 1 eV from the vacuum level, suggesting that all of these materials are easily oxidized, perhaps negatively impacting their reliability.

Here, TSPO1 is used as an exciton/hole blocking layer since its HOMO energy extends well below that of 5.0–5.3 eV of the hole conducting $Ir(pmp)_3$ isomers. However, in a device where both the host and dopant have small differences in energy between the LUMO

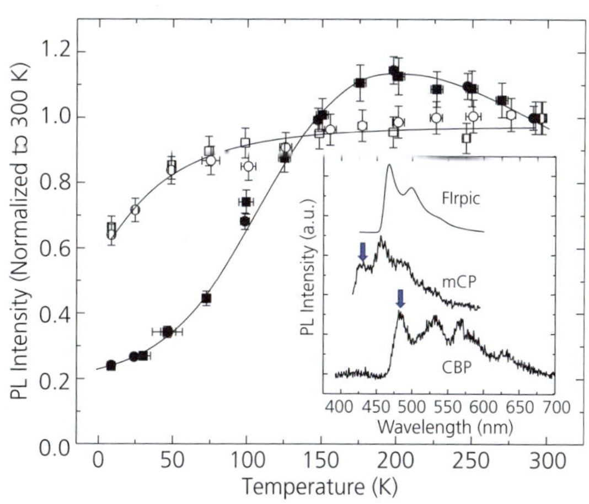

Figure 6.52 PL intensity vs. temperature of FIrpic in mCP (open symbols) and CBP (closed symbols). *Inset*: Triplet emission spectra of FIrpic, mCP, and CBP. Since the latter two molecules are fluorophores, the weak triplet emission spectra are obtained at 10K to freeze out non-radiative recombination channels. Arrows indicate the 0–0, $T_1 \rightarrow S_0$ transitions. (Holmes et al., 2003b).

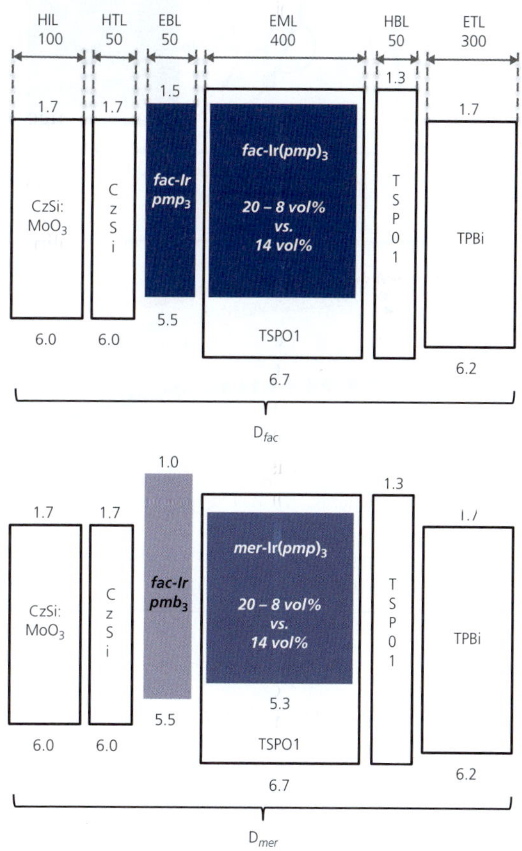

Figure 6.53 Energy level diagrams of a deep blue (a) *fac*- and (b) *mer*-Ir $(pmp)_3$ PHOLEDs. These devices are labeled D_{fac} and D_{mer}, respectively. Energies relative to vacuum are shown adjacent to the appropriate level for all materials (Lee et al., 2016).

and the vacuum level, EBLs are also required. The dopant itself, *fac*-Ir(pmp)$_3$ is, therefore, used as the electron blocker in the *fac*-isomer emitting device (called D_{fac}), whereas an even smaller LUMO energy blocker, *fac*-Ir(pmb)$_3$ is used as an EBL for the meridional device (called D_{mer}).

The EL spectra and quantum efficiencies of the two devices are shown in Figs. 6.54a and b, respectively. As expected for systems with strong spin–orbit coupling, the ^{3}MLCT-rich *mer*-Ir(pmp)$_3$ spectral peak is significantly red-shifted compared with the *fac*-isomer. Also, the quantum efficiency is ~50% higher for *mer*- than for *fac*-Ir(pmp)$_3$ due to

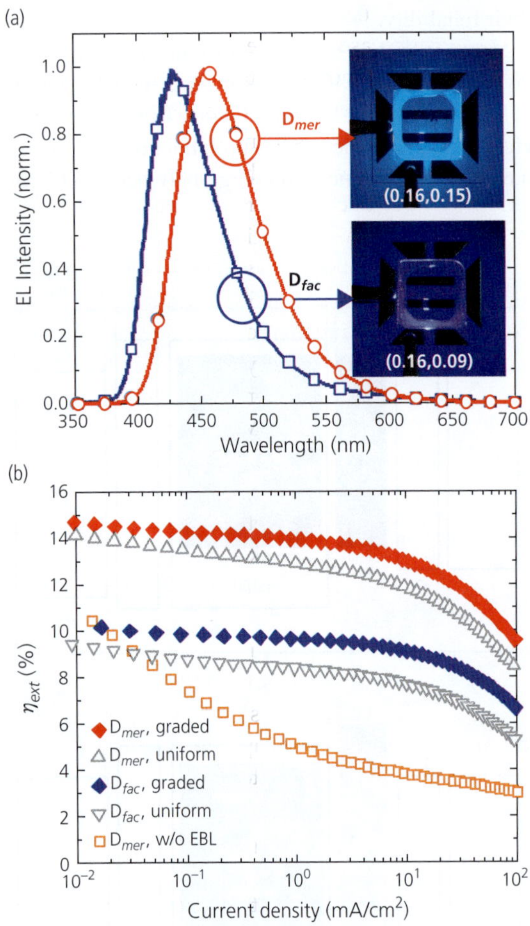

Figure 6.54 (a) Normalized EL spectra of *mer*- and *fac*-Ir(pmp)$_3$ PHOLEDs (labeled D_{mer}, D_{fac}, respectively). Inset: Photographic images of the deep blue emission from these PHOLEDs along with their CIE coordinates (in parentheses). (b) External quantum efficiency vs. current density for D_{fac} and D_{mer} with the dopant graded or uniformly doped throughout the EML. Squares are data for D_{mer} without an EBL (Lee et al., 2016).

its higher radiative rate. Note that there are two D_{mer} and D_{fac} devices shown: one each with uniformly doped EMLs, and the other with the doping graded from 8 vol.% to 20% to more evenly distribute the exciton formation zone across the EML, thus reducing the rate of TTA (Section 6.3.5). As discussed above, the dopant is a conductor whose hole mobility is larger than the electron mobility in the host (TSPO1). Decreasing the Ir(pmp)$_3$ concentration by grading moves the recombination zone away from the HTL/EML interface toward the ETL, thus reducing the opportunity for annihilation, and hence increasing efficiency.

Finally, we point out that D_{mer} in the absence of an EBL has a comparatively low quantum efficiency that continuously decreases from its value at $j \to 0$. Clearly, while charge blocking is challenging in deep blue emitting devices, it is nevertheless essential for achieving high efficiency at high brightness.

An example green emitting PHOLED employing a doped p-i-n structure is shown in Fig. 6.55. The energy level diagram is shown in the inset. The device consists of an ITO anode, and a 50 nm thick layer of m-MTDATA doped p-type with 2 mol. % F_4-TCNQ that serves as both HIL and HTL. This is followed by a 10 nm thick, wide energy gap Ir(ppz)$_3$ EBL that also confines excitons within the 5 nm thick, 7 wt% Ir(ppy)$_3$ in CBP EML. Here, Ir(ppz)$_3$ has a very shallow LUMO energy that provides a barrier to electron conduction, yet its HOMO is aligned with that of m-MTDATA to allow efficient hole injection into the EML. Furthermore, its wide energy gap prevents excitons from quenching in the doped HTL. The ETL/exciton blocking layer is comprised of a 55 nm thick Bphen layer that was either undoped, or doped with LiF or Li to

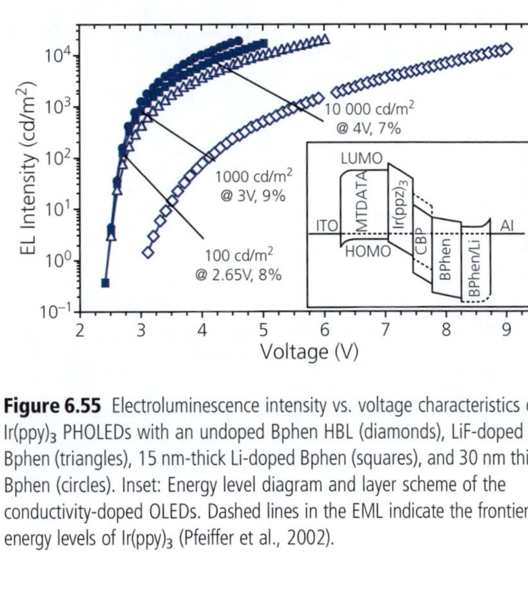

Figure 6.55 Electroluminescence intensity vs. voltage characteristics of Ir(ppy)$_3$ PHOLEDs with an undoped Bphen HBL (diamonds), LiF-doped Bphen (triangles), 15 nm-thick Li-doped Bphen (squares), and 30 nm thick Bphen (circles). Inset: Energy level diagram and layer scheme of the conductivity-doped OLEDs. Dashed lines in the EML indicate the frontier energy levels of Ir(ppy)$_3$ (Pfeiffer et al., 2002).

increase conductivity. This is capped by a Li-doped Bphen EIL with an electron concentration of $\sim 10^{18}$ cm^{-3}, and finally an Al cathode. The total thickness of the ETL and EIL combination is 55 nm.

To understand the effectiveness of n-type doping of the ETL on the operating voltage, three devices were compared with their EL intensity vs. voltage behavior shown. The undoped ETL showed considerably higher operating voltage (i.e. the voltage to achieve a given luminance) than devices where the ETL was doped with either LiF or Li. For example, the undoped device achieves a luminance of 10,000 cd/m^2 at 8.5 V, compared to only 4 V for devices with Li-doped Bphen layers (Pfeiffer et al., 2002).

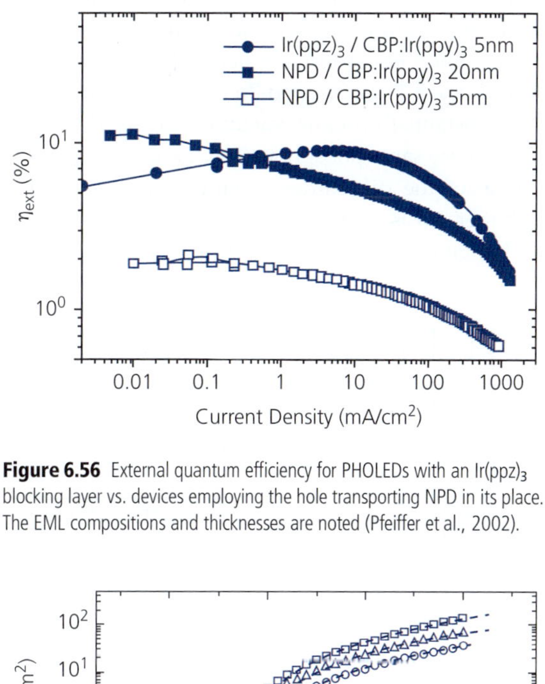

Figure 6.56 External quantum efficiency for PHOLEDs with an $Ir(ppz)_3$ blocking layer vs. devices employing the hole transporting NPD in its place. The EML compositions and thicknesses are noted (Pfeiffer et al., 2002).

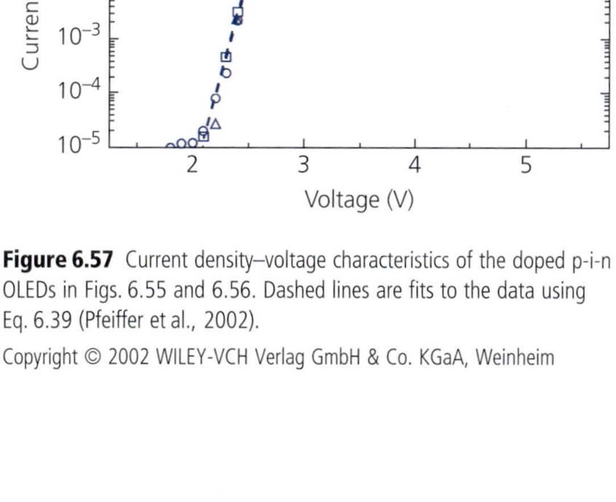

Figure 6.57 Current density–voltage characteristics of the doped p-i-n OLEDs in Figs. 6.55 and 6.56. Dashed lines are fits to the data using Eq. 6.39 (Pfeiffer et al., 2002).

The $Ir(ppz)_3$ HBL increases the quantum efficiency, particularly for very thin EMLs where exciton leakage into the doped HTLs results in quenching, or alternatively, very low emission intensity in the HTL due to electron leakage. Figure 6.56 shows the external quantum efficiency for a 5 nm thick EML device with the $Ir(ppz)_3$ blocking layer, compared to a device lacking this layer and another with a thicker (20 nm) EML. In the devices without $Ir(ppz)_3$, the EBL was replaced with an undoped hole transporting but non-blocking NPD layer to maintain a constant device thickness. If there is unity quantum efficiency, the light output should be directly proportional to the current, that is, every electron injected results in the emission of a photon. In Fig. 6.56, this is clearly not the case: often there is a decrease in efficiency at low current due to traps that lead to non-radiative exciton recombination. Once the traps are filled at higher current (i.e. the recombination rate is less than the charge injection rate), the efficiency increases. At the highest currents, exciton annihilation with charges and/or other excitons leads to a further decrease in efficiency. These processes will be discussed in subsequent sections: here we point out that the devices in Fig. 6.56 suffer from both trap recombination and annihilation.

The increase in efficiency for the 5 nm thick EML with the blocking layer compared to one without is dramatic. The thicker EML results in the same high efficiency as the thin, blocked device but only at very low currents. There, the hole concentration is sufficiently low for complete recombination of all injected holes within the thick emission zone. However, charge balance is reduced at higher currents due to leakage into the HTL.

Interestingly, the p-i-n current density–voltage (j–V) characteristics differ from trap-limited space charge conduction commonly observed for simple bilayer OLEDs considered in Chapter 4. In Figure 6.57 we show fits of the j–V characteristics to a Shockley-type diode equation:

$$j = j_0\left(\exp\left(q\{V - [j/j_T]^{1/m}\}/nk_BT\right) - 1\right), \tag{6.39}$$

where, following our notation in Chapter 4, j_0 is the saturation current, and n is the ideality factor. The term in square brackets is the voltage drop outside of the junction due to space charge limited current, where $j_T = (9/8)(\varepsilon_r\varepsilon_0\mu/d^3)$. Here, μ is the mobility, d is the thickness of the undoped EML, and m is the characteristic trap exponent. Then $m = 2$ for trap free transport, and $n = 2$ for generation and recombination via mid-energy-gap states.

The fits to Eq. 6.39 shown by dashed lines in Fig. 6.57 extend over seven decades using $m = 1.7$

and $n = 2.1$. This implies that at high voltages, the current is dominated by trap-free space charge limited injection, while at low currents (in the exponential region of the *j–V* characteristics), the dominant conduction mechanism is diffusion across the anisotype HJ limited by generation-recombination within the energy gap. That is, the doped device behaves analogously to an inorganic p-n junction diode, rather than as a dielectric OLED simply limited by trap-dominated space charge limited currents, as in the device in Fig. 6.22. The fits for the different samples are due to differences in the doping of the ETL. This reflects that the doped Bphen HTL (Sample C) has a thinner space charge region (i.e. smaller *d*) than that doped only half way across (Sample B), or one in which LiF leads to only a very thin doped region near the cathode (Sample A).

6.3.3.2 Polymer PHOLEDs

There are three strategies for realizing high efficiency polymer PHOLEDs. The most actively pursued is blending small molecule phosphorescent complexes into a polymer host. The polymer host transports the electrons and holes to the doped region, followed by energy transfer from the polymer to the phosphor molecule. In a second copolymerization strategy, the phosphor moiety is part of the polymer backbone. The third approach incorporates the triplet-emitting transition metal complex as a pendant group attached to the polymer chain. The highest efficiency is achieved in multilayer combinations comprising an HIL and active region adjacent to the anode contact, followed by vacuum deposited small molecule EBL and ETL layers capped by a metal cathode (Han et al., 2015).

The most commonly used polymer host is the non-conjugated PVK due to its high singlet and triplet energies that enable efficient energy transfer for chromophores spanning the visible spectrum. The effectiveness of multilayer structures, particularly the use of an exciton blocking layer in PVK:Ir(ppy)$_3$ PHOLEDs is illustrated by comparing devices with and without an EBL/ETL bilayer deposited on the polymer blend EML. For example, a 70 nm thick layer of PVK:Ir(ppy)$_3$ (3.5%) blend sandwiched between an ITO anode and a Mg:Ag cathode was directly compared to a device comprising a 70 nm PVK$_{0.9}$PBD$_{0.1}$:Ir(ppy)$_3$ (3.5%)/5 nm TPBI/30 nm Alq$_3$ PHOLED with contacts similar to that used in the single layer device (Vaeth and Tang, 2002). Here, PBD is blended with PVK due to its superior electron conducting properties. Furthermore, TPBi is a vacuum-deposited, wide energy gap blocking layer. Finally, Alq$_3$ is the ETL, also deposited in vacuum. The single layer device had an external quantum efficiency of approximately 2.5% at an operating voltage of 15.1 V, whereas the device with the EBL had $\eta_{ext} = 6.7\%$ at 15.9 V, increasing to 8.5% at somewhat lower voltages. Thus, blocking layers can improve exciton confinement in both polymer and small molecule PHOLEDs.

The ability of PVK to excite triplet emission across the visible spectrum is demonstrated by blending with the Ir(III) dopants in Fig. 6.58a. The spectra of the devices shown in Fig. 6.58b span the visible spectrum from the sky blue when doped with 10 wt% FIrpic, 5 wt% Irppy$_3$ in the green, 5 wt% Bt$_2$Ir(acac) in the yellow, and 5–10 wt% Btp$_2$Ir(acac) in the deep red. Longer wavelength emission is accessed by increasing the conjugation length of the ligand. Blue emission from FIrpic is achieved by attaching electron-withdrawing fluorine atoms to the phenyl group. The peak external efficiencies of the devices were 1.2%, 5.2%, 2.0%, and 2.1%, respectively, which are substantially below vacuum deposited devices using these same phosphors (Kawamura et al., 2002).

One of the difficulties in employing conjugated polymer hosts is that they tend to have a large exchange energy (~0.5–0.7 eV) that appears to be

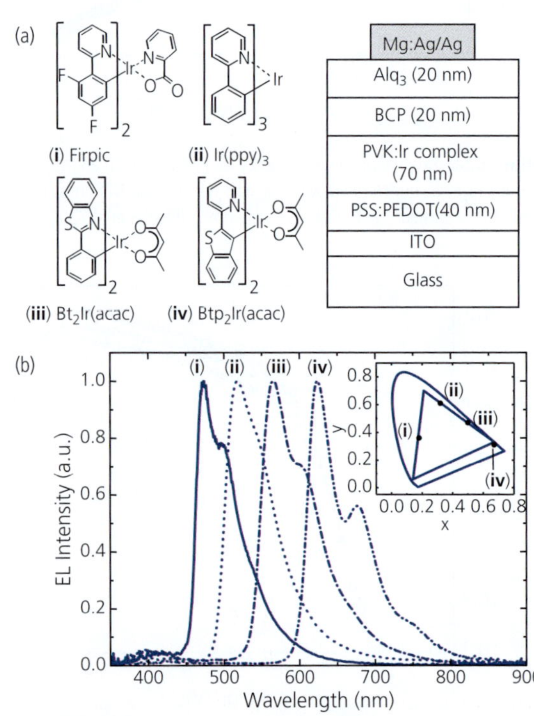

Figure 6.58 (a) Molecular structures of four Ir(III) complexes numbered **i–iv** used in the device structure on the right. (b) Spectra from the PVK-based PHOLEDs in (a) along with their CIE chromaticity coordinates (inset) (Kawamura et al., 2002).

independent of polymer composition or chain length. The resulting low triplet energy makes it difficult or even impossible to transfer triplets from the host to a comparatively high energy green or blue emitting phosphor. However, when conjugated polyfluorenes are used as hosts for deep red emitting phosphors, the PHOLED performance is comparable with non-conjugated PVK (Jiang et al., 2004).

A striking example of the material-independent exchange energy is shown in Fig. 6.59a for a series of conjugated polymers including PPP, ladder type LPPP, phenyl-substituted PPV, polyfluorenes, as well as polymers with Pt in the main chain (Köhler and Beljonne, 2004). It is remarkable that the exchange energy, $\Delta E_{ST} \approx 0.7$ eV, which is the difference between the $S_1 \rightarrow S_0$ and $T_1 \rightarrow S_0$ transitions, is independent of the polymer composition. To

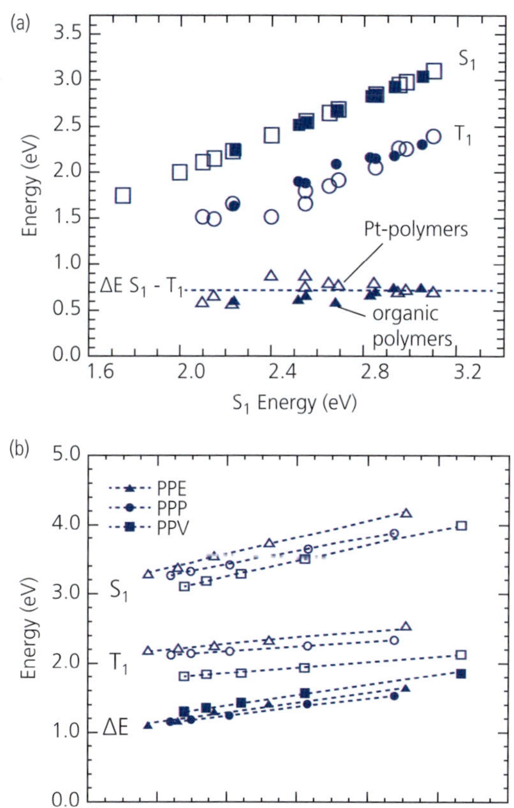

Figure 6.59 (a) The lowest singlet (S_1) and triplet (T_1) energies vs. S_1 energy for a variety of conjugated Pt-containing and organic conjugated polymers. Also shown is their energy difference of $\Delta E_{ST} \approx 0.7$ eV for all materials investigated. (b) Calculated transition energies for three conjugated molecules as a function of (oligomer length)$^{-1}$ (Köhler and Beljonne, 2004).

understand this phenomenon, Hartree–Fock semi-empirical intermediate neglect of differential overlap calculations (Section 3.4.2) were used to estimate the excited state energies of oligomers of varying length of three molecules, namely PPP, PPV, and PPE. The theoretical results in Fig. 6.59b are consistent with experiment. That is, while the energies of S_1 and T_1 decrease only weakly with increasing oligomer length due to the increase in conjugation length, and is independent of polymer composition. The residual dependence of ΔE_{ST} on oligomer length is due to the limited extent of the singlet and triplet states. Torsional twists in the chain along the polymer backbone increase the dependence of S_1 energy on length, while the far more compact triplet state does not have a length dependence (Köhler and Beljonne, 2004). This limitation in tuning ΔE_{ST} is lifted to a significant degree in small molecules, where the exciton radius, r_{eh}, can be engineered by structurally separating the electron and hole on electron withdrawing and donating groups within a single molecule. Indeed, ΔE_{ST} can be reduced to nearly zero in small molecules optimized for delayed fluorescence (see Sections 3.7.5 and 6.3.4).

High efficiency polymer PHOLED emission across the visible spectrum has been achieved using the hybrid solution process-vacuum deposited device in Fig. 6.60. The emissive material consists of a vinyl backbone onto which are appended both a hole transporting carbazole unit, as well as 0.1–2.0 mol. % of either a red, green or blue phosphorescent (RPP, GPP, BPP, respectively) functional group. The Ir(III) complexes are red emitting $Btp_2Ir(acac)$, green $Irppy_2(acac)$ or blue FIrpic. Electron conduction is enhanced by also blending in 30 wt% OXD-7 into the 50 nm thick emissive layer. Importantly, the efficiency is increased by 20% in the red to nearly 100% in the blue by including the vacuum deposited BAlq hole blocking layer (Tokito et al., 2003a, 2003b).

The emission spectra of the three different devices are shown in Fig. 6.61, with the devices exhibiting $\eta_{ext} = 6.6\%$ (R), 11% (G), and 6.9% (B). Appending the emissive group onto a polymer backbone, along with functionalization of the polymer and blend to balance carrier injection can be used to precisely control the emissive dopant concentration while preventing aggregation. However, the molecular synthesis routes leading to these structures are more complex than in simple blends, and hence may increase the cost of the materials.

Finally, we discuss copolymers that combine a transporting moiety and the triplet-emitting center

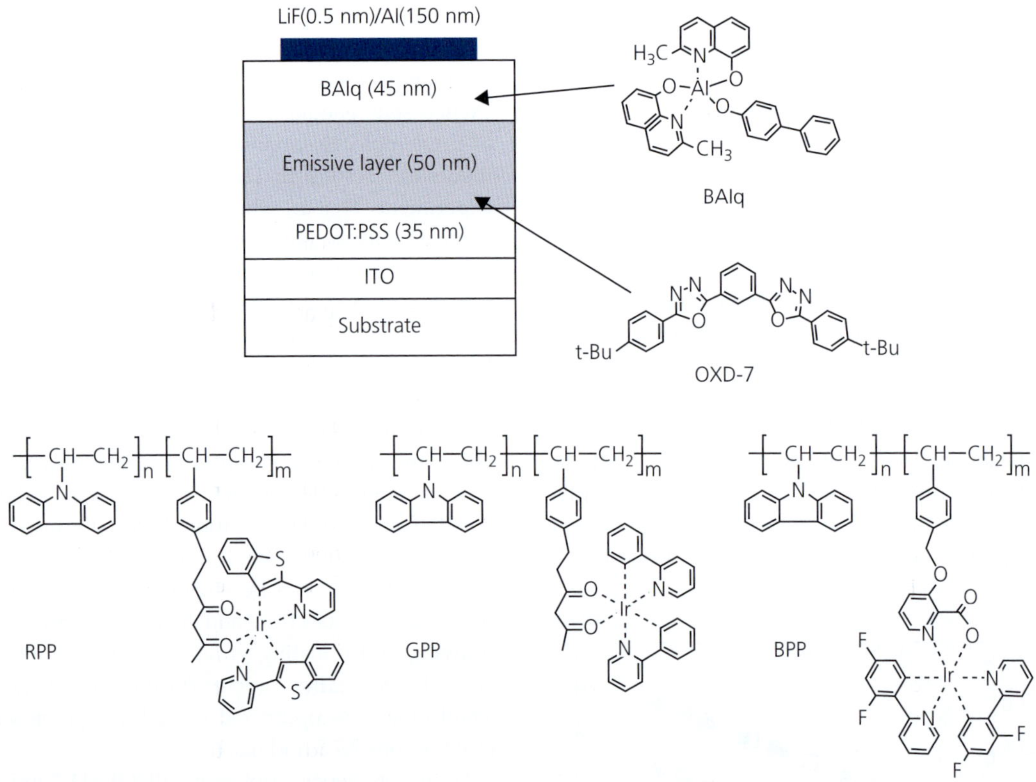

Figure 6.60 Red, green, and blue device structures and chemical structural formulae of materials used in the blocking layer (BAlq) and in the emissive layer (OXD-7), and the red (RPP), green (GPP), or blue (BPP) phosphorescent polymer materials (Tokito et al., 2003a).

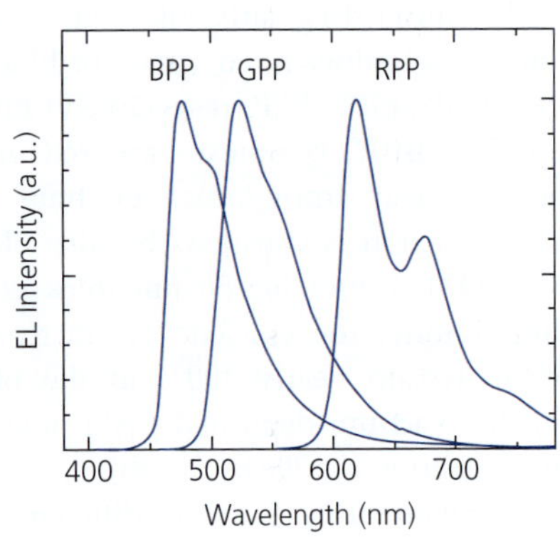

Figure 6.61 Emission spectra of the devices in Fig. 6.60 (Tokito et al., 2003a).

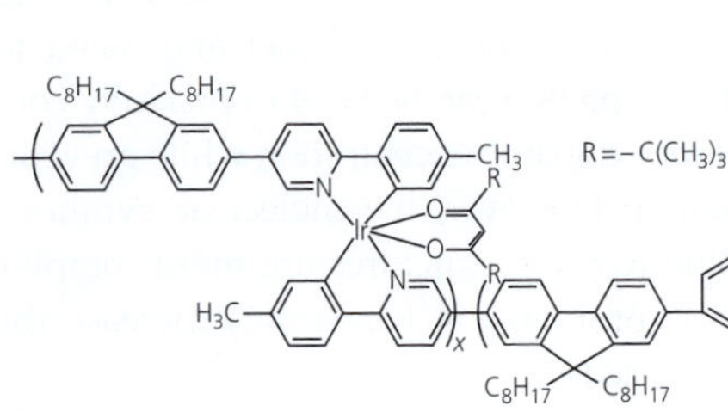

Scheme 1.

within the chain itself. Such a bis-chelating Ir(III) copolymer is shown in **Scheme 1** (Zhen et al., 2005). These polymers consist of a polyfluorene-chelating unit combined with a fluorine-*alt*-carbazole (PFCz) unit. Energy transfer from the fluorine to the emissive complex is nearly complete, even at complex concentrations of only 4%. Indeed, it was observed that intrachain energy transfer is more efficient than transfer to small molecules in blended PHOLEDs with comparable concentrations of the same Ir complex. Interestingly, the peak wavelength of the copolymer is shifted to 575 nm from 550 nm for the analogous blended device. This results from the extended conjugation of the pyridine group due to its linkage with the neighboring fluorenes.

A PHOLED with the structure ITO/PEDOT:PSS/PVK/copolymer/Ba/Al, where $(x, y) = (0.04, 0.96)$ in Scheme 1, had a peak quantum efficiency of 4.1% along with a somewhat elevated voltage of 13 V at 32 mA/cm^2.

In summary, numerous strategies have been devised to achieve efficient polymer PHOLEDs based on both Pt(II) and Ir(III) complexes. Generally, the highest quantum efficiencies and lowest voltages are obtained by blending a small molecule triplet-emitting molecule along with a charge conducting polymer host that transfers energy to the emitter. High efficiency can also be obtained by combining polymeric emitting regions with a vacuum deposited EBL and ETL. In spite of the considerable effort that has been devoted to developing solution processed PHOLEDs, the highest performances across the visible spectrum have been achieved using small molecule devices that benefit from the freedom to deposit high purity layers of precisely controlled thicknesses, and that are optimized to perform their separate functions within the device.

6.3.3.3 Triplet emitting complexes

Substantial efforts have focussed on developing high efficiency triplet emitting, transition metal complexes across the visible spectrum. The most successful materials are based on Ir(III) complexes, although efficient emission has also been achieved using Pt(II)-, Os(III)-, Au(III)-, and Cu(II)-centered molecules. In Fig. 6.62 and Table 6.7 we provide a sampling of Ir, Pt, Cu, and Os complexes, including their efficiencies and color coordinates. Further examples of heavy metal phosphor complexes not found in Table 6.7 will be provided in the following sections as their uses in specific device examples are introduced. For a more complete discussion of such complexes, the reader is referred to several reviews (Lamansky et al., 2001, Thompson et al., 2007, Yam and Wong, 2011, Choy et al., 2014, Jou et al., 2015, Li et al., 2016b).

With few exceptions (mostly the Cu complexes with comparatively low spin orbit coupling due to the low atomic number, $Z = 29$), the efficiencies are $\geq$20%. As apparent from in Eq. 6.1, their 100% emission efficiency results in a potential efficiency of 20% when OLEDs are grown on glass. Reports of higher efficiencies are almost certainly the result of light outcoupling enhancements, either due to a modified device structure or due to alignment of TDMs parallel to the substrate (see Section 6.6).

As noted above, the emission is red-shifted by increasing the conjugation length of the ligands. Hence, red emitting phosphors often feature an extra phenyl group fused to either the phenyl or pyrazole of the phenylpyridine ligand, whereas blue emitting phosphors contain electron-withdrawing halides or cyano groups. The molecules in Table 6.7 span the visible from the deep blue to the deep red. Even 8.5% efficient infrared PHOLEDs with a narrow peak at $\lambda =$ 772 nm have been demonstrated based on Pt(II)-tetraphenyltetrabenzoporphyrin doped at 4 wt% in Alq_3 (Sun et al., 2007).

In Fig. 6.63 we show EL images from a few, high efficiency ($\eta_{int} \sim 100\%$) Ir-complex based PHOLEDs that span the visible spectrum. Also shown are the coordinates of these emitters on the accompanying CIE chromaticity chart. Apparently, even this limited selection of electrophosphorescent OLEDs covers the sRGB and NTSC standard color gamuts indicated by the two triangles in the chart.

6.3.4 Thermally assisted delayed fluorescence

An alternative way to exploit high efficiency triplet emission is known as thermally assisted delayed fluorescence (TADF). This process, described in Section 3.7.5, entails the use of molecules whose singlet–triplet exchange energy splitting, ΔE_{ST}, is comparable to or less than the thermal energy, $k_B T$. From perturbation theory, the intersystem crossing rate for $S_1 \leftarrow T_1$ is inversely proportional to ΔE_{ST}, with rate $k_{ISC,T} \sim \left|\frac{\langle S_1|H_{SO}|T_1\rangle}{\Delta E_{ST}}\right|^2$ (cf. Eq. 3.189). Hence, the endothermic intersystem energy transfer $S_1 \leftarrow T_1$ can be nearly 100% efficient provided that competing non-radiative rates are less than $k_{ISC,T}$. This transition, therefore, leads to delayed fluorescence. The $S_1 \leftarrow T_1$ process is sometimes referred to as reverse intersystem crossing (RISC), but in fact it is simply just an endothermic variation of the exothermic ISC process.

The energetic processes leading to efficient TADF are shown in Fig. 6.67. Efficient radiative transitions from the singlet state require that the natural rate of fluorescence, k_F, exceed all other rates. In particular, $k_F >> k_{ISC,S}, k_{ISC,T}, k_{ph}$, where $k_{ISC,S}$ corresponds to the rate of the exothermic ISC process. Since TADF employs molecules with a small ΔE_{ST} that lack a heavy metal atom, the phosphorescent emission rate, k_{Ph}, is expected to be smaller than the other competing rates except for non-radiative transitions from T_1 at rate k_{nrT}. Also, for high efficiency, it follows that the non-radiative transition rate of $S_1 \rightarrow S_0$ must be $k_{nrS} \ll k_F$.

Minimization of ΔE_{ST} requires that the HOMO/LUMO overlap within the emissive molecule also be small. This can be achieved with dipolar molecules comprising spatially separated donor and acceptor groups whose interactions are reduced by a twist along the molecular backbone. That is, the small exchange energies of TADF molecules results from reduced *intramolecular charge transfer* between the donor and acceptor groups. The same feature leads to a broadened, featureless emission spectrum characteristic of charge transfer states. This strategy has led to the development of a series of dopants with

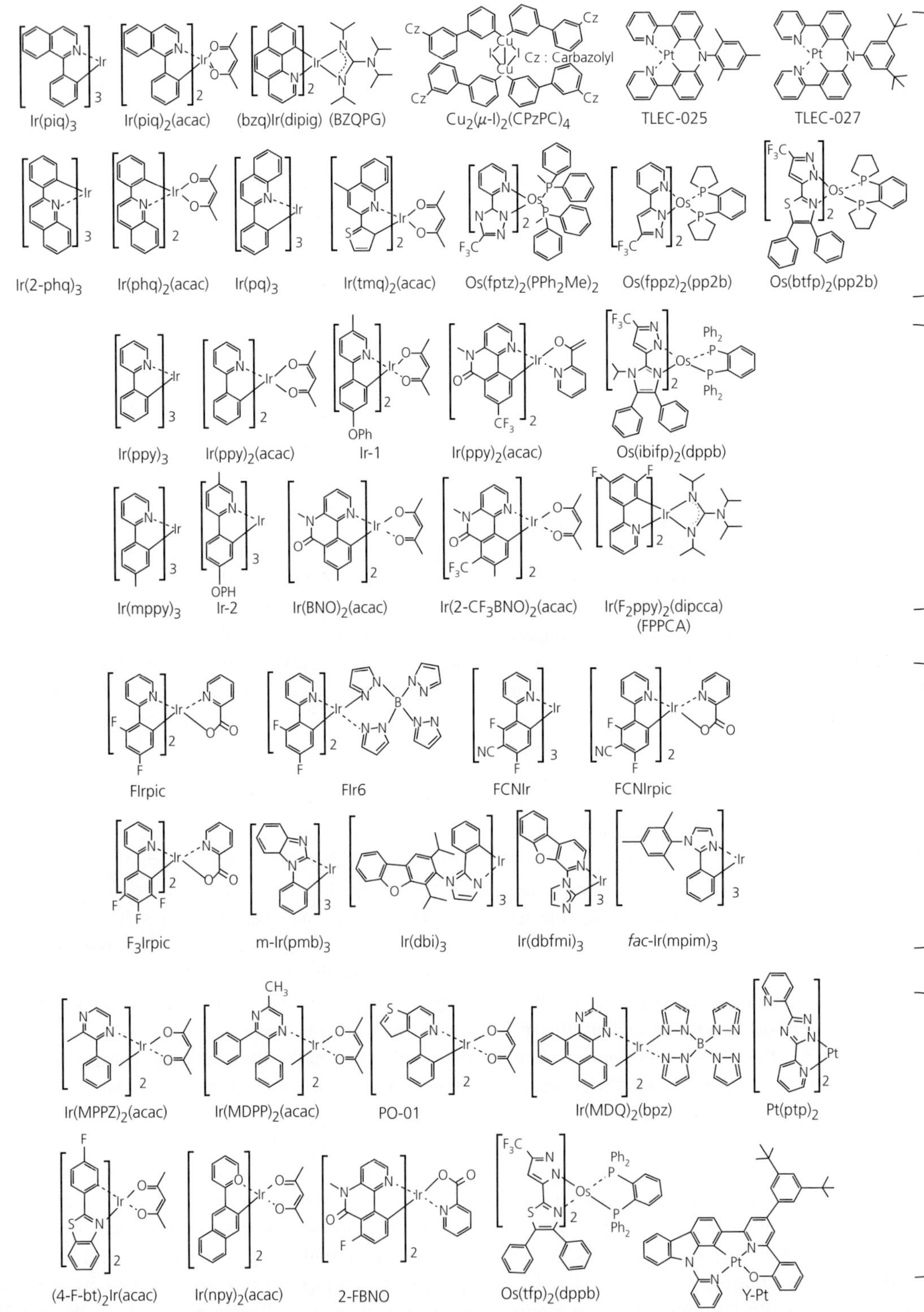

Figure 6.62 Example transition metal complexes used in red, green, blue, and yellow emitting PHOLEDs (Jou et al., 2015).

Table 6.7 Performance of PHOLEDs employing the triplet emitting complexes in Fig. 6.62 and their hosts. References with further information for each device are in Jou et al. (2015)

Emitter[a]	Host[b]	η_c (cd A)$^{-1}$	η_p (lm W^{-1})	η_{ext} (%)	CIE$_{xy}$/λ_{EL} (nm)
$Ir(piq)_3$	CzPPQCz	27.5	24.1	21.4	(0.67, 0.33)/—
$Ir(pq)_3$	CzPPQ	75.8	68.1	25.6	(0.55, 0.45)/—
$Ir(piq)_2(acac)$	BIQF	28.5	24.9	24.4	(0.68, 0.32)/623
$Ir(2\text{-}phq)_3$	TCTA/TPBi	47.4	47.2	20.3	(0.57, 0.43)/—
$Ir(phq)_2(acac)$	$Bebq_2$	26.53	29.58	21.0	(0.62, 0.37)/—
$Ir(tmq)_2(acac)$	BIQMCz	32.4	30.9	24.9	—/618
FPPCA/FPPCA + BZQPG/BZQPG/BZQPG + CBP		—	76.8	27.9	(0.57, 0.41)/—
TLEC-025	$Bebq_2$	—	20.7[a]	18.5[a]	(0.66, 0.34)/—
TLEC-027	$Bebq_2$	—	25.2[a]	18.2[a]	(0.66, 0.34)/—
$Os(fptz)_2(PPh_2Me)_2$	POAPF	34.5	43.9	18.4	
$Os(fppz)_2(pp2b)$	CzPhO	34.8	45.2	14.3	(0.54, 0.46)/—
$Os(btfp)_2(pp2b)$	TCTA	18.3	17.6	15.6	(0.63, 0.37)/—
$Cu2(\mu\text{-}I)_2(CPzPC)_4$	—	15.9	14.0	6.6	—/590
$Ir(ppy)_3$	CBP	105	133	29	(0.32, 0.64)/514
$Ir(mppy)_3$	CBP + PBD + TPD	69.0	60.0	22.0	—/510
$Ir(ppy)_2(acac)$	TCTA + B3PYMPM	106	127.3	30.2	—/520
$(BNO)_2Ir(acac)$	CBP	89.1[a]	—	—	(0.22, 0.51)/—
$(2\text{-}CF_3BNO)_2Ir(acac)$	CBP	94.5[b]	69.3[b]	25.7[b]	(0.28, 0.64)/—
$(3\text{-}CF_3BNO)_2Ir(prz)$	CBP	95.6[b]	60.8[b]	23.8[b]	(0.34, 0.60)/—
$Os(ibifp)_2(dppb)$	TCTA	48.3	50.9	15.6	(0.30, 0.53)/—
Ir-1	CBP	84.6	77.6	24.5	(0.26, 0.63)/504
Ir-2	CBP	76.2	72.8	22.5	(0.29, 0.59)/500
FIrpic	CbBPCb	53.6	50.6	30.1	(0.14, 0.30)/—
FIr6	mCP + UGH3	39.5	39.2	22.9	(0.16, 0.29)/—
FCNIr	PPO2	21.1	16.6	18.4	(0.14, 0.15)/—
FCNIrpic	mCPPO1	31.0	29.8	25.1	(0.14, 0.18)/—
F_3Irpic	UGH2	37.6	33.7	14.3	—/484
$m\text{-}Ir(pmb)_3$	UGH2	—	1.7	5.8	(0.17, 0.06)/—
$Ir(dbfmi)_3$	PO9	28.6	35.9	18.6	(0.15, 0.19)/—
$Ir(dbi)_3$	mCP	61.5	43.7	23.1	(0.19, 0.44)/
$fac\text{-}Ir\text{-}(mpim)_3$	TCTA	72.9	92.3	32.2	—/474
$Ir(MPPZ)_2(acac)$	CBP	37.3	20.3	13.2	—/580
$(4\text{-}F\text{-}bt)_2Ir(acac)$	CBP	52.4	29.4	17.2	(0.48, 0.51)/—
$Ir(MDPP)_2(acac)$	NPB	—	9.9	6.2	(0.52, 48)/—
$Ir(npy)_2(acac)$	mCP	33.6	21.8	10.5	(0.45, 0.54)/—
PO-01	CBP	83.0[a]	82.0[a]	26.0[a]	(0.49, 0.51)/—
$(2\text{-}BNO)_2Ir(prz)$	TCTA	75.9[a]	84.8[a]	22.1[a]	(0.45, 0.54)/—
$Ir(MDQ)_2(Bpz)$	CBP	77.4	50.7	21.5	(0.38, 0.60)/—
$Os(tfp)_2(dppb)$	TCTA	42.1	47.2	15.7	(0.49, 0.48)/—
$Pt(ptp)_2$	—	—	79.2[a]	20.5[a]	(0.49, 0.50)/—
Y-Pt	mCP	74.9	52.1	—	(0.52, 0.47)/—

[a] At 100 cd m^{-2}.
[b] At 1000 cd m^{-2}, / for the double emissive layer, and + the mixed host emissive layer.

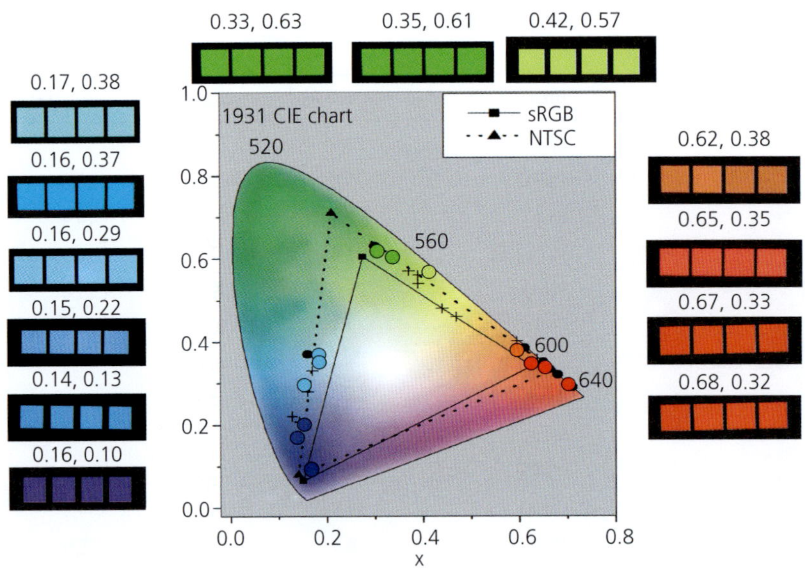

Figure 6.63 Photographic EL images of Ir-based PHOLEDs and their color coordinates plotted in the 1931 CIE chromaticity chart (circles). Also indicated are the sRGB and NTSC color spaces. Courtesy, Universal Display Corp.

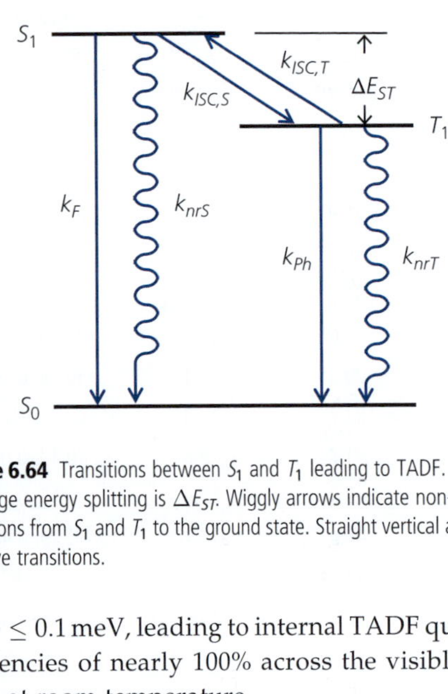

Figure 6.64 Transitions between S_1 and T_1 leading to TADF. The exchange energy splitting is ΔE_{ST}. Wiggly arrows indicate non-radiative transitions from S_1 and T_1 to the ground state. Straight vertical arrows are radiative transitions.

$\Delta E_{ST} \leq 0.1$ meV, leading to internal TADF quantum efficiencies of nearly 100% across the visible spectrum at room temperature.

An efficient molecular motif is based on a carbazolyl dicyanobenzene structure, where the carbazole is an electron donating group and the dicyanobenzene is an electron withdrawing acceptor (Uoyama et al., 2012). Several examples of these compounds and their PL spectra are shown in Fig. 6.65. The molecular structural formulae show a pronounced twist between the carbazole and the dicyanobenzene groups that minimize orbital overlaps, and hence results in a small exchange energy. In Fig. 3.87 we showed that the lowest unoccupied natural transition orbital is centered on the cyanobenzene, whereas the highest occupied transition orbital centers on the carbazole groups in 4CzIPN, with similar orbital configurations expected for analogous molecules in this series. A consequence of the minimal overlap of the transition orbitals is a featureless, broad excimer-like luminescence spectrum. Hence, TADF molecules tend to have non-saturated emission spectra compared with conventional fluorescent or phosphorescent lumophores.

Figure 6.66 shows emission transients of a TADF-based OLED consisting of 5 wt% 4CzIPN in CBP, forming the emissive layer. Prompt fluorescence with a natural lifetime of $\tau_p = 18$ ns, followed by TADF with $\tau_{TADF} = 5.1$ µs at 300 K is shown in Fig. 6.66a. The transient lifetime increases as temperature is decreased due to thermally activated energy transfer, as implied in Fig. 6.64. The spectra for both the prompt and delayed fluorescence signals in Fig. 6.66b have peaks at a wavelength of 507 nm, which supports the conclusion that the delayed signal is due to the same $S_1 \to S_0$ transition as the prompt fluorescence emission.

The TADF efficiency is the integral of both the prompt and delayed emission spectra. That is, from Eq. 6.1,

$$\eta_{ext} = \eta_{int}\eta_{out} = (\eta_F + \eta_{DF})\eta_{out}, \qquad (6.40)$$

where for consistency, we use the formal definition of rapid emission as fluorescence (with efficiency η_F),

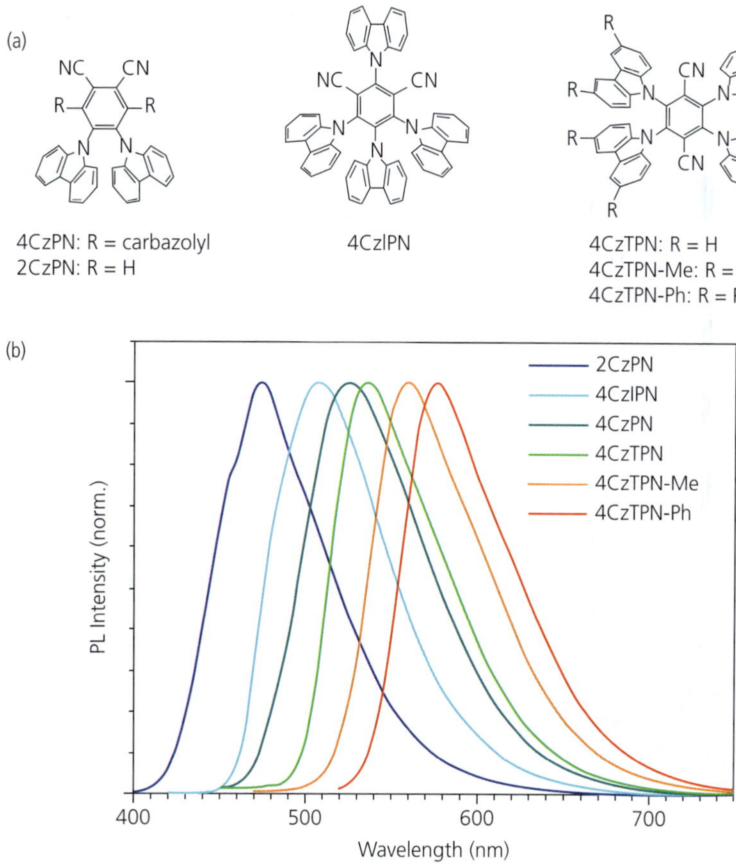

Figure 6.65 (a) Carbazolyl dicyanobenzene molecules and (b) their characteristically featureless photoluminescence spectra. Functionalizing the molecular species leads to tuning of the spectral peak from the blue into the red (Uoyama et al., 2012).

and delayed fluorescence (with efficiency η_{DF}). Then it follows:

$$\eta_{\text{int}} = \gamma\Big[\chi_S\phi_F + \Big(\chi_S\phi_{ISC,S} + \chi_T\Big)\phi_{ISC,T}\Big]. \tag{6.41}$$

Here, γ is the charge balance factor, ϕ_F is the fluorescence quantum yield, and $\phi_{ISC,S}$ and $\phi_{ISC,T}$ are the quantum yields for producing triplets and singlets, respectively (cf. Fig. 6.64). Hence, the first term is simply η_F and the second is the efficiency for emission via TADF, namely η_{DF}.

Figure 6.66c shows the PL quantum efficiency for both the fluorescent and TADF emission, along with the combined spectra that corresponds to the integrated area under the transient PL data. There is almost no temperature dependence of the prompt spectral intensity, as expected for direct fluorescence with rate k_F. However, the TADF signal appears to decrease by approximately 30% from room temperature to $T < 10$ K, indicating a weak temperature dependence for this endothermic process. The temperature dependent emission rate is understood if we assume that T_1 and S_1 are in thermal equilibrium (Kirchhoff et al., 1983). The ratio of concentrations of excitations in state S_1 to T_1 (corresponding to $[S_1]$ and $[T_1]$, respectively), is given by their respective ISC rates, that is,

$$K = [S_1]/[T_1] = k_{ISC,T}/k_{ISC,S}. \tag{6.42}$$

This relationship is valid if the rates are greater than the competing rates in Fig. 6.64 to allow for equilibrium between the two states to be established. That is, we assume that and $k_{ISC,S} \gg k_{Ph} + k_{nrT}$ and $k_{ISC,S} \gg k_F + k_{nrS}$. Then it follows that the quantum yields from T_1 and S_1 are

$$\phi_{Ph} = \frac{k_{Ph}}{(k_F + k_{nrS})K + k_{Ph} + k_{nrT}} \tag{6.43a}$$

and

$$\phi_F = \frac{k_F K}{(k_F + k_{nrS})K + k_{Ph} + k_{nrT}}. \tag{6.43b}$$

Note that ϕ_{Ph} and ϕ_F are the radiative quantum yields, whereas $\phi_{ISC,T}$ and $\phi_{ISC,S}$ in Eq. 6.41 are the yields (i.e. the probabilities) of intersystem crossing for $S_1 \leftarrow T_1$ and $S_1 \rightarrow T_1$, respectively. Then, the rate for delayed fluorescence is given by

$$k_{TADF} = \frac{(k_F + k_{nrS})K + k_{Ph} + k_{nrT}}{1+K}. \tag{6.44}$$

In Section 3.7.5 we found the RISC rate is proportional to the Boltzmann factor, $\exp(-\Delta E_{ST}/k_B T)$, which weights the relative filling of T_1 and S_1 at thermal equilibrium. Since the degeneracy of the triplet is three times that of the singlet, and assuming that the zero field splitting for the triplet state is negligible, then we can write the relative concentrations of the triplet and singlet states, K, as

$$K = \frac{k_{ISC,T}}{k_{ISC,S}} = \frac{1}{3}\exp(-\Delta E_{ST}/k_B T), \tag{6.45}$$

which finally yields the expression for the rate of population of triplet states by ISC:

$$k_{ISC,T} = \frac{k_{ISC,S}}{3}\exp(-\Delta E_{ST}/k_B T). \tag{6.46}$$

(Compare this with Eq. 3.208, with $k^{\infty}_{ISC,T} = k_{ISC,S}/3$.)

By combining Eqs. 6.42, 6.44, and 6.45, we can obtain $k_{ISC,T}$ by making the reasonable assumptions that $k_{Ph} + k_{nrT} \ll (k_F + k_{nrS})$ and $K \lesssim 1$ in efficient TADF systems. Then,

$$k_{ISC,T} = \frac{k_{TADF}k_{ISC}}{k_F + k_{nrS} - k_{TADF}} \approx \frac{k_{TADF}k_{ISC}}{k_F}, \tag{6.47}$$

where on the right hand expression we have made the further assumption that k_{nrS} and k_{TADF} are small compared to the prompt fluorescence rate, k_F. Since both k_{ISC} and k_F are largely temperature independent, the activation energy taken from the slope of the k_{TADF} vs. $1/T$ gives a small exchange energy splitting via Eqs. 6.46 and 6.47. For 4CzIPN, for example, it was found that $\Delta E_{ST} = 83$ meV (Uoyama et al., 2012). This is more than three times the thermal energy at room temperature. From the intercept of a plot of $k_{ISC,T}$ vs. $1/T$, at $T \rightarrow \infty$ we obtain $k_{ISC,T} = k_{ISC,S}/3$.

Although the singlet–triplet splitting is larger than $k_B T$, endothermic transfer $S_1 \leftarrow T_1$ can, nevertheless, be efficient in the absence of parasitic decay channels. In Fig. 6.67 we show η_{ext} for three OLEDs with EMLs

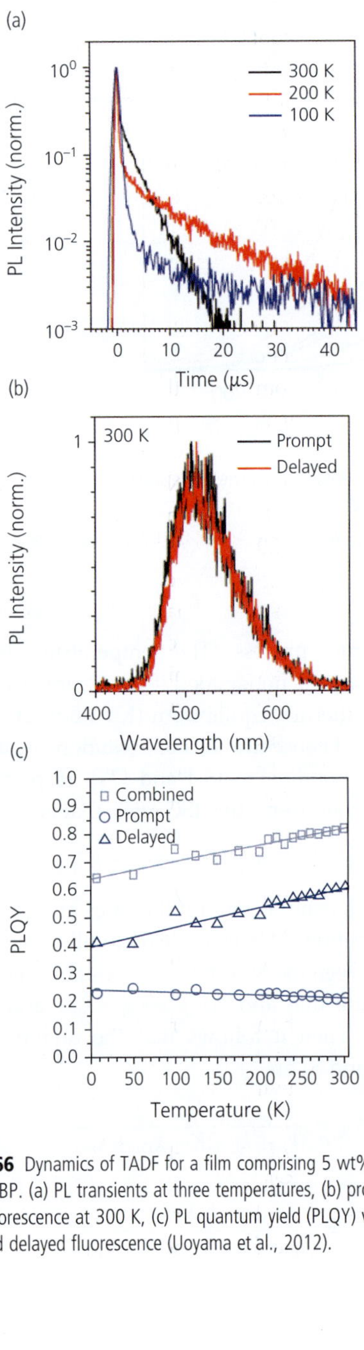

Figure 6.66 Dynamics of TADF for a film comprising 5 wt% 4CzIPN doped in CBP. (a) PL transients at three temperatures, (b) prompt and delayed fluorescence at 300 K, (c) PL quantum yield (PLQY) vs. T for prompt and delayed fluorescence (Uoyama et al., 2012).

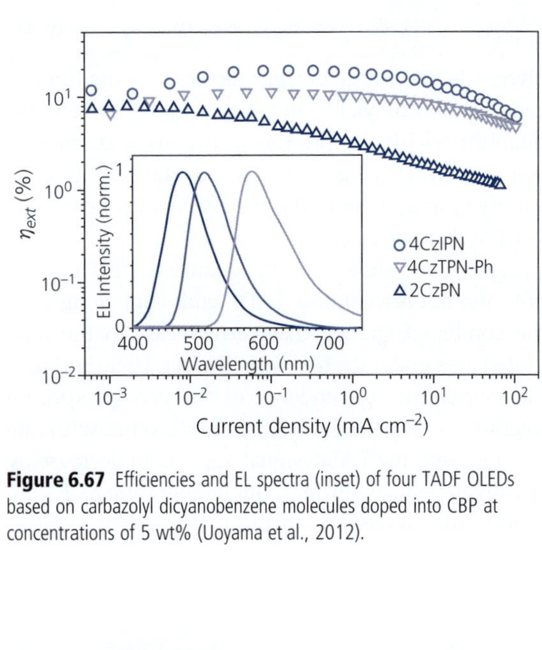

Figure 6.67 Efficiencies and EL spectra (inset) of four TADF OLEDs based on carbazolyl dicyanobenzene molecules doped into CBP at concentrations of 5 wt% (Uoyama et al., 2012).

comprising 15 nm 4CzIPN or 4CzTPN-Ph (5 wt%): CBP for green and yellow emitting devices, respectively. For cyan OLEDs, the EML was 20 nm 2CzPN(5 wt%):PPT:CBP. These structures are similar to those employed in PHOLEDs, consisting of a layering scheme that follows: anode/HTL/EML/HBL/ETL/EIL/cathode. Since triplets have a far longer lifetimes than singlets, it is important to confine them with an HBL/EBL around the EML to ensure that they are harvested by the TADF chromophores. Furthermore, the triplet and singlet energies of the host should exceed that of the guest to ensure efficient energy transfer. As in PHOLEDs, this transfer can occur by exchange (Dexter), dipole coupling (Förster) or even direct charge transfer to the guest from the adjacent ETL and/or HTL layers.

The green emitting 4CzIPN-based device has a peak room temperature efficiency of $\eta_{ext} = 19.3\%$, corresponding to $\eta_{int} \simeq 100\%$, with $\eta_{ext} = 11 \pm 1\%$ for the yellow, and $8 \pm 1\%$ for the cyan devices (Uoyama et al., 2012). Since these are inherently phosphorescent devices (i.e. emission rates are controlled by the triplet lifetime, $1/k_{ph}$), they are capable of harvesting both triplets and singlets similar to heavy metal triplet emitters.

The maximum rate for delayed fluorescent emission is given by Eq. 6.44. Hence, for $k_F \sim 10^8\ \mathrm{s}^{-1}$ and $\Delta E_{ST} > 0.07$ eV, then $K \approx 0.025$, yielding

$$\tau_{TADF} = k_{TADF}^{-1} = \frac{1+K}{(k_F + k_{nrS})K} \approx 400\mathrm{ns}. \qquad (6.48)$$

This is apparently a *lower* limit to the emission time for non-metalorganic TADF molecules, with high performance materials demonstrated that exhibit $\tau_{TADF} \sim$ 2–100 μs (Zhang et al., 2012, 2015, Wang et al., 2015, Wang et al., 2016).

Since their introduction in OLEDs, TADF dopant materials based on carbazolyl, Cu and Au complexes have produced devices with $\eta_{\mathrm{int}} \simeq 100\%$ and spectra spanning the entire visible spectrum. These molecules consistently follow the motif of electron donating and withdrawing groups at opposite ends of a twisted backbone, thus limiting the orbital overlap between these groups. Examples of a few such organic fluorophores listed in Table 6.8 provide their τ_{TADF} in solution or solid matrices, peak emission wavelengths and maximum efficiencies of OLEDs in which they are used. The delay time scales with the exchange energy, as expected. The longest $\tau_{TADF} = 86$ μs is found for the deep red fluorophore, TPA-DCPP (compound **1R**) with an $\Delta E_{ST} = 130$ meV, and the shortest is for DMAC-BP (**7G**) with $\tau_{TADF} = 2.7$ μs and a $\Delta E_{ST} = 70$ meV. It is indeed remarkable that OLEDs employing molecules with $\tau_{TADF} > 50$ μs have $\eta_{ext} >$ 10%, indicative the presence of very limited non-radiative recombination pathways.

In some cases, both doped and undoped EMLs have been shown to have nearly 100% internal quantum efficiency. The green emitting devices (**4G** and **7G**) have nearly lossless endothermic transfer from the triplet to the singlet state, followed by fluorescence whose rate, k_F, is significantly larger than competing non-radiative processes, at rate k_{nrS}.

One variation on these molecular motifs is the deep blue "twin emitter" design of 44TCzPN (**9B**) (Kim et al., 2016b). Generally, the carbazole fluorophores are connected via a phenyl group (e.g. **3Y**, **4G**) that is incapable of controlling the degree of conjugation between these units. In the twin emitter design, interconnection is via the pair of carbazoles that controls the angle between the end groups (also carbazoles), thereby decreasing conjugation. This strategy appears to eliminate the red shift, leading to deep blue emission at high OLED quantum efficiency. Indeed, **9B** has a rather short τ_{TADF} even with its large exchange energy of 210 meV.

Up to this point, only organic molecules have been discussed. In Sections 3.6.1 and 3.7.5 we introduced metal–organic complexes that use a Cu linker between the donor and acceptor moieties. These moieties show characteristic, structureless emission that is tuned from the blue to the NIR spectral regions, often with nearly 100% internal quantum efficiency (Wallesch et al., 2014, Yersin et al., 2017, Shi et al., 2019). Similarly Ag and Au can link the donor acceptor groups with similar characteristics to those obtained for the organic complexes in Table 6.8. Figure 6.68a shows the emission spectra of OLEDs employing two Au(III) aryl emitting TADF compounds whose molecular structural formulae are also provided. The device structure was: ITO/PEDOT:PSS/PVK:OXD-7:Au(III) complex (80:20)/TPBi/LiF/Al, where the bottom layers are deposited from solutions, whereas the TPBi blocker and cathode contacts are deposited by VTE. As for the carbazolyl and Cu complexes, the emission spectra are broad and featureless extending from the blue into the NIR at 700 nm. The peak emission intensity is at 509 nm and 500 nm, with the shorter wavelength corresponding to the fluorinated compound that has stronger electron-withdrawing character than the compound lacking the four F atoms.

The structure of the non-fluorinated compound is shown in Fig. 6.68b. The twist in the molecular structure across the Au linker provides isolation between the orbitals on the donor and acceptor moieties. This

Table 6.8 Representative fluorescent emitters and hosts used in TADF OLEDs

No.	Emitting polymer	Conductive host	λ_{max} (nm)	η_{ext} (%)	τ_{TADF} (μs)	CIE coord.	Ref.
1R	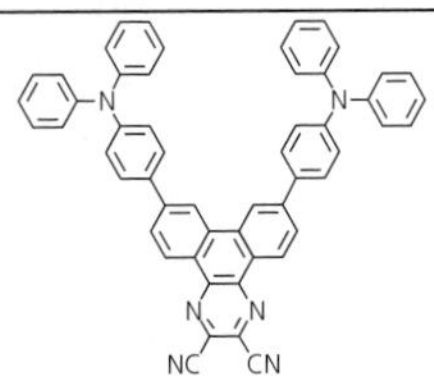 TPA-DCPP	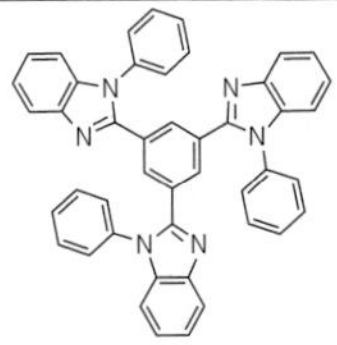 TPBi	668	9.8	86	0.68, 0.32	(Wang et al., 2015)
2Y	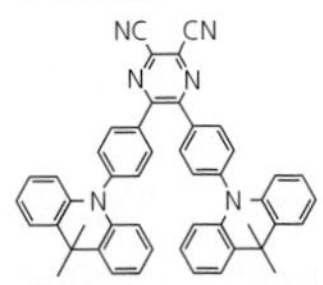 PyCN-ACR	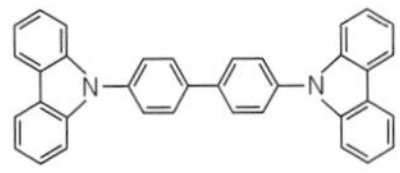 CBP	575	15.6	4.6		(Cai et al., 2016)
3Y	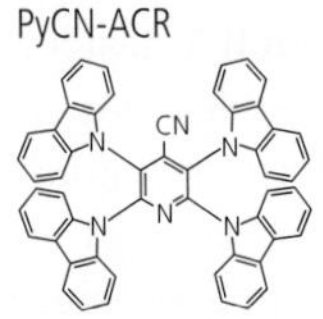 4CzCNPy	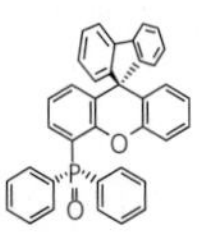 SFXSPO	560	19.7	14.2	0.31, 0.59	(Li et al., 2016a)
4G	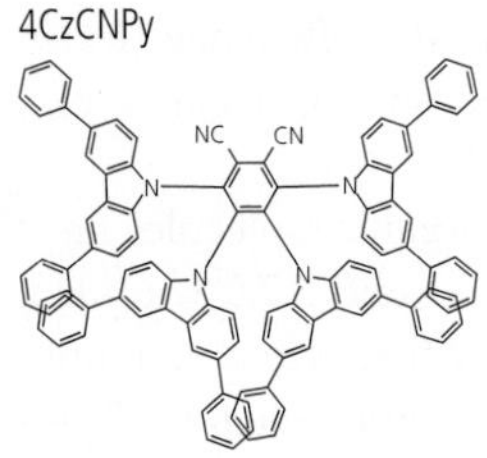 4CzPNPh	SFXSPO	516	22.5	5.1	0.48, 0.51	(Li et al., 2016a)
5G	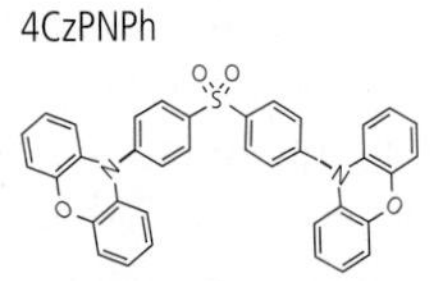 PXZ-DPS	CBP	507	17.5	2.6		(Zhang et al., 2014a)
6G	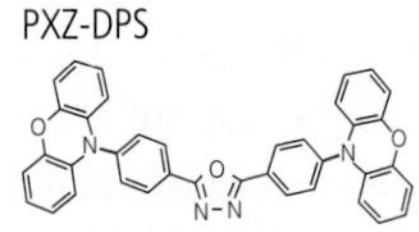 bis-PXZ-OXD	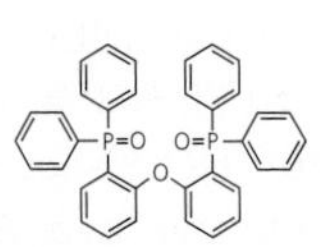 DPEPO	508	14.9	73		(Tanaka et al., 2015)
7G	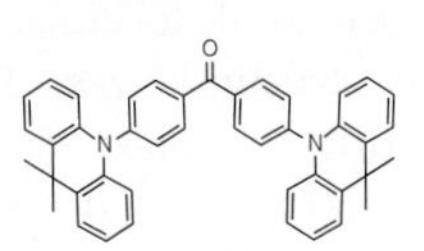 DMAC-BP	None	510	18.9	2.7	0.26, 0.55	(Zhang et al., 2015)
8B	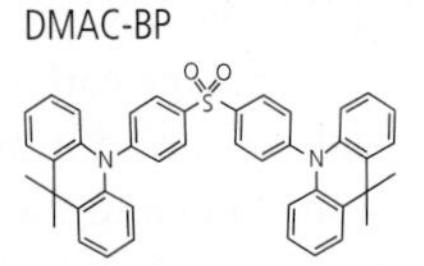 DMAC-DPS	None	481	14.4	3.0	0.17, 0.30	(Zhang et al., 2015)
9B	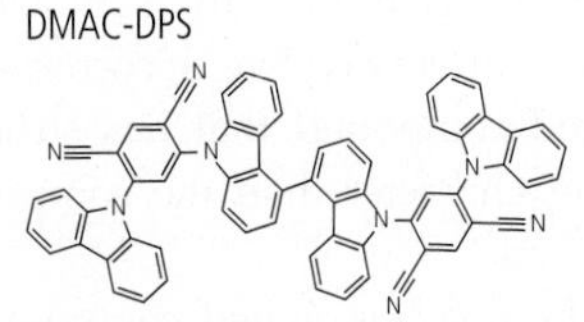 44TCzPN	DPEPO	414	19.5	4.2	0.16, 0.23	(Kim et al., 2016b)

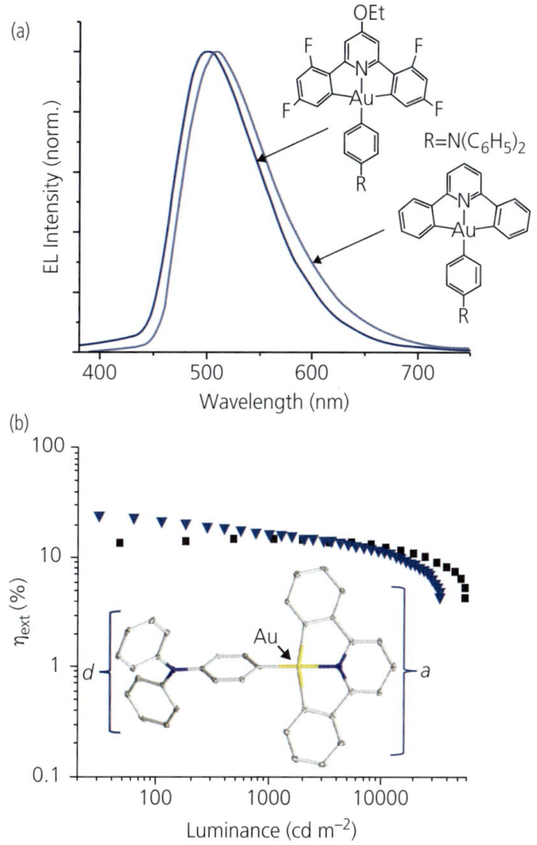

Figure 6.68 (a) Electroluminescence spectra of two TADF OLEDs employing blue Au-aryl emitters, whose structural formulae are indicated. (b) External quantum efficiency vs. luminance for the emitters in (a). Also shown is the molecular structure of the lower of the two compounds in (a) indicating the locations of the Au linker and the donor (*d*) and acceptor (*a*) moieties. After To et al. (2017).

twist leads to a reduced HOMO-LUMO overlap, and hence a reduced exchange energy, giving Au(III)-aryl compounds their TADF character. The temperature dependence of the emission lifetime of this Au complex following an analysis used for the Cu compounds in Eq. 3.212 gives $\Delta E_{ST} = 26$ meV, with 39 meV for the fluorinated complex in Fig. 6.68a (To et al., 2017). Notably, metalorganic TADF molecules often have radiative lifetimes $< 1 \mu s$, in contrast to the generally far longer lifetimes (typically > 5 μs) for pure organic compounds.

It is found in Fig. 6.68b that at 1000 cd/m^2, external efficiencies of 14.7% and 16.5% are obtained for the non-fluorinated and fluorinated complexes, respectively, with corresponding peak efficiencies at low luminance of 14.8% and 23.8%. These high room temperature efficiencies are due to the very small exchange energies, and hence small TADF lifetimes of these molecules.

The broad and unsaturated emission of TADF compounds results from the lack of intramolecular overlap of the donor-acceptor moieties, giving the emission a charge transfer character. More saturated emission can be achieved using a TADF *molecular sensitizer* (or *assistant molecule*) co-doped with a conventional fluorophore with its reduced, excitonic emission linewidths (Nakanotani et al., 2014). If the singlet level of the sensitizing molecule is higher than that of the fluorophore, and if there is spectral overlap between the singlet emission spectrum of the sensitizer and the absorption spectrum of the fluorescent emitter, then efficient Förster energy transfer can occur from the TADF molecule to the fluorescent emitter. Sensitized TADF is analogous to phosphor-sensitized fluorescence discussed in Section 6.3.1 (cf. Figs. 6.30 and 6.31) that was similarly developed to access the exceptionally large palette of highly efficient fluorescent dopants in OLEDs (Baldo et al., 2000b, Kanno et al., 2006c, Chang et al., 2007) The mechanism of charge to exciton to dopant transfer is shown in Fig. 6.69. Note that sensitized TADF has the added benefit of accessing the far larger number of fluorescent dyes that are available as compared with TADF compounds, thus providing greater flexibility in choosing an emission color to satisfy a particular application need.

For sensitized TADF to be efficient, there must be 100% triplet-to-singlet up-conversion on the sensitizer, followed by equally efficient singlet transfer from the sensitizer to the fluorescent emitter. We have seen that the first step in this energy cascade can be achieved using TADF molecules with a small exchange energy splitting. We have also shown in Sections 6.3.2, 6.3.3, and Chapter 3, that FRET can also be highly efficient when the acceptor and donor are in resonance, and within a Förster radius, R_0.

The down-conversion process in Fig. 6.69 starts by the formation of an injected electron and hole into an exciton. This can occur on the host molecule, or directly on the TADF assistant. The exciton on the assistant is excited to S_1 and then subsequently exothermically transferred by FRET onto the emitting fluorophore. Figure 6.70a shows the PL spectrum of the blue TADF-assisted system consisting of a film using the TADF assistant ACRSA in DPEPO as host, along with the absorption and PL spectra of the TBPe dopant in a solution of CH_2Cl_2. The overlap between the ACRSA emission and TBPe absorption spectra yields a transfer radius is $R_0 = 2.2$ nm. The energies of the S_1 and T_1 states are 3.50 and 3.00 eV for DPEPO, and 2.55 eV and 2.52 eV for ACRSA, respectively, and S_1 for the emitter is 2.69 eV. Since ACRSA

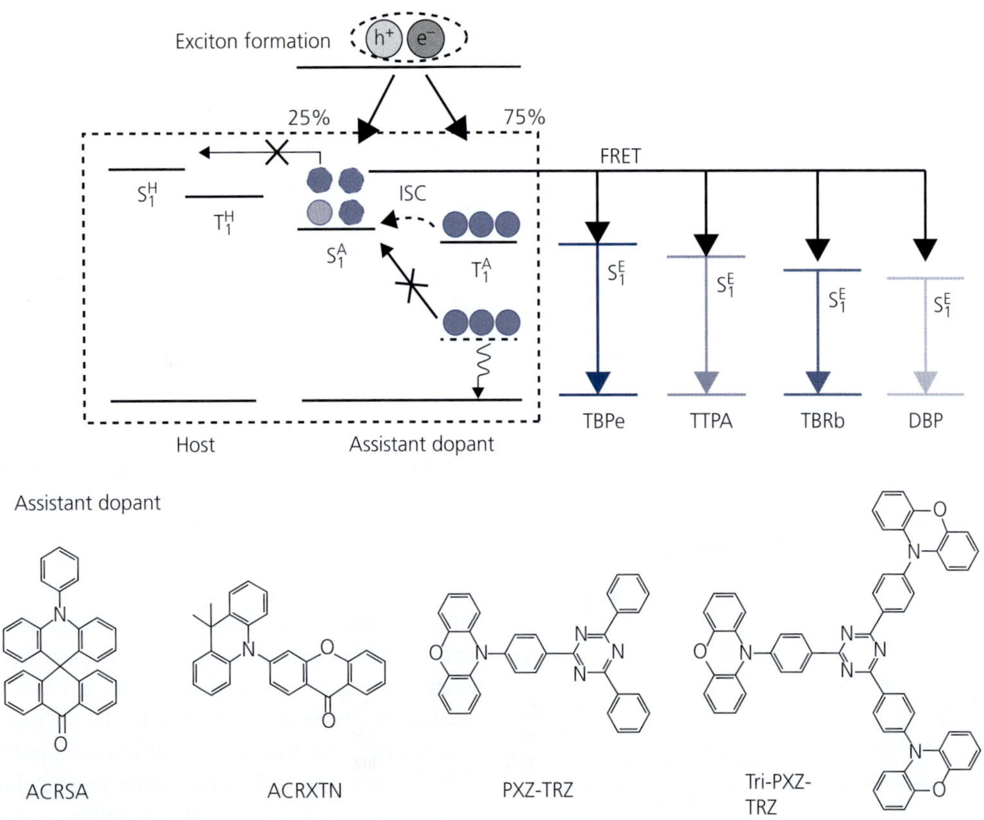

Figure 6.69 TADF sensitized fluorescence. A host transfers charge to a TADF assistant molecule (dashed box) doped with a fluorophore. Exothermic FRET transfer of the exciton from the assistant to the fluorophore results in emission within the visible or NIR spectrum, depending on the excitonic energy of the fluorophore. Molecular formulae of several assistant TADF dopants are shown at the bottom (Nakanotani et al., 2014).

is doped at 15 wt% into DPEPO, both singlets and triplets are efficiently transferred by exchange and/or FRET to the assistant. In contrast, the emitter molecule, TBPe, is doped at only 1wt% into the host-assistant blend, ensuring that only FRET of the singlets in the assistant is active. This avoids direct transfer of triplets from the assistant to the emitter molecule, which would lead to their loss. The effect of the assistant is apparent in Fig. 6.70b. The external quantum efficiency of the device with the TADF sensitizer has a maximum of 13.4%, falling to 8.7% at 1000 cd/m^2, with cyan CIE coordinates of (0.17, 0.30). This is compared with a maximum efficiency of only 1.5% for the TPBe device lacking the sensitizer.

The high energy spectra in Fig. 6.70a point to one of the challenges of the sensitization strategy used in both PHOLEDs and TADF OLEDs: deep blue fluorescent OLEDs require a sensitizer molecule of even higher (ultraviolet) energy emission to allow for efficient exothermic transfer to the fluorophore, and perhaps even higher energy for the host excited states. As will be shown in Section 6.7, high energy emitting materials have considerably shorter operational lifetimes than their lower energy (e.g. green or red) emitting counterparts. Hence, it is problematic to realize long-lived devices with emission in the deep blue.

The spectra of sensitized OLEDs using the various materials combinations in Fig. 6.69 are shown in Fig. 6.70c. Unlike TADF emission, the vibronic structure of the emitting fluorophore is apparent, resulting in spectra that are comparatively narrow. All materials combinations have the appropriate energetic arrangement of $S_1^H, T_1^H > S_1^A, T_1^A \gtrsim S_1^F$, where superscripts H, A, and F correspond to the host, assistant and fluorescent emitter molecules, respectively. The peak efficiencies thus obtained using 1 wt% of the fluorophore in 15–50 wt% assistant doped into their respective hosts, are 8.7%, 11.7%, 17.2%, and 10.9% for the blue, green, yellow, and red OLEDs, respectively. The efficiencies are consistently lower than optimized TADF devices due to energy transfer losses incurred at each of the

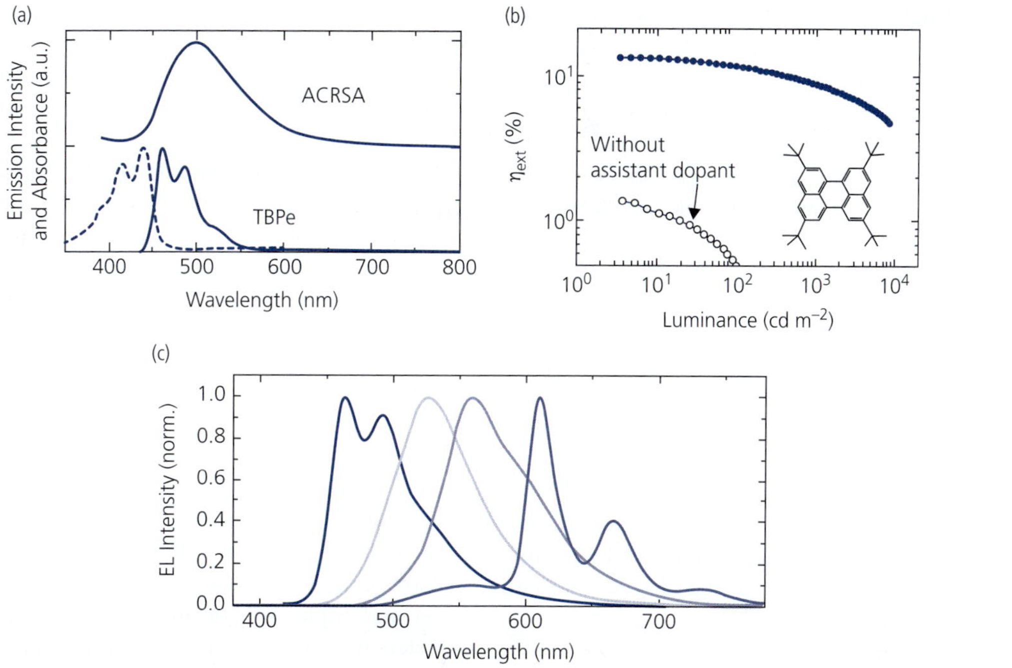

Figure 6.70 (a) Emission spectra of the TADF sensitizer, ACRSA doped at 15 wt% into a DPEPO host, and the absorption (dashed line) and emission (lower solid line) spectra of TBPe in a 10 μM/L solution of CH_2Cl_2. (b) External quantum efficiency of a DPEPO:TBPe (1 wt%) EML OLED with and without 15 wt% doping with ACRSA. Inset: Chemical structural formula of TBPe. (c) Spectra of the blue, green, orange, and red fluorescent dopants in Fig. 6.69 (Nakanotani et al., 2014).

steps in the cascade from host to assistant to emitter, as well as to the less than unity PLQY of the various molecules employed. This is particularly clear in the red-emitting device in Fig. 6.70c where incomplete transfer results in a short wavelength tail extending from $\lambda = 500$ to 590 nm.

6.3.5 Exciton annihilation and management in OLEDs

At high current, the efficiency of all OLEDs decreases (see, for example, Figs. 6.54 and 6.56). There are three primary sources for this phenomenon: saturation of dopant molecular sites with excited states, loss of charge balance in the EML, and exciton annihilation, all of which increase with current (Giebink and Forrest, 2008). Exciton annihilation with other excitons and/or polarons is particularly important, and was introduced in Section 3.10. The annihilation reactions that result in decreased efficiency at high current are the following:

Fluorescent OLEDs

$$S+T \xrightarrow{k_{STA}} T^n + S_0 \xrightarrow{k_{Tn}} T + S_0 \quad \text{(STA)} \tag{6.49a}$$

$$S+P \xrightarrow{k_{SPA}} P^n + S_0 \xrightarrow{k_{Pn}} P + S_0 \quad \text{(SPA)} \tag{6.49b}$$

$$S+S \xrightarrow{k_{SSA}} S^n + S_0 \xrightarrow{k_{Sn}} S + S_0 \quad \text{(SSA)} \tag{6.49c}$$

Phosphorescent and TADF OLEDs

$$T+T \xrightarrow{k_{TTA-S}} S^n + S_0 \xrightarrow{k_{Sn}} S + S_0 \quad \text{(TTA}-\text{S)} \tag{6.50}$$

$$T+T \xrightarrow{k_{TTA-T}} T^n + S_0 \xrightarrow{k_{Tn}} T + S_0 \quad \text{(TTA}-\text{T)} \tag{6.51a}$$

$$T+P \xrightarrow{k_{TPA}} P^n + S_0 \xrightarrow{k_{Pn}} P + S_0 \quad \text{(TPA)} \tag{6.51b}$$

Here, the annihilation processes are defined as in Chapter 3: singlet–triplet (STA), singlet–polaron (SPA), singlet–singlet (SSA), triplet–triplet that produces singlets or triplets (TTA-S or TTA-T, respectively), and triplet–polaron (TPA). Note that Eq. 6.50 gives rise to delayed fluorescence, and thus leads to *increased* efficiency in singlet emitting OLEDs.

These processes require the collision of two particles (either an exciton or a polaron) at a common molecular site. The collision results in the delivery of the energy from the emitting exciton to a second species (exciton or polaron), which is promoted to a higher energy (hot) state, denoted as S^n, P^n, or T^n, where $n > 1$. Non-radiative relaxation then returns the hot state back to its lowest energy level where either a further annihilation event, or a transition to the ground state occurs. The rates (k) of the corresponding processes are also defined in Eqs. 6.49–6.51. Since the same triplet population leads to singlet or triplet annihilation, the rate constants governing the relative contributions of processes in Eqs. 6.50 and 6.51a are related via $k_{TTA-S} = \alpha k_{TT}$ and $k_{TTA-T} = (1-\alpha)k_{TT}$, respectively, where α is the fraction of annihilation events that follow TTA-S *vs.* k_{TT}, which is the total TTA rate constant. Since these processes depend on exciton concentration, then TTA-S occurs at rate $\alpha k_{TT}T^2$ and TTA-T at rate and $(1-\alpha)k_{TT}T^2$.

With these processes and rates, we can write a generalized rate equation for the dynamics of the exciton population in the presence of both optical and electrical excitation at position x within the emission zone of the OLED, viz.

$$\frac{dN(x,t)}{dt} = G(x,j) - N(x,t)[k_N + K(x,j)], \qquad (6.52)$$

where $G(x, j)$ is the electrical generation rate at current density j, $k_N = k_F$ or k_{Ph} for a singlet or triplet exciton, respectively, and the exciton concentration, $N(x, t)$ is either the singlet ($S(x, t)$) or triplet ($T(x, t)$) density. These can be either optically or electrically generated or both, such that $N(x,t) = N_{EL}(x,t) + N_{PL}(x,t)$. Finally, $K(x, j)$ is the aggregate rate due to quenching by all possible mono- and bimolecular processes, and can be written (Giebink and Forrest, 2008)

$$K(x,j) = k_n n(x,j) + k_p p(x,j) + k_{NNE}N_{EL}(x,j) + k_{NNP}N_{PL}(x) + \zeta(F) + \ldots \qquad (6.53)$$

In this expression, the generalized rates due to electron or hole quenching are k_n and k_p with corresponding electron and hole densities, $n(x)$ and $p(x)$. The second two terms are the generalized bimolecular exciton recombination rates due to electrical (k_{NNE}) or optical (k_{NNP}) excitation. Finally, $\zeta(F)$ is exciton quenching rate due to the local electric field $F(j)$. As we will see below, most practical situations require that we consider only the few processes that are the most significant to the exciton dynamics in a particular OLED.

6.3.5.1 Exciton annihilation, triplet fusion, and singlet fission in fluorescent OLEDs

From Eqs. 6.49a and 6.50, we see that triplet annihilation can lead to either a decrease or an increase in the singlet density (Zhang and Forrest, 2012). Process TTA-S results in delayed fluorescent emission that adds to OLED efficiency by creating additional singlets from non-radiative triplets. Several reports (Kondakov et al., 2009, Yokoyama et al., 2011, Wallikewitz et al., 2012) show $\alpha > 0$, resulting in fluorescent OLEDs whose internal efficiency exceeds the 25% spin-statistical limit due to this process.

The rate of STA is $k_{STA}ST$. Thus, using Eqs. 6.52 and 6.53 we can describe the singlet and triplet density dynamics by

$$\frac{dS}{dt} = \gamma(j)\frac{j}{4qd} - k_F S - k_{STA}ST + \alpha k_{TT}T^2, \qquad (6.54)$$

$$\frac{dT}{dt} = \gamma(j)\frac{3j}{4qd} - k_{Ph}T - (1+\alpha)k_{TT}T^2, \qquad (6.55)$$

where $\gamma(j)$ is the current-dependent charge balance factor, and d is the EML thickness. The ¼ and ¾ terms in these expressions reflect the statistics of singlet and triplet spin formation. In steady state, the external quantum efficiency is calculated for $S(t=\infty, J)$ to give (Zhang and Forrest, 2012)

$$\eta_{ext}(j) = \eta_{out}\phi_S \frac{k_F S(t=\infty, j)}{j/qd}. \qquad (6.56)$$

Rubrene is a particularly interesting fluorescent molecule since its singlet energy ($E_S \sim 2.2$ eV) is twice that of the triplet ($E_T \sim 1.1$ eV), thereby enabling efficient, resonant TTA-S (Kondakov et al., 2009). When rubrene is doped with a fluorophore with emission energy, E_F, such that $E_T < E_F < E_S$ (e.g. DBP with $E_F = 2.0$ eV), two triplets produced on rubrene can undergo TTA-T (a process called *triplet fusion*), to produce one rubrene singlet. Subsequently, the singlets can transfer via FRET to the fluorophore followed by radiative emission. Unlike TADF, this route to singlet generation is lossy, since it takes two excited states (triplets) to generate a radiative singlet, see Fig. 6.71. Given the dual generation processes of direct singlet excitation and singlet production via triplet fusion, the OLED efficiency can exceed that determined by spin statistics, where only one singlet is generated per three triplets. Figure 6.71 also shows the competing process of singlet fission, whereby one singlet can produce two triplets, following: $S \xrightarrow{k_{S2T}} 2T$ under the condition that $E_S \geq 2E_T$. This process competes with radiative emission from DBP. Due to the exothermic transfer from rubrene to DBP, it is likely that $k_F >> k_{S2T}$, and hence the latter process is largely inactive.

These several processes have been identified in a 1 vol.% DBP:rubrene OLED with a peak $\eta_{ext} = 6.7\%$ at $j = 75$ mA/cm^2 (Fig. 6.72), clearly exceeding the spin-

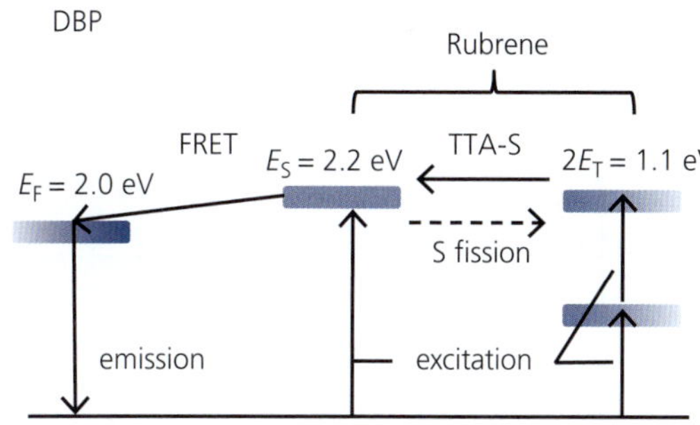

Figure 6.71 Emission from a DBP:rubrene OLED due to a combination of direct electrical excitation of rubrene singlets and triplets, along with triplet fusion (right) resulting in additional rubrene singlets.

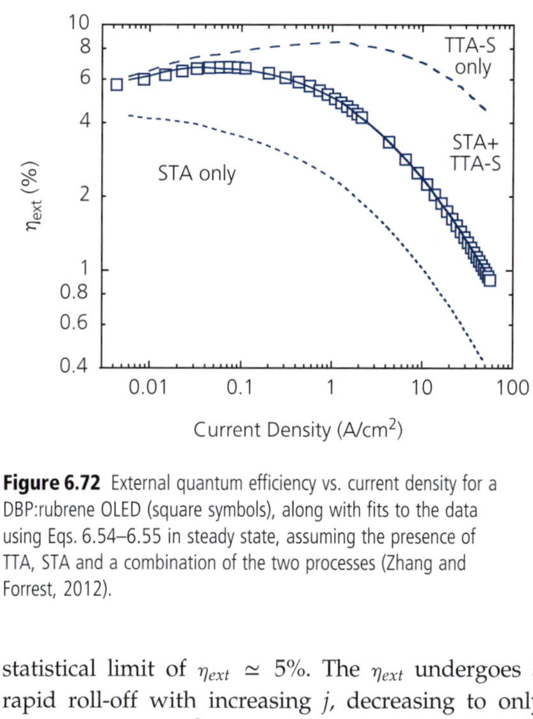

Figure 6.72 External quantum efficiency vs. current density for a DBP:rubrene OLED (square symbols), along with fits to the data using Eqs. 6.54–6.55 in steady state, assuming the presence of TTA, STA and a combination of the two processes (Zhang and Forrest, 2012).

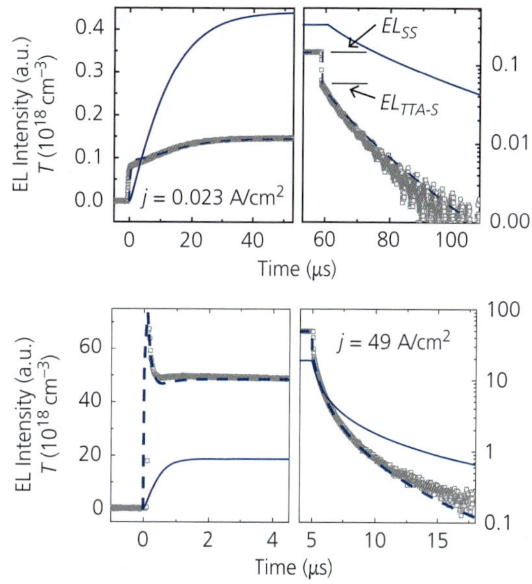

Figure 6.73 Transient response of DBP:rubrene OLED EL intensity at pulse turn-on (left panels) and turn-off (right panels) at two different current densities. Calculated triplet densities (solid lines) and singlet densities (dashed lines) are compared with the data. The steady state EL_{SS} and delayed EL_{TTA-S} due to triplet fusion are shown (Zhang and Forrest, 2012).

statistical limit of $\eta_{ext} \simeq 5\%$. The η_{ext} undergoes a rapid roll-off with increasing j, decreasing to only 0.9% at 57 A/cm^2. The EL spectrum shows pure DBP emission with a peak wavelength of λ= 610 nm, and with almost no change with j, indicating that FRET is efficient (Zhang and Forrest, 2012). A solution to the steady state rate equations that considers both triplet fusion and STA (Eqs. 6.54–6.56), yields a fit to the data (solid line) for $k_{STA} = 7.1 \times 10^{-11}$ cm^{-3} s^{-1} and $k_{TTA-S} = 6.0 \times 10^{-14}$ cm^{-3} s^{-1}. Eliminating one or the other of the processes, however, does not adequately describe the observed dependence of η_{ext} on j.

The theory also requires charge balance. The fits are reasonable assuming $\gamma(j) = 1$ for $j > 1$ A/cm^2, after which $\gamma(j)$ decreases due to charge leakage through the emissive layer, and field-induced exciton dissociation (Giebink and Forrest, 2008). The leakage is evident in the emission spectra for OLEDs that show a noticeable increase in the contribution from rubrene (at λ= 550 nm) at $j > 60$ A/cm^2. Exciton–polaron annihilation is also possible in fluorescent OLEDs; however, it is not needed to achieve the fit in Fig. 6.72.

The time dependent solutions to Eqs. 6.54 and 6.55 that include both STA and TTA-S with rate constants used in Fig. 6.72 also accurately represent the data obtained for both the turn-on and turn-off transients of the EL intensity, shown in Fig. 6.73. At low j, the EL intensity gradually increases to steady state over tens of microseconds. This is consistent with TTA-S since the triplet density slowly increases over this timescale. In contrast, at high j, the EL intensity rapidly peaks, and then gradually decays to a steady-state intensity approximately 70% of its peak value. This is a signature of STA since the singlet density decreases with increasing triplet density due to increased annihilation (Zhang et al., 2010). The EL turn-off transients for all current densities show delayed fluorescence (right hand graphs in Fig. 6.73) due to TTA-S (Kondakov et al., 2009, Zhang et al., 2010). The steady-state EL intensity, EL_{SS}, and the initial delayed fluorescence intensity, EL_{TTA-S} are related by the ratio, $\rho(j) = EL_{TTA-S}/EL_{SS}$ that characterizes the fraction of TTA-generated-emission relative to the total emission intensity.

The question now arises: what is the maximum quantum efficiency that can be achieved by delayed fluorescence generated by TTA-S? Using Eqs. 6.54 and 6.55, we can write $\rho(j)$ as

$$\rho(j) = \frac{3ak_{TTA-S}T^2}{k_{Ph}T + (1+a)k_{TTA-S}T^2 + 3ak_{TTA-S}T^2}. \quad (6.57)$$

Since the phosphorescent rate is very small compared to the other rates; then $\lim_{j\to\infty} \rho = 3a/(4a+1)$. The upper limit of the internal quantum efficiency is $\eta_{int,max} = 25\%/(1-\rho) = 25\% \times (4a+1)/(a+1)$. When $a = 1$, $\eta_{int,max} = 62.5\%$; a limit can only be reached for $k_{TTA\text{-}S}T >> k_{Ph}$ with negligible STA. This limit is considerably lower than 100% that is accessible by triplet emission via phosphorescence or TADF. Indeed, the highest observed internal quantum efficiency via delayed fluorescence is ~50–55% (Kondakov et al., 2009, Zhang and Forrest, 2012).

It is apparent from the foregoing that TTA-S can add to the device efficiency at low j, but at high j it must compete with losses due to STA, where the build-up of triplets results in quenching of the high density of singlets. To find the j where the increase in efficiency is overcome by losses, we define the effective triplet current density as

$$j_T = \frac{4qd}{\gamma(j)}\left(ak_{TT}T^2 - k_{STA}ST\right), \quad (6.58)$$

which allows us to rewrite Eq. 6.54 as

$$\frac{dS}{dt} = \gamma(j)\frac{j}{4qd}\left[1 + \frac{j_T}{j}\right] - k_F S. \quad (6.59)$$

Now, j_T can be either positive or negative, depending on whether TTA-S or STA dominates. The calculated j_T/j for the DBP:rubrene OLED is shown in Fig. 6.74. With increasing j, j_T/j is initially positive, and increases due to TTA-S to a peak of 0.53 at $j = 50$ mA/cm^2; and then decreases due to the increased rate of STA. A critical current density, j_C can be defined when $j_T/j = 0$ where the effects of TTA and STA are equal. Above $j_C = 2.2$ A/cm^2 triplets lead to a net decrease in quantum efficiency.

Singlet fission can result up to 125% internal efficiency if the triplets are harvested by a molecule with a large phosphorescent efficiency, such as a heavy metal complex. An energy level diagram showing singlet fission yielding two triplets per singlet is shown in Fig. 6.75. We assume that the host molecule can engage in efficient singlet fission, that is, $E_{HS1} \geq 2E_{HT1}$, where the substrate H refers to the host. A guest phosphor whose triplet energy is $E_{GT1} \leq E_{HT1}$ is doped into the host at a concentration that

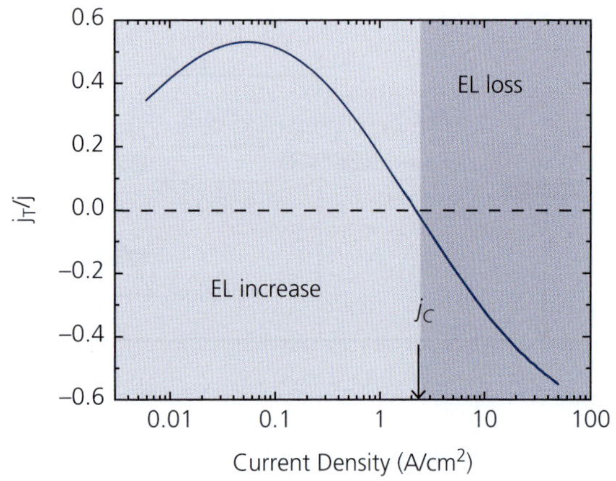

Figure 6.74 Effective triplet normalized current density, j_T vs. j for a 1% vol. DBP:rubrene OLED. The critical current (j_C) where the increase in quantum efficiency due to TTA-S is exceeded by loss due to STA is shown.

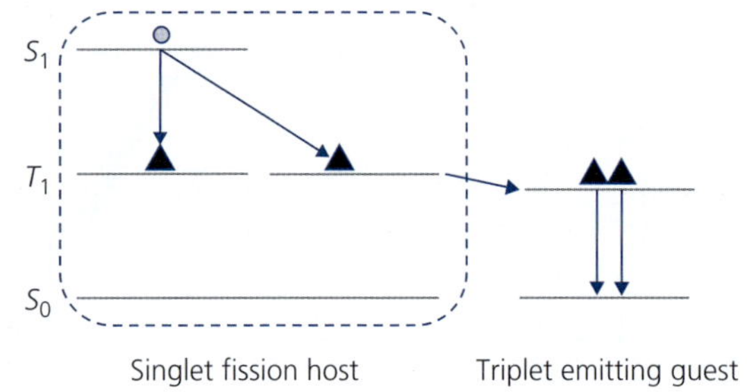

Figure 6.75 Energy level diagram of the emission region of an OLED undergoing the process of singlet fission yielding two emissive triplets (triangles) for every singlet (circle).

permits Dexter transfer of triplets from the host to the guest molecule. The host is electrically excited, creating the usual 1:3 singlet:triplet ratio. The triplets are transferred directly to the guest, accounting for 75% of the excited states. On the other hand, the singlet population undergoes fission to create two triplets for each singlet. These excess triplets then transfer to the phosphor, which add to the emission intensity. In principle, therefore, there are 125 emitting triplets generated for each 100 electron–hole pairs injected into the OLED EML.

There is somewhat weak evidence that this process has been observed in the rubrene:Erq_3 host/guest system (Nagata et al., 2018). The guest is a rare earth phosphor. Unlike heavy metal electrophosphorescent complexes, the triplet transition occurs directly between atomic orbitals on the rare earth, rather than via ligand transitions. For Er, the intra-4f transition $^4I_{13/2} \rightarrow {}^4I_{15/2}$ ($^4I_{15/2}$ corresponds to the ground state) has a NIR emission energy of 0.80 eV. This is below the rubrene triplet at 1.14 eV, which allows for efficient exothermic transfer to the Er ion. The 8-hyrdoxyquinoline ligand singlet and triplet energies on Erq_3 are 2.85 eV and 2.23 eV, respectively, and hence do not participate in the emission process. While rare earth molecules undergo triplet emission with narrow emission spectra, coupling to the atomic orbitals is weak, and hence such devices have not been found useful for OLED applications.

To determine whether or not singlet fission to the Er complex, the change in PL intensity, ΔI_{PL}, vs. magnetic field was measured for a 2 mol. % Erq_3:rubrene film. As we will discuss in detail in Section 7.6, the magnetic field, B, decreases the number of correlated triplet pairs with singlet character, denoted by $^1(TT)$. Thus, the magnetic field enhances the rate of the reaction: $S_1 \rightleftharpoons {}^1(TT) \rightleftharpoons T_1 + T_1$ (Piland et al., 2013). As the field increases, there should be a decrease in the number of singlet fission products. It is observed that the fraction of light emitted in the NIR by the Er-complex with increasing magnetic field decreases by 9.5% at $B = 0.4$T from its value at $B = 0$ due to a decrease in the production of $^1(TT)$ precursor states. Also, note that the fraction of PL in the visible from rubrene fluorescence increases (+11.3%) due to a concomitant increase in singlets that are not lost to triplets.

The emission spectrum from an OLED employing an Erq_3:rubrene EML is identical to the narrow emission spectrum centered at 1.55 μm characteristic of the Er $^4I_{13/2} \rightarrow {}^4I_{15/2}$ transition. Quantitative analysis of the dependence of the intensity dependence on B yields a singlet fission yield under electrical excitation of 51.6%. However, the efficiency of the OLED is very low (i.e. far from the predicted 125%), which is attributed to the PLQY of only $\Phi_{PL} \sim 10^{-3}$ for Erq_3.

6.3.5.2 Reducing efficiency roll-off via triplet management

Triplet-induced quenching in OLEDs (Eq. 6.49a) can be reduced or even eliminated by blending a "triplet manager" molecule into the EML to divert triplets from the emitting guest molecule. The process of triplet management is illustrated in Fig. 6.76. If the T manager has a higher S and lower T energy than the guest, it facilitates Förster transfer of singlets to the guest, and Dexter transfer of triplets from the guest to the manager, thereby potentially eliminating STA on the guest. This strategy was demonstrated in OLEDs with EMLs comprising an Alq_3 host doped with 2 vol. % of DCM2 along with various concentrations ($0 \leq x \leq 70$ vol.%) of ADN for triplet management. The triplet energies of Alq_3 ($E_{T1} = 2$ eV), DCM2 (1.74 eV), and ADN (1.69 eV) are consistent with the requirements in Fig. 6.76 (Zhang et al., 2011). Since

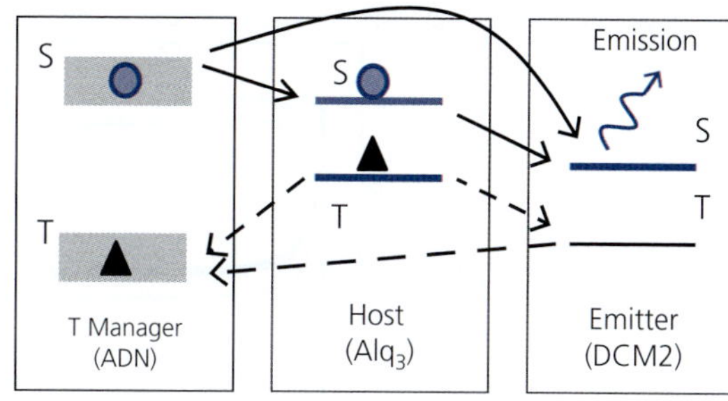

Figure 6.76 Concept of a triplet managed fluorescent OLED EML. The triplet and singlet levels and the transitions between them are shown. The manager is co-doped with the emitter (guest) and host molecules in the EML. Its triplet energy is lower than that of the host and the fluorophore to capture triplets before they can undergo STA with host and emitter singlets. Förster transfers are shown as solid arrows, and exchange (Dexter) transfers by dashed arrows.

triplet transfer occurs primarily by exchange interactions, the manager concentration must be high (> 10%) for this scheme to be effective. Transfer of singlets is via Förster transfer, requiring dopant concentrations of only a few percent, which is sufficiently low to avoid concentration quenching.

Figure 6.77a shows the transient EL for a series of managed OLEDs with the structure of: ITO/35 nm NPD/25 nm-thick EML/25 nm BCP ETL/0.8 nm LiF/100 nm Al with different ADN fractions, x, in the EML. For the unmanaged (control) device ($x = 0$), the EL rapidly reaches a peak and then decays to a steady state intensity at ~50% of its initial value. This decrease in efficiency is due to STA as the triplet population builds up over several milliseconds (see Fig. 6.73, lower left) (Zhang et al., 2010). When the guest triplet population approaches steady state, STA stabilizes at a reduced equilibrium singlet population. In the presence of the T manager, the magnitude of the EL transient decay reduces with increasing x, and disappears entirely at $x \geq 50\%$, indicating the complete suppression of STA. Figure 6.77b shows the EL intensity peak-to-steady-state ratio, $\rho(j)$. We find that $\rho(j)$ decreases with increased manager concentration, and $\rho(j) = 1$ for $x \geq 50\%$, corresponding to an absence of STA.

Two annihilation processes dominate energy transfer in triplet-managed OLEDs, namely guest STA, $S + T_G \rightarrow S_0 + T_G$, and host-to-guest triplet Dexter transfer, $T_H \rightarrow T_G$. The EL transient follows the singlet population whose dynamics are described by (Zhang et al., 2011)

$$\frac{dS(t)}{dt} = -(k_F + k_{STA}T_G(t))S(t) + \gamma\frac{1}{4}\frac{j}{qd}, \quad (6.60)$$

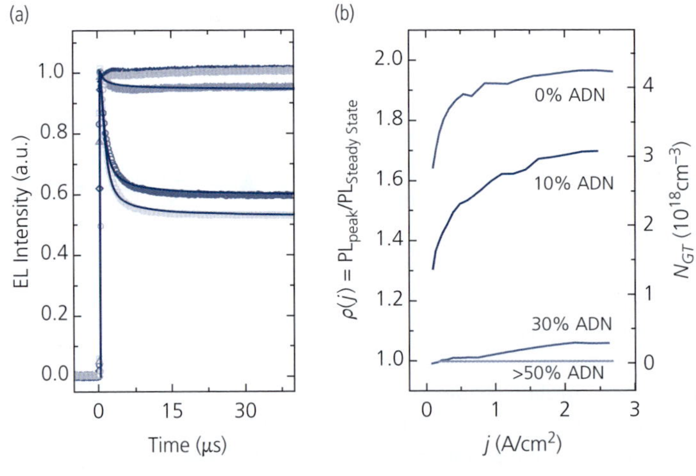

Figure 6.77 (a) Triplet managed transients for an OLED with an Alq_3 host, 2 vol.% DCM2 and an ADN manager at an injection current of $j = 2$ A/cm^2. The transients, from top to bottom, have an ADN concentration of 70% (yellow), 50% (green), 30%, 10%, 0%. Solid lines are fits the STA theory in Eqs. 6.60–6.62. (b) Peak PL to steady state PL intensities as functions of ADN concentration vs. current density. Also shown is the guest triplet concentration, N_{GT} (Zhang et al., 2011).

$$\frac{dT_H(t)}{dt} = -k_{HG}\exp\left(-\frac{2}{L}\sqrt[3]{\frac{1}{N_{GT}(j) - T_G(t)}}\right)T_H(t) + \gamma\frac{3}{4}\frac{j}{qd}, \quad (6.61)$$

and

$$\frac{dT_G(t)}{dt} = k_{HG}\exp\left(-\frac{2}{L}\sqrt[3]{\frac{1}{N_{GT}(j) - T_G(t)}}\right)T_H(t), \quad (6.62)$$

where we assume γ is independent of current, k_{HG} and L are the host-to-guest triplet transfer rate coefficient and molecular van der Waals radius, respectively, and $N_{GT}(j)$ is the saturation guest T population. That is, when $T_G(t) \rightarrow N_{GT}(j)$, the transfer stops and $dT_G/dt = 0$. The exponential factors are due to transfer of excitons between molecules consistent with the Dexter process. Eventually, T_H saturates due to TTA and natural decay, but these processes are slower ($\sim$1 ms) than the time scale for EL decay, and are thus neglected.

Assuming the boundary condition that $T_G = 0$ at $t = 0$, and $T_G = N_{GT}$ as $t \rightarrow \infty$ in Eq. 6.61, we obtain

$$\rho(j) = 1 + \frac{k_{STA}N_{GT}(j)}{k_F}. \quad (6.63)$$

The external quantum efficiency roll-off is then given by $\frac{\eta_{ext}(j)}{\eta_0} = \frac{\gamma(j)}{\rho(j)}$, where η_0 is the external efficiency for $\gamma = 1$, and $\gamma(j)$ includes the current dependent charge balance along with all other loss factors under electrical excitation, such as SPA.

For a 2 vol.% DCM2 concentration, $L \approx 3$ nm as determined from the average distance between dopant molecules. With the constraint on $N_{GT}(j)$ given by Eq. 6.63 and $\gamma = 1$, the EL transients are fit by Eqs. 6.60–6.62 as shown by the solid lines in Fig. 6.77a. From the fits, we obtain $k_{STA} = 1.5 \times 10^{-10}$ cm^3 s^{-1} independent of host and manager concentrations, and $k_{HG} = 2.5 \times 10^7$ s^{-1} for $x = 0$ and 10%, and $k_{HG} = 3 \times 10^8$ s^{-1} for $x = 30\%$ ADN. Also, N_{GT} is obtained using Eq. 6.63, with results indicated on the right hand side of Fig. 6.77b. The introduction of the triplet manager substantially reduces N_{GT}. For example, at $j = 0.5$–2.5 A/cm^2, N_{GT} is reduced from 4×10^{18} cm^{-3} ($x = 0$) to $<3 \times 10^{17}$ cm^{-3} ($x = 30\%$), and to 0 cm^{-3} at $x \geq 50\%$. The manager effectively eliminates the nearly 50% decline in brightness (and efficiency) observed in its absence. This strategy not only emphasizes the importance of STA in fluorescent OLEDs, but it also shows a path to achieve very high brightness via the intentional reduction in the triplet population. We will show in Sections 6.7 and 6.8 that triplet management can also result in an increased PHOLED lifetime and continuous wave (CW) operation of optically pumped organic lasers.

6.3.5.3 Efficiency roll-off in PHOLEDs

A defining characteristic of phosphorescent emission is its long triplet decay time ($\sim$500 ns to 10's of μs) compared with fluorescence (1–10 ns) While this is the origin of the high efficiency in PHOLEDs and TADF OLEDs, it also provides an extended period in which a triplet can interact with another excited state or polaron. Given the lack of a strong absorption into the triplet manifold and the non-radiative nature of the host triplet, energy transfer occurs via hopping or direct charge trapping in phosphorescent devices. This requires a phosphor doping density that is significantly greater than used in fluorescent devices, where energy transfer is primarily via long-range FRET. This also results in an increased probability for triplet–triplet and triplet–polaron encounters on the dopant molecules, or in dopant molecular clusters at high current densities, leading to additional channels for non-radiative recombination at high current.

In Section 3.9 we showed that k_{TTA} is related to the triplet diffusion constant, D, via

$$k_{TTA} = 8\pi D R_c \tag{6.64}$$

where R_c is the exciton capture radius. The factor of two difference between this expression and Eq. 3.261 is that here we consider the collision rate between two identical (i.e. triplet–triplet) particles. It has been argued that exciton annihilation can occur in a single step via FRET (Staroske et al., 2007, Zhang et al., 2016a), or via multiple exchange transfers from molecule to molecule (Zhang and Forrest, 2013), as illustrated in Fig. 6.78. Following the treatment in Section 3.9, we obtain the diffusion constants for these two different processes:

$$D_{FRET} = a^2 k_{Ph} \left(\frac{R_0}{a}\right)^6 \tag{6.65}$$

and

$$D_{ex} = a^2 K J_\lambda \exp(-2a/L), \tag{6.66}$$

where $a = 1/\sqrt[3]{CN}$ is nominally the distance between dopant molecules determined by the doping fraction, C, and the molecular density, N. Also, R_0 is the Förster transfer radius, K is a constant, and J_λ is the overlap integral between the emission and absorption spectra of the donor and acceptor molecules, respectively (see Section 3.8). By measuring the dependence of k_{TTA} vs. a for PtOEP:CBP, Ir(ppy)$_3$:CBP, and Ir(ppy)$_2$acac:CBP guest–host systems in Fig. 6.79, the analysis using Eqs. 6.64–6.66 accurately fits the Dexter process (straight lines) compared to FRET (dashed lines), suggesting that triplet diffusion in is predominantly due to multi-step transfer. Note that all of these systems are characterized by strongly confined triplets on the guest molecules whose triplet energies are significantly less than that of the host (e.g. the Ir-complexes both have $E_{T1} = 2.4$ eV). On the right hand axis in Fig. 6.79, we show the intermolecular hopping rate, $k_H = 6k_{TTA}/(8\pi a^2)$. The number of steps taken prior to an annihilation reaction is N = k_H/k_{Ph}, which gives N $\sim$ 3–30 (Zhang and Forrest, 2013).

The efficiency roll-off in PHOLEDs at high current density is thus due to a combination of: (i) TTA yielding singlets and triplets, (ii) TPA (Baldo et al., 2000a), and (iii) loss of charge balance (Reineke et al., 2007, Giebink and Forrest, 2008). It has also been suggested (Kalinowski et al., 2002) that loss of triplets in PHOLEDs can arise from field-induced quenching due to field ionization (Onsager, 1938). Given the compact nature of triplets, this mechanism is unlikely to be important.

Focusing our attention on the primary mechanism of TTA, we write the respective dynamical equation:

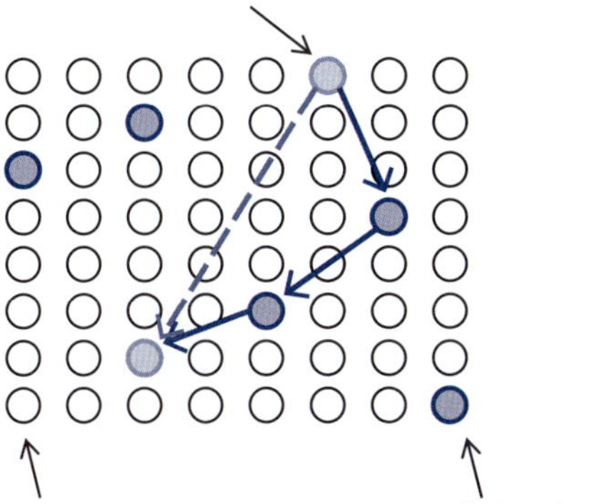

Figure 6.78 Single step (dashed arrow) FRET vs. multistep (solid arrows) exchange-mediated diffusion resulting in triplet–triplet annihilation.

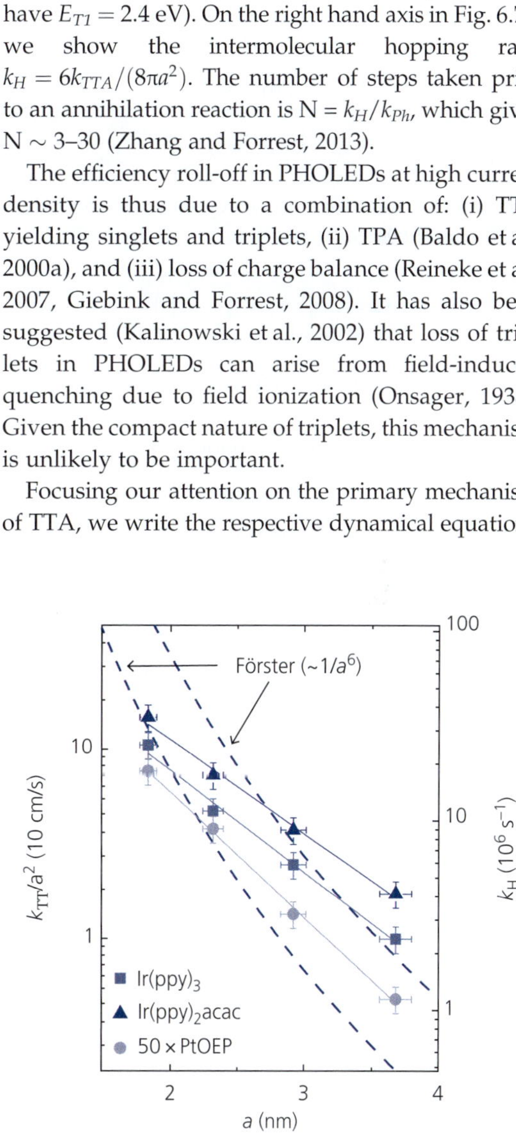

Figure 6.79 Triplet–triplet annihilation rate, k_{TTA}/a^2, vs. the distance between dopant molecules, a, for three phosphors doped into CBP. Also shown (right hand axis) is the intermolecular hopping rate, k_H (Zhang and Forrest, 2013).

$$\frac{dT}{dt} = \frac{\gamma(j)j}{qd} - k_{Ph}T - \frac{1}{2}k_{TTA}T^2. \quad (6.67)$$

The factor of ½ accounts for the loss of one of two triplets in TTA (Chandrasekhar, 1943). Following the end of the excitation pulse (i.e. $j = 0$), the solution to Eq. 6.68 is

$$T(t) = \frac{T(0)}{\left(1 + T(0)\frac{k_{TTA}}{2k_{Ph}}\right)\exp(k_{Ph}t) - T(0)\frac{k_{TTA}}{k_{Ph}}}. \quad (6.68)$$

Assuming that the luminescence intensity (L) is linearly proportional to the concentration of excited states, that is, $L(t) \propto k_{Ph}T$, then the time evolution of phosphorescent emission referenced to its value at $t = 0$ is obtained from Eq. 6.67:

$$L(t) = \frac{L(0)}{\left(1 + T(0)\frac{k_{TTA}}{2k_{Ph}}\right)\exp(k_{Ph}t) - T(0)\frac{k_{TTA}}{2k_{Ph}}}. \quad (6.69)$$

The external quantum efficiency can also be calculated from the steady state solution of Eq. 6.67:

$$\frac{\eta_{ext}}{\eta_0} = \frac{j_0}{4j}\left(\sqrt{1 + 8\frac{j}{j_0}} - 1\right), \quad (6.70)$$

where η_0 is the quantum efficiency in the absence of TTA (measured at $j \to 0$), and the roll-off current, j_0, is defined at $\eta_{ext} = \frac{1}{2}\eta_0$. Then it follows that:

$$j_0 = \frac{2k_{Ph}^2 qd}{k_{TTA}}. \quad (6.71)$$

The importance of TTA in determining the roll-off in efficiency of PHOLEDs has been widely observed. For example, in Fig. 6.80a we show the transient PL transient data of an 8 vol.% PtOEP:Alq_3 film. The triplet density, $T(0)$, obtained from the absorption coefficient of Alq_3 (Garbuzov et al., 1996) and the optical intensity, are provided. The increased curvature with triplet density shows the quadratic dependence of TTA. Solutions to Eq. 6.69 provide fits to the data, yielding $k_{Ph} = 3.1 \times 10^4\ s^{-1}$ and $k_{TTA} = 1.2 \times 10^{-13}\ cm^3\ s^{-1}$, shown by the solid lines in the figure (Baldo et al., 2000a).

This same materials system was used in a PHOLED with a 10 nm PtOEP:Alq_3 EML, where the PtOEP doping concentration was 1, 8, and 16 vol.%. The solid lines in Fig. 6.80b are steady state solutions to Eq. 6.67, yielding the roll-off current densities, j_0, shown for each PtOEP concentration. Apparently, j_0 increases with concentration due to triplet diffusion. As the dopant concentration increases, triplets diffusing on the host can more easily find a dopant molecule and subsequently radiatively recombine, leading to reduced collisions with other triplets at high current. The value of j_0 that is inferred from k_{TTA} obtained from transient analysis (Eq. 6.71, dashed line in Fig. 6.80b) is in reasonable agreement with that obtained in steady state for the sample with 8 vol.% PtOEP, providing further support for the dominance of TTA at high j.

The theory in Eq. 6.67 neglects host triplets. However, at the highest current densities and lowest dopant concentrations, the exceptionally long-lived triplets on the fluorescent host molecules will themselves annihilate. To take this into account, terms corresponding to guest–host and host-host triplet annihilation of the form $\frac{1}{2}k_{GH}T_GT_H$ and $\frac{1}{2}k_{HH}T_H^2$, respectively, must be included. Here, k_{GH} and k_{HH} are the

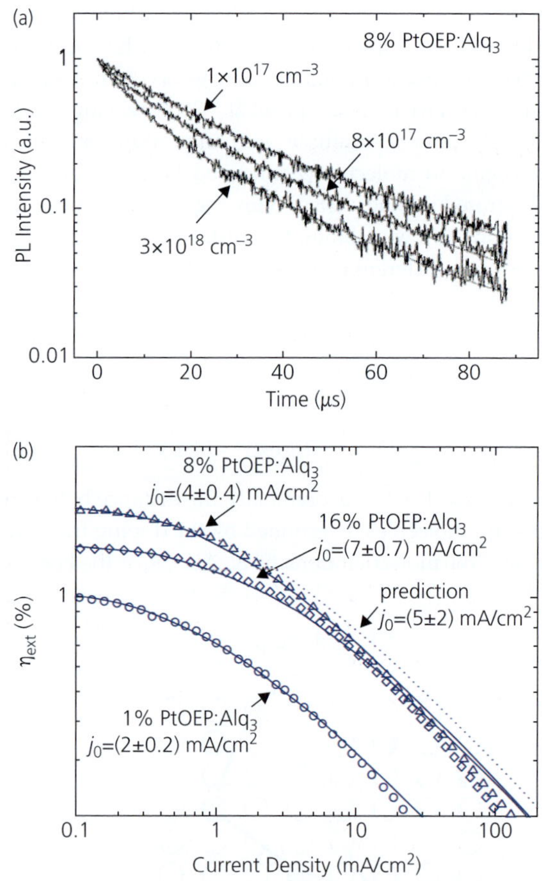

Figure 6.80 (a) Transient PL of a 8 vol.% PtOEP:Alq_3 film excited by 500 ps pulses at λ= 337 nm. The triplet densities at t = 0 are indicated for each transient (see text). (b) Steady state external quantum efficiencies (data points) of PHOLEDs employing PtOEP:Alq_3 EMLs with various PtOEP concentrations. Also shown are fits to steady state theory (solid lines) with the corresponding roll-off current densities, j_0, indicated. Also shown is a fit (dashed line) to the 8 vol.% PtOEP:Alq_3 EML PHOLED using k_{TTA} obtained from the transient data in (a) (Baldo et al., 2000a).

respective guest–host and host-host triplet annihilation rates (Baldo et al., 2000a).

One further high triplet process that must be considered in PHOLEDs is TPA. In this case, the generalized dynamical Eqs. 6.52 and 6.53 reduce to

$$\frac{dT}{dt} = -k_{Ph} - (k_{TPAn}n + k_{TPAp}p)T + \frac{\gamma(j)j}{qd}, \qquad (6.72)$$

where $k_{TPAn,p}$ is the TPA rate for electrons, holes with densities n, p, respectively. Now, the current in OLEDs typically follows $j \propto V^{m+1}$. Hence, solutions to Eq. 6.72 are

$$\frac{\eta_{ext}(j)}{\eta_0} = \frac{1}{1 + \alpha j^{1/(m+1)}}, \qquad (6.73)$$

where we have assumed that γ and α are constants independent of j (Baldo et al., 2000a, Reineke et al., 2007).

The importance of TPA has been investigated by observing the PL intensity due to optical pumping of the EML as a function of current in single carrier (i.e. electron or hole) devices. Then, from changes in the transient decay rate, the magnitude of k_{TPA} can be determined (Baldo et al., 2000a, Reineke et al., 2007, Giebink and Forrest, 2008). This method has been applied to study TPA in electron-only devices with the structure: ITO/45 nm Alq_3/10 nm 6 vol.% PtOEP: Alq_3/45 nm Alq_3/Mg:Ag cathode. The PL transients were obtained by exciting the devices with a pulsed N_2 laser at λ= 337 nm, with results shown in Fig. 6.81a. The transient solutions to Eq. 6.72 with $p = 0$ and $m = 1$ in Eq. 6.74 are shown by the dashed lines that provide a convincing fit to the data. The decay rate increases with j, characteristic of enhanced TPA at high electron densities. Values of $k_{TPA}n$ ranging from 4 to 35 μs^{-1} as inferred from these fits are provided in Fig. 6.81b. The lifetime is decreased to half its initial value at $j_0 = 200$ mA/cm^2. This is compared with $j_0 \simeq 4$ mA/cm^2 due to TTA. Hence, while TPA is present in the PtOEP:Alq_3 system, its contribution to the observed efficiency roll-off is insignificant at even the highest currents. This same conclusion has been reached for Ir-based devices (Reineke et al., 2007, Giebink and Forrest, 2008).

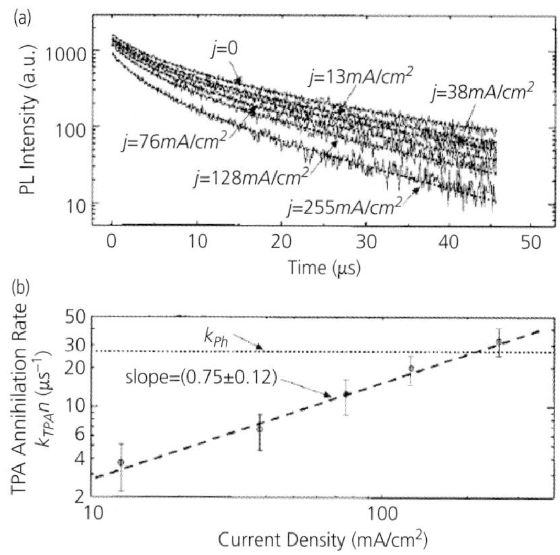

Figure 6.81 (a) Photoluminescence intensity vs. time and drive current of a 6 vol.% PtOEP:Alq_3 electron-only device. Lines fit to the data follow Eq. 6.73. (b) TPA annihilation rate vs. current density for the device in (a). The EL lifetime, k_{Ph}, is reduced to 50% of its zero current at $j = j_0 = 200$ mA/cm^2 (Baldo et al., 2000a).

6.4 OLED displays

Mobile applications such as smart phones and tablets are now dominantly featuring OLED displays due to their high contrast, high efficiency (leading to long battery lifetimes), thinness, and optical design simplicity (which leads to lower costs). As we have discussed in Section 6.1, OLEDs are also moving rapidly into the TV market for many of these same reasons. In this section, we discuss the elements of an OLED display, and the demands their designs place on the pixel architecture itself.

A schematic design of OLED displays is illustrated in Fig. 6.82. There are two OLED display architectures. For mobile applications, red, green and blue OLED sub-pixels are positioned side-by-side above the substrate plane to give the appearance of full color rendition to the viewer who, in the far field, cannot resolve the individual pixels (typically ~10–30 μm in diameter). Each sub-pixel is a top emitting device using a transparent cathode that rests on its own driver transistor circuit to provide rapid, individual addressing of the OLEDs. A circular polarizer with a low reflection coating is located on the display surface to eliminate back reflections. The second architecture employed in many television sets comprises three white OLED (WOLED) sub-pixels positioned beneath a set of color filters (CF). The color filters produce the R, G, and B output needed to generate the full-color image. Often, a fourth white emitting OLED is used along with the R, G, and B pixels to enhance luminosity of a white display field, enabling intense high definition images. Since these displays are viewed from a distance, the pixel pitch and dimensions (>50 μm) are more relaxed than for mobile displays. The color-filtered designs in Fig. 6.82b are inherently less efficient than the RGB architecture in Fig. 6.82a since the unwanted portions of the spectrum are removed from the white emission spectrum, whereas the RGB display pixel delivers only the color needed to achieve the desired color coordinates of the pixel triad.

The relative simplicity of the OLED display is in striking contrast to a LCD shown in Fig. 6.83. The LCD is based on a white backlight that distributes light across the entire image field. Interposed between the backlight and the observer are several layers needed to create the image: a polarizer is followed by the

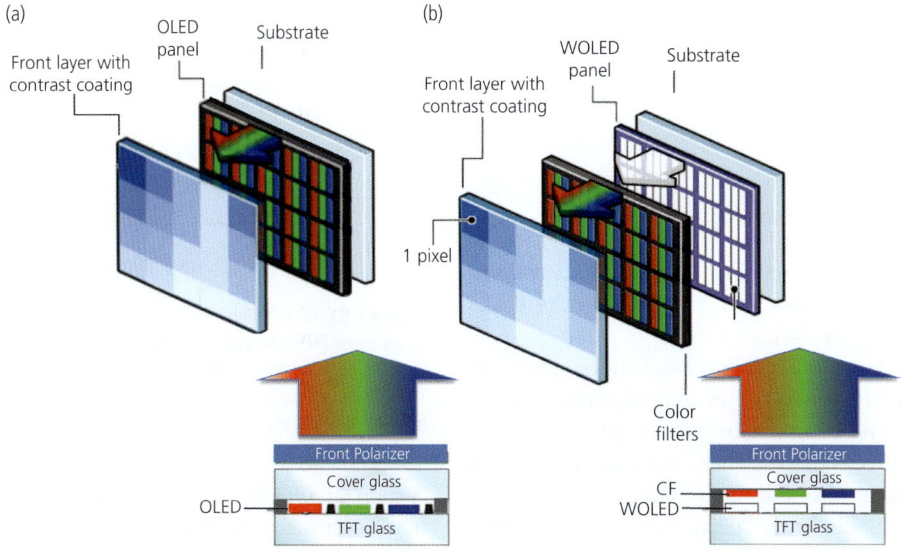

Figure 6.82 Schematic illustration of two types of OLED displays. (a) Full color RGB OLED pixel display used predominantly in mobile applications, and (b) white OLED (WOLED) pixels + color filters used in televisions. The front contrast layer includes a circular polarizer and an additional coating to minimize reflections. Insets: Cross-sections showing the essential details of the full color pixels used in these displays. TFT = thin film transistor drive electronics, CF = color filter. (www.flatpanelshd.com)

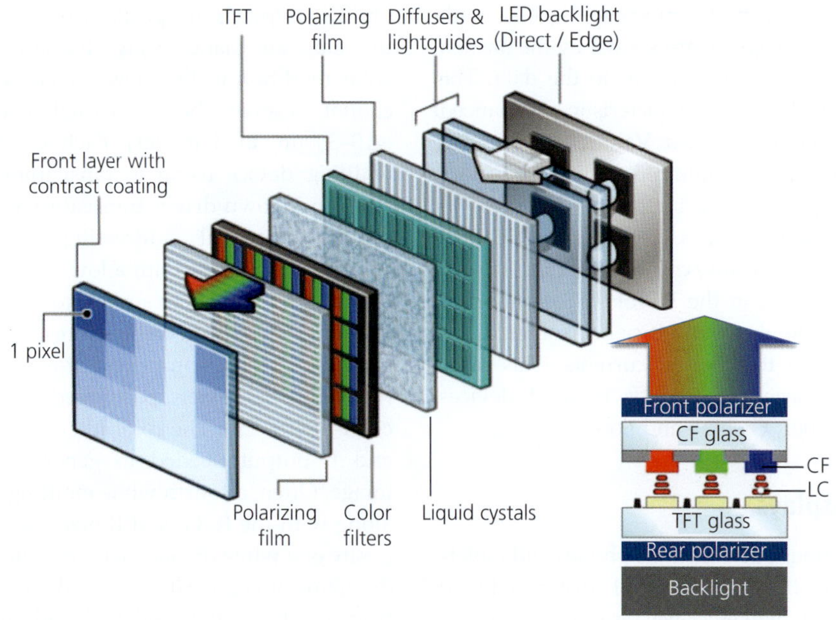

Figure 6.83 Schematic illustration of a liquid crystal (LC) display using an LED backlight. Inset: Cross-section showing the essential details of the full color pixels used in the display. TFT = thin film transistor drive electronics. (www.flatpanelshd.com)

LC light valve array that switches the light on and off by application of an electric field. The field rotates the LC molecules, which in turn rotate the polarization vector of the incident light. Similar to an OLED display, the LC array is addressed by a TFT backplane positioned behind each light valve. Next is the color filter array that admits only R, G, or B light from each of the LC devices. This is followed by a second polarizer oriented perpendicular to the first, and finally the front glass with an antireflection coating to provide a high contrast image. Great care is needed to generate a uniform backlight that prevents visible differences in intensity, contrast, shadowing, etc. (known as display *mura*). This is done by using either a compact fluorescent tube, or an array of LEDs located around the display periphery. Its light output is then evenly distributed across the entire panel via diffusers and light guides. Alternatively, an array of LEDs is placed on the backplane itself to locally illuminate a few thousand pixels at an intensity that ensures image consistency under a variety of room lighting situations. There are several optical elements used in light distribution not shown in Fig. 6.83 including a diffuser, reflector, light guides, etc.

The LCD is inherently different than an OLED display in that an LCD is a color-subtractive device, whereas an OLED is a color-additive device. Color subtraction results in the waste of energy since light is required to illuminate a pixel even when it is turned off. Also, since the LC light valve itself is not 100% opaque when off, the color gamut as well as the contrast of the LCD is considerably less than an OLED display. One means to improve LCD efficiency and color gamut is to use a sheet of quantum dots (QDs) in front of a blue LED backlight. The QD diameters are varied to down-convert the blue light to a narrow green and red emission spectrum. The larger QDs emit in the red, whereas the smaller in the green. Hence, the spectrally broad emission of a white LED or fluorescent backlight is replaced by more saturated and efficient R, G, and B sources whose colors are once again selected by the LC arrays. This improved display is called QD-LED LCD. We note that the speed of response and viewing angle dependence of LCDs remain unchanged by the presence of QD *color-changing media*.

A qualitative comparison of the performance LCDs, QD-LED and OLED displays is provided in Table 6.9. Display pixels are required to deliver a maximum brightness of 100 nits. Given the several optical elements with only limited transparency, approximately 5% of the light from the LCD backlight reaches the viewer. In contrast, 30–40% of the light emitted by the OLED is viewed, resulting in a considerable savings in power consumption. The reduced complexity and thinness of the OLED display allows for flexible or foldable form factors that are unachievable via LCD technology. Indeed, the fewer layers and optical elements of the OLED display should also lead to lower cost, although to date LCD displays are less expensive due to the much longer experience gained in their manufacture. As we have discussed, OLEDs can be switched on or off in a few microseconds, whereas switching the LC molecules requires their physical rotation, which takes place over several milliseconds. Hence, fast moving images are blurred or *tunnel* (i.e. rapidly moving objects briefly vanish), which is completely avoided in OLED displays. The polarized light in an LCD also results in variations in contrast and brightness with viewing angle.

Table 6.9 Qualitative comparison of features of LCD and OLED displays

Feature	LCD	QD-LED LCD	OLED
Thickness	>2.5 cm	>2.5 cm	$\leq$ 3 mm
Power[a]	1	0.7	0.3
Life span	50,000 h	50,000 h	40,000 h
Viewing angle[b]	100–150°	100–150°	170°
Contrast ratio[c]	5000:1	5000:1	10^6:1
Response speed	1–5 ms	1–5 ms	1–10 μs
Form factor	flat	flat	flat, curved, flexible

[a] Power consumption is approximate and normalized to that of an LCD assuming all displays have the same area. The OLED display assumes that R and G colors are produced by phosphorescence, and B by fluorescence.

[b] Viewing angle in LCD and QD-LED LCD displays is superior along the horizontal compared with the vertical directions. OLEDs are nearly ideal Lambertian sources, and hence have no angular viewing dependence.

[c] Contrasts ratios of OLEDs are limited by front surface reflections.

Importantly, the exceptionally high on/off ratio of an OLED display compared with an LCD gives the images a crisp, high contrast, and even 3D appearance.

Finally, we note that both standard and high definition (4K and 8K) LCD and OLED displays are commercialized, with the latter displays now available up to 77″ along the diagonal. Given the ability to achieve extremely small OLED pixels coupled to their fast response time, OLEDs are now entering the virtual and augmented reality equipment markets. The performance objectives for OLEDs used in displays are:

(i) *High external quantum efficiency* for all R, G, and B sub-pixel OLEDs used to achieve the desired color gamut at maximum display intensity (typically ~100 nits) is required to reduce the total display current, and hence increase the battery life in mobile applications. While 100 nits is the intensity needed for a white display field, we will show in the following section that the actual device intensity needs to be significantly higher than this when its limited *aperture ratio* (i.e. the area of a pixel that emits light vs. the total pixel footprint) and the front surface polarizer losses are considered.

(ii) *Minimal angle dependence of intensity and wavelength*, that is, they should be close to ideal Lambertian sources to avoid color shifts and display dimming with angle that are common in LCDs.

(iii) *Low voltage operation at display intensities* to be compatible with low power drive electronics. Combined with a high quantum efficiency, this implies a high display power efficiency.

(iv) *Saturated color emission in blue, green, and red* permitting the largest possible color gamut. This implies narrow natural emission linewidths for the R, G and B sub-pixels.

(v) *Very long operational lifetimes* with stable color and intensities extending to >10,000 h at display brightness.

(vi) *Top emission* is important for high-resolution displays used in small appliances for mobile applications. This allows for a high pixel pitch when combined with an opaque transistor driver backplane.

(vii) *Rapid temporal response* to create realistic, fast moving images. Pixel switching times of ~1 ms are generally adequate.

(viii) *Low cost materials and fabrication processes*. To date, most displays are fabricated on glass and plastic substrates that can be costly. Also, the organic materials should be easily accessible via simple and efficient synthetic processes of only a few steps. Synthetic accessibility has a far greater impact on cost than does that of the materials themselves. For example, Ir or Pt in phosphors have only a minimal impact on the total display cost if the emitting metal complexes are readily synthesized. Also, fabrication costs can be excessive when using shadow masks and similar "indirect" patterning processes. Direct printing via OVJP or ink jet printing, for example, hold promise for lowering costs in future generation OLED displays. Patterning must also have a sufficient resolution for the particular display application: ~10–30 μm for mobile appliances, and 50–100 μm for high definition TVs.

In the following section, we will discuss a few of the most important considerations that displays require in the design of the OLED itself.

6.4.1 Top emitting OLEDs for displays

Example sub-pixel structures used in displays are illustrated in Fig. 6.84. The OLED is often constructed directly onto the source or drain contact of the driver TFT that is fabricated on the substrate surface. The array of TFT drivers is called the *backplane*. The backplane is typically based on *low temperature poly-Si* (LTPS) technology. Recently, considerable progress has been made in metal oxide transistor technology, which is attractive for OLED displays due to its low temperature processing. Metal oxide electronics may eventually overtake LTPS backplanes, particularly in flexible displays where the substrate is a temperature-sensitive plastic foil. The backplane circuit comprises a multi-transistor current driver (e.g. a four transistor current mirror or a voltage stabilization circuit, Sasaoka et al., 2001) used to provide a high on/off contrast ratio (i.e. high *image dynamic range*), fast pixel switching, and temperature and compensation to prevent color shifts with prolonged operation of the display (Tsujimura, 2012).

To save space and achieve a high pixel pitch, top emitting devices are placed directly onto the TFT surface (Fig. 6.84a). This architecture is particularly relevant for small mobile displays, or displays employed in virtual reality appliances that require very high resolution. The architecture features a transparent or semitransparent cathode on top of a conventional OLED stack consisting of an anode on the bottom and cathode on top, or inverted OLED whose anode and cathode are reversed. There are two choices to achieve an efficient transparent cathode: use a very thin, semitransparent low work function metal, or employ a transparent metal oxide such as ITO. Whereas the former approach may have high resistance or a discontinuous contact, the high work functions of metal oxides can create an energy barrier to electron injection that can reduce efficiency. In Fig. 6.84b, we see that a bottom emitting pixel requires that the OLED be fabricated over a window in the backplane. High resolution may not always be required, making this architecture desirable for many applications.

Table 6.10 shows the approximate R, G, and B pixel brightness needed for a tablet display (Samsung S8/4 at 100% duty cycle) used at a maximum viewing brightness of 400 nits, assuming a front polarizer loss of 50%. Interestingly, the low aperture ratio and polarizer losses require that the actual pixel brightness is >10 times the intensity that is viewed. Also note that the green pixel is operated at ~10 times the luminance of the blue pixel, and nearly three times that of the red. This allows for uniform aging of all pixels to avoid color shifts with display use, since the green PHOLED lives far longer than the blue fluorescent OLED (see Section 6.7). It is worth noting, too, that the intensities required in mobile appliances such as tablets are considerably higher than for TVs and monitors, since the former are designed for use in bright outdoor environments, whereas larger, stationary displays are ordinarily viewed indoors.

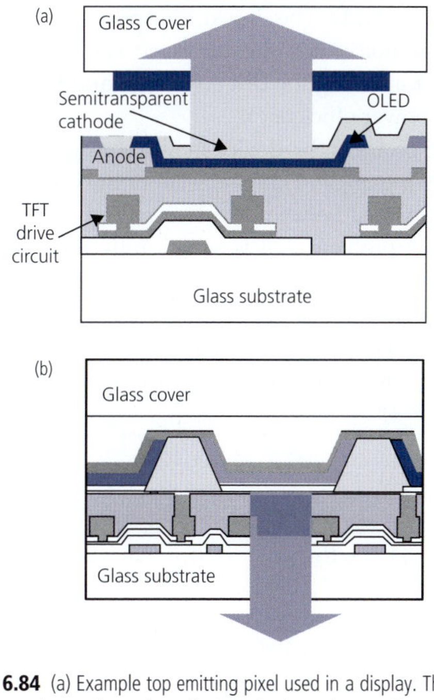

Figure 6.84 (a) Example top emitting pixel used in a display. The OLED is constructed onto the surface of a TFT circuit, and uses either a fully transparent metal oxide or semitransparent thin metal cathode at the emitting surface. A color filter may or may not be used for modifying the pixel spectrum. (b) A bottom emitting display pixel.

Table 6.10 Approximate pixel aperture ratios and brightnesses used in a mobile tablet display (C. W. Tang, private communication, 2017)

Color	Aperture ratio (%)	Screen lum. (nits)	Pixel lum. (nits)
Red	6.7	100	2990
Green	9.7	285	5880
Blue	8.4	25	600
White	24.8	400	3225

In Chapter 4 we showed that efficient electron injection requires that the cathode work function align with the LUMO of the organic layer that it contacts. Alternatively, defects at the surface of the semiconductor can act as an energy "ladder" to assist electrons in surmounting the cathode-organic barrier. This is illustrated for an Alq_3-based device with an ITO cathode sputtered onto a CuPc buffer in Fig. 6.85 (Parthasarathy et al., 1998). To avoid damage to the underlying organic induced during ITO sputter deposition, a very low power (5–10 W for the initial layer) is used. This results in a growth rate of <0.1 Å/s for the initial few monolayers. The growth rate is then increased to achieve the final thickness. Some degree of damage to the interface is necessary since the ionization potential of ITO is too large for electron

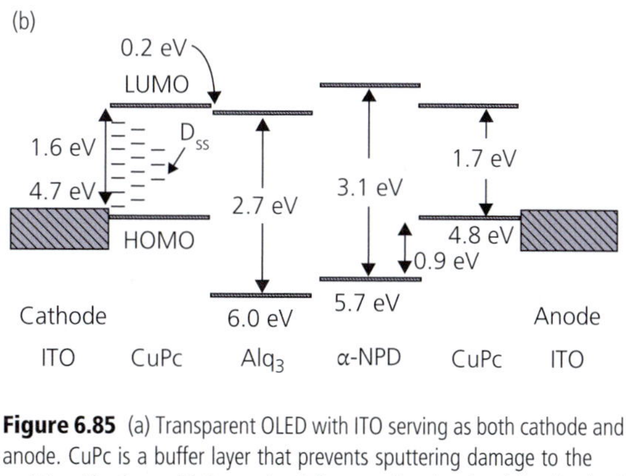

Figure 6.85 (a) Transparent OLED with ITO serving as both cathode and anode. CuPc is a buffer layer that prevents sputtering damage to the emitting Alq_3 layer. (b) Energy level diagram showing the defect levels of density D_{ss} in the CuPc buffer (Parthasarathy et al., 1998).

injection. Hence a thin (3–6 nm) and relatively robust CuPc interface buffer layer is grown on the Alq_3 EML surface. The buffer incurs limited damage that provides defect levels, or energy steps that lower the injection barrier (Parthasarathy et al., 1998).

A feature of a top emitting OLED is that the anode contact on the substrate can also be made transparent using a second metal oxide contact such as ITO or indium zinc oxide (IZO). This type of device is very nearly totally transparent, thereby enabling transparent OLED (TOLED) displays and windows, shown in Figs. 6.1 and 6.2 (Bulovic et al., 1996).

Doping the electron and hole transport layers in an inverted n-i-p PHOLED with transparent cathode and anode contacts results in considerably lower voltage operation compared with transparent fluorescent OLEDs. The device in Fig. 6.86 features a 15 nm thick Li-doped Bphen EIL deposited onto an ITO bottom cathode, followed by a 20 nm thick Bphen ETL/EBL, a 10 nm thick EML consisting of CBP doped with $Ir(ppy)_3$, a 15 nm thick HBL comprising $Ir(ppz)_3$, a 200 nm thick F_4-TCNQ doped m-MTDATA HTL, and a 20 nm thick CuPc HIL capped with a 60 nm thick ITO anode (Pfeiffer et al., 2003). The Li-doped BPhen increases the cathode electron injection efficiency while also absorbing sputtering damage during ITO deposition. This device is 80% transparent at the peak emission wavelength of $Ir(ppy)_3$ of $\lambda = 510$ nm, and is >65% transparent across the entire visible spectrum. The voltage required to achieve a given luminance is reduced by ~1.5–2 V compared to a conventional bottom emitting OLED due to the doping of the transport layers. The peak quantum efficiency of the n-i-p OLED is 7.5% at a luminance of 1000 cd/m^2, which is less than that for a conventional device with $\eta_{ext} = 9\%$ at this brightness, indicative of a

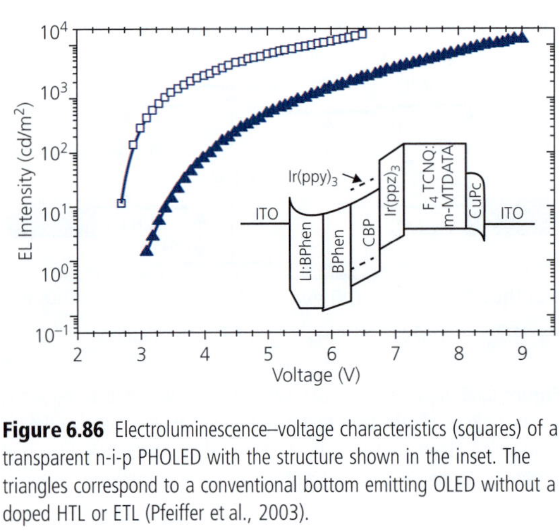

Figure 6.86 Electroluminescence–voltage characteristics (squares) of a transparent n-i-p PHOLED with the structure shown in the inset. The triangles correspond to a conventional bottom emitting OLED without a doped HTL or ETL (Pfeiffer et al., 2003).

lower injection efficiency of the low work function metal oxide cathode.

Higher efficiencies have been achieved by replacing the ITO cathode with a thin, semitransparent metal contact. Combining a semitransparent metal contact with doped charge transport layers whose thickness can be varied without significantly changing the operating voltage, can be used to adjust the optical field within the device to optimize the light outcoupling efficiency (Lin et al., 2006, Leem et al., 2009, Hofmann et al., 2010). For example, $\eta_{ext} = 29\%$ is achieved by combining these features in a top emitting device with a thick metal anode on the substrate surface and a thin, semitransparent Ag cathode capped with an α-NPD antireflection layer (Hofmann et al., 2010). The red-emitting device shown in the inset of Fig. 6.87 comprises an $Ir(MDQ)_2(acac)$ phosphor doped at 10 vol.% in α-NPD. The doped transport layers provide flexible choices for the dielectric stack thicknesses without affecting the drive voltage (and hence luminance power efficiency).

The spectra of this device show a strong dependence on viewing angle. This is a result of the optical microcavity formed between the two reflective metal electrodes; a problem that is significantly reduced when both contacts use more transparent metal oxides such as in Fig. 6.86. The spectra are modeled assuming a weak Fabry–Pérot microcavity, which yields the following angle-dependent intensity spectrum (Deppe et al., 1994):

$$I(\lambda,\theta)/I_0(\lambda) = \frac{T_2[1 + R_1 + 2\sqrt{R_1}\cos(-\phi_1 + 4\pi n_{EML} z_{EML}\cos\theta_{EML}/\lambda)]}{\left(1-\sqrt{R_1R_2}\right)^2 + 4\sqrt{R_1R_2}\sin^2(\Delta\phi/2)}, \tag{6.74}$$

where T and R are the transmissivity and reflectivity of the rear anode mirror (subscript 1) and the front semitransparent cathode (subscript 2), z_{EML} is the position of the EML relative to the anode, and θ_{EML} is the angle with index of refraction, n_{EML}. Also, ϕ_1, ϕ_2, and $\Delta\phi$ are the phase shifts at the mirror, the semitransparent cathode, and the total phase shift accumulated by a photon after one transit of the cavity. The total phase shift due to the accumulation of phase shifts within each layer, i, in the stack is

$$\Delta\phi = -\phi_1 - \phi_2 + \sum_i \frac{4\pi n_i d_i \cos\theta_i}{\lambda}. \tag{6.75}$$

Here, n_i and d_i are the refractive index and thickness of the ith layer, respectively, and θ_i is the angle of propagation. The total optical cavity length is then (Lee et al., 2006)

$$L_{opt} = \frac{\lambda}{4\pi}(|\phi_1| + |\phi_1|) + \sum_i n_i d_i \approx \frac{\lambda}{4\pi}(|\phi_1| + |\phi_1|) + nL. \tag{6.76}$$

At normal incidence, the numerator in Eq. 6.74 has a maximum when the emitting molecule lies precisely at the antinode of the optical field in the dielectric stack, in which case $nz/\lambda = (2m + 1)/2$, where $m = 0, 1, 2\ldots$ is the optical order of the cavity, and z is the distance to the anode. The denominator in Eq. 6.74 is determined by the interference within the layers. Thus, the thickness of the doped HTL is chose such that the EML is located at the antinode of the optical field.

One other spectral feature that is affected by the microcavity is the emission full width at half maximum (FWHM):

$$\frac{\Delta\lambda}{\lambda} = \frac{\lambda}{2nL_{cav}}\left[\frac{1 - \sqrt{R_1R_2}}{\pi(R_1R_2)^{1/4}}\right]. \tag{6.77}$$

That is, as the reflectivity increases, the spectral line narrows; an effect that can be significant in OLEDs with both top and bottom reflective metal contacts. The allowed cavity modes are enhanced at the expense of modes that are not resonantly coupled to the cavity. Now the integrated emission intensity enhancement depends on the natural linewidth of the emission source ($\Delta\lambda_n$) and the cavity spectral width in Eq. 6.77, viz.

$$I_{\text{int}} = I(\lambda_0)\sqrt{\pi \ln 2}(\Delta\lambda/\Delta\lambda_n) \tag{6.78}$$

for a Gaussian lineshape. The resonant enhancement factor is (Deppe et al., 1994)

$$I(\lambda_0) = \frac{\xi}{2}\frac{(1 + \sqrt{R_1})^2(1 - R_2)}{(1 - \sqrt{R_1R_2})^2}\left(\frac{\tau_{cav}}{\tau_n}\right), \tag{6.79}$$

where $\xi = 0$, 1, or 2 if the emitting dipoles are located at a cavity node, if they are distributed over several optical nodes, or if they are located at an antinode, respectively. Also, τ_{cav} is the lifetime of the dipole in the cavity that is typically longer than the natural lifetime, τ_n, due to the *Purcell effect* (see below). However, in moderately weak cavities using emitters with broad spectral lifetimes, $\tau_{cav} \simeq \tau_n$. Indeed, even in relatively strong cavities that comprise cathodes with reflectivities approaching 70% (and hence a greatly attenuated emission intensity according to Eq. 6.79), the spectrum is only slightly narrowed from its natural linewidth, although even a minor narrowing can result in a noticeable angular dependence of the spectrum (Lee et al., 2006).

The spectra in Fig. 6.87 shows pronounced microcavity-induced shifts. The peak wavelength at normal incidence is 665 nm at normal incidence, moving to 610 nm at $\theta = 50°$. This is accompanied by intensity attenuation as the viewing angle moves away from the cavity resonance. Furthermore, the spectral narrowing is evident when compared to the PL spectrum of the emitter, $Ir(MDQ)_2(acac)$ (dashed line). The cavity linewidth at normal incidence is $\Delta\lambda =$ 60 nm compared to a natural linewidth of $\Delta\lambda_n = 90$ nm. The first order cavity in Fig. 6.87 has a peak quantum efficiency of 29%, falling to 17% and 12% for

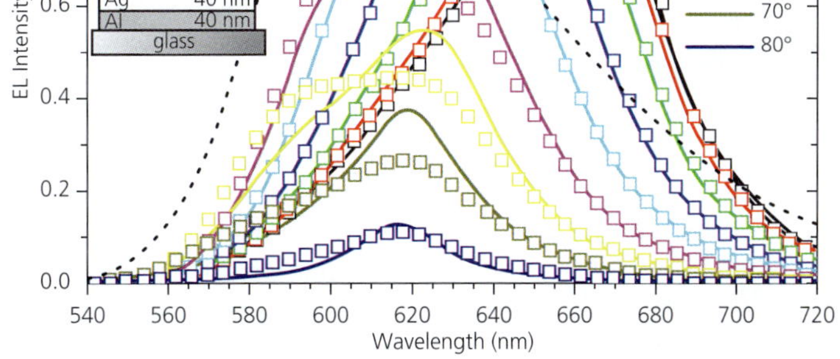

Figure 6.87 Emission intensity vs. wavelength and viewing angle (θ) for the top emitting PHOLED shown in the inset, whose emission region is indicated by the red dashed line. NDP-2 is a proprietary acceptor molecule. The data are shown by the lines, and simulations by the data points. Also shown (dashed line) is the PL spectrum from the phosphor, $Ir(MDQ)_2(acac)$ (Hofmann et al., 2010).

second and third order cavities obtained by increasing the doped HTL thickness.

The calculated output intensities as a function of the ETL and HTL, and cathode thickness (and hence reflectivity, R_1) for these PHOLEDs are shown in Fig. 6.88 (Hofmann et al., 2010). The integrated intensity decreases as the optical cavity thickness is increased to accept higher order optical modes. Similarly, as the optical order is increased, the maximum output is achieved with thinner and less reflective top contacts.

To summarize, we have described two strategies for achieving top emission. The less reflective transparent oxide cathode produces a very weak microcavity, and hence the impact on the OLED optical output intensity and spectral content is minimal, although this is at the expense of a reduced (by ~20%) output efficiency. We will show in Section 6.6, however, that optimal designs can eliminate these losses by using improved cathode materials and sputtering conditions. In contrast, semitransparent metal cathodes can produce a higher output intensity compared to conventional devices. This is due to microcavity resonances that dramatically shift the OLED color

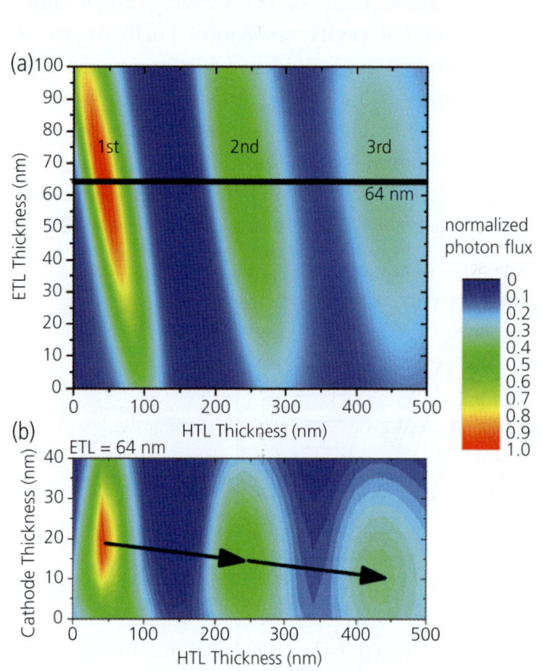

Figure 6.88 Calculation of the output intensity ETL versus (a) HTL and (b) cathode thickness for first, second, and third order cavities with structures analogous to that in Fig. 6.87. In (a) the line at 64 nm corresponds to fabricated devices. In (b), the peak intensity shifts to thinner cathodes as the optical order increases. The integrated intensity also decreases with order (Hofmann et al., 2010).

Reprinted from Hofmann, S., Thomschke, M., Freitag, P., Furno, M., Lüssem, B. & Leo, K. 2010. Top-emitting organic light-emitting diodes: Influence of cavity design. Applied Physics Letters, 97, 253308, with the permission of AIP Publishing.

coordinates with viewing angle. Careful cavity design and control of layer thicknesses is imperative to assure the device-to-device variations are minimized.

Whether transparent oxide or semitransparent thin metal cathodes are used, both structures present challenges to processing and contact resistance. That is, the metal oxide resistivity is considerably larger than for metals, yet very thin metal films are also resistive and can be discontinuous. Many top emitting OLED displays use semitransparent metal contacts largely due to their simplicity and compatibility with other deposition processes used in display manufacture.

6.4.2 Full color displays and pixelation

In the context of Fig. 6.82 we noted that full color images require dividing each pixel into three (R, G, B) sub-pixels that can be independently tuned to achieve the desired luminosity and color coordinates. Each sub-pixel, in turn, requires its own TFT drive circuit. Several different schemes have been devised to pixelate the display, two of which are illustrated in Fig. 6.82: positioning of the triad of sub-pixels side-by-side to achieve a brilliant, large color gamut display, and the use of white emitting OLED (WOLED) pixels positioned beneath R, G and B color filters. Several other methods of pixelation are shown in Fig. 6.89. Bottom emitting pixels are shown, although each configuration can be inverted for top emission.

Beyond the RGB and WOLED + color filter pixelation schemes, two others warrant our attention. In Fig. 6.89c the RGB triad comprises three blue OLEDs, two of them positioned over fluorescent *color changing media* (CCM) (Niko et al., 1997, Sokolik et al., 2001). In contrast to color filters that simply eliminate the unwanted spectral content and thus lead to efficiency losses, CCM are films that luminesce with near unity PL quantum yield when pumped by the blue OLED. Fluorophores such as coumarin for green emission, and laser dyes of lumogen and DCM have been employed to generate the green and red pixels, respectively. Colloids containing QDs of different diameters is another type of CCM. Direct pumping of the red CCM by a blue source is inefficient. Hence, it is usually combined with the green fluorophore, such that there is a cascade from blue to green to red.

While this scheme is more efficient than simple color filtering of white light, it remains considerably less efficient that direct RGB emission in Fig. 6.89a. The internal efficiency of the CCM (which ignores outcoupling and spin dependent losses) is calculated using

$$\eta_{int,CCM} = \eta_{ext,OLED}(1 - R)\phi_{CCM}\exp(-\alpha_{CCM}d_{CCM}), \tag{6.80}$$

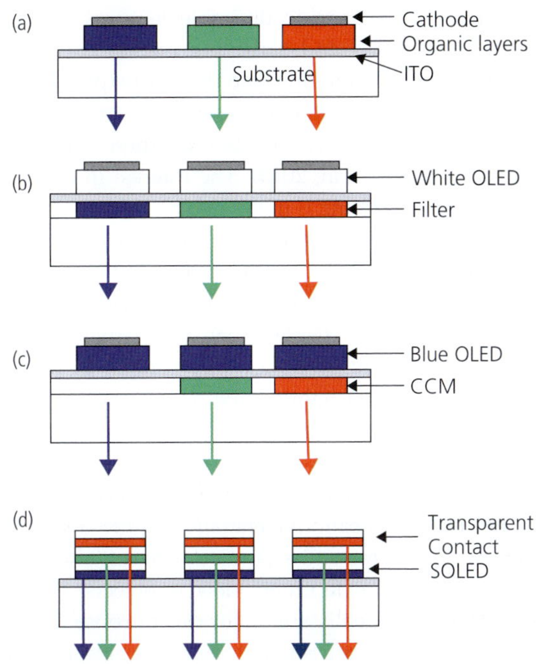

Figure 6.89 Several methods of pixelation in full color displays. (a) RGB pixels positioned side-by-side, (b) individually addressable WOLEDs positioned above RGB filters, (c) blue OLEDs positioned above R and G emitting color changing media (CCM), and (d) stacked RGB OLEDs (SOLEDs).

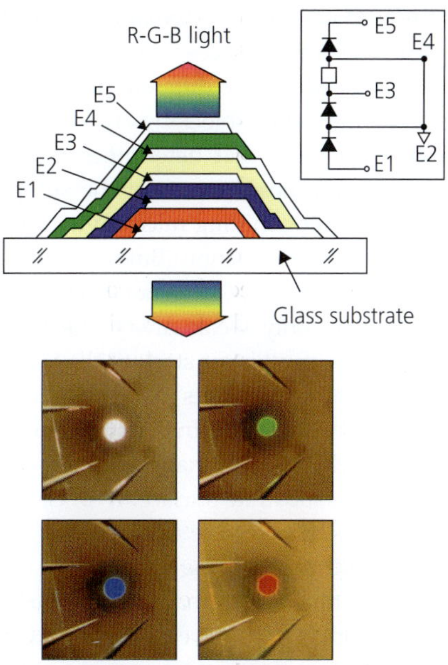

Figure 6.90 Cross-section of a SOLED with its biasing scheme shown in inset. Also shown are images of R, G, B, and white emission achieved by separate biasing of the various elements in the stack. White is achieved by simultaneously turning on all three pixels (Parthasarathy et al., 1999).

where $\eta_{ext,OLED}$ is the external efficiency of the blue OLED, R is the reflectivity of the CCM, ϕ_{CCM} its PLQY, α_{CCM} is its absorption coefficient, and d_{CCM} its thickness. Similar to filtering schemes, unabsorbed light constitutes both leakage and loss that can result in shifting of color coordinates. Furthermore, we note that there is a power loss associated with down-conversion since a fraction of energy of the blue photon is lost in generating green or red fluorescence. That is, the optical output power of the CCM sub-pixel is reduced by the ratio of the peak OLED and CCM wavelengths. That is,

$$\eta_{P,CCM} = \eta_{P,OLED}(\lambda_{OLED}/\lambda_{CCM}). \tag{6.81}$$

These losses are also present in filtered schemes in Fig. 6.89b. Since the CCM is an active light emitting layer, it is subject to long term degradation experienced by the OLED itself. In spite of its many advantages, CCM pixelation is not widely employed due to the added complexity and requirements for stable and efficient fluorophores that strongly absorb in the blue OLED spectral region. In contrast, QD-based CCMs are currently finding use in large OLED displays.

Direct emission can also be achieved by vertically stacking the sub-pixels, as shown in Fig. 6.89d. The stacked OLED (SOLED) vertically positions the separately optimized sub-pixels, separated by transparent contacts that allow for individual addressing (Shen et al., 1997). A detail of a SOLED pixel is shown in Fig. 6.90 along with images of the SOLED biased to emit R, G, B, and white (Parthasarathy et al., 1999). This scheme has advantages for high resolution displays since it forms a compact triad in the vertical dimension, thus minimizing the pixel footprint. Each sub-pixel can, in principle, achieve 100% internal efficiency using phosphors or TADF lumophores provided that the intervening electrodes are optically lossless. Metal oxide and thin metal intermediate electrodes are employed for this purpose. In Section 6.5 we show that SOLEDs are attractive for high intensity lighting applications, albeit in those devices there is no need to contact the intermediate electrodes to separately tune elements in the stack.

Although there are several advantages to this architecture, bringing the contacts out from the individual layers onto the substrate without shorting can be challenging, and can occupy a large peripheral area around the base of the stack that increases the effective pixel size. The addressing scheme for stacked contacts in Fig. 6.90 differs from that used in laterally separated, individually addressed pixels. To simplify biasing, an insulating layer (shown by a small box in the inset) is inserted between contacts E3 and E4. This avoids the need to reverse the polarity of the contacts or inverting one of the devices in the stack, and hence eliminates the need for two different pixel addressing circuits when the current blocking layer is absent.

A design constraint that determines the stacking order and the thickness of each of the sub-pixels is the existence of color-shifting microcavity effects discussed in the previous section (Bulovic et al., 1998a). While the metal-free, ITO interconnection electrodes in Fig. 6.90 show only weak cavity effects, placing them in the deep, multilayer stack of the SOLED can result in significant color shifts that reduce the accessible color gamut of the pixel. In the device in Fig. 6.90, the optimal stacking order that minimizes cavity effects is, from the substrate surface, R–B–G. Also, improved color coordinates are achieved by adjusting the cavity length through the appropriate choice of transport layer thicknesses that is made possible by conductivity doping of the layers. The SOLED can use either top or bottom transparent contacts, or both, for use in transparent or top emitting displays.

6.5 OLED lighting and lighting devices

Approximately 25% of the total industrial electricity use in the US is devoted to powering illumination sources, and yet a vast majority of those sources are highly inefficient (~5%) and short-lifetime (800–1000 h) incandescent light bulbs; a technology that is approaching its 150th anniversary. Hence, considerable effort has been devoted over the last few decades to develop alternative sources that can provide pleasant illumination for building interiors as well as for signage and street lighting. Currently, inorganic LEDs provide desirable features of efficiency, brightness and satisfying colors, and as a result are rapidly moving into the marketplace. The ultrahigh efficiency and tunable color emission of OLEDs has also drawn attention for these same applications (Kassakian, 2013). The interest in OLEDs for solid state lighting is motivated by their low form factor including bendability and conformability, and the ability to precisely tune their emission to achieve desired CRI and CCT values. The development of WOLEDs used in displays has also driven rapid progress in lighting applications. Unlike displays, however, WOLED lighting fixtures do not require high resolution patterning of R, G and B emitting pixels, thus leading to a significant cost savings and simplification of their manufacturing process. Interior lighting applications generally require a highly uniform, glare-free emission that casts soft shadows to provide some depth to viewed objects. OLED illumination is ideally suited for this purpose. However, they are less attractive for outdoor lighting due to their relatively low intensities.

Table 6.11 provides a comparison between OLED and other lighting sources. Currently, OLED efficiencies are as high as 150 lm/W, which is second only to inorganic LEDs. By combining multiple emitting phosphors, a high CRI at the desired CCT.

Importantly, the surface temperature of OLED fixtures is only a few degrees above ambient (Qi and Forrest, 2011, Bergemann et al., 2012, Levermore et al., 2012). This feature is critically important when considering installation in buildings since this eliminates the necessity to dissipate heat generated in its fixture, or *luminaire*. A thermal image of a high intensity green

Table 6.11 Comparison of the performance of lighting sources

Feature	Incandescent	Fluorescent	LED	OLED
Efficacy	17 lm/W	40–90 lm/W	80–90 lm/W, white 65 lm/W, warm white 240 lm/W, lab demo	150 lm/W, lab demos
CRI	100	80–85	80, white 90, warm white	Up to 95
Form factor	Heat generating	Long or compact gas filled glass tube	Point source high intensity lamp	Large area thin diffuse source, flexible, transparent
Bulb temp.	3000°C	50°C	150°C	30°C
Toxicity	None	Mercury	Low	Low
Lifetime	0.8 kh	20 kh	50 kh	40 kh
Dimmable	Yes, but much lower efficacy	Yes, efficiency decreases	Yes, efficiency increases	Yes, efficiency increases
Cost ($/klm)	0.7	3.00 (CFL)	3.00	~100

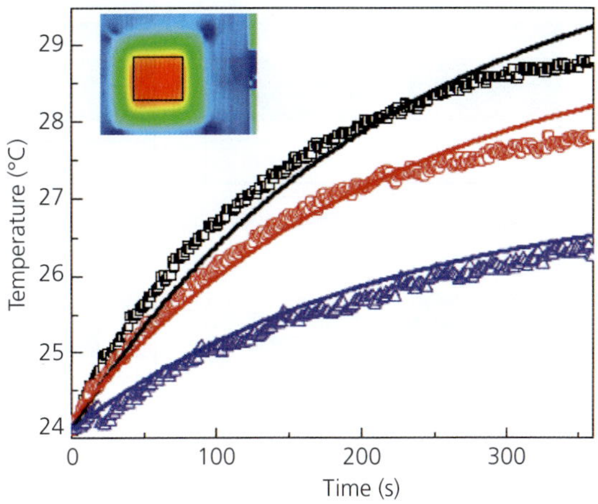

Figure 6.91 Temperature increase of a 25 cm^2 green emitting PHOLED in a sealed glass package following the onset of current steps with steady state values of (from bottom to top) 2, 3, and 4 mA/cm^2. The lines are fits using a multilayer transfer-matrix approach. The largest impact on the peak temperature is the air gap between the OLED and the top glass package lid. Inset: Infrared thermal image of the device in equilibrium at 3 mA/cm^2. The ambient (blue color) is at 24°C and the peak temperature (red) is 28.2°C (Bergemann et al., 2012).

PHOLED is shown in Fig. 6.91 along with the thermal transients at different drive currents. The theoretical fits to the data (lines) are based on a transfer matrix approach using Laplace transforms (Qi and Forrest, 2011, Bergemann et al., 2012) to account for the multiple layers comprising the packaged device, and their thermal losses (including convection, conduction and radiation). Even at very high currents, the maximum temperature change is only $\Delta T \sim 5°C$, which is in striking contrast to the temperature at the surface of an inorganic LED (~125°C). Interestingly, the temperature of the OLED is primarily determined by the package configuration. A small air gap between the OLED and the lid in Fig. 6.91 results in reduced heat conduction away from the source, accounting for a significant fraction of the ΔT observed.

The low temperature of the OLED is due to a combination of its high luminous efficiency and large area. As a distributed lighting source, its thermal and optical power densities are several orders of magnitude lower than in inorganic devices. For example, a 1 mm^2 LED area can provide >1 klm light output, whereas an OLED requires ~1 m^2 to deliver the same luminance.

The large area format of OLED lighting has several advantages as well as some notable disadvantages. We have seen in Chapter 5 that OLEDs can be deposited on lightweight flexible plastic or metal foils, allowing them to be adapted to many different building configurations. This feature alone provides architectural advantages when designing lighting fixtures that are installed in modern structures. Furthermore, unlike other lighting technologies, *the OLED itself is the luminaire*: it does not need to be mounted into reflecting cones or other fixtures that reduce glare and distribute the light. To provide similar areal lighting, LED point sources must be mounted on a light distribution system. For example, diffusers can add weight, complexity, and cost to an otherwise simple LED source. On the other hand, large OLED fixtures can provide uniform lighting where shadows cast are too soft, leading to reduced depth perception of the lighted space. And the large substrate areas lead to high cost and potentially low manufacturing yield. Hence, the current price of OLED lighting is ~ US \$100/klm, which is at least 20 times higher than LED lighting and 200 times that of incandescence. These costs are expected to be reduced as the demand for OLED lighting increases, but at present they are too high for widespread deployment.

The lifetime of WOLEDs is somewhat less than that of LEDs (40 kh vs. > 50 kh, respectively), which is limited by the lifetime of the blue emitting component (Section 6.7). Nevertheless, WOLED lifetime is sufficient for many indoor illumination applications. And finally, the materials and processes used in fabricating OLED fixtures have relatively low environmental impact. Unlike Hg-containing fluorescent lighting, there are no significant amounts of toxic materials used in OLED fixtures, permitting disposal without special handling.

Hence, OLEDs present an attractive source of efficient solid state-lighting (Kassakian, 2013). As we will see in the following sections, numerous device structures have been developed to improve OLEDs for illumination applications. When considering possible device architectures, the following criteria should be met:

i. *High efficiency*: Ultimately, the value of a light source is determined by its luminous power efficiency, or efficacy (Section 6.2.3). For this reason, PHOLEDs are suitable for all lighting applications with their potential for $\eta_{int} = 100\%$. Doped charge conductive layers are also useful for reducing voltage, and hence power loss. We will show that, in principle, hybrid blue fluorescent/green + red phosphorescent WOLEDs can also obtain unity internal efficiency.
ii. *High brightness*: The surface brightness of an OLED lighting fixture should be ~3000–5000 cd/m^2. Higher brightness leads to glare, and lower brightness requires excessively large areas to produce the desired level of illumination. Both brightness and

efficiency also demand that effective light outcoupling methods be used to minimize optical losses. This topic is the subject of Section 6.6.

iii. *High CRI at the desired CCT*: A *CRI* > 80, and preferably >90, is needed for aesthetically pleasing lighting appliances. CCTs of ~3000–4000 K are also desirable. Due to the relatively narrow emission spectra of OLEDs, this generally requires that the EML comprise three or more lumophores that span the visible spectrum to result in acceptable color renditions while also achieving high efficiency.

iv. *Color and intensity stability*: The lighting source should be free of color drift over its operational lifetime. Current standards also demand that the source should not lose more than 30% of its initial intensity (known as T70, i.e. the time to reach 70% of its initial luminance) in times less than 10 kh.

v. *Low cost*: Minimizing OLED complexity and fabrication processes should lead to an increased manufacturing yield and reduced materials cost.

There are a great many approaches to achieving these objectives. Figure 6.92 shows six different example WOLED designs. There are many variations for each of these species, some of which are detailed in the ensuing discussions. The wide choice of structure speaks to the inherent flexibility available in the design of organic electronic devices. Organic thin film devices allow for layering schemes that combine numerous functional materials into a single structure without the need for lattice matching to a particular substrate. While considerable early work focused on fluorescence (Kido et al., 1995), the need for high efficiency compels us to focus our discussion on structures that are capable of achieving unity internal efficiency. All such devices comprise at least one phosphor or TADF material. Similarly, we will not discuss color down conversion via CCM used in displays (Lee et al., 2011a). Our treatment of the topic of WOLED lighting sources is meant to provide insights into how devices are designed and optimized for illumination applications. The reader interested in more in-depth discussions of this topic is referred to *Further reading* and the extensive scientific literature.

6.5.1 Multicolor blended EMLs

Figure 6.92a shows a WOLED with a single EML comprising three or four lumophores separated into different sectors in a common EML. These can be phosphors emitting in the red, green and blue to provide

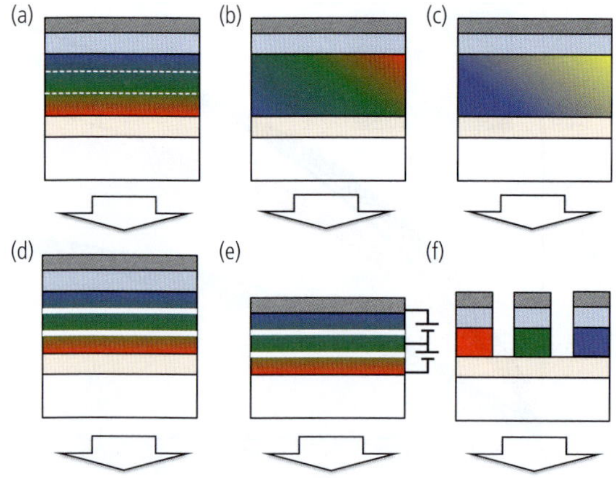

Figure 6.92 Several different WOLED architectures. The EMLs are in color, the transport layers (e.g. the HTL, ETL, etc.) are shown in orange and light blue, and the cathode is gray. Most layers and details are omitted for clarity. (a) RGB lumophores spatially separated in a single EML (emission zones delineated by dashed lines), (b) blended EML consisting of 3 or more lumophores, (c) two color blended EML consisting of two lumophores or a phosphor that emits from the blue monomer + red/yellow excimer state, (d) stacked WOLED consisting of any combination of R, G, B, or white emitting zones separated by transparent charge generation layers (white stripes), (e) vertically stacked WOLED with each luminous region separately addressable to tune color, and (f) horizontally striped WOLED with each sub-pixel separately addressed to tune color.

a high CRI along with a high efficacy. Even higher CRIs are achieved by including a yellow emitting phosphor into the EML. Alternatively, a fluorescent blue emitter can be combined with red and green phosphors to achieve a *hybrid fluorescent/phosphorescent WOLED*.

The earliest example of an all-phosphor WOLED was based on this structure comprised the cyan-emitting FIrpic (λ_{peak} = 472 nm), yellow-emitting $Bt_2Ir(acac)$ (λ_{peak} = 563 nm), and red-emitting $Btp_2Ir(acac)$ (λ_{peak} = 620 nm), doped into a primarily electron conducting CBP host (D'Andrade et al., 2002b). Blending different phosphors into a common transport layer presents several challenges. Since triplets are readily transferred from the blue to the yellow to the red phosphors, their relative positions, doping concentrations, and the thicknesses of zones within the EML determine the spectral balance. Since the exciton formation in this structure is primarily at the HTL/EML interface, and FIrpic is the least efficient of the three phosphors, the highest efficiency is obtained using an EML comprising CBP doped with a 20 nm thick 6 wt% FIrpic region at the HTL/EML interface, followed by a 2 nm thick 8 wt% $Btp_2Ir(acac)$ region, and topped with a 2 nm thick 8 wt% $Bt_2Ir(acac)$ region. The exciton blocking ETL comprises BCP, followed by the metal cathode. The triplet exciton transfer length is 8.3 nm in CBP. This distance allows triplet excitons to drift across the entire 24 nm thick EML while being sequentially transferred to each of the dopants in the desired proportions to result in efficient light emission of the appropriate white color balance. This phosphorescent WOLED had a maximum $\eta_{ext} = 5.2\%$, a power efficiency of 6.4 lm/W and a $CRI = 83$. The CIE coordinates at (0.37, 0.40) place it near to the Planckian locus (Fig. 6.13) with CCT $\simeq$ 4000 K. A problem encountered by combining three lumophores in a single layer is that the color balance shifts with intensity, that is, as the current increases, the peak emission can move from one part of the EML toward another, thus resulting in a variation in the balance between blue *vs.* yellow *vs.* red emission.

To more completely separate the dopants and to control color drifts with current, a different host for each emission color can be used. An example is shown in Fig. 6.93, where 20 wt% of the blue dopant, FIr6, is doped into UGH2, followed by 2 wt% $Ir(ppy)_3$ (green) in mCP and 4 wt% PQIr (red) in TCTA. The

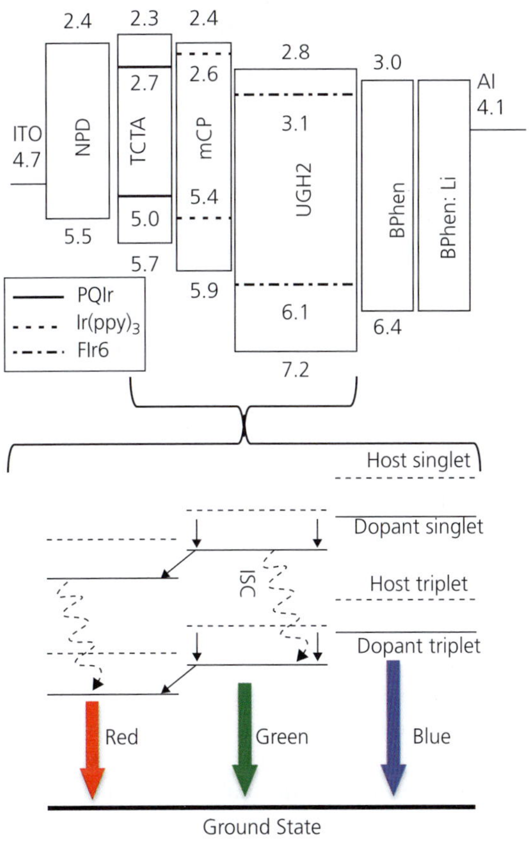

Figure 6.93 Energy level diagram of a triple dopant/triple host phosphorescent WOLED. The top panel shows the energy level scheme of the diode, with the transitions in the EML in the bottom panel. HOMO and LUMO energies (in eV) for all materials are shown in the top diagram. From Sun and Forrest (2008b).

exciton formation zone is distributed between the blue and green emitting regions of the EML. The small energy steps between each of the layers impede electron and hole transfer, retaining the desired amount of charge within each layer. Combining this balance with the amount and position of phosphor dopants results in the desired spectrum. In this architecture, the red phosphor, PQIr is excited via Dexter transfer from excitons originating on $Ir(ppy)_3$, whereas those generated on FIr6 are trapped.

This device exhibits near unity internal quantum efficiency, largely attributed to the generation of all excitons on the dopants themselves, and the high radiative efficiency and energy transfer to the several phosphors. This device has an external efficiency of 23%, along with a power efficiency of 37 lm/W at 500 cd/m^2, falling only slightly from its low brightness peak efficiency of $\eta_{ext} = 28\%$. At 10 mA/cm^2 (~20 lm/W), the drive voltage is 5.0 V, indicating that the small energy barriers between dopants do not appreciably decrease η_P. This device has $CRI = 81$ at coordinates (0.37,0.41), corresponding to a $CCT \simeq 4000$ K at 1 mA/cm^2. Indeed, the CRI and CIE coordinates shift only by ~2–3% even as *j* increases by a factor of 100. Hence, this scheme is effective in maintaining color stability at all brightness levels (Sun and Forrest, 2007, 2008b).

High efficiencies have also been achieved with a simplification to the structure in Fig. 6.93 by doping two phosphors (20 wt% FIrpic and 8 wt% $Ir(ppy)_3$) into a single, 12 nm thick TPBi host, and the third phosphor (10 wt% $Ir(MDQ)_2(acac)$) doped into an even wider gap 8 nm thick TCTA host. Unlike the device in Fig. 6.93 whose spectrum is controlled by energy barriers between layers, the two-host structure is controlled by diffusion across undoped regions of the hosts to limit the rate of energy transfer. Thus, excitons are generated in the interface region (dashed box, Fig. 6.94) primarily on FIrpic and TCTA. The resonant HOMO levels of both materials encourage hole transfer across the HJ, whereas electrons are blocked from entering the red-doped zone. The red dopant is spaced away from the blue doped section at the junction to prevent energy transfer that would substantially decrease the blue spectral content with increasing current. Hence, the primary mechanism for populating the red emission zone is via diffusion from FIrpic at rate $k_{b\text{-}r}$. Further, a portion of excitons diffuse from this zone to the green-doped region, which again is offset by a 2 nm thick TPBi diffusion zone. This allows host-dopant Dexter transfer at rate $k_{b\text{-}g}$.

Since the emission is controlled by diffusion rather than energy barriers, there is little or no spectral shift from $L = 100$ to 5000 cd/m^2. These devices exhibited

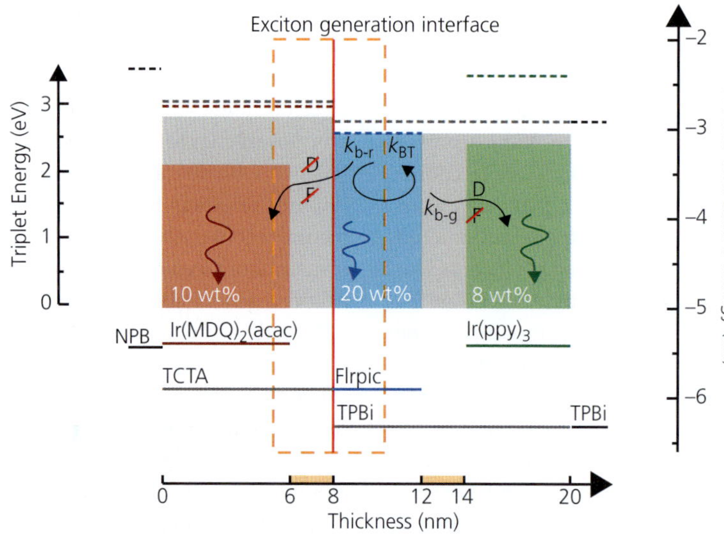

Figure 6.94 Energy level scheme of a high efficiency triple doped EML comprising two hosts. The TCTA and TPBi hosts and phosphors used are indicated, along with their relative positions within the 20 nm thick EML. Active energy transfer processes are denoted **D** for Dexter and **F** for FRET. These processes are crossed out if inactive. Also, the several transfer rates, *k*, are shown (Reineke et al., 2009).

an efficiency (using a hemispherical lens to extract the glass substrate modes) of $\eta_{ext} = 24\%$ and $\eta_P = 55$ lm/W at 1000 cd/m^2. At this luminance, $CRI = 80$ and the chromaticity coordinates are (0.45, 0.47) which gives it a yellow-green cast. The spectral shift away from the Planckian locus at $CCT \simeq 3000$ K is one reason for its high power efficiency. The efficiency was increased further via the use of conductivity dopants in both the NPB HTL and the BPhen ETL. Even higher efficiencies are achieved with enhanced light outcoupling schemes, to be discussed in Section 6.6 (Reineke et al., 2009).

The devices thus far have been characterized by thick EMLs (20–25 nm) and color rendering indices that appear to have a maximum value of 80, with most structures achieving a maximum external efficiency (without using additional outcoupling optics) of ~20–25%. To achieve the combination of both very high CRI and quantum efficiency comparable to the best available lighting sources, four dopants may be required. That is, including a yellow phosphor can pull the coordinates toward the Planckian locus when added to the red, green, and blue emitting sections. The device in Fig. 6.95a shows an EML comprising a blue emitting FIrpic layer on the EML side of the ETL/EML interface where exciton formation occurs (Chang et al., 2013b). Excitons generated on the TPBi ETL transfer to the common CBP host and are subsequently transferred to FIrpic. Both hosts are chosen for their high triplet energies, leading to efficient energy transfer to FIrpic in the exciton formation region. Those excitons not harvested in FIrpic continue to diffuse through the 33.5 nm thick EML, entering layers with dopants with a lower triplet energy than in the previous layer. As in the devices already discussed, the thickness of each layer and the phosphor doping concentration determines the emission intensity as the excitons move along the energy cascade from high to low energy via: CBP→FIrpic→Ir(ppy)$_3$→Bt$_2$Ir(acac)→Ir(MDQ)$_2$(acac), shown in the energy level diagram in Fig. 6.95b No intermediate exciton transport layers such as those in Fig. 6.94 are used. This allows for lossless energy harvesting by the lower energy emitting species.

To facilitate energy transfer to the lowest energy phosphors of Bt$_2$Ir(acac) and Ir(MDQ)$_2$(acac), the yellow and red emitting zones are co-doped with Ir(ppy)$_3$. This allows for very efficient transfer from CBP to Ir(ppy)$_3$ due to the energetic resonance between these two species, and then sequentially from Ir(ppy)$_3$ to the redder dopants. Complete transfer is inferred from the lack of Ir(ppy)$_3$ emission in PHOLEDs where only Ir(ppy)$_3$ and the long wavelength emitters are co-doped into the same EML.

The spectral intensity of the device with only the blue dopant, the B+G dopants, the B+G+Y dopants and then finally all four dopants in Fig. 6.99c show that by increasing the number of dopants, the spectrum naturally broadens, and importantly, η_{ext} at $L = 1000$ cd/m^2 increases from 8.6% to 19.2%. Hence, all four dopants in a single CBP host are needed to harvest ~100% of the generated excited states. The very high internal efficiency of the device, along with the use of four phosphor dopants leads to $\eta_P = 61.7$ lm/W, and $\eta_{ext} = 41.5\%$ using a hemispherical lens to extract glass substrate modes. The $CRI = 83$ with chromaticity coordinates of (0.44, 0.45) at $L = 1000$ cd/m^2, correspond to a warm white $CCT = 3332$ K. High performance is obtained at the expense of device complexity, which may lead to high cost.

The three dopants can also be blended within the same zone of a single host EML (Fig. 6.92b). As in other designs, the various doping concentrations must be carefully balanced to obtain the desired emission spectrum and efficiency. The energy level diagram in Fig. 6.96 shows an example of this structure comprising the blue, green and red phosphors, FIr6,

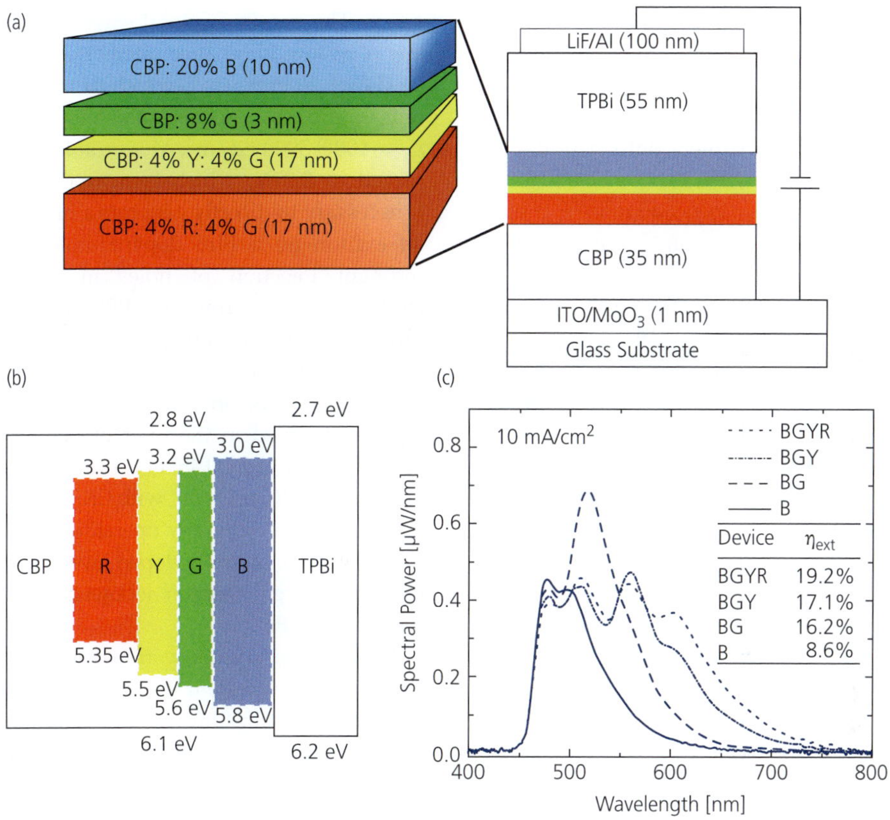

Figure 6.95 Four-dopant, single-host high CRI WOLED. (a) Schematic of the structure with dopant concentrations in CBP shown. (b) Energy level diagram of the WOLED showing the $B \rightarrow G \rightarrow Y \rightarrow R$ energy cascade. (c) Spectral output powers of OLEDs with different numbers of dopants in the EML. Inset: External quantum efficiency at L=1000 cd/m^2 of the devices in (a) and (b). (Chang et al., 2013b).

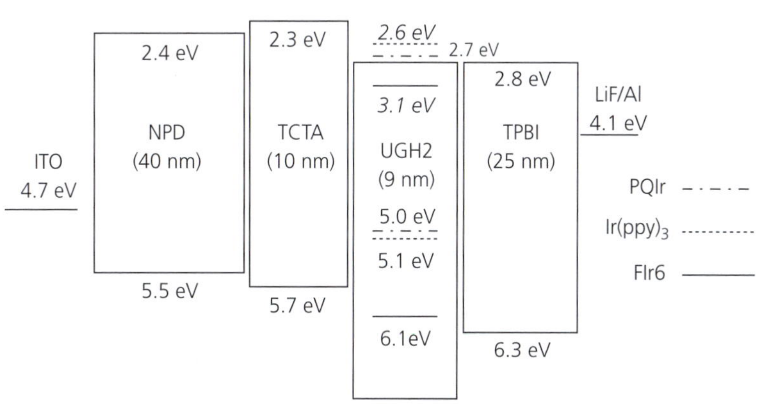

Figure 6.96 Energy level diagram of a three phosphor WOLED blended into a single thin (9 nm) ultrawide energy gap host (UGH2) (D'Andrade et al., 2004).

$Ir(ppy)_3$ and PQIr *simultaneously* co-doped at 20 wt%, 0.5 wt%, and 2 wt%, respectively, into the same UHG2 host. Given its wide energy gap, UGH2 does not carry charge; FIr6 thus serves as both an electron and hole conductor. Its energy levels are aligned to the LUMO of the TPBi ETL/EBL and the HOMO of the TCTA HTL, allowing for direct electron and hole injection onto FIr6. This avoids the need for energy transfer from host to guest, and the accompanying exchange energy losses incurred forming singlets on the host that subsequently transfer to the triplet of the blue dopant.

Direct exciton formation on FIr6 is followed by transfer of some of the triplets onto $Ir(ppy)_3$ and PQIr in proportions determined by their relative concentrations. Given the high sensitivity of the eye to green, the $Ir(ppy)_3$ concentration is kept relatively low to limit its emission intensity. The performance of this device is summarized in Fig. 6.97a. The power efficiencies in both the forward viewing direction, and the total efficiency are indicated. Also shown in the forward direction are the currents at which the WOLED reaches 100, 1000, and 10^4 cd/m^2 luminance. The maximum total external efficiency is 42 lm/W at low luminance intensity, decreasing to 10 lm/W at 20 mA/cm^2, which is approximately 10 times the current needed to achieve a luminance of L_O = 1000–3000 cd/m^2. Measurable emission is observed at 3.3 V, which is near the estimated 3.0 V necessary to directly generate a blue emitting triplet exciton on a FIr6 dopant molecule, assuming resonant injection of holes in TCTA into the HOMO, and electrons in TPBI into the LUMO of FIr6. The voltage required to excite green and red emission from $Ir(ppy)_3$ and PQIr, respectively, is <3.0 V. Thus, the emission color shifts with increasing voltage, as observed in the inset of Fig. 6.97a. The CIE coordinates of (0.43, 0.45) at 0.1 mA/cm^2 blue-shift to (0.38, 0.45) at 10 mA/cm^2, with $CRI = 80$ ($CCT \simeq 4100$ K). The exceptionally low operating voltage is due to the use of thin, multiply doped layers along with near resonant charge injection followed by direct triplet formation on FIr6.

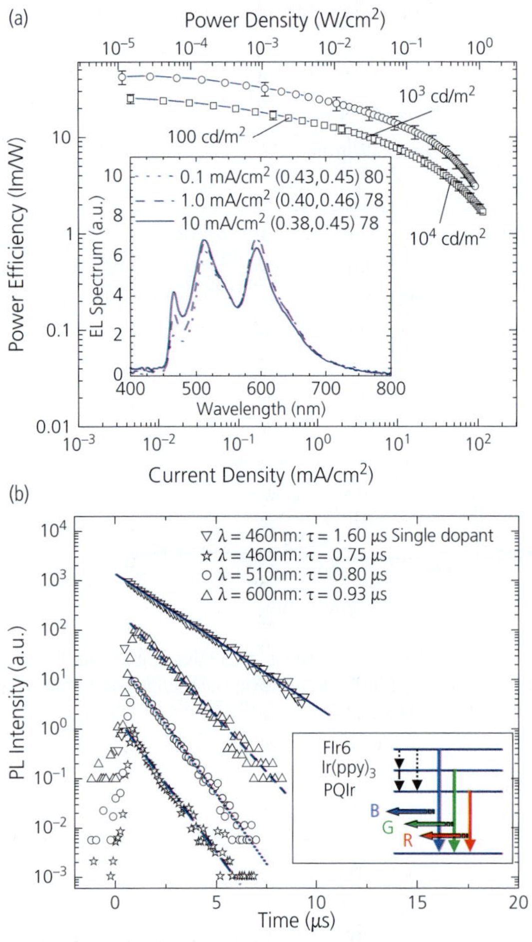

Figure 6.97 (a) Power efficiency vs. current density of the triple-doped WOLED in Fig. 6.96 in the forward viewing direction (squares), and the total emitted power (circles). Inset: EL spectra at three different current densities, along with the CIE coordinates and CRI values. (b) PL transient response of a FIr6 doped film (inverse triangles), and a triple-doped film taken at the wavelengths indicated in the legend. The characteristic phosphorescent decay times (τ) obtained from the fits indicated by lines for each spectral component. Inset: Relaxation process of the triple-doped film (D'Andrade et al., 2004).

To understand energy transfer between the dopants, the PL transient decay was studied in optically pumped thin-films of UGH2 doped at 20 wt% with FIr6, and of the triple-doped EML. Given its high doping concentration, only FIr6 is excited by the $\lambda = 337$ nm pump. The energy transfer channels originating from the excitation of FIr6 to the various dopants are depicted in Fig. 6.97b, inset. Energy transfer from FIr6 to both $Ir(ppy)_3$ and PQIr is shown by dotted lines. Although energy transfer from $Ir(ppy)_3$ to PQIr is possible, the low concentration of both species makes this unlikely. Solid lines show radiative relaxation to the ground state. The lifetime of FIr6 decreases from τ = 1.60 μs in a UGH2:FIr6 film, to 0.75 μs in the triple doped system.

The lifetimes of $Ir(ppy)_3$ (Baldo and Forrest, 2000) and PQIr (Lamansky et al., 2001) are τ = 0.8 μs and 2.0 μs (D'Andrade et al., 2004). It was found that τ = 0.80 μs, and 0.93 μs for the peak emission features at λ = 510 nm and 600 nm. Hence, the lifetime decrease in the triple doped film suggests that FIr6 emission is quenched by energy transfer to the green and red dopants. Thus, the luminescence process starts with exciton formation on FIr6, followed by direct transfer to the lower energy phosphors to obtain balanced white emission. The lifetime of the coupled system is determined by the dopant with the shortest natural phosphorescent decay rate (i.e. $Ir(ppy)_3$).

Given the large area required to provide the total luminance necessary for interior lighting, solution processing may be a viable path to low cost. As we have shown, high efficiency along with acceptable color content to achieve a pleasant color balance also requires multiple dopants arranged to avoid spectral shifts with brightness. Previously, we showed that solution processes generally impose a trade-off between complexity and performance in OLEDs for displays. The same appears to be the case for WOLED lighting.

A high performance solution processed WOLED consists of the B, G, R phosphors 15% FIrpic, 1% $Ir(ppy)_3$, and 0.6% $Ir(MDQ)_2(acac)$ used with a mixed host comprising the hole conducting moieties, TCTA and TAPC. The 40 nm thick EML was sandwiched between PEDOT:PSS HIL and a vacuum deposited TmPyPB ETL, capped with a LiF/Al cathode. Here, TmPyBP, TCTA and TAPC are all chosen for their high triplet energies of 2.78, 2.78, and 3.5 eV, respectively, compared to FIrpic (2.65 eV). The TCTA was blended at 25% concentration into TAPC due to its resonant LUMO energy (2.3 eV measured from vacuum) alignment with FIrpic, whereas the HOMO of TAPC (5.3 eV) is aligned with PEDOT:PSS (5.1 eV). Hence, this blend comprises three emitters with efficient energy transfer from the host while also supporting both electron and hole transport. As a

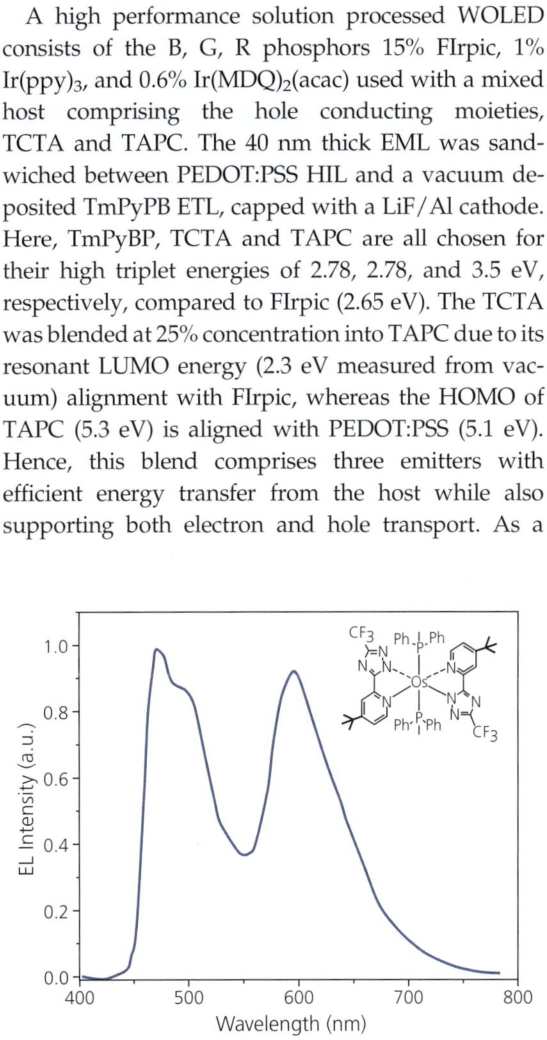

Figure 6.98 Emission spectra for a solution-processed WOLED. The CIE color coordinates of (0.38,0.38) are relatively independent of voltage, and hence current. Inset: Chemical structural formula of the orange emitting phosphor, Os-O. After Huang et al. (2009).

result, the solution processed EML WOLED has $\eta_P = 23.3$ lm/W, $V = 4.6$V, $CRI = 73$, and $CIE = (0.35, 0.41)$ corresponding to $CCT \simeq 4700$ K at $L = 1000$ cd/m^2 (Fu et al., 2012). This performance is competitive with analogous vacuum-deposited blended EML WOLEDs.

An example polymer-based blended two-phosphor EML WOLED is shown in Fig. 6.98 (Huang et al., 2009). This device consists of a 30 nm thick PEDOT:PSS hole injection and transport layer. The electron and hole transporting EML comprise a 70 nm thick blend of the small molecule phosphors, 7 wt% blue emitting FIrpic and 0.25 wt% orange emitting Os-complex, Os-O (see inset, Fig. 6.98), co-doped into PVK mixed with 30 wt% OXD-7. Only a very small amount of Os-O is required since its lower triplet energy makes it an efficient exciton trap. Together, the two dopants emit broadly across the visible to provide white light. The optimized device exhibited $\eta_P = 18.5$ lm/W and $\eta_{ext} = 12.6\%$.

As noted previously, a white light source can be obtained using only two color emitters. This is illustrated for two pairs of illuminants (cyan-red or blue-yellow) in the chromaticity space in Fig. 6.99. As long as the line connecting the CIE coordinates of the illuminant pairs passes through the Planckian locus, white emission will be perceived. However, as apparent from Fig. 6.98, only two emitters result in weak green emission due to the relative narrowness of exciton spectra. Hence, while the source will *appear* white, its CRI will be lower than required for a high quality illumination source. The device in Fig. 6.98 has CIE coordinates of (0.38, 0.38), but the $CRI < 70$ results from spectral "holes" in the green (Su et al., 2008, Wang et al., 2009a, Wu et al., 2009). While substituting two phosphors for three lends simplicity to the WOLED design and also to high efficacy, the low color purity would render objects using this source with an odd appearance.

Thus far, we have shown two types of blended EMLs; those with dopants segregated into zones as in Fig. 6.92a, and those that are fully blended (Fig. 6.92b). The zone/fully blended architectures can also be combined. This strategy has been applied to a device in which a separate, 4 nm thick exciton generating zone comprising only 8 wt% FIrpic is placed adjacent to an 11 nm thick zone containing a blend of the green emitting 8 wt% $Ir(ppy)_3$ and the red phosphor, 1 wt%, $Ir(PPQ)_2(acac)$ doped into a common, high triplet energy (2.9 eV) mCP host. As in other such designs, the highest triplet energy phosphor, FIrpic, is located at the exciton formation

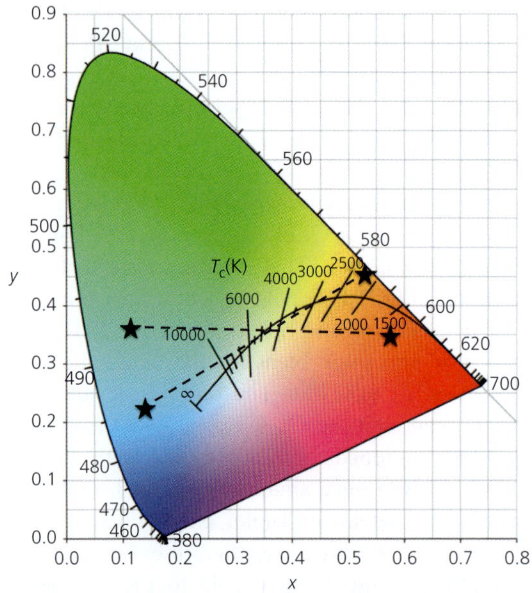

Figure 6.99 1931 chromaticity chart showing two pairs of illuminants indicated by stars. The straight lines connecting these pairs intersect the Planckian locus at a $CCT = 4600$ K.

zone near the TAZ-based ETL/EML interface. The co-doping of the RG zone results in color that is determined by the energy transfer efficiency from the relatively high concentration of G to R dopants, almost eliminating the color dependence on current. For example, it is found that the color coordinate shift over a luminance variation from 500 to 10,000 cd/m^2 of the RG-B device is $\Delta CIE = (0.00, 0.00)$. This is compared to (+0.01, −0.01) for a PHOLED comprising the same dopants in an EML of equal thickness (15 nm) but separated into three discrete R-G-B zones. The peak performance of the two-zone, three dopant WOLED is $\eta_{ext} = 19.1\%$, $\eta_P = 37.3$ lm/W, $CIE = (0.39, 0.42)$ corresponding to $CCT \simeq 4000$ K, which is competitive with the three zone device using the same materials combinations (Wang et al., 2009b).

6.5.2 Hybrid fluorescent/phosphorescent WOLEDs

Among the many properties required for OLED lighting, achieving high power efficiency and long operational lifetime are the most important. Unfortunately, obtaining blue PHOLEDs with long lifetime has been problematic (see Section 6.7), leading to alternative architectures that replace the blue phosphorescent unit with a more stable, but efficient fluorescent substitute. We have shown that OLED displays employ fluorescent blue OLEDs to minimize the loss of reliability encountered in using blue PHOLEDs, albeit at a significant loss in efficiency for this color pixel. However, this compromise is unacceptable in WOLEDs that must rely on 100% internal efficiency to be competitive with existing lighting sources.

The spectrum of white emission is approximately 25% blue, with the remainder in the green to the red spectral regions. This fortuitous division in spectral content suggests a means for achieving satisfactory white emission with $\eta_{int} = 100\%$ by using a blue fluorophore to harvest the 25% electrically generated singlets, and green and red phosphors to harvest the remaining 75% triplets. Harvesting singlets directly on a fluorescent molecule has several advantages: it eliminates the need for a host with a triplet energy that is greater than that of the blue phosphor. Further, it eliminates exchange energy losses in converting singlets to high energy (blue emitting) triplets, thus reducing the PHOLED drive voltage. This results in an increase in power efficiency needed for lighting. And finally, blue singlet emitting materials have considerably longer operational lifetimes than blue phosphors.

The first demonstration of a hybrid fluorescent/phosphorescent WOLED employed a blue fluorophore placed directly into the exciton generation regions of the EML. The red and green phosphors were located within a triplet diffusion length of their origin, but sufficiently far from the blue fluorophore to prevent direct energy transfer. The independent harvesting of singlets and triplets is illustrated by the energy level diagram in Fig. 6.100a (Sun et al., 2006). The process starts with electrical generation of excitons in the host in a ratio of $\chi_S = 0.25$ and $\chi_T = 0.75$. The singlets transfer via FRET to the blue fluorophore, BCzVBi, which is doped at 5 wt% in CBP to minimize direct Dexter transfer of triplets from host to dopant. A thin (6 nm) CBP spacer that is thinner than the triplet diffusion length separates the blue fluorescent/exciton generation region from green (5 wt% $Ir(ppy)_3$) and red (4 wt% PQIr) phosphorescent zones that harvest the triplets via Dexter transfer from the CBP host (see inset, Fig. 6.100b for the device structure). In this demonstration, the BCzVBi has a lower triplet energy than either of the phosphors or CBP, and hence transfer to the BCzVBi triplet is a significant loss channel.

By omitting the phosphors and measuring the blue emission for devices with only the blue fluorophore doped near the ETL or HTL interface, it was determined that all excitons were formed within ~5 nm of both interfaces. The device spectrum in Fig. 6.100 at $j = 100$ mA/cm^2 is fit using

$$\eta_{ext} = (1 - \chi_{trap})\eta_B + [(1 - \chi_{trap})\chi_T + \chi_{trap}]\eta_{GR}, \quad (6.82)$$

where χ_{trap} is the ratio of excitons formed by charge trapping directly on the phosphors to those diffusing from the generating interfaces, η_B is the efficiency of blue emission (which includes the probability of singlet formation), and η_{GR} is the efficiency of the red and green phosphorescence.

Analysis of the spectra in Fig. 6.100b (with layer thicknesses $b_1 = 15$ nm, $b_2 = 10$ nm, $r = 8$ nm, and $g = 12$ nm, along with a 20 nm thick BPhen ETL) indicate that $\chi_{trap} = 20\%$, with the remaining triplets diffusing through the fluorescent and host layers to reach the phosphors. Note that there is only a small color shift when the current varies by over a factor of 100, indicating a reasonably stable balance of exciton distribution in this diffusion-controlled device. The direct exciton formation via trapping results in a decrease in blue emission, which ultimately limits the device efficiency (Sun et al., 2006). Indeed, the low efficiency of the blue fluorophore coupled to exciton losses ultimately leads to a WOLED efficiency of $\eta_{ext} = 18.7\%$ and $\eta_P = 23.8$ lm/W at 500 cd/m^2, along with $CRI = 85$ at $CIE = (0.40, 0.41)$, corresponding to $CCT = 3750$ K.

A device similar to that in Fig. 6.100 eliminates the zone containing the red phosphor, which is replaced by the red fluorophore, DCJTB. This is lightly doped (~0.08 wt%) into the 15 nm thick, green emitting zone with 8 wt% Ir(ppy)$_3$ in CBP (Kanno et al., 2006c). Thus, red emission from the fluorophore is due to phosphor-sensitized fluorescence in the green emitting zone (cf. Fig. 6.30). Triplets generated at the HTL/EML and ETL/EML interfaces drift to the green-red phosphorescent/fluorescent zone across two, 2 nm thick undoped CBP host spacer layers. Triplets are partially harvested by the phosphor, where a fraction is Förster-transferred to the singlet state of red-emitting DCJTB. The blue spectrum is emitted from singlets transferred directly to BCzVBi from the CBP at the edges of the EML. Since all the singlets are harvested by the blue dopant, and triplets by the green phosphor, this two-dopant structure still has the potential of $\eta_{int} = 100\%$ even though two fluorescent dopants are used.

Evidence for sensitized fluorescence is shown by the transient EL response in the green and red spectral regions in Fig. 6.101, inset. The rapid (~1 ns) fluorescence of DCJTB takes several hundred nanoseconds since its excitation is via transfer from the Ir (ppy)$_3$ triplet manifold. Even though the fluorescent/phosphorescent/fluorescent architecture has the potential for unity quantum efficiency, there are several losses using the particular dopants chosen in this particular device. While the spectrum is independent of current in Fig. 6.101 due to diffusion-controlled transfer, the quantum and power efficiencies are approximately 20% lower than the fluorescent/phosphorescent device in Fig. 6.100 due to losses in the sensitized fluorescent transfer. The CRI of the device is 79 with CIE coordinates of approximately (0.38,0.42).

There are several variations to the hybrid fluorescent/phosphorescent structure that can improve its performance (Rosenow et al., 2010). Nearly unity internal efficiency has been achieved with the structure whose energy level diagram is shown in Fig. 6.102a (Sun et al., 2014a). A common CBP host is used with excitons formed near its interface with the TmPyPb ETL and at the TCTA HTL. The blue fluorescent

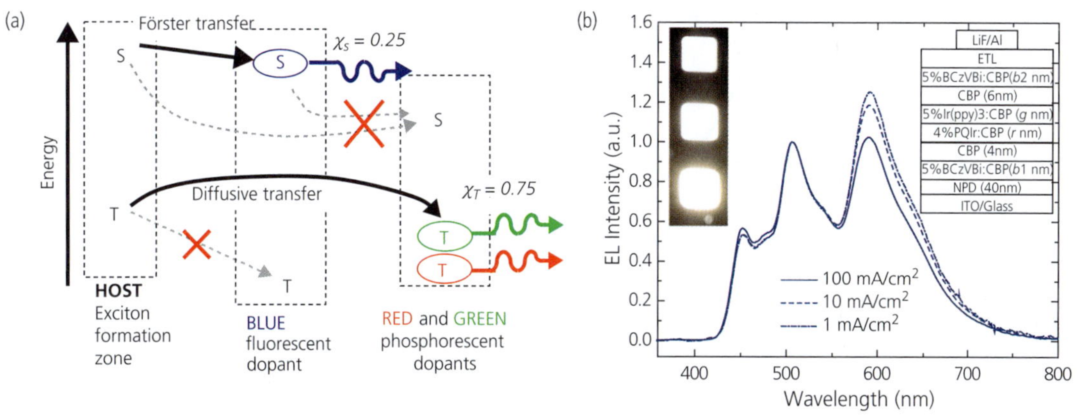

Figure 6.100 Hybrid fluorescent/phosphorescent WOLED. (a) Energy level scheme showing Dexter (dashed lines) and Förster (solid lines) transfer pathways. χ_S and χ_T are the singlet and triplet formation ratios, respectively. (b) Spectra of the hybrid WOLED whose structure is shown in the right inset, at several current densities. Left inset: Images of the WOLED emission at these same current densities (Sun et al., 2006).

4P-NPD triplet energy (2.3 eV) is lower than that of the green phosphor, $Ir(ppy)_2(acac)$ at 2.4 eV. Hence the green phosphor is positioned near to the TCTA with a 2 nm thick spacer between this material and the red emitting $Ir(MDQ)_2(acac)$, whose triplet energy (2.0 eV) is less than that of 4P-NPD. Since the singlet exciton diffusion length is considerably shorter than the 8 nm thick 4P-NPD, there is efficient triplet transfer between these latter two materials, while the singlets are immediately harvested by the fluorophore. The spacer discourages transfer of $Ir(ppy)_2(acac)$ triplets to the red phosphor, thus maintaining the desired color balance.

The spectrum in Fig. 6.102b shows significant color shifts with brightness attributed to a shift of the exciton generation zone from the ETL toward the HTL interface with increasing drive current (note that the green intensity stays constant in these normalized spectra). A forward viewing external quantum efficiency of 21.2% indicates nearly unity internal efficiency. Furthermore, the maximum $\eta_P = 40.7$ lm/W decreases only weakly with intensity. The high power efficiency is achieved by n-type doping the ETL with 4% Cs_2CO_3 to minimize the drive voltage. The $CRI = 85$ and $CIE = (0.42, 0.44)$ corresponding to $CCT = 3544$ K (Sun et al., 2014a).

Following similar design concepts to that in Fig. 6.102, but including a yellow emitting phosphor in a four-color fluorescent/phosphorescent WOLED, a very high $CRI = 94$ was achieved along with $\eta_{ext} = 14.3\%$ and $\eta_P = 26.2$ lm/W (Sun et al., 2014b). Hence, hybrid devices have been shown to combine high quantum efficiency with good color quality, while potentially eliminating the short operational lifetime of all-phosphorescent WOLEDs.

The hybrid approach has also been demonstrated, albeit with less success, in polymers. Solution processing creates difficulties in the control of emission color via a zone-doped structure such as exploited in vacuum-deposited small molecule WOLEDs. A method that takes advantage of the synthetic diversity of polymers is to copolymerize red, green, and blue emitting moieties along the molecular backbone. In **Scheme 2** we show a polymer with segments that are incorporated into the feed stock at concentrations that lead to the desired color balance (Jiang et al., 2006). In this copolymer, green fluorescent benzothiadiazole units are incorporated into a blue emitting polyfluorene backbone. Then a red emitting, pendant Ir phosphor complex fills out the spectrum. Copolymer WOLEDs based on these three lumophores have not shown particularly high efficiencies, although this approach may

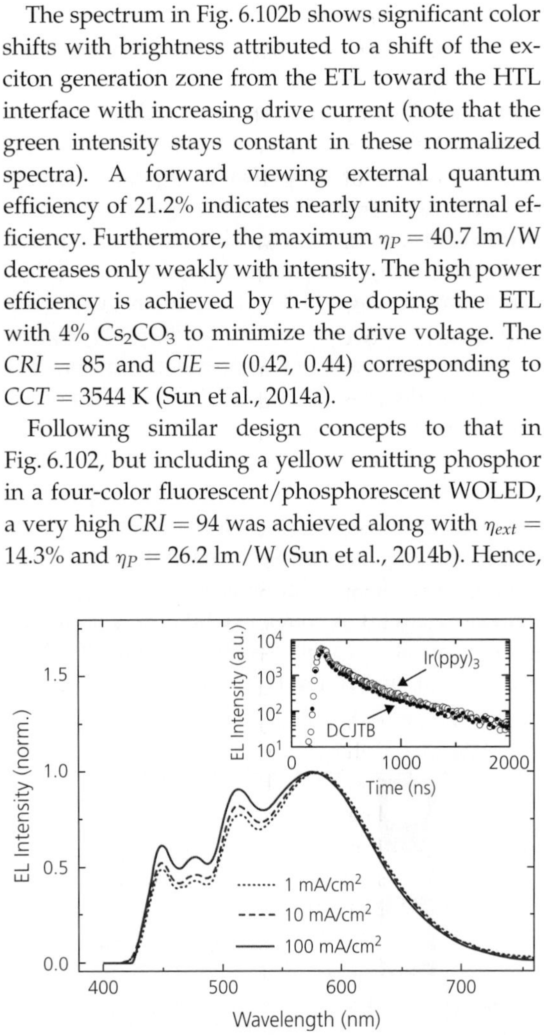

Figure 6.101 EL spectrum of a blue BCzVBi fluorescent/green $Ir(ppy)_3$ + red DCJTB phosphorescent/fluorescent WOLED at several current densities. Inset: EL transient for the green and red spectral components (Kanno et al., 2006c).

Reprinted from Kanno, H., Sun, Y. & Forrest, S. R. 2006c. White organic light-emitting device based on a compound fluorescent-phosphor-sensitized-fluorescent emission layer. Applied Physics Letters, 89, 143516, with the permission of AIP Publishing.

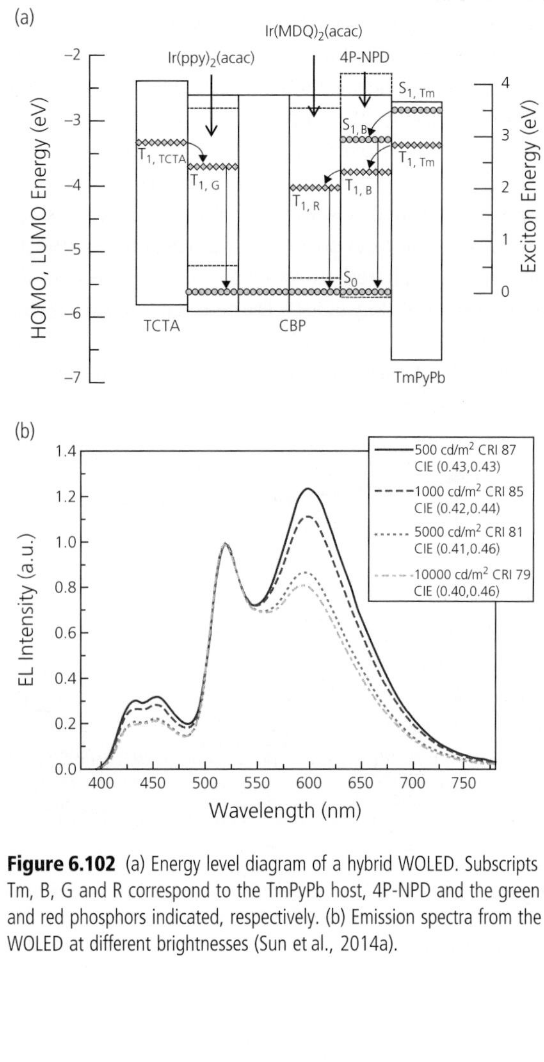

Figure 6.102 (a) Energy level diagram of a hybrid WOLED. Subscripts Tm, B, G and R correspond to the TmPyPb host, 4P-NPD and the green and red phosphors indicated, respectively. (b) Emission spectra from the WOLED at different brightnesses (Sun et al., 2014a).

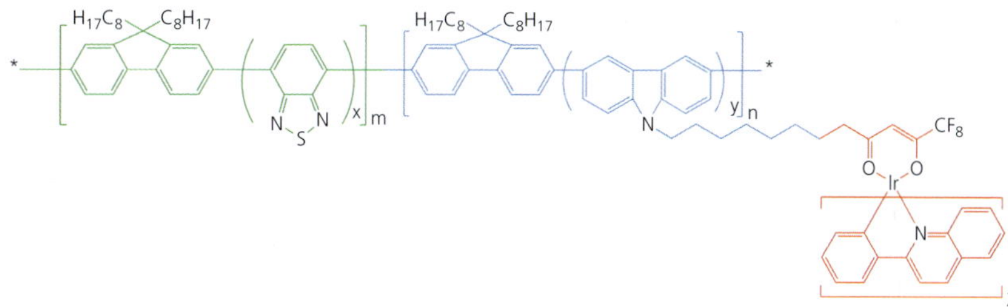

Scheme 2. Left to right: green, blue, and red emitting units.

provide a path to achieving high efficiency with improved emitting units and charge conduction characteristics in a simplified WOLED structure.

6.5.3 WOLEDs based on excimer emission

In the preceding sections, we have seen that producing high quality white light can entail complex engineering of the EML that typically contains three (or even more) luminescent zones, each with a phosphor or fluorophore emitting in a separate region of the visible spectrum. In this context, in Section 3.6.3 we showed that the emission spectra of excimers and exciplexes are broad and featureless. Hence, it is reasonable to expect that such excited state species might be a route to creating a broad, white spectrum. The blue emitting square planar Pt phosphorescent molecules can efficiently emit from exciplexes due to coupling within the molecular stacks, thus leading to broad white light emission (Adamovich et al., 2002). In particular, the monomeric (blue) emission of a single lumophore can be combined with broad green-red excimer emission from the same molecular species to result in a single-dopant WOLED. By carefully controlling the doping concentration of the phosphors in the host, a fraction of the molecules will form excimer emitting clusters, whereas the remainder will be isolated within the blend to emit from their native blue monomer state.

Monomer/excimer WOLEDs are attractive due to their inherent simplicity. While they have demonstrated high efficiencies, one drawback is that the emission spectrum is difficult to precisely control since it is a sensitive function of both doping concentration as well as the position of the exciton formation and light emitting zones within the device (D'Andrade and Forrest, 2004, Fleetham and Li, 2014). Hence, most such devices have $CRI \leq 70$, which is below that required for high quality lighting.

The earliest demonstration of a phosphorescent monomer/excimer WOLED employed a series of blue emitting square planar Pt molecules comprising fluorine electron withdrawing groups (Adamovich et al., 2002). At high concentrations, the relatively strong Pt··Pt bonds form oligomers in the ground state. Their close π–π stacking gives rise to broad and unstructured excimer emission.

An example of interactions between Pt molecules leading to monomer/excimer WOLED spectra is illustrated in Fig. 6.103. The binuclear Pt interactions result in triplet metal–metal ligand charge transfer states (^{3}MMLCT) that are destabilized compared to the ^{3}MLCT of the monomer. Thus, the dz^2 and Pt-π levels are split due to interactions between Pt atoms on neighboring molecules. This leads to additional, lower energy and broadened, non-bonding σ^* transitions. The combination of monomer ^{3}MLCT and excimer ^{3}MMLCT transitions results in the broad emission spectrum of Fig. 6.103b. In Sections 3.6.3 and 3.7.3 we provide a detailed description of the excimer and spin–orbit interactions in dihedral and octahedral transition metal complexes.

The EL spectra in Fig. 6.104 are from PHOLEDs whose EML comprises a tridentate square planar Pt (II) molecule, Pt-17, mixed at several different concentrations in the host, 26mCPy (Bakken et al., 2012). At only 2 wt% Pt-17, the spectrum is dominated by monomer emission between $\lambda = 440$ nm and 530 nm, with a weak tail extending to longer wavelengths due to excimer emission from a few molecular aggregates. As the Pt-17 concentration is increased to $> 18\%$, the emission is dominated by excimer transitions, resulting in a shift from blue to orange, with broad peak at 600 nm. Due to non-radiative coupling from within the aggregates, however, the peak external efficiency decreases from $\eta_{ext} = 13\%$ at 2 wt% concentration, to $\eta_{ext} = 8\%$ when the concentration is increased to 26 wt%. Hence, while excimer transitions can lead to broad spectral emission, at the same time they lead to a decrease in quantum efficiency.

The highest performance WOLEDs reported using the structures in Fig. 6.104 with a 40 nm thick 18 wt% Pt-17:(1:1 TAPC:PO15) EML had a forward viewing

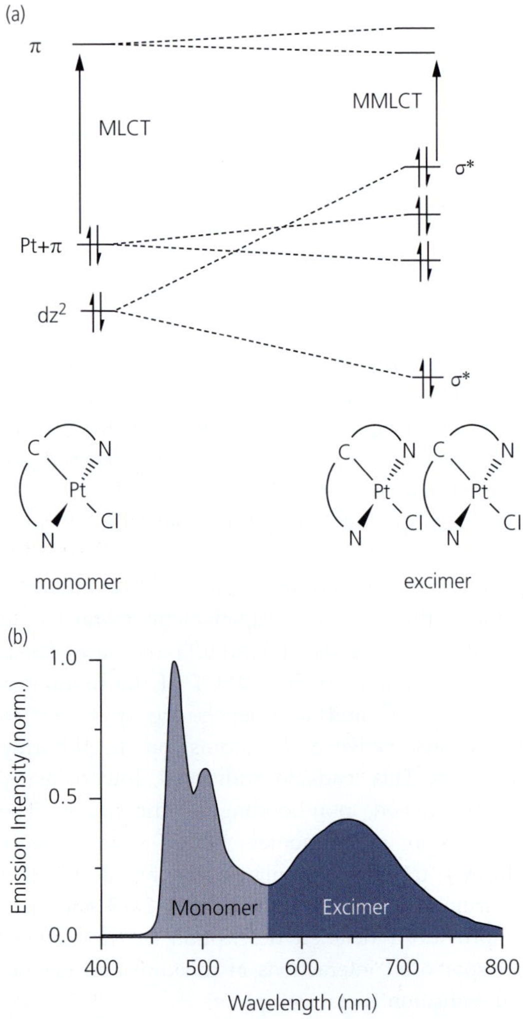

Figure 6.103 (a) Energy level scheme for the ^{3}MLCT state of a dihedral Pt monomer, and the broadened ^{3}MMLCT excimer state of a pair of coupled molecules. Molecular structures for the monomer and the Pt ··· Pt dimer are illustrated. (b) Spectrum from a film containing a mixture of monomer and excimer species (Fleetham and Li, 2014).

$\eta_{ext} = 15.7\%$ with $\eta_P = 27.3$ lm/W, $CRI = 80$, and $CIE = (0.37,0.40)$, corresponding to $CCT \simeq 4600$ K at a luminance of approximately 500 cd/m^2.

The design of monomer/excimer WOLEDs depends on the control of the dynamics leading to the generation of excimers in a molecular mixture. The mechanism by which the energy is transferred to the excimer state is different for neat and doped films. In neat films, excitons are first generated on the monomer, and are subsequently transferred to the excimer state. The formation rate of optically generated triplet excimers from adjacent excited and unexcited monomers follows:

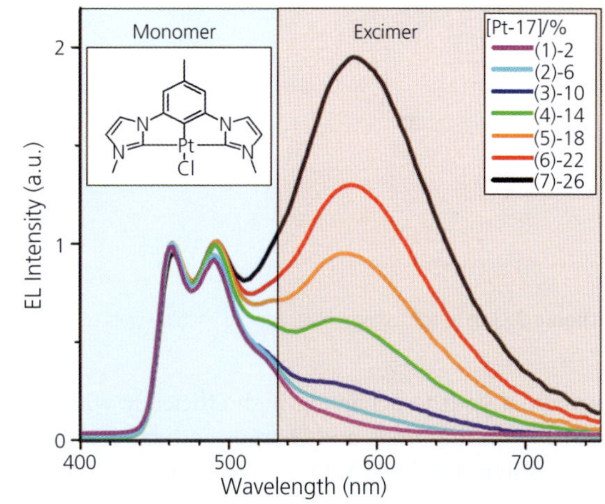

Figure 6.104 EL spectra normalized to the peak monomer intensity of a PHOLED with the structure: ITO:PEDOT:PSS/NPD (30 nm)/TAPC (10 nm)/*x* % Pt-17:26mCPy (25 nm)/PO15 (20 nm)/LiF/Al. The spectral regions for monomer and excimer emission are shown. Inset: Molecular structural formula of Pt-17 (Bakken et al., 2012).

$$^3M^* + {}^1M \xrightarrow{k_X} {}^3E^* \xrightarrow{k_{XP}} h\nu + 2S_0, \tag{6.83}$$

where k_X is the rate of formation of the triplet excimer, k_{XP} is the rate of excimer phosphorescence, $^3M^*$ is the excited monomer triplet state, ^1M is the ground state, and $^3E^*$ is the excited triplet excimer state in the solid film. If the formation is via electrical injection, a neutral charge transfer complex is formed by the combination of an anion and cation precursor pair:

$$M^- + M^+ \xrightarrow{k_{ACX}} {}^3E^* \xrightarrow{k_{XP}} h\nu + 2S_0, \tag{6.84}$$

where k_{ACX} is the rate of excimer formation from the charge transfer couple, M^- and M^+. Following analyses similar to that developed for exciton dynamics in Section 6.3.5, the rate equations for excimer formation in a neat film are (D'Andrade and Forrest, 2003)

$$\frac{d^3M^*}{dt} = -k_{Ph}({}^3M^*) - k_{nr}^M({}^3M^*) - k_X({}^1MM^*) + G$$
$$\frac{d^3E^*}{dt} = k_X({}^1M{}^3M^*) - k_{XP}({}^3E^*) - k_{nr}^E({}^3E^*) \tag{6.85}$$

where k_{nr}^M and k_{nr}^E are the non-radiative decay rates of the excited monomer and excimer, respectively, and G is the generation rate of triplet monomers due to the intersystem crossing of photogenerated monomer singlet states on the phosphor. As previously (cf. Eqs. 6.54 and 6.55), in these rate equations, M and

E represent monomer and excimer *concentrations* (which differs from their representations as states in Eqs. 6.83 and 6.84). The solutions to Eq. 6.85 at $G = 0$ are

$$\begin{aligned} {}^3M^*(t) &= A_1 \exp(-k_1 t) \\ {}^3E^*(t) &= A_2 \exp(-k_1 t) + A_3 \exp(-k_2 t) \end{aligned} \tag{6.86}$$

where A_1, A_2, and A_3 are determined by initial conditions, and the decay rates, k_1 and k_2, are given by

$$\begin{aligned} k_1 &= k_{Ph} + k_{nr}^M + k_X(^1M) \\ k_2 &= k_{XP} + k_{nr}^E \end{aligned} \tag{6.87}$$

Triplet energy transfer between adjacent molecules is by thermally activated diffusion with an activation energy of E_a, giving (Baldo and Forrest, 2000)

$$k_X(^1M) = k_X^\infty \exp[-E_a/k_B T], \tag{6.88}$$

where k_X^∞ is the monomer–excimer transfer rate at $T \to \infty$. When $E_a \ll k_B T$, excimers are formed by exothermic transfer of the excited state from the monomer, whereas for $E_a > k_B T$, the formation is endothermic. Hence, the relative emission intensities from excimer to monomer should change significantly with temperature, and the lifetime of the monomer should follow Eq. 6.87. Equation 6.88 also implies that the formation of triplet excimers is monomer-concentration dependent.

From the foregoing, we can find the lifetime of the monomer exciton vs. temperature, T, using

$$\frac{k_1(T)}{k_1(T_0)} = \frac{k_{Ph} + k_{nr}^M + k_X(^1M)}{\left(k_{Ph} + k_{nr}^M + k_X^\infty \exp[-E_a/kT]\right)}, \tag{6.89}$$

with steady state solutions to Eq. 6.85 of

$${}^3M^* = \frac{G}{k_{Ph} + k_{nr}^M + k_X(^1M)}, \tag{6.90}$$

$${}^3E^* = \frac{k_X(^1M)}{k_{Ph} + k_{nr}^M + k_X(^1M)} \times \left[\frac{G}{k_{XP} + k_{nr}^E}\right]. \tag{6.91}$$

The monomer emission intensity is proportional to $k_{Ph}(^3M^*)$, and follows (D'Andrade and Forrest, 2003)

$$L^M(T) = \frac{A_4\left(k_{Ph} + k_{nr}^M\right)}{\left(k_{Ph} + k_{nr}^M + k_X^\infty \exp[-E_a/k_B T]\right)}, \tag{6.92}$$

where A_4 is a constant. Given that and $k_r^E\left[^3E_0^*\right]$ is proportional to the excimer emission intensity, and assuming that the PL efficiency of the excimer is temperature independent, the emission intensity of the excimer is

$$L^X(T) = \frac{A_5 \exp[-E_a/k_B T]}{\left(k_{Ph} + k_{nr}^M + k_X^\infty \exp[E_a/k_B T]\right)}, \tag{6.93}$$

where A_5 is a constant. Note that the low efficiency of excimer diffusion due to exciton self-trapping reduces quenching from TTA compared with monomer-based electrophosphorescent systems.

In the solid state, monomers must be nearest neighbors to form an excimer. We have seen that solid films of square planar molecules with close π–π ligand contact should promote the formation of excimer species, even in lightly doped films. Further, the rate equations above suggest that the ratio of excimer-to-monomer emission depends on the monomer concentration.

Varying the doping concentration of an excimer-forming molecule thus controls the excimer-to-monomer emission intensity ratio. The PL and absorption of the host and guest molecules, respectively, must overlap for efficient Dexter and Förster energy transfer to occur between the two species (see Section 3.8). The rate equations for excimer formation processes in a *doped* system are thus

$$\begin{aligned} \frac{d^1H^*}{dt} &= -k_{nr}^H(^1H^*) - k_{ET}(^1M^1H^*) + G \\ \frac{d^3M^*}{dt} &= -k_{Ph}(^3M^*) - k_{nr}^M(^3M^*) - Qk_X(^1M^3M^*) \\ &\quad + k_{ET}(^1M^1H^*) \\ \frac{d^3E^*}{dt} &= Qk_X(^1M^3M^*) - k_r^E(^3E^*) - k_{nr}^E(^3E^*) \end{aligned} \tag{6.94}$$

where k_{nr}^H is the non-radiative decay rate of the excited host, ^1H is the singlet host state concentration, k_{ET} is the rate at which the exciton is transferred from the host to the dopant molecule, and Q is the probability that two dopant molecules are nearest neighbors. At doping levels where all host emission is quenched, the radiative decay rate of the excited host state is assumed to be negligible compared to the transfer rate of the exciton from the host to the dopant molecule.

For neat films, $Q = 1$, and for doped films, $Q = 1 - (1 - f)^C$, where C is the number of possible lattice positions around one molecule that can lead to the formation of an excimeric pair (i.e. the coordination number), and f is the mole fraction of the dopant. This assumes that dopant molecules are randomly dispersed without a tendency to form clusters. Apparently, $Q = Cf$ for small dopant concentrations. The excimer and monomer emission intensities are then:

$$L^M = \frac{k_{Ph}}{k_{Ph} + k_{nr}^M + Qk_X(^1M)} \cdot \left[\frac{k_{ET}(^1M)G}{k_{nr}^H + k_{ET}(^1M)}\right]$$

$$L^X = \frac{k_r^E}{k_r^E + k_{nr}^E} \cdot \left[\frac{Qk_X(^1M)}{k_r^M + k_{nr}^M + Qk_X(^1M)}\right] \cdot \frac{k_{ET}(^1M)G}{k_{nr}^H + k_{ET}(^1M)}$$

$$\frac{L^X}{L^M} = \frac{k_r^E}{k_r^E + k_{nr}^E} \cdot \left[\frac{Qk_X(^1M)}{k_{Ph}}\right] \qquad (6.95)$$

The ratio L^X/L^M varies as Q^1M, and for small dopant concentrations, Q varies linearly with concentration. The excimer-to-monomer emission intensity ratio should, therefore, exhibit a quadratic dependence with ^1M.

This model has been applied to understand the dynamics of the monomer–excimer emitting phosphor, FPt1 (D'Andrade et al., 2002a, D'Andrade and Forrest, 2003), shown in the inset, Fig. 6.105. The PL spectra of a neat thin film of FPt1 are shown as a function of temperature, along with the EL spectrum of a PHOLED with a 30 nm thick, neat FPt1 EML. At low temperatures, the excimer emission at 2.07 eV is completely dominated by monomer emission with a clearly resolved vibronic progression whose corresponding transitions are labeled. This contrasts with the highest temperature spectrum where only weak excimer emission is observed.

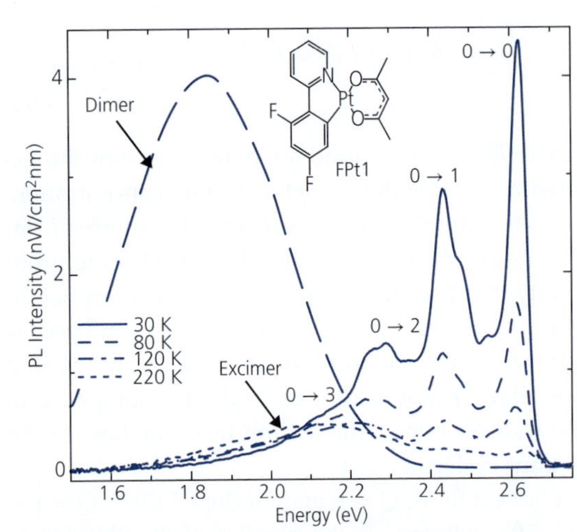

Figure 6.105 PL spectra of a planar FPt1 molecule (molecular structural formula in the inset) at several temperatures. The monomer emission vibronic transitions from the triplet to the ground state are labeled, as well as the excimer emission which is most prominent at 220 K. Also shown is the room temperature dimer EL spectrum from an FPt1 PHOLED. The EL intensity is not normalized to the PL intensity shown on the ordinate (D'Andrade and Forrest, 2003).

The room temperature PHOLED EL spectrum shows none of the features found in the PL spectrum. Instead, only a broad peak, red-shifted to 1.84 eV associated with dimer emission is observed. Recall from Section 3.6 that excimers and dimers differ in that the former has no bound ground state. It results from an excitation shared by two nearest neighbor molecules that form a new, coupled eigenstate. In contrast, the dimer is stable in both the ground and the excited state, and results from the weak Pt···Pt interactions between adjacent molecules. These weak bonds are considerably more stable than the excimer interactions, thus accounting for the significant red shift between the excimer and dimer spectra.

The excimer dynamics are obtained from the temperature dependence of the relative monomer and excimer PL intensities, as shown in Fig. 6.106. The data show the intensities of the highest energy monomer feature in Fig. 6.105 (i.e. the $^3M^*(0{\to}0)$ transition) and the excimer ($^3E^*$) as a function of temperature. The monomer intensity increases as temperature is

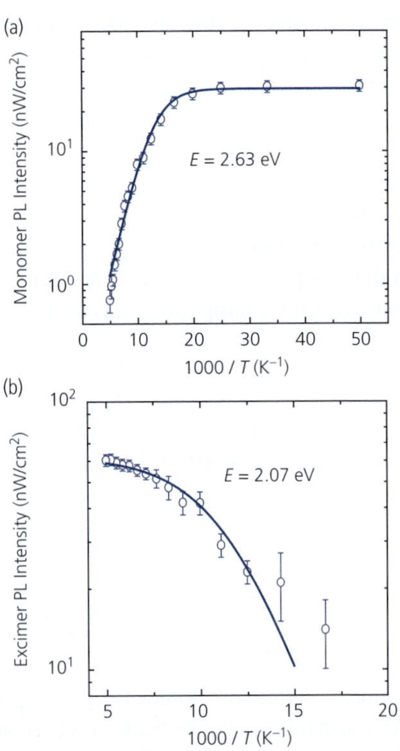

Figure 6.106 (a) Integrated intensity of the $^3M^*(0{\to}0)$ monomer emission spectrum of a FPt1-based PHOLED, and (b) the $^3E^*$ spectrum vs. inverse temperature. Lines are fits to Eqs. 6.92 and 6.93 in steady state (D'Andrade and Forrest, 2003).

reduced, until it saturates at $T < 40$ K. At the same time, the excimer peak decreases with temperature until its intensity is overwhelmed by the monomer emission, again at approximately 40 K.

The seemingly opposite response to temperature of the excimer and monomer intensities suggests that these processes are linked, that is, the monomer excitation transfers more rapidly to the excimer state as temperature increases. The foregoing analysis leading to the temperature dependences of the luminance in Eqs. 6.92 and 6.93, therefore, should give us insight into the energy transfer process. Indeed, the solid lines in Fig. 6.106 are fits to these equations, yielding activation energies of $E_a = 35$ meV for $L^M(T)$, and 39 meV for $L^X(T)$. Interestingly, even the magnitudes of these temperature variations are approximately equal, with $k_X^\infty/(k_{Ph} + k_{nr}^M) \approx 185$ in both cases. This strongly suggests that the two states are coupled. Following the model of excimer-monomer coupling presented in Figs. 3.44 and 3.45 (Birks, 1975, Nasu, 1987), there is apparently an activation energy barrier between monomer and excimer manifolds, as shown in Fig. 6.107. As triplets on individual monomers diffuse toward each other, they encounter a barrier due to Pauli exclusion and Coulomb repulsion of their orbitals. At a critical approach distance, their wavefunctions are shared to form an excimer state that is stabilized relative to the individual monomers. This results in a red shift of the emission energy as large as 560 meV. Analyses of the data are consistent with an energy barrier between these two states of 37 meV that inhibits excimer formation as temperature is decreased. As noted above, a more stable Pt···Pt dimer state is identified in the EL spectra that results in an even greater red shift, to 1.84 eV.

A demonstration of the control of Pt···Pt dimerization, and hence the ^{3}MMLCT interactions is based on a series of binuclear Pt complexes shown in Fig 6.108, where the relative opening angle between adjacent ligands is varied by modifications of the steric bulk of the bridging pyrazolate (Ma et al., 2005, 2006). This series of molecules was introduced in Section 3.7.3 (Fig. 3.79) in the context of spin–orbit coupling. The complexes have the general formula C^NPt(μ-pz$'$)$_2$PtC^N (here, C^N = 2-(2,4-difluorophenyl)pyridyl, pz$'$ for compound **1** = μ-pyrazolate, **2** = 3-methyl-5-*tert*-butylpyrazolate, and **3** = 3,5-bis(*tert*-butyl) pyrazolate). The opening angle determines the Pt···Pt bond length, and hence the strength of the ^{3}MMLCT transition. As the Pt atoms are spaced more closely together, the dimer becomes increasingly stable, resulting in a larger red shift as seen in both the WOLED

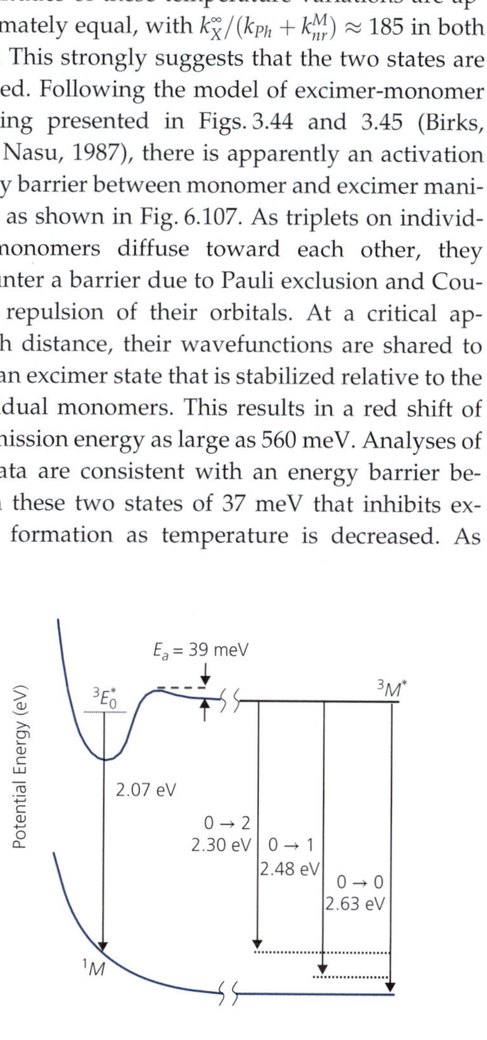

Figure 6.107 Energetic landscape of FPt1 showing the excimer and monomer transition energies, as well as the 39 meV activation energy barrier separating the states (D'Andrade and Forrest, 2003).

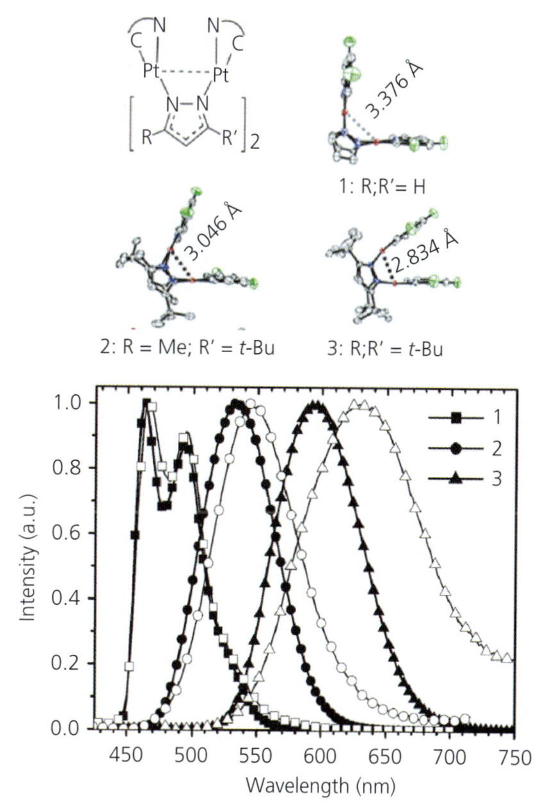

Figure 6.108 (from top left) Molecular structural formulae of the three binuclear Pt complexes shown. The Pt···Pt separation distances shown are adjusted by varying the steric bulk of the functional groups on the bridging pyrazolate: Me = methyl; *t*-Bu = *tert*-butyl. The lower figure shows that the EL spectra for PHOLEDs (filled symbols) using dopants **1**, **2**, and **3** are blue shifted compared to their PL spectra in solid solutions of polystyrene (open symbols) (Ma et al., 2006).

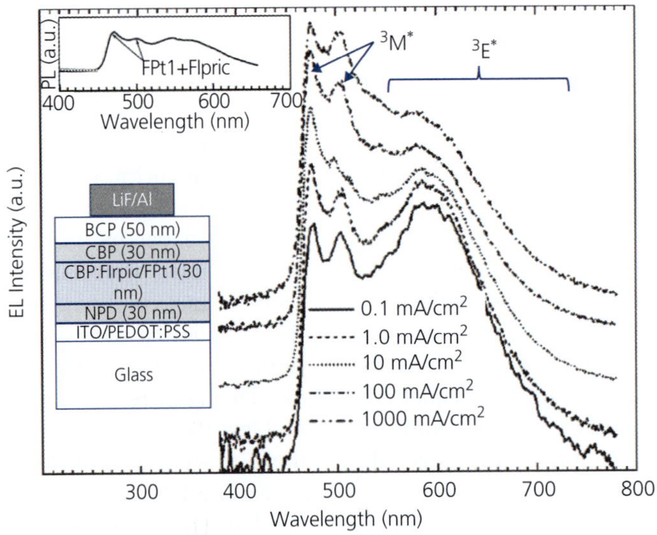

Figure 6.109 EL spectra of an FPt1+FIrpic OLED at several current densities. The monomer and excimer contributions to the spectra are noted. Upper inset: PL spectrum of the CBP + 6 wt% FPt1 + 6 wt% FIrpic blend. Lower inset: Schematic of the PHOLED structure. After D'Andrade et al. (2002a).

EL and PL spectra of these compounds in Fig. 6.108. The PL spectra are taken with the molecules mixed into a large energy gap polystyrene matrix. In the PHOLEDs, the complexes are doped at 8 wt% in mCP in an EML sandwiched between an NPD HTL, and a BCP EBL/Alq_3 ETL cap. When the ligand angle is 90° for compound **1**, there is very little Pt–Pt interaction, resulting in a nearly unchanged spectrum from the monomer ^{3}LC/^{3}MLCT transition of the square planar (C^N)Pt complex itself. In contrast, **2** and **3** have considerably stronger Pt dimer interactions, resulting in increasingly larger red shifts of up to 200 nm in solid solution as the bond distance decreases to 2.834 Å. The differences in EL and PL spectra are attributed to different couplings to the host. The orbital interactions between **2** and **3** and mCP result in a hypsochromic shift compared to polystyrene, where such interactions are completely absent. The shifts are a result of the solid-state solvation polarization interactions discussed in Section 3.6.7. Further, compound **1** shows no such shift since the monomer excited state is confined to the ligands themselves.

While spectral shifts and broadening are precisely controlled by the ligand bulk, it does not result in sufficient spectral broadening to generate efficient white emission. In fact, only compound **1** generates excimer broadening in the presence of monomer emission at high doping concentrations (30–70%) due to the exposure of its ligands that allows for monomer coupling to nearby molecules. Unfortunately, the efficiencies of these white PHOLEDs are < 4%. When **3** is used as a dopant in **1**, the monomer emission from the latter compound is mixed with the orange-red spectrum of the latter, giving broad red emission with 5.4% efficiency (Ma et al., 2006).

The foregoing analysis of excimers is general, and can be used to understand and ultimately optimize excimer-based PHOLED emission in neat systems, as well as in the often more efficient guest–host systems that can precisely control the excimer-to-monomer intensity ratio. The phosphor, FPt1, itself has proven to yield modestly efficient WOLEDs when combined with an additional blue phosphor, FIrpic, in a CBP host. The FIrpic is included to shift the spectrum toward the green to compensate for the strong red shift of FPt1 needed to achieve high efficiency. The PL spectra of the CBP + 6 wt% FPt1 + 6 wt% FIrpic blend is shown in Fig. 6.109, upper inset. The monomer emission features are apparent at 450–525 nm, whereas the broad excimer emission is seen at longer wavelengths. The WOLED EL spectra at several current densities employing this double-doped EML are provided in Fig. 6.109. Again, both the monomer $^3M^*$ and excimer $^3E^*$ features are readily resolved. The devices had slightly green CIE coordinates of (0.33, 0.44), resulting in $CRI = 78$ and $CCT \approx 5500$ K. However, as is often the case with excimer-monomer devices, the forward viewing efficiencies were only $\eta_{ext} = 4\%$ and $\eta_P = 4.4$ lm/W (D'Andrade et al., 2002a).

Among the most efficient single dopant monomer–excimer devices are those based on a symmetric tetradentate square planar Pt (Pt707) doped at 14 wt% into mCBP (Li et al., 2014). The monomer-to-excimer ratio is controlled by the concentration, as is the efficiency.

The devices exhibit $\eta_{ext} = 17.7\%$ and $\eta_P = 24.0$ lm/W at 1000 cd/m^2, making them competitive with many multi-monomer emitting phosphor devices. Further, their CIE coordinates are (0.37, 0.42) (corresponding to $CCT \simeq 4550$ K), although the low $CRI = 70$ makes these devices unsuitable for high quality lighting applications.

6.5.4 WOLEDs employing TADF emission

In this section we discuss two WOLEDs based on TADF that have been engineered to achieve high efficiency. The first uses a blue TADF emitter that transfers a portion of its energy to conventional yellow and red emitting fluorescent materials (Higuchi et al., 2015). The second is based on a blue emitting TADF emitter used in combination with a yellow Ir-phosphor molecule (Cho et al., 2015). Both structures share many of the same operational principles as hybrid fluorescent/phosphorescent and phosphor-sensitized fluorescent WOLEDs in Section 6.5.2.

The hybrid TADF/sensitized fluorescent device EML energy scheme is shown in Fig. 6.110. The EML consists of an electron and hole conducting blue emitting TADF molecule, DMACDPS, located in the exciton generation region. Its peak wavelength is 480 nm, and when used alone as an emitting molecule, the blue OLED has an external efficiency of 15%. In the WOLED, the blue emitting/exciton formation zone is separated by a 2 nm thick mCP spacer, chosen for its high triplet energy to avoid recombination of the TADF molecular triplets. At the opposite side of the spacer, the mCP is doped at 1 wt% with red emitting DBP and 10 wt% yellow emitting TTPA. The spacer prevents direct singlet transfer to these molecules, which would inevitably result in a red OLED based solely on DBP emission. Rather, the spacer requires energy transfer from DMACDPS to TTPA to DBP. By maintaining a very low DBP concentration, a balance of blue, yellow and red emission is established.

The excitation processes of the various lumophores are verified by the transient response of the WOLED in Fig. 6.111a. Whereas the fluorescent emission from TTPA and DBP occurs on a timescale of 1–10 ns, all three WOLED spectra are governed by the phosphorescent lifetime of DMACDPS of $\tau \approx 4$ μs. Thus, the luminescence persists until all blue triplets have been converted to singlets, and subsequently radiatively recombine, or transfer to the fluorophores that emit at longer wavelengths.

The WOLED emission spectrum is relatively independent of current, as shown in Fig. 6.111b. This indicates that the process in Fig. 6.110 is indeed limited by FRET-based diffusion. While good spectral coverage across the visible is not achieved (hence resulting in a low CRI), the CIE coordinates are unchanged over a decade variation in current. The maximum

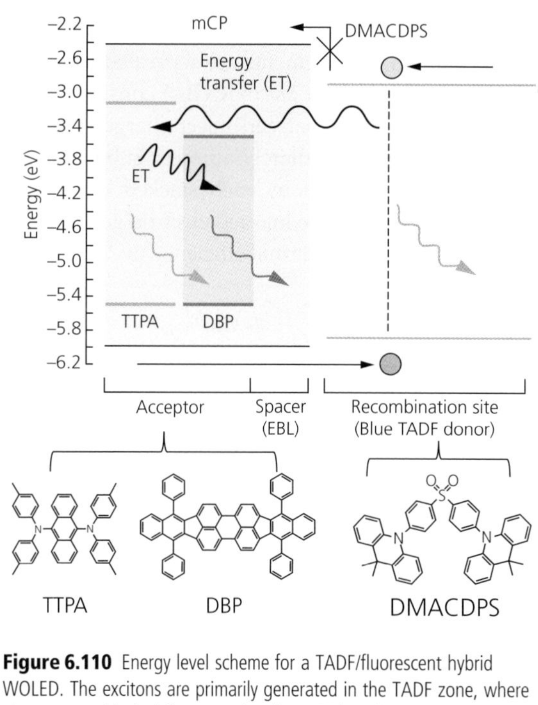

Figure 6.110 Energy level scheme for a TADF/fluorescent hybrid WOLED. The excitons are primarily generated in the TADF zone, where electrons are blocked from entering the mCP host by a 0.5 eV energy barrier. A spacer layer separates the fluorophores, TTPA and DBP, from the exciton formation zone. Energy transfer (bold wiggly arrow) and emission (gray wiggly arrows) are shown (Higuchi et al., 2015).

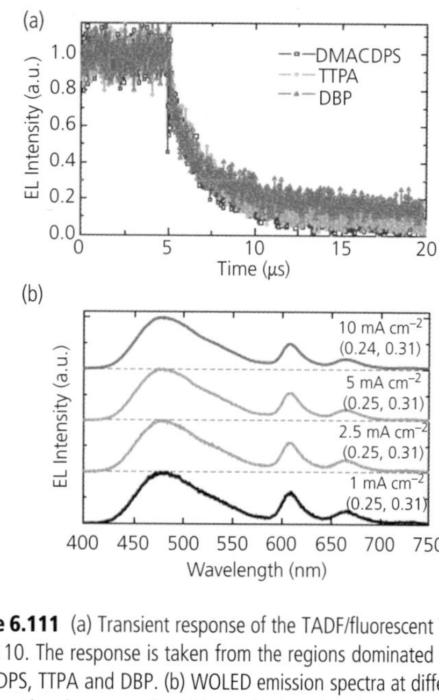

Figure 6.111 (a) Transient response of the TADF/fluorescent WOLED of Fig. 6.110. The response is taken from the regions dominated by DMACDPS, TTPA and DBP. (b) WOLED emission spectra at different drive currents. Also shown are the CIE coordinates of each spectrum (Higuchi et al., 2015).

external quantum efficiency is 12.1% and $\eta_P = 22.0$ lm/W, with CIE = (0.25, 0.31) at 1000 cd/m^2, resulting in a green-blue cast (Higuchi et al., 2015).

An alternative two-color WOLED uses a TADF molecule that serves the dual purpose of a blue emitter, and as a host to a high efficiency yellow Ir-based phosphor (Cho et al., 2015). The very high efficiency of both the blue TADF molecule, DCzIPN at 16.4% and the yellow Ir complex, PO-01, results in a high WOLED efficiency as well. The orange and blue emitting sections in the EML are separated to prevent rapid energy transfer from DCzIPN to PO-01 in the mCP host. Excitons are generated at both the ETL/EML and HTL/EML interfaces. A fraction is harvested on the TADF emitter, and the remainder diffuse to PO-01. The WOLED has a maximum external efficiency of 21% achieved at 100 cd/m^2 with CIE color coordinates of (0.31, 0.33), which is near to the isoenergetic point of (0.33, 0.33). However, the color coordinates are intensity dependent with the greatest blue spectral content at an intermediate luminance of 1000 cd/m^2 due to shifts in the position of the emission zone due to exciton formation moving between the ETL to the HTL sides of the EML. Also, as is characteristic of two-color WOLEDs, there is only weak emission in the green. Thus, while the CIE coordinates are nearly ideal, the CRI is too low for high quality illumination sources (Cho et al., 2015).

Given the results in this section, it is unclear whether TADF-based WOLEDs or those employing excimer/exciplex emission will have the high performance necessary to be useful in general lighting applications. However, the simplicity of single or even double dopant WOLEDs is an attractive means for lowering cost.

6.5.5 Stacked and striped phosphorescent WOLEDs

We found in Section 6.4 that RGB OLEDs can be arranged in a stack with each color element individually addressed to achieve any desired color, including white with a bias-selectable CRI and CCT (Shen et al., 1997, Gu et al., 1999, Parthasarathy et al., 1999). Alternatively, the potential of the internal contacts can be allowed to float, as in Fig. 6.92d. In this configuration, both high efficiency and brightness are achieved, although dynamic color tuning is not possible.

The operating principle of the stacked WOLED is shown in Fig. 6.112. The SOLED consists of N emitting sub-elements comprising individual OLEDs that emit either the same or different colors. That is, the design of each element in the stack can be

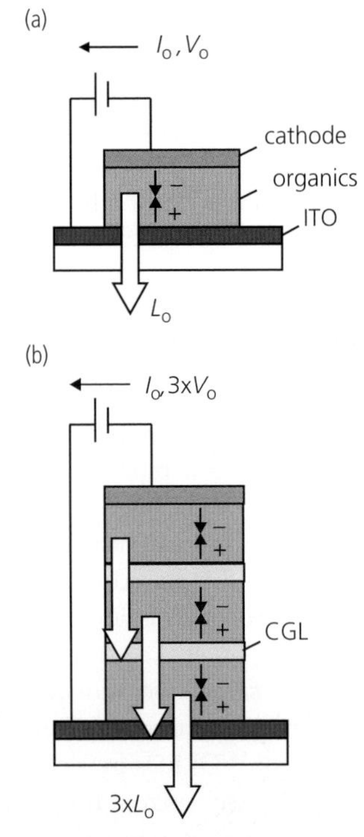

Figure 6.112 Comparison between (a) a single element OLED and (b) a stacked OLED, or SOLED. For a single element OLED, a current, I_0, at voltage, V_0 generates luminance, L_0. For a three sub-element SOLED, I_0 generates three times the luminance, $3L_0$, but requires voltage, $3V_0$.

independently designed and optimized to serve a particular function. The elements are separated by transparent *charge generation layers* (CGLs) consisting of a conducting medium that can inject charge of either polarity into the immediately adjacent sub-elements. The luminance emitted by each stacked element is additive. Hence, a single injected electron generates N photons, with a total external efficiency of:

$$\eta_{ext} = \sum_{i=1}^{N} T_{CGL,i}\eta_{int,i}\eta_{out,i}, \tag{6.96}$$

where $T_{CGL,i}$ is the transmissivity of the CGL and the transparent contact of the ith element, and $\eta_{int,i}$ is the internal quantum efficiency. Here $i = 1$ is for the bottom sub-element, and $i = N$ for the top element in Fig. 6.112b. Interestingly, there are different outcoupling efficiencies for each member of the stack since the optical field is different depending on position (see Section 6.6) (Bulovic et al., 1998a). We will show that the CGL transmissivity can approach 100%, allowing, in principle, the external quantum efficiency of a SOLED to exceed 100%.

The increased luminance comes at the expense of increased voltage. Thus, a current, I_0, at voltage, V_0, delivers a luminous intensity, L_0, for a single element

OLED. All other factors being equal, when placed in a stack of N elements, I_0 results in NL_0, but at voltage NV_0. Thus, the power efficiency remains unchanged. This is an obvious restatement of the conservation of energy. The disadvantage of higher operational voltage and increased layer complexity notwithstanding, there are compelling advantages of using SOLEDs in lighting and displays:

(i) The additive luminance leads to very high brightness at low current, particularly useful in lighting applications.
(ii) Since a higher luminance is achieved at lower current densities, and since the operational lifetime of an OLED depends on current, the SOLED architecture results in a longer useful device life (see Section 6.7) (Zhang et al., 2014b).
(iii) Lower current at a particular brightness leads to reduced operating temperature, which also extends device lifetime (Adamovich et al., 2012, Levermore et al., 2012).
(iv) Reduced current operation reduces annihilation-based losses, resulting at higher efficiency at an equivalent current (and hence higher brightness) than a single element device.
(v) The SOLED architecture allows for a flexible approach to device design. For example, white emission can be achieved by stacking individual white emitting devices (Kanno et al., 2006a, 2006b, Adamovich et al., 2012). Alternatively, red, green, and blue PHOLEDs (Liao et al., 2004, Qi et al., 2008, Chen et al., 2011), or any combination of a multicolor and monochromatic elements, can be stacked (Lee et al., 2008, Chen et al., 2011).
(vi) The SOLED thickness is $\sim Nd$, where d is the thickness of each sub-element comprising the stack. The added thickness can result in reduced shorts between anode and cathode, and hence an increased manufacturing yield, although internal shorts between elements is not expected to differ from single element OLEDs.

Since many approaches to achieving white emission already discussed are compatible with the SOLED architecture, a primary focus of SOLED research has been to develop transparent and electrically lossless CGLs. The earliest devices employed ITO, which stacks multiple TOLEDs in an integrated structure (Shen et al., 1997, Gu et al., 1999, Parthasarathy et al., 1999, Matsumoto et al., 2003). The energetic sputtering process is helpful in creating defects that promote electron injection into the ETL. However, as discussed in Section 6.4, significant damage and shorting must be avoided by using a very low sputtering power, and the resulting slow deposition rate is inconvenient for rapid, high volume manufacturing.

To avoid sputtering, compound CGLs comprising vacuum thermally deposited metal oxides such as V_2O_3 (Matsumoto et al., 2003) or MoO_3 (Kanno et al., 2006a, 2006b, Lee et al., 2008) used in conjunction with a doped organic electron injection layer have shown good performance. A comparison of the transmittance of these two oxides is provided in Fig. 6.113. For broad spectral bandwidth WOLEDs, MoO_3 is a better choice due to its high transmittance from 85% in the deep blue, to >90% in the red.

A device structure comprising two RGB white emitting sub-pixel elements along with a 10 nm MoO_3/20 nm BPhen:Li (1:1 molar ratio) CGL is shown in Fig. 6.114. The very deep MoO_3 LUMO (-6.35 eV below the vacuum level (Bao et al., 2010)) allows for rapid and efficient electron–hole recombination at its interface with the NPD HTL of the upper element. Similarly, the Li-doped BPhen in the CGL supplies electrons for injection into the lower element. The residual positive charge in the n-type Li-doped layer captures excess electrons in the MoO_3, thus completing the circuit with charges injected from the anode and cathode. While there is a large barrier between the MoO_3 LUMO and the Li^+ level, the excess charge induces energy level bending that allows for charge transfer until the Li and MoO_3 are once more neutral (Lee et al., 2012).

The function of the CGL can be understood in terms of the energy level alignments shown in Fig. 6.115. The CGL in the device in Fig. 6.114 is illustrated in Fig. 6.115a. MoO_3 has a deep LUMO that aligns with

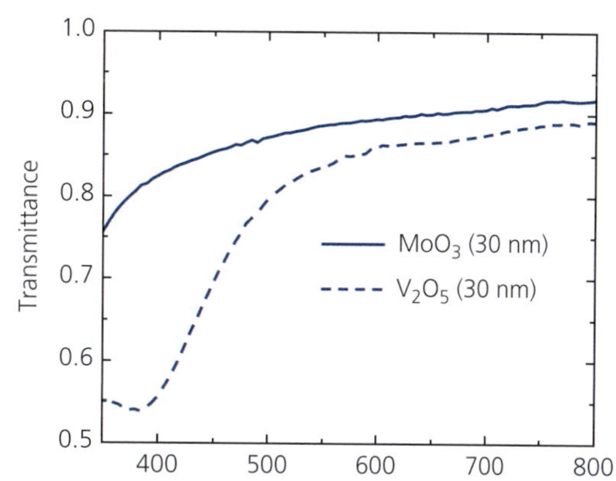

Figure 6.113 Transmittance of 30 nm thick thermally evaporated layers of MoO_3 and V_2O_3 (Kanno et al., 2006a).

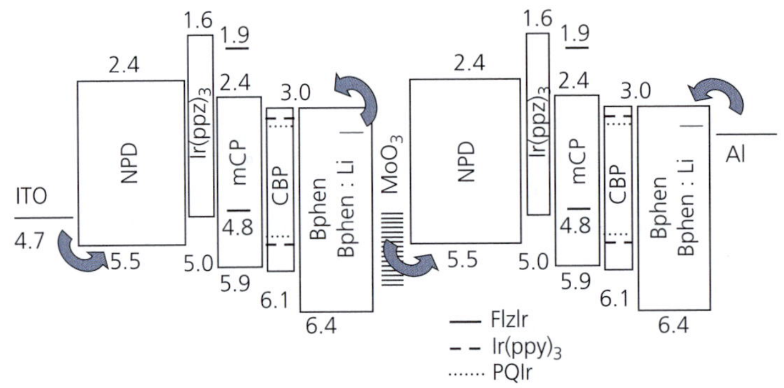

Figure 6.114 Energy level diagram of a dual element WOLED using a compound MoO_3/BPhen:Li hole/electron injecting CGL. Numbers indicate HOMO and LUMO energies relative to vacuum of the various materials in eV. Arrows show the direction of charge injection (Kanno et al., 2006a).

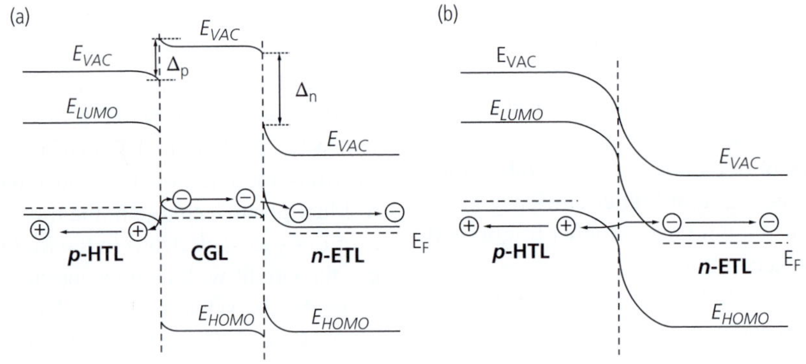

Figure 6.115 Two types of CGLs used in multi-stage WOLEDs. (a) Charge generation via interface recombination at a CGL with a deep LUMO level. (b) Tunnel-junction CGL between a heavily doped p-HTL and n-ETL.

the n-doped ETL, forming an interface dipole of energy, Δ_n. The LUMO alignment to the p-HTL permits efficient electron–hole recombination. The dipole plus energy level bending near the interface due to dopant charge depletion (i.e. electron injection from Li:BPhen into the adjacent OLED ETL) subsequently allows for electron-injection into the ETL of the second OLED in the stack. Alternatively, a highly doped p-n organic junction induces significant interface energy level bending, leading to tunneling of holes and electrons across the interface region. This case, shown in Fig. 6.115b, will be discussed further below.

Returning to the device in Fig. 6.114, each sub-element has the following structure: NPD (40 nm)/ $Ir(ppz)_3$ (10 nm)/mCP:10 wt% FlzIr (20 nm)/CPB:3 wt% $Ir(ppy)_3$:10 wt% PQIr/Bphen(20nm)/Li:BPhen (1:1 molar ratio) (20 nm). The EML is comprised of two sections: one with green and red phosphors in a CBP host, and the other with the blue phosphor (FlzIr) doped into mCP. Both elements use a combination EBL (BPhen) and hole blocking layer ($Ir(ppz)_3$), the latter needed due to the very high LUMO energy of the blue dopant.

The stacked WOLED performance is shown in Fig. 6.116 (Kanno et al., 2006a). The peak η_{ext} moves monotonically to higher currents, and hence higher brightness, with the number of stages. As a result, the current density required to achieve a given luminance decreases as the number of sub-elements increases from 1, to 2, to 3. Both effects are due to reduced annihilation since the EML is effectively "spread out" among the multiple elements. Furthermore, the peak efficiency is approximately a linear function of the number of stacked elements. Thus, the peak forward viewing efficiencies from the glass surface are η_{ext} = 13.6, 24.7, and 34.9% for 1-, 2-, and 3-element SOLEDs. These efficiencies correspond to 1.8 and 2.6 times the efficiency of a single element device for the 2- and 3-element SOLEDs, suggesting that losses at each CGL are only $\sim$ 10%. As noted, conservation of energy requires that the power efficiency of multi-element devices not exceed that of a single stage OLED. This is clearly the case in Fig. 6.116a, where the power efficiency of each device is approximately equal.

In Fig. 6.116b we show the *j*–*V* characteristics for the same three SOLEDs whose characteristics are

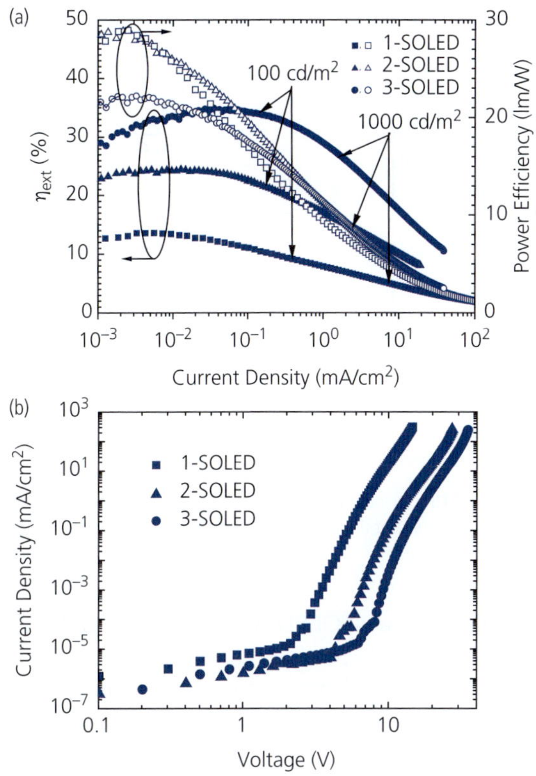

Figure 6.116 (a) External quantum and power efficiencies of 1-, 2-, and 3-unit SOLEDs with the structure in Fig. 6.114. The currents needed to achieve luminances of 100 and 1000 cd/m^2 are indicated by arrows. (b) Current density vs. voltage characteristics of the devices in (a) (Kanno et al., 2006a).

Table 6.12 Performance characteristics of the white phosphorescent SOLEDs in Figs. 6.114 and 6.116 (Kanno et al., 2006a)

Device	η_{ext} (max) [%][a]	η_P (max) [lm/W][a]	Volt.[b]	CIE coord.[b]	CRI[b]
1-SOLED	13.6	28.9	9.2	(0.36, 0.46)	62
2-SOLED	24.7	28.9	18.0	(0.39, 0.45)	64
3-SOLED	34.9	22.7	24.1	(0.35, 0.44)	66

[a] Efficiencies measured in the forward viewing direction.
[b] Voltage, CIE, and CRI all measured at $j = 10$ mA/cm^2.

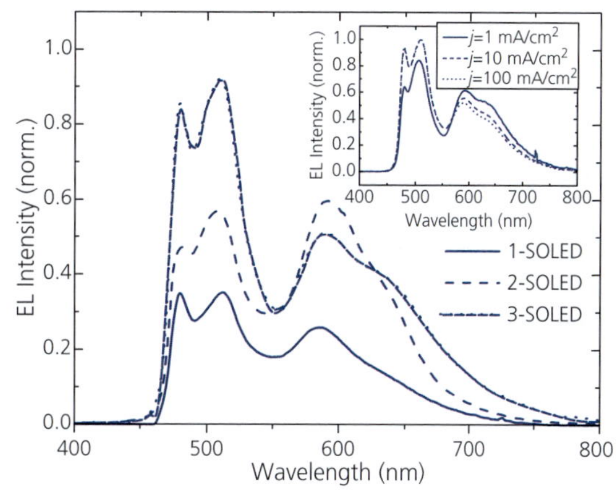

Figure 6.117 EL spectra for 1-, 2-, and 3-element SOLEDs in Figs. 6.114 and 6.116. Inset: The spectra of the 3-SOLED vs. current density (Kanno et al., 2006a).

summarized in Table 6.12. Since the applied voltage is approximately an integer multiple of the number of identical elements, we infer that the charge injection from the CGLs is nearly lossless. Indeed, the charge injection of the CGLs appears to be even more efficient than for the Al and ITO cathode and anode.

As noted previously, SOLEDs can exhibit noticeable microcavity effects due to the numerous dielectric layers comprising the stacks. These effects can be ameliorated by varying the thickness of the layers, thereby changing the optical path lengths (and hence the optical intensity profiles) within the stacks. The influence of stacking on the EL spectra of the devices in Figs. 6.114 and 6.116 is shown in Fig. 6.117. As elements are added to the stack, the peak emission moves from the blue, to the green, and then back to the blue for 1-, 2-, and 3-element devices, respectively. Since the emission zone in each element is also current dependent, the spectra vary with current, as shown in the inset, Fig. 6.117.

The MoO_3/BPhen:Li compound CGL has been adapted to many different stacked structures. For example, a significantly improved color rendition has been achieved by stacking fluorescent/phosphorescent white elements having structures and materials sets similar to that in Fig. 6.100. A 3-SOLED device with such a hybrid sub-element had a total efficiency of $\eta_{ext} = 57\%$, $\eta_P = 22$ lm/W and CIE coordinates of (0.38, 0.44), corresponding to $CCT \simeq 4100$ K with $CRI = 82$ at a luminance of 1000 cd/m^2. This device showed an almost optically and electrically lossless CGL. The peak external efficiencies were 1.9 and 2.85 times larger for the 2- and 3-SOLEDs compared to the single element devices, corresponding to only a 5% loss at each stage. Similarly, the change in voltage drop across the device also scaled nearly linearly with the number of elements (Kanno et al., 2006b).

A different architectural approach using the MoO_3/BPhen:Li CGL places three different color WOLEDs in a stack, as shown in Fig. 6.118a. One element (nearest the cathode) has a red-emitting EML consisting of PQIr doped CBP, followed by the green emitting $Ir(ppy)_3$:CBP element, and a blue emitting FIr6:UGH2 element (Qi et al., 2008). This is an "RGB" structure,

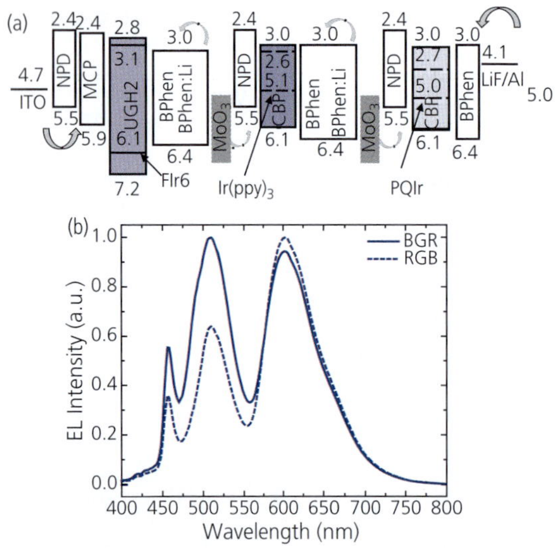

Figure 6.118 (a) Energy level structure of an RGB phosphorescent WOLED. (b) Spectra depending on ordering of the sub-elements. BGR or RGB corresponds to the pixel ordering from cathode to anode (Qi et al., 2008).

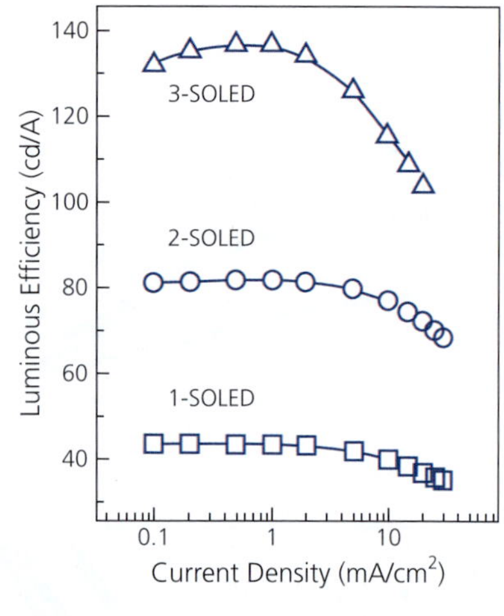

Figure 6.119 Luminous efficiency vs. current for 1-, 2-, and 3-element green phosphorescent ($Ir(ppy)_3$-based) SOLEDs with fully organic CGLs (Liao et al., 2004).

where the first letter corresponds to the element nearest the cathode. The sub-pixel ordering is critical in managing microcavity effects to obtain the desired output spectrum. In Fig. 6.118, the color balance of the BGR SOLED is improved compared to the RGB SOLED, with CRI values for the two devices of 79 and 63, respectively. Due to the larger content of green and red emission, however, the RGB device has the highest quantum and power efficiencies of $\eta_{ext} = 36\%$, and $\eta_P =$ 21 lm/W and CIE coordinates of (0.45, 0.36), corresponding to $CCT \simeq 2450$ K with $CRI = 63$. This is a very warm white, with characteristics that fall short of acceptable for lighting applications, although further optimization is undoubtedly possible.

Lithium used in the CGLs in Fig. 6.118 can diffuse through the layers as the device ages. For this reason, other n-type dopants such as Cs_2CO_3 have been used as the electron donor (Lee et al., 2008), as well as devices whose CGL comprises a combination of strong n-type donors such as LiF and Ca in mixtures with thin cathode metals Al and Ag (Sun et al., 2005).

Metal oxide/Li-doped organic CGLs are simply p-n junctions that efficiently inject electrons and holes into the adjacent OLED elements. Fully organic doped p-n tunnel junctions (Fig. 6.115b) have also been found to be an effective CGL architecture. The first such example used an n-type layer consisting of 1.2 vol.% Li doped into electron conducting Alq_3 or TPBI, combined with a p-type, α-NPD layer doped with 1 vol.% $FeCl_3$ in both fluorescent and phosphorescent stacked devices (Liao et al., 2004). The organic junctions were used in green monochrome $Ir(ppy)_3$-based SOLEDs, with results in Fig. 6.119. The structure of the 3-SOLED was ITO/SA/EML/CGL/EML/CGL/EML/SC/Mg:Ag, with the 2-SOLED and 1-SOLED (single element) structures eliminating one or two EML/CGL units, respectively. Here, the separate units are EML = 30 nm NPD/20 nm 5 vol.% $Ir(ppy)_3$:CBP/20 nm TPBI; an anode side optical spacer unit to control microcavity emission spectra of SA = 60 nm NPD, and a cathode side optical spacer of SC = 40 nm TPBI. Also, CGL = Alq_3:Li/NPD:$FeCl_3$. The CGLs are nearly lossless, with luminous efficiencies of 43.5, 81.7, and 136.3 cd/A at $j = 1$ mA/cm^2 for the 1-, 2-, and 3-unit devices respectively. This corresponds to efficiency ratios of 1.9 and 3.1 for the 2-SOLED and 3-SOLED compared to the single element device. The voltages also nearly exactly double, and increase by 150% for comparable, fluorescent 2- and 3-element devices compared to the conventional OLED.

An organic that has similar frontier orbital energy alignments as MoO_3 is HATCN, with a LUMO energy of −6.0 eV (from the vacuum level) and a HOMO at −9.9 eV (Kim et al., 2009). This can be used in an all-organic CGL with a multilayer structure of p-HTL/HATCN/n-ETL, which is analogous to charge injecting junctions using MoO_3 (Liao et al., 2008, Lee et al., 2012).

A CGL that is a hybrid combination of organic and inorganic compounds has been demonstrated in tandem white phosphorescent SOLEDs comprising a combination of a blue element stacked with a red/green emitting element (see Fig. 6.120). The green-red section employed $Ir(MDQ)_2(acac)$ and $Ir(ppy)_2(acac)$ for red and green, and FIrpic for the blue emitting element. The CGL *p-n* junction comprised a p-MoO_3 doped TCTA layer, and an n-Li_2CO_3 doped TPBi layer. These were separated by an interfacial region consisting of a type II pentacene donor/C_{60} acceptor HJ. The rationale for the interface zone is its ability to transfer charge across the donor/acceptor junction. This results in a reservoir of holes on the p-side and electrons on the n-side that greatly increase the conductivity, and the opportunity for charges to recombine within the CGL even at high current densities. A consequence is that there is very little dependence of quantum efficiency on luminance up to >5000 cd/m^2 (Chen et al., 2011).

The effectiveness of the hybrid HJ was assessed by comparing the performance of the device in Fig. 6.124 with an identical two stage SOLED lacking the junction. The pentacene/C_{60} CGL device had a voltage of 6.0 and 6.9 V at 100 and 1000 cd/m^2, respectively, which is significantly lower with 6.7 and 8.3 V for the device lacking the layer. This has a significant impact on the power efficiency of the devices. At 1000 cd/m^2, the device with the hybrid interface junction had $\eta_{ext} =$ 45%, $\eta_P = 45$ lm/W and CIE coordinates of (0.39, 0.44), corresponding to $CCT \simeq 4000$ K with $CRI = 78$. These metrics are only slightly degraded from their values at ten times lower luminance (Chen et al., 2011).

We previously noted that color tunability afforded by a device with individual contacts to each element in the stack can be used to change the CRI, CCT, and color coordinates of a light source simply by changing the current drive across each of the stacked elements. This has numerous advantages, in that room lighting often has different requirements for different occasions or times of day. For example, to complement the human circadian rhythm, the workplace can benefit from lighting fixtures with a higher red content (i.e. lower CCT) in the mornings and evenings, and a cooler, bluer hue during mid-day hours. Also, color-variable lighting can set a mood or highlight objects within a space. For all of these reasons, color tunability afforded by OLED lighting makes it uniquely attractive to the consumer.

A difficulty in realizing color-tunable SOLEDs is fabricating and accessing low resistance contacts between the stacked elements. Illumination sources, however, are customarily viewed in the far field, obviating the need for the fabrication of small pixels that eases access to the interior contacts. Alternatively, the R, G, and B elements can be formed into long, laterally spaced stripes, as in Fig. 6.92f. This considerably eases both patterning and contacting of the individual color emission regions. Further, both stacked and striped PHOLEDs can achieve 100% internal efficiency in a white device, since the individual R, G, and B PHOLEDs can be separately optimized. While these approaches are more complex (and hence costlier) than two terminal WOLEDs, they provide the highest potential efficiency, and a wide range of color tunability.

A striped RGB WOLED has been fabricated by direct printing using organic vapor jet printing as described in Section 5.6.6 (Arnold et al., 2008). The device architecture and fabrication process are shown schematically in Fig. 6.121. Starting at the

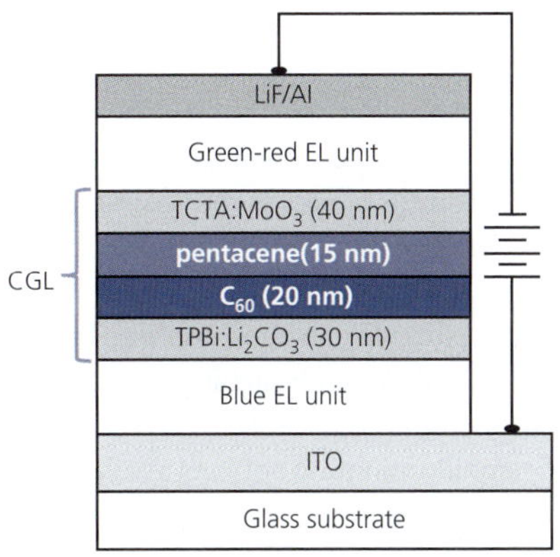

Figure 6.120 A phosphorescent WOLED comprising a stack of green-red and blue emitting elements, along with a hybrid organic/inorganic compound CGL (Chen et al., 2011).

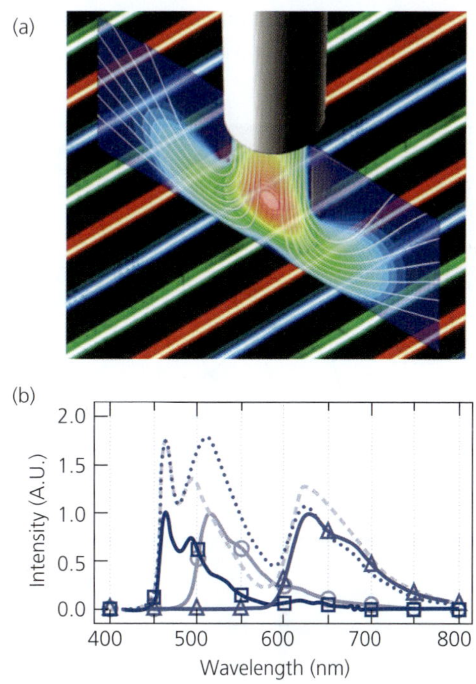

Figure 6.121 (a) Conceptual view of direct printing of closely spaced R, G, and B PHOLED stripes by organic vapor jet printing to create a tunable color illumination source. The nozzle and its organic vapor flow field are shown above the stipes. (b) Spectra of the three PHOLEDs (solid lines), including two different white spectra obtained by differentially tuning the emission intensities from the PHOLEDs (Arnold et al., 2008).

ITO-coated glass substrate, the NPD HTL is blanket-deposited by vacuum thermal evaporation across the entire substrate. Then the separate EMLs are sequentially printed by volatilizing the dopants and hosts in a dilute, hot N_2 gas stream and carried through a nozzle that directs the dopant/host mixture onto a substrate that is linearly translated at a rate ranging from 3 to 15 mm/s. The substrate is returned to the vacuum deposition chamber where the BPhen EBL, Alq_3 ETL and LiF/Al cathodes are deposited through striped shadow masks to pattern the PHOLEDs. The R, G, and B EMLs consisted of the commercial, red emitting Ir-dopant RD-15, green emitting $Ir(ppy)_3$ doped in CBP, and blue emitting FIr6 doped into mCP within the OVJP print-head apparatus.

The emission spectra of the three parallel striped PHOLEDs are shown in Fig. 6.121b, along with white spectra with color coordinates of (0.33, 0.33) (dashed line) and (0.40, 0.31) (dotted line) accessed by differentially tuning the current of the individual stripes. The source appears as a single hue (including white) in the far field, or in the near field by placing the stripes behind a ground glass light diffuser. For the devices shown, the peak efficiency of the source at coordinates of (0.40, 0.31) was $\eta_{ext} = 12.1\%$, $\eta_P = 11.9$ lm/W. These values are comparable to those obtained using analogous vacuum-deposited PHOLED stripes (Arnold et al., 2008).

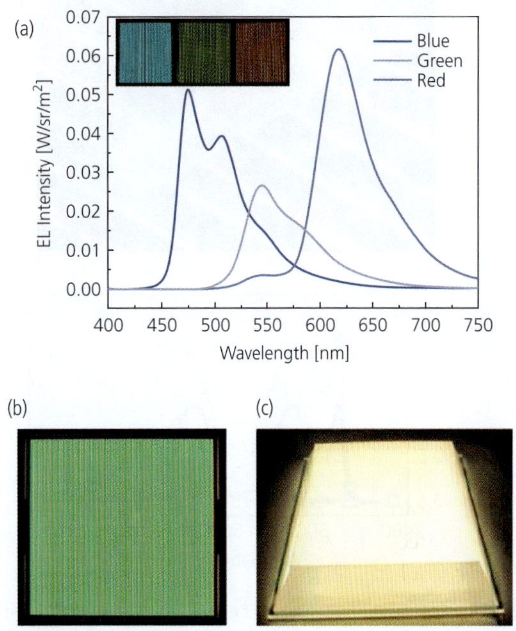

Figure 6.122 (a) Emission of the separate R, G, and B striped PHOLEDs of the 15 cm × 15 cm panel shown in the inset, with each of the individual striped PHOLEDs turned on independently. (b) All green pixels turned on simultaneously. (c) The lighting panel using a thick light extraction block plus color diffuser to give a uniform white appearance (Weaver et al., 2014).

A similar striped device was fabricated using high efficiency phosphorescent dopants deposited entirely by VTE. The 15 cm × 15 cm panel in Fig. 6.122a employed a thick, light-extracting block and diffuser. Figure 6.122b shows only the green stripes turned on. Without the light extraction block, the striped WOLED had $\eta_P = 41$ lm/W at 1000 cd/m^2, with color coordinates of (0.476, 0.433), $CRI = 90$, and $CCT = 2700$ K. The light extraction block in Fig. 6.122c increased the output efficiency by a factor of 1.53, leading to an efficacy of 63 lm/W at 1000 cd/m^2, with drive voltages of 3.0 V, 2.7 V, and 4.2 V for the R, G, and B stripes, respectively (Weaver et al., 2014). These results suggest that striped PHOLED designs are suited for producing both highly efficient and attractive color tunable lighting sources.

6.6 Light outcoupling

In the foregoing sections, we have seen that the internal efficiency of PHOLEDs can be as high as 100%. However, the external efficiency on flat glass substrates is limited to approximately 20% due to TIR. The fraction of light emitted from the dielectric substrate surface from a Lambertian source is $\sim 1/2n^2 \simeq 0.2$, where $n = 1.5$ is the refractive index of glass (Greenham et al., 1994). Hence, there is considerable benefit to be gained if the light trapped within the OLED is extracted. There are four primary sources of loss illustrated in Fig. 6.123 that decrease η_{out} in Eq. 6.1: (i) *substrate modes*, due to light trapped in the substrate from TIR at the substrate/air interface, (ii) *waveguide modes* due to launching of light parallel to the substrate plane within the relatively high index of refraction layers comprising the device and the transparent anode, (iii) *surface plasmon polariton modes* (SPP) excited in the conductive contacts, and (iv) *lossy modes* due to material and contact absorption.

In this section, we discuss strategies to extract light from each of these loss channels. We will find that extracting substrate modes is the easiest, and can lead to a 100% increase in the outcoupled light intensity, followed by waveguide modes, and the most difficult to outcouple, surface plasmon modes. If outcoupled, η_{out} between 70–80% can be achieved which, combined with their potential $\eta_{int} = 100\%$, can make OLEDs the most efficient illumination sources available.

There are numerous solutions to outcoupling the trapped modes from OLEDs. We point out, however, that all *practical* solutions (i.e. those that are

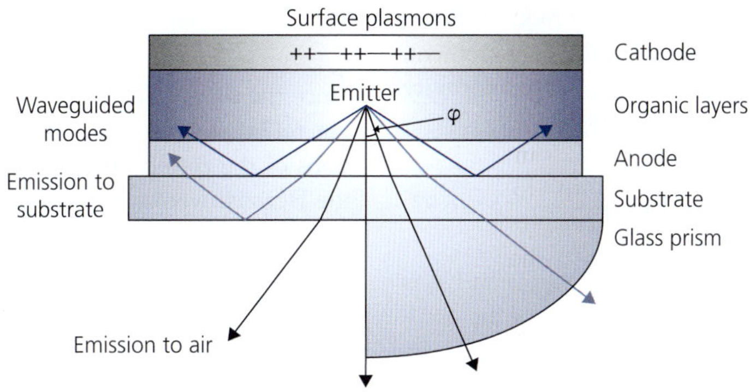

Figure 6.123 Optical paths launched by an emitting molecule within an OLED. Shown on the right bottom of the substrate is a glass lens or prism used to outcouple substrate modes that would otherwise be totally internally reflected (Brütting et al., 2013).

acceptable for incorporation in large area devices such as displays or lighting) should have the following properties:

(i) The wavelength and intensity of the OLED should not be affected by the outcoupling scheme, and in particular, should not be a function of viewing angle.
(ii) The solution should be minimally invasive of the OLED structure itself. That is, the ideal outcoupling scheme should not require changes to the OLED which is assumed to have been separately optimized to achieve maximum internal and luminous efficiencies, with the desired color coordinates and brightness. Placing additional constraints on the OLED design by an outcoupling scheme can possibly result in a decrease in performance.
(iii) The solution should be inexpensive, allowing for minimal cost impact when used over the large areas required by both displays and lighting applications.

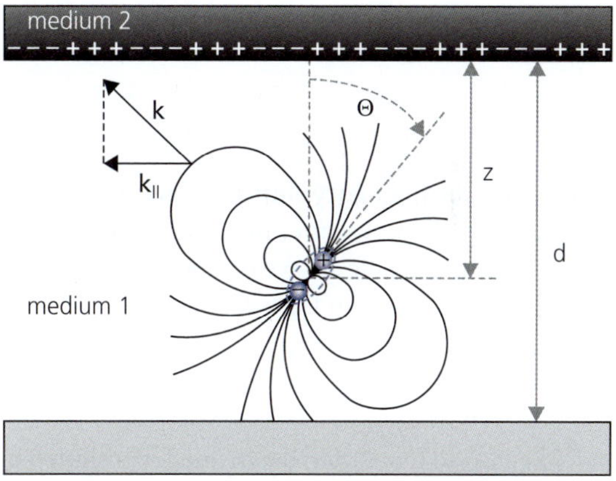

Figure 6.124 Illustration of a radiating molecular dipole at angle Θ in the optical cavity of thickness d formed by an OLED. The direction of propagation is $\mathbf{k}$, with in-plane component $\mathbf{k}_{||}$. A mirror forms the top cavity surface and a semi-reflecting layer forms the bottom surface (Brütting et al., 2013).

6.6.1 Theory of outcoupling

In its simplest terms, we view an OLED as an optical cavity comprising a dielectric in which there is an emitting molecule (medium 1) bounded on one surface by a reflective, metal contact (medium 2). Of course, the OLED is considerably more complicated than this simple two layer picture, but the bilayer embodies the essential physics needed to calculate the output spectrum and angular power distribution. Including the additional layers that more accurately represent an OLED is a simple matter that will be described later.

In Fig. 6.128 we show a radiating dipole (i.e. the emitter molecule) at angle Θ and position z in an optical microcavity of thickness d formed by the OLED. The dipole radiation field results in propagation whose directions are defined by the vector $\mathbf{k}$, with component $\mathbf{k}_{||}$ along the substrate plane, and $\mathbf{k}_{\perp}$ perpendicular to the plane. When the dipole oscillates near an interface between two materials, it transfers energy to that interface (i.e. it excites a plasmon), resulting in a reflected electric field, modifying both the dipole frequency and lifetime. The interaction of the dipole and its field within a cavity is known as the Purcell effect. The reflected electric field, $\mathbf{F}_R$, is strongest near a highly reflective medium (i.e. at the interface with a metal mirror versus at a dielectric discontinuity). The field modifies the OLED emission. There are two related situations to consider: one in which the interface is weakly reflecting in an OLED with two transparent contacts, and the second more common scenario where one surface is a reflective metal cathode or anode contact.

Following the procedure of *Chance, Prock, and Silbey* (CPS), the equation of motion for a dipole of moment, **p**, oscillating at natural frequency, ω, in a microcavity is given by (Chance et al., 1974, 1978)

$$\frac{d^2\mathbf{p}}{dt^2}+\omega^2\mathbf{p}=\frac{q^2}{m^*}\mathbf{F}_R-\frac{1}{\tau_0}\frac{d\mathbf{p}}{dt}, \tag{6.97}$$

where m^* is the effective mass of the dipole and τ_0 is the natural damping (i.e. radiative) time of the oscillation in the absence of the mirror. Now both the dipole and the field oscillate at the same frequency, and have the same vector dependence since one is the image of the other. Equation 6.97 has the following solutions:

$$\mathbf{p}=\mathbf{p}_0\exp(-i(\omega+\Delta\omega)t-t/2\tau) \tag{6.98a}$$

and

$$\mathbf{F}_R=\mathbf{F}_0\exp(-i(\omega+\Delta\omega)t-t/2\tau), \tag{6.98b}$$

where $\Delta\omega$ is the frequency shift and τ is the lifetime in the presence of the mirror. In the limit of $\Delta\omega\rightarrow 0$, which is characteristic of weak microcavities, then substituting Eq. 6.98 into Eq. 6.97 gives

$$\frac{1}{\tau}=\frac{1}{\tau_0}+\left[\frac{q^2}{2p_0m^*\omega}\right]\mathrm{Im}(F_0) \tag{6.99a}$$

and, likewise, to first order, the frequency is shifted by

$$\Delta\omega\approx\frac{1}{8\omega\tau^2}-\frac{1}{4\omega_0\tau_0\tau}-\frac{q^2}{2m^*\omega_0p_0}\mathrm{Re}(F_0). \tag{6.99b}$$

Generally, the frequency shift is sufficiently small that it is ignored in treatments of the weak microcavity effects typically encountered in OLEDs. From these two expressions, we see that the radiative rate is related to the out-of-phase electric field, and the frequency shift to the in-phase components of the reflected field.

The total field within medium 1 is the superposition of the dipole field and that reflected from the mirror. The field is calculated by considering a dipole in a homogeneous dielectric medium with permittivity, $\varepsilon_0\varepsilon_r$. Then the *Hertzian vector* is given by

$$\boldsymbol{\Pi}=\frac{\mathbf{p}_0\exp(ik_1r)}{4\pi\varepsilon_0\varepsilon_r r}, \tag{6.100}$$

where r is the distance from the dipole to the point of observation, and k_1 is the component of the wavevector parallel to $\hat{\mathbf{r}}$ in medium 1. Then, the electric field is given by

$$\mathbf{F}=k_1^2\Pi+\nabla\nabla\cdot\boldsymbol{\Pi} \tag{6.101}$$

and the magnetic field by

$$\mathbf{H}=-i\omega\nabla\times\boldsymbol{\Pi}. \tag{6.102}$$

Since the dipole field is cylindrically symmetric (see Fig. 6.124), we can solve this in terms of Bessel functions of the zeroth order, J_0. Then the expression for the Hertzian vector is (Wasey and Barnes, 2000)

$$\boldsymbol{\Pi}=\frac{i\mathbf{p}_0}{4\pi\varepsilon_0\varepsilon_r}\int_0^\infty\frac{u}{l}J(u\rho)\exp(il|z|)du, \tag{6.103}$$

where the integral is over the variable, $u=k_\|/k_1$, $k_\|$ is the in-plane wavevector, and $l=(k_1^2-u^2)^{1/2}$ is the wavevector for propagation perpendicular (i.e. in the $\hat{\mathbf{z}}$-direction) to the plane of the cavity (k_z). For this expression, we use the cylindrical coordinate system, (ρ, ϕ, z). Over the domain, $0\leq u\leq 1$, energy is radiated into the half-space below the mirror. In a multilayer system, waveguide modes can also be excited for particular angles of emission. At $u>1$, non-radiative, evanescent SPP modes are excited, and when u becomes large at small separations between the dipole and medium 2, lossy non-radiative surface waves are generated.

With the modified emission rate, we can proceed with our determination of the reflected field. An isotropically oriented dipole (which is not always the case for emitting molecules in OLEDs, see Section 6.6.5) can have its components decomposed into directions parallel (||) and (⊥) perpendicular to the plane, yielding

$$\frac{1}{\tau_{\|,\perp}}=\frac{1}{\tau_0}(1-\eta_{int}^0z_{\|,\perp}), \tag{6.104}$$

where η_{int}^0 is the internal quantum yield of the dipole in medium 1 without the metal mirror. The parameters $z_{\|,\perp}$ are given by (Barnes, 1998)

$$z_\perp=1-\frac{3}{2}\mathrm{Im}\int_0^\infty\frac{u^3}{l_1}(1-r_{1,2}^p)\exp(-i\beta)du \tag{6.105a}$$

and

$$z_\|=1-\frac{3}{4}\mathrm{Im}\int_0^\infty\frac{u}{l_1}\left[(1+r_{1,2}^s)-(1-u^2)(1+r_{1,2}^p)\right]\times\exp(-i\beta)du, \tag{6.105b}$$

where $r_{1,2}^{s,p}$ are the Fresnel reflection coefficients for s- and p-polarized light at the interface between media 1 and 2, l_1 is the value of l in medium 1, and β is the phase difference between the emitted and reflected plane waves. This treatment can be extended to consider multiple layers based on transfer-matrix or dyadic Green's function formalisms to account for the reflection at each interface, or depending on the value of u, excitation of guided modes within a particular medium (Chance et al., 1978, Celebi et al., 2007, Setz et al., 2011).

Equation 6.104 gives the decay rate due to a dipole oriented either parallel or perpendicular to the mirror. For an isotropic dipole orientation, the total observed damping rate is simply

$$\frac{1}{\tau} = \frac{2}{3\tau_{\parallel}} + \frac{1}{3\tau_{\perp}}, \tag{6.106}$$

where there are two possible in-plane (x, y), and one out-of-plane (z) orientation.

The modification of the emission rate with distance can be dramatic, as shown by the data points for the normalized PL lifetime for $Ir(ppy)_3$ emission in the PHOLED in Fig. 6.125 (Setz et al., 2011). Depending the distance as determined by a dielectric spacer to the contact (in this case the ETL), the radiative lifetime is considerably shorter for dipoles near the mirror surface, to $\tau > \tau_0$ at ~150 nm distant from the mirror. For this device, the minimum distance from the emitting molecule to the cathode is 10 nm, determined by the hole blocking layer thickness. However, as the dipole distance is decreased, then $\tau \rightarrow 0$ (Amos and Barnes, 1997). The oscillations, which arise from the phase differences between the superposed dipole and reflected fields, are damped as the distance from the dipole to the mirror increases. The peaks occur at field maxima, that is, where the reflected and the emitted fields are in phase. The damping results since the reflected field intensity also decreases with distance. At very small dipole-to-mirror distances, however, the lifetime vanishes due to efficient excitation of SPPs in the metal. The solid lines in Fig. 6.125 are fits to the data based on CPS theory. Each line corresponds to a different emitter efficiency, η_{int}^0. As the emitting molecule becomes increasingly efficient, the oscillations are stronger due to the increased interference from the superimposed reflected field. The best fit to the data corresponds to $\eta_{int}^0 = 0.5$, where the CPS theory accurately approximates the observed optical emission.

An SPP mode is schematically illustrated in Fig. 6.126. Light incident on the metal couples to the charge at its surface, inducing a charge density wave that is trapped at the metal/dielectric interface. The resulting p-polarized mode has wavelength, λ_{SPP}. The dispersion relationship of these modes is

$$k_{SPP}/k_0 = \sqrt{\frac{\varepsilon_1(\omega)\varepsilon_2(\omega)}{\varepsilon_1(\omega) + \varepsilon_2(\omega)}}, \tag{6.107}$$

where $\varepsilon_{1,2}(\omega)$ is the complex relative dielectric constant of medium 1, 2 at frequency $\omega = k_{\parallel}/c$. The field from the SPP evanescently penetrates a short distance, δ_2, into the metal mirror. Resistance in the metal leads

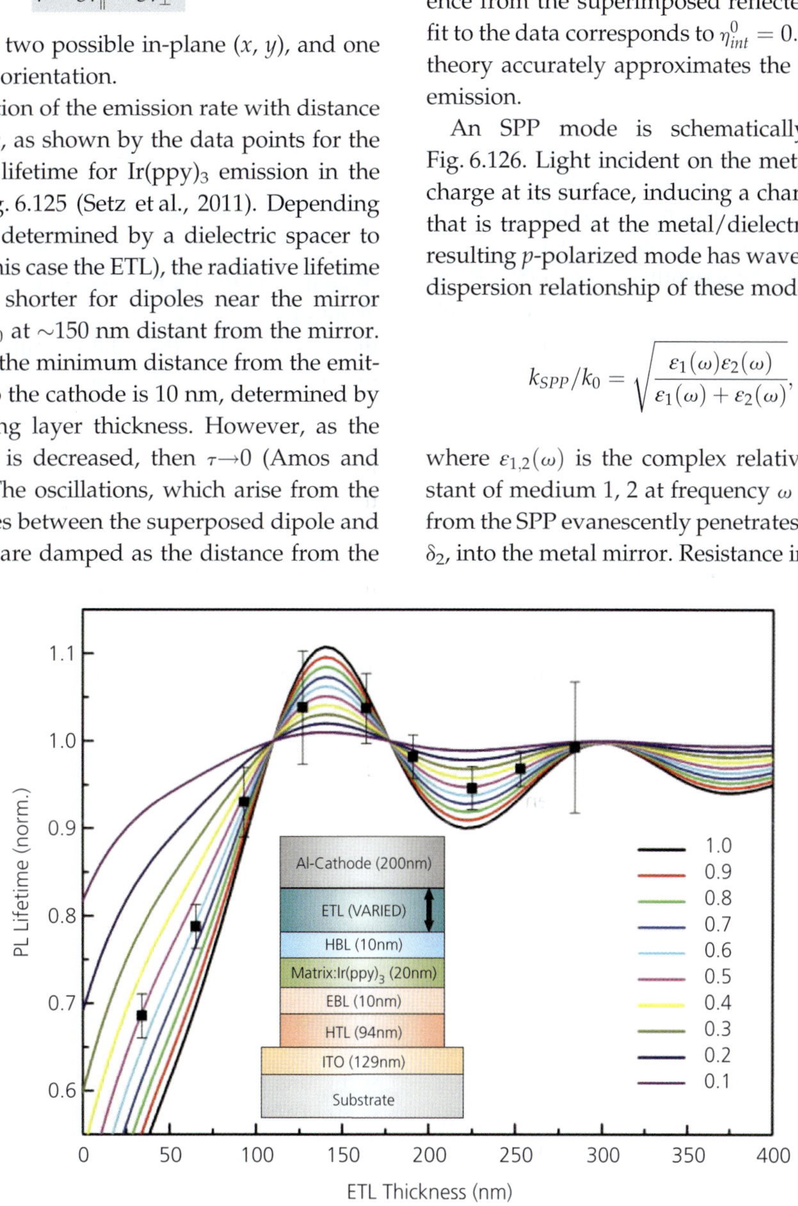

Figure 6.125 Oscillations in the lifetime of $Ir(ppy)_3$ emission calculated (lines) as a function of spacer (i.e. ETL) thickness between the emission layer and the Al cathode for the device structure shown in the inset. Measurements shown by squares. Each curve corresponds to a different internal efficiency, η_{int}^0, for $Ir(ppy)_3$ emission. The best fit to the optical model corresponds to $\eta_{int}^0 = 0.5$. After Setz et al. (2011) and Brütting et al. (2013).

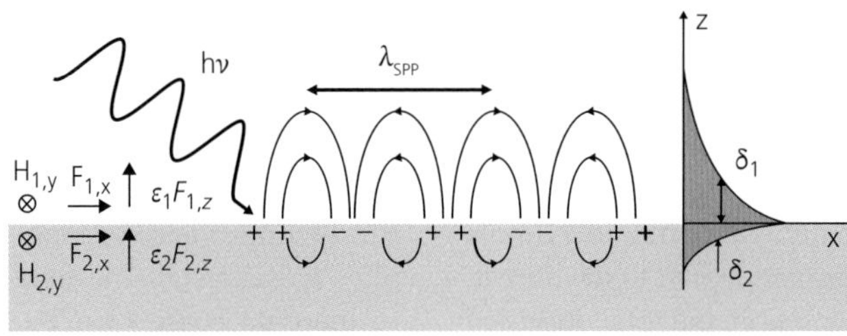

Figure 6.126 Schematic illustration of a surface plasmon polariton excited by dipole emission at energy $h\nu$. The SPP wavelength is λ_{SPP}, and its penetration depth is δ_2 into the metal mirror (medium 2) and δ_1 into the dielectric (medium 1). The permittivities of the two materials are shown, with $\varepsilon_1 = \varepsilon_0\varepsilon_{r1}$, for example.

to losses. As the distance from the radiative dipole to the metal decreases, the coupling increases, resulting in a significantly increased loss, as observed in Fig. 6.125. However, the field can penetrate into the metal layer if its thickness is $d \ll \delta_2$, and hence can be extracted from the opposite (unilluminated) metal surface.

With the solutions for the electric field in Eqs. 6.100–6.103, the Poynting vector, $\mathbf{S} = \mathbf{F} \times \mathbf{H}$ yields the energy flux, from which we obtain the total (integrated) power emitted by the dipole at wavelength λ, viz.

$$P_{tot}(\lambda) = \frac{\tau_0}{\tau} = \left[1 - \eta_{int}^0(\lambda)\right] + \eta_{int}^0(\lambda)\int_0^\infty P(k_{\|}, \lambda, z)dk_{\|}. \tag{6.108}$$

The first term in Eq. 6.108 is for transverse magnetic (TM, i.e. p-polarized) radiation emitted by dipoles oriented parallel to the substrate plane, and the second term includes both TM and transverse electric (TE, i.e. s-polarized) radiation emitted by perpendicular dipoles.

Thus, the power is equal the ratio of the rates for emission with and without the mirror. The rate within the cavity is $1/\tau = 1/\tau_r' + 1/\tau_{nr}$, where τ_r' is the modified lifetime in the absence of non-radiative relaxation at rate τ_{nr}, and $1/\tau_0 = 1/\tau_r + 1/\tau_{nr}$ is the analogous expression in the absence of the mirror.

If the emission spectrum of the dipoles is $S(\lambda)$, the total emitted power is found by integrating over λ:

$$P_{tot} = 1 + \eta_{int}^0\left\{\int_0^\infty\int_0^\infty S(\lambda)P(k_{\|}, \lambda, z)dk_{\|}d\lambda - 1\right\}. \tag{6.109}$$

More simply, the integral is the *Purcell factor*, *PF*, giving

$$P_{tot} = 1 + \eta_{int}^0(PF - 1). \tag{6.110}$$

This modifies the cavity lifetime: $1/\tau_r' = PF/\tau_r$, that is, the Purcell factor is the ratio of the radiative rate in a cavity to the natural lifetime of the emitter. Since the efficiency is given by $\eta_{int}^0 = \tau_{nr}/(\tau_r + \tau_{nr})$ with a similar expression for η_{int}, then Eqs. 6.108–6.110 yield the quantum efficiency of the dipole in the cavity:

$$\frac{\eta_{int}}{\eta_{int}^0} = F\frac{\tau}{\tau_0} = \frac{PF}{1 + \eta_{int}^0(PF - 1)}. \tag{6.111}$$

In analyzing the OLED spectrum, it is useful to decompose the power dissipated into its various modes, as well as the total output power as determined using Eq. 6.108. This can be accomplished by calculating the fraction of optical power emitted into each channel (i.e. the various waveguide and substrate modes) bounded by the allowed wavevectors, k_1 and k_2:

$$P_{mod} = \eta_{int}^0\int_0^\infty\int_{k_1}^2 S(\lambda)P(k_{\|}, \lambda, z)dk_{\|}d\lambda. \tag{6.112}$$

Then it is straightforward to determine the total external quantum efficiency from the microcavity via

$$\eta_{ext} = \eta_{int}\eta_{out} = \frac{P_{air}}{P_{tot}}, \tag{6.113}$$

where P_{air} corresponds to modes that escape the cavity within the domain $0 \le k_{\|} \le k_0$, where k_0 is the wavevector in free space.

The results of a calculation of the emitted power intensity vs. $k_{\|}/k$ as a function of ETL thickness (and hence the distance from the emitting molecules from the reflective metal cathode) for an OLED using this formalism is shown in Fig. 6.127 (Chance et al., 1978, Neyts, 1998, Celebi et al., 2007). The in-plane wavevector provides the power coupled separately into SPP, waveguide, substrate, and air (radiative) modes. Note that the actual power radiated into air modes is reduced by Fresnel reflections for s-polarized waves at the substrate (glass, with index $n_s = 1.5$)/air interface, given by

$$r_{sa}^{s} = \frac{n_s \cos\theta_s - n_a \cos\theta_a}{n_s \cos\theta_s + n_a \cos\theta_a}, \tag{6.114}$$

where the subscripts s and a refer to the substrate and air, respectively, with $n_a = 1$. For p-polarization:

$$r_{sa}^{p} = \frac{n_a \cos\theta_s - n_s \cos\theta_a}{n_a \cos\theta_s + n_s \cos\theta_a}. \tag{6.115}$$

The reflectivity of the substrate is $R_{sa}^{s,p} = |r_{sa}^{s,p}|^2$.

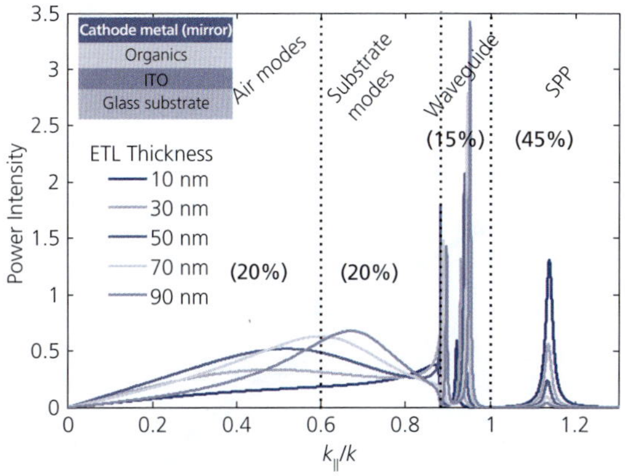

Figure 6.127 Calculated power at $\lambda = 510$ nm vs. in-plane wavevector for the structure shown in the inset. The refractive index of the organic layers is $n = 1.7$, ITO $n = 1.85$, and for the Al cathode, the complex index is $\tilde{n} = 0.9 + i6.9$. The detailed structure is glass substrate/ITO 70 nm/HTL 40 nm, EML 15 nm, and the ETL thickness is varied to space the dipole at different distances from the cathode. All emitting molecules are assumed to be at the EML/HTL interface.

As the spacer layer (i.e. the ETL) thickness between the emitter and the cathode increases, coupling to non-radiative SPP modes decreases. A much weaker modal dependence can also be found by varying the HTL and ITO anode thicknesses, although this does not significantly affect the excitation efficiency of SPPs. The waveguide, comprising the organic OLED stack and the ITO, supports at least two (TM, TE) modes. The number of modes depends on the total thickness of the organic layers, where more waveguide modes are expected in thick, stacked WOLEDs, for example.

The area under the various peaks is the integrated power within that particular mode. The broad SPP modes represents up to 45% of the power, whereas the very narrow waveguide modes trap only 15% of the power, with an additional 20% trapped within the substrate. As we show below, propagation in the substrate and waveguide modes is not lossy, and hence can be outcoupled if the in-plane propagating wave can be directed into out-of-plane radiative modes.

The calculated integrated powers of several WOLED optical modes as a function of ETL thickness (d_{ETL}) are decomposed in Fig. 6.128. The EML is separated into three, stacked R, G, and B emitting zones. The lines show the total emission from each zone, and the color bands show their average. We find that the air modes are approximately equal at 20% of the emission at the first and second cavity maxima at $d_{ETL} = 60$ nm and 220 nm. The substrate and waveguide mode

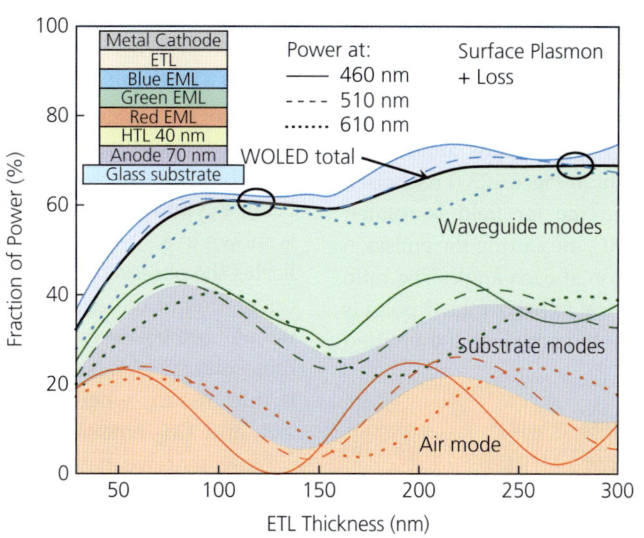

Figure 6.128 Power dissipation vs. ETL thickness for the WOLED structure in the inset. The lines explicitly show the power for three different wavelengths corresponding to the peaks in the blue, green and red spectra of the phosphor dopants, FIrpic, Irppy$_3$ and PQIr, respectively. The bands give the total output power that is the average of the combined WOLED spectrum in the ratio of B:G:R of 0.45:0.63:1 to yield $CRI = 0.91$. The black line is the total power coupled into the three lossless propagating modes. Optical constants for each layer are as in Fig. 6.127. The emitter molecules are placed in the center of each EML whose thicknesses are 10 nm. The ellipses note the ETL thicknesses where the intensities of all three wavelengths are equal, providing an undistorted white color.

contributions are small when $d_{ETL}\rightarrow 0$, since a large fraction of the light is coupled into a combination of SPP and lossy metal modes as the dipoles approach the cathode. These losses are smallest for the blue emitters located at the largest optical distance (defined by the wavelength, λ_0/n) from the mirror. As d_{ETL} increases, the waveguide (comprising the ITO, ETL, and the EMLs) can support a larger number of modes. Without outcoupling, an increasing amount of power, therefore, is trapped in the waveguide, as indicated by its increase at the expense of the other contributions. Finally, as the ETL thickness increases, coupling to SPPs and lossy metal modes decreases as predicted by CPS theory. In all cases the metal cathode contributes the largest loss, which is never less than 30% of the total, even at very large d_{ETL}. Also, such a large d_{ETL} may be impractical due to the series resistance that thick layers contribute, reducing the luminous power efficiency more rapidly than can be compensated by the increased external quantum efficiency.

Interestingly, there are at least two values of d_{ETL} where the total power from the three non-lossy modes is equal across the visible spectrum. These points at 120 nm and 275 nm are indicated by ellipses in Fig. 6.128. At these "isoenergetic" points, the emitted color is the same as that of the dipoles in a completely non-lossy, dielectric medium. It is at these thicknesses that the WOLED has a $CRI = 0.91$ with an emission that is native to the molecules themselves. Interestingly, good color balance occurs at points where the air mode is near its minimum. By moving away from the isoenergetic points, the ratio of R to G to B is changed, thereby changing the WOLED *CRI*, CIE coordinates, and *CCT*.

The air and substrate optical modes can be conveniently measured using a cylindrical lens attached to the substrate, as on the right in Fig. 6.123 (Frischeisen et al., 2010). A detector in the far field is rotated through angles from 0 to 90°, measuring the emission intensity and its polarization at each angle. The ratio of light emitted by the vertical to horizontal dipoles is then obtained from the polarization of the emitted radiation following:

$$\Theta = \frac{TM_{\parallel}}{TE_{\perp} + TM_{\perp} + TM_{\parallel}}, \tag{6.116}$$

leading to $\Theta = 0.33$ for random alignment and $\Theta = 0$ for all molecular TDMs oriented parallel to the substrate plane (i.e. dipole *emission* is in the perpendicular direction, cf. Eq. 6.105) (Schmidt et al., 2011). The emission from horizontally aligned TDMs relative to the substrate plane is thus decomposed into transverse electric ($TE_{\perp}$) and magnetic ($TM_{\perp}$) modes

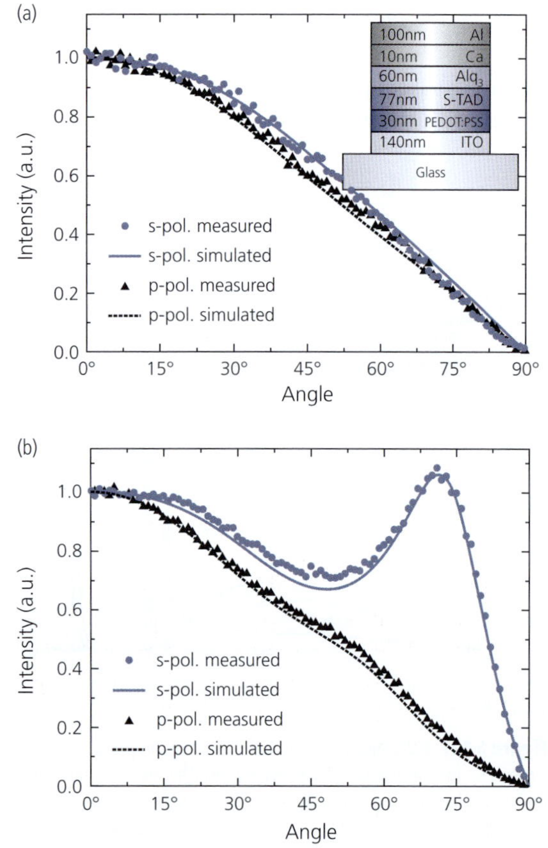

Figure 6.129 *s*- and *p*-polarized emission from the device in the inset (a) without and (b) with a cylindrical lens used for extracting substrate modes, as illustrated in Fig. 6.123. Data are represented by points, theory by lines (Brütting et al., 2013).

using a polarization filter, whereas the vertically aligned TDM emits into the ($TM_{\parallel}$) mode.

A fit of the foregoing optical analysis to the angular emission intensity is shown in Fig. 6.129a for the OLED in the inset. The emission profile obtained using a cylindrical lens attached to the glass substrate is shown in 6.129b. The optical model accurately replicates the observed emission pattern. In this case the orientation of the Alq_3 dipoles is isotropic, although we will see below that this is not the case for many molecular species. Note that the *s*-polarized emission is due to $TE_{\perp}$, whereas *p*-polarization collects both $TM_{\parallel}$ and $TM_{\perp}$ emission.

Now that we have developed an accurate and predictive optical model based on microcavity effects in OLEDs, we are prepared to discuss strategies used in efficiently outcoupling light from substrate and waveguide modes, as well as means for minimizing or even eliminating coupling to lossy metal contact modes. In keeping with our guiding principles for effective and practical outcoupling schemes, we will limit the following discussion to those techniques that are (i) easily and cost-effectively applied over large areas, (ii) are largely wavelength- and angle-independent, and (iii) minimally intrude into the OLED structure itself. The principal methods are roughly categorized into those that outcouple substrate, waveguide and SPP modes, although several methods can extract more than a single type of emission. The following sections are divided by the primary extraction mechanism of a given approach.

6.6.2 Substrate mode outcoupling

Efficient substrate light outcoupling methods seek to reduce TIR at the substrate-air interface. These modes account for 20% of the power trapped in the substrate of bottom emitting devices such as in Figs. 6.127 and 6.128. One simple extraction method is to roughen the emitting surface to create a region whose index is intermediate between the substrate (typically glass or plastic) and air. Roughening on a scale of the emitting wavelength provides an anti-reflection coating as well as a layer that isotropically scatters internal reflections, thereby giving reflections from the metal cathode additional opportunities to be incident at this interface within the TIR cone. This angle is defined by Snell's law as $\sin\theta_{TIR} = 1/n_s$ where θ_{TIR} is the half angle for TIR. Reflections from the cathode result in ~5–10% loss on each bounce. If the textured surface is far from the emission zone (due to an optically thick substrate), it creates a diffuse image of a pixel, although this would not necessarily be a disadvantage to a white illumination fixture. Roughening results in ~20% increase in external efficiency (Zhou et al., 2011), representing only a small fraction of the total power in the substrate modes.

The scattering layer can be placed immediately beneath the transparent anode with significantly improved outcoupling compared with substrate back surface roughening. Methods of surface roughening include creating voids beneath the transparent anode (Fig. 6.130a) or embedding high refractive index scattering inhomogeneities in a low index matrix, also beneath the anode (Fig. 6.130b). The rough surface does not support waveguiding, and hence scatters light into the substrate. The texture in Fig. 6.130a can be produced by sandblasting, etching or grinding. The uneven surface shown is coated with a 50 nm thick layer of high index (n = 2.1) Ta_2O_5 that spreads the guided modes into the roughened regions. The surface is subsequently planarized with a 200 nm thick layer of the low index polymer to prevent the surface irregularities from penetrating into the OLED active region. It is likely that internally reflected emission also scatters

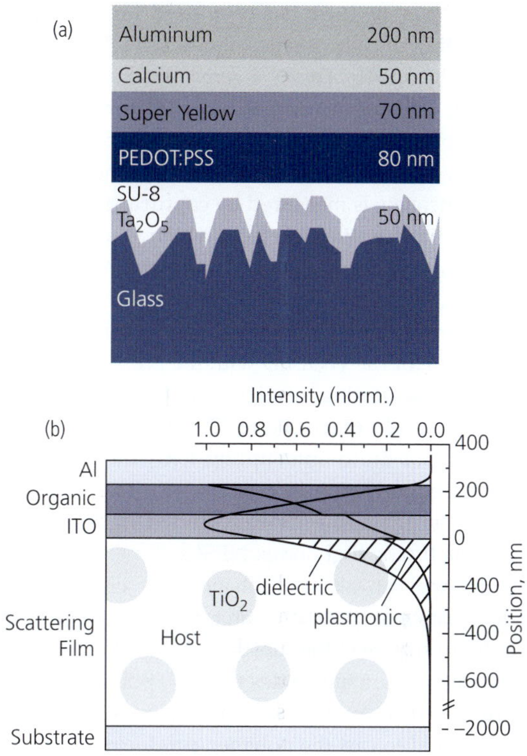

Figure 6.130 (a) Random roughness of a high index scattering layer below a transparent anode (Riedel et al., 2010). (b) Embedded high index TiO_2 nanoparticle scattering layer in the same location. The lines represent dielectric and plasmonic mode intensity distributions (Chang et al., 2013a)

off of these features, whose scale is on the order of the wavelength of light (e.g. sandblasting results in a root mean square (rms) roughness of 200 nm, and a peak-to valley height of 900 nm; Riedel et al., 2010), giving the trapped radiation multiple opportunities to couple into air modes. The best improvement of outcoupling of $< 40\%$ was achieved using sandblasting. This falls short of extracting all of the substrate modes, which would approximately double the light output (see Fig. 6.128).

Significant improvements in outcoupling are achieved dispersions of TiO_2 ($n = 2.2$) nanoparticles (with mean diameters of 240 nm) in a low index ($n = 1.5$) polymer positioned immediately below the transparent anode, as shown in Fig. 6.130b (Chang et al., 2013a). This dispersion has an effective index that approaches the weighted average of the volume concentrations of the spheres and matrix. As in the case of the roughened surfaces, the nanoparticle scattering layer (NPSL) primarily re-scatters internally reflected light, while also redirecting substrate modes to $\theta < \theta_{TIR}$. The power distribution of these two modes is shown in Fig. 6.130b.

The NPSL results in haze and image blurring (see Fig. 6.131a), making this suitable for lighting but not for displays. The layer has been used for increasing outcoupling of a three-color stacked WOLED structure in Fig. 6.131b. This is a semitransparent device since the cathode comprises only 20 nm thick Ag layer capped with 50 nm thick layer of Spiro-TAD that acts as an anti-reflection coating. The total external efficiency is measured by capturing light from both top band bottom surfaces which is in the ratio: $L_{bottom}/L_{top} = 1.5{:}1$. Devices without the NPSL have $L_{bottom}/L_{top} = 1.9{:}1$. The WOLED efficiency was improved from 22% (32 lm/W) for the control, to 33% (46 lm/W) for the WOLED with the NPSL. This suggests that scattering results in a 50% increase in substrate mode outcoupling. By applying a hemispherical substrate outcoupling lens, the efficiency is increased further to 46% (62 lm/W). This indicates incomplete scattering of waveguided light into the substrate by the high index NPSL.

Scattering can also be enhanced by creating low index voids in a film that is attached onto the bottom (emitting) surface of the substrate (Koh et al., 2015). To create the voids, a high index ($n > 1.7$) polyimide layer is immersed in an *antisolvent* that is miscible in the solvent used to cast the polymer, producing a porous structure that scatters the substrate modes as shown in Fig. 6.132. The ~2 μm thick film contains a high density of cavities a few microns in diameter. The films are slightly hazy, similar to the high index sub-anode film of Fig. 6.131. This scheme increases outcoupling by nearly 80%, which corresponds to nearly half of the trapped emission compared to a similar device on a flat glass substrate. Furthermore, the emission is nearly perfectly Lambertian.

We have seen that the light trapped in substrate modes is almost entirely extracted using a large spherical or cylindrical lens (Madigan et al., 2000). The diameter of the lens, however, must be even larger than the OLED itself, which adds considerable thickness to a device. For displays, substrate thicknesses of only 0.3–0.5 mm are desirable, thus severely limiting the utility of this approach. Furthermore, the single lens must be centered over the emitting area, extending 30–45° beyond the edge of the pixel to fully capture all of the light. This solution can be simplified for lighting by replacing the lens with a large trapezoidal outcoupling block (D'Andrade and Brown, 2006, Horst, 2007), or with trapezoids whose minor surface is equal to the pixel size itself (Gu et al., 1997b). However, none of these methods are practical given their complexity (and hence high cost), bulkiness or weight.

Microlens arrays (MLAs) have proven to be a simple and effective means for outcoupling a majority of the substrate modes (Möller and Forrest, 2001, Sun and Forrest, 2006, 2008a). The microlenses themselves are much smaller than the pixel size, and are placed on the far (emitting) surface of the substrate. They can also be attached to the surface of top emitting OLEDs with equal effect. The enhanced outcoupling results from presenting a curved surface whose index is approximately matched to the substrate, as shown in Fig. 6.133a. Emitted light incident on the microlens

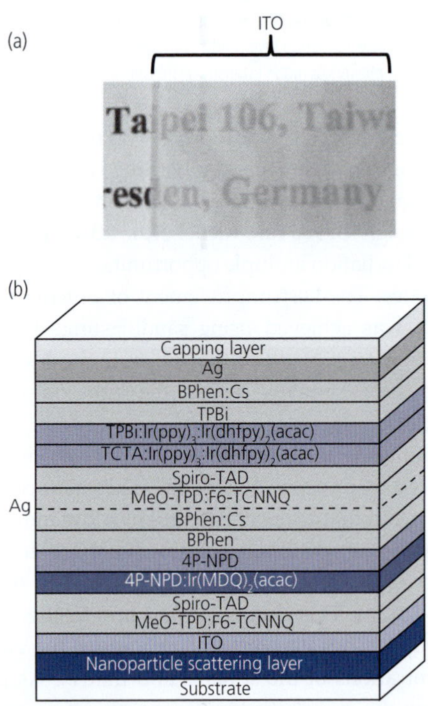

Figure 6.131 (a) Translucent nanoparticle scattering layer on a glass substrate. The film scatters light from the printed sheet, leading to some image blurring. (b) Transparent stacked WOLED with a NPSL used to increase light outcoupling. The phosphors used for light emission are red = $Ir(MDQ)_2(acac)$, green = blends of $Ir(ppy)_3$ and $Ir(dhfpy)_2(acac)$, and blue emission uses the fluorophore 4P-NPD (Chang et al., 2013a).

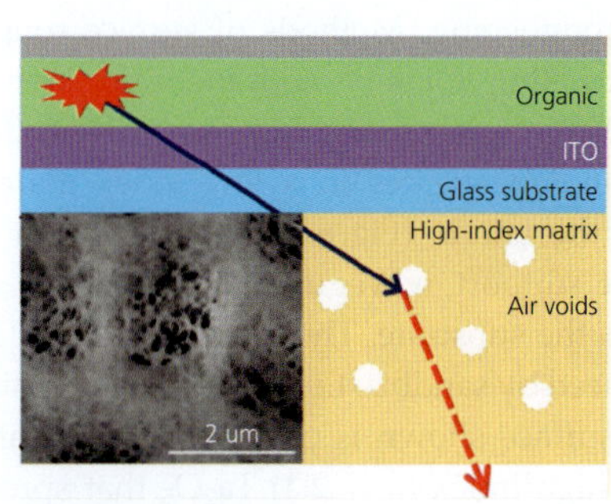

Figure 6.132 Schematic of scattering from voids in a film attached to the emitting glass substrate surface of an OLED. The micrograph at the left shows the blister-like cavities in the high index polyimide film (Koh et al., 2015).

hemisphere at angles $> \theta_{TIR}$ on a flat surface finds itself at $< \theta_{TIR}$ at the curved lens/air interface, and hence can escape. Additionally, light emitted at $\theta >> \theta_{TIR}$ may be back scattered into the device, is once more reflected by the metal contact, and then can exit the substrate as shown in Fig. 6.133a. This path leads to some loss on reflection, hence the MLAs are $< 100\%$ efficient.

An example array consisting of hexagonally close packed, 10 μm diameter polydimethylsiloxane (PDMS) hemispheres is shown in Fig. 6.133b. A disadvantage of MLAs is that they slightly blur the image. Figure 6.133c shows the image of a printed sheet partially covered by a MLA on a 0.35 mm thick glass substrate. Image blurring is apparent for 5 pt. type and below, although the lenses used to produce the images had a muffin-tin rather than spherical shape that has superior performance (Möller and Forrest, 2001).

An attractive feature of MLAs is the ease (and hence low cost) with which they can be produced in large sheets in batch or roll-to-roll processes. A reusable mold is fabricated as shown in Fig. 6.134a on either a plate or a roller. (1) A photoresist mask (hatched layer in the figure) with small ~1 μm diameter circular openings on a hexagonal lattice with a 10 μm pitch is applied to the surface of a glass sheet. (2) The glass is isotropically etched using an HF solution to form hemispherical openings that touch each other on their peripheries. (3) The mold is then pressed into a soft PDMS film which is subsequently thermally cured, and (4) the mold is removed, leaving the MLA film for subsequent attachment to the OLED substrate surface (Sun and Forrest, 2006).

The emission spectrum from the fluorescent/phosphorescent WOLED in Fig. 6.100 (Sun et al., 2006) is nearly independent of viewing angle, as shown in Fig. 6.134b, and the power distribution in the far field is approximately Lambertian (see inset), although the emission at wide viewing angles is enhanced compared with a conventional WOLED on a flat glass substrate. It is found that the PDMS arrays result in a 1.7 times enhancement in outcoupling efficiency compared to a flat glass substrate (Sun and Forrest, 2008a). This is close to the calculated

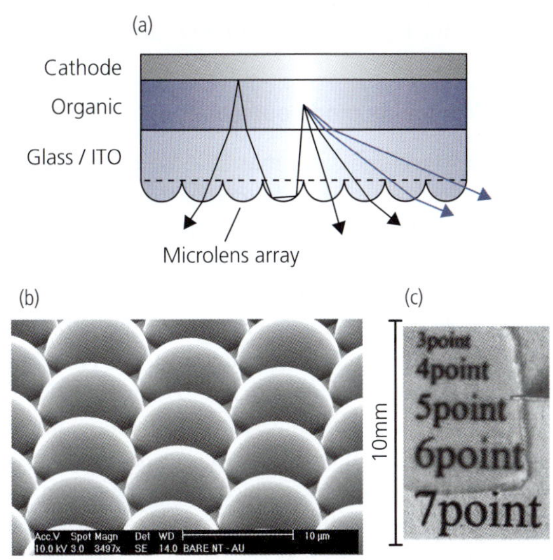

Figure 6.133 (a) Principle of operation of the microlens array. Blue rays correspond to substrate angles $\theta > \theta_{TIR}$, whereas black rays are at $\theta < \theta_{TIR}$ (Brütting et al., 2013). (b) Micrograph of a hexagonal array of 10 μm diameter hemispherical PDMS microlenses (Sun and Forrest, 2008a). (c) A film of 10 μm diameter muffin-tin shaped microlenses on top of a printed sheet showing minor image blurring (Möller and Forrest, 2001).

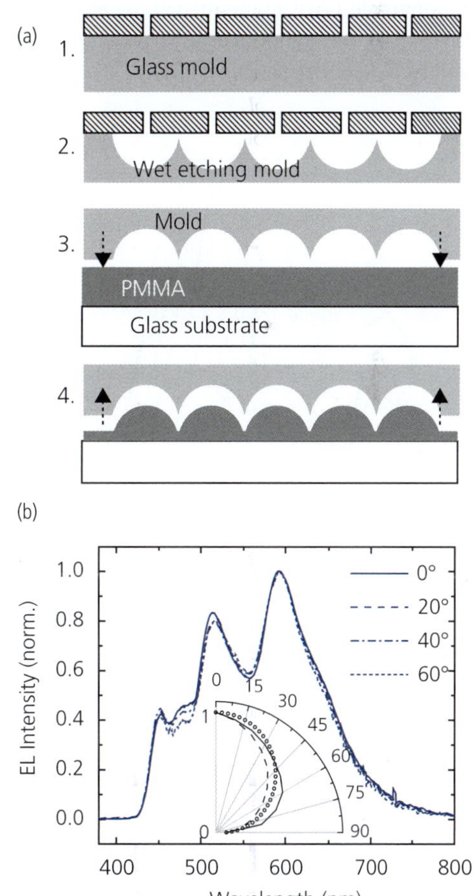

Figure 6.134 (a) Process for forming polymer microlens arrays. (b) Emission spectrum from a fluorescent/phosphorescent WOLED as a function of viewing angle. Inset: Polar emission pattern for a device with (solid line) and without (dashed line) a microlens array (Sun and Forrest, 2006).

maximum enhancement of ~1.8 times for this lens and substrate combination.

MLAs attached to the surface of a top emitting OLED significantly improve outcoupling while eliminating image blurring (Yang et al., 2007, Thomschke et al., 2012). One feature of the top surface MLA is that it acts as an effective light diffusing medium, thus reducing microcavity effects arising from the metal contacts on both top and bottom OLED surfaces. Optical outcoupling and light diffusion has been demonstrated with a tandem WOLED comprising a thick bottom Ag contact, and a semitransparent cathode using a 15 nm thick Ag film. The bottom element in the stack comprised a 10 nm thick yellow emitting EML with 8 wt% $Ir(ppy)_3$ in TCTA. This combination assists energy transfer to a yellow 1 wt% $Ir(dhfpy)_2$(acac) phosphor. The top blue/red 4P-NPD/Ir $(MDQ)_2$(acac) EML was separated from the bottom device with a CGL comprising an ultrathin (0.5 nm) Ag nanoparticle layer embedded in n- and p-doped conducting layers. Without the MLA, the spectrum showed two distinct peaks (see Fig. 6.135a). Since the

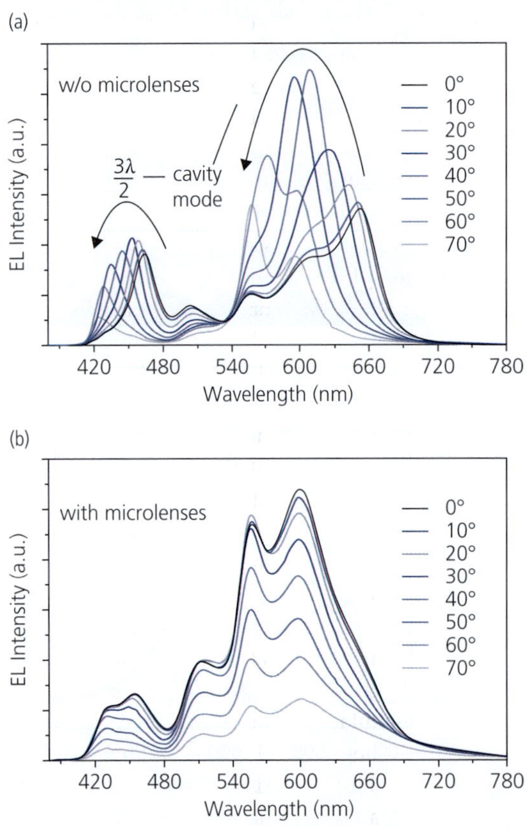

Figure 6.135 The spectra and viewing angle dependence of a tandem, blue/red + yellow top emitting WOLED (a) without and (b) with a high index polymer microlens array attached to the top emitting surface (Thomschke et al., 2012).

blue and red lumophores are co-doped into a single layer, the blue emission has a peak at the $m = 1$ (corresponding to $3\lambda/2$) optical order centered at 460 nm (normal incidence). The red is at the $m = 0$ order centered at $\lambda = 660$ nm. The yellow emitting element also is at the $m = 0$ mode at approximately 600 nm. The positions of the two EMLs within the cavity are adjusted by the thicknesses of the doped HTL and ETL in the blue/red element. There is a strong microcavity resonance associated with this device, as apparent from the spectral dependence on viewing angle in Fig. 6.135a (Thomschke et al., 2012).

A high index ($n = 1.71$) MLA with a 10 μm pitch was attached via lamination to the top contact with an intervening 200 nm thick NPD layer with approximately the same index of refraction as the MLA. Two effects are observed in Fig. 6.135b. The first is the almost total absence of spectral shifts with viewing angle; the microcavity effects have essentially been eliminated. Furthermore, the clear articulation of the cavity modes at $3\lambda/2$ and λ disappear due to $m = 0$ and 1 modal mixing. The net effect is a high $CRI = 86$ at $CIE = (0.516, 0.420)$ for a device with an external efficiency of 22.8% and $\eta_P = 25.2$ lm/W. This is nearly a 1.4 times increase in efficiency compared to a device without the microlenses. This demonstration provides evidence that MLAs are effective for extracting both waveguide and substrate modes, whether they are placed on the top or substrate OLED surface.

6.6.3 Waveguide mode outcoupling

Modal analysis indicates that 20–30% of the OLED emission excites waveguide modes supported in the organic and ITO layers. Due to the small but non-negligible loss in the ITO with complex index of refraction of $\tilde{n} = 1.95 + i0.0035$ (Chang et al., 2013a), these modes must be outcoupled into the substrate to limit attenuation. Thick organic and/or ITO layers can support high order TM and TE modes. Thus, waveguide mode trapping is more severe in multilayer WOLEDs than in thinner, monochrome pixels.

Unlike substrate modes, outcoupling of waveguide modes requires structures that are in close proximity to, or even within the OLED active layers themselves. Many solutions are complex, and can reduce device performance and manufacturing yield or increase fabrication costs. It is not possible to directly measure the contribution of waveguide modes to the total power since they propagate parallel to the substrate plane. Indeed, SPPs also propagate in the in-plane direction, and may at least partially be outcoupled using the same strategies as those employed to extract guided

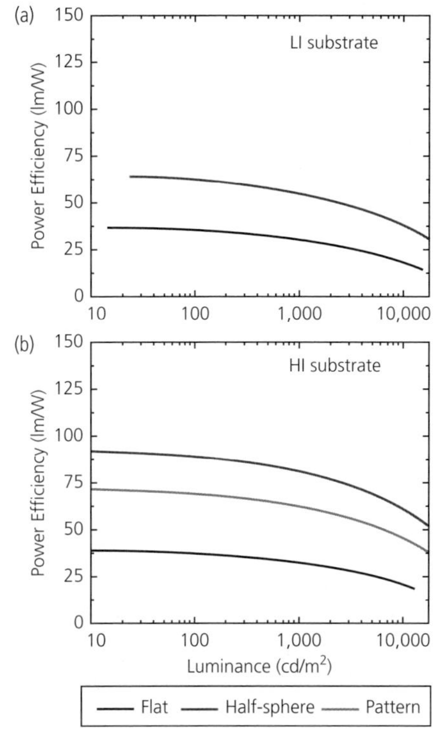

Figure 6.136 Power efficiency vs. luminance for the WOLED in Fig. 6.94 on a (a) low index (LI) and (b) high index (HI) glass substrate. The black line is with no substrate mode outcoupling, dark blue is with index-matching fluid and a hemispherical lens, and light blue is with an array of pyramidal structures cut into the emitting substrate surface (Reineke et al., 2009).

Reprinted by permission from Springer Nature, *Nature*, 459, 234, Reineke, S., Lindner, F., Schwartz, G., Seidler, N., Walzer, K., Lüssem, B. & Leo, K., White organic light-emitting diodes with fluorescent tube efficiency. Copyright 2009.

modes. However, the SPP mode attenuation length is far shorter due to the large contribution of the imaginary part of the complex index of refraction by the metal layer.

The outcoupling of guided modes requires scattering into the substrate, followed by a second substrate light extraction scheme described in the previous section. A straightforward means for outcoupling guided modes is to use high index of refraction substrates instead of glass. If the substrate index is greater than that of the organic layers (typically $n \sim 1.7$), the guided modes evanescently couple into the substrate (Nakamura et al., 2005, Gaertner and Greiner, 2008, Mladenovski et al., 2009), as shown for WOLEDs using high index MLAs in Fig. 6.135. The emitting layer of the OLED should be sufficiently far from the cathode to avoid excitation of lossy or SPP modes in the metal contact. Unfortunately, high index glasses are generally more costly than conventional substrate glass.

Enhanced waveguide mode outcoupling using high index N-LAF 21 Schott glass substrates ($n = 1.78$ compared to $n \simeq 1.7$ for the organic active waveguide layers) is observed for the WOLED structure in Fig. 6.94, with results in Fig. 6.136. Absent substrate mode outcoupling enhancements (black lines), the peak power efficiency is approximately the same for the high and low index substrates, since in the former case, the increased outcoupling of waveguide modes is offset by the increased TIR at the substrate/air interface. Using *index-matching fluid* (IMF) between the substrate and a spherical lens extracts all of the substrate modes. Using this combination of extraction schemes, the power efficiency at 1000 cd/m^2 (dark blue line), increases by greater than two, to 81 lm/W for the high index substrate, but only to 55 lm/W for the conventional glass substrate. The high index substrate was further modified by cutting an array of large pyramids (0.55 mm pitch, 90° angle) into its emitting surface, which has effects similar to using microlenses. That is, the angled surfaces effectively increase the escape cone of light trapped by TIR on flat surfaces (Nakamura et al., 2005, Gaertner and Greiner, 2008). The enhancement leads to 63 lm/W, which is a significant improvement over that obtained for the low index substrate with complete substrate mode outcoupling. The spectra of both devices are nearly identical, since extracting waveguide modes does not change the device optical properties as long as the high index medium is non-absorbing.

Gratings or two-dimensional (2D) photonic band gap structures placed within or near to the OLED active region offer additional opportunities for waveguide mode outcoupling (Lee et al., 2003, Do et al., 2004, Ishihara et al., 2007, Zhou et al., 2014a). A grating placed near the emitting dipole scatters the propagating mode at an angle that is a function of wavelength. While a grating can efficiently scatter guided modes into the substrate, its energy dispersive nature requires that the outcoupled radiation pass through a diffuser to remove chromatic and angular dependences. Microlenses, roughened surfaces, or multilayer coatings can all serve this purpose.

Gratings comprise wavelength-scale features patterned over large substrate areas, resulting in potentially high manufacturing costs. Patterning can use nanoimprint lithography (Ishihara et al., 2007, Zhou et al., 2014a, 2014b) or, as in Fig. 6.137a, interference lithography (Matterson et al., 2001, Bocksrocker et al., 2012). The latter entails the interference of two beams split from the same laser source, and projected onto the substrate to expose a layer of photoresist (Fig. 6.137b) with a grating pattern of the appropriate period. Next, a high index layer (e.g. TiO_2 with $n = 2.2$) is evaporated onto the photoresist surface (Fig. 6.137c), which is subsequently lifted off to leave behind the high index grating (Fig. 6.137d and e). Since the grating lies in only one direction, at most it can only outcouple 50% of the light emitted by vertically aligned dipoles that radiate isotropically within the plane. This problem is eliminated using 2D gratings, or photonic crystals that are produced by rotating the substrate by 90° and repeating the exposure and development steps (Lee et al., 2003, Do et al., 2004, Ishihara et al., 2007).

The grating is planarized with PEDOT:PSS and then the WOLED is completed by depositing the organic layers and cathode contact (Fig. 6.137f). Figure 6.137g shows a cross-section of the somewhat nonplanar device and grating. This minor surface ripple is actually helpful in outcoupling light, since $k_{||}$ waveguide modes scatter at each encounter with the undulations (Koo et al., 2010). In addition, the ripples reduce the efficiency of coupling to SPP modes, and can scatter them into the waveguide and substrate modes (Hobson et al., 2002).

Modes that are coupled into the substrate either escape or are reflected back to the grating at the substrate/air interface. The backward travelling wave modes can recouple into the waveguide. The depth of the grating determines its scattering efficiency. In the example of Fig. 6.137, it was found that the 15 nm deep grating was less likely to recouple back-scattered light than one with 35 nm depth. To eliminate chromatic and angular dispersion and minimize recoupling light into the waveguide, an MLA can be deposited on the substrate surface. The combination of microlenses formed by thermally reflowing an array of polymer pillars and a 15 nm deep grating resulted in a factor of four improvement in outcoupling (104% enhancement from the gratings and 94% from the microlens diffuser) compared to an identical WOLED on a flat glass substrate. Furthermore, the emission was found to be approximately Lambertian across the visible spectrum when the grating was paired with the lens array (Bocksrocker et al., 2012).

Random wavelength-scale buckling of a film deposited onto the flat substrate surface that provides a rippled surface on which the OLED layers are deposited can be used to avoid high resolution grating patterning over large substrate areas (Koo et al., 2010, Lee et al., 2014). The advantages of such corrugations are two-fold: they do not require photolithography over larger areas, and the nanostructures are isotropic, which means that they are equally effective in outcoupling light propagating in either the x or y in-plane directions. The buckling in Fig. 6.138 is produced by

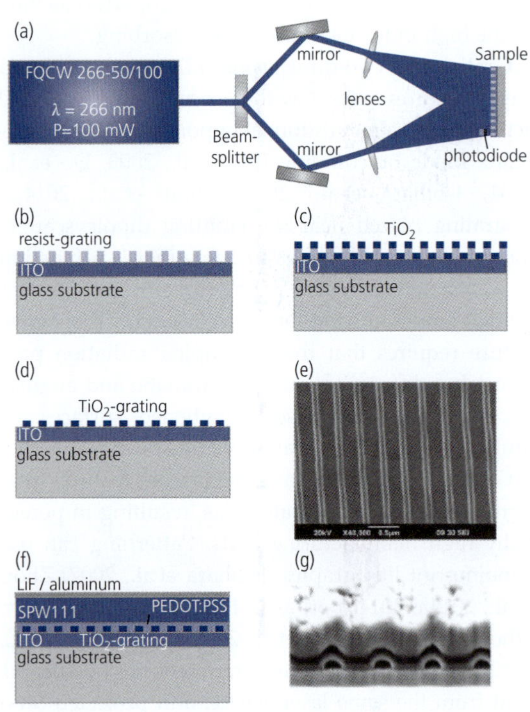

Figure 6.137 Fabrication steps of a TiO_2 grating within the anode contact of a WOLED for extracting waveguide modes. See text for a description of the procedure (Bocksrocker et al., 2012).

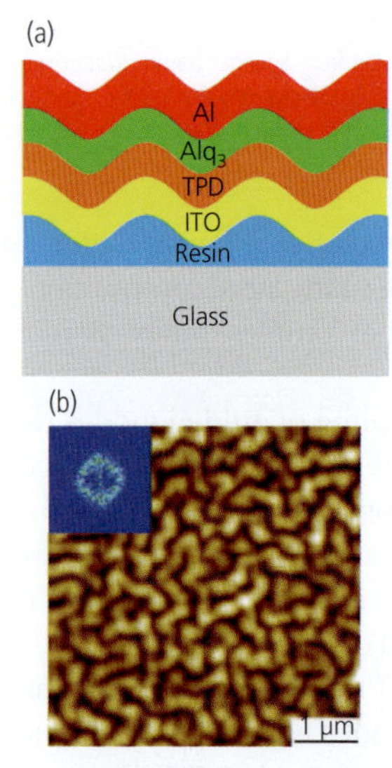

Figure 6.138 (a) Schematic cross-section of an OLED fabricated on the surface of a buckled, UV curable resin. (b) Atomic force micrograph of a 5 µm × 5 µm area of a buckled surface. The inset shows a fast Fourier transform of the patterns indicating an effective buckling period of ~400 nm. Layer thicknesses: Al = 150 nm; Alq_3 = 30 nm; TPD = 40 nm; ITO = 120 nm (Koo et al., 2010).

depositing a 10 nm thick Al film on a PDMS layer heated to 100°C. After deposition, the bilayer cools. The buckles are formed to relieve strain induced by the differences in thermal expansion between PDMS and Al. The depth of the corrugations is $D \approx \Lambda\sigma^{1/2}$, where σ is the compressive stress exerted by the metal on the polymer, and Λ is the periodicity of the features. A decrease in Λ results in a proportionate decrease in the corrugation depth, and consequently the outcoupling efficiency is also decreased. The depth is increased by making successive, buckled film masters. The second master is created by spinning a layer of PDMS onto the original Al/PDMS bilayer, and repeating the process to obtain single-, double- and triple-cast masters. To form the devices in Figure 6.138a, a UV-curable resin is spun on to the glass substrate, and then the PDMS master is pressed into its surface, thereby transferring the buckled pattern. This is followed by deposition of the OLED layers.

This process results in D from 40 to 70 nm, leading to a characteristic buckling length of $\Lambda = 400$ nm. Outcoupling enhancements of ~2–4 times (with the larger enhancements at longer wavelengths) compared to a conventional, Alq_3 device are observed. The ripples outcouple waveguide as well as a fraction of the SPP modes, due to the rapid change in waveguide and contact angles relative to the substrate plane. A disadvantage of this scheme is that it disturbs the planar morphology of the OLED structure. As a result, the current at a given voltage is increased compared with a planar device due to local thinning of the active region (Koo et al., 2010).

A method for waveguide mode outcoupling that does not require high resolution lithography, and hence is wavelength and viewing angle independent is the use of *low index grids* (LIGs) in the OLED active region, as illustrated in Fig. 6.139a (Sun and Forrest, 2008a). A region with an index lower than the organic layers is placed within the active region, thereby scattering light into the substrate modes. The grid structure is 2D, leading to outcoupling of all in-plane modes, independent of propagation direction. The primary requirement for the grid periodicity is that the low index regions are sufficiently closely spaced to avoid excessive waveguide mode attenuation, but are far enough apart to prevent chromatic dispersion. Beyond this, there are no conditions on grid shape. Since the low index dielectric region is placed within the OLED active layers, the regions over the grid lines are non-emissive. This reduces the total luminance but not the luminescent efficiency, since both the electrically active current and light emitting device areas are similarly reduced. Typical grid dimensions are a dielectric region width of $w_{LIG} = 1$ μm and a spacing of $w_{org} = 6$ μm, requiring an increase in pixel size of ~25% to achieve the same luminance as a device lacking the grid.

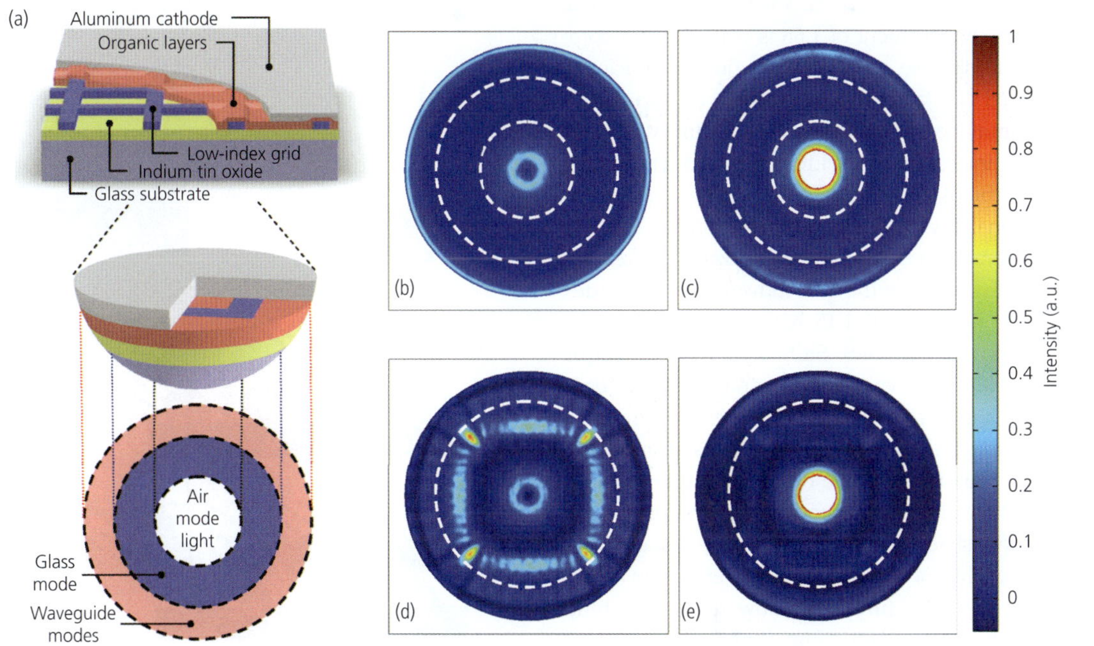

Figure 6.139 Low index grid (LIG) outcoupling structure. (a) Schematic of the device as well as the half-sphere used for calculating the total output power. (b)–(e) Simulated power intensities through the curved surface of the simulation domain for vertical and horizontal emitting dipole orientations, with and without the LIG. (b) Vertical, and (c) horizontal dipole, no LIG. (d) vertical and (e) horizontal dipole with LIG. In (c) and (e) the full range of values is not shown. The vast majority of light is in the white center of the disk. Dashed lines show the azimuthal angles of $\theta = 30°$ and 60° (Slootsky and Forrest, 2009).

Full-wave modal analysis of the guided light shows the effectiveness of a silica grid ($n_{GRID} = 1.45$) in coupling vertical dipole emission into the azimuthal direction. Polar plots of the emission intensity calculated for vertically and horizontally aligned dipoles in a conventional OLED are shown in Fig. 6.139b and c, respectively (Slootsky and Forrest, 2009). Emission from vertical dipoles is weak, with a considerable amount of equatorial propagation along the waveguide longitude at an angle of $\theta = 90°$. Emission from horizontal dipoles, on the other hand, is almost completely along the azimuth, as expected. The LIG placed within the organic ($n = 1.75$) and above the ITO ($n = 1.8$) guiding layers efficiently scatters the vertical dipole emission from the first grid cell (Fig. 6.139d) while leaving the horizontal dipole emission largely unperturbed (Fig. 6.139e). The dielectric discontinuities scatter waveguide light into the substrate with such high efficiency that light scattered from second nearest neighbor cells contribute only negligibly to the total emission intensity.

The LIG concept was demonstrated using the fluorescent/phosphorescent WOLED structure in Fig. 6.100 along with a 100 nm thick silica LIG with dimensions of $w_{LIG} = 1$ µm and $w_{org} = 6$ µm. The LIG device exhibits an enhancement of 1.32 in outcoupling compared to a conventional device, which exceeds that predicted from modal calculations since the SPP modes are also partially scattered by the uneven waveguiding region. Using a MLA with the conventional device, an enhancement of 1.68 is realized, and combining both of these coupling strategies, the efficiency is increased by a factor of 2.3 to achieve $\eta_{ext} = 34\%$ and a peak luminance power efficiency of 68 lm/W. Due to the large dimensions of the LIG, the broad white emission spectrum is unchanged from that of the analogous conventional device (Sun and Forrest, 2008a).

Considerable benefits can be realized by reducing the refractive index of the LIG. Indeed, very low index LIGs with $n_{LIG} = 1.10$–1.15 have been achieved by depositing 100 nm thick SiO_2 at a highly oblique angle (80–85°) relative to the substrate plane. *Oblique angle* or *glancing incidence deposition* results in nucleation of the film at random positions around the substrate. Due to the angle of the substrate plane to the direction of the evaporated flux, the nucleation centers cast shadows on regions more distant from the source. This increases the rate of deposition on the islands which then further shadow the area behind them, leading to a porous film similar to an aerogel (Xi et al., 2005, Barranco et al., 2016).

The oblique angle deposition process used in optical coatings for both OLEDs and OPVs can be understood from the illustration in Fig. 6.140. A point source (Fig. 6.140a) results in more defined shadowing, and hence narrower pillars than an extended source due to the range of arrival angles between α_1 and α_2 in Fig. 6.140b. The pillar tilt angle, β, is determined by the angle of the substrate to the source, θ, and α. The index of the resulting film equals the volume average of the indices of the deposit and the voids. Results of the e-beam deposition of low index SiO_2 at $\alpha = 80$–$85°$ and $\theta = 0$ forming the LIG is shown in Fig. 6.141. The top image (Fig. 6.141a) shows a highly porous film, comprised of densely packed pillars of nearly the same diameter. The side view in Fig. 6.141b shows that the pillars have a kink midway along their length. This results since the substrate was rotated around an angle of $\phi = 180°$ after the first ~50 nm of SiO_2 was deposited. The index of refraction of the film was $n = 1.10$–1.15. Using porous grids with the same dimensions as the device in Fig. 6.139, outcoupling enhancements of approximately a factor of

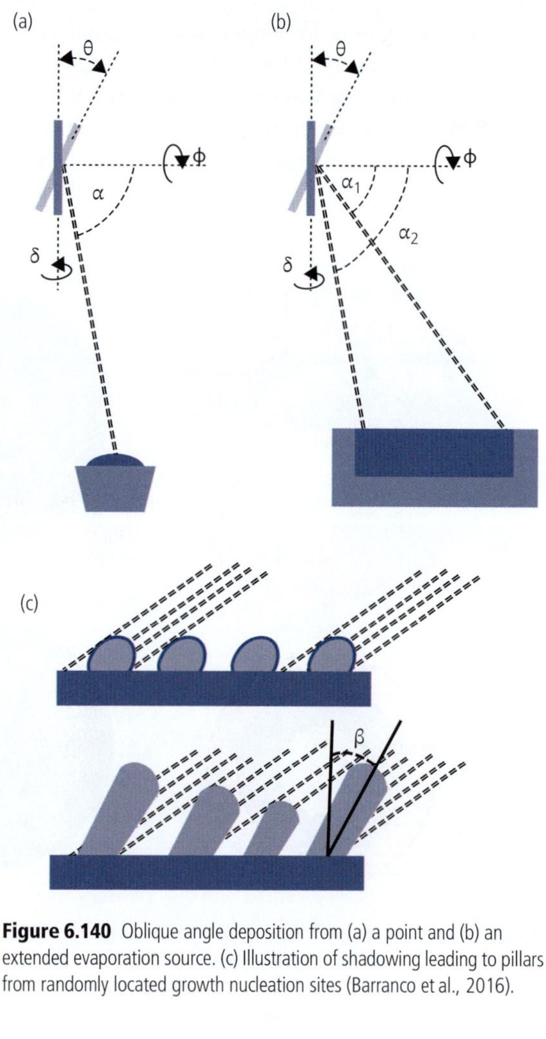

Figure 6.140 Oblique angle deposition from (a) a point and (b) an extended evaporation source. (c) Illustration of shadowing leading to pillars from randomly located growth nucleation sites (Barranco et al., 2016).

two compared to conventional devices on flat substrates were observed (Slootsky and Forrest, 2010).

The structure of Fig. 6.139 penetrates the device active region, which can lead to losses in manufacturing yield due to shorts. Alternatively, the grid can be located beneath the ITO anode and the OLED layers, as shown in Fig. 6.142a. The grid can have either a higher or lower index than the medium in which it is embedded, since light is scattered as a result of the index discontinuity that it encounters as it propagates in-plane (Qu et al., 2015). The evanescent tail of the guided modes penetrates beyond the ITO into the grid region in Fig. 6.142b, and then is scattered into substrate and air modes. With an index difference between the "host" and the grid of 2.2, the enhancement in outcoupled radiation can be as high as 1.4, decreasing with the index contrast. This 2D structure is effective in scattering all in-plane modes, and its efficiency is largely independent of the shape or the size of the grid. The sub-anode grid dimensions are limited by the same factors as for the LIG: it must be small enough to minimize propagation losses prior to encountering a grid line, and it must be sufficiently large (i.e. >1 μm) to avoid chromatic dispersion accompanying a grating or photonic crystal structure. Since the grid is located close to the EML, there is no image blurring due to the "near field" scattering of light.

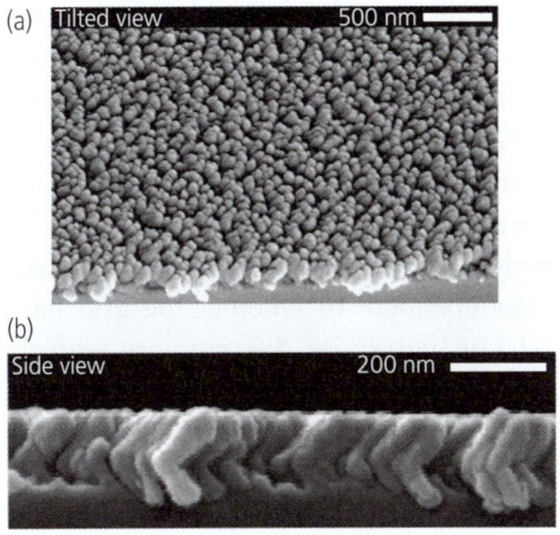

Figure 6.141 (a) Top and (b) side view SEM images of SiO_2 forming a low index grid used in OLED outcoupling (Slootsky and Forrest, 2010).

Reprinted with permission from Slootsky, M. & Forrest, S.R., 2010, The Optical Society.

Sub-anode grids have been fabricated using either low index (air) or high index TiO_2 grid lines, with similar results. Fabrication of the TiO_2 grid is shown in Fig. 6.143. The buried grid has the advantage of not interfering with the OLED structure. However, a consequence of the resulting planar OLED is that the grid only scatters waveguide modes. The SPP modes, whose evanescent fields do not extend as far as the grid, remain unaffected, and their losses remain significant. The fraction of power in each mode as a function of ETL thickness in a PHOLED whose structure is: 70 nm ITO/2 nm MoO_3/40 nm CBP/15 nm 15 wt% $Ir(ppy)_3$ doped in CBP/65 nm TPBi ETL/1 nm LiF/Al is shown in Fig. 6.144. This device is deposited

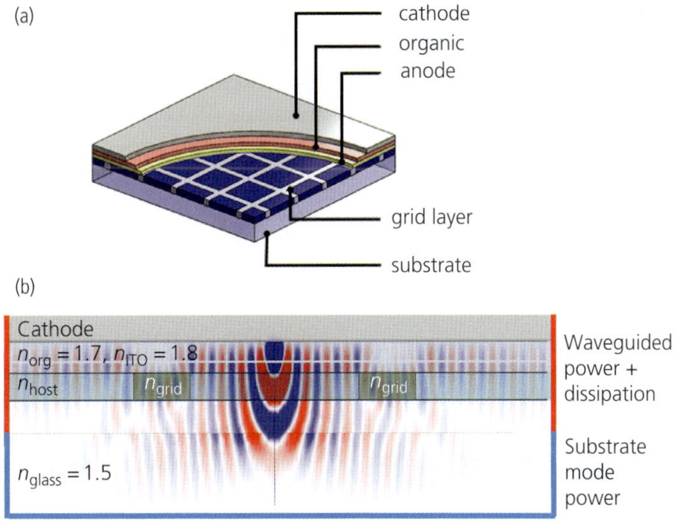

Figure 6.142 (a) Schematic cut-away view of an OLED with a sub-anode grid (Qu et al., 2015). (b) Calculated wavefronts of guided light in the ITO and organic layers. Scattering of the wavefront from the index contrast presented by the grid results in an increase in substrate modes at the expense of waveguide modes.

(a) Qu, Y., Slootsky, M. & Forrest, S. R. 2015. Enhanced light extraction from organic light-emitting devices using a sub-anode grid. *Nature Photon.*, 9, 758

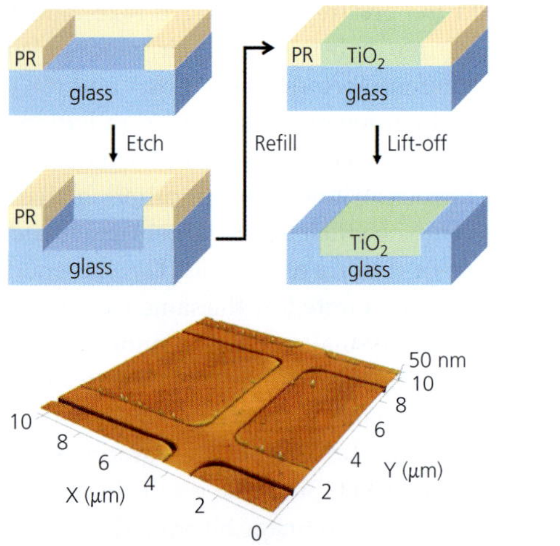

Figure 6.143 Fabrication sequence for the sub-anode grid substrate. Starting at top left: Photoresist (PR) of the grid pattern is deposited and exposed to define the grid lines. The glass substrate is etched and TiO_2 is thermally deposited to refill the openings. The grid is completed by lifting off the excess TiO_2 with the PR. The OLED including the ITO anode is deposited onto this modified substrate. The AFM image at the bottom shows the grid with a line width of 1 μm and a line spacing of 6 μm. The grid is slightly higher (by 8 nm) than the surrounding substrate (Qu et al., 2015).

Qu, Y., Slootsky, M. & Forrest, S. R. 2015. Enhanced light extraction from organic light-emitting devices using a sub-anode grid. *Nature Photon.*, 9, 758

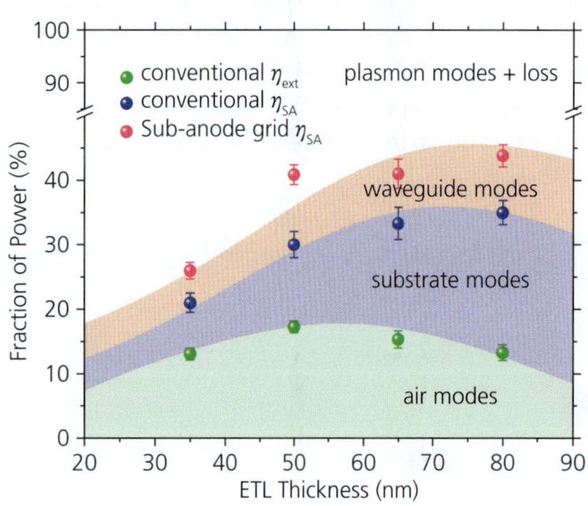

Figure 6.144 Power coupled into each mode of an $Ir(ppy)_3$ PHOLED vs. ETL thickness. Measurements of contributions from air and substrate modes are based on conventional PHOLEDs without, and with index-matching fluid between the substrate and detector surfaces, respectively. The waveguide modes are measured using an analogous PHOLED with a sub-anode grid, also using IMF (Qu et al., 2015).

Qu, Y., Slootsky, M. & Forrest, S. R. 2015. Enhanced light extraction from organic light-emitting devices using a sub-anode grid. *Nature Photon.*, 9, 758

onto a 340 nm thick hexagonal TiO_2 grid, with grid line width of 1 μm and a hexagon side dimension of 3 μm. The calculation uses the Green's function analysis based on CPS theory in Section 6.6.1.

The power in the individual modes in Fig. 6.144 is measured as follows: (i) Air modes employ a conventional device and all of the radiation in the forward viewing direction is measured with a photodetector in the far field. (ii) Substrate modes are assigned to the difference in the output from the device in (i) and for that same device placed in contact with the photodetector using a substrate IMF to outcouple all of the internally trapped modes. (iii) The device with an identical structure to (i) and (ii), except that it includes a sub-anode grid, is placed in contact with the detector using an IMF. The difference in output between (ii) and (iii) is assigned to waveguide modes. (iv) Plasmonic, internal efficiency and metal losses are assigned to the difference between (iii) and 100%. The calculations match well with the data (points in Fig. 6.144) for several different ETL thicknesses. Indeed, the sub-anode grid apparently extracts nearly 100% of the waveguide modes, resulting in an enhancement of $\sim$ 1.45 times compared to the conventional device. This enhancement is independent of whether the grid has a low index (air) or high index discontinuity with the substrate, or whether it is a square or a hexagon. Furthermore, no systematic changes in the emission spectra are observed for any of these device variations, and the output intensity profile is found to be a nearly ideal Lambertian (Qu et al., 2015). However, approximately 50% of the power is lost to SPP and other sources, emphasizing that reduction of coupling to SPPs is essential in achieving high efficiency.

Another approach to outcoupling waveguide modes is to fabricate the thin OLED directly onto the flat surface of a high index film shaped into an MLA, and then attaching the entire unit onto the surface of the glass substrate. This so-called built-up light extraction substrate (BLES) is shown schematically in Fig. 6.145. The device comprises a high refractive index ($n = 1.77$ at a wavelength of 550 nm) PEN film with a 10 μm diameter MLA with the same high index. This is attached to a glass substrate with a small air gap between the array and the glass. The planar surface of the PEN serves as a substrate for the growth of the OLED, including the transparent anode. In effect, this is simply an MLA mounted on a high index substrate, then placed on a glass substrate to enable further encapsulation. The calculated light extraction efficiency is plotted vs. residual OLED loss in comparison to a MLA on a low index (glass) substrate. The efficiency is enhanced by the BLES, particularly as the OLED losses due to absorption, SPP modes, etc. are reduced.

To test the efficacy of the waveguide mode extraction using the BLES, 1 cm^2 two unit (one comprising a red-green emitting, and the second a blue emitting sub-element) white PHOLEDs were grown onto the BLES surface, and then mounted onto the glass substrate (Yamae et al., 2013). The BLES device had an efficacy of 114 lm/W and an operational lifetime of 10^5 h to half its initial luminance of 1000 cd/m^2 due to the encapsulation provided by the glass slab (Komoda et al., 2012). The emission has $CRI = 80$ and $CCT = 2550$ K giving it a reddish cast. It was estimated that 50% of the light was extracted via this strategy. Note that the emission pattern in Fig. 6.145, inset, is non-Lambertian due to the tendency for MLAs to preferentially scatter into near in-plane angles (Möller and Forrest, 2001).

An improvement on the BLES that extracts nearly 100% of the trapped light in the active region into the air and substrate modes embeds the MLA completely within the glass substrate immediately beneath the bottom, transparent electrode (Qu et al., 2018). The sub-electrode microlens array (SEMLA) is shown in Fig. 6.146a. It comprises a high index ($n_{\mathrm{SEMLA}} = 1.8$) planar spacer layer on top of a hexagonal close-packed array of 10 μm diameter hemispherical lenses. The SEMLA surface is smooth to eliminate optical scattering, and does not intrude into the device active region. Figure 6.146b shows a glass substrate containing the SEMLA structure (highlighted by dashed lines) on a printed background, with no apparent impact on image sharpness. The inset shows a microscopic image of a section of a glass substrate covered by the SEMLA. The substrate is patterned with a square grid with 100 μm period and 20 μm wide metal gridlines. The SEMLA has negligible impact on image resolution when magnified at this level.

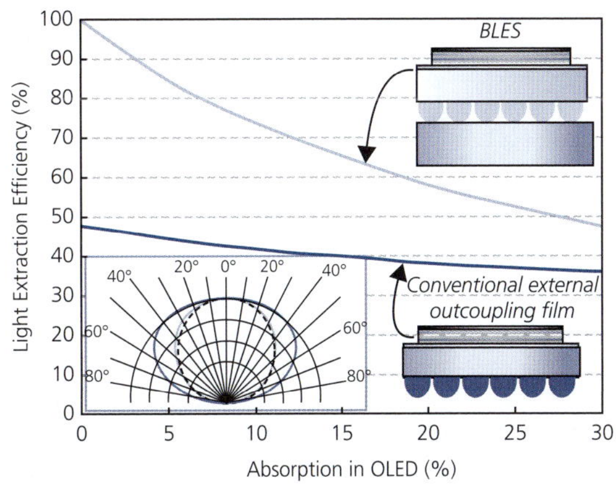

Figure 6.145 Light extraction efficiency vs. excess loss in the OLED for a built-up light extraction (BLES) microlens architecture and a conventional device on a glass substrate with a microlens array. Inset: Radiation pattern of the BLES (red line), a Lambertian source (black line) and a conventional OLED (green line) (Yamae et al., 2013).

The high refractive index of the SEMLA extracts waveguide modes from the organic and ITO anode layers. Figures 6.147a and b show the calculated optical power distribution in devices with and without the SEMLA based on Green's function analysis, assuming that the spacer layer is semi-infinite. The device structure used in the calculation is: 40 nm ITO anode/40 nm HTL/20 nm EML/ETL/Al cathode. The waveguide modes are reduced to almost zero for an ETL thickness < 70 nm. Indeed, the SEMLA extracts all radiated optical power except for the SPP modes. The optical power exits into the glass substrate from the high index hemispherical MLA. The light extraction from the SEMLA into glass ($n_{\mathrm{glass}} = 1.45$) is more efficient than from an external MLA ($n_{\mathrm{MLA}} = 1.4$–1.5) into air ($n_{\mathrm{air}} = 1$) due to reduced reflection at the lens/glass interface with its larger critical angle. Air gaps between the substrate and the MLA as in Fig. 6.145 increase reflections, and hence are avoided.

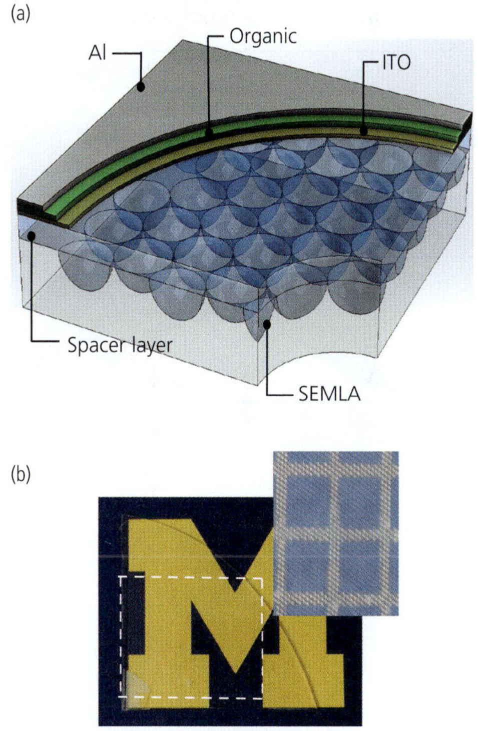

Figure 6.146 (a) Schematic illustration of devices on a sub-electrode microlens array (SEMLA) substrate. (b) Image of a SEMLA substrate on a printed image, with the microlens side facing the image. The white dashed lines enclose the SEMLA structure area. Inset: A microscopic image of a SEMLA on top of a substrate patterned with a square grid with a 100 μm pitch and 20 μm wide gridlines. Use of the logo is by permission (Qu et al., 2018).

Both green emitting PHOLEDs and WOLEDs were fabricated on SEMLA substrates, using an additional MLA attached to the bottom of the substrates to extract substrate modes. The performance characteristics of the green PHOLEDs with different substrates are shown in Fig. 6.148a. At 100 cd/m^2, the SEMLA alone enhances η_{ext} from 25% for a PHOLED on a flat glass substrate, to 30%, representing improvement by an outcoupling enhancement factor of $EF = 1.2$. The outcoupling is further enhanced to $EF = 1.9$ to $\eta_{ext} = 47\%$ using an external MLA. Using IMF at the glass-air interface in conjunction with the SEMLA extracts up to 70% (an $EF = 2.6$) of the total generated photons, compared with 51% for conventional glass, and 60% for sapphire substrates whose refractive index is similar to that of the SEMLA ($n_{sap} = 1.77$). The most efficient device using the SEMLA reaches $\eta_{ext} = 70\%$. Similar enhancements are achieved for a WOLED fabricated on the SEMLA surface.

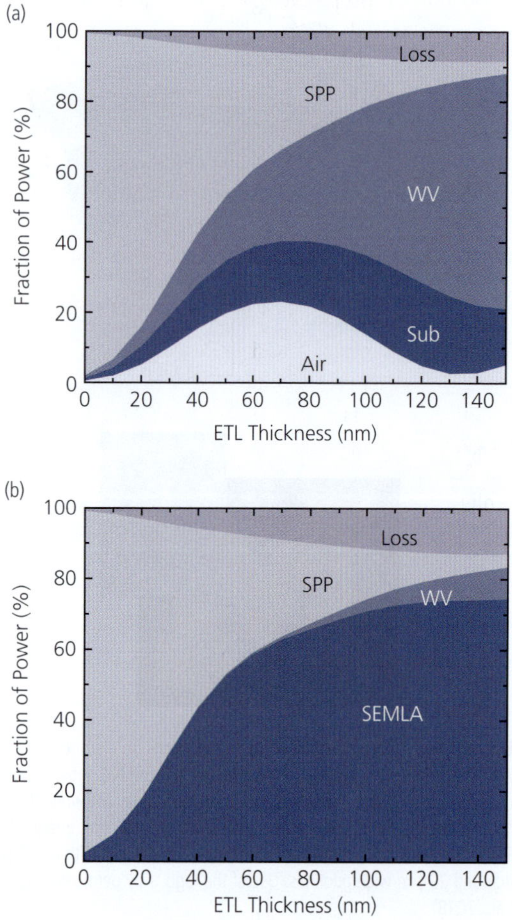

Figure 6.147 Optical power modal analysis vs. electron transport layer (ETL) thickness on (a) a conventional glass substrate and (b) a SEMLA substrate (Qu et al., 2018).

The SEMLA eliminates microcavity effects, and hence does not induce spectral shifts even at viewing angles ~60°. This is inferred from the angular intensity profile of the conventional device in Fig. 6.148b,

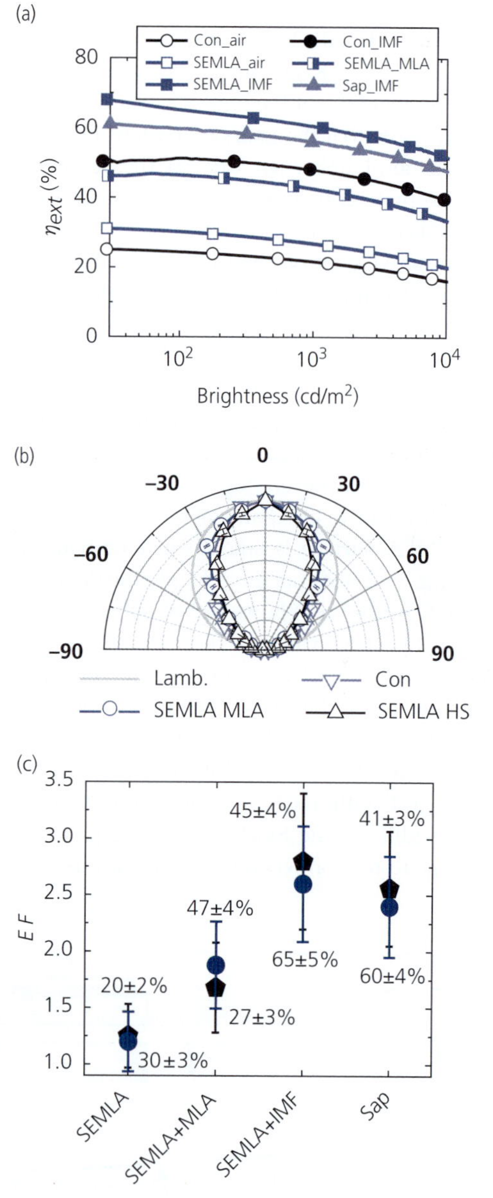

Figure 6.148 (a) Measured η_{ext} vs. luminance of green emitting PHOLEDs on conventional flat glass (Con), SEMLA and sapphire (Sap) substrates. Measurements with no additional outcoupling at the substrate and air interface (air), with an external microlens array (MLA), and with index matching fluid (IMF) are indicated. (b) The angular luminosity distribution of WOLEDs on different substrates. Lamb. Indicates an ideal Lambertian emission pattern. (c) Enhancement factor (*EF*) of green and white PHOLEDs on different substrates compared with conventional flat substrates with identical device structures. The η_{ext} are indicated. The circles and the pentagons denote the green and white PHOLEDs, respectively (Qu et al., 2018).

which is narrower than a Lambertian source, whereas the SEMLA-modified substrates show higher intensities at large viewing angles, bringing them closer to a Lambertian pattern. The similar enhancement factors of light extraction from both the green and white PHOLEDs are summarized in Fig. 6.148c, indicating that the SEMLA performs the same, independent of emission wavelength. The MLA outcouples $\sim$ 70% photons from the substrate and air modes combined. Outcoupling using an IMF or high index sapphire substrate in Fig. 6.148c shows that $\sim$20% photons remain trapped in the substrate using the SEMLA in combination with the MLA. Solutions to improve extraction from substrate modes are, therefore, necessary to fully exploit the advantages of efficient waveguide and surface plasmon mode outcoupling structures.

6.6.4 Surface plasmon polariton mode outcoupling

Both SPP and waveguide modes can be coupled into glass modes by surface roughness near to the metal surface (Hobson et al., 2002). For example, patterning a grating onto the thin metal electrode is one means to extract SPP modes from the OLED surface (Gifford and Hall, 2002, Yates et al., 2006). As previously, the grating can be formed by interference or nanoimprint lithography on the substrate surface, followed by deposition of the several OLED layers and metal. If the metal cathode is semitransparent, with a grating periodicity, Λ, then in-plane modes can outcouple the SPP mode at both the dielectric interface and at the metal/air interface if it satisfies the Bragg condition:

$$k_{||}(\lambda) = \pm k_{SPP}(\lambda) \pm \frac{2\pi}{\Lambda}, \qquad (6.117)$$

where k_{SPP} is the wavevector of the SPP mode at the metal-air interface (see Eq. 6.107). Modes satisfying Eq. 6.117 radiate from the top metal surface as long as the metal is thin enough to be semitransparent, that is, it is less than or on the order of the skin depth of the metal (Gifford and Hall, 2002). The signs in Eq. 6.117 are due to the two counter-propagating modes at each wavelength due to the mirror symmetry of the grating. Hence, the grating produces a splitting of the emission spectrum into both a red- and blue-shifted mode.

Since the SPP modes are lossy, their outcoupling efficiency is limited. Furthermore, this is another solution that disrupts the planar OLED structure. Alternatively, employing a very thick ETL avoids exciting SPP and metal loss modes altogether, although increasing the layer thickness at some point leads to high device resistance. Replacing the metal contact with a less conductive transparent oxide almost completely eliminates both SPP and lossy metal modes. This solution has been implemented by growing a transparent PHOLED onto the surface of a planarized reflective Ag mirror beneath the WOLED bottom contact, as shown in Fig. 6.149. Employing a scattering structure within the substrate while retaining a planar surface allows freedom for optimizing the scattering layer dimensions without affecting the OLED electrical properties. The reflector is a patterned grid of raised rectangles coated with Ag. Similar to the LIG, the periodicity is on the order of several wavelengths to avoid angle- and wavelength-dependent effects. A planarizing dielectric layer fills in the depressions and extends above the raised portions of the grid on which the electrodes and organic layers are deposited. The thick and thin spacer regions couple to different optical modes propagating beneath the transparent electrode (Qu et al., 2017).

The simulated modal power distributions vs. wavelength of the thick and thin regions between the grid surface and the transparent OLED bottom contact are shown in Fig. 6.150a and b (Gontijo et al., 1999). The calculations assume an OLED whose active region comprises a 130-nm-thick organic layer (n_{org} = 1.8) sandwiched between 80-nm-thick transparent electrodes (50 nm IZO and 30 nm MoO_3, $n_{IZO} = n_{MoO3}$ = 2), with the EML placed 60 nm above the bottom electrode. A randomly oriented green emitting molecular dipole is placed at the position of the EML. The spacer refractive index is n = 1.5. The optical outcoupling efficiency is >20% when spacer layer thicknesses are

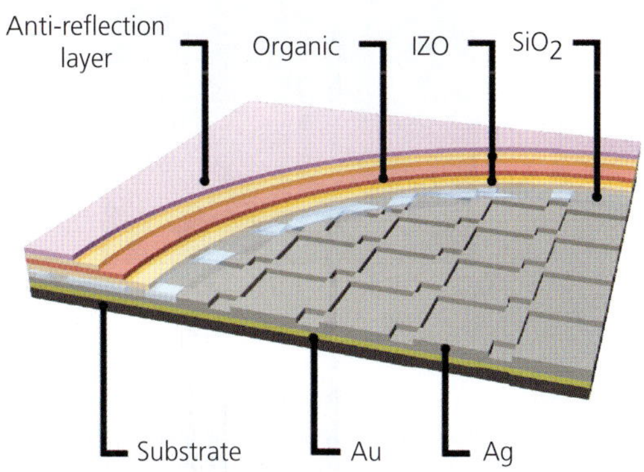

Figure 6.149 Schematic cross-sectional view of a transparent OLED grown on a reflective sub-electrode grid (Qu et al., 2017).

65 and 245 nm in the thin and thick dielectric regions beneath the OLED, respectively.

The in-plane wavevector, $k_{\parallel}$, corresponds to light propagating in the organic layers with refractive index n_{org}, and the total wavevector is k. Modes with $u = k_{\parallel}/k < n_{\text{air}}/n_{\text{org}}$ are radiative, $n_{air}/n_{org} < u < 1$ are waveguided, and modes with $u > 1$ are lost to SPPs. For the thin cavity regions (Fig. 5.150a), there are four waveguide modes, and a weak SPP mode. The thick cavity region (Fig. 5.150b) adds three more waveguide modes but has no SPP mode. Figure 6.150c shows the power distribution of a conventional top emitting OLED with the structure Ag/ 90 nm organic layers/20 nm top semitransparent Ag layer, with a 20 nm EML centered in the organic active region. This structure does not support any waveguide modes, but instead has two, intense SPP modes. The first SPP mode at $u \approx 0.6$ and $\lambda \approx 540$ nm propagates in the top thin Ag film. According to Eq. 6.107 and given $n_{\text{air}} < n_{\text{org}}$, this SPP mode lies at $u < 1$ rather than $u > 1$ for organic/metal SSPs. The second SPP mode at $u > 1.5$ is along the Ag/ organic interface lies at. Thus, compared to conventional OLEDs, metal-coated grid devices successfully suppress SPP modes while coupling power into the waveguide modes.

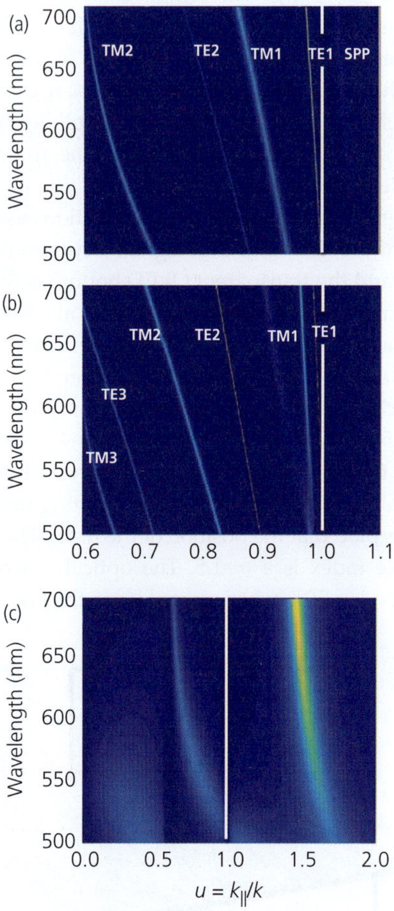

Figure 6.150 The modal power distributions within the cavities found between the raised and depressed portions of the grid and the OLED in Fig. 6.149. Here, $k_{\|}$ is the in-plane wavevector for light within the organic layers with refractive index, n_{org}. The total wavevector is k. The waveguide and SPP modes (at $u > n_{\text{air}}/n_{\text{org}}$), of the cavities over the (a) gridlines and (b) depressions. (c) The power distribution of a conventional top emitting OLED optimized over the same spectral range. The vertical line corresponds to $u = k_{\|}/k = 1$. At higher $k_{\|}$, the waves are evanescent and lossy (Qu et al., 2017).

The scattering by the grid is a consequence of the mismatch between the waveguide modes supported by the two cavities above the grid lines and depressions, which can be estimated by the overlap of the wavevectors of these modes. The TE_1 and TM_1 modes in the thin SiO_2 cavity regions (Fig. 6.150a) have some overlap with the TE_1 and TM_1 modes in the thick dielectric regions (Fig. 6.150b), and thus can propagate in plane. The modes that are not aligned between cavities are completely scattered into the substrate by the grid.

The sub-electrode reflecting grid for an OLED with the structure: glass substrate/50 nm IZO/30 nm MoO_3/30 nm BPhen: Li (molar 1:1)/30 nm BPhen/ 30 nm 8 vol.% $Ir(ppy)_3$ in CBP/40 nm TAPC/30 nm MoO_3/50 nm IZO comprised an array of 1 μm × 3 μm raised rectangles surrounding 3 μm × 3 μm square depressions. The spacer thicknesses varied between 245 nm and 65 nm. The design is forgiving in that there is no pronounced sensitivity to either grid shape (it can be ordered, or random), or dimensions, as long as the feature sizes are larger than λ and less than δ_{SPP}. As with the non-reflecting sub-anode grid structure, there is no image blurring due to the proximity of the mirror to the EML.

In Fig. 6.151 we show that η_{ext} is increased from 20% to 30% using the metallic scattering grid compared to a similar PHOLED fabricated on a substrate with a mirror but no grid. The angular intensity profiles of both devices are broadened from a Lambertian function. The peak intensity of the flat mirror device is normal to the surface, whereas the grid OLED intensity has a maximum at 20° from normal. This is due to microcavity effects that are inevitably encountered in top emitting OLEDs.

Although a significant efficiency enhancement is obtained using the reflecting grid, 60% of the optical power is still lost. In addition to the limited grid scattering efficiency due to the spacer thicknesses used, the scattered light incurs losses at each reflection from the mirror surface. A diffuser or MLA comprised of high refractive index materials added to the

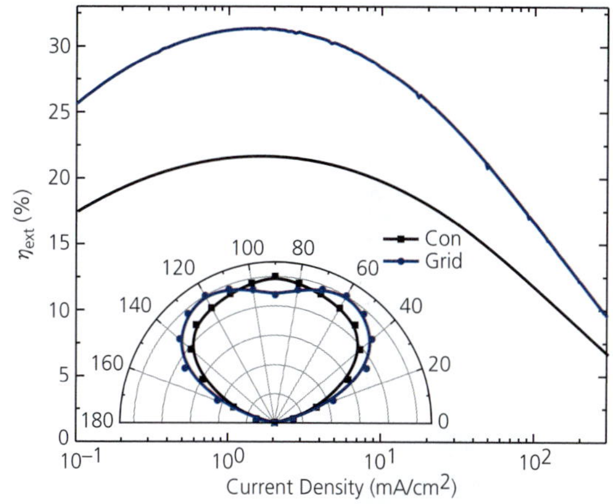

Figure 6.151 External quantum efficiencies of a top emitting PHOLED with a sub-anode mirror separated from the PHOLED by a 65 nm thick SiO_2 spacer (black), and a metal-coated grid device (blue). Inset: Angular intensity profiles of the conventional and metal coated grid PHOLEDs (Qu et al., 2017).

top surface of the transparent OLEDs can reduce cavity resonances. Green's function analysis shows that a MLA on the device emitting surface could further improve the efficiency by >30% (Qu et al., 2017).

A final, potentially very low cost example that completely eliminates SPP modes replaces the bottom metal electrode in top-emitting WOLEDs with the ITO anode deposited on a highly reflective, diffuse dielectric such as used in the interior surfaces of integrating spheres. The structure design is illustrated in Fig. 6.152a. The rough dielectric reflector surface efficiently scatters the waveguided light emitted from a transparent OLED deposited on a planarizing, high refractive index (n_P =1.8) polymer slab that extracts modes trapped in the OLED, while presenting a flat surface on which the OLED layers are grown. Unlike the other methods described in this section, the diffuser-based PHOLED requires no additional light extraction such as index matching layers or MLAs, making this a very simple design.

As illustrated in Fig. 6.152a, light emitted from the OLED exits either from the top, transparent ITO contact, or enters the waveguide layer where it propagates until it is incident on the rough dielectric reflector surface. There, the light is scattered into a Lambertian profile, and light within the TIR exit cone is emitted into the viewing direction. Light incident at angles greater than θ_{TIR} at the polymer–air interface is returned to the diffuse reflector where it is scattered once again. This repeats until the light is either absorbed or scattered into the viewing direction. Since the light requires several bounces before exiting the device, the diffuser must have a reflectivity that exceeds that of the metal, and losses in the transparent anode and polymer waveguide layer must also be small. Here, PTFE (Teflon®) is well suited for this purpose due to its nearly 99% reflectance across the visible.

The simplicity of fabrication is illustrated by the process sequence in Fig. 6.152b. The PTFE reflector is coated with the polymer waveguide on its surface. Next, a PDMS stamp attached onto a glass handle is pressed onto the polymer for planarization, during which the polymer was cured by exposure to UV light prior to stamp removal. The resulting polymer waveguide layer was 0.25 mm thick. Following this procedure, the ITO anode, the OLED layers, and the transparent, top ITO cathode are fabricated by conventional methods.

For a lossless reflector and waveguide layer, the ratio of incident to scattered light power towards the viewing direction from a single diffuse reflection (R_S) is determined using Snell's law, viz. $R_S = (n_{air}/n_P)^2$. Then, the ratio of the light intensity extracted into the viewing direction to that confined within waveguide layer (η_D) following the path in Fig. 6.152a is (Kim et al., 2018)

$$\eta_D = R_S + (1 - R_S)\cdot R_S + (1 - R_S)^2\cdot R_S + \ldots$$
$$= \sum_{n=0}^{\infty} (1 - R_S)^n\cdot R_S = 1. \quad (6.118)$$

It follows that total outcoupling efficiency (η_{out}) of the OLED is

$$\eta_{out} = \eta_{TA} + \eta_D\eta_S, \quad (6.119)$$

where η_{TA} is the fraction of light emitted from the top surface, and η_S is the fraction of light coupled into the slab waveguide. A diffuse reflecting, a green emitting PHOLED showed a peak η_{ext} = 37%, whereas for comparison, a device using only a metal mirror had η_{ext} = 15%. This substantial difference is due to the much larger losses introduced due to reflections from a metal mirror than a dielectric reflector. From Eq. 6.119, the light coupled into the waveguide layer shows $\eta_{out} = \eta_{int}\eta_S\eta_D$ = 30% and 8% for the diffuser and metal mirror structures, respectively, resulting in a 3.8-fold increase in extraction into the viewing direction via diffuse scattering.

The diffuser increases the *étendue* of the OLED, thus introducing emission outside of the PHOLED active area defined by the device contacts. Figure 6.153a shows the normalized radiated power intensity assuming a polymer waveguide loss of α = 0.4 mm^{-1}, a thickness of 240 μm, and a (1 mm)^2 device area. An image of the emission

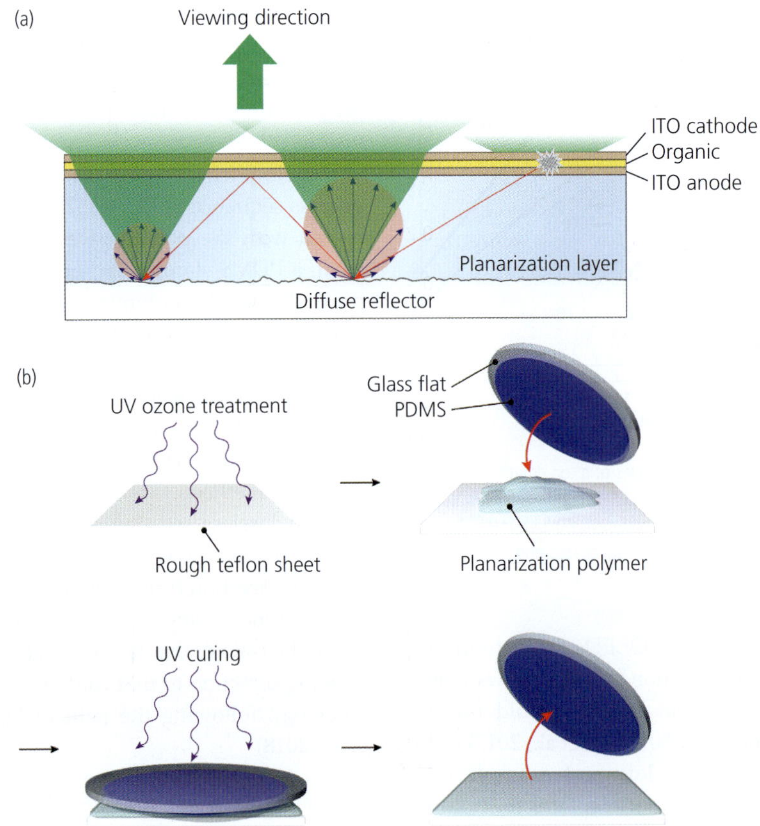

Figure 6.152 (a) Schematic illustration of the transparent OLED on the surface of a polymer waveguide/planarization layer on a diffuse dielectric reflector. The path of light emission from a region within the OLED (star) shows multiple reflections from the diffuse reflecting surface prior to escape from the top surface of the device. Direct emission from the surface is also shown at right. At each bounce, the light direction is randomized, resulting in a Lambertian emission profile (red ovals), some of which is reflected into the optical escape cone. (b) Fabrication sequence of the diffuser and polymer waveguide/planarization layer. After Kim et al. (2018).

from the PHOLED is shown in Fig. 6.153b, right, inset. The diffuse reflector isotropically redistributes the power, thus showing a nearly circular emission pattern. The measured radial intensity profile of the PHOLED following the trajectory of the horizontal dashed line in the simulated 2D intensity map is shown at the right in Fig. 6.153a. Integrating the radiated power outside the device active area indicates that approximately 55% of the total emission is emitted beyond the contact periphery. Given the broad dispersion of the light beneath the OLED area, this method results in considerable image blurring, making it useful only for lighting applications.

For large lighting fixtures, the fraction of emission in the tails vs. directly under the active area decreases exponentially with OLED diameter. Hence, in an extended lighting source, all the light eventually emerges via the top transparent OLED surface. The integrated modal power of air and substrate modes after each reflection is given in Fig. 6.153c, showing that ~90% of the light is emitted (including the direct emission from the top surface) after five bounces from the diffuse reflecting layer.

6.6.5 Outcoupling via molecular alignment

We illustrated several strategies in the previous sections that are effective in outcoupling up to 70% of the emission that is otherwise lost to surface plasmon and waveguide modes. A completely different approach to avoiding coupling to these loss channels is to align the TDMs of the emitting molecules parallel to the substrate plane. In this orientation, the emission is directed almost entirely perpendicular to the emitting surface, thus only exciting air and substrate modes. Equation 6.116 indicates that the degree of outcoupling is controlled by the ratio of TDM alignment in the vertical to the horizontal direction. That is, for $\Theta = 0$,

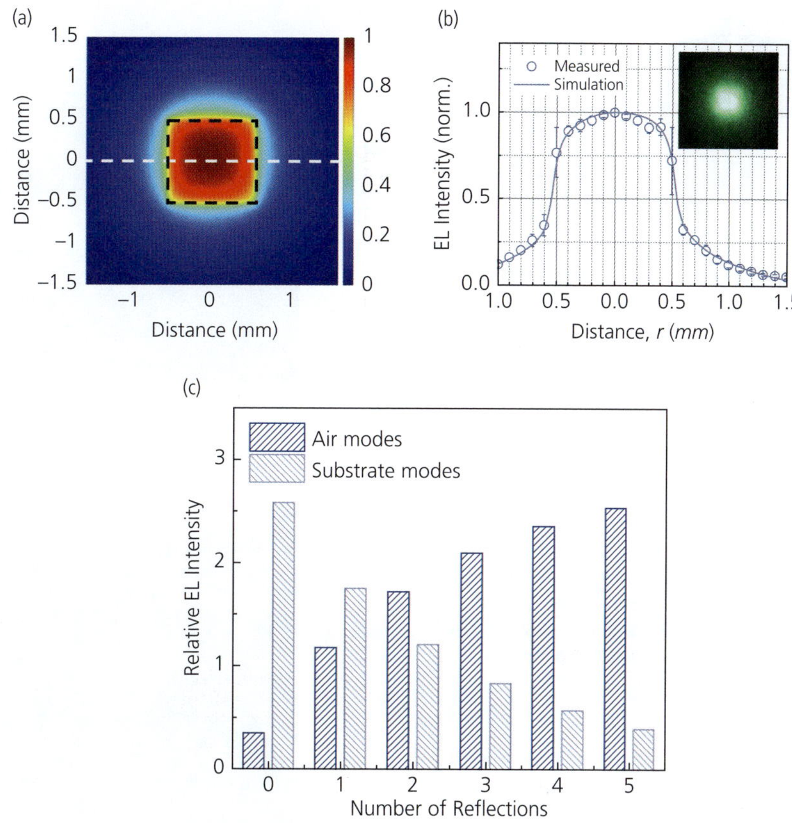

Figure 6.153 (a) Simulated radiant power for a PHOLED on a diffuse reflector substrate. (*left*) The black dashed line indicates the device active area. (right) Measured (circles) radial peak intensity profile along the white dashed line bisecting the PHOLED in (a) along with the ray tracing fit (line). Inset: Photograph of emission from the green, diffuse PHOLED. (c) Distribution of modal power in air and substrate modes following each reflection (Kim et al., 2018).

radiation is emitted normal to the substrate plane due to complete horizontal dipole alignment. The dipole radiation field due to vertical TDM orientation is shown in Fig. 6.154a. From electrostatics, the field intensity of a vertical dipole follows $\sin^2 \alpha$ (Rao, 1994), where α is the angle between the dipole vector and field direction (Rao, 1994). Thus, perfect dipolar orientation parallel to the substrate will excite both waveguide and SPP modes. These contributions will be far smaller than if the dipoles are randomly or vertically oriented.

The fraction of power emitted into each mode with different orientations for a green emitting device with $\eta_{ext} = 1$ and the structure of: glass substrate/ 140 nm ITO/77 nm HTL/60 nm EML/d_{ETL} ETL/Ca: Al cathode are calculated using CPS theory with result in Fig. 6.155, where d_{ETL} is the ETL thickness. Complete horizontal dipole orientation (Fig. 6.155a) results in plasmonic loss only for $d_{ETL} < 50$ nm. At larger ETL thicknesses where the metal cathode is in the weak fringe field of the dipole, the plasmonic losses nearly vanish. Of course, with increasing ETL thickness, an increasing amount of energy is dissipated into waveguide modes. Horizontally oriented dipoles emit only ~20% into air and another 40% into the substrate modes. The opposite situation obtains for predominantly vertical orientation in Fig. 6.155c. Here, plasmonic modes are dominant even for d_{ETL} as large as 150 nm, resulting in almost no emission into either air or even substrate modes. However, once the ETL is sufficiently thick, excitation of SPP modes is weakened and becomes negligible, yet the dominance of waveguide modes becomes even stronger than for random (Fig. 6.155b) or horizontal dipolar orientations, since the field intensity for vertical dipoles is almost entirely in-plane.

The external efficiencies of PHOLEDs as a function of molecular TDM orientation are summarized in Fig. 6.156. For this calculation, it is assumed that the 30 nm thick EML is spaced 40 nm from an Al cathode. The device is on a flat glass substrate with no further outcoupling employed. The PL quantum yield of an Ir(ppy)$_2$acac molecule is $\phi_{PL} = 0.94$, corresponding to

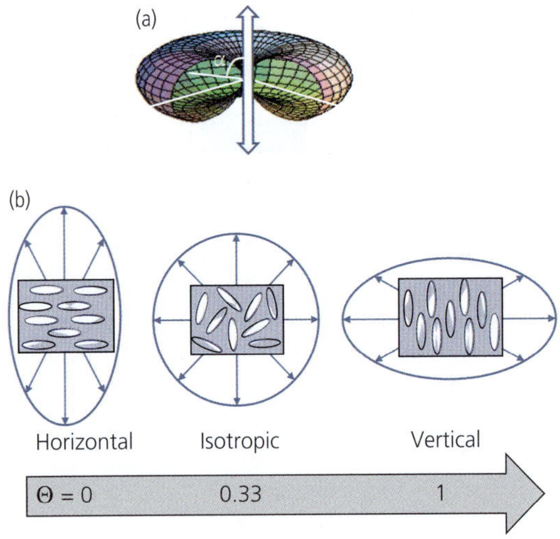

Figure 6.154 (a) Radiation field intensity distribution for a molecule with a vertical TDM (block arrow). (b) Schematic of the radiation fields from perfect horizontal (Θ = 0) to vertical (Θ = 1) TDM alignment. The TDMs are illustrated by ellipses, and the substrate plane is assumed to be parallel to the gray arrow at bottom (Courtesy, M. E. Thompson, 2018).

the horizontal dashed line. This analysis shows that for perfect horizontal TDM alignment and $\phi_{PL} = 1$, a maximum external quantum efficiency of 46% is possible.

The TDM orientation for the PHOLED was measured using the PL intensity vs. angle with results in Fig. 6.157. The fits employ independent ellipsometric measurements of the ordinary and extraordinary refractive indices (n_o and n_e, respectively) of host matrix (TCTA:B3PYMPM in this case). Then, the angle-dependent effective refractive index is

$$\frac{1}{n_{eff}^2} = \frac{\cos^2\theta}{n_o^2} + \frac{\sin^2\theta}{n_e^2}. \tag{6.120}$$

The best fit to the data yields an alignment principally in the horizontal direction, with $\Theta = 0.23$. Referring to Fig. 6.156, this should result in a PHOLED with $\eta_{ext} = 29.7\%$, which is consistent with the measured value of 30.2% (Kim et al., 2013b).

Preferential molecular alignment parallel to the substrate plane resulting in increased quantum efficiency has been observed in a wide range of fluorescent and phosphorescent materials. Early on, it was found that spin-cast polymer chains can align in the plane, resulting in increased outcoupling (McBranch et al., 1995). Linearly extended vacuum deposited small molecules also show a tendency to lie with their TDM parallel to the substrate (Frischeisen et al., 2010, Frischeisen et al., 2011), as do square planar Pt complexes (Wang et al., 2014a, Kim et al., 2016a, Ly et al., 2016). Also, the heteroleptic molecules, $(C{\wedge}N)_2IrX$, where X is the ancillary ligand show alignment parallel to the substrate, independent of the host in which they are doped (Schmidt et al., 2011, Graf et al., 2014, Kim et al., 2014a, Kim et al., 2014b, Jurow

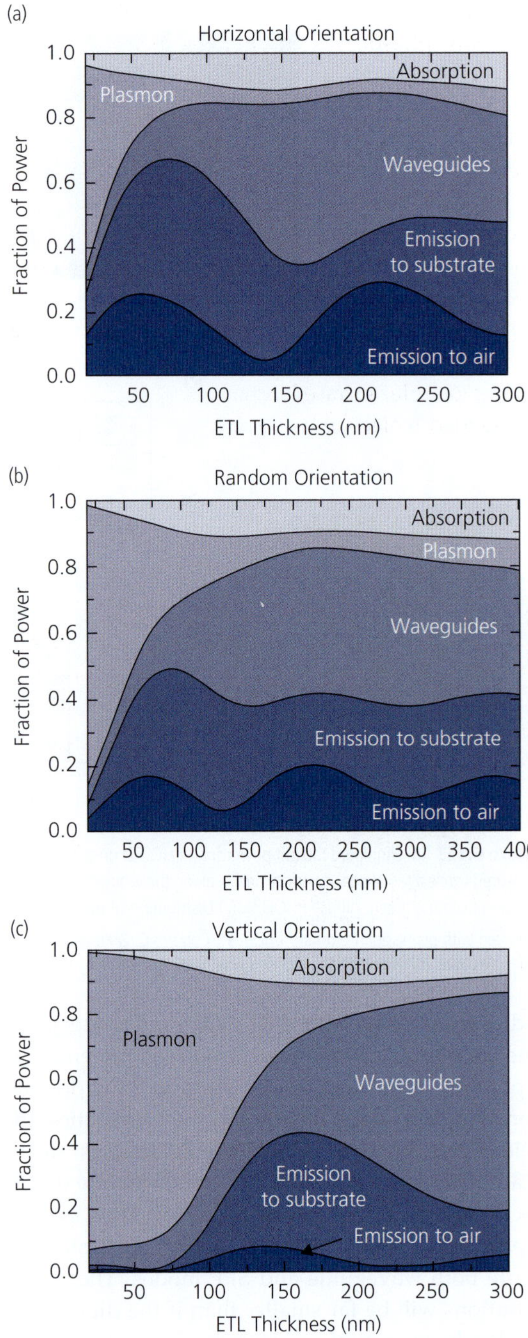

Figure 6.155 Relative fraction of emitted power for different dipole orientations and the contribution from each OLED mode. It is assumed that $\eta_{ext} = 1$ (Brütting et al., 2013).

et al., 2016). In contrast, small molecular weight homoleptic phosphorescent dopants such as $Ir(ppy)_3$ whose dipole moments are isotropically distributed, show no tendency for preferred alignment (Graf et al., 2014, Kim et al., 2014b).

Several models have been advanced to explain the driving force that leads to molecular alignment. We have seen in Chapter 5 that ostensibly planar molecules grow into two preferred orientations relative to the substrate: flat lying or standing up. The different growth habits are a function of molecule/surface vs. molecule/molecule bond energies. That is, a strong molecule/surface binding due, for example to an extended molecular π-system, leads to a flat-lying motif. However, very strong intermolecular π–π interactions may draw the molecules into standing perpendicular to the substrate to minimize energy. The preferred stacking habit is often determined solely by these interactions, or they can, to some degree, be controlled by the thermodynamic conditions used during growth. For example, the substrate temperature can be increased to drive vertical vs. horizontal alignment (Komino et al., 2014, Kim et al., 2019). Alternatively, spin-casting rather than vacuum deposition of films can lead to isotropic TDM orientations (Lampe et al., 2016).

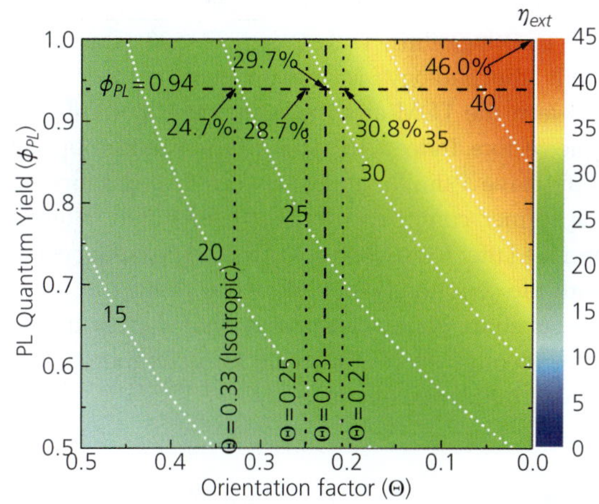

Figure 6.156 External PHOLED quantum efficiency dependence on dipole orientation factor, Θ, and PL quantum yield. For complete horizontal dipole alignment (Θ = 0), a maximum η_{ext} = 46% is possible for a device on a flat glass substrate (Kim et al., 2013b).

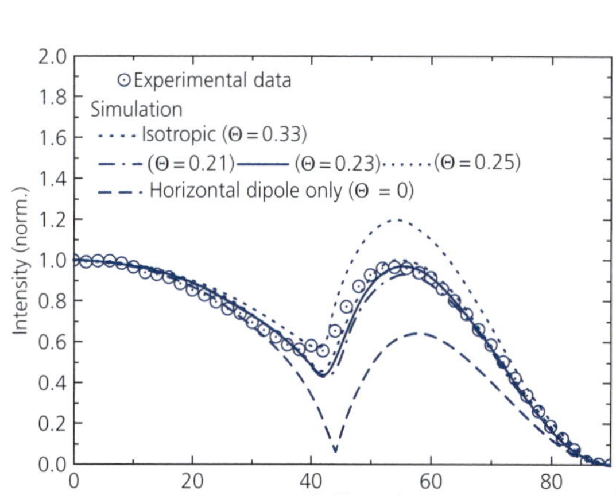

Figure 6.157 PL intensity vs. angle for an EML comprising 8 vol.% $Ir(ppy)_2(acac)$ in a 1:1 TCTA:B3PYMPM co-host. The data (circles) are fit (lines) to a dipole alignment factor of Θ = 0.23. Also shown are expected angular dependences for Θ = 0 and 0.33 (Kim et al., 2013b).

The film growth habits leading to TDM orientation has been determined for the homoleptic 2-pyrazinyl pyrazolate Pt(II) complex, $Pt(fprpz)_2$, using *grazing incidence wide angle x-ray scattering* (GIWAXS), with results given in Fig. 6.158. Strong intermolecular interactions result in a vertical molecular orientation relative to a fused silica substrate. This is induced by the ^{3}MMLCT interactions whose excimer TDM lies in the substrate plane between adjacent molecules in a vertical stack. The resulting excimer emission has a peak wavelength of 740 nm. The horizontal alignment factor of Θ = 0.87 in the PHOLEDs with neat EMLs indicates near perfect horizontal alignment of the TDM, leading to η_{ext} = 24% (Ly et al., 2016).

In neat films, planar molecular orientations appear to be driven by molecule/interface energies and growth kinetics. In contrast, orientation is largely independent of the host material for a number of homo- and heteroleptic pseudo-octahedral Ir-complexes. Table 6.13 lists a number of such complexes, all of

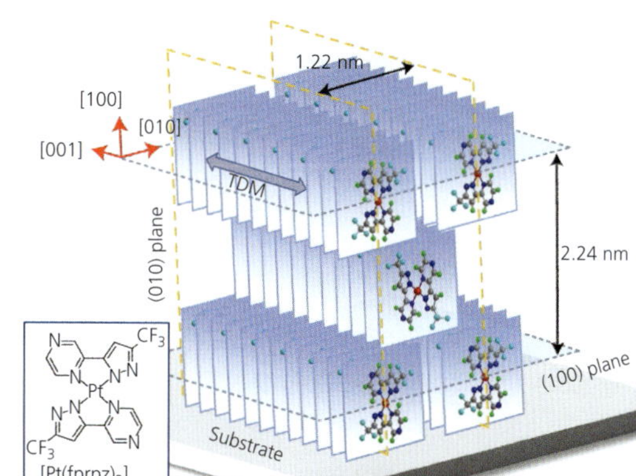

Figure 6.158 Stacking habit of $Pt(fprpz)_2$ on a fused silica substrate. The TDM orientation of the excimer state is parallel to the substrate (blue block arrow). Inset: Molecular structural formula of $Pt(fprpz)_2$ (Ly et al., 2016).

which are approximately spherically shaped. However, the TDM vectors of molecules containing an aliphatic acetyl acetate (acac) group lie preferentially parallel to the substrate plane. Those without such a group (the three Ir complexes near the bottom of the Table) appear to have no preferred orientation.

It is noteworthy that the horizontal alignment factor is, thus, nearly independent of the static dipole moment of the molecules, or the host matrix shape or composition. Several theories have been advanced as to the source of this alignment. One model suggests that high molecular dipole moments lead to electrostatically driven aggregation of dopants, which tends to randomize the orientation (Graf et al., 2014). This potentially explains why $Ir(ppy)_3$, with its comparatively large static dipole moment of 6.40 D exhibits a nearly isotropic TDM in the solid state. However, this apparently contradicts the observed preferred horizontal orientations of $Ir(bppo)_2(acac)$ (bppo = benzopyranopyridinone) and $Ir(tBuCN\text{-}F)_2(acac)$ with even higher dipole moments. It has been suggested that the large static dipole can lead to electrostatic guest–host interactions that align the aggregates at the growth interface. While host molecules are often approximately planar with a high aspect ratio, they have a very small or no dipole moment. Once again, Table 6.13 shows no apparent correlation between host or dopant composition, which brings this model into doubt (Kim et al., 2014a).

To identify the molecular properties that drive orientation, the angular emission patterns of Ir-complexes with coumarin-based bppo ligands shown in Fig. 6.159 were measured. The carbonyl (C=O) groups are dipolar, allowing the molecules to cluster when doped at 20 vol.% in a CBP host. Their ground state (static) dipole moments range from 6.18 D for $Ir(bppo)_2(acac)$ to 8.44 D for $Ir(ppy)(bppo)_2$. The broadened and quenched spectra with increasing doping concentration from 6 to 20vol.% indicate that aggregates are, in fact, formed.

Interestingly, only the PL spectra of $Ir(bppo)_2(acac)$ show preferential horizontal alignment with $\Theta = 0.22$. The other two bppo-based molecules form isotropically emitting films, with $\Theta = 0.32$–0.33. This alignment morphology is independent of the angle between the TDM and the permanent molecular dipole vectors. Indeed, $Ir(bppo)_2Ir(acac)$ and $Ir(bppo)_2Ir(ppy)$ have a similar angle between these vectors, yet the former compound forms preferentially aligned, and the latter forms isotropic films when co-deposited with the common host, CBP, whose long axis lies in the substrate plane. Hence, it is unlikely that dipole-dependent guest–host interactions determine the degree of horizontal alignment.

Table 6.13 Horizontal alignment factors and static dipole moments of several Ir-phosphors in various host matrices (Jurow et al., 2016)

Emitter molecule	Host	Orientation (Θ)	Static dipole moment (D)	Reference
$Ir(dhfpy)_2(acac)$	NPD	0.25	1.16	(Graf et al., 2014)
$Ir(ppy)_2(acac)$	CBP	0.23	1.66	(Graf et al., 2014)
	TCTA/B3PYMPM	0.24		(Kim et al., 2014b)
$Ir(ppy)_2(tmd)$	TCTA/B3PYMPM	0.22	1.72	(Kim et al., 2014b)
$Ir(MDQ)_2(acac)$	NPD	0.24	1.75	(Graf et al., 2014)
	NPD/B3PYMPM	0.20		(Kim et al., 2014a)
$Ir(bt)_2(acac)$	BPhen	0.22	1.76	(Graf et al., 2014)
$Ir(chpy)_3$	NPD	0.23	2.02	(Graf et al., 2014)
$Ir(mphmq)_2(tmd)$	NPD/B3PYMPM	0.18	2.42	(Kim et al., 2014a)
$Ir(mphq)_2(acac)$	NPD/B3PYMPM	0.23	2.52	(Kim et al., 2014a)
$Ir(phq)_3$	NPD/B3PYMPM	0.30	4.63	(Kim et al., 2014a)
$Ir(piq)_3$	NPD	0.22	5.20	(Graf et al., 2014)
$Ir(bppo)_2(acac)$	CBP	0.22	6.18	(Jurow et al., 2016)
$Ir(ppy)_3$	CBP	0.31	6.40	(Graf et al., 2014)
	TCTA/B3PYMPM	0.33		(Kim et al., 2014b)
$Ir(ppy)_2(bppo)$	CBP	0.32	8.25	(Jurow et al., 2016)
$Ir(ppy)(bppo)_2$	CBP	0.33	8.44	(Jurow et al., 2016)
$Ir(tBuCN\text{-}F)_2(acac)$	pBCb2Cz/TSPO1	0.23	10.59	(Lee et al., 2015a)

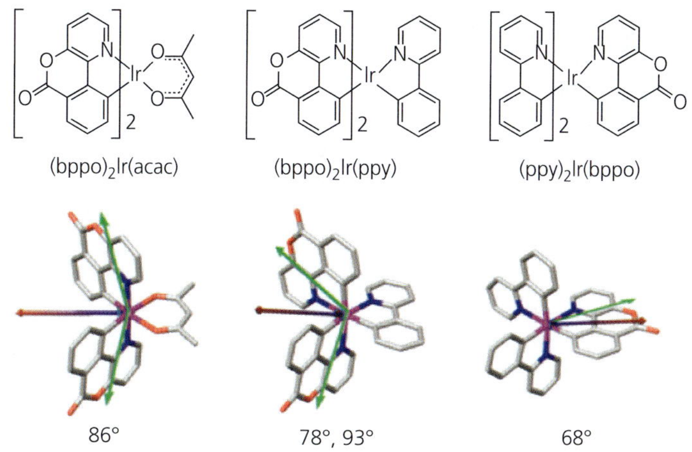

Figure 6.159 Three heteroleptic Ir-complexes, each with a different number of emitting ligands employing the coumarin-based ligand, bppo. The TDM vectors are shown by the green arrows, and the red arrows are the permanent (static) dipole moments. The numbers below each structure give the angle between the TDM and static dipole vectors (Jurow et al., 2016).

It has also been proposed that hosts with low *glass transition temperatures* (T_g) may form clusters during or after deposition (Dalal et al., 2015). The glass transition temperature is where the amorphous material makes a transition from a hard glassy state to a more flexible (rubbery) viscous state. The dependence of molecular orientation on host composition, and the process used in depositing the films for a series of heteroleptic Ir-complexes containing aliphatic acetylacetonate ancillary ligands as well as the homoleptic Ir(ppy)$_3$ has been investigated, with results summarized in Fig. 6.160 (Lampe et al., 2016). The different hosts have a range of glass transition temperatures (and hence propensity for crystallization). They were co-deposited either from vacuum, or spin-casting with four Ir-complexes. Spiro2-CBP and PMMA have considerably higher T_g than either CBP or NPB, which are known to crystallize at temperatures as low as 50°C. With the exception of Ir(ppy)$_3$, vacuum deposition results in preferred horizontal alignment in all hosts tested (open symbols). The alignment following spin casting, however does not show this preference. Solution processed guest–host blends result in an isotropic orientation of the TDM vectors, and in a few cases there is even a slight preference for vertical dipole alignment.

A consistent explanation of these phenomena is offered in Fig. 6.161. Deposition of a CBP:Ir(ppy)$_2$(acac) layer occurs at the vacuum/solid interface of the film. At the growth interface, the aliphatic (acac) groups are presumably forced to orient toward vacuum while the phenylpyridine groups highlighted in red lie along the molecular C2 axis extending from the Ir core. The molecules thus aligned adhere to the solid surface, thereby creating a preferred in-plane orientation of the TDM that is perpendicular to the C2 axis. In contrast, CBP arrives without a preferred orientation This relative orientation of vectors is shown for Ir(bppo)$_2$(acac) in Fig. 6.159. The orientation is presumably driven by a surface energy mechanism similar to that of molecules with hydrophobic and hydrophilic end groups in Langmuir-Blodgett thin film deposition (see Section 5.4.2). In this picture, the molecules deposited from solution in Fig. 6.161 lack a defined vacuum/solid interface, preventing them from finding a preferred emitter or host orientation (Jurow et al., 2016). While this is a plausible explanation for the horizontal alignment observed for Ir-complexes containing ancillary aliphatic groups grown by VTE, to our knowledge there is not yet a quantitative energetic model to support it.

The methods of alignment discussed above are thermodynamically driven, with the outcome dependent on the energetic relationship between the molecule and substrate surface. In Section 5.4.2, we showed that the energy topography of the surface can be modified by pretreatment and by templating. The various morphologies that can be achieved during growth on various surfaces are illustrated in Fig. 6.162. In the foregoing discussion we showed that blends of the guest and host can be random, or the host itself can drive the orientation of the dopants within the blend to maximize horizontal TDM alignment. The most directed approach is to initiate

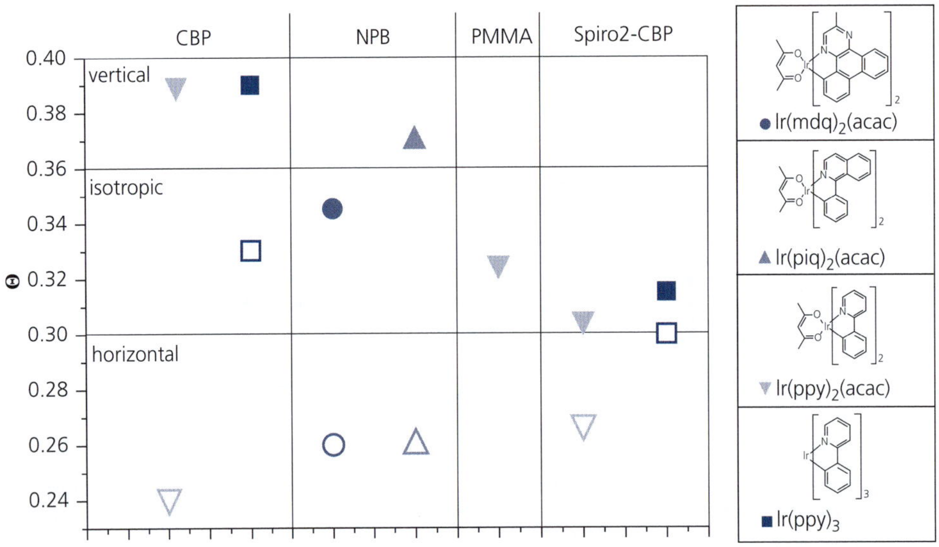

Figure 6.160 Horizontal TDM alignment factor for various host and dopant combinations. Filled symbols correspond to samples cast from solution, open symbols to co-deposition via vacuum thermal evaporation (Lampe et al., 2016).

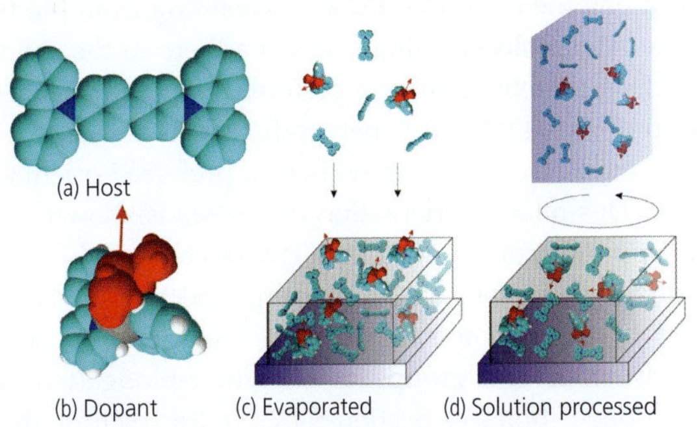

Figure 6.161 Molecular structures of the (a) CBP host and (b) Ir $(ppy)_2(acac)$. The acac group is highlighted in red, and the C2 axis vector between the Ir atom and this aliphatic group is shown (arrow). (c) Growth of a CBP:$Ir(ppy)_2(acac)$ via vacuum thermal evaporation. At the growth interface, the acac groups orient toward vacuum, and the phenylpyridine groups adhere to the solid surface creating a preferred in-plane orientation of the TDM located perpendicular to the C2 axis (cf. Fig. 6.159). (d) Solution-processed films have no vacuum/solid interface, and hence randomize the TDM orientation (Lampe et al., 2016).

growth of the guest/host system on a pre-deposited *template layer*, thus ensuring that the desired alignment is forced on the subsequently deposited structure (right hand side of the figure).

An example of forced alignment by deposition on a structural template is shown for the square planar Pt complex, PtD shown in the inset of Fig. 6.163. The PtD was deposited by VTE on a bilayer of 1.5 nm PTCDA/ 1.5 nm NTCDA on sapphire. As shown previously, PTCDA self-organizes into planar stacks with monolayer flatness across a large area. However, the energy gap of PTCDA is small (2.2 eV), and hence it can quench excitons in the layer deposited on its surface. Thus, the wide energy gap NTCDA is deposited on the PTCDA film, thereby transferring the template structure to the subsequently deposited Pt-complex. The templated PtD shows a decreased horizontal orientation ($\Theta_{hor} = 0.33 \pm 0.01$) of the PtD dimer whose TDM extends between Pt centers of adjacent molecules, compared to deposition on a bare fused silica substrate with $\Theta_{hor} = 0.91 \pm 0.01$ (see Fig. 6.163a). Thus, the horizontal TDM for the neat film corresponds to PtD molecules orientated with their planes perpendicular to the substrate surface. To verify this spatial relationship, an X-ray pole figure of the (200) plane is provided in Fig. 6.163b. The film deposited directly on the sapphire substrate in shows the (200) diffraction peak ($2\theta = 8.2$–$8.5°$) at a radial angle of $\psi = 0°$, suggesting that (200) plane lies parallel to the substrate. Note that the (200) plane is perpendicular to the molecular stacking axis along the (020) plane. There is a striking contrast between the TDM orientations with and without the exogenous template, from parallel to nearly completely vertical relative to the substrate plane, respectively (Kim et al., 2019).

In summary, molecular alignment appears to be an effective means to improve optical outcoupling by decreasing the efficiency of excitation of waveguide and SPP modes using many octahedral or dihedral phosphorescent complexes, as well as high aspect

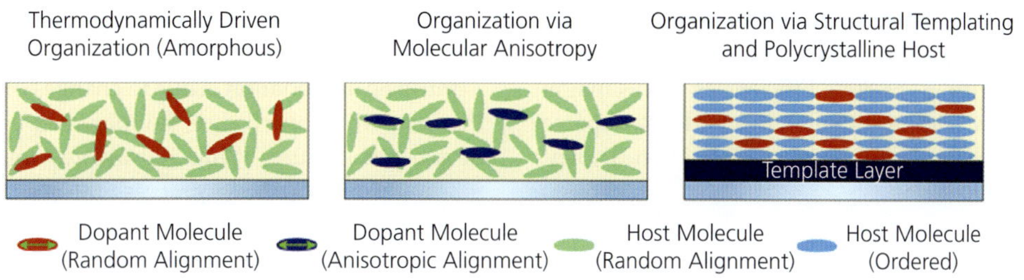

Figure 6.162 Film morphologies attained during growth. From left to right, amorphous organization of emitting (dopant) and host molecules, dopant molecules aligned due to interactions with the host, and controlled orientation of both dopant and host by growth on a pre-deposited, ordered template layer (Kim et al., 2019).

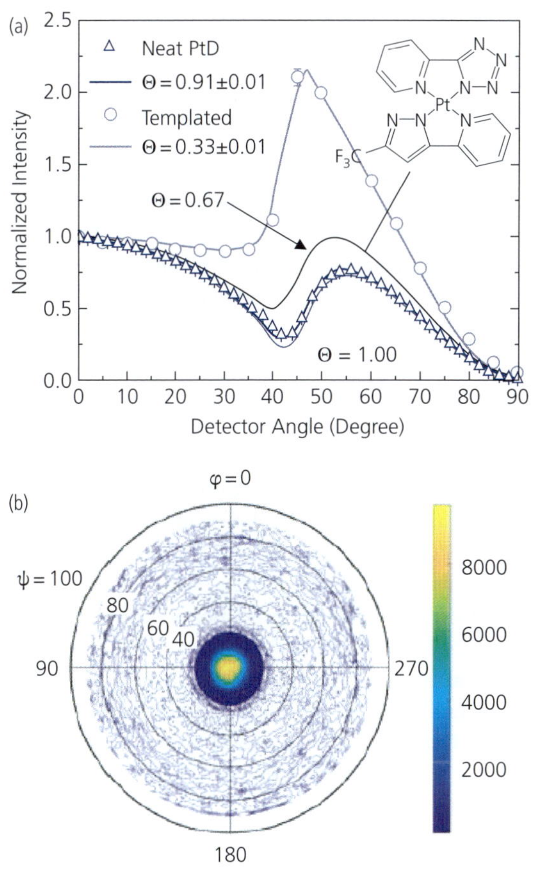

Figure 6.163 Control of the PtD transition dipole moment via templating. (a) Angle-dependent *p*-polarized photoluminescence of a neat PtD film grown on a sapphire substrate and on a pre-deposited template layer. Inset: Molecular structural formula of PtD. (b) X-ray pole figure of the (200) plane ($2\theta = 8.4°$) for a neat PtD film on sapphire. The color bar at right gives the number of counts, that is, the intensity of the X-ray signal (Kim et al., 2019).

ratio planar molecules used in fluorescent and TADF-based OLEDs. Alignment is controlled to a limited degree by the conditions used during growth (Ràfols-Ribé et al., 2018), material glass transition temperature, spin casting, substrate structure, templating, etc., although the primary driving force for alignment is the molecular structure itself. The coupling to only substrate and air modes is limited by the spatially distributed emission pattern of the dipoles. Nevertheless, efficiencies in the absence of additional substrate outcoupling schemes can potentially be as high as 45%. The primary disadvantage of dipole alignment is that it requires that the molecular design and deposition process promote TDM alignment in the desired direction. Given the numerous other requirements placed on high performance molecules used in optimized display and lighting applications, also demanding that materials meet this additional design criterion might not always be possible.

6.6.6 Summary and prospects

In this section we have reviewed several criteria that should be met for effective outcoupling, and have shown a few ingenious approaches toward achieving that goal. We have found that extracting substrate modes is achieved using a dense array of microlenses that can be fabricated by molding polymers, followed by their attachment to the substrate. Extracting waveguide and SPP modes is considerably more challenging. Yet, the use of modified substrates (e.g. sub-anode grids), and transparent contacts are effective since they can eliminate exciting modes in the first place. Extracting ~70% of the trapped light has been achieved, although not without adding complexity to the OLED design. Finally, we have shown that alignment of molecular transition dipoles parallel to the substrate plane can eliminate most waveguide and SPP modes, particularly for planar Pt-based phosphors and fluorescent molecules, as well as some Ir-complexes with non-emissive ancillary ligands.

Based on these findings, it appears likely that practical outcoupling schemes that outcouple 80% of the emitting power are possible. Since phosphorescent complexes have shown up to 100% internal efficiency, achieving this goal will result in display and lighting appliances with $\eta_{ext} = 80\%$. Efficiencies of this magnitude are now leading to new, and perhaps unprecedented applications in efficient information displays and lighting.

6.7 Reliability of organic light emitters

Since the invention of efficient OLEDs, questions have been raised about their potential for practical use given the nearly universal observation up to that time that organic electronic device properties were unstable, that is, they rapidly change over time and often suffer from catastrophic failure soon after fabrication. This lack of reliability makes such devices unsuitable for practical applications, where the user of a smart phone, tablet or TV monitor reasonably expects their appliance to work trouble-free over many years. For this reason, quantifying and increasing the reliability of OLEDs has been the focus of considerable research over several decades. Today, OLEDs are beginning to dominate the entire display industry due, among other things, to the progress made in extending their operational lifetimes.

In this section, we review the methodologies employed in quantifying OLED reliability, and discuss metrics that must be met to qualify devices for display and lighting applications. This discussion lays out general techniques that are also useful for characterizing other electronic devices including OPVs and OTFTs in Chapters 7 and 8. Once the methods have been established, we will discuss the major failure mechanisms that are active in OLEDs. Device aging, or failure, falls generally into two categories: failure due to *extrinsic* or *intrinsic* mechanisms. Extrinsic causes for degradation are those whose origins are not an inherent property of the material or the device itself. Examples include catastrophic failure due to shorts via dust particles left on the substrate, material impurities or film inhomogeneities introduced during film preparation. In Chapter 5 we showed that extrinsic impurities have many deleterious effects on device operation, such as reduction in luminescence from non-radiative recombination, reduction in exciton and charge diffusion lengths, and most assuredly on device stability. Also, exposure to environmental contaminants such as moisture or oxygen in poorly or unpackaged devices falls into this category. In the following discussion, we generally assume that extrinsic sources of failure have been eliminated in environmentally protected devices that have been fabricated by stable and proven processes, using materials that have been highly purified prior to use.

Intrinsic degradation is due to changes over time of the inherent properties of the materials used or the device architecture. While not always easily distinguished from extrinsic process, intrinsic processes include: (i) Degradation of contacts and heterointerfaces that changes the charge density and balance within the OLED structure, (ii) thermally induced morphological instabilities, (iii) dissociation and fragmentation of chemical species within the OLED, (iv) energy-driven degradation of blue phosphorescent and TADF emitters, and (v) exposure to environmental agents in OLEDs employing semipermeable (e.g. plastic) substrates.

6.7.1 Quantifying OLED long term performance and reliability

The lifetime of electronic devices is determined by first setting the criteria for failure. In OLEDs, this is typically defined as the time, t, it takes for the luminance ($L(t)$) to reach a predetermined value lower than its initial luminance (L_0), when the device is deemed to no longer be useful (i.e. it has failed). Then its *time to failure* or *lifetime* is denoted Tx, or alternatively, LTx, where x is 100 times the ratio $L(t)/L_0$ that corresponds to device failure. For example, if $L/L_0 = 0.5$ is set as the failure criterion, then the time t' that it takes to reach this luminance corresponds to $LT50 = t'$. To put this value in context, the time to failure must be specified along with the test conditions that might affect its value, including L_0, temperature, T, ambient properties (e.g. humidity), etc.

The intended application sets the conditions that define failure. For example, displays are operated at $\sim$100 cd/m^2, although we have shown in Table 6.10 that the actual pixel luminance on the display front plane can be nearly 60 times that value due to the small aperture ratio and differential aging of the R, G and B pixels. Furthermore, the eye is highly sensitive to small luminance differences between pixels. Hence, display pixels are commonly tested at $L_0 = 1000$ cd/m^2, and are assumed to fail once 5–10% of the initial luminance is lost. That is, *LT95* or *LT90* provides the failure criterion for information display pixels. The industrial metric for the lifetime of lighting sources, on the other hand, not only employs a different luminance and luminance loss criterion, but also a shift in color coordinates. Typically, OLED lighting sources are tested at $L_0 = 3000$ cd/m^2, and are assumed to have failed at *LT70*, with a maximum color shift of $\Delta u = \Delta v \approx 0.03$.

Since lifetime depends on multiple factors that vary from device to device, meaningful failure analysis must be done on a statistically significant population of devices, with the ultimate objective of predicting the operational lifetime of fresh devices manufactured by the same process as the test population. The measured lifetime that characterizes the population is specified as the *mean time to failure*, or *MTTF*, of members of that population. Other important quantities to measure include the *failure rate*, which is the number of failures per unit time, and *B(X) life* which is the time at which the probability for failure has reached *X*%. The statistical analysis of the failure of a population of devices independent of the specifics of the technology employed is known as a *Weibull analysis* (Weibull, 1939).

Several functions have been used to empirically model $L(t)$ to predict the lifetime for a particular device architecture, materials set, and fabrication process. Given their parametric nature, the best models accurately fit the luminance for times relevant for practical device operation. These functions include the *stretched exponential*:

$$L(t)/L_0 = \exp\left[-(t/\tau)^{\beta}\right], \tag{6.121}$$

where τ is the characteristic luminance decay time and β is the dispersion factor that determines the shape (or "speed") of the exponential decay (Ishii, 2003, Féry et al., 2005). This is a particularly simple functional form, requiring only two fitting parameters. Furthermore, the value of β is largely independent of L_0, as shown by the data in Fig. 6.164 for a green emitting OLED fit using $\beta = 0.53$ for L_0 from 100 to 800 cd/m^2.

There is no theoretical framework with which to predict the stretched exponential. If we assume that the luminance degradation arises from destruction of emitting molecules via reaction with other species such as excited states or polarons (Aziz et al., 1999), then the luminance depends on the concentration of surviving emissive molecules at time t. In this simple picture, we expect that $L(t)$ would follow Eq. 6.121 with $\beta = 1$, that is, a simple exponential. Since typically $\beta < 1$ as in the population in Fig. 6.164, the rate of degradation must depend on the depletion of more than one species. That is, it might be catalyzed by the addition of an excited state of a nearby emissive molecule, or an impurity. The number of nearby pairs, therefore, rapidly decreases at the early stages of life, slowing as the EML is left only with lone molecules that have no opportunity to react with appropriate species in their immediate neighborhood.

For long-lived devices, it is often inconvenient to wait until failure under the anticipated operating conditions to determine the LTx of the population. Hence, the use of a functional form such as Eq. 6.121 is essential, but its accuracy in predicting LTx depends on the accuracy with which τ and β can be determined over the time the device is under test. Hence, we often quote the *extrapolated luminous lifetime*, which is the LTx determined by extrapolating the actual measured time to the calculated lifetime of the population using a characteristic degradation function as the stretched exponential.

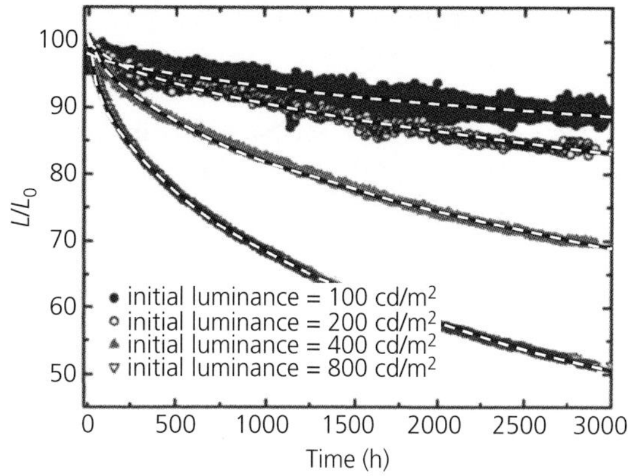

Figure 6.164 Luminance decay of an OLED for several values of L_0 shown in legend. Data are shown by points, and fits (dahed lines) to a stretched exponential (Eq. 6.121) with $\beta = 0.53$ for various τ are shown by lines (Féry et al., 2005).

An alternative means for decreasing the time under test is to use *accelerated aging* methodologies. For example, for lighting applications, we require L_0 = 1000–3000 cd/m^2 and $LT70$ = 50,000 h. For displays, the criteria are even more stringent: typically L_0 = 600–6000 cd/m^2 (depending on the pixel color, see Table 6.10) and $LT90$ = 10,000–30,000 h, since only a few percent of differential aging between color pixels is easily discerned as a shift in hue (Laaperi, 2008). Hence, it is impractical to wait for such long times to directly measure the MTTF. There are two principal means for accelerating the aging of OLEDs: by increasing intensity (which, equivalently, corresponds to an increase in drive current), or by increasing the temperature.

Once again, the aging factors are empirically determined by measuring the changes in the slope of $L(t)$ as either intensity or temperature conditions are varied. It has been found that the lifetime depends on the initial luminance following:

$$LTx(L_0) = LTx(L_{0tst})\cdot\left[\frac{L_{0tst}}{L_0}\right]^n, \tag{6.122}$$

where L_{0tst} is the initial luminance of the device under accelerated test conditions, which is greater than its value under normal operating conditions (L_0), and n is

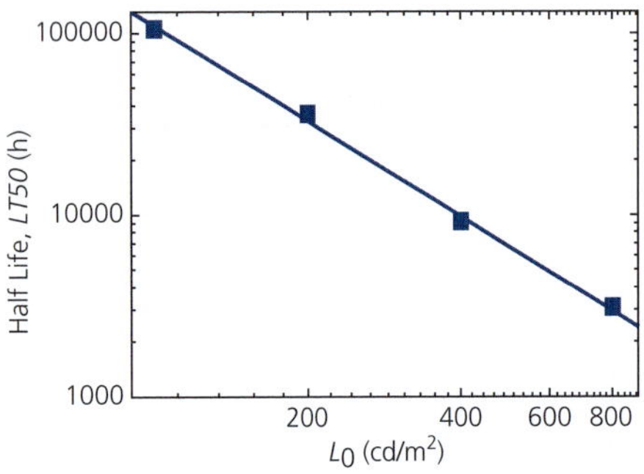

Figure 6.165 Accelerated aging of the green OLED in Fig. 6.164 via increasing initial luminance, L_0. The data are shown by squares, and the power law fit using Eq. 6.122 with $n = 1.7$ is shown by the line (Féry et al., 2005).

the acceleration factor that varies between 1.5 and 2.5 (Féry et al., 2005, Orselli et al., 2012, Zhang et al., 2014b). A plot of *LT50* vs. L_0 for the device in Fig. 6.164 is shown in Fig. 6.165. The data (squares) are fit by the straight line with $n = 1.7$. Note, that since $L \propto j$, we conclude that lifetime is determined by the total amount of charge that has passed through the OLED, and since $n > 1$, increasing the initial charge density results in more rapid generation of non-radiative defects than at lower charge densities. This is consistent with the model that aging is catalyzed by multiple excitations on a single molecule, or between neighboring molecules. Our discussion of energy-related degradation in Section 6.7.4 lends further support to this hypothesis.

Increasing the operating temperature can also accelerate aging. We might assume that the probability for destroying an emitting molecule is thermally activated, that is, the probability is $P = \exp(-\Delta E_A/k_BT)$, where ΔE_A is the activation energy for the formation of non-radiative molecular species. As noted above, the probability is increased by the presence of other emitting molecules. Hence the energy for defect formation decreases as the number of emitting molecules, n_e, decreases (Féry et al., 2005):

$$\Delta E_A(n_e) = \Delta E_{A0} - K(n_{e0} - n_e), \tag{6.123}$$

where K is a constant, and n_{e0} and ΔE_{A0} are the number of emitting molecules and the defect activation energy at $t = 0$. Then the rate of defect formation is

$$\frac{dn_e}{dt} = \frac{1}{\tau_r}\exp(-\Delta E_{\rm A}(n_e)/k_BT) - K' n_e, \tag{6.124}$$

where K' is chosen to meet the boundary condition that $dn_e/dt = 0$ at $t \to \infty$, and τ_r is the radiative emission time for the molecules that competes with defect formation. Solving Eq. 6.123 using Eq. 6.124 yields a thermally activated luminance decay time

$$\tau = \frac{\tau_r}{n_{e0}}\exp\left[\left(\Delta E_{A0} - K n_{e0}^2\right)/k_BT\right]. \tag{6.125}$$

Hence, the luminance time constant, τ, decreases with increasing T. From Eq. 6.121, this suggests that increasing temperature will accelerate the degradation in luminance. The dependence of the activation energy on the number of emitting centers in Eq. 6.125 thus yields the stretched exponential behavior of the luminance.

We have previously noted that, lacking an analytical theory for device degradation, the fits to $L(t)$ are inherently empirical. Hence, other expressions have been used to approximate this decay, but none with as few parameters as Eq. 6.121 (Tsujimura et al., 2012, Yoshioka et al., 2014). In fact, the precision of the extrapolated lifetime using this simple expression has not always been acceptable (Meerheim et al., 2006, Orselli et al., 2012). To increase the accuracy over a long test period, a sum of exponentials can be employed, viz. (Yoshioka et al., 2014)

$$L(t)/L_0 = \lambda\exp(-t/\tau_1) + (1-\lambda)\exp(-t/\tau_2), \tag{6.126}$$

where λ, τ_1 and τ_2 are fitting parameters. The first term expresses the rapid initial decay, and the second fits the extended long-term decay. Hence, $\tau_1 \ll \tau_2$ and λ depends on the relative magnitudes of these two aging regimes. Since most of the luminance loss occurs over the long term in reasonably high performance OLEDs, it is safely assumed that $\lambda \ll 1$. In fact, the physical assumptions leading to both Eqs. 6.121 and 6.126 are similar; that initial decay is more rapid than long term luminance loss due to a an initially accelerated depletion of emitting molecules.

The use of Eq. 6.126 entails the separate fits to the early and late stages of luminance decay, with $L(t)$ vs. j shown in Fig. 6.166 for a PHOLED with the structure: glass substrate/150 nm ITO/60 nm HATCN/30 nm $Ir(ppy)_3$ (6%):CBP/10 nm BAlq/30 nm Alq_3/1.6 nm LiF/150 nm Al. The sum of exponentials provides an accurate fit to the data at all current densities, with a reasonable expectation for predicting *LT50* even at the lowest current densities examined. The rather small impact of early aging is observed at the lowest current densities ($m \leq 2$), where adjustments at $t < 200$ h were needed to assure that $L(t)/L_0 = 1$ at $t = 0$.

The thermal activation of degradation following Eq. 6.125 is shown in Fig. 6.167 for the same green PHOLED. We observe that the data for the long term rate is thermally activated, independent of

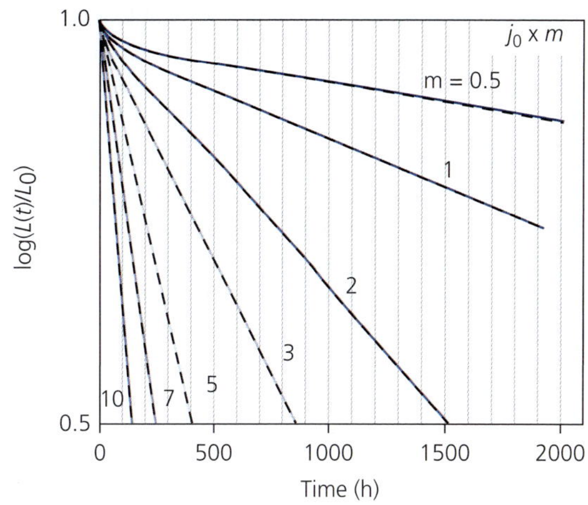

Figure 6.166 Luminance vs. time of a green PHOLED at several different current densities at $T = 25°C$. Here, $j_0 = 6$ mA/cm^2 corresponds to $L_0 = 1800$ cd/m^2. The data are shown by solid lines, and the fits to the sum of exponentials (Eq. 6.125) are dashed lines (Yoshioka et al., 2014).

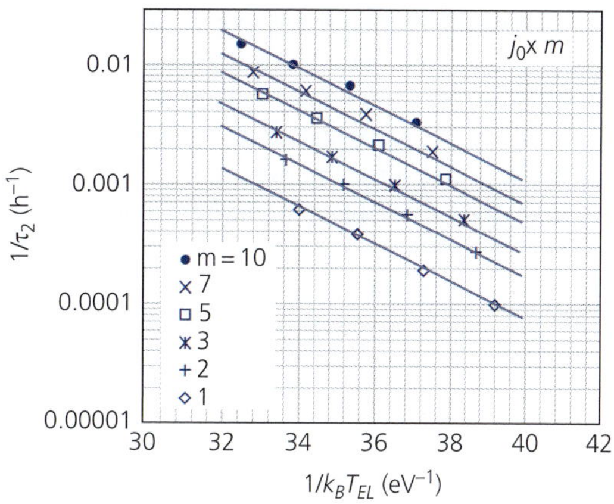

Figure 6.167 Thermal activation of τ_2 in Eq. 6.126 at several current densities. Here, $j_0 = 6$ mA/cm^2 corresponding to $L_0 = 1800$ cd/m^2. T_{EL} is the actual temperature of the organic layers as determined from the temperature dependent j–V characteristics. The data are shown by points, and the fits to Eq. 6.125 (assuming $Kn_{e0}^2 \to 0$) are lines (Yoshioka et al., 2014).

current density. The activation energy, which is also independent of j, is $\Delta E_{A0} = 0.34 \pm 0.02$eV. The thermally activated luminance decay suggests that increasing temperature can predictably accelerate the OLED aging. However, it is important that a single value of ΔE_{A0} is associated with each failure mechanism. Hence, the aging temperature should not be so high as to introduce new degradation channels that are inactive under normal device operating conditions (i.e. employing $T > T_g$, the glass transition temperature of the layers).

Combining the results of the data in Figs. 6.166 and 6.167, along with the sum of exponentials empirical analysis, we arrive at an expression for the long-term decay time constant of

$$\frac{1}{\tau_2} = K'' j^{\alpha} \exp(-\Delta E_{A0}/k_B T), \tag{6.127}$$

where K'' is a fitting parameter. The universality of Eq. 6.127 is illustrated by a plot of all the lifetime, current and temperature data of Figs. 6.166 and 6.167, in Fig. 6.168 using $\alpha = 1.20$ and $\Delta E_{A0} = 0.34$ eV. The observation that all the data fall on a single line implies that ΔE_{A0} and α are independent of the test conditions. Furthermore, these data support the conclusion that the luminance depends on the total amount of charge that has flowed through the device at a given time during its life cycle.

The test conditions used in obtaining reproducible and accurate data for the long term reliability of the devices must be carefully controlled. As noted previously, to achieve long lifetimes and reliable data

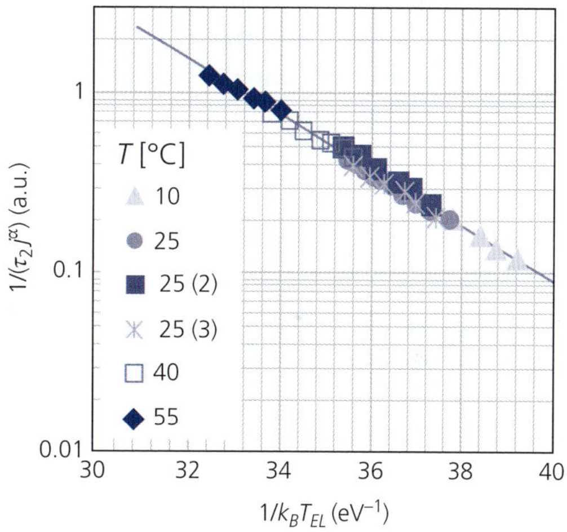

Figure 6.168 Universal dependence of the luminance loss of green PHOLEDs on current and ambient temperature (T) following Eq. 6.127 (Yoshioka et al., 2014).

sets, the materials used must be purified, and the fabrication conditions are sufficiently stable to ensure there is little variation between device characteristics made on different processing cycles. Absent this stability of the fabrication process, the aging tests are likely to be tainted by minor, and relatively uninteresting variations between device populations whose presence will obscure the underlying sources of failure and result in errors in the extrapolated device lifetime.

Beyond process stability and materials purity, long-term reliability tests should have the following features:

(i) A statistically significant population of devices should be examined to ensure reliable determination of *LTx*. Since the random uncertainty of a parameter in a population of *N* devices is proportional to $\sqrt{N}$, a minimum number of devices for a trustworthy sampling typically ranges from 10 to 100.

(ii) Care must be taken to ensure the test parameters (e.g. j, T) and measurement systems are accurate and stable over long test periods of typically 100–10,000 h.

(iii) The devices must be maintained in a controlled environment for the duration of the test. These parameters include humidity, temperature and ambient light. It is common that the OLEDs are contained in packages that are epoxy sealed in ultrahigh purity nitrogen (<1 ppm O_2 or H_2O), along with a desiccant to scavenge residual H_2O incorporated in the organic films and package prior to encapsulation. A typical OLED package used in lifetime and in practical applications is illustrated in Fig. 6.169. The packages are sealed with a UV cured epoxy applied around the periphery of the lid and substrate. The lid is often either metal or a second thin glass sheet. A desiccant such as BaO is deposited on the inner surface of the lid.

6.7.2 Degradation due to contacts and interfaces

The quality of the contacts and heterointerfaces within the device structure can have significant effects on the OLED lifetime. This failure mode can result from defects, impurities, charge, and excited state-induced chemical degradation of the layers. And when the interface presents a significant barrier to charge transport it can induce thermally-activated morphological and chemical changes. Many of these effects are of exogenous origin, and hence can be reduced or eliminated by using only stable materials, and the OLED fabrication processes are carefully controlled to prevent defects and impurities from entering the device active area.

Dark spot defects visible at the cathode contact have been identified as a significant source of failure (Burrows et al., 1994). The formation of dark spots was microscopically investigated in conventional Alq_3/TPD devices. The devices exhibit considerable non-uniformities in EL intensity across their diameters (Fig. 6.170a and b). Several bright spots observed when the device is first turned on in air eventually transition to dark spots. Also, small dark spots grow, and even a few dark spots become intensely, albeit briefly luminescent with time. These changes are attributed to film non-uniformities, or to dust or other defects residing on the substrate that create pinholes. The electric field at the edges of the non-uniform region is higher than elsewhere, forming areas where the current (and hence the EL) is higher. These quickly carbonize, resulting in regions of relatively low emission intensity that continue to grow over time until the entire device area is consumed, rendering the OLED entirely non-emissive.

Packaged devices aged for several hundred hours show numerous pinholes apparent in the top electrode surface. When viewed from the bottom, the pinholes are still visible, which indicates that they penetrate the entire film. Furthermore, they are surrounded by discolored and non-emissive halos (Fig. 6.170c and d). These regions are possibly due to de-wetting (i.e. blistering) of the organic from the metal cathode. As in the EL images, the pinholes are likely due to defects in the film, electrode or ITO, creating regions of high electric field that burn out over time. It has been proposed that the discolored areas may be due to galvanic corrosion of Mg, resulting in the evolution of $Mg(OH)_2$ gas that forces the

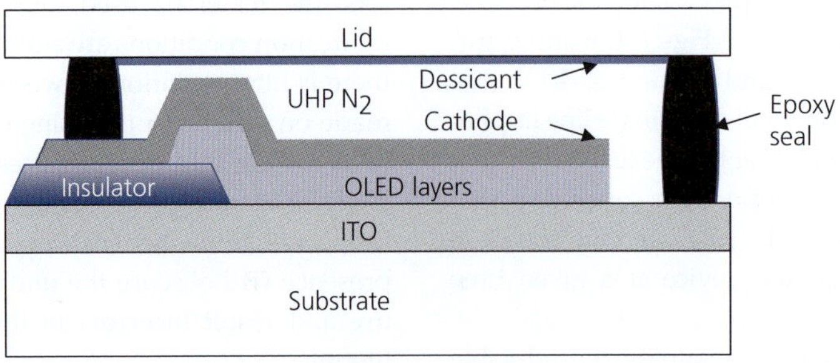

Figure 6.169 Example OLED packaging scheme. The insulator (often SiN_x or SiO_2) is used to prevent the metal cathode and ITO anode from shorting. The desiccant (e.g. BaO) is pre-deposited onto the glass or metal lid. Ultrahigh purity (UHP) N_2 or other inert gas fills the gap between lid and substrate. The package is sealed with UV curable epoxy applied around its periphery.

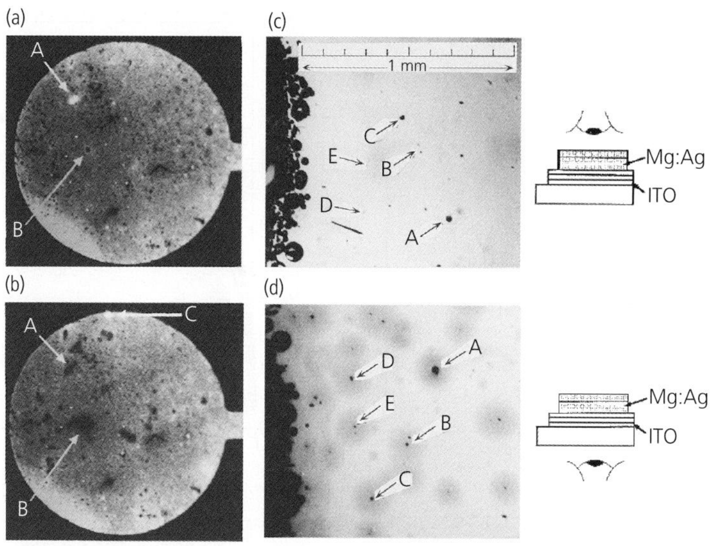

Figure 6.170 (a) Relative EL intensity image of an as-grown unpackaged Alq_3/TPD OLED as viewed via the ITO contact. (b) Image taken after operation for 4 min in air, showing that one defect (A) transitions from a high intensity spot to a dark spot, and another that begins as a small dark spot (B) has grown. An edge defect (C) transitions from dark to light. (c) Micrographs near the edge of the cathode contact of an encapsulated OLED after several hundred hours of operation as viewed from the top and (d) bottom surfaces of the device as shown in diagrams to the right (Burrows et al., 1994).

organic to delaminate from the electrode around the pinholes, subsequently admitting water into the active region (Aziz et al., 1998).

Metal cathodes other than Mg:Ag have shown similar effects of electrolysis with water. For example, Alq_3/TPD OLEDs using Al contacts develop nonemissive, delaminated dome-like bubbles after operation in air. Mass spectrometry analysis indicates that these delaminated areas are filled with oxygen and hydrogen as products of electrolytic reactions between water and Al (Do et al., 1996). The ITO anode, too, can impact device lifetime. Plasma and other surface treatments clean the surface, and reduce the injection barrier to holes (and hence reduce the electric field across the anode/organic interface) leading to improved OLED operational lifetime (Wu et al., 1997b). Indeed, residual water in PEDOT:PSS anodes can lead to rapid degradation, similar to the effects found when cathodes are exposed to humid environments (Fehse et al., 2008).

Organic–organic interfaces have also been implicated in device degradation. Specifically, interactions between the high density of hole polarons and excitons at HTL/EML interfaces lead to a rise in operating voltage that accompanies a decrease in luminance. Following the method of Giebink et al. (2009) described in the following section, a hole-only conducting device was used to investigate changes in the interfaces of a 5 nm thick layer of a TPBi ETL sandwiched between two 20 nm thick CBP HTLs, as shown in the inset of Fig. 6.171a. The voltage rise across devices (ΔV) operated in nitrogen was monitored as a function of time under 0.5 mW/cm^2 illumination at $\lambda = 365$ nm (denoted by **L**), under a constant current density of 20 mA/cm^2 (**j**), and under their combination (**L+j**). Excitons are only generated in TPBi at this wavelength. Using these various conditions, the separate effects of current and illumination are determined.

In Fig. 6.171a, there is a minor increase in voltage over time for conditions **L**, a significantly higher increase with **j**, and a higher increase when both excitations are simultaneously applied, **L+j**. Notably, this combined excitation results in a significant increase in voltage rise compared to the sum of the voltage rises from separate application of these stimuli (solid line corresponding to $\Sigma(\mathbf{L},\mathbf{j})$). In contrast, in a device lacking the HTL/ETL interface, there is no enhanced degradation from the combination of current and illumination, as shown in Fig. 6.171b. From this we infer that excitons generated in CBP interact with holes in TPBi at their interface, inducing the voltage change.

The polaron/exciton induced degradation was further examined by placing a very thin red fluorescent DCJTB layer doped at 4% in CBP in the proximity of the interface. This layer decreases the exciton lifetime from ~1.3 ns in CBP, to 0.5 ns in the doped zone. If excitons play a role in degradation, then decreasing their natural decay rate should result in a decreased impact on the operating voltage over

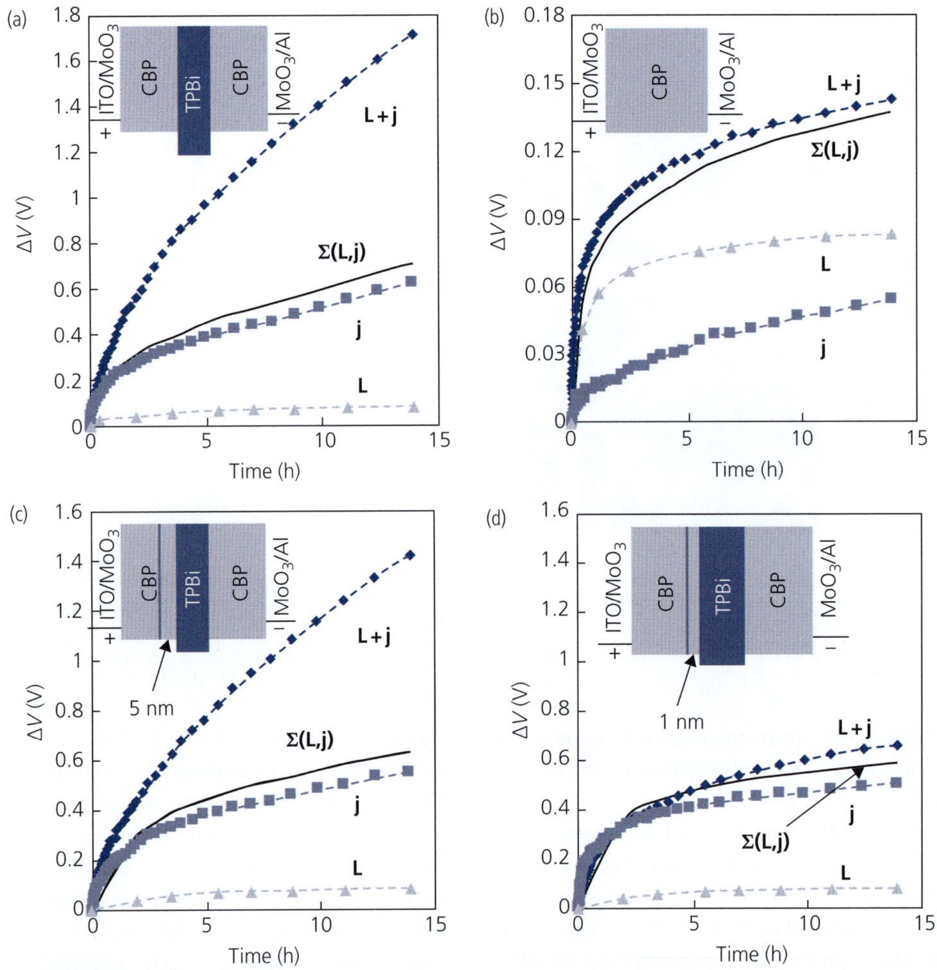

Figure 6.171 Voltage rise (ΔV) vs operation time for hole only devices in the presence of current (**j**), illumination at $\lambda = 365$ nm (**L**) and the presence of both (**L+j**). The data are shown by points, and the solid line is the sum of the **j** and **L** data, denoted Σ(**L,j**). In (c) and (d), a thin layer of the red emitting fluorophore, DCJTB is placed at (c) 5 nm and (d) 1 nm distance from the interface between CBP and TPBi. Insets: Device structures used in the corresponding tests. After Wang and Aziz (2013).

time. That is indeed the case in Fig. 6.171c and d, where placing the DCJTB layer within a Förster transfer radius of the interface results in a smaller increase in ΔV than when it is placed farther away. When placed within 1 nm of the interface (Fig. 6.171d), the sum of the individual impacts of current and illumination, Σ(**L,j**), is nearly the same as their combination, **L+j**.

The coexistence of a high density of excitons and polarons in OLEDs can result in interactions that lead to physical aggregation of the host (CBP) molecules (Wang et al., 2014b). Aggregates may introduce mid-energy gap interfacial defects that lead to increased voltage and decreased luminance. The formation of interface traps was identified by thermally stimulated current measurements of TADF-based OLEDs with ultrathin (~2–3 nm) dipolar Liq layers located at the EML/HBL or HBL/ETL interface (Tsang et al., 2016). These morphological effects appear to be more prevalent in wider energy gap hosts (Zhang and Aziz, 2016). Attribution to morphological changes is based on an absence of new chemical species formed during aging, or to thermal effects that would introduce spectral changes for the **j** devices that are also found in the **L+j** devices. A long wavelength spectral feature due to aggregation (and thus excimer formation) in an OLED comprising these two materials was observed only in the **L+j** case (Wang et al., 2014b). Unfortunately, no explanation as to why excitons and polarons induce aggregation of host molecules has been proposed, nor is there direct morphological evidence for the existence of such aggregates.

An alternative explanation for these effects is that a decrease in energy barriers at the interfaces reduces the charge pile-up, thus reducing chemical reactions catalyzed by cationic or anionic species. The effectiveness of reducing energy barriers is tested by blending the electron and hole transporting materials in the EML. Schematic illustrations of three different compositionally graded HJ EMLs are shown in Fig. 6.172a. The conventional abrupt heterostructure consists of an ETL doped with the fluorophore adjacent to the HTL. Holes, electrons, and excitons pile up at the HJ between the layers, with the charge density depending exponentially on the size of the HOMO and LUMO energy offsets at the interface. In the graded mixed HJ, the dopant resides in an area where the HTL and ETL are mixed. This allows for considerable penetration of both electrons and holes in the exciton formation region, reducing the peak density of these species, but not the total number of excitons generated. The third device is a uniformly mixed HTL-ETL-chromophore device that spreads the polarons and excitons evenly throughout the blended EML.

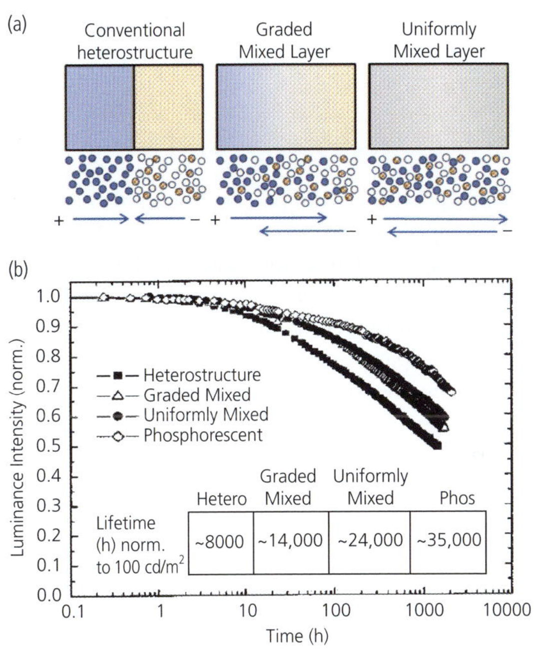

Figure 6.172 (a) Schematic of three types of HJ OLEDs: a conventional bilayer HJ, a graded-mixed HJ where the ETL and HTL are blended along with the dopant across the EML, and a uniformly mixed co-host device. The "range" of hole and electron injection is illustrated by the arrows beneath each junction. (b) Luminance as a function of aging time for packaged devices using C545T as the fluorescent dopant with the structures in (a), or an abrupt HJ $Ir(ppy)_3$ PHOLED. The LT_{50} normalized to $L_0 = 100$ cd/m^2 for each device is shown in the inset table. After Chwang et al. (2002).

The lifetimes of three green emitting HJ OLEDs in Fig. 6.172 based on C545T-doped fluorescent OLEDs employing NPD as the hole transporting, and Alq_3 as the electron transporting hosts were compared to an $Ir(ppy)_3$:CBP abrupt HJ PHOLED. The most efficient fluorescent device had an abrupt HJ with an efficiency of 15.5 cd/A, followed by the graded mixed device at 9.5 cd/A, and finally the graded junction device at 8 cd/A. The lower efficiency of the graded devices results from reduced charge confinement, and hence a broader emission zone (Chwang et al., 2002).

The reduction of local densities of polarons and excitons leads to a considerable increase in operational lifetime (see Fig. 6.172b). Consistent with the hypothesis that charged and excited molecules lead to degradation, the abrupt junction device shows a lifetime of $LT50 = 8000$ h (normalized to $L_0 = 100$ cd/m^2), compared with 14,000 h for the graded mixed structure, and 24,000 h for the uniformly graded device. Interestingly, the PHOLED has the longest lifetime with $LT50 = 35{,}000$ h. This is due to its requiring a significantly lower current to achieve the same luminance as the fluorescent devices (cf. Eq. 6.127), along with the more rapid harvesting of triplet states that linger in fluorescent devices (Burrows et al., 2000). Reactive triplets either on the host or dopant in fluorescent devices have lifetimes of seconds or longer, during which time the molecules comprising the EML can undergo chemical dissociation. In PHOLEDs, the triplet is harvested by the dopant and radiatively recombines on timescales of a few hundred nanoseconds to a few microseconds, drastically reducing the time available for excited state reactions. Hence, it is reasonable to assume that the vastly shortened triplet lifetimes in PHOLEDs vs. that in fluorescent devices should also yield lifetime improvements. We will show in Section 6.7.4 that this is indeed true for all but blue emitting PHOLEDs where energy-driven aging mechanisms are active.

6.7.3 Degradation due to thermal effects and chemical decomposition

At this point it should be apparent that degradation is most likely due to a combination of effects. That is, assigning device aging simply to aggregation from a combination of polarons and excitons, or to layer delamination in the absence of thermal effects, environmental exposure, or energy-induced molecular decomposition generally provides a far from complete picture of all the mechanisms that shorten device operational lifetime.

In this context, we showed in Section 6.7.1 that the decrease in luminance is accelerated at elevated temperatures, with an activation energy, ΔE_A. Thermal effects are an excellent example of how a single environmental variable can lead to numerous degradation pathways by accelerating morphological changes, degradation of molecular species, interdiffusion of constituents across interfaces, layer delamination, and so on. Primarily, elevated temperatures have two effects: disrupting the morphology of the as-deposited films that is particularly significant for low T_g materials, and decomposition of the constituent molecules.

Exceeding T_g of the materials comprising the OLED can lead to their crystallization on re-cooling. The crystallized areas disrupt the lamellar film structure, leading to shorts across the EML. For sufficiently low T_g, this process can occur even at room temperature, and can be accelerated by local Joule heating during device operation. For most devices, Joule heating (with dissipated power density of jV) is small, although this does not avoid the problem of crystallization when the ambient temperature is increased. Hence, finding materials sets with high T_g is of great importance (O'Brien et al., 1998). While the device temperature rise may only be a few degrees above ambient, the local temperature at energy barriers between layers, or within a particular resistive layer itself, may be considerably higher.

The surface temperatures of several WOLED panels measured using a *forward looking infrared* (FLIR) *thermal imaging* camera are shown in Fig. 6.173. The two 15×15 cm^2 phosphorescent panels have efficacies at $L_0 = 3000$ cd/m^2 of 49 lm/W, whereas the 8×8 cm^2 fluorescent panel has an efficacy of 16 lm/W. The luminous emittance plotted in Fig. 6.173 is the efficacy times the area filled by the emitting surface. The latter is known as the *aperture ratio*. The surface temperatures of the PHOLED panels are considerably lower than the fluorescent panel. At $L_0 = 3000$ cd/m^2, the surface temperatures are 29.4°C and 27.2°C for the panel in the center inset photo (panel 1) and the one on the right (panel 2), respectively. The difference is a result of the different aperture ratios for the two panels. In contrast, the lower efficiency fluorescent panel has a surface temperature of 40.5°C (Levermore et al., 2012).

The relationship between operating temperature and lifetime is shown in Fig. 6.174, with a thermal image of the surface of one half of panel 2 in the absence of heat sinking. Consistent with the discussion in Section 6.7.1, there is a pronounced

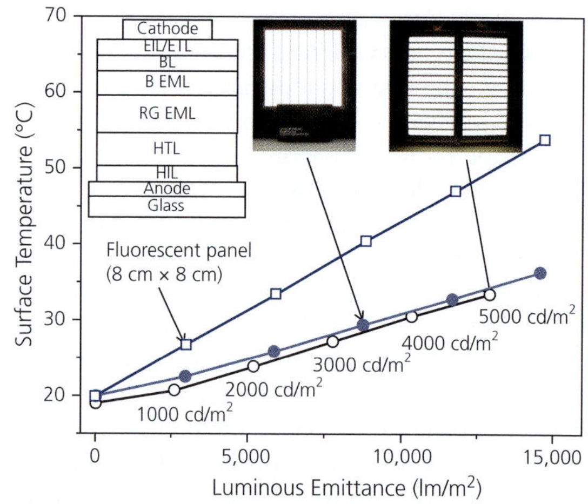

Figure 6.173 Temperature of white light emitting panels vs. luminance emittance. The PHOLED panels are 15×15 cm^2, and have the structure and layouts shown in the insets. The fluorescent panel is 8×8 cm^2 (Levermore et al., 2012).

dependence of panel lifetime on temperature. At 3000 cd/m^2, the activation energy of the lifetime is $E_A = 0.40$ eV, leading to a 1.65-fold increase in lifetime per 10°C reduction in temperature. The cause of this degradation is not apparent from these data alone.

The relationship between device temperature and current is

$$jV - P_{opt} = c_{sp}\rho L\frac{\partial T}{\partial t} - \kappa_T \nabla T, \qquad (6.128)$$

where P_{opt} is the optical power emitted by the OLED, c_{sp} is the specific heat capacity (with units of [J/kg K]), ρ is the mass density of the solid, L is the layer thickness, and κ_T is the thermal conductivity ([W/m K]). The terms on the left hand side of Eq. 6.128 give the power dissipated via Joule heating, which is decreased by the fraction of power carried by outcoupled photons, the first term on the right is the transient temperature rise, and the last term is due to thermal conductivity via the substrate or convection. We assume that the temperature rise is sufficiently small that radiative losses (equal to $\varepsilon_{rad}\sigma_{SB}T^4$, where ε_{rad} is the emissivity and $\sigma_{SB} = 5.67 \times 10^{-8}$ W m^{-2} K^{-4} is the Stefan–Boltzmann constant) are small. However, this is not always the case, with a more precise approach to calculating the temperature of multilayer OLEDs to be found elsewhere (Qi and Forrest, 2011, Bergemann et al., 2012).

Since the primary heat loss in steady state is via the substrate, the choice of substrate materials is of vital importance. Figure 6.175 shows thermal images as a function of time for top emitting devices with the structure: substrate/300 nm Ag/4 nm MoO_3/54 nm

NPD/53 nm Alq_3/1 nm Yb/20 nm Ag/45 nm Alq_3 driven at $L_0 = 10{,}000$ cd/m^2. The top Alq_3 layer serves as an antireflection coating to improve optical outcoupling (Chung et al., 2009). Three different substrates were used: glass (with $\kappa_T = 1$ W/m K), polymer planarized SUS stainless steel (16 W/m K) and Si (150 mW/m K). The substrate conductivity plays a significant role in the steady state temperature (64.5, 40.5, and 21.4°C, respectively) as well as the lifetime. With $L_0 = 10{,}000$ cd/m^2, the *LT80* = 31 h for glass, 96 h for SUS stainless, and 198 h for Si, corresponding to $\Delta E_A = 0.39$ eV.

The temperature can be further reduced through the use of pulsed rather than continuous current drive. The pulsed duty cycle should be sufficiently small to allow for thermal recovery with time constant $\tau_T = c_{sp}\rho L^2/\kappa_T$ between pulses (Cao et al., 2015). Temperature is also reduced by decreasing the contact (e.g. ITO/HTL) energy barriers or by employing doped transport layers to decrease series resistance, and thus Joule heating (Meerheim et al., 2006).

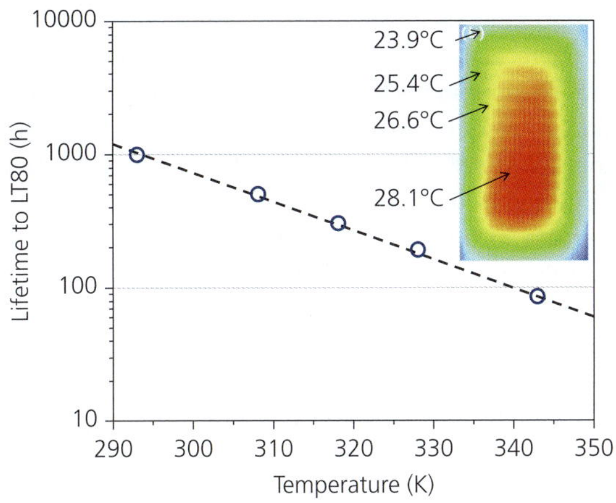

Figure 6.174 Panel lifetime to *LT80* vs operating temperature. Inset: FLIR surface image of the panel at 3000 cd/m^2 after 10 min of operation (Levermore et al., 2012).

The morphological damage from heating can be manifested by local crystallization or film expansion. When the films are heated to $T > T_g$, the entire structure undergoes expansion, leading to strain between layers, and possibly delamination. A dramatic demonstration of the dependence of strain with temperature in Fig. 6.176a obtained via X-ray beam specular reflection from the surface of an initially planar TPD/Alq_3 heterostructure on a Si substrate. The initial molecularly flat surface results in an extended series of intensity oscillations due to Fabry–Pérot modes from X-ray reflection at each interface within in the structure. The film at $T = 91$°C has fewer resolvable oscillations than at lower temperatures, indicating a reduction in flatness.

A Fourier transform of these oscillations yields the layer thicknesses (Fenter et al., 1997). The thermal expansion of the two layers plus that of the entire film is shown in Fig. 6.176b. The thickness is unchanged up to 83°C, at which point the Alq_3 thickness decreases by 5 Å while the TPD film increases by 11 Å, leading to a net film expansion of 6 Å. At 93°C, all three layers show a rapid increase in thickness, accompanied by a decrease in intensity of the Fourier transform peaks (Fig. 6.176c). This results from film roughening: as the film thickness varies across the sample, beam scattering increases, resulting in destructive interference of the reflected X-ray wavefront.

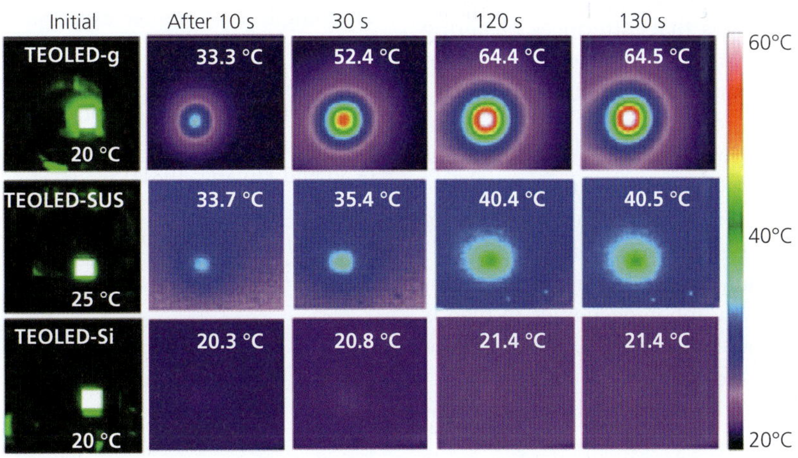

Figure 6.175 Visible (left hand column) and FLIR images of top emitting fluorescent OLEDs (TE-OLEDs) as a function of time after turn-on for devices made (top to bottom) on glass, SUS, and Si substrates. The initial luminance is 10,000 cd/m^2 (Chung et al., 2009).

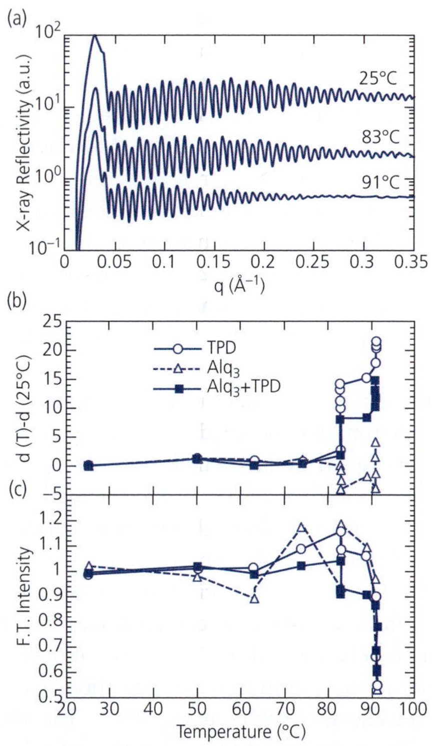

Figure 6.176 (a) X-ray reflectivity data of a 29.6 nm TPD/51.8 nm Alq_3 heterojunction on a Si substrate at three different temperatures. *q* is the X-ray momentum transfer which is a function of scattering angle. (b) Thickness (*d*) expansion (in Å) and (c) Fourier transform intensity of the oscillations for the heterostructure and layers vs. temperature. Here, $d(T)$ is the layer thickness at temperature, *T* (Fenter et al., 1997).

This reduces the intensity and the regularity of the oscillations.

These data, along with the absence of X-ray evidence for crystallization, suggest that the TPD film undergoes expansion once its glass transition temperature of 60°C is exceeded. Expansion results in strain between the films that introduces blistering and other damage that propagates to the heterointerface. Thermal expansion induced strain is also likely to introduce defects that lead to non-radiative recombination of charges and excitons, and hence a reduced OLED luminance. While the effects revealed in Fig. 6.176 occur by uniform heating, similar effects can occur on a local level due to current "hot spots" arising from non-uniformities in growth, inhomogeneities in blends of more than one molecular species, contacts, substrate irregularities, etc.

Thermal effects can be minimized by using materials with a high T_g. Mixing low and high T_g components in a single layer increases T_g above that of the the lowest component. In effect, the doping of the film by a high T_g species acts to "pin" the morphology in its as-grown state. This method for increasing T_g is useful when a low T_g material has a desired optoelectronic property (e.g. a large energy gap in a blocking layer) but has insufficient morphological stability needed for a long device lifetime. The glass transition temperature for the mixture is

$$\frac{1}{T_{gM}} = \frac{w_A}{T_{gA}} + \frac{w_B}{T_{gB}}, \tag{6.129}$$

where $w_{A,B}$ is the weight percent of components A, B, respectively (Rudin and Choi, 2013). Component mixing has been used to significantly extend the operational lifetime of $Ir(ppy)_3$:CBP PHOLEDs by blending the blocker BPhen ($T_g = 62°C$) with BAlq ($T_g = 92°C$) (D'Andrade et al., 2003).

Simultaneous exposure to heat and current generates defects in the bulk due to molecular fragmentation. The fragments form exciton quenching as well as charge recombination sites (Scholz et al., 2007, Kondakov, 2008, Kondakov, 2015). Trap generation is inferred from an increase in operating voltage and a decrease in luminance efficiency. We can determine the relationship between the voltage rise and luminance loss over time by assuming that traps are formed at rate K_X. Then the density of fixed charge trapped at position x, is $Q(x)$. Following the treatment of Giebink et al. (2009), we can write Poisson's equation as

$$-\frac{d^2V}{dx^2} = \frac{dF}{dx} = \frac{q}{\varepsilon_r\varepsilon_0}(n(x) + n_t(x) + \zeta Q(x)). \tag{6.130}$$

Here, the trapped electron density is $n_t(x)$, and the fixed charge due to defects is treated as a perturbation of strength, ζ. A similar expression applies for hole densities $p >> n$.

Following the treatment in Section 4.4.2 for an exponential trap distribution in the limit of $n \ll n_t$, we obtain a solution of the form: $V \cong V_0 + \zeta V_1$, where (Kao and Hwang, 1981)

$$V_0 = \frac{l}{2l+1}\left[\frac{l+1}{l}d\right]^{2l+1/l+1}\left(\frac{qH_t}{\varepsilon_r\varepsilon_0 N_{LUMO}^{1/l}}\right)^{l/l+1}\left(\frac{j}{q\mu}\right)^{1/l+1}. \tag{6.131}$$

Here, d is the film thickness and μ is the electron mobility. The characteristic trap temperature is T_c, where $l = T_c/T$, and N_{LUMO} and H_t are the densities of LUMO and trap states at the transport level, respectively. When an a.c. voltage is applied whose modulation rate exceeds the characteristic trapping frequency, the small-signal capacitance is given by

$$C = \frac{dQ_{tot}}{dV} = \frac{1}{\mu}\frac{d}{dV}\left(\frac{j}{F}\right), \quad (6.132)$$

where Q_{tot} is the total mobile charge. Then from Eq. 6.131,

$$C_0 = q\frac{N_{LUMO}}{H_t^l}\left(\frac{l(2l+1)}{(l+1)^2}\frac{\varepsilon_r\varepsilon_0}{q}\right)^l \frac{l}{(2l-1)}\frac{V^{l-1}}{d^{2l-1}}. \quad (6.133)$$

Adding the area-specific geometric capacitance, $C_{geom} = \varepsilon/d$ to C_0 gives the total device capacitance.

We start with the *ansatz* that defect formation requires the presence of a polaron on the molecule, forming an anionic or cationic state. Then the rate of change of the defect density is:

$$\frac{dQ}{dt} = K_X(n + n_t). \quad (6.134)$$

Defects can be formed on both the ground and excited states. Thus, K_X is written as

$$K_X = K_{X1} + K_{X2}G, \quad (6.135)$$

where K_{X1} is due to ground state degradation while K_{X2} describes degradation of the excited state, such as via exciton–polaron annihilation. Hence, K_{X2} is only present for molecules subject to an excitation at rate, G.

Equation 6.130 can be solved to first order in the perturbation ζ under the assumption that $n \ll n_t$. Then, the voltage at constant j increases with time according to

$$V(t) \approx V_0 + \zeta V_1(t) = V_0\left(1 + \frac{\zeta t l K_X}{t K_X + l + 1}\right). \quad (6.136)$$

Similarly, the capacitance decreases over time:

$$C(t) \approx C_0\left(1 + \frac{\zeta t l K_X}{t K_X + l + 1}\right)^{-1}. \quad (6.137)$$

Temporal changes in voltage, $\Delta V = V(t) - V_0$, and capacitance, $\Delta C = C(t) - C_0$, are thus nonlinear at a constant j since polaron-induced defect states increase the voltage, which in turn increases the polaron density.

A schematic diagram of an experimental setup used to determine the role of defects in the OLED voltage and luminance over time is shown in Fig. 6.177a. In a demonstration of the method, four equivalent electron-only test structures (eTST) shown in Fig. 6.177b were simultaneously monitored to separately understand influences of polarons and excitons. The test conditions for each of the devices are the same as in Fig. 6.171: constant intensity optical excitation (condition **L**), constant current (**j**), and the combination of the two (**L+j**). A device that is exposed to neither current nor illumination is used as a control (**C**). Each device is monitored over an extended time interval by periodically sampling both the voltage at constant j, and its capacitance-voltage characteristics. Optical excitation at $\lambda = 442$ nm is only absorbed by Ir$(ppy)_3$, thereby eliminating exciton generation on mCBP for the device in Fig. 6.177b (Holzer et al., 2005).

The eTST device consist of the 15 wt% Ir$(ppy)_3$:mCBP EMLs sandwiched between electron conducting BAlq layers. Analogous hole-only devices locate the EML between NPD layers. To isolate effects of degradation in the BAlq buffer, a control device (eCTL) lacking the EML was fabricated, shown in Fig. 6.177b. Extrinsic factors such as degradation due to environmental exposure are identified using the control device, eCTL, and the **C** test condition.

Example $\Delta C = C(t) - C(t=0)$ data collected at 12 V for the electron-only devices in Fig. 6.177b at optical intensities of 1.1 and 19.5mW/cm^2 are shown in Fig. 6.178. Electron current alone (condition **j**, open squares) reduces the capacitance of eTST in Fig. 6.178a. A further decrease in capacitance occurs for the combined condition, **L+j** (open circles), with negligible deviation from condition **j** at low pump intensities, to progressively larger shifts as intensity, I, is increased. No changes are observed in any of the eCTL devices (see Fig. 6.178b), or for eTST at any intensity, **L**, or **C** conditions. Fits to the data using Eq. 6.136 assuming, $\mu = 2.2 \times 10^{-7}$ cm^2/V s, $l = 5.5$, $H_t = 1.3 \times 10^{18}$ cm^{-3} (obtained from the j–V characteristics), $N_{LUMO} = 10^{23}$ cm^{-3}, and $\zeta = 0.05$ are shown by the solid lines in Fig. 6.178a.

The data obtained in this study yield a rate of defect formation for eTST devices of $K_X(\mathrm{s}^{-1}) \approx 3.6 \times 10^{-5} + 5.2 \times 10^{-7} I(\mathrm{mW/cm}^2)$. This linear relationship with a non-zero intercept at $I \to 0$ suggests degradation by polaron interactions (condition **j**). Since the LUMO of Ir$(ppy)_3$ is above that of mCBP, electron conduction through the Ir$(ppy)_3$:mCBP matrix occurs via the host. Thus, mCBP anions appear to be unstable, and become more so upon optical excitation.

From Eq. 6.137, the linear relationship for eTST can be rewritten:

$$K_X = K_{X1} + K_{X2}\left[\frac{I\tau_{Ph}(1 - \exp(-2\alpha_{EML}x_{EML}))}{x_{EML}E_{ph}}\right], \quad (6.138)$$

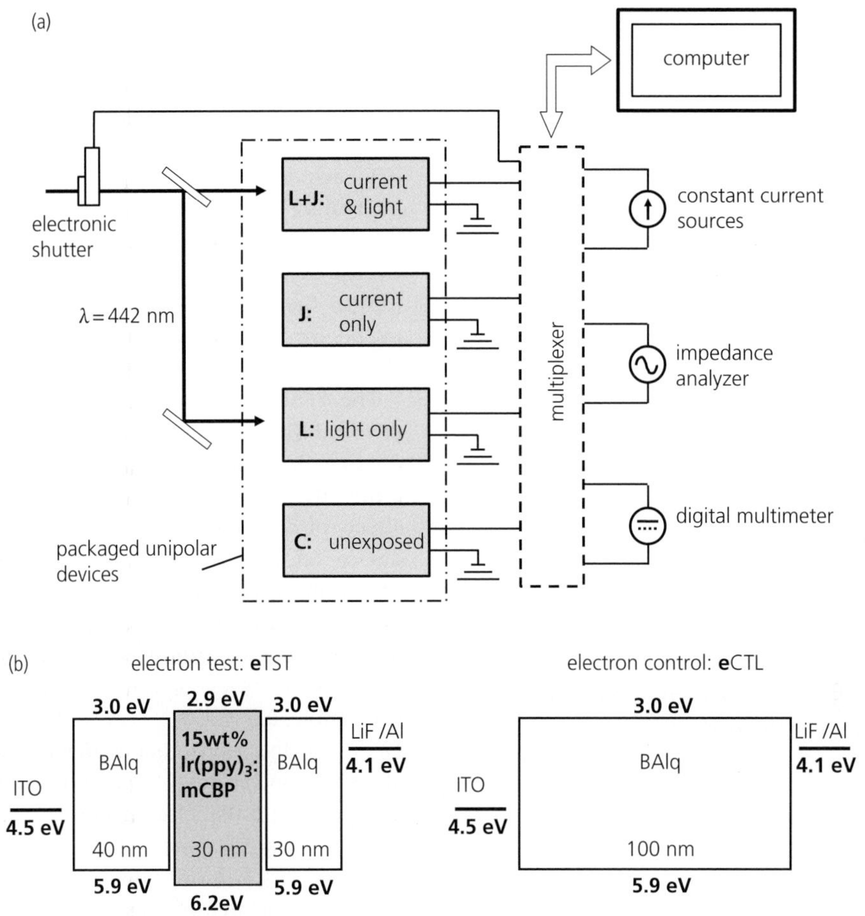

Figure 6.177 (a) Test setup for measuring degradation mechanisms in OLEDs under different test conditions noted in text. (b) Example electron-only (eTST) PHOLED test structure along with its electron-only control (eCTL) (Giebink et al., 2009).

where the quantity in brackets is the density of $Ir(ppy)_3$ excitons generated at intensity, I. Thus, K_{X2} is the rate coefficient due to the combined probability of annihilation with $Ir(ppy)_3$ excitons followed by dissociation of the excited polaron state. Also, τ_{Ph} is the $Ir(ppy)_3$ phosphorescent lifetime, x_{EML} is the width of the $Ir(ppy)_3$-doped layer, α_{EML} is its absorption coefficient, and the photon energy is E_{ph}. Taking $\alpha_{EML} = 4 \times 10^3$ cm^{-1} at $\lambda = 442$nm (Holzer et al., 2005). we obtain $K_{X2} = (3\pm1) \times 10^{-20}$cm^{3}s^{-1}, which is consistent with the empirical relationship given above. These results indicate luminance loss in an $Ir(ppy)_3$:mCBP OLED is predominantly due to non-radiative electron–hole recombination at defect sites formed by the interaction of an electron polaron on the host mCBP and an exciton on $Ir(ppy)_3$. The defect sites are responsible for the increase in voltage and decrease in capacitance and luminance with device operation (Giebink et al., 2009).

The preceding analysis indicates that generation of deep levels over time is one mechanism responsible for device aging. Yet it does not provide direct information about the *chemical origin* of those defects. One possible source is molecular fragmentation following reactions between radical and neutral species, perhaps resulting in spontaneous dimerization. Molecular fragments are identified using *laser desorption/ionization time-of-flight mass spectroscopy* (LDI-TOF-MS). A schematic diagram of the spectrometer is shown in Fig. 6.179. A high energy (UV) laser beam ionizes and desorbs molecules at the target. The ionized species are accelerated and selected by the ratio of mass-to-charge (m/z) in the precursor selector. Unwanted ionic species are eliminated by collisions with non-reactive (e.g. He) and/or reactive (e.g. H_2) gases in the collision cell. The remaining molecules are accelerated once more, and subsequently selected in the mass spectrometer analyzer. The measurement provides the relationship between ionic species concentration vs. mass. Analyzing constituent materials before and after device aging can be used to identify molecular fragments and compounds produced under stress during OLED operation (Meerheim et al., 2008, Moraes et al., 2011). Given the high intensity and energy of the laser desorption beam, a high density of molecular excited states are generated during the measurement. This results in exciton annihilation whose impact on defect generation can also be studied (Sandanayaka et al., 2015, Lee et al., 2017).

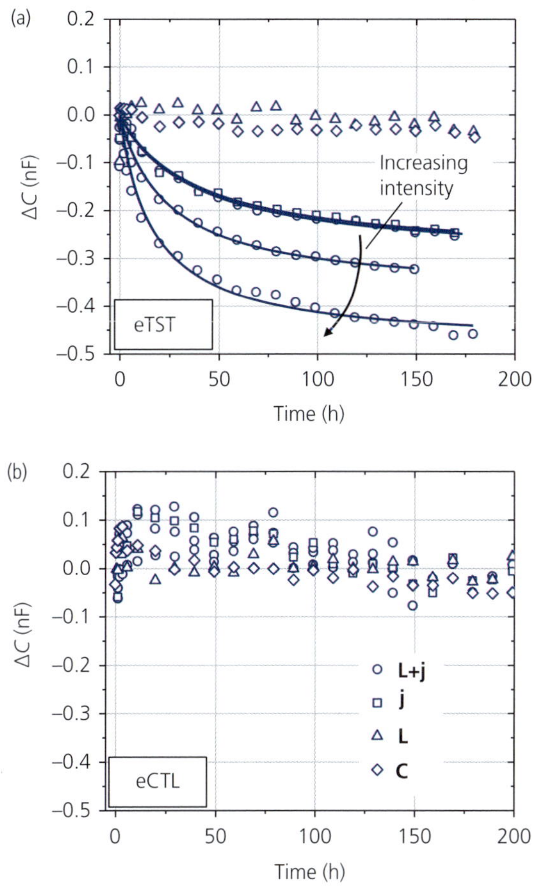

Figure 6.178 (a) Change in capacitance for (a) eTST and (b) eCTL devices in Fig. 6.177b as a function of operating time. Solid lines are fits to the data following Eq. 6.137 (Giebink et al., 2009).

Figure 6.180 shows the LDI-TOF-MS intensity spectra as a function of *m/z* for both a fresh and aged $Ir(ppy)_3$:TCTA PHOLED whose structure is shown in Fig. 6.180a, inset. Two identical devices were fabricated and encapsulated. The fresh device served as a control, whereas the aged device was driven at 2.5 mA/cm^2 (corresponding to $L_0 = 1012$ cd/m^2), for 215 h, at which point it had lost 7% of its initial luminance. The packages were then opened, and the top contacts were peeled off to obtain the mass spectra.

The control device spectra exhibit signals from all of the components comprising the OLED. After aging, several new peaks emerge, notably those at or near $m/z = 663$, 795, and 833. The larger mass species is assigned to a dissociative reaction of $Ir(ppy)_3$ to form a reversible pair of fragments, $[Ir(ppy)_2]^+ + [ppy]^-$. The heavier fragment reacts with BPhen at the EML/HBL interface to form $Ir(ppy)_2$ (BPhen) with $m/z = 833.2$. The sequence of proposed complexation reactions is shown in Fig. 6.186b and c. The features at $m/z = 663$ and 675 are identified with BPhen dimers associated with a proton or Cs^+ ion, respectively. Furthermore, it is proposed that the Ir $(ppy)_2$(BPhen) complex results in energy gap states at the interface that lead to recombination and accumulation of charge (Moraes et al., 2011). Hence, while

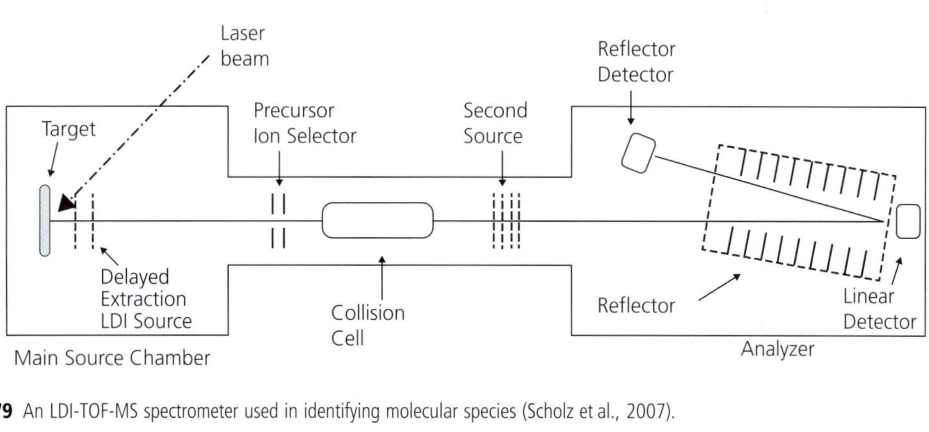

Figure 6.179 An LDI-TOF-MS spectrometer used in identifying molecular species (Scholz et al., 2007).

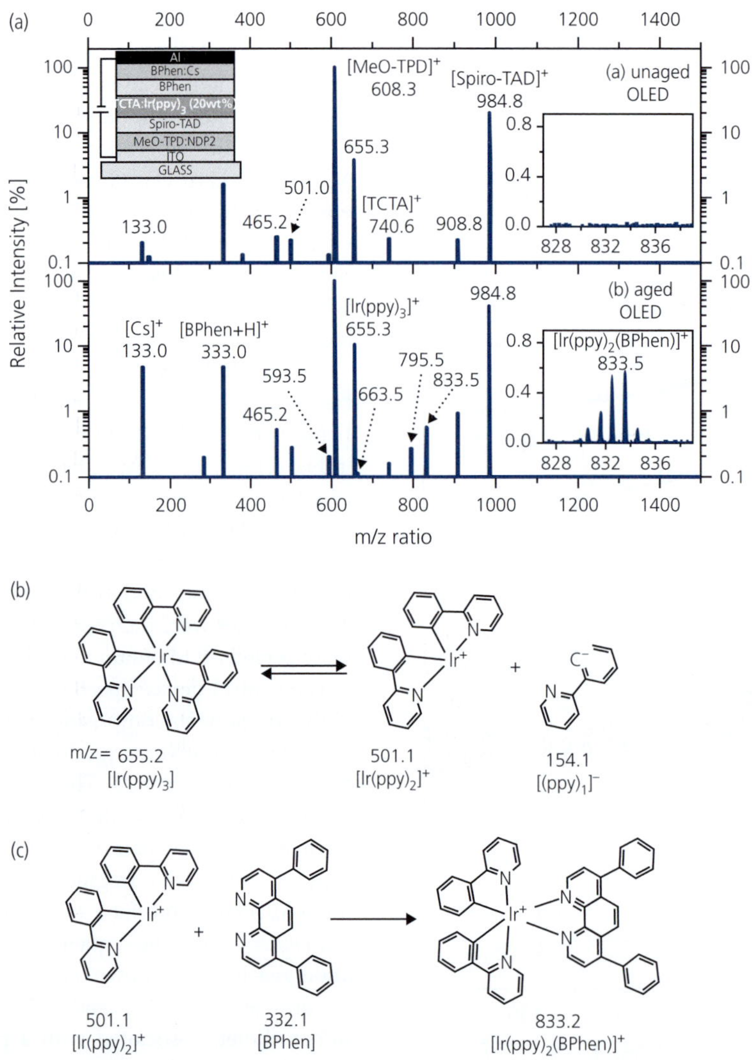

Figure 6.180 (a) Mass spectra of an unaged and aged PHOLED with the structure shown in the left-hand inset. The right insets show a detail of the mass to charge ratio (*m/z*) between 828 and 838. (b) Proposed fragments of $Ir(ppy)_3$ and their masses generated on aging. (c) Proposed complexation between $[Ir(ppy)_2]^+$ and BPhen in the HBL (Moraes et al., 2011).

these investigations provide evidence for interface reactions as well as defects formed within the HBL, they do not rule out defects that occur within the bulk of the EML, nor do they associate a particular reaction product to a defect state that is responsible for the drop in luminescence.

A study of aged $Ir(ppy)_3$-based PHOLEDs using a CBP host has also found numerous radical and ionic species that are fragmentation products from CBP. The species found after aging were isolated using high performance liquid chromatography (HPLC) and analyzed by nuclear magnetic resonance (NMR) spectroscopy. Calculations of the energy levels of the anions, cations and radicals led to two conclusions: several electronic states developed within the CBP energy gap, but they had very low oscillator strengths and hence were non-radiative. It is reasonable, therefore, to assume that these degradation products may lead to the loss of luminance and increased voltage of the aged devices, yet no direct correlation was reported between a particular species and the changes in device performance (Kondakov et al., 2007).

The inferential connections between reactions at interfaces, as well as within the bulk layers themselves discussed in this section, and the resulting reduced performance is striking. We will show in the following section how molecular fragments can be associated with specific defect energy levels in aged OLEDs.

6.7.4 Lifetime of blue PHOLEDs: energy-driven degradation

While it is not possible to consistently compare the intrinsic lifetime of OLEDs comprising different materials systems and with different emission processes, it is clear that even the most optimized blue emitting devices have far shorter lifetimes than either red or green emitting OLEDs. Indeed, the blue emitter in OLED displays and lighting is the major limiting factor in the useable lifetime of such appliances. With this caveat, a comparison of lifetime of several OLEDs emitting at different wavelengths is provided in Table 6.14. Here, the lifetimes of PHOLEDs, fluorescent OLEDs, and TADF-based OLEDs taken from different laboratories are listed. Each of these devices comprises a different materials set and device architecture, so finding a consistent set of data is not possible. Nevertheless, there is a rough correlation between the energy of emission and device stability; the lower photon energy (red, green) emitting devices exhibit longer lifetimes than blue emitting devices. This is particularly pronounced for PHOLEDs and TADF emitters, with a weaker correspondence for fluorescent OLEDs. Furthermore, the development of TADF emitters lags far behind that of phosphorescent and fluorescent molecular complexes. Anticipated advances in TADF molecular engineering will undoubtedly lead to significant improvements in these materials as well.

A compilation of blue PHOLED lifetimes is given in Table 6.15, roughly ordered by their CIE coordinates from deep blue to greenish-blue (i.e. cyan). The combination of longest lifetime and shortest wavelength emission is found for $Ir(dmp)_3$; a material that will be discussed further below. However, even in this case the lifetime is insufficient to qualify for either lighting or display applications.

Since *deep* blue emitters are required in displays, it has been necessary to pair the less efficient but longer-lived fluorescent blue OLEDs with red and green phosphors to achieve a suitably long lifetime. In addition, red and green PHOLEDs have considerably longer lifetimes than their fluorescent counterparts. This leads to image "burn in" in displays that is a consequence of loss of blue intensity as the screen ages. As noted previously, phosphor dopants rapidly harvest both singlet and triplet excitons. Since material degradation is catalyzed by the combined presence of an excited state and a charge, rapid elimination of the exciton should increase device lifetime. In analogous fluorescent OLEDs, the triplets have no

Table 6.14 Example lifetimes of PHOLEDs, fluorescent and TADF OLEDs normalized to $L_0 = 1000$ cd/m^2

Color	CIE coord.	Lum. eff. (lm/W)	*LT95* (h)	*LT50* (h)
PHOLEDs[a]				
Deep red	(0.69, 0.31)	17	14,000	250,000
Red	(0.64, 0.36)	30	50,000	900,000
Yellow	(0.44, 0.54)	81	85,000	1,450,000
Green	(0.31, 0.63)	85	18,000	400,000
Light blue	(0.18, 0.42)	50	700	20,000
Fluorescent OLEDs[b]				
Red	(0.67, 0.33)	11		160,000
Green	(0.29, 0.64)	37		200,000
Deep blue	(0.14, 0.12)	9.9		11,000
TADF OLEDs				
Green[c]	(0.34, 0.58)	15	1380[e]	
Light blue[d]	(0.18,0.34)	17		40[f]

[a] Source: website, Universal Display Corp.
[b] Source: website, Idemitsu Kosan.
[c] Source: Tsang et al. (2016).
[d] Source: Cui et al. (2016).
[e] *LT90*.
[f] $L_0 = 500$ cd/m^2 normalized to $L_0 = 1000$ cd/m^2 using $n = 1.7$ acceleration factor in Eq. 6.122.

quantum mechanically allowed transition path to the ground state, and hence they linger for many seconds or longer, possibly negatively impacting their lifetime.

The anomalously short lifetimes of blue OLEDs and PHOLEDs leads one to conclude that aging is intrinsically linked to the energy of the excited state. Indeed, it can be argued that extending the blue PHOLED lifetime to several tens of thousands of hours is the most important and difficult problem confronting the field of organic electronics today. Controlling intrinsic failure mechanisms in OLEDs should also deepen our understanding of failure mechanisms in organic electronic devices in general, ultimately leading to their acceptance in numerous other practical applications.

To understand the source of blue OLED failure, we first need to understand what is different for these vs. longer wavelength emitting devices:

(i) Lack of a large palette of hosts, charge transport, and blocking layers. There are very few host materials whose triplet energies are higher than that of the dopant. Blue photon energies are typically >2.9 eV, which requires triplet energies at least that high. Since the triplet energy is less than for the singlet by the exchange energy (ranging from 0.1 to 0.5 eV), the host singlet energy is pushed into the ultraviolet. To date, the most successfully employed hosts have a biphenyl carbazole structure such as mCBP. Similarly, electron and hole blocking layers must have very low LUMO energies (relative to vacuum). This results in materials sets that are easily oxidized and whose ability to balance charges within the EML is limited.

(ii) The photon energy is sufficient to break many atomic bonds found in organic materials. In Table 6.16 we show a few bond energies (BE) found in common organic optoelectronic materials. Clearly, low energy bonds (e.g. O–O, N–O) must be avoided since energy dissipation on such bonds results in dissociation. However, almost all molecules such as CBP have at least one C–N bond, which is close to the photon energy from deep blue emitters. And as we show in the ensuing discussion, two particle collisions such as TTA or TPA momentarily promote the molecular excited state to double the photon energy, that is, as high as 5.5–6 eV. No bond can withstand such a high energy excitation, but this assumes that all

Table 6.15 Lifetime comparisons of selected blue and cyan PHOLEDs

Phosphor	CIE	η_{ext} (%)	L_0 (cd/m^2)	*LT80* (h)	Ref.
FIrpic & Pyrene-CN	(0.15, 0.13)	~2.5	500	~1	(Lee et al., 2015b)
Ir(dmp)$_3$	(0.15, 0.25)	N/A	1880	75[a]	(Yamamoto et al., 2011)
Ir(dmp)$_3$	(0.16, 0.30)	9.6	1000	334	(Lee et al., 2017)
Ir(dmp)$_3$	(0.16, 0.31)	9.5	1000	213	(Zhang et al., 2014b)
PtNON	(0.17, 0.32)	9.1	1000	335[a]	(Fleetham et al., 2016)
FIrpic	(0.16, 0.36)	12.6	1000	0.1	(Zhuang et al., 2013)
Ir(ipripmi)	(0.17, 0.37)	~14	1000[b]	24	(Klubek et al., 2014)
Ir(iprpmi)$_3$	(0.17, 0.40)	19.6	1000	~20	(Zhang et al., 2016b)
Ir(dbi)$_3$	(0.19, 0.44)	23.0	1000	15	(Oh et al., 2016)
Ir(dbi)$_3$	(0.19, 0.44)	4	1000	154	(Jeon and Lee, 2015)

[a] Extrapolated using an acceleration factor of $n = 1.7$ (cf. Eq. 6.122).
[b] Estimated.

Table 6.16 Bond energies (BE) of several atomic species commonly found in organic semiconductors

Bond	BE (eV)	Bond	BE (eV)
C–C	3.64	N–N	1.69
C–H	4.28	N–O	2.08
C–O	3.71	N–H	4.05
C–N	3.04	O–O	1.51
C–F	5.03	H–H	4.52

the excess energy from this reaction is dumped into a single bond, which is not always true. The deeper blue emitters have even higher energy photons than blue-green PHOLEDs, leading to ever shorter lifetimes. Doubling of green (2.3 eV) and red (2.0 eV) photon energies is far less destructive of the molecular structure.

(iii) The triplet radiative lifetime tends to be longer (2–5 μs) in blue emitting phosphors than in longer wavelength complexes (≤2 μs) due to reduced ^{3}MLCT or ^{3}LC coupling. The longer lifetime leads to a higher probability for collision with another triplet or polaron to produce a "hot" excited state. This problem is exacerbated in TADF-based diodes due to their typically longer radiative lifetimes (2–20 μs). This is reflected in the data in Table 6.14. By the same argument, the short lifetime of the singlet state (~1–10 ns) is responsible for the longer lifetimes of fluorescent blue OLEDs.

In Section 3.7.3, we discussed the energetic landscape of metalorganic phosphors. There, we found that the principal routes to high efficiency phosphorescence is via ^{3}MCLT and ^{3}LC transitions. In this context, blue emission can be accessed by chemical modifications that either destabilize the LUMO, or stabilize the HOMO, as shown in Fig. 6.181a. Destabilization of the LUMO implies moving the LUMO closer to the vacuum level. This creates a vulnerability to oxidative dissociation of the molecules. As the T_1 energy increases, the energy difference between T_1 and the metal centered state, ^{3}MC decreases. In Chapter 3, we showed that ^{3}MC is an antibonding, σ^* orbital, and further, due to its t_{2g} symmetry, the $t_{2g} \rightarrow e_g$ excited to ground state transitions are not allowed due to parity violations. Therefore, ^{3}MC is a dark, or non-radiative state.

Examples of LUMO destabilization for tris-bidentate molecules starting from Ir(ppy)$_3$ are shown in Fig. 6.182. For Ir(ppy)$_3$, the activation energy between the T_1 and ^{3}MC state is 0.5 eV. Hence, the latter state is thermally inaccessible and does not engage in molecular fragmentation. Assuming that TTA or TPA supplies the excess energy necessary for molecular dissociation, the hot (i.e. multiply excited) state will have at least twice the triplet energy, that is, $2E_{T1}$. Now, the free energy from the ground state, G_0, to the dissociative state, G_{diss} is the energy of reaction, viz. $\Delta G_{react} = G_{diss} - G_0$. Thus, dissociation requires that $\Delta G_{react} > 2E_{T1}$. From DFT calculations, it is found that $\Delta G_{react} \approx 2E_{T1} \approx 5.0$eV (Jacquemin and Escudero, 2017). Since not all doubly excited states immediately localize all of their energy on a single bond, it is unlikely that Ir(ppy)$_3$ dissociates either by hot excited states or by thermal activation to the ^{3}MC state.

The circumstances are considerably different for the larger energy gap Ir(ppz)$_3$. In this case, the ^{3}MC state

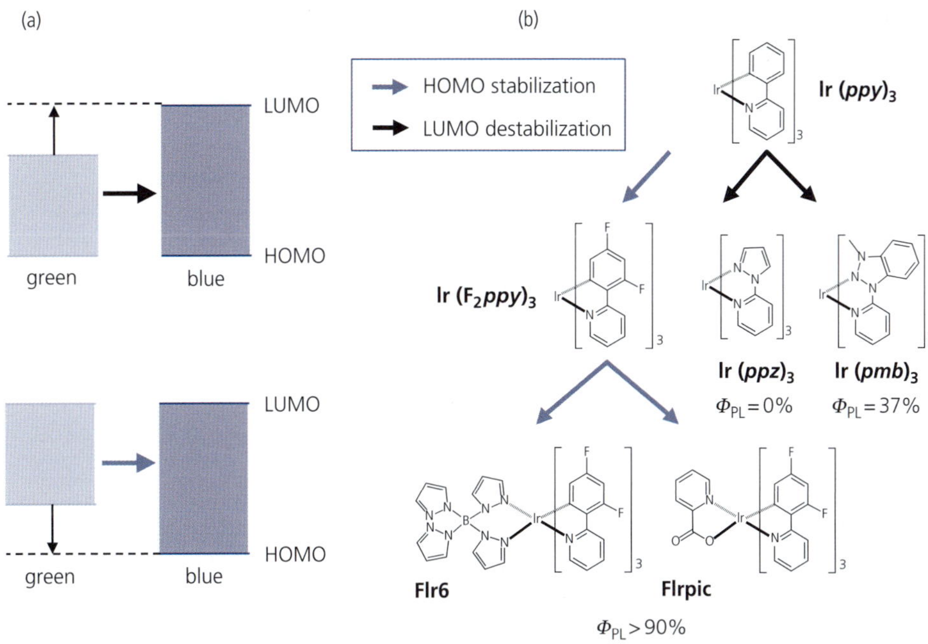

Figure 6.181 Pathways to stable blue phosphor complexes. (a) (top) LUMO destabilization and (bottom) HOMO stabilization. (b) Example blue emitting Ir-complexes, starting with the archetype green emitting Ir(ppy)$_3$.

is thermally accessible to T_1, with an activation energy of only 0.2 eV. Due to the ease of access to ^{3}MC, the molecule is non-radiative, that is, the PLQY is $\Phi_{PL} = 0$. Furthermore, $\Delta G_{react} = 4.18$eV, which is considerably smaller than $2E_{T1} = 6.0$ eV. This can result in the rupture of the N_1–C bond, forming the five coordinated trigonal bipyramidal (TBP) structure in Fig. 6.182, center. The energy difference between TBP and ^{3}MC of 0.4 eV is also too large for the molecule to recover by healing of the bond rupture. For all of these reasons, LUMO destabilization by this chemical route is not a viable approach to achieve long-lived blue phosphor.

The third molecule to consider is Ir(pmb)$_3$, shown in Figs. 6.181b and 6.182. Like Ir(ppz)$_3$, T_1 and ^{3}MC are separated by 0.2 eV. This leads to Φ_{PL} of only 37%. However, since ^{3}MC and TBP are degenerate, bond rupture can result in reversibility of the dissociation reaction and hence a longer operational lifetime of PHOLEDs employing this species. Furthermore, it has been suggested that tris-bidentate N-heterocyclic carbene (NHC) molecules such as Ir(pmb)$_3$ can have a longer lifetime as well as a high Φ_{PL} since $\Delta G_{react} = 5.77$eV compared to $2E_{T1} = 6.0$ eV for the dissociative reaction $\mathrm{Ir(pmb)_3} \rightarrow [\mathrm{Ir(pmb)_2}]^+ + [\mathrm{pmb}]^-$. Unfortunately, this competes unfavorably with $\mathrm{Ir(pmb)_3} \rightarrow \mathrm{Ir(pmb)_2}\,(C_6H_5)^+\,(C_8N_2H_7)^-$ where $\Delta G_{react} = 4.60$eV (Jacquemin and Escudero, 2017). Hence, this particular NHC still results in unstable PHOLEDs, although NHCs with different energetic manifolds may be found that do not undergo this type of low energy ligand fragmentation.

The second approach to achieving blue is via HOMO stabilization. This is accomplished by replacing the luminescent moiety with an ancillary ligand such as acetylacetone (acac), picolinate (pic), or pyrazolyl (ppz). This raises the ^{1}MLCT state without affecting the ^{3}LC. From Eq. 3.185 we see that the triplet is stabilized by the singlet via

$$E'(^3LC) = E(^3LC) + \frac{|\langle ^1MLCT|H_{SO}|^3LC\rangle|^2}{E(^3LC) - E(^1MLCT)}. \quad (6.139)$$

The increase in ^{1}MLCT thus results in an increase in ^{3}LC, thereby shifting the emission toward the blue. The attachment of ancillary, non-radiative ligands has proven to be a viable approach to achieving greenish-blue (cyan) emission, illustrated by the two examples of FIr6 and FIrpic in Fig. 6.181. However, this entails an Ir-N double bond, or even the weaker Ir–O bond for pic-complexes. Furthermore, the addition of an electron-withdrawing F-atom can also lead to molecular instabilities. Hence, HOMO stabilization has, to this point, been unsuccessful in leading to stable, deep blue emission.

The energetics of molecular dissociation in blue TADF and metal complexes are shown in Fig. 6.183. The excited state in one molecule (which for the purposes of this discussion will be assumed to be a long-lived triplet) can collide with an excited state in a second molecule (Fig. 6.183a) or a polaron (Fig. 6.183b). The collision results in the non-radiative transition of the exciton on the first molecule to its ground state, along with promotion of the second excited state or polaron to a very high, hot energy state. Although this highly

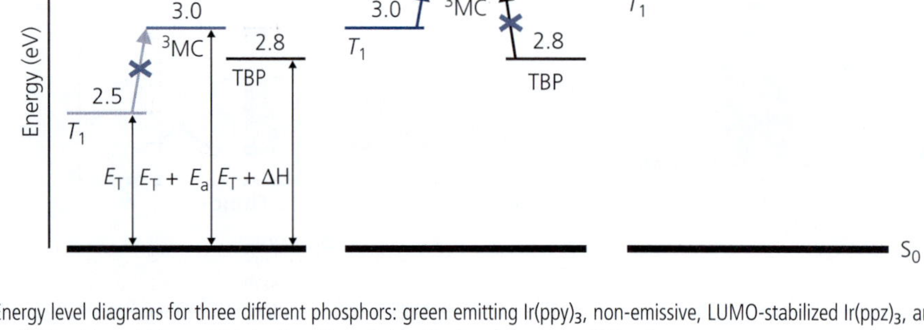

Figure 6.182 Energy level diagrams for three different phosphors: green emitting Ir(ppy)$_3$, non-emissive, LUMO-stabilized Ir(ppz)$_3$, and emissive LUMO-stabilized Ir(pmb)$_3$. TBP = trigonal bipyramidal structure illustrated by the bond-cleavage of Ir(ppz)$_3$.

excited state is expected to be very short lived due to coupling to numerous lower energy states, occasionally a dissociative reaction (to state D) will occur that results in the destruction of the excited molecule. This is analogous to non-radiative *Auger* processes in inorganic semiconductor lasers and LEDs (Coldren and Corzine, 1995).

The rich energy landscape of phosphors and hosts is exemplified by the excited state energy manifolds in Fig. 6.184 for two phosphors, $Ir(pmp)_3$ and $Ir(dmp)_3$, and the host mCBP. Although the energies shown are only a sampling of the entire manifold, it is readily apparent that there are numerous hot excited states accessible to Auger recombination in Fig. 6.183. Interestingly, while the host mCBP has a high triplet state suitable for blue PHOLEDs, its singlet is inherently unstable with a dissociative state energy, $D_1 = 3.6$ eV $< S_1 = 3.7$ eV (Lee et al., 2017).

Defects formed by triplet annihilation result in both a decrease in luminance and an increase in operating voltage. Hence, it is reasonable to assume that the annihilation reactions result in defect states in the host and/or the dopant. The states may be mid-energy gap traps that quench excitons in the EML. The number of defects formed increases with time, thereby decreasing luminance and increasing voltage over time. This situation is shown schematically in Fig. 6.185 for a single, discrete, deep defect state at energy, E_t, that lies within the energy gap of the host. Luminescence quenching by defects occurs when energy is transferred to the trap state by a Förster or Dexter transition from the guest or host molecule.

To quantify the effects of trap states on device operation over time, we assume that only hole traps with E_t are formed. In practice, however, defects may have either or both their HOMO and LUMO levels, or singly occupied molecular orbitals within the host energy gap, and hence can act as traps to either electrons or holes (Kondakov et al., 2007). The defect state itself may not lead directly to a quenching transition,

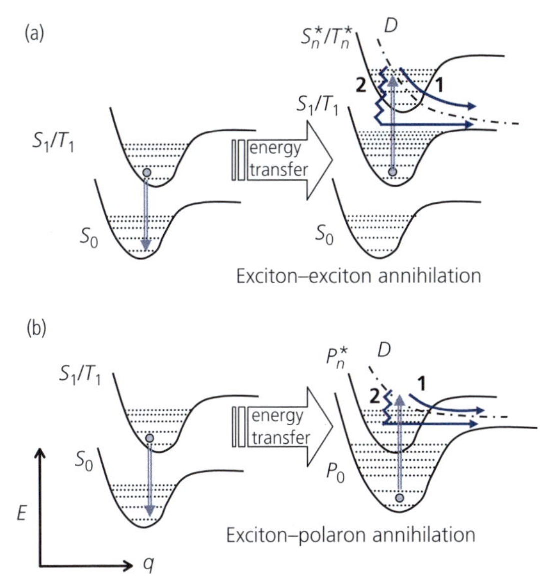

Figure 6.183 Configuration space diagrams showing how two particle collisions result in molecular dissociation reactions (D = predissociative state). (a) Annihilation between excitons on adjacent molecules results in the non-radiative transition to the ground state of the first molecule, and promotion of the lowest excited state of the adjacent molecule to a higher manifold (denoted S_n^*/T_n^*). (b) As (a) except for exciton–polaron (denoted P_0, P_n^*) collisions (Giebink et al., 2008).

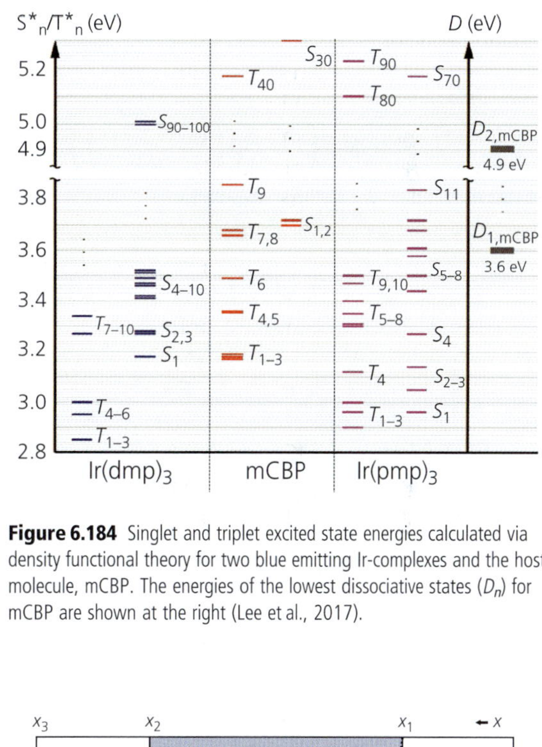

Figure 6.184 Singlet and triplet excited state energies calculated via density functional theory for two blue emitting Ir-complexes and the host molecule, mCBP. The energies of the lowest dissociative states (D_n) for mCBP are shown at the right (Lee et al., 2017).

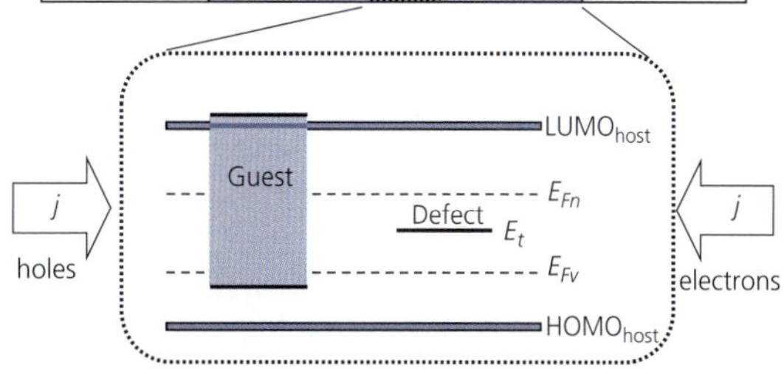

Figure 6.185 Schematic energy level scheme of a PHOLED. A detail of the guest blue phosphor/host EML recombination zone of width d_{rec} shows the guest (emitter) and host energy levels, along with the approximately mid-energy gap defect at energy E_t.

but becomes active once it is occupied with a trapped polaron.

The evolution of the energy-driven formation of midgap defects in the EML is determined by assuming that radiative decay decreases exponentially with characteristic length d_{rec} from the edge of the EML where excitons are formed, as in Fig. 6.185 (Tutis et al., 2003). For simplicity, we further assume that equal numbers of electrons and holes enter the recombination zone, that is, the charge balance efficiency is unity. Excitons either form directly on guest molecules, or they rapidly transfer from the host due to their high doping concentration.

Following the treatment of Giebink et al. (2008), these assumptions lead to rate equations for hole (p), electron (n), and exciton (N) densities in the recombination zone of

$$\frac{dp(x,t,t')}{dt} = \frac{j}{qd_{rec}\left(1-\exp\left(-\frac{(x_2-x_1)}{d_{rec}}\right)\right)} \times \exp\left(-\frac{(x-x_1)}{d_{rec}}\right) - \gamma n(x,t,t')p(x,t,t') - \sigma v_{th}[f_D(E_t)]Q(x,t')p(x,t,t'), \quad (6.140)$$

$$\frac{dn(x,t,t')}{dt} = \frac{j}{qd_{rec}\left(1-\exp\left(-\frac{(x_2-x_1)}{d_{rec}}\right)\right)} \times \exp\left(-\frac{(x-x_1)}{d_{rec}}\right) - \gamma n(x,t,t')p(x,t,t') - \tilde{\gamma}[1-f_D(E_t)]Q(x,t')n(x,t,t'), \quad (6.141)$$

$$\frac{dN(x,t,t')}{dt} = \gamma n(x,t,t')p(x,t,t') - \left(\frac{1}{\tau_{Ph}} + k_{DR}Q(x,t')\right)N(x,t,t'). \quad (6.142)$$

The electron, hole, and exciton densities (Eqs. 6.140, 6.141, and 6.142, respectively) depend on the timescale of transport and energy level transitions, t (on the order of microseconds), as well as on that of degradation, t' (hours), due to formation of defects of density $Q(x, t')$. The electron and hole densities are also functions of the current density, j. Excitons are formed at the Langevin rate of $\gamma = q(\mu_n + \mu_p)/(\varepsilon_r\varepsilon_0)$ (see Section 4.5.4) and decay with natural lifetime, τ_{Ph}. The hole and electron mobilities in the doped emissive layer are μ_p and μ_n, respectively.

In Eq. 6.140, holes with thermal velocity, v_{th} $\sim 10^7$cm/s, trap at defect sites of energy, E_t, and capture cross-section, σ. The Fermi factor, $f_D(E_t) = [\exp(E_t - E_{Fv}) + 1]^{-1}$, gives the probability that the hole trap is empty, where E_{fv} is the hole quasi-Fermi energy. Electrons in Eq. 6.141 non-radiatively recombine at a rate proportional to the trapped hole density, $Q(x,t')$ $[1 - f_D(E_t)]$, and the reduced Langevin coefficient, $\tilde{\gamma} = q\mu_n/(\varepsilon_r\varepsilon_0)$, since trapped holes are assumed to be immobile. Quenching of excitons by defects occurs at the bimolecular rate given by k_{DR} in Eq. 6.142.

There are four possible routes leading to defect generation expressed by the following:

$$\frac{dQ(x,t')}{dt'} = \begin{cases} K_X n(x,t'), K_X p(x,t') & (6.143a) \\ K_X N(x,t') & (6.143b) \\ K_X N^2(x,t') & (6.143c) \\ K_X N(x,t')n(x,t'), K_X N(x,t')p(x,t') & (6.143d) \end{cases}$$

Here, the defect formation rate is K_X, whose dimensions are consistent with the order of the reaction (i.e. [s^{-1}] for first order, and [$cm^3\,s^{-1}$] for second order processes). In Eq. 6.143a, only the presence of an electron or hole (i.e. a polaron) leads to molecular degradation, while in Eq. 6.143b excitons are responsible. More likely, the higher energy second-order processes of exciton–exciton annihilation in Eq. 6.143c and exciton–polaron annihilation in Eq. 6.143d will lead to luminance loss and increased voltage. Note that this theory is similar to the defect generation model used to describe the voltage rise in devices exposed to current and illumination in Section 6.7.3. The principal difference is that here we consider only those processes where excess energy dissipation leads to bond breaking reactions specifically within the EML.

On the short timescale, t, Eqs. 6.140–6.143 are at steady state and can be solved to yield $N(x,t')$. The resulting coupled differential equations containing $N(x,t')$ and $Q(x,t')$ are solved numerically to arrive at the normalized luminescence:

$$L_{norm}(t') = \frac{\int_{x_1}^{x_2} N(x,t')dx}{\int_{x_1}^{x_2} N(x,0)dx}, \quad (6.144)$$

and the defect formation rate per exciton, averaged over the recombination zone:

$$F_X(t') = \frac{1}{d_{rec}}\int_{x_1}^{x_2} \frac{1}{N(x,t')}\frac{dQ(x,t')}{dt'}dx. \quad (6.145)$$

Here, the integration limits, x_1 and x_2, are defined in Fig. 6.185.

The density of trapped charge increases with defect density following $\rho_t(x, t') = qQ(x, t')\ [1 - f_D(E_t)]$. Assuming that the increase in ρ_t is offset by an equal density of opposing charge at the cathode, and that the free charge distribution under steady-state operation is unperturbed, then the voltage rise is

$$\Delta V(t') \approx \int_0^{x_3} x\rho_t(x, t')dx. \tag{6.146}$$

Non-radiative recombination via defects should also decrease the radiative lifetime of the phosphor via: $1/\tau_{Ph} = 1/\tau_{Ph0} + 1/\tau_{nr}$, where τ_{Ph0} is the natural lifetime for phosphorescent emission. Hence, changes in the PL decay rate vs. t' are also due to trap formation.

To determine which of the intrinsic processes in Eq. 6.143 dominate, the long-term changes in luminance and voltage can be measured for a population of OLEDs whose fabrication process is stable, and the materials are as free from impurities as possible. The most active aging mechanism represented in Eq. 6.143 is the one that yields the best fit to the $L_{norm}(t')$ and $\Delta V(t')$ data.

Figure 6.186 shows the application of this procedure to a population of packaged PHOLEDs whose 30 nm thick EML comprises 9 wt% Ir(dmp)$_3$ doped into mCBP. The full device structure is: ITO/30 nm α-NPD/EML/5 nm mCBP/40 nm Alq$_3$/0.8 nm LiF/100 nm Al (Giebink et al., 2008). The solid lines show the fits to TTA- and TPA-induced degradation, corresponding to Eqs. 6.143c and d, respectively. The fits use the following model-dependent parameters: d_{rec} = 10 nm and 8 nm, $k_{DR} = 4 \times 10^{-12}$ cm^3/s and 5×10^{-12} cm^3/s, and $K_X = 1.7 \times 10^{-22}$ cm^3/s and 7×10^{-24} cm^3/s for TTA and TPA, respectively. Similar fits attempted for the various other monomolecular processes in Eq. 6.143 did not provide reasonable matches to the data. The fits, therefore, suggest that the most likely contribution to both the luminance and voltage changes over time is TPA, although contributions from TTA cannot be ruled out.

The relative contributions to luminance loss from exciton quenching and non-radiative recombination inferred from the PL transients vs. time are estimated to be about 70% and 30% respectively. The rate of exciton decay through quenching is $-k_{DR}QN$ (see Eq. 6.142), whereas the loss of

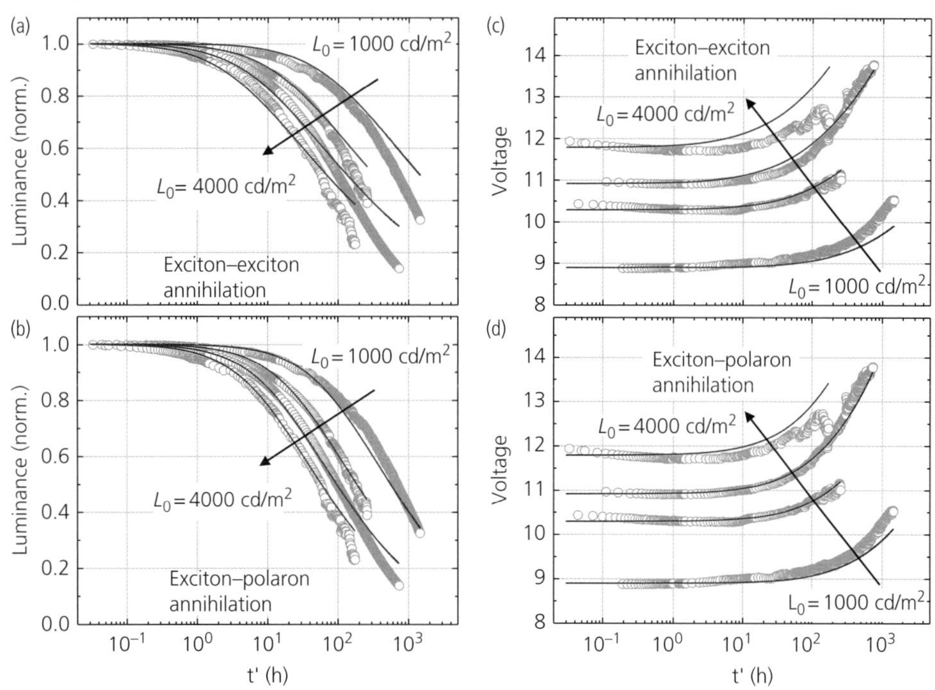

Figure 6.186 Luminance vs. time measured (points) for the blue PHOLED described in text, and fit to (a) TTA and (b) TPA models. These same processes are used to fit the voltage rise vs. time data in (c) and (d). The various curves correspond to initial luminances of 1000, 2000, 3000, and 4000 cd/m^2 (Giebink et al., 2008).

excitons to trapped charge from Eq. 6.141 is given by $-\tilde{\gamma}[1-f_D]Qn$. The 70/30 split is then the ratio of these two rates.

Now, the average defect density is given by

$$Q_{AVG}(t') = \frac{1}{d_{rec}} \int Q(x,t')dx. \qquad (6.147)$$

From the fits to the exciton–polaron model in Fig. 6.186, it is found that a defect density of 10^{18} cm^{-3}, which is ~ 0.1% of the molecular density, leads to >50% loss in luminescence. Furthermore, the rate of defect formation across the EML at 1000 cd/m^2 is $F_X \approx 0.04\ h^{-1}$, which is equivalent to a defect formation rate of ~$6 \times 10^{11}\ cm^{-3}\ s^{-1}$.

The probability that an exciton–polaron encounter leads to formation of a defect can now be estimated. Assuming that if a polaron comes within the exciton encounter radius, r, then the number of encounters per unit volume is $Nn\left(\frac{4}{3}\pi r^3\right)$, each of which has probability, P, for resulting in a defect. The number of defects produced during τ_{Ph} is $\Delta Q = K_X Nn\tau_{Ph} = PNn\left(\frac{4}{3}\pi r^3\right)$. Thus, the probability that an encounter leads to the formation of a defect is

$$P_{TPA} = \frac{3K_X\tau_{Ph}}{4\pi r^3}. \qquad (6.148)$$

Using $K_X = 7 \times 10^{-24}\ cm^{-3}\ s^{-1}$ and $\tau = 1.1\ \mu s$ for $Ir(dmp)_3$, and assuming nearest molecular neighbor interactions (corresponding to r~1 nm), then $P \approx 2 \times 10^{-9}$, or roughly 1 in 5×10^8 encounters result in molecular dissociation. Note that in a cavity with a Purcell factor, $PF > 1$, the probability of defect formation is decreased by this same amount since the lifetime is reduced according to $\tau'_{ph} \rightarrow \tau_{ph}/PF$ (cf. Eq. 6.110*ff*). Similarly, if TTA is the primary source of degradation, two excitons are required, and the probability for a destructive event is reduced by $P_{TTA} \rightarrow \tau'^2_{Ph} \sim \tau^2_{Ph}/PF^2$.

The exceptionally low rate of defect formation suggests that numerous pathways exist for the hot polaron to non-destructively relax back to T_1. In Fig. 6.184, we showed a near continuum of accessible hot excited states. Transitions among these states leads to extremely fast rates of internal conversion (~picoseconds) that compete with dissociation. Even with this remarkably low dissociation rate, the device has an $LT50 = 700$ h with $L_0 = 1000\ cd/m^2$ due to the very high exciton and polaron densities required to achieve this luminance.

From the foregoing, we conclude that strategies for extending blue PHOLED lifetime must involve lowering the densities of excitons and polarons. From the fits in Fig. 6.186, the device lifetime can more than double as the EML thickness becomes large ($d_{rec} \rightarrow \infty$). However, there is little improvement in $LT80$ for $d_{rec} > 30$ nm. Since voltage increases rapidly with EML thickness, 30 nm is near the upper limit for power efficient OLEDs. Alternatively, control of electron and hole mobilities in the EML, as well as the strategic placement of energy barriers in the device can lead to a more uniform distribution of excited states within the recombination zone.

Engineering host and guest molecules to lower the annihilation probability also leads to increased operating lifetime, as well as improved efficiency at high brightness (Reineke et al., 2007). Since the probability for defect formation is proportional to τ_{Ph} (Eq. 6.148), one of the most effective methods for decreasing the probability of defect production is to decrease τ_{Ph}. However, as seen in Section 3.7.3, the natural lifetime is intrinsically limited by the magnitude of zero-field splitting to ~500 ns for metalorganic complexes. Furthermore, blue emitting complexes typically have reduced metal–ligand interactions, leading to $\tau_{Ph} > 2\ \mu s$, setting a limit on the leverage that changes in this parameter can have on improving the lifetime of blue PHOLEDs.

Note that the foregoing analysis is independent of the source of the luminescence. Hence, it is equally applicable to TADF emitters. Indeed, the exceptionally long radiative lifetimes of 2–20 μs of non-metalorganic TADF molecules can account for the relatively poor reliabilities observed in OLEDs employing these materials (see Table 6.14).

The prediction that reducing the exciton density within the EML results in a concomitant reduction in the defect formation rate has been directly tested by grading the dopant concentration in the EMLs of blue PHOLEDs (Zhang et al., 2014b). In $Ir(dmp)_3$:mCBP devices such as in Fig. 6.186, the $Ir(dmp)_3$ preferentially conducts holes and mCBP conducts electrons in the EML, as shown in Fig. 6.187a. Once the carriers encounter each other on adjacent molecules, an exciplex is briefly formed, which then transfers to a triplet exciton on $Ir(dmp)_3$, resulting in blue emission. Since the hole mobility on the Ir-complex is higher than electrons on mCBP, the electron-polarons and excitons in a uniformly doped emission region pile up near the ETL/HBL interface. This high density of charge encourages TPA and TTA, accelerating aging.

The exciton formation zone was uniformly spread across the entire EML by linearly grading the hole-conducting dopant from 18% at the EML/HTL interface to 8% at the EML/HBL interface. The reduction in $Ir(dmp)_3$ concentration leads to a reduced hole polaron hopping efficiency (and hence mobility), shifting the exciton formation zone away from the HBL toward the center of the EML.

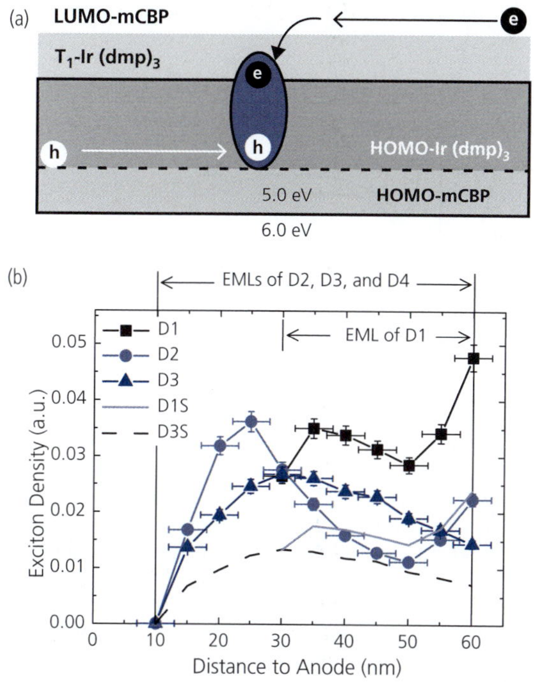

Figure 6.187 (a) Schematic of the energetics and charge conduction levels in the Ir(dmp)$_3$:mCBP EML. Holes are conducted on the Ir(dmp)$_3$ HOMO, and electrons on the mCBP LUMO. (b) Exciton profiles in 5 different devices described in text (Zhang et al., 2014b).

To determine the spatial distribution of excitons in the EML in both graded and ungraded PHOLEDs, thin (1.5 nm), red emitting 2 vol.% PQIr "sensing" layers were doped in the EMLs of several otherwise identical devices, but spaced at 5 nm intervals in each PHOLED. The relative intensity of red emission from each of the positions within the EML gives the relative exciton density at that position, thus mapping the exciton spatial distributions. Since the thickness of the sensing layer is much less than that of the EML, its presence does not significantly perturb the voltage or exciton distribution in the EML from that in the absence of the sensor layers. This procedure is used to map the exciton distribution in three different device architectures. (i) D1—a uniformly doped PHOLED with a conventional structure that includes an HTL. The EML is uniformly doped comprising 13 vol.% Ir(dmp)$_3$:mCBP. (ii) D2—same as D1 except the HTL (in this case, NPD) is omitted and the EML thickness is increased to 50 nm to ensure that the Ir(dmp)$_3$ is an efficient hole conductor, and (iii) D3—same as D2 except the Ir(dmp)$_3$ concentration is also linearly graded from 18 vol.% to 8 vol.%. Two stacked devices were included in the study. D1S, which is a double stack of D1 with 70 nm thick 2 vol.% Li-doped Alq_3 interconnecting charge generation layer between stacked elements, and D3S which is a double stack of D3 elements with the same interconnecting layer as in D1S. Stacked devices have an exciton concentration that is reduced by the number of stacked elements compared to single element devices (Adamovich et al., 2012).

The measured exciton distributions are shown in Fig. 6.187b. In D1, significant exciton accumulation occurs at the EML/HBL interface. In D2 that lacks an HTL, the exciton formation zone is broadened due to the increase in the EML thickness. As a result, the exciton density at the HIL/EML interface is reduced due to a correspondingly low concentration of holes. The reduced hole transport efficiency of these devices leads to electron penetration deep into the EML, resulting in a peak exciton density near the HIL/EML interface. In contrast, efficient hole transport near the HIL/EML interface and the gradual hole blocking in the EML in D3 lead to a more uniform exciton distribution (and hence higher η_{ext} at high current due to reduced TTA) compared to both D1 and D2. Figure 6.187b also shows the inferred exciton density profiles of D1S and D3S, estimated at 50% of the exciton density in D1 and D3.

The lifetime data of D1, D3, and D3S are shown in Fig. 6.188, along with fits to the TPA model and empirical model of Eq. 6.121. Consistent with the TPA analysis, the lifetime of D3 is approximately three times longer than D1 due to the extended exciton recombination zone, and is ten times longer when two D3 elements are stacked.

The results from this series of devices along with the lifetime fitting parameters of $k_{DR} \simeq 1\times\ 10^{-11}\mathrm{cm^3s^{-1}}$ and $K_X \simeq 7 \times 10^{-24}\mathrm{cm^3s^{-1}}$ from the TPA model are summarized in Table 6.17 (Zhang et al., 2014b). The values from the fits are consistent with those in Fig. 6.186 using this same materials system. Furthermore, the efficiency of D3 is higher than that of D1 and D2. This is also expected for devices in which bimolecular annihilation is reduced in EMLs that have a lower local exciton density. Finally, the voltage at $L_0 = 1000\ \mathrm{cd/m^2}$ is higher for D3 (7.7 V) than for D1 (6.9 V) which is the result of the larger EML thickness of the former device.

Reducing the exciton density via grading confirms the TPA mechanism, however, it is unclear that this strategy alone can result in blue PHOLEDs with sufficiently long lifetimes to be practical. To achieve this latter objective, hot excited states must be entirely eliminated before they can damage the emitting molecules. This can be partially accomplished by including a third molecular species; a *hot excited state manager molecule* into the phosphor-doped EML. The

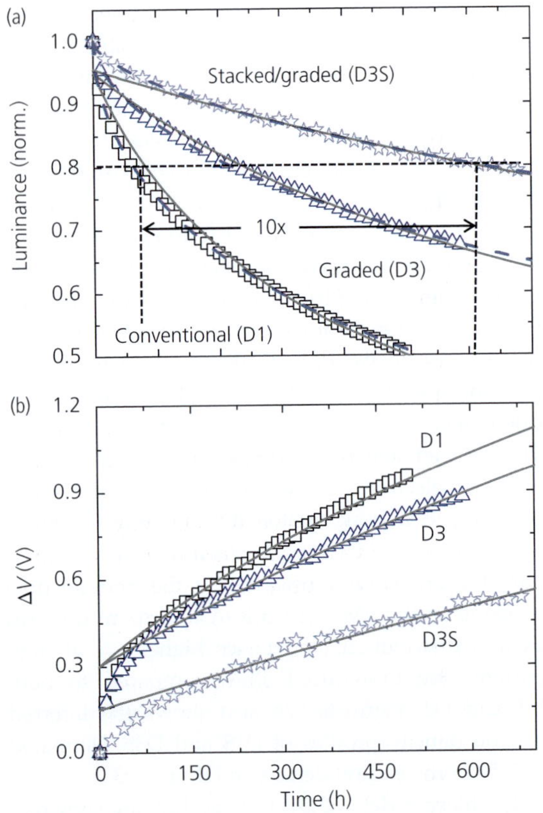

Figure 6.188 (a) Blue PHOLED luminance vs. aging time for three different devices described in text, and (b) the voltage rise vs. time for these same devices. The solid lines are fits to the TPA degradation model using values listed in Table 6.17, and the dashed lines are fits to the stretched exponential in Eq. 6.121 (Zhang et al., 2014b).

Table 6.17 Characteristics at $L_0 = 1000$ cd/m^2 for blue emitting PHOLEDs

Device	η_{ext} (%)	j (mA/cm^2)	V (V)	*LT80* (h)	*LT50* (h)
D1	8.5	6.2	6.9	56	510
D2	8.5	6.2	8.7	-	-
D3	9.5	5.7	7.7	213	1500[a]
D1S	16.0	2.8	12.9	-	-
D3S	18.0	2.9	14.3	616	3500[a]

[a] Estimated extrapolations from the TPA model.

manager dissipates the excess energy generated in TTA and TPA reactions prior to localization on a single molecular bond.

The Jablonski diagram in Fig. 6.189 illustrates the process of hot excited state management. By introducing a manager whose excited state energies S_M/T_M are greater than that of the dopant, excitons formed on, or transferred to the manager can be returned to

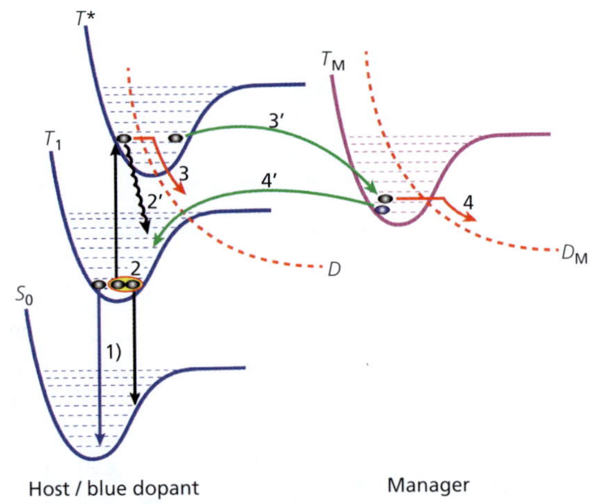

Figure 6.189 Jablonski diagram of an EML containing the manager. Here, S_0 is the ground state, T_1 is the lowest energy triplet state, and T^* is a hot triplet manifold of the dopant or host. D represents the dissociative state accessed via the predissociative potential in the EML. T_M is the lowest triplet state of the manager. Possible energy-transfer pathways are numbered as follows: (1) radiative recombination, (2) TTA resulting in excitation to T^*, (2′) internal conversion and vibrational relaxation, (3) and (4) dissociative reactions leading to molecular dissociation, (3′) exothermic energy transfer for transitions to the manager, and (4′) back transfer from the manager to T_1 (Lee et al., 2017).

the dopant for emission. Exothermic transfer from S_n^*/T_n^* to S_M/T_M is also allowed, and damage to these molecules via dissociative reactions (process 3) is reduced provided that the rate for $S_n^*/T_n^* \rightarrow S_M/T_M$ is comparable or higher than $S_n^*/T_n^* \rightarrow D$, where D is the predissociative state, via process 3′. A transferred singlet undergoes vibrational relaxation and subsequently transfers back to the lowest dopant singlet state. Alternatively, the thermalized singlet state intersystem crosses to the triplet state via $S_M \rightarrow T_M$, which subsequently transfers back to the dopant or host ($T_M \rightarrow T_1$) via process 4′. This leads to radiative recombination (process 1), or is recycled back to S_n^*/T_n^*. It is also possible that the high energy T_M state can result in dissociation of the manager itself via $T_M \rightarrow D_M$ (process 4), that is, where the manager serves as a sacrificial additive to the EML. Process 4 is not optimal since the number of effective managers decreases over time, providing less protection for the host and dopant. Even in this case, however, the manager can still increase device stability.

From the foregoing, three primary criteria must be met for effective design of the manager. (i) The exciton energy of the manager should be higher than lowest exciton states (S_1/T_1) of the dopant; (ii) the rate of transfer to the manager (process 3′) must be comparable to or higher than that for dissociation (process 3); and (iii) the manager should be sufficiently stable such that it does not degrade on a timescale short compared to that of the unmanaged device (process 4).

The principle of hot excited state management in Ir(dmp)$_3$:mCBP based PHOLEDs was investigated by including the manager, *mer*-Ir(pmp)$_3$, in the EML. The manager triplet energy is at $E_{T1} = 3.1$ eV compared with the dopant at $E_{T1} = 2.8$ eV. Hence, at least criterion (i) is satisfied (Lee et al., 2017). The devices tested consisted of a 50 nm thick EML that was graded from 15 vol.% Ir(dmp)$_3$:mCBP at the HTL/EML interface, to 5 vol.% Ir(dmp)$_3$:mCBP at the EML/HBL interface. Six different structures were compared: M0 has 3 vol.% of the manager molecule doped uniformly across the EML, whereas M1–M5 have a 10 nm zone of the EML doped with 3 vol.% *mer*-Ir(pmp)$_3$ doped in 10 nm wide zones starting at the HTL/EML interface (M1), and ending in the last 10 nm wide zone at the EML/HBL interface (M5). The PHOLEDs were otherwise analogous to D1 in Fig. 6.187, except that the HTL comprised CPD.

No emission was observed from the high energy manager, indicating that there was complete energy transfer from *mer*-Ir(pmp)$_3$ to Ir(dmp)$_3$. The luminance and voltage changes with operating time are fit using a TTA model modified from that in Eq. 6.140–6.142 (lines, Fig. 6.190a) that includes voltage increases due to deep traps formed in both the EML and the HTL, and luminance loss only due to a trap within the EML itself.

The exciton distribution is determined by doping 1.5 Å thick red phosphor slabs at successively distant positions from the HTL/EML interface, and then measuring the relative red emission intensities from these layers. Grading results in a reasonably uniform exciton distribution, see Fig. 6.180b. Interestingly, the *LT90* and *LT80* lifetimes of the devices M1–M5 qualitative match the exciton density profile in Fig. 6.190b, that is, the lifetime is longest when the manager is placed in the zone with highest exciton density (M3). This is expected if the manager is in fact eliminating excitations from the dopant and host prior to their damaging the active molecules in the emitting zone.

The performance of all devices is summarized in Table 6.18 and Fig. 6.190c. The improvement in lifetime at *LT90* is 5.2-fold compared to that of a conventional, ungraded EML device (CONV), falling to 3.5-fold at *LT80*. A similar decrease is seen for the graded (GRAD) device whose structure is similar to that in Fig. 6.187. This is attributed to depletion of the manager due to dissociative reactions, corresponding to process 4 in Fig. 6.189. While the improvements are insufficient for displays, hot excited state management appears to provide a strategy for extending the lifetime of blue PHOLEDs provided stable manager molecules can be developed that promote extremely rapid energy transfer from the dopant and host. Notably, the drive voltages of the managed devices are higher than for GRAD and CONV. This is due to the lower hole conductivity of *mer*-Ir(pmp)$_3$ compared with Ir(dmp)$_3$. Increased voltage leads to increased Joule heating, and hence a relative reduction in lifetime. Local doping of the manager in M3, therefore provides an improved lifetime compared with uniformly doped M0.

Since the fits in Fig. 6.190a suggest that traps in the HTL are partially responsible for the voltage rise, it is reasonable to assume that degradation of the blocking and transport layers in PHOLEDs may also occur over time. We have already noted that LUMO energies of layers used in blue PHOLED lie close to the vacuum level, and thus are more easily oxidized than materials used in longer wavelength emitting devices. Hence, loss of charge balance may also contribute to the loss of luminance in blue PHOLEDs. The contributions from the transport and blocking layers have been independently measured by placing thin red phosphorescent and fluorescent sensing layers in the ETL and HTL in the CONV devices in Fig. 6.190. The emission from these sensing layers as a function of time is used to determine whether the charge balance or exciton leakage from the EML is also changing. The phosphor dopant, PtOEP in the ETL and HTL is sensitive to triplet leakage, and the DCM2 fluorophore placed in the ETL is sensitive to loss of charge balance, since electrons and holes in the Alq$_3$ ETL transfer to the DCM2 and result in red emission. The 0.5 Å thick sensor layers were placed within 5 nm of the EML or HBL, as shown in Fig. 6.191a (Coburn and Forrest, 2017).

The blue spectral intensities in Fig. 6.191b and c decrease with operating time as the device ages. However, there are only very small changes to the PtOEP emission on the HTL and ETL sides of the device, indicating a minor increase in triplet exciton leakage from the EML. In contrast, there is no change in DCM2 emission since the HBL does not undergo significant changes in its ability to block holes during the timescale of the experiments. These experiments find that no more than 3% of the luminance loss and voltage rise is due to changes in charge or exciton confinement within the EML. Hence, the degradation of the PHOLEDs is indeed almost entirely due to energy-induced degradation of the molecular constituents in the EML.

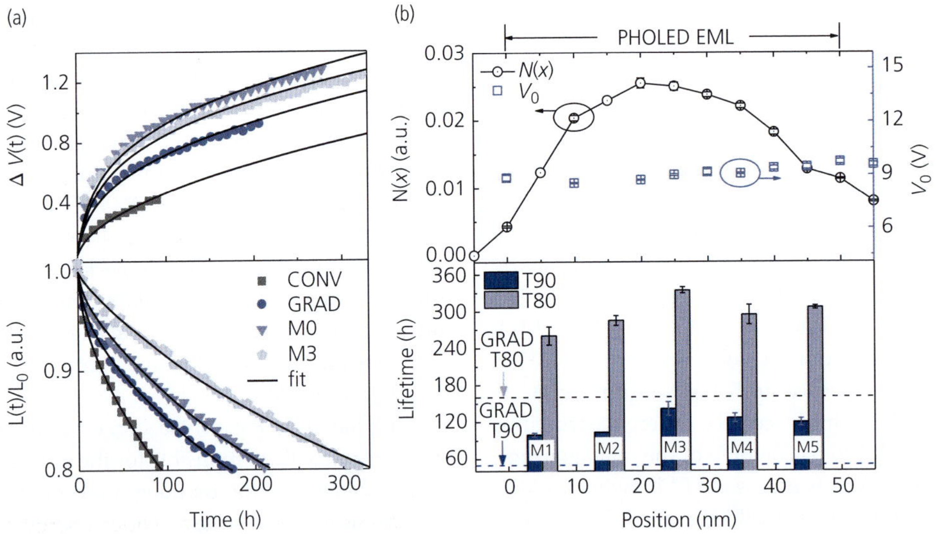

Figure 6.190 (a) Lifetime characteristics of the conventional (CONV), linearly graded (GRAD), and managed PHOLEDs M0 and M3. Top and bottom plots show the time evolution of the operating voltage change, $\Delta V(t) = V(t) - V_0$, and the normalized luminance degradation, $L(t)/L_0$, respectively. Solid lines are fits based on the TTA/TPA model. (b) (*Top*) Exciton density profile, $N(x)$, at $L_0 = 1000$ cd/m^2 of the EML vs. position, x, from the HTL/EML interface, and operating voltages obtained using sensing layers. (*Bottom*) Lifetimes (*LT90* and *LT80*) of managed devices (M1–M5) as functions of the position of the manager in the EML. *LT90* and *LT80* of the managed devices are compared with those of GRAD (dotted lines). The variation in lifetimes qualitatively follows the exciton density profile (Lee et al., 2017).

Table 6.18 Results for hot excited state managed (M0–M5) and unmanaged (CONV and GRAD) devices (Lee et al., 2017)

Device	j_0[a]	η_{ext}[a]	V_0[a]	*LT90*	*LT80*	ΔV(*LT90*)	ΔV(*LT80*)
	[mA/cm^2]	[%]	[V]	[h]	[h]	[V]	[V]
CONV	6.7	8.0	6.6	27	93	0.3	0.4
GRAD	5.7	8.9	8.0	47	173	0.6	0.9
M0	5.5	9.4	9.2	71	226	0.9	1.2
M1	5.4	9.5	8.8	99	260	1.2	1.6
M2	5.4	9.3	8.9	103	285	0.7	1.0
M3	5.3	9.6	9.0	141	334	1.1	1.5
				5.2×[b] 3.0×[c]	3.5×[b] 1.9×[c]		
M4	5.2	9.6	8.6	126	294	1.0	1.3
M5	5.1	9.9	8.6	119	306	0.9	1.2

[a] Parameters are measured at $L_0 = 1000$ cd/m^2.
[b] Increase in lifetime compared to the conventional, ungraded device (CONV) lifetime.
[c] Increase in lifetime compared to the graded device (GRAD) lifetime.

The TTA/TPA model is based solely on the assumption that deep levels are generated by destructive, high-energy reactions with the emitting and host molecules, but it does not identify the chemical source of the defects. These chemical origins have been studied for the graded device, D3, in Figs. 6.187 and 6.188 using LDI-TOF-MS (see Section 6.7.3). The mass spectra for fresh (unaged) devices, and for packaged devices aged to *LT40* starting at $L_0 = 3000$ cd/m^2 are shown in Fig. 6.192a. The aged device spectra exhibit peaks from the undamaged dopant, Ir(dmp)$_3$, the host, mCBP, and the Alq$_3$ ETL molecules. But after aging, several additional peaks appear. These are associated with fragments of the dopant (DX), host (HX) and ETL (EX) molecules, where X identifies a particular fragment species.

The proposed fragmentation products of Ir(dmp)$_3$ and mCBP are shown in Fig. 6.192b. The bond rupture sites on the parent molecules are indicated by wiggly lines. D1 is the most massive fragment resulting from

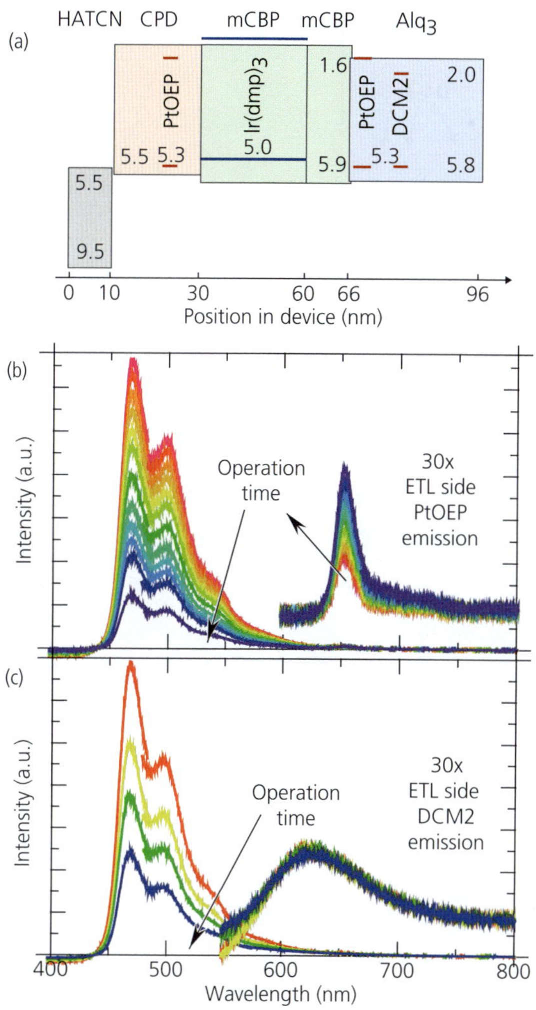

Figure 6.191 (a) Energy level diagram of a blue PHOLED showing materials used in each layer, along with the placement of the DCM2 and PtOEP sensing layers in the Alq_3 ETL, and PtOEP layer within the CPD HTL. (b) Spectra of the blue PHOLED as a function of operating time along with the increasing ETL-side emission spectrum from PtOEP. The spectral evolutions are shown periodically along the directions of the arrows between 0 and 214 h of operation at $j_0 = 30$ mA/cm^2 ($L_0 \sim 5000$ cd/m^2). (c) As in (b) but showing the DCM2 singlet emission from 0 to 109 h (Coburn and Forrest, 2017).

Reprinted figure with permission from Coburn, C. & Forrest, S. R., *Phys. Rev. Applied Letters*, 7, 041002, 2017. Copyright 2017 by the American Physical Society.

cleavage of an entire ligand, which is labelled D4 with $m/z = 334$. Other products include a 2-coordinated Ir atom containing only a single ligand (D2) and a pair of ligands (D3) with twice the mass of D1. Several other products, D4–D7 are due to loss of only a few low mass groups from one or more of the ligands themselves. Similarly, mCBP, which we have already shown is unstable when excited to its S1 state, dissociates into several smaller species, labeled H1–H3.

While a large number of fragments are identified, the question remains as to whether, or what fragments actually produce traps that can affect the luminance and voltage of the PHOLED over time. For this purpose, the energy levels of the various decay products have been calculated using TD-DFT, with results provided in Fig. 6.193. The calculations suggest that the D6, H1, H3, and H4 radicals have deeper LUMOs than mCBP. These energy levels are available to trap electrons, and thus form recombination centers that lead to a drop in luminescence and an increase in voltage. The D7 radical can quench excitons since it has lower triplet energy than $Ir(dmp)_3$. This fragment species, however, does not have an impact on charge trapping due its wide energy gap. Further, mCBP forms neutral fragments (i.e. carbazole, phenyl-carbazole, etc.) upon proton abstraction. These have wider energy gaps and higher E_T than $Ir(dmp)_3$, and hence they, too, are inactive in affecting voltage and luminescence over time.

Figure 6.193 shows that the D2, D4, and H1–H4 cations have deeper LUMO energies than the HOMO of the dopant. These positively charged fragments can attract electrons to become neutral species as shown, and hence do not persist as electronically active states. This leaves the anion series of D1 and H1–H4. Electron transfer from these states will also be rapid, converting them to neutral species. Thus, only the neutral fragments, D6, H1, H3, and H4, appear to be responsible for device aging. The calculations also show that these species have lower E_T than $Ir(dmp)_3$, suggesting that they are likely sources of exciton quenching that is absent from the triplet annihilation model.

While deep blue emitters are required for displays, only cyan emission is needed in an illumination source to achieve an acceptable CCT and CRI. Coupled to the relaxed criterion of *LT70* (compared to *LT90–LT95* for display pixels) the lifetime of WOLEDs nearly meets many of the requirements demanded for interior lighting. An example structure of a high brightness, long-lived all-phosphorescent stacked WOLED consisting of a single blue emitting element and one, two, or three red-green (R-G) elements (devices D3–D5, respectively) is shown in Fig. 6.194a. Since only ~25% of the light emitted by an illumination source is blue, only a single blue element is required to achieve a high total brightness. The WOLED employs a blue element whose graded, Ir(dmp):mCBP EML center section is doped at 3 vol.% with an $Ir(pmp)_3$ manager molecule, along with a R-G bilayer comprising 25 nm thick and 10 nm thick 10% PQIr:8% $Ir(5\text{-ph-ppy})_3$ in mCBP (R-G) layer that is separated by a 3 nm thick BAlq spacer from a pure red emitting 5 nm thick 10% PQIr:BAlq layer. The composition of each of the

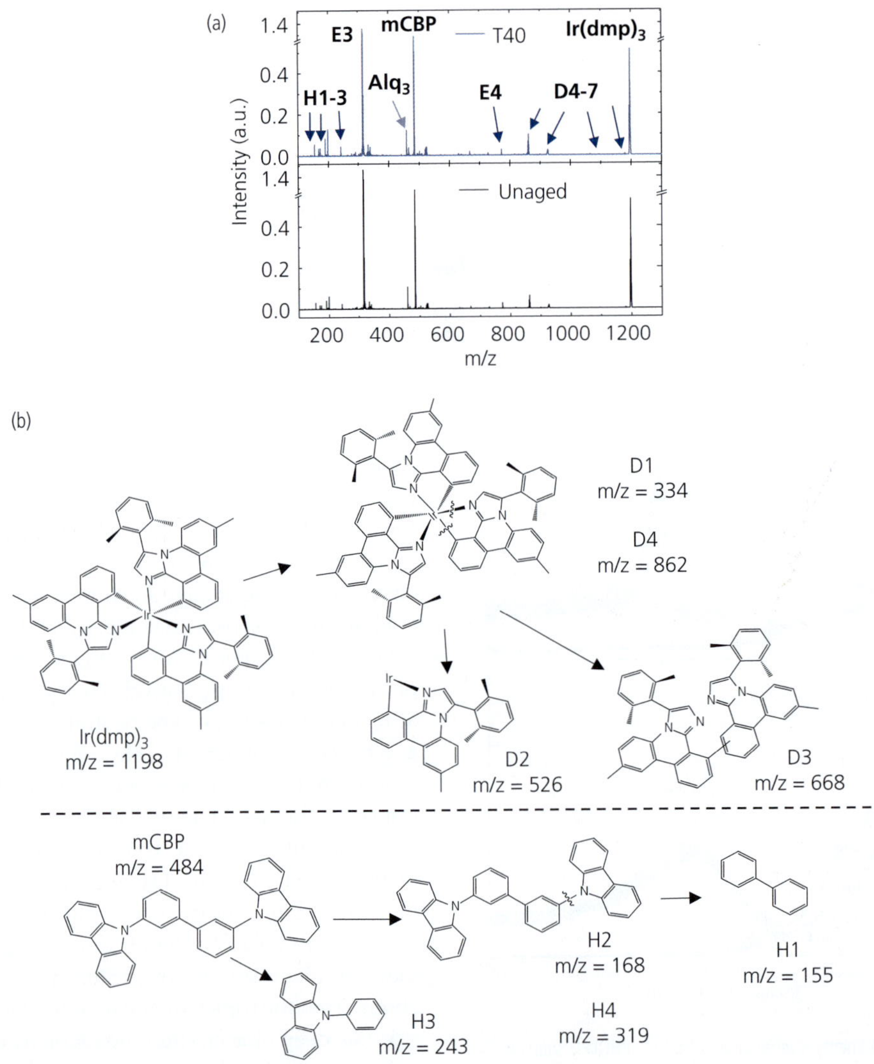

Figure 6.192 (a) LDI-TOF-MS spectra for aged and unaged graded EML PHOLEDs with the D3 structure. Here, H = host, D = dopant, and E = ETL fragments. The unaged molecular features are also labelled. (b) Proposed molecular fragmentation routes for dopant and host molecules corresponding to the *m/z* peaks in (a). The wiggly lines indicate possible bond rupture sites (Jeong et al., 2019).

elements is provided for this 48-layer phosphorescent diode in Fig. 6.194a. The layers are optimized to achieve the desired color coordinates and brightness demanded of high quality lighting sources. The 3-R-G element device, D5, exhibits a maximum brightness of ~200,000 cd/m^2, η_{ext} = 170%, a luminous power efficiency at L_0 = 1000 cd/m^2 of 50 lm/W using substrate outcoupling index matching fluid, a *CCT* = 3000 K and *CRI* = 89.

The *LT70* of devices D3–D5 as a function of initial luminance is shown in Fig. 6.194b. The vertical dashed line corresponds to L_0 = 3000 cd/m^2, which is the nominal operating point of an illumination source. The LT70 of D5 using IMF outcoupling is approximately 14,000 h, which increases to 80,000 h at L_0 = 1000 cd/m^2. Using fewer R-G elements results in a small decrease in lifetime since these layers must be deliver a large fraction of the emission brightness, and thus undergo the highest stress. The color shifts during aging are criteria that are as important as luminance loss in setting the failure criteria of an illumination source. The device in Fig. 6.194 in fact redshifts with age, due to the more rapid differential aging of the blue emitting element, as expected. At *LT70*, the color coordinates of D5 change as follows: $\Delta CCT = -360$ K, $\Delta CRI = -0.8$, and $\Delta u, \Delta v = 0.03, 0$. While all of these performance characteristics indicate that all-phosphorescent stacked PHOLEDs can meet

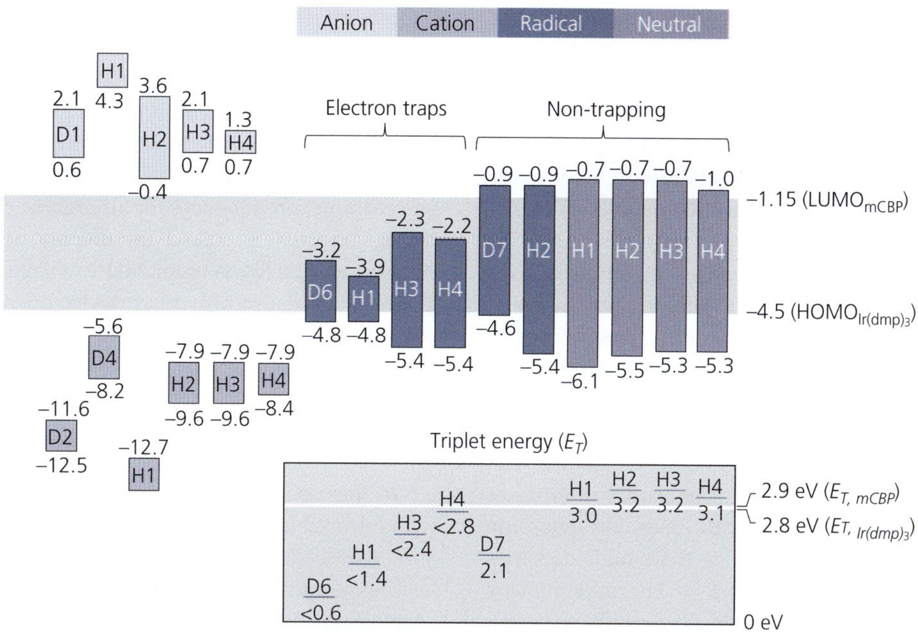

Figure 6.193 Energy level diagram of the fragments produced on PHOLED aging. The broad gray shaded region shows the LUMO energy of mCBP and the HOMO energy of $Ir(dmp)_3$. Anion, cation, radical, and neutral species are displayed as shown in the legend at top. Triplet energies of the parent as well as neutral and radical fragment molecules are given at the bottom right (Jeong et al., 2019).

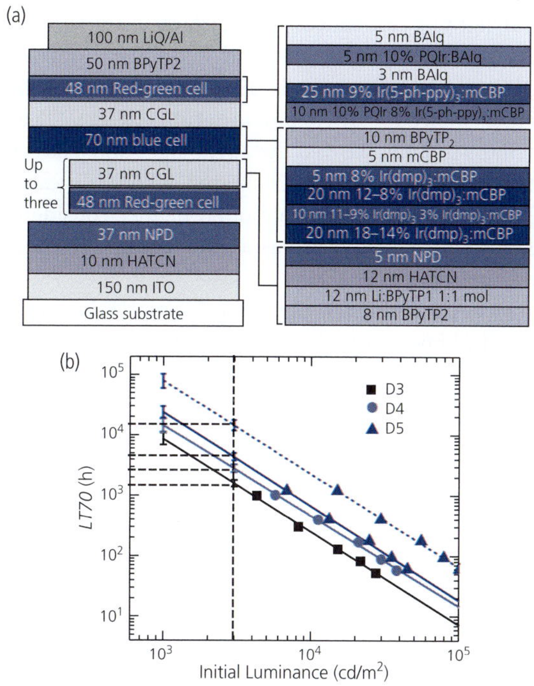

Figure 6.194 (a) Stacked, all-phosphorescent WOLEDs. SWOLEDs D3, D4, and D5 have one, two, and three CGL/red-green element pairs below the blue emitting element, respectively. The total number of layers in D5 is 48. (b) Extrapolation of *LT70* vs. initial brightness. Error bars represent one standard deviation from a fit to the data using the relation: $LT70 \times L_0^n = const$, with $n = 1.50$–1.54. The diagonal dashed line corresponds to the initial luminance obtained using index matching fluid for substrate mode outcoupling. The vertical dashed line corresponds to an initial luminance of 3000 nits (Coburn et al., 2017).

many of the criteria required of efficient and bright white illumination sources, energy-driven degradation of the blue element still presents the limiting factor in their lifetime.

To summarize this discussion, a combination of TTA and TPA initiated reactions is the source of intrinsic losses in luminance and operating voltage of blue light emitting devices. The energies produced are at least double that of the molecular bond energies. Hence, energy-driven aging becomes increasingly important as the emission wavelength is reduced. To mitigate these effects, decreasing the exciton and polaron densities in the EML can significantly increase blue PHOLED lifetime by decreasing the probability for annihilation events. An alternative route to an extending the PHOLED lifetime is to incorporate hot excited state managers in the EML to siphon off the excess energy induced by TPA and TTA.

There is one other important route to increasing the blue PHOLED lifetime, and that is by decreasing the exciton recombination lifetime itself (see Eq. 6.148). Shorter exciton lifetimes reduce the opportunity for annihilation reactions, since this also reduces the exciton concentration in the EML. This is the primary reason that blue fluorescent OLEDs have significantly longer operational lifetimes than PHOLEDs—the lifetime of the singlet state is often $< 10^{-2}$ times that of triplets. As a result, fluorescent blue OLEDs have lifetimes at least 10 times greater than phosphorescent blue devices, see Table 6.14.

We have seen in Section 3.7.3 that improved ^{3}MLCT results in a decreased τ_{Ph}, but a few hundred nanoseconds appears to be a minimum attainable value in transition metal complexes (Yersin et al., 2011). Due to the ligand structure of blue PHOLEDs with their strongly electron withdrawing nature, the spin–orbit coupling, and hence the degree of ^{3}MLCT is reduced, leading to $\tau_{Ph} > 2$ μs. Cavity effects (i.e. via the Purcell effect), or rapid extraction of energy via surface plasmons or other "external" approaches to decreasing the exciton lifetime offer possibilities to reducing exciton recombination times, and hence increasing device stability.

It is worth reiterating that the problem, and the solution for extended blue lifetime is equally relevant for both transition metal phosphor complexes and TADF molecules. In both cases, the emission is driven by triplets, and it is their high densities and long lifetimes that lead to energy-enhanced aging. Notably, TADF molecules, with their reduced spin–orbit coupling, typically have significantly longer exciton recombination times than metal-based phosphor complexes. Hence, it is not surprising that TADF-based OLEDs have shorter lifetimes than those based on Ir(dmp)$_3$ and Pt-based blue complexes (see Tables 6.14 and 6.15) (Nakanotani et al., 2013). Thus, while TADF molecules are still in the early stages of development, it seems unlikely that their lifetimes will exceed those of the metal–organic complexes. Nevertheless, managers, EML grading, reducing the excited state lifetime, and other such strategies that have proven beneficial to blue PHOLEDs should serve equally well for TADF-based blue emitting OLEDs.

Finally, efficient blue emission can also be accessed via phosphor- or TADF-sensitized fluorescence (Baldo et al., 2000b, Nakanotani et al., 2014). At first glance, this would appear to be a promising route to achieving the efficiency of triplet emission and the lifetime of fluorescent OLEDs. Since the energy transfer process to the blue fluorophore from the initiating phosphor or TADF molecule is exothermic, this requires that the initial state be at a higher energy. That is, the initiating molecule must emit in the UV to excite a deep blue emitting fluorophore. Furthermore, the recombination time in sensitized systems is determined by the initiating molecule, not the fluorophore (Baldo et al., 2000b). We have seen that the operating lifetime decreases very rapidly with increased energy, and that even accessing the deep blue with long lived OLEDs has not yet been possible. Hence, it has proven to be more difficult to achieve a long-lived blue sensitized fluorescent device than one that directly emits from the phosphor or TADF complex itself.

We end this section where we began. Achieving deep blue PHOLEDs with both a high efficiency and stability is arguably the most challenging high value problem confronting organic electronics today, with far-reaching consequences for all organic devices if it can be adequately solved for display applications. Whichever strategy is employed to increase lifetime, it must not compromise other device operating characteristics of high efficiency, low operating voltage and color purity.

6.7.5 OLED lifetime on flexible plastic substrates

Up to this point, we have been concerned primarily with intrinsic luminance loss due to chemical degradation that leads to morphological changes from Joule heating, exciton–polaron interactions in the bulk and at interfaces, and so on. However, there is one case where special attention needs to be paid to extrinsic loss when the OLEDs are deposited onto flexible plastic substrates. A significant advantage of OLEDs is their flexibility, opening the door to numerous exciting applications such as wearable or rollable displays and lighting appliances. The earliest example of an OLED on a plastic substrate employed polymer OLEDs (Gustafsson et al., 1992), soon followed by small molecule devices (Gu et al., 1997a). These demonstrations showed the advantages and also several of the problems encountered when fabricating OLEDs on plastic foils, although neither considered environmental exposure as a potential source of device lifetime.

More recently, the Samsung Galaxy Edge series smart phones employ flexible displays mounted behind the front glass panel. Furthermore, flexible substrates allow for rapid, low cost and high volume production of OLED appliances. As we found in Section 5.7, plastic foils such as PET, PEN, PES, PC, and PI have been used as substrates, but they have oxygen with *water vapor transmission rates* of $WVTR \sim$ 1–0.1 g/m^2/day. However, $WVTR < 10^{-5}$ g/m^2/day is required for long term, stable OLED operation (see Section 5.8). Alternative flexible substrates can be employed, such as metal (Wu et al., 1997a, Cheon et al., 2006) or very thin glass foils, but these materials bring their own challenges such as unacceptable levels of surface roughness, fragility and cost (Bhandari and Gaspar, 2015).

In Section 5.8, we discussed processes for depositing encapsulating layers on plastic films using multiple polymer/inorganic layer coatings (Moro et al.,

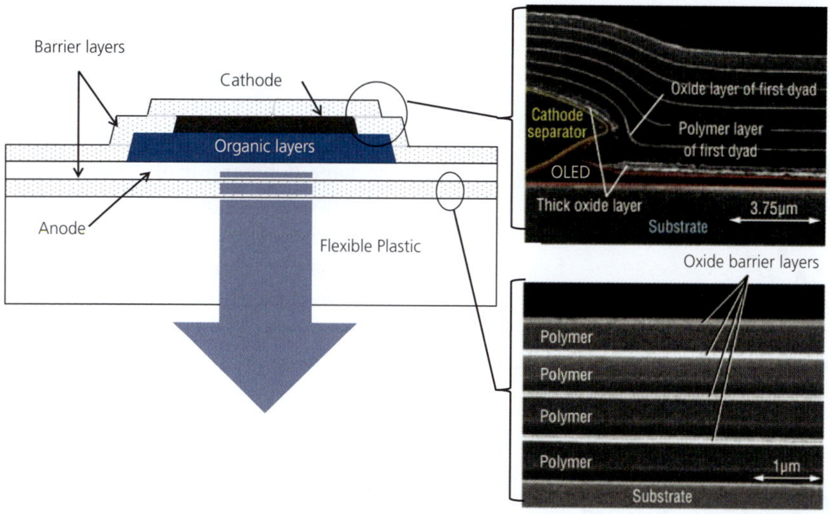

Figure 6.195 Schematic of an OLED encapsulated with barrier layer films on the top of a plastic substrate, as well as conformally onto the OLED surface. To the right (*top*) is a scanning electron micrograph of a conformal multilayer coating on the OLED surface, and (*bottom*) on the substrate. Multilayer coatings can be deposited by CVD or ALD. Images from Suen and Chu (2008).

2004), or more precise and conformal *atomic layer deposition* (ALD). For flexible OLED applications, the barrier coating needs to cover both the bottom and top device surfaces, as shown in Fig. 6.195.

Effective encapsulating layers must have the following properties:

(i) *Low moisture and oxygen permeation rates*: A *WVTR* $< 10^{-6}$ g/m^2/day is preferred, with 10^{-5} g/m^2/day considered to be the minimum acceptable rate (see Fig. 5.90).
(ii) *Flexibility*: They must survive repetitive bends over radii of typically a few centimeters during the projected lifetime of the devices. Depending on the application, this can be many thousands to hundreds of thousands of flexes.
(iii) *Conformability*: The coatings should be capable of conforming to surface features, and be tolerant to morphological defects such as dust. This is shown by the ALD multilayer film deposited onto the surface of an OLED in Fig. 6.195 (top right image) where a smooth transition covers the 100 nm thick cathode and the organic layers without a break in the multilayer barrier structure.
(iv) *Throughput and low cost*: Low cost demands high product throughput and yield. However, ALD growth is slow, resulting in a long takt time in fabricating the substrates. Low cost also implies that low energy processes are combined with low cost materials. To date, barrier-coated plastic substrates are more costly to produce than metal foils, but they also have compelling advantages such as smoothness and transparency that may ultimately be more important than cost alone.
(v) *Non-destructive processing*: The deposition of the encapsulation directly onto the OLED surface must be achieved at sufficiently low temperature, and cannot use materials or chemistries that lead to deterioration of the underlying active device layers.
(vi) *Small coefficient of thermal expansion*: The expansion coefficient should be sufficiently small such that stress from the substrate or coating does not degrade device performance as the temperature is varied over the full range needed in device fabrication and operation.
(vii) *Transparency*: For bottom emitting devices, the substrate must be >90% transparent across the visible spectrum. Metal foils and other opaque substrates require top emitting architectures.
(viii) *High Young's modulus*: In roll-to-roll manufacturing, the substrate (often called the *web*) should be resistant to stretching or other geometric distortion as it is translated from the feed to the take-up reel. Distortions result in misalignment with masks, uneven pixel spacing, etc.

The permeability of plastic to moisture can promote the formation of dark spot defects, and hence a loss of emissive area over time. In Fig. 6.196a we show that the efficiency of fresh and aged Ir(ppy)$_3$:CBP PHOLEDs with $L_0 = 600$ cd/m^2 results in $LT50 = 2500$ h.

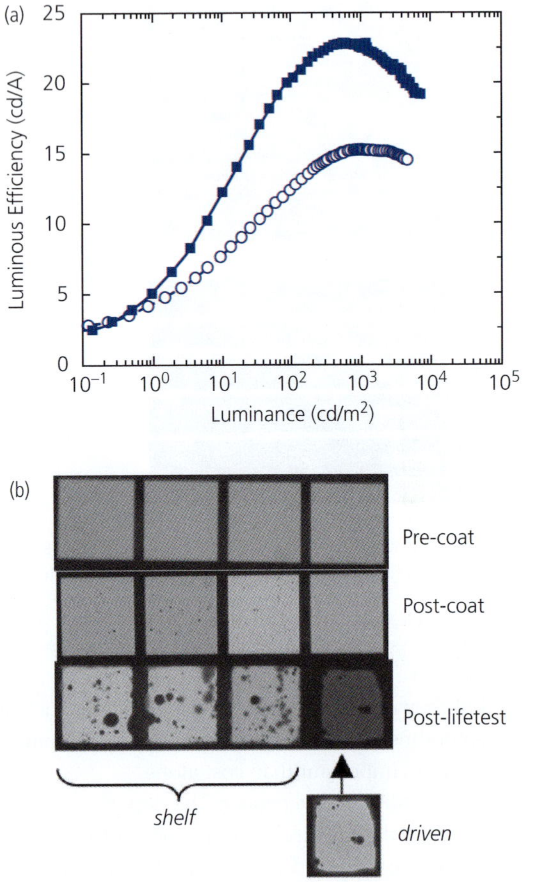

Figure 6.196 (a) The efficiency before (squares) and after (circles) life test of an $Ir(ppy)_3$:CBP PHOLED encapsulated in an organic/inorganic barrier layer film. (b) Images of an array of 5 mm² PHOLEDs before coating by the top barrier layer encapsulation (*top row*), after coating (*middle row*), and post life test (*bottom row*). The pixel on the far right was aged at $L_0 = 600$ cd/m² until $LT50 = 2500$ h (Chwang et al., 2003).

These devices were deposited on a PET substrate, with a barrier coating comprising four *dyads* (or layer pairs) comprising 5–7 μm thick pairs of reactively sputtered, 10–30 nm thick Al_2O_3 and polyacrylate films (Weaver et al., 2002). The polyacrylate itself was deposited by flash evaporation of the acrylate monomer followed by cross-linking via exposure to ultraviolet light. A similar multilayer structure was deposited onto the top surface of the PHOLED, completing the encapsulation. These barrier layer structures delay, but do not halt the ingress of environmental contaminants. The laminate presents a long circuitous path that water and oxygen molecules must take along the polyacrylate layer before finding a defect that allows the contaminant to penetrate the inorganic barrier. After sufficient time has elapsed, these contaminants find such a path, and arrive at the PHOLED active layers (Graff et al., 2004). It is estimated that multilayer film stacks have $WVTR = 4 \times 10^{-6}$ g/m²/day (Weaver et al., 2002).

Using multilayer coatings, the encapsulated device has $LT50 = 2500$ h, compared to 9000 h for glass encapsulation that included a desiccant within the package (Chwang et al., 2003). The source of the luminance loss is apparent from the pixel images in Fig. 6.196b. The dark spots appear over a large fraction of the emitting surface after many hours of storage (denoted "shelf" in the image), or operation ("driven"). Interestingly, the same dark spots are visible in the stored devices as after operation. In both cases, they grow considerably after aging. The dark spots are likely due to particulates that are present during PHOLED film deposition, thus having an origin similar to the dark spots in devices on glass in Fig. 6.170.

Highly conformal coatings of precise thickness can be applied by ALD. The conformality is useful in protecting the OLEDs from penetration by dust particulates or rough surfaces. Figure 6.197a is a micrograph of a top OLED surface encapsulation comprising five pairs of 3 nm Al_2O_3/3 nm ZnO ALD layers separated by 120 nm layer of silica nanoparticles embedded in an organic–inorganic, sol–gel nanocomposite (Jeong et al., 2016). The ALD process is at 70°C using trimethylaluminum and H_2O as precursors for Al_2O_3, and diethylzinc and H_2O for ZnO. The deposition time of each layer is 0.2 s, followed by a 10 s purge. The nanocomposite was spin-coated to achieve the desired thickness. The bottom emitting Alq_3-based OLEDs were deposited onto PET substrates pre-coated with 2.5 dyads of the ALD/nanocomposite stacks. While the deposition was slow, the devices themselves were robust with $WTVR = 1.91 \times 10^{-5}$ g/m²/day as a result of the lack of defects in the ALD layers. After 1000 bends over a 3 cm radius, the devices showed no degradation of their current-voltage or luminance-voltage characteristics (Fig. 6.197b). Furthermore, this nanostratified/nanocomposite structure was exposed for 30 days at 30°C and 90% relative humidity without the OLEDs developing significant dark spot defects (Fig. 6.197c).

Due to their many unique features, flexible OLEDs on plastic substrates remain a subject of intense investigation. While their properties are inferior to analogous devices on glass, they already have operational lifetimes that are acceptable for many applications. It is anticipated that flexible OLED technology will continue to improve, and that their characteristics will eventually equal those of devices on conventional substrates, albeit possibly at a higher cost.

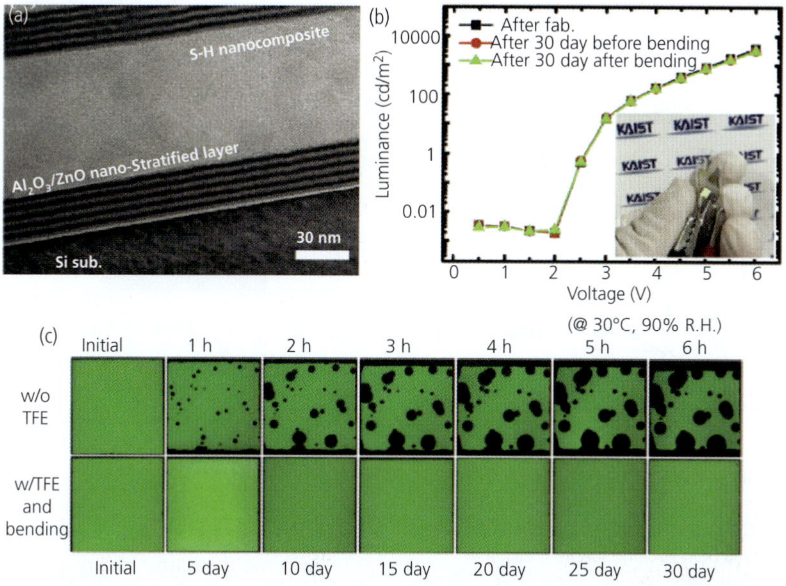

Figure 6.197 (a) High resolution transmission electron micrograph of a nanostratified barrier layer used as thin film encapsulation for a fluorescent OLED. (b) Luminance–voltage characteristics immediately after fabrication, after 30 days exposure to 30°C and 90% relative humidity, and an aged device after 1000 bends over a radius of 3 cm. Inset: Photograph of the FOLED. (c) Luminescence images taken at $L_0 = 400$ cd/m^2 for an unencapsulated and an encapsulated device vs. aging time under the conditions in (b). No aging results under continuous operation are available (Jeong et al., 2016).

6.8 Organic semiconductor lasers

Organic semiconductor lasers (OSLs) have been pursued since their first demonstrations in the mid-1990s in the hopes that they would eventually lead to a new generation of intense optical sources that could be tailored to emit at desired wavelengths by drawing on the same enormous palette of materials developed for OLEDs. And while organic lasers do indeed have many remarkable properties such as emission over a wide range of wavelengths spanning the visible to the NIR, nearly temperature-independent operating characteristics, ability for growth on any smooth substrate (including flexible plastic foils) without materials constraints set by lattice matching, etc., to date only optically pumped lasing using organic semiconductors have been convincingly achieved. In this section we will discuss the principles of organic lasers, some of their unique operating characteristics, and the prospects for electrically pumped lasing. Indeed, optically pumped lasers in themselves hold great interest due to the insights they provide into the properties of organics under intense optical or electrical excitation. The scientific rewards for their study, coupled to the ease with which they can be fabricated and tested, makes the investigation of organic lasers an ongoing source of fascination for scientists and engineers alike.

A laser is an *optical resonator* that comprises three components: a *gain medium*, a low loss optical cavity, and a pump source (electrical or optical). The pump intensity must be sufficient to drive the gain medium into an excited state population inversion, leading to stimulated emission. Organic materials are particularly suited as gain media since they can have very high fluorescent yields, and their Franck–Condon shifts between the absorption and emission spectra lead to exceptionally low material absorption losses. Energy transfer from the pump-absorbing host to an emitting guest molecule results in a further bathochromic shift that reduces absorption losses to near zero. The four-level energy scheme for a laser based on Förster energy transfer from an optically pumped Alq_3 host to a DCM guest is shown in Fig. 6.198. Many such guest–host schemes have been developed for optically pumped lasing based on both small molecule and polymer active media (Hide et al., 1996, Kozlov et al., 1997a, 1998, Vasdekis et al., 2006).

Lasing, along with *amplified spontaneous emission* (ASE) that is easily confused with lasing, has also been observed in single component systems, where the molecular species used to absorb the pump radiation also serves as the laser gain medium (Tessler et al., 1996). However, such single component gain media are not as efficient, and have higher thresholds than four level guest–host systems.

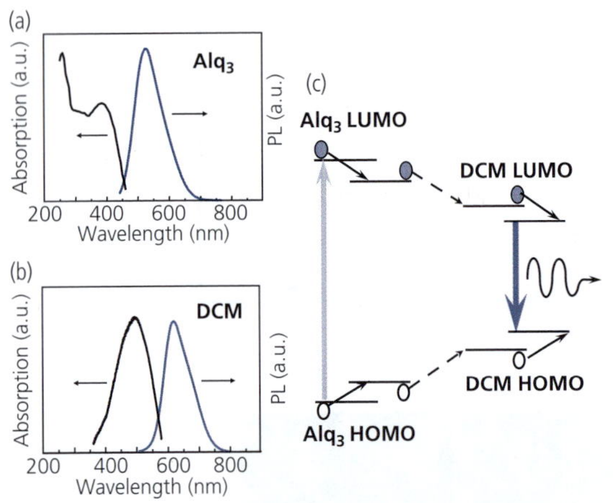

Figure 6.198 The absorption and emission spectra of (a) Alq_3 and (b) DCM. The overlap of the Alq_3 emission and DCM absorption spectra lead to efficient Förster energy transfer from the host to guest. The almost complete separation of the red DCM emission spectrum from the absorption of the two species results in very low optical losses. (c) Energy scheme of the Alq_3:DCM system leading to red laser emission (Kozlov et al., 1998).

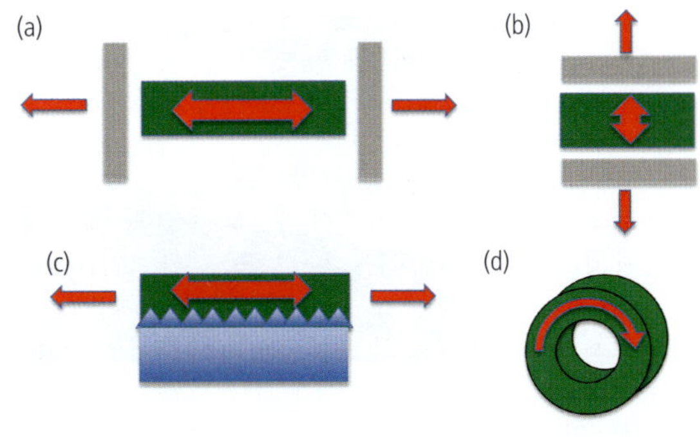

Figure 6.199 Resonators used in lasers. Green regions are the gain media, gray are the mirrors, blue is the substrate, and the optical mode propagation directions are shown in red. (a) Fabry–Pérot or longitudinal laser, (b) vertical cavity surface emitting laser, (c) distributed feedback laser, and (d) ring resonator laser.

The second element of the laser is the resonant cavity that provides optical feedback via the use of high reflectivity mirrors bounding the gain medium. The resonator extends the photon lifetime within the cavity, thereby building up the photon density to create a population inversion (i.e. a non-equilibrium concentration of lasing molecules in their excited states). Numerous resonator geometries have been demonstrated, with a few such designs illustrated in Fig. 6.199.

(i) The most common and simplest form of resonator is the Fabry–Pérot resonator, or longitudinal cavity, in Fig. 6.199a, where the gain medium is sandwiched between two high reflectivity mirrors (Kozlov et al., 1997a). A high mirror reflectivity decreases the pump threshold required for population inversion. Leakage of a fraction of the light through the mirrors results in the observed laser emission. The advantage of this structure is that the overlap between the optical modes and the gain medium is large, leading to high gain and hence high brightness and low threshold. Often, the mirrors are simply cleaved facets of the gain medium whose reflectance is determined by the difference in index of refraction between the semiconductor and air. The reflectivity at normal incidence is given by (cf. Eq. 6.114)

$$R_{ga} = \left|\frac{n_g - n_a}{n_g - n_a}\right|^2, \tag{6.149}$$

where n_g and n_a are the refractive indexes of the gain medium and air, respectively. The two mirrors may have different reflectivities, in which case most of the light is emitted from the lowest reflectivity mirror.

(ii) A *vertical cavity surface emitting laser* (VCSEL) is shown in Fig. 6.199b. This too is a Fabry–Pérot resonator, where the propagation is transverse to the thin film sandwiched between mirrors. At least one mirror must be semitransparent (Hide et al., 1996, Bulovic et al., 1998b). Due to the limited overlap between the optical mode and the gain medium, VCSELs require exceptionally high mirror reflectivity and high gain. An advantage of the VCSEL is its compactness (it is only as thick as the gain medium film), and its cylindrical symmetry. This can result in a symmetric, conical output beam that is easily coupled to external optical systems such as fibers, lenses, etc. that commonly also have cylindrical symmetry. If the mirrors are multilayer dielectric stacks whose reflectance stop band is at the same wavelength as the peak of the gain spectrum (approximately coincident with the PL spectral peak of the emitting species), the external mirrors are known as *distributed Bragg reflectors* (DBRs). Thus, VCSEL is a DBR laser with a single optical mode whose wavelength is determined by the periodicity (Λ) of the dielectric layers comprising the mirror stack:

$$\Lambda = \frac{m\lambda_0}{4}\left(\frac{1}{\bar{n}_1} + \frac{1}{\bar{n}_2}\right) \approx \frac{m\lambda_0}{2\bar{n}}, \tag{6.150}$$

where $\bar{n}_1$ and $\bar{n}_2$ are the effective refrative indexes of the two media in the reflectors, $\bar{n} = (\bar{n}_1 + \bar{n}_2)/2$, and λ_0 is the emission wavelength. Also, $m = \pm 1$,

±2,...is the grating order ($m = 1$ is a first order grating, 2 is a second order grating, and so on).

(iii) When the reflective grating is placed either within or in close proximity to the gain medium as in Fig. 6.199c, the structure is known as a *distributed feedback* (DFB) *laser* (McGehee et al., 1998, Kozlov et al., 2000, Karnutsch et al., 2006). This is distinct from a DBR laser where the grating is located outside of the cavity as discussed in the context of VCSELs. External gratings can also be used in longitudinal geometries (Vasdekis et al., 2006). The DFB grating condition is similar to Eq. 6.150.

The symmetrical, counter-propagating optical fields within the gain medium create a degeneracy resulting in two output modes, separated by the cavity *free spectral range*:

$$\Delta\lambda = \frac{\lambda_0^2}{2\bar{n}L}, \qquad (6.151)$$

where L is the cavity length. The cavity symmetry can be lifted to create only a single mode output by "breaking" the grating into two sections of unequal length. This can also be done by inserting a spacer of length, $\lambda/4$ within the grating. The spacer creates a phase delay for the counter-propagating modes that favors one over the other. DFB resonators are a common means for providing high reflectivity, and hence a low lasing threshold.

(iv) An approach that differs from these one dimensional architectures is the *ring resonator* in Fig. 6.199d. The resonator can be formed by depositing the organic gain medium around a cylindrical core such as an optical fiber (Frolov et al., 1997, Schülzgen et al., 1998). The mode is trapped by circulation within the ring by TIR due to the higher index of the gain medium compared to the surrounding air. Only modes whose wavelengths that are integer multiples of the ring circumference constructively interfere, and hence have the possibility for lasing. That is, the modes must satisfy the following condition:

$$\lambda_0 = \frac{2\pi r\bar{n}}{m}, \qquad (6.152)$$

where r is the ring radius. The light is outcoupled from the ring due to scattering from imperfections around its periphery. However, this does not result in a usable output that is efficiently coupled into an external optical system. A method for more focused outcoupling is to bring a waveguide into close proximity, and within the plane of the ring. The evanescent optical field from the mode trapped within the ring evanescently couples into the waveguide at the point of closest approach. The advantage of the ring resonator is that it provides for a maximal overlap between the optical mode and the gain medium as the mode circulates around the ring periphery, and can have an extremely large quality, or *Q-factor* ($\sim 10^9$) compared with other cavities described, where Q ranges from 10 to 10^4. For an emission line of width $\Delta\lambda$, the Q-factor is:

$$Q = \frac{\Delta\lambda}{\lambda_0}. \qquad (6.153)$$

The final element comprising the laser is the excitation source. To date, all OSLs have been excited by a short wavelength optical pump, such as a N_2, He–Cd laser, or InGaN diode laser. The pump beam is typically projected onto the laser cavity as a stripe that establishes a well-defined region of high optical gain, forming a so-called *gain-guided laser*, that is, the lasing mode is also confined to the region being pumped where there is a population inversion of excited states. In longitudinal (e.g. Fabry–Pérot or DFB) lasers, the optical pump beam illuminates a stripe along the length of the active medium defining the gain region shown in Fig. 6.200a. The intensity is sufficient to induce a population inversion, and hence this is the region where the stimulated emission intensity is higher than the losses, resulting in mode propagation along the stripe. The mode is said to be gain-guided within the stripe since the high excited state population induces transparency. Hence, both the optical mode and the inverted excited state population coexist within the stripe, maximizing the gain.

Reflective mirrors, cleaved facets, or gratings at both ends of the stripe provide the surfaces that define the oscillator dimensions. Surface emitting lasers are

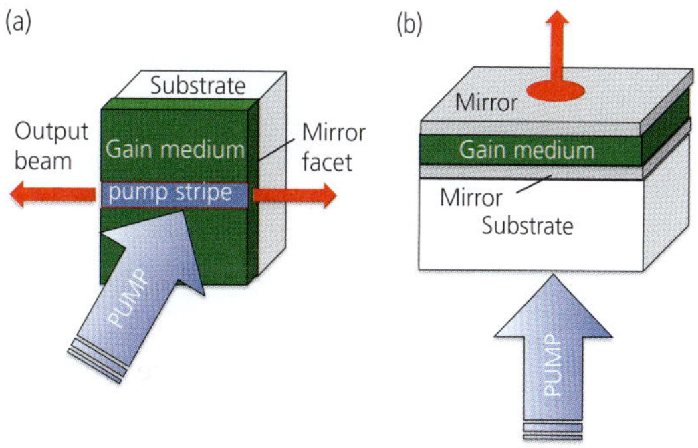

Figure 6.200 Optical pumping schemes for (a) longitudinal and (b) surface emitting lasers.

pumped either from the bottom or top surfaces to create a high gain region within the active medium; see Fig. 6.200b. Depending on whether or not the two surface mirrors have the same reflectivity, the output beam can be preferentially emitted from the top or bottom surface. Ring resonator lasers are typically pumped by directing the exciting laser beam at the gain medium along the ring axis.

In a slab laser, a waveguide is formed transverse to the laser cavity by the pump itself (Fig. 6.201a). As gain continues to increase to the point of transparency of the laser medium, the index of refraction decreases. This causes a break-up (*mode repulsion*) of the single transverse mode, thus reducing the overlap between the optical field and the gain region. Mode repulsion is prevented by employing the *index-guided ridge waveguide* structure in Fig. 6.201b. A shallow ridge in the gain medium is optically pumped. Because the gain medium has a higher index than the surrounding air, the effective index sampled by the mode is higher in this region than elsewhere. The ridge, therefore, provides additional confinement that stabilizes the transverse mode even at high pump powers. Since high efficiency, low threshold lasing requires confinement of the optical mode in the direction perpendicular to the waveguide plane, the gain medium must have a higher index of refraction than either the substrate or the air.

There has been considerable confusion in the literature in identifying a true laser source. Often *superluminescence* or ASE are mistaken for laser emission (Denton et al., 1997). Here, ASE refers to the presence of gain in a medium in the absence of a resonant cavity that produces stimulated emission via photon recycling. Hence it is useful to enumerate those properties of the emitted light that should be observed before the device can be unambiguously identified as a laser oscillator.

(i) *Lasers exhibit a clear threshold delineating regions of spontaneous and stimulated emission.* The threshold is identified by an abrupt increase in slope efficiency, which is the ratio of the output intensity to the pump energy (or in the case of an electrically pumped device, the drive current). Examples of the threshold and high slope efficiency above threshold are illustrated for Alq_3: DCM (2.5%) double and single heterostructure, Fabry–Pérot lasers on InP substrates in Fig. 6.202. Below threshold, the spontaneous emission efficiency is only ~1%, whereas it rises abruptly once lasing is reached. The higher refractive index of the DCM doped layer (with $n = 1.78$) compared to the Alq_3 cladding layer ($n = 1.72$) in the double heterostructure in the upper left of Fig. 6.202 provides optical confinement within the doped layer (Kozlov et al., 1998). The lasing mode is trapped in the thin, high refractive index active region resulting in a significant decrease in threshold and increase in slope efficiency compared with the single heterostructure in the lower right of the figure. Note that it is customary to display the light output vs. excitation intensity on a linear plot where the threshold can be accurately determined from the point of intersection between the spontaneous and stimulated emission regimes. Displaying these data on semi-log or log-log plots can result in an S-shaped curve that can lead one to conclude there is a lasing threshold, yet a straight line with an intercept at the origin

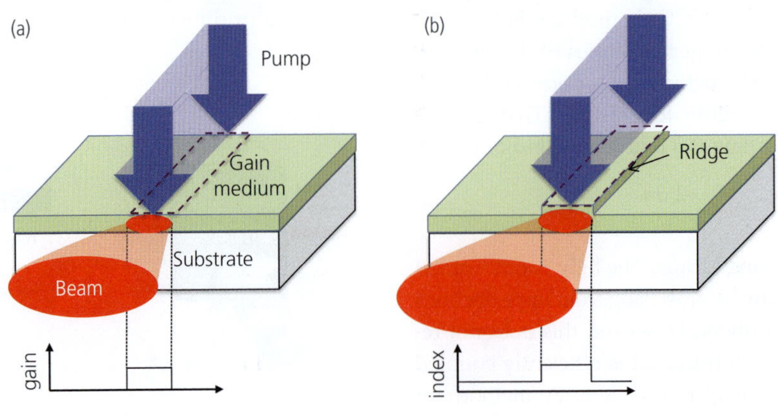

Figure 6.201 (a) Gain-guided and (b) index-guided Fabry–Pérot laser structures. Qualitative plots of the gain and index of refraction along the transverse axis are shown below each structure. The dashed line is the approximate zone being pumped. To achieve optical confinement in the vertical direction, the gain medium must have a higher index than either air or the substrate.

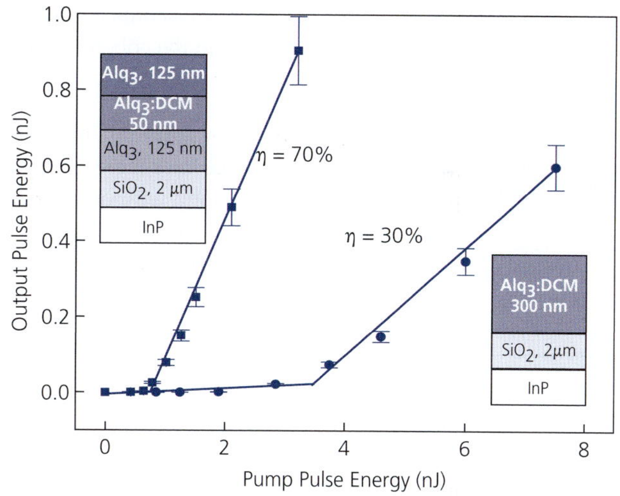

Figure 6.202 Light output vs. input pump energy for two Alq_3:DCM Fabry–Pérot lasers. Upper left: Double heterostructure laser with a slope efficiency of $\eta = 70\%$ and a threshold energy of 1 nJ. Lower right: Single heterostructure with a slope efficiency of 30% and a threshold of 3.5 nJ. The pump is a N_2 laser with a wavelength of 337 nm and a pulse length of 500 ps (Kozlov et al., 1997a).

will also show an S-shaped dependence if the abscissa is a logarithmic scale.

(ii) *Laser linewidths are substantially narrower that spontaneous emission sources.* A transition from a broad spontaneous PL spectrum to a single or multimode spectrum at threshold is an essential feature of lasing. Narrowing above threshold is shown in Fig. 6.203 for an optically pumped polymer-based DFB laser. The active lasing region consists of the semiconducting polymer, BeEH-PPV. The 0.15–0.35 μm thick polymer was deposited onto the surface of a 1 μm thick SiO_2 layer on whose surface was etched a grating with $\Lambda = 170$ to 185 nm (see inset) (McGehee et al., 1998). The spontaneous emission spectrum below threshold is characteristically broad (dashed line), but once the pump energy exceeds threshold there is a dramatic narrowing (solid line). Examination of the emission in the inset shows two modes offset by approximately $\Delta\lambda = 2$ nm arising from the pair of resonances from counter-propagating waves supported by the grating. The differences in mode intensities arises since the short wavelength mode is better aligned to the gain peak of the medium, which is located at the peak of the spontaneous (PL) emission spectrum.

Caution must be used in associating spectral narrowing with gain, however, Microcavity effects, and propagation in waveguides near their modal cutoff can also result in narrow spectra even when only spontaneous emission is present (Yokoyama et al., 2008). Modal narrowing in the absence of gain is easily identified, however, since

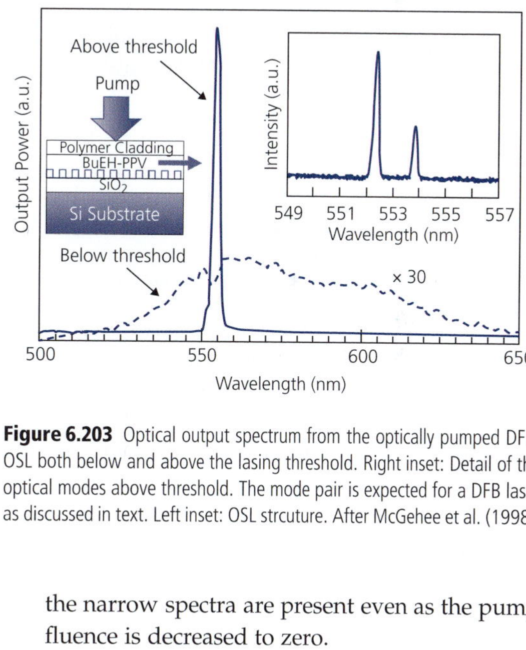

Figure 6.203 Optical output spectrum from the optically pumped DFB OSL both below and above the lasing threshold. Right inset: Detail of the optical modes above threshold. The mode pair is expected for a DFB laser as discussed in text. Left inset: OSL strcuture. After McGehee et al. (1998).

the narrow spectra are present even as the pump fluence is decreased to zero.

Lacking a wavelength-selective element such as a grating, numerous modes are supported within the gain envelope. This is shown by the multimode spectrum of the Fabry–Pérot laser in Fig. 6.204. The cavity is bounded by organic facets cleaved at both ends of a $L = 500$ μm long slab of Alq_3:DCM (2.5%). The modes are evenly spaced at 2 Å, with a mode width of 0.6 Å. The free spectral range given by Eq. 6.151 determines the mode spacing.

(iii) *Lasers are spatially coherent light sources.* This is identified by a diffraction-limited emission

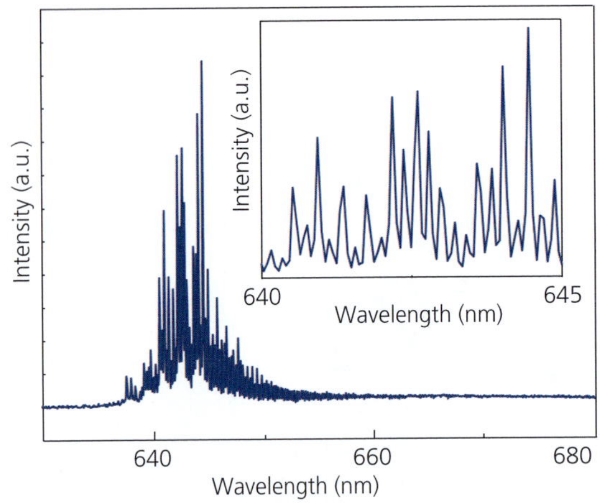

Figure 6.204 Multimode spectrum of a 500 μm long Alq_3:DCM (2.5%) single heterostructure OSL in Fig. 6.202. The modes entirely fill the gain spectrum of DCM defined by its broad PL spectrum, with the highest intensity modes at the gain peak (Kozlov et al., 1998).

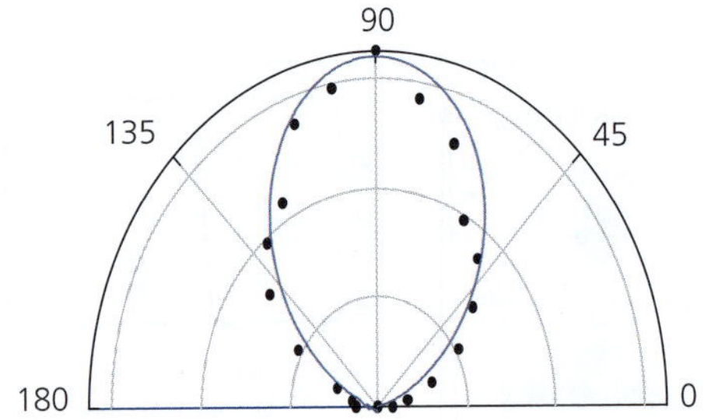

Figure 6.205 Intensity of polarized light emission as a function of the angle, α, of a linear polarizer relative to the normal to the substrate plane of the single heterostructure OSL in Fig. 6.202. The line follows $\sin^2\alpha$, as expected for TE-polarized emission (Kozlov et al., 1997a).

beam, or in the case of a multimode laser, a speckle pattern in the laser output. Coherence can also be identified by polarization of the output beam as shown in Fig. 6.205 for the single heterostructure OSL in Fig. 6.202. The beam is polarized in the plane of the active medium, which is expected for TE emission. The line in Fig. 6.205 is a fit to the TE polarized beam intensity that follows $I(\alpha) = I(\pi/2)\sin^2\alpha$, where α is the angle between the polarizer axis and the normal to the substrate plane, and $I(\alpha)$ is the intensity at α. The existence of a well-defined beam is clear evidence for lasing. While spontaneous emission can result in a spectrally broad, Lambertian pattern, lasers are characterized by narrow and directional transverse mode emission.

(iv) *Lasers are temporally coherent sources.* Temporal coherence is the property that two wave sources have the same period, frequency and waveform. Thus, a purely monochromatic source is temporally coherent and can interfere with itself. If a source emits a range of frequencies of width Δf, then the coherence time, τ_C, is

$$\tau_C \Delta f \leq 1. \tag{6.154}$$

Thus, the coherence time decreases as the source becomes increasingly polychromatic: lasers have a long temporal coherence, whereas white light sources are incoherent. The temporal coherence is typically measured using a Michelson or Mach–Zehnder interferometer.

6.8.1 Theory of lasing in optically pumped OSLs

Since laser emission is due to inversion of the singlet exciton population, all sources of singlet generation and losses must be included in the rate equations that describe laser action. Previously (Section 6.3.5) we have shown that singlets are primarily lost to SSA and STA in the absence of an electric current. The slow buildup of triplets during optical pumping due to the presence of weak spin–orbit coupling even in fluorescent materials has been found to turn off laser emission after only a few tens or hundreds of nanoseconds following the onset of the optical pump. Furthermore, triplet-induced absorption in the cavity can reduce the net optical gain (Kozlov et al., 2000, Baldo et al., 2002, Giebink and Forrest, 2009, Lehnhardt et al., 2010). Including all significant cavity losses, the coupled rate equations that describe laser transient behavior under optical pumping are as follows (Zhang and Forrest, 2011):

$$\frac{dS}{dt} = \frac{\eta_p I_p}{h\nu_p d} - k_S S - k_{ISC} S - k_{ST} S T_G - \gamma \frac{c}{\bar{n}} P, \tag{6.155}$$

$$\frac{dT_H}{dt} = k_{ISC} S - k_{HG} \exp\left(-\frac{2}{R_0}\sqrt[3]{\frac{1}{N_0 - T_G}}\right) T_H, \tag{6.156}$$

$$\frac{dT_G}{dt} = k_{HG} \exp\left(-\frac{2}{R_0}\sqrt[3]{\frac{1}{N_0 - T_G}}\right) T_H, \tag{6.157}$$

$$\frac{dP}{dt} = \left(\Gamma\gamma - \alpha_{CAV} - \frac{1}{L} ln(R) - \Gamma\sigma_{TT} T_G\right) c\bar{n} P + \Gamma\beta k_S S, \tag{6.158}$$

where S, T_H, T_G, and P are guest singlet, host triplet, guest triplet, and lasing mode photon densities, respectively, η_p is the fraction of the pump emission absorbed by the organic film, I_p is the pump intensity, $h\nu_p$ is the pump photon energy, k_S is the guest singlet natural decay rate, k_{ISC} is the host ISC rate, k_{ST} is the guest STA rate, $\gamma = \sigma_{stim}S$ is the gain, is σ_{stim} the stimulated emission cross-section, and $\bar{n}$ is the effective refractive index of the dielectric cladding layer and gain medium waveguide (Chuang, 2009). The optical confinement factor, Γ, is the ratio of the optical mode intensity confined within the gain region of thickness, d, to the total intensity of the propagating mode. Also, k_{HG} is the host–guest Dexter transfer coefficient where we assume that triplets are transferred by exchange interactions, R_0 is the guest–host van der Waals radius (~ 1 nm), $N_0 > T_G$ is the guest triplet saturation population (i.e. the point at which the triplet loss rate equals its generation rate as determined from the saturation of PL quenching, as we have observed for OLEDs in Section 6.3.5), α_{CAV} is the cavity loss without contributions from triplet absorption given by α_{TT}, and σ_{TT} is the guest triplet absorption cross-section. The losses from the optical cavity with mirrors with reflectivities of r_1 and r_2 gives a total reflectivity of $R = r_1 r_2$. The mirror loss is given by the logarithmic term in Eq. 6.158 for a cavity of length, L. That is, the mirror losses decrease with increasing cavity length. Finally, β is the spontaneous emission factor, which is equal to the inverse of the total number of modes within the PL (i.e. the gain) spectrum of the active medium (Coldren and Corzine, 1995). For spontaneous emission, $P = 0$, and the intensity is proportional to S.

The exponential functions in Eqs. 6.156 and 6.157 are due to Dexter transfer that decays exponentially with molecular distance, R_0. The simultaneous solutions of this system of coupled equations provide the transient response of an OSL both below and above the lasing threshold.

Lasing is achieved when the net gain, $g(t)$, is equal to the loss. This is the point when the laser medium has reached *transparency*. Thus, from Eq. 6.161, the lasing condition is given by

$$g(t) = \Gamma\sigma_{stim}S(t) - \alpha_{CAV} - \frac{1}{L}\ln(R) - \Gamma\sigma_{TT}T_G(t) = 0. \tag{6.159}$$

This equation can be solved to determine the threshold singlet population, $S_{TH}(t)$. Surprisingly, OSLs exhibit two distinct threshold singlet populations. The CW threshold population (S_{CW}) must be high enough to overcome triplet losses as $T_G(t)$ increases with time. In contrast, the lasing population under pulsed excitation is $S_{PS} < S_{CW}$, since for short pumping durations the triplet population is too small to affect the cavity gain. Then, ignoring mirror losses, as $t \to 0$, the triplet loss $\Gamma\sigma_{TT}T_G \ll \alpha_{CAV}$, and thus Eq. 6.159 yields $S_{PS} = \alpha_{CAV}/(\Gamma\sigma_{stim})$. With time, T_G increases, which increases the associated loss until $\Gamma\sigma_{TT}T_G > \alpha_{CAV}$. Finally, T_G reaches its saturation density, N_0, at which point the triplet loss can no longer increase, and $S_{CW} = (\alpha_{CAV} + \Gamma\sigma_{TT}N_0)/(\Gamma\sigma_{stim})$.

The CW threshold is approximately

$$\begin{aligned} I_{CW}(N_0) &= h\nu_p d(k_S + k_{ISC} + k_{ST}N_0) \\ &\times (\alpha_{CAV} - \frac{1}{L}\ln(R) + \Gamma\sigma_{TT}N_0)/(\eta\Gamma\sigma_{stim}), \end{aligned} \tag{6.160}$$

compared to the pulse threshold $I_{PS} = I_{CW}(N_0 \approx 0)$. Thus, I_{CW} is a *quadratic* function of guest triplet saturation population. That is, $k_{ST}N_0$ is due to STA that reduces the gain; and $\Gamma\sigma_{TT}N_0$ is due to triplet absorption that leads to loss.

The presence of two thresholds appears to be a unique feature of OSLs, and explains why CW lasing was not observed for 15 years after the first demonstrations of laser action in an organic semiconductor (Zhang and Forrest, 2011).

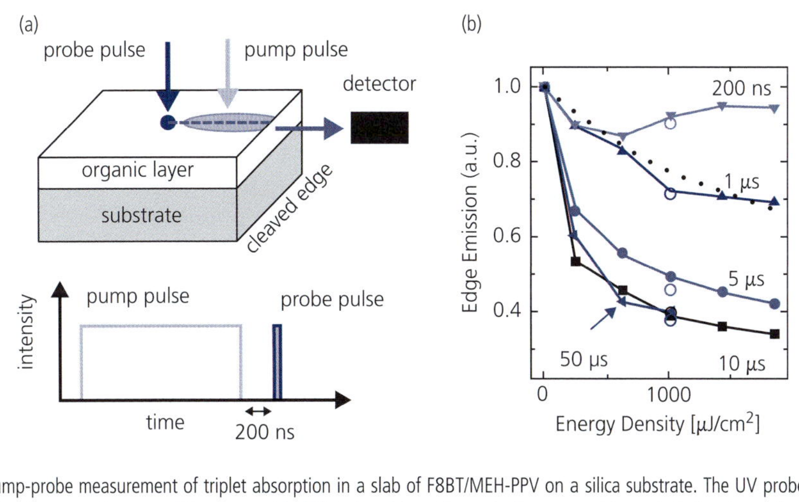

Figure 6.206 (a) Pump-probe measurement of triplet absorption in a slab of F8BT/MEH-PPV on a silica substrate. The UV probe beam is delayed by 200 nm from the end of the pump beam, and the waveguided PL induced from the probe is measured by a photodetector at the edge of the slab. (b) Edge emission intensity vs. pump energy density and duration (Lehnhardt et al., 2010).

The importance of triplet absorption has been measured in the polymer blend F8BT/MEH-PPV (16 wt%) using the pump-probe arrangement in Fig. 6.206a. A waveguide comprising a slab of active material is pumped in the UV to create a large density of singlets along with a much smaller density of triplets formed via weak ISC. After a delay of 200 ns during which time the singlets completely recombine, a second UV probe pulse creates a second population of singlets whose radiation propagates through the now triplet-rich, previously pumped region. By measuring the intensity of the emission as a function of distance from the waveguide edge, the triplet absorption coefficient is determined. The long interaction length between the waveguided fluorescence and the residual triplet population minimizes measurement inaccuracies (Lehnhardt et al., 2010).

The edge emission intensity vs. pump energy density and duration is shown in Fig. 6.206b. The intensity falls to an asymptotic value once the saturation triplet density, N_0, is reached. Furthermore, as the pump duration and energy density increase, the triplet density also increases, as apparent from the larger decrease in edge emission intensity. A fit to the data (dashed line, Fig. 6.206b) using a theory similar to that in Eqs. 6.155–6.158 yields a spectrally broad triplet absorption loss of $\alpha_{TA} \sim 30\ \text{cm}^{-1}$.

The laser rate equations have been used to analyze a gain-guided DFB laser that included a triplet manager analogous to managed OLEDs with reduced TTA discussed in Section 6.3.5. The OSL comprised a 200 nm thick gain layer containing the manager, ADN, co-deposited into the guest–host gain medium consisting of 2 vol.% DCM2 in Alq_3. The manager is included to remove triplets generated by the pump, thereby reducing N_0 and extending the lasing duration. As noted in Section 6.3.5 (cf. Fig. 6.76), ADN has lower triplet ($E_T = 1.69$ eV) and higher singlet energy ($E_S = 2.83$ eV) than Alq_3 ($E_T = 1.99$ eV and $E_S = 2.38$ eV). Furthermore, $E_S = 2.03$ eV and $E_T = 1.74$ eV for DCM2. The manager concentration in Alq_3(98 $- x$ vol.%):DCM2(2 vol.%) is x vol.% ADN ($x = 0$, 10, 30, 50, 70, 100). The same films were deposited on gratings with a period of $\Lambda = 430$ nm and a 50 nm depth on the SiO_2-on-Si substrate to form the DFB OSLs. Output from a 0.6 W pump laser diode at wavelength $\lambda = 405$ nm was focused to a 150 μm × 250 μm spot on the thin film.

Results from pumping OSLs with different concentrations of managers along with fits to Eqs. 6.155–6.158 are shown in Fig. 6.207. The theory and measurement agree well for both the unmanaged laser ($x = 0$), as well as lasers with increasing amounts of the manager, ADN (Fig. 6.207a). As in the case of the managed OLED, the initial drop in PL intensity is considerably reduced with increasing concentrations of the manager until there is no apparent loss from TTA at $x = 100$. The fits to the lasing transients in

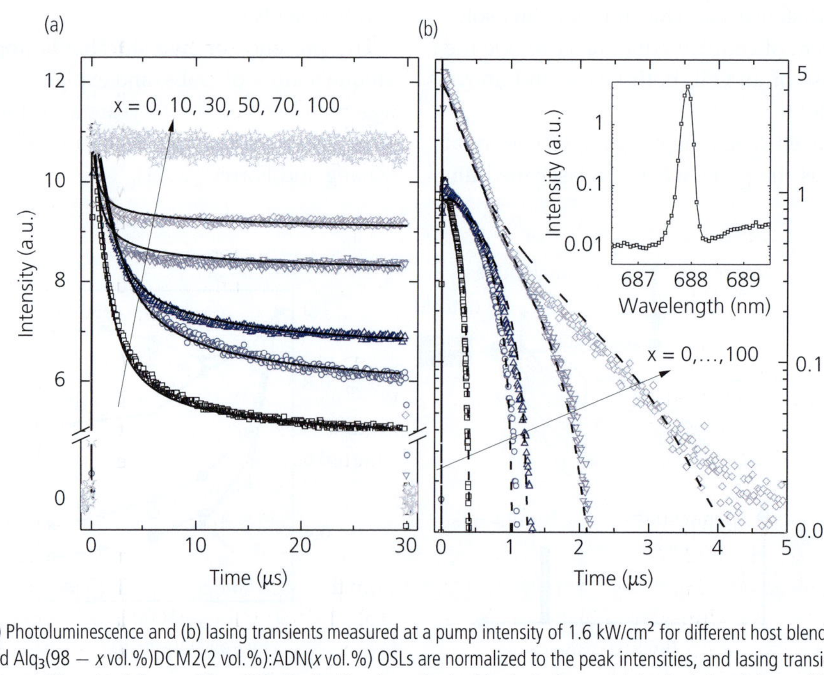

Figure 6.207 (a) Photoluminescence and (b) lasing transients measured at a pump intensity of 1.6 kW/cm^2 for different host blends. The PL transients of the triplet-managed Alq_3(98 $- x$ vol.%)DCM2(2 vol.%):ADN(x vol.%) OSLs are normalized to the peak intensities, and lasing transients are normalized to unity for $x = 0$, 10, and 30, and to 5 for $x = 50$ and 70. The fits (lines) are obtained by the lasing model that includes triplet annihilation and absorption, using parameters summarized in Table 6.19. Inset: Single mode lasing spectrum of an $x = 70$ OSL (Zhang and Forrest, 2011).

Table 6.19 Parameters for PL and lasing transients fits, and the corresponding, measured pulsed (I_{PS}) and calculated CW (I_{CW}) lasing threshold intensities for Alq_3(98 − x vol.%):DCM2(2 vol.%):ADN(x vol.%) OSLs (Zhang and Forrest, 2011)

Device (x vol.%)	I_{PS}[a] (kW/cm^2)	k_{HG}[a] ($10^{10}\ s^{-1}$)	k_{ISC}[a] ($10^7\ s^{-1}$)	N_0[a] ($10^{18}/cm^3$)	σ_{TT}[b] ($10^{-17}\ cm^2$)	σ_{stim}[a] ($10^{-16}\ cm^2$)	I_{CW} (kW/cm^2)
0	0.93	4.0	3.3	5.0	4.0	1.9	32
10	0.75	3.5	2.6	3.9	3.8	2.0	19
30	0.72	13	2.3	2.8	3.6	2.4	8.8
50	0.45	3.0×10^3	1.7	1.5	4.3	2.1	3.7
70	0.43	5.0×10^5	1.3	0.92	4.1	2.3	2.2

[a] Parameters from measurement.
[b] Parameters from fits to data.

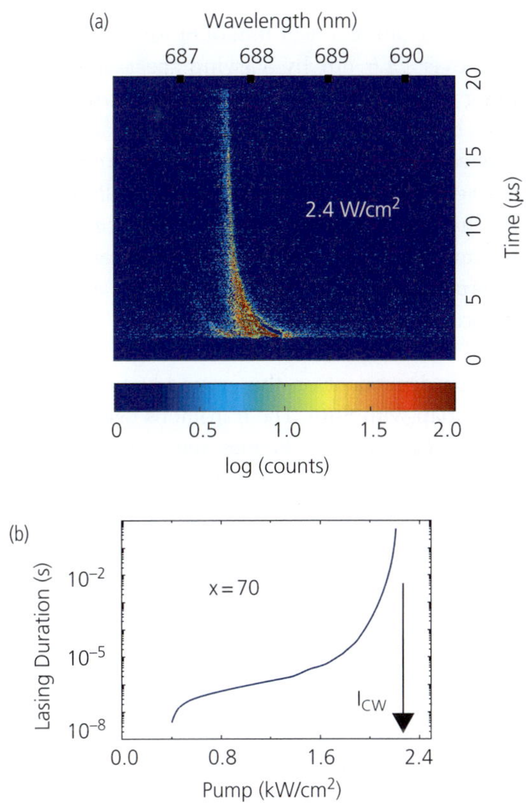

Figure 6.208 (a) Streak camera image of laser emission for the triplet-managed OSL in Fig. 6.207 measured at a pump intensity 2.4 kW/cm^2. This shows a CW lasing duration of >18 μs lasing duration. (b) Simulated lasing duration evolution with increasing pump power for a triplet managed OSL (Zhang and Forrest, 2011).

Fig. 6.207b are more revealing. In the absence of the manager, triplets rapidly build up, resulting in a lasing duration of only 400 ns. However, as the manager is increased to 70%, the duration is increased by a factor of ten, to 4.5 μs. At all concentrations, the data are accurately fit to the model described above, using parameters listed in Table 6.19.

From the fits, we see that in unmanaged devices, the pulsed threshold pump fluence is more than 30 times less than its CW value. Indeed, $I_{CW} = 32$ kW/cm^2 is well above the damage threshold of this (and most) organic semiconductor systems. However, at $x = 70$, the CW threshold drops to only 2.2 kW/cm^2, which should not induce immediate damage to the organics. Figure 6.208a shows a streak camera image of CW lasing achieved at $I_{CW} = 2.4$ kW/cm^2. The lasing lasts as long as the pump beam is applied, and in fact only ceases after 100 μs once the high excitation density damages the gain region. Thus, CW lasing is achieved once the triplet population is sufficiently reduced. The lasing duration vs. pump power is calculated for this same $x = 70$ managed laser, indicating that the lasing duration becomes infinite once I_{CW} is exceeded.

6.8.2 Distinguishing characteristics of OSLs

We noted at the beginning of our discussion that OSLs are interesting due to many characteristics that distinguish them from conventional, inorganic laser diodes. In the following, we discuss a few of these unique characteristics that have provided motivation to seek means for electrical pumping, and in some cases provide sufficient cause for using OSLs in optically pumped applications as well.

(i) *Access to a broad and tunable range of emission wavelengths across the visible and NIR.* The appeal of organic materials is their versatility. By simple modifications of chemical structure, a nearly unlimited range of properties can be accessed. From the earliest demonstrations of both polymer and small molecule-based lasers, researchers demonstrated OSLs with emission from the deep blue to the NIR (Hide et al., 1996, Kozlov et al., 1998), with example lasing spectra in Fig. 6.209 for six different small molecule dopants and two hosts.

Another feature of organics is their broad gain spectrum, as determined by the width of the PL

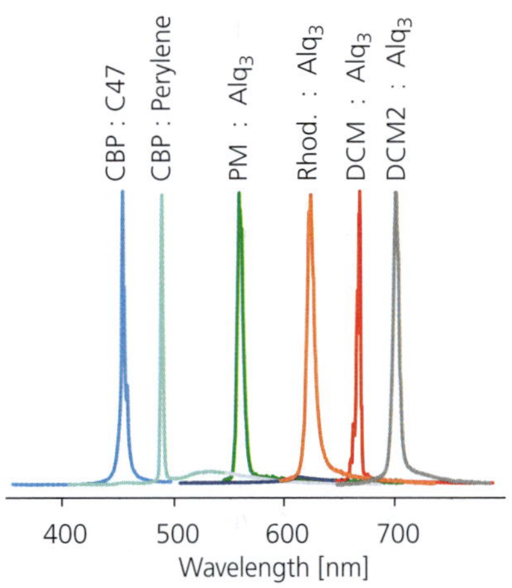

Figure 6.209 Lasing wavelengths of several host–guest systems. The host materials are CBP and Alq_3. Dopants are Rhod = Rhodamine 6G, PM = pyromethane, C47 = coumarin 47. After Kozlov et al. (1998).

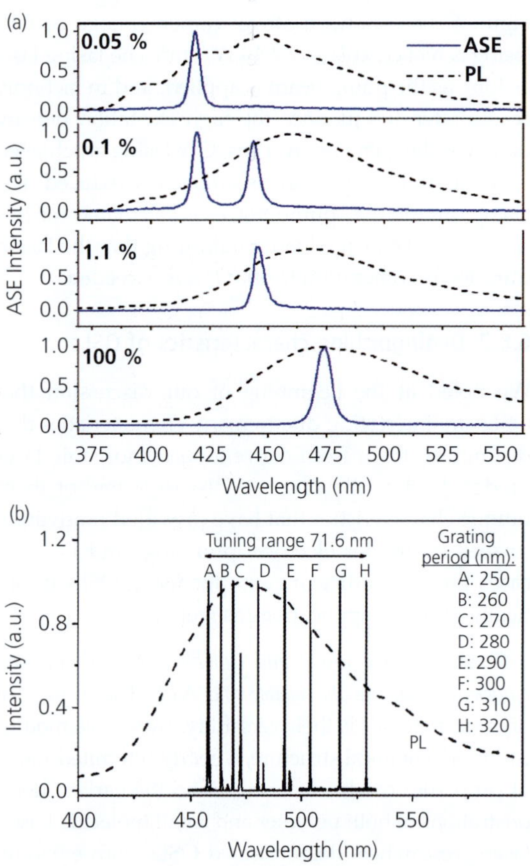

Figure 6.210 Extraordinary tuning range of a small molecule guest–host system. (a) Amplified spontaneous emission spectra of the guest, Spiro-DPVBi with concentration (wt.%) in Spiro-BTBi. (b) Lasing spectra for a neat guest using second order DFB gratings of different periods to tune the wavelength. The total tuning range is 71.6 nm (Schneider et al., 2004).

emission peak. This allows for an exceptionally broad selection of lasing wavelengths in even a single molecular system. An example of this tuning range is shown in Fig. 6.210. The optically pumped ASE spectra of the guest, Spiro-DPVBi, in the host, Spiro-BTBi, red shifts with guest concentration, from 415 nm at 0.05% to 475 nm at 100% (Fig. 6.210a). The red shift is likely due to dielectric polarization, via the SSSE discussed in Section 3.6.7 (Bulovic et al., 1999). The accessible wavelength range is determined by the broad PL spectrum shown by the dashed line. Indeed, SSSE red shifts in the lasing spectrum itself have been observed over a remarkable range of 35 nm by varying the DCM concentration from 0.05–10% in Alq_3:DCM Fabry-Pérot lasers (Kozlov et al., 1997b). The DFB lasing spectrum of a neat film of Spiro-DPVBi can be tuned over 71.6 nm by varying the second order grating period on a Si/SiO_2 substrate (Fig. 6.210b).

The broad gain spectrum of OSLs can also be exploited to select the lasing wavelength from among the several vibronic modes of a single molecular species. This can be understood from the electronic state diagram in Fig. 6.211. In this example, optical pumping results in the transition: $S_1(v_e = 2) \leftarrow S_0(v_g = 0)$. Emission occurs from the $S_1(0)$ state to any one of the ground state vibronics, with the strength of a given transition dependent on the ground state wavefunction overlap with $S_1(0)$. The transitions between energies: $E_1 \rightarrow E_4 \rightarrow E_3 \rightarrow E_2$ defines the four-level lasing system, with the gain envelope extending across the

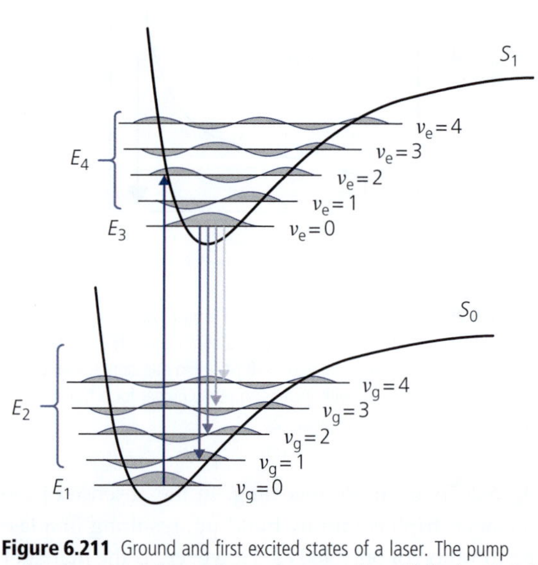

Figure 6.211 Ground and first excited states of a laser. The pump excites the molecule into its S_1 state (gray arrow), and following Kasha's rule, emission occurs from $S_1(v_e = 0)$ to any one of several vibronic states (v_g = 0–4) within the S_0 manifold (green arrows). Due to lack of orbital overlap between $S_1(0)$ and $S_0(0)$, the oscillator strength is very small (Dong et al., 2018).

accessible ground state vibronics for which the oscillator strength for a particular transition is significant (in this case, from $v_g = 1$ to 4). Ordinarily, lasing occurs between $S_1(0)$ and $S_0(1)$. However, the other ground state vibronics can be selected by increasing their oscillator strength relative to the $S_1(0) \rightarrow S_0(1)$ transition. The selection of a given transition can thus be used to "digitally tune" the lasing wavelength.

Selection of a particular vibronic transition has been demonstrated for nanocrystals of the green emitting thiophene, BP1T, by doping with low concentrations of the absorber BP3T. The molecular structures of these complexes are shown in Fig. 6.212. The doped, 20 µm hexagonal crystallites were grown by OVPD in an Ar carrier gas on quartz substrates. Since the excited state dipole moments are oriented perpendicular to the (001) crystal plane that lies parallel to the substrate surface, emission is preferentially within the plane, and is incident to the facets. The crystallites thus form ring (or *whispering gallery*) lasers, with cavity length L equal to the sum of the lengths of the 6 facets.

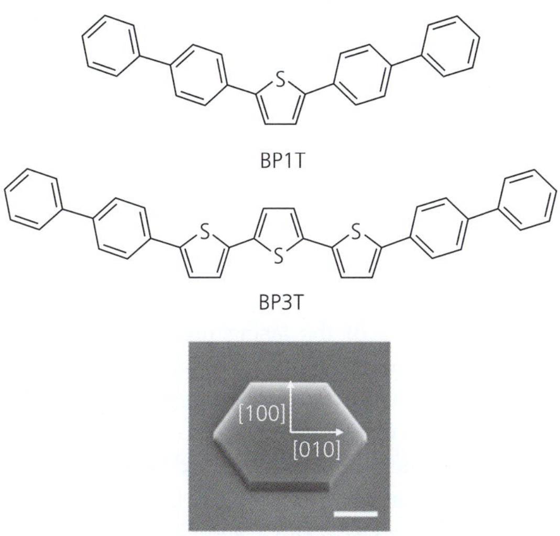

Figure 6.212 Molecular structural formulae of the luminescent BP1T, and the UV–green absorber, BP3T. The electron micrograph shows a single BP1T nanocrystal along with two crystalline axes indicated. The scale bar is 5 µm (Dong et al., 2018).

The BP1T emission and BP3T absorption spectra are shown in Fig. 6.213a. The low concentrations of the dopant are easily controlled by OVPD by placing both the dopant and host in a temperature gradient along the length of a horizontal tube furnace. Each species is located at the optimal point to control its individual evaporation rate. From the figure, we see that the long wavelength absorption tail of BP3T extends into the high energy limb of the BP1T emission spectrum. As the dopant concentration is increased,

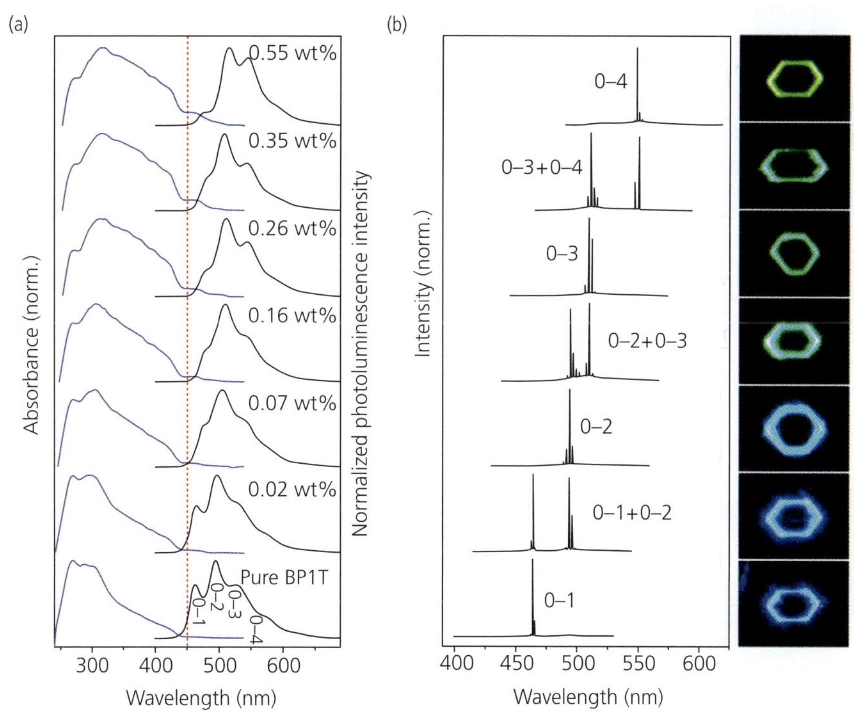

Figure 6.213 (a) Absorption and PL spectra of BP3T and BP1T, respectively, as functions of BP3T concentration. The transitions of the BP1T vibronic progression are shown for the pure BP1T crystal. (b) Lasing spectrum vs. BP3T concentration shown in (a). The various transitions are indicated. Left: Micrographs of the laser emission from a BP1T crystal, varying from blue (bottom) to green (top) (Dong et al., 2018).

the higher energy BP1T emission is reduced, effectively reducing the strength of the 0–1 transition, moving to the 0–4 transition at a BP3T concentration of 0.55 wt%. This, in turn selects the lasing mode, as shown in Fig. 6.213b. The pure crystal emits only from the 0–1 transition in the blue, at a wavelength of $\lambda = 464$ nm. The emission pattern is shown to the right, indicating that the crystal facets indeed form mirrors that confine the optical mode within the hexagonal resonant cavity. As the dopant concentration increases to 0.07 wt%, lasing "digitally" shifts to the 0–2 transition at 495 nm, at 0.26 wt% to the 0–3 mode, and finally at 0.55 wt%, emission is solely from the 0–4 transition in the green, at 550 nm. Interestingly, at intermediate concentrations of BP3T, laser emission from two modes are simultaneously observed. These lasers exhibit multimode emission, indicating that there may be more than one optical path that circulates within the hexagonal cavity at one time.

Selecting the lasing wavelength by absorption has the adverse effect of increasing the threshold pump power, which increases by approximately 20% from the undoped to the 0.55 wt% sample. This roughly corresponds to the increased absorption of BP3T (Dong et al., 2018). The ability to select the lasing wavelength from the transitions from within the vibronic progression of a single molecular species is due to the very broad gain spectrum of organic molecules.

One further exploitation of the broad gain spectrum is an electrically tunable organic VCSEL (OVCSEL) in Fig. 6.214a. The DCM (2.5%):Alq_3 gain medium is separated from a dielectric DBR stack by a 1 μm thick air gap. The top reflector of this bottom emitting device is a 100 nm thick layer of Ag deposited onto the gain medium surface. By applying a voltage between the Au top contact and the ITO layer that underlies the dielectric mirror stack, the cavity length is reduced via electrostatic attraction. This results in a shift in resonant wavelength of $\Delta\lambda > 10$ nm at 20 V, as shown in Fig. 6.214b.

From these few examples, we see that the exceptionally broad gain spectrum of OSLs can lead to wavelength versatility over a large range.

(ii) *Temperature independent gain characteristics*. Inorganic semiconductor laser characteristics such as the threshold current, slope efficiency and emission wavelength show significant temperature dependencies. This arises from the broad densities of states of electrons and holes. As temperature increases, the Fermi energy moves toward the band edge due to the increase in free charge density. This, in turn, increases the threshold current needed to reach transparency. The threshold current (or excitation density) in inorganic laser diodes follows the empirical relationship: $j_{th} = j_0 \exp(T/T_0)$ where T_0 is the characteristic temperature of the lasing medium and cavity architecture. For group III–V semiconductor lasers such as those based on InP or GaAs, $T_0 \sim 100$–150 K (Coldren and Corzine, 1995). To reduce the effects of temperature, the confinement of the density of states in engineered quantum well and QD structures are widely employed (Arakawa and Sakaki, 1982). We have shown in Part I that the energetics of organic semiconductors are inherently different from their inorganic analogues. Disorder leads to a lack of well-developed bands except in only a handful of molecular systems. Thus, a strong energy gap dependence on temperature in inorganic semiconductors is absent in organics. Indeed, in organic host–guest systems, excitations are localized on the guest molecule after Förster transfer from the host. Thus, we can think of such excited, widely dispersed fluorescent molecules as natural quantum dots. These molecular systems thus have very weak threshold power dependence on temperature.

The temperature dependent characteristics for an Alq_3:DCM (5%) Fabry–Pérot laser are shown in Fig. 6.215, and are compared with expectations for a

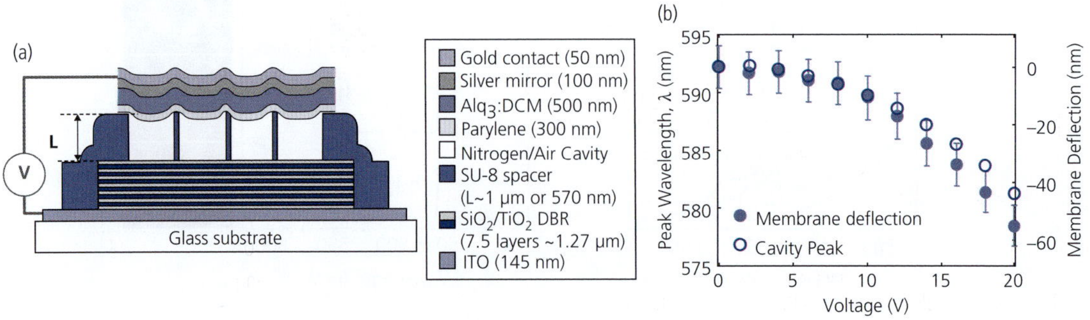

Figure 6.214 (a) Schematic diagram of a mechanically tunable OVCSEL. (b) Optically pumped OVCSEL wavelength and mirror deflection distance as functions of applied voltage between the Au and ITO contacts in (a) (Chang et al., 2014).

conventional GaAs-based semiconductor laser diode (dashed lines). As expected for a quantum-confined state, there is only a small increase in threshold pump energy with temperature from 0°C to 150°C for the organic laser. Its characteristic temperature of $T_0 =$ 1000 K is nearly an order of magnitude higher than for the inorganic device. The slope efficiency of the organic laser does not change over this same temperature range, whereas many inorganic laser diodes do not even reach threshold at the highest temperatures used in this experiment (Kozlov et al., 1997b). Moreover, the wavelength is also relatively temperature independent shifting by only a few nanometers over this broad temperature range. In contrast, the inorganic laser wavelength is expected to depend on its energy gap, which follows the empirical relationship: $E_g(T) = E_g(0) - \alpha T^2/(T+\beta)$, with $\alpha = 5.4 \times 10^{-4}$ eV/K and $\beta = 204$ K for GaAs (Sze, 1981). The small red shift of the organic is possibly due thermal expansion of the Fabry–Pérot cavity at high temperatures.

The extraordinary and intrinsic temperature independence of OSLs provides a compelling case for their use as wavelength standards whose intensities and wavelengths are independent of large temperature swings. The access to a broad range of emission wavelengths further suggests their suitability for a diversity of sensing, timing and communications applications.

(iii) *Ultra-narrow lasing linewidths.* In addition to stable wavelength operation, the linewidths of OSLs can be extraordinarily narrow. Indeed, Fourier transform limited linewidths have been observed for small molecule OVCSELs, as shown in Fig. 6.216. This device comprises a single metal mirror and a DBR stack with a reflectivity of $R_1 = 0.995$. This gives a microcavity Q of

$$Q = \frac{\lambda_0}{\Delta\lambda} = \frac{2\pi}{\lambda_0} \frac{2\bar{n}L}{ln(1/R_1R_2)} \approx 420, \tag{6.161}$$

where $R_2 = 0.91$ is the Ag mirror reflectivity and $\bar{n}L = 3.5\lambda$ is the effective optical cavity length. The measured linewidth above threshold is only $\delta\lambda = 0.2$ Å (Bulovic et al., 1998b)

The Fourier transform limited beam width is determined by the laser bandwidth. That is, high bandwidth operation transforms into a spectrally broader line. The temporal response of the OVCSEL is shown in Fig. 6.217. The pulse response is characterized by a series of *relaxation oscillations* of period $T = 100$ ps and a duration of $\Delta t = 40$ ps. Relaxation oscillations arise from a mismatch between the spontaneous emission lifetime of the laser medium, τ, and the photon lifetime in the cavity, t_C, which is simply the photon round trip time divided by the probability of escape per round trip. The relaxation oscillation frequency is given by

$$f_{ro} = \frac{1}{T} = \frac{1}{2\pi}\sqrt{\frac{(\rho-1)}{t_{C\tau}} - \left(\frac{\rho}{2\tau}\right)}. \tag{6.162}$$

Here, ρ is called the pump parameter, which is the ratio of the pump power to its value at threshold (i.e. $\rho = E/E_{TH}$ in Fig. 6.217). The Fourier transform time of 40 ps results in the observed linewidth of $\delta\lambda = 0.2$ Å, indicating that this device has reached its theoretical bandwidth limit.

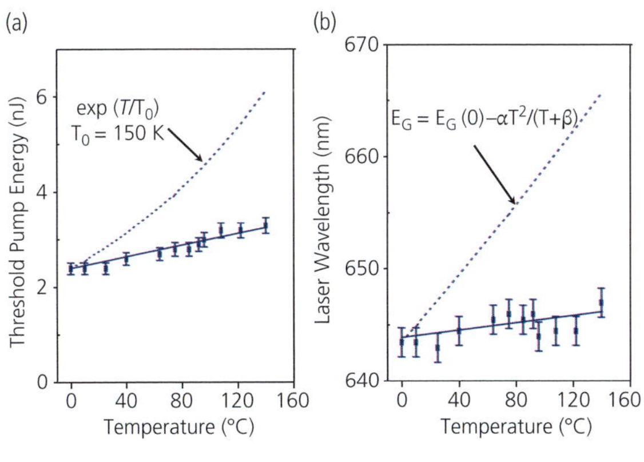

Figure 6.215 (a) Threshold pump energy for an Alq$_3$:DCM Fabry–Perot laser vs. temperature. The dashed line is characteristic of GaAs lasers with $T_0 = 150$ K, and the solid line is a fit to the data for the OSL with $T_0 =$ 1000 K. (b) Laser wavelength vs. temperature. The dashed line shows the change in GaAs energy gap with temperature, and the data show the dependence for the organic laser (Kozlov et al., 1997b).

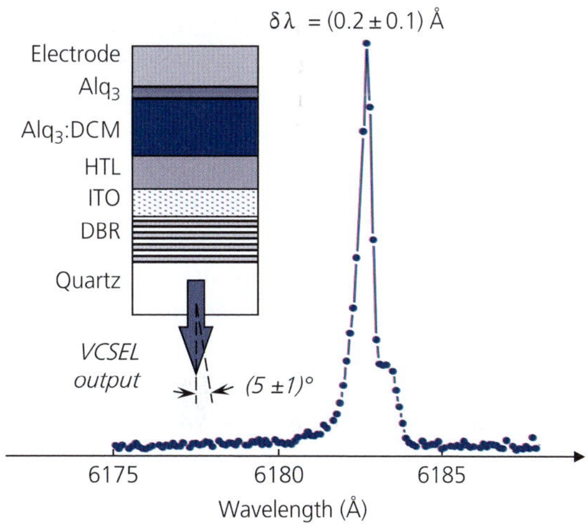

Figure 6.216 Single mode laser emission spectrum of the OVCSEL structure shown in the inset. The exit beam angle half-width is 5°, and the laser linewidth is 0.2 Å (Bulovic et al., 1998b)

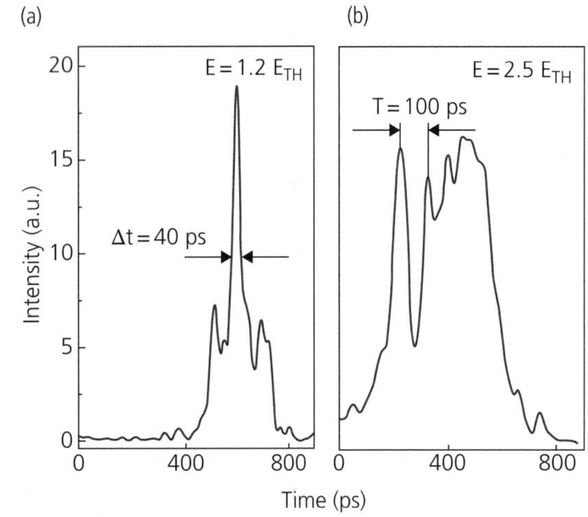

Figure 6.217 Temporal response of the OVCSEL in Fig. 6.216 at two different pump excitation energies above threshold, E_{TH}. The relaxation oscillations have a period of $T = 100$ ps and a duration of $\Delta t = 40$ ps (Bulovic et al., 1998b).

Given this result, we are in the position to estimate the maximum bandwidth (or minimum linewidth) that can be achieved with an organic laser. This ultimately depends on the value of the threshold gain, the size of the cavity, and its *Q*-factor (Henry, 1986). Taking these factors into account, the power–bandwidth product for OSLs is $P\Delta f = 1$ GHz mW, which is comparable to that achieved using inorganic semiconductor lasers (Bulovic et al., 1998b)

(iv) *Compatibility with flexible substrates.* A common property of all organic electronic devices is that they can be grown on almost any flat substrate without regard to its composition or mechanical properties. This property also applies to OSLs. Almost as soon as the first demonstrations of OSLs were announced, researchers reported OSLs fabricated on a variety of flexible substrates (Berggren et al., 1998, Kallinger et al., 1998, Kozlov et al., 1998, Herrnsdorf et al., 2010). The challenge is to fabricate high quality gratings that result in a low threshold and a well-defined output beam, even when the substrate is flexed. Single mode DFB lasers have been demonstrated by spin coating a photoactive polymer, or vacuum deposition of small molecule guest–host systems onto photopatterned or pressure-embossed gratings in plastic substrates (Berggren et al., 1998, Kallinger et al., 1998). To date, however, the lowest pump thresholds (~2.7 kW/cm^2) (Herrnsdorf et al., 2010) on flexible substrates are almost two orders of magnitude higher than lasers on flat substrates (Xia et al., 2009). In spite of these deficiencies, the allure of fabricating arrays of multicolor organic lasers on flexible and lightweight substrates has led to considerable developments of these structures.

6.8.3 Achieving electrically pumped OSLs

In Section 6.8.1 we showed that even the small number of residual triplets that are generated by optical pumping are sufficient to terminate lasing emission after a few hundred nanoseconds. The losses due to triplets are associated with triplet absorption and STA. These losses are enormously increased in electrically pumped devices, where 75% of the injected charges result in triplets. For example, in the Alq$_3$:DCM Fabry–Pérot laser, it was found that the threshold gain corresponded to a cavity loss of $\alpha_{CAV} \approx 30$ cm^{-1}. Including mirror losses (with $R = 7\%$) of $\alpha_M \approx 25$ cm^{-1}, the total loss at threshold was estimated to be $\alpha_T \approx 55$ cm^{-1}. However, at $j = 100$ A/cm^2, the loss increases nearly ten-fold, to $\alpha_T \approx 400$ cm^{-1} (Kozlov et al., 2000). Clearly, electrical pumping can induce very large losses that must be overcome by a commensurate amount of gain, presenting a daunting challenge to reaching threshold. We have already seen in Section 6.3.5 that triplet losses in high intensity OLEDs leads to a significant drop in efficiency as the triplet population increases over time.

Laser diodes are characterized by two features. There must be electrical confinement within the gain medium, that is, the charge balance factor defined for OLEDs must be $\gamma \sim 1$ to ensure a very high density of excited states leading to population inversion in the active material. And there must be optical confinement within the gain region to ensure maximal overlap between the optical field and the charge/exciton distribution. Without both of these factors, it is unlikely that electrically pumped lasing can be achieved. The longitudinal laser structure in Fig. 6.218 possesses these characteristics, but it also illustrates the difficulties in reducing losses. Optical confinement is provided by the refractive index of the active region that is presumably higher than that of the surrounding HTL and ETL. Transparent contacts can be used to avoid the very high losses due to absorption and SPPs excited in the metal contacts. These losses are exacerbated by low index contrast between organic layers that lead to only weak optical confinement. To minimize mirror losses, a DBR or a DFB structure is required.

An advantage of the longitudinal arrangement is the extended gain region. However, contact losses are considerably easier to manage in a VCSEL where the optical field has nodes at the contact surfaces. This must be traded off against a shorter gain region, and the need to make Ohmic contact to top and bottom surfaces that may also contain dielectric DBR mirror stacks.

The laser equations for electrical pumping are identical to those used for optical pumping (Eq. 6.155–6.158) except that the optical generation terms must be modified to account for electrical injection at current, j. Thus, we rewrite these expressions as

$$\frac{dS}{dt} = \frac{\chi_S j}{qd} - k_S S - k_{ISC} S - k_{ST} S T_G + \frac{1}{2} k_{TT} T_H^2 - \gamma c \bar{n} P \tag{6.163}$$

and

$$\frac{dT_H}{dt} = \frac{(1-\chi_S)j}{qd} + k_{ISC} S - k_{TT} T_H^2 - k_{HG} \exp\left(-\frac{2}{R_0}\sqrt[3]{\frac{1}{N_0 - T_G}}\right) T_H. \tag{6.164}$$

Here, we assume that the triplets are formed only on the host, and χ_S is the singlet generation probability. Also, we have included the TTA rate, k_{TT}, to account for triplet interactions only on host molecules since the density of guests, and hence guest triplets, is expected to be considerably smaller. For electrical pumping, N_0 approaches the total host concentration. Direct triplet formation on the guest requires analogous straightforward modifications of Eq. 6.158. The threshold condition in Eq. 6.159 also requires modifications consistent with the above expressions.

There have been only a few studies on the impact of triplet losses on the threshold currents of OSLs (Baldo et al., 2002, Giebink and Forrest, 2009). Figure 6.219a shows the results of one such theoretical analysis that summarizes the challenges in achieving lasing via current injection. This study considers losses due to mirrors, and optical and polaron absorption. For fluorescent materials, the upper limit to efficiency is determined by $\chi_S = ¼$, although the external efficiency will almost certainly be less than 25% for most lasing systems. Even in this best case, the total cavity losses are *optimistically* on the order of 500 cm^{-1}, leading to a minimum threshold current density of $\sim$10 kA/cm^2, which is well beyond the steady-state damage threshold (10–100 A/cm^2) of most organic semiconductor materials. While $j > 12$ kA/cm^2 has been achieved in very small (25 μm) CuPc thin film devices mounted on heat sinks sandwiched between ITO and Mg:Ag electrodes, this current is still at the lower limit expected for lasing in an OLED. Moreover, an Alq_3:NPD device was electrically pumped with $j > 500$ A/cm^2, reaching a maximum brightness of 1.5×10^6 cd/m^2, yet no mention was made of observations of spectral

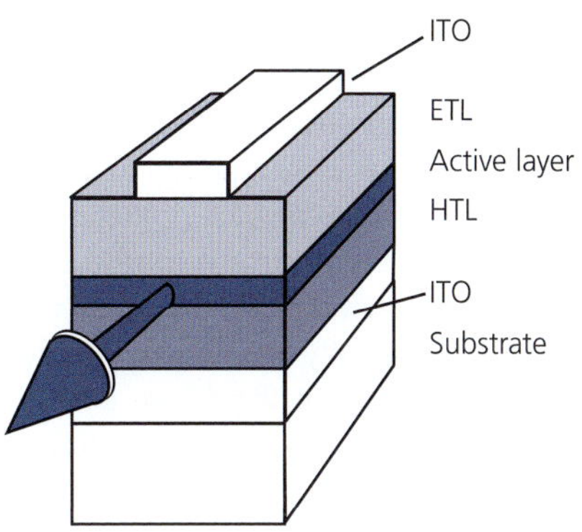

Figure 6.218 Schematic illustration of an electrically pumped organic semiconductor laser.

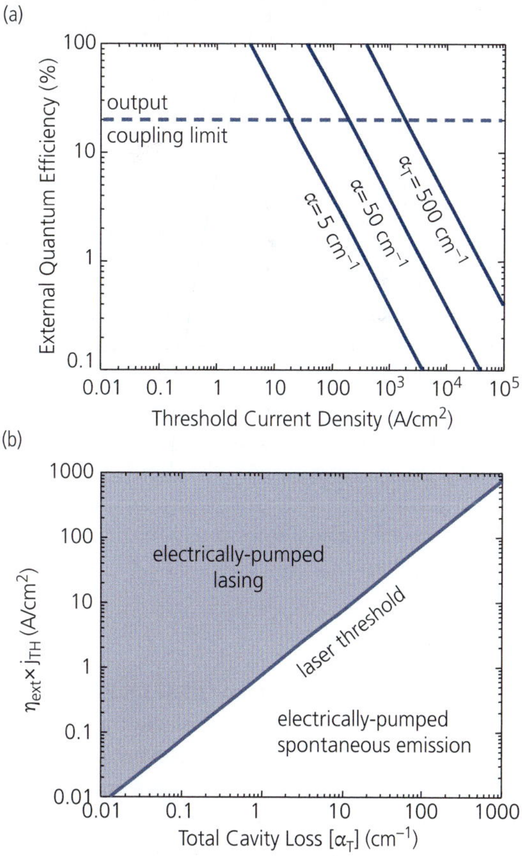

Figure 6.219 Calculated relationship between (a) external quantum efficiency, η_{ext}, and threshold current density, j_{TH}, and its dependence on total cavity loss. (b) The product $\eta_{ext} j_{TH}$ vs total cavity loss (Baldo et al., 2002).

narrowing or a superlinear *j–V* characteristic indicative of ASE or superluminescence (Yamamoto et al., 2005).

Finally, it is useful to consider the product of external quantum efficiency and threshold current density to determine the combined effects of materials and excitation. This is plotted in Fig. 6.219b. There, the regions between spontaneous emission and lasing are delineated as functions of cavity loss.

Since $\chi_S = ¼$ under electrical pumping leads to an extremely rapid build-up of triplets, the laser emission transient will be extremely short (~ns) in the absence of management strategies. Indeed, if the cavity losses are too high, lasing will not be reached at *any* value of *j* (Giebink and Forrest, 2009). Furthermore, even for low cavity losses, the device capacitance may lead to a turn-on time constant due to device charging that is too long to avoid the build-up of triplet losses.

In view of the foregoing considerations, it appears unlikely that a "conventional" OLED-like structure such as that in Fig. 6.218 will lead to electrically pumped OSLs. However, electrically pumped lasing may yet be achieved in structures where the gain region and that used for electrical excitation are separated. For example, it has been shown that a vertical cavity polymer lasing region can exhibit laser emission when integrated with a high intensity, pulsed InGaN LED pump. Using 36 ns pulses, a threshold was reached at an LED drive current of 156 A, as shown in Fig. 6.220 (Yang et al., 2008). This hybrid organic–inorganic laser suggests that spatially separating these functions may indeed eventually lead to an all-organic laser. But it appears that this will occur only when the effects of triplet annihilation and absorption are eliminated from the lasing region. Should that be accomplished, the rapid degradation of organic materials under intense excitation found in laser cavities still constitutes a significant barrier to their practicality.

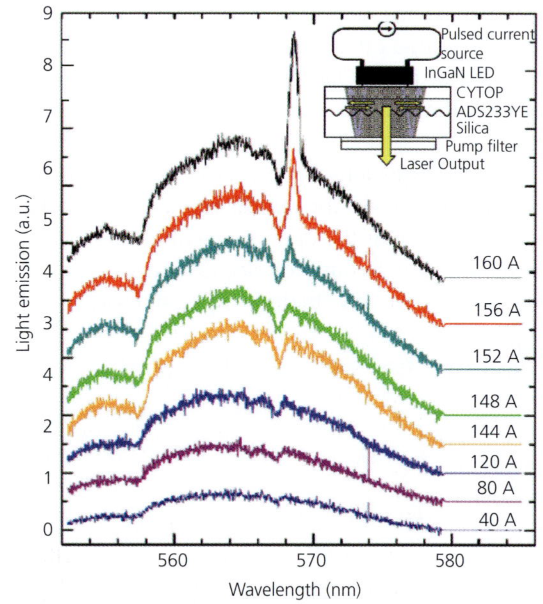

Figure 6.220 Externally, InGaN LED pumped polymer laser spectrum as a function of LED drive current. Lasing threshold is at 156 A. The DFB structure using a CYTOP™ fluoropolymer buffer layer and the luminescent molecule ADS233YE is shown in the right inset. The DFB grating period is 350 nm and its depth is 70nm (Yang et al., 2008).

6.9 Summing up

In this chapter, we have found that OLEDs are the driving force behind the organic electronics "revolution." They have many attractive features that make them compelling solutions for displays and lighting, including high brightness and efficiency, access to a wide color gamut, flexible substrates, etc. The list of attributes is extensive, and has led to a rapidly growing industry with OLED displays becoming dominant in mobile applications. These displays have also made significant inroads into the television and monitor markets. The prospects for OLED lighting are also becoming increasingly bright!

While organic light emission science and technology has made enormous strides since the invention of the low voltage, bilayer OLED in the mid-1980s, there are still major challenges that must be overcome for this technology to reach its full potential. In the chapter, we have discussed many of those challenges, which we summarize here.

(i) *Phosphorescent blue OLED lifetime*: We noted that achieving acceptably long lifetimes for blue PHOLEDs is arguably one of the most difficult yet most important challenges facing the field of organic electronics today. The increase in display and lighting efficiencies that would accompany replacing the longer-lived blue fluorescent devices is approximately 25%, leading to a substantial boost in mobile display battery lifetimes. Yet, we have seen that the lifetime is intrinsically limited by energy-driven molecular dissociation. The degradation can partially be mitigated by decreasing the exciton density in the device emission zone, decreasing the emission lifetime of the phosphor (or TADF) molecule, or by siphoning off excess energy of hot excited states using managers. These methods have not yet led to sufficiently long deep blue PHOLED lifetimes to bring them anywhere near a practical solution for displays. Armed with insights into the chemical origins of the electronic defects, it may be possible to engineer even more robust molecular materials systems that will eventually lead to acceptable performance.

(ii) *Outcoupling all of the photons*. Modern PHOLEDs and TADF molecules can have 100% internal quantum efficiencies, yet most of the light (as much as 80%) is trapped within the device. We have found that the trapped light is launched into substrate, waveguide, SPP, and lossy modes. Numerous creative schemes have been demonstrated to extract each of these modes from OLEDs, with maximum efficiencies approaching ~80%. For a solution to be practical, it must allow for the emission to be independent of wavelength and viewing angle, it should not "disturb" the construction of the OLED itself, nor put demands on the materials sets employed, degrade the OLED performance in any way, and it should be very low cost when applied over large substrate areas. Clearly, much progress is still needed to fully solve this high value problem.

(iii) *Low cost, high resolution patterning for high definition and large area displays*. While this was primarily a topic of Chapter 5, the ubiquity of OLED displays in applications from virtual reality to large screen televisions requires that considerable progress must still be made to achieve low cost, high yield pixel patterning. Currently, displays use fine metal shadow masks that require constant cleaning and are a challenge to implement over large areas. High speed direct patterning, either by ink or vapor jet, appear to be far better suited for pixelation at "any scale, over any area," as required for these applications.

(iv) *Long lifetime, flexible encapsulation*. Already, flexible, ultrathin and lightweight OLED displays are appearing in the marketplace due to their compelling form factors. Indeed, flexibility appears to be a unique feature of organic display and lighting technology. However, we have shown that the water and oxygen permeability of plastic substrates is unacceptably high. This problem has been partially solved by using inorganic moisture and oxygen barriers deposited by ALD or plasma chemical vapor deposition onto the plastic, and directly within the OLED encapsulating layer itself. These coatings are not completely effective, nor are they low cost. However, achievement of long-lived OLEDs on plastic substrates leads to applications that extend beyond just their attractive performance characteristics. High speed patterning along with manufacturing by roll-to-roll processes require flexible substrates. Their achievement would ultimately result in realizing the dream of printing optoelectronic devices "on demand" and "by the kilometer."

(v) *Electrically pumped OSLs*. While optically pumped OSLs were initially demonstrated more than two decades ago, electrical pumping has yet to be convincingly achieved. We have shown that, like long lifetime blue PHOLEDs, there are intrinsic limitations to realizing such a device. The losses due to triplets often overwhelm the gain achieved by the fluorophores in the active media. Thus, electrically pumped organic laser diode designs should separate the zones where excited states are created via current injection, and the light generation (i.e. the exciton recombination) region. Ultimately the solution to this problem will be found since the interest in OSLs is driven by their wavelength versatility, temperature independent operating characteristics, and compatibility with plastic substrates.

These are only five of the many remaining challenges facing OLEDs today. Nevertheless, OLEDs are, and remain likely to continue to be well into the future, the most successful organic electronic device technology. Indeed, their development and continuous improvement over the three decades since their invention are already leading the way toward realizing numerous other practical organic electronic devices such as detectors, solar cells and transistors, to be discussed in the following chapters.

Further reading

1. W. L. Barnes, 1998. Fluorescence near interfaces: the role of photonic mode density. *J. Mod. Opt.*, 45, 661.
2. W. Brütting and C. Adachi (Eds.), 2012. *Physics of Organic Semiconductors*. Weinheim, Germany: Wiley-VCH.
3. W. Brütting, J. Frischeisen, T. D. Schmidt, B. J. Scholz and C. Mayr, 2013. Device efficiency of organic light-emitting diodes: progress by improved light outcoupling. *Phys. Stat. Sol. (a)*, 210, 44.
4. M. Bukshtab, 2012. *Applied Photometry, Radiometry and Measurements of Optical Losses*. Dordrecht: Springer.
5. L. A. Coldren and S. W. Corzine, 1995. *Diode Lasers and Photonic Integrated Circuits*. New York: Wiley-Interscience.
6. S. R. Forrest, D. D. C. Bradley, and M. E. Thompson, 2003. Measuring the efficiency of organic light-emitting devices. *Adv. Mater.*, 15, 1043.

7. D. J. Gaspar and E. Plikarov (Eds.), 2015. *OLED Fundamentals: Materials, Devices, and Processing of Organic Light-Emitting Diodes*. Boca Raton, FL: CRC Press.
8. Z. R. Li (Ed.), 2015. *Organic Light-Emitting Materials and Devices*. Boca Raton, FL: CRC Press.
9. S. Reineke, M. Thomschke, B. Lüssem, and K. Leo, 2013. White organic light-emitting diodes: status and perspective. *Rev. Mod. Phys.*, 85, 1245.
10. S. Scholz, D. Kondakov, B. Lüssem, and K. Leo, 2015. Degradation mechanisms and reactions in organic light-emitting devices. *Chem. Rev.*, 115, 8449.
11. T. Tsujimura, 2012. *OLED Display Fundamentals and Applications*. Hoboken, NJ: John Wiley & Sons.

Problems

1. From the three PHOLED spectra in Fig. 6.15a:
   (a) Calculate the color coordinates for each spectrum to verify the gamut shown in Fig. 6.15b.
   (b) All spectra are approximately equally broad and unsaturated. Why does the red spectrum appear nearly saturated in the CIE chart, appearing closest to the color locus?
   (c) Assuming that the peak intensities of the three colors are in a ratio of ¼, 1, ½ in blue, green, and red, what is the CRI and CCT that can be achieved by the combination of these three color elements? Plot this point on the 1931 CIE chromaticity chart.
2. Figure P6.2 shows the emission spectrum of the white OLED used in LG's 55" TV.
   (a) From this spectrum calculate the CRI, CCT, and (u, v) 1964 CIE color coordinates of the WOLED source.
   (b) Would this WOLED be considered a good room illumination source? Explain.
3. Figure 6.111 shows the spectrum at four different current densities from a two-color WOLED based on a combination of TADF and Ir-phosphor emitting molecules.
   (a) Calculate the CRI, CCT, and color coordinates at the current densities shown.
   (b) Plot the coordinates on the 1931 CIE chromaticity chart. What are the Δu, Δv distances to the Planckian locus?
4. Figure P.6.4 shows the *j*–*V* and *L*–*V* characteristics of a red emitting OLED, along with its emission spectrum in the inset.
   (a) What are the CIE coordinates of this device?
   (b) Plot the photometric characteristics of luminous efficiency and luminous power efficiency as functions of *j*.
   (c) Plot the equivalent radiometric units as those in (b).
   (d) Is this a PHOLED or fluorescent OLED? Explain your answer.

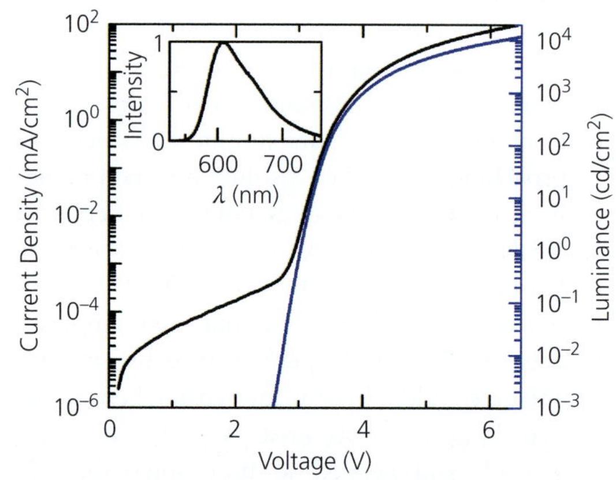

Figure P6.4 Performance characteristics of a red OLED.

5. Figure P.6.5 shows the emission spectrum (blue) of a host, and the absorption spectrum (red) of a guest molecular pair for an infrared emitting OLED. The effective radius of the host is r_H and the guest is r_G
   (a) Plot the dependence of the Förster transfer radius, R_0, and the FRET transfer efficiency vs. guest concentration, N_G.
   (b) At what concentration is the efficiency maximized for a fluorescence lifetime of the guest, τ_D? Assume that the non-radiative concentration quenching rate is given by $k_{cc} = k_{cc0}\exp(-r/(r_G + r_H))$, where r is the molecular separation.
   (c) Repeat part (b) for Dexter transfer.

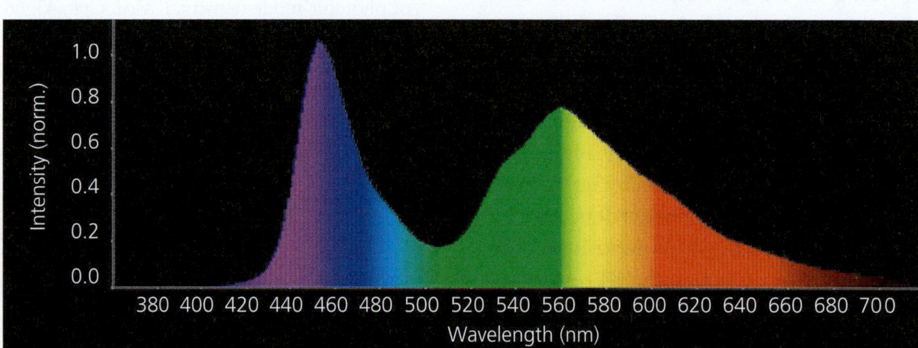

Figure P6.2 Normalized emission spectrum of a WOLED.

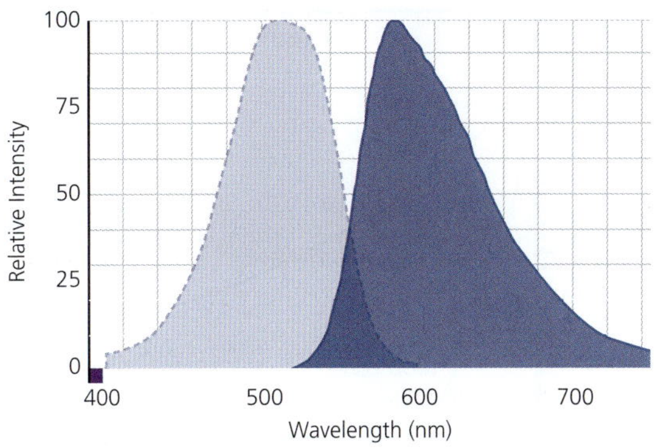

Figure P6.5 The emission (blue) and absorption (red) spectra of a guest–host molecular pair.

6. Analysis of delayed fluorescence in Section 6.3.5 is based on the assumption that singlet and triplet excitons are electrically generated according to random statistics. Here we explore situations where this might not be the case.
   (a) Write the coupled rate equations for triplet and singlet concentrations in OLEDs with singlet–triplet statistics of $\chi_{ST} = 0, 0.1, 0.25$, and 0.5.
   (b) Plot the ratio of delayed fluorescence to direct fluorescence vs. j assuming that $k_{TTA-S}T = k_F = 100k_{Ph}$ as a function of χ_{ST} over the same domain as in (a).
   (c) What is the maximum quantum efficiency that can be attained in each case of χ_{ST} as in (a)?
   (d) Plot the effective triplet current density (j_T in Eq. 6.58) vs. j for a device with an EML thickness of $d = 50$ nm and $\gamma = 1$, independent of the j vs. χ_{ST} with the values in (a).
7. The excimer dynamics are described by the rate equations in Section 6.5.3 for a guest–host system.
   (a) Plot $M^*(t)$, $E^*(t)$, and $H^*(t)$ with the following initial conditions: $t = 0$, $G = 0$, ${}^3M^*(0) = (0.001)^1M(0)$, $f = 0.1$, and $k_{Ph} = 0.01k_{ET} = 0.1k_r \ll k_X$ in the absence of other non-radiative recombination processes.
   (b) Under these conditions, sketch the time evolution of the spectra for the FPt1+FIrpic diode in Fig. P6.7.

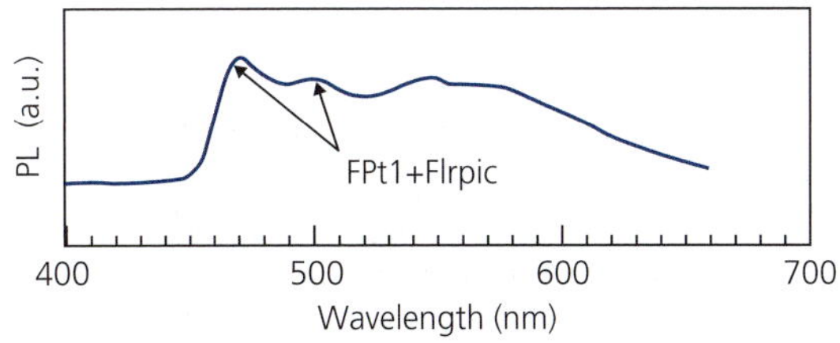

Figure P6.7 Blue monomer/excimer emission.

8. The *transfer matrix approach* is a simple and accurate means for calculating wave propagation through multilayer media, such as that encountered in OLEDs and OPVs. There are many elementary texts and sources for learning about this powerful method of field calculation (see, for example, Heavens, 1991). Briefly, the electric field incident on a medium along the propagation direction, z, is written $F(z) = F_r e^{ikz} + F_l e^{-ikz}$, where subscripts r, l refer to the right, left propagating waves, respectively, with wavevector $k = 2\pi\,\tilde{n}/\lambda_0$, where $\tilde{n}$ is the complex index of refraction of the medium, and λ_0 is the wavelength in free space. For the purposes of this discussion, we will only consider plane waves incident normally on a stratified medium lying in the x-y direction. Then, after propagating through a layer of thickness d, the field and its derivative, $F_z'(d) = dF\big/dz\big|_{z=d}$

$$\begin{pmatrix} F(z+d) \\ F_z'(z+d) \end{pmatrix} = M \begin{pmatrix} F(z) \\ F_z'(z) \end{pmatrix} \tag{6.P8a}$$

where

$$M = \begin{pmatrix} \cos kd & \frac{1}{k}\sin kd \\ -k\sin kd & \cos kd \end{pmatrix} \tag{6.P8b}$$

is called the transfer matrix. For a series of N layers, and with the ith layer with thickness and wavenumber (d_i, k_i), then,

$$\begin{pmatrix} F(z+L) \\ F_z'(z+L) \end{pmatrix} = \prod_{i=1}^{N} M_i \begin{pmatrix} F(z) \\ F_z'(z) \end{pmatrix}, \tag{6.P8c}$$

where $L = \sum_{i=1}^{N} d_i$ is the total thickness of the stack.
   (a) Generalize Eq. 6.P8b for a plane wave arriving at any angle, θ, relative to the normal at $\theta = 0$.
   (b) Write expressions for the reflectivity and transmissivity of the stack in terms of the amplitude of a wave normally incident from the $-z$ direction.
   (c) Now consider the three-layer medium in Fig. P6.8. Calculate the field exiting from the soda-lime glass substrate into air, that is, calculate the transmissivity of the stack for light normally incident on the surface with wavevector k_0.
9. Since the device in Fig. P6.8 is approximately what an OLED structure might look like, we would like to enhance outcoupling from the glass substrate using an anti-reflection coating (ARC).

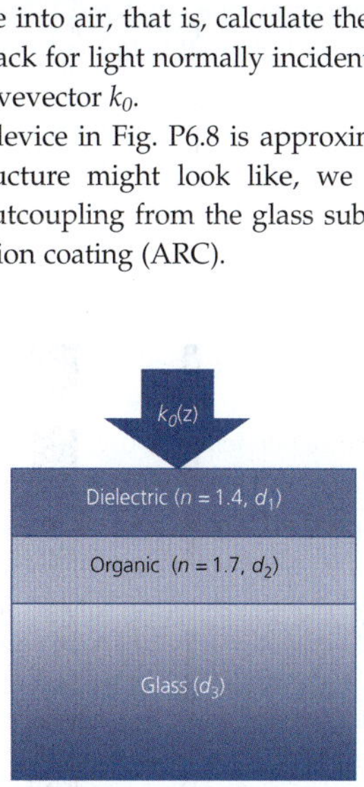

Figure P6.8 Three-layer medium bounded by air, with an electric field propagating in the z-direction with wavevector $k_0(z)$.

(a) We want to use a two-layer ARC at the glass/air interface comprising TiO_2 and silicon dioxide with indexes of refraction of $n_{TiO2} = 2.61$, and $n_{SiNx} = 1.45$. What are the optimal thicknesses of the individual layers in the bilayer coating to maximize the field intensity transmitted into the glass substrate? In what order should these two materials be placed on the glass surface?

(b) Very low index *aerogels* have also been used as antireflection coatings. What is the transmittance of the stack in the figure for such a porous, lossless silica medium with $n = 1.15$ and $d = 0.5$ mm?

10. Microlens arrays have been shown to outcouple light by increasing the light cone angle that can escape from the glass substrate into the air modes.
   (a) Assuming that the EML emits isotropically into 4π solid angle, plot the approximate outcoupling efficiency as a function of microlens diameter, $d_{\mu l}$, from a bottom emitting OLED (index $n_{OLED} = 1.7$) with a perfectly reflective top contact, and where the EML is located sufficiently far from the emitting dipole layer so as not to generate SPP modes. The glass substrate has a thickness d_{sub}. Assume that the microlens array (index $n_{\mu l} < n_{OLED}$) forms a two dimensional, closely packed hexagonal lattice, and that the OLED diameter, d_{OLED}, is much smaller than d_{sub}. (Hint: Ray tracing is a good method to solving this problem.)
   (b) Plot the outcoupling enhancement fraction for the device with the microlens array compared to one with a flat glass substrate using the same geometry as in (a)?
   (c) Plot the dependence of outcoupling efficiency on $n_{\mu l}$ for a fixed $d_{\mu l} \ll d_{OLED}$
11. Calculate the approximate polar emission profile (i.e. intensity vs. angle, θ) of a microlens array comprising 10 μm diameter close-packed hemispherical lenses with indexes of refraction of $n = 1.3$ and 1.7 from:
   (a) A bottom emitting OLED on a 0.3 mm thick glass substrate.
   (b) A top emitting OLED with a transparent top cathode.
   (c) What is the deviation from Lambertian for the two devices in (a) and (b)? That is, calculate the fraction of the total emission that falls outside a perfect Lambertian emitter.
   (d) What effect will the differences with a Lambertian source have on the image resolution and other appearances of the devices?
12. Calculate the dependence of dipole field intensity as a function of angle for a molecule with a transition dipole moment aligned 30° with respect to the substrate plane.
13. Figure P6.13 shows the luminance lifetime data for an OLED driven at the current densities of 10, 20, 30, 50, and 100 mA/cm². The initial luminances, L_0, are provided in the figure. Assume that the emission spectrum of this device is the same as $Ir(ppy)_3$ (although do not assume the device is $Ir(ppy)_3$).
   (a) What are the device efficiencies (η_L and η_{ext}) at $t = 0$ at each current? Is this likely to be a fluorescent OLED or a PHOLED?

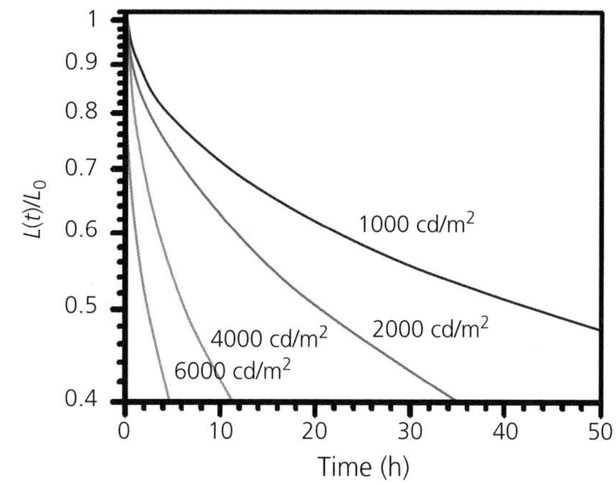

Figure P6.13 Aging data for an OLED. Initial luminances are indicated.

   (b) Fit the curves to the stretched exponential and the sum of exponentials expressions in Section 6.7.1. What do best fits yield for the time constants and acceleration factors τ, τ_1, τ_2, β, and n? (Since lifetime analyses are based on statistical approaches, the best fits should be selected by a minimization of mean square deviations, χ^2, of the data from the lifetime expressions).
   (c) Write a general expression for $L(t)$ in terms of L_0 and j that best predicts the lifetime of the devices. What is *LT80* and j_0 for this device at $L_0 = 100$ cd/m²?
14. For the blue PHOLED structure that generated the data and fits in Fig. 6.192, and assuming only TPA is active in aging,
   (a) Plot the *LT70* of the device as a function of recombination zone thickness, from $d_{rec} = 5$ nm to 100 nm.
   (b) Assuming that the voltage of the device increases as $\frac{dV}{d(d_{rec})} = Ad_{rec}^2$ where $A = 0.001\mathrm{V/nm}^3$, what is the optimal EML thickness that gives the largest $LT70\mathrm{x}\eta_P$ figure of merit?
15. We have seen that the OLED lifetime is thermally activated, that is, as temperature increases by 10°C, the lifetime decreases by approximately a factor of 1.6–1.7. Hence, it is important to understand the change in temperature with drive current in both display and lighting appliances. Consider a green OLED ($\lambda_{max} = 520$ nm) with an internal quantum efficiency of 50% comprised of a 100 nm thick Al cathode, a 100 nm thick organic active region, and a glass substrate of thickness, d_{sub}. Assume the top and bottom surfaces are maintained at the ambient temperature, T_0. The materials parameters for these several layers are given in Table 6P.15.
   (a) Based on the power balance Eq. 6.128, plot the steady state OLED active region temperature over the domain $0 < j < 100$ mA/cm² for $d_{sub} = 0.1$, 0.3, 1, and 5 mm?
   (b) Calculate the junction temperature in (a) allowing the top contact to reach its own equilibrium temperature.

Table 6P.15 Thermal parameters used in Problem 6.15

Material	Heat capacity (J/kg K)	Density (kg/m^3)	Thermal conductivity (W/m K)
Al	900	0.0027	237
Organic	1500	0.0017	0.2
Glass	840	0.0025	1.1

(c) Assume that a current step, j, is applied to the device at $t = 0$. What is the equilibration time constant as a function of organic layer thickness, $d_{org} = 10$–200 nm?

16. An organic semiconductor laser has mirror reflectivities of $R_1 = R_2 = 0.6$, a radiative lifetime of 5 ns, an effective index of $\bar{n} = 1.7$, and a wavelength of $\lambda_0 = 600$ nm.
    (a) Plot the transform-limited bandwidth vs. the cavity of length L in the domain of 50 μm to 500 μm at twice the threshold.
    (b) Plot the laser linewidth over the same domain in L as (a).

References

Adachi, C., Baldo, M. A., Thompson, M. E. & Forrest, S. R. 2001a. Nearly 100% internal phosphorescence efficiency in an organic light emitting device. *J. Appl. Phys.*, 90, 5048.

Adachi, C., Kwong, R. C., Djurovich, P., Adamovich, V., Baldo, M. A., Thompson, M. E. & Forrest, S. R. 2001b. Endothermic energy transfer: a mechanism for generating very efficient high energy phosphorescent emission in organic materials. *Appl. Phys. Lett.*, 79, 2082.

Adachi, C., Tsutsui, T. & Saito, S. 1990. Confinement of charge carriers and molecular excitons within 5-nm-thick emitter layer in organic electroluminescent devices with a double heterostructure. *Appl. Phys. Lett.*, 57, 531.

Adachi, C., Tsutsui, T. & Saito, S. 1991. Electroluminescent mechanism of organic multilayer thin film devices. *Optoelectron. Devices Technol.*, 6, 25.

Adamovich, V., Brooks, J., Tamayo, A., Alexander, A. M., Djurovich, P. I., D'Andrade, B. W., Adachi, C., Forrest, S. R. & Thompson, M. E. 2002. High efficiency single dopant white electrophosphorescent light emitting diodes. *New J. Chem.*, 26, 1171.

Adamovich, V. I., Levermore, P. A., Xu, X., Dyatkin, A. B., Elshenawy, Z., Weaver, M. S. & Brown, J. J. 2012. High-performance phosphorescent white-stacked organic light-emitting devices for solid-state lighting. *J. Photonics Energy*, 2, 021202.

Amos, R. M. & Barnes, W. L. 1997. Modification of the spontaneous emission rate of Eu^{3+} ions close to a thin metal mirror. *Phys. Rev. B*, 55, 7249.

Arakawa, Y. & Sakaki, H. 1982. Multidimensional quantum well laser and temperature dependence of its threshold current. *Appl. Phys. Lett.*, 40, 939.

Arnold, M. S., McGraw, G. C., Lunt, R. R. & Forrest, S. R. 2008. Direct vapor jet printing of three color segment organic light emitting devices for white light illumination. *Appl. Phys. Lett.*, 92, 053301.

Aziz, H., Popovic, Z., Tripp, C. P., Hu, N.-X. & Hor, A.-M. 1998. Degradation processes at the cathode/organic interface in organic light emitting devices with Mg:Ag cathodes. *Appl. Phys. Lett.*, 72, 2642.

Aziz, H., Popovic, Z. D., Hu, N.-X., Hor, A.-M. & Xu, G. 1999. Degradation mechanism of small molecule-based organic light-emitting devices. *Science*, 283, 1900.

Bakken, N., Wang, Z. & Li, J. 2012. Highly efficient white organic light emitting device using a single emitter. *J. Photonics Energy*, 2, 021203.

Baldo, M. A. & Forrest, S. R. 2000. Transient analysis of organic electrophosphorescence. I. Transient analysis of triplet energy transfer. *Phys. Rev. B*, 62, 10958.

Baldo, M. A., Adachi, C. & Forrest, S. R. 2000a. Transient analysis of organic electrophosphorescence. II. Transient analysis of triplet–triplet annihilation. *Phys. Rev. B*, 62, 10967.

Baldo, M. A., Holmes, R. J. & Forrest, S. R. 2002. Prospects for electrically pumped organic lasers. *Phys. Rev. B*, 66, 035321.

Baldo, M. A., Lemansky, S., Burrows, P. E., Thompson, M. E. & Forrest, S. R. 1999a. Very high efficiency green organic light emitting devices based on electrophosphorescence. *Appl. Phys. Lett.*, 76, 4.

Baldo, M. A., O'Brien, D. F., Thompson, M. E. & Forrest, S. R. 1999b. The excitonic singlet–triplet ratio in a semiconducting organic thin film. *Phys. Rev. B*, 60, 14422.

Baldo, M. A., O'Brien, D. F., You, Y., Shoustikov, A., Thompson, M. E. & Forrest, S. R. 1998. High efficiency phosphorescent emission from organic electroluminescent devices. *Nature*, 395, 151.

Baldo, M. A., Thompson, M. E. & Forrest, S. R. 2000b. High-efficiency fluorescent organic light-emitting devices using a phosphorescent sensitizer. *Nature*, 403, 750.

Bao, Q. Y., Yang, J. P., Li, Y. Q. & Tanga, J. X. 2010. Electronic structures of MoO3-based charge generation layer for tandem organic light-emitting diodes. *Appl. Phys. Lett.*, 97, 063303.

Barnes, W. L. 1998. Fluorescence near interfaces: the role of photonic mode density. *J. Mod. Opt.*, 45, 661.

Barranco, A., Borras, A., Gonzalez-Elipe, A. R. & Palmero, A. 2016. Perspectives on oblique angle deposition of thin films: From fundamentals to devices. *Prog. Mater. Sci.*, 76, 59.

Bergemann, K. J., Krasny, R. & Forrest, S. R. 2012. Thermal properties of organic light-emitting diodes. *Org. Electron.*, 13, 1565.

Berggren, M., Dodabalapur, A., Slusher, R. E., Timko, A. & Nalamasu, O. 1998. Organic solid-state lasers with imprinted gratings on plastic substrates. *Appl. Phys. Lett.*, 72, 410.

Bernanose, A. 1955. Electroluminescence of organic compounds. *Br. J. Appl. Phys.*, 6, S54.

Bernanose, A. & Vouaux, P. 1953. Électroluminescence organique: étude du mode d'émission. *J. Chim. Phys.*, 50, 64.

Bhandari, A. & Gaspar, D. J. 2015. Substrates. In: Gaspar, D. J. & Polikarpov, E. (eds.) *OLED Fundamentals: Materials, Devices, and Processing of Organc Light-Emitting Diodes*. Boca Raton, FL: CRC Press.

Birks, J. B. 1975. Excimers. *Rep. Prog. Phys.*, 38, 903.

Blom, P. W. M., de Jong, M. J. M. & Vleggaar, J. J. M. 1996. Electron and hole transport in poly(*p*-phenylene vinylene) devices. *Appl. Phys. Lett.*, 68, 3308.

Bocksrocker, T., Preinfalk, J. B., Asche-Tauscher, J., Pargner, A., Eschenbaum, C., Maier-Flaig, F. & Lemmer, U. 2012. White organic light emitting diodes with enhanced internal and external outcoupling for ultra-efficient light extraction and Lambertian emission. *Opt. Express*, 20, A932.

Brütting, W. & Frischeisen, J. 2012. Device efficiency of organic light-emitting diodes. In: Brütting, W. & Adachi, C. (eds.) *Physics of Organic Semiconductors*. Weinheim, Germany: Wiley-VCH.

Brütting, W., Frischeisen, J., Schmidt, T. D., Scholz, B. J. & Mayr, C. 2013. Device efficiency of organic light-emitting diodes: progress by improved light outcoupling. *Phys. Stat. Sol. (a)*, 210, 44.

Bulović, V., Deshpande, R., Thompson, M. E. & Forrest, S. R. 1999. Tuning the color emission of thin film molecular organic light emitting devices by the solid state solvation effect. *Chem. Phys. Lett.*, 308, 317.

Bulović, V., Gu, G., Burrows, P. E., Thompson, M. E. & Forrest, S. R. 1996. Transparent light emitting devices. *Nature*, 380, 29.

Bulović, V., Khalfin, V. B., Gu, G., Burrows, P. E., Garbuzov, D. Z. & Forrest, S. R. 1998a. Weak microcavity effects in organic light emitting devices. *Phys. Rev. B*, 58, 3730.

Bulović, V., Kozlov, V. G., Khalfin, V. B. & Forrest, S. R. 1998b. Transform-limited, narrow-linewidth lasing action in organic semiconductor microcavities. *Science*, 279, 553.

Bulović, V., Shoustikov, A., Baldo, M. A., Bose, E., Kozlov, V. G., Thompson, M. E. & Forrest, S. R. 1998c. Bright, saturated red-to-yellow organic light-emitting devices based on polarization-induced spectral shifts. *Chem. Phys. Lett.*, 287, 455.

Burroughes, J. H., Bradley, D. D. C., Brown, A. R., Marks, R. N., Mackay, K., Friend, R. H., Burns, P. L. & Holmes, A. B. 1990. Light-emitting diodes based on conjugated polymers. *Nature*, 347, 539.

Burrows, P. E., Bulovic, V., Forrest, S. R., Sapochak, L. S., McCarty, D. M. & Thompson, M. E. 1994. Reliability and degradation of organic light emitting devices. *Appl. Phys. Lett.*, 65, 2922.

Burrows, P. E., Forrest, S. R., Zhou, T. X. & Michalsk, L. 2000. Operating lifetime of phosphorescent organic light emitting devices. *Appl. Phys. Lett.*, 76, 2493.

Burrows, P. E., Shen, Z., Bulovic, V., McCarty, D. M., Forrest, S. R., Cronin, J. A. & Thompson, M. E. 1996. The relationship between electroluminescence and current transport in organic heterojunction light emitting devices. *J. Appl. Phys.*, 79, 7991.

Cai, X., Li, X., Xie, G., He, Z., Gao, K., Liu, K., Chen, D., Cao, Y. & Su, S.-J. 2016. 'Rate-limited effect' of reverse intersystem crossing process: the key for tuning thermally activated delayed fluorescence lifetime and efficiency roll-off of organic light emitting diodes. *Chem. Scie.*, 7, 4264.

Cao, X. A., Li, X. M., Yang, R. Y. & Zhou, Y. M. 2015. Effects of localized heating at heterointerfaces on the reliability of organic light-emitting diodes. *IEEE Electron. Dev. Lett.*, 36, 847.

Cao, Y., Parker, I. D., Yu, G., Zhang, C. & Heeger, A. J. 1999. Improved quantum efficiency for electroluminescence in semiconducting polymers. *Nature*, 397, 414.

Celebi, K., Heidel, T. D. & Baldo, M. A. 2007. Simplified calculation of dipole energy transport in a multilayer stack using dyadic Green's functions. *Opt. Express*, 15, 1762.

Chance, R. R., Prock, A. & Silbey, R. 1974. Lifetime of an emitting molecule near a partially reflecting surface. *J. Chem. Phys.*, 60, 2744.

Chance, R. R., Prock, A. & Silbey, R. 1978. Molecular fluorescence and energy transfer near interfaces. In: Prigogine, I. & Rice, S. A. (eds.) *Advances in Chemical Physics*. New York: John Wiley and Sons.

Chandrasekhar, S. 1943. Stochastic problems in physics and astronomy. *Rev. Mod. Phys.*, 15, 1.

Chang, C.-H., Lu, Y.-J., Liu, C.-C., Yeh, Y.-H. & Wu, C.-C. 2007. Efficient white OLEDs employing phosphorescent sensitization. *J. Display Technol.*, 3, 193.

Chang, H.-W., Lee, J., Hofmann, S., Hyun, K. Y., Muller-Meskamp, L., Lussem, B., C.-C. Wu, Leo, K. & Gather, M. C. 2013a. Nanoparticle based scattering layers for optical efficiency enhancement of organic light-emitting diodes and organic solar cells. *J. Appl. Phys.*, 113, 204502.

Chang, W., Akselrod, G. M. & Bulović, V. 2015. Solid-state solvation and enhanced exciton diffusion in doped organic thin films under mechanical pressure. *ACS Nano*, 9, 4412.

Chang, W., Wang, A., Murarka, A., Akselrod, G. M., Packard, C., Lang, J. H. & Bulović, V. 2014. Electrically tunable organic vertical-cavity surface-emitting laser. *Appl. Phys. Lett.*, 105, 073303.

Chang, Y.-L., Song, Y., Wang, Z., Helander, M. G., Qiu, J., Chai, L., Liu, Z., Scholes, G. D. & Lu, Z. 2013b. Highly efficient warm white organic light-emitting diodes by triplet exciton conversion. *Adv. Funct. Mater.*, 23, 705–712.

Chang, Y. J. & Chow, T. J. 2011. Highly efficient red fluorescent dyes for organic light-emitting diodes. *J. Mater. Chem.*, 21, 3091.

Chen, C.-T. 2004. Evolution of red organic light-emitting diodes: materials and devices. *Chem. Mater.*, 16, 4389.

Chen, Y., Chen, J., Ma, D., Yan, D. & Wang, L. 2011. Tandem white phosphorescent organic light-emitting diodes based on interface-modified C_{60}/pentacene organic heterojunction as charge generation layer. *Appl. Phys. Lett.*, 99, 103304.

Cheng, Y.-J., Liu, M. S., Zhang, Y., Niu, Y., Huang, F., Ka, J.-W., Yip, H.-L., Tian, Y. & Jen, A. K. Y. 2008. Thermally cross-linkable hole-transporting materials on conducting polymer: synthesis, characterization, and applications for polymer light-emitting devices. *Chem. Mater.*, 20, 413.

Cheon, J. H., Choi, J. H., Hur, J. H., Jang, J., Shin, H. S., Jeong, J. K., Mo, Y. G. & Chung, H. K. 2006. Active-matrix OLED on bendable metal foil. *IEEE Trans. Electron Devices*, 53, 1273.

Cho, Y. J., Yook, K. S. & Lee, J. Y. 2015. Cool and warm hybrid white organic light-emitting diode with blue delayed fluorescent emitter both as blue emitter and triplet host. *Sci. Rep.*, 5, 7859.

Choy, W. C. H., Chan, W. K. & Yuan, Y. 2014. Recent advances in transition metal complexes and light-management engineering in organic optoelectronic devices. *Adv. Mater.*, 26, 5368.

Chuang, S. L. 2009. *Physics of Photonic Devices*. Hoboken, NJ: John Wiley & Sons.

Chung, S., Lee, J.-H., Jeong, J., Kim, J.-J. & Hong, Y. 2009. Substrate thermal conductivity effects on heat dissipation and lifetime of organic light-emitting diodes. *Appl. Phys. Lett.*, 94, 253302.

Chwang, A. B., Kwong, R. C. & Brown, J. J. 2002. Graded mixed-layer organic light-emitting devices. *Appl. Phys. Lett.*, 80, 725.

Chwang, A. B., Rothman, M. A., Mao, S. Y., Hewitt, R. H., Weaver, M. S., Silvernail, J. A., Rajan, K., Hack, M., Brown, J. J., Chu, X., Moro, L., Krajewski, T. & Rutherford, N. 2003. Thin film encapsulated flexible organic electroluminescent displays. *Appl. Phys. Lett.*, 83, 413.

Coburn, C. & Forrest, S. R. 2017. Effects of charge balance and exciton confinement on the operational lifetime of blue phosphorescent organic light-emitting diodes. *Phys. Rev. Appl, Lett*, 7, 041002.

Coburn, C., Jeong, C. & Forrest, S. R. 2017. Reliable, all-phosphorescent stacked white organic light emitting devices with a high color rendering index. *ACS Photonics*, 5, 630.

Coldren, L. A. & Corzine, S. W. 1995. *Diode Lasers and Photonic Integrated Circuits*. New York: Wiley-Interscience.

Cui, L.-S., Deng, Y.-L., Tsang, D. P.-K., Jiang, Z.-Q., Zhang, Q., Liao, L.-S. & Adachi, C. 2016. Controlling synergistic oxidation processes for efficient and stable blue thermally activated delayed fluorescence devices. *Adv. Mater.*, 28, 7620.

D'Andrade, B. W. & Brown, J. J. 2006. Organic light-emitting device luminaire for illumination applications. *Appl. Phys. Lett.*, 88, 192908.

D'Andrade, B. W. & Forrest, S. R. 2003. Formation of triplet excimers and dimers in amorphous organic thin films and light emitting devices. *Chem. Phys.*, 286, 321.

D'Andrade, B. & Forrest, S. R. 2004. White organic light emitting devices for solid-state lighting. *Adv. Mater.*, 16, 1585.

D'Andrade, B. W., Baldo, M. A., Adachi, C., Brooks, J., Thompson, M. E. & Forrest, S. R. 2001. High-efficiency yellow double-doped organic light-emitting devices based on phosphor-sensitized fluorescence. *Appl. Phys. Lett.*, 79, 1045.

D'Andrade, B. W., Brooks, J., Adamovich, V., Thompson, M. E. & Forrest, S. R. 2002a. White light emission using triplet excimers in electrophosphorescent organic light-emitting devices. *Adv. Mater.*, 14, 1032.

D'Andrade, B. W., Forrest, S. R. & Chwang, A. B. 2003. Operational stability of electrophosphorescent devices containing p- and n-doped transport layers. *Appl. Phys. Lett.*, 83, 3858.

D'Andrade, B. W., Holmes, R. & Forrest, S. R. 2004. Efficient organic electrophosphorescent white light emitting device with a triple-doped emissive layer. *Adv. Mater.*, 16, 624.

D'Andrade, B. W., Thompson, M. E. & Forrest, S. R. 2002b. Controlling exciton diffusion in multilayer white phosphorescent organic light emitting devices. *Adv. Mater.*, 14, 147.

Dalal, S. S., Walters, D. M., Lyubimov, I., de Pablo, J. J. & Ediger, M. D. 2015. Tunable molecular orientation and elevated thermal stability of vapor-deposited organic semiconductors. *Proc. Natl. Acad. Sci.*, 112, 4227.

Denton, G. J., Tessler, N., Stevens, M. A. & Friend, R. H. 1997. Spectral narrowing in optically pumped poly (*p*-phenylenevinylene) films. *Adv. Mater.*, 9, 547.

Deppe, D. G., Lei, C., Lin, C. C. & Huffaker, D. L. 1994. Spontaneous emission from planar microstructures. *J. Mod. Opt.*, 41, 325.

Difley, S., Beljonne, D. & vanVoorhis, T. 2008. On the singlet–triplet splitting of geminate electron–hole pairs in organic semiconductors. *J. Am. Chem. Soc.*, 130, 3420.

Do, L.-M., Oyamada, M., Koike, A., Han, E.-M., Yamamoto, N. & Fujihira, M. 1996. Morphological change in the degradation of Al electrode surfaces of electroluminescent devices by fluorescence microscopy and AFM. *Thin Solid Films*, 273, 209–213.

Do, Y. R., Kim, Y.-C., Song, Y.-W. & Lee, Y.-H. 2004. Enhanced light extraction efficiency from organic light emitting diodes by insertion of a two-dimensional photonic crystal structure. *J. Appl. Phys.*, 96, 7629.

Dong, H., Zhang, C., Liu, Y., Yan, Y., Hu, F. & Zhao, Y. S. 2018. Organic microcrystal vibronic lasers with full-spectrum tunable output beyond the franck–condon principle. *Angew. Chem.*, 130, 3162.

Engmann, S., Barito, A. J., Bittle, E. G., Giebink, N. C., Richter, L. J. & Gundlach, D. J. 2019. Higher order effects in organic LEDs with sub-bandgap turn-on. *Nat. Commun.*, 10, 227.

Fehse, K., Meerheim, R., Walzer, K., Leo, K., Lövenich, W. & Elschner, A. 2008. Lifetime of organic light emitting diodes on polymer anodes. *Appl. Phys. Lett.*, 93, 083303.

Fenter, P., Schreiber, F., Bulović, V. & Forrest, S. R. 1997. Thermally induced failure mechanisms of organic light emitting device structures probed by X-ray specular reflectivity. *Chem. Phys. Lett.*, 277, 521.

Féry, C., Racine, B., Vaufrey, D., Doyeux, H. & Cinà, S. 2005. Physical mechanism responsible for the stretched exponential decay behavior of aging organic light-emitting diodes. *Appl. Phys. Lett.*, 87, 213502.

Fleetham, T. & Li, J. 2014. Recent advances in white organic light-emitting diodes employing a single-emissive material. *J. Photonics Energy*, 4, 040991.

Fleetham, T. B., Huang, L., Klimes, K., Brooks, J. & Li, J. 2016. Tetradentate Pt(II) complexes with 6-membered chelate rings: a new route for stable and efficient blue organic light emitting diodes. *Chem. Mater.*, 28, 3276.

Forrest, S. R., Bradley, D. D. C. & Thompson, M. E. 2003. Measuring the efficiency of organic light-emitting devices. *Adv. Mater.*, 15, 1043.

Förster, T. 1959. 10th Spiers Memorial Lecture: transfer mechanisms of electronic excitations. *Disc. Faraday Soc.*, 27, 7.

Frischeisen, J., Yokoyama, D., Adachi, C. & Brütting, W. 2010. Determination of molecular dipole orientation in doped fluorescent organic thin films by photoluminescence measurements. *Appl. Phys. Lett.*, 96, 073302.

Frischeisen, J., Yokoyama, D., Endo, A., Adachi, C. & Brütting, W. 2011. Increased light outcoupling efficiency in dye-doped small molecule organic light-emitting diodes with horizontally aligned emitters. *Org. Electron.*, 12, 809.

Frolov, S. V., Shkunov, M., Vardeny, Z. V. & Yoshino, K. 1997. Ring microlasers from conducting polymers. *Phys. Rev. B*, 56, R4363.

Fu, Q., Chen, J., Shi, C. & Ma, D. 2012. Solution-processed small molecules as mixed host for highly efficient blue and white phosphorescent organic light-emitting diodes. *ACS Appl. Mater. Interfaces*, 4, 6579.

Fujikawa, H. 2002. Efficient organic organic light-emitting diodes doped with coumarin. *R&D Rev. Toyota CRDL*, 37, 71.

Gaertner, G. & Greiner, H. 2008. Light extraction from OLEDS with (high) index matched glass substrates. *Proc. SPIE*, 6999, 69992T.

Garbuzov, D. Z., Bulović, V., Burrows, P. E. & Forrest, S. R. 1996. Photoluminescence efficiency and absorption of aluminum-tris-quinolate (Alq_3) thin films. *Chem. Phys. Lett.*, 249, 433.

Gaspar, D. J. & Plikarov, E. (Eds.) 2015. *OLED Fundamentals: Materials, Devices, and Processing of Organic Light-Emitting Diodes*. Boca Raton, LA: CRC Press.

Giebink, N. C. & Forrest, S. R. 2008. Quantum efficiency roll-off at high brightness in fluorescent and phosphorescent organic light emitting diodes. *Phys. Rev. B*, 77, 235215.

Giebink, N. C. & Forrest, S. R. 2009. Temporal response of optically pumped organic semiconductor lasers and its implication for reaching threshold under electrical excitation. *Phys. Rev. B*, 79, 073302.

Giebink, N. C., D'Andrade, B. W., Weaver, M. S., Brown, J. J. & Forrest, S. R. 2009. Direct evidence for degradation of polaron excited states in organic light emitting diodes. *J. Appl. Phys.*, 105, 124514.

Giebink, N. C., D'Andrade, B. W., Weaver, M. S., Mackenzie, P. B., Brown, J. J., Thompson, M. E. & Forrest, S. R. 2008. Intrinsic luminance loss in phosphorescent small-molecule organic light emitting devices due to bimolecular annihilation reactions. *J. Appl. Phys.*, 103, 044509.

Gifford, D. K. & Hall, D. G. 2002. Emission through one of two metal electrodes of an organic light-emitting diode via surface-plasmon cross coupling. *Appl. Phys. Lett.*, 81, 4315.

Gong, X. 2013. Polymer light-emitting diodes (PLEDs): devices and materials. In: Hu, W. (ed.) *Organic Optoelectronics*. 1st ed. Weinheim, Germany: Wiley-VCH Verlag.

Gong, X., Robinson, M. R., Ostrowski, J. C., Moses, D., Bazan, G. C. & Heeger, A. J. 2002. High efficiency polymer based electrophosphorescent devices. *Adv. Mater.*, 14, 581.

Gontijo, I., Boroditsky, M., Yablonovitch, E., Keller, S., Mishra, U. K. & DenBaars, S. P. 1999. Coupling of InGaN quantum-well photoluminescence to silver surface plasmons. *Phys. Rev. B*, 60, 11564.

Graf, A., Liehm, P., Murawski, C., Hofmann, S., Leo, K. & Gather, M. C. 2014. Correlating the transition dipole moment orientation of phosphorescent emitter molecules in OLEDs with basic material properties. *J. Mater. Chem. C*, 2, 10298.

Graff, G. L., Williford, R. E. & Burrows, P. E. 2004. Mechanisms of vapor permeation through multilayer barrier films: lag time versus equilibrium permeation. *J. Appl. Phys.*, 96, 1849.

Green, A. P., Butler, K. T. & Buckley, A. R. 2013. Tuning of the emission energy of fluorophores using solid state solvation for efficient luminescent solar concentrators. *Appl. Phys. Lett.*, 102, 133501.

Greenham, N. C., Friend, R. H. & Bradley, D. D. C. 1994. Angular dependence of the emission from a conjugated polymer light-emitting diode: implications for efficiency calculations. *Adv. Mater.*, 6, 491.

Gu, G., Burrows, P. E., Venkatesh, S., Forrest, S. R. & Thompson, M. E. 1997a. Vacuum-deposited, nonpolymeric flexible organic light emitting devices. *Opt. Lett.*, 22, 172.

Gu, G., Garbuzov, D. Z., Burrows, P. E., Venkatesh, S., Forrest, S. R. & Thompson, M. E. 1997b. High-external--quantum-efficiency organic light-emitting devices. *Opt. Lett.*, 22, 396.

Gu, G., Parthasarathy, G., Burrows, P. E., Tian, P., Hill, I. G., Kahn, A. & Forrest, S. R. 1999. Transparent stacked organic light emitting devices. I. Design principles and transparent compound electrodes. *J. Appl. Phys.*, 86, 4067.

Gustafsson, G., Cao, Y., Treacy, G., Klavetter, F., Colaneri, N. & Heeger, A. 1992. Flexible light emitting diode. *Nature*, 357, 477.

Han, X., Perepichka, D. F., Meng, H. & Ling, M. M. 2015. Phosphorescent polymer light-emitting diodes. In: Li, Z., Li, Z. R. & Meng, H. (eds.) *Organic Light Emitting Materials and Devices*. 2nd ed. Boca Raton, FL: CRC Press.

Heavens, O. S. 1991. *Optical Properties of Thin Solid Films*. London: Courier Corp.

Helfrich, W. & Schneider, W. G. 1965. Recombination radiation in anthracene crystals. *Phys. Rev. Lett.*, 14, 229.

Henry, C. 1986. Theory of spontaneous emission noise in open resonators and its application to lasers and optical amplifiers. *J. Lightwave Technol.*, 4, 288.

Herrnsdorf, J., Guilhabert, B., Chen, Y., Kanibolotsky, A. L., Mackintosh, A. R., Pethrick, R. A., Skabara, P. J., Gu, E., Laurand, N. & Dawson, M. D. 2010. Flexible blue-emitting encapsulated organic semiconductor DFB laser. *Opt. Express*, 18, 25535.

Hide, F., Diaz-Garcia, M., Schwartz, B. J., Andersson, M. R., Pei, Q. & Heeger, A. J. 1996. Semiconducting polymers: a new class of solid-state laser materials. *Science*, 273, 1833.

Higuchi, T., Nakanotani, H. & Adachi, C. 2015. High-efficiency white organic light-emitting diodes based on a blue thermally activated delayed fluorescent emitter combined with green and red fluorescent emitters. *Adv. Mater.*, 27, 2019.

Ho, S., Liu, S., Chen, Y. & So, F. 2015. Review of recent progress in multilayer solution-processed organic light-emitting diodes. *J. Photonics Energy*, 5, 057611.

Hobson, P. A., Wedge, S., Wasey, J. A. E., Sage, I. & Barnes, W. L. 2002. Surface plasmon mediate emission from organic light-emitting diodes. *Adv. Mater.*, 14, 1393.

Hofmann, S., Thomschke, M., Freitag, P., Furno, M., Lüssem, B. & Leo, K. 2010. Top-emitting organic light-emitting diodes: influence of cavity design. *Appl. Phys. Lett.*, 97, 253308.

Holmes, R. J., D'Andrade, B. W., Forrest, S. R., Ren, X., Li, J. & Thompson, M. E. 2003a. Efficient, deep-blue organic electrophosphorescence by guest charge trapping. *Appl. Phys. Lett.*, 83, 3818.

Holmes, R. J., Forrest, S. R., Tung, Y.-J., Kwong, R. C., Brown, J. J., Garon, S. & Thompson, M. E. 2003b. Blue organic electrophosphorescence using exothermic host–guest energy transfer. *Appl. Phys. Lett.*, 82, 2422.

Holzer, W., Penzkofer, A. & Tsuboi, T. 2005. Absorption and emission spectroscopic characterization of $Ir(ppy)_3$. *Chem. Phys.*, 308, 93.

Horst, G. 2007. Light Extraction from organic light emitting diode substrates: simulation and experiment. *Jpn. J. Appl. Phys.*, 46, 4125.

Hou, Q., Zhou, Q., Zhang, Y., Yang, W., Yang, R. & Cao, Y. 2004. Synthesis and electroluminescent properties of high-efficiency saturated red emitter based on copolymers from fluorene and 4,7-di(4-hexylthien-2-yl)-2,1,3-benzothiadiazole. *Macromolecules*, 37, 6299.

Huang, F., Shih, P.-I., Shu, C.-F., Chi, Y. & Jen, A. K.-Y. 2009. Highly efficient polymer white-light-emitting diodes based on lithium salts doped electron transporting layer. *Adv. Mater.*, 21, 361.

Ishihara, K., Fujita, M., Matsubara, I., Asano, T., Noda, S., Ohata, H., Hirasawa, A., Nakada, H. & Shimoji, N. 2007. Organic light-emitting diodes with photonic crystals on glass substrate fabricated by nanoimprint lithography. *Appl. Phys. Lett.*, 90, 111114.

Ishii, M. 2003. Luminance decay mechanisms in organic light-emitting diodes. *R&D Rev. Toyota CRDL*, 38.

Jacquemin, D. & Escudero, D. 2017. The short device lifetimes of blue PhOLEDs: insights into the photostability of blue Ir (III) complexes. *Chem. Sci.*, 8, 7844.

Jeon, S. K. & Lee, J. Y. 2015. Four times lifetime improvement of blue phosphorescent organic light-emitting diodes by managing recombination zone. *Org. Electron.*, 27, 202.

Jeon, Y.-M., Lee, J.-Y., Kim, J.-W., Lee, C.-W. & Gong, M.-S. 2010. Deep-blue OLEDs using novel efficient spiro-type dopant materials. *Org. Electron.*, 11, 1844.

Jeong, C., Coburn, C., Idris, M., Li, Y., Djurovich, P. I., Thompson, M. E. & Forrest, S. R. 2019. Understanding molecular fragmentation in blue phosphorescent organic light-emitting devices. *Org. Electron.*, 64, 15.

Jeong, E. G., Han, Y. C., Im, H.-G. & Bae, B.-S. 2016. Highly reliable hybrid nano-stratified moisture barrier for encapsulating flexible OLEDs. *Org. Electron.*, 33, 150.

Jeong, H., Shin, H., Lee, J., Kim, B., Park, Y.-I., Yook, K. S., An, B.-K. & Park, J. 2015. Recent progress in the use of fluorescent and phosphorescent organic compounds for organic light emitting diode lighting. *J. Photonics Energy*, 5, 057608.

Jiang, C., Yang, W., Peng, J., Xiao, S. & Cao, Y. 2004. High-efficiency, saturated red-phosphorescent polymer light emitting diodes based on conjugated and non-conjugated polymers doped with an Ir complex. *Adv. Mater.*, 16, 537.

Jiang, J. X., Xu, Y. H., Yang, W., Guan, R., Liu, Z. Q., Zhen, H. Y. & Cao, Y. 2006. High-efficiency white-light-emitting devices from a single polymer by mixing singlet and triplet emission. *Adv. Mater.*, 18, 1769.

Jou, J.-H., Kumar, S., Agrawal, A., Li, T.-H. & Sahoo, S. 2015. Approaches for fabricating high efficiency organic light emitting diodes. *J. Mater. Chem. C*, 3, 2974.

Jurow, M. J., Mayr, C., Schmidt, T. D., Lampe, T., Djurovich, P. I., Brutting, W. & Thompson, M. E. 2016. Understanding and predicting the orientation of heteroleptic phosphors in organic light-emitting materials. *Nat. Mater*, 15, 85.

Kadashchuk, A., Vakhnin, A., Blonski, I., Beljonne, D., Shuai, Z., Brédas, J. L., Arkhipov, V. I., Heremans, P., Emelianova, E. V. & Bässler, H. 2004. Singlet–triplet splitting of geminate electron–hole pairs in conjugated polymers. *Phys. Rev. Lett.*, 93, 066803.

Kalinowski, J., Stampor, W., Mężyk, J., Cocchi, M., Virgili, D., Fattori, V. & DiMarco, P. 2002. Quenching effects in organic electrophosphorescence. *Phys. Rev. B*, 66, 235321.

Kallinger, C., Hilmer, M., Haugeneder, A., Perner, M., Spirkl, W., Lemmer, U., Feldmann, J., Scherf, U., Müllen, K., Gombert, A. & Wittmer, V. 1998. A flexible conjugated polymer laser. *Adv. Mater.*, 10, 920.

Kanno, H., Holmes, R. J., Sun, Y. & Forrest, S. R. 2006a. White stacked electrophosphorescent organic light-emitting devices employing MoO_3 as a charge generation layer. *Adv. Mater.*, 18, 339.

Kanno, H., Sun, Y. & Forrest, S. R. 2006b. Stacked white organic light emitting devices based on a combination of fluorescent and phosphorescent emitters. *Appl. Phys. Lett.*, 89, 023503.

Kanno, H., Sun, Y. & Forrest, S. R. 2006c. White organic light-emitting device based on a compound fluorescent-phosphor-sensitized-fluorescent emission layer. *Appl. Phys. Lett.*, 89, 143516.

Kao, K. C. & Hwang, W. 1981. *Current Transport in Solids*. Oxford: Pergamon.

Karnutsch, C., Gýrtner, C., Haug, V., Lemmer, U., Farrell, T., Nehls, B. S., Scherf, U., Wang, J., Weimann, T., Heliotis, G., Pflumm, C., deMello, J. C. & Bradley, D. D. C. 2006. Low threshold blue conjugated polymer lasers with first- and second-order distributed feedback. *Appl. Phys.* 89, 201108.

Kassakian, J. (Ed.) 2013. *Assessment of Advanced Solid State Lighting*. Washington, DC: National Academies Press.

Kawamura, Y., Yanagida, S. & Forrest, S. R. 2002. Energy transfer in polymer electrophosphorescent light emitting devices with single and multiple doped luminescent layers. *J. Appl. Phys.*, 92, 87.

Kido, J., Kimura, M. & Nagai, K. 1995. Multilayer white light-emitting organic electroluminescent device. *Science*, 267, 1332.

Kim, J., Qu, Y., Coburn, C. & Forrest, S. R. 2018. Efficient outcoupling of organic light-emitting devices using a light-scattering dielectric layer. *ACS Photonics*, 5, 3315.

Kim, J.-S., Friend, R. H., Grizzi, I. & Burroughes, J. H. 2005. Spin-cast thin semiconducting polymer interlayer for improving device efficiency of polymer light-emitting diodes. *Appl Phys Lett.*, 87, 023506.

Kim, J.-S., Ho, P. K. H., Greenham, N. C. & Friend, R. H. 2000. Electroluminescence emission pattern of organic light-emitting diodes: implications for device efficiency calculations. *J. Appl. Phys.*, 88, 1073.

Kim, J. C., Batagoda, T., Lee, J., Sylvinson, D., Ding, K., Saris, P., Kaipa, U., Oswald, I. W. H., Omary, M. A., Thompson, M. E. & Forrest, S. R. 2019. Systematic control of the orientation of organic phosphorescent Pt complexes in thin films for increased optical outcoupling. *Adv. Mater.*, 31, 1900921.

Kim, K.-H., Lee, S., Moon, C.-K., Kim, S.-Y., Park, Y.-S., Lee, J.-H., Lee, J. W., Huh, J., You, Y. & Kim, J.-J. 2014a. Phosphorescent dye-based supramolecules for high-efficiency organic light-emitting diodes. *Nat. Commun.*, 5, 4769.

Kim, K.-H., Liao, J.-L., Lee, S. W., Sim, B., Moon, C.-K., Lee, G.-H., Kim, H. J., Chi, Y. & Kim, J.-J. 2016a. Crystal organic light-emitting diodes with perfectly oriented non-doped Pt-based emitting layer. *Adv. Mater.*, 28, 2526.

Kim, K.-H., Moon, C.-K., Lee, J.-H., Kim, S.-Y. & Kim, J.-J. 2014b. Highly efficient organic light-emitting diodes with phosphorescent emitters having high quantum yield and horizontal orientation of transition dipole moments. *Adv. Mater.*, 26, 3844.

Kim, M., Jeon, S. K., Hwang, S.-H., Lee, S.-S., Yu, E. & Lee, J. Y. 2016b. Highly efficient and color tunable thermally activated delayed fluorescent emitters using a 'twin emitter' molecular design. *Chem. Commun.*, 52, 339.

Kim, R., Lee, S., Kim, K.-H., Lee, Y.-J., Kwon, S.-K., Kim, J.-J. & Kim, Y.-H. 2013a. Extremely deep blue and highly efficient non-doped organic light emitting diodes using an asymmetric anthracene derivative with a xylene unit. *Chem. Commun.*, 49, 4664.

Kim, S.-Y., Jeong, W.-I., Mayr, C., Park, Y.-S., Kim, K.-H., Lee, J.-H., Moon, C.-K., Brütting, W. & Kim, J.-J. 2013b. Organic light-emitting diodes with 30% external quantum efficiency based on a horizontally oriented emitter. *Adv. Funct. Mater.*, 23, 3896.

Kim, S. O., Lee, K. H., Kim, G. Y., Seo, J. H., Kim, Y. K. & Yoon, S. S. 2010. A highly efficient deep blue fluorescent OLED based on diphenylaminofluorenylstyrene-containing emitting materials. *Synth. Met.*, 160, 1259.

Kim, Y.-K., Kim, J. W. & Park, Y. 2009. Energy level alignment at a charge generation interface between 4,4′-bis(*N*-phenyl-1-naphthylamino)biphenyl and 1,4,5,8,9,11-hexaazatriphenylene-hexacarbonitrile. *Appl. Phys. Lett.*, 94, 063305.

King, S. M., Cass, M., Pintani, M., Coward, C., Dias, F. B., Monkman, A. P. & Roberts, M. 2011. The contribution of triplet–triplet annihilation to the lifetime and efficiency of fluorescent polymer organic light emitting diodes. *J. Appl. Phys.*, 109, 074502.

Kirchhoff, J. R., Gamache, R. E., Blaskie, M. W., Del Paggio, A. A., Lengel, R. K. & McMillin, D. R. 1983. Temperature dependence of luminescence from $Cu(NN)^{2+}$ systems in fluid solution. Evidence for the participation of two excited states. *Inorg. Chem.*, 22, 2380.

Klubek, K. P., Dong, S.-C., Liao, L.-S., Tang, C. W. & Rothberg, L. J. 2014. Investigating blue phosphorescent iridium cyclometalated dopant with phenyl-imidazole ligands. *Org. Electron.*, 15, 3127.

Koh, T.-W., Spechler, J. A., Lee, K. M., Arnold, C. B. & Rand, B. P. 2015. Enhanced outcoupling in organic light-emitting diodes via a high-index contrast scattering layer. *ACS Photonics*, 2, 1366.

Köhler, A. & Beljonne, D. 2004. The singlet–triplet exchange energy in conjugated polymers. *Adv. Funct. Mater.*, 14, 11.

Komino, T., Tanaka, H. & Adachi, C. 2014. Selectively controlled orientational order in linear-shaped thermally activated delayed fluorescent dopants. *Chem. Mater.*, 26, 3665.

Komoda, T., Yamae, K., Kittichungchit, V., Tsuji, H. & Ide, N. 2012. Extremely high performance white OLEDs for lighting. *SID Symp. Dig. Tech. Pap.*, 43, 613.

Kondakov, D. Y. 2008. Role of chemical reactions of arylamine hole transport materials in operational degradation of organic light-emitting diodes. *J. Appl. Phys.*, 104, 084520.

Kondakov, D. Y. 2015. Device degradation. In: Gaspar, D. J. & Plikarov, E. (eds.) *OLED Fundamentals: Materials, Devices, and Processing of Organic Light-Emitting Diodes*. Boca Raton, LA: CRC Press.

Kondakov, D. Y., Lenhart, W. C. & Nichols, W. F. 2007. Operational degradation of organic light-emitting diodes: mechanism and identification of chemical products. *J. Appl. Phys.*, 101, 024512.

Kondakov, D. Y., Pawlik, T. D., Hatwar, T. K. & Spindler, J. P. 2009. Triplet annihilation exceeding spin statistical limit in highly efficient fluorescent organic light-emitting diodes. *J. Appl. Phys.*, 106, 124510.

Koo, W. H., Jeong, S. M., Araoka, F., Ishikawa, K., Nishimura, S., Toyooka, T. & Takezoe, H. 2010. Light extraction from organic light-emitting diodes enhanced by spontaneously formed buckles. *Nat. Photonics*, 4, 222.

Kozlov, V., Parthasarathy, G., Burrows, P. E., Khalfin, V., Wang, J., Chou, S. & Forrest, S. 2000. Structures for organic diode lasers and optical properties of organic semiconductors under intense optical and electrical excitations. *IEEE J. Quantum Electron.*, 36, 18.

Kozlov, V. G., Bulovic, V., Burrows, P. E., Baldo, M., Khalfin, V. B., Parthasarathy, G. & Forrest, S. R. 1998. Study of lasing action based on Forster energy transfer in optically pumped organic semiconductor thin films. *J. Appl. Phys.*, 84, 4096.

Kozlov, V. G., Bulovic, V., Burrows, P. E. & Forrest, S. R. 1997a. Laser action in organic semiconductor waveguide and double heterostructure devices. *Nature*, 389, 352.

Kozlov, V. G., Bulovic, V. & Forrest, S. R. 1997b. Temperature independent performance of organic semiconductor lasers. *Appl. Phys. Lett.*, 71, 2575.

Kulkarni, A. P., Tonzola, C. J., Babel, A. & Jenekhe, S. A. 2004. Electron transport materials for organic light-emitting diodes. *Chem. Mater.*, 16, 4556.

Kuo, C.-J., Li, T.-Y., Lien, C.-C., Liu, C.-H., Wu, F.-I. & Huang, M.-J. 2009. Bis(phenanthroimidazolyl)biphenyl derivatives as saturated blue emitters for electroluminescent devices. *J. Mater. Chem.*, 19, 1865.

Laaperi, A. 2008. OLED lifetime issues from a mobile-phone-industry point of view. *J. Soc. Inf. Display*, 16, 1125.

Lamansky, S., Djurovich, P., Murphy, D., Abdel-Razzaq, F., Lee, H.-E., Adach, C., P. E. Burrows, Forrest, S. R. & Thompson, M. E. 2001. Highly phosphorescent bis-cyclometalated iridium complexes: synthesis, photophysical characterization, and use in organic light emitting diodes. *J. Am. Chem. Soc.*, 123, 4304.

Lampe, T., Schmidt, T. D., Jurow, M. J., Djurovich, P. I., Thompson, M. E. & Brütting, W. 2016. Dependence of phosphorescent emitter orientation on deposition technique in doped organic films. *Chem. Mater.*, 28, 712.

Lee, B. R., Jung, E. D., Park, J. S., Nam, Y. S., Min, S. H., Kim, B.-S., Lee, K.-M., Jeong, J.-R., Friend, R. H., Kim, J.-S., Kim, S. O. & Song, M. H. 2014. Highly efficient inverted polymer light-emitting diodes using surface modifications of ZnO layer. *Nat. Commun.*, 5, 4840.

Lee, J., Chen, H.-F., Batagoda, T., Coburn, C., Djurovich, P. I., Thompson, M. E. & Forrest, S. R. 2016. Deep blue phosphorescent organic light-emitting diodes with very high brightness and efficiency. *Nat. Mater.*, 14, 92.

Lee, J., Chopra, N., Bera, D., Maslov, S., Eom, S.-H., Zheng, Y., Holloway, P., Xue, J. & So, F. 2011a. Down-conversion white organic light-emitting diodes using microcavity structure. *Adv. Energy Mater.*, 1, 174.

Lee, J., Jeong, C., Batagoda, T., Coburn, C., Thompson, M. E. & Forrest, S. R. 2017. Hot excited state management for long-lived blue phosphorescent organic light-emitting diodes. *Nat. Commun.*, 8, 15566.

Lee, J.-H., Chen, K.-Y., Hsiao, C.-C., Chen, H.-C., Chang, C.-H., Kiang, Y.-W. & Yang, C. C. 2006. Radiation simulations of top-emitting organic light-emitting devices with two- and three-microcavity structures. *J. Display Technol.*, 2, 130.

Lee, J.-H., Sarada, G., Moon, C.-K., Cho, W., Kim, K.-H., Park, Y. G., Lee, J. Y., Jin, S.-H. & Kim, J.-J. 2015a. Finely tuned blue iridium complexes with varying horizontal emission dipole ratios and quantum yields for phosphorescent organic light-emitting diodes. *Adv. Op. Mater.*, 3, 211.

Lee, J. S. 2017. Lifetime and efficiency of blue phosphorescent organic light-emitting diodes. Ph.D. thesis, University of Michigan.

Lee, K. H., Kang, L. K., Lee, J. Y., Kang, S., Jeon, S. O., Yook, K. S., Lee, J. Y. & Yoon, S. S. 2010a. Molecular engineering of blue fluorescent molecules based on silicon end-capped diphenylaminofluorene derivatives for efficient organic light-emitting materials. *Adv. Funct. Mater.*, 20, 1345.

Lee, K. H., Kim, S. M., Kim, J. Y., Kim, Y. K. & Yoon, S. S. 2010b. Red fluorescent organic light-emitting diodes using modified pyran-containing DCJTB derivatives. *Bull. Korean Chem. Soc.*, 31, 2884.

Lee, K. H., Kwon, Y. S., Lee, J. Y., Kang, S., Yook, K. S., Jeon, S. O., Lee, J. Y. & Yoon, S. S. 2011b. Highly efficient blue organic light-emitting diodes based on 2-(diphenylamino) fluoren-7-ylvinylarene derivatives that bear a tert-butyl group. *Chem. Eur. J.*, 17, 12994.

Lee, M.-T., Yen, C.-K., Yang, W.-P., Chen, H.-H., Liao, C.-H., Tsai, C.-H. & Chen, C. H. 2004. Efficient green coumarin dopants for organic light-emitting devices. *Org. Lett.*, 6, 1241.

Lee, S., Lee, J.-H., Lee, J.-H. & Kim, J.-J. 2012. The mechanism of charge generation in charge-generation units composed of p-doped hole-transporting layer/HATCN/n-doped electron-transporting layers. *Adv. Funct. Mater.*, 22, 855.

Lee, S. E., Lee, H. W., Lee, J. W., Hwang, K. M., Park, S. N., Yoon, S. S. & Kim, Y. K. 2015b. Optimization of hybrid blue organic light-emitting diodes based on singlet and triplet exciton diffusion length. *Jpn. J. Appl. Phys.*, 54, 06FG09.

Lee, T.-W., Noh, T., Choi, B.-K., Kim, M.-S., Shin, D. W. & Kido, J. 2008. High-efficiency stacked white organic light-emitting diodes. *Appl. Phys. Lett.*, 92, 043301.

Lee, Y.-J., Kim, S.-H., Huh, J., Kim, G.-H., Lee, Y.-H., Cho, S.-H., Kim, Y.-C. & Do, Y. R. 2003. A high-extraction-efficiency nanopatterned organic light-emitting diode. *Appl. Phys. Lett.*, 82, 3779.

Leem, D.-S., Kim, S.-Y., Lee, J.-H. & Kim, J.-J. 2009. High efficiency p-i-n top-emitting organic light-emitting diodes with a nearly Lambertian emission pattern. *J. Appl. Phys.*, 106, 063114.

Lehnhardt, M., Riedl, T., Weimann, T. & Kowalsky, W. 2010. Impact of triplet absorption and triplet-singlet annihilation on the dynamics of optically pumped organic solid-state lasers. *Phys. Rev. B*, 81, 165206.

Levermore, P. A., Dyatkin, A. B., Elshenawy, Z. M., Pang, H., Silvernail, J., Krall, E., Kwong, R., Weaver, M. S., Ma, R., Brown, J. J., Qi, X. & Forrest, S. R. 2012. Phosphorescent OLEDs for high efficacy long lifetime solid state lighting. *J. Photonics Energy*, 2, 021205

Li, G., Fleetham, T. & Li, J. 2014. Efficient and stable white organic light-emitting diodes employing a single emitter. *Adv. Mater.*, 26, 2931.

Li, J., Ding, D., Tao, Y., Y, W., Chen, R., Xie, L., Huang, W. & Xu, H. 2016a. A significantly twisted spirocyclic phosphine oxide as a universal host for high efficiency full-color thermally activated delayed fluorescence diodes. *Adv. Mater.*, 28, 3122.

Li, K., Tong, G. S. M., Wan, Q., Cheng, G., Tong, W.-Y., Ang, W.-H., Kwong, W.-L. & Che, C.-M. 2016b. Highly phosphorescent platinum(II) emitters: photophysics, materials and biological applications. *Chem. Sci.*, 7, 1653.

Li, W., Liu, D., Shen, F., Ma, D., Wang, Z., Feng, T., Xu, Y., Yang, B. & Ma, Y. 2012. A twisting donor–acceptor molecule with an intercrossed excited state for highly efficient, deep-blue electroluminescence. *Adv. Funct. Mater.*, 22, 2797.

Liao, L. S., Klubek, K. P. & C. W. Tang, C. W. 2004. High-efficiency tandem organic light-emitting diodes. *Appl. Phys. Lett.*, 84, 167.

Liao, L. S., Slusarek, W. K., Hatwar, T. K., Ricks, M. L. & Comfort, D. L. 2008. Tandem organic light-emitting diode using hexaazatriphenylene hexacarbonitrile in the intermediate connector. *Adv. Mater.*, 20, 324.

Lin, C.-L., Cho, T.-Y., Chang, C.-H. & Wu, C.-C. 2006. Enhancing light outcoupling of organic light-emitting devices by locating emitters around the second antinode of the reflective metal electrode. *Appl. Phys. Lett.*, 88, 081114.

Liu, T.-H., Iou, C.-Y., Wen, S.-W. & Chen, C.-H. 2003. 4-(Dicyanomethylene)-2-t-butyl-6-(1,1,7,1-tetramethyljulolidyl-9-enyl)-4H-pyran doped red emitters in organic light emitting devices. *Thin Solid Films*, 441, 223.

Lunt, R. R., Giebink, N. C., Belak, A. A., Benziger, J. B. & Forrest, S. R. 2009. Exciton diffusion lengths of organic semiconductor thin films measured by spectrally resolved photoluminescence quenching. *J. Appl. Phys.*, 105, 053711.

Ly, K. T., Chen-Cheng, R. W., Lin, H.-W., Shiau, Y.-J., Liu, S.-H., P.-T. Chou, Tsao, C.-S., Huang, Y.-C. & Chi, Y. 2016. Near-infrared organic light-emitting diodes with very high external quantum efficiency and radiance. *Nat. Photonics*, 11, 63.

Ma, B., Djurovich, P. I., Garon, S., Alleyne, B. & Thompson, M. E. 2006. Platinum binuclear complexes as phosphorescent dopants for monochromatic and white organic light-emitting diodes. *Adv. Funct. Mater.*, 16, 2438.

Ma, B., Djurovich, P. I., Yousufuddin, M., Bau, R. & Thompson, M. E. 2005. Synthetic control of Pt • • • Pt separation and photophysics of binuclear Pt complexes. *J. Am. Chem. Soc.*, 127, 28.

MacAdam, D. L. 1942. Visual sensitivities to color differences in daylight. *J. Opt. Soc. Am.*, 32, 247.

Madigan, C. F., Lu, M.-H. & Sturm, J. C. 2000. Improvement of output coupling efficiency of organic light-emitting diodes by backside substrate modification. *Appl. Phys. Lett.*, 76, 1650.

Matsumoto, T., Nakada, T., Endo, J., Mori, K., Kawamura, N., Yokoi, A. & Kido, J. 2003. Multiphoton organic EL device having charge generation layer. *SID Symp. Dig. Tech. Pap.*, 34, 979.

Matterson, B. J., Lupton, J. M., Safonov, A. F., Salt, M. G., Barnes, W. L. & Samuel, I. D. W. 2001. Increased efficiency and controlled light output from a microstructured light-emitting diode. *Adv. Mater.*, 13, 123.

McClure, D. S. 1949. Selection rules for singlet–triplet perturbations in polyatomic molecules. *J. Chem. Phys.*, 17, 665.

McClure, D. S. 1952. Spin–orbit interaction in aromatic molecules. *J. Chem. Phys.*, 20, 682.

McBranch, D., Campbell, I. H., Smith, D. L. & Ferraris, J. P. 1995. Optical determination of chain orientation in electroluminescent polymer films. *Appl. Phys. Lett.*, 66, 1175.

McGehee, M. D., Dıaz-Garcıa, M., Hide, F., Gupta, R., Miller, E. K., Moses, D. & Heeger, A. 1998. Semiconducting polymer distributed feedback lasers. *Appl. Phys. Lett.*, 72, 1536.

McNeill, C. R. & Greenham, N. C. 2009. Conjugated-polymer blends for optoelectronics. *Adv. Mater.*, 21, 3840.

Meerheim, R., Scholz, S., Olthof, S., Schwartz, G., Reineke, S., Walzer, K. & Leo, K. 2008. Influence of charge balance and exciton distribution on efficiency and lifetime of phosphorescent organic light-emitting devices. *J. Appl. Phys.*, 104, 014510.

Meerheim, R., Walzer, K., Pfeiffer, M. & Leo, K. 2006. Ultrastable and efficient red organic light emitting diodes with doped transport layers. *Appl. Phys. Lett.*, 89, 061111.

Mladenovski, S., Neyts, K., Pavicic, D., Werner, A. & Rothe, C. 2009. Exceptionally efficient organic light emitting devices using high refractive index substrates. *Opt. Express*, 17, 7562.

Möller, S. & Forrest, S. R. 2001. Improved light outcoupling in organic light emitting devices employing ordered microlens arrays. *J. Appl. Phys.*, 91, 3324.

Moraes, I. R. d., Scholz, S., Lüssem, B. & Leo, K. 2011. Role of oxygen-bonds in the degradation process of phosphorescent organic light emitting diodes. *Appl. Phys. Lett.*, 99, 053302.

Moro, L., Krajewski, T. A., Rutherford, N. M., Philips, O., Visser, R. J., Gross, M., Bennett, W. D. & Graff, G. 2004. Process and design of a multilayer thin film encapsulation of passive matrix OLED displays. *Proc. SPIE*, 5214, 83.

Morteani, A. C., Dhoot, A. S., Kim, J. S., Silva, C., Greenham, N. C., Murphy, C., Moons, E., Ciná, S., Burroughes, J. H. & Friend, R. H. 2003. Barrier-free electron–hole capture in polymer blend heterojunction light-emitting diodes. *Adv. Mater.*, 15, 1708.

Nagata, R., Nakanotani, H., Potscavage Jr, W. J. & Adachi, C. 2018. Exploiting singlet fission in organic light-emitting diodes. *Adv. Mater.*, 30, 1801484.

Nakamura, T., Tsutsumi, N., Juni, N. & Fujii, H. 2005. Thin-film waveguiding mode light extraction in organic electroluminescent device using high refractive index substrate. *J. Appl. Phys.*, 97, 054505.

Nakanotani, H., Higuchi, T., Furukawa, T., Masui, K., Morimoto, K., Numata, M., Tanaka, H., Sagara, Y., Yasuda, T. & Adachi, C. 2014. High-efficiency organic light-emitting diodes with fluorescent emitters. *Nat. Commun.*, 5, 4016.

Nakanotani, H., Masui, K., Nishide, J., Shibata, T. & Adachi, C. 2013. Promising operational stability of high-efficiency organic light-emitting diodes based on thermally activated delayed fluorescence. *Sci. Rep.*, 3, 2127.

Nasu, K. 1987. Tunneling and relaxation from the free state to the self-localized state of the exciton: nonadiabatic multi-steptheory. *J. Lumin.*, 38, 90.

Neyts, K. A. 1998. Simulation of light emission from thin film microcavities. *J. Opt. Soc. Am. A*, 15, 962.

Niko, A., Tasch, S., Meghdadi, F., Brandstätter, C. & Leising, G. 1997. Red-green-blue emission of parahexaphenyl devices with color-converting media. *J. Appl. Phys.*, 82, 4177.

O'Brien, D. F., Burrows, P. E., Forrest, S. R., Koene, B. E., Loy, D. E. & Thompson, M. E. 1998. Hole transporting materials with high glass transition temperatures for use in organic light emitting devices. *Adv. Mater.*, 10, 1108.

O'Brien, D. F., Baldo, M. A., Thompson, M. E. & Forrest, S. R. 1999. Improved energy transfer in electrophosphorescent devices. *Appl. Phys. Lett.*, 74, 442.

Oh, C. S., Choi, J. M. & Lee, J. Y. 2016. Chemical bond stabilization and exciton management by CN modified host material for improved efficiency and lifetime in blue phosphorescent organic light-emitting diodes. *Adv. Opt. Mater.*, 4, 1281.

Okumoto, K., Kanno, H., Hamada, Y., Takahashi, H. & Shibata, K. 2006a. High efficiency red organic light-emitting devices using tetraphenyldibenzoperiflanthene-doped rubrene as an emitting layer. *Appl. Phys. Lett.*, 89, 013502.

Okumoto, K., Kanno, H., Hamada, Y., Takahashi, H. & Shibata, K. 2006b. Green fluorescent organic light-emitting device with external quantum efficiency of nearly 10%. *Appl. Phys. Lett.*, 89, 063504.

Onsager, L. 1938. Initial recombination of ions. *Phys. Rev.*, 54, 554.

Orselli, E., Maunoury, J., Bascour, D. & Catinat, J.-P. 2012. Orange phosphorescent organic light-emitting diodes with high operational stability. *Org. Electron.*, 13, 1506.

Parthasarathy, G., Burrows, P. E., Khalfin, V., Kozlov, V. G. & Forrest, S. R. 1998. A metal-free cathode for organic semiconductor devices. *Appl. Phys. Lett.*, 72, 2138.

Parthasarathy, G., Gu, G. & Forrest, S. R. 1999. A full-color transparent metal free stacked organic light emitting device with simplified pixel biasing. *Adv. Mater.*, 11, 907.

Perepichka, I. F., Perepichka, D. F., Meng, H. & Wudl, F. 2005. Light-emitting polythiophenes. *Adv. Mater.*, 17, 2281.

Pfeiffer, M., Forrest, S. R., Leo, K. & Thompson, M. E. 2002. Electrophosphoresent p-i-n organic light emitting devices for very high efficiency flat panel displays. *Adv. Mater.*, 14, 1633.

Pfeiffer, M., Forrest, S. R., Zhou, X. & Leo, K. 2003. A low drive voltage, transparent, metal-free n-i-p electrophosphorescent light emitting diode. *Org. Electron.*, 4, 21.

Piland, G. B., Burdett, J. J., Kurunthu, D. & Bardeen, C. J. 2013. Magnetic field effects on singlet fission and fluorescence decay dynamics in amorphous rubrene. *J. Phys. Chem. C*, 117, 1224.

Pope, M., Kallmann, H. & Magnante, P. 1963. Electroluminescence in organic crystals. *J. Chem. Phys.*, 38, 2042.

Qi, X. & Forrest, S. R. 2011. Thermal analysis of high intensity organic light-emitting diodes based on a transmission matrix approach. *J. Appl. Phys.*, 110, 124516.

Qi, X., Slootsky, M. & Forrest, S. R. 2008. Stacked white organic light emitting devices consisting of separate red, green and blue elements. *Appl. Phys. Lett.*, 93, 1933206.

Qu, Y., Coburn, C., Fan, D. & Forrest, S. R. 2017. Elimination of plasmon losses and enhanced light extraction of top-emitting organic light-emitting devices using a reflective subelectrode grid. *ACS Photonics*, 4, 363.

Qu, Y., Kim, J., Coburn, C. & Forrest, S. R. 2018. Efficient, nonintrusive outcoupling in organic light emitting devices using embedded microlens arrays. *ACS Photonics*, 5, 2453.

Qu, Y., Slootsky, M. & Forrest, S. R. 2015. Enhanced light extraction from organic light-emitting devices using a sub-anode grid. *Nat. Photonics*, 9, 758.

Ràfols-Ribé, J., Will, P.-A., Hänisch, C., Gonzalez-Silveira, M., Lenk, S., Rodríguez-Viejo, J. & Reineke, S. 2018. High-performance organic light-emitting diodes comprising ultrastable glass layers. *Sci. Adv.*, 4, eaar8332.

Rao, N. N. 1994. *Elements of Engineering Electromagnetics*. Englewood Cliffs, NJ: Prentice-Hall.

Reineke, S. & Baldo, M. A. 2012. Recent progress in the understanding of exciton dynamics within phosphorescent OLEDs. *Phys. Stat. Sol. (a)*, 209, 2341.

Reineke, S., Lindner, F., Schwartz, G., Seidler, N., Walzer, K., Lüssem, B. & Leo, K. 2009. White organic light-emitting diodes with fluorescent tube efficiency. *Nature*, 459, 234.

Reineke, S., Thomschke, M., Lüssem, B. & Leo, K. 2013. White organic light-emitting diodes: status and perspective. *Rev. Mod. Phys.*, 85, 1245.

Reineke, S., Walzer, K. & Leo, K. 2007. Triplet-exciton quenching in organic phosphorscent light-emitting diodes with Ir-based emitters. *Phys. Rev. B*, 75, 125328.

Reufer, M., Walter, M. J., Lagoudakis, P. G., Hummel, B., Kolb, J. S., Roskos, H. G., Scherf, U. & Lupton, J. M. 2005. Spin-conserving carrier recombination in conjugated polymers. *Nat. Mater.*, 4, 340.

Riedel, B., Kaiser, I., Hauss, J., Lemmer, U. & Gerken, M. 2010. Improving the outcoupling efficiency of indium-tin--oxide-free organic light-emitting diodes via rough internal interfaces. *Opt. Express*, 18, A631.

Rosenow, T. C., Furno, M., Reineke, S., Olthof, S., Lüssem, B. & Leo, K. 2010. Highly efficient white organic light-emitting diodes based on fluorescent blue emitters. *J. Appl. Phys.*, 108, 113113.

Rudin, A. & Choi, P. 2013. *The Elements of Polymer Science and Engineering*. Waltham, MA: Academic Press.

Sandanayaka, A. S. D., Matsushima, T. & Adachi, C. 2015. Degradation mechanisms of organic light-emitting diodes based on thermally activated delayed fluorescence molecules. *J. Phys. Chem. C*, 119, 23845.

Sasaoka, T., Sekiya, M., Yumoto, A., Yamada, J., Hirano, T., Iwase, Y., Yamada, T., Ishibashi, T., Mori, T., Asano, M., Tamura, S. & Urabe, T. 2001. 24.4L: A 13.0-inch AM-OLED display with top emitting structure and adaptive current mode programmed pixel circuit (TAC). *SID Symp. Dig. Tech. Pap.*, 32, 384.

Schmidt, T. D., Setz, D. S., Flämmich, M., Frischeisen, J., Michaelis, D., Krummacher, B. C., Danz, N. & Brütting, W. 2011. Evidence for non-isotropic emitter orientation in a red phosphorescent organic light-emitting diode and its implications for determining the emitter's radiative quantum efficiency. *Appl. Phys. Lett.*, 99, 163302.

Schneider, D., Rabe, T., Riedl, T., Dobbertin, T., Kröger, M., Becker, E., Johannes, H.-H., Kowalsky, W., Weimann, T., Wang, J. & Hinze, P. 2004. Laser threshold reduction in an all-spiro guest–host system. *Appl. Phys. Lett.*, 85, 1659.

Scholz, S., Corten, C., Walzer, K., Kuckling, D. & Leo, K. 2007. Photochemical reactions in organic semiconductor thin films. *Org. Electron.*, 8, 709.

Schülzgen, A., Spiegelberg, C., Morrell, M. M., Mendes, S. B., Allemand, P.-M., Kawabe, Y., Kuwata-Gonokami, M., Honkanen, S., Fallahi, M., Kippelen, B. & Peyghambarian, N. 1998. Light amplification and laser emission in conjugated polymers. *Opt. Eng.*, 37, 1149.

Segal, M., Baldo, M. A., Holmes, R. J., Forrest, S. R. & Soos, Z. G. 2003. Excitonic singlet–triplet ratios in molecular and polymeric organic materials. *Phys. Rev. B*, 68, 075211.

Segal, M., Singh, M., Rivoire, K., Dilfey, S., Van Voorhis, T. & Baldo, M. A. 2007. Extrafluorescent electroluminescence in organic light-emitting devices. *Nat. Mater.*, 6, 374.

Sekine, C., Tsubata, Y., Yamada, T., Kitano, M. & Doi, S. 2014. Recent progress of high performance polymer OLED and OPV materials for organic printed electronics. *Sci. Technol. Adv. Mater.*, 15, 034203.

Setz, D. S., Schmidt, T. D., Flämmich, M., Nowy, S., Frischeisen, J., Krumm, B. C., Dobbertin, T., Heuser, K., Michaelis, D., Danz, N., Brütting, W. & Winnacker, A. 2011. Comprehensive efficiency analysis of organic light-emitting devices. *J. Photonics Energy*, 1, 011006.

Shen, Z., Burrows, P. E., Bulovic, V., Forrest, S. R. & Thompson, M. E. 1997. Three color, tunable organic light emitting device. *Science*, 276, 2009.

Shi, J. & Tang, C. W. 1997. Doped organic electroluminescent devices with improved stability. *Appl. Phys. Lett.*, 70, 1665.

Shi, S., Jung, M. C., Coburn, C., A. Tadle, D. Sylvinson, Djurovich, P. I., Forrest, S. R. & Thompson, M. E. 2019. Highly efficient photo- and electroluminescence from two coordinate Cu(I) complexes featuring non-conventional N heterocyclic carbenes. *J. Am. Chem. Soc.*, 141, 3576.

Shirota, Y. & Kageyama, H. 2007. Charge carrier transporting molecular materials and their applications in devices. *Chem. Rev.*, 107, 953.

Shu, C.-F., Dodda, R., Wu, F.-I., Liu, M. S. & Jen, A. K.-Y. 2003. Highly efficient blue-light-emitting diodes from polyfluorene containing bipolar pendant groups. *Macromolecules*, 36, 6698.

Slootsky, M. & Forrest, S. R. 2009. Full-wave simulation of enhanced outcoupling of organic light emitting devices with an embedded low-index grid. *Appl. Phys. Lett.*, 94, 163302.

Slootsky, M. & Forrest, S. R. 2010. Enhancing waveguided light extraction in organic light emitting devices using an ultra-low-index grid. *Opt. Lett.*, 35, 1052.

Sokolik, I., Howard, W. E., Ghosh, A. P., Shershukov, V. M., Skripkina, V. T., Tolmachev, A. V., Voronkina, N. I., Dudkin, V. A., Pukha, V. E. & Pishkin, O. S. 2001. P-46: new color changing materials for OLED microdisplays. *SID Symp. Dig. Tech. Pap.*, 32, 727.

Staroske, W., Pfeiffer, M., Leo, K. & Hoffmann, M. 2007. Single-step triplet–triplet annihilation: an intrinsic limit for high brightness efficiency of phosphorescent organic light emitting devices. *Phys. Rev. Lett.*, 98, 197402.

Su, S.-J., Gonmori, E., Sasabe, H. & Kido, J. 2008. Highly efficient organic blue-and white-light-emitting devices having a carrier- and exciton-confining structure for reduced efficiency roll-off. *Adv. Mater.*, 20, 4189.

Suen, C.-S. & Chu, X. 2008. Multilayer thin film barrier for protection of flex-electronics. *Solid State Technol.*, 51(3), 36.

Suh, M., Bailey, J., Kim, S. W., Kim, K., Yun, D.-J., Jung, Y., Hamilton, I., Chander, N., Wang, X., Bradley, D. D. C., Jeon, D. Y. & Kim, J.-S. 2015. High-efficiency polymer LEDs with fast response times fabricated via selection of electron-injecting conjugated polyelectrolyte backbone structure. *ACS Appl. Mater. Interfaces*, 7, 26566.

Sun, J. X., Zhu, X. L., Peng, H. J., Wong, M. & Kwok, H. S. 2005. Effective intermediate layers for highly efficient stacked organic light-emitting devices. *Appl. Phys. Lett.*, 87, 093504.

Sun, N., Wang, Q., Zhao, Y., Yang, D., Zhao, F., Chen, J. & Ma, D. 2014a. A hybrid white organic light-emitting diode with above 20% external quantum efficiency and extremely low efficiency roll-off. *J. Mater. Chem. C*, 2, 7494.

Sun, N., Zhao, Y., Zhao, F., Chen, Y., Yang, D., Chen, J. & Ma, D. 2014b. A white organic light-emitting diode with ultra-high color rendering index, high efficiency and extremely low efficiency roll-off. *Appl. Phys. Lett.*, 150, 013303.

Sun, Y. & Forrest, S. R. 2006. Organic light emitting devices with enhanced out-coupling via microlenses fabricated by imprint lithography. *J. Appl. Phys.*, 100, 073106.

Sun, Y. & Forrest, S. R. 2007. High-efficiency white organic light emitting devices with three separate phosphorescent emission layers. *Appl. Phys. Lett.*, 91, 263503.

Sun, Y. & Forrest, S. R. 2008a. Enhanced light outcoupling of organic light emitting devices using embedded low-index grids. *Nat. Photonics*, 2, 483.

Sun, Y. & Forrest, S. R. 2008b. Multiple exciton generation regions in phosphorescent white organic light emitting devices. *Org. Electron.*, 9, 994.

Sun, Y., Borek, C., Hanson, K., Djurovich, P. I., Brooks, J., Brown, J. J., Thompson, M. E. & Forrest, S. R. 2007. Photophysics of Pt-porphyrin electrophosphorescent devices emitting in the near infrared. *Appl. Phys. Lett.*, 90, 213503.

Sun, Y., Giebink, N., Kanno, H., Ma, B., Thompson, M. E. & Forrest, S. R. 2006. Management of singlet and triplet excitons for efficient white organic light emitting devices. *Nature*, 440, 908.

Swanson, L. S., Shinar, J., Brown, A. R., Bradley, D. D. C., Friend, R. H., Burn, P. L., Kraft, A. & Holmes, A. B. 1993. Electroluminescence-, conductivity-, and photoconductivity-detected magnetic resonance study of poly(*p*-phenylenevinylene)-based light emitting diodes. *Synth. Met.*, 55, 241.

Sze, S. M. 1981. *Physics of Semiconductor Devices*. New York: Wiley-Interscience.

Tanaka, H., Shizu, K., Lee, J. & Adachi, C. 2015. Effect of atom substitution in chalcogenodiazole-containing thermally activated delayed fluorescence emitters on radiationless transition. *J. Phys. Chem. C*, 119, 2948.

Tang, C. W. & VanSlyke, S. A. 1987. Organic electroluminescent devices. *Appl. Phys. Lett.*, 51, 913.

Tang, C. W., VanSlyke, S. A. & Chen, C. H. 1989. Electroluminescence of doped organic thin films. *J. Appl. Phys.*, 65, 3610.

Tessler, N., Denton, G. & Friend, R. 1996. Lasing from conjugated-polymer microcavities. *Nature*, 382, 695.

Thompson, M. E., Djurovich, P. E., Barlow, S. & Marder, S. 2007. Organometallic complexes for optoelectronic applications. In: Mingos, M. and Crabtree, R. (eds.) *Comprehensive Organometallic Chemistry III*. Elsevier Science.

Thomschke, M., Reineke, S., Lüssem, B. & Leo, K. 2012. Highly efficient white top-emitting organic light-emitting diodes comprising laminated microlens films. *Nano Lett.*, 12, 424.

To, W.-P., Zhou, D., Tong, G. S. M., Cheng, G., Yang, C. & Che, C.-M. 2017. Highly luminescent pincer gold(III) aryl emitters: thermally activated delayed fluorescence and solution-processed OLEDs. *Angew. Chem.*, 129, 14224.

Tokito, S., Suzuki, M. & Sato, F. 2003a. Improvement of emission efficiency in polymer light-emitting devices based on phosphorescent polymers. *Thin Solid Films*, 445, 353.

Tokito, S., Suzuki, M., Sato, F., Kamachi, M. & Shirane, K. 2003b. High-efficiency phosphorescent polymer light-emitting devices. *Org. Electron.*, 4, 105.

Tokito, S., Taga, Y. & Tsutsui, T. 1997. Strongly modified emission from organic electroluminescent device with a microcavity. *Synth. Met.*, 91, 49.

Tokito, S., Weifurtner, K.-H., Fujikawa, H., Tsutsui, T. & Taga, Y. 2001. Acene containing polyfluorenes for red, green, and blue emission in organic light-emitting diodes. In: *Organic Light-Emitting Materials and Devices IV* (Vol. 4105, pp. 69–74). International Society for Optics and Photonics.

Tonzola, C. J., Alam, M. M., Kaminsky, W. & Jenekhe, S. A. 2003. New n-type organic semiconductors: synthesis, single crystal structures, cyclic voltammetry, photophysics, electron transport, and electroluminescence of a series of diphenylanthrazolines. *J. Am. Chem. Soc.*, 125, 13548.

Tsang, D. P.-K., Matsushima, T. & Adachi, C. 2016. Operational stability enhancement in organic light-emitting diodes with ultrathin Liq interlayers. *Sci. Rep.*, 6, 22463.

Tsujimura, T. 2012. *OLED Display Fundamentals and Applications*. Hoboken, NJ: John Wiley & Sons.

Tsujimura, T., Furukawa, K., Ii, H., Kashiwagi, H., Miyoshi, M., Mano, S., Araki, H. & Ezaki, A. 2012. Commercialization of world's first all-phosphorescent OLED product for lighting application. *SID Symp. Dig. Tech. Pap.*, 43, 605.

Tutis, E., Berner, D. & Zuppiroli, L. 2003. Internal electric field and charge distribution in multilayer organic light-emitting diodes. *J. Appl. Phys.*, 93, 4594.

Uoyama, H., Goushi, K., Shizu, K., Nomura, H. & Adachi, C. 2012. Highly efficient organic light-emitting diodes from delayed fluorescence. *Nature*, 492, 234.

Vaeth, K. M. & Tang, C. W. 2002. Light-emitting diodes based on phosphorescent guest/polymeric host systems. *J. Appl. Phys.*, 92, 3447.

Vasdekis, A. E., Tsiminis, G., Ribierre, J. C., O'Faolain, L., Krauss, T. F., Turnbull, G. A. & Samuel, I. D. W. 2006. Diode pumped distributed Bragg reflector lasers based on a dye-to-polymer energy transfer blend. *Opt. Express*, 14, 9211.

Wallesch, M., Volz, D., Zink, D. M., Schepers, U., Nieger, M., Baumann, T. & Bräse, S. 2014. Bright coppertunities: multinuclear Cu I complexes with N–P ligands and their applications. *Chem. Eur. J.*, 20, 6578.

Wallikewitz, B. H., Kabra, D., Gélinas, S. & Friend, R. H. 2012. Triplet dynamics in fluorescent polymer light-emitting diodes. *Phys. Rev. B*, 85, 045209.

Wang, C., Li, X., Pan, Y., Zhang, S., Yao, L., Bai, Q., Li, W., Lu, P., Yang, B., Su, S. & Ma, Y. 2016. Highly efficient nondoped green organic light-emitting diodes with combination of high photoluminescence and high exciton utilization. *ACS Appl. Mater. Interfaces*, 8, 3041.

Wang, Q. & Aziz, H. 2013. Degradation of organic/organic interfaces in organic light-emitting devices due to polaron-exciton interactions. *ACS Appl. Mater. Interfaces*, 5, 8733.

Wang, Q., Ding, J., Ma, D., Cheng, Y., Wang, L., Jing, X. & Wang, F. 2009a. Harvesting excitons via two parallel channels for efficient white organic LEDs with nearly 100% internal quantum efficiency: fabrication and emission-mechanism analysis. *Adv. Funct. Mater.*, 19, 84.

Wang, Q., Ding, J., Ma, D., Cheng, Y., Wang, L. & Wang, F. 2009b. Manipulating charges and excitons within a single-host system to accomplish efficiency/CRI/color-stability trade-off for high performance OWLEDs. *Adv. Mater.*, 21, 2397.

Wang, Q., Oswald, I. W. H., Yang, X., Zhou, G., Jia, H., Qiao, Q., Chen, Y., Hoshikawa-Halbert, J. & Gnade, B. E. 2014a. A non-doped phosphorescent organic light-emitting device with above 31% external quantum efficiency. *Adv. Mater.*, 26, 8107.

Wang, Q., Sun, B. & Aziz, H. 2014b. Exciton–polaron-induced aggregation of wide-bandgap materials and its implication on the electroluminescence stability of phosphorescent organic light-emitting devices. *Adv. Funct. Mater.*, 24, 2975.

Wang, S., Yan, X., Cheng, Z., Zhang, H., Liu, Y. & Wang, Y. 2015. Highly efficient near-infrared delayed fluorescence organic light emitting diodes using a phenanthrene-based charge-transfer compound. *Angew. Chem. Int. Ed.*, 54, 13068.

Wasey, J. A. E. & Barnes, W. L. 2000. Efficiency of spontaneous emission from planar microcavities. *J. Mod. Opt.*, 47, 725.

Weaver, M., Michalski, L., Rajan, K., Rothman, M., Silvernail, J., Brown, J. J., Burrows, P. E., Graff, G. L., Gross, M. E. & Martin, P. M. 2002. Organic light-emitting devices with extended operating lifetimes on plastic substrates. *Appl. Phys. Lett.*, 81, 2929.

Weaver, M. S., Xu, X., Pang, H., Ma, R., Brown, J. J. & Lu, M.-H. 2014. Color tunable phosphorescent white OLED lighting panel. *SID Symp. Dig. Tech. Pap.*, 45, 672.

Wei, B., Liu, J.-Z., Zhang, Y., Zhang, J.-H., Peng, H.-N., Fan, H.-L., He, Y.-B. & Gao, X.-C. 2010. Stable, glassy, and versatile binaphthalene derivatives capable of efficient hole transport, hosting, and deep-blue light emission. *Adv. Funct. Mater.*, 20, 2448.

Weibull, W. 1939. A statistical theory of the strength of materials. *R. Swedish Acad. Eng. Sci. Proc.*, 151, 1.

Wen, S.-W., Lee, M.-T. & Chen, C. H. 2005. Recent development of blue fluorescent OLED materials and devices. *J. Display Technol.*, 1, 90.

Wilson, J. S., Dhoot, A. S., Seeley, A. J. A. B., Khan, M. S., Kohler, A. & Friend, R. H. 2001. Spin-dependent exciton formation in π-conjugated compounds. *Nature*, 413, 828.

Wohlgenannt, M. & Vardeny, Z. V. 2003. Spin-dependent exciton formation rates in π-conjugated materials. *J. Phys.: Condens. Matter*, 15, R83.

Wohlgenannt, M., Tandon, K., Mazumdar, S., Ramasesha, S. & Vardeny, Z. V. 2001. Formation cross-sections of singlet and triplet excitons in π-conjugated polymers. *Nature*, 409, 494.

Wu, C., Theiuss, S., Gu, G., Lu, M., Sturm, J., Wagner, S. & Forrest, S. 1997a. Integration of organic LEDs and amorphous Si TFTs onto flexible and lightweight metal foil substrates. *IEEE Electron Device Lett.*, 18, 609.

Wu, C. C., Wu, C. I., Sturm, J. C. & Kahn, A. 1997b. Surface modification of indium tin oxide by plasma treatment: an effective method to improve the efficiency, brightness, and reliability of organic light emitting devices. *Appl. Phys. Lett.*, 70, 1348.

Wu, H., Zhou, G., Zou, J., Ho, C.-L., Wong, W.-Y., Yang, W., Peng, J. & Cao, Y. 2009. Efficient polymer white-light emitting devices for solid-state lighting. *Adv. Mater.*, 21, 4181.

Wu, K. C., Ku, P. J., Lin, C. S., Shih, H. T., Wu, F. I., Huang, M. J., Lin, J. J., Chen, I. C. & Cheng, C. H. 2008. The photophysical properties of dipyrenylbenzenes and their application as exceedingly efficient blue emitters for electroluminescent devices. *Adv. Funct. Mater.*, 18, 67.

Wu, Y., Zhou, Y. C., Wu, H. R., Zhan, Y. Q., Zhou, J., Zhang, S. T., Zhao, J. M., Wang, Z. J., X. M. Ding & Hou, X. Y. 2005. Metal-induced photoluminescence quenching of tri-(8-hydroxyquinoline) aluminum. *Appl. Phys. Lett.*, 87, 044104.

Xi, J.-Q., Kim, J. K. & Schubert, E. F. 2005. Silica nanorod-array films with very low refractive indices. *Nano Lett.*, 5, 1385.

Xia, L., Xue, Y., Xiong, K., Cai, C., Peng, Z., Wu, Y., Li, Y., Miao, J., Chen, D., Hu, Z., Wang, J., Peng, X., Mo, Y. & Hou, L. 2015. Highly improved efficiency of deep-blue fluorescent polymer light-emitting device based on a novel hole interface modifier with 1,3,5-triazine core. *ACS Appl Mater Interfaces*, 7, 26405.

Xia, R., Lai, W.-Y., Levermore, P. A., Huang, W. & Bradley, D. D. C. 2009. Low-threshold distributed-feedback lasers based on pyrene-cored starburst molecules with 1,3,6,8-attached oligo(9,9-dialkylfluorene) arms. *Adv. Funct. Mater.*, 19, 2844.

Yam, V. W.-W. & Wong, K. M.-C. 2011. Luminescent metal complexes of d6, d8 and d10 transition metal centres. *Chem. Commun.*, 47, 11579.

Yamae, K., Tsuji, H., Kittichungchit, V., Ide, N. & Komoda, T. 2013. Highly efficient white OLEDs with over 100 lm/W for general lighting. *SID Symp. Dig. Tech.Pap.*, 44, 916.

Yamamoto, H., Brooks, J., Weaver, M. S., Brown, J. J., Murakami, T. & Murata, H. 2011. Improved initial drop in operational lifetime of blue phosphorescent organic light emitting device fabricated under ultra high vacuum condition. *Appl. Phys. Lett.*, 99, 033301.

Yamamoto, H., Kasajima, H., Yokoyama, W., Sasabe, H. & Adachi, C. 2005. Extremely-high-density carrier injection and transport over 12,000 A/cm^2 into organic thin films. *Appl. Phys. Lett.*, 86, 083502.

Yang, C.-J., Liu, S.-H., Hsieh, H.-H., Liu, C.-C., Cho, T.-Y. & Wu, C.-C. 2007. Microcavity top-emitting organic light-emitting devices integrated with microlens arrays: simultaneous enhancement of quantum efficiency, cd/A efficiency, color performances, and image resolution. *Appl. Phys. Lett.*, 91, 253508.

Yang, Y., Turnbull, G. A. & Samuel, I. D. W. 2008. Hybrid optoelectronics: a polymer laser pumped by a nitride light-emitting diode. *Appl. Phys.* .92, 163306.

Yates, C. J., Samuel, I. D. W., Burn, P. L., Wedge, S. & Barnes, W. L. 2006. Surface plasmon-polariton mediated emission from phosphorescent dendrimer light-emitting diodes. *Appl. Phys. Lett.*, 88, 161105.

Yersin, H., Czerwieniec, R., Shafikov, M. Z. & Suleymanova, A. F. 2017. TADF material design: photophysical background and case studies focusing on Cu I and Ag I complexes. *ChemPhysChem*, 18, 3508.

Yersin, H., Rausch, A. F., Czerwieniec, R., Hofbeck, T. & Fischer, T. 2011. The triplet state of organo-transition metal compounds. Triplet harvesting and singlet harvesting for efficient OLEDs. *Coord. Chem. Rev.*, 255, 2622.

Yokoyama, D., Moriwake, M. & Adachi, C. 2008. Spectrally narrow emissions at cutoff wavelength from edges of optically and electrically pumped anisotropic organic films. *J. Appl. Phys.*, 103, 123104.

Yokoyama, D., Park, Y., Kim, B., Kim, S., Pu, Y.-J., Kido, J. & Park, J. 2011. Dual efficiency enhancement by delayed fluorescence and dipole orientation in high-efficiency fluorescent organic light-emitting diodes. *Appl. Phys. Lett.*, 99, 123303.

Yoshioka, T., Sugimoto, K., Katagi, K., Kitago, Y., Tajima, M., Miyaguchi, S., Tsutsui, T., Iwasaki, R. & Furukawa, Y. 2014. An improved method for lifetime prediction based on decoupling of the joule self-heating effect from coulombic degradation in accelerated aging tests of OLEDs. *SID Symp. Dig. Tech. Pap.*, 45, 642.

Zhang, L., van Eersel, H., Bobbert, P. A. & Coehoorn, R. 2016a. Clarifying the mechanism of triplet–triplet annihilation in phosphorescent organic host–guest systems: a combined experimental and simulation study. *Chem. Phys. Lett.*, 652, 142.

Zhang, L., Zhang, Y.-X., Hu, Y., Shi, X.-B., Jiang, Z.-Q., Wang, Z.-K. & Liao, L.-S. 2016b. Highly efficient blue phosphorescent organic light-emitting diodes employing a host material with small bandgap. *ACS Appl. Mater. Interfaces*, 8, 16186.

Zhang, Q., Komino, T., Huang, S., Matsunami, S., Goushi, K. & Adachi, C. 2012. Triplet exciton confinement in green organic light-emitting diodes containing luminescent charge-transfer Cu(I) complexes. *Adv. Funct. Mater.*, 22, 2327.

Zhang, Q., Li, B., Huang, S., Nomura, H., Tanaka, H. & Adachi, C. 2014a. Efficient blue organic light-emitting diodes employing thermally activated delayed fluorescence. *Nat. Photonics*, 8, 326.

Zhang, Q., Tsang, D., Kuwabara, H., Hatae, Y., Li, B., Takahashi, T., Lee, S. Y., Yasuda, T. & Adachi, C. 2015. Nearly 100% internal quantum efficiency in undoped electroluminescent devices employing pure organic emitters. *Adv. Mater.*, 27, 2096.

Zhang, Y. & Aziz, H. 2016. Influence of the guest on aggregation of the host by exciton–polaron interactions and its effects on the stability of phosphorescent organic light-emitting devices. *ACS Appl. Mater. Interfaces*, 8, 14088.

Zhang, Y. & Forrest, S. R. 2011. Existence of continuous-wave threshold for organic semiconductor lasers. *Phys. Rev. B*, 84, 241301(R).

Zhang, Y. & Forrest, S. R. 2012. Triplets contribute to both an increase and loss in fluorescent yield in organic light emitting diode. *Phys. Rev. Lett.*, 108, 267404.

Zhang, Y. & Forrest, S. R. 2013. Triplet diffusion leads to triplet–triplet annihilation in organic phosphorescent emitters. *Chem. Phys. Lett.*, 590, 106.

Zhang, Y., Lee, J. & Forrest, S. R. 2014b. Ten-fold increase in the lifetime of blue phosphorescent organic light emitting diodes. *Nat. Commun.*, 5, 5008.

Zhang, Y., Slootsky, M. & Forrest, S. R. 2011. Enhanced efficiency in high-brightness fluorescent organic light emitting diodes through triplet management. *Appl. Phys. Lett.*, 99, 223303.

Zhang, Y., Whited, M., Thompson, M. E. & Forrest, S. R. 2010. Singlet–triplet quenching in high intensity fluorescent organic light emitting diodes. *Chem. Phys. Lett.*, 495, 161.

Zhen, H., Jiang, C., Yang, W., Jiang, J., Huang, F. & Cao, Y. 2005. Synthesis and properties of electrophosphorescent chelating polymers with iridium complexes in the conjugated backbone. *Chem. Eur. J.*, 11, 5007.

Zheng, C.-J., Zhao, W.-M., Wang, Z.-Q., Huang, D., Ye, J., Ou, X.-M., Zhang, X.-H., Lee, C.-S. & Lee, S.-T. 2010. Highly efficient non-doped deep-blue organic light-emitting diodes based on anthracene derivatives. *J. Mater. Chem.*, 20, 1560.

Zhou, J., Ai, N., Wang, L., Zheng, H., Luo, C., Jiang, Z., Yu, S., Cao, Y. & Wang, J. 2011. Roughening the white OLED substrate's surface through sandblasting to improve the external quantum efficiency. *Org. Electron.*, 12, 648.

Zhou, L., Jiang, X., Li, Y., Shi, A., Chen, J., Ou, Q., Liu, H. & Tang, J. 2014a. Light extraction of trapped optical modes in polymer light-emitting diodes with nanoimprinted double-pattern gratings. *ACS Appl. Mater. Interfaces*, 6, 18139.

Zhou, L., Ou, Q.-D., Chen, J.-D., Shen, S., Tang, J.-X., Li, Y.-Q. & Lee, S.-T. 2014b. Light manipulation for organic optoelectronics using bio-inspired moth's eye nanostructures. *Sci. Rep.*, 4, 4040.

Zhuang, J., Li, W., Su, W., Liu, Y., Shen, Q., Liao, L. & Zhou, M. 2013. Highly efficient phosphorescent organic light-emitting diodes using a homoleptic iridium (III) complex as a sky-blue dopant. *Org. Electron.*, 14, 2596.

CHAPTER 7

Organic light detectors

"I'd put my money on the sun and solar energy. What a source of power! I hope we don't have to wait until oil and coal run out before we tackle that."

Thomas A. Edison, Inventor, 1931

What prophetic words from Thomas Edison, 23 years before the invention of the solar cell! Yet, here we are in the twenty-first century, and solar power generation has finally taken root as a significant (but far from dominant) alternative to fossil fuels. Our question in this chapter, however, is not whether solar power will play a significant role in the production of clean, abundant, and inexpensive renewable energy (it already does), but whether organic thin film materials, with their ultra-lightweight form factor along with their ability to be deposited onto virtually any reasonably smooth surface including flexible plastic and metal foils, will contribute to the explosively growing field of solar electricity production. But solar cells are not the only organic detector technology. Solar cells, also known as organic photovoltaics (OPVs), are a member of a broader class of photodetectors that includes organic photodiodes, photoconductors, and phototransistors. It is the former two types of detectors that are the subject of this chapter. Phototransistors, which are a combination of the photodetection and gain/switching elements, are treated in Chapter 8. As always, our discussion relies heavily on the foundations laid in Part I on the processes of photogeneration and charge conduction in organic materials.

To fully understand the process of photogeneration of charge, we begin with a discussion of the fundamentals of the organic photodetection process. The types of organic detectors will then be described in broad terms, before we engage in a detailed discussion of photodiodes. Their performance characteristics, including spectral sensitivity (i.e. their external quantum efficiency spectrum), noise, and bandwidth characteristics form the basis of this part of the chapter. This sets the groundwork for discussing specific photodetector architectures and applications.

While developments in photoconductors and photodiodes have led to impressive device performance characteristics, the vast majority of research has focused on solar cells since these are inherently large area devices; an application for which organic electronics is optimally suited to fill. Hence, the bulk of this chapter will focus on organic solar cells, principally in what differentiates them from incumbent technologies, and the mechanisms that govern their performance. After discussing the basics of solar cell operation including the thermodynamic limits to organic solar cell efficiency, we will describe the various types and structures of devices developed to reach the highest possible efficiencies. Indeed, progress in OPVs has largely been paced by developments in materials and device architectures. Hence, we will introduce several of the most interesting materials that have been developed for use in the various device regions including the active, blocking, and transport layers, the morphologies of the thin films that lead to efficient exciton dissociation and charge collection, and the device architectures that provide the highest power conversion efficiencies. Efficient solar

Organic Electronics: Foundations to Applications. Stephen R. Forrest. © Stephen R. Forrest 2020.
Published in 2020 by Oxford University Press. DOI: 10.1093/oso/9780198529729.001.0001

power harvesting requires specialized devices, with the highest performance often achieved with multijunction cells. The design of such cells can be quite complex due to the dependence of efficiency on balancing the charge generation and extraction in multiple stacked elements. And a transparent charge recombination zone must exist between each element. The design principles of multijunction cells will therefore be discussed in detail.

Beyond bulk and mixed heterojunctions, there are many other architectures that have been explored. These include cascade and blended donor or acceptor cells, ternary mixtures used to achieve broad spectral coverage, hybrid organic/inorganic quantum dot semiconductor structures, and structures exploiting singlet fission. This latter process involves the formation of two low energy triplet excitons for each high energy singlet generated on photon absorption. These interesting structures and phenomena introduce important physical and device concepts that may eventually bear fruit.

The chapter closes with a discussion of device reliability and the potential for large-scale, low cost manufacture. Indeed, it is these two features that will ultimately determine whether or not OPVs become a practical source of solar energy harvesting. It will become apparent that, given some additional progress in achieving low cost and high efficiency, OPVs are indeed well-positioned to become a vital source of electricity derived from the nearly limitless supply of solar energy arriving every day on the Earth's surface. Given that providing an ubiquitous source of clean energy to fulfill humankind's needs over the long term is imperative, we must continue to work to improve, and to provide solar energy via large-area organic electronic technology that is so well suited to the purpose.

7.1 Operating principles of organic photodetectors

A photodetector is a transducer that converts photons to an electric current. There are numerous types of photodetectors ranging from bolometers to avalanche photodiodes. In this chapter, however, we focus our attention on only the primary types of detectors that employ organic electronic active materials. These respond in the spectral region ranging from the ultraviolet (with wavelengths $\lambda > 300$ nm) into the near infrared (NIR, at $\lambda < 1500$ nm). The NIR wavelengths of interest are also known as the *short wavelength IR*, or *SWIR* band. Within these visible to near-visible spectral regions, organic electronic devices fall into three classes: *photoconductors*, *photodiodes*, and *phototransistors*. The principal difference between them is that the photodiode comprises an internal junction that actively separates photogenerated excitons, while the photoconductor and phototransistor lack a junction. The latter two devices exhibit *photocurrent gain* (i.e. more than one electron is produced per incident photon) although their bandwidth is usually far lower than the photodetector that has an internal field to rapidly sweep out charge. To understand phototransistors, knowledge of transistor operation is required, and hence is treated separately in the following chapter.

Photocurrent in organic semiconductor devices is generated by absorption of a photon of energy $h\nu$, greater than the HOMO–LUMO gap energy of the semiconductor, E_G. Here, h is Planck's constant, and ν is the frequency of the incident radiation. While this fundamental absorption process is dominant in most devices, there can be other, lower energy sites that absorb photons with $h\nu < E_G$. For example, charge transfer (CT) states at the heterojunction formed between a donor and acceptor in a photodiode (Section 4.7.2), or at the semiconductor/metal contacts can add to the absorption, creating a tail that extends deep into the energy gap. We will find that CT state absorption plays a vital role in the efficiency and photodetection process at photodiode junctions.

7.1.1 Photoconductivity and photoconductors

7.1.1.1 General considerations

A schematic illustration of an in-plane photoconductor is shown in Fig. 7.1. Photoconductivity is a

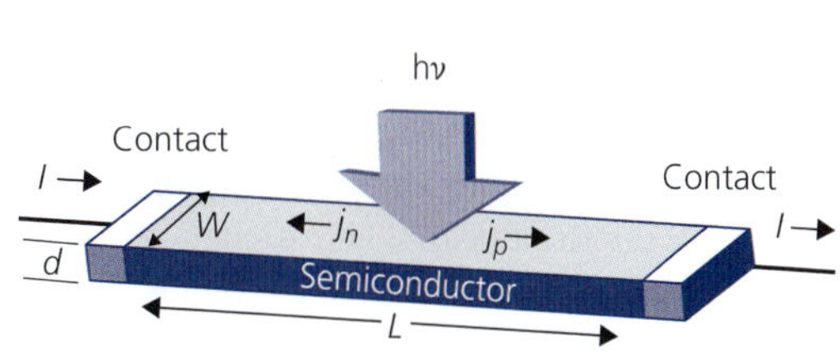

Figure 7.1 Schematic illustration of a photoconductor showing the electron (j_n) and hole (j_p) current densities generated by an incident photon of energy $h\nu$. The current in the external circuit is I.

fundamental property of all semiconductors, and hence this property was demonstrated long before the first junction photodiodes were reported. Photoconductivity in triphenylmethane under pressure was observed in 1930 (Petrikaln, 1930), which predated the first clear identification of an organic semiconductor, violanthrone (see Chapter 1) (Akamatu and Inokuchi, 1950). The first systematic studies of photoconductivity in organics were apparently based on the small molecule rhodamine B (Nelson, 1954).

As illustrated, the photoconductor is simply a slab of semiconductor that is sandwiched between ohmic metal contacts. Light incident on the semiconductor generates excitons that dissociate into free electrons and holes. These carriers subsequently drift to opposite contacts due to a small electric field placed across the semiconductor by an externally applied voltage. The source of charge can be either via excitation across the energy gap to generate an intrinsic photocurrent, or via excitation from deep trap levels to a conduction (HOMO or LUMO) energy level. This latter process is extrinsic, in that it requires a defect or dopant that is not native to the semiconductor itself. In the following discussion we will consider only intrinsic processes, since the effects of traps can greatly complicate (and potentially obscure) the analysis of this deceptively simple process.

Rose first elegantly elucidated the general problem of photoconductivity in inorganic semiconductors (Rose, 1955a, 1955b, 1963). The conductivity of the semiconductor is

$$\sigma = q\left(\mu_n n + \mu_p p\right), \tag{7.1}$$

where μ_n and μ_p are the bulk electron and hole mobilities and n and p are the total densities of electrons and holes. In the presence of illumination, we have $n = n_{ph} + n_0$ and $p = p_{ph} + p_0$, where n_{ph} (p_{ph}) and n_0 (p_0) are the densities of photogenerated and background free electrons (holes), respectively. The background free charge densities are subject to the law of mass action:

$$n_0 p_0 = n_i^2, \tag{7.2}$$

where n_i is the intrinsic carrier concentration. In the absence of background doping, $n_0 = p_0 = n_i$. Also, conservation of charge requires that $n_{ph} = p_{ph}$. In an intrinsic inorganic semiconductor where excitation is simply between the valence and conduction bands, Rose asserts that "*only the life time of a free carrier describes the essential performance and contains the essential physics of a photoconductor*" (Rose, 1955a). Thus, we can write for the carrier generation rate, G_{ph}, under illumination:

$$G_{ph} = k_D n_{ph} = \frac{\eta_{ext}(P_{inc}\lambda/hc)}{dWL}, \tag{7.3}$$

where $k_D = 1/\tau_D$ is the rate of photogeneration of carriers equal to the reciprocal of its lifetime, τ_D, P_{inc} is the incident optical power, c is the speed of light, d is the thickness of the semiconductor, L is its length, and W is its width (cf. Fig. 7.1). Also, η_{ext} is the *external quantum efficiency* of the photoconductor, which is equal to the total number of photons incident to the electrons generated. In Eq. 7.3 we assume that all photons are absorbed in the semiconductor, that is, $d \gg 1/\alpha(\lambda)$, where α is the optical absorption length at wavelength, λ.

From the above, we obtain the total photocurrent density (in one dimension):

$$\begin{aligned} j_{ph} &= \sigma F = q n_{ph}\left(\mu_n + \mu_p\right)\frac{V_a}{L} \\ &= q\frac{\eta_{ext}(P_{inc}\lambda/hc)}{k_D}\left(\mu_n + \mu_p\right)\frac{V_a}{dWL^2}, \end{aligned} \tag{7.4}$$

where V_a is the voltage across the sample of length, L. Also, F is the electric field, which in a resistor is $F = V_a/L$. By the same reasoning, the dark current for an intrinsic semiconductor is given by

$$j_D = q n_i\left(\mu_n + \mu_p\right)\frac{V_a}{L}, \tag{7.5}$$

giving a total current of

$$j_T = j_{ph} + j_D = q\left(n_{ph} + n_i\right)\left(\mu_n + \mu_p\right)\frac{V_a}{L}. \tag{7.6}$$

The foregoing discussion is based on the premise that a photoconductor is a simple resistor, following Ohm's law in Eqs. 7.4–7.6. Barriers due to non-Ohmic contacts, traps, or morphological irregularities within the bulk of the photoconductor can result in significant deviations from the ideal case considered here. The existence of energy barriers results in the local depletion of charge, creating a deviation from the linear field approximation of $F = V_a/L$. In situations that exist at high voltage or intensity, or for photoconductors with poor contacts, therefore, our analysis breaks down and a model that includes extrinsic factors is required.

7.1.1.2 Gain and bandwidth

Returning to the photocurrent in Eq. 7.4, and defining $j' = q\eta_{ext}(P_{inc}\lambda/hc)/dW$, then the *photocurrent gain* is given by

$$g = \frac{j_{ph}}{j'} = \tau_D\left(\mu_n + \mu_p\right)\frac{V_a}{L^2}. \tag{7.7}$$

That is, the photoconductor has gain whereby one photon can generate more than one electron. Then, in some cases, $g\eta_{EQE} > 1$. Indeed, g has been observed for many inorganic photoconductors to be as high as 10^6.

The carrier velocity is related to mobility via $v = \mu F$, and since the carrier transit time is $t_{tr,n(p)} = L/v_{n(p)} = L^2/\left(\mu_{n(p)} V_a\right)$ for electrons (holes), then it follows that

$$g = \tau_D\left(\frac{1}{t_{tr,n}} + \frac{1}{t_{tr,p}}\right). \tag{7.8}$$

Thus, we have the remarkable result that the gain is equal to the ratio of the carrier lifetime to its transit time. Indeed, if the mobility of one carrier type is substantially larger than the other, then the gain is simply the ratio of the lifetime to the transit time of that charge.

Photoconductors operate in quasi-equilibrium, that is, in the regime where the applied voltage is low enough to avoid injection of charge from contacts. Hence all regions of the device are undepleted, and conduction is strictly ohmic, as stated in in Eq. 7.4. If a charge is swept out of the sample by the applied voltage in time, $t_{tr} < \tau_D$, the semiconductor is forced out of equilibrium until a like charge is injected from the opposite contact. This process of charge transit followed by injection to re-establish equilibrium repeats itself until the excess charge recombines, and the sample is once more charge-neutral. During time, τ_D, g charges are injected following the absorption of a single photon, thus giving rise to photoconductive gain.

The time to reestablish equilibrium is τ_D, and hence the photogenerated charge has a transient response to a light pulse following

$$n_{ph}(t) = n_{ph0}\exp(-t/\tau_D). \tag{7.9}$$

This leads to a bandwidth of $\Delta f = 1/2\pi\tau_D$, or a *gain-bandwidth product* of

$$g\Delta f = \frac{1}{2\pi}\left(\frac{1}{t_{tr,n}} + \frac{1}{t_{tr,p}}\right), \tag{7.10}$$

which is the inverse of the transit time of the charges. The gain-bandwidth product expresses a fundamental trade-off of all photodetectors that determines its viability for a particular application. Unfortunately, photoconductors generally have a slow recovery time as well as a long transit time, leading to a small $g\Delta f$ compared to photodiodes that operate under reverse bias.

7.1.1.3 Photogeneration in excitonic photoconductors

Thus far we have considered the simple case of an inorganic semiconductor device in the absence of traps. An organic photoconductor is substantially more difficult to analyze since the fundamental excitation does not directly generate free change in the bands, but rather it proceeds via the intermediate generation of an exciton. In Chapter 4.7.2, we analyzed charge generation at an organic type II heterojunction. There, excitons drift to a donor–acceptor junction where they separate into a more loosely bound PP state that eventually dissociates into free charge. A photoconductor lacks a heterojunction, and hence the exciton must dissociate directly into free charges, an often inefficient process that lacks an energetic driving force. Of course, excitons may diffuse to a contact that can lead to efficient charge generation, or it can dissociate at a defect within the semiconductor bulk. Notwithstanding these secondary processes with their own inherent inefficiencies (i.e. the exciton may have to diffuse long distances before finding a dissociation site), organic photoconductors consequently have considerably lower efficiencies than do organic photodiodes and solar cells.

The energetics of exciton dissociation is illustrated in Fig. 7.2. This scenario shows that a photogenerated exciton current, j_X, is generated by absorption of a photon of energy $h\nu$ greater than the HOMO-LUMO gap of the semiconductor. The exciton then can recombine at rate k_D, or it can dissociate into a free electron–hole pair at rate k_{diss}. Recombination of free electrons and holes at rate k_{rec} leads to a loss in

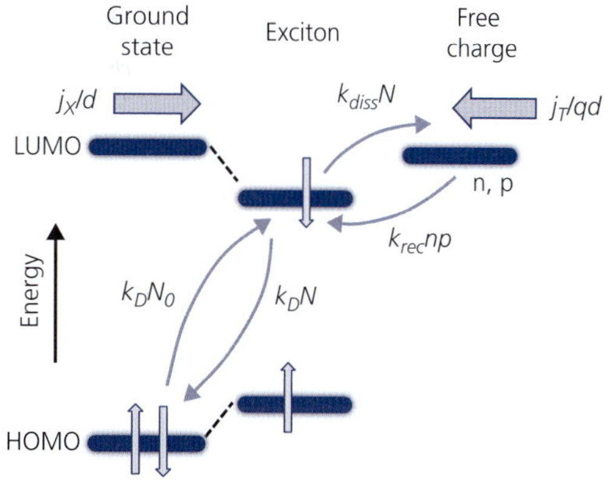

Figure 7.2 Schematic of the energetics of free charge generation in an organic photoconductor. Rates are denoted by k, the exciton density is N, the free charge densities are n, p, the total current is j_T and the exciton current is j_X. The vertical arrows indicate electrons and their spins.

the photocurrent, $j_{ph} = j_T - j_D$. Excitons can also be thermally generated at rate k_D. Once again ignoring complications due to recombination at traps, the steady state rate equations governing photogeneration are

$$\frac{j_{ph}}{dq} - k_{rec}\left(n^2 - n_i^2\right) + k_{diss}(N - N_0) = 0 \qquad (7.11)$$

and

$$\frac{j_X}{d} - k_D(N - N_0) - k_{diss}N + k_{rec}n^2 = 0. \qquad (7.12)$$

Here, N is the exciton density with value N_0 at equilibrium (i.e. in the absence of illumination), and we assume that the material is undoped, that is, $n_0 = p_0 = n_i$. Now, when $j_X = 0$, then $N = N_0$ and from Eq. 7.12 we obtain

$$\frac{k_{rec}n_i}{k_{diss}} = \frac{N_0}{n_i}, \qquad (7.13)$$

which is a consequence of detailed balance.

We can define the *internal quantum efficiency* as the ratio of the number of *electrons generated per photon absorbed* in the device active region, which differs from η_{ext} by ignoring reflections from the surface or absorption in electronically inactive areas of the sample. This can be expressed as the ratio of rates:

$$\eta_{int} = \frac{k_{diss}}{k_{diss} + k_D}. \qquad (7.14)$$

Then it follows that

$$\eta_{int}k_D = (1 - \eta_{int})k_{diss}. \qquad (7.15)$$

Finally, as in Chapter 4, the Langevin recombination rate between free carriers is given by

$$k_{rec} = \frac{q}{\varepsilon}\left(\mu_e + \mu_p\right), \qquad (7.16)$$

where $\varepsilon = \varepsilon_r\varepsilon_0$ is the semiconductor permittivity, ε_r is the relative dielectric constant, and ε_0 is the permittivity of free space. Using Eq. 7.5, we can replace the photocurrent with

$$j_{ph} = k_{rec}n_{ph}D, \qquad (7.17)$$

where $D = \varepsilon F$ is the electric displacement field. Substituting for N by using Eq. 7.11 in Eq. 7.12 gives

$$j_{ph} = dq(1 - \eta_{int})\left[\frac{j_{ph}^2}{k_{rec}D^2} + \frac{2n_i j_{ph}}{D}\right] - q_{\eta_{int}}j_X. \qquad (7.18)$$

The quadratic term in the photocurrent results from the two step recombination process via the exciton that does not exist in non-excitonic, inorganic semiconductors. This expression, unfortunately, does not have a tidy solution. It is helpful at this point to define the recombination current as

$$j_{rec} = \frac{k_{rec}D^2}{dq(1 - \eta_{int})}, \qquad (7.19)$$

which represents the reverse current due to Langevin recombination of photogenerated electrons and holes back to the bound exciton state. Then, the number of photogenerated charges from Eq. 7.18 is given by

$$n_{ph} = \frac{j_{rec}}{k_{rec}D}\left\{1 - \frac{2dqn_i}{D}(1 - \eta_{int}) - \left[\left(1 - \frac{2dqn_i}{D}(1 - \eta_{int})\right)^2 + 4q\eta_{int}\frac{j_X}{j_{rec}}\right]^{\frac{1}{2}}\right\}. \qquad (7.20)$$

For unipolar conduction in a doped semiconductor, n_i is replaced by the majority background carrier density. For example, for electrons, $n_0 \gg n_i$, and n_0 is substituted in Eq. 7.20. For bipolar conduction, the definition of j_{rec} must be adjusted accordingly.

There are a couple of notable aspects to this otherwise ungainly expression. First, if $\eta_{int} = 0$, this implies that $j_X = 0$, in which case n_{ph} also vanishes. On the contrary, the expression is undefined as $\eta_{int} \to 1$, since in this limit, j_{rec} blows up. Outside of these limits, we see that n_{ph} is, to first order, linear in D^{-1} (and therefore L/V_a) consistent with Ohm's law, although the detailed dependence of n_{ph} on V_a departs significantly from linearity as the terms in D^{-2} become large at very low electric fields. However, this only occurs when the background charge concentration, n_0 is also large. Indeed, for an intrinsic semiconductor or when n_0 is small, the expression to first order simplifies to

$$n_{ph} = \frac{j_{rec}}{2k_{rec}D}\left\{1 - \left[1 + 4q\eta_{int}\frac{j_X}{j_{rec}}\right]^{\frac{1}{2}}\right\} \approx q\eta_{int}\frac{j_X}{k_{rec}D}. \qquad (7.21)$$

The last approximation on the right is valid at low light levels (i.e. j_X small). In this limit, the expression leads to the obvious conclusion that in the absence of a significant background charge density and recombination, $j_{ph} \approx q\eta_{int}j_X$.

We emphasize that the treatment assumes that the photoconductor is in equilibrium, that is, the dark current is conducted by charge native to the device, whether it be due to intrinsic thermal generation (with density n_i), or due to extrinsic charge introduced via impurity doping (e.g. with density n_0). As seen in Section 4.4.2, charge injection whereby $n_0 \to n_0 + n_{inj}$, with n_{inj} equal to the injected charge density, results in a significant departure from ohmic conduction used in the derivation of Eq. 7.21.

The intermediate generation of an exciton also results in a slower response to an optical impulse compared to an inorganic semiconductor since the exciton diffusion time (τ_{ED}) must be added to the carrier lifetime in Eq. 7.8. That is the excitonic photoconductor has a bandwidth of

$$\Delta f = \frac{1}{2\pi}\left(\frac{1}{\tau_D + \tau_{ED}}\right) \approx \frac{1}{2\pi\tau_D}, \tag{7.22}$$

where the term on the far right hand side is the expression for the bandwidth of a conventional photoconductor, since in general $\tau_D \gg \tau_{ED}$ (i.e. $\tau_D \sim$ microseconds, whereas $\tau_{ED} \sim$ nanoseconds). If this inequality is not valid, $g\Delta f$ in Eq. 7.10 must be modified accordingly.

7.1.1.4 Quantum efficiency and responsivity

Equation 7.3 was used to define the external quantum efficiency in terms of the charge generation rate, assuming that all photons are absorbed within a thickness, d. This is rarely the case. Indeed, in organics, the lateral photoconductor configuration is the exception rather than the rule due to the very short exciton and charge diffusion lengths, and low charge mobilities in these inherently disordered materials. Hence, many photoconductors have a vertical geometry, that is similar to OLEDs and photodiodes (see Fig. 7.3). An *antireflection coating* (ARC) is often deposited on the light incident surface to reduce or eliminate Fresnel reflections, increasing the photon intensity arriving at the absorbing photoconductive layer. The photoconductive layer itself is sandwiched between a transparent and a reflective metal contact, and hence light undergoes two passes through the

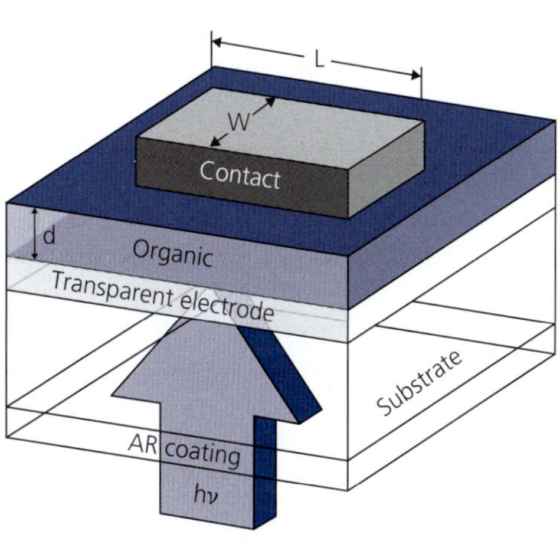

Figure 7.3 Schematic illustration of a vertical photoconductor.

organic layer. Note that the convention used in specifying device dimensions is different from that used in lateral geometry devices in Fig. 7.1. In the vertical geometry, the length of the photoconductor is defined by the layer thickness, d. Hence, the device area is LW, whereas in the lateral geometry it is dW.

The flux light at position x within a film follows the *Beer–Lambert law*, that is, $\Phi(x) = \Phi_0 \exp(-\alpha(\lambda)x)$, where Φ_0 is the incident flux intensity. Thus, after propagating a distance $2d$, and ignoring losses due to reflection at the metal contact and at the dielectric boundaries within the multilayer structure, the total number of absorbed photons is

$$\begin{aligned} G_{ph} &= \frac{(1-R)(P_{inc}\lambda/hc)}{dWL} \\ &\quad \times \left[\int_0^{2d} \exp(-\alpha(\lambda)x)dx \Big/ \int_0^{\infty} \exp(-\alpha(\lambda)x)dx\right] \\ &= \frac{(1-R)(P_{inc}\lambda/hc)}{dWL}[1-\exp(-2\alpha(\lambda)d)] \end{aligned} \tag{7.23}$$

where R is the reflectivity of the incident surface. Comparing this with Eq. 7.3 gives

$$\begin{aligned} \eta_{ext}(\lambda) &= (1-R)\eta_{int}(\lambda)\eta_A(\lambda) \\ &= (1-R)\left(\frac{k_{diss}}{k_{diss}+k_D}\right)[1-\exp(-2\alpha(\lambda)d)], \end{aligned} \tag{7.24}$$

where η_A is the absorption efficiency in the active region. This expression gives the quantitative relationship between the internal and external quantum efficiencies. That is, η_{int} includes only those photons that reach the active semiconductor region. We emphasize that Eq. 7.24 is approximate in that it replaces the actual optical field intensity distribution within the multilayer device stack with a simple exponentially decaying flux density. However, we have seen in the previous chapters that devices as in Fig. 7.3 form weak microcavities that can substantially impact the photon distribution within the active layer. A more precise calculation of the optical field distribution and its relation to detector efficiency is found in Section 7.1.2.

A representative compilation of the absorption coefficients of the molecular donors DBP, DPSQ, and DTDCPB, and the fullerene acceptors C_{60} and C_{70} are compared to the common inorganic semiconductors, Si, Ge and, GaAs, in Table 7.1. The absorption coefficients for the organics are at least one order of magnitude higher than the inorganics. In particular, the organics have $\alpha > 10^5$ cm^{-1}, compared to the indirect band gap Si and Ge with $\alpha = 10^2$–10^3 cm^{-1}, and the

Table 7.1 Absorption coefficients (α) and refractive indexes (n) for representative organic and inorganic photodetector materials

Material	λ_{max}[a] (nm)	α (cm^{-1})	n[b] (at λ_{max})
Si	800	3×10^2	3.42
Ge	1300	7×10^3	3.68
GaAs	800	1.3×10^4	4.1
C_{60}	345	3.20×10^5	2.14
C_{70}	385	2.01×10^5	1.99
DBP	605	2.70×10^5	2.5
DPSQ	690	3.23×10^5	1.9
DTDCPB	590	1.09×10^5	1.75

[a] λ_{max} for inorganic semiconductors taken at a point approximately 0.2 eV above the band edge.
[b] gn for inorganic semiconductors taken approximately 0.2 eV below the band edge.

direct gap GaAs at 10^4 cm^{-1}. The higher absorption coefficient is due to the molecular (i.e. non-chemical bonded) nature of organics leading to high oscillator strength for $S_1 \leftarrow S_0$ transitions. This is compared to transitions in inorganics where the oscillator strength is distributed over the high density of states within the broad energy bands. The consequence of the higher absorption coefficient is the ability to use thinner layers to absorb the incident light in an organic photodetector.

The reflectivity at the interface between two non-absorbing layers with indexes of refraction of n_1 and n_2 at normal incidence is

$$R = \frac{(n_1 - n_2)^2}{(n_1 + n_2)^2}. \tag{7.25}$$

The more general case of incidence as a function of angle relative to the substrate plane is discussed further in Section 7.2.2. Generally, illumination is incident from free space with $n_1 = 1$, and $n_2 = n_{sub}$ is the refractive index of the substrate. Then,

$$R = \frac{(1 - n_{sub})^2}{(1 + n_{sub})^2}. \tag{7.26}$$

Equation 7.24 requires knowledge of the rates of dissociation and recombination. It is, therefore, more convenient to express the external efficiency in terms of the measured photocurrent density, j_{ph}. Then, using Eq. 7.3 it is apparent that

$$g\eta_{ext} = \frac{j_{ph}A}{q(P_{inc}\lambda/hc)}, \tag{7.27}$$

where $A = WL$ is the sample area. This is the most common and useful expression for calculating the change generation efficiency using all easily measurable quantities without knowing the details of the detection process. However, in a photoconductor or for that matter any detector that exhibits gain, a measurement of the photocurrent does not directly provide η_{ext}. Rather, we can only measure the *gain-efficiency product*, $g\eta_{ext}$, since the total number of electrons per photon is the observable quantity.

Finally, it is useful to define the detector *responsivity*, $\mathcal{R}$, that is, the current generated per incident unit of optical power. From Eq. 7.27,

$$\mathcal{R} = \frac{j_{ph}A}{P_{inc}} = qg\eta_{ext}\left(\frac{\lambda}{hc}\right)[\text{A/W}]. \tag{7.28}$$

Thus, the responsivity increases with wavelength since the photodetector is essentially a photon counter. Absorption of low energy photons (corresponding to longer wavelengths) generates the same number of photons as higher energy photons as long as all the photons have energy that exceeds the energy gap. Since power decreases with increasing wavelength, the responsivity correspondingly increases.

7.1.1.5 Noise, detectivity, and dynamic range

There are three principal noise sources for all detectors: *shot noise*, *thermal* (or *Johnson*) *noise*, and *1/f noise* (Rose, 1955a, 1963, van der Ziel, 1986). Noise has its origins in the uncertainty in the random arrival of charge at an electrode. The noise currents, therefore, can be determined using statistical analysis. Since the more rigorous approaches are beyond the scope of this book, here we will use a more intuitive method introduced by Albert Rose to derive the expressions for calculating noise currents in photodetectors (Rose, 1963). As we will see in Chapter 8, many of these same expressions are also used to quantify transistor noise.

The noise current is the product of the root mean square (rms) fluctuation in the average number of electrons arriving at an electrode per time interval, τ, and the amount of charge contributed (ζq) on each arrival, as illustrated in Fig. 7.4. If $\langle n \rangle$ is the average number of charges, then the rms fluctuation is $\langle n \rangle^{1/2}$. Thus, the rms noise current is

$$\langle i_n^2 \rangle^{1/2} = \frac{\langle n \rangle^{1/2}}{\tau}\zeta q, \tag{7.29}$$

yielding a mean square fluctuation of

$$\langle i_n^2 \rangle = \frac{\langle n \rangle}{\tau^2}(\zeta q)^2 = \frac{q i_T \zeta}{\tau}. \tag{7.30}$$

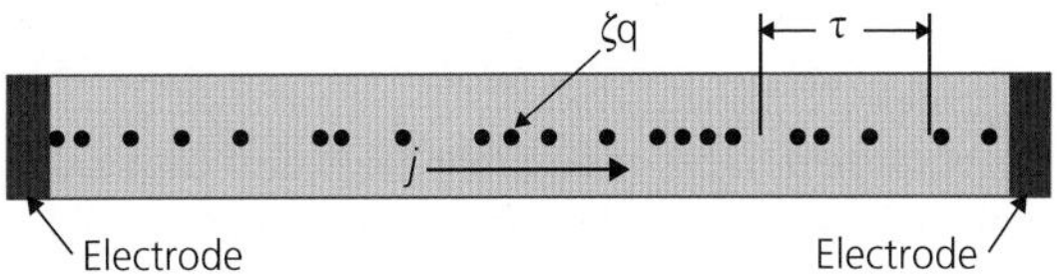

Figure 7.4 Illustration of charges carrying current, j, in a photoconductor arriving at an electrode. The amount of charge delivered per event is ζq, and the measurement interval is τ.

Here, $i_T = q\langle n\rangle/\tau$ is the mean total current. The time interval is related to the bandwidth, Δf, via $\tau = 1/2\Delta f$. In traversing the distance between electrodes, carriers undergo a gain, $\zeta = g$. This results in a mean square shot noise current of

$$\langle i_s^2\rangle = 4qgi_T\Delta f. \tag{7.31}$$

The extra factor of 2 arises since the charges in a photoconductor can either be generated (i.e. they add to the current) or recombine (they reduce the current). However, since both are random processes, they contribute equally to the current fluctuation and hence must be included in Eq. 7.31.

The situation is somewhat different when $V_a \to 0$. By illuminating or heating the semiconductor, we generate charges that recombine after traveling a distance equal to their diffusion length, L_D. Then each generated charge contributes, on average, qL_D/L to the total current, where L is the length of the photoconductor in the lateral geometry in Fig. 7.1. Then, from Eq. 7.30, the total noise current in the frequency interval, Δf, is

$$\langle i_n^2\rangle = 4\frac{\langle n\rangle}{\tau}\left(q\frac{L_D}{L}\right)^2\Delta f. \tag{7.32}$$

Again, we must include contributions from charge diffusing either toward or away from an electrode, accounting for the extra factor of 2. The diffusion length is related to the carrier lifetime via: $L_D^2 = D\tau$, and from the Einstein relation in thermal equilibrium, $\frac{D}{\mu_n} = \frac{k_BT}{q}$, we obtain

$$L_D^2 = \frac{k_BT\mu_n\tau}{q}. \tag{7.33}$$

For simplicity, we assume that the hole mobility is smaller than the electron mobility, and hence can be ignored. Substituting Eq. 7.33 into 7.32, and relating the average number of charges to their density via $\langle n\rangle = nLA$, then mean square thermal or Johnson noise current is

$$\langle i_{th}^2\rangle = \frac{4k_BT\Delta f}{R_{PC}}, \tag{7.34}$$

where $R_{PC} = L/\sigma A$ is the series resistance of the photoconductor under illumination.

It is interesting to note the voltage at which the shot and thermal noise contributions in photoconductors are equal. In the shot noise limit, the gain is $g = \tau/t_{tr} = \tau\mu F/L$. Recall, the thermal noise was calculated assuming $V_a \to 0$. In this case, the gain is $g_0 = L_D/L$; that is additional carriers are drawn into the photoconductor to maintain equilibrium until the diffusing charge arrives at the contacts. In many organic photoconductors, $g_0 < 1$, whereas $g > 1$. Writing g_0 in terms of the mobility, using Eq. 7.33 we obtain

$$g_0 = \frac{k_BT\mu_n\tau}{qL_DL}. \tag{7.35}$$

Setting $g = g_0$, then the thermal and shot noise contributions are equal at:

$$F_{diff} = \frac{k_BT}{qL_D}. \tag{7.36}$$

Here, F_{diff} is the *diffusion field*, above which shot noise dominates thermal noise. Indeed, F_{diff} is a convenient means to identify the transition between diffusion (i.e. equilibrium) and drift in a photoconductor.

Thus far we have considered a trap-free photoconductor. Both static and dynamic disorder makes this the exception rather than the rule in organic semiconductors. As a charge traverses the conductor, it can be trapped after a time, τ_t in an energy level within the HOMO–LUMO gap, where it subsequently recombines. The lifetime of the carrier fluctuates due to randomness of the trapping process, and this in turn introduces randomness in the charge arrival time at the electrodes. Taking the example where $\tau_t \ll \tau$, then the gain is reduced to: $g_t = \tau_t/t_{tr} = \tau_t\mu F/L$, and the bandwidth becomes $\Delta f_t = 1/2\tau_t$. The shot noise in this instance is, from Eq. 7.30,

$$\langle i_{st}^2\rangle = 4qg_ti_T\Delta f_t. \tag{7.37}$$

The spectral densities, that is, the noise per unit bandwidth, $\langle i_n^2\rangle/\Delta f$, of both shot and thermal noise is bandwidth-independent. Thus, both of these are white noise sources. An additional and important source of noise is $1/f$ or *flicker noise*. This is not a white noise source since its spectral density follows

$$\langle i_f^2\rangle = \frac{\kappa\Delta f}{f^\alpha}, \tag{7.38}$$

where $\alpha \geq 1$ depends on the specific nature of the noise source and κ is a constant. The origins of flicker noise are not as well understood as thermal or shot noise, although it is often attributed to poor contacts or trapping at defects within the semiconductor. Its functional dependence indicates that $1/f$ noise dominates at low frequencies, but then is overcome by white noise sources as the frequency is increased.

An illustration of the several noise contributions and their frequency domains is shown in Fig. 7.5. Note the frequency independence of the white noise sources compared to the $1/f$ noise spectrum.

The total noise current of the photoconductor is found by adding all of the mean square currents, viz.

$$\langle i_n^2 \rangle = \langle i_s^2 \rangle + \langle i_{st}^2 \rangle + \langle i_{th}^2 \rangle + \langle i_f^2 \rangle. \quad (7.39)$$

For a time-averaged incident optical power (assuming a sinusoidal signal of frequency $\ll \Delta f$) of $\bar{P}_{inc} = P_{inc}/2$, the time averaged signal current is, from Eq. 7.27,

$$\bar{i}_{ph} = qg\eta_{ext}\left(\bar{P}_{inc}\lambda/hc\right), \quad (7.40)$$

leading to a signal-to-noise power ratio (which is proportional to i^2) of:

$$\frac{S}{N} = \frac{\langle i_{ph}^2 \rangle}{\langle i_n^2 \rangle}. \quad (7.41)$$

In Eq. 7.40 we assume a sinusoidal input power of $P(t) = P_{inc}e^{-j\omega t}$, where the modulation frequency is $\omega \ll 2\pi\Delta f = 1/\tau_D$. The *noise equivalent power*, *NEP*, is defined as the rms optical power required to achieve $S/N = 1$ per unit bandwidth. That is,

$$NEP = \frac{\sqrt{\langle i_n^2 \rangle/\Delta f}}{\mathcal{R}}. \quad (7.42)$$

Then the detectivity is simply the inverse of the *NEP*, that is,

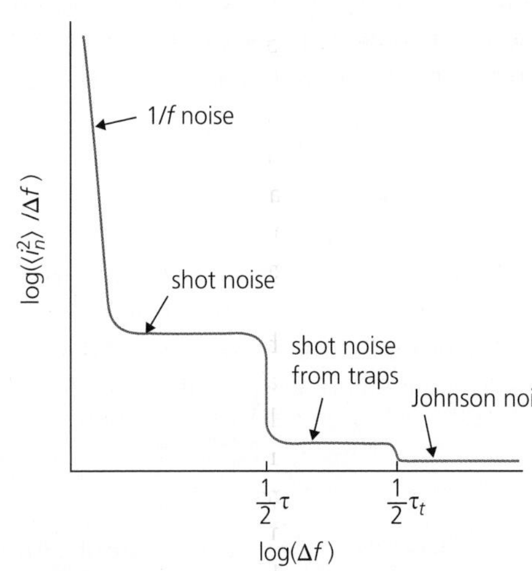

Figure 7.5 Mean square current noise spectral density in a photoconductor.

$$D = \frac{1}{NEP}. \quad (7.43)$$

To compare devices of different areas, it is convenient to normalize D to the device area and bandwidth. For this purpose, we define use the *specific detectivity*, given by

$$D^* = \frac{\sqrt{A\Delta f}}{NEP} = \mathcal{R}\sqrt{\frac{A\Delta f}{\langle i_n^2 \rangle}}, \text{ [Jones]} \quad (7.44)$$

where the units of Jones are [cm Hz$^{1/2}$/W]. Then, for a trap-free photoconductor operating near $V_a = 0$, $\langle i_n^2 \rangle \approx \frac{4k_BT\Delta f}{R_{PC}} + \frac{\Delta f}{f^\alpha}$ we obtain

$$D^* = q\eta_{ext}(\lambda/hc)\sqrt{\frac{A}{(4k_BT)/R_{PC} + \kappa/f^\alpha}}. \quad (7.45)$$

At higher voltages, or in the presence of traps, the detectivity is dominated by thermal or trap shot noise, which all decrease with the square root of the noise current.

One final quantity defining detector performance is its *dynamic range*, *DR*, that is, the range over which the detector response is linear. If we assume that i_{ph}^{max} is the maximum photocurrent that can be sourced by the photoconductor before there is a noticeable departure from linear behavior (i.e. η_{ext} decreases from its maximum value), then the dynamic range is defined as:

$$DR = 20log_{10}\left(\frac{i_{ph}^{max}}{i_{ph}^{min}}\right) \text{ [dB]} \quad (7.46)$$

where i_{ph}^{min} is the minimum photocurrent that delivers the least acceptable *S/N*. The maximum photocurrent is sometimes defined as the point at which i_{ph} departs from its linear value by -1 dB (corresponding to i'_{ph}). Often this decrease in efficiency is due to voltage drops across the contacts or other regions of the photodetector structure, which introduce a parasitic resistance, $R_{contact}$, in series with the photoconductive semiconductor. Then from Eq. 7.41 and assuming $S/N > 1$, $i_{ph}^{min} = \langle i_n^2 \rangle^{1/2}$:

$$DR = 10log_{10}\left(\frac{i_{ph}'^2}{\langle i_n^2 \rangle}\right), \quad (7.47)$$

where the noise current is given by the appropriate expressions.

7.1.2 Photodiodes and solar cells

The primary difference between a photodiode and a photoconductor is that the photodiode has a rectifying junction and is operated away from equilibrium due to a combination of internal and applied electric fields. We also distinguish between a photodiode operated under reverse bias and at zero bias. A reverse bias is applied to increase the charge collection efficiency beyond that attained with only the internal field due to the junction, as well as to increase the detector bandwidth by accelerating photogenerated charges toward their respective contacts. This mode of operation is known as the *photodetector mode*. Under reverse bias, the detector is a shot-noise-limited current source that can be sensitive to very small photocurrents for a small j_D. Alternatively, the detector can be operated at $V_a = 0$, relying only on the internal field to separate and sweep out the charges. This results in slower response and often lower efficiency, although the biasing scheme complexity is reduced from that needed to apply an external reverse bias. This is known as the *photovoltaic mode* of operation, and is universally used for photodiodes employed as solar cells (Section 7.3), but is also often used for photodetector sensor applications.

The organic photodiode (OPD) comprises a donor/acceptor (D-A) type II heterojunction (HJ) that separates a photogenerated exciton into a polaron pair, which subsequently dissociates into a free electron and hole. The multistep photogeneration process in the HJ introduced in Section 4.7.2 is illustrated in Fig. 7.6. The process is initiated by absorption of a photon in either the donor or acceptor (step **1**, Fig. 7.6). The photon generates an exciton that diffuses toward the D-A junction (step **2**). At the HJ, the electron transfers from the donor to the acceptor, and the hole remains in the donor (step **3**). If absorption occurs in the acceptor, the electron stays within that layer while the hole transfers to the donor. The driving force for the transfer is the offset in energies between the LUMOs and HOMOs of the two layers, whose magnitude is determined by the energy gaps and the HOMO–LUMO energy difference, ΔE_{HL}. The new state originating from exciton dissociation is the *polaron pair*, or CT state (shown prior to relaxation in step **3**). Processes **1–3** are collectively known as *photoinduced charge transfer*. Finally, the electron and hole are dissociated from the bound PP state via thermal excitation. The free polarons drift away from the junction and are collected in the external circuit (step **4**).

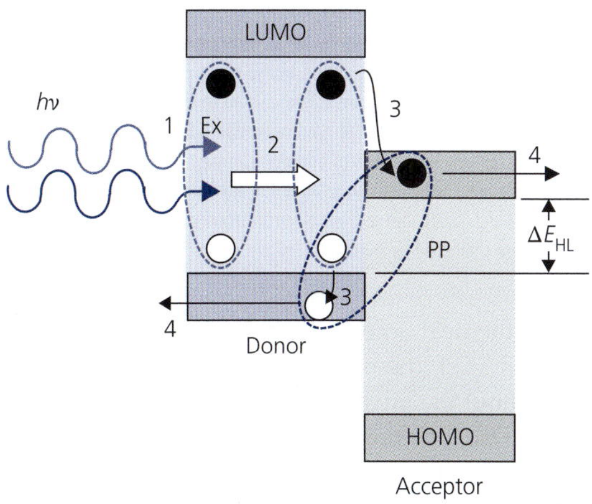

Figure 7.6 Illustration of the photogeneration process in a D-A heterojunction. Ex = exciton, PP = polaron pair. The dashed lines indicate Coulomb binding between electrons (solid circles) and holes (open circles).

The energy level diagram in Fig. 7.6 represents the equilibrium condition where $V_a = 0$. This is known as *short circuit operation*. Photodiodes are often reverse-biased to achieve both high bandwidth and high efficiency. Under reverse bias, the bands are tilted downward from left to right, thus resulting in efficient hole collection at the anode contact, and electron collection and at the cathode.

Figure 7.6 implies that internal quantum efficiency for photogeneration is the product of the efficiencies of each of the four steps:

$$\eta_{ext} = \eta_A \eta_{int} = \eta_A \eta_{ED} \eta_{CT} \eta_{CC}, \tag{7.48}$$

where η_{A} is the absorption efficiency (step **1**), η_{ED} is the exciton diffusion efficiency (step **2**), η_{CT} is the charge transfer efficiency (step **3**), and η_{CC} is the charge collection efficiency (step **4**). Since each of these processes has $\eta < 1$ (particularly η_{ED} and η_{CT}) the total efficiency can be small. We will find, however, that control of the morphology of the D-A junction that removes some of the limitations of a planar HJ illustrated in Fig. 7.6 can result in an internal efficiency that approaches unity.

Photogeneration in organics differs from the process in inorganic semiconductors, in that the latter does not proceed via the generation and dissociation of tightly bound excitons. In inorganic semiconductors, absorption results in the direct promotion of free carriers to the conduction and valence bands that are subsequently collected at the adjacent contacts. Thus, $\eta_{int} = \eta_A \eta_{CC}$, thereby eliminating the two most inefficient steps required in organic detectors (i.e. it eliminates η_{ED} and η_{CT}). Band-to-band excitation also simplifies the device structure, which does not require a HJ to provide the necessary energy for exciton dissociation.

7.1.2.1 Photodiode and solar cell architectures and energetics

Figure 7.7 provides an archetype layering scheme of an OPD, although actual structures can be significantly more complex than illustrated. The substrate-illuminated device begins with a transparent anode, typically ITO. This is followed by a transparent anode buffer layer that reduces the barrier to charge injection. Typical materials used for this purpose are PEDOT:PSS or MoO_3. Next is the organic active region. We will find in subsequent sections that this can be a bilayer D-A HJ, a bulk HJ, or a homogeneously mixed HJ. Often neat planar layers are included in the active region along with the blended film to increase absorption and/or ensure that there are no continuous pathways for charge to transit from cathode to anode that create shorts. The active region is often capped by a transparent exciton blocking/electron conducting layer (EBL). The wide energy gap EBL has multiple purposes (Peumans and Forrest, 2001, Peumans et al., 2003b). Energy barriers formed at the LUMO and HOMO edges and its wide energy gap prevents excitons generated in the active region from diffusing to, and recombining at the cathode. It is a transparent spacer that ensures that the incident optical field is maximized within the active region. And finally, the EBL serves as a buffer that prevents damage to the active region during deposition of the metal cathode.

There are several variants to this basic design. A transparent metal oxide cathode can be employed if device transparency at wavelengths other than those absorbed in the active region is required for use as an energy-generating window, for example. Furthermore, top-illuminated detectors are enabled using a metal contact on the substrate and a transparent top contact. Top illumination is often more convenient for integration with backplane circuits and for use in modules. Also, inverted structures reverse the positions of the cathode and anode from that shown in Fig. 7.7 to improve electrode performance for some organic active layer combinations.

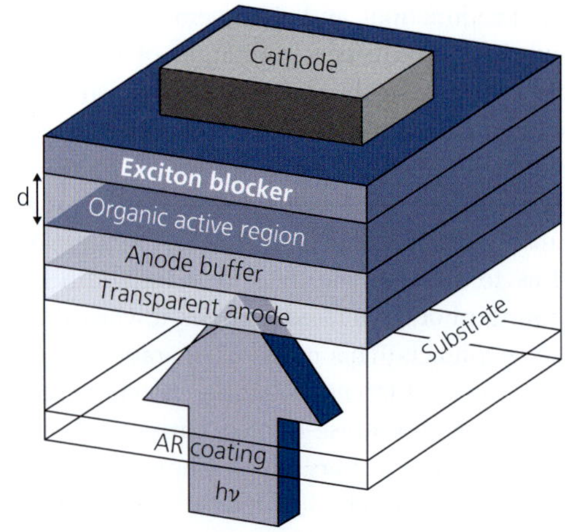

Figure 7.7 Illustration of an archetype, substrate-illuminated organic photodiode.

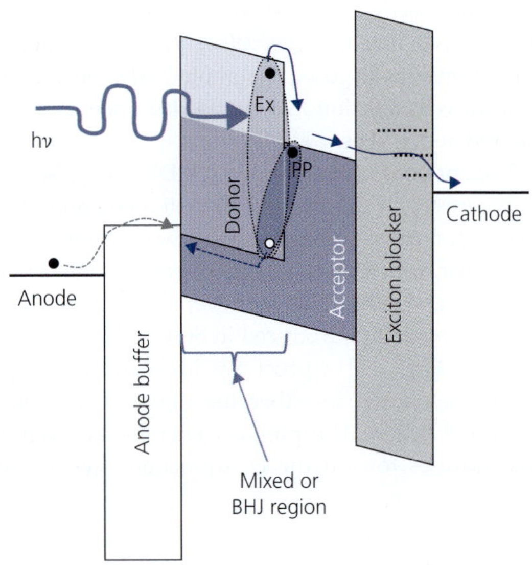

Figure 7.8 Energy level diagram of an archetype OPD illuminated via the anode. In this diode, the anode buffer has a deep LUMO level that promotes transport of electrons (filled circles) to the mixed D-A layer where they can recombine with photogenerated holes (open circle). The process of exciton (Ex) diffusion and polaron pair (PP) formation prior to dissociation are illustrated. Horizontal dashed lines in the exciton blocker indicate defect levels that enable electron transport to the cathode. Excitons are indicated by ovals.

The energy level diagram of the notional OPD in Fig. 7.7 is shown in Fig. 7.8 to illustrate the functions of the various layers. Starting from the anode side, light is absorbed in the organic active region, in this case comprising a mixed D-A region with an additional region formed by a layer of the neat acceptor material. In fact, light can be absorbed across the entire active region although we show here only processes that arise from absorption in the mixed region itself. While this device has a conventional planar-mixed architecture used in small molecule OPVs, the neat acceptor region is often absent. The excitons (Ex) dissociate into polaron pairs (PPs) throughout the mixed or BHJ region, or they diffuse to the D-A junction in planar structures. The PPs then dissociate (or recombine) to generate free electron polarons in the acceptor, and hole polarons in the donor that move along continuous percolating paths in the mixed or BHJ region toward their respective electrodes.

The anode buffer is a transparent layer that is analogous to the hole injection layer in OLEDs (Section 6.1). As shown in the figure, the buffer has a very deep LUMO energy characteristic of MoO_3 or HAT-CN. It transports electrons from the anode contact to its interface with the donor layer where they recombine with photogenerated holes, thus "completing the circuit." Alternatively, the buffer can have a very shallow HOMO energy that aligns with the donor HOMO. In this modality, holes are transported from the donor, through the buffer, and to the anode.

There are several different designs for the EBL that will be discussed in Section 7.4. The embodiment in Fig. 7.8 comprises a wide energy gap organic layer that results in transparency. While the wide energy gap provides for efficient blocking of excitons from reaching the cathode, it likewise confines electrons within the acceptor. Electronic defect levels (illustrated by dashed lines in the energy gap) may be generated during deposition of the cathode contact. These defects provide sites for electron hopping, thus allowing their transport to the cathode. The presence of such defects depends on the blocking material used, that is, it must be susceptible to damage inflicted by energetic metal atoms. The defects must also extend across the layer at sufficiently high densities to create low resistance conducting paths. This condition limits the thickness of the blocking layer to a range that may not be optimal to ensure that the optical field is also maximized within the active layer. Hence, several alternative blocking layer architectures have been demonstrated that to not depend on damage from the cathode. Due to the important role that EBLs play in high efficiency detection, alternative schemes to that in Fig. 7.8 are discussed in Section 7.4.1.

In contrast to photodiodes used in sensing applications that are optimized to detect a comparatively narrow spectral band, a solar cell must maximize its power output by harvesting photons generated across a broad range of the solar spectrum. Excitonic absorption linewidths are insufficiently broad to harvest sunlight across the entire visible and near IR spectra, hence both the donor and acceptor layers should be comprised of materials whose absorption bands do not entirely overlap. It is rarely the case that even two materials are sufficient to achieve optimal solar coverage that should extend over a wavelength range of at least 400 nm $< \lambda <$ 1100 nm (corresponding to photon energies 1.1 eV $< E_{ph} <$ 3.1 eV). Furthermore, a single junction cell must absorb photons from the blue into the NIR to provide good coverage of the solar spectrum. This implies that only one, or at most two (donor and acceptor) narrow energy gap materials comprise the cell. This results in a low open circuit voltage determined by the smallest energy gaps employed. Shorter wavelength photons thermalize after absorption in the narrow gap region, thus surrendering much of their energy to heat. Thermalization losses are reduced by combining multiple cells, some absorbing at short wavelengths while others absorbing deeper into the red. For these reasons, the highest efficiency cells are configured into stacks of subcells, each comprising donor and acceptor materials that absorb in different, partially overlapping absorption bands (Hiramoto et al., 1990, Yakimov and Forrest, 2002, Peumans et al., 2003b, Ameri et al., 2009, Che et al., 2014, 2018b).

A schematic illustration of a multijunction cell is shown in Fig. 7.9. The cell consists of N subcells (SC-1 to SC-N) separated by nearly electrically and optically lossless *charge recombination zones* (CRZs). The CRZ may consist of an exciton blocking layer, a charge recombination region and an anode buffer that share an interface between the two adjacent subcells, or it may consist of only a single layer that serves all of these purposes. An equivalent circuit of the stack is also shown in Fig. 7.9. Each of the subcells is a separate, current-generating diode. If each of the CRZs is electrically lossless, then the total open circuit voltage (see the following section) is equal to the sum of the voltages across each of the subcells. The maximum output voltage is thus the sum of the individual open circuit voltages (V_{OC}), viz.

$$V_{OC,T} = \sum_{i=1}^{N} V_{OC,i}. \tag{7.49}$$

In a bottom illuminated device, the top-most cell (SC-1) is known as the *back cell,* and SC-N upon which the light is initially incident is the *front cell.* Since the cells are in series, the current across each of the cells must be equal. Hence, the photocurrent, j_{ph}, is limited by the subcell generating the lowest photocurrent. By convention, in the photovoltaic mode, the photocurrent is positive.

The CRZs must be transparent such that the light arriving via the substrate (or via transparent contact in a top illuminated device) is not absorbed before arriving at one of the subcells. The purpose of the CRZ is to provide sites for charge recombination between each of the cells. For the example in Fig. 7.9, a photon absorbed in SC-N results in a hole collected at the anode and an electron that drifts toward the CRZ separating this subcell from SC-(N-1). Similarly, the hole generated in SC-(N-1) drifts

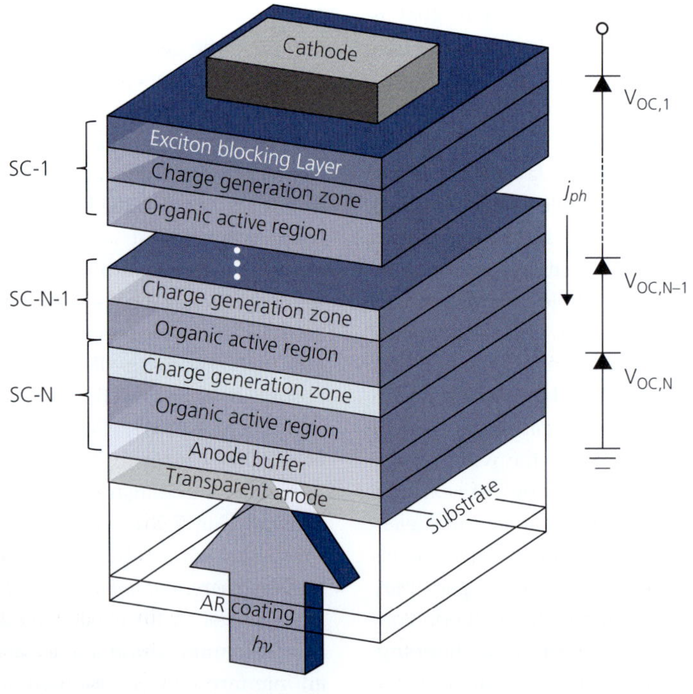

Figure 7.9 Schematic illustration of an archetype, multijunction photodiode or solar cell. Its equivalent circuit is shown on the right, with j_{ph} the photocurrent and $V_{OC,i}$ the open circuit voltage of the *i*th subcell (SC-i).

toward that same CRZ whereas its electron drifts toward the next subcell (SC-(N-2)) in the stack. The electron and hole arriving at the CRZ between the two lower subcells recombine without loss within this layer. This is analogous to charge generation layers (CGLs) used in stacked OLEDs. Their different nomenclatures suggest that the opposite processes occur in these two types of zones found between stacked elements, yet in fact their operating principles are essentially identical.

The spectral separation in a dual junction (tandem) solar cell is shown in Fig. 7.10a. This arrangement places the long wavelength absorbing subcell adjacent to the anode, and the short wavelength subcell nearest to the cathode. In Section 7.1.2, we show that this is the optimal arrangement for multijunction organic cells since the intensity of short wavelength illumination is highest near to the reflecting cathode, where there is a node in the standing wave amplitude. This is similar to inorganic cells where the high-energy absorbing cell is placed in the front cell position. The onset of absorption in an inorganic semiconductor is at the gap energy, but due to their very broad bands, absorption extends deep into the blue from the longest wavelength onset. Hence, a narrow band gap, red absorbing subcell must be placed at the back position, otherwise it blocks the higher energy photons from being absorbed in a short wavelength (large band gap) absorbing cell placed behind it.

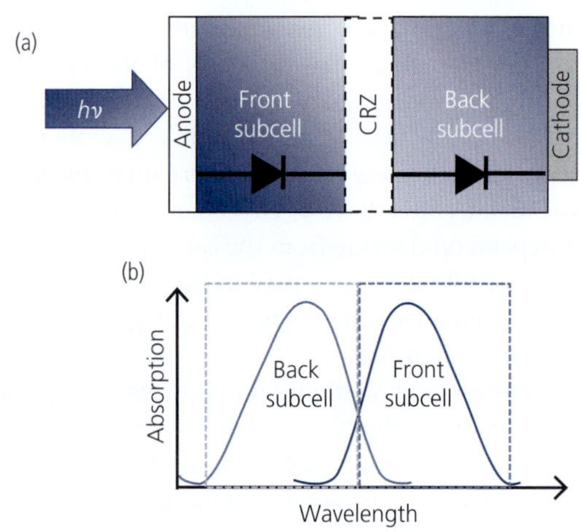

Figure 7.10 Each subcell in a multijunction OPV absorbs a different part of the solar spectrum. (a) A schematic illustration of a tandem cell with a red absorbing front subcell and a blue-green absorbing back subcell. (b) The excitonic absorption spectra are sufficiently narrow to only harvest photons over a limited spectral range.

In contrast, organic excitonic absorption bands are relatively narrow, as illustrated in Fig. 7.10b. Hence, the subcells do not require this same ordering of absorption bands, giving an increased flexibility in cell design. The ordering of organic absorbing cells can thus be optimized to maximize absorption of specific wavelengths, depending on the field profile within the multilayer stack without being overly concerned about whether light from a front cell will be blocked from the cell behind it (see Section 7.1.2). Of course, this assumes that the absorption spectra of the various subcells do not significantly overlap, as is the case for the cell in Fig. 7.10.

Energy level diagrams for a series-stacked tandem photodiode in equilibrium and under illumination are shown in Figs. 7.11a and b, respectively (the anode buffer and exciton blocking layers are omitted for clarity). The recombination of photogenerated electrons and holes within the CRZ is illustrated in Fig. 7.11b. The CRZ itself may be a thin, semitransparent metal such as Ag, or a transparent metal oxide such as MoO_3 or ZnO with a large electron affinity. The composition of this and other layers are discussed further in Section 7.4. Without the CRZ, the electrons and holes build up at the D-A interface between SC-1 and SC-2. This results in a potential that tilts the bands to reduce $V_{OC,T}$ back towards its equilibrium value of $V_{OC,T} = 0$, at which point the device no longer conducts. Note that the energy level diagrams imply a planar cell architecture. That is, the donor and acceptor layers are continuous layers meeting at a single, flat interface. However, interdigitated, mixed, and bulk HJs all share the same energetics as illustrated here, since the donor and acceptor components in these mixed morphologies still contact the anode and cathode, respectively (see Section 7.1.2.3).

Finally, it is worth mentioning that the series connected stack, while being the most frequently implemented, is not the only possible architecture. The requirement for balanced charge generation in all the subcells can be removed by connecting the subcells in parallel, as shown in Fig. 7.12. The disadvantages of this device are twofold. First, parallel connection requires contact to the transparent, low resistance electrodes between each subcell in the stack. The electrodes no longer serve as recombination zones since charges of similar sign are incident on both electrode surfaces. In practice, it is difficult to simultaneously achieve very low resistance and optically lossless contacts in a thin film between the active organics. Fabricating devices with transparent electrodes on separate substrates and then mechanically sandwiching the devices together is one route to obtain contacts to interior electrodes. In addition, the two different substrates can form the base and lid of a package if the entire unit is sealed around its periphery.

The second disadvantage of the parallel configuration is that the voltage at the output is equal to

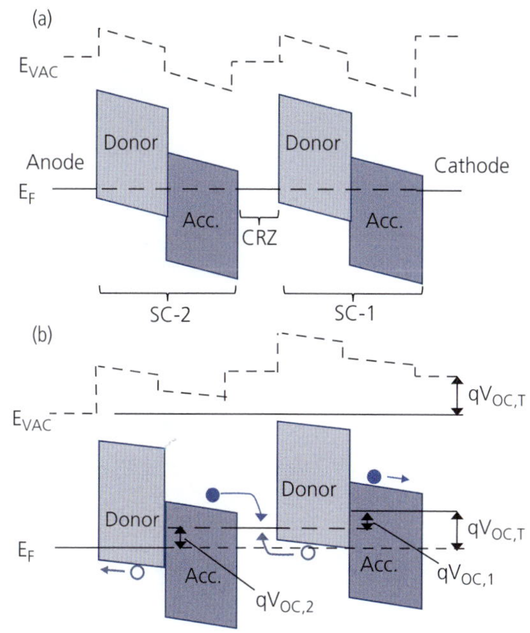

Figure 7.11 Energy level diagrams of a tandem OPV cell in (a) equilibrium and (b) under illumination. For simplicity, the presence of dipoles at organic interfaces is ignored. Electrons are shown as closed circles, and holes as open circles.

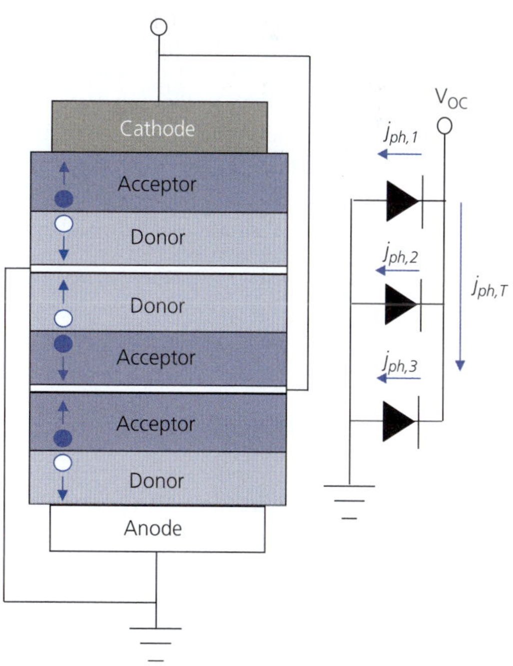

Figure 7.12 Schematic illustration and equivalent circuit of a parallel connected stacked photodiode or photovoltaic cell. Electrons are closed circles, holes are open circles. The total photocurrent, $j_{ph,T}$, is the sum of the separate photocurrents, $j_{ph,i}$, generated in the several subcells.

the minimum of all the voltages of the constituent subcells, and the photocurrent is the sum of the photocurrents from each of the cells (see the equivalent circuit in Fig. 7.12) following

$$j_{ph,T} = \sum_{i=1}^{N} j_{ph,i}. \tag{7.50}$$

As a result, the cell is a high current/low voltage element. This combination is susceptible to efficiency loss due to its internal series resistance—an effect that is minimized in the series stack in Fig. 7.9.

Hybrid combinations of series and parallel cells may ultimately provide the highest efficiency along with a reduced need for perfect current or voltage balance. For example, two low V_{OC} (i.e. long wavelength) cells can be stacked in series, and then connected in parallel to a high V_{OC} cell that absorbs at shorter wavelengths (see Fig. 7.13a). If the sum of the two low voltage cells is approximately equal to that of the high voltage cell, the voltage loss, which is the difference between the voltage generated by the single junction and the series double junction cell, can be minimized as shown by the equivalent circuit in Fig. 7.13b.

The series tandem can be fabricated on one transparent substrate (e.g. glass) in a conventional bottom illuminated, transparent top contact configuration, while the short wavelength device is fabricated in a top-illuminated configuration on a second substrate, which can either be transparent or opaque. The contacts from both are brought to the substrate edges and separately accessed after the substrates are bonded to form a sealed package. An illustration of the assembly procedure of the two substrates and their associated OPVs into a single, sealed package is shown in Fig. 7.14.

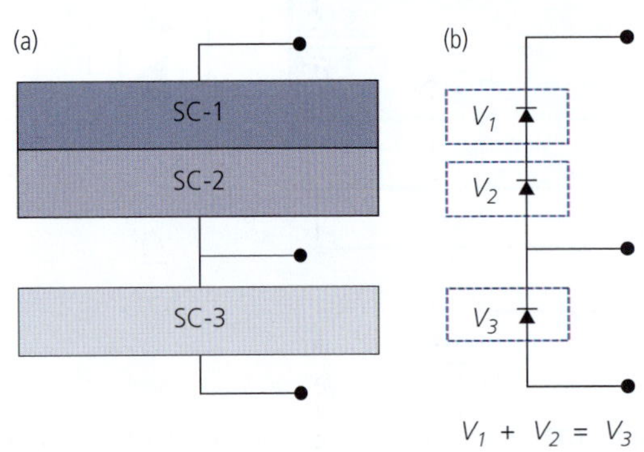

Figure 7.13 (a) Series-parallel OPV schematic structure and (b) equivalent circuit.

7.1.2.2 *j–V* characteristics in the dark and under illumination

In Section 4.7.2, we used a theory based on PP production mediated by excitons to derive the current in an organic HJ in the absence of traps (Giebink et al., 2010b). This analysis showed that the *j-V* characteristics were similar, but not identical to the Shockley ideal diode equation, viz. (see Eq. 4.282)

$$j = j_{0s}\left[\exp(q(V_a - jAR_{ser})/k_BT) - \frac{k_{PPd}}{k_{PPd,eq}}\right] - j_{ph}, \tag{7.51}$$

where

$$j_{0s} = qa_0k_{rec}N_{HOMO}N_{LUMO}(1 - \eta_{PPd})\exp(-\Delta E_{HL}/k_BT) \tag{7.52}$$

is the saturation current density, a_0 is the PP radius calculated using the Bohr model of the hydrogen atom and the dielectric constants of the contacting D and A materials, N_{HOMO} and N_{LUMO} are the densities of states of the corresponding frontier orbitals,

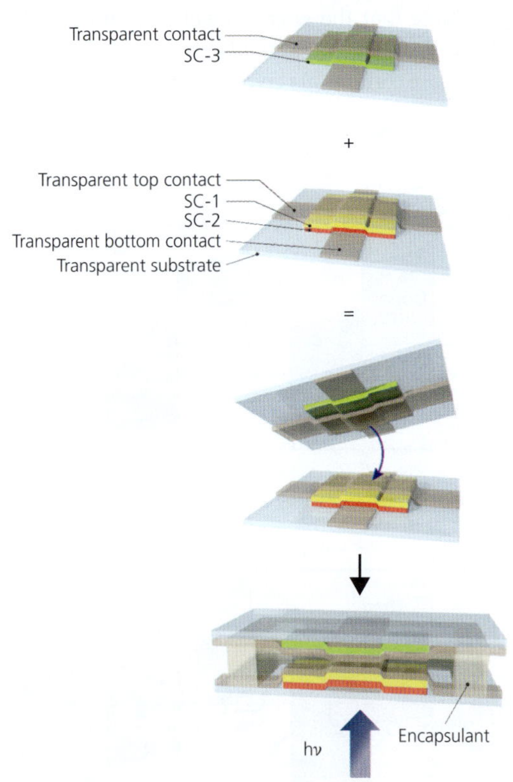

Figure 7.14 Illustration of the assembly of OPVs on separate substrates into a single, packaged device. From top: The single and double junction cells are separately fabricated, assembled face to face and then encapsulated to seal the devices from the environment.

and k_{PPd} and $k_{PPd,eq}$ are the PP dissociation rates at voltages V_a and 0, respectively. These rates are analogous to k_{diss} for the photoconductor in Fig. 7.2 (cf. Fig. 4.103). Also included is the series resistance, R_{ser}, which combines contact and internal layer resistances into a single quantity. The voltage drop across this resistance is $V_{ser} = jAR_{ser}$. Finally,

$$\eta_{PPd} = \frac{k_{PPd}}{k_{PPd} + k_{PPr}} \tag{7.53}$$

is the PP dissociation efficiency, where k_{PPr} is the rate of recombination of PPs. Hence, the photocurrent is $j_{ph} = q\eta_{PPd} j_X$, again analogous to the process in Fig. 7.2. In the more common situation where traps are present, we have shown that (Eq. 4.314):

$$j = j_{0D}\left[\exp(q(V_a - jAR_{ser})/n_D k_B T) - \frac{k_{PPd}}{k_{PPd0}}\right] + j_{0A}\left[\exp(q(V_a - jAR_{ser})/n_A k_B T) - \frac{k_{PPd}}{k_{PPd0}}\right] - j_{ph}, \tag{7.54}$$

where j_{0D} and j_{0A} are the saturation current densities due to traps on the donor and acceptor sides of the HJ, respectively (and hence depend on the respective trapped charge densities, among other factors found in Eq. 7.52), and n_D and n_A are the ideality factors, that define the voltage division between the donor and acceptor sides of the HJ. For $n_D, n_A \sim 1$, diffusion in the absence of traps is the dominant recombination process, and when $n_D, n_A \geq 2$, recombination via traps on the donor or acceptor side of the HJ is present.

For simplicity, we can assume that the HJ is approximately symmetrical, in that the density of trap states is equal on both the donor and acceptor sides, and that $N_S = \sqrt{N_{LUMO} N_{HOMO}}$ is the joint density of states. It is straightforward to show for this case that there is only a single ideality factor, $n_S = n_A = n_D$. Then Eqs. 7.51 and 7.54 reduce to (Giebink et al., 2010b)

$$j = j_0\left[\exp(q(V_a - jAR_{ser})/n_S k_B T) - \frac{k_{PPd}}{k_{PPd,eq}}\right] - j_{ph} \tag{7.55}$$

and the saturation current becomes

$$j_0 = qa_0 k_{rec} N_S^2 (1 - \eta_{PPd}) \exp(-\Delta E_{HL}/k_B T). \tag{7.56}$$

In addition to the series resistance, shunts across the junction must be included. The leakage through the shunt adds to j_0, and is accounted for by adding a parallel resistance, R_{shunt}. The equivalent circuit for an OPD or OPV containing all the circuit elements that serve as proxies to describe numerous complex physical phenomena is shown in Fig. 7.15. Including R_{shunt} into the "master equation" Eq. 7.55 then gives

$$j = j_0\left[\exp(q(V_a - jAR_{ser})/n_S k_B T) - \frac{k_{PPd}}{k_{PPd,eq}}\right] + \frac{V_a - jAR_{ser}}{R_{shunt}} - j_{ph} \tag{7.57}$$

The j–V characteristics of an archetype photodiode are shown in Fig. 7.16, which are similar to those shown on a semi-log scale in Fig. 4.113. The linear scale is helpful in revealing several parameters useful to understanding photodetector operation. The photodetector and photovoltaic modes correspond to the characteristics in the third and fourth quadrants, respectively. A photodetector used in the photovoltaic mode at $V_a = 0$ V yields the detector short circuit current, j_{SC}. Solar cells, however, are operated at the

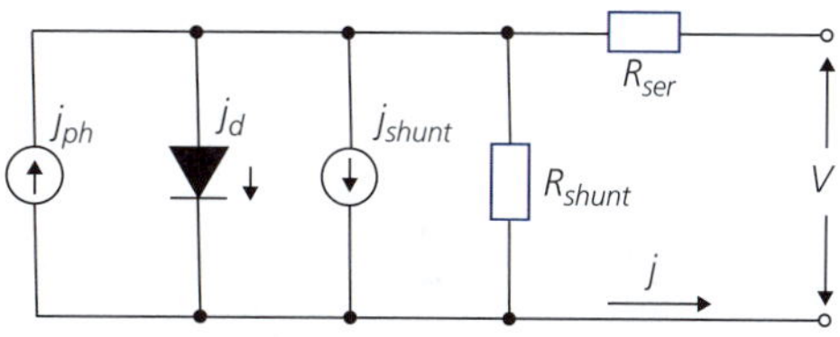

Figure 7.15 Equivalent circuit for an organic photodiode.

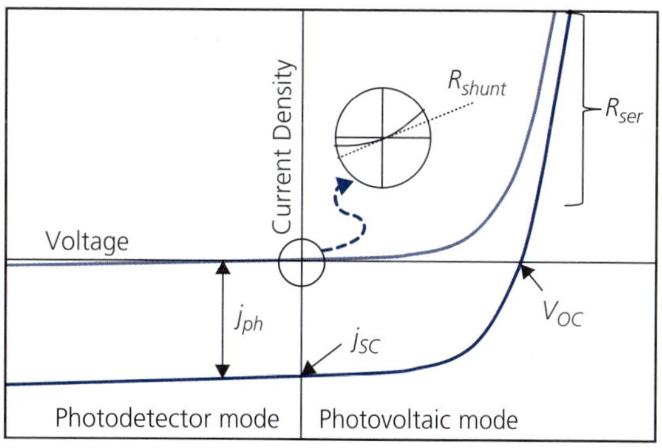

Figure 7.16 Current–voltage characteristics of an organic HJ diode in the dark (light blue line) and under illumination (dark blue line), generating a photocurrent density, j_{ph}. R_{shunt} is the shunt resistance determined by the slope of the j–V curve at $V_a = 0$ V, which is magnified in the inset. Also shown are the open circuit voltage (V_{OC}) and the short circuit current density (j_{SC}). The region dominated by series resistance is in the first quadrant. The photodetector and photovoltaic modes of operation are indicated.

maximum power generation point, which is intermediate between $V_a = 0V$ and V_{OC}.

At low forward bias, the current increases exponentially with V_a, with exponential slope equal to $q/n_S k_B T$. While Eq. 7.54 predicts two slopes since in general, $n_A \neq n_D$, it is also true that only one trapping process will dominate at, and be observable in a linear plot such as Fig. 7.16. As forward bias is increased, V_{ser}, the voltage drop across the series resistance, becomes significant and the slope of the *j-V* characteristic is linear, following Ohm's law determined by R_{ser}.

Under reverse bias, the dark current increases slowly with V_a due to increased PP ionization where, $k_{PPd}(V_a) > k_{PPd}(0) = k_{PPd0}$ as approximated by Onsager–Braun theory (Eq. 4.305). As we will find in Sec. 7.2, other factors such as generation-recombination of charge and photoconductivity can also contribute to the slope in reverse bias. At large reverse bias, the slope increases at a greater rate than shown in Fig. 7.16. This is the region of *junction breakdown*. Breakdown can have several sources, including *tunneling* of charge at high fields across the HJ, *avalanche breakdown* due to *impact ionization* of high energy carriers with valence electrons on the molecules in the layers, *microplasmas*, and *thermal runaway* (Sze, 1981). Whereas tunneling and avalanche breakdown may be reversible (i.e. the *j–V* characteristic at lower voltage is recovered without hysteresis), microplasmas and thermal effects permanently damage the junction. Furthermore, due to the low charge mobilities, it is unlikely that impact ionization leading to carrier multiplication (and hence photocurrent gain) can occur in OPDs. Hence, the most prevalent causes of breakdown under reverse bias are thermal effects or shorting via microplasmas. Tunneling breakdown may also occur in some situations, although it is rare given the relatively small extent of the electron wavefunctions in molecular solids. Tunneling breakdown is characterized by an approximately exponential increase in dark current with voltage, and only a minor dependence on temperature (Sze, 1981).

The slope of the *j–V* characteristic at $V_a = 0$ gives the shunt resistance:

$$R_{shunt} = \frac{1}{A} \frac{dV_a}{dj}\bigg|_{V_a=0}, \tag{7.58}$$

illustrated by the diagonal line in the inset. The shunt resistance is a function of charge leakage around the junction, and is important in devices that are operated near 0 V. For the trap free D-A HJ (corresponding to $n_S = 1$), and ignoring the small voltage dependence of k_{PPd} as $V_a \to 0$ (Braun, 1984, Peumans and Forrest, 2004, Giebink et al., 2010b), then from Eq. 7.55,

$$R_{shunt} = R_{ser} + \frac{k_B T}{qA}\left(\frac{1}{j + j_0 + j_{ph}}\right). \tag{7.59}$$

That is, at $V_a = 0$ and in the dark, the shunt resistance is simply equal to the series plus the *thermal resistance*, $R_{th} = k_B T/(qAj_0)$. This latter term is due to recombination across the HJ. For a one-sided junction (i.e. where either j_{0D} or j_{0A} dominates), the thermal resistance is increased by the ideality factor, n_D or n_A. Under illumination, the shunt resistance increases to $R_{th} = k_B T/qA(j_0 + j_{ph})$. There are often parasitic resistances that can contribute to R_{shunt}. For example, defects at the rectifying HJ can result in leakage that is not accounted for by the ideal expressions for *j vs. V*. The shunt resistance in this case is found by combining the current through the parasitic resistance, R_p, in parallel with the expression in Eq. 7.55. Indeed, in such situations, it is not uncommon for $R_p \ll R_{ser} + R_{th}$. Thus, R_{shunt} and its deviation from the ideal Eq. 7.59 is used to diagnose the quality of the rectifying junction.

Under illumination, the *j–V* characteristics undergo a nearly rigid shift by an amount equal to $-j_{ph}$. The curve crosses the $j = 0$ axis at the open circuit voltage, V_{OC}, and at $V_a = 0$ the short circuit current is $j_{SC} = j_{ph}$. According to Eq. 7.6, $j_{ph} = j_T - j_0$ (here we replace j_D with j_0 from the diode equation). The photocurrent contribution to the *j–V* characteristics is indicated in Fig. 7.16. We show below that application of reverse bias can result in an increase in j_{ph} that is larger than the increase in j_0 due to contributions from $\eta_{CC}(V)$, since a large applied voltage can efficiently sweep out photogenerated electrons and holes before they recombine.

7.1.2.3 Active region morphologies

Up until this point, we considered only a simple photoactive region consisting of the donor layer in contact with the acceptor to form a planar interface. However, in both OPDs and OPVs, considerably higher efficiencies are achieved using irregular or interdigitated interfaces presenting a large area for exciton dissociation.

Photogeneration occurs in both the acceptor and donor regions, within a depletion region of width w_T, and in the undepleted regions lying remote from the HJ. However, we have shown in Chapter 4 that unless they are intentionally doped, organic materials have few background charges, in which case the active region is fully depleted. There are three principal morphologies employed in organic detectors that maximize efficiency and bandwidth. They are the planar heterojunction that was discussed in reference to Fig. 7.6, the random *bulk heterojunction* (BHJ), and the more ordered *controlled BHJ*. An illustrative comparison between the planar and BHJ architectures is shown in Fig. 7.17.

In the planar junction, light may be absorbed at distances much larger than L_D from the point of dissociation. This leads to a trade-off between η_{ED} and η_A in Eq. 7.48 since $\alpha \ll 1/L_D$. Therefore, efficient detectors employ a morphology where the D and A regions form an interdigitated junction with a large surface area that completely fills the active region volume (Halls et al., 1995, Yu et al., 1995). This is the BHJ architecture in Figs. 7.17b and c that was originally inspired by the large area junction of a *dye sensitized solar cell* (DSSC). In a DSSC, a rough TiO_2 surface is coated with a sensitizing dye, and the entire structure is immersed in an electrolyte (O'Regan and Grätzel, 1991). The large area of the TiO_2 surface enables the efficient harvesting of photons without creating resistive bottlenecks to charge generated at the sensitizer/TiO_2 interface.

The idealized BHJ in Fig. 7.17b shows a perfectly ordered, interdigitated network of A and D materials. The network optimizes the photodetector by orthogonalizing the directions for light absorption and exciton diffusion. That is, light is absorbed along the direction parallel to the interdigitation which can extend for distances $\gg 1/\alpha$. However, the width of the fingers should be $\leq L_D$ to assure that excitons reach a heterointerface where they dissociate prior to their recombination. This geometric optimization breaks the fundamental trade-off between η_{ED} and η_A, allowing both to simultaneously approach unity.

Achieving the ideal, interdigitated BHJ with the appropriate dimensions requires nanometer control of the morphology. In Section 5.4.2. we showed that management of strain via organic vapor phase deposition (OVPD) of a CuPc/PTCBI HJ achieved this precisely controlled structure (Yang et al., 2005a). The resulting solar cell comprised 35 nm long by 30 nm wide nanocrystalline pillars of CuPc covered by blanket deposition of the PTCBI acceptor. These dimensions are close to the exciton diffusion length in the CuPc. The growth of the BHJ with this precise geometric and morphological control led to an approximately 2.5-fold increase in the OPV cell quantum efficiency compared to that of an analogous planar HJ made from these same materials.

Conventionally, BHJ structures are obtained by blending two, partially miscible polymers in a solution (Yu et al., 1995). This forms a thermodynamically driven, random D-A network shown in Fig. 7.17c. While the blend provides a large surface area junction, it contains numerous defects such as islands of one material, or *cul de sacs* that prevent charge extraction to the electrodes. A layer of the unblended donor (next to the anode) or acceptor (adjacent to the cathode) is required to prevent shorting by direct bridging by one or both of these materials across the entire active volume of the device. The BHJ shown in Fig. 7.17c features a zone of neat acceptor near to the anode.

Many variations on the BHJ morphology have been demonstrated in both polymer and small molecular weight materials systems. A notable and successful variant is the mixed HJ, or the *planar-mixed HJ* (PM-HJ) shown schematically in Fig. 7.18 (Uchida et al.,

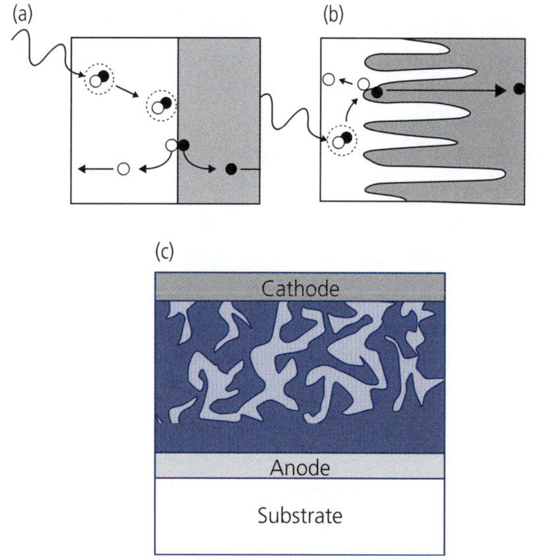

Figure 7.17 Illustration of photodetection in an idealized (a) planar and (b) bulk heterojunction (BHJ) diode. The acceptor regions are shaded. Holes are represented by open circles, electrons by filled circles, excitons are shown by dotted circles, and the D-A junction is the boundary with the shaded region. (c) A BHJ diode with blended donor and acceptor regions forming a random network. The acceptor regions are in light blue. The neat, donor region (dark blue) adjacent to the anode prevents shorting of the acceptor from cathode to anode.

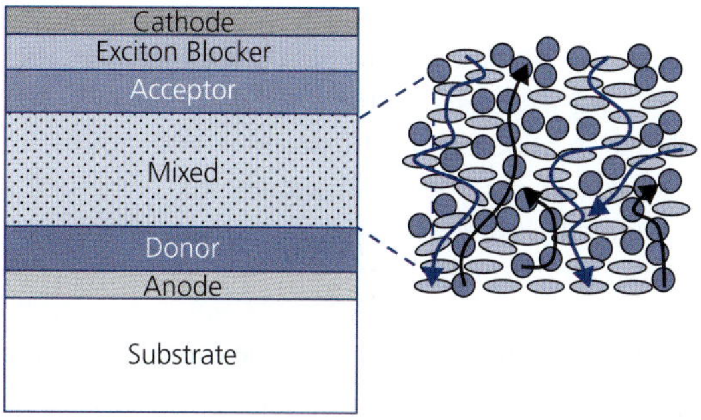

Figure 7.18 Illustration of the planar-mixed HJ (PM-HJ) structure. The mixed region is shown in detail on the right, where the acceptor molecules are dark, shaded circles and the donor molecules are light, open ellipses. Percolating pathways for electrons (black lines) and holes (blue lines) are shown.

2004, Xue et al., 2005). This small-molecule-based device comprises a region with a blend of donor and acceptor molecules deposited via co-deposition in vacuum or OVPD. The mixed layer is sandwiched between neat, planar layers of donor molecules adjacent to the anode, and a neat acceptor adjacent to a transparent and electron conducting/exciton blocking layer (see Section 7.4.1). A structural variant employs one or sometimes even no neat layers. Excitons in the mixed layer are generated on either a donor or acceptor molecule. The exciton rapidly dissociates into a PP since acceptor and donor molecules are in contact everywhere within the mixture. This is an improvement over the BHJ since there is neither need, nor opportunity for exciton diffusion over even short distances. This leads to $\eta_{ED} = 1$ for the PM-HJ.

While exciton dissociation occurs with unity efficiency in the mixed region, there remains two losses that must be overcome. The first is due to absorption in the neat, planar layers. The excitons generated in the donor and acceptor slabs surrounding the blend must diffuse to the mixed region to undergo dissociation. While the neat layers increase the optical path length and hence the number of excitons generated, as well as preventing shorts from forming between the cathode and anode by the mixed layer, they should be kept at thicknesses $< L_D$ to minimize exciton recombination. The second requirement is that percolating pathways for electron and hole conduction on the acceptor and donor molecules, respectively, must exist within the mixed layer to ensure that the charges do not recombine prior to collection. If conducting paths from the excitonic point of origin to a charge-collecting electrode are incomplete, charge will build up until the potential within the mixed layer that opposes the external bias is sufficiently large to shut off further charge collection. This is controlled by the morphology the blend, and the ratio of the particular donor and acceptor molecules used. Further, their steric and electrostatic characteristics must be considered in choosing the deposition and other processes used in their fabrication to ensure that *bicontinuous*, low resistance pathways for charge conduction are formed.

The disruption of the conductive pathways, and the disorder induced by the close proximity of donor and acceptor molecules, results in a reduced charge mobility compared with that in neat layers of the same materials. To quantify the conductive properties of the blend, we define the *carrier drift length*, $l = \mu\tau F$ as the distance a charge can travel before recombination. Here, τ is the carrier lifetime. A high charge collection efficiency, η_{CC}, requires that l be comparable to, or less than the mixed-layer thickness. Charge transport in the homogeneous layers is considered to be lossless since photogenerated holes and electrons are spatially separated, except when V_a approaches the built-in junction potential. The collection efficiency can therefore be written (Xue et al., 2005)

$$\eta_{CC} = \frac{L_C}{x_M}(1 - \exp(-x_M/L_C)), \tag{7.60}$$

where the charge collection length, L_C, is the sum of the hole and electron drift lengths prior to recombination, and x_M is the thickness of the mixed layer (Crandall, 1983). From this expression we find that $\eta_{CC} = 1$ for $x_M \ll L_C$, and $\eta_{CC} = L_C/x_M$ for $x_M \gg L_C$, suggesting that the mixed layer should be thinner than the collection length to obtain a high efficiency. Furthermore, η_{CC} is a function of the applied voltage through the field dependence of L_C:

$$L_C = \left(\tau_{Dp}\mu_p + \tau_{Dn}\mu_n\right)(V_a - V_{bi})/x_T, \tag{7.61}$$

where τ_{Dp} and τ_{Dn} are the hole and electron lifetimes within the mixed layer, and x_T is the thickness of all layers between the anode and cathode contacts. Hence, η_{CC} increases with the applied voltage, which efficiently separates charges immediately upon their photogeneration. To take full advantage of the mixed molecular HJ structure, it is important to have high carrier mobilities in a thin mixed layer such that L_C is comparable to the optical absorption length.

It should be apparent that the morphologies of the absorbing regions of planar, BHJ and PM-HJ devices are determining factors in the detector efficiency. Ordered and interconnected regions lead to higher charge mobilities and longer diffusion lengths, increasing both η_{ED} and η_{CC}, and hence η_{ext} in the photodiode. For this reason, considerable effort has been directed towards controlling the film morphologies in organic photodetectors. This topic is discussed further below.

7.1.2.4 Bandwidth

The energy level diagram of a generalized symmetric, p-intrinsic (i)-n excitonic HJ under reverse bias is shown in Fig. 7.19. Photons can be incident from either the right or the left, depending on which of the contacts is transparent. For the purpose of this discussion, we assume that light is incident via a transparent anode contact on the donor, and that the junction is at $x = 0$. Absorption occurs in both the donor and acceptor, generating carriers across the entire device active region, depending on the local flux density,

$\Phi(x,\lambda)$. There is minimal charge recombination within the depletion region following exciton dissociation since electrons and holes are rapidly swept to the contacts by the applied and internal fields (resulting from V_a and the built-in voltage, V_{bi}, respectively). Hence, unlike a photoconductor, a photodetector has a gain, $g = 1$.

The response time of the photodetector is limited by four factors: (i) the time for exciton diffusion from the point of generation to the HJ, τ_{ED}, (ii) charge drift time across the depletion region of total width $w_T = w_D + w_A$, (iii) charge diffusion time through the undepleted regions of the detector at $w_A < x < -w_D$, and (iv) the time required to charge the device capacitance, C_j.

Charges within the depletion region drift in the junction field resulting in a transit time, $t_{tr} = t_{tr,n} + t_{tr,p}$, where the electron and hole transit times ($t_{tr,n}$ and $t_{tr,p}$, respectively) depend on the electric field and the charge mobilities in the acceptor and donor layers. These times are typically less than the diffusion time, τ_{ED}. The rate at which charges diffuse through undepleted regions of the photodiode depends on their diffusion constants. The undepleted region transit time ($t_{tr,un}$) also depends on whether or not there is a drift component to the transit time equal to

$$t_{tr,unp} = (x_D - w_D)^2 \Big/ \left(\mu_p \cdot \delta V_a\right) \tag{7.62}$$

for the example of hole drift in the donor. Here, δV_a is the residual voltage across the undepleted layer, and x_D is the total donor layer thickness. A similar expression obtains for electrons in the undepleted acceptor region. It is also apparent that the drift time depends on the total layer thicknesses, x_D or x_A. For very thick layers, with $w_T \ll x_D + x_A$, we would expect $t_{tr} < t_{tr,unp} + t_{tr,unn}$, in which case the device will behave more like a photoconductor than a photodiode. It will exhibit photoconductive gain, whose bandwidth depends on the minority carrier lifetime. This device design is clearly undesirable, and hence will not be considered further. Hence, we assume there is little background charge as is typical of most OPDs, such that the undepleted regions are much thinner than w_T. Then, the time for charge transport across the undepleted zones of the detector is less than either t_{tr} or τ_{ED}, and can be neglected.

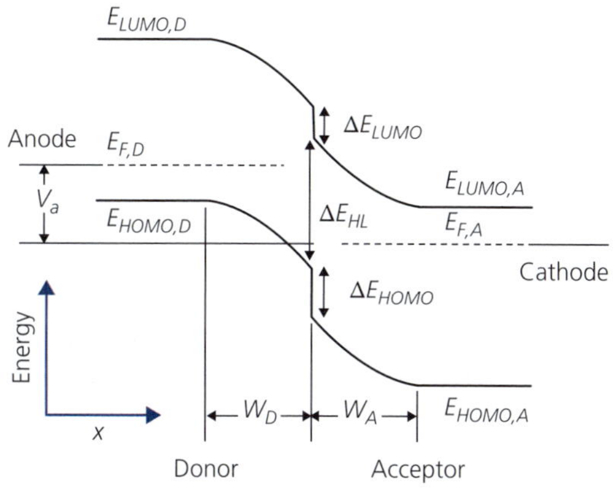

Figure 7.19 Energy level diagram of a D-A HJ detector under reverse bias (V_a), defining energies and dimensions.

Charging of the junction capacitance, C_j, places a further time constraint on detector response. Here, C_j is the series capacitance of the two depleted regions:

$$C_j = \frac{C_D C_A}{C_D + C_A} = \varepsilon_0 A \frac{\varepsilon_D \varepsilon_A}{(w_A \varepsilon_D + w_D \varepsilon_A)}. \tag{7.63}$$

This is in addition to the parasitic capacitances from the contact and other peripheral contributing capacitances, C_p, to give the total device capacitance of $C_T = C_j + C_p$. Since achieving a short response time requires thin layers, and $w_T \to x_D + x_A$, then in most situations $C_T \approx C_j$. The total resistance, R_T, that determines the RC-limited response time of $\tau_{RC} = R_T C_T$ is inferred from the small signal equivalent diode circuit in Fig. 7.20 to include capacitance and external loads. The photocurrent and noise current sources are in parallel with all circuit elements except the device series resistance. Thus, $R_T = \left(R_{ser} + R_j\right) \| (R_L \| R_{in})$ where R_L is the load resistance of the detection circuit, and R_{in} is the effective input resistance of the following current amplifier. For a very low dark current device, it is reasonable to expect that the junction resistance as determined by the diode equation, $R_j = i_T / V_a \gg R_{ser}$, except at high reverse voltages where the device undergoes reverse breakdown. Similarly, it follows that $R_j \gg R_L, R_{in}$, thus giving

$$\tau_{RC} = (R_L \| R_{in}) C_T. \tag{7.64}$$

The foregoing discussion provides the bandwidth for an organic HJ detector of:

$$\Delta f = \frac{1}{2\pi}\left(\frac{1}{t_{tr}} + \frac{1}{\tau_{ED}} + \frac{1}{\tau_{RC}}\right) \tag{7.65}$$

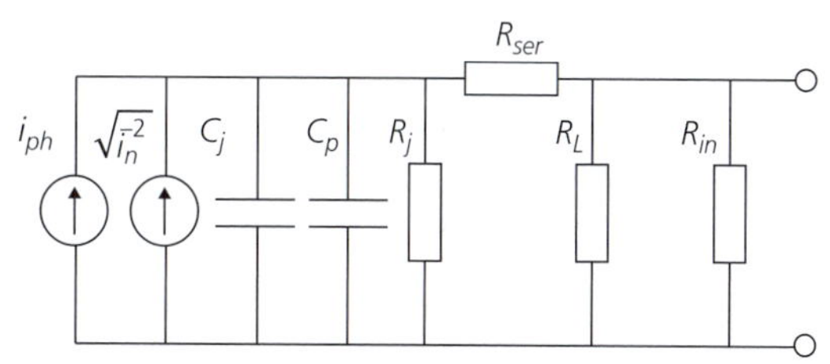

Figure 7.20 Small signal equivalent circuit of a photodiode.

and, since the gain of the OPD is $g = 1$, we have a gain-bandwidth product of

$$g\Delta f = \Delta f \tag{7.66}$$

(cf. Eq. 7.10). We assume that the transit time through undepleted regions is small compared to the other terms in Eq. 7.65. In many detectors, however, the three terms in Eq. 7.65 are comparable. We will find in Section 7.2 that diode designs have been devised to maximize detector bandwidth that eliminate one or more of the time constants, resulting in bandwidths approaching 1 GHz (Peumans et al., 2000b).

7.1.2.5 Noise and dynamic range

Since $n, p \to 0$ in the depletion region in reverse-biased photodiodes, the probability of trapping and detrapping of carriers is negligible. Hence $1/f$ noise is not a dominant source of noise. In contrast, noise from charge *generation* can be significant, leading to shot noise given by

$$\langle i_{sh}^2 \rangle = 2q(i_D + i_{ph})\Delta f = 2qi_T\Delta f, \tag{7.67}$$

which differs from the expression for photoconductors (Eq. 7.37) by a factor of 2 since, in the absence of charge in the depletion region, recombination is minimal. Also, for OPDs, $g = 1$. The other source of noise is the thermal, or Johnson noise, which from Eq. 7.34,

$$\langle i_{th}^2 \rangle = \frac{4k_BT\Delta f}{(R_{ser} + R_j)\|(R_L\|R_{in})} \approx \frac{4k_BT\Delta f}{R_L\|R_{in}}. \tag{7.68}$$

Due to the high junction resistance, the thermal noise is generally dominated by the impedance of the external circuit.

The total noise of a photodiode is typically much smaller than for a photoconductor due to the low dark resistance (and hence high i_D) of the latter. Further, the Johnson noise of the photoconductor cannot be smaller than that of a photodiode since, even for the best case of a very high dark resistance, a photoconductor still must be used at the input of a detection circuit whose load and input resistances are, in principle, equal to those used for a photodiode.

The *S/N*, *NEP*, and D^* then follow directly from our treatment for photoconductors in Sec. 7.1.1. The specific detectivity is:

$$D^* = q\eta_{ext}(\lambda/hc)\sqrt{\frac{A}{2qi_T + 4k_BT/R_L\|R_{in}}}. \tag{7.69}$$

From the foregoing, we infer that D^* is much larger for a photodiode due to its lower noise compared to the photoconductor. Similarly, from Eq. 7.47 we infer that the *DR* of the photodiode is also larger due to its reduced *NEP*. Series resistance from contacts limits the maximum photocurrent, although undepleted donor and acceptor regions in OPDs can also contribute to R_{ser}, placing further limits on i_{ph}, and hence *DR*.

7.1.3 Comparison of photoconductors and photodiodes

The simplicity of photoconductors led to their early demonstration and development, long preceding the demonstration of a HJ OPD that showed an asymmetric, rectifying *j–V* characteristic (Tang, 1986). However, as we have found in this section, the bandwidth and noise performance of reverse-biased photodiodes far exceeds that of photoconductors. For these reasons, photodiodes are preferred for almost all sensitive detection applications. While photoconductors can source photocurrent, they are incapable of providing power to an external load. In contrast, photodiodes operating in the fourth quadrant of the *j–V* characteristics (where $V > 0$ and $j < 0$, and vice versa) in the so-called photovoltaic mode are useful as a power generating solar cell. Due to the potential low cost and ability to deploy such devices on large, flexible substrates, OPVs have been the subject of considerable research by the organic electronics community, and hence will be discussed at length in subsequent sections. We summarize the foregoing discussion by comparing the characteristics of photoconductors and OPDs operated in the detection (in contrast to the photovoltaic) mode in Table 7.2.

7.1.4 Modeling efficiency: optical and charge generation in OPDs and OPVs

According to Eq. 7.48, the product of the efficiencies of four processes leading to charge generation determines the quantum efficiency. Two of the processes, namely absorption (η_A) and exciton diffusion (η_{ED}), are coupled since the diffusion efficiency depends on the distance that the exciton is generated from a D-A junction, and this in turn depends on the distribution of the optical field within the device. In describing the optical field of a photoconductor, we used the simple model of an exponentially decaying flux density with distance (the Beer–Lambert law) within the absorbing semiconductor. Such a treatment is valid only in the case of a single layer structure, whereas multilayered OPDs form weak microcavities. Indeed, in Chapter 6 we showed that the optical field distribution in multilayer OLEDs was more complex than predicted

Table 7.2 Comparison of operating parameters of photoconductors and photodiodes

Parameter	Photoconductor	Photodiode
Operating voltage	Near equilibrium ($V_a \to 0$)	Reverse bias
Photocurrent gain (g)	τ/t_r (1–10^6)	1
η_{int}	$k_{diss}/(k_{diss}+k_D)$	$k_{PPd}/(k_{PPd}+k_{PPr})$
η_{ext}	$\dfrac{j_{ph}A}{qg(P_{inc}\lambda/hc)}$	$\dfrac{j_{ph}A}{q(P_{inc}\lambda/hc)}$
Responsivity	$qg\eta_{ext}(\lambda/hc)$	$q\eta_{ext}(\lambda/hc)$
Bandwidth (Δf)	$1/2\pi\tau_D$	$1/2\pi t_{tr}$
Gain-bandwidth product ($g\Delta f$)	$1/2\pi t_{tr}$	$1/2\pi t_{tr}$
$\bar{i}_n^{\,2}/\Delta f$	$(4k_BT)/R_{PC}+\kappa/f^{\alpha}$	$2qi_T+4k_BT/R_L\Vert R_{in}$
Specific detectivity (D^*)	$q\eta_{ext}(\lambda/hc)\sqrt{\dfrac{A}{(4k_BT)/R_{PC}+\kappa/f^{\alpha}}}$	$q\eta_{ext}(\lambda/hc)\sqrt{\dfrac{A}{2qi_T+4k_BT/R_L\Vert R_{in}}}$

by such a simple considerations, and thus optical outcoupling of emitted radiation could only be understood and ultimately controlled by a detailed knowledge of the complex indices of refraction of all the layers comprising the device. This problem is even more complex in OPDs (and by extension, OPVs) since the product of $\eta_A\eta_{ED}$ ultimately determines η_{int}.

The absorption and exciton diffusion efficiencies in the photodetector are functions of the optical properties of the materials employed, excitation wavelength λ, layer thicknesses, and the layer order. To evaluate $\eta_A\eta_{ED}$, the optical electric field amplitude $\mathbf{F}(x)$ is calculated as a function of position in the generalized multilayer structure in Fig. 7.21 following the treatment of Heavens (1991) and later adapted to solve the specific situation of organic detectors (Pettersson et al., 1999, Peumans et al., 2003b). Generally, we assume that the materials comprising the device are homogeneous and isotropic, and the jth layer of a stack consisting of m layers between a semi-infinite substrate (layer $j = 0$) and air ($j = m+1$) is thus described by a complex index of refraction, viz.

$$\tilde{n}_j = n_j + i\kappa_j, \qquad (7.70)$$

where n_j and κ_j are the real and imaginary parts of the refractive index, respectively. In the harmonic field approximation at frequency ω, we can write

$$\mathbf{F}_j(x,t) = \mathbf{F}_{0j}\exp\left(-\xi_j x - \omega t\right)\hat{\mathbf{z}}, \qquad (7.71)$$

where $\xi_j = (2\pi/\lambda)\,\tilde{n}_j$ is the complex propagation constant. Since we are only concerned with the spatial distribution of the field, we can omit the time dependence of $\mathbf{F}$. Furthermore, since we assume that the

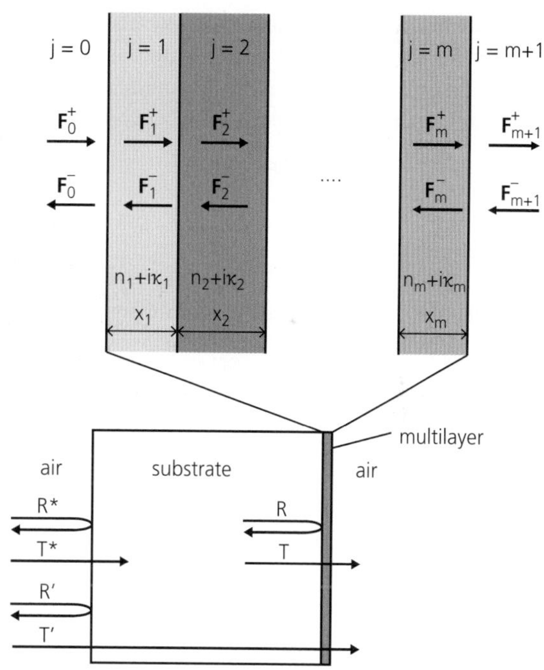

Figure 7.21 Schematic of a multilayer stack with light incident in the $+x$ direction from the left. Superscripts + and − refer to propagation in the $+x$ and $-x$ directions, respectively.

structure is semi-infinite in the plane normal to the propagation direction, x, we do not need to distinguish between the polarization directions of the field. In the following treatment, we use the convention that a + (−) superscript represents propagation in the $+x$ ($-x$) direction. The thickness of the jth layer is x_j, and light is incident on the substrate in the $+x$ direction. The interfaces between layers are assumed flat on the scale of the wavelength.

In the *transfer matrix formalism*, propagation of light is described by 2×2 matrices subject to the continuity of the tangential component of the electric field at an interface. At the interface between layers j and k, the optical field is described by the interface matrix I_{jk}:

$$\begin{bmatrix} \mathbf{F}_j^+ \\ \mathbf{F}_j^- \end{bmatrix} = I_{jk} \begin{bmatrix} \mathbf{F}_k^+ \\ \mathbf{F}_k^- \end{bmatrix} = \frac{1}{t_{jk}} \begin{bmatrix} 1 & r_{jk} \\ r_{jk} & 1 \end{bmatrix} \begin{bmatrix} \mathbf{F}_k^+ \\ \mathbf{F}_k^- \end{bmatrix}, \quad (7.72)$$

where r_{jk} and t_{jk} are the complex Fresnel reflection and transmission coefficients, respectively, given by

$$r_{jk} = \frac{(\tilde{n}_j - \tilde{n}_k)}{(\tilde{n}_j + \tilde{n}_k)} \quad (7.73a)$$

and

$$t_{jk} = \frac{2\,\tilde{n}_j}{(\tilde{n}_j + \tilde{n}_k)}. \quad (7.73b)$$

The treatment thus far has considered reflection and transmission at each interface. We also need to account for the phase shift accumulated as the field propagates through a layer of thickness x_j. This is expressed in terms of the phase matrix given by

$$L_j = \begin{bmatrix} e^{-i\xi_j x_j} & 0 \\ 0 & e^{i\xi_j x_j} \end{bmatrix}, \quad (7.74)$$

where $\xi_j x_j$ is the phase shift.

Thus, the field incident on the layer stack from the substrate, $j = 0$, is related to the transmitted field at $j = m+1$ by the scattering matrix, $\mathbf{S}$:

$$\begin{bmatrix} \mathbf{F}_{m+1}^+ \\ 0 \end{bmatrix} = \mathbf{S} \begin{bmatrix} \mathbf{F}_0^+ \\ \mathbf{F}_0^- \end{bmatrix} = \prod_{l=1}^{m} I_{(l-1)l} L_l \begin{bmatrix} \mathbf{F}_0^+ \\ \mathbf{F}_0^- \end{bmatrix}. \quad (7.75)$$

Here, $\mathbf{F}_{m+1}^- = 0$ implies that there is no reflected wave at the end of the multilayer stack. The convenience of the transfer matrix method is that the field exiting from an arbitrary layer system is found simply by calculating the Fresnel coefficients at each interface and the phase change within each layer, and then taking the products of their respective 2×2 matrices with the incident field. Then, with

$$\mathbf{S} = \begin{bmatrix} S_{11} & S_{12} \\ S_{21} & S_{22} \end{bmatrix}, \quad (7.76)$$

we obtain the total reflection (r) and transmission (t) coefficients of the stack:

$$r = \frac{\mathbf{F}_0^-}{\mathbf{F}_0^+} = \frac{S_{21}}{S_{11}} \quad (7.77a)$$

and

$$t = \frac{\mathbf{F}_{m+1}^+}{\mathbf{F}_0^+} = \frac{1}{S_{11}}. \quad (7.77b)$$

The absorption efficiency of the multilayer structure is equal to the fraction of the light neither transmitted nor reflected, which is given by

$$\eta_A = 1 - T - R = 1 - \frac{|t|^2 \tilde{n}_{m+1}}{\tilde{n}_0} - |r|^2, \quad (7.78)$$

where T and R are the transmissivity and reflectivity, respectively. Thus far, we have assumed that the substrate is infinitely thick. In fact it is only optically thick, that is, $x_{sub} \gg \lambda$, and is typically ~0.1–1 mm. Hence, the effect of the substrate is included by correcting T and R for reflections at the air/substrate and substrate/multilayer interfaces rather than by including it directly in the transfer matrix calculation. This gives substrate reflectivities and transmissivities for light incident from air with $\tilde{n} = 1$ of (cf. Eq. 7.26):

$$R_{sub} = \frac{(1 - n_{sub})^2}{(1 + n_{sub})^2} \quad (7.79a)$$

and

$$T_{sub} = \frac{4}{(1 + n_{sub})^2}, \quad (7.79b)$$

where we have made the reasonable assumption that the substrate is non-absorbing (i.e. $\tilde{n}_{sub} = n_{sub}$). This gives a total reflectivity and transmissivity of the multilayer structure, including the substrate, of

$$R' = \frac{R_{sub} + R}{1 + R_{sub} R} \quad (7.80a)$$

and

$$T' = \frac{T_{sub} T}{1 + R_{sub} R}, \quad (7.80b)$$

yielding the total absorption efficiency of $\eta_A = 1 - T' - R'$.

Since we are ultimately only interested in absorption within the active region (presuming that other layers do not generate excitons that can reach a D-A interface, and hence cannot contribute to the photocurrent), we need to calculate the *local* electric field within the layers comprising the dielectric stack. Let us assume that we are interested in the field within layer, j. Then we can separate the system into three parts: those layers in the regions $0 < x < x'$, the layer j itself, and those at $x > x' + x_j$, where x' is the sum of the thicknesses of the layers x_0 to x_{j-1}. Then the scattering matrix of the entire structure is given by

$$\mathbf{S} = \mathbf{S}_j^- L_j \mathbf{S}_j^+, \quad (7.81)$$

with

$$\mathbf{S}_j^- = \left(\prod_{l=1}^{j-1} I_{(l-1)l} L_l\right) I_{(j-1)j} \tag{7.82a}$$

and

$$\mathbf{S}_j^+ = \left(\prod_{l=j+1}^{m} I_{(l-1)l} L_l\right) I_{m(m+1)}. \tag{7.82b}$$

The electric field propagating in the positive direction in layer j at the left interface is related to the incident plane wave via

$$t_j^+ = \frac{\mathbf{F}_j^+}{\mathbf{F}_0^+} = \frac{S_{j11}^+}{S_{j11}^+ S_{j11}^- e^{-i\xi_j x_j} + S_{j21}^+ S_{j12}^- e^{i\xi_j x_j}}, \tag{7.83}$$

and similarly for the electric field propagating in the negative direction at the left interface:

$$t_j^- = \frac{\mathbf{F}_j^-}{\mathbf{F}_0^+} = t_j^+ \frac{S_{j21}^+}{S_{j11}^+} e^{i2\xi_j x_j}. \tag{7.84}$$

Finally, we arrive at the total electric field at an arbitrary position inside layer j as the linear superposition of the fields propagating in opposite directions, and whose intensities are modified by the phase at x:

$$\frac{\mathbf{F}_j(x)}{\mathbf{F}_0^+} = t_j^+(x) + t_j^-(x) = t_j^+ e^{i\xi_j x} + t_j^- e^{-i\xi_j x}. \tag{7.85}$$

The time-averaged *Poynting vector* gives the power as a function of position, which for the harmonic field in Eq. 7.71 yields the local generation rate:

$$G_j(x) = \frac{2\pi\varepsilon_0 \kappa_j n_j}{h} |\mathbf{F}_j(x)|^2, \tag{7.86}$$

which is related to $\eta_A P_{inc}$ via Eq. 7.3. This differs from the simple model of absorption calculated using the Beer–Lambert relationship since it considers reflection and absorption within each layer of the dielectric stack in the OPD.

To calculate η_{ext}, we must also consider exciton diffusion, that is, we need to calculate the probability, η_{ED}, that an exciton generated at x will diffuse to a D-A junction where charge transfer followed by carrier generation can occur. The continuum steady state exciton diffusion equation for the jth layer is, therefore,

$$L_{Dj}^2 \frac{\partial^2 N_j(x)}{\partial x^2} - N_j(x) + \tau_{Dj} G_j(x) = 0, \tag{7.87}$$

where $\tau_{Dj} = L_{Dj}^2 / D_j$ is the exciton diffusion lifetime in layer j, D_j is its diffusion coefficient, and $N_j(x)$ is the exciton density.

Equation 7.87 is solved using two possible boundary conditions: (i) ideal non-quenching interfaces, and (ii) quenching interfaces with an infinite surface recombination velocity. The first boundary condition, corresponding to $\partial N_j(x)/\partial x = 0$, applies to excitons in a narrow energy gap organic semiconductor at a heterointerface whose LUMO and HOMO levels are nested within the LUMO and HOMO of the contacting material (type I HJ), prohibiting exciton dissociation by via charge transfer. This case is clearly undesirable for OPDs. The second boundary condition, corresponds to $N_j(x) = 0$, and applies to type II D-A HJs where all excitons dissociate by CT and contribute to the photocurrent, as well as to organic/metal contacts where excitons non-radiatively recombine without contributing to the photocurrent. The photocurrent can then be calculated from the exciton diffusion current evaluated at the D-A interface:

$$j_{ph,j} = qD_j \left|\frac{\partial N_j(x)}{\partial x}\right|_{x=x_{HJ}}, \tag{7.88}$$

where x_{HJ} is the position of the heterointerface. Then, the external efficiency from layer j leading to charge transfer is given by the ratio of the photocurrent to the incident photon flux:

$$\eta_{ext,j} = 2h \frac{j_{ph,j}}{q\varepsilon_0 \lambda |\mathbf{F}_0^+|^2}, \tag{7.89}$$

where we have assumed that $\eta_{CT}\eta_{CC} \approx 1$, which is often the case for the most efficient D-A HJ photodetectors. Now, we have seen that $G_j \propto |\mathbf{F}_0^+|^2$, and thus $j_{ph,j} \propto |\mathbf{F}_0^+|^2$. Hence the external quantum efficiency is independent of the incident optical power in the linear range of OPD operation, that is, where series resistance or other second order effects such as charge annihilation are small.

The external quantum efficiency in stratified structures is determined by first calculating $|\mathbf{F}_j(x)|^2$ throughout the multilayer as shown in Fig. 7.22a for the archetype device: ITO anode/32 nm PEDOT:PSS hole injection layer (HIL)/15 nm CuPc donor/35 nm C_{60} acceptor/16 nm BCP exciton blocking layer (EBL)/100 nm Al cathode at $\lambda = 600$ nm. Optical constants for each layer are provided in the table in Fig. 7.23. The local generation rate is calculated using the electric field and the optical properties of the active layers, as shown in Fig. 7.22b. Finally, using L_D determined from independent measurements such as spectrally resolved photoluminescent quenching (see Section 3.9, Lunt et al., 2009 and Bergemann and Forrest, 2011) the diffusion equation is solved to

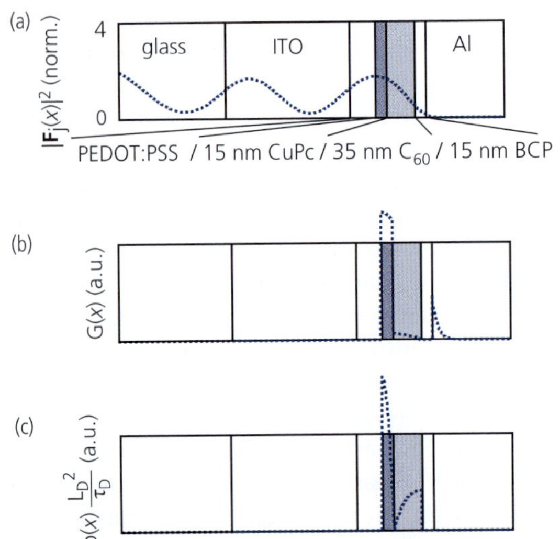

Figure 7.22 Steps used in calculating the external quantum efficiency of an archetype organic photodiode. (a) The OPD structure along with the normalized intensity distribution calculated using the optical constants for each layer provided in Fig. 7.23. (b) The optical absorption profile at $\lambda = 600$ nm, and (c) the exciton profile calculated using the method described in text. All calculations are shown as dashed lines (Peumans et al., 2003b).

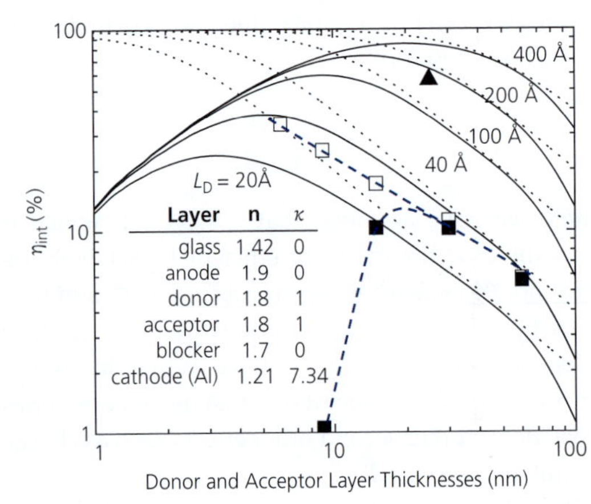

Figure 7.23 Internal quantum efficiency vs. donor and acceptor layer thicknesses for a symmetric organic OPD whose materials parameters are provided in the table. The structure is on a glass substrate coated with a 150 nm thick anode, donor and acceptor layers, a 10 nm thick, transparent exciton blocking layer, and a 150 nm thick Al cathode. The solid lines are calculated for various L_D. The dotted lines assume no absorption at the metal cathode. The open and filled squares correspond to measurements of devices described in text, and the solid triangle is for a glass/ITO/PEDOT:PSS/CuPc/C_{60}/BCP/Ag OPD (Peumans et al., 2003b).

obtain $N(x)$ in Fig. 7.22c. Then, the photocurrent implied from Eq. 7.86 yields η_{ext} (Peumans et al., 2003b).

Extending the analysis for the specific structure in Fig. 7.22, we provide a guide in Fig. 7.23 for designing bilayer OPDs and OPV cells, where η_{int} is used as a performance parameter, since it is a measure of the exciton harvesting efficiency. In these calculations, the wavelength is fixed at $\lambda = 620$ nm. The symmetric device structure is: glass/150 nm anode/x nm donor/x nm acceptor/10 nm spacer/150 nm cathode. For the donor and acceptor layers, $\kappa = 1.0$ was used, which corresponds to $\alpha = 2\pi\kappa/\lambda = 1.0 \times 10^5$ cm^{-1}, which is typical for $\pi^* \leftarrow \pi$ transitions (see Table 7.1). Since this device is symmetrical, the same L_D was used for both the donor and acceptor layers. Of course, accurate modeling of a realistic detector requires precise knowledge of the layer thicknesses, the optical properties (n and κ), and L_D of all materials used. The solid lines are model calculations of η_{int} *vs.* the donor and acceptor layer thicknesses assuming L_D = 2–40 nm. The dotted lines represent a simple exciton diffusion model that does not account for interference or absorption in regions where $x > L_D$, or at metal contacts.

The maximum η_{int} is obtained for donor and acceptor layer thicknesses $\sim L_D$. Thicker layers suffer from absorption in regions far away from the D-A interface. For layers thinner than L_D, absorption in the metal cathode decreases η_{int}, causing a large deviation from the simple model (dotted lines). This latter effect poses a limit to the achievable η_{int} for a given L_D. The data points in the figure show measurements of devices. The solid squares correspond to glass/ITO/CuPc/PTCBI/Ag, and the open squares are for a similar device except that it contains a transparent BCP exciton blocking layer between the cathode and the acceptor. Also shown (triangle) is the result for a device that replaces the acceptor, PTCBI, with C_{60}. The fits to the PTCBI device indicate that its diffusion length is approximately 3 nm. However, we note that the efficiency decreases dramatically for an acceptor layer thickness <10 nm for the device lacking the EBL. This is due to exciton quenching at the cathode when the thickness is less than $\sim 2L_D$. In contrast, the device with the EBL shows an increasing η_{int} as $x_A \rightarrow L_D$. Also, the longer diffusion length of C_{60} results in a concomitant increase in the internal efficiency. In all cases, the optical model yields an accurate value for the device efficiency.

The transfer matrix approach has been used in devices with intentionally strong microcavities that allow for tuning the absorption spectra to enable wavelength-selective detection. Such a spectrally selective OPD has been demonstrated with the following structure (see Fig. 7.24a, An et al., 2009): Al anode (50 nm)/MoO_3 transparent optical spacer (110–220 nm)/CuPc or α-NPD donor (20 nm)/C_{60} acceptor (40 nm)/BCP EBL (7 nm)/Ag cathode

(40 nm). The strong microcavity results from the dielectric layers sandwiched between the reflective Al anode and the semitransparent, reflective Ag cathode through which the illumination is incident. The MoO_3 anode buffer layer thickness is adjusted from 110 nm to 220 nm to tune the cavity to the desired wavelength (An et al., 2009).

Figure 7.24a shows the calculated optical intensity distribution for $\lambda = 520$ nm and 430 nm. The peak intensity within the CuPc/C_{60} active region is at 520 nm, whereas the intensity is weaker at 430 nm. Hence, it is expected that the OPD will have a greater sensitivity at the longer wavelengths. The transfer matrix formalism coupled to the exciton diffusion equation using measured optical constants and L_D has been applied to devices with different MoO_3 thicknesses. Atomic force microscope (AFM) images of the films showed an rms roughness of 14.5 nm for Al and 7.3 nm for Ag. Hence, it is reasonable to expect that films deposited onto the Al surface will replicate at least some fraction of the roughness of the base layer. The calculations, therefore, were modified from those in the foregoing discussion to account for rough interfaces that were generated during layer deposition.

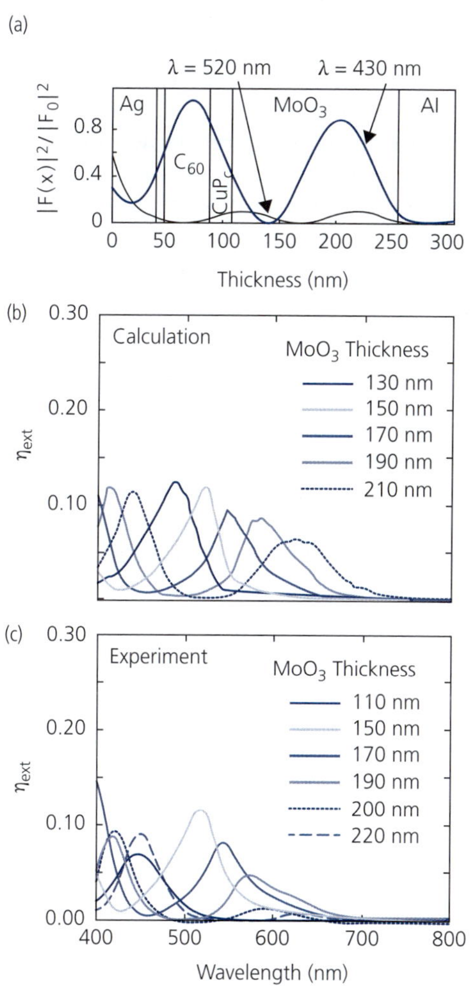

Figure 7.24 (a) Sample OPD structure with a MoO_3 optical spacer/anode buffer layer thickness of 150 nm, and the optical field distribution at two wavelengths of 520 nm (in resonance with the CuPc/C_{60} active region) and 430 nm (off resonance). (b) Simulated and (c) measured external quantum efficiency spectra for devices in (a) for varying MoO_3 spacer thicknesses. Devices with 200 nm and 220 nm thick MoO_3 thicknesses use CuPc instead of α-NPD (An et al., 2009).

Rough surfaces are a common feature of many multilayer organic thin films, and hence it is worthwhile to consider this situation in some detail. The reflection and transmission of a plane wave from a rough surface can be understood from Fig. 7.25a. Rays of wavelength, λ_0, incident from a medium with index of refraction n_0 at angle β are reflected from surface irregularities of height, Δh. The reflected light undergoes a position-dependent phase shift of

$$\Delta\Phi_R = \frac{4\pi\Delta h n_0}{\lambda_0}\cos\beta. \tag{7.90}$$

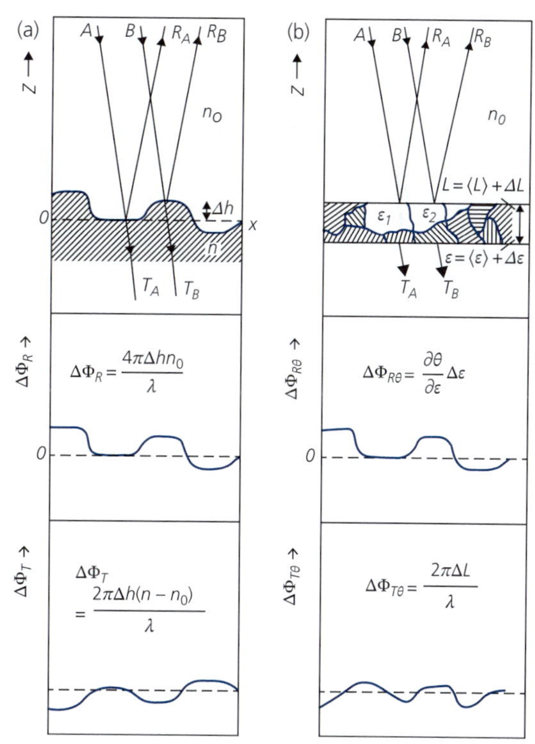

Figure 7.25 (a) Reflection and transmission of a pair of rays, A, B, that are incident at angle β to the surface normal. The middle and lower panels show variations in phase vs. position for the reflected (*R*) and transmitted (*T*) waves. (b) As in (a) for a heterogeneous medium with domains of different dielectric constants, ε_1 and ε_2 (Filiński, 1972).

For a surface with a Gaussian distribution of surface roughness, the reflection coefficient is thus modified (Filiński, 1972):

$$R'(\beta) = R_s(\beta)\exp\left(-\Delta\Phi_R^2\right) = R_s(\beta)\exp\left[-\left(\frac{4\pi\sigma n_0}{\lambda_0}\cos\beta\right)^2\right], \tag{7.91}$$

where the rms surface roughness is given by

$$\sigma = \left(\sum \Delta h^2\right)^{1/2}. \tag{7.92}$$

Here, R_s is the reflection coefficient obtained for a specularly smooth surface. Using similar reasoning for a transmitted wave (see Fig. 7.25a), we have a phase shift of

$$\Delta\Phi_T = \frac{2\pi\Delta h}{\lambda_0}(n - n_0)\cos\beta \tag{7.93}$$

to yield a modified transmission coefficient of

$$\begin{aligned} T'(\beta) &= T_s(\beta)\exp\left(-\Delta\Phi_T^2\right) \\ &= (1 - R_s(\beta))\exp\left[-\left(\frac{2\pi\sigma(n - n_0)}{\lambda_0}\cos\beta\right)^2\right]. \end{aligned} \tag{7.94}$$

Film composition heterogeneity, that is, where the composition of the medium is spatially variant such as in a D-A blend or BHJ, can also be present. Spatial variations in dielectric properties due to domains with different crystal orientation, densities, or even from a high, spatially varying impurity concentration may exist. This introduces a heterogeneous dielectric medium that comprises domains of differing dielectric constants, ε_1, ε_2,..., as shown in Fig. 7.25b. Rays traveling through different regions of the film will therefore undergo phase shifts similar to that experienced from a rough surface. Then, $\Delta\Phi_\theta$ is the rms fluctuation in phase due to a distribution of dielectric constants, viz.

$$\Delta\Phi_\theta = \frac{2\pi L}{\lambda}\left[\sum\left(\frac{\Delta n}{n^2}\right)^2\right]^{1/2}, \tag{7.95}$$

where L is the total layer thickness in Fig. 7.25b. In absorptive media, the phase shift due to variations in both the real and imaginary indexes of refraction (Δn) are included. Then in analogy to our treatment for roughness, we modify the reflection coefficient by:

$$R'_\theta = R_s\exp\left(-\Delta\Phi_\theta^2\right). \tag{7.96}$$

A similar expression can be derived for the transmission coefficient, T'_θ. Finally, both surface roughness and material inhomogeneities are incorporated into the expressions used in the transfer matrices by adding their respective contributions to the phase shifts encountered in propagating through the ith layer.

Returning to Fig. 7.24b, the external quantum efficiency is calculated for devices with various MoO_3 buffer layer thicknesses, incorporating the effects of surface roughness imparted by the Al anode. The strong microcavity arising from the active semiconductor layers sandwiched between two reflective metal contacts results in narrow peaks, each at a different wavelength depending on the buffer layer thickness. For the thickest buffer layers, there are two maxima, one at short and the other at longer wavelength. These dual peaks correspond to two intensity maxima, one at $\lambda/4n$ and the second at 3 $\lambda/4n$ lying within the D-A layers. The experimental results in Fig. 7.24c are reasonably close to predictions, although there are some notable differences, particularly at long wavelengths. This difference is due to inaccuracies in the layer refractive indexes which were taken from literature values rather than being directly measured (An et al., 2009). Note that the devices with MoO_3 thicknesses of 200 nm and 220 nm (dashed lines) used α-NPD in place of CuPc as the donor to enhance the short wavelength absorption.

One feature of microcavities discussed in Chapter 6 is their sensitivity to light incidence angle and polarization. This is indeed the case for the devices in Fig. 7.24, where wavelength peak shifts as large as 40–50 nm are observed as the angle is varied from 0° (normal incidence) to 60° (An et al., 2009). This effect is undesirable if the incidence angle is not fixed, whereas it can be used to detect the incidence angle if the source wavelength is known.

7.1.5 Modeling efficiency: dependence on film morphology

Thus far we have considered only charge generation and the optical fields in planar, bilayer heterojunctions (PHJs). However, we have shown that efficient detectors use BHJs or mixed HJs. The characteristics of such entangled structures (cf. Fig. 7.17) must be accurately modeled to optimize materials and architectures for improved performance. Developing morphological models is more straightforward for small molecules deposited from the vapor phase than for solution processed films. In the former case, the dynamics of domain formation is based only on the intermolecular interaction energies at the substrate surface, along with the particular kinetics of the growth process that involves non-interacting particles in the vapor phase. More complex solution phase interactions along with undefined molecular shapes govern polymer morphologies. Achieving a desired morphology is ultimately an experimental process with a good degree of trial and error. A few such processes and the resulting morphologies achieved for films deposited from solution are described in Section 7.4.2. Here we focus on the less complicated problem of modeling mixtures deposited from the vapor phase (i.e. by either VTE or OVPD). This treatment provides a generalized methodology for understanding how the foregoing discussion of PHJs informs the more complex treatment of blends.

The basic elements for "constructing" a given small molecule morphology is to assume that the blends consist of solids predominantly bonded by van der Waals forces. Then the structure can be modeled using the atom–atom potential model developed in Section 2.5. In the following, we describe a general *dynamic Monte Carlo* (DMC) energy-based simulation method introduced by Adams et al. (1993) and then adapted to the specific problem of mixed small molecule HJ domain formation during growth (Watkins et al., 2005, Yang and Forrest, 2008). Variations on this *molecular dynamics simulation* procedure have been used to understand morphologies in OPVs, OLEDs, transistors, and numerous other thin film structures.

We start by assuming the organic molecules are arranged on a cubic (Watkins et al., 2005, Yang and Forrest, 2008) or a face-centered cubic lattice to account for molecular asymmetries, allowing extended molecules to occupy two lattice sites (Ding et al., 2018). The lattice constant is a, which is on the order of a molecular diameter (typically ~1 nm). The lattice sizes discussed below are $(100a)^2$ in the substrate (x–y) plane, with varying depth in the substrate normal (z) direction. Periodic boundary conditions are applied. The bottom and top of the lattice are assumed to contact the anode and the exciton blocking layer, respectively.

Phase separation from a *homogeneously* mixed active region is generated using an entropy-driven site spin-exchange Ising model (Adams et al., 1993) whose parameters are adjusted to reproduce the observed grain size and distribution for binary blends. Lattice sites are allowed to exchange their spins between all 26 proximate neighbors (6 in the <100> direction, 12 in the <110> direction, and 8 in the <111> direction). Since the van der Waals interaction between like molecules is generally larger than the interaction between different molecules, the change in system energy that ensues from the exchange of two neighboring molecules (spins) is represented using the Ising Hamiltonian for the energy of site, i:

$$\varepsilon_i = -\frac{J}{2}\sum_j \left(\delta_{S_i,S_j} - 1\right), \tag{7.97}$$

where δ_{S_i,S_j} is the Kronecker delta for spins occupying sites i and j. The coupling between sites is given by J, which is arbitrarily set to +1. The summation is over first and second nearest neighbor sites, with the energetic contribution decreased by $\sqrt{2}$ for the latter. Exchange attempts between sites i and j are accepted with the probability $P(\Delta\varepsilon)$. This is thermally activated according to (Kawasaki, 1972)

$$P(\Delta\varepsilon) = \frac{\exp(-\Delta\varepsilon/k_BT)}{1+\exp(-\Delta\varepsilon/k_BT)}, \tag{7.98}$$

where the energy difference between sites i, j is $\Delta\varepsilon$. Here, T is in units of J/k_B. This procedure leads to large-scale phase separation occurring in a system with limited miscibility. The accepted spin exchanges tend to form aggregates of like molecules by lowering the total system energy, while the grain size is controlled by the number of executed spin exchanges.

The growth of more complex D-A nanocrystalline networks is modeled assuming growth by alternate deposition of ultrathin layers of D and A molecules (Yang et al., 2005b, Ding et al., 2018), taking into account gas phase transport, surface diffusion, and evaporation back into the gas phase. The structural morphology is simulated based on the atom–atom potential method. Recall from Section 2.5 that the total bond potential between neighboring molecules i, j is $U(R_{ij}) = \frac{1}{2}\sum_{m,n}\phi_{m,n}(R_{ij})$. Here, $R_{ij} = |\mathbf{R}_i - \mathbf{R}_j|$ is the relative molecular positions and $\phi_{m,n}(R_{ij})$ is the potential between the mth and nth atoms in different molecules, i and j. Both the London and Buckingham potentials are convenient for large-scale atom–atom modeling of organics. The energy of molecule i, is then obtained by summing over all neighboring molecules j.

It is assumed that the organic molecules are randomly generated in the gas phase at a distance far from the substrate. Once incident on the solid surface, they diffuse from site i to unoccupied site j, at a rate:

$$w_{i\to j} = w_0 f(\Delta\Phi), \tag{7.99}$$

where w_0 is the hopping attempt frequency, and $\Delta\Phi = \Phi_j - \Phi_i$ is the difference in energy of the molecule occupying site i or j. Then,

$$f(\Delta\Phi) = \begin{cases} \exp(-\Delta\Phi/k_BT) & \text{for } \Delta\Phi \geq 0 \\ 1 & \text{for } \Delta\Phi < 0 \end{cases} \tag{7.100}$$

A molecule on the surface can also re-evaporate back into the gas phase where the destination site, j, is above the solid surface, with $\Phi_i = 0$. The dominance of bond energies between like molecules over those between dissimilar molecules results in the formation of large, homogeneous domains.

Morphologies are "constructed" by assuming that the first homogeneous layer consists of D molecules. Following that, a thin A layer is deposited, forming aggregates that do not provide full coverage of the underlying D layer. A second, thin D layer is then deposited to partially cover the A layer, but it also directly contacts the first, continuous D layer. Several such thin, discontinuous D-A layer pairs are sequentially deposited, followed by a thick and continuous capping layer.

To connect the morphology to the charge generation efficiency, exciton and carrier transport within the simulated morphology must also be modeled. We found in Chapter 4 that short-range hopping between localized states provides a reasonable picture of the transport in disordered organic systems. For this purpose, the *first reaction method* (FRM) using the DMC algorithm is applied. In FRM, the probability of an allowed process (e.g. exciton site hopping) is calculated following the first time the exciton or charge appears, and is updated following exciton dissociation, quenching, bimolecular recombination or charge collection. Compared to a full DMC calculation in which the probability is recalculated after each change of system configuration, FRM significantly reduces computation time (Watkins et al., 2005). The FRM calculation is reasonably accurate as long as the particle densities are small, where interparticle interactions can be neglected. This is generally the case for detectors and solar cells under modest illumination intensities. The FRM simulation generates excitons at rates consistent with the incident light intensity and spectrum, and tracks their diffusion to the D-A HJ where they dissociate into free carriers.

Förster energy transfer governs exciton diffusion in materials commonly used for detection. Hence, the hopping time, τ_{EX}, depends on R_{ij}^6 and the difference in site energies, ΔE_{ij}, viz. (Scheidler et al., 1996, Yang and Forrest, 2008)

$$\tau_{EX} = \tau_D \left(\frac{R_{ij}}{a}\right)^6 \frac{log(X)}{f(\Delta E_{ij})}. \tag{7.101}$$

Here, τ_D is the exciton lifetime, and $0 < X < 1$ is a random number. The hopping probability is based on the Miller–Abrahams model in Section 4.3.2:

$$f(\Delta E_{ij}) = \begin{cases} \exp(-\Delta E_{ij}/k_BT) & \text{for } \Delta E_{ij} \geq 0 \\ 1 & \text{for } \Delta E_{ij} < 0 \end{cases}. \tag{7.102}$$

Onsager–Braun theory (Braun, 1984) is used to describe exciton dissociation, where the generated carrier gains energy equal to the difference between the D-A energy offset and the exciton binding energy.

Finally, electron transport is calculated based on effective-medium theory (cf. Fishchuck et al., 2002 and Section 4.3.2). This yields the carrier hopping time:

$$\tau_{ch} = \frac{qa^2\exp(4a[R_{ij} - a])}{6k_BT\mu_{n(p)}} \frac{1}{f(\Delta E_{ij})}, \tag{7.103}$$

where $\mu_{n(p)}$ is the electron (hole) mobility.

This analysis has been used to understand photogeneration in CuPc/C_{60} D-A HJs with six different morphologies: a bilayer, planar HJ (PHJ), a homogeneous D-A mixture, a PM-HJ, an ordered BHJ consisting of pillars with diameters greater than L_D, arranged in a chessboard pattern bounded on top and bottom with a D and A layer, and a random nanocrystalline network. The PHJ cell consists of two homogeneous layers separately comprised of D and A molecules. The PM-HJ cell is comprised of a D-A mixed layer sandwiched between top and bottom continuous homogeneous layers (Yang and Forrest, 2008).

The morphological evolution of an initially homogeneously mixed, 60 nm thick CuPc:C_{60} film that is subsequently annealed by calculating the structure after a specified number of DMC steps (i.e. "spin exchanges") is provided in Table 7.3. The average domain size is expressed in units of lattice constants, [a], and the specific interface area in units of [a^2]. The specific interface area is the ratio of the total calculated D-A interface area to its projection onto the substrate plane. The homogeneous mixture has a domain size of a, equal to the volume occupied by a single molecule, and a specific interface area $>60a^2$. As annealing proceeds to 100 steps, the domain volume grows by segregation of the components, finally leading to a domain size of $10a$ and a much smaller projected area of $6a^2$.

When the top surface is unconstrained (i.e. it is free to develop 3D structure), the simulations predict roughening during annealing that increases with layer thickness, consistent with phenomena observed for thermally annealed CuPc/PTCBI mixed junctions (Peumans et al., 2003a). Roughening is shown in Fig. 7.26 for the simulated growth of a sample consisting of a stack of alternating CuPc

Table 7.3 Average domain sizes, specific interface areas and cell efficiencies of mixed films before and after annealing for different numbers of DMC steps for a $100\times100\times60$ nm^3 lattice. The shaded areas are one of the two components (D or A), and the "empty" areas are occupied by the opposite component (Yang and Forrest, 2008)

DMC steps	**0**	**1**	**10**	**50**	**100**
Morphology			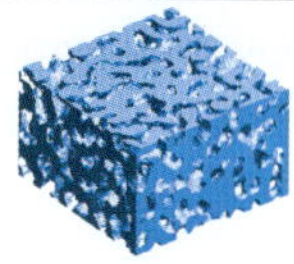		
Domain size [a]	1.0	2.0	4.9	7.9	10.3
Specific interface area [a^2]	61.2	29.7	12.2	7.6	6.0
η_{CC}	0.31	0.34	0.40	0.44	0.43
η_{ED}	1.00	1.0	0.97	0.93	0.87
η_{int}	0.31	0.34	0.39	0.41	0.37

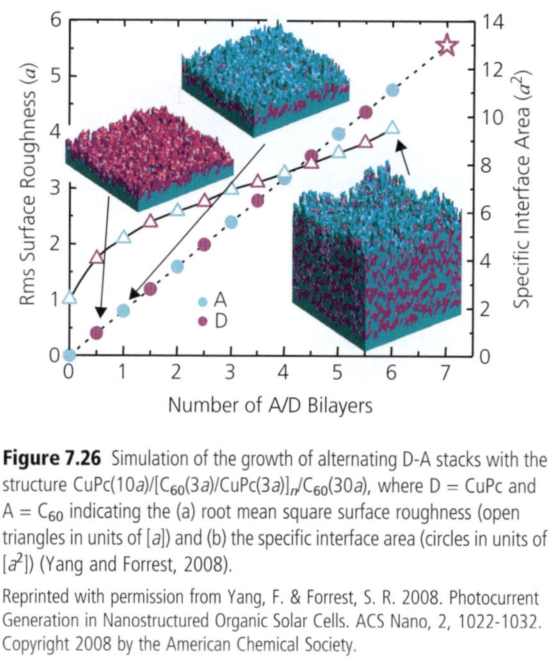

Figure 7.26 Simulation of the growth of alternating D-A stacks with the structure CuPc(10a)/[C_{60}(3a)/CuPc(3a)]$_n$/C_{60}(30a), where D = CuPc and A = C_{60} indicating the (a) root mean square surface roughness (open triangles in units of [a]) and (b) the specific interface area (circles in units of [a^2]) (Yang and Forrest, 2008).

and C_{60} layers as follows: CuPc(10a)/[C_{60}(3a)/CuPc (3a)]$_n$/C_{60}(30a), where n is the number of bilayers. Surface roughening is a general property of nanocrystalline growth, although its magnitude depends on the particular materials systems and growth dynamics involved.

By combining the results of morphological simulations with the charge and excitonic dynamics of Eqs. 7.101–7.103, we can determine the internal efficiency of the mixed or BHJ device. Moreover, the external efficiency of the detector is found by including the physical dimensions of the device with the optical constants (n, κ) of the layer stack. In Table 7.3, for example, we use the exciton diffusion and charge collection efficiencies to arrive at the internal quantum efficiency (recall $\eta_{int} = \eta_{ED}\eta_{CT}\eta_{CC}$) as a function of morphology obtained through mixed HJ annealing. In these calculations, it has been assumed that the charge transfer efficiency, η_{CT}, is unity. As we have seen in this and previous chapters, however, this is not generally true, particularly for junctions where ΔE_{HL} is small. Nevertheless, it is straightforward to include this factor into the calculation of η_{int} based on discussions from Chapter 4 and Section 7.4.

Simulated efficiencies of five CuPc/C_{60} morphologies are shown in Fig. 7.27. For these calculations the following materials parameters were employed for CuPc and C_{60}, respectively: $L_D = 15$ nm, 40 nm; $\tau_D =$ 10 ns, 1 μs, and $\mu_p = 7 \times 10^{-4}$ cm^2/V s and $\mu_n = 5 \times 10^{-2}$ cm^2/V s. As expected, the exciton diffusion efficiency of the planar HJ decreases exponentially with d, although its charge collection efficiency is always close to 100%. The opposite holds true for the homogenous mixture, were charge collection is impeded by the lack of percolating conductive paths, yet every photogenerated exciton has a neighboring counter molecule that results in $\eta_{ED} = 1$. The planar mixed device shares properties of both of these extremes. Unsurprisingly, the external quantum efficiency (and hence j_{ph}) is maximized for an ideal chessboard morphology, or for a nanocrystalline network. We have shown in Section 5.4.2 that a chessboard structure can approximately be achieved by the strained growth of CuPc nanopillars on ITO using OVPD (Yang et al., 2005a). Nanocrystalline networks, although more random than chessboards are more readily achieved during growth and yet deliver nearly equivalent performance. Finally, the analysis provides information on the optimal active layer thicknesses for a given materials system. It is apparent that the efficiency saturates at thicknesses between 60 nm and 80 nm, which indeed is close to that used in OPVs comprising small molecule active regions.

It is encouraging that experimental results summarized in Table 7.4 are in agreement with the simulations. In particular, the agreement between the transfer matrix calculations of the optical field intensity, and the exciton diffusion theory has proven this analysis to be invaluable in accurately predicting the design and performance of a wide range of highly complex planar, bulk HJ, mixed HJ, and multi-element stacked organic detectors.

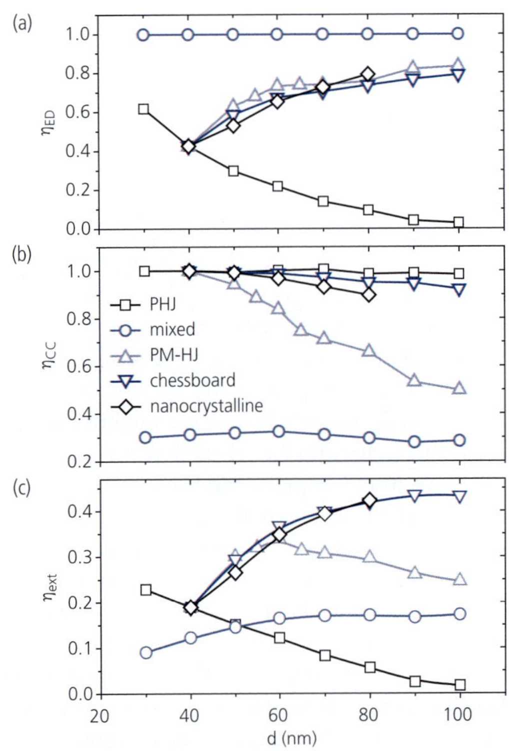

Figure 7.27 Simulations of five CuPc/C_{60} HJ morphologies yielding the (a) exciton diffusion, (b) charge collection and (c) external quantum efficiencies as functions of the active layer thickness, *d*. The PHJ cell has the structure of CuPc(d_0 nm)/C_{60}(d_0 +20 nm), where $2d_0 + 20$ nm $= d$. The PM-HJ, chessboard structures comprises a 10 nm thick bottom CuPc layer and a 30 nm thick top C_{60} layer. The nanocrystalline structure is CuPc (10 nm)/[C_{60}(3 nm)/CuPc(3 nm)]$_n$/C_{60}(30 nm), where $n = 0, 2, 4, 6$, and 8. The square pillar width in the chessboard structure is 5 nm (Yang and Forrest, 2008).

7.2 Organic photoconductors and photodiodes: properties and examples

Having considered the theoretical underpinnings of organic photodetection, in this section we give examples of how this property is implemented in devices. We also introduce several examples of photodetector applications that benefit from the unique attributes of organic semiconductors, particularly their flexibility and conformability. We will show that very high sensitivity (i.e. low noise) and bandwidth can be achieved using OPDs demonstrated to span the electromagnetic spectrum from X-rays to the NIR. We begin in Section 7.2.1 with a brief discussion of photoconductive detectors, and then in Section 7.2.2 we move to descriptions of the many and varied implementations of OPDs. The discussion of an organic device that is a hybrid between a detector and a transistor is the organic phototransistor, or OPT. A discussion of OPTs is found in Section 8.5.

7.2.1 Photoconductors

We have found in Chapter 4 that a wealth of knowledge about charge transport and photogeneration in organic semiconductors has been acquired via the study of their photoconductive properties (Borsenberger et al., 1978, Karl et al., 1991, Bässler, 1993). However, there are far fewer examples of organic photoconductors used in device applications. A notable exception is their early application in *electrophotography* (also known as *xerography*). In its most elementary embodiment, the electrophotographic process requires a large photoconductive sheet mounted on a roller with a contact on its surface. The sheet is charged by the application of voltage to the roller electrode. When illuminated, charges are generated and subsequently separated

Table 7.4 Comparison between simulated and observed η_{ext} for various CuPc/C_{60} solar cells under 1 sun intensity, AM1.5G illumination

HJ structure	Simulated η_{ext}	Measured η_{ext}	Ref.
PHJ:	0.13	0.14	(Yang et al., 2007)
CuPc(20nm)/C_{60}(40nm)		0.15	(Peumans and Forrest, 2001, Fujishima et al., 2009)
Mixed HJ:	0.16	0.16	(Sullivan et al., 2004)
CuPc:C_{60}(1:1, 60 nm)			
Nanocrystalline network HJ:	0.42	0.43	(Yang et al., 2007)
CuPc(10 nm)/[C_{60}(3 nm)/CuPc(3 nm)]$_6$/C_{60}(30 nm)			

by the resulting field. Charges migrate to the surface of the sheet, neutralizing the photogenerated charge in the illuminated areas. Thus, the sheet carries an electrostatic, latent image that is subsequently decorated by colored particles charged opposite to the latent charge on the surface. This converts the electrostatic image to a permanent image once the particles are "fixed" to the surface (i.e. via heating). For a comprehensive description of organic photoconductors used in electrophotography, the reader is referred to reviews by Law (1993) and Weiss and Abkowitz (2010). It is worth noting that electrophotography provided the materials, and in many ways inspired early successful demonstrations of efficient OLEDs and OPVs at Kodak in the 1980s (Tang, 1986, Tang and VanSlyke, 1987), and that ultimately led to our current "organic electronic revolution."

The electrophotographic configuration differs from that of the discrete organic photoconductive (OPC) device that converts light to a photocurrent. Due to their inferior performance (i.e. low bandwidth and high noise) compared to photodiodes, there are few reports of OPCs developed for sensitive detection purposes. However, these deficiencies are partially counterbalanced by photocurrent gain, and the relative simplicity of the OPC.

There is considerable confusion that leads to misidentification of photoconductors as photodiodes (Aihara et al., 2003, Campbell and Crone, 2007, 2009, Hokuto et al., 2011). It is therefore important to once more articulate their distinct attributes. (i) A photoconductor does not have a junction. This results in symmetrical, Ohmic characteristics under an applied voltage of both polarities. That is, $j \propto V$ over the full range of normal operation. A diode-like, rectifying characteristic indicates the presence of a junction arising from the presence of two different semiconductors (i.e. a HJ), or at the contacts (a Schottky barrier). Indeed, asymmetrical rectifying characteristics are an unmistakable feature of an OPD. (ii) Photocurrent gain is observed in an OPC but not in an OPD. In Section 7.1, we noted that the unambiguous identification of gain can be problematic, given that the photocurrent is proportional to $g\eta_{ext}$. If η_{ext} is small, the product $g\eta_{ext}$ may be less than unity. Hence, to ensure that there is photoconductive gain, a comparison of the device response with a calibrated detection standard with a known efficiency must yield $g\eta_{ext} > 1$. Otherwise it is possible to assume that the external efficiency is very small, resulting in a gain that is mistakenly inferred to be large, and still arrive at $g\eta_{ext} < 1$. (iii) The photocurrent is proportional to V_a in the OPC, whereas in an OPD, j_{ph} is independent, or at best weakly dependent on voltage. Recall from Eq. 7.7 that the OPC photocurrent gain follows $g \propto V_a$. Hence, $j_{ph} \propto g\eta_{ext} \propto V_a$. In contrast, $g = 1$ at all voltages for an OPD. (iv) Lacking a rectifying junction, an OPC does not exhibit a photovoltaic effect (cf. Fig. 7.17). That is, its j-V characteristics intercept the origin both in the dark and under illumination. Thus, an OPC cannot be used as a photovoltaic power source. By meeting all four of these performance criteria, an OPC can be clearly distinguished from a photodiode.

A device with all of the attributes of a photoconductor is the interdigitated structure in Fig. 7.28a. The OPC uses a conventional lateral geometry where the semiconductor (a 200 nm thick pentacene layer) is deposited by VTE onto the surface of a 3×3 mm^2 interdigitated Cr/Au electrode pattern, with an electrode "finger" width of 10 µm separated by 20 µm long photoconductive channels. Light is incident onto the electrodes via the glass substrate. The I–V characteristics in the dark and under illumination are shown in Fig. 7.28b. The characteristics are symmetrical, since both electrodes are comprised of the same metal (and hence the same work function). The small nonlinearity in the curves is an indication of a residual extrinsic injection from the contacts. However, this small deviation from Ohm's law is insufficient to significantly affect the photoconductive nature of the device. Under illumination, the increase

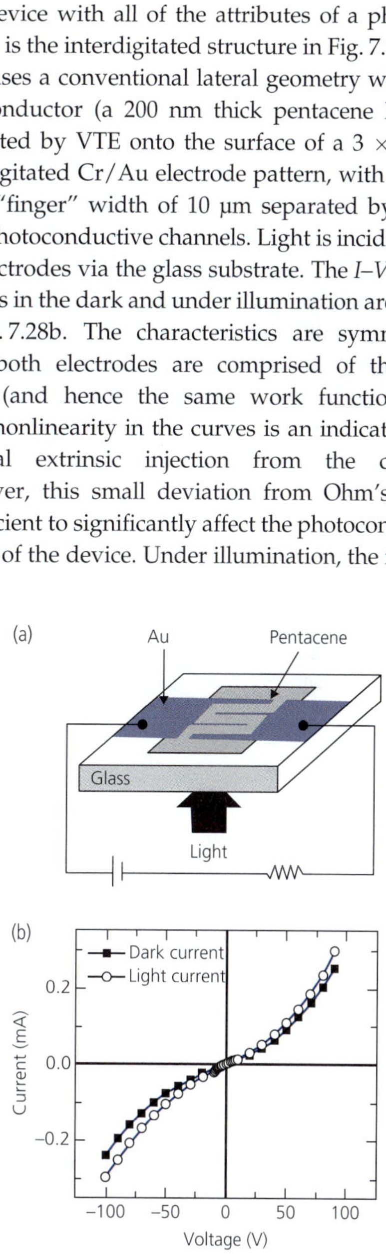

Figure 7.28 (a) Schematic illustration of a pentacene OPC and its test circuit. (b) Current–voltage characteristics of the OPC in the dark and under 0.5 mW illumination at $\lambda = 632$ nm (Gao and Hegmann, 2008).

in current is due to the addition of a photoconductive current, i_{ph}. Note that even in the presence of a photocurrent, the *I-V* curve still intercepts the origin, indicating a lack of a photovoltaic effect, which is consistent with the absence of a rectifying junction.

The operating characteristics of the pentacene OPC are shown in Fig. 7.29. The photocurrent is linear in voltage to $V_a \sim 60$ V (Fig. 7.29a). At higher voltages, the photocurrent increases superlinearly, possibly due to trap filling or heating. Nevertheless, the adherence to Ohm's law over a wide range of voltage is clear.

Figure 7.29b shows the responsivity of the OPC compared with a Si photodiode used to determine the OPC efficiency. Also shown is a line corresponding to a unity external efficiency. Due to the dependence of $\mathcal{R}$ on incident wavelength, the external efficiency has a slope that increases with λ. Importantly, the OPC shows a responsivity that significantly exceeds that for $\eta_{ext} = 1$, providing unambiguous evidence for the presence of photocurrent gain. From the OPC responsivity and analysis of the photocurrent spectra with voltage, it is found that $g = 16$. The gain decreases with increasing optical input power, following a sublinear dependence of $g = \tau_t \mu_e V_a / L^2 [1 + (P/P_0)^n]$, where $n = 0.65$. This dependence is attributed to shallow electron traps, with a trap residence time, τ_t. As power increases, the traps become saturated with charge, and the transit time decreases, resulting in the concomitant decrease in gain.

The presence of a rectifying junction does not rule out the possibility that photoconductivity may also contribute to the photoresponse of a photodiode. Indeed, photogeneration in regions beyond the depleted zones has been quantified for planar phthalocyanine donor/C_{60} acceptor HJ diodes (Jeong et al., 2012, Renshaw et al., 2012). This combination of effects complicates the analysis of the *j-V* characteristics beyond that used to describe the ideal situation where all the charge is generated by excitons arriving and dissociating at the D-A HJ. The effect of photoconductivity on the photocurrent characteristics of the D-A HJ photodiode are illustrated in Fig. 7.30a for a device with structure: ITO/SubPc (13 nm)/C_{60}

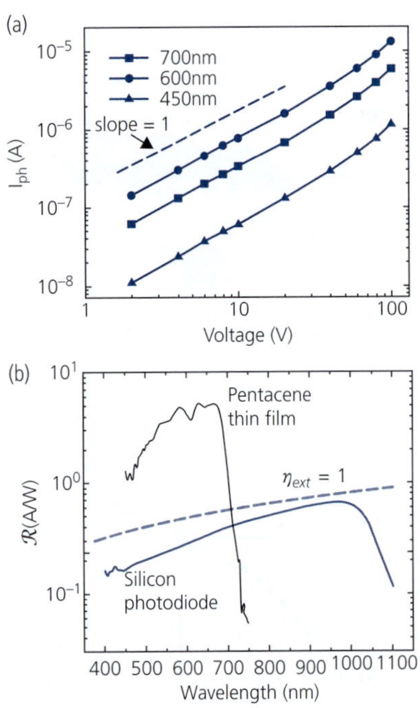

Figure 7.29 (a) Photocurrent–voltage characteristics of the OPC in Fig. 7.28 at three different wavelengths shown in the legend. The dashed line indicates a unity slope. (b) Responsivity vs. wavelength of the OPC, also showing that of a Si photodiode used to determine the OPC efficiency. Dashed line shows the response expected for a unity external quantum efficiency (Gao and Hegmann, 2008).

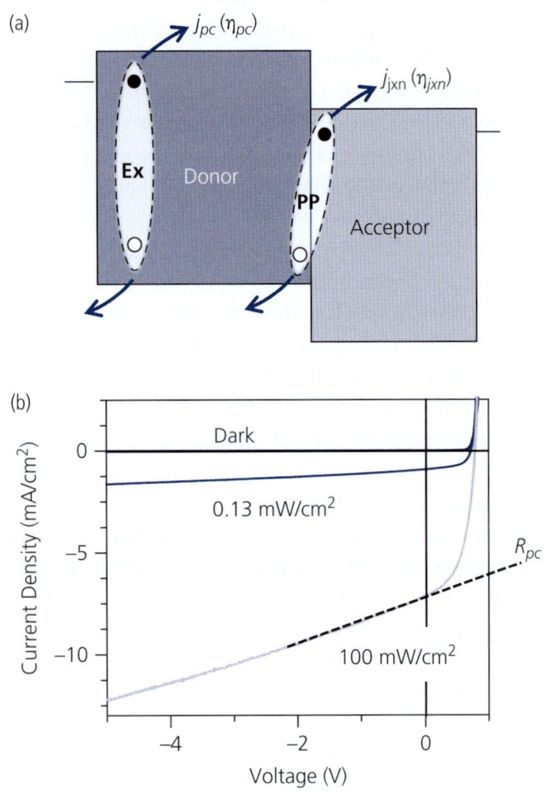

Figure 7.30 (a) Energy level diagram of a HJ photodiode showing two sources of photocurrent: polaron pair (PP) dissociation at the HJ leading to a slope in the reverse-biased photocurrent, j_{jxn}, and exciton (Ex) dissociation in the bulk of the donor or acceptor layer leading to a photoconductive current, j_{pc}. (b) Bipolar *j–V* characteristics of a SubPc (donor)/C_{60} (acceptor) planar HJ photodiode in the dark and under white light illumination at two different incident powers. The slope of the dashed line extrapolation from the linear reverse characteristics gives the photoresistance, R_{pc}. Data from Renshaw et al. (2012).

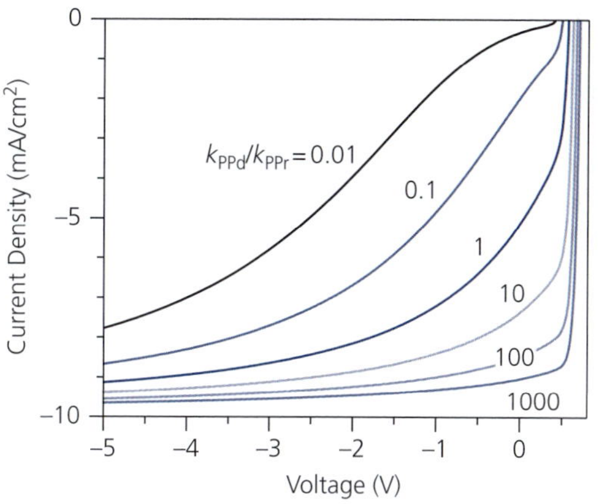

Figure 7.31 Calculated j–V characteristics for an archetype planar HJ OPD under reverse bias up to −5 V under 1 sun (1 kW/m²) illumination. The curves correspond to different ratios of the PP dissociation and recombination rates, k_{PPd}, k_{PPr}, respectively. There is no combination of rates that results in a linear region characteristic of photoconductivity as in Fig. 7.30 (Renshaw et al., 2012).

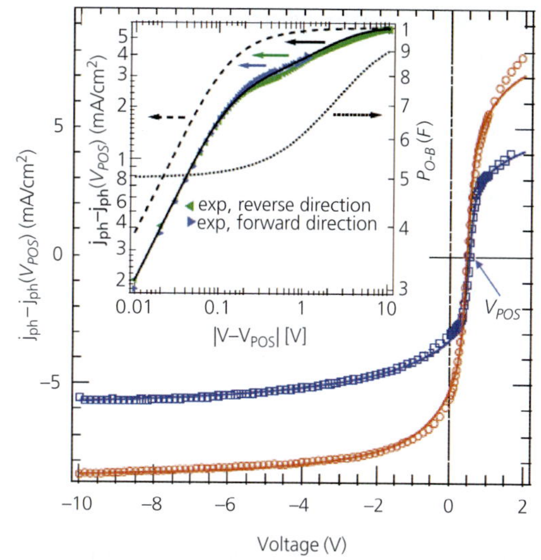

Figure 7.32 Current–voltage characteristics of P3HT:PC_{61}BM BHJ photodiode under 1 sun intensity illumination from a white emitting LED. V_{POS} is the voltage at the symmetry point in the j-V characteristics. Data are points for Ag (circles) and Ca/Al (squares) cathodes, and lines are fits to the data assuming O-B and Sokel–Hughes processes. Inset: Experimental j–V characteristics on a log-log scale of the Ca/Al BHJ, as well as fits to O-B (dotted line), S-H (dashed line), and their combination (solid line). Axis on the right is the O-B dissociation probability (Limpinsel et al., 2010).

(40 nm)/BCP (8 nm)/Ag (100 nm). The two sources of photocurrent are direct exciton dissociation in the bulk with efficiency, η_{pc}, and exciton migration to the D-A junction with efficiency, η_{jxn}. The photoconductive contribution, according to Eq. 7.4, should scale linearly with voltage. In Fig. 7.30b, we show the bipolar j-V characteristics in the dark and under white light illumination. Over a wide range of reverse bias, the current under illumination linearly increases, along with a current contribution due to a photoconductance, R_{pc}^{-1}.

We have found in Section 4.7.2 (cf. Eq. 4.303 and 4.305) that the voltage dependence of the junction current under reverse bias is due to field-induced PP dissociation; an effect observed for both planar and bulk HJs (Mihailetchi et al., 2004, Giebink et al., 2010a, Giebink et al., 2010b, Limpinsel et al., 2010). Contrary to the simple linear dependence on voltage predicted for photoconductivity, field-induced PP dissociation increases approximately exponentially, and can be mistaken for photoconductivity (Braun, 1984). In Fig. 7.31, the reverse j-V characteristics are plotted for an ideal HJ diode in the absence of photocurrent, but for varying ratios of k_{PPd} to k_{PPr}. Over this extended range of reverse bias, there is no region where the curves are linear as in Fig. 7.30 due to the dominance of Onsager–Braun (O-B) dissociation, with its approximately exponential voltage dependence.

PP dissociation via O-B resulting in j-V characteristics similar to those predicted in Fig. 7.31 has been observed in PC_{61}BM acceptor-based BHJ diodes in Fig. 7.32. The characteristics are symmetric about voltage, V_{POS}, between 0.52 and 0.64 V. This is the built-in potential along with a contribution from differences in the contact work functions (either Ca/Al or Ag cathodes, and an ITO anode). The field at the contacts partially depletes the semiconductor, and hence reduces the internal electric field to $F = |V_a - V_{POS}|/d$, where d is the BHJ film thickness. The photogeneration process is due to two processes in these diodes: excitons can either migrate to a P3HT (D)-PC_{61}BM (A) HJ where they form a PP. Alternatively, the large work function/LUMO offsets can separate the PPs at the cathode, yielding a constant photocurrent offset of j_{cont}.

In the bulk, the field-assisted PP dissociation efficiency determined from O-B theory is given by:

$$P_{O-B}(F) = \eta_{PPd}(F) = \frac{k_{PPd}(F)}{k_{PPd}(F) + k_{Prr}} \tag{7.104}$$

Dissociation is followed by charge collection, whose efficiency, η_{CC}, is also dependent on the electric field. That is, the charges can recombine due to collisions with counter charges (e.g. photogenerated electrons with holes) in the entangled BHJ regions where high densities of charges find themselves in close proximity under intense illumination. Charge recombination in insulators prior to collection leads to a collection efficiency of (Sokel and Hughes, 1982)

$$\eta_{CC}(V) = \frac{\exp(qV/k_BT)+1}{\exp(qV/k_BT)-1} - \left(\frac{2k_BT}{qV}\right) \quad (7.105)$$

The voltage dependences of both PP dissociation via O-B in Eq. 4.305 and collection via Eq. 7.105 lead to a photocurrent in both the forward and reverse directions that is described by combining Eqs. 7.104 and 7.105 to obtain (Limpinsel et al., 2010)

$$\begin{aligned} j_{ph} &= qGdP_{O-B}(F)\eta_{CC}(V_a - V_{POS}) + j_{cont} \\ &= qGd\eta_{PPd}(F)\eta_{CC}(V_a - V_{POS}) + j_{cont}. \end{aligned} \quad (7.106)$$

As previously, G is the exciton generation rate.

Results of fitting Eq. 7.106 to the j–V data of the P3HT:PC$_{61}$BM cell are shown by the solid lines in Fig. 7.32. The photocurrent offset due to photogeneration at the contacts leads to a higher j_{cont} for the device with a Ag compared to a Ca/Al cathode. The various contributions to the photocurrent from field-dependent PP dissociation, and from recombination in the bulk, are plotted in the inset for the Ca/Al cathode. Note that near V_{POS}, the j-V response is linear and is indistinguishable from photoconductivity.

Photoconductivity can, indeed, dominate the current in reverse bias (e.g. Fig. 7.30). In this case, we can extend the analysis for an ideal HJ in the presence of traps to obtain

$$\begin{aligned} j = &\, j_{0D}\left[\exp(q(V_a - jAR_{ser})/n_Dk_BT) - \frac{k_{PPd}}{k_{PPd0}}\right] \\ &+ j_{0A}\left[\exp(q(V_a - jAR_{ser})/n_Ak_BT) - \frac{k_{PPd}}{k_{PPd0}}\right] \\ &- j_{jxn} + j_{pc} \end{aligned} \quad (7.107)$$

where, using Eq. 7.4, we obtain the photoconductive contribution to the current in terms of its charge generation efficiency, η_{pc} (Renshaw et al., 2012):

$$\begin{aligned} j_{pc} &= S_{pc}(V_a - jR_{ser} - V_{bi}) \\ &= q\eta_{pc}G(\tau_e\mu_e + \tau_h\mu_h)(V_a - jR_{ser} - V_{bi})/d. \end{aligned} \quad (7.108)$$

The j–V characteristics can be fit by Eq. 7.107, provided that an independent measurement of the photoconductance, $S_{pc} = 1/R_{pc}$, is separately obtained for the D and A layers of the HJ. This is found by measuring the photoconductivities of single layer devices comprising either the D or A layer sandwiched between ohmic electrodes. Figure 7.33 shows one such measurement for an ITO (150 nm)/SubPc (60 nm)/Ag (100 nm) OPC at several different simulated solar illumination intensities. The symmetric characteristics increase linearly for both positive and

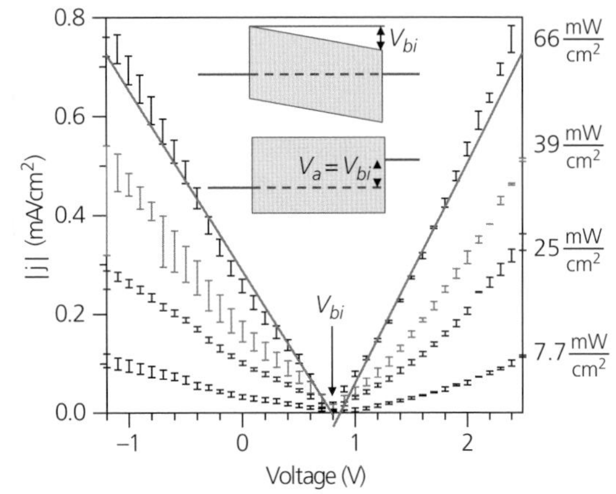

Figure 7.33 Response of an ITO/SubPc/Ag photoconductor to different intensities of simulated solar illumination. The solid line is a linear fit to the data. Inset: Energy level diagrams of the PC at (top) equilibrium and (bottom) flat band. Adapted from Renshaw et al. (2012).

negative V_a, as expected for a photoconductor in the absence of space charge effects or energy level bending. The photocurrent is zero at $V_a = V_{bi} = 0.8$ eV, which is equal to the work function differences of the cathode and anode contacts. Unlike the BHJ in Fig. 7.32, there is no evidence for energy level bending at the contacts. The energy level diagrams for this device at both equilibrium and at flat band, where the absence of an electric field results in $j_{pc} = 0$, are shown in the inset. From these and similar data obtained from a C$_{60}$ OPC, it is found that $S_{pc} = 0.40$ mA/V cm^2 at a light intensity of 66 mW/cm^2 for SubPc and $S_{pc} = 0.82$ mA/V cm^2 at 100 mW/cm^2 for C$_{60}$.

Photoconductivity measurements were used to fit the planar SubPc/C$_{60}$ HJ characteristics in Fig. 7.30, using $S_{pc} = 0.90$ mA/V cm^2 at 1 sun (100 mW/cm^2) intensity, white light illumination. The fits also assume $j_{0A} = 1.0$ pA/cm^2, $j_{0D} = 47$ nA/cm^2, $R_{ser} = 3.7\ \Omega$ cm^2, and ideality factors of $n_A = 1.6$, and $n_D = 5.9$. These parameters, along with $V_{bi} = 0.8$ V then provide a simulation of the illuminated j–V data shown by the solid line in Fig. 7.34. The dashed line only includes effects of $k_{PPd}(V_a)$ from Onsager–Braun dissociation that does not account for the observed reverse-biased voltage dependence. By including the additional current due to photoconductivity, the simulation and the device data are in agreement over the entire range of measurement, indicating efficient PP dissociation and the importance of photoconductivity in these archetype OPV structures. The presence of photoconductance results in an increase in j_{sc} by 15% in the SubPc/C$_{60}$ devices. Hence, in

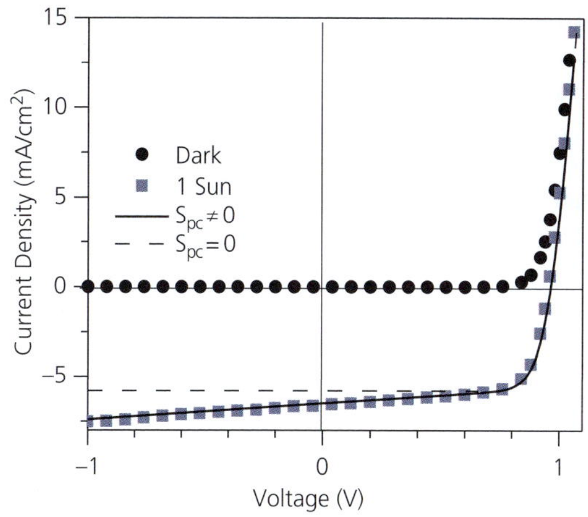

Figure 7.34 Current–voltage characteristics in the dark (circles) and under 1 sun intensity, simulated solar illumination (squares) of the D-A HJ in Fig. 7.30. The dashed line is a fit to Eq. 7.55 in the absence of photocurrent (S_{pc} = 0), and the solid line is includes S_{pc} as independently measured for SubPc and C_{60} photoconductors in Fig. 7.33 (Renshaw et al., 2012).

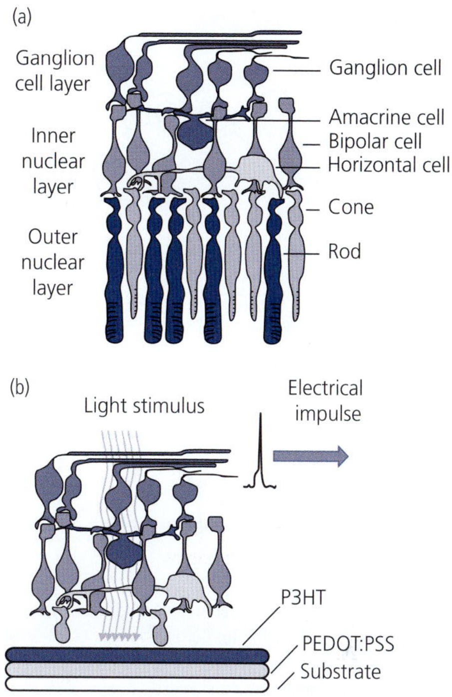

Figure 7.35 (a) Normal retina structure showing light-receptive rod and cone cells connected to the ganglion cell layer that delivers electrical impulses to the brain when activated by light. (b) A degenerated retinal structure where the photoreceptive cells are replaced by a P3HT photoconductor. Illumination of the OPC generates charges that stimulate the nerve network, analogous to direct stimulation of the rods and cones. Adapted from Ghezzi et al. (2013).

planar HJs, this effect cannot be ignored, although in bulk and mixed HJs where there are no undepleted photoactive regions, the contributions of photoconductivity are insignificant.

Photoconductors have seen few photonic device applications. One notable exception is the use of P3HT photoconductors as photoreceptors in retinal implants. The polymer is deposited on an aqueous PEDOT:PSS electrode on a 30 μm thick silk fibroin substrate, illustrated in Fig. 7.35. The unit was implanted into a rat retina whose photoreceptive rod and cone cells had degenerated from retinitis pigmentosa (Ghezzi et al., 2013). The ionic, aqueous environment within the retina creates the second contact that interfaces with the inner nuclear layer. When illuminated, photogenerated electrons stimulate the nerve cells that contact the photoconductor, sending an electrical impulse to the brain via the ganglion network. Thus, retinal photosensitivity was restored (Maya-Vetencourt et al., 2017). The biocompatibility, flexibility, and functionality of organic electronics make this technology attractive for such medical prosthetics.

7.2.2 Photodiodes

An optimized photodetector should be designed to have a high bandwidth, low dark current (and hence low noise), and most importantly, high external quantum efficiency. This combination of attributes requires that the materials, layering scheme and layer morphology be selected and controlled to achieve the required performance goals. In this section, we will primarily focus on photodiodes for signal detection, and leave the discussion of solar cells to Section 7.3. However, since many of the attributes of OPDs are shared with OPVs, it is not possible to completely separate the discussion of devices that are optimized for both applications. Hence, to fully understand the operation of an OPV, we must first take a close look at detectors used as optical sensors.

7.2.2.1 Photodiode materials

The first bilayer OPD with a high rectification ratio (i.e. the ratio of current in the forward to that in the reverse biased direction at the same $|V_a|$) comprised the donor, CuPc, and the acceptor, PTCBI. The j–V characteristics obtained in the dark and under 75 mW/cm^2, AM2 simulated solar illumination (see Section 7.3) are shown in Fig. 7.36. The device area as determined by the Ag cathode was 0.1 cm^2. The donor, acceptor, and cathode layers were deposited by VTE. The HJ formed between the D and

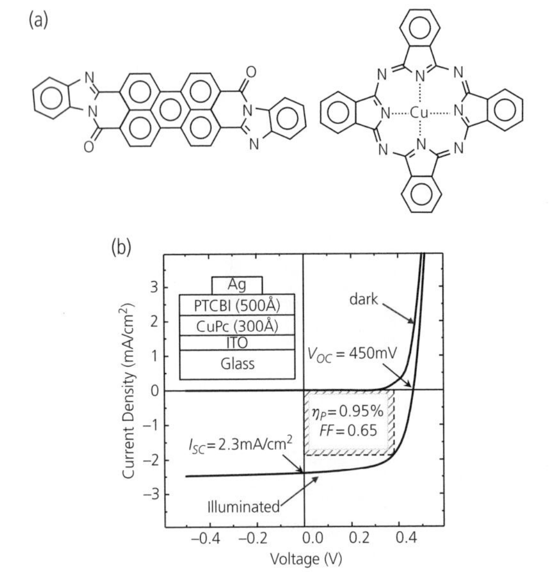

Figure 7.36 (a) Molecular structural formulae of the acceptor, PTCBI (left), and the donor, CuPc (right). (b) Current–voltage characteristics of the OPV under AM2 white light illumination of 75 mW/cm^2. The short circuit current (I_{SC}) and open circuit voltage (V_{OC}) are shown, along with the power conversion efficiency (η_P) and fill factor (FF) at the maximum power point. Inset: Device structure. Adapted from Tang (1986).

A molecules produces an apparently exponential turn-on in the forward direction, and a low dark current of 5–10 nA/cm^2 under reverse bias The ideality factor, determined from the forward-biased characteristics fit to the Shockley diode equation, is n = 1.5–1.7 (Tang, 1986). The operation of this OPD is consistent with discussions both in Chapter 4 and Section 7.1, that is, excitons generated within the D and A layers drift to the HJ where they undergo dissociation. The liberated electrons flow in the acceptor to the cathode, and the holes transport through the donor to the anode.

Since this demonstration, many other donor and acceptor materials have been identified that have led to higher optical response and improved sensitivity. The donors are chosen to absorb in the wavelength range of interest while ensuring that the positions of the frontier HOMO and LUMO levels create a type II (staggered) HJ with the acceptor. The energy level offsets between the HOMO and LUMO levels of the D and A pair must be sufficiently large to promote efficient PP formation, followed by their dissociation into free charge. This requires HOMO and LUMO energy offsets of 0.2–1 eV.

There are several reviews of materials used in OPDs and OPCs (Dong et al., 2012, Baeg et al., 2013, Jansen-van Vuuren et al., 2016). We provide a few selected examples of donors and acceptors employed in photodiodes for light sensing applications, with further example materials with particular applications to solar energy conversion discussed in Section 7.4.4. Figure 7.37 shows several example OPD donor materials and their respective absorption bands. Molecules that span the spectrum from the UV for solar blind detection, all the way to the SWIR band between 0.9 and 1.7 μm wavelength, have been developed for use in high efficiency detectors. The most commonly employed donor is P3HT due to its compatibility in forming bulk HJs with PCBM. With an energy gap of 1.9 eV, it absorbs from the light blue into the red. Similarly, MEH-PPV is a soluble, red-to-green absorbing material that can be paired with an appropriate acceptor to access the middle of the visible spectrum (Halls et al., 1995, Yu et al., 1995). Several polymers used in OPDs that absorb farther into the red and NIR have also been demonstrated, such as PDDTT (Gong et al., 2009) and PCPDTBT (Peet et al., 2007) in Fig. 7.37 that have long wavelength absorption cutoffs at 1300 nm and 950 nm, respectively.

There are numerous donors based on small molecules that span the spectrum from the UV into the SWIR region. One particularly broad class of molecules is the metal phthalocyanines, MPcs. Their high thermal stability, ease of vacuum thermal sublimation, reasonably good hole mobilities, and diversity of optical absorption spectra resulting from the large choice of metal centers (most notably Zn, Cu, Sn, and Pb) makes them attractive for detection. Metal Pcs have been used in detectors sensitive from the green into the NIR (Tang, 1986, Peumans and Forrest, 2001, Bailey-Salzman et al., 2006, Pfuetzner et al., 2009). Porphyrins, whose properties are closely related to the Pcs, have also been employed for optical sensing. A notable example is porphyrin tapes, where two or more porphyrin rings are fused to increase the electron conjugation length, thus pushing absorption to wavelengths as long as 1400 nm (Zimmerman et al., 2010a). The linear polyacene, pentacene, and polyacene derivatives such as rubrene have also been used due to their high hole mobilities (and hence high operating bandwidths when employed in OPDs) and absorption in the visible (Tsai et al., 2009, Yang et al., 2013). Many of the donors used for detection in the blue and UV have their origins as hole conducting hosts and transport layers. Hosts require large energy gaps to effectively transfer excited states to lower energy dopant molecules. Furthermore, the high mobilities required for hole transport make them suitable as donors. Two examples in Fig. 7.37 are m-MTDATA with a long wavelength cutoff of 375 nm, NSN with a cutoff of

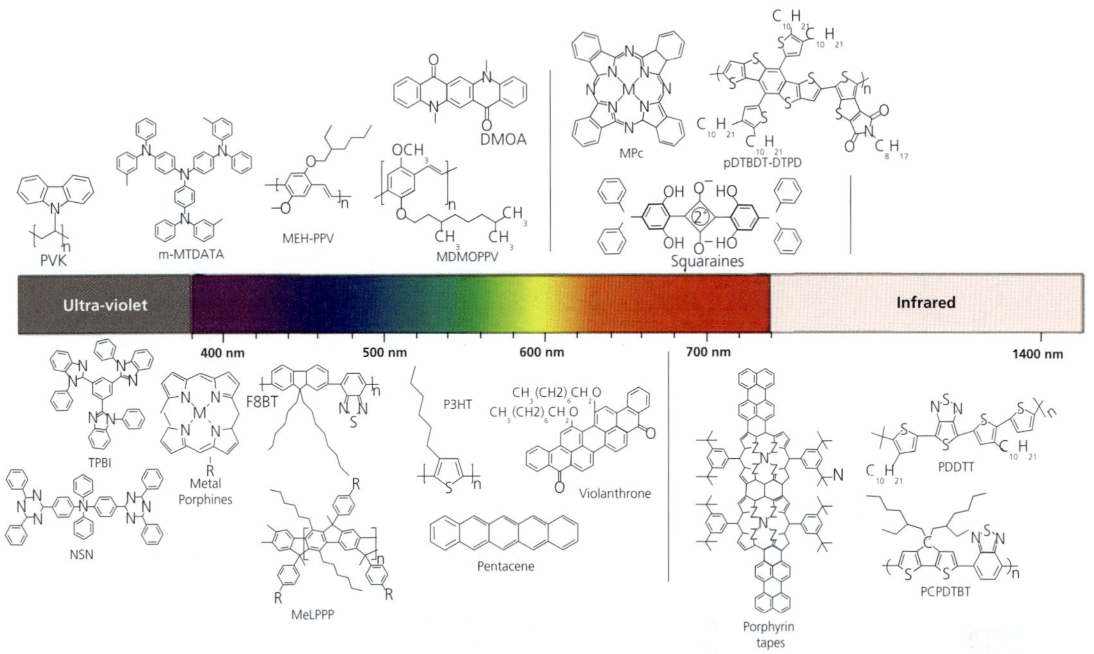

Figure 7.37 Example donor molecules employed in OPDs are shown, along with the spectral range of their absorption from the ultraviolet into the short wavelength infrared (SWIR).

310 nm, and TPBi (cutoff of ~400 nm) (Yan et al., 2009). When combined with a UV absorbing acceptor such as the polymer, PVK, the detectors are useful in environments with large background solar irradiation. This allows for so-called *solar-blind detection*, which is used to detect deep UV light that is emitted, for example, by the flame from a missile or jet engine. This emission must often be detectable in bright sunlight, or when viewed while backlit by the sun (Li et al., 2011).

In contrast to donors, comparatively few acceptors have been developed. An acceptor must have good electron conductivity along with appropriately positioned HOMO and LUMO levels relative to the donor for efficient exciton dissociation. In subsequent sections, we will find that the advancement of solar cell efficiency has primarily been paced by the identification of new, efficient acceptor molecules. The rectifier in Fig. 7.36 uses a perylene-based acceptor due its large LUMO energy. However, solar cell power conversion efficiencies employing thermally deposited perylenes has been limited to only ~1%. In the 1990s, the fullerene, C_{60} (Yu et al., 1994) and its soluble derivative $PC_{61}BM$ (Yu et al., 1995) were successfully used in efficient solution-based detectors, and this was extended to small molecule based OPDs and OPVs in 2001 (Peumans and Forrest, 2001). More recently, C_{70} and its analog, $PC_{71}BM$, have been favored due to their broader spectral absorption than their C_{60}-based analogs. Functionalization with silyl groups in SIMEF raises the LUMO energy to −3.7 eV compared to −3.8 eV for $PC_{61}BM$, increasing ΔE_{HL}, and hence the open circuit voltage of OPDs operated in the photovoltaic mode. In addition, its favorable packing morphology is helpful for promoting electron transport (Matsuo et al., 2009). The fullerene, ICBA, also has frontier orbital energies that can provide a slightly modified optical absorption and energy offsets from $PC_{61}BM$. It also has improved solubility that broadens applications accessible to the fullerene family. Several of example fullerenes are shown in Fig. 7.38.

Fullerene acceptors led to the first solution-processed BHJ (Yu et al., 1995) and mixed HJ structures (Xue et al., 2005). Their spherical symmetry leads to rapid and efficient PP production in planar as well as blended D-A films. The success of fullerene acceptors in achieving high OPD and OPV efficiencies was unmatched by other molecules until the middle of the second decade of the twenty-first century when a new class of non-fullerene, non-perylene acceptor molecules (called *non-fullerene acceptors*, NFAs) emerged, an advance that has led to a renewed burst of development of high efficiency OPDs (see Section 7.4.4).

A limitation of the fullerene family is apparent from the very few modifications that have been introduced to tailor their energetic and/or physical properties. Indeed, the molecules in Fig. 7.38 represent

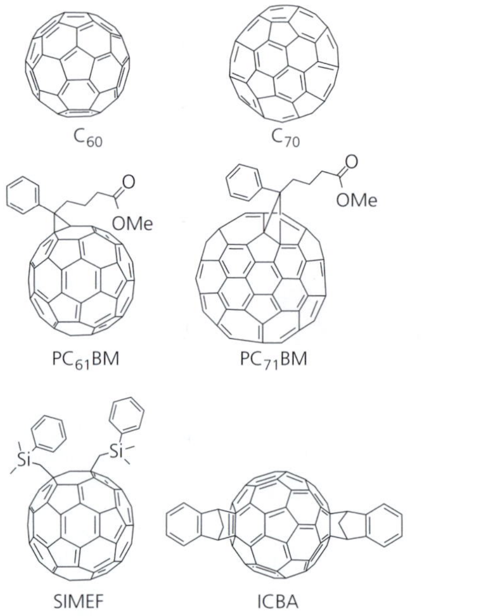

Figure 7.38 Fullerenes used in organic photodiodes. C_{60} and C_{70} are employed in small molecule devices. The remaining fullerenes are used in solution-based devices due to their solubility arising from attaching the pendant groups to the fullerene cores.

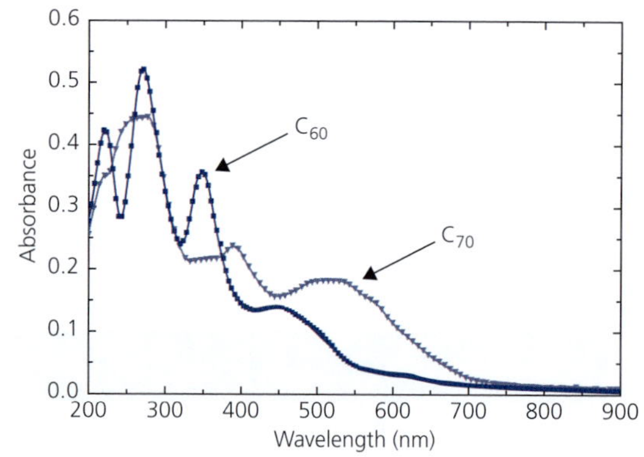

Figure 7.39 Absorbance spectra of 20 nm thick C_{60} and C_{70} films (Pfuetzner et al., 2009).

almost all examples that have been used in high efficiency detectors.

The absorbance spectra of C_{60} and C_{71} are given in Fig. 7.39. *Absorbance* is defined as the logarithm of the transmittance of a substance, that is,

$$A = -log_{10}T = \int_0^d \alpha_{10}(x)dx, \tag{7.109}$$

where the film thickness is d and α_{10} is the *decadic attenuation coefficient* given by $\alpha_{10} = \alpha/log10$. The higher order fullerene, C_{70}, is preferable for most OPD and OPV applications due to its absorption into the NIR, whereas C_{60} only absorbs weakly from the yellow to red due to low absorbance of the CT state between 550 and 650 nm.

The squaraines (SQs) are solution processed small molecules that can be functionalized by capping end groups that shift their absorption spectra from the visible to the NIR. For example, the modified SQs such as the diphenyl moiety in Fig. 7.37 absorbs out to 850 nm (Binda et al., 2009, Wei et al., 2011b). An example glycolic chain-substituted SQ (GlySQ) blended with $PC_{61}BM$ has demonstrated high efficiency as a photodetector due to its low dark current. The detector dark current was reduced by interposing a 30 nm thick layer of a cross-linked and insoluble, wide energy gap MEH-PPV anode buffer layer between the ITO anode and the BHJ active region to prevent electron injection under reverse bias, as shown in Fig. 7.40. Dark currents of 2 nA/cm^2 at 1 V reverse bias are indicative of the importance of anode buffer layers, to be discussed further in Section 7.4.1 in the context of OPVs. The specific detectivity at this voltage for a 4 mm^2 detector is $D^* = 3.4 \times 10^{12}$ Jones vs. 5.9×10^{11} Jones for devices with and without the buffer, respectively. The 3 dB roll off frequency is 1 MHz, indicating that parasitic resistance from the electron-blocking layer does not degrade the detector dynamic response.

Demonstrating devices with high quantum efficiencies at wavelengths extending into the NIR and beyond is particularly difficult to achieve due to the energy gap law (see Section 3.6.1), which states that the rate of non-radiative exciton recombination increases as the energy gap is reduced. Hence, narrow energy gap materials needed for NIR detection are vulnerable to dissipative reactions that ultimately result in reduced diffusion lengths, and hence lower efficiencies. In addition, narrow energy gaps lead to increased dark currents due to thermal charge generation from HOMO to LUMO. In spite of these challenges, considerable success has been achieved in demonstrating efficient NIR photodetectors using optimized materials.

An efficient long wavelength organic detector is based on solution-processed small molecular weight porphyrin tapes comprising two Zn-metallated porphyrins, triply-linked at the meso–meso and both β–β positions shown in Fig. 7.41. Long wavelengths are accessed by molecular linking to extend the electron conjugation length, thereby leading to a reduced energy gap compared to the core porphyrin molecule itself (Tsuda and Osuka, 2001, Duncan et al., 2006). The Zn porphyrins with 3,5-di-*tert*-butylphenyl

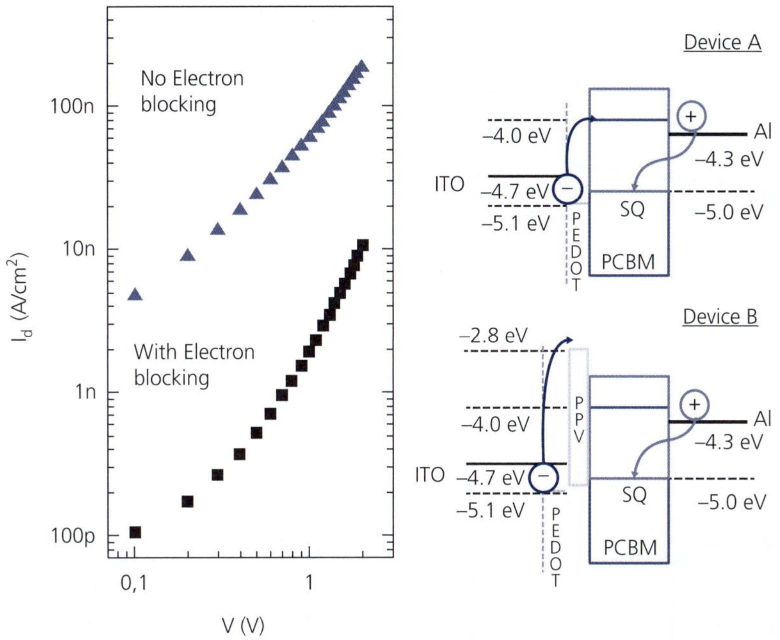

Figure 7.40 Reverse dark current characteristics of a GlySQ:$PC_{61}BM$ OPD in presence and absence of an MEH-PPV electron blocking layer at the anode. Energy level diagrams for the devices are shown at right. Dashed lines give the energies of the corresponding organic and ITO layers (Binda et al., 2011).

side groups in Fig. 7.41 are functionalized by using three different terminations: Psub with singly bonded pyrene, Pfused with a doubly-bonded pyrene, and Ppery with a doubly fused perylene. The conjugation length is shortest for Psub, and longest for Ppery. Forming the additional bond to the pyrene end group reduces the energy gap of Psub by 0.20 eV. The absorption peaks are at 1150 nm (Psub), 1400 nm (Pfused), and 1485 nm (Ppery). The compounds absorb throughout the visible and into the NIR, with intense Q-band absorption peaks between $\lambda = 1050$ nm and 1350 nm (Zimmerman et al., 2010a, 2011).

For Psub:PCBM:4,4′-bipyridyl (Bipy) additive (1:0.1:1), a maximum $D^* = 8.8 \times 10^{11}$ Jones and $\eta_{ext} = 3.6\%$ at $\lambda = 1150$ nm was observed, whereas a Pfused:PCBM:Bipy (1:0.75:1) OPD had an order of magnitude lower D^*, but with $\eta_{ext} = 10.5\%$ at $\lambda = 1400$ nm. The *j–V* characteristics, spectrally resolved external quantum efficiencies, and specific detectivities for Psub:PCBM:Bipy and Pfused:PCBM:Bipy devices are shown in Fig. 7.42. Devices using films of neat perylene-fused porphyrin dimers (not shown) achieved an $\eta_{ext} = 9.1\%$ at $\lambda = 1485$ nm (Zimmerman et al., 2011).

The detectivities achieved using the porphyrin tapes are significantly less than for other SWIR-sensitive detectors such as those based on InGaAs ($\sim 10^{13}$ Jones), but are comparable to cooled PbS detectors (Webster, 1999, Rogalski, 2002). The temporal response of the Psub-based OPDs to pulsed optical excitation using a 50 Ω load yielded a 3 dB roll-off frequency of 56 MHz at 0 V. The *RC* time constant-limited frequency of the OPD is ~200 MHz, indicating that the response is limited by charge transit times or internal parasitic resistances (Zimmerman et al., 2010a).

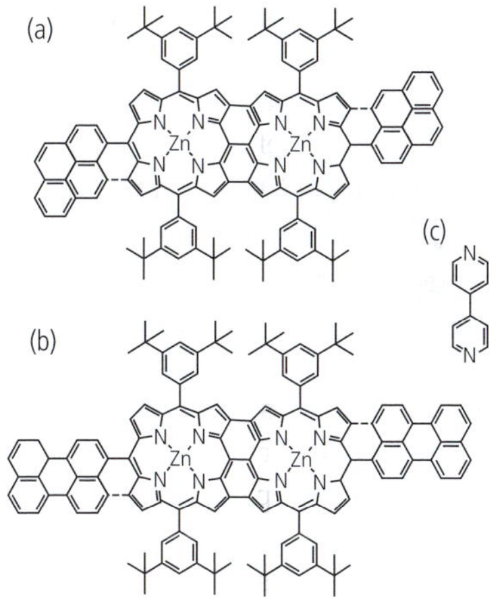

Figure 7.41 Fused Zn porphyrin donors with (a) singly (Psub) and doubly fused (Pfused) pyrene, and (b) a doubly fused perylene (Ppery). (c) 4,4′-Bipyridyl (Bipy) additive (Zimmerman et al., 2011).

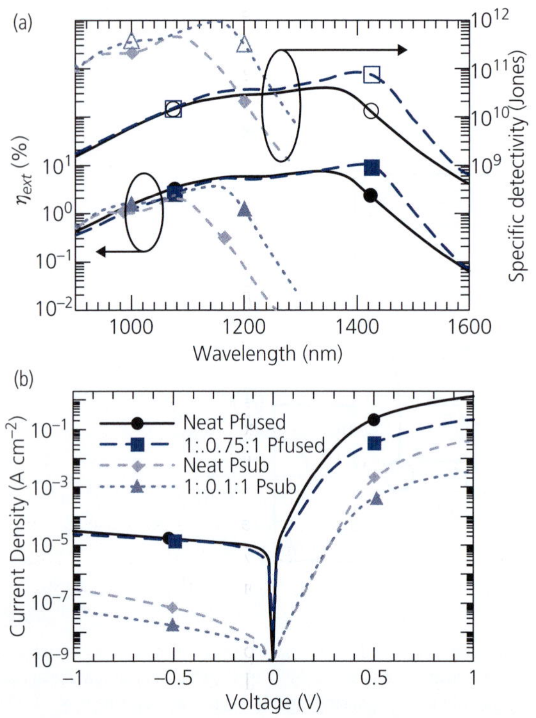

Figure 7.42 (a) External quantum efficiency and specific detectivity spectra of devices fabricated from Psub and Pfused compounds in Fig. 7.41 with different additive concentrations. (b) The *j–V* characteristics of the devices in (a). The molar ratios shown are for porphyrin:$PC_{61}BM$:Bipy (Zimmerman et al., 2011).

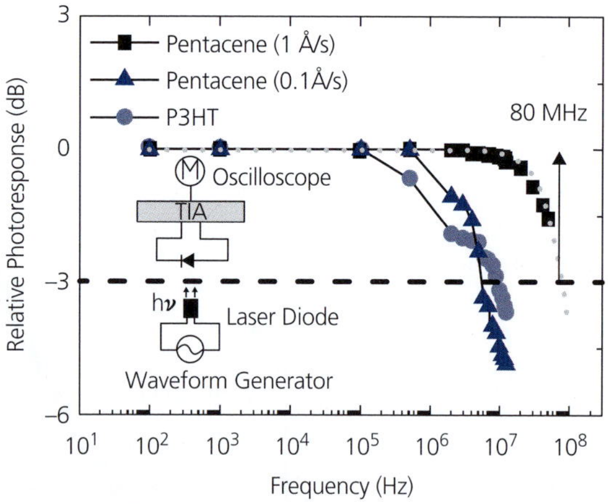

Figure 7.43 Frequency response of donor/C_{60} OPDs with the donors indicated in the legend. The dashed line follows $P = P_0/(1 + f/f_{3dB})$ to obtain an extrapolated bandwidth of 80 MHz for the pentacene (1 Å/s) device. Inset: Test system used to measure bandwidth. TIA = transimpedance amplifier. The system bandwidth is 50 MHz, thus requiring the extrapolation to the −3 dB intensity point of the highest bandwidth device. After Tsai et al. (2009).

There are other approaches to achieving long wavelength absorption such as employing the donor, BDN, that absorbs to ~1500 nm (Dalgleish et al., 2012), or ester functionalized PTT with absorption to 950 nm (Yao et al., 2007). However, the bandwidth and dark currents (and hence noise) of OPDs based on these thiol-based molecules do not achieve the high performance of OPDs employing porphyrin tapes.

7.2.2.2 High bandwidth OPDs

There have been numerous demonstrations of high bandwidth OPDs, ranging from several tens of MHz (Morimune et al., 2006, Tsai et al., 2009, Zimmerman et al., 2010a, Yang et al., 2013) to 430 MHz for multilayer OPDs absorbing across the visible spectrum (Peumans et al., 2000b, 2003b). As noted in Section 7.1.2, the bandwidth of a detector is limited by a combination of the exciton diffusion time, the charge extraction, or transit time in the depletion region, charge diffusion across undepleted regions, and the *RC* time constant due to detector series resistance and capacitance. The transit times, in turn, can be influenced by charge trapping at defects in both the depleted and undepleted regions of the OPD.

The bandwidth of a transit-time-limited device is determined by the hole mobility in the donor and the electron mobility in the acceptor. The mobilities, in turn, are affected by the presence of static disorder, see Section 4.3.2. The effects of disorder on detector bandwidth have been explored by comparing the bandwidths of C_{60} devices comprising either a highly disordered P3HT donor layer, or with a pentacene donor thermally deposited in vacuum at various rates (Tsai et al., 2009). The responses of three, 4 mm^2 devices to a pulsed laser diode at a wavelength of 633 nm are shown in Fig. 7.43. The frontier energies of both P3HT and pentacene are nearly equal, thereby ensuring that differences in bandwidths between devices are due to morphology determined by the deposition rate rather than energetics that might affect the PP dissociation rate. The mobilities of the various layers, measured from their space charge limited *j–V* characteristics were 1.9×10^{-3}, 6.0×10^{-5}, 2.3×10^{-5}, and 2.8×10^{-4} cm^2/V s for C_{60}, P3HT, and pentacene deposited at 0.1 Å/s and at 1 Å/s, respectively.

The high mobility sample (pentacene, 1 Å/s) has an eight-fold higher roll-off frequency of $\Delta f = 80$ MHz compared to the other, more disordered devices. Here, Δf is measured at the frequency where they photoresponse drops by 3 dB from its value at low frequency. This difference is roughly equal to the corresponding differences in mobilities between the donor films. All device responses are limited by the hole transit time across the donor layers.

A high bandwidth OPD with $\Delta f = 430$ MHz is based on alternating D-A multilayer structure that places interfaces in close proximity to the points of exciton generation, resulting in $\tau_{ED} \to 0$ (see the energy level diagram in the inset, Fig. 7.44). The device consists of a stack of ultrathin (ranging from 1 to 16 nm), alternating PTCBI acceptor and CuPc donor layers, sandwiched between an ITO anode and a BCP exciton blocking layer, and capped by a Ag cathode. To maintain sharp interfaces, the multilayer stack was deposited by the UHV process of organic molecular beam deposition (Section 5.4.2). The sharp interfaces require an external electric field to efficiently extract the photogenerated holes and electrons trapped by the energy barriers to electron and hole transport formed by the HOMO and LUMO energy offsets at each heterointerface.

The room temperature *j–V* characteristics for devices with different individual layer thicknesses but with the same total thickness of 32 nm, are shown in Fig. 7.44. For comparison, data for a mixed-layer device (dashed line) is also shown. The differences between the mixed and the multilayer *j–V* characteristics suggest different transport mechanisms; the multilayers are dominated by tunneling and the mixed HJ by charge diffusion (Peumans et al., 2003b). The dark current density at a fixed reverse voltage decreases by over an order of magnitude as the thickness of the individual layers is reduced from 16 to 0.5 nm.

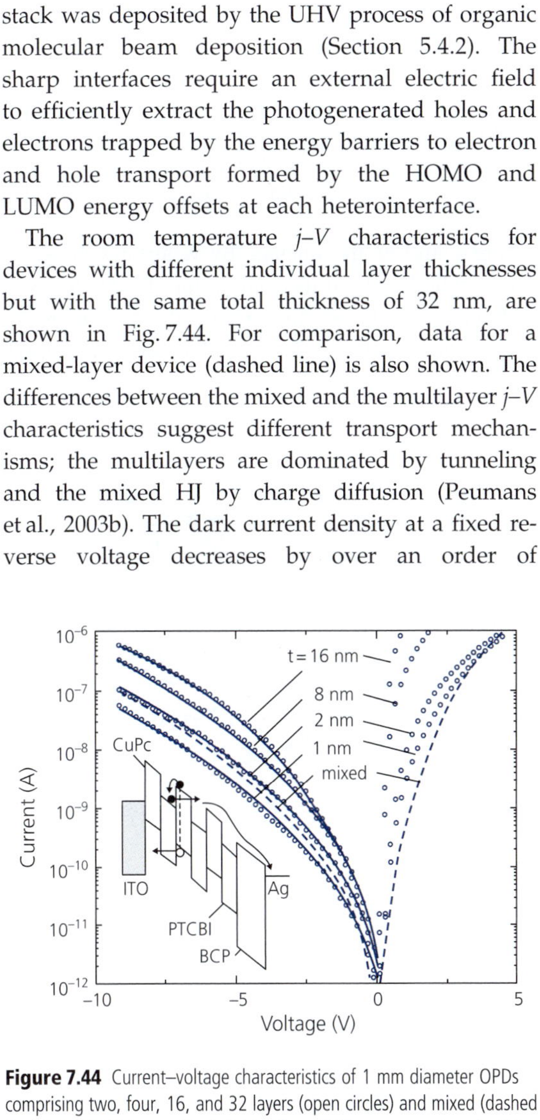

Figure 7.44 Current–voltage characteristics of 1 mm diameter OPDs comprising two, four, 16, and 32 layers (open circles) and mixed (dashed line) CuPc/PTCBI layers. The individual layer thicknesses are given by *t*. The number of layers is adjusted to reach a total absorption region thickness of 32 nm. Fits of the data to theory based on Frenkel–Poole emission are shown by solid lines. Inset: Schematic energy level diagram of the alternating donor–acceptor multilayer photodetector device under reverse bias. The process of exciton dissociation, followed by carrier separation via tunneling, is illustrated (Peumans et al., 2000b).

The *j-V* characteristics are understood by assuming that electrons and holes trapped in the acceptor and donor regions escape via *Frenkel–Poole emission* (see Section 4.3.2). This tunneling mechanism illustrated in Fig. 7.44 follows

$$j(F) = AF\exp\left(\frac{q}{rk_BT}\sqrt{\frac{qF}{\pi\varepsilon_r\varepsilon_0}}\right), \qquad (7.110)$$

where A is a constant and $1 \le r \le 2$ is determined by the degree of impurity charge compensation ($r = 2$ for a fully compensated semiconductor). The solid lines in Fig. 7.44 show fits to this expression using w_T inferred from capacitance–voltage data. The dark *j–V* characteristics can be accurately fit over 5–6 orders of magnitude in *j* using Eq. 7.110. These data suggest that the PTCBI layers donate their electrons, and hence compensate all the charged acceptor levels in the adjacent CuPc layers up to a distance 10 nm from the heterointerface, thus fully depleting the multilayer stack.

The external quantum efficiency of a $t = 5$ Å × 32 alternating layer photodetector as a function of the applied bias is shown in Fig. 7.45. The absorption efficiency is also shown (dashed line). The similar spectral shapes of $\eta_{ext}(\lambda)$ and $\eta_A(\lambda)$ confirm that charge generation takes place via exciton dissociation. Under short circuit conditions, the peak $\eta_{ext}(\lambda) = 1\%$, increasing $\eta_{ext}(\lambda) > 70\%$ between wavelengths of 540 nm to 700 nm at −10 V. This indicates efficient exciton dissociation at the D-A interfaces followed by fast and complete charge separation and transport.

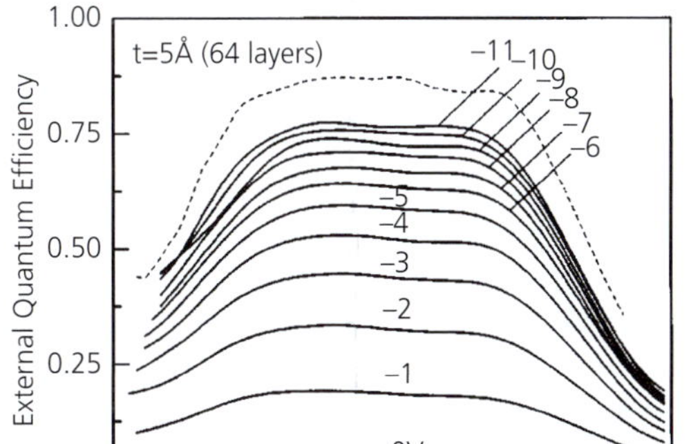

Figure 7.45 External quantum (solid lines) and absorption efficiencies (dashed line) of a 5 Å × 32 period PTCBI/CuPc OPD vs. wavelength and reverse bias (Peumans et al., 2003b).

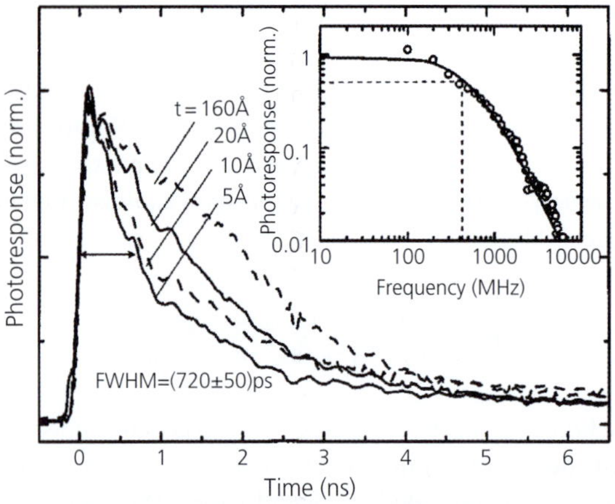

Figure 7.46 Temporal response of 0.1 mm diameter CuPc/PTCBI OPDs with varying individual layer thicknesses (t) with a 320 Å total active layer thickness. Inset: Fourier transform of the temporal response of a 32-period device normalized to dc operation. The 3 dB bandwidth of the detector is $\Delta f = 430$ MHz (Peumans et al., 2000b).

In spite of the many energy barriers present in the multilayer structures, carrier transport is rapid due to tunneling through the thin barriers. This is supported by measurements of the temporal response of 0.1 mm diameter devices under excitation with short (~1 ps) optical pulses at $\lambda = 670$ nm and -9 V, shown in Fig. 7.46. The response times decrease with increasing number of interfaces, which is contrary to expectations if charge percolation is by successive trapping and hopping, as typical for mixed junctions. Devices incorporating thinner layers exhibit a faster response due to a reduction in the exciton diffusion lifetime. The response time of the bilayer device (t = 160 Å) is limited by the intrinsic exciton lifetime, which is 1.8 ns in a neat PTCBI. The highest bandwidth is achieved with the device with t = 5 Å, which allows for rapid and efficient tunneling of electrons and holes throughout the device. This OPD exhibits a full width at half maximum (FWHM) response of 720 ps, corresponding to $\Delta f = 430$ MHz. Interestingly, the combination of high quantum efficiency in conjunction with the high bandwidth makes the performance of this OPD comparable or even superior to Si p-i-n photodiodes, although the leakage currents, and hence the detectivity of the OPD is inferior to those latter PDs.

7.2.2.3 OPDs based on nanotubes and quantum dots

A means for extending OPD sensitivity into the NIR while avoiding rapid recombination in narrow energy gap organics is to combine long wavelength absorbing *inorganic* species in an organic matrix. Two inorganic materials that have been exploited for this purpose are semiconducting single-walled carbon nanotubes (SWNTs) and compound semiconducting quantum dots (QDs).

Carbon nanotubes (CNTs) can form very small energy gap semiconductor or metallic solids, depending on their diameters and chiralities. Provided that the metallic SWNTs are removed, light absorbed by the semiconducting nanotubes can generate low energy excitons that migrate to a type II HJ formed with an organic in a composite, or colloid, where they undergo charge transfer and generation similar to conventional organic D-A HJs.

We characterize the CNT structure by its *chirality* vector, $\mathbf{C}_h$, shown in Fig. 7.47a. For a graphene sheet with unit vectors $\mathbf{a}_1$ and $\mathbf{a}_2$:

$$\mathbf{C}_h = n\mathbf{a}_1 + m\mathbf{a}_2, \tag{7.111}$$

where n and m are integers. By wrapping the sheet into a cylinder about its axial vector, $\mathbf{T}$, nanotubes of different chiralities are achieved. Examples of the *armchair*, *chiral*, and *zigzag* structures are shown

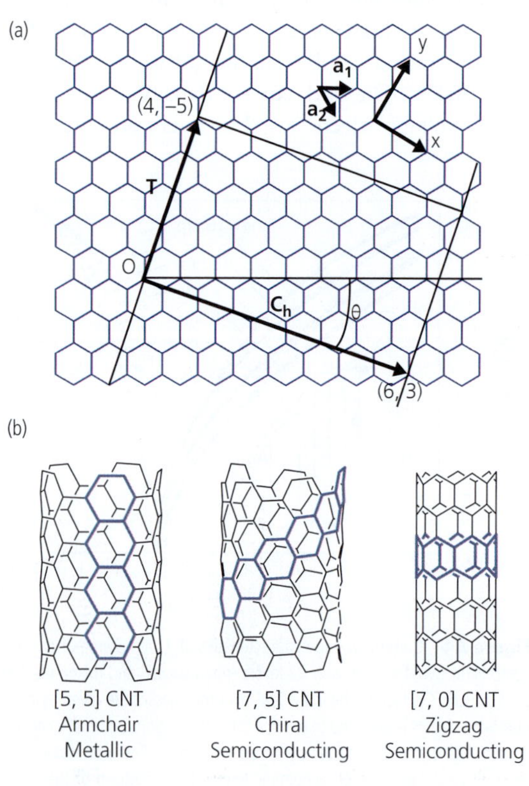

Figure 7.47 (a) Carbon nanotube naming scheme based on graphene unit cell vectors, $\mathbf{a}_1$ and $\mathbf{a}_2$. $\mathbf{T}$ denotes the tube axis. (b) Three example chiralities of SWNTs.

in Fig. 7.47b. The different chiralities result in strikingly different conductive properties. If $n = m$, the tube forms an armchair structure that is metallic. If $n - m$ is a multiple of 3 with $n \neq m$ and $nm \neq 0$, the band gap is very small, that is, the tube is quasimetallic, although semimetallic CNTs do not exist. Otherwise the tube is a semiconductor whose energy gap is varied both by its chirality and diameter.

Since an OPD requires that the composite be semiconducting, the metallic CNTs, as well as multiple walled tubes and bundles need to be removed from the source material stock. As discussed in Section 5.3.4, metallic and other undesired fractions can be separated using ultracentrifugation. This is a time consuming process, yet it has resulted in detectors with sensitivities deep into the IR.

The first report of a composite detector employing a SWNT–polycarbonate blend was a photoconductor in the lateral geometry of Fig. 7.1. This device was sensitive in both the long wavelength IR and in the visible. It had a long recovery time constant (~100 s) after removal of the IR excitation, and showed only a few percent change in conductivity in the dark vs. under white illumination of 0.7 mW/cm^2 (Pradhan et al., 2008).

This demonstration was followed by a CNT–polymer/organic semiconductor type II HJ OPD with the structure shown in Fig. 7.48a. The SWNTs with diameters ranging from 0.7 to 1.1 nm were purified via ultracentrifugation, and then wrapped by immersion in a solution with a semiconducting polymer, either MDMO-PPV or P3HT. Polymer wrapping prevents the CNTs from contacting each other and forming bundles. The CNT solution was applied by doctor blade onto an ITO coated glass substrate, and then capped either with a neat C_{60} layer or a SnPc:C_{60} (1:3) D-A blend that was employed to enhance the photodiode visible-NIR response. The layer was coated with a C_{60} layer, a BCP blocking layer and a Ag cathode.

The energy level diagram of the MDMO-PPV:CNT/C_{60} device in Fig. 7.48b is useful to understand the photogeneration mechanism. Photons absorbed directly in the CNT form excitons that subsequently migrate to the HJ with the C_{60} where they dissociate. When combined with C_{60}, the CNT acts as a donor. If the C_{60} is blended with SnPc, shorter wavelength photons not absorbed by the CNT are absorbed by the SnPc, which is also a donor. Including SnPc fills the hole in the spectrum from approximately 500 nm to 800 nm where the SWNTs are only weakly absorbing.

A scanning electron micrograph showing the morphology of the MDMO-PPV:CNT blend is

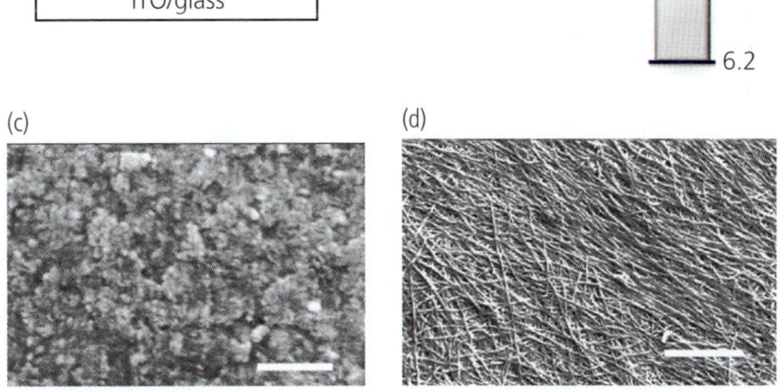

Figure 7.48 (a) Layer scheme of a composite SWNT–polymer/organic HJ OPD. The diagram represents three different devices: in two of the devices, the semiconducting nanotubes were wrapped in the polymer MDMO-PPV, or P3HT. In one device, the composite was contacted directly by a C_{60} acceptor layer, and in another, it contacted a SnPc:C_{60} blend. (b) Energy level scheme of the MDMO-PPV:CNT/C_{60} OPD. Scanning electron micrograph images of a film of (c) nanotube/MDMO-PPV composite and (d) bare nanotubes. The scale bars = 500 nm (Arnold et al., 2009).

provided in Fig. 7.48c. The film has an rms surface roughness of 4 nm, compared to the very irregular surface of a mat of bare CNTs in Fig. 7.48d, with a roughness of 20 nm. Hence, the polymer wrapping has the added effect of planarizing the film to prevent shorts from the CNT to the cathode.

The performances of several CNT device structures are shown in Fig. 7.49. Absorption by the population of semiconducting CNTs having easily identifiable chiralities (indexed adjacent to each peak) results in a rich responsivity spectrum ranging from 550 nm (due to the E22 absorption), to >1500 nm by the E11 band. The responsivity at $\lambda = 1155$ nm at -0.7 V is 23 mA/W, corresponding to $\eta_{ext} = 2.3\%$. For MDMO-PPV/CNT layer thicknesses of 14 nm, $\eta_{int} > 25\%$ between $\lambda = 1000$ and 1400 nm. The responsivity significantly decreases at $\lambda > 1450$ nm due to of a lack of absorption since there are no CNTs with diameters >1.1 nm.

The temporal response of a 0.3 mm diameter MDMO-PPV:CNT device was measured at 0 V and $\lambda = 847$ nm using a laser pulse width of ~1 ns. The rise time, limited by the RC time constant of the diode was $\tau_{RC} = 1.84$ ns. The OPD response had a time of $\tau = 7.16$ ns limited by charge drift. This corresponds to a 3 dB detector bandwidth of 31 MHz.

The *j–V* characteristics of a P3HT:CNT (1:3) OPD are shown in Fig. 7.49b, and its resulting specific detectivity is in Fig. 7.49c. A fit to the *j–V* data (solid line, Fig. 7.49b) gives an ideality factor of 1.34. Based on these data, the CNT:P3HT/C_{60}:SnPc photodetector at 0 V exhibits $D^* > 10^{10}$ Jones from $\lambda < 400$ nm to $\lambda = 1450$ nm, with a peak of $\sim 3 \times 10^{11}$ Jones at 750 nm due to intense absorption by the SnPc, and 7×10^{10} Jones at 1200 nm and 1300 nm due to the SWNT blend (Arnold et al., 2009).

An analogous device employing poly(9,9-dichlorofluorene) wrapping to selectively disperse only CNTs with chiralities (7,5), (7,6), (8,6), (8,7), and (9,7) had response between 1000 nm and 1350 nm due to the narrower population of nanotubes (Bindl et al., 2010). This chiral selectivity all but eliminated the undesirable metallic CNTs, while permitting optimization of the overlap of the optical field with the device active region. As a result, the external quantum efficiency at 1200 nm reached 12.9%, corresponding to $\eta_{int} = 91\%$. However, the reverse dark

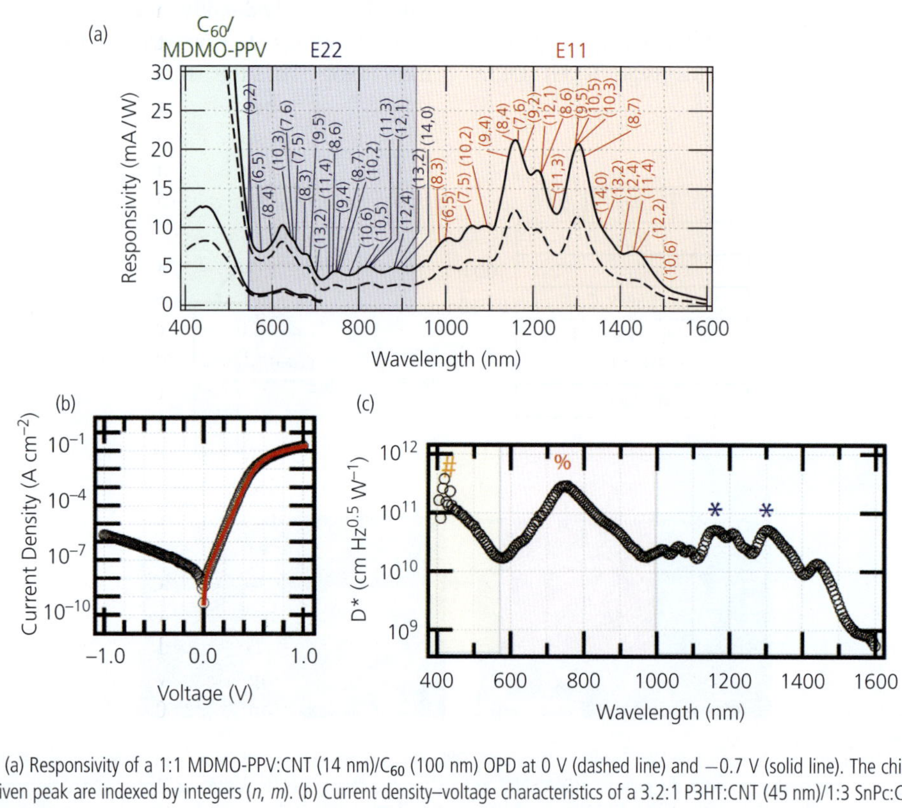

Figure 7.49 (a) Responsivity of a 1:1 MDMO-PPV:CNT (14 nm)/C_{60} (100 nm) OPD at 0 V (dashed line) and -0.7 V (solid line). The chiralities of CNTs leading to a given peak are indexed by integers (*n*, *m*). (b) Current density–voltage characteristics of a 3.2:1 P3HT:CNT (45 nm)/1:3 SnPc:C_{60} (10 nm)/C_{60} (100 nm) OPD. The solid line is a fit to the data using the ideal diode equation. (c) The specific detectivity of the device in (b). Responsivities due to absorption in the P3HT and C_{60}, SnPc, and the SWNTs are labeled by "#", "%," and "*", respectively (Arnold et al., 2009).

current was ~10 mA/cm^2, which is approximately 10^5 times higher than the P3HT-based device.

Colloids consisting of small bandgap compound semiconductor QDs are an alternative route to accessing the infrared. The exciton energy of a quantum-confined state within a nanoscale particle (or dot) depends on the particle diameter. The exciton energy in a sphere of radius, r_0, is found by solving Schrodinger's equation using the following Hamiltonian:

$$H = -\frac{\hbar^2}{2m_e^*}\nabla_e^2 - \frac{\hbar^2}{2m_h^*}\nabla_h^2 - \frac{q^2}{4\pi\varepsilon_0\varepsilon_{QD}|\mathbf{r}_e - \mathbf{r}_h|} + V(r_e, r_h), \quad (7.112)$$

where the boundary potential at the periphery is given by

$$V(r_e, r_h) = \begin{cases} 0 & \text{for } r_e, r_h < r_0 \\ \infty & \text{elsewhere} \end{cases}. \quad (7.113)$$

Here, $\mathbf{r}_e$ and $\mathbf{r}_h$ are the positions of the electron and hole, respectively, within the QD, and ε_{QD} is its relative dielectric constant. The first two terms in Eq. 7.112 are the kinetic energies of the electron and hole, respectively, the third term is their binding energy, and $V(r_e, r_h)$ is the potential that binds the charges to the QD.

Since QDs are nanocrystalline solids, they are often facetted, and hence the spherical dot approximation is not always valid, although it is sufficient to understand the factors determining the QD optical properties. The lowest lying (1s) Wannier–Mott exciton for a QD in a dielectric medium can be approximated by a hydrogenic state. Using time independent perturbation theory, the solutions to Eq. 7.112 yield (Brus, 1984)

$$E = \frac{\hbar^2\pi^2}{2r_0^2 m_r^*} - \frac{1.8q^2}{4\pi\varepsilon_0\varepsilon_{QD} r_0}, \quad (7.114)$$

where $1/m_r^* = 1/m_e^* + 1/m_h^*$ is the reduced effective mass. The first term is the kinetic energy of the quantum confined 1s state. For higher order (principal quantum number, $n > 1$) excited states, this energy takes the familiar form for a particle in a box, viz. $E_n = \hbar^2\pi^2 n^2/2r_0^2 m_r^*$. The second term is the Coulomb potential energy of the electron–hole pair.

An interesting analogy connects QD and doped donor–acceptor organic systems. Indeed, in Section 6.8 we showed that a fluorescent dopant molecule in a large energy gap host results in laser performance analogous to that of QD semiconductor lasers. The latter devices show weak temperature-dependent spectral and threshold characteristics due to the confinement of the luminescent state by the barriers formed by the wide energy gap semiconductor surrounding the narrow energy gap QD. The density of states of the dot is comparatively narrow relative to the bulk, leading to reduced thermal broadening of the electron and hole populations, and hence the reduced temperature dependence of the lasing threshold. Similarly, the excitons confined on isolated organic dopant molecules are also relatively immune from thermal broadening effects due to their spatial and energetic localization.

Equation 7.114 provides the dependence of E on particle radius and effective mass. Thus, the energy increases as r_0^2 and m_r^* decrease. This dependence on diameter is shown for several inorganic QDs in Fig. 7.50. The lowest energy absorption transitions ($1s_e \leftarrow 1s_h$ corresponding the 1s transitions from the hole to electron state, respectively) calculated in Eq. 7.114 are shown in the inset. Dots with a larger diameter have a lower energy than smaller dots (the latter are on the right). A composite incorporating narrow band gap QDs such as PbS, InSb and PbSe of the appropriate radii should, therefore result in absorption all the way into the mid-IR (the so-called *mid-wavelength IR*, MWIR, band from 3 to 5 μm).

Considerable advances have been made in achieving populations of QDs with a narrow distribution of diameters about a mean, target value. These so-called *monodisperse* populations can be achieved by ultracentrifugation and sedimentation of a relatively broad starting population of diameters (see Section 5.3.4). Alternatively, adding precise amounts of precursors during synthesis can also produce a monodisperse diameter population whose diameters are controlled by depletion of the precursors in solution which terminates dot growth. Micrographs of a synthetically generated, monodisperse population of PbS nanocrystals with an absorption peak at 1440 nm are shown in Fig. 7.51, with a high resolution image of a single dot showing its crystal structure, along with a selective area electron diffraction (SAED) pattern of the nanocrystals in the inset. The nanocrystals are synthesized by mixing a solution of PbO and oleic acid with TMS and octadecene in a 2:1 Pb/S ratio. The reaction runs to completion when all of the TMS is consumed. The resulting size dispersion is 10–15%, yielding narrow absorption peaks that can be tuned from $\lambda = 800$ nm to 1800 nm. The dot lattice constant is 5.9 Å oriented along the [100] direction, which is the same as that observed for bulk PbS. The colloid in Fig. 7.51 shows that the dots are separated by the oleic acid ligands, thus preventing clustering and fusing of neighboring dots into larger particles with lower energy

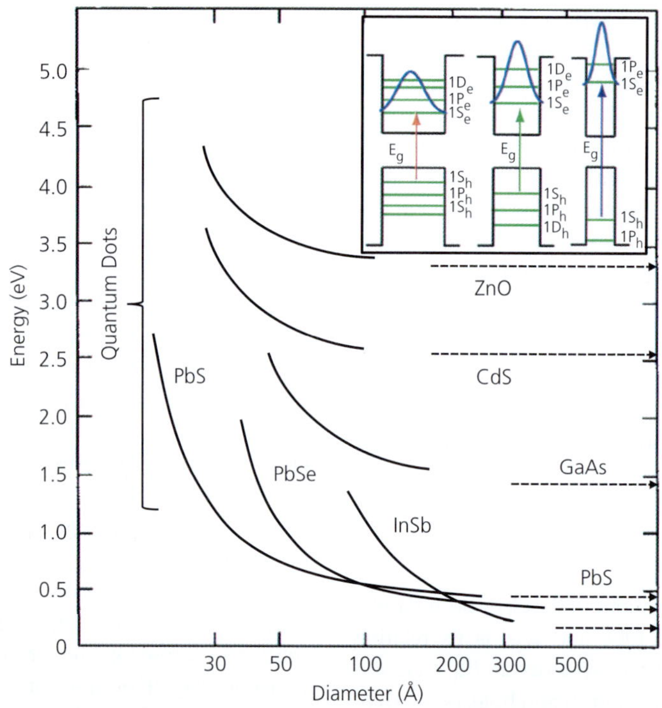

Figure 7.50 Lowest excited state energies of spherical semiconductor quantum dots of different compositions. The dashed lines show the band gap energies of the bulk semiconductors, which is approached at large diameters. Data from Brus (1984) and Kang and Wise (1997). Inset: Transitions between the $1s_h$ and $1s_e$ for three QDs of different diameters, increasing from left (lowest energy transition) to right (highest energy transition). Higher order 1p and 1d states are also shown.

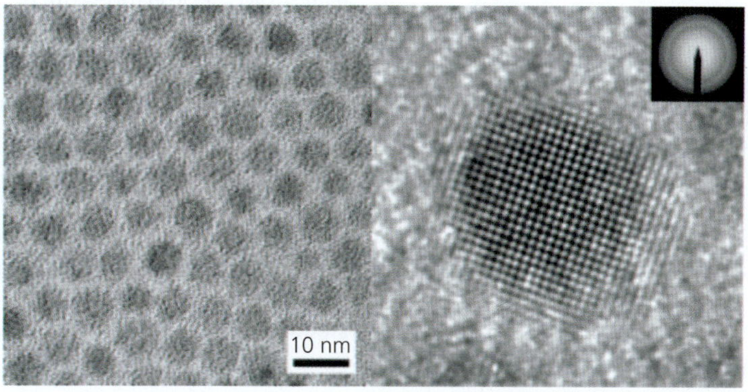

Figure 7.51 (a) High resolution transmission electron micrograph of a colloid of PbS nanocrystals. (b) Electron micrograph of a single nanocrystal showing its crystal structure. Note its spherical shape. Inset: A selected area electron diffraction (SAED) pattern of the nanocrystals showing rings that correspond to the bulk lattice spacing of PbS (Hines and Scholes, 2003).

transitions. Hence, the ligands serve two purposes: to control the growth of the dot population, and to form a solid colloidal dispersion that isolates individual dots.

The precision with which diameters and absorption spectra can be controlled is illustrated in Fig. 7.52a for populations of PbSe QDs. The radii are synthetically tuned from 3.3 nm to 8.1 nm. Evidence for the monodispersity of the populations is provided by the narrow spectral peaks corresponding to each diameter, tuned from approximately 1200 nm to 2100 nm. Interestingly, the spectra also exhibit higher order ($n = 2, 3$) transitions at shorter wavelengths, providing further evidence for monodispersity. The dependence of the lowest transition energy vs. dot radius in Fig. 7.52b shows a fit

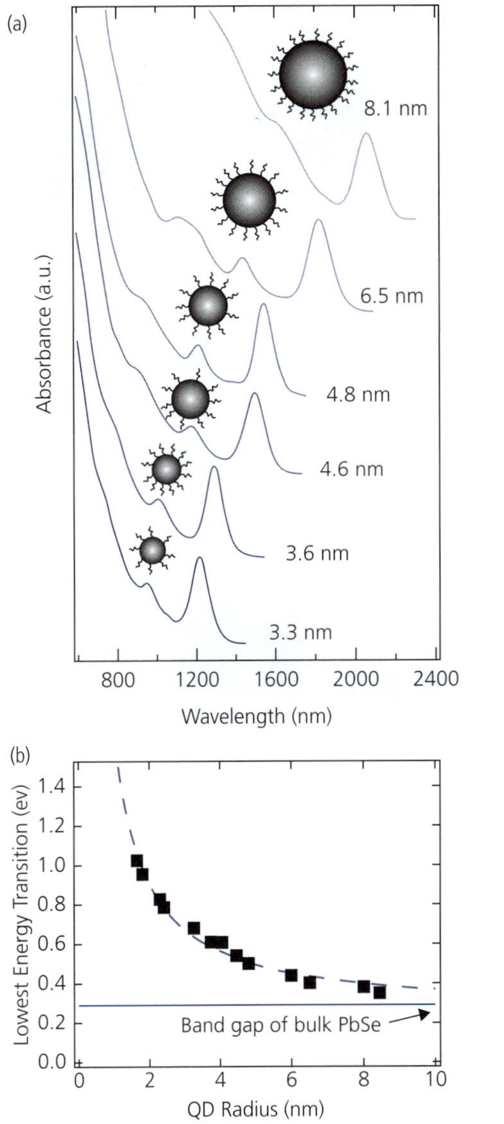

Figure 7.52 (a) Absorption spectra of monodisperse populations of PbSe quantum dots showing the dot with its isolating ligand structure. The corresponding dot radius for each spectrum is shown. (b) Lowest energy absorption transition vs radius of PbSe QDs. The dashed line is a fit to the data using Eq. 7.114 (Semonin et al., 2012).

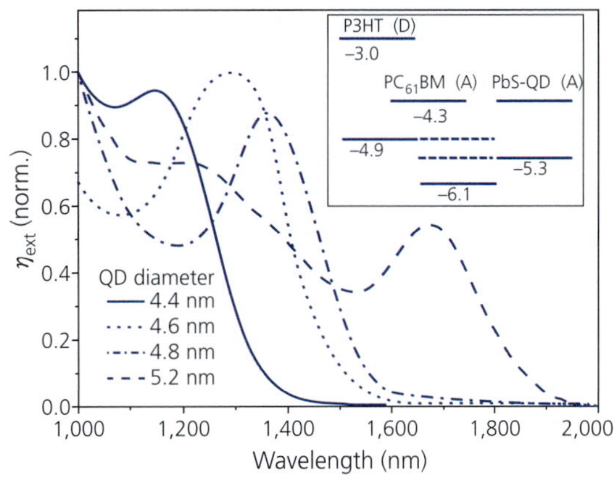

Figure 7.53 Normalized external quantum efficiency vs. wavelength for P3HT:$PC_{61}BM$:PbS-QD OPDs for different diameter QD populations. The curve for the 4.4 nm diameter dots corresponds to a maximum efficiency of ~15%. Inset: HOMO and LUMO energies (in eV) in the OPD active regions (Rauch et al., 2009).

(dashed line) that is based on quantum confinement following Eq. 7.114.

Rectifying QD photodiode arrays with sensitivity to $\lambda \leq 1900$ nm integrated with a CMOS passive pixel sensor (see Section 8.9.2) and a readout integrated circuit (ROIC) have been demonstrated. The external quantum efficiencies of the QD OPDs are shown in Fig. 7.53 (Rauch et al., 2009). The IR-imaging circuit comprised a 256 × 256 *focal plane array* (FPA) with a pixel pitch of 154 μm. The devices were a monodisperse population of PbS/oleic acid QDs blended in a P3HT donor and $PC_{61}BM$ acceptor BHJ active region. The 100–250 nm thick BHJ P3HT:$PC_{61}BM$:PbS-QD (1:1:4) absorbing region was applied by a doctor blade onto the ITO/PEDOT:PSS bottom anode that made direct contact to the drain of the transistors in the address array, and was capped by an Al cathode.

The energy level diagram of the QD OPD is shown in Fig. 7.53, inset. Both the PbS QDs and $PC_{61}BM$ are acceptors, and P3HT is the donor. Current leakage from each of the constituents, and hence at longer wavelengths (i.e. larger radii dots), the dark current is prohibitively large for sensitive IR detection. Devices employing 4.5 nm diameter dots with a long wavelength cutoff at approximately 1450 nm had a dark current of 10 nA/m^2 at −1 V. Combined with a peak external quantum efficiency of 16.5% at 1220 nm, this corresponds to $D^* = 2.3 \times 10^9$ Jones, which falls short of devices employing porphyrin tapes or CNTs. An unexpectedly large increase of η_{ext} with reverse bias is due to photoconductive gain in the blended active region. The imaging FPA had $f_{3dB} = 2.5$ kHz, which was limited by the device capacitance. This bandwidth can be increased by reducing the pixel pitch or by modifying the device thickness and other architectural details.

7.2.2.4 Photodiode applications

Photodiodes have found an enormous range of applications for sensing, imaging, and communications. In this context, OPDs are a relatively new entrant into the photodetector family of devices, and hence the numbers of sensing applications that have been demonstrated are few by comparison to those employing inorganic semiconductors. Nevertheless, the very broad wavelength range accessible to OPDs due to the variety of functional molecules that are synthetically accessible, make this class of detectors adaptable to detection from the UV to the MWIR. But "competing with silicon" is never a good idea unless there is a clear advantage to be gained by using a new technology. As we have already pointed out, organic electronic devices have features that are difficult and often impossible to access using their inorganic counterparts due to their inherent flexibility, stretchability, ultralight weight, narrow absorption spectral widths (leading to selective OPD transparency and detection), low temperature processing, and deposition that can lead to integration of devices on virtually any large area, flat substrate. In addition, one feature that we have not discussed up to now is the radiation hardness of organic thin films (Li and Yang, 2008, Kumar et al., 2011, Paternò et al., 2017, Barbé et al., 2018). The very small molecular weight atoms comprising organic devices make them relatively transparent to X-ray and gamma radiation. Hence, they may be useful for applications in space or in other high-energy radiation environments. In this regard, X-ray detection by combining scintillators with large area OPDs has been demonstrated, although this application domain remains in its infancy (Arca et al., 2013).

In this section, we consider only a few of the many sensing applications where OPDs have been employed. Our examples are chosen based on the unique materials and/or architectural features that organics provide. First, we describe OPDs used in full color imaging. Next, we show that the conformability of organics enables their application in hemispherical FPAs (HFPAs) that mimic the form, size, and several optical features of the human eye. This is followed by descriptions of position-sensitive detectors used for measuring both macroscopic and nanoscopic positions of illumination sources. We conclude with descriptions of an OPD integrated with an OLED to form an optoelectronic logical switching element and an organic charge coupled device imager. On occasion, OPDs have also been suggested for use in optical communications. We will not consider this application since there are many alternatives to OPDs that deliver considerably higher performance. This application space is adequately served by incumbent NIR detector technologies (e.g. based on Si and InGaAs), leaving only very specialized niches that may eventually be filled by OPDs.

Producing a full color image using an FPA is analogous to the challenge of providing a full color display

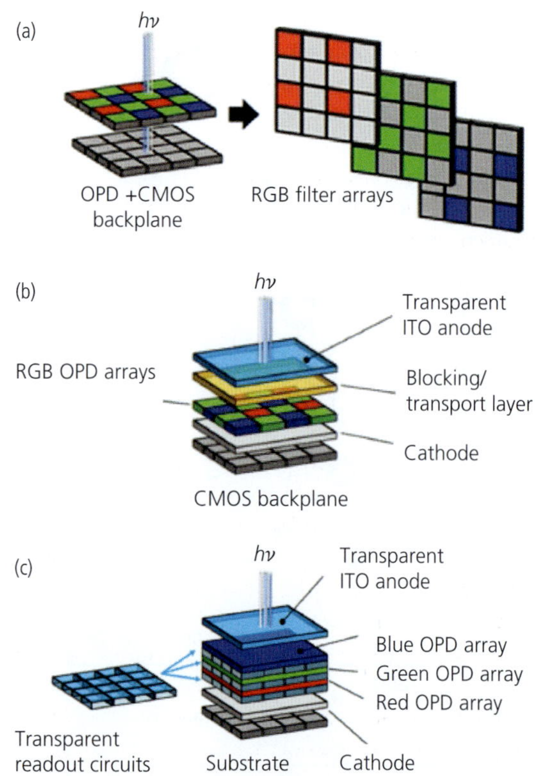

Figure 7.54 Three different two-dimensional (2D) full color imaging array architectures. (a) An OPD array combined with an RGB filter array. (b) R, G, and B sensitive, narrow spectral bandwidth OPD pixel arrays integrated with a CMOS ROIC. (c) Stacked R, G, and B narrow spectral bandwidth OPD pixel arrays. After Jansen-van Vuuren et al. (2016). © 2016 WILEY-VCH Verlag GmbH & Co. KGaA, Weinheim

image, and hence the architectures used in both applications are similar. While there are numerous optical methods for producing full color images using dispersive elements such as lenses, gratings and prisms, here we discuss how images are formed by the planar integration of devices and filters in a thin film construction. Figure 7.54a shows a broad spectral bandwidth OPD array integrated onto a CMOS pixel sensor/ROIC backplane. Color separation is provided by placing a red-green-blue (RGB) color filter mask above the OPDs, thus limiting the spectral window of each detector. The rectangular close-packing of the filter indicates that one of the color triad (green in this example) will have two pixels per array "unit cell", while the other two colors are each detected by only a single pixel. This inherently limits the imager resolution, requiring four detectors per full-color image pixel. Note that a similar duplication of the green pixel is often found in OLED displays (cf. Fig. 5.58). The filter architecture is analogous to OLED displays employing white emitting pixels placed beneath RGB filter arrays, or blue OLEDs used to pump R and G fluorescent down-converting "color changing" layers (see Section 6.4.2).

A second design is to use narrowly spectrally sensitive OPD arrays, with one OPD that detects in the blue, another in the green and a third in the red. The architecture in Fig. 7.54b is analogous to RGB emitting OLED displays employed in mobile applications. The OPDs are deposited directly onto the CMOS backplane readout array. Finally, in Fig. 7.54c we show a stacked architecture, where the light is incident on vertically disposed and separated R, G and B OPD arrays (Hokuto et al., 2007, Hokuto et al., 2011). Each color detecting array is integrated with a transparent ROIC within the stack. Alternatively, a single ROIC backplane can be located beneath the stacked OPD arrays, which is then vertically connected using vias to their corresponding transistor address circuits. The array has at least three times the resolution of the laterally disposed pixels in Fig. 7.54a and b. This advantage must be weighed against the disadvantage of increased fabrication and interconnection complexity, as well as the demands placed on the lens system to achieve a high image-plane depth of field.

An early demonstration of a full color imager was a one-dimensional scanning array of BHJ OPDs (Yu et al., 1998, 2000). The broad spectral bandwidth linear array comprises 102, ITO/150–180 nm thick P3OT:$PC_{61}BM$/Al OPDs on glass or flexible Mylar substrates. The array had a resolution of 40 dpi, with a total length of 64.8 mm. Images were obtained by sequentially interposing R, G, and B filters between the array and the object, and then scanning the image.

One-dimensional scanners capture images one line at a time, which is slow and only produces a blur-free image of fixed objects. These problems are eliminated with 2D CMOS FPAs commonly employed in digital cameras. A stacked, 2D imaging array is shown in Fig. 7.55a. Transparent ZnO thin film transistors (TFTs) are used as the address and readout electronics. The 128 × 96 arrays comprised small molecule-based OPDs with a 100 × 100 μm^2 pixel size, and with an image capture rate of 10 frames per second. The structure of the pixels is detailed in Fig. 7.55b. The ZnO transistors have a bottom gate, top source and drain configuration, comprising a SiO (50 nm)/SiN (100 nm) gate insulator, and a 45 nm thick ZnO channel. The OPDs are deposited onto the surface of an isolating, 200 nm thick SiN_x interlayer, and are connected to the drain via ITO and a 100 nm thick MoW signal line. A second, transparent ITO counter electrode caps the device. Light is incident onto the OPD via the substrate. A photomicrograph of a section of the ZnO TFT array is shown in Fig. 7.55c.

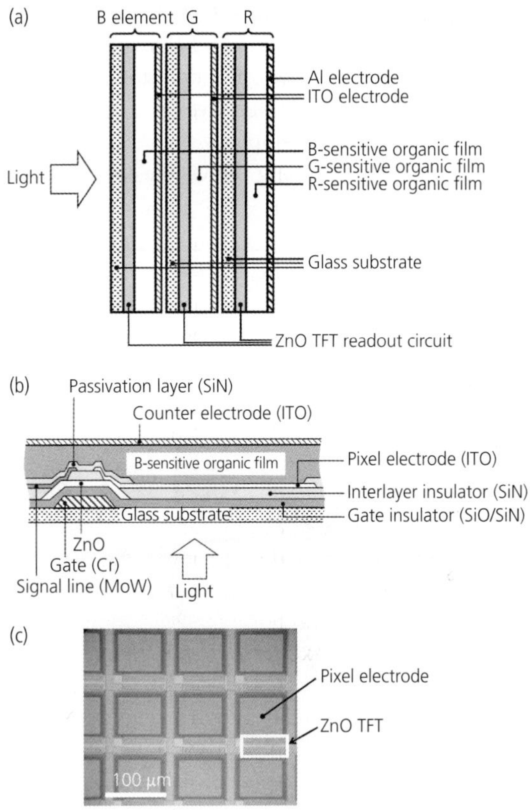

Figure 7.55 (a) Schematic of an RGB imaging system where each color OPD along with its readout circuitry is fabricated on separate substrates and aligned in a sandwich structure. (b) Cross-section schematic of a blue-sensitive pixel in the array. (c) Micrograph of a section of the ZnO thin film transistor array (Hokuto et al., 2011).

The detectors themselves consisted of the following photoactive regions: (i) B—coumarin 6 donor/C_{60} acceptor, (ii) G—DMQA donor/Py-PTC acceptor, and (iii) ZnPc+TiOPc donor/Alq_3 acceptor. ITO was used for the B and G counter electrodes, whereas Al was used for the R-sensitive device. Images of the three devices are shown in Fig. 7.56, inset, whose colors are the complement of their absorption spectra. Thus, the blue element appears yellow, the green is magenta, and the red is cyan. The external quantum efficiencies of the three sub-pixels are 7.6% at 10 V reverse bias for blue, 7.0% at 3 V for green, and 18% at 15 V for red. Their dark currents ranged from approximately 10 nA/cm^2 for blue to 0.5 $\mu A/cm^2$ at 15 V for red. These device characteristics are adequate for producing a full color image, albeit with some defects (i.e. line-outs) present that presumably can be eliminated with improved process control.

A hybrid combination of a CMOS imager and a green sensitive OPD array has been demonstrated by stacking the FPA, filters, and detectors, forming a hybrid of all three architectures in Fig. 7.54. The 5 Mpx CMOS array employs two Si PDs with 50–60% efficiency spanning the visible spectrum (see Fig. 7.57a). Blue and red color filters are deposited onto the pixelated Si PD and backplane surface. The Si electronics are planarized using chemomechanical polishing and covered with a semitransparent green-sensitive OPD. The flat surface is subsequently coated with sputter-deposited ITO that is connected to the underlying metal FPA circuit electrode. The OPD comprised a MoO_x:Al (5nm) cathode, DM-2,9-DMQA donor:SubPc acceptor (1:1, 90 nm) active region, and a MoO_x (10 nm)/ITO anode. A schematic illustration of the three-color pixel with a pitch of 1.4 μm is shown in Fig. 7.57b (Lim et al., 2015).

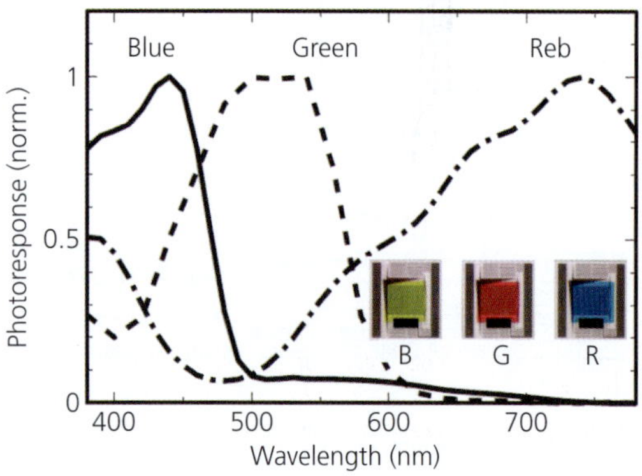

Figure 7.56 Normalized spectral response of the three-color OPDs used in the arrays in Fig. 7.55. Inset: Photographs of the three pixels (Hokuto et al., 2011).

The efficiencies of the two filtered Si PDs and the green OPD are approximately equal across the visible, as shown in Fig. 7.57a. The specific detectivity at the operating point of -3V reverse bias is 2×10^{12} Jones and the roll off frequency is $\Delta f = 76.7$ kHz. A 5MPx image of a color sample panel obtained using this hybrid Si/organic PD FPA is shown in Fig. 7.57c. Good color balance is obtained due to the nearly equal quantum efficiencies of the three RGB sub-pixels.

Flat FPAs used in cameras do not present an optimal form factor for imaging systems since the focal planes of spherical lenses are themselves spherical.

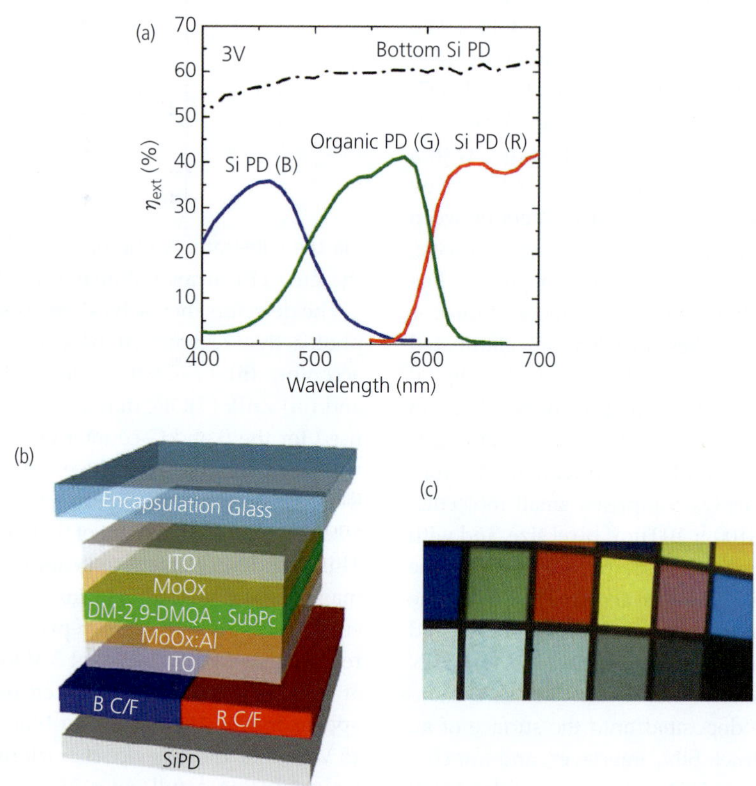

Figure 7.57 (a) Photoresponse of a hybrid Si/green OPD FPA showing the filtered and unfiltered Si PD responses. (b) Structure of the three-color-sensing pixel where the Si PDs are located beneath red and blue color filters (CF), and the green OPD covers the entire pixel surface. (c) Images of a color sample acquired with the 5 Mpx sensor array (Lim et al., 2015).

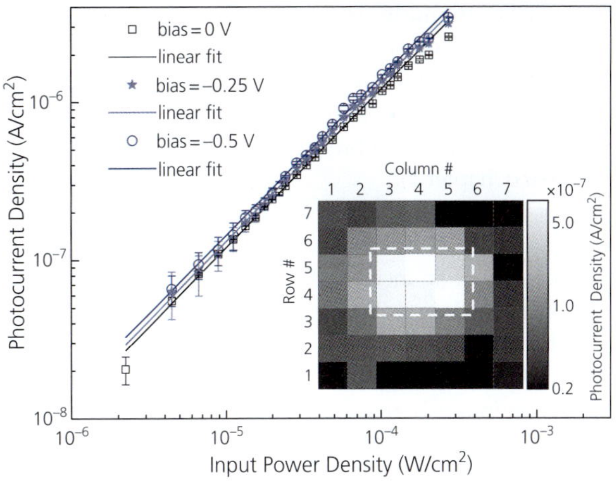

Figure 7.58 Photocurrent density vs. input power density of a (200 μm)^2 photodetector in a 20 × 20 array under various biases. Linear fits at each bias are shown as solid lines. The dynamic range, determined by the 1 dB compression point and minimum detectable power level, is approximately 20 dB, giving a 7-bit resolution. Inset: Grayscale image of a rectangular illuminated area (dashed line) on the hemispherical FPA, corresponding to a 2 × 3 block of photodetectors. Photocurrents of pixels in the 7 × 7 block containing the illuminated (at 633 nm) pixels show contrast with unilluminated pixels (Xu et al., 2008).

Thus, focusing an image onto a planar FPA requires a heavy, narrow field of view (FOV), high f number multi-element lens system to reduce image aberrations and chromatic dispersion. The planar form factor is a consequence of the limitations of photolithographic processing on inorganic semiconductor substrates. In contrast, the approximately spherical retina of the human eye provides a nearly ideal imaging system that employs a relatively simple, single element convex lens. Thus, the eye is compact, lightweight, has a very wide FOV nearly free of image distortion, and has a very low f/number.

In Section 5.6.4 we described the steps used in fabrication of a retina-like, HFPA that exploits the flexibility and conformability of organic electronics. The architectural challenges of mounting devices on the HFPA surface are met using arrays of OPDs on thermoformable substrates such as PET (Xu et al., 2008, 2009). The Au anode (10 nm)/CuPc (50 nm)/C_{60} (50 nm)/BCP (10 nm)/Ag (20 nm) OPDs were formed into a 100 × 100 passive matrix array of (40 μm)^2 OPDs on a 1 cm radius hemispherical surface. The external detector quantum efficiency had a maximum of 12.6% at $\lambda = 640$ nm. This was limited primarily by absorption in the thin Au anode oriented toward the incident illumination, and was increased to 20.5% at 605 nm when the Au was replaced by ITO in a strain-free stamp transfer process (Xu et al., 2009).

A grayscale image produced by the pixel photocurrents due to placing a square aperture in the focal plane of a 20 × 20 × (200 μm)^2 array is shown in the inset of Fig. 7.58. Apparently, several of the pixels lying beyond the aperture boundary (dashed line) also generate current that blurs the image. These are known as *sneak currents* in the passive matrix architecture arising from photocurrents within an illuminated row and column that add to the dark currents of the unilluminated detectors. This effect is illustrated in Fig. 7.59. The sneak currents can be reduced, but not eliminated by floating, rather than grounding the unaddressed row and column lines.

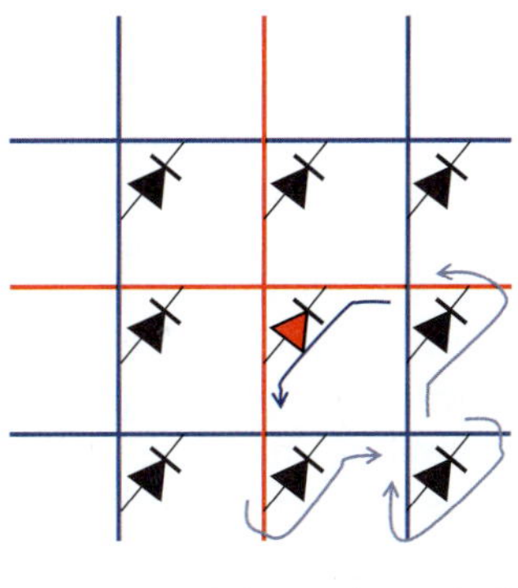

Figure 7.59 A passive matrix diode array showing the illuminated diode (red) and the selected row and column address lines. The photocurrent is shown by the black dashed line. Several parasitic sneak current paths are shown by blue dashed lines.

The photodetector specific detectivity is $D^* = 1.25 \times 10^{11}$ Jones for the devices with the ITO anode. The detectivity in the array, and hence its dynamic range, is determined by shot noise from the sneak currents. Hence, the shot noise increases as $\sqrt{N}$, where N is the number of diodes within a row that are addressed simultaneously during read out. The detectivity can be substantially increased if each photodiode is accompanied by a transistor address circuit that interrogates only one detector at a time while eliminating contributions from unilluminated pixels (see Chapter 8).

Position sensitive detectors (PSDs) are an important class of sensors that use the lateral photoeffect (Wallmark, 1957) to detect the location of an incident light beam. PSDs are commonly used in robotic vision, machine tool alignment, and guidance systems. The advantage of thin film PSDs over photodetector arrays is their ability to continuously track a signal with the resolution limited only by the light beam size, and not by detector size or pixel pitch. Furthermore, a thin film PSD requires only two outputs, whereas arrays require address lines to each detector. Since PSDs must track a beam over a large distance, the adaptability of organics to large, flexible substrates makes them well suited to this application (Rand et al., 2003, Kabra et al., 2004, Cabanillas–Gonzalez et al., 2011).

An example PSD comprises a long, transparent ITO strip deposited on a glass substrate used as both the anode for the photodetector and a resistive layer (see Fig. 7.60, upper inset). Light incident via the substrate generates a photocurrent, I_{ph}, that is collected at two, separated anodes. The total dark current for the device, I_D, is equally split between the two electrodes that are connected to an external circuit. The photocurrent contributions I_1 and I_2 are independently detected (see equivalent circuit, Fig. 7.60, lower inset) with magnitudes (Rand et al., 2003)

$$I_1 = \frac{I_D}{2} + I_{ph}\left(\frac{R_2}{R_1 + R_2}\right) \tag{7.115a}$$

and

$$I_2 = \frac{I_D}{2} + I_{ph}\left(\frac{R_1}{R_1 + R_2}\right), \tag{7.115b}$$

where R_1 and R_2 are the resistances of the ITO strip from the position of the incident light beam to the adjacent electrode. The currents detected at the two electrodes give the location, x_L, of the incident beam

$$x_L = \frac{(I_1 - I_2)L}{2(I_1 + I_2)}, \tag{7.116}$$

where L is the length of the device active region. Ideally, the ratio of currents should be a linear function of the light beam position. In practice, however, nonlinearities are observed that are quantified using: $\delta = 2\sigma/\Delta I_{FS}$ where σ is the rms deviation from a least squares fit to the current data, and ΔI_{FS} is the total swing in current for a beam scanned across total distance, L. Another characteristic is the fractional error in measuring the beam position given by, $\Delta l/L$, where Δl is the difference between the actual and calculated positions.

The bilayer PSD structure in Fig. 7.60 employed a 3 cm × 1 mm resistive ITO anode strip. The device is sensitive to wavelengths from 450 nm to 750 nm, with a peak efficiency of 12% at −2 V reverse bias. The data in Fig. 7.60 show the response of the PSD to a 180 μm diameter beam for different reverse biases ranging from 0 to −2 V. It is found that x_L is a linear function of beam position with a spatial resolution of 20 μm, and an error of $\Delta l/L = 0.05$ at −2 V. All other voltages show an error less than 1%. The table in Fig. 7.60

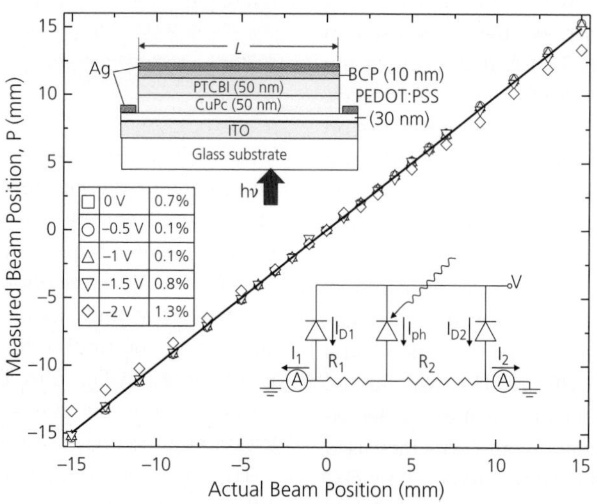

Figure 7.60 Measured vs. actual beam positions for the position sensitive OPD, shown in upper inset. Lower inset shows the equivalent circuit of the OPD. The table provides a legend and the positional error, δ, of the beam over the OPD length, L. Adapted from Rand et al. (2003).

gives δ vs. reverse bias as well as for different optical powers at −1 V.

The tracking speed is one further criterion that characterizes the PSD. The capacitance of the PSD is 15.3 nF at −1 V bias, resulting in $\tau = RC = 9.2$ μs for the ITO strip with a resistance of 600 Ω. Accurate and stable tracking can be obtained for a beam velocity (v_{max}) less than the resolution divided by the response time. In this case, $v_{max} = 20$ μm/9.2 μs = 2.2 m/s.

A 2D PSD can be realized by placing pairs of contacts orthogonal to each other along the device periphery. As the device area increases to accommodate larger ranges of beam scans, the dark current also increases, reducing the sensitivity and scan rate.

There are several alternative architectures that can be implemented to locate the beam position. For example, the absorption spectrum can be spatially varied by grading the donor and acceptor thicknesses between the contacts. The position is determined from the spectral response at x_L (Cabanillas–Gonzalez et al., 2011). This approach requires the use of a donor and acceptor with distinct absorption spectra, precise linear grading of the bilayer along the PSD length, the use of at least two wavelengths to obtain the relative absorption in the donor vs. the acceptor layer, and layers whose thicknesses are not significantly greater than the exciton diffusion length in each layer. This latter constraint ensures that excitons generated far from the D-A junction do not recombine before arriving at the HJ. While this architecture may be attractive for limited applications, it is not suited for general position sensitive detection.

The ability to deposit organics onto virtually any surface has enabled a high spatial resolution scanning probe PSD (An et al., 2008). The scanning PSD is located on the vertex of a conventional, Si AFM tip by forming a 500 nm diameter platform using focused ion beam (FIB) milling (see Fig. 7.61a). The OPD cathode comprising Al on a thin Ni adhesion layer was deposited onto the tip surface, making contact to the Si. The entire surface was coated with perylene, and the tip face was milled to open a 500 nm window that defined the detector active area. This was followed by the deposition of a C_{60}/α-NPD bilayer active region coated by a blocking layer and anode metallization. External contact to the diode is made via the Si tip and the outer metal coating remote from the tip itself.

The tip is placed in contact with a metallized glass substrate with an 8 × 20 μm aperture illuminated from its reverse surface (Fig. 7.61a). A topographic scan of the aperture surface shown in Fig. 7.61b is acquired from the AFM as the tip is translated along the aperture surface. Simultaneously, an optical scan is obtained by measuring the OPD current vs. position (Fig. 7.61c). Line scans with micron-scale resolution of the topography and the photocurrent along the dashed lines in Fig. 7.61b and c are shown in Fig. 7.61d and e, respectively (An et al., 2008).

Another application for an OPD is in the bistable optical latch shown schematically in Fig. 7.62a (Xue and Forrest, 2002). The green emitting PHOLED has a 12 nm thick semitransparent Mg–Ag/ITO cathode that serves as the anode of the eight-period, multilayer

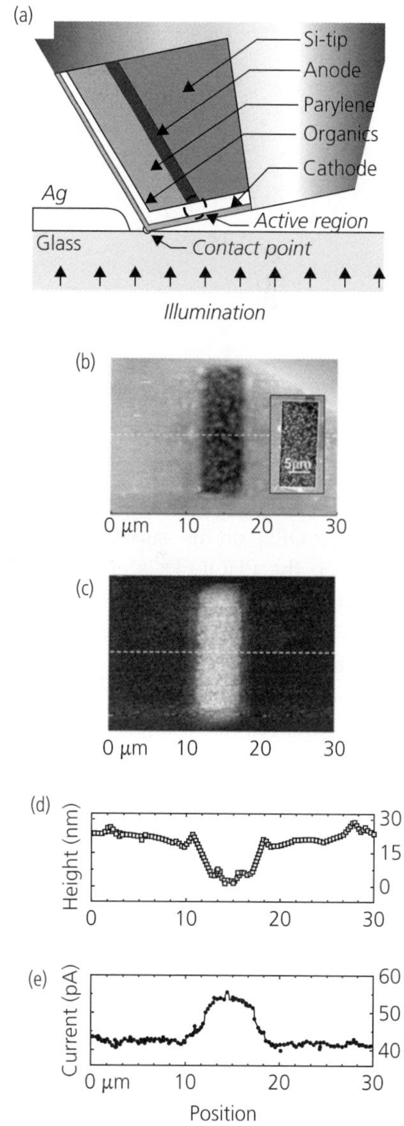

Figure 7.61 (a) Schematic illustration of a microdetector at the end of a flattened AFM tip. (b) Topographic scan of an aperture in a Ag film on glass. Inset: Scanning electron micrograph of the 8 × 20 μm aperture. (c) Optical image of the aperture using the micro-OPD illuminated from the reverse side of the glass. (d) Topographic scan and (e) photocurrent scan of the aperture along the dashed lines in (b) and (c) (An et al., 2008).

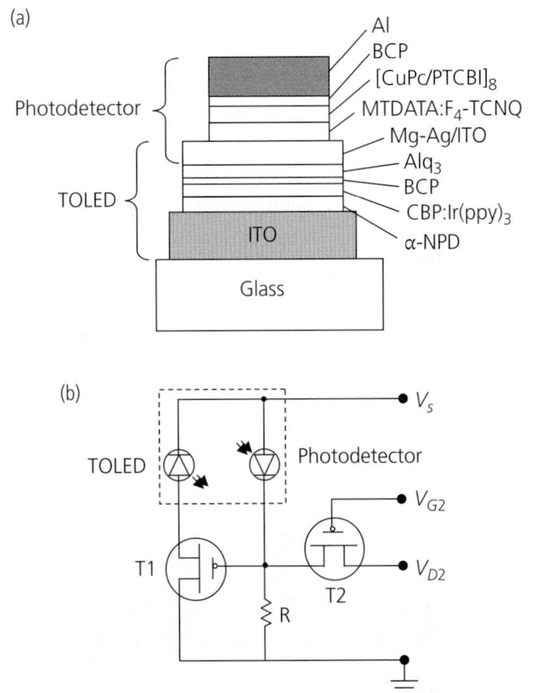

Figure 7.62 (a) Schematic cross-section of a transparent OLED integrated with an OPD in an optical bistable latch. (b) The integrated latch (dashed box) and external circuit used to create bistable operation (Xue and Forrest, 2002).

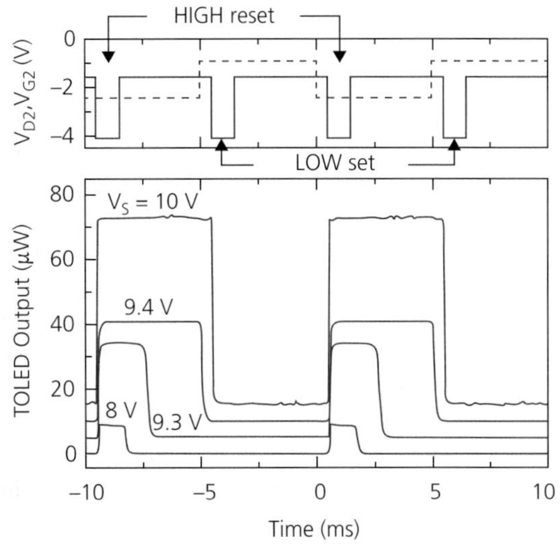

Figure 7.63 Operation of the TOLED/OPD bistable switch in Fig. 7.62. The upper panel shows the waveforms of the switching signals V_{D2} (dashed line) and V_{G2} (solid line). The lower panel shows the TOLED emission power vs. V_S (shifted vertically for clarity) (Xue and Forrest, 2002).

$[\text{CuPc/PTCBI}]_8$ OPD on its surface (cf. device in Fig. 7.44). When the PHOLED is energized, light is emitted through the substrate into the viewing direction. Light is also emitted from its top surface, and absorbed in the adjacent OPD. The resulting photocurrent provides positive feedback via an external load, R, at the drain of transistor, T1 (Fig. 7.62b). The circuit latches the transparent PHOLED in the emitting state until a blanking voltage, V_{G2}, is applied to the gate of T2. Thus there are two possible states of the circuit. In the LOW state, the PHOLED does not emit light, so that the current passing through the photodetector, is solely its dark current. By choosing an appropriate value of the load resistor, R, then $V_{T1} < V_{G1}^L < 0$, where V_{T1} is the threshold voltage of T1, $V_{G1} = -|I_{PD}| \cdot R$ is the gate voltage of T1, and the superscript "L" represents the LOW state. Hence, T1 remains off to maintain the LOW state of the PHOLED. In the HIGH state, the PHOLED emits green light, generating a photocurrent. When the photodetector current is sufficiently high at $V_{g1} < V_{T1} < 0$ (V_{g1} is the gate voltage of T1), T1 is turned on and the PHOLED continues to emit. Transistor, T2, therefore, toggles the switch between HIGH and LOW.

Optical bistability is shown in Fig. 7.63, where V_S is varied from 0 to -10 V. The input of the drain and gate voltages, V_{D2} and V_{G2}, are shown in the upper panel. Here V_{G2} (solid line) is delayed from V_{D2} (dashed line) by 0.5 ms, and has a pulse width of 1 ms. During the pulse window of V_{G2}, T2 is turned on, setting V_{G1} to -0.95 V or -2.45 V, thereby switching the device to LOW, or reset to HIGH, respectively. The waveforms of the PHOLED bottom emission intensity at different V_S are shown in Fig. 7.63. At $V_S = -8$ V, the PHOLED is on during the reset window. However, the HIGH state is not stable, and the PHOLED turns off when T2 is switched off. As V_S is increased, the PHOLED remains on for a brief period after T2 is switched off. The stable HIGH state is achieved for $V_S = -10$ V, that is, the PHOLED is turned on at the onset of the reset window. It remains on until the onset of the set window, at which time it is turned off and remains off until the next reset pulse. The reset or set windows can be as narrow as 60 ns to make the device switch between the two stable states. The bandwidth of the circuit is $\Delta f_{3dB} = 25$ kHz limited by the external components (Xue and Forrest, 2002).

This integrated switch has potential applications in displays, where static image regions can be set, allowing for low bandwidth circuits that need to refresh only those pixels whose image content changes between frames. Also, it can simulate paper whereby an image can be generated and then maintained for long periods without need for refreshing the data. Finally, given its digital response to set and reset pulses, the switch is a building block for photonic logic circuits. Taken alone, the integrated PHOLED/photodetector can be used in linear circuit applications such as in automatic display brightness control and sensing.

One last example application is an organic charge coupled device, or OCCD. Charge coupled devices transport charge packets near a semiconductor-insulator interface by sequentially passing them along between closely spaced electrodes in a voltage-controlled "bucket brigade" fashion. Since their introduction (Amelio et al., 1970, Boyle and Smith, 1970), CCDs have been successfully applied to high sensitivity optical imaging as a result of their favorable signal-to-noise ratio, high quantum efficiency, linearity, and speed (Janesick et al., 1987, Tiffenberg et al., 2017). In Section 4.5.5, we showed that charge transport can occur across distances of several centimeters in organic thin film heterostructures using energetic barriers to isolate electron and hole polarons (Burlingame et al., 2018). This striking phenomenon has led to the demonstration of an OCCD where electron packets delivered from an array of OPDs are transferred using a linear, three-phase shift register. Reaching the end of the register, the packets are read out, providing information about the intensity and spatial distribution of the incident light. In contrast to the operation of conventional CCDs that attract minority carriers to depletion regions at the semiconductor-insulator interface, the electrodes of the n-type organic semiconductor shift register are negatively biased to block charge transport via repulsive potential barriers along its length (Coburn et al., 2019).

A four-pixel OCCD is shown schematically in Fig. 7.64a. The materials used in the OPD array and electron-transporting channel are identical to those shown in Fig. 4.77. Thirteen metal electrodes are photolithographically patterned on a glass substrate and coated with Al_2O_3 – 11 electrodes separated by 50 μm define the shift register, and one each forms the guard and shutter electrodes. Four photodetectors with DTDCPB:C_{70} HJ active regions are deposited on the anode, each connected to the nearest shift register electrode by a 20 nm thick C_{60} electron-transporting channel. Inside the shift register, the channel runs in a vertical stripe that extends past the guard electrode and terminates under the Ag cathode. The three shift register potential phases are denoted Φ_1–Φ_3. The electrode nearest the guard electrode is Φ_1, followed by Φ_2, Φ_3, Φ_1, etc. A photograph of the device is shown in Fig 7.64b.

Photocurrent is collected between the anode and cathode in response to illumination of the photodiodes. This current is cut off by applying < -8 V at any one of the shutter, guard, or Φ_1 electrodes since they bisect the channels connecting the OPDs to the cathode. Charge transport is effected by first biasing

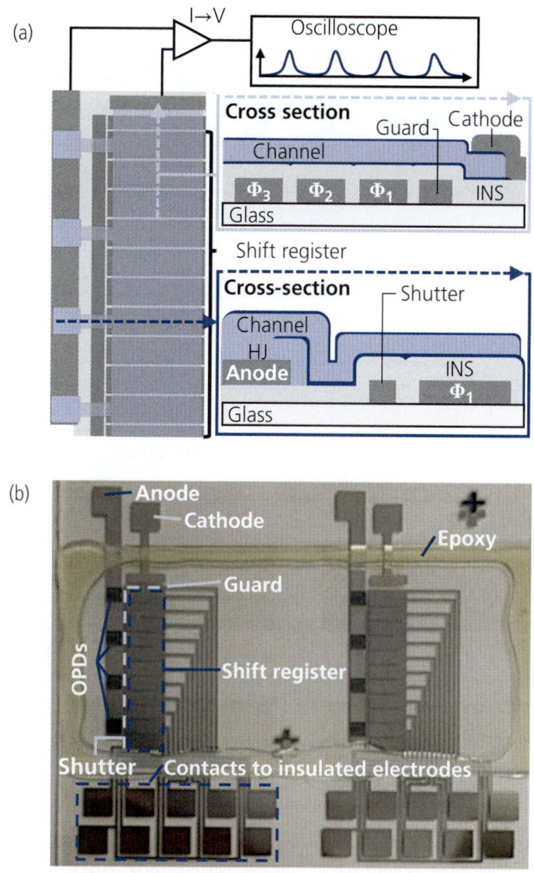

Figure 7.64 (a) Top and cross-section schematic views of the OCCD, and the measurement circuit. The guard, shutter, and Φ_3 have a single contact pad each, while Φ_1 and Φ_2 electrodes have multiple pads that are connected externally. (b) Photograph of two, 1.8 cm long OCCDs located side by side. The contact pads and traces, alignment marks, and encapsulation cover slide attached to the substrate using UV cured epoxy are visible (Coburn et al., 2019).

Φ_2, Φ_3, and shutter electrodes to -10 V, and $V_{guard} = -8.75$ V. This expels charge from the channel above the biased electrodes, leaving background charge packets confined above Φ_1. The shift register is clocked using a phased sequence of three potentials, advancing the charge packets toward the cathode following the scheme in Fig. 7.65a. Because $V_{guard} > -10$ V, electrons diffuse towards the cathode when expelled from above the adjacent Φ_1 electrode, preventing them from flowing back into the shift register. After expelling background charge, $V_{shutter}$ swings to -9.6V, which blocks the background (dark) charge from leaking into the shift register. Then one HJ is illuminated, generating charges and raising the potential of electrons in the channel. This increases the rate of electrons leaking from the illuminated HJ over the shutter. Following exposure, $V_{shutter}$ is switched back to -12 V and the shift register is emptied by a

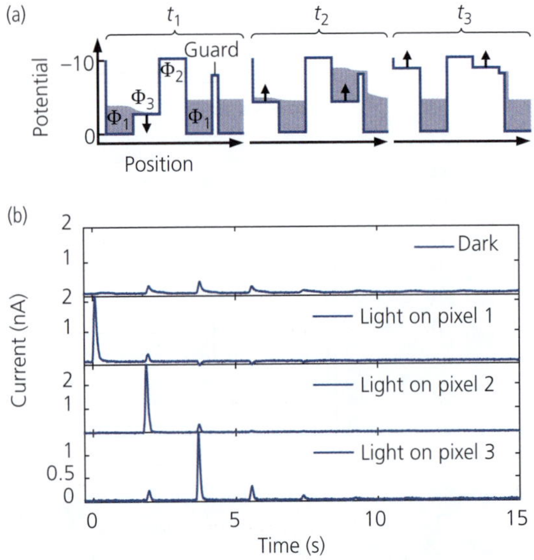

Figure 7.65 (a) The potential sequence for electrodes at the end of the shift register. The electrode potentials are shown at times t_1–t_3. High potentials repel the electron charge packets, indicated as shaded areas. (b) Measured signal for dark and single pixel illumination exposures. Here, $t = 0$ corresponds to the first time Φ_1 swings negative following exposure.

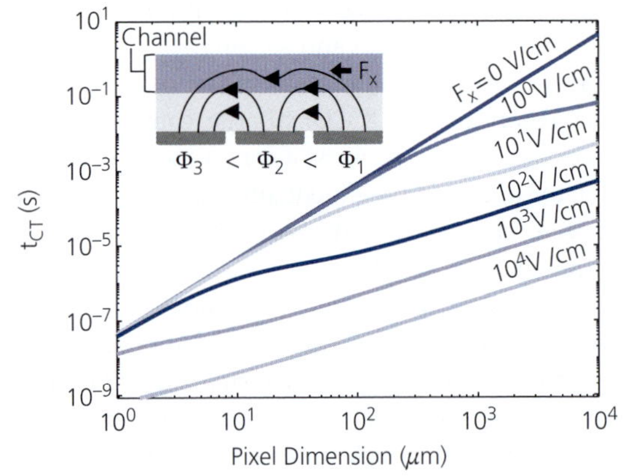

Figure 7.66 Simulated charge transfer time calculated from the solution to the drift-diffusion equation vs. electrode size and fringe field, F_x. The transfer time, t_{CT}, is the time at which 99.99% of the initial uniform charge distribution has transferred to the adjacent electrode. Inset: Illustration of the fringe fields for an electrode-channel spacing similar to the electrode width. The fields are established by applying voltages such that $\Phi_1 > \Phi_2 > \Phi_3$. Lines represent electric field lines in the direction of the arrows (Coburn et al., 2019).

series of eight clock cycles. Figure 7.65b shows the measured current in the dark and with illumination focused onto one pixel per measurement, resulting in a distinct current pulse observed at a time delay corresponding to the position of the illuminated pixel.

Using majority carrier electrons to store information contrasts with the operation of inorganic CCDs that transfer minority carriers. Using majority carriers precludes using attractive potentials (positive electrode bias, in the case of electrons) to confine charge packets because no depletion region would form. Also, the response of the OCCD is quite slow compared to inorganic CCDs, primarily due to its dependence on charge diffusion blocked by the shift register potentials. The speed of response can be significantly increased, however, by applying an electric field along the transport direction. This is shown via a calculation of the time required for 99.99% charge transfer between adjacent electrodes, t_{CT}, using the drift-diffusion equation as a function of electric field, with results in Fig. 7.66. The readout time for a $N \times N$ matrix of pixels ($N \gg 1$) is $t_{RO} \approx t_{CT}(3 \times N)^2 / N_{amp}$, where N_{amp} is the number of readout amplifiers. In the diffusion limit, a 500×500 sensor with $N_{amp} = 4$ requires 6–7 μm diameter pixels to achieve $t_{RO} < 1$ s, while 10 μm pixels and $N_{amp} = 500$ (1 per column) yields $t_{RO} < 20$ ms. Introducing drift transport significantly decreases t_{RO}. Drift can be increased by fringing fields as shown in Fig. 7.66, inset.

Despite the inherently lower charge mobility in organic compared to inorganic semiconductors, OCCDs may find uses in space applications, where their light weight (Cardinaletti et al., 2018, Schreurs et al., 2018) and comparatively low sensitivity to cosmic radiation (Kumar et al., 2010, Paternò et al., 2017) can be beneficial. They can function as an element in ultralight optical imaging systems when used with thin film Fresnel lenses (O'Neill et al., 2003), or as lightweight memory devices. Additionally, stacked, semitransparent OCCDs with individual pixel triads absorbing in the red, green and blue spectral regions can produce color images.

7.3 Organic solar cells

Whenever there is significant opportunity to provide a social benefit that may also be financially rewarding, there are many contenders to capture the new market. The solar energy field is a singular example where benefits to the commercial sector as well as to the social welfare of humanity, abound. This has led to considerable competition attention and investment in next generation solar cells that deliver advantages beyond today's dominant source of solar power generation: crystalline and polycrystalline Si photovoltaics. It is important, therefore, to clearly establish the benefits afforded by new approaches such as OPVs for generating low cost and environmentally friendly electricity from solar energy. In the following, we attempt to lay out important considerations for new solar harvesting technologies that will clarify the advantages, and potential application spaces that are available to OPV technology.

(i) *Cost*: The cost of the solar cell is wrapped up in all of its performance parameters, from efficiency to operational lifetime, to the bill of materials and manufacturing yield. Silicon solar provides a module cost of < \$0.50/Wp, (here Wp is the peak wattage output of the module). Added to this are *soft costs* (e.g. costs of financing, taxation, installation, etc.), inverters, support equipment, and so on. A disadvantage of Si, however, is in the transportation, installation, and support of heavy, fragile, glass-like panels. In contrast, OPVs can be fabricated on lightweight and flexible substrates that reduce the requirement for costly and heavy support structures. The cells can be applied to windows, rooftops, curved automobile surfaces, etc. in the form of tapes, panels, or adhesive appliqués. Furthermore, roll-to-roll (R2R) fabrication of high efficiency single and multijunction cells on flexible substrates has a realistic opportunity to dramatically reduce cost, as OPVs are produced "by the kilometer" rather than traditional semiconductors fabricated in small batches of wafers.

(ii) *Efficiency*: Silicon module power conversion efficiencies (*PCE*) exceed 20%, with even higher efficiencies (nearly 29%) achieved for GaAs-based cells (Green et al., 2017). A sampling of efficiencies of several solar cell technologies is provided in Table 7.5. Among these examples, organic solar cells apparently have the lowest efficiencies. However, there are many other criteria beyond efficiency that must be met before a technology gains popular acceptance, including cost, reliability, environmental friendliness, and large-scale manufacturability. For example, comparison of the costs of Si and GaAs technologies is an illustration of how efficiency is only one important parameter to consider. Multijunction, group III–V semiconductor cells have $PCE \sim 46\%$ under intense solar concentration, but are prohibitively expensive for terrestrial commoditization. In many applications, efficiencies at or around 15% may be adequate, provided that other factors such as lower cost support equipment, longer lifetime, etc. compensate for the lower efficiency. OPV efficiencies are now poised to exceed 17% with the introduction of NFAs. Yet this is still significantly below the ~25% thermodynamic efficiency limit for OPV single junction cells (Giebink et al., 2011), with even higher efficiencies projected for multijunction cells. Hence, it is reasonable to assume that with continued development of OPV materials and device architectures, efficiencies will eventually exceed 20%.

(iii) *Reliability*: Solar cell standards demand 20–25 year module lifetimes. Both Si and GaAs technologies meet this requirement. As noted above, reliability also ultimately factors into the total module cost of ownership, where a longer operational lifetime implies a longer period between panel disposal and replacement. Reduced manufacturing costs or higher efficiencies may reduce this requirement to some extent, but reliability must be high to achieve market acceptance. OPVs have been making substantial progress in this domain as well. Operational stability under accepted operating environmental conditions is still in its early stages, but lifetimes of some high efficiency cells are now well in excess of 20 years. Considerable strides are being made to understand

Table 7.5 Example solar cell efficiencies

Cell type	Configuration	Max. PCE[a]	Ref.[b]
Inorganic, thermodynamic limit	Single junction	31%	(Shockley and Queisser, 1961)
OPV, thermodynamic limit	Single junction	22–27%	(Giebink et al., 2011)
Multijunction III–V cell	Concentrated	46%	(Green et al., 2017)
GaAs cell	Single junction	28.8%	(Green et al., 2017)
Crystalline Si cell	Si single junction	26.3%	(Green et al., 2017)
Crystalline Si cell	Module	24.0%	(Green et al., 2017)
CIGS	Thin film	24.4%	(Green et al., 2017)
Perovskite	Thin film	19.7%	(Green et al., 2017)
Organic cell	Single junction	15.7%	(Yuan et al., 2019)
Organic cell	Multijunction[c]	15.4%	(Li et al., 2018)

[a] Measured under AM1.5G spectrum at 1 kW/m^2, except for the concentrated multijunction III–V cell measured at 508 suns intensity.
[b] Green et al. (2017) is a summary compilation of results. References to specific data quoted can be found in the compilation.
[c] Although higher efficiencies have been reported, multijunction cells are notoriously difficult to measure accurately. This is the highest known reliable value to date.

and reduce failure mechanisms. Based on experience in the now ubiquitous OLED display industry, the concerns for OPV lifetime are dissipating, yet considerable work remains for this technology to achieve widespread acceptance.

(iv) *Environmental friendliness*: The complete absence, or at least minimization of toxic materials used in both manufacturing and in the completed panels is essential. The ubiquitous deployment of solar panels magnifies even a small level of toxicity into a major environmental hazard. Here OPV excels beyond all other incumbents. The active materials are similar to dyes used in clothing, ink jet printers, displays (LCD and OLED), paints, and so on. In short, they are already widespread, and many materials are environmentally benign. The most problematic components are the inorganic materials used such as indium tin oxide. Heavy metals (e.g. Pb, Se, and Cd) and toxic and reactive chemicals (e.g. halide salts) are also largely absent in OPVs, and the synthetic processes are similar to those used in preparing dyes. Even less solvent-intensive methods are used in the production of fullerenes. These processes and their products tend to be low cost and importantly, a very low energy investment is required in the production of both the constituent materials as well as in the fabrication of the cells themselves. That is, the energy required for OPV production is rapidly returned early in their operational life.

(v) *New capabilities*. One factor in choosing a new technological paradigm is the features that it offers that cannot be easily replicated by existing solutions. In this regard, OPV technology has demonstrated its ability for efficient energy production even if the cells are transparent, or semitransparent in the visible. This is a result of the ability to tailor OPV materials sets that absorb primarily in the NIR between wavelengths of 650 nm to 1100 nm, thus capturing a significant amount of the solar power that is invisible to the human eye. This makes OPVs particularly attractive for coatings on windows or on building facades (so-called *building integrated photovoltaics* and *building applied photovoltaics*, or BIPV and BAPV, respectively). This feature distinguishes OPVs from conventional Si, GaAs, and other thin film inorganic technologies and provides unique opportunities for their early and rapid adoption by large, unserved markets when combined with other attributes of high efficiency and reliability, and low-cost manufacture.

(vi) *Leveraging existing industries and capabilities*: Starting up an entirely new manufacturing process with new materials where there is no significant existing technical base is prohibitively expensive. The Si solar industry took advantage of, and leveraged the integrated circuit knowledge base and industrial capabilities for producing raw materials before it gained a sufficiently strong foothold to become self-supporting and sustainable on its own. Likewise, OPVs are building on the enormous economic base and manufacturing industry of OLED displays and lighting. The semiconductor equipment infrastructure industry for OLEDs alone is $\sim$ \$US 20 bn/year. Much of this same equipment and expertise can be readily transferred to the OPV industry. Similarly, the OLED display panel industry represents a $>$ \$US 20 bn market, with $>$2 bn OLED smartphone displays, more than a million TV sets, and a high number of electronic tablets that use large display panels in the hands of consumers today. Hence, the infrastructure and knowhow for manufacturing OPV panels is in place, and growing daily. Major OLED manufacturers are currently developing OPVs, recognizing the synergies that exist for rapid scaling into a major industry.

(vii) *Synergies with other industries, and workforce development*: The lessons learned in fabricating OPVs (e.g. by R2R production) can be rapidly translated into the R2R manufacture of low cost OLED lighting, and indeed can also be translated into the manufacture of inorganic semiconductor appliances. The work force skill and expertise developed in exploring OPV technology has rapidly adapted to solve problems in these other industries. The intellectual curiosity and imaginative versatility one gains in studying organic electronics can lead to unanticipated advances in organic as well as inorganic semiconductor device research and technology.

In the remainder of this section, we explore many potential advantages of OPVs by describing their underlying physics and technology. Indeed, while OLEDs are the only clear example of an organic electronic technology making a large commercial impact, OPVs stand poised to be the second example owing to their inherent adaptability to large area, low cost, flexible devices that offer competitive performance.

7.3.1 Solar cell basics

A solar cell converts solar irradiance into electrical energy. In its simplest terms, a solar cell is a photodiode that is operated in the fourth quadrant of its j–V characteristics, that is, it is operated in the photovoltaic mode. It is only in this quadrant that power can be produced. A solar cell must be optimized to capture the maximum solar irradiation with minimum loss, whereas a photodiode is tuned to have low noise and high sensitivity within a narrow spectral band. As such, the materials, the details of the active region, the methods used for coupling broad solar radiation vs. a narrower wavelength band into the solar cell, and the contacts must be optimized for power rather than current generation.

Solar cell, or photovoltaic cell, operation can be understood from its j–V characteristics in Fig. 7.67. The open circuit voltage (V_{OC}) and short circuit current density (j_{SC}) are defined by the intercepts of the j–V characteristics with the V and j axes, respectively. Here, $j_{SC} = j_T - j_D$, where j_T is the total current and j_D is the dark current. Typically, $j_{SC} \gg j_D$ such that it is simply equal to the photocurrent, j_{ph} at $V = 0$. In the fourth quadrant, the power density generated is $P = -j_{ph}V < 0$. In the first and third quadrants, $P > 0$ and hence power is dissipated. By convention, in solar cells the power generated is taken as positive. The maximum power generated by the cell (called the *maximum power point*, or MPP) is defined by the largest area rectangle (shaded area) whose corner opposite the origin intersects with the j–V curve. Using our sign convention, the maximum power density generated is given by

$$P_M = j_M V_M, \tag{7.117}$$

where j_M and V_M are the corresponding current and voltage at the MPP. It is convenient to define the cell figure of merit, the *fill factor*, FF, as the ratio of the areas of the rectangles defined by j_{SC} and V_{OC} to that at P_M, viz.

$$FF = \frac{j_M V_M}{j_{SC} V_{OC}}, \tag{7.118}$$

which, according to Eq. 7.55 for an organic HJ and the Shockley ideal diode equation (Eq. 4.282) for inorganic junctions, is always less than unity. Then the power conversion efficiency (PCE or η_P) is the ratio of the electrical power generated by the cell to the optical power density incident from the sun, P_{opt}. Thus,

$$\eta_P = \frac{j_M V_M}{P_{opt}} = \frac{FF(j_{SC} V_{OC})}{P_{opt}}. \tag{7.119}$$

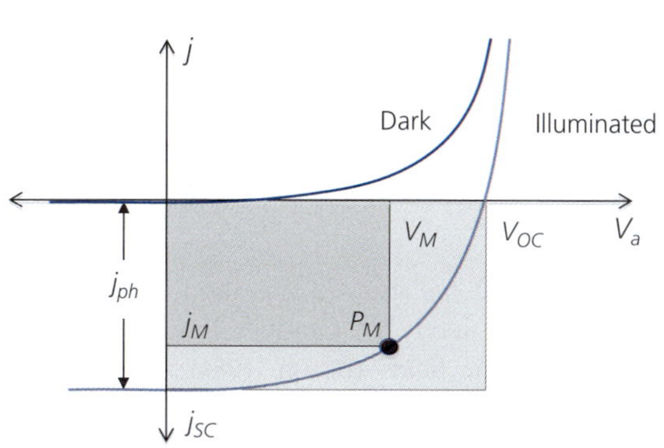

Figure 7.67 Current–voltage characteristics of a solar cell in the fourth quadrant. Variables are defined in text.

Clearly, we seek designs that have the highest fill factor, that is, that conform most closely to a diode whose ideality factor, $n \approx 1$. This requires a minimum series resistance and a maximum shunt resistance that approaches the thermal resistance of the junction (see Section 7.1.2).

The short circuit current density depends on the efficiency of the solar cell and its overlap with the solar spectrum:

$$j_{SC} = q \int \eta_{ext}(\lambda)\phi_{sun}(\lambda)d\lambda, \tag{7.120}$$

where $\phi_{sun}(\lambda)$ is the solar spectral flux density. The reference solar spectrum standards are set by the American Section of the International Association of Testing Materials (ASTM). The flux at the Earth's surface is a function of the spectral properties of the intervening atmosphere, or the *air mass* (AM) that acts as a filter to solar radiation. Referring to Fig. 7.68, the irradiance arriving at the top of the atmosphere from a solar black body source with a temperature of approximately 5800 K is known as the AM0 reference spectrum, ASTM-490-00, with an intensity of 1.353 kW/m^2. This spectrum is primarily used by the aerospace community for determining satellite solar panel efficiencies. The AM1 spectrum corresponds to light impinging normal to the surface at the zenith ($z = 0°$, i.e. at normal incidence). This spectrum is observed within the Earth's equatorial regions and has an intensity of 1 kW/m^2. The most commonly used spectral standard is that found in the temperate regions, where the intensity is 9638 W/m^2, but for convenience is often simply taken as 1 kW/m^2. The spectrum is different from that at the equator, since light travels through a deeper blanket of atmosphere. The reference spectrum (ASTM G-173-03) is known as the *global* AM1.5, or AM1.5G spectrum. This reference is an average of collection of the solar irradiance from a network of observation points across the United States. It assumes incidence

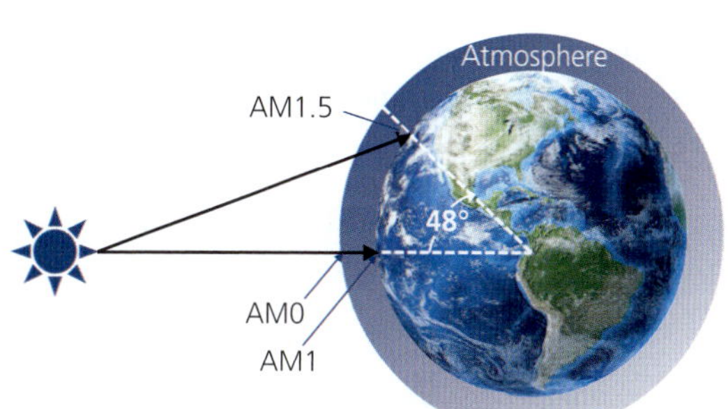

Figure 7.68 Solar irradiance on the Earth and the corresponding air mass designations.

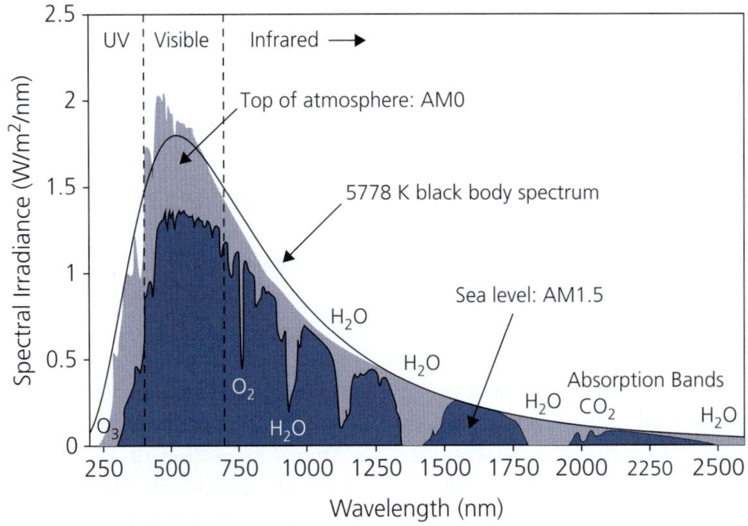

Figure 7.69 Spectra for several different illumination conditions.

on a tilted surface at 37°, and a zenith at $z = 48°$ facing due south.

Different illumination conditions, along with the corresponding spectral irradiance intensities are shown in Fig. 7.69. The AM1.5G spectrum shows the strong effects of atmospheric absorption, particularly by H_2O across the IR. The measurement of solar cell efficiency begins with this understanding of reference spectra. However, the process of standardized quantification of cell efficiency is made more complicated due to the absence of such spectra in the laboratory where the solar cell is evaluated; a topic that is covered in Section 7.3.3.

An approximate relation between solar intensity and the air mass at sea level with the sun at the zenith is given by

$$I = 1.1 \cdot I_0 [0.7]^{AM^{0.678}}. \tag{7.121}$$

Here, $I_0 = 1.353\ \mathrm{kW/m^2}$ corresponds to the AM0 intensity at the top of the Earth's atmosphere, and 1.1 assumes that there is an additional 10% power due to diffuse scattering.

Another quantity in Eq. 7.119 that determines the power conversion efficiency is V_{OC}. From Eq. 7.55, setting $j = 0$, we obtain

$$V_{OC} = \frac{n_S k_B T}{q} log\left(\frac{j_{ph}}{j_0} + \frac{k_{PPd}}{k_{PPd,eq}}\right) \approx \frac{n_S k_B T}{q} log\left(\frac{j_{SC}}{j_0} + 1\right). \tag{7.122}$$

In the expression to the right, we take $j_{ph} = j_{SC}$ at $V_a = 0$ for convenience, and that ionization of PPs is small, such that $k_{PPd} = k_{PPd,eq}$. This is rarely, if ever the case for OPVs, and hence this latter expression can lead to errors. In any case, V_{OC} increases logarithmically with the photocurrent, and with decreasing dark saturation current.

Plots of the fourth quadrant j–V characteristics for several different shunt and series resistances and their implication on the maximum power generated by a solar cell (P_M) are shown in Fig. 7.70a and b. In describing solar cell characteristics, the fourth quadrant is often "flipped" into the first quadrant, plotting both j_{ph} and V as positive quantities. The generated power has a maximum at P_M, falling off as V_a deviates from this point. The power curves, therefore, are the locus of areas of the rectangles within each point of the j–V characteristic. A decreasing R_{shunt} decreases the curvature at the MPP, which depresses both j_M and V_M. Similarly, an increasing series resistance decreases the forward-biased slope, which results in a decreased V_M. The fill factor is strongly influenced by both the series and shunt resistances.

Now, from Chapter 4, and using Eq. 7.56 for j_0, V_{OC} for a symmetric junction is

$$qV_{OC} = \Delta E_{HL} - n_S k_B T ln\left[\left(\frac{k_{rec} k_{PPr}}{k_{PPd} + k_{PPr}}\right) \frac{q a_0 N_S^2}{j_{ph}}\right]. \tag{7.123}$$

In Section 4.7.2, we showed that the PP dissociation rate is determined by Onsager–Braun ionization at small electric fields (Braun, 1984). Then, $k_{PPd} \to \frac{3}{4\pi a_0^3} k_{rec} \exp(-E_B/k_B T)$, where E_B is the PP binding energy and a_0 is its radius (approximately equal to

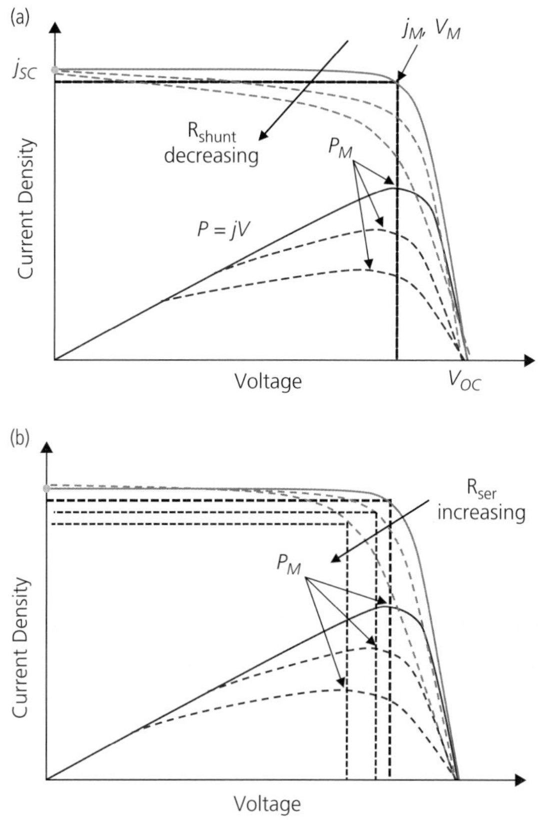

Figure 7.70 Power-generating quadrant of a solar cell. The ideal j–V characteristics (grey solid lines) and with (a) decreasing shunt and (b) increasing series resistances (grey dashed lines) are shown. The blue lines (solid and dashed) are the corresponding power output curves equal to jV. The maximum power outputs, P_M, at the MPPs are indicated.

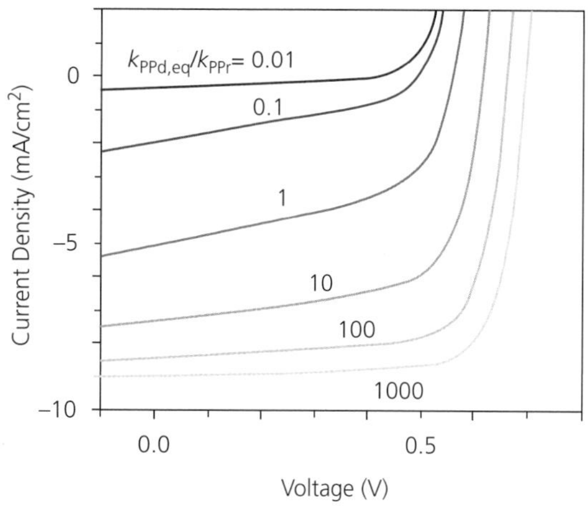

Figure 7.71 Calculated j–V characteristics in the fourth quadrant for an archetype planar HJ photodiode in reverse bias ($V_a < 0$) under AM1.5G solar illumination at an intensity of 100 mW/cm^2. The curves correspond to different ratios of the PP dissociation and recombination rates, k_{PPd} and k_{PPr}, respectively (Renshaw et al., 2012).

the molecular separation distance at the HJ). In the limit of very high illumination intensity or low temperature, this expression leads to a maximum open circuit voltage of

$$qV_{OC}^{max} \to \Delta E_{HL} - E_B \tag{7.124}$$

(cf. Eq. 4.308). Hence, an essential challenge for optimizing OPV efficiency is to reduce E_B. We will show that NFAs are effective in reducing this loss channel compared with conventional fullerene-based OPVs.

Equation 7.123 shows that V_{OC} is dependent on several factors. Notably, there is a competition between PP recombination vs. dissociation, expressed by the recombination efficiency of

$$\eta_{PPr} = \frac{k_{PPr}}{k_{PPr} + k_{PPd}}. \tag{7.125}$$

When the dissociation rate is increased relative to recombination (i.e. $k_{PPd} > k_{PPr}$), the losses are decreased, leading to an increase in V_{OC}. The trade-off is illustrated in Fig. 7.71 where the j–V characteristics are plotted for an ideal HJ diode for varying ratios of k_{PPd} to k_{PPr} (Renshaw et al., 2012). Interestingly, the decrease in FF and j_{SC} with decreasing k_{PPd}/k_{PPr} is even larger than the decrease in V_{OC}. This is due to curvature in the j–V characteristics as a result of field-induced PP ionization and current losses from recombination. Hence, not only is it essential to reduce k_{PPr}, but also to increase a_0 to achieve high efficiency. We will show in the following sections that these factors are controlled by molecular design as well as the morphologies of the donor and acceptor phases within mixed and BHJ active regions.

The maximum power is found by multiplying the current in Eq. 7.55 by V_a, and then setting the first derivative to zero. Once P_M is found, then from Eq. 7.117, it can be shown that

$$V_M = V_{OC} - \frac{n_S k_B T}{q} log\left(\frac{qV_M}{n_S k_B T} - \chi(V_M)\right), \tag{7.126}$$

where $\chi(V_M) = \frac{k_{PPd}(V_M)}{k_{PPd,eq}}$ and we assume for convenience that the change in χ is small and nearly constant with V_a. Now, setting $\beta = (j_{SC} + j_M)/j_{SC}$, then at the maximum power

$$\beta_M = \left[\chi(V_M) + log\left(\beta_M \frac{j_{SC} + j_D}{j_D}\right)\right]^{-1}. \tag{7.127}$$

Here, we have replaced the saturation current, j_0, with j_D, which includes all parasitic sources that contribute to the dark current. The FF in the absence of series resistance, and for $R_{shunt} \to \infty$, can now be expressed in terms of β_M as

$$FF(R_{shunt}, R_{ser}) = FF(\infty, 0) = \frac{(\chi(V_M) - \beta_M) log[\beta_M(j_{SC} + j_D)/j_D]}{log[(j_{SC} + j_D)/j_D]}. \quad (7.128)$$

Hence, the fill factor depends on both the dark and the short circuit currents (and hence the incident power). An increase in power results in an increase in j_{SC}, which in turn decreases β_M. Ultimately, this increases FF. An increasing dark current, j_D, has the opposite effect; its presence results in a decrease in FF, as implied by the results in Fig. 7.71 where $j_0 \propto (1 - k_{PPd,eq}/k_{PPr})$ for a large PP recombination rate.

A realistic model of cell performance must include the dependence of FF on the parasitic cell resistances. For this, we define the *power loss penalty* (f_{PLP}), which is the fractional decrease in power from its value in the absence of series resistance, and an infinite shunt resistance. For a finite shunt resistance R_{shunt}, and $R_{ser} \to 0$, then,

$$f_{PLP} = \frac{V_M^2}{R_{shunt}} \Big/ j_M V_M = \frac{V_M}{j_M R_{shunt}} \approx \frac{V_{OC}}{j_{SC} R_{shunt}}. \quad (7.129)$$

Similarly, for a nonzero R_{ser} and $R_{shunt} \to \infty$,

$$f_{PLP} = \frac{j_M^2 R_{ser}}{j_M V_M} = \frac{j_M R_{ser}}{V_M} \approx \frac{j_{SC} R_{ser}}{V_{OC}}. \quad (7.130)$$

Combining the results of Eqs. 7.129 and 7.130 we have

$$FF(R_{shunt}, R_{ser}) = FF(\infty, 0)(1 - f_{PLP}) = FF(\infty, 0)\left(1 - \frac{V_{OC}}{j_{SC} R_{shunt}} - \frac{j_{SC} R_{ser}}{V_{OC}}\right). \quad (7.131)$$

Figure 7.72 shows a plot of FF as a function of j_{SC} for a solar cell with several values of R_{ser} and R_{shunt}. The increase of FF with increasing current, and its decrease in the presence of parasitic resistances is apparent. Note that this plot extends beyond one sun intensity, showing the FF for concentrated illumination. The concentration factor of a solar cell is equal to

$$CF = \frac{P_{inc}}{1\ \text{kW}-\text{m}^{-2}}. \quad (7.132)$$

7.3.2 Thermodynamic limits to OPV efficiency

Before embarking on an exploration of means for achieving high performance OPVs, we first calculate the fundamental limits to their efficiency. The *Shockley-Queisser* (SQ) *thermodynamic efficiency limit* (Shockley and Queisser, 1961) was derived for inorganic solar cells using the principle of detailed balance (Shockley and van Roosbroeck, 1954), which requires that the absorbed and emitted radiation fluxes be equal for a cell in thermal equilibrium. The SQ treatment is based on the premise that photon interactions with the cell occur via the generation

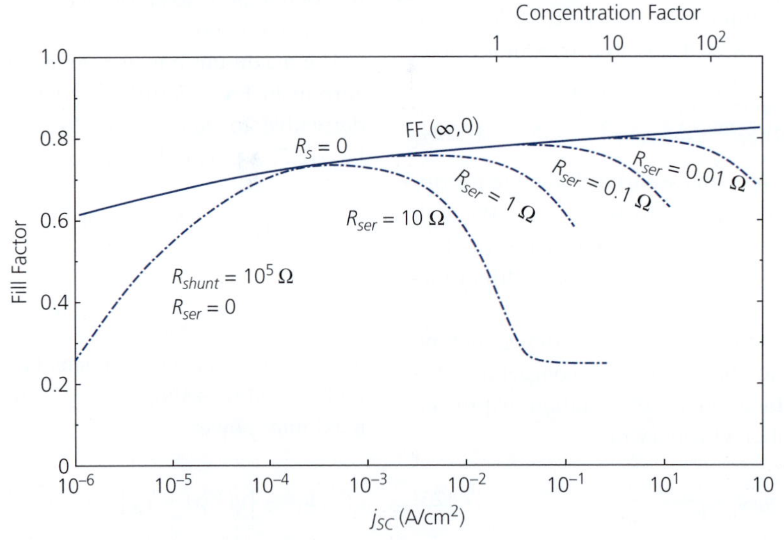

Figure 7.72 Fill factor vs. short circuit current density for several values of series and shunt resistances, R_{ser} and R_{shunt}. The cell has the following properties: j_{SC} = 1 nA/cm^2, n_S = 1, active area = 1 cm^2. $R_{shunt} = \infty$ except for the dashed curve. The concentration ratio is defined as the intensity of the light relative to 1 sun (1 kW/m^2). After Bube and Fahrenbruch (1981).

and recombination of free electron–hole pairs. This is a good assumption for inorganic semiconductor solar cells because photon absorption leads to the direct generation of free carriers. As a result, the chemical potential, μ_{ph}, of emitted photons is the same as that of the electron and hole quasi-Fermi level splitting, which for an ideal cell is equal to the voltage difference between its contacts. That is, $\mu_{ph} = V_a$, and the emitted light intensity is increased by $\exp\left(q\mu_{ph}/k_BT\right)$ relative to equilibrium (Luque et al., 2002, Luque and Marti, 2003, Würfel and Würfel, 2009).

The situation is fundamentally different in organic semiconductors, where the primary excitations are tightly bound Frenkel and charge transfer excitons with binding energies ranging from 0.3 to 1.2 eV. Thus, photons interact only with the bound exciton population, and hence the photon chemical potential is that of the exciton population, not that of the free electrons and holes. The efficiency limit of an excitonic solar cell consequently depends on both the optical energy gap and the exciton and PP (or CT) binding energies, which is in stark contrast to the SQ limit for inorganic cells, which applies to the special case where these binding energies can be neglected.

Charge generation in organic semiconductors via exciton and PP dissociation leads to the diode equations in materials containing, or are free of traps in Sections 4.7.2 and 7.2. It is a multistep process that begins by photon absorption that generates Frenkel excitons within either the donor or acceptor layer. On reaching the D-A HJ, the exciton dissociates into a charge transfer state whose Gibb's free energy is less than that of the excitonic precursor by ΔG_{CT} (see Fig. 7.73). This state also leads to increased radiative and/or non-radiative recombination. The last step in the charge generation process is the liberation of the CT state binding energy, E_B, to generate free charges.

The thermodynamically limited efficiency can be obtained using the detailed balanced approach of Shockley and Queisser, although this generally requires the implicit and artificial inclusion of losses arising from formation of, and radiation by the intermediate CT state (Kirchartz et al., 2009). Here, we present a treatment by Giebink et al. (2011) based on the Second Law of Thermodynamics where the inclusion of the free energy loss, ΔG_{CT}, is an explicit element of the model. We make the following assumptions. (i) The cell consists of only a single junction. Stacked, multijunction cells can exceed the limits calculated here, as we will show in subsequent sections. Nevertheless, single elements within such stacked structures are still limited by these results. (ii) Each photon can only generate a single exciton. Multiexciton generation processes such as singlet fission are not included. (iii) Every incident photon with energy greater than the optical energy gap, E_{opt}, of

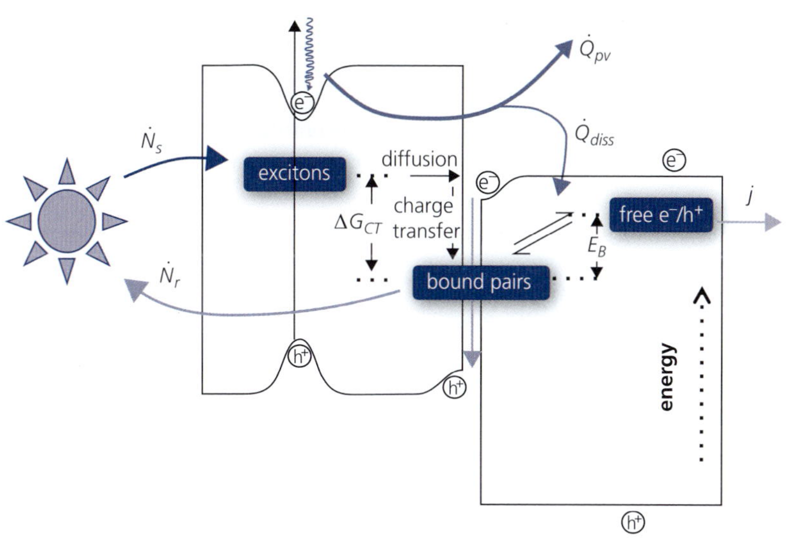

Figure 7.73 The photogeneration of free charge in an excitonic solar cell in the radiative limit. The incident solar photon flux ($\dot{N}_s$) is fully absorbed by the cell, forming bound excitons. Radiative exciton recombination produces the luminescent photon flux ($\dot{N}_r$) emitted from the cell. Alternatively, excitons dissociate at the D-A heterointerface with a loss in free energy of ΔG_{CT}, to form polaron pairs with binding energy, E_B. Subsequent formation of free carriers requires heat input ($\dot{Q}_{diss}$) to overcome E_B, subtracting from the heat produced in the photovoltaic process, $\dot{Q}_{pv}$. The current from the free electrons (e^-) and holes (h^+) is given by j. In an ideal OPV, recombination only occurs from PP states since all excitons reach the HJ (Giebink et al., 2011).

either the donor or acceptor is absorbed ($\eta_A = 1$). (iv) A photogenerated exciton ultimately arrives at the HJ with unity efficiency, that is, $\eta_{ED} = 1$. (v) All photogenerated free charges, n, p, are collected at their corresponding contacts ($\eta_{CC} = 1$). Finally, (vi) the solar radiation at the Earth's surface is that of a perfect black body with a surface temperature of $T_s = 5778$ K. Thus, effects due to atmospheric absorption that define the AM1.5G spectrum in Fig. 7.69 are ignored.

The exciton dissociation process must be accompanied by a net decrease in Gibb's free energy, whose change is given by (Lewis and Randall, 1961, Zemansky and Dittman, 1981, Giebink et al., 2011)

$$dG = \mu dN = dE - TdS + \mathcal{P}d\mathcal{V} < 0. \tag{7.133}$$

Here, N is the particle number, E is the internal energy, S is the entropy, $\mathcal{P}$ is the pressure, and $\mathcal{V}$ is the volume. Since the particle flux is equal to the cell current, $I = jA$, and for a constant volume, the Gibb's free energy flux is (Luque and Marti, 2003)

$$\Delta\dot{G}_{diss} = (\mu_{eh} - \mu_{PP})I = (E_{eh} - E_{PP})I/q - T_a\Delta\dot{S}_{diss}, \tag{7.134}$$

where T_a is the cell temperature, and the subscripts PP and eh refer to the bound polaron pair and free electron/hole pairs, respectively. Since the entropy change is equal to the heat supplied to the reaction, $T_a\Delta\dot{S}_{diss} = \dot{Q}_{diss}$, and since $E_B = E_{eh} - E_{PP}$, we have

$$q(\mu_{PP} - \mu_{eh}) = -E_B + q\dot{Q}_{diss}/I > 0. \tag{7.135}$$

Note that the difference between the exciton binding energy, E_X, and that of the PP is the free energy decrease, ΔG_{CT}. That is,

$$E_{PP} = E_X + \Delta G_{CT}, \tag{7.136}$$

where $\Delta G_{CT} < 0$.

According to Eq. 7.135, photocurrent is generated if the entropy increase upon dissociation (i.e. heat input, $\dot{Q}_{diss}$ in Fig. 7.73) fully compensates for E_{PP}. This process, therefore, results in cooling of the OPV. Hence, when $\dot{Q}_{diss} > (I/q)E_B$, the heat flux is sufficient to drive exciton dissociation, and the exciton chemical potential can, in principle, equal the cell voltage which simply recovers the SQ result (Shockley and Queisser, 1961, Luque and Marti, 2003). However, when $\dot{Q}_{diss} < (I/q)E_B$, the heat input is insufficient and μ_{PP} must exceed V_a by the amount required to compensate for this deficiency.

The total luminescent photon flux radiated from the cell, assuming that it is a black body, is determined by integrating the flux at each energy:

$$\Phi(\mu, E_{opt}, T, f_\omega) = \frac{2f_\omega}{h^3c^2}\int_{E_{opt}}^{\infty}\frac{E^2}{\exp[(E - q\mu)/k_BT_a] - 1}dE = \frac{2f_\omega}{h^3c^2}\int_{E_{opt}}^{\infty}\phi(E, T_a)dE, \tag{7.137}$$

where $E = hc/\lambda$ is the photon energy and f_ω is the *étendue* of the source. The lower integration limit, E_{opt}, is the minimum exciton absorption energy of the cell (i.e. its optical energy gap, see below), and $\phi(E)$ is the photon flux between E and $E + dE$. The possible increase of μ relative to V_a as discussed above increases the radiative recombination rate compared to an inorganic solar cell, and leads to a maximum efficiency that is lower than that given by the SQ limit. Physically, the increase in exciton chemical potential indicates that the exciton density must increase relative to that of free electrons and holes to compensate for the large exciton binding energy and small internal electric field, and thereby maintain a total dissociation flux that is capable of supplying the cell photocurrent.

The étendue is the geometrical extent of the source (sun) as seen from the system (i.e. the solar cell). That is, f_ω is the product of the area of the solar cell and the solid angle subtended by the source at the solar cell at the Earth's surface. The angular acceptance of the cell is π. If the half angle subtended by the sun is θ_{sun}, then,

$$f_\omega = \pi sin^2\theta_{sun} = \pi sin^2(0.267°) = 5.52 \times 10^{-5}. \tag{7.138}$$

The entropy generated by charge generation is given by (Luque and Marti, 1997, 2003, Luque et al., 2002)

$$T_a\dot{S}_{PV} = \dot{Q}_{PV} = -IV_a + (\dot{E}_s - T_a\dot{S}_s) - (\dot{E}_r - T_a\dot{S}_r) = \int_{E_{opt}}^{\infty}[q(\mu_s - V_a)\phi_s(E) - q(\mu_{PP} - V_a)\phi_r(E) + (\dot{\omega}_s - \dot{\omega}_r)]dE, \tag{7.139}$$

where the subscripts s and r correspond to thermodynamic fluxes of radiation from the sun and cell, respectively, and $\dot{Q}_{PV}$ is the heat from the excess energy used to generate an exciton with energy, E_{opt} (see Fig. 7.73). In Eq. 7.139, $\dot{\omega}$ is the spectral energy flux. Also, we take $\mu_s = E(1\text{–}T_a/T_s)$ as the chemical potential equivalent of the sun at T_s (Luque and Marti, 2003).

The first line of Eq. 7.139 shows that the entropy generation is equal to the difference between the

Helmholtz free energy flux of incoming solar radiation and that of the outgoing luminescent radiation, less the work done by the cell (Luque and Marti, 1997). Since only radiative recombination occurs in an ideal cell, that is, $I = q(\dot{N}_s - \dot{N}_r)$, then from Eqs. 7.135, 7.137, and 7.139 we obtain

$$E_B \int_{E_{opt}}^{\infty} (\phi_s - \phi_r) dE = \int_{E_{opt}}^{\infty} q(\mu_s - V_a)\phi_s dE - \int_{E_{opt}}^{\infty} q(\mu_{PP} - V_a)\phi_r dE + \int_{E_{opt}}^{\infty} (\dot{\omega}_S - \dot{\omega}_r) dE, \quad (7.140)$$

where $\phi_s(E)$, $\phi_r(E)$, $\dot{\omega}_r$, and $\dot{\omega}_S$ are evaluated by assuming that the only radiative species is the PP state, since all excitons reach the HJ where these states are formed. Equation 7.140 allows for an implicit solution of $\mu_{PP} = V_a$ (Giebink et al., 2011).

The PP or CT state can also absorb photons, as required by detailed balance, and as observed experimentally (see Section 4.7.2) (Vandewal et al., 2009, 2010). Since the PP is an extended interface state, both its oscillator strength and density are considerably smaller than that of the bulk exciton. Ideally, the bulk absorbance of the cell, $\alpha(E)$, is unity above the exciton energy gap. However, in OPVs, $\alpha(E)$ is significantly lower in the region between E_{PP} and the exciton binding energy, E_X. From these assumptions, the current for the HJ cell is

$$j_{SC} = q \int_{E_{PP}}^{\infty} \alpha(E)(\phi_s(E) - \phi_r(E)) dE, \quad (7.141)$$

where

$$\alpha(E) = \begin{cases} 0 & \text{for} \quad E < E_{PP} \\ \alpha_{PP} & \text{for} \quad E_{PP} < E < E_X \\ 1 & \text{for} \quad E > E_X \end{cases}, \quad (7.142)$$

and the domain of the CT absorption and emission region is $E_{PP} - E_X = \Delta G_{CT}$ (cf. Eq. 7.136). The ratio between absorption by the exciton and the PP state is typically $a_{PP} = 10^{-3}$ (Vandewal et al., 2009, 2010). Equations 7.141 and 7.142 imply that the larger energy gap of bulk materials dominates in absorption (at $E > E_X$), whereas the smaller binding energy of the pairs dominates the emission (at $E_X > E > E_{PP}$). Consequently, the radiative recombination rate is increased, since it occurs via low energy bound pair states. This formulation of the absorption coefficient has also been approximated by a Gaussian distribution for $\alpha(E)$, which follows the approximately Gaussian tail of CT emission in polymer solar cells (Kirchartz et al., 2009).

Once a solution for $\mu_{PP} = V_a$ is found from Eqs. 7.140–7.142, then using the j–V characteristics for an ideal excitonic HJ (e.g. Eq. 7.57 with $n_S = 1$, $R_{ser} = 0$) and $T_s = 5778$ K, the extracted power density is

$$P_{pv} = V_a j(V_a). \quad (7.143)$$

Now the solar power density, or solar irradiance, is given by Planck's black body law:

$$P_{sun} = \frac{2f_\omega}{h^3c^2} \int_0^{\infty} \phi(E, T_s) E dE = \frac{2f_\omega}{h^3c^2} \int_0^{\infty} \frac{E^2}{\exp(E/k_B T_s) - 1} E dE. \quad (7.144)$$

Then it follows that

$$\eta_P(V_a) = \frac{V_a j(V_a)}{P_{sun}}, \quad (7.145)$$

which has a maximum at

$$\frac{d}{dV_a}(V_a j(V_a)) = 0. \quad (7.146)$$

The energy loss that ultimately determines the cell efficiency limit is a function of the saturation current (j_0 in Eq. 7.56), which is an exponential function of the energy level offset at the D-A HJ, and the PP dissociation and recombination rates (found in η_{PPd}) that are implicit in the foregoing analysis.

Figure 7.74 shows the cell efficiency vs. the organic absorber optical energy gap for several values of the free energy loss for charge transfer, $-\Delta G_{CT}$, calculated using Eqs. 7.141 and 7.142 assuming $T_s = 5778$ K and $T_a = 300$ K. The calculations assume that the integrated solar intensity at the Earth's surface is equal to 1 kW/cm^2. The SQ limit is shown for $-\Delta G_{CT} \to 0$, although such a minimal free energy difference, in practice, produces negligible power since there is no driving force for excitons to form bound PPs. As $-\Delta G_{CT}$ increases, the theoretical efficiency peak monotonically decreases and shifts toward higher optical gap energies.

The peak efficiency depends on the CT state absorbance, a_{PP}. Equation 7.141 shows that the SQ limit of $\eta_P = 31\%$ is recovered when either $a_{PP} \to 0$ or $a_{PP} \to 1$. However, $a_{PP} > 0$ with its maximum value limited by the low oscillator strength of the PP state. In the physically relevant range of $10^{-4} < a_{PP} < 10^{-2}$, the

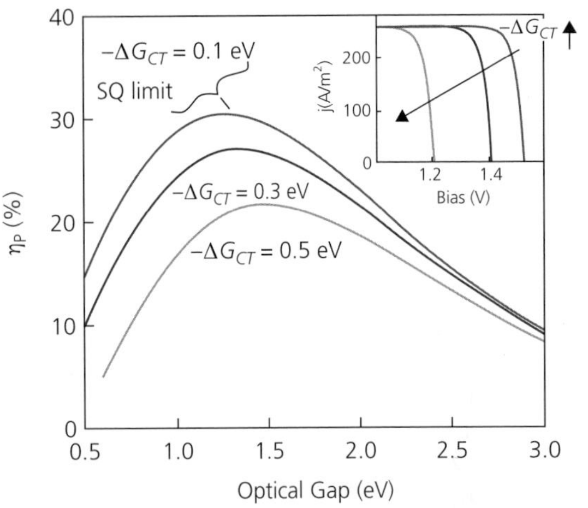

Figure 7.74 Maximum efficiency for an ideal single junction OPV cell as a function of exciton optical energy gap and several values of the free energy for charge transfer at the HJ, $-\Delta G_{CT}$. Increased recombination from the lower energy bound pair states leads to reduced efficiency due to the decrease in V_{OC}, as shown in the inset for a cell with $E_X = 1.8$ eV (Giebink et al., 2011).

efficiency decreases for increasing a_{PP}. Although the efficiency limits are based on values chosen for a_{PP} as obtained from experiment, the conclusion is general: *any heterojunction OPV cell where exciton dissociation results in a lower energy state with incomplete coupling to the radiation field, necessarily operates with an efficiency below the SQ limit* (Street, 2008).

A practical lower limit of $-\Delta G_{CT} \sim 0.2$ eV is required for efficient charge transfer, leading to photocurrent generation (Rand et al., 2007). However, we will find in the following sections that $-\Delta G_{CT}$ between 0.4 eV and 0.8 eV are common. Taking $-\Delta G_{CT} = 0.3$ to 0.5 eV, as typical of high efficiency OPVs (Rand et al., 2007, Veldman et al., 2009), a maximum single junction solar cell efficiency of $\eta_P =$ 27–22% at energy gaps of 1.35 eV and 1.5 eV, respectively, is obtained from Fig. 7.71.

The inset of Fig. 7.71 shows that the decrease in efficiency is entirely due to a decrease in V_{OC}, and is not affected by j_{SC} due to our assumptions that all photons absorbed ultimately result in PP formation. Since the small PP binding energy does not limit dissociation, then $V_a = \mu_{PP}$. In this case, the maximum open circuit voltage is $qV_{OC}^{max} = \Delta E_{HL} - E_B = E_{PP}$ which, as we have shown in our derivation of the excitonic HJ diode equations, occurs in the high intensity limit (e.g. under optical concentration), or as $T_a \to 0$ (Rand et al., 2007, Giebink et al., 2010b, Vandewal et al., 2010).

It should now be apparent that PP absorption and radiative emission introduce fundamental limitations to the efficiency of OPVs. Materials with a high external radiative quantum efficiency (and hence a low recombination efficiency) will produce a higher efficiency solar cell, independent of whether that cell is based on organic semiconductors where charge generation is mediated by an intermediate PP state, or in inorganic semiconductors that generate charge directly from band-to-band transitions (Miller et al., 2012).

Including the effects of non-radiative processes that often dominate recombination in practical organic solar cells suppresses V_{OC} from the limiting case treated above. Decreasing the electronic interaction between the D and A molecules across the HJ reduces both the oscillator strength of the bound pair (i.e. its coupling to the radiation field via a_{PP}) and the orbital overlap that facilitates non-radiative decay, thus increasing V_{OC}. This has been observed in experiments where the degree of π–π intermolecular orbital overlap between donor and acceptor molecules is varied (Perez et al., 2009, Zimmerman et al., 2012), suggesting this is a useful design strategy for increasing cell efficiency.

From the foregoing discussion and that of Miller et al. (2012), the dependence of η_{ext} on the observed (vs. the theoretical maximum) open circuit voltage can be analytically determined. Equation 7.122 shows that the open circuit voltage is dependent on the logarithm of the ratio of j_{SC} to j_0 (or j_D). For a cell with external quantum efficiency η_{ext}, a CT state electroluminescence quantum efficiency of η_{CT} due to electrical injection of charges (which is proportional to the number of radiative CT states), and an ambient (at T_a) black body flux of ϕ_{BB}, we can write (Rau, 2007)

$$j_{SC} = q\int_0^\infty \eta_{ext}(E)\phi_s(E, T_s)dE \tag{7.147}$$

and the dark current is

$$j_0 = \frac{q}{\eta_{CT}}\int_0^\infty \eta_{CT}(E)\phi_{BB}(E, T_a)dE. \tag{7.148}$$

When $\eta_{CT} = 1$, all the recombination is radiative (the *radiative limit*), and $j_0 = j_0^{rad}$. This results in a minimum j_0 at a maximum voltage (sometimes referred to as the SQ voltage) $V_{OC}^{max} = V_{OC}^{rad}$, where

$$V_{OC}^{rad} = \frac{k_BT}{q}ln\left(\frac{j_{SC}}{j_0^{rad}} + \frac{k_{PPd}}{k_{PPd,eq}}\right)$$
$$= \frac{k_BT}{q}ln\left(\frac{\int_0^\infty \eta_{ext}(E)\phi_s(E, T_s)dE}{\int_0^\infty \eta_{ext}(E)\phi_{BB}(E, T_a)dE} + \frac{k_{PPd}}{k_{PPd,eq}}\right). \tag{7.149}$$

Now, in the absence of a PP state, $a_{PP} = 1$ at $E > E_G$, and is zero elsewhere. This condition yields the dark current in the SQ limit from Eq. 7.148 by setting $\eta_{EL} = 1$:

$$j_0^{SQ} = q\int_{E_{opt}}^{\infty} \eta_{ext}(E)\phi_{BB}(E, T_a)dE, \quad (7.150)$$

where $\eta_{ext} = 1$ for $E > E_X$, and $\eta_{ext} = 0$ otherwise. The maximum or SQ-limited voltage is, therefore,

$$V_{OC}^{max} = \frac{k_B T}{q} ln\left(\frac{j_{SC}}{j_0^{SQ}} + \frac{k_{PPd}}{k_{PPd,eq}}\right) = \frac{k_B T}{q} ln\left(\frac{\int_0^{\infty} \eta_{ext}(E)\phi_s(E, T_s)dE}{\int_{E_G}^{\infty} \eta_{ext}(E)\phi_{BB}(E, T_a)dE} + \frac{k_{PPd}}{k_{PPd,eq}}\right). \quad (7.151)$$

To account for the formation of the PP at the organic HJ, we must include the effects of PP absorption using the expression for $\alpha(E)$ in Eq. 7.142. This absorption incurs a loss compared to the S-Q voltage of

$$\Delta V_{OC}^{PP} = V_{OC}^{max} - V_{OC}^{rad}. \quad (7.152)$$

Finally, in a practical diode, we must consider that CT state radiation is not 100% efficient, that is, where $\eta_{CT} < 1$. Then, there is an additional voltage loss of

$$\Delta V_{OC}^{nr} = V_{OC}^{rad} - V_{OC} = -m\left(\frac{k_B T}{q}\right) log(\eta_{CT}) \quad (7.153a)$$

or

$$\Delta V_{OC}^{nr} = -m\left(\frac{k_B T}{q}\right) log\left(\frac{\eta_{EL}}{\gamma \chi_{em}}\right) \quad (7.153b)$$

where m is a parameter that accounts for non-idealities such as charge that recombines in regions remote from the HJ including interfaces and contacts, η_{EL} is the measured external radiative quantum efficiency, γ is the charge balance factor and χ_{em} is the ratio of emissive CT states to the number of injected electron-hole pairs (e.g. in a conventional OLED where there is significant exchange energy splitting between ^{3}CT and ^{1}CT, $\chi_{em} = ¼$, see Chapter 6.2.1). In Eq. 7.153b, we write the internal CT electroluminescence efficiency, η_{CT}, in terms of the measurable quantity, η_{EL}, taking into account effects of charge balance (which can be strongly dependent on voltage and current) and emissive CT exciton fraction following Eq. 6.1.

Combining the above results, the open circuit voltage is

$$qV_{OC} = E_X^{min} - E_{loss}, \quad (7.154)$$

where the minimum exciton energy is E_X^{min} and (Liu et al., 2016a)

$$\begin{aligned} E_{loss} &= \left(E_X^{min} - qV_{OC}^{max}\right) + \left(qV_{OC}^{max} - qV_{OC}^{rad}\right) + \left(qV_{OC}^{rad} - qV_{OC}\right) \\ &= \left(E_X^{min} - qV_{OC}^{max}\right) + q\Delta V_{OC}^{PP} + q\Delta V_{OC}^{nr}. \end{aligned} \quad (7.155)$$

In an ideal diode, E_{loss} is the *sum* of the free energy losses for both the D and A materials, ΔG_{CT}, which is the second term in the expression on the right hand side of Eq. 7.155 (cf. Eq. 7.136). Including the effects of non-radiative recombination and other non-idealities gives the third term. Then, the efficiency limit in Fig. 7.74 is found by replacing ΔG_{CT} with E_{loss}.

The fundamental differences and similarities between inorganic and organic semiconductor junctions are accounted for by all three terms on the right in Eq. 7.155. (i) In the first term, the band gap energy of an inorganic is replaced by the excitonic absorption in organics. (ii) Although below band gap absorption in inorganics must be considered in calculating its thermodynamic limit, the spread in energies of PP states can be very large, making the second term more significant in organic compared with inorganic HJs. (iii) For both systems, a high external radiative quantum efficiency results in a reduced energy loss. Unfortunately, the radiative efficiency tends to be very low ($<10^{-3}$) for materials used in OPVs. This introduces losses in the third term that are considerably greater than in inorganics. However, this constraint provides guidance in the design of molecules with much higher η_{EL} for use in high efficiency OPV active regions.

Accurate determination of E_X^{min}, which is equivalent to E_{opt}, can be complicated by the existence of strong aggregate or CT absorption below the exciton energy. We showed in Section 3.5.3 that the energy gap can be assigned to the energy at the crossing point between the zero phonon energies of the absorption and emission spectra (the so-called 0–0 crossing point energy, E_{00}). An method often used to determine E_{opt} particularly in cases where the emission spectrum is unavailable or weak, is to linearly extrapolate along the low energy tail of the exciton absorption from its peak to its intercept with the energy axis, as shown in Fig. 7.75 for the NFA molecule, TT-FIC. The molecule features a rigid conjugated backbone that promotes π-stacking, which gives rise to intense aggregate absorption at wavelengths > 950 nm. Obtaining E_{opt} from the intercept of the tail with the abcissa can underestimate its value. In contrast, extrapolating from the exciton peak as shown in Fig. 7.75 (dashed line)

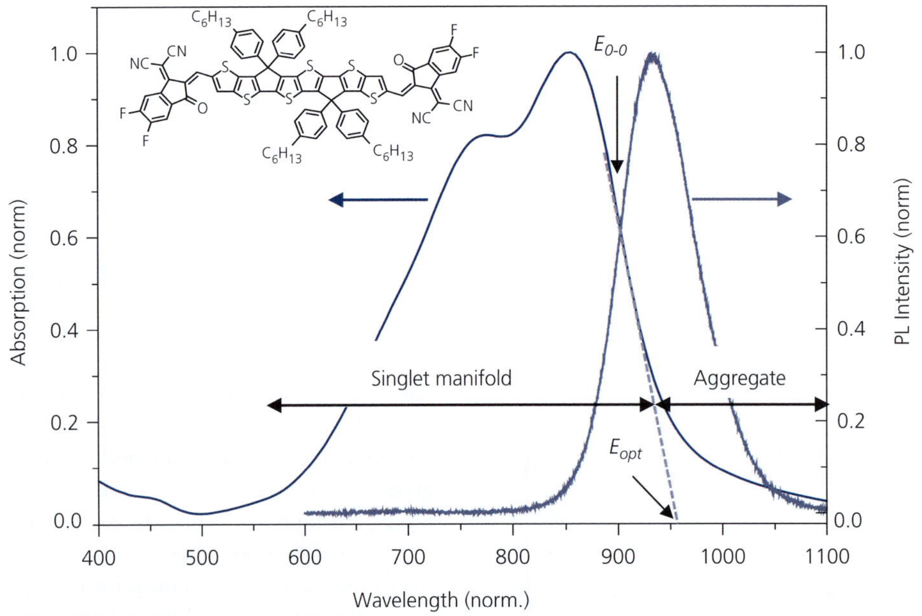

Figure 7.75 Absorption and exciton emission spectra of TT-FIC, whose molecular formula is shown in the inset. The 0–0 crossing, E_{00}, and optical energy gap found from extrapolation of the exciton peak, E_{opt}, are indicated. The long absorption tail starting at ~925 nm is due to aggregates. The lowest singlet state absorption range is also indicated.

gives a value closer to E_G, but at smaller energies than the 0–0 crossing. The difference between $E_{0\text{-}0}$ and E_{opt} estimated by extrapolation is ~70 meV. In spite of possible inaccuracies, the extrapolation method is straightforward and has the advantage of avoiding errors incurred by using the weak, but extended aggreagate tail.

We now calculate the short circuit current using the AM1.5G reference spectrum, $E_{REF}(\lambda)$ in Fig. 7.69 to find η_P for a more realistic (i.e. non-ideal) OPV. The j_{SC} is obtained by integrating the OPV responsivity given by $R(\lambda) = q\eta_{ext}(\lambda)[\lambda/hc]$, and $E_{REF}(\lambda)$. As in the calculations above, we assume that η_{ext} is constant across the visible spectrum up to E_X, the low energy onset of the absorbing material. It is likely that in a high performance bulk or mixed HJ cell, that η_{ED}, $\eta_{CC} \approx 1$. With these assumptions along with Eq. 7.154, we can estimate the power conversion efficiency using several device characteristics that are within reasonable reach using current technology. An example efficiency calculation shown in Fig. 7.76, assuming benchmark values of $\eta_{ext} = 90\%$ (vs. $\eta_{ext} = 100\%$ in Fig. 7.74) and $FF = 0.75$. For fullerene-based OPVs, E_{loss} is between 0.5–0.7 eV, whereas for NFAs (see Section 7.4.3), E_{loss} typically lies between 0.4–0.6 eV. The ripples in the curves in Fig. 7.76 arise from using a reference AM1.5G spectrum in contrast to the

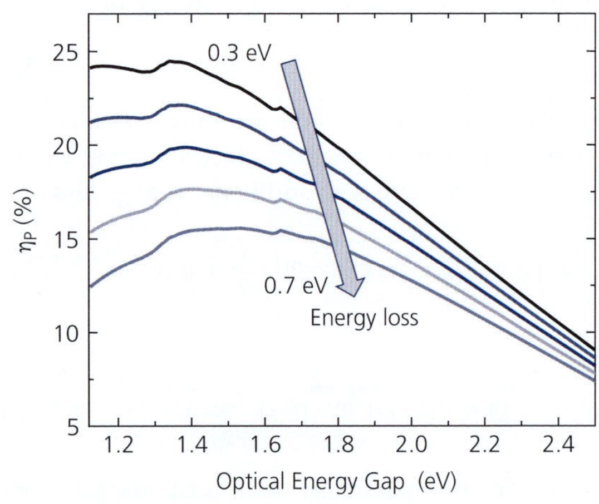

Figure 7.76 Approximate power conversion efficiency vs. long wavelength absorption onset for an OPV with $\eta_{ext} = 90\%$ and $FF = 0.75$ for several values of E_{loss}. The reference spectrum is AM1.5G (Che, 2018).

smooth black body radiation function employed for Fig. 7.74. As an example, we assume that $E_X^{min} = 1.4$ eV (corresponding to $\lambda \approx 900$ nm) and $E_{loss} = 0.5$ eV. Then we obtain a power conversion efficiency of 19.9% for a single junction cell, which falls somewhat short of 22% for $\Delta G_{CT} = E_{loss} = 0.5$ eV in Fig, 7.74. In both cases, however, the peak efficiency is achieved for an optical gap of approximately 1.4 eV, revealing a practical guideline in the design of light absorbing materials for use in OPVs.

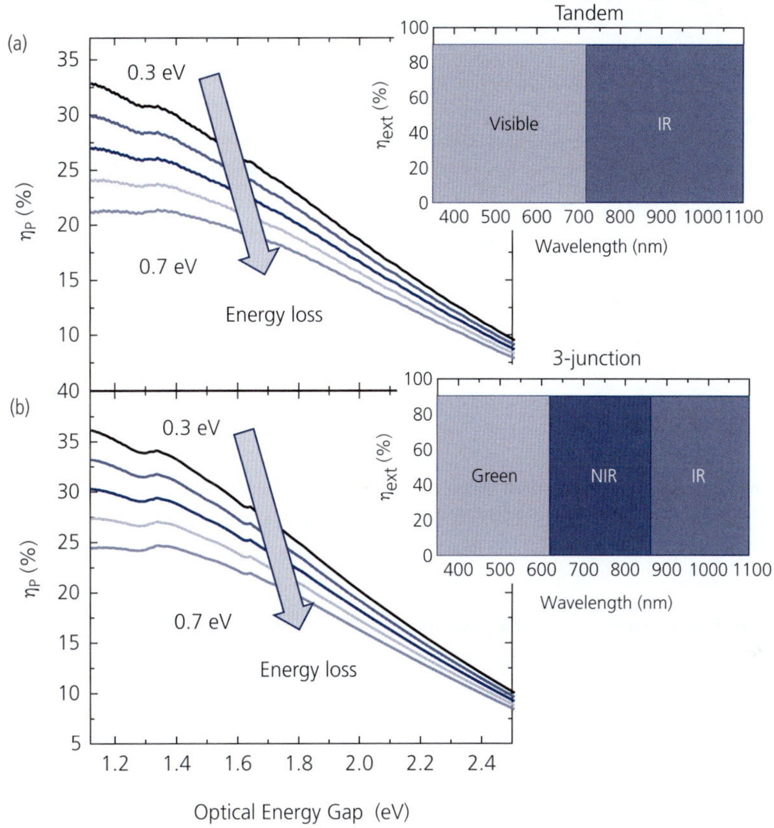

Figure 7.77 Power conversion efficiency vs. long wavelength absorption onset for the reddest absorbing subcell in (a) tandem and (b) triple junction, series-connected OPVs with $\eta_{ext} = 90\%$ and $FF = 0.75$. The energy loss, E_{loss}, of each subcell is assumed to be equal. The insets show the external quantum efficiencies assumed for each subcell, for an OPV whose reddest subcell has $E_G = 1.1$ eV. The reference spectrum is AM1.5G (Che, 2018).

Using this same approach, we can calculate the realistically achievable efficiencies of series-connected multijunction cells, assuming that the cell comprises materials with completely non-overlapping absorption spectra. Current continuity requires that each subcell generate the same photocurrent, which determines the absorption band that can be occupied by each subcell, from 350 nm up to the optical gap of the lowest energy-absorbing subcell. For simplicity, we assume that the individual subcells equally divide the absorption spectrum. Results of these calculations are shown in Fig. 7.77, where once again we use $\eta_{ext} = 90\%$, $FF = 0.75$ and E_{loss} is the same for each cell in the stack. We also assume that there are no optical or electrical losses introduced by the CRZs. That is, the total cell V_{OC} is the sum of V_{OC} of the individual subcells, see Eq. 7.49. The efficiencies in the stacked cells are considerably higher than for the single junction cells, even when single junction cells operate at the thermodynamic limit. The tandem cell (again taking $E_{loss} = 0.5$ eV) gives $\eta_P = 27\%$ and the triple junction device yields $\eta_P = 29\%$. While the efficiency increases slowly from $E_G = E_X^{min} < 1.4$ eV to $E_X^{min} = 1.1$ eV, there is little advantage gained by including materials with longer wavelengths in the reddest absorbing subcell.

The improvement between one, two, and three subcells suggests that the efficiency can be further increased by simply increasing the depth of the stack. Figure 7.78 shows a plot of η_P vs. E_{loss} for cells comprising as many as six series-stacked subcells. While there is a monotonic increase in efficiency with the number of cells, the increase becomes negligible for greater than five subcells. The reduced benefit of adding more cells arises from the diminished current contributed by each successive cell that absorbs an ever-smaller slice of the solar spectrum.

Note that these calculations only provide guidance to expectations for multijunction cell efficiencies. In fact, the actual efficiencies will be even lower than shown in Figs. 7.76–7.78 since η_{ext} is not expected to be a step function with abrupt cutoffs at high and low energies, and each subcell will undoubtedly have

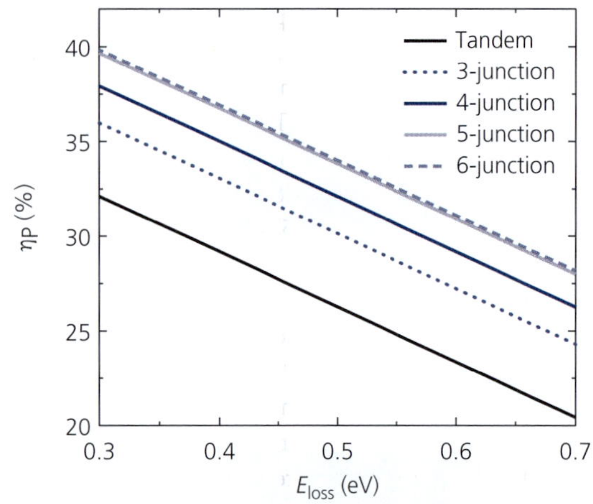

Figure 7.78 Power conversion efficiency vs. energy loss for series-connected multijunction cells with a long wavelength cutoff energy of $E_G = 1.1$ eV (Che, 2018).
Reprinted with permission.

different E_{loss}, FF, and maximum η_{ext}. Also, no assumptions have been made about the match of the optical field with the absorption spectrum of a particular subcell; it is only assumed that these factors combine to result in $\eta_{ext} = 90\%$ across each absorption band. Nevertheless, these calculations provide insights into the factors that govern, and ultimately limit practical OPV efficiencies.

7.3.3 Measuring solar cell efficiency

Accurate quantification of solar cell performance is essential for making valid comparisons of results from different laboratories, and for qualification of devices for commercial production and use. The fundamental challenge of measuring efficiencies of cells based on technologies that employ new, unproven materials sets and device architectures is that the laboratory devices are often small (1–10 mm^2 is typical) and unstable. Complicating matters further is that the measurement of solar cell power conversion efficiency is considerably more involved than simply measuring the external quantum efficiency of a photodetector. The elements required for measuring η_P are as follows:

(i) *Calibrated reference detector*. The reference should cover at least the same spectral range as the test device. Reference detector calibrations should be traceable to a qualified standards laboratory such as the US National Institute of Standards and Technology (NIST).

(ii) *Solar reference spectrum*. The standard for AM1.5G illumination is currently ASTM G-173-03.

(iii) *Solar simulator lamp with an appropriate optical filter*. Xenon lamps are most frequently employed due to their long operational lifetime, and coverage of the visible and NIR spectra that approximately corresponds to the solar spectrum. The output of the simulator should be adjusted using the appropriate optical filter (e.g. AM1.5G) to match the reference spectrum as closely as possible. The output illumination should be uniform across the entire test and reference cell areas, and its intensity should reach, at a minimum, 1 sun (1 kW/m^2). Higher power is desirable for full characterization of the cell performance.

(iv) *Test cell*. This is the device to be characterized. Its area should be as large as possible. However, the highest efficiency laboratory "hero" results typically use small cells to avoid loss due to parasitic resistances. Such small cells can lead to inaccuracies in determining the device active area, whose uncertainty is a function of the square of the cell linear dimensions.

(v) *Optical mask*. This is used to shade the cell periphery from the incident light to avoid collection of charge beyond the presumed active area, as defined by the region of overlap between anode and cathode contacts (Kim et al., 2008, Burlingame et al., 2018).

(vi) *Monochromator*. To assure accuracy, the cell efficiency should be measured by two different methods: (1) using a solar simulator to measure j_{SC}, FF, and V_{OC} under prescribed measurement conditions, and (2) by measuring η_{ext} vs. λ. The integral of the quantum efficiency spectrum convolved with the reference spectrum is then used to calculate j_{SC}. The integral method is generally more accurate than that obtained with a solar simulator, and hence should always be employed as a check of the other measurement. For this purpose, and for quantifying the simulated spectra and the response of the reference detector, a monochromator is required.

Ultimately the complexity of measuring η_P arises from inaccuracies in determining the differences between the lamp (i.e. solar simulator) and the solar spectra, and the effects that those differences have on determining j_{SC}. The FF and V_{OC} are easily and accurately extracted from the j-V characteristics in the 4th quadrant. Since V_{OC} is only logarithmically dependent on j_{SC} (Eq. 7.122), and since it is straightforward to set the incident power accurately with a calibrated detector, the open circuit voltage can be obtained to within 1%.

Methods for accurately measuring the spectrally dependent photocurrent have been described previously (Seaman, 1982, Emery and Osterwald, 1986, Osterwald, 1986, Shrotriya et al., 2006). Here we outline the measurement process, particularly with reference to difficulties encountered with small diameter OPVs. If followed carefully, this procedure should yield the power conversion efficiency to within a few percent (relative). To assure the highest confidence in the outcome, an independent laboratory should also duplicate the measurements. Several precision measurement laboratories such as the National Renewable Energy Laboratory in (Golden, CO, USA), Fraunhofer Institute for Solar Energy Systems (Fraunhofer ISE, Freiburg, Germany), and the Tokyo Laboratory Japan Electrical Safety & Environment Technology Laboratories (JET) are three such examples.

The devices should be measured under *standard reporting conditions* (SRC) which are $P_{inc} = 1\ \mathrm{kW/m^2}$, AM1.5G reference spectrum, and a cell temperature of 25°C. The importance of calibrating the detector and source lamp spectral responses can be understood from the sample data in Fig. 7.79 for the test and reference cells, the Xe source lamp with an AM1.5G filter, and the reference spectrum. The test cell comprises the structure: glass/ITO/MoO_3 (10 nm)/DTDCPB:C_{70} (1:1, 80 nm) HJ/BPhen:C_{60} (8 nm)/BPhen (5 nm)/Ag (100 nm) (Griffith et al., 2015). The reference is a KG5-filtered Si photodiode whose calibration is traceable to a NIST standard. The *relative*, or *normalized*, spectral responsivities of the OPV and the Si reference detector are $S_T(\lambda)$ and $S_{REF}(\lambda)$, respectively in Fig. 7.79a. This is not an optimal situation since $S_{REF}(\lambda)$ does not fully overlap $S_T(\lambda)$ at long wavelengths, leading to a significant difference in photocurrent generated by the two diodes at 800 nm $> \lambda >$ 600 nm, shown on the right in Fig. 7.79a. Nevertheless, the responses are sufficiently similar to compensate for their mismatch, as shown below. Similarly, the normalized irradiance spectra of the solar simulator source ($E_{SIM}(\lambda)$) and the reference ($E_{REF}(\lambda)$) are shown in Fig. 7.79b. Again, there are areas of the spectra where the overlap is poor, particularly at

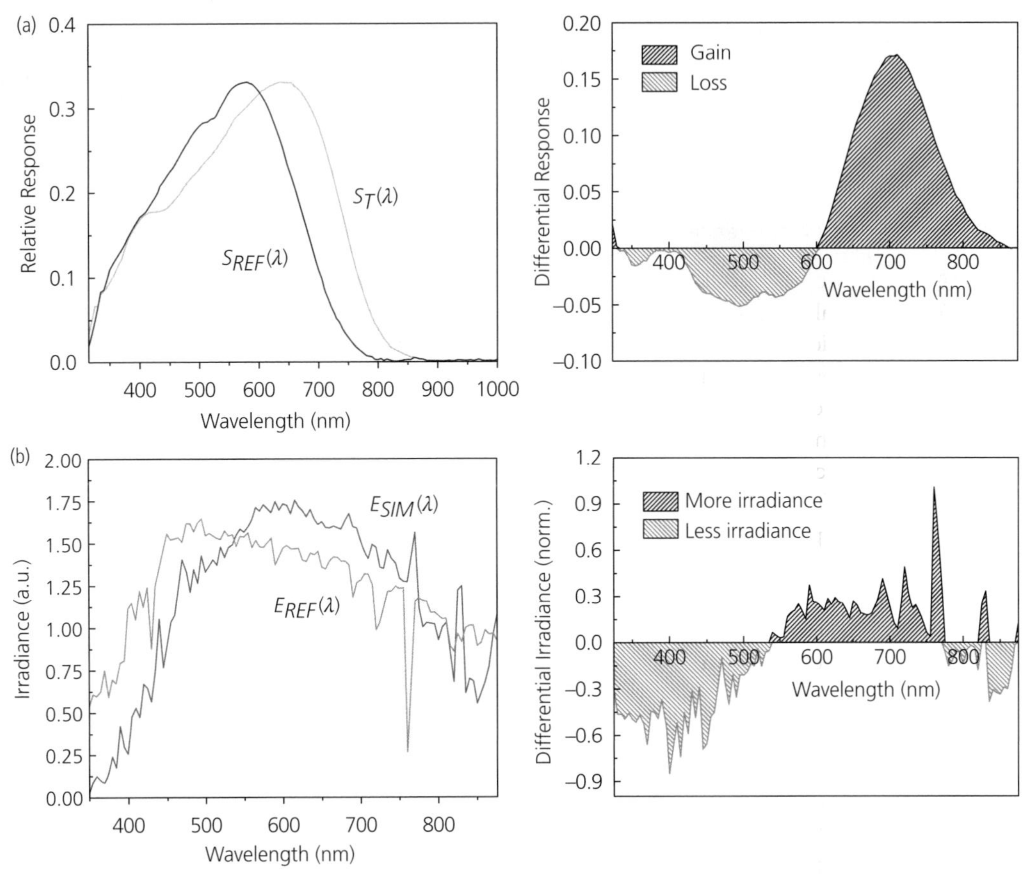

Figure 7.79 (a) Relative detector responses (*S*) and (b) irradiance (*E*) of the AM1.5G filtered Xe source, and reference AM1.5G spectra. The subscript *REF* is the reference Si detector and reference solar spectrum, T is the detector under test, and *SIM* is the lamp spectrum. The differential detector response and spectral irradiance are shown to the right of the corresponding data.

$\lambda < 450$ nm. The fractional error in the measured photocurrent using these two spectra is

$$f_{error} = \frac{j_{SIM}^T}{j_{REF}^T}, \tag{7.156}$$

where j_{SIM}^T and j_{REF}^T are the short circuit currents for the test device illuminated by the simulated and reference spectra, respectively. The currents are expressed by integrals over the two spectra:

$$j_{SIM}^T = \int_{\lambda_1}^{\lambda_2} E_{SIM}(\lambda) S_T(\lambda) d\lambda \tag{7.157a}$$

and

$$j_{REF}^T = \int_{\lambda_1}^{\lambda_2} E_{REF}(\lambda) S_T(\lambda) d\lambda, \tag{7.157b}$$

where the limits of integration, λ_1 and λ_2, should extend across the entire range of response of the detector under test.

The products $E_{REF,SIM}(\lambda)S_T(\lambda)$ of the spectra in Fig. 7.79 are presented in Fig. 7.80a, along with the difference in the two spectra in Fig. 7.80b. The differences are indeed quite large, even though the reference and simulated spectra in Fig. 7.80 overlap reasonably well. The currents obtained using Eq. 7.157 are $j_{SIM}^T = 16.3\ \text{mA/ cm}^2$, whereas $j_{REF}^T = 14.6\ \text{mA/ cm}^2$. This is nearly a 12% overestimation of the short circuit current using the simulated spectrum, which would lead to a similar overestimation of η_P. The discrepancy arises primarily from the differences between the OPV response and that of the Si detector with a filter that cuts off at a shorter wavelength. However, there are many detectors to choose from.

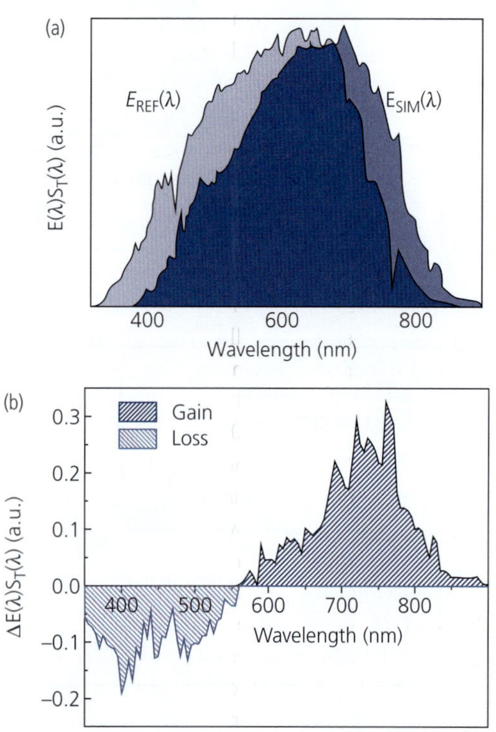

Figure 7.80 (a) The irradiance–OPV responsivity product obtained using a reference (E_{REF}) and a simulated (E_{SIM}) AM1.5G spectra with the device data in Fig. 7.79. These products are the integrands in Eq. 7.158. (b) Differential values of the integrands for the two measurements in (a).

Sample spectra for Si detectors with different filter types are provided in Fig. 7.81. The most complete overlap with the OPV spectrum in Fig. 7.79 is achieved with a KG3 or perhaps KG2 filtered Si detector.

By replacing $S_T(\lambda)$ with $S_{REF}(\lambda)$ in Eq. 7.157, analogous expressions exist for the reference cell current under simulated and reference illumination (j_{SIM}^{REF} and j_{REF}^{REF}, respectively). Substituting these expressions into Eq. 7.156, we obtain a measurement error of

$$f_{error} = \frac{\int_{\lambda_1}^{\lambda_2} E_{SIM}(\lambda) S_T(\lambda) d\lambda \int_{\lambda_1}^{\lambda_2} E_{REF}(\lambda) S_{REF}(\lambda) d\lambda}{\int_{\lambda_1}^{\lambda_2} E_{REF}(\lambda) S_T(\lambda) d\lambda \int_{\lambda_1}^{\lambda_2} E_{SIM}(\lambda) S_{REF}(\lambda) d\lambda} \left(\frac{j_{SIM}^{REF}}{j_{REF}^{REF}} \right). \tag{7.158}$$

The differential responses and spectra can be compensated to obtain an accurate match of the reference and source spectra by calculating the *spectral mismatch correction factor*, *M*, given by

$$M = \frac{j_{SIM}^T j_{REF}^{REF}}{j_{REF}^T j_{SIM}^{REF}} = \frac{\int_{\lambda_1}^{\lambda_2} E_{SIM}(\lambda) S_T(\lambda) d\lambda \int_{\lambda_1}^{\lambda_2} E_{REF}(\lambda) S_{REF}(\lambda) d\lambda}{\int_{\lambda_1}^{\lambda_2} E_{REF}(\lambda) S_T(\lambda) d\lambda \int_{\lambda_1}^{\lambda_2} E_{SIM}(\lambda) S_{REF}(\lambda) d\lambda}, \tag{7.159}$$

such that $f_{error} = M(j_{SIM}^{REF}/j_{REF}^{REF})$. Note that this correction factor has several interesting properties. For example, if the simulated and reference spectra are identical, $M = 1$ and no correction is required. Similarly, if the responsivities of the test and reference detectors are equal, $M = 1$ and once again, no correction is required. Thus, for accurate measurements, M should be as close to unity as possible, and preferably the difference should be $< 10\%$. This is achieved by making the simulator and solar reference spectra, or the reference and test detector responsivity spectra as similar as possible.

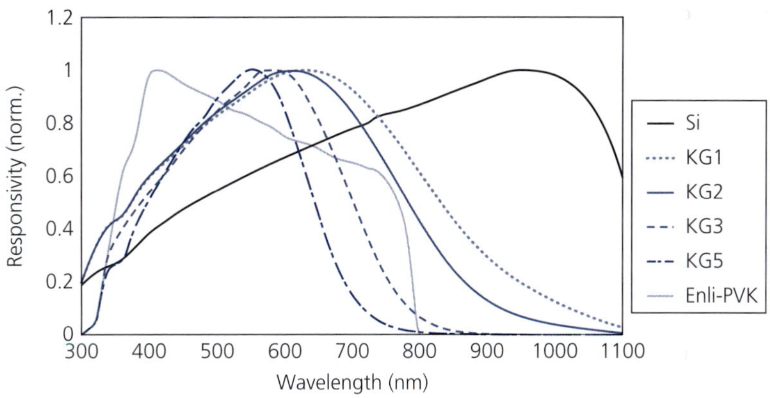

Figure 7.81 Relative Si detector responsivities using several different front-surface filters.

Using Eqs. 7.157–7.159 in Eq. 7.156, we obtain the calibrated short circuit current of the test cell for the reference spectrum:

$$j_{REF}^{T} = \frac{j_{REF}^{REF} \cdot j_{SIM}^{T}}{M \cdot j_{SIM}^{REF}}. \tag{7.160}$$

Measurement of j_{REF}^{T} begins by calibrating the integrated power of the simulator to the total solar irradiance, E_{REF}, under SRC. This is done by setting $j_{SIM}^{REF} = j_{REF}^{REF}$. Furthermore, the simulator spectrum, $E_{SIM}(\lambda)$, must be obtained using the monochromator system. Now, the *effective irradiance* of the solar simulator is only that fraction of total power that overlaps the response of the reference detector, and is given by

$$E_{SIM}^{eff} = M \cdot j_{SIM}^{REF}/CN, \tag{7.161}$$

where CN is the calibration number, which is the ratio of the short circuit current of the reference to the total irradiance, E_{TOT}. By setting the simulator intensity to $E_{SIM}^{eff} = E_{REF}$, then $j_{REF}^{REF} = CN \cdot E_{REF}$ and

$$\frac{j_{SIM}^{REF}}{j_{REF}^{REF}} = \frac{E_{SIM}^{eff}}{M \cdot E_{REF}}. \tag{7.162}$$

Returning to our device in Fig. 7.78, the first ratio on the right in Eq. 7.159 taken from data in Fig. 7.79 is 16.3/14.6 = 1.116. Balancing the power such that $E_{SIM}^{eff} = E_{REF}$ yields an adjustment of 12.2 mA/cm^2 for the reference current using the simulated spectrum, and 11.4 mA/cm^2 for the reference spectrum. By making the lamp intensity the same as E_{REF}, however, the second ratio on the right hand side of Eq. 7.159 is equal to unity and is ignored. In this example, therefore, we have $M = 1.116$, which is larger than desired for an accurate measurement.

The test device is mounted directly beneath a mask of known area that is smaller than that of the solar cell to underfill the device area. Then under short circuit conditions, we obtain j_{SIM}^{T}. The mask should then be removed. The fourth quadrant j–V characteristics measured at E_{REF} then yield V_{OC} and FF. Using these measured values and Eq. 7.160, we obtain the calibrated test cell response.

A cross-check of the efficiency can be obtained by measuring the absolute detector responsivity, $S_T(\lambda)$, using illumination from a monochromator that is varied from λ_1 to λ_2, and calculating j_{REF}^{T} from Eq. 7.157b. This measurement should be done while the device is flooded by simulator illumination of ~1 sun intensity to avoid errors due to recombination at low light levels, or resistive losses at higher intensities. Also, the monochromator output intensity should be calibrated at each wavelength using the reference detector. Finally, as in the case of the j–V measurement, the monochromatic optical beam should underfill the device area to ensure that all incident photons are detected. If the measurements are accurate, the short circuit current calculated from Eq. 7.157b should agree with that of Eq. 7.160 to within 5%.

Underfilling the device active area by using a mask or a focused light source is crucial for obtaining accurate results, especially with small diameter OPVs whose peripheral area is a substantial fraction of the total device area. In some OPV structures, light sensitivity extends beyond the area defined by the contacts, which leads to errors in precisely determining device active area when calculating j_{SC}. A striking example of peripheral charge generation is shown in Fig. 7.82 for a planar-mixed small molecule OPV whose structure is shown schematically in the inset. The donor for this experiment is DTDCBP that was used in the devices in Figs. 7.78 and 7.79. Illumination

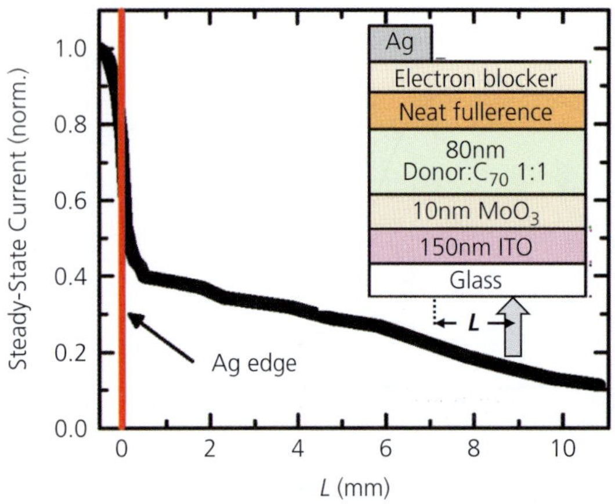

Figure 7.82 Normalized steady state current *vs.* distance from the cathode, *L*, of a <40 μm diameter focused beam at a wavelength of 633 nm. The red vertical line shows the edge of the cathode contact. Inset: OPV structure used in the measurement (Burlingame et al., 2018).

Burlingame, Q., Coburn, C., Che, X., Panda, A., Qu, Y. & Forrest, S. R. 2018. Centimetre-scale electron diffusion in photoactive organic heterostructures. *Nature*, 10.1038/nature25148

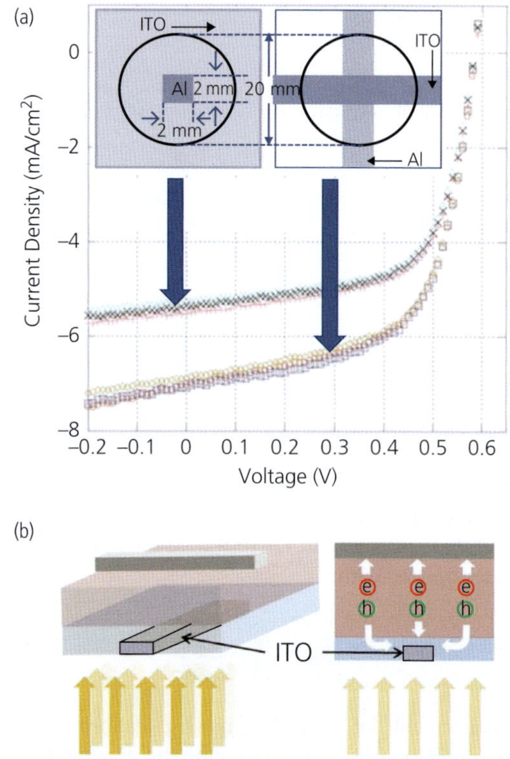

Figure 7.83 (a) Current–voltage characteristics of two (2 mm)^2 P3HT:$PC_{61}BM$ OPVs using different contact arrangements. The region of the 20 μm diameter light spot illuminating the device is outlined by the black circle. Inset: The contact arrangements are the island (left) and crossbar geometry (right). (b) Model for explaining the large current observed for the crossbar geometry, where the PEDOT:PSS layer (blue) overcoats the ITO contact, providing a larger effective anode for charge collection along the length of the perpendicular Al cathode (gray). The device active region is shown in red (Kim et al., 2008).

from an optical fiber placed in close proximity to the glass substrate is projected into the active region of the OPV. The current is measured as the fiber is slowly scanned across the diode diameter defined by the Ag cathode, and well beyond its periphery to a distance, *L*. Excitons generated in the mixed layer dissociate into holes that are immediately collected in the underlying ITO anode, but electrons must enter the 10 nm thick neat fullerene (C_{60}) layer where they are trapped between the BCP electron blocking layer and the donor. The photocurrent collection is efficient, even when the optical beam is located 1 cm from the Ag cathode (Burlingame et al., 2018). This is an extreme case of edge current collection that depends on the unconventional choice of donors and details of the architecture employed. If one attempted to measure the efficiency of the device with overfilled illumination that extended well beyond the Ag cathode, the current would be far greater than anticipated. This would lead to a substantial overestimation of efficiency, since practical solar cells are large area devices where peripheral currents, even as large as in this example, do not significantly contribute to the total cell current.

The cell layout (i.e. the contact arrangement) also can affect the magnitude of the peripheral currents in small area devices. For example, the *j–V* characteristics from ITO/PEDOT:PSS (65 nm)/P3HT:$PC_{61}BM$ (1:1, 150 nm)/LiF (1 nm)/Al (100 nm) BHJ OPVs with two different electrode arrangements are illustrated in Fig. 7.83 (Kim et al., 2008). Both devices have overlapping cathode/anode contact areas of (2 mm)^2, but in one device the Al cathode is an isolated island on the large area active layer and ITO anode, whereas in the second OPV the cathode and anode form an overlapping crossbar pattern. A large optical beam overfills the contacted areas with AM1.5G, 1 sun illumination. The short circuit currents obtained from these otherwise identical devices are significantly different; the island layout gives j_{SC} = 5.5 mA/cm^2, whereas the crossbar gives 7 mA/cm^2 (Fig. 7.83a). The difference is attributed to excess edge currents in the latter device, illustrated in Fig. 7.83b. The conducting PEDOT:PSS layer overlaps the ITO stripe, providing a route for the collection of photogenerated holes along the orthogonal Al stripe, thus providing a larger effective area than the overlap of the two stripes. Edge collection is considerably reduced when the cathode is only an island, and can be further reduced or even eliminated if the PEDOT:PSS layer is omitted. In either case, this points to the large errors that can be incurred when the illumination overfills the contact area of an OPV.

The most effective means to reduce edge effects is to increase the device area. However, this comes at the price of reduced efficiency due to increased series resistance of the ITO anode. Hence, the areas of cells developed to demonstrate the highest efficiencies are typically minimized to eliminate extrinsic resistances that decrease the *FF*, and hence both V_M and I_M (Green, 2017). The effect of contact size for single and multijunction cells was systematically compared using devices with generic structures shown schematically in Fig. 7.9. Specifically, single, tandem, triple and four junction OPV cells based on the donors, DTDCTB and DBP, were combined with C_{60} and C_{70} acceptors. A DBP:C_{70} planar-mixed HJ was employed as a blue-green absorbing subcell (SC-1 in Fig. 7.9) while DTDCTB:C_{60} PM-HJ primarily absorbs in the NIR (SC-2). The stacked subcells are interconnected with a CRZ comprising a BPhen:C_{60} mixed layer (1:1, 5 nm)/Ag (0.1 nm)/MoO_3 (5 nm) (see Section 7.5.2). Four-junction cells comprised (from the cathode): Ag (100 nm)/BPhen (7 nm)/SC-1/CRZ/SC-2/CRZ/SC-1/CRZ/SC-2/MoO_3 (10 nm)/ITO. For triple, tandem and single junction cells, the subcells and CRZs closest to the anode were eliminated according to the number of stages (e.g. a tandem had active regions of SC-1/CRZ/SC-2) (Xiao et al., 2015).

The *FF* and η_P vs. cell area are shown in Fig. 7.84a and b, respectively, for single, double, triple, and quadruple junction cells. The j_{SC} and V_{OC} are unaffected by device area over the limited range explored, consistent with the effects of increased series resistance inferred from Fig. 7.70. The performance characteristics of each cell type as area, $A \to 0$ are provided in Table 7.6. Among various devices, single junction cells exhibit the largest drop in η_P of 17% as area increases from 1 mm^2 to 1 cm^2, whereas tandem, triple and four junction cells suffer reductions of 8.5%, 6%, and 4%, respectively. The reduced loss in efficiency for multijunction cells is primarily due to their increased V_{OC} and reduced j_{SC} with each additional sub-element. That is, the incremental power loss due to series resistance, $\Delta P_{loss} = I^2 R_{ser}$, decreases with the square of the cell current.

The specific series resistance ($R_{ser}A$) of the OPV is given by

$$R_{ser}A = \frac{R_{\square}LA}{W} + \sum_{N} (\rho_O t_O + r_{int}), \tag{7.163}$$

where the sum is over each element in the stacked device consisting of N =1, 2, 3, or 4 sub-elements, $R_{\square}$ is the sheet resistance of the ITO anode, L and W are the length and width of the ITO contact, ρ_O is the resistivity of the organic layer, t_O is its thickness, and r_{int} is the specific interface resistance within the cell stack. The first term on the right is due to the ITO anode resistance, while the terms in the summation are due to the organic layer and interface resistances. This expression is simplified by assuming that the total internal junction resistance, r_{jn}, of each subcell is the same. Then,

$$R_{ser}A = \frac{R_{\square}LA}{W} + Nr_{jn}. \tag{7.164}$$

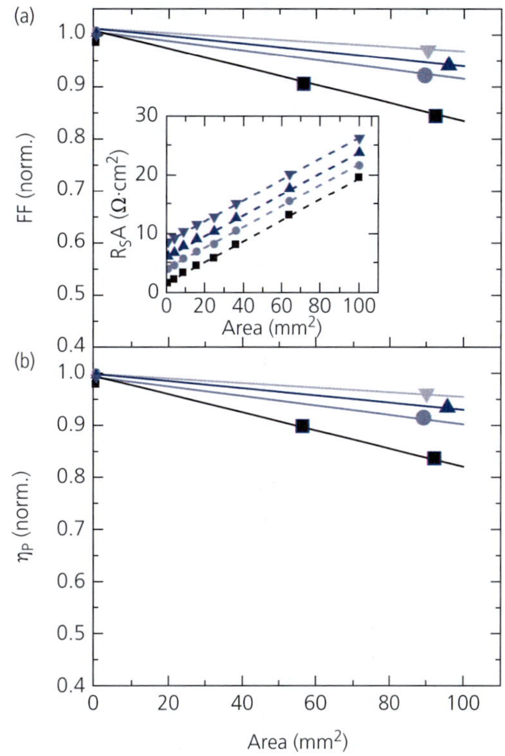

Figure 7.84 (a) Normalized fill factor and (b) power conversion efficiency vs. area of single (squares), tandem (circles), triple (up triangles), and quadruple junction (down triangles) OPVs whose structures are described in text. Inset: Series resistance vs. area. Adapted from Xiao et al. (2015).

Table 7.6 Performance of 1 mm^2 single, tandem, triple, and quadruple junction OPV cells (Xiao et al., 2015)

No. junctions	j_{SC} (mA/cm²)	V_{OC} (V)	*FF*	η_P (%)
Single	11.3	0.92	0.59	6.1
Tandem	7.5	1.70	0.58	7.4
Triple	5.3	2.59	0.58	8.0
Quadruple	4.7	3.34	0.57	8.9

As shown in the inset of Fig. 7.84a, $R_{ser}A$ increases linearly with device area, as expected. The slopes of the lines give an average junction resistance $\bar{r}_{jn} = 2.1\ \Omega\ \text{cm}^2$. For the 1 mm^2 single junction cell, $R_{\square}A = 0.015\ \Omega\ \text{cm}^2$, which is small compared with r_{jn}. However, for the 1 cm^2 single junction cell, $R_{\square}A$ increases to 18 $\Omega\ \text{cm}^2$, which now dominates. Multijunction cells show similar trends. However, since the total resistance of the organic layers is higher for multijunction than for single junction OPVs due to the increased number of layers and interfaces, scaling of resistance with area is not as pronounced as for single junction OPVs, and hence the ITO sheet resistance plays a relatively small role in the latter cells compared to contributions from r_{jn}. For large area cells, the sheet resistance dominates, leading to a significant drop in η_P.

The results of this analysis for the normalized fill factor of $FF(\infty, R_{ser}A)/FF(\infty, 0)$ *vs.* j_{SC} and V_{OC} for 1 cm^2 cells are shown in Fig. 7.85. Consistent with our previous discussion, *FF* is maximized as j_{SC} is reduced and V_{OC} is increased in multijunction cells.

Measurement of η_P for multijunction cells is considerably more challenging than for single junction cells since the subcells cannot be independently contacted, and therefore Eq. 7.159 can no longer be used to calculate the spectral mismatch factor. The total current in a series-connected cell is limited by the subcell that passes the lowest current. Therefore, the current at a given wavelength generated by measuring the quantum efficiency of one subcell is inherently ambiguous due to current simultaneously generated by the other subcells in the stack. To reduce the error in j_{REF}^T, therefore, a reference detector whose efficiency spectrum most closely matches that of the test device, that is, $j_{SIM}^{REF} \approx j_{SIM}^T$, can be employed along with a simulator whose spectrum approximates that of the reference AM1.5G spectrum. This, however, provides no more than a loose approximation of efficiency that does not substitute for a precise measurement.

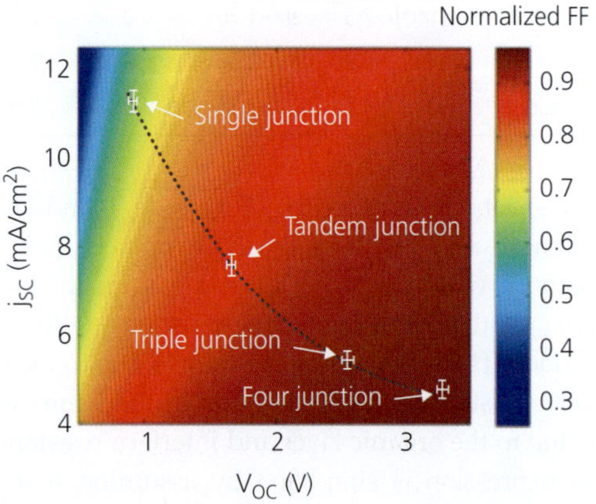

Figure 7.85 Calculated normalized $FF = FF(\infty, R_{ser})$ as a function of j_{SC} and V_{OC}. The stars represent the measured coordinates of OPV cells whose properties are listed in Table 7.6. The dashed line is a guide to the eye (Xiao et al., 2015).

Inaccuracies can be reduced by measuring the cell efficiency out-of-doors on a clear day, making sure to measure the solar spectrum and intensity under which the cell is being tested. However accurate this test may be, it is not a practical, solution since the solar irradiance depends on many factors including latitude, atmospheric conditions, time of day, etc. Hence, such a measurement does not conform to a standard by which other cells, measured elsewhere and at different times can be compared. A standardized laboratory method for multijunction cell efficiency measurements has been implemented at NREL using the so-called one sun multisource solar simulator (OSMSS) shown in Fig. 7.86a (Moriarty et al., 2012). This apparatus consists of nine spectral channels whose intensities can be independently varied to create a 1 sun intensity spectrum that reasonably approximates that of the reference spectrum. The procedure for finding this spectrum for a particular test subcell is accomplished by independently exciting each subcell via optical biasing. This requires the illumination of subcells not being tested by wavelengths not absorbed by the subject subcells. Optical biasing thus avoids the other devices in the stack from limiting the total current. Then, as before, the current of the ith cell in the stack is

$$j_{SIM,i}^T = \int_{\lambda_1}^{\lambda_2} E_{SIM}(\lambda) S_T^i(\lambda) d\lambda. \tag{7.165}$$

The ratio, R_{ij}, of currents between any combination of two subcells i, j, is then adjusted to

$$R_{ij} = \frac{\int_{\lambda_1}^{\lambda_2} E_{REF}(\lambda) S_T^i(\lambda) d\lambda \int_{\lambda_1}^{\lambda_2} E_{SIM}(\lambda) S_T^j(\lambda) d\lambda}{\int_{\lambda_1}^{\lambda_2} E_{REF}(\lambda) S_T^j(\lambda) d\lambda \int_{\lambda_1}^{\lambda_2} E_{SIM}(\lambda) S_T^i(\lambda) d\lambda} = 1. \tag{7.166}$$

Once the spectrum is constructed by iteratively setting $R_{ij} = 1$ for all i, j (usually at an intensity within each band of ~0.5 sun for that band), the full intensity is set to 1 sun using a reference detector. The power conversion efficiency is then obtained by dividing j_{SIM}^T by the mismatch factor which is the same for the N subcells by following the procedure that sets all

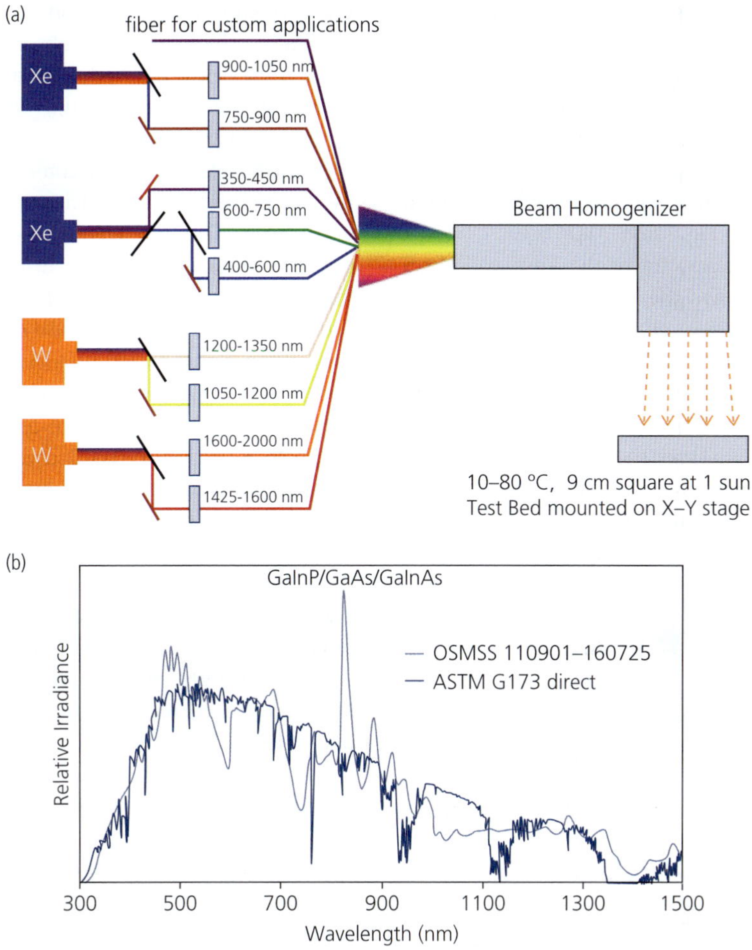

Figure 7.86 (a) Schematic of the one sun multisource solar simulator (OSMSS). (b) The spectrum constructed for a GaInP/GaAs/GaAs triple junction cell using the OSMSS vs. the reference AM1.5G ASTM G173 direct spectrum (Moriarty et al., 2012).

$R_{ij} = 1$. The complex OSMSS system can be replaced by a single, intense light source with a series of filters that are switched in and out to measure the currents of the individual subcell pairs.

An example spectrum constructed in this manner for a triple junction GaInP/GaAs/GaInAs cell is shown in comparison to an AM1.5G reference spectrum in Fig. 7.86b. The spectral agreement is sufficient that the photocurrents generated in each subcell were within 1% of their value under the reference.

As we will find in subsequent sections, multijunction OPVs can deliver superior performance to single junction cells. As such, it is not practical to send all experimental cells to standards laboratories whose measurements are more suitable for qualifying solar cells for commercial use. Indeed, sending model cells to such laboratories is both time consuming and often unnecessary. Nor can every laboratory afford a complex system such as the OSMSS, nor have the expertise to operate it properly. For these reasons, one requires a method that provides a reliable efficiency measurement that is simple, inexpensive and accurate. The light biasing setup illustrated in Fig. 7.87a provides a means for such an efficiency measurement.

The measurement procedure is as follows for the example tandem cell whose individual subcell spectra are shown in Fig. 7.87b. The device should comprise subcells whose individual spectra have at least one region that does not overlap the spectra of other subcells in the stack. Then, a high intensity monochromatic light source operating in the transmission window of the subcell under test is used to selectively "turn on" the other cells such they deliver a current close to j_{SC} under 1 sun illumination. Then a weak, chopped light source is passed through a monochromator to excite the subcell under test to obtain its current at each wavelength, and hence its efficiency spectrum, $\eta_{ext}(\lambda)$. To avoid current generated from the bias light source due to residual absorption by the subcell under test, the monochromatic light must be synchronously detected at the frequency of the chopper. This procedure is reproduced to capture $\eta_{ext}(\lambda)$ for each cell in the stack. The placement of a red and blue bias light for the example tandem spectrum is shown in Fig. 7.87b.

The overlap of the spectra of the long and short wavelength cells allows the short wavelength bias illumination at 530 nm to effectively "turn on" both cells, while also enabling a measurement of $\eta_{ext}(\lambda)$ of the long wavelength cell. In contrast, the short wavelength subcell does not absorb the long wavelength bias beam at 780 nm. Thus, the efficiency measured for the short wavelength cell must be corrected for very low intensity excitation, since in some OPVs the subcell efficiency varies with intensity, decreasing as $P_{inc} \rightarrow 1$ sun. This correction is easily obtained by fabricating single junction subcells that are similar to those used in the stacked OPV and measuring η_{ext}(-P_{inc}) for each at 1 sun, AM1.5G simulated illumination.

A further complication common to OPVs is that the charge extraction efficiency increases with reverse bias. The j–V characteristics for a strongly optically biased and a current-limiting subcell in a tandem are illustrated in Fig. 7.88 (dotted line). These conditions show that the current-limiting subcell is reverse biased at $-V_0$ (dashed line), yielding a higher j_{ph} than its own j_{SC}. Hence, the measured $\eta_{ext}(\lambda)$ must be corrected to account for the overestimation by Δj_{ph}. This is accomplished by forward biasing the tandem by $+V_0$ during the efficiency measurement. The amount of forward bias to apply to arrive at $\Delta j_{ph} = 0$ is once again obtained from the response of discrete cells engineered to closely resemble the subcells in the multijunction device (Gilot et al., 2010a, 2010b).

Once the light bias measurements are obtained, and suitable corrections for voltage and intensity-dependent efficiency have been applied, the total quantum efficiency of the tandem is obtained from the sum of the external efficiencies of the constituent subcells. Then, integrating over the reference spectrum, j^T_{REF} is obtained without requiring a measurement of the spectral mismatch factor. Consequently, this provides η_P once V_{OC} and FF are measured under SRC. While it is increasingly complicated to obtain accurate efficiencies of devices with more than two subcells, the approach is general and can in principle be extended to

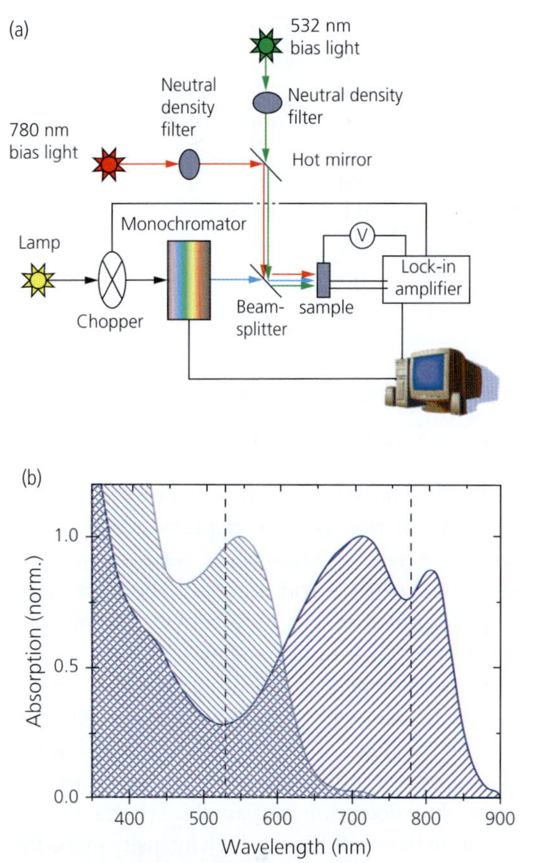

Figure 7.87 (a) Experimental setup for measuring the efficiency of subcells in a tandem OPV using the light biasing method. Example spectra of the subcells in a tandem OPV. The dashed lines are the wavelengths of the bias illumination sources. Bias illumination at 530 nm is used during measurement of the red absorbing subcell, and bias at 780 nm is used during measurement of the blue-green cell (Gilot et al., 2010b).

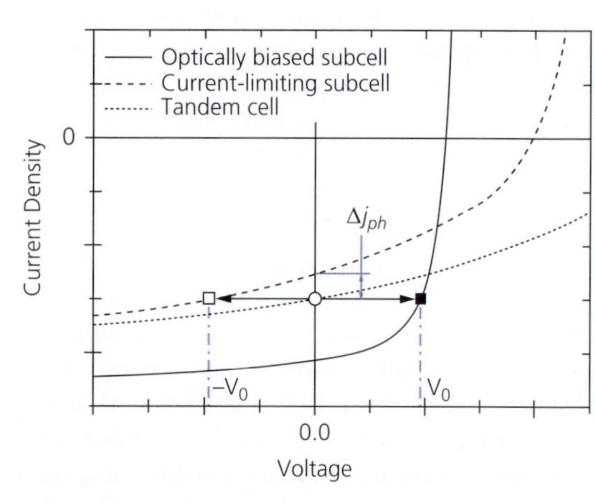

Figure 7.88 Current–voltage characteristics of a tandem OPV, and the current-limiting, and the optically biased subcells. Under short circuit conditions, $V_0 = -V_0$ such that the total tandem cell bias is $V = 0$ V. Adapted from Gilot et al. (2010b).

multijunction cells with any number of stacked elements (Di Carlo Rasi et al., 2017).

7.4 Architectures, morphologies, and materials for OPVs

Three critical factors that determine the efficiency of an organic solar cell are the device layering scheme, or architecture, the morphologies of the layers with particular focus on the optically active and exciton blocking regions, and the materials used in all of the layers. Having laid the general groundwork for understanding the operation, quantification of performance, and limits to the efficiency of OPVs, we are prepared to look at specific methods to attain high efficiency and long lifetime by optimizing the several cell attributes.

We begin this section with a brief history of OPVs that have led to current devices with operational lifetimes >20 years and efficiencies >16%. The advances in OPVs can be divided into six periods, each marked by discoveries in architecture, layer morphology, and materials. These innovations, or stepping stones in OPV development, are the demonstration of (i) Schottky barrier solar cells (Chamberlain, 1983), (ii) bilayer junction solar cells (Tang, 1986), (iii) multijunction cells (Hiramoto et al., 1990) with subcells separated by optically lossless CRZs (Yakimov and Forrest, 2002), (iv) the BHJ OPV by both solution (Halls et al., 1995, Yu et al., 1995) and vacuum processing (Xue et al., 2005), (v) fullerene acceptors in both solution (Yu et al., 1995) and vacuum deposited (Peumans and Forrest, 2001) OPVs, and (vi) high efficiency, non-fullerene-based OPVs (Cnops et al., 2014). While these are largely technological advances, OPVs are second only to OLEDs in igniting the revolution in our understanding of the physics and chemistry of organic electronic materials.

The earliest OPVs comprised only a single organic semiconductor sandwiched between two metal electrodes with different work functions that create an internal field. One of the contacts is non-Ohmic, thereby creating a rectifying Schottky barrier diode. Excitons that drift to the blocking electrode undergo dissociation, and the generated free charges are collected (Chamberlain, 1983). The highest efficiencies were achieved using the metal-free phthalocyanine, H_2Pc, or merocyanine dyes as the active layer due to their stability and relatively broad spectral coverage (Morel et al., 1978, Chamberlain and Clooney, 1979). The disadvantage of this architecture is the thin photoactive region that is confined near to the dissociating electrode, the lack of a large Schottky energy barrier limiting the open circuit voltage while introducing large background currents, and the narrow spectral range spanned by only a single chromophore. Hence, these devices exhibited power conversion efficiencies <0.1% and fill factors of ~0.25 were not uncommon ($FF = 0.25$ corresponds to a linear dependence of j vs. V in the fourth quadrant, where $j_M V_M = \left[\frac{1}{2} j_{SC}\right]\left[\frac{1}{2} V_{OC}\right]$).

Most, if not all of the deficiencies of the metal–organic OPV were eliminated by the introduction of the bilayer OPV comprising a PTCBI acceptor and CuPc donor layers (Tang, 1986). The device shown in Fig. 7.36 employs a type II HJ to efficiently dissociate excitons. This device represents a transformative advance in organic solar harvesting devices. The OPV showed a $V_{OC} = 0.45$ V, $j_{SC} = 2.3$ mA/cm^2, and $FF = 0.65$ under AM2 (75 mW/cm^2) illumination, producing $\eta_P = 0.95\%$. Even more remarkable than the relatively high efficiency was the close to ideal diode rectifying characteristic of the diode with very low leakage current; characteristics that are essential to reaching high performance. Lastly, the materials employed are stable if kept in an oxygen- and moisture-free environment.

The next significant advance was the demonstration of a multijunction cell that delivers increased power at low current and high voltage, in contrast to a single junction cell that sources low voltage at high current. The challenge to successful multijunction cell operation is to create a zone for efficient charge recombination between the front and back cells (see Section 7.1.2). The first attempt used a thin

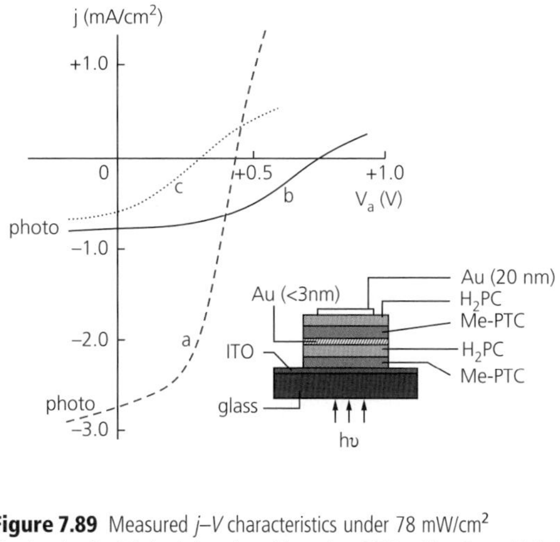

Figure 7.89 Measured *j*–*V* characteristics under 78 mW/cm^2 illumination for (a) single junction, (b) tandem OPVs with a 2 nm thick Au charge recombination layer, and (c) a tandem cell without the Au layer. Inset: Cell structure. Adapted from Hiramoto et al. (1990).

(~2 nm) Au sheet between the subcells, as shown in Fig. 7.89 (Hiramoto et al., 1990). Both the front and back cells used Me-PTC as the acceptor, and the donor in both cells was H_2Pc. While the Au recombination layer increased the cell voltage, it unfortunately formed a rectifying contact with the H_2Pc layer in the front cell. Furthermore, the Au layer was sufficiently thick to be optically absorbing. For these reasons, the voltage was increased by approximately 80% from the single junction cell, but the current was reduced by nearly a factor of three. Hence, η_P was lower than for the single junction cell.

Problems with the device in Fig. 7.89 were resolved by depositing an ultrathin layer (0.5 nm) of Ag in place of the thicker Au. The thin Ag is discontinuous, only forming sub-nanometer clusters while still being completely transparent (Yakimov and Forrest, 2002). The nanoclusters provided efficient charge recombination sites with no barrier formation with the organic films. The introduction of the electrically lossless Ag nanoparticle CRZ led to the stacking of up to five identical PTCBI/CuPc planar subcells whose V_{OC} increased linearly up to three subcells (Fig. 7.90a). Furthermore, η_P of the stacked cells increased superlinearly between a single junction and tandem cell, and the devices remained efficient even under intense illumination of 10 suns (1 W/cm^2). This indicated that the CRZs are also optically lossless. Figure 7.90b shows that increasing the number of cells beyond two leads to a decrease in efficiency since the front-most cells absorb much of the light required in the back cells, thus limiting the maximum attainable j_{SC}. While the fabrication of multijunction cells is more challenging than for single junction OPVs, the introduction of loss-free CRZs provided motivation to combine multiple subcells, each absorbing in different spectral region to reduce thermalization losses, increase voltage, and ultimately increase power conversion efficiency.

The devices described thus far comprise a planar HJ structure. However, as already shown, the exciton diffusion length is typically considerably smaller than the optical absorption length, leading to a low η_{ED} in Eq. 7.48. This efficiency bottleneck was eliminated by the introduction of the BHJ and mixed HJ architectures, where excitons dissociate throughout the active region volume (Halls et al., 1995, Yu et al., 1995, Xue et al., 2005). This eliminates exciton diffusion, resulting in $\eta_{ED} = 1$, although optimization of layer morphology is essential in preventing a reduction in the charge collection efficiency due to disorder within the absorbing region. A discussion of these morphologies is found in Section 7.4.2.

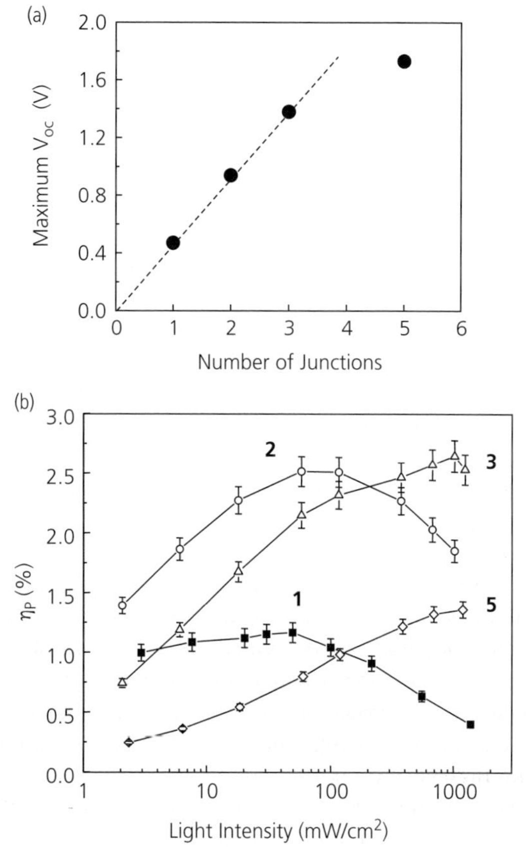

Figure 7.90 (a) Maximum V_{OC} vs. number of PTCBI/CuPc junctions in multilayer OPVs. Each subcell is separated by a 5 Å thick Ag charge recombination layer. (b) Power conversion efficiency of the cells vs. light intensity from 0.01 to 10 sun, AM1.5 solar spectrum. The number of subcells in the stacked OPV are indicated (Yakimov and Forrest, 2002).

The introduction of the BHJ was contemporaneous with a second important advance: the use of fullerenes (specifically $PC_{61}BM$) as an acceptor molecule in a photodiode operating in the photovoltaic mode (Yu et al., 1995). Rapid donor–acceptor charge transfer, along with the high spatial symmetry of the spherical or nearly spherical fullerenes led to a remarkable improvement in the charge transfer efficiency and electron conductivity (and thus an increase in η_{CC}) compared to previously employed perylene-based acceptors. Following the introduction of solution processable $PC_{61}BM$, the first solar cells based on vapor-deposited C_{60} were employed in a so-called *double heterostructure* OPV with an efficiency of 3.6% (Peumans and Forrest, 2001). Using planar-mixed active regions, fullerene-based OPV efficiencies increased to 5% (Xue et al., 2005).

Significant improvements in device performance have been accompanied by innovations in materials and device architectures (i.e. the planar HJ, BHJ, and

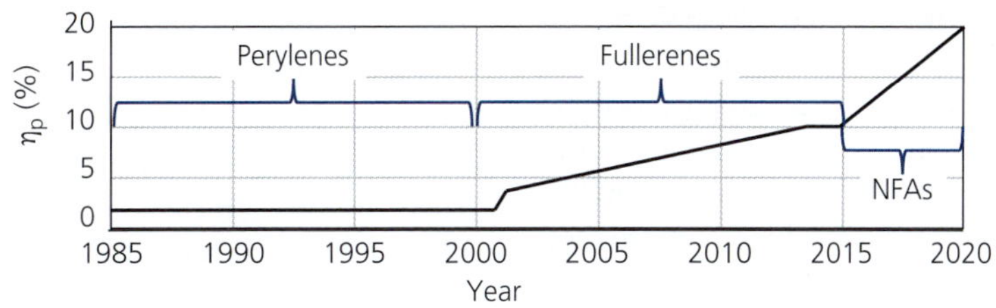

Figure 7.91 Rapid increases in cell efficiency have been paced by the introduction of new acceptor molecules grouped into three major categories: perylenes, fullerenes, and NFAs.

the double HJ that included an exciton blocking layer; Xue et al., 2005). In this context, introducing new acceptors, starting with the perylenes, then the fullerenes, and most recently the NFAs have led to concomitant increases in OPV efficiencies. In fact, any material can be used as an acceptor provided that it forms a type II HJ with the selected donor with sufficiently large energy offsets (~0.5 eV) to drive exciton dissociation. A suitable acceptor must also have high electron mobility and a LUMO energy that approximately matches the cathode metal work function. We have seen in Chapter 4 that materials with high electron mobilities are uncommon, which is one reason that the fullerenes dominated OPV designs for over a decade.

The importance of the development of new acceptors is illustrated in the OPV efficiency timeline in Fig. 7.91. We have seen in foregoing sections that the diode efficiency depends on the product of the absorption, electron diffusion, charge transfer, and charge collection efficiencies. All of these factors depend on a combination of materials properties (i.e. absorption coefficient and HOMO–LUMO energy offsets at the D-A interface), as well as layer morphology. Indeed, exciton diffusion and charge mobility intimately depend on the film structure at the nanometer scale. The fullerenes improved the coverage of the solar spectrum due to their broad absorption compared to the perylenes, and their spherical or near-spherical symmetries allow close packing, which in turn increases the exciton diffusion length and mobility.

Because of these attributes, the fullerenes supported an extended run of efficiency increases, although the rate of improvement slowed in the late 2010s, primarily due to the large energy loss from the large fullerene exciton binding energy. Also, while spectrally broad, fullerene absorption does not extend into the NIR. The acceptor palette expanded when a high efficiency non-fullerene cell based on halogenated subphthalocyanines was demonstrated. Recent innovations in NFAs

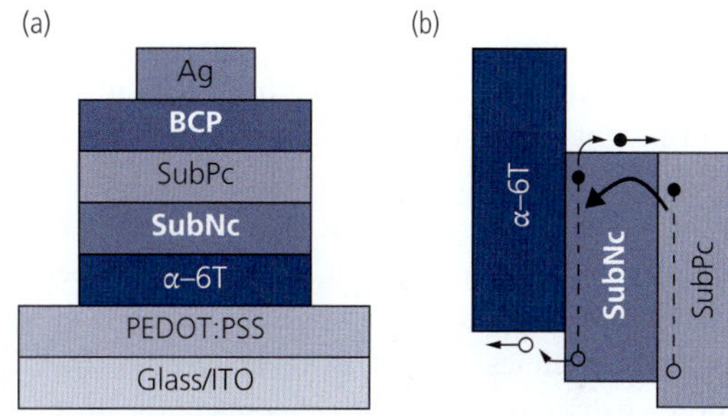

Figure 7.92 (a) Structure of the NFA-based OPV cell. (b) Proposed energy level diagram showing an exciton cascade that transports excitons within the acceptor to the dissociating HJ between α-6T and SubNc (Cnops et al., 2014).

have resulted in single junction cell efficiencies exceeding 16% (Yuan et al., 2019).

An early high efficiency cell containing only non-fullerene-based materials is shown in Fig. 7.92. This device employed the low ionization energy α-6T donor forming a planar HJ with a bilayer of the NFAs, SubNc, and SubPc. The trilayer cell provides a site for the absorption of high energy photons in the 18 nm thick SubPc, and lower energy photons in the 12 nm thick SubNc layer. Exciton transport from SubNc to SubPc is via Förster resonant energy transfer, and subsequently drift to the dissociating interface with the α-6T.

The use of these three materials provides broad spectral coverage shown in Fig. 7.93a, extending from <400 nm due to the α-6T, to 750 nm, which is the long-wavelength cutoff of SubNc. The j–V characteristics of several different combinations of acceptors, including C_{60}, SubPc, SubNc, and SubPc/SubNc are shown in Fig. 7.93b. The larger spectral coverage of the NFAs results in a substantial increase in j_{SC}. Importantly, the V_{OC} increases from 0.42 V for C_{60} to 1.09 V for SubPc. The voltage increase is due to the larger HOMO–LUMO offset energies between the subphthalocyanines and α-6T, and to the lower exciton binding energies of the NFAs. The highest efficiency achieved is with the SubPc/SubNc device, with $V_{OC} = 0.96$ V, $j_{SC} = 14.55$ mA/cm^2, $FF = 0.61$, resulting in $\eta_P = 8.4\%$ (Cnops et al., 2014). Since this demonstration, advancements in developing NFAs with absorption deeper into the NIR and with smaller binding energies have continued apace. A more thorough discussion of NFAs and solar cell architectures developed to exploit their properties is found in Section 7.4.4.

7.4.1 Architectural elements of high performance OPVs

The basic architectural elements of conventional single junction cells include, starting from the substrate, a transparent anode contact, anode buffer, optically absorbing region, an exciton blocking/electron transporting cathode buffer, and the cathode contact. Multijunction cells also include a CRZ between subcells. Inverted and transparent cells contain similar elements but sometimes stacked in reverse order. In this section we discuss each of the elements in the order in which they are stacked in a conventional OPV structure, starting from the substrate.

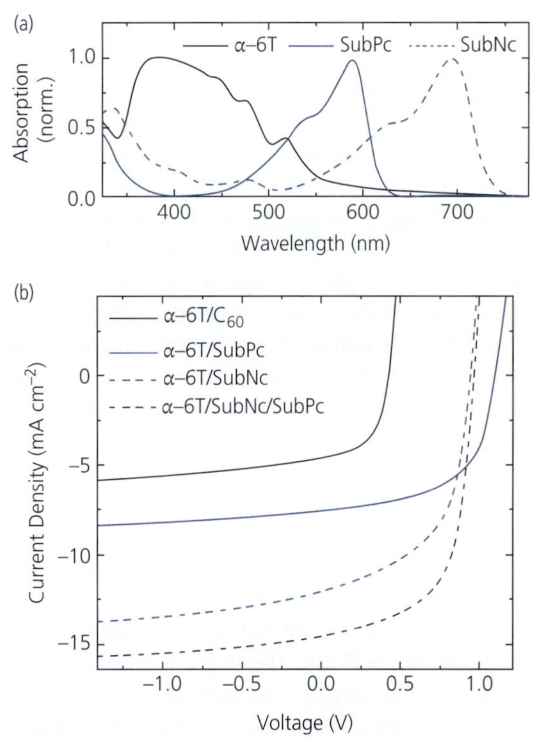

Figure 7.93 (a) Normalized absorption spectra of the materials employed in the device in Fig. 7.92. (b) Measured j–V characteristics of several devices using fullerene and non-fullerene acceptors under simulated, 1 sun intensity solar illumination (Cnops et al., 2014).

7.4.1.1 Anode buffers

The anode buffer reduces injection barriers to holes exiting the donor, allowing for low resistance transport to the high work function anode (typically ITO) contact. There are two principal architectures that accomplish this purpose: the high LUMO energy, or large work function electron injecting buffer in Fig. 7.94a, or the low HOMO energy, large energy gap (and hence transparent) hole transporting buffer in Fig. 7.94b. These are Type 1 and Type 2 buffers, respectively. Note that interface dipoles can exist between the donor, anode, and the anode buffer. These substantially alter the simple pictures in Fig. 7.94, and indeed their presence can ultimately determine the efficiency of charge injection altogether (Kanai et al., 2010). The origins and effects of interface dipoles have been discussed in Section 4.6.

Type 1 buffers inject electrons that recombine with photogenerated holes in the donor. Commonly, the anode is the p-type ITO. The energy level alignments with the buffer depend on processing conditions that can change the stoichiometry, and hence the work function and ionization potential of the anode. ITO typically has a valence band maximum energy of −4.7 eV and an energy gap of 3.75 eV (Méndez-Pinzón et al., 2010). A wide energy gap organic material that satisfies the Type 1 criterion is HAT-CN with an ionization energy of −9.44 eV, energy gap of 4.24 eV, and a Fermi energy located only 0.26 eV below the LUMO (Oh et al., 2017). The LUMO energy, therefore, is positioned ~0.5 eV *below* that of ITO, allowing for barrier-free injection of electrons from the buffer LUMO into ITO.

In contrast, Type 2 buffers have an ionization energy comparable to that of the donor and the anode contact, allowing for efficient hole conduction.

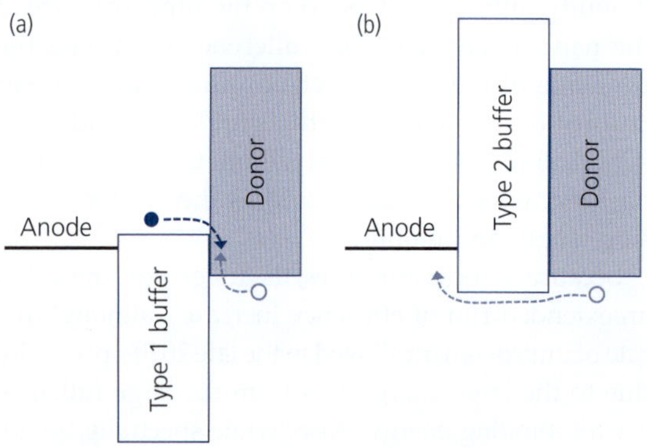

Figure 7.94 (a) Type 1 electron injecting and (b) Type 2 hole conducting anode buffers that connects the anode contact with the donor in the active region. Interface dipoles that may exist depending on materials combinations used, are omitted.

An example of a Type 2 buffer is PEDOT:PSS (Zhang et al., 2007). Like ITO, the energy levels and their alignments are dependent on the precise mixtures, and preparation of this material. Using photoelectron spectroscopic probes, the ionization energy of p-type PEDOT:PSS is 4.9 eV, and the energy gap is ~1.6 eV which places its HOMO level at only ~0.2 eV below that of the valence band maximum of ITO (Ratcliff et al., 2012). Its narrow energy gap makes this material weakly absorbing in the visible. Thus only very thin layers can be employed in efficient OPVs.

Transparent metal oxides (TCOs, or MO_x) have been extensively exploited as anode buffers (Han et al., 2009, Kanai et al., 2010, Dasgupta et al., 2013, White et al., 2016, Guo et al., 2017). This class of materials has a broad range of energy gaps and Fermi energies, yet they generally serve as reasonably efficient hole buffers. The most often used oxide is MoO_3, which has a deep conduction band minimum of close to 7 eV, thus forming a Type 1 buffer. Numerous models have been advanced to explain its apparent ability to transport holes, which is opposite to the electron conducting mechanism in Fig. 7.94a. The models impute the presence of very large interface dipoles, or transport through mid-energy gap states due to oxygen vacancies that give rise to a defect-induced hole transport band (Kanai et al., 2010, Dasgupta et al., 2013). However, these scenarios are unlikely since the hole conductivity within an impurity band with a limited density of states would presumably be far less than the electron conductivity in the conduction band, as both would serve the same purpose of transporting charge from the donor to the anode.

The ability of TCOs to conduct holes to the anode is explained by the differences in HOMO energies between several molecular solids such as α-NPD, 2T-NATA, and CBP and the work function, ϕ_M, of a range of TCOs. As shown in Fig. 7.95, the energy difference fits on a universal curve when plotted against $\phi_M - IE_{org}$. Here, IE_{org} is the ionization energy of the organic overlayer measured by ultraviolet photoelectron spectroscopy (see Appendix C). The offset energy between the Fermi energy of the metal oxide and the donor HOMO, ΔE_H, increases linearly for $\phi_M - IE_{org} < 0$, until $\phi_M - IE_{org} = 0$, at which point the HOMO energy is pinned at $\Delta E_H \approx 0.3$ eV. Increasing ϕ_M further does not result in a change in the HOMO energy relative to the metal oxide. This trend indicates that details of the organic, and the electronic structure or the reconstruction of the MO_x surface, has little if anything to do with the energy level alignments.

The apparently universal relationship results from filling of oxygen vacancies in the transition metal oxide via charge exchange with the organic layer on its surface. The electronic structure of transition metal oxides is primarily determined by filling their *d*-band orbitals, as illustrated in Fig. 7.96a. The metal oxides can be broadly separated into four classes. Class 1 oxides

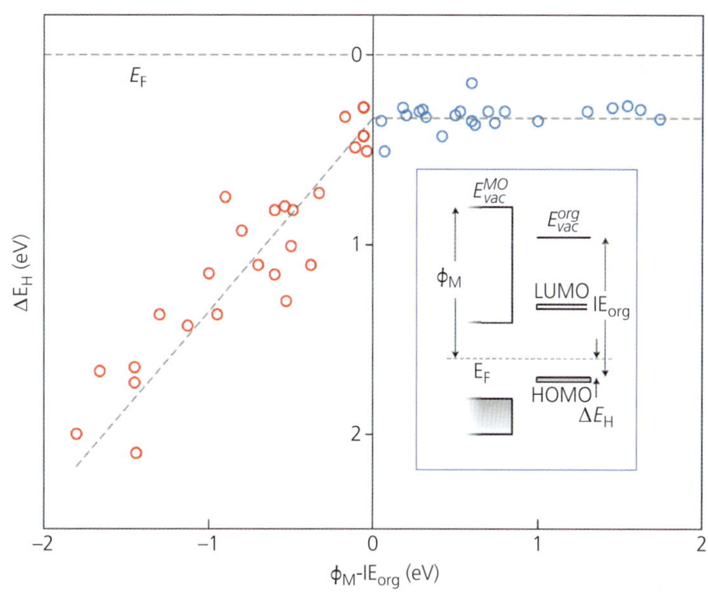

Figure 7.95 HOMO offset energy, ΔE_H, vs. the difference in metal oxide work function, ϕ_M, and organic ionization energy, IE_{org}, for a variety of metal oxides and organic semiconductors. The Fermi energy, E_F, is shown at $\Delta E_H = 0$. Inset: Definition of the variables used in the plot (Greiner et al., 2012).

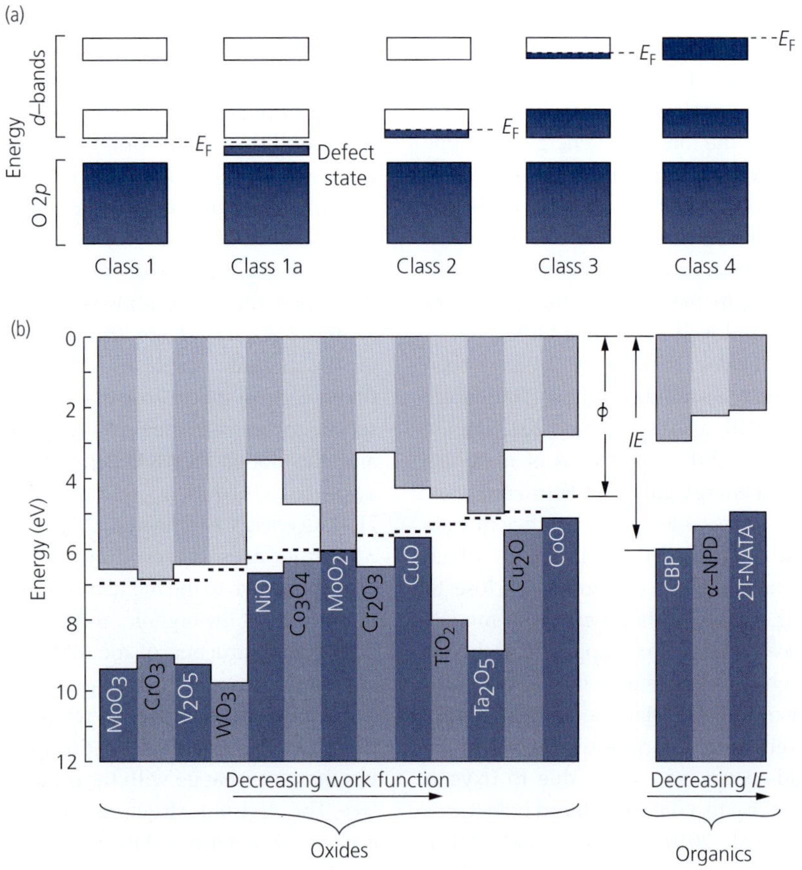

Figure 7.96 (a) Schematic illustration of the energy band structure of transition metal oxides with different *d*-band occupancies. Also shown are the O 2p orbitals. (b) Energy gaps and Fermi energies of metal oxides whose energies are plotted in Fig. 7.95, along with the organics used to form the heterojunctions (Greiner et al., 2012).

(e.g. V_2O_5, MoO_3, WO_3, TiO_2, Ta_2O_5) are d^0 insulators in their native state since they have empty conduction levels. Since these materials generally are non-stoichiometric compounds due to a high density of oxygen vacancies, the *d*-bands are only partially occupied. The vacancies form defect states below the Fermi energy, giving them n-type character (Class 1a). Class 2 metal oxides have partially filled *d*-bands with a low density of electrons at energies higher than d^0, giving them metallic character. Examples of Class 2 materials are WO_2, TiO, MoO_2, and Cr_2O_3. Class 3 oxides have a large number of electrons in these energy bands, filling them to even higher energies (examples include Co_3O_4, CoO, NiO, CuO). Finally, oxides whose bands are completely filled (known as the d^{10} oxides) form Class 4 semiconductors (Ag_2O, Cu_2O).

The relative positions of the MO_x Fermi energies to those of the organics are illustrated in Fig. 7.96b. In spite of the variety of electronic structures, energy gaps and compositions of the oxides, the Fermi energies all lie within a narrow range from approximately 4.5 eV for CoO to 7 eV for MoO_3. Many of the oxides are suitable as Type 1 (e.g. MoO_3, WO_3, etc.) or Type 2 (Cu_2O, NiO, etc.) buffers, depending on the organic used as the donor.

The dependence of ΔE_H on $\phi_M - IE_{org}$ and the source of the Fermi level pinning in Fig. 7.95 are explained as follows: As the HOMO energy of the organic moves closer to the vacuum level, E_{vac}^{org}, than E_F of the MO_x, there is electron transfer from the molecular layer to the oxide. The number of cation molecules, M^+, formed is given by Fermi statistics, viz. (Greiner et al., 2012)

$$[M^+] = \frac{[M^0]}{1 + g\exp(E_{form}/k_B T)}, \tag{7.167}$$

where $[M^+]$ is the concentration of ionized molecules, $[M^0]$ is the total concentration of molecules, g is the degeneracy factor which is equal to the total number of equivalent charge states of the molecule, and E_{form} is the formation energy of the charged molecular state. Once the Fermi energy equals, or exceeds the HOMO energy, charge transfer no longer occurs, and the number of ionized molecules becomes saturated. Increasing the Fermi energy further, therefore, does not lead to a concomitant increase in $[M^+]$.

From this analysis, the HOMO energy relative to E_F is given by

$$\Delta E_H = IE_{org} - \alpha\phi + \Delta + \frac{q^2 \rho d}{2\varepsilon_0} \sum_{i=0}^{\infty} \frac{1}{1 + g\exp\left(E^i_{form}/k_B T\right)}, \tag{7.168}$$

where α is a constant dependent on the degree of screening of the surface potential by the molecular layer, Δ is the interface dipole energy, ρ is the area density of molecules in each layer, and d is the distance between the molecular plane and the substrate. The sum is over the layers from the MO_x surface into the bulk, each with different formation energies, E^i_{form}, due to the decreasing relaxation energy of the molecules located at increasing distances from the interface. That is, the formation energy for a molecular ion, M^+, in layer i is $E^i_{form} = IE_{org} - \phi_M - iE_{relax}$, where E_{relax} is the energy relaxation between each layer.

The fit to the data by Eq. 7.168 gives evidence for charge transfer. This universal dependence also rules out the possibility that the energy level offsets for MO_x-based anode buffers depends on details of surface chemistry, conduction via defect states, etc. Indeed, the phenomenon is neither process nor surface-chemistry dependent, making the metal oxides ideal candidates for anode buffers.

7.4.1.2 Active regions: cascade and ternary blend OPVs

In Section 7.1.2 we introduced the generic structures of the planar HJ, the BHJ, the planar-mixed device, and the so-called "controlled" HJ comprising a nearly ideal morphology comprising finger-like protrusions of the donor into the acceptor region with diameters on the scale of the exciton diffusion length, and finger lengths comparable to the optical absorption length. Beyond these basic structures, many alternatives have been demonstrated that are designed to increase η_{ext}, V_{OC}, and *FF* to overcome the characteristically low mobility and exciton diffusion lengths of organic thin films.

There are two variants of these structures known as *cascade* and *ternary blend* OPVs. We have already introduced an acceptor cascade OPV in Fig. 7.92, utilizing a three layer, planar, thin film active region comprising a donor layer along with two acceptor layers whose energy gaps increase with distance from the D-A HJ (Cnops et al., 2014). The second variation of the cascade is one that blends either two donors and one acceptor molecule, or vice versa, into a single bulk or mixed HJ structure. These architectures will be separately discussed, since their operating principles differ in a few important respects, although in both cases their purpose is to extend the coverage of the solar spectrum by increasing the number of absorbing molecular species within the active region.

Often, properties of a specific donor material that favorably impact one performance metric, negatively impact another. For example, molecular geometries that frustrate intermolecular electronic interactions at the D-A junction can result in an increased open circuit voltage (cf. Section 4.7.2). In contrast, exciton diffusion and charge transport are generally enhanced by strong interactions within the layers (Servet et al., 1994, Coropceanu et al., 2007). The competition between conflicting requirements can be resolved by combining several donor or acceptor materials, each optimized for a different function, within the active region. The cascade structure in Fig. 7.97a offers this multifunctional design strategy by partitioning the donor region into a multilayer stack of complementary materials, D_1, D_2, and D_3, with successively larger energy gaps according to their distance to the D-A HJ. The progression in energy gaps allows for directed transfer of excitons generated in large energy gap layers to the lower energy gap layers, until the excited state ultimately arrives in D_1 where it can diffuse to the HJ. Here, D_3 is designed to prevent exciton diffusion to the anode while suppressing dark current and absorbing the highest energy photons. D_1 absorbs the longest wavelength light while also facilitating exciton transfer to the D-A interface. Also, D_1 should maximize V_{OC} by choosing a material with the largest possible HOMO–LUMO offset energy that permits efficient exciton dissociation. The intermediate donor, D_2 is used to complete the coverage of the solar spectrum. In the design in Fig. 7.97a, the HOMO energies of all donors are aligned such that holes generated at the HJ are transported to the anode without encountering a significant energy barrier that would decrease η_{CC}.

The energy gap ordering of $D_3 > D_2 > D_1$ for the series Dpt, tetracene (Ttn), and rubrene (Rbn) is shown in Fig. 7.97b. Due to the spectral overlap between the emission and absorption spectra of molecules in adjacent layers, excitations in Dpt Förster transfer to Ttn, and likewise, Ttn excitations transfer to Rbn. A summary of the device performances for several devices employing these donors contacting a 40 nm thick C_{60} acceptor in a planar junction OPV are provided in Table 7.7. The V_{OC} of the cascade device is 0.74 V, which is noticeably lower than that of the single donor Rbn OPV. This is attributed to singlet fission

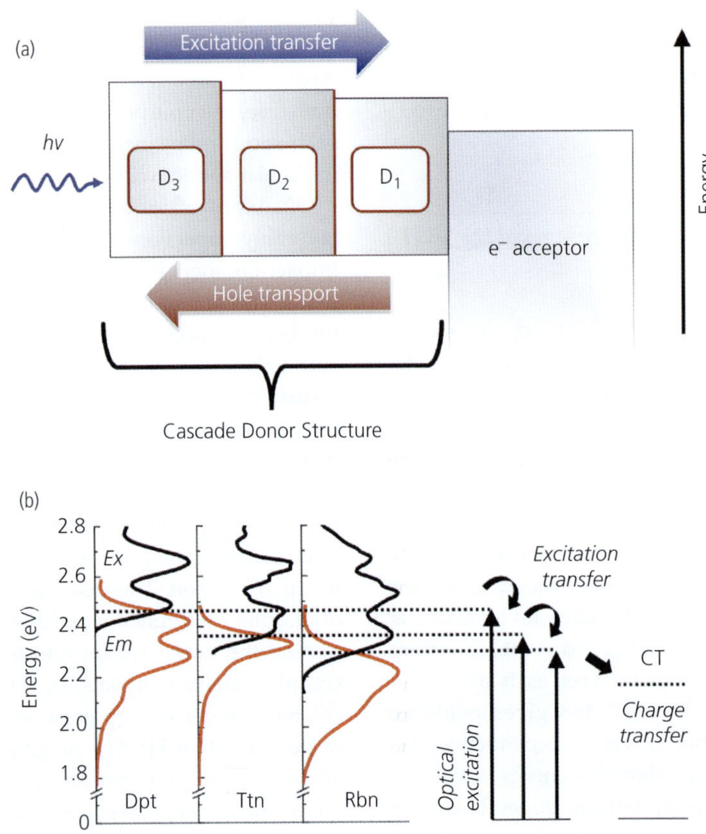

Figure 7.97 (a) Multi-donor (D_1, D_2, D_3) energy cascade OPV. (b) Excitation (Ex) and emission (Em) spectra of the three donors, Dpt, Ttn, and Rbn, along with transitions obtained from optical absorption and charge transfer at the HJ. The excited state energy cascade progression is shown by the successively lower energies from Dpt to Rbn (Schlenker et al., 2011).

Table 7.7 Comparison of performance of cascades and single donor devices using materials in Fig. 7.94 (Schlenker et al., 2011)

Donor[a]	j_{SC}[b] (mA/cm^2)	V_{OC} (V)	*FF*	η_P (%)
Dpt (10 nm)	1.1	0.65	0.41	0.30
Ttn (45 nm)	2.9	0.54	0.54	0.83
Rbn (20 nm)[c]	2.4	0.92	0.43	0.92
Dpt (10 nm)/Ttn (35 nm)	1.4	0.79	0.37	0.41
Dpt (5 nm)/Ttn (20 nm)/Rbn (20 nm)	2.6	0.74	0.50	0.95

[a] All devices employed a 40 nm C_{60} acceptor, 10 nm BCP blocking layer, and an Al cathode.
[b] Measurements made under 1 sun intensity, AM1.5G simulated illumination.
[c] Data from Perez et al. (2009).

that lowers the photovoltage, and perhaps leads to inefficient energy transfer between the various layers in the cascade (Schlenker et al., 2011). Nonetheless, the cascade device performance (including *FF* and j_{SC}, and consequently, η_P) is improved compared to any of the single donor combinations studied.

A marked improvement in donor cascade devices was achieved using the triple donor stack comprising Dpt (10 nm)/Rbn (10 nm)/DBP (10 nm) along with a C_{60} (35 nm) acceptor followed by a C_{60}:Bphen (1:1, 8nm)/Bphen (5 nm) exciton blocking and electron filtering buffer layer (see Section 7.4.1). The optical gaps of Dpt (D_1), Rbn (D_2), and DBP (D_3) are 2.5eV, 2.2 eV, and 2.0 eV, respectively, whereas the HOMO energy of each donor is at 5.4 eV referenced to vacuum. Hence, this triad of materials satisfies the architectural requirements of a cascade in that the energy gaps allow for transfer from $D_1 \rightarrow D_2 \rightarrow D_3$, whereas the HOMO energy alignment creates a path for efficient hole extraction to the anode (Griffith and Forrest, 2014).

The absorption and emission spectra of the various donor constituents used in conjunction with their exciton diffusion lengths of $L_D = 30$ nm, 35 nm, 20 nm, and 28 nm for Dpt, Rbn, DBP, and C_{60}, respectively, provide a quantitative determination of the contribution to the total short circuit current of the cascade

Table 7.8 Short-circuit current density contributions to the total cascade OPV current calculated for each organic active layer (Griffith and Forrest, 2014)

Donor	j_{SC} (Dpt)[a]	j_{SC} (Rbn)	j_{SC} (DBP)	j_{SC} (C_{60})	Total j_{SC}
Dpt/Rbn/DBP	0.13 (1.8%)	0.11 (1.5%)	3.88 (51.0%)	3.48 (45.7%)	7.61
Rbn/DBP	—	0.12 (1.7%)	3.63 (50.7%)	3.41 (47.6%)	7.16
Dpt/Rbn	0.15 (3.8%)	0.20 (5.0%)	—	3.57 (91.2%)	3.92
DBP	—	—	2.01 (33.9%)	3.92 (66.1%)	5.93
Rbn	—	0.20 (6.1%)	—	3.14 (93.9%)	3.34
Dpt	0.21 (8.4%)	—	—	2.28 (91.6%)	2.49

[a] Relative percentage contributions of a given material to j_{SC} are in parentheses. All values in mA/cm^2.

device from each layer. This analysis assumes that absorption in each layer is followed by diffusion to the interface with the next layer. The exciton is then transferred via FRET to the following layer, and the process continues until exciton dissociation occurs at the DBP/C_{60} HJ. Calculated estimates of the j_{SC} contributions from each layer are summarized in Table 7.8. In single donor devices, the preponderance of the short circuit current is generated in the C_{60} layer due to its long diffusion length and its broad spectral absorption. However, in the cascades (with the exception of the Dpt/Rbn double cascade), photocurrent generation is more balanced. Indeed, the triple cascade device shows that the current is primarily generated in the donors, although only a small fraction is due to transfer from Dpt to Rbn, and Rbn to DBP, which is attributed to the low absorption coefficients of Dpt and Rbn.

Although the currents contributed by D_1 and D_2 are not substantial, the power conversion efficiency of the triple cascade (7.1%) is markedly higher than any of the single donor devices, where the largest efficiency of 4.3% is obtained for the DBP/C_{60} bilayer OPV (see Table 7.9). Exciton blocking by large energy gap Dpt and Rbn layers is likely responsible for this significant improvement in η_P, which from Table 7.9 appears to arise primarily from an increase in j_{SC}. That is, excitons generated in DBP are prevented from diffusing to the anode where they are quenched.

These early demonstrations indicate that cascades can significantly improve OPV efficiency if the various layers are selected for high absorption coefficients, with limited overlap of absorption between layers. Several materials with approximately equal HOMO energies that might prove useful in donor cascades are provided in Fig. 7.98. Using these and many other materials available, there appears to be no limit to the number of donors or acceptors that can be employed in an efficient cascade as long as the design criteria of descending energy gaps and nearly equal transport level energies are met.

Table 7.9 Performance of 3, 2, and 1 donor cascade OPVs (Griffith and Forrest, 2014)

Donor	V_{OC} (V)	j_{SC}[a] (mA/cm^2)	*FF*	η_P (%)
Dpt/Rbn/DBP	0.94	10.6	0.71	7.1
Rbn/DBP	0.94	9.3	0.68	6.0
Dpt/Rbn	0.90	3.9	0.52	1.9
DBP	0.92	6.7	0.69	4.3
Rbn	0.89	3.8	0.47	1.6
Dpt	0.80	2.7	0.39	0.8

[a] Values measured 1 sun AM 1.5G simulated illumination.

The NFA cascade in Fig. 7.92 provides an example of how to balance contributions from each of the layers. In Fig. 7.93, the absorption from the donor (α-6T) and the two acceptors (SubNc, acceptor A_1, and SubPc, acceptor A_2) have only a limited spectral overlap. Contrary to the donor cascades, the acceptor LUMO energies must be approximately matched. There is a LUMO barrier of 0. 23 eV between SubPc and SubNc, which may result in a small loss in electron collection efficiency at the cathode (Endres et al., 2016). Nevertheless, this materials combination results in a broad spectral response, extending from 350 nm to nearly 750 nm with an average of $\eta_{ext} \sim 60\%$. As a result, the efficiency of the double acceptor cascade is $\eta_P = 8.4\%$, whereas the D/A_1 device has $\eta_P = 6.0\%$, and D/A_2 gives 4.7% (Cnops et al., 2014).

Thus far we have considered cascades where excitons are funneled to a single, dissociating D-A HJ. The lack of differences in HOMO energies in the donor cascades prevents dissociation within the donor region itself. Intermediate charge generation is possible using an energy staircase at both the HOMO and LUMO energies, as shown in Fig. 7.99. This results in multiple exciton dissociation sites within the active region that eliminate the need for energy

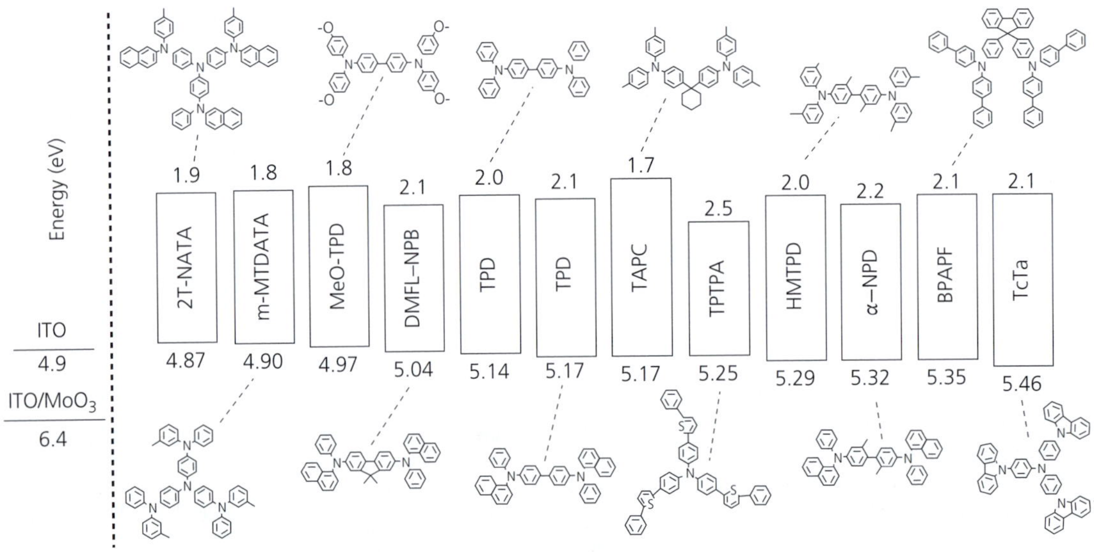

Figure 7.98 Molecules and their HOMO and LUMO energies useful in donor cascades (Barito et al., 2014).

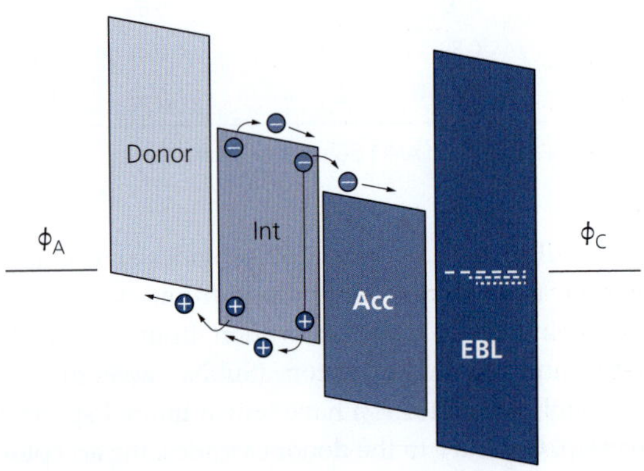

Figure 7.99 Proposed equilibrium energy level diagram of a multijunction cascade OPV consisting of a donor layer, an intermediate layer (Int), an acceptor (Acc), and an exciton blocking layer (EBL).

transfer along the entire cascade to the HJ (Hong et al., 2009, Heidel et al., 2011, Barito et al., 2014). The use of multiple HJs eliminates the trade-off between a short diffusion length and the longer optical absorption length, that often limits the efficiency in bilayer planar HJ cells.

The photodetection process in the multijunction cascade is shown in Fig. 7.99. There are two sites for exciton dissociation, one at the donor/intermediate layer (Int), and one at the Int/acceptor (Acc) HJ. Both HJs must have sufficiently large HOMO–LUMO energy offsets to induce charge transfer in the forward direction. In fact, the two junctions serve as two photocurrent generators in series. The interlayer serves as both the acceptor for the left-most junction as well as the donor to the right-most junction. This shared purpose also affects the total device voltage.

The benefit of this design is that the multiple layers provide charge generation sites within a diffusion length of the photon absorption site. Also, the layer compositions can be chosen to give broad spectral coverage. Beyond having complementary absorption spectra, efficient multistep devices must satisfy two further design criteria that are not easily met. As noted, the HOMO–LUMO offsets at each HJ must be sufficiently large to promote efficient exciton dissociation. Also, current continuity requires that the electron and hole mobilities of Int be approximately equal to avoid electron or hole pile-up at the junction. Charge pile-up creates an internal voltage that bends the energy levels opposite to that in Fig. 7.99, which eventually stops the current flow, to significantly reduce efficiency.

The *j–V* characteristics of a multistep α-NPD/SubPc/C_{60} cascade device shown in Fig. 7.100 exhibit the benefits and challenges of the architecture. The open circuit voltage is approximately equal to that of a SubPc/C_{60} OPV, but considerably lower than the α-NPD/SubPc device. The multistep cascade features an inflection in the characteristics near V_{OC} that substantially decreases *FF* from 0.62 for the bilayer lacking the donor, to 0.49 for the multistep OPV (Barito et al., 2013). The inflection is characteristic of two energy barriers that can be modeled by the back-to-back diode circuit shown in the inset. The intermediate SubPc layer has a significantly higher hole than electron mobility. This leads to pile-up of

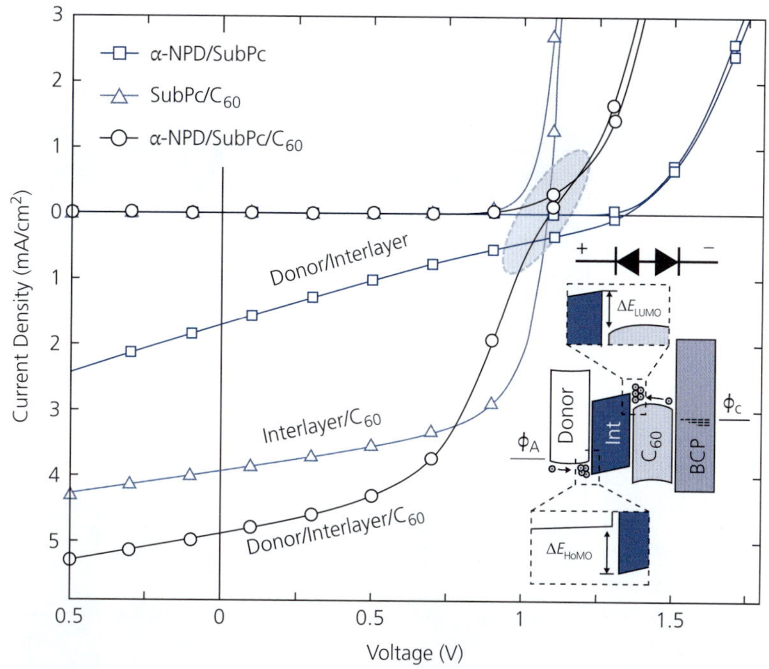

Figure 7.100 Current density–voltage characteristics of a multistep cascade device at 1 sun, AM1.5 simulated solar illumination, and in the dark. Individual bilayer devices comprised of the same materials are also shown. The layer thicknesses and compositions are α-NPD (donor, 5 nm), SubPc (Int, 13 nm), C_{60} (Acc, 36 nm). The active layers are sandwiched between an ITO/MoO_3 anode and a 10 nm thick BCP EBL and an Al cathode. The oval highlights the inflection in the cascade device characteristics. After Barito et al. (2013). Inset: Proposed energy level diagram at open circuit (bottom) along with an equivalent circuit (top). After Barito et al. (2014).

electrons at the C_{60}/Int interface, and holes at the donor/Int interface that, in turn, reduces the junction potential, thereby reversing the internal electric field. The reduction in current produces the inflection. This problem has been observed to a greater or lesser extent in other such cascade devices (Heidel et al., 2011, Barito et al., 2014). The inflection is expected to be more pronounced with increasing light intensity, and hence increased charge accumulation at the junctions. Thus, the OPV can switch from a conventional single diode-like characteristic at low light intensity with the two junctions serving as current generators, to a device at high intensity with two opposing junctions that effectively shut off the current flow.

The characteristics in Fig. 7.100 reflect the existence of internal energy barriers within the cascade that can only be reduced by a careful choice of materials. This choice, however, may ultimately result in compromises in other materials properties that are required to achieve high performance. Consequently, multistep cascade devices have not offered improved performance over BHJ, mixed HJ or exciton cascade devices.

Planar junction architectures of three or more components can be replaced by ternary *blend* BHJs of those same components to simplify the active region structure. The blended cells can consist of two donors and an acceptor (i.e. D_1D_2A), or two acceptors with a single donor (DA_1A_2) (Ameri et al., 2013, Fu et al., 2018). The blends are generally cast from solution, although they can also be realized by co-deposition by VTE (Rand et al., 2005b, Yang et al., 2008). The exciton generation and charge transport processes are analogous to those found for exciton and multijunction cascades, although the complete intermixing of three components can lead to parallel processes in contrast to serial transport experienced using cascades of individual layers to a single dissociating junction. Unlike cascades, the blended ternary BHJ performance critically depends on morphology. The ability to extract charge through complex networks comprising three or more constituents amplifies morphological challenges already confronted in binary BHJs, and in small molecule mixed regions.

The possible routes to photogeneration in D_1D_2A ternary cells are illustrated in Fig. 7.101. The three constituents are represented by their individual frontier energy diagrams, although they are homogeneously intermixed in the actual structures. Figure 7.101a shows parallel charge transfer, where excitons are generated in both D_1 and D_2 (as well as

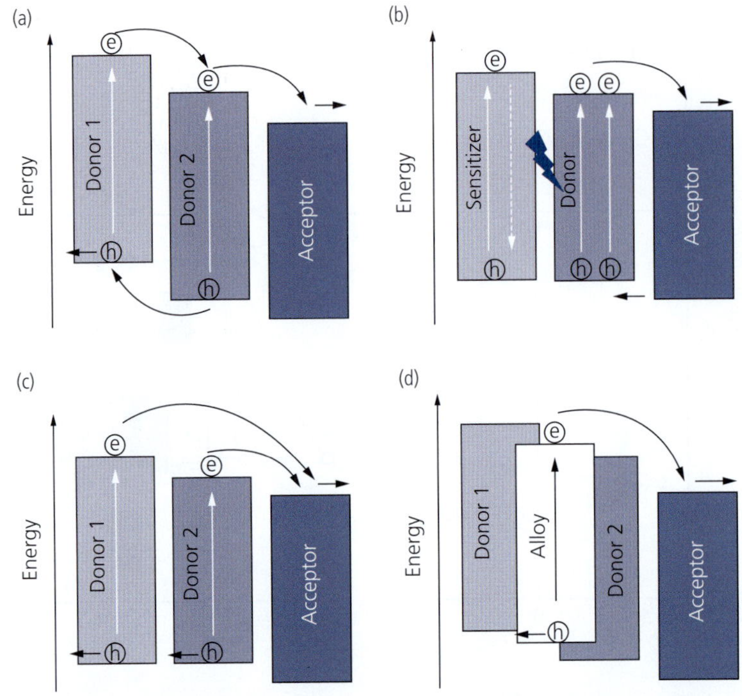

Figure 7.101 Possible photocurrent generation routes of D_1D_2A ternary blend BHJs. (a) Parallel charge generation, (b) FRET transfer from a sensitizer (D_1) to a donor (D_2), (c) parallel exciton dissociation, and (d) dissociation at A in a "molecular alloy" formed by mixing D_1 and D_2 with A.

in A, but not shown). Electrons generated by dissociation at the D_1/D_2 HJ diffuse through D_2 and onward to the acceptor. Similarly, holes generated between D_1 and D_2 transfer to D_1 for conduction to the anode, and electrons diffuse through A to the cathode. Of course, dissociation also occurs at the D_2/A interface, given that these are homogeneous blends of all three constituents. Direct charge transfer occurs when there is weak or no overlap between the emission spectrum of D_1 and the absorption of D_2, thus reducing the probability for FRET. The blend operates on the same principle as the multi-junction cascade in Fig. 7.99.

Energy transfer (Fig. 7.101b) is another potential route to charge generation that is active when there is significant overlap between the emission spectrum of D_1 (the "sensitizer") and the absorption spectrum of D_2 (Honda et al., 2011, Huang et al., 2013). This mechanism is identical to that for the energy cascade in Fig. 7.97, except that the active region morphology is a blend as opposed to discrete layers. The wide energy gap material transfers photogenerated excitons to the lower energy gap donor, where they subsequently dissociate at the HJ between D_2 and A. Also, direct exciton generation is possible in D_2 itself.

The mechanism in Fig. 7.101c models the ternary blend cell as two parallel, rectifying junctions comprising D_1A and D_2A. Both junctions independently generate excitons by absorption of light. The excitons transport though the donors where they dissociate into free charges at both D-A junctions. Subsequently, the holes are transported to the anode within the donors in which they originated. This system does not undergo significant energy transfer between donors as in Fig. 7.101b, nor do the donors share conductive pathways for charges generated on the other components in the blend.

The so-called "alloy" model in Fig. 7.101d is qualitatively different than the other three mechanisms. In this picture, a new material, or *molecular alloy* is formed by mixing of several materials in a common active region. The alloy thus formed has emergent properties that are not characteristic of the original materials of which it is comprised.

Parallel charge generation has been proposed for the D_1D_2A ternary mixture of P3HT, PCPDTBT and $PC_{61}BM$, whose energy levels are shown in Fig. 7.102a (Koppe et al., 2010). Since in the blend, all materials contact each other, several transfer paths ($D_1 \rightarrow D_2 \rightarrow A$, $D_1 \rightarrow A$, and $D_2 \rightarrow A$) are possible. Also, the diagram suggests that FRET from D_1 to D_2 is not active in this materials combination due to their poor spectral overlap.

Photoinduced absorption (PA) was used to study parallel charge transport in the device. The method uses a

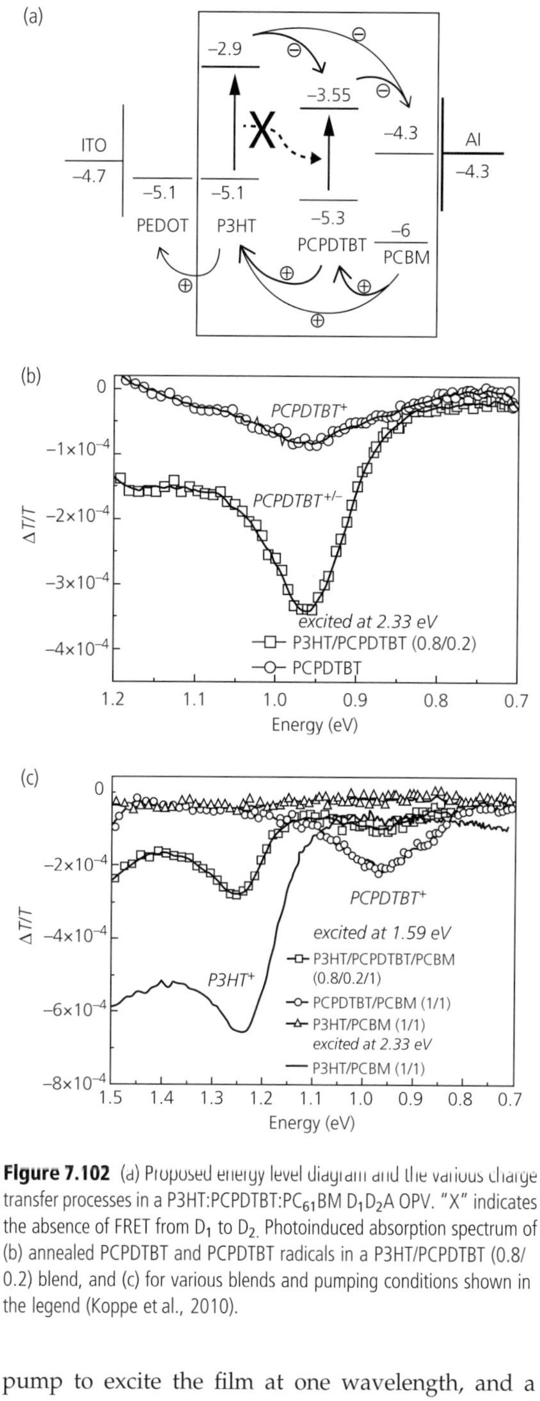

Figure 7.102 (a) Proposed energy level diagram and the various charge transfer processes in a P3HT:PCPDTBT:$PC_{61}BM$ D_1D_2A OPV. "X" indicates the absence of FRET from D_1 to D_2. Photoinduced absorption spectrum of (b) annealed PCPDTBT and PCPDTBT radicals in a P3HT/PCPDTBT (0.8/0.2) blend, and (c) for various blends and pumping conditions shown in the legend (Koppe et al., 2010).

pump to excite the film at one wavelength, and a probe to measure the resulting change in transmission ($\Delta T/T$). The PA spectra are shown in Fig. 7.102b and c for two different excitation energies. The PCPDTBT film absorbs at 1.59 eV, where P3HT is transparent. Exciting the films at 2.33 eV where both films absorb (Fig. 7.102b) generates a radical cation, $PCPDTBT^{+}$, leading to an absorption peak at 0.96 eV. The larger peak for the P3HT:PCPDCTBT (1:1) indicates that photogenerated electrons efficiently transport from P3HT to PCPDCTBT, as noted by the bold line between their LUMOs in Fig. 7.102a. In Fig. 7.102c, excitation at 2.33 eV of the P3HT:$PC_{61}BM$ (1:1) junction results in hole polaron absorption in P3HT at 1.25 eV. This signal disappears when pumped at 1.59 eV where P3HT is nearly transparent. In contrast, pumping the PCPDTBT:$PC_{61}BM$ (1:1) film at 1.59 eV gives rise to a secondary hole polaron peak in P3HT at ~0.95 eV, suggesting that holes are transported directly from $PC_{61}BM$ to PCPDTBT. The strong P3HT polaron absorption band found in the double donor structure of P3HT:PCPDTBT:$PC_{61}BM$ is due to efficient hole transport from the low to the high energy gap donors. These data provide evidence that the primary routes for charge generation are due to exciton dissociation at both the D_1A and D_2A interfaces, and then conduction of both electrons and holes through the entire D-A network to the electrodes. Lack of evidence for energy transfer strengthens these conclusions (Koppe et al., 2010).

The photoluminescence lifetime of D_1D_2A blends was analyzed to confirm that FRET leads to the energy transfer from D_1 to D_2 according to Fig. 7.101b for a D_1 = P3HT, D_2 = SQ, and A = $PC_{61}BM$ blended ternary OPV. In Section 3.8.1 we found that energy transfer is a process that competes with the natural fluorescent decay of the molecule. That is, the radiative decay time is

$$\tau_{tot} = \frac{\tau_{ET}\tau_F}{\tau_{ET} + \tau_F}. \tag{7.169}$$

Here, τ_{ET} is the FRET transfer time and τ_F is the natural fluorescence decay time. It was found that τ_{tot} decreased from 223 ps for 0% SQ in P3HT, to 52.4 ps for 1 wt% SQ, to 9.9 ps for 5 wt% SQ. This was one of several pieces of data suggesting that FRET was the most active process for this particular materials system (Huang et al., 2013).

The ternary open circuit voltage, V_{OCT}, for the parallel circuit model of Fig. 7.101c is analyzed using the j–V characteristics in Fig. 7.103. Here, we separately plot the characteristics for the two, independent parallel "subcells," D_1A and D_2A, that coexist in the blend. From Kirchoff's Law, the open circuit voltage of D_1D_2A corresponds to the point where the $j_1 = -j_2$, such that the total short current is $j_{SCT} = 0$. The V_{OCT} of the network is calculated using this condition, along with $V_1 = V_2$ in Eq. 7.122. The solution is complicated unless we assume that the ideality factors for the two junctions are approximately equal, that is $n_{ST} = n_{S1} \approx n_{S2}$. Then, in the approximation that $k_{PPd}(V) = k_{PPd,eq}$:

$$V_{OCT} = \frac{n_{ST}k_BT}{q} log\left(\frac{j_{SCT}}{j_{0T}} + 1\right), \tag{7.170}$$

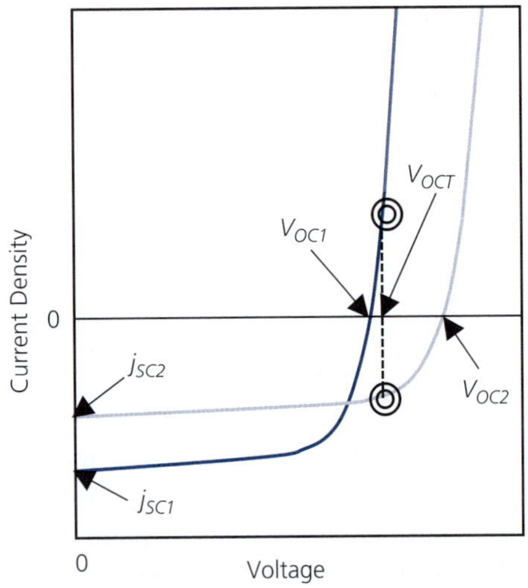

Figure 7.103 Example *j*–*V* characteristics of two parallel "subcells" in a ternary blend such as in Fig. 7.101c. The open circuit voltage of the ternary blend, V_{OCT}, is at the point where the current from subcell 1 (dark blue line) is equal and opposite to that of subcell 2 (light blue line).

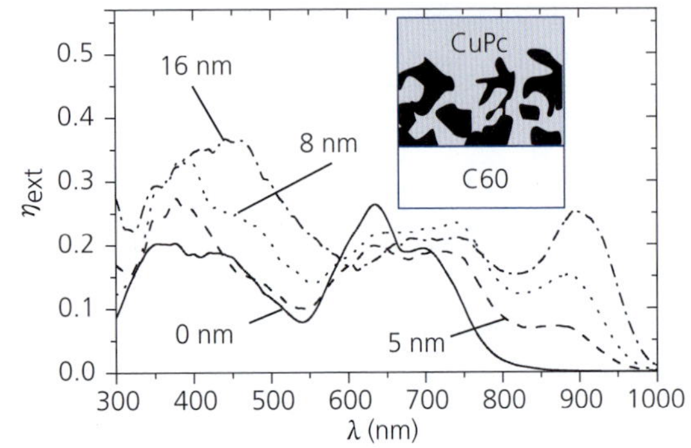

Figure 7.104 External quantum efficiency spectra for CuPc (15 nm)/SnPc (*x* nm indicated)/C_{60} ternary OPVs grown by organic vapor phase deposition. Inset: Illustration of the junction showing the discontinuous SnPc (black) and CuPc (light blue) donor regions (Yang et al., 2008).

where the saturation currents are given by $j_{0T} = j_{01} + j_{02}$, and we assume that $j_{SC} \gg j_0$ for both subcells.

It is apparent from Fig. 7.103 and Eq. 7.170 that V_{OCT} must always lie between the upper and lower limits set by V_{OC1} and V_{OC2}. This conclusion is consistent with voltage measurements of ternary devices, and is independent of which model in Fig. 7.101 is active. An additional implication of Kirchhoff's law is that the less efficient junction limits the current from the more efficient junction to ensure that voltages from both junctions are equal at the diode terminals.

A ternary cell based on vapor co-deposited D_1D_2A HJ (D_1 = CuPc, D_2 = SnPc, A = C_{60}) formed a parallel junction similar to that in Fig. 7.101c (Rand et al., 2005b). The parallel network results from incomplete wetting of the CuPc by a thin (5 nm) layer of SnPc, over-coated by the thick (40 nm) C_{60} acceptor. In subsequent work, OVPD of the layers in a ITO/SnPc (*x* nm)/CuPc (40 nm)/BCP (10 nm)/Ag OPV resulted in formation of a more controlled network of D_1, D_2, and A (Yang et al., 2008). The solar coverage provided by this ternary combination is significantly expanded by the addition of even a thin SnPc layer due to its strong absorption into the NIR, as shown by the η_{ext} spectra shown in Fig. 7.104. At a SnPc thickness of $x = 0$ nm (corresponding to D_1A), the absorption extends only to 750 nm. However, a peak emerges at 900 nm with the addition of SnPc, providing spectral coverage from 300 nm to 1000 nm for the ternary OPV.

The inset in Fig. 7.104 shows an illustration of the morphology achieved during growth. The SnPc forms

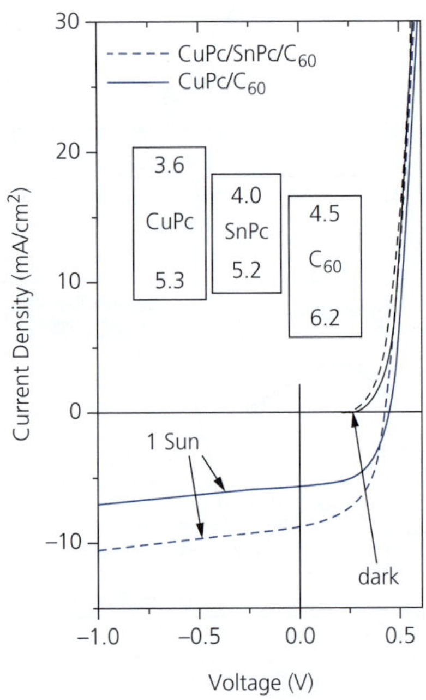

Figure 7.105 Current density–voltage characteristics of OPVs grown by OVPD in the dark, and under 1 sun, AM1.5G simulated illumination. The active regions comprise CuPc (15 nm)/C_{60} (40 nm) and CuPc (15 nm)/SnPc (16 nm)/C_{60} (40 nm) for the binary and ternary cells, respectively. Inset: Energy level diagram of the ternary active region (Yang et al., 2008).

crystalline clusters that segregate into separate junctions with the C_{60}. Gaps between the islands allow for parallel, direct contact between the acceptor and the second donor, CuPc. The *j*–*V* characteristics of this device are shown in Fig. 7.105 along with a binary CuPc/C_{60} OPV for comparison. The j_{SC} is increased by approximately 40% in the ternary compared to the binary cell due to the additional NIR absorption of the SnPc. Furthermore, the existence of a parallel conduction network is inferred since V_{OC} = 0.44 V is the same as that for the binary device. The V_{OC} of an analogous NPD/CuPc/C_{60} device was also 0.44 V, whereas V_{OC} of the NPD/C_{60} binary was 0.87 V. This suggests that the V_{OC} is "pinned" to the lowest voltage in the parallel network, although we have seen that this is not always the case. In the junction in Fig. 7.104, the HOMO energies of both SnPc and CuPc are similar, resulting in a similar ΔE_{HL}, and hence V_{OC} should be independent of any a particular network model. The conclusion to be drawn is that in this device, V_{OC} is determined not by the junction energies, but rather is due to rapid charge recombination at the interface.

A solution-based system that shows parallel diode-like behavior consists of a blend of DTffBT (D_1), DTPyT (D_2,), and $PC_{61}BM$ as the acceptor. The parallel BHJ (PBHJ) results are compared with those of separate diodes comprising only D_1A and D_2A "subcells" in Fig. 7.106. The absorption spectrum of the PBHJ OPV is the sum of the absorption from the

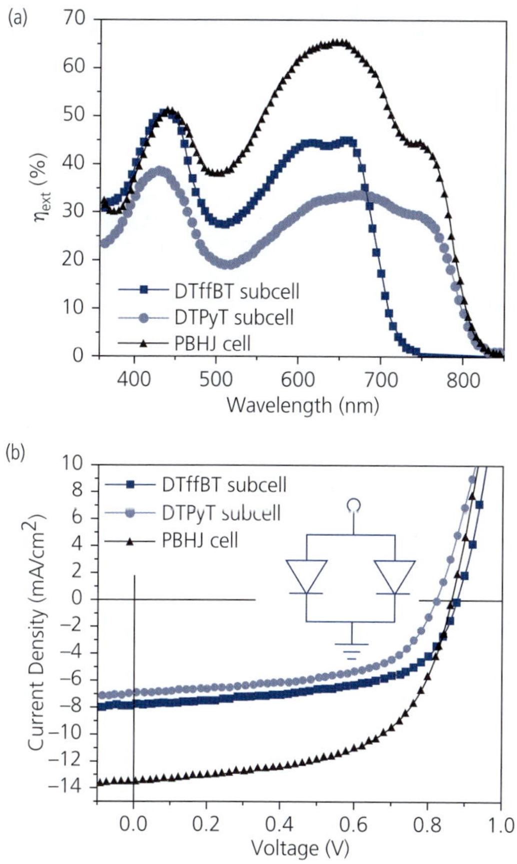

Figure 7.106 (a) External quantum efficiency spectra, and (b) current density-voltage characteristics of a D_1D_2A parallel BHJ ternary OPV consisting of the donors DTffBT and DPTyP with $PC_{61}BM$ as acceptor. PBHJ is the parallel BHJ device. *Inset*: Parallel junction equivalent circuit (Yang et al., 2012a).

two constituent subcells. Likewise, the external quantum efficiency spectra in Fig. 7.106a are also a superposition of D_1A and D_2A at wavelengths where the spectra from both subcells overlap. At long wavelengths where only DTPyT absorbs, the efficiency is larger than for the DTPyT:$PC_{61}BM$ OPV. It is speculated that this is a result of reduced resistance of the hole current due to the presence of additional parallel conductive paths afforded by DTffBT in the ternary blend (Yang et al., 2012a). Increased charge collection due to improved conduction should also result in lower series resistance. Figure 7.106b does indeed show a higher slope in the forward biased characteristics compared with the individual subcell characteristics that support this hypothesis. Direct absorption into the CT state between the two donors (at energy ~ 1.6 eV) may also contribute to the long wavelength quantum efficiency.

It is observed that the $V_{OC} = 0.87$ V of the PBHJ is close to, but less than the $V_{OC} = 0.89$ V of the larger energy gap donor, DTffBT ($E_G = 1.7$ eV), while it is considerably larger than $V_{OC} = 0.83$ V for the smaller band gap DTPyT ($E_G = 1.5$ eV).

The alloy model in Fig. 7.101d qualitatively differs from the other models. It suggests the ternary blend forms a molecular alloy between the donors in a D_1D_2A, or the acceptors in DA_1A_2 blend. In fact, molecular alloys have been proposed over several decades to describe electronic, optical and physical properties of mixtures that are not found in their individual molecular components (Bellows and Prasad, 1977, Engler et al., 1977, Rai and George, 1992, Sturz et al., 2004, Mogeritsch et al., 2009, Murata et al., 2010, Dou et al., 2019). Before one proposes the existence of such a material, however, the concept of an alloy must be clearly defined in the context of organic materials, and what makes such a substance distinct from a simple blend of multiple components commonly found in almost all organic photonic devices.

An alloy is a mixture of two atoms or compounds (most often metals such as brass or stainless steel, or compound semiconductors such as GaAlAs or InGaAsP) that share *chemical* bonds to form a new material whose electronic or physical properties differ from those of the pure substances of which it is comprised. Furthermore, once an alloy is formed, its thermal properties (e.g. the melting temperature of a eutectic) may change substantially from the pure solids of which it is comprised, making recovery of the initial components problematic. In contrast, the constituents in a blend of a small molecule donor and acceptor may be separated by raising the temperature at or near the sublimation point of the lowest temperature constituent. However, a tightly bound charge transfer salt such as TTF:TCNQ may not be so easily separated once the compound solid is formed. Although the individual constituents in this case might not separate on heating, the spectrum of the complex is, nevertheless, only a superposition of the spectra of the cationic and anionic species in the solid.

But what about molecular blends in ternary cells, or even in doped emission layers of OLEDs? Again, one would expect that an alloy formed from a blend would possess emergent properties not found in any of the individual constituents. It is not a simple matter to envision how an alloy would emerge from a mixture of van der Waals bonded materials. Cross-linking between molecules, electronic orbital coordination via hydrogen bonding, etc. may result in the formation of a molecular alloy, or it may simply form a different molecule (e.g. bond linkage between molecules A and B results in a new compound C, corresponding to an oligomer or even a polymer, but not an alloy).

The formation of a molecular alloy for the DA_1A_2 ternary BHJ comprising P3HT/ICBA/$PC_{61}BM$ blend has been proposed based on the photocurrent spectra of the CT and exciton states as functions of ICBA fraction in $PC_{61}BM$ shown in Fig. 7.107a (Street et al., 2013). The long wavelength CT photocurrent spectrum (which is proportional to the absorption spectrum) monotonically shifts to larger energies with ICBA fraction. Yet the exciton-induced photocurrent spectra between 1.6 eV and 1.8 eV are a simple linear superposition of the spectra of the two constituent acceptors. This indicates that the exciton energies are unperturbed from those of the individual molecules, even when those molecules are blended. It was argued that changes in the exciton spectrum of an alloy would be small due to their localized nature.

On the other hand, CT states extend across the D-A HJ, and hence are more strongly perturbed when combined into an alloy. The CT energies and V_{OC} for the P3HT/ICBA/$PC_{61}BM$ OPVs measured at two different relative values of photocurrent (PC = 0.01 and 0.1, indicated by the horizontal dashed lines in Fig. 7.107a) are plotted in Fig. 7.107b. The CT energy, which in the ideal case is equal to ΔE_{HL}, increases with ICBA fraction. Since V_{OC} is also a linear function of ΔE_{HL} (cf. Eq. 7.123), the open circuit voltage of the alloy, V_{OCA}, should follow *Vegard's law*:

$$V_{OCA} = (1-x)V_{OC1} + xV_{OC2} - bx(1-x), \quad (7.171)$$

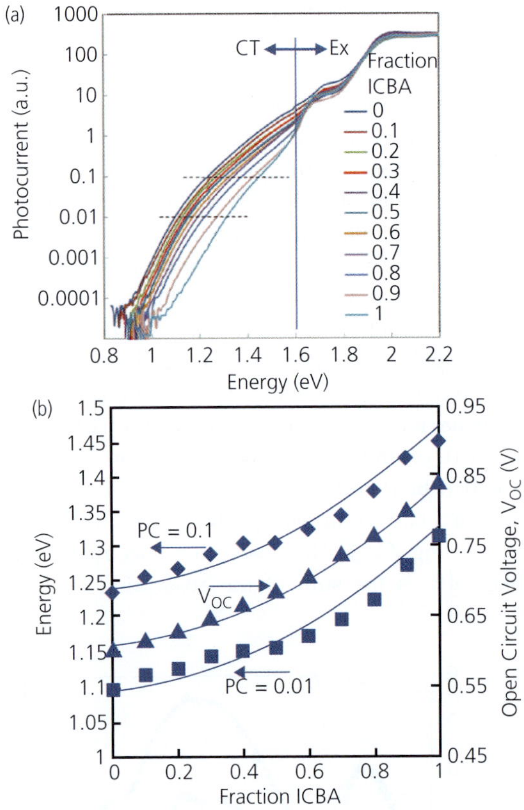

Figure 7.107 (a) Photocurrent spectrum of the DA_1A_2 ternary OPV as a function of fraction of A_2 = ICBA mixed with $A_1 = PC_{61}BM$. The ranges of CT and exciton (Ex) response are indicated. (b) CT energy for two different photocurrents (PC = 0.01 and 0.1) corresponding to the dashed lines in (a). Also shown is the open circuit voltage (Street et al., 2013).

where $b = 0.18$ is the bowing parameter and x is the ratio of ICBA to $PC_{61}BM$. The line is a fit to V_{OCA} and CT spectral peak positions vs. x, indicating that these two parameters have the same dependence on x. It is notable that the range spanned by the alloy voltage is $V_{OC1} < V_{OCA} < V_{OC2}$, which is similar to that of all the other ternary BHJs discussed, independent of the transport model invoked.

While the trends in Fig. 7.107 may imply the existence of an alloy, they fall short of providing conclusive evidence. Indeed, the nearly linear interpolation of V_{OCA} in Eq. 7.171 (albeit with a small amount of quadratic bowing) suggests that the photocurrent due to CT states is also a linear superposition of the two independent CT states for DA_1 and DA_2, much like that found for the exciton spectra. Furthermore, the CT spectra are insufficiently distinct from the excitonic tails to accurately fit their energies. Hence, more convincing evidence should be provided by observation of the transient and photoluminescence spectra of the CT state whose features cannot be explained by a linear superposition of independent CT states.

The most direct evidence for electronic interactions could come from the CT state electroluminescence spectra arising from forward current injection into the active regions of the two binary junctions compared to that of the ternary. These spectra are shown in Fig. 7.108a for the same P3HT:ICBA:PC_{61}BM system for several different ICBA:PC_{61}BM blend ratios. There is a monotonic shift of the EL peak from that of ICBA at 980 nm, to PC_{61}BM at 1100 nm as the PC_{61}BM fraction is increased from 0 to 1. The peaks at $\lambda <$ 850 nm are excitonic features from the three constituents. The peak positions of the CT spectra are easily resolved compared to the absorption data in Fig. 7.107a. The EL features are fit to a series of Gaussians, where only the relative intensities of the two CT states were varied linearly with concentration, that is, all the peak widths and positions remain fixed in each of the blends. An example deconvolution of the spectrum of a P3HT:ICBA:PC_{61}BM (1:0.5:0.5) blend into its separate component peaks is shown in Fig. 7.108b.

A plot of the ratio of ICBA to the combined CT peak intensities vs. PC_{61}BM fraction is shown in Fig. 7.109a. All blends fit a linear function with a unity slope. This provides clear evidence that the CT feature is a simple linear superposition of the individual CT states of P3HT:ICBA and P3HT:PC_{61}BM HJs. Indeed, this linear superposition of Gaussian peaks also accurately replicates the bowing observed in the CT state energy and V_{OC} of the ternary blends, as shown in Fig. 7.109b. Note that the CT radiative lifetime of 0.95 ns is the same for all the blends, as well as for the individual P3HT:ICBA and P3HT:PC_{61}BM junctions. The HOMO energies of the blends were also unaffected by the blend ratio. Photoelectron spectroscopy measurements indicate that the HOMO energies of the blends are pinned to that of ICBA which is slightly higher than for PC_{61}BM, thus forming a hole trap (Huang et al., 2019).

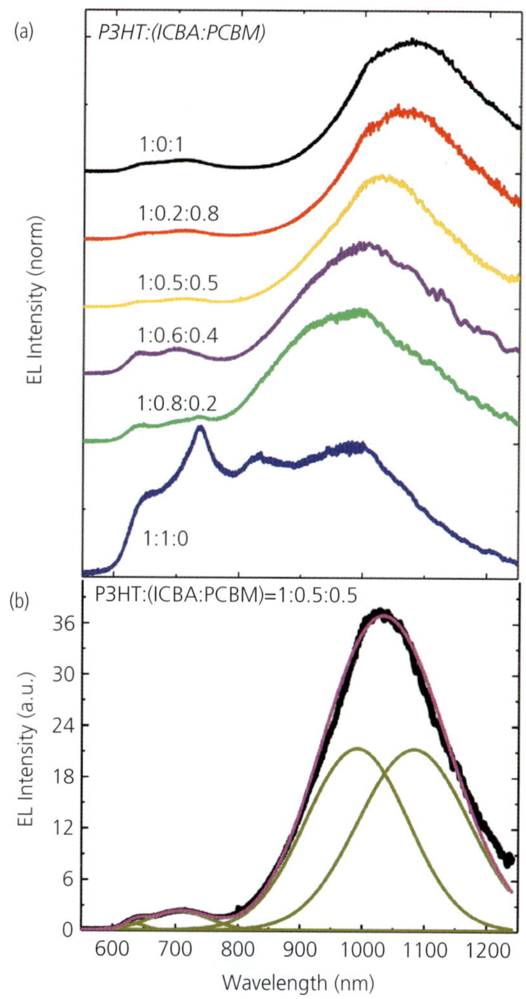

Figure 7.108 (a) CT state EL intensity spectra for different P3HT:ICBA:PC_{61}BM blend ratios. (b) Example fitting procedure to the various excitonic and CT features for the 1:0.5:0.5 blend. The lower (green) curves show the relative heights and widths of the constituent CT states, and the upper (red) curve fit to the data is their sum (Huang et al., 2019).

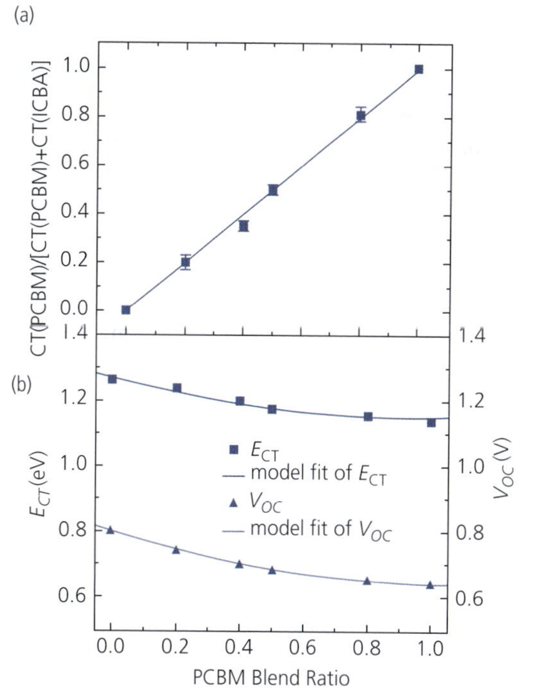

Figure 7.109 (a) Ratio of the ICBA to the total integrated CT state intensity vs. PC_{61}BM/ICBA ratio. (b) Charge transfer exciton peak energy and V_{OC} vs. PC_{61}BM/ICBA ratio. Lines are fits to a linear superposition of the CT features in Fig. 7.108 (Huang et al., 2019).

From these data, no new, "alloy" state is found in this particular system. The question remains whether or not alloys are formed in other materials systems. More broadly, there remains little or no conclusive evidence for the existence of an alloy whose optoelectronic properties are not adequately described by a simple blend comprising multiple, independent constituent molecules.

Extensive reviews of device performance and materials used in DA_1A_2 and D_1D_2A ternary BHJs can be found in the literature (Ameri et al., 2013, Fu et al., 2018). Here, we highlight three example high performance ternary OPVs.

A DA_1A_2 cell comprising a combination of the donor, PBDB-T with the acceptors IT-M, (A_1) and the bis-adduct of $PC_{71}BM$ (A_2) in the ratio of PBDB-T:IT-M:bis-$PC_{71}BM$ (1:1:0.2) is shown in Fig. 7.110. The blend was annealed for 10 min at 100°C prior to the deposition of the PFN-Br exciton blocking layer. A summary of the performances of the ternary as well as the separate binary D-A combination BHJ OPVs is provided in Table 7.10. The ternary device achieves both a higher *FF* and j_{SC} than either binary device, leading to a maximum power conversion efficiency of 12.2%, compared to 10.8% for the DA_1 and 6.25% for the DA_2 OPVs. Consistent with previous discussions, the open circuit voltage of the ternary device is 0.952 V, lying between $V_{OC1} = 0.937$ V and $V_{OC2} = 1.02$ V. The higher *FF* suggests improved electron conductivity in the network comprising two acceptors, and hence reduced series resistance. This is supported by the energy level diagram of the materials in Fig. 7.110 where the optical LUMO energies of bis-$PC_{71}BM$ and IT-M are nearly aligned to promote barrier-free electron transport to the cathode, consistent with the parallel dissociation model of Fig. 7.101c. The reduced open circuit voltage of the ternary compared to DA_2 is compensated by the increased *FF* and j_{SC}.

An example D_1D_2A OPV comprises the spirobifluorene (SF) symmetrically disposed between two 1,3-benzothiazole (BT) units capped with (3-ethylhexyl-4-oxothiazolidine-2-yl)dimalononitrile (RCN) in the acceptor, SFBRCN, combined with the donors PTB7-Th (D_1) and PBDB-T (D_2) (see Fig. 7.111). One of the donors (PBDB-T) also serves as the donor in the DA_1A_2 device in Fig. 7.110. Similar to the device in Fig. 7.92, donors and acceptors can be interchanged since their function is determined by their energy levels relative to the other materials employed in the HJ. The performance of the device along with its separately characterized D and A components are summarized in Table 7.11.

An alternative to the blended and layered multi-junction active region is an amalgamation of the two, whereby two independently fabricated D-A junctions are placed in a series stack. After, attachment, their interface is blended to avoid the use of a CRZ that is otherwise required to interconnect subcells in a tandem architecture. In Fig. 7.112 we show the complementary absorption of the two different HJ OPVs comprising PTB7-Th:$PC_{71}BM$ (BHJ1) and PDPP3T: $PC_{71}BM$ (BHJ2) employed in the stacked "bi-ternary"

Table 7.10 Performance characteristics of a ternary DA_1A_2 OPV and its binary components (Zhao et al., 2017b)

$D:A_1:A_2$ ratio	V_{OC} (V)	j_{SC}[a] (mA/cm^2)	*FF*	η_P (max) (%)	η_P (ave)[b] (%)
1:1:0	0.937	16.7	0.69	10.80	10.45
1:1:0.2	0.952	17.4	0.74	12.20	11.75
1:0:1	1.02	10.6	0.58	6.25	5.86

[a] Cell area = 4 mm^2, measured under 1 sun, AM1.5G simulated illumination.
[b] Sample size = 100 diodes.

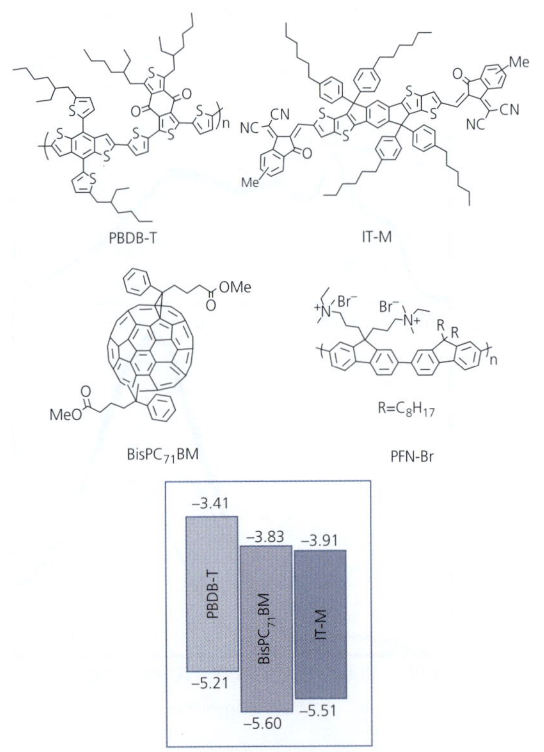

Figure 7.110 Molecular structural formulae of materials used in a DA_1A_2 OPV. Inset: HOMO energies measured by ultraviolet photoelectron spectroscopy. Presumably the LUMO energies are the sum of the HOMO energies and the optical energy gap of each material (Zhao et al., 2017b).

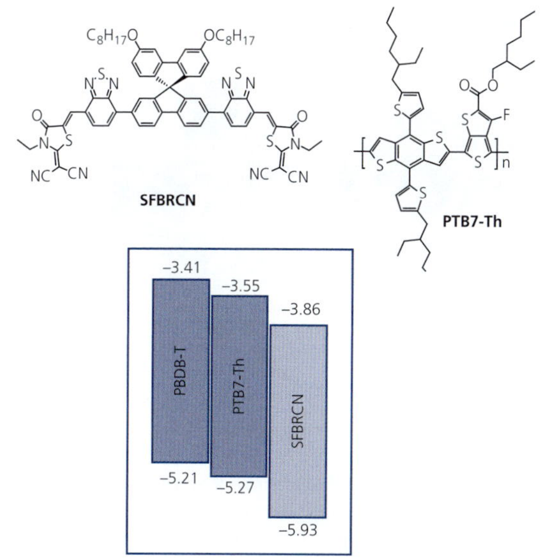

Figure 7.111 Molecular structural formulae of materials used in the D_1D_2A OPV. Inset: Energy level diagram of the active region (Xu et al., 2017).

Table 7.11 Performance characteristics of a ternary D_1D_2A OPV and its binary components (Xu et al., 2017)

D_1:D_2:A ratio	V_{OC} (V)	J_{SC}[a] (mA/cm^2)	*FF*	η_P (max) (%)	η_P (ave)[b] (%)
0:1:1	0.97	13.7	0.71	9.39	9.17
0.7:1:1	0.93	17.9	0.74	12.27	11.89
1:0:1	0.91	17.1	0.65	10.10	9.87

[a] Cell area = 4 mm^2 measured under 1 sun, AM1.5G simulated illumination.
[b] Sample size = 20 diodes.

OPV (Huang et al., 2017). The marginal overlap of the absorption spectra between 700–800 nm makes this a nearly ideal combination for efficient coverage of the solar spectrum up to a wavelength of 900 nm.

The two BHJs are combined by physical bonding through the process illustrated in Fig. 7.113. Each active layer is cast from solution onto separate substrates. The 200 nm thick BHJ1 is deposited onto a ZnO_x buffer on the surface of the ITO anode from a solution containing 1,8-diiodooctane. The second substrate comprises the 50 nm thick PDPP3T:$PC_{71}BM$ BHJ2 active region using a polyurethane-coated polycarbonate (PUA/PC) substrate. When the two BHJs are brought into contact and pressure is applied to the PUA/PC stamp at 75°C for 2 min, the active regions bond. The stamp is peeled away, and methanol is used to remove residual solvent in BHJ1. The device is completed by vacuum deposition of the cathode.

Time of flight secondary ion mass spectroscopy (TOF-SIMS) was used to determine the degree of intermixing of the two BHJs at the interface. In TOF-SIMS, the film is slowly removed by sputtering while tracking the relative F, N, and Zn mass spectral intensities to obtain a depth profile of the relative concentrations of PTB7-Th, PDPP3T, and ZnO_x, respectively. The spectra in Fig. 7.114a show a 25 nm thick intermixed zone (corresponding to a sputtering time of 100 s). Immediately following contact, the BHJs form a sharp interface (upper left, Fig. 7.114b). Soaking the structure in methanol results in dissolution of the as-deposited layers, forcing some intermixing of the two donor polymers at the junction. When the methanol is removed by

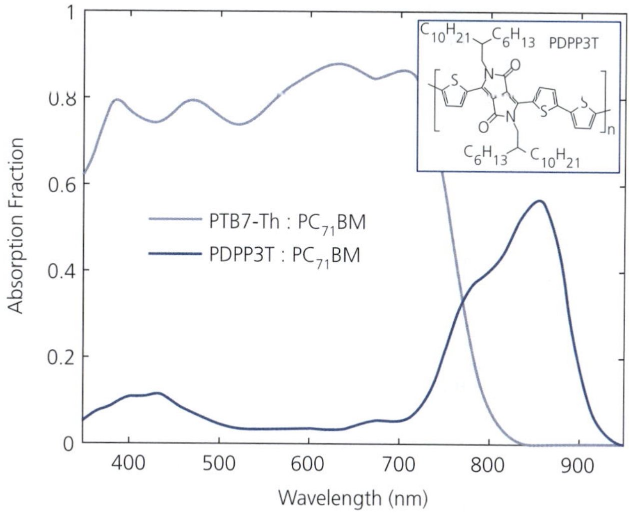

Figure 7.112 Absorption spectra of two separate binary cells whose active regions are in contact to form a bi-ternary OPV. Inset: Molecular structural formula of PDPP3T used as the donor in the BHJ2 subcell. The structural formula of PTB7-Th is given in Fig. 7.111 (Huang et al., 2017).

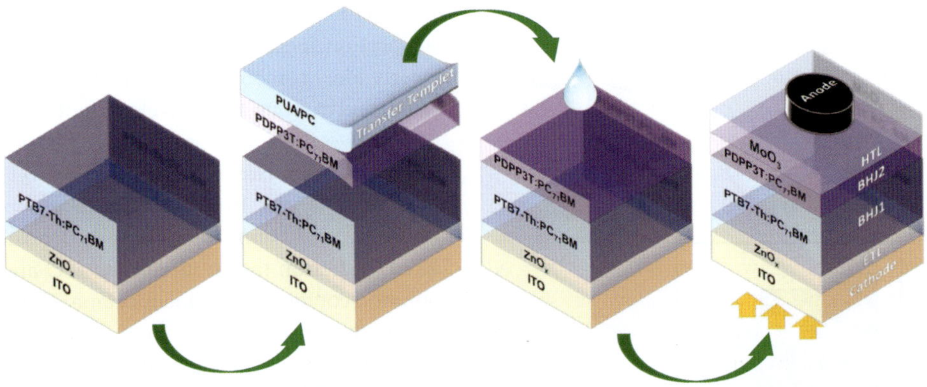

Figure 7.113 Fabrication of the stamp-bonded bi-ternary OPV. From left: BHJ1 is fabricated on an ITO-coated glass substrate. This is coated by a ZnO_x anode buffer and the PTB7-Th:$PC_{71}BM$ active region. The PDPP3T:$PC_{71}BM$ BHJ2 active region is deposited onto a PUA/PC elastomeric stamp that is peeled off once the two subcells are bonded by pressure welding. In the third step, methanol is used to soak the junction to remove the residual solvent used in the BHJ1 solution. Finally, the MoO_3 electron buffer and Ag cathode are deposited in vacuum (Huang et al., 2017).

Table 7.12 Performance characteristics of a bi-ternary (BHJ1/2) OPV and its binary components (Huang et al., 2017)

Cell[a]	V_{OC} (V)	j_{SC}[b] (mA/cm^2)	*FF*	η_P (max) (%)	η_P (ave)[c] (%)
BHJ1	0.81	18.5	0.70	10.5	10.3
BHJ2	0.69	7.4	0.71	3.6	3.5
BHJ1/2	0.77	23.8	0.67	12.3	11.9

[a] Active region thicknesses: BHJ1 = 200 nm, BHJ2 = 50 nm with the same respective thickness for BHJ1/2.
[b] Cell area = 3.14 mm^2, measurement under 1 sun, AM1.5G simulated illumination.
[c] Sample size = 20 diodes.

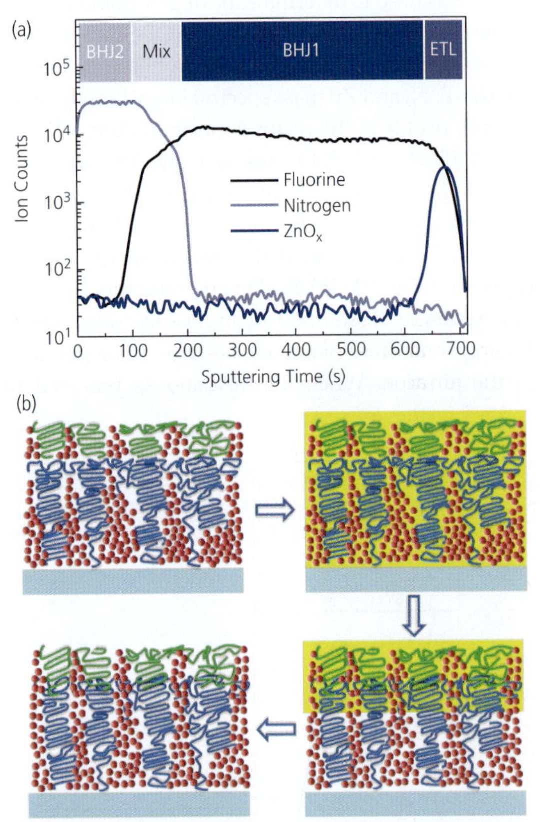

Figure 7.114 (a) TOF-SIMS signal intensity vs. sputtering time for the methanol-treated bi-ternary cell in Fig. 7.113. The data show significant blending of PDPP3T (correlated with the N signal) and PTB7-Th (F signal) in the interface region. (b) Illustration of the development of the interdiffused interface morphology by application of methanol (yellow). The polymer chains are color coded as follows: green = PDPP3T, blue = PTB7-Th, and red = PC71BM (Huang et al., 2017).

drying, it leaves a continuous, hole conducting path along the blended polymer interface. Due to the nearly spherical symmetry of $PC_{71}BM$, the acceptor regions interconnect by close molecular packing that forms $PC_{71}BM$ nanocrystals (Huang et al., 2017). A similar vacuum deposited architecture has also been reported (Li et al., 2019).

The characteristics of the bi-ternary (BHJ1/2) device and its constituent BHJ1 and BHJ2 subcells are provided in Table 7.12. As in the other multilayer cascade and ternary architectures, V_{OC} of the three-component cell is between the limits set by BHJ1 and BHJ2.

In this section, we have seen that there are several ways to combine three or more molecular constituents into an OPV to increase coverage of the solar spectrum. These active region architectures can either consist of a series of layers, or can form a blend in a ternary (or perhaps higher order) BHJ. Several models have been advanced to describe the photogeneration process in ternary blend cells. The distinction between these models is sometimes unclear, since multiple processes may be active in such complex blends. Nevertheless, a common feature of the blended junctions is that their open circuit voltages lie within the range bounded by the of the constituent junctions, although the short circuit current can be significantly greater due to the improved coverage of the solar spectrum by three or more constituents vs. conventional binary OPVs.

While the requirements placed on the materials used in such devices are stringent (e.g. alignment of HOMO and/or LUMO levels, overlap of emission and absorption spectra for efficient energy transfer, etc.), the increased efficiency combined with the relative simplicity of the structures has resulted in considerable interest in on both cascade and ternary blend OPVs.

7.4.1.3 Exciton blocking layers

A significant improvement in cell performance was realized by introduction of the *exciton blocking layer* (EBL), also known as the *cathode buffer layer*, interposed between the active region and the cathode (Peumans et al., 2000a, Peumans and Forrest, 2001). The electron-conducting, transparent EBL serves multiple purposes:

(i) The layer prevents excitons generated in the acceptor layer from migrating to the cathode contact where they are quenched without generating photocurrent. This is particularly important for bilayer and planar-mixed HJ active regions where the exciton diffusion length can exceed that of the layer thickness.

(ii) The optical field has a node at the metal cathode surface. The wide energy gap EBL, therefore, provides a transparent spacer whose thickness is adjusted to locate the active layers at a distance $\sim\lambda/4n$ from the cathode. At this position, the optical intensity is maximized, thus maximizing η_{ext}.

(iii) The layer serves as a buffer between the cathode and acceptor, preventing damage to the active region induced during metal deposition, thus eliminating potential exciton quenching sites or electrical shorts across the junction.

The EBL in an OPV was inspired by, and serves several of the same functions as the hole blocking/exciton blocking layer in OLEDs. In the latter structure, the HBL/EBL prevents triplet exciton quenching at the cathode, as well as providing an optical spacer to minimize optical microcavity effects (Baldo et al., 1998, Baldo et al., 1999).

Since the initial demonstration of a so-called double heterostructure OPV employing a BCP EBL (Peumans

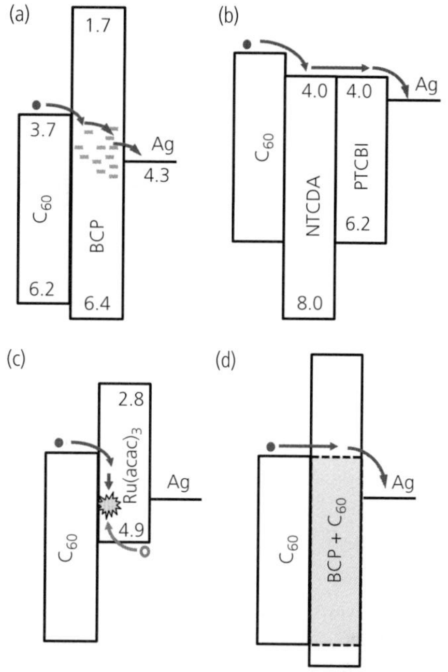

Figure 7.115 Different exciton blocking layer schemes and the mechanisms for extracting electrons for the example of a C_{60} acceptor layer. (a) Trap assisted electron transport, (b) transport along the EBL LUMO, (c) e-h recombination at the acceptor interface, and (d) electron filtering along percolating conducting paths in a doped, wide energy gap layer.

et al., 2000a), several other blocking layer concepts shown in Fig. 7.115 have been demonstrated. The illustrations assume a C_{60} acceptor layer for energy reference, although these configurations are adaptable to all acceptors with the proper choice of blocking layer materials. Each design relies on a large energy gap EBL to ensure its transparency and to avoid exciton transfer from the acceptor. The differences between EBLs are the mechanisms for electron transport. Figure 7.115a shows the wide energy gap BCP that conducts photogenerated electrons from the acceptor via a high density of defect states introduced during cathode deposition (Peumans and Forrest, 2001). Since the defects penetrate only $\sim$10 nm into the BCP, the thickness of the buffer may be too small to ensure an optical intensity maximum at the D-A HJ. Nevertheless, EBLs employing defect conduction in wide energy gap buffers are still the most frequently employed in OPVs due to its simplicity.

Figure 7.115b employs a pair of acceptor-like materials whose LUMO energies align with that of the acceptor. The specific bilayer shown is comprised of the wide energy gap NTCDA adjacent to a second electron conducting layer of PTCBI. The PTCBI makes ohmic contact to Ag but absorbs in the red. To minimize optical losses, the PTCBI layer is kept thin, whereas the wider energy gap NTCDA is used as the transparent optical spacer. The LUMO energy alignments result in low resistance electron conduction to the cathode (Lassiter et al., 2011).

An alternative to employing electron transport is to use a wide energy gap EBL with a small ionization potential, as shown in Fig. 7.115c. Electrons generated in the acceptor recombine at the EBL/acceptor interface with holes conducted via the HOMO level of the EBL. In the figure, we show the mechanism of hole conduction in a Ru(acac)$_3$ EBL from the cathode to C_{60} (Rand et al., 2005a).

Figure 7.115d shows what is perhaps the most effective means for satisfying all of the criteria of a high performance EBL. This so-called electron-filtering exciton blocking layer, comprises a mixture of C_{60} in a wide energy gap matrix (Bartynski et al., 2013, Xiao et al., 2014). The spherical symmetry of the C_{60} molecules allows them pack into continuous percolating electron conduction paths that completely penetrate the transparent exciton-blocking matrix in which they are blended. The LUMO energy of C_{60} aligns with that of the C_{60} (or C_{70}) acceptor, providing an energy-barrier-free electron conducting path to the cathode. And while C_{60} absorbs in the visible, the absorption is due to a charge transfer transition that is largely eliminated at the low concentrations required for conduction. Hence, this composite is both electrically conducting and optically transparent.

We now provide further detail of the conduction mechanisms and optical properties of these four EBL architectures, and the OPV performance enhancements that result from their use.

The defect-based EBL in Fig. 7.115a was first included in a conventional CuPc/PTCBI cell where the 10 nm thick BCP blocking layer was positioned between the Ag cathode and the perylene-based acceptor (inset, Fig. 7.116). The 10 nm thickness is approximately equal to the penetration depth of damage from the cathode deposition; larger thicknesses result in a high cell series resistance due to the lack of conducting paths for electrons extending from acceptor to cathode. A device with 30 nm CuPc lacking an EBL has a peak external efficiency of ~8% at $\lambda = 540$ nm for a 30 nm PTCBI thick layer (dashed-dotted line, closed squares, Fig. 7.116). The external efficiency drops dramatically for thinner PTCBI layers due to exciton quenching at the cathode, as well as a vanishing optical field within the thin active region placed in such close proximity to the metal contact. For the double heterostructure comprising the EBL, η_{ext} continues to increase with

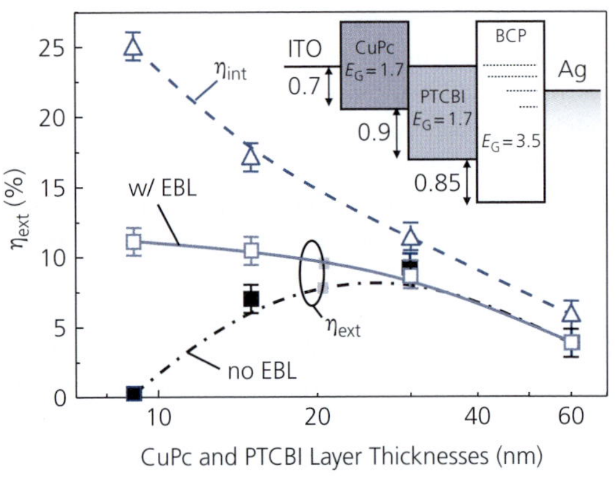

Figure 7.116 Quantum efficiency vs. active layer thicknesses for a CuPc donor/PTCBI acceptor double heterostructure OPV employing a 10 nm thick BCP EBL. The closed and open squares are the external efficiencies for devices with a 30 nm thick CuPc, and varying PTCBI thickness. The open triangles are calculated for devices with equal donor and acceptor layer thicknesses. Inset: Energy level diagram for the OPVs. All numbers are in eV. Dashed lines in the BCP represent defect levels that permit electron conduction. After Peumans et al. (2000a).

decreasing PTCBI thickness until it reaches a maximum of $\eta_{ext} = 11\%$ at approximately 10 nm, due to a higher optical field in the active region and reduced exciton quenching at the cathode.

The device efficiency improves even as the PTCBI thickness is reduced, which in turn reduces the cell series resistance. Hence, devices operated at illumination intensities as high as 6 suns (6 kW/m^2) without a reduction in efficiency.

The fullerene C_{60} was first introduced in an OPV that employed a 12 nm thick BCP EBL. The large exciton diffusion length of C_{60} (~8–10 nm) compared to PTCBI (3 nm) necessitates the use of the blocking layer to prevent exciton quenching at the cathode. The optimized double heterostructure comprised ITO/PEDOT:PSS/20 nm CuPc/40 nm C_{60}/12 nm BCP/Ag, achieving $\eta_P = 3.6\%$ at 1 sun, AM1.5G simulated illumination (Peumans and Forrest, 2001). Contrary to PTCBI-based devices, η_P peaks at 1 sun intensity, and falls off at both higher and lower solar fluence. Due to its higher efficiency, the C_{60} cell delivers at least three times the current as the PTCBI-based cell. Hence, the former suffers loss at high intensity due to the cell series resistance. Resistance also leads to a roll-off in *FF*, from a peak of approximately 0.54 at 1 sun, to 0.32 at 10 suns.

Damage-induced defect conduction constrains the thickness of the wide energy gap matrix to the penetration depth of the damage caused by deposition of the metal cathode. The penetration depth is dependent on both the materials and the deposition processes, but typically the layer thickness can be no greater than 10 nm if high resistance to electron conduction is to be avoided.

Doping the wide energy gap material with a second constituent (e.g. PTCBI) stabilizes its morphology while increasing its conductivity (Peumans et al., 2000a). Alternatively, doping the layer with metal by co-evaporation can also provide a high conductivity even for thick layers. It has been found, for example, that BCP co-evaporated with 20% Yb leads to EBLs as thick as 40 nm without a noticeable loss in *FF* and j_{SC} in CuPc/C_{60} OPVs. It is unclear, however, whether the metal doping creates conductive percolating paths, or that it simply extends the depth of damage that ordinarily stops near the surface when induced during cathode deposition. Furthermore, the presence of metal atoms can increase absorption by the EBL, although this was not a particularly significant effect for the Yb: BCP layer (Chan et al., 2006).

The buffer in Fig. 7.115b employs materials whose electron conduction levels align with that of the acceptor. Figure 7.117 compares the performances of several different buffer layer configurations used in conjunction with a 15 nm 1-NPSQ/40 nm C_{60} planar

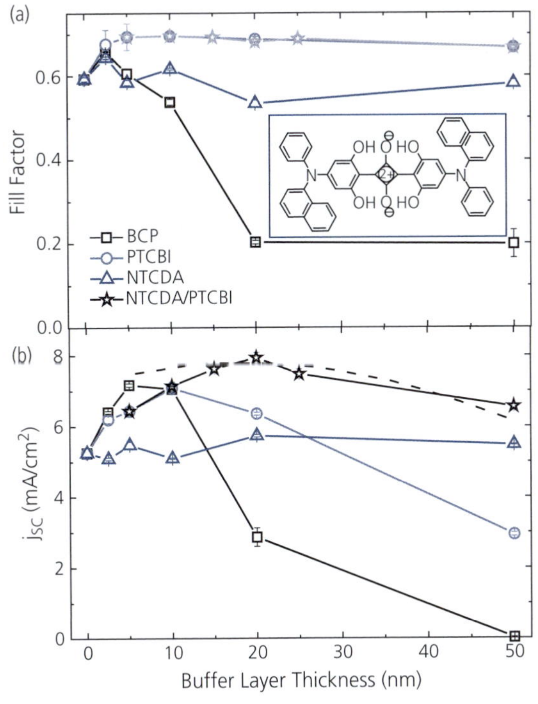

Figure 7.117 (a) Fill factor and (b) short circuit current density vs. buffer layer thickness for 1-NPSQ/C_{60} OPVs. The buffer layers used are indicated in the legends. The dashed line is j_{SC} calculated for the NTCDA/PTCBI buffer due to the optical field intensity in the active region. Inset in (a): Molecular formula of 1-NPSQ (Lassiter et al., 2011).

junction. The buffer layer consists of neat BCP, PTCBI or NTCDA, or 15 nm NTCDA/5 nm PTCBI. PTCBI has a low electron injection barrier with Ag, although it absorbs in the red. Furthermore, PTCBI is a robust molecule that is resistant to damage due to cathode deposition, which is not the case for either the wide energy gap (4.0 eV), and therefore transparent NTCDA or BCP. Hence, the combination of a thin PTCBI layer used to absorb damage and to present a low energy barrier to electron injection, and a wide gap NTCDA layer that acts as an optical spacer results in an efficient EBL structure.

In Fig. 7.117a, we see that *FF* is maximized between 0.60 and 0.70 for the several buffers tested. As expected, *FF* for BCP rapidly decreases at thicknesses greater than 5–10 nm, whereas the other buffers show no such decrease up to 50 nm. However, the *FF* of the NTCDA buffer ($FF = 0.60$) does not achieve the high values of either the neat PTCBI buffer ($FF = 0.70$), or the compound NTCDA/ PTCBI EBL ($FF = 0.68$).

The highest j_{SC} (Fig. 7.117b) is achieved for the compound buffer due to its combination of low energy barrier to electron injection, and transparency. The j_{SC} of the neat PTCBI buffer device decreases with thickness due to its absorption in the red and yellow, and the neat NTCDA layer never reaches the high j_{SC} of the compound buffer due to the barrier formed with Ag. The dashed line is a calculation of j_{SC} for the NTCDA/PTCBI buffer that indicates an optimum optical spacer thickness of ~20 nm, which is consistent with measurement (Lassiter et al., 2011).

The coupling between layer thickness and conductivity can also be reduced by employing an EBL with a shallow HOMO energy that approximately aligns with the LUMO of the acceptor. This results in so-called reciprocal carrier collection, illustrated in Fig. 7.115c, whereby holes are transported from the cathode to the acceptor/blocking layer (A/B) interface where they recombine with photogenerated electrons. The process was originally demonstrated using a Ru(acac)$_3$ buffer, which has a d^5 open shell that permits hole transport. The OPV structure: ITO/CuPc (20 nm)/C_{60} (40 nm)/ Ru(acac)$_3$ (20 nm)/Ag is shown in Fig. 7.118a.

Comparisons of device performance using the Ru(acac)$_3$ buffer with that of a device with no EBL, and one using BCP, are shown in Fig. 7.118b and c. The responsivity (j_{SC}/P_{inc}) at 1 sun, AM1.5G simulated illumination shows that the thickness dependence of the BCP-based device is eliminated using the Ru(acac)$_3$ buffer. On the other hand, removing the buffer entirely results in a significant decrease in response. The responsivity of the OPV with the

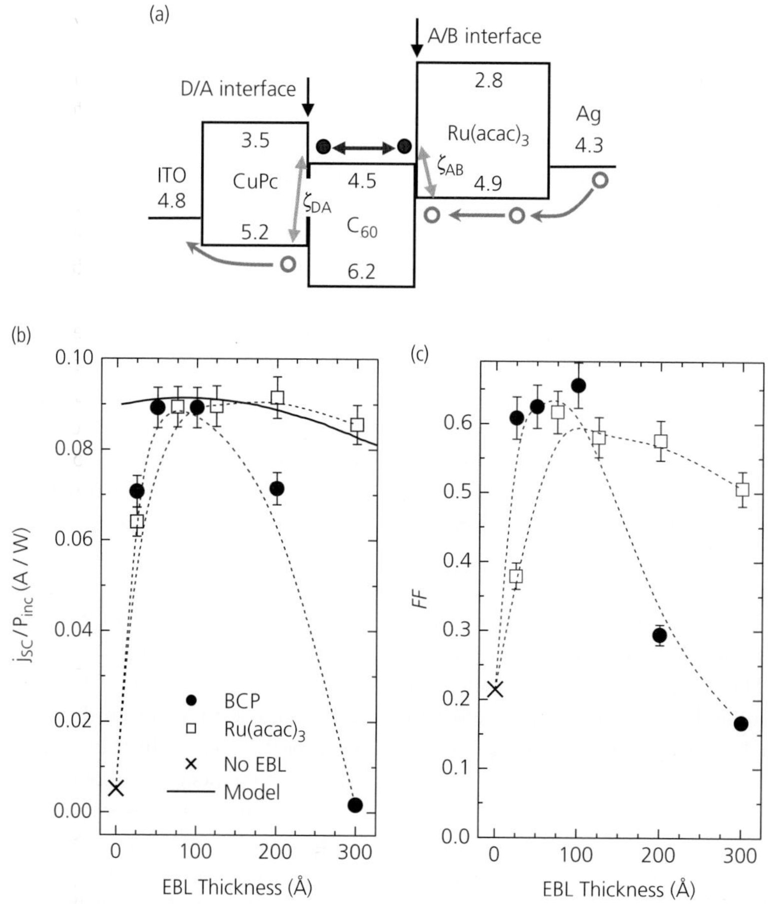

Figure 7.118 (a) Energy level diagram of an OPV with a charge reciprocating EBL. Electrons are filled circles, holes are open circles. (b) Solar responsivity and (c) fill factor vs. EBL thickness, comparing device performance for various EBL structures. The solid line in (b) is a fit a model that includes the optical field and exciton diffusion length in the active region. Dashed lines are guides to the eye (Rand et al., 2005a).

$Ru(acac)_3$ buffer follows the optical absorption/diffusion model (solid line, Fig. 7.118b) of Section 7.1.2. The loss in response with thickness of the device employing BCP is due to its high series resistance, which is supported by a similar sharp drop in *FF*, which is absent for the $Ru(acac)_3$ buffer (see Fig. 7.118c) (Rand et al., 2005a).

The reciprocal OPV structure has two energy barriers: one at the D/A interface for exciton dissociation, and the second at the acceptor/EBL junction. Hence, the charge density within the acceptor depends on the relative magnitudes of the dissociation rate of PPs at the D/A junction, and the recombination rate at the acceptor/buffer (A/B) junction. Figure 7.100 illustrates that two opposing junctions can be approximately modeled by two, back-to-back rectifiers. This generates in an "S" shaped j–V characteristic that produces a dramatic loss in efficiency, especially at high illumination intensities where electrons pile up within the acceptor and ultimately recombine without being collected. The response of the device in Fig. 7.118 thus depends on the magnitude of the A/B HOMO–LUMO offset energy, $\Delta E_{HL,AB}$. As this energy increases, the recombination rate also increases, leading to lower efficiency at high intensity.

A quantitative analysis can be used as a guide for choosing materials for a reciprocating EBL. The device is a combination of two PP recombination sites at the D/A and A/B interfaces that are coupled by current continuity, that is, $j_{DA} = j_{AB}$, and voltage $V_a = V_{DA} + V_{AB}$. Then it can be shown that the current for a symmetric D/A junction is (Renshaw et al., 2011)

$$j = \frac{\left[j_{0S,DA}\left(\exp(qV_a/n_S k_B T) - k_{PPd,DA}/k_{PPd,eq,DA}\right) - q\eta_{PPd} j_X\right]}{\left[1 + \left(j_{0S,DA}/j_{0S,AB}\right)\exp(qV_a/n_S k_B T)\right]}. \tag{7.172}$$

Series and shunt resistances are ignored for simplicity. Also, the subscripts identify the variable associated with either the D/A or A/B junction.

Reciprocal charge collection at the A/B interfaces of ITO/CuPc (40 nm)/C_{60} (40 nm)/RuL_3 (20 nm)/Ag OPVs has been studied to determine the dependence of the j–V characteristics on $\Delta E_{HL,AB}$. The family of RuL_3 compounds is shown in Fig. 7.119. In these molecules, L is the ligand in the tridentate complex.

The j–V characteristics of the RuL_3 buffers OPVs under 1 sun, AM1.5G simulated illumination are provided in the linear and semi-log plots in Figs. 7.120a and b, respectively. With the exception of Ru(acac)$_3$, to a greater or lesser extent, all of the characteristics have an "S" shape whereby the current increase falls sharply away from exponential near V_{OC}. It is presumed that the j–V characteristics of Ru(acac)$_3$ also undergo an inflection, but at voltages higher than shown in the figure. The lines in Fig. 7.120 are fits to the data using Eq. 7.172, from which we find that $j_{0s,AB}$ increases by more than three orders of magnitude, from 6.4 μA/cm^2 for Ru(fhba)$_3$ to $\geq$ 10 mA/cm^2 for Ru(acac)$_3$.

These results can be understood from the energetics of the A/B junction. The frontier orbital energies of C_{60} and the various RuL_3 compounds are shown in Fig. 7.121a. The offset energy at the C_{60}/RuL_3 A/B interface increases from 1.1 eV for Ru(acac)$_3$, to a maximum of 2.64 eV for Ru(fhba)$_3$. These

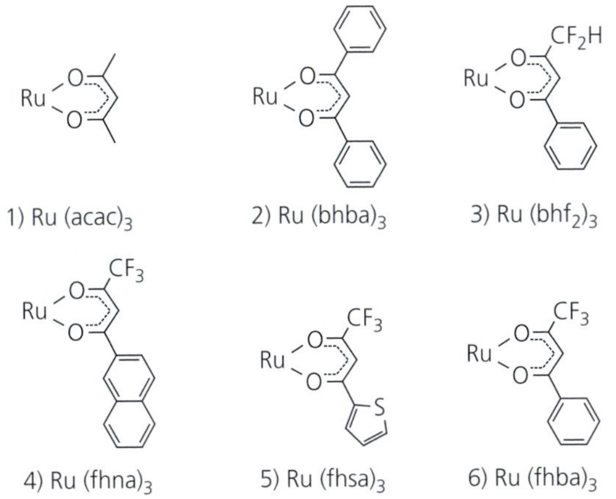

Figure 7.119 Molecular structural formulae of RuL_3 complexes used for EBLs in CuPc/C_{60} OPVs (Renshaw et al., 2011).

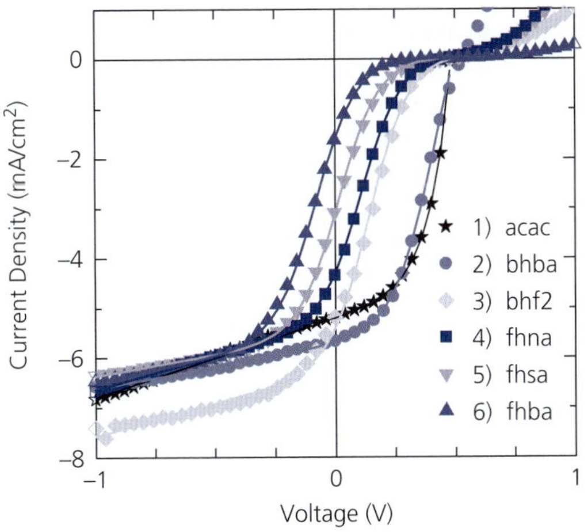

Figure 7.120 The j–V characteristics of CuPc/C_{60} OPVs using the RuL_3 compounds in Fig. 7.119 as EBLs. The lines are fits to the data (points) using Eq. 7.172 (Renshaw et al., 2011).

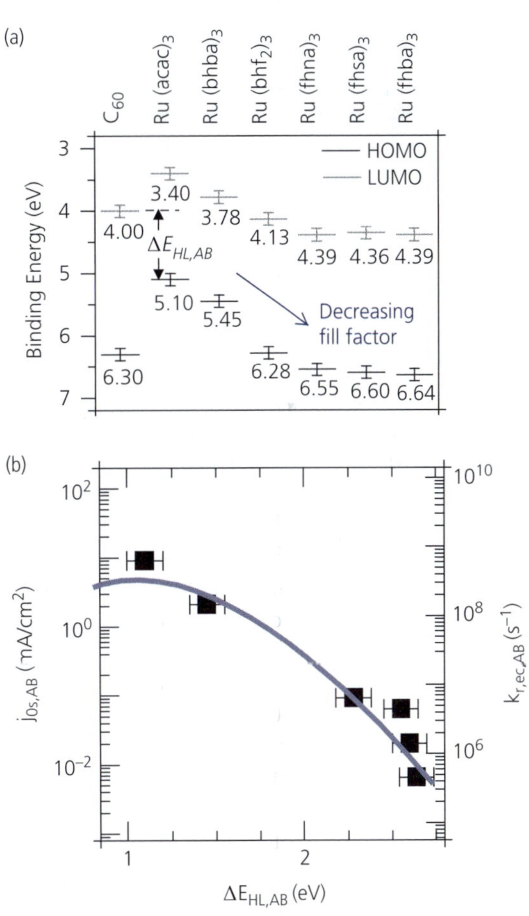

Figure 7.121 (a) Frontier orbital energies of C_{60} and the RuL_3 compounds in Fig. 7.119. The HOMO energies are obtained from ultraviolet photoelectron spectroscopy, and the LUMO energies are the HOMO plus the optical energy gap of the compounds. $\Delta E_{HL,AB}$ for C_{60}/Ru(acac)$_3$ is shown by example. (b) Saturation current for the A/B junction obtained from the fits to the j–V data in Fig. 7.120, vs. $\Delta E_{HL,AB}$ obtained from (a). The right axis is the electron–hole recombination rate at the A/B interface (Renshaw et al., 2011).

energies are then used along with fits to the j–V characteristics to obtain the saturation current of the A/B junction vs. HOMO–LUMO offset energy in Fig. 7.121b. The $j_{0s,AB}$ for Ru(acac)$_3$ is a lower limit due to a lack of deviation from exponential current for this complex.

Equation 7.56 shows that $j_{s0,AB} \propto k_{rec,AB}$. The recombination rates can thus be calculated from the saturation currents for the various complexes, with results shown on the right axis of Fig. 7.121b. The correspondence between $k_{rec,AB}$ (and thus $j_{s0,AB}$) and offset energy is described by Marcus theory (cf. Eq. 4.91):

$$k_{rec,AB} = k_{rec0,AB}\exp\left(-\left(\lambda_{reorg} - \Delta E_{HL,AB}\right)^2/4\lambda_{reorg}k_BT\right), \tag{7.173}$$

where λ_{reorg} is the molecular reorganization energy. Reciprocal charge recombination at the acceptor/EBL interface occurs in the inversion regime, in which case $\Delta E_{HL,AB} > \lambda_{reorg}$. A fit to the recombination rates is shown in Fig. 7.121b (solid line), using $\lambda_{reorg} = 0.6$ eV, and $k_{rec0,AB}$ is chosen for the best correspondence with the data (Closs et al., 1986). The fit shows a slope reversal at $\Delta E_{HL,AB} \approx 1$ eV as the rate enters the normal Marcus transfer region.

The implication of this analysis is that the energy offset at the A/B interface must be minimized for the EBL to prevent charge pile-up at high illumination intensities. Of the compounds tested, only Ru(acac)$_3$ with $\Delta E_{HL,AB} \approx 1.1$ eV is suitable for this purpose. Nevertheless, the benefit of using hole conducting EBLs is their ability to simultaneously achieve high conductivity and high transparency over a large range of thicknesses.

Electron-filtering EBLs (Fig. 7.115d) offer a particularly effective means for achieving the combination of high electron conductivity and transparency. While it was first employed in fullerene-based cells, it can also be applied to OPVs with NFAs since it only relies on the approximate alignment of the LUMO of C_{60} to that of the acceptor material (Bartynski et al., 2013, Bergemann et al., 2015). The EBL employs C_{60} as a dopant in a wide energy gap matrix. Its spherical symmetry allows it to closely pack, forming extended, high conductivity percolating paths for electrons between the acceptor and cathode when doped even at low concentrations within BCP. Furthermore, the absorption feature of C_{60} centered at a wavelength of 450 nm results from a CT state between molecules. As we have shown in Section 3.6.6, the CT state absorption decreases approximately quadratically with dilution. Hence, as the C_{60} concentration is reduced, the EBL becomes increasingly transparent, resulting in a buffer

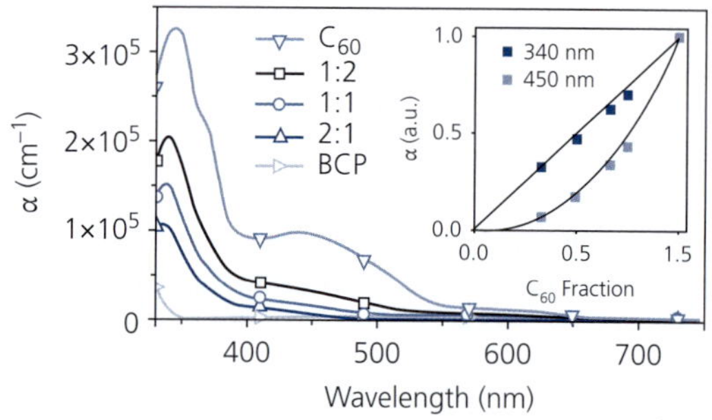

Figure 7.122 Absorption coefficient of films of BCP mixed with C_{60} at the different volume fractions. Inset: The dependence of the absorption coefficient on C_{60} fraction at two different wavelengths corresponding to singlet absorption at 340 nm and CT state absorption at 450 nm (Bartynski et al., 2013).

that has the desired combination of high transparency and conductivity over a large range of doped layer thicknesses.

The absorption spectra as functions of C_{60} fraction when co-deposited in vacuum with BCP are shown in Fig. 7.122. The CT peak at 450 nm is found to decrease more rapidly than the singlet absorption peak at shorter wavelengths. The dependence of these two absorption features on C_{60} volume fraction, y, is plotted in the inset. The peak at 340 nm decreases linearly with C_{60} fraction as expected for an excitation that is confined to a single molecule. However, the long wavelength feature increases as $y^{2.7}$, affirming its multi-molecular origin. These data suggest, therefore, that as long is y is sufficiently small, the EBL absorption is also low. The question remains whether the conductivity at small y is adequate to support charge conduction at high illumination intensities.

This question is addressed using the device in Fig. 7.123a where a 10 nm thick BCP: C_{60} 1:1 blocking layer is sandwiched between two neat C_{60} layers, one of thickness (40 – x) nm adjacent to a PTCBI EBL, and the other of thickness, x, forming a D-A junction with the donor, DPSQ (Wang et al., 2011). The corresponding j–V characteristics are shown in Fig. 7.123b. We observe that j_{SC} increases linearly with x, as expected for an increasing acceptor layer thickness (for $x < L_D \sim 40$ nm for C_{60}) at the D-A junction. These data also suggest that excitons generated in the C_{60} layer adjacent to the cathode do not contribute to the photocurrent, since they lack a D-A junction and are blocked by the intervening BCP:C_{60} (1:1) layer. Further, $FF = 0.72$ for all the devices, independent of x. Thus, electrons generated at the D-A junction are easily transported to the cathode through the BCP:C_{60} layer.

The electron filtering and exciton blocking properties of BCP:C_{60} (z:y) buffer were analyzed using a

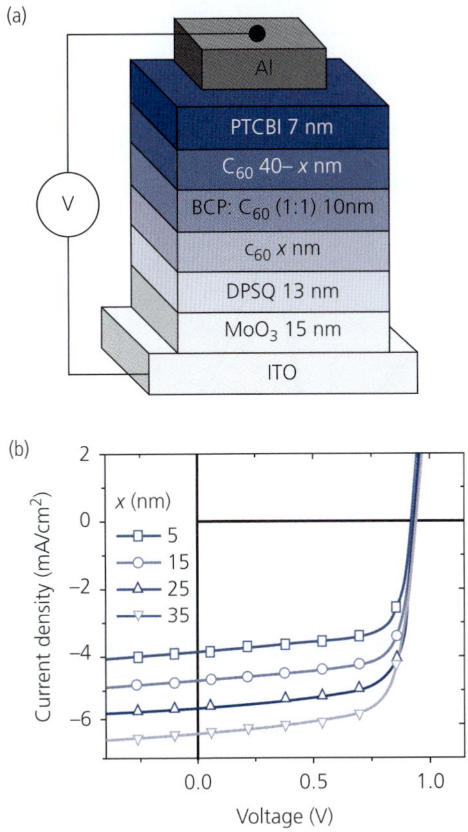

Figure 7.123 (a) Device used to study the electron conducting and exciton blocking characteristics of a BCP:C_{60} (1:1) EBL. (b) The *j–V* characteristics of the device in (a) for varying C_{60} acceptor layer thicknesses (Bartynski et al., 2013).

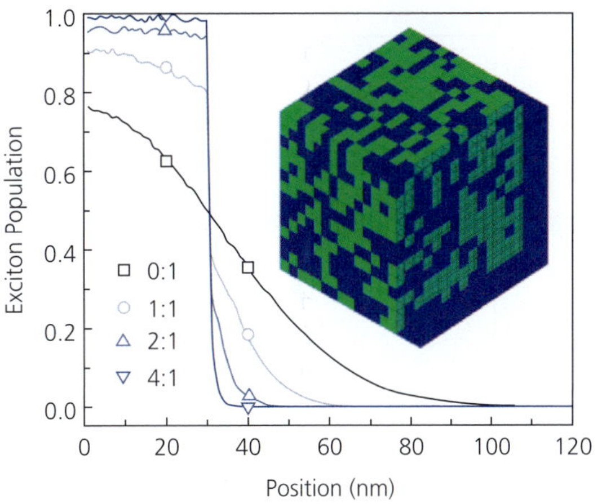

Figure 7.124 Monte Carlo simulation of the exciton blocking properties of BCP:C_{60} (*z*:*y*) buffer layers. The cube shows an illustration of a 1:1 mixed layer used, where the different colored blocks correspond to randomly distributed BCP and C_{60} molecules (Bergemann et al., 2015).

combination of simulation and experiment. Monte Carlo simulations of the EBLs are shown in Fig. 7.124. The simulations assume a random distribution of available (C_{60}) and unavailable (BCP) sites for exciton transport. The excitons move via hopping on a cubic lattice to nearest neighbor sites in a random walk. Any exciton that attempts to move onto a BCP site remains stationary for that step. The BCP:C_{60} interface reflects excitons back into the acceptor if no available site exists in the blocking layer. The plot shows that a neat C_{60} EBL ($z = 0, y = 1$) has no blocking capability, with 50% of the excitons entering the EBL from the acceptor. However, as the BCP fraction increases, the layer becomes increasingly effective at preventing excitons from entering the EBL. The blocking efficiency (i.e. the ratio of excitons reflected back into the acceptor to those entering into the EBL) increases from 81% at $z/y = 1$, to 98% at $z/y = 4$.

The electron transport properties were also examined by measuring the layer resistivity as a function of C_{60} concentration in the EBL. The resistivity only slowly increases from $z/y = 0.6$ to 0.9, and then increases abruptly to the insulating state at $z/y = 1$. Similarly, DBP/C_{60} OPVs with a 10 nm thick BCP:C_{60} buffer shows an efficiency of 6.6% until $z/y = 0.9$, above which η_P decreased to 5.9%. Phase segregation of C_{60}, if it is present at all, is not observed using current-sensitive AFM. Hence, conduction occurs along C_{60} nanowires once the concentration of C_{60} drops below 50%.

Finally, exciton blockers have been used on the donor side of the junction, although this architecture is uncommon since the exciton diffusion lengths in donors are generally considerably smaller (~5 nm) than for the fullerenes, and the use of mixed or bulk HJs reduces the diffusion lengths even further. Also, positioning a transparent layer near the anode only weakly affects the optical field distribution. Nevertheless, the insertion of a wide energy gap exciton blocker between the anode buffer and the donor can increase cell efficiency in some instances.

Figure 7.125a shows the external quantum efficiency of a ITO/PEDOT:PSS/TPTPA (x nm)/DBP (y nm)/ C_{60} (50 nm)/BCP (10 nm)/Al OPV, where a constant thickness of the TPTPA anode blocking layer and DBP donor was $x + y = 20$ nm. Here, TPTPA is suitable as a blocking layer with an energy gap of 2.7 eV compared to only 1.9 eV for the donor, leading to absorption at $\lambda < 450$ nm. The η_{ext} spectra are taken for devices with different TPTPA thicknesses, viz. device 1: $x = 0$ (no blocking layer), 2: $x = 1$ nm, 3: $x = 3$ nm, 4: $x = 10$ nm, and 5: $x = 15$ nm. The quantum efficiency is primarily affected by DBP absorption, since only excitons absorbed within the donor are impacted by the presence

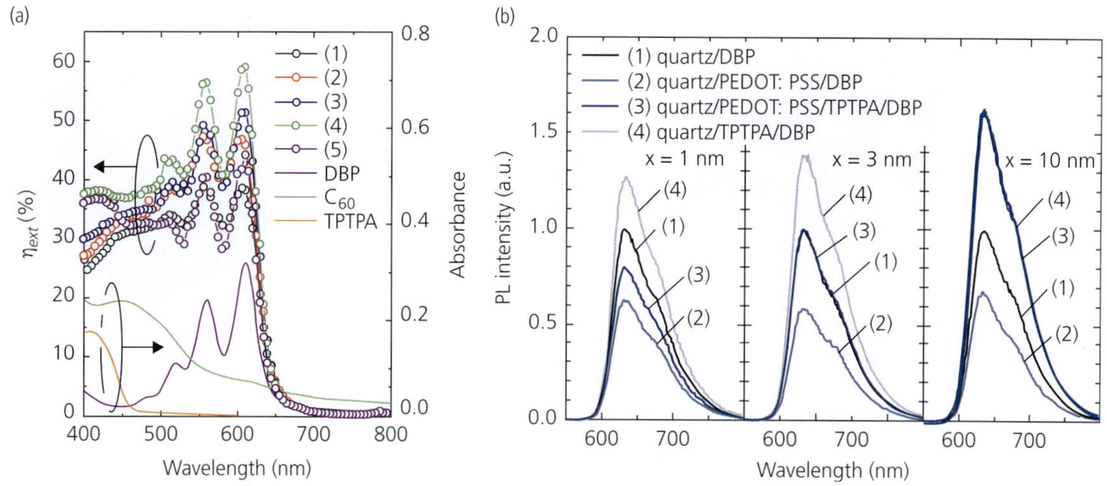

Figure 7.125 (a) External quantum efficiency and absorbance spectra of the devices (1–5, see text) and layers employed in OPVs using anode EBLs. The spectra correspond to different TPTPA EBL thicknesses, *x*. (b) Photoluminescence intensity spectra of DBP films used in the layers shown in the legend (Hirade and Adachi, 2011).

of the donor-side EBL. In fact, the largest enhancement is achieved for device 4, where $x = y = 10$ nm. The efficiency decreases for thinner EBLs that are less effective in confining excitons within the donor. For thicker TPTPA, the donor absorption decreases (Hirade and Adachi, 2011).

The DBP thin film PL intensity spectra in Fig. 7.125b provide evidence for the blocking properties of the EBL. The spectra from four structures are compared for different TPTPA thicknesses. In film 1, the DBP was deposited directly on quartz, which provides a blocking, non-quenching surface. Film 2 is quartz/PEDOT:PSS/DBP, film 3 is quartz/PEDOT:PSS/TPTPA/DBP, and film 4 is quartz/TPTPA/DBP. The highest intensity for all TPTPA thicknesses is film 4, indicating that TPTPA forms an interface with DBP that exhibits even less quenching than the nominally non-quenching quartz substrate. On the other hand, the lowest intensity is observed for film 2 where DBP is deposited directly on PEDOT:PSS which forms an efficient quenching interface. Film 3, which is the structure employed in the OPV cells, is as effective as film 4 in blocking excitons in the absence of quenching, but only when the TPTPA is 10 nm thick as is necessary to completely block exciton transport to the anode buffer.

An OPV with $x = 0$ nm showed $\eta_P = 3.9\%$, whereas it increased to 5.0% for $x = 10$ nm. Also, the *FF* increased from 0.72 to 0.77 over this same range of thicknesses. As noted previously, the absorption in the donor layer decreases with increasing *x*, resulting in a concomitant decrease in device performance (Hirade and Adachi, 2011).

7.4.2 OPV transparency and flexibility, and the role of contacts

The narrow excitonic absorption bands of organics compared to the very broad absorption of inorganic semiconductors requires the inclusion of three or even more organic materials to achieve full spectral coverage across the visible and into the NIR. For example, in Fig. 7.126 we show the absorption spectra of six homologous NFA molecules used in high efficiency OPVs. Each of the spectra is rigidochromically shifted by approximately 80 meV, with a half width of ∼150–250 nm. The complexity inherent in requiring several molecules to cover the solar spectrum is offset by an opportunity to employ OPVs as power generating windows in building applied or building integrated photovoltaic applications. The examples in Fig. 7.126 indicate that this collection of molecular species leaves transparency windows across broad swatches of the visible and NIR spectra. By choosing organic materials that selectively absorb in the NIR, one can create a cell that has a tint, or even a neutral density semitransparent window while also providing an efficient power source, provided that the anode and cathode contacts are also semitransparent.

To quantify the appearance and efficiency of power generating windows, we introduce two parameters: the *average photopic transmission, APT* (or variously the *average visible transmission, AVT*) and the *light utilization efficiency, LUE* (Traverse et al., 2017). Using the same approach for windows as taken to quantify illumination sources, the perceived color of sunlight

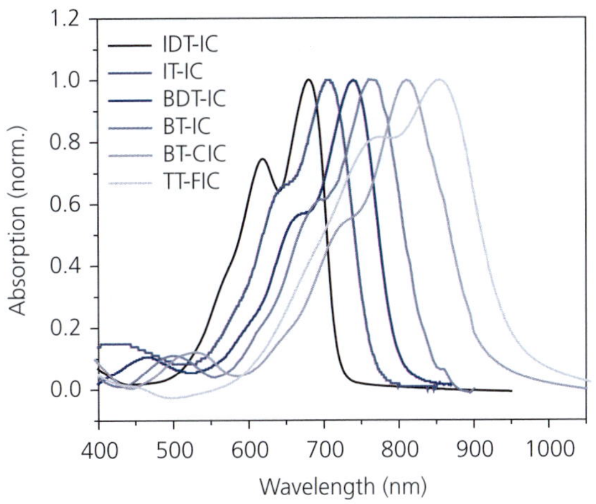

Figure 7.126 Absorption spectra from six NFAs used in high efficiency OPVs. The molecular formulae and other details of these compounds are provided in Section 7.4.4.

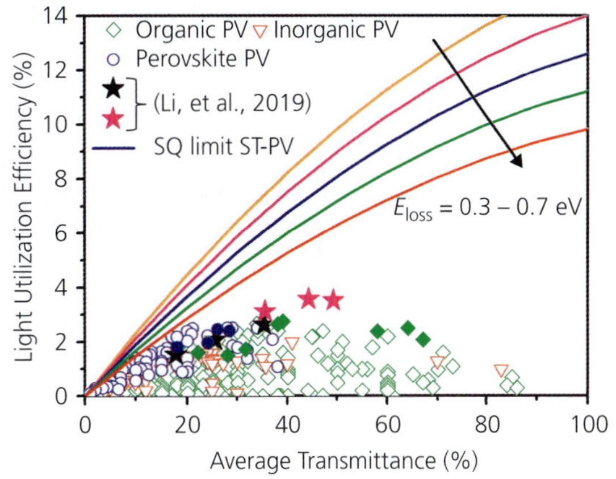

Figure 7.127 Light utilization efficiencies (*LUE*) vs. *APT* for several inorganic, perovskite, and OPV cells. The lines are the thermodynamically limited *LUE* curves for a range of energy losses found in OPVs. The stars refer to OPVs from Li et al. (2019). Adapted from (Traverse et al., 2017).

through a window (which during daytime hours is an illumination source) is expressed by the window's chromaticity coordinates (cf. Section 6.2.3). The *APT* is the integrated transmission spectrum, $T(\lambda)$, weighted by the *photopic response* of the human eye, $V(\lambda)$, and is given by

$$APT = \frac{\int_{visible} T(\lambda)E_{REF}(\lambda)V(\lambda)d\lambda}{\int_{visible} E_{REF}(\lambda)V(\lambda)d\lambda}, \tag{7.174}$$

where $E_{REF}(\lambda)$ is the reference solar spectrum.

The power conversion efficiency is also needed to fully quantify the performance of a window. For this purpose, we define the *LUE*:

$$LUE = APT \times \eta_P. \tag{7.175}$$

Thus, $LUE = 0$ for a totally opaque cell (where $APT = 0$), although it can be as high as 20% for a cell that is 100% transparent to the human eye. For example, with $\eta_P = 15\%$ and $APT = 50\%$, then $LUE = 7.5\%$.

Transparency can be achieved by two approaches. Opaque cells based on inorganic semiconductors such as Si and GaAs can be separated into thin strips or sections, thus admitting light through the gaps between the individual cells. This provides an optical neutral density appearance. However, the fabrication process of such a window that entails precise placement and interconnection of the cell segments can be costly. An attractive and far less complex approach is offered by organic cells that can be transparent across the visible, yet highly absorbing in the NIR using appropriate materials and device structures. The window color can also be selected simply by changing the optical coatings of the OPVs, or selecting materials that remove desired spectral regions.

A compilation of *LUE* values for inorganic, perovskite and OPVs are provided by the data points in Fig. 7.127. Devices with the highest *LUE* are based on organic thin film cells, even though the absolute efficiencies of many inorganic cells are higher. The maximum $LUE = 3.5\%$ in this compilation is expected to increase as new device architectures, optical filtering schemes and organic materials are developed. One note of caution in reading these data is that not all cells are neutral density. That is, the light transmitted may have its color coordinates significantly shifted away from a standard solar illuminant with $CCT = 5800$ K. Hence, objects illuminated by the sun through such a window may not appear in their true colors (i.e. $CRI < 1$). This is particularly true for the organic and perovskite cells that can have noticeable tints. If the coloration is too strong, the cells no longer perform as windows, but rather as power generating optical filters that do not serve the purpose of building integrated appliances.

The lines in Fig. 7.127 are calculations based on the thermodynamically limited response of OPVs according to the theory in Section 7.3.2. The curves correspond to different values of energy loss typical of OPVs, assuming that the devices absorb in the UV and NIR (with energy gaps that yield the thermodynamically limited output power), yet are semitransparent in the visible. We observe that the *LUE* increases with *AVT* as a greater fraction of energy is harvested from the UV and NIR. These curves have a maximum at $LUE = \eta_P \rightarrow \eta_P^{max}$ and $AVT = 100\%$, where η_P^{max} is the thermodynamically limited power conversion efficiency in the absence of contributions from the visible spectrum.

We can use the same analysis to estimate the limits to the power conversion efficiency attainable by semi-transparent, multijunction OPVs. The efficiency limit for a series, and a series-parallel connection of three subcell devices is shown in Fig. 7.129a and b, respectively. The series-connected triple junction cell comprises a visible absorbing subcell whose external quantum efficiency is $\eta_{ext,vis} = [1 - VT]$, where we assume that the transparency window lies between 350 nm $< \lambda <$ 625 nm. Here, VT is the visible transmission, which differs from AVT in that it is the arithmetic mean of the transmission within a specified optical domain. This is unrelated to the perceived appearance of the solar cell. A $VT = 50\%$, therefore, does not necessarily have a neutral color. The NIR and IR subcells in Fig. 7.128 assume $\eta_{ext,IR} = 90\%$. The subcell absorption bandwidths are adjusted such that the current along the stack is balanced. For $E_{loss} = 0.5$ eV, we calculate $\eta_P = 19\%$ and $LUE = 9.5\%$. That is, the triple junction design achieves nearly the same high efficiency when it is 50% transparent in the visible, with that of completely absorbing single junction cell in Fig. 7.74.

An alternative multijunction architecture is the series/parallel connection of Fig. 7.129b. Parallel circuits require voltage balance between sources. However, the operational voltage of NIR absorbing subcells is lower than from visible absorbing cells due to their smaller energy gaps. Voltage balance can be achieved by placing two long wavelength series-connected subcells in parallel with a visible absorbing cell. The series subcell stack must be current matched, whereas the MPP voltage optimally follows $V_{M,IR} = V_{M,IR1} + V_{M,IR2} = V_{M,vis}$. Here, $V_{M,IR}$ is the total IR tandem voltage, equal to the sum of the two IR subcells, $V_{M,IR1}$ and $V_{M,IR2}$, which in turn is assumed to equal the voltage of the visible absorbing subcell, $V_{M,vis}$.

The matched, multijunction cell efficiency *vs.* transparency is provided in Fig. 7.128b. Again, taking $VT = 50\%$ and $E_{loss} = 0.5$ eV, we obtain $\eta_P = 19\%$ which is approximately that of the series stacked cell. Importantly, the dependence of the efficiency on transparency is less pronounced for the series-parallel OPV. Presumably, this results since the visible absorbing cell voltage is only weakly dependent on intensity. Voltage balance is maintained even though the current generated is significantly reduced by the increased transparency. In contrast, a fully series-connected cell is current-limited by the visible absorbing subcell that is made increasingly transparent via thinning of its active region, yielding its stronger dependence on VT.

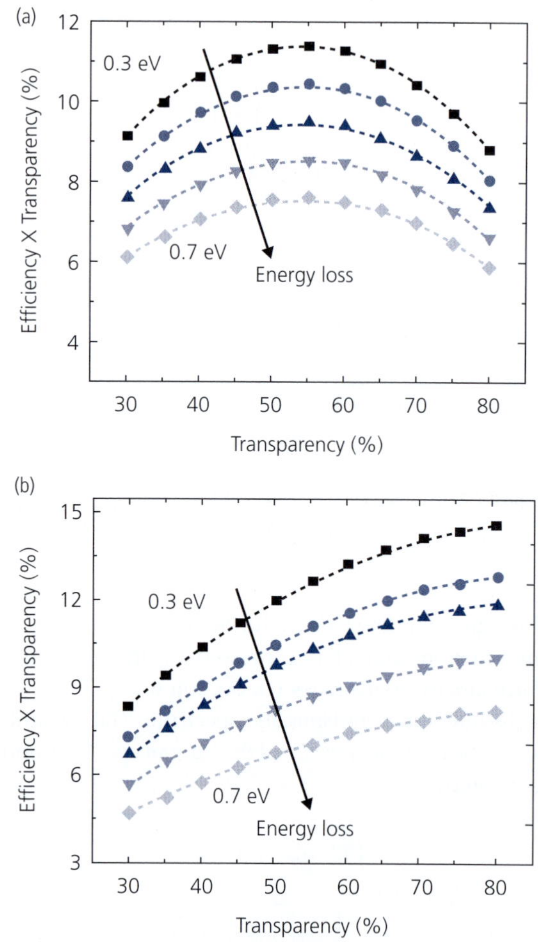

Figure 7.128 Power conversion efficiency–average transparency product vs. cell transparency for (a) series and (b) series/parallel connected triple junction OPVs. Adapted from Che (2018).

Semitransparent cells require optically transmissive and low resistance contacts. To ensure low injection barriers to electrons into the acceptor, the cathode metals must have low work functions. Similarly, anodes have high work functions. Furthermore, since solar cells cover large surface areas, the contacts must have low sheet resistance. A requirement of all materials used in solar cells including the contacts, is low cost and the ability to be applied over very large substrate areas. Typically, thermally evaporated Al (with a work function of $q\phi_m$ = 4.3 eV) and Ag (4.2 eV) are preferred, although Ag readily diffuses into the organics, creating damage and even shorts across the cell. Indium tin oxide and other metal oxides are almost universally used as anode materials due to their high work functions, relatively low resistance and high transparency. However, the sheet resistance of ITO is several orders of magnitude higher than metals of comparable thickness, and the sputtering process and materials costs used in their manufacture can be prohibitively high. Lastly, metal oxide films are brittle, preventing them from flexing during use or manufacture.

A straightforward approach to achieving contact semitransparency is to use a metal whose thickness is less than, or on the order the optical skin depth (see Section 6.2.2),

$$\delta = \sqrt{2/\omega\mu_r\mu_0\sigma_{dc}}. \tag{7.176}$$

Here, δ is the characteristic distance for electric field attenuation through the metal, viz. $\mathbf{F}(\mathbf{r}) \propto \exp(-\hat{\mathbf{n}}\cdot\mathbf{r}/\delta)$. Also, ω is the frequency of the incident radiation, μ_0 is the magnetic permeability of free space and μ_r is its relative permeability, σ_{dc} is its d.c. conductivity, $\hat{\mathbf{n}}$ is the unit vector normal to the metal surface, and $\hat{\mathbf{r}}$ is the propagation direction. For most metals, $\delta \sim 50$ nm which sets an upper limit to electrode thickness when used in semitransparent cells. However, since σ_{dc} for a contact metal is much higher than ITO, the thin metal contact generally has a lower resistance than the anode. Electromagnetic theory gives the dependence of the transmittance of a conductor of thickness, d, on conductivity via (Dressel and Grüner, 2002)

$$T = \frac{1}{\left(1+\dfrac{Z_0\sigma_{ac}d}{2}\right)^2} = \frac{1}{\left(1+\dfrac{Z_0\sigma_{ac}}{2R_\square\sigma_{dc}}\right)^2}, \tag{7.177}$$

where σ_{ac} is the frequency dependent a.c. conductivity, $R_\square = 1/(\sigma_{dc}d)$ is the contact sheet resistance, and $Z_0 = \sqrt{\mu_0/\varepsilon_0} = 377\ \Omega$ is the impedance of free space.

An example of the dependence of the transparency of OPVs on the thicknesses of Ag cathodes on a structure comprising ITO/ZnO (25 nm)/PCE-10:BT-CIC (130 nm)/MoO_3 (15 nm), is provided in Table 7.13 (Li et al., 2017c). The devices show a broad transparency window between 550 nm to 900 nm, with a bluish tint. The *VT* varies from 43% for a Ag thickness of 10 nm, to 26% at 20 nm. As the device becomes increasingly transparent, the reflection from the cathode into the thin active region is decreased, eliminating the double pass of the resulting standing wave. Hence, for a device with a 100 nm thick, reflective cathode, $\eta_P = 11.2\%$, decreasing to 7.1% for the device with a 10 nm thick cathode. As expected, the series and sheet resistances also increase with decreasing cathode thickness.

The inverse relationship between transparency and sheet resistance of a flexible PEDOT:PSS anode inferred from Eq. 7.177 is shown in Fig. 7.129. The film thickness is varied by sequentially spin coating layers of PEDOT:PSS containing 6 vol.% ethylene glycol, and annealing at 120°C between each deposition step. The solid line is a fit to $R_\square = 1/(\sigma_{dc}d)$, with $\sigma_{dc} = 1200$ S/cm. The film with a sheet resistance of 75 Ω/sq. has a transmissivity of ~ 80%. This is 50–100 times higher than ITO, which itself is unsuitably high for use in large modules (Kim et al., 2011).

A comparison of the transmissivities weighted by the solar spectrum between 400 nm and 800 nm for several different anode materials is provided in Fig. 7.130. To reduce the sheet resistance of PEDOT:PSS or ITO without significantly affecting transparency, the film can be deposited onto the surface of a thin and widely spaced metal grid. A Ag grid with a period of 400 nm and a gridline width of 40 nm is approximately 90% transparent yet gives a sheet resistance of ~ 1 Ω/sq., as shown by the solid line in Fig. 7.130. However, high resolution photolithography required to produce the fine grid features is costly. An alternative is to use a micron-scale, screen-printed grid covered by a thicker layer of PEDOT:PSS, although this configuration is inferior to the transparency and conductivity provided by ITO (Aernouts et al., 2004).

Random, crystalline Ag nanowire meshes may be useful replacements to a patterned grid (Lee et al.,

Table 7.13 Performance of PCE-10/BT-CIC OPVs employing Ag cathodes of different thicknesses (Li et al., 2017c)

Ag thickness (nm)	J_{SC}[a] (mA/cm^2)	V_{oc} (V)	*FF* (%)	η_P (%)	*VT* (%)	R_{ser} (Ω cm^2)	$R_\square$ (Ω/sq.)
10	15.8	0.68	66.2	7.1	43	2.1	28.0
15	17.0	0.68	67.1	7.7	33	1.7	4.3
20	18.0	0.68	67.5	8.2	26	1.7	2.4
100	22.5	0.70	71.0	11.2	0	—	—

[a] Obtained at 1 sun, AM1.5G simulated solar illumination.

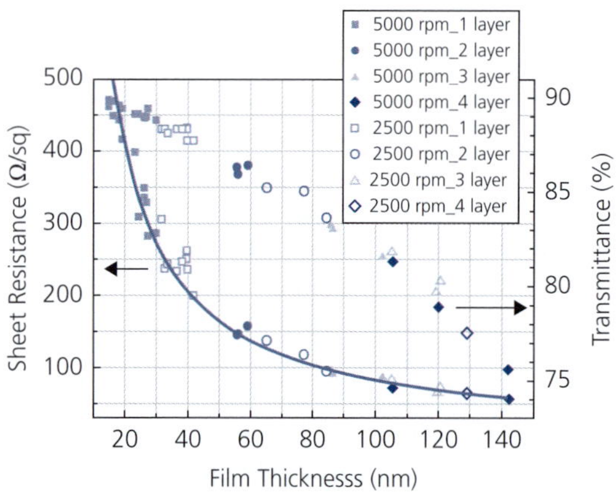

Figure 7.129 Sheet resistance and transmittance vs. PEDOT:PSS film thickness (*d*), controlled by adjusting the spin velocity and number of layers. The line is a fit to $R_{\square} = 1/(\sigma_{dc}d)$ (Kim et al., 2011).

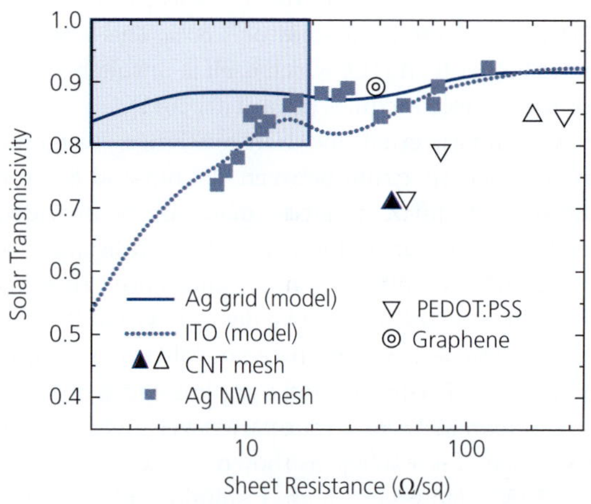

Figure 7.130 Transmissivity weighted by the AM1.5G spectrum *vs.* sheet resistance for several different OPV anode materials. Very few materials are suitable for use in OPVs, whose preferred properties lie within the shaded box. After Lee et al. (2008). Data for PEDOT:PSS from Kim et al. (2011), carbon nanotubes (CNT) from Rowell et al. (2006) and Van De Lagemaat et al. (2006) (open and filled triangles, respectively), and four-monolayer HNO_3-doped graphene from Bae et al. (2010).

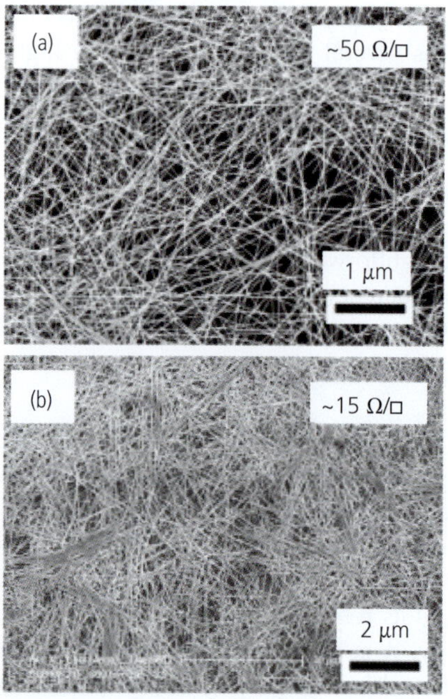

Figure 7.131 SEM images of Ag nanowire meshes with sheet resistances of (a) 50 Ω/sq. and (b) 15 Ω/sq. (Hu et al., 2010).

2008, Hu et al., 2010). Nanowires that are ~10 μm long by 100 nm in diameter grown from solution are shown in Fig. 7.131. As-deposited wires are coated with a thin insulating barrier that results in a low conductivity mat. Annealing at 200°C fuses the wires that significantly reduce the film resistance to 8 Ω/sq. (filled squares, Fig. 7.131) (Kim et al., 2011).

An increase in nanowire density leads to an increase in conductivity, but also in a loss of transmittance. The transmission (*T*) of the 50 Ω/sq. film is ~85% (Fig. 7.131a) whereas for the 15 Ω/sq. sample, $T \sim$ 65% (Fig. 7.131b) (Hu et al., 2010). Prior to fabrication of the OPVs, the rough surface of the meshes is planarized to prevent shorting across the thin device active region. This can be done by coating the mats with a thin PEDOT:PSS layer that also provides a uniformly conductive surface. ZnO nanoparticles with diameters between 3–5 nm synthesized in solution have also been used to fill the gaps between the nanowires, resulting in P3HT:PC_{61}BM BHJ OPVs with $\eta_P = 3.9\%$ and $FF = 0.68$; values that are marginally superior to those achieved using an ITO anode with a ZnO nanoparticle treated surface, where $\eta_P = 3.5\%$ (Ajuria et al., 2012).

Nanowire mats can also form semitransparent top contacts (Chen et al., 2012a). A Ag nanowire composite cathode was deposited onto the surface of a 20 nm thick TiO_2 nanoparticle cathode interfacial layer on the OPV structure in Fig. 7.132a. The OPV BHJ consists of the donor, PBDTT-DPP blended with PC_{61}BM, on an ITO/PEDOT:PSS anode. The BHJ active region strongly absorbs in the UV and NIR, with only 30–40% absorption in the visible. The nanowires are spray-coated and fused using alcohol-based solvents onto the TiO_2 interlayer. The last step fills in the gaps between the nanowires by spreading a conducting ITO nanoparticle film suspended in a polymer binder. This creates a relatively low

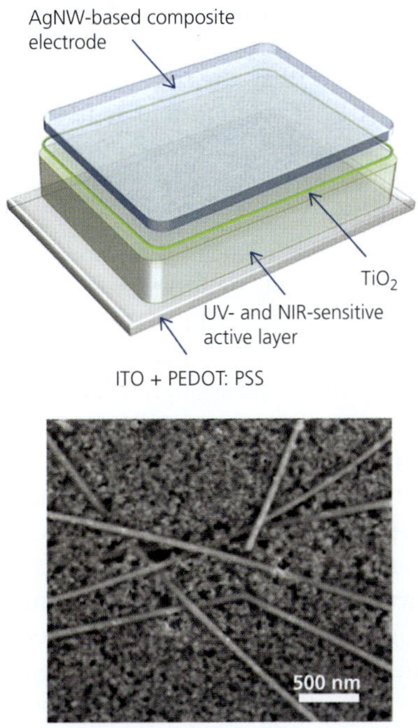

Figure 7.132 (a) Schematic illustration of a semitransparent OPV employing a Ag nanowire (Ag NW) composite top cathode and an ITO-PEDOT:PSS anode. (b) Scanning electron microscope image of the underside of the top electrode showing the Ag NWs and the ITO nanoparticle filler. This image was taken after the electrode was peeled off of the organic layers (Chen et al., 2012a).

resistance path between the point of charge injection onto the high conductivity Ag mesh. A micrograph of the underside of the contact in Fig. 7.132b shows that the film exposes nanowires to the underlying layers while creating a continuous film that contacts the OPV layers below. The contact has $AVT = 87\%$ and a sheet resistance of 30 Ω/sq (Chen et al., 2012a). Organic solar cells employing this cathode show $\eta_P = 4.0\%$ and 3.8% under 1 sun AM1.5G spectral illumination incident from the anode and cathode sides, respectively. The *FF* of the transparent device is 0.57 compared with 0.60 for the conventional device due to increased resistance from the Ag nanowire cathode contact.

CNTs can also form flexible conductive mats. Indeed, Fig. 7.48b shows that its low HOMO energy gives it donor-like characteristics, and thus is suitable as an anode contact. Single walled nanotubes prepared by arc discharge can have transparencies as high as 90% for 30 nm thick films, but with a correspondingly high sheet resistance of 200 Ω/sq. (Fig. 7.130). As discussed previously, a random population of nanotubes contains approximately 1/3 metallic and 2/3 semiconducting tubes, but only the former contribute to the conductivity. This leads to a relatively high sheet resistance that is comparable, or even inferior to that attained by PEDOT:PSS (Hu et al., 2004, Rowell et al., 2006). The resistance is reduced by annealing at ~100°C to remove surfactant and improve contact between overlapping tubes. Gaps between the tubes can be filled, and the surface planarized by coating with a layer of PEDOT:PSS (Rowell et al., 2006, Van De Lagemaat et al., 2006).

Lastly, graphene has been explored for flexible and transparent electrodes. Due to the delocalized electron wavefunctions along the graphene sheet, its in-plane conductivity and carrier mobility is very high (up to 200,000 cm^2/V s), and as a monolayer it is transparent. Indeed, HNO_3-doped p-type, four monolayer thick graphene sheets have been produced in a continuous R2R process with a sheet resistance of 30 Ω/sq. and a transparency of ~ 90%. Monolayers are successively transferred to obtain multilayer films with the desired combination of resistance and transparency (Bae et al., 2010).

The performance of graphene anodes adapted for use in OPVs is not quite as high. For example, three monolayers of graphene were successively transferred to a quartz substrate and then coated with a planarizing layer of PEDOT:PSS. The doping process results in an 80% transparent film with a sheet resistance of ~100 Ω/sq (Kim et al., 2010). Combining this anode with 21 nm CuPc/40 nm C_{60}/9.3 nm BCP double HJ device with a Ag cathode resulted in $\eta_P = 1.6\%$ and $FF = 0.42$, compared to $\eta_P = 1.8\%$ and $FF = 0.56$ for an analogous device employing an ITO/PEDOT:PSS anode (Park et al., 2010a). The reduced *FF* for the graphene-based OPV is a result of its higher anode injection barrier and sheet resistance. Single monolayers of graphene coated with PEDOT:PSS have been employed as a top, transparent contact on P3HT/PC_{61}BM OPVs employing an ITO/ZnO anode. The sheet resistance was ~500 Ω/sq., which results in a low fill factor of ~0.4 (Liu et al., 2011).

The recurring theme of the foregoing discussion is that potentially lower cost and flexible contacts can be achieved using metal nanowires, PEDOT:PSS, nanotubes or graphene, although they generally deliver inferior performance compared to those using brittle, transparent metal oxide contacts.

The loss in efficiency that accompanies transparency can be partially recovered by recycling the NIR photons back into the active region, thereby doubling the absorption length, while the visible photons experience only a single pass as they transit across the cell. This is accomplished by appropriate combination of an ARC and a mirror in the NIR (Koeppe et al., 2009, Lunt and Bulovic, 2011, Betancur et al., 2013). This takes advantage of the narrow spectra of excitonic materials that absorb primarily in the non-visible spectral regions. To selectively transmit multiple spectral regions, a single ARC or *distributed Bragg reflector* (DBR) is insufficient. On the other hand, *aperiodic dielectric reflectors* (ADRs) comprising multiple layers with different thicknesses and indexes of refraction, can achieve this purpose (Sergeant et al., 2009).

Since the sun is a broad spectral source, the ARCs and ADRs must also have broad spectral transmission or reflection. This is a more complex problem than designing a coating for a relatively narrow spectral source such as an OLED or OPD. Determining the layer thicknesses starts by choosing the number of layers and the refractive indices of the materials to be used in the coating, and the target output spectrum. This is the inverse solution of the conventional design approach that starts with the structure and optical properties of the multilayer stack, and then using transfer matrix formalism (Section 7.1.4) to calculate its transmission, absorption and reflection coefficients. An example of a broad band, aperiodic dielectric stack comprising six periods of SiN_x and SiO_2 of varying thicknesses is shown in Fig. 7.133. The solution is found by using a *genetic algorithm* to iteratively approach the target spectrum (shown at right). The algorithm iteratively varies the thicknesses of the layers until the output spectrum most closely approaches that of the target. A difficulty inherent in this approach is that the algorithm may become trapped in a local solution that is not optimal. To avoid this possibility, the procedure is attempted multiple times, each with a randomly chosen set of initial conditions.

The absorption efficiency for the aperiodic ARC in Fig. 7.133 is reasonably uniform across the absorption spectrum of the CuPc:PTCBI (1:1) OPV active region for which it is designed. Indeed, the electric field intensity appears to be strongly confined to the organic multilayer, as shown in Fig. 7.133a. The average absorption efficiency is 26.6%, compared with 19.8% without the dielectric stack (Peumans, 2004). A more uniform spectrum may be achieved by increasing the number of layers in the stack.

The structure in Fig. 7.134 is an example device employing an ADR designed to reflect UV and NIR radiation to create a standing wave within the PTB7:PC_{71}BM active region of an OPV, while being transmissive to visible radiation. The dielectrics employed are LiF ($n = 1.39$) and MoO_3 ($n = 2.52$) forming a stack on a semitransparent, 10 nm thick Ag cathode.

The calculated *APT* and j_{SC} comparing an OPV with only a thin Ag layer (called Tn-Ag), and one with an ADR on the surface of the Tn-Ag (ADR-Tn-Ag) vs. the number of layers in the dielectric stack is

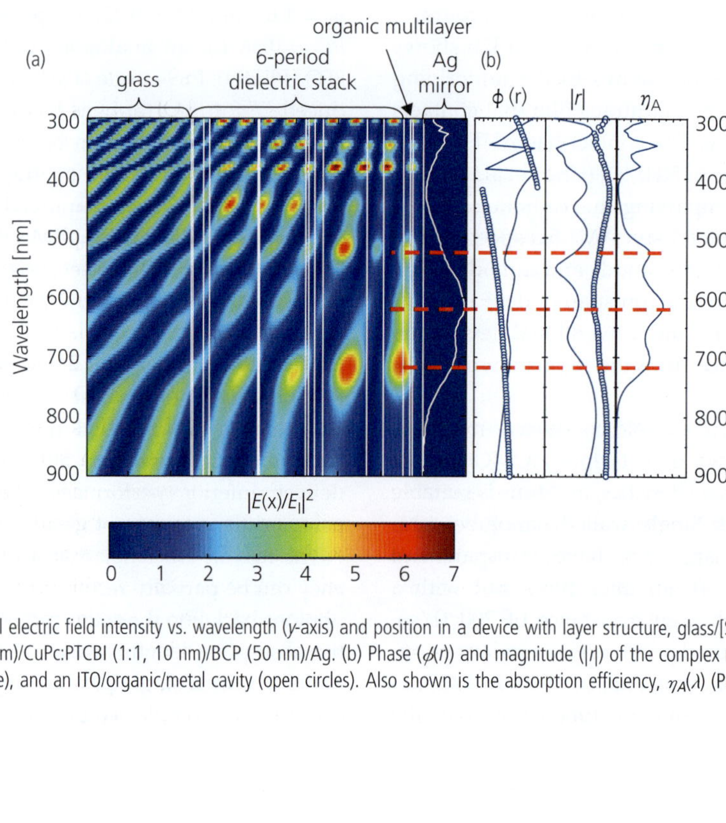

Figure 7.133 (a) Optical electric field intensity vs. wavelength (y-axis) and position in a device with layer structure, glass/$[SiN_x/SiO_2]_6$/ITO (15 nm)/PEDOT:PSS (32 nm)/CuPc:PTCBI (1:1, 10 nm)/BCP (50 nm)/Ag. (b) Phase ($\phi(r)$) and magnitude ($|r|$) of the complex reflection coefficient, r, of the coating (solid line), and an ITO/organic/metal cavity (open circles). Also shown is the absorption efficiency, $\eta_A(\lambda)$ (Peumans, 2004). Reprinted with permission.

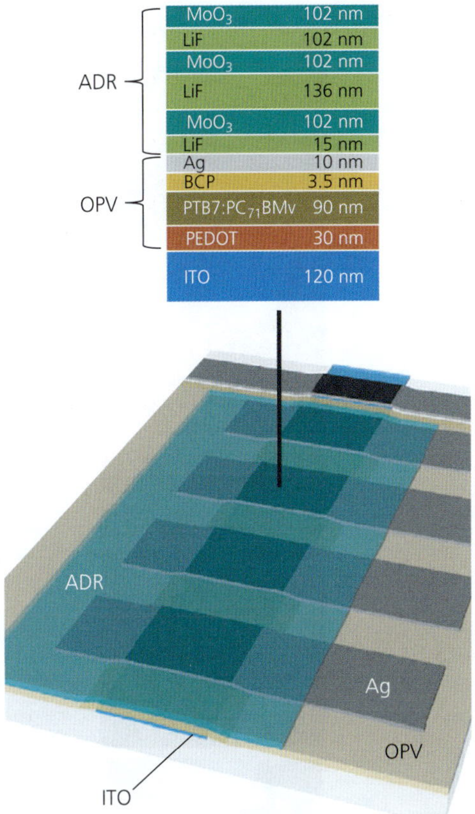

Figure 7.134 Structure (top) and layout of an array of OPVs employing an aperiodic dielectric reflector (ADR) that reflects UV and NIR radiation back into the active region to maximize absorption, while transmitting in the visible. After Betancur et al. (2013).

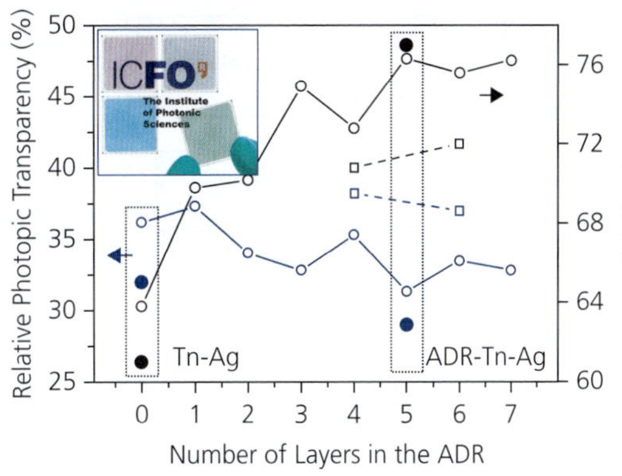

Figure 7.135 Relative photopic transparency and short circuit current for a PTB7:PC_{71}BM OPV vs. the number of layers in the ADR of Fig. 7.134. Comparison of these parameters is between a device with an ADR on a thin Ag electrode (ADR-Tn-Ag), and a similar device lacking the ADR (Tn-Ag). The measured device performances are given by solid circles, and the calculated performances by open circles. Open squares are calculations for an optimal periodic Bragg reflector. Inset: Transparent films optimized for different transmission spectra placed on the logo of the Institut de Ciencies Fotoniques (ICFO) (Betancur et al., 2013).

Table 7.14 Performance of PTB7:PC_{71}BM OPVs with different cathode designs (Betancur et al., 2013)

Cathode	j_{SC} (mA/cm^2)	η_P (%)	*APT* (%)
Tk-Ag	14.0	7.3	0
Tn-Ag	8.5	3.6	32
ADR-Tn-Ag	10.7	5.2	29

shown in Fig. 7.135. This result is compared with that for an optimized, periodic DBR comprising four and six alternating layers (squares). The plot shows that adding layers past four or five has a diminishing impact. The stack used for the OPVs comprises five layers, providing an *APT* only 5% less than for the Tn-Ag sample (open blue circles), but with a short circuit current nearly 20% higher (open black circles). The DBR gives a higher transparency than the ADR, but at the cost of a lower j_{SC}. The measured performance of the OPVs (solid circles) is close to that obtained via calculation for both Tn-Ag and ADR-Tn-Ag. The inset, Fig. 7.135, shows the appearance of OPVs optimized to transmit different colors, from red to blue to neutral. Generally, building architects favor neutral density coatings on windows, which is accessible via the appropriate dielectric stack designs.

The power conversion efficiencies and *APT*s of the variously coated OPVs are summarized in Table 7.14. There are minor differences in *FF* and V_{OC} attributed to damage incurred during dielectric stack deposition. The largest differences are found in j_{SC} and η_P where the device with the aperiodic grating has a 20% higher j_{SC} and a 40% higher η_P compared with the Tn-Ag device, although the visible transparencies are nearly equal (Betancur et al., 2013).

We found in Section 6.2.2 that replacing the semitransparent metal with ITO can also result in a large *APT*, although there is a possibility that its deposition can damage the underlying organic layers. Furthermore, the large work function of ITO makes it an inefficient electron injecting contact that is overcome, ironically, by the formation of electron-conducting defect states induced by this damage. A buffer layer is often employed between the ITO cathode and the device active regions to eliminate the formation of exciton quenching defects in the light absorbing layers (Parthasarathy et al., 1998).

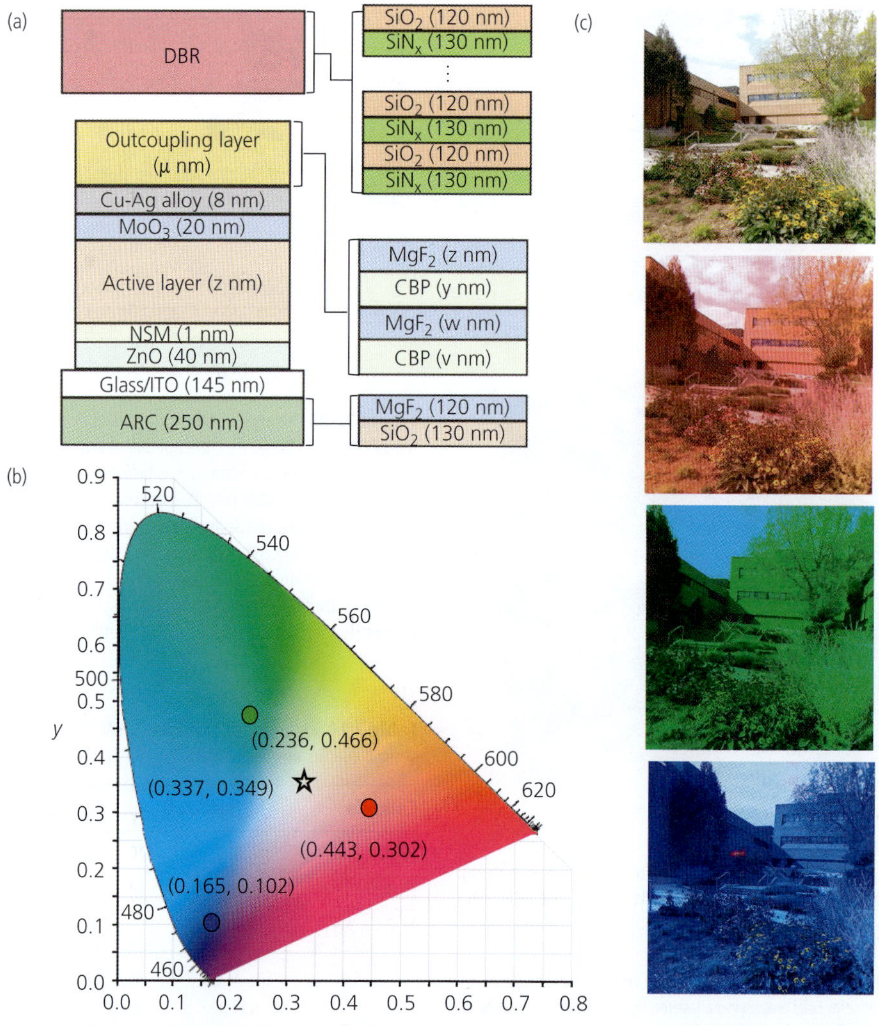

Figure 7.136 (a) General structure of the semitransparent OPV with an NIR reflecting DBR, an ARC on the bottom surface, and an optical outcoupling layer on the top surface. (b) Chromaticity coordinates for the four OPVs in (c) with different outcoupling and active layer thicknesses (Li et al., 2019).

While the narrow spectral width of NIR absorbing materials can be exploited to achieve both high efficiency and transparency when combined with a semitransparent cathode, their spectral overlap with the visible spectrum can lead to unwanted coloration of the cells. Hence, it is essential to also engineer the outcoupled spectrum of the devices to achieve the desired hue. In the previous example, most of the attention is directed at improving the coupling of the broad, incident spectrum into the solar cell active region. The appearance and efficiency of the cell can simultaneously be tailored by applying a second coating on the exit surface, called an *outcoupling* (OC) layer. By combining the elements of materials sets featuring narrow absorption spectra with multilayer ARC and OC films, as well as a DBR used for NIR light recycling, a high *LUE* is achieved for the device in Fig. 7.136a. Starting from the top surface, a DBR comprising 12.5 alternating layers of plasma enhanced CVD SiN_x and SiO_2 is deposited on a separate glass substrate. This is attached to the surface of the OPV that is fabricated on a second glass substrate (see also, Lunt and Bulovic, 2011). The device itself has several innovations. The anode buffer consists of a ZnO film coated with a 1 nm thick NSM hole injection layer that minimizes charge trapping at the ZnO surface. The active region comprises a solution-deposited PCE-10:BT-CIC:TT-FIC (1:1.25:0.5) ternary blend that is capped with a MoO_3 cathode buffer and a semitransparent, sputtered 8 nm thick 2% Cu-Ag alloy film (Cho et al., 2017). The alloy has a resistance of 11.3 $\Omega/sq.$, which is comparable to a vacuum deposited Ag film of equal thickness, although the 2% Cu-Ag is >20% more transmissive across the visible. Most importantly, the very thin alloy film is continuous and smooth compared to Ag, which tends to form clusters that results in a rough surface. The final layer is a two period, CBP/MgF_2 light outcoupling layer that reduces trapping of visible radiation within the cell. Unlike the ARC, the outcoupling layer has a narrower spectral bandwidth, and thus controls the tint of the cell. Images of an outdoor scene in daylight taken through several different cells with OC and active layer thicknesses tuned to transmit different wavelengths are shown in Fig. 7.136c. The CIE coordinates of light from an AM1.5G simulator viewed through the cells are indicated by the data points in Fig. 7.136b. The neutral density cell (top image) has coordinates of (0.337, 0.349), that is, it lies close to the isoenergetic chromaticity point of (0.33, 0.33) at 5000 K (see Fig. 6.13). Similar to a light source, the window appearance is defined by its color temperature (here approximately 5000 K), and its color rendering index. In this case, $CRI = 87$, indicating that objects illuminated through the OPV will appear in approximately their natural colors. The other cells in Fig. 7.136 strongly filter the light, and as a result have a deeply colored appearance.

The performance characteristics of these cells are summarized in Table 7.15. Included in this compilation is a second neutral density cell with CIE coordinates of (0.280,0.335), exhibiting a slightly green cast. Importantly, the highest performance neutral density OPV has $LUE = 3.56\%$ and $\eta_P = 8.0\%$, giving photopic transparency of nearly 50%. From Fig. 7.127, this is nearly double the *LUE* of inorganic and perovskite cells, suggesting that transparent windows provide a unique opportunity for OPV applications, particularly when coupled to their ability to be deployed on flexible thin films that can be easily attached within the pocket between the outer and inner panes of insulated windows commonly found in commercial and residential buildings.

Table 7.15 Operating characteristics of multi-colored semitransparent OPVs (Li et al., 2019)

Device	j_{SC}[a] (mA/cm^2)	V_{OC} (V)	*FF*	η_P[b] (%)	*AVT* (%)	*LUE* (%)
Blue	18.6	0.67	0.71	8.9	4.0	0.36
Green	18.1	0.68	0.71	8.7	17.0	1.48
Red	17.3	0.68	0.70	8.3	5.9	0.49
Neutral[c]	13.2	0.65	0.67	5.8	44.3	2.57
	16.2	0.68	0.72	8.0	44.2	3.56

[a] The j_{SC} values are calculated from the integral of the *EQE* spectrum.
[b] η_P calculated based on measurement of eight devices under 1 sun, AM1.5G reference spectrum along with j_{SC}.
[c] The first entry has CIE coordinates of (0.337, 0.349), the second has (0.280, 0.335).

7.4.3 Bulk and mixed heterojunction morphologies

Bulk and mixed HJs are nanostructured aggregates of donor and acceptor molecules that present a large interface area where excitons can dissociate without having to diffuse long distances from their origin. This structure eliminates the trade-off between the relatively short exciton diffusion length and the long optical absorption length that limits the efficiency of planar HJ devices. However, the inherent disorganization of these random structures can impede charge conduction to the contacts due to the circuitous and often discontinuous paths between the charge point of origin to its respective contact. To maximize the probability of exciton dissociation and minimize resistance to charge transport, the morphology of the mixed regions must be controlled with nanoscale precision. We have seen in Chapter 5 that morphology is determined by numerous factors, including the method of deposition, post-deposition annealing, the ratio of donor to acceptor molecules in the blended junction, and by the steric properties of the donor and acceptor molecules. All of these factors become increasingly complex to manage as the number of constituent molecules forming the active region is increased. It is no surprise, then, that controlling and understanding the morphologies of ternary active regions is considerably more difficult than for binary OPVs.

In this section we discuss a few of the frequently used methods for controlling morphologies to maximize OPV efficiencies. We start by considering guidance provided by the ideal diode theory of Section 4.7.2 to understand the optimal morphologies needed for high efficiency. Then we provide examples of how to achieve those structures based on discussions in Section 5.5 on solvent and thermal annealing of both small molecule and polymer blends.

7.4.3.1 Dependence of efficiency on morphology: theoretical perspectives

The ideal diode equation for excitonic HJs provides insights into how morphology impacts efficiency. The external efficiency in Eq. 7.48 depends on the product of the efficiencies of processes leading to charge generation. The objective of the BHJ is to reduce the distance between the point of photon absorption to the site of exciton dissociation at the D-A interface, such that the exciton diffusion efficiency, $\eta_{ED} \rightarrow 1$. The blended active region, unfortunately, also creates a long and tortuous route for the charges generated at the heterointerface that increases the probability for Langevin or bimolecular recombination. This results in a charge collection efficiency, $\eta_{CC} < 1$. The optimal morphology, therefore, must simultaneously maximize both η_{ED} and η_{CC}.

The morphology also controls the ability for the PP to dissociate into free charges. The complex process of free charge generation at the heterointerface discussed in Section 4.7.2 is folded into the charge transfer efficiency, η_{CT}. Recall that the current of an excitonic HJ given by Eq. 7.55 is a function of two morphology-dependent terms: the saturation current and the photocurrent (j_0 and j_{ph}, respectively). That is, $j_0 \propto 1 - \eta_{PPd}$ and $j_{ph} \propto \eta_{PPd}$, where η_{PPd} is the PP dissociation efficiency given by $\eta_{PPd} = k_{PPd}/(k_{PPd} + k_{PPr})$ (cf. Eq. 7.53). Thus, morphologies that lead to $k_{PPd} \gg k_{PPr}$ maximize j_{ph} (and hence j_{SC}), while simultaneously minimizing the dark current, j_0, both of which lead to an increase in V_{OC} (Eq. 7.122) Moreover, Onsager–Braun theory shows that exciton dissociation into free polarons is enhanced by an applied electric field, that is, $k_{PPd} = k_{PPd}(F)$. Then, from Eq. 7.123 we find that $V_{OC} = V_{OC}^0 + \frac{n_S k_B T}{q} log(\eta_{PPd})$, where we take $k_{PPd} \propto k_{rec}$ at low voltage, and V_{OC}^0 is the open circuit voltage in the absence of recombination, that is, k_{PPr} or k_{rec} are negligibly small.

A consequence of this analysis is that there is a significant benefit to maximizing the rate of PP dissociation vs. recombination. This is apparent from the plot of the calculated j–V characteristics of an ideal diode in Fig. 7.71 for various ratios of k_{PPd}/k_{PPr}. All three critical parameters governing the power conversion efficiency, that is, V_{OC}, j_{SC}, and FF, are significantly increased as k_{PPd}/k_{PPr} is increased from 0.01 to 10^3.

The ideal morphology is, therefore, one that minimizes k_{PPr} by decreasing the electronic coupling between D and A molecules at the HJ. This is achieved by using compounds that are sterically hindered from close packing, thus limiting the orbital overlap with neighboring molecules across the HJ. Unfortunately, steric hindrance can also lead to poor coupling between like molecules. That is, it can reduce either or both of the D–D or A–A molecular couplings. This increases the resistance to charge transport along percolating paths from the point of exciton dissociation to the electrodes, thereby decreasing η_{CC} and increasing cell series resistance (and thus decreasing FF). Hence, the preferred morphology is one that achieves a delicate balance between order needed for low resistance conduction, and disorder that promotes exciton dissociation very near to their point of origin.

Example HJ morphologies and their potential influence on k_{PPr} are illustrated in Fig. 7.137 for a model interface between C_{60} and a planar donor molecule (Wang et al., 2011). Figure 7.137a shows an ordered arrangement of donor molecules with their planar surface adjacent to the C_{60} layer. Presumably, the π-orbitals are parallel to the planar face and hence overlap with that on the C_{60} molecules. This leads to strong interactions between electronic systems, thus increasing k_{PPr} at the expense of k_{PPd}. Figure 7.137b shows an edge-on arrangement. Here the π-systems extend between adjacent donor molecules, but do not significantly overlap with those of C_{60}. In this situation, we anticipate that k_{PPr} is small. However, conduction of photogenerated holes to layers below the interface is problematic given the presumed edge-on relationship of the donor molecules in the underlying bulk layer. The optimal situation is shown in Fig. 7.137c comprising an interface where the D and A molecules interdiffuse, leading to limited

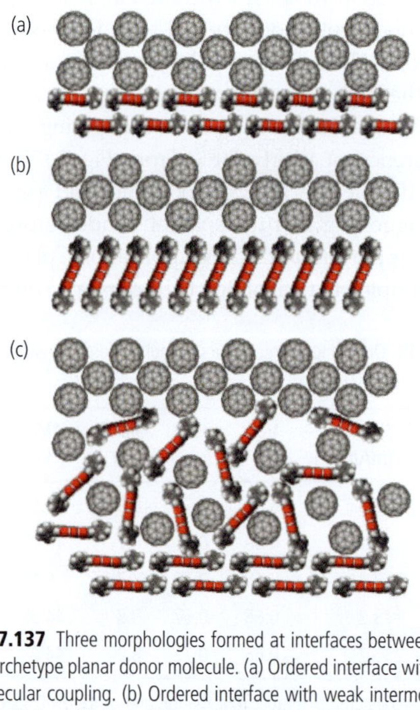

Figure 7.137 Three morphologies formed at interfaces between C_{60} and an archetype planar donor molecule. (a) Ordered interface with strong intermolecular coupling. (b) Ordered interface with weak intermolecular coupling. (c) Interdiffusion of D and A molecules at the interface with a combination of couplings. Ordered structures lie beyond the interface region.

disorder. In the bulk, away from the interface, both the D and A molecules form ordered structures that support a high conductivity for both holes and electrons, respectively. While there are pairs of D and A molecules within the interface region that may have strong π–π interactions, on average those interactions are considerably smaller than in the ordered structure of Fig. 7.137a. This structure achieves the necessary balance of a high k_{PPd}/k_{PPr} as well as a high conductivity through the layer bulk (and hence a high *FF* and j_{SC}).

Proposing desired morphologies, that is, disorder at interfaces and order within the bulk, is one thing, but achieving them and showing that they are effective in increasing efficiency is a more daunting challenge. A series of experiments that were directed at independently tailoring the morphologies at the interface and in the bulk of bilayer DPSQ/C_{60} HJ OPVs endeavored to directly test the validity of these concepts (Zimmerman et al., 2012). The OPV structure is shown in Fig. 7.138 along with the molecular structural formula for the SQ donor, DPSQ. The donor is spun on a MoO_3 surface from a chloroform (CF) solution. In the "as-cast" (AC) device, the C_{60} layer was subsequently thermally evaporated onto the DPSQ surface, followed by the PTCBI EBL and the Ag cathode. In the "pre-C_{60}" device, immediately following DPSQ deposition, the donor layer was solvent vapor annealed for 10 min in a sealed vial containing DCM (Wei et al., 2011a). Solvent vapor annealing (SVA) entails the infusion of solvent molecules that re-dissolve the film, allowing the molecules to readjust to a lower energy, crystalline state. Solvent annealing promotes mass transport over large distances, resulting in the growth of nanocrystalline domains. The annealing time is chosen to optimize both efficiency and yield. Subsequent to the annealing step, the remainder of the layers used in the as-cast device in Fig. 7.139 were deposited. The "post-C_{60}" process employed SVA after the C_{60} deposition but before completing the device.

Morphologies of AC and pre-C_{60} films studied using X-ray diffraction data shown in Fig. 7.139a.

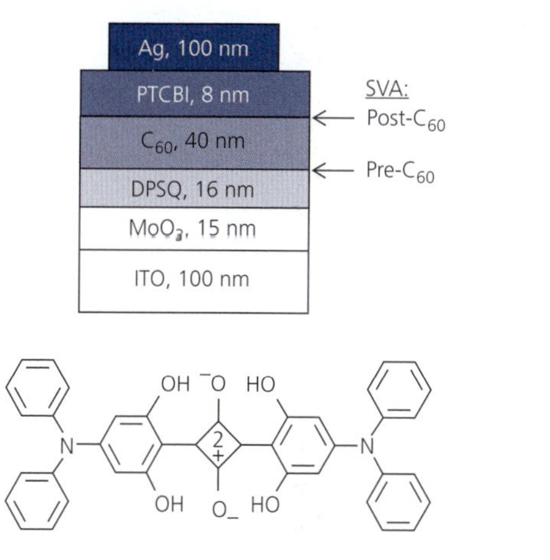

Figure 7.138 The structure used to control interface and bulk film morphologies. Also shown is the molecular structural formula of DPSQ. Three different devices were tested: "as-cast" where the entire device was constructed without annealing, "pre-C_{60}" where the DPSQ was solvent vapor annealed prior to deposition of the remaining layers of the device, and "post-C_{60}" where the device was completed up to the deposition of C_{60} prior to SVA, followed by deposition of the PTCBI exciton blocking layer and cathode.

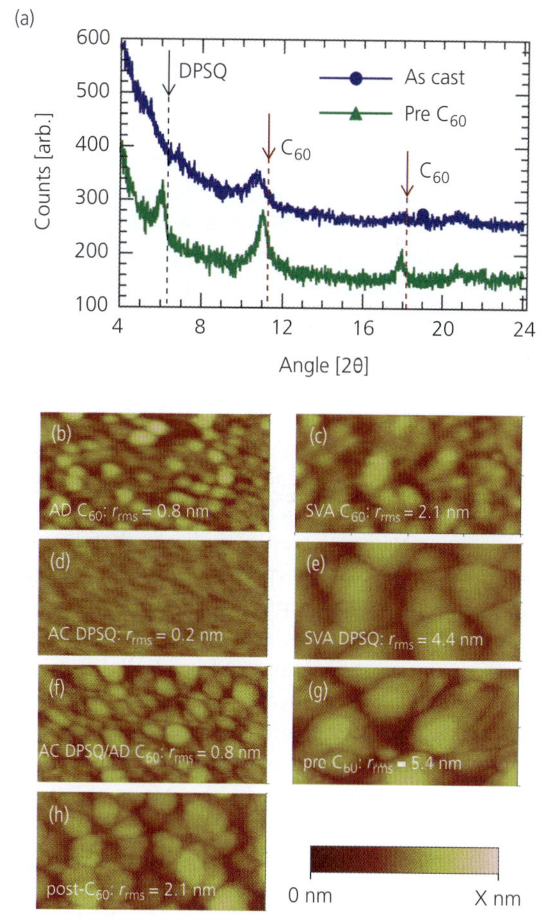

Figure 7.139 (a) X-ray diffraction patterns identifying molecular peaks in both as cast (AC) and pre-C_{60} thin films. (b)–(h) AFM images of the surfaces of as deposited (AD) and SVA films of C_{60}, AC and SVA films of DPSQ, and AC DPSQ/C_{60}, pre- and post-C_{60} films. The root mean square roughness, r_{rms}, is shown in each image. The scale bar at lower right corresponds to (b) $X=10$ nm, (c, f, h) $X=20$ nm, (d) $X=5$ nm, and (e, g) $X=50$ nm. All fields of view are 1 × 0. 5 μm (Zimmerman et al., 2012).

Reprinted with permission from Zimmerman, J. D., Xiao, X., Renshaw, C. K., Wang, S., Diev, V. V., Thompson, M. E. & Forrest, S. R. 2012. Independent Control of Bulk and Interfacial Morphologies of Small Molecular Weight Organic Heterojunction Solar Cells. Nano Letters, 12, 4366–4371. Copyright 2012 by the American Chemical Society.

The AC film shows no DPSQ feature, although two weak peaks associated with C_{60} indicate limited crystallization. In contrast, the pre-C_{60} pattern shows that both DPSQ and C_{60} exhibit long range order. Apparently, crystallization of DPSQ provides a template on which the C_{60} can form crystallites on a larger scale than when deposited onto amorphous AC DPSQ films. The post-C_{60} structure also shows C_{60} crystalline order although there is no corresponding peak for DPSQ. Nevertheless, absorption spectra of the film have a pronounced dimer peak due to DPSQ, indicating the presence of at least small ordered domains arising from this treatment.

Film surface morphologies also show evidence for DPSQ crystallization due to SVA, see Figs. 7.139b–h. As-deposited C_{60} and AC DPSQ surfaces are smooth, with rms roughnesses of 0.8 nm and 0.2 nm, respectively, which is indicative (although not proof) of an amorphous film. However, when exposed to SVA, the C_{60} and DPSQ roughness increases to 2.1 nm and 4.4 nm due to formation of nanocrystallites. This roughness appears in the pre-C_{60} sample, although the post-C_{60} roughness is similar in scale to that of the SVA C_{60} sample. Thus, it appears that the C_{60} layer prevents the underlying DPSQ surface from roughening, even though it undergoes a degree of internal crystallization. The C_{60} cap layer preserves the interface disorder while allowing the molecules within the film bulk to adjust during annealing (Peumans et al., 2003a).

The morphological studies indicate that the AC sample is completely amorphous, both at the interface and within the bulk. The pre-C_{60} sample is ordered in both the D and A layers, and at their interface. The post-C_{60} film most likely has ordered bulk layers, but the presence of the C_{60} cap locks in disorder at its interface with DPSQ. These conclusions are supported by the *j*–*V* characteristics in Fig. 7.140. We make the following observations. (i) The AC sample has a high $V_{OC} = 0.94$ V, but $j_{SC} = 5.3$ mA/cm^2 is relatively low. The high V_{OC} suggests that k_{PPd}/k_{PPr} is large (cf. Eq. 7.123) due to disorder, but it is difficult to extract charge through the amorphous bulk. (ii) The pre-C_{60} has a high degree of order in the bulk, leading to an increase in j_{SC}, but the order at the interface increases molecular coupling between C_{60} and DPSQ, reducing k_{PPd}/k_{PPr}, and hence $V_{OC} = 0.86$ V. (iii) The ordered bulk of post-C_{60} results in the highest $j_{SC} = 7.0$ mA/cm^2, while disorder locked in at the as-grown interface maintains a high k_{PPd}/k_{PPr}, resulting in V_{OC} similar to the disordered AC device.

The interplay between interface and bulk morphologies is summarized in Fig. 7.141. The results provide compelling evidence that disorder at the interface leads to reduced intermolecular electronic coupling, and thus an increased V_{OC}, but the disorder in the bulk leads to a decreased charge extraction, and hence a low j_{SC}. The ideal diode analysis in Section 4.7.2, points to the essential role of morphology in determining OPV performance. While these results are obtained for a bilayer, small molecule diode comprising both solution and vapor deposited molecular species, the conclusions also apply to polymer-based devices. In the following section, we will show several approaches for achieving optimized morphologies in both solution and vapor processed cells. The precise control exhibited for the DPSQ/C_{60} system is by no means easy to achieve with other

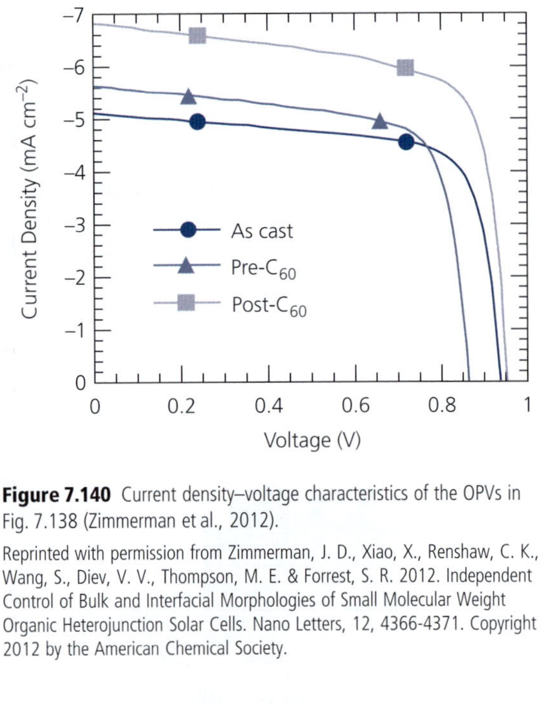

Figure 7.140 Current density–voltage characteristics of the OPVs in Fig. 7.138 (Zimmerman et al., 2012).

Reprinted with permission from Zimmerman, J. D., Xiao, X., Renshaw, C. K., Wang, S., Diev, V. V., Thompson, M. E. & Forrest, S. R. 2012. Independent Control of Bulk and Interfacial Morphologies of Small Molecular Weight Organic Heterojunction Solar Cells. Nano Letters, 12, 4366-4371. Copyright 2012 by the American Chemical Society.

C_{60} / DPSQ	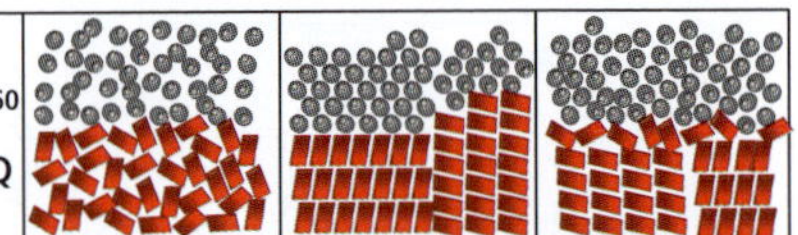		
	As cast	**Pre C_{60}**	**Post C_{60}**
Bulk DPSQ	Amorphous	Ordered	Mod. order
Bulk C_{60}	Weakorder	Ordered	Weak order
Interface	Disordered	Ordered	Disordered
Surface	Smooth	Rough	Smooth
k_{PPr}	Low	High	Low
V_{OC}	High	Low	High
J_{SC}	Low	Moderate	High

Figure 7.141 Summary of results obtained from analysis of results for devices and films subjected to SVA treatment.

materials or processes, although it is necessary for reaching a high OPV efficiency.

In addition to influencing charge transfer, the degree of interface order impacts the energy loss associated with charge transfer. Indeed, this is implied in the foregoing experiments, where we have seen that k_{PPr} increases with order, suggesting that the CT energy is lower (and consequently the energy loss larger) for more ordered interfaces due to stronger intermolecular interactions across the HJ. The relationship between interface order and the CT energy is illustrated in Fig. 7.142a. The D-A junction comprising, in this example, a donor polymer and a fullerene acceptor, can assume several morphologies whose CT energies are, in descending order: (i) fully disordered, (ii) ordered on the donor side, disordered on the other, (iii) disordered on the donor side, ordered on the other, and (iv) ordered on both sides of the D-A junction. This picture is strikingly similar to Fig. 7.141 used to describe CT at small molecule HJs.

The CT state energies in ternary blend OPVs comprising a mixture of amorphous (AnE-PV*ba*) and semicrystalline (AnE-PV*ab*) donor polymers with $PC_{61}BM$ were analyzed to understand whether the energy loss picture in Fig. 7.142a is valid. Here, AnE-PV*ab* is an anthracene-containing poly(*p*-phenyleneethynylene)-*alt*-poly(*p*-phenylenevinylene) copolymer (AnE-PV), with octyloxy side-chains at the

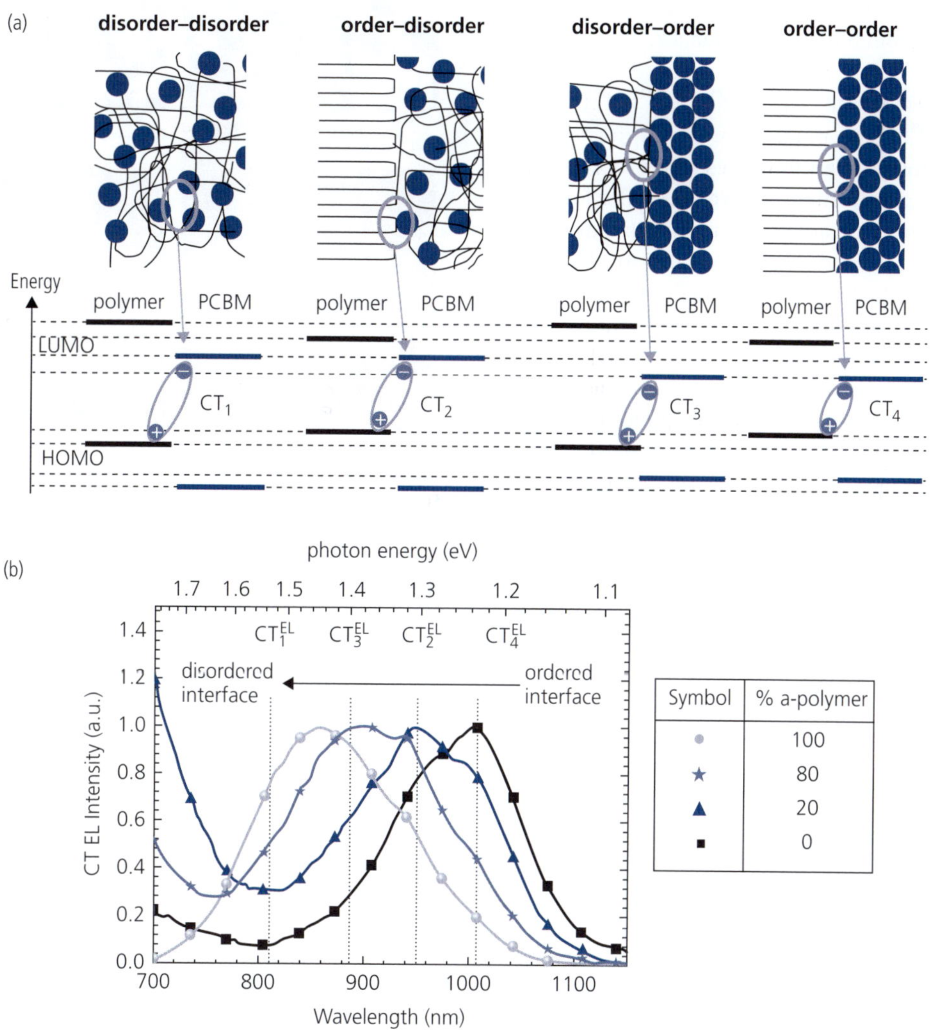

Figure 7.142 (a) Illustration of the relationship between interface morphology and the CT energy for (from left) fully disordered interface; order on one side, disorder on the other; disorder/order; and a fully ordered interface. For this example, the donor is a polymer (lines) and the acceptor is $PC_{61}BM$. (b) CT electroluminescence spectra obtained by fitting the full EL data to a Gaussian-Lorentzian function for different concentrations of an amorphous donor polymer (AnE-PV*ba*) in a ternary blend also comprising the semicrystalline donor, AnE-PV*ab*, and $PC_{61}BM$. After Kästner et al. (2017).

PPE side, and 2-ethylhexyloxy side-chains at the PPV-side. Amorphous AnE-PVba comprises AnE-PV with 2-ethylhexyloxy side-chains at the PPE side and octyloxy side-chains at the PPV. The CT EL spectra obtained by current injection, vs. concentration of the amorphous phase of the donor polymer were measured and fit to a Gaussian–Lorentzian line shape function to extract the EL peak energies. The results of the fits are shown in Fig. 7.142b. The highest energy CT state, CT_1 is at $E_{CT1} = 1.513$ eV, and is found in the fully disordered OPV. As the ratio of the disordered phase, AnEPV*ba*, to the ordered AnEPV*ab*, donor is increased, the CT state energies decrease from $E_{CT1} = 1.513$ eV to $E_{CT3} = 1.386$ eV to $E_{CT2} = 1.308$ eV to $E_{CT4} = 1.233$ eV, consistent with the picture in Fig. 7.142a. This is similar to the DPSQ/C_{60} system in Fig. 7.141, and consistent with predictions from theory in Section 4.7.2.

7.4.3.2 Optimizing morphology during deposition and processing

Here we provide a few examples where morphological control leads to high performance. The following discussion is not meant to be an exhaustive treatment of a very large field, but rather an introduction to those morphologies that are beneficial, and some of the characteristics that they introduce into the active regions of OPVs.

Structural annealing. Annealing an as-deposited film adds energy to the system, allowing molecules that are in a metastable configuration (e.g. in an amorphous structure) to reorganize into a more stable, lower energy state. This creates nano- or microcrystalline domains, driving towards a large surface area D-A junction. The domains themselves have low resistance due to high charge mobilities compared to amorphous regions. Intimate packing of the crystalline and amorphous phases can lead to a higher optical density, and hence higher external quantum efficiency. These effects ultimately increase j_{SC} and *FF*, although increased interfacial order can have the unintended consequence of a reduced V_{OC}.

The solvents used in casting the materials into films, as well as the duration and temperature used, have a significant impact on BHJ morphology (Li et al., 2005, Mori et al., 2011, Mori et al., 2015). An example of such effects is observed in the AFM images Fig. 7.143 of non-fullerene P3HT:PF12TBT (1:1) BHJs that are solution cast from three solvents with widely differing boiling points (BP), *o*-dichlorobenzene (DCB, BP = 180°C), chlorobenzene (CB, BP = 131°C), and CF (BP = 61.2°C). The large-scale surface morphology depends on solvent, with

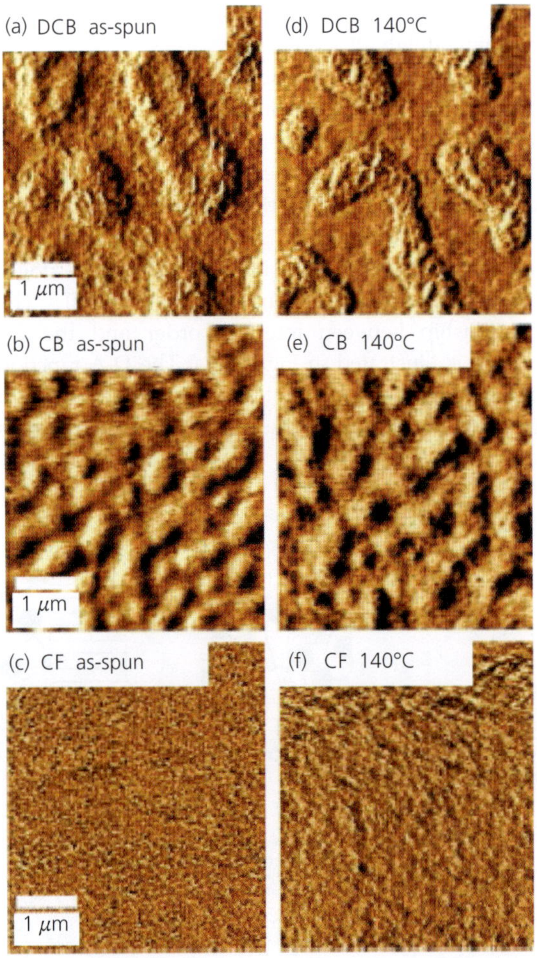

Figure 7.143 AFM images of both (a–c) as-spin-cast, and (d–f) thermally annealed P3HT:PF12TBT film surfaces using different solvents (DCB, CB, CF) during casting. Annealing conditions are 140°C for 10 min (Mori et al., 2011).

the largest phase separation of the donor and acceptors occurring for the highest BP solvents. Spin-casting in the low boiling point CF, results in an apparently homogeneous blend without evidence for phase separation. Hence, there is a monotonic dependence of domain size on solvent boiling point temperature. Thermal annealing does not noticeably change the surface morphology of the as-cast films. The high BP solvents take longer to evaporate, giving the D and A molecular species more time to phase separate.

A qualitatively different picture emerges when we examine the performance of OPVs comprising these same active regions after thermal annealing in

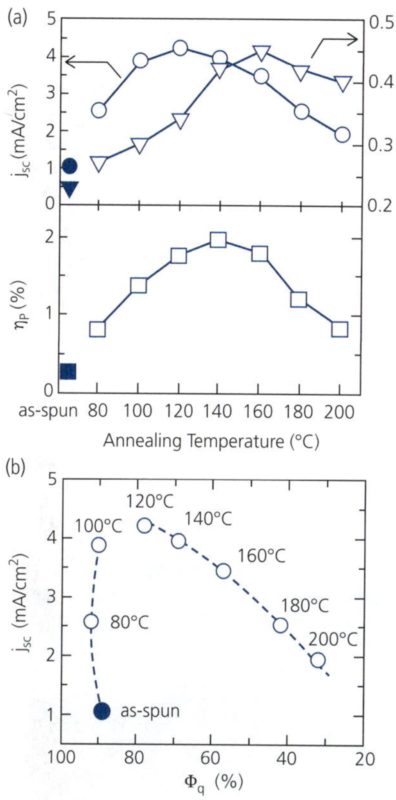

Figure 7.144 The dependence of (a) performance of P3HT/PF12TBT BHJ OPVs vs. thermal annealing temperature for 10 min in N_2. All devices are spin-cast from a CF solution. (b) j_{SC} vs. the PL quenching efficiency, Φ_q, vs. annealing temperature (Mori et al., 2011).

N_2 at various temperatures. The devices were spin-cast from CF with the structure: ITO/PEDOT: PSS (40 nm)/P3HT:PF12TBT (1:1, 70 nm)/LiF (1 nm)/Al. Figure 7.144a shows j_{SC}, FF, and η_P of the devices vs. annealing temperature, T_{anneal}. Figure 7.144b shows j_{SC} vs. the PL quenching efficiency, $\Phi_q = 1 - PL_{blend}/PL_{neat}$, and T_{anneal}. Here, PL_{neat} is the PL intensity of a neat film of PF12TBT, and PL_{blend} is the intensity of the P3HT:PF12TBT blend. We expect that excitons are immediately dissociated followed their generation, leading to $\Phi_q \to 1$, in a perfectly homogenous mixture of D and A molecules. On the other hand, the excitons radiatively recombine prior to dissociation in domains whose size is much greater than the diffusion length, leading to $\Phi_q \to 0$. Thus, Φ_q gives a qualitative measure of the ratio of grain size to L_D, the latter of which is on the order of a few tens of nanometers.

From Fig. 7.144b, we find that j_{SC} is at a minimum for the as-spun device, increasing to a peak at $T_{anneal} =$ 120°C, and decreasing as temperature is further increased. This suggests that the as-spun film is homogeneous with very small, or nonexistent PF12TBT domains, resulting in $\Phi_q \approx 1$. The donor and acceptor components are so thoroughly intermixed that there are few continuous paths to extract charges once the excitons dissociate at the domain boundaries. This is confirmed by the small j_{SC} and FF for the as-spun films. As T_{anneal} is increased, the molecules reorganize into larger domains that form a *bicontinuous network* that extends to the OPV electrodes, yet the domain sizes are still less than L_D. At $T_{anneal} = 120°C$, the domains are $\sim L_D$, leading to a maximum j_{SC}. Interestingly, FF continues to increase as T_{anneal} is further increased. Thus, the PF12TBT domains continue to grow, leading to improved electron conduction (and presumably better hole conduction on P3HT). However, the excitons may not have an equally large probability to dissociate before recombining in the larger domains, and hence j_{SC} decreases. This results in a peak power conversion efficiency and quenching ratio at $T_{anneal} = 140°C$.

The electron mobility in PF12TBT and the hole mobility in neat P3HT, and in P3HT:PF12TBT (1:1) blends are plotted vs. T_{anneal} in Fig. 7.145. The mobilities are extracted from the j–V characteristics in the space-charge limited conduction regimes of hole- and electron-only devices (Section 4.4.2). The neat films have significantly higher mobilities than the blends. However, as the films are annealed, both electron and hole mobilities increase significantly, reaching a maximum at T_{anneal} between 120–140°C. These trends track those of the FF in Fig. 7.144, due to the interrelationships between mobility, FF, and R_{ser}.

This study clarifies the different effects that film deposition and post-deposition annealing have on film morphology, and consequently device performance. The choice of solvents and other deposition conditions creates a global morphology that may, or may not be optimized for both exciton dissociation and charge collection. On the other hand, annealing can provide additional energy that allows the molecules to seek a configuration closer to equilibrium, forming nanodomains of D and A molecules whose size is controlled by the annealing conditions. If the process is driven too far, the domains become so extensive that the device performance once more is reduced. This suggests that the optimum BHJ morphology is metastable, with obvious implications concerning the long-term performance of the devices themselves (see Section 7.8).

SVA can also create internal order. The small solvent molecules diffuse into the film, preferentially dissolving one of the components, which is then free to reorganize into phase-separated domains. The

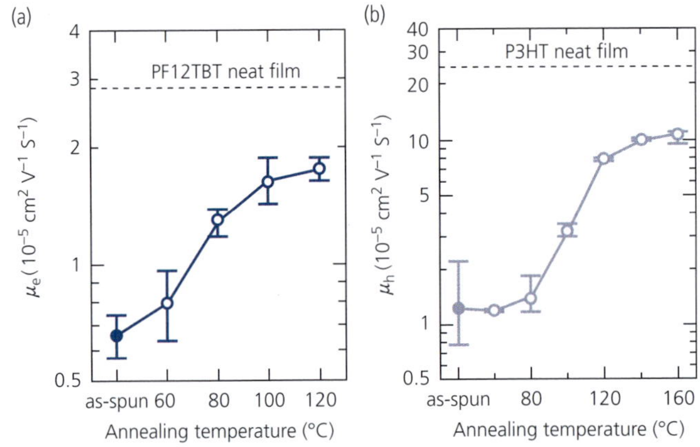

Figure 7.145 (a) Electron and (b) hole mobilities vs. T_{anneal} for P3HT:PF12TBT (1:1) films (Mori et al., 2015).

archetype polymer-based P3HT:$PC_{61}BM$ 1:1 blend morphology was studied by spin casting from a DCB solution onto a glass substrate pre-coated with ITO and 70 nm PEDOT:PSS. The samples were thermally annealed for 20 min at 130°C. Alternatively, the film was solvent annealed by drying at room temperature in a covered container for 3 h. Electron diffraction patterns of 100 nm thick P3HT:$PC_{61}BM$ as-deposited films, and films subjected to SVA and TA are shown in Fig. 7.146. The outer ring corresponding to the (020) crystalline plane of as-deposited P3HT is considerably weaker than for either the TA or SVA samples. In contrast, the $PC_{61}BM$ ring is broad, suggesting a nanocrystalline morphology. Moreover, the $PC_{61}BM$ diffraction intensity is not significantly affected by either annealing procedure. We conclude that annealing tends to organize the P3HT into crystalline domains, but has a much smaller effect on the $PC_{61}BM$ morphology. These smaller molecules evidently intercalate between the P3HT fibrils.

The relative influence of thermal vs. solvent annealing on the microstructure of P3HT:$PC_{61}BM$ blends was also studied by 3D electron tomography. Electron tomography has been developed primarily for examining biological specimens (Frank, 1992). *Electron tomography*, or *transmission electron microtomography* (TEMT), constructs a 3D image of a sample by combining a series of 2D TEM images taken at closely spaced angles relative to the incident electron beam. The images are taken over a large observation angle, preferably the 180° half space above the sample plane. Fourier analysis is then used to construct a volume image from a composite of the 2D images (Jinnai et al., 2004).

The 3D morphologies of the blends in Fig. 7.146 are revealed in the 3D electron tomographs in Fig. 7.147.

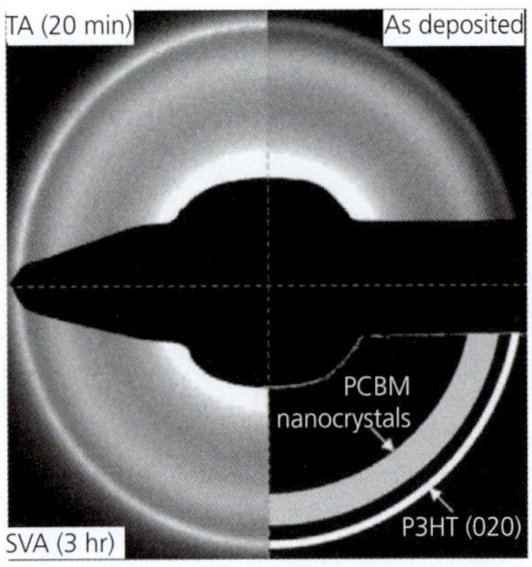

Figure 7.146 Electron diffraction patterns of as-deposited, thermally annealed (TA), and solvent vapor annealed P3HT:$PC_{61}BM$ (1:1) BHJ thin films. The fourth quadrant is used to identify the diffraction features (van Bavel et al., 2009).

The P3HT fibrils are very short and disorganized in the as-deposited sample. Further, there does not appear to be a significant change in their density as the film thickness is traversed from top to bottom. The volume image shows a relatively featureless film structure throughout. Imaging the TA and SVA samples slice by slice through the film tells a different story. The P3HT crystallites are several microns long, typically ~15 nm wide and only a few nanometers thick. Furthermore, they appear to form a denser network near the bottom of the film vs. toward the top. There are no remarkable differences in the appearance of either the TA or SVA samples. The differences with the as-deposited sample are particularly striking. The P3HT volume fraction increases from ~25% near the top of the sample, to 45% near the bottom. From this we infer that the $PC_{61}BM$ had the opposite concentration gradient, with the film being richer in $PC_{61}BM$ near the top. This component segregation is important for OPV performance, since it explains why the cathode and anode are not shorted, as would be expected for a homogeneous, bicontinuous network of D and A regions spanning the entire active layer thickness.

Solvent additives. The addition of small molecular weight bipyridine or $PC_{61}BM$ increases coupling between porphyrin molecules to form strongly coupled macromolecular assemblies with high conductivity (see Section 7.2.2). Alternatively, additives can be used to selectively dissolve one of the molecular constituents in a blend. If the evaporation rate of the additive is less than that of the primary solvent, the species soluble in the additive reorganizes into a more ordered state. Numerous additive systems have been investigated to improve the nanocrystalline morphology of polymer-based BHJ systems (Peet et al., 2007, Lou et al., 2011, Sharenko et al., 2014).

The morphological evolution of films of a donor copolymer of diketopyrrolopyrolle and quarterthiophene (pDPP) mixed with $PC_{71}BM$ (1:1) has been tracked in real time by *in situ grazing incidence X-ray diffraction* (GIXD), and *grazing incidence small angle X-ray scattering* (GISAXS). Glancing incidence X-ray analysis has proven to be an essential tool for revealing both the molecular as well as the macrostructure of organic thin films, whether they are based on polymer or small molecules. The mixture was dissolved in a solution of CF and DCB. CF is a good solvent for pDPP, which is not the case for the high boiling point, DCB, which therefore selectively solubilizes $PC_{71}BM$. Organic solar cells were fabricated by spin casting the D-A mixture onto an ITO/PEDOT:PSS coated glass substrate to form a 120–140 nm thick BHJ active layer capped with a LiF (1 nm)/Al (100 nm) cathode (Liu et al., 2012). A summary of OPV performance obtained using different CF:DCB additive concentrations is provided in Table 7.16. The additive results in significant improvements in efficiency, primarily a result of short circuit current. The best performance was for the device with a CF:DCB (4:1 vol. ratio), which presumably leads to the optimum nanomorphology.

The GIXD patterns as a function of drying time at room temperature for this system are shown in Fig. 7.148a. The highest peak corresponds to the (100) crystal plane of $PC_{71}BM$. It grows slowly after solvent exposure, then accelerates in intensity, and finally saturates into a stable morphology. The peak intensity is plotted in Fig. 7.148b. As drying proceeds, the film weight decreases rapidly within the first 100 s due to evaporation of CF, and then slowly continues to decrease as the less volatile DCB additive slowly leaves the film.

The size of the crystalline clusters can be estimated from their *radius of gyration* (R_g). This quantity is determined by tracking the intensity of the GISAXS diffraction at small scattering angles, that is, small q, following

$$I(q) \simeq I(0)\exp\left(-q^2R_g^2/3\right), \qquad (7.178)$$

where $I(q)$ is the X-ray intensity at q. Equation 7.178 is known as the *Guinier approximation* that applies to approximately spherical clusters (Guinier and Fournet, 1955, Putnam, 2016). In Fig. 7.148b, we

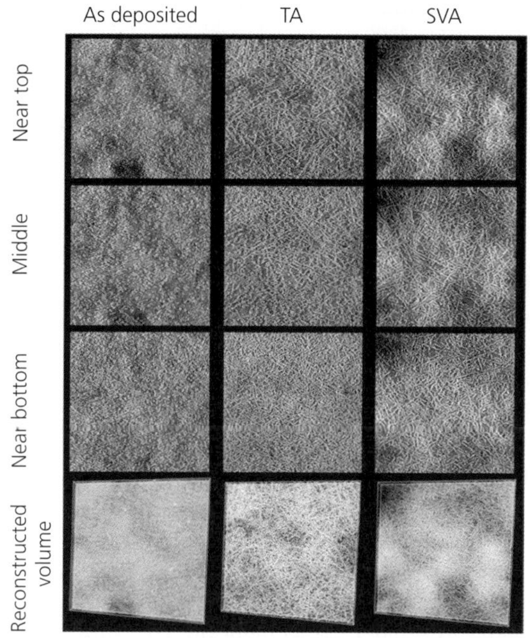

Figure 7.147 Electron tomographs of P3HT:$PC_{61}BM$ (1:1) blends that are freshly deposited, thermally and solvent vapor annealed. The images from electron tomography from near the bottom, in the middle and near the top of the layer are assembled, along with images taken at other angles (and thus depths) to reconstruct the volume images at the bottom. The field of view is 1.7 μm × 1.7 μm × 100 nm thick (van Bavel et al., 2009).

Table 7.16 Performance characteristics of pDPP:$PC_{71}BM$ (1:1) OPVs for different primary solvent, CF, to additive, DCB, ratios (Liu et al., 2012)

CF:DCB	V_{OC} (V)	j_{SC} (mA/cm^2)	*FF*	η_P (%)
1:0	0.62	2.8	61	1.1
16:1	0.61	12.9	52	4.2
4:1	0.63	14.8	60	5.6

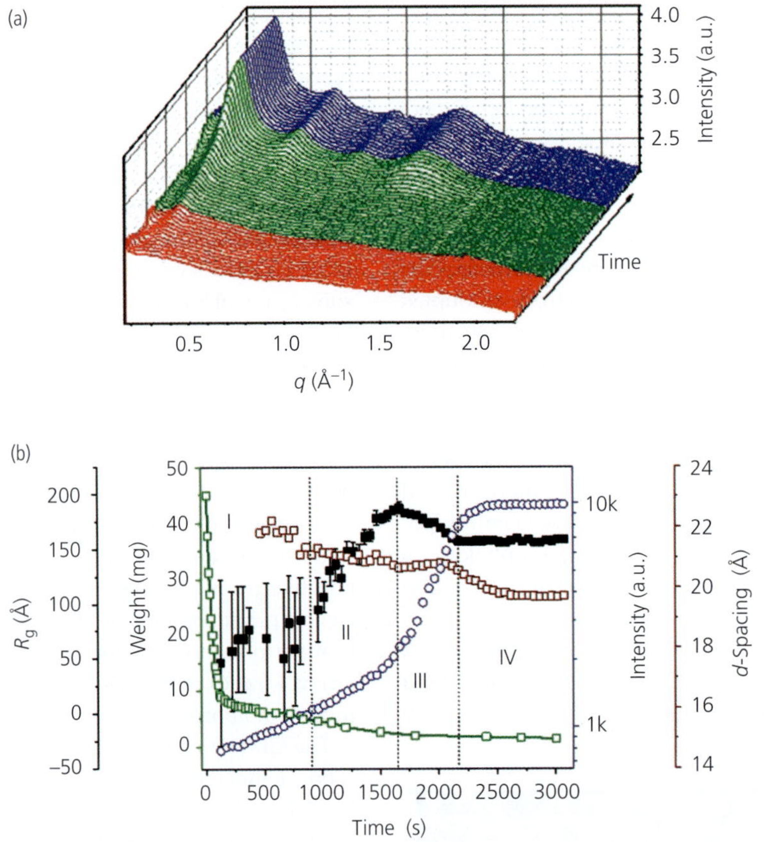

Figure 7.148 (a) GIXD diffraction pattern tracking the structural evolution during drying of a pDPP:$PC_{71}BM$ BHJ film infused with a CF:DCB (1:4) solvent additive. The highest peak corresponds to the (100) crystal plane of $PC_{71}BM$. The next two adjacent satellite peaks correspond to reflections from the (200) and (300) planes. (b) From X-ray and other data, the domain radius of gyration (R_g, solid squares), film weight (green open squares), (100) diffraction peak intensity (blue open circles) and crystal *d*-spacing (red squares) are obtained as a function of film drying time. The morphological changes are divided into four periods (see Fig. 7.150) (Liu et al., 2012).

find that $R_g \approx 50$ Å for the as-deposited film. As the film dries, the crystallites grow until R_g peaks at 200 Å after 1600 s, and then decreases slightly to a saturation value of 150 Å. The *d*-spacing steadily declines by ~2 Å during drying, reaching a steady state value of 20 Å.

These data give a detailed picture of the crystallization process summarized by the illustrations in Fig. 7.149. Initially (zone I, Figs. 7.148 and 7.149), there is considerable solvent infused that creates a nearly homogeneous blend of pDPP and $PC_{71}BM$. As the CF evaporates, the pDPP begins to crystallize, forming fibrils. The loss of solvent volume results in film compaction and a concomitant loss of weight. However, the more slowly evaporating additive, DCB, keeps the $PC_{71}BM$ in solution and mobile within the film (zone II), during which pDPP continues to crystallize forming fibrils of increasing length. The remaining additive evaporates during zone III, causing crystalline $PC_{71}BM$ domains to form in the regions between those of pDPP. The decrease in *d*-spacing over time results from the evaporation of residual solvent in the interstices between the pDPP chains. Finally, all of the primary solvent and additive has evaporated, leaving a stable, compact and ordered structure that has a higher conductivity and optical density than the as-deposited film (zone IV).

The various solution-based processes such as slow drying, thermal annealing, and the use of additives are qualitatively similar in that they use time and energy to drive a non-equilibrium structure toward a lower energy, and stable morphology.

Achieving nanocrystalline thin films during vapor deposition. Thus far we have focused on post-deposition processes to achieve a desired nanocrystalline morphology in solution-based thin films. In Section 5.4.2 we have shown that the morphology of small molecule thin films can be controlled during both VTE and OVPD growth. Here, we focus on the development of nanocrystalline structures during growth in DBP:

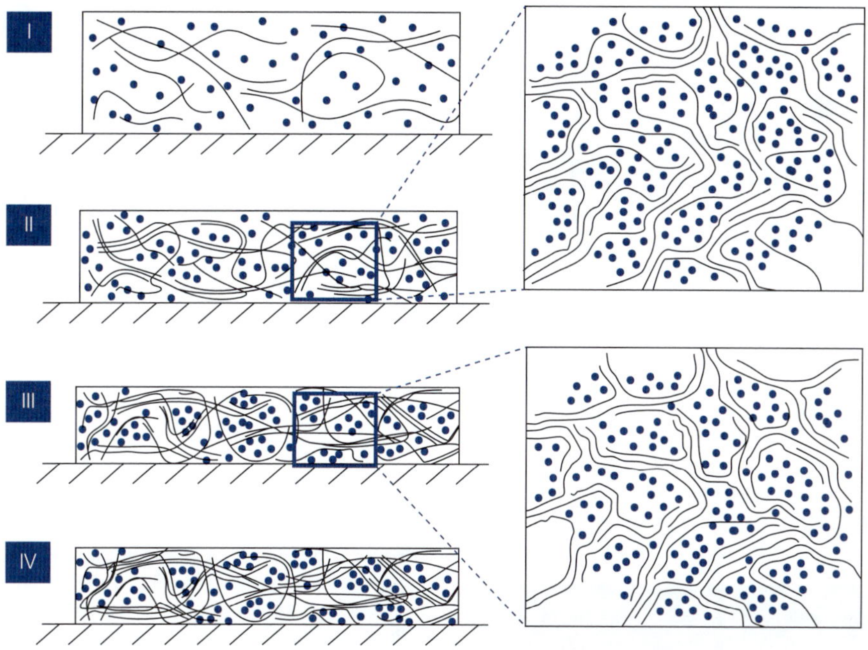

Figure 7.149 Conceptual illustration of the morphological evolution of the pDPP:$PC_{71}BM$ BHJ film inferred from the data in Fig. 7.148. Here, I–IV correspond to each of the zones in Fig. 7.148b. Lines = pDPP, dots = $PC_{71}BM$ (Liu et al., 2012).

C_{70} mixed HJs. This system is of particular interest due to several remarkable properties that lead to high efficiency photon-to-charge conversion. Like the P3HT:PCBM system, DBP:C_{70} mixed HJs have been thoroughly studied due both to its high efficiency, and the understanding that it has brought to the relationship between nanostructure, optical absorption properties, and charge generation (Fujishima et al., 2009, Hirade and Adachi, 2011, Yokoyama et al., 2012, Xiao et al., 2013). One particularly notable feature is that when used in a mixed HJ, the optimum fraction of DBP to C_{70} is much less than 1:1. In Table 7.17 we show that when 40 nm thick co-deposited DBP:C_{70} blends on a ITO/MoO_3-coated glass substrate are capped with a 8 nm thick Bphen EBL and an Al cathode, the highest power conversion efficiencies and *FF* are achieved at DBP fractions of 10–12%. (Xiao et al., 2013) When the mixed HJ is capped with a neat C_{70} layer to form a planar-mixed structure, the power conversion efficiency increases from 4.7%, to 6.4%.

To explain the high efficiencies achieved with such a dilute donor concentration, it was proposed that the DBP and C_{70} form separate domains during VTE growth, as illustrated in Fig. 7.150. The domains are sufficiently extensive to allow for conduction of both electrons and holes along uninterrupted paths to their respective electrodes. The microstructure of blends with various C_{70} concentrations were examined by TEM, with a sampling of images shown in Fig. 7.151. The nanocrystalline domain size for neat C_{70} is 20–50 nm, but it decreases rapidly with increasing DBP concentration. This suggests that C_{70} domains are replaced by DBP nanocrystals or by amorphous regions, neither of which are visible in the images.

Table 7.17 Device performance for various DBP:C_{70} blend ratios (Xiao et al., 2013)

Blend ratio	V_{OC} (V)	j_{SC} (mA/cm^2)	*FF*	η_P (%)
1:4	0.85	9.1	0.50	3.9
1:6	0.87	9.5	0.53	4.4
1:8	0.89	9.7	0.55	4.7
1:10	0.90	9.3	0.55	4.5

An interesting feature of the DBP-C_{70} CT state noted in Section 4.7.2 is its energy dependence on fullerene concentration. Since photogeneration of charges occurs via the CT state, the dependence of the CT state properties on concentration determines the OPV photogeneration efficiency. The EL spectra of the CT spectra of C_{70} in blends with various C_{70} concentrations are shown in Fig. 7.152. The

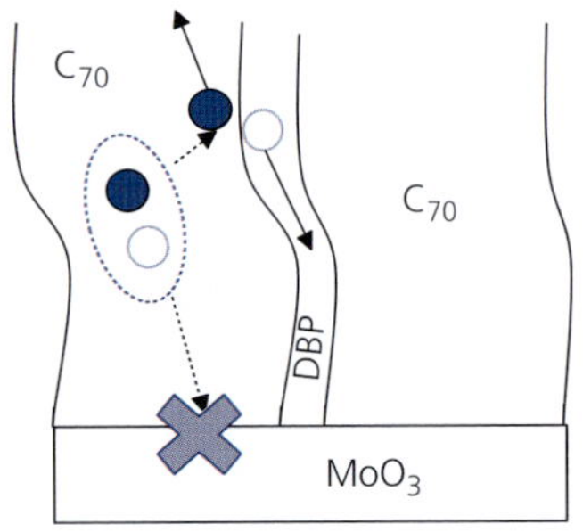

Figure 7.150 Proposed morphology of a dilute DBP:C_{70} mixed HJ in the region of the anode buffer. Due to the strong intramolecular interactions within DBP and C_{70}, they segregate into continuous domains. For example, an exciton (dashed oval) generated in C_{70} drifts to an interface with DBP where it dissociates into electrons (blue circles) and holes (open circles). The domains provide continuous paths to the electrodes for both carrier types. The "X" indicates that excitons migrating to the MoO_3 buffer are quenched.

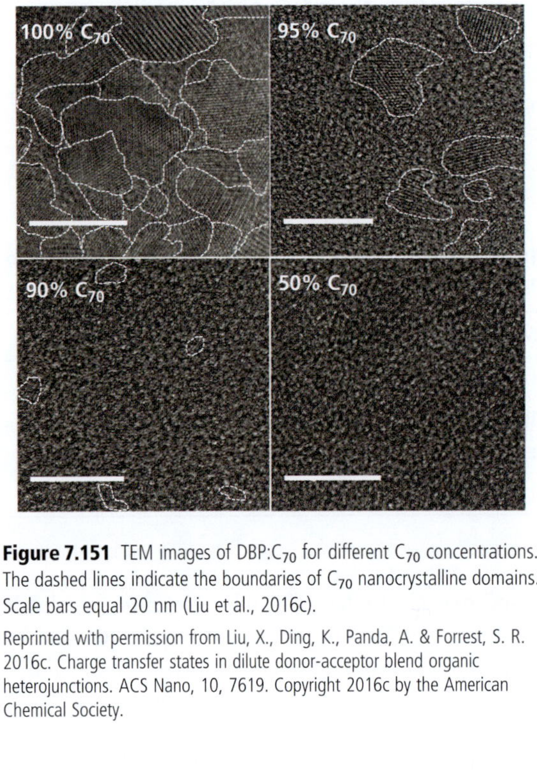

Figure 7.151 TEM images of DBP:C_{70} for different C_{70} concentrations. The dashed lines indicate the boundaries of C_{70} nanocrystalline domains. Scale bars equal 20 nm (Liu et al., 2016c).

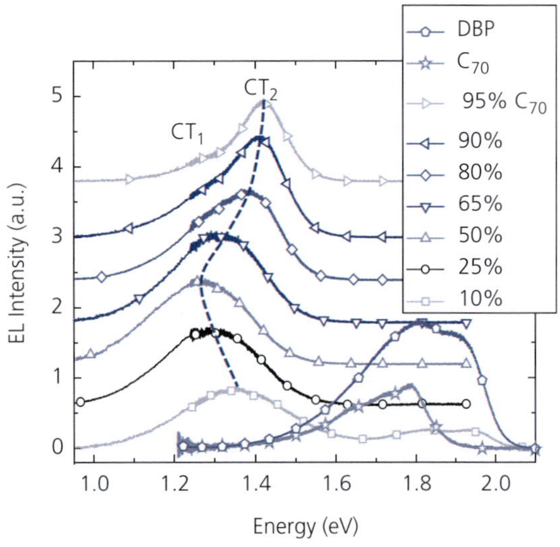

Figure 7.152 Electroluminescence spectra of DBP:C_{70} blends for various C_{70} concentrations ranging from 0 (neat DBP) to 100% (neat C_{70}). The excitonic spectra of DBP and C_{70} are also shown at bottom (Liu et al., 2016c).

EL spectra are obtained by directly forming the CT state via charge injection under forward bias. The PL spectra provided in Fig. 4.108 show a similar dependence of CT peak position on DBP concentration. The singlet emission from DBP and C_{70} in both PL and EL are at higher energies, and hence are well resolved from the CT emission. The two CT states (CT_1 and CT_2) are apparent in both spectra. Interestingly, while CT_1 does not depend on blend ratio, CT_2 is strongly influenced by this parameter, initially decreasing in energy as C_{70} concentration is increased from 10% to 50%, and then increasing as the concentration is further increased.

The change of dielectric constant for different blend ratios has typically been invoked to explain the shift in E_{CT} (Loi et al., 2007, Piersimoni et al., 2011, Bernardo et al., 2014). However, the optical dielectric constant of DBP:C_{70} blend increases monotonically with C_{70} concentration. This fails to explain the trend of CT_2 that has an energy minimum at 50% C_{70}, and a change in slope toward an apparent asymptote at $>$ 80% C_{70} concentration.

An alternative explanation is that the exciton binding energy responds to changes in CT state localization that depends on the size of the C_{70} nanocrystals or DBP aggregates. As the C_{70} concentration increases to $>$ 50%, the electron on a C_{70} nanocrystal bound to the hole on a DBP molecule at the domain boundary is delocalized over several C_{70} molecules, leading to a decrease in the binding energy and therefore an increase of CT state energy. Similarly, as the C_{70} concentration decreases below 50%, the DBP aggregate size increases, resulting in increased hole delocalization on the other side of the boundary.

This scenario has been modeled for CT states in blends with $>$ 50% C_{70} where the electron is confined in approximately spherical crystalline domains with average radii, $<r_0>$. An electron within the domain is confined by a potential barrier $\phi_b = 0.5$ eV on its boundary, which corresponds to the difference between the LUMO energies of DBP and C_{70} (Xiao et al., 2013). The electron Hamiltonian is

$$H = -\frac{\hbar}{2m_e^*}\nabla^2 + V(\mathbf{r}),$$

$$V(\mathbf{r}) = \begin{cases} -\dfrac{q^2}{4\pi\varepsilon_0\varepsilon_{r,C_{70}}|\mathbf{r}-\mathbf{a}|} & (|\mathbf{r}-\mathbf{r_0}|<r_0) \\ -\dfrac{q^2}{4\pi\varepsilon_0\varepsilon_{r,DBP}|\mathbf{r}-\mathbf{a}|}+\phi_b & (|\mathbf{r}-\mathbf{r_0}|>r_0) \end{cases}, \quad (7.179)$$

where m_e^* is the electron effective mass and $|\mathbf{r}-\mathbf{a}|$ is the distance between the hole and the electron. The relative dielectric constants are $\varepsilon_{r,C_{70}} = 4.60$ and $\varepsilon_{r,DBP} = 3.92$ at $\lambda = 900$ nm.

The domain radii are found via x-ray diffraction, where the (111) C_{70} peak intensity and width are tracked vs. blend ratio. The mean radius of the domains is obtained from *Scherrer analysis* of the diffraction peak FWHM, given by

$$\langle r_0 \rangle \simeq \frac{K\lambda_X}{\Delta\theta\cos\theta}, \quad (7.180)$$

where K is a constant close to unity that depends on the shape of the crystallites, λ_X is the X-ray wavelength, θ is the Bragg diffraction angle, and $\Delta\theta$ is the FWHM of the diffraction line after removing broadening due to instrumental resolution limits. X-ray data yield mean radii ranging from 1 nm for 50% C_{70}, to approximately 6 nm for neat C_{70} films.

The energy eigenvalues of the Hamiltonian are obtained from Eq. 7.179, yielding the energy of the Coulombically bound electron states. The energy difference between the bound and unbound ($V(\mathbf{r}) = 0$) states gives the CT binding energy, shown in Fig. 7.153a. The electron density of a Coulombically bound state for a C_{70} domain radius $<r_0> = 2$ nm is shown in the inset. The calculated binding energies fit reasonably well with the measured CT energy, E_{CT}, after including the 1.52 eV energy difference between the DBP HOMO and the C_{70} LUMO.

The CT_2 lifetime is also dependent on the C_{70} domain size, with the lifetime increasing with $<r_0>$, as shown in Fig. 7.153b. Quantum confinement in the C_{70} domains with decreasing $<r_0>$ leads to an increased overlap of the electron and hole probability densities, resulting in a concomitant increase in the CT_2 recombination rate. To model this dependence, the overlap of the probability density of the excited ($|\psi_{CT}|^2$) and ground states ($|\psi_G|^2$), that is, $J = \int |\psi_{CT}|^2|\psi_G|^2 dr^3$, are calculated, where ψ_G is the electron wavefunction at the DBP HOMO obtained from time dependent density functional theory. The

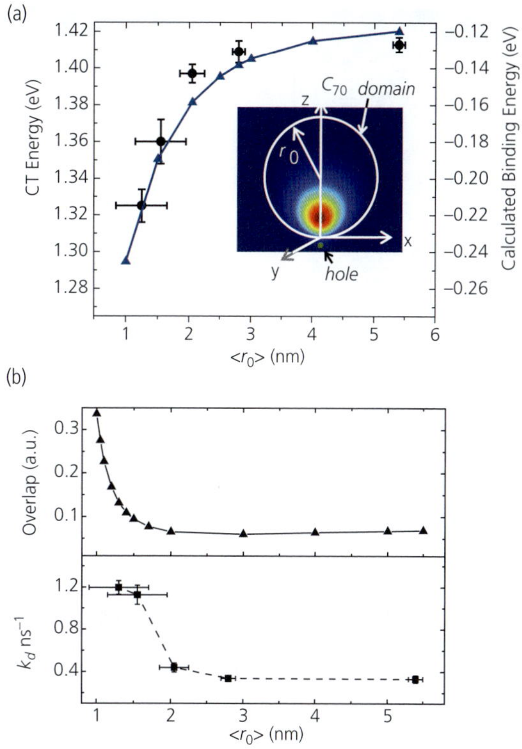

Figure 7.153 (a) Measured peak energies of CT_2 states (data points) and their calculated binding energies (line) as functions of C_{70} domain radius, $<r_0>$. Inset: Calculated electron density distribution for a domain with $<r_0> = 2$ nm. (b) (Upper panel) Calculated probability density overlap integral vs. $<r_0>$. (Lower panel) Measured recombination rate, k_d, for CT_2 vs. $<r_0>$ (Liu et al., 2016c).

overlap integral, J, vs. C_{70} domain radius is shown in Fig. 7.153b, upper panel, which qualitatively agrees with the trend observed for the decay rate of CT_2, in the lower panel.

The model was further tested by measuring the CT energies for nanostructures where the domain sizes were precisely controlled during growth. Transmission electron micrographs of ultrathin layer, alternating layers of DBP and C_{70} grown by VTE are shown in Fig. 7.154. To enhance image contrast, the DBP is functionalized by attaching two Br molecules. It is noteworthy that continuous layers as thin as 2 nm are achieved using VTE. The CT_2 energy shift with layer thickness is in quantitative agreement with the data in Fig. 7.152 and 7.153, confirming the confinement of electrons and holes within C_{70} and DBP of limited scale and dimension (Ding et al., 2018).

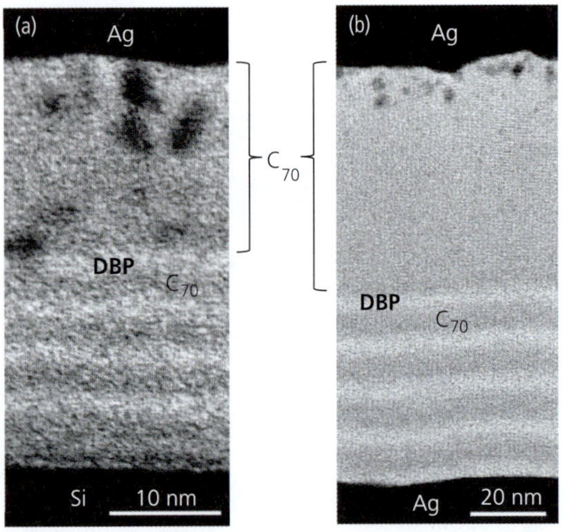

Figure 7.154 (a) Bright field cross-sectional scanning transmission electron microscopic image of five pairs of 2-nm-thick alternating Br_2-DBP/C_{70} layers capped by a 20-nm-thick C_{70} layer and a Ag protection layer. (b) Image of five pairs of 5-nm-thick alternating layers capped by a 20-nm-thick C_{70} layer and a Ag layer. The dark spots within the C_{70} cap are Ag nanoparticles that diffused into the organic (Ding et al., 2018).

The quantum confinement of electrons thus results in a change in the CT state energy and lifetime. This, in turn affects the OPV efficiency due to the competition between the PP dissociation and its recombination rate (see Fig. 7.71). These results once more show the central role played by morphology in OPV active regions, and the necessity for its control either during or after film growth.

Thus far, we have found that details of the nanomorphology affect the energetics of the solid, but we need to clarify why mixtures with the donor diluted to only 10% of the total has the highest η_P. This implies that the absorption and charge collection efficiencies remain high even for such low donor concentrations. Dilute mixtures have also resulted in high efficiencies in polymer blends (Cates Miller et al., 2012). The DMC model in Section 7.1.2 has been used to simulate the morphologies of dilute DBP:C_{70} blends to answer this question (Ding et al., 2018). The extended planar core of DBP (see inset, Fig. 7.155) encourages π–π stacking between molecules that excludes C_{70} molecules to the regions between the stacks. Holes resulting from exciton dissociation are extracted through the resulting contiguous, percolating paths of DBP molecules. The calculated 2D density of percolating paths for hole transport in a simulated solid grown on the anode contact by co-deposition of DBP and C_{70} is given by

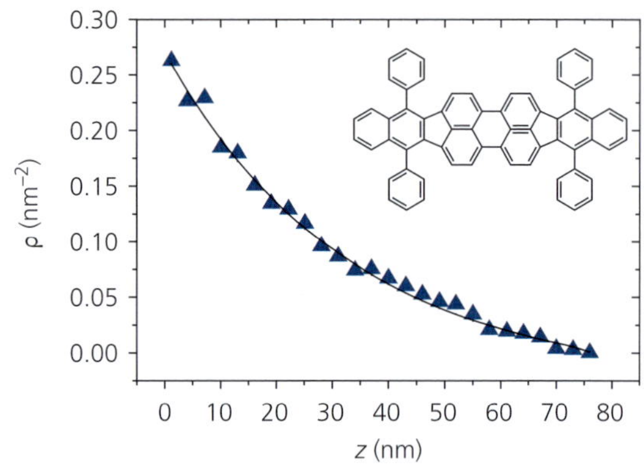

Figure 7.155 Simulated 2D density of DBP percolating paths, ρ, in the substrate plane vs. distance from the anode contact, z (triangles). The line is an empirical exponential fit to the data following Eq. 7.182. Inset: Molecular structural formula of DBP. After Ding et al. (2018).

$$\rho(z) = N_p(z)/A. \tag{7.181}$$

Here, $N_p(z)$ is the number of DBP molecules formed into continuous regions extending a distance z from the anode, and A is the anode contact area. Figure 7.155 shows the 2D density of percolating paths in the substrate plane in a blend with 10% DBP versus the distance from the anode. Empirically, the simulation follows

$$\rho(z) = \rho_0(z)e^{-z/\zeta}, \tag{7.182}$$

shown by the line in Fig. 7.155. Here, ρ_0 is the 2D density of percolating paths originating at the anode. From the fit (solid line), $\zeta = 33$ nm is the characteristic percolation length. At a mixed active region thickness of $d = 50$ nm that is typical for an efficient OPV cell, then $\rho \approx 0.05$ nm$^{\times 2}$ which corresponds to a mean distance between percolating paths of 4.5 nm. This is comparable the C_{70} exciton diffusion length (~4–8 nm), suggesting that the holes can be extracted throughout the active layer with a DBP concentration of only 10 mol.%. Thus, π–π stacking of DBP results in a high density of continuous paths to the anode from DBP, and the cathode from the C_{70} regions. Hence, the crystal packing characteristics of this system lead to $\eta_{CC} \to 1$ as long as the thickness of the active region does not exceed the mean percolation length of the comprising molecular species.

A similar situation has been modeled for the packing of PBTTT and $PC_{71}BM$. Molecular mechanics modeling shows that the $PC_{71}BM$ intercalates the PBTTT network forming a scaffold comprising the polymer backbone and side chains. These organized structures support electron and hole conduction over long distances from their point of origin, and yet place the

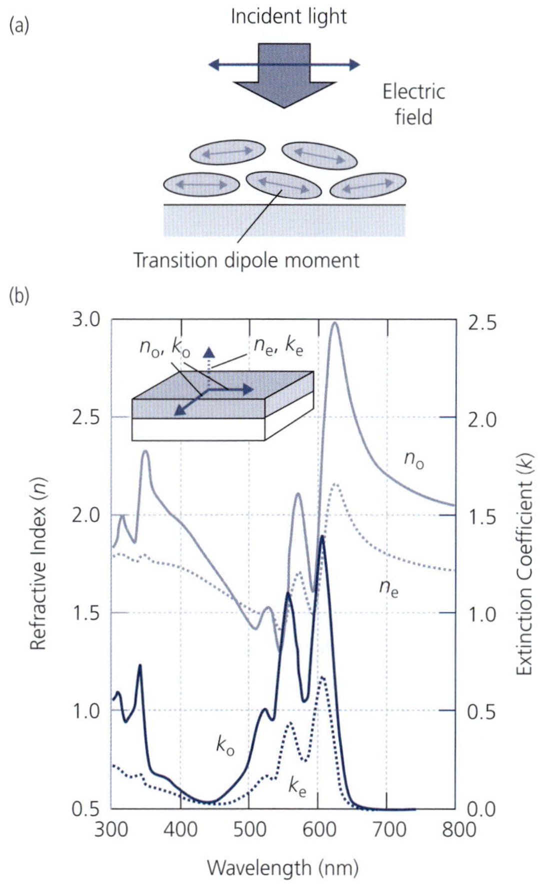

Figure 7.156 (a) The DBP molecules have a large in-plane transition dipole moment that preferentially aligns parallel to the substrate plane. (b) The index of refraction and extinction coefficients to the extraordinary (*e*) and ordinary (*o*) incident wave exhibit a large anisotropy, or film birefringence (Yokoyama et al., 2012).

D and A molecules in intimate proximity that ensures efficient exciton dissociation (Cates Miller et al., 2012).

Finally, we must determine why the external efficiency, and hence η_A, is so high even though only 10% of the DBP molecules contribute to the absorption in the dilute blends. The answer is again found to result from local π–π stacking of DBP. The large conjugated electron system in the plane of the molecule results in a transition dipole moment that preferentially orients parallel to the substrate, shown in Fig. 7.156a. Thus, the extinction coefficient for the ordinary wave is nearly triple that of the extraordinary wave (Fig. 7.156b). Since the molecules preferentially orient parallel to the substrates during growth, the species are able to efficiently absorb light even though the material is present in relatively small concentrations. Further, only those percolating paths connected to the anode and oriented perpendicular to the substrate plane are electrically active. This fortuitously creates a morphology whereby only those molecules in the most favorable orientation for optical absorption participate in the photogeneration process. Indeed, the DBP thin film absorption coefficient is $\alpha = 3.4 \times 10^5\ \mathrm{cm}^{-1}$ at a wavelength of 560 nm, and $4.3 \times 10^5\ \mathrm{cm}^{-1}$ at 610 nm. This is compared to the broader but weaker absorption of C_{70}, with $\alpha = 6 \times 10^4\ \mathrm{cm}^{-1}$ from 400 nm to 650 nm.

The conclusion from this section is that morphology can be precisely controlled by the steric aspects of the D and A molecules, and conditions used both during and after growth. While each system is different due to different molecular shapes and bonding properties, numerous tools exist for guiding growth to achieve morphologies that can lead to high solar cell power conversion efficiencies. These processing methods are augmented by precise techniques for characterizing structure, ranging from electron microscopy to X-ray diffraction, and by accurate computational tools that connect structure with the microscopic nature of charge and exciton transport. Taken together, the quest for high efficiency OPV materials and structures has vastly improved our understanding and control of organic thin films for a variety of applications, while at the same time providing a detailed view into the fundamental physics underpinning these materials.

7.4.4 Materials optimized for use in OPVs

The range of donor and acceptor materials available for use in both small molecule and polymer cells is enormous, and grows every day. In selecting materials, one needs to focus on those properties that lead to stable, low cost and efficient solar energy harvesting. The properties of an optimal donor–acceptor combination can be summarized as follows:

(i) A high extinction coefficient across the desired spectrum. For single junction cells, absorption should span the range that leads to thermodynamically limited efficiency (Section 7.3.2). The multijunction cell should have the same property, although, individual D-A subcells should have only moderate spectral overlap with each other to achieve current balance in series-connected architectures. The trade-off between energy gap (and ultimately V_{OC}) and power conversion efficiency suggests that the optimal energy gap for a single junction cell needs to be ~1.4 eV (or a long wavelength absorption cutoff of 900 nm. But the multijunction cell is not bound by this photon energy thermalization limit. The maximum total, series-

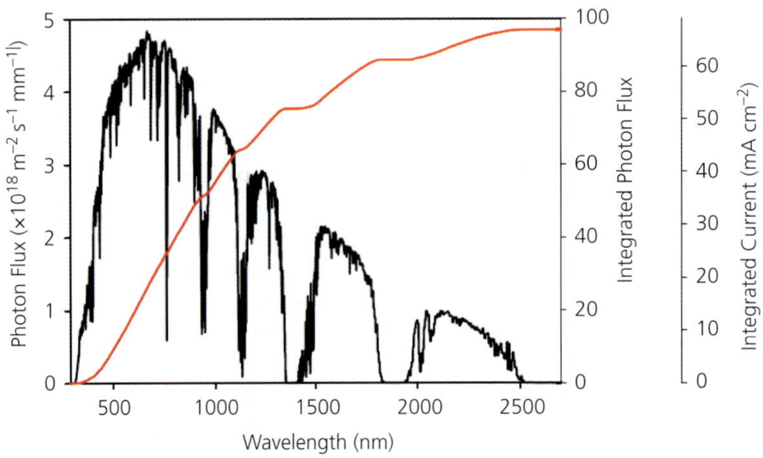

Figure 7.157 Photon flux from an AM1.5G, 1 sun intensity spectrum (black line) and the integrated photon flux and OPV current (assuming $\eta_{ext} = 1$) vs. wavelength.

connected cell V_{OC} is simply the sum of the voltages of the subcells, whereas the current of the cell is limited by the smallest current generated by a subcell in the stack. To understand the limit to the spectral range that can practically be used in solar energy harvesting, the AM1.5G reference spectrum is plotted in Fig. 7.157 along with the integrated photon flux from $\lambda = 280$ nm, at 1 sun (1 kW/m^2) intensity. Also shown is the integrated current density, assuming $\eta_{ext} = 1$. The integrated photon flux at an energy gap cutoff wavelength of 1100 nm is 63% of the total solar emission, and the current is 43% of the possible total. This implies that the subcells in a multijunction OPV must have an increasingly broad spectrum to balance currents generated at shorter wavelengths. But a broader range of wavelengths implies that V_{OC} of the long wavelength subcells is also reduced since its energy gap (and hence ΔE_{HL}) is only as large as the lowest energy photon absorbed. This, in turn, limits the power delivered by the longer wavelength subcells. Hence, there is little to be gained by extending the wavelength range beyond 1200–1400 nm, even in multijunction cells. Also, the energy gap law (Section 3.6.1) implies that non-radiative recombination exponentially increases with decreasing energy gap. Thus, it becomes difficult if not impossible to have high η_{ext} for long wavelength absorbing cells, placing an additional constraint on finding materials combinations that absorb at $\lambda > 1400$ nm.

(ii) A high V_{OC} requires a high ΔE_{HL} (see Eq. 7.123). Optimized OPVs maximize this offset, while also maintaining a sufficient LUMO and/or HOMO energy offset to drive efficient charge transfer while avoiding back transfer. As a rule of thumb, the total energy offset between the D and A frontier orbitals should be $\Delta E_{LUMO} + \Delta E_{HOMO} \sim 0.5$ eV, with neither offset being <0.1 eV (or approximately $4k_BT$ at room temperature).

(iii) The energy loss, E_{loss}, which is the energy difference between the energy gap (defined approximately as the exciton low energy cutoff) and qV_{OC}, needs to be minimized. It is due to several factors, including the free energy of CT state formation, and the magnitudes of the radiative and non-radiative losses (cf. Section 7.3.2). We find in Section 7.4.4 that NFAs can have significantly smaller E_{loss} compared to fullerene-based cells.

(iv) The materials must be easy to deposit from either the vapor or solution phase, to achieve D-A mixed or BHJ morphologies that lead to high efficiency. That is, deposition and post-processing must result in a nanomorphology that results in $\eta_{ED} \approx \eta_{CC} \to 1$ while also achieving a low cell series resistance.

(v) The materials must be inexpensive. This requires the use of low cost precursors, and even more importantly, a simple route to synthesis of bulk quantities of the active materials. The ultimate metric for the value of a solar cell technology is the cost per watt of the modules. A considerable cost of manufacture of large area solar cells is tied up in materials cost, supply and materials use efficiency in cell manufacturing.

We discussed several donor materials and fullerene derivatives in the context of OPDs in Section 7.2.2, and in subsequent sections. Rather than revisit those molecules, many of which have also found use in efficient OPVs, here we focus on materials that have primarily been optimized for solar energy harvesting applications. As in the chapters concerning light emitting devices and transistors, the discussion is by no means comprehensive. Rather, it provides a sample of archetype materials and the principles used in their design. There are numerous reviews that treat these subjects more comprehensively to which the reader is directed (Dong et al., 2012, Burkhart and Thompson, 2014, Benten et al., 2016, Collins et al., 2017, Fu et al., 2018, Hou et al., 2018).

7.4.4.1 Donors

The design challenge of solution processed donors is to find structures that can readily crystallize, and yet are soluble in solutions also containing the acceptor molecules. These are somewhat conflicting requirements, since the tendency to crystallize requires structures with low solubility. Solubilization generally entails attachment of alkyl groups on the conjugated molecular backbone that sterically hinder π-stacking of neighboring molecules into a nanocrystalline morphology.

The design of most high performance donor and acceptor molecules is to symmetrically position alternating donor (d) and acceptor (a) units along the molecular backbone around a central core. The discussions in this and the following sections use lower case "d" and "a" to connote electron-donating and electron-accepting subgroups within a molecule. Thus the molecular designs typically follow an a-d-a, a-d′-d-d′-a, a-π-d-π-a, or similar architecture, as schematically illustrated in Fig. 7.158. Here, d′ is simply a different electron-donating unit than d, and π is representative of a linkage that extends the conjugation between the functional units comprising the molecule. It is also generally true that the π-linkages may have donor- or acceptor-like properties, but it is their function in creating an extended electron conjugation that dominates.

The advantage of this design is that molecules are combinations of "molecular modules" that simplifies

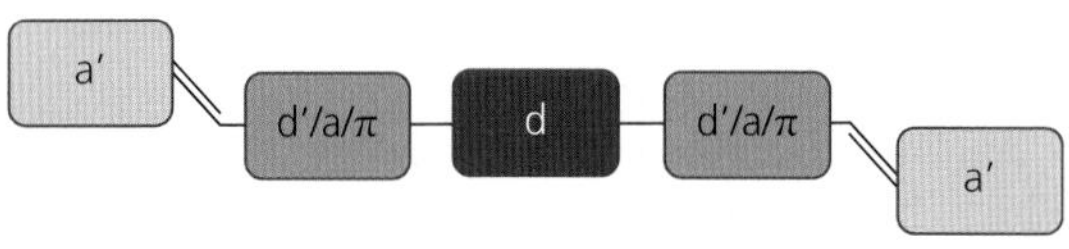

Figure 7.158 Modular, axisymmetric architecture commonly used in the design of high performance donors and non-fullerene acceptors. Here, "d" and "a" represent donor and acceptor units, and "π" is a conjugated bridge between units.

both their design and synthesis. Once the backbone structure is fixed, the acceptor end units are attached. This lends itself to an iterative approach to finding the structure best suited to solution processing into the desired morphology, and with the appropriate electronic properties (e.g. its HOMO energy and optical energy gap). The conjugated backbone has other advantages as well. It is generally rigid and planar, which allows the molecules to form closely packed conductive stacks within a BHJ. Furthermore, the extended electron system reduces barriers to intramolecular charge transfer (ICT) between electron rich and electron poor segments along the molecular framework. By distributing the electron density based on principles also used for TADF molecules in OLEDs (Section 6.3.4), the exciton binding energy is reduced. This in turn reduces the energy gap to enable long wavelength absorption. Finally, the extended conjugation reduces the driving force to generate free carriers at the D-A HJ, thus minimizing the energy loss, and hence increasing the V_{OC}.

Figure 7.159 provides the molecular structural formulae of several polycyclic monomeric d and a units commonly found in donor and acceptor molecules. The planar moieties forming the rigid conjugated backbone lead to ordered π-stacking arrangements. To ensure that the materials are soluble, extended alkane chains or other sterically bulky units are attached to either or both the donor and acceptor groups. The presence of the strategically positioned alkane does not necessarily hinder the ability of the materials to form nanocrystalline domains during deposition or post-annealing of the films.

With rare exception, almost all donor molecules and NFAs comprise a thiophene group. These multipurpose molecular units are used as connectors between d and a functional groups, and, as illustrated in Fig. 7.159, they serve as constituents in those same groups. They can also aid in planarizing the molecules to encourage π-stacking arrangements. As π-bridges, they can extend the conjugation length within a molecule, and are also helpful in controlling the morphology of the films. Ultimately, inclusion of thiophenes creates the molecular modularity that leads to simplified synthetic approaches to extended d–a structures.

Table 7.18 shows a compilation of several archetype polymer donors combined with fullerenes in BHJ OPVs. With only the exception of P3HT:$IC_{70}BA$ and PBnDT-DTffBT:$PC_{61}BM$ (1:1), the polymers are blended with $PC_{71}BM$ due to its broad absorption spectrum extending into the red. Since all the donors are paired with fullerenes, this table can be used for comparison

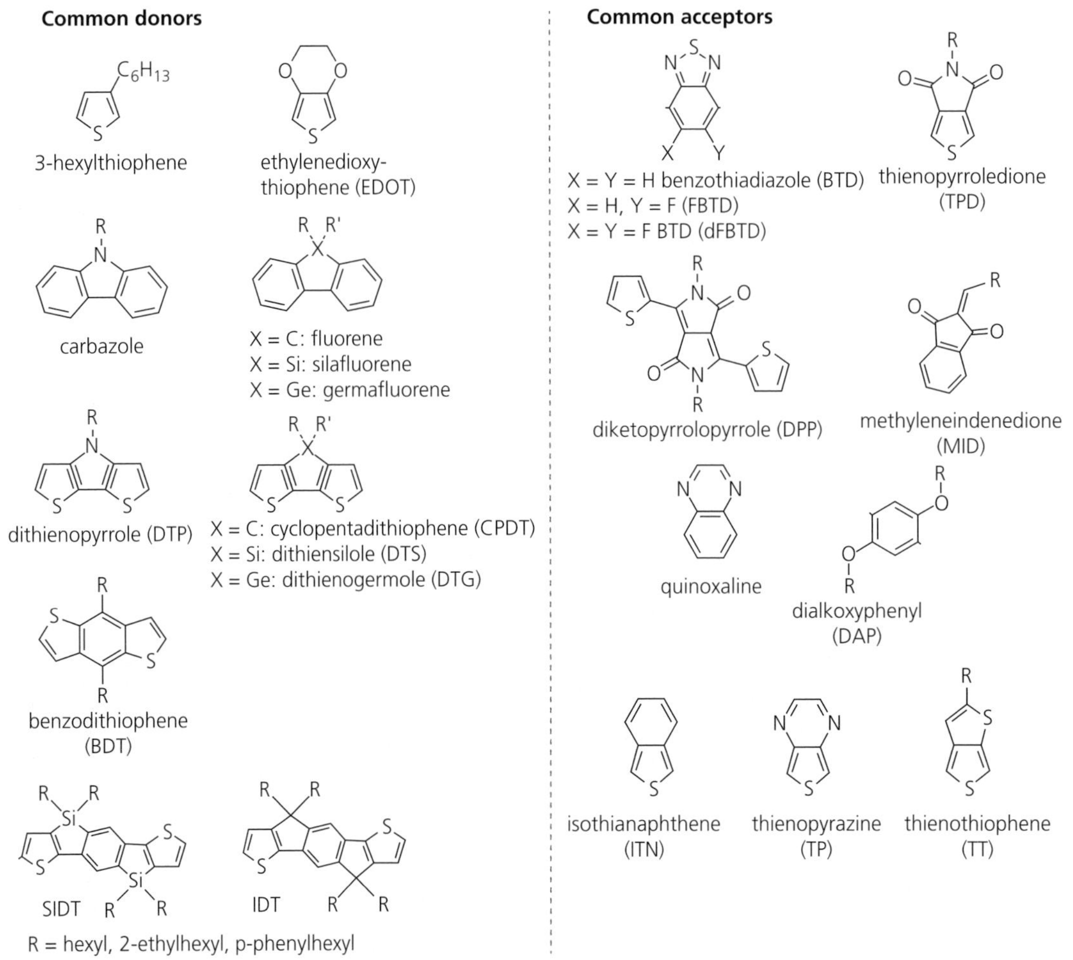

Figure 7.159 Structural formulae of monomer units commonly found in solution processed donor and acceptor compounds.

of their various characteristics, rather than on the specifics of different D-A combinations. With the exception of the homopolymer P3HT, the compounds are copolymers comprising alternating d/a linkages. P3HT is included here since it has found extensive use in OPV and OPD applications due to the relatively high V_{OC} achieved when combined with fullerenes, and its absorption into the red. Most significant is its ability to form nanocrystalline BHJ networks as discussed previously. The hexyl groups on each segment of the thiophene backbone can alternate regularly in a head-to-tail arrangement, or they can fall in a random head-to-head, tail-to-tail or head-to-tail pattern. We will show in Chapter 8 that the regular head-to-tail arrangements along all segments of the polymer backbone is known as *regioregularity*. This property allows for close and organized packing that leads to a high film mobility, and hence a high FF and j_{SC}. In contrast, the random arrangement results in a *regiorandom* polymer that is incapable of crystal formation.

An example of a strong acceptor moiety is benzothiadiazole (BTD) found in PBnDT-DTffBT, as well as in PCDTBT and the naphtho[1,2-c:5,6-c]bis [1,2,5]thiadiazole-containing PBDT-DTNT. Here, PCDTBT combines the BTD unit with a carbazole donor that increases E_{HOMO} to -5.5 eV compared to -5.25 eV for P3HT. This, in turn, increases V_{OC} by 40 meV. The PBnDT-DTffBT copolymer employs a difluorinated BTD unit that increases its electron-withdrawing properties to increase E_{HOMO} even further to -5.54 eV. Also, the film-forming properties of the latter molecule are inferior to PCDTBT as evidenced by the lower $FF = 0.61$ compared to 0.66 for the non-fluorinated species.

Another BTD-based polymer is PBDT-DTNT, which has the highest efficiency of 8.4%. In this molecule, the BTD is replaced by the fused ring naphthobisthiadiazole unit. This has a slightly stronger electron-withdrawing character than the single ring BTD, and its larger conjugated structure may

Table 7.18A Performance of selected polymer donors paired with fullerene acceptors

Polymer		Acc.	E_{HOMO} (eV)	E_{opt} (eV)	j_{SC} (mA/cm^2)	V_{OC} (V)	FF	$\eta_{P,max}$ (%)	Ref.
A	P3HT (1:1)	$IC_{70}BA$	−5.25	1.9	11.35	0.87	0.75	7.40	(Guo et al., 2012)
B	PCDTBT (1:4)	$PC_{71}BM$	−5.5	1.9	11.95	0.91	0.66	7.2	(Sun et al., 2011)
C	PBnDT-DTffBT (1:1)	$PC_{61}BM$	−5.54	1.7	12.91	0.91	0.61	7.2	(Zhou et al., 2011)
D	PBDT-DTNT (1:1.5)	$PC_{71}BM$	−5.19	1.6	17.4	0.75	0.61	8.6	(Yang et al., 2012b)
E	PBDTT-DPP (1:2)	$PC_{71}BM$	−5.35	1.44	14.0	0.73	0.65	6.6	(Dou et al., 2012)
F	PDTG-TPD (1:1.5)	$PC_{71}BM$	−5.6	1.69	14.4	0.86	0.69	8.5	(Amb et al., 2011, Small et al., 2012)
G	PIDT-phanQ (1:3)	$PC_{71}BM$	−5.28	1.65	11.41	0.88	0.66	6.63	(Zhang et al., 2011, Li et al., 2012)
H	PTB7 (1:1.5)	$PC_{71}BM$	−5.15	1.84	17.5	0.75	0.70	9.2	(He et al., 2012)
I	PBDTDTT-S-T (1:1)	$PC_{71}BM$	−5.04	1.65	16.35	0.69	0.66	7.81	(Huang et al., 2012)

Notes: (i) OPV performance characteristics are obtained at 1 sun intensity, AM1.5G illumination, except for PBDT-DTNT which as obtained at 0.95 sun intensity.
(ii) The optical energy gap obtained from extrapolation of the long wavelength absorption tail to its intercept with the energy axis. Extended weak absorption due to CT states is ignored.
(iii) The maximum (not average) power conversion efficiency for a population of OPVs, $\eta_{P,max}$, is quoted.
(iv) The j_{SC}, V_{OC}, and FF quoted for PBDT-DTNT and PBDTDTT-S-T are for the average device performance.

Table 7.18B Performance of selected polymer donors paired with fullerene acceptors

Polymer

A

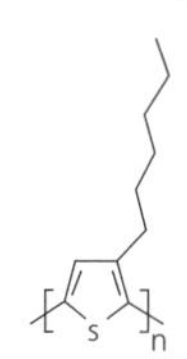

B

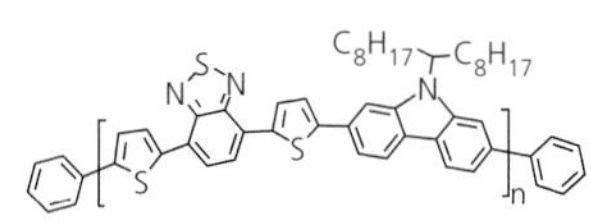

C

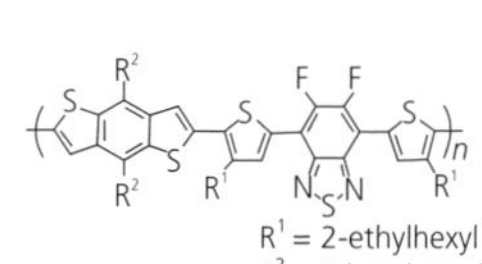

D

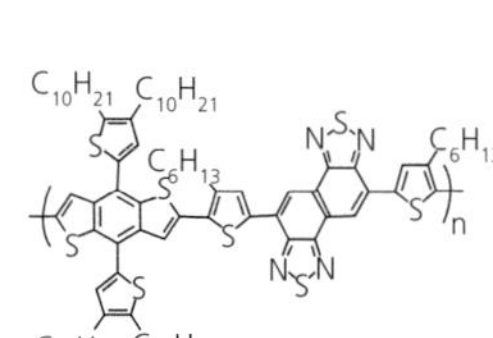

(*Continued*)

Table 7.18B Continued

Polymer	
E	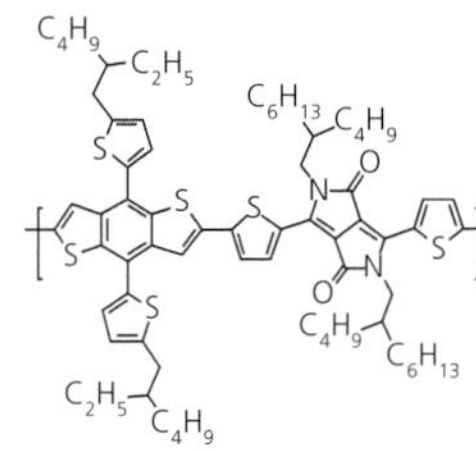
F	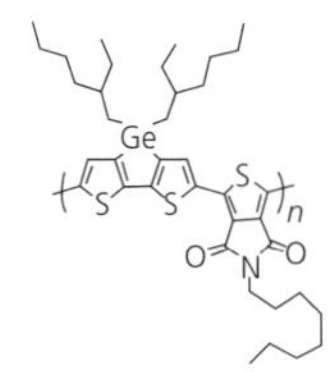
G	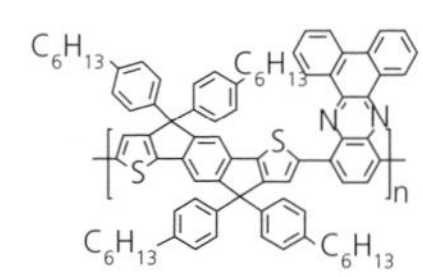
H	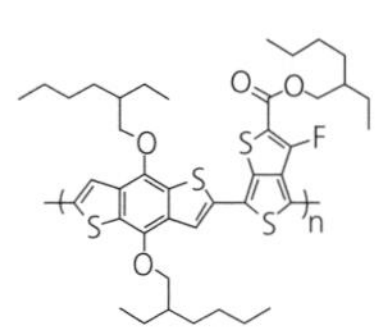
I	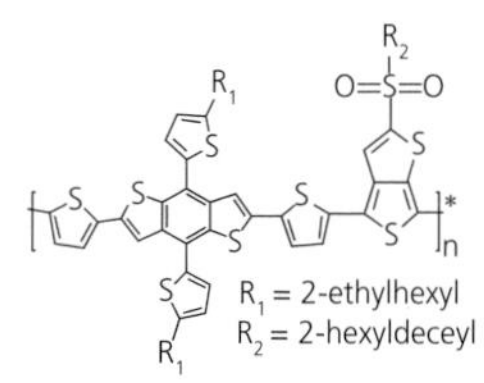

rigidify the molecular backbone to some degree. This leads to improved hole conductivity, as well as a decrease in the energy gap. Indeed, the small optical energy gap of 1.6 eV is primarily responsible for the comparatively large j_{SC} = 17.4 mA/cm^2 that compensates for the lower V_{OC} compared to the other BTD-based molecules. This example reinforces the importance of extending the absorption of molecules into the NIR to achieve high power conversion efficiencies.

Substituting thienopyrroledione (TPD) for BTD results in a somewhat weaker electron-withdrawing character in the next molecule in the series, PDTG-TPD. The molecule has high solubility provided by the alkane chains, while molecular packing is encouraged by the presence of the Ge bridge in the dithienogermole (DTG) unit. We note that a similar silole moiety has been used in place of DTG with similar results (Chu et al., 2012). The energy gap of PDTG-TPD is rather large at 1.69 eV due to the weak acceptor nature of the TPD relative to BTD. This is offset by the deep HOMO of −5.6 eV that leads to a relatively large V_{OC}. On balance, these several factors lead to a single junction efficiency of 8.5%.

PBDTT-DPP employs a diketopyrrolopyrrole (DPP) acceptor group linked to the benzodithiophene

(BDT) donor. While DPP is a strong electron acceptor, BDT is a comparatively weak donor, both of which act to lower the energy gap while maintaining a deep E_{HOMO}. In principle, this should result in a high short circuit current and large V_{OC}. The PDBTT-DPP exhibits a low energy gap of 1.44 eV (corresponding to $\lambda = 1.16$ μm), ultimately achieving an efficiency of 6.6%. Note that the ethylhexyl and butyloctyl groups may hinder the packing, resulting in a *FF* of only 0.65.

The next molecule in the family is a representative quinoxaline d/a copolymer, PIDT-phanQ. The quinoxaline is a very strong acceptor that, when copolymerized with the indacenodithiophene group, results in an optical energy gap of 1.65 eV. An alternative approach is to employ thienothiophene (TT) functionalized with a strongly electron-withdrawing fluorine molecule to convert the TT into a strong acceptor group, as found in PTB7. The F atom avoids steric hindrance to packing while at the same time it lowers E_{HOMO}. The result is the highest $\eta_P = 9.2\%$ of the compounds listed, which results from a combination of large $FF = 0.70$ and $j_{SC} = 17.5$ mA/cm^2, both indicative of a favorable BHJ morphology, and a moderate $V_{OC} = 0.75$ V.

The last donor polymer is another TT-based copolymer PBTDTTT-S-T containing the weak electron-withdrawing potential of TT strengthened by functionalization with a sulfonyl group. Unfortunately, the sulfonyl results in disordered packing, but this deficiency is rectified by including the thiophene side groups to the BDT donor that enforce the π-stacking arrangement needed for a high hole mobility. The resulting OPV has $\eta_P = 7.81\%$, along with a moderate energy gap of 1.65 eV that allows for absorption into the NIR.

The molecules in Table 7.18 are representative of classes of some of the most efficient polymer donors reported in OPVs. We have seen that the copolymers share common properties of combining donor and acceptor units along the backbone to reduce exciton binding, thus promoting efficient charge transfer to the acceptor molecule. In addition, the large alkyl chains appended to either one or both of the d and a groups engender a high solubility. A plethora of modifications have been engineered that are analogous to these structures that optimize the film forming or energetic properties to achieve even higher performance than reported here. One of the obvious and recurring benefits of organic electronics exemplified by this list of compounds is the nearly limitless space available for optimization that is possible by iterative optimization of chemical structures.

The design of small molecules follows that of polymers, with the same combinations of d and a groups defined in Fig. 7.159. A sample of such small molecule donors is provided in Table 7.19. The combination of multiple d, a and π units results in extended molecular structures that cannot be deposited via thermal evaporation, which is unlike the case of many of the donors discussed in the context of OLEDs and OPDs. Nevertheless, solution processed small molecules can also result in high efficiency if the appropriate post-deposition annealing processes are employed.

The first example donor in Table 7.19 is an oligomer comprising five linked thiophenes, DRCN5T. The acceptor groups are based on rhodanine that is made into a strong electron- withdrawing moiety by its CN appendages. Interestingly, it was found that the series of compounds, DRCN*n*T have improved morphologies when *n* is an odd number of thiophenes, presumably due to closer π–π stacking of thiophene rings on neighboring molecules (Kan et al., 2015). In addition, the HOMO energies of this series increases monotonically, from −5.34 eV for $n = 4$ to −5.02 eV for $n = 8$. The optimum OPV performance is obtained by creating a nanocrystalline network through thermal annealing followed by SVA in CF. A maximum power conversion efficiency of 10.1% is achieved even though the optical gap is comparatively large at 1.60 eV. Indeed, with only the exception of the porphyrin-based donor, DPPEZnP-TEH, the optical gaps of these molecules lie between 1.6 eV and 1.8 eV, ultimately limiting the efficiencies attained.

The next two molecules, DR3T-SBDT and DR3T-BDTT, contain BDT donor groups, flanked on both sides by three axisymmetric thiophenes, and capped by rhodanine end groups lacking the strong electron-withdrawing cyano moieties of DRCN5T. The difference between the DR3T- structures is the additional thiophene of DR3T-BDTT that replaces the S terminations in DR3T-SBDT to increase the propensity to π-stack. The fill factor of this molecule is 0.76 compared with 0.73 for DR3T-SBDT due to improved hole conductivity. As in the case of DRCN5T, DR3T-SBDT also underwent thermal and solvent annealing to reach its highest efficiency of 9.95%.

The molecule p-DTG(FBT_2Th_2)$_2$ has a dithienogermole donor flanked by strongly electron-withdrawing fluorinated BTD (FBTD in Fig. 7.159) groups bridged by thiols to form an extended d′-a-

Table 7.19 Performance of selected solution-processed small molecule donors paired with $PC_{71}BM$ acceptors

Donor		E_{HOMO} (eV)	E_{opt} (eV)	j_{SC} (mA/cm^2)	V_{OC} (V)	*FF*	η_{max} (%)	Ref.
A	DRCN5T (1:0.8)	−5.22	1.60	15.66	0.92	0.68	10.1	(Kan et al., 2015)
B	DR3T-SBDT (1:0.8)	−5.07	1.62	14.45	0.91	0.73	9.95	(Kan et al., 2014)
C	DR3T-BDTT (1:0.8)	−5.51	1.77	14.21	0.89	0.76	9.58	(Li et al., 2015)
D	p-DTG $(FBT_2Th_2)_2$ (1:1)	−5.17	1.62	15.7	0.78	0.73	9.1	(Gupta et al., 2016)
E	BIT6F (1:3)	−5.32	1.79	13.39	0.89	0.75	9.09	(Wang et al., 2016)
F	OPD2FBT2-1-MID (1:0.7)	−5.28	1.75	13.5	0.94	0.71	9.25	(Yuan et al., 2016)
G	NDPPFBT (1:2)	−5.25	1.56	10.71	0.88	0.72	7.00	(Wang et al., 2015)
H	DPPEZnP-THE (1:1.2)	−5.14	1.37	16.76	0.78	0.62	8.08	(Gao et al., 2015)

Notes:

(i) OPV characteristics obtained at 1 sun intensity, AM1.5G illumination.
(ii) The maximum (not average) power conversion efficiency for a population of OPVs $\eta_{P,max}$ is quoted.
(iii) Blend ratios correspond to (D:A).

Table 7.19B Performance of selected solution-processed small molecule donors paired with $PC_{71}BM$ acceptors

Donor	
A	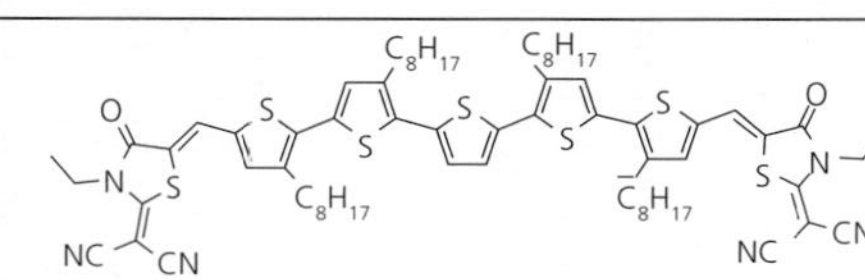
B	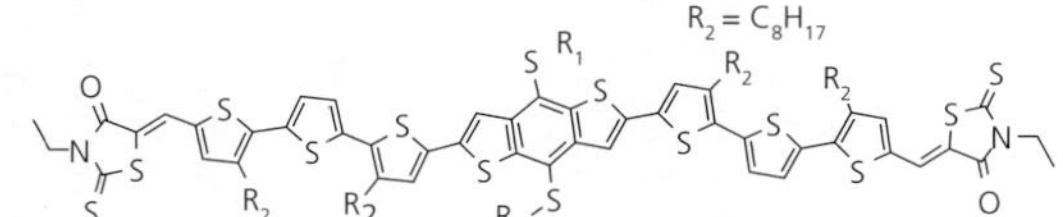
C	R₁ = 2-ethylhexyl R₂ = C₈H₁₇ 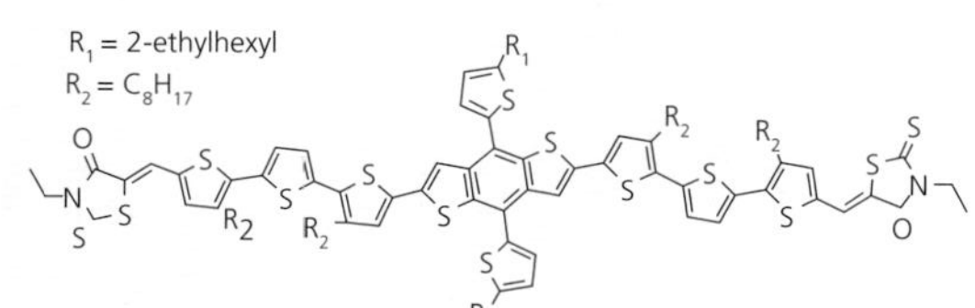

(Continued)

Table 7.19B Continued

Donor	
D	R_1 = 2-ethylhexyl R_2 = C_6H_{17} 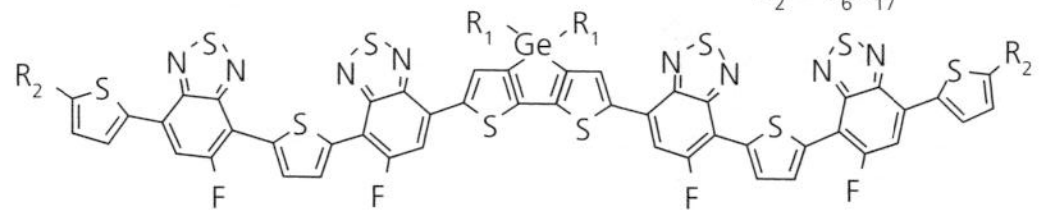
E	R_1 = 2-ethylhexyl R_2 = C_6H_{13} 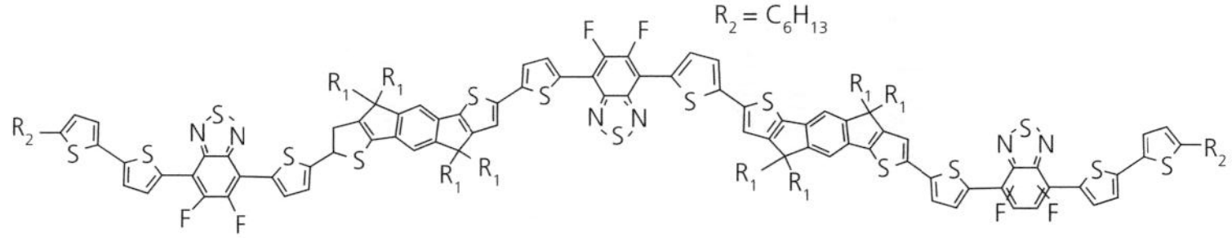
F	R = 2-butyloctyl 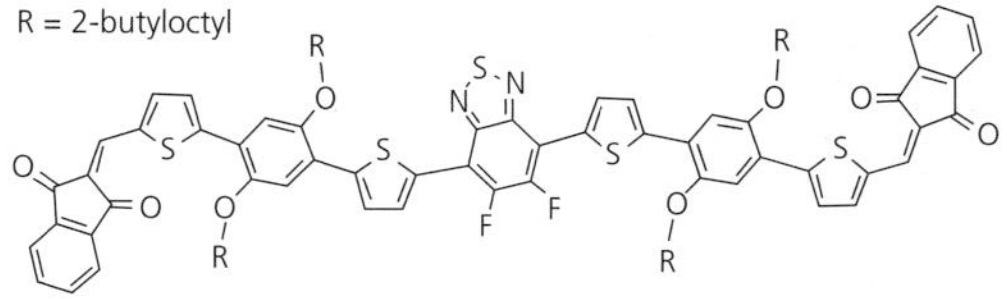
G	R_1 = 2-ethylhexyl R_2 = C_6H_{13} 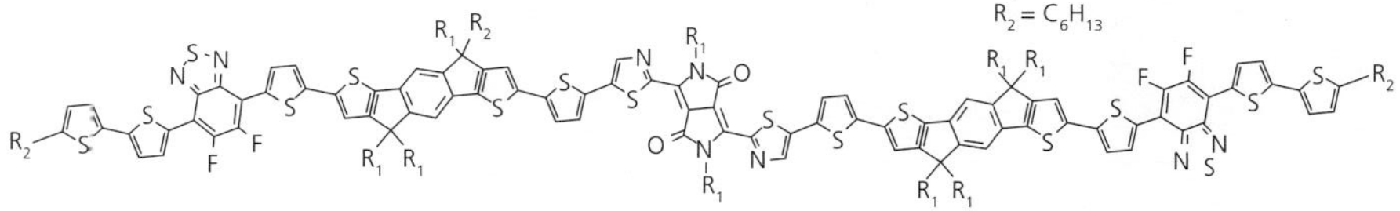
H	R =2-ethylhexyl 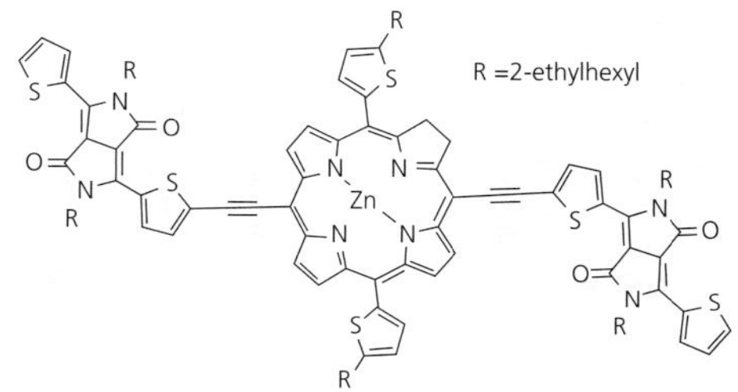

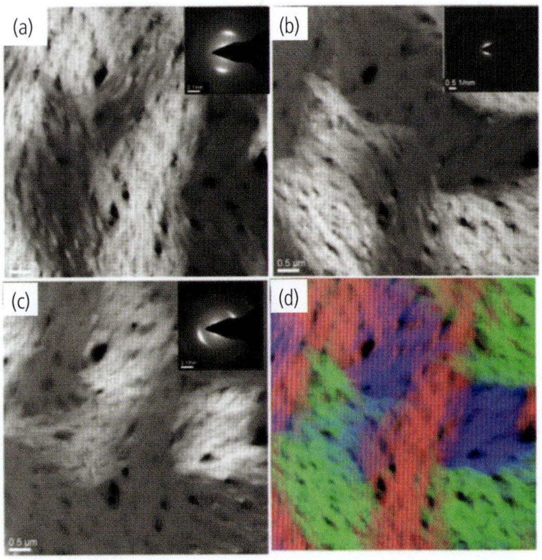

Figure 7.160 (a–c) Dark field electron micrographs of a p-DTG(FBT_2Th_2)$_2$ thin film taken at angles of (a) 60°, (b) 120°, and (c) 180° showing a network of partially oriented nanocrystalline fibrils at each angle, corresponding to different depths within the film. (d) A false color composite image of (a–c) to illustrate the orientation of the nanocrystals vs. film depth. Insets: Selected area electron diffraction (SAED) patterns of the films. The two diffraction intensity rings correspond to a *d*-spacing of ~3.5 Å (Gupta et al., 2016).

Reproduced by permission of The Royal Society of Chemistry.

d′-a-d-a-d′-a-d′ network where d = DTG, a = thiophene (Th), and d′ = FBDT. This molecule was spin-coated with $PC_{71}BM$ from a CB solution with 5% CF and 0.4% DIO as an additive. The film was thermally annealed to achieve an efficiency of 9.1%. Even with this extended structure, $FF = 0.73$ and $j_{SC} = 15.7$ mA/cm^2 are competitive with the more compact structures in Table 7.19.

The remarkably lacey nanostructure formed by films of p-DTG(FBT_2Th_2)$_2$ are shown by the dark field TEM images taken at different angles (60°, 120°, and 180°) in Fig. 7.160a–c, with the composite image shown in Fig. 7.161d. The SAED patterns for these images in the insets indicate an interplanar spacing of 3.5 Å. The composite image shows a mat with fibrils oriented along various preferred directions, depending on their depth from the surface.

An even more extended structure is that of BIT6F, comprising a central diflourobenzothiadiazole (dFBTD) acceptor core symmetrically flanked by Th and indacenodithiophene (IDT) groups to form a molecule with an optical energy gap of nearly 1.8 eV. As in the case of DRCN*x*T, molecules within the series of BIT*x*F exhibited progressively larger HOMO energies of −5.33 eV for $x = 4$, to −5.28 eV for $x = 10$. Similarly, the optical energy gaps decreased from 1.81 to 1.77 eV for this same range of the number of fluorine atoms. No significant differences in energies were observed for $x > 10$, suggesting that this number of fluorine atoms did not result in a further extension of the conjugation length of the molecule. This may be limited by twists and bends that accumulate with molecular length. By combining thermal annealing and SVA, significant film ordering is achieved.

OPD2FBT2-1-MID has a dFBTD core surrounded by dialkoxyphenyl (DAP) acceptors, and capped by a methyleneidenedione (MID) acceptor unit. Four axisymmetrically disposed thiophene groups provide the donor character along the backbone. Even for such an extended oligomeric backbone, DFT calculations show that the molecule is remarkably flat, allowing for aggregation in closely spaced π–π stacks. Ordered films leading to high efficiency OPVs after SVA in CF are obtained. The use of the 0.8% 2-chlorophenyl additive mitigates the lack of molecular mobility due to its large size, allowing the compounds to organize during annealing.

The most spatially extended oligomer, NDPPFBT, has an axisymmetric d′-a′-d-a-d-a′-d′ structure, where d′ = dithiophene, a′ = dFBTD, d = IDT linked by a Th group to a′, and the central acceptor a = diketopyrrolopyrrole (DPP). Its extensive conjugation length coupled with a strong acceptor core moiety leads to reduced energy loss, and consequently a high $V_{OC} = 0.88$ V (Wang et al., 2015). Smooth films are achieved by SVA in tetrahydrofuran for 60 s. It is among the longest wavelength absorbing donors of those in Table 7.19, with a cutoff at $\lambda = 800$ nm.

A substantially different approach to achieving ICT is illustrated by DPPEZnP-TEH that employs a Zn porphyrin donor as the d-group, linked to two DPP acceptors via alkyne groups in an a-d-a architecture. We have shown in Section 7.2 that porphyrins (and their phthalocyanine analogs) are stable donor molecules used in bilayer and mixed OPDs. In DPPEZnP-TEH, the ZnP unit results in the smallest energy gap (1.37 eV, with a wavelength cutoff of ~900 nm) of the compounds in Table 7.19, making this a candidate for NIR absorption. Use in an OPV gives the highest short circuit current of 16.76 mA/cm^2. Unfortunately, the molecular design raises the HOMO to −5.14 eV, which in turn, reduces V_{OC}. Finally, the strong tendency to form aggregates that is promoted by the discotic porphyrin structure may ultimately limit *FF* to 0.62. To achieve the desired morphology, the blend films with $PC_{71}BM$ were thermally annealed in the presence of a 1% pyridine additive. In an OPV employing a PFN blocking layer, $\eta_{max} = 8.08\%$ was achieved using this donor.

Up to this point, we have only considered solution processed small molecule and planar donors that feature ICT between d and a groups using the building blocks in Fig. 7.159. These molecules contain thiophene, or thiol-containing groups to enhance planarity and intermolecular π-stacking. However, we have previously shown that the SQ compound, DPSQ, can also achieve a morphology that leads to nearly ideal structures for charge transfer at the HJ. It is instructive to consider the generalized class of functionalized SQ donors to provide deeper insights into how molecular structure drives thin film morphology, and ultimately device performance.

SQ dyes are intensely fluorescent, and likewise have intense absorption in the red and NIR. They feature a four-membered, electron deficient aromatic ring, and are readily functionalized with aryl and alkane groups to modify their film forming and optical properties. The tendency of the end groups to twist out of the SQ plane sterically hinders the molecules from forming ordered domains. Hence, initial work with SQ-based dyes resulted in low $\eta_P \sim 1.2$–2%, which resulted from a low $FF \sim 0.35$ and $V_{OC} < 0.6$ V when blended with $PC_{61}BM$ (Silvestri et al., 2008, Bagnis et al., 2010). Nevertheless, their ease of functionalization has led to the development of a series of *N,N*-diarylanilionsquaraine derivatives with end groups that drive their π–π stacking, and hence in increase *FF* in fullerene-based OPVs (Wang et al., 2011). Example functionalized squaraines (*f*SQs) synthesized to determine the link between molecular structure, film morphology, and OPV performance are shown in Fig. 7.161 (Wei et al., 2011b). There are three species of functionalized molecules: the "parent" SQ that features the sterically limiting butyl groups; three axisymmetric SQs, namely 1-NPSQ with naphthyl plus aryl functional units, DPSQ with an arylamine, and PSQ with symmetric propylamino groups; and the asymmetric molecules, DPASQ and ASSQ, functionalized with either butyl or amine units.

The butyl end-groups of the parent SQ sterically limit close packing with adjacent molecules, that in turn limits the film hole conductivity, and exciton diffusion length to only $L_D = 1.6$ nm. This is compared with the asymmetric SQs with $L_D = 11$ nm and the symmetric compounds with $L_D \sim 3$ nm. Replacement of the end groups with planar aryl moieties exerts control over the crystalline morphology by influencing the intermolecular contact distance while tuning the optical absorption spectrum and energy levels of the films. The arylamine substitutions enhance π-stacking, which results in an increased thin film density. Table 7.20 shows that the film densities

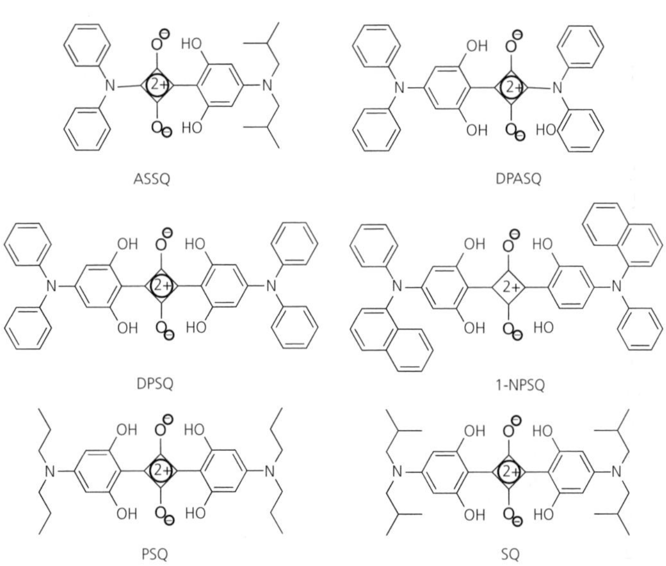

Figure 7.161 Molecular structural formula of six functionalized squaraines.

vary from 1.45 g/cm^3 for DPASQ, to 1.27 g/cm^3 for ASSQ. The asymmetric compounds also have considerably larger optical energy gaps of 1.9–2.1 eV compared to the symmetric *f*SQs whose energy gaps all lie between 1.5 and 1.6 eV, making the latter suitable for detection into the NIR.

The central SQ unit, combined with the symmetrically disposed functional end groups, creates a tendency for the films to crystallize. With the exception of PSQ, all *f*SQs form amorphous films when cast from CF solutions, yet when they are thermally annealed between 70°C and 90°C, the symmetric *f*SQs readily crystallize, which is apparent from the SAED patterns for 1-NPSQ, DPSQ, and SQ in Fig. 7.162. However, the asymmetric compounds do not crystallize due to their lack of inversion symmetry, although they may nevertheless form small domains that are not found in the unannealed films. The case of PSQ differs, in that it shows weak crystallinity both before and after annealing (Fig. 7.163b and e), whereas DPSQ shows the most extensive order apparent from by the well-defined diffraction spots in Fig. 7.162d.

Bilayer HJ OPVs comprising the structure: ITO/MoO_3 (8 nm)/*f*SQ (8.5 nm)/C_{60}(40 nm)/PTCBI (8 nm)/Ag (100 nm) were fabricated, with characteristics listed in Table 7.20. The *FF* scales monotonically with film density, ρ, as shown in Fig. 7.163. The highest $FF = 0.73$ and $\rho = 1.45$ g/cm^2 for DPASQ, falls to 0.63 and 1.27 g/cm^2 for ASSQ. This trend indicates that the intermolecular stacking distance is reduced by the arylamine functional appendages, which in turn

Table 7.20 Physical properties of *f*SQ films, and optoelectronic characteristics of *f*SQ/C_{60} OPVs (Wei et al., 2011b)

Molecule	E_{HOMO}[a] (eV)	E_{opt} (eV)	ρ(g/cm^3)	V_{oc}[b] (V)	j_{SC} (mA/cm^2)	*FF*	η_P (%)	R_{ser} (Ω cm^2)
DPASQ	5.4	1.9	1.45	1.0	5.5	0.73	4.0	3.8
DPSQ	5.3	1.5	1.39	0.91	7.2	0.70	4.8	1.7
1-NPSQ	5.3	1.6	1.35	0.90	6.9	0.71	4.3	9.3
PSQ	5.2	1.5	1.34	0.68	9.5	0.69	4.6	3.6
SQ	5.1	1.6	1.29	0.79	8.0	0.68	4.4	4.7
ASSQ	5.3	2.1	1.27	0.92	6.3	0.63	3.7	41.2

[a] Measured by ultraviolet photoelectron spectroscopy.
[b] All OPV characteristics measured under 1 sun, AM1.5G simulated illumination. All films used in OPVs were thermally annealed at $T = 70$–90°C.

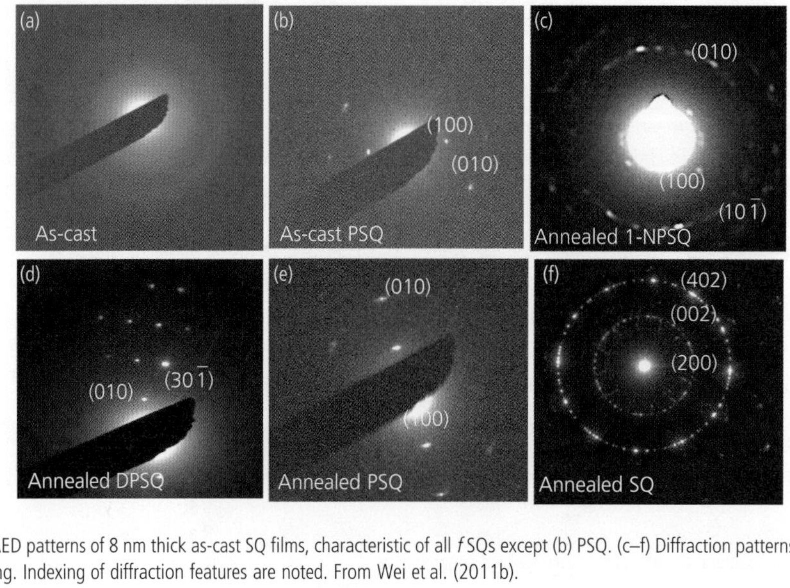

Figure 7.162 (a) SAED patterns of 8 nm thick as-cast SQ films, characteristic of all *f*SQs except (b) PSQ. (c–f) Diffraction patterns of representative *f*SQs after thermal annealing. Indexing of diffraction features are noted. From Wei et al. (2011b).

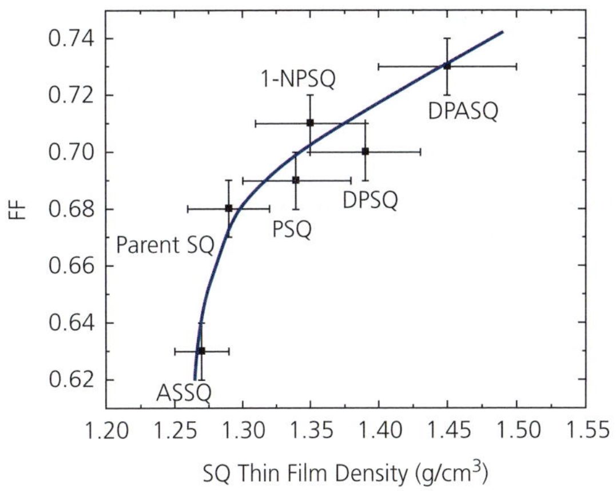

Figure 7.163 Relationship between FF and film density for the *f*SQ family of compounds (Wei et al., 2011b).

Reprinted with permission from Wei, G., Xiao, X., Wang, S., Sun, K., Bergemann, K. J., Thompson, M. E. & Forrest, S. R. 2011b. Functionalized squaraine donors for nanocrystalline organic photovoltaics. ACS Nano, 6, 972-978. Copyright 2011 by the American Chemical Society.

results in improved molecular density, and ultimately in a higher hole mobility (which is indirectly inferred from the increased *FF*). The highest performance ($\eta_P =$ 4.8%) is achieved for the symmetric DPSQ with its two phenyls on each end of the SQ core. This is higher than for DPASQ (4.0%) although it has a higher ρ and *FF*, but at the cost of a lower j_{SC} arising from a larger optical energy gap.

Currently, there are fewer vapor-deposited small molecular weight donors yielding OPVs with efficiency as high as the solution-processed donors in Tables 7.18 and 7.19. A notable exception are the highly dipolar d-a-a′ molecules comprising a strong electron donating ditolylaminothienyl group connected to electron-withdrawing dicyanovinylene via a second acceptor, BTD (Chen et al., 2012b). These molecules have ground state dipole moments of 10–15 D, causing them to crystallize with π–π stacking distances of 3.4–3.5 Å. Their intermolecular dipole attraction is sufficiently strong that annealing does not improve order in the blends comprising d-a-a′ structures (Che et al., 2018a). Two isomers of asymmetric heterotetracenes with different thienyl ring arrangements, antiBTDC and synBTDC, are shown in Fig. 7.164a. These molecules employ a terminal benzene ring fused with the electron-donating dithieno[3,2-b:2′,3′-d]pyrrole (DTP) donor unit to yield a coplanar structure.

The electronic density plots in Fig. 7.164b show the HOMOs are delocalized over the molecular backbone while the LUMOs are localized on the electron-deficient benzothiadiazole-dicyanovinylene fragments. The lowest energy transitions of antiBTDC

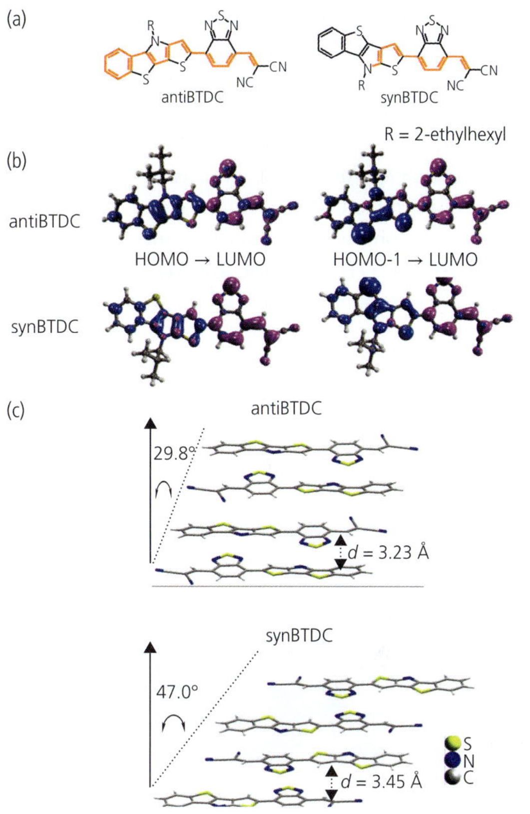

Figure 7.164 (a) Molecular structural formulae of antiBTDC and synBTDC. The colored paths trace the molecular orbital conjugation. Note the longer conjugation length of antiBTDC compared with synBTDC. (b) Charge densities of the lowest energy singlet excitations, $S_1 \leftarrow S_0$. Blue denotes occupied and purple symbolizes unoccupied molecular orbitals. (c) Crystal packing configurations of antiBTDC and synBTDC. Side chains are omitted for clarity (Che et al., 2016).

and synBTDC comprise both π–π* and ICT character that ultimately reduce the energy loss when combined with C_{70} in a mixed HJ. The MO spatial overlap for $S_1 \leftarrow S_0$ is calculated using DFT to be 43% and 27% for antiBTDC and synBTDC, respectively. This reflects the longer electron conjugation of antiBTDC, indicated by the colored bonds in Fig. 7.164a. The internal dipole moments of the ground and excited states are μ_g = 15.42 D and μ_e = 16.48 D, respectively, for antiBTDC, compared with synBTDC where μ_g = 14.84 D and μ_e = 13.77 D.

The large ground state dipole moments lead to an antiparallel molecular arrangement along the long backbone axes, as shown by the crystal packing configurations in Figs. 7.164c. The smaller intermolecular distance, enforced by the larger dipole moment and the longer conjugation length of antiBTDC results in improved characteristics when the isomers are employed in OPV cells whose structure is shown in Fig. 7.165, inset. The longer conjugation length results in absorption considerably deeper into the NIR, with a cutoff wavelength of 800 nm compared with 750 nm for synBTDC.

The characteristics of the devices employing these isomers in mixed HJs with C_{70} are provided in Table 7.21, along with the performance of analogous d-a-a′ donors, DTDCPB and iBuBTDC. Among this group of compounds, DTDCPB has the highest $\eta_P = 9.6\%$, largely as a result of its close interplanar spacing of ca. 3.5 Å (Chen et al., 2012b) that leads to a high fill factor of 0.67 in the blended films. The advantageous performance characteristics of DTDCPB has led to its use in tandem cells discussed in Section 7.5.

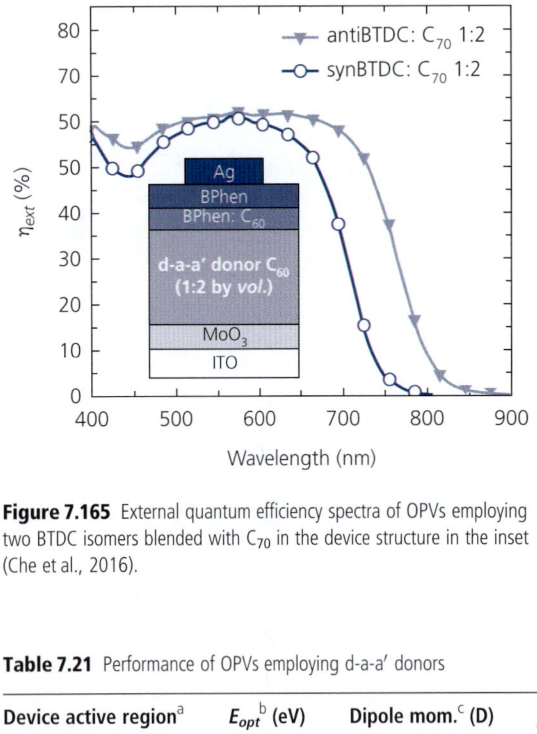

Figure 7.165 External quantum efficiency spectra of OPVs employing two BTDC isomers blended with C_{70} in the device structure in the inset (Che et al., 2016).

Two-dimensional (2D) GIXD reveals the morphologies of the thin films and provides an understanding the origin of the high *FF* of these dipolar donors. In Figs. 7.166a and b, we compare the GIXD diffraction patterns of iBuBTDC, antiBTDC, and C_{70}. The strongest diffraction signal in the in-plane direction (corresponding to high diffraction intensity shown by the yellow regions at large q_{xy} and low q_z scattering parameters) with almost no out-of-plane component (indicated by the low intensity broad blue pattern in q_z and low q_{xy}) is observed for antiBTDC, suggesting a preferred edge-on orientation. In contrast, iBuBTDC shows an intense signal at all q_z. The higher ratio of face-on π–π stacking in the iBuBTDC film leads to efficient charge transport perpendicular to the substrate, resulting in the highest *FF* compared to antiBTDC (see Table 7.21). The diffraction patterns of the d-a-a′ species differ considerably from C_{70}, which shows maxima due to scattering from several different molecular planes, consistent with its tendency to crystallize into an organized fcc lattice. The line-cut patterns shown in Fig. 7.166b emphasize the in-plane and out-of-plane stacking habits of these three compounds. The higher intensities correspond to higher organization along a given film axis.

An interesting series of vacuum deposited donors is based on a conjugated backbone comprising oligothiophenes attached at the β-position, and symmetrically capped with strongly electron-withdrawing dicyanovinyl (DCV) groups, shown in the inset, Fig. 7.167b. These DCV*n*T compounds reveal the relationship between the length of the conjugated thiophene oligomer and the HOMO and LUMO energies. The peak molar extinction

Table 7.21 Performance of OPVs employing d-a-a′ donors

Device active region[a]	E_{opt}[b] (eV)	Dipole mom.[c] (D)	j_{SC}[d] (mA/cm^2)	V_{OC} (V)	*FF*	η_P (%)	Ref.
iBuBTDC:C_{70} (1:3, 80 nm)	1.53	>10	16.5	0.94	0.60	9.3	(Che et al., 2018a)
antiBTDC:C_{70} (1:2, 60 nm)	1.55	15.4	14.2	0.91	0.56	7.2	(Che et al., 2016)
synBTDC:C_{70} (1:2, 50 nm)	1.65	14.8	11.6	1.01	0.52	6.1	(Che et al., 2016)
DTDCPB:C_{70} (1:1, 80 nm)	1.75	12.0	15.8	0.92	0.67	9.6	(Griffith et al., 2015)

[a] Cathode buffers employed Bphen:C_{60} (1:1) electron filtering layers.
[b] Obtained from the long wavelength absorption cutoff.
[c] Ground state dipole moments obtained from DFT calculations.
[d] Electrical characteristics measured at 1 sun, AM1.5G simulated illumination.

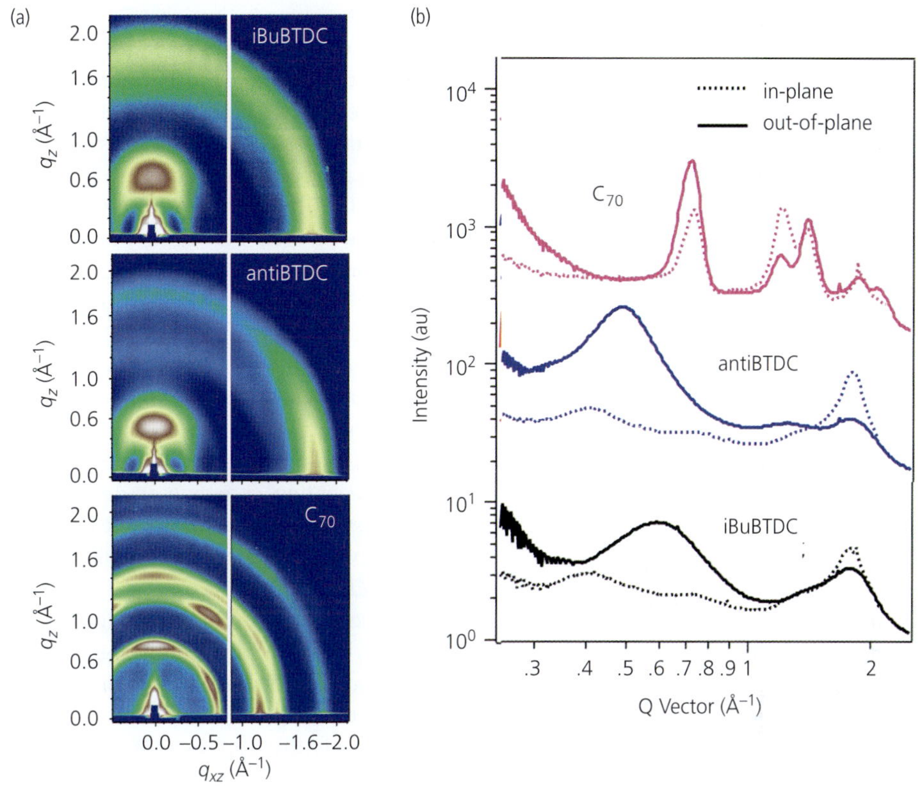

Figure 7.166 (a) 2D grazing incidence X-ray diffraction (GIXD) scattering patterns of vacuum deposited donor and acceptor thin films. Blue = low intensity, yellow-red = high intensity. (b) Corresponding intensity profiles along the lines in (a) (Che et al., 2018a).

coefficients of compounds $n = 1$–5 (Fig. 7.167a) monotonically shift to longer wavelengths with an increasing number of thiophenes. The spectra lose their articulated vibronic features with increasing n, presumably due to increasing torsional disorder with molecular length. This suggests that increasing n to even larger numbers of thiophene units will result in ever smaller spectral shifts, since bends and kinks in the chains ultimately limit the extent of the excited state supported by the molecules (cf. Section 3.5.3).

The increased conjugation length with n is apparent in the HOMO and LUMO energies obtained via cyclic voltammetry, as shown in Fig. 7.167b. The high E_{LUMO} results from attachment of the strongly electron-withdrawing DCV units, which is advantageous for increasing V_{OC} in OPVs using a fullerene acceptor. The figure shows a monotonic decrease in E_{HOMO} with n, along with a concomitant decrease in E_G as the conjugation length is increased.

This molecular series illustrates the systematic relationship between molecular structure and energetics. The highest efficiency p-i-n OPVs employ DCV5T with an active region thickness of 40 nm. The operating characteristics under approximately 1 sun intensity, AM1.5G illumination are $j_{SC} = 11.1$ mA/cm^2, $V_{OC} = 09.7$, $FF = 0.49$ leading to $\eta_P = 5.2\%$ (Fitzner et al., 2011).

7.4.4.2 Non-fullerene acceptors

The earliest bilayer cell employed the non-fullerene, perylene-based imidazole, PTCBI, as the acceptor (Tang, 1986). The limited efficiency resulting from $E_{loss} = E_{opt} - qV_{OC} > 1$ eV (i.e. the difference between the exciton energy and qV_{OC}) led to their eventual replacement by fullerenes (Yu et al., 1995, Peumans and Forrest, 2001). Nevertheless, perylenediimides (PDIs) of which PTCBI is an example, have several properties that are advantageous for their use as acceptors. Primarily, their large LUMO energies allow them to be paired with a wide range of donor molecules. Also, their planar, extended conjugated electron systems promote close π-stacking, giving a high electron mobility. Contravening the advantage of

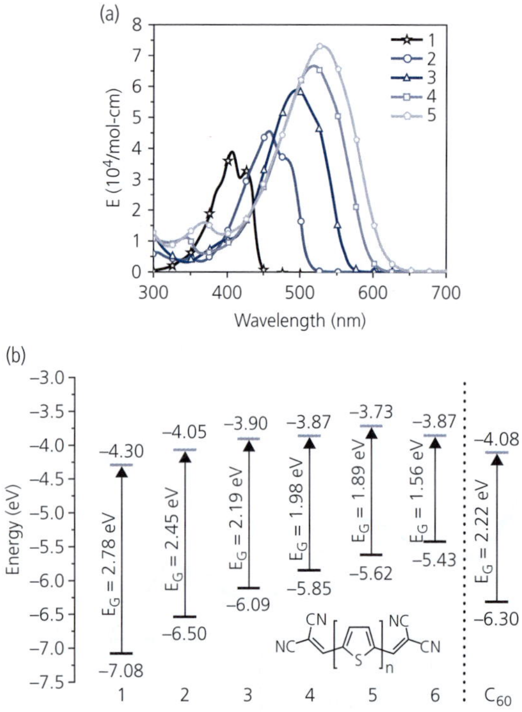

Figure 7.167 (a) Molar extinction coefficient vs. wavelength for DCV*n*T donors dissolved in DCM, where $n = 1$–5. Compound $n = 6$ is absent due to its low solubility in DCM. (b) HOMO and LUMO energies for DCV*n*T donors whose molecular structural formulae are shown in the inset. The energies of C_{60} are shown for reference (Fitzner et al., 2011).

stacking is their tendency to aggregate. This prevents high efficiency in mixed or bulk HJ structures, where homogeneous blending with the acceptor molecule is required. Also, as shown in Chapter 3, close packing leads to strong coupling of the excited states with the intermolecular phonon population, which in turn can lead to rapid non-radiative recombination, and hence a reduced diffusion length.

To take advantage of the benefits presented by PDIs while minimizing their disadvantages, a large family of solution-processable PDI-based oligomers have been introduced (Li et al., 2017b), with some of the highest efficiency materials paired with polymer donors listed in Table 7.22. These PDIs have a very low solubility, which can be enhanced by attaching alkanes to the peripheral imide groups.

The energy loss provides a measure of the efficiency for charge transfer from A to D. In the table, E_{opt} is listed for the acceptor in the OPV. The E_{loss} of these molecules is nearly 1 eV, with the notable exception of SF-PDI$_2$ comprising two PDIs linked by a spirobifluorene-2,7-diyl at the bay positions (Yan et al., 2013). We introduced the P3TEA:SF-PDI$_2$ D-A system in Section 4.7.2 due to the apparent absence of energy loss in transferring from the exciton to the CT state. Furthermore, the electroluminescence efficiency of the PDI is $\sim 0.5 \times 10^{-4}$, resulting in a voltage loss of 0.26 V. This, combined with 3 ps electron transfer time results in a high $V_{OC} = 1.11$ eV (Liu et al., 2016a).

The Se-bridged PDI dimer, SdiPBI-Se, exhibits a maximum power conversion efficiency of 8.42%, which is in part due to the large $FF = 0.70$. Including Se in the bay positions of the PDIs has several effects. Selenium is a large atom that allows for facile ionization, thus raising the LUMO of the molecule and enhancing its electron-accepting character. It introduces a twist between the perylene bisimide (PBI) pair, preventing excessive aggregation. It may also enhance the electron mobility due to improved orbital overlap with neighboring molecules, thus leading to a higher *FF*.

The next two compounds are the trimers, TPH and TPH-Se. These triperylene hexaimides form tridentate structures with their rylene units severely twisted out of plane, preventing aggregation. Here, rylene generally applies to a dye compound comprising naphthalene units linked at their peri-positions, resulting in compounds such as perylenes, naphthalene diimides, and so forth. The large conjugated systems on the so-called propeller units promote strong π–π interactions between molecules that result in a high *FF* and j_{SC}. As in SdiPDI-Se, the Se-annulated TPH-Se has a higher LUMO energy than TPH, which is inferred from its reduced E_{HOMO}, but with a similar energy gap of 1.93 eV. As a result, the efficiency is improved from 8.28% for TPH to 9.28% for TPH-Se.

The last PDI in Table 7.22 is an electron-withdrawing tetramer linked by a benzodithiophene, electron-rich core. The molecule, TPB, has the conventional a-d-a architecture that leads to a reduced exciton binding energy. This is reflected in a low $E_{loss} = 0.79$ eV. It also has the lowest optical energy gap of the compounds listed, possibly due to the extended electron conjugation of the tetramer. However, this same attribute leads to a reduced $FF = 0.58$, which ultimately leads to a lower $\eta_P = 8.47\%$.

Many if not all the compounds in Table 7.22, and in other PDI-based OPVs, lack solubility from the rylene moieties, thus requiring additional annealing or solvent additives to achieve the target morphologies. While the performance of these compounds are remarkably high, they fall short of those achieved with the fullerenes. The large conjugated electron

Table 7.22A Performance of selected perylenediimide-based NFA:donor OPVs

Acceptor		E_{HOMO} (eV)	E_{opt} (eV)	E_{loss} (eV)	j_{SC} (mA/cm^2)	V_{OC} (V)	*FF*	η_{max} (%)	Ref.
A	P3TEA (1:1.5)	−5.71	1.72	0.61	13.27	1.11	0.64	9.50	(Liu et al., 2016a)
B	PDBT-T1 (1:1)	−6.09	1.93	0.98	12.48	0.95	0.70	8.42	(Meng et al., 2015)
C	PDBT-T1 (1:1)	−6.02	1.93	0.96	12.40	0.97	0.70	8.28	(Meng et al., 2016)
D	PDBT-T1 (1:1)	−5.97	1.93	0.93	12.72	1.00	0.72	9.28	(Meng et al., 2016)
E	PTB7-Th (1:1)	−5.71	1.58	0.79	17.90	0.79	0.58	8.47	(Wu et al., 2016)

Notes: (i) All OPV characteristics obtained at 1 sun, AM1.5G simulated solar illumination.
(ii) D:A blend ratios shown in parentheses in donor column.

Table 7.22B Performance of selected perylenediimide-based NFA:donor OPVs

	Acceptor		Donor
A	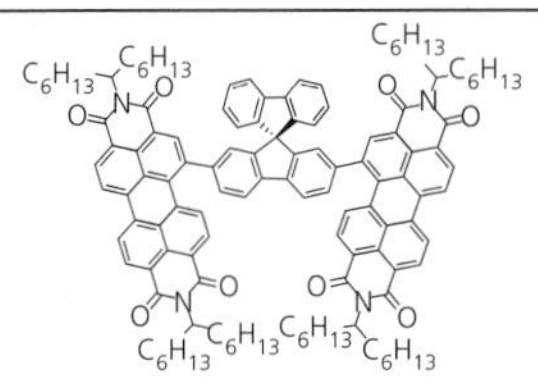	SF-PDI$_2$	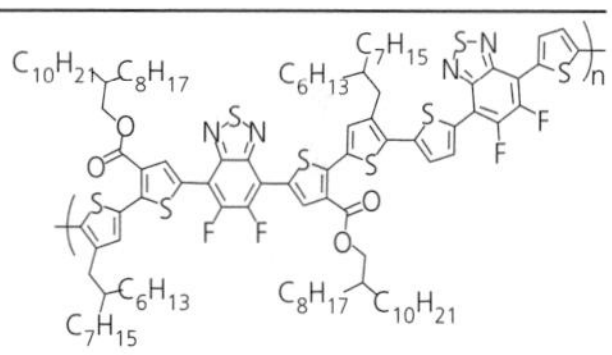
B	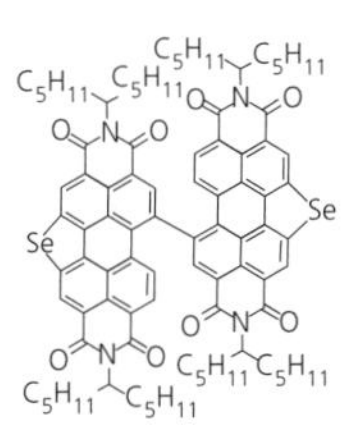	SdiPBI-Se	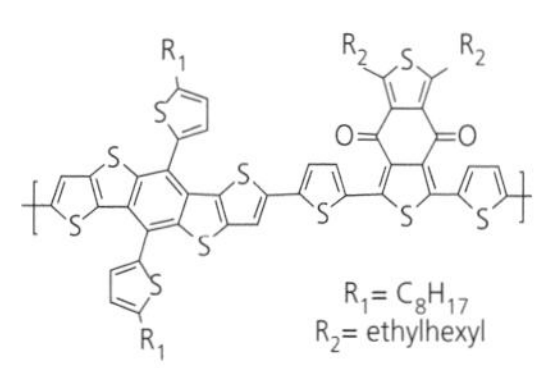
C	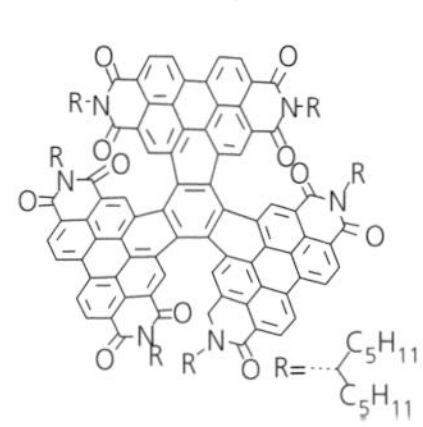	TPH	
D	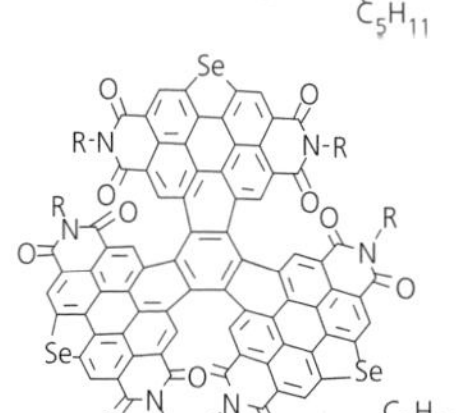	TPH-Se	
E	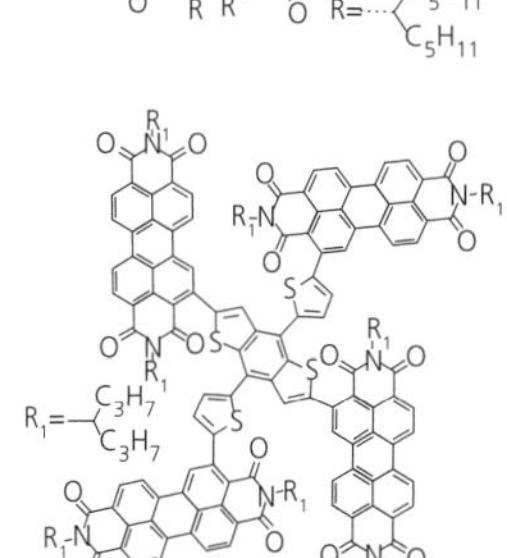	TPB	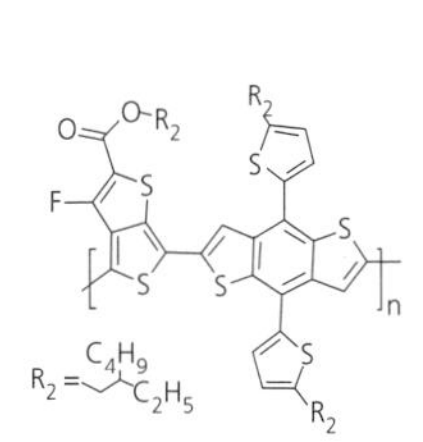

system on the fullerene periphery results in very rapid (hundreds of femtoseconds) charge transfer, and hence high η_{CT}, and a concomitant reduction in E_{loss} to 0.6–0.8 eV. Moreover, the near spherical symmetries of both C_{60} and C_{70} lead to close packing into nanocrystalline domains. This promotes a high electron conductivity and hence a high *FF* and low series resistance. All of these factors have led to the widespread use of fullerenes as acceptors in OPVs.

Fullerenes, however, also have significant disadvantages. For example, while E_{loss} is less than in the PDIs, its magnitude ultimately limits cell efficiency. Furthermore, as in Section 7.2.2, their absorption spectra are confined to the visible, with absorption coefficients that are often smaller than more tightly π-stacked compounds such as planar donors in Section 7.4.4.1. To avoid these deficiencies, a family of thiophene-based NFAs has emerged with E_{loss} approaching 0 V in some cases, and with energy gaps that allow for absorption into the NIR. The design flexibility afforded by this molecular family has led to a new generation of OPVs with single junction efficiencies > 15%.

The general architecture of this class of calamitic mesogenic a-d-a NFAs is similar to that followed by donors (Table 7.19), and is illustrated in Fig. 7.168. A *calamitic mesogen* is a molecule that has liquid crystalline properties (i.e. it is *mesogenic* such that, depending on temperature, it can behave as a solid or liquid), and it has a rigid core that allows it to organize into layered stacks (i.e. it is calamitic). The alternation of electron rich thiophene donor units and electron-withdrawing 2-(3-oxo-2,3-dihydroinden-1-ylidene)malononitrile moieties allows for facile ICT. The tendency to π-stack competes with the need for solubility provided by alkane groups attached along its length. All high performance NFAs have

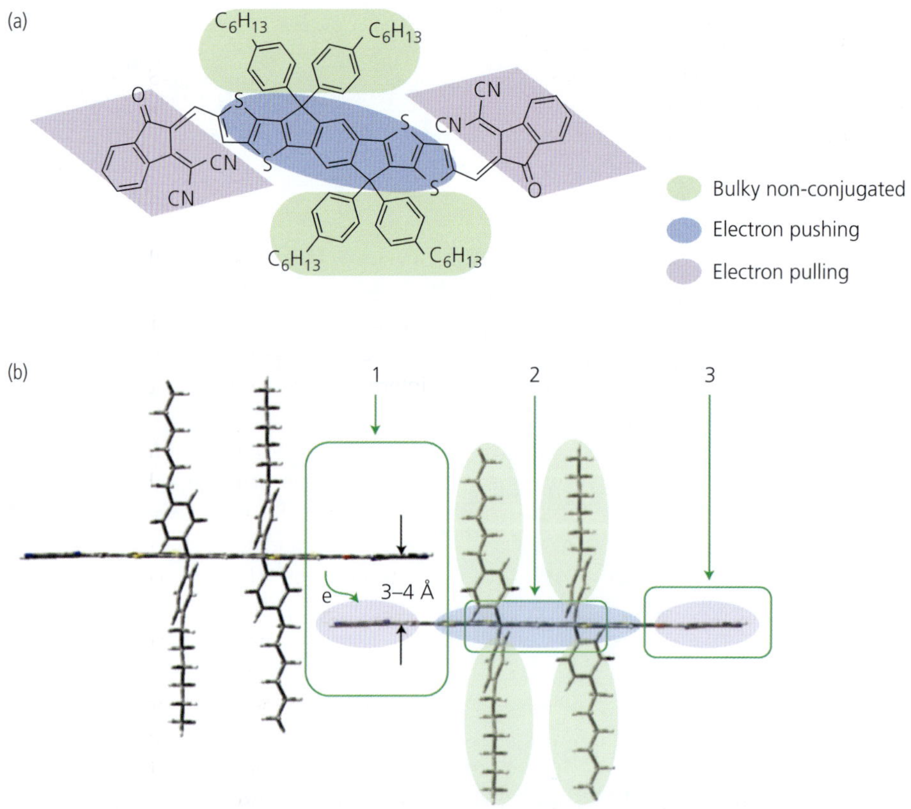

Figure 7.168 (a) Molecular structure of the NFA, IT-IC, showing its electron push–pull architecture. (b) The packing motif of IT-IC with each molecular sub-unit performing a different function: (1) intermolecular overlap results in efficient electron conduction between members of the stack; (2) the donor backbone is protected by the alkyl groups also used to increase solubility; and (3) the a-unit that lacks shielding by alkyls to promote close π-stacking with neighbors. After Hou et al. (2018).

low energy gaps, and contain thiol groups as π-bridges or donor units.

Figure 7.168 illustrates how the different molecular units are designed to affect both morphology and charge transfer in the archetype NFA, IT-IC (see spectrum, Fig. 7.126). The molecule comprises a seven-ring fused donor core (indacenodithieno[3,2-b]thiophene, IT), terminated with 2-(3-oxo-2,3-dihydroinden-1-ylidene)malononitrile acceptor units. The core is shielded from close packing with other, similar molecules by the alkane chains that afford the needed molecular solubility. However, the alkanes do not impede molecular stacking of the acceptor end groups, whose electron-withdrawing character is enhanced by the CN groups (Fig. 7.168a). This permits intermolecular stacking to within 3–4 Å, and hence a high electron conductivity along the stacks. The rigid, planar backbone (Fig. 7.168b) enables efficient ITC necessary to reduce E_{opt} and E_{loss} when IT-IC is employed in OPVs. IT-IC has a long wavelength cutoff of 925 nm corresponding to $E_{opt} = 1.35$ eV, making it suitable for NIR sensitive OPVs. Thus, the extended backbone with the bulky central donor unit protected by the alkane chains creates the needed delicate balance between solubility and close packing in a slip-stack molecular structure. The OPVs consisting of ITO/PEDOT:PSS/PTB7-Th:ITIC (1:1.3)/Al thermally annealed for 15 min at 150°C exhibited $\eta_P = 6.8\%$. While this is still lower than obtained when PTB7-Th is blended with either $PC_{71}BM$ ($\eta_P = 7.5\%$) (Lin et al., 2015), or using a ZnO/C_{60} cathode buffer with $PC_{71}BM$ (9.35%) (Liao et al., 2013), we will show in the ensuing discussion that the calamitic NFAs can lead to exceptionally high single junction efficiencies when combined with the appropriate donor. Importantly, the narrow absorption spectra make these molecules particularly well-suited for visibly transparent OPV applications.

A theme throughout our discussion of OPV materials has been that energy loss sets a fundamental limit to the open circuit voltage and hence the efficiency of OPVs. This loss can be as large as 0.7 eV for fullerene acceptors, although NFAs with the a-d-a motif of the archetype IT-IC, and indeed in the d-a-d donors reduce this to ≤ 0.6 eV. Understanding the molecular origins of energy losses can be used to devise design rules that lead to even higher efficiency CT. In the following we describe an analytical framework of Liu and co-workers that connects molecular design to E_{loss} (Liu et al., 2019).

In Section 7.3.2 we showed that the driving force required to overcome the binding energy is supplied by the offset of the LUMO and HOMO energies, ΔE_{LUMO} and ΔE_{HOMO}. As a result, the dissociation process incurs an energy loss during charge transfer:

$$\Delta E_{CT} = E_X - E_{CT} \approx \Delta E_{HOMO(LUMO)} - E_B, \quad (7.183)$$

where, as previously, E_X is the energy difference between S_1 and S_0, and E_B is the exciton binding energy. Additional losses arise from both radiative (ΔE_r) and non-radiative (ΔE_{nr}) recombination at the HJ (Menke et al., 2018):

$$\Delta E_{rec} = E_{CT} - qV_{OC} = \Delta E_r + \Delta E_{nr}. \quad (7.184)$$

Therefore, the total energy loss at the donor–acceptor HJ is (cf. Eq. 7.155)

$$E_{loss} = \Delta E_{CT} + \Delta E_{rec} = E_X - qV_{OC}. \quad (7.185)$$

From semi-classical Marcus theory (see Section 4.3.2), the transfer rate from an acceptor exciton (A^*) to a CT state (A^-/D^+) illustrated in Fig. 7.169a is given by (cf. Eq. 4.89)

$$k_{CT} = \frac{J^2}{\hbar}\left(\frac{\pi}{\lambda_{reorg}k_BT}\right)\sum_{n'}\frac{S^{n'}}{n'!}\exp(-S) \times \exp\left[\frac{-\left(\Delta E_{CT} + \lambda_{reorg} + n'\hbar\langle\omega\rangle\right)}{4\lambda_{reorg}k_BT}\right], \quad (7.186)$$

where, as previously, J is the electronic coupling between the acceptor excited state (ψ_{A^*D}) and CT ($\psi_{A^-D^+}$) wavefunctions, λ_{reorg} is the intermolecular reorganization energy, and S is the Huang–Rhys factor for intermolecular electron–phonon coupling. The intramolecular phonon modes are replaced by a single mode with mean energy $\hbar\langle\omega_I\rangle$, typically corresponding to the dominant (C—C) bond stretch with $\hbar\langle\omega_I\rangle \approx 0.17$ eV. Also, the Gibbs free energy is given by ΔE_{CT}.

The CT process between the initial A^*/D initial state to the A^-/D^+ final state and the definition of ΔE_{CT} are illustrated in Fig. 7.169a. Quantum mechanical tunneling mediates transfer through the barrier separating the initial and final states. Following Kasha's rule, the transition occurs from the lowest vibrational state ($n = 0$) of the excited acceptor to the n' vibrational state in the CT (A^-/D^+) manifold. The Frank–Condon (FC) integral of the tunneling transition is then simplified to $\exp(-S)\frac{S^{n'}}{n'!}$. The reorganization energy, λ, during charge transfer is the sum of *intra*molecular (λ_I) and *inter*molecular (λ_O) contributions. Intermolecular reorganization is due to electronic polarization, and consequently lattice

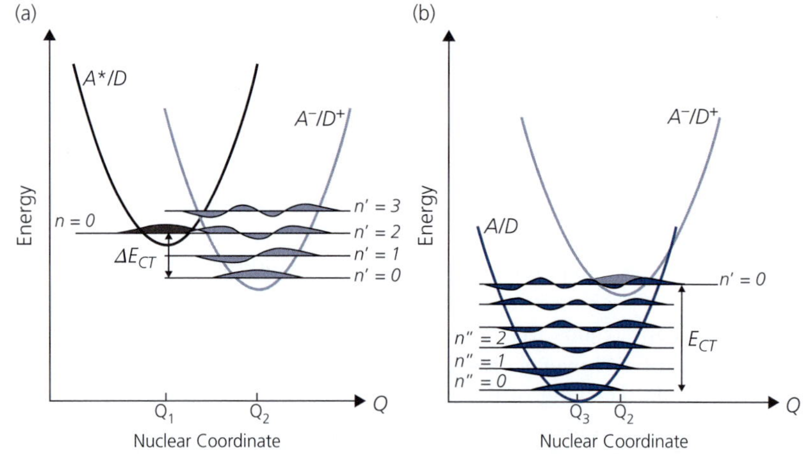

Figure 7.169 Energy diagrams of (a) charge transfer from the acceptor exciton (A^*) to the charge transfer (CT) state (A^-/D^+), and (b) non-radiative recombination from the CT state (A^-/D^+) to the ground state (A/D). The Gibbs free energies are ΔE_{CT} in (a) and E_{CT} in (b). Symmetric parabolic potentials are assumed for the initial and final states. Equilibrium configurations of A^*/D, A^-/D^+, and A/D manifolds are indicated by Q_1, Q_2, and Q_3, respectively (Liu et al., 2019).

relaxation of the surrounding medium. Assuming a spherically symmetric charge distribution on the donor and acceptor, and treating the medium as an isotropic dielectric continuum, λ_O is found from Eq. 4.61.

Non-radiative recombination energy losses (ΔE_{nr}) occur through coupling from CT (A^-/D^+) to the ground state (A/D), as illustrated in Fig. 7.169b. The transition rate, k_{nr}, can be described by Eq. 7.186 after making the following substitutions. The Gibbs free energy is now E_{CT} instead of ΔE_{CT}, and $J \to J'$, $S \to S'$, and $\lambda_{reorg} \to \lambda'_{reorg}$ to account for the electronic coupling between the CT and the ground state, which may or may not differ from those in Eq. 7.186 that describes coupling between A^* and CT (cf. Fig. 7.169). Clearly, k_{nr} increases with decreasing E_{CT} in accordance with the energy gap law. For non-radiative recombination, the energy loss is (cf. Eq. 7.153) (Rau, 2007)

$$\Delta E_{nr} = -mk_B T \ln\left(\frac{1}{\eta_{CT}}\right) = -mk_B T \ln\left(\frac{k_r + k_{nr}}{k_r}\right). \tag{7.187}$$

To achieve low energy loss yet efficient charge transfer, E_B and ΔE_{HOMO} must be minimized simultaneously (cf. Eq. 7.183). Secondly, decreasing both intra- and intermolecular electron–phonon coupling leads to an increased k_{CT} at a minimum cost to ΔE_{CT}, as well as a decrease in k_{nr}, and consequently, ΔE_{nr}.

Energy loss is a function of E_B, S, and λ_{reorg}, which in turn are specific to a particular molecular structure. The exciton binding energies have been calculated using DFT for both fullerenes and NFAs. The exciton binding energy is $E_B = (E_- + E_+) - (E_0 + E_X)$, where

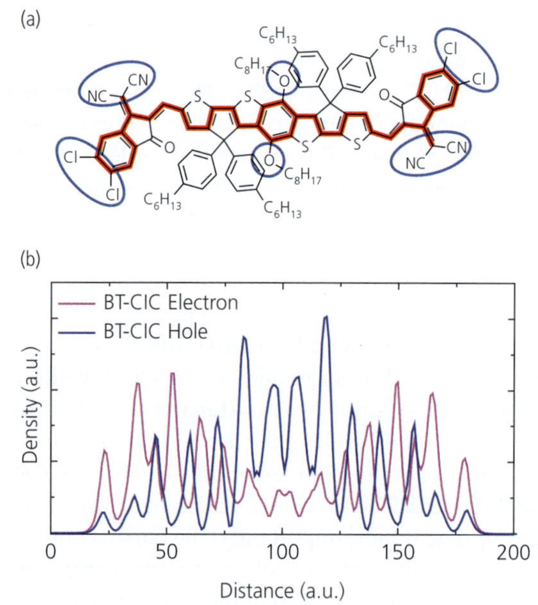

Figure 7.170 (a) Molecular structure of BT-ClC. The red bold line traces the electron conjugation path comprising alternating C—C single and double bonds. Blue circles indicate the electron-rich (i.e. oxy-) and deficient (i.e. chloro- and cyano-) moieties. (b) The electron and hole density distributions along the molecular length of BT-ClC calculated using DFT (Liu et al., 2019).

E_- and E_+ are the optimized energies of radical anions and cations of the molecule, respectively, and E_0 is the ground state energy. Figure 7.170a shows the molecular structure of the example NFA, BT-ClC. The electron and hole distributions along the molecular backbone of the excited state are shown in

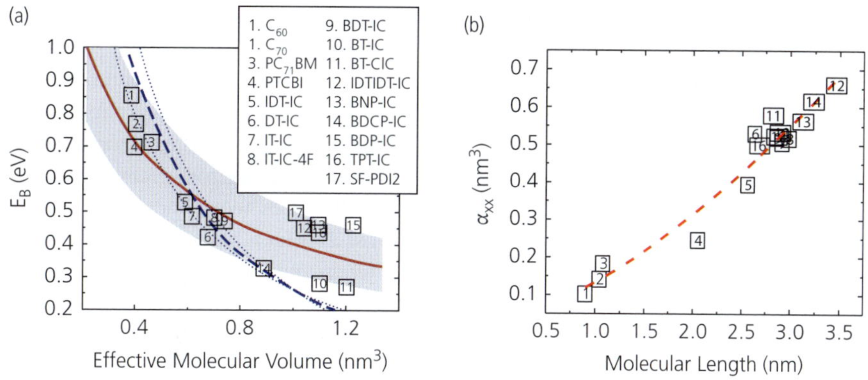

Figure 7.171 (a) Exciton binding energy calculated via DFT vs. effective molecular volume for several acceptor molecules indicated in legend. The solid line is a fit assuming electron confinement within the effective molecular volume for each molecule. The calculation assumes a dielectric parameter, $f = 5.4$. The shaded area is the 95% confidence band. The dashed line assumes f depends on molecular polarizability from (b). The confidence limits of the calculations are shown by dotted lines. (b) Calculated molecular polarizability along the molecular longitudinal axis, α_{xx}, vs. the effective molecular length l. The dashed line is the fit using the empirical power-law described in text (Liu et al., 2019).

Fig. 7.170b. The conjugated electron system is indicated by the bold red line bridging the electron-donating thiophene chain to the electron deficient cyano and Cl groups. This results in symmetric ICT from the middle to the two ends of the optically excited molecule, which features electron rich regions near the end caps, and an increased hole concentration toward the center of the backbone. The relative segregation of electron and hole densities results in the reduced E_B characteristic of a-d-a acceptors and d-a-d donors, resulting in their reduced energy loss according to Eq. 7.185 compared to the more compact fullerenes. The rigid coplanar structure of a-d-a NFAs ensures that the electron wavefunction is delocalized over the entire molecular backbone, unlike polymers that can twist at C—C single bonds that ultimately terminates the extent of the excited state (see Section 3.6).

For a given offset of the donor and acceptor frontier orbital energies, ΔE_{CT}, is maximized by minimizing the acceptor or donor exciton binding energy, which in turn increases k_{CT} in the Marcus non-inverted regime. The combined influences of molecular size, conjugation length, and strength of the electron donating and withdrawing groups determines the magnitude of ICT in the acceptors (or donors), and this ultimately determines E_B. The effects of molecular size on E_B can be quantitatively estimated assuming that the electron wavefunction for a particle in a three dimensional "molecular box" is bounded by the extent of the electron and hole distributions (see Fig. 7.170b). The wavefunctions are $\Psi_{n_x,n_y,n_z} = \sqrt{\frac{8}{v}}\sin\left(\frac{n_x\pi x}{l}\right)\sin\left(\frac{n_x\pi y}{w}\right)\sin\left(\frac{n_z\pi z}{h}\right)$, where n_x, n_y, n_z are quantum numbers. This can be used to calculate E_B vs. effective molecular volume $v = lwh$. The effective molecular length l, width w, and height h are defined by a box whose boundaries are set at the point that the electron density falls to 0.02 au^{-3}, a metric commonly used to define the iso-surface of electron distributions and the size of a molecule (Frisch et al., 2009).

Figure 7.171 shows the calculated E_B for a population of fullerenes and NFAs vs. v. For this analysis, an average molecular height of $h = 0.3$ nm and width of $w = 0.8$ nm was assumed, with l varied according to the linear dimension of the molecule. The binding energy monotonically decreases, with v independent of the details of the molecular structure due to increased separation between the electron and hole densities (Fig. 7.170b). The solid line is the binding energy obtained from the hole and electron wavefunctions, $\Psi_{0,0,0}$ and $\Psi_{1,0,0}$, respectively, found from solutions to Schrödinger's equation. Thus,

$$E_B = q^2\int\frac{|\Psi_{0,0,0}(\mathbf{r}_h)|^2|\Psi_{1,0,0}(\mathbf{r}_e)|^2}{4\pi\varepsilon_0 f|\mathbf{r}_h - \mathbf{r}_e|}d^3\mathbf{r}_h d^3\mathbf{r}_e, \qquad (7.188)$$

where the r_h and r_e are the position vectors of the hole and electron, respectively, and f is a parameter that accounts for the relative dielectric constant of the molecule and the relaxation energy of the ionic species. Remarkably, the solid line corresponds to only a single value of $f = 5.4$ that approximately fits the entire population of acceptors, ranging from highly symmetric fullerenes to extended and planar NFAs. The deviations of the data from the solid line are due to details of the molecular structures that impact the molecular polarizabilities and relative dielectric constants in the solid.

Since the dielectric constant is related to the polarizability, α_{xx}, we can interpret the analysis leading to Eq. 7.188 by calculating the polarizability vs. molecular length l. The polarizability follows an empirical power-law dependence (Van Dyck et al., 2017):

$$\alpha_{xx} = \alpha_0 + \alpha_1 l^n, \tag{7.189}$$

where α_0 is the polarizability of terminal bonds such as C–H, and α_1 is related to the polarizability of conjugated or non-conjugated chains of length l. A fit to the data in Fig. 7.171b gives $\alpha_0 = 0.035\ \text{nm}^3$, $\alpha_1 = 0.098\ \text{nm}^{3\text{-n}}$, and $n = 1.51$. The weak, superlinear increase of polarizability ($n > 1$) is due to the contributions of conjugation and ICT. The relationship between the dielectric constant of the medium and the polarizabilities of each constituent, α_i, is given by the *Clausius–Mossotti relationship*:

$$\frac{\varepsilon - 1}{\varepsilon + 2} = \frac{4\pi}{3}\sum_i \mathrm{N}_i \alpha_i, \tag{7.190}$$

where the molecular packing density is N_i. The effect of the dielectric constant change on E_B is estimated using Eqs. 7.189 and 7.190 to replace constant f of Eq. 7.188, and assuming an isotropically polarizable medium comprising randomly oriented molecules with $N = 1.3\ \text{nm}^{-3}$. This yields the blue dashed line in Fig. 7.171a. The fit closely follows the data for molecules **1** to **11** compared to that using a constant f, although there are noticable departures from the data arising from approximations made to the molecular structure and the relaxation of ionic species. The less accurate fits to molecules **12**, **13**, and **15–17** are attributed to molecular bends that disturb the electron conjugation, thereby increasing E_B.

Achieving a small ΔE_{CT} and ΔE_{nr} requires simultaneously reducing both intra- and intermolecular electron–phonon coupling. According to Eq. 4.61, the increased dielectric constant of NFAs reduces the intermolecular coupling, λ_0. The intramolecular coupling constant, S', also changes with the size and rigidity of the molecules, that is, they scale with the increased FC shift that is twice the total reorganization energy, $E_{FC} \approx 2\lambda_{CT}$. This yields $\lambda_{CT} = 0.11$ eV for IT-IC/PBDB-T HJs compared to $\lambda_{CT} = 0.24$ eV for $PC_{71}BM$/PBDB-T, resulting in a higher k_{nr} for the latter system. Thus, $\eta_{CT} = \dfrac{k_r}{k_r + k_{nr}}$ is higher in the IT-IC/-PBDB-T HJ. Indeed, the non-radiative recombination loss using Eq. 7.188 and η_{EL} of IT-IC/PBDB-T and $PC_{71}BM$/PBDB-T HJs result in a 0.05 ± 0.01 eV smaller ΔE_{nr} in the IT-IC system, leading to the generally higher power conversion efficiencies observed for NFA vs. fullerene based OPVs. The presence of nonbonding orbitals in the cyano, chloro or oxy groups introduce lone pair electrons that do not participate in bonding, and thus avoid electron–phonon coupling. At the same time, the cyano and chloro groups attract electrons and reduce the charge density along the backbones, effectively reducing electron–phonon coupling.

Finally, it is worth noting that the treatment assumed thus far is based on semi-classical Marcus transfer, where only direct coupling of the CT to the ground state is assumed. A more complete theory must take into consideration coupling of all the excited states, and most importantly, the coupling of the lowest excited (S_1) state to the CT manifold. This is illustrated in Fig. 7.172, that explicitly includes the strength of the coupling of the S_1 to the CT state ($t_{LE\text{-}CT}$) in addition to the coupling of CT to the ground state ($t_{CT\text{-}G}$) considered above. These multiple contributions can be quantified using Marcus–Jortner electron transfer theory (Marcus and Sutin, 1985, Bixon and Jortner, 1999). The potential energy surface of the transition from S_1 to a CT state is illustrated in Fig. 7.172 assuming a simple harmonic oscillator model. The dashed lines at the crossing points present the adiabatic surface, while solid lines are non-adiabatic. In the case of non-adiabatic transitions within the Born–Oppenheimer approximation, the CT rate constant from Fermi's golden rule is (Bixon and Jortner, 1999)

$$\begin{aligned} k_{CT} &= \frac{2\pi}{\hbar}|\langle\psi_{S1}|H|\psi_{CT}\rangle|^2|\langle v_{S1}|v_{CT}\rangle|\rho(E_{CT}) \\ &= \frac{2\pi}{\hbar} t_{LE-CT}^{\ 2}(FCWD). \end{aligned} \tag{7.191}$$

Here, $t_{LE-CT} = \langle\psi_{S1}\,|\,\mathrm{H}\,|\,\psi_{CT}\rangle$ is the matrix element that mixes the wavefunctions of CT (ψ_{CT}) and local excited

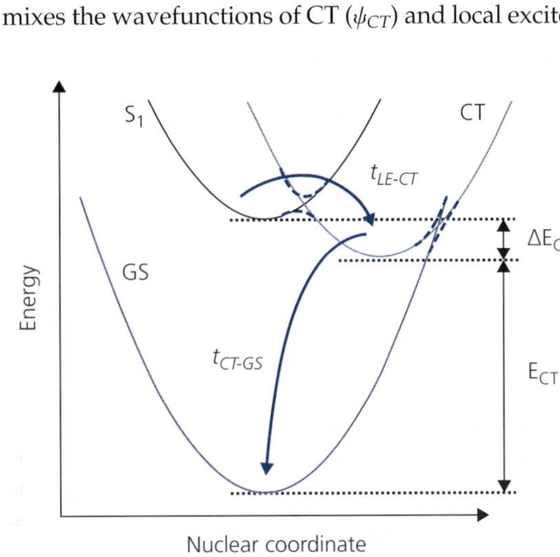

Figure 7.172 Coupling of the CT state to both the ground (GS) and lowest excited state (S_1). These transition and their corresponding strengths (denoted $t_{CT\text{-}GS}$ and $t_{LE\text{-}CT}$, respectively) must be considered to fully account for energy losses in the charge generation process.

state (ψ_{S1}) of acceptor or donor, H is the interaction Hamiltonian, $\rho(E_{CT})$ is the density of CT states, and $FCWD$ denotes the Franck–Condon factor weighted density of states. In a semi-classical approximation at room temperature, Eq. 7.191 gives

$$k_{CT} = \frac{2\pi}{\hbar} t_{LE-CT}{}^2 FCWD(-\Delta G_{LE-CT}) = \frac{2\pi}{\hbar} t_{LE-CT}{}^2 \frac{1}{\sqrt{4\pi\lambda_0 k_B T}} \times \left[\sum_{n'} \exp(-S)\frac{S^{n'}}{n'!}\exp\left(-\frac{(-\Delta G_{LE-CT} + \lambda_0 + n'\hbar\langle\omega_l\rangle)^2}{4\lambda_0 k_B T}\right)\right], \quad (7.192)$$

where the Gibbs free energy change ΔG_{LE-CT} is the sum of energy offset, ΔE_{CT}, and the entropy change. The $FCWD$ factor is assumed to be the sum of vibrational overlap integrals between the lowest ($n = 0$) vibrational state of S_1 and the state n' of the CT manifold. Each transition from $n = 0$ to n' in the sum, contributed by low frequency intermolecular modes ($\hbar\omega_0 \ll k_BT$) in the medium, is adequately described by the classical picture by an Arrhenius-type exponential term.

Recently, derivations of the matrix element between S_0 and hybridized CT-S_1 states have been developed (Chen et al., 2018, Eisner et al., 2019). That analysis finds that k_{nrad} decreases with decreasing energy offset, while k_{rad} stays approximately constant when hybridization is absent. Hybridization increases both k_{rad} and k_{nrad}, but has a stronger impact on k_{rad} than on k_{nrad}, giving rise to the decrease in $\Delta V_{OC,nr}$. While this approach is formally correct, its application depends on knowing the coupling strengths between the states involved. However, the experimental evidence thus far is insufficiently detailed to justify invoking the more complex interactions involved in the three-state formalism.

The foregoing analysis suggests molecular design strategies that can further reduce energy loss. Currently, a-d-a-type NFAs are inherently symmetric, and hence possess a nearly zero dipole moment. However, asymmetric electron-withdrawing end groups in a-d-a molecules provide freedom to tune the dipole moment and further reduce E_B, S, and λ_0 by changing the effective intramolecular electron–hole separation and the molecular packing in bulk HJs. Additionally, the relationship between molecular dimension, rigidity and energy loss points to the benefits of increasing the molecular volume. However, there is a limit to the extent to which the length of the molecular backbone can be increased without bending, and hence terminating the effective conjugation length. Extending the conjugation into two and three dimensions can further reduce E_B, S, and λ_O, as compared to 1D conjugation used up to now. This may also enhance intermolecular π–π interactions along all directions, leading to increased charge carrier mobilities. Indeed, early demonstrations of increasing the molecular volume in two dimensions based on dithienopicenocarbazole-based a-d-a-type (Yao et al., 2018) or spiro-fused PDI NFAs (Gao et al., 2017) provide an illustration of the benefit of this strategy.

Table 7.23 lists several different solution-processed a-d-a NFAs that are chosen to demonstrate the facility with which the modular design of molecules in Fig. 7.168 can lead to systematic variations in electronic properties. The first six entries in the table, namely IDT-IC, IT-IC, BDT-IC, BT-IC, BT-CIC, and TT-FIC, correspond to the materials whose absorption spectra are found in Fig. 7.127. While some of the molecules have efficiencies exceeding 11%, the most interesting aspect is the nearly rigidochromic shift of the spectra, with optical energy gaps monotonically increasing from 1.28 eV for TT-FIC, to 1.73 eV for IDT-IC. This is accompanied by a concomitant, near monotonic decrease in E_{HOMO} from −5.43 eV to −5.61 eV, respectively. This trend is depicted in the energy level diagram in Fig. 7.173. The energy losses, which are determined using the E_{opt} of the acceptors, depend on the specific choice of donor. The minimum value of $E_{loss} = 0.56$ eV is found for the J71:BT-IC blends, which is smaller than obtained with any of the PDI-based acceptors in Table 7.22.

We can qualitatively understand the source of the decreasing energy by examining the structures of the six NFAs themselves. The planar, electron-donating backbone of each molecule comprises a series of fused rings, each contributing to the conjugation length of the electron system. The d-unit is capped at each end by a 2-(3-oxo-2,3-dihydroinden-1-ylidene) malononitrile a-unit, and in the case of BT-CIC and TT-FIC, the a-units are functionalized with Cl and F, which further increases their electron-withdrawing ability. Then, starting with the smallest energy gap IDT-IC comprising five fused rings (containing two Th), we progress to lower E_{opt} with IT-IC (6 rings, 4 Th), BDT-IC (7 rings, 4 Th), BT-IC (3 fused rings disposed on opposite sides of a alkoxy group), BT-CIC (as in BT-IC but with strongly electron-withdrawing Cl functionalized a-units), and TT-FIC with eight rings having an even more electron deficient F-functionalization. The strong donating ability of the alkoxy group in BDT-IC also results in a somewhat higher E_{HOMO}, evident in Fig. 7.173. Hence, this series shows a progressively increasing conjugation length leading to decreasing energy gaps. These modifications tend to lower the E_{HOMO}, and thus E_{LUMO}. Finally, the planar donor core of the molecules promotes ordered π-stacking, giving $FF > 0.7$ in almost all cases.

Table 7.23A Performance of selected calamitic NFA:donor OPVs

Acceptor		E_{HOMO} (eV)	E_{opt} (eV)	E_{loss} (eV)	j_{SC} (mA/cm^2)	V_{OC} (V)	*FF*	η_{max} (%)	Ref.
A	PTB7-Th (1:1.5)	−5.61	1.73	0.90	9.53	0.83	0.40	3.05	(Li et al., 2016)
B	J71 (1:1)	−5.58	1.59	0.64	17.32	0.94	0.70	11.4	(Lin et al., 2015, Bin et al., 2016, Sun et al., 2018a)
C	J71 (1:1.2)	−5.51	1.53	0.61	17.3	0.92	0.66	10.5	(Li et al., 2017e)
D	J71 (1:1.5)	−5.32	1.46	0.56	17.8	0.90	0.66	10.5	(Li et al., 2017d)
E	PCE-10 (1:1.5)	−5.49	1.35	0.65	22.5	0.70	0.71	11.2	(Li et al., 2017c)
F	PTB7-Th (1:1.7)	−5.43	1.28	0.61	25.6	0.67	0.72	10.9	(Dai et al., 2018)
G	PBDB-T-SF (1:1)	−5.66	1.48	0.60	20.5	0.88	0.72	13.1	(Zhao et al., 2017a)
H	PBDB-T (1:1)	−5.42	1.39	0.54	21.6	0.85	0.72	13.1	(Sun et al., 2018b)
I	PFBDB-T (1:1.25)	−5.63	1.52	0.58	19.6	0.94	0.72	13.2	(Fei et al., 2018)
J	PM6 (1:1.2)	−5.65	1.33	0.50	25.3	0.83	74.8	15.7	(Yuan et al., 2019)

Notes: (i) All OPV characteristics obtained at 1 sun, AM1.5G simulated solar illumination.
(ii) D:A blend ratios shown in parentheses in donor column OPV efficiencies obtained have opened the path toward eventually achieving near thermodynamically limited performance.

Table 7.23B Performance of selected calamitic NFA:donor OPVs

	Acceptor		Donor
A	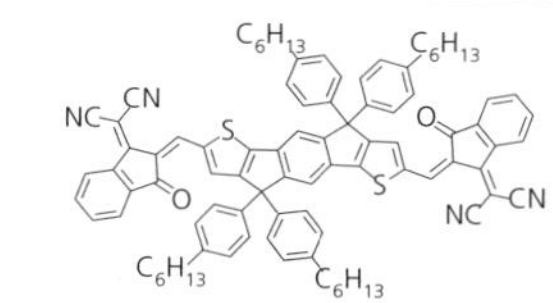	IDT-IC	
B	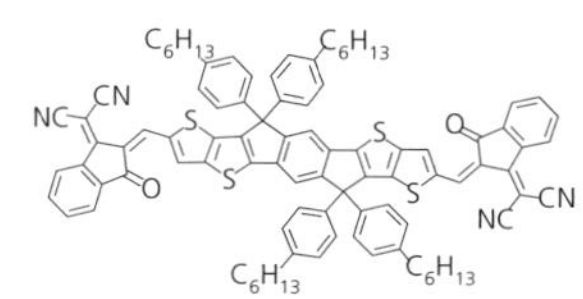	IT-IC	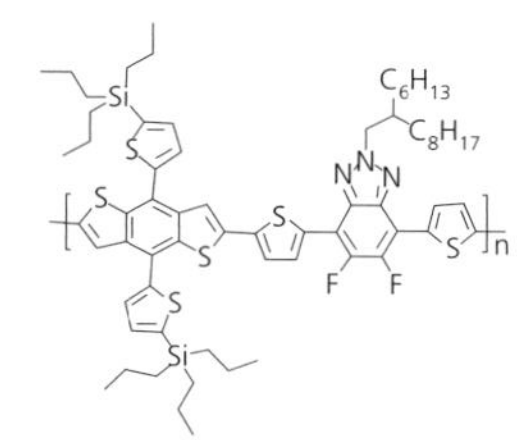
C	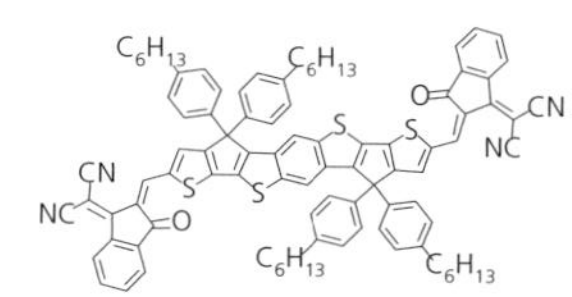	BDT-IC	
D	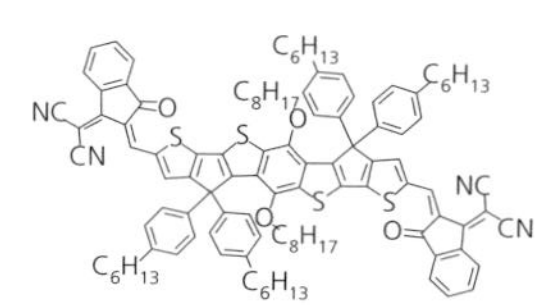	BT-IC	
E	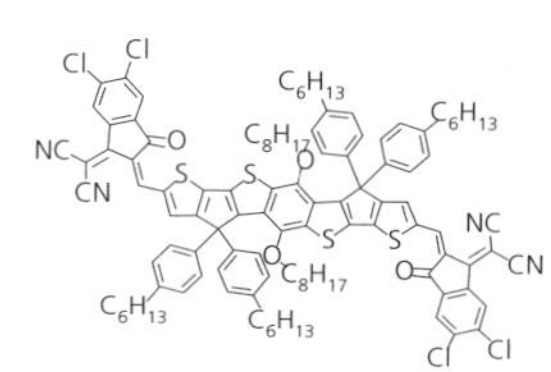	BT-ClC	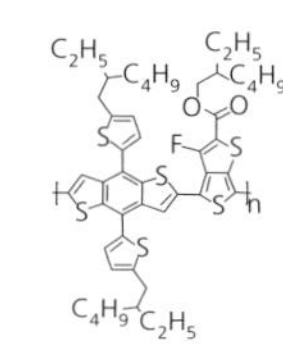

(Continued)

Table 7.23B Continued

	Acceptor		Donor
F	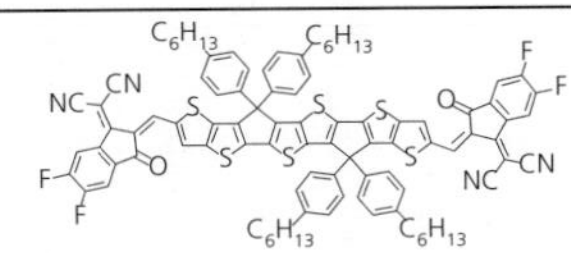	TT-FIC	
G	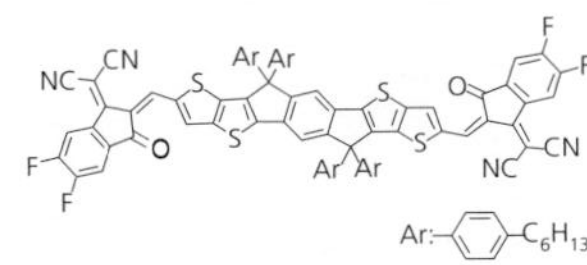	IT-4F	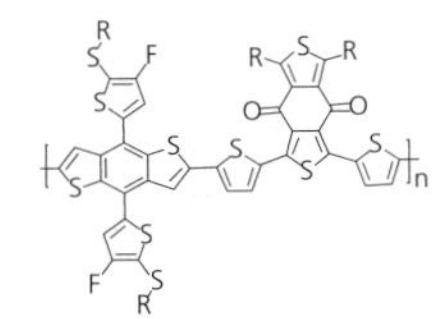
H	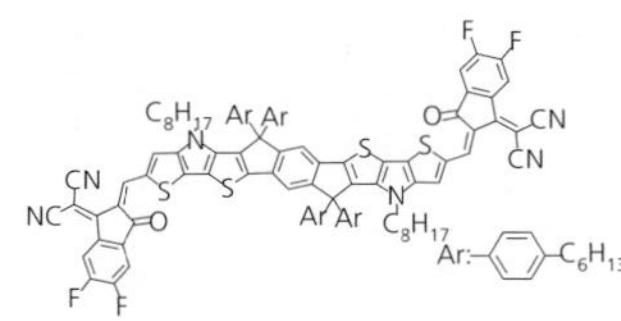	INPIC-4F	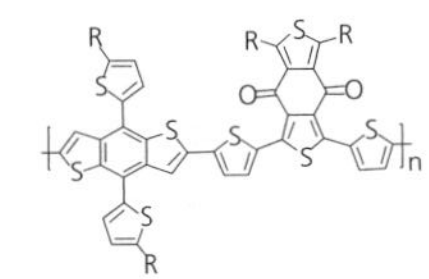
I	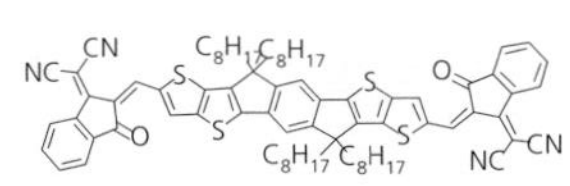	C_8-IT-IC	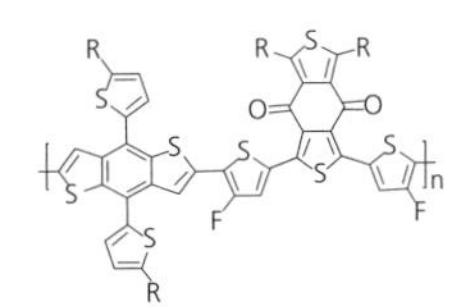
J	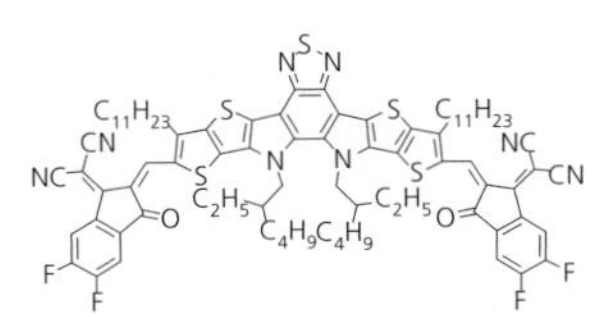	Y6	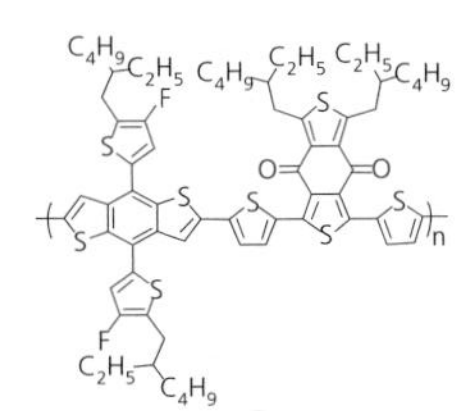

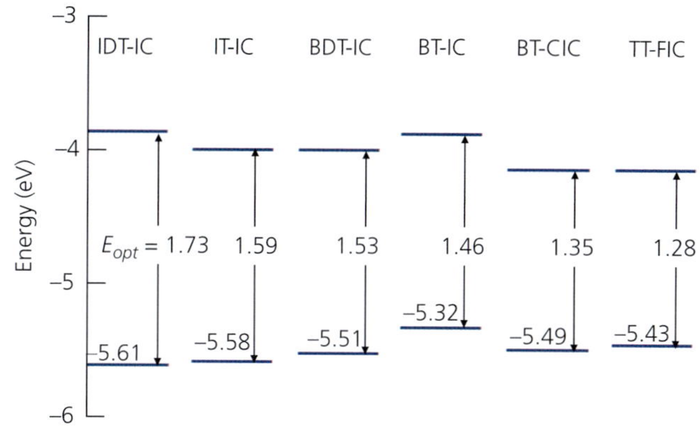

Figure 7.173 Progression in the optical energy gap and HOMO energy for the series of NFAs in Fig. 7.126. E_{opt} is the energy for the $S_1 \leftarrow S_0$ transition.

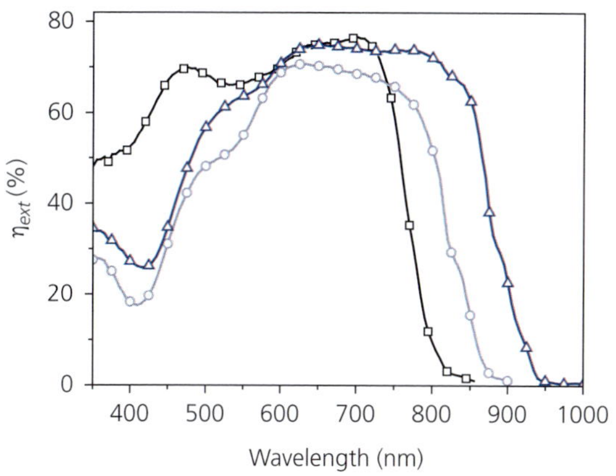

Figure 7.174 External efficiency spectra of BHJ OPVs employing a PCE-10 donor and a BT-IC (circles), BT-CIC (triangles), or $PC_{71}BM$ (squares) (Li et al., 2017c).

The absorption of these molecules is quite strong, with $\alpha \sim 10^5\ \text{cm}^{-1}$. However, since their absorption is narrow and generally centered in the NIR, a donor with absorption in the visible is needed to achieve more complete coverage of the solar spectrum when visible transparency is not required. Examples of the external quantum efficiency spectra of BT-CIC and BT-CIC OPVs with the structure ITO/ZnO(25 nm)/PCE-10:NFA (1:1.5, 130 nm)/MoO_3 (15 nm)/Ag are shown in Fig. 7.174. For comparison, η_{ext} for a PCE-10:$PC_{71}BM$ OPV with the same blend proportion and thickness is also provided. Remarkably, the OPVs based on the NFAs have η_{ext} approaching 80% over much of the NIR, with the longest wavelength response, and highest $\eta_P = 11.2\%$ observed for BT-CIC. The short wavelength response of the $PC_{71}BM$-based OPV is the highest of the group, since the fullerene has a higher absorption coefficient than either of the NFAs, and the donor PCE-10 also has limited absorption in the blue spectral region.

The next four NFAs all have $\eta_P > 13\%$. The acceptor, C_8-IT-IC, is a modified IT-IC whose hexyl side chains are replaced by octyl units. Given the marginal differences in the energetics of these two compounds, the improved morphology obtained by the extended octyl chains apparently leads to the higher efficiency of C_8-ITIC. In fact, the highest efficiency achieved for both materials employed a PFBDB-T donor blend that was annealed at 160°C to improve its nanocrystalline morphology.

The IT-4F acceptor has a strongly electron-withdrawing F-functionalized a-capping unit. This increases both the HOMO and LUMO energies compared to its non-fluorinated analog, IT-IC. While this reduces the CT energy somewhat, and hence reduces V_{OC}, (0.88 V vs. 0.94 V for IT-IC), the smaller optical energy gap leads to a higher current of $j_{SC} = 20.5\ \text{mA/cm}^2$ compared with 17.32 mA/cm². Taken together, $\eta_P = 13.1\%$ is obtained in a D-A BHJ device comprising the fluorinated donor, PBDB-T. An analogous fluorinated acceptor, INPIC-4F achieves a similarly high power conversion efficiency, indicating the effectiveness of employing strongly electron-withdrawing a-groups along with an extended, electron-rich 5,5,12,12-tetrakis (4-hexylphenyl)-indacenobis(dithieno[3,2-b:2′,3′-d]pyrrol) (INP) donor core.

The highest efficiency is obtained with the NFA, Y6 used in combination with the donor, PM6 (Zhang et al., 2015) in a 1:1.2 (w/w) D:A ratio. This molecule is also built around a BT core, with strong electron-withdrawing F-functionalized acceptor capping groups. The low optical energy gap of 1.33 eV coupled with the exceptionally low energy loss of 0.5 eV leads to a relatively high V_{OC}. The low energy loss may be a result of a twist about the molecular core enforced by avoidance of the alkyl groups attached to the central N-atoms. This, in turn further separates the electron and hole densities, leading to a reduced E_B. The efficiency reported is 15.7% for both conventional and inverted OPVs.

7.4.4.3 Materials used in non-fullerene ternary OPVs

In Section 7.4.1, we introduced the concept of the ternary OPV where the active region consists of either a donor and two acceptors or vice versa. Adding a third component into the BHJ can increase the solar coverage, which often leads to higher single junction OPV efficiencies. In Table 7.24 we summarize the characteristics of several high efficiency ternary solar cells. Each of these devices comprises at least one NFA, often combined with a fullerene acceptor. Since all a-d-a NFAs require solution processing, both the donor and secondary acceptor additions to the blend are also solution-processed. Furthermore, the ternary OPVs in Table 7.24 have a DA_1A_2 architecture. The structural formulae of compounds listed in the table and not referred to elsewhere in this chapter are provided in Fig. 7.175.

Ternary device performance is particularly sensitive to details in film morphology due to the need to sim-

Table 7.24 Performance characteristics of selected DA_1A_2 OPV cells

Materials	Blend ratio	j_{SC} (mA/cm^2)	V_{OC1} (V)	V_{OC2} (V)	V_{OCT} (V)	*FF*	η_{max} (%)	Ref.
PBDB-T:IT-M:bis- $PC_{71}BM$	1:0.2:1	17.4	0.94	1.02	0.95	0.74	12.2	(Zhao et al., 2017b)
PBDB-T:ITIC:N2200	1:0.9:0.1	16.9	0.92	0.90	0.93	0.73	11.4	(An et al., 2018)
DTRB-T:IDIC: $PC_{71}BM$	1:0.5:0.5	15.5	0.99	0.99	0.99	0.68	10.5	(Zhang et al., 2017)
PDBT-T1:$PC_{71}BM$:ITIC-Th	1:0.5:0.5	15.5	0.92	0.95	0.93	0.71	10.2	(Liu et al., 2017)
PDBT-T1-SdiPDI-Se:ITIC-Th	1:0.5:0.5	15.4	0.95	0.92	0.94	0.71	10.3	(Liu et al., 2016b)
PTB7-Th:IDT-2BR:PDI-2DTT	1:1:0.01	14.5	1.05	—	1.03	0.65	9.7	(Cheng et al., 2017)

Notes: (i) The materials and blend ratio sequences follow the order $D:A_1:A_2$.
(ii) V_{OC1}, V_{OC2}, and V_{OCT} correspond to the open circuit voltages of DA_1, DA_2, and DA_1A_2, respectively.
(iii) Performance characteristics are obtained at 1 sun, AM1.5G simulated illumination.

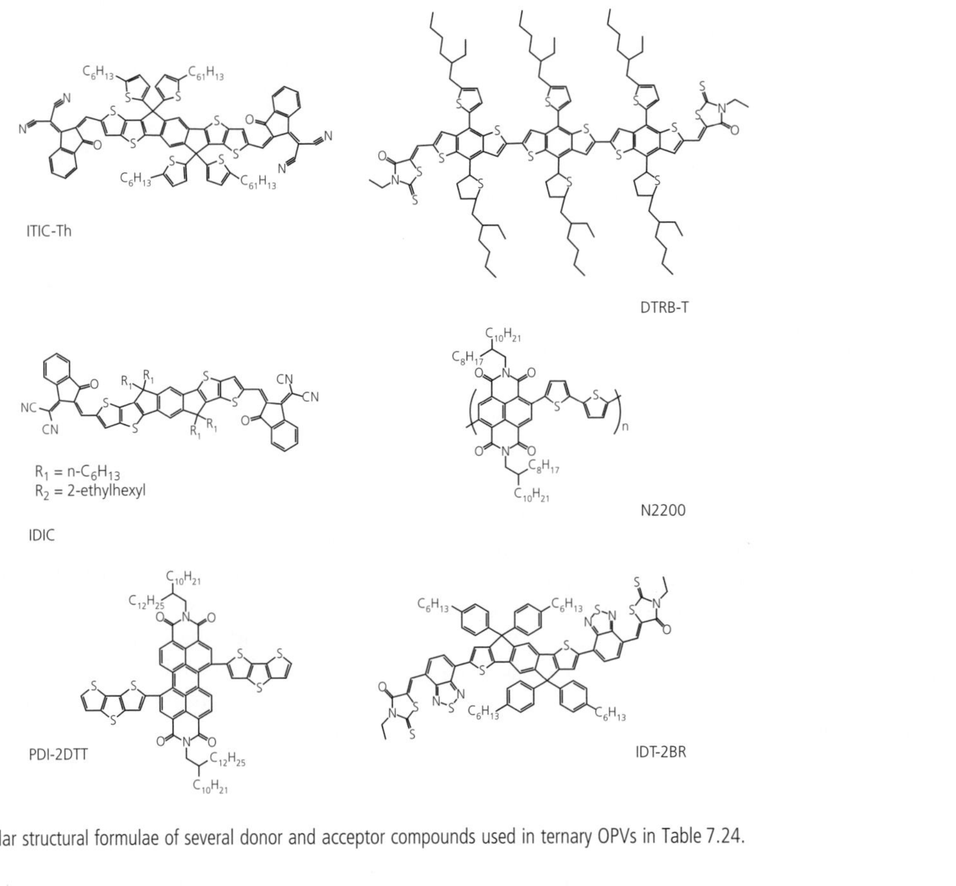

Figure 7.175 Molecular structural formulae of several donor and acceptor compounds used in ternary OPVs in Table 7.24.

ultaneously balance the local concentrations of the three molecular components. With one exception (PTB7-Th:IDT-2BR:PDI-2DTT), the ternary mixtures consist of high concentrations of both acceptors, which in total equal the concentration of the donor. A second observation is that the open circuit voltage of the ternary OPV, V_{OCT}, falls within the limits set by the individual binary devices, V_{OC1} and V_{OC2}. Although the PBDB-T:ITIC:N2200 appears to violate this "rule," V_{OCT} exceeds V_{OC1} by just slightly more than 10 meV, and V_{OC2} is also close to V_{OC1}. This apparent anomaly has been attributed to the improved morphology of the ternary, which also resulted in a significantly increased $FF = 0.73$ and decreased series resistance $R_{ser} = 3.5\ \Omega\ \mathrm{cm}^2$ for the blend containing 10% of the poly(naphthalene bithiophene), N2200, compared to the binary DA_1 (0.68, 3.8 Ω cm^2) and DA_2 (0.73, 5.3 Ω cm^2) devices (An et al., 2018). We showed in Section 7.4.1 that this finding is consistent with all four models used to describe ternary device operation, absent other extrinsic factors such as morphology, junction non-idealities, etc.

An interesting exception to the devices in Table 7.24 employs a PTB7-Th:IDT-2BR:PDI-2DTT active region. The highest efficiency device using this blend was reached when only 1% of the acceptor, PDI-2DTT, is included. Such a low concentration should not be thought of as a third ternary component, but rather as a dopant in a binary BHJ (Cheng et al., 2017). The rationale for the addition of PDI-2DTT is that the HOMO and LUMO offset energies between PTB7-Th and the primary a-d-a NFA, IDT-2BR, of 32 meV and 10 meV, respectively, are too small to efficiently drive charge transfer at the HJ. Doping with PDI-2DTT increases the offset energies with the donor to 40 meV and 31 meV, respectively, which is sufficient to prevent back transfer of dissociated charges into a bound CT state. At concentrations of $\leq 1\%$, the films show a powder texture with the molecular lamellar spacing consistent with that of PTB7-Th:IDT-2BR. However, at higher PDI-2DTT concentrations, numerous new distinct crystalline rings are observed in the GISAXS diffraction patterns. Apparently, this more crystalline morphology decreases the hole and electron mobilities by a factor of three from the undoped DA_1 film, and the *FF* is also reduced. Clearly, only a small addition of A_2 is advantageous for device performance.

7.5 Multijunction cells

The multijunction cell offers several advantages over single junction OPVs, many of which were touched on in Section 7.1.2 and subsequent sections. Briefly, these advantages include:

(i) *Increased coverage of the solar spectrum by combining subcells whose absorption spectra do not entirely overlap*. Given the narrow absorption spectra characteristic of excitonic materials, the case for employing multijunction cells for achieving high efficiency is compelling.

(ii) *Reduced thermalization losses in series-connected stacks*. The energy difference between the incident photon and the single junction cell energy gap is lost to heat, thus setting a thermodynamic limit to the maximum power that the cell can generate (see Section 7.3.2). By combining cells of different energy gaps, thermalization losses due to high energy photons absorbed by large energy gap subcells are reduced. Similarly, low energy photons are absorbed by subcells with smaller energy gaps. Thus, a multijunction cell can, in principle, exceed the thermodynamically limited efficiency of the single junction cell.

(iii) *Increased terminal voltage*. The V_{OC} of a series-connected multijunction cell is the sum of the V_{OC} of the subcells in the stack. Additionally, the multijunction cell current is limited by the subcell generating the lowest current. As a result, the series-connected stack is a "high voltage, low current" OPV. Consequently, tandem cells are less vulnerable to resistive losses that equal I^2R_{ser}.

(iv) *Reduced current increases cell efficiency*. Due to the generally high internal resistance of an OPV, the *FF* of a stacked cell can be higher than that of its constituent subcells under an equivalent illumination intensity. The *FF* is a function of R_{ser} (see Fig. 7.70). When the incident light is absorbed in different subcells in the stack, the resulting reduction in cell current results in an increased *FF*, and hence η_P.

(v) *Reduced current reduces the cell vulnerability to current-induced failure mechanisms*.

Subcells connected in parallel were also considered in Section 7.1.2. While parallel-connected multijunction cells do not require current balance, they nevertheless require matched operating voltages for each parallel element. In contrast to a series cell, parallel-connected cells can be "high current, low voltage" devices. For this reason, they lack many of the advantages of the series multijunction OPV, and hence will not be a focus of further discussion.

The ease of combining numerous materials and multiple layers in organic devices, and the benefits accrued by stacking subcells absorbing in different spectral regions, makes the series-connected multijunction architecture suited for achieving high efficiency and reliable OPVs. In this section we focus on three topics: (i) the design principles leading to cell optimization, (ii) the design of optically and electrically lossless CRZs between subcells, and (iii) examples of multijunction cell architectures used to understand how materials and processing problems are solved to achieve high efficiency.

7.5.1 Design principles of optimized multijunction OPVs

To develop a systematic approach to multijunction cell design, we consider the operating point of a series tandem (i.e. dual junction) cell, and then generalize the method to N junctions. Figure 7.176 illustrates the j–V characteristics of two cells comprising a tandem. Since the central terminals of tandems cannot be accessed, these characteristics correspond to single junction OPVs whose characteristics are close, but not identical to those expected when similar devices are used as subcells in the completed device. Subcell 1 has a low $V_{OC,1}$, whereas subcell 2 has the smaller $j_{SC,2}$ and lower fill factor. Presuming that the CRZ connecting the subcells is lossless, then the tandem meets two conditions:

$$V_{OC,T} = V_{OC,1} + V_{OC,2} \tag{7.193a}$$

and

$$j_{SC,T} = j_1 = j_2. \tag{7.193b}$$

This second condition corresponds to the horizontal dashed line in Fig. 7.176 where the current in a series-connected cell is balanced, that is, the current delivered by each subcell is equal. This requires that the voltage across the two subcells at short circuit meet the condition: $V_T = V_1 + V_2 = 0 \Rightarrow V_1 = -V_2$.

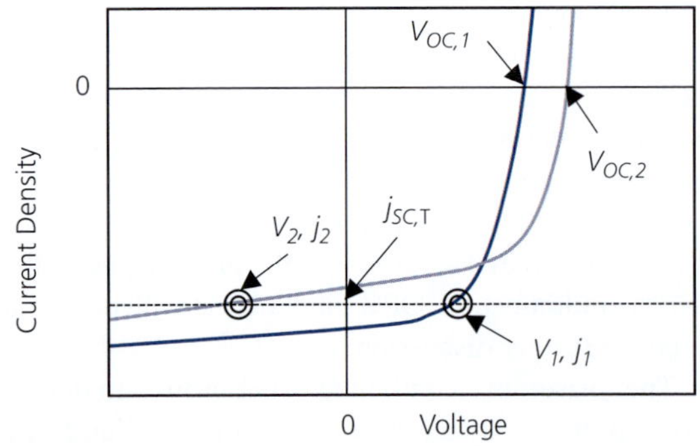

Figure 7.176 Current–voltage characteristics of two subcells and the operating current of a series-connected tandem solar cell comprising these subcells.

A similar analysis for a parallel tandem is provided in Fig. 7.103 in the context of the operation of a ternary BHJ. For a parallel diode connection, the operating points lie along vertical, constant voltage lines, and the currents are additive (the converse of Eq. 7.193).

The performance of the multijunction cell is modelled by constructing its j–V characteristics from those of the discrete devices, with the goal of finding j_M, V_M at the MPP, and hence η_P. This must be done using numerical methods along with the conditions of Eq. 7.193 at each (V, j) coordinate. Care must be taken such that the spacer and active layers in the discrete cells result in an optical field that is close to that experienced when they are placed in the stack. We start by defining $\Delta\eta_P$ as the power conversion efficiency penalty at the N-junction cell MPP relative to its value if each individual subcell in the stack were operating at its own MPP. That is,

$$\Delta\eta_P = 1 - \frac{j_{M,T}V_{M,T}}{\sum_{i=1}^{N} j_{M,i}V_{M,i}}, \tag{7.194}$$

where $j_{M,i}$ is the current density at MPP of subcell $i = 1, 2 \ldots N$ (as numbered from the reflective contact), and $V_{M,i}$ is the voltage at the MPP of the ith subcell, which is defined as

$$MPP_i = j_{M,i}V_{M,i} = FF_i j_{SC,i}V_{OC,i}. \tag{7.195}$$

Setting $\Delta\eta_P = 0$, we obtain $j_{M,T}V_{M-T} = j_{M,T}\sum_{i=1}^{N} V_{M,i}$.

Thus, if all subcells operate near to their individual $(V_{M,-i}, j_{M,i})$, then $\Delta\eta_P$ is minimized. Thus, a criterion to achieve the maximum power conversion efficiency, $\eta_{P,T}$, for a multijunction cell is to match $j_{M,i}$ for each subcell.

The foregoing analysis has been used to project the performance of a series tandem OPV comprised of two different sets of subcells whose single junction performances are listed in Table 7.25. One tandem (T1), combines cells *i* and *ii* with nearly equal fill factors, while the second tandem (T2) is comprised of cells *iii* and *iv* with widely different *FF*s. The different situations should lead to different MPPs and introduce significantly different power efficiency penalties when used in the tandem cells. Note that these data are obtained from four different, small molecule OPVs whose j–V characteristics and solar spectral responses were measured and fit to the diode theory in Section 7.1.2. Furthermore, the optical properties of the various compounds were measured to determine the response of the tandem cells in which they were placed. For reference, the cells used in the models are (*i*) DBP/C_{70} PM-HJ, (*ii*) DPASQ:DPSQ (4:6)/C_{70} bilayer, (*iii*) SubPc/C_{70} graded HJ, and (*iv*) DPSQ/C_{70} bilayer HJ (Lassiter et al., 2013a).

Figure 7.177a shows contour plots of the performance calculated for T1 whose front subcell, that is, SC-2 in Fig. 7.9, is cell *i*. When the *FF* of the two subcells are approximately matched, we see that FF_T is closer to the higher of the two individual fill factors ($FF_b =$ 0.66) when the subcell j_{SC} are matched. Also, the power penalty is very small if the individual subcells operate at approximately the same j_{SC} when used in the tandem. The relationship between $\Delta\eta_P$ and $j_{SC,1}$ and $j_{SC,2}$ becomes apparent for T1 in Fig. 7.177b. Here, the differences between cell *i* and SC-2, for example, are due to the position of cell *i* in the stack, the overlap of its spectrum with SC-1 (using the materials in cell *ii*), and the thicknesses and other parameters chosen for all of the layers comprising the tandem OPV. Thus, for devices in Fig. 7.177b (top and bottom panels) with very different j_{SC}, a power penalty of $> 6\%$ is incurred compared to the situation where $j_{SC,1} = j_{SC,2}$, which is approximately achieved for the scenario in Fig. 7.177b, center panel. As predicted above, when $P_M = j_M V_M$ is approximately equal for each subcell, then the power delivered by the tandem is given by $P_{M,T} = j_M(V_{M,1} + V_{M,2})$.

The penalties incurred by using significantly mismatched subcells *iii* and *iv* in T2, are considerably larger. In this case, FF_T trends towards that of the

Table 7.25 Example discrete cell characteristics used in Fig. 7.178 (Lassiter et al., 2013a)

Parameter	Cell *i*	Cell *ii*	Cell *iii*	Cell *iv*
V_{OC} (V)	0.89	0.96	1.04	0.94
FF	0.61	0.66	0.48	0.71
j_{SC} (mA/cm^2)	9.4	7.7	8.5	6.1
η_P (%)	5.0	4.9	4.3	4.1

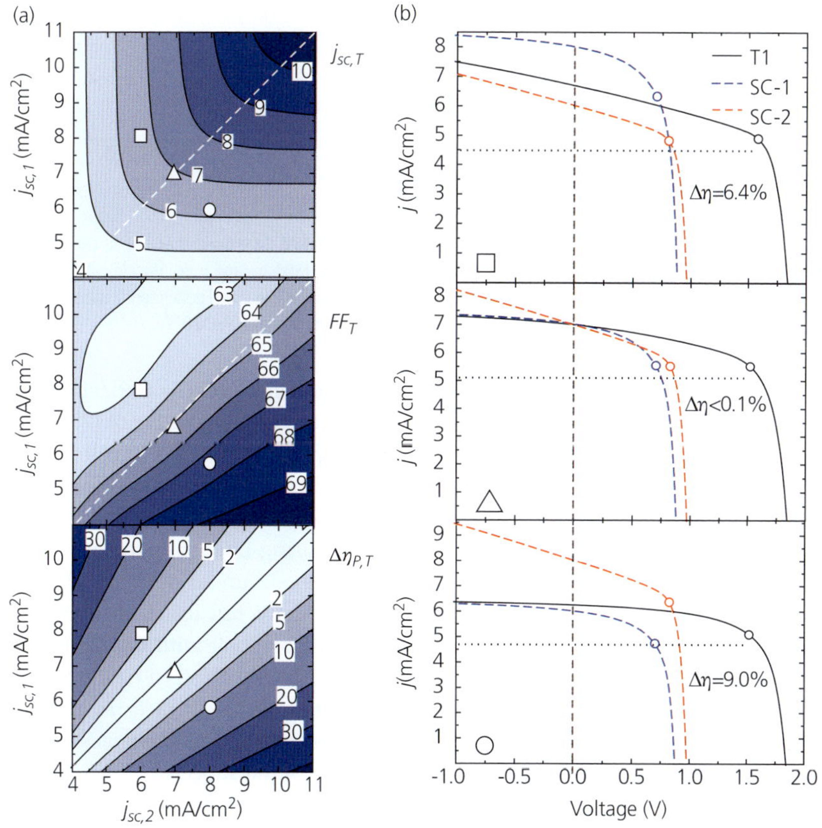

Figure 7.177 (a) From top to bottom: $j_{SC,T}$, FF_T, and $\Delta\eta_{P,T}$ for tandem T1 comprising cells *i* and *ii* in Table 7.25. The diagonal lines are for the optimal case of $j_{SC,2} = j_{SC,1}$. (b) Three *j-V* characteristics for the tandem T1 modelled in (a) employing front and back cells SC-2 and SC-1, respectively. The small circles on the characteristics indicate the MPPs of each device. The symbols in the lower left correspond to those shown in (a). After Lassiter et al. (2013a).

subcell with the lowest current, that is, when $j_{SC,1} \gg j_{SC,2}$, then $FF_T \to FF_{SC,2}$, and when $j_{SC,2} \gg j_{SC,1}$, then $FF_T \to FF_{SC,1}$. When the short circuit currents of the two subcells are equal, then $FF_T = (FF_{SC,1} + FF_{SC,2})/2$. Finally, the power penalties are also more severe when the individual fill factors and currents are increasingly asymmetric. The power penalty is then no longer centered along the diagonal in Fig. 7.177a, but rather $\Delta\eta_P$ it is larger when $j_{SC,2} > j_{SC,1}$. Hence, the tandem reaches the highest performance only when the constituent single junction cells have closely matched characteristics in the stacked architecture.

The angle-dependent sensitivity of thin, organic devices arises from optical interference between the stacks comprising several layers with different optical constants. We showed in Section 6.6 that interference significantly affects the OLED emission spectrum. The same angle-dependence affects the total diurnal energy generated by a solar cell, whose efficiency may vary with the incidence angle. The variation of the efficiency vs. incident angle of AM1.5G simulated illumination of a DBP:C_{70} single junction cell, a DTDCPB:C_{70} front subcell/DBP:C_{70} back subcell tandem, and four-junction cells comprising a double stack of the tandem structure is shown in Fig. 7.178. Details of the various devices are provided in Section 7.5.3. Also plotted is cosθ which is proportional to the expected cell angular response as a function of angle from the normal (at $\theta = 0$) in the absence of microcavity effects. The single and tandem junction cells closely follow the cosine law. Only the four-junction cell, comprising a stack of over 20 layers and a thickness of 250 nm, shows a noticeable angle-dependence at $\theta > 45°$. While the efficiency at higher angles is marginally lower, the total energy available for harvesting at such oblique angles also decreases. Hence, the penalty incurred due to angle dependent interference within a deeply stacked multijunction cell is small compared to the advantage of its significantly higher efficiency at normal incidence.

The angle dependence of the current in even deeply stacked OPVs is significantly less pronounced than the angle-dependent emission observed in stacked OLEDs. This is a result of the "damping" effect of absorption in OPVs that is absent in light emitters. As light propagates within the multijunction stack it is partially absorbed by each subcell, attenuating the magnitude of the optical interference.

To understand where the angle dependence originates in the stacks, light biasing described in Section 7.3.3 was used to separately determine the quantum efficiencies vs. angle for both the DTDCPB:

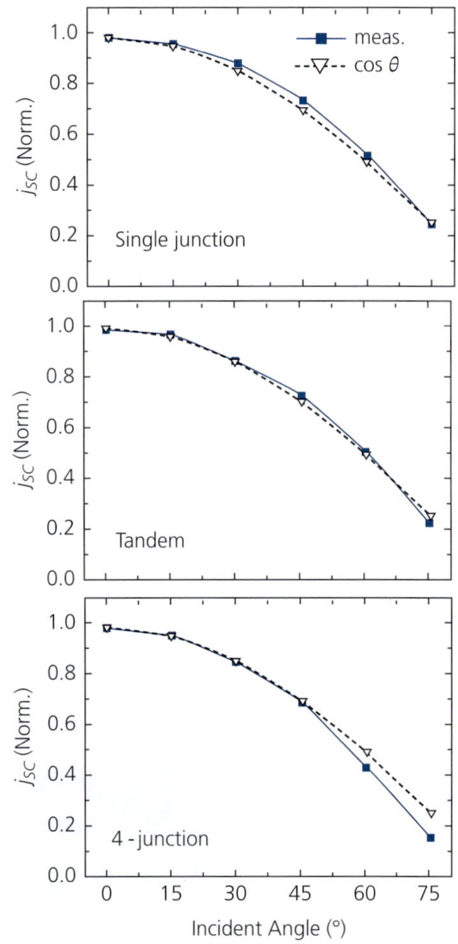

Figure 7.178 Measured short circuit current vs. angle of incidence, θ, for single, tandem, and four-junction OPVs described in Section 7.5.3. Also shown is the cos θ dependence expected for a device with no microcavity effects (Che, 2018).

C_{70} front and DBP:C_{70} back subcells. The experimental setup is shown in Fig. 7.179a. The η_{ext} of the long wavelength DTDCPB:C_{70} cell is obtained by providing light bias from an intense, 505 nm wavelength LED, whereas the short wavelength DBP:C_{70} cell is biased using LED emission at 735 nm. The η_{ext} of the separate cells under normally incident monochromatic light delivered using an optical fiber is shown in Fig. 7.179b, along with the light biasing wavelengths (dashed lines). This is a nearly ideal system to obtain accurate efficiency measurements of the separate components due to their lack of significant spectral overlap of the subcells.

The angle of the monochromatic light source is varied from $\theta = 0$ to 60°, with results for the two cells shown in Fig. 7.179c and d. Almost all angle dependence arises from the short wavelength absorbing back subcell (SC-1). This is consistent with

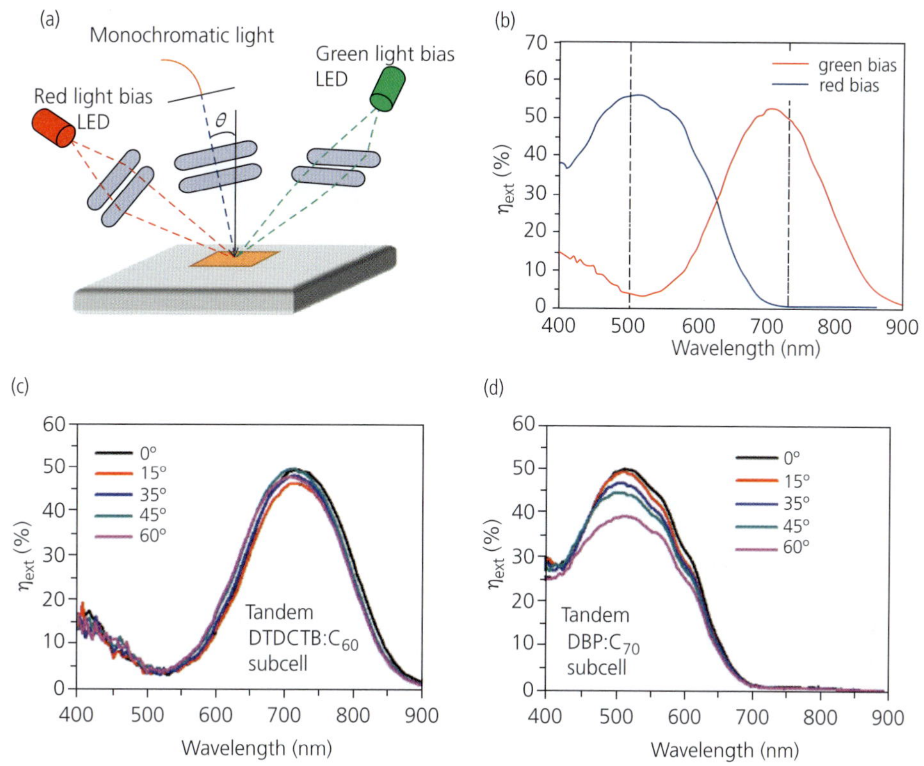

Figure 7.179 (a) Experimental setup for measuring the angle-dependent external quantum efficiencies of separate subcells in a tandem OPV. (b) The η_{ext} of the separate subcells measured at normal incidence with light bias wavelengths shown by the dashed lines, and η_{ext} for the (c) front and (d) back subcells vs. angle, θ. After Che (2018).

optical models, whereby short wavelength light in a microcavity has a longer optical path length (equal to $nd/\lambda\cos\theta$, where d is the layer thickness and n is its index of refraction) than light at longer wavelengths. For this device, there is a 15% decrease in η_{ext} of the DBP:C_{70} cell as θ is varied from 0° to 60°, which accounts for nearly all of the deviation of the cell response in Fig. 7.178 over this angular range.

7.5.2 Charge recombination zones

The CRZ is interposed between each subcell in a multijunction OPV. It serves a purpose similar to a tunnel junction between subcells in inorganic multijunctions; that is, it provides a site for electrons in one subcell to recombine with holes in the adjacent subcell to "complete the circuit." In its absence, the photogenerated electrons and holes cannot reach the opposing electrodes, creating a buildup of charge and, consequently an opposing internal potential that halts the current flow. The ideal CRZ should be electrically lossless, that is, it must provide a site with 100% recombination efficiency. It must also be optically lossless to avoid absorption outside of the OPV active region.

The first demonstration of a CRZ was a thin (3 nm) Au layer placed between two identical H_2Pc-MePTC subcells (Fig. 7.89) (Hiramoto et al., 1990). Energy barriers between the Au thin film and the subcells leads to a less than unity recombination efficiency, and the relatively thick layer also introduces significant optical losses (Hadipour et al., 2006). This situation was improved by replacing Au with an ultrathin (1–5 Å) Ag layer that is both optically and electrically lossless (Yakimov and Forrest, 2002, Che et al., 2014).

The submonolayer Ag CRZ is orders of magnitude thinner than the optical skin depth (Eq. 7.176), making the layer nearly transparent. Its effectiveness as a site for charge recombination is understood in the context of the TEM images in Fig. 7.180. The thin Ag deposit does not consistently wet the surface; rather it forms a widely dispersed population of Ag clusters on the organic surface that range from < 1 nm to 7–8 nm diameter (Fig. 7.180a). The cross-sectional image of a CuPc/PTCBI tandem cell in Fig. 7.180b shows that the clusters do not form into a 2D layer, but rather diffuse into the organic film, creating a band of nanoparticles (NPs).

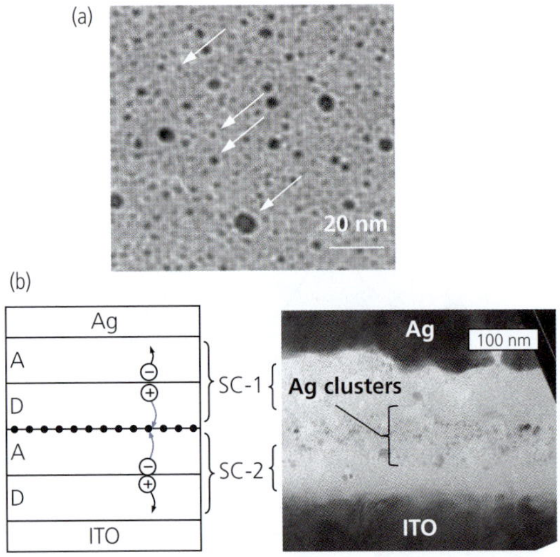

Figure 7.180 (a) Transmission electron microscope images of a PTCBI surface on which Ag nanoparticles (mean Ag film thickness, 5 Å) are deposited. Arrows point to Ag clusters of different diameters (Peumans et al., 2003b). (b) *Left*: Schematic diagram of a tandem cell with a Ag NP CRZ (dots) between subcells 1 and 2 (SC-1, SC-2, respectively). *Right*: TEM image of a thinned cross-section of the tandem cell. Ag NPs are shown as dark clusters (Rand et al., 2004).

The NP layer provides for highly efficient charge recombination without optical losses. When an electron from an adjacent acceptor layer is at distance r from a neutral Ag NP metal sphere, it is attracted by its mirror image with a force of $q^2/(4\pi\varepsilon_r\varepsilon_0 r^2)$. Once the electron binds to the sphere, the NP is negatively charged, and subsequently attracts a hole from the adjacent donor layer with the same force. The electron and hole recombine, and the process begins once more starting with the neutral NP. The binding energy of an electron to the NP is $E_B = q^2/(8\pi\varepsilon_r\varepsilon_0 r)$. At room temperature, and at a distance, $r = 4$ nm, $E_B = 2.4k_BT$. From the small field approximation of Onsager theory, the capture probability of the charge is given by (Onsager, 1938, Tachiya, 1988)

$$P_{cap} = \exp\left(-\frac{r_c}{r}\right)\left(1 + \frac{qr_c}{2k_BT}F\right), \tag{7.196}$$

where the capture radius is $r_c = q^2/(4\pi\varepsilon_r\varepsilon_0 k_BT)$. For $F = 0$, $P_{cap} \to 0.99$ (see Section 4.7.2), making the NPs extremely efficient recombination centers even though they cover only a small fraction of the organic surface.

Closely packed metallic NPs also affect the incident illumination by supporting plasmons that reradiate into the adjacent layers. Silver has a pronounced plasmonic absorption feature centered at approximately $\lambda = 440$ nm (Rand et al., 2004). The plasmonic field is

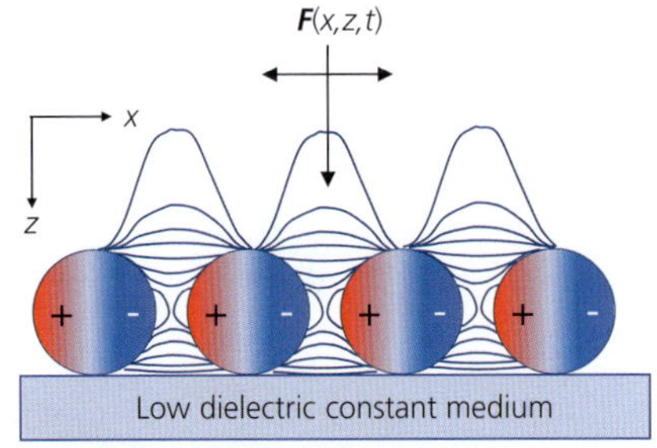

Figure 7.181 Instantaneous field lines due to plasmons in metal spheres induced by an incident field, *F*(*x*,*z*,*t*). The metal spheres lie on a half plane occupied by a low dielectric medium (e.g. glass), and the upper half-plane is filled by a higher dielectric organic film resulting in an asymmetric field profile. The instantaneous induced charge distribution is shown by "+" and "−."

depicted in Fig. 7.181, where the normally incident electromagnetic field induces an instantaneous redistribution of charge on the surfaces of the spheres. The charge density wave reradiates in a pattern determined by the surrounding dielectric medium. In the illustration, the spheres are located on the surface of a low index medium (e.g. glass), and are buried in a higher index dielectric such as an organic semiconductor. The NPs act as an antenna array that "focuses" the field intensity within a nearby junction, thus increasing absorption. The field intensification is responsible for the increased efficiency of multijunction cells in Fig. 7.90b, where the dual junction cell has a power conversion efficiency > 2.5 times that of the single junction cell. This 20% enhancement is due to the increased field arising from plasmons supported by the Ag NP layer.

A simulation of the electric field enhancement due to a Ag NP layer embedded in an organic film, and sandwiched between ITO and Al is shown in Fig. 7.182a (Peumans, 2004). The inset shows an increase of 100–1000 times the incident field. Indeed, the field intensity remains high in a zone whose width is 10–20 nm, which overlaps the D and A layers on either side. Figure 7.182b shows the measured absorbance for a Ag NP layer on quartz, a PTCBI layer on quartz, and a quartz/NP/PTCBI structure. The absorbance of the latter structure (black line) is nearly double the sum (dashed line) of the contributions of just the NPs and PTCBI, accounting for the increased efficiency of the tandem cell compared to the single junction cell in Fig. 7.90.

The most common CRZ used for solution processed multijunction cells comprises a layer of ZnO NPs (Gilot et al., 2007, Janssen et al., 2007, Li et al., 2013). The metal oxide NPs were initially

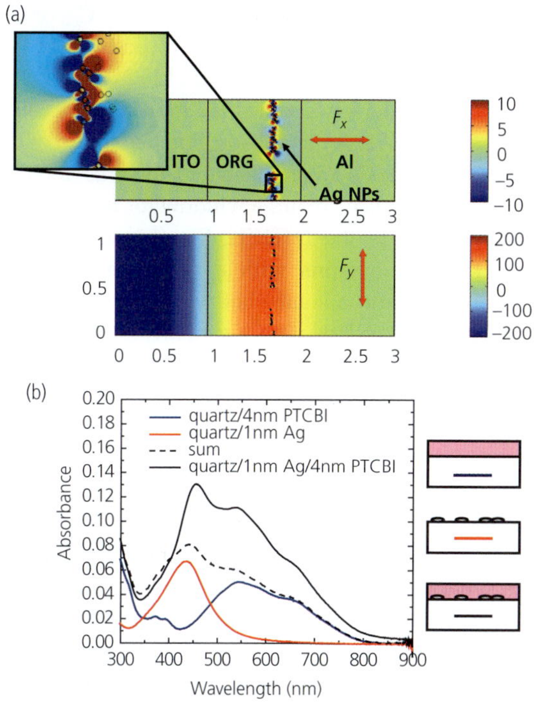

Figure 7.182 (a) Calculated relative perpendicular (top) and parallel (bottom) electric field intensities in an ITO/organic/Al structure. The field strengths are relative, with red being the most intense. Scales on the plot axes are in 100 nm units. Inset: A detail of a section of the NP layer. Circles indicate positions of the layers. (b) Absorbance of the quartz/NP/PTCBI structure (black line) is greater than the sum (dashed line) of the quartz/NP (red line) and quartz/PTCBI (blue line) layers. After Peumans (2004).

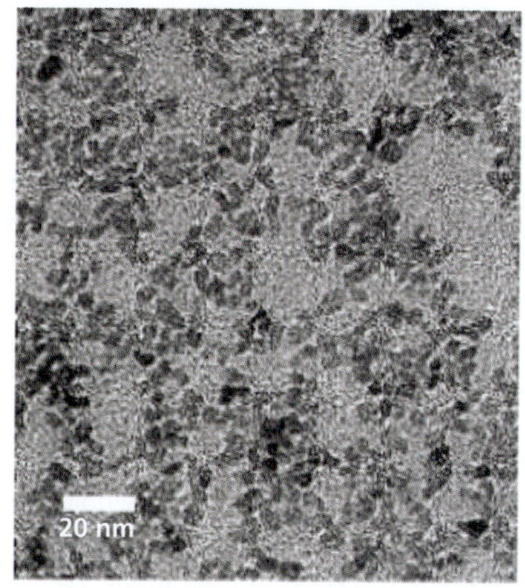

Figure 7.183 TEM image of a monodisperse film of 5 nm diameter ZnO NPs (Beek et al., 2005).

introduced as an acceptor in BHJ cells employing MDMO-PPV as the donor (Beek et al., 2005). However, an electron-transporting ZnO NP layer, in combination with a pH-neutral PEDOT:PSS hole transporting layer, forms an efficient and transparent CRZ that can be cast from solution.

The monodisperse particle populations are synthesized by a combination of hydrolysis and condensation of a Zn acetate dehydrate in a KOH: methanol solution. A TEM image of a solution cast ZnO NP film is shown in Fig. 7.183. Layer thicknesses of 10–30 nm provide optimal interconnection between subcells with minimal optical losses (Gilot et al., 2007). To produce a more continuous network and drive off excess solution, the NPs are often annealed at elevated temperatures (e.g. 150°C for 10 min.), although this may damage the underlying organic layers (Hau et al., 2010, Kouijzer et al., 2012).

The ZnO NPs can result in a Schottky barrier with the adjacent layers, which creates an internal series resistance within the multijunction cell. A *self-assembled monolayer* (SAM) cast onto the NP layer surface is one means for changing the surface energy with the objective of decreasing the barrier to charge injection from the active region into the CRZ. A SAM is a space-filling, ultrathin (e.g. a single molecular layer) film of molecules. When attached to a surface, it can change the film morphology and/or energetics. A SAM comprising pyrrolidine-functionalized C_{60} molecules was used to improve injection from the compound ZnO NP/PEDOT:PSS CRZ in multijunction polymer cells (Zhang et al., 2006). Following the deposition of the NP layer, the SAM is deposited from a solution of CB and THF, and then excess molecules unattached to the ZnO surface are removed by rinsing in a pure THF solution (Hau et al., 2008). When applied to single, double and triple junction OPVs based on P3HT:PC_{61}BM BHJs, the SAM improves the *FF* (e.g. from 0.49 to 0.61 for single junction cells without and with the SAM, and 0.49 to 0.55 for tandem cells) due to a significant reduction in barrier resistance at the BHJ/CRZ interface (Hau et al., 2010).

7.5.3 Example multijunction OPV structures and performances

Multijunction OPVs have enjoyed continuous improvement since their introduction in the 1990s, to the point where efficiencies of over 15% have been achieved. The narrow absorption spectrum characteristic of excitonic materials makes it probable that multijunction cells, with their combination of several absorbing materials that achieve a broader spectral coverage than single junction cells will ultimately become the structure that achieves the highest efficiencies. The lower current and higher voltages offered by series-connected cells also leads to reduced resistive losses when used in large area modules. In this section we provide examples of tandem and triple junction cells as a means to point out the benefits of, and challenges in realizing multijunction organic solar cells.

Vapor deposited multijunction solar cells. Vapor deposited materials easily form, multilayer complex devices since dry processing does not affect the morphology of previously deposited layers, regardless of their composition or conditions used in their deposition. Unfortunately, there are only a limited number of high efficiency donors and NFAs that can be thermally deposited without decomposition. Nevertheless, multijunction OPVs with at least some of the subcells deposited from the vapor phase have achieved among the highest OPV efficiencies to date (Che et al., 2018b, Li et al., 2018).

Vacuum-deposited tandem, triple, and quadruple junction cells are shown in Fig. 7.184a. A DTDCPB: C_{60} (1:1)/C_{60} (5 nm) PM-HJ cell (SC-2) used to absorb primarily in the infrared is paired with a DPB:C_{70} (10:1)/C_{70} (5 nm) PM-HJ cell (SC-1) that absorbs broadly across the visible. The CRZs interposed between the subcells comprise a BPhen:C_{60} (1:1) (5 nm) electron filtering buffer adjacent to the neat C_{60} or C_{70} layer of the corresponding cell, followed by a 0.1 nm thick Ag NP layer, and a 5 nm thick MoO_3 spacer. The optical field intensity profile for a four-junction cell is shown in Fig. 7.184b. The absorption maximum of the DPB:C_{70} (10:1)/C_{70} (5 nm) active region is centered at $\lambda = 500$ nm, which creates a window for transmitting NIR photons to the DTDCPB:C_{60} (1:1)/C_{60} (5 nm) PM-HJ active regions with a peak absorption at 650–700 nm, and a somewhat weaker absorption at 400 nm that overlaps with a similarly weak absorption in the DBP-based subcell. Importantly, the BPhen:C_{60} (1:1) (5 nm)/Ag (0.1 nm)/MoO_3 (5 nm) CRZs are nearly completely transparent across the visible and NIR, as shown by the solid blue coloration of those layers in the figure.

The CRZ is also nearly electrically lossless. In Table 7.26, we see that the sum of the V_{OC} values for the separately fabricated subcells is 0.82 + 0.90 V = 1.72 V. This is identical to the measured V_{OC} of the tandem. The quadruple OPV should have double the voltage of the tandem, that is, 3.44 V. The actual voltage is only slightly lower than this at 3.38 V. The residual voltage loss may be due to losses from lower illumination intensity of the individual subcells in a four-junction stack. Recall that $qV_{OC} \propto k_B T \log(j_{SC}/j_0 + 1)$. Since j_{SC} is lower for multijunction than for single junction cells, we expect that V_{OC} will also be less than that of individual subcells

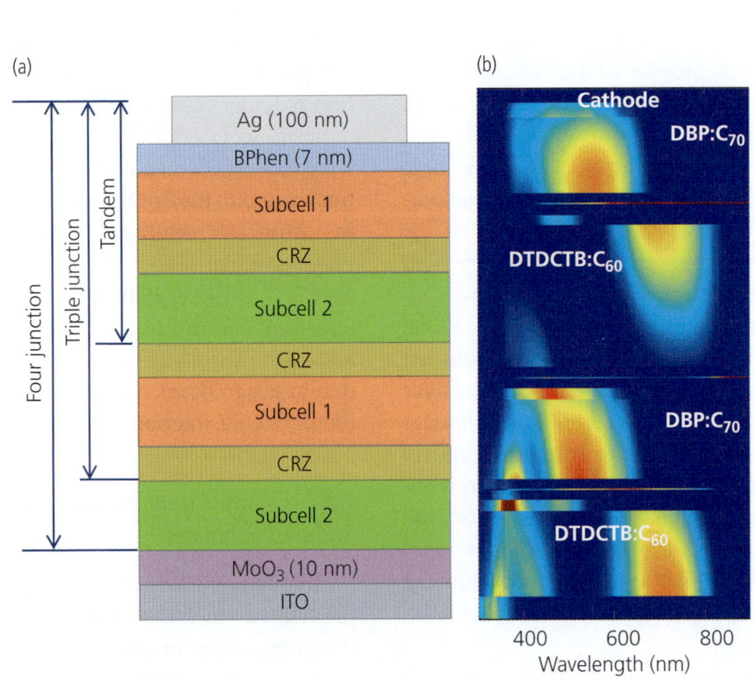

Figure 7.184 (a) Vacuum-deposited multijunction OPV structures. The active regions are Subcell 1: DTDCTB:C_{60} (1:1) and DBP:C_{70} (10:1). (b) Optical field distribution calculated for a quad-junction cell. Red indicates the highest absorption, and blue the lowest. After Che et al. (2014) and Che (2018).

exposed to full 1 sun intensity. The drop in V_{OC} is partially compensated by an increased FF at the reduced j_{SC} (cf. Fig. 7.72). That is, FF is limited to some extent by cell series resistance that becomes less significant as the intensity per subcell decreases with number of junctions in the stack.

The power conversion efficiencies summarized in Table 7.26 show that the tandem does not have double the efficiency of the individual subcells. The subcells in the stack have different (thinner) layer thicknesses to ensure current balance. Nevertheless, the tandem cell has $\eta_P = 10\%$, which is substantially higher than either of the optimized single junction cells of 5–6%. Increasing the number of stacked elements to three and four results in a continuously decreasing advantage, with $\eta_P = 11.1\%$ and 12.6%, respectively. Note, however, that these efficiencies may be approximately 10% higher than their actual values as the measurements were made without a mask on very small (1 mm diameter) devices. Thus, edge currents result in an artificially higher j_{SC} when the samples are flooded with illumination (Che et al., 2014).

Table 7.26 Measured subcells and tandem, triple, and quadruple junction cell performances (Che et al., 2014, Che, 2018)

Cells[a]	j_{SC} (mA/cm^2)	V_{OC} (V)	FF	η_P (%)
SC-1	11.3	0.90	0.61	6.2
SC-2	10.4	0.82	0.59	5.1
Tandem	9.9	1.72	0.59	10.0
Triple	7.3	2.58	0.59	11.1
Quadruple	6.4	3.38	0.58	12.6

[a] SC-2 refers to a PM-HJ DTDCTB:C_{60} subcell nearest the anode in the tandem; SC-1 refers to the PM-HJ DBP:C_{70} subcell near the cathode in the tandem; "tandem" employs a BPhen:C_{60} interconnecting buffer layer; "triple" employs a second DBP:C_{70} subcell adjacent to the substrate in additional to the tandem structure. Quadruple employs a final DTDCTB:C_{60} subcell nearest the cathode.

We have seen that the highest efficiency single junction OPVs employ solution-processed small molecule NFAs. Hence, to employ these materials in tandems requires that at least one subcell be deposited from solution. A tandem device employing a combination of vapor deposited, fullerene-based subcell (SC-2) and a second, NIR absorbing NFA-based subcell (SC-1) is shown in Fig. 7.185 (Che et al., 2018b). The challenge in fabricating this device is to avoid damage to the vapor deposited SC-2 during the solution deposition of SC-1 onto its surface with only a thin, and potentially porous CRZ separating the stacked elements.

The problem of eroding the underlying layers is solved by using an acidic (pH = 1–2) PEDOT:PSS layer cast onto the surface of the Ag NP/BPhen:C_{60} CRZ. Both the Bphen:C_{60}/Ag NP layer and PEDOT:PSS are hydrophilic, leading to uniform wetting by the PEDOT:PSS. Pinholes, or other physical defects through the CRZ may expose the underlying DTDCPB:C_{70} (1:2) active region, but the PEDOT:PSS does not penetrate into the pre-deposited active layer. The hydrophobic PCE-10:BT-CIC (1:1.5) BHJ

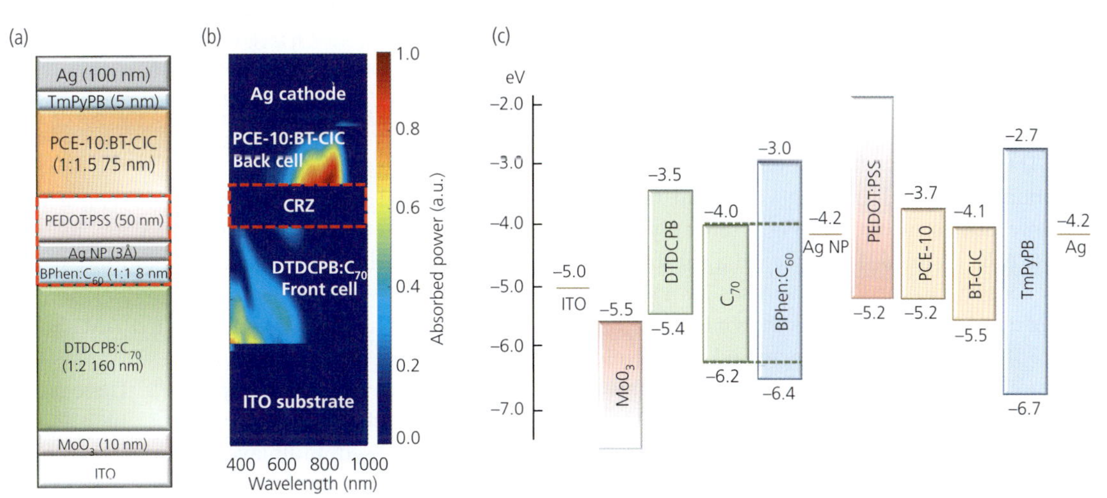

Figure 7.185 (a) Architecture and (b) calculated absorbed power distribution in the solution/vapor deposited fullerene/NFA-based tandem OPV. (c) Equilibrium energy level diagram of the cell in (a). The energy levels of C_{60} in the electron-filtering BPhen:C_{60} layer are indicated by dashed lines. The buffer layer used for the back cell (SC-1) is the wide energy gap, vapor-deposited TmPyPB (Che et al., 2018b).

Che, X., Li, Y., Qu, Y. & Forrest, S. R. 2018b. High Fabrication Yield Organic Tandem Photovoltaics Combining Vacuum- and Solution-Processed Subcells with 15% Efficiency. *Nature Energy*, 3, 422 non-fullerene active region is cast from a CF/CB solution (10:1 by vol.) which prevents it from fully wetting, and hence dissolving the aqueous PEDOT: PSS layer underneath, thereby preventing damage to the underlying layers. The device is completed by the vacuum deposition of a 5 nm thick TmPyPB exciton blocking buffer layer. Using the combination of hydrophobic and hydrophilic materials results in >95% fabrication yield of devices ranging in areas from 2 mm^2 to 1 cm^2. In addition, the performance characteristics of the devices are stable over storage periods of several months, indicating that the layers do not host residual solvents that may continue to degrade the layers in the completed device (Che et al., 2018a).

The optical intensity distribution within the tandem cell shows nearly complete absorption in the visible by the vacuum-deposited DTDCPB:C_{70} mixed junction, with the NIR illumination out to 900 nm absorbed in the PCE-10:BT-CIC BHJ non-fullerene active region. The 2 mm^2 antireflection coated (ARC) devices showed an efficiency of up to 15% under 1 sun, AM1.5G simulated illumination, decreasing to 11.5% for 1 cm^2, non-ARC devices. The drop in efficiency for the larger devices is primarily a result of the series resistance of the ITO anode contact to SC-2 (see Section 7.3.3).

The efficiency of devices without the ARC is approximately 14.3%. From Eq. 7.26, we find that Fresnel reflections at the glass surface result in a 4% loss for a glass refractive index of 1.5. A low index, bilayer ARC comprising 120 nm MgF_2 (with index $n = 1.38$) and a low index SiO_2 ($n = 1.12$) reduce the reflections to $< 1\%$. We have shown in Section 6.6.3 that thermally depositing SiO_2 at a large oblique angle relative to the source results in a porous layer with an index that is the weighted mean of the volume of air and SiO_2 within the layer. Oblique incidence of the vapor results in random nucleation of silica across the substrate. The islands thus formed then shadow further deposition of material, creating voids that make an open, low index structure (see Fig. 7.186a). By depositing the ARC on the distal surface of a substrate, the optical coupling to the OPV increases by 4% across much of the visible spectrum, as shown in Fig. 7.186b. In Fig. 7.186c we find that the increased optical intensity results in a concomitant increase in j_{SC}, thus increasing the maximum η_P from 14.3% to 15%.

This same solution/vapor deposited device structure was demonstrated, except where the solution-processed binary layer was replaced by a broader NIR absorbing DA_1A_2 ternary subcell. The subcell comprised a blend of PCE-10:BT-CIC:TT-FIC (1:1.25:0.5), with absorption extending to 1000 nm. By including an ARC, The device efficiency using light biasing to obtain the quantum efficiencies of both the front and back subcells, yielding a nearly constant $\eta_{ext} = 80\%$ from 380 nm – 900 nm. Including an ARC on the bottom glass substrate surface resulted in $\eta_P = 15.4\%$, with $V_{OC} = 1.56$V, $FF = 0.71$, and $j_{SC} =$ 13.8 mA/cm^2 under 1 sun, simulated AM1.5G illumination (Li et al., 2018).

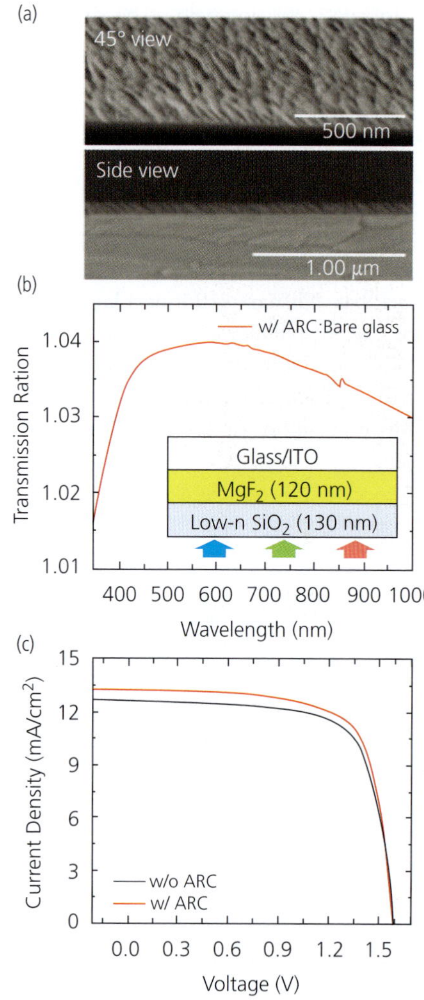

Figure 7.186 (a) Two SEM views of the surface of the SiO_2 ARC layer thermally deposited at glancing angle of 85° from normal. (b) Ratio of transmitted light through a glass substrate with a bilayer ARC to that of an uncoated surface. (c) Fourth quadrant *j–V* characteristics of the tandem OPV in Fig. 7.185 with and without an ARC (Che et al., 2018b).

Solution-processed multijunction cells. There are two principal methods for achieving high quality solution-processed multilayer structures: using orthogonal solvents to deposit successive layers, and inserting an insoluble and relatively thick CRZ between the subcells. In either case, the options for achieving optimized optical absorption and electrical performance are limited by the inherent constraints on the choice of materials, solvents and desired morphologies.

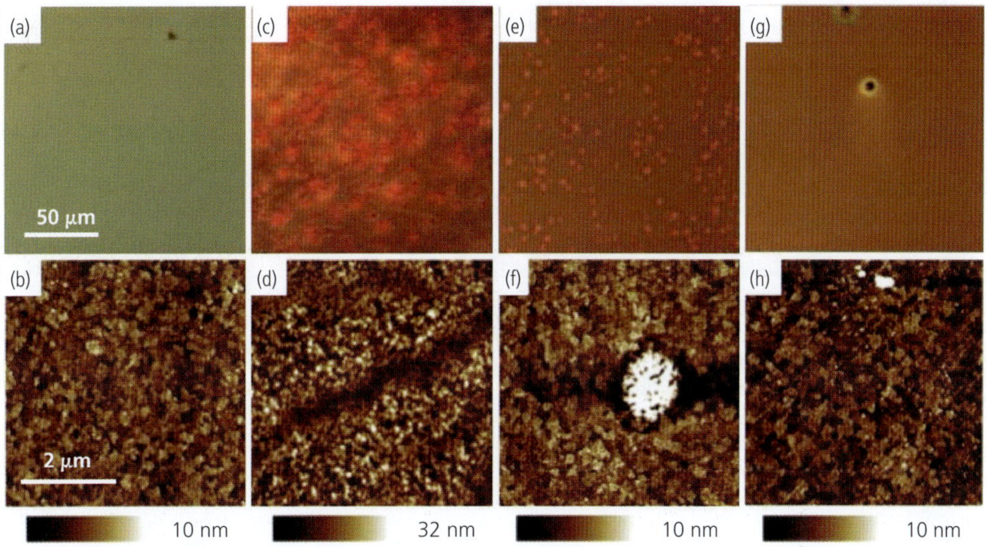

Figure 7.187 Optical (top row) and atomic force (bottom row) micrographs for samples processed using: (a, b) SVA, no THF exposure; (c, d) as-cast, 5 nm MoO_3, THF exposure; (e, f) SVA, 5 nm MoO_3 layer, THF exposure; (g, h) SVA, 20 nm MoO_3, THF exposure. The vertical scales for the bottom row of micrographs are indicated beneath each image (Lassiter et al., 2013b).

As discussed in Section 5.4.2, only three orthogonal (i.e. non-interacting) solvent classes are employed in organic electronic device fabrication: aqueous, organic, and the less frequently used fluorinated solvents. Hence, using orthogonal solvent systems to produce deeply stacked multijunction OPVs offers only limited opportunities to achieve the desired structures. One example of an orthogonal solution-processed, *f*SQ-based tandem cell used a combination of CF and THF solutions in the following structure: glass/ITO/25 nm MoO_3/15 nm blended *f*SQ (CF)/10 nm C_{70}/5 nm PTCBI/0.1 nm Ag/7 nm MoO_3/15 nm blended *f*SQ (THF)/28 nm C_{60}/5 nm PTCBI/100 nm Ag (Lassiter et al., 2013b). The layers were solvent vapor annealed after deposition of each PTCBI layer. The *f*SQ layers comprised a DPASQ:DPSQ (4:6 by vol.) blend. The neat C_{70} and other layers in the stack (e.g. MoO_3, PTCBI and Ag) were vacuum deposited after the active, blended layers were spun-on from their respective solutions. The SVA was accomplished by exposure of the layers to a DCM vapor to achieve a dense, nanocrystalline morphology of the *f*SQ blend. The annealed morphology reduces the layer porosity, improving its resistance to penetration by the solvents applied during subsequent processing. The MoO_3 and C_{70} layers on the surface of the front subcell (SC-2) provide additional barriers that impede penetration of the THF used in deposition of SC-1 into underlying layers.

Damage incurred by spinning THF on the top of the base structure of: glass/ITO/25 nm MoO_3/15 nm blended *f*SQ (spun on using CF)/10 nm C_{70}/5 nm PTCBI/0.1 nm Ag/x MoO_3, where $x = 5$ or 20 nm is shown in the micrographs in Fig. 7.187. Figures 7.187a and b correspond to a SVA sample with a MoO_3 thickness of $x = 5$ nm. The layer was not exposed to THF, resulting in a smooth surface morphology, with a rms roughness of 1.5 nm. Figures 7.187c and d show as-cast films with $x = 5$ nm, where THF was spun on the top of the same structure. The films significantly roughen to 4.9 nm due to re-dissolution of the underlying *f*SQ layers. In Figs. 7.187e and f, the layers were solvent vapor annealed after deposition of PTCBI, followed by exposure to THF. The film contains ~1 μm-diameter by 10 nm-high protrusions covering 10% of its surface due to swelling from THF penetration into the underlying layers. As the MoO_3 thickness is increased to $x = 20$ nm, the density of protrusions decreases, and are nearly eliminated (Fig. 7.87g and h). Tandem cells processed using these structures and procedures had $V_{OC} = 1.78$ V, $FF = 0.67$, and $\eta_P = 6.2\%$ (Lassiter et al., 2013b).

In this example, the devices areas were only 1 mm^2. It is unclear, therefore, what yield can be achieved by scaling the OPVs to larger areas. Their vulnerability to pinholes, along with the limited selection of available orthogonal solvent chemistries, has focused work on solution-processed tandem designs that employ relatively thick ZnO NP/PEDOT:PSS CRZs that have a low density of pinholes and other physical defects.

Fullerene-based tandem and triple junction cells based entirely on solution processed layers (with the

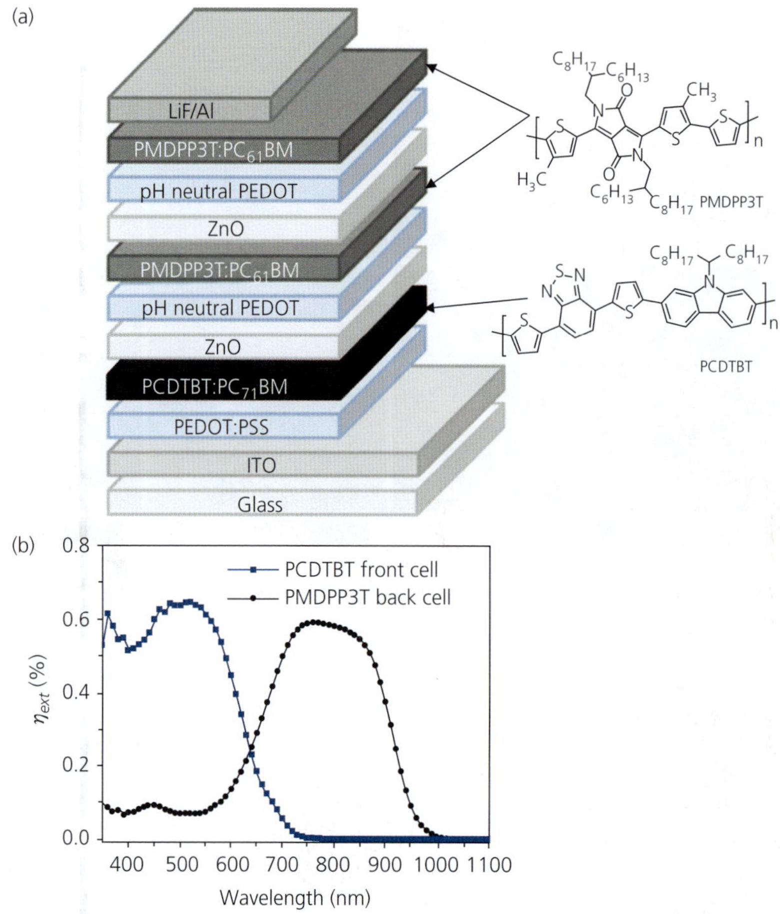

Figure 7.188 (a) Schematic illustration of a solution deposited triple junction cell. The short-wavelength absorbing PCDTBT-based SC-3 and the longer-wavelength absorbing PMDPP3T cells, SC-1 and SC-2 are shown along with the active polymer molecular structures. (b) External quantum efficiencies of the subcells employed in a tandem cell that contains only a single PMDPP3T-based element (Li et al., 2013).

exception of the cathode and anode contacts) have been demonstrated with at least a 50% increase in efficiency over optimized single junction cells that are used with the stacked devices. The structure of the triple junction cell is shown in Fig. 7.188a, along with the materials comprising both the long and short wavelength subcells. The short wavelength absorbing SC-3 positioned adjacent to the anode is a BHJ comprising a PCDTBT:$PC_{71}BM$ (1:4) blend. This is combined with a PMDPP3T:$PC_{61}BM$ (1:3) BHJ that absorbs at wavelengths > 900 nm. The triple junction analog of the tandem employs two PMDPP3T:$PC_{61}BM$ elements, SC-1 and SC-2.

The external efficiency spectra of the two subcells in the tandem are independently obtained by appropriate light and electrical biasing (Section 7.3.3), with the results shown in Fig. 7.188b. The peak efficiencies of the two cells is ~60%, which is nearly identical to the efficiencies obtained in the discrete cells as a consequence of the complementary spectra of the subcells employed.

Table 7.27 Performance characteristics of single, tandem, and triple junction, solution-processed fullerene-based cells (Li et al., 2013)

Cells	j_{SC} (mA/cm^2)	V_{OC} (V)	*FF*	η_P (%)
PCDTBT:$PC_{71}BM$	9.76	0.87	0.56	4.73
PMDPP3T:$PC_{61}BM$	15.3	0.61	0.65	6.00
Tandem	9.58	1.49	0.62	8.90
Triple	7.34	2.09	0.63	9.64

Measurements made at 1 sun intensity, AM1.5G simulated illumination without solar spectral mismatch correction.

A summary of the performance of the single, tandem and triple junction cells is provided in Table 7.27. The tandem cell efficiency is significantly higher than that of the individual cells, with $V_{OC,T} \approx V_{OC,1} + V_{OC,2}$, that indicates that the CRZ has very low

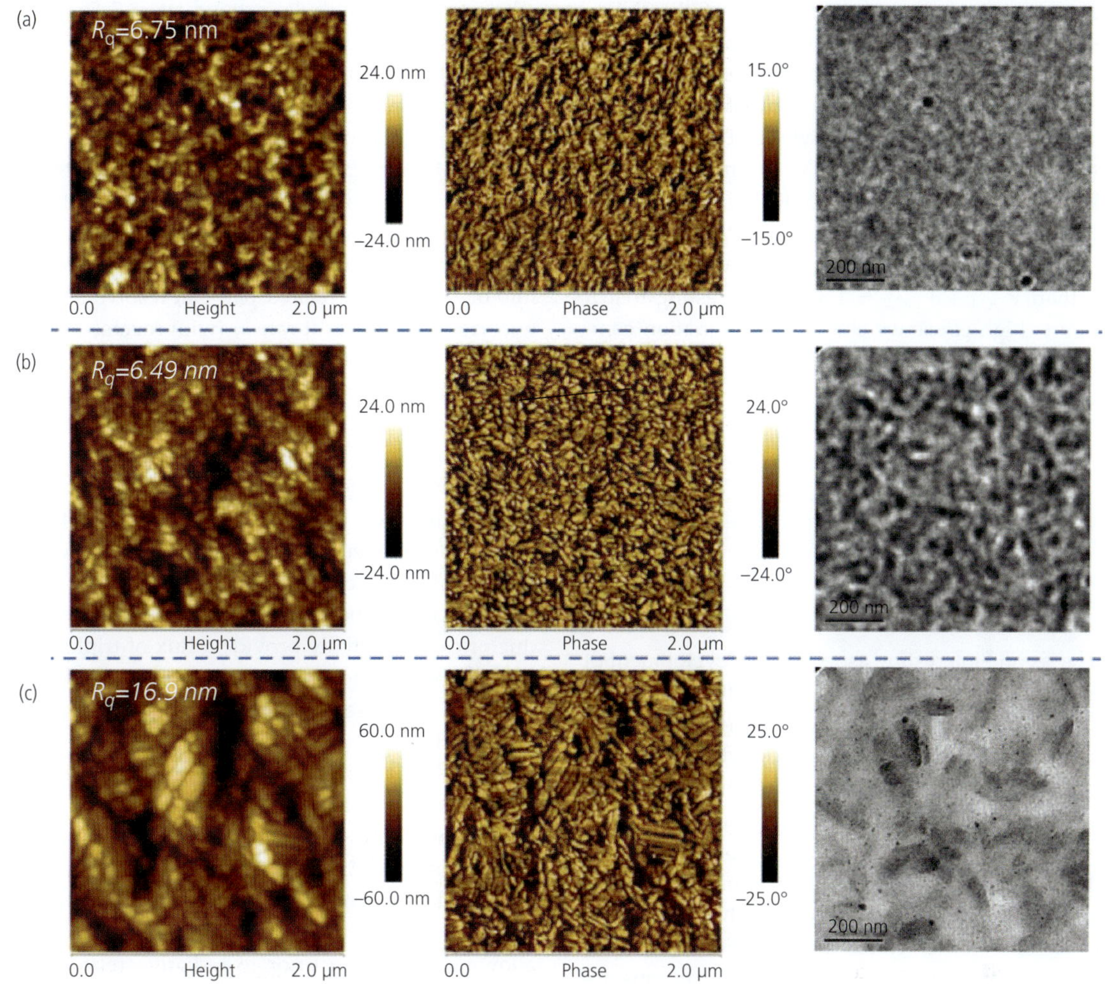

Figure 7.189 Micrographs of a PBDTTT-E-T:IEICO blend film (a) as-deposited, (b) after 1 min and (c) after 4 min SVA in a chlorobenzene solution. From left to right: AFM topography, AFM phase, and TEM images (Cui et al., 2017).

electrical loss. The tandem cell efficiency is limited by the unbalanced current generation of the short and long wavelength elements. That is, the current from the long wavelength SC-2 is larger than from SC-3. The balance is improved by splitting the long wavelength cell into two subcells, SC-1 and SC-2 in a 1+2 triple junction design. While this decreases the current generated by an individual PMDPP3T:PC_{61}BM BHJ, the efficiency is increased by the increases in V_{OC} and FF, and the incrementally higher current generated by SC-3. As a result, the triple junction cell has the highest power conversion efficiency of 9.64%.

The reduced energy loss and improved NIR response of NFA-based OPVs leads to significantly improved efficiency when these acceptors are incorporated into multijunction cells (Cui et al., 2017, Qin et al., 2017, Cheng et al., 2018). However, while NFAs are useful for NIR absorption, there are few examples of these molecules that have sufficiently large energy gaps for absorption in the visible. An exception is ITCC-M used in a 1:1 mixture with the donor PBDB-T, a large energy gap derivative of ITIC. This species has an absorption peak at 600 nm ($E_G = 1.68$ eV), which is blue-shifted by approximately 50 nm from that of IT-IC. The narrow energy gap cell consists of a blend of the donor polymer, PBDTTT-E-T (Huo et al., 2011), with the acceptor, IEICO, the latter molecule with $E_G = 1.36$ eV. This combination of active materials results in a fullerene-free tandem OPV with $\eta_P = 13.0\%$ (Cui et al., 2017).

There are two interesting features of this NFA-based tandem cell. The first is the replacement of the PEDOT:PSS hole conducting component of the ZnO NP CRZ with a 6 nm thick film of the strongly electron-accepting polyelectrolyte, PCP-Na (**Scheme 1**) (Cui et al., 2016). The benefit of this material is that it is more transparent in the NIR than PEDOT:PSS, although it has a relatively strong absorption between 400–500 nm. The tandem $V_{OC,T} = 1.79$ V, which is 2% less than $V_{OC,1} + V_{OC,2} = 1.82$ V, suggesting that it supports high efficiency charge recombination comparable to PEDOT:PSS.

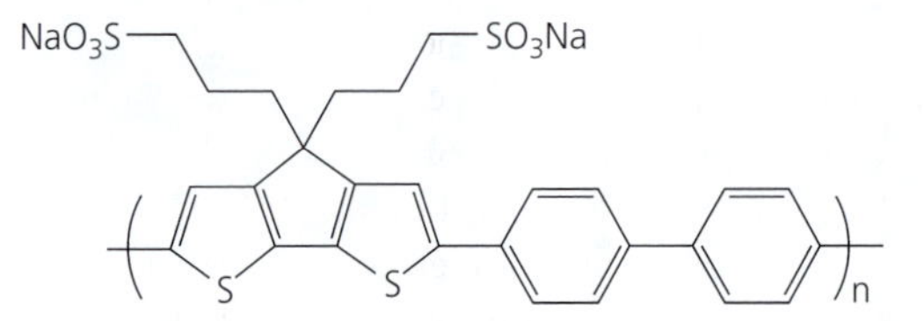

Scheme 1

The second interesting aspect is the development of the morphology of the PBDTTT-E-T:IEICO blend due to annealing in a CB solution. The micrographs in Fig. 7.189a shows that the as-deposited films have a surface rms roughness of 6.75 nm, which is unchanged after SVA for 1 min (Fig. 7.189b). However, the TEM images show the development of a fibrous structure indicative of small, nanocrystalline domains due to the aggregation of IEICO molecules. As the exposure to the CB solution is extended to 4 min (Fig. 7.189c), the surface exhibits a distinct and larger-scale, fibrous domain structure. The highest performance device is obtained for cells annealed for only 1 min, which results in the optimal phase separation between the donor polymer and the small molecule NFA.

7.6 Singlet fission

Multijunction cells are only one way to reduce thermalization losses and incomplete solar spectral coverage in single active layer solar cells. Another method to reduce losses incorporates *singlet fission* (SF) into the solar energy harvesting process (Dexter, 1979). SF is a *multi-exciton generation* (MEG) process that occurs when the singlet energy, E_S, is at least twice that of the triplet, that is, $E_S \geq 2E_T$. Then, it is possible for a photogenerated singlet exciton to split into two triplets via the reaction (Johnson and Merrifield, 1970, Smith and Michl, 2010):

$$S_0 + S_0 + h\nu \rightarrow S_1 + S_0 \underset{k'_{fis}}{\overset{k_{fis}}{\rightleftharpoons}} {}^1(TT) \underset{k'_T}{\overset{k_T}{\rightleftharpoons}} T_1 + T_1, \tag{7.197}$$

where S_0 is the molecular ground state, S_1 is the first excited singlet, and T_1 is the first excited triplet state. The third step in the reaction results in the formation of a *correlated triplet state*, $^1(TT)$, with singlet multiplicity (i.e. total spin quantum number $S = 0$), followed by separation into two uncorrelated triplets, T_1, that diffuse in the medium. The reaction in Eq. 7.197 is reversible, with k_{fis} equal to the rate of triplet pair formation (with the opposite reaction rate of k'_{fis}). The correlated pair can resolve into two, uncorrelated triplets at rate k_T (with reverse rate k'_T).

If the triplets are harvested at a type II HJ, the SF process can, in principle, double the internal quantum yield, that is, the number of excitons generated per incident photon. The benefit of this process is that triplets have longer diffusion lengths than singlets, and the diffusion length is often longer than the optical absorption length (Dexter, 1979). Hence, coating a layer of organic SF material (i.e. a singlet fission sensitizer) onto the surface of an inorganic semiconductor can potentially provide a flux of triplet states that dissociate into charges at the organic/inorganic interface. Likewise, this process can be active at fully organic HJs.

Figure 7.190 shows an illustration of the energetics leading to the reaction in Eq. 7.197. Fission involves the formation of two triplets on adjacent molecules in a radiationless transition analogous to internal conversion. Efficient fission requires that the two molecules that host the triplets be strongly coupled, allowing for fission prior to radiative relaxation of the singlet. Fission is very fast (~ps), and hence competes efficiently with fluorescence that occurs on a nanosecond timescale. The various combinations of spins of the excited states can produce three possible, nearly degenerate correlated triplet pair states: the singlet $^1(TT)$ ($S = 0$), triplet $^3(TT)$ ($S = 1$), and quintet $^5(TT)$ ($S = 2$). The multiplicity of each of the states implies that there are up to nine possible outcomes for triplet fission from S_1: one from the singlet, three from

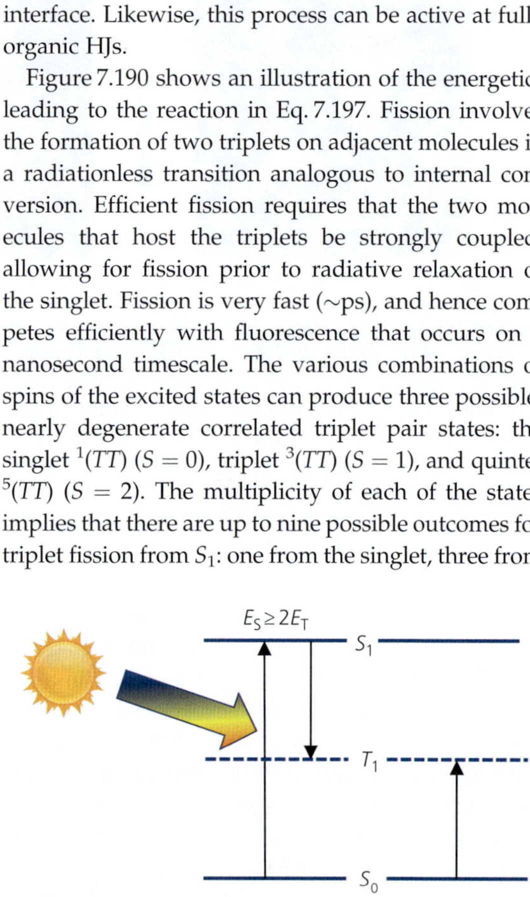

Figure 7.190 Singlet fission into two triplets. The singlet and triplet energies are E_S and E_T, respectively.

the triplet, and five from the quintet state. Only one of the states ($^1(TT)$) has singlet character, and hence its formation is spin-conserving. Nevertheless, all nine states can partially mix at zero magnetic field. Their near degeneracy lifts the prohibition against. violation of total spin and symmetry conservation (Smith and Michl, 2010). This allows all three combinations to originate from $^1(TT)$, particularly in systems where the coupled molecules are the same, that is, in homo-fission systems most commonly active in OPVs. In contrast to intersystem crossing that couples singlets and triplets via spin-orbit interactions, SF is governed by spin dipole-dipole interactions that sensitively depend on the molecular shape and the spatial relationship between neighboring molecules that enhance the dipole coupling matrix elements.

The sequence of steps leading to photocurrent generation is illustrated in Fig. 7.191 for pentacene. Initially, a photon is absorbed, generating a singlet state that is delocalized over two or more pentacene molecules. The singlet subsequently forms a correlated triplet pair on neighboring pentacene molecules at rate, k_{fis}, or if there is an acceptor molecule such as C_{60} nearby, it will form a CT state that dissociates into electrons and holes. Presumably, the correlated pair is formed via a briefly transient excimer intermediate state with a lifetime of a few hundred femtoseconds to picoseconds. The triplet pair then separates into two uncorrelated triplets at rate k_T that diffuse to the acceptor molecule and dissociate into free charges.

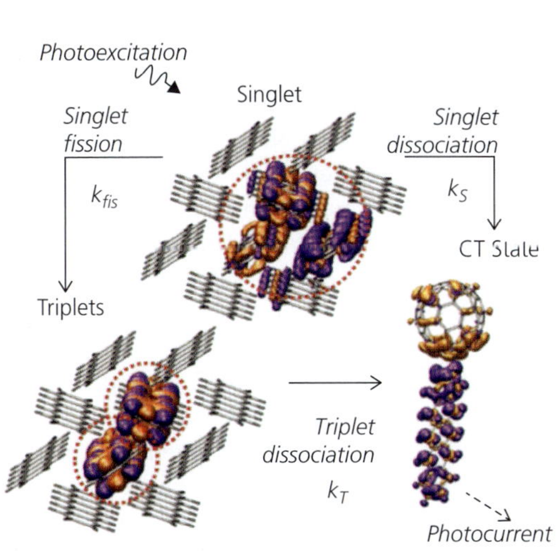

Figure 7.191 Steps from photoexcitation of a pair of pentacene molecules, to singlet fission, ending with dissociation into an electron–hole polaron pair at the interface with an acceptor molecule (in this example, C_{60}). The orange and purple regions indicate electronic orbitals, and the dashed circles indicate the location of singlet and triplet excited states. Rates (k_{fis}, k_S, k_T) for each of the reactions are indicated (Congreve et al., 2013).

The rates of forward and reverse fission depend on the multiplicity of states available for both reactions (9 and 2, respectively). In thermal equilibrium, the rates are related via a Boltzmann distribution, viz. (Merrifield et al., 1969, Groff et al., 1970)

$$\frac{k'_{fis}}{k_{fis}} = \frac{9}{2}\exp[-(2E_T - E_S)/k_B T]. \qquad (7.198)$$

Hence, k_{fis} increases with the number of available states, which in turn should result in a decrease in direct fluorescence intensity. Equation 7.198 shows that the SF efficiency is thermally activated, as has indeed been found for the external quantum efficiency of tetracene/CuPc/C_{60} detectors (Jadhav et al., 2011). In tetracene, SF is marginally endothermic, with $E_S = 2.36$ eV and $E_T = 1.2$ eV.

Simply generating triplets via fission cannot increase the power output of a solar cell since, while the current can be doubled, the voltage is reduced by at least 50%. However, the power output of a solar cell can be increased if a SF sensitizing layer contacts a low energy gap D-A HJ that also harvests photons into the NIR. The triplets generated from absorption of high energy photons by the sensitizer transfer to the HJ where they subsequently dissociate into free polarons in the usual manner. This process is illustrated by energy level diagram in Fig. 7.192 for a conceptual SF-sensitized OPV (Jadhav et al., 2011,

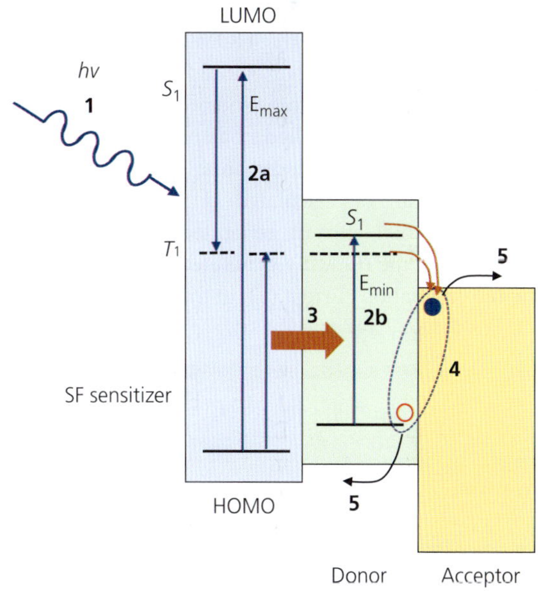

Figure 7.192 Energy level diagram of a conceptual singlet fission-sensitized OPV showing the steps (**1**–**5**) leading to charge generation. The dashed oval line shows the PP state which is a precursor to the free electron (solid circle) and hole (open circle). The block arrow (**3**) indicates exciton transfer to the HJ.

Wu et al., 2014). The steps for charge generation are: (**1**) photon absorption in the sensitizer and the HJ layers. Singlets are generated in the sensitizer (**2a**) and in the HJ (showing only absorption by the donor for clarity, **2b**) due to absorption of short and long wavelength light, respectively. No triplets are excited in either material since these are optically dark states. That is, due to spin selection rules they cannot be directly generated via absorption. The narrow excitonic absorption bands of organics limit the minimum and maximum photon energies that are absorbed. The minimum energy absorbed, E_{min}, is determined by the optical energy gaps of the materials comprising the D-A junction, which also determines the open circuit voltage of the device. The maximum photon energy (E_{max}) is extended to the highest absorption energy of the sensitizer. Over the spectral bandwidth of the sensitizer, two triplets are generated for each absorbed photon. (**3**) Triplet exciton fission products diffuse and transfer to triplet states in the D-A HJ materials via the Dexter process (see Section 3.8.2). This introduces a requirement that the T_1 state of the SF sensitizer be nearly equal to, or greater than T_1 of the D or A material that it contacts. Endothermic transfer is also possible if E_T of the sensitizer is within a few k_BT of the E_T of the HJ materials. The generated triplet and singlet excitons arrive at the HJ where they form a PP, or CT state (**4**), and subsequently dissociate into free charges (**5**). Note that the HOMO energies of the sensitizer and the donor should nearly align to prevent trapping of photogenerated holes.

Following the detailed balance treatment in Section 7.3.2, the thermodynamic limits to an ideal SF device can be calculated (Hanna and Nozik, 2006). The total current generated is equal to the difference of the optically generated current (j_{ph}) and the recombination current (j_0). In the absence of losses (cf. Eqs. 7.137, 7.147 and 7.148),

$$j_{ph} = q \int_{E_G}^{E_{max}} \eta_{QY}(E)\phi_s(E)dE \qquad (7.199)$$

and

$$j_0 = \frac{2\pi q}{c^2h^3} \int_{E_G}^{\infty} \frac{\eta_{QY}(E)E^2}{\exp\left(\left(E - q\eta_{QY}(E)V\right)/k_BT\right) - 1} dE, \qquad (7.200)$$

with

$$j_T = j_{ph} - j_0. \qquad (7.201)$$

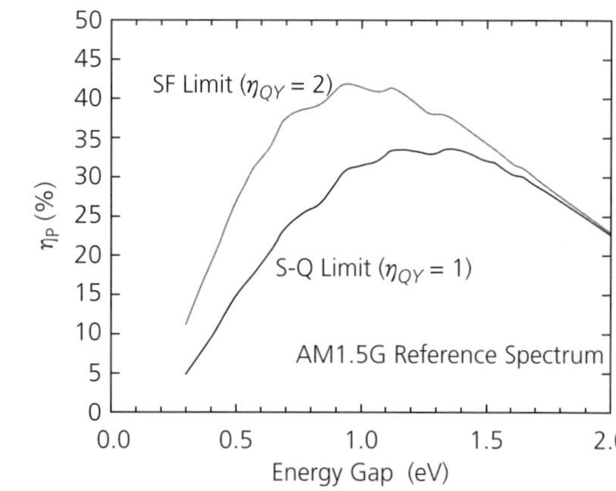

Figure 7.193 Thermodynamic power conversion efficiency limit under AM1.5G illumination in the absence ($\eta_{QY} = 1$) and presence ($\eta_{QY} = 2$) of singlet fission. Here, we assume the material energy gap is the triplet energy. Adapted from Hanna and Nozik (2006).

Here, η_{QY} is the quantum yield for exciton-to-charge generation. For lossless SF, then $\eta_{QY} = 1$ at $E_T < E < 2E_T$, and $\eta_{QY} = 2$ at $E > 2E_T$. In this simple case we thus assume step-wise response to the incident photon energy.

The thermodynamic efficiency limit for SF OPVs using these assumptions is plotted in Fig. 7.193 assuming an AM1.5G reference spectrum. The SQ limit of 33.7% corresponds to $\eta_{QY} = 1$. Note that this is different than the limit in Section 7.3.2 (~31%) since that earlier calculation assumed an ideal black body source, which neglects the details of the AM1.5G spectrum. For a SF-sensitized device, $\eta_{QY} = 2$ and the efficiency limit increases dramatically to 41.9% while the energy gap at the peak efficiency red-shifts to 0.93 eV from 1.34 eV. Note that these calculations assume $E_{loss} = 0$. However, for organics, $E_{loss} = 0.3$–0.7 eV, and hence the maximum efficiency must be decreased accordingly (cf. Fig. 7.74).

Exploitation of the excited state multiplication process sensitively depends on several factors. As noted, the condition of $E_S \geq 2E_T$ must be satisfied. To reach maximum efficiency, E_S should exceed 2 eV to achieve an optimal triplet energy of ~1 eV, according to Fig. 7.193. Furthermore, the interaction requires two strongly coupled molecules, which depends on molecular shapes and their relative conformations to ensure appropriate orbital overlap in the solid state that leads to efficient correlated triplet pair production. The coupling cannot be so strong, however, that it prevents triplets from diffusing to the D-A junction. That is, strongly coupled triplets may not separate, in which case $k'_{fis} > k_{fis}$ in Eq. 7.194. Finally, if the formation of uncorrelated triplets, $^1(TT) \rightarrow T_1 + T_1$, is slow, the reverse reaction back to S_1 is again favored. The trapped triplet may annihilate with polarons generated by the dissociation of excitons at the HJ, or even with other triplets. Beyond these fundamental considerations, the energetic alignment of the SF sensitizer with the donor, large triplet diffusion lengths and efficient energy transfer create additional constraints to SF in OPVs.

The observation of SF can be traced back to spectroscopic measurements made on the polyacenes, anthracene and tetracene (Singh et al., 1965, Swenberg and Stacy, 1968, Geacintov et al., 1969, Johnson and Merrifield, 1970). The transitions are *endergonic*, meaning that they occur most efficiently from the first vibronic excited state, and hence the probability for SF increases with temperature (Singh et al., 1965). The presence of SF in the polyacenes was inferred from the low PL efficiency of tetracene, which was attributed to strong singlet quenching by the nearly resonant generation of two triplets (Swenberg and Stacy, 1968). Since then, numerous materials have been developed and the photophysical processes leading to SF have been studied from both practical and fundamental perspectives (Pope and Swenberg, 1982, Zimmerman et al., 2010b, Chan et al., 2012, Yost et al., 2014, Busby et al., 2015, Ito et al., 2018). Given the many factors that must be avoided to achieve high performance devices, there have been few if any reports where SF has led to improvement in the photogeneration efficiency. In this context, two methods have been used to exploit SF in detectors: employing a SF sensitizer (i) in an all-organic OPV as in Fig. 7.192 or (ii) in an organic-semiconductor QD, hybrid HJ.

7.6.1 Fully organic singlet fission OPVs

An all-organic, pentacene-based SF OPV has been demonstrated with an external quantum efficiency >100%. The structure for the device active region is shown in the inset of Fig. 7.194. It consists of an 8 nm thick P3HT EBL on a PEDOT-PSS/ITO anode, followed by a 15 nm thick pentacene (Pn) SF/donor layer, a 35 nm thick C_{60} acceptor capped by a BCP blocking layer immediately beneath the Ag cathode. Both Pn and P3HT generate singlets. The P3HT singlets transfer to the Pn where fission occurs, resulting in potentially twice as many excitons that contribute to the photocurrent. A MgF_2 ARC is applied to the substrate to minimize reflective losses at the glass/air interface.

Lacking P3HT, the peak external quantum efficiency is 24%. However, with all three layers, η_{ext} =

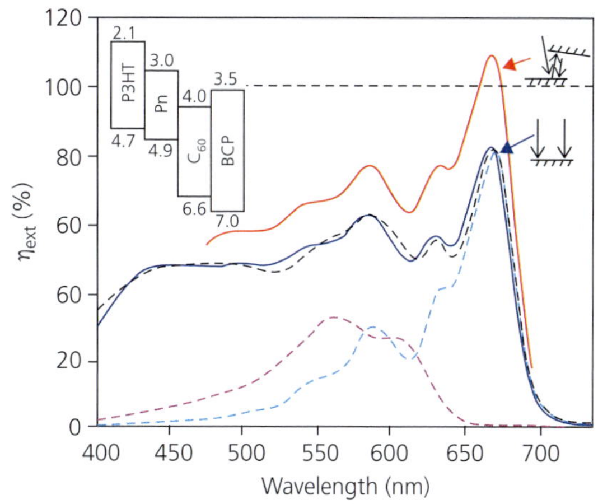

Figure 7.194 External quantum efficiency of an organic SF photovoltaic, with (red line) and without (blue line) optical trapping. The efficiencies calculated based on optical modeling of the thin films are shown for pentacence (blue dashes) and P3HT (purple dashes). The model applied to the full device structure with SF is shown by black dashes. Inset: upper left is the energy level diagram of the SF OPV. Upper right is the mirror arrangement for light trapping and for direct, one pass light incidence (Congreve et al., 2013).

82% at $\lambda = 670$ nm. A very thin (15 nm) Pn layer is employed to minimize losses in exciton diffusion from the donor to the D-A heterointerface. As a result, only 49% of the incident light at 670 nm is absorbed. From this, along with modeling of the optical intensity distribution within the OPV microcavity, the internal quantum efficiency is calculated to be 160% for absorption in the Pn, and 150% for absorption in P3HT. These > 100% efficiencies are not possible without the presence of multiple exciton generation per incident photon.

The external quantum efficiency is increased using light trapping via illuminating the OPV through an aperture that has a reflecting surface (see Section 7.7). Thus, light that is not absorbed in the first pass through the active region is reflected off of the Ag cathode, returns to the aperture/mirror and has a second opportunity for absorption. Using the arrangement shown in the upper right inset in Fig. 7.194, the peak external quantum efficiency increases to 109%. The power conversion efficiency of the SF device under 1 sun, simulated AM1.5G illumination is 1.8% (Congreve et al., 2013), which is not significantly different from 1.5% obtained using conventional Pn/C_{60} OPVs (Yoo et al., 2004).

The high detector quantum efficiency provides inferential substantiation for SF. A more direct signature is the photocurrent and/or fluorescence response to an applied magnetic field. In a magnetic field (B), Zeeman splitting of the states lifts their degeneracy. Under a weak magnetic field, the number of triplet states sharing some singlet character increases from three to six, and then decreases to two at high magnetic field (Groff et al., 1970). Thus, at low fields, the photocurrent decreases due to enhanced singlet quenching with increasing B when SF is present.

Writing the dependence of the fission rate on B as: $k_{fis}(B) = \chi(B)k_{fis}^0$, the fractional change in photocurrent with B is given by (Congreve et al., 2013)

$$\delta(B) = \frac{j(B) - j(0)}{j(0)} = \frac{k_S k_{fis}^0(\chi(B) - 1)}{\left(2k_{fis}^0 + k_S\right)\left(\chi(B)k_{fis}^0 + k_S\right)}, \tag{7.202}$$

where, as above, k_S is the fluorescence decay rate. Here, $\chi(B)$ contains the functional dependence of the fission rate on field. Now δ is maximized at $k_S = \sqrt{2\chi(B)}k_{fis}^0$, which from Eq. 7.202 leads to an expression for the triplet yield of

$$\eta_{fis}(B) = \frac{2k_{fis}^0}{k_{fis}^0 + k_S} = \frac{(1 - \delta(B))\chi - 1 \pm \sqrt{[\delta(B)(\chi + 2) - \chi + 1]^2 - 8\delta^2(B)\chi}}{(\delta(B) + 1)(\chi - 1)}. \tag{7.203}$$

The pentacene device using either PTCBI or C_{60} as the acceptor, and with varying pentacene thicknesses were placed in a low magnetic field (0.4 T). The resulting changes in photocurrent, triplet yield, and η_{int} are shown in Fig. 7.195. The photocurrent change increases with Pn layer thickness, until approximately 1 nm (see Fig. 7.195a), after which it begins to decrease. Equation 7.202 is used to extract the triplet yield in Fig. 7.195b. At thicknesses of > 8nm, the yield approaches 200%, that is, every singlet generated in Pn results in the formation of a triplet pair. Finally, using the values of δ, and η_{ext} and optical modeling, the internal quantum efficiency reaches a maximum of η_{int} = 160% at a Pn thickness of 10 nm (Fig. 7.195c). The roll-off at larger thickness is due to the generation of triplets that recombine prior to charge separation at the HJ due to their limited diffusion length in Pn. The maximum η_{int} = 200% is not achieved due to recombination and other losses of the triplets. The results are independent of the acceptor used (C_{60} or PTCBI), indicating that the effects are entirely due to photogeneration within the donor layer, consistent with the SF process. Similar magnetic field dependences are observed for tetracene-based devices (Wu et al., 2014).

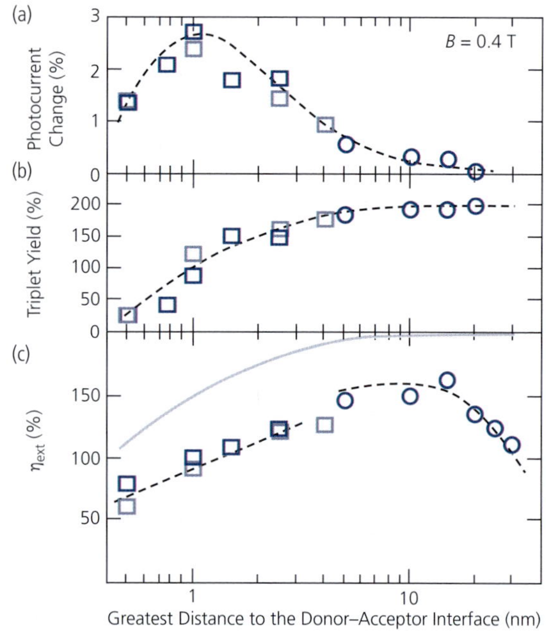

Figure 7.195 (a): Photocurrent change ($-\delta$), triplet yield (η_{fis}), and η_{int} for P3HT/Pn/C_{60} (circles) or PTCBI (squares) devices in Fig. 7.194 vs. Pn layer thickness. Dashed lines are guides to the eye. Solid line is the maximum achievable η_{int} as determined from the measured η_{ext} and optical modeling of the thin device structure (Congreve et al., 2013).

7.6.2 Hybrid organic/QD singlet fission OPVs

An alternative design for harvesting triplets generated by SF is the hybrid organic–inorganic QD structure shown in Fig. 7.196 (Ehrler et al., 2012, Yang et al., 2015). This is similar to the organic–Si detector initially proposed by Dexter (1979). If transfer across the organic/inorganic HJ is efficient (see Section 4.7.3), this may ultimately be superior to SF at fully organic HJs (Tabachnyk et al., 2014, Thompson et al., 2014). The device in Fig. 7.196a comprises a 16 nm thick TIPS-Pn SF sensitizing layer solution-cast onto a pre-deposited layer of either PbSe or PbS QDs of various diameters. The active region is sandwiched between a TiO_2/ITO cathode and a Au anode. Photocurrent is due to direct singlet exciton generation via absorption in the QDs, or triplet generation via fission following absorption by the TIPS-Pn S_1 state (dashed line, Fig. 7.196b). Up to 85% of the illumination intensity incident on the QD layer is absorbed, which limits the contribution due to SF to the remaining 15% absorbed in the TIPS-Pn.

The external quantum efficiency spectrum contributed by only the TIPS-Pn is extracted from the calculated light fraction (via transfer matrix methods) absorbed by this layer, with results in Fig. 7.196b for

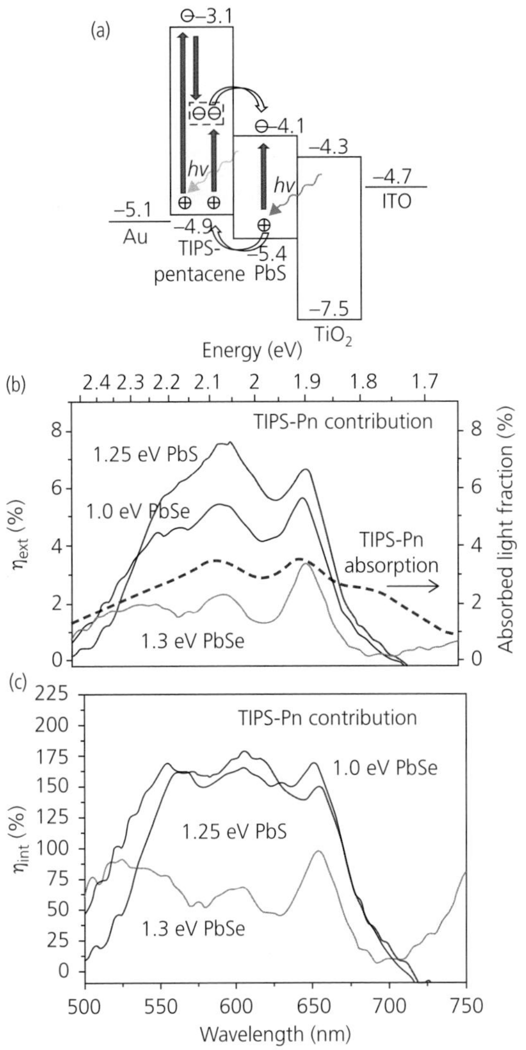

Figure 7.196 (a) Device structure for a SF organic–QD solar cell. (b) Contribution to the external quantum efficiency spectrum from TIPS-Pn of the device in (a) employing either PbSe or PbS quantum dots of different diameters (and hence absorption energies). The calculated fraction of absorbed light in the TIPS-Pn layer is also shown. (c) Internal quantum efficiencies for the devices in (b) (Yang et al., 2015).

several different diameter PbSe and PbS QDs. As shown in Section 7.2.2, the transition energy (given in the figures) is directly related to the dot diameter: the energy increases with decreasing dot size. Notably, the PbSe QD absorption at 1.3 eV is greater than the T_1 energy of TIPS-Pn. In this instance, therefore, the peak η_{ext} is only 3%, or roughly half that of the 1.35 eV PbS and 1.0 eV PbSe QDs where the corresponding $T_1(Pn) \rightarrow 1S_e(QD)$ transition is exothermic. Here $1S_e$ is the lowest electronic excited state of the QD (cf. Fig. 7.50). Accounting for the limited absorption in the TIPS-Pn layer, the internal quantum efficiency spectrum is shown in Fig. 7.196c. It is inferred that a maximum $\eta_{int} = 170\%$ is obtained for the low energy gap QDs, and the PbSe cell with a 1.3 eV energy has $\eta_{int} = 80\%$, presumably due solely to direct S_1 transfer from TIPS-Pn to the QD. The presence of SF was confirmed using B-field measurements as previously described for the Pn-C_{60} devices. When illuminated with a simulated solar spectrum (conditions not specified), the TIPS-Pn/PbS cell had $\eta_P = 4.8\%$, $FF = 0.41$, and $V_{OC} = 0.59$ V (Yang et al., 2015). The voltage increases with QD bandgap, consistent with its increased HOMO–LUMO offset energy with the organic.

It remains unclear whether OPVs employing SF will ultimately have improved power conversion efficiencies compared to single or multijunction devices. While the process has been realized in both organic and colloidal QD systems, the most useful application may yet be that originally envisioned by Dexter, where the efficiency of conventional Si cells is enhanced by depositing an organic fission layer onto its surface. This sensitizing layer converts singlets generated by high energy (blue) photons into a pair of triplets, which augments the photocurrent generated by the inorganic cell (Dexter, 1979, Einzinger et al., 2019). Many challenges must still be overcome to find appropriate materials, morphologies and device structures before SF can be effectively exploited in solar energy harvesting applications.

7.7 Light trapping and concentration

OPV cells are thin film devices comprising a substrate, multiple organic and metal layers that form optical microcavities. As such, OPVs trap light. As we have shown in Section 7.1.2, the light reflected by the metallic contact supports a standing wave that has more than one opportunity to be absorbed within the device active region. By the same process, the optical field can suffer losses on reflection and scattering from the several interfaces in the structure. However, the weak microcavity of the thin film OPV is usually insufficient to result in complete absorption of the incident solar radiation. In Sections 7.4.2 and 7.5.2, we introduced several methods that improve the absorption via optical trapping that include application of ARCs, outcoupling layers, ADRs, and plasmonic field focusing from nanoparticles in the CRZs of multijunction cells. Yet, even these light coupling schemes fall short of increasing the absorption efficiency to ~100% achieved by conventional solar cell materials and structures. In this section, we describe additional light trapping schemes used to increase η_{ext}.

The basic strategy of all light trapping or in-coupling schemes is the same: to increase the optical path length (and hence, the photon lifetime) within the OPV. This allows for increased light absorption in the thin active layers. The requirements that are placed on all practical schemes are nearly identical to those needed for white organic light emitting devices. These are:

(i) *Low cost over large areas*. The primary application of organic electronic appliances is for large area and low cost devices. Solar energy harvesting is a prime example of such an application, since a solar cell is only effective if it captures a large amount of sunlight whose power density is only 1 kW/m^2. This demand for low cost electricity places stringent demands on the light trapping scheme, since it must not add significantly to the solar module cost.

(ii) *Insensitive to wavelength and incidence angle*. The useable solar spectrum extends from the near UV (< 400 nm) to the NIR (>1.1 µm). In addition, the sun follows a *diurnal solar path* (the day arc), requiring that the time available for power generation is maximized, extending from solar angles as low as 10–20° above the horizon near dawn and sunset, to the azimuth at midday. Non-tracking solar modules offer the lowest cost. Furthermore, many solar trackers are unsuitable for use in building integrated or building applied PV. These conditions demand an optical trapping scheme whose efficiency is relatively independent of the angle of incidence of the solar flux.

(iii) *Non-invasive of the OPV active region*. Great care is taken to maximize the efficiency of solar cells by a judicious choice of materials, processing methods to achieve the desired morphology, and multilayer device architecture. Optical trapping designs are generally "added on" to the device or to its substrate after device fabrication is completed. Hence, it is beneficial to avoid schemes that impact the design, or intrude upon the structure of the OPV. To the extent that this is possible, the method should be general, such that it can be applied to any number of device types, independent of the details of their structure or film properties. This non-invasive property creates significant added value to a particular light coupling approach.

The effects of light trapping are illustrated for light incident on a rough surface in Fig. 7.197. The light enters a medium with index of refraction, n, and is absorbed in the device active region before being specularly reflected off of the back surface reflector (e.g. a metal cathode). The unabsorbed light once more scatters from the top surface or escapes. This process occurs until all of the incident photons are

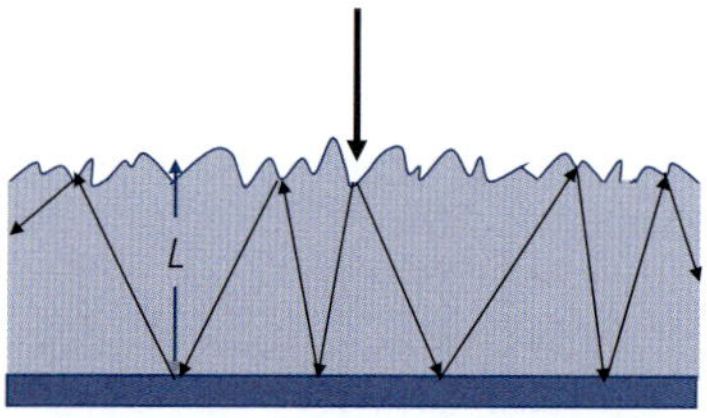

Figure 7.197 Light incident on a rough surface of a sample of thickness, L, is scattered and reflected off of a back reflecting surface in a thin film solar cell.

either absorbed or escape back toward the direction of illumination. Assuming that the scattered light intensity distribution over time is identical to its distribution in phase space, the distribution is said to be ergodic, and hence we can treat the scattering using statistical ray optics (Yablonovitch, 1982). Then, the intensity of radiation from a black body within a medium at frequency, ω, within a solid angle $d\Omega$ is

$$I(\omega) = \frac{n^2}{4\pi^3} \frac{\omega^2}{c^2} \frac{\hbar\omega}{\exp(\hbar\omega/k_B T) - 1} d\Omega d\omega. \tag{7.204}$$

This expression implies that the intensity within the medium (I_{med}) is increased by n^2 compared to the incident light intensity, I_{inc}, viz.

$$I_{med}(\omega) = n^2 I_{inc}(\omega). \tag{7.205}$$

For simplicity we assume that n is frequency-independent, and varies only slowly with position. The back surface reflector (BSR) in the solar cell then doubles this intensity to

$$I_{med}(\omega) = 2n^2 I_{inc}(\omega). \tag{7.206}$$

For a sample of thickness L, and absorption coefficient α, the fraction of incident light that is absorbed is obtained by integrating over the optical paths within the volume, viz. (Yablonovitch, 1982)

$$f = \frac{2\alpha L A_{inc} I_{med}}{A_{inc} I_{inc}} = 4n^2 \alpha L, \tag{7.207}$$

where we assume that α is wavelength independent, and that the dependence of both I_{med} and I_{inc} on ω are the same. Here, A_{inc} is the area illuminated by the incident radiation. This is an important result since it implies that the absorption in the film is increased by a factor of $4n^2$ due to light trapped within its volume. In organics, $n \sim 1.5$–2, resulting in an increase in absorption by 9–16 simply by using an efficient light trapping scheme. This increase provides flexibility in the choice of materials and structures where the charge extraction and cell series resistance can be reduced by using very thin cells that still absorb most of the incident radiation. Importantly, the high volume intensity also leads to an increased open circuit voltage according to Eq. 7.122. This, in turn increases the cell power conversion efficiency (Eq. 7.119).

If the incident angle is restricted to the half-angle, θ, the enhancement factor is similarly increased to $4n^2/\sin^2\theta$ (Campbell and Green, 1986). This restriction can be implemented by using solar concentrators, lenses and other external means, as will be shown in the examples discussed below.

The absorbing medium in many thin film cells based on inorganic semiconductors fills the entire scattering volume as in Fig. 7.197. However, we have seen that the active region thicknesses in OPVs are typically $L \ll \lambda$. In this case, the modal power can be considerably enhanced even over that predicted by Eq. 7.207. For example, by replacing the BSR with a grating of period, Λ, the incident radiation can be redirected into the relatively high index absorption region surrounded by low index media (i.e. a glass substrate). The modes confined within the waveguiding active region are subsequently absorbed, generating photocurrent. An illustration of the geometry for a waveguide employing a grating on one surface is shown in Fig. 7.198a. The in-plane modes have propagation vector $k_{\|} \leq \omega/c$. Figure 7.198b is the absorption enhancement factor for a square grating coupler obtained using statistical coupled mode theory (Yu et al., 2010). The enhancement is as large as $12n^2$ using a sub-wavelength periodicity of $\Lambda/\lambda \leq 1$. Numerous modes exist at larger grating periods, although these do not result in enhancement $> 4n^2$ obtained in a non-wavelength dispersive structure. Thus, there is considerable potential benefit to be gained by employing sub-wavelength structures within the OPV, although these benefits may be outweighed by the increased complexity and cost required in their fabrication.

In the remainder of this section, we introduce several light trapping solutions demonstrated in OPVs. Since light trapping and concentration schemes employing surface plasmons from metal nanoparticles in tandem CRZs, ARCs, and ADR stacks have already been considered, we do not discuss them further here.

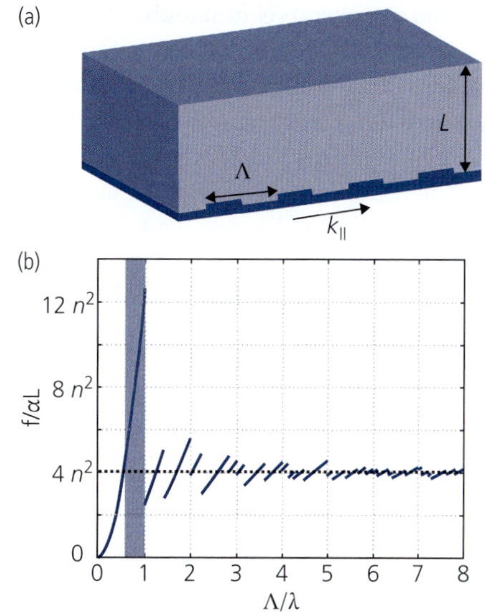

Figure 7.198 (a) Modal propagation with in-plane wavevector $k_{\|}$ in a guided wave structure of thickness, L, and back surface grating of periodicity, Λ. (b) Absorption enhancement factor vs. Λ/λ. The shaded region shows an absorption enhancement greater than the thick structure with a roughened surface whose enhancement is $4n^2$ (dashed line) (Yu et al., 2010).

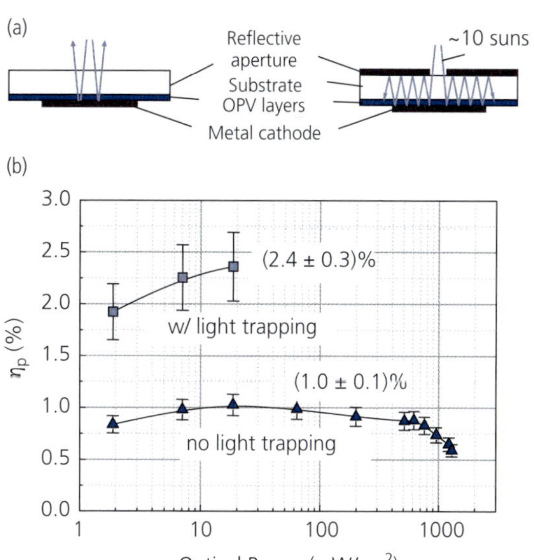

Figure 7.199 (a) A conventional OPV structure (left), and one with light concentrated through an aperture that reflects light trapped within the substrate. The optimum light intensity concentration at the aperture is equal to the ratio of the reflective OPV cathode to the aperture area. (b) The efficiency of a CuPc/PTCBI OPV with no light trapping, and with 10 suns intensity focused through a reflective aperture. Adapted from Peumans et al. (2000a).

7.7.1 Light trapping using reflective apertures

The high quantum efficiencies for the SF device in Fig. 7.194 were obtained using a narrow reflective aperture, whose principle of operation of this method is illustrated in Fig. 7.199a. In a conventional device, the light is incident on the flat, distal surface of the OPV, and is reflected once from the cathode. The thin layers do not absorb all of the incident radiation, which is subsequently lost to radiation back toward the emitter (i.e. the sun). A mirror containing an aperture can accept concentrated radiation at a limited angle, θ. The light that is reflected from the cathode and not absorbed in its second pass through the active region is then once more reflected from the mirror, providing additional opportunities for reabsorption. If we ignore mirror losses, the potential increase in photocurrent can be as high as $4n^2/\sin^2\theta$. Effective light trapping occurs for an aperture diameter, d, less than the total mirror diameter, D. Then the incident light must be concentrated by a factor of $(D/d)^2$ to result in a net increase in absorption.

The benefit of employing the light trap has been realized using an ITO/CuPc (15 nm)/PTCBI (6 nm)/ BCP:PTCBI (30 nm)/Ag OPV with results shown in Fig. 7.199b (Peumans et al., 2000a). We have seen in Section 7.4.1 that this device structure is capable of achieving high efficiency even at very high intensities. Indeed, the maximum efficiency of the conventional structure is $\eta_P = 1\%$ up to intensities of at least 6 suns. The use of a 10 sun concentrating aperture increases the power conversion efficiency to $\eta_P = 2.4\%$ due to the increased opportunity for absorption enabled by reflections from the mirror surface.

Practical implementation of this scheme requires concentration of the incident radiation using mini-*Winston collectors* or *compound parabolic concentrators* (CPCs) that can be made from a metallized thermoformed plastic sheet, a polymer poured into a suitably shaped mold, etc. (Winston and Hinterberger, 1975). The example shown in Fig. 7.200a is a 10× concentrator with an aperture of 10 μm, and a concentrator opening of 100 μm. The reflectivity of the mirror surface at the distal side of the glass substrate, the cathode, and on the interior of the CPCs is $R = 0.95$, typical of a Ag mirror. Ray tracing simulations show that the maximum intensity at the cell assuming a hexagonal arrangement of concentrators never locally exceeds 1.4 suns.

The light trapping efficiency, η_{LT}, is equal to the ratio of the light absorbed in the OPV using the 10× CPCs to that in the absence of the CPCs. This efficiency is plotted vs. the total absorption efficiency (η_A) of the active region in Fig. 7.200b. The benefit of the light trap increases as η_A decreases, with the advantage disappearing as $\eta_A \rightarrow 1$. For nearly transparent layers with $\eta_A = 0.01$, the advantage increases to $\eta_{LT} > 15$. Note that concentration with mini-CPCs requires solar tracking throughout the day and year (Lee et al., 2015). This increases the cost and size of the solar cells, potentially eliminating advantages gained in their use. Furthermore, solar tracking is not always practical in applications such as BIPV.

Microlens arrays (MLAs) that are focused into an array of apertures can also serve as light traps, as shown in Fig. 7.201. The incident illumination is focused by the microlenses onto the small diameter apertures. By displacing the OPV below the aperture plane, the light passing through the aperture defocuses, and the non-absorbed radiation reflected back from the electrode returns to the mirror surface where only the fraction of light directly incident on the aperture escapes. The remainder then has a second chance to be absorbed. This architecture depends on alignment of the aperture array to the lens focal points and that the light is normally incident to the substrate plane. This demands precise solar tracking to maintain the focal points in the center of each aperture, regardless of incidence angle. Also, the

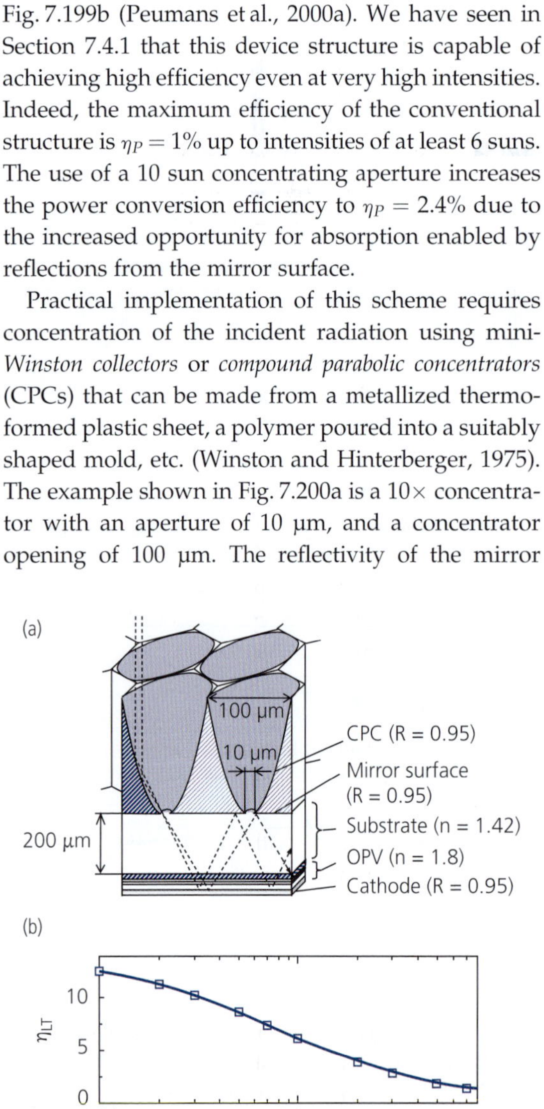

Figure 7.200 (a) Design of a 10× mini-compound parabolic concentrator attached to a mirrored aperture surface on the substrate of an OPV. The reflectivity (R) of each surface is indicated. (b) Calculated light trapping efficiency vs. absorption efficiency of the OPV layers (Peumans, 2004).

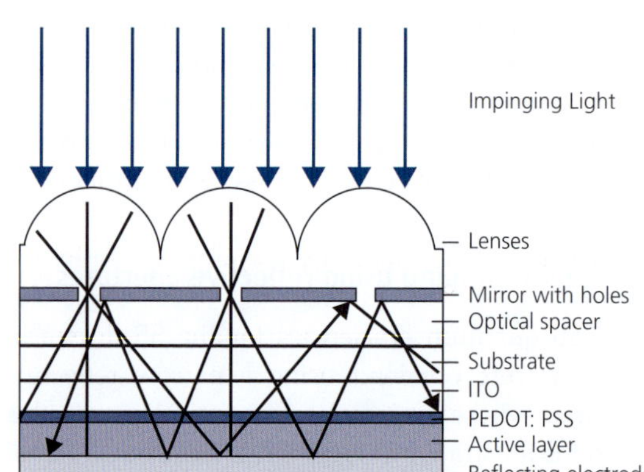

Figure 7.201 Microlens array concentrators. The lens focal points are at the aperture array located at the surface of an optical spacer layer (e.g. a glass substrate) above the OPV active layers (Tvingstedt et al., 2008).

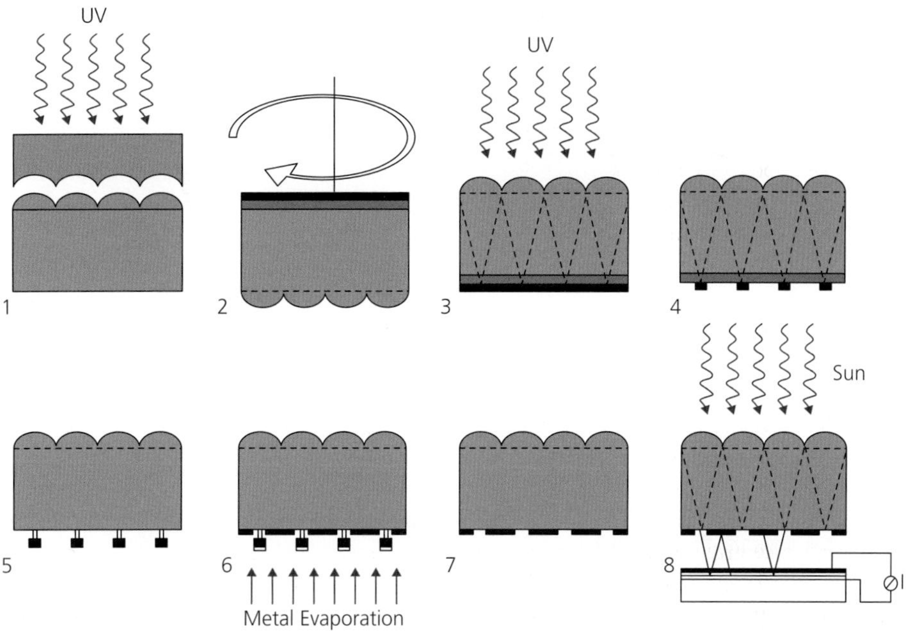

Figure 7.202 Example process for fabricating a microlens array that is self-aligned with the aperture array (Tvingstedt et al., 2008).

focal plane of the hemispherical lenses describes an arc such that off-normal illumination will both move the focal point as well as elevate it above the aperture, causing defocusing and a loss of illumination intensity in the OPV active region.

Self-alignment of the focal points to the apertures is achieved by using the MLA itself to pattern the aperture array (see the process in Fig. 7.202). In step **1**, the polymer MLA (lens radius, r) is formed by embossing the polymer layer with a glass mold comprising an array of hemispherical depressions, and then curing by exposure to UV light. Alternatively, as shown in Section 6.6.2, a thermally cured polymer can be poured into the mold and then peeled off and attached to the substrate (Möller and Forrest, 2001, Sun and Forrest, 2006). Next (step **2**), a bilayer photoresist is applied to the opposite side of the substrate whose thickness is equal to the lens focal length of $f_{lens} = r[n_{lens}/(n_{lens}+1)]$, and is exposed (step **3**) through the lenses themselves using UV light. The index of the lens (n_{lens}) is assumed to match that of the substrate, although suitable adjustments can be made if this is not the case. The use of the lens array to expose the photoresist ensures accurate self-alignment. The photoresist is then developed (steps **4** and **5**) to form an undercut pattern due to the different development rates of the two resists comprising the bilayer. The reflective metal layer is deposited (step **6**), and excess metal is lifted off by dissolving the photoresist (step **7**), thus leaving behind the aligned aperture array. The entire assembly can be attached to, or integrated with the OPV, as shown in step **8**.

A lensed aperture OPV has been demonstrated employing 200 μm diameter microlenses with 90% transmission in the forward direction (microlenses facing the illumination source), and 15% in the opposite direction. Combining this with a 30 nm thick fluorene copolymer donor (APFO Green-9):$PC_{71}BM$ (1:3) OPV increased the cell efficiency by 25%. Most of the improvement came from increased absorption in the wavelength range from 500–750 nm where the BHJ absorbed <50%, whereas the system was less effective at shorter wavelengths where absorption reaches 90% (Tvingstedt et al., 2008). This coupling improvement can possibly be increased using a larger lens radius-to-aperture ratio, although doing so requires an improved optical quality of the lenses and more precise tracking of the light source throughout the day.

7.7.2 V-traps

A V-trap is the combination of two opposing OPVs forming a V-shaped substrate symmetrically disposed about an opening angle, 2α (see Fig. 7.203a). For only partially absorbing active regions, the incident light undergoes several bounces before exiting the V, thus providing a longer optical path length that increases with decreasing α. Fabrication of the trap is straightforward. In Fig. 7.203b, the substrate is prepared by beveling one edge to match it with an identical substrate. This sets the opening angle. The

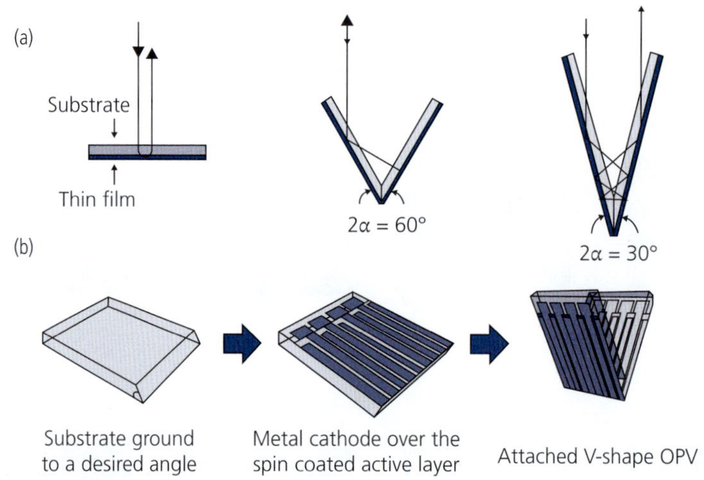

Figure 7.203 (a) Schematic illustration of V-trap OPVs with different opening angles, 2α. (b) Assembly of the V-trap (Kim et al., 2013). Reprinted with permission from Kim, 2013, The Optical Society.

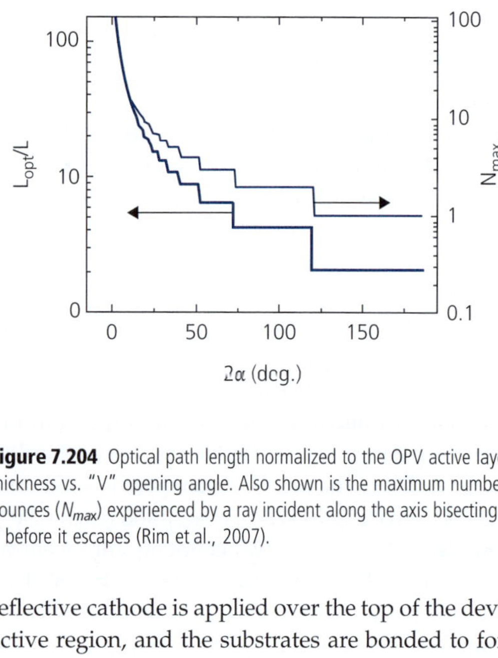

Figure 7.204 Optical path length normalized to the OPV active layer thickness vs. "V" opening angle. Also shown is the maximum number of bounces (N_{max}) experienced by a ray incident along the axis bisecting the V before it escapes (Rim et al., 2007).

reflective cathode is applied over the top of the device active region, and the substrates are bonded to form the V with the semi-transparent contacts facing the inner side of the trap.

The number of bounces, N_{max}, increases as the opening angle is decreased, with the highest density of illumination found near the vertex. A plot of the optical path length, L_{opt}, normalized to the active layer thickness, L, along with N_{max} vs. 2α is shown in Fig. 7.204. The functions are step-wise in 2α, with $N_{\max} \to 1$ as $2\alpha \to 180°$, corresponding to a planar substrate. In contrast, when the sides become parallel, we have $N_{\max} \to \infty$, leading to an infinite optical path length. Of course, the optical path length and number of bounces only reaches these values if the absorption on each bounce is negligible, suggesting that there is an optimum angle that leads to the largest combination of opening (and hence maximum solar acceptance), and photocurrent. Furthermore, the length of the cell increases with α. This introduces another trade-off between module depth, OPV size, and optical path length. An important distinction between this trapping scheme and that using apertures is that the V-trap is not a concentrator. Hence, in principle it does not require solar tracking as long as the aperture is sufficiently open (i.e. $2\alpha \sim \pi/2$) to capture most of the sunlight generated during the day. The planar-V architecture should be oriented in an east-west direction to capture sunlight during most of the day, as long as the opening angle is not too shallow as to create shadowing in the early morning and late afternoon hours.

V-traps have demonstrated increased efficiency when identical cells are used on both legs, with the largest enhancements corresponding to cells with the lowest absorption, or in spectral regions where the cells are most transparent (Rim et al., 2007, Tvingstedt et al., 2007). Alternatively, V-traps provide opportunities to combine different cells on the opposing planes, forming either parallel- or series-connected tandem OPVs. This is shown in Fig. 7.205, inset, for a device comprising a green absorbing APFO-G9:$PC_{61}BM$ BHJ cell on one arm (Inganäs et al., 2004) and a red absorbing APFO3:$PC_{61}BM$ BHJ on the other (Tvingstedt et al., 2007). The absorbance spectra of the two discrete, planar cells do not significantly overlap (Fig. 7.205a), allowing the two sections to be either connected in series or parallel while maintaining current and voltage balance, respectively. The spectra of the tandem cells folded at

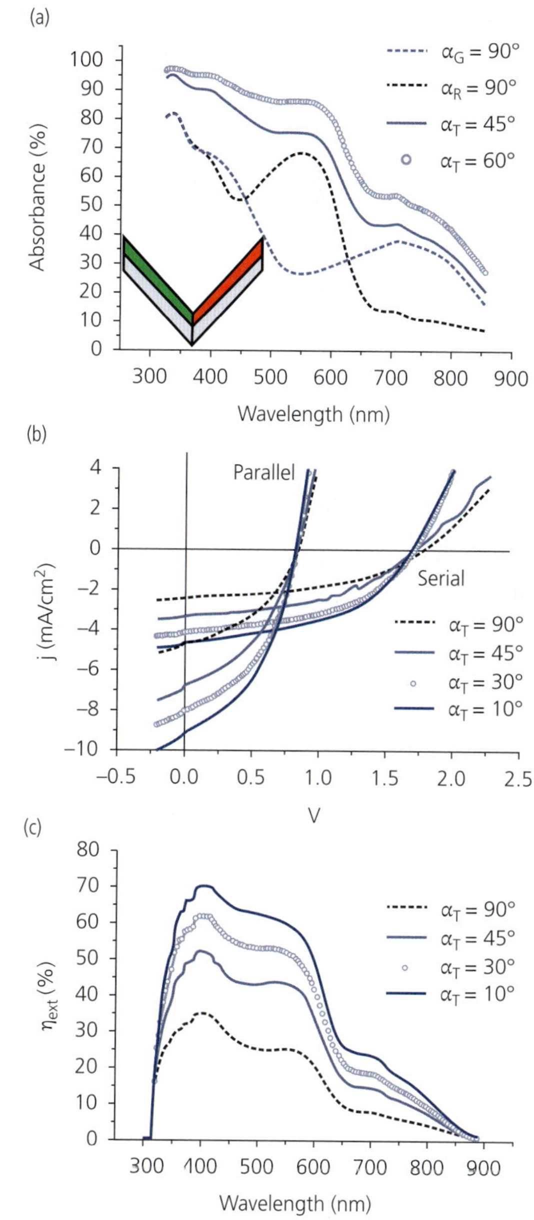

Figure 7.205 (a) Absorbance, (b) j–V characteristics, and (c) external quantum efficiency spectra of planar green and red absorbing OPVs (α_G, $\alpha_R = 90°$, respectively), and tandem cells at different V-trap opening angles (α_T). Inset in (a): V-trap structure with $\alpha_T = 45°$, with the green absorbing cell on the left and the red cell on the right (Tvingstedt et al., 2007).

either 45° or 60° are broadened, and are a superposition of spectral features of both subcells. Note that the smaller opening angle results in increased absorption due to the longer effective optical path length.

The series and parallel-connected tandem cell j–V characteristics under 1 sun illumination are shown in Fig. 7.205b. The series-connected cell has an increase in η_P from 2.0% to 3.7% on decreasing the opening angle from 90° to 10°. The parallel junction performs somewhat better since current balance is not required, increasing from 3.3% to 4.4%. Finally, the external quantum efficiency in Fig. 7.205c for the parallel-connected cell exhibits the broad response characteristics of both the long and short wavelength arms of the V. Indeed, the efficiency is more than doubled for the cell with a 10° opening angle compared to a conventional planar tandem cell.

An additional degree of freedom offered by V-traps with different cell types on each arm is that the orientation toward the sun can be optimized to generate the greatest total diurnal energy harvesting. Since the red content of sunlight at Earth's surface is larger in the morning and early evening, it is advantageous to orient the long wavelength-absorbing cell on the upper half of the V toward the south (when in the northern hemisphere). Simulations of total daily energy harvesting have found that the most efficient combination of OPVs employing V-traps is using a parallel connection to avoid problems with current balance during the course of a day. Except for large tilt angles where the cosine losses are significant, the diurnal energy harvesting for a V-trap with a module opening angle of $\alpha = 70°$ can be as much as 1.5 times greater than for an equivalent flat cell (Andersson et al., 2011).

7.7.3 Nanoscale dielectric scatterers

Surface roughening leads to light scattering, which in turns increases the optical path length within the OPV active region. In Section 6.6.2 we found that 240 nm diameter TiO_2 spheres embedded in a transparent organic film beneath the OLED increases the outcoupling by scattering light from the active region into the substrate escape cone. Likewise, such a tactic can be used to scatter light within an OPV, also resulting in enhanced efficiency (Adams et al., 1993, Tang et al., 2013). The high dielectric constant of the TiO_2 within a low index matrix can effectively randomize the ray angles entering the active region.

Improved performance of both single and tandem junction polymer OPVs has been demonstrated by embedding 25–35 nm diameter TiO_2 NPs in a PDMS film, and then pressing the solution between two glass slides to form a 300 μm thick, flexible, diffuse reflector that is attached to the OPV (Tang et al., 2013). The visible reflectivity of the films is varied from near 100% for a TiO_2:PDMS weight ratio of 2:1 (the high reflective, HR, film), to ~85% for a ratio of 0.1:1 (the low reflective, LR, film). The ratio can be adjusted to full transparency as the TiO_2 concentration is decreased. Absorption in the TiO_2 reduces the reflectance to near zero at $\lambda < 400$ nm. The films can be attached after the OPV fabrication is complete, their response is relatively wavelength independent, and they are lightweight and flexible. As such, scattering layers satisfy many of the requirements of practical light traps.

The external efficiency of a semitransparent (ST) OPV comprising a TQ1:$PC_{71}BM$ (2:5) active region sandwiched between an ITO/PFPA-1 bottom cathode on glass (Tang et al., 2012) and a PEDOT:PSS anode is shown in Fig. 7.206. The efficiency has a maximum of 45% at 380 nm, and a second peak of 38% at 475 nm. Using a Ag mirror placed parallel to the glass substrate, the efficiency at 475 nm increases to 52%, whereas with the HR scattering film, it reaches 70%. The increase obtained by using a mirror *vs*. the dielectric film is due to the Lambertian scattering profile of the latter, leading to the optical field intensification in the in-plane direction as inferred from Fig. 7.197. The enhancement factor, which is the ratio of the external efficiency of the ST device with the scattering layer to that lacking the layer is also shown in Fig. 7.206. There is no enhancement at short wavelengths due to absorption by the TiO_2, while at longer wavelengths where the devices are most transparent, the enhancement factor is >2.5. Absorption within the layers or contacts decreases the enhancement by attenuating the intensity of the light during multiple passes through the OPV.

Scattering layers with adjustable transparency and reflectance can improve current balance in series-connected multijunction cells. An example tandem structure is shown in Fig. 7.207, which is a combination of two, semitransparent single junction OPVs grown on different substrates. The OPV on substrate 2 employs a high reflectance scattering layer that sets up a standing wave within the stack. A second, lower reflectance scattering layer is positioned between the cells. This limits the light penetrating to the upper cell while increasing the reflection into the lower cell. Using the appropriate spectral response and transparency of the LR layer, the current generated in the two cells can be precisely balanced, leading to a higher overall efficiency. In a parallel-connected tandem, only one HR layer is required since current balance is not essential.

A cell using the LR and HR TiO_2/PDMS scattering layers was assembled to implement this strategy. It combined a short wavelength absorbing TQ1:$PC_{71}BM$ top subcell and a second P3TI:$PC_{71}BM$ (2:3) longer wavelength absorbing subcell adjacent to the HR scattering layer. The device with the LR layer interposed between subcells exhibited $\eta_P = 6.9\%$, but without this balancing layer, the efficiency fell to 6.0% (Tang et al., 2013).

7.7.4 Scattering via gratings and textured surfaces

Gratings are effective in coupling incident radiation into waveguide modes (cf. Fig. 7.198), and hence can significantly increase absorption. The gratings can be

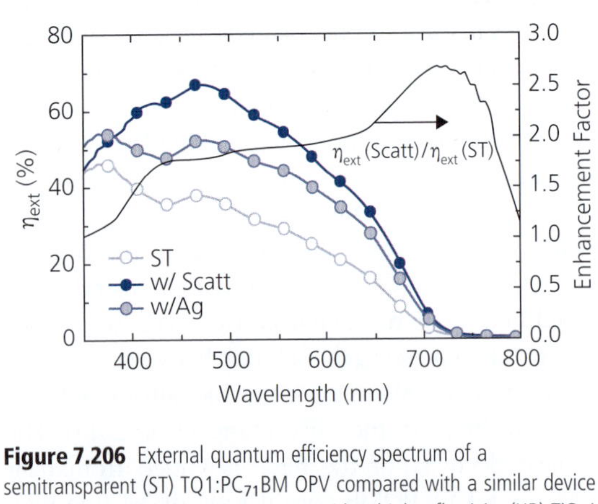

Figure 7.206 External quantum efficiency spectrum of a semitransparent (ST) TQ1:$PC_{71}BM$ OPV compared with a similar device positioned parallel to a Ag mirror, or with a high reflectivity (HR) TiO_2/PDMS scattering layer attached to the glass substrate. Also shown is the enhancement ratio in η_{ext} of the device with the scattering layer to the ST device (Tang et al., 2013).

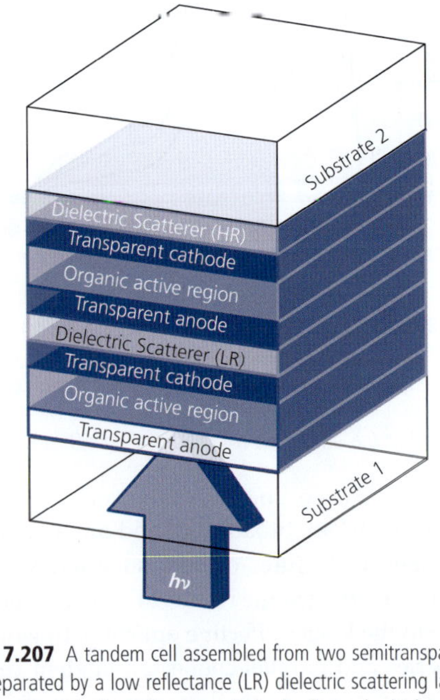

Figure 7.207 A tandem cell assembled from two semitransparent OPVs separated by a low reflectance (LR) dielectric scattering layer to balance currents in the series connected subcells, and a high reflectance (HR) scattering layer to increase the optical path length within the device.

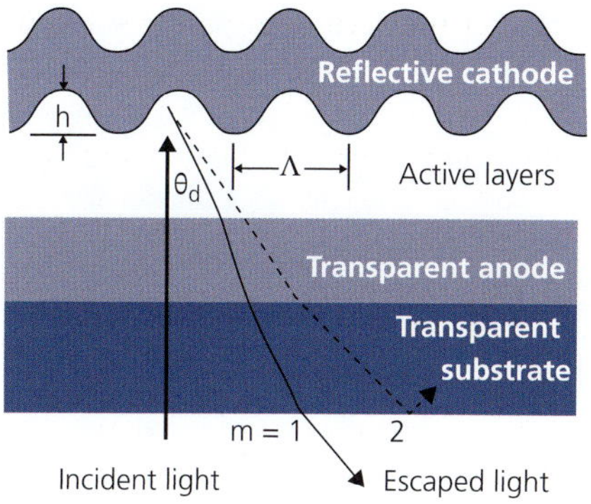

Figure 7.208 Illustration of grating coupling of incident light into the active layers of an OPV. Grating orders $m = 1$ and 2, and the diffraction angle relative to the substrate normal, θ_d, are shown.

placed either beneath or on the surface of the detector, but in either case, they must be in close proximity to efficiently couple light into the absorptive active region. Also, 2D gratings are more efficient than 1D gratings since they provide coupling independent of the relative orientation of the sun to the solar module. An illustration of an OPV with a grating on its reflective (metal) cathode surface is shown in Fig. 7.208. The grating period is Λ, with depth, h. The diffraction condition for the grating is

$$m\lambda = n_{act}\Lambda(\sin\theta_i + \sin\theta_d), \tag{7.208}$$

where $m = 0, \pm1, \pm2\ldots$ is the diffraction order, θ_i is the angle of incidence, and θ_d is the diffraction angle. Two values of m are shown in the figure ($m = 0$ is omitted since it corresponds to specular reflection at $\theta_i = -\theta_d$). For solar radiation in the spectral range of 300 nm $< \lambda$ $<$ 500 nm, Eq. 7.208 is satisfied for $m = 0, \pm1, \pm2$ at a grating period of $\Lambda = 500$ nm and $n = 2$, whereas when 500 nm $< \lambda <$ 700 nm, then $m = 0, \pm1$. Note that beyond redirecting light into the waveguide modes, the diffraction angle for some radiation can also exceed the angle for total internal reflection (TIR) at the substrate–air interface (i.e. at $\sin\theta_d \geq 1/n_{sub}$, where n_{sub} is the substrate index). This reflects the mode back through the active layers, allowing a second opportunity for absorption.

An example soft imprint stamping process (see Section 5.6.4) used for forming gratings on the surface of a P3HT:PC_{61}BM OPV active region is illustrated in Fig. 7.209. The first step is to fabricate the 1D and 2D grating stamps by patterning a photosensitive polymer using optical exposure via a holographic interference pattern with a period, Λ, equal to that required for the grating. Alternatively, an inorganic master template can be coated with a suitable photoresist and similarly exposed. The master is then processed to create indentations that replicate the grating pattern. For a 2D grating, the master is exposed twice, the second exposure made after rotating by 90° followed by development. The master can be directly used as the printing tool, or it can serve as a template by coating it with an elastomer that is cured, and then removed from the master to form the stamp. The advantage of using a replica stamp is that its composition can be made compatible with the processing of the OPV grating. PDMS is a suitable option since it can be easily peeled off from the active device layers without damage owing to its high elasticity (Stolz Roman et al., 2000, Niggemann et al., 2004, Na et al., 2008).

Next, the active layer is spun onto the substrate. Prior to curing, the stamp is pressed onto its surface (Fig. 7.209a), followed by detachment, and then coating the surface with the metal cathode. The surface and the active region are thus left with an impression of the grating pattern as defined by the master. AFM images of 1D and 2D grating patterns on the surface of ITO/P3HT:PC_{61}BM/Ca/Al OPVs using an azo-polymer master and PDMS stamp are shown in Fig. 7.209b. Patterns of similar dimensions are replicated on the master and the stamp, indicating the high fidelity of the multiple transfer and stamping processes (Na et al., 2008).

The fourth quadrant j–V characteristics of devices with 1D and 2D gratings are compared with an analogous, conventional (reference) OPV in Fig. 7.210. The inset shows the corresponding external quantum efficiencies for these same diodes. The j_{SC} is increased for the grating devices compared to the reference OPV due to the increase in η_{ext}, with a larger enhancement achieved with the 2D grating. The power conversion efficiencies are $\eta_P = 3.6\%$, 4.1% and 4.3% for the reference, 1D and 2D devices, respectively. One of the drawbacks that decreases the advantage of this design is having the metal cathode in such close proximity to the waveguiding active region. We have shown in Section 6.6.4 that optical modes efficiently couple strongly to lossy plasmon modes in metal layers in the near field. This loss channel significantly reduces the light intensity in the OPV active region, and hence reduces the photocurrent.

Gratings rely on wavelength-scale features to couple incident plane waves into guided modes. While in principle these structures can lead to an increase of $12n^2$ in intensity within the waveguide, more easily fabricated micron-scale textures can also be beneficial, although they lead to a somewhat reduced optical coupling. An advantage of larger scale

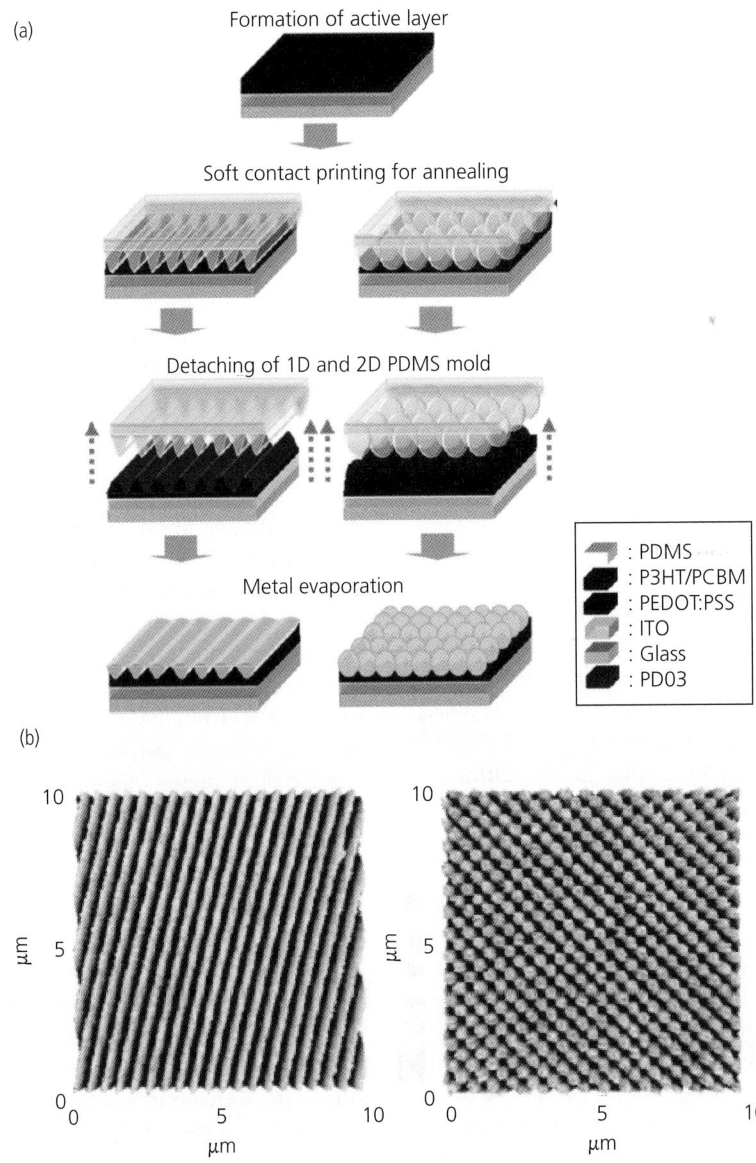

Figure 7.209 (a) Soft lithography stamping process for forming 1D and 2D gratings in a solution-cast OPV active region. (b) AFM images of (*left*) a P3HT:$PC_{61}BM$ OPV surface with a 1D grating with $\Lambda = 500$nm, $h = 20$ nm, and (*right*) a 2D grating with the same dimensions (Na et al., 2008).

features are their reduced sensitivity to wavelength and incidence angle. Numerous texturing schemes have been explored, including *echelle reflectors* (Tvingstedt et al., 2012), multifaceted prismatic retro-reflectors (Esiner et al., 2013), MLAs (Myers et al., 2012), and substrates with sharp creases and folds (Kim et al., 2012).

An example of a reflector with a large feature size is the echelle prism reflector in Fig. 7.211. An echelle grating is one with coarsely defined grooves that are useful in scattering near-normal incidence light into the in-plane direction. The reflector shown comprises a series of right, triangular ridges standing 40 μm from base to apex, with a pitch of 55 μm. It is formed by micro-embossing an acrylic resin substrate, and is then coated with a Ag reflective layer. The prism array is attached to the back surface of a semitransparent OPV. It functions by scattering unabsorbed light into angles greater than TIR. By doing so, it traps the light within the substrate plane until it is finally absorbed. The asymmetric shape of the prisms is chosen to prevent scattering back into the incident

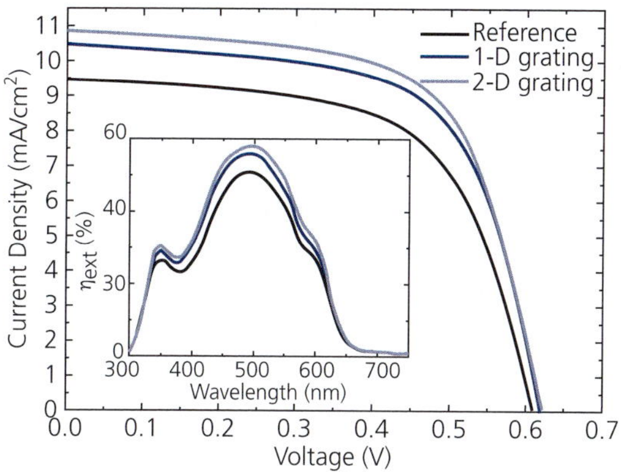

Figure 7.210 Fourth quadrant *j–V* characteristics of a reference P3HT: $PC_{61}BM$ OPV on a conventional substrate, and similar devices with 1D and 2D surface gratings. Inset: External quantum efficiency spectra of the devices (Na et al., 2008).

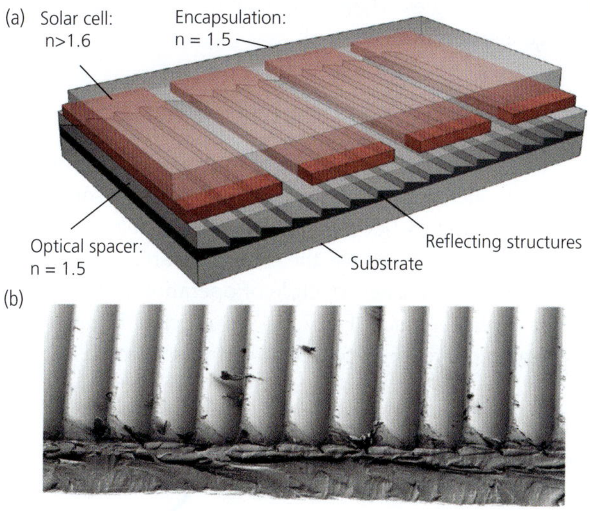

Figure 7.211 (a) An array of four OPVs with a reflecting echelle light trap attached to the substrate. Semitransparent top electrodes are shown in red. (b) A SEM image of the echelle reflector. The prism pitch is 55 μm, and the height is 40 μm (Tvingstedt et al., 2012).

direction, as would happen for equilateral ridges. The shape is not unique; there are any number of macroscopic, asymmetric retroreflectors that can serve the same purpose. However, a triangular shape is simple to produce and uniformly coat with a reflecting layer using vacuum deposition.

The external quantum efficiency of a semitransparent TQ1:$PC_{71}BM$ OPV compared with that using the prism reflector, or a flat Ag mirror is provided in Fig. 7.212a. The increased efficiency using the prism reflector compared to that using a flat mirror is significant, particularly at long wavelengths where the active region is most transparent. This has been a common theme of all light trapping methods considered; the enhancement is greatest when the light requires a long optical path through thin active regions to be fully absorbed. For this combination of light traps and OPV active region materials, the ratio of η_{ext} of the echelle reflector to the planar mirror reaches 1.5 at $\lambda = 700$ nm, falling to 1.0 at $\lambda < 400$ nm where the active region absorption is quite strong. Under illumination from an AM1.5 simulated solar spectrum, the increase in j_{SC} is 24% over the device with the planar mirror.

The efficacy of many of the strategies described thus far is sensitive to incidence angle, and hence they require solar tracking throughout the day to fully exploit their advantages. Large-scale texturing, on the other hand, avoids this problem. For example, in Fig. 7.212b we show the short circuit current of the device with a flat mirror vs. one with the prism grating as a function of angle to a normally incident light

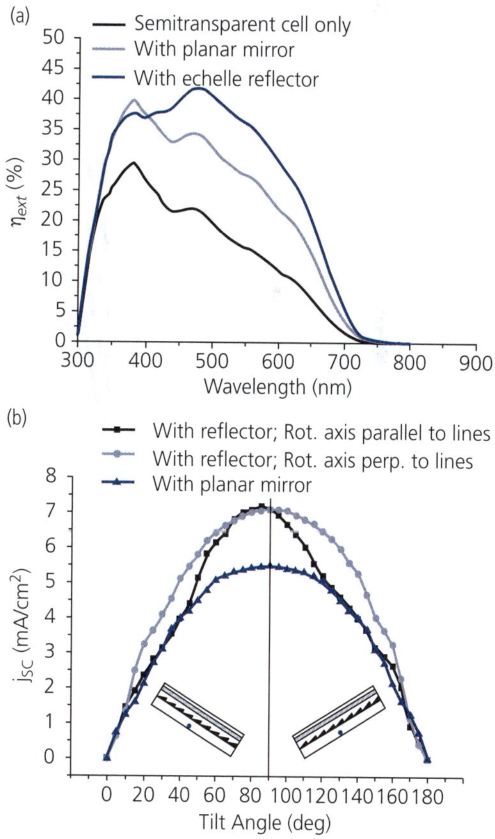

Figure 7.212 (a) External quantum efficiency spectrum of a semitransparent TQ1:$PC_{71}BM$ OPV without a reflector, with a Ag mirror, and with an echelle reflector attached to its distal surface. (b) Short circuit current of the OPV with a flat Ag mirror and with an echelle reflector rotated parallel and perpendicular to the structure ridge lines relative to a fixed, normally incident light source (Tvingstedt et al., 2012).

source, thereby replicating the response to the diurnal solar path. The amount of energy harvested in a day (which is proportional to the areas under the curves) depends on the relative orientation of the blaze lines and the axis of rotation, but in either axial orientation, it is larger than that harvested using a planar mirror. The largest energy benefit is accrued when the prism ridges are aligned perpendicular to the rotation axis, which is equivalent to aligning the prism lines along the east–west direction.

MLAs attached to the substrate are another means for using microscale texture to redirect light into angles greater than that required for TIR. The effects on the incident illumination using a MLA are illustrated in Fig. 7.213a. The lens array refracts the incoming light into directions that increase the optical path length within the active region. To exploit their benefits, the active layer position and thickness must be adjusted relative to that of a device where MLAs are absent. Thus, ray tracing combined with transfer matrix modeling is necessary to find the optimum device design (Myers et al., 2012).

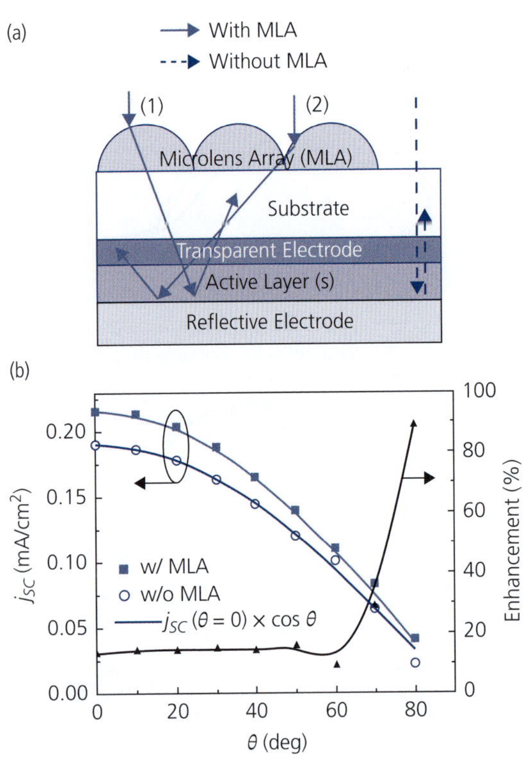

Figure 7.213 (a) Illustration of a device with a MLA, showing redirection of light from the curved lens surface. Dashed arrows show light paths on a conventional, flat substrate. (b) Angular dependence of j_{SC} for a SubPc/C_{60} bilayer OPV with and without a MLA. The ratio of the currents is the enhancement factor, shown at right (Myers et al., 2012).

Figure 7.213b shows that the MLA generates an increased photocurrent at all incidence angles, eliminating the need for solar tracking. In this demonstration, a 100 μm diameter, close-packed array of polymer microlenses with index of refraction n = 1.56 were fabricated using a PDMS mold. The mold itself was fabricated by spreading the PDMS solution over a raft of closely packed polystyrene microspheres. Once cured, the microspheres are lifted off using scotch tape. The polymer employed for the array was then poured into the mold, and UV-cured. The efficiency enhancement for a variety of solar cell materials and structures is ~15–20% compared to similar devices lacking the lens arrays. This enhancement is shown for the diurnal energy harvesting of a SubPc (12 nm)/C_{60} (80 nm)/BCP OPV in Fig. 7.213b. The enhancement compared to devices without the array is approximately 15%, significantly increasing at high angles where, unfortunately, the total energy generation is quite small.

7.7.5 Luminescent solar concentrators

An interesting application of the properties of organics to increase the performance of *inorganic* solar cells is via the use of *organic luminescent solar collectors* (OLSCs) that reduce the solar cell area, and hence module cost. The principle of operation of the OLSC is explained using Fig. 7.214. An efficient luminescent organic dye is dispersed in a transparent plate, for example, glass or plastic. The dye molecules absorb

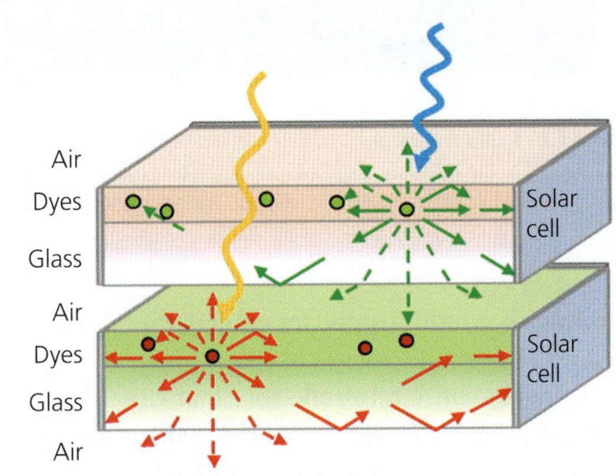

Figure 7.214 Configuration of two, stacked organic luminescent solar concentrators. Light is incident on the concentrators and is absorbed by fluorescent organic molecules in either a thin film (shown) or dispersed in the bulk of the collector plate. The molecules re-emit, with a large fraction of the emitted light trapped in the plate by TIR. The emitted photons are waveguided toward the periphery where they are absorbed and detected by inorganic solar cells. Stacking two such concentrators, one absorbing at short and the other at long wavelengths can be used to cover the solar spectrum (Currie et al., 2008).

light incident on the large, flat surface of the plate, and then re-emit light at their longer fluorescent wavelengths into 4π solid angle. Light emitted within the escape cone leave the plate from the top and bottom surfaces, whereas other photons are trapped by TIR and propagate along its length. Solar cells positioned along the periphery of the plate absorb the waveguided photons. By stacking two plates as shown, short wavelength radiation can be absorbed in one plate, and long wavelength radiation in the other, thus filling in the gaps in the absorption spectra of the luminescent dyes at the expense of doubling the solar cell area required at the edges of the waveguide slabs. The *geometric gain*, G, is the ratio of the concentrator area to that of the PV cells. Since there are propagation losses inherent in this approach, we define the *flux gain*, F, which is G corrected for loss.

There are several benefits and shortcomings of using OLSCs. The primary benefit is that this arrangement does not require solar tracking. The only optical coupling loss is due to cos θ_i, where θ_i is the incident solar angle from normal. If the concentrator cost is less than that of the solar cell, which is most likely the case, then the solar cell cost is reduced by F. Shortcomings of OLSCs include incomplete absorption over the relevant solar spectral band, a limited PL quantum yield (PLQY) of the dyes, and reabsorption of the emitted photons by other dye molecules in the waveguide (Hermann, 1982). The first problem is mitigated by making the concentrator suitably thick, and by using a high concentration of the luminescent molecules. The second is circumvented by using high efficiency lumophores, although as we have seen in Chapters 3 and 6, a high density of dopants generally leads to reductions in PQLY due to concentration quenching. The last challenge is met by selecting molecules with large Stokes shifts, that is, where the emission spectrum is significantly red-shifted from the absorption tail. Even when these conditions are met, the geometric gain in single dye systems is only between 3 and 4 (Batchelder et al., 1981). The stability of organic dyes, particularly in the presence of UV solar irradiation, is also a concern, although significant advances in developing highly stable lumophores for OLEDs have considerably reduced this problem.

Reabsorption losses can be minimized in host-guest luminescent systems, where the host molecule absorbs the incident radiation, and then transfers the excited state to the lumophore with nearly unity efficiency by FRET. The host absorption spectrum can be adjusted to cover greater portions of the solar spectrum, it can result in highly efficient PLQY of the lumophore, and can also shift the absorption of the lumophore to regions of the spectrum that only weakly absorb the re-emitted photons. For example, using metal–organic phosphors as the luminescent molecule, the PLQY in doped systems can approach 100%, but the phosphors are very weakly absorbing due to the disallowed $T_1 \leftarrow S_0$ transition, even in the presence of strong spin-orbit coupling (Section 3.7.3).

Doped OLSCs have been demonstrated using both fluorescent and phosphorescent molecular guest–host systems (Currie et al., 2008). Example absorption and emission spectra for the fluorescent laser dye, DCJTB, and the phosphor, Pt(TPBP), are shown in Fig. 7.215a and b, respectively. The dashed lines are the spectra due to a 5.7 μm thick film of Alq_3:DCTJB (2%) on a glass substrate. The ratio of the peak luminescence intensity to the absorbance of the film at that same wavelength is the *self-absorption ratio*, S. Thus, S is an effective measure of how lossy the dye system is due to self-absorption of its emitted spectrum. For this blend, $S = 80$. Self-absorption is considerably reduced by including a sensitizer molecule (rubrene) that increases the Förster transfer efficiency from the wide energy gap Alq_3 to the red emitting DCJTB. The presence of rubrene in a blend of Alq_3: rubrene (30%):DCJTB (1%) reduces the dopant concentration, and hence its absorbance, resulting in $S = 220$ (dashed lines). The greatest luminescence to absorption advantage is obtained using the very weakly absorbing phosphor, Pt(TPBP). In Fig. 7.215b, we show $S = 500$ for a 5.8 μm thick Alq_3DCJTB (2%):Pt (TPBP) (4%) film.

The performance of the OLSCs has been measured using single as well as double waveguide

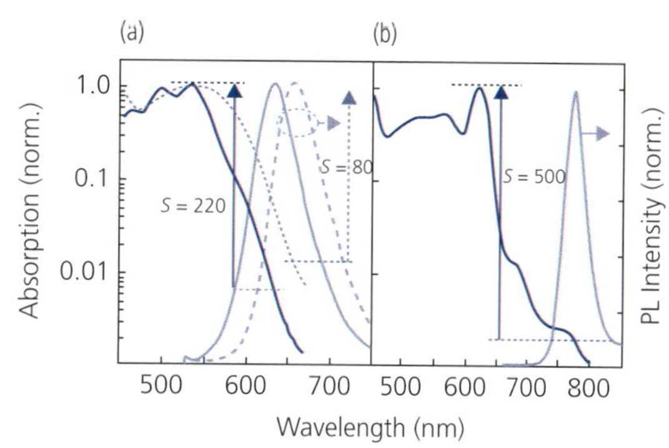

Figure 7.215 Absorption and emission spectra of an Alq_3:DCTJB (dashed line) and an Alq_3:rubrene:DCTJB (solid line) film, and (b) an Alq_3: DCTJB:Pt(TPBP) film. S is the self-absorption ratio equal to the normalized peak luminescence intensity to the absorption at that wavelength (Currie et al., 2008).

configurations. The efficiency obtained using the films in Fig. 7.215a and b ranges between 30% for the Alq_3:DCJTB (2%):Pt(TPBP) (4%) film, to 50% for both DCJTB blends. Approximately 20% of the losses result from re-emission into the concentrator escape cone, and the remainder from reabsorption. Combining the phosphorescent and rubrene-based fluorescent OLSCs in tandem with the red emitting phosphor on the bottom, the concentrator efficiency increased to 60% with emission extending to $\lambda =$ 675 nm.

Crystalline Si solar cells have been mounted onto one edge of a 100 × 100 × 1 mm glass substrate with the external quantum efficiency and flux gains plotted vs. geometric gain in Fig. 7.216a and b, respectively. The geometric gain is varied by scanning a focused incident light beam normal to the edge supporting the solar cell. The lines show fits based on optical modeling of the losses and the film optical properties. The efficiency falls off rapidly with G for films where reabsorption is strongest (i.e. for the DCTJB and rubrene:DCTJB films), providing the least benefit from concentration. Almost no losses are experienced by the Pt(TPBP) film due to its very low self-absorption. The flux gain, Fig. 7.219b rises rapidly until self-absorption dominates. As previously, the phosphor-doped film shows the possibility for high F along with large geometric gains, potentially leading to very low cost solar modules. However, $F \sim 10$ was the maximum measured in the best cases, which is nevertheless a considerable improvement over demonstrations employing single, relatively highly absorbing fluorescent dye compounds (Slooff et al., 2008). While considerable work has been focused over the last several years on exploiting the fluorescent concentrator concept, the efficiencies still remain too low to make them cost competitive with conventional Si cells that are now commercialized for < US \$0.50/Wp.

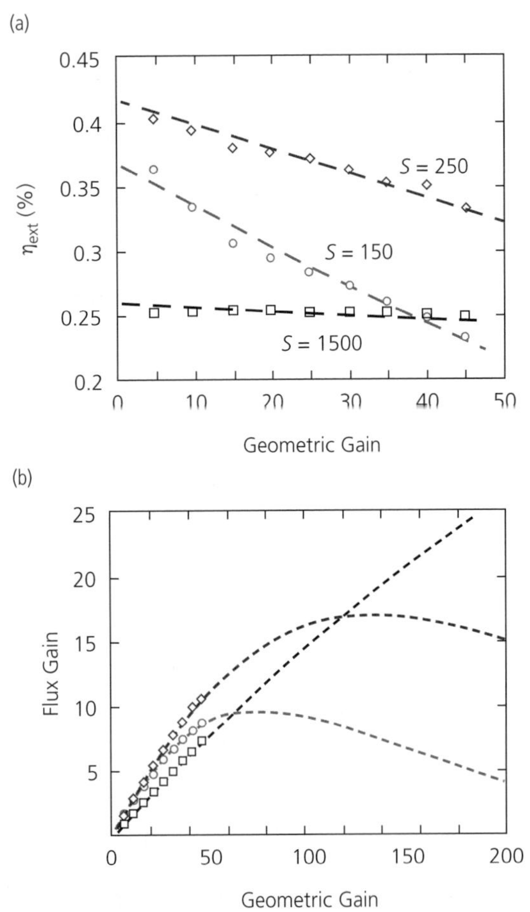

Figure 7.216 (a) External quantum efficiency and (b) flux gain vs. geometric gain for the Pt(TPBP) (black line), rubrene:DCTJB (dark blue), and DJTCB (light blue) OLSCs. Data in (a) are measurements using a Si solar cell on one edge of the 100 × 100 × 1 mm plate with the luminescent film on its surface. Lines are fits based on optical modeling (Currie et al., 2008).

We found in Section 6.6.5 that the emission from lumophores is most intense perpendicular to the excited state dipole moment. Hence, the 20% loss incurred due to emission within the escape cone of the slab can be reduced if molecules are intentionally aligned with their dipole moments oriented normal to the plate surface. This orientation, however, also results in increased absorption since the ground and excited state dipole moments are often approximately aligned. Placing a diffuser in front of an oriented film can randomize the light input wavevector. By using the rod-shaped, green-emitting Coumarin 6 molecules dispersed in a liquid crystal scaffold on a glass substrate, some degree of vertical alignment is achieved, increasing the light trapping efficiency to 81% from 66% for an OLSC where the dye molecules are randomly dispersed in a transparent polymer matrix (Mulder et al., 2010).

A potentially interesting variant on the OLSCs are those that absorb only in the UV or NIR spectral regions. This results in visible transparency, which may find applications as glass on windows, building facades, etc. While the idea is attractive (Yang and Lunt, 2017), there is a lack of NIR-absorbing and emitting lumophores with reasonably high efficiencies. We showed in Section 3.6 that the efficiency of organic materials falls off rapidly with energy gap (the energy gap law), making it unlikely that NIR OLSCs can achieve performance metrics necessary for practical applications. For example, OLSCs employing cyanine dyes have only achieved $\eta_P = 0.4\%$ using edge-mounted Si cells (Zhao et al., 2014, Banal et al.,

2015). While UV-absorbing organic materials may provide higher efficiencies, their stability to short wavelength solar radiation introduces additional concerns.

7.8 Reliability of organic photovoltaics

All solar cell technologies must possess three cardinal attributes: (i) low manufacturing cost at high yield, (ii) high efficiency, and (iii) long-term operational reliability. In fact, all of these attributes can be reduced to the primary consideration of low cost, since they all act in concert to deliver a higher output power using panels with the smallest area footprint at an acceptably low level of maintenance. In this section, we are concerned with understanding the factors that deliver long-term operational stability. In Section 6.7 we introduced the basic methodologies used to determine the reliability of semiconductor devices. Lifetime test methods used to qualify OPVs must account for the particularly challenging set of conditions to which solar cells are exposed that are not experienced by OLEDs or transistors. For example, OPVs are used out-of-doors, bathed in intense broadband solar irradiation that extends from the UV to the IR. The ambient temperatures can range from $-40°C$ to $100°C$, and the humidity from $\sim 20\%$ to 100%. To meet the standards of established technologies, the lifetime of OPVs must be at least 20 years as determined by the point at which their power conversion efficiency decreases by 20% from its initial value (also known as *T80*, see Section 6.7). Other extreme environmental factors such as exposure to dust, hail, and wind loading, can apply significant stress to the OPV module. At this point, there are few OPVs that have been deployed in outdoor environments, since the technology is still in its infancy. If successful, however, OPVs may find widespread, and perhaps ubiquitous use for power generation on buildings, mobile appliances, and other applications where they have the potential to provide significant value over incumbent technologies. For this to occur, long-term OPV reliability must be established.

The primary objective of reliability testing is to identify significant failure mechanisms, and their origins. None of these mechanisms can be defined unless high purity and stable materials sets are developed, and packaging schemes are implemented that prevent extrinsic, short-term failure. Once the primary degradation routes are understood and resolved, qualification of devices and modules can be undertaken by exposure to the elements that possibly introduce new, extrinsic routes to failure. As with OLEDs, we attribute *intrinsic failure* mechanisms to aging of the molecular constituents in the layer stack, contact degradation, changes in morphology, and device architecture. *Extrinsic failure* mechanisms arise from package failure, ingress of contaminants (e.g. O_2 and H_2O) into the package, impurities within the materials left during processing, delamination of contacts from the package, and so on.

Solar cell operational standards require a *T80* of 10–30 years, a metric that is dependent on the application. Here, Tx is the time for the power conversion efficiency to drop to $x\%$ from its initial value. It is impractical to qualify a new technology over even a small fraction of these times. Hence, accelerated aging protocols need to be established to accurately extrapolate the performance loss in the short term into the long term. We found in Section 6.7 that degradation in OLEDs can be predictably accelerated at elevated temperatures and drive currents. Accelerated aging of OPVs, with their use in out-of-doors venues, involve parameters that can be applied in the laboratory, but that also include aging tests that simulate the environment that is endured during actual field use. These environments can have widely varying demands, depending where the panels are deployed (e.g. in the arctic or equatorial regions, on the sea shore, or in the desert). This places demands on developing relevant and standardized testing protocols that are appropriate across technology platforms and their field of use.

Given this backdrop, efforts have been dedicated to defining standardized test conditions that are suitable for both laboratory, and for consumer-ready, qualification of OPVs. Recommended test protocols have been defined through a consensus study by the OPV community via the International Summit on OPV Stability (ISOS). The test protocols fall into three categories; basic (Level 1), intermediate (Level 2), and advanced (Level 3). The lowest level is meant for research laboratories with only modest access to necessary test equipment. It provides guidelines for those labs such that results can be widely understood and reproduced by independent investigators. Level 2 increases the comprehensiveness and standardization of the testing protocols that can lead to a unambiguous comparisons by different laboratories. Level 3 is close to, but not identical with standardized tests for market-ready products. Each level also has a different standard for shelf-life testing in the dark (D-1–D-3), outdoors (O-1–O-3), laboratory weather testing (L-1–L-3), thermal cycling (T-1–T-3), and solar-thermal humidity cycling (LT-1–LT-3). The conditions used in each of these testing regimes are summarized in Table 7.28. Further details can be found in the consensus report itself (Reese et al., 2011).

Table 7.28 Test protocols for OPVs at different stages of development. (Reese et al., 2011)

Three levels						
Basic (Level 1)			"Hand held" measurements using the simplest equipment and few conditions			
Intermediate (Level 2)			Fixed conditions and protocols suited for most labs			
Advanced (Level 3)			Standardized tests applied in certified labs. Extended range of parameters to monitor, etc.			
Test type	**Dark**			**Outdoor**		
Test ID	ISOS-D-1 Shelf	ISOS-D-2 High temp. storage	ISOS-D-3 Damp heat	ISOS-O-1 Outdoor	ISOS-O-2 Outdoor	ISOS-O-3 Outdoor
Light source	None	None	None	Sunlight	Sunlight	Sunlight
Temp.[a]	Ambient	65/85 °C	65/85 °C	Ambient	Ambient	Ambient
Relative humidity (R.H.)[a]	Ambient	Ambient (low)	85%	Ambient	Ambient	Ambient
Environment[a]	Ambient	Oven	Env. chamber	Outdoor	Outdoor	Outdoor
Characterization light source	Solar simulator or sunlight	Solar simulator	Solar simulator	Solar simulator	Sunlight	Sunlight and solar simulator
Load[b]	Open circuit	Open circuit	Open circuit	MPP or open circuit	MPP or open circuit	MPP
Test type	**Laboratory weathering testing**			**Thermal cycling**		
Test ID	ISOS-L-1 Laboratory weathering	ISOS-L-2 Laboratory weathering	ISOS-L-3 Laboratory weathering	ISOS-T-1 Thermal cycling	ISOS-T-2 Thermal cycling	ISOS- T-3 Thermal cycling
Light source	Simulator	Simulator	Simulator	None	None	None
Temp.[a]	Ambient	65/85 °C	65/85 °C	Between room temp. and 65/85 °C	Between room temp. and 65/85 °C	−40 to +85 °C
Relative humidity (R.H.)[a]	Ambient	Ambient	Near 50%	Ambient	Ambient	Near 55%
Environment/ setup	Light only	Light and Temp.	Light, Temp. and R.H.	Hot plate/oven	Oven/env. chamb.	Env. chamb.
Characterization light source	Solar simulator	Solar simulator	Solar simulator	Solar simulator or sunlight	Solar simulator	Solar simulator
Load[b]	MPP or open circuit	MPP or open circuit	MPP	Open circuit	Open circuit	Open circuit
Test type	**Solar-thermal-humidity Cycling**					
Test ID	ISOS-LT-1 solar-thermal cycling		ISOS-LT-2 solar-thermal-humidity cycling		ISOS-LT-3 solar-thermal-humidity-freeze cycling	
Light source	Simulator		Simulator		Simulator	
Temp.	Linear or step ramping between room temp. and 65 °C		Linear ramping between 5 and 65 °C		Linear ramping between −25 and 65 °C	
Relative humidity (R.H.)	Monitored, uncontrolled		Monitored, controlled at 50% beyond 40 °C		Monitored, controlled at 50% beyond 40 °C	
Environment/ setup	Weathering chamber		Env. chamb. with sun simulation		Env. chamb. with sun simulation and freezing	
Characterization light source	Solar simulator		Solar simulator		Solar simulator	
Load[b]	MPP or open circuit		MPP or open circuit		MPP or open circuit	

[a] The ambient conditions are defined as 23 °C/50%RH in general, and 27 °C/65%RH accepted in tropical countries according to ISO 291(2008): Plastics—Standard atmospheres for conditioning and testing.

[b] Open circuit refers to a simply disconnected device or device connected to a sourcemeter set to 0 current.

Almost, but not all OPV tests have thus far used Level 1 protocols. With the increased performance of OPVs, both in terms of their electrical characteristics and reliability, there have been more frequent reports of Level 1 to Level 2, Level 2, and even field tests. This is an indication that OPV technology is maturing, and moving toward practical implementation.

Similar to OLEDs, the aging of OPVs often follows a two-stage trajectory illustrated in Fig. 7.217. The plot shows the dependence of a measured parameter, P (e.g. V_{OC}, j_{SC}, FF, η_P, etc.) on operating time. The first stage of rapid but limited decrease is known as the *burn-in* period. During this decline, from the initial value of the test parameter (i.e. P_0 at T_0, or alternatively, P_{100} at *T100*) the device undergoes adjustments to a condition that was not fully established during fabrication. That is, burn-in may result from early stage morphological annealing induced by current generation, uptake of residual impurities in the film or package that cause finite degradation until the impurities are exhausted, contact annealing and so on. It is possible, although not necessarily the case that the measured parameter will decrease to even less than 80% of P_0, corresponding to P_{80} at *T80*. At some time, T_S, the characteristics take on a second, stabilized slope corresponding to long-term aging. As long as P_S is not significantly less than P_0, this long-term aging defines the useful term of operation of the device. Determining T_S80, that is, the usable lifetime of the solar cell during its stable phase of operation, is the objective of most reliability experiments. However,

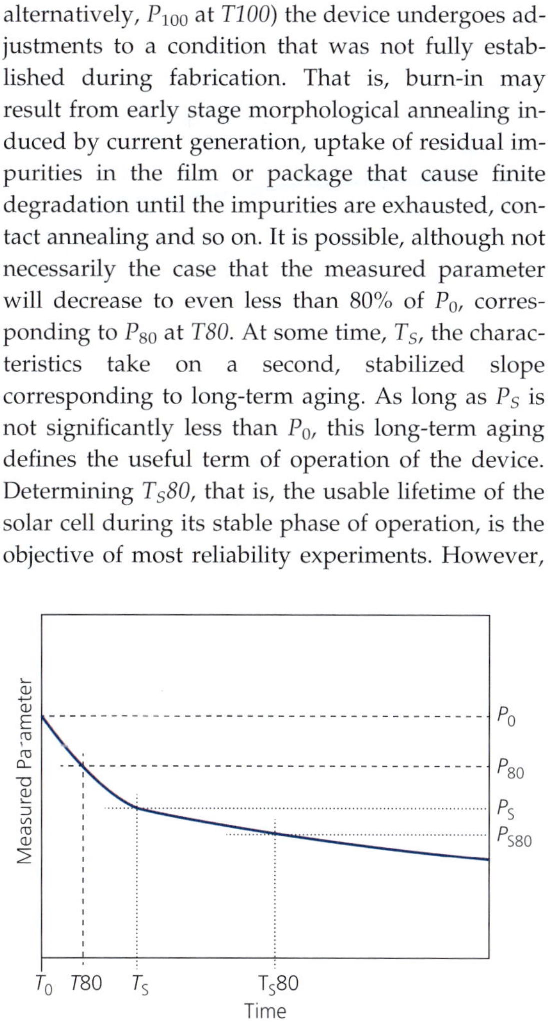

Figure 7.217 Characteristic parameter aging trajectory for an OPV. The initial value of a test parameter is P_0 obtained at time T_0. There are two time constants for aging: Early rapid decay (called burn-in) that occurs until the onset of the second, more stable degradation at time, T_S. (with parameter value P_S). *T80* and T_S80 are the times that the measured parameter is decreased to 80% of its initial value to P_{80} and P_{S80}, respectively.

since T_S80 in practical cells is too long to measure in most laboratory settings. Hence, the aging process must be accelerated. This is accomplished by either elevating the cell temperature and/or the light intensity (which can also increase cell temperature and current) during the aging cycle. Here, temperature and intensity are two potential acceleration factors. For useful analysis, however, temperature and intensity must be analyzed independently, which can be done by aggressive cooling of the illuminated solar cell, or measuring the temperature dependence of degradation of the test parameter, and then subsequently measuring its intensity dependence at a fixed temperature. The independently observed thermal degradation is then used to separate out the contributions due only to high illumination intensity.

Accelerated aging must not be so harsh (i.e. by exposure to an excessively high temperature or light intensity) that aging processes not encountered under normal operation are introduced. This results in extrinsic degradation that may not provide an accurate prediction of cell lifetime during standard operating conditions.

The two-phase aging can be modeled using a *sum of exponentials* of the form (Section 6.7):

$$P(t) = P_0\exp(-t/\tau_1) + P_{ex}\exp(-t/\tau_2), \tag{7.209}$$

where τ_1 is the characteristic burn-in time constant, and τ_2 is the long-term degradation time constant. Now, $\tau_1 \ll \tau_2$ as is often the case, which allows for accurate determination of the coefficient, P_{ex}. A representative set of aging data is shown in Fig. 7.218. At the upper left, we find that $FF(t)$ undergoes rapid change during burn-in, and thereafter is constant. From the FF data, therefore, we obtain the time constant, τ_1. Note that the identification of T_S can be somewhat arbitrary, except in cases such as this one where $\tau_1 \ll \tau_2$. The short circuit current, however, does not show changes during burn-in, hence giving a value of τ_2 as well as an extrapolated value of $j_{SC,ex}$ at $t \to 0$. The open circuit voltage is apparently unchanged from its initial value of V_{OC}. The product of all of these curves results in the time evolution of the power conversion efficiency shown at the lower right. For this device, the functional form of the efficiency is

$$\begin{aligned}\eta_P(t) &= \eta_{P0}\exp\left[-t\left(\frac{1}{\tau_1}+\frac{1}{\tau_2}\right)\right] + \eta_{P,ex}\exp\left[-\frac{t}{\tau_2}\right]\\ &\approx \eta_{P0}\exp\left[-\frac{t}{\tau_1}\right] + \eta_{P,ex}\exp\left[-\frac{t}{\tau_2}\right]\end{aligned} \tag{7.210}$$

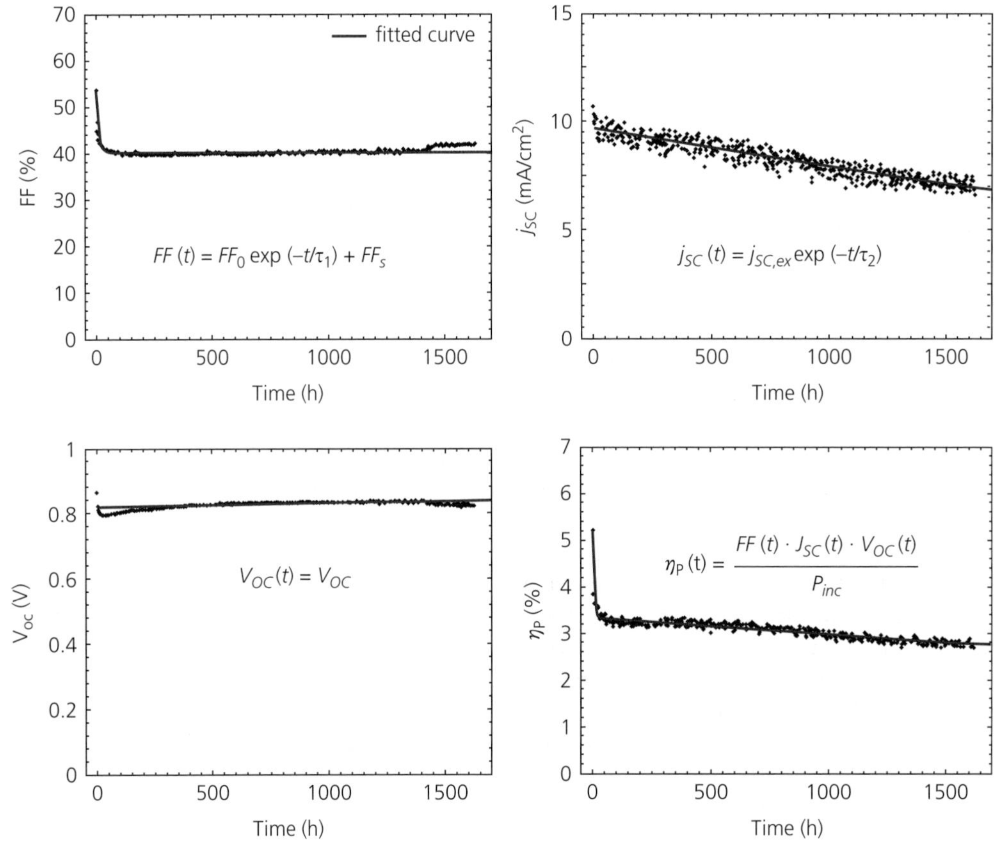

Figure 7.218 Aging of several OPV operating parameters and empirical fits (lines) to the data (points). The functional form of the fits to *FF*, j_{SC}, and V_{OC} are indicated. Adapted from Roesch et al. (2015).

since $\tau_1 \ll \tau_2$. The constants η_{P0} and $\eta_{P,ex}$ are easily obtained from the functions from which they are derived, or directly from the plot in Fig. 7.218.

Note that while an exponential decrease in performance is suggestive of a process related to chemical degradation, the long-term data in Fig. 7.218 can arguably be just as well fit by a linear empirical relationship, viz.

$$P(t) = P_0 \exp(-t/\tau_1) + P_{ex}(1 - t/\tau_2). \tag{7.211}$$

An additional functional form, often used in fitting the lifetime response of OLEDs, and that also has been applied to small molecule OPVs is the *stretched exponential*:

$$P(t) = P_0 \exp\left[-(t/\tau_1)^{\beta}\right], \tag{7.212}$$

where $\beta \leq 1$ is a fitting parameter. This functional form does not distinguish between initial burn-in and long-term degradation, and hence is more appropriate for fitting those degradation processes that proceed monotonically from *T100* to *T0*.

Once the various lifetime parameters in Eqs. 7.209, 7.211, or 7.212 are determined, it is then possible to extrapolate the efficiency to T_S*80*. As in previous discussions on OLED reliability, there is an element of faith that is needed to assume that there are only two principal aging mechanisms captured by the time constants, τ_1 and τ_2, and that a third set of parameters are not needed to extend the extrapolation to even longer times when additional failure mechanisms possibly emerge. In addition, catastrophic failure due to package leakage or breakage, or migration of a metal contact into the OPV active region, etc., may also lead to an early death of the devices that is not accounted for by the time constants, τ_1 and τ_2. We will return to these extrinsic but nevertheless important failure mechanisms in subsequent sections.

Ultimately, the value of a cell is determined by the total energy that it delivers during its usable life. This is known as the *lifetime energy yield*, or E_{80}, and is calculated using (Roesch et al., 2015)

$$E_{80} = \int_{t=0}^{T_{S80}} \eta_P(t) P_{inc} dt, \tag{7.213}$$

where P_{inc} is generally taken to equal 1 sun intensity (i.e. 1 kW/m^2). The E_{80} calculated for the device in Fig. 7.218 is presented in Fig. 7.219, yielding a $T_S80 =$ 1930 h and $E_{80} \approx 58$ kWh/m^2.

There is no need to accelerate the aging of the cell in Figs. 7.218 and 7.219 since the degradation is sufficiently fast that the experiment is completed within less than 1500 h, or 2 months. However, it would not be possible to obtain the T_S80 for more durable cells, in which case the aging process needs to be accelerated by increasing temperature and/or intensity. If we assume that the post-burn-in decay of the parameter, P, is due to a single kinetically-driven process, for example, molecular fragmentation or reactions with other species in the active region, it should follow an Arrhenius behavior with activation energy, E_a. Then, the degradation rate of parameter P is:

$$k_{deg} = 1/\tau = k_0 \exp(-E_a/k_B T), \tag{7.214}$$

where k_0 is a constant. If we assume that the decrease in performance is a function of the incident power, then,

$$k_{deg} = k_0 \left(I_{test}/I_{ref} \right)^{\gamma}, \tag{7.215}$$

where γ is an empirical constant, I_{test} is the power used during the aging tests, and I_{ref} is that used during normal operation (typically 1 sun). Then we can define the *aging acceleration factor* between two different powers and temperatures (I_1, I_2 and T_1, T_2, respectively) using

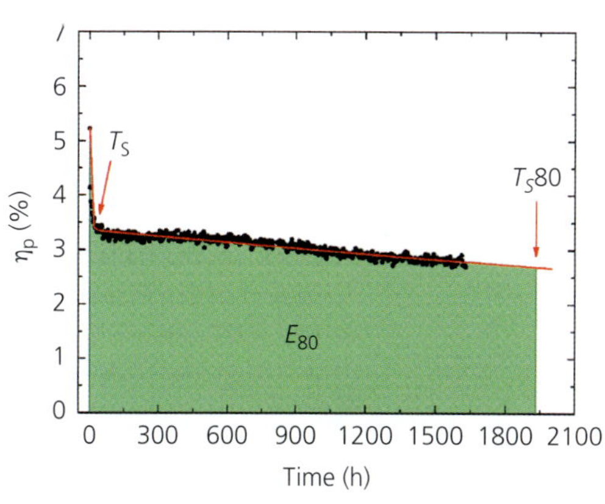

Figure 7.219 The total energy harvested during the cell lifetime is the product of the area of the shaded region and $P_{inc} = 1$ kW/cm^2. Data are for the cell in Fig. 7.218 (Roesch et al., 2015).

$$A = \left(\frac{I_1}{I_2} \right)^{\gamma} \exp\left[-\frac{E_a}{k_B} \left(\frac{1}{T_1} - \frac{1}{T_2} \right) \right]. \tag{7.216}$$

The procedure for finding γ and E_a is similar to methods discussed for OLEDs. For example, if the decay constant is dominated by long-term degradation, then $k_{deg} = 1/\tau_2$ is measured at several different temperatures (Schuller et al., 2004). An Arrhenius plot of k_{deg} vs. $1/T$ yields E_a, from which the temperature dependence of A is found. This, in turn, yields τ in the relevant operating regime and temperature. This procedure has been applied to a DBP:C_{70} (1:8) OPV employing a Bphen:C_{60}/BPhen exciton blocking buffer layer, with results for *FF*(*t*) at several temperatures given in Fig. 7.220a. The data are fit by the biexponential in Eq. 7.209 from which the rate constants, k_1 and

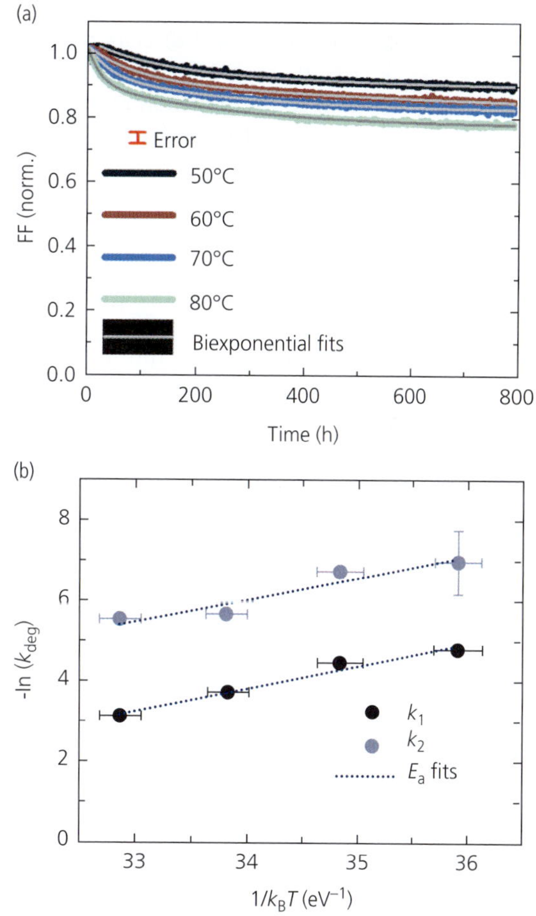

Figure 7.220 (a) Time dependence vs. temperature of the normalized *FF* of a DBP:C_{70} (1:8) OPV employing a BPhen:C_{60}/BPhen buffer layer. Lines are biexponential fits to the data (points). (b) Arrhenius plot of the fast (k_1) and slow (k_2) degradation rates obtained from the fits in (a). Lines are activation energy fits according to Eq. 7.214 yielding $E_a = 0.53$–0.56 eV (Burlingame et al., 2016).

k_2 are extracted at each temperature. These are plotted in Fig. 7.220b as a function of $1/k_BT$ to obtain nearly equal activation energies ($E_a \sim 0.55$ eV) for both burn-in and long term operation. While it is unusual for the activation energy to be the same over both time scales, the apparent coincidence arises since both are affected by BPhen crystallization; the short term loss in *FF* is due to increased shunt currents across the junction induced at high temperature, and the longer term loss arises from increased recombination due to slow intrusion of BPhen into the active region (Burlingame et al., 2016).

Similarly, the acceleration factor, γ, is found by measuring the rate of degradation of the OPV parameters at several different intensities. To separately measure mechanisms affected only by intensity, the temperature dependence must first be eliminated by cooling the devices, or by ensuring that the temperature dependence of the lifetime is minimal.

The degradation of DBP:C_{70} (1:8) cells comprising a TPBi:C_{60} (1:1, 10 nm)/TPBi (3 nm) buffer layer has been measured as a function of intensity. Thermal stability is ensured by replacing the low glass transition temperature BPhen ($T_g = 61$°C) blocking layer with TPBi, that has a substantially higher $T_g = 120$°C.

It is challenging to achieve high intensity exposure using a standard solar simulator employing a Xe lamp. Intensity-dependent aging experiments, therefore, often use extremely high brightness (up to 70 suns) LED light sources. The LED intensity does not fade over the time scale of the experiment, although it has a significantly different spectrum from that of the sun. The spectra from cool-white and UV emitting LEDs are shown in Fig. 7.221a superimposed on the AM1.5G reference and the spectrum from a Xe-arc lamp conventionally used to simulate solar radiation. The most damaging spectral region for organics is the UV, which is not represented by the broadly emitting LEDs. Hence, the DBP:C_{70} devices are also exposed to irradiation from a separate UV-emitting LED with a peak wavelength of 365 nm. The absorption of the OPV active region is minimal beyond 700 nm, thus the lack of NIR spectral content does not impact the measured lifetimes. Since active cooling is employed, the temperature of the devices remains well below T_g of the exciton blocking material (see Fig. 7.221b). Indeed, the devices exhibited no temperature-dependent aging up to $T = 130$°C (Burlingame et al., 2016).

It is remarkable in itself that the OPVs can tolerate illumination at such the extreme intensities. The device characteristics up to concentrated *outdoor* solar irradiances of 27 suns are provided in Fig. 7.222. Since DBP:C_{70} OPVs typically have efficiencies ~8%, the current at 37 suns is >0.3 A/cm^2. The *j–V* characteristics (Fig. 7.222a) show that at the highest intensity, the forward and reverse currents are approximately equal, achieving a maximum of ~0.12 A/cm^2. The symmetry of the curves indicates that the current is limited by the internal cell series resistance.

The slope of the short circuit current in Fig. 7.222b is relatively constant until the highest intensity, indicating that the quantum efficiency is constant over this range of intensities. However, in Fig. 7.222c, a roll-off in V_{OC} from its expected logarithmic dependence on irradiance is observed, due to increased recombination from the accumulation of a very large density of charge in the mixed active region. Finally, *FF* is significantly decreased at high intensity (Fig. 7.222d), again due to cell series resistance. This results in a decrease in η_P from 6.5% at 1 sun, to 3% at 27 suns.

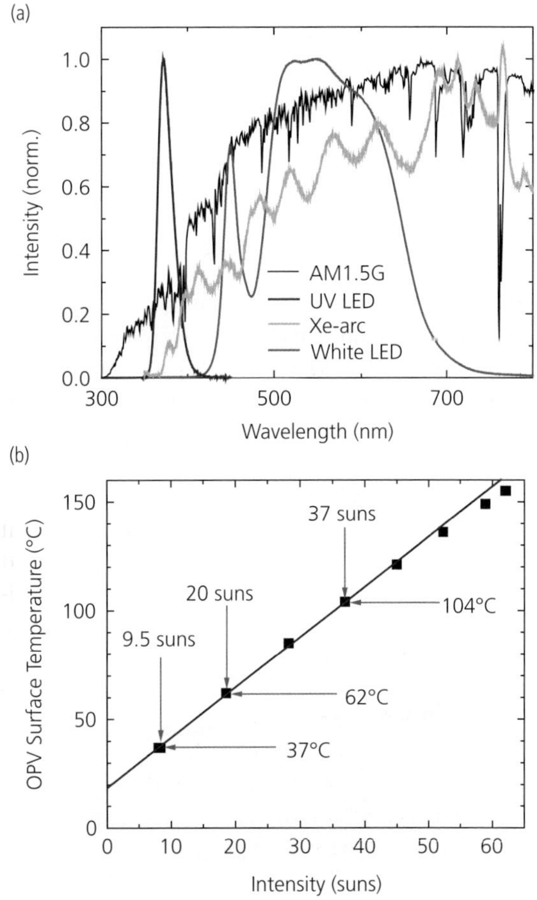

Figure 7.221 White and UV LED spectra used in OPV aging. For comparison, the Xe-arc lamp and AM1.5G reference spectra are also shown. (b) Illumination intensity from a white LED vs. temperature on the OPV surface. The three aging conditions used for OPVs in Fig. 7.223 are indicated (Burlingame et al., 2019).

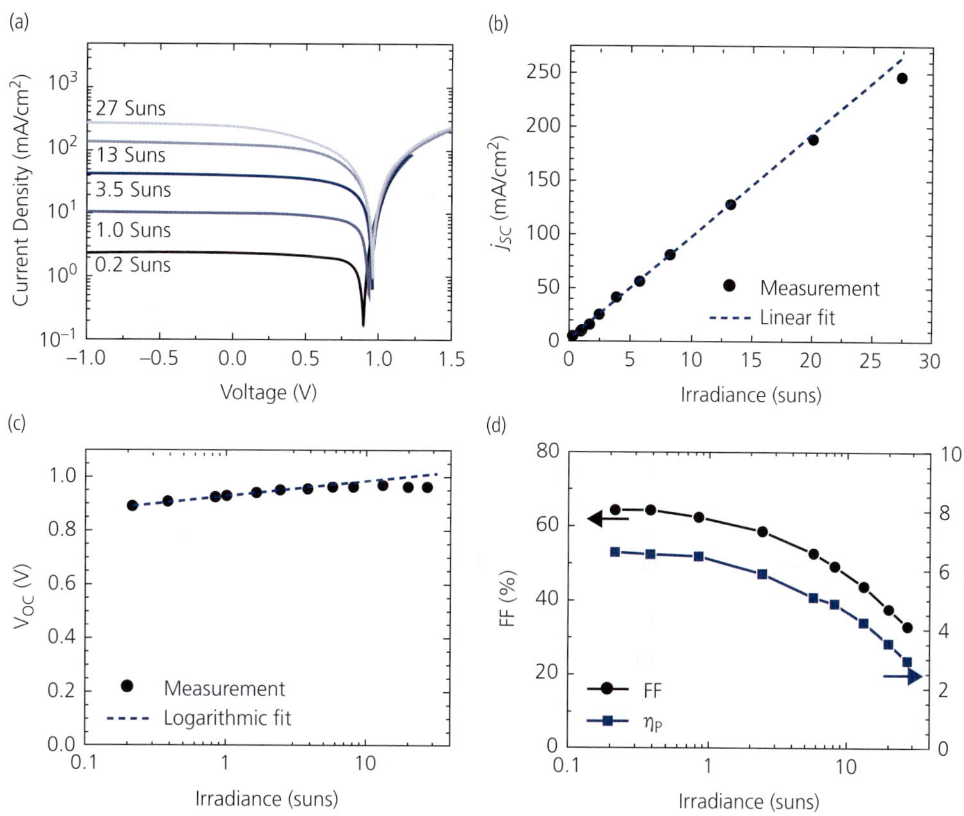

Figure 7.222 Performance of CBP:C_{70} OPVs vs. incident concentrated solar power intensity. (a) Current-voltage characteristics, (b) short circuit current, (c) open circuit voltage, and (d) *FF* and power conversion efficiency (Burlingame et al., 2017).

Figure 7.223 shows the performance characteristics of a population of TPBi-based OPVs aged at several different illumination intensities. This population shows no appreciable early stage (i.e. burn-in) degradation (<3–5%) in any of its characteristics, and hence only long-term aging data are shown. There is some deviation in j_{SC} and *FF* (Fig. 7.223b) at 36.9 suns (36.9 kW/cm^2) from the data taken at lower intensities due to degradation of the ITO and MoO_3 anode buffer. That is, the reduction in efficiency of the highest intensity data is primarily due to loss of transparency of these metal oxides. This is accompanied by an equally small loss in charge extraction efficiency, as determined by measuring the change in slope of the reverse-biased j_{ph}-V characteristics both before and after aging (Burlingame et al., 2019). The change in power conversion efficiency is accurately fit by the stretched exponential function in Eq. 7.212 using $\beta = 0.28$ and $\tau = 1.1 \times 10^{10}$ h. Surprisingly, no degradation is experienced up to 20 suns equivalent UV illumination (inset, Fig. 7.223a), attesting to the remarkable stability of the materials employed.

Figure 7.224a shows k_{deg} vs. I_{test}. Fitting the data with Eq. 7.216 yields $\gamma = 2.5$. The equivalent lifetime under 1 sun (I_{ref}) exposure can be extrapolated using γ for the OPVs aged at each intensity. As shown in Fig. 7.224b, the η_P at all intensities follows a similar trend that results in a lifetime of $T80 = 4.9 \times 10^7$ h. This extremely long lifetime is similar to operating lifetimes that have been reported for phosphorescent organic light emitting devices operated at 10 mA/cm^2 (Burrows et al., 2000), which is comparable with 12.5 mA/cm^2 generated at 1 sun intensity by the OPVs.

To convert the OPV lifetimes into outdoor lifetime projections, we assume an average of 5 kWh/m^2 of sunlight per day (e.g. 5 h of equivalent AM1.5G radiation), although value depends on location (e.g. 4.2 kWh/m^2/day in Detroit, MI to 6.5 kWh/m^2/day in Phoenix, AZ) (Habte et al., 2017). Therefore, the extrapolated *T80* corresponds to $> 2.7 \times 10^4$ yr, outdoor equivalent. While such lifetimes are not practically achievable given the presence of extrinsic failure mechanisms such as contact weathering, package failure, and the possibility of late-onset degradation, the extrapolation nevertheless suggests that the *intrinsic*

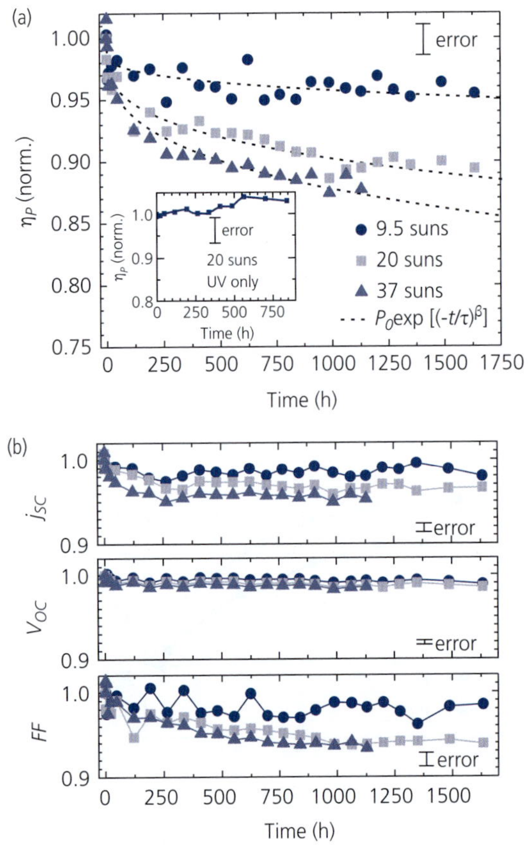

Figure 7.223 Measured DBP:C_{70} (1:8) OPV characteristics at various white LED intensities, from 1 to 37 suns. Full OPV performance data are found in (a) and (b). Dashed lines in (a) are fits to the stretched exponential function in the legend. Inset in (a): Power conversion efficiency vs. aging time under 20 suns equivalent illumination in the UV (Burlingame et al., 2019).

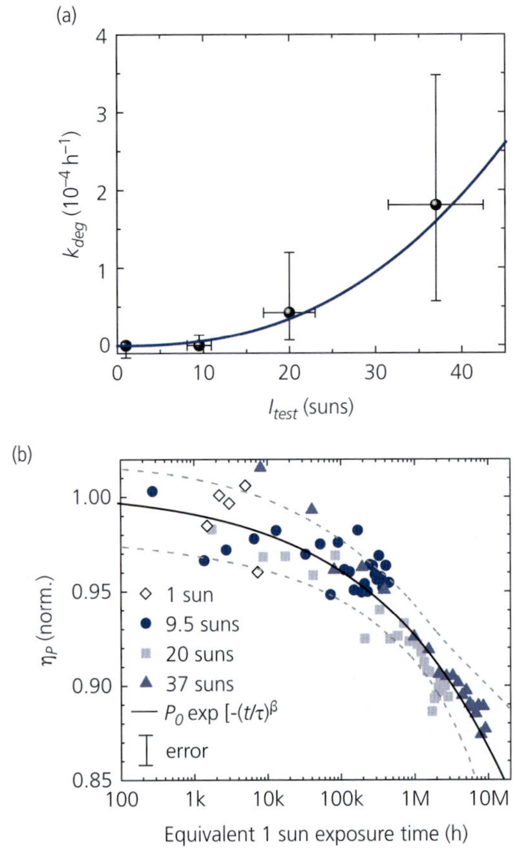

Figure 7.224 Degradation rate, k_{deg}, vs. aging intensity, I_{test}, for OPVs under 1 sun AM1.5G simulated, and 9.5, 20, and 37 suns equivalent illumination from white LEDs. The best fit (solid line) to Eq. 7.216 using the data in Fig. 7.223 is shown with an acceleration factor of $\gamma = 2.5$. (b) η_P vs. equivalent 1 sun exposure time, defined as time [h] × intensity [suns$^{2.5}$], for OPVs under all illumination conditions employed. The data are fit to a stretched exponential to estimate the *T80* lifetime with the best fit shown as solid line while upper and lower 95% confidence bounds as dashed lines, respectively (Burlingame et al., 2019).

stability of OPVs in an O_2- and H_2O-free environment can substantially exceed the requirements of even the most demanding applications.

The remarkable stability observed in the OPVs is due to a combination of the photochemical stability of the constituent materials, the morphological and thermal stability of the blended organic layers, the rapid exciton dissociation provided by the blended HJ architecture, and the high purity of the thermally-evaporated organics. The only chemical weak point in either C_{70} or DBP appears to be the single carbon-carbon bond in DBP that connects its perylene core to its peripheral phenyl groups. However, blending DBP into C_{70} compensates for this weakness as no DBP fragments are observed in the mass spectra of aged DBP:C_{70} films (Burlingame et al., 2019). As we found in Section 6.7.4, long-lived excitons can dissociate chemical bonds, a process exacerbated by multi-exciton annihilation interactions that result in highly energetic exciton species that are destructive to organic materials. The blended HJs based on co-evaporation of the donor and acceptor materials rapidly quench excitons via charge transfer, thereby preventing the excitons from dissipating energy and cleaving chemical bonds. Thermal stability of all the materials, and the blend morphology play important roles in the reliability of this particular materials system.

Having described the methodologies commonly employed to quantify aging of OPVs, the remainder of this section presents a few of the mechanisms leading to performance degradation due to materials, contacts, architectures and encapsulation. Only the latter source is extrinsic, although it is no less important than intrinsic degradation routes since the reliability of the fielded

OPV module ultimately depends on all of these factors, regardless of origin. Several reviews on this topic present many detailed discussions beyond the few cases presented here (Grossiord et al., 2012, Jørgensen et al., 2012, Voroshazi, 2014, Cheng and Zhan, 2016).

7.8.1 Materials and morphological degradation

While there are several sources for the decrease of OPV performance over time, none is more fundamental than the degradation of the organic molecules comprising the layer structure itself. Many organic molecules are vulnerable to photooxidative reactions that cause bleaching of their absorption bands after exposure to light and air over a period of time (ranging from minutes to days). Bleaching of several copolymers employing donor–acceptor conjugated backbones in air under 1 sun, AM1.5G illumination (30% relative humidity, *RH*) has been systematically investigated to determine which moieties are most susceptible to damage. Several commonly used donor and acceptor groups employed in block copolymer donors are listed in Fig. 7.225 (Manceau et al., 2011). The stabilities of these building blocks (arbitrarily defined as the time for the optical density of the film to decrease by ~50% from its as-deposited value) range from a few hours for fluorene to thousands of hours for thiophene.

Several factors that affect stability can be summarized as follows: (i) Polycyclic aromatic units within the backbone exhibit high stability. This is consistent with findings discussed in previous chapters where molecules comprising C=C double bonds are not susceptible to photophysical damage. (ii) Conjugated thiophene-containing structures such as P3HT tend to have far longer lifetimes than, for example, fluorenes with their weak quaternary C bonds. Indeed, the fluorenes appear to be the least stable of the family of molecular units shown. (iii) Molecules with exocyclic double bonds (e.g. vinyl groups in MEH-PPV) are

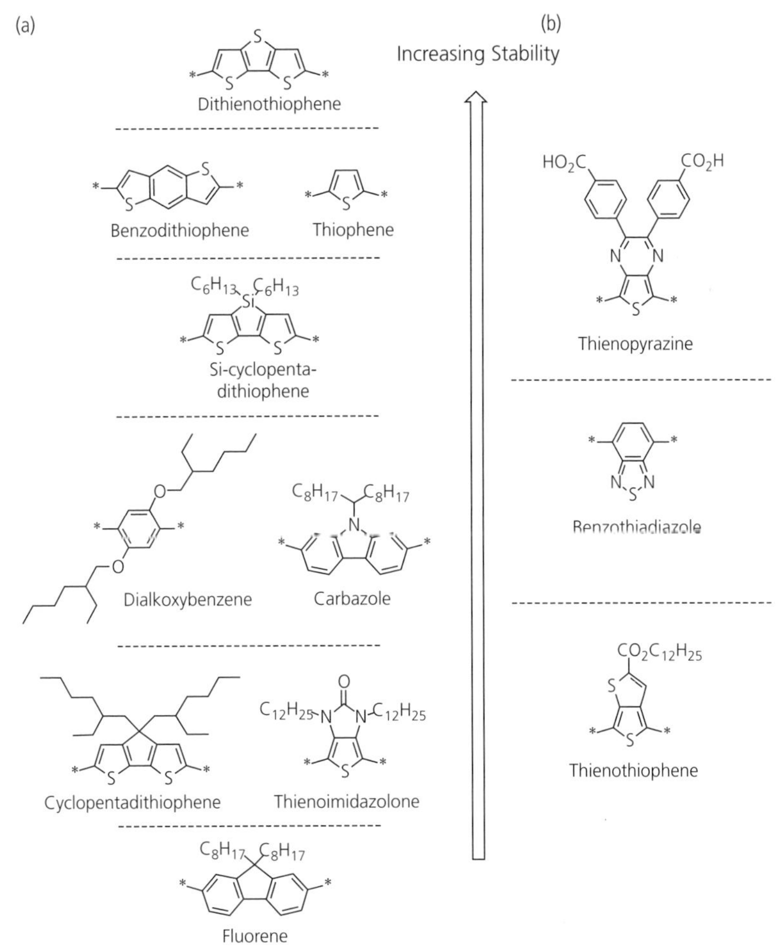

Figure 7.225 Relative photooxidative stabilities of a selection of (a) donor and (b) acceptor moieties employed in block copolymers used as donors (Manceau et al., 2011).

susceptible to oxidative cleavage. The vinylene bonds are vulnerable to side chain radicals cleaved during aging. (iii) The composition of the side chains can determine their ability to form radicals during photo-oxidation. (iv) Bond energies lower than C=C, such as C—N and C—O, are readily cleaved in the presence of photoexcitation and oxygen. This is similar to what is commonly observed in OLEDs, where molecules with such bonds are more rapidly degraded than those without.

There are two principal methods to reduce or even eliminate photooxidative bleaching. The first, obvious route is to eliminate oxygen by encapsulating the OPV in a package that prevents ingress of atmospheric contaminants. The second is to use a mixed or bulk HJ that decreases the lifetime of the excited state from nanoseconds typical of singlets, to picoseconds—a time determined by charge transfer from the donor to acceptor in the blends. Elimination of excited state interactions has already been discussed in the context of long-lived DBP:C_{70} OPVs. In contrast, neat films of SubPc exposed to 1 sun, AM1.5G illumination in air lose approximately 40% of their absorbance in only 2 h (see Fig. 7.226a). The three exposed phenyl rings of the molecule that are vulnerable to reductive reactions with radical oxygen or water in the atmosphere may be one source of this effect. This is supported by Fourier Transform IR (FTIR) spectra of SubPc films exposed to light and air, where a C=O spectral peak emerges over time (Burlingame et al., 2015). However, when blended with C_{60} in a 1:4 mixture (Fig. 7.226b), the change in absorbance is significantly reduced, even in air. Charge transfer to C_{60} results in a decrease in the SubPc singlet exciton lifetime, decreasing the probability that the excited state undergoes photooxidative bleaching. When the films are

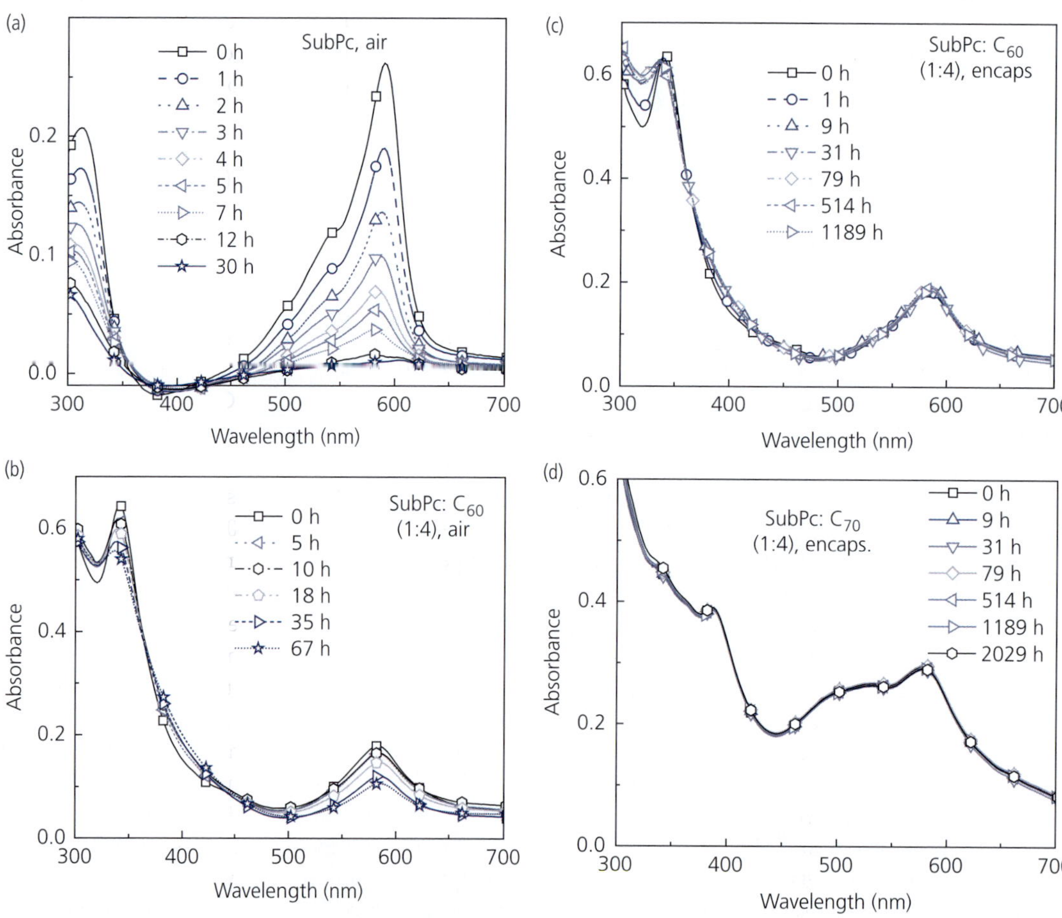

Figure 7.226 Absorbance spectra of SubPc and SubPc:fullerene blend films vs. time of exposure to (a, b) air and (c, d) encapsulated in a N_2-filled, epoxy-sealed glass package. The donor:acceptor ratio and species are indicated in each figure. Note the difference in time scales in the legends (Wang et al., 2014).

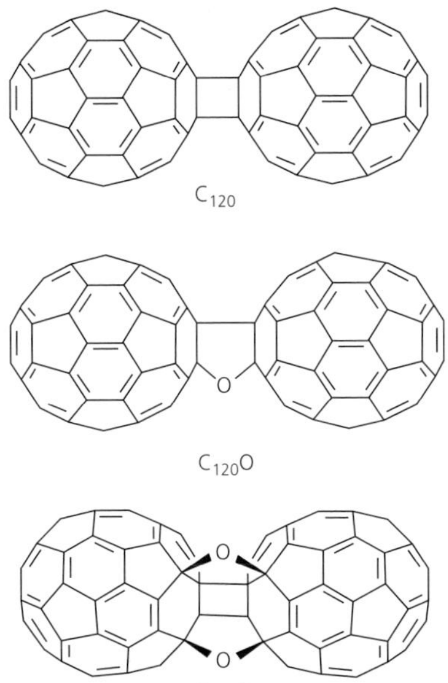

Figure 7.227 Possible C_{60} oligomers formed during exposure to air and/or illumination (Lebedkin et al., 1998).

encapsulated in an ultrapure N_2 atmosphere, oxygen is eliminated and the absorbance is stable over time, with the exception of a small decrease in C_{60} absorption at $\lambda = 350$ nm, as seen in Fig. 7.226c.

The change in C_{60} absorption has been attributed to photodimerization of adjacent C_{60} molecules. The spherical shape of the molecule puts its bonds under strain. Since two adjacent molecules are separated in the film by ~4.2 Å, the molecular bonds can break to form a bridge, resulting in a 2+2 cycloaddition between C=C double bonds. This results in a C_{120} dimer, or even higher order oligomers. Further, FTIR spectra of the films exposed to air suggest the formation of C_{60}-O_x-C_{60} oligomers (Wang et al., 2014). The oligomeric reductions of C_{60} have been identified by Raman spectroscopy, and are shown in Fig. 7.230 (Lebedkin et al., 1998). The ability of C_{70} to photopolymerize is significantly less, since it has only 10 C=C bonds, compared with 30 in C_{60}. Hence, there are fewer equivalent orientations that C_{70} can take that result in photooxidation. Also, given its lower symmetry, not all C_{70} packing arrangements lead to dimerization. The stability of SubPc:C_{70} (1:4) films is inferred from the lack of changes in absorbance over 2000 h, in Fig. 7.226d.

Photo-oligomerization results in a decreased exciton lifetime due to recombination at chemical impurities within the film. This, in turn reduces the exciton diffusion length, ultimately resulting in a decrease in η_{ext} in bilayer structures. Recall from Section 7.1.4 that the process of exciton diffusion to the HJ is described in steady state by

$$D\frac{\partial^2 N(x)}{\partial x^2} - k_{ex}N(x) + G(x) = 0, \tag{7.217}$$

where D is the exciton diffusivity, $N(x)$ is their density as a function of position, x, from the HJ, k_{ex} is the total exciton recombination rate, and $G(x)$ is the generation rate (cf. Eq. 7.87). During device operation, a fraction of non-radiative exciton recombination events results in the formation of C_{60} oligomers. The fraction of monomers remaining at time, t, is then

$$\Gamma(x,t) = \frac{[M(x,t)]}{[M_0]} = (1-\Gamma_\infty)\exp[-k_F N(x,t)t] + \Gamma_\infty, \tag{7.218}$$

where $[M(x, t)]$ is the monomer concentration at time t, $[M_0]$ is the initial monomer density, k_F is the oligomer formation rate, and Γ is the percentage of monomers that are unable to oligomerize due to their particular energetic and chemical environments. Including the effects of C_{60} oligomerization, Eq. 7.217 becomes

$$D\frac{\partial^2 N(x,t)}{\partial x^2} - \left[\left(k_{ex,1} + k_F\right)\Gamma(x,t) + k_{ex,n}(1-\Gamma(x,t))\right] \times N(x,t) + G(x,t) = 0, \tag{7.219}$$

where $k_{ex,1}$ is the exciton recombination rate on C_{60} monomers and $k_{ex,n}$ is the rate on the oligomers.

The dependence of external quantum efficiency on aging time for the SubPc/C_{60} HJ OPV is shown in Fig. 7.228a, along with a schematic of the device structure in the inset, Fig. 7.228b. The spectral changes are only evident in the portions of the spectra related to C_{60}; the SubPc peak at 600 nm remains nearly unchanged. The most pronounced decrease is centered at 450 nm, corresponding to the C_{60} CT state (see Section 7.4.1). The disappearance of this feature is consistent with oligomerization: if two adjacent C_{60} molecules share a CT state, this state will vanish once the molecules form a dimer.

The measured η_{ext} at 450 nm *vs.* aging time is shown in Fig. 7.228b. Burn-in results in a decrease in efficiency by nearly 50% after 5 h, which is followed by a much slower, exponential roll-off. The fit to the data using the quantitative analysis of Eq. 7.219 is shown by the line, using a C_{60} singlet recombination rate of $10^6\ s^{-1}$ (Kuhnke et al., 1997), $k_{ex,n} = 6 \times 10^6\ s^{-1}$, and an oligomerization rate of 80 s^{-1}. Further, it is assumed that the asymptotic concentration of monomers is $\Gamma_\infty \to 0$, that is, there is complete oligomerization of

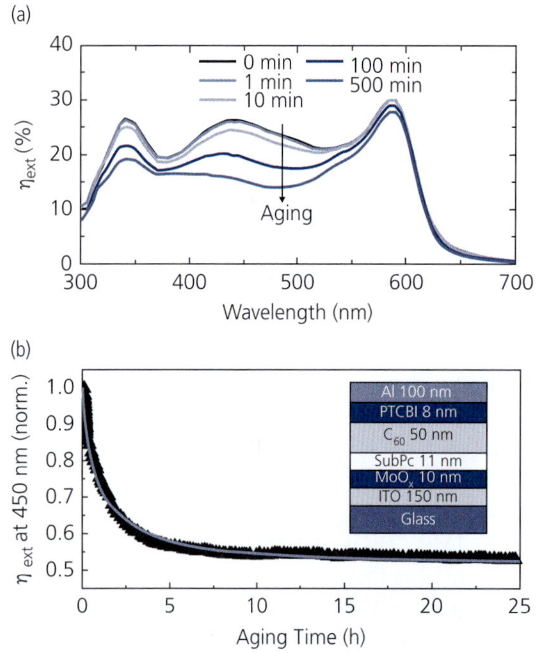

Figure 7.228 (a) External quantum efficiency spectra of a bilayer SubPc:C_{60} OPV as a function of aging time. (b) Normalized efficiency (points) along with a fit to the data (line) assuming reduced exciton diffusion length due to C_{60} photo-polymerization. Inset: Schematic of the OPV structure (Burlingame et al., 2015).

the C_{60} over time (Burlingame et al., 2015). Note that the functional form of the analysis is strikingly similar to the sum of exponentials, with the burn-in rate controlled by that for oligomerization.

Laser desorption induced time of flight mass spectroscopy was used to determine the relative concentration of molecular species and reaction products in the aged films. From this study, it was determined that $\Gamma_\infty = 10.4\%$ was reached during aging. Complete oligomerization, in practice is not possible since individual molecules become isolated from each other during the formation of intervening oligomers. In any case, the formation of a high concentration of oligomers with considerably shorter exciton lifetimes than in monomers results in a concomitant decrease in diffusion length. In a bilayer structure, this decreases the external efficiency accordingly. Once the monomer population is effectively exhausted, the OPV efficiency decline translates into a slow aging process having a different origin.

Beyond photo-polymerization, the chemical changes during burn-in may have contributions from complexation with the donor, or morphological changes within the blends. For example, reactions between the donor polymer, PCDTBT, and $PC_{71}BM$ have been implicated in 40% loss of efficiency during burn-in. Comparisons of encapsulated (oxygen-free) devices using this BHJ blend employing a conventional PEDOT:PSS anode with Al/Ca cathode structure, or an inverted ZnO NP cathode with a MoO_x/Ag anode, showed the same efficiency loss before leveling off into a more gentle aging rate with an extrapolated $T80 = 7$ years (Mateker et al., 2015). With no apparent dependence on contact arrangement, morphology or exposure to oxygen, it is assumed that the aging is due to changes in the bulk as a result of increased resistance from recombination in the active region. An additional observation is that regiorandom and amorphous PCDTBT films degrade more rapidly than regioregular and more crystalline films, presumably due to the higher density of the latter morphologies that impedes intercalation by residual impurities left after film preparation.

Self-terminating chemical reactions must be responsible for burn-in since they occur over only a limited duration. The C_{60} photo-oligomerization in SubPc/C_{60} OPVs is one such self-terminating process. A combination of FTIR and electron spin resonance has shown that the early stage degradation of PCDTBT:$PC_{71}BM$ BHJs is due to a combination of complexation with the fullerene, and polymerization (Tournebize et al., 2014). The weak C—N bond between the carbazole group and its octane side chain in PCDTBT cleaves under illumination following the reaction pathway in Fig. 7.229. It is notable that even after 1500 h exposure to light in vacuum, the film absorbance does not change significantly, eliminating the possibility for photooxidation. The resulting carbazole radicals can undergo trapping by the fullerene. Scavenging of the N—$CH(C_8H_{17})_2$ radicals by cross-linking with $PC_{71}BM$ effectively terminates the reaction. The result is a stable film chemistry and morphology whose properties are unchanged by further exposure to light over exceptionally long periods, as found by the long-term stability of OPVs employing this materials combination. No such termination occurs in the pure polymer films, where formation of long chain polycarbazole moieties continues to completion.

The foregoing has shown that fullerenes are implicated in several burn-in mechanisms, namely dimerization and complexation. Furthermore, its small diameter and symmetrical shape can lead to time-dependent morphological changes in blends with donors that also change the properties of the active region. Considerable care is used to achieve the optimum morphology in BHJs that promotes efficient exciton dissociation as well as low resistance to

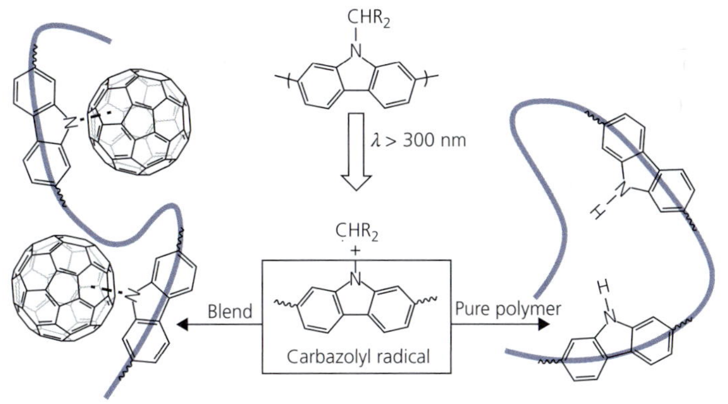

Figure 7.229 Scission of the carbazole group in PCDTBT in the presence of light results in the production of carbazolyl and CHR_2 ($R = C_8H_{17}$) radicals. These can be trapped by the fullerene in the BHJ blend (left), or in the pure polymer, forming polycarbazoles indicated by the blue lines (right) (Tournebize et al., 2014).

charge extraction. While the former requires an intimate and nearly homogeneous blend of donor and acceptor molecules, the latter property is optimized by the development of longer-range structural order within the bicontinuous D-A network. These apparently conflicting purposes and morphologies have led to the development of mixtures containing both amorphous and nanocrystalline, phase-separated regions. The separation of two solids from a blend into coexisting phases is known as *spinodal decomposition*. This is accomplished through the choice of molecular structures, the use of additives, thermal annealing, etc. The resulting morphologies are metastable. The addition of heat or illumination, therefore, can supply sufficient energy to cause structural reorganization. The concomitant rapid change in OPV characteristics due to spinodal decompositon has been observed in, for example, PffBT4T-2OD (also known as PCE11):$PC_{61}BM$ (1:1.2) active region OPVs (Li et al., 2017a). The separation of two solids from a blend into two coexisting phases is known as spinodal decomposition. This process appears to be active in the PCE11:$PC_{61}BM$ system.

The effects considered thus far are associated with fullerene acceptors. By replacing fullerenes with NFAs, dimerization, complexation, and morphological changes experienced by fullerenes are eliminated. This has been found to be the case for P3HT:IDTBR (**Scheme 2**) D-A OPVs. The aging data along with those of a P3HT:$PC_{61}BM$ BHJ are shown in Fig. 7.230. Both devices are fabricated by doctor-blade application of the active material blends in ambient. They form an inverted architecture with a ITO/ZnO bottom cathode, and a PEDOT:PSS/Ag

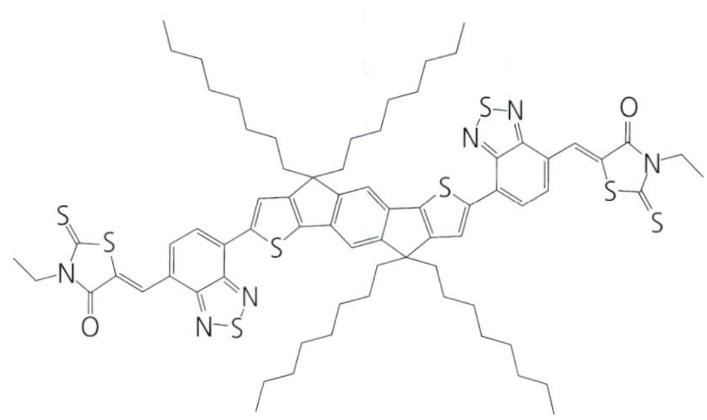

Scheme 2

anode (Gasparini et al., 2017). The burn-in transient that results in a 30% reduction in η_P of the $PC_{61}BM$-based OPV during the initial 500 h of operation is absent in the IDTBR-based device. Almost all of the change responsible for the loss in $PC_{61}BM$ OPV efficiency is due to a reduction in j_{SC}, that we have previously shown is due to photo-oligomerization. While there is no burn-in loss in the NFAs, both devices suffer from a long term monotonic decay in *FF* and j_{SC}, resulting in *T80* $\sim$ 5000 h for the NFA-based cell. Hence, reducing burn-in loss does not guarantee that the reliability of the devices will meet the demands required of practical solar cells.

7.8.2 Contacts and other interfaces

Contact stability is a key determinant of device operational lifetime. This has been shown to be true for OLEDs, and it is equally true for OPVs. Contacts must form a very low and stable injection barrier where the cell series resistance does not increase over time. The contact should not undergo reactions with the organic

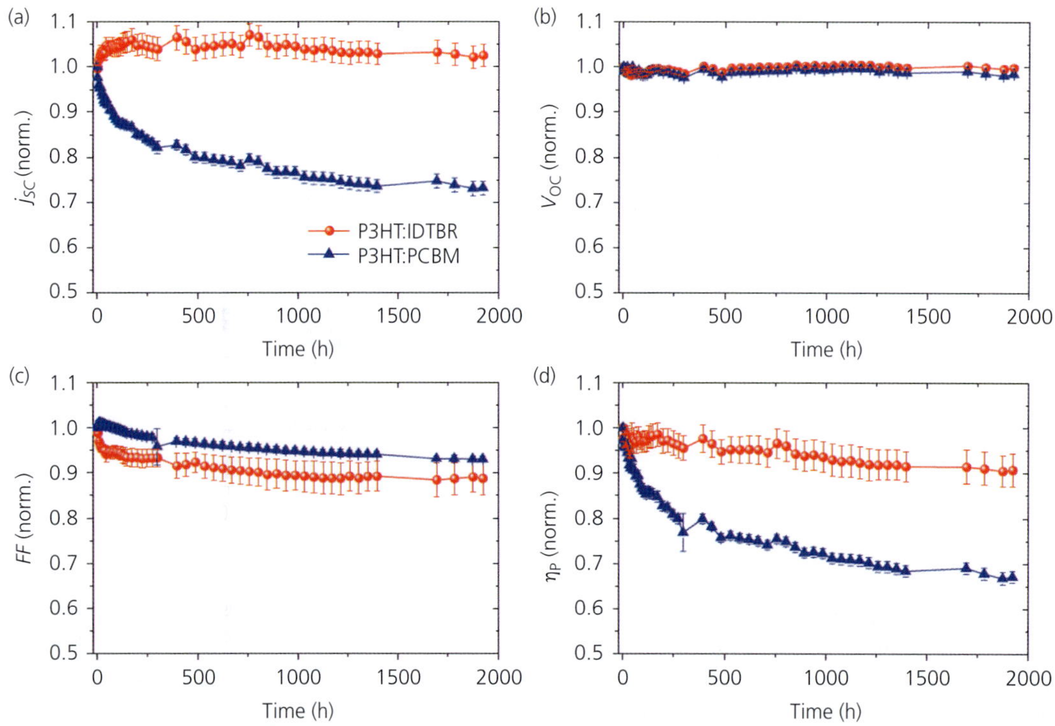

Figure 7.230 Comparison of (a) j_{SC}, (b) V_{OC}, (c) FF, and (d) η_P during aging of P3HT:PC_{61}BM and P3HT:IDTBR BHJ OPVs (Gasparini et al., 2017).

layers, it should be free of pinholes, physically robust (i.e. it should adhere strongly to the organic thin films), and should be resistant to change when exposed to trace contaminants remaining within the package following device fabrication. Since contact quality depends on so many factors, it is difficult to separate contact-related aging phenomena that are intrinsic (e.g. due to metal–organic interactions) or extrinsic (e.g. arising from contamination or physical defects). Hence, in studying OPV aging, its environment must be tightly controlled to enable a clear understanding of the mechanisms for degradation.

Contact diagnostics are similar to those employed in OLEDs (cf. Section 6.7.2). These include pulling or peeling off the contact using scotch tape, microscopic imaging to identify physical defects, thermal imaging to locate hot spots due to current crowding, and EL and PL imaging of the OPV active region. Dark regions under forward bias indicate a high rate of nonradiative recombination that do not generate photocurrent when reverse biased. Thus, monitoring the EL intensity over time provides a quantitative measurement of how the active OPV area is evolving. If the ratio of the time dependent photocurrent to EL intensity, $EL(t)$, given by $R(t) = j_{SC}(t)/EL(t)$, is constant, it is likely that the loss of contact area is the source of degradation in the OPV. In contrast, when $R(t) < 1$, then there are additional sources of EL loss due to recombination in the bulk of the active region beyond a simple reduction in contact area (Seeland et al., 2011). The photoluminescence intensity is also a function of bulk recombination, and hence PL imaging of the OPV can be used along with EL to separate contact from trap-related recombination within the BHJ.

Figure 7.231 shows a plot of η_P vs. the residual active area of a p-i-i ZnPc: C_{60} cell with the structure shown in the inset after extended exposure to simulated solar illumination. The device is capped by a degassed, 175 μm thick PET film attached to the glass substrate by a bead of UV-cured epoxy along its edges. As discussed in Section 6.7.5, PET and similar polymer films do not form a substantial barrier to atmospheric constituents, which undoubtedly account for many of the defects that grow during exposure. The residual area over time is determined from the total EL intensity obtained under forward bias. An image of an aged contact is shown in the top inset, Fig. 7.231. The linear relationship between η_P and the active area indicates that $R(t) = 1$. That is, the degradation in efficiency is due to the loss of active area, possibly due to infiltration of water through pinholes in the Al contact.

Example EL and thermal images of 0.5 cm^2 contacts to PCDTBT:PC_{61}BM OPVs after aging are shown in

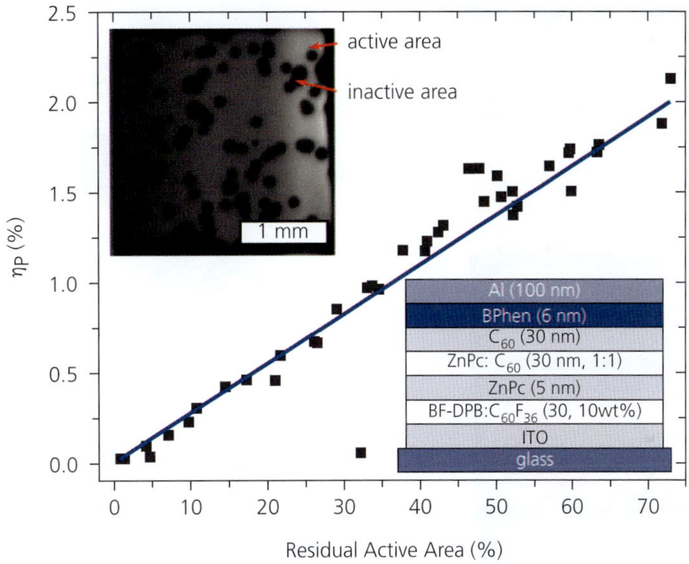

Figure 7.231 OPV efficiency vs. active area determined via EL imaging under 1 sun intensity white illumination. Top inset: Example contact image showing dark spots. Bottom inset: p-i-i OPV structure used in the experiments. The BF-DPB layer is p-doped with $C_{60}F_{36}$. Adapted from Klumbies et al. (2014).

Fig. 7.232. Immediately following fabrication in a controlled, clean environment, the devices show featureless contact images, with no evidence for pinholes or hot spots that would arise from current nonuniformities (Seeland et al., 2011). However, numerous features emerge over time. They appear whether the devices are stored in the dark or operated at elevated temperatures. The integrity of the package and the choice of contact metal impact the degree and rapidity of defect formation. The packages comprised two glass layers bonded at their periphery using thermo-setting epoxy. It is found that a bare Al cathode exposed to air during aging exhibits several pinholes (dark spots, Fig. 7.232a) and hot areas (white regions, Fig. 7.232b), along with variations in EL intensity across the device. In contrast, packaged OPVs employing TiO_x/Al cathodes are almost featureless after aging (Fig. 7.232c and d). The pinholes tend to grow with time since they allow water and oxygen to enter the active region, as well as by oxidizing the Al contact. This promotes photobleaching of the active region. This increases the contact resistance around the perimeter of the pinhole, causing an increase in local temperature and metal delamination. Preventing exposure to atmosphere in the packaged devices eliminates many of these problems. Interestingly, the devices close to the edges of the package (i.e. near the epoxy seal) had a noticeably higher defect density than devices near the center. This suggests the solar cells are vulnerable to environmental exposure; presumably devices near the edge are contaminated by diffusion from water and oxygen sooner than are devices deeper within the package. Indeed, the devices near the edge may themselves getter oxygen and water, fixing it at locations distant from the other, longer-lived OPVs.

The OPV characteristics do not age nearly as fast for the TiO_x/Al cathodes as for the bare Al cathodes, even when packaged. Finally, TiO_x/Ag contacts show a much higher concentration of non-uniformities than do the Al contacts. Silver rapidly migrates through the organic layers, creating current shunts that render large areas of the device inactive (Roesch et al., 2013, Song et al., 2015). The shunts result in the large bright regions across the device area in Fig. 7.232e and f.

The TiO_x buffer placed immediately beneath the metal contact serves multiple purposes illustrated in Fig. 7.233. As described in Section 7.4.1, the buffer acts as an optical spacer, an exciton blocker and electron transporting medium, and it protects the active region from damage during metal contact deposition. The electron-conducting TiO_x extends the device lifetime since it is considerably denser than the organic layers, and as such it forms a barrier to oxygen and water. Furthermore, since the layer is oxygen-deficient, it can scavenge oxygen from both the active region as well as from the interface with the metal contact (Lee et al., 2007). The metal oxide layer deposition from solution is illustrated in the right inset, Fig. 7.233. A precursor of Ti(IV) isopropoxide (TiOR = $Ti[OCH(CH_3)_2]_4$) is

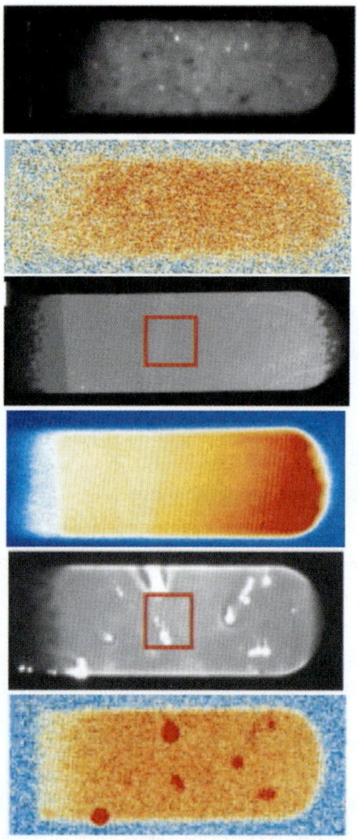

Figure 7.232 Thermal (grayscale) and electroluminescent (color) images of 0.5 cm^2 PCDTBT:PC$_{61}$BM OPV contacts following aging under different conditions. (a, b) Unpackaged devices employing TiO$_x$/Al contacts aged for 1100 h at 1 sun intensity, (c, d) Packaged devices employing TiO$_x$/Al contacts aged for 1150 h at 1 sun intensity. (e, f) Packaged devices employing TiO$_x$/Ag contacts aged for 1150 h at 1 sun intensity. The temperature is highest in the white regions of the thermal images. The EL images range from red (most intense) to blue (no EL). Red boxes in (c) and (e) indicate regions where η_{ext} of the devices were measured (Roesch et al., 2013).

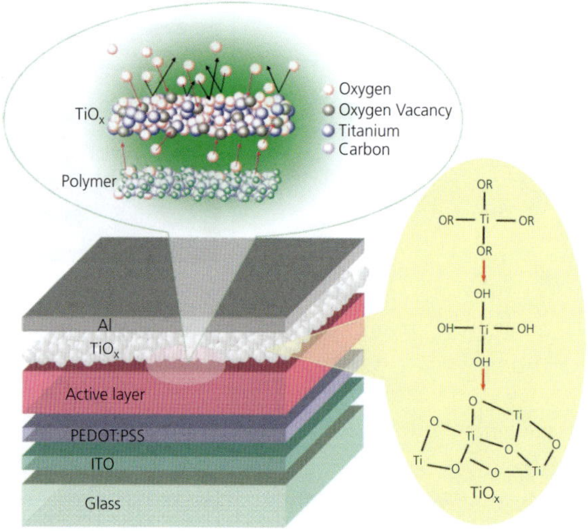

Figure 7.233 Structure of a polymer OPV employing a TiO$_x$ cathode buffer layer. The top inset illustrates the ability of TiO$_x$ to form a barrier that prevents atmospheric contaminants from entering the active region, and also to scavenge O$_2$ at TiO$_x$ vacancies. The right inset shows the low temperature sol-gel process used to synthesize the TiO$_x$ layer (Lee et al., 2007).

hydrolyzed by mixing with 1-methoxyethanol and ethanolamine. After spin casting the solution onto the OPV surface, it is annealed in air at 80°C for 10 min to form a 30–40 nm thick film.

The ability of TiO$_x$ to act as a barrier and as an oxygen getter is inferred from the data in Fig. 7.234, where TiO$_x$ was used to protect polyfluorene (PF) films that are highly vulnerable to photooxidation (cf. Fig. 7.225). The samples prepared for this test were: glass/PF, glass/TiO$_x$/PF, glass/PF/TiO$_x$, and glass/TiO$_x$/PF/TiO$_x$. Figure 7.234a, upper panel, shows the PL intensity of PF for the freshly prepared samples. The bare PF film shows the emergence of a strong, secondary emission peak at 550 nm after exposure to air at 150°C (Fig. 7.234a, lower panel). The peak intensity is dramatically reduced, or even eliminated for the glass/PF/TiO$_x$, and glass/TiO$_x$/PF/TiO$_x$ samples. Hence, the TiO$_x$ cap retards oxidation. The peak intensity for the glass/TiO$_x$/PF sample is below that of the bare PF film. Thus, even when oxygen permeates the PF, it can be captured and rendered inactive by TiO$_x$ buried beneath its surface. These findings are reinforced by the XPS spectra of the O$_{1s}$ peak normalized to C$_{1s}$ in Fig. 7.234b for bare and capped PF films after 48 h exposure at high temperature in air. The spectra were obtained after removal of the cap. They show that there is very little oxygen in the capped film compared to the bare PF, indicative of the barrier-forming properties of the buffer.

When the active region is heated above the glass transition temperature of a constituent material, the blend can locally separate, with one or more molecular constituents migrating toward the contacts. In some cases this introduces a barrier to charge extraction at the electrodes. Figure 7.235 shows the aging schedule of two PC$_{61}$BM-based BHJ devices in the conventional, top cathode configuration. One device employs the donor P3HT with a $T_g = 56°C$, and the other uses PCDTBT with $T_g = 125°C$ and 130°C for the thin film and bulk, respectively. In acquiring the data, the devices are brought to temperature and held for 3 h before the temperature is raised by another 10°C,

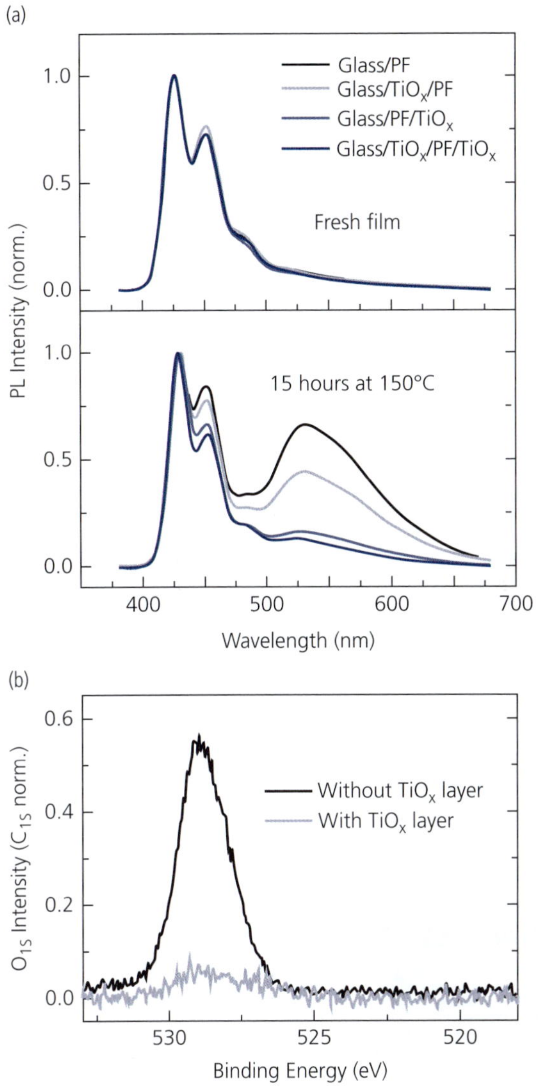

Figure 7.234 (a) The PL intensity of as-deposited polyfluorene (PF) films and after 15 h exposure to air at 150°C. The legend shows the various film structures with and without a TiO_x layer. (b) XPS spectrum of the O_{1s} emission intensity of a bare PF film, and one capped by TiO_x after 48 h exposure to air at 150°C (Lee et al., 2007).

all the while monitoring V_{OC}. When $T \geq 57$°C for the P3HT-based OPV, there is a noticable decrease in V_{OC} during the holding period. This is observed at 127°C for the device comprising PCDTBT. The changes in V_{OC}, therefore, occur at or very near to the T_g of the respective materials. By peeling off the cathode using scotch tape following the tests and then replacing it, the original device characteristics are recovered. The decrease in V_{OC} at T_g appears to be indpendent of cathode material, including MoO_3/Ag and PEIE/Ag.

These findings suggest that the donor migrates to the cathode and forms a very thin barrier to electron transport from the acceptor. However, this inference is not directly supported by spectroscopic evidence. It is notable that the changes in V_{OC} are not encountered in inverted devices, where the cathode is on the substrate surface. In that configuration, the donor would selectively migrate toward the top surface that is capped by the anode, thus eliminating the formation of the energy barrier to electrons at the cathode (Sachs-Quintana et al., 2014).

To summarize this discussion, the aging characteristics of OPVs are strongly influenced by the contacts, that can be protected by encapsulation impermeable to atmospheric exposure, particularly to O_2 and H_2O. There are numerous pathways to failure originating from the choice of active region and contact metals, the chemical interactions at the interfaces between them, and their reaction with contaminants. Furthermore, combining materials with substantially different bulk moduli and coefficients of thermal expansion creates strain that results in morphological defects such as voids and fracture induced by changes in temperature or mechanical flexing. The effects of stress on various device regions are illustrated in Fig. 7.236. The value of organic electronic devices, and particularly OPVs, is partially due to their ability to be scaled to very large sizes, and to be fabricated on flexible, lightweight and often stretchable surfaces (Lipomi and Bao, 2011, Savagatrup et al., 2015). These substrates can introduce significant stresses on organic films and their encapsulants that are not ordinarily experienced by brittle, inorganic semiconductors.

7.8.3 Encapsulation

While we have seen that materials, morphologies, device architectures, interfaces and contacts all play roles in determining reliability of OPV modules, ultimately, these features are vulnerable to deterioration if exposed to the environment. The modules must last for at least 20 years in environments that range from bitter cold (−40°C) to extremely hot (100°C), and from arid to tropical. The modules may be subjected to bombardment by sand and hail, and severe wind loads as well as heavy coatings of snow. The list of environmental hazards is extensive, placing extreme demands on organic thin film solar cells that are comprised of what are generally believed to be rather fragile, carbon-based semiconductor materials. Thus, cell and module packaging must withstand the most adverse circumstances while also being low cost.

Fortunately, a great deal of experience has been gained in packaging from the manufacture of

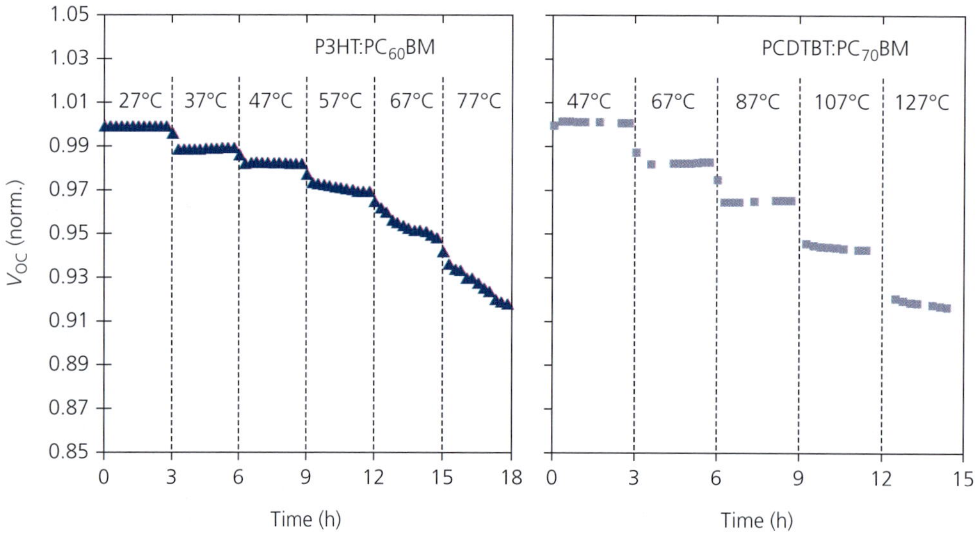

Figure 7.235 Time dependence of V_{OC} using a step-wise increase in temperature of two BHJ OPVs employing a Ca (7 nm)/Al (150 nm) top cathode, and a ITO/PEDOT:PSS anode aged in N_2 (Mateker and McGehee, 2017).

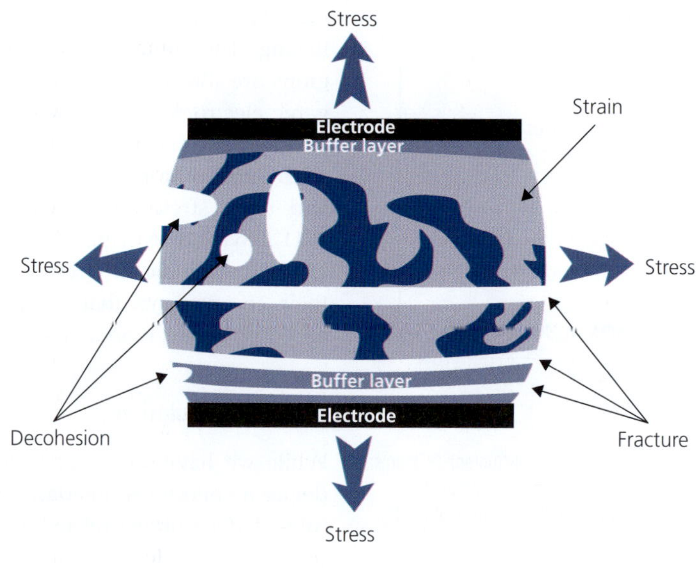

Figure 7.236 Schematic of possible strains and failure modes introduced in an OPV subjected to stress, either from mechanical, optical, or thermal sources (Cheng and Zhan, 2016).

long-lived OLED displays and lighting appliances. Of course, the environments encountered by OLEDs are not nearly as challenging as for solar cells, but flexible and lightweight encapsulation schemes, along with sophisticated methods for their evaluation have been developed. Many of these solutions are described in Section 6.7.5. We extend that discussion here, focusing on technologies adapted specifically to packaging and large-scale manufacturing of OPVs.

Adverse environmental exposure can lead to catastrophic package failure. Figure 7.237 shows the failure of a UV-cured, epoxy sealed glass package used to encapsulate small molecule OPVs in Fig. 7.220. The devices were continuously aged under open circuit conditions out-of-doors at Sde Boqer in the Negev Desert, Israel. The data from four devices shows a spread in diurnal efficiencies with the devices held at a fixed angle of 30° relative to the horizon. Rainfall on day 108 damaged the package seal, initiating a rapid decline in efficiency. This was due to a similarly rapid reduction in j_{SC}, with V_{OC} and FF largely unaffected by the package

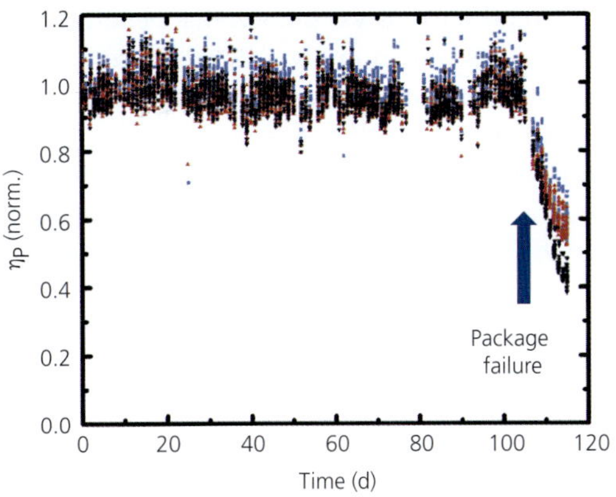

Figure 7.237 Normalized η_P for a population of four DBP:$PC_{71}BM$ OPVs vs. time spent outdoors. Package failure occurred on day 108. After Burlingame et al. (2016).

failure. The vulnerability to variable weather events is but one of many challenges that must be overcome by environmentally robust, yet inexpensive encapsulation technologies.

Flexible packages must have at least one transparent surface. This can be provided by a plastic foil with an inorganic barrier layer that retards or eliminates penetration into the active device region (Dennler et al., 2006). In Sections 5.8 and 6.7.5, we discussed the process of contaminant ingress through multilayer barrier coatings embedded in polymer substrates. By interleaving impermeable, thin inorganic layers between organic layers, the contaminant must diffuse along an extended, circuitous path between defects in successive layers before it enters the cavity containing the active device (Graff et al., 2004). While the multilayer cannot prevent the eventual entry of contaminants since there is always a finite density of defects in every film, the time needed to reach the device may be sufficiently long to enable the intended application. An alternative, impermeable flexible encapsulant is very thin glass, although this is quite fragile and possibly unsuitable for solar cell applications. The second surface may or may not need to be transparent, and hence a second plastic barrier, or metal foil, respectively, can be employed for this purpose.

A coating is characterized by its *water vapor transmission rate* (*WVTR*) and *oxygen transmission rate* (*OTR*) (see Section 5.8). Suitable coatings for long-lived operation needed in OPVs have $WVTR \sim 10^{-5}$–10^{-6} g/m^2-d and $OTR < 10^{-1}$ cm^3/m^2 d atm. The *WVTR* of a package can be determined using a Ca sensor layer. A water sensitive test coupon comprises a thin Ca layer deposited on a glass substrate. In the example device in Fig. 7.238, top inset, the active region (here represented by a single C_{60} layer, lower

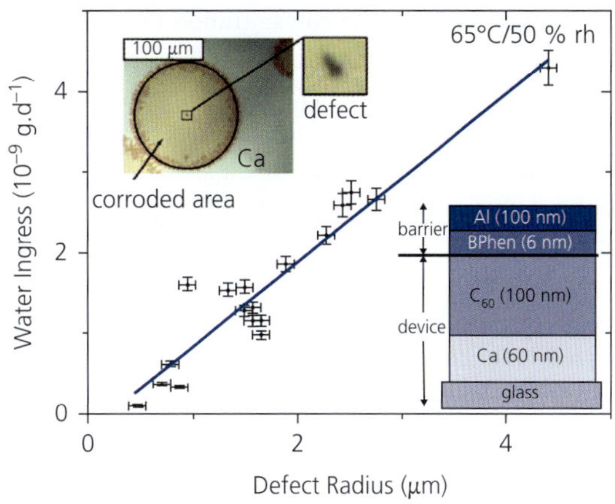

Figure 7.238 Water vapor transmission rate vs. defect radius. Upper inset: Micrograph of a defect in the center of a corroded area. The defect itself is much smaller than the yellow corroded area. Lower inset: Test coupon used to measure *WVTR*. After Klumbies et al. (2014).

inset), followed by the BPhen exciton blocker and Al contact are deposited on the Ca surface. Calcium corrodes by contact with two water molecules to form $Ca(OH)_2$ (Cros et al., 2006). The complete hydrolysis requires 1.4 g(H_2O)/cm^3(Ca). Thus, the 60 nm thick layer is converted to $Ca(OH)_2$ by 8.4 μg/cm^2 of H_2O. Since Ca is a conducting, opaque metal, and $Ca(OH)_2$ is a transparent insulator, water penetration results in significant changes in both its optical and electrical characteristics.

The Ca test was used to investigate the barrier formed by the Al/Bphen layer of the devices in Fig. 7.231 when exposed to 50% relative humidity (*RH*) at 65°C. Figure 7.238, upper inset, shows a micrograph of the Ca layer (as viewed through the glass substrate) in the region surrounding a small defect in the Al cathode. The defect is apparently a pinhole located at the center of the corroded area that allows water to penetrate to the Ca layer, locally converting it to $Ca(OH)_2$. The corroded area grows as the water content increases. Interestingly, the amount of water penetrating to the test layer is proportional to its radius, not the area of the defect. This results since the water molecules must diffuse laterally to reach areas not directly beneath the defect itself. The extent of the diffusion region, therefore, depends on the length of the defect periphery. Another finding is that the amount of water in the OPV active region at *T50* is 2.0 μg/cm^2, independent of the exposure of the devices to a combination of temperatures between 38°C and 65°C and relative humidity between 11% and 90%. This is unsurprising since the damage to a constant volume of active material by H_2O is ultimately determined only by the concentration of H_2O within that volume (Klumbies et al., 2014).

Table 7.29 shows the *WVTR* of several different encapsulant lids that are epoxy-sealed to a glass substrate carrying the devices in Fig. 7.231. The device without an encapsulant only has BPhen/Al cathode as a barrier, whereas there are several different package lids employing different combinations of PET and reactively sputtered zinc-tin-oxide (ZTO). The triple barrier layer comprises two, 100 nm thick ZTO layers forming a sandwich with a 500 nm $SiO_xC_yH_z$ plasma-deposited polymer layer. The package employing a glass lid has the lowest *WVTR* due entirely to water penetration of the thin UV-cured epoxy bead around its periphery.

A population of devices encapsulated by these different lids were exposed to 1 sun intensity illumination at $RH = 5.5\%$ using ISOS-L-1 conditions (i.e. at open circuit, see Table 7.28). The normalized short circuit current vs. total water penetration for the devices is shown in Fig. 7.239a. There, $j_{SC}(t)$ follows the stretched exponential of Eq. 7.212, indicating that there is no discernable difference between burn-in and long term degradation: the device performance monotonically depends on the amount of water to which it is exposed. In the figure, the *T50* of each device was different; curves to the left have lower *WVTR*, and hence have a larger *T50*, than curves to the right.

With only the exception of the glass lid device, the shape of the curves is approximately the same, with almost no change in j_{SC} until a threshold is reached, after which the device rapidly degrades. This suggests that once a threshold concentration of H_2O has entered the active region, rapid degradation follows. The water volume that is required to reach *T50* for each encapsulation is indicated in the plots. Degradation occurs at smaller water ingress as the *WVTR* of a package is reduced. For both the glass and 245 nm thick ZTO-coated PET, the amount of water penetrating to the device is insufficient to lead to complete degradation (i.e. to *T0*) over the 1150 h duration of the experiment. The thick, single layer ZTO is more effective than even the three-layer ZTO/organic/ZTO barrier. Presumably, when the ZTO reaches a critical thickness, the total defect area drops dramatically (da Silva Sobrinho et al., 2000). Nevertheless, some formation of defects is unavoidable. Thus, as expected, the best performance is obtained for the fully glass encapsulated OPV.

Table 7.29 Water vapor transmission rates into the devices in Fig. 7.231 using different package lids (Hermenau et al., 2012)

Encapsulation	*WVTR* (g/m² d)
None	6.9
125 μm PET	0.43
125 μm PET + 100 nm ZTO	0.02
125 μm PET + triple barrier layer	0.0014
75 μm PET + 245 nm ZTO	3.7×10^{-4}
Glass	3.9×10^{-5}

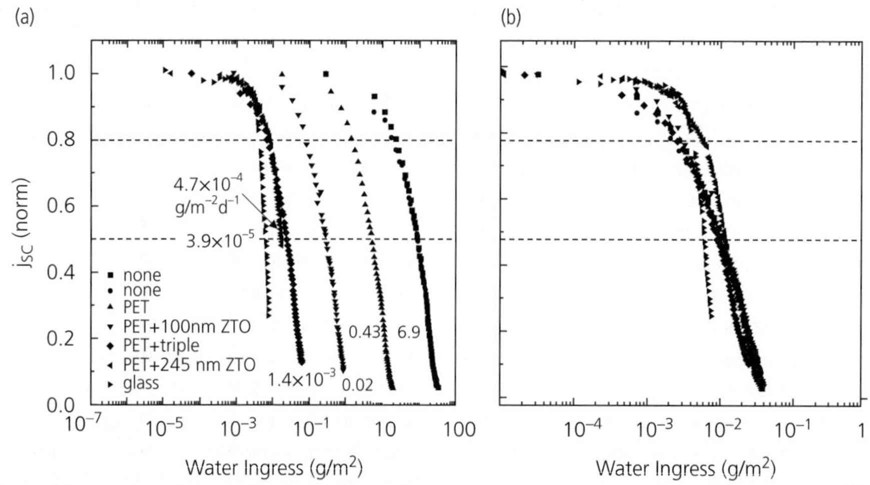

Figure 7.239 (a) Normalized j_{SC} vs. total amount of water penetration into the packages encapsulated using a glass substrate and the lids in Table 7.29. The number associated with each curve is the total amount of water transmission at *T50*. The devices are the same as in Fig. 7.231. (b) The data in (a) adjusted for the effects of the water penetration of the BPhen/Al cathode such that only water penetration through the encapsulant is counted (Hermenau et al., 2012).

The result in Fig. 7.239a is counterintuitive. Why would a lower *WVTR* require less water to degrade the device (albeit over a longer time) than if the barrier to water penetration is reduced (i.e. in a device with no encapsulation)? This apparent contradiction is understood if there is more than the single barrier to water. One barrier (the encapsulation) is far more resistive to water penetration than the secondary barrier (the cathode). The time to penetrate the lid is much longer than to overcome the 100 nm thick Al cathode, and hence the former dominates the time to device failure. Once the moisture enters the package, however, it rapidly moves into the device active volume, since the cathode presents only a small additional time delay. Since *WVTR* is a measure of the conductance of the barrier to water, the total resistance to water penetration of the two barriers is simply: $1/WVTR_{total} = 1/WVTR_{lid} + 1/WVTR_{Al}$. Using $WVTR_{Al} = 8 \times 10^{-4}$ g/m d, the data in Fig. 7.239a are replotted to account only for the transmission of the lid, $WVTR_{lid}$, as shown in Fig. 7.239b, where all the curves now overlap (Hermenau et al., 2012). This corroborates the hypothesis that the devices fail once a fixed amount of water has entered the package.

Encapsulations employing a plastic lid or substrate still fall well short of the performance of glass packages, with none showing the endurance required for operation longer than a few weeks or months when tested using ISOS-D-3, L-2, and L-3 conditions (Corazza et al., 2014, Voroshazi, 2014). Furthermore, sensitivity to UV light of many polymers results in their discoloration and embrittlement, leading to additional routes to package and device failure. We have seen that even the epoxy seal can fail catastrophically when exposed to adverse environments commonly encountered out-of-doors, and has a *WVTR* that in itself is too high for exterior use. Until these problems are solved, it is expected that OPVs will not be found in the most demanding conditions now dominated by commodity Si solar cells. There are many large niche applications, however, that do not require such a high degree of robustness. As discussed in previous sections, building integrated PV such as power generating windows, and even in some building applied PV applications, the harsh conditions faced by large power generating plants can be avoided. These niches are potentially enormous, and the low cost, lightweight, flexibility and most importantly, the semitransparency of OPVs are possibly the best choice for filling these needs.

7.9 Scaling up to modules

The primary challenges in scaling to modules are the reduction of series resistance, and the adaptability of the process to volume, low-cost and high yield manufacturing. We have briefly touched on both of these topics in previous sections. Series resistance primarily originates from the transparent contact. In Section 7.4.2 we have shown that there are several approaches to reducing series resistance while maintaining transparency, but none supersede the performance of ITO due to its combination of high transparency and conductivity. Yet ITO substantially limits the ability to scale the devices to modules due to both its high sheet resistance and cost. Furthermore, ITO is a brittle metal oxide, limiting the bending radius that can be tolerated before cracking. While deposition of OPVs on flat glass substrates may provide a solution for specialized applications, a considerable focus of research has been on developing continuous roll processes. While this is the path to realizing low cost, volume manufacturing, the flexible packaging required for R2R production are not sufficiently robust to withstand harsh exterior environments over long periods.

Here we discuss contact designs used for large modules, and some of the features of current R2R manufacturing. For this latter topic, we rely on the discussions of both Chapters 5 and 6, while in the following we describe only those issues that are specific to the scalable manufacture of organic solar cell modules in a continuous web environment. Batch manufacturing on large glass plates is a well-developed technology for the display industry.

In the context of multijunction cells in Sections 7.1.2 and 7.5, the power losses due to cell series resistance follow

$$\Delta P_{loss} = I^2 R_{ser}. \tag{7.220}$$

Thus, series-connected cells in modules are preferred over parallel-connected cells to minimize losses. An illustration of a typical cell arrangement in a module is shown in Fig. 7.240. The module consists of a 5 × 5 array of individual cells in a series-parallel circuit. The series strings are along the columns, and are connected in parallel by the bus bars at the front and back of the module. Sub-electrode metal contacts overcoat the exposed ITO edges to minimize contact series resistance (Park et al., 2010b). The use of a combination of series and parallel connections prevents the voltage from getting so high that arcing, or difficulties in interfacing

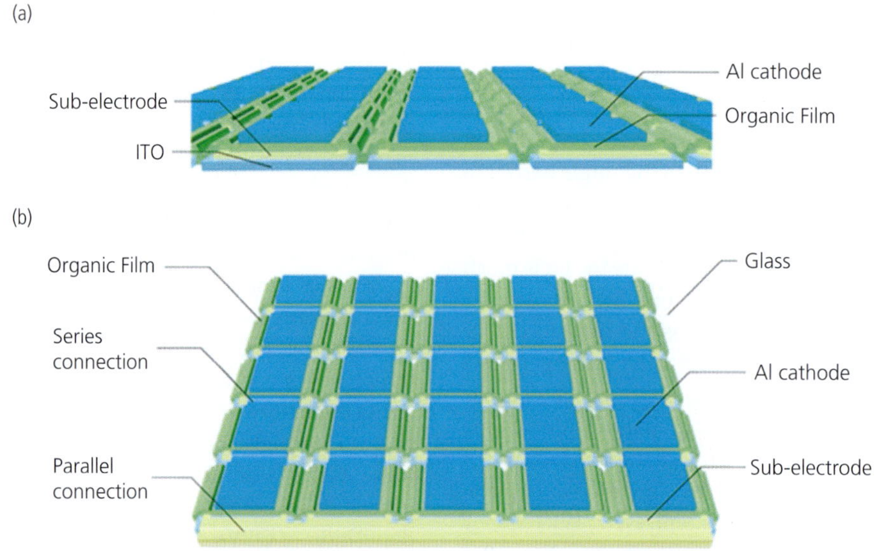

Figure 7.240 (a) Edge and (b) top views of an OPV module connected in a series-parallel circuit (Xiao et al., 2015).

with the control circuitry, become problematic. Thus, for a single cell operating at its MPP voltage, V_M, the 25 OPV array output voltage is $5V_M$.

The series circuit is realized by connecting the anode of one OPV with the cathode of the adjacent device, as shown in detail in Fig. 7.241. The step made by the cathode from the bottom ITO contact on one cell to the top of the organic layer creates optically inactive regions that allow for alignment tolerances, and provide a sufficiently large anode-cathode contact to minimize series resistance. The aperture ratio, or *geometric fill factor*, is $GFF = A_{cell}/A_{tot}$, where A_{cell} is the active area of the cell, and A_{tot} is the total footprint occupied by each solar cell including inactive regions. For a cell of width, W, and length, L, then from Fig. 7.241 we have $GFF = L/(L + \Delta L)$, where ΔL is the distance between cell active regions along the x-direction. The module power conversion efficiency is

$$\eta_{P,module} = \eta_{P,cell} \cdot GFF = \eta_{P,cell} \frac{L}{L + \Delta L}. \qquad (7.221)$$

One of the challenges in module design and fabrication is to minimize ΔL.

Power loss due to series resistance of an individual cell was calculated in Section 7.3.3. The analysis for a module extends that treatment to account for additional losses due to the layout of the series-connected array. In addition to the series resistance of the contacts of the individual cells, power losses in the module also must include contributions from the anode

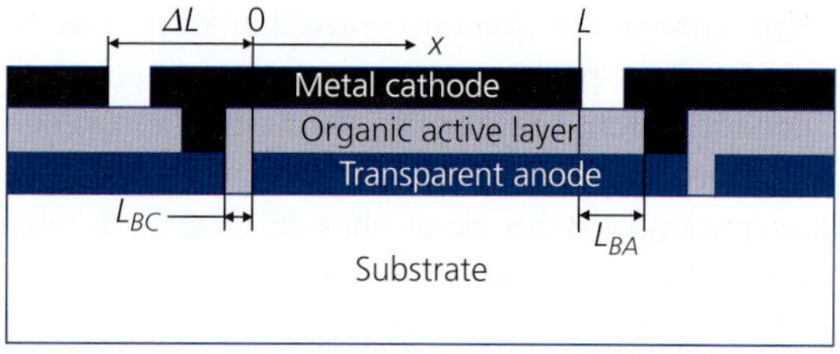

Figure 7.241 Schematic of the series connection of an OPV of active region length, L, to its neighbors. Note how the cathode must be deposited to connect the bottom transparent anode to the top of the organic active region.

and cathode bridges of lengths L_{BA} and L_{BC}, respectively, and the contact resistance of the cathode-anode contact (Hoppe et al., 2012).

If we consider the cell itself to be a distributed source generating an incremental current at point x (see Fig. 7.241) of $dI(V,x) = j(V)Wdx$ where W is the contact width, then the total current at x is $I(V,x) = j(V)Wx$. It follows that the incremental change in resistance is $dR = R_{\square}dx/W$, yielding a total power loss from the contact sheet resistance, $R_{\square}$, of

$$\Delta P_{sheet} = \frac{R_{\square}}{W} \int_0^L I(V)^2 dx = \frac{R_{\square}}{W} \int_0^L [j(V)Wx]^2 dx = I(V)^2 \left[\frac{R_{\square}L}{3W}\right]. \qquad (7.222)$$

Thus the total sheet resistance from the contact of length L is

$$R_{sheet} = \frac{R_{\square}L}{3W}. \tag{7.223}$$

A similar analysis yields the bridge resistances of

$$R_{BA,C} = R_{\square A,C}\frac{\Delta L}{3W}, \tag{7.224}$$

where it is assumed that both the anode and cathode (subscripts A and C, respectively) bridge lengths are equal, but their sheet resistances, $R_{\square A,C}$, are not. The contact between the anode and cathode also contributes resistance,

$$R_C = \rho_C \frac{3}{W\Delta L}, \tag{7.225}$$

where ρ_C is the contact resistivity.

These resistances can be independently measured on the substrates used in the modules. Similarly, the device properties are derived from their $j(V)$ characteristics. From the latter, $\eta_{P,cell}$ is calculated in the usual manner. Finally, the module efficiency is found from Eq. 7.221.

This analysis is used to optimize the layout by minimizing both ΔL and $R_{\square}$ through appropriate materials choices, device structure, etc. It has been applied to P3HT:PC$_{61}$BM solar cells to evaluate the performance of different anodes (ITO, PEDOT:PSS) and substrates (PET and glass). Table 7.30 lists measured values of $R_{\square}$ and ρ_C. With these parameters in hand, along with the characteristics of the individual OPV cells, the module efficiency is calculated as a function of L and ΔL assuming ITO bridge lengths of $L_{BA} = \Delta L/3$ and $W = 5$ cm. The results are plotted for R_{ser} (which is the sum of Eqs. 7.223–7.224), FF and η_P in Fig. 7.242. The result is for ITO on glass, with a maximum individual cell efficiency of 4%. The reduction in η_P with L and ΔL is due to the reduction in FF, since both j_{SC} and V_{OC} are largely unaffected by the series resistance. The R_{ser} is almost entirely dominated by the ITO sheet resistance, increasing with the length of the anode bridge and the cell. It is noted that the cell efficiency in the module drops by over 30% as the cell length is increased to 2 cm, and the distance between cells to 0.5 cm. Not surprisingly in view of Table 7.30, the erosion in efficiency is far more pronounced for ITO on PET, and by using PEDOT:PSS to replace ITO (Hoppe et al., 2012).

Figure 7.243 shows data from a module comprising a 5 × 5 array of 1 cm^2 tandem cells in Section 7.3.3 (see Fig. 7.84 and 7.240). The dominant contribution to the series resistance from ITO is reduced using a Ag sub-electrode bus bar around 95% of the device perimeter (Park et al., 2010b), thereby reducing the total contact resistance to only 30% of that for a 1 cm^2 cell without sub-electrodes. Furthermore, a sub-electrode is employed around each individual cell in the array.

Table 7.30 Sheet and contact resistivities for common materials used in OPV modules (Hoppe et al., 2012)

Material	Layer thickness	$R_{\square}$ (Ω/sq.)
Al	135	0.16
ITO on glass	150	12.5
ITO on PET	150	50
PEDOT:PSS PH1000	150	100
Material		**ρ_C with Al (mΩ cm^2)**
ITO on glass		11
ITO on PET		17
PEDOT:PSS PH1000		200

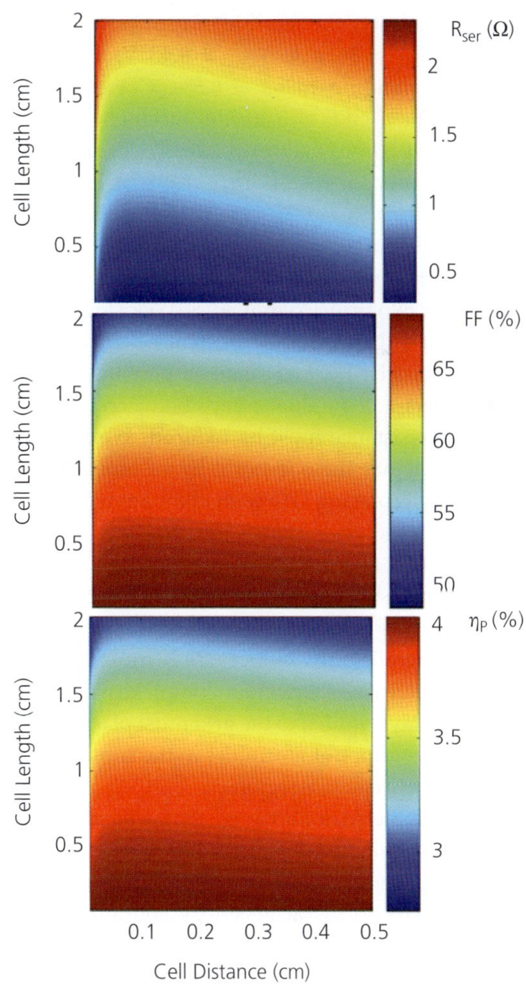

Figure 7.242 Efficiency of P3HT:PC$_{61}$BM solar modules as functions of individual cell lengths, L, and separation distances, ΔL, assuming L_{BA} = 1/3ΔL (Hoppe et al., 2012).

A photograph of the module fabricated by shadow mask patterning on a glass substrate is shown in the inset of Fig. 7.243a. For a column of five discrete cells connected in series, V_{OC} = 8.45 V, which is within 0.5% of the sum of V_{OC} for 5 individual cells, while the short-circuit current of a column is 7.5 mA, the same as for the total discrete cell currents. The *I-V* characteristics of these components measured under 1 sun, AM1.5G simulated illumination are shown in Fig. 7.243a. The characteristics indicate that resistive losses are minimal in the module compared to that of the individual cells due to a combination of the use of tandem cells, sub-electrodes, and an optimized circuit layout. The difference between the experimental and fitted *I–V* characteristics for the module is due to differences in individual device performances from variations in film thickness and contact resistances across the module. Under simulated AM1.5G illumination at one sun intensity, the module output power is 162 mW with I_{SC} = 36 mA, V_{OC} = 8.45 V, and FF = 53%, corresponding to η_P = 6.5%. Correcting the efficiency for a $GFF \approx 0.46$, the module efficiency is $\eta_{P,module}$ = 3.0% (Xiao et al., 2015). The vacuum deposition process results in a 100% discrete cell yield, as shown in Fig. 7.243b. The efficiency of the discrete cells range from 6.7% to 7.2%, with a variation less than 10% across the module.

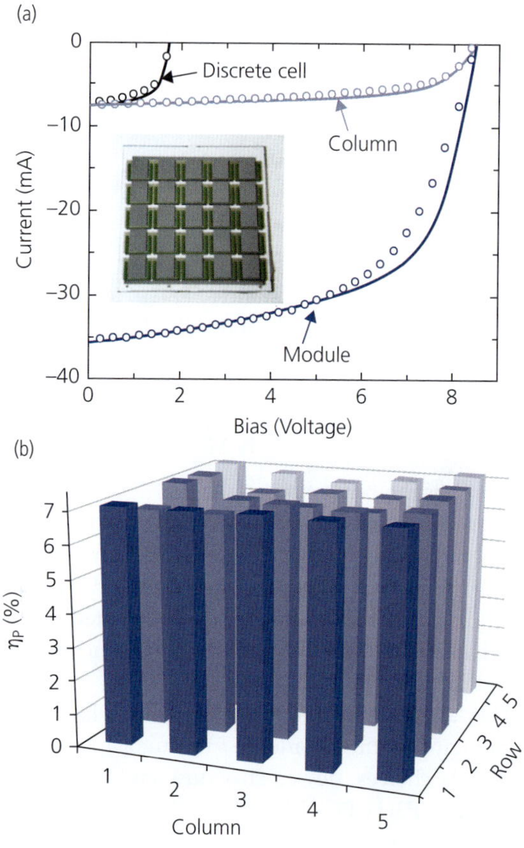

Figure 7.243 (a) Experimental (circles) and calculated (solid lines) *I–V* characteristics of a discrete, 1 cm^2 tandem OPV on the module (black), a column of 5 cells connected in series (light blue), and the entire module (dark blue) under 1 sun illumination. Inset: Photo of the OPV module. (b) Efficiency distribution of discrete OPV cells across the module (Xiao et al., 2015).

There are two principal remedies to closing the gap between the cell and module efficiencies. The individual cells must be kept as small as possible without unduly affecting *GFF* due to gaps between cells, and the anode resistance must be reduced. We found in Section 7.4.1 that there is not a lot of improvement expected by replacing ITO with alternative materials such as Ag nanowires. Hence, an attractive approach used in OLEDs and in OPVs is to embed a thin metal grid (laid out as streets and avenues, or honeycombs) in the ITO or PEDOT:PSS layer (Zhou et al., 2014). This can significantly reduce sheet resistance, although it increases fabrication complexity since the transparent layer added to the grid must planarize the surface to avoid non-uniformities in the subsequently deposited device active regions.

Large volume manufacturing that will ultimately make OPVs a viable technology has been extensively explored for solution-processed materials. Rapid module production is achieved by continuous R2R thin film deposition on flexible plastic (generally PEN or PET) substrates. Several solution-based R2R deposition methods have already been discussed in Chapter 5, and are reviewed elsewhere (Krebs, 2009). Patterning of the electrode on the substrate is the most challenging step, since the other layers including the organic and counter-electrodes are deposited over the full surface. The final step is encapsulation, which also must be done in the R2R environment to avoid slowing the rate of material output. Indeed, continuous deployment of reels of substrate material requires that all layer depositions and patterns, independent of material or film thickness, must occur at the same rate, again to avoid slowing or intermittently halting the fabrication process.

Several example electrode patterning processes are shown schematically in Fig. 7.244. Transparent electrode processes are provided in Fig. 7.244a–c, and opaque substrates used for devices with transparent top contacts are in Fig. 7.244d and e. The layers are applied in ambient on a continuous, 305 mm wide PET web. The most common anode is ITO, which is also the most difficult to pattern. The substrate in Fig. 7.244a is supplied with a full surface

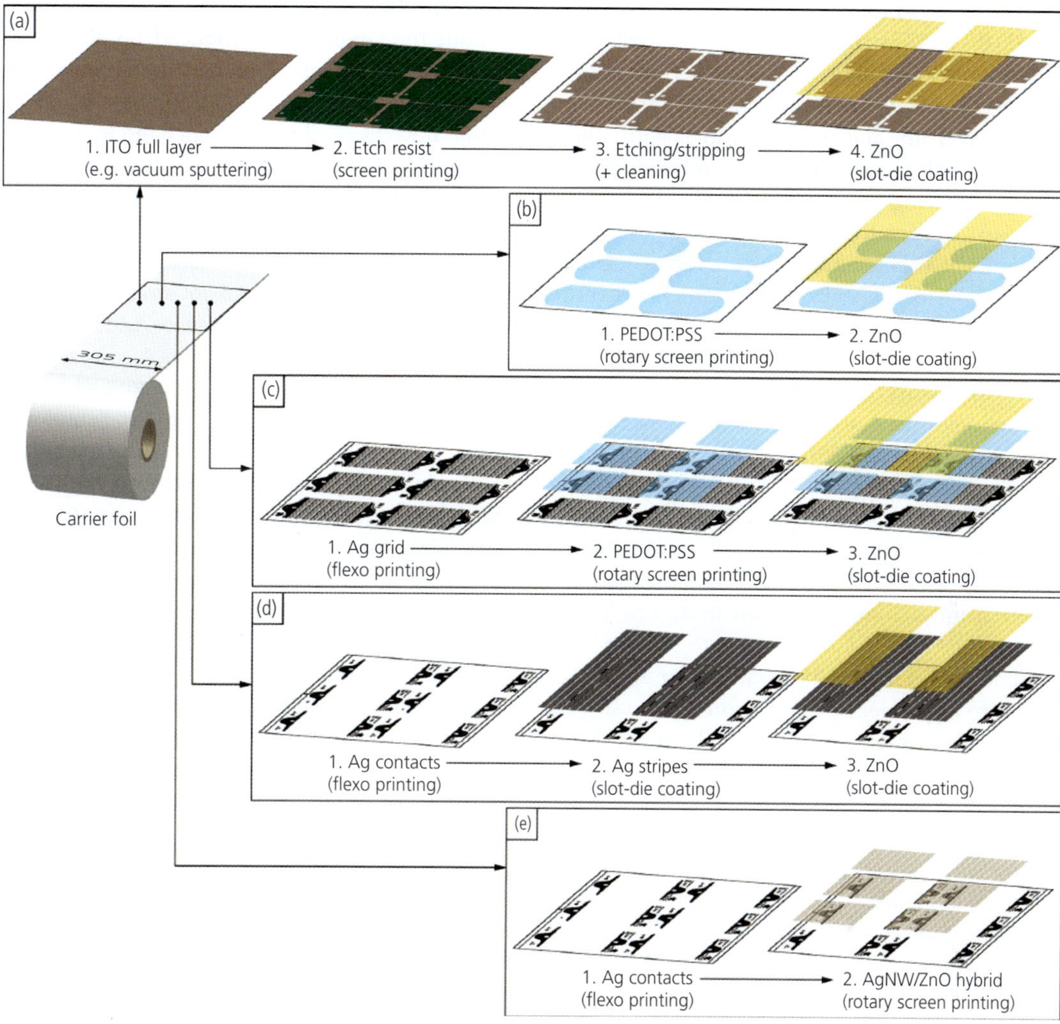

Figure 7.244 Several processes for patterning either (a–c) transparent or (d, e) opaque reflective electrodes. The demonstrations of these methods are on a 305 mm wide PET web. Details of each process are provided in the text (Hösel et al., 2014).

coating of sputter-deposited ITO. Screen printing is used to apply an etch resist, which is stripped by wet chemical or dry processing, and then the remaining resist is removed, also via plasma processing. The ZnO cathode layer is applied from solution via slot-die coating. This, and other coating and patterning methods are described in Section 5.6. A simpler process employs only two layers (Fig. 7.244b). Here, PEDOT:PSS is applied via rotary screen printing, which comprises a cylinder with a screen whose pattern matches that to be transferred. The solution is continuously pushed through the screen with a squeegee, after which the film is dried. Again, the ZnO is applied by slot-die coating, with the pattern slightly offset from the underlying PEDOT:PSS to allow for contact to this layer. Figure 7.244c shows the patterning of a Ag NP grid cathode. It begins by flexo-printing of the Ag NP or Ag nanowire solution. Flexo-printing employs a soft stamp whose raised areas define the pattern. Ink, in this case the Ag NP solution, is continuously applied to the stamp roller, and is transferred to the substrate as it is rolled through (Krebs, 2009). Then the PEDOT:PSS and ZnO layers are added as before.

Reflective substrate coatings start, as shown in Fig. 7.244d by application of Ag contacts on the edge of the sheet using flexo-printed Ag NPs. The cathode stripes are subsequently patterned using slot-die coating, and completed by a layer of ZnO. A simplified version of this process replaces the Ag stripes with a composite solution of Ag nanowires (NW), plus ZnO using rotary screen printing.

A further method of patterning that can be integrated into the R2R system is *laser scribing*. *Laser thermal ablation* can selectively remove layers at several stages of the fabrication sequence, depending on the layer absorption spectrum and the laser wavelength. This process can result in cuts only a few microns wide, potentially achieving $GFF > 98\%$. Laser ablation has been used to make small (3–10 devices) modules with a 500 kHz train of 350 fs pulses at peak energies of 16 μJ and 6.4 μJ at the fundamental wavelength of 1040 nm and the first harmonic at 520 nm, respectively. The two wavelengths are selectively absorbed within different layers of the structure. It is estimated that the laser can produce modules at a rate of 1 m/s in a volume manufacturing facility (Li et al., 2014).

Photographs of the results of each of the printing processes in Fig. 7.244 are shown in Fig. 7.245. There are six modules on each 12″ × 12″ sheet. Solution-processed R2R coating proceeds at rates that vary from < 1 m/min to up to 100 m/min. The layer thickness control and patterning resolution tend to decline as the translation speed is brought into the upper range. As noted previously, the printing processes must be capable of dispensing adequate amounts of materials at the requisite web translation speed, and the equipment must be kept free of excess material accumulation and clogging during the manufacturing cycle. The rate of cell fabrication is limited by the slowest step in the continuous chain.

A compilation of performance of single devices and modules fabricated by the methods in Fig. 7.244 is provided in Table 7.31. The devices employ various transparent conductive electrodes and donor polymers. The active regions are BHJ blend films with $PC_{61}BM$ as the acceptor. The performance of both the single cells produced on a laboratory scale "mini-roll coater", as well as the larger modules fabricated on the 305 mm system fall short of similar devices produced by batch mode processing. There are many reasons for the drop in efficiency, many of which have already been touched on. The increased contact resistance, loss of active area between cells, compromises made in choice of printable electrode materials and substrates, all conspire to decrease solar cell performance. Additionally, the R2R process proceeds in ambient. A consistent finding for all semiconductor devices, not just organics, is that ultrahigh purity source materials and an absence of contamination in the processing environment are essential ingredients for achieving high performance and long-term device stability. Exposing active materials to the atmosphere, which in itself is a poorly defined environment that changes from day to day, initiates the photooxidation process before the devices are encapsulated. The substrates, too, need to be baked out prior to use to eliminate moisture and other contamination that can outgas into the encapsulated device. Ultimately, this demand for high purity and avoidance of environmental contamination must be included at all stages of device fabrication up to, and including encapsulation.

There is considerably less information available for R2R processes based on vapor deposition of small molecules. There are a handful industry reports of production of OPVs by multiple, interconnected vacuum deposition chambers in a single R2R system. Active area efficiencies of 7.7% over areas of approximately 4600 cm^2 have been reported with a $GFF = 66\%$ (Uhrich et al., 2017).

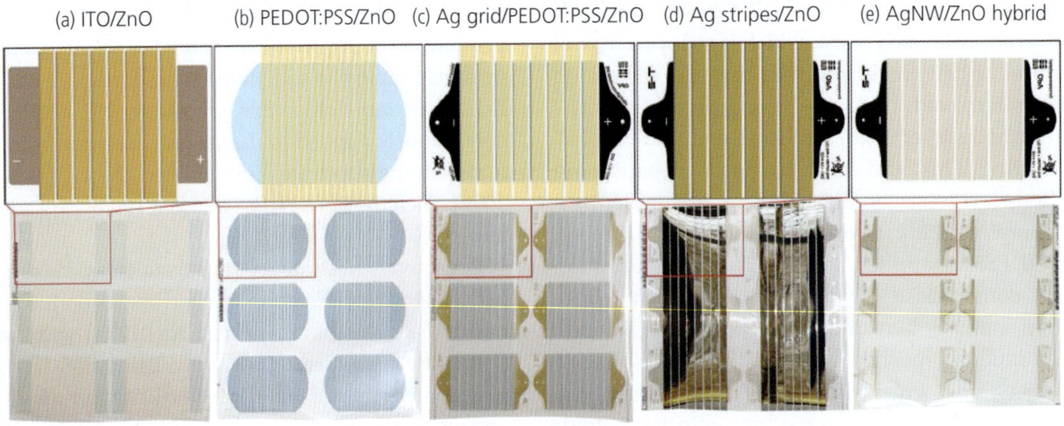

Figure 7.245 Images of 12″ × 12″ motifs (bottom row) and details of modules on the motifs (top row) produced in ambient using the processes in Fig. 7.244 (Hösel et al., 2014).

Table 7.31 Performance of OPV cells and modules fabricated by R2R processing in ambient based on patterned electrode and donor polymer compositions. Each device employs a $PC_{61}BM$ acceptor (Hösel et al., 2014)

Patterned electrode	Donor polymer	V_{OC} (V)	j_{SC} (mA/cm^2)	*FF*	η_P (%)	Area (cm^2)
Single cells, mini-rollcoater processed						
PEDOT:PSS	P3HT	0.52	6.35	0.48	1.61	0.2
PEDOT:PSS	PBDTthd-DTBT[a]	0.69	7.64	0.53	2.82	0.4
Ag grid/PEDOT:PSS	P3HT	0.53	7.98	0.52	2.24	0.7
Ag	P3HT	0.55	7.67	0.57	2.45	1.4
Ag NW	P3HT	0.52	9.6	0.55	2.75	0.7
Ag NW	PBDTthd-DTBTf[a]	0.71	11.01	0.50	3.9	0.7
ITO	PDTSTTz-4[b]	0.67	10.46	0.47	3.29	0.8
Modules, R2R processed						
PEDOT:PSS	P3HT	8.88	13	0.48	1.83	30 (16 cells)
Ag grid/PEDOT:PSS	P3HT	4.2	41	0.60	1.82	57 (8 cells)
Ag NW	P3HT	4.19	63.7	0.53	2.46	57 8 cells)

[a] Carlé et al. (2014).
[b] Helgesen et al. (2013).

An adaptation of R2R deposition is the integration of two different vapor deposition technologies into a single, laboratory system. The system combines high vacuum thermal evaporation with organic vapor phase deposition. This takes advantage of the versatility of VTE to grow complex layer systems, and the morphological control and very high deposition rates suitable for mass production using OVPD. A schematic illustration of the integrated system is shown in Fig. 7.246.

The system connects the VTE and OVPD systems via a translation chamber comprising a series of rollers that transports a 75 mm-wide flexible, tempered steel web. The use of steel eliminates problems of registration due to stretching, warping and heating of thermally sensitive substrates such as PET. This allows for attachment of a polymer or thin glass substrate onto the web without concern for its thermal or mechanical properties, nor whether it might be damaged during translation. The web is held stationary, or translated at up to 2.5 cm/s speed during deposition in either growth chamber. An actively cooled pusher arm is extended to shift the web from the translation chamber that contains the roller mechanism, into the OVPD growth position. A continuous coolant flow within the arm maintains the substrate at low temperature while in the hot OVPD chamber. The OVPD chamber itself comprises a cylindrical quartz reactor enclosed in a horizontal furnace. Quartz barrels contain the source boats. The organic source materials are volatilized by the flow of hot inert (N_2) gas that carries the molecules downstream where they are adsorbed on the cooled web film.

During VTE growth, the gate valve between the OVPD chamber and the translation chamber is closed, and the VTE chamber is pumped to 10^{-7} torr. Before OVPD operation, the gate valve is opened and the pressure throughout the tool is equalized to the OVPD base pressure of 10^{-2} torr. Shadow masks at both the VTE and OVPD growth positions allow for contact and film patterning.

This system has demonstrated an rms film roughness of only 0.40 nm even at growth rates of approximately 16 Å/s at translation speeds of 1.7 cm/s. Both single junction and tandem OPVs have been grown in this system, requiring web translation between the VTE and OVPD chambers twice for the tandem cells. A comparison of several devices (both single and tandem junction) grown while stationary and while translating at 0.3 cm/s is provided in Table 7.32. There is no significant difference between devices grown by R2R with the PET substrate either translating during growth, or halted at the growth position before moving on to the next. Also, the device characteristics are comparable to those grown by conventional VTE. All devices, however, are somewhat inferior to those grown on glass substrates due to the lower sheet resistance of ITO on glass compared with PET.

While this is only an early demonstration of R2R growth using vapor deposited small molecules in an integrated system, it is encouraging that very

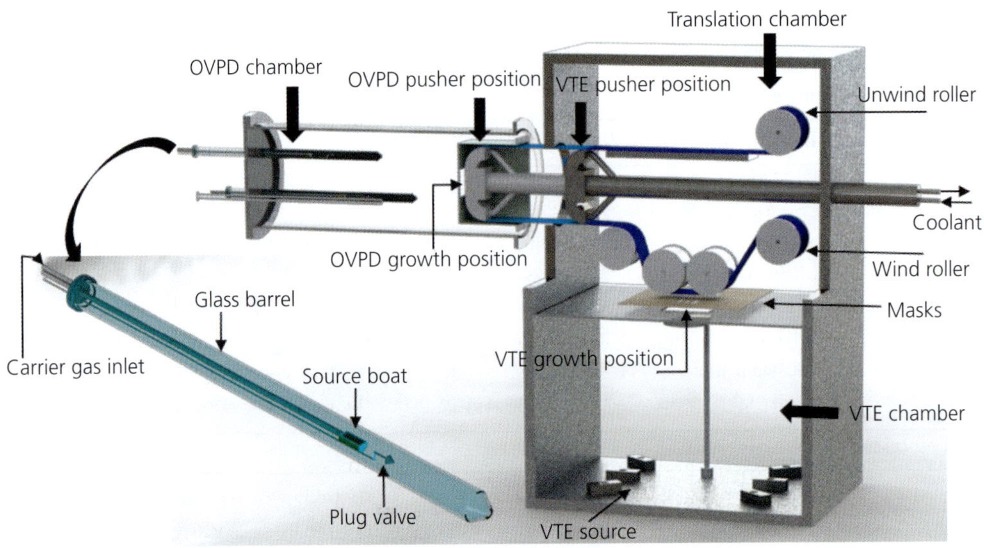

Figure 7.246 Cross-section schematic of the roll-to-roll (R2R) system integrating VTE with OVPD. A magnified view of OVPD source barrel containing the organic source boat and plug valve is shown (Qu and Forrest, 2018).

Table 7.32 Device performance of single and tandem junction OPVs fabricated by conventional VTE growth, and in the R2R system combining VTE and OVPD (Qu and Forrest, 2018)

Substrate	Dep.	No. of junctions	Trans. speed (mm/s)	j_{sc} (mA/cm^2)	V_{oc} (V)	*FF*	η_P (%)
Glass	VTE	1	0	15.1	0.92	0.70	9.7
PET	VTE	1	0	14.7	0.92	0.65	8.8
PET	R2R	1	0	14.5	0.92	0.65	8.6
PET	R2R	1	3	14.4	0.92	0.64	8.5
PET	R2R	2	0	8.6	1.77	0.58	8.9

different growth processes can be integrated in a single deposition tool. Combining solution and vacuum deposition in a continuous web-based growth line provides additional opportunities to realize high efficiency devices that are hybrid combinations of several different processes (Griffith et al., 2016). Indeed, we have seen that the combination of solution and vapor processed subcells in a tandem architecture can lead to high fabrication yields and efficiencies (Che et al., 2018b). Thus far, however, hybrid solution and vapor-grown systems have not enabled the continuous throughput of substrates as has been done in fully solution or vapor deposition R2R systems. It is possible to extend the laboratory system in Fig. 7.246 into a continuous process (as opposed to one where the substrate moves back and forth between growth chambers). Further, it is feasible to integrate high vacuum and solution-processed deposition systems using differentially pumped plenum chambers between system modules operating at different pressures. However, significant challenges related to device performance, encapsulation and contact resistance must first be overcome before large volume manufacturing of OPVs via R2R technologies becomes practical.

7.10 Summing up

In this chapter, we have shown that organic thin film materials are ideally suited for high performance optical detection in applications requiring flexibility, light weight, and low cost. Intense and narrow exciton spectra enable thin organic active regions that only absorb in selected wavelength bands. This is distinct from the behavior of inorganic semiconductors whose broad energy bands absorb from the band edge, deep into the shorter wavelength spectral regions. The narrow absorption spectra of organics opens up new possibilities for use of OPDs as wavelength selective sensors designed to capture only those regions of the spectrum needed for a given application, while otherwise being transparent. OPDs have other characteristics that are superior to their inorganic counterparts. Their thin active regions lead to short charge transit times that allow for very high bandwidth, high quantum efficiency operation. And the ability to deposit films on a variety of flexible, large area substrates makes them useful as position sensors. The range of applications that can benefit from the unique attributes of OPDs is limitless.

While OPDs have these distinctive features, a preponderance of effort in exploiting the optical detection properties of organics has focused on their use as solar cells. Their selective absorption spectra makes them attractive for application to power-generating windows that are semitransparent in the visible, yet absorb strongly in the UV and NIR. Thus, high efficiency coupled to semitransparency, long lifetime, flexibility and low cost on large substrates has driven intense worldwide interest in OPVs as a potentially important contributor to solving humankind's clean energy supply crisis. Of no less importance is the environmental sustainability of OPVs. The materials used in the cells and modules are often recyclable, and do not present undue toxic hazards that are frequently drawbacks of other solar technologies. Use of toxic materials in solar panels creates a disposal challenge at the end of their useful life; concerns that are largely absent in OPVs. The very low temperatures used in OPV fabrication compared to those required for thin film inorganic semiconductor processes coupled with the relatively low cost of source materials, suggests that OPVs will eventually be cost competitive with Si (Mulligan et al., 2015, Gambhir et al., 2016).

The efficiencies of OPVs are also fast becoming competitive with far more mature solar cell technologies, and their long term stability also appears to be reaching toward the 10–20 year benchmark required for low cost solar energy generation. All of these features suggest that there is indeed a very promising future for OPVs. To date, however, this promise has not been realized. The fabrication of solar cell modules based on either solution or vapor deposited materials has yet to yield performance that is near to that obtained on the laboratory bench. This is due to the many compromises that are made in the choice of materials, processes, and encapsulants when moving to large volume manufacturing. Many years of research and development still need to be invested to completely overcome these hurdles. Moreover, this is not the type of research that can be done in small laboratories. It will take significant investment by industry and/or governments to resolve many of the most challenging manufacturing issues. The investments needed are of a scale similar to those made in moving OLEDs from laboratory curiosities into today's massive organic global display market. The outstanding, differentiating attributes of OPVs makes these investments almost inevitable. We just do not know when.

Finally, and perhaps most important of all, is that research on OPVs and OPDs has revealed the richness of the phenomena involved in photoinduced charge generation in disordered media. The multistep process that starts with the absorption of a photon to generate an exciton, followed by its diffusion to an energy discontinuity at the donor–acceptor HJ, leading to efficient charge transfer, and finally the generation of a free PP have been a subject of deep fascination since long before the first efficient bilayer OPV was demonstrated. The linkages found between the physical processes of optical absorption, energy transport and charge generation with materials composition and thin film morphologies, have impacts on our understanding of disordered solids that go far beyond their particular applications in photodetection. Indeed, many of the insights developed in understanding light generation in OLEDs have furthered our understanding of the complementary process of light detection in organic detectors. It is this foundation of knowledge that will ultimately open doors to devices and applications that have yet to be imagined. Thus, for practical and fundamental reasons alike, the pursuit of high efficiency OPDs and OPVs has created a watershed of knowledge that will continue to grow in the years ahead, all the while bringing this exciting technology closer to a transformative new generation of devices that will help to provide abundant, clean energy to fill humankind's ever increasing demands.

Further reading

M. Born and E. Wolf, 1980. *Principles of Optics*. Oxford: Pergamon.

C. Deibel and V. Dyakonov, 2010. Polymer–fullerene bulk heterojunction solar cells. *Rep. Prog. Phys.*, 73, 096401.

A. L. Fahrenbruch and R. H. Bube, 1983.*Fundamentals of Solar Cells*. New York: Academic Press.

O. S. Heavens, 1991. *Optical Properties of Thin Solid Films*. New York: Dover.

C. Krebs (Ed.), 2009. Special issue on processing and preparation of polymer and organic solar cells. *Solar Energy Mater. Solar Cells*, 93, 393–538.

J. C. MacKay, 2009. *Sustainable Energy—Without the Hot Air.* Cambridge: UIT Cambridge, Ltd.

J. Nelson, 2003. *The Physics of Solar Cells.* London: Imperial College Press.

B. P. Rand and H. Richter, 2014. *Organic Solar Cells: Fundamentals, Devices and Upscaling.* Singapore: Pan Stanford Publishing.

A. Rose, 1963. *Concepts in Photoconductivity and Allied Problems.* New York: Interscience.

Sze, S. M. 1981. *Physics of Semiconductor Devices.* New York: Wiley-Interscience.

A. van der Ziel, 1986. *Noise in Solid State Devices and Circuits.* New York: John Wiley and Sons.

Problems

1. In Section 7.1 we showed that photogeneration in an excitonic photoconductor results in a charge density given by Eq. 7.20, viz.:

$$n_{ph} = \frac{j_{rec}}{k_{rec}D}\left\{1 - \frac{2dqn_i}{D}\left(1-\eta_{IQE}\right) - \left[\left(1 - \frac{2dqn_i}{D}\left(1-\eta_{IQE}\right)\right)^2 + 4q\eta_{IQE}\frac{j_X}{j_{rec}}\right]^{\frac{1}{2}}\right\}.$$

   (a) Derive this expression.
   (b) Derive an expression for n_{ph} in a doped semiconductor with $n_0 > p_0$ and $\mu_e \approx \mu_p$
   (p) Plot n_{ph} *vs.* D and for n_0/p_0 from 1 to 10^5.

2. The expressions such as in Problem 1 are for the charge density in steady state.
   (a) Set up the time dependent rate equations for $n_{ph}(t)$ assuming that the illumination is a harmonic function at frequency ω. Assume that the semiconductor is intrinsic.
   (b) What is the characteristic time constant for the return to equilibrium of the photoconductor for a step function optical pulse that terminates abruptly at $t = 0$? Plot the decay transient. What variables is it most sensitive to?
   (c) What is the gain–bandwidth product of the photoconductor in (a) and (b)?

3. The expressions for noise in photoconductors in Section 7.1.1 assume a single charge carrier (electrons) and deep traps that act as recombination centers. Derive expressions for the shot and Johnson noise for two cases:
   (a) The electron and hole mobilities are equal.
   (b) The traps are shallow, that is, they reside close to a frontier energy level, at or just below the Fermi energy.

4. For a symmetric heterojunction, that is, where the trap density of states (H_{TA}, H_{TD} with characteristic trap temperatures of T_{tA}, T_{tD}) are equal on each side of the junction,
   (a) Derive the ideal diode equation for a symmetric heterojunction, including an expression for the saturation current, j_{sym0}, assuming that the HOMO and LUMO densities of states are equal to N_S.
   (b) Show that the open circuit voltage for the symmetric junction is given by

$$qV_{OC} = \Delta E_{HL} - n_S k_B T \ln\left[\left(\frac{k_{rec}k_{PPr}}{k_{PPd}+k_{PPr}}\right)\frac{qa_0 N_S^2}{j_{ph}}\right].$$

5. The transfer matrix method in Section 7.1.4 was developed for light incident normal to the substrate.
   (a) Derive the scattering and phase matrices, I_{jk} and L_j, respectively, for light incident at an angle, θ, from normal.
   (b) Plot the total light absorbed in a layer of thickness 10 nm sandwiched between an Al contact and a 0.1 mm thick glass substrate as a function of angle, θ. The layer has $\tilde{n} = 1.7 + i1.0$ at the wavelength of $\lambda = 500$ nm.

6. Derive the generalized transfer matrices I_{jk} and L_j for:
   (a) Layers with an rms surface roughness, σ.
   (b) Layers with an rms index inhomogeneity, Δn.
   (c) Layers with both surface roughness and index inhomogeneities as in (a) and (b).

7. Plot the first, second, and third order transitions vs. diameter for quantum dots in air composed of:
   (a) InP
   (b) InAs
   (c) In Fig. 7.52, verify that the three peaks correspond to the 1S–1S, 1P–1P, and 1D–1D transitions for a spherical PbSe dot of diameter $2\rho = 6.5$ nm. The effective masses of electrons and holes in PbSe are approximately $m_e^* = 0.12m_0$ and $m_h^* = 0.07m_0$, respectively. (See Fig. 7.50 for the definition of the transitions.)

8. By integrating over the relevant half space, show that the étendue is given by $f_\omega = \pi \sin^2\theta_{sun}$.

9. Calculate the approximate spectral mismatch factor for the OPV in Fig. 7.79a (denoted S_T) using:
   (a) A Si detector with a KG3 filter (see Fig. 7.81).
   (b) A Si detector with a KG5 filter.
   (c) Assuming that the device has $j_{SC} = 15\ \text{mA/cm}^2$, $FF = 0.72$, and $V_{OC} = 0.92$V at SRC measured under a solar simulator whose spectrum is given in Fig. 7.79b using a KG3 filter, what is its power conversion efficiency?
   (d) For the device in (c), what is its peak external quantum efficiency and responsivity?

10. The *j*–*V* characteristics of inorganic and organic junctions are governed by two different equations; the Shockley equation for the former and the excitonic ideal diode equation for the latter. The power conversion efficiency of a photodiode operated in the photovoltaic mode is proportional to the fill factor, $FF = \frac{V_M I_M}{V_{OC} I_{SC}}$, where V_M and I_M are the voltage and current at the maximum power point, and V_{OC} and I_{SC} are the open circuit voltage and short circuit current taken from the *j*–*V* data in the fourth quadrant. Assume a saturation current density of j_{sat}.
   (a) Write an approximate expression for the maximum fill factor expected for a Shockley diode as a function of ideality factor, n.

(b) Plot the maximum FF vs. n over the domain $n = [1,3]$.
(c) As in part b, for an excitonic diode over $n = [1,4]$, assume that the polaron pair dissociation efficiency is approximately given by Frenkel-Poole emission. That is, $k_{PPd} = K F \exp\left(\frac{\sqrt{q^3 F/\pi\varepsilon}}{k_B T}\right)$. Here, K is a constant, and the electric field is given by $F = |V - V_{bi}|/d$, where V_{bi} is the built-in potential and d is the total HJ layer thickness. Which device (organic or inorganic) has a higher FF and thus power conversion efficiency? For simplicity, assume only one type of charge trap exists in the non-ideal (n >1) case.

11. Rewrite Problem 10a for:
    (a) A device with series resistance, R_{ser}.
    (b) Now recalculate FF vs. R_{ser} for $n = 2$.
    (c) What is the difference in shape of the reverse j-V characteristics of an excitonic HJ assuming Onsager–Braun or Frenkel–Poole emission theory for k_{PPd}. Be quantitative (i.e. calculate these curves and plot on a normalized scale of j/j_0 over a voltage domain of 0 to −5 V). Consider both the voltage and temperature dependences.
12. In Section 7.3.1, we considered the effects of series and shunt resistance on solar cell efficiency. Starting with the ideal diode equation of excitonic junctions,
    (a) Derive an expression for V_M in Eq. 7.126, including the effects of R_{ser} and R_{shunt}.
    (b) Plot in the j–V characteristics in the limit of $R_{ser} = 0$ and $R_{shunt} = \infty$ in the 4th quadrant for $k_{PPd}/k_{PPr} = 10$ and 100, with $n_S = 1.5$ and $V_{OC} = 1$ V, and $j_{SC}/j_0 = 10^2$ and 10^4 with $j_{SC} = 10$ mA/cm^2. How do these differ from the curves in Fig. 7.71?
    (c) Derive the expression for FF in Eq. 7.131.
13. A D-A heterojunction detector has $\Delta E_{HL} = 0.3$ eV. Assuming equal masses for electrons and holes, and a density of states of $N = 10^{21}$ cm^{-3} eV^{-1} for both the donor and acceptor, for a device whose characteristics are dominated by generation and recombination in the dark:
    (a) Plot the fourth quadrant j–V characteristics of the OPV in the dark and under 1 sun illumination assuming a uniform $\eta_{ext} = 70\%$ from $\lambda = 400$ nm to 1 μm, and $\eta_{ext} = 0$ elsewhere. Assume the work functions of the cathode and anode contacts are equal.
    (b) Now assume that there is a second internal junction within the device whose polarity is opposite to the first junction (perhaps due to a second HJ, or a poor contact). Assuming the saturation current of this second junction is $j_{02} = 0.1 j_{01}$, where $j_{0(1,2)}$ are the saturation currents of the respective junctions, plot the j–V characteristics under the same illumination conditions as in (a).
14. Figure 7.100 suggests that the j–V characteristics of a double junction formed by a multistep cascade have an "S" shaped appearance that can be modeled as two back-to-back diodes.
    (a) Draw the full equivalent circuit of the double junction, assuming that one junction has saturation current, j_{01} and the other has current j_{02}. Each has a shunt resistance, R_{shunt1} and R_{shunt2} and the internal resistance of the layer connecting the two junctions is R_{ser}. Make sure you include an element for the photocurrent, j_{ph}.
    (b) Show that the j–V characteristics are given by:

$$j = j_{01}[\exp(qV_1/n_1 k_B T) - 1] + \frac{V_1}{R_{shunt1}} - j_{ph}$$
$$= -j_{02}[\exp(qV_2/n_2 k_B T) - 1] + \frac{V_2}{R_{shunt2}}.$$

For this expression, we assume that $V_a = V_1 + V_2 + jR_{ser}$, and that only junction 1 generates photocurrent. Here, subscripts 1 and 2 refer to the different junctions.

    (c) In the special case where $jR_{shunt2} >> V_2$, show that:

$$j = j_{01}\left[\left(1 - \frac{j}{j_{02}}\right)^{n_2/n_1} \exp(qV_a/n_1 k_B T) - 1\right] + \frac{V_1}{R_{shunt1}} + \frac{n_2 k_B T}{q R_{shunt1}} ln\left(1 - \frac{j}{j_{02}}\right) - j_{ph}$$

    (d) Letting $R_{shunt1} \to \infty$ plot the j-V characteristics over the domain V = (−1,1) V of a device with $j_{01} = 10$ μA/cm^2, $j_{02} = 0.1$ mA/cm^2, $j_{ph} = 5$ mA/cm^2, and $n_1 = n_2 = 2$. Does it show an "S" like characteristic? Why or why not?
15. Consider a series-connected multijunction OPV with three different absorbing cells with Gaussian absorption profiles centered at 500 nm, 750 nm, and 900 nm. The charge recombination zone is both electrically and optically lossless, and the saturation current is 10^{-6} A/cm^2 for each subcell. Assuming that the peak external quantum efficiencies are 65%, 80%, and 90% with FWHMs of 50 nm, 100 nm, and 125 nm, respectively, then calculate:
    (a) The efficiency of the triple junction cell assuming that the optical fields are centered within the appropriate subcell, assuming $FF = 70\%$ and $E_{loss} = 0.5$ eV for each subcell.
    (b) From front to back, what is the optimal position for each cell?
    (c) What is the optimal thickness of each subcell to maximize light absorption? For this, assume that each device is a mixed or bulk heterojunction structure.
16. In Fig. P7.16 are shown the fourth quadrant j–V characteristics under 1 sun, AM1.5G illumination of three subcells used in a multijunction cell.
    (a) Make a table that lists the j_{SC}, V_{OC}, FF, j_M, V_M, and P_M of each cell.
    (b) Consider a tandem cell comprised of Subcells A and B. Construct the j–V characteristics of the tandem and put its j_{SC}, V_{OC}, FF, j_M, V_M, and P_M data in your table. Also list the power penalty, $\Delta\eta_P$ for the tandem. Which cell limits the performance?
    (c) Repeat (b) for a triple junction cell comprised of subcells A, B, and C.

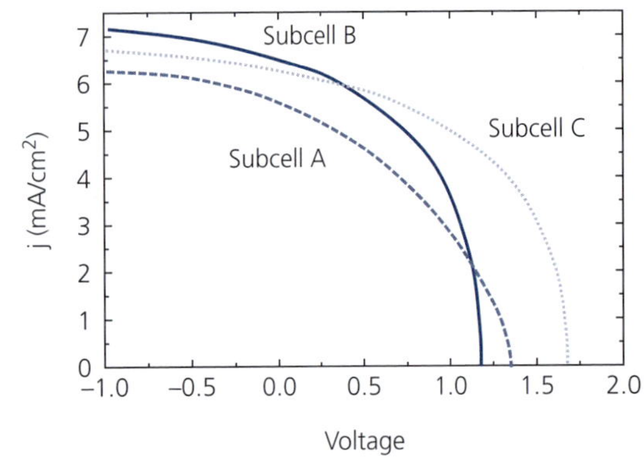

Figure P7.16 The *j*–*V* characteristics of three subcells used in multijunction OPVs.

17. Singlet fission results in the generation of three possible correlated triplet pair states, $^1(TT)$, $^3(TT)$, and $^5(TT)$.
    (a) Using bracket notation, write down the spin wavefunctions for the 9 possible triplet pair states. Clearly define the meaning of all of the notations used.
    (b) Derive Eqs. 7.202 and 7.203 for the change in photocurrent and fission efficiency as functions of magnetic field.
    (c) Plot the functions in (b) vs. magnetic field, from $B = 0$ to the saturation field (i.e. where δ and η become approximately independent of B).

References

Adams, C. D., Srolovitz, D. J. & Atzmon, M. 1993. Monte Carlo simulation of phase separation during thin-film codeposition. *J. Appl. Phys.*, 74, 1707.

Aernouts, T., Vanlaeke, P., Geens, W., Poortmans, J., Heremans, P., Borghs, S., Mertens, R., Andriessen, R. & Leenders, L. 2004. Printable anodes for flexible organic solar cell modules. *Thin Solid Films*, 451–452, 22.

Aihara, S., Hirano, Y., Tajima, T., Tanioka, K., Abe, M., Saito, N., Kamata, N. & Terunuma, D. 2003. Wavelength selectivities of organic photoconductive films: dye-doped polysilanes and zinc phthalocyanine/tris-8-hydroxyquinoline aluminum double layer. *Appl. Phys. Lett.*, 82, 511.

Ajuria, J., Ugarte, I., Cambarau, W., Etxebarria, I., Tena-Zaera, R. & Pacios, R. 2012. Insights on the working principles of flexible and efficient ITO-free organic solar cells based on solution processed Ag nanowire electrodes. *Solar Energy Mater. Solar Cells*, 102, 148.

Akamatu, H. & Inokuchi, H. 1950. On the electrical conductivity of violanthrone, iso-violanthrone, and pyranthrone. *J. Chem. Phys.*, 18, 810.

Amb, C. M., Chen, S., Graham, K. R., Subbiah, J., Small, C. E., So, F. & Reynolds, J. R. 2011. Dithienogermole as a fused electron donor in bulk heterojunction solar cells. *J. Am. Chem. Soc.*, 133, 10062.

Amelio, G. F., Tompsett, M. F. & Smith, G. E. 1970. Experimental verification of charge coupled device concept. *Bell Syst. Tech. J.*, 49, 593.

Ameri, T., Dennler, G., Lungenschmied, C. & Brabec, C. J. 2009. Organic tandem solar cells: a review. *Energy Envrion. Sci.*, 2, 347.

Ameri, T., Khoram, P., Min, J. & Brabec, C. J. 2013. Organic ternary solar cells: a review. *Adv. Mater.*, 25, 4245.

An, K. H., O'Connor, B., Pipe, K. P. & Shtein, M. 2009. Organic photodetector with spectral response tunable across the visible spectrum by means of internal optical microcavity. *Org. Electron.*, 10, 1152.

An, K. H., O'Connor, B., Pipe, K. P., Zhao, Y. & Shtein, M. 2008. Scanning optical probe microscopy with submicrometer resolution using an organic photodetector. *Appl. Phys. Lett.*, 93, 033311.

An, Q., Zhang, F., Gao, W., Sun, Q., Zhang, M., Yang, C. & Zhang, J. 2018. High-efficiency and air stable fullerene-free ternary organic solar cells. *Nano Energy*, 45, 177.

Andersson, B. V., Wuerfel, U. & Inganäs, O. 2011. Full day modelling of V-shaped organic solar cell. *Solar Energy*, 85, 1257.

Arca, F., Kohlstädt, E., Tedde, S. F., Lugli, P. & Hayden, O. 2013. Large active area organic photodiodes for short-pulse X-ray detection. *IEEE Trans Electron Devices*, 60, 1663.

Arnold, M. S., Zimmerman, J. D., Renshaw, C. K., Xu, X., Lunt, R. R., Austin, C. M. & Forrest, S. R. 2009. Broad spectral response using carbon nanotube/organic semiconductor photodetectors. *Nano Lett.*, 9, 3354.

Bae, S., Kim, H., Lee, Y., Xu, X., Park, J.-S., Zheng, Y., Balakrishnan, J., Lei, T., Kim, H. R. & Song, Y. I. 2010. Roll-to-roll production of 30-inch graphene films for transparent electrodes. *Nat. Nanotechnol.*, 5, 574.

Baeg, K.-J., Binda, M., Natali, D., Caironi, M. & Noh, Y.-Y. 2013. Organic light detectors: photodiodes and phototransistors. *Adv. Mater.*, 25, 4267.

Bagnis, D., Beverina, L., Huang, H., Silvestri, F., Yao, Y., Yan, H., Pagani, G. A., Marks, T. J. & Facchetti, A. 2010. Marked alkyl- vs alkenyl-substitutent effects on squaraine dye solid-state structure, carrier mobility, and bulk-heterojunction solar cell efficiency. *J. Am. Chem. Soc.*, 132, 4074.

Bailey-Salzman, R., Rand, B. P. & Forrest, S. R. 2006. Semi-transparent organic photovoltaic cells. *Appl. Phys. Lett.*, 88, 233502.

Baldo, M. A., Lemansky, S., Burrows, P. E., Thompson, M. E. & Forrest, S. R. 1999. Very high efficiency green organic light emitting devices based on electrophosphorescence. *Appl. Phys. Lett.*, 76, 4.

Baldo, M. A., O'Brien, D. F., You, Y., Shoustikov, A., Thompson, M. E. & Forrest, S. R. 1998. High efficiency phosphorescent emission from organic electroluminescent devices. *Nature*, 395, 151.

Banal, J. L., White, J. M., Lam, T. W., Blakers, A. W., Ghiggino, K. P. & Wong, W. W. H. 2015. A transparent planar concentrator using aggregates of gem-pyrene ethenes. *Adv. Energy Mater.*, 5, 1500818.

Barbé, J., Lee, H. K. H., Toyota, H., Hirose, K., Sato, S.-i., Ohshima, T., Heasman, K. C. & Tsoi, W. C. 2018. Characterization of stability of benchmark organic photovoltaic films after proton and electron bombardments. *Appl. Phys. Lett.*, 113, 183301.

Barito, A., Sykes, M. E., Bilby, D., Amonoo, J., Jin, Y., Morris, S. E., Green, P. F., Kim, J. & Shtein, M. 2013. Recovering lost excitons in organic photovoltaics using a transparent dissociation layer. *J. Appl. Phys.*, 113, 203110.

Barito, A., Sykes, M. E., Huang, B., Bilby, D., Frieberg, B., Kim, J., Green, P. F. & Shtein, M. 2014. Universal Design Principles for Cascade Heterojunction Solar Cells with High Fill Factors and Internal Quantum Efficiencies Approaching 100%. *Adv. Energy Mater.*, 4, 1400216.

Bartynski, A. N., Trinh, C., Panda, A., Bergemann, K., Lassiter, B. E., Zimmerman, J. D., Forrest, S. R. & Thompson, M. E. 2013. A fullerene-based organic exciton blocking layer with high electron conductivity. *Nano Lett.*, 13, 3315.

Bässler, H. 1993. Charge transport in disordered organic photoconductors. *Phys. Stat. Sol.*, 175, 15.

Batchelder, J., Zewail, A. & Cole, T. 1981. Luminescent solar concentrators. 2. Experimental and theoretical analysis of their possible efficiencies. *Appl.* .20, 3733.

Beek, W. J., Wienk, M. M., Kemerink, M., Yang, X. & Janssen, R. A. 2005. Hybrid zinc oxide conjugated polymer bulk heterojunction solar cells. *J. Phys. Chem. B*, 109, 9505.

Bellows, J. C. & Prasad, P. N. 1977. Molecular motions and lattice stability of a disordered organic alloy: binary solid solutions of 1,4-dihalonaphthalenes. *J. Chem. Phys.*, 67, 5802.

Benten, H., Mori, D., Ohkita, H. & Ito, S. 2016. Recent research progress of polymer donor/polymer acceptor blend solar cells. *J. Mater. Chem. A*, 4, 5340.

Bergemann, K. J., Amonoo, J. A., Song, B., Green, P. F. & Forrest, S. R. 2015. Surprisingly high conductivity and efficient exciton blocking in fullerene/wide-energy-gap small molecule mixtures. *Nano Lett.*, 15, 3994.

Bergemann, K. J. & Forrest, S. R. 2011. Measurement of exciton diffusion lengths in optically thin organic films. *Appl. Phys. Lett.*, 99, 274.

Bernardo, B., Cheyns, D., Verreet, B., Schaller, R. D., Rand, B. P. & Giebink, N. C. 2014. Delocalization and dielectric screening of charge transfer states in organic photovoltaic cells. *Nat. Commun.*, 5, 3245.

Betancur, R., Romero-Gomez, P., Martinez-Otero, A., Elias, X., Maymó, M. & Martorell, J. 2013. Transparent polymer solar cells employing a layered light-trapping architecture. *Nat. Photonics*, 7, 995.

Bin, H., Gao, L., Zhang, Z.-G., Yang, Y., Zhang, Y., Zhang, C., Chen, S., Xue, L., Yang, C., Xiao, M. & Li, Y. 2016. 11.4% Efficiency non-fullerene polymer solar cells with trialkylsilyl substituted 2D-conjugated polymer as donor. *Nat. Commun.*, 7, 13651.

Binda, M., Agostinelli, T., Caironi, M., Natali, D., Sampietro, M., Beverina, L., Ruffo, R. & Silvestri, F. 2009. Fast and air stable near-infrared organic detector based on squaraine dyes. *Org. Electron.*, 10, 1314.

Binda, M., Iacchetti, A., Natali, D., Beverina, L., Sassi, M. & Sampietro, M. 2011. High detectivity squaraine-based near infrared photodetector with nA/cm^2 dark current. *Appl. Phys. Lett.*, 98, 073303.

Bindl, D. J., Wu, M.-Y., Prehn, F. C. & Arnold, M. S. 2010. Efficiently harvesting excitons from electronic type-controlled semiconducting carbon nanotube films. *Nano Lett.*, 11, 455.

Bixon, M. & Jortner, J. 1999. Electron transfer—from isolated molecules to biomolecules. *Adv. Chem. Phys.*,106, 35.

Borsenberger, P. M., Chowdry, A., Hoesterey, D. C. & Mey, W. 1978. An aggregate organic photoconductor. II. Photoconduction properties. *J. Appl. Phys.*, 49, 5555.

Boyle, W. S. & Smith, G. E. 1970. Charge coupled semiconductor devices. *Bell Syst. Tech. J.*, 49, 587.

Braun, C. L. 1984. Electric field assisted dissociation of charge transfer states as a mechanism of photocarrier production. *J. Chem. Phys.*, 80, 4157.

Brus, L. E. 1984. Electron–electron and electron–hole interactions in small semiconductor crystallites: the size dependence of the lowest excited electronic state. *J. Chem. Phys.*, 80, 4403.

Bube, R. H. & Fahrenbruch, A. L. 1981. Photovoltaic effect. *Adv. Electron. Electron Phys.*, 56, 163.

Burkhart, B. & Thompson, B. C. 2014. Solution-processed donors. In: Rand, B. P. & Richter, H. (eds.) *Organic Solar Cells: Fundamentals, Devices, and Upscaling*. Singapore: Pan Stanford Publishing.

Burlingame, Q., Coburn, C., Che, X., Panda, A., Qu, Y. & Forrest, S. R. 2018. Centimetre-scale electron diffusion in photoactive organic heterostructures. *Nature*, 554, 77.

Burlingame, Q., Song, B., Ciammaruchi, L., Zanotti, G., Hankett, J., Chen, Z., Katz, E. A. & Forrest, S. R. 2016. Reliability of small molecule organic photovoltaics with electron-filtering compound buffer layers. *Adv. Energy Mater.*, 6, 1601094.

Burlingame, Q., Tong, X., Hankett, J., Slootsky, M., Chen, Z. & Forrest, S. R. 2015. Photochemical origins of burn-in degradation in small molecular weight organic photovoltaic cells. *Energy Environ. Sci.*, 8, 1005.

Burlingame, Q., Huang, X., Liu, X., Jeong, C., Coburn, C. & Forrest, S. R. 2019. Intrinsically stable organic solar cells under high intensity illumination. *Nature*, 573, 394.

Burlingame, Q., Zanotti, G., Ciammaruchi, L., Katz, E. A. & Forrest, S. R. 2017. Outdoor operation of small-molecule organic photovoltaics. *Org. Electron.*, 41, 274.

Burrows, P. E., Forrest, S. R., Zhou, T. X. & Michalsk, L. 2000. Operating lifetime of phosphorescent organic light emitting devices. *Appl. Phys. Lett.*, 76, 2493.

Busby, E., Xia, J., Wu, Q., Low, J. Z., Song, R., Miller, J. R., Zhu, X. Y., Campos, Luis M. & Sfeir, M. Y. 2015. A design strategy for intramolecular singlet fission mediated by charge-transfer states in donor–acceptor organic materials. *Nat. Mater.*, 14, 426.

Cabanillas-Gonzalez, J., Peña–Rodríguez, O., Lopez, I. S., Schmidt, M., Alonso, M. I., Goñi, A. R. & Campoy-Quiles, M. 2011. Organic position sensitive photodetectors based on lateral donor–acceptor concentration gradients. *Appl. Phys. Lett.*, 99, 103305.

Campbell, I. H. & Crone, B. K. 2007. Bulk photoconductive gain in poly(phenylene vinylene) based diodes. *J. Appl. Phys.*, 101, 024502.

Campbell, I. H. & Crone, B. K. 2009. A near infrared organic photodiode with gain at low bias voltage. *Appl. Phys. Lett.*, 95, 263302.

Campbell, P. & Green, M. A. 1986. The limiting efficiency of silicon solar cells under concentrated sunlight. *IEEE Trans. Electron Devices*, 33, 234.

Cardinaletti, I., Vangerven, T., Nagels, S., Cornelissen, R., Schreurs, D., Hruby, J., Vodnik, J., Devisscher, D., Kesters, J., D'Haen, J., Franquet, A., Spampinato, V., Conard, T., Maes, W., Deferme, W. & Manca, J. V. 2018. Organic and perovskite solar cells for space applications. *Solar Energy Mater. Solar Cells*, 182, 121.

Carlé, J. E., Helgesen, M., Zawacka, N. K., Madsen, M. V., Bundgaard, E. & Krebs, F. C. 2014. A comparative study of fluorine substituents for enhanced stability of flexible and ITO-free high-performance polymer solar cells. *J. Polym. Sci. Part B: Polym. Phys.*, 52, 893.

Cates Miller, N., Cho, E., Junk Matthias, J. N., Gysel, R., Risko, C., Kim, D., Sweetnam, S., Miller Chad, E., Richter Lee, J., Kline, R. J., Heeney, M., McCulloch, I., Amassian, A., Acevedo-Feliz, D., Knox, C., Hansen Michael, R., Dudenko, D., Chmelka Bradley, F., Toney Michael, F., Brédas, J. L. & McGehee Michael, D. 2012. Use of X-ray diffraction, molecular simulations, and spectroscopy to determine the molecular packing in a polymer-fullerene bimolecular crystal. *Adv. Mater.*, 24, 6071.

Chamberlain, G. 1983. Organic solar cells: a review. *Solar Cells*, 8, 47.

Chamberlain, G. A. & Clooney, P. J. 1979. Photoelectric properties of aluminium/copper phthalocyanine/gold photovoltaic cells. *Chem. Phys. Lett.*, 66, 88.

Chan, M. Y., Lai, S. L., Lau, K. M., Lee, C. S. & Lee, S. T. 2006. Application of metal-doped organic layer both as exciton blocker and optical spacer for organic photovoltaic devices. *Appl. Phys. Lett.*, 89, 163515.

Chan, W.-L., Ligges, M. & Zhu, X. Y. 2012. The energy barrier in singlet fission can be overcome through coherent coupling and entropic gain. *Nat. Chem.*, 4, 840.

Che, X. 2018. *High efficiency single and multijunction photovoltaics*. PhD thesis, University of Michigan.

Che, X., Chung, C.-L., Liu, X., Chou, S.-H., Y.-H. Liu, Wong, K.-T. & Forrest, S. R. 2016. Regioisomeric effects of donor–acceptor–acceptor′ small-molecule donors on the open circuit voltage of organic photovoltaics. *Adv. Mater.*, 28, 8248.

Che, X., Chung, C. L., Hsu, C. C., Liu, F., Wong, K. T. & Forrest, S. R. 2018a. Donor–acceptor–acceptor's molecules for vacuum-deposited organic photovoltaics with efficiency exceeding 9%. *Adv. Energy Mater.*, 8, 1703603.

Che, X., Li, Y., Qu, Y. & Forrest, S. R. 2018b. High fabrication yield organic tandem photovoltaics combining vacuum- and solution-processed subcells with 15% efficiency. *Nat. Energy*, 3, 422.

Che, X., Xiao, X., Zimmerman, J. D., Fan, D. & Forrest, S. R. 2014. High efficiency vacuum deposited small molecule organic tandem and triple-junction photovoltaic cells. *Adv. Energy Mater.*, 4, 568.

Chen, C.-C., Dou, L., Zhu, R., Chung, C.-H., Song, T.-B., Zheng, Y. B., Hawks, S., Li, G., Weiss, P. S. & Yang, Y. 2012a. Visibly transparent polymer solar cells produced by solution processing. *ACS Nano*, 6, 7185.

Chen, X.-K., Coropceanu, V. & Breédas, J.-L. 2018. Assessing the nature of the charge-transfer electronic states in organic solar cells. *Nat. Commun*, 9, 5295.

Chen, Y. H., Lin, L. Y., Lu, C. W., Lin, F., Huang, Z. Y., Lin, H. W., Wang, P. H., Liu, Y. H., Wong, K. T., Wen, J., Miller, D. J. & Darling, S. B. 2012b. Vacuum-deposited small-molecule organic solar cells with high power conversion efficiencies by judicious molecular design and device optimization. *J. Am. Chem. Soc.*, 134, 13616.

Cheng, P., Li, G., Zhan, X. & Yang, Y. 2018. Next-generation organic photovoltaics based on non-fullerene acceptors. *Nat. Photonics*, 12, 131.

Cheng, P. & Zhan, X. 2016. Stability of organic solar cells: challenges and strategies. *Chem. Soc. Rev.*, 45, 2544.

Cheng, P., Zhang, M., Lau, T. K., Wu, Y., Jia, B., Wang, J., Yan, C., Qin, M., Lu, X. & Zhan, X. 2017. Realizing small energy loss of 0.55 eV, high open-circuit voltage >1 V and high efficiency >10% in fullerene-free polymer solar cells via energy driver. *Adv. Mater.*, 29, 1605216.

Cho, K.-S., Kim, E., Kim, D.-W. & Kim, H.-K. 2017. Highly flexible and semi-transparent Ag–Cu alloy electrodes for high performance flexible thin film heaters. *RSC Adv.*, 7, 45484.

Chu, T. Y., Lu, J., Beaupré, S., Zhang, Y., Pouliot, J. R., Zhou, J., Najari, A., Leclerc, M. & Tao, Y. 2012. Effects of the molecular weight and the side-chain length on the photovoltaic performance of dithienosilole/thienopyrrolodione copolymers. *Adv. Funct.Mater.*, 22, 2345.

Closs, G., Calcaterra, L., Green, N., Penfield, K. & Miller, J. 1986. Distance, stereoelectronic effects, and the Marcus inverted region in intramolecular electron transfer in organic radical anions. *J. Phys. Chem.*, 90, 3673.

Cnops, K., Rand, B. P., Cheyns, D., Verreet, B., Empl, M. A. & Heremans, P. 2014. 8.4% efficient fullerene-free organic solar cells exploiting long-range exciton energy transfer. *Nat. Commun.*, 5, 3406.

Coburn, C., Fan, D. & Forrest, S. R. 2019. An organic charge-coupled device. *ACS Photonics*, 6, 2090.

Collins, S. D., Ran, N. A., Heiber, M. C. & Nguyen, T. Q. 2017. Small is powerful: recent progress in solution-processed small molecule solar cells. *Adv. Energy Mater.*, 7, 1602242.

Congreve, D. N., Lee, J., Thompson, N. J., Hontz, E., Yost, S. R., Reusswig, P. D., Bahlke, M. E., Reineke, S., Van Voorhis, T. & Baldo, M. A. 2013. External quantum efficiency above 100% in a singlet-exciton-fission-based organic photovoltaic cell. *Science*, 340, 334.

Corazza, M., Krebs, F. C. & Gevorgyan, S. A. 2014. Predicting, categorizing and intercomparing the lifetime of OPVs for different ageing tests. *Solar Energy Mater. Solar Cells*, 130, 99.

Coropceanu, V., Cornil, J., Filho, D. A. d. S., Olivier, Y., Silbey, R. & Brédas, J.-L. 2007. Charge transport in organic semiconductors. *Chem. Rev.*, 107, 926.

Crandall, R. S. 1983. Modeling of thin film solar cells: uniform field approximation. *J. Appl. Phys.*, 54, 7176.

Cros, S., Firon, M., Lenfant, S., Trouslard, P. & Beck, L. 2006. Study of thin calcium electrode degradation by ion beam analysis. *Nucl. Instrum. Methods Phys. Res. B*, 251, 257.

Cui, Y., Xu, B., Yang, B., Yao, H., Li, S. & Hou, J. 2016. A novel pH neutral self-doped polymer for anode interfacial layer in efficient polymer solar cells. *Macromolecules*, 49, 8126.

Cui, Y., Yao, H., Gao, B., Qin, Y., Zhang, S., Yang, B., He, C., Xu, B. & Hou, J. 2017. Fine-tuned photoactive and interconnection layers for achieving over 13% efficiency in a fullerene-free tandem organic solar cell. *J. Am. Chem. Soc.*, 139, 7302.

Currie, M. J., Mapel, J. K., Heidel, T. D., Goffri, S. & Baldo, M. A. 2008. High-efficiency organic solar concentrators for photovoltaics. *Science*, 321, 226.

da Silva Sobrinho, A. S., Czeremuszkin, G., Latrèche, M. & Wertheimer, M. R. 2000. Defect-permeation correlation for ultrathin transparent barrier coatings on polymers. *J. Vacuum Sci. Technol. A*, 18, 149.

Dai, S., Li, T., Wang, W., Xiao, Y., Lau, T. K., Li, Z., Liu, K., Lu, X. & Zhan, X. 2018. Enhancing the performance of polymer solar cells via core engineering of nir-absorbing electron acceptors. *Adv. Mater.*, 30, 1706571.

Dalgleish, S., Matsushita, M. M., Hu, L., Li, B., Yoshikawa, H. & Awaga, K. 2012. Utilizing photocurrent transients for dithiolene-based photodetection: stepwise improvements at communications relevant wavelengths. *J. Am. Chem. Soc.*, 134, 12742.

Dasgupta, B., Goh, W. P., Z. E. Ooi, Wong, L. M., Jiang, C. Y., Ren, Y., Tok, E. S., Pan, J., Zhang, J. & Chiam, S. Y. 2013. Enhanced extraction rates through gap states of molybdenum oxide anode buffer. *J. Phys. Chem. C*, 117, 9206.

Dennler, G., Lungenschmied, C., Neugebauer, H., Sariciftci, N. S., Latrèche, M., Czeremuszkin, G. & Wertheimer, M. R. 2006. A new encapsulation solution for flexible organic solar cells. *Thin Solid Films*, 511, 349.

Dexter, D. L. 1979. Two ideas on energy transfer phenomena: Ion-pair effects involving the OH stretching mode, and sensitization of photovoltaic cells. *J. Lumin.*, 18–19, 779.

Di Carlo Rasi, D., Hendriks, K. H., Wienk, M. M. & Janssen, R. A. 2017. Accurate characterization of triple-junction polymer solar cells. *Adv. Energy Mater.*, 7, 1701664.

Ding, K., Liu, X. & Forrest, S. R. 2018. Charge transfer and collection in dilute organic donor–acceptor heterojunction blends. *Nano Lett.*, 18, 3180.

Dong, H., Zhu, H., Gong, X. & Hu, W. 2012. Organic photoresponse materials and devices. *Chem. Soc. Rev.*, 41, 1754.

Dou, J.-H., Yu, Z.-A., Zhang, J., Zheng, Y.-Q., Yao, Z.-F., Tu, Z., Wang, X., Huang, S., Liu, C. & Sun, J. 2019. Organic semiconducting alloys with tunable energy levels. *J. Am. Chem. Soc.*, 141, 6561.

Dou, L., Gao, J., Richard, E., You, J., Chen, C.-C., Cha, K. C., He, Y., Li, G. & Yang, Y. 2012. Systematic investigation of benzodithiophene-and diketopyrrolopyrrole-based low-bandgap polymers designed for single junction and tandem polymer solar cells. *J. Am. Chem. Soc.*, 134, 10071.

Dressel, M. & Grüner, G. 2002. *Electrodynamics of Solids: Optical Properties of Electrons in Matter*. Cambridge: Cambridge University Press.

Duncan, T. V., Susumu, K., Sinks, L. E. & Therien, M. J. 2006. Exceptional near-infrared fluorescence quantum yields and excited-state absorptivity of highly conjugated porphyrin arrays. *J. Am. Chem. Soc.*, 128, 9000.

Ehrler, B., Wilson, M. W. B., Rao, A., Friend, R. H. & Greenham, N. C. 2012. Singlet exciton fission-sensitized infrared quantum dot solar cells. *Nano Lett.*, 12, 1053.

Einzinger, M., Wu, T., Kompalla, J. F., Smith, H. L., Perkinson, C. F., Nienhaus, L., Wieghold, S., Congreve, D. N., Kahn, A., Bawendi, M. G. & Baldo, M. A. 2019. Sensitization of silicon by singlet exciton fission in tetracene. *Nature*, 571, 90.

Eisner, F. D., Azzouzi, M., Fei, Z., Hou, X., Anthopoulos, T. D., Dennis, T. J. S., Heeney, M. & Nelson, J. 2019. Hybridization of local exciton and charge-transfer states reduces nonradiative voltage losses in organic solar cells. *J. Am. Chem. Soc.*, 141, 6362.

Emery, K. A. & Osterwald, C. R. 1986. Solar cell efficiency measurements. *Solar Cells*, 17, 253.

Endres, J., Pelczer, I., Rand, B. P. & Kahn, A. 2016. Determination of energy level alignment within an energy cascade organic solar cell. *Chem. Mater.*, 28, 794.

Engler, E. M., Scott, B. A., Etemad, S., Penney, T. & Patel, V. V. 1977. Organic alloys: synthesis and properties of solid solutions of tetraselenafulvalene-tetracyano-p-quinodimethane (TSeF-TCNQ) and tetrathiafulvalene-tetracyano-p-quinodimethane (TTF-TCNQ). *J. Am. Chem. Soc.*, 99, 5909.

Esiner, S., Bus, T., Wienk Martijn, M., Hermans, K. & Janssen René, A. J. 2013. Quantification and validation of the efficiency enhancement reached by application of a retroreflective light trapping texture on a polymer solar cell. *Adv. Energy Mater.*, 3, 1013.

Fei, Z., Eisner Flurin, D., Jiao, X., Azzouzi, M., Röhr Jason, A., Han, Y., Shahid, M., Chesman Anthony, S. R., Easton Christopher, D., McNeill Christopher, R., Anthopoulos Thomas, D., Nelson, J. & Heeney, M. 2018. An alkylated indacenodithieno[3,2-b]thiophene-based nonfullerene acceptor with high crystallinity exhibiting single junction solar cell efficiencies greater than 13% with low voltage losses. *Adv. Mater.*, 30, 1705209.

Filiński, I. 1972. The effects of sample imperfections on optical spectra. *Phys. Stat. Sol. (b)*, 49, 577.

Fishchuck, I. I., Hertel, D., Bässler, H. & Kadashchuk, A. K. 2002. Effective-medium theory of hopping charge-carrier transport in weakly disordered organic solids. *Phys. Rev. B*, 65, 125201.

Fitzner, R., Reinold, E., Mishra, A., Mena-Osteritz, E., Ziehlke, H., Körner, C., Leo, K., Riede, M., Weil, M., Tsaryova, O., Weiß, A., Uhrich, C., Pfeiffer, M. & Bäuerle, P. 2011. Dicyanovinyl-substituted oligothiophenes: structure-property relationships and application in vacuum-processed small molecule organic solar cells. *Adv. Funct. Mater.*, 21, 897.

Frank, J. 1992. *Electron Tomography*. New York: Springer.

Frisch, M. J., Trucks, G. W., Schlegel, H. B., Scuseria, G. E., Robb, M. A., Cheeseman, G., Scalmani, J. R., Barone, V., Mennucci, B., Petersson, G. A., Nakatsuji, H., Caricato, M., Li, X., Hratchian, H. P., Izmaylov, A. F., Bloino, J., Zheng,

G., Sonnenberg, J. L., Hada, M. & Ehara, M. 2009. Gaussian 09. Revision E.01 ed.

Fu, H., Wang, Z. & Sun, Y. 2018. Advances in non-fullerene acceptor based ternary organic solar cells. *Solar RRL*, 2, 1700158.

Fujishima, D., Kanno, H., Kinoshita, T., Maruyama, E., Tanaka, M., Shirakawa, M. & Shibata, K. 2009. Organic thin-film solar cell employing a novel electron-donor material. *Solar Energy Mater. Solar Cells*, 93, 1029.

Gambhir, A., Sandwell, P. & Nelson, J. 2016. The future costs of OPV—a bottom-up model of material and manufacturing costs with uncertainty analysis. *Solar Energy Mater. Solar Cells*, 156, 49.

Gao, G., Liang, N., Geng, H., Jiang, W., Fu, H., Feng, J., Hou, J., Feng, X. & Wang, Z. 2017. Spiro-fused perylene diimide arrays. *J. Am. Chem. Soc.*, 139, 15914.

Gao, J. & Hegmann, F. A. 2008. Bulk photoconductive gain in pentacene thin films. *Appl. Phys. Lett.*, 93, 223306.

Gao, K., Li, L., Lai, T., Xiao, L., Huang, Y., Huang, F., Peng, J., Cao, Y., Liu, F. & Russell, T. P. 2015. Deep absorbing porphyrin small molecule for high-performance organic solar cells with very low energy losses. *J. Am. Chem. Soc.*, 137, 7282.

Gasparini, N., Salvador, M., Strohm, S., Heumueller, T., Levchuk, I., Wadsworth, A., Bannock, J. H., de Mello, J. C., Egelhaaf, H. J. & Baran, D. 2017. Burn-in free nonfullerene-based organic solar cells. *Adv. Energy Mater.*, 7, 1700770.

Geacintov, N., Pope, M. & Vogel, F. 1969. Effect of magnetic field on the fluorescence of tetracene crystals: exciton fission. *Phys. Rev. Lett.*, 22, 593.

Ghezzi, D., Antognazza, M. R., Maccarone, R., Bellani, S., Lanzarini, E., Martino, N., Mete, M., Pertile, G., Bisti, S. & Lanzani, G. 2013. A polymer optoelectronic interface restores light sensitivity in blind rat retinas. *Nat. Photonics*, 7, 400.

Giebink, N. C., Lassiter, B. E., Wiederrecht, G. P., Wasielewski, M. R. & Forrest, S. R. 2010a. The ideal diode equation for organic heterojunctions. II. The role of polaron pair recombination. *Phys. Rev. B*, 82, 155306.

Giebink, N. C., Wiederrecht, G. P., Wasielewski, M. R. & Forrest, S. R. 2010b. The ideal diode equation for organic heterojunctions. I. Derivation and application. *Phys. Rev. B*, 82, 155305.

Giebink, N. C., Wiederrecht, G. P., Wasielewski, M. R. & Forrest, S. R. 2011. The thermodynamic efficiency limit of excitonic solar cells. *Phys. Rev. B*, 83, 195326.

Gilot, J., Wienk, M. M. & Janssen, R. A. 2010a. Optimizing polymer tandem solar cells. *Adv. Mater.*, 22, E67.

Gilot, J., Wienk, M. M. & Janssen, R. A. J. 2007. Double and triple junction polymer solar cells processed from solution. *Appl. Phys. Lett.*, 90, 143512.

Gilot, J., Wienk, M. M. & Janssen, R. A. J. 2010b. Measuring the external quantum efficiency of two-terminal polymer tandem solar cells. *Adv. Funct. Mater.*, 20, 3904.

Gong, X., Tong, M., Xia, Y., Cai, W., Moon, J. S., Cao, Y., Yu, G., Shieh, C.-L., Nilsson, B. & Heeger, A. J. 2009. High-detectivity polymer photodetectors with spectral response from 300 nm to 1450 nm. *Science*, 325, 1665.

Graff, G. L., Williford, R. E. & Burrows, P. E. 2004. Mechanisms of vapor permeation through multilayer barrier films: Lag time versus equilibrium permeation *J. Appl. Phys.*, 96, 1849.

Green, M. A. 2017. Corrigendum to 'Solar cell efficiency tables (version 49)'[Prog. Photovolt: Res. Appl. 2017; 25:3–13]. *Prog. Photovolt.: Res. Appl.*, 25, 333.

Green, M. A., Emery, K., Hishikawa, Y., Warta, W., Dunlop, E. D., Levi, D. H. & Ho-Baillie, A. W. Y. 2017. Solar cell efficiency tables (version 49). *Prog. Photovolt.: Res. Appl.*, 25, 3.

Greiner, M. T., Helander, M. G., Tang, W.-M., Wang, Z.-B., Qiu, J. & Lu, Z.-H. 2012. Universal energy-level alignment of molecules on metal oxides. *Nat. Mater.*, 11, 76.

Griffith, M. J., Cooling, N. A., Vaughan, B., Elkington, D. C., Hart, A. S., Lyons, A. G., Quereshi, S., Belcher, W. J. & Dastoor, P. C. 2016. Combining printing, coating, and vacuum deposition on the roll-to-roll scale: a hybrid organic photovoltaics fabrication. *IEEE J. Sel. Top. Quantum Electron.*, 22, 112.

Griffith, O. L. & Forrest, S. R. 2014. Exciton management in organic photovoltaic multi-donor energy cascades. *Nano Lett.*, 14, 2353.

Griffith, O. L., Liu, X., Amonoo, J. A., Djurovich, P. I., Thompson, M. E., Green, P. F. & Forrest, S. R. 2015. Charge transport and exciton dissociation in organic solar cells consisting of dipolar donors mixed with C_{70}. *Phys. Rev. B*, 92, 085404.

Groff, R., Avakian, P. & Merrifield, R. 1970. Coexistence of exciton fission and fusion in tetracene crystals. *Phys. Rev. B*, 1, 815.

Grossiord, N., Kroon, J. M., Andriessen, R. & Blom, P. W. M. 2012. Degradation mechanisms in organic photovoltaic devices. *Org. Electron.*, 13, 432.

Guinier, A. & Fournet, G. 1955. *Small-Angle Scattering of X rays*. New York: John Wiley & Sons

Guo, X., Cui, C., Zhang, M., Huo, L., Huang, Y., Hou, J. & Li, Y. 2012. High efficiency polymer solar cells based on poly (3-hexylthiophene)/indene-C_{70} bisadduct with solvent additive. *Energy Envrion. Sci.*, 5, 7943.

Guo, Y., Lei, H., Xiong, L., Li, B. & Fang, G. 2017. Organic solar cells based on a Cu_2O/FBT-TH4 anode buffer layer with enhanced power conversion efficiency and ambient stability. *J. Mater. Chem. C*, 5, 8033.

Gupta, V., Lai, L. F., Datt, R., Chand, S., Heeger, A. J., Bazan, G. C. & Singh, S. P. 2016. Dithienogermole-based solution-processed molecular solar cells with efficiency over 9%. *Chem. Commun.*, 52, 8596.

Habte, A., Sengupta, M. & Lopez, A. 2017. Evaluation of the National Solar Radiation Database (NSRDB): 1998–2015. Golden, CO: National Renewable Energy Lab.

Hadipour, A., deBoer, B., Wildeman, J., Kooistra, F. B., Hummelen, J. C., Turbiez, M. G. R., Wienk, M. M., Janssen, R. A. J. & Blom, P. W. M. 2006. Solution-processed organic tandem cells. *Adv. Funct. Mater.*, 16, 1897.

Halls, J. J. M., Walsh, C. A., Greenham, N. C., Marseglia, E. A., Friend, R. H., Moratti, S. C. & Holmes, A. B. 1995. Efficient photodiodes from interpenetrating polymer networks. *Nature*, 376, 498.

Han, S., Shin, W. S., Seo, M., Gupta, D., Moon, S.-J. & Yoo, S. 2009. Improving performance of organic solar cells using amorphous tungsten oxides as an interfacial buffer layer on transparent anodes. *Org. Electron.*, 10, 791.

Hanna, M. C. & Nozik, A. J. 2006. Solar conversion efficiency of photovoltaic and photoelectrolysis cells with carrier multiplication absorbers. *J. Appl. Phys.*, 100, 074510.

Hau, S. K., Yip, H.-L., Chen, K.-S., Zou, J. & Jen, A. K.-Y. 2010. Solution processed inverted tandem polymer solar cells with self-assembled monolayer modified interfacial layers. *Appl. Phys. Lett.*, 97, 253307.

Hau, S. K., Yip, H.-L., Ma, H. & Jen, A. K.-Y. 2008. High performance ambient processed inverted polymer solar cells through interfacial modification with a fullerene self-assembled monolayer. *Appl. Phys. Lett.*, 93, 233304.

He, Z., Zhong, C., Su, S., Xu, M., Wu, H. & Cao, Y. 2012. Enhanced power-conversion efficiency in polymer solar cells using an inverted device structure. *Nat. Photonics*, 6, 591.

Heavens, O. S. 1991. *Optical Properties of Thin Solid Films*. New York: Dover.

Heidel, T. D., Hochbaum, D., Sussman, J. M., Singh, V., Bahlke, M. E., Hiromi, I., Lee, J. & Baldo, M. A. 2011. Reducing recombination losses in planar organic photovoltaic cells using multiple step charge separation. *J. Appl. Phys.*, 109, 104502.

Helgesen, M., Carlé, J. E. & Krebs, F. C. 2013. Slot-die coating of a high performance copolymer in a readily scalable roll process for polymer solar cells. *Adv. Energy Mater.*, 3, 1664.

Hermann, A. M. 1982. Luminescent solar concentrators—a review. *Solar Energy*, 29, 323.

Hermenau, M., Schubert, S., Klumbies, H., Fahlteich, J., Müller-Meskamp, L., Leo, K. & Riede, M. 2012. The effect of barrier performance on the lifetime of small-molecule organic solar cells. *Solar Energy Mater. Solar Cells*, 97, 102.

Hines, M. A. & Scholes, G. D. 2003. Colloidal PbS nanocrystals with size-tunable near-infrared emission: observation of post-synthesis self-narrowing of the particle size distribution. *Adv. Mater.*, 15, 1844.

Hirade, M. & Adachi, C. 2011. Small molecular organic photovoltaic cells with exciton blocking layer at anode interface for improved device performance. *Appl. Phys. Lett.*, 99, 153302.

Hiramoto, M., Suezaki, M. & Yokoyama, M. 1990. Effect of thin gold interstitial layer on the photovoltaic properties of tandem organic solar cell. *Chem. Lett.*, 19, 327.

Hokuto, S., Satoshi, A., Toshihisa, W., Hiroshi, O., Misao, K. & Norifumi, E. 2007. Color sensors with three vertically stacked organic photodetectors. *Jpn. J. Appl. Phys.*, 46, L1240.

Hokuto, S., Satoshi, A., Toshihisa, W., Hiroshi, O., Toshikatsu, S., Misao, K., Norifumi, E., Takahiro, H., Tokiyoshi, M., Mamoru, F. & Takashi, H. 2011. A 128×96 pixel stack-type color image sensor: stack of individual blue-, green-, and red-sensitive organic photoconductive films integrated with a ZnO thin film transistor readout circuit. *Jpn. J. Appl. Phys.*, 50, 024103.

Honda, S., Yokoya, S., Ohkita, H., Benten, H. & Ito, S. 2011. Light-harvesting mechanism in polymer/fullerene/dye ternary blends studied by transient absorption spectroscopy. *J. Phys. Chem. C*, 115, 11306.

Hong, Z. R., Lessmann, R., Maennig, B., Huang, Q., Harada, K., Riede, M. & Leo, K. 2009. Antenna effects and improved efficiency in multiple heterojunction photovoltaic cells based on pentacene, zinc phthalocyanine, and C_{60}. *J. Appl. Phys.*, 106, 064511.

Hoppe, H., Seeland, M. & Muhsin, B. 2012. Optimal geometric design of monolithic thin-film solar modules: Architecture of polymer solar cells. *Solar Energy Mater. Solar Cells*, 97, 119.

Hösel, M., Angmo, D., Søndergaard, R. R., dos Reis Benatto, G. A., Carlé, J. E., Jørgensen, M. & Krebs, F. C. 2014. High-volume processed, ITO-free superstrates and substrates for roll-to-roll development of organic electronics. *Adv. Sci.*, 1, 1400002.

Hou, J., Inganäs, O., Friend, R. H. & Gao, F. 2018. Organic solar cells based on non-fullerene acceptors. *Nat. Mater.*, 17, 119.

Hu, L., Hecht, D. & Grüner, G. 2004. Percolation in transparent and conducting carbon nanotube networks. *Nano Lett.*, 4, 2513.

Hu, L., Kim, H. S., Lee, J.-Y., Peumans, P. & Cui, Y. 2010. Scalable coating and properties of transparent, flexible, silver nanowire electrodes. *ACS Nano*, 4, 2955.

Huang, J., Wang, H., Yan, K., Zhang, X., Chen, H., Li, C. Z. & Yu, J. 2017. Highly efficient organic solar cells consisting of double bulk heterojunction layers. *Adv. Mater.*, 29, 1606729.

Huang, J.-S., Goh, T., Li, X., Sfeir, M. Y., Bielinski, E. A., Tomasulo, S., Lee, M. L., Hazari, N. & Taylor, A. D. 2013. Polymer bulk heterojunction solar cells employing Förster resonance energy transfer. *Nat. Photonics*, 7, 479.

Huang, X., Liu, X., Ding, K. and Forrest, S. R. 2020. Is there such a thing as a molecular organic alloy?. *Mater. Horiz.* 7, 244.

Huang, Y., Guo, X., Liu, F., Huo, L., Chen, Y., Russell, T. P., Han, C. C., Li, Y. & Hou, J. 2012. Improving the ordering and photovoltaic properties by extending π-conjugated area of electron-donating units in polymers with D-A structure. *Adv. Mater.*, 24, 3383.

Huo, L., Zhang, S., Guo, X., Xu, F., Li, Y. & Hou, J. 2011. Replacing alkoxy groups with alkylthienyl groups: a feasible approach to improve the properties of photovoltaic polymers. *Angew. Chem. Int. Ed.*, 50, 9697.

Inganäs, O., Svensson, M., Zhang, F., Gadisa, A., Persson, N. K., Wang, X. & Andersson, M. R. 2004. Low bandgap alternating polyfluorene copolymers in plastic photodiodes and solar cells. *Appl. Phys. A*, 79, 31.

Ito, S., Nagami, T. & Nakano, M. 2018. Molecular design for efficient singlet fission. *J. Photochem. Photobiol. C*, 34, 85.

Jadhav, P. J., Mohanty, A., Sussman, J., Lee, J. & Baldo, M. A. 2011. Singlet exciton fission in nanostructured organic solar cells. *Nano Lett.*, 11, 1495.

Janesick, J. R., Elliott, T., Collins, S., Blouke, M. M. & Freeman, J. 1987. Scientific charge-coupled-devices. *Opt. Eng.*, 26, 692.

Jansen-van Vuuren, R. D., Armin, A., Pandey, A. K., Burn, P. L. & Meredith, P. 2016. Organic photodiodes: the future of full color detection and image sensing. *Adv. Mater.*, 28, 4766.

Janssen, A. G. F., Riedl, T., Hamwi, S., Johannes, H. H. & Kowalsky, W. 2007. Highly efficient organic tandem solar cells using an improved connecting architecture. *Appl. Phys. Lett.*, 91, 073519.

Jeong, W. I., Lee, Y. E., Shim, H. S., Kim, T. M., Kim, S. Y. & Kim, J. J. 2012. Photoconductivity of C_{60} as an origin of bias-dependent photocurrent in organic photovoltaics. *Adv. Funct. Mater.*, 22, 3089.

Jinnai, H., Nishikawa, Y., Ikehara, T. & Nishi, T. 2004. Emerging technologies for the 3D analysis of polymer structures. In: Fatkullin, N., Ikehara, T., Jinnai, H., Kawata, S., Kimmich, R., Nishi, T., Nishikawa, Y. & Sun, H. B. (eds.) *NMR • 3D Analysis • Photopolymerization*. Berlin: Springer.

Johnson, R. C. & Merrifield, R. E. 1970. Effects of magnetic fields on the mutual annihilation of triplet excitons in anthracene crystals. *Phys. Rev. B*, 1, 896.

Jørgensen, M., Norrman, K., Gevorgyan, S. A., Tromholt, T., Andreasen, B. & Krebs, F. C. 2012. Stability of polymer solar cells. *Adv. Mater.*, 24, 580.

Kabra, D., Singh, T. B. & Narayan, K. S. 2004. Semiconducting-polymer-based position-sensitive detectors. *Appl. Phys. Lett.*, 85, 5073.

Kan, B., Li, M., Zhang, Q., Liu, F., Wan, X., Wang, Y., Ni, W., Long, G., Yang, X., Feng, H., Zuo, Y., Zhang, M., Huang, F., Cao, Y., Russell, T. P. & Chen, Y. 2015. A series of simple oligomer-like small molecules based on oligothiophenes for solution-processed solar cells with high efficiency. *J. Am. Chem. Soc.*, 137, 3886.

Kan, B., Zhang, Q., Li, M., Wan, X., Ni, W., Long, G., Wang, Y., Yang, X., Feng, H. & Chen, Y. 2014. Solution-processed organic solar cells based on dialkylthiol-substituted benzodithiophene unit with efficiency near 10%. *J. Am. Chem. Soc.*, 136, 15529.

Kanai, K., Koizumi, K., Ouchi, S., Tsukamoto, Y., Sakanoue, K., Ouchi, Y. & Seki, K. 2010. Electronic structure of anode interface with molybdenum oxide buffer layer. *Org. Electron.*, 11, 188.

Kang, I. & Wise, F. W. 1997. Electronic structure and optical properties of PbS and PbSe quantum dots. *J. Opt. Soc. Am. B*, 14, 1632.

Karl, N., Marktanner, J., Stehle, R. & Warta, W. 1991. High field saturation of charge carrier drift velocities in ultrapurified organic photoconductors. *Synth. Met.*, 41–43, 2473.

Kästner, C., Vandewal, K., Egbe, D. A. M. & Hoppe, H. 2017. Revelation of interfacial energetics in organic multiheterojunctions. *Adv. Sci.*, 4, 1600331.

Kawasaki, K. 1972. Kinetics of Ising models. In: Domb, C. & Green, M. S. (eds.) *Phase Transitions and Critical Phenomena*. New York: Academic.

Kim, J. B., Kim, P., Pégard, N. C., Oh, S. J., Kagan, C. R., Fleischer, J. W., Stone, H. A. & Loo, Y.-L. 2012. Wrinkles and deep folds as photonic structures in photovoltaics. *Nat. Photonics*, 6, 327.

Kim, K. K., Alfonso, R., Yumeng, S., Hyesung, P., Li, L.-J., Lee, Y. H. & Kong, J. 2010. Enhancing the conductivity of transparent graphene films via doping. *Nanotechnology*, 21, 285205.

Kim, M.-S., Kang, M.-G., Guo, L. J. & Kim, J. 2008. Choice of electrode geometry for accurate measurement of organic photovoltaic cell performance. *Appl. Phys. Lett.*, 92, 114.

Kim, S. J., Margulis, G. Y., Rim, S.-B., Brongersma, M. L., McGehee, M. D. & Peumans, P. 2013. Geometric light trapping with a V-trap for efficient organic solar cells. *Opt. Express*, 21, A305.

Kim, Y. H., Sachse, C., Machala Michael, L., May, C., Müller-Meskamp, L. & Leo, K. 2011. Highly conductive PEDOT: PSS electrode with optimized solvent and thermal post-treatment for ITO-free organic solar cells. *Adv. Funct. Mater.*, 21, 1076.

Kirchartz, T., Taretto, K. & Rau, U. 2009. Efficiency limits of organic bulk heterojunction solar cells. *J. Phys. Chem. C*, 113, 17958.

Klumbies, H., Karl, M., Hermenau, M., Rösch, R., Seeland, M., Hoppe, H., Müller-Meskamp, L. & Leo, K. 2014. Water ingress into and climate dependent lifetime of organic photovoltaic cells investigated by calcium corrosion tests. *Solar Energy Mater. Solar Cells*, 120, 685.

Koeppe, R., Hoeglinger, D., Troshin Pavel, A., Lyubovskaya Rimma, N., Razumov Vladimir, F. & Sariciftci, N. S. 2009. Organic solar cells with semitransparent metal back contacts for power window applications. *ChemSusChem*, 2, 309.

Koppe, M., Egelhaaf, H. J., Dennler, G., Scharber, M. C., Brabec, C. J., Schilinsky, P. & Hoth, C. N. 2010. Near IR sensitization of organic bulk heterojunction solar cells: towards optimization of the spectral response of organic solar cells. *Adv. Funct. Mater.*, 20, 338.

Kouijzer, S., Esiner, S., Frijters, C. H., Turbiez, M., Wienk, M. M. & Janssen, R. A. J. 2012. Efficient inverted tandem polymer solar cells with a solution-processed recombination layer. *Adv. Energy Mater.*, 2, 945.

Krebs, F. C. 2009. Fabrication and processing of polymer solar cells: a review of printing and coating techniques. *Solar Energy Mater. Solar Cells*, 93, 394.

Kuhnke, K., Becker, R., Epple, M. & Kern, K. 1997. C_{60} exciton quenching near metal surfaces. *Phys. Rev. Lett.*, 79, 3246.

Kumar, A., Devine, R., Mayberry, C., Lei, B., Li, G. & Yang, Y. 2010. Origin of radiation-induced degradation in polymer solar cells. *Adv. Funct. Mater.*, 20, 2729.

Kumar, A., Rosen, N., Devine, R. & Yang, Y. 2011. Interface design to improve stability of polymer solar cells for potential space applications. *Energy Envrion. Sci.*, 4, 4917.

Lassiter, B. E., Kyle Renshaw, C. & Forrest, S. R. 2013a. Understanding tandem organic photovoltaic cell performance. *J. Appl. Phys.*, 113, 214505.

Lassiter, B. E., Wei, G., Wang, S., Zimmerman, J. D., Diev, V. V., Thompson, M. E. & Forrest, S. R. 2011. Organic photovoltaics incorporating electron conducting exciton blocking layers. *Appl. Phys. Lett.*, 98, 112.

Lassiter, B. E., Zimmerman, J. D. & Forrest, S. R. 2013b. Tandem organic photovoltaics incorporating two solution-processed small molecule donor layers. *J. Appl. Phys.*, 103, 123305.

Law, K. Y. 1993. Organic photoconductive materials: recent trends and developments. *Chem. Rev.*, 93, 449.

Lebedkin, S., Gromov, A., Giesa, S., Gleiter, R., Renker, B., Rietschel, H. & Krätschmer, W. 1998. Raman scattering study of C_{120}, a C_{60} dimer. *Chem. Phys. Lett.*, 285, 210.

Lee, J.-Y., Connor, S. T., Cui, Y. & Peumans, P. 2008. Solution-processed metal nanowire mesh transparent electrodes. *Nano Lett*, 8, 689.

Lee, K., Kim, J. Y., Park, S. H., Kim, S. H., Cho, S. & Heeger, A. J. 2007. Air-stable polymer electronic devices. *Adv. Mater.*, 19, 2445.

Lee, K., Lee, J., Mazor, B. & Forrest, S. R. 2015. Transforming the cost of solar-to-electrical energy conversion: Integrating thin-film GaAs solar cells with non-tracking mini-concentrators. *Light: Sci. Appl.*, 4, e288.

Lewis, G. N. & Randall, M. 1961. *Thermodynamics*. New York: McGraw-Hill.

Li, C.-Z., Chueh, C.-C., Yip, H.-L., O'Malley, K. M., Chen, W.-C. & Jen, A. K.-Y. 2012. Effective interfacial layer to enhance efficiency of polymer solar cells via solution-processed fullerene-surfactants. *J. Mater. Chem.*, 22, 8574.

Li, G., Shrotriya, V., Huang, J., Yao, Y., Moriarty, T., Emery, K. & Yang, Y. 2005. High-efficiency solution processable polymer photovoltaic cells by self-organization of polymer blends. *Nat. Mater.*, 4, 864.

Li, G. & Yang, Y. 2008. Radiation induced damage and recovery in poly(3-hexylthiophene) based polymer solar cells. *Nanotechnology*, 19, 424014.

Li, H.-G., Wu, G., Chen, H.-Z. & Wang, M. 2011. Spectral response tuning and realization of quasi-solar-blind detection in organic ultraviolet photodetectors. *Org. Electron.*, 12, 70.

Li, M., Liu, F., Wan, X., Ni, W., Kan, B., Feng, H., Zhang, Q., Yang, X., Wang, Y. & Zhang, Y. 2015. Subtle balance between length scale of phase separation and domain purification in small-molecule bulk-heterojunction blends under solvent vapor treatment. *Adv. Mater.*, 27, 6296.

Li, N., Kubis, P., Forberich, K., Ameri, T., Krebs, F. C. & Brabec, C. J. 2014. Towards large-scale production of solution-processed organic tandem modules based on ternary composites: design of the intermediate layer, device optimization and laser based module processing. *Solar Energy Mater. Solar Cells*, 120, 701.

Li, N., Perea, J. D., Kassar, T., Richter, M., Heumueller, T., Matt, G. J., Hou, Y., Güldal, N. S., Chen, H. & Chen, S. 2017a. Abnormal strong burn-in degradation of highly efficient polymer solar cells caused by spinodal donor–acceptor demixing. *Nat. Commun.*, 8, 14541.

Li, W., Furlan, A., Hendriks, K. H., Wienk, M. M. & Janssen, R. A. J. 2013. Efficient tandem and triple-junction polymer solar cells. *J. Am. Chem. Soc.*, 135, 5529.

Li, W., Yao, H., Zhang, H., Li, S. & Hou, J. 2017b. Potential of nonfullerene small molecules with high photovoltaic performance. *Chem. Asian J.*, 12, 2160.

Li, Y., Ji, C., Qu, Y., Huang, X., Hou, S., Li, C.-Z., Liao, L.-S., Guo, L. J. & Forrest, S. R. High efficiency semi-transparent organic photovoltaics. 46th IEEE PVSC, June 16–21, 2019, Chicago, IL. Paper 484.

Li, Y., Lin, J.-D., Che, X., Qu, Y., Liu, F., Liao, L.-S. & Forrest, S. R. 2017c. High efficiency near-infrared and semitransparent non-fullerene acceptor organic photovoltaic cells. *J. Am. Chem. Soc.*, 139, 17114.

Li, Y., Lin, J. D., Liu, X., Qu, Y., Wu, F. P., Liu, F., Jiang, Z. Q. & Forrest, S. R. 2018. Near-infrared ternary tandem solar cells. *Adv. Mater.*, 30, 1804416.

Li, Y., Liu, X., Wu, F.-P., Zhou, Y., Jiang, Z.-Q., Song, B., Xia, Y., Zhang, Z.-G., Gao, F. & Inganäs, O. 2016. Non-fullerene acceptor with low energy loss and high external quantum efficiency: towards high performance polymer solar cells. *J. Mater. Chem. A*, 4, 5890.

Li, Y., Zhong, L., Gautam, B., Bin, H.-J., Lin, J.-D., Wu, F.-P., Zhang, Z., Jiang, Z.-Q., Zhang, Z.-G. & Gundogdu, K. 2017d. A near-infrared non-fullerene electron acceptor for high performance polymer solar cells. *Energy Envrion. Sci.*, 10, 1610.

Li, Y., Zhong, L., Lin, J. D., Wu, F. P., Bin, H. J., Zhang, Z., Xu, L., Jiang, Z. Q., Zhang, Z. G., Liu, F. & Russell Thomas, P. 2017e. Isomeric effects of solution processed ladder-type non-fullerene electron acceptors. *Solar RRL*, 1, 1700107.

Li, Y., Sheriff Jr, H.K., Liu, X., Wang, C.K., Ding, K., Han, H., Wong, K.T. & Forrest, S.R. 2019. Vacuum-deposited biternary organic photovoltaics. *J. Am. Chem. Soc.*, 141, 18204.

Liao, S. H., Jhuo, H. J., Cheng, Y. S. & Chen, S. A. 2013. Fullerene derivative-doped zinc oxide nanofilm as the cathode of inverted polymer solar cells with low-bandgap polymer (PTB7-Th) for high performance. *Adv. Mater.*, 25, 4766.

Lim, S.-J., Leem, D.-S., Park, K.-B., Kim, K.-S., Sul, S., Na, K., Lee, G. H., Heo, C.-J., Lee, K.-H. & Bulliard, X. 2015. Organic-on-silicon complementary metal–oxide–semiconductor colour image sensors. *Sci. Rep.*, 5, 7708.

Limpinsel, M., Wagenpfahl, A., Mingebach, M., Deibel, C. & Dyakonov, V. 2010. Photocurrent in bulk heterojunction solar cells. *Phys. Rev. B*, 81, 085203.

Lin, Y., Wang, J., Zhang, Z. G., Bai, H., Li, Y., Zhu, D. & Zhan, X. 2015. An electron acceptor challenging fullerenes for efficient polymer solar cells. *Adv. Mater.*, 27, 1170.

Lipomi, D. J. & Bao, Z. 2011. Stretchable, elastic materials and devices for solar energy conversion. *Energy Envrion. Sci.*, 4, 3314.

Liu, F., Gu, Y., Wang, C., Zhao, W., Chen, D., Briseno Alejandro, L. & Russell Thomas, P. 2012. Efficient polymer solar cells based on a low bandgap semi-crystalline DPP polymer-PCBM blends. *Adv. Mater.*, 24, 3947.

Liu, J., Chen, S., Qian, D., Gautam, B., Yang, G., Zhao, J., Bergqvist, J., Zhang, F., Ma, W. & Ade, H. 2016a. Fast charge separation in a non-fullerene organic solar cell with a small driving force. *Nat. Energy*, 1, 16089.

Liu, T., Guo, Y., Yi, Y., Huo, L., Xue, X., Sun, X., Fu, H., Xiong, W., Meng, D., Wang, Z., Liu, F., Russell Thomas, P. & Sun, Y. 2016b. Ternary organic solar cells based on two compatible nonfullerene acceptors with power conversion efficiency >10%. *Adv. Mater.*, 28, 10008.

Liu, T., Xue, X., Huo, L., Sun, X., An, Q., Zhang, F., Russell, T. P., Liu, F. & Sun, Y. 2017. Highly efficient parallel-like ternary organic solar cells. *Chem. Mater.*, 29, 2914.

Liu, X., Ding, K., Panda, A. & Forrest, S. R. 2016c. Charge transfer states in dilute donor–acceptor blend organic heterojunctions. *ACS Nano*, 10, 7619.

Liu, X., Li, Y., Ding, K. & Forrest, S. R. 2019. Energy loss in organic photovoltaics: Nonfullerene versus fullerene acceptors. *Phys. Rev. Appl.*, 11, 024060.

Liu, Z., Li, J., Sun, Z.-H., Tai, G., Lau, S.-P. & Yan, F. 2011. The application of highly doped single-layer graphene as the top electrodes of semitransparent organic solar cells. *ACS Nano*, 6, 810.

Loi, M. A., Toffanin, S., Muccini, M., Forster, M., Scherf, U. & Scharber, M. 2007. Charge transfer excitons in bulk heterojunctions of a polyfluorene copolymer and a fullerene derivative. *Adv. Funct. Mater.*, 17, 2111.

Lou, S. J., Szarko, J. M., Xu, T., Yu, L., Marks, T. J. & Chen, L. X. 2011. Effects of additives on the morphology of solution phase aggregates formed by active layer components of high-efficiency organic solar cells. *J. Am. Chem. Soc.*, 133, 20661.

Lunt, R. R. & Bulovic, V. 2011. Transparent, near-infrared organic photovoltaic solar cells for window and energy-scavenging applications. *Appl. Phys. Lett.*, 98, 113305.

Lunt, R. R., Giebink, N. C., Belak, A. A., Benziger, J. B. & Forrest, S. R. 2009. Exciton diffusion lengths of organic semiconductor thin films measured by spectrally resolved photoluminescence quenching. *J. Appl. Phys.*, 105, 053711.

Luque, A. & Marti, A. 1997. Entropy production in photovoltaic conversion. *Phys. Rev. B*, 55, 6994.

Luque, A. & Marti, A. 2003. Theoretical limits of photovoltaic conversion. In: Luque, A. & Hegedus, S. (eds.) *Handbook of Photovoltaic Science and Engineering*. Chichester, UK: Wiley.

Luque, A., Marti, A. & Cuadra, L. 2002. Thermodynamics of solar energy conversion in novel structures. *Physica E*, 14, 107.

Manceau, M., Bundgaard, E., Carlé, J. E., Hagemann, O., Helgesen, M., Søndergaard, R., Jørgensen, M. & Krebs, F. C. 2011. Photochemical stability of π-conjugated polymers for polymer solar cells: a rule of thumb. *J. Mater. Chem.*, 21, 4132.

Marcus, R. A. & Sutin, N. 1985. Electron transfers in chemistry and biology. *Biochim. Biophys. Acta Rev.Bioenergetics*, 811, 265.

Mateker, W. R. & McGehee, M. D. 2017. Progress in understanding degradation mechanisms and improving stability in organic photovoltaics. *Adv. Mater.*, 29, 1603940.

Mateker, W. R., Sachs-Quintana, I. T., Burkhard, G. F., Cheacharoen, R. & McGehee, M. D. 2015. Minimal long-term intrinsic degradation observed in a polymer solar cell illuminated in an oxygen-free environment. *Chem. Mater.*, 27, 404.

Matsuo, Y., Sato, Y., Niinomi, T., Soga, I., Tanaka, H. & Nakamura, E. 2009. Columnar structure in bulk heterojunction in solution-processable three-layered pin organic photovoltaic devices using tetrabenzoporphyrin precursor and silylmethyl [60] fullerene. *J. Am. Chem. Soc.*, 131, 16048.

Maya-Vetencourt, J. F., Ghezzi, D., Antognazza, M. R., Colombo, E., Mete, M., Feyen, P., Desii, A., Buschiazzo, A., Di Paolo, M., Di Marco, S., Ticconi, F., Emionite, L., Shmal, D., Marini, C., Donelli, I., Freddi, G., Maccarone, R., Bisti, S., Sambuceti, G., Pertile, G., Lanzani, G. & Benfenati, F. 2017. A fully organic retinal prosthesis restores vision in a rat model of degenerative blindness. *Nat. Mater.*, 16, 681.

Méndez-Pinzón, H. A., Pardo-Pardo, D. R., Cuéllar-Alvarado, J. P., Salcedo-Reyes, J. C., Vera, R. & Páez-Sierra, B. A. 2010. Analysis of the current-voltage characteristics of polymer-based organic light-emitting diodes (OLEDs) deposited by spin coating. *Univ. Sci.*, 15, 68.

Meng, D., Fu, H., Xiao, C., Meng, X., Winands, T., Ma, W., Wei, W., Fan, B., Huo, L., Doltsinis, N. L., Li, Y., Sun, Y. & Wang, Z. 2016. Three-bladed rylene propellers with three-dimensional network assembly for organic electronics. *J. Am. Chem. Soc.*, 138, 10184.

Meng, D., Sun, D., Zhong, C., Liu, T., Fan, B., Huo, L., Li, Y., Jiang, W., Choi, H., Kim, T., Kim, J. Y., Sun, Y., Wang, Z. & Heeger, A. J. 2015. High-performance solution-processed non-fullerene organic solar cells based on selenophene-containing perylene bisimide acceptor. *J. Am. Chem. Soc.*, 138, 375.

Menke, S. M., Ran, N. A., Bazan, G. C. & Friend, R. H. 2018. Understanding energy loss in organic solar cells: toward a new efficiency regime. *Joule*, 2, 25.

Merrifield, R., Avakian, P. & Groff, R. 1969. Fission of singlet excitons into pairs of triplet excitons in tetracene crystals. *Chem. Phys. Lett.*, 3, 386.

Mihailetchi, V., Koster, L., Hummelen, J. & Blom, P. 2004. Photocurrent generation in polymer-fullerene bulk heterojunctions. *Phys. Rev. Lett.*, 93, 216601.

Miller, O. D., Yablonovitch, E. & Kurtz, S. R. 2012. Strong internal and external luminescence as solar cells approach the Shockley–Queisser limit. *IEEE J. Photovolt.*, 2, 303.

Mogeritsch, J., Ludwig, A., Eck, S., Grasser, M. & McKay, B. 2009. Thermal stability of a binary non-faceted/non-faceted peritectic organic alloy at elevated temperatures. *Scripta Mater.*, 60, 882.

Möller, S. & Forrest, S. R. 2001. Improved light outcoupling in organic light emitting devices employing ordered microlens arrays. *J. Appl. Phys.*, 91, 3324.

Morel, D. L., Ghosh, A. K., Feng, T., Stogryn, E. L., Purwin, P. E., Shaw, R. F. & Fishman, C. 1978. High-efficiency organic solar cells. *Appl. Phys. Lett.*, 32, 495.

Mori, D., Benten, H., Kosaka, J., Ohkita, H., Ito, S. & Miyake, K. 2011. Polymer/polymer blend solar cells with 2.0% efficiency developed by thermal purification of nanoscale-phase-separated morphology. *ACS Appl. Mater. Interfaces*, 3, 2924.

Mori, D., Benten, H., Ohkita, H. & Ito, S. 2015. Morphology-limited free carrier generation in donor/acceptor polymer blend solar cells composed of poly(3-hexylthiophene) and fluorene-based copolymer. *Adv. Energy Mater.*, 5, 1500304.

Moriarty, T., Jablonski, J. & Emery, K. Algorithm for building a spectrum for NREL's one-sun multi-source simulator. 38th IEEE Photovoltaic Specialists Conference (PVSC), 2012, 001291.

Morimune, T., Kajii, H. & Ohmori, Y. 2006. Photoresponse properties of a high-speed organic photodetector based on copper–phthalocyanine under red light illumination. *IEEE Photonics Technol. Lett.*, 18, 2662.

Mulder, C. L., Reusswig, P. D., Velázquez, A. M., Kim, H., Rotschild, C. & Baldo, M. A. 2010. Dye alignment in luminescent solar concentrators. I. Vertical alignment for improved waveguide coupling. *Opt. Express*, 18, A79.

Mulligan, C. J., Bilen, C., Zhou, X., Belcher, W. J. & Dastoor, P. C. 2015. Levelised cost of electricity for organic photovoltaics. *Solar Energy Mater. Solar Cells*, 133, 26.

Murata, T., Shao, X., Nakano, Y., Yamochi, H., Uruichi, M., Yakushi, K., Saito, G. & Tanaka, K. 2010. Tuning of multi-instabilities in organic alloy, $[(EDO\text{-}TTF)_{1-x}(MeEDO\text{-}TTF)_x]$ 2PF6. *Chem. Mater.*, 22, 3121.

Myers, J. D., Cao, W., Cassidy, V., Eom, S.-H., Zhou, R., Yang, L., You, W. & Xue, J. 2012. A universal optical approach to enhancing efficiency of organic-based photovoltaic devices. *Energy Envrion. Sci.*, 5, 6900.

Na, S.-I., Kim, S.-S., Jo, J., Oh, S.-H., Kim, J. & Kim, D.-Y. 2008. Efficient polymer solar cells with surface relief gratings fabricated by simple soft lithography. *Adv. Funct. Mater.*, 18, 3956.

Nelson, R. C. 1954. Organic photoconductors. I. The kinetics of photoconductivity in rhodamine B. *J. Chem. Phys.*, 22, 885.

Niggemann, M., Glatthaar, M., Gombert, A., Hinsch, A. & Wittwer, V. 2004. Diffraction gratings and buried nano-electrodes—architectures for organic solar cells. *Thin Solid Films*, 451–452, 619.

O'Neill, M. J., Piszczor, M. F., Eskenazi, M. I., McDanal, A. J., George, P. J., Botke, M. M., Brandhorst, H. W., Edwards, D. L. & Hoppe, D. T. 2003. Ultra-light stretched Fresnel lens solar concentrator for space power applications. *Proc. SPIE*, 5179, 116.

O'Regan, B. & Grätzel, M. 1991. A low-cost, high-efficiency solar cell based on dye-sensitized colloidal TiO_2 films. *Nature*, 353, 737.

Oh, E., Park, S., Jeong, J., Kang, S. J., Lee, H. & Yi, Y. 2017. Energy level alignment at the interface of NPB/HAT-CN/graphene for flexible organic light-emitting diodes. *Chem. Phys. Lett.*, 668, 64.

Onsager, L. 1938. Initial recombination of ions. *Phys. Rev.*, 54, 554.

Osterwald, C. R. 1986. Translation of device performance measurements to reference conditions. *Solar Cells*, 18, 269.

Park, H., Rowehl, J. A., Kim, K. K., Bulovic, V. & Kong, J. 2010a. Doped graphene electrodes for organic solar cells. *Nanotechnology*, 21, 505204.

Park, S.-Y., Jeong, W.-I., Kim, D.-G., Kim, J.-K., Lim, D. C., Kim, J. H., Kim, J.-J. & Kang, J.-W. 2010b. Large-area organic solar cells with metal subelectrode on indium tin oxide anode. *Appl. Phys. Lett.*, 96, 173301.

Parthasarathy, G., Burrows, P. E., Khalfin, V., Kozlov, V. G. & Forrest, S. R. 1998. A metal-free cathode for organic semiconductor devices. *Appl. Phys. Lett.*, 72, 2138.

Paternò, G. M., Robbiano, V., Fraser, K. J., Frost, C., García Sakai, V. & Cacialli, F. 2017. Neutron radiation tolerance of two benchmark thiophene-based conjugated polymers: the importance of crystallinity for organic avionics. *Sci. Rep.*, 7, 41013.

Peet, J., Kim, J. Y., Coates, N. E., Ma, W. L., Moses, D., Heeger, A. J. & Bazan, G. C. 2007. Efficiency enhancement in low-bandgap polymer solar cells by processing with alkane dithiols. *Nat. Mater.*, 6, 497.

Perez, M. D., Borek, C., Forrest, S. R. & Thompson, M. E. 2009. Molecular and morphological influences on the open circuit voltages of organic photovoltaic devices. *J. Am. Chem. Soc.*, 131, 9281.

Petrikaln, A. 1930. Über den Einfluss des Gasdrucks auf die lichtelektrische Leitfähigkeit organisher Farbstoffe. *Z. Phys. Chem.*, B10, 9.

Pettersson, L. A. A., Roman, L. S. & Inganäs, O. 1999. Modeling photocurrent action spectra of photovoltaic devices based on organic thin films. *J. Appl. Phys.*, 86, 487.

Peumans, P. 2004. *Organic thin-film photodiodes*. PhD thesis, Princeton University.

Peumans, P. & Forrest, S. R. 2001. Very high efficiency double heterostructure copper phthalocyanine/C_{60} photovoltaic cells. *Appl. Phys. Lett.*, 79, 126.

Peumans, P. & Forrest, S. R. 2004. Separation of geminate charge-pairs at donor–acceptor interfaces in disordered solids. *Chem. Phys. Lett.*, 398, 27.

Peumans, P., Bulovic, V. & Forrest, S. R. 2000a. Efficient photon harvesting at high optical intensities in ultrathin organic double heterostructure photovoltaic diodes. *Appl. Phys. Lett.*, 76, 2650.

Peumans, P., Bulovic, V. & Forrest, S. R. 2000b. Efficient, high-bandwidth organic multilayer photodetectors. *Appl. Phys. Lett.*, 76, 3855.

Peumans, P., Uchida, S. & Forrest, S. R. 2003a. Efficient bulk heterojunction photovoltaic cells based on small molecular weight organic thin films. *Nature*, 425, 158.

Peumans, P., Yakimov, A. & Forrest, S. R. 2003b. Small molecular weight organic thin-film photodetectors and solar cells. *J. Appl. Phys.*, 93, 3693.

Pfuetzner, S., Meiss, J., Petrich, A., Riede, M. & Leo, K. 2009. Improved bulk heterojunction organic solar cells employing C_{70} fullerenes. *Appl. Phys. Lett.*, 94, 145.

Piersimoni, F., Chambon, S., Vandewal, K., Mens, R., Boonen, T., Gadisa, A., Izquierdo, M., Filippone, S., Ruttens, B., D'Haen, J., Martin, N., Lutsen, L., Vanderzande, D., Adriaensens, P. & Manca, J. V. 2011. Influence of fullerene ordering on the energy of the charge-transfer state and open-circuit voltage in polymer:fullerene solar cells. *J. Phys. Chem. C*, 115, 10873.

Pope, M. & Swenberg, C. E. 1982. *Electronic Processes in Organic Crystals*. Oxford: Clarendon Press.

Pradhan, B., Setyowati, K., Liu, H., Waldeck, D. H. & Chen, J. 2008. Carbon nanotube– polymer nanocomposite infrared sensor. *Nano Lett.*, 8, 1142.

Putnam, C. D. 2016. Guinier peak analysis for visual and automated inspection of small-angle X-ray scattering data. *J. Appl. Crystallogr.*, 49, 1412.

Qin, Y., Chen, Y., Cui, Y., Zhang, S., Yao, H., Huang, J., Li, W., Zheng, Z. & Hou, J. 2017. Achieving 12.8% efficiency by simultaneously improving open-circuit voltage and short-circuit current density in tandem organic solar cells. *Adv. Mater.*, 29, 1606340.

Qu, B. & Forrest, S. R. 2018. Continuous roll-to-roll fabrication of organic photovoltaic cells via interconnected high-vacuum and low-pressure organic vapor phase deposition systems. *Appl. Phys. Lett.*, 113, 053302.

Rai, U. & George, S. 1992. Physicochemical studies on organic eutectics and the 1:1 addition compound: benzidine-α-naphthol system. *J. Mater. Sci.*, 27, 711.

Rand, B. P., Burk, D. P. & Forrest, S. R. 2007. Offset energies at organic semiconductor heterojunctions and their influence on the open-circuit voltage of thin-film solar cells. *Phys. Rev. B*, 75, 115327.

Rand, B. P., Li, J., Xue, J., Holmes, R. J., Thompson, M. E. & Forrest, S. R. 2005a. Organic double-heterostructure photovoltaic cells employing thick tris(acetylacetonato)ruthenium(III) exciton-blocking layers. *Adv. Mater.*, 17, 2714.

Rand, B. P., Peumans, P. & Forrest, S. R. 2004. Long-range absorption enhancement in organic tandem thin-film solar cells containing silver nanoclusters. *J. Appl. Phys.*, 96, 7519.

Rand, B. P., Xue, J., Lange, M. & Forrest, S. R. 2003. Thin-film organic position sensitive detectors. *IEEE Photonics Technol. Lett.*, 15, 1279.

Rand, B. P., Xue, J., Yang, F. & Forrest, S. R. 2005b. Organic solar cells with sensitivity extending into the near infrared. *Appl. Phys. Lett.*, 87, 233508.

Ratcliff, E. L., Meyer, J., Steirer, K. X., Armstrong, N. R., Olson, D. & Kahn, A. 2012. Energy level alignment in PCDTBT: $PC_{70}BM$ solar cells: solution processed NiO_x for improved hole collection and efficiency. *Org. Electron.*, 13, 744.

Rau, U. 2007. Reciprocity relation between photovoltaic quantum efficiency and electroluminescent emission of solar cells. *Phys. Rev. B*, 76, 085303.

Rauch, T., Böberl, M., Tedde, S. F., Fürst, J., Kovalenko, M. V., Hesser, G., Lemmer, U., Heiss, W. & Hayden, O. 2009. Near-infrared imaging with quantum-dot-sensitized organic photodiodes. *Nat. Photonics*, 3, 332.

Reese, M. O., Gevorgyan, S. A., Jørgensen, M., Bundgaard, E., Kurtz, S. R., Ginley, D. S., Olson, D. C., Lloyd, M. T., Morvillo, P., Katz, E. A., Elschner, A., Haillant, O., Currier, T. R., Shrotriya, V., Hermenau, M., Riede, M., R. Kirov, K., Trimmel, G., Rath, T., Inganäs, O., Zhang, F., Andersson, M., Tvingstedt, K., Lira-Cantu, M., Laird, D., McGuiness, C., Gowrisanker, S., Pannone, M., Xiao, M., Hauch, J., Steim, R., DeLongchamp, D. M., Rösch, R., Hoppe, H., Espinosa, N., Urbina, A., Yaman-Uzunoglu, G., Bonekamp, J.-B., van Breemen, A. J. J. M., Girotto, C., Voroshazi, E. & Krebs, F. C. 2011. Consensus stability testing protocols for organic photovoltaic materials and devices. *Solar Energy Mater. Solar Cells*, 95, 1253.

Renshaw, C. K., Schlenker, C. W., Thompson, M. E. & Forrest, S. R. 2011. Reciprocal carrier collection in organic photovoltaics. *Phys. Rev. B*, 84, 045315.

Renshaw, C. K., Zimmerman, J. D., Lassiter, B. E. & Forrest, S. R. 2012. Photoconductivity in donor–acceptor heterojunction organic photovoltaics. *Phys. Rev. B*, 86, 085324.

Rim, S.-B., Zhao, S., Scully, S. R., McGehee, M. D. & Peumans, P. 2007. An effective light trapping configuration for thin-film solar cells. *Appl. Phys. Lett.*, 91, 243501.

Roesch, R., Eberhardt, K.-R., Engmann, S., Gobsch, G. & Hoppe, H. 2013. Polymer solar cells with enhanced lifetime by improved electrode stability and sealing. *Solar Energy Mater. Solar Cells*, 117, 59.

Roesch, R., Faber, T., von Hauff, E., Brown Thomas, M., Lira-Cantu, M. & Hoppe, H. 2015. Procedures and practices for evaluating thin-film solar cell stability. *Adv. Energy Mater.*, 5, 1501407.

Rogalski, A. 2002. Infrared detectors: an overview. *Infrared Phys. Technol.*, 43, 187.

Rose, A. 1955a. Performance of photoconductors. *Proc. IRE*, 43, 1850.

Rose, A. 1955b. Space-charge-limited currents in solids. *Phys. Rev.*, 97, 1538.

Rose, A. 1963. *Concepts in Photoconductivity and Allied Problems*. New York: Interscience.

Rowell, M. W., Topinka, M. A., McGehee, M. D., Prall, H.-J., Dennler, G., Sariciftci, N. S., Hu, L. & Gruner, G. 2006. Organic solar cells with carbon nanotube network electrodes. *Appl. Phys. Lett.*, 88, 233506.

Sachs-Quintana, I. T., Heumüller, T., Mateker William, R., Orozco Darian, E., Cheacharoen, R., Sweetnam, S., Brabec Christoph, J. & McGehee Michael, D. 2014. Electron barrier formation at the organic-back contact interface is the first step in thermal degradation of polymer solar cells. *Adv. Funct. Mater.*, 24, 3978.

Savagatrup, S., Printz, A. D., O'Connor, T. F., Zaretski, A. V., Rodriquez, D., Sawyer, E. J., Rajan, K. M., Acosta, R. I., Root, S. E. & Lipomi, D. J. 2015. Mechanical degradation and stability of organic solar cells: molecular and microstructural determinants. *Energy Envrion. Sci.*, 8, 55.

Scheidler, M., Lemmer, U., Kersting, R., Karg, S., Riess, W., Cleve, B., Mahrt, R. F., Kurz, H., Bässler, H., Göbel, E. O. & Thomas, P. 1996. Monte Carlo study of picosecond exciton relaxation and dissociation in poly(phenylenevinylene). *Phys. Rev. B*, 54, 5536.

Schlenker, C. W., Barlier, V. S., Chin, S. W., Whited, M. T., McAnally, R. E., Forrest, S. R. & Thompson, M. E. 2011. Cascade organic solar cells. *Chem. Mater.*, 23, 4132.

Schreurs, D., Nagels, S., Cardinaletti, I., Vangerven, T., Cornelissen, R., Vodnik, J., Hruby, J., Deferme, W. & Manca, J. V. 2018. Methodology of the first combined in-flight and ex situ stability assessment of organic-based solar cells for space applications. *J. Mater. Res.*, 33, 1841.

Schuller, S., Schilinsky, P., Hauch, J. & Brabec, C. 2004. Determination of the degradation constant of bulk heterojunction solar cells by accelerated lifetime measurements. *Appl. Phys. A*, 79, 37.

Seaman, C. H. 1982. Calibration of solar cells by the reference cell method: the spectral mismatch problem. *Solar Energy*, 29, 291.

Seeland, M., Rösch, R. & Hoppe, H. 2011. Luminescence imaging of polymer solar cells: Visualization of progressing degradation. *J. Appl. Phys.*, 109, 064513.

Semonin, O. E., Luther, J. M. & Beard, M. C. 2012. Quantum dots for nextgeneration photovoltaics. *Mater. Today*, 15, 508.

Sergeant, N. P., Pincon, O., Agrawal, M. & Peumans, P. 2009. Design of wide-angle solar-selective absorbers using aperiodic metal-dielectric stacks. *Opt. Express*, 17, 22800.

Servet, B., Horowitz, G., Ries, S., Lagorsse, O., Alnot, P., Yassar, A., Deloffre, F., Srivastava, P. & Hajlaoui, R. 1994. Polymorphism and charge transport in vacuum-evaporated sexithiophene films. *Chem. Mater.*, 6, 1809.

Sharenko, A., Gehrig, D., Laquai, F. d. r. & Nguyen, T.-Q. 2014. The effect of solvent additive on the charge generation and photovoltaic performance of a solution-processed small molecule: perylene diimide bulk heterojunction solar cell. *Chem. Mater.*, 26, 4109.

Shockley, W. & Queisser, H. J. 1961. Detailed balance limit of efficiency of p-n junction solar cells. *J. Appl. Phys.*, 32, 510.

Shockley, W. & van Roosbroeck, W. 1954. Photon-radiative recombination of electrons and holes in germanium. *Phys. Rev.*, 94, 1558.

Shrotriya, V., Li, G., Yao, Y., Moriarty, T., Emery, K. & Yang, Y. 2006. Accurate measurement and characterization of organic solar cells. *Adv. Funct. Mater.*, 16, 2016.

Silvestri, F., Irwin, M. D., Beverina, L., Facchetti, A., Pagani, G. A. & Marks, T. J. 2008. Efficient squaraine-based solution processable bulk-heterojunction solar cells. *J. Am. Chem. Soc.*, 130, 17640.

Singh, S., Jones, W., Siebrand, W., Stoicheff, B. & Schneider, W. 1965. Laser generation of excitons and fluorescence in anthracene crystals. *J. Chem. Phys.*, 42, 330.

Slooff, L. H., Bende, E. E., Burgers, A. R., Budel, T., Pravetton, i. M., Kenny, R. P., Dunlop, E. D. & Büchtemann, A. 2008. A luminescent solar concentrator with 7.1% power conversion efficiency. *Phys. Stat. Sol. Rapid Res. Lett.*, 2, 257.

Small, C. E., Chen, S., Subbiah, J., Amb, C. M., Tsang, S.-W., Lai, T.-H., Reynolds, J. R. & So, F. 2012. High-efficiency inverted dithienogermole–thienopyrrolodione-based polymer solar cells. *Nat. Photonics*, 6, 115.

Smith, M. B. & Michl, J. 2010. Singlet Fission. *Chemical Reviews*, 110, 6891–6936.

Sokel, R. & Hughes, R. C. 1982. Numerical analysis of transient photoconductivity in insulators. *J. Appl. Phys.*, 53, 7414.

Song, B., Burlingame, Q. C., Lee, K. & Forrest, S. R. 2015. Reliability of mixed-heterojunction organic photovoltaics grown via organic vapor phase deposition. *Adv. Energy Mater.*, 5, 1401952.

Stolz Roman, L., Inganäs, O., Granlund, T., Nyberg, T., Svensson, M., Andersson, M. R. & Hummelen, J. C. 2000. Trapping light in polymer photodiodes with soft embossed gratings. *Adv. Mater.*, 12, 189.

Street, R. 2008. Carrier mobility, structural order, and solar cell efficiency of organic heterojunction devices. *Appl. Phys. Lett.*, 93, 361.

Street, R. A., Davies, D., Khlyabich, P. P., Burkhart, B. & Thompson, B. C. 2013. Origin of the tunable open-circuit voltage in ternary blend bulk heterojunction organic solar cells. *J. Am. Chem. Soc.*, 135, 986.

Sturz, L., Witusiewicz, V. T., Hecht, U. & Rex, S. 2004. Organic alloy systems suitable for the investigation of regular binary and ternary eutectic growth. *J. Cryst. Growth*, 270, 273.

Sullivan, P., Heutz, S., Schultes, S. & Jones, T. 2004. Influence of codeposition on the performance of CuPc–C_{60} heterojunction photovoltaic devices. *Appl. Phys. Lett.*, 84, 1210.

Sun, C., Pan, F., Bin, H., Zhang, J., Xue, L., Qiu, B., Wei, Z., Zhang, Z.-G. & Li, Y. 2018a. A low cost and high performance polymer donor material for polymer solar cells. *Nat. Commun.*, 9, 743.

Sun, J., Ma, X., Zhang, Z., Yu, J., Zhou, J., Yin, X., Yang, L., Geng, R., Zhu, R., Zhang, F. & Tang, W. 2018b. Dithieno [3,2-b:2′,3′-d]pyrrol fused nonfullerene acceptors enabling over 13% efficiency for organic solar cells. *Adv. Mater.*, 30, 1707150.

Sun, Y. & Forrest, S. R. 2006. Organic light emitting devices with enhanced out-coupling via microlenses fabricated by imprint lithography. *J. Appl. Phys.*, 100, 073106.

Sun, Y., Takacs, C. J., Cowan, S. R., Seo, J. H., Gong, X., Roy, A. & Heeger, A. J. 2011. Efficient, air-stable bulk heterojunction polymer solar cells using MoO_x as the anode interfacial layer. *Adv. Mater.*, 23, 2226.

Swenberg, C. & Stacy, W. 1968. Bimolecular radiationless transitions in crystalline tetracene. *Chem. Phys. Lett.*, 2, 327.

Sze, S. M. 1981. *Physics of Semiconductor Devices*. New York: Wiley-Interscience.

Tabachnyk, M., Ehrler, B., Gélinas, S., Böhm, M. L., Walker, B. J., Musselman, K. P., Greenham, N. C., Friend, R. H. & Rao, A. 2014. Resonant energy transfer of triplet excitons from pentacene to PbSe nanocrystals. *Nat. Mater.*, 13, 1033.

Tachiya, M. 1988. Breakdown of the Onsager theory of geminate ion recombination. *J. Chem. Phys.*, 89, 6929.

Tang, C. W. 1986. Two-layer organic photovoltaic cell. *Appl. Phys. Lett.*, 48, 183.

Tang, C. W. & VanSlyke, S. A. 1987. Organic Electroluminescent Devices. *Appl. Phys. Lett.*, 51, 913.

Tang, Z., Andersson, L. M., George, Z., Vandewal, K., Tvingstedt, K., Heriksson, P., Kroon, R., Andersson, M. R. & Inganäs, O. 2012. Interlayer for modified cathode in highly efficient inverted ITO-free organic solar cells. *Adv. Mater.*, 24, 554.

Tang, Z., Elfwing, A., Bergqvist, J., Tress, W. & Inganäs, O. 2013. Light trapping with dielectric scatterers in single- and tandem-junction organic solar cells. *Adv. Energy Mater.*, 3, 1606.

Thompson, N. J., Wilson, M. W. B., Congreve, D. N., Brown, P. R., Scherer, J. M., Bischof, Thomas S., Wu, M., Geva, N., Welborn, M., Voorhis, T. V., Bulović, V., Bawendi, M. G. & Baldo, Marc A. 2014. Energy harvesting of non-emissive triplet excitons in tetracene by emissive PbS nanocrystals. *Nat. Mater.*, 13, 1039.

Tiffenberg, J., Sofo-Haro, M., Drlica-Wagner, A., Essig, R., Guardincerri, Y., Holland, S., Volansky, T. & Yu, T. T. 2017. Single-electron and single-photon sensitivity with a silicon skipper CCD. *Phys. Rev. Letters*, 119, 131802.

Tournebize, A., Rivaton, A., Gardette, J. L., Lombard, C., Pépin-Donat, B., Beaupré, S. & Leclerc, M. 2014. How photoinduced crosslinking under operating conditions can reduce PCDTBT-based solar cell efficiency and then stabilize it. *Adv. Energy Mater.*, 4, 1301530.

Traverse, C. J., Pandey, R., Barr, M. C. & Lunt, R. R. 2017. Emergence of highly transparent photovoltaics for distributed applications. *Nat. Energy*, 2, 849.

Tsai, W.-W., Chao, Y.-C., Chen, E.-C., Zan, H.-W., Meng, H.-F. & Hsu, C.-S. 2009. Increasing organic vertical carrier mobility for the application of high speed bilayered organic photodetector. *Appl. Phys. Lett.*, 95, 213308.

Tsuda, A. & Osuka, A. 2001. Fully conjugated porphyrin tapes with electronic absorption bands that reach into infrared. *Science*, 293, 79.

Tvingstedt, K., Andersson, V., Zhang, F. & Inganäs, O. 2007. Folded reflective tandem polymer solar cell doubles efficiency. *Appl. Phys. Lett.*, 91, 123514.

Tvingstedt, K., Dal Zilio, S., Inganäs, O. & Tormen, M. 2008. Trapping light with micro lenses in thin film organic photovoltaic cells. *Opt.Express*, 16, 21608.

Tvingstedt, K., Tang, Z. & Inganäs, O. 2012. Light trapping with total internal reflection and transparent electrodes in organic photovoltaic devices. *Appl. Phys. Lett.*, 101, 163902.

Uchida, S., Xue, J., Rand, B. P. & Forrest, S. R. 2004. Organic small molecule solar cells with a homogeneously mixed copper phthalocyanine: C_{60} active layer. *Appl. Phys. Lett.*, 84, 4218.

Uhrich, C., Weiß, A. & Pfeiffer, M. 2017. Roll-to-roll production of organic solar cells. In *Organic, Hybrid, and Perovskite Photovoltaics XVIII. Proc. SPIE* 10363, 103630F.

van Bavel, S. S., Sourty, E., With, G. d. & Loos, J. 2009. Three-dimensional nanoscale organization of bulk heterojunction polymer solar cells. *Nano Lett.*, 9, 507.

Van De Lagemaat, J., Barnes, T. M., Rumbles, G., Shaheen, S. E., Coutts, T. J., Weeks, C., Levitsky, I., Peltola, J. & Glatkowski, P. 2006. Organic solar cells with carbon nanotubes replacing In_2O_3:Sn as the transparent electrode. *Appl. Phys. Lett.*, 88, 233503.

van der Ziel, A. 1986. *Noise in Solid State Devices and Circuits*. New York: John Wiley and Sons.

Van Dyck, C., Marks, T. J. & Ratner, M. A. 2017. Chain length dependence of the dielectric constant and polarizability in conjugated organic thin films. *ACS Nano*, 11, 5970.

Vandewal, K., Tvingstedt, K., Gadisa, A., Inganäs, O. & Manca, J. V. 2009. On the origin of the open-circuit voltage of polymer–fullerene solar cells. *Nat. Mater.*, 8, 904.

Vandewal, K., Tvingstedt, K., Gadisa, A., Inganäs, O. & Manca, J. V. 2010. Relating the open-circuit voltage to interface molecular properties of donor: acceptor bulk heterojunction solar cells. *Phys. Rev. B*, 81, 125204.

Veldman, D., Meskers, S. C. J. & Janssen, R. A. J. 2009. The energy of charge-transfer states in electron donor–acceptor blends: insight into the energy losses in organic solar cells. *Adv. Funct. Mater.*, 19, 1939.

Voroshazi, E. 2014. Stability of organic photovoltaic cells: failure mechanisms and operational stability. In: Rand, B. P. & Richter, H. (eds.) *Organic Solar Cells: Fundamentals, Devices and Upscaling*. Singapore: Pan Stanford Publishing.

Wallmark, J. T. 1957. A new semiconductor photocell using lateral photoeffect. *Proc. IRE*, 45, 474.

Wang, J.-L., Liu, K.-K., Yan, J., Wu, Z., Liu, F., Xiao, F., Chang, Z.-F., Wu, H.-B., Cao, Y. & Russell, T. P. 2016. Series of multifluorine substituted oligomers for organic solar cells with efficiency over 9% and fill factor of 0.77 by combination thermal and solvent vapor annealing. *J. Am. Chem. Soc.*, 138, 7687.

Wang, J.-L., Wu, Z., Miao, J.-S., Liu, K.-K., Chang, Z.-F., Zhang, R.-B., Wu, H.-B. & Cao, Y. 2015. Solution-processed diketopyrrolopyrrole-containing small-molecule organic solar cells with 7.0% efficiency: in-depth investigation on the effects of structure modification and solvent vapor annealing. *Chem. Mater.*, 27, 4338.

Wang, N., Tong, X., Burlingame, Q., Yu, J. & Forrest, S. R. 2014. Photodegradation of small-molecule organic photovoltaics. *Solar Energy Mater. Solar Cells*, 125, 170.

Wang, S., Hall, L., Diev, V. V., Haiges, R., Wei, G., Xiao, X., Djurovich, P. I., Forrest, S. R. & Thompson, M. E. 2011. *N*, *N*-Diaryl anilinosquaraines and their application to organic photovoltaics. *Chem. Mater.*, 23, 4789.

Watkins, P. K., Walker, A. B. & Verschoor, G. L. 2005. Dynamical Monte Carlo modelling of organic solar cells: the dependence of internal quantum efficiency on morphology. *Nano Lett.*, 5, 1814.

Webster, J. G. 1999. *The Measurement, Instrumentation, and Sensors Handbook*. Boca Raton, FL: CRC Press.

Wei, G., Sun, K., Wang, S., Thompson, M. E. & Forrest, S. R. 2011a. Solvent annealed crystalline squaraine: $PC_{70}BM$ (1:6) solar cells. *Adv. Energy Mater.*, 1, 184.

Wei, G., Xiao, X., Wang, S., Sun, K., Bergemann, K. J., Thompson, M. E. & Forrest, S. R. 2011b. Functionalized squaraine donors for nanocrystalline organic photovoltaics. *ACS Nano*, 6, 972.

Weiss, D. S. & Abkowitz, M. 2010. Advances in organic photoconductor technology. *Chem. Rev.*, 110, 479.

White, R. T., Thibau, E. S. & Lu, Z.-H. 2016. Interface structure of MoO_3 on organic semiconductors. *Sci. Rep.*, 6, 21109.

Winston, R. & Hinterberger, H. 1975. Principles of cylindrical concentrators for solar energy. *Solar Energy*, 17, 255.

Wu, Q., Zhao, D., Schneider, A. M., Chen, W. & Yu, L. 2016. Covalently bound clusters of alpha-substituted PDI: rival electron acceptors to fullerene for organic solar cells. *J. Am. Chem. Soc.*, 138, 7248.

Wu, T. C., Thompson, N. J., Congreve, D. N., Hontz, E., Yost, S. R., Van Voorhis, T. & Baldo, M. A. 2014. Singlet fission efficiency in tetracene-based organic solar cells. *Appl. Phys. Lett.*, 104, 193901.

Würfel, P. & Würfel, U. 2009. *Physics of Solar Cells: From Basic Principles to Advanced Concepts*. Weinheim, Germany: Wiley-VCH.

Xiao, X., Bergemann, K. J., Zimmerman, J. D., Lee, K. & Forrest, S. R. 2014. Small molecule planar-mixed heterojunction photovoltaic cell with a fullerene-based electron filtering buffer. *Adv. Energy Mater.*, 4, 1301557.

Xiao, X., Lee, K. & Forrest, S. R. 2015. Scalability of multijunction organic solar cells for large area organic solar modules. *Appl. Phys. Lett.*, 106, 213301.

Xiao, X., Zimmerman, J. D., Lassiter, B. E., Bergemann, K. J. & Forrest, S. R. 2013. A hybrid planar-mixed tetraphenyldibenzoperiflanthene/C_{70} photovoltaic cell. *Appl. Phys. Lett.*, 102, 0733302.

Xu, X., Bi, Z., Ma, W., Wang, Z., Choy, W. C. H., Wu, W., Zhang, G., Li, Y. & Peng, Q. 2017. Highly efficient ternary-blend polymer solar cells enabled by a nonfullerene acceptor and two polymer donors with a broad composition tolerance. *Adv. Mater.*, 29, 1704271.

Xu, X., Davanco, M., X. Qi & Forrest, S. R. 2008. Direct transfer patterning on three dimensionally deformed surfaces at micrometer resolutions and its application to hemispherical focal plane detector arrays. *Org. Electron.*, 9, 1122.

Xu, X., Mihnev, M., Taylor, A. & Forrest, S. R. 2009. Organic photodetector arrays with indium tin oxide electrodes patterned using directly transferred metal masks. *Appl. Phys. Lett.*, 94, 043313.

Xue, J. & Forrest, S. R. 2002. Organic optical bistable switch. *Appl. Phys. Lett.*, 82, 136.

Xue, J., Rand, B. P., Uchida, S. & Forrest, S. R. 2005. A hybrid planar-mixed molecular heterojunction photovoltaic cell. *Adv. Mater.*, 17, 66.

Yablonovitch, E. 1982. Statistical ray optics. *J. Opt. Soc. Am.*, 72, 899.

Yakimov, A. & Forrest, S. R. 2002. High photovoltage multiple-heterojunction organic solar cells incorporating interfacial metallic clusters. *Appl. Phys. Lett.*, 80, 1667.

Yan, F., Liu, H., Li, W., Chu, B., Su, Z., Zhang, G., Chen, Y., Zhu, J., Yang, D. & Wang, J. 2009. Double wavelength ultraviolet light sensitive organic photodetector. *Appl. Phys. Lett.*, 95, 253308.

Yan, Q., Zhou, Y., Zheng, Y.-Q., Pei, J. & Zhao, D. 2013. Towards rational design of organic electron acceptors for photovoltaics: a study based on perylenediimide derivatives. *Chem. Sci.*, 4, 4389.

Yang, C. & Lunt, R. R. 2017. Limits of visibly transparent luminescent solar concentrators. *Adv. Opt. Mater.*, 5, 1600851.

Yang, D., Zhou, X. & Ma, D. 2013. Fast response organic photodetectors with high detectivity based on rubrene and C_{60}. *Org. Electron.*, 14, 3019.

Yang, F. & Forrest, S. R. 2008. Photocurrent generation in nanostructured organic solar cells. *ACS Nano*, 2, 1022.

Yang, F., Lunt, R. R. & Forrest, S. R. 2008. Simultaneous heterojunction organic solar cells with broad spectral sensitivity. *Appl. Phys. Lett.*, 92, 38.

Yang, F., Shtein, M. & Forrest, S. R. 2005a. Controlled growth of a molecular bulk heterojunction photovoltaic cell. *Nat. Mater.*, 4, 39.

Yang, F., Shtein, M. & Forrest, S. R. 2005b. Morphology control and material mixing by high-temperature organic vapor-phase deposition and its application to thin-film solar cells. *J. Appl. Phys.*, 98, 014906.

Yang, F., Sun, K. & Forrest, S. R. 2007. Efficient solar cells using all-organic nanocrystalline networks. *Adv. Mater.*, 19, 4166.

Yang, L., Tabachnyk, M., Bayliss, S. L., Böhm, M. L., Broch, K., Greenham, N. C., Friend, R. H. & Ehrler, B. 2015. Solution-processable singlet fission photovoltaic devices. *Nano Lett.*, 15, 354.

Yang, L., Zhou, H., Price, S. C. & You, W. 2012a. Parallel-like bulk heterojunction polymer solar cells. *J. Am. Chem. Soc.*, 134, 5432.

Yang, T., Wang, M., Duan, C., Hu, X., Huang, L., Peng, J., Huang, F. & Gong, X. 2012b. Inverted polymer solar cells with 8.4% efficiency by conjugated polyelectrolyte. *Energy Envrion. Sci.*, 5, 8208.

Yao, Y., Liang, Y., Shrotriya, V., Xiao, S., Yu, L. & Yang, Y. 2007. Plastic near-infrared photodetectors utilizing low band gap polymer. *Adv. Mater.*, 19, 3979.

Yao, Z., Liao, X., Gao, K., Lin, F., Xu, X., Shi, X., Zuo, L., Liu, F., Chen, Y. & Jen, A. K. Y. 2018. Dithienopicenocarbazole-based acceptors for efficient organic solar cells with optoelectronic response over 1000 nm and an extremely low energy loss. *J. Am. Chem. Soc.*, 140, 2054.

Yokoyama, D., Qiang Wang, Z., Pu, Y.-J., Kobayashi, K., Kido, J. & Hong, Z. 2012. High-efficiency simple planar heterojunction organic thin-film photovoltaics with horizontally oriented amorphous donors. *Solar Energy Mater. Solar Cells*, 98, 472.

Yoo, S., Domercq, B. & Kippelen, B. 2004. Efficient thin-film organic solar cells based on pentacene/C_{60} heterojunctions. *Appl. Phys. Lett.*, 85, 5427.

Yost, S. R., Lee, J., Wilson, M. W., Wu, T., McMahon, D. P., Parkhurst, R. R., Thompson, N. J., Congreve, D. N., Rao, A. & Johnson, K. 2014. A transferable model for singlet-fission kinetics. *Nat. Chem.*, 6, 492.

Yu, G., Gao, J., Hummelen, J., Wudl, F. & Heeger, A. J. 1995. Polymer photovoltaic cells: enhanced efficiencies via a network of internal donor–acceptor heterojunctions. *Science*, 270, 1789.

Yu, G., Pakbaz, K. & Heeger, A. J. 1994. Semiconducting polymer diodes: large size, low cost photodetectors with excellent visible-ultraviolet sensitivity. *Appl. Phys. Lett.*, 64, 3422.

Yu, G., Srdanov, G., Wang, J., Wang, H., Cao, Y. & Heeger, A. J. 2000. Large area, full-color, digital image sensors made with semiconducting polymers. *Synth. Met.*, 111–112, 133.

Yu, G., Wang, J., McElvain, J. & Heeger, A. J. 1998. Large-area, full-color image sensors made with semiconducting polymers. *Adv. Mater.*, 10, 1431.

Yu, Z., Raman, A. & Fan, S. 2010. Fundamental limit of nanophotonic light trapping in solar cells. *Proc. Natl. Acad. Sci.*, 107, 17491.

Yuan, J., Zhang, Y., Zhou, L., Zhang, G., Yip, H.-L., Lau, T.-K., Lu, X., Zhu, C., Peng, H., Johnson, P. A., Leclerc, M., Cao, Y., Ulanski, J., Li, Y. & Zou, Y. 2019. Single-junction organic solar cell with over 15% efficiency using fused-ring acceptor with electron-deficient core. *Joule*, 3, 1140.

Yuan, L., Lu, K., Xia, B., Zhang, J., Wang, Z., Wang, Z., Deng, D., Fang, J., Zhu, L. & Wei, Z. 2016. Acceptor end-capped oligomeric conjugated molecules with broadened absorption and enhanced extinction coefficients for high-efficiency organic solar cells. *Adv. Mater.*, 28, 5980.

Zemansky, M. & Dittman, R. 1981. *Heat and Thermodynamics*. New York: McGraw-Hill.

Zhang, F., Vollmer, A., Zhang, J., Xu, Z., Rabe, J. & Koch, N. 2007. Energy level alignment and morphology of interfaces between molecular and polymeric organic semiconductors. *Org. Electron.*, 8, 606.

Zhang, H., Wang, X., Yang, L., Zhang, S., Zhang, Y., He, C., Ma, W. & Hou, J. 2017. Improved domain size and purity enables efficient all-small-molecule ternary solar cells. *Adv. Mater.*, 29, 1703777.

Zhang, J., Morton, J. J. L., Sambrook, M. R., Porfyrakis, K., Ardavan, A. & Briggs, G. A. D. 2006. The effects of a pyrrolidine functional group on the magnetic properties of N@C60. *Chem. Phys. Lett.*, 432, 523.

Zhang, M., Guo, X., Ma, W., Ade, H. & Hou, J. 2015. A large-bandgap conjugated polymer for versatile photovoltaic applications with high performance. *Adv. Mater.*, 27, 4655.

Zhang, Y., Zou, J., Yip, H.-L., Chen, K.-S., Zeigler, D. F., Sun, Y. & Jen, A. K. Y. 2011. Indacenodithiophene and quinoxaline-based conjugated polymers for highly efficient polymer solar cells. *Chem. Mater.*, 23, 2289.

Zhao, W., Li, S., Yao, H., Zhang, S., Zhang, Y., Yang, B. & Hou, J. 2017a. Molecular optimization enables over 13% efficiency in organic solar cells. *J. Am. Chem. Soc.*, 139, 7148.

Zhao, W., Li, S., Zhang, S., Liu, X. & Hou, J. 2017b. Ternary polymer solar cells based on two acceptors and one donor for achieving 12.2% efficiency. *Adv. Mater.*, 29, 1604059.

Zhao, Y., Meek, G. A., Levine, B. G. & Lunt, R. R. 2014. Near-infrared harvesting transparent luminescent solar concentrators. *Adv. Opt. Mater.*, 2, 606.

Zhou, H., Yang, L., Stuart, A. C., Price, S. C., Liu, S. & You, W. 2011. Development of fluorinated benzothiadiazole as a structural unit for a polymer solar cell of 7% efficiency. *Angew. Chem.*, 123, 3051.

Zhou, L., Xiang, H.-Y., Shen, S., Li, Y.-Q., Chen, J.-D., Xie, H.-J., Goldthorpe, I. A., Chen, L.-S., Lee, S.-T. & Tang, J.-X. 2014. High-performance flexible organic light-emitting diodes using embedded silver network transparent electrodes. *ACS Nano*, 8, 12796.

Zimmerman, J. D., Diev, V. V., Hanson, K., Lunt, R. R., Yu, E. K., Thompson, M. E. & Forrest, S. R. 2010a. Porphyrin-tape/C_{60} organic photodetectors with 6.5% external quantum efficiency in the near infrared. *Adv. Mater.*, 22, 2780.

Zimmerman, J. D., Xiao, X., Renshaw, C. K., Wang, S., Diev, V. V., Thompson, M. E. & Forrest, S. R. 2012. Independent control of bulk and interfacial morphologies of small molecular weight organic heterojunction solar cells. *Nano Lett.*, 12, 4366.

Zimmerman, J. D., Yu, E. K., Diev, V. V., Hanson, K., Thompson, M. E. & Forrest, S. R. 2011. Use of additives in porphyrin-tape/C60 near-infrared photodetectors. *Org. Electron.*, 12, 869.

Zimmerman, P. M., Zhang, Z. & Musgrave, C. B. 2010b. Singlet fission in pentacene through multi-exciton quantum states. *Nat. Chem.*, 2, 648.

CHAPTER 8

Organic thin film transistors

"God made the bulk; the surface was invented by the devil."

Wolfgang Pauli, Nobel Laureate in Physics

Organic materials are excitonic materials. In this regard, they would seem to be ideally suited for optoelectronic devices, such as OLEDs, OPVs, and photodetectors. But what about their uses as purely electronic materials? Specifically, do organics hold promise as a platform for thin film transistors useful solely for electronic applications? In this chapter we address just this question.

As the epigram to this chapter implies, organic thin film transistors (OTFTs) differ from the other devices that we have considered in that they are controlled by the surface, rather than the film bulk. By this we mean that the charge flows between source and drain contacts along the organic semiconductor/dielectric interface. From many decades of engineering of conventional semiconductors, it is indisputable that surface physics, and particularly surface charge transport is far more complex than analogous phenomena in the bulk. The complications are inherent in the 2D nature of interfaces, introducing defects and lattice reconstruction that occurs at surfaces. Hence, the understanding and optimization of OTFTs necessarily implies that we examine the complex microscopic properties of organic semiconductor/gate insulator interfaces.

The chapter opens by introducing basic transistor architectures and the theory of their operation. This discussion relies on the understanding of charge transport in organics that was developed in Chapter 4, and is now extended to consider interface transport. Next, we elucidate all of the many forms that OTFTs take, since the positions of the gate, source and drain contacts relative to each other and to the films on which they are deposited determine the device attributes. Also, we take this opportunity to discuss several different variants, such as split and dual gate OTFTs, and a different type of transistor altogether: the vertical transistor that is analogous to a vacuum tube triode. Unlike the OTFT, vertical transistor properties depend on the film bulk rather than its surface, making it more akin to the devices in the previous two chapters.

Our discussion of transistor operating principles and architectures is followed by consideration of the phototransistor. While this hybrid between a transistor and a photodetector can exhibit optical responsivity considerably greater than a photodiode owing to its inherent gain, enhanced photosensitivity comes at the expense of a smaller bandwidth of operation and a somewhat more complex structure. However, its ability to be integrated into circuits with increased functionality beyond that of a simple photodetector can make it attractive for specialized applications where bandwidth is not a dominant concern. The treatment of phototransistors necessarily draws heavily on discussions and definitions used in understanding photodetectors. Hence, the reader is referred to Chapter 7 to fully appreciate the phototransistor discussion presented here.

The performance of organic transistors, like all electronic devices, depends almost entirely on the materials used. The OTFTs are operated with few exceptions in the accumulation mode, whereby charges are drawn

Organic Electronics: Foundations to Applications. Stephen R. Forrest. © Stephen R. Forrest 2020.
Published in 2020 by Oxford University Press. DOI: 10.1093/oso/9780198529729.001.0001

into the channel by the potentials determined by the gate and drain voltages. The carrier type can be either electrons or holes, depending on the metal/channel work functions, and which carrier type has the highest mobility. We find that nearly equal performance has been obtained for n- and p-channel devices, and indeed ambipolar transistors supporting both carrier types have been demonstrated and employed in high performance circuits. We will discuss the introduction of materials with high mobilities and low voltages that have been employed in complementary logic circuits and oscillators, and that operate in the kilohertz to megahertz regimes with acceptable noise margins.

No discussion of devices is complete without considering stability, that is, the ability of the OTFTs to maintain high performance over long periods of operation, in adverse environments and under stressful conditions. Many of the electrical characteristics of OTFTs, such as threshold voltage, drain current, field effect mobility, and output conductance, drift over time. In most cases, exposure of the gate insulator, the channel, and the insulator/channel interface to a humid environment is responsible for many, but not all, of the device instabilities. Encapsulation can delay, but not completely eliminate water from entering the device active regions, hence providing only a limited degree of device stability over time.

We close out our discussion of OTFTs with a description of a few of the impressive applications and circuits that have been demonstrated. In the relatively short time since their introduction, the performance of OTFTs has surpassed even that of amorphous Si devices in terms of mobility, stability, and operating voltage. However, to be adopted in practical applications, we must clearly define the competitive advantage that OTFTs offer over conventional thin film solutions such as amorphous or polycrystalline Si, or metal oxide transistors. Without such clarity of purpose, OTFTs simply become yet another "technological solution looking for a problem!" In this context, OTFTs have a considerable potential to occupy new application spaces that are inaccessible to incumbent inorganic transistor solutions as a result of their ability to be fabricated on ultrathin (~1 μm) substrates over very large areas by volume, roll-to-roll (R2R) fabrication processes. In our discussion, we therefore focus on applications that require large area circuits such as for tactile, gas and optical sensing, and for the emerging area of imperceptible, biocompatible devices for medical diagnostics, entertainment, and robotics. In spite of this promise, we must keep in mind that OTFTs have yet to be adopted on any significant scale by the electronics manufacturing industry; a situation different from that experienced by OLEDs, and to a much lesser extent by organic solar cells.

Ultimately, the widespread adoption of organic transistor technology remains uncertain. This is not a question that can or should be answered in this book: that is left to the marketplace and the consumer to decide. Independent of that question, however, there is little doubt that research and engineering of OTFTs has led to a deeper understanding of charge transport in organics, and to the development of highly sophisticated small molecule and polymer conductive media. This in itself has led to improvements in optoelectronic devices such as OLEDs and OPVs that rely on highly conductive charge transport layers to achieve high performance.

For the ensuing discussions, we assume that the reader has at least a passing acquaintance with the basics of transistor design and operation. There are many excellent texts available that can provide that background knowledge, as well as affording a deeper appreciation for this very broad topic than can be presented here. Some of these references are found in *Further reading*.

8.1 Thin film transistor basics

The organic thin film transistor (OTFT) is a *metal–insulator–semiconductor field effect transistor*, or MISFET, whose basic configuration is shown in Fig. 8.1. It is a three terminal device comprising a source (S) and a drain (D) contact made to a thin film semiconducting channel. An insulating layer (the gate insulator) is in contact with at least one surface of the semiconductor. The gate contact (G) is located on the opposite side of the insulator, bridging the gap between S and D. Figure 8.1 also defines several important geometric factors: the gate width (W) is typically much larger than its length (L). The channel thickness is d, and the insulator thickness is t_i. Charge is introduced into the channel by injection from the source contact due to a potential difference from the source to the gate, V_{GS}. Then conduction in the channel occurs by applying a voltage between source and drain, V_{DS}. Since the amount of charge injected into the channel depends on the gate voltage, the source-drain current, I_D, is modulated by the magnitude of V_{GS}. Thus, field effect transistors are essentially gate-controlled resistors.

The sign of the charge that carries the drain current determines whether the OTFT is a p- or n-channel device. In most organics, the channel itself is undoped, although it may have a residual concentration of holes or electrons that determine the position of the Fermi energy at equilibrium (see Chapter 4). This is a different operating regime than for an inorganic MISFET where the channel region is doped n- or p-type, and hence the semiconductor bulk, or body, supplies the carriers at the insulator/semiconductor interface. For example, the high hole mobility in pentacene makes this material useful in p-channel OTFTs. A negative voltage at the gate draws holes to the semiconductor/insulator interface, while a negative voltage at the drain leads to conduction within the channel. These are the same bias conditions used for an inorganic p-channel MISFET. The density of charge in the channel is largely unrelated to the background charge concentration. Indeed, in OTFTs it is important that the background charge concentration be very small such that, at $V_{GS} = 0$ (corresponding to the transistor "off" state), the channel leakage is also very small. Hence, the OTFT operates in the charge *accumulation mode*. It is normally off, that is, in the absence of a negative gate voltage there is ideally no charge in the channel, and hence no current can flow from source to drain. A normally off transistor is known as an *enhancement mode* device.

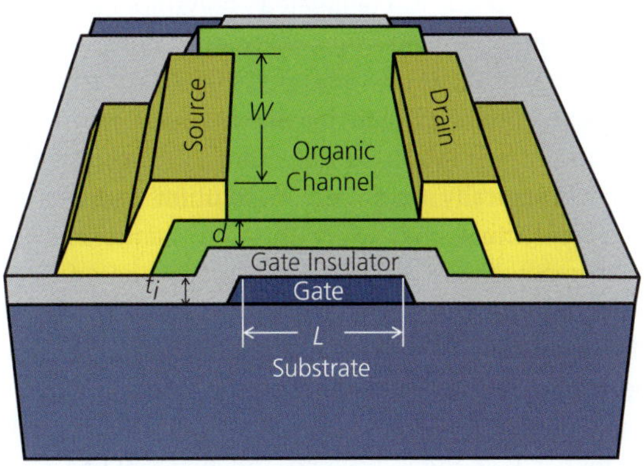

Figure 8.1 Schematic perspective of an organic thin film transistor.

The device pictured in Fig. 8.1 has a bottom gate-top source-drain contact configuration, or briefly, it is a *bottom-gate/top-contact* (BG/TC) transistor. However, as shown in Fig. 8.2, many other possible architectures realized by placing the gate and channel contacts on either the same or opposing sides of the channel semiconductor. The BG/TC transistor is possibly the easiest to fabricate, and hence is the most common architecture since it allows the S and D contacts to be deposited through a shadow mask. We have found in Chapter 4 that deposition of metals onto organics can damage the metal/organic interface, introducing traps that can lower the injection barrier, and hence the contact resistance. Another commonly employed architecture is the bottom gate/bottom contact (BG/BC) configuration. This

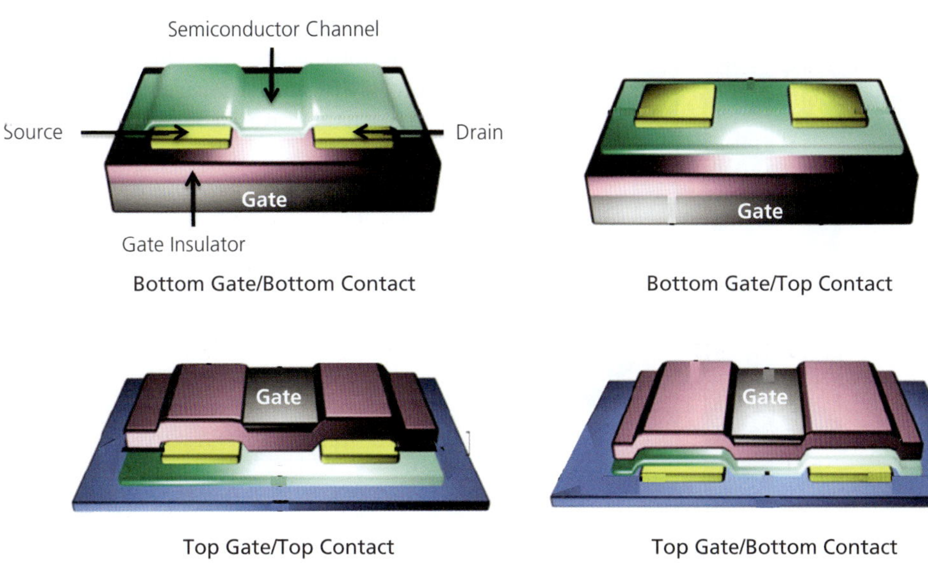

Figure 8.2 Possible OTFT configurations. Clockwise from top left: bottom gate/bottom contact (BG/BC), BG/TC, TG/BC, and TG/TC.

places all contacts on the same surface of the semiconductor, which allows for photolithographic pattering to achieve very short gate geometries. The gate contact is accessed by extending it beyond and parallel to the source and drain contacts. The selection of a particular architecture is ultimately based on the choice of materials, the patterning method, and the demands placed by the circuit on its intended application (e.g. high current, high bandwidth, low noise, etc.).

8.2 A brief history of OTFTs

We begin our tour through the history of OTFTs with the invention of the field effect transistor itself. This can be traced to a patent by Lilienfeld who proposed, but did not demonstrate a "device for controlling electric current" employing CuS as the channel material, and Al_2O_3 as the gate insulator (Lilienfeld, 1933). The first realization of a thin film transistor grown on an insulating substrate employed a vapor deposited microcrystalline film of CdS as the semiconductor, Au as the gate electrode, and SiO as the gate insulator. This TG/BC structure was fabricated on a glass substrate. As shown in Fig. 8.3, it bears all of the attributes of modern OTFTs (Weimer, 1962).

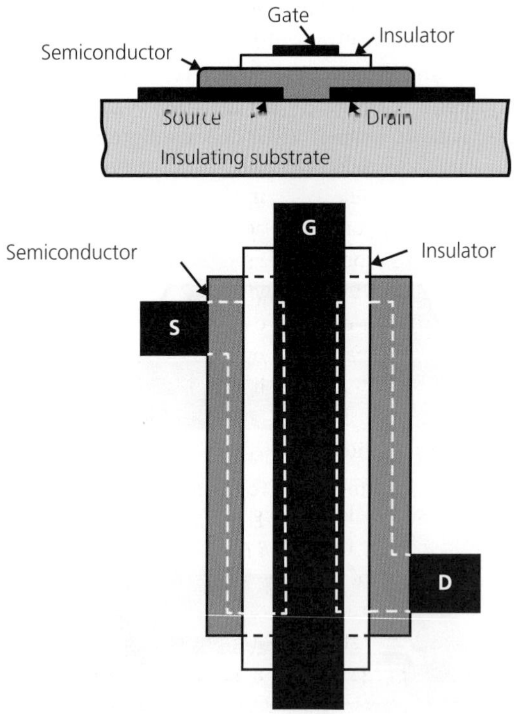

Figure 8.3 Microcrystalline CdS TFT on a glass substrate in cross-sectional (top) and plan (bottom) views. Note how the gate contact overlaps the source and drain to minimize channel resistance (Weimer, 1962).

The door to OTFTs was opened by depositing a merocyanine channel on interdigitated S and D contacts patterned onto a 300 nm thick SiO_2 gate insulator. This combination of materials used a 1 Ω cm p-Si substrate that served as the gate electrode. The BG/BC configuration is shown in Fig. 8.4 (Kudo et al., 1984). The device dimensions were $L = 20$ μm, $W = 7$ mm, and $d = 40$ nm. The device did not enter the saturation regime, nor did the channel pinch off, even at +15 V, indicating excessive background conduction of the merocyanine dye. A small hole *field effect mobility* of $\sim 10^{-5}$ cm^2/V s was extracted from the device characteristics in the unsaturated, or linear regime. Hence, while this device has the elements of an OTFT, it did not show clear transistor operation.

Apparently, the first report of an OTFT that showed all of the characteristics of a functioning transistor exhibiting both linear and saturated drain current characteristics, used a conjugated polythiophene thin film channel in a BG/BC configuration (see Fig. 8.5) (Tsumura et al., 1986). The polythiophene film was deposited on a SiO_2 gate insulator ($t_i = 300$ nm) on a 4–8 Ω-cm n-Si substrate gate contact. The film was electrochemically de-doped in solution to ensure a non-conducting channel in the "off" state. The enhancement mode OTFT had dimensions of $L = 10$ μm, $W = 2$ mm, and $d = 140$ nm. The drain-current characteristics have three characteristics that are expected of all practical enhancement mode FETs: (i) a linear onset of I_D vs. V_{DS} extending up to $V_{DS} \simeq V_{GS}$. (ii) At higher V_{DS}, the I_D becomes independent of V_{DS} as the transistor enters the saturation regime at pinch-off. (iii) The drain current, $I_D \to 0$ as $V_{GS} \to 0$. A hole mobility of $\sim 10^{-5}$ cm^2/V s is inferred from analysis of these characteristics.

Within three years, a vacuum-deposited (vs. solution processed) small molecule thin film of α-sexithiophene (α-6T) OTFT was demonstrated (Horowitz et al.,

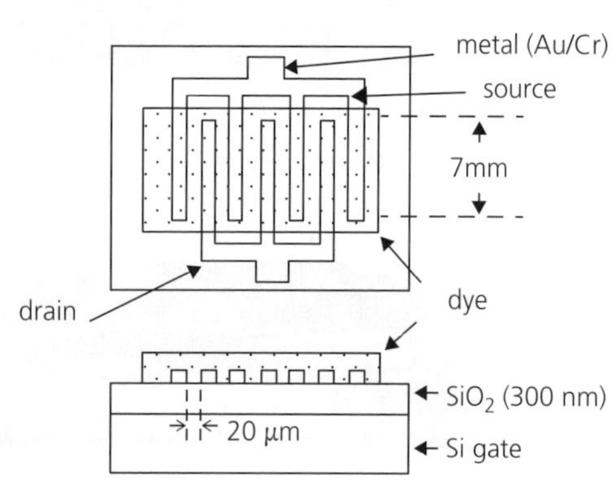

Figure 8.4 Plan (top) and cross-sectional views of the merocyanine OTFT (Kudo et al., 1984).

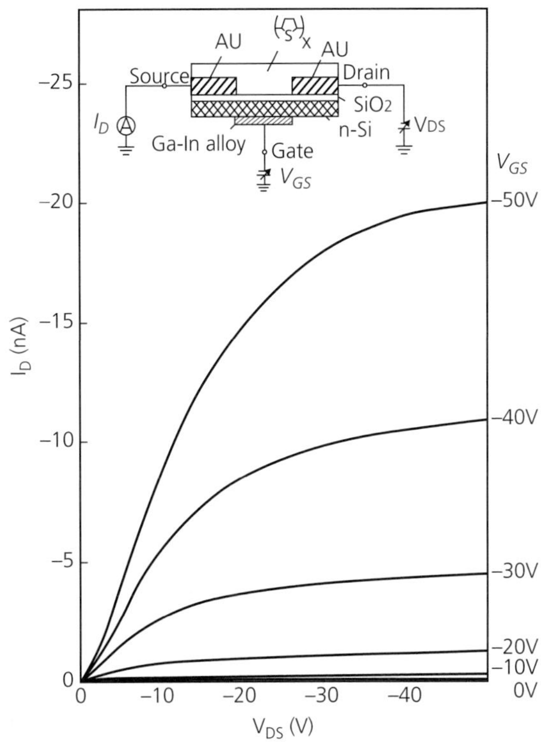

Figure 8.5 Polythiophene p-channel OTFT drain current characteristics showing linear and saturation regions of operation, as well as a negligible "off" current at $V_{GS} = 0$ (Tsumura et al., 1988).

1989). The devices were in a BG/TC configuration, again using n-Si as the gate contact and substrate, and a 300 nm thick SiO_2 gate oxide. The OTFT had dimensions of $L = 90$ μm, $W = 310$ μm, and $d = 70$ nm. The as-deposited films exhibited excessive channel conductivity, and hence required de-doping by heating in air for 3 h at 120°C after deposition. Following de-doping, the enhancement mode transistors exhibited a saturation-regime mobility of $\sim 10^{-4}$ cm^2/V s, which was the highest reported up to that time.

The next significant advance in OTFTs was made by replacing the Si gate coated with an SiO_2 gate insulator by a metal electrode and an organic dielectric, cyanoethylpullulan (CYEPL) in the structure shown in Fig. 8.6, inset (Garnier et al., 1990). The BG/TC structure starts with a 25 μm thick, flexible poly(parabanic acid) (PPA) substrate. A 1 mm wide by 20 mm long Au gate contact stripe is then vacuum deposited, followed by solution drop deposition of CYEPL. Depositing an α-6T channel and Au S and D electrodes completed the $L = 50$ μm, $W = 5$ mm, and $d = 50$ nm OTFT. This structure has several characteristics that distinguished it from previous work. A metal strip replaced the conducting substrate used previously as the gate contact. This allows for applying different gate biases to different transistors integrated on the same substrate, which is required in almost all electronic circuit designs. Further, the use of a plastic substrate allows for bending and twisting without changing the device characteristics, as shown in Fig. 8.6. This takes advantage of the inherent flexibility of van der Waals bonded solids, arguably one of the most important characteristics of organic electronic devices. And finally, the high quality of the organic insulator/channel interface resulted in an unprecedented high hole field effect mobility of 0.43 cm^2/V-s. This represents at least a 100-fold improvement over previously demonstrated devices, and indeed made this OTFT competitive with amorphous Si thin film transistors then in use in LCD back-planes. The high mobility also leads to a low $V_{GS} \sim 10$ V compared with previous devices where $V_{GS} \sim 50$ V was common (cf. Fig. 8.5). This early, high performance OTFT considerably heightened interest in the technology, and subsequently led to its further rapid development.

As seen in Chapter 4, charge mobility in organics is strongly dependent on disorder. Dynamic disorder that arises from coupling of phonons to the molecular solid presents an intrinsic limitation to the bulk mobility of 1–5 cm^2/V s in molecular solids. Static disorder due to structural defects or impurities, however, has an exogenous origin, and can reduce the charge mobility even further. This led, in subsequent years to, finding deposition conditions and pure materials that generated more ordered thin films, and even single crystals to achieve high field effect mobilities. We will show below, however, that field effect mobilities obtained in single crystal OTFTs are not significantly higher than in polycrystalline organic transistors. While single crystal OTFTs have been helpful in understanding the limits to performance of organic electronic devices, practical devices must be manufactured over large substrate areas using conventional thin film deposition from the solution or vapor phases.

The introduction of pentacene, which readily forms crystalline regions under a wide range of deposition conditions, has led to significant increases in field effect mobilities in OTFTs (Lin et al., 1997a). Indeed, a dependence of mobility and transistor on-off current ratio (i.e. the ratio of the drain current in the transistor "on" state in saturation to the "off" state where no charge is accumulated under the gate, see Section 8.3.2) is directly traced to both pentacene nanomorphology and the purity of the source material. When deposited onto a substrate at elevated temperatures (60–120°C), pentacene molecules tend

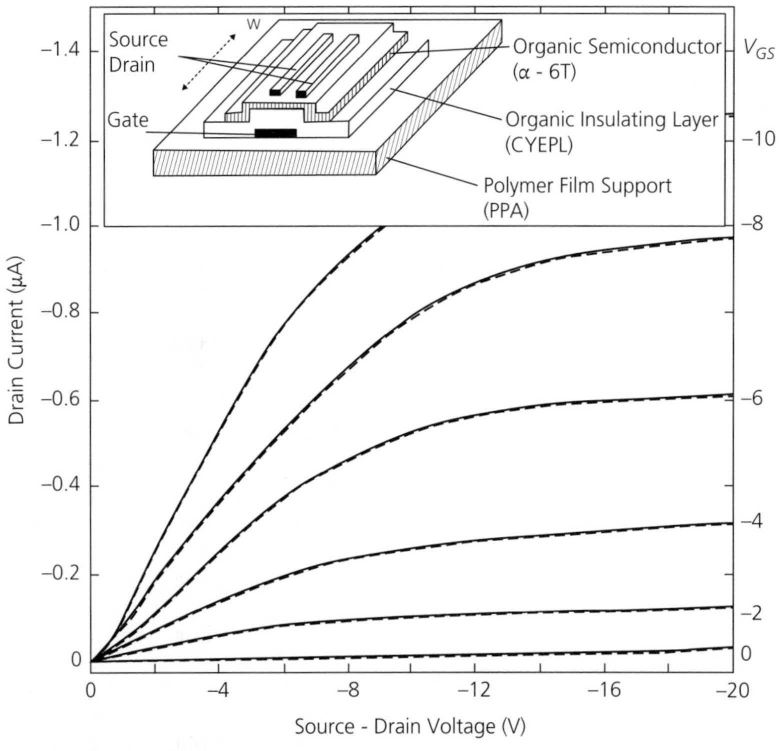

Figure 8.6 Drain current characteristics of a BG/TC OTFT obtained in air. The solid lines are for an as-fabricated device, and the dashed lines are the transfer characteristics after bending and twisting of the device on its flexible substrate. Inset: Perspective illustration of the OTFT featuring a polymer substrate, and small molecule organic insulator and channel layers (Garnier et al., 1990).

to stand up, forming π-stacks along the in-plane direction. This orientation results in a high hopping mobility between molecules along the channel. The pentacene BG/TC OTFT employing a SiO_2 insulator and a doped Si substrate gate contact showed a hole mobility of 0.7 cm^2/V s. Importantly, the on-off current ratio of the transistor was increased from previous reports of from $\sim 10^3$–10^4 to 10^8. A high on-off current ratio is essential for obtaining large display backplane arrays that deliver a suitably high contrast between the pixel on and off states, as well as a large noise margin in logic circuits (see Section 8.3.4).

High mobility was attributed to nanocrystalline order achieved via vacuum deposition on heated substrates, and the on-off ratio to the high purity (and hence a low background charge density) of the material that reduced the off-state leakage current. This result was improved by functionalizing the SiO_2 surface with a self-assembled, space-filling monolayer of octadecyltrichlorosilane (OTS) prior to deposition of a 30 nm thick pentacene channel onto the heated (90°C) substrate. The OTS reduces the surface energy relative to pentacene. Growth on OTS produces dendritic crystals separated by grain boundaries that are at a lower density and perhaps smaller than achieved by pentacene growth directly on SiO_2. The gaps are filled in by subsequent deposition of an additional 20 nm of pentacene after the substrate is cooled to room temperature. This BG/BC SiO_2 gate insulator OTFT showed a mobility of 1.5 cm^2/V-s and an on-off drain current ratio of 10^8 (Lin et al., 1997b).

The dependence of mobility on surface and bulk morphologies was demonstrated for pentacene-based OTFTs grown by organic vapor phase deposited (OVPD, see Section 5.4.2) on a SiO_2 insulator, with and without an OTS monolayer. Growth by OVPD allows for considerable control over morphology via variations in the chamber temperature and pressure, source temperature and gas flow rates. Hence, it provides an ideal platform for studying the effects of morphology on device parameters. Figure 8.7 shows scanning electron micrographs of 70–100 nm thick pentacene films grown by OVPD on a thermally grown SiO_2 on an n-Si substrate, with and without an OTS layer pre-deposited onto the oxide surface (Shtein et al., 2002). The crystallite size increases from 0.5 to >5 μm as equilibrium growth conditions are approached at high substrate temperatures, high chamber pressures and low growth rates. Under these

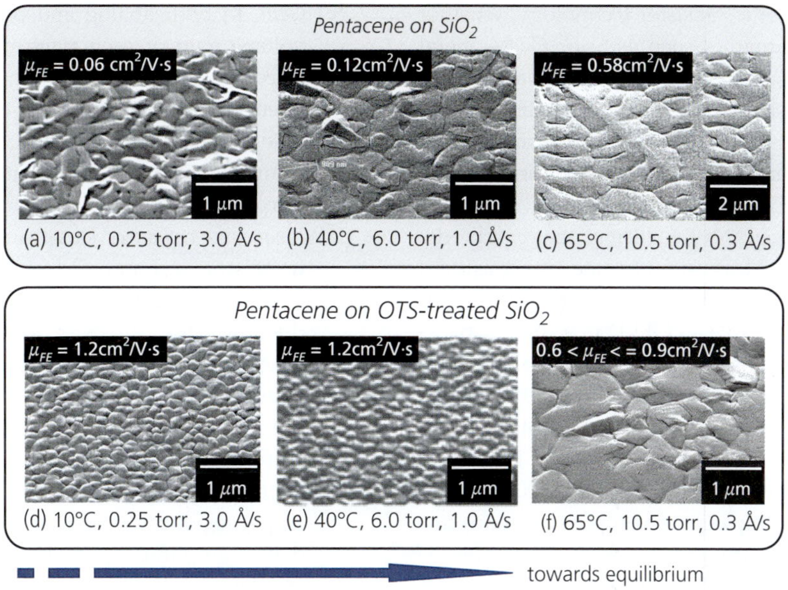

Figure 8.7 Scanning electron micrographs of pentacene thin films grown by OVPD on (a–c) thermally grown SiO_2 and (d–f) OTS pretreated SiO_2. The conditions for OVPD growth for each film (left to right) are substrate temperature, chamber pressure and growth rate. The films on the right are grown closer to thermodynamic equilibrium than on the left (Shtein et al., 2002).

conditions, arriving molecules have more freedom to move along the surface, where they find low energy attachment sites before being buried by subsequently arriving molecules. The crystallite boundaries are readily apparent in the images of Fig. 8.7a–c.

OVPD growth was used in fabricating BG/TC OTFTs whose resulting hole field effect mobilities are shown in the figure. Increasing the grain size results in at least a tenfold increase in mobility due to the reduction in the number of grains, and possibly the width of the grain boundaries. In addition, the purified pentacene channels yielded a drain current on-off ratio of $\sim 10^5$. However, even the largest crystallites grown near equilibrium do not extend all the way across the gate length of 45 μm. If this were the case, even higher mobilities might be achieved.

Growth on an OTS-functionalized surface results in a considerably smaller-scale surface texture but higher mobilities. Indeed, non-equilibrium growth results in mobilities of up to 1.6 cm^2/V s compared to the device grown near equilibrium with a field-effect mobility of μ_{FE} = 0.6–0.9 cm^2/V-s. Both of these mobilities are higher than for films grown on bare SiO_2. The on-off ratio also increased to $\sim 10^7$ compared to the bare SiO_2 devices. Apparently, the textures on the film surfaces in Fig. 8.7d and 8.7e are not due to grain boundaries. X-ray diffraction analysis of the films shows that the OTS forces the first few monolayers of pentacene to lie flat on the substrate plane, as well as creating denser films with fewer grain boundaries. The flat-lying pentacene may lead to improved order in the subsequently deposited layers that ultimately transport charge within the channel.

Higher mobility OTFTs employed organic polymer gate insulators with a vacuum deposited pentacene channel. The BG/TC transistors used a Si gate contact and substrate, followed by either cross-linked PVP or PVP-co-HEM. The pentacene channel and contacts were deposited on the polymer surface. Transistors with dimensions of L = 130 μm and W = 170 μm, yielded a hole field effect mobility of approximately $\mu_{FE} = 3$ cm^2/V s for both insulators as compared to similar devices employing SiO_2 (Klauk et al., 2002).

In this brief history, we have only discussed high mobility in hole conducting channels. In Section 8.6 we discuss several polymers and small molecules leading to both hole and electron dominated OTFTs.

8.3 Operating principles and definitions

The previous sections have described several prominent features of OTFTs. We have provided the basic layouts and several operating parameters such as the field effect mobility and on-off current ratio that characterize the most basic device attributes. In this section, we develop the theory of operation of the thin film transistor. The treatment provides the relationships between device geometry, materials characteristics and equivalent circuit performance in both transient and steady-state modes of operation.

The usefulness of a transistor technology, whether it is based on inorganics such as Si or GaAs, metal oxide transistors, nanostructures such as carbon nanotubes (CNTs), or OTFTs, depends entirely on the application it is intended to serve. The application places several demands on the transistor, including:

(i) *Current capacity*. Is the current that can be delivered by the transistor per gate area consistent with the needs of the load? For example, for the OTFT to drive an OLED in a display backplane, the transistor circuit must have a footprint no greater than that of the emitting device ($\leq 10^4\ \mu m^2$) while delivering sufficient current to achieve the required display brightness (≥ 1 mA/cm^2 for a luminance $L > 1000$ cd/m^2). The relationship between output (i.e. drain) current, and device and materials parameters is therefore a central consideration in transistor design.

(ii) *On/off current ratio*. When used as a switch, and even in analog circuit applications, the drain current in the on state must be several orders of magnitude higher than in the off state. For digital circuits, this ensures reliable, low noise operation of a logic circuit consisting of a cascading chain of gates. That is, the transistor must be capable of distinguishing between a high vs. a low input state without error, to ensure that its output responds accordingly for the next gate in the chain. The distinction between on and off (or logical high and low states) is quantified by the circuit *noise margin* (Section 8.3.4). Further, the off state is determined by the channel leakage current. This sets the *noise floor* in both analog and digital circuits below which an output signal cannot be reliably detected.

(iii) *Operating voltage*. Most low power dissipation circuits today operate at ~5 V or less. The supply voltage can be higher than this, although the transistor can be assigned only a portion of the total available voltage to allow for other circuit elements to function.

(iv) *Uniformity*. Device turn-on, or threshold voltage, must be confined to an acceptably small variance from device to device in large circuits, otherwise the margin for failure that depends on triggering logic gates becomes too small for reliable operation. The primary quantities that determine the operating point are the transistor *threshold voltage, sub-threshold slope*, and *transconductance*.

(v) *Gain*. In both analog and digital circuits, the voltage gain of each stage must be sufficient to compensate for losses, and to drive subsequent circuits. The transistor transconductance, as determined by the charge mobility in the channel, must be sufficiently high for this purpose.

(vi) *Frequency response*. Determined by both the transistor geometry (e.g. gate area and hence gate capacitance) and the transconductance, the transistor must operate at frequencies sufficient to deliver the required circuit bandwidth.

(vii) *Noise*. The transistor noise in an analog circuit is amplified at each stage, and can eventually overwhelm the small input signal. For example, the input current noise of a transimpedance amplifier must be smaller than the smallest detectable signal of a photodiode designed to detect a small photon fluence. In a digital circuit, the output noise current of a transistor should not be so high as to mistakenly trip the logic state of the following gate. The *dynamic range* of the amplifier is the ratio of the lowest detectable signal determined by the input transistor noise floor, to the maximum signal before the amplifier deviates significantly from a linear response.

(viii) *Stability*. The transistor must be stable over time, otherwise at some point the output characteristics drift outside of the margins where the circuit can reliably operate.

In this section, we establish a theoretical framework for determining whether or not a particular transistor architecture or material meets the demands of its target application. Our discussion begins at the most general level, considering all operational modes of MISFETs. However, as previously noted, OTFTs primarily (but not exclusively) operate in the accumulation-enhancement regime, and hence the discussions primarily focus on this common situation. Within this framework, we identify the applications that are most suitably addressed by OTFTs, and which can be better served by alternative transistor technologies.

8.3.1 Metal–insulator–semiconductor contacts

The key to understanding OTFT operation is the *metal gate–insulator–semiconductor* (MIS) capacitor. In Fig. 8.8 we show the energy level diagram for a MIS capacitor comprising a metal contact and

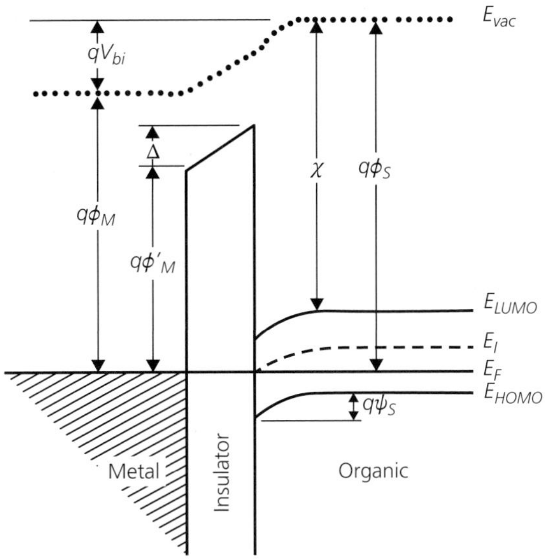

Figure 8.8 Equilibrium energy level diagram of a metal–insulator–organic contact where the organic is p-type.

semiconductor with work functions, ϕ_M and ϕ_S, respectively. In this example, $\phi_M < \phi_S$. When the metal, insulator and semiconductor are brought into contact, charge flows until equilibrium is established. Then the Fermi energy, E_F, across the device is level, resulting in bending of E_{LUMO}, E_{HOMO}, and the vacuum level, E_{VAC}, near the dielectric/semiconductor interface. This induces a dipole of energy, Δ, in the insulator, and depletion of holes within the semiconductor. From inspection of this diagram:

$$\phi_{MS} = \phi_M - \phi_S = \phi_M - \frac{1}{q}(\chi + (E_{LUMO} - E_F)) = qV_{bi}, \tag{8.1}$$

where χ is the electron affinity of the semiconductor and V_{bi} is the *built-in potential*. It is convenient to define the surface potential,

$$q\psi_S = q\phi_{MS} - \Delta = qV_{bi} - \Delta, \tag{8.2}$$

as the band bending at the semiconductor/insulator interface. In the special case where $\phi_M = \phi_S$, $V_{bi} = 0$ and the bands are flat throughout the semiconductor channel. This is known as the *flat-band condition*. The device in Fig. 8.8 achieves flat-band when the potential applied to the metal (i.e. the gate potential in a transistor) is $V_{GS} = \psi_S$.

In MIS systems, there are several sources of trapped charge both within the insulator and at the interfaces. Specifically, trapped charges per area due to fixed (ΔQ_F) and mobile (ΔQ_M) ions in the insulator bulk, and interface traps (ΔQ_{IT}), contribute to a change in gate voltage. The presence of charge traps in equilibrium also contributes to the energy level bending, and hence affects the surface potential in the channel. Defining the sum of all charges trapped at defects both within the bulk and at the surfaces of the insulator as $\Delta Q_i = \Delta Q_F + \Delta Q_M + \Delta Q_{IT}$ (each term of which can be either positive or negative, depending on the polarity of the trap at equilibrium), the built-in potential becomes

$$V'_{bi} = V_{bi} + \Delta V_{bi} = \phi_{MS} + \frac{\Delta Q_i}{C_i}, \tag{8.3}$$

where the insulator capacitance per area in the absence of traps is

$$C_i = \frac{\varepsilon_i \varepsilon_0}{t_i}. \tag{8.4}$$

Here, both Q and C are normalized to the contact area, and ε_i and ε_0 are the relative dielectric constant of the insulator and the permittivity of free space, respectively.

Control of the channel charge and potential by the gate voltage in an MIS capacitor is shown in Fig. 8.9, assuming $\phi_M = \phi_S$ and $\Delta Q_i = 0$. The energy level diagrams are shown at the top of the figure, and the charge distributions are shown at the bottom. Generally, the charge density in the organic is small, but rarely is it zero. In the MIS capacitor, we assume the organic is infinitely thick with a background hole density, $p > 0$.

At $V_{GS} = 0$, the device is at thermal equilibrium and the bands are flat. At $V_{GS} < 0$, holes accumulate at the surface at density $+Q_S$. This is the *accumulation regime*. Under a small positive bias less than the threshold voltage, that is, at $V_{GS} < V_T$, charge is depleted from the insulator/semiconductor interface, leaving only fixed bulk charge, $-Q_B$, due to the residual holes in the channel. At $V_{GS} > V_T$ the Fermi level in the semiconductor moves above mid-gap and the minority charge concentration at the surface surpasses the majority carrier concentration. In this case, the semiconductor is said to be *inverted*, where the number of electrons, $n > p$. Note that this description is general. However, due to the low background charge concentrations in undoped organics, typically only the accumulation regime is accessed by applying a negative gate voltage.

The inversion condition occurs when the free charge at the interface (i.e. electrons in Fig. 8.9) equals the holes in the bulk, and therefore creates a neutrality condition. This is the depletion-inversion transition point, and from inspection of Fig. 8.8, occurs when $q\psi_S = 2(E_F - E_I)_{bulk}$ where the last term refers to the values of E_F and E_I in the channel bulk. Here, E_I is the

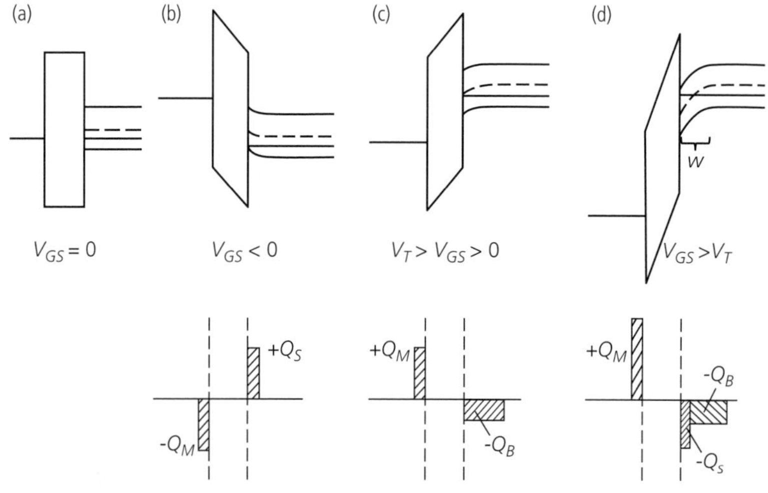

Figure 8.9 Gate voltage control of hole concentration at the organic/semiconductor interface in an MIS capacitor. (a) Flat band condition (thermal equilibrium) where $\phi_M = \phi_S$. (b) Accumulation, (c) depletion, (d) inversion at $V_{GS} > V_T$, the threshold voltage. Diagrams at bottom show the charge at each side of the insulator.

intrinsic level energy. The gate voltage at the transition point is the threshold voltage, V_T, and is defined by satisfying the condition for the total surface charge:

$$Q_S = -C_i(V_{GS} - V_T) = 0. \tag{8.5}$$

In the absence of channel doping, then

$$V_T = \frac{2}{q}(E_F - E_I)_{bulk} = 2\psi_B, \tag{8.6}$$

where ψ_B is the Fermi energy relative to the intrinsic level in the bulk of the semiconductor. The charge density in the bulk of the semiconductor is given by (see Eq. 4.266*ff.*)

$$Q_B = -qN_A w = -\sqrt{2qN_A\varepsilon_S\varepsilon_0(2\psi_B)}, \tag{8.7}$$

where N_A is the density of residual acceptors in the bulk. The total threshold voltage is thus

$$V_T = 2\left(\psi_B + \frac{\sqrt{qN_A\varepsilon_S\varepsilon_0\psi_B}}{C_i}\right) \approx 2\psi_B + \frac{qN_Ad}{C_i}. \tag{8.8}$$

The expression to the far right is valid for a lightly doped organic of thickness d that is completely depleted at $V_{GS} = 0$. This situation is the most commonly encountered in organic semiconductors.

A qualitative plot of the surface charge, $\log(|Q_S|)$, vs. the surface energy $q\psi_S$ for an MIS capacitor is provided in Fig. 8.10. This plot shows how the gate potential controls the charge density at the semiconductor/insulator interface. The electron surface charge density for an MIS capacitor with an electron-rich semiconductor bulk follows

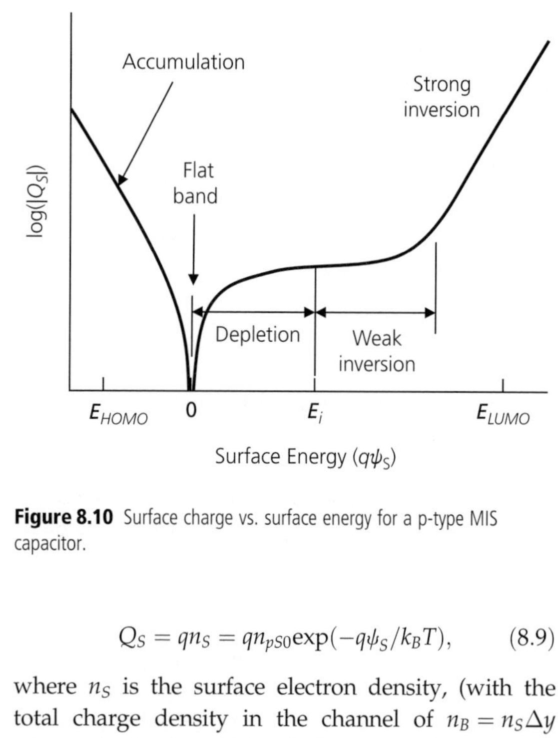

Figure 8.10 Surface charge vs. surface energy for a p-type MIS capacitor.

$$Q_S = qn_S = qn_{pS0}\exp(-q\psi_S/k_BT), \tag{8.9}$$

where n_S is the surface electron density, (with the total charge density in the channel of $n_B = n_S\Delta y$ where Δy is the width of the charge layer at the insulator/semiconductor interface), and n_{pS0} is the electron surface charge density at equilibrium. For an undoped semiconductor, Q_S increases exponentially in both the accumulation and strong inversion regimes. In weak inversion, the charge density increases only slowly with ψ_S, whose exact functional dependence is found from a solution to Poisson's equation in the channel (Sze, 1981). Importantly, at flat band, the semiconductor is in equilibrium where the number of thermally generated holes and electrons are equal, thus resulting in the absence of a net charge at the interface.

The capacitance of the gate under accumulation is simply the geometric capacitance of the insulator in Eq. 8.4. At high frequencies where the surface charge cannot respond to the a.c. signal, C_i is added in series to the capacitance of the depletion region in the channel bulk, C_B. Then, the total capacitance is

$$C_T = \frac{\varepsilon_S \varepsilon_i \varepsilon_0}{\varepsilon_S t_i + \varepsilon_i w}, \tag{8.10}$$

where ε_S is the semiconductor dielectric constant, and w is the depletion region width. As is often the case in organics, the background charge density is small, and the thin semiconductor is completely depleted ($w \to d$). At low frequency, the interface charge follows the variation in voltage, and the capacitance returns to C_i once inversion is reached.

8.3.2 OTFT operation

Figure 8.9 illustrates several bias regimes at a MIS junction. Since organics have a low background charge density, charge is drawn into the channel via injection from the S contact by applying a bias, $|V_{GS}| > |V_T|$. Furthermore, the channel is "normally off" when no charge exists at $V_{GS} = 0$. This corresponds to enhancement mode operation. For the unusual case where the channel is heavily doped, it can be conducting at $V_{GS} = 0$, in which case it is "normally on." This is the depletion mode of operation, which is rarely exploited in OTFTs due to the difficulty in obtaining sufficiently high free charge concentrations via doping.

In the accumulation-enhancement mode, at $|V_{GS}| > |V_T|$, injected charge in the channel diffuses along the insulator interface to the drain. Figure 8.11 shows the OTFT along the length of the gate at different V_{DS}. The illustration shows the extent of the conduction region from $V_{DS} = 0$ (Fig. 8.11a), to $V_{DS} = V_{DSsat}$ (Fig. 8.11b), that is, the point at which the drain potential leads to a pinch-off of the charged layer at the drain. At small V_{GS}, therefore, the channel current is in the linear regime when I_D vs. V_{DS} approximately follows Ohm's law. In this regime, the channel resistance is relatively independent of V_{DS}. At $|V_{DS}| = |V_{DSat}|$, the current demanded by the drain potential can no longer be supplied by the source without increasing V_{GS} needed to increase the charge density at the insulator interface. This creates a depletion region between the pinch-off point and the source, preventing the drain current from increasing further.

Thus, when $|V_{DS}| > |V_{DSat}|$, the free charge recedes a distance, Δx, from the drain (Fig. 8.11c). The drain current is pinned at I_{Dsat} assuming there is no series resistance contributed by the depleted gate region. Since the drain current is constant at higher voltages, this is known as the *saturation regime*. The various current regimes along with the I_D vs. V_{DS} characteristics of an n-channel OTFT (where both V_{GS} and V_{DS} are positive) are shown in Fig. 8.12.

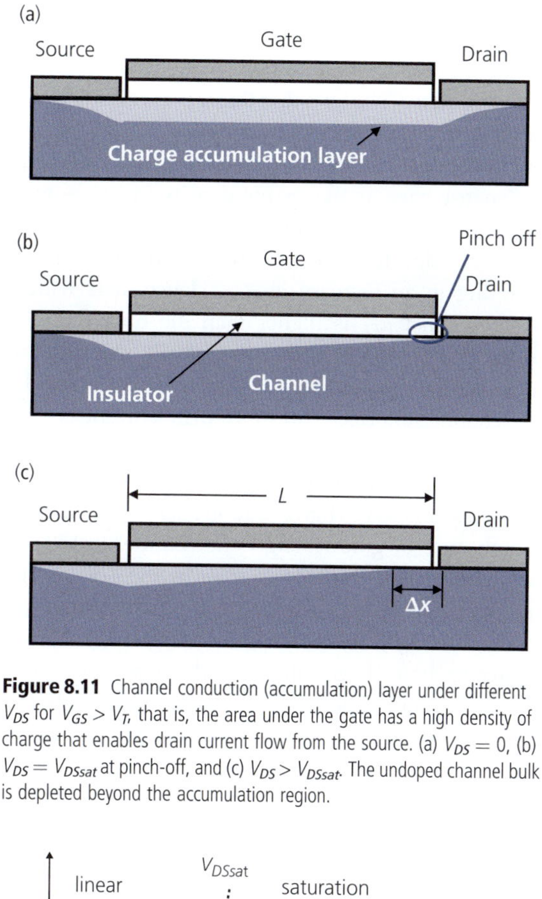

Figure 8.11 Channel conduction (accumulation) layer under different V_{DS} for $V_{GS} > V_T$, that is, the area under the gate has a high density of charge that enables drain current flow from the source. (a) $V_{DS} = 0$, (b) $V_{DS} = V_{DSsat}$ at pinch-off, and (c) $V_{DS} > V_{DSsat}$. The undoped channel bulk is depleted beyond the accumulation region.

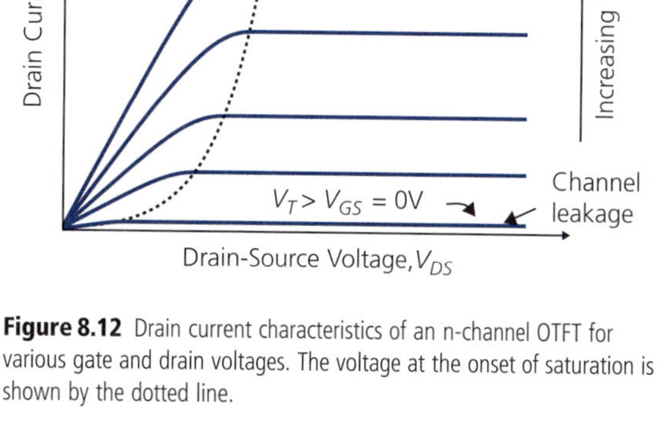

Figure 8.12 Drain current characteristics of an n-channel OTFT for various gate and drain voltages. The voltage at the onset of saturation is shown by the dotted line.

The OTFT operation can be modeled using the distributed resistive/capacitance network in Fig. 8.13. The equivalent circuit approximates the channel by a series of N incremental resistances, δr. Similarly, the gate dielectric consists of a parallel network of incremental insulator capacitances, δC_i. The voltage at point x from the source along the channel of length, L, is $V(x)$. For simplicity, we ignore the effects of S and D contact resistances, which can be significant (see below).

For calculating device characteristics, we make several assumptions. (i) The charge mobility at the interface, that is, the field effect mobility, μ_{FE}, is independent of electric field under all bias conditions. We have found in Chapter 4 that this is in fact not the case in disordered organic solids where a density of conductive states extends from the HOMO or LUMO into the energy gap. As the charge density increases with V_{GS} and V_{DS}, the filling of these tail states pushes the mobility edge toward the frontier orbital energies, resulting in an increase in μ_{FE}. Hence, it is widely observed that μ_{FE} is higher than the bulk mobility of the channel material that is typically measured at much lower charge densities than exist at the insulator/channel interface. (ii) Only drift current is considered; diffusion current contributions are ignored. (iii) The channel is lightly doped, or undoped, that is, $E_F \approx E_I$ at $V_{DS} = 0$. This condition also implies that, in the absence of gate leakage, the channel leakage is negligible when the transistor is in the "off" state. (iv) The electric field normal to the gate contact (in the y-direction) is considerably larger than from source to drain. This is the *gradual channel approximation* that applies when $t_i \ll L$, which is almost always true for OTFTs. (v) We start by assuming ambipolar conduction in the channel (Schmechel et al., 2005). Ambipolar implies that both electrons and holes can transport current. Except in special cases where the material has a nearly equal electron and hole field effect mobility, that is, where $\mu_{FEn} \approx \mu_{FEp}$, or in blends of electron and hole conducting constituents, the OTFT shows unipolar conduction where I_D is carried by only a single charge type. Once the full ambipolar treatment is developed, it is readily simplified to model unipolar operation.

In an ambipolar OTFT, the total charge at the insulator/channel interface at x is

$$Q_S(x) = q(p_S(x) - n_S(x)) = C_i(V(x) - V_{GS}). \quad (8.11)$$

Here, $p_S(x)$ and $n_S(x)$ are the respective surface charge concentrations (cf. Eq. 8.9). At this point, we neglect the threshold voltage and assume that $|V_{GS}| \geq |V_T|$. This simplification allows us to consider a different threshold for electrons (V_{Tn}) and holes (V_{Tp}); factors that will be included below.

Since the channel in the linear regime behaves as a resistor with total resistance R_{tot}, then from Fig. 8.13 it follows that

$$V(x) = V_{DS}\frac{R(x)}{R_{tot}}, \quad (8.12)$$

where the resistance at x as measured from ground (i.e. from the source contact) is

$$R(x) = \int_0^x \frac{dx}{qW\left[\mu_{FEp}p_S(x) + \mu_{FEn}n_S(x)\right]} \quad (8.13)$$

and $R_{tot} = R(L)$. Substituting Eqs. 8.12 and 8.13 into Eq. 8.11 yields the change in surface charge vs. distance along the channel:

$$\frac{dQ_S(x)}{dx} = \frac{C_iV_{DS}}{R_{tot}}\frac{1}{qW\left[\mu_{FEp}p_S(x) + \mu_{FEn}n_S(x)\right]}. \quad (8.14)$$

Figure 8.14 illustrates the accumulation region for holes injected from the source and electrons from the drain (i.e. $V_{DS} < 0$). To good approximation, we can assume that all electrons and holes that meet at point x_0 recombine. Hence, we have the following conditions for the charge distributions:

$$\begin{aligned} Q_S(x) &= qp_S(x) \quad & 0 \leq x < x_0 \\ Q_S(x) &= -qn_S(x) \quad & x_0 \leq x < L \end{aligned} \quad (8.15)$$

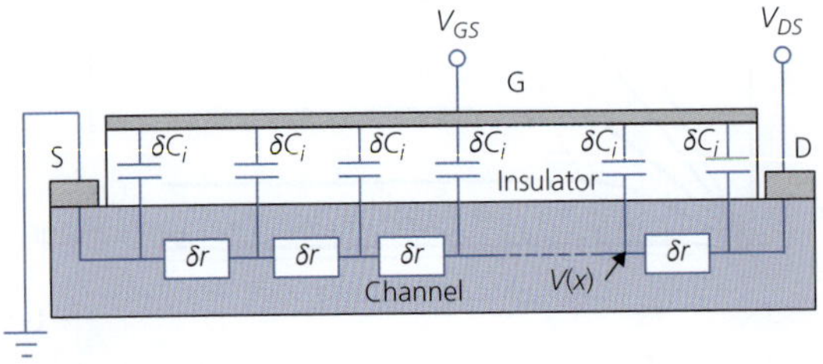

Figure 8.13 Equivalent circuit used in calculating the large signal OTFT operating characteristics.

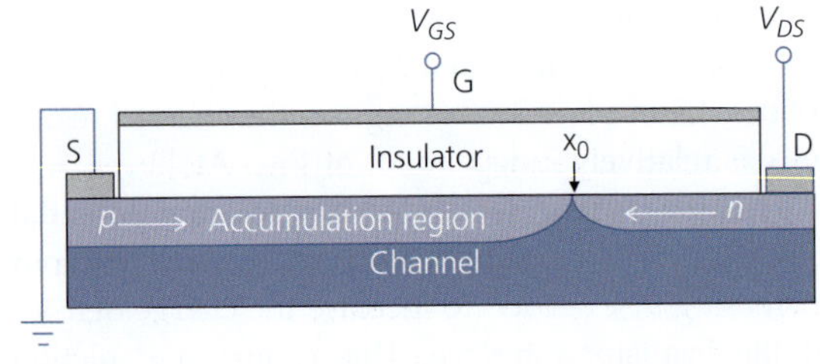

Figure 8.14 Ambipolar OTFT showing the accumulation region under bipolar charge injection.

That is, the accumulation region is divided into two zones, one that conducts only holes and the other, electrons. With this, we can simplify Eq. 8.14 in the region $x < x_0$:

$$\frac{dQ_S(x)}{dx} = \frac{C_i V_{DS}}{R_{tot}} \frac{1}{qW\mu_{FEp} Q_S}, \tag{8.16}$$

with an analogous relationship for electrons at $x \geq x_0$. Integrating Eq. 8.16 up to the point x gives

$$|Q_S(x)| = \sqrt{(C_i V_{GS})^2 \pm \frac{2C_i}{\mu_{FE}} \frac{V_{DS}}{R_{tot}} x}, \tag{8.17}$$

where the "+" sign refers to holes (in which case $\mu_{FE} = \mu_{FEp}$), and "−" to electrons (where $\mu_{FE} = \mu_{FEn}$). Since $V(L) = V_{DS}$, and using Eqs. 8.11 and 8.17 we can eliminate $Q_S(L)$:

$$R_{tot} = \frac{2L}{\mu_{FE} W C_i |2V_{GS} - V_{DS}|}. \tag{8.18}$$

Finally, we calculate the drain current characteristics using Ohm's law, $I_D = V_{DS}/R_{tot}$, along with our solution for R_{tot} in Eq. 8.18. First, let us consider the case of a unipolar transistor where $\mu_{FE} = \mu_{FEn} >> \mu_{FEp}$. Then electrons populate the inversion region when V_{GS}, $V_{DS} > 0$. Including the threshold voltage, V_T, we obtain in the linear region:

$$I_D = \frac{W}{L} \mu_{FEn} C_i \left(V_{GS} - V_T - \frac{V_{DS}}{2} \right) V_{DS}, \quad \begin{pmatrix} 0 \leq V_{DS} \leq V_{DSsat} \\ V_{GS} > V_T \end{pmatrix}. \tag{8.19}$$

Equation 8.19 describes the unipolar drain current characteristics of the OTFT in the linear regime at $V_{DS} < V_{GS} - V_T$. In this regime, the drain current follows Ohm's law, that is, $I_D \sim (V_{GS} - V_T)V_{DS} \propto V_{DS}$.

The saturation current at $V_{DS} > V_{DSsat} = V_{GS} - V_T$ is fixed at its value at V_{DSsat}. Since the conduction is unipolar, saturation occurs when $x_0 = L$ and

$$I_{Dsat} = \frac{W}{2L} \mu_{FEn} C_i (V_{GS} - V_T)^2, \quad \begin{pmatrix} V_{DS} > V_{DSsat} \\ V_{GS} > V_T \end{pmatrix}. \tag{8.20}$$

The saturation current scales linearly with the dielectric capacitance, and not surprisingly, the mobility. The capacitance is proportional to the ratio ε_r/t_i. Therefore, the choice of both the organic semiconductor and dielectric determines the magnitude of the current, and ultimately the gain that the transistor can source.

Two further quantities are useful in characterizing transistor performance. The first is the *transconductance*, g_m (units of [A/V] = [S]), which quantifies the modulation of the drain current due to changes in V_{GS}. The transconductance is related to the transistor gain. Thus, for a constant V_{DS},

$$g_m = \left. \frac{\partial I_D}{\partial V_{GS}} \right|_{V_{DS}} = \begin{cases} \frac{W}{L} \mu_{FEn} C_i V_{DS} & V_{DS} \leq V_{DSsat} \\ \frac{W}{L} \mu_{FEn} C_i (V_{GS} - V_T) & V_{DS} > V_{DSsat} \end{cases}. \tag{8.21}$$

As noted, high gain is also dependent on materials, requiring a high C_i and μ_{FE}. The second quantity is the *output*, or *drain conductance*, g_D, equal to the slope of the I_D vs. V_{DS} characteristics. The output conductance ultimately determines the output impedance of the transistor, and hence the maximum circuit gain that can be achieved. In the absence of drain or source contact resistances:

$$g_D = \left. \frac{\partial I_D}{\partial V_{DS}} \right|_{V_{GS}} = \begin{cases} \frac{W}{L} \mu_{FEn} C_i (V_{GS} - V_T - V_{DS}) & V_{DS} \leq V_{DSsat} \\ 0 & V_{DS} > V_{DSsat} \end{cases}. \tag{8.22}$$

In realistic devices, it is not possible to have a vanishing output conductance in saturation. Diffusion, generation-recombination, and other parasitic currents such as gate insulator leakage contribute to the channel current originating within the transistor depletion region. Leakage depends on both V_{GS} and V_{DS}, and hence $g_D \to g_{Dsat}(V_{GS}, V_{DS})$ at $V_{DS} > V_{DSsat}$, where g_{Dsat} is the output conductance in the saturation region. Furthermore, the source (r_S) and drain (r_D) resistances differ from zero. Their inclusion in the analysis leads to expressions for the *measured values* of g_m and g_D (g'_m and g'_D, respectively) of (Chou and Antoniadis, 1987)

$$g'_m = \frac{g_m}{1 + r_S g_m} \tag{8.23}$$

and

$$g'_D = \frac{g_D}{1 + (r_S + r_D) g_D}. \tag{8.24}$$

Thus, for $r_S \to 0$, then $g'_m \to g_m$. In contrast, in the saturation regime for large leakage (corresponding to g_{Dsat} large), then $g'_D = (r_S + r_D)^{-1}$, that is, the output conductance is limited by sum of the source and drain resistances that includes a combination of both undepleted channel and contact resistances.

From the expressions for I_D in both the linear and saturation modes, we can extract several important features of the OTFT. The dependence of I_D on V_{GS} at fixed V_{DS} (known as the transistor *transfer characteristics*) in these two modes is shown in Fig. 8.15. According to Eq. 8.19, I_D in the linear regime at fixed V_{DS} is proportional to V_{GS} with an intercept at $V_{GS} = V_T$, and a slope equal to g_m (cf. Eq. 8.21). The field effect mobility, therefore, is extracted from g_m. Similarly, in the saturation regime (Eq. 8.20), a plot of $\sqrt{I_{Dsat}}$ vs. V_{GS} yields $\sqrt{\mu_{FEn}}$, with an intercept of V_T. In practice, the mobility should be obtained from the saturation rather than the linear regime data, since the latter may be diluted by series resistance and channel leakage that are less significant as the current is increased into saturation. Deviations at high I_D in both plots are due to series resistance, field-dependent mobility and other non-ideal effects.

The value of μ_{FE} can alternatively obtained from a plot of g_D vs. V_{GS} at $V_{DS} = 0$. From Eq. 8.22, the slope is $\frac{W}{L}\mu_{FEn}C_i$ and the intercept at $V_{GS} = 0$ gives V_T. This measurement is reliable only if the interface charge density is small, since charging or discharging of traps as the Fermi energy is swept through the energy gap with V_{GS} reduces the slope, affecting the mobility measurement.

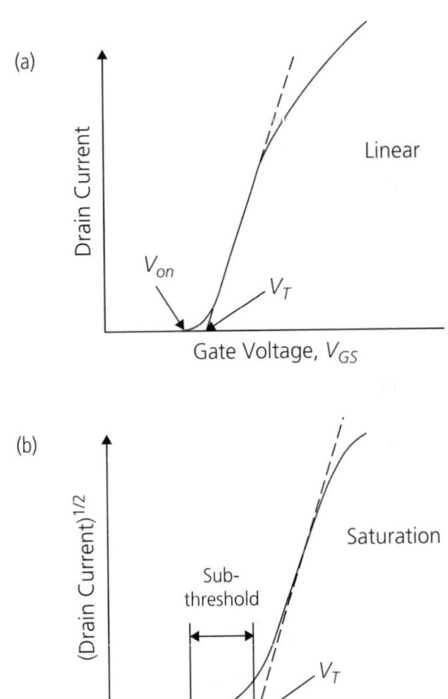

Figure 8.15 Drain current vs. V_{GS} in the (a) linear and (b) saturation regimes of an enhancement mode OTFT. Also shown are the threshold voltage, V_T, and linear extrapolations to the characteristics (dashed line).

We have noted previously in this discussion as well as in Section 4.4.3, that the field effect mobility is not the same as the bulk mobility for several reasons that should be apparent from the foregoing analysis. The field effect mobility is associated with conduction along the organic/insulator interface, and hence is influenced by the presence of charge trapped at the interface. It is also influenced by contact resistances (Eqs. 8.23 and 8.24), the exceptionally high density of charge in the OTFT that leads to filling of the Gaussian density of states (DOS), and static disorder within the channel. This has led to numerous reports of unphysically high single crystal mobilities extracted from the transistor transconductance.

The accuracy of the measurements have been studied using rubrene single crystal OTFTs whose structure is shown in Fig. 8.16a, and a photograph is shown in Fig. 8.16b. The rubrene crystal was grown by physical vapor deposition in flowing Ar in a heated reactor tube. The crystal was attached to the SiO_2 gate dielectric pretreated with OTS, used to improve its adherence to the dielectric. The $I_{Dsat}^{1/2}$ vs. V_{GS} characteristics in the saturation regime ($V_{DS} = -20$ V) are shown in Fig. 8.16c. There is a noticeable kink in the characteristics at $V_{GS} = -6$ V, which differs from expectations from the theory of Eq. 8.20. The mobility extracted from the data at $V_{GS} < -6$ V gives $\mu_{FEp} = 0.9\ cm^2/V\,s$, which is more than six times lower than $6\ cm^2/V\,s$ at $-6\ V < V_{GS} < 0$. A survey of the literature shows that the discontinuity in the transfer characteristics can result in differences of greater than a factor of ten in the field effect mobilities measured using the d.c. output characteristics. This finding is independent of whether the device is based on single crystals or polymers, and for both carrier types (Bittle et al., 2016, Choi et al., 2018).

Figure 8.17 shows fits to the I_D vs. V_{DS} characteristics in both the linear and saturation regimes of transistors employing the two different values of the field effect mobility from Fig. 8.16. The fits to the lower mobility obtained at a more negative gate voltage (Fig. 8.17a) are considerably more faithful to the data than using the higher mobility, suggesting that this latter value is significantly overestimated. Analysis of the transistor performance based on voltage-dependent mobility measurements and admittance spectroscopy indicates that the transition from the high to low mobility regimes is accompanied by an abrupt decrease in contact resistance. At low gate bias,

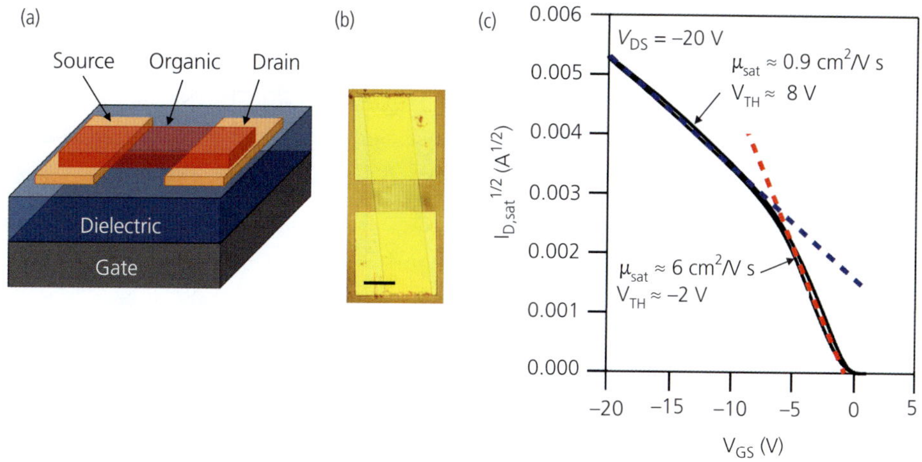

Figure 8.16 (a) Schematic illustration of a single crystal OTFT. (b) Micrograph of a single crystal of rubrene overlaying the S and D and G (not shown) contacts forming the single crystal OTFT. The scale bar is 100 μm. (c) The transfer characteristics of the rubrene transistor in the saturation regime at $V_{DS} = -20$V. After Bittle et al. (2016).

the contact resistance is dominated by thermionic emission of charge at the source contact, and is approximately equal to the channel resistance. As the gate voltage is increased, a high density of charge accumulates under the contact, reducing the injection barrier by the onset of field emission into the channel. With further increases in gate voltage, the resistance remains constant, at value of more than two orders of magnitude smaller than the channel resistance. In this case, the contact resistance no longer significantly impacts the mobility measurement. Indeed, when contact resistance effects are included, the mobility is independent of voltage equal to it value at $V_{GS} < -6$ V. The anomalous differences between the field effect and bulk mobilities are thereby reduced, although these two quantities characterize two very different conduction regimes. Hence, it should be apparent that measurements of very high mobilities from the d.c. transistor characteristics should be treated with considerable caution (Choi et al., 2018).

Returning to the more general solution for ambipolar conduction, the recombination point, x_0, is determined by modeling the accumulation region as a voltage divider comprising two resistances, R_p and R_n (Schmechel et al., 2005):

$$R_p = \frac{2x_0}{\mu_{FEp} W C_i |V_{GS}|},\qquad R_n = \frac{2(L-x_0)}{\mu_{FEn} W C_i |V_{DS}-V_{GS}|} \tag{8.25}$$

with a total channel resistance of $R_{tot} = R_p + R_n$ and $|V_{GS}|/R_p = |V_{DS} - V_{GS}|/R_n$. Thus,

$$x_0 = \frac{L V_{GS}^2}{V_{GS}^2 + \frac{\mu_{FEn}}{\mu_{FEp}}(V_{DS}-V_{GS})^2}. \tag{8.26}$$

As expected, x_0 depends on the relative magnitudes of μ_{FEn} and μ_{FEp}. Here we have assumed that holes are injected at $x = 0$ and electrons at $x = L$. As μ_{FEn} increases relative to μ_{FEp}, x_0 decreases until $\mu_{FEp} \to 0$, corresponding to $x_0 \to 0$, that is, the electron-rich accumulation region extends across the entire channel.

A plot of the charge density distribution for various ratios of μ_{FEp}/μ_{FEn} is given in Fig. 8.18. When the mobilities are equal, $x_0 = L/2$, but as μ_{FEp} increases, the point where the charge density vanishes moves toward the drain. Unipolar conduction corresponds to $\mu_{FEp}/\mu_{FEn} = 100$.

With expressions for R_n, R_p, and x_0, we can calculate a general solution to the drain currents for ambipolar OTFTs. The characteristics are similar to those for a unipolar device in the linear regime in Eq. 8.19, although at higher drain voltages ($V_{DS} \geq V_{GS}$), the I_D vs. V_{DS} characteristics follow

$$I_{Dsat} = \frac{WC_i}{2L}\left[\mu_{FEn}(V_{GS}-V'_T)^2 + \mu_{FEp}(V_{DS}-V_{GS}+V'_T)^2\right],\quad \begin{pmatrix} V_{GS} < V_{DS} \leq V_{DSsat} \\ V_{GS} > V'_T \end{pmatrix}. \tag{8.27}$$

There are several differences between this expression and that for unipolar conduction in Eq. 8.20. For $\mu_{FEn} >> \mu_{FEp}$, the conduction is in fact unipolar, and Eq. 8.27 is the same as Eq. 8.20. Further, the threshold voltage V_T is replaced by V'_T, allowing for the

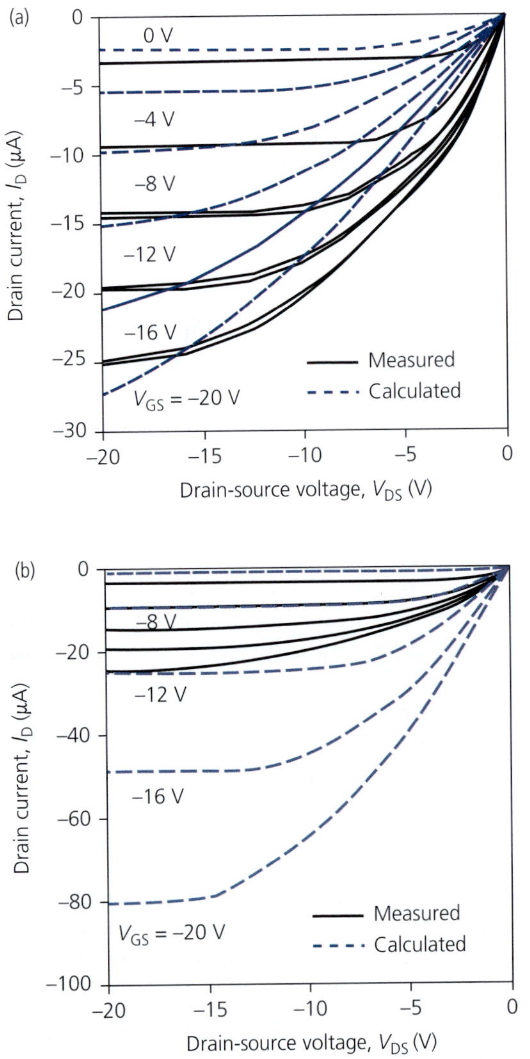

Figure 8.17 Measured and simulated drain current characteristics using the mobility obtained at (a) high and (b) low negative gate voltages from Fig. 8.16 (Bittle et al., 2016).

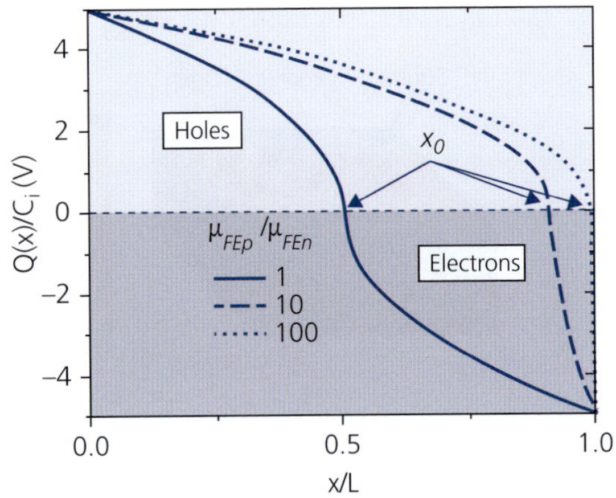

Figure 8.18 Charge density and location of the recombination point, x_0 vs. the relative electron and hole field effect mobilities, μ_{FEp} and μ_{FEn}, respectively. Here, $V_{GS} = -15$ V, $V_{DS} = -10$ V, and $V_T = 0$ V. After Schmechel et al. (2005).

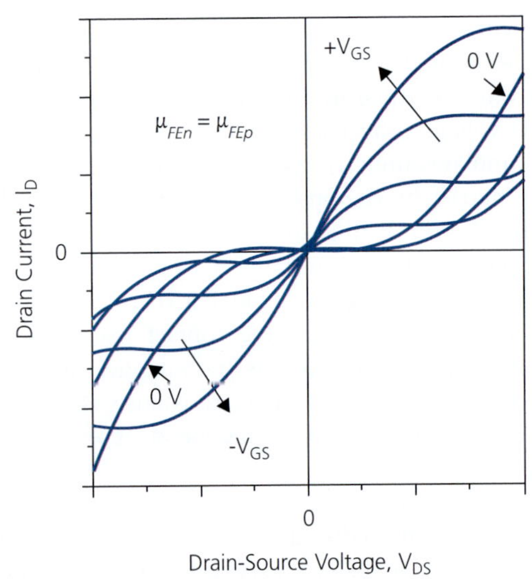

Figure 8.19 Ambipolar OTFT drain current characteristics for $\mu_{FEp} = \mu_{FEn}$. The threshold voltage for both is $V_T = 0$ (Zaumseil and Sirringhaus, 2007).

possibility of two thresholds; one for electrons and the other for holes. This is discussed further below. And finally, there is no current saturation regime. The last term on the right indicates that the current increases quadratically with V_{DS} when $V_{DS} > V_{GS}$.

The drain current characteristics for an ambipolar OTFT with $\mu_{FEn} = \mu_{FEp}$ and $V_T' = 0$ V are illustrated in Fig. 8.19. The onset of the $I_D \propto V_{DS}^2$ dependence is at $V_{DS} = V_{GS}$, as predicted by Eq. 8.27. Hence, a notable characteristic of the curves is that they cross over each other and exhibit saturated behavior over a larger range of V_{DS} with increasing V_{GS}. Furthermore, they are expected to lose their symmetry about $V_{DS} = 0$ when the electron and hole mobilities and threshold voltages significantly differ.

Ambipolar transistors have a comparatively low output conductance, g_D at low V_{GS}, making them less useful than unipolar OTFTs as voltage amplifiers. The transconductance of ambipolar devices is inferred from the I_D vs. V_{GS} characteristics in Fig. 8.20. When there is a large asymmetry in the mobilities, the transfer characteristics are similar to those in Fig. 8.15. However, as the difference in mobilities is reduced, an inflection in the drain current is observed near $V_{GS} = 0$ V, which becomes more pronounced with increasing μ_{FEp}. Inspection of Eq. 8.27 shows that at small V_{GS}, the second term dominates and the slope inverts.

The drain current expression assumes a single threshold voltage, V_T'. In most devices, however, different threshold voltages, V_{Tn} and V_{Tp}, are required to achieve electron or hole accumulation, respectively. We now write the explicit expressions for I_D for the example of $V_{Tp} \leq V_{Tn}$. The ambipolar transistor has three regimes of operation. In the linear regime,

$$I_D = \frac{WC_i}{L}\mu_{FEn}\left(V_{GS} - V_{Tn} - \frac{V_{DS}}{2}\right)V_{DS}, \quad \begin{pmatrix} 0 \leq V_{DS} \leq V_{DSsat} \\ V_{GS} > V_T \end{pmatrix}. \tag{8.28}$$

Beyond the linear regime, the transistor enters the saturation regime, where

$$I_D = \frac{WC_i}{2L}\mu_{FEn}(V_{GS} - V_{Tn})^2, \quad \begin{pmatrix} V_{DS} \geq V_{GS} - V_{Tn} \\ V_{DS} \leq V_{GS} - V_{Tp} \end{pmatrix}. \tag{8.29}$$

Finally, in the ambipolar (quadratic) regime, we have

$$I_D = \frac{WC_i}{2L}\left\{\mu_{FEn}(V_{GS} - V_{Tn})^2 + \mu_{FEp}\left(V_{DS} - V_{GS} + V_{Tp}\right)^2\right\}, \quad V_{DS} \geq V_{GS} - V_{Tp} \geq V_{GS} - V_{Tn}. \tag{8.30}$$

Analogous expressions are obtained for $\mu_{FEn} < \mu_{FEp}$ and $V_{Tp} > V_{Tn}$.

These drain current characteristics are easily reduced to the equivalent unipolar expressions by simply allowing the mobility of the less mobile carrier to vanish. Since nearly all high performance OTFTs operate under unipolar injection, for clarity we will confine the rest of our discussion here to unipolar devices. Example ambipolar transistors, and their application to electronic circuits is found in Section 8.4.4.

In Fig. 8.15, the drain current deviates from zero at $V_{GS} < V_T$. A high performance transistor is characterized by a small drain current below threshold, as well as an onset (called the *switch-on voltage*, V_{on}) very near to V_T. An abrupt current onset enables a small voltage swing to switch the OTFT from the off to the on state. The rate of change of the current at $V_{on} < V_{GS} < V_T$ is known as the *sub-threshold slope*, defined as

$$S = \frac{\partial V_{GS}}{\partial(\log_{10} I_D)}. \tag{8.31}$$

For example, for a Si MOSFET $S \sim 60$ meV/decade of drain current, and for a-Si TFTs, $S = 0.3$–1.5 V/decade. This larger value for amorphous Si is likely due to states that extend into the band gap (tail states) due to structural disorder or defects. A similar situation exists in organics, where both dynamic and static disorder lead to a high density of tail states. It is therefore unsurprising that S is large, and for OTFTs it can be as high as 4–5 V/decade (Lin et al., 1997b), although with improved crystalline order in the channel and a reduced interface charge density of states (DOS), $S = 0.5$ V/decade has been obtained for pentacene/SiO_2 OTFTs (Klauk et al., 2002).

The sub-threshold current is due to thermally excited charge that surmounts the energy barrier between the injecting source contact and the gate, as illustrated in Fig. 8.21 for an OTFT in weak and strong accumulation. Figure 8.21a shows depletion region (shaded area), where there is a barrier with a

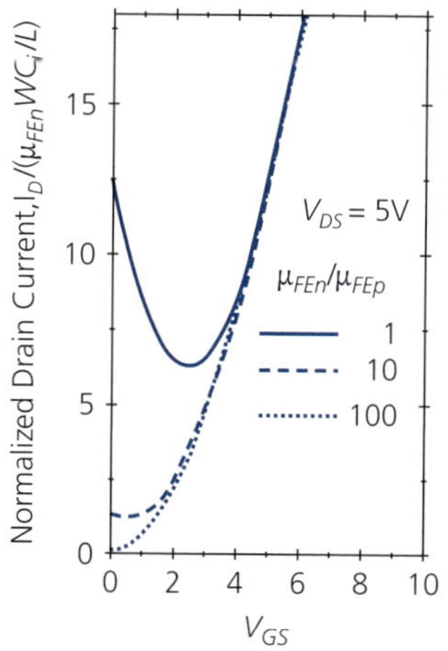

Figure 8.20 Transfer characteristics of an ambipolar OTFT in the saturation regime for various ratios of electron to hole mobility (Schmechel et al., 2005).

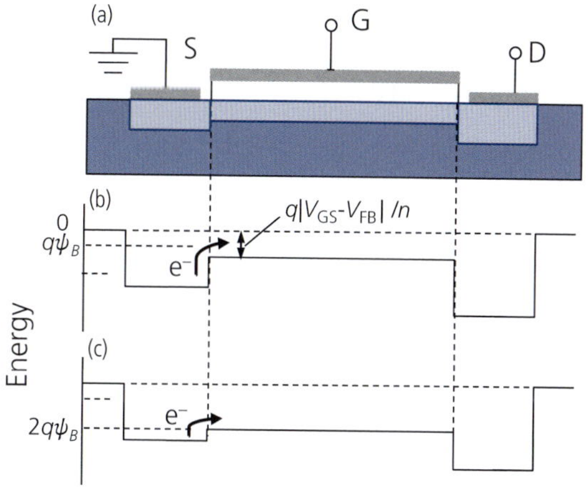

Figure 8.21 Energetic landscape under sub-threshold conditions. (a) Profile of the OTFT. The lightly shaded area under the gate is the depletion region. (b) Energy along the channel under weak accumulation, and (c) energy under strong accumulation. The energy barrier height between source and gate is qV_{GS}/n.

built-in potential at each contact. In Fig. 8.21b, the energy barrier between source and gate decreases with gate voltage as $q|V_{GS} - V_{FB}|/n$, where V_{FB} is the flat band voltage ($V_{FB} = 0$ in Fig. 8.21b), and n is the ideality factor for the injection current. We have shown previously that, in the absence of trapped charge, $V_{FB} = \Delta/q$. At $V_{GS} < V_T$, charge surmounts the barrier by thermionic emission, resulting in a drain leakage current of

$$I_D = I_{D0}\exp(q|V_{GS} - V_{FB}|/nk_BT) = I'_{D0}\exp(qV_{GS}/nk_BT), \tag{8.32}$$

where I_{D0} and I'_{D0} are constants. That is, I_D increases exponentially as the barrier height decreases. In Fig. 8.21c, the barrier is significantly reduced near threshold as $V_{GS} \to 2\psi_B$, improving charge injection into the channel.

The expressions developed thus far, and in particular those leading to Eqs. 8.20 and 8.32, are based on treatments of conventional transistors that operate in the inversion mode. Since OTFTs operate in the accumulation mode, the relationships between the threshold voltage, and the turn-on voltage, V_{on} (and hence, V_{FB}) are simply empirical models. Nevertheless, Eq. 8.32 is in good agreement with OTFT operation in the sub-threshold regime. And while V_T extracted from these characteristics is not identically equivalent to that obtained for inversion-mode FETs, it is nevertheless useful for describing OTFT performance.

The ideality factor is the fraction of the gate voltage dropped across the semiconductor, that is, $n = V_{GS}/\psi_S$. This is calculated by inspection of the gate equivalent circuit in Fig. 8.22. In the presence of traps, the circuit consists of the series/parallel combination of the capacitances of the insulator, depletion region (C_B), and interface traps, C_{IT}. The voltage divider thus yields

$$n = 1 + \frac{C_B + C_{IT}}{C_i} \approx 1 + \frac{C_{IT}}{C_i}, \tag{8.33}$$

where, on the right, we assume that $C_B \to 0$ in an organic where the dopant concentration is very small. The interface capacitance is $C_{IT} = qN_{IT}$, where N_{IT} is the interface charge DOS. Then we obtain $n \geq 1$.

Using Eq. 8.32 in Eq. 8.31, we obtain

$$S = 2.3\frac{nk_BT}{q}. \tag{8.34}$$

For an ideal MISFET, $n = 1$ and thus the smallest possible sub-threshold slope at room temperature is $S = 60$ mV/decade. The trapped charge density is obtained from S using Eq. 8.33.

The sub-threshold characteristics of a DNTT, p-channel OTFT are shown in Fig. 8.23. The device has a BG/TC configuration. The AlO_x gate dielectric is a 3.6 nm layer grown by oxygen plasma vapor deposition (PVD). Prior to vacuum depositing DNTT, the insulator surface is treated with a solution of alkylphosphonic acid that forms a *self-assembled monolayer* (SAM) that modifies the surface energy, and/or surface trap density. Further discussion of the use of SAMs in influencing OTFT characteristics is found in Section 8.6.1.

Figure 8.23 is a semi-log plot from which the sub-threshold slope of $S = 80$ mV/decade is obtained from the exponential characteristics in both the linear and saturation regimes at $V_{DS} = -0.1$ V and -1.5 V, respectively. The threshold at $V_{GS} = V_T = -1.4$ V is obtained from the transition of the current from an exponential to a linear increase in I_D at $V_{GS} = V_T$. The value of S is remarkably close to the theoretical minimum of 60 meV/decade for $n \approx 1$ (Eq. 8.34). An interface trap density of $N_{IT} = 2 \times 10^{12}$ V^{-1} cm^{-2} is inferred from the measured n and using Eq. 8.33.

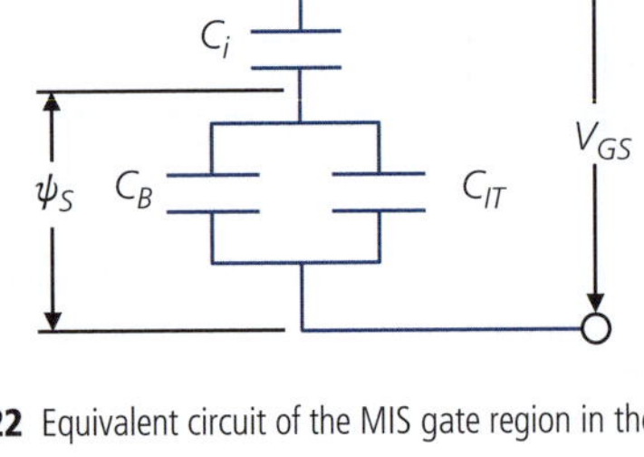

Figure 8.22 Equivalent circuit of the MIS gate region in the presence of interface traps.

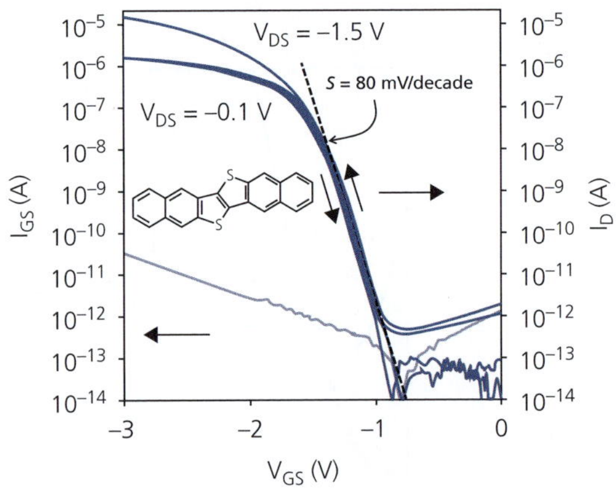

Figure 8.23 Sub-threshold drain and gate leakage current characteristics of a DNTT, p-channel OTFT with dimensions of $L = 10$ μm and $W = 100$ μm. The characteristics in the linear and saturation regimes at $V_{DS} = -0.1$ and -1.5 V, respectively, are shown. The sub-threshold slope (dashed line) is obtained at low I_D. Inset: Molecular structural formula of DNTT (Klauk, 2010).

At $V_{GS} < V_T$, the drain current is due to emission over the source contact-channel barrier, recombination within the channel, shunt and other parasitic leakage paths, and leakage through the gate dielectric. The sum of the leakage currents determines the off current of the transistor. The gate leakage current is also plotted in Fig. 8.23, and is ~1 pA in the saturation regime. The channel resistances, $r_D + r_S$, limit the on current. The transistor performance as a switch depends on the ratio of on-to-off current ratio, $I_{on}/I_{off} = 10^6 - 10^7$, and by S which sets the minimum voltage needed to swing the transistor between these logic states.

8.3.3 Frequency response

The equivalent circuit for the small signal a.c. response of an OTFT is shown in Fig. 8.24. It is customary to use lower case variables (e.g. i_D vs. I_D, v_{DS} vs. V_{DS}, etc.) to denote the small signal values of the corresponding parameters. Now the small signal drain current is

$$i_D(v_{DS}, v_{GS}) = I_D(V_{DS} + v_{DS}, V_{GS} + v_{GS}) - I_D(V_{DS}, V_{GS}). \tag{8.35}$$

This expression is expanded in a Taylor series to give, to first order,

$$i_D = \left.\frac{\partial I_D}{\partial V_{DS}}\right|_{V_{GS}} v_{DS} + \left.\frac{\partial I_D}{\partial V_{GS}}\right|_{V_{DS}} v_{GS}, \tag{8.36}$$

which, using the definitions in Eqs. 8.21 and 8.22, yields

$$i_D = g_D v_{DS} + g_m v_{GS}. \tag{8.37}$$

Both the gate-to-source (C_{GS}) and gate-to-drain (C_{GD}) capacitances contribute to the a.c. input impedance These capacitances are the sum of the geometric dielectric capacitance (Eq. 8.4) and the parasitic gate-source and gate-drain capacitances that result from overlap between the gate and the channel contacts. Under a.c. gate voltage modulation, the reactance from these capacitances shunts the gate voltage. The total equivalent gate capacitance is, therefore,

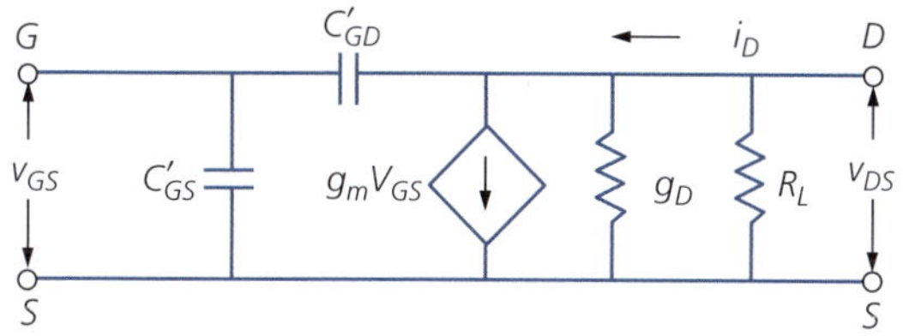

Figure 8.24 OTFT small signal a.c. equivalent circuit. Here, C'_{GS} and C'_{GD} are the total gate-source and gate-drain capacitances, respectively.

$$C_G = C_{GS} + C_{GD}(1 + A_v), \tag{8.38}$$

where A_v is the voltage gain. In an a.c. circuit, the external load resistance, R_L, appears across the drain, and hence A_v is simply

$$A_v = \frac{\partial v_{DS}}{\partial v_{GS}} = \frac{\partial v_{DS}}{\partial i_D}\frac{\partial i_D}{\partial v_{GS}} = \left(R_L \| \frac{1}{g'_D}\right) g_m, \tag{8.39}$$

where, in the a.c. equivalent circuit of a common source amplifier, the transistor output resistance, $1/g'_D$, is in parallel with R_L. For circuits with moderate voltage gain, $R_L < 1/g'_D$, leading to $R_L \| \frac{1}{g'_D} \to R_L$. The drain capacitance in Eq. 8.38 appears at the OTFT input, and hence A_v amplifies its effect on the output impedance. The total drain capacitance

$$C_M = C_{GD}(1 + A_v) \tag{8.40}$$

is known as the *Miller capacitance*, and its amplification is known as the *Miller effect*. The difficulty in precise patterning and alignment of the gate within the channel can result in a significant dependence of the OTFT a.c. response on C_{GD}, and therefore the parasitic capacitances of the source and drain to the gate cannot be ignored.

For a sinusoidal input angular frequency, ω, the small signal input current is purely reactive:

$$i_{GS} = WLC_G \left.\frac{\partial v_{GS}}{\partial t}\right|_{V_{DS}} = j\omega(WLC_G)v_{GS} = j2\pi f(WLC_G)v_{GS}, \tag{8.41}$$

where f is the modulation frequency of the gate. Also, $j = \sqrt{-1}$ indicates that the gate current and voltage are phase-shifted by $\pi/2$. Similarly, the current at the drain is, from Eq. 8.21,

$$i_D \simeq g_m v_{GS}, \tag{8.42}$$

resulting in a current gain of

$$\left|\frac{i_D}{i_G}\right| = \frac{g_m}{2\pi f(WLC_G)}. \tag{8.43}$$

Since the reactive current i_G is linear in f, whereas i_D is frequency independent, we define the maximum frequency of practical operation as the *cutoff*, or *transfer frequency*, f_T, which is defined at the point where $|i_D/i_G| = 1$. Thus,

$$f_T = \frac{g_m}{2\pi WLC_G} = \frac{g_m}{2\pi WL(C_{GS} + C_M)}. \tag{8.44}$$

Apparently, f_T decreases approximately inversely with A_v in high gain transistor circuits, with the maximum frequency response at $A_v \sim 1$. This is an expression of the gain-bandwidth product introduced in the context of photodetectors in Section 7.1.1.

The geometric relationship between capacitances and the electrode configuration that is helpful in analyzing a BG/TC OTFT is shown in Fig. 8.25a. For convenience, we assume a symmetrical device where the insulator and parasitic contact capacitances (C_i and C_C, respectively) are shared equally between the source and drain. The *transfer length*, L_T, is the overlap of the drain and source contacts with the channel that participates in current injection into the channel. The d.c. equivalent circuit in Fig. 8.25b shows that $C_{GS} = C_{GD}$ for this particular device geometry.

Using Eq. 8.21 in the saturation regime, we obtain

$$f_T = \frac{g_m}{2\pi W(L+\Delta L)C_i} = \frac{\mu_{FEac}(V_{GS}-V_T)}{2\pi L(L+\Delta L)}. \quad (8.45)$$

Here, we assume that the *total* overlap of both the gate and drain contacts with the source is ΔL. Then the total insulator geometric capacitance is increased by $L + \Delta L$. We have shown above that the current is limited by both the contact and channel resistances. At high frequencies, the contact resistance can significantly impact the frequency response as well. Indeed, in Eqs. 8.23 and 8.24 we found that g_m decreases with increasing resistance. Since μ_{FE} is determined from a measurement of the transconductance, this implies that the measured mobility, and hence f_T, are also dependent on resistance. To calculate contact resistance effects, we replace V_{DS} in the linear regime with $V_{DS} - I_D R_C$, where R_C is the total contact resistance that is the sum of the contributions from both source and drain (Benor and Knipp, 2008). Then, from Eq. 8.19, we have

$$I_D = \frac{W}{L}\mu_{FE0}C_G\left(V_{GS}-V_T-\frac{1}{2}(V_{DS}-I_DR_C)\right) \times (V_{DS}-I_DR_C), \quad (8.46)$$

where the insulator capacitance is replaced by the total gate capacitance, $C_G = C_{GS} + C_{GD}$, and μ_{FE0} is the channel mobility in the absence of contact resistance. For small V_{DS}, Eq. 8.46 simplifies to

$$I_D = \frac{W\mu_{FE0}C_G(V_{GS}-V_T)V_{DS}}{L+W\mu_{FE0}C_G(V_{GS}-V_T)R_C}. \quad (8.47)$$

Comparison with Eq. 8.19 suggests that the effective mobility measured in an OTFT whose contact resistance is comparable to that of the channel resistance is expressed as

$$\mu_{FEeff} = \frac{\mu_{FE0}L}{L+W\mu_{FE0}C_G(V_{GS}-V_T)R_C}. \quad (8.48)$$

An analogous expression can be derived for the effective mobility in the saturation region starting with Eq. 8.20. In the presence of both parasitic capacitance and resistance, the foregoing yields

$$f_T = \frac{\mu_{FE0}(V_{GS}-V_T)}{2\pi L(L+\Delta L)}\left[\frac{1}{1+W\mu_{FE0}C_G(V_{GS}-V_T)R_C/L}\right]. \quad (8.49)$$

Now $f_T \sim (V_{GS}-V_T)/L^2$, and decreases with both R_C and C_G. Thus, as the voltage increases, so does the charge density in the channel, allowing for more rapid charging and discharging of the gate capacitance in response to a small signal input.

The cutoff frequency vs. channel length is calculated for BG/TC DH7T OTFTs in Fig. 8.26, showing the individual contributions from contact resistance and overlap capacitance. The organic channel is deposited onto a SiO_2 dielectric first coated with an HMDS SAM. The transfer length, $L_T = 2$–3 μm, and $\Delta L = 5$ μm were obtained for the photolithographically patterned gate with lengths as small as 50 nm. The devices have a mobility of $\mu_{FE0} = 0.02$ cm^2/V s, and the fits to the data (points) including all parasitic effects including overlap capacitances, yields a contact resistance of $R_CW = 3.5$ kΩ cm (Hoppe et al., 2010).

A maximum frequency response of 200 kHz is obtained at $L = 1$ μm. As the gate length decreases, the contact resistance exceeds the channel length, limiting

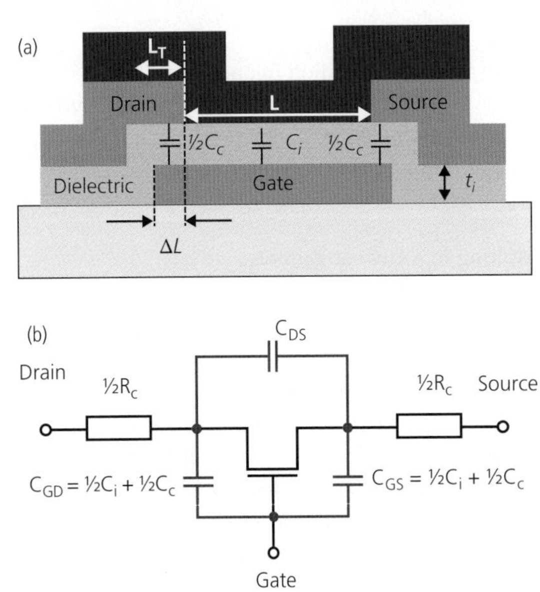

Figure 8.25 (a) Cross-sectional view of a BG/TC OTFT showing the various capacitive circuit elements and geometric factors; (b) d.c. equivalent circuit of the OTFT. The contact capacitance and resistance are C_c and R_c, respectively.

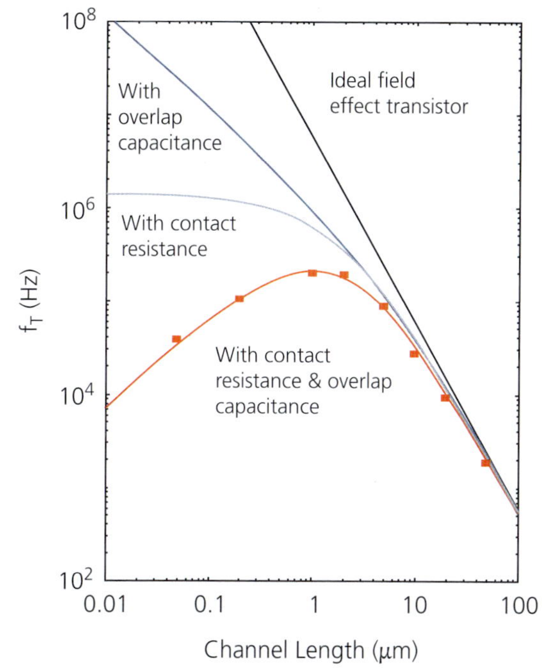

Figure 8.26 Cutoff frequency vs channel length and the separate contributions to the response due to overlap capacitance and contact resistance for BG/TC DH7T OTFT. The fit (red line) to the data (points) is obtained using $R_cW = 3.5$ kΩ cm (Hoppe et al., 2010).

f_T. Indeed, Eqs. 8.48 and 8.49 indicate that the frequency response, even for very high mobility materials, is ultimately limited by contact resistance which, for OTFTs, is rarely below 100 Ω-cm. Thus the prospect for achieving $f_T > 100$ MHz when μ_{FE0} is as high as 10 cm^2/V s and $V_{DS} \sim 10$ V, is unlikely unless a considerable reduction in R_CW below current values of several hundred Ω cm are realized (Klauk, 2018).

Finally, it is important to distinguish the mobility measured under a.c. and d.c. conditions. The mobilities thus obtained are not generally equal, since the small signal response under a.c. modulation may be limited, in part, by the trapping and detrapping of charge that follows the input signal. The a.c. mobility is loosely analogous to a time of flight measurement of the bulk mobility (μ_{TOF}, see Section 4.4.3), except that the values may significantly differ, since μ_{FEac} is influenced by the morphology and charge density at the insulator/semiconductor interface, whereas μ_{TOF} is a property of the bulk.

To understand the scale of μ_{FEac} in OTFTs, we return to the device characteristics in Fig. 8.23. In that example, $C_i = 0.8$ μF/cm^2, $\Delta L = 10$ μm, giving a total input capacitance of 25 pF. A transconductance of $g_m = 15$ μS was measured from the saturation characteristics at $V_{DS} = -3$ V, yielding $f_T \sim 100$ kHz (Klauk, 2010). The best determination of f_T, however, is by direct measurement rather than via inference from the transconductance in Eq. 8.44. That is, g_m is obtained from the d.c. transistor characteristics, which may not accurately extrapolate into the a.c. domain.

A typical test setup for measuring f_T is shown in Fig. 8.27. The operating point of the device under test (DUT) is set by the two power supplies, V_{GS} and V_{DD}. Both a d.c. bias and the oscillating test signal ($v_{GS}(t)$) are applied to the gate through a bias tee. The tee capacitively couples the a.c. signal to the gate, which is isolated from ground by the inductive choke between the V_{GS} supply and the gate. The gate and drain a.c. currents are measured using current amplifiers and a fast digital oscilloscope or spectrum analyzer. The a.c. input signal is typically a sine wave, although a fast pulse, approaching that of a delta function can be used in conjunction with a spectrum analyzer. The fast pulse has a uniform frequency spectrum, and hence its roll-off at high frequency to the point corresponding to $|i_G(t)/i_D(t)| = 1$ from its d.c. value provides a direct measurement of f_T. The disadvantage of the fast-pulse technique is that the pulse magnitude must be sufficient to detect the current signal across the entire spectrum up to f_T.

Results of frequency response measurements on an n-channel, C_{60} OTFT are shown in Fig. 8.28a. The device has a BG/TC configuration, with L varied from 2 μm to 10 μm. Overlap between the gate and the output contacts of $\Delta L = 1$–2 μm gives rise to a parasitic capacitance, C_p. The other transistor dimensions are $W = 1$ mm, $d = 100$ nm, and $t_i = 200$ nm. The data are obtained in the saturation regime with $V_{DS} = 25$ V and $V_{GS} = 20$ V, leading to a linear increase in i_{GS} with f that is consistent with reactive coupling through C_G in Eq. 8.41. In contrast, i_D is relatively frequency independent until $f = f_T = 20$ MHz. The frequency dependence of i_D is more pronounced as L increases due to the increased parasitic C_{GD}. The steady-state calculations of f_T based on g_m, and measurement are shown in Fig. 8.28b (dashed line and data, respectively). Here, g_m is obtained using an electron mobility of μ_{FEn} between 0.86 and 1.1 cm^2/V s, which is a fitting parameter applied to achieve agreement with the measured cutoff frequency. The difference between theory and experiment is largest at large f_T, corresponding to a short gate length. This is attributed to a parasitic capacitance, C_p, not entirely accounted for by the measured gate/drain overlap. A fit to the data assuming an average value for C_p for all gate lengths is shown by the solid line in Fig. 8.28b (Kitamura and Arakawa,

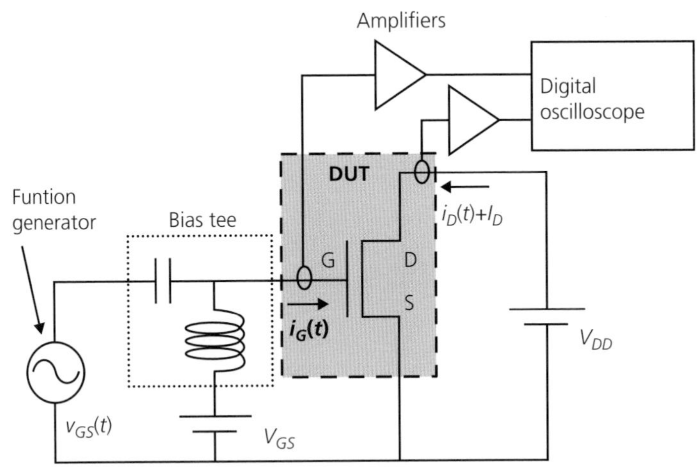

Figure 8.27 Setup for testing the frequency response of a transistor. The shaded region highlights the device under test (DUT).

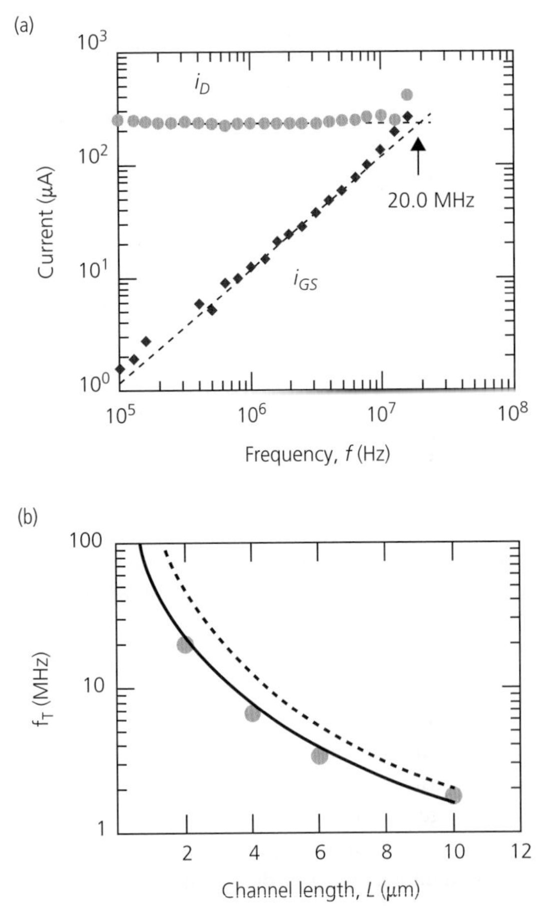

Figure 8.28 (a) Small signal drain and gate currents vs. frequency for an *n*-channel C_{60} OTFT. The crossing point corresponds to $f_T = 20$ MHz for a gate length of $L \approx 2$ μm. Dashed lines are fits to the data. (b) Cutoff frequency vs. gate length. The dashed line is a calculation using the transconductance (Eq. 8.44), and the solid line includes a parasitic capacitance that provides the best fit to the data (Kitamura and Arakawa, 2009).

2009). If the overlap between input and output electrodes is decreased using self-aligned patterning and gate lengths of $L \to 1$ μm, and a decreased contact resistance, f_T may exceed 50 MHz.

A dual transistor, unipolar inverter consists of a load (T1) and a drive (T2) transistor in a common source (i.e. source ground) configuration. The load comprises a transistor whose gate and drain are shorted. This active load is in saturation since $V_{GS} = V_{DS} = V_{DD}$, the supply voltage. The drive transistor source is grounded, and the output is across the load, and is supplied at the drain of T2. The approximate voltage gain of the stage is

$$A_v = \frac{\partial V_{OUT}}{\partial V_{IN}} = -\frac{g_{m2}}{g_{m1}} = -\sqrt{\frac{(W_2/L_2)}{(W_1/L_1)}}, \qquad (8.50)$$

where the last expression follows from Eq. 8.21 for g_m in saturation. Here, $g_{m1,2}$ refers to transistors T1, T2 in Fig. 8.29, respectively. For the data in Fig. 8.29b, $A_v = -2$. Each stage must have $|A_v| > 1$ to fully regenerate the signal from the previous stage. The inverter has an approximate voltage swing of $\Delta V_{IN} = 0.7$ V. For linear operation, the gate voltage of T2 is biased midway within its linear regime, that is, at $V_{IN} \approx -1.7$ V. Importantly, the circuit voltage never swings completely between 0V and $V_{DD} = -4$ V, indicating that even in the quiescent, switched state, the circuit dissipates power. We show below that this limits the number of unipolar stages due to the maximum current that can be sourced, and by circuit heating. Quiescent power dissipation is eliminated via the use of complementary logic.

A *ring oscillator* circuit is another means to determine the transistor frequency response. A ring

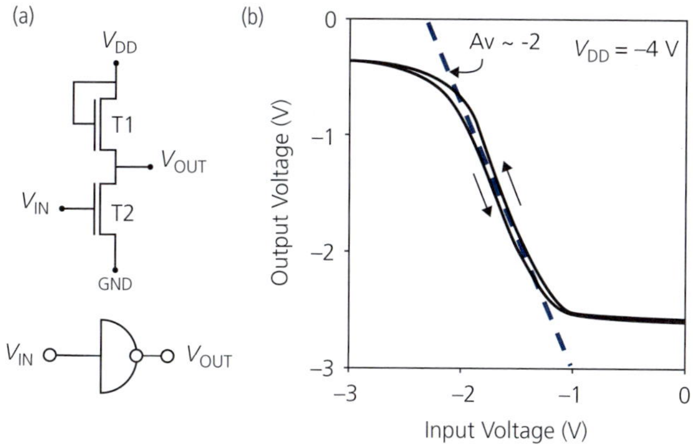

Figure 8.29 (a) Unipolar voltage inverting amplifier circuit consisting of a load (T1) and a drive transistor (T2). The logical symbol for the inverter is shown at the bottom. (b) Inverter voltage transfer characteristics comprising transistors similar to that in Fig. 8.23 shows a small hysteresis (arrows point in direction of the voltage sweep). The dashed line is the slope between the amplifier on and off states in the range $-2\text{ V} < V_{IN} < -1.3$ V. The voltage gain is $A_v = \partial V_{OUT}/\partial V_{IN} = -2$. After Klauk (2010).

oscillator comprises a series connection of an odd number of identical inverter circuits (Fig. 8.29a), with the output of the last inverter tied back to the input of the first inverter in the series. Each inverter input is out of phase by π from its output. Hence, the first inverter switches the second in the opposite sense, propagating through all the stages. The output of the last stage is "high" when the input of the first stage is "low." The feedback loop thus flips the first stage back into its "low" state. Thus, the circuit oscillates (or *rings*) continuously at frequency $2f_{osc} = 1/N\tau_{delay}$, accumulating a delay per stage of $\tau_{delay} = 1/2Nf_{osc}$. The factor of 2 accounts for two transistor state changes per cycle. This number is not identical to f_T since each inverter output drives the input of the following stage. The interconnections between stages and their associated parasitic capacitances give $f_{delay} = \left(2\tau_{delay}\right)^{-1} < f_T$, although the gate delay thus obtained is nevertheless useful for characterizing the transistor small signal response in an actual circuit.

The logical circuit schematic and a micrograph of a 7-stage ring oscillator are shown in Fig. 8.30. The device has an output stage that buffers the oscillator from the subsequent load. The transistors use a small molecule/polymer blend for the channel, where the small molecule is a good hole conductor ($\mu_{FEp} \sim 0.1\ \text{cm}^2/\text{V s}$), and the polymer is used to provide a uniform film morphology. The gate lengths and widths for the BG/BC configuration are $L = 1.5$ μm for both T1 and T2, and $W_1 = 0.2$ mm and $W_2 = 2$ mm (giving $A_v = -3.2$ per stage). The gate dielectric is PVP. A maximum $f_{osc} = 100.2$ kHz is measured, giving a stage delay of $\tau_{delay} = 712$ ns at $V_{DD} = -120$ V (Smith et al., 2008). This corresponds to a cutoff frequency that

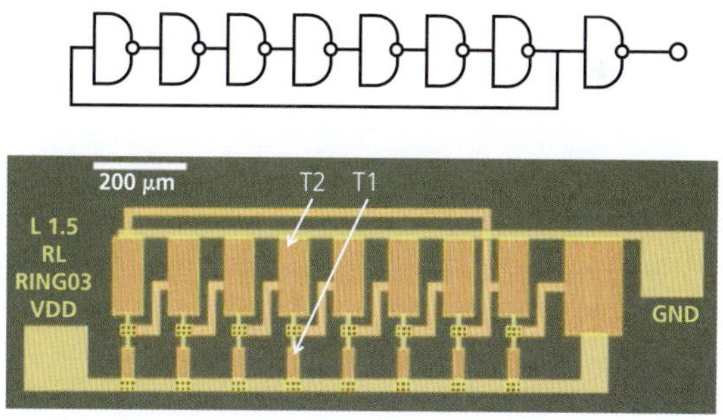

Figure 8.30 Logic diagram (top) and layout of a seven-stage ring oscillator comprising a blended small molecule/polymer channel. The feedback line from the 7th to the first stage is shown at the top of the micrograph. Note the asymmetry in sizes between the load (T1) and drive (T2) transistors in each stage. The last stage is used for driving an external load. White scale bar is 200 μm. The output inverter driver is also shown in the logic diagram, top. After Smith et al. (2008).

is more than ten times smaller than the device in Fig. 8.28, as a result of the ten times lower mobility of the channel layer compared with C_{60}.

The oscillator in Fig. 8.30 has an exceptionally high V_{DD}. Increasing the supply voltage decreases the gate charging time, thus decreasing the gate delay. This is illustrated in Fig. 8.31 for the signal delay measured for 11-stage ring oscillators employing BG/TC transistors. The OTFTs comprised a 20 nm thick layer of the vacuum deposited small molecule, C_{10}-DNTT, with $\mu_{FEp} = 1.2\ \text{cm}^2/\text{V s}$. An AlO_x/SAM insulating layer is identical to that used in the devices in Fig. 8.23. The thin insulator layer and the substantial overlap between the gate and output electrodes of $\Delta L = 5$–40 μm ultimately limits f_{delay}. However, a benefit of the thin insulator is the low OTFT operating voltage used (Zschieschang et al., 2013). To minimize

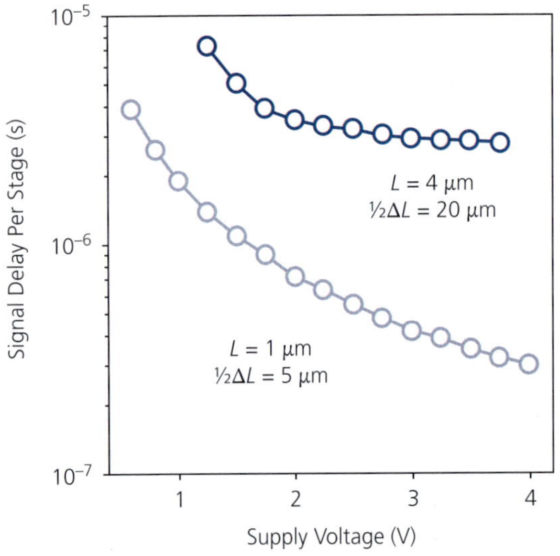

Figure 8.31 Gate delay time per stage for C_{10}-DNTT p-channel OTFTs with two different gate lengths. The circuit is an 11-stage ring oscillator (Zschieschang et al., 2013).

gate capacitance, the widths are only 72 µm and 24 µm for T1 and T2, respectively. As expected, the shorter gates have lower capacitance, and hence a smaller τ_{delay}, which also decreases with increasing V_{DD}. The smallest delay time measured is 300 ns at $V_{DD} = 4$V. From the measured $g_m = 1.2$ mS/mm and $C_G = 90$ pF/mm, the cutoff frequency is $f_T \approx 1/2\tau_{delay} = 2$ MHz. This is reasonably close to the measured $f_{delay} = 1.6$ MHz, indicating that the effects of parasitic capacitance are small.

8.3.4 Transistor noise

The usefulness of a transistor in both analog and digital circuits is often determined by its noise characteristics. In an analog application such as a linear amplifier, the noise of the input transistor stage determines the smallest detectable signal by the subsequent circuitry. Hence, the input noise determines the amplifier dynamic range, which is the ratio of the largest signal that still results in a linear response, to the smallest signal set by the noise floor. In a digital circuit, the *noise margin* defines the maximum allowable noise output of one stage that still correctly triggers the logic state of the following stage (Hauser, 1993). Hence, lower gate noise results in a larger noise margin, and hence more error-free operation of the circuit.

Thin film transistors have several noise sources, each contributing a *mean square noise current spectral density* of $S_{In} = \langle i_n^2 \rangle$. The total noise spectral density, S_{Itot}, is that of all sources added in quadrature. Thus,

$$S_{Itot} = \sum_n \langle i_n^2 \rangle. \tag{8.51}$$

The determination of the separate noise sources in Eq. 8.51 follows an analysis developed for photodetectors in Section 7.1.2.

When the voltage between drain and source is zero for a channel in accumulation, the channel performs as a resistor with conductance, g_0. The noise from a resistor in equilibrium is due to the random arrival of charge at an electrode. It is variously known as *thermal, Johnson*, or *Nyquist noise*, and has a spectral density of

$$S_{th} = \langle i_{th}^2 \rangle = 4k_B T g_0. \tag{8.52}$$

The frequency independence of i_{th}^2 suggests that the noise current increases linearly with the bandwidth of the measurement window, and hence it is a *white noise* source. In saturation, $V_{DS} \neq 0$, and hence the OTFT channel is no longer a homogenous resistor, that is, the conductance varies along the channel. To good approximation, it can be shown that the thermal noise current in saturation is given by (Gray and Meyer, 1984)

$$S_{th} = \langle i_{th}^2 \rangle = \frac{8}{3} k_B T g_m, \tag{8.53}$$

which is commonly used to model Si MOSFETs. Now the channel current is coupled back to the gate via C_G. This, in turn generates a temporal gate potential resulting in a spectrally dependent current noise:

$$\langle i_G^2 \rangle = \frac{4}{5} k_B T g_m \left(\frac{f}{f_T} \right)^2. \tag{8.54}$$

Comparing this to Eq. 8.53, $i_G^2 < i_{th}^2$ unless $f > f_T$, and hence the gate plays an insignificant role in the overall transistor noise.

We have shown in the previous section that OTFTs operate at relatively low frequencies, with $f_T < 10$ MHz. In this domain, the dominant noise source is known as $1/f$, or *flicker noise*. There are many potential sources for $1/f$ noise described by as many models. Nevertheless, the noise behavior generally can be fit to Hooge's empirical formula (Hooge et al., 1981):

$$\frac{S_f}{I_D^2} = \frac{\langle i_f^2 \rangle}{I_D^2} = \frac{\alpha}{N f^\gamma}, \tag{8.55}$$

where α is the *Hooge parameter*, N is the number of carriers in the channel, and $\gamma \sim 1$. Unlike the other white noise sources, $1/f$ noise is frequency dependent. Flicker noise has variously been attributed to random fluctuations in the charge mobility as a function of gate voltage, trapping and detrapping of charges from the insulator/channel interface and from within the bulk of the insulator into the channel, trapping within the channel itself, and parasitic resistances that shunt or lie in series with the channel. Indeed, the magnitude of S_f, and even its slope, γ, depend on the choice of materials, physical dimensions of the OTFT, processes used in fabrication, and so on. The details of the various physical origins are largely contained in the Hooge parameter, α, which is orders of magnitude higher in organics than in Si MOSFETs. To a much lesser extent, these processes also affect γ (Vandamme et al., 2002, Marinov and Deen, 2015).

An example noise measurement circuit schematic is shown in Fig. 8.32. The DUT is biased to its operating point. The drain current is input to a high gain transimpedance amplifier, whose output is connected to an electronic spectrum analyzer. The system should be calibrated with a known noise source. For example, a resistor is a thermal white noise source (S_{th}) with mean square spectral density of $4k_BT/R_{TEST}$ (cf. Eq. 8.52). A resistance value (R_{TEST}) can be chosen that doubles the noise at the spectrum analyzer, in which case the test resistor noise is equal to that of the DUT.

The noise measurement for a diF-TESADT, p-channel OTFT operated in its linear regime is shown in Fig. 8.33. The device has a BG/TC configuration with a 200 nm thick SiO_2 insulating layer, yielding $V_T = 5$ V. Analysis of the drain current characteristics gives $\mu_{FEp} = 0.3$ cm^2/V s (Jurchescu et al., 2008). The noise spectrum has an approximately

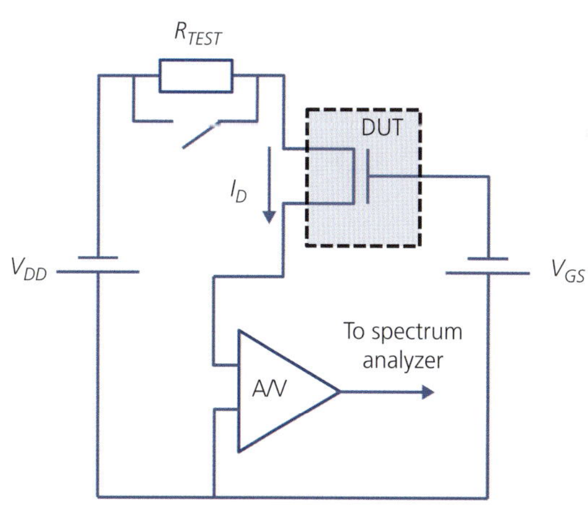

Figure 8.32 Example test setup for measuring transistor channel noise current. The device under test (DUT) is biased at its operating point, and the noise current is amplified (after removal of the d.c. component of I_D) by a low noise transimpedance amplifier (A/V). The output is connected to a spectrum analyzer. The test resistor, R_{TEST}, provides a noise source of known amplitude.

$1/f^\gamma$ dependence over more than two decades in frequency, with a slope of $\gamma = 1.05$.

The origins of the flicker noise in Fig. 8.33 were studied by measuring the hole mobility and noise spectral density as a function of the microcrystalline morphology of the semiconductor in the channel. The size and extent of crystallite formation is controlled by using different surface treatments of the Au S and D contacts, as well as by various treatments of the oxide. The type 1 device has no surface treatments resulting in a very high density of small crystallites. This poorly ordered region is apparent from the uniform coloration of the channel in Fig. 8.34a. For type 2 devices, the contacts are treated with a pentafluorobenzenethiol solution. This resulted in a more ordered, large grain polycrystalline zone extending approximately 10 μm from each electrode into the channel. When the gate length exceeded 20 μm, its central region reverted to the poorly ordered morphology of a type 1 device. Likewise, the type 3 device had type 2-treated contacts, but with an HMDS-functionalized oxide dielectric surface. This allowed for an extension of the large-scale crystallinity to distances of $L \sim 50$ μm.

Figure 8.34b shows that there is a continuous decrease in μ_{FEp} with L for all three transistor types. The rate of decrease significantly increases when the morphology transitions from highly ordered, to a mixture, or ordered with disordered regions in type 2 and type 3 devices. Indeed, the more ordered channels have higher mobilities (0.13 and 0.24 cm^2/V s for type 2 and 3, respectively) as $L \to 10$ μm than the more poorly ordered type 1 device (0.0022 cm^2/V s). The noise, S_f, at short gate lengths decreases monotonically with increasing L, since the fluctuations decrease as the fraction of the ordered region to the total gate length increases (see Fig. 8.34c). Notably, the transitions to lower mobility at $L \sim 25$ μm for type 2, and 50 μm for type 3 OTFTs are accompanied by abrupt increases in $1/f$ noise. From the magnitude of S_f/I_D^2, Hooge parameters at $L = 10$ μm of $\alpha = 50$, 3, and 1.5 are obtained for type 1, 2, and 3 OTFTs, respectively. We infer, therefore, that changes in S_f follow changes in mobility, which in turn are related to the extent of the crystalline order. Hence, $1/f$ noise appears to be due to scattering or charge trapping at grain boundaries. That is, while increased scattering at boundaries decreases mobility, the noise itself may result from random trapping and detrapping of charge at these same sites. Unfortunately, the data in Fig. 8.34 do not distinguish between these two scenarios, which may both be active.

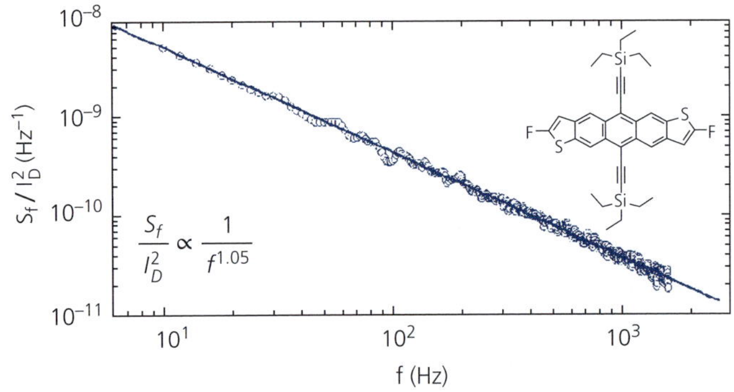

Figure 8.33 Noise spectrum of a diF-TESADT OTFT with channel dimensions of $W = 270$ μm and $L = 10$ μm. The bias conditions are $V_{GS} - V_T = -20$ V and $V_{DS} = -4$ V. Inset: Molecular structural formula of diF-TESADT (Jurchescu et al., 2008).

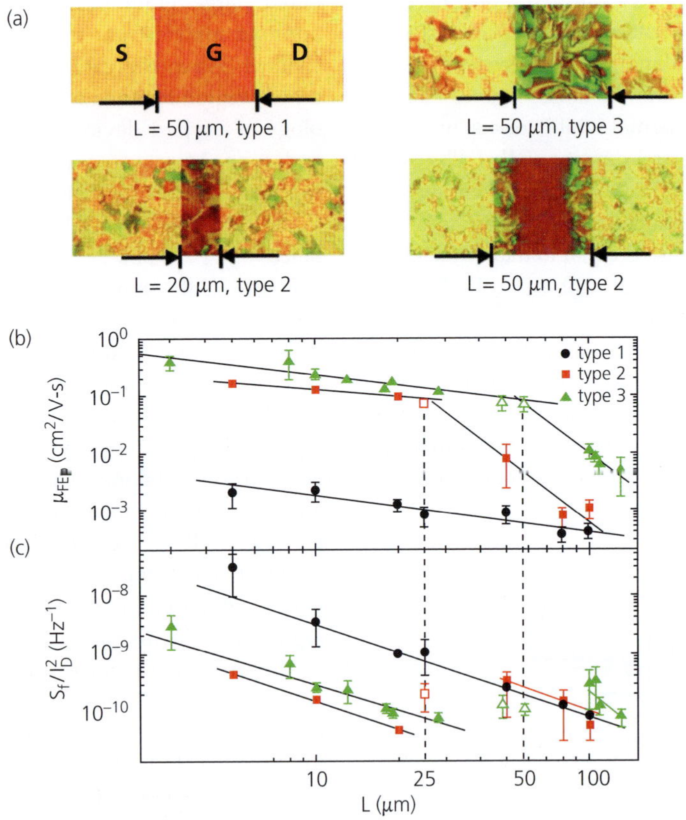

Figure 8.34 (a) Microscope images of diF-TESADT channel OTFTs with different surface treatments of the source and drain electrodes for type 1, 2, and 3 structures. (b) Hole field effect mobility and (c) $1/f$ noise spectrum vs. gate length (L). Dashed lines indicate abrupt transitions in μ_{FEp} (Jurchescu et al., 2008).

Typically, S_f is measured in the linear regime of OTFT operation. The charge density in the channel is thus: $Q = \pm qN = WLC_i(V_{GS} - V_T)$, such that

$$\frac{S_f}{I_D^2} = \frac{q\alpha}{WLC_i(V_{GS} - V_T)f^\gamma}. \tag{8.56}$$

Thus, the noise current scales inversely with gate area (WL), and voltage, V_{GS}. Indeed, the relationship of $S_f/I_D^2 \propto 1/L$ is apparent in Fig. 8.34c. A similar length-dependent scaling along with $S_f/I_D^2 \propto 1/V_{GS}$ has been found for PTV, p-channel OTFTs (Vandamme et al., 2002).

The universality of the Hooge empirical relationship is shown by a compilation of the noise spectra measured from a large number of both small molecule and polymer OTFTs in Fig. 8.35. Regardless of

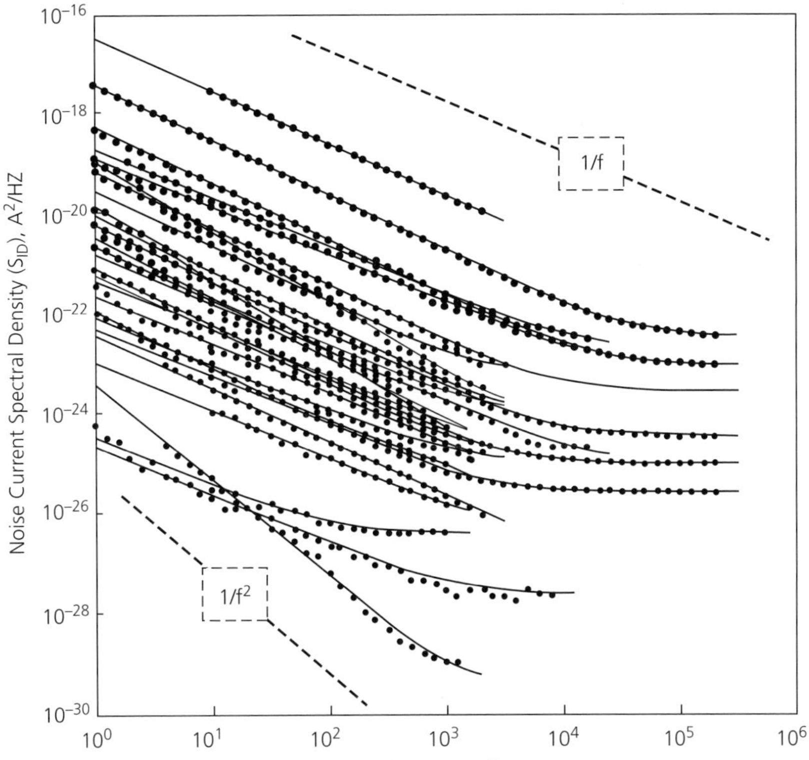

Figure 8.35 Noise spectral densities for a variety of polymer and small molecule OTFTs (data points). Dashed lines show power laws for $S_f \sim 1/f$ and $1/f^2$. Solid lines show fits to $S_f(1Hz)/f + S_{white}$ (Marinov and Deen, 2015).

the fabrication conditions, materials, device geometry, and other laboratory conditions used during measurement, with only one exception, the spectral density data (points in the figure) follow $S_f(1\text{Hz})/f + S_{white}$ (solid lines). The white noise component (S_{white}) provides an asymptotic value of S_f at high frequencies for many of the data sets. White noise has two possible origins: it represents either the noise floor of the measurement system, or it is due to channel noise given by Eq. 8.52. It is indeed remarkable that the dominant noise source in OTFTs is nearly independent of the details of the device or fabrication conditions. Statistical analyses of this population gives $\gamma = 1.1 \pm 0.2$, with a Hooge parameter that has a statistical spread with a standard deviation of $\sigma \sim 9$ dB_{10}. This is no larger than that for inorganic CMOS and bipolar transistors. However, the magnitude of the flicker noise in OTFTs is ~3000 times higher than for inorganic devices (Marinov and Deen, 2015).

In addition to random fluctuations in the noise current, variations in V_T due to small differences in the insulator/channel interface defect density or distribution, film morphology from process variations during fabrication, unintentional doping or impurities in the channel, etc. determine whether or not a circuit consisting of a series of stages can consistently operate in a fault-free manner. The transistor noise margin provides a quantitative metric that characterizes the ability of a transistor to operate reliably (i.e. its "robustness" to variations from device to device) in a circuit. The noise margin is the maximum variation tolerated at the output of a logic gate that results in error-free switching at the input of the following gate in a circuit.

The noise margin can be understood in the context of an infinite chain of inverters in Fig. 8.36a. The challenge is to determine if the voltage noise at the input of an inverter rises to a level that inadvertently swings its output to a state (e.g. from high, H, to low, L) that is different from that required by the previous stage (e.g. H). Figure 8.36b shows the equivalent circuit of the logic chain that can be modeled as a cross-coupled inverter pair with noise sources δV_a and δV_b. In the equivalent, an H output from amplifier A_a drives the output of amplifier A_b to L. This is identical to each pair of gates in the infinite circuit in Fig. 8.36a.

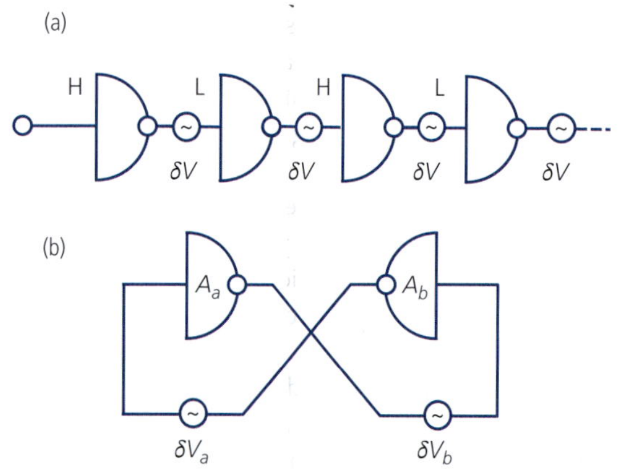

Figure 8.36 (a) An infinite chain of inverters with the high (H) and low (L) inputs at alternate stages. The noise voltage input at each stage contributed by the previous stage is δV. (b) Equivalent circuit of the infinite chain. Each stage has a voltage gain of A.

The transfer function of the cross-coupled pair is shown in Fig. 8.37. The curve corresponding to $V_{out} = f(V_{in})$ is the transfer characteristic of an inverter (e.g. with its circuit shown schematically with its transfer function in Fig. 8.29). Its cross-coupled pair is shown as the mirror image, $V_{in} = f^{-1}(V_{out})$, where V_{out} is the stage output and V_{in} is its input voltage. The circles in the figure indicate the voltage at which the drive transistor transitions from the linear to the saturation regime ($V_{in,a}$) with the load in saturation, or when the load transistor moves from the saturation to the linear regime ($V_{in,b}$) with the drive transistor in saturation.

The inverter operates error free if the input voltage is greater than V_{IH}, which corresponds to the minimum voltage needed to assure the inverter will switch to its output low voltage, V_{OL}. Similarly, a low input voltage must be less than V_{IL} to drive the output into its high state at V_{OH} (Hauser, 1993). These operating conditions are summarized as follows:

$$\begin{gathered} V_{in} \leq V_{IL} \Rightarrow V_{out} \geq V_{OH} \\ V_{in} \geq V_{IH} \Rightarrow V_{out} \leq V_{OL} \\ V_{IH} > V_{IL} \end{gathered} \tag{8.57}$$

We define the noise margins for the high and low states as

$$\begin{aligned} NM_H &= V_{OH} - V_{IH} \\ NM_L &= V_{IL} - V_{OL} \end{aligned} . \tag{8.58}$$

The noise margins are shown graphically by the solid rectangles in Fig. 8.37 for a cross-coupled inverter. For the most robust operation, the area within the rectangles should be maximized. Comparison with Fig. 8.37 suggests there are two ways to increase

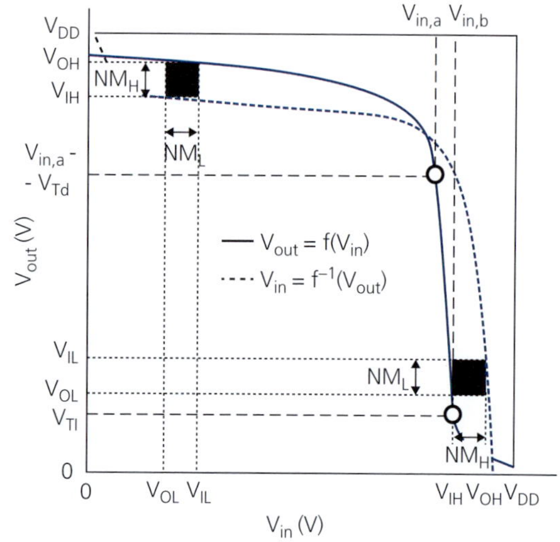

Figure 8.37 Illustration of the noise margin (*NM*) of a series of inverters as in Fig. 8.36 (De Vusser et al., 2006).

NM: decreasing the stage gain, which decreases the slope of the transfer characteristics, $A_{vI} = \partial V_{OUT}/\partial V_{IN}$, and shifting the threshold voltage of the drive and load transistors to move the transfer characteristic in a direction that increases the separation between input and output states in Eq. 8.57.

8.4 Alternative thin film transistor architectures

Figure 8.2 illustrates various common OTFT configurations. However, several other transistor designs have been explored due to potential applications not served by the conventional, lateral architecture. These include dual gate OTFTs, depletion mode, and vertical channel transistors. In this section we describe several of these alternative designs and what they offer that differs from those already discussed.

8.4.1 Dual gate OTFTs

The dual gate OTFT shown in Fig. 8.38 consists of the channel sandwiched between two insulators with opposing gate contacts, called the bottom and top gates. Voltage applied to the bottom gate changes the potential across the channel, which in turn changes the operating point (i.e. V_T) of the transistor. The device can be thought of as two parallel transistors sharing a common channel. For example, consider a p-channel OTFT operating in the accumulation-saturation regime. Then, the top gate potential is $V_{GT} - V_{TT} < 0$, where V_{GT} is the top gate-source voltage, and V_{TT} is the top gate threshold voltage, both of which are < 0V. Application of a small positive potential to the bottom gate (V_{GB}) partially depletes the channel. This reduces the charge at the top of the channel (the top channel) by the ratio of the bottom-to-top capacitances, C_B and C_T, respectively. The effect is a shift of the top channel threshold voltage by:

$$\Delta V_{TT} = \frac{C_B}{C_T} V_{GB}. \tag{8.59}$$

Conversely, if the bottom gate is negatively biased, the top gate threshold is shifted in the opposite direction. At some point, $V_{GB} - V_{TB} < 0$, where V_{TB} is the threshold voltage of the bottom gate, leading to accumulation at the bottom channel. Since the source and drain contacts of both the OTFTs are shared, this additional accumulation layer adds to the total drain current. Increasing $|V_{GB}|$ higher than $|V_{GT}|$ eventually leads to a shift of the accumulation layer to the bottom channel while completely depleting the top channel.

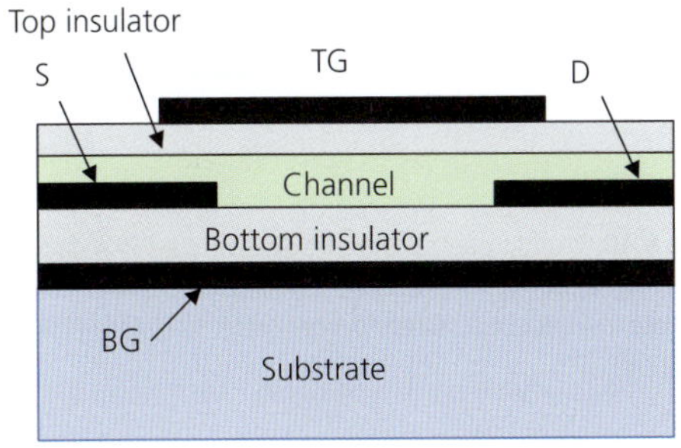

Figure 8.38 Schematic cross-section of a dual gate OTFT.

The second gate can be used to control the threshold voltage over a substantial range, which increases the noise margin in large integrated circuits. Indeed, very large circuits including a 99-stage ring oscillator and a 64-bit radio frequency identification (RFID) transponder have been demonstrated using dual gate OTFT technology (Myny et al., 2011). The design is appealing due to its simplicity, and its compatibility with methods used in fabricating large and complex circuits. The top insulator can also serve as an encapsulation layer that can retard degradation in performance over time. Requiring a more complex biasing scheme consisting of four rather than three terminals to each transistor partially offsets benefits gained by its fabrication simplicity.

Dual gate transistors have been implemented using several different small molecule and polymer semiconductor channels (Cui and Liang, 2005, Gelinck et al., 2005, Morana et al., 2005, Maddalena et al., 2008, Brondijk et al., 2012). Example transfer characteristics of a dual gate PTAA, p-channel OTFT are shown in Fig. 8.39. The bottom gate-substrate/insulator combination is p-Si/1 μm SiO_2 functionalized with HMDS before the 80 nm thick PTAA channel is deposited. The top gate/insulator comprises a Au contact on 400 nm thick PIBM. At $V_{GT} = 0$, the device behaves as a single gate transistor with $V_{TB} = 0$ V. As the top gate is positively biased, the channel depletes, resulting in a threshold shift towards more negative V_{GB}. The additional potential is required to restore the charge under the bottom gate to its value at $V_{GT} = 0$. Conversely, as the top gate is negatively biased, the

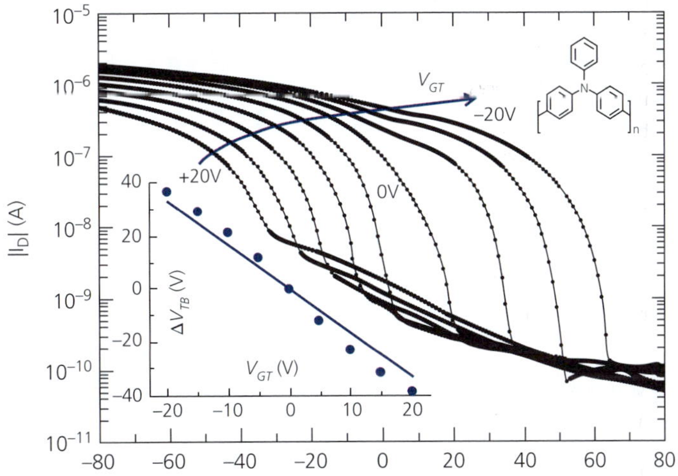

Figure 8.39 Drain current vs. bottom gate bias (V_{GB}) characteristics for a PTAA dual gate transistor with $W = 1$ cm by $L = 10$ μm, operating in saturation. The curves correspond to various top gate biases (V_{GT}) in 5V steps. Bottom inset: Threshold voltage shift of the bottom gate vs. V_{GT}. Top inset: Molecular structural formula of PTAA (Spijkman et al., 2008).

top channel enters the accumulation mode, and the transfer characteristics require more positive bias to maintain the same drain current (i.e. the charge in the bottom channel is reduced, thereby decreasing I_D). The linear dependence of the threshold shift on V_{GT} predicted by Eq. 8.59 is shown in the lower inset. The threshold voltage is varied over approximately 80 V, with a 40 V swing in V_{GT} (Spijkman et al., 2008).

The change in slope at $V_{GT} < 0$V and $V_{GB} > 0$ is due to increasing depletion of the channel with V_{GB}, as the conducting region shifts toward the top gate. The charge density in the top accumulation region is much larger than the background concentration in the channel. Hence, once the channel is depleted all the way to the top accumulation region, an additional bottom gate voltage must be applied to completely turn off the drain current. Hence, sweeping charge from the undoped channel region takes less voltage than doing so in the accumulation region, resulting in the change in slope, indicated by the kink at bottom gate biases between 10 V and 20 V. The channel leakage current is also affected by changes in V_{GT}. When the channel is strongly depleted at $V_{GT} > 0$, the OTFT leakage is significant (see the rising current following the steep drop in drain current), and is equal to the top channel current.

Increased noise margins are achieved using inverters comprising the dual gate OTFT in Fig. 8.39. The inverter circuit and its transfer characteristics are shown in Fig. 8.40. Due to the ability to shift the threshold voltage, the noise margin of the inverter is dramatically improved from that obtained using a single gate inverter. The dual gate $NM = 5.9$ V is achieved at $V_{GT,driver} = 50$ V and $V_{GT,load} = -10$ V, compared to the single gate inverter with $NM = 0.5$ V (Spijkman et al., 2008).

Increasing NM using the dual gate design leads to reduced errors in circuit operation compared to single gate devices. This was necessary for the realization of the RFID transponder shown in Fig. 8.41 (Myny

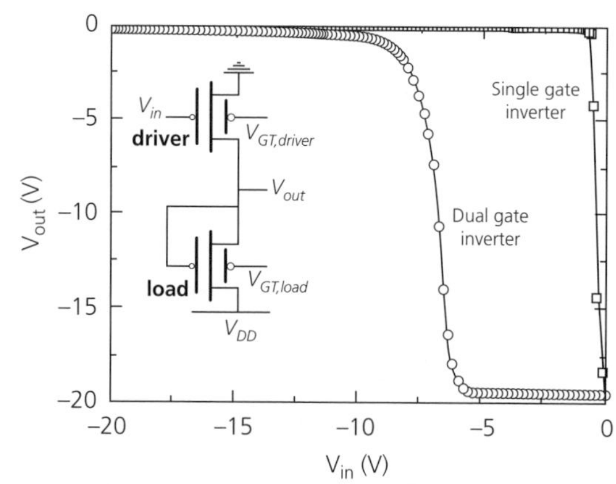

Figure 8.40 Inverter transfer characteristics of the PTAA dual gate OTFTs in Fig. 8.39. Also shown are the characteristics for a comparable single gate inverter circuit. Inset: Inverter circuit schematic (Spijkman et al., 2008).

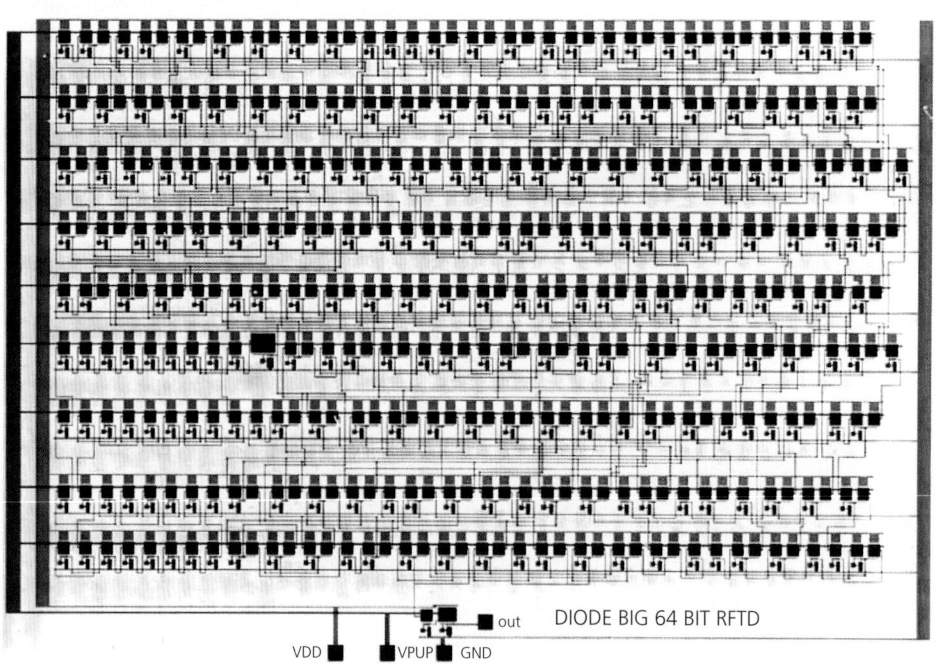

Figure 8.41 Photographic image of a 64-bit RFID transponder operating at 4.3 kb/s using dual gate inverter logic (Myny et al., 2011). Copyright 2011 IEEE. Reprinted, with permission, from IEEE Journal of Solid-State Circuits, 46, 1223–30.

et al., 2011). The dual gate allows for flexibility in design of logic circuits that require both depletion and/or accumulation mode operation, level shifting, etc. The 4.3 kb/s circuit employs spin-cast pentacene dual gate transistors with a polyvinylphenol insulator for both top and bottom dielectrics (Gelinck et al., 2005). While dual gate transistors add some complexity that may not be necessary in simple analog circuits, they may well become indispensible in more sophisticated multistage digital circuits given their ability to precisely adjust the operating points of devices that otherwise may be subject to variations in materials properties and fabrication processes.

8.4.2 Doped channel OTFTs

We have noted that conventional OTFT channels are undoped. Nevertheless, unintentional doping can be a source of channel leakage that limits the on/off ratio, and also affects the threshold voltage. In contrast, intentional doping of the channel can compensate charges due to ionized impurities and defect states that lead to shifts in V_T (Abe et al., 2005, Lee and Chen, 2011, Hählen et al., 2012, Olthof et al., 2012, Hein et al., 2014, Lüssem et al., 2016), while it can significantly reduce channel contact resistance (Minari et al., 2007, Tiwari et al., 2010, Wakatsuki et al., 2011, Günther et al., 2016). It has been suggested that doping can lead to depletion and inversion mode operation (Lüssem et al., 2013), although convincing demonstrations of such OTFTs are absent.

The dependence of V_T on the purity of the channel semiconductor, and the impact of doping on the charge mobility has been investigated for BG/TC C_{60} n-channel OTFTs with dimensions $W = 1.2$ mm, $L = 100$ µm, and a SiO_2 dielectric with $t_i = 200$ nm. The channel source material was either 99.9% pure C_{60} evaporated as-received from the supplier or after purification by vacuum thermal gradient sublimation (see Section 5.3.2). The channels were doped by co-evaporation with the n-type dimer dopant, $[RuCP^*(mes)]_2$, at molar ratios (MR) between 1.8×10^{-4} to 1.1×10^{-3}. The dopant molecules react to form two $[RuCP^*(mes)]^+$ cations that provide negative polarons to compensate positive charges from traps within the C_{60}.

The drain current transfer characteristics of the OTFTs employing as-received C_{60} with and without doping are shown in Fig. 8.42a. The threshold voltage is a sensitive function of MR, shifting from $V_T = 17.9$ V for the undoped channel, to only 4.7 V for $MR = 1.1 \times 10^{-3}$ (see Fig. 8.42b). The change of

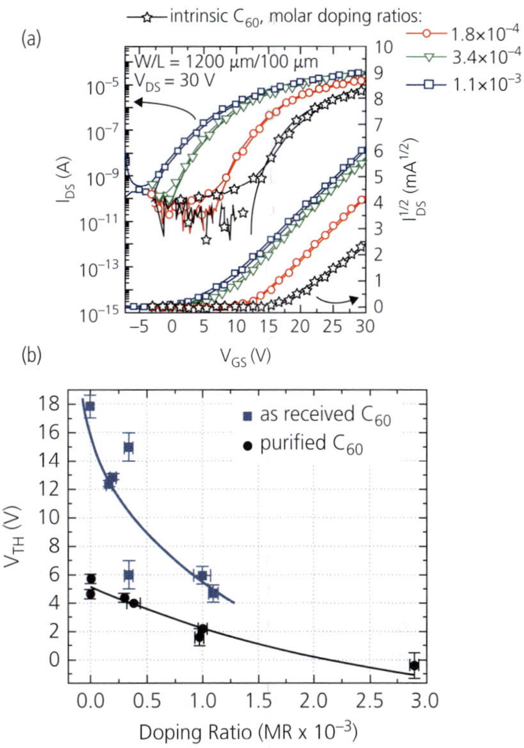

Figure 8.42 (a) Drain current transfer characteristics of n-channel OTFTs using as-received C_{60} channels that are undoped (labeled intrinsic), and with different molar ratios (MR) of the acceptor, $[RuCP^*(mes)]^+$. (b) Threshold voltages of as-received and purified C_{60} OTFTs vs. dopant MR (Olthof et al., 2012).

the threshold on doping concentration decreases as MR increases, setting an upper limit to the density of background charges. Similarly, the electron mobility increases from ~0.4 cm^2/V s for the undoped to 0.6 cm^2/V s for the doped samples, also suggesting that the charge density in the channel increases with MR. In Section 4.3.2, we found that mobility increases with filling of the DOS, pushing the mobility edge closer to the LUMO. Hence, compensation of residual positive charge in the channel by doping increases the electron density, and hence μ_{FEn}. Note also that the I_{on}/I_{off} is not strongly dependent on MR, increasing by only a factor of 10–30 over the entire range of MR studied.

The effects on V_T are even more dramatic for channels employing pre-purified C_{60}. The V_T for the purified source material is 4.7 V, equivalent to that for the heavily compensated channel for the as-received C_{60}. Addition of the dopant is effective in neutralizing adventitious impurities in the as-received C_{60}. A second conclusion, also discussed in Section 5.2, is that material purification is essential to achieving high performance electronic and photonic devices (Bailey-Salzman et al., 2006). Additional doping continues to reduce V_T to as low as ~0 V at $MR = 2.9 \times 10^{-3}$ (Fig. 8.42b). However, at this high doping concentration, I_{on}/I_{off} is reduced to 70 due to the high electron concentration in the channel. A second indication of the improved purity of the C_{60} is the relatively high mobility of ~1.6–1.7 cm^2/V s for the intrinsic and doped OTFTs.

Transistor performance is influenced by the total output resistance, which is the series combination of the channel resistance, $R_{ch} = 1/g'_D$ (cf. Eqs. 8.22 and 8.24), the depleted semiconductor under source and drain contacts, R_{bulk}, and the resistance of the contact itself to the channel layer, R_{int} (see Fig. 8.43a). Ignoring channel leakage, in the linear regime of operation, the total on resistance of the OTFT above threshold is

$$R_{ON} = R_{bulk} + R_{\text{int}} + \frac{1}{g_D} = R_C + \frac{L}{\mu_{FE} W C_i (V_{GS} - V_T)}. \tag{8.60}$$

To separately determine R_C and R_{ch}, either the gate length or voltage is varied while measuring I_D *vs.* V_{DS}.

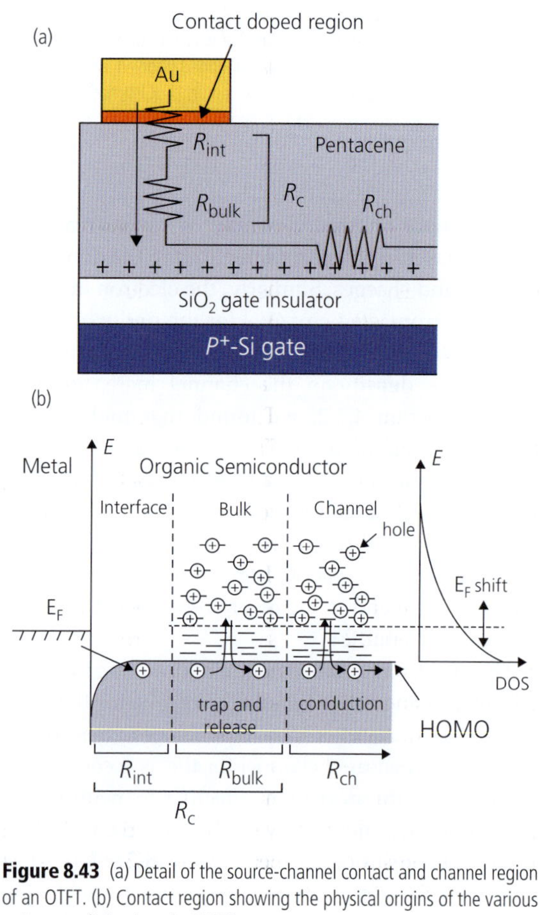

Figure 8.43 (a) Detail of the source-channel contact and channel region of an OTFT. (b) Contact region showing the physical origins of the various resistances (Minari et al., 2007).

The slope gives R_{ch} and extrapolating the data to $L \to 0$ gives R_C.

As the contact and other parasitic channel resistances increase, the measured output conductance decreases (see Eq. 8.24), which limits the gain in a common source inverting amplifier. Doping under the gate (Minari et al., 2007) or inducing damage from deposition of the metal contacts are means for reducing R_{int}. Doping, therefore, results in an output conductance that is limited by the depleted bulk resistance, and R_{ch}. The bulk resistance is found to follow the *Meyer–Neldel rule* (MNR) that asserts that conductivity is thermally activated:

$$\sigma = A \exp(-\Delta E_a / k_B T), \tag{8.61}$$

with a prefactor of

$$A = \sigma_0 \exp(\Delta E_a / k_B T_0). \tag{8.62}$$

Here, σ_0 is a constant and ΔE_a is the activation energy. The MNR is found to be broadly applicable to disordered solids, and specifically to organic semiconductors (Kemeny and Rosenberg, 1970, Roberts, 1971, Gutmann et al., 1983, Metselaar and Oversluizen, 1984). There is no concise derivation of Eqs. 8.61 and 8.62, although the MNR has been attributed to thermal activation over the interface, a Schottky barrier between the metal and the semiconductor, thermally activated tunneling due to small polaron hopping between molecules (Kemeny and Rosenberg, 1970), or an exponentially decreasing DOS into the energy gap of the disordered semiconductor (Roberts, 1971). For this latter case, T_0 is a characteristic temperature determined by the slope of the DOS, as illustrated in Fig. 8.43b. This is similar to T_0 employed to model the current-voltage characteristics in disordered organic media discussed in Section 4.4.2.

Since $R_{bulk} \propto 1/\sigma$, we have from Eqs. 8.60 and 8.61:

$$R_{bulk} = R_{bulk0} \exp(-\Delta E_a / k_B T_0) \exp(\Delta E_a / k_B T), \tag{8.63}$$

where $R_{bulk,0}$ is a temperature-independent prefactor.

The validity of MNR has been explored in BG/TC pentacene OTFTs where the p-channel is doped with a 3 nm thick layer of the acceptor, F_4TCNQ, located immediately beneath the Au source and drain contacts; see Fig. 8.43a. With the doped layer, the interface resistance, R_{int} is three times less than in devices where it is absent. For doped contacts, $R_{int} \ll R_{bulk}$. The contact resistance vs. T at several gate voltages $V_{GS} > V_T$ extracted from the OTFT in the linear regime (V_{DS} small) is shown in Fig. 8.44. A feature of Eq. 8.63 is that the resistance is thermally activated. The data taken at various V_{GS} intersect at a single point corresponding

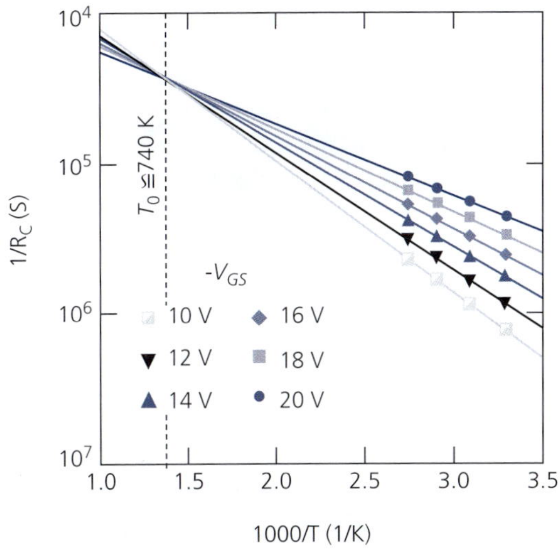

Figure 8.44 Contact resistance vs. 1000/T for several gate voltages for the device in Fig. 8.43a with $W \approx 1$ mm, $L = 20$ μm, and $t_i = 200$ nm SiO_2. The resistances are thermally activated, crossing at $T_0 = 740$ K, corresponding to an activation energy of $\Delta E_a = 64$ meV. After Minari et al. (2007).

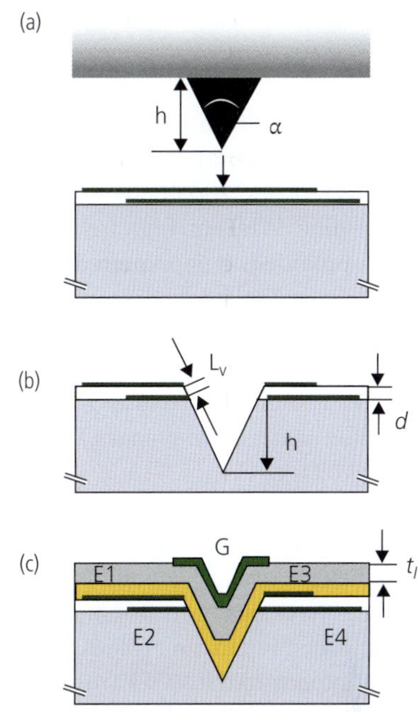

Figure 8.45 Fabrication sequence for a vertical organic field effect transistor (VOFET). (a) Source and drain contacts (black) separated by a polymer insulator (gray) of thickness d are deposited onto a plastic substrate. (b) It is cut by a wedge-shaped punch, leaving a "V"-shaped notch in the layers and substrate. (c) The polymer channel layer (orange) is deposited, followed by the gate insulator and the top gate contact (Stutzmann et al., 2003).

to T_0, indicating that the approximation of $R_C \cong R_{bulk}$ is valid. The data yield an activation energy of $\Delta E_a = k_B T_0 = 64$ meV (Minari et al., 2007). This is double that of the channel conductance, suggesting that hopping between molecules stacked end-to-end in the bulk, and parallel to the substrate plane requires less energy compared to that needed to overcome the contact barrier.

The thermal activation of R_{bulk} inferred from Fig. 8.44 indicates the bulk conductivity is ultimately limited by charge trapping in an exponential DOS. A similar situation results in an increase in channel conductivity with V_{GS} as the Fermi energy approaches the HOMO energy. The injection process is illustrated schematically in Fig. 8.43b.

While doping of the contacts and the bulk are both means for improving the output characteristics of an OTFT, it can also increase the channel conductance to the point that I_{on}/I_{off} is significantly reduced (Lim et al., 2007, Lee and Chen, 2011, Olthof et al., 2012).

8.4.3 Vertical organic field effect transistors

The current sourced by the OTFT and its frequency response are limited by the channel length, L. Large gate lengths require large V_{DS} for the device to achieve saturation. In many of the examples above, V_{DS} between 20 and 100 V are common. Orienting the gate along the vertical direction, perpendicular to the substrate plane, in principle, eliminates this shortcoming. In the *vertical organic field effect transistor* (VOFET), the gate length is determined by the channel layer thickness (a few hundred nanometers), which in principle should lead to high frequency operation. In this geometry the current is determined by the device diameter, which can be considerably higher than in lateral OTFTs with a similar device footprint. An example VOFET and its fabrication process is illustrated in Fig. 8.45. The source and drain contacts are deposited on a PET substrate, and are separated from each other by a 700 nm thick insulating layer of PVP. This trilayer structure is embossed using a V-shaped metal wedge of height $h = 10$ μm, and with angle $\alpha = 70°$. The wedge cuts a channel through the metal contacts and the substrate. The p-channel material, F8T2, is deposited from solution. The gate insulator is a second PVP layer, followed by the gate contact electrode.

The VOFET can be operated employing two geometries. By using electrodes E1 and E3 as source and drain, a conventional planar OTFT is accessed with $L = 12$–17 μm as defined by the depth of the embossed pattern. In the second geometry, E1 and E2 or E3 and E4 are the source and drain electrodes, and in this case the gate length is only $L_v \approx 900$ nm.

The μ_{FE} measured for both device geometries is in the range 2–3 × 10^{-3} cm^2/V s, consistent with conventional F8T2 OTFTs. The devices did not turn off entirely at $V_{GS} = 0$, which limited the on/off ratio to only ~10^3. Also, the VOFET (E1–E2) did not exhibit saturated characteristics: apparently it only operated in the linear regime. While no high frequency characteristics were provided, it is expected that the frequency response is limited by the parasitic drain-source capacitance that results from substantial overlap of the contacts separated by only a thin layer of insulating PVP. The drain current of $I_D = 6$ μA is 20 times larger than that for the planar OTFT, although the operating voltages ($V_{GS} = V_{DS} = -20$ V) are approximately the same for both geometries. Hence, while this architecture has attributes of high drain current, low voltage and potentially high frequency operation, these have not been demonstrated by the devices in Fig. 8.45 (Stutzmann et al., 2003).

A second approach to the VOFET uses a *permeable source electrode* illustrated in Fig. 8.46a (Ben-Sasson et al., 2009). The gate electrode is positioned beneath the gate dielectric, while the source and drain electrodes are vertically stacked with the channel layer sandwiched between them. A Schottky barrier between the source electrode and the channel or a thin oxide coating prevents leakage when $V_{GS} = V_{DS} = 0$. As gate bias is applied, the barrier is lowered, and current is injected into the channel, creating a conductive path from source to drain.

The permeable source contact pores are approximately equal to the organic channel thickness. This is done by depositing a rough and very thin metal layer onto the gate dielectric (Ma and Yang, 2004) that is almost certainly discontinuous, containing a distribution of fissure sizes and spacings. Alternatively, the permeable source is fabricated by depositing a copolymer where one component is dissolved away with a material-selective solvent, leaving behind the remaining copolymer that forms micropillars of the desired height and spacing. The micropillars are coated with a thin gate metal. The pillars are subsequently removed by immersing in a second solvent, only leaving metal between the pillars. An image of a permeable, 10-nm-thick Au source electrode is shown in Fig. 8.46b.

The operating principle of the n-channel permeable source transistor is understood from the simulated charge densities at several V_{DS} (with $V_{GS} = 5$ V) shown in Fig. 8.47. At $V_{DS} = 0$, no charge flows from source to drain due to the Schottky barrier or oxide layer between permeable source layer and the semiconductor channel (Fig. 8.47a). A charge

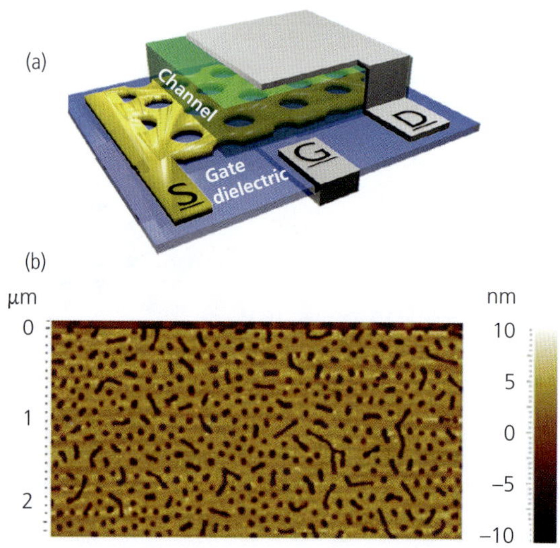

Figure 8.46 (a) Schematic of a permeable source VOFET (Ben-Sasson et al., 2012). (b) Atomic force microscope image of a 10 nm thick Au grid patterned via lift-off used as the source electrode (Ben-Sasson et al., 2009).

accumulation layer forms at the gate dielectric interface within the pores due to the positive gate potential. As the drain potential is made positive ($V_{DS} = 5$V, Fig. 8.47b), electrons are drawn into the channel due to field lowering of the source contact energy barrier. Current between source and drain increases linearly with voltage, and the transistor enters the linear regime. As the drain voltage is increased, the demand for charge at the drain exceeds the supply from the source. The current is limited by space charge injection, and hence is no longer linearly dependent on voltage. The current is higher near the edges of the source electrode, that is, at the point of injection, and lower toward the center, as shown in Fig. 8.47c. The output current henceforward increases only slowly with V_{DS} as the transistor characteristics become saturated.

The drain current can be increased by depositing a layer of semiconductor onto the gate insulator prior to deposition of the permeable S contact. This effectively increases the area of the contact by allowing charge extraction from the entire pore area rather than only from its periphery. Increasing the layer thickness beneath S increases the area of this "virtual contact." Calculated output characteristics for a PTCDI-C_8 channel transistor assuming $\mu_n = 0.1$ cm^2/V-s for 10 nm and 20 nm thick buried source layers are shown in Fig. 8.48. There are several features of note from these simulations. The first is the very narrow linear region, that gives way to saturation at $V_{DS} \ll V_{GS}$. Also, the current is high compared to a conventional OTFT due to the very short channel length,

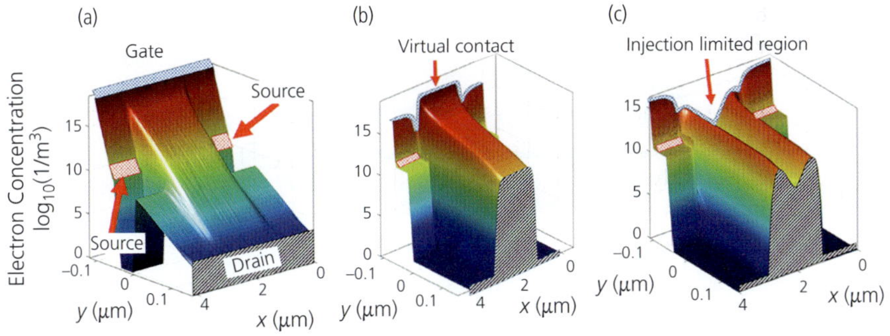

Figure 8.47 Calculated charge density in an n-channel permeable source transistor at operating points of $V_{GS} = 5$ V and V_{DS} of (a) 0 V, (b) 5 V, and (c) 8 V. The barrier height at the source/channel contact is 0.7 eV (Sheleg et al., 2017).

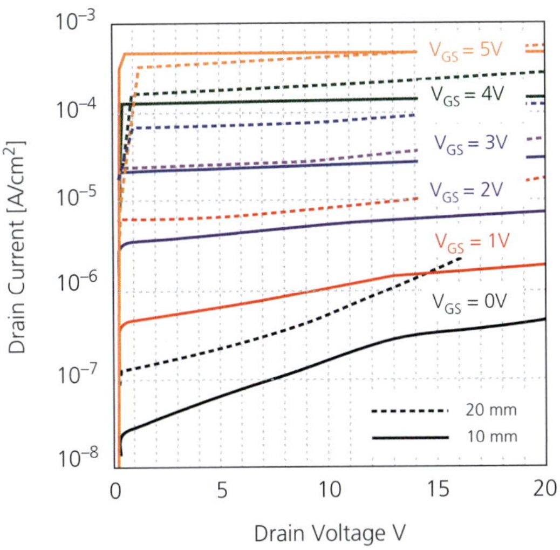

Figure 8.48 Calculated characteristics of a PTCDI-C_8 VOFET with a 10 nm or 20 nm thick buried source contact semiconductor layer. The channel layer thickness is 150 nm, and the SiO_2 gate insulator is 40 nm thick (Sheleg et al., 2017).

which is equal to the device thickness of 150 nm. For example, drain currents of up to 4 A/cm^2 are more than sufficient to drive an OLED (Ma and Yang, 2004), although the cutoff frequencies are limited by the high gate-source and drain-source capacitances due to complete overlap of the S and D contacts with the gate. As expected, the current is increased with the thickness of the buried source film, since this increases the cross-sectional area effective in charge injection. Finally, the output conductance of vertical transistors tends to be high due to thermionic emission across the source/channel barrier (Ma and Yang, 2004, Ben-Sasson et al., 2009, 2012). This limits the amount of gain that can be achieved, and hence the noise margin of circuits employing the VOFET may be small. High g_D also reduces the on/off ratio, although $I_{on}/I_{off} > 10^6$ has been achieved.

A *permeable base transistor* is an alternative vertical geometry that provides high current gain (Fujimoto et al., 2005, Fischer et al., 2012, Klinger et al., 2017). Its operating principle is similar to a vacuum tube triode or a bipolar junction transistor as opposed to a gate-voltage-controlled FET. An illustration of a permeable base transistor is shown in Fig. 8.49a. The device consists of an organic semiconductor sandwiched between ohmic emitter and collector electrodes. The current is controlled by interposing a very thin, permeable metal sheet (in this case, Al) near the center of the organic thin film. The base contact metal is oxidized in air for several minutes at elevated temperatures. This creates an insulating barrier around the electrode to prevent the flow of base current. It also may create pinholes due to crystallization of the organic that protrudes through fissures in the metal layer. The penetration of the organic permits unimpeded charge transport from emitter to collector (Fischer et al., 2012). An alternative explanation for the base current is hot electron transport across the base metal, that is, high energy electrons are emitted from the C_{60} into the base whose thickness is on the order of or less than their mean free path of ~50 nm. Hence, electrons can penetrate the base without loss until the thickness exceeds this limit (Fujimoto et al., 2005).

The *common emitter current gain* is $h_{FE} = \partial I_C/\partial I_B$, where I_C is the collector and I_B is the base current. The device in Fig. 8.49 shows a maximum of $h_{FE} = 10^5$ at a base-emitter voltage of $V_{BE} = 1$V; a value that is comparable to high performance inorganic heterojunction (HJ) bipolar transistors (Streetman and Banerjee, 2006). Furthermore, the *base transport factor* is given by: $\alpha_{CE} = \partial I_C/\partial I_E$, where I_E is the emitter current. Hence, α_{CE} is a measure of the loss of current as it transits the base electrode. Since $I_E = I_C + I_B$, then

$$h_{FE} = \frac{\alpha_{CE}}{1 - \alpha_{CE}}. \tag{8.64}$$

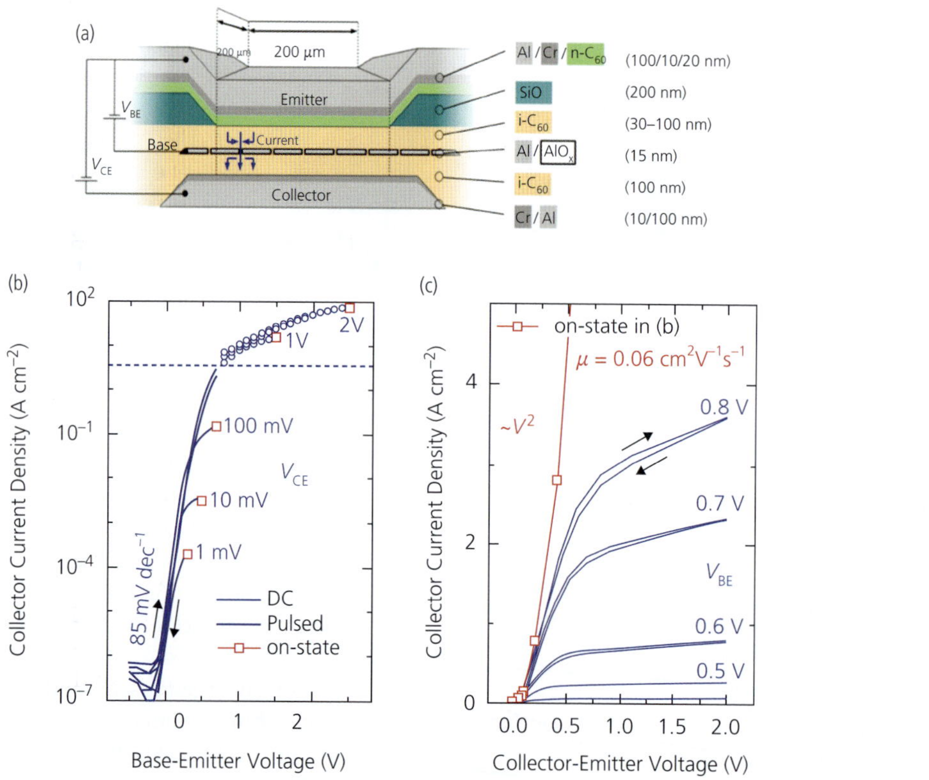

Figure 8.49 (a) Schematic view of a permeable base transistor based on undoped (intrinsic) i-C_{60}, and an oxidized Al base layer. (b) Transfer and (c) output characteristics of the transistor. The squares in (b) indicate the transition from the linear to saturation regime. High current density data are taken in pulsed mode to minimize Joule heating. In (c) the linear regime is dominated by space charge limited transport, yielding an electron mobility of 0.06 cm^2/V s (Klinger et al., 2017).

Thus, for the transistor in Fig. 8.49, we obtain $\alpha_{CE} = 99.999\%$, indicating near lossless transport through the permeable base.

From the transfer characteristics in Fig. 8.49b, the device delivers a current density of 75 A/cm^2 at $V_{CE} = 2$ V, and $V_{BE} = 2.5$ V under pulsed operation. This is increased to >1 kA/cm^2 by increasing the emitter area, thus reducing limitations due to current injection. This is orders of magnitude higher than the current that is sourced by conventional, lateral geometry OTFTs. Other characteristics of the permeable base transistor are $I_{on}/I_{off} = 10^8$ and $S = 85$ mV/decade; the latter parameter nearly equal to the theoretical minimum of 60 mV/decade (cf. Eq. 8.34). The output characteristics in Fig. 8.49c indicate that the linear regime is dominated by space-charge-limited injection from the emitter, following its characteristic $I_C \sim V_{CE}^2$ behavior (see Section 4.4.2). From these data, the electron mobility in the vacuum-deposited i-C_{60} layers is $\mu_n = 0.06$ cm^2/V s. Note that this is bulk and not a field effect mobility measured in the lateral OTFT configuration. As is the case for all triode architectures, the output conductance is comparatively high, leading to a potentially low voltage gain when used in common-emitter circuits.

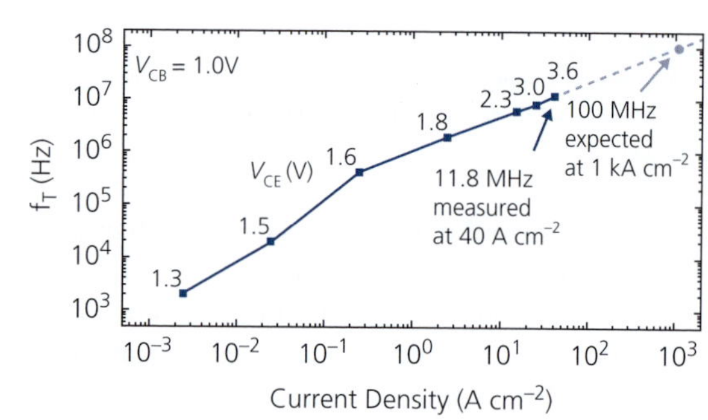

Figure 8.50 Cutoff frequency vs. I_{CE} and V_{CE} of the transistor in Fig. 8.49 (Klinger et al., 2017).

Given the short emitter-collector distance, we expect that the transit time across the transistor bulk will be short, even though the mobility is low. Indeed, this is found to be the case from the data for the transfer frequency vs. I_C in Fig. 8.50. A $f_T = 11.8$ MHz is obtained at 40 A/cm^2, which is approximately half of that obtained in a C_{60} OTFT with $L = 2$ μm (Fig. 8.26). The projected frequency response is 100 MHz at $I_C = 1$ kA/cm^2. However, this projection does not account for contact resistance effects that become increasingly dominant at high currents and frequency.

There are several variants to vertical transistors beyond those discussed here, including static induction transistors (Kudo et al., 1998) and devices employing a variety of other permeable base and source concepts. A common attribute of these alternative structures is that their output currents fail to enter saturation, they require very precise control of fabrication parameters, and have high output capacitance, and hence presumably low cutoff frequencies. Countering these disadvantages is their ability to source high currents at low voltages. Nevertheless, conventional single and dual gate OTFTs dominate applications comprising integrated circuits from simple inverters to ring oscillators, shift registers and other digital and analog circuits.

8.4.4 Complementary logic and ambipolar transistors

Unipolar conduction in the channel is limited by several factors, including the contact work functions or metal/insulator barriers that preferentially inject one carrier type, or asymmetries in mobilities between holes and electrons that leads to low resistance transport for one or the other, but not both carriers. However, some channels can have balanced hole and electron mobilities, and contacts can be engineered to inject electrons and holes with equal efficiency. In such cases, reversing the polarities of V_{DS} and V_{GS} can result in the conduction of either charge, but in opposite directions. *Ambipolar transistors*, or integrated combinations of *n*- and *p*-channel transistors, enable *complementary logic* circuits that form the foundation of all modern electronics (known as complementary MOS, or CMOS). A complementary logic inverter circuit is shown in Fig. 8.51a. An n-channel and p-channel FET are connected in series, or for ambipolar devices, a single OTFT type is used in both positions. When V_{in} is in the logical "high," or H state (i.e. $V_{in} = V_{DD}$), the p-channel OTFT is off, and the *n*-OTFT is on (i.e. conducting), and hence $V_{out} = 0$, that is, in the "low" or L state. When $V_{in} = 0$, the p-channel OTFT turns on with the n-channel off, and $V_{out} = V_{DD}$. Thus, the circuit output switches its logical state to H. The transfer characteristic of this inverter is shown in Fig. 8.51b. The advantage of this circuit compared to the unipolar inverter in Fig. 8.29, and the reasons for its widespread adoption in electronic logic circuits, is its low static power dissipation. Since one of the transistors is always off, the circuit is non-conducting (i.e. $I_D = 0$) except during the brief time that it switches its logic state. Furthermore, given the very low I_{off} of enhancement mode transistors, the circuit has a high noise margin that leads to robust, error-free operation in integrated circuits.

An integrated combination of n- and p-channel OTFTs with nearly balanced electronic properties is illustrated in Fig. 8.52. The fabrication of this circuit on a flexible PEN substrate involves five vacuum depositions through a series of shadow masks to pattern (i) the Al gate contacts, (ii) the gate interconnection vias followed by deposition of the gate insulator consisting of ~3 nm thick Al_2O_3 whose surface is functionalized by a 25%/75% mixture of an alkyl/fluoroalkyl phosphonic acid SAM, resulting in a total insulator thickness of 5.7 nm. The SAM lowers the threshold voltage (see Section 8.6.1), (iii) the 12 nm thick p-channel semiconductor comprising DPh-BTBT, (iv) the 15 nm thick n-channel semiconductor, N1100, and finally (v) the

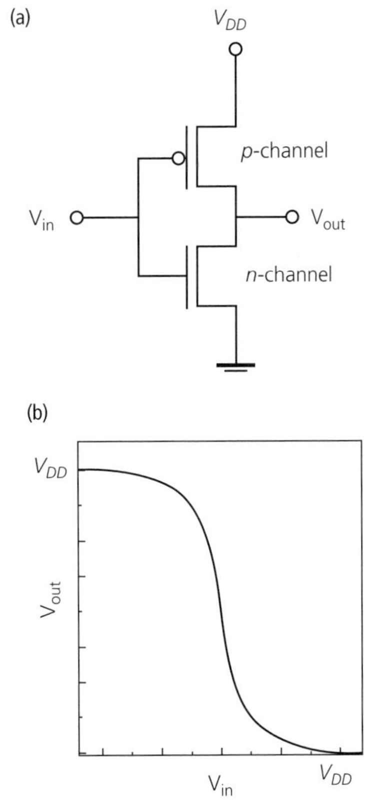

Figure 8.51 (a) Complementary inverter circuit comprising series-connected p- and n-channel OTFTs. Note the different symbols used for the complementary transistors. (b) Transfer characteristic of the inverter circuit.

Au source and drain contacts. The need to align the several masks results in a relaxed gate dimension of L= 40 μm with gate-to-source and gate-to-drain overlaps of 80 μm each (Zschieschang et al., 2017).

The characteristics of the n- and p-channel OTFTs with $W = 1$ mm are nearly balanced, as shown in Fig. 8.53a and b. Fits to the drain current transfer characteristics in the saturation region show $|V_T| <$ 0.5 V, and nearly equal mobilities and sub-threshold slopes for the complementary pair. The transfer characteristics of the circuit in Fig. 8.53c show logic state switching at $V_{in} = 0.5$ V, and a maximum switching current of 1 nA. The static power dissipation of the circuit is <1 pW. This is in marked contrast to the unipolar inverter that draws significant current even in the quiescent mode, and points to the low-power advantage of complementary logic. Furthermore, this circuit shows a maximum voltage gain of $A_v = 160$. One disadvantage of very low voltage, and hence low power, is that the gate takes longer to charge during switching. Hence, the cutoff frequency obtained from a five-stage ring oscillator with a $\tau_{delay} = 20$ ms is only $f_T = 1/(2\tau_{delay}) = 25$ Hz. As the supply voltage is increased along with the static and switching powers, the cutoff frequency also increases.

While the performance of the complementary circuit can be modified for a particular application, circuit fabrication can be complex and lead to compromises in design (e.g. large alignment tolerances are required for positioning multiple shadow masks without error). An alternative approach is to use dual ambipolar OTFTs in place of separate n- and p-channel transistors. This eliminates the need for separate deposition steps for electron and hole conducting channels. There are three different architectures for achieving ambipolar OTFT operation. (i) Use a bilayer channel, with one layer having a high electron mobility and the other a high hole mobility (Kang et al., 2005, Kuwahara et al., 2005, Wang et al.,

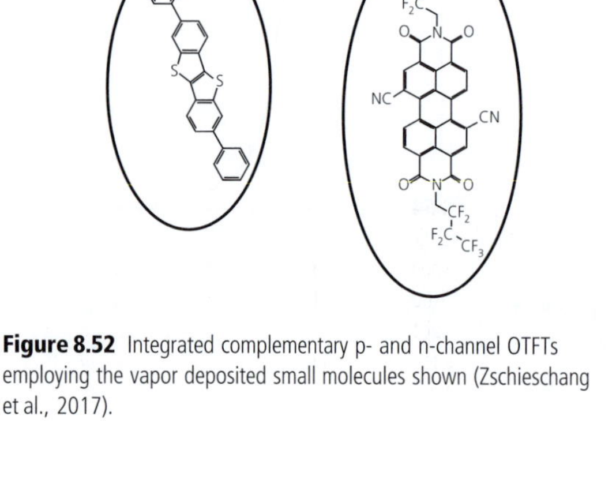

Figure 8.52 Integrated complementary p- and n-channel OTFTs employing the vapor deposited small molecules shown (Zschieschang et al., 2017).

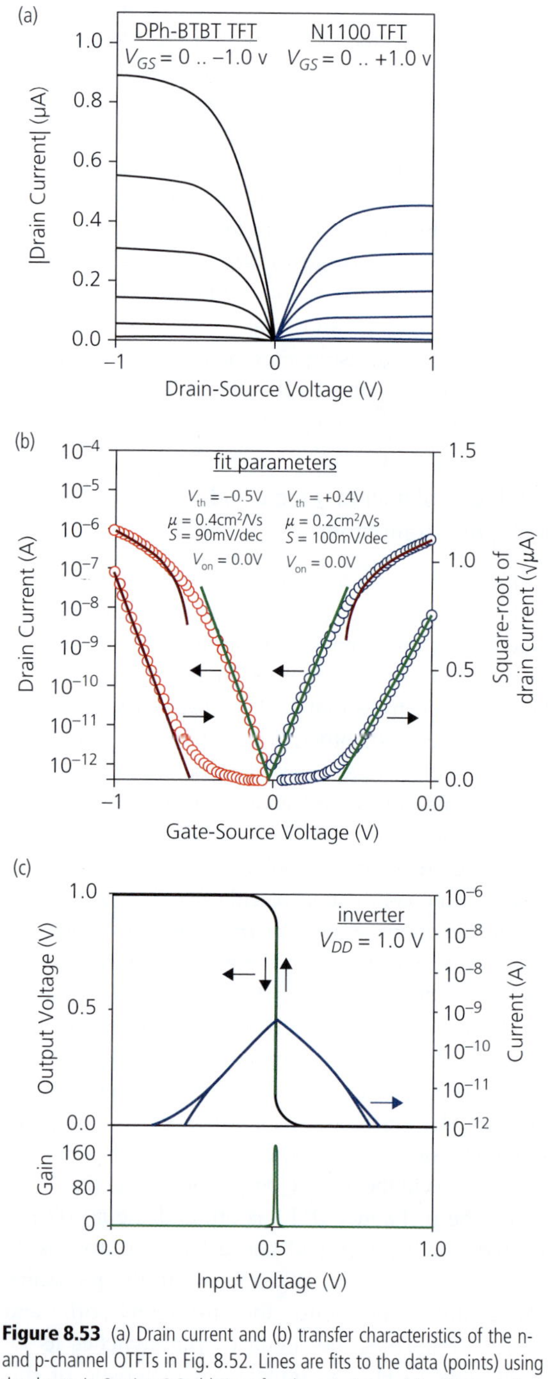

Figure 8.53 (a) Drain current and (b) transfer characteristics of the n- and p-channel OTFTs in Fig. 8.52. Lines are fits to the data (points) using the theory in Section 8.3. (c) Transfer characteristics of the integrated complementary inverter showing the gain (lower panel) and drain current during switching (Zschieschang et al., 2017).

2006). Since two different layers are deposited sequentially, this design preferentially employs vacuum deposited small molecules, where dissolution of pre-deposited, thin layers is not a concern. (ii) Use a single layer whose source and drain contacts have small barriers to electron and hole injection. It is preferable to use different metals for source and drain (one with large work function for hole injection, and the other with a small work function for electron injection) (Schmechel et al., 2005). Alternatively, a narrow energy gap channel semiconductor can be used with a small injection barrier to either carrier. This promotes ambipolar charge injection at the cost of increased channel leakage due to thermal generation (see Chapter 4). (iii) Use a blend of a hole and electron conducting organic semiconductor channel material (Meijer et al., 2003). Given the ease with which materials can be blended in solution, polymers are well suited for this purpose. In the following we provide examples of each of these architectures, with a further discussion of potential materials combinations that support ambipolar conduction in Section 8.6.

In Fig. 8.54a we show the bipolar output characteristics of a bilayer of the electron conducting $F_{16}CuPc$,

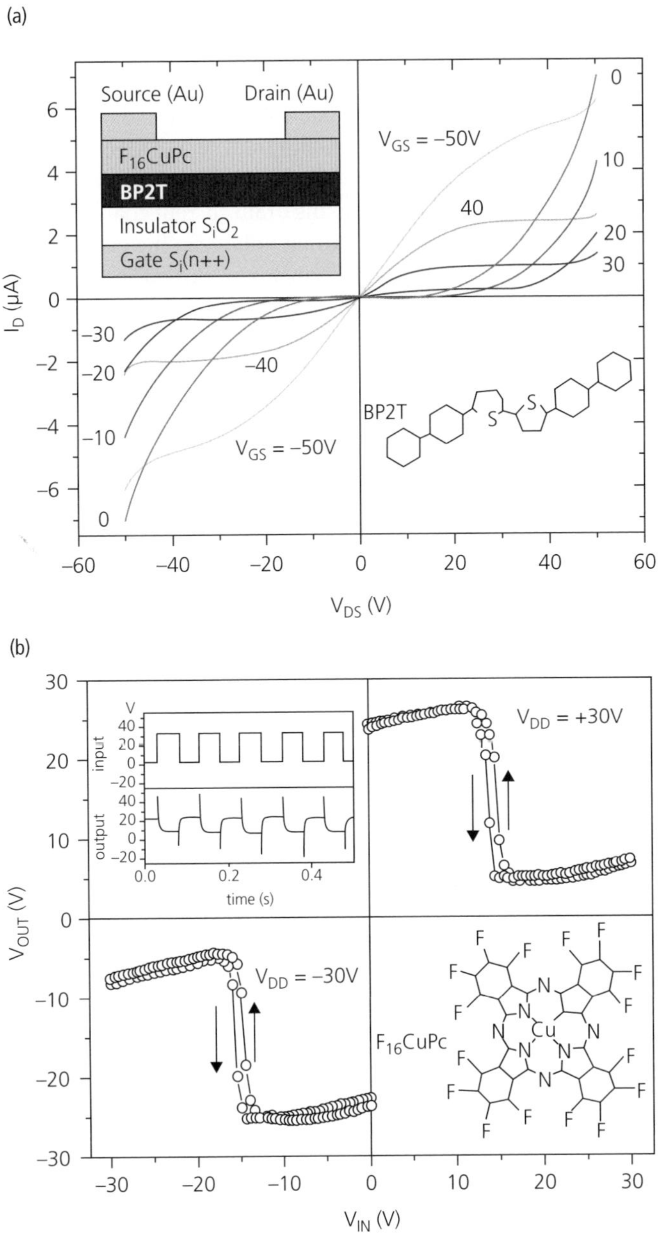

Figure 8.54 (a) Ambipolar drain current characteristics of the bilayer OTFT with $L = 180$ μm, $W = 1350$ μm and $t_i = 150$ nm shown in the inset, upper left. (b) Transfer characteristics of a complementary inverter circuit. Inset: Pulse response of the inverter. Insets, 4th quadrants: Molecular structural formulae of the molecules used in the channel (Wang et al., 2006).

and hole conducting BP2T whose molecular structural formulae are shown in the inset. The LUMO energy of 4.61 eV for $F_{16}CuPc$ and the HOMO energy of 5.24 eV for BP2T allow for reasonably efficient and symmetrical injection of electrons and holes from the Au source and drain contacts (the Au work function is 5.1 eV), respectively. The BP2T and $F_{16}CuPc$ layers are sequentially deposited to thicknesses of 3 nm and 20 nm on a thermally grown SiO_2 gate insulator, respectively, and then Au for the S and D contacts is deposited, all by vacuum thermal evaporation.

The drain current characteristics in Fig. 8.54a are remarkably symmetrical at $\pm V_{DS}$ due to the near equal electron and hole mobilities of 0.036 and 0.04 cm^2/V s, and threshold voltages of $V_{Tn} = 13$ V and $V_{Tp} = -10$ V. The characteristics are similar to those determined using the analysis in Section 8.3.2, and illustrated in Fig. 8.19. The drain current shows $I_D \propto V_{DS}^2$ ambipolar behavior for small V_{GS}, reverting to the unipolar saturation regime as the gate voltage is increased. Implementation in a complementary OTFT inverter shows abrupt switching at $V_{in} = \pm 15V$ for $V_{DD} = \pm 30V$ (Fig. 8.54b), with a voltage gain of 13. Switching response to a square V_{in} pulse stream is shown in the inset.

A second architecture for ambipolar operation uses a single channel material but with different S and D contacts; one with a large work function for hole injection, and the second with a smaller work function for electron injection. One example uses a pentacene channel that can transport both carriers, albeit with different mobilities of $\mu_{FEp} \approx 0.2$ cm^2/V s and $\mu_{FEn} \approx 0.1$ cm^2/V s. A BG/TC ambipolar pentacene OTFT using a thermally grown SiO_2 gate dielectric has bipolar drain current characteristics shown in Fig. 8.55. The electron and hole injecting channel contacts are Ca and Au, respectively. The fits to Eqs. 8.19 and 8.27 assume the mobilities and threshold voltages are dependent on gate voltage. In addition to differences in mobilities, the threshold voltages vary from $V_{Tn} \approx 18$ V to $V_{Tp} \approx -35$ V. The observed asymmetry may be due to differences in injection efficiencies from the Ca and Au contacts.

Finally, ambipolar transistors using blends of the electron conducting $PC_{61}BM$ and the hole conducting OC_1C_{10}-PPV have been demonstrated (Meijer et al., 2003). A challenge with blended channels is efficient injection into the constituents from the source and drain contacts. In this particular example, energy barriers exceeding 0.8 V are overcome, in part, by varying the gate potential. Significant hysteresis is observed in the drain current characteristics when sweeping the drain voltage in the positive and negative directions, which is reflected in hysteresis in a complementary inverter circuit incorporating these devices. The hysteresis may arise from charge trapping and detrapping within the blend network, at the contacts, etc. In general, the blended channel ambipolar OTFTs appear to have inferior performance to bilayer and single component ambipolar conducting, small molecule and polymer channel devices.

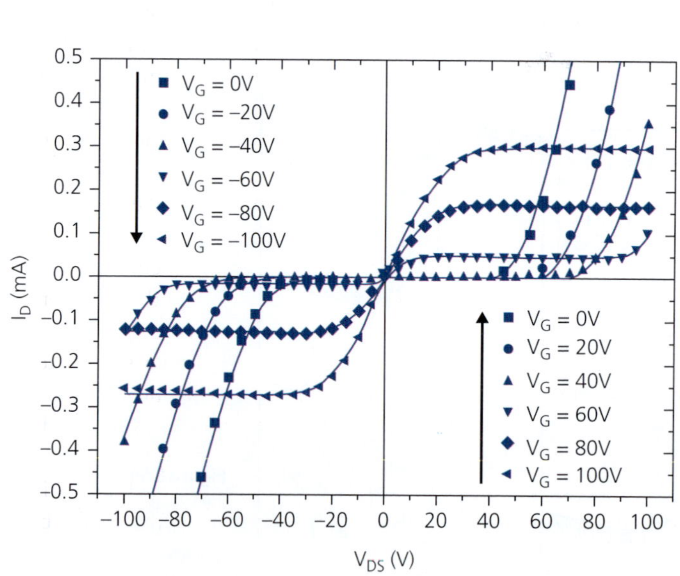

Figure 8.55 Drain current characteristics of an ambipolar pentacene OTFT with electron and hole injecting channel contacts. The transistor dimensions are $L = 70$ μm, $W = 8$ mm, and $t_i = 200$ nm. Lines are fits to theory in Section 8.3 (Schmechel et al., 2005).

8.4.5 Split gate transistors

Ambipolar transistors have a disadvantage of a low I_{on}/I_{off} since there is no bias condition whereby the channel is completely free of conduction from electrons or holes. This results in a high off-state power dissipation and a low noise margin. A means of limiting leakage is to use a split gate electrode (Hsu et al., 2010), as shown in Fig. 8.56a (Yoo et al., 2017). This BG/BC device consists of a main and a side (or control) gate beneath the drain contact separated by a small gap, allowing for their independent bias. When the gates are shorted, the device behaves as an ambipolar OTFT providing that an ambipolar conducting channel is used. When the side and main gates are independently biased, the side gate raises or lowers the potential barrier beneath the source electrode. When the control voltage $V_{CS} < 0$, holes are drawn into the channel at the exclusion of electrons, and then the transistor behaves as a p-channel OTFT. Conversely, when $V_{CS} > 0$, n-channel operation is induced. Hence, adjustment of the side gate voltage for different transistors within a circuit enables complementary transistor operation without compromising I_{on}/I_{off}.

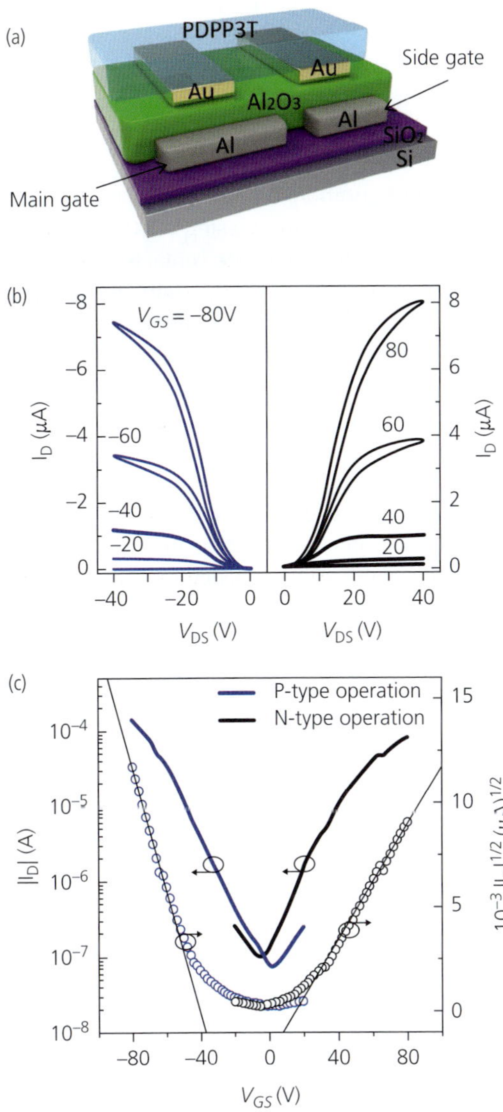

Figure 8.56 (a) Schematic view of a split gate OTFT. (b) Unipolar drain current characteristics for both p- and n-channel conduction (left and right panels, respectively) accomplished by control of the side gate bias. The OTFT dimensions are $L = 6$ μm, $W = 810$ μm, with a gate electrode gap of 1 μm. $V_{GS} = 0$ for the lowest curves in both panels. (c) Transfer characteristics of the spit gate OTFT. Solid lines are fits to the $\sqrt{I_D}$ vs. V_{GS} characteristics (Yoo et al., 2017).

The drain current characteristics exhibiting complementary unipolar operation of a split gate OTFT employing the ambipolar conducting channel material, PDPP3T, is shown in Fig. 8.56b. The transistor features a $t_i = 200$ nm thick Al_2O_3 gate insulator deposited by ALD, along with a spin-coated PDPP3T channel vacuum annealed at 180°C.

The magnitude of the hysteresis in the drain current induced by changing the V_{DS} sweep direction once it has reached its maximum value (±40 V), depends on how fast the voltage is scanned. Hysteresis is a common feature of many OTFTs, and results from trapping and detrapping of charge from the semiconductor/insulator interface, or from within the semiconductor bulk. Changing the occupancy of traps shifts the surface potential, ψ_S, which has an effect similar to changing the gate bias, V_{GS}. The change in I_D at a fixed V_{DS} is known as *drain current drift*. Its presence can significantly reduce the noise margin of digital circuits over time, and can also adversely affect the operating point of analog circuits.

Nearly symmetrical characteristics are achieved for both p- and n-channel operation (left and right panels, respectively, in Fig. 8.56c), with $I_{on}/I_{off} = 10^3$. While this is not particularly high, it is an improvement of between 10 and 100 over conventional ambipolar OTFTs. Indeed, $I_{on}/I_{off} = 10^7$ has been achieved for OTFTs with two split gates (a trigate architecture), where charge injection is controlled by side gates under both the source and the drain (Uemura et al., 2014, Torricelli et al., 2016).

The transfer characteristics in Fig. 8.56c show nearly ideal symmetrical behavior when used in either the p- or n-channel modes of operation. The threshold voltages in both cases are $V_{Tp} \approx V_{Tn} \approx 0$, which is optimal for complementary transistor operation. The hole an electron mobilities are also comparable, with $\mu_{FEp} = 0.026$ cm^2/V s and $\mu_{FEn} = 0.012$ cm^2/V s.

The cutoff frequencies of split gate OTFTs are higher than for conventional, single gate designs due to their reduced contact resistance and higher charge concentration at the insulator interface at low gate voltages. This can be understood as follows: The main gate is placed between two split side gates buried within the insulator, as in Fig. 8.57a, eliminating the gap between the channel contacts and the gate that often exists in a conventional device. Thus, the charge density at $V_{CS} \neq 0$ can be high even at $V_{GS} = 0$. In Section 8.3.2, we found that the total channel resistance is given by: $R_{tot} = 1/g_D + r_S + r_D = R_{ch} + R_C$, where $R_{ch} = 1/g_D$ and R_C is the sum of the contact resistances (cf. Fig. 8.25). The split gate draws charge into the channel even at $V_{GS} = 0$, thereby greatly decreasing the contact barriers that contribute to R_C.

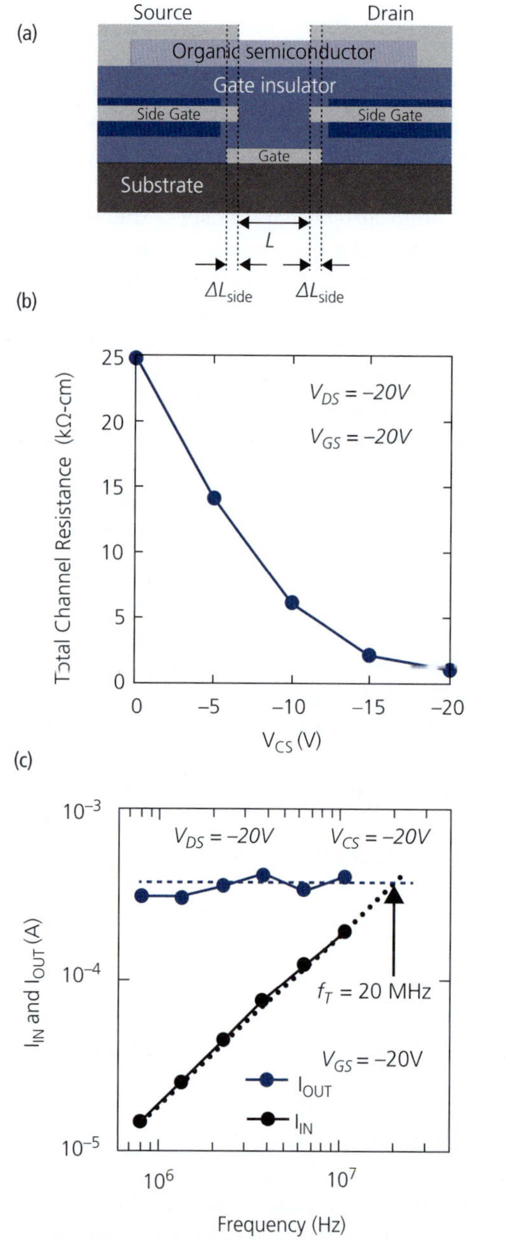

Figure 8.57 (a) Schematic illustration of a trigate OTFT with a buried split gate. (b) Channel resistance vs. side gate control voltage, V_{CS} for a C_{10}-DNTT/F_6TNAP OTFT with $W = 3$ mm, $L = 2.5$ μm, and $\Delta L_{side} = 500$ nm. (c) Frequency response of the trigate OTFT (Uemura et al., 2014).

The reduction in channel resistance with V_{DS} is shown for a buried split gate OTFT in Fig. 8.57b. This device comprises the hole conducting semiconductor, C_{10}-DNTT, doped with the acceptor, F_6TNAP, beneath the S and D contacts to reduce $R_{contact}$ (Uemura et al., 2014). To create the buried split gate configuration, the Al_2O_3 insulator is deposited by ALD at 150°C in two steps: 200 nm followed by deposition of the split gate contacts, and capped with 100 nm that is in contact with the vacuum-deposited channel layer. The second insulator surface is functionalized with an n-tetradecylphosphonic acid SAM prior to C_{10}-DNTT deposition to result in $\mu_{FEp} = 0.4$ cm^2/V s.

The contact resistance in the saturation regime decreases from 20 kΩ cm at $V_{CS} = 0$ corresponding to that of a conventional OTFT, to 1.2 kΩ cm at $V_{CS} = -20$ V. This dramatic reduction in R_{tot} results in a considerably enhanced frequency response, which is inversely proportional to the time to charge the gate capacitance, $\tau = R_{tot}C_{tot}$. This device exhibits $f_T = 20$ MHz (Fig. 8.57c); a significant improvement over conventional OTFTs.

The foregoing results suggest that split gate OTFTs can have a high frequency response and improved ambipolar conducting properties compared with single gate structures. The question remains whether the higher fabrication complexity counters the benefits of the multiple gate design. High gain, high frequency complementary logic circuits would appear to benefit from the split gate architecture, and indeed in some cases they are enabled by it.

8.5 Phototransistors

An organic phototransistor (OPT) is a thin film transistor with light incident on the organic semiconducting channel. Since the potential at the organic/insulator interface is a sensitive function of the gate voltage, optically generated charges shift the Fermi energy at that interface, causing a concomitant and potentially large change in the drain current. Thus, the phototransistor is a device that can have a high optoelectronic gain, albeit at a reduced bandwidth due to limitations set by their gain-bandwidth product (see Section 7.1.1). In this section, we consider the principal modes of operation of the phototransistor, and provide a few example OPTs. Since an OPT is a hybrid between a photodetector and a transistor, a complete understanding of its operation requires a knowledge of conventional organic photodiode and photoconductor behavior. A few of the important expressions derived in Chapter 7 for OPDs will be reproduced here for convenience, with more complete discussions of their origins found in that chapter.

An illustration of a top illuminated, BG/TC OPT is shown in Fig. 8.58. The light is incident on the channel between source and drain contacts. If a transparent gate contact is used, illumination can be incident via the substrate. Using a metal contact in the TG configuration is also possible. Indeed, all the various iterations of OTFT architectures in Section 8.1 have also been applied to OPTs.

The phototransistor can operate in either the photovoltaic or the photoconductive mode. In the photovoltaic mode, the charge concentration at the insulator/channel interface is increased under illumination due to an optically-induced change in the gate voltage. This change, called the *photovoltage*, V_{ph}, is determined using

$$\Delta C_{GS} = \frac{\partial Q}{\partial V_{GS}} = q^2 \frac{\partial n_s}{\partial E_F}, \tag{8.65}$$

where ΔC_{GS} is the change in gate capacitance due to a change in surface charge concentration $\Delta Q = \Delta n_s$. The photovoltage, V_{ph}, is simply equal to the change in the Fermi energy at the interface, viz. (Romero et al., 1996)

$$qV_{ph} = \Delta E_F. \tag{8.66}$$

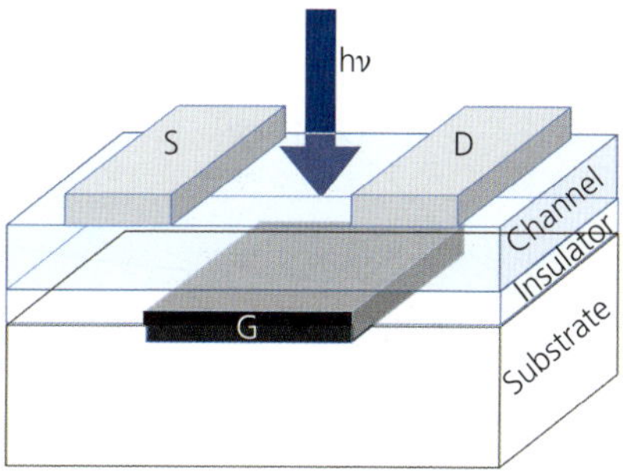

Figure 8.58 Schematic illustration of a BG/TC, top illuminated organic phototransistor.

The total interface charge density in the presence of illumination, n_T, is related to that in the dark, n_0, by the energy level bending, ΔE_F, via Boltzmann statistics:

$$n_T = n_0 \exp(q\Delta E_F / k_B T). \tag{8.67}$$

The charge density is related to the total channel current via:

$$n_T = n_{ph} + n_0 = \frac{(j_{ph} + j_D)}{q\mu F}, \tag{8.68}$$

where j_{ph} and j_D are the induced photocurrent and the drain current in the dark, respectively. We assume for simplicity that the mobility is field independent, which is valid if the photocurrent does not significantly exceed the dark current. Then, combining Eqs. 8.66–8.68, we obtain

$$V_{ph} = \frac{nk_B T}{q} \ln\left(\frac{j_{DT}}{j_D}\right) = \frac{nk_B T}{q} \ln\left(1 + \frac{j_{ph}}{j_D}\right), \tag{8.69}$$

where the total drain current is $j_{DT} = j_{ph} + j_D$. Also, the factor $n \le 1$ accounts for the decrease in energy level bending as the photocurrent is increased (Chen, 1981). That is, at low power, we expect $n = 1$. However, at high power the bending becomes sublinear with j_{ph} as ΔE_F approaches the nearest frontier orbital energy.

From Section 7.1, the photocurrent is related to the incident optical power per area, P_{inc}, via $j_{ph} = \eta_{ext} q\lambda P_{inc}/hc$, leading to

$$V_{ph} = \frac{nk_B T}{q} \ln\left(1 + \frac{q\eta_{ext}\lambda P_{inc}}{j_D hc}\right), \tag{8.70}$$

where h is Planck's constant and η_{ext} is the external quantum efficiency for charge generation (see Section 7.1). The external efficiency is given by

$$\eta_{ext} = \frac{hc}{q\lambda P_{opt}}(I_{DT} - I_D), \tag{8.71}$$

where the total optical power is $P_{opt} = WLP_{inc}$ and $I_{DT} = WLj_{DT}$ is the total drain current under illumination.

The relationship in Eq. 8.70 was originally derived in the context of high mobility, modulation doped inorganic transistors (Chen, 1981, Takanashi et al., 1998, 1999). However, the derivation does not specifically apply to a particular device geometry; rather it relies only on the presence of a sheet density of charge confined to a surface by surrounding energy barriers where Boltzmann statistics apply. Hence, it should also be valid for organic OPTs where a large sheet charge exists in the accumulation layer at the organic/gate insulator interface.

Finally, the photocurrent in the photovoltaic mode is related to the photovoltage by the transconductance, viz.

$$I_{ph} = g_m V_{ph} = \frac{g_m n k_B T}{q} \ln\left(1 + \frac{q\eta_{ext}\lambda P_{inc}}{j_D hc}\right). \tag{8.72}$$

The photocurrent, therefore, is linear in P_{inc} at low power, and saturates once the incident power generates a photocurrent larger than the background drain current, that is, when $\eta_{ext} P_{inc} > \frac{j_D hc}{q\lambda}$.

In the photoconductive mode, illumination generates electrons and holes within the accumulation region. The lack of a junction in the channel results in an increase in its photoconductivity, and hence an increase in the drain current. From Eq. 7.4, we obtain

$$I_{ph} = q\eta_{ext}\mu_{FE} d \frac{(P_{inc}\lambda/hc)}{k_D} \frac{V_a}{L}, \tag{8.73}$$

where d is the thickness of the channel, which in this approximation is less than the exciton diffusion length, L_D. As in Chapter 7, k_D is the lifetime of the exciton. Generally, the photoconductive current is considerably less than the photovoltaic current since the latter is amplified by the transconductance. That is, the photoconductive current is $I_{ph} \sim P_{inc}$, whereas the photovoltaic current follows $I_{ph} \sim g_m P_{inc}$ at small incident powers. Photoconductivity is generally only observed near $V_{GS} = 0$ where $g_m \to 0$, and the photovoltaic effect vanishes.

Charge generation in the unipolar channel suggests that photocurrent effects are small, given the absence of an exciton-dissociating donor–acceptor HJ. Yet, this contradicts the observation of a very large responsivity of $>10^4$ A/W that has been observed (Guo et al., 2010). While the photogeneration mechanism is not well understood, several hypotheses for the large photocurrent have been proposed. For example, exciton dissociation may occur via field ionization at the insulator-semiconductor interface (Guo et al., 2009). However, the electric field strengths rarely exceed 5 x 10^5 V/cm at this junction. We have seen in Section 4.7.2, Fig. 4.104 that Onsager–Braun theory predicts a probability for dissociation of $\sim 10^{-4}$ for a 0.7 nm exciton radius at this field, which is too small to account for the very large photocurrents observed. An alternative mechanism is that excitons may be quenched by defect states at the interface, trapping one carrier type and releasing the other charge into the channel. The change in trapped surface charge induces a change in threshold voltage, thus increasing the current according to Eq. 8.72 (Wang et al., 2010). The change in threshold voltage is expected to have a far greater effect on the drain current than field ionization.

Representative drain current characteristics in the dark and under white light illumination for a P3HT/SiO_2 BG/TC OPT are shown in Fig. 8.59. The saturated dark characteristics in Fig. 8.59a yield a mobility of $\mu_{FEp} = 0.07$ cm^2/V s. There is a small curvature in the linear regime due to source contact resistance. Under white light illumination and $V_{GS} = 0$ V in Fig. 8.59b, the photocurrent overwhelms the < 1 μA dark drain current, and increases with illumination intensity. The photocurrent also increases with drain current. In the saturation regime, at $V_{GS} = -40$ V, the drain current is ~ 100 μA at $V_{DS} = -60$ V, compared to 20 μA at $V_{GS} = 0$ V (Pal et al., 2010). The polymer absorption extends from 2.0 eV to >3.5 eV, thus resulting in photocurrent generation across much of the visible spectrum. Often, OPTs are characterized using unspecified white illumination making it

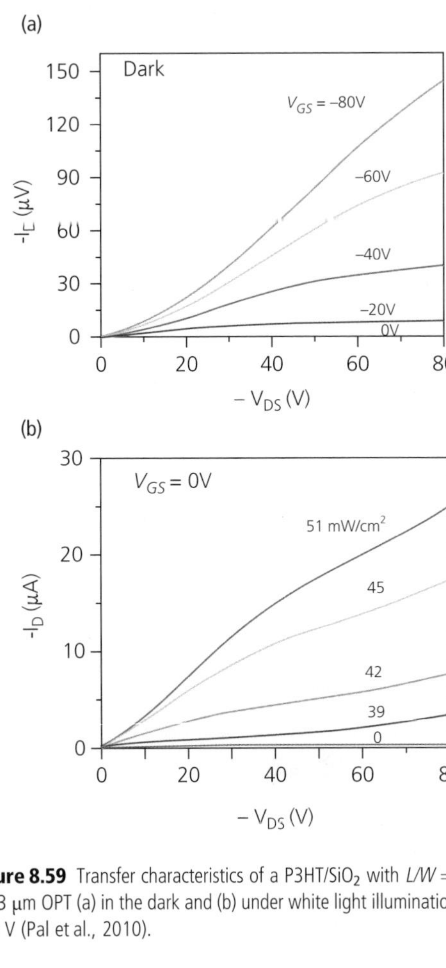

Figure 8.59 Transfer characteristics of a P3HT/SiO_2 with $L/W = 200$ μm/3 μm OPT (a) in the dark and (b) under white light illumination at $V_{GS} = 0$ V (Pal et al., 2010).

difficult to compare characteristics taken in different laboratories. As a rule, it is more useful to use monochromatic light sources that coincide with the peak absorption of the semiconductor.

The characteristics in the dark and under illumination for this OPT are shown in Fig. 8.60a for $V_{DS} = -60$ V. The photoresponse ratio of $P = I_{ph}/I_D$, has a peak of 3800 at $V_{GS} = -12$ V. Note that under illumination, the channel current does not switch off, but rather saturates at a photocurrent of $I_{ph} = 4$ µA. A plot of $\sqrt{|I_D|}$ vs. V_{GS} in Fig. 8.60b shows that slopes in the dark and under illumination are approximately equal, indicating that the field effect is independent of illumination. This eliminates the possibility that photogenerated minority carriers (electrons) or light-assisted trapping or detrapping at the interface affect the hole mobility (Wang et al., 2010). As in Fig. 8.60a, illumination induces a photocurrent offset that prevents the transistor from fully switching off, even for $V_{GS} > 0$. Further, the illuminated characteristics indicate a threshold voltage shift to +12 V, compared to −13 V in the dark. This ΔV_T is equal to the photovoltage given in Eq. 8.70. An increase in the sub-threshold slope for the illuminated device is also apparent from the increased linearity of the data near $V_{GS} = 0$ V.

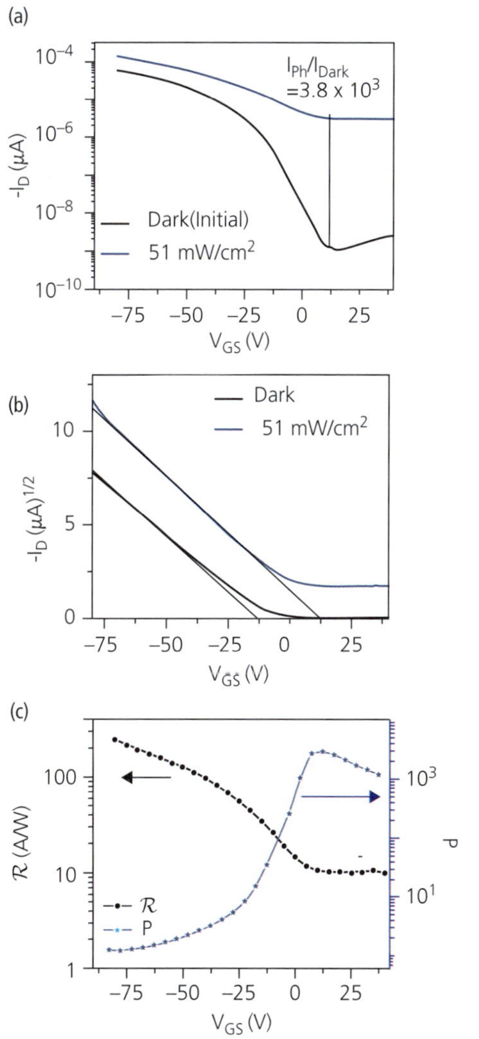

Figure 8.60 (a, b) Transfer characteristics of the P3HT/SiO_2 OPT in Fig. 8.59 in the dark and under illumination for $V_{DS} = -60$ V. (c) Responsivity and photoresponse ratio, $P = I_{ph}/I_D$ vs. V_{GS}, at $V_{DS} = -60$ V (Pal et al., 2010).

Finally, the responsivity and photoresponse are provided in Fig. 8.60c. A maximum responsivity of $\mathcal{R} = 250$ A/W is observed in saturation at $V_{GS} = -75$ V and $V_{DS} = -60$ V. Due to the large I_D at this voltage, the photoresponse is relatively small. Nevertheless, as we have seen in Section 7.2, this is ~100 times the responsivity of a photodiode. Hence, we expect that the bandwidth of the OPT will be at most, 100 times lower than an OPD of comparable dimensions. It is uncommon to find OPTs whose bandwidths exceed a few kilohertz (Baeg et al., 2013).

The photocurrent and photovoltaic regimes of operation have been explored for small molecule-based BPTT/SiO_2 BG/TC OPTs, with results shown in Fig. 8.61 (Noh et al., 2005). The device exhibited $\mu_{FEp} = 0.082$ cm^2/V s and a threshold voltage of $V_T = -29$ V. The maximum responsivity of the OPTs is 82 A/W at an intensity of 1.55 mW/cm^2 at $\lambda = 340$ nm, with a photoresponse of 2×10^5. Fits to Eqs. 8.70 and 8.73 (lines, Fig. 8.61a) are obtained in the transistor on and off states, corresponding to $V_{GS} = -50$ V and the V_{GS} at the minimum I_D, respectively. Fits affirm that the photoconductive current dominates when the OPT is in the off state, whereas it follows a logarithmic dependence on optical power when the device is in the on state. Also, the change in the threshold voltage, $\Delta V_T = V_{ph}$, also follows the theoretical trend predicted by Eq. 8.72, as shown in Fig. 8.61b.

Ambipolar OPTs that consist of either bilayer or blended channels can increase the charge generation efficiency by combining materials that form a type II, exciton dissociating interface (Marjanović et al., 2006, Anthopoulos, 2007, Labram et al., 2010). An example of a bilayer OPT is shown schematically in Fig. 8.62a, inset. The device comprises a bottom Al gate contact deposited on a glass substrate. The Al is exposed to an O_2 plasma to enable growth of a 5–10 nm thick AlO_x insulator, which is then capped by dipping in a octadecylphosphonic acid solution to deposit a 2.3 nm thick SAM. Next, a 100 nm thick pentacene donor channel is deposited by VTE, followed by Au S and

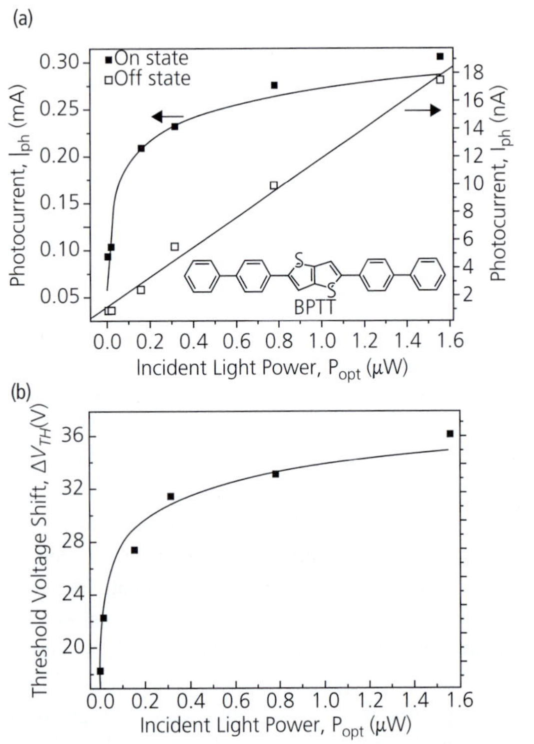

Figure 8.61 (a) Photocurrent vs. total incident optical power $P_{opt} = WLP_{inc}$ at $V_{GS} = -50$ V and $\lambda = 340$ nm (closed squares, on state) and at the minimum I_D (open squares, off state) at $V_{DS} = -50$ V for a BPTT OPT with $W/L = 5$ mm/20 μm. Inset: Molecular structural formula of BPTT used as the channel semiconductor. (b) Threshold voltage shift in the on state. Lines are fits to Eqs. 8.70–8.73 (Noh et al., 2005).

D contacts, and then capped with drop-cast $PC_{61}BM$ in CB, see inset, Fig. 8.62b. Exposing the pentacene to the $PC_{61}BM$ solution roughens its surface, reducing its mobility to $\mu_{FEp} = 7 \times 10^{-4}$ cm^2/V s (compared to 0.1–1 cm^2/V s typically observed). The electron mobility of the $PC_{61}BM$ layer is $\mu_{FEn} = 0.2$ cm^2/V s. This difference results in asymmetric transfer characteristics shown in Fig. 8.62a. The threshold voltages are $V_{Tn} = 0.7$ V and $V_{Tp} = -1.4$ V.

The donor–acceptor HJ gives a significant ambipolar photoresponse when illuminated at $\lambda = 469$ nm (2.64 eV), as shown in Fig. 8.62b. Light of this wavelength is absorbed in both layers, although it is primarily absorbed in the top $PC_{61}BM$ layer on which it is incident. The n-channel threshold voltage shifts by $\Delta V_{Tn} = -0.2$ V for an illumination intensity of 3.2 mW/cm^2, while $\Delta V_{Tp} = 0$ V. This difference is possibly due to charge trapping at the roughened pentacene/$PC_{61}BM$ interface. Importantly, I_{off} also increases with illumination intensity due to the large density of photogenerated charges at the D-A junction.

The photocurrent efficiency *vs.* illumination intensity at $V_{DS} = -0.3$ V is shown in Fig. 8.63. The maximum external efficiency is only 0.8% obtained at the lowest power. As the input power increases, the efficiency decreases. This is a common feature of OPTs, and is due to a saturation in $\Delta V_T = V_{ph}$ as the density of photogenerated charges increases (see Eq. 8.72). That is, recombination of photogenerated charges

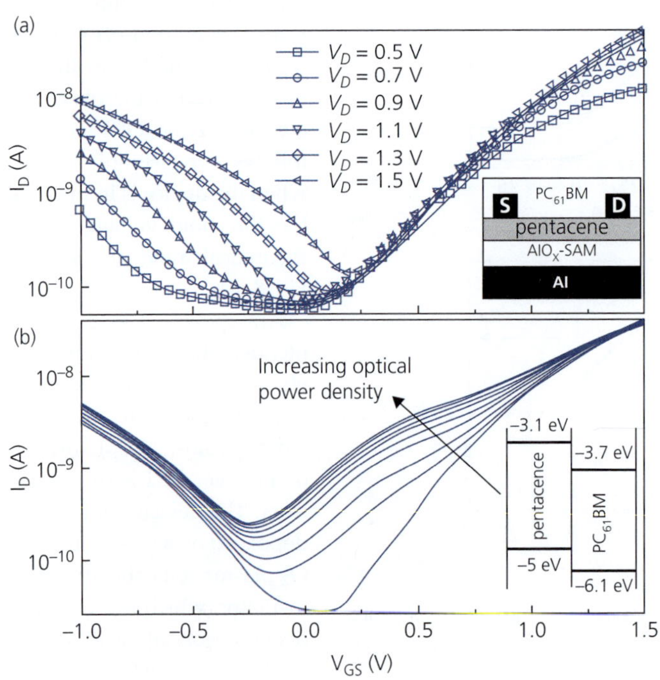

Figure 8.62 (a) Dark ambipolar transfer characteristics of the pentacene/$PC_{61}BM$ transistor shown in the inset. (b) Characteristics at $V_{DS} = -1$ V under $\lambda = 469$ nm illumination from a blue LED. Inset: Energy level diagram showing the exciton separating HJ (Labram et al., 2010).

increases with the number of electron-hole pairs, thus decreasing their ability to increase the channel current at high illumination intensities (Wasapinyokul et al., 2009). This imposes a limit to the linear dynamic range of OPTs, which may be increased somewhat by increasing V_{DS} to more efficiently sweep out photogenerated charges.

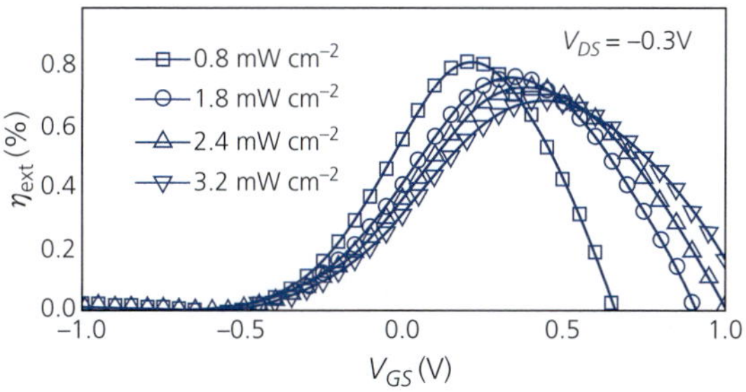

Figure 8.63 External quantum efficiency vs. V_{GS} of the device in Fig. 8.62 for various input optical powers and $V_{DS} = -0.3$ V (Labram et al., 2010).

We have shown that the bandwidth of OPTs is also limited by their gain. This is in addition to other factors that limit response such as gate and drain capacitances, charge mobility, etc., see Section 8.3.3. The rise and fall time transient response to a pulsed LED source of the OPT in Fig. 8.62 with $W/L = 1$ mm/ 50 μm are shown in Fig. 8.64a. The total rise and fall time is $t_r + t_f = 435$ ms, corresponding to a bandwidth of $\Delta f = 1/2(t_r + t_f) \approx 1$ Hz. This is quite slow compared to most OPTs, and is most probably due to charge trapping at the highly disrupted pentacene/ $PC_{61}BM$ interface.

Finally, low voltage ambipolar OPTs have also been used as complementary transistors in the inverter circuit shown in Fig. 8.64b, inset. In the dark, the circuit shows a maximum gain of $A_v = 35$ at a supply voltage of $V_{DD} = 2.5$ V, and an input voltage of $V_{IN} = 0.7$ V. The gain decreases with increasing channel illumination intensity, accompanied by a voltage shift of approximately -0.2 V at a power of 1.2 mW/cm^2. The operating point shift is due to a

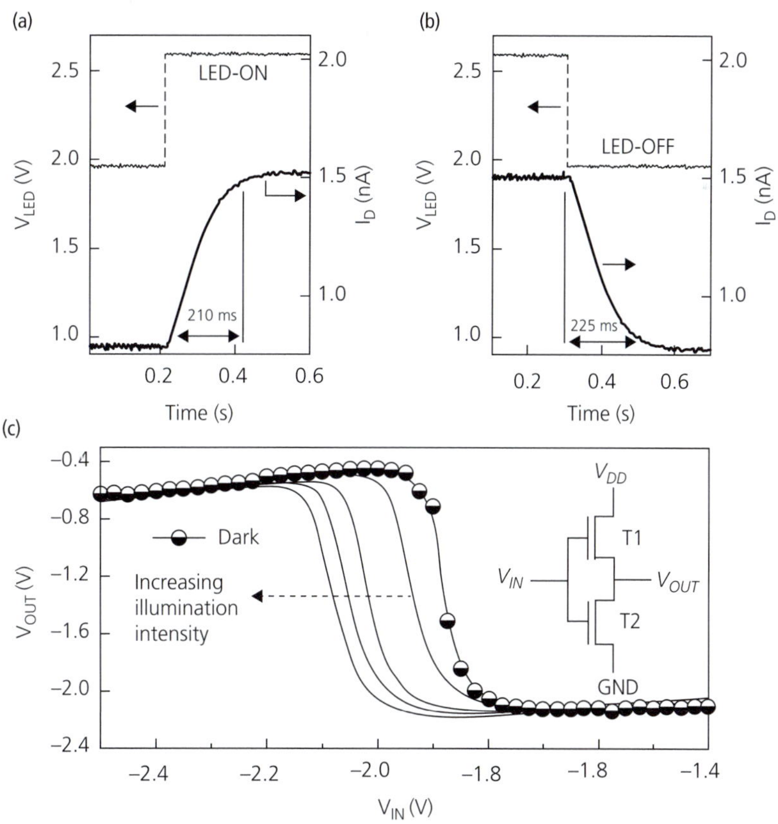

Figure 8.64 (a) Rise and (b) fall times of the OPT in Fig. 8.62 to a fast optical pulse whose waveforms are shown in the upper halves of the panels. V_{LED} is the voltage applied to the InGaN LED source. (c) Transfer characteristic of a complimentary ambipolar OPT pair in the dark and under illumination up to 1.2 mW/cm^2. Inset: Schematic of the complementary inverter circuit (Labram et al., 2010).

similar shift observed for the threshold voltage seen in Fig. 8.62b. Presumably, the channels of both transistors, T1 and T2 are illuminated simultaneously and at the same power. Differential illumination of the transistors can provide optical logic functionality, where the circuit output voltage, V_{OUT}, depends on whether one or both transistors are illuminated, and whether or not the intensity during illumination is the same (Anthopoulos, 2007).

In summary, an OPT is capable of very high photosensitivity (and hence responsivity) due to the strong dependence of the gate photovoltage on illumination intensity. However, this advantage is counterbalanced by its slow response to changes in optical signals, confining them to applications where bandwidths less than a few kilohertz are acceptable. Also, the devices are more complex than organic photodiodes discussed in Chapter 7. Nevertheless, their integration into complex circuits may provide advantages in specialized applications where optical signal processing is combined with detection.

8.6 OTFT materials

The performance of an OTFT is determined by several factors, including the organic semiconductor used for the channel and the dielectric, the deposition conditions for the channel layer that leads to a particular morphology, the quality of the contacts to the source and drain, and the gate and channel geometries. Both small molecules and polymers have been successfully used in OTFTs, and considerable effort has been directed toward morphology control to achieve high mobilities. These methods include deposition of large crystalline domains from the vapor or solution phases, applying strain to the as-deposited films to assist in driving their orientation into a preferred configuration, and orientation of single crystals. One of the most attractive features of organics is that they lend themselves to a nearly limitless modification of properties through variations in their composition and structure.

Figure 8.7 gives an example of how transistor performance is linked to film morphology. Unlike OLEDs that require amorphous emission regions, or blended OPVs requiring nanocrystalline domains distributed throughout the active region, OTFTs benefit from ordered crystal morphologies with polycrystalline grains extending across the entire channel. Single crystals exhibit the highest mobilities since charges are neither scattered nor trapped at grain boundaries, or in disordered regions (Reese and Bao, 2007, Liu et al., 2009a). Indeed, field effect mobilities can be several orders of magnitude higher for single crystals than for disordered films (Sundar et al., 2004, Stingelin-Stutzmann et al., 2005, Seo et al., 2006). Unfortunately, there has been no convincing demonstration that single crystals can be reproducibly deposited over large areas, and with mobilities that exceed those of thin films deposited from the vapor or solution phases. In addition to achieving crystalline growth, in Chapter 4 we found that organic crystals are characterized by highly anisotropic conductivities. Hence, the crystals need to align with their high mobility axes parallel to the S-D conducting direction. For these reasons, effort has been directed to finding materials and scalable process conditions that lead to thin film morphologies beneficial to high mobilities and on/off ratios.

Small molecule organics can take on crystal habits ranging from herringbone to lamellar π-stacking arrangements that depend on the molecular

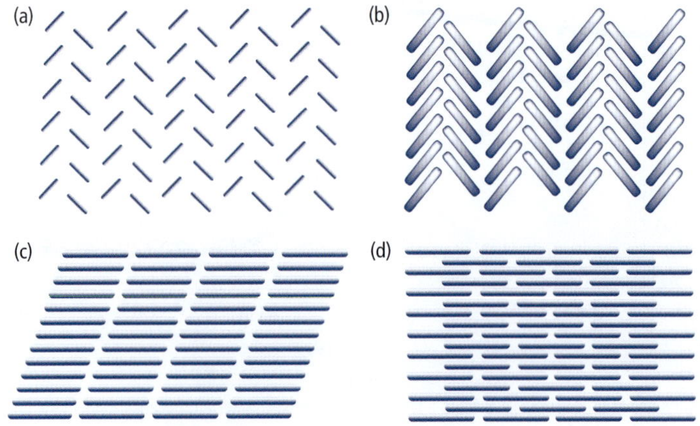

Figure 8.65 Common small molecular crystal stacking habits found in transistor thin films. (a) Face-to-edge herringbone structure lacking strong π–π stacking. (b) Closely packed, co-facial herringbone structure with strong π–π stacking. (c) Slip-stack co-facial –π stacking. (d) Lamellar, brick-like π–π stacking.

structure and growth process; several possible stacking habits are illustrated in Fig. 8.65 (also see Chapters 2 and 5). Unlike vertical device architectures found in OLEDs and OPVs, OTFTs require the high conductivity direction parallel to the substrate along the S-D direction. Often, the organic crystal forms the herringbone motif in Fig. 8.65a and b. While these structures do not have the most advantageous π-orbital overlap between molecules within the plane, they have the advantage of a reasonably isotropic conductivity, and hence the in plane orientation relative to the S–D axis becomes less important. The open, face-to-edge structure in Fig. 8.65a results in the lowest conductivity due to the relative absence of intermolecular orbital overlap.

All else being equal, the planar, or flat-lying habits of Fig. 8.65c and d should have a higher conductivity perpendicular to the molecular plane than the herringbone structures. Both the slip-stack and lamellar configurations can have tight intermolecular spacing and a low hopping energy barrier between adjacent molecules. The two different structures are driven by intermolecular van der Waals forces that determine the lowest energy configuration. These motifs are often found in polyaromatic hydrocarbon molecules such as naphthalene, anthracene or tetracene. Unfortunately, often the same forces that lead to lamellar stacking may also cause the molecules to lie flat on the substrate (e.g. PTCDA). Then the high conductivity direction lies perpendicular, and not parallel to the substrate as required in OTFTs. The high mobility that arises from parallel stacking is of no use in this configuration.

The orientation relative to the substrate plane is due to a balance of forces between the molecule and substrate, and between the molecules themselves. Hence, pentacene molecular stacks orient with their molecular axes perpendicular to an SiO_2 gate insulator, allowing the lamella to π-stack along the S–D direction giving a high in-plane mobility. In the following sections we will find that planar molecules are often functionalized using alkane "tails" to force them into the appropriate orientation relative to the substrate. Also, as noted in Section 8.3, a pre-deposited SAM can functionalize the surface to drive the orientation of subsequently deposited molecules. These strategies reinforce the conclusion that realization of high performance OTFTs requires exploitation and control of the interplay between the molecular structure, the insulator-organic interface morphology and surface energies, and the deposition and post-deposition process conditions. As a consequence, seeking high channel mobilities, low contact resistances and large I_{on}/I_{off} has become a very rich subject for investigation.

The situation for polymer films is analogous to that of small molecular weight materials. That is, achieving high in-plane mobility depends on the ability of the polymers to form crystals with the optimal interchain π-coupling direction positioned in the channel plane. The situation with a common p-channel polymer, P3HT, is shown in Fig. 8.66a. The extended polymer chain creates even larger anisotropies in electronic properties than do many small molecules. The highest conductivity direction is along the length of the chain, where each segment is attached to the next with a conjugated, covalent bond (Venkateshvaran et al., 2014). Interchain π–π stacking also promotes conduction. In the case of P3HT, close packing along this second dimension is promoted by the unhybridized *p*-orbitals of the thienylene groups. The least conductive direction is along the alkyl chains (the vertical direction in Fig. 8.66a).

The most advantageous orientation is for the molecular lamella to couple with the π-stacking direction along the channel, as shown in Fig. 8.66b. Clearly, the out-of-plane stacking arrangement in Fig. 8.66c requires conduction edge-wise between molecules. The insulating nature of alkanes and lack of intermolecular π-orbital overlap prohibits a high mobility in this case.

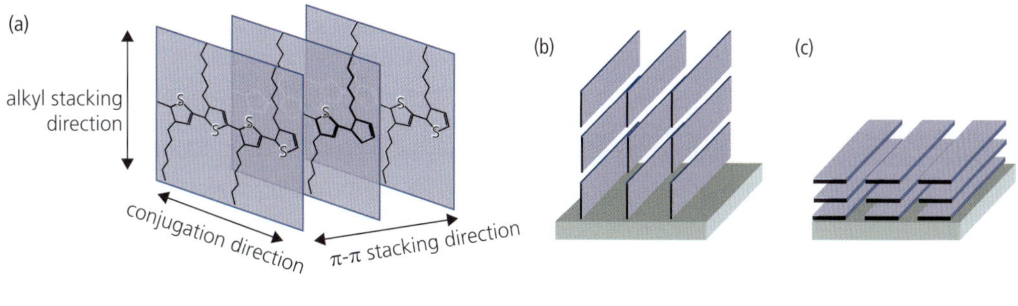

Figure 8.66 (a) P3HT molecular structure, showing the anisotropic orientation of the π–π stacking, conjugation and alkyl stacking directions. Molecular planes are π-stacked (b) parallel and (c) perpendicular to the substrate plane. The former orientation is most useful for OTFT applications (Salleo, 2007).

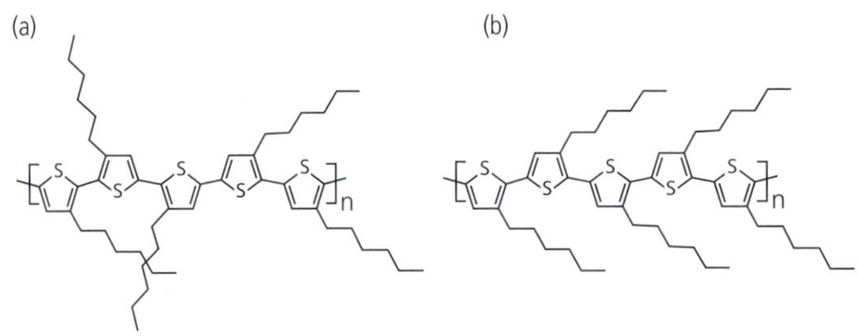

Figure 8.67 (a) Regiorandom and (b) regioregular P3HT molecules.

The charge mobility and the ability for polymers to crystallize into ordered structures is complicated by several factors that are not present in small molecule systems. These include regioregularity, molecular weight, rapidity with which the polymers are deposited, and post deposition annealing used to establish equilibrium crystalline morphologies. *Regioregularity* is the property whereby the alkyl moieties on each thienylene unit are attached at the same location. This is understood by inspection of Fig. 8.67. The irregular structure in Fig. 8.67a shows a random attachment of the alkyl groups on the both sides of the conjugated backbone. Due to steric interference between alkyls, the thienylenes along the backbone are twisted out of the molecular plane. This prevents strong π–π interactions, and at the same time breaks the conjugation along the backbone, such that the mobility in the two most favorable directions is substantially reduced. When the alkyl groups are on the same side, they are known as head-to-head or tail-to-tail arrangements (Fig. 8.67b). The uniform head-to-tail attachment maintains the planarity of the thienylene cores, as required to achieve the ordered structures in Fig. 8.66.

The molecular weight (M_n) and processing conditions of the polymer films determine the degree of order as well as the relative orientation to the substrate plane. For example, low molecular weight films prepared by slow casting from a dilute solution result in the high mobility configuration of Fig. 8.66b, whereas high molecular weight films prefer the flat-lying and low mobility arrangement in Fig. 8.66c, even though the conjugation length increases with M_n (Sirringhaus et al., 1999).

The dramatic dependence of morphology on M_n is shown by the atomic force microscope (AFM) images in Fig. 8.68 of regioregular P3HT thin films spin-cast on an HMDS-functionalized SiO_2 surface on Si. A rod-like morphology is observed for $M_n = 3.2$ kDa, whereas the $M_n = 31.1$ kDa film exhibits nanoparticle (NP) clusters. The lower molecular weight film is found to have a higher degree of crystalline order than that

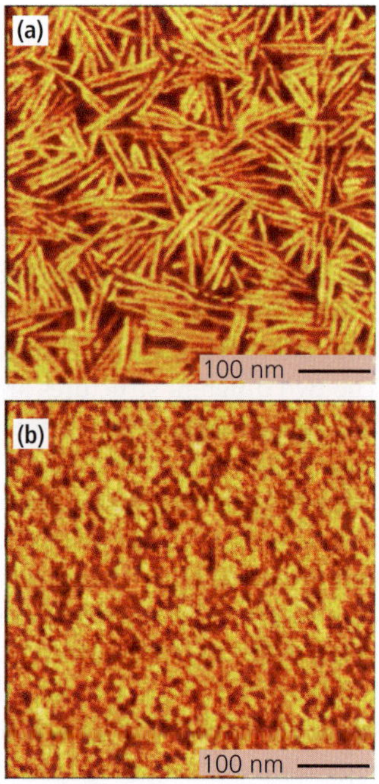

Figure 8.68 AFM images of regioregular P3HT films on HMDS coated SiO_2. (a) $M_n = 3.2$ kDa, (b) 31.1 kDa (Kline et al., 2003).

with larger molecular weights (Kline et al., 2003, Verilhac et al., 2006). Further, the rods form into extended bunches if the low M_n regioregular P3HT is slowly deposited by dip coating vs. rapid spin-cast deposition (Verilhac et al., 2006). Slow crystallization allows the films to approach their equilibrium morphologies with lamella oriented with their intermolecular π-stacks in the plane, although the morphological dependence on M_n is stronger than it is on the rate of deposition. The rate is slowest for dip coating, less slow for drop-casting, and fastest for spin-coating. Thus, shorter chains are more likely to form ordered domains separated by grain boundaries between segments, independent of the rate of casting. Indeed, both a low molecular weight and slow crystallization lead to a lower mobility, as reflected in the data in

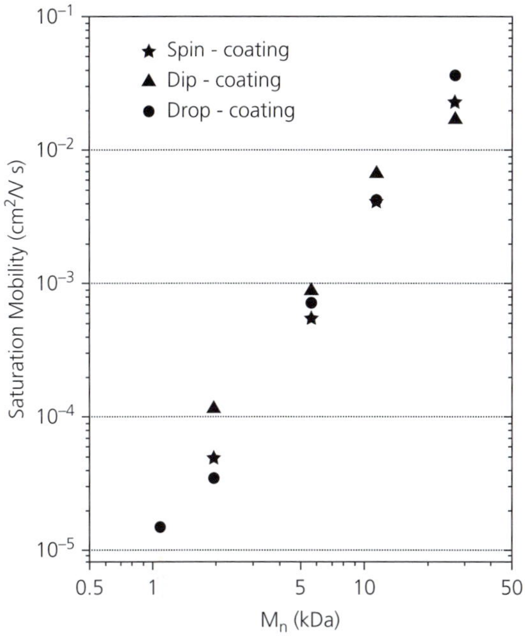

Figure 8.69 Dependence of regioregular P3HT μ_{FEp} on M_n and film coating method on OTFTs with an SiO_2 gate insulator (Verilhac et al., 2006).

Reprinted from Synthetic Metals, 156, Verilhac, J.-M., LeBlevennec, G., Djurado, D., Rieutord, F., Chouiki, M., Travers, J.-P. & Pron, A., Effect of macromolecular parameters and processing conditions on supramolecular organisation, morphology and electrical transport properties in thin layers of regioregular poly (3-hexylthiophene). 815-823, Copyright 2006 with permission from Elsevier.

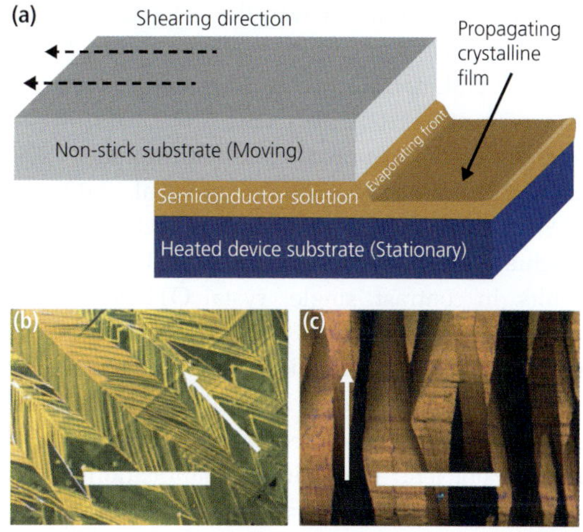

Figure 8.70 (a) Solution shearing employs two planar slabs translated relative to each other, with a solution of the semiconductor sandwiched between them. Oriented films of the small molecules (b) TMS-4T and (c) dbo-P2TP prepared by solution shearing along the directions of the arrows. The films are shown using polarization optical microscopy. The scale bars are 1 mm (Liu et al., 2009a).

Fig. 8.69. The extended chains formed as M_n increases are likely to have many points of contact with adjacent chains, leading to efficient charge hopping with a reduced probability for trapping at grain boundaries or at macroscopic boundaries between the chains (Kline et al., 2003).

Another means for imposing order along the channel is via *solution shearing* (Becerril et al., 2008). This process, illustrated in Fig. 8.70, is similar to preparing films using a *doctor blade*. A dilute solution of the semiconductor is deposited onto the target substrate. A second planar slab with a non-wettable surface is placed onto the surface of the solution, and is drawn parallel to the substrate while the entire system is gently heated. The solvent at the liquid front evaporates as the moving slab is translated, leaving behind crystalline precipitates that orient along the drawing direction. These crystallites are nucleation sites for additional material deposition, resulting in a preferred polycrystalline orientation also parallel to the drawing direction. Polarization microscope images of solution sheared films of the small molecules trimethylsilyl-substituted quarterthiophene (TMS-4T) and dbo-P2TP are shown in Fig. 8.70b and c, respectively. There is a clear morphological dependence on the shearing direction (indicated by the arrows), resulting in large-scale orientation of molecules and high conductivity along the channel. Dip coating can also result in directional crystallite formation, which may be in part responsible for improved performance using that technique.

Molecular structure is another factor that determines molecular arrangement. For example, interchain van der Waals forces in the planar conjugated F8BT copolymer has a preference for the flat-lying (tile) morphology of Fig. 8.66c (Donley et al., 2005). As a consequence, electron field effect mobilities of $\sim 10^{-3}$ cm^2/V s are observed, in comparison to upstanding molecular structures such as P3HT, whose mobilities are a factor of 100 or more larger.

Below, we describe approaches to surface functionalization using SAMs, and then present several small molecules and polymers used in p-channel, n-channel, and ambipolar OTFTs. Many p- and n-channel materials often show ambipolar conductivity, depending on the contacts employed. If the hole and electron mobilities of a material are significantly different, it is discussed in the sections focused on unipolar transport.

In the same spirit that we discussed OLED and OPV materials in Chapters 6 and 7, the example materials should provide an appreciation of the challenges that are faced by the materials scientist and chemist in developing high performance compounds for transistor applications. Given the necessarily cursory nature of our treatment, the reader is referred to extensive literature from which our examples are drawn (Newman et al., 2004, Facchetti, 2007, Murphy and Fréchet, 2007, Zaumseil and Sirringhaus, 2007, Klauk, 2010, Wang et al., 2011, Zhao et al., 2013).

We will also include some further techniques used to achieve morphologies that yield the best transistor results. In contrast, single crystal OTFTs and their associated materials will not be discussed. While these devices hold scientific interest, the value of the OTFTs is in their use in integrated circuits. Single crystal transistors often exhibit high mobilities, yet they do not lend themselves to fabrication beyond one or only a very few transistors on a substrate due to the special, and non-scalable handling necessitated in their fabrication (Reese and Bao, 2007).

8.6.1 Self-assembled monolayers

A SAM is a space-filling layer of surface molecules that serve two primary purposes. The most common is to functionalize the insulator surface to increase the mobility and decrease V_T. The application of SAMs can also result in improvements in sub-threshold slope and on-off ratio. Their second use is in bottom contact OTFTs, where the channel semiconductor is deposited onto the source and drain contacts. SAMs have been developed to increase adhesion that decreases contact resistances, decrease the metal/semiconductor barrier height, and nucleates crystalline growth that can spread into the channel, thus increasing mobility.

An example SAM is OTS, which is a precursor to functionalizing SiO_2 gate insulators. Packing into a space-filling monolayer begins by hyroxylating the insulator surface by exposure to UV-ozone, as in Fig. 8.71a. This creates a hydrophilic surface, which is transformed to hydrophobic by subsequent adsorption of the OTS. Next, the Cl atoms are washed away and are replaced with Si–O bonds to the SiO_2 surface. In Fig. 8.71b and c, the distal end of the alkane chain can be oxidized by further exposure to UV-ozone to controllably adjust the hydrophobicity of the surface. Alternatively, the end of the alkane chains can be functionalized to create nucleation sites for subsequently deposited channel molecules, including attachment of hydroxyl, anthryl, or other groups, depending on the desired morphological outcome (DiBenedetto et al., 2009).

Chlorosilanes such as ODTS and HMDS are members of a large class of SAM precursors. Another class of stable SAMs useful for modifying Al_2O_3 and other metal oxide surfaces are the anthryl-terminated phosphonic acids (PAs) in Fig. 8.72 (Ma et al., 2012). We show how the several molecular elements in the SAM are designed to perform different functions, leading to the desired structure/property relationships in the channel layer. The precursor molecule is the anthryl-terminated phosphonic acid containing a

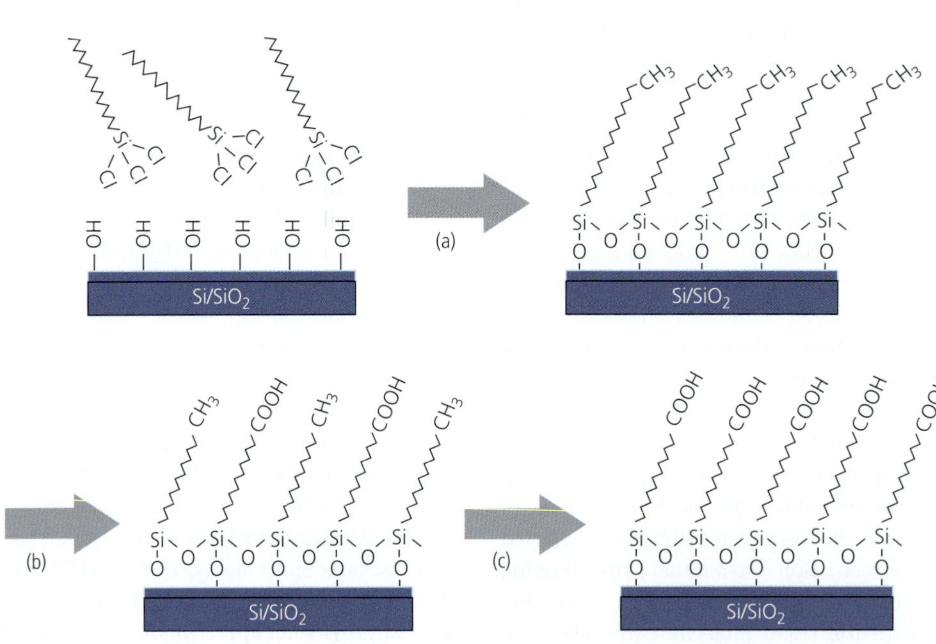

Figure 8.71 (a) Attachment of OTS onto a SiO_2 surface forming a close-packed, space filling monolayer. (b) Partial and (c) full oxidation of the distal end of the alkane chain to functionalize it for subsequent deposition of the channel molecules (Son et al., 2008).

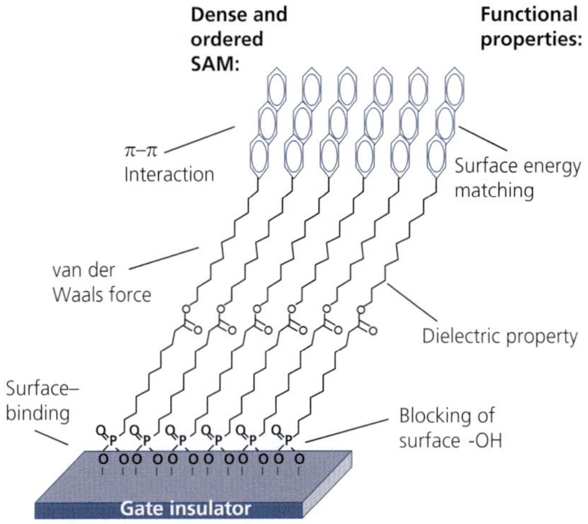

Figure 8.72 Surface functionalization using a π-σ-phosphonic acid SAM employing an anthryl end group, with the function of each component shown (Ma et al., 2012).

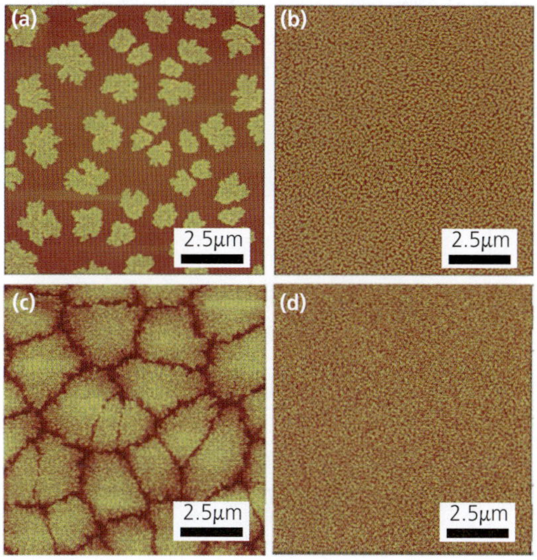

Figure 8.73 AFM images of ultrathin films of C_{60} on (a) ODPA on HfO_2, (b) bare HfO_2. (c) 50 nm thick C_{60} on ODPA/HfO_2 and (d) on bare HfO_2 (Acton et al., 2008).

combination of π and σ bonds. The molecule (or its analogs) can be applied by either vacuum or solution deposition. It attaches to the oxide via a reaction that replaces the surface hydroxyl group with an insulator P–O bond to the oxide (e.g. Al_2O_3 or HfO_2). The attachment of this space-filling monolayer blocks the surface states, while the aliphatic and anthryl groups close pack. The ensemble is stabilized by strong intermolecular π–π interactions between anthracene terminations. The anthracene groups tailor the surface energy and morphology such that the subsequently deposited channel molecules (typically a linear polyacene such as tetracene or pentacene) adhere in an orientation that promotes conductivity in the in-plane direction. Close packing of the SAM promotes alignment of the channel molecules in a layer that is largely free of defects, and that templates the surface to subsequently promote nearly commensurate or quasiepitaxial growth (Section 2.6). The alkane chains themselves are insulating and dipolar, which can have a significant impact on the surface potential, ψ_S, and hence V_T and μ_{FE}. As with the chlorosilanes, different functionalizing end groups can be tailored for the purpose of modifying different insulator surfaces and channel morphologies.

The modification of the morphology of C_{60} films on a HfO_2 surface in the absence or presence of the π-σ-PA, octadecylphosphonic acid (ODPA), is shown in the AFM images in Fig. 8.73. To prepare the surface, the HfO_2-coated substrates are immersed in a solution of 0.1 mM ODPA mixed in a 1:1 solution of THF and ethanol at 70°C where they remain for 12 h, followed by rinsing in solvents and then dried. The addition of the SAM has a profound effect on the subsequently vacuum deposited C_{60} film morphology. That is, from the nucleation (Figs. 8.73a and d) to the bulk film (Figs. 8.73c and d) stages, the presence of ODPA results in larger grain sizes (~2.3 μm, average *vs.* 0.1 μm on the SAM and on bare HfO_2 surfaces, respectively). This is attributed to more efficient diffusion of C_{60} during film deposition on the SAM. The larger grain size results in a higher mobility compared to the unfunctionalized oxide surface. Thus, $\mu_{FEn} = 0.28$ vs. 0.097, and $I_{on}/I_{off} = 10^5$ vs. 10^3, respectively (Acton et al., 2008).

Various explanations have been advanced to explain how SAMs on the gate insulator modify transistor characteristics, including the control of morphology by influencing the surface energy, and hence the surface wettability (Janssen et al., 2006). Close-packing can create a template for ordering the deposited film. And the interface potential itself can be influenced by passivation (i.e. neutralizing) of traps at the insulator surface, or it can introduce a dipole that shifts the energy levels (see Section 8.3).

Surface dipoles introduced by a series of organosilane SAMs with different molecular dipole moments on an SiO_2 gate insulator alters the charge mobilities of both pentacene and C_{60} OTFTs (Kobayashi et al., 2004). The molecules used are based on fluoroalkyl-, alkl-, and aminoalkylsilane (denoted F, CH_3, and NH_2, respectively). The molecular formula and packing of the F-SAM is shown in Fig. 8.74a. The drain current transfer characteristics for BG/TC pentacene OTFTs employing these SAMs, as well as an untreated SiO_2 insulator are shown in Fig. 8.74b. The most dipolar species, the F-SAM, results in the largest shift in V_T to +17 V, with V_T monotonically decreasing with dipole moment to a minimum of $V_T = -11$ V for both the untreated and NH_2-SAM (with molecular dipole moments of $p = 0$ and 0.60 D, respectively) functionalized surfaces.

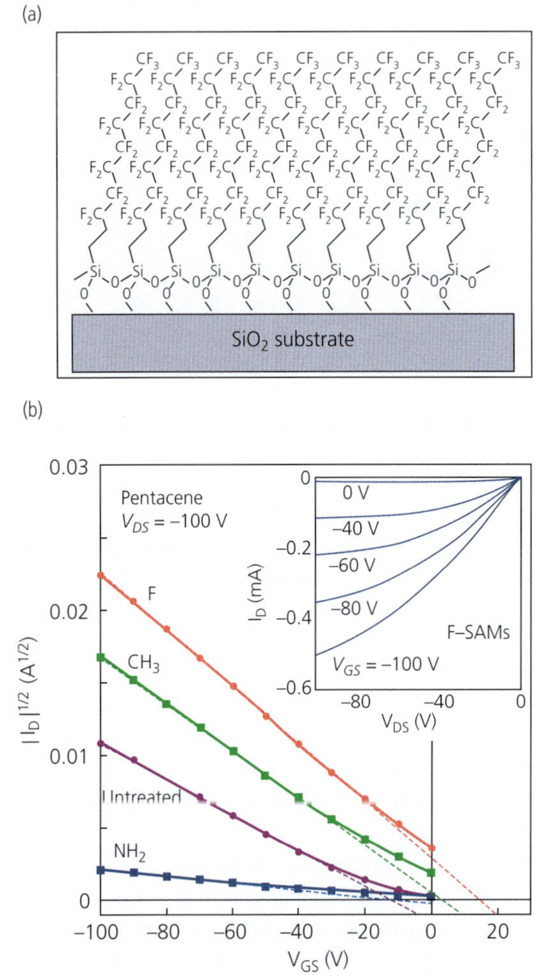

Figure 8.74 (a) Schematic of a closely packed F-SAM comprising $(CF_3)(CF_2)_7(CH_2)_2Si(OC_2H_5)_3$ on an SiO_2 surface. (b) Transfer characteristics of a SiO_2 gate insulator/pentacene OTFT employing an untreated surface and F-, CH_3-, and NH_2-SAM layers described in text. Inset: Drain current characteristics of the pentacene OTFT employing the F-SAM (Kobayashi et al., 2004).

The relative surface energies of the SAM surfaces are determined by the water contact angles (θ) listed in Table 8.1. In Section 5.6.5, we showed that the surface energy between the solid and the ambient is given by the Young equation (Israelachvili, 2011):

$$\sigma_{sg} = \sigma_{sl} + \sigma_{lg} \cos \theta, \tag{8.74}$$

where the interfacial surface energy between the liquid and the solid is σ_{sl}, and between the liquid and the ambient is σ_{lg}. Larger contact angles (i.e. for F-SAM and CH_3-SAM) correspond to a greater hydrophobicity, and hence a higher liquid–solid surface energy. The pronounced dependence of mobility and V_T on SAM composition for both pentacene p-channel OTFTs in Fig. 8.74, and C_{60} n-channel OTFTs (not shown) indicate that the SAM dipole moment controls the semiconductor/insulator interface potential. This, in turn, determines the charge type in the channel at $V_{GS} = 0$ V. The majority carriers at the interface in equilibrium are electrons, with $V_T > 0$ for F-SAM and CH_3-SAM devices, whereas holes populate the untreated OTFT interface. It is not a simple matter to isolate the origin of the dependence of the OTFT on a single factor such as molecular dipole moment due to the complex interrelationships between chemistry, morphology and trap structure.

Earlier in this section, we noted that SAMs have been applied to BC OTFTs to seed ordered growth, or to increase the adhesion of the channel layer to the source and drain contacts. As in the case of insulator/semiconductor junctions, the dipole moment at metal/semiconductor contacts also influences the injection barrier height (see Section 4.6). All of these factors suggest that SAM functionalization of the S and D surfaces can affect both the contact and channel resistances. Evaporated Au electrodes are favored for S and D contacts since the films are uniformly flat and continuous. A thin layer of Cr or Ti

Table 8.1 Pentacene OTFT with different SAMs on SiO_2 gate insulators (Kobayashi et al., 2004)

SAM [Nomenclature]	H_2O contact angle (°)	Dipole moment[a] (D)	μ_{FEp} (cm^2/V s)	V_T
$(CF_3)(CF_2)_7(CH_2)_2Si(OC_2H_5)_3$ [**F**]	108.8	2.3	0.20	17
$(CH_3)(CH_2)_7(CH_2)_2Si(OC_2H_5)_3$ [**CH_3**]	105.1	0.95	0.13	5.0
Untreated	24–35[b]	0	0.086	−11
$(NH_2)(CH_2)_3Si(OC_2H_5)_3$ [**NH_2**]	36.7	0.60	0.0024	−11

[a] Total dipole moment.

[b] Considerable variability in this measurement found in different laboratories.

deposited prior to the thick Au contact improves adhesion to an underlying oxide since both of these refractory metals form strong metal-oxygen bonds. The top surface of the Au is functionalized using one of any number of thiols that can form a SAM that attaches to the Au via the Au–S bond (Hu et al., 2005, Marmont et al., 2008). Examples of a few thiolated molecules used in SAMs are provided in Fig. 8.75. The most commonly used species are the alkane-thiols, although the alkane chain is non-conducting. It can be replaced with conducting analogs exemplified by the oligo (phenyleneethynylene) thiols or the oligo(phenylenevinylene) thiols (DiBenedetto et al., 2009).

The efficacy of using thiols on Au contacts is illustrated in Fig. 8.76a. The OTFT is a BG/BC architecture employing an OTS-treated SiO_2 on Si gate insulator and contact. The channel comprising PQT-12 is shown in the inset of Fig. 8.76a. Due to the bottom contact arrangement, application of PQT- 12 on the untreated Au surfaces leads to a large channel series resistance as evidenced by the nonlinear rise in I_D at low V_{DS} (highlighted by the oval). When the Au contacts are immersed in a solution of 1-octanethiol ($C_8H_{17}SH$) in ethanol for 24 h, and then rinsed in solvents prior to spin-coating with PQT-12, the resistance decreases, and the low voltage drain current characteristics become linear, as shown in Fig. 8.76b. The SAM initiates polycrystalline growth in an otherwise amorphous film (see AFM images in Fig. 8.76c and d). This not only reduces series resistance, but also results in a nearly ten-fold increase in mobility, from $\mu_{FEp} = 0.015$ cm^2/V s for the untreated, vs. 0.09 cm^2/V s for the SAM-functionalized S and D contact surfaces. The increase in mobility may be a function of increased film conductivity or decreased contact resistance (cf.

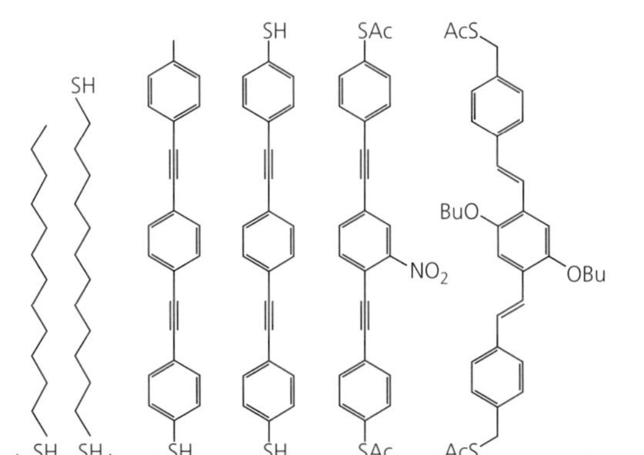

Figure 8.75 Example thiols used as SAMs for functionalizing Au contact surfaces.

Eq. 8.23) due to improved wetting of the Au surface. This, in turn, reduces the injection barrier to the polythiophene channel layer (Cai et al., 2008).

8.6.2 Small molecules for p-channel transistors

We now introduce several important small molecules developed for p-channel OTFTs. As discussed in Chapter 4, there is no intrinsic reason why hole mobilities are higher than that for electrons. The family of p-channel materials is, nevertheless, more extensive than for other conductivity types. The molecules selected typically have field effect mobilities > 0.5 cm^2/V s which is the nominal minimum value for practical applications. In a few instances, lower mobility materials are discussed to illustrate a point or introduce a family of structures that are of more general interest as OTFT channel materials. The potential of a transistor, however, is not simply a function of mobility. As shown in previous sections, the threshold voltage should be close to zero, the contact resistance low, and the sub-threshold slope should approach the theoretical minimum of 60 meV/decade. Transistors with all of these qualities ensure high bandwidth, low voltage operation and a high noise margin when employed in circuits. The materials cited fulfill one but generally not all of these criteria.

The molecular structural formulae for several p-channel materials are shown in Figs. 8.77 and 8.78. In Table 8.2 we summarize the properties of OTFTs made with the species in both figures, along with relevant aspects of device architecture and process conditions. Channel layers deposited by either vapor or solution phase deposition are represented.

Compounds **1–3** are representative fused ring aromatics. The molecules tend to form lamellar stacks in either a brick- or slipped-stack configuration due to interactions between their extended π-orbital systems. The molecules are insoluble in organic or aqueous solvents, requiring VTE deposition onto OTS-treated substrates to achieve the highest performance. They tend to form polycrystalline morphologies that lead to mobilities as high as 5 cm^2/V s in the case of pentacene (**2**). That result was taken from a statistical study of a large sample of devices that underscored the importance of materials purity (as achieved by thermal gradient sublimation and deposition in high vacuum), and process control as a cornerstone of high performance OTFTs (Kelley et al., 2003b).

It is tempting to assume that increasing the number of fused rings in the linear polyacenes will result in more ordered π-stacking. However, as we found in Chapter 3, extending the molecule increases its

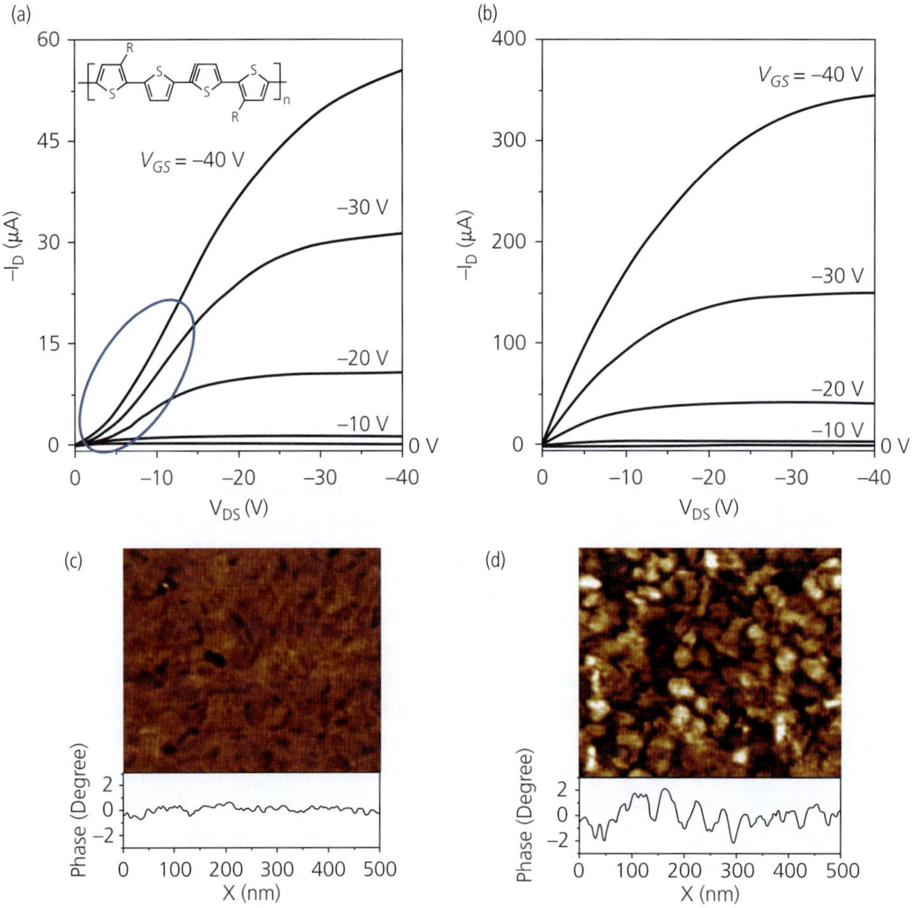

Figure 8.76 (a) Drain current characteristics of a BG/TBC p-channel OTFT with $L = 20$ μm and $W = 9$ mm employing the polythiophene, PQT-12 ($R = C_{12}H_{25}$ in inset) with untreated Au S and D contacts. The oval highlights the nonlinear characteristics at low V_{DS}. (b) An analogous device to (a) but with the S and D contacts pre-treated with a 1-octanethiol SAM. (c) AFM image of the untreated substrate (lower panel is the AFM image phase indicating relative surface roughness). (d) Image of the SAM-treated substrate (Cai et al., 2008).

conjugation length. This, in turn, results in a reduced energy gap, which can increase the off current (and hence decrease I_{on}/I_{off}), as well as decrease device stability. Indeed, the longer molecules may not necessarily form deep, ordered stacks compared with molecules with five or fewer rings. Hence, few investigations of linear polycyclic aromatics have gone beyond pentacene. However, bianthracene (**4**) and other anthracene and pentacene derivatives (**5**, **6**) have shown good stacking behavior along with concomitantly high μ_{FEp} and I_{on}/I_{off}. These molecules can be grown in an up-standing orientation when deposited on SAM-modified insulator surfaces, leading to enhanced π-stacking and higher stabilities than their un-modified core moieties.

Chapter 5.4.2 discussed OTFTs employing rubrene (**7**). Single crystals of this molecule form a co-facial herringbone structure (Fig. 8.65b), although films are most often amorphous with very low mobilities. To achieve mobilities approaching 1 cm^2/V s, the molecules were crystallized by dissolving in the glassy vitrifying species, 5,12-diphenylanthracene. This creates a hypereutectic that grows into extended crystallites rather than in a completely disordered film. The vitrifying mixture is blended into an ultrahigh molecular weight polystyrene matrix to provide mechanical stability and flexibility needed for use in circuits on bendable substrates. The polycrystallinity of the films containing all three components is apparent in the polarized optical micrograph in Fig. 8.79 (Stingelin-Stutzmann et al., 2005).

The series of compounds **8**–**10** are solution-processable variants of pentacene or tetracene, some with propyl (Pr) and others with ethyl (Et) groups appended to the perimeter (peri) groups surrounding the molecular core to improve solubility. The most fully studied molecule in this series is TIPS-pentacene, **8**. It forms lamellar structures with mobilities of

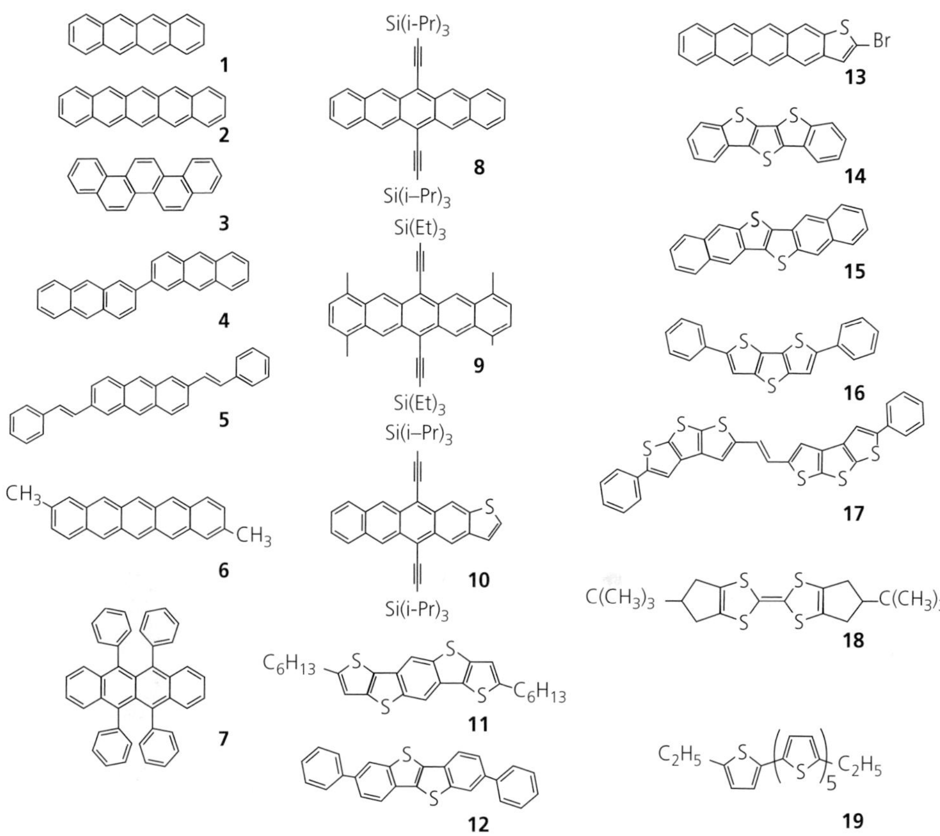

Figure 8.77 Example small molecules used in p-channel OTFTs and listed in Table 8.2.

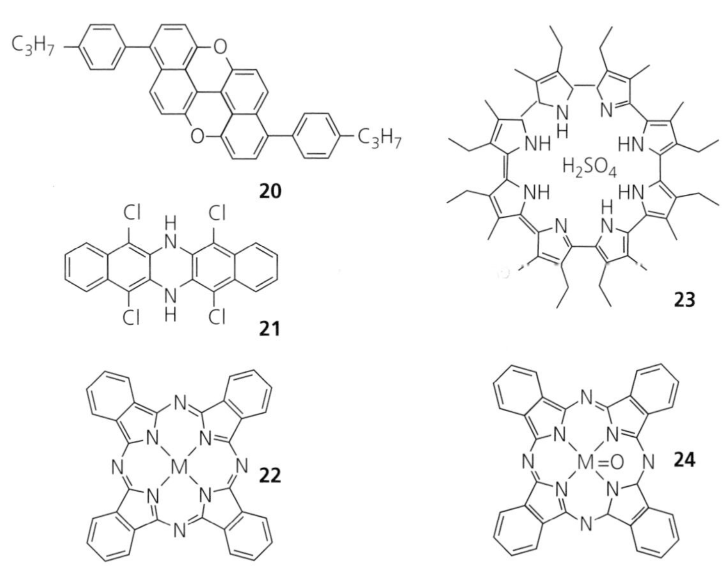

Figure 8.78 Additional small molecules used in p-channel OTFTs and listed in Table 8.2.

$\mu_{FEp} = 1.8\ \text{cm}^2/\text{V s}$ when cast from solution. The solubility and crystal-forming ability is increased further with the tetramethylated variation, **9**.

Strong S-S interactions between adjacent molecules have made various thiophene derivatives a focus of considerable attention for use in OTFTs, several of which **10–19** are shown in Fig. 8.77. Attaching alkyl chains (e.g. **11**) leads to good film forming characteristics via dip or spin coating, and improved intermolecular π–π stacking is made possible by

Table 8.2 Example small molecules and device structures employed in *p*-channel OTFTs

Mol.	Structure	Insulator	μ_{FEp} (cm²/V s)	V_T (V)	I_{on}/I_{off}	Comments	Ref.
1	BG/BC	OTS/SiO_2	0.12	−3	10^7		(Gundlach et al., 2002)
2	BG/TC	AMS[a]/Al_2O_3	5.1	−7	10^6		(Kelley et al., 2003b)
3	BG/TC	SiO_2	1.75	−80	10^4	O_2 expos. 70 h.	(Okamoto et al., 2008)
4	BG/BC	OTS/SiO_2	1.0	−55	10^7		(Kumaki et al., 2008)
5	BG/TC	OTS/SiO_2	1.3	−16	10^7	$T_{sub} = 80°C$	(Klauk et al., 2007)
6	BG/TC	P6D/Al_2O_3[b]	2.5	−3.8	10^6		(Kelley et al., 2003a)
7	BG/BC	PVP	0.7	0	10^6	Vitrified mixture w. DPA	(Stingelin-Stutzmann et al., 2005)
8	BG/BC	HMDS/SiO_2	≤1.8	−3.4	10^8	Drop cast	(Park et al., 2007)
9	BG/TC	OTS/SiO_2	2.5	−20	10^6		(Llorente et al., 2009)
10	BG/TC	OTS/SiO_2	0.78	−5	10^9		(Tang et al., 2008b)
11	BG/TC	SiO_2	1.7	−39	10^7	Dip coated	(Gao et al., 2009)
12	BG/TC	ODTS/SiO_2	1.5–3.9	−15	10^6–10^8	Device variations	(Izawa et al., 2008)
13	BG/TC	OTS/SiO_2	0.79	−11	10^7		(Tang et al., 2008a)
14	BG/TC	OTS/SiO_2	0.51	−67	10^6		(Gao et al., 2007)
15	BG/TC	OTS/SiO_2	2.9	−11	10^7	$T_{sub} = 60°C$	(Yamamoto and Takimiya, 2007)
16	BG/TC	SiO_2	0.42	−23.4	10^7	$T_{sub} = 70°C$	(Sun et al., 2006)
17	BG/TC	OTS/SiO_2	2.0	−31	10^8	$T_{sub} = 100/27°C$[c]	(Zhang et al., 2009)
18	BG/TC	OTS/SiO_2	0.98	−2.2	10^6	$T_{sub} = 50°C$	(Kanno et al., 2009)
19	BG/TC	PVP	1.1		10^4		(Halik et al., 2003)
20	BG/TC	OTS/PVP	0.6–0.8	4	10^5		(Kobayashi et al., 2009)
21	BG/TC	HMDS/SiO_2	0.3	−20	10^7	$T_{sub} = 150°C$	(Drolet et al., 2005)
22	BG/SC[d]	TaO_2	0.11	−8.9	10^5	$T_{sub} = 150°C$; M = Cu	(Zhang et al., 2004)
23	BG/BC	SiO_2	0.68	0	10^5	30 layer L-B dep.[e]	(Xu et al., 2005)
24	BG/TC	OTS/SiO_2	3.3	−17.8	10^7	$T_{sub} = 120°C$; M = Ti	(Li et al., 2007)

[a] AMS is a 10 nm thick layer on 150 nm sputtered Al_2O_3.
[b] P6D SAM on the Al_2O_3 surface.
[c] Two step channel deposition done at two different T_{sub} at rates of 0.1 then increased to 0.4–0.6 Å/s.
[d] SC = sandwiched contacts between a 30 nm thick CuPc channel layer and a 50 nm thick CoPc layer that aids in charge injection.
[e] L-B = channel layer consists of 30 layers deposited by the Langmuir–Blodgett method described in Section 5.4.2.

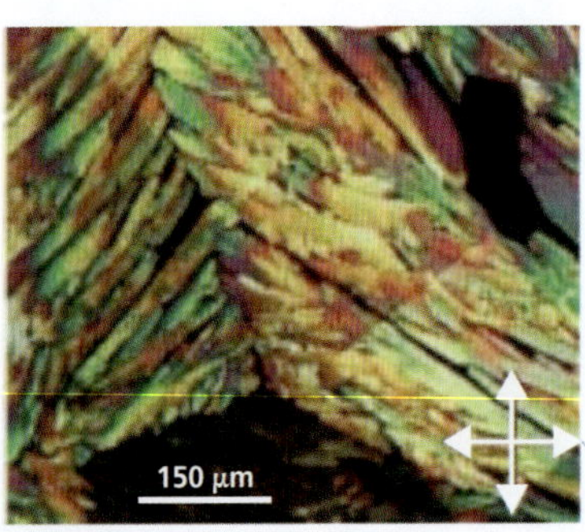

Figure 8.79 Polarized optical micrograph of a polycrystalline rubrene:DPA: polystyrene hypereutectic film formed at a temperature $T_{eutectic} < T < T_{melt}$. The needle-like features are the individual crystalline domains. The arrows show the polarization directions viewed (Stingelin-Stutzmann et al., 2005).

end-capping the molecules with planar phenyl groups, as in the case of DPh-BTBT, **12**, where high mobilities (1.5–3.9 cm²/V s) along with very high $I_{on}/I_{off} = 10^8$ are achieved. Indeed, this compound, along with **15** (DNTT, see Fig. 8.23) have shown $V_T \sim 0$ V, making them suitable for low power circuit applications (Zschieschang et al., 2017). Indeed, nearly all of the fused thienoacenes shown have high mobilities, and in most cases suitably low V_T due to their ability to form ordered stacks. Many of these moieties (e.g. **15**–**18**) can be vacuum deposited.

The oligothiophenes (α-6T, **19**, being an example) have attracted considerable attention due to the ease of functionalization, especially when compared to phenyl-based compounds, or to the ability to controllably adjust the number of thiophene units. As the number of links in the oligomer increase, the molecular solubility decreases. Unfortunately, the stability of these molecules is not high, nor is their mobility. For **19**, attaching alkyl chains at the end positions increases the oligomer solubility. Finally, the rigid (HMTTF) molecule readily forms crystallites with hole mobilities as high as 3.6 cm^2/V s, albeit with a low ($\sim 10^3$) on-off current ratio (Kanno et al., 2009). Hence, the best combination of characteristics is for device using **18** in Table 8.2.

An advantage of many of the compounds in Fig. 8.78 is their stabilities. The *peri*-xanthenoxanthene, **20**, has shown exceptional stability to air exposure, even at temperatures as high as 150°C. The molecule has been adapted for vacuum and solution processing (with and without the alkane moiety). It features a nanocrystalline structure necessary for high mobility. The thin film mobilities obtained via vacuum and solution processing on a PVP gate dielectric are $\mu_{FEp} = 0.8$ and 0.4 cm^2/V s, and the threshold voltages are $V_T = 0$ V and 4 V, respectively (Kobayashi et al., 2009). Hence, as is often the case, improved performance is achieved by vacuum deposition as a result of its ability to produce a more highly ordered film. The halogen-substituted azaacene, **21**, also leads to advantageous π-stacking, and hence high mobility due to intermolecular Cl–Cl interactions.

Lastly, we turn to the phthalocyanines (Pcs) and porphyrins (**22–24**). Phthalocyanines are particularly stable compounds that have found ubiquitous applications as dyes. Their stability naturally leads to interest in their suitability for electronic devices. Most metallated phthalocyanines (MPc, **22**), including CuPc, form into a range of stacking habits, depending on deposition conditions, although the slip-stack is the most common (Fig. 8.65c). To be useful in OTFTs, the stacks must tilt edge-on relative to the substrate, which depends on the relative energies between molecules and substrate, and among the molecules themselves (Yook et al., 2011). Unfortunately, the MPcs have not shown high mobilities, and hence are not often employed in OTFTs. An exception is the metal oxide Pcs (**24**) such as TiOPc. Unlike the MPcs, MOPcs are non-planar, forming a pyramidal structure leading to very tightly spaced molecules (~3.15 Å) with significant π–π overlaps (Li et al., 2007). The result is higher mobilities than the MPcs and that are, in a few cases (including VOPc, Li et al., 2008) competitive with many of the thiophene and acene-based compounds. An analogous cyclic molecular family to the Pcs is the porphyrins (**23**, for example). Langmuir–Blodgett techniques (Section 5.4.2) have been used to deposit 30 monolayers of the acid-based cyclo[8]pyrrole to achieve a modest mobility and $V_T = 0$ V (Xu et al., 2005).

8.6.3 Small molecules for n-channel transistors

The selection of high performance n-channel materials is more limited than for p-channel semiconductors, possibly due to their reduced air stability, and also due in part to the shorter time in their development. In Section 8.2 we showed that the early OTFT demonstrations were almost entirely focused on p-channel devices due to the immediate availability of established, high hole mobility materials such as the thiophenes and the linear polyacenes. The n-channel materials generally are based on molecules with strong electron withdrawing groups via attachment of amino, cyano, or halogen moieties, or molecular cores with large HOMO energies such as the naphthalene- or perylenetetracarboxylic diimide families. The deep energy levels create low electron injection barriers from the source contact.

Many materials used for *n*-channel transistors have nearly comparable hole and electron mobilities. In that case, they can be employed in ambipolar OTFTs whose characteristics are primarily determined by the S and D contact barriers, see Fig. 8.80. That is, a high work function (ϕ_m) metal results in a low HOMO energy (E_{HOMO})/work function offset allowing for efficient hole injection into the channel. This combination is preferable for p-channel contacts. Likewise, a low work function metal is employed for n-channel operation. As seen in Chapter 4, the Fermi energy of the organic is often determined by the injected, rather than the intrinsic charge, which justifies the relative position of the metal and semiconductor energies in

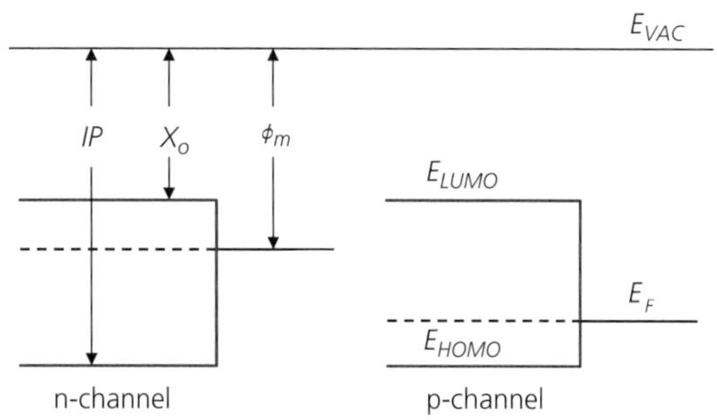

Figure 8.80 Channel contact energies favorable for n- and p-channel OTFTs. IP = ionization potential, χ_O = electron affinity of the organic, and ϕ_m = contact metal work function. Energy level bending on the organic side of the contact is omitted for clarity.

Fig. 8.80 for ambipolar conducting organics. Reducing the energy gap of the organic is an alternative means to reducing the injection barrier, although this can lead to higher output conductance and a lower I_{on}/I_{off}.

Example materials primarily employed in n-channel devices are shown in Fig. 8.81, with detailed OTFT characteristics listed in Table 8.3. Perfluoropentacene (**25**) is an example of a halogen-containing linear polyacene, where the electron-withdrawing F atoms replace H at the pentacene *peri* positions. The electron mobility of $\mu_{FEn} = 0.11$ cm^2/V s is considerably less than the core, hole-conducting pentacene (**2**) with $\mu_{FEp} = 5.1$ cm^2/V s. Fluorination has also been employed to improve electron conduction in the phthalocyanines, with the example of $SnCl_2F_{16}Pc$ (**26**) yielding $\mu_{FEn} = 0.3$ cm^2/V s, compared with the best *p*-channel TiOPc, (**24**) of $\mu_{FEp} = 3.3$ cm^2/V s. Compound **26** forms a tightly packed co-facial, polycrystalline herringbone structure (Fig. 8.65b) with grain sizes ~1 μm. Like most of the Pcs, this compound is reasonably air stable.

Several halogenated analogs of p-channel materials shown are **27** – **29**. Notably, the tetrahalogenated TTF derivative, **27**, is analogous to the p-conducting compound **18**. Both are planar, lending themselves to compact and ordered π-stacking arrangements. One of the highest electron mobilities is achieved with **28** (and its analogous thiophene derivative, **29**), with $\mu_{FEn} = 1.83$ cm^2/V s. This molecule, too, forms a nearly planar motif leading to tight π-π stacking. Unfortunately, V_T approaches 78 V, leading to an excessive voltage swing that makes it unusable for circuit applications.

An example non-halogenated molecular structure is DCMT, **30**, where the four cyanomethylenes are electron-withdrawing. Interestingly, low work function Mg contacts were used in place of the conventional Au contacts to achieve improved electron injection, leading to a mobility of 0.2 cm^2/V s. Oligothiophenes are also good hole conductors, but when halogenated (**31**), they result in electron-conducting semiconductors useful in n-channel transistors.

Naphthalene- and perylenetetracarboxylic diimides (compounds **32**–**34**) have comparatively large E_{HOMO}, making them excellent building blocks for n-channel devices. Their stability and the ease with which they can be synthetically modified has led to their extensive use in OTFTs. The central aromatic

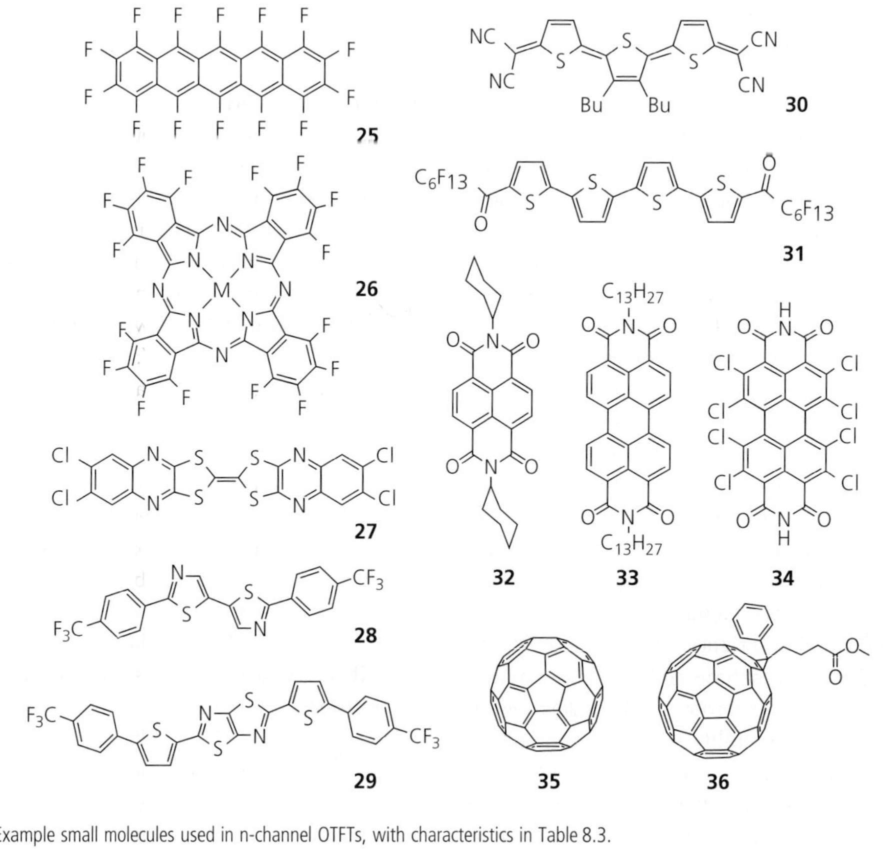

Figure 8.81 Example small molecules used in n-channel OTFTs, with characteristics in Table 8.3.

Table 8.3 Example small molecules and device structures employed in n-channel OTFTs

Mol.	Structure	Insulator	μ_{FEn} (cm^2/V s)	V_T (V)	I_{on}/I_{off}	Comments	Ref.
25	BG/TC	OTS/SiO_2	0.11	0	10^5	$T_{sub} = 50°C$	(Sakamoto et al., 2004)
26	BG/TC	p-6P[a]/SiO_2	0.3	27	10^7	M = $SnCl_2$	(Song et al., 2008)
27	BG/TC	HMDS/SiO_2	0.11	50	10^7	Meas. at 10^{-5} Pa	(Naraso et al., 2006)
28	BG/TC	OTS/SiO_2	1.83	78	10^4		(Ando et al., 2005)
29	BG/TC	ODTS/SiO_2	1.2	67	10^7		(Kumaki et al., 2007)
30	BG/BC	SiO_2	0.2	11	10^6	$T_{sub} = 130°C$	(Chesterfield et al., 2003)
31	BG/TC	PS[b]/SiO_2	1.7	24	10^9		(Yoon et al., 2006)
32	BG/TC	OTS/SiO_2	6.2	58	10^9	Ar atmos.[c]	(Shukla et al., 2008)
33	BG/TC	SiO_2	2.1	60	10^5	140°C anneal	(Tatemichi et al., 2006)
34	BG/TC	OTS/SiO_2	0.82	28	10^8		(Gsänger et al., 2010)
35	BG/BC	HMDS/SiO_2	3	17	10^6	$T_{sub} = 120°C$	(Kitamura et al., 2008)
36	BG/TC	PVA	0.2	−20	10^3		(Singh et al., 2005)

[a] p6P is a 3.5 nm thick SAM deposited from solution.
[b] PS is 24 nm thick with $M_w = 280$k.
[c] Tested under an Ar atmosphere, showing large hysteresis in the forward and backward sweep directions. In air, μ_{FE} was reduced to 0.41 cm^2/V s.

cores are functionalized with carbon rings (**32**) or alkane groups (**33**) to increase their solubility, and to encourage π-stacking of the conjugated planar perylene or naphthalene cores along the in-plane direction. Indeed, one of the advantages of these structures is their large and extended conjugated π-systems that lead to the formation of very deep, ordered stacks. The cyclohexyl substituted naphthalene compound (**32**) shows a remarkably high electron mobility of 6.2 cm^2/V s in an Ar atmosphere. Unfortunately, this particular compound is unstable, showing vastly different (and lower) performance in oxygen and air. Even in Ar, the device drain current characteristics showed significant hysteresis, depending on whether V_{DS} is swept from 0 to $+V_{DS}$ or vice versa. The tridecyl-functionalized perylenetetracarboxylic diimide (**33**), on the other hand, is quite stable in air, and has $\mu_{FEn} = 2.1$ cm^2/V s. When the alkane is replaced with C_3F_7, very high mobility ($\mu_{FEn} = 1.42$ cm^2/V s) was observed for air-stable BG/TC OTFTs, comprising a solution-deposited OTS SAM on SiO_2 gate insulator. This mobility is double that achieved via vacuum deposition of the SAM (Schmidt et al., 2009). Solution processing via substrate immersion in an OTS solvent solution results in improved order, leading to larger crystallite formation within the transistor plane. The most stable compound, albeit with the lowest mobility of these three planar compounds is 1,2,5,6,7,8,11,12-octachloroperylene-3,4,9,10-tetracarboxylic diimide (**34**). The OTFTs have a stable μ_{FEn} and I_{on}/I_{off} after exposure to air for 20 months. This is attributed to their deep E_{LUMO} (and hence low probability for reduction), and the exceptionally close intermolecular packing promoted by Cl–Cl interactions at distances as small as 3.25 Å (Gsänger et al., 2010).

The last two *n*-channel materials are the vacuum-deposited fullerene, C_{60}, (**35**) and its solution deposited analog, $PC_{61}BM$ (**36**) that are extensively used in organic detectors (see Chapter 7). The spherical symmetry of C_{60} promotes intimate close-packed arrangements when deposited on almost any surface, and in all three orthogonal directions. The resulting strong π-coupling between adjacent molecules has made this material desirable for *n*-channel FETs, with mobilities in excess of 3 cm^2/V s and with relatively low threshold voltages of $V_T \sim 10$ V. In contrast, the butyric acid methyl ester functionalization of $PC_{61}BM$ that leads to its solubility, also lends steric bulk to the molecules that prohibits its close packing and structural ordering. Hence, solution-processed fullerene-based OTFTs have characteristics that are inferior to those of their vacuum-deposited analogs. Unlike the perylene and naphthalene-based compounds, fullerenes are not easily functionalized. The higher order fullerenes such as C_{70} that lack the symmetry of C_{60}, exhibit field effect mobilities approximately three orders of magnitude smaller than the latter molecule (Haddon, 1996). An additional disadvantage of fullerene-based OTFTs is that their operating characteristics are not as stable as the perylene- and naphthalene-based compounds.

8.6.4 Polymers for p-channel transistors

Example polymers developed for *p*-channel OTFTs are shown in Fig. 8.82. Properties of OTFTs using these molecules are provided in Table 8.4. The molecules fall into two general families; those based on the regioregular polythiophenes (**37–40**), and thiophene copolymers with rigid, fused molecular backbones (**41**, **42**). Molecular design entails the incorporation of high performance monomers, such as those listed in Table 8.2, into a high molecular weight, soluble chain that can be crystallized into a structure with favorable intermolecular π–π coupling such as in Fig. 8.66b.

Oligothiophenes and their derivatives are often employed in small molecule TFTs (**19**), and indeed are among the first materials to demonstrate organic transistor behavior (see Section 8.2). The thiophene core is easily functionalized to achieve a desired transistor property. Likewise, the thiophene core is employed in numerous *p*-channel polymers in Fig. 8.82. Typically, the performance of polymeric materials is inferior to that of small molecules due to the usual difficulties in achieving high purity, a monodisperse molecular weight distribution, and the needed level of crystalline order over long channels. This is the case in comparing **37**, which is P3HT in Section 8.6.1, and PQT-12 (compound **38**) that have mobilities approximately five times lower than small molecule oligothiophenes (**19**). Annealing polymer films can result in improved crystallinity, and hence improved charge transport. For example, polythiophenes with crystalline nano-domains form after annealing at elevated temperatures, resulting in significant improvements in μ_{FEp} (Ong et al., 2005).

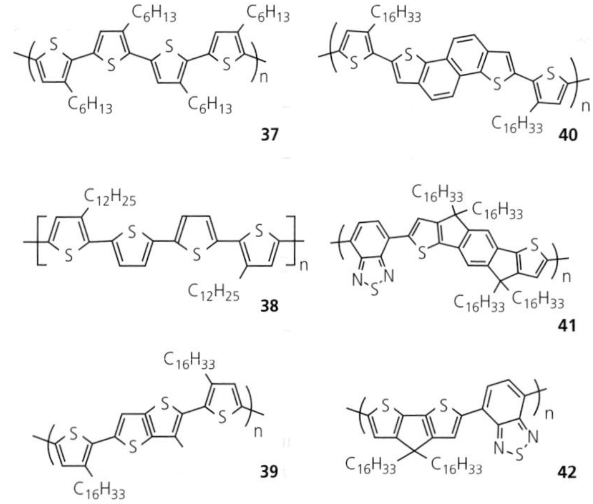

Figure 8.82 Example polymers used in p-channel OTFTs, with characteristics in Table 8.4.

High thiophene mobilities have been achieved with the PB16TTT block copolymer (**39**), whose fused rings promote planar stacking, particularly when deposited on SiO_2 surfaces functionalized with fluoroalkyl-based SAMs that induce ordered π–π coupling along the in-plane direction. The fused rings localize the electron density, preventing it from spreading along the chain. Field effect hole mobilities of 1 cm^2/V s along with a low $V_T = -8$ V and $I_{on}/I_{off} = 10^6$ have been achieved. The extended fused rings of **40**, lead to a performance similar to PB16TTT.

The rigid conjugated backbones of polythiophene copolymers, **41** and **42**, show exceptionally high mobilities that are comparable to several high performance small molecule materials in Table 8.2. Compound **41** (IDT-BT) is a fused ladder indacenodithiophene (IDT) alternating copolymer with benzothiadiazole (BT). Its molecular structure encourages ordered crystal formation, avoiding the twists that occur in the other polythiophene molecules shown in Fig. 8.82. Similarly, the copolymer, **42**, is found to order on PTES-treated SiO_2 gate insulators with their long axes parallel to the substrate, which is perpendicular to the dip coating direction. Both materials have mobilities >1 cm^2/V s, albeit with somewhat low on-off current ratios.

8.6.5 Polymers for n-channel transistors

There have been only a handful of demonstrations of high electron mobility polymers useful in unipolar n-channel OTFTs. Many n-channel polymer devices have unstable characteristics, with reports providing only incomplete information about the details of fabrication, or other important transistor operating characteristics. Several of the highest performance materials useful in *n*-channel devices are ambipolar conductors: these compounds will be discussed in Section 8.6.6.

Examples of n-channel polymers are provided in Fig. 8.83 with device characteristics summarized in Table 8.5. Of these materials, thorough device characterization of a population of devices using a variety of dielectric layers and substrates has only been provided for **44** (Yan et al., 2009).

The intermolecular hydrogen bonding of the electron-deficient diketopyrrolopyrrole-diketopyrrolopyrrole copolymer (DPP-DPP, **43**) leads to close π–π stacking, and the alkyl chains provide for molecular solubility. For these reasons, DPP-DPP has a high mobility coupled to a low V_T compared with other *n*-channel polymers listed in Table 8.5. The $\mu_{FEn} = 3$ cm^2/V s is the best result obtained among the devices tested (Kanimozhi et al., 2012).

Table 8.4 Example polymers and device structures employed in p-channel OTFTs

Mol.	Structure	Insulator	μ_{FEp} (cm²/V s)	V_T (V)	I_{on}/I_{off}	Comments	Ref.
37	BG/BC	HMDS/SiO_2	0.18	23	400	T_{an} = 160°C, 3 min, N_2[a]	(Wang et al., 2003)
38	BG/TC	OTS/SiO_2	0.2	0	10^7–10^8	T_{an} = 145°C, 10–20 min	(Ong et al., 2005)
39	BG/TC	FOTS[b]/SiO_2	1.0	−8	10^6		(Umeda et al., 2009)
40	BG/TC	HMDS/SiO_2	0.54	−15	10^7	T_{an} = 150°C	(Osaka et al., 2010)
41	TG/BC	CYTOP[c]	1.2	−25	10^4	T_{an} = 100°C	(Zhang et al., 2010)
42	BG/TC	PTES/SiO_2	1.4	8	10^5	T_{an} = 200°C, 1 h, N_2	(Tsao et al., 2009)

[a] T_{an} = annealed at 160°C for 3 min in N_2 atmosphere to increase I_{on}/I_{off}.
[b] FOTS = tridecafluorooctyltrichlorosilane SAM.
[c] CYTOP = fluorocarbon polymer from Asahi Glass, Co.

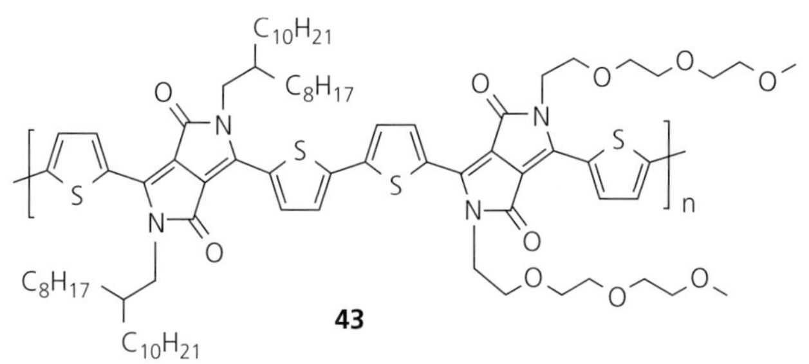

43

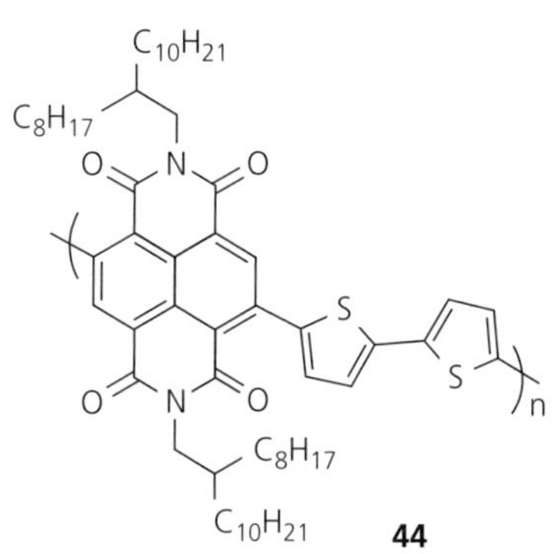

44

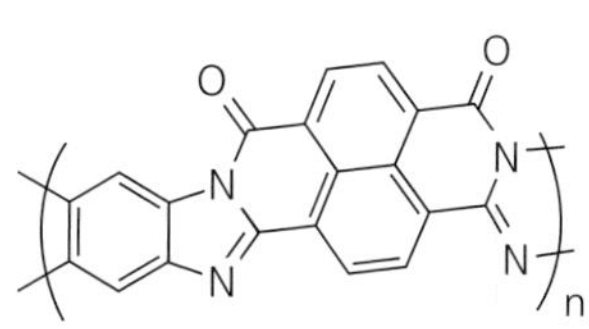

45

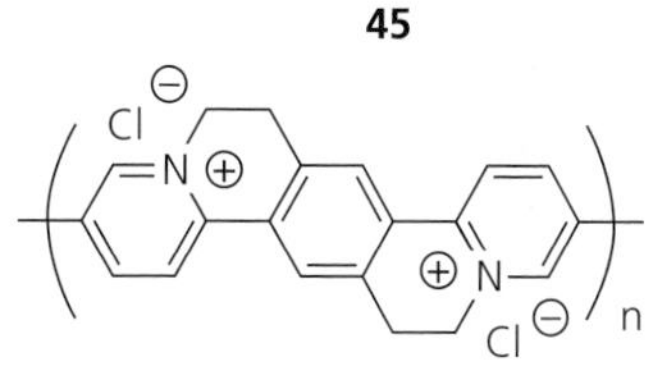

46

Figure 8.83 Example polymers used in n-channel OTFTs, with characteristics in Table 8.5.

Table 8.5 Example polymer materials and device structures employed in *n*-channel OTFTs

Mol.	Structure	Insulator	μ_{FEn} (cm²/V s)	V_T (V)	I_{on}/I_{off}	Comments	Ref.
43	TG/BC	CYTOP	3	−8	10^4		(Kanimozhi et al., 2012)
44	TG/BC	D2200[a]	0.2–0.85	5–10	10^6–10^7		(Yan et al., 2009)
45	BG/BC	HMDS/SiO_2	0.03–0.1	25	10^6	T_{an} = 100°C	(Babel and Jenekhe, 2003)
46	BG/BC	HMDS/SiO_2	3.4	8.4	20	Unstable	(Izuhara and Swager, 2009)

[a] D2200 is a polyolefin-polyacrylate from ActivInk, with $\varepsilon_r = 3.2$.

The next two compounds (**44** and **45**) use the high ionization energy conjugated naphthalene molecule as their core moiety, largely due to its electronegativity and its ability to form intermolecular π-stacks. The alkyl chains in **44** also lead to good solubility, and align the molecules in a configuration similar to P3HT in Fig. 8.66a. OTFTs based on this molecule were fabricated into TG/BC devices on glass, PEN and PET substrates using a number of polymer dielectrics (only D2200 is listed but other

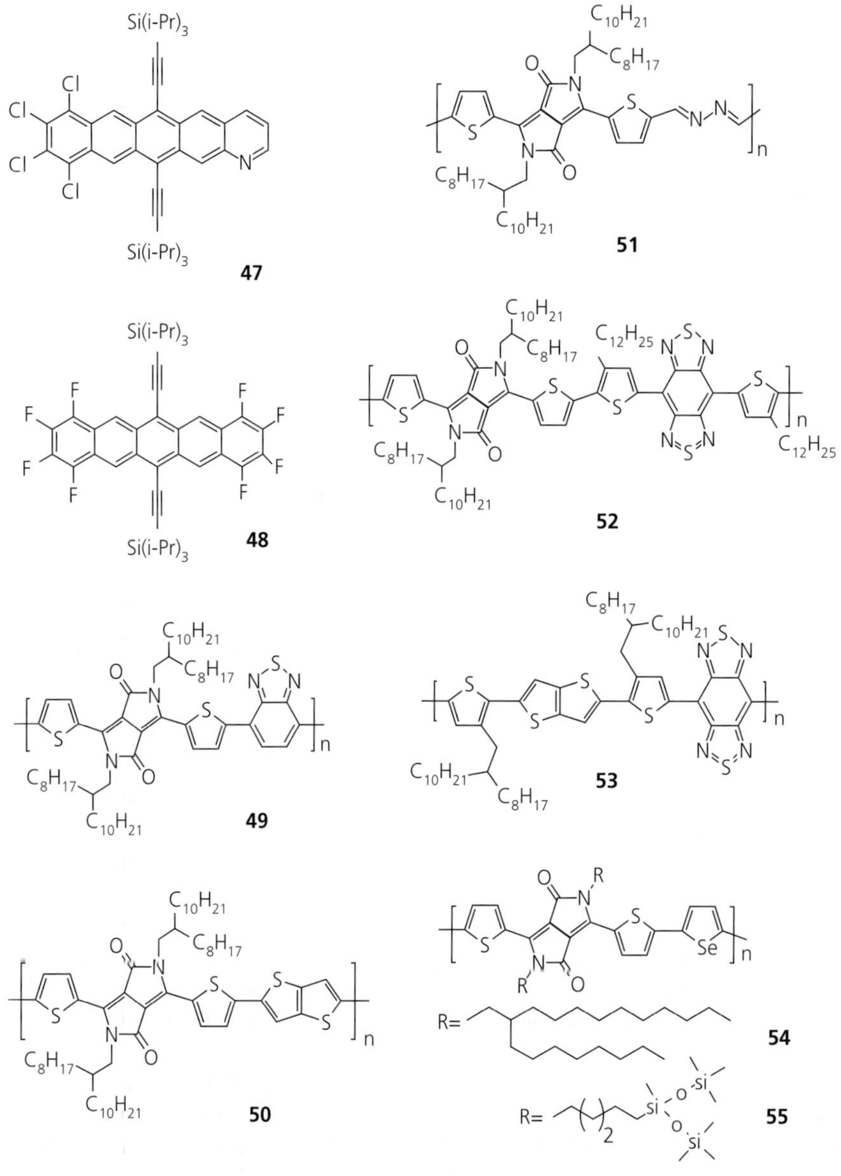

Figure 8.84 Example molecules used in ambipolar channel OTFTs, with characteristics in Table 8.6.

insulators including CYTOP, polystyrene, polymethyl methacrylate, etc. were tested). The device properties are reproducible and stable when exposed to humid air environments. Furthermore, inverter circuits with $A_v = 26$–29 were fabricated by gravure printing of an n-channel device with **44** paired with a P3HT (**37**) p-channel OTFT.

The poly(pyridinium phenylene) compound, **46**, is interesting in that it is water soluble. Electron-deficient pyridinium rings lower the LUMO energy. The planar close-stacking motif leads to an exceptionally high mobility of $\mu_{FEn} = 3.4$ cm^2/V s, although the drain current characteristics were irregular and unstable. Also, $I_{on}/I_{off} = 20$, indicating substantial channel leakage. Hence, while some of these characteristics are intriguing, it is not of practical use at this time.

8.6.6 Materials for ambipolar transistors

Several representative materials with nearly balanced electron and hole conducting properties for ambipolar transistor applications are shown in Fig. 8.84, with the corresponding OTFT characteristics summarized in Table 8.6. The example materials

Table 8.6 Example materials and device structures employed in ambipolar OTFTs

Mol.	Structure	Insulator	μ_{FEp} (cm^2/V s)	μ_{FEn} (cm^2/V s)	V_{Tp}[a] (V)	V_{Tn} (V)	$(I_{on}/I_{off})_p$	$(I_{on}/I_{off})_n$	Comments	Ref.
47	BG/TC	OTMS/ SiO_2	0.18	0.20	−40	+44	N/A	N/A		(Zeng et al., 2013)
48	BG/TC	OTS/SiO_2	0.33	0.41	−15	+50	200	10^6	$T_{sub} = 60°C$	(Tang et al., 2009)
49	BG/TC	OTS/SiO_2	0.35	0.40	−24	+40	90	25	$T_{an} = 200°C$	(Sonar et al., 2010)
50	TG/BC	PMMA	1.36	1.56	−16	0	10^5–10^6	10^5–10^6	$T_{an} = 320°C$	(Chen et al., 2012)
51	TG/BC	CYTOP	0.36	0.41	−10	+8	10^7	10^6	$T_{an} = 200°° C$	(Hong et al., 2012)
52	BG/BC	DTS'SiO_2	1.17	1.32	−10	+40	600	200	$T_{an} = 240°C$	(Yuen et al., 2011)
53	BG/BC	DTS/SiO_2	1.0	0.7	+10	+33	~100	~50	$T_{an} = 240°C$	(Fan et al., 2012)
54	BG/TC	OTS/SiO_2	2.48	0.43	−5	+70	10^5	10^3	Sol. sheared[b] $T_{an} = 220°C$	(Lee et al., 2012a)
55	BG/TC	OTS/SiO_2	8.84	4.34	0	+75	10^4	50	Sol. sheared $T_{an} = 220°C$	(Lee et al., 2013)

[a] $V_{Tp,n}$ extracted from data where possible if not explicitly stated. Values may also represent $V_{onp,n}$ in some cases.
[b] Solution sheared to align molecules.

are selected due to their nearly equal electron and hole mobilities (necessary for high noise margin, low power dissipation complementary logic, see Section 8.4.4), and their beneficial properties for use in OTFTs.

A key attribute of an ambipolar material is a narrow energy gap that reduces the barrier for either hole or electron injection from the source contact. This, unfortunately, also leads to increased off current. In Table 8.6, I_{on}/I_{off} of most ambipolar OTFTs is $\leq 10^3$, which is 10^3–10^4 times smaller than for unipolar OTFTs. A contact that is equally efficient for injecting electrons and holes is required. In all of the examples provided, Au is used due to its good film forming properties and a work function that approximately centers its Fermi energy within the ambipolar material energy gap.

Only two small molecule examples (**47, 48**) are provided; both are analogs of TIPS-pentacene. The molecules contain halogens (Cl and F, respectively) that lower the LUMO energies due to their electron-withdrawing natures. The lower LUMO also results in improved air stability, and their planar cores promote π-stacking. While these molecules provide reasonably balanced electron and hole mobilities, their mobilities are <0.5 cm^2/V s, which is considerably less than almost all the ambipolar polymers (**49–55**). And as is the case for the polymers, the threshold voltages are large, particularly for n-channel conduction.

The next four polymers (**49–52**) feature an electron-deficient DPP moiety that promotes π-stacking. The alkyl groups result in a high solubility. Hence, when assembled in a copolymer with an electron donating group (such as DPPT-TT, which comprises a thieno[3,2-*b*]thiophene along with two thiophene repeat units in **50**), the resulting donor–acceptor motif reduces the energy gap. Compounds **50** and **52** are of particular note due to mobilities that reach to >1.5 cm^2/V s. Indeed, **50** also has an unusually low V_{Tn} and V_{Tp} along with high on-off current ratios characteristic of unipolar compounds.

The energetics of DPPT-TT in Fig. 8.85 is instructive for understanding n-channel and ambipolar OTFTs. The energy gap of the molecule is E_G = 1.26 eV, which results in a comparatively lower energy barrier for both electron and hole injection from Au. The preparation of the contacts also impacts the injection barrier. The contacts were prepared by solvent cleaning and exposure to an O_2 plasma. For the device in Table 8.6, the solvent-cleaned contacts give nearly equal hole and electron injection barriers of approximately 0.6 to 0.7 eV, respectively. When the somewhat weaker benzothiadiazole acceptor in benzobisthiadiazole bithiophene-thienothiophene replaces DPP (PBBTTT, **53**), the LUMO and HOMO energies are −3.80 eV and −4.36 eV, respectively, leading to an energy gap of only 0.56 eV. This produces a large I_{off}, and hence an unacceptably low $I_{on}/I_{off} \sim 10$–100.

The DPP-based copolymers that replace one of the thiophene units with a selenophene (**54, 55**), and the alkyl group with a solubilizing siloxane, results in the highest mobilities of 8.84 and 4.34 cm^2/V s for holes and electrons in **55**, respectively. The three-dimensional (3D) conductivity was enhanced by limiting the length of the solubilizing side chains. This resulted in dense packing when the hexyl was replaced by a pentyl unit (Lee et al., 2013).

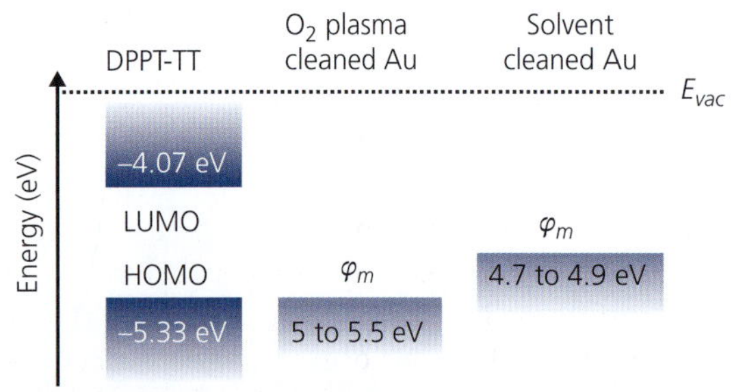

Figure 8.85 Energetics of DPPT-TT and the work functions of Au contacts prepared by O_2 plasma, or solvent cleaning (Chen et al., 2012).

We have noted that the nanocrystalline morphology of solution-processed molecules depends on the rate and method with which the materials are deposited on the gate insulator. For example, a slower rate of deposition via dip coating, vs. drop or spin casting, can result in higher crystalline order, as observed for compounds **54** and **55**.

Solution shearing is one method that produces the unusually high mobilities observed for both the dithienyl-diketopyrrolopyrrole-selenophene copolymer, PTDPPSe with the long, branched 2-octyldodecyl side chain, (**54**) and a hexyl substituted PTDPPSe-Si analogous to (**55**). Atomic force micrographs of the surfaces of these films are shown in Fig. 8.86 (Lee et al., 2012a). While the morphological distinction between drop-cast and solution-sheared films is not as dramatic as implied by the optical micrographs in Fig. 8.70, the differences are nevertheless

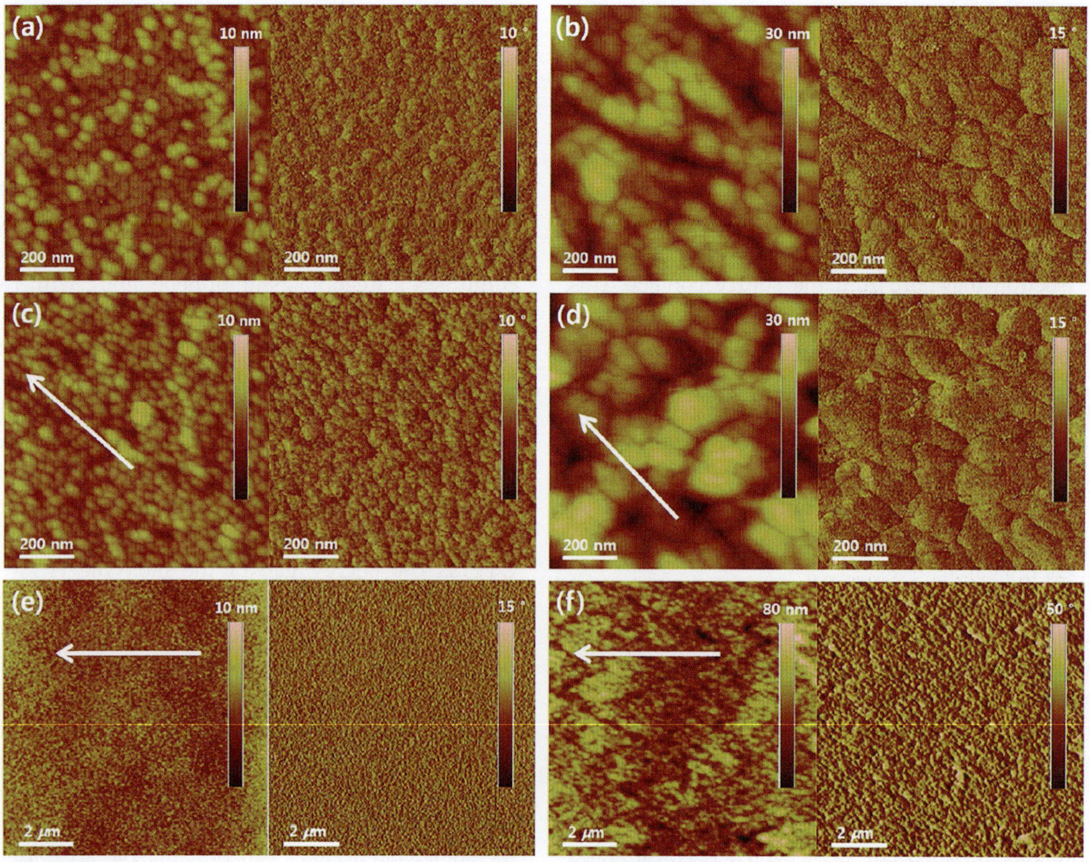

Figure 8.86 AFM images of (a, c, e) PTDPPSe and (b, d, f) PTDPPSe-Si thin films. (a, b) Drop cast and (c–f) solution sheared films. Arrows indicate the shear direction (Lee et al., 2012a).

apparent. Directionality is visible for the sheared films in Fig. 8.86c and f. It is noteworthy that the hole mobilities are more than twice that for similarly annealed (at 220°C) spin and drop-cast films of **55**, and the electron mobilities are larger by a factor of nearly three. Even higher mobilities (>8 cm^2/V s) have been obtained for solution-sheared TIPS-pentacene that results in millimeter wide by centimeter long crystallites (Diao et al., 2013). These results suggest that entrainment of the crystallites along the in-plane S–D direction can significantly improve performance.

8.7 Material deposition and transistor patterning

Section 8.6 emphasized the need for achieving long range order in the in-plane direction. In this context, several different deposition and post-deposition techniques for achieving these desired morphologies were described. A slow deposition process such as drop-casting results in a morphology close to an equilibrium structure. Rapid solvent evaporation in spin-casting leads to a disordered thin film. We also showed that solution shearing induces crystallization along the shear direction via the controlled, slow evaporation of the solvent at the liquid/thin film interface. Nearly all OTFTs benefit from deposition on heated substrates, or by post-deposition thermal annealing of the semiconductor film. Annealing drives the as-deposited morphology toward equilibrium, which results in increased conductivity (and hence mobility) along preferred crystalline directions (cf. Chapter 5.5). Exposure of a solution-deposited film to a solvent vapor also promotes morphological reorganization, again resulting in improved nano- and microcrystalline order (Lee et al., 2009).

Here, we introduce several additional methods for driving thin film morphologies toward a high mobility, equilibrium structure (Liu et al., 2009a, Klauk, 2010, Wang et al., 2011). Our discussion builds on the more general deposition processes discussed at length in Chapter 5. Since patterning and deposition are central elements of the OTFT fabrication process (Sirringhaus et al., 2006), we also include a discussion on patterning processes that result in thin dielectrics and short channel lengths.

8.7.1 Molecular alignment via dip coating and associated techniques

In dip coating, the substrate is slowly withdrawn from a saturated solution of the target molecule (Tsao et al., 2009, Jang et al., 2012). This is analogous to solution shearing, except that the substrate/liquid/air boundary constitutes free, unconstrained interfaces. Precipitation of the crystallites occurs at the liquid/air boundary due to fast evaporation of a volatile solvent. Initially the nucleation of crystallites is random. However, as the substrate is extracted from the solution, the growth is driven along the pulling direction, creating extended crystallites. This is analogous to the Bridgeman method used to grow bulk crystals, whereby slowly withdrawing a seed crystal from the melt creates a growth interface at the liquid surface (see Section 5.4.1). Figure 8.87a shows the dip coating process that generates an array of parallel crystal strips that nucleate at the liquid/air boundary. The growth morphology depends on a balance between substrate pull rate, solution temperature, and molecular concentration. When the substrate is translated at high speed, the crystal thickness and width is decreased while the space between the strips is increased. The strips do not grow into each other due to local depletion of the solute, and the preference for the molecules to crystallize on the already-deposited film. Similarly, increasing the solution evaporation rate by increasing temperature or using a lower boiling point solvent has the same effect as increasing the pull rate. An AFM image of a TIPS-pentacene film grown by pulling from a DCM solution at 150 μm/s is shown in Fig. 8.87b. The lower panel indicates a film thickness of 80-90 μm, and a crystal strip width of ~10 μm (Jang et al., 2012).

The dependence of the optimum pull rate vs. temperature for TIPS-pentacene in solutions of DCM, CF, and toluene is shown in Fig. 8.87c. Also shown are polarization optical micrographs of the resulting films obtained at several growth rates and temperatures. The optimum growth rate and conditions are defined as those that yield the highest hole field effect mobility. Micrographs of the corresponding optimal film morphologies are outlined in colored boxes. As noted, the optimal growth rate increases with temperature and decreasing solution boiling point (the b.p. for DCM, CF, and toluene are 40.0, 60.4, and 110.6°C, respectively). At rates ≥200 μm/s in DCM, the rate of substrate translation exceeds the growth rate and consequently, the strips become discontinuous. In the least volatile toluene solution, the films at low temperatures or slow growth rates no longer produce well-defined and uniform strips, since the nearly equilibrium conditions drive the growth of a three dimensional bulk structure.

The dependence of stacking order on growth rate for TIPS-pentacene in DCM is quantitatively understood from the glancing incidence X-ray diffraction patterns in Fig. 8.88a. Only peaks along the in-

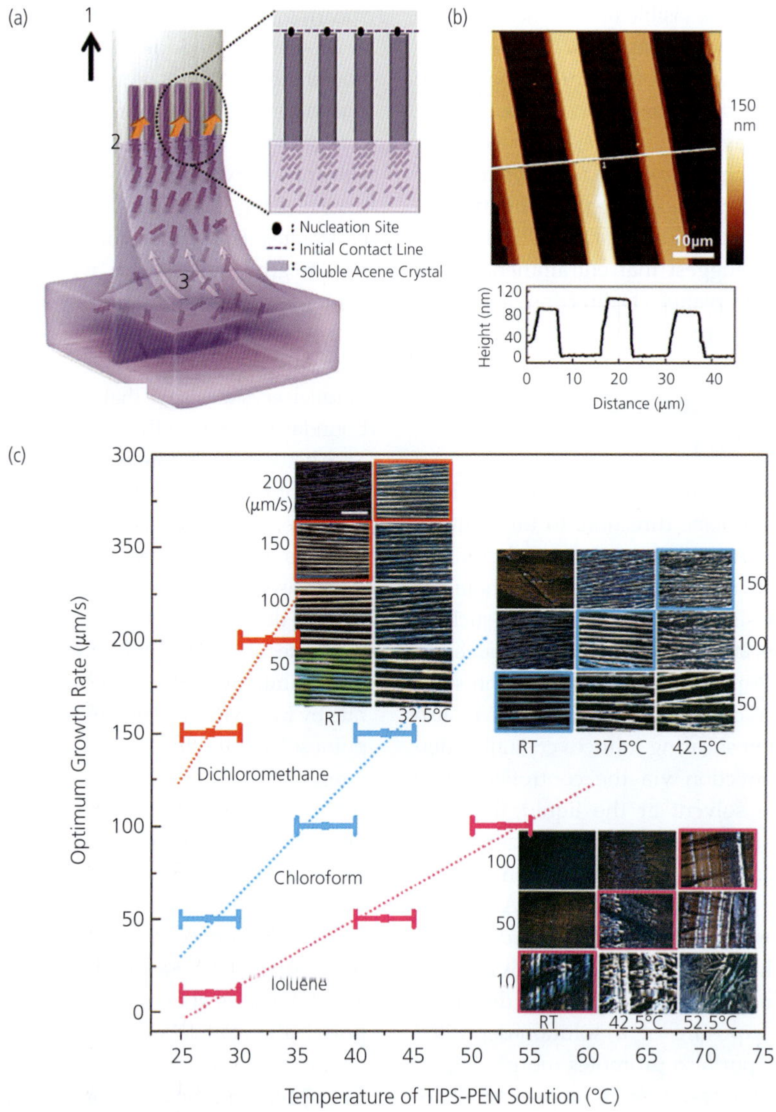

Figure 8.87 (a) Illustration of the mechanism of crystal growth via dip coating. The substrate (**1**) is pulled from solution (**3**), forming crystalline strips (**2**) along the pulling direction. (b) AFM image of TIPS-pentacene strips pulled from a DCM solution. Below is shown the thickness profile of the crystallites. (c) Optimum pull rate vs. temperature for TIPS-pentacene in different solutions. The insets are polarization optical micrographs of growths at different temperatures and pull rates showing the crystal strips (Jang et al., 2012).

plane direction are observed, corresponding to an intermolecular spacing of $d = 7.74$ Å. This is consistent with the intermolecular π-stacking distance along the b-axis (Anthony et al., 2001). The weakest alignment is found for the sample drawn at 10 μm/s. The highest X-ray intensity, and hence order, is found for the 150 μm/s sample. (The film drawn at 300 μm/s is discontinuous). The low resistivity directions are found along the a- and b-axes: hence alignment of these axes along the channel as inferred from the X-ray data (Fig. 8.88b) should lead to the highest mobilities (Anthony et al., 2001).

The alignment produced by dip coating of a TIPS-pentacene/DCM solution has a significant impact on OTFT performance when the direction of growth is along the channel direction. A comparison of the characteristics of BG/TC transistors with an ALD-grown Al_2O_3 dielectric at several different dip coating rates at room temperature is given in Table 8.7. The film deposited at 150 μm/s results in the highest μ_{FEp}, but at the expense of a high V_T and a lower I_{on}/I_{off} than at other rates. The film grown closest to equilibrium corresponds to the slowest growth rate of 10 μm/s. This leads to thick films whose mobilities are two

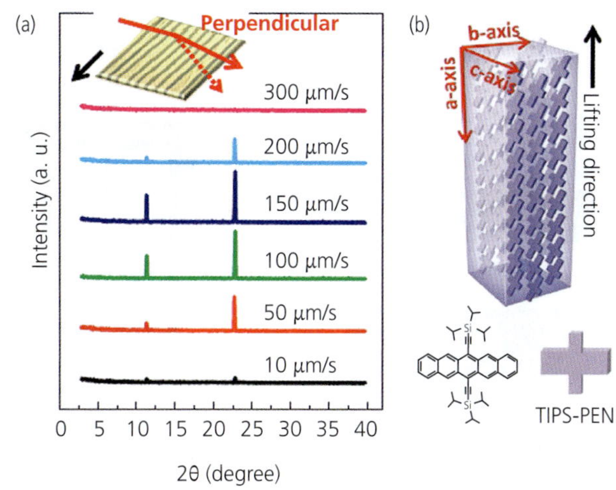

Figure 8.88 (a) Glancing incidence X-ray diffraction patterns of crystals produced at different dip-coating rates from a solution of TIPS-pentacene in DCM. Inset: Configuration of the X-ray measurement where the incident beam is perpendicular to the pull direction, and at a 0.15° angle relative to the substrate plane. (b) Schematic of the TIPS-pentacene crystal. The molecules are represented by the crosses shown in the lower diagram (Jang et al., 2012).

Table 8.7 OTFT characteristics for several dip-coating rates for TIPS-pentacene in DCM (Jang et al., 2012)

Rate (μm/s)	μ_{FEp} (cm^2/V s)	V_T (V)	I_{on}/I_{off}	S (V/decade)
10	0.005	−7.4	48	6.4
50	0.13	−8.0	8.5×10^4	1.89
100	0.6–1.0	−7.6	1.6×10^6	1.66
150	0.85–1.3	−10.6	7.6×10^5	1.3
200	0.05	−10.2	4×10^4	1.54

decades lower than the more organized crystal structure obtained at 150 μm/s.

Zone-casting shown in Fig. 8.89a is a variant of dip coating. The solution containing the semiconductor solute is cast from a wide nozzle onto a moving substrate at a controlled rate (Pisula et al., 2005, Miskiewicz et al., 2006, Mas-Torrent et al., 2008). Crystals nucleate at the initial nozzle position. As the solvent evaporates, the molecules grow along the substrate translation direction. Films of TTF-4SC18 (see Fig. 8.89a, inset) dissolved in toluene are grown by zone-casting at a substrate temperature of 80°C. This is sufficient to volatilize the solvent, generating structures shown in Fig. 8.89b–d. The molecules stand with their alkyl groups oriented toward the substrate, forming π–π coupled stacks between adjacent, planar TTF units. This places the high mobility direction along the substrate plane, as illustrated in Fig. 8.89a. The optical micrograph in Fig. 8.89b shows alignment of long thin crystallites along the substrate translation direction. The molecular stacking axis is longitudinally oriented along the growth direction. The AFM image suggest that the surface morphology is terraced, with a roughness of approximately 30 nm (Fig. 8.89c) (Miskiewicz et al., 2006).

Bottom gate/top contact OTFTs were fabricated with the channel oriented parallel (Fig. 8.89d) and perpendicular to the casting direction. The gate dielectric comprises plasma enhanced, sequential depositions of 100 nm Si_3N_4 followed by 50 nm SiO_2. With a gate of $L = 50$ μm oriented parallel to the casting direction, $\mu_{FEp} = 0.015\ cm^2/V\ s$, $V_T = -6$ V to −10 V, and $I_{on}/I_{off} = 10^4$. When oriented perpendicular, these parameters are $0.0017\ cm^2/V\ s$, −15 V to −120 V, and 10^2–10^4, respectively. The reduced performance is a result of misorientation of the π–π stacks perpendicular to the transport direction, leading to a large number of crystalline grain boundaries that interrupt hole transport.

Zone casting, solution shearing, and dip coating rely on ordered crystallization via controlled evaporation. The actual morphology of the resulting film is ultimately determined by the likelihood that the material itself will self-assemble into π-stacks in the in-plane direction. Another method to develop crystalline order is *solvent exchange* illustrated in Fig. 8.90a. The molecular species is dissolved in a solvent where it has a high solubility (e.g. TIPS-pentacene in toluene). A small amount of this solution is introduced into a large volume of a poor solvent (e.g. acetonitrile). As the solution evaporates, the molecules precipitate from solution onto the substrate. The molecules stack into ribbons, driven by their own intermolecular interactions. The microribbons

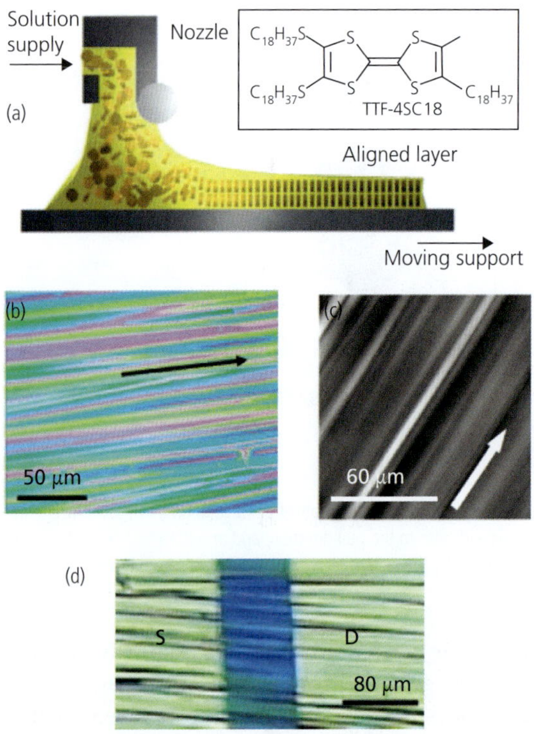

Figure 8.89 (a) Schematic of the zone casting process (Pisula et al., 2005). Inset: Molecular structural formula of TTF-4SC18 used in (b)–(d). (b) Optical micrograph and (c) AFM image of a TTF-4SC18 film grown by zone casting from a toluene solution. The direction of substrate translation is indicated by arrows. The maximum- to-minimum height in the image is ~30 nm. (d) The S and D contacts of a BG/TC OTFT are shown with the channel. Crystallites are aligned parallel to the translation direction (Miskiewicz et al., 2006).

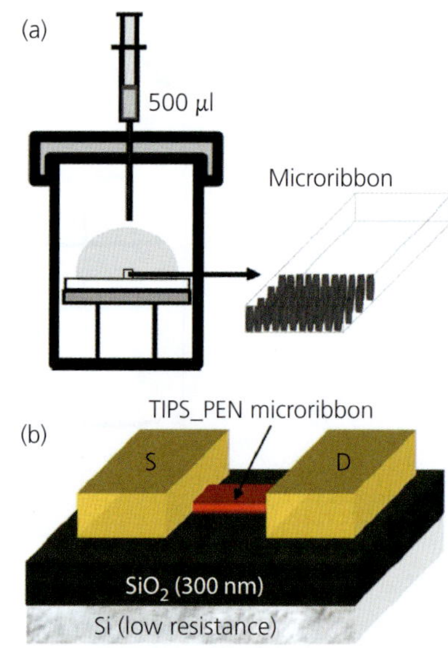

Figure 8.90 (a) Schematic of solvent exchange crystallization. A small volume of solution is dropped onto a substrate via syringe within a larger volume containing a second solvent. The molecular material has a reduced solubility in the second (shown by the shaded hemisphere) *vs.* the first solvent. The solvent is allowed to slowly evaporate, leaving molecular stacks forming microribbons. (b) BG/TC OTFT with a TIPS-pentacene microribbon spanning the gap from S to D (Kim et al., 2007).

are coated with the S and D contacts to form the BG/TC OTFT in Fig. 8.90b. Unlike zone casting and related techniques, solvent exchange results in randomly positioned microribbons on the substrate. Indeed, neighboring ribbons may even overlap. This requires placement of each ribbon across individual transistor channels. In the following section, we will show that assembly of the macrostructure (i.e. the ribbons) may be oriented by pre-patterning "entrainment" structures on the target substrate. Field effect mobilities for OTFTs in Fig. 8.90b are as high as 1.42 cm^2/V s ($I_{on}/I_{off} = 10^4$) compared to 0.17 cm^2/V s for TIPS-pentacene transistors deposited by conventional solution coating (Kim et al., 2007).

One final precipitation scheme to consider is *controlled drying* of a semiconductor-containing solution applied via inkjet printing. The dynamics of droplet formation and evaporation illustrated in Fig. 8.91 are also discussed in Section 5.6.5. In a heterogeneous solution, evaporative flow drives the solution to the droplet edges, which is countered by *Marangoni flow* that is determined by the surface energy of the droplet. This leaves the precipitation of solute at the edges of the droplet as it shrinks. Provided the molecules tend to form π-stacks, they self-assemble into a "coffee ring" with the high conductivity direction oriented in a radial direction from the injection point. To exploit the radial crystalline orientation, the source and drain contacts need to be configured into concentric circles (Lim et al., 2008).

As with other precipitation methods, the rate of evaporation controls the ability of the solute to form an organized, equilibrium structure. Since it is not possible to precisely control the evaporation rate of a single component solvent, a second, high boiling point (b.p.) solvent is introduced. The second solvent counters the outward convective flow of the low b.p. solvent via Marangoni flow. The balance of these two flows is controlled by the solvent mixture, delivering the molecules to the droplet periphery at a rate that is optimal for self-assembly.

8.7.2 Molecular orientation on surface-modified substrates

In Section 5.4.2, we showed that molecules can align to corrugations in the topography of the surface energy at the substrate when deposited under

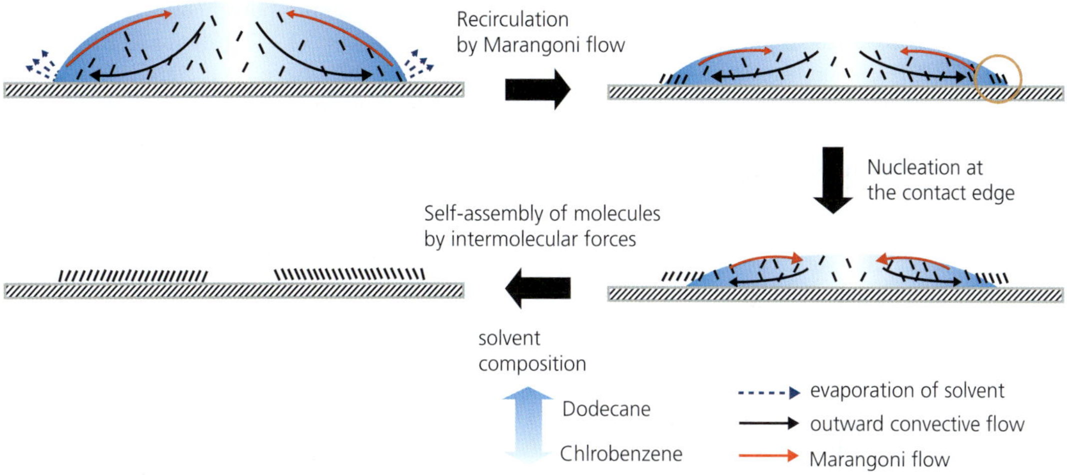

Figure 8.91 Formation of self-assembled π-stacked films of an organic semiconductor delivered to a substrate by inkjet printing (Lim et al., 2008).

appropriate conditions. This was discussed in the context of quasiepitaxy, whereby the substrate is prepared to create a low surface energy direction, followed by deposition of a layer whose comparatively stronger intermolecular binding forces drive self-assembly initiated by the substrate.

Numerous surface preparations have been developed that establish self-ordering in transistors (Liu et al., 2009a). These have resulted in significant anisotropies in charge mobility in the in-plane direction. For example, *rubbing* the substrate surface prior to deposition can influence alignment of liquid crystal molecules (Ishihara et al., 1989, Depp and Howard, 1993). The process uses a polymer coating that is mechanically rubbed, or brushed along a given direction. Subsequently deposited molecules assemble into anisotropically ordered morphologies induced by energetic corrugations impressed on the alignment layer. Rubbing has also been used to force alignment of organic semiconductors deposited in the channel of OTFTs to increase mobility (Sirringhaus et al., 2000b, Chen et al., 2001). Figure 8.92 shows the morphology of a 50 nm thick film of $F_{16}CuPc$ (compound **26**, M = Cu, Fig. 8.79), vacuum deposited onto a 5 nm thick rubbed film of the same composition on a SiO_2 layer on Si. The pre-deposited layer is rubbed at room temperature using a PTFE (or Teflon®) bar that is repeatedly and gently slid along the substrate approximately ten times. The striations in the AFM images have widths of 50–100 nm, and lengths of approximately 10μm along the rubbing direction. The root-mean-square (rms) roughness of the ridges is 2.5 nm. Crystallographic and optical analysis suggests that these ridges are thin crystallites of molecular

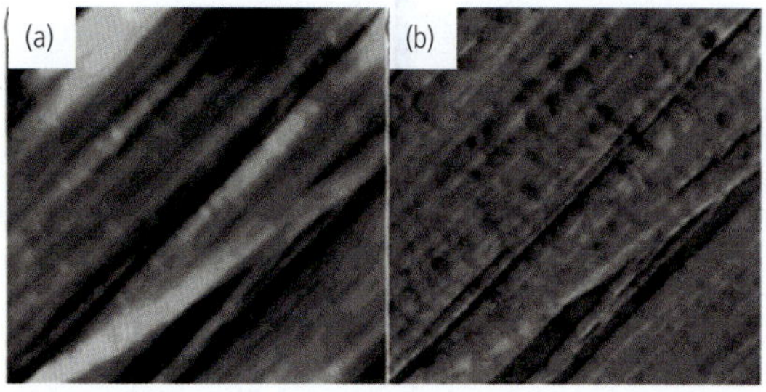

Figure 8.92 (a) Height and (b) phase contrast AFM images of a 50 nm thick $F_{16}CuPc$ film on a 5 nm thick, pre-rubbed template layer of the same composition (Chen et al., 2001).

stacks of $F_{16}CuPc$ that preferably align along the rubbing direction.

Top contact, bottom gate OTFTs using pentacene films grown on a 5 nm thick rubbed layer of the same composition on a SiO_2 gate dielectric show large asymmetries in hole mobility, supporting the conclusion that rubbing entrains subsequently deposited material to π-stack along the rubbing direction. The mobility in channels oriented parallel to the rubbing direction is $\mu_{FEp} = 0.24$–$0.25\ \mathrm{cm^2/V\ s}$. The mobility anisotropy ratio is $\mu_{FEp\|}/\mu_{FEp\perp} > 40$, where the subscripts $\|$ and $\perp$ refer directions parallel and perpendicular to the rubbing direction respectively (Chen et al., 2001).

A more forceful means for aligning molecules is to initially form a highly directional template on the substrate surface, followed by deposition either from solution or vapor, of the semiconductor film. *Friction transfer deposition* is one such templating method (Nagamatsu et al., 2003, 2004, Li et al., 2005, Dong et al., 2008). The PTFE alignment layer illustrated in Fig. 8.93a is deposited by applying a constant pressure to a PTFE rod or pellet, with the substrate heated to 280°C. At this temperature, the PTFE softens while being translated along the substrate. This leaves behind PTFE chains aligned along the drawing direction. Next, a rigid-rod polymer such as PPE with thioacetyl end groups (TA-PPE) in a THF solution is drop-cast onto the PTFE surface. The polymer orients along the PTFE chain direction as the solvent evaporates, apparent from the striated films shown in Figs. 8.93b and c. The photoconductivity of the aligned TA-PPE measured after depositing S and D contacts on the film surface increases by 20 times compared with unaligned films (Dong et al., 2008). This technique has also been used to align polymers in OTFTs, creating conductive anisotropies as large as $\sigma_{\parallel}/\sigma_{\perp} = 100$ (Li et al., 2005).

Other substrate modification techniques that coerce the directionality of subsequently deposited films include deposition on structured surfaces by *nanoimprinting* (Zheng et al., 2007), and by exposure of a photosensitive layer to linearly polarized light. This latter *photoalignment* technique has been adapted from methods used to orient molecules in LCDs (Ichimura, 2000). A 100 nm thick polyimide layer deposited on a SiO_2 gate dielectric acted as a growth template of the 100 nm thick pentacene channel by exposing the polyimide with polarized UV light. The BG/TC OTFT exhibited a maximum hole mobility of 0.75 cm^2/V s with $\mu_{FEp\parallel}/\mu_{FEp\perp} = 13 - 25$ (Chou and Cheng, 2004).

Small molecules have been aligned by deposition on a thin templating layer that self-assembles during deposition due to strong intermolecular van der Waals forces (Wang et al., 2007, Lassiter et al., 2010, Yook et al., 2011). For example, a 2 nm thick layer of p6P deposited by ultrahigh vacuum organic molecular beam deposition onto a SiO_2 substrate at 180°C forms ~2 μm diameter nanocrystalline islands (Fig. 8.94a). Next, ZnPc is deposited onto the p6P surface islands. The ZnPc molecules stand upright on the oligophenylene, p6P surface, with their crystalline direction imposed by the underlying p6P template lattice orientation (Fig. 8.94b and c) (Wang et al., 2007). In contrast, ZnPc deposited onto bare SiO_2 substrates showed small domain sizes that are comparatively rougher, and randomly oriented relative to each other, see Fig. 8.94d.

While this example shows that the ZnPc domains have random in-plane orientation with grain sizes of ~10 μm, the oligophenylene forces the ZnPc to stand up rather than lie flat on the substrate. This orientation improves in-plane intermolecular π–π coupling, and hence the mobility along the channel. Indeed, ZnPc initiated on p6P films on BG/TC OTFTs result in $\mu_{FEp} = 0.32$ cm^2/V s compared to only 0.022 cm^2/V s for films grown on SiO_2 without the template. It is likely that a combination of the templating layer deposited on a directionally modified substrate surface

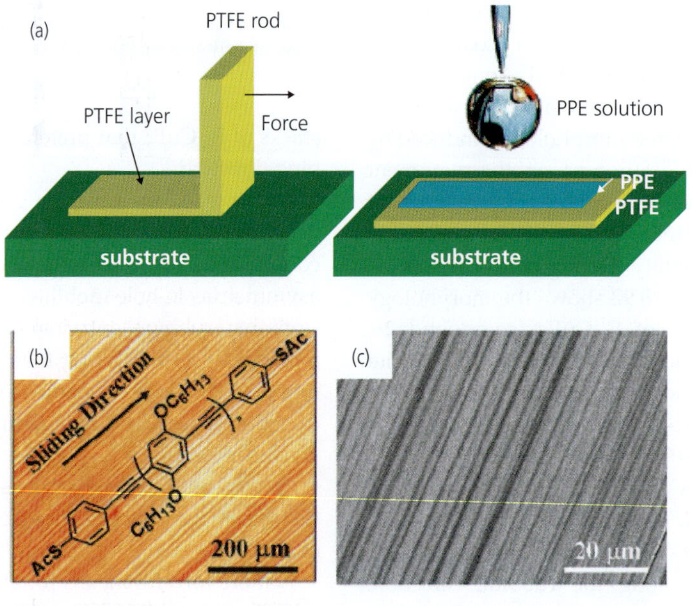

Figure 8.93 (a) Friction transfer of a PTFE template layer (left) followed by drop casting of a solution of the semiconductor, PPE, whose molecular structural formula is shown in (b). (b) Polarized optical microscope and (c) SEM images of the PPE film using the friction transfer method (Dong et al., 2008).

(e.g. via friction transfer, rubbing, etc.) would result in further orientation of the π-stacks along the channel from source to drain.

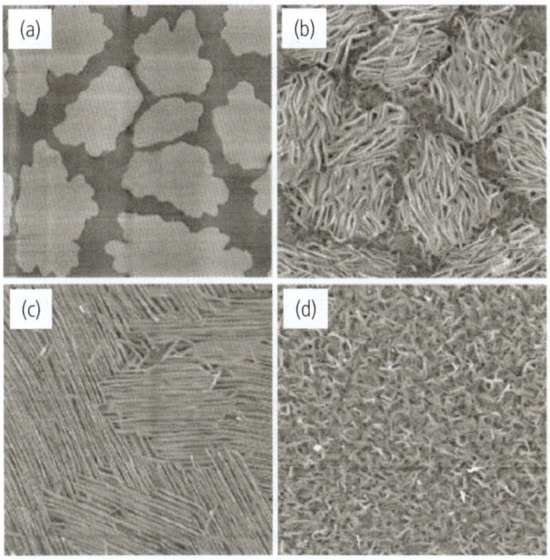

Figure 8.94 AFM image of a (10 μm)^2 region of a (a) 2 nm p6P, (b) 3 nm ZnPc on 2 nm p6P, (c) 30 nm ZnPc on 2 nm p6P, and (d) 30 nm ZnPc film. The films are on SiO_2 (Wang et al., 2007).

8.7.3 Direct printing of OTFT contacts

A discussion of numerous printing techniques of organics and their contacts has already been provided in Section 5.6. Several of these methods, such as gravure and flexographic printing have been employed to pattern the S, D and G contacts of OTFTs (Sirringhaus et al., 2006, Yan et al., 2009). Here, we discuss two methods that are particularly promising for contact patterning due to their potential for high resolution, and ability to print over large areas: inkjet and microcontact printing.

Well-defined inkjet printed contacts can be realized by functionalizing the substrate surface into regions that are variously hydrophobic and hydrophilic. Then, droplets dispensed onto the surface form regions whose boundaries are defined by the difference in energy of the solution and the surface on which it is deposited. Figure 8.95a illustrates how droplets are confined by variations in substrate surface energy to form OTFT contacts. The first step involves deposition of a 50 nm thick hydrophobic layer of PI on a glass substrate surface. After pattering, the surface is made hydrophilic by exposure to an O_2 plasma that also removes PI from the source and drain regions. Next, a conducting PEDOT:PSS layer is dispensed from an inkjet nozzle near to the PI barriers. Due to

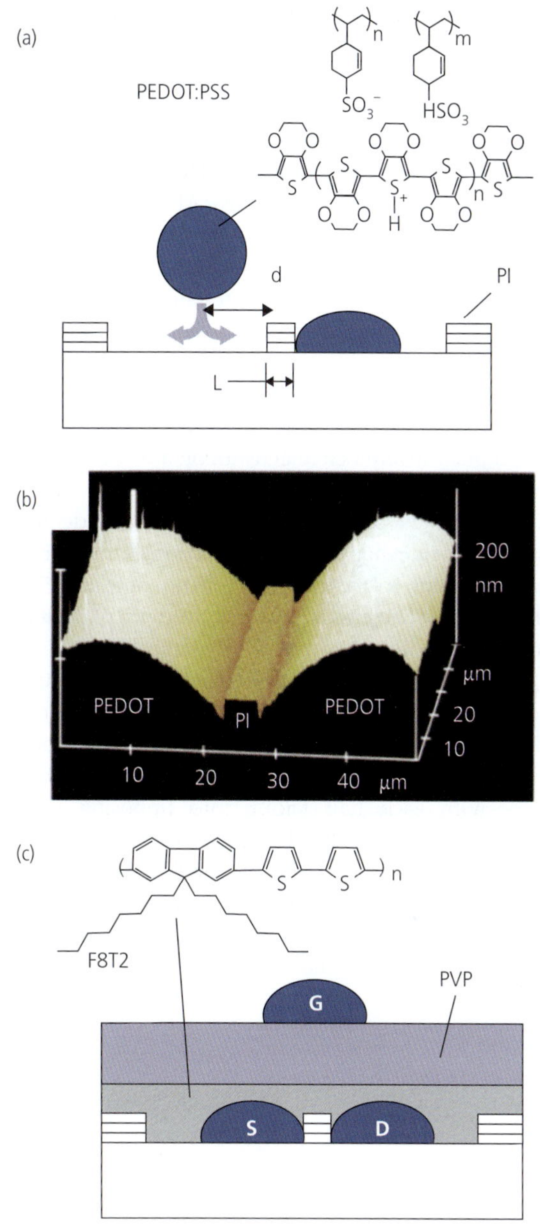

Figure 8.95 Inkjet printing of contacts. (a) Droplets of conductive, aqueous PEDOT:PSS solution are dispensed onto a hydrophilic glass substrate with a patterned hydrophobic polyimide (PI) barrier that defines the source and drain contacts separated by the channel of length, *L*. (b) AFM image of the S and D contacts. (c) Completion of the OTFT by deposition of the F8T2 channel and PVP gate insulator, followed by a second inkjet printed PEDOT:PSS gate contact (Sirringhaus et al., 2000a).

the low surface energy of the substrate, the droplet flows until it comes into contact with the barrier, whose width, *L*, defines the channel length. Since the PI surface is hydrophobic, the droplet does not overtop the barrier, forming a pattern whose resolution is limited by photolithography. Droplets are continually ejected onto the substrate surface as the inkjet nozzle is translated parallel to the barrier, defining the channel width. The rate of droplet dispensing is set by the inkjet nozzle speed to write a continuous conducting contact without necks or breaks.

The effects of the non-wetting barrier are visualized by the droplet shape in the AFM image of the contacts in Fig. 8.95b. The OTFT is completed by depositing F8T2 in a xylene solution (see Fig. 8.95c). This solvent is orthogonal to the aqueous PEDOT:PSS solution, thus preventing re-dissolving of the S and D contacts. Finally, the PVP gate insulator is applied, followed by a second inkjet printing of the PEDOT:PSS gate electrode.

Unipolar inverter circuits featuring a pair of printed F8T2 transistors with $L = 5$ μm, interconnected by inkjet- printed conducting vias have been demonstrated (Sirringhaus et al., 2000a). The vias are realized by dispensing a droplet of isopropanol onto the PVP, thereby dissolving it and allowing the drain of the load to connect to the source of the drive transistor.

It may not be necessary to dispense two droplets to separately define the S and D contacts. A single, larger droplet can alternatively be dispensed directly onto the center of the barrier. The droplet then splits into two, with each half falling into opposing wells (Sirringhaus et al., 2006). Furthermore, controlling the volume of the droplet consisting of Ag nanoparticles suspended in a tetradecane solution to less than 1 fl has led to pattern resolutions of <1 μm, which is ultimately limited by the diameter of the droplets. The Ag particles form into conductive S and D electrodes when fused at 130°C for 1 h in N_2. An illustration of BG/TC OTFT employing printed Ag NP S and D contacts is shown in Fig. 8.96a, with a micrograph of the top of the device shown in Fig. 8.96b. The thickness, and hence the resistivity of the lines is determined by the number of printer passes. As the number of passes increases from 1 to 20, the thickness increases from 30 nm to 600 nm, and the resistivity approaches that of bulk Ag. The dependence of resistivity on the number of passes is shown in Fig. 8.96c. Due to the very thin AlO_x gate insulator functionalized by an n-octadecylphosphonic acid SAM, the operating voltage of the OTFT is <3 V, making the devices useful for high gain, low power circuits operating at ~1 MHz (Sekitani et al., 2008). A position-sensitive pressure sensor discussed in Section 8.9 has been realized with integrated arrays of such printed transistors (Noguchi et al., 2006).

Contact printing of electrodes is an alternative to inkjet printing. In Section 5.6.4, we introduced the concept of cold-welding as a method for transferring very high resolution metal or organic patterns from a stamp to a receiving surface to fabricate OLEDs, detectors and other organic electronic devices (Kim et al., 2000, Kim and Forrest, 2003, Xu et al., 2008). A variation of the stamping process has been used to both functionalize and pattern the gates of BG/TC OTFTs.

The patterning process is illustrated in Fig. 8.97 (Zschieschang et al., 2008). The fabrication of the OTFT begins by coating the substrate (in this

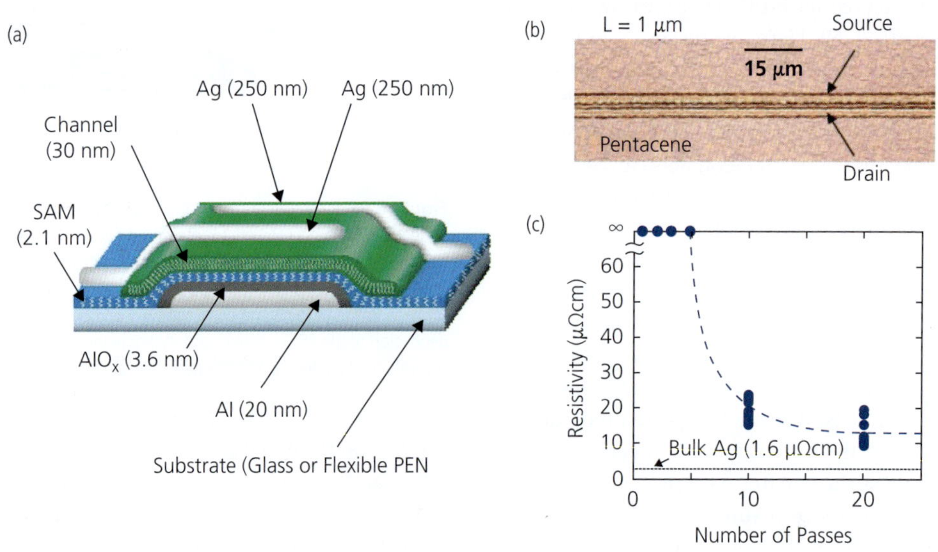

Figure 8.96 High resolution inkjet-printed pentacene OTFT. (a) Schematic of the OTFT with thicknesses of each layer indicated. (b) Micrograph of the OTFT channel region with $L = 1$ μm. (c) Resistivity of the inkjet printed S and D contacts vs. number of printing passes (Sekitani et al., 2008).

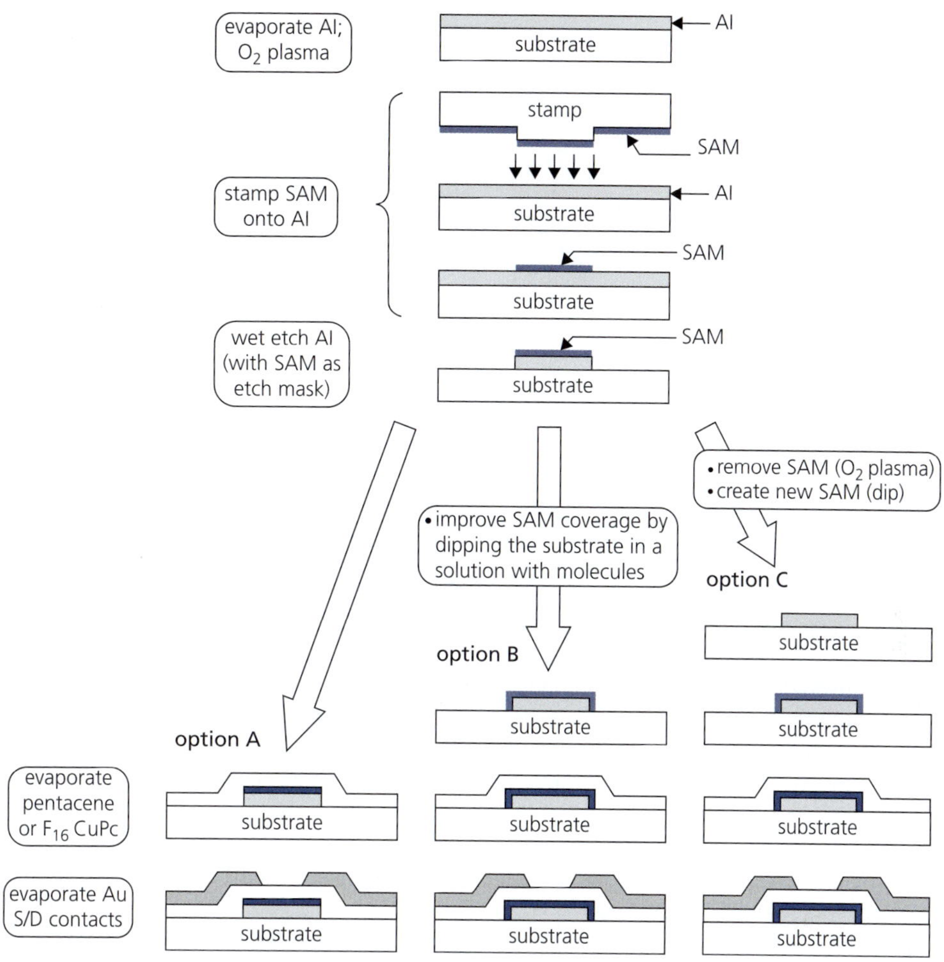

Figure 8.97 Three microcontact approaches for using an SAM to both pattern the oxidized gate contact, and functionalize its surface (Zschieschang et al., 2008).

example, glass) with a layer of Al. The substrate is exposed to an O_2 plasma to form a 3.6 nm thick oxide surface that serves as the gate insulator in the BG/TC OTFT. Next, a PDMS elastomeric stamp with the gate patterns occupying raised portions is formed by photolithographic patterning followed by etching recesses in a master (typically Si or glass). The master is filled with PDMS, which is subsequently cured and peeled from the master. The stamp is dipped in a solution containing the appropriate functionalizing SAM such as *n*-octadecylphosphonic acid, which attaches to the PDMS surface. The stamp is pressed against the AlO_x surface on the substrate, leaving behind the 2.1 nm thick SAM, which is stabilized by heating and drying. Areas covered by the SAM are protected from the wet acid etchant used to remove the Al/AlO_x in the exposed regions. At this point, there are three options for completing the device. **A**: vacuum deposit the organic channel layer, followed by deposition of the source and drain electrodes that are patterned using a shadow mask. **B**: The substrate is dipped in a solution of phosphonic acid to improve coverage along the exposed edges of the Al gate to prevent leakage into the channel. This step also repairs damage to the top gate surface SAM incurred on wet etching and handling of the sample during the gate etch step. The remainder of the process is as in option **A**. **C**: Assuming that the gate etching step irreparably damages the SAM, it is completely removed in an O_2 plasma, and the substrate is dipped in the SAM solution. Subsequent steps then are once more identical to option **A**. The disadvantages of options **B** and **C** are that they are slightly more complex than **A**, but this is counterbalanced by the

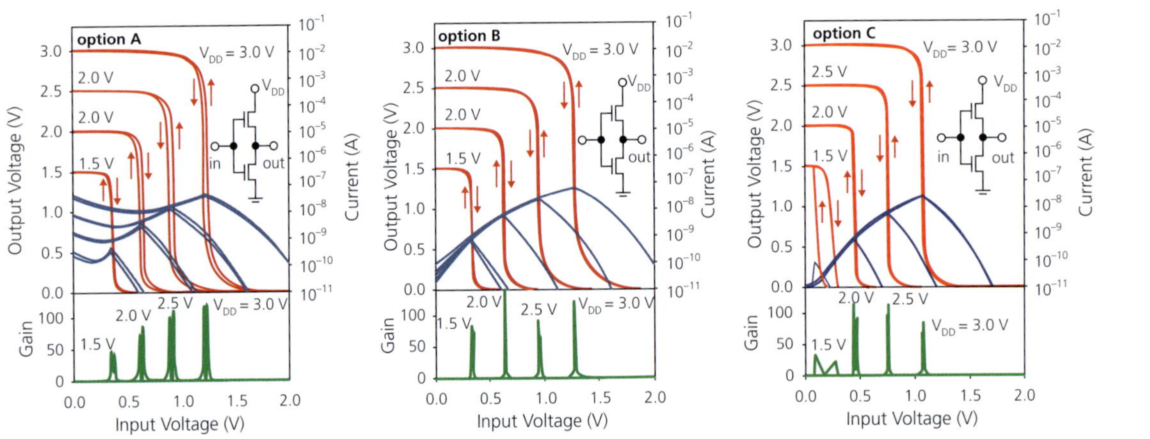

Figure 8.98 Complementary inverter transfer characteristics based on pentacene p-channel and $F_{16}CuPc$ n-channel OTFTs fabricated by the three options in Fig. 8.97. The transfer characteristics shown by red lines are referred to the left axes, and the gate leakage current shown by blue lines refer to the right axes (Zschieschang et al., 2008).

advantage of improved device performance and yield due to reduced gate leakage.

This process has been implemented to fabricate complementary inverters based on pentacene p-channel and $F_{16}CuPc$ n-channel OTFTs. The field effect mobilities were $\mu_{FEp} = 0.4$ cm^2/V s and $\mu_{FEn} = 0.01$ cm^2/V s, respectively. Importantly, the gate leakage current, I_{GS}, decreased from 10^{-8} A for option **A**, to $<10^{-10}$ A for **B**, to $<10^{-11}$ A for **C**. This results in significantly lower power dissipation in the inverter quiescent mode. With a peak $I_D = 1$ μA, the ratio $I_D/I_{GS} = 100$, 10^4, and $>10^5$, for options **A**, **B**, and **C**, respectively (Zschieschang et al., 2008).

The dependence of the inverter transfer characteristics on process procedure is shown in Fig. 8.98. Owing to the very thin gate insulator comprising the AlO_x/SAM combination, the operating voltages are <3.0 V. The gate leakage (blue lines) as $V_{in} \to 0$ V is related to the quiescent circuit power dissipation. The repair and replacement of the SAM following gate etching has a dramatic effect on reducing the leakage from ~40 nA, to 110 pA, to 12 pA for circuits fabricated by options **A**, **B**, and **C**, respectively. This corresponds to respective power dissipations of 114 nW, 330 pW, and 36 pW, a substantial difference when employed in large integrated circuits where each gate contributes to the total circuit power. Other performance parameters of the inverters such as voltage gain and operating voltage are relatively independent of process procedure.

The stamping process has numerous variants, as described in Section 5.6.4. The apparent ease with which direct printing can produce high performance, low power dissipation circuits is emblematic of one of the signature advantages of organic electronics. The adaptability of processing techniques that are largely inaccessible to inorganic semiconductors allows for their implementation over very large areas, and they may even be adapted to continuous R2R manufacturing processes. Ultimately, this can lead to low cost, provided that the performance of the devices thus produced is sufficient to meet the demands of their intended applications.

8.8 Reliability of organic transistors and circuits

Integrated circuits are expected to provide reliable operation from days to years, while being exposed to environments that may require hermetic packaging However, the stability requirements for OTFTs may not always be as demanding as for other organic electronic devices such as OLEDs and OPVs, especially where the circuit is employed for chemical sensing, or in one-time use, disposable medical applications.

Unstable device operation is identified by time-dependent changes in the threshold voltage, subthreshold slope, transistor "off" current, channel mobility, output (or channel) conductance, gate leakage, and hysteresis in the drain current as V_{DS} is swept across the transistor operating range in the positive vs. the negative directions, etc. If we rule out catastrophic failure due to shorts or dielectric breakdown, the most common sources of device aging are time-dependent charge trapping at interfaces and grain boundaries due to morphological changes in the thin films, ingress of contaminants from the environment, decomposition of the organic semiconductors, migration of ions within the dielectric and into the channel, and deterioration of the source and drain contacts, also potentially due to material decomposition or even electromigration of the S and D metals into the organic layer.

We have seen that the noise margin determines the circuit reliability, that is, larger noise margins in a logic stage result in a greater probability that the subsequent stage will correctly identify its input as a high or low state. The gains and threshold voltages of the individual transistor stages, in turn, determine the noise margin. Clearly, the threshold voltage must be stable, and within a minimal spread of values for a large integrated circuit to operate properly. Hence a preponderance of work on understanding and improving OTFT reliability has focused on mechanisms that affect V_T. In the following sections, we describe methodologies used to quantify and understand changes in V_T and other operating characteristics, and several theories as to the source of these phenomena.

8.8.1 Tracking and understanding threshold voltage shifts

The lifetime of the transistor is quantified in terms of its duration of use (*operating lifetime*) and storage (*shelf life*). The device is under far higher stress during operation, and hence this is the most important mode of aging. However, the devices must be stable even when not in operation to ensure that the circuit is fully operational when it is eventually "plugged in." The latter failure mode occurs on a much longer time scale than the former, and is almost always due to the use of unstable organic molecules or exposure to contaminants during storage or fabrication. For this discussion, we assume that the shelf life of the device is long, and is not of primary concern since its origins are almost always extrinsic.

Intrinsic aging processes that limit the operating lifetime are not easily eliminated by improved processing or packaging. They are studied by applying a d.c. or a.c. signal to the gate while measuring the transfer characteristics over an extended period. Degradation can be accelerated by using a high voltage stress, elevated temperature, or large drain current, although test conditions are generally those used under normal operating conditions (Lee et al., 2012b, Jia et al., 2018).

The drift of the transfer characteristics (or the drain current at fixed V_{GS}) over an extended period of a PTAA BG/TC transistor employing an HMDS-coated SiO_2 gate insulator are shown in Fig. 8.99a. The characteristics are obtained by applying a constant stress of $V_{DS} = -3$ V and $V_{GS} = -20$ V while periodically halting the test to measure the I_D-V_{GS} characteristics (Sharma et al., 2009). Several time-dependent changes are noticed in these characteristics. The threshold voltage, which is estimated by the extrapolation of

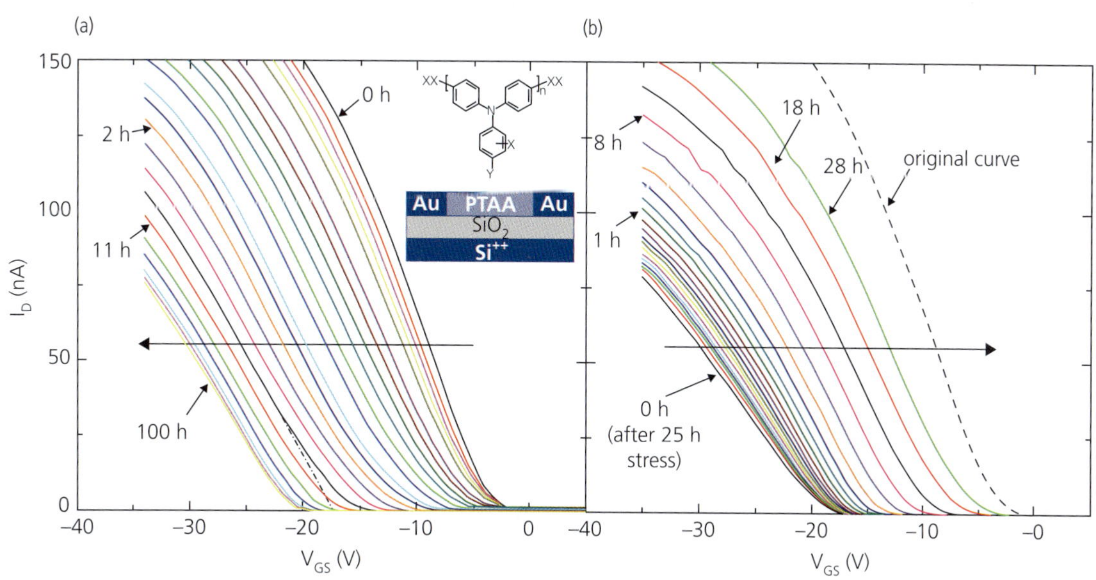

Figure 8.99 (a) Transfer characteristics of the device in the inset as a function of time. The stress conditions used during aging are $V_{DS} = -3$ V and $V_{GS} = -20$ V. (b) After the stress is applied for 25 h and then removed, the transfer characteristics slowly return toward their original values. The original transfer curve (dashed line) is not entirely recovered within the time of the test (Sharma et al., 2010).

the linear region of the characteristics to $I_D = 0$ (dashed line and arrow, Fig. 8.99a), shifts to larger negative voltage (i.e. in the direction of the applied gate potential) for this p-channel device. The threshold voltage appears to reach an asymptote after ~100 h at the gate stress bias of −20 V. Also, the slope of the I_D–V_{GS} characteristics decreases with aging, indicating a decrease in the mobility and sub-threshold slope, S. Once the stress is removed after its application for 25 h, the drift in the transfer characteristics reverses direction, returning toward their original values (dashed line, Fig. 8.99b). However, the characteristics rarely recover completely, showing a permanent offset in V_T from its original, as-grown value.

The drift in V_{GS} has been attributed to defects that gradually migrate into the channel, or reside at the organic/insulator interface. This induces in a change in the fixed charge concentration in the dielectric bulk or at the interface that reduces the field in the channel for a given V_{GS}. This, in turn, reduces the drain current (Sharma et al., 2009, 2010). The shift in threshold voltage is given by

$$\Delta V_T(t) = (V_T(t) - V_T(0)) = \frac{q\Delta N_{tr}(t)}{C_i}, \tag{8.75}$$

where $\Delta N_{tr}(t)$ is the time-dependent change in insulator and/or interface charge concentration.

The drift has been fit to a model used to describe metastable defect formation in hydrogenated amorphous Si transistors that show similar long-term behavior (Jackson et al., 1989). While the mechanism of hydrogen defect formation is clearly not active in organic transistors, the model is based on the existence of an exponential distribution of trap energies that results in dispersive diffusion of defects over time, similar to that observed for the charge mobility in the presence of an exponential density of defect states (see Section 4.3.2). The diffusion constant follows

$$D(t) = D_{00}(\nu t)^{m-1}, \tag{8.76}$$

where D_{00} is a constant, ν is the attempt frequency for diffusion (i.e. the frequency with which a defect attempts to move from site i to $i\pm1$ in the solid), and $m = T/T_0$. The characteristic temperature, T_0, defines the exponential defect distribution (cf. Eq. 4.140). That is, the defect distribution follows, $h_{tr}(E) = h_{tr0}\exp(-E/E_T)$, where h_{tr0} is a constant, E_T is the characteristic defect energy, and E is the energy.

The change in defect density with time is then (Jackson et al., 1989)

$$\frac{d\Delta N_{tr}}{dt} = -AD(t)\Delta N_{tr} = -AD_{00}(\nu t)^{m-1}\Delta N_{tr}, \tag{8.77}$$

where A is a constant. The solution to this equation is the familiar *stretched exponential* that provides a good fit to the time rate of change of other electronic devices such as OLEDs and OPVs. That is, from Eq. 8.77 we obtain

$$\Delta N_{tr}(t) = \Delta N_{tr}(0)\exp\left(-\frac{t}{\tau}\right)^m, \tag{8.78}$$

which yields a time dependent threshold voltage of

$$\Delta V_T(t) = \Delta V_T(\infty)\left(1 - \exp\left(-\frac{t}{\tau}\right)^m\right). \tag{8.79}$$

The maximum threshold voltage shift at $t = \infty$ cannot exceed the gate voltage, thus, $\Delta V_T(\infty) \to V_{GS}$. The time constant for ΔV_T is given by

$$\tau = (2\pi\nu)^{-1}\exp(E_T/k_BT), \tag{8.80}$$

which has activation energy:

$$E_T = k_BT_0 log\left(\frac{2\pi\nu m}{AD_{00}}\right). \tag{8.81}$$

Hence, the rate of threshold voltage shift is thermally activated with energy, E_T. The activation energy is the energy required for the defects to form or migrate into the channel, and thus provide a site for charge trapping.

The defect diffusion model has been applied to PTAA OTFTs in Fig. 8.99 with results shown in Fig. 8.100 (Mathijssen et al., 2007). The thermal activation of the characteristic time constant for ΔV_T is obtained from the fit to stretched exponential data following Eq. 8.79 at several different temperatures in Fig. 8.100a. The fit yields a time constant of $\tau = 10^7$ s (approximately 2800 h) at room temperature, and $m = 0.3$ using an attempt frequency of $\nu = 1$ kHz. The thermal activation energy for defect migration is $E_T = 0.6$ eV, which differs significantly from $E_T \sim 1$ eV for more thermally active hydrogen defects in a-Si (Deane et al., 1998). In Fig. 8.100a, inset, the trap exponent, m, is plotted vs. temperature. As noted above, the model predicts that $m \propto T$, which is indeed found for the PTAA transistors. The slope gives $T_0 = 900$ K, which is comparable to $T_0 = 720$ K for a-Si.

The stretched exponential fit to the threshold voltage shift provides an accurate representation of the data (see Fig. 8.100b). The applied voltage stress is $V_{GS} = -20$ V at $T = 140$°C. The parameters used to fit the data are $m = 0.43$ and $\tau = 10^4$ s. Data similar to those in Fig. 8.100 are also obtained during recovery. The T_0 is the same as obtained under stress, since the trap density distribution does not change with time. However, the activation energy during recovery is 0.3 eV, or only half that during stress. Likewise the attempt frequency is only 4 Hz, compared to 1 kHz (Mathijssen et al.,

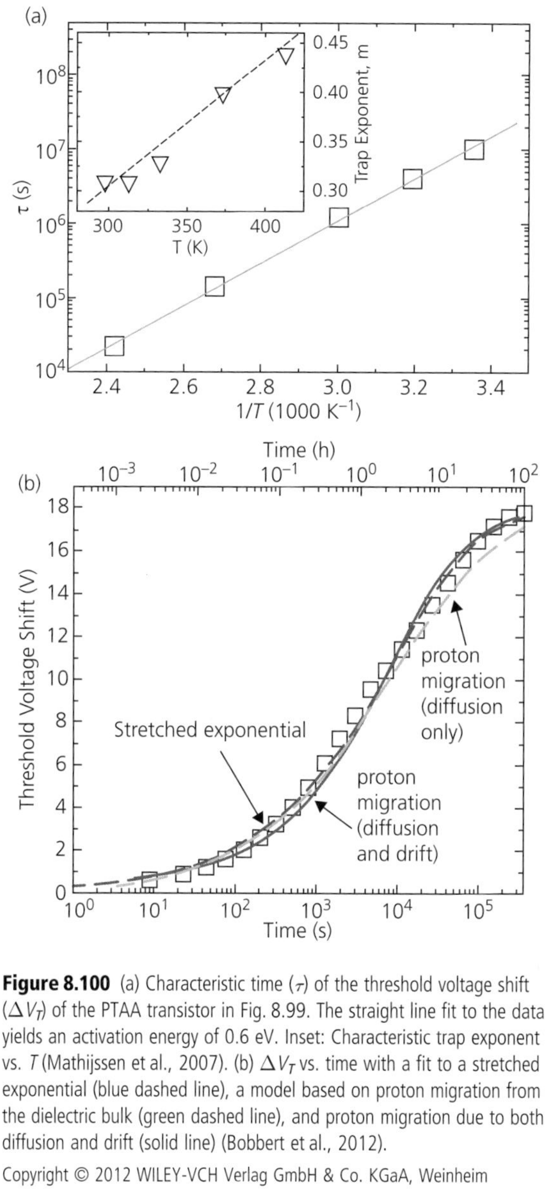

Figure 8.100 (a) Characteristic time (τ) of the threshold voltage shift (ΔV_T) of the PTAA transistor in Fig. 8.99. The straight line fit to the data yields an activation energy of 0.6 eV. Inset: Characteristic trap exponent vs. T (Mathijssen et al., 2007). (b) ΔV_T vs. time with a fit to a stretched exponential (blue dashed line), a model based on proton migration from the dielectric bulk (green dashed line), and proton migration due to both diffusion and drift (solid line) (Bobbert et al., 2012).

2007). These differences suggest that trapped charge relaxation occurs once stress is removed, resulting in a much longer time for the charge to be neutralized.

The thermal activation of the aging time, τ, suggests that degradation can be accelerated at elevated temperatures. For example, in pentacene films with $\tau_0 = 10^{-8}$ s and $E_T = 0.67$ eV, aging for 30 min at 80°C is equivalent to aging for 2 days at room temperature (Lee et al., 2012b).

Figure 8.101 illustrates how the pinch-off voltage (and analogously, the threshold voltage) depends on charge trapped in the dielectric. Prior to the application of stress (Fig. 8.101a), the threshold voltage of the p-channel OTFT is positive due to negatively trapped charges within the dielectric, or at the interface. At equilibrium (i.e. all the contacts are grounded), the number of holes in the channel exactly compensates the immobile trapped charges. This causes the potential at the dielectric/channel interface to vanish, that is, the surface potential is $V_S = 0$. For drain current to flow, a negative V_{GS} is applied to increase the mobile holes to a concentration that exceeds the immobile negative charge in the gate insulator. The injected holes are trapped, resulting in a positively charged trap concentration that exceeds that of the residual trapped electrons. In this case, $V_S > 0$ at equilibrium, and V_S and V_T become negative, as shown in Fig. 8.101b.

Charge trapping has been observed along the OTFT channel by measuring the spatial dependence of the surface potential using a modification of AFM, known as *scanning Kelvin probe microscopy* (SKPM) (Nonnenmacher et al., 1991, Bürgi et al., 2002), shown schematically in Fig. 8.102a. In SKPM, the potential between a conducting AFM tip and the sample surface is $V_{tip}(x) = \Delta\phi_{TS} + \psi_S(x)$, where $\Delta\phi_{TS}$ is the difference in work functions between the probe tip and the sample, and $\psi_S(x)$ is the local value of the surface potential. In the AFM non-contact mode, a small vibration of the tip at frequency, ω, induces a change in capacitance, ΔC. This, in turn induces a current:

$$i(t) = \omega \Delta C V_{tip} \cos(\omega t). \tag{8.82}$$

In SKPM, the voltage across the tip is set to $i = 0$, and then the force between tip and sample at a position x is inferred from the voltage-induced oscillation of the tip as it is translated along the sample surface. The oscillations, which yield $\psi_S(x)$, are detected using a beam of light that is reflected from the top surface of the probe (Martin et al., 1988).

The potential along the channel of HMDS-functionalized SiO_2/PTAA OTFTs was measured by first stressing the transistor, measuring $\psi_S(x)$ along the channel by SKPM, and then peeling away (i.e. exfoliating) the PTAA channel. Then $\psi_S(x)$ was measured once more. The organic was removed by attaching a piece of adhesive tape to its surface, and mechanically lifting it off from the dielectric. As shown in Fig. 8.102, initially V_P, the pinch-off voltage, is positive due to excess negative charge in the SiO_2. As a consequence, the surface potential after PTAA removal is negative (data point 1, Fig. 8.102b and c). With the channel semiconductor in place, holes are drawn in from the source and drain to compensate the negative charge, resulting in $\psi_S(x) = 0$ across the entire channel length (point 2). After stress, V_P becomes negative since the surface potential is positive due to filling of surface traps with positive charge drawn in from the channel contacts (point 3). This immobile positive charge

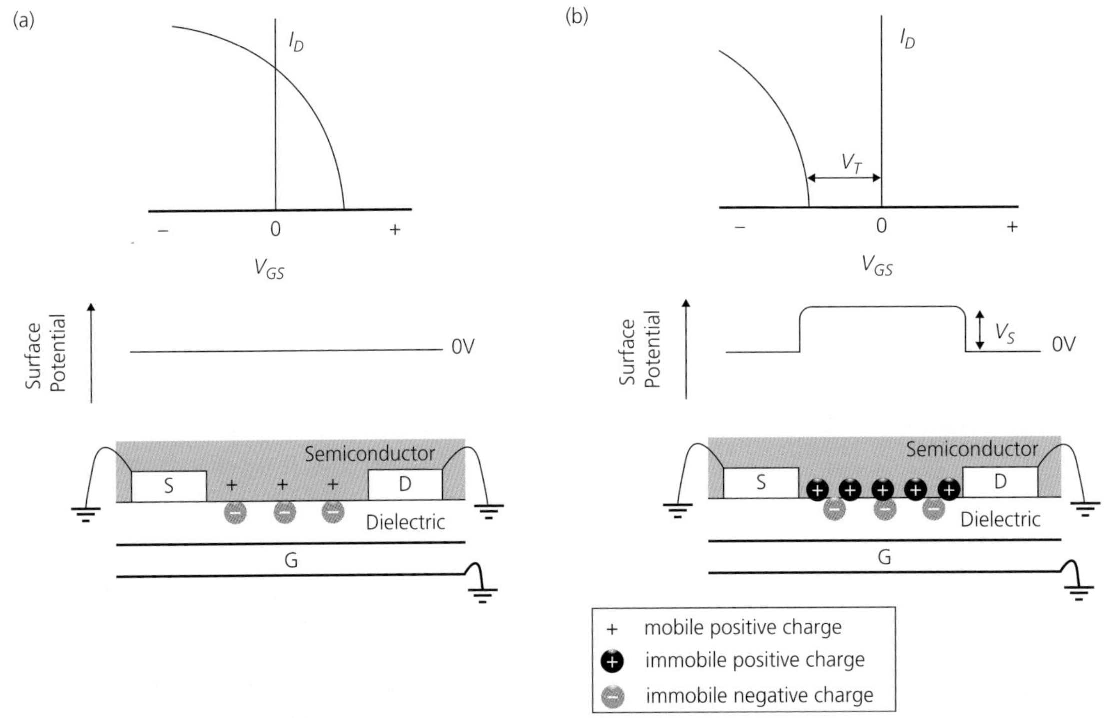

Figure 8.101 Schematic illustration of the transistor transfer characteristics (top panels) and how the charge distribution in the dielectric (bottom panels) influences V_T. The center panels illustrate the potential at the organic/dielectric interface (the surface potential, V_S) when all OTFT contacts are grounded. (a) Before application of stress, negative charge is trapped in the dielectric. (b) Following stress, injected holes become trapped and the V_T shifts to positive V_{GS}. Adapted from Mathijssen et al. (2010).

remains even when the PTAA is attached. Since the charge is not removed along with the channel semiconductor, it must be present in the insulator itself. Charge residing in the gate insulator or at the insulator/semiconductor surface gives rise to the drift in V_T with time. Similar phenomena, but with opposite polarities have also been observed in n-channel OTFTs employing polymer (Chua et al., 2005, Barra et al., 2010) or vacuum-deposited perylene derivative-based channels on an HMDS-functionalized SiO_2 gate dielectric (Mathijssen et al., 2010).

8.8.2 Effects of water on transistor stability

The source of trapped charge is obviously different than hydrogenation of a-Si implicated in the threshold voltage drift in those devices. The long durations observed in the voltage drifts suggest that the effects are due to ionic motion within the gate insulator. Adsorbed water on SiO_2 is known to increase its surface conductivity. Large concentrations of electrons and holes at the SiO_2 surface near the source and drain electrodes produces redox reactions that introduce mobile charge species, via

$$2H_2O \leftrightarrow 4H^+ + O_2 + 4e^- \tag{8.83}$$

That is, water electrolysis results in the production of protons and electrons on the insulator surface. The charges are mobile and, hence, are easily neutralized by charges within the accumulation layer. A slower diffusion of charge into the insulator, therefore, must also be active. It has been proposed that the electrons in the redox reaction Eq. 8.83 are neutralized by holes in a p-channel OTFT, leaving an excess of protons (Sharma et al., 2009, 2010). The surface protons subsequently and reversibly migrate into the insulator on a time scale consistent with that observed in the drift of V_T.

The volume concentration of protons in the dielectric bulk should be linearly proportional to the areal hole concentration, viz.

$$[H^+] = a[h^+], \tag{8.84}$$

where a is a constant. Hence, a higher drain current results in a higher $[H^+]$ concentration, leading to a more rapid and larger drain current drift over time. A fit to $\Delta V_T(t)$ in Fig. 8.100b that takes only proton diffusion into account (green dashed line) results in noticeable differences with the data, suggesting that this simple model is insufficient to describe charge

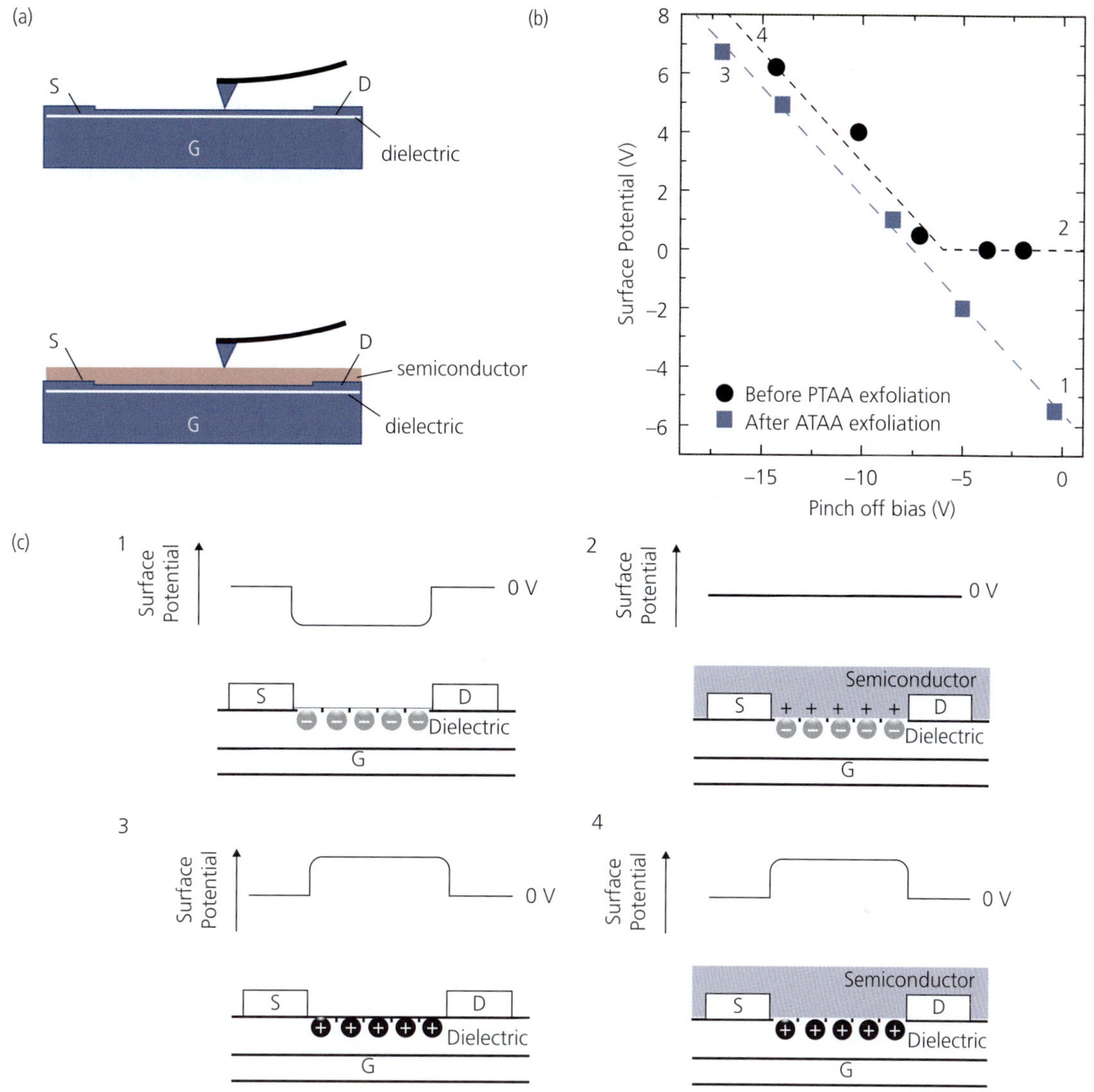

Figure 8.102 (a) Schematic of the scanning Kelvin probe microscope (SKPM) experiment for measuring the local surface potential on the dielectric (top) and on the organic semiconductor channel of an OTFT. (b) Surface potential in the channel of a HMDS-SiO_2/PTAA OTFT measured by SKPM before and after removal of the PTAA from the dielectric surface. The numbers (1–4) correspond to surface potentials measured for otherwise identical OTFTs with, and without PTAA after periods of stress ranging from no stress (points 1, 2), to 2 h of stress at $V_{GS} = -60$ V (points 3, 4). (c) Schematic of the measured surface potentials due to the corresponding charge distributions within the channel (Mathijssen et al., 2010).

diffusion. The fit to the stretched exponential function is more accurate when the presence of the electric field under the gate is included (solid line). The fit employs only two parameters, $\alpha = 2.2\ \text{nm}^{-1}$ and $D = 1.6 \times 10^{-19}\ \text{cm}^2/\text{s}$, from which one infers that, after a stress of 4×10^5 s, the protons penetrate into the insulator to a distance of 30 nm (Sharma et al., 2009, Bobbert et al., 2012). The protons in the dielectric contribute to an electric field at the semiconductor/insulator interface that results in charge scattering. Consequently, the charge mobility is reduced along with the drift in V_T, both leading to a reduction of I_D and S with time.

Note that the threshold recovery also follows a stretched exponential, albeit with a different time constant as noted previously. During recovery, $[h^+] \to 0$. Since the proton concentration is in equilibrium at the insulator surface, protons reverse their migration out of the dielectric where they react with solvated oxygen to form holes once the stress is removed (cf. the redox reaction, Eq. 8.83). This causes V_T to relax back toward its initial value.

The model is specific to SiO_2 gate dielectrics independent of the organic semiconductor used as the channel material. The time constant for the stretched exponential is given by

$$\tau = \frac{1}{\pi a^2 D}, \tag{8.85}$$

where a and D are properties only of the dielectric (Sharma et al., 2010). Similar mechanisms may also be active in other dielectrics since the redox reaction only requires the presence of water.

The effects of water on drift during bias stress have been inferred from the dependence of the drain current vs. temperature for several BG/BC OTFTs employing different channel materials (poly (3-sexithiophene), polyarylamine, and sexithiophene) on an HMDS-treated SiO_2 gate insulator. The devices were exposed to a humid environment for 24 h following fabrication, and then excess moisture was removed by exposure to vacuum for 20 min prior to electrical characterization. All devices showed that the drain current increases monotonically with temperature, independent of the channel material used, until approximately $T = 200$ K. At this point, the drain current dips and rises once again until approximately 273 K when a second, smaller decrease in I_D is observed (see Fig. 8.103). The as-grown devices (i.e. those not exposed to the humid environment) did not show either of these features;

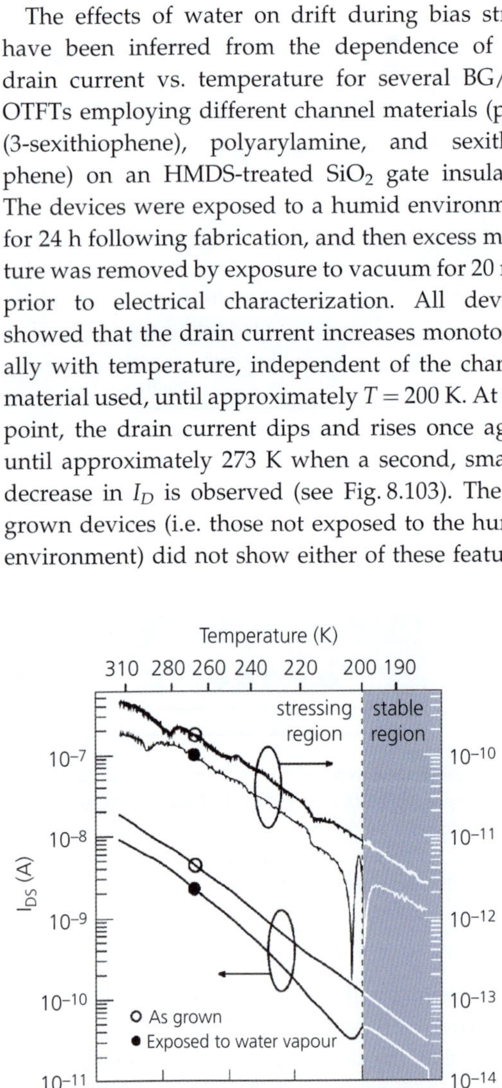

Figure 8.103 The dependence of drain current on temperature for a BG/BC poly(arylamine) OTFT using an HMDS-functionalized SiO_2 gate insulator. The as grown sample is compared to one intentionally exposed for 24 h to water vapor, and then exposed to high vacuum prior to measurement. The derivative of the drain current vs. temperature is also shown to emphasize changes at $T = 273$ K and 200–220 K (Gomes et al., 2006).

I_D simply increases with temperature across the entire range tested (Gomes et al., 2006). Water in confined environments (e.g. water molecules dispersed at the semiconductor/insulator interface, or in the insulator bulk) are supercooled in the liquid phase below ~273 K. In this environment, water transitions to the solid phase only when $T = 200$ K. The additional feature at 273 K corresponds to freezing of water that does not undergo a supercooled phase transition. Only liquid water can undergo the reaction in Eq. 8.83 that results in protonation of the insulator.

The drain current unexposed to water is higher than in its presence, and shows none of the anomalies of the latter devices. Furthermore, the fact that these features are found in each of the organic channel layers whether they are vacuum deposited (in the case of sexithiophene) or cast from solution (for the polymer channels) suggests that the water-related defects are either attached to the insulator surface, or diffuse into its bulk. These experiments show that cooling discontinuously changes the drain current at precisely those temperatures at which water undergoes a liquid–solid phase transition, reinforcing the conclusion that water plays an important role in transistor instability (Gomes et al., 2006).

We have shown in the foregoing discussion how protons generated from a water redox reaction, and that subsequently diffuse into the gate insulator at a rate largely independent of the presence of a SAM (Chua et al., 2005), can cause threshold voltage drifts by electric field screening in the channel. Another possible effect of water arises from its dipole moment of $p = 1.85$ D. We showed in Section 4.6.3 that dipoles at the surface of an organic can affect the density of transport states (Baldo and Forrest, 2001). In the context of an OTFT, the dipolar water molecules adsorbed onto the dielectric surface induce a broadening of the transport DOS within the channel, as illustrated in Fig. 8.104a (Richards et al., 2008). The H_2O dipole alignment at the interface is disordered. This creates a distribution of local electric fields experienced by organic molecules near the interface. This, in turn, broadens the charge carrier DOS in the channel, as apparent from the calculated profiles in Fig. 8.104b (Richards et al., 2008), with the broadening extending several nanometers into the charge accumulation region (Sharma et al., 2011).

The Gaussian disorder model in Chapter 4 provides an empirical dependence of mobility on the width in the DOS ($\sigma_{GDM}(z)$) at distance, z, from the dielectric surface. That is (Novikov et al., 1998, Hertel and Bässler, 2008),

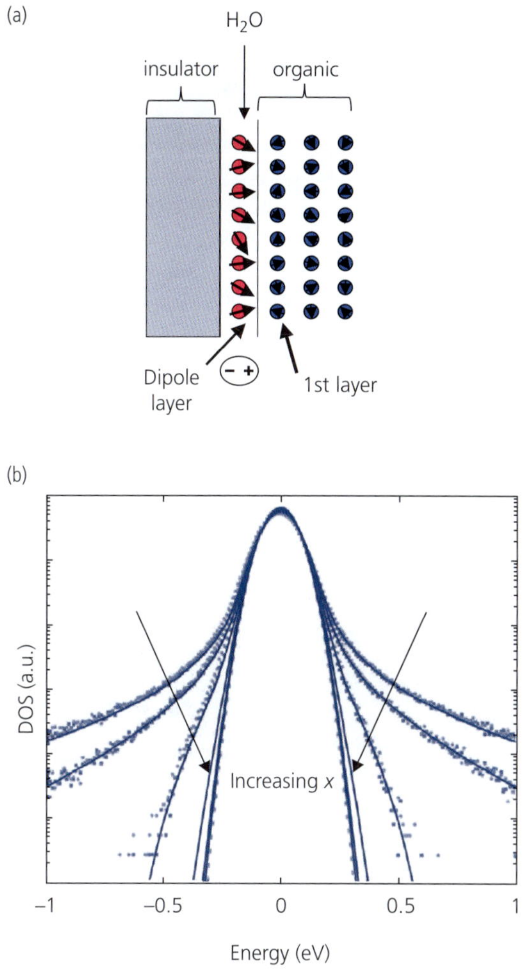

Figure 8.104 (a) Schematic illustration of surface dipoles at the insulator/organic interface. (b) Calculated density of states (DOS) in a PTAA channel due to a surface dipole of moment $p = 2$ D as a function of distance from the interface (x) ranging from 0 to 5 Å. Adapted from Richards et al. (2008).

$$\mu(z) = \mu_0 \exp\left(-\left(\frac{3\sigma_{GDM}(z)}{5k_BT}\right)^2\right). \tag{8.86}$$

Here μ_0 is the mobility at $T \to \infty$. The field effect mobility, μ_{FE}, is then

$$\mu_{FE} = \int_0^{z_{\max}} \mu(z)p(z)dz, \tag{8.87}$$

where $z_{\max}$ (V_{DS}, V_{GS}) is the width of the accumulation region and $p(z)$ is the hole concentration. Equation 8.86 shows that μ is a decreasing function of the width of the Gaussian DOS. Hence, the presence of water (or any dipolar contaminant) at the dielectric surface results in a decrease in mobility, and hence a decrease in I_D. With increased exposure time during device operation, the amount of water adsorbed at the gate dielectric/organic interface is also expected to increase, resulting in a shift in V_T.

The decrease in mobility with dipole strength has been observed by comparing μ_{FEp} obtained for PTAA OTFTs using either a high ($\varepsilon_i = 3.8$ for PMMA) or low dielectric constant ($\varepsilon_i = 2.1$–2.3 for a fluorene-thiophene copolymer, poly(9,9-dioctylfluorene-*co*-bithiophene)) gate insulator. The mobilities for the low ε_i insulators were found to exceed the PMMA-based devices by nearly an order of magnitude. The higher ε_i corresponds to a larger polarizability, and hence a higher dipole moment (Veres et al., 2003). The trend in μ_{FEp} for these devices is, therefore, consistent with the prediction of Eqs. 8.86 and 8.87.

While the change in mobility due to water adsorption must affect the mobility as described, the time constant of the shifts in V_T should be relatively short, determined by the rate of water adsorption and desorption. This is inconsistent with the very long time constants (10^4–10^7 s) associated with ΔV_T in Figs. 8.99 and 8.100. Hence, proton diffusion is more likely than water adsorption to be the source of the long term drifts in current and voltage. However, it is also clear that dipolar broadening of the DOS can indeed affect the transistor transfer characteristics. This can lead to hysteresis in I_D when sweeping V_{DS} in the positive *vs.* the negative direction observed in many OTFTs after exposure to atmosphere. Hysteresis may also be due to trapping of charge within the near-interface region (Noh et al., 2006, Uemura et al., 2014).

The time dependent drifts in the drain characteristics can be reduced by employing a bilayer dielectric, whereby the drift due to charge trapping in one insulator interface is compensated at the second interface by a mechanism in the opposite direction. This has been demonstrated in OTFTs employing channels that are microcrystalline (μc) blends of TIPS-pentacene or dif-TES-ADT and PTAA. The gate dielectric comprises 35 nm CTYOP directly layered onto the semiconductor channel, followed by a layer of Al_2O_3/HfO_2 of approximately the same total thickness. The secondary dielectric is deposited in five alternating layers of the two constituents by ALD. Its ΔV_T approximately compensates the drift due to the CYTOP insulator, although the mechanism for these two competing processes was not explained. Nevertheless, the devices showed minimal drifts even when immersed in water or air. Indeed, the minor drifts encountered by direct immersion in water were largely reversed when the devices were placed in vacuum at 100°C for 16 h (Jia et al., 2018).

The deleterious effects of water and other environmental contaminants can also be mitigated, at least to some degree, by device packaging, similar to that employed to extend the lifetime of OPVs and OLEDs. The primary goal of encapsulation is to passivate the semiconductor, that is, to make it less vulnerable to changes when operated in ambient. A passivating coating should have the following properties: (i) it should prevent ingress of contaminants into the device active region, (ii) its application should be non-destructive to the device on which it is deposited, and (iii) it should prevent or retard degradation at elevated temperatures by preventing the organics from vaporizing. (iv) Since OTFTs are often used in circuits that need to conform to an irregular substrate or are bent during operation, encapsulations may also need to allow for device flexibility.

A BG/BC pentacene OTFT with a flexible polyether sulfone substrate has been coated with a polymer passivating layer to decrease time-dependent drifts in I_D, μ_{FE}, and V_T (Han et al., 2006). The passivating layer consists of a 500 nm thick dichromated PVA film, capped with a 1 μm thick photoinitialized acrylate, see inset, Fig. 8.105a. After patterning the acrylate layer, the devices were rinsed in water and then baked in vacuum at 80°C for 4 h. Following this treatment, the devices showed $\mu_{FEp} = 0.80\ \mathrm{cm}^2/\mathrm{V\ s}$ and $V_T = -9.2$ V.

The drain current and V_T characteristics vs. time are shown in Fig. 8.105a and b, respectively. The I_D data were collected in the linear regime at $V_{GS} = -30$ V and $V_{DS} = -1$ V. The empirical dependences of I_D and V_T are shown by the straight line fits to the data using

$$I_D(t) = I_{D0} \exp(-\gamma t) \tag{8.88a}$$

and

$$V_T(t) = V_{T0} \exp(-\beta t), \tag{8.88b}$$

where I_{D0} and V_{T0} are the values of drain current and threshold voltage at $t = 0$, and γ and β are degradation rate constants. The passivated devices have considerably lower degradation rates than unpassivated ones, suggesting that the bilayer coating is effective in reducing drifts in the operating characteristics. Interestingly, while the slopes of the I_D data are both negative (i.e. I_D decreases with time) the rates of change in V_T are opposite for the passivated and unpassivated OTFTs. This is attributed to negatively charged traps due to water at grain boundaries in the pentacene film, requiring increasing the gate voltage over time in the unpassivated devices. The opposite slope for passivated devices is presumably due to oxygen

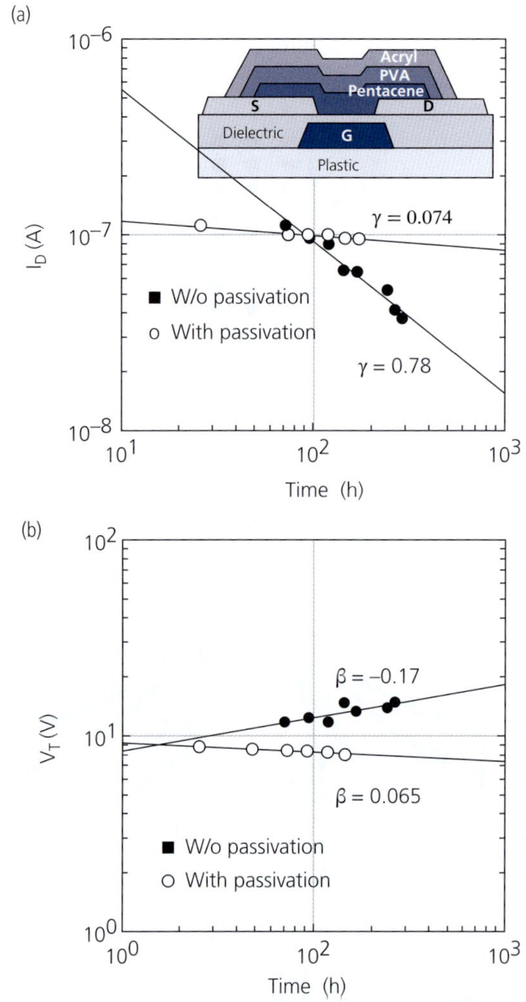

Figure 8.105 (a) Drain current vs. time for both a passivated (i.e. coated) and uncoated pentacene OTFTs. (b) Threshold voltage vs. time for the devices in (a). Lines are fits to Eqs. 8.88a and b. Adapted from Han et al. (2006).

entering the film, which forms a negatively charged oxygen ion (Han et al., 2006). Regardless of the origins of the effects, coating the devices is effective in extending their operational lifetime by reducing their exposure to the environment.

To summarize this section, ionic migration that is widely observed in conventional MOSFETs and MISFETs is undoubtedly also active in OTFTs. When stressed by an external potential, there is a change in the total electric field in the channel. This, in turn, causes both the threshold voltage and current to change with time, ultimately leading to circuit failure as the changes exceed the acceptable noise margin. Importantly, the mechanistic studies described highlight the vulnerability of OTFTs to exposure to moisture and other atmospheric contaminants, which must be considered when adapting this technology for a particular application such as displays, RFIDs, or sensing.

8.8.3 Other sources of instabilities in OTFT performance

Thus far, we have focused almost entirely on the effects of moisture on the transistor characteristics. Here we will describe a few of the other possible causes that reduce operational lifetimes. This should give a view of a few of the more important effects among a complex range of phenomena.

Gate leakage: Recall that the gate leakage must be minimized to reduce the off current, and hence the power dissipation in complementary logic circuits. In addition, very thin insulator layers are required to achieve low voltage operation. But thin insulators often have a tendency to have pinholes, and can undergo dielectric breakdown at 1–10 MV/cm, leading to catastrophic failure (Halik et al., 2004).

Figure 8.106 shows the leakage of a polystyrene gate insulator on a pentacene channel for a device in its off state ($V_{GS} = 0$ V and $V_{DS} = -30$ V) in air over a period of weeks. The drain current tracks the gate leakage. That is, I_D increases at the same rate as I_{GS}, although the current polarities are opposite. After annealing in air at 70°C for 5 min, both I_{GS} and I_D are decreased by the same amount. Analysis of the *C–V* characteristics of a metal–polystyrene MIS capacitor indicates that aging results in an increase in positive trapped charge within the insulator. The positive charge shifts the gate potential, resulting in the observed changes in I_D. The positive charge arises

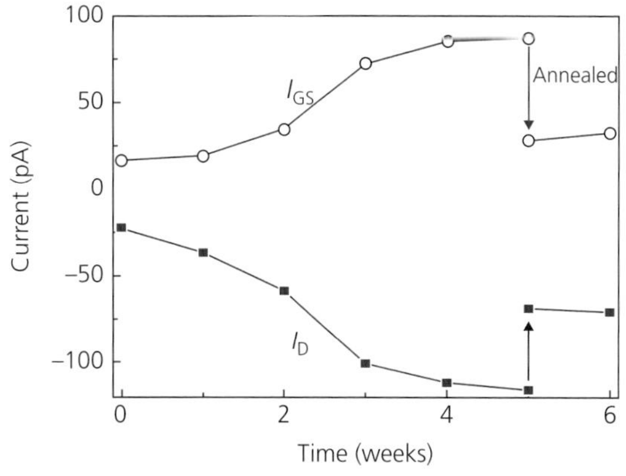

Figure 8.106 Gate and drain current vs. time of a BG/TC pentacene OTFT in the off state with a polystyrene gate dielectric. The aging is in air at $T = 24$°C and 42% relative humidity. Annealing conditions are 70°C in air for 5 min (Hong et al., 2008).

from diffusion of moisture into the bulk, which is removed by high temperature annealing that restores the original gate and drain leakage currents (Hong et al., 2008). This is consistent with the proton diffusion model in Section 8.8.2.

Interestingly, gate leakage also accelerates the aging of single crystal tetracene OTFTs in the on state, even when it is several orders of magnitude smaller than the drain current (De Boer et al., 2005). While no definitive reason has been given for this effect, it was suggested that high energy electrons accelerated through the high dielectric constant metal oxide dielectric induce damage in the tetracene near the dielectric/organic interface.

Bulk defects: Charges trapped within the channel can result in changes in current, just as the charges at the insulator surface. Both result in a local change in potential that influences the charge density in the accumulation region. Indeed, many reports suggest that charge trapping in the semiconductor bulk can result in threshold and drain current drifts (Salleo and Street, 2003, Chang and Subramanian, 2006), and in the hysteresis in the transistor output characteristics observed during positive and negative sweeps of V_{GS} and/or V_{DS} (Ucurum et al., 2008). Unfortunately, it is difficult to definitively determine the location of the traps. Hence, most reports of bulk trapping are based on inferential analyses. An exception is obtained using spatially resolved SKPM, where the location of the charge trapping sites can be directly correlated with grain boundaries between domains within the channel semiconductor thin film (Tello et al., 2008, Hallam et al., 2009). Figure 8.107 provides one such direct study of charge trapping in the bulk of a p-channel pBTTT on an untreated SiO_2 gate dielectric in a BG/TC OTFT. The polymer is known to form a "ribbon" morphology when annealed at 260°C.

To study charge trapping, the OTFT was stressed at $V_{GS} = -80$ V for 60 min in ultrahigh vacuum, with the intention of filling deep traps within the film. During stress, the drain current decreased by >50% resulting from a shift in V_T. Once the stress was removed ($V_{GS} = V_{DS} = 0$ V), I_D recovered to 80% of its original value over the ensuing 4–6 h due to trapped charge release. The average surface potential in the channel following stress is shown in Fig. 8.107a. The change in surface potential of 8.5 V corresponds to a total surface trapped charge density of 6×10^{11} cm^{-2} for an insulator capacitance of 11 nF/cm^2. The surface potential returns to its original value prior to stress (horizontal dashed line) over approximately 6 h. Also shown is the average relative amplitude of the potential corrugations across the channel before

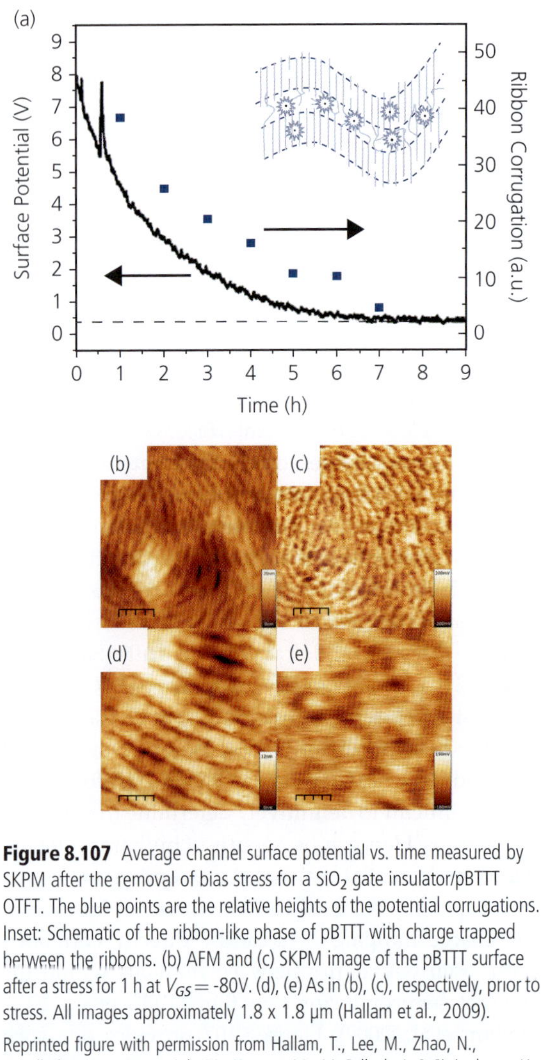

Figure 8.107 Average channel surface potential vs. time measured by SKPM after the removal of bias stress for a SiO_2 gate insulator/pBTTT OTFT. The blue points are the relative heights of the potential corrugations. Inset: Schematic of the ribbon-like phase of pBTTT with charge trapped between the ribbons. (b) AFM and (c) SKPM image of the pBTTT surface after a stress for 1 h at V_{GS} = -80V. (d), (e) As in (b), (c), respectively, prior to stress. All images approximately 1.8 x 1.8 μm (Hallam et al., 2009).

Reprinted figure with permission from Hallam, T., Lee, M., Zhao, N., Nandhakumar, I., Kemerink, M., Heeney, M., McCulloch, I. & Sirringhaus, H., Physical Review Letters, 103, 256803. Copyright 2009 by the American Physical Society.

and after stress (Figs. 8.107c and e). These corrugations decrease in amplitude during recovery; they tend to diminish over the same period that the surface potential relaxes to its pre-stress value.

The topographical AFM images in Figs. 8.107b and d show the ribbon-like domains. The potential corrugations of the stressed film (Fig. 8.107c) match the topographic pattern. The white areas of the SKPM image are more positive than the shaded regions. The correspondence between the AFM topographic data and the SKPM derived potential indicate that the trapped charge resides at the grain boundaries between the individual ribbons. The relative lack of contrast of the unstressed film (Fig. 8.107e) indicates the absence of trapped positive charge in regions between the domains (Hallam et al., 2009).

The surface potentials of the OTFTs were measured during operation to further understand the nature of bulk charge trapping. Figure 8.108 (top scan) shows the topographic roughness of the films. The ribbon domains are at the high points of the profile, and the grain boundaries are at the valleys. The SKPM profiles acquired during operation in the linear and saturation regimes (depending on V_{DS}) are shown in the lower figure. Within the grains, the potential is relatively flat, corresponding to a very small internal electric field. In contrast, there is a peak in the potential at each grain boundary, indicating a high density of positive charge associated with trapping at these locations. Note that even at $V_{DS} = 0$ V, the peaks exist. This suggests that the potential maxima between grains result from immobile charge and not current, since at $V_{DS} = 0$, I_D also vanishes.

Surface potential measurements have also implicated structural boundaries in vacuum deposited pentacene films as trapping sites leading to long-term changes in transistor characteristics, particularly as the film thickness is reduced to the point where it incompletely covers the insulator (Tello et al., 2008). Finally, trapping within the channel bulk is inferred from the dependence of drain current under bias stress on pentacene layer thickness. The thickness dependence indicates the bulk is the source of traps, as opposed to the surface (Chang and Subramanian, 2006).

While the SKPM measurements, and the observed thickness dependence of drifts in V_T and I_D on channel layer thickness point to the importance of trapping in the bulk semiconductor layer, the origins of the trapping sites remain unclear, and indeed may not be the same for all semiconductors. For example, reduced orbital overlap between molecules across the grain boundaries can be a source of levels deep in the energy gap that inhibit charge transport between grains, and hence result in accumulation of charge in the channel. Alternatively, the grain boundaries may provide sites for the ingress of impurities (e.g. H_2O) that also can trap charge. Independent of the mechanism for the formation of traps, their presence in the bulk has been established as an important cause of stress-induced changes in device performance.

Contact degradation: Contact resistance may change as defects form in the channel, or from *electromigration* of metal into the channel at high drain currents. This generally manifests itself as a decrease in channel output conductance, as well as an apparent decrease in field effect mobility (cf. Eqs. 8.23 and 8.24). The rate and magnitude of the changes depend on whether a top or bottom contact configuration is

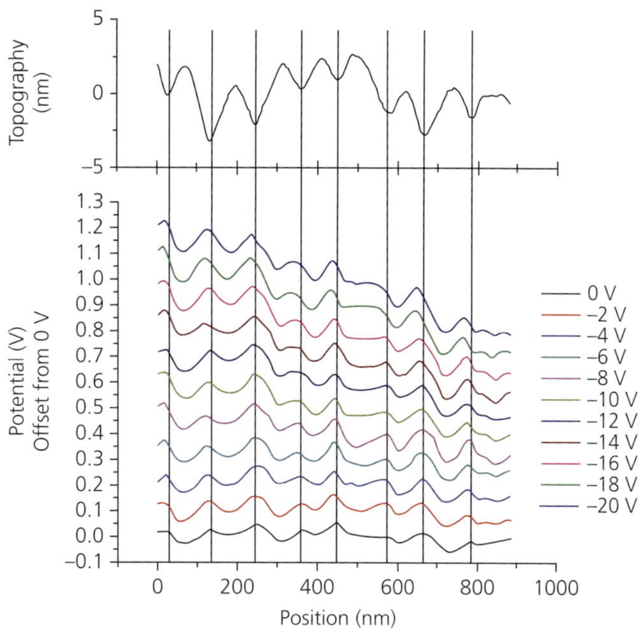

Figure 8.108 Surface topography acquired via AFM (top) and the surface potentials (bottom) measured by SKPM along the 20 μm long channel of the pBTTT OTFT in Fig. 8.107 as a function of V_{DS} (legend) at $V_{GS} = -30$ V (Hallam et al., 2009).

employed. In TC OTFTs, the source and drain metals contact a large channel surface area. Since the electric field lines extend to the opposite (dielectric) surface of the channel, injection can occur over most of the metal contact area. The injection area also increases with the amount of overlap between the S and D contacts and the gate, which extends the accumulation region well into the channel contact area. In contrast, bottom contacts result in preferential injection from the edge of the metal layer into the very thin, charge accumulation layer. The differences in contact areas are reflected in the contact resistances: 1–2 MΩ cm vs. 0.2–0.5 MΩ cm for the BC vs. TC devices, respectively (Richards and Sirringhaus, 2008). The edge-on contact of the BC configuration also results in several orders of magnitude higher current *densities* needed to achieve the same drain current compared to the top contact structure. This is more likely to cause local heating and the formation of defects in the channel than at lower currents, as we have seen for OLEDs and OPVs. The stress-induced defects lead a reduction of contact area that ultimately results in shorting or reduced injection efficiency over time.

The output characteristics of BG/TC and BG/BC F8T2 transistors are shown in Fig. 8.109a and b, respectively. Although the devices have different gate dielectrics (polystyrene for the BG/TC, and SiO_2 for the BG/BC device), the current in the BC OTFT is several times lower than in the TC case. This, presumably, is due to the larger contact area in the latter structure. Also, there is hysteresis for the BC OTFT that is noticeably absent in the TC device. This suggests a more stable injection efficiency for the S and D contacts of the TC transistor.

Gate bias stress of the BG/BC OTFT applied for several minutes induces a noticeable reduction in the drain current, and an increase in V_T and source resistance, with a much smaller fractional increase in drain resistance. The degradation of the contacts becomes important only when the contact resistance equals, or exceeds that of the channel. The channel resistance in accumulation is given by $R_{ch} = L/Wpq\mu_{FE_p}$ for a p-channel device. This sets a minimum channel length for a contact resistance of R_C of $L_{min} = pq\mu_{FEp}R_C$. Thus, when $R_C > WR_{ch}$, the contact resistance dominates the output characteristics. Potentiometry suggests that injection from the source is limited by bulk ohmic contact resistance along with injection into the depletion region at the contact, whereas R_C alone determines the drain resistance (Bürgi et al., 2003). Hence, it is the change in source (vs. the drain) resistance over time that is responsible for the contact-limited operational lifetime in some OTFTs.

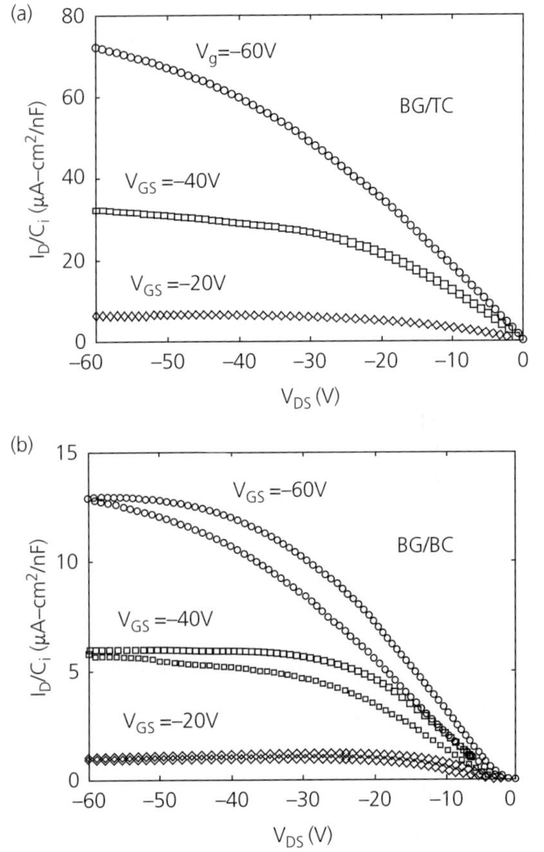

Figure 8.109 Drain current characteristics of OTFTs employing a p-channel F8T2 polymer channel on a (a) polystyrene gate insulator in a BG/TC configuration, and (b) SiO_2 gate insulator in a BG/BC configuration. The current is normalized to the gate capacitance for $L = 20$ μm and $W = 1$ cm (Richards and Sirringhaus, 2008).

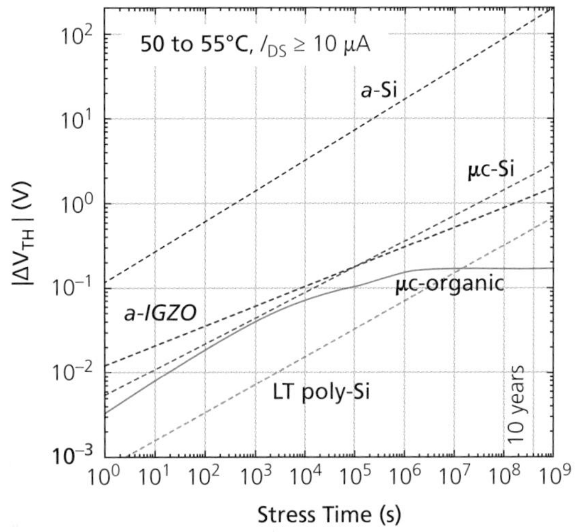

Figure 8.110 Comparison of the change in threshold voltage under bias stress at elevated temperatures for several different thin film transistor technologies (Jia et al., 2018).

8.8.4 Achieving reliable OTFTs

We have shown that OTFTs are susceptible to bias-stress-induced changes to the threshold voltage, drain current, mobility, contact resistances, and so on. Mobile charge in the gate insulator, at the insulator/channel interface, and within the channel itself and at the contacts all contribute to aging, with water being the most deleterious impurity observed in almost all OTFTs.

The most effective means extending OTFT lifetime, therefore, is to prevent or delay water entry into the channel region. This can be accomplished by packaging or encapsulation. However, one of the most attractive features of OTFTs is their flexible and/or conformable form factor. Flexible encapsulation as in Fig. 8.105 is only marginally effective in retarding the aging process. Beyond using improved encapsulation, there have been reports of both *n*- and *p*-channel organic semiconductors and composites that are comparatively immune to contamination from ambient water or oxygen (Roberts et al., 2008, Lei et al., 2011, Kuribara et al., 2012, Feng et al., 2016). A combination of all of these strategies can significantly extend the OTFT stability.

Ultimately, the reliability that is achieved must meet the demands of the target application. Since thin film transistor technology serves a wide range of applications, from driving display elements, to chemical and medical sensing, to RFID transmit and receive circuits, it is useful to compare the stabilities of several TFT materials systems. Figure 8.110 shows the change in threshold voltage over time for TFT technologies compared with microcrystalline (μc)-OTFTs with a TIPS-pentacene:PTAA channel and a bilayer CYTOP/Al_2O_3/HfO_2 gate insulator (Jia et al., 2018). The inorganic transistors based on a-Si, amorphous indium gallium zinc oxide (a-IGZO), μc-Si, and *low temperature polycrystalline Si* (LTPS) all show long term drifts that are on the order of, or larger than the organic device (Arai and Sasaoka, 2011). Care must be taken in making such comparisons, however, since the choice of devices is necessarily selective, and their performance characteristics vary over a wide range. Nevertheless, we can conclude that OTFTs can have competitive reliability and electrical performance compared to many inorganic thin film technologies, and indeed may have operational lifetimes well over 10 years in some cases.

8.9 Organic transistor circuit applications

When considering the possible applications of a new technology, we must first understand the incumbent technologies that stand to be replaced or augmented. We must ask, what is compelling about the new technology, and what new applications can it fill that otherwise are not served by existing solutions? In the case of OLEDs and OPVs the answer is clear. OLED displays offer high efficiency, a large color gamut, and flexibility that are unique and not well served by existing LCD or quantum dot/LED/LCD (QLED) displays. Likewise OPVs can be efficient and can serve the building integrated solar markets in ways that cannot be replicated using Si or compound semiconductor solar energy harvesting devices. The case for organic transistors is less clear. There is little doubt that enormous progress has been made in improving OTFT performance and reliability, and numerous examples of ICs such as logic circuits, RFIDs, ring oscillators, etc. have been reported (Crone et al., 2000, H. Klauk, 2006, Gelinck et al., 2010, Taylor, 2016). Yet, the electronics market is completely dominated by Si ICs that are low cost, high performance, extremely sophisticated and complex, reliable and adaptable to an almost limitless range of applications. Hence, prior to discussing particular applications where OTFTs may find use, we list several attributes of the technology that distinguish it from its competitors, including those based on crystalline, polycrystalline, and amorphous inorganic semiconductor, and metal oxide thin film transistors.

- *Adaptability to large area circuits*: Almost all applications envisioned for OTFTs entail their use as active matrix pixel address arrays in sensing and display applications. Their appeal in this regard is the same as for almost all other organic electronic devices: the ability to deposit materials over large areas on any reasonably flat substrate. Hence, OTFTs are ideally suited for use in pixel address circuits. The competition in this space is thin film Si (LTPS) and metal oxide transistors that can also be deposited over large areas.
- *Flexibility, stretchability, and conformability*: A distinguishing feature of OTFTs is their ability to be deposited at room temperature. This allows for the use of plastic substrates that are flexible and conformable, and uniquely, on stretchable surfaces. Whereas other transistor technologies (most notably metal oxide TFTs) can also be deposited at low temperature, an apparently unique feature of OTFTs is their compatibility with ultrathin ($\sim$1 μm) substrates that opens completely new application areas, particularly in wearable electronics.
- *Potential low cost when fabricated over large areas*: In Section 5.7, we have seen that room temperature R2R deposition on flexible plastic or metal foil substrates is a path to very low manufacturing costs. Thus, large area arrays of pixel address transistors for detection, sensing, and displays can be manufactured at very large volumes using R2R methods. This can potentially lead to significant cost reductions compared to technologies based on conventional, inorganic semiconductors that are more suitably processed in batches.

The competitive edge of OTFTs, therefore, lies in delivering large area electronic circuits on flexible substrates at very low cost. While this provides a motivation for developing OTFT applications requiring these attributes, it is equally important to understand where OTFTs are deficient. This chapter has focused on the performance of OTFTs, and the following properties of these devices have shown to present some limitations compared to inorganic transistor technologies:

- *Low output current*: The low output conductance of an OTFT (which is a result of their low field effect mobilities and large contact resistances) limits the current that they can deliver compared to many inorganic transistors. For this reason, an OTFT is more suitable as a voltage than a current source. While organic transistors have been proposed and even demonstrated for OLED pixel drivers (Yagi et al., 2008, Noda et al., 2011), a large, in-plane OTFT footprint is necessary to provide the current that can source sufficient current to attain high OLED brightness. Hence, while using OTFTs for OLED backplanes is possible, it is not optimal for realizing high resolution and intense OLED displays. As shown in Section 8.4.3, vertical transistors may be suitable for this purpose. In general, however, OTFTs are more useful as voltage sources for field-driven displays.
- *Limited stability*: Drain current drift during operation leads to a decrease in noise margin and hence instabilities in integrated circuit operation. OTFTs share this deficiency with a-Si TFTs, although drift can be lower in polycrystalline Si transistors. Very long term operation (>10 years) typical of inorganic semiconductor transistors have not been reported for OTFTs, although as we have seen in Fig. 8.110, a few OTFT designs are showing promise of reliabilities competitive with inorganic TFTs. In this context, many sensors, such as those found in medical diagnostics, do not need to be used over long periods, and hence may provide an application domain that can take advantage of the low cost and conformability of OTFTs.

- *Low cutoff frequency*: OTFTs have 3 dB bandwidths typically less than a few MHz. This places another constraint on the operating space available. Radio frequency circuits (e.g. RFIDs operating well into the MHz range), and fast logic circuits are beyond the reach of almost all OTFT technologies. However, many display and medical sensing applications do not require circuits with high bandwidth capabilities, providing a potentially rich domain for their application.

In the remainder of this section we provide three potential application areas for OTFT-based electronics: voltage-driven displays, tactile, chemical and other sensor applications, and "imperceptible" devices for medical electronic sensing. All of these applications have the common needs for large area and flexible circuits. The demand for device performance and stability varies, depending on the application. There are of course, other emerging areas where OTFTs are finding use at the laboratory level, but these three examples should serve to illustrate the potential utility of the technology.

8.9.1 Flexible display backplanes

Organic transistor arrays are well-suited for information display backplanes to activate voltage-driven pixels (Gelinck et al., 2006, Yoneya et al., 2006). These displays include LCDs and electrophoretic displays. While a-Si TFTs can serve as the pixel driver, their fabrication requires high temperature processing that is incompatible with lightweight, flexible plastic substrates that are often used for this application. A 320 × 240 pixel, 4.7" quarter video graphics array (QVGA) electrophoretic display driven by a solution-processed pentacene OTFT backplane is shown in Fig. 8.111a (Gelinck et al., 2006). The backplane is fabricated on a 25 μm thick PEN substrate, with a single BG/BC address transistor per pixel. The backplane is fabricated by attaching the substrate to a rigid "handle," and then patterning the contacts and the semiconductor layer via conventional photolithography. The solution-processable, 100 nm thick pentacene, and the 350 nm thick PVP gate insulator are deposited by spin-coating. Following completion of the OTFTs, a second PVP field insulator is spin-coated onto the transistor surface to separate the Au pixel electrode from the semiconductor surface while defining the pixel dimensions (see Fig. 8.111b). Once the backplane fabrication is complete, it is peeled off of the handle.

The display front plane comprises a sheet of microcapsules filled with reflective, electronegative white particles and less electronegative black particles suspended in a transparent fluid (Comiskey et al., 1998). The top surface of the sheet is coated with an ITO counter electrode, resulting in a total display-plus-backplane thickness of 100 μm. Applying a potential across the microcapsules causes the white and black particles (typically 1–5 μm in diameter) to migrate to the opposite top and bottom surfaces of the capsule. When the white particles are at the top, the display is more reflective than when the black particles are on top. Switching is achieved via a reversal of the pixel electrode potential.

Physical and electrical characteristics of the display are provided in Table 8.8. The properties of electrophoretic displays, including low frame rates, modest drive voltages, high on/off current ratio, bendability over a radius of only 7.5 mm, and ultralight weight (1.6 g) are enabled by the properties of the backplane OTFTs. Patterning the gates with $L = 5$ μm uses stamp transfer, or transfer from a rigid handle. The pixel size is considerably larger than the total area defined by the OTFT gate dimensions, resulting in a display with a very high aperture ratio (97.5%),

(a) (b)

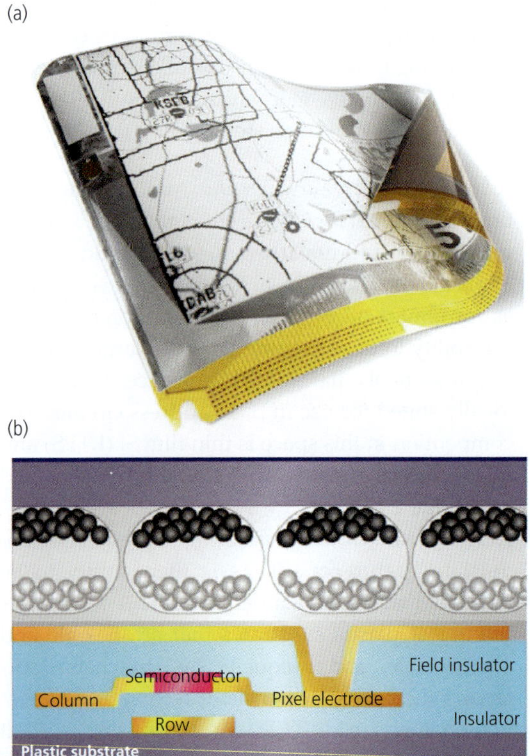

Figure 8.111 (a) Photograph of a flexible, 320 × 240 pixel, 4.7 inch QVGA electrophoretic display with an OTFT address backplane. (b) Schematic illustration of the backplane transistor and the electrophoretic display elements. The total thickness of the integrated structure is 100 μm (Gelinck et al., 2006).

Table 8.8 Characteristics of the electrophoretic display in Fig. 8.111 (Gelinck et al., 2006)

Parameter	Value
Backplane OTFTs	
L/W	5 μm/140 μm
μ_{FE}	0.01 cm^2/V s
I_{on}/I_{off}	1 μA/6 pA
V_T	0.2 V
Backplane characteristics	
Row voltages	±25 V
Column voltages	−15 V/0 V/+15 V
Common electrode voltage	4 V
Display	
Panel size	4.7" (72 × 96 mm^2)
Pixel size	300 × 300 μm^2
Display bend radius	7.5 mm
Resolution	320 × 240 (85 ppi)
Aperture ratio	97.5%
Thickness	100 μm
Frame rate	50 Hz
Image update time	1.5 s
No. gray levels	4
Contrast	10:1
Maximum reflection	35–40%
Weight	1.6 g

defined as the fraction of area occupied by the display device to the total area of the pixel.

8.9.2 Sensors and sensor arrays

Organic transistors have been used in numerous sensing applications. One example is the detector *focal plane array* (FPA). In the previous chapter, we showed that passive matrix arrays can be shaped into hemispheres to create an imaging system with the size, relative optical simplicity and some critical functionalities of the human eye (Xu et al., 2008, 2009). Passive matrix addressing, however, has the disadvantage of generating large background, or sneak currents (see Section 7.2.2). A passive matrix is comprised of N rows and M columns of detectors that are addressed by an array of electrodes (busses) connected to all of the detector cathodes, and an orthogonal array of column electrodes connected to all of the anodes. Then, if the reverse dark current of each device is i_D, the total dark current measured between each pair of column and row busses is Ni_D. Since the mean square shot noise current per unit bandwidth is $\langle i_S^2 \rangle = 2qNi_D$, the added shot noise from the array decreases the sensitivity as the FPA size increases.

To reduce the dark leakage currents in passive matrix FPAs, it is conventional practice to use an active matrix backplane that places one or more transistors at each detector as the pixel addressing element. The circuit may consist of five or more transistors as well as a charge storage capacitor to serve the multiple purposes of pixel addressing, pre-amplification and impedance matching. While the more complex circuit (known as an *active pixel sensor*) can result in very high bandwidth, ultralow noise operation, the size of the transistor circuits may ultimately limit the *pixel pitch*, that is, the number of detectors per length along a column or row (Bigas et al., 2006). Given the limited current sourcing capacity of an OTFT, the transistors must be considerably larger that their inorganic counterparts to achieve a large FPA dynamic range, DR, given by

$$DR = 20\log_{10}\left(\frac{I_{max}}{I_{min}}\right) \text{ (dB)}, \tag{8.89}$$

where I_{max} is the maximum photocurrent prior to deviating from a linear detector response (and hence constant efficiency), due to series resistance, heating, etc. The I_{max} is identified when the current falls below a predetermined linearity criterion, for example, when the difference between the responsivity of the pixel falls 1 dB below its value at lower input optical powers. Also, I_{min} is the photocurrent required to give a minimum acceptable *signal-to-noise ratio* (SNR). The I_{min} is the current *noise floor* determined by the detector and input transistor.

To achieve a very high pixel pitch (corresponding to a high image resolution) using OTFTs, a single transistor *passive pixel sensor* (PPS) provides a useful compromise (Renshaw et al., 2010, Tong and Forrest, 2011). An example pixel structure along with its circuit schematic is shown in Fig. 8.112. The photodetector is a bilayer consisting of the donor/acceptor pair of 90 nm thick SubPc/50 nm thick C_{60} with a 30 nm thick MoO_3 buffer. The device has an ITO anode and an Al cathode connected to a Au drain contact. The BG/TC OTFT consists of a parylene gate insulator and a 50 nm thick pentacene channel. The bias between the Al gate and the Au source addresses the pixel (Fig. 8.112, inset). All layers are deposited in vacuum and defined by shadow masks. The detector dimensions are 100 μm × 800 μm, and the OTFT dimensions are $W/L = 500$ μm/30 μm.

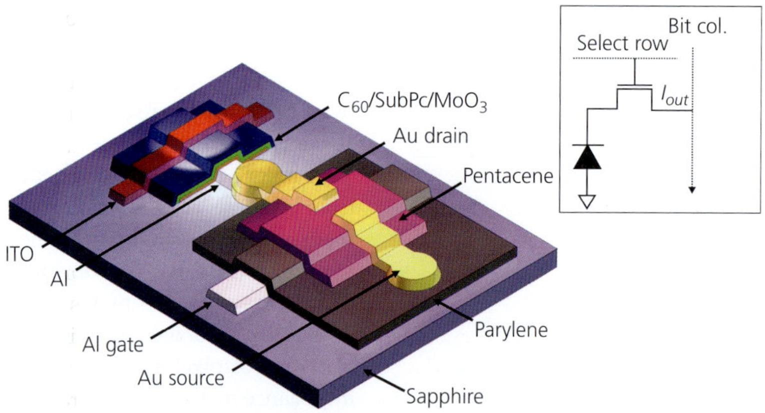

Figure 8.112 Perspective illustration of an organic passive pixel sensor (PPS). Inset: Circuit diagram of the PPS (Tong and Forrest, 2011).

The response of the photocurrent (i.e. I_{out} in Fig. 8.112) is linear over four decades, showing a departure from linearity at $I_{max} = 6$ μA. Also, the minimum detectable photocurrent is $I_{\min} = 300$ pA (see Fig. 8.113a). These detection limits yield $DR = 20\log\left(I_{max}/I_{min}\right) = 86$ dB.

The external quantum efficiency of the detector is 18.2% at a wavelength of 580 nm, and the speed of response of the circuit, shown in Fig. 8.113b yields a pixel readout time of ~400 μs. This is limited by the OTFT output conductance, given the relatively low $\mu_{FEp} = 0.09\ \mathrm{cm^2/V\ s}$. The on-off current and particularly the dark leakage of this device ultimately determines the ability of the PPS to fully turn off and contribute only negligible noise when used in an array. Here, $I_{on}/I_{off} = 10^5$, with a pixel off-current of $I_{out} = 31$ pA (Tong and Forrest, 2011). The pixel is too large for use in high resolution FPAs, and the small aperture ratio leads to a reduced *SNR*. Nevertheless, the circuit shows that even a rather modest performance OTFT can serve as an effective switch in an FPA. The transistor can be made smaller and the aperture ratio can approach unity by vertically integrating the detector and OTFT, and by using more precise patterning via photolithography or direct pixel printing.

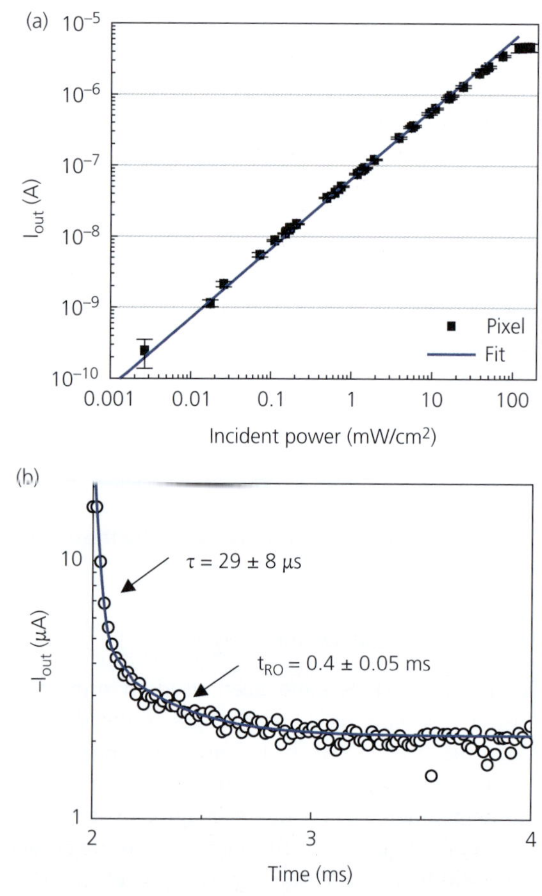

Figure 8.113 (a) Dynamic range of the passive pixel sensor in Fig. 8.112 showing the output current vs. incident optical power at a wavelength of 580 nm. The line shows the linear response of the PPS. (b) Temporal response of the PPS. Fit to a biexponential (line) gives an RC time constant of $\tau = 29$ μs and a readout time of $t_{RO} = 400$ μs (Tong and Forrest, 2011).

Tactile sensors, with uses in robotics, personal entertainment, pressure sensing, etc., present significant integration and materials challenges when implemented in active matrices. A tactile sensor can comprise a transducer whose resistance changes with pressure. The sensor arrays often must be flexible, and even conformal to fit a human extremity or other irregular surface. To obtain spatial information, the sensor can be connected to the drain of an OTFT in a cross-point array similar to the PPS in Fig. 8.112, inset, except where the photodiode is replaced with the pressure-sensitive element.

The fabrication sequence used to produce one such tactile sensor array is shown in Fig. 8.114. Figure 8.114a shows the flexible PEN base film with via holes cut by laser drilling, and then back-filled by thermally evaporated gate and via contacts on the top and bottom film surfaces. In this example, patterning is based on shadow masks, although inkjet printing has also been used (Noguchi et al., 2006). A 500 nm thick polyimide layer is deposited and thermally cured (Fig. 8.114b), and laser drilled to access the metal pad connected to the sensor electrode. The 50 nm thick pentacene channel is deposited in Fig. 8.114c, followed by deposition of the source and drain electrodes. The source makes contact to the back side sensor electrode via the two, laser-drilled vias. Finally, a pressure variable resistive film with a Cu electrode is laminated to a polyimide film, and is subsequently attached to the back surface of the base film (Fig. 8.114d).

The pressure-sensitive transducer is a colloid of conducting graphite particles in a PDMS film. Application of pressure compresses the film. This, in turn, increases the total contact area between the particles, decreasing the resistance between the Cu plane and the pixel transistor source electrode. Since the source electrode defines the pressure sensitive area, the Cu counter electrode covers the entire surface of the array. This is similar to the architecture used in the electrophoretic display described in Section 8.9.1.

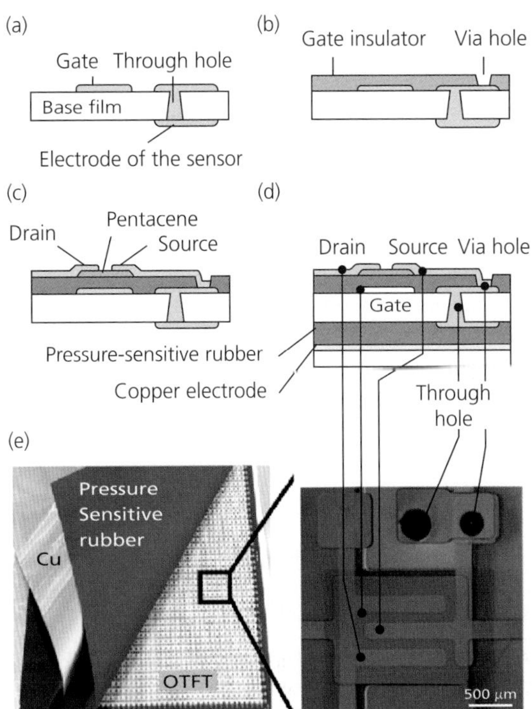

Figure 8.114 (a–d) Fabrication sequence for a pressure sensor array with an OTFT pixel address backplane. (e) View of the array showing a detailed image of an individual pixel (right). The scale bar is 500 μm. Adapted from Someya et al. (2004) and Noguchi et al. (2006).

The 32 × 32 pixel array fabricated using the process sequence in Fig. 8.114 has a sensor pitch of 2.54 mm, which is equal to the dimension of the transistor in Fig. 8.114e. The OTFT characteristics are provided in Fig. 8.115a, with $\mu_{FEp} = 1.4$ cm^2/V s with $I_{on}/I_{off} = 10^6$. These characteristics lead to the transfer characteristics in Fig. 8.115b. The drain current is determined by the resistance of the pressure transducer for $V_{DS} = -20$ V: it varies from 10 MΩ at pressure, $P = 0$, to 1 kΩ at $P = 30$ kPa. Pixelation allows pressure contact patterns to be reproduced to within the spatial resolution of the transistors. Also, there is only a small change in characteristics as the array is bent over radii as small as 10 mm. Changes in response are observed when the bend radius is reduced to 2 mm, although the array remains functional even in that case (Someya et al., 2004).

A type of temperature sensor is the *thermistor*. In array format, the sensors provide spatially resolved information that can have applications to health monitoring, robotics, and monitoring of local environments. A 16 × 16 flexible thermistor array on a 12 μm thick PEN substrate is shown schematically in

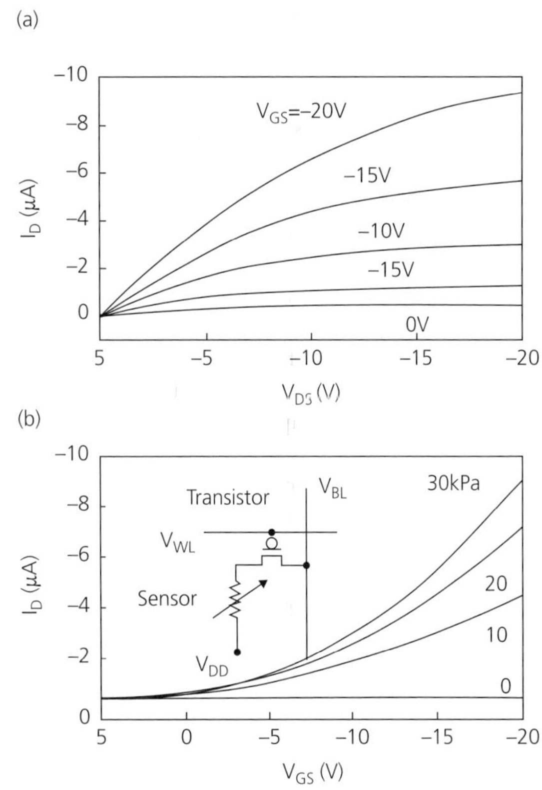

Figure 8.115 (a) Drain current characteristics for the transistors in the sensor array in Fig. 8.114. (b) The I_D vs. V_{GS} of a pixel on the array as a function of pressure. Inset: Pixel circuit schematic (Someya et al., 2004).

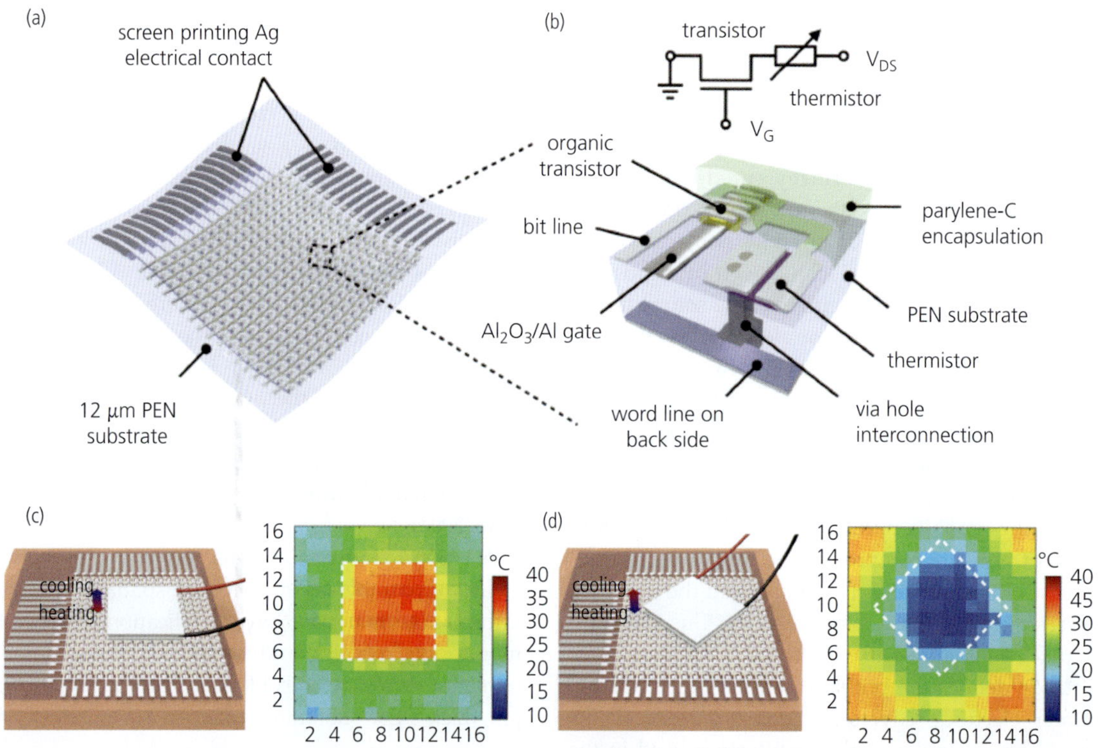

Figure 8.116 (a) Illustration of the 16 × 16 temperature sensor array with (b) an illustration of an individual sensing pixel and its circuit schematic. By placing a thermoelectric heater onto the array, both (c) heating and (d) cooling thermal images are produced, as shown to the right of each illustration. The thermal scale in (c) is 10–40°C and in (d) is 30–70°C. Blue indicates the coolest regions, red are the hottest (Ren et al., 2016).

Fig. 8.116a and b. The thermistor is a temperature-variable resistor connected to the OTFT drain contact. As temperature increases, the resistance decreases, resulting in an increased I_D. For accurate thermometry, the characteristics of the transistor itself must not be temperature sensitive. The array in Fig. 8.116 operated between temperatures of 20°C and 100°C, a range over which the transfer characteristics of the BG/TC DNTT transistors (employing Al gate contacts and oxidized Al/octadecylphosphonic acid SAM gate insulator) in Fig. 8.116b were thermally stable (Ren et al., 2016).

The thermistor itself comprises a layer of 5 nm Ag NPs sandwiched between two, 20 nm thick pentacene thin films. The temperature coefficient of resistance of a transducer comprising 3 nm Ag NPs is thermally activated, as shown in Fig. 8.117a. We have shown in Chapter 7 that ultrathin Ag layers deposited onto organics do not wet the organic layer, but rather forms NPs that are uniformly dispersed over the organic surface (and may also penetrate a considerable distance into the organic film). The NPs are too dispersed to contact each other, and thus the enhanced conductivity and significant increase in thermal activation energy compared with neat pentacene is due to NP-induced trap states in the pentacene (Fig. 8.117b and c). The traps lie above the pentacene transport level in the HOMO tail states. The traps may result from damage induced in the pentacene from impact of the high kinetic energy, evaporated Ag atoms during deposition (see Section 4.6.3).

The energy difference between the states is the activation energy, E_A. Then the conductivity is thermally activated following: $\sigma = \sigma_0 \exp(-E_A/k_BT)$, where σ_0 is the conductivity at $T \to \infty$. It follows that the *temperature coefficient of resistance* of the thermistor is

$$TCR = \frac{1}{R}\frac{dR}{dT} = -\frac{E_A}{k_BT^2}. \tag{8.90}$$

The devices in Fig. 8.117a have $E_A = 332$ meV, compared to a pentacene film with 157 meV. Further, we observe that the conductivity decreases by 300 times as the temperature is decreased from room temperature to 220 K. Hence, the thermistor provides a very sensitive element with a temperature resolution of 0.4°C and $TCR = 0.044/°C$ between 20°C and 100°C.

The DNTT transistors used in the array have performance similar to those discussed previously (cf. Fig. 8.23) in that they exhibit both low voltage ($V_{GS} \sim 4$ V) and low gate leakage (50 pA), required for high

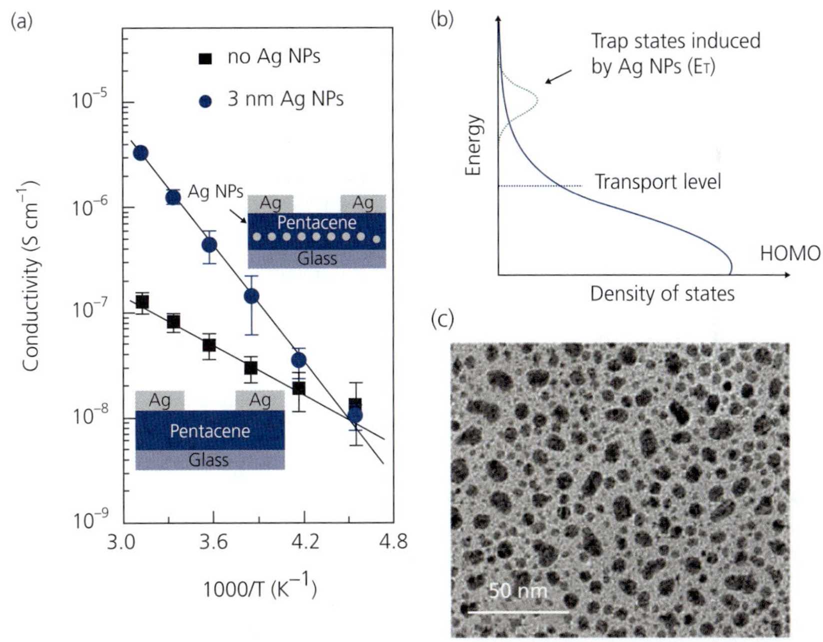

Figure 8.117 (a) Thermal activation of pentacene and a pentacene layer with 3 nm (average) Ag NPs deposited near the center of the pentacene film. Schematics of the two resistors are shown as insets. (b) Proposed DOS of the pentacene channel leading to thermal activation of the conductivity. (c) Transmission electron micrograph of the Ag NPs on pentacene (Ren et al., 2013).

sensitivity thermal monitoring. Also, by encapsulating the transistors with 6 μm parylene, the electronics are located very near the *neutral strain point* (see Section 8.9.3). Thus, the arrays were bent 10^4 times over a radius of 3.2 mm without changes in their operating characteristics (Ren et al., 2016). Imaging of both the hot and cold surfaces of a 2×2 cm^2 Peltier heater is shown in Fig. 8.116c and d, respectively. The spreading of the thermal image due to thermal conduction ultimately limits the spatial resolution of this device (Fig. 8.116d). However, the resolution can be improved by using less thermally conducting substrates or encapsulation layers.

The sensitivity of an OTFT to environmental factors discussed in Section 8.8 has led to their use for sensing a variety of analytes, ranging from water, oxygen, nitrous oxide, and organic solvents (Crone et al., 2001, Mabeck and Malliaras, 2006, Roberts et al., 2008, Marinelli et al., 2009, Zhang et al., 2015, Feng et al., 2016). This might seem a case of "turning lemons into lemonade" by exploiting the unwanted sensitivity of OTFTs to the environment into an advantage. Hence it is useful to clarify the characteristics of practical chemical sensors. They are: (i) *Selectivity* (or *specificity*). The sensor must be able to distinguish between analytes within the test environment. (ii) *Sensitivity*. The sensor must respond with an easily detectable SNR to the target analyte at the lowest possible concentrations. (iii) *Linearity*. A linear response is desirable to provide an output signal that is directly proportional to the analyte concentration over a wide dynamic range. (iv) *Response time*. This should be as short as needed, particularly when detecting toxic or corrosive substances that must be rapidly identified to protect life and equipment in time to reduce or eliminate the source of exposure. (v) *Reversibility*. Following the termination of exposure to the subject analyte, the sensor should rapidly recover to its original, pre-exposure condition, and hence be ready for reuse. (vi) *Reproducibility*. The sensor itself should have predictable and reproducible characteristics to avoid extensive calibration, or recalibration between uses.

An OTFT sensor uses a bottom gate configuration. This gives the analyte access to the sensitive channel and dielectric zones of the device. The target analyte creates or neutralizes traps within these regions or at interfaces, resulting in a change in operating parameters. Most commonly, changes in drain current are monitored over time as a signature of the presence or absence of a particular species. A high performance sensor generates a time-dependent change in drain current, $\Delta I_D(t)$, that is easily measured even at trace concentrations of the target analyte. Specificity and speed of response is achieved by the choice of semiconductor and/or dielectric, or by functionalizing the semiconductor surface with compounds that selectively react with the target molecules.

A demonstration of a sensor family that provides repeatable, selective, and rapid response to exposure to organic solvents is based on a series of thiophenes functionalized with increasingly longer end groups. The channel semiconductors are: α-sexithiophene (α-6T), dihexyl α-6T (DH α-6T), didodecyl α-6T (DD α-6T), and dioctadecyl α-6T (DOD α-6T) used in BG/TC OTFTs with an SiO_2 dielectric. The change in drain current in saturation, $\Delta I_D(t)$, of a DH α-6T sensor before, during, and after 72 consecutive, 5 s exposures to 1-hexanol is shown in Fig. 8.118. The recovery period is 1 min, which can be accelerated by an electrical bias applied after each sweep. The drain current recovers to within 2% of its initial, unexposed value after reach cycle (Crone et al., 2001). Analysis of the DD α-6T OTFT transfer characteristics due to exposure to 1-octanol indicates that $\Delta I_D(t)$ is in response to a change in V_T, whereas μ_{FEp} is unaffected. Exposure to ketones and nitriles produce changes in both parameters. These results indicate selective detection of molecules that variously affect charge trapping in the bulk (and hence V_T and μ_{FE}), or at the interface (which primarily affects V_T).

The selectivity "map" of several OTFTs vs. analyte is provided in Fig. 8.119. From the diversity of responses, an array of OTFTs with different channel materials can be used to develop a "fingerprint" for each analyte, thus creating a sensing system that can readily distinguish between substances to which it is exposed.

The sensitivity and selectivity to trace amounts of three different impurities of a BG/BC p-channel OTFT comprising a D_{3A} film spin-coated onto a SiO_2 gate dielectric (Marinelli et al., 2009) is illustrated in Fig. 8.120. The response and recovery are reproducible, and occur over times of 50–100 s. The sensitivity of the detector shows that NO_2 concentrations as low as 500 ppb are readily detected, whereas there is a lower sensitivity of 1 ppm for NO. The device is also selective, as there are no changes observed from exposure to CO. However, the sensor lacks the ability to distinguish between the different nitrides. Without prior knowledge of the analyte concentration, distinguishing between exposure to NO_2 or NO is not possible.

Analysis of the transistor transfer characteristics indicate that the changes in drain current are the result of changes in μ_{FEp}, whereas V_T is almost unchanged for all concentrations of NO_2. The change in mobility indicates that the analyte affects the conductive properties of the channel semiconductor, or the dielectric/semiconductor interface. The nitrides have a strong electron accepting nature. Introduction of these impurities in the channel can neutralize negatively charged traps, shifting the threshold voltage to more positive values. This, in turn, is equivalent to increasing the applied field in the channel,

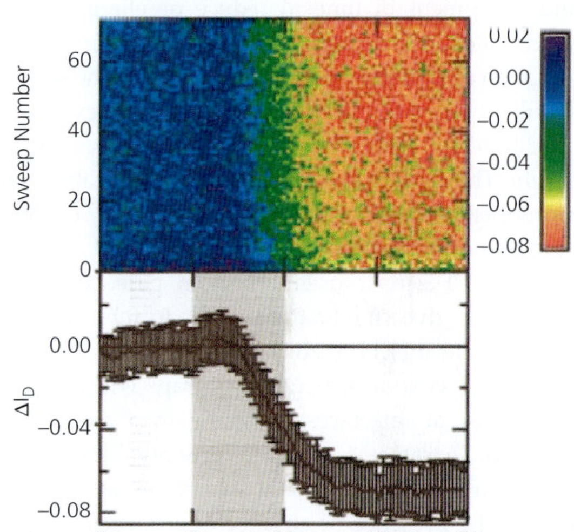

Figure 8.118 Change in I_D of a DH α-6T OTFT exposed for 5 s (shaded area, lower panel) to 1-hexanol vapor. The upper panel shows the fractional change in current (scale bar on right) for 72 exposures. The lower panel shows the change and deviations of the response from the data in the upper panel. The 1 min recovery between each sweep is not shown (Crone et al., 2001).

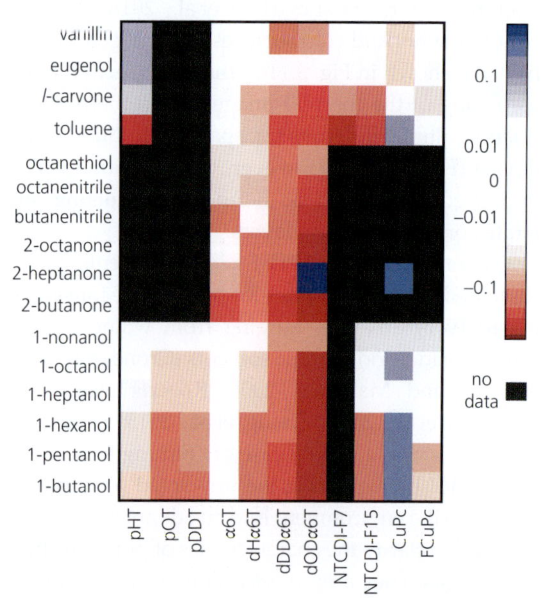

Figure 8.119 The fractional change in I_D for 16 analytes (vertical axis) exposed to OTFTs with 11 different channel materials (horizontal axis). A positive change in current is shown in blue, a negative shift is shown in red, and black indicates no data are available (Crone et al., 2001).

leading to a higher drain current at a fixed V_{GS}, as observed.

The ability to deposit OTFTs on virtually any reasonably flat surface provides an opportunity for OTFT sensors to find applications that are inaccessible to competing technologies. For example, high yield OTFT arrays and circuits deposited on banknotes have been shown to have performance characteristics similar to those deposited on more conventional plastic and glass substrates (Zschieschang et al., 2011). This allows for the direct integration of organic transistor-based RFIDs onto paper currency as an anti-counterfeit measure. While this solution has yet to be adopted, and indeed may ultimately prove to be impractical, demonstrations such as this can open up new possibilities for applications of organic electronic circuits and devices.

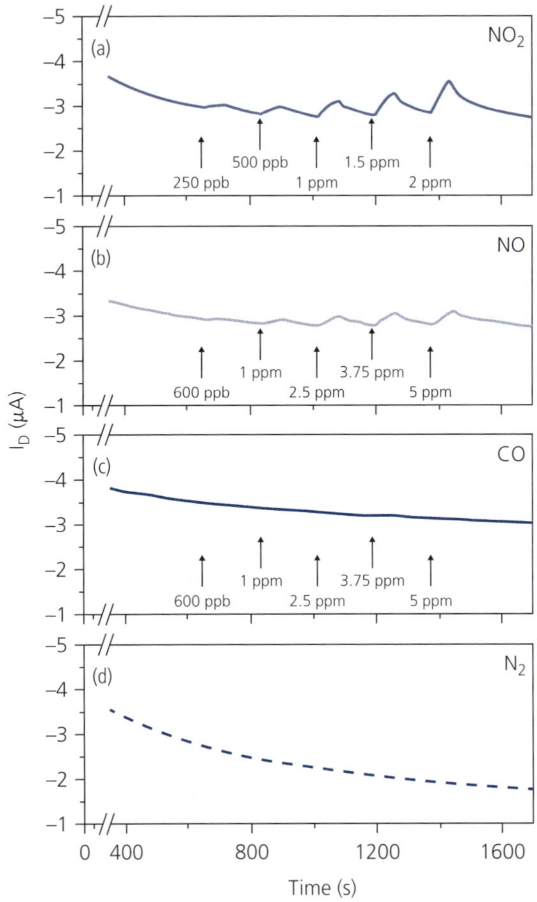

Figure 8.120 (a–c) Drain current vs. exposure of a D_{3A}/SiO_2 OTFT at $V_{GS} = -40$ V to three different analytes. The concentrations of the analyte vapors are indicated. (d) The drift due to operation in N_2 is used to remove the baseline response from the data in (a-c) (Marinelli et al., 2009).

8.9.3 Ultrathin, stretchable, and biocompatible electronics

A most compelling application with the potential to make a significant impact in the fields of medicine, robotics, and entertainment has been termed *imperceptible electronics* (Kaltenbrunner et al., 2013). This concept applies to electronics that are so thin, lightweight, and conformable that they cannot be felt when worn on the body, nor do they disturb the organism to which they are attached (e.g. to internal organs, limbs, etc.). Imperceptibility is achieved by fabricating devices on ultrathin and stretchable substrates, such that they conform to the natural topography of skin, the tissue of internal organs, etc., even as these tissues are naturally stretched or bent *in vivo* (Sekitani et al., 2010). Images in Fig, 8.121 show electronic circuits on ultrathin and lightweight substrates that can be bent over very small radii, or even crumpled without suffering losses in performance.

Achieving these goals requires more than just being able to handle such thin and easily distorted substrates. Bending over small radii requires that the device avoid excessive strain that can damage the active electronic layers. Also, long-term operation

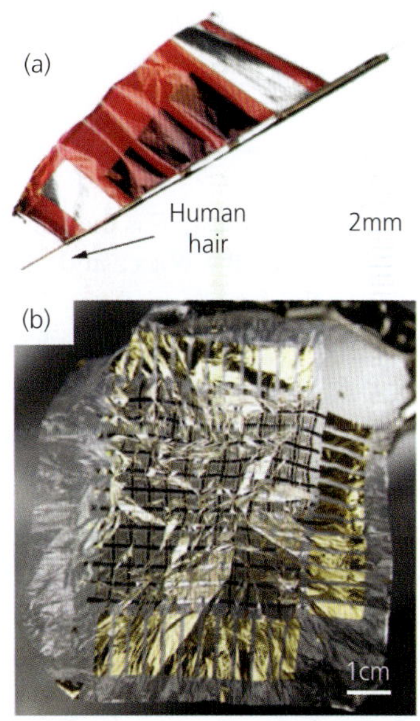

Figure 8.121 Examples of ultrathin imperceptible organic electronics. (a) A polymer, P3HT/PCBM solar cell on a 1.4 μm thick PET substrate wrapped around a human hair with a 35 μm radius (Kaltenbrunner et al., 2012). (b) A crumpled 12 × 12 matrix of tactile sensor pixels on a 1.2 μm thick PEN substrate (Kaltenbrunner et al., 2013).

requires encapsulation that does not impede flexibility. The examples in Fig. 8.121 are flexible, but not conformable: a property that requires stretching of the substrate. This places additional constraints, such as the ability for the interconnects to remain conductive even when significantly strained.

Encapsulation can be used to eliminate, or greatly reduce the strain on flexing if the active electronics are located at the neutral strain point in the multilayer package. This is illustrated in Fig. 8.122 for a film is bent over a radius, R. The bottom surface is under a negative compressive strain, ε_B, and the top surface is under a positive tensile strain. The *axial strain*, ε_A, or *proportional deformation*, is constant along the entire film. Thus, there is a location (the neutral strain point) within the film where the strain, $\varepsilon_B = 0$. For a homogeneous film, this point lies at its center. For the general case where the *Young's modulus* of the substrate is Y_{sub}, and that of an encapsulating film is Y_e, then the neutral point for substrate and encapsulation layer thicknesses of d_{sub} and d_e, respectively, are related via (Suo et al., 1999)

$$\frac{d_{sub}}{d_e} = \sqrt{\frac{Y_e}{Y_{sub}}}. \tag{8.91}$$

The Young's modulus is the ratio of stress to strain, that is, it measures the extension or shrinkage of a material under tensile or compressive stress. Thus,

$$Y = \frac{\sigma(\varepsilon)}{\varepsilon} = \frac{\mathcal{F}L_0}{A\Delta L}, \tag{8.92}$$

where σ is the uniaxial stress, $\mathcal{F}$ is the tensile force, L_0 is the initial length of the object and A is its area perpendicular to the force, and $\Delta L = L - L_0$ is the length change. When the active electronics are placed at the point of zero strain, it can be bent over very

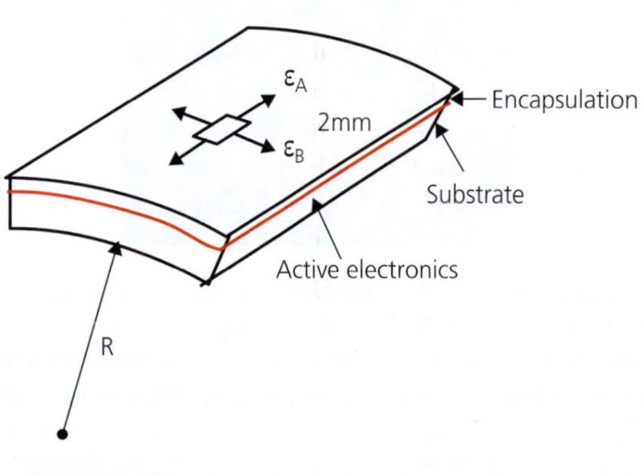

Figure 8.122 Active electronics on a thin substrate and encapsulated by a top thin film is bent over radius R, inducing an axial (ε_A) and bending strain (ε_B).

sharp radii, which is required for ultrathin substrates as in Fig. 8.121.

Elastic conductors are another essential component of bendable and conformable electronics. The interconnects between active devices must remain conductive, ideally without a significant change in resistivity as they are distorted. One method to achieve a stretchable interconnect is to pattern a metal meander, coil or flexible bridge between active circuit elements that expands as the elements are separated, for example on the surface of a plane that is distorted into a 3D shape (Huang et al., 2007, Gonzalez et al., 2008, Ko et al., 2008). While this may accommodate very large strains, it involves complicated processes that may violate the condition for low cost. An alternative is to introduce bundles of conducting carbon nanotubes (CNTs) or metal nanowires into a prestrained (i.e. stretched) elastomer such as PDMS (Lee et al., 2012c, Sekitani and Someya, 2012). The contacts between the long conducting tubes and wires form percolating paths along the length of the composite. The stress applied during film formation is relaxed once the curing process is complete, allowing the film to shrink. It can then be repeatedly stretched to its initial length, and then relaxed without a change in the resistance.

An example of a conducting film comprising a matrix of Ag nanowires in an elastomeric matrix that can withstand strains as high as 900% without failure is shown in Fig. 8.123. The nanowires are grown by a solution process that results in a very high aspect ratio: the diameters range from 100 to 150 nm over lengths of 100–200 μm. The nanowire films achieve 90–96% transparency and a resistivity from 9–70 Ω/sq. (Lee et al., 2012c). Improved contact between nanowires (and hence a lower resistivity and higher tolerance to strain without failure) is obtained by thermally annealing the composite during curing. Results of the prestrained films and the dependence of resistance on the amount of strain applied during fabrication are shown in Fig. 8.123. Strains as high as 460% are applied without failure (indicated by an abrupt and very large increase in R/R_0, where R_0 is the resistance at zero strain). No significant changes in resistance are observed after 200 stretching cycles. The figure also shows illustrations of the nanowire composite under several different strain conditions. On the bottom, left hand side, the prestrained film is in its relaxed state, where the Ag nanowires (orange region) form a dense conducting network. As the film is stretched, the nanowire network spreads out, until at the

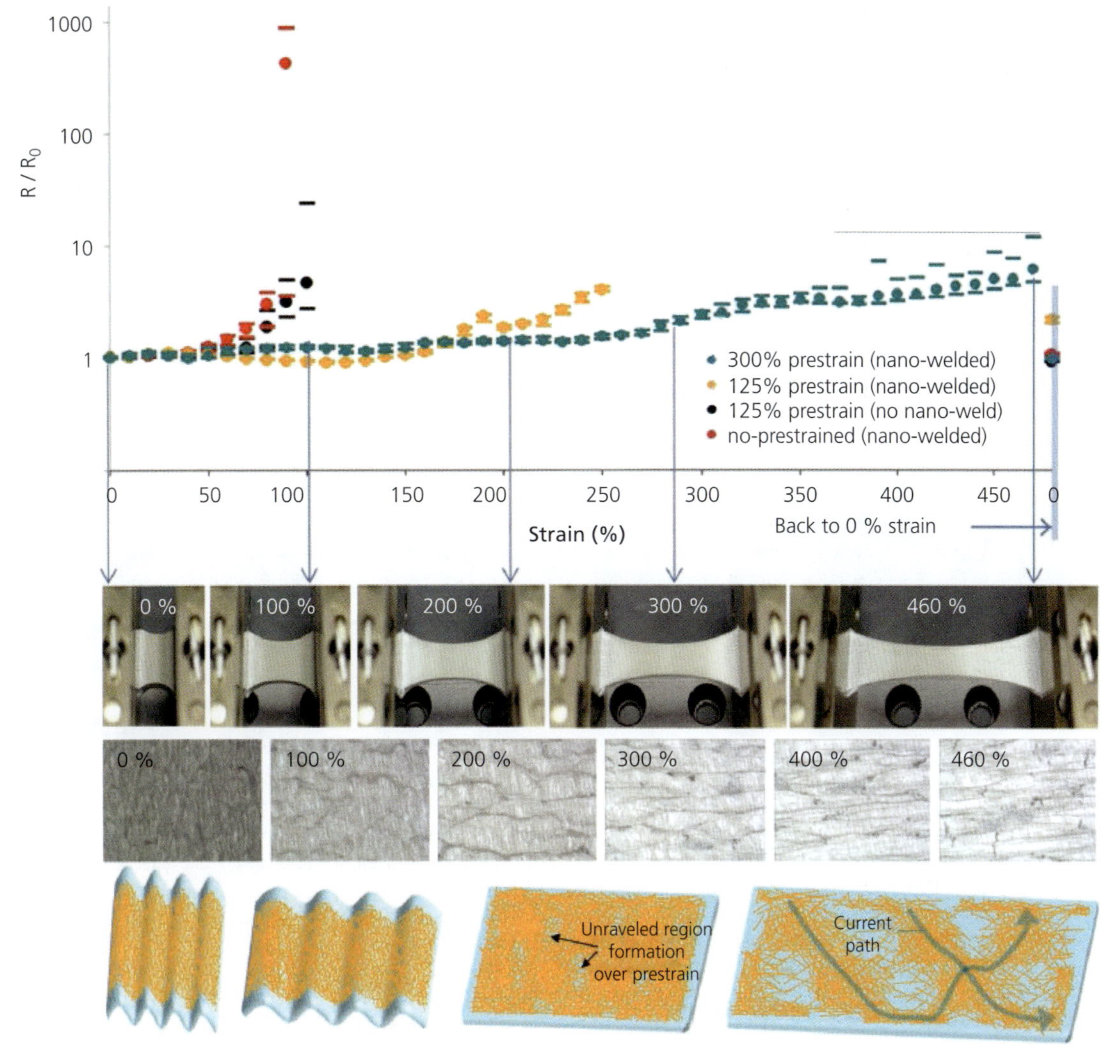

Figure 8.123 Top graph shows the resistance (R) of pre-strained Ag nanowire films normalized to their values under zero strain (R_0). The photos show the macroscopic and microscopic views of the strained films, and the illustrations at bottom show the conductive pathways in unstrained (left), and highly strained (right) films (Lee et al., 2012c).

highest strains applied, the network opens up, but percolating conducting pathways (arrows) remain. The resistance in this case is somewhat higher than at zero strain. From the top graph, however, we see that it has only increased ten-fold even at $\varepsilon = 460\%$. At higher strains, the networks break apart and the resistance abruptly increases.

There are many different approaches to achieving very high conductivity using mechanical meshes, nanowires, CNTs, etc. For a more complete description of these techniques that are a central element of highly flexible and conformable electronics, the reader is referred to the extensive literature on this subject (Hammock et al., 2013, Lipomi and Bao, 2017).

Transistors in tactile matrix arrays such as that in Figs. 8.114 and 8.121b represent early demonstrations of an imperceptible electronic application that use prestrained transistors such as in Fig. 8.124. A DNTT transistor is fabricated on the neutral plane sandwiched between a 1.2 μm thick PEN substrate and a perylene encapsulation layer of approximately the same thickness (Fig. 8.124a). Since the Young's moduli of the two materials are approximately equal, Eq. 8.91 indicates that the neutral strain point for the electronics is in the center of the sandwich. The structure of the BG/TC transistor comprises a 25 nm thick Al gate, a 19 nm thick potentiostatically anodized AlO_x gate insulator, an n-tetradecyl phosphonic acid

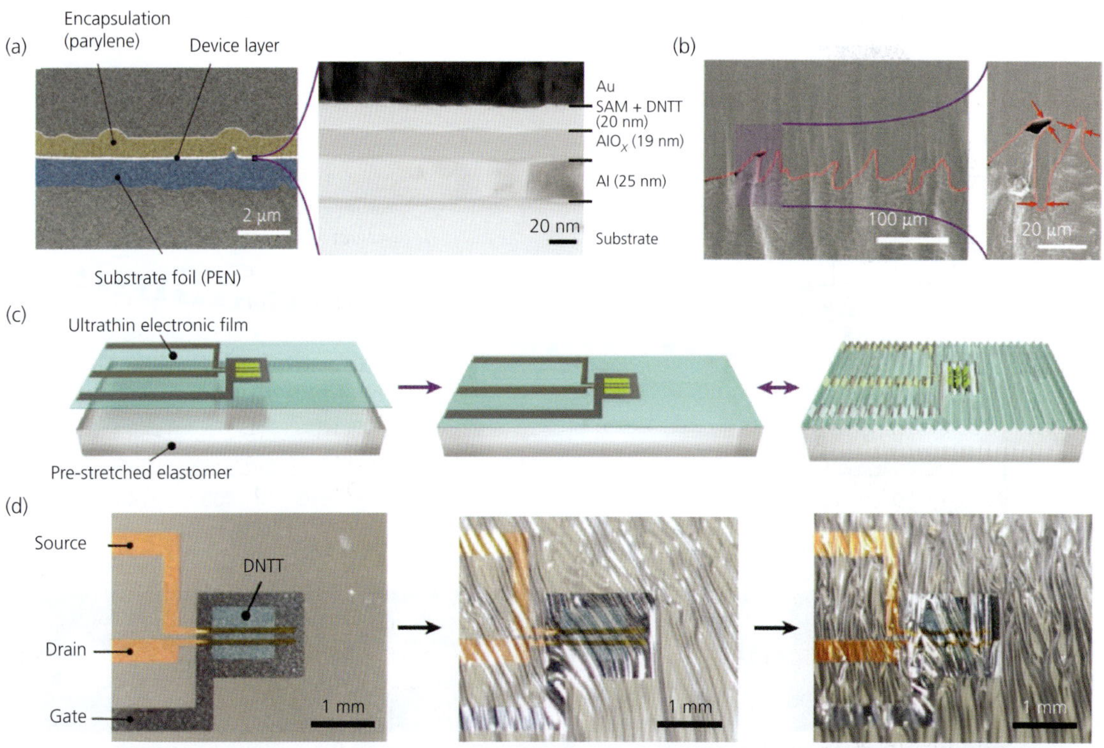

Figure 8.124 (a) A thin film transistor used in the tactile sensor backplane array of Fig. 8.114b. It is fabricated in the neutral plane between the 1.2 µm thick PEN substrate and an equally thick perylene encapsulation. (b) Cross-section of the film after it is compressed following the procedure in (c). The red line is the transistor plane. (c) Process for wrinkling the transistor in (a) by placing the fabricated, ultrathin OTFT on a pre-stretched substrate and then allowing it to relax. (d) Micrographs of the transistor before and after relaxation (Kaltenbrunner et al., 2013).

SAM, and a p-channel layer comprised of DNTT. The source and drain metallization used VTE grown Au. The OTFTs showed $\mu_{FEp} = 3\ \text{cm}^2/\text{V s}$.

The cross-sectional SEM image of the transistors in Fig. 8.124b after wrinkling shows folds with radii as small as 5 µm within the device area. The folds are formed by fabricating the transistors on a flat, rigid handle. The devices are then peeled off from the handle and attached to a stretched, elastomeric substrate. The substrate is allowed to relax, crumpling the transistor, creating deep folds as shown in Fig. 8.124b and d. Interestingly, even with such deep and sharp creases, there is no perceptible change in I_D, V_T, and μ_{FEp} up to tensile strains of 233%, and compressive strains of 80%. Radial compressive strains of 40% were also achieved without changes in performance. The devices were cycled through compressive and tensile motions up to 200 times, also without significant changes in device characteristics (Kaltenbrunner et al., 2013). This demonstrates that fabrication at the neutral strain position enables electronic devices that are nearly immune from flexing and stretching.

A potential application of electronics on such thin substrates is the *in vivo* and real time monitoring of a beating heart by wrapping it in a biocompatible array of electrical sensors (Sekitani et al., 2016, Park et al., 2018). The sensor inputs are the nerve impulses that trigger the contraction of the heart muscle. These impulses, however, are only 1–2 mV. An integrated voltage amplifier at each pixel is needed to generate an electrocardiogram with an SNR that is easily monitored without error.

The circuit schematic of the heart sensor is shown in Fig. 8.125a. A biocompatible multiwalled CNT gel electrode is placed in contact with the heart tissue to be monitored. The electrode connects to one plate of the input capacitor that, in turn, is connected to the gate of the input transistor of a

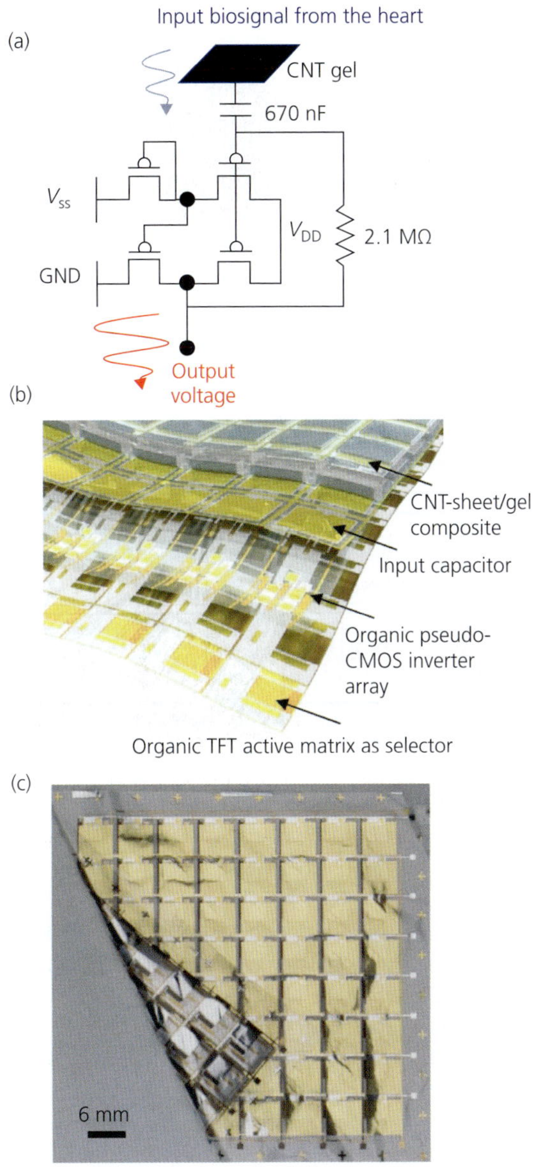

Figure 8.125 (a) Amplifier circuit consisting of a carbon nanotube (CNT) gel contact, input capacitor and voltage amplifier circuit. (b) Schematic illustration of the multilayer, ultrathin amplifier array. (c) Image of the completed array (Sekitani et al., 2016).

single stage voltage amplifier circuit. This differentiating voltage amplifier enables detection of the small voltage pulses transmitted from the nervous system encasing the heart muscle. The success of this application relies on two key features: (i) biocompatibility of the electrode in contact with the heart, and (ii) ultralight weight and flexibility that prevents irritation, and otherwise does not affect the normal function of the organ.

The electrode itself consists of a stretchable and soft aqueous polyrotaxane-based gel with a moveable cyclodextrin/polyethylene cross-linker mixed with multiwall CNTs. The sheet in the array shown in Fig. 8.125b and c is produced by drop casting the gel onto a PTFE sheet. The array is attached to the capacitor sheet comprising a SAM/AlO_x dielectric sandwiched between Au electrodes. The active devices are p-channel DNTT OTFTs employing structures similar to those used in the tactile sensor arrays in Fig. 8.114. In this case, the gate dielectric is a 4 nm thick layer of AlO_x formed by oxygen plasma treatment of the Al gate electrode. This is followed by deposition of a 2 nm thick *n*-octadecylphosphonic acid SAM prior to deposition of the DNTT and the Au source and drain electrodes. The entire array has a 3 dB frequency roll-off of ~20 Hz, as determined by the input capacitor. The transistors themselves show f_T = 42.7 kHz. Due to the use of the ultrathin (1.2 μm) PEN substrate, transistor operation is unaffected by bends around a 50 μm radius.

Electrocardiograms taken by wrapping the circuit around a rat's heart (Fig. 8.126a) are shown in Fig. 8.126b. The gel electrode in contact with the heart had a thickness of 1 mm and an outer dimension of 6 × 6 mm^2. The electrical input of 1.2 mV was amplified 200 times to result in a 220 mV output signal, increasing the SNR from 0.53 to 64 before and after amplification.

Biocompatibility also requires that the devices withstand exposure to extreme conditions found in autoclave sterilization. This entails exposure to steam at temperatures exceeding 120°C. Experiments on DNTT transistors have shown that they are sufficiently robust to withstand such treatments for up to 30 min without inducing large changes in their operating characteristics. To withstand high humidity and temperature, the devices are fabricated on moisture-impermeable and rigid Si substrates. Furthermore, the top parylene encapsulation layer is replaced by a trilayer comprised of 300 nm parylene, followed by 200 nm thick Au, and a cap of a second 1 μm thick parylene layer. The initial parylene layer protects the OTFT surface from the Au coating. However, the parylene water vapor transmission rate (see Section 5.8) is too high to prevent degradation at elevated temperature and humidity, and hence the Au layer is inserted as a moisture barrier. The parylene cap provides the final, biocompatible coating. The trilayer, along with the environmentally impervious Si substrate,

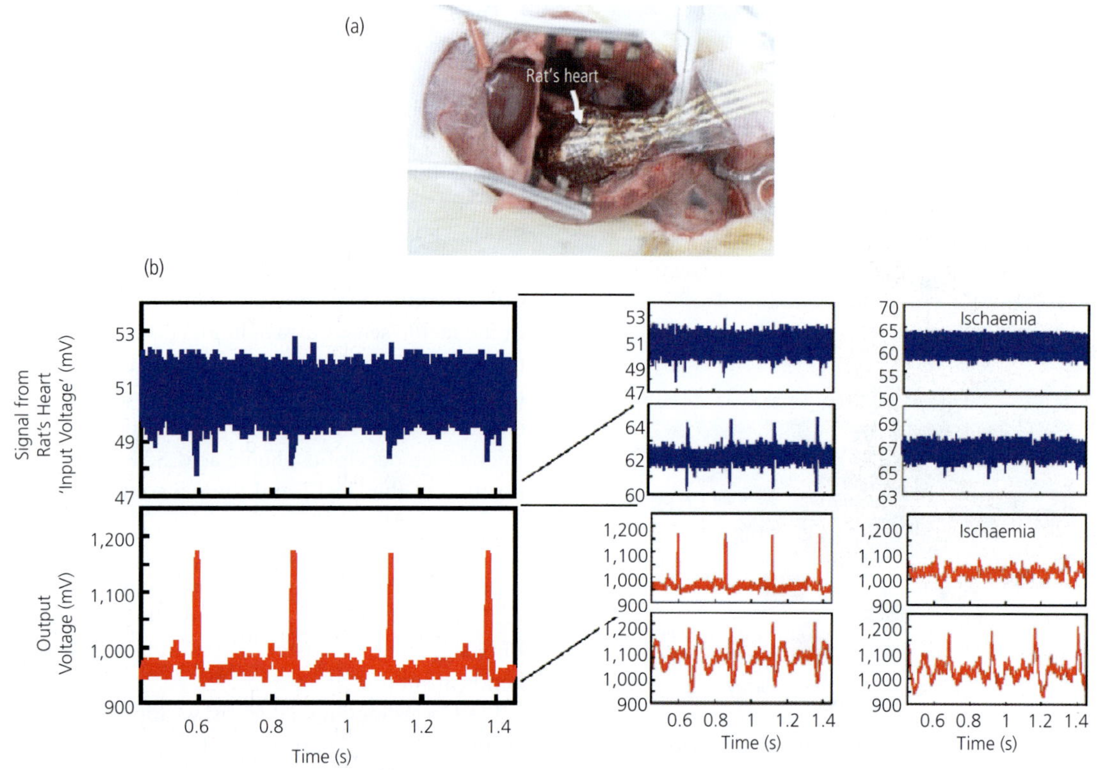

Figure 8.126 (a) Photograph of the electrocardio transducer attached to a rat's heart. (b) Electrocardiograms of the unamplified (blue) and amplified (red) heart impulses for different input conditions, including the ischemic state under myocardial infarction (Sekitani et al., 2016).

leads to stable OTFT performance up to 150°C (Kuribara et al., 2012).

One additional feature that can be added to make the electronic device precisely and softly conform to a surface is to mount the circuits on a polymer with shape memory. *Shape memory* is the property where a deformed material can return to its original 3D conformation once the stimulus that has shaped it is removed. Thus, the material can "remember" its original form. *Shape memory polymers* (SMPs) can therefore be used to conform to surfaces by heating above their glass transition temperatures, T_g, whence they are "pretrained" to fit into a particular shape. Reheating the material causes it to assume its original conformation (Liu et al., 2009b, Nair et al., 2010). For use with living organisms, the T_g of the SMP must be sufficiently low to prevent injury during deformation.

An example of a 25 μm thick thiolene/acrylate SMP on a parylene-coated DNTT OTFT on a 1.4 μm thick PEN substrate is shown in Fig. 8.127a. The system is initially formed into a 2.5 mm radius helix by post-cure annealing above T_g. It is then passed through a slot in a thermal barrier at $T > T_g = 70$°C. This causes it to reform into its original helical shape where it wraps itself around a cylinder of similar dimension (Reeder et al., 2014). The transistor characteristics are shown in Fig. 8.127b when the OTFT is in its original configuration, in its flattened state, and when the helix is recovered by heating. There is a small shift of V_T to more negative voltages with deformation, but no apparent change in μ_{FEp} as inferred from the similar slopes of the curves.

Conformation of the SMP OTFT circuit *in vivo* in Fig. 8.128 shows its biocompatibility and its possible utility for monitoring biological functions in real time. The device is implanted into the living tissue of a rat (Fig. 8.128a) where it is shaped due to a rise in temperature in its subcutaneous location. After 24 h (Fig. 8.128b) the device has conformed to the tissue shape. Once removed, it retains its initial, heat-induced conformation (Fig. 8.128c).

These several demonstrations, while still at early stages, point to the promise of utilizing imperceptibly thin electronic circuits for medical purposes. It is indeed remarkable that organic devices can meet many of the most rigorous demands of sterilization and biocompatibility that are pillars of modern medical practice. These versatile attributes, therefore, may lead to the opening up of the exceedingly broad applications space of biological and medical sensing that cannot readily be accessed using conventional, inorganic semiconductor-based electronics.

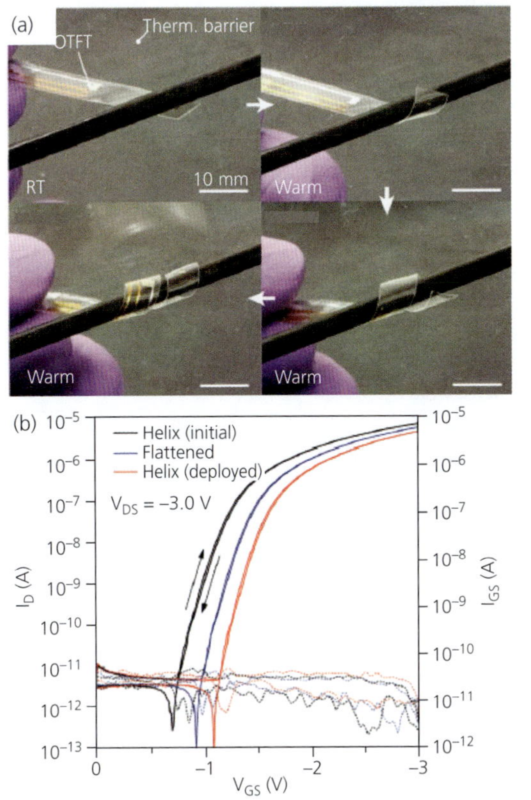

Figure 8.127 (a) Images of a SMP OTFT undergoing thermal relaxation (clockwise from top left) into a 2.5 mm radius helix when inserted through a thermal barrier. The far side of the barrier is at room temperature (RT). (b) Transistor transfer characteristics when the DNTT OTFT is in its original, flattened and deployed helical shapes (Reeder et al., 2014).

8.10 Summing up

In this chapter we described the remarkable advances made in OTFTs over the more than three decades since their first demonstrations. The operating voltages have dropped by an order of magnitude, and the mobilities have increased from $<0.01\ cm^2/V\ s$ until today, when they are routinely $>1\ cm^2/V\ s$. The OTFTs operate in the accumulation mode, which is distinct from conventional CMOS that operates in the depletion or inversion regimes. Materials for both n- and p-channel operation enable complementary logic with nearly identical characteristics for transistors based on these different carrier types. And we have seen that the channel as well as the gate dielectric layers can be deposited with monolayer control. Finally, many of the most important materials have shown good stability when operated in low humidity environments.

Similarly significant advances in device architecture have complemented advances in materials. Large area circuits with hundreds of devices for RFIDs, oscillators, and array backplanes employ OTFTs that are lithographically defined on substrates ranging from glass to ultrathin plastic. Split gate, dual gate OTFTs, and high performance vertical transistors have been used in a variety of circuit applications. Photosensitive transistors, or phototransistors, offer responsivities that are significantly higher than HJ photodiodes due to inherent transistor gain. Unfortunately, higher responsivities lead to a concomitant reduction in bandwidth.

As in all photonic and electronic devices, reliability during long term operation is essential. Considerable work, therefore, has been devoted to understanding how the operating characteristics change during extended use. Most of the work on device reliability has focused on drifts in V_T, I_D, and μ_{FE} that are accelerated under bias-stress conditions. The most pervasive

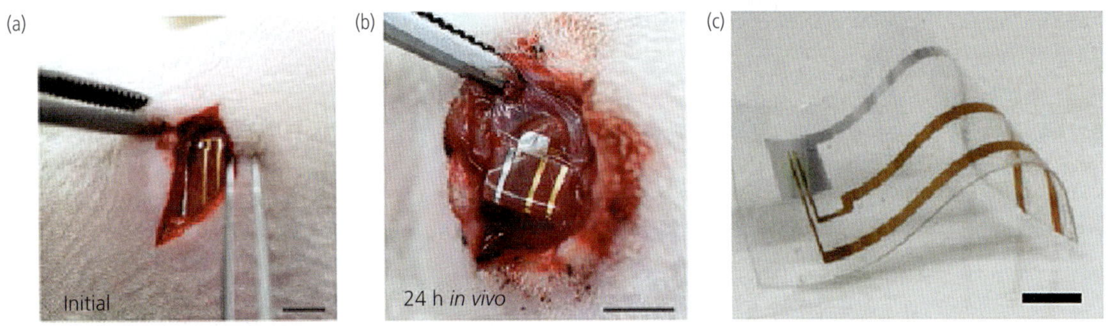

Figure 8.128 *In vivo* conformation of an SMP OTFT to the tissue of a rat. (a) As initially implanted, (b) after 24 h, and (c) after removal from the organism (Reeder et al., 2014).

contaminant leading to changes is moisture. The dipolar water molecule can invade the dielectric, attach itself to the dielectric/semiconductor interface, or bind to structural defects within the semiconductor itself. Water contamination, and contamination by other impurities effectively screens the electric field induced by the gate voltage, and hence shifts the V_T required to achieve a given charge concentration in the channel. As in the case of OPVs and OLEDs, encapsulation of the transistor circuits can provide an effective barrier to contamination, ultimately delaying degradation.

While there have been significant and impressive advances in OTFTs, there is as yet no large-scale application that has adopted this new transistor platform. Should such an application arise, it will exploit the unique features of OTFTs that are common to all organic electronic devices: flexibility, conformability, ultralight weight, and potentially low cost when deployed over large areas. We have shown that OTFTs provide unique capabilities in tactile sensing, identification of analytes through gas sensing, and as photodetector FPA pixel addressing circuits. Importantly, the ability to fabricate very high performance circuits on ultrathin substrates has enabled a new genus of biocompatible, "imperceptible" electronic appliances. Early demonstrations of their application to monitoring organ activity suggest that OTFT circuits may find unique opportunities in medical monitoring, imaging, entertainment, and robotics. It remains to be seen which of these applications will first emerge. But there is little doubt that the unique form factors and relatively high performance of OTFTs will enable numerous applications currently not adequately served by conventional inorganic semiconductor-based electronics.

Further reading

1. S. A. Di Benedetto, A. Facchetti, M. A. Ratner, and T. J. Marks, 2009. Molecular self-assembled monolayers and multilayers for organic and unconventional inorganic thin-film transistor applications. *Adv. Mater.*, 21, 1407.
2. C. R. Kagan and P. Andry (Eds.), 2003. *Thin-Film Transistors*. New York: Marcel Dekker, Inc.
3. H. Klauk, 2010. Organic thin-film transistors. *Chem. Soc. Rev.*, 39, 2643.
4. H. Klauk (Ed.), 2006. *Organic Electronics, Materials, Manufacturing and Applications*. Weinheim, Germany: Wiley-VCH Verlag.
5. D. J. Lipomi and Z. Bao, 2017. Stretchable and ultraflexible organic electronics. *MRS Bull.*, 42, 93.
6. R. F. Pierret, 1996. *Semiconductor Device Fundamentals*. Reading, MA: Addison Wesley Longman.
7. A. Rose, 1963. *Concepts in Photoconductivity and Allied Problems*. New York: Interscience.
8. B. G. Streetman and S. K. Banerjee, 2006. *Solid State Electronic Devices*. Upper Saddle River, NJ: Prentice Hall.
9. S. M. Sze, 1981. *Physics of Semiconductor Devices*. New York: Wiley-Interscience.
10. C. Wang, H. Dong, W. Hu, Y. Liu, and D. Zhu, 2011. Semiconducting π-conjugated systems in field-effect transistors: a material odyssey of organic electronics. *Chem. Rev.*, 112, 2208.
11. J. Zaumseil and H. Sirringhaus, 2007. Electron and ambipolar transport in organic field-effect transistors. *Chem. Rev.*, 107, 1296.

Problems

1. Write expressions for the high and low frequency capacitance-voltage characteristics for an n-channel metal/insulator/organic semiconductor device. Assume the dielectric constant of the insulator is ε_i and that of the organic are ε_S, work functions are ϕ_M and ϕ_S, respectively, a bulk charge concentration of N_D, and for a surface state density of:
   (a) $N_{ss} = 0$.
   (b) N_{ss} large, and negative.
   (c) Assuming the trapping rate is 10^{-4} s, plot the characteristics for $\phi_M = 4.5$ eV, $\phi_S = 4.8$ eV, $\varepsilon_i = 3.5$, $\varepsilon_S = 3.2$, $N_D = 10^{15}$ cm^{-3}, and $N_{SS} = 0$ at a capacitance measurement frequency of 20 kHz. Indicate V_{FB} on the plot.
   (d) As in (c) but $N_{SS} = -5 \times 10^{12}$ cm^{-2}.
2. Consider a p-channel OTFT with dimensions $L = 50$ μm, $W = 5$ mm, and $t_i = 50$ nm for the gate length, width, and insulator thickness, respectively. For $\mu_{FE} = 1$ cm^2/V s and a threshold voltage of $V_T = 2.5$ V using an SiO_2 gate insulator and $V_{GS} = 0$ to -40 V,
   (a) Plot the drain current vs. drain-source voltage characteristics of this device.
   (b) What is the background doping concentration of the channel?
   (c) What is the surface potential at $V_{GS} = 0$?
3. The equivalent circuit of a p-channel OTFT in Problem 2 is shown in Fig. P8.3. Here, v'_G is the gate voltage in the absence of external resistance, R_S, and i_D is the small signal drain current. Based on this equivalent circuit,

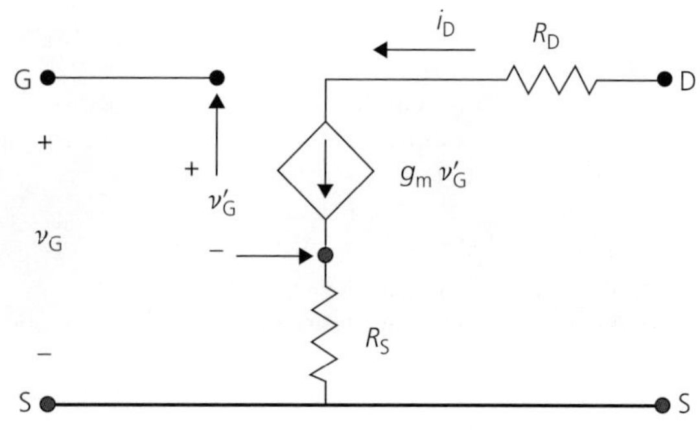

Figure P8.3 Equivalent circuit of an OTFT.

(a) Show that the measured transconductance and drain (or output) conductance are given by Eqs. 8.23 and 8.24, respectively.
(b) For the device in Fig. P8.3, show the drain current characteristics in the presence of $R_D = 5\ \Omega$ and $R_S = 1\ \Omega$.

4. Equation 8.48 shows the effective field effect mobility, μ_{FEeff}, in the presence of contact resistance derived in the linear regime. Derive an expression for μ_{FEeff} in the saturation regime.
5. Derive a general expression for the Johnson noise channel current for an OTFT in saturation. Show that to good approximation, the noise current is given by $S_{th} = i_{th}^2 = \frac{8}{3}(k_B T)g_m$.
6. In an ambipolar OTFT, the voltage ($V(x)$) varies along the channel from source (at $x = 0$) to drain (at $x = L$). Assuming that there is 100% recombination of electrons and holes at the point x_0 along the channel,
   (a) Derive an expression for the voltage $V(x)$ vs. x/L along the channel for a threshold voltage of $V_{Tn} = V_{Tp} = V_T = 0.1V_{GS}$ and $V_{DS}/V_{GS} = 0.5, 1, 5$.
   (b) Plot the normalized drain current $\left(I_D/\left(\mu_{FEp}WLC_i\right)\right)$ characteristics in the linear, saturation and ambipolar regimes of operation of an ambipolar OTFT for $\mu_{FEp} = 10\mu_{FEn}$ for $V_{DS}/V_{GS} = 0.5, 1, 5, 10$.
   (c) Derive an expression for the transconductance in an ambipolar OTFT in each of the three regimes of operation. Plot the normalized transconductance vs. V_{GS} for each of the voltage conditions in part (b).
7. Design a complementary inverter logic circuit comprising a p-channel organic phototransistor based on the P3HT transistor shown in Fig. 8.59.
   (a) Select an n-channel material from Sections 8.6.3 or 8.6.5 to form a reasonable complement to P3HT. Decide on the appropriate W/L and dielectric, and calculate and plot the resulting drain current characteristics.
   (b) In combination with the n-channel device, assume that the channel of the *p*-channel device is illuminated at energies slightly above its energy gap. Determine the logical output function for the light ON and OFF states at the input of the illuminated organic phototransistor.
   (c) Calculate and plot the transfer characteristics of the circuit as a function of the supply voltage, V_{DD}, the input bias voltage, V_{IN}, and the optical intensity illuminating the OPT.
8. In Chapter 6 we showed that OLEDs need to be driven at 10 mA/cm^2 to achieve very high brightness often required in displays and lighting. The question is whether or not OTFTs can serve as backplane pixel address transistors in displays. We will simplify this consideration by assuming only a single OTFT is required to address each pixel, although in fact it takes more than one transistor per pixel in practical applications to fully control the OLED brightness over the period of its useful life in a display.
   (a) Assuming that the pixel dimensions are (50 μm)^2 and that the OLED occupies as much area as the OTFT (e.g. it is layered onto the transistor surface), what is the current that can be delivered to an OLED in a common source configuration with $\mu_{FEp} = 1$ cm^2/V s assuming a power supply voltage of $V_{DD} = -10$ V and $V_{GS} = -5$ V? The SiO_2 gate dielectric has thickness, $t_i = 100$ nm and $L = 10$ μm.
   (b) Plot the drain current density vs. gate capacitance per area for the supply voltage in (a). What is the highest practical drain current density that can be achieved using OTFTs? Is it adequate for driving the OLEDs? Is there another thin film technology that would be more practical?
9. Figure 8.109b shows a polymer OTFT with hysteresis in its drain current characteristics. Assume that these characteristics are obtained at a voltage sweep rate of 500 mV/s. The dielectric thickness is 300 nm and the S and D contact thicknesses are also 300 nm.
   (a) Calculate the total trapped charge observed in these characteristics.
   (b) If the positive voltage sweep results in larger I_D than for the negative voltage sweep, what is the sign of the trapped charge?
   (c) What is the approximate electric field near the S and D contacts at $V_G = -40$ V? What does this imply about the position of the trap level in the organic energy gap (i.e. are they hole or electron traps)?
10. A detector with a quantum efficiency at low optical input powers ($P_{in} \approx 0$) of η_{EQE0}, and dark current density of I_{dark}, is connected to the input of the OTFT in Fig. 8.98, option **B**.
   (a) What is the total rms noise current vs. input voltage of a common source single stage amplifier for this circuit?
   (b) For an output conductance of the input OTFT of g_D and a diode series resistance of R_{ser}, what is the maximum linear input optical power range, assuming that the maximum linear power corresponds to a reduction in response to $\Delta\eta_{EQE}(P_{in})/\eta_{EQE0}$ of – 1dB?
   (c) What is the dynamic range of this circuit, assuming that the minimum detectable signal requires SNR = 2?
   (d) Assuming the device values in Table P8.10, what is the *DR*? How many bits of grey scale of an incident light beam can be obtained from this circuit?

Table P8.10 Circuit parameter values

Parameter	Value
I_{dark}	1 nA
g_D	1 μS
R_{ser}	100 Ω
$\Delta\eta_{EQE}$	1 dB

11. (a) Calculate the expected gate leakage for the transistor in Fig. 8.98, option **B** assuming Frenkel–Poole emission is its dominant source for an 3.6 nm thick AlO_x gate insulator coated by a 2.1 nm thick SAM (cf. Fig. 8.96).
   (b) How closely does this agree with the data shown from the complementary inverter pair, assuming both the drive and load transistors have the same W and L?

12. Figure 8.53b shows the transfer characteristics of n- and p-channel OTFTs used in a complementary inverter. The transistor dimensions are $W/L = 25$, $L = 40$ μm, $t_i = 5.7$ nm SAM/AlO_x gate insulator, and a unit-area capacitance of 700 nF/cm^2.
    (a) Calculate f_T for both transistors at $V_{GS} = \pm 0.5$ V.
    (b) Calculate f_T for both transistors at $V_{GS} = \pm 1$ V.
    (c) How does the calculated f_T compare with that obtained from the gate delay of 20 ms for a 5 stage ring oscillator? Explain the sources of the difference.
13. Using ambipolar compound **50** in Table 8.6:
    (a) Specify the W/L, L, and t_i of a complementary transistor inverter comprising a drive and load with $A_v = 25$. The gate dielectric is Al_2O_3.
    (b) Calculate the drain current and transfer characteristics of both the n- and p-channel transistors.
    (c) Calculate g_m and g_D of the transistors in (a) in both the linear and saturation regimes.
    (d) As in (a), design the physical characteristics of the inverter with a 3 dB roll-off frequency of 5 MHz.
    (e) Describe the fabrication sequence to make the transistor pair in (a).
14. For a complementary inverter using transistor pair in Fig. 8.53, graphically determine the noise margin by plotting V_{out} *vs.* V_{in} for the load and drive transistors.
15. We have shown in Section 8.9.3 tactile arrays based on transducers whose resistance is varied by the application of pressure. An alternative transducer that can be very sensitive to pressure changes is one based on capacitance.
    (a) Using the transistor whose characteristics are given in Fig. 8.23, design a capacitive pressure transducer pixel based on a PDMS dielectric. The Young's modulus is 2 MPa and the compressive modulus is 130 MPa. Make sure that you specify the pixel footprint (i.e. area) and the a.c. circuit used at each pixel.
    (b) What is the minimum deflection your pixel can detect at a measurement frequency of 1 MHz and an rms measurement signal amplitude of 10 meV?
    (c) What is the pressure that corresponds to the deflection in (b)?

References

Abe, Y., Hasegawa, T., Takahashi, Y., Yamada, T. & Tokura, Y. 2005. Control of threshold voltage in pentacene thin-film transistors using carrier doping at the charge-transfer interface with organic acceptors. *Appl. Phys. Lett.*, 87, 153506.

Acton, O., Ting, G., Ma, H. & Jen, A. K.-Y. 2008. Low-voltage high-performance C_{60} thin film transistors via low-surface-energy phosphonic acid monolayer/hafnium oxide hybrid dielectric. *Appl. Phys. Lett.*, 93, 311.

Ando, S., Murakami, R., Nishida, J.-i., Tada, H., Inoue, Y., Tokito, S. & Yamashita, Y. 2005. n-Type organic field-effect transistors with very high electron mobility based on thiazole oligomers with trifluoromethylphenyl groups. *J. Am. Chem. Soc.*, 127, 14996.

Anthony, J. E., Brooks, J. S., Eaton, D. L. & Parkin, S. R. 2001. Functionalized pentacene: improved electronic properties from control of solid-state order. *J. Am. Chem. Soc.*, 123, 9482.

Anthopoulos, T. D. 2007. Electro-optical circuits based on light-sensing ambipolar organic field-effect transistors. *Appl. Phys. Lett.*, 91, 113513.

Arai, T. & Sasaoka, T. 2011. Emergent oxide TFT technologies for next-generation AM-OLED displays. *SID Symp. Dig. Tech. Pap.*, 49, 710.

Babel, A. & Jenekhe, S. A. 2003. High electron mobility in ladder polymer field-effect transistors. *J. Am. Chem. Soc.*, 125, 13656.

Baeg, K.-J., Binda, M., Natali, D., Caironi, M. & Noh, Y.-Y. 2013. Organic light detectors: photodiodes and phototransistors. *Adv. Mater.*, 25, 4267.

Bailey-Salzman, R., Rand, B. P. & Forrest, S. R. 2006. Semi-transparent organic photovoltaic cells. *Appl. Phys. Lett.*, 88, 233502.

Baldo, M. A. & Forrest, S. R. 2001. Interface-limited injection in amorphous organic semiconductors. *Phys. Rev. B*, 64, 085201.

Barra, M., Girolamo, F. D., Chiarella, F., Salluzzo, M., Chen, Z., Facchetti, A., Anderson, L. & Cassinese, A. 2010. Transport property and charge trap comparison for N-channel perylene diimide transistors with different air-stability. *J. Phys. Chem. C*, 114, 20387.

Becerril, H. A., Roberts, M. E., Liu, Z., Locklin, J. & Bao, Z. 2008. High-performance organic thin-film transistors through solution-sheared deposition of small-molecule organic semiconductors. *Adv. Mater.*, 20, 2588.

Ben-Sasson, A. J., Avnon, E., Ploshnik, E., Globerman, O., Shenhar, R., Frey, G. L. & Tessler, N. 2009. Patterned electrode vertical field effect transistor fabricated using block copolymer nanotemplates. *Appl. Phys. Lett.*, 95, 302.

Ben-Sasson, A. J., Chen, Z., Facchetti, A. & Tessler, N. 2012. Solution-processed ambipolar vertical organic field effect transistor. *Appl. Phys. Lett.*, 100, 263306.

Benor, A. & Knipp, D. 2008. Contact effects in organic thin film transistors with printed electrodes. *Org. Electron.*, 9, 209.

Bigas, M., Cabruja, E., Forest, J. & Salvi, J. 2006. Review of CMOS image sensors. *Microelectron. J.*, 37, 433.

Bittle, E. G., Basham, J. I., Jackson, T. N., Jurchescu, O. D. & Gundlach, D. J. 2016. Mobility overestimation due to gated contacts in organic field-effect transistors. *Nat. Commun.*, 7, 10908.

Bobbert, P. A., Sharma, A., Mathijssen, S. G. J., Kemerink, M. & de Leeuw, D. M. 2012. Operational stability of organic field-effect transistors. *Adv. Mater.*, 24, 1146.

Brondijk, J., Spijkman, M., Torricelli, F., Blom, P. & De Leeuw, D. 2012. Charge transport in dual-gate organic field-effect transistors. *Appl. Phys. Lett.*, 100, 20.

Bürgi, L., Richards, T. J., Friend, R. H. & Sirringhaus, H. 2003. Close look at charge carrier injection in polymer field-effect transistors. *J. Appl. Phys.*, 94, 6129.

Bürgi, L., Sirringhaus, H. & Friend, R. 2002. Noncontact potentiometry of polymer field-effect transistors. *Appl. Phys. Lett.*, 80, 2913.

Cai, Q. J., Chan-Park, M. B., Lu, Z. S., Li, C. M. & Ong, B. S. 2008. Bottom-contact poly(3,3‴-didodecylquaterthiophene) thin-film transistors with gold source-drain electrodes modified by alkanethiol monolayers. *Langmuir*, 24, 11889.

Chang, J. B. & Subramanian, V. 2006. Effect of active layer thickness on bias stress effect in pentacene thin-film transistors. *Appl. Phys. Lett.*, 88, 233513.

Chen, C. Y. 1981. Theory of a modulated barrier photodiode. *Appl. Phys. Lett.*, 39, 979.

Chen, X. L., Lovinger, A. J., Bao, Z. & Sapjeta, J. 2001. Morphological and transistor studies of organic molecular semiconductors with anisotropic electrical characteristics. *Chem. Mater.*, 13, 1341.

Chen, Z., Lee, M. J., Shahid Ashraf, R., Gu, Y., Albert-Seifried, S., Meedom Nielsen, M., Schroeder, B., Anthopoulos, T. D., Heeney, M., McCulloch, I. & Sirringhaus, H. 2012. High-performance ambipolar diketopyrrolopyrrole-thieno[3, 2-b]thiophene copolymer field-effect transistors with balanced hole and electron mobilities. *Adv. Mater.*, 24, 647.

Chesterfield, R. J., Newman, C. R., Pappenfus, T. M., Ewbank, P. C., Haukaas, M. H., Mann, K. R., Miller, L. L. & Frisbie, C. D. 2003. High electron mobility and ambipolar transport in organic thin-film transistors based on a π-stacking quinoidal terthiophene. *Adv. Mater.*, 15, 1278.

Choi, H. H., Cho, K., Frisbie, C. D., Sirringhaus, H. & Podzorov, V. 2017. Critical assessment of charge mobility extraction in FETs. *Nat. Mater.*, 17, 2.

Chou, S. Y. & Antoniadis, D. A. 1987. Relationship between measured and intrinsic transconductances of FETs. *IEEE Trans. Electron. Dev. Lett.*, ED-34, 448.

Chou, W. Y. & Cheng, H. L. 2004. An orientation-controlled pentacene film aligned by photoaligned polyimide for organic thin-film transistor applications. *Adv. Funct. Mater.*, 14, 811.

Chua, L.-L., Zaumseil, J., Chang, J.-F., Ou, E. C.-W., Ho, P. K.-H., Sirringhaus, H. & Friend, R. H. 2005. General observation of n-type field-effect behaviour in organic semiconductors. *Nature*, 434, 194.

Comiskey, B., Albert, J., Yoshizawa, H. & Jacobson, J. 1998. An electrophoretic ink for all-printed reflective electronic displays. *Nature*, 394, 253.

Crone, B., Dodabalapur, A., Gelperin, A., Torsi, L., Katz, H., Lovinger, A. & Bao, Z. 2001. Electronic sensing of vapors with organic transistors. *Appl. Phys. Lett.*, 78, 2229.

Crone, B., Dodabalapur, A., Lin, Y.-Y., Filas, R., Bao, Z., LaDuca, A., Sarpeshkar, R., Katz, H. & Li, W. 2000. Large-scale complementary integrated circuits based on organic transistors. *Nature*, 403, 521.

Cui, T. & Liang, G. 2005. Dual-gate pentacene organic field-effect transistors based on a nanoassembled SiO_2 nanoparticle thin film as the gate dielectric layer. *Appl. Phys. Lett.*, 86, 064102.

De Boer, R., Iosad, N., Stassen, A., Klapwijk, T. & Morpurgo, A. 2005. Influence of the gate leakage current on the stability of organic single-crystal field-effect transistors. *Appl. Phys. Lett.*, 86, 032103.

De Vusser, S., Genoe, J. & Heremans, P. 2006. Influence of transistor parameters on the noise margin of organic digital circuits. *IEEE Trans. Electron Devices*, 53, 601.

Deane, S., Wehrspohn, R. & Powell, M. 1998. Unification of the time and temperature dependence of dangling-bond-defect creation and removal in amorphous-silicon thin-film transistors. *Phys. Rev. B*, 58, 12625.

Depp, S. W. & Howard, W. E. 1993. Flat-panel displays. *Sci. Am.*, 268, 90.

Diao, Y., Tee, B. C. K., Giri, G., Xu, J., Kim, D. H., Becerril, H. A., Stoltenberg, R. M., Lee, T. H., Xue, G., Mannsfeld, S. C. B. & Bao, Z. 2013. Solution coating of large-area organic semiconductor thin films with aligned single-crystalline domains. *Nat. Mater.*, 12, 665.

DiBenedetto, S. A., Facchetti, A., Ratner, M. A. & Marks, T. J. 2009. Molecular self-assembled monolayers and multilayers for organic and unconventional inorganic thin-film transistor applications. *Adv. Mater.*, 21, 1407.

Dong, H., Li, H., Wang, E., Wei, Z., Xu, W., Hu, W. & Yan, S. 2008. Ordering rigid rod conjugated polymer molecules for high performance photoswitchers. *Langmuir*, 24, 13241.

Donley, C. L., Zaumseil, J., Andreasen, J. W., Nielsen, M. M., Sirringhaus, H., Friend, R. H. & Kim, J.-S. 2005. Effects of packing structure on the optoelectronic and charge transport properties in poly(9,9-di-*n*-octylfluorene-alt-benzothiadiazole). *J. Am. Chem. Soc.*, 127, 12890.

Drolet, N., Morin, J. F., Leclerc, N., Wakim, S., Tao, Y. & Leclerc, M. 2005. 2,7-Carbazolenevinylene-based oligomer thin-film transistors: high mobility through structural ordering. *Adv. Funct. Mater.*, 15, 1671.

Facchetti, A. 2007. Semiconductors for organic transistors. *Mater. Today*, 10, 28.

Fan, J., Yuen, J. D., Wang, M., Seifter, J., Seo, J. H., Mohebbi, A. R., Zakhidov, D., Heeger, A. & Wudl, F. 2012. High-performance ambipolar transistors and inverters from an ultralow bandgap polymer. *Adv. Mater.*, 24, 2186.

Feng, L., Tang, W., Zhao, J., Yang, R., Hu, W., Li, Q., Wang, R. & Guo, X. 2016. Unencapsulated air-stable organic field effect transistor by all solution processes for low power vapor sensing. *Sci. Rep.*, 6, 20671.

Fischer, A., Scholz, R., Leo, K. & Lüssem, B. 2012. An all C_{60} vertical transistor for high frequency and high current density applications. *Appl. Phys. Lett.*, 101, 213303.

Fujimoto, S.-y., Nakayama, K.-i. & Yokoyama, M. 2005. Fabrication of a vertical-type organic transistor with a planar metal base. *Appl. Phys. Lett.*, 87, 133503.

Gao, J. H., Li, R. J., Li, L. Q., Meng, Q., Jiang, H., Li, H. X. & Hu, W. P. 2007. High-Performance field-effect transistor based on dibenzo[d,d′]thieno[3,2-b;4,5-b′]dithiophene, an easily synthesized semiconductor with high ionization potential. *Adv. Mater.*, 19, 3008.

Gao, P., Beckmann, D., Tsao, H. N., Feng, X., Enkelmann, V., Baumgarten, M., Pisula, W. & Müllen, K. 2009. Dithieno[2,3-d;2′,3′-d′]benzo[1,2-b;4,5-b′]dithiophene (DTBDT) as semiconductor for high-performance, solution-processed organic field-effect transistors. *Adv. Mater.*, 21, 213.

Garnier, F., Horowitz, G., Peng, X. & Fichou, D. 1990. An all-organic 'soft' thin film transistor with very high carrier mobility. *Adv. Mater.*, 2, 592.

Gelinck, G., Heremans, P., Nomoto, K. & Anthopoulos, T. D. 2010. Organic transistors in optical displays and microelectronic applications. *Adv. Mater.*, 22, 3778.

Gelinck, G., Huitema, H., Mil, M. v., Veenendaal, E. v., Lieshout, P., Touwslager, F., Patry, S., Sohn, S., Whitesides, T. & McCreary, M. 2006. A rollable, organic electrophoretic QVGA display with field-shielded pixel architecture. *J. Soc. Inf. Display*, 14, 113.

Gelinck, G. H., Veenendaal, E. v. & Coehoorn, R. 2005. Dual-gate organic thin-film transistors. *Appl. Phys. Lett.*, 87, 073508.

Gomes, H. L., Stallinga, P., Cölle, M., De Leeuw, D. & Biscarini, F. 2006. Electrical instabilities in organic semiconductors caused by trapped supercooled water. *Appl. Phys. Lett.*, 88, 082101.

Gonzalez, M., Axisa, F., Bulcke, M. V., Brosteaux, D., Vandevelde, B. & Vanfleteren, J. 2008. Design of metal interconnects for stretchable electronic circuits. *Microelectron. Reliability*, 48, 825.

Gray, P. R. & Meyer, R. G. 1984. *Analysis and Design of Analog Integrated Circuits*. New York: Wiley.

Gsänger, M., Oh, J. H., Könemann, M., Höffken, H. W., Krause, A.-M., Bao, Z. & Würthner, F. 2010. A crystal-engineered hydrogen-bonded octachloroperylene diimide with a twisted core: an n-channel organic semiconductor. *Angew. Chem.*, 122, 752.

Gundlach, D. J., Nichols, J. A., Zhou, L. & Jackson, T. N. 2002. Thin-film transistors based on well-ordered thermally evaporated naphthacene films. *Appl. Phys. Lett.*, 80, 2925.

Günther, A. A., Sawatzki, M., Formánek, P., Kasemann, D. & Leo, K. 2016. Contact doping for vertical organic field-effect transistors. *Adv. Funct. Mater.*, 26, 768.

Guo, Y., Du, C., Di, C.-a., Zheng, J., Sun, X., Wen, Y., Zhang, L., Wu, W., Yu, G. & Liu, Y. 2009. Field dependent and high light sensitive organic phototransistors based on linear asymmetric organic semiconductor. *Appl. Phys. Lett.*, 94, 143303.

Guo, Y., Du, C., Yu, G., Di, C. a., Jiang, S., Xi, H., Zheng, J., Yan, S., Yu, C. & Hu, W. 2010. High-performance phototransistors based on organic microribbons prepared by a solution self-assembly process. *Adv. Funct. Mater.*, 20, 1019.

Gutmann, F., Keyzer, H., Lyons, L. E. & Somoano, R. B. 1983. *Organic Semiconductors, Part B*. Malabar, FL: Robert E. Krieger.

Haddon, R. C. 1996. C_{70} thin film transistors. *J. Am. Chem. Soc.*, 118, 3041.

Hählen, T., Vanoni, C., Wäckerlin, C., Jung, T. A. & Tsujino, S. 2012. Surface doping in pentacene thin-film transistors with few monolayer thick channels. *Appl. Phys. Lett.*, 101, 033305.

Halik, M., Klauk, H., Zschieschang, U., Schmid, G., Dehm, C., Schütz, M., Maisch, S., Effenberger, F., Brunnbauer, M. & Stellacci, F. 2004. Low-voltage organic transistors with an amorphous molecular gate dielectric. *Nature*, 431, 963.

Halik, M., Klauk, H., Zschieschang, U., Schmid, G., Ponomarenko, S., Kirchmeyer, S. & Weber, W. 2003. Relationship between molecular structure and electrical performance of oligothiophene organic thin film transistors. *Adv. Mater.*, 15, 917.

Hallam, T., Lee, M., Zhao, N., Nandhakumar, I., Kemerink, M., Heeney, M., McCulloch, I. & Sirringhaus, H. 2009. Local charge trapping in conjugated polymers resolved by scanning Kelvin probe microscopy. *Phys. Rev. Lett.*, 103, 256803.

Hammock, M. L., Chortos, A., Tee, B. C. K., Tok, J. B. H. & Bao, Z. 2013. The evolution of electronic skin (e-skin): a brief history, design considerations, and recent progress. *Adv. Mater.*, 25, 5997.

Han, S. H., Kim, J. H., Jang, J., Cho, S. M., Oh, M. H., Lee, S. H. & Choo, D. J. 2006. Lifetime of organic thin-film transistors with organic passivation layers. *Appl. Phys. Lett.*, 88, 073519.

Hauser, J. R. 1993. Noise margin criteria for digital logic circuits. *IEEE Trans. Educ.*, 36, 363.

Hein, M. P., Zakhidov, A. A., Lüssem, B., Jankowski, J., Tietze, M. L., Riede, M. K. & Leo, K. 2014. Molecular doping for control of gate bias stress in organic thin film transistors. *Appl. Phys. Lett.*, 104, 013507.

Hertel, D. & Bässler, H. 2008. Photoconduction in amorphous organic solids. *ChemPhysChem*, 9, 666.

Hong, S., Choi, J. & Kim, Y. 2008. Degraded OFF-state current of organic thin-film transistor and annealing effect. *IEEE Trans. Electron Devices*, 55, 3602.

Hong, W., Sun, B., Aziz, H., Park, W.-T., Noh, Y.-Y. & Li, Y. 2012. A conjugated polyazine containing diketopyrrolopyrrole for ambipolar organic thin film transistors. *Chem. Commun.*, 48, 8413.

Hooge, F. N., Kleinpenning, T. G. M. & Vandamme, L. K. J. 1981. Experimental studies on 1/f noise. *Rep. Prog. Phys.*, 44, 479.

Hoppe, A., Knipp, D., Gburek, B., Benor, A., Marinkovic, M. & Wagner, V. 2010. Scaling limits of organic thin film transistors. *Org. Electron.*, 11, 626.

Horowitz, G., Fichou, D., Peng, X., Xu, Z. & Garnier, F. 1989. A field-effect transistor based on conjugated alpha-sexithienyl. *Solid State Commun.*, 72, 381.

Hsu, B. B. Y., Namdas, E. B., Yuen, J. D., Cho, S., Samuel, I. D. W. & Heeger, A. J. 2010. Split-gate organic field effect transistors: control over charge injection and transport. *Adv. Mater.*, 22, 4649.

Hu, W. S., Tao, Y. T., Hsu, Y. J., Wei, D. H. & Wu, Y. S. 2005. Molecular orientation of evaporated pentacene films on gold: alignment effect of self-assembled monolayer. *Langmuir*, 21, 2260.

Huang, K., Dinyari, R., Lanzara, G., Kim, J. Y., Feng, J., Vancura, C., Chang, F. K. & Peumans, P. An approach to cost-effective, robust, large-area electronics using monolithic silicon. 2007 IEEE International Electron Devices Meeting, 10–12 Dec 2007, 217.

Ichimura, K. 2000. Photoalignment of liquid-crystal systems. *Chem. Rev.*, 100, 1847.

Ishihara, S., Wakemoto, H., Nakazima, K. & Matsuo, Y. 1989. The effect of rubbed polymer films on the liquid crystal alignment. *Liquid Cryst.*, 4, 669.

Israelachvili, J. N. 2011. *Intermolecular and Surface Forces*. Waltham, MA: Academic Press.

Izawa, T., Miyazaki, E. & Takimiya, K. 2008. Molecular ordering of high-performance soluble molecular semiconductors and re-evaluation of their field-effect transistor characteristics. *Adv. Mater.*, 20, 3388.

Izuhara, D. & Swager, T. M. 2009. Poly(pyridinium phenylene)s: water-soluble n-type polymers. *J. Am. Chem. Soc.*, 131, 17724.

Jackson, W. B., Marshall, J. M. & Moyer, M. D. 1989. Role of hydrogen in the formation of metastable defects in hydrogenated amorphous silicon. *Phys. Rev. B*, 39, 1164.

Jang, J., Nam, S., Im, K., Hur, J., Cha, S. N., Kim, J., Son, H. B., Suh, H., Loth, M. A. & Anthony, J. E. 2012. Highly crystalline soluble acene crystal arrays for organic transistors: mechanism of crystal growth during dip-coating. *Adv. Funct. Mater.*, 22, 1005.

Janssen, D., De Palma, R., Verlaak, S., Heremans, P. & Dehaen, W. 2006. Static solvent contact angle measurements, surface free energy and wettability determination of various self-assembled monolayers on silicon dioxide. *Thin Solid Films*, 515, 1433.

Jia, X., Fuentes-Hernandez, C., Wang, C.-Y., Park, Y. & Kippelen, B. 2018. Stable organic thin-film transistors. *Sci. Adv.*, 4, eaao1705.

Jurchescu, O. D., Hamadani, B. H., Xiong, H. D., Park, S. K., Subramanian, S., Zimmerman, N. M., Anthony, J. E., Jackson, T. N. & Gundlach, D. J. 2008. Correlation between microstructure, electronic properties and flicker noise in organic thin film transistors. *Appl. Phys. Lett.*, 92, 132103.

Kaltenbrunner, M., Sekitani, T., Reeder, J., Yokota, T., Kuribara, K., Tokuhara, T., Drack, M., Schwödiauer, R., Graz, I., Bauer-Gogonea, S., Bauer, S. & Someya, T. 2013. An ultra-lightweight design for imperceptible plastic electronics. *Nature*, 499, 458.

Kaltenbrunner, M., White, M. S., Głowacki, E. D., Sekitani, T., Someya, T., Sariciftci, N. S. & Bauer, S. 2012. Ultrathin and lightweight organic solar cells with high flexibility. *Nat. Commun.*, 3, 770.

Kang, S., Yi, Y., Kim, C., Cho, K., Seo, J., Noh, M., Jeong, K., Yoo, K.-H. & Whang, C. 2005. Ambipolar organic thin-film transistors using C_{60}/pentacene structure: characterization of electronic structure and device property. *Appl. Phys. Lett.*, 87, 233502.

Kanimozhi, C., Yaacobi-Gross, N., Chou, K. W., Amassian, A., Anthopoulos, T. D. & Patil, S. 2012. Diketopyrrolopyrrole–diketopyrrolopyrrole-based conjugated copolymer for high-mobility organic field-effect transistors. *J. Am. Chem. Soc.*, 134, 16532.

Kanno, M., Bando, Y., Shirahata, T., Inoue, J.-i., Wada, H. & Mori, T. 2009. Stabilization of organic field-effect transistors in hexamethylenetetrathiafulvalene derivatives substituted by bulky alkyl groups. *J. Mater. Chem.*, 19, 6548.

Kelley, T. W., Boardman, L. D., Dunbar, T. D., Muyres, D. V., Pellerite, M. J. & Smith, T. P. 2003a. High-performance OTFTs using surface-modified alumina dielectrics. *J. Phys. Chem. B*, 107, 5877.

Kelley, T. W., Muyres, D. V., Baude, P. F., Smith, T. P. & Jones, T. D. 2003b. High performance organic thin film transistors. *MRS Online Proceedings Library Archive*, 771.

Kemeny, G. & Rosenberg, B. 1970. Theory of the pre-exponential factor in organic semiconductors. *J. Chem. Phys.*, 52, 4151.

Kim, C., Burrows, P. E. & Forrest, S. R. 2000. Micropatterning of organic electronic devices by cold-welding. *Science*, 288, 831.

Kim, C. & Forrest, S. R. 2003. Fabrication of organic light-emitting devices by low pressure cold welding. *Adv. Mater.*, 15, 541.

Kim, D. H., Lee, D. Y., Lee, H. S., Lee, W. H., Kim, Y. H., Han, J. I. & Cho, K. 2007. High-mobility organic transistors based on single-crystalline microribbons of triisopropylsilylethynyl pentacene via solution-phase self-assembly. *Adv. Mater.*, 19, 678.

Kitamura, M., Aomori, S., Na, J. H. & Arakawa, Y. 2008. Bottom-contact fullerene C_{60} thin-film transistors with high field-effect mobilities. *Appl. Phys. Lett.*, 93, 033313.

Kitamura, M. & Arakawa, Y. 2009. Current-gain cutoff frequencies above 10 MHz for organic thin-film transistors with high mobility and low parastic capacitance. *Appl. Phys. Lett.*, 95, 023503.

Klauk, H. (Ed.) 2006. *Organic Electronics: Materials, Manufacturing and Applications*. Berlin: Wiley-VCH Verlag.

Klauk, H. 2010. Organic thin-film transistors. *Chem. Soc. Rev.*, 39, 2643.

Klauk, H. 2018. Will we see gigahertz organic transistors? *Adv. Electron. Mater.*, 4, 1700474.

Klauk, H., Halik, M., Zschieschang, U., Schmid, G., Radlik, W. & Weber, W. 2002. High-mobility polymer gate dielectric pentacene thin film transistors. *J. Appl. Phys.*, 92, 5259.

Klauk, H., Zschieschang, U., Weitz, R. T., Meng, H., Sun, F., Nunes, G., Keys, D. E., Fincher, C. R. & Xiang, Z. 2007. Organic transistors based on di(phenylvinyl)anthracene: performance and stability. *Adv. Mater.*, 19, 3882.

Kline, R. J., McGehee, M. D., Kadnikova, E. N., Liu, J. & Fréchet, J. M. J. 2003. Controlling the field-effect mobility of regioregular polythiophene by changing the molecular weight. *Adv. Mater.*, 15, 1519.

Klinger, M. P., Fischer, A., Kaschura, F., Widmer, J., Kheradmand-Boroujeni, B., Ellinger, F. & Leo, K. 2017. Organic power electronics: transistor operation in the kA/cm^2 regime. *Sci. Rep.*, 7, 44713.

Ko, H. C., Stoykovich, M. P., Song, J., Malyarchuk, V., Choi, W. M., Yu, C.-J., Geddes Iii, J. B., Xiao, J., Wang, S., Huang, Y. & Rogers, J. A. 2008. A hemispherical electronic eye camera based on compressible silicon optoelectronics. *Nature*, 454, 748.

Kobayashi, N., Sasaki, M. & Nomoto, K. 2009. Stable peri-xanthenoxanthene thin-film transistors with efficient carrier injection. *Chem. Mater.*, 21, 552.

Kobayashi, S., Nishikawa, T., Takenobu, T., Mori, S., Shimoda, T., Mitani, T., Shimotani, H., Yoshimoto, N., Ogawa, S. & Iwasa, Y. 2004. Control of carrier density by self-assembled monolayers in organic field-effect transistors. *Nat. Mater.*, 3, 317.

Kudo, K., Xing Wang, D., Iizuka, M., Kuniyoshi, S. & Tanaka, K. 1998. Schottky gate static induction transistor using copper phthalocyanine films. *Thin Solid Films*, 331, 51.

Kudo, K., Yamashina, M. & Moriizumi, T. 1984. Field effect measurement of organic dye films. *Jpn. J. Appl. Phys.*, 23, 130.

Kumaki, D., Ando, S., Shimono, S., Yamashita, Y., Umeda, T. & Tokito, S. 2007. Significant improvement of electron mobility in organic thin-film transistors based on thiazolothiazole derivative by employing self-assembled monolayer. *Appl. Phys. Lett.*, 90, 053506.

Kumaki, D., Umeda, T., Suzuki, T. & Tokito, S. 2008. High-mobility bottom-contact thin-film transistor based on anthracene oligomer. *Org. Electron.*, 9, 921.

Kuribara, K., Wang, H., Uchiyama, N., Fukuda, K., Yokota, T., Zschieschang, U., Jaye, C., Fischer, D., Klauk, H., Yamamoto, T., Takimiya, K., Ikeda, M., Kuwabara, H., Sekitani, T., Loo, Y.-L. & Someya, T. 2012. Organic transistors with high thermal stability for medical applications. *Nat. Commun.*, 3, 723.

Kuwahara, E., Kusai, H., Nagano, T., Takayanagi, T. & Kubozono, Y. 2005. Fabrication of a logic gate circuit based on ambipolar field-effect transistors with thin films of C_{60} and pentacene. *Chem. Phys. Lett.*, 413, 379.

Labram, J. G., Wöbkenberg, P. H., Bradley, D. D. C. & Anthopoulos, T. D. 2010. Low-voltage ambipolar phototransistors based on a pentacene/PC61BM heterostructure and a self-assembled nano-dielectric. *Org. Electron.*, 11, 1250.

Lassiter, B. E., Lunt, R. R., Renshaw, C. K. & Forrest, S. R. 2010. Structural templating of multiple polycrystalline layers in organic photovoltaic cells. *Opt. Express*, 18, A444.

Lee, C.-T. & Chen, H.-C. 2011. Performance improvement mechanisms of organic thin-film transistors using MoO_x-doped pentacene as channel layer. *Org. Electron.*, 12, 1852.

Lee, J., Han, A.-R., Kim, J., Kim, Y., Oh, J. H. & Yang, C. 2012a. Solution-processable ambipolar diketopyrrolopyrrole–selenophene polymer with unprecedentedly high hole and electron mobilities. *J. Am. Chem. Soc.*, 134, 20713.

Lee, J., Han, A.-R., Yu, H., Shin, T. J., Yang, C. & Oh, J. H. 2013. Boosting the ambipolar performance of solution-processable polymer semiconductors via hybrid side-chain engineering. *J. Am. Chem. Soc.*, 135, 9540.

Lee, K., Weis, M., Manaka, T. & Iwamoto, M. 2012b. Conservation of the injection and transit time ratio in organic field-effect transistors: a thermally accelerated aging study. *J. Appl. Phys.*, 111, 104505.

Lee, P., Lee, J., Lee, H., Yeo, J., Hong, S., Nam, K. H., Lee, D., Lee, S. S. & Ko, S. H. 2012c. Highly stretchable and highly conductive metal electrode by very long metal nanowire percolation network. *Adv. Mater.*, 24, 3326.

Lee, S. S., Kim, C. S., Gomez, E. D., Purushothaman, B., Toney, M. F., Wang, C., Hexemer, A., Anthony, J. E. & Loo, Y.-L. 2009. Controlling nucleation and crystallization in solution-processed organic semiconductors for thin-film transistors. *Adv. Mater.*, 21, 3605.

Lei, T., Cao, Y., Fan, Y., Liu, C.-J., Yuan, S.-C. & Pei, J. 2011. High-performance air-stable organic field-effect transistors: isoindigo-based conjugated polymers. *J. Am. Chem. Soc.*, 133, 6099.

Li, L., Tang, Q., Li, H. & Hu, W. 2008. Molecular orientation and interface compatibility for high performance organic thin film transistor based on vanadyl phthalocyanine. *J. Phys. Chem. B*, 112, 10405.

Li, L., Tang, Q., Li, H., Yang, X., Hu, W., Song, Y., Shuai, Z., Xu, W., Liu, Y. & Zhu, D. 2007. An ultra closely π-stacked organic semiconductor for high performance field-effect transistors. *Adv. Mater.*, 19, 2613.

Li, S. P., Newsome, C. J., Russell, D. M., Kugler, T., Ishida, M. & Shimoda, T. 2005. Friction transfer deposition of ordered conjugated polymer nanowires and transistor fabrication. *Appl. Phys. Lett.*, 87, 062101.

Lilienfeld, J. E. 1933. *Device for controlling electric current.* US Pat. 1,900,018.

Lim, E., Jung, B.-J., Chikamatsu, M., Azumi, R., Yoshida, Y., Yase, K., Do, L.-M. & Shim, H.-K. 2007. Doping effect of solution-processed thin-film transistors based on polyfluorene. *J. Mater. Chem.*, 17, 1416.

Lim, J. A., Lee, W. H., Lee, H. S., Lee, J. H., Park, Y. D. & Cho, K. 2008. Self-organization of ink-jet-printed triisopropylsilylethynyl pentacene via evaporation-induced flows in a drying droplet. *Adv. Funct. Mater.*, 18, 229.

Lin, Y.-Y., Gundlach, D., Nelson, S. & Jackson, T. 1997a. Stacked pentacene layer organic thin-film transistors with improved characteristics. *IEEE Electron Device Lett.*, 18, 606.

Lin, Y.-Y., Gundlach, D. J., Nelson, S. F. & Jackson, T. N. 1997b. Pentacene-based organic thin film transistors. *IEEE Trans. Electron. Dev. Lett.*, 44, 1325.

Lipomi, D. J. & Bao, Z. 2017. Stretchable and ultraflexible organic electronics. *MRS Bull.*, 42, 93.

Liu, S., Wang, W. M., Briseno, A. L., Mannsfeld, S. C. B. & Bao, Z. 2009a. Controlled deposition of crystalline organic semiconductors for field-effect-transistor applications. *Adv. Mater.*, 21, 1217.

Liu, Y., Lv, H., Lan, X., Leng, J. & Du, S. 2009b. Review of electro-active shape-memory polymer composite. *Compos. Sci. Technol.*, 69, 2064.

Llorente, G. R., Dufourg-Madec, M.-B., Crouch, D. J., Pritchard, R. G., Ogier, S. & Yeates, S. G. 2009. High performance, acene-based organic thin film transistors. *Chem. Commun.*, 3059.

Lüssem, B., Tietze, M. L., Kleemann, H., Hoßbach, C., Bartha, J. W., Zakhidov, A. & Leo, K. 2013. Doped organic transistors operating in the inversion and depletion regime. *Nat. Commun.*, 4, 2775.

Lüssem, B. r., Keum, C.-M., Kasemann, D., Naab, B., Bao, Z. & Leo, K. 2016. Doped organic transistors. *Chem. Rev.*, 116, 13714.

Ma, H., Acton, O., Hutchins, D. O., Cernetic, N. & Jen, A. K.-Y. 2012. Multifunctional phosphonic acid self-assembled monolayers on metal oxides as dielectrics, interface modification layers and semiconductors for low-voltage high-performance organic field-effect transistors. *Physi. Chem. Chem. Phys.*, 14, 14110.

Ma, L. & Yang, Y. 2004. Unique architecture and concept for high-performance organic transistors. *Appl. Phys. Lett.*, 85, 5084.

Mabeck, J. T. & Malliaras, G. G. 2006. Chemical and biological sensors based on organic thin-film transistors. *Anal. Bioanal. Chem.*, 384, 343.

Maddalena, F., Spijkman, M., Brondijk, J., Fonteijn, P., Brouwer, F., Hummelen, J., De Leeuw, D., Blom, P. & De Boer, B. 2008. Device characteristics of polymer dual-gate field-effect transistors. *Org. Electron.*, 9, 839.

Marinelli, F., Dell'Aquila, A., Torsi, L., Tey, J., Suranna, G. P., Mastrorilli, P., Romanazzi, G., Nobile, C. F., Mhaisalkar, S. G., Cioffi, N. & Palmisano, F. 2009. An organic field effect transistor as a selective NO_x sensor operated at room temperature. *Sens. Actuators B: Chem.*, 140, 445.

Marinov, O. & Deen, M. J. 2015. Low-frequency noise in organic transistors. 2015 International Conference on Noise and Fluctuations, 1.

Marjanović, N., Singh, T. B., Dennler, G., Günes, S., Neugebauer, H., Sariciftci, N. S., Schwödiauer, R. & Bauer, S. 2006. Photoresponse of organic field-effect transistors based on conjugated polymer/fullerene blends. *Org. Electron.*, 7, 188.

Marmont, P., Battaglini, N., Lang, P., Horowitz, G., Hwang, J., Kahn, A., Amato, C. & Calas, P. 2008. Improving charge injection in organic thin-film transistors with thiol-based self-assembled monolayers. *Org. Electron.*, 9, 419.

Martin, Y., Abraham, D. W. & Wickramasinghe, H. K. 1988. High-resolution capacitance measurement and potentiometry by force microscopy. *Appl. Phys. Lett.*, 52, 1103.

Mas-Torrent, M., Masirek, S., Hadley, P., Crivillers, N., Oxtoby, N. S., Reuter, P., Veciana, J., Rovira, C. & Tracz, A. 2008. Organic field-effect transistors (OFETs) of highly oriented films of dithiophene-tetrathiafulvalene prepared by zone casting. *Org. Electron.*, 9, 143.

Mathijssen, S. G., Cölle, M., Gomes, H., Smits, E. C., de Boer, B., McCulloch, I., Bobbert, P. A. & de Leeuw, D. M. 2007. Dynamics of threshold voltage shifts in organic and amorphous silicon field-effect transistors. *Adv. Mater.*, 19, 2785.

Mathijssen, S. G., Spijkman, M. J., Andringa, A. M., van Hal, P. A., McCulloch, I., Kemerink, M., Janssen, R. A. & de Leeuw, D. M. 2010. Revealing buried interfaces to understand the origins of threshold voltage shifts in organic field-effect transistors. *Adv. Mater.*, 22, 5105.

Meijer, E., De Leeuw, D., Setayesh, S., Van Veenendaal, E., Huisman, B.-H., Blom, P., Hummelen, J., Scherf, U. & Klapwijk, T. 2003. Solution-processed ambipolar organic field-effect transistors and inverters. *Nat. Mater.*, 2, 678.

Metselaar, R. & Oversluizen, G. 1984. The Meyer–Nedel rule in semiconductors. *J. Solid State Chem.*, 55, 320.

Minari, T., Miyadera, T., Tsukagoshi, K., Aoyagi, Y. & Ito, H. 2007. Charge injection process in organic field-effect transistors. *Appl. Phys. Lett.*, 91, 053508.

Miskiewicz, P., Mas-Torrent, M., Jung, J., Kotarba, S., Glowacki, I., Gomar-Nadal, E., Amabilino, D. B., Veciana, J., Krause, B., Carbone, D., Rovira, C. & Ulanski, J. 2006. Efficient high area OFETs by solution based processing of a π-electron rich donor. *Chem. Mater.*, 18, 4724.

Morana, M., Bret, G. & Brabec, C. 2005. Double-gate organic field-effect transistor. *Appl. Phys. Lett.*, 87, 153511.

Murphy, A. R. & Fréchet, J. M. 2007. Organic semiconducting oligomers for use in thin film transistors. *Chem. Rev.*, 107, 1066.

Myny, K., Beenhakkers, M. J., van Aerle, N. A., Gelinck, G. H., Genoe, J., Dehaene, W. & Heremans, P. 2011. Unipolar organic transistor circuits made robust by dual-gate technology. *IEEE J. Solid-State Circuits*, 46, 1223.

Nagamatsu, S., Takashima, W., Kaneto, K., Yoshida, Y., Tanigaki, N. & Yase, K. 2004. Polymer field-effect transistors by a drawing method. *Appl. Phys. Lett.*, 84, 4608.

Nagamatsu, S., Takashima, W., Kaneto, K., Yoshida, Y., Tanigaki, N., Yase, K. & Omote, K. 2003. Backbone arrangement in 'friction-transferred' regioregular poly(3-alkylthiophene)s. *Macromolecules*, 36, 5252.

Nair, D. P., Cramer, N. B., Scott, T. F., Bowman, C. N. & Shandas, R. 2010. Photopolymerized thiol-ene systems as shape memory polymers. *Polymer*, 51, 4383.

Naraso, Nishida, J.-i., Kumaki, D., Tokito, S. & Yamashita, Y. 2006. High performance n- and p-type field-effect transistors based on tetrathiafulvalene derivatives. *J. Am. Chem. Soc.*, 128, 9598.

Newman, C. R., Frisbie, C. D., da Silva Filho, D. A., Brédas, J.-L., Ewbank, P. C. & Mann, K. R. 2004. Introduction to organic thin film transistors and design of n-channel organic semiconductors. *Chem. Mater.*, 16, 4436.

Noda, M., Kobayashi, N., Katsuhara, M., Yumoto, A., Ushikura, S., Yasuda, R., Hirai, N., Yukawa, G., Yagi, I., Nomoto, K. & Urabe, T. 2011. An OTFT-driven rollable OLED display. *J. Soc. Inf. Display*, 19, 316.

Noguchi, Y., Sekitani, T. & Someya, T. 2006. Organic-transistor-based flexible pressure sensors using ink-jet-printed electrodes and gate dielectric layers. *Appl. Phys. Lett.*, 89, 253507.

Noh, Y.-Y., Kim, D.-Y., Yoshida, Y., Yase, K., Jung, B.-J., Lim, E. & Shim, H.-K. 2005. High-photosensitivity p-channel organic phototransistors based on a biphenyl end-capped fused bithiophene oligomer. *Appl. Phys. Lett.*, 86, 043501.

Noh, Y. H., Young Park, S., Seo, S.-M. & Lee, H. H. 2006. Root cause of hysteresis in organic thin film transistor with polymer dielectric. *Org. Electron.*, 7, 271.

Nonnenmacher, M., o'Boyle, M. & Wickramasinghe, H. K. 1991. Kelvin probe force microscopy. *Appl. Phys. Lett.*, 58, 2921.

Novikov, S. V., Dunlap, D. H., Kenkre, V. M., Parris, P. E. & Vannikov, A. V. 1998. Essential role of correlations in governing charge transport in disordered organic materials. *Phys. Rev. Lett.*, 81, 4472.

Okamoto, H., Kawasaki, N., Kaji, Y., Kubozono, Y., Fujiwara, A. & Yamaji, M. 2008. Air-assisted high-performance field-effect transistor with thin films of picene. *J. Am. Chem. Soc.*, 130, 10470.

Olthof, S., Singh, S., Mohapatra, S. K., Barlow, S., Marder, S. R., Kippelen, B. & Kahn, A. 2012. Passivation of trap states in unpurified and purified C_{60} and the influence on organic field-effect transistor performance. *Appl. Phys. Lett.*, 101, 253303.

Ong, B. S., Wu, Y., Liu, P. & Gardner, S. 2005. Structurally ordered polythiophene nanoparticles for high-performance organic thin-film transistors. *Adv. Mater.*, 17, 1141.

Osaka, I., Abe, T., Shinamura, S., Miyazaki, E. & Takimiya, K. 2010. High-mobility semiconducting naphthodithiophene copolymers. *J. Am. Chem. Soc.*, 132, 5000.

Pal, T., Arif, M. & Khondaker, S. I. 2010. High performance organic phototransistor based on regioregular poly(3-hexylthiophene). *Nanotechnology*, 21, 325201.

Park, S., Heo, S. W., Lee, W., Inoue, D., Jiang, Z., Yu, K., Jinno, H., Hashizume, D., Sekino, M. & Yokota, T. 2018. Self-powered ultra-flexible electronics via nano-grating-patterned organic photovoltaics. *Nature*, 561, 516.

Park, S. K., Jackson, T. N., Anthony, J. E. & Mourey, D. A. 2007. High mobility solution processed 6,13-bis (triisopropyl-silylethynyl)pentacene organic thin film transistors. *Appl. Phys. Lett.*, 91, 063514.

Pisula, W., Menon, A., Stepputat, M., Lieberwirth, I., Kolb, U., Tracz, A., Sirringhaus, H., Pakula, T. & Müllen, K. 2005. A zone-casting technique for device fabrication of field-effect transistors based on discotic hexa-peri-hexabenzocoronene. *Adv. Mater.*, 17, 684.

Reeder, J., Kaltenbrunner, M., Ware, T., Arreaga-Salas, D., Avendano-Bolivar, A., Yokota, T., Inoue, Y., Sekino, M., Voit, W., Sekitani, T. & Someya, T. 2014. Mechanically adaptive organic transistors for implantable electronics. *Adv. Mater.*, 26, 4967.

Reese, C. & Bao, Z. 2007. Organic single-crystal field-effect transistors. *Mater. Today*, 10, 20.

Ren, X., Chan, P. K., Lu, J., Huang, B. & Leung, D. C. 2013. High dynamic range organic temperature sensor. *Adv. Mater.*, 25, 1291.

Ren, X., Pei, K., Peng, B., Zhang, Z., Wang, Z., Wang, X. & Chan, P. K. 2016. A low-operating-power and flexible active-matrix organic-transistor temperature-sensor array. *Adv. Mater.*, 28, 4832.

Renshaw, C. K., Xu, X. & Forrest, S. R. 2010. A monolithically integrated organic photodetector and thin film transistor. *Org. Electron.*, 11, 175.

Richards, T., Bird, M. & Sirringhaus, H. 2008. A quantitative analytical model for static dipolar disorder broadening of the density of states at organic heterointerfaces. *J. Chem. Phys.*, 128, 234905.

Richards, T. & Sirringhaus, H. 2008. Bias-stress induced contact and channel degradation in staggered and coplanar organic field-effect transistors. *Appl. Phys. Lett.*, 92, 023512.

Roberts, G. G. 1971. Thermally assisted tunnelling and pseudointrinsic conduction: two mechanisms to explain the Meyer–Neldel rule. *J. Phys. C: Solid State Phys.*, 4, 3167.

Roberts, M. E., Mannsfeld, S. C., Queraltó, N., Reese, C., Locklin, J., Knoll, W. & Bao, Z. 2008. Water-stable organic transistors and their application in chemical and biological sensors. *Proc. Natl. Acad. Sci.*, 105, 12134.

Romero, M. A., Martinez, M. A. G. & Herczfeld, P. R. 1996. An analytical model for the photodetection mechanisms in high-electron mobility transistors. *IEEE Trans. Microwave Theory Tech.*, 44, 2279.

Sakamoto, Y., Suzuki, T., Kobayashi, M., Gao, Y., Fukai, Y., Inoue, Y., Sato, F. & Tokito, S. 2004. Perfluoropentacene: high-performance p–n junctions and complementary circuits with pentacene. *J. Am. Chem. Soc.*, 126, 8138.

Salleo, A. 2007. Charge transport in polymeric transistors. *Mater. Today*, 10, 38.

Salleo, A. & Street, R. 2003. Light-induced bias stress reversal in polyfluorene thin-film transistors. *J. Appl. Phys.*, 94, 471.

Schmechel, R., Ahles, M. & von Seggern, H. 2005. A pentacene ambipolar transistor: experiment and theory. *J. Appl. Phys.*, 98, 084511.

Schmidt, R., Oh, J. H., Sun, Y.-S., Deppisch, M., Krause, A.-M., Radacki, K., Braunschweig, H., Könemann, M., Erk, P., Bao, Z. & Würthner, F. 2009. High-performance air-stable n-channel organic thin film transistors based on halogenated perylene bisimide semiconductors. *J. Am. Chem. Soc.*, 131, 6215.

Sekitani, T., Noguchi, Y., Zschieschang, U., Klauk, H. & Someya, T. 2008. Organic transistors manufactured using inkjet technology with subfemtoliter accuracy. *Proc. Natl. Acad. Sci.*, 105, 4976.

Sekitani, T. & Someya, T. 2012. Stretchable organic integrated circuits for large-area electronic skin surfaces. *MRS Bull.*, 37, 236.

Sekitani, T., Yokota, T., Kuribara, K., Kaltenbrunner, M., Fukushima, T., Inoue, Y., Sekino, M., Isoyama, T., Abe, Y. & Onodera, H. 2016. Ultraflexible organic amplifier with biocompatible gel electrodes. *Nat. Commun.*, 7, 11425.

Sekitani, T., Zschieschang, U., Klauk, H. & Someya, T. 2010. Flexible organic transistors and circuits with extreme bending stability. *Nat. Mater.*, 9, 1015.

Seo, J. H., Park, D. S., Cho, S. W., Kim, C. Y., Jang, W. C., Whang, C. N., Yoo, K.-H., Chang, G. S., Pedersen, T. & Moewes, A. 2006. Buffer layer effect on the structural and electrical properties of rubrene-based organic thin-film transistors. *Appl. Phys. Lett.*, 89, 163505.

Sharma, A., Janssen, N., Mathijssen, S., de Leeuw, D., Kemerink, M. & Bobbert, P. 2011. Effect of Coulomb scattering from trapped charges on the mobility in an organic field-effect transistor. *Phys. Rev. B*, 83, 125310.

Sharma, A., Mathijssen, S., Kemerink, M., De Leeuw, D. & Bobbert, P. 2009. Proton migration mechanism for the instability of organic field-effect transistors. *Appl. Phys. Lett.*, 95, 331.

Sharma, A., Mathijssen, S., Smits, E., Kemerink, M., De Leeuw, D. & Bobbert, P. 2010. Proton migration mechanism for operational instabilities in organic field-effect transistors. *Phys. Rev. B*, 82, 075322.

Sheleg, G., Greenman, M., Lussem, B. & Tessler, N. 2017. Removing the current-limit of vertical organic field effect transistors. *J. Appl. Phys.*, 122, 195502.

Shtein, M., Mapel, J., Benziger, J. B. & Forrest, S. R. 2002. Effects of film morphology and gate dielectric surface preparation on the electrical characteristics of organic-vapor-phase-deposited pentacene thin-film transistors. *Appl. Phys. Lett.*, 81, 268.

Shukla, D., Nelson, S. F., Freeman, D. C., Rajeswaran, M., Ahearn, W. G., Meyer, D. M. & Carey, J. T. 2008. Thin-film morphology control in naphthalene-diimide-based semiconductors: high mobility n-type semiconductor for organic thin-film transistors. *Chem. Mater.*, 20, 7486.

Singh, T. B., Marjanović, N., Stadler, P., Auinger, M., Matt, G. J., Günes, S., Sariciftci, N. S., Schwödiauer, R. & Bauer, S. 2005. Fabrication and characterization of solution-processed methanofullerene-based organic field-effect transistors. *J. Appl. Phys.*, 97, 083714.

Sirringhaus, H., Brown, P. J., Friend, R. H. & Nielsen, M. M. 1999. Two-dimensional charge transport in self-organized, high-mobility conjugated polymers. *Nature*, 401, 685.

Sirringhaus, H., Kawase, T., Friend, R. H., Shimoda, T., Inbasekaran, M., Wu, W. & Woo, E. P. 2000a. High-resolution inkjet printing of all-polymer transistor circuits. *Science*, 290, 2123.

Sirringhaus, H., Sele, C. W., vonWerne, T. & Ramsdale, C. 2006. Manufacturing of organic transistor circuits by solution-based printing. In: Klauk, H. (ed.) *Organic Electronics, Materials, Manufacturing and Applications*. Berlin: Wiley-VCH Verlag.

Sirringhaus, H., Wilson, R., Friend, R., Inbasekaran, M., Wu, W., Woo, E., Grell, M. & Bradley, D. 2000b. Mobility enhancement in conjugated polymer field-effect transistors through chain alignment in a liquid-crystalline phase. *Appl. Phys. Lett.*, 77, 406.

Smith, J., Hamilton, R., Heeney, M., Leeuw, D. M. d., Cantatore, E., Anthony, J. E., McCulloch, I., Bradley, D. D. C. & Anthopoulos, T. D. 2008. High-performance organic integrated circuits based on solution processable polymer-small molecule blends. *Appl. Phys. Lett.*, 93, 253301.

Someya, T., Sekitani, T., Iba, S., Kato, Y., Kawaguchi, H. & Sakurai, T. 2004. A large-area, flexible pressure sensor matrix with organic field-effect transistors for artificial skin applications. *Proc. Natl. Acad. Sci.*, 101, 9966.

Son, Y., Kim, C., Yang, D. H. & Ahn, D. J. 2008. Spreading of an inkjet droplet on a solid surface with a controlled contact angle at low Weber and Reynolds numbers. *Langmuir*, 24, 2900.

Sonar, P., Singh, S. P., Li, Y., Soh, M. S. & Dodabalapur, A. 2010. A low-bandgap diketopyrrolopyrrole-benzothiadiazole-based copolymer for high-mobility ambipolar organic thin-film transistors. *Adv. Mater.*, 22, 5409.

Song, D., Wang, H., Zhu, F., Yang, J., Tian, H., Geng, Y. & Yan, D. 2008. Phthalocyanato tin(IV) dichloride: an air-stable, high-performance, n-type organic semiconductor with a high field-effect electron mobility. *Adv. Mater.*, 20, 2142.

Spijkman, M., Smits, E. C. P., Blom, P. W. M., Leeuw, D. M. d., Côme, Y. B. S., Setayesh, S. & Cantatore, E. 2008. Increasing the noise margin in organic circuits using dual gate field-effect transistors. *Appl. Phys. Lett.*, 92, 143304.

Stingelin-Stutzmann, N., Smits, E. C. P., Wondergem, H., Tanase, C., Blom, P. W. M., Smith, P. & de Leeuw, D. 2005. Organic thin-film electronics from vitreous solution-processed rubrene hypereutectics. *Nat. Mater.*, 4, 601.

Streetman, B. G. & Banerjee, S. K. 2006. *Solid State Electronic Devices*. Upper Saddle River, NJ: Prentice Hall.

Stutzmann, N., Friend, R. H. & Sirringhaus, H. 2003. Self-aligned, vertical-channel, polymer field-effect transistors. *Science*, 299, 1881.

Sun, Y. M., Ma, Y. Q., Liu, Y. Q., Lin, Y. Y., Wang, Z. Y., Wang, Y., Di, C. A., Xiao, K., Chen, X. M., Qiu, W. F., Zhang, B., Yu, G., Hu, W. P. & Zhu, D. B. 2006. High-performance and stable organic thin-film transistors based on fused thiophenes. *Adv. Funct. Mater.*, 16, 426.

Sundar, V. C., Zaumseil, J., Podzorov, V., Menard, E., Willett, R. L., Someya, T., Gershenson, M. E. & Rogers, J. A. 2004. Elastomeric transistor stamps: reversible probing of charge transport in organic crystals. *Science*, 303, 1644.

Suo, Z., Ma, E., Gleskova, H. & Wagner, S. 1999. Mechanics of rollable and foldable film-on-foil electronics. *Appl. Phys. Lett.*, 74, 1177.

Sze, S. M. 1981. *Physics of Semiconductor Devices*. New York: Wiley-Interscience.

Takanashi, Y., Takahata, K. & Muramoto, Y. 1998. Characteristics of InAlAs/InGaAs high electron mobility transistors under 1.3-μm laser illumination. *IEEE Electron Device Lett.*, 19, 472.

Takanashi, Y., Takahata, K. & Muramoto, Y. 1999. Characteristics of InAlAs/InGaAs high-electron-mobility transistors under illumination with modulated light. *IEEE Trans. Electron Devices*, 46, 2271.

Tang, M. L., Oh, J. H., Reichardt, A. D. & Bao, Z. 2009. Chlorination: a general route toward electron transport in organic semiconductors. *J. Am. Chem. Soc.*, 131, 3733.

Tang, M. L., Reichardt, A. D., Okamoto, T., Miyaki, N. & Bao, Z. 2008a. Functionalized asymmetric linear acenes for high-performance organic semiconductors. *Adv. Funct. Mater.*, 18, 1579.

Tang, M. L., Reichardt, A. D., Siegrist, T., Mannsfeld, S. C. B. & Bao, Z. 2008b. Trialkylsilylethynyl-functionalized tetraceno[2,3-b]thiophene and anthra[2,3-b]thiophene organic transistors. *Chem. Mater.*, 20, 4669.

Tatemichi, S., Ichikawa, M., Koyama, T. & Taniguchi, Y. 2006. High mobility n-type thin-film transistors based on *N,N′*-ditridecylperylene diimide with thermal treatments. *Appl. Phys. Lett.*, 89, 112108.

Taylor, D. M. 2016. Progress in organic integrated circuit manufacture. *Jpn. J. Appl. Phys.*, 55, 02BA01.

Tello, M., Chiesa, M., Duffy, C. M. & Sirringhaus, H. 2008. Charge trapping in intergrain regions of pentacene thin film transistors. *Adv. Funct. Mater.*, 18, 3907.

Tiwari, S. P., Potscavage, W. J., Sajoto, T., Barlow, S., Marder, S. R. & Kippelen, B. 2010. Pentacene organic field-effect transistors with doped electrode-semiconductor contacts. *Org. Electron.*, 11, 860.

Tong, X. & Forrest, S. R. 2011. An integrated organic passive pixel sensor. *Org. Electron.*, 12, 1822.

Torricelli, F., Ghittorelli, M., Smits, E. C. P., Roelofs, C. W. S., Janssen, R. A. J., Gelinck, G. H., Kovács-Vajna, Z. M. & Cantatore, E. 2016. Ambipolar organic tri-gate transistor for low-power complementary electronics. *Adv. Mater.*, 28, 284.

Tsao, H. N., Cho, D., Andreasen, J. W., Rouhanipour, A., Breiby, D. W., Pisula, W. & Müllen, K. 2009. The influence of morphology on high-performance polymer field-effect transistors. *Adv. Mater.*, 21, 209.

Tsumura, A., Koezuka, H. & Ando, T. 1986. Macromolecular electronic device: field-effect transistor with a polythiophene thin film. *Appl. Phys. Lett.*, 49, 1210.

Tsumura, A., Koezuka, H. & Ando, T. 1988. Polythiophene field-effect transistor: its characteristics and operation mechanism. *Synth. Met.*, 25, 11.

Ucurum, C., Goebel, H., Yildirim, F. A., Bauhofer, W. & Krautschneider, W. 2008. Hole trap related hysteresis in pentacene field-effect transistors. *J. Appl. Phys.*, 104, 084501.

Uemura, T., Matsumoto, T., Miyake, K., Uno, M., Ohnishi, S., Kato, T., Katayama, M., Shinamura, S., Hamada, M. & Kang, M. J. 2014. Split-gate organic field-effect transistors for high-speed operation. *Adv. Mater.*, 26, 2983.

Umeda, T., Kumaki, D. & Tokito, S. 2009. Surface-energy-dependent field-effect mobilities up to 1 cm2/Vs for polymer thin-film transistor. *J. Appl. Phys.*, 105, 024516.

Vandamme, L. K. J., Feyaerts, R., Trefán, G. & Detcheverry, C. 2002. 1/f noise in pentacene and poly-thienylene vinylene thin film transistors. *J. Appl. Phys.*, 91, 719.

Venkateshvaran, D., Nikolka, M., Sadhanala, A., Lemaur, V., Zelazny, M., Kepa, M., Hurhangee, M., Kronemeijer, A. J., Pecunia, V. & Nasrallah, I. 2014. Approaching disorder-free transport in high-mobility conjugated polymers. *Nature*, 515, 384.

Veres, J., Ogier, S. D., Leeming, S. W., Cupertino, D. C. & Mohialdin Khaffaf, S. 2003. Low-k insulators as the choice of dielectrics in organic field-effect transistors. *Adv. Funct. Mater.*, 13, 199.

Verilhac, J.-M., LeBlevennec, G., Djurado, D., Rieutord, F., Chouiki, M., Travers, J.-P. & Pron, A. 2006. Effect of macromolecular parameters and processing conditions on supramolecular organisation, morphology and electrical transport properties in thin layers of regioregular poly (3-hexylthiophene). *Synth. Met.*, 156, 815.

Wakatsuki, Y., Noda, K., Wada, Y., Toyabe, T. & Matsushige, K. 2011. Molecular doping effect in bottom-gate, bottom-contact pentacene thin-film transistors. *J. Appl. Phys.*, 110, 054505.

Wang, C., Dong, H., Hu, W., Liu, Y. & Zhu, D. 2011. Semiconducting π-conjugated systems in field-effect transistors: a material odyssey of organic electronics. *Chem. Rev.*, 112, 2208.

Wang, G., Swensen, J., Moses, D. & Heeger, A. J. 2003. Increased mobility from regioregular poly(3-hexylthiophene) field-effect transistors. *J. Appl. Phys.*, 93, 6137.

Wang, H., Wang, J., Yan, X., Shi, J., Tian, H., Geng, Y. & Yan, D. 2006. Ambipolar organic field-effect transistors with air stability, high mobility, and balanced transport. *Appl. Phys. Lett.*, 88, 133508.

Wang, H., Zhu, F., Yang, J., Geng, Y. & Yan, D. 2007. Weak epitaxy growth affording high-mobility thin films of disk-like organic semiconductors. *Adv. Mater.*, 19, 2168.

Wang, X., Wasapinyokul, K., De Tan, W., Rawcliffe, R., Campbell, A. J. & Bradley, D. D. C. 2010. Device physics of highly sensitive thin film polyfluorene copolymer organic phototransistors. *J. Appl. Phys.*, 107, 024509.

Wasapinyokul, K., Milne, W. & Chu, D. 2009. Photoresponse and saturation behavior of organic thin film transistors. *J. Appl. Phys.*, 105, 024509.

Weimer, P. K. 1962. The TFT a new thin-film transistor. *Proc. IRE*, 50, 1462.

Xu, H., Yu, G., Xu, W., Xu, Y., Cui, G., Zhang, D., Liu, Y. & Zhu, D. 2005. High-performance field-effect transistors based on langmuir–blodgett films of cyclo[8]pyrrole. *Langmuir*, 21, 5391.

Xu, X., Davanco, M., X. Qi & Forrest, S. R. 2008. Direct transfer patterning on three dimensionally deformed surfaces at micrometer resolutions and its application to hemispherical focal plane detector arrays. *Org. Electron.*, 9, 1122.

Xu, X., Mihnev, M., Taylor, A. & Forrest, S. R. 2009. Organic photodetector arrays with indium tin oxide electrodes patterned using directly transferred metal masks. *Appl. Phys. Lett.*, 94, 043313.

Yagi, I., Hirai, N., Miyamoto, Y., Noda, M., Imaoka, A., Yoneya, N., Nomoto, K., Kasahara, J., Yumoto, A. & Urabe, T. 2008. A flexible full-color AMOLED display driven by OTFTs. *J. Soc. Inf. Display*, 16, 15.

Yamamoto, T. & Takimiya, K. 2007. Facile synthesis of highly π-extended heteroarenes, dinaphtho[2,3-b:2′,3′-f] chalcogenopheno[3,2-b]chalcogenophenes, and their application to field-effect transistors. *J. Am. Chem. Soc.*, 129, 2224.

Yan, H., Chen, Z., Zheng, Y., Newman, C., Quinn, J. R., Dötz, F., Kastler, M. & Facchetti, A. 2009. A high-mobility electron-transporting polymer for printed transistors. *Nature*, 457, 679.

Yoneya, N., Kimura, N., Hirai, N., Yagi, I., Noda, M., Nomoto, K., Wada, M., Kasahara, J., Tsukagoshi, K. & Aoyagi, Y. 2006. All-organic TFT-driven QQVGA active-matrix polymer-dispersed LCD with solution-processed insulator, electrodes, and wires. *SID Symp. Dig. Tech. Pap.*, 37, 123.

Yoo, H., Ghittorelli, M., Lee, D.-K., Smits, E. C., Gelinck, G. H., Ahn, H., Lee, H.-K., Torricelli, F. & Kim, J.-J. 2017. Balancing hole and electron conduction in ambipolar split-gate thin-film Transistors. *Sci. Rep.*, 7, 5015.

Yook, K. S., Doo Chin, B., Yeob Lee, J., Lassiter, B. E. & Forrest, S. R. 2011. Vertical orientation of copper phthalocyanine in organic solar cells using a small molecular weight organic templating layer. *Appl. Phys. Lett.*, 99, 150.

Yoon, M.-H., Kim, C., Facchetti, A. & Marks, T. J. 2006. Gate dielectric chemical structure–organic field-effect transistor performance correlations for electron, hole, and ambipolar organic semiconductors. *J. Am. Chem. Soc.*, 128, 12851.

Yuen, J. D., Fan, J., Seifter, J., Lim, B., Hufschmid, R., Heeger, A. J. & Wudl, F. 2011. High performance weak donor–acceptor polymers in thin film transistors: effect of the acceptor on electronic properties, ambipolar conductivity, mobility, and thermal stability. *J. Am. Chem. Soc.*, 133, 20799.

Zaumseil, J. & Sirringhaus, H. 2007. Electron and ambipolar transport in organic field-effect transistors. *Chem. Rev.*, 107, 1296.

Zeng, W.-J., Zhou, X.-Y., Pan, X.-J., Song, C.-L. & Zhang, H.-L. 2013. High performance CMOS-like inverter based on an ambipolar organic semiconductor and low cost metals. *AIP Adv.*, 3, 012101.

Zhang, C., Chen, P. & Hu, W. 2015. Organic field-effect transistor-based gas sensors. *Chemi. Soc. Rev.*, 44, 2087.

Zhang, J., Wang, J., Wang, H. & Yan, D. 2004. Organic thin-film transistors in sandwich configuration. *Appl. Phys. Lett.*, 84, 142.

Zhang, L., Tan, L., Wang, Z., Hu, W. & Zhu, D. 2009. High-performance, stable organic field-effect transistors based on trans-1,2-(dithieno[2,3-b:3′,2′-d]thiophene)ethene. *Chem. Mater.*, 21, 1993.

Zhang, W., Smith, J., Watkins, S. E., Gysel, R., McGehee, M., Salleo, A., Kirkpatrick, J., Ashraf, S., Anthopoulos, T. & Heeney, M. 2010. Indacenodithiophene semiconducting polymers for high-performance, air-stable transistors. *J. Am. Chem. Soc.*, 132, 11437.

Zhao, Y., Guo, Y. & Liu, Y. 2013. 25th anniversary article: Recent advances in n-type and ambipolar organic field-effect transistors. *Adv. Mater.*, 25, 5372.

Zheng, Z., Yim, K.-H., Saifullah, M. S. M., Welland, M. E., Friend, R. H., Kim, J.-S. & Huck, W. T. S. 2007. Uniaxial alignment of liquid-crystalline conjugated polymers by nanoconfinement. *Nano Lett.*, 7, 987.

Zschieschang, U., Bader, V. P. & Klauk, H. 2017. Below-one-volt organic thin-film transistors with large on/off current ratios. *Org. Electron.*, 49, 179.

Zschieschang, U., Halik, M. & Klauk, H. 2008. Microcontact-printed self-assembled monolayers as ultrathin gate dielectrics in organic thin-film transistors and complementary circuits. *Langmuir*, 24, 1665.

Zschieschang, U., Hofmockel, R., Rödel, R., Kraft, U., Kang, M. J., Takimiya, K., Zaki, T., Letzkus, F., Butschke, J. & Richter, H. 2013. Megahertz operation of flexible low-voltage organic thin-film transistors. *Org. Electron.*, 14, 1516.

Zschieschang, U., Yamamoto, T., Takimiya, K., Kuwabara, H., Ikeda, M., Sekitani, T., Someya, T. & Klauk, H. 2011. Organic electronics on banknotes. *Adv. Mater.*, 23, 654.

CHAPTER 9

Expect the unexpected: more possibilities for organic electronics

"If you don't know where you are going, you might wind up someplace else."

Yogi Berra, Catcher for the New York Yankees from 1946 to 1963.

The possibilities for organic semiconductors seem limitless. With each year, researchers discover new phenomena, and engineers invent new devices. In Chapter 1 we enumerated many common myths surrounding organic materials and devices. Many of those myths suggested that there were limitations that were later found not to be so at all. Indeed, it seems that every property of inorganic semiconductors has its analog in organics, even though their materials properties are fundamentally different. Yet it is not the similarities between these systems that make organics interesting. Their value lies in the new opportunities for discovery and innovation that they offer. The purpose of this final chapter is to open a window to the many surprises organics have yet to reveal. And perhaps this chapter will give you new ideas and paths to explore that continue to exploit the "organic difference."

Thus far in Part II, we have focused on three broad classes of devices and their applications: light emitters, photodetectors, and transistors. There are several subclasses of devices within each of these categories. For example, photodetectors include photoconductors, photodiodes, phototransistors, solar cells, and many different specialized optical sensors. The emphasis on these three general device classes is justified by their attracting the preponderance of attention by scientists, engineers and industries driven by their commercial potential. But a fascinating aspect of organics that has given impetus for their continued exploration is the unparalleled versatility in their properties compared to other materials systems. The versatility is enabled by a nearly infinite palette of materials. The entire scope of their application space is only limited by our imaginations.

Here, we introduce a few devices and scientific topics that do not fit neatly into those covered in other chapters. Yet they illustrate the breadth of the field, and some of the devices or their descendants may eventually find applications at scales that will become comparable to those for emitters, detectors and transistors. Given the abundance of research on novel devices, this could be a discussion without end. Hence, only a few examples are introduced. The choices of topics are based on their potential for making a substantial impact on the future of the field, or because they reveal a unique property of organics that has not been adequately considered in previous chapters. Even so, very large and important scientific and application domains have been omitted. Perhaps the most significant omission is that of waveguide-based optical devices. Organics are exceptionally well suited for use as waveguides. These are not included since passive waveguides do not employ an active electronic or excitonic property of the materials. Similarly, waveguide devices employing nonlinear optical effects such as optical switching, optical mixing, etc. rarely rely on an active electronic property such as electrical conduction or optical generation. This topic, therefore, falls outside of the boundaries that define organic electronics, where we apply the working definition of applications that require current for activation, or their characteristics are governed by excitonic effects. We

Organic Electronics: Foundations to Applications. Stephen R. Forrest. © Stephen R. Forrest 2020.
Published in 2020 by Oxford University Press. DOI: 10.1093/oso/9780198529729.001.0001

have also chosen to omit this large class of materials and devices in deference to other excellent texts available that specifically treat this subject. Other topics that have been omitted include a vast and ever growing array of electronic sensors such as distance and motion detectors, bioelectronic sensors, and so on. The list is indeed very long. The interested reader is also encouraged to explore these topics that are well represented in the scientific literature. There simply is insufficient space or time to cover everything in this one volume.

The topics selected for inclusion in this chapter are (i) light emitting electrochemical cells, (ii) microcavity exciton polaritons, (iii) thermoelectricity, (iv) organic memories, (v) organic/two-dimensional (2D) junctions, and (vi) single molecule devices. Each of these devices and/or phenomena has features that are peculiar to organic semiconductors. For example, light emitting electrochemical cells exploit the property of ionic conduction in active electronic devices. Exciton polaritons provide an understanding of organic molecules in resonant optical cavities. This leads to phenomena such as ultrastrong optical coupling, room temperature polariton lasing, and Bose–Einstein condensation. Thermoelectricity and electronic memories represent two broad and very promising application areas where organic materials can offer advantages over inorganics. And finally, limited dimensional systems open entirely new possibilities for organics to be productively combined with ultrahigh mobility 2D inorganic films, or in single molecule devices that may enable next generation (i.e. beyond Moore's law) electronics used in artificial intelligence, quantum computing and information processing, and molecular sensing.

The topics addressed, therefore, provide a limited but nevertheless important window into what this field may have in store for the future. As the title of this chapter suggests, in the field of organics, it is always a good idea to expect the unexpected.

9.1 Light emitting electrochemical cells

The conventional organic light emitting device consists of one or more neutral organic species comprising the emission layer (EML). A different approach to achieving electroluminescence is to combine the chromophore with an ionic medium, or salt. The resulting device is known as a *light emitting electrochemical cell* (LEC). The LEC was initially demonstrated using a 1:1 blend of the polymers, MEH-PPV and PEO, doped with the salt, CF_3SO_3Li (Pei et al., 1995). This is known as a polymer LEC, or pLEC. Soon after, a small molecular weight LEC (SMLEC) was demonstrated based on the ionic ruthenium polypyridyl complex, $Ru(phen)_3^{2+}[2Cl^-]$ (Lee et al., 1996).

A distinctive feature of the LEC is that it is an ambipolar device: the cathode and anode contacts can be interchanged. Likewise, it consists of only a single layer that is ionically doped via charge injection from the contacts. Another feature is that the energy barriers between the contacts and the active region are reduced by charge injection. This allows for the use of relatively high work function contacts that are resistant to degradation when exposed to the ambient, thus eliminating the need for potentially costly packaging required to protect conventional OLEDs. Another distinguishing attribute is that the devices are oriented within the substrate plane, with anode and cathode positioned laterally as in a photoconductor. Thus, the devices and their emission zones can be large; on the order of millimeters vs. active regions in OLEDs that are only tens of nanometers thick.

These several advantages are outweighed by several disadvantages that have prevented LECs from becoming a substitute for OLEDs. Chief among the disadvantages is that they take a long time (hundreds of microseconds to tens of minutes) for the light to turn on to full intensity. This eliminates the possibility of their use in conventional information displays, although the faster devices may be suitably fast for lighting. Since the conductivity is based on ionic transport, the device performance (i.e. its conductivity and luminous efficiency) is highly temperature dependent: elevated temperatures result in faster response, higher conductivity and efficiency. Cooling such devices even slightly below room temperature can deactivate almost all of their electronic properties. The light output reaches a peak, and then slowly decreases as the lateral dimension of the light emission zone shrinks with the continued lateral growth of the ionically doped conductive regions where excitons are quenched. Finally, the efficiencies of LECs are well below that of conventional polymer and small molecule OLEDs, even though phosphorescent transition metal complexes have been employed as the active organic material (van Reenen et al., 2013).

A schematic of an LEC is provided in Fig. 9.1. It comprises two metal electrodes of the same composition. One is the anode and the other is the cathode, depending on the polarity of the externally applied voltage. The luminescent polymer mixed with a salt provides ionic conductivity. Typically, the active layer is from 100 nm to 300 nm thick. The simplicity of this structure is one of its attractions; it requires none of the complex layer schemes demanded of high performance OLEDs.

Two explanations have been advanced that describe the properties of LECs: the *electrodynamic* (ED) model, and the *electrochemical doping* (ECD) model. These two mechanisms are illustrated in Fig. 9.2. Apparently both can be active in the same device, depending on the bias voltage and whether the contacts are blocking or non-blocking (Meier et al., 2014).

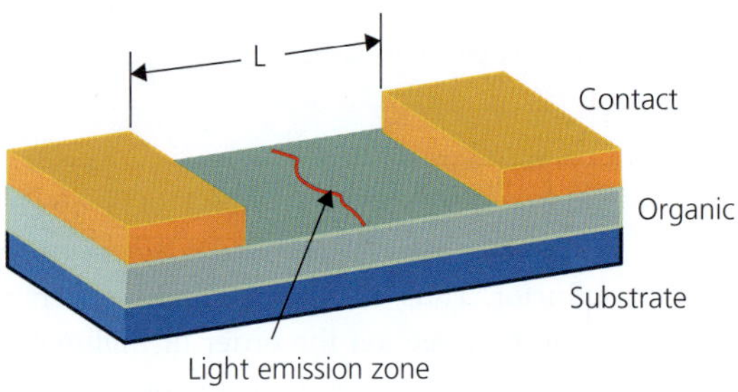

Figure 9.1 Schematic of a light emitting electrochemical cell of length, *L*. Light emission is from an irregular, undoped stripe located between the metal contacts.

In the ED model, nearly all of the potential is dropped across the ionically doped layers (EDLs) at the anode and cathode. These regions contain a high concentration of uncompensated anions (at the anode) and cations (at the cathode) that are attracted to these two regions by the applied voltage. The concentration of ionic species is sufficient to lower the injection barriers, making ohmic contact to the organic layer. Between the contact zones, the ionic species are compensated by injected free charges. This results in an extended charge-neutral region where oppositely charged free polarons combine to form radiative excitons. As soon as a recombination event occurs, additional free charge is drawn in from the contacts to maintain charge neutrality in the central, compensated region of the device.

In the ECD model, electrons are drawn in from the EDL at the cathode to compensate the cations, forming a conductive "n-doped" region. Similarly, oppositely charged ions and polarons injected from the anode generate a "p-doped" region. In either ionically doped region, the potential drop is minimal, resulting in diffusion-limited charge transport. An intrinsic, emissive charge recombination zone forms at the interface between the doped regions. Thus, the EDL model invokes the formation of a p-i-n type diode.

The applied voltage results in a redistribution of the ionic species within the EDL layers that range in thickness from 2 nm to 6 nm. The time to establish the zones, and for the charges to redistribute within

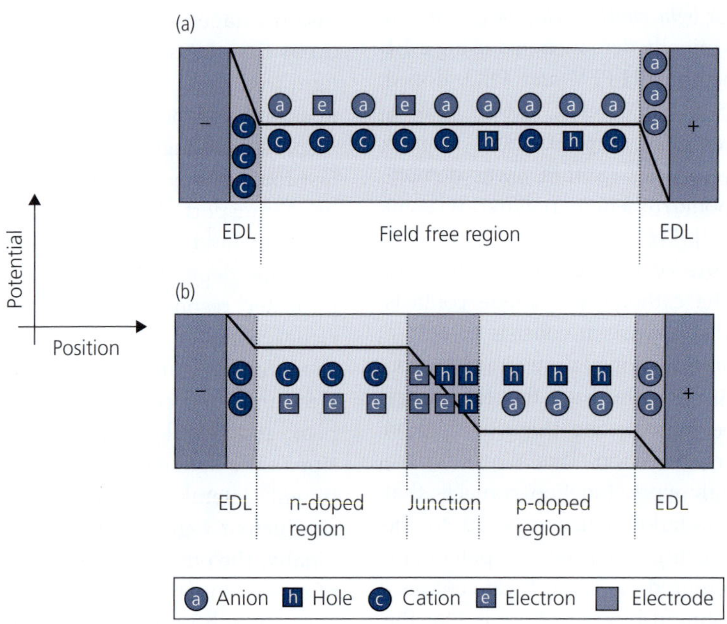

Figure 9.2 The distribution of free polarons and ions in an LEC according to (a) the electrodynamic (ED) and (b) the electrochemical doping (ECD) models. The black lines show the potential drops across the device active regions (van Reenen et al., 2010).

the bulk of the organic layer, is ultimately a function of the ionic mobility. This results in a slow, temperature-dependent response of both pLEC and SMLEC devices. Once the highly conductive regions are formed, the intervening undoped region consists of uncompensated charges that are able to recombine and emit light. The position of the emission region depends on the ratio of mobilities of the cations and anions. The conductive zone increases in width with the continued injection of free polarons over time. This causes the compensated regions to grow at the expense of the uncompensated region, thus resulting in the shrinkage of the recombination zone, and a concomitant decrease in emission intensity. The rate of advance of the doping fronts is determined by the magnitude of the applied voltage, the ionic mobilities, and the thickness of the organic layer.

The various operating regimes have been modeled to determine the ranges of validity of the ED and EDL pictures, and under which circumstances these models apply (van Reenen et al., 2010, Mills and Lonergan, 2012). The potential distribution for a pLEC comprising an active region, with and without blocking contacts, is shown in Fig. 9.3a. In both cases, there is a large voltage drop across the very narrow EDL near the cathode and anode contacts. For the blocking contact, there is almost no potential drop across the intervening polymer layer. This corresponds to the ED model of Fig. 9.2a where the entire active region has a distribution of cations and anions that are compensated by free polarons of opposite polarity. In contrast, a p-i-n junction forms when Ohmic contacts are used that do not block charge injection. Then, the voltage drop across the heavily doped p and n regions vanishes, and appears only across the junction and at the doped regions near the contacts. This corresponds to the ECD regime of operation in Fig. 9.2b.

The free charge and ion distributions for pLECs with blocking and non-blocking contacts are shown in Fig. 9.3b and c, respectively. For blocking contacts, the ionic and free charge distributions lie directly on top of each other, indicating complete compensation that produces a charge-neutral region across the semiconductor. These distributions change near the contacts where the anion and cation densities are large, but the free charge densities fall toward zero. The recombination zone is relatively large and near the center of the active region between the contacts. It is in this region that the anion and cation densities are small.

The electrochemically doped regime is established by using non-blocking contacts, as shown in Fig. 9.3c. The compensation between ionic and mobile free charge is no longer complete in either the p- or n-doped regions since the cations and anions are not completely separated, contrary to the case of blocking contacts. This results in a recombination zone that is narrower in the ECD than the ED regime.

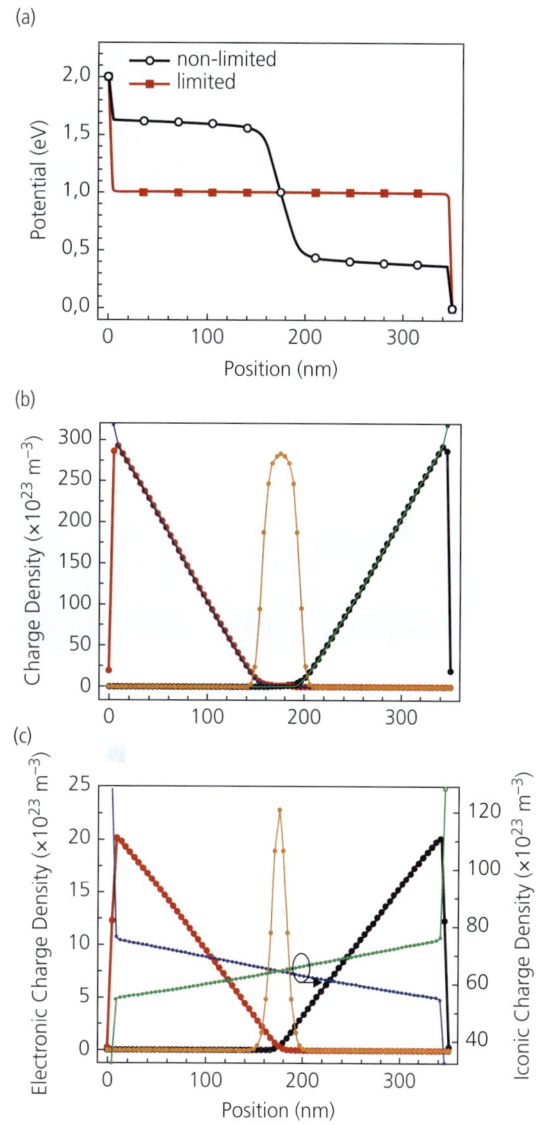

Figure 9.3 (a) Simulated charge distribution for injection limiting (blocking) vs. non-limiting (i.e. Ohmic) LEC contacts. The charge distribution for (b) limiting and (c) non-limiting contacts. The orange line shows the recombination profile, whereas the red, black, green, and blue lines show the electron, hole, anion, and cation densities, respectively (van Reenen et al., 2010).

The validity of this analysis has been tested by measuring the potential across the active region of a pLEC with *scanning Kelvin probe microscopy* (SKPM, see

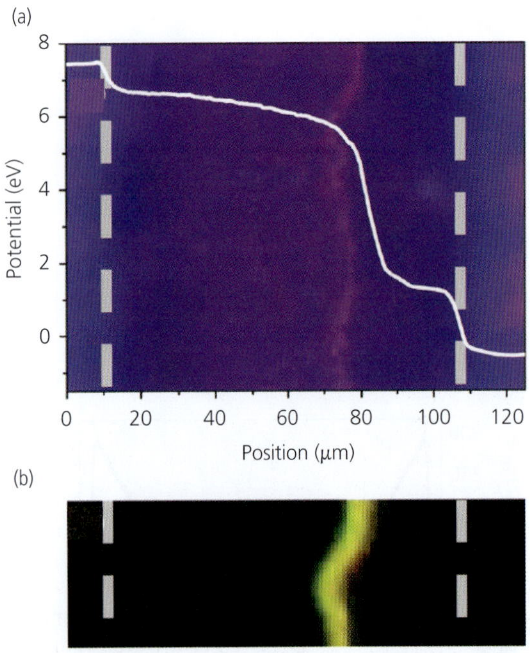

Figure 9.4 (a) SKPM map of the potential across a pLEC device with non-blocking contacts, at an applied voltage of 8 V. The region between the dashed lines is the polymer, and the contacts are outside of the lines. (b) Electroluminescence image under steady state operation at 8 V (van Reenen et al., 2010).

Section 8.8.1). The device in Fig. 9.4 comprises Au/SY-PPV:PEO:$K^+CF_3SO_3^-$ (1:1.35:0.25)/Au. Here, SY-PPV is a PPV copolymer known as "super yellow," and potassium triflate, $K^+CF_3SO_3^-$, is the salt (see Fig. 9.5). The separation between contacts is 100 μm. Since Au forms non-blocking contacts, the SKPM image in Fig. 9.4a corresponds to the ECD regime. Indeed, the potential distribution across the device is consistent with expectations from Fig. 9.2b. There is little or no voltage drop across the doped regions, whereas voltage only drops near the anode and cathode (at the left and right of the SKPM map, respectively) and across the undoped junction region. Likewise, the light emission zone is located in the junction region (Fig. 9.4b), where charges can recombine without quenching. Note the thinness and irregular shape of the recombination zone resulting from local variations in ionic transport within the organic semiconductor. It is not centered between contacts due to differences in the cation and anion mobilities: in this instance the anion mobility is the smaller of the two, resulting in recombination closer to the cathode.

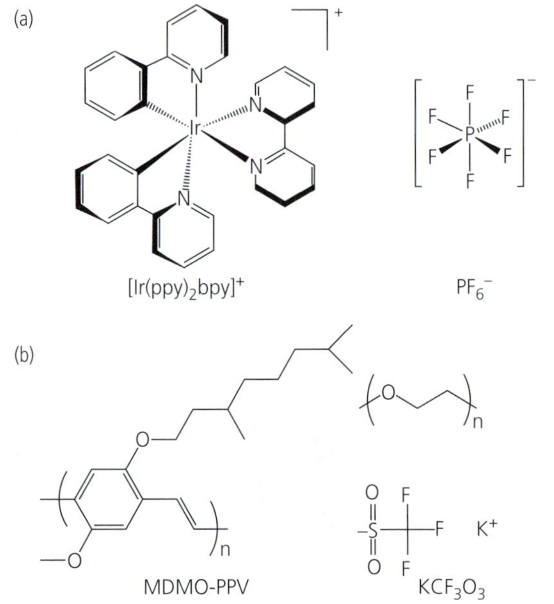

Figure 9.5 Archetype materials systems used in (a) SMLEC and (b) pLEC devices.

Introducing physically small anion–cation pairs increases their mobilities. Two archetype systems are shown in Fig. 9.5. The small molecule Ir complex in Fig. 9.5a comprises the cation-anion couple $[Ir(ppy)_2bpy]^+[PF_6]^-$ in a 1:1 blend. The transition metal complex serves both as a phosphorescent emitting molecule as well as a cation. The polymer system comprises an MDMO-PPV fluorophore blended with PEO that improves the ionic mobility. The KCF_3SO_3 forms a blend with the polymer mixture in the ratio MDMO-PPV:PEO:KF_3SO_3 = 1:1.35:0.25.

The transient response at different temperatures of a MEH-PPV:PEO:KCF_3SO_3 (10:5:0.1.2) pLEC is shown in Fig. 9.6a. The 700 μm long device is imaged using a detector positioned at the opposite end of a scanning, single mode optical pickup fiber whose diameter is sufficiently small to resolve the spatial distribution of the photoluminescence (PL) and electroluminescence (EL) along the length of the channel, as shown in Fig. 9.6b. Heating to 360 K while biased at 20 V activates the electroluminescence. After heating, the device is cooled to 170 K, freezing the emission and doped zones in place. At this temperature, the behavior of these regions can be studied over long periods without change. Note that the current drops by at least one order of magnitude as the sample is cooled due to the decreased electron and hole mobilities, whereas the ionic mobilities nearly vanish.

The PL and EL maps of the active region during the high temperature activation cycle are shown in a series

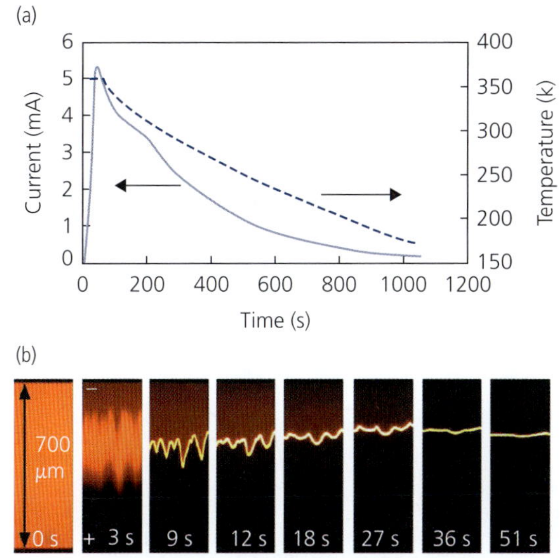

Figure 9.6 (a) Dependence of current on time as a MEH-PPV:PEO: KCF_3SO_3 pLEC is cooled from 360 K to 170 K. (b) The device is "activated" during the first 51 s at 360 K before cooling. The images are of the photoluminescence zone. The line appearing at 9 s is due to electroluminescence from the cell. The cathode is at the top of the images, and the anode at the bottom. After AlTal and Gao (2016).

of images in Fig. 9.6b. Initially, there is no electronic doping. Thus, the film is compensated, leading to uniform PL across the sample. Once a voltage is applied, charges are injected from the contacts, and the anions and cations drift toward the oppositely charged contacts. This quenches the luminescence, thereby narrowing the region where PL is observed. After 9 s, fully doped n- and p-regions are established, at which point the PL intensity drops to near zero while a line of intense EL is observed near the cathode. The p-doped region is dark, totally quenching all PL. The offset of the EL emission toward the cathode is due to the higher mobility of cations than anions. As time progresses, the jaggedness of the EL emission line decreases, and its intensity decreases due to narrowing of the i-region as discussed above (AlTal and Gao, 2016). The response in Fig. 9.6 shows that changes in the luminescence profile take at least 1 min to stabilize, even at 360 K. Simply cooling the samples to only a few degrees below room temperature can result in response times of hours (van Reenen et al., 2013). Qualitatively similar time-dependent behavior is observed for both pLECs and SMLECs, although the quantitative sensitivities to time and temperature may vary widely due to the different dynamics of the ionic and polaronic species in these different materials systems.

Many advantages to LECs have already been mentioned, the most notable being their reduced sensitivity to environmental exposure. However, their performance is substantially inferior to even modestly performing OLEDs. One of the highest performance LECs uses the oligoether, TMPE-OH as an electron transporter containing the salt, $LiCF_3SO_3$, along with the light emitting SY-PPV. The device turn-on times are ~10–2000 s to reach a luminance of 300 cd/m^2, depending on details of the composition of the electron transporter. The maximum luminance efficiency is 14 cd/A with an operational lifetime of approximately 350 h (Mindemark et al., 2016).

Electrophosphorescent SMLECs employing the Ir-based cyan emitting molecule-anion pair $[Ir(dfppz)_2(dtb\text{-}bpy)]^+[PF_6]^-$ blended with the near infrared (NIR) emitting cationic charge trapping DOTCI, achieved an external quantum efficiency of 12.8%. A maximum brightness of 16.6 cd/m^2 is obtained after 62 min following application of 3.3 V, although after 45 min the luminescence intensity decayed to half that value. This slow response is also obtained in a device with a vertical (vs. lateral) architecture of ITO/PEDOT:PSS/emissive layer (200 nm)/Ag. While the results are inferior to those of conventional small molecule PHOLEDs using the analogous, neutral Ir complex, the relatively high efficiency reported for the LEC is primarily a result of improved carrier mobility, and hence charge balance from the addition of DOTCI. Note that while the fluorescent emission of DOTCI is in the NIR, it is only doped at 0.01 wt% in the Ir salt complex. This prevents energy transfer, and thus there is no NIR emission from DOTCI at 700 nm (Liao et al., 2011).

9.2 Microcavity organic exciton polaritons

A *polariton* is a quantum mechanical quasiparticle that is the superposition of a photon that is *strongly coupled* to a material state (e.g. an exciton) in a resonant optical cavity. The polariton is unique in that it contains attributes of both states of which it is comprised. Thus, like a photon, it is a boson governed by Bose–Einstein statistics. The most immediate outcome of this property is that a population of polaritons can occupy a single state, forming a *Bose–Einstein condensate* (BEC) that exhibits coherent, laser-like emission, superfluidity, vortices, and strong nonlinear optical effects such as all-optical switching. On the other hand, its matter-like properties give the polariton mass, although it is far lighter than the excitons of which it is comprised. Thus, an exciton-polariton can diffuse farther and faster than the bare excitonic state.

Figure 9.7 is an illustration of the polaritonic state. A microcavity with distance, L, between a pair of mirrors has a resonance at frequency, ω_{cav}, given by

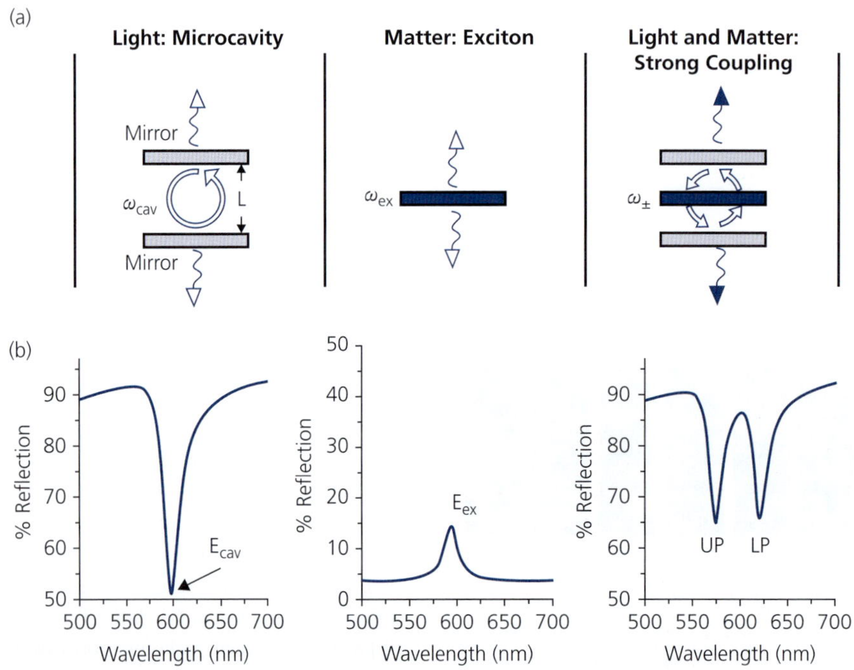

Figure 9.7 Conceptual illustration of the physics of microcavity exciton polaritons (upper panels). A photon cavity mode, combined with an exciton mode, forms a new quasiparticle known as an exciton-polariton in the strong coupling regime. Resonances in the reflectivity for each situation are shown in the lower panels. Here, $E_{cav} = \hbar\omega_{cav}$ is the energy and frequency of the cavity mode, $E_{ex} = \hbar\omega_{ex}$ is for the exciton mode, and UP and LP refer to the upper and lower polariton branches, respectively After Tischler et al. (2007).

the well-known Fabry–Pérot expression $k_z = N\pi/L_{eff}$. Here, N is the integer mode number, L_{eff} is the effective cavity length which is the sum of the separations between the mirrors and the penetration of the optical field into the mirrors, and k_z is the out-of-plane wavevector. An example reflection spectrum of the bare cavity is shown in the left-hand lower panel in Fig. 9.7. This cavity has a stop band at 600 nm.

Since the in-plane wavevector, $k_{\|}$, is unconstrained by energy barriers at the mirrors, the cavity photon dispersion is given by

$$E_{cav} = \hbar\omega_{cav} = \hbar(c/n)\sqrt{k_z^2 + k_{\|}^2} = \hbar(c/n)\sqrt{(N\pi/L)^2 + k_{\|}^2}. \quad (9.1)$$

Here, $\hbar$ is Planck's constant divided by 2π, c is the speed of light, and n is the refractive index within the cavity. The blue dashed line in Fig. 9.8 shows the cavity mode dispersion spectrum. The photon outside of the cavity follows a linear dispersion, $E_{vac} = \hbar ck$, shown by the diagonal lines in the figure. Within the cavity at small $k_{\|}$, the curvature of the cavity mode implies an effective mass for the photon equal to:

$$m_{cav} = \frac{\pi\hbar Nn}{Lc}. \quad (9.2)$$

Typically, $m_{cav} \approx 10^{-5} m_0$, where m_0 is the electron rest mass.

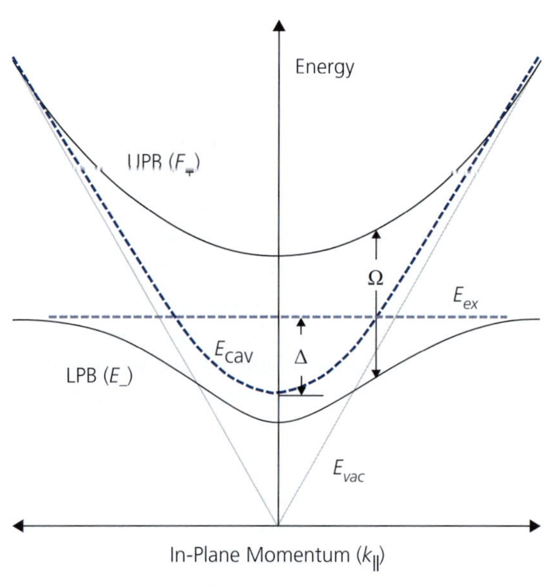

Figure 9.8 Dispersion relation of a polariton showing the upper and lower polariton branches (UPB at E_+ and LPB at E_-, respectively). Also the various energies are defined including the Rabi splitting energy, Ω, and the cavity detuning, $\Delta < 0$.

The exciton is a property of the material filling the cavity, with a binding energy of E_{ex}. The frequency of the exciton is largely unchanged when the material is placed inside the cavity except at large $k_{\|}$. The reflectivity of an excitonic material is shown in Fig. 9.7, center lower panel, and its lack of dispersion is shown by the flat dashed line in Fig 9.8.

For the condition where ω_{ex} is resonant with ω_{cav}, a new state is formed that is a superposition of two coupled oscillators (the photon and the exciton, see Fig. 9.7, right panel). The new state, sharing characteristics of both the cavity photon and the exciton, is the polariton. As with all coupled oscillator pairs, the polariton has two normal modes, or resonances that are separated into the more exciton-like *lower polariton branch* (LPB) and the photon-like *upper polariton branch* (UPB). The splitting energy at the anti-crossing between the LPB and UPB is known as the *Rabi energy*, Ω (see Fig. 9.8). This is linearly proportional to the exciton–photon dipole interaction strength, g_0. The coupling is divided between the two normal modes, and hence is equal to half of the splitting between branches.

The plot in Fig. 9.8 shows negative detuning, Δ, between the cavity mode and the exciton, that is,

$$\Delta = E_{cav}(k_{\parallel} = 0) - E_{ex}(k_{\parallel} = 0) < 0. \tag{9.3}$$

The Rabi splitting is found at the value of $k_{\parallel}$ where the cavity and exciton energies are equal. The exciton effective mass is also significantly reduced due to its admixture with the photon mode. The mass is obtained starting with (Deng et al., 2010)

$$\alpha = \frac{1}{2}\left(1 + \frac{\Delta E(k_{\parallel})}{\sqrt{\Delta E(k_{\parallel})^2 + \Omega^2}}\right) \tag{9.4a}$$

and

$$\beta = \frac{1}{2}\left(1 - \frac{\Delta E(k_{\parallel})}{\sqrt{\Delta E(k_{\parallel})^2 + \Omega^2}}\right), \tag{9.4b}$$

where $\Delta E(k_{\parallel}) = E_{ex}(k_{\parallel}) - E_{cav}(k_{\parallel})$. For $\Delta E{=}0$ (e.g. at $k_{\parallel} = 0$ when $\Delta E(0) = \Delta = 0$), then $\alpha = \beta = \frac{1}{2}$, and the polariton consists of half photon and half exciton content in both the lower and upper polariton branches. The variables, α and β are the absolute squares of the so-called *Hopfield coefficients*. The effective mass of the polariton in both the lower and upper branches are related to these coefficients by

$$\frac{1}{m_{LPB}} = \frac{\alpha}{m_{ex}} + \frac{\beta}{m_{cav}} \tag{9.5a}$$

and

$$\frac{1}{m_{UPB}} = \frac{\beta}{m_{ex}} + \frac{\alpha}{m_{cav}}. \tag{9.5b}$$

With $m_{cav} \approx 10^{-5} m_0$, and the reduced mass of the exciton in organic compounds of $m_{ex} \geq m_0$, we see that the polariton effective mass is nearly equal to the cavity photon mass in both branches. This results in a longer exciton lifetime due to a reduced interaction between electron and hole owing to the larger exciton radius (see Eq. 3.142), and a much larger exciton diffusivity. This suggests that exciton polaritons based on organic thin films can have vastly different properties than a free exciton, at least near $k_{\parallel} = 0$. At higher in-plane momenta, however, the LPB flattens and takes on principally excitonic character, in which case $\beta \to 0$, $\alpha \to 1$, and $m_{LPB} \to m_{ex}$. Similarly, the upper branch becomes purely photon-like.

The energies of two coupled harmonic oscillators are found from solutions to the eigenvalue equation:

$$\begin{aligned} H\Psi_{p,k} &= \begin{pmatrix} E_{cav} - i\gamma_{cav}(k_{\parallel}) & \Omega/2 \\ \Omega/2 & E_{ex} - i\gamma_{ex}(k_{\parallel}) \end{pmatrix} \Psi_{p,k} \\ &= E_{LPB,UPB}\Psi_{p,k}, \end{aligned} \tag{9.6}$$

which yields the following solutions for the upper and lower branches:

$$\begin{aligned} E_{LPB,UPB}(k_{\parallel}) = \frac{1}{2}\Big\{ & E_{ex} + E_{cav} + i(\gamma_{cav} + \gamma_{ex}) \\ & \pm \left[\Omega^2 + (E_{ex} + E_{cav} + i(\gamma_{cav} - \gamma_{ex}))^2\right]^{1/2} \Big\}. \end{aligned} \tag{9.7}$$

Equation 9.7 is the general solution to the dispersion relationship comprising the two branches in Fig. 9.8. This solution only applies to a coupled binary oscillator system. However, excitonic states in organics show strong vibronic character, often exhibiting 0–0, 0–1, 0–2, etc. modes that, if contained within the stop band of a microcavity, can also couple to the polariton state, generating additional branches in the dispersion relationship. Then, depending on the number of coupled modes, the 2×2 matrix of Eq. 9.6 needs to be expanded accordingly (Kéna-Cohen and Forrest, 2008).

9.2.1 Cavity designs

In Eq. 9.7, γ_{cav} and γ_{ex} are damping terms that add to the homogeneously broadened cavity and exciton linewidths, respectively. Cavities with mirrors whose reflectivities are less than unity leak photons, thus decreasing the photon lifetime. Escaping photons couple to the external vacuum field where they can be observed, and thus provide information about the polariton population. The *quality factor* of a cavity comprising two *distributed Bragg reflectors* (DBRs) with reflectivities of R_1 and R_2 is

$$Q = \frac{E_{cav}}{\gamma_{cav}} \approx \frac{\pi (R_1 R_2)^{1/4}}{1 - (R_1 R_2)^{1/2}}. \quad (9.8)$$

A cavity with a high Q has highly reflective mirrors ($R_1, R_2 \to 1$), allowing for hundreds to thousands of photon cycles within the cavity before escape. The high Q results in a narrow linewidth, and hence a concomitantly longer photon lifetime of $\tau_{cav} \approx \hbar/\gamma_{cav}$.

Two types of cavities that are most commonly used in organic polaritonic devices are shown in Fig. 9.9. Two DBRs can form a symmetric cavity that sandwiches the organic at the central antinode of the electric field (Fig. 9.9a). Alternatively, one DBR is replaced with a metal mirror in an asymmetric arrangement (Fig. 9.9b). This structure generally has a much lower $Q \sim$ 50–100, while the double DBR cavity can have $Q \sim$ 500–5000. If both mirrors are replaced by metals, with one being ultrathin and hence semitransparent, the Q is reduced to ~10. Such cavities are known to result in ultrastrong coupling where Ω is an appreciable fraction of the exciton energy. This is considered further in Section 9.2.5.

Finally, it is notable that one of the mirrors can be eliminated completely in a one-sided, or "open" dielectric structure shown in Fig. 9.10a. The optical mode is trapped in the relatively high index DBR and the organic film on its surface, penetrating into air with an evanescent traveling wave, known as a Bloch surface wave (BSW) (Lerario et al., 2017). The BSW is excited by resonant coupling to the optical pump using a prism attached to the glass substrate (known as the *Kretschmann configuration*). This all-dielectric cavity can have a very high Q-factor due to the use of a lossless, low index dielectric (air) on one surface.

A compilation of cavity types, their Q-factors, cavity volumes, polariton velocities, and propagation lengths is found in Table 9.1. The table points to the important characteristics of polaritons compared with bare organic materials. Excitations in high Q cavities result in a dramatic reduction in exciton effective mass that consequently leads to extraordinarily long diffusion lengths (~100 μm). We have seen in previous chapters that a long diffusion length can be utilized to great advantage in many photonic devices including OLEDs, OPDs, and OPVs.

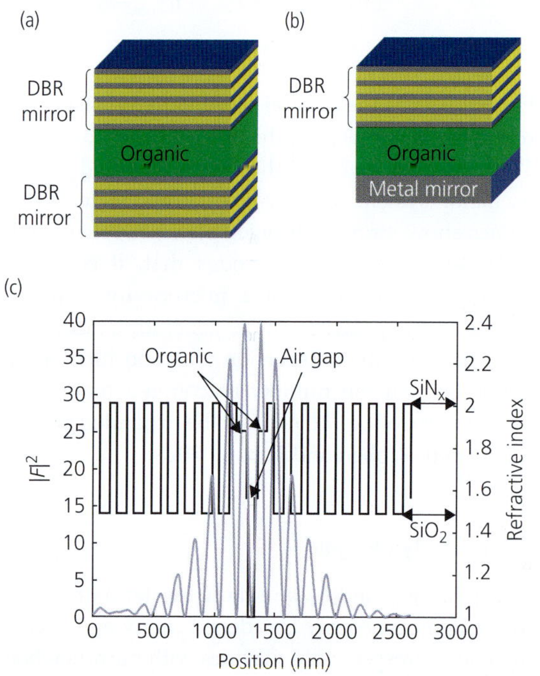

Figure 9.9 Two example cavities for observing polaritons. (a) A high Q cavity with two DBR mirrors, and (b) a lower Q cavity with one metal and one DBR mirror. (c) Optical field intensity ($|F|^2$) and index profiles in an archetype double DBR structure. The organic layer is split into two sections with an air gap in between. The optical field has antinodes in the organic and within the gap.

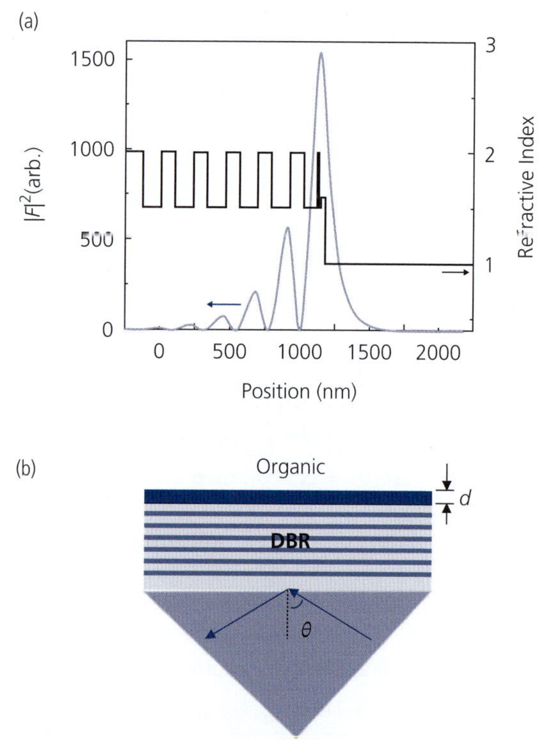

Figure 9.10 (a) Calculated optical field intensity distribution in a one-sided optical structure whose index profile is shown to the right. The DBR consists of six pairs of alternating layers of 135 nm thick SiO_2 and 101 nm thick SiN_x on glass. It is capped by an organic film of thickness, d. (b) Experimental configuration for exciting Bloch surface waves as shown in (a).

The optical field intensity profile for a symmetric DBR cavity is shown in Fig. 9.9c, superimposed on the index of refraction profile of the SiO_2/SiN_x mirror stacks. The organic film itself is divided into two sections, each attached to a different mirror surface, with a gap between them. This arrangement results in field antinodes within the organics as well as in the gap, maximizing interactions between the field and the excitonic material layers. Note that very little field leakage beyond the DBR results in a long τ_{cav}, and hence a high Q.

The radiative width of the exciton line is a materials property that depends on the exciton oscillator strength, f_{ex} (see Section 3.6.1). Thus,

$$\gamma_{ex} = \frac{\pi}{n} \frac{q^2}{4\pi\varepsilon_0} \frac{\hbar}{m_{ex}c} f_{ex}, \tag{9.9}$$

where ε_0 is the permittivity of free space and q is the electron charge. The interaction between photon and exciton is determined by the magnitude of f_{ex}. The linewidth increases linearly with the oscillator strength, which is much larger for organic Frenkel excitons compared with dielectrically screened, and therefore weakly bound Wannier–Mott (W-M) states in inorganic semiconductors.

Taking line broadening into consideration, the Rabi splitting energy is given by (Zhu et al., 1990)

$$\Omega = \sqrt{[aL\mathcal{F}\gamma_{ex}\gamma_{cav}/\pi] - \left[(\gamma_{ex} - \gamma_{cav})^2/4\right]}. \tag{9.10}$$

Here, $\mathcal{F}$ is the cavity finesse and α is the absorption coefficient of the organic, which like γ_{ex}, is proportional to the oscillator strength. Thus the Rabi splitting increases with γ_{ex} and γ_{cav}, and thus f_{ex} and Q, up to the point where the linewidth becomes so large that the second term in Eq. 9.10 is dominated by cavity losses. Thus, the strong coupling regime corresponds to the condition where Ω is greater than the combined widths of the cavity and exciton modes.

A comparison of semiconductors commonly used for studying polaritons is provided in Table 9.2. The benefit of using organics is primarily due to their large exciton binding energy. Thus, organic polaritons

Table 9.1 Comparison of cavity types and the resulting organic polariton characteristics

Cavity type	Q	Vol.[a] (μm^3)	Ω (meV)	τ_{cav} (ps)	Velocity (μm/ps)	Prop. length (μm)	Ref.[b]
Metal-org-metal	10–30	~1	>300	0.05–3	0.2–0.4	1	(Rozenman et al., 2017)
Metal-org-DBR	50	~3	~100	—	—	—	(Hobson et al., 2002)
DBR-org-DBR	800	~10	80–700	—	1–2	3	(Kéna-Cohen et al., 2008, Guillet and Brimont, 2016)
DBR-org	7000	—	50–200	1	120–150	120–300	(Lerario et al., 2017)
Plasmon-org[c]	10	10^{-7}	80–200	—	—	—	(Baranov et al., 2017)

[a] Vol. = cavity volume.
[b] Some references provide reviews by multiple groups. Many additional details can be found in those reviews.
[c] Plasmons are supported in the near field in a metal cavity surrounding the organic. They are lossy due to finite resistance of metals, thus yielding relatively low-Q cavities.

Table 9.2 Comparison of materials that support exciton-polaritons

Characteristic	Zinc blende compound semiconductors	Wurtzite compound semiconductors	Organics[a]
Examples	GaAs, InP, CdTe	GaN, ZnO	Cyanine dyes, PAHs, Pcs
ε_{opt}[b]	10–15	4–6	3–4
Exciton[c]	W-M	W-M	Frenkel
E_B	5–20 meV	20–60 meV	0.5–1 eV
γ_{ex}	0.1–1 meV	5–10 meV	25–100 meV
Rabi splitting	3–25 meV	30–150 meV	25–250 meV
Relaxation path	Acoustic phonons, ex-ex scattering	Acoustic, LO phonons, ex-ex scattering	Optical phonons, radiative

[a] PAH = polyaromatic hydrocarbon, Pc = phthalocyanine.
[b] ε_{opt} is the dielectric constant at optical frequencies.
[c] W-M = Wannier–Mott exciton.

are stable at room temperature. Combined with their large oscillator strengths, unique polariton phenomena such as Bose–Einstein condensation and polariton lasing can be observed in ambient laboratory conditions. These physical properties are also reflected in the large Rabi splitting energies of organics compared to either high or low dielectric constant inorganic semiconductors. Thus, organic semiconductors offer a realistic route to the exploitation of polariton phenomena in devices that are operated at, or even well above room temperature.

9.2.2 Optically pumped organic polaritons

Optical pumping of the microcavity is generally done at high $k_{\|}$, as shown in Fig. 9.11. This creates an exciton reservoir that continually replenishes polaritons that are lost due to photon leakage from the cavity. At high pump intensities, a BEC may form at $k_{\|} = 0$. The density at the LPB minimum however, is reduced by the exciton bottleneck at the inflection in the dispersion of the LPB. The large phonon energies characteristic of intramolecular vibrations partially avoids the formation of the bottleneck. This is in contrast to inorganic semiconductors where the phonon energies are at least an order of magnitude smaller, inhibiting transfer to $k_{\|} = 0$ of the LPB. Thus, strong exciton–exciton and exciton–phonon scattering in organics reduces the in-plane momentum, allowing the condensate to form.

The exciton–exciton scattering rate in a microcavity is (Tassone and Yamamoto, 1999)

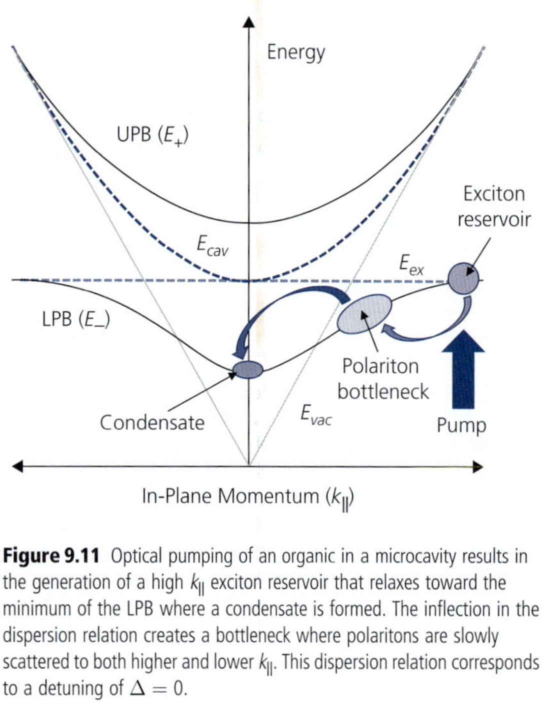

Figure 9.11 Optical pumping of an organic in a microcavity results in the generation of a high $k_{\|}$ exciton reservoir that relaxes toward the minimum of the LPB where a condensate is formed. The inflection in the dispersion relation creates a bottleneck where polaritons are slowly scattered to both higher and lower $k_{\|}$. This dispersion relation corresponds to a detuning of $\Delta = 0$.

$$\Gamma_{ex-ex} = \frac{\pi^2}{2}\left(n_{ex}m_{ex}a_0^4\right)\frac{E_B^2}{\hbar^3}, \tag{9.11}$$

where a_0 is the excitonic Bohr radius, E_B is the exciton binding energy, and n_{ex} is the exciton density per unit area in the microcavity. We have shown that a_0 for Frenkel excitons is ~10% of the radius for W-M states in inorganics (Eq. 3.142), but E_B and m_{ex} are 10–100 times that for inorganics. Also, n_{ex} can be 100 times larger for organics due to the large density of states (DOS). Taking this into consideration, $\Gamma_{\text{ex-ex}} \sim 10^2$–$10^4$ times higher for organics than for inorganic semiconductors, thereby reducing the bottleneck to condensation by an approximately equivalent amount.

The foregoing analysis suggests that narrow, homogeneously broadened exciton linewidths are necessary to observe strong coupling, which would perhaps rule out the possibility of such an observation in organics whose vibronically broadened states result in a linewidth greater than Ω. However, the first report of strong coupling in an organic employed the porphyrin, 4TBPPZn (Lidzey et al., 1998). Porphyrins are characterized by their narrow Q-band absorption in the visible and high luminescence efficiency. The room temperature Rabi splitting showed $\Omega = 160$ meV due to the ~10^3 times higher f_{ex} for the organic compared with inorganic InGaAs quantum wells, where strong coupling is observed only at very low temperatures. Another approach to achieving comparatively narrow linewidths is to use ordered organic layers known as *J-aggregates*. Comprised of cyanine dyes, these ultrathin films feature sharp absorption bands suitable for strong coupling. The molecules are precipitated from solution onto a substrate to form a an ordered morphology. When sandwiched between a dielectric DBR and a Ag mirror, Rabi splittings as large as 180 meV are observed (Lidzey et al., 1999).

Not long after these demonstrations, strong coupling was also observed in vacuum-deposited thin films with structural order extending possibly to only nanocrystalline dimensions. The demonstration employed a 60 nm thick layer of NTCDA sandwiched between DBR and metal mirrors. It was striking that both the 0–0 and 0–1 vibronic modes were hybridized within the cavity stop band, resulting in two exciton–phonon coupled states that effectively transferred energy between the LPB, the middle polariton branch (MPB), and the UPB. The Rabi splittings observed were exceptionally large: $\Omega_{0\text{-}0} = 280$ meV for the 0–0 exciton-polariton, and $\Omega_{0\text{-}1} = 120$ meV for the 0–1 state (Holmes and Forrest, 2004). Even though the absorption lines of amorphous NTCDA films are broad, polaritons can be observed when the vibronic relaxation time is longer than the Rabi lifetime, $\Omega/\hbar$.

This is equivalent to the condition that the Rabi splitting should be on the order of, or greater than the homogeneously broadened exciton linewidth (Skolnick et al., 1998, Holmes and Forrest, 2004).

An example of hybridization of the multiple vibronics of polycrystalline tetracene illustrates how strong coupling is quantified and analyzed, and illustrates the robustness of organic exciton-polariton states. Figure 9.12 shows the absorption spectrum of a vacuum deposited tetracene film. It features an extended progression of vibrational modes from 0–0 to 0–4, with the higher modes becoming less distinct due to a combination of decreasing intensity and line broadening. Interestingly, the Davydov components of the 0–0 mode (p-band) along the **a** and **b** crystalline axes are resolved using light that is selectifvely polarized along these axes. Recall from Section 3.6.4 that Davydov splitting results from Coulomb interactions between the two tetracene molecules within each unit cell. Although the domains with diameters ranging from 1–10 μm are axially symmetric in the substrate plane, the observation of Davydov modes is due to ordered stacking in the out-of-plane direction typical of many planar aromatic hydrocarbons.

To observe strong coupling, tetracene films were deposited onto an Al back surface mirror, and then capped by a semitransparent, 35 nm thick Ag top mirror through which the cavity is illuminated. The use of two metal mirrors results in $Q \sim 26$. Reflectivity spectra taken in 5° increments are provided in Fig. 9.13a. The spectra show a series of minima corresponding to the LPBs of both Davydov components. Two branches that are hybrids between the LPBs and the higher order vibronics are also visible. The UPB, with its photon-like behavior, is clearly resolved at approximately 2.8–3.0 eV.

Simulations of the dispersion data require that the crystalline anisotropy be taken into account. This is done using the dielectric tensor of tetracene, viz.

$$\varepsilon_{ij}(\omega) = \begin{pmatrix} \varepsilon_a(\omega)\cos^2\varphi + \varepsilon_b(\omega)\sin^2\varphi & \cos\varphi\sin\varphi(\varepsilon_a(\omega) - \varepsilon_b(\omega)) & 0 \\ \cos\varphi\sin\varphi(\varepsilon_a(\omega) - \varepsilon_b(\omega)) & \varepsilon_a(\omega)\sin^2\varphi + \varepsilon_b(\omega)\cos^2\varphi & 0 \\ 0 & 0 & \varepsilon_c \end{pmatrix}. \tag{9.12}$$

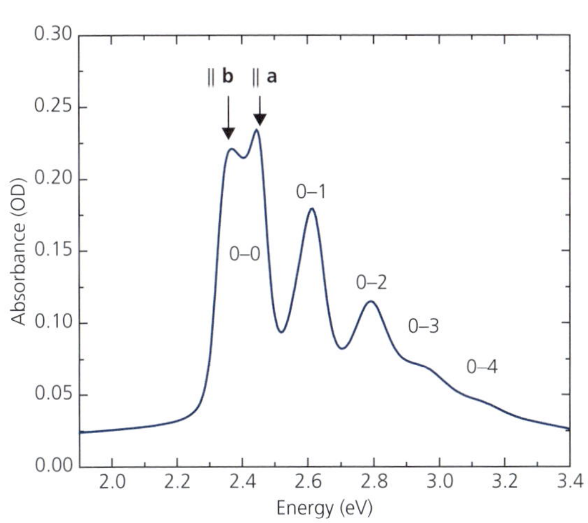

Figure 9.12 Absorption spectrum of a 140 nm thick, vacuum deposited polycrystalline tetracene film showing its vibronic progression. The 0–0 peak is split into its Davydov components parallel to the **a** and **b** crystalline axes.

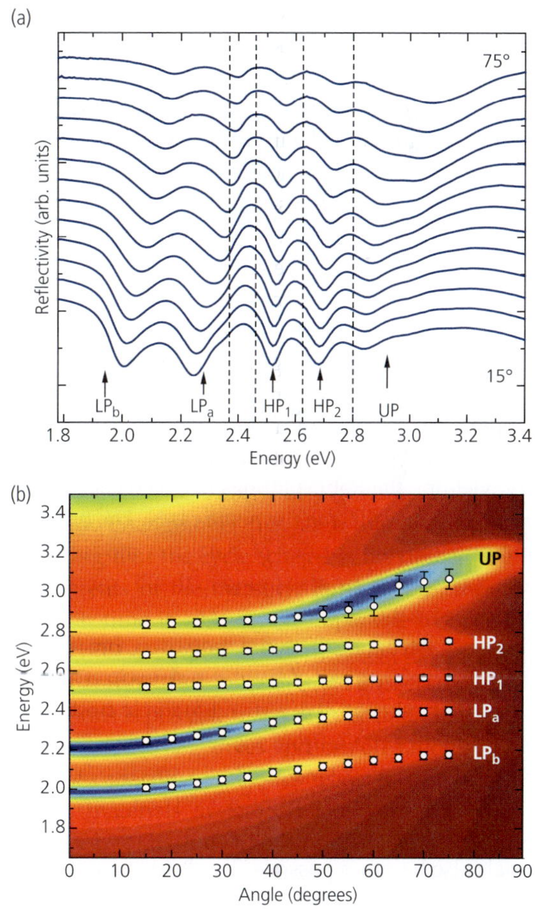

Figure 9.13 (a) Angle-dependent reflectivity spectra of a microcavity containing a 160 nm thick tetracene film. The spectra are taken in 5° increments. The dotted lines indicate the energies of the bare resonances observed in the absence of a cavity. Positions of the two lower polariton branches (LPBs) from the Davydov splitting of the 0–0 exciton, hybrid branches (HP) and upper branch (UP) are indicated. (b) Color contour plot of the simulated average *s*-polarized reflectivity of the tetracene microcavities. The data points are taken from the plot in (a) (Kéna-Cohen and Forrest, 2008).

Here, φ is the angle of the principal crystalline axis relative to the laboratory frame x-axis, ω is the light frequency, and ε_a, ε_b, and ε_c are the relative dielectric constants along **a**, **b**, and **c** directions of the triclinic tetracene unit cell. Also, the in-plane momentum is obtained from a projection of the angle of the incident light to the sample, θ. Equation 9.1 gives the cavity photon energy, from which we obtain the dispersion relation:

$$E(k) = E_0\sqrt{1 + \hbar^2 c^2 k_{\|}^2 / n_{eff}^2 E_0^2}, \tag{9.13}$$

where $E_0 = hc/n_{eff}L$ is the energy at $k_{\|} = 0$, and n_{eff} is the effective index of refraction of the uncoupled, non-absorbing cavity taking into account field penetration into the mirrors. Now (Skolnick et al., 1998),

$$k_{\|} = \frac{E(k)}{\hbar c}\sin\theta. \tag{9.14}$$

Substituting Eq. 9.14 into Eq. 9.13 gives the dependence of energy on the light incident angle:

$$E(\theta) = E_0\left(1 - \frac{\sin^2\theta}{n_{eff}^2}\right)^{-1/2}. \tag{9.15}$$

Solutions to the 5×5 eigenvalue equation (cf. Eq. 9.6) are solved for a five-level (see Fig. 9.12) coupled oscillator. The results are provided in the color map in Fig. 9.13b for the two LPBs, the hybrid branches coupled to the 0–1 and 0–2 vibronics, and the UPB. The data (points) are accurately fit by this theory, yielding the Rabi splittings between each pair of bands of nearly 400 meV in some cases. The branches and anticrossing features of polariton modes in the reflection spectra are also found in the PL spectrum of many materials.

Energy exchange between polariton branches is due to hybridization of cavity and exciton modes. Or, as we have seen, it can also result in hybridization of multiple vibronics from the same molecules. Likewise, if more than one excitonic species (e.g. two organics) with near resonant excited states are blended into a cavity, energy exchange between these materials also becomes possible. This leads to many new possibilities for devices, and for exploiting emergent energy transport phenomena.

An example of the utility of strong coupling can be found in the use of hybridized vibrational modes of different molecules to catalyze reactions between them. Strong coupling between carbonyl, C=O stretching modes of PMMA and DMF in Fig. 9.14 is possible due to their very narrow transmission spectral linewidths and near resonance at 1729 cm^{-1} and 1662 cm^{-1}. The IR states exhibit strong coupling by sandwiching a blend of PMMA and DMF between two DBRs consisting of Ge ($n = 4.01$) and ZnS ($n = 2.24$), with a cavity resonance at $E_{cav} = 0.21$ eV that couples to both states. The DBR materials are chosen for their transparency in the near IR, and the ease of film growth by vacuum thermal evaporation. The appropriate thickness of each layer in the DBR is $\lambda/4n$. The transmission spectra of the blend result in three polariton branches corresponding to hybridization of the photon and the two IR modes of PMMA and DMF.

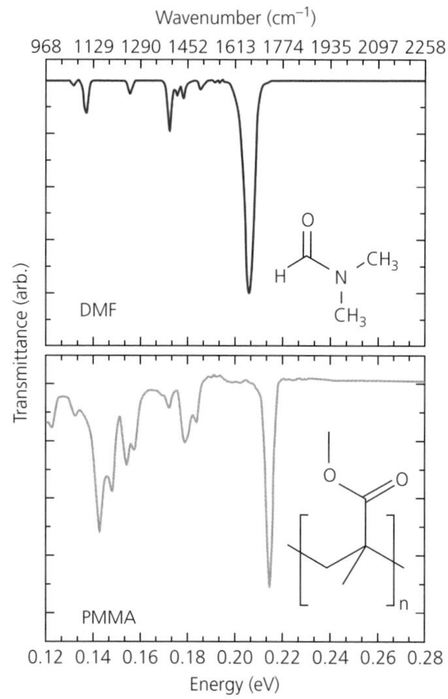

Figure 9.14 Infrared transmission spectra of DMF and PMMA. Molecular formulae of the molecules are shown in the insets (Muallem et al., 2016).

The measured dispersion relationships are shown in Fig. 9.15a. The simulated dispersion curves are shown by solid lines. These curves are obtained by solutions to the 3×3 secular matrix of Eq. 9.6 assuming $\gamma = 0$. The polariton state mixing coefficients, α, β, and γ, are the new eigenvectors of the strongly coupled system, and give the relative contributions from the DMF, and PMMA vibronic states, and the cavity mode, respectively, in each of the three branches. Thus α, β, and γ, are found from solutions to

$$\begin{bmatrix} E_{cav} & V_{DMF} & V_{PMMA} \\ V_{DMF} & E_{DMF} & 0 \\ V_{PMMA} & 0 & E_{PMMA} \end{bmatrix}\begin{bmatrix}\alpha \\ \beta \\ \gamma\end{bmatrix} = E\begin{bmatrix}\alpha \\ \beta \\ \gamma\end{bmatrix}. \tag{9.16}$$

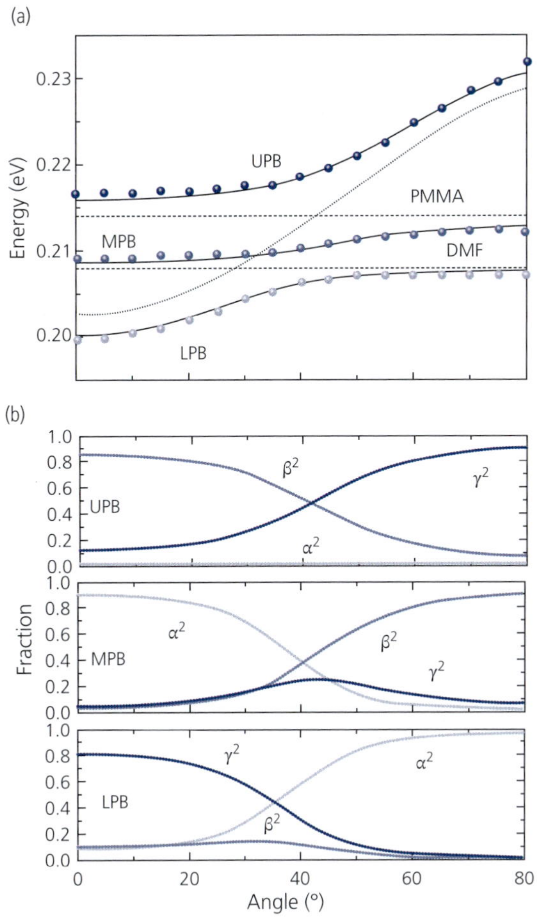

Figure 9.15 (a) Measured (points) and calculated (solid lines) infrared dispersion curves of a PMMA:DMF microcavity showing the upper, middle, and lower (UPB, MPB, and LPB, respectively) polariton branches, giving evidence for strong coupling of carbonyl modes in the two materials. Dashed lines correspond to the empty cavity mode, and the PMMA and DMF vibrational states. (b) Solutions to the energy eigenvalue equation give the Hopfield coefficients that show the relative contributions from the DMF, PMMA, and photon modes (α^2, β^2, γ^2, respectively) in the various branches (Muallem et al., 2016).

Here, the two interaction potentials are (V_{DMF}, V_{PMMA}), and the uncoupled exciton energies are (E_F, $E_{W\text{-}M}$). The potentials are related to the Rabi frequencies by $V = \Omega/2$. The calculated mixing coefficients for each branch are shown in Fig. 9.15b.

The UPB is principally photon-like (γ) at high angles, whereas the PMMA contribution (β) is strongest at low angles. There is almost no contribution from DMF (α) in the upper branch. The situation is reversed for the LPB, where near $k_{\|} = 0$, the contribution from DMF is small compared to that of the photon mode. At high angles, the relative contributions are reversed, and the branch becomes almost completely excitonic. The MPB is most interesting, since it consists of contributions from both molecular states, with only a minor contribution from both cavity mode. However, at approximately 40°, $\alpha^2 = \beta^2$ indicating complete mixing and energy transfer between PMMA and DMF. The PMMA content is highest at large $k_{\|}$, and the DMF contribution is dominant as $k_{\|} \to 0$. The cavity mode contribution is substantial in the transition region where there is complete mixing between the DMF, PMMA and photon modes (Muallem et al., 2016). One anticipates that chemical interactions can be enhanced from this mixing within the MPB. Indeed, the control of the kinetic coupling between spiropyran and merocyanine derivatives by coupling to cavity modes has been reported, opening the path to using quantum electrodynamic effects to tune the rate of chemical reactions between species whose vibronic states lie within the cavity resonance (Hutchison et al., 2012).

9.2.3 Hybrid organic/inorganic semiconductor polaritons

Energy transfer from excitonic to conventional semiconductors is achieved by hybridization of Frenkel and W-M excitons in the same resonant cavity. The first observation of this hybridization was achieved by combining W-M excitons in red emitting InGaP multiple quantum wells (QWs) that are resonant with the Soret band of TPP, using the structure in Fig. 9.16a. The bottom mirror is a DBR composed of 24 pairs of undoped $Al_{0.46}Ga_{0.54}As$ ($n = 3.45$) and AlAs ($n = 3.04$), grown by gas source molecular beam epitaxy on a GaAs substrate. The nine, 12 nm thick $In_{0.52}Ga_{0.48}P$ QWs between 12.5 nm thick $Al_{0.4}Ga_{0.6}In_{0.5}P$ barriers are similarly grown to form the inorganic active region. A 320 nm thick TPP layer is deposited onto the 75 nm thick AlGaInP spacer by VTE to produce a nanocrystalline film. Finally, the TPP is capped with a 5.5 pair DBR consisting of sputter-deposited TiO_2 ($n = 2.30$) and SiO_2 ($n = 1.46$).

The index of refraction profile and the resulting electric field within the cavity is shown in Fig. 9.16b. Several field antinodes coincide with the strategically placed, three QW groups spaced by AlGaInP. Also, the TPP is sufficiently thick to contain several field maxima, ensuring strong coupling to both the inorganic and organic regions of the cavity. Strong coupling is observed at both 100 K and 4 K (Holmes et al., 2006). The reduced temperature is necessary to ensure the stability of the lower binding energy, W-M state. The dispersion relationships shown in Fig. 9.17a taken from the reflectivity spectra at $T = 100$ K show hybridization of the W-M and Frenkel excitons in the

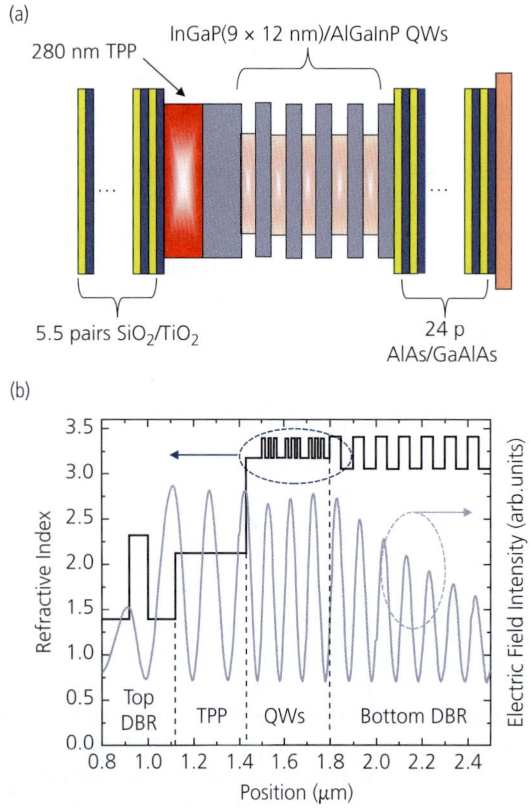

Figure 9.16 (a) Hybrid organic/inorganic polariton cavity comprising a 24-period AlAs/GaAlAs DBR, 9-period InGaP/AlGaInP quantum wells/Al GaInP spacer/TPP/5.5-period SiO_2/TiO_2 DBR. (b) Index of refraction and electric field profiles across the cavity.

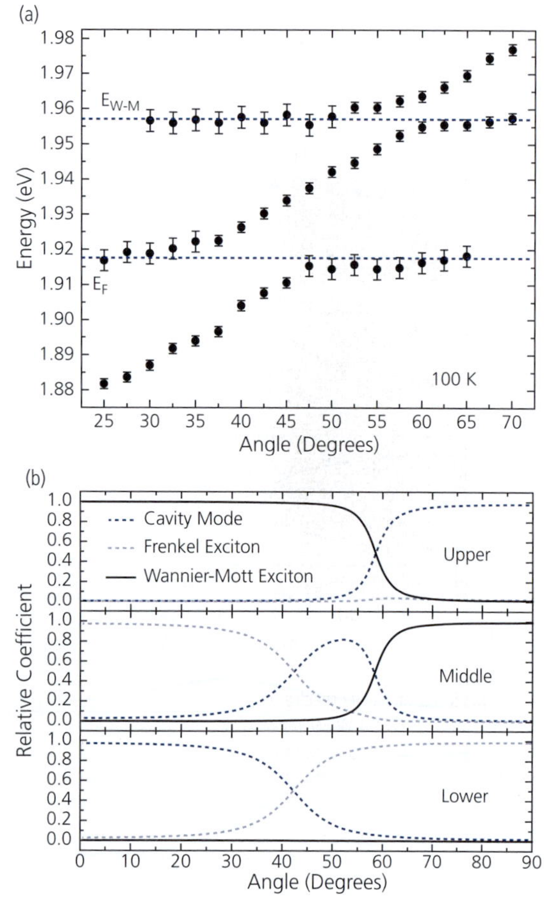

Figure 9.17 (a) Measured (points) and calculated (lines) dispersion relations showing strong coupling between the Frenkel (F) and Wannier-Mott (W-M) states of the structure in Fig. 9.16. Also shown are the respective uncoupled exciton energies, E_F and $E_{W\text{-}M}$. (b) Mixing coefficients for the various modes in the three polariton branches. After Holmes et al. (2006).

MPB, with mixing coefficients in Fig. 9.17b confirming the extent of the hybridization. Due to the energetic separation of the uncoupled excitonic states, there is only approximately 20% Frenkel and W-M character at 55°. Pronounced and rapid energy transfer between these two very different excitonic states is promoted by strong microcavity coupling analogous to the formation of a hybrid charge transfer exciton (HCTE) at an organic/inorganic heterojunction (cf. Section 4.7.3).

Room temperature W-M/Frenkel hybridization has been achieved between ZnO and NTCDA. ZnO has a strongly bound (and hence thermally stable) W-M state with $E_B = 60$ meV. The cavity structure is shown in Fig. 9.18, inset, with the dispersion relation obtained from its reflectivity spectra shown in the figure. An interesting feature of this cavity is that the NTCDA 0–1 vibronic is in near perfect resonance with the ZnO exciton. Fits to the data using an approach similar to Eq. 9.16 yield a Rabi splitting energy of $\Omega_2 = 320$ meV between the middle and upper branches. The lower branch splitting is roughly half that value. The resonance between the ZnO and NTCDA 0–1 state leads to strong mixing, with $\alpha^2 = \beta^2 = 0.4$ at an angle of 30°, suggesting strong energy transfer even at room temperature (Slootsky et al., 2014).

9.2.4 Polariton lasers and Bose–Einstein condensation

A particularly attractive feature of polaritons is their ability to form a BEC in the LPB. Emission from the condensate is coherent, and unlike a laser, there is no requirement for population inversion. Hence, while it shares many of the characteristics of conventional lasers, emission from the equilibrium condensate does not require a threshold pump intensity as long as the LPB population can be supplied from the exciton reservoir at a rate that is greater than or equal to the polariton decay rate itself. Formation of a condensate at the polariton ground state energy is inhibited by a strong polariton bottleneck, shown in Fig. 9.11. However, we have already noted that the bottleneck is considerably reduced in organics compared with inorganics due to their high phonon energies and strong exciton–exciton scattering.

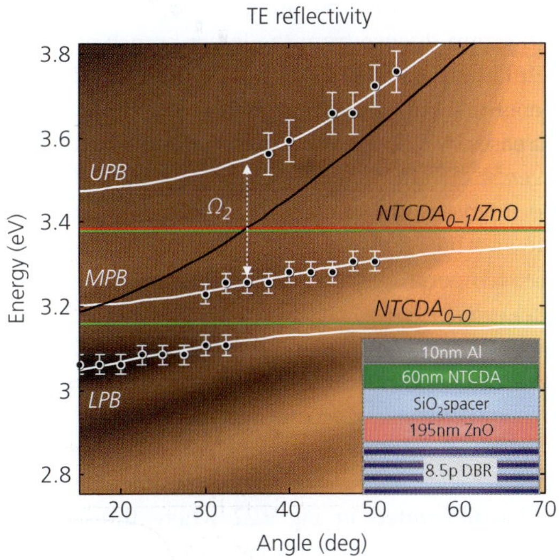

Figure 9.18 TE reflectivity spectra (points) and fits to the dispersion relations of a ZnO/NTCDA W-M/Frenkel hybridized polariton cavity (white lines). The uncoupled exciton states are shown by the horizontal dashed lines. The NTCDA 0–1 vibronic is resonant with the ZnO W-M state. The cavity mode is shown by the curved black line. Inset: Structure used in the measurements (Slootsky et al., 2014).

Room temperature polariton lasing in anthracene (Kéna-Cohen and Forrest, 2010) contrasts with lasing in inorganics at low temperature needed to stabilize the W-M exciton (Deng et al., 2003). Anthracene single crystals have narrow linewidths that are suitable for both strong coupling and polariton lasing. (Kéna-Cohen et al., 2008) The cavity structure is shown in the inset of Fig. 9.19. Two DBRs are deposited onto separate quartz substrates. Gold strips are then deposited on the DBRs whose combined thickness is equal to the cavity length, *L*. The substrates are brought into contact, and the Au strips cold-weld bond by applying pressure (Section 5.6.4) to form the empty cavity. An open end of the cavity is immersed into a molten solution of anthracene that is drawn into the gaps between Au strips by capillary action (Section 5.4.1). Nucleation at the strip edges results in the growth of continuous, single crystals, as shown in the upper micrograph in the left inset of Fig. 9.19. In contrast, a film of anthracene grown on the substrate without the

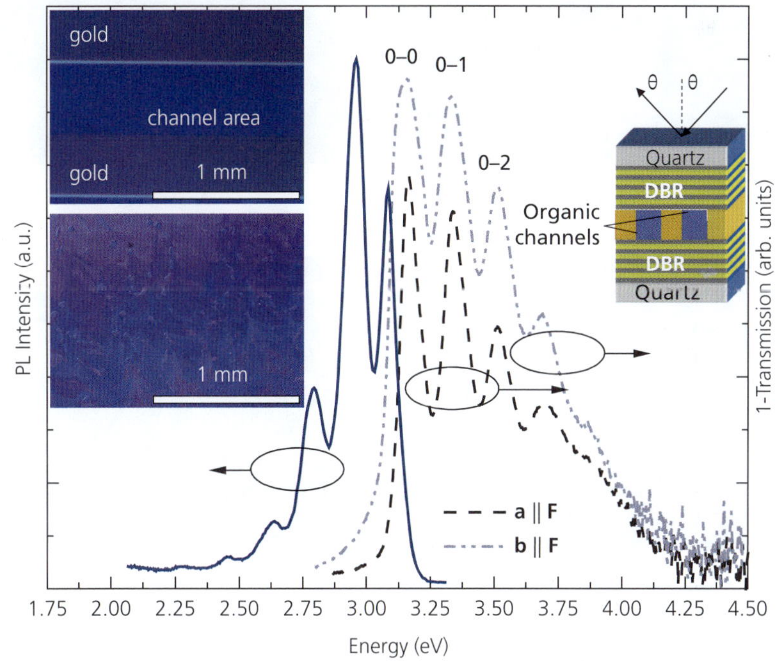

Figure 9.19 Right: Inverse transmission (1-T) spectrum of a 120 nm thick anthracene single crystal channel for electric field polarizations along the **a** and **b** crystal axes. Left: Photoluminescence spectrum of the anthracene single crystal. The vibronic spectra mirror the absorption spectrum, with the highest emission energy peak narrowed by self-absorption. Right inset: Schematic of the microcavity. Left insets: Phase contrast micrographs of the channel area within the cavity (top) and a thicker area far away from the gold channels where multiple domains are observed (bottom) (Kéna-Cohen et al., 2008).

constrained dimensions of the cavity forms a rough, polycrystalline film, as shown in the lower micrograph. The spectra of the crystalline films shows Davydov splitting between light polarized along the **a** and **b** crystal axes, similar to that found for tetracene in Fig. 9.12. The narrow linewidths are suitable for demonstrating strong coupling. Thus, the growth of single crystals within a precisely fabricated high-Q ($\approx$600) cavity are essential steps to realizing polariton lasing at thresholds well below that assumed for conventional anthracene lasers (Kéna-Cohen and Forrest, 2010).

The dispersion relations for cavity luminescence at room temperature and 77 K are shown for the lower polariton **b** polarized branch (LP_b) in Fig. 9.20. A signature of condensation is the abrupt and significant narrowing of the angular dispersion near the ground polariton state as the pump power is increased above threshold. Furthermore, the luminescence stays within the LP_b branch rather than moving into the cavity mode, as would be expected for conventional lasing.

The transition to polariton lasing is accompanied by a rapid increase in the slope of the output intensity vs. pump fluence, as well as a narrowing of the output emission spectrum. The trends in output intensity, spectral linewidth and angular extent of emission with pump fluence from the lower branch at ~120 K are plotted in Fig. 9.21. The lasing threshold is at a fluence of 17 nJ, at which point the spectral linewidth decreases by a factor or 2 and the angular extent of the lower branch luminescence decreases from 40° to 7°. The intensity dispersion in Fig. 9.20 approximately follows a Bose-Einstein distribution with a characteristic temperature of 326 K when plotted vs. $E(k_{\parallel}) - E_0$, suggesting that the condensate is near equilibrium. Also, the PL lifetime dramatically decreases from 1 ns below threshold, to $<$ 30 ps above threshold (Kéna-Cohen and Forrest, 2010).

Unlike a conventional laser, the polariton laser characteristics are strongly temperature dependent. This is illustrated in Fig. 9.22 where threshold vs. temperature of a polariton laser is compared with a

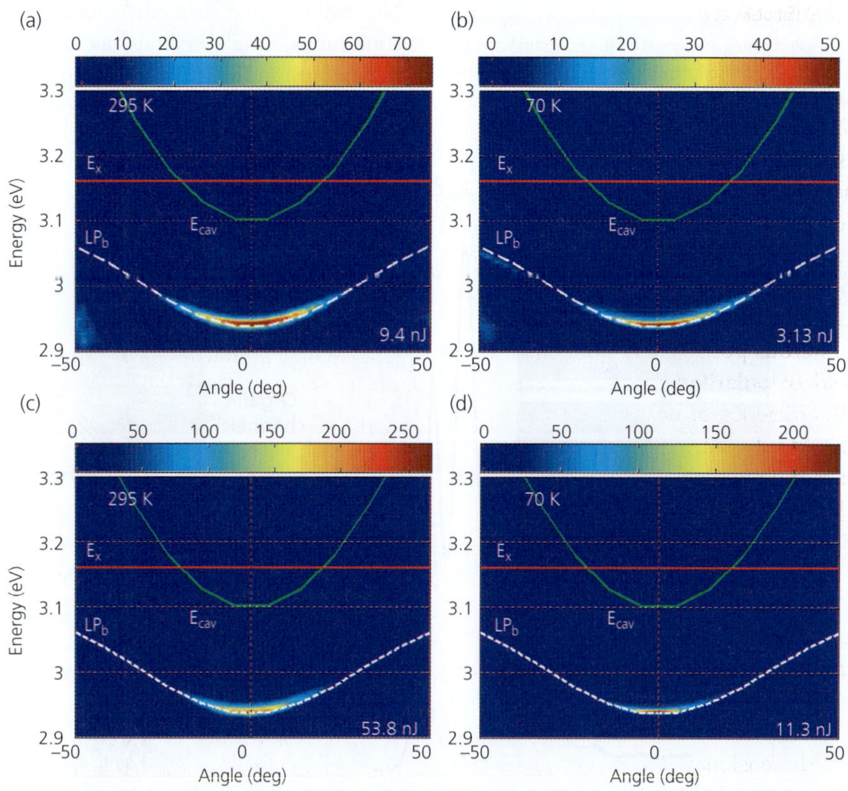

Figure 9.20 Angle-resolved polariton photoluminescence showing dispersion in LP_b below (a) and (b), and above (c) and (d) the lasing threshold at room temperature and 70 K. The dispersion fit to a coupled-oscillator model (dashed line) remains unchanged, indicating strong coupling both below and above threshold. The cavity photon (green line) and the lowest anthracene exciton absorption (red line) are detuned by $\Delta = -60$ meV. Also, the pump fluences at $\lambda = 360$ nm are indicated (Slootsky et al., 2012).

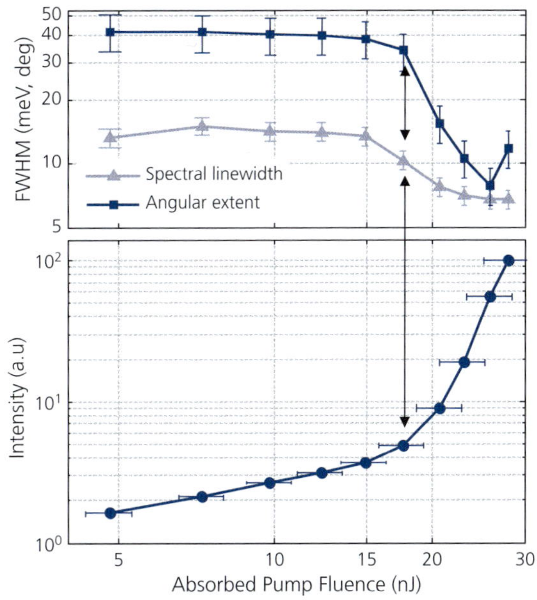

Figure 9.21 Integrated intensity (circles), spectral linewidth (triangles), and angular extent of polariton emission (squares) at 120 K as functions of absorbed pump fluence. All three parameters exhibit a transition at the lasing threshold at 17 nJ (Slootsky et al., 2012).

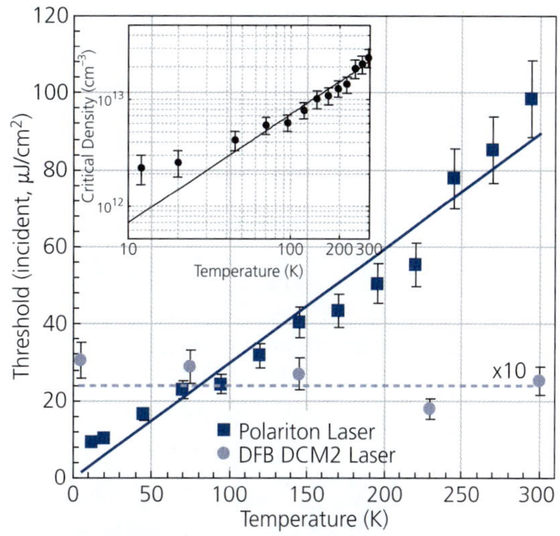

Figure 9.22 Temperature-dependent threshold for the anthracene polariton laser (squares) and a conventional organic distributed feedback laser (circles). The conventional laser threshold is temperature independent, in marked contrast to the polariton laser. Lines are a guide to the eye. Inset: Calculated polariton population assuming 50% absorption of incident light and a polariton quantum efficiency of 10^{-6} (circles), and the critical density for condensation in the thermodynamic limit (line) as functions of temperature (Slootsky et al., 2012).

conventional, low threshold laser employing an Alq_3 host doped at 2 wt% with the red-emitting fluorophore, DCM2, described in Section 6.8. The conventional laser has a threshold that is approximately ten times lower than the polariton laser. Furthermore, its threshold and other lasing characteristics are temperature independent owing to the "quantum dot" nature of isolated fluorescent molecules in a wide energy gap host matrix that contribute to the laser emission.

There are numerous possible explanations for the higher threshold of polariton lasers, including the much lower PL efficiency of neat anthracene (~0.6) compared to DCM2 doped in a host matrix (~0.9), and temperature-independent losses in the anthracene cavity. Note that the increase in polariton lasing threshold with temperature is opposite to that expected for condensation limited by a bottleneck between the ground state and the exciton reservoir, where lack of thermally-assisted scattering results in an increase in threshold as temperature is decreased.

The critical density for BEC is calculated by assuming a finite two-dimensional system approximately equal to the pump spot of radius, R, using the Bose–Einstein distribution function:

$$N_c = \frac{1}{\pi R^2} \sum_{k_{\|} \geq 2\pi/R} \left\{ \exp\left[\left(E(k_{\|}) - E(0) - \mu \right) / k_B T \right] - 1 \right\}^{-1}. \tag{9.17}$$

The chemical potential, μ, approaches 0, allowing for an unlimited number of bosons in excess of the critical density, N_c, to populate the ground state. Substituting the LP_b dispersion and evaluating the sum in Eq. 9.17 gives the critical density as a function of temperature, as shown by the line in the inset in Fig. 9.22. The calculated linear regime for anthracene extends above room temperature to nearly 400 K due to strong light-matter interactions compared to that of analogous inorganic systems.

The polariton density is $N = 0.5 \times \eta \times N_{inc}/V$ for a PL quantum efficiency, η, and an estimated 50% absorption of incident light by the active medium of volume, V. Also, N_{inc} is the number of incident photons. Comparing the experimental threshold pump fluence with the critical BEC density in the thermodynamic limit in Eq. 9.17 yields a quantum efficiency of only $\eta \approx 10^{-6}$. The discrepancy between N_c (which assumes an infinite particle lifetime) and the pump power required for lasing arises from the existence of loss pathways for the relaxation of excitons from the reservoir to the bottom of the polariton branch.

The strength of exciton–photon interactions in organics has led to observations of BEC in the disordered polymer, MeLPPP, sandwiched between two Ta_2O_3/SiO_2 DBRs. This ladder-type polymer has a relatively narrow emission spectrum with a peak at 2.68 eV (Plumhof et al., 2014). The device exhibits another interesting feature of polariton lasers: spectral blue shifts with pump fluence shown in Fig. 9.23a–c. There are several notable characteristics of these data. The pronounced decrease in the extent of in-plane momentum above threshold is accompanied by symmetric lasing intensity maxima about $k_{\parallel} = 0$. This is due to incomplete thermalization from the polariton bottleneck between the exciton reservoir and the LPB ground state. This, in turn, results in an increase in threshold as temperature is decreased, opposite to that of the crystalline anthracene laser. Furthermore, laser emission (not shown) is blue-shifted from the polariton emission by approximately 10 meV, although it remains >20 meV below the cavity mode energy. Note that a second population of non-lasing polaritons is also apparent in Fig. 9.23c at lower energies, possibly due to the second higher vibronic state of the polymer that also is coupled to the cavity. The blue shift with pump fluence is due to repulsive polaron interactions, and is opposite to the red shift experienced by conventional lasers that are pulled toward regions of higher gain as the transparency increases above threshold. It is notable that neither the disordered polymer nor crystalline anthracene polariton lasers make a transition to conventional lasing. This differs from inorganic lasers, possibly since the conventional lasing threshold may exceed the damage limit of the organics.

An additional feature of a polariton laser is spatial beam coherence, illustrated by the interferograms in Fig. 9.24 for a thermally evaporated, amorphous TDAF layer sandwiched between six-period, Ta_2O_3/ SiO_2 DBRs. The cavity $Q \sim 600$ leads to Rabi splitting energies of 580 meV for a detuning of $\Delta \sim 400$ meV.

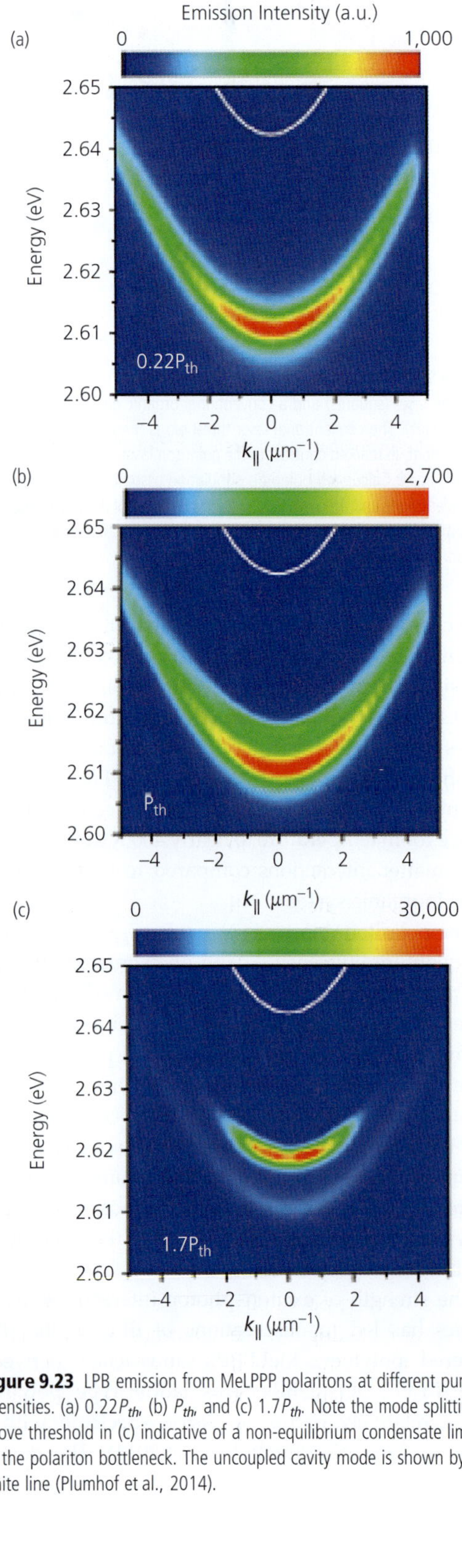

Figure 9.23 LPB emission from MeLPPP polaritons at different pump intensities. (a) $0.22P_{th}$, (b) P_{th}, and (c) $1.7P_{th}$. Note the mode splitting above threshold in (c) indicative of a non-equilibrium condensate limited by the polariton bottleneck. The uncoupled cavity mode is shown by the white line (Plumhof et al., 2014).

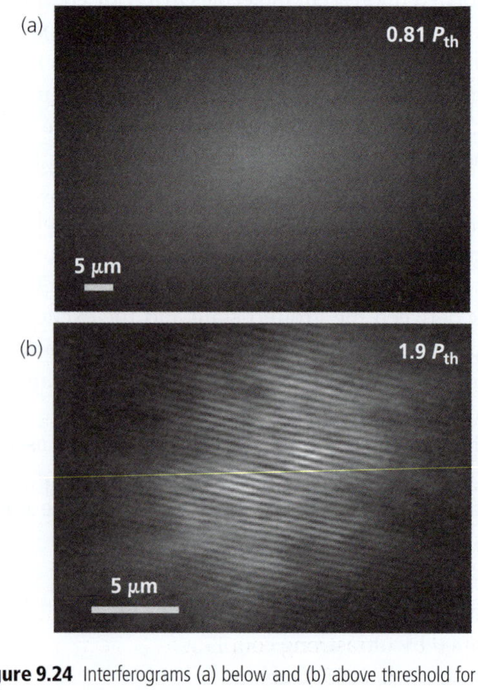

Figure 9.24 Interferograms (a) below and (b) above threshold for a TDAF polariton laser (Daskalakis et al., 2014).

Below threshold, no fringes appear (Fig. 9.24a), indicating a total absence of spatial coherence. At $1.9P_{th}$, ordered interference fringes are visible across the entire emitting area of the laser, indicative of coherent emission. Also, the emission is TM polarized, independent of the pump beam polarization (Daskalakis et al., 2014). The presence of fringes and polarized emission both point to spatial and temporal coherence.

We can draw several conclusions from the foregoing observations of polariton lasing. The first is that it can be observed for both crystalline and amorphous films as long as strong coupling is present. As noted above, this is possible when the exciton and cavity linewidths are less than the Rabi splitting that, unlike conventional semiconductors, can be very large. Further, polariton lasing exhibits features that distinguish it from conventional lasing, such as a blue shift in emission above threshold, and in some cases an increase in threshold as temperature is decreased. Polariton lasing has a threshold far below that of its conventional lasing threshold. The existence of a low threshold is a consequence of the reduced polariton bottleneck whereby supply of polaritons from the exciton reservoir exceeds losses due to a combination of non-radiative and radiative processes. To date, the thresholds remain too high to enable electrically pumped organic polariton lasing at room temperature. Cooling is not expected to improve this situation. Finally, like a conventional laser, the emission from a polariton laser is polarized, and has both temporal and spatial coherence.

9.2.5 Ultrastrong coupling and polaritonic OLEDs and OPDs

In spite of the high thresholds for BEC formation, a step toward realizing electrically pumped polaritonic lasers has been made by the demonstration of a polaritonic OLEDs and organic photodetectors. Both device types employ a pair of metal mirrors, resulting in *ultrastrong coupling* between the exciton and photon modes. Here, ultrastrong coupling is defined as the condition where the Rabi splitting is an appreciable fraction of the exciton energy, that is, when the nominal condition that $\Omega \geq 0.1E_{ex}$ is met. The simple dipolar interactions employed in analyzing strongly coupled polariton dynamics must be modified to include nonlinear effects such as antiresonant couplings in the ultrastrong coupling regime (Ciuti et al., 2005, Kéna-Cohen et al., 2013).

Small molecule based OLEDs whose spectra are modified by ultrastrong coupling have been reported (Tischler et al., 2005, Gubbin et al., 2014, Genco et al., 2018). Ultrastrong coupling has resulted in a large

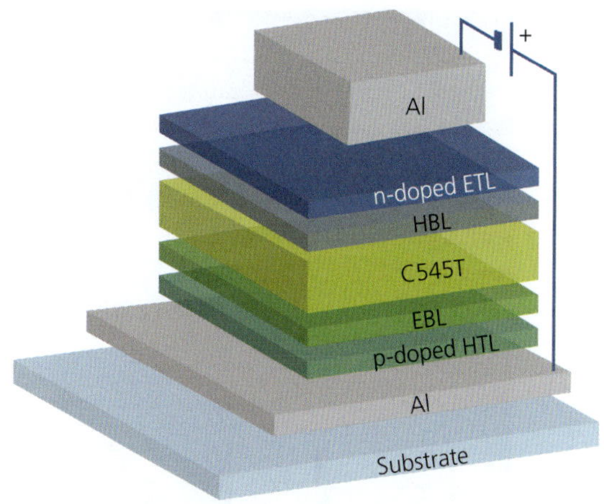

Figure 9.25 Polaritonic OLED structure exhibiting ultrastrong coupling via both electrical and optical pumping. The cavity is formed by the two Al mirrors, and the EML comprises a 55 nm thick neat C545T fluorescent layer (Genco et al., 2018).

exciton to polariton conversion efficiency approaching 25% in the device structure in Fig. 9.25. This device comprises a 55 nm thick C545T green fluorescent emitting layer (EML). The amount of detuning is adjusted by varying the thickness of the EML as well as the surrounding charge injecting layers from a total thickness of from 110–130 nm. Ultrastrong coupling is achieved by sandwiching this conventional OLED structure between two Al electrodes, with emission emerging from the bottom, 25 nm thick semitransparent anode.

Ultrastrong coupling was verified by measuring the reflectance due to optical pumping. There the ratio of the Rabi splitting energy to that of the bare exciton was a remarkable $G = 0.92\text{ eV}/2.5\text{ eV} = 0.37$, which is well beyond the nominal threshold of 0.1 for ultrastrong coupling. Emission intensity vs. angle from OLEDs with cavity detunings of 26 meV (A1) and −283 meV (A2) are shown in Fig. 9.26. The emission follows the expected dispersion curve of the LPB, since the polaritons relax into the lowest available energy state more rapidly than their recombination rate. Also, this dispersion shows that the emission is due to coupling to the electrically generated excitons. Note that the emission from cavity A2 with the large detuning is not centered at the LP branch minimum, but rather has the highest intensity at approximately 40° from normal. This is due to the polariton bottleneck, preventing polaritons that are pumped at high angles from relaxing into the minimum energy state at 0°. The external quantum efficiency of polariton emission is 0.2%, which corresponds to an approximately 25% coupling of exciton to polariton state emission. Also, the OLEDs exhibit a maximum intensity of 700 cd/m^2.

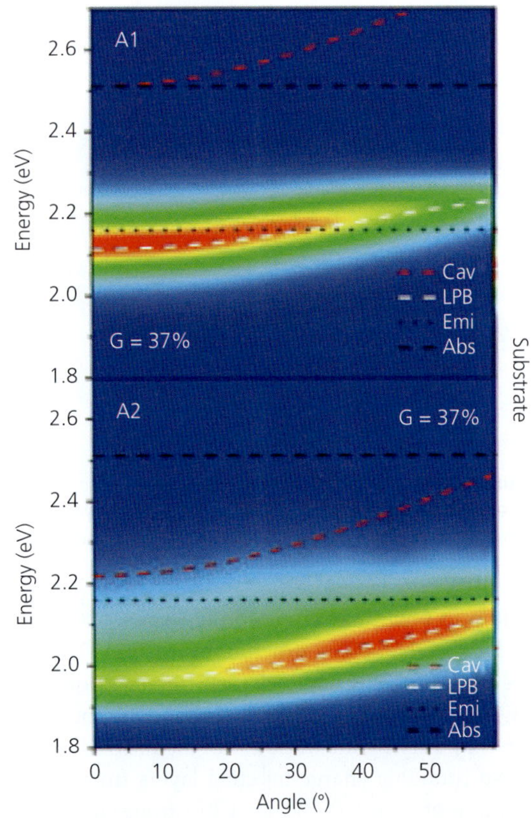

Figure 9.26 Polariton OLED emission from the LP branch for two different cavity detunings: A1 = 26 meV, and A2 = −283 meV. The bare exciton absorption (black dashed line), the photon dispersion (red dashed line), the exciton emission (black dotted line), and the LP branch dispersion (white dashed line) energies are also shown. The color map spans from highest (red) to lowest (blue) emission intensity (Genco et al., 2018).

Ultrastrong coupling has been exploited in a polaritonic organic photodiode (OPD), where splitting between the branches results in a large red shift of the LPB. This permits photodetection at much longer wavelengths than for the same materials used in an uncoupled detector. Conceptually, this can be understood by the illustration in Fig. 9.27. Like the polaritonic OLED, the active region is sandwiched between metal mirrors to create a strong coupling environment. The organic active region consists of a donor–acceptor junction that can dissociate the polaritonic state, thereby generating photocurrent. Ultrastrong coupling splits the uncoupled excitonic absorption spectrum into widely separated upper and lower branches. The LPB is significantly red-shifted, with a nearly equal blue shift of the UPB. This, in principle can result in absorption deep into the IR that is typically poorly accessed by the bare organic films due to rapid recombination of IR excitons as limited by the energy gap law (see Section 3.6.1).

A NIR absorbing OPD employed a structure analogous to that in Fig. 9.27. The structure is a planar-mixed HJ (see Section 7.1.2): Ag anode (75 nm)/HATCN (15 nm)/CuPc (15 nm)/SnPc:C_{60} (0.73:0.27 by vol., 110 nm)/C_{60} (30 nm)/Bphen (15 nm)/LiF (1 nm)/Ag cathode (25 nm). The absorption of the SnPc donor has a peak at 850 nm, with a tail extending to 1 μm. The absorption and external quantum efficiency (η_{ext}) spectra of the device are provided by the color maps in Fig. 9.28a and b, respectively. Remarkably, a very large Rabi splitting of $\Omega = 0.70$ eV is 42% of the uncoupled exciton wavelength. The wavelength shift of the LPB is as large as 0.44 eV, and the blue shift of the UPB is 0.36 eV. The η_{ext} spectrum follows that of

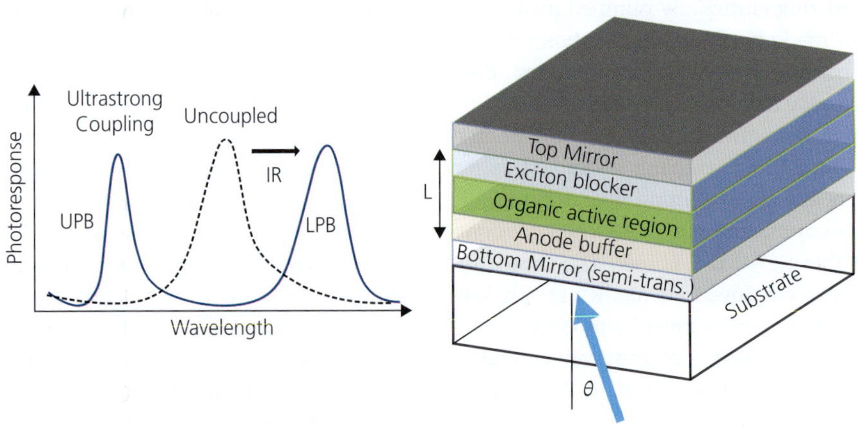

Figure 9.27 (right) Schematic of a strongly coupled OPD in a cavity comprising two metal mirrors. The thin lower mirror is semi-transparent. (left) The spectrum of an uncoupled exciton is split by a significant fraction of the exciton energy by ultrastrong coupling. The lower branch is shifted into the infrared (IR).

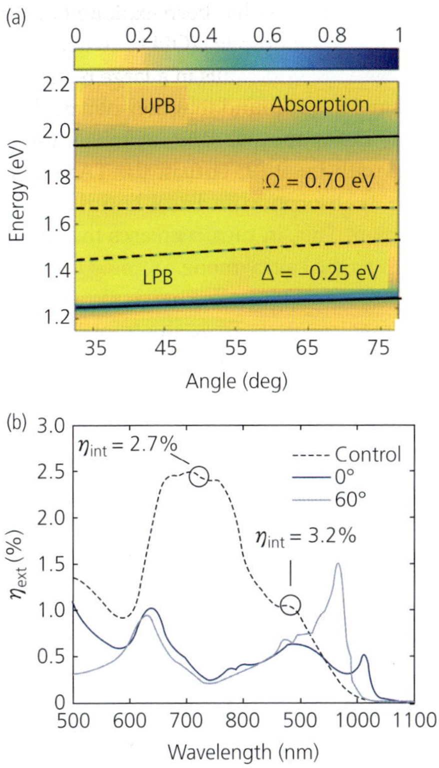

Figure 9.28 (a) Absorption dispersion for an ultrastrong coupled SnPc: C_{60} OPD. Solid lines are fits to coupled oscillator theory assuming a Rabi splitting of 0.70 eV and a detuning of −0.25 eV. The dashed lines are the bare exciton (horizontal) and cavity modes (diagonal). (b) OPDs with weak coupling (control, dashed line) and strong coupling at two different incident angles (Eizner et al., 2018).

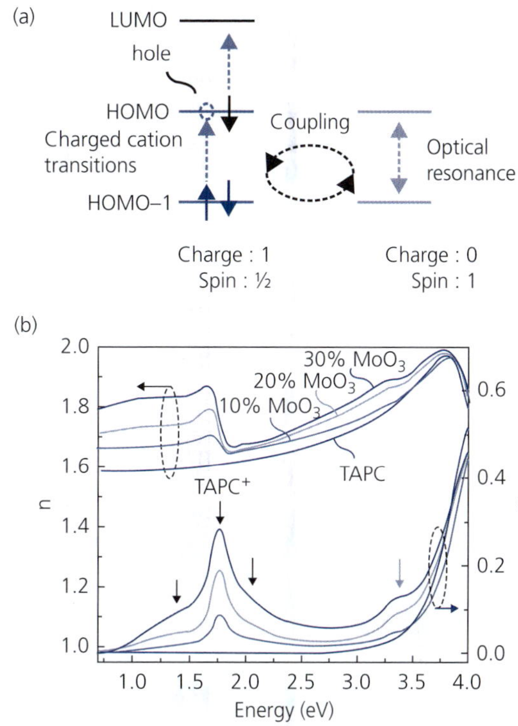

Figure 9.29 (a) Energy scheme for a hole polaron-polariton on a cationic molecule. The hole in the HOMO level is coupled to the electron in HOMO-1 via resonance with the cavity field. (b) Index of refraction and extinction coefficient of neutral (blue vertical arrow) and cationic (black arrows) TAPC as a function of MoO_3 doping concentration (Cheng et al., 2018).

the absorption in Fig. 9.28a. Additional spectral details are resolved in the internal (η_{int}) and external quantum efficiency plots in Fig. 9.28b. The strongly coupled spectra at 0° and 60° reveal the two peaks at short and long wavelengths. The external efficiency at long wavelength is considerably higher than for an analogous, uncoupled device (dashed line), employing a non-reflective ITO anode. In particular, the peaks at 970 nm and 1 μm for incident angles of 60° and 0°, respectively, are considerably more intense than for the conventional OPD. The peak external quantum efficiencies at both short and long wavelengths of the strongly coupled devices, however, fall short of those for conventional excitonic OPDs (Eizner et al., 2018).

Finally, strong coupling between photons and charges can generate a polaron-polariton that is qualitatively different from exciton-polaritons. The former couples a spin-½ particle (a polaron) with a spin-1 photon, resulting in a quasiparticle that follows Fermi statistics. This is in contrast to the exciton-polaron which is a boson. The charged polaron-polariton can respond to an external electric field, and is similar to a *trion*, which is a weakly bound charge-exciton state found in inorganic semiconductors at low temperatures. The large Rabi splitting of the polaron-polariton is a consequence of the intense, narrow spectral linewidths of anionic and cationic radical organic species, making them stable even at room temperature.

An energy level diagram of a cationic polaron-polariton is shown in Fig. 9.29a. A missing electron in the HOMO (i.e. a hole) is bound to an electron in the HOMO-1 state, and is coupled to a resonant cavity photon to form the quasiparticle. Analogously, a molecular anion radical with an unpaired electron in the LUMO couples to the LUMO+1 or other higher LUMO states. To prepare the cationic states, the hole conducting TAPC is doped with MoO_3 whose large LUMO energy serves as an electron acceptor. Electrons are transferred from the HOMO of TAPC to MoO_3. Increasing the dopant concentration increases the density of $TAPC^+$ cations.

Evidence for charge transfer is inferred from the spectra in Fig. 9.29b. The extinction peak (with attenuation coefficient, κ) at 1.77 eV and its sidebands only exists in the presence of the dopant. Therefore, these features are assigned to a singly charged radical cation species. The neutral exciton is found at approximately 3.5 eV. The increase in κ is accompanied by a drop in the real part of the index of refraction, n. This is confirmed by direct optical measurements of the cationic species in solution (Cheng et al., 2018).

Films of 10 wt% MoO_3:TAPC were sandwiched between a reflective, 100 nm thick Ag mirror and a semitransparent, 17 nm Ag mirror that form a strongly coupled cavity. Reflectivity spectra are shown in Fig. 9.30a along with the calculated spectra assuming a coupled oscillator pair. The $TAPC^+$ reflectivity peak at 1.77 eV is split, resolving the anticrossing between the two polaron branches with a Rabi energy of 320 meV. The magnitude of the splitting is indicative of ultrastrong coupling found in cavities bound by metal mirrors.

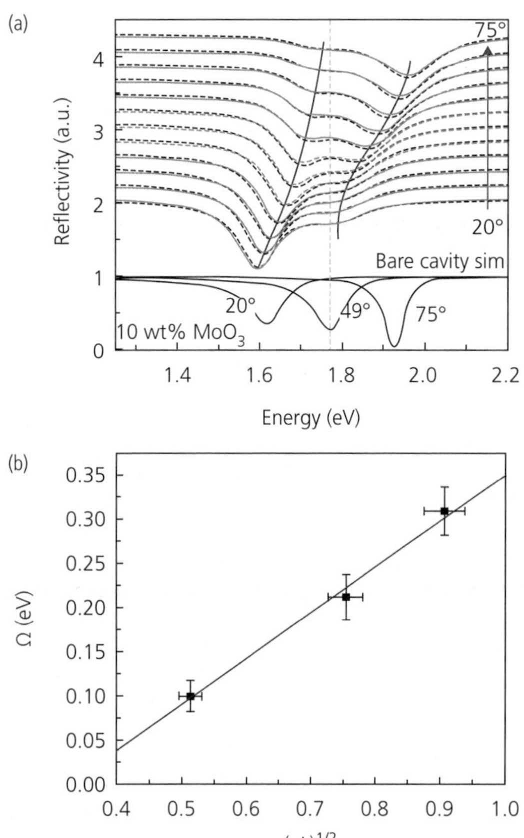

Figure 9.30 (a) Measured reflectivity spectra (blue lines) of a 10 wt% MoO_3:TAPC film in an optical microcavity. Fits to the data using a coupled oscillator model are shown by the dashed lines that highlight the anticrossing of the LPB and UPB. The lower spectra show the calculated uncoupled exciton spectra at several angles. (b) Rabi splitting energy vs. optical absorbance (Cheng et al., 2018).

Interestingly, it is found that the splitting is related to the film absorbance via $\Omega \propto (aL)^{1/2}$ (see Fig. 9.30b), consistent with Eq. 9.10. That is, the $TAPC^+$ density, and hence its absorption increases linearly with MoO_3 concentration. This, in turn, increases the coupling of the $TAPC^+$ to the cavity mode.

Since the polariton carries charge, it can be electrostatically manipulated. Thus, the transport of polarons from source to drain in a resonant cavity transistor can be influenced by charge injection and the gate potential. A remarkable aspect of polaron-polaritons lies in their room temperature stability that apparently is unique to organic materials with their low dielectric constants, and consequently large binding energies of the molecular excited states.

The foregoing devices experience ultrastrong coupling using cavities comprising two metal mirrors. However, ultrastrong coupling can be observed for any cavity that provides sufficient optical feedback. An all-dielectric structure that supports BSWs has demonstrated this degree of coupling in DBP thin films on the surface of a DBR pre-deposited on a fused silica substrate, see Fig. 9.10. The DBP absorption spectrum in Fig. 9.31a shows four well-resolved peaks corresponding to the 0–0, 0–1, 0–2, and 0–3 vibronics in the S_1 manifold. Fits to the coupled oscillator formalism for a $d = 10$ nm thick film are shown in Fig. 9.31b. All three of the dominant vibronics are coupled to the BSWs, yielding a Rabi splitting of 215 meV, which is at the lower edge of the ultrastrong coupling regime.

The coupling increases with organic film thickness. This is anticipated since the evanescent field that leaks out of the DBR overlaps more of the organic film as its thickness increases. However, the effect saturates due to the exponential decay of the field intensity with distance from the DBR surface. In the constant field approximation, the Rabi splitting energy is given by $\Omega(d) = A\sqrt{d - d_0}$, where A is a constant and d_0 is a distance that scales with the line broadening parameter, γ. This expression results in the dashed line fit in Fig. 9.31c, clearly overestimating the coupling strength at $d > 15$ nm. To correct for reduced field interactions, the exponentially decreasing field amplitude over characteristic length, l, must be integrated across the film thickness. Then, the Rabi splitting follows (Hou et al., 2019)

$$\Omega(d) \propto \mu\sqrt{N\bar{F}} = \mu\sqrt{N}\int_0^d \frac{e^{-x/l}}{d}dx, \tag{9.18}$$

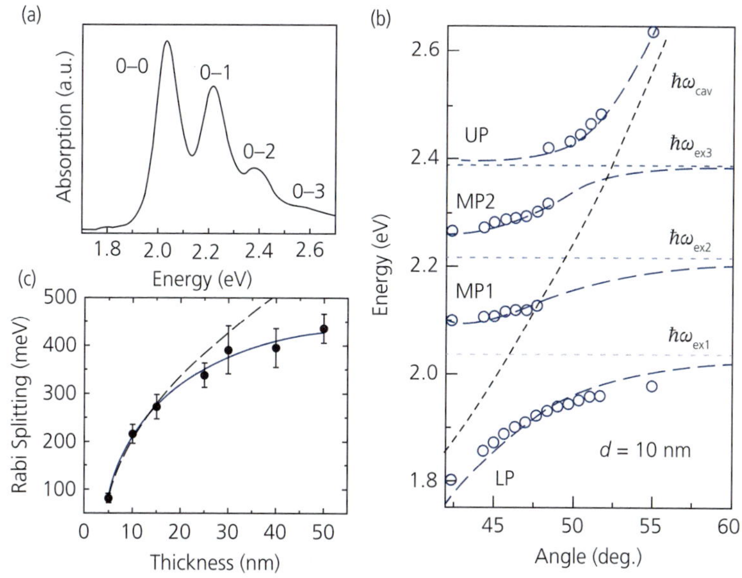

Figure 9.31 (a) DBP absorption spectrum. (b) Reflectivity spectrum of a 10 nm thick DBP film in an all-dielectric cavity, showing coupling of the three lowest vibronics in the strong coupling regime. The exciton energies of ω_{ex1}, ω_{ex2}, and ω_{ex3} correspond to the 0–0, 0–1, and 0–2 transitions, respectively. The cavity mode is also shown. (c) Rabi splitting vs. DBP thickness, *d*. The onset of ultrastrong coupling occurs when $\Omega \approx 0.1E_{ex} \approx 215$ meV corresponding to $d \gtrsim 10$ nm. The dashed and solid lines are discussed in text (Hou et al., 2019).

where $\bar{F}$ is the mean electric field, N is the number of oscillators that is proportional to d, and μ is molecular transition dipole moment. As shown by the solid line in Fig. 9.31c, this expression no longer overestimates the degree of coupling. Although the effect saturates with film thickness, significant coupling is achieved with the all-dielectric cavity, reaching $\Omega > 0.2E_{ex}$ at film thicknesses of 50 nm.

While the results of the active polaritonic OLEDs and OPDs are promising, their performances still fall short compared with conventional devices. Nevertheless, as noted above, the stable room temperature polariton phenomena in organic semiconductors open doors to many new potential nonlinear optical applications, ranging from electrically pumped polariton lasers, to all-optical logic gates and IR detectors (Sanvitto and Kéna-Cohen, 2016, Snoke and Keeling, 2017). The list of properties accessed via strong coupling, and of their concomitant optical phenomena yet to be discovered, is long and growing. For example, early demonstrations of using polaritons for the control of chemical reactions may eventually lead to new chemistries based on the interaction of light and matter. The insights that polaritonic phenomena provide into the understanding, and ultimately the control of energy transfer between both homogeneous (organic/organic) and heterogeneous (organic/inorganic) materials systems provides compelling reasons to continue their investigation and exploitation.

9.3 Organic thermoelectricity

Thermoelectric materials provide a simple and potentially low cost route to temperature control, measurement, and thermal energy conversion to electricity. However, many inorganic materials used in efficient thermoelectric devices are often difficult or costly to synthesize, are comprised of difficult to obtain materials, and may be brittle, preventing their being shaped into desired form factors. The need for efficient thermoelectrics useful in a wide range of thermal generating and harvesting applications has therefore focused attention on the use of organic thin films that can be made from earth-abundant materials at low cost, while having adaptable form factors via fabrication on large area, flexible substrates.

The processes of thermal generation (i.e. heating and refrigeration) and energy harvesting are complementary, and hence can be understood from a single analytical perspective. The *Seebeck effect* is when a temperature difference across a semiconductor results in a voltage difference that can deliver power to a load (energy harvesting). The converse, *Peltier effect*, is when current across a semiconductor develops a temperature difference, enabling heating or cooling, depending on the direction of current flow.

These two effects are conveniently understood with reference to the thermocouple in Fig. 9.32, where two different materials (**a** and **b**) form a pair of parallel

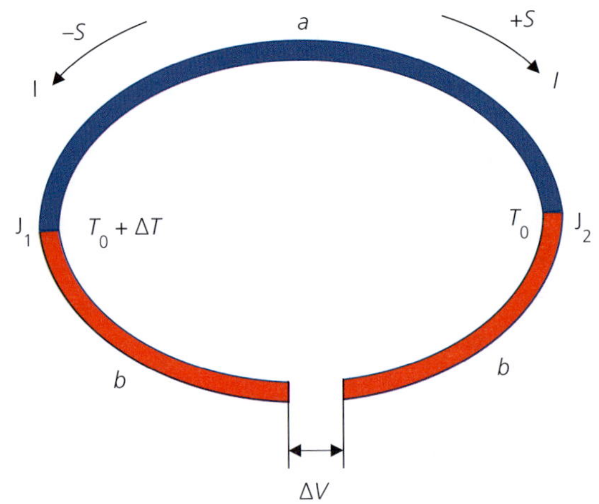

Figure 9.32 Schematic of two dissimilar materials, **a**, **b**, forming a thermocouple. The thermocouple comprises a pair of junctions, J_1, J_2, at different temperatures. This induces a voltage drop, ΔV, and hence a current flow, I, in an external load. A clockwise current flow results from a material with a positive Seebeck coefficient, S. Otherwise S is negative.

junctions, J_1 and J_2. If the two junctions are at different temperatures, T_0 and $T_0 + \Delta T$, a voltage is developed across the gap in material **b**, whose magnitude is given by the *Seebeck coefficient*:

$$S_{ab} = \Delta V / \Delta T. \tag{9.19}$$

In this discussion, we use S_{ab} for the Seebeck coefficient, although α_{ab} is also used. If the current flows from the hot to the cold junction ($\Delta T > 0$), then $S > 0$, whereas if it flows in the reverse direction, $S < 0$. The direction of current flow depends on whether the majority carriers are electrons or holes. The Seebeck effect is thus also useful for determining the majority carrier type of a semiconductor.

The converse phenomenon of heating or refrigeration of a junction via the passage of current, I, is characterized by the *Peltier coefficient*:

$$\pi_{ab} = I/\dot{Q}_{th}, \tag{9.20}$$

where $\dot{Q}_{th}$ is the rate of cooling or heating. That is, one of the junctions is heated at rate $\dot{Q}_{th} > 0$, while the other is cooled at rate $\dot{Q}_{th} < 0$. The Peltier effect is most frequently used for thermoelectric heaters and coolers. The Seebeck and Peltier coefficients are related via the *Kelvin relationship*:

$$S_{ab} = \pi_{ab}/\Delta T. \tag{9.21}$$

The magnitude of the thermoelectric effect of a particular material is characterized by the dimensionless figure of merit, z:

$$zT = \frac{S^2 \sigma T}{\kappa}, \tag{9.22}$$

where σ is the electrical conductivity and κ is its thermal conductivity. Here, z is an intensive parameter that is defined by the material characteristics alone. However, to characterize a thermoelectric device with resistance, R, and thermal conductance K, use of the extensive parameter:

$$ZT = \frac{S^2 T}{R\mathrm{K}} \tag{9.23}$$

is more appropriate.

Finally, the *power factor* of the material is

$$PF = S^2 \sigma. \tag{9.24}$$

Equation 9.22 shows that a high zT is achieved for materials with a high electrical and a low thermal conductivity. High electrical conductivity is required to reduce internal power loss, I^2R. On the other hand, κ must be small such that the device can support a large thermal gradient across a small distance so as not to thermally short the hot to the cold end. Shorting results in a decreased ΔT across the thermocouple.

The thermal conductivity is due to the sum of the conductivity from the lattice, κ_{lat}, and the electron population, κ_{el}. Heat transport by phonons is expressed using

$$\kappa_{lat} = \frac{1}{3} C_v v \lambda_{ph}, \tag{9.25}$$

where C_v is the heat capacity per volume, v is the velocity of sound, and λ_{ph} is the phonon mean free path. For conducting polymers, the lattice contribution to the thermal conductivity is 0.1–1 W m^{-1} K^{-1}. The electronic contribution is related to the electrical conductivity via the *Weidemann–Franz law*, viz.

$$\kappa_{el} = L\sigma T, \tag{9.26}$$

where $L = (\pi^2/3)(k_B/q)$ is the *Lorenz factor*. Thus, κ_{el} is proportional to the electrical conductivity, leading to a tradeoff in finding σ that maximizes zT. In organic semiconductors, $\kappa_{el} \ll \kappa_{lat}$, and often can be neglected.

The Seebeck coefficient for a material is ultimately determined by the DOS, $g(E)$, available for conduction by electrons and/or holes. Recall from Section 4.3.2 that the HOMO and LUMO densities of states can be approximated by Gaussian distributions about the frontier energy levels, E_{HOMO} and E_{LUMO}, respectively. We have shown that conduction is affected by both static and dynamic disorder, which are characteristics of both polymers and small molecules. Thus, the picture is made more complete by inclusion of a separate Gaussian DOS centered at energy E_T near the Fermi energy. An example DOS for an archetype p-type material is shown in Fig. 9.33. A similar trap state distribution can also exist at the LUMO edge, and in fact this most likely is the case in most organics. The conductivity of carriers via either conduction or valence states then can be written

$$\sigma = \int \sigma(E)dE = q\int g(E)\mu(E)f(E)[1-f(E)]dE, \quad (9.27)$$

where $\mu(E)$ is the energy-dependent mobility, and the terms including the Fermi function, $f(E)$, account for the probability that a charge can move from an occupied to an unoccupied state. This analysis suggests that the fraction of electrons between E and $E + dE$ contributing to conduction is given by $\sigma(E)dE/\sigma$, in which case we can calculate the Seebeck coefficient for metallic materials (where $\kappa_{lat} < \kappa_{el} \propto \sigma$) as follows (Fritzsche, 1971, Mott and Davis, 1979):

$$S = -\frac{k_B}{q}\int\left(\frac{E-E_F}{k_BT}\right)\frac{\sigma(E)}{\sigma}dE. \quad (9.28)$$

This relationship is valid for materials where the conduction takes place at or near the Fermi energy, E_F, and σ is a slowly varying function about E_F. Note that in p- and n-type materials, the conduction level energies, E, are less than or greater than E_F. Hence, the sign of the integrand, and consequently, S, changes for the different materials types. By convention, $S < 0$ for n-type, and $S > 0$ for p-type semiconductors.

In n- or p-type organics, the charges are primarily transported within either the HOMO or the LUMO level state distributions where E may significantly differ from E_F, making Eq. 9.28 incomplete (Kaiser, 1989). In disordered organic materials, electron–phonon interactions must be considered, but are

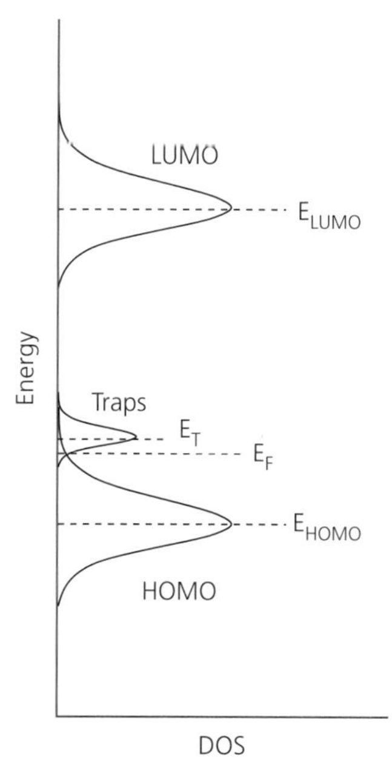

Figure 9.33 Density of states (DOS) at the HOMO and LUMO energies, along with a trap density centered at E_T.

absent in this expression. It is likely that organics are single energy band conductors, that is, either holes or electrons dominate the conductive properties of a particular material. If transport occurs by three-dimensional (3D) *variable range hopping*, then $\sigma = \sigma_0 \exp\left[-(T_0/T)^{1/4}\right]$ where σ_0 and T_0 are constants. In this case, Eq. 9.28 becomes (Mott and Davis, 1979)

$$S = \frac{k_B^2}{2q}(T_0T)^{1/2}\frac{d\ln D(E)}{dE}\bigg|_{E=E_F}, \quad (9.29)$$

leading to the more realistic approximation used for degenerate semiconductors (Cutler et al., 1964):

$$S = \left(\frac{2k_Bm^*T}{3\hbar^2}\right)\left(\frac{\pi}{3n}\right)^{2/3}\left(\frac{k_B}{q}\right). \quad (9.30)$$

This is known as the *Mott formula*. From the foregoing, a thermoelectric device must balance the electrical and thermal conduction properties to maximize its efficiency. The tradeoffs are qualitatively illustrated for insulators, semiconductors, and metals in Fig. 9.34. In the lower panel, we show that the electron contribution to the thermal conductivity is important only in metals, and in very highly doped (i.e. degenerate) semiconductors, whereas for most semiconductors and insulators, κ_{lat} is dominant. The decrease in S due to an increasing σ is counteracted by the low κ of insulators. Taking these factors into consideration, the power factor peaks for highly doped semiconductors. However, this applies only to materials systems where σ and κ are linked. One advantage of organic materials is the ease with which doping and thermal conductivity can be independently controlled. We will find below that composites consisting of carbon nanotubes (CNTs) dispersed in polymer matrices, or ionic doping of PEDOT are two strategies used to optimize this tradeoff.

An additional figure of merit for a *thermoelectric generator* (TEG) is its maximum efficiency, η_{TEG}, which is equal to the ratio of the energy supplied to the load to the heat absorbed at the hot junction. The thermodynamic limit of η_{TEG} is the *Carnot efficiency*, $\eta_C = \Delta T/T_H = (T_H - T_C)/T_H < 1$, where T_H (T_C) is the temperature of the hot (cold) junction. Then, it can be shown for a generator with figure of merit, Z (Rowe, 2005),

$$\eta_{TEG} = \eta_C\frac{\sqrt{1+Z\bar{T}}-1}{\sqrt{1+Z\bar{T}}+T_C/T_H}, \quad (9.31)$$

where $\bar{T} = (T_H + T_C)/2$ is the average temperature.

Organic semiconductors are useful for thermal power generation at temperatures that are tolerable for these materials, that is, at $T \leq 150°C$. Fortunately this range represents the largest fraction of waste heat generated in common household and building environments. Inorganics, however, can be used where more extreme temperatures are produced, such as scavenging heat from automobile exhausts, incinerators, and in the production of materials such as Si, steel, etc. Taking T_C = 300 K and T_H = 450 K as reference temperatures, the Carnot efficiency is η_C = 0.33. Of course, η_C increases with T_H, ultimately limiting the maximum η_{TEG} that can be reached at the lower end of the temperature range accessible to organics. Figure 9.35 shows the dependence of η_{TEG} and η_C on T_H for several values of Z in the temperature range typically accessed by organic TEGs. The highest $ZT \sim 1$–2 is achieved with inorganic alloys such as Bi_2Te_3 and SiGe (Snyder and Toberer, 2011), which is at least a factor of four higher than reported for organic TEGs (Russ et al., 2016). Also, Z is roughly 50% that of z in the most efficient, practical thermoelectric appliances. Thus, for an organic TEG operating at 450 K, we find an upper practical value of $Z \sim 0.8 \times 10^{-3}$ K^{-1}, giving $\eta_{TEG} \sim 2.2\%$ at $T_C = 300$ K.

Analogously, the maximum efficiency for cooling is

$$\eta_{COP} = \frac{T_C}{\eta_C T_H} \frac{\sqrt{1+Z\bar{T}} - T_H/T_C}{\sqrt{1+Z\bar{T}}+1}, \qquad (9.32)$$

which is also known as the maximum *coefficient of performance*. This corresponds to a maximum temperature difference of $\Delta T_{\max} = \frac{1}{2} Z T_C^2$. A plot of the theoretical maximum temperature for a generator with figure of merit Z, and with $T_H = 298$ K is shown in Fig. 9.36. Using the performance values given above, a refrigeration module employing organic thin films is expected to have $\Delta T_{max} \sim 50$ K.

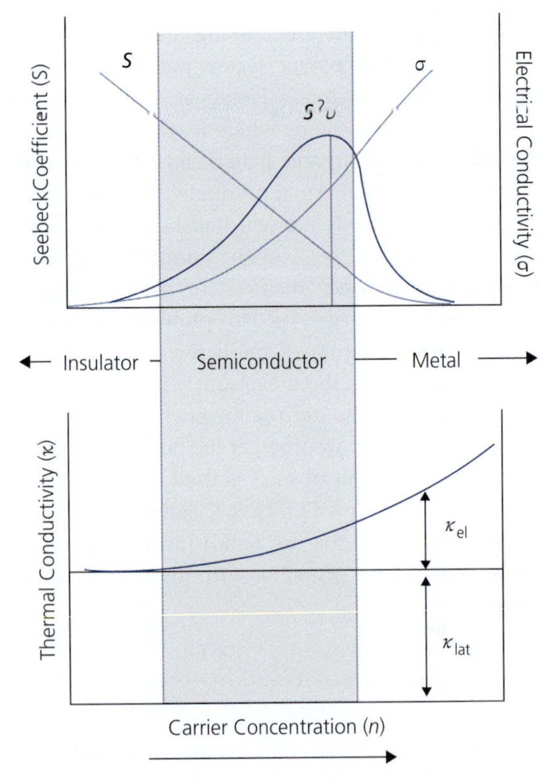

Figure 9.34 Tradeoffs between the Seebeck coefficient, electrical conductivity, thermal conductivity, and power factor vs. carrier concentration in thermoelectric materials, ranging from insulators to metals.

9.3.1 Organic thermoelectric module architectures

Figure 9.37 provides a compilation of representative ZT_{max} values versus the year of their demonstration for several inorganic and organic thermoelectric materials. The inorganic materials undergo a significant increase in ZT_{max} when nanostructured such as with quantum wells or nanowires, to break the strict dependence between the electrical and thermal conductivities of the materials (Hicks and Dresselhaus, 1993). The highest values are obtained for Pb-based

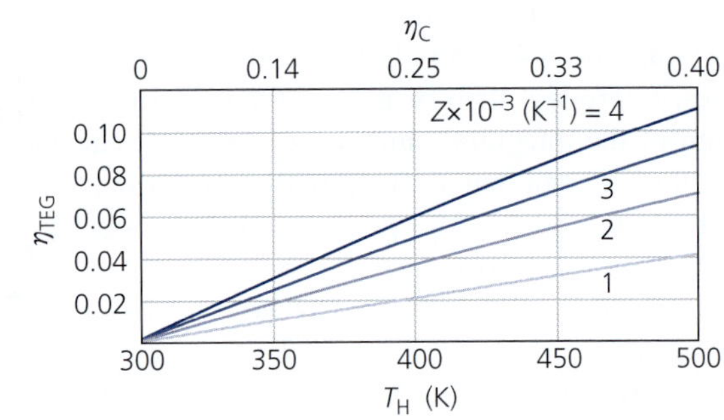

Figure 9.35 Thermoelectric generation efficiency (η_{TEG}) vs. hot junction temperature (T_H) for several Z values. Also shown is the Carnot efficiency (η_C) for $T_C = 300$ K.

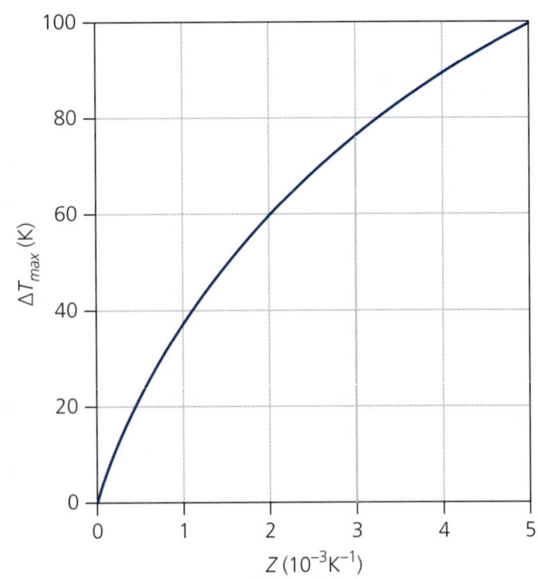

Figure 9.36 Maximum temperature difference of a thermoelectric refrigerator vs. Z at $T_H = 298$ K.

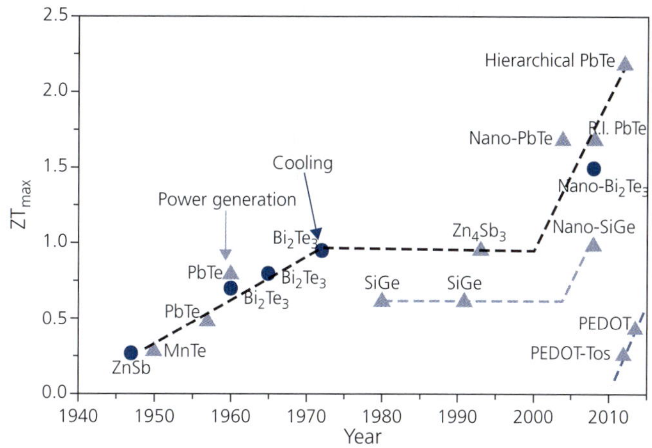

Figure 9.37 ZT_{max} of representative thermoelectric materials vs. the year of their demonstration used in power generation (triangles) and refrigeration (circles) applications. The inflection in ZT_{max} around the year 2000 separates values obtained for inorganic bulk and nanostructured materials. The best organic values are obtained for PEDOT with a tosylate anion (PEDOT-Tos) (Bubnova et al., 2011) and PEDOT itself (Kim et al., 2013b). Adapted from Heremans et al. (2013).

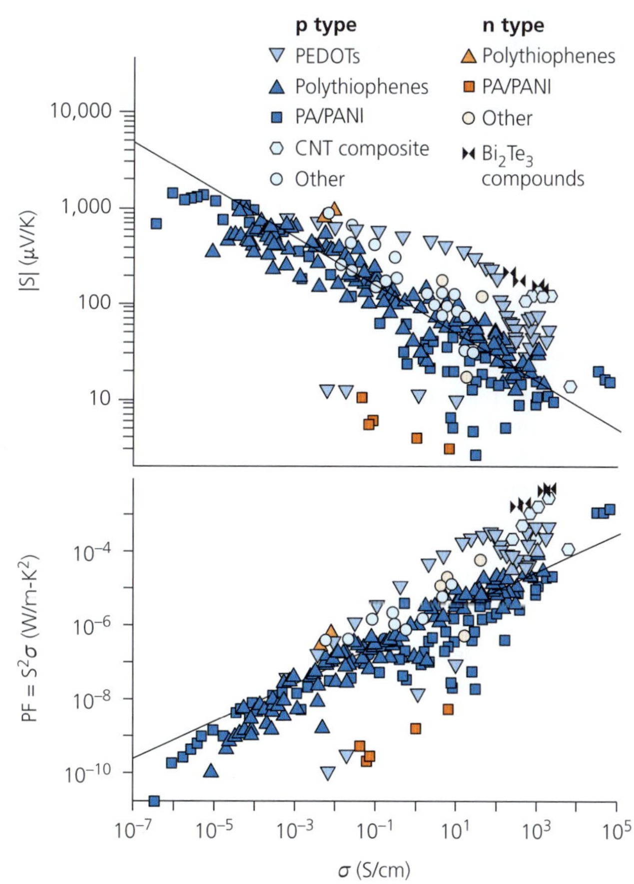

Figure 9.38 Seebeck coefficients and power factors for a variety of organic thermoelectric materials. Included for comparison is the high performance inorganic alloy, Bi_2Te_3. Solid lines correspond to the empirical relationships: $S \propto \sigma^{-\frac{1}{4}}$ and $PF \propto \sigma^{\frac{1}{2}}$. References for each material are found in Russ et al. (2016).

II–VI semiconductors, with $ZT_{max} > 2$. Nanostructuring of organic materials has also led to higher thermoelectric figures of merit, as discussed below. To date, the highest ZT_{max} for organics hovers at just less than 0.5, with the best results obtained for PEDOT and related polymers (Bubnova et al., 2011, Kim et al., 2013b).

Selected values of S and PF for a population of both n- and p-type organic thermoelectric materials are shown in Fig. 9.38. The various points correspond to data taken from the literature, with specific references provided in Russ et al. (2016). The data roughly follow the empirical relationships $S \propto \sigma^{-1/4}$ and $PF = S^2\sigma \propto \sigma^{1/2}$ shown by the lines. While there is considerable scatter of the data about the trend lines, it is nevertheless remarkable that the data follow these trends over approximately 11 decades in conductivity, and are independent of whether the material is n- or p-type. The origin of these empirical relationships is as yet unclear, and does not appear to be well fit by variable range hopping (cf. Eq. 9.29) or conduction along the mobility edge. It has been suggested that this trend may be coincidental, in that it may exist for high conductivity (and hence degenerately doped) organics that blends smoothly into a similar functionality for low doped, semiconducting-to-insulating materials (Glaudell et al., 2015, Russ et al., 2016). In any case, it is clear that the relationship between conductivity and the thermoelectric properties of organics is robust, and hence can be a useful predictive tool for selecting materials for specific applications.

The performance of the inorganic bismuth chalcogenide, Bi_2Te_3, is shown for comparison (black, opposing triangles). While this material has the highest *PF*, CNT composites, polyanylenes (PANI), polyacetylenes (PA), and PEDOT-based compounds are competitive, reaching a maximum $PF \sim 3 \times 10^{-3}$ W m^{-1} K^{-2}. Also, due to the generally higher hole compared to electron mobilities in organics, the thermoelectric properties of p-type organic thin films (blue data points) also tend to be superior to n-type materials (orange points).

To scavenge a meaningful amount of energy, or to cool or heat a load with a significant thermal mass, the thermoelectric elements must be arranged in a module comprising a parallel array of individual thermocouples or heaters, as shown in Fig. 9.39. The two types of elements employ either bulk or thin film architectures (Fig. 9.39a and b, respectively). The hot, heat-rejecting side is isolated from the heat-absorbing (cold) side by thermoelectric materials that have a low thermal but high electrical conductivity.

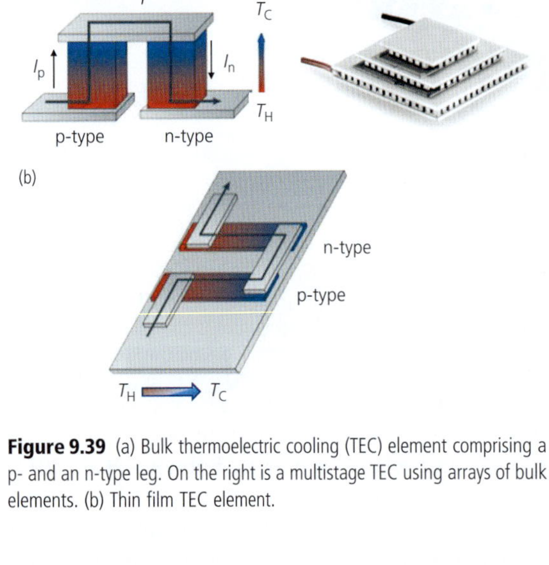

Figure 9.39 (a) Bulk thermoelectric cooling (TEC) element comprising a p- and an n-type leg. On the right is a multistage TEC using arrays of bulk elements. (b) Thin film TEC element.

A module such as shown on the right in Fig. 9.39a consists of an electrical series connection of individual elements, with each element absorbing and rejecting heat in parallel. Since n- and p-type materials have Seebeck coefficients of opposite sign, using a combination of materials with opposite majority carrier types results in effective thermal isolation of the hot and cold surfaces while producing equal electron and hole currents (I_n and I_p). If the values of σ and κ are identical for materials of both carrier types (which is almost never the case), then $z = Z$.

Stacking multiple thermoelectric coolers (TEC) or heat generators (TEG), with each subsequent cooler module smaller than the one beneath it, results in an increased ΔT between the top and bottom surfaces. The stack should be mounted on an efficient heat sink to dispose of waste heat when used as a TEC, or to increase the highest possible temperature of a TEG. The diminishing area for each module compensates for the less than unity efficiency of the Peltier effect.

While bulk elements are readily achieved using inorganics, organic materials are more adaptable to a thin film form factor, making the geometry in Fig. 9.39b preferable. The thin film TEC or TEG can be mounted on a flexible and/or conformable sheet. This allows for roll-to-roll manufacturing of large area modules, which can lower cost and increase production throughput. A disadvantage of thin film TE devices is that thermal conduction through the substrate can thermally short-circuit the device. Using low thermal conductivity substrates is, therefore, imperative.

The thin film architecture also places constraints on its orientation to achieve intimate contact with the thermal load. As in the case of bulk elements, using a combination of *n*- and *p*-type legs connected in series increases the total current generated per device area, while providing thermal isolation of the cold and hot regions of the load. Matching hole and electron currents is accomplished in the in-plane configuration by adjusting the leg widths (w) via printing. For example, if $I_n = xI_p$ in a cooler whose legs are of equal width, then the currents are balanced when $x = w_p/w_n$, where w_p and w_n are the widths of the p- and n-type legs, respectively. Note too, that the thermal properties of the n- and p-type regions must be matched to achieve the highest performance. Thermal and electrical matching conditions may not always result in the same leg geometries, in which case tradeoffs are inevitable.

An example configuration for thin film thermoelectrics uses a corrugated structure as shown in Fig. 9.40a, originally demonstrated for Ag/Ni TEG modules. The p- and n-type organic materials along with their series/parallel electrical connections are printed onto a thermoformable (e.g. PET) base film. The array is attached only to the bottom substrate and the top lid (also PET) along the interconnect lines. Heating the package results in shrinkage of the lid and base, causing the TEG array to buckle into the third dimension. This allows for contact between the hot and cold surfaces that are separated only by the thin films themselves and the resulting air gaps. The detail in Fig. 9.40b shows that the n- and p-type stripe widths can be independently adjusted to match the currents generated in each material. This is one of many possible configurations. Depending on the application (e.g. for thermometry on an irregular skin surface), the 2D planar configuration may in fact be preferable to 3D arrangements.

9.3.2 Example devices

The thermoelectric performance, including relevant materials parameters for many organic thin films and composites optimized for their thermoelectric properties, can be found elsewhere (Chen et al., 2015b, Toshima, 2017). Among the various materials used as thermoelectrics, PEDOT has shown the highest performance (Bubnova et al., 2013, Kim et al., 2013b). As

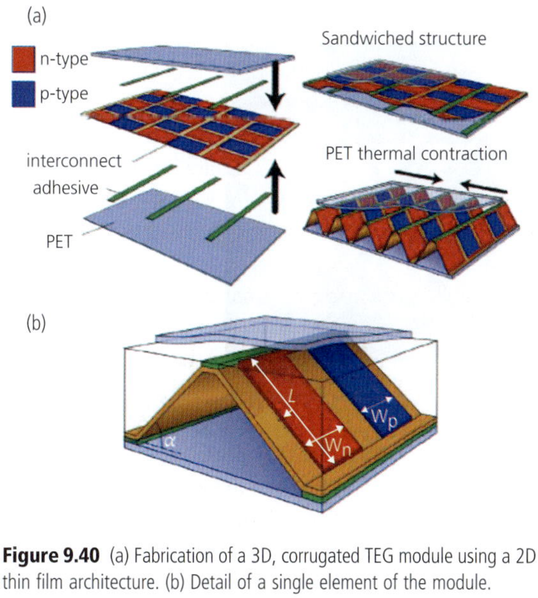

Figure 9.40 (a) Fabrication of a 3D, corrugated TEG module using a 2D thin film architecture. (b) Detail of a single element of the module. Adapted from Sun et al. (2015).

we have shown in the preceding discussion, organics have strong electron–phonon interactions that decouple the electrical from the thermal conductivity. The challenge for organics, then, is to achieve a sufficiently high electrical conductivity in both p- and n-type materials to avoid dissipative I^2R losses. In this regard, the conductivity of PEDOT can be controlled over several orders of magnitude by changing its ratio with an ionic species such as PSS. Since PSS is hydrophilic and PEDOT is hydrophobic, they can be readily separated using organic solvents.

Dedoping, which increases the electrical resistance, has been achieved by thermal annealing films of PEDOT:PSS spun onto glass from solutions mixed in either ethylene glycol (EG) or DMSO. Figure 9.41 shows the electrical and thermal properties of the cast films that are dedoped by subsequent immersion in an EG solution for extended periods; the doping is decreased with the duration that the mixture spends in solution. The electrical conductivity in Fig. 9.41a increases with time until approximately 100 min, at which point it reaches a peak value close to 1000 S/cm. After 200–300 min, the change in conductivity decreases due to completion of the dedoping process. Concurrently, the thermal conductivity measured perpendicular to the film plane ($\kappa_\perp$), which is ~60% of the in-plane conductivity ($\kappa_\parallel$), decreases monotonically with time, as shown in Fig. 9.41b. The similarity of the dependence of σ and κ on annealing time indicates that the electrons in PEDOT:PSS partially contribute to the film thermal conductivity.

The Seebeck coefficient, S, is found by measuring the change in voltage due to a temperature difference across the sample after accounting for the substrate thermal conductivity (Fig. 9.41c). Combining measurements of S and σ gives the thermal power in Fig. 9.41d, and the figure of merit, ZT. It is observed that PF is maximized for the film with maximum σ, achieving $S^2\sigma = 500\ \mu\text{W m}^{-1}\text{ K}^{-2}$, with a corresponding $ZT = 0.42$. As shown in Fig. 9.37, this value is less than the best inorganic alloy thermoelectric materials, although it has the advantages of using earth-abundant source materials to produce flexible thin films.

An approach to segregating the thermal and electrical properties is to create a nanostructured composite that has regions of high electrical conductivity percolating through a thermally insulating, dielectric matrix. Such composites are achieved by blending conductive materials such as metal nanoparticles (NPs) with a conducting organic or CNTs in an insulating polymer. Figure 9.42a illustrates the random

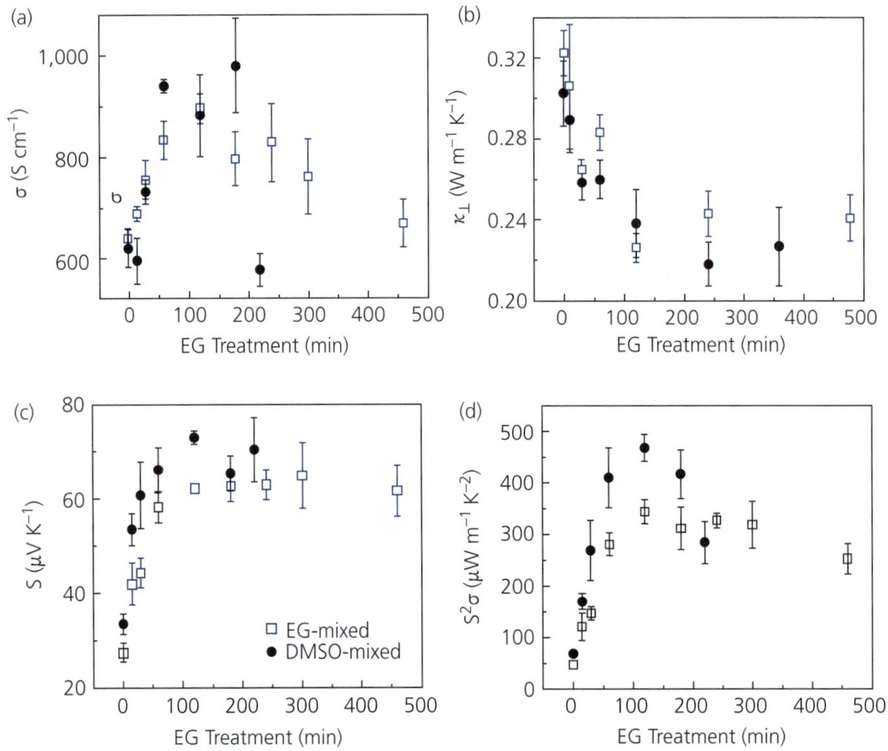

Figure 9.41 Characteristics of thermally annealed PEDOT:PSS films mixed in either EG or DMSO vs. exposure time in an EG solution. (a) Electrical conductivity, (b) thermal conductivity perpendicular to the film plane, (c) Seebeck coefficient, and (d) power factor (Kim et al., 2013b).

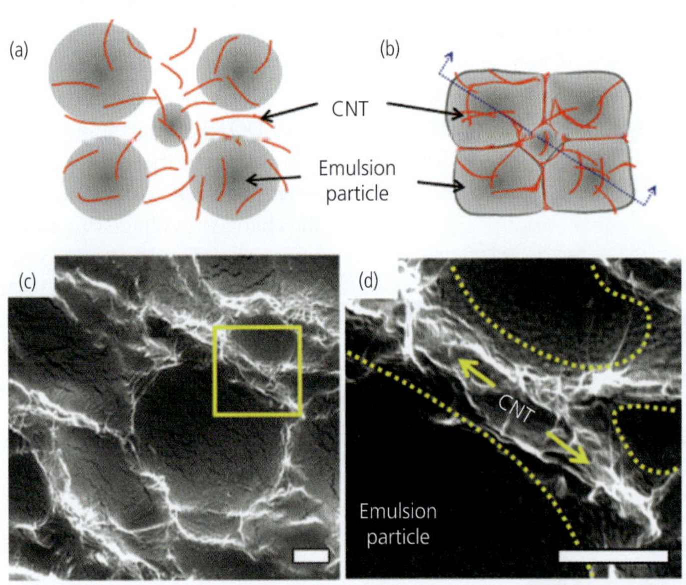

Figure 9.42 Illustration of (a) a solution of a polymer nanoparticle emulsion mixed with carbon nanotubes (CNTs) and (b) the material after solidification. Electron micrographs of the dried film containing 5 wt% CNTs imaged using a scanning electron microscope at (c) wide angle and (d) at higher magnification of the region in the yellow box in (c). The scale bars correspond to 1 μm (Yu et al., 2008).

distribution of CNTs in a solution containing a polymer emulsion. An *emulsion* is a stable suspension of particles or liquid in a solution of different composition (e.g. water). When the emulsion is dried, a film is formed that excludes the CNTs from the particle volumes, forming conducting paths within the interstices along the boundaries between the particles (see Fig. 9.42b). In this way, narrow, electrically conducting paths are formed along the CNTs that are separated by regions that inhibit thermal transport.

This film morphology is produced by blending 1–20 wt% of a distribution of single, double, and triple-walled CNTs into a PVA homopolymer emulsion that comprised 55 wt% solids in water. The PVA particle diameter distribution varies from 0.14 μm to 3.5 μm. The CNTs are dispersed in the matrix by first adding gum arabic before blending into the PVA. The solid composites are then dried in ambient for 2 days, followed by vacuum for 24 h to generate the 3D network in Fig. 9.42b. A microscope image of the resulting network is shown in Fig. 9.42c. The CNT clusters lie within the excluded volumes between PVA particles, forming what is apparently a continuous network. Indeed, the detail image in Fig. 9.42d shows a high density of tubes that forms electrically conducting paths that percolate throughout the film volume.

The thermoelectric performance of the composite is shown in Fig. 9.43. The electrical conductivity of the networks increases over several decades as the CNT concentration increases from 1 wt% to 20 wt%, indicating that the density of percolating pathways increases with the volume of nanotubes. It reaches a maximum of 50 S/cm at 20 wt% CNT, which is considerably less than for PEDOT:PSS, possibly due to resistance between the nanotubes within the conducting paths. We showed in Section 7.4.2 that casting CNTs from polymer suspensions can result in resistive contacts between overlapping nanotubes. In contrast to the electrical conductivity, the thermal conductivity (Fig. 9.43, inset) is largely unaffected by the presence of the nanotubes. The Seebeck coefficient, too, is relatively constant with CNT concentration even up to 20 wt%. At 20 wt%, a maximum $zT = 0.006$ is obtained at 300 K, limited primarily by the low conductivity of the CNT regions (Yu et al., 2008).

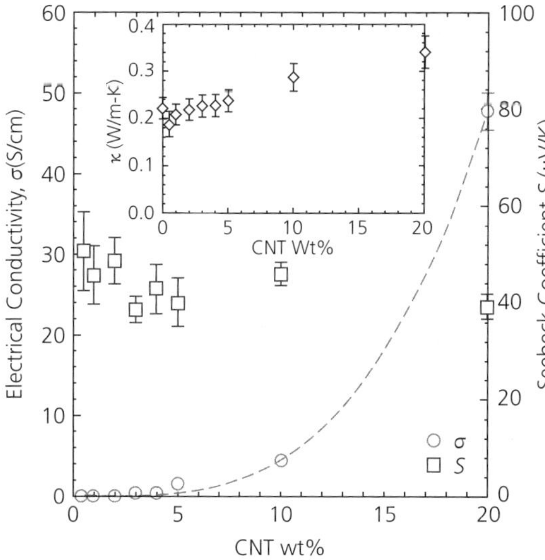

Figure 9.43 Electrical conductivity and Seebeck coefficient vs. CNT concentration in the PVA emulsions in Fig. 9.42. Inset: Thermal conductivity vs. CNT concentration. Adapted from Yu et al. (2008).

The conductivity of CNTs is improved by including an ionic species within the film bulk. Table 9.3 shows results obtained by including the cationic polymer, *n*-PETT, into a mixture of the insulating PVC and a dispersion of CNTs. The *n*-PETT forms 10–50 nm diameter NPs that are dispersed with CNTs to form an electrically conducting but thermally insulating three-component network. The *n*-PETT is an n-type material, and thus leads to a negative Seebeck coefficient even when dispersed by itself in PVC. However, the conductivity of this material is quite low compared to the CNT and CNT/PVC films. Notably, the high hole conductivity of CNTs leads to a positive *S*. The benefit of using *n*-PETT to electrically interconnect the CNTs is apparent from the *n*-PETT:CNT:PVC (10:8:3) composite that shows an electrical conductivity of 630 S/cm, which is almost equal to that of a sheet of pure CNTs. The thermal conductivity of this three component film is only 0.07 W m^{-1} K^{-1}, which is approximately the same as that of CNTs diluted in PVC. As a result, a peak $zT = 0.31$ at 340 K is achieved, which is nearly that of dedoped PEDOT:PSS.

A five-leg thermocouple comprising composite *n*-PETT:CNT:PVC films printed on a polyimide substrate is shown in the inset of Fig. 9.44. This represents one of the few device-scale demonstrations of a thermoelectric device based entirely on organic semiconductors. The $t = 5.7$ μm thick by $w = 4$ mm legs are series-connected by Ag electrodes in a module comprising five, $L = 5$ mm long elements. A temperature difference of $\Delta T = 100$ K perpendicular to the module results in the voltage and power-

Table 9.3 Thermoelectric characteristics of mixed films at 340 K (Toshima et al., 2015)

Film	Form	S (μV/K)	σ (S/cm)	PF (μW m^{-1} K^{-2})
n-PETT	Block	−39.6	0.036	0.006
n-PETT:PVC (1:3)	Film	−32.9	0.004	0.0005
CNT	Sheet	38.2	691	101
CNT:PVC (8:13)	Film	31.9	23.2	2.3
n-PETT:CNT:PVC (10:8:3)	Film	30.5	630	58.6

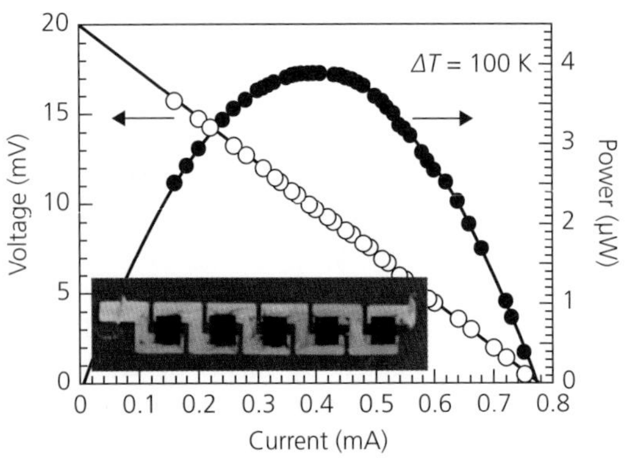

Figure 9.44 Voltage– and power–current characteristics of a five-leg, series-connected thermocouple comprising an *n*-PETT:CNT:PVC film with Ag bridging electrodes. The characteristics are obtained for a temperature difference of $\Delta T = 100$ K. Inset: Photograph of the thermocouple on a polyimide substrate. The height of a semiconductor element is 5 mm (Toshima et al., 2015).

current characteristics obtained under a varying load of 0–100 Ω shown in Fig. 9.44. The temperature difference, and hence the module efficiency is ultimately limited by the thermal conductivity of the Ag interconnects and the polyimide substrate. Extrapolation of the voltage to $I = 0$ gives an open circuit voltage of 19.9 mV. The internal series resistance obtained from the slope is $R_{ser} = 25.6\ \Omega$. The peak power, which is equal to $P_{\max} = (\frac{1}{2}V_O)^2/R_{ser} = (NS\Delta T)^2/4[NL/(\sigma wt)]$, is 3.88 μW, where $N = 5$ is the number of thermoelectric elements.

The foregoing discussion suggests that organic thermoelectric properties can approach those of inorganic alloys. The highest performances are achieved using high conductivity PEDOT derivatives, as well as with polymer composites with CNTs, metallic (Yoshida and Toshima, 2014), and even thermoelectric alloys such as Bi_2Te_3 and related compounds (Toshima and Ichikawa, 2015). While the substantial benefits of organics include ease of processing using low cost, flexible, earth-abundant source materials, several challenges remain to be solved before they find widespread application. For many organic TEG and TEC configurations, substrate thermal conductivity remains a limitation. Also, the lack of thermal stability of organic materials, particularly at temperatures >200°C, limits their use at high temperatures where the Carnot efficiency is large. One further issue that needs attention is the inherent anisotropy of organic materials. For example, the in-plane thermal conductivity, $\kappa_{\|}$, can differ from the out-of-plane value, $\kappa_{\perp}$, by at least a factor of 2 (Kim et al., 2013b). Similar anisotropies in the electrical conductivity are also common. Hence, it is not always clear that the high zT values reported will be experienced in devices assuming a perhaps unfavorable orientation, leading to a much smaller ZT than anticipated. And finally, while the highest performance is obtained from p-type organic materials, there are very few, if any high performance n-type materials (see Fig. 9.38). Ultimately, this may require modules with metal interconnects between elements, or a significant asymmetry in device dimensions for the n- and p-type legs. Hence, materials development merits further attention before such devices find practical deployment.

9.4 Memories

Organic materials offer the possibility of delivering very inexpensive memory devices on flexible substrates. Applications where this may be an attractive form factor include RFIDs, and buffers for data acquisition from sensors used in medicine, robotics, and imaging. Several generic types of organic memories including *static* and *dynamic random access memories* (SRAM and DRAM) and *read-only memories* (ROM) have been demonstrated. A RAM is where the access to all memory elements (i.e. bits) is rapid and independent of the position of the bit in the memory. This is in contrast to time-accessed or serially addressed memories such as magnetic tapes. A *dynamic memory* is one where the contents can be changed on demand, whereas in a ROM, the contents are permanently written to contain vital information that does not change with time, such as the basic input–output system (BIOS) used to launch a computer operating system. A ROM is a *non-volatile memory* that preserves its contents even in the absence of power. Similarly, RAMs can be either volatile or non-volatile, depending on the character of the memory element. While a ROM is essential for all computation applications, non-volatile RAM is the most versatile and ubiquitous. It is the basis of NAND flash memories that pervade modern computational systems. Retention times of 10 years and read–write cycle reliability into the thousands are desirable for conventional memories, although this is dependent on the specific demands of a given application. The state of the memory should be retained over operating temperatures from −40°C to +85°C.

Organic memories are arranged in an array of metallic column and row interconnects separated by the memory material that changes its conductive state when a bit is written or erased. This so-called *cross-point array* minimizes the area occupied by each active memory element. Packing of arrays in a 3D configuration is the predominant form of high density NAND flash, since its bit density per circuit area is multiplied by the number of 2D array layers.

The most common and simple mechanism for storing information is using an organic material or device whose resistance (i.e. its memory state) is changed by several decades on application of a voltage during the write, or "set" cycle. The memory location is then read by measuring the magnitude of the current (and thus the resistance) through the device at a voltage much lower than that needed to write the bit. The resistance change can be permanent (i.e. where the memory element is a *fuse* that is blown once it is written), or it can be reversible by application of a second write pulse.

The read and write cycles of a volatile resistive memory are shown in Figs. 9.45a and b for unipolar and bipolar switching, respectively. Initially, the memory device is highly resistive and is in the OFF state (which may represent either a logical "1" or "0"). When queried by a small read voltage, the current is small due to the high resistance of the element. When the voltage is driven to the set voltage, V_1, the resistance is reduced, giving a constant current (CC in Fig. 9.45) equivalent to the ON state. To reset the device, the current in in the ON state is increased until its value at "reset" is reached, corresponding to voltage V_2. This returns the element to its high resistance, OFF state. To non-destructively read the state of the memory, the voltage swing is maintained in the range $0 < V < V_2$. The *memory window*, ΔV, is the voltage difference needed to switch between logic states, that is, $\Delta V = |V_1 - V_2|$. The operation is unipolar since the voltage swings for SET and RESET are either both positive or negative. A bipolar device undergoes the same sequence of steps for reading and writing, except that the voltage polarities used for set and reset are opposite to each other, as shown in Fig. 9.45b. The choice of unipolar or bipolar switching depends on the circuit architecture and availability of the appropriate voltages.

Figure 9.45 points to a common feature of all memory elements. There must be hysteresis in the *I–V* characteristics, leading to an open loop that clearly distinguishes between the ON and OFF states (with corresponding currents of I_{on} and I_{off}, respectively). The magnitude of I_{on}/I_{off} ultimately limits the size of a practical memory array. To achieve a moderate to large memory size, $I_{on}/I_{off} = 10^5$–10^6 is required (Prime and Paul, 2009). Also, the read cycle must be non-destructive of the memory contents, otherwise the contents must be refreshed by delivering a second set or reset signal after the memory is queried for a certain number of cycles. Read and write times limit

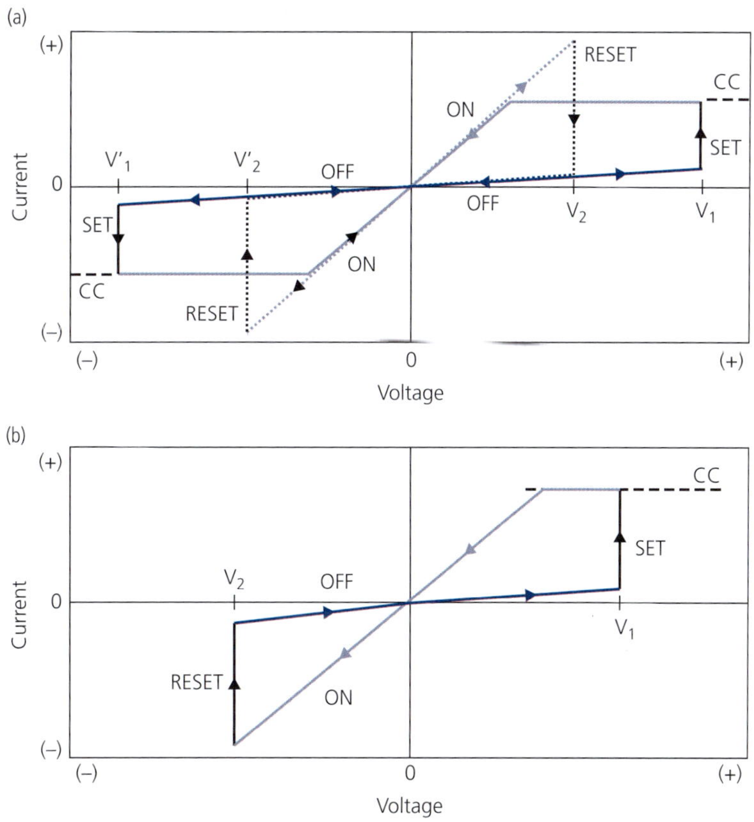

Figure 9.45 (a) Unipolar resistive memory switching. CC = constant current used in the read cycle. (b) Bipolar memory switching (Heremans et al., 2011).

the memory access rate, and are typically on the order of several microseconds for high speed NAND flash.

9.4.1 Memory arrays

All digital memories comprise arrays of elements. Leakage currents from neighboring, unaddressed memory cells can lead to errors in identifying whether the contents of the target element is a **0** or a **1**. Similarly, leakage can result in errors in writing data to the desired location. The source of this failure is the accumulation of *sneak currents*, which were discussed in the context of photodiode arrays in Section 7.2.2. This is illustrated in the fuse memory array in Fig. 9.46a. A logical **0** for this device is contained in a low resistance memory element, whereas a **1** is stored by an open element, or "blown fuse." Matrix memories are read row by row. The logic state of each memory cell is determined by scanning through the several columns. To address the target element at (b,x) in Fig. 9.46a, a voltage is applied between row b and column x, and the current I_{on} is read at the row output. However, unblown fuses at (b,y), (c,x), and (c,y) provide a secondary path, adding to the total current. Thus, if element (b,x) is in the **0** state, a current will still be read out at row b due to the sneak currents through this secondary path, mistaking the logic state of (b,x) for a **0**. Any number of paths on a large scale array thus makes the error-free readout of memory unlikely. Similar ambiguities exist in writing the memory. In spite of its shortcomings, a three stage, 3D stacked organic resistive array has been reported in an architecture that is presumably inspired by 3D NAND flash memories based on Si (Song et al., 2010). No data were provided on the array performance other than that of individual memory elements. This makes it impossible to evaluate the contrast between the **1** and **0** states, although according to Fig. 9.46a, this limits its practicality as a memory circuit.

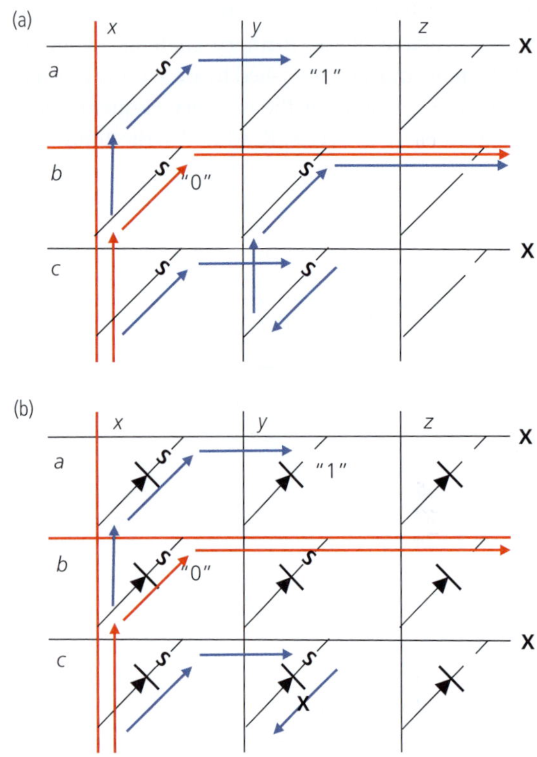

Figure 9.46 (a) A cross-point memory array comprising fuses (**S**) that represent a logical **0** in the low resistance mode, and a **1** when open (i.e. a blown fuse). The circuit reading the memory connects the column on the far left with the second row (highlighted in red). Unconnected rows are marked **X**. Red arrows connect rows and columns through the addressed fuse at (b,x). In this example, a parasitic current (blue arrows) through the unblown fuses at (b,y), (c,x) and (c,y) contribute parasitic current through (b,x) in the **0** state. (b) The same array except where a nonlinear circuit element (a diode) accompanies each fuse. The sneak current through (c,y) is significantly reduced through the reverse-biased diode at (c,y).

Sneak currents are eliminated, or at least significantly reduced by including a non-linear circuit element such as a diode or an access transistor at each cross-point, as shown in Fig. 9.46b. The sneak current at point (c,y) encounters a reverse-biased diode, thus breaking its path to readout line b. However, reverse leakage currents in the diode, or parasitic source-drain currents in the access transistor add from each column and row, potentially resulting in a current that is sufficient to be mistaken for a logical **1**, even if such a state is in error. The accumulation of leakage currents, therefore, also limits the size of the array.

An alternative nonlinear element is the access transistor in Fig. 9.47. This configuration employs an additional interconnect, called the "word line," that switches an access transistor channel between the conducting and non-conducting states, thereby isolating

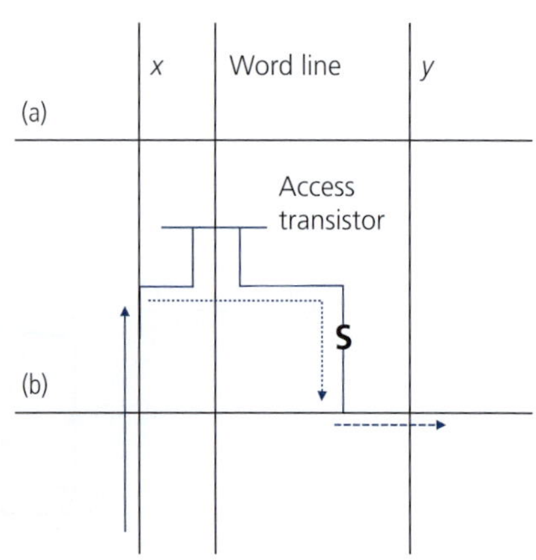

Figure 9.47 One cell of a memory array employing an access transistor with a resistive memory element, **S**.

the desired pixel. Here, the channel leakage sets a limit to the array size that gives an acceptable on/off current ratio for error-free read and write operations. A common means to increase the memory size in either diode or transistor arrays without incurring readout errors, while simultaneously decreasing the memory element access time is to divide the array into blocks, and then using multiple read/write electronics in parallel to address the various memory sectors.

Organic memory elements are configured into several different classes, depending on the phenomenon that is exploited to store data. The bit can be stored by a change in resistance as discussed above, or in capacitance. The resistance changes are induced by application of a switching voltage that results in a chemical change (e.g. reduction or oxidation) of the organic molecules comprising the element, charge transfer to trapping sites that change the potential between electrodes, or electromigration of metal filaments that short the electrodes. Alternatively, an organic ferroelectric material has two stable dipolar polarization states that can either point along, or opposite to the direction of the applied field. The dielectric properties of the ferroelectric can be switched by an applied voltage, inducing a change in the device capacitance.

9.4.2 WORM memories

A resistive *write once, read many times* (WORM) memory array based on PEDOT:PSS fuses that are integrated in series with an amorphous Si p-i-n junctions is shown in Fig. 9.48. The device consists of an array of thin film diodes pre-deposited onto a flexible, stainless steel substrate. The Al diode cathodes are arranged in columns. The p-Si regions form one contact of a ~50 nm thick film of PEDOT:PSS (1:2.5) spun onto the diode, and then cured at elevated temperature. The Au row contacts are vacuum deposited through a shadow mask to form 17 μm^2 row and column overlaps that define each memory cell. The 16 × 16 array in Fig. 9.48b is fabricated using cold-welded rows and columns applied with a PDMS stamp (see Section 5.6.4).

The asymmetric I–V characteristics of one element are shown in Fig. 9.49a, along with the switching cycle in Fig. 9.49b. The forward current of the Si diode (open circles) reaches 100 μA at a voltage of 1.5 V, whereas under reverse bias, the leakage is only 2.3 nA at −1.5 V, yielding a rectifying ratio of ~10^5. Due to the series resistance of the thin organic film, the forward current is reduced to 400 nA for the memory element. To switch the fuse state from "closed" to "open," the voltage is ramped at 1.0 V/s from 0 V → 10 V → 0 V. During each voltage step of duration $t_d = 10$ ms, the voltage is applied for $\tau = 4$ ms (a 40% duty cycle); see inset, Fig. 9.49b. Ramping the voltage to 8 V opens the fuse, reducing the current to a background leakage of ≤ 10 pA at all voltages applied. Hence, the contrast between a **1** and **0** is $I_{on}/I_{off} \sim 10^4$ at a forward bias of 1.5 V. The memory elements irreversibly and reproducibly switch on the application of the voltage.

The film undergoes a transition to its high resistivity state by traversing two distinct regions: **A** from $V = 0$

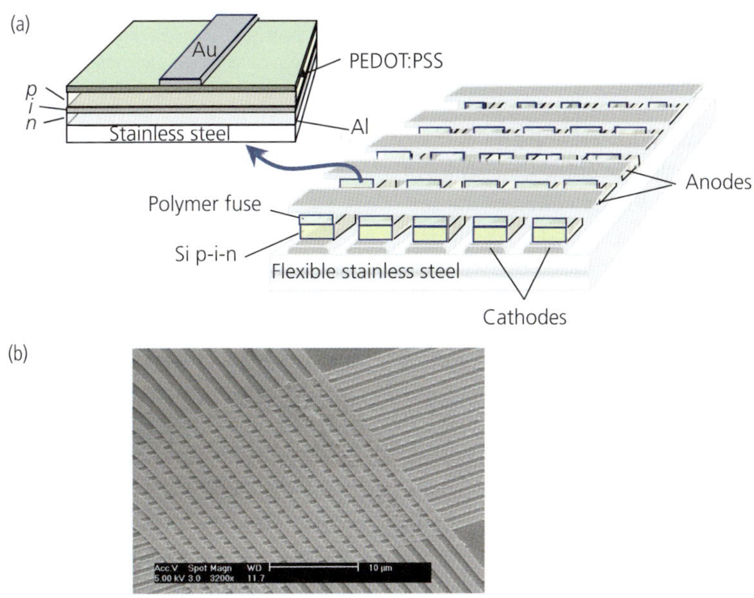

Figure 9.48 (a) PEDOT:PSS fuse/Si p-i-n diode WORM memory array with a detail of a single memory element at the upper left. (b) SEM image of a WORM memory with only PEDOT:PSS fuses between row and column array electrodes fabricated by cold-weld electrode stamp transfer. The scale bar is 10 μm. Adapted from Möller et al. (2003).

First publication in Nature, 2003.

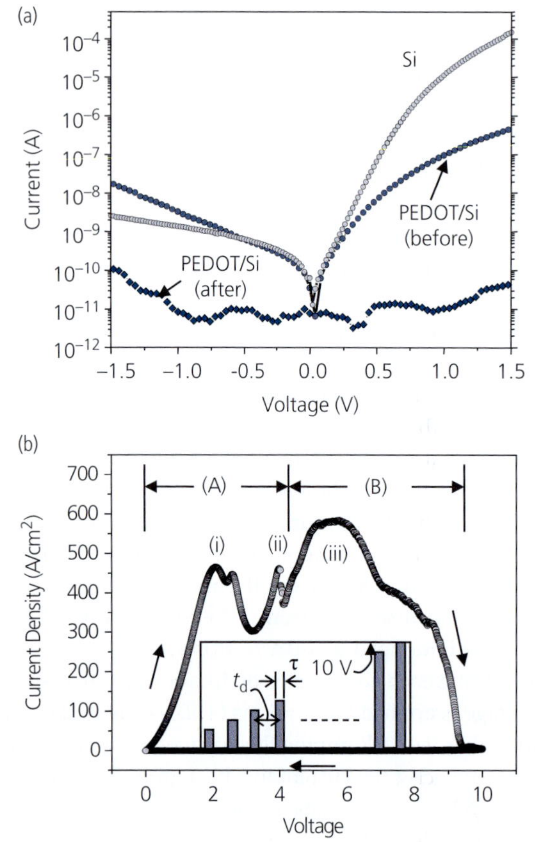

Figure 9.49 (a) Current–voltage characteristics of the PEDOT:PSS/Si p-i-n diode before and after switching. (b) The switching process of the memory element. Two regions are observed during application of a ramp of switching voltage pulses. Region **A** results in small, semi-reversible increases in resistance, whereas **B** results in an irreversible, 10^4 increase in resistance. Features of note are labeled (i), (ii), and (iii). Inset: Switching pulse sequence (Möller et al., 2003).

First publication in Nature, 2003.

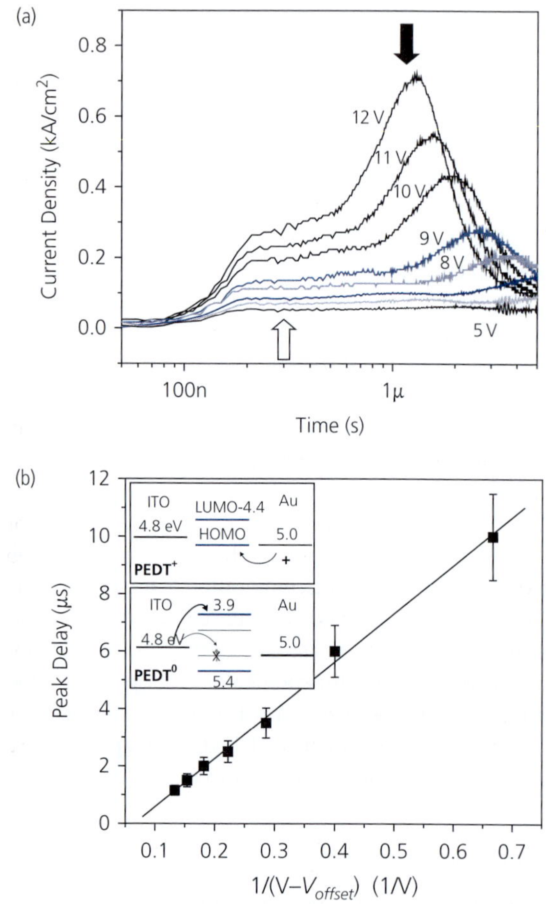

Figure 9.50 (a) Transient response of the current density across a 55 nm thick PEDOT:PSS film vs. voltage during a >10 μs voltage pulse with a rise time of 100 ns. The hollow arrow shows the region where no changes in conductivity are observed. The solid arrow indicates the current peak at which there is a permanent drop in conductivity, as apparent from the decrease in current density following the peak. (b) Peak transient delay vs. $1/(V - V_{offset})$, where $V_{offset} = 4.5$ V is required for electron injection. The straight line fit to the data yields an electron mobility of $\mu_n = (1.7 \pm 0.4) \times 10^{-6}$ cm²/V s. Inset: Energy level diagrams of PEDOT in the oxidized (upper panel) and neutral (lower panel) states, along with the electrode arrangement (Möller et al., 2003).

to 4 V and **B** at $V \geq 4$ V. At the lowest voltages (0–2 V), the current density is reversible over the time scales of the experiment (minutes). At higher voltages, the current reaches a first (i), and then a second peak (ii). The permanent conductivity changes induced by sweeps of up to 4 V are small. The magnitude of the change depends on the pulse duration, τ. The second, sharp peak (ii) that appears for lower voltage sweeps, is reduced in magnitude and broadens as τ is increased from 0.5 ms to 4 ms. However, once the film is biased into region **B**, a qualitatively different process is initiated, resulting in a dramatic and permanent decrease in conductivity. Indeed, once the last broad peak (iii) appears, there is a precipitous decrease in conductivity by several orders of magnitude, and the fuse is permanently opened (Möller et al., 2003).

Figure 9.50a shows the transient response of a 55 nm thick PEDOT film sandwiched between Au and ITO electrodes at several different voltages. The response at >5 V is separated into two regions: a plateau (hollow arrow), followed by a current peak (solid arrow). The delay between the pulse onset and the peak decreases with increasing voltage, reaching a current density of 700 A/cm² after 1 μs for a pulse height of 12 V. The current peak corresponds to initiation of the permanent change in film conductivity in region **B**. The peak delay from the leading edge of the voltage pulse is linearly dependent on $1/(V - V_{offset})$, where $V_{offset} =$ 4.5 V is the voltage drop across the Au contact necessary to inject an appreciable concentration of electrons.

The transient response is due to of double carrier injection (Hack and Street, 1992). In PEDOT, the current is primarily carried by holes (Johansson et al., 2002). However, at $V > V_{offset}$, electrons are injected, thereby neutralizing some of the hole space charge. This, in turn, increases the internal electric field within the polymer, leading to a further increase in hole current. This phenomenon of "double injection gain" is similar to current gain in a photoconductor (see Section 7.1.1), and leads to the current peak observed. At higher pulse voltages, the peak delay decreases due to the larger density of injected electrons. Although the current is controlled by the presence of electrons, they do not contribute appreciably to the film conductivity.

Since PSS is electrically insulating, conduction occurs by hole transport along low resistivity percolation pathways formed by $PEDOT^+$ chains throughout the film. The energy levels of $PEDOT^+$ responsible for charge transport are shown in the top inset, Fig. 9.50b (Xing et al., 1997, Greczynski et al., 2001, Heuer et al., 2002). Holes are injected from both Au and ITO into the HOMO of $PEDOT^+$ at 5.0 eV. The LUMO is 0.6 eV above the HOMO, presenting a barrier to electron injection. To promote electron injection, $PEDOT^+$ is electrochemically reduced to $PEDOT^0$, and the energy level scheme changes to that in Fig. 9.50b, lower inset. In neutral $PEDOT^0$, electrons can only occupy the LUMO at 3.9 eV. Hence, there is rapid, sequential oxidation and reduction by the successive transfer of holes and electrons into the polymer.

The reduction of $PEDT^+$ continues with increasing current until a maximum number of chains in the neutral $PEDT^0$ state are achieved at voltages >2 V but still within region **A**. At $V_{offset} > 4.5$ V, electron injection leads to process **B** where a permanent decrease in film conductivity by up to a factor of 10^3 occurs. The temperature rise of ~200°C at current densities of ~1kA/cm^2 dedopes the $PEDOT^+$ (Pei et al., 1994). Thermal effects can lead to segregation of the PEDOT and PSS components, creating additional and permanent energy barriers at the contacts that can greatly decrease the conductance between electrodes (Xu et al., 2006).

An $N \times N$ resistive matrix memory is addressed row-by-row by measuring the current flowing from N columns to the selected row. Hence, the maximum number of pixels in a row that can be accommodated without leading to read errors (i.e. detecting that a pixel is written when in fact it is not) is $N = mod\,|I_{F1}/I_{F0}|/S_B \sim 10^3$, assuming the signal-to-background current ratio required to reliably distinguish a **1** from a **0** is $S_B = 10$. Here, I_{F1} and I_{F0} are the forward biased currents for the two logical states shown in Fig. 9.49a. Assuming that the current of a written pixel is $<10^{-4}$ of the forward-biased current of a single element, this suggests that a 1 Mb memory is achievable with these devices.

9.4.3 Reversible resistive memories

Conductive filament growth has been used for achieving non-volatile, reversible memory elements. Analogous to the WORM, the formation of conducting metallic filaments is thermally activated. Switching between the conducting and insulating states has been attributed to the formation of conductive salts between a metal electrode and an organic film. For example, the conducting Cu^+TCNQ^- salt can be neutralized into a non-conducting Cu + TCNQ state by reversing the polarity between Cu contacts to the TCNQ thin film. Alternatively, the non-conducting state is associated with the formation of a thin, insulating oxide at the counter electrode to the Cu (Billen et al., 2007). The conductivity of a Cu/CuTCNQ/Al memory cross point unexposed to air is in its conducting, or *low resistance state* (LRS) that cannot be switched to a *high resistance state* (HRS) by applying a voltage. When exposed to air, the HRS is the native state, which switches to a LRS by applying either positive or negative 5–10 V.

The model proposed for filamentary conduction is illustrated in Fig. 9.51a, with the reverse process of filament dissolution in Fig. 9.51b. Switching from HRS to LRS occurs via the reduction of Cu from the anode to Cu^+, or Cu^{2+} from Cu^+, with $TCNQ^-$ in the film bulk (processes i, ii, and iii in Fig. 9.51a, respectively). The Cu cations migrate toward the oxidized cathode where they penetrate the imperfect, very thin insulating coating. Oxidation of the Cu^+ at the cathode surface produces a metallic Cu filament that, over time and under sufficient bias, shorts the oxide, dramatically decreasing the resistance. Reversing the contact polarity results in the opposite reaction, whereby neutral Cu in the filaments is reduced, dissolving the filaments and driving the element from LRS back to HRS. The thermal image in Fig. 9.51c shows hot spots that likely arise from conduction along the metallic filaments from the top Al contact.

Note that voltage applied to query the memory state can reverse either of these electrochemical processes, ultimately destroying the information held within the element. The CuTCNQ system can be used for ~10^4 cycles before a refresh voltage pulse is required (Kever et al., 2006). A drawback of this memory mechanism is its sensitivity to the fabrication process and ambient conditions, making it somewhat irreproducible and unreliable when fabricated in

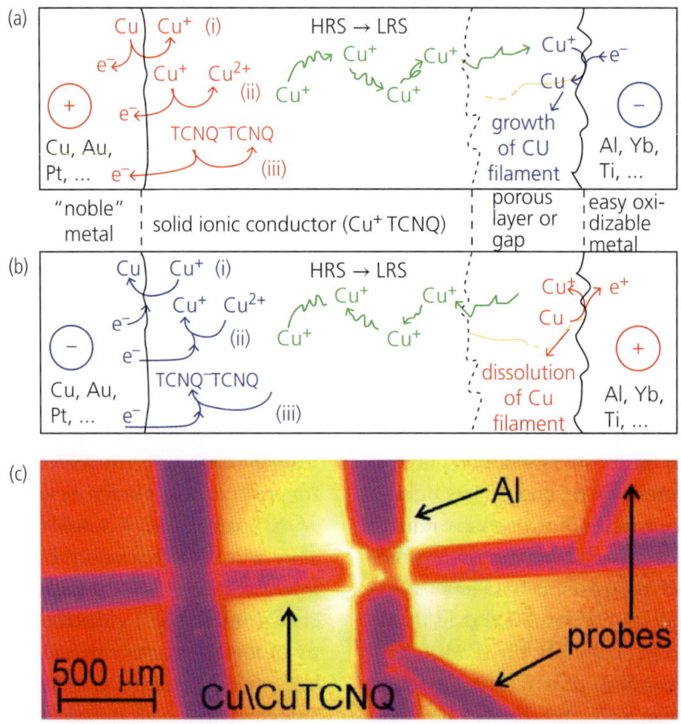

Figure 9.51 (a) Proposed electrochemical process resulting in the formation of conductive Cu filaments (gold color) that penetrate the thin, porous oxide formed on the metal contact. This leads to transformation from the high resistance (HRS) to the low resistance (LRS) state. (b) Dissolution of the filaments due to oxidation of Cu. Oxidation is shown in red, reduction in blue, Cu+ transport in green. (c) Infrared (thermal) image of a Cu/CuTCNQ/Al memory cross point. Yellow areas are hottest, blue are coldest. Note the hot (pinhole) spot within the central cross point (Billen et al., 2007).

different labs, or when used under less than perfectly controlled conditions. This sensitivity is expected for a process that depends on the random formation of pinholes whose density and dimensions are subject to variability. These variations can be significantly reduced by intentionally growing thin oxide layers using chemical vapor deposition (CVD) or atomic layer deposition (Muller et al., 2009). Of greater concern is that $I_{on}/I_{off} \sim$ 10–100 is too small for use in large-scale memory arrays.

Metal NP dispersions within an organic matrix have also exhibited reversible resistance over several orders of magnitude, along with very fast switching between the LRS and the HRS. This resistive memory device does not require the adventitious and somewhat uncontrolled formation and dissolution of filaments, and hence it can potentially result in less process-dependent, more predictable performance. Illustrations of two different NP resistive memory architectures are provided in Fig. 9.52a and b. The NPs can be intentionally deposited in a narrow zone or layer (Ma et al., 2002, Yang et al., 2006), or they can be homogeneously dispersed throughout the bulk of the resistive zone (Tseng et al., 2005, Lin et al., 2007b).

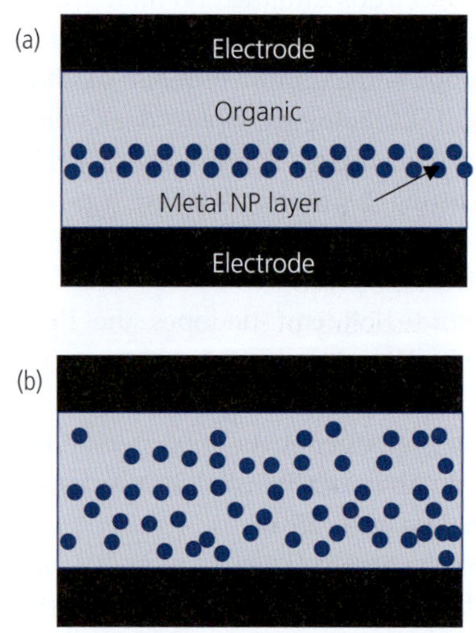

Figure 9.52 Resistive memory devices employing (a) a layer of metal nanoparticles (NPs) embedded in a conducting organic matrix sandwiched between two metal electrodes. (b) A homogeneous blend of NPs and organic.

The layered structure (Fig. 9.52a) is conveniently fabricated by vacuum thermal evaporation of a combination of thin (~50 nm) small molecule organic layers that sandwich a thin (~20 nm), non-wetting metal layer. The blend memory cell comprises a conducting polymer mixed in solution with the metal NPs and cast onto the substrate. Both devices show nearly identical performance, and are largely independent of the composition of the NPs and the conducting organic thin films. Applying a voltage larger than a threshold, V_1, between the counter electrodes induces an abrupt increase in current. The LRS has a conductivity that is 100–10^4 higher than the HRS. Applying a reverse potential returns the resistor to its HRS.

Results from a metal NP/polymer blend memory element are shown in Fig. 9.53. The device comprises a 70 nm thick Au NP/PANI film spin coated onto an Al electrode, and then coated with a vacuum deposited Al counter-electrode. An image of a region of the Au NP/PANI nanofiber film is provided in Fig. 9.53a. There is a high density of NPs that condense within a fibrous conducting network. Switching is observed for NP diameters < 20 nm, whereas at larger diameters the device switches into its LRS only once. Three switching cycles of the device are shown in Fig. 9.53b. In sweep A, the bias is ramped from 0 to +4 V. At $V_1 = 3$ V, the current abruptly jumps from 100 nA to 100 μA. At higher voltages, there is a weak decrease in current due to a negative differential resistance. The voltage is then decreased from +4 V back to 0 V (sweep B). The device remains in the LRS during this sweep, with the current falling at the same rate as when the device is in the HRS. The LRS is returned to the HRS (i.e. the contents of the memory are erased) by applying a reverse bias of −5 V. It can be turned on once again by following the same voltage sweep as used initially (sweep C). It is noteworthy that the LRS and HRS during this and subsequent sweeps are nearly identical. The response to a sequence of read, write and erase voltage pulses is shown in Fig. 9.53c, showing the reproducibility of the switching characteristics over several cycles (Tseng et al., 2005).

The data are retained with no significant change in conductivity for over 3 days even if small voltage stresses are periodically applied. Importantly, the switching time is ~ 25 ns, which is limited by the resistance-capacitance (RC) charging time of the resistor. Note that similar qualitative and quantitative behavior is obtained for the layer structure (Fig. 9.52a) employing a 20 nm thick Al layer sandwiched between two 50 nm thick AIDCN organic thin films (Ma et al., 2002). Microscopic images of Al layer indicate its coalescence into NPs embedded into the organics due to dewetting during deposition (Yang et al., 2006). We have seen in Section 7.5.2 that thin Ag layers deposited onto organic surfaces can diffuse deep into the film bulk, and hence the structure may

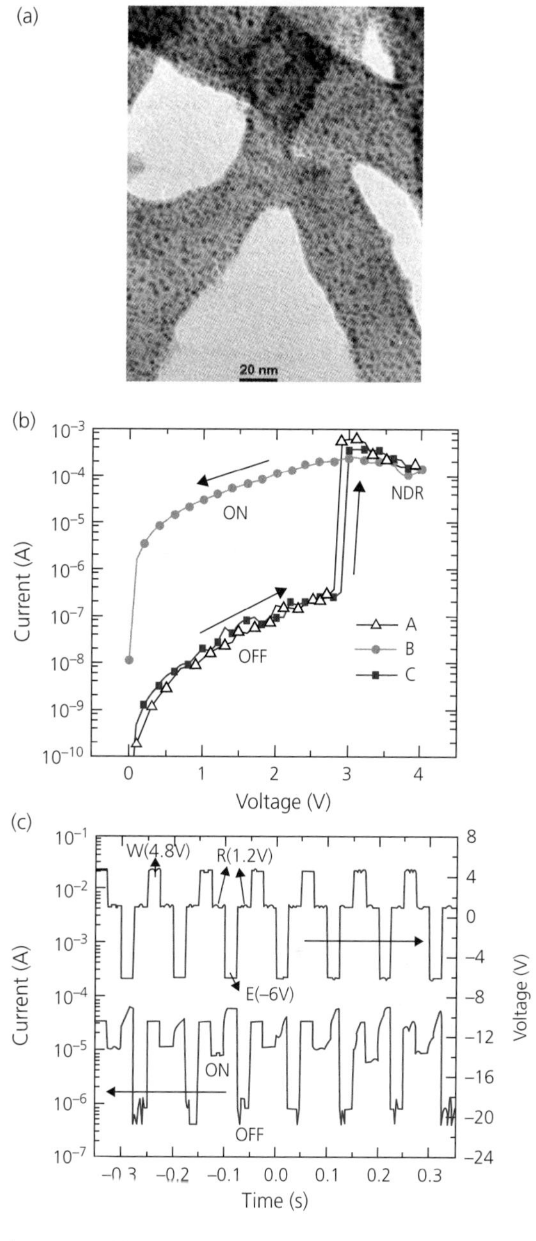

Figure 9.53 (a) Transmission electron micrograph of a section of a Au NP/PANI blend. The nanofiber regions contain a high density of NPs visible as black dots. (b) Switching characteristics of the blend showing an approximately 1000 times increase in current from the HRS to the LRS. Current switching occurs at $V_1 = 3$ V. The memory contents are erased at $V_2 = -5$ V. There is a small negative differential resistance (NDR) above V_2. (c) Top: Sequence of write (W), read (R), and erase (E) voltage pulses. Bottom: Current response to the voltage cycle (Tseng et al., 2005).

actually be a more homogeneous dispersion of NPs than suggested in Fig. 9.52a.

The nearly identical resistance switching properties of devices containing an apparently uniform metallic layer and one in which NPs are dispersed within the polymer, and the independence of the characteristics to the choice of metal, suggest that the same mechanism is responsible for switching in both. The source of the switching has been attributed to several factors, including tunneling of charge into the metal NPs (Bozano et al., 2005) and trap filling that creates large changes in the space charge limited current (Lin et al., 2007a). However, neither of these mechanisms appears to explain the long retention time of the memory state, nor the very fast switching time of <25 ns.

A plausible suggestion for conductivity switching is analogous to that used to describe switching of PEDOT:PSS in the WORM memory, and in CuTCNQ in filamentary memories. Prior to switching, the Au NP and the PANI are both neutral, and hence the film conductivity is low. When the voltage exceeds V_1, there is partial charge transfer from PANI to Au, as illustrated in **Scheme 1**. As in the case of the Cu^+TCNQ^- complex, NP^-PANI^+ becomes conductive. To reverse the process, a negative voltage is applied to neutralize the two components of the film. The charge transfer process is extremely rapid, consistent with the fast switching time of the devices. Further, the transferred charge is stable due to the large energy barrier between the metal and the polymer, making the relaxation time (i.e. memory state decay) long compared to that expected for trapping at deep levels within a semiconductor energy gap. The existence of a charge transfer state has been affirmed using X-ray photoelectron spectroscopy, where it was found that the N_{1s} state of the PANI is shifted in the composite compared to a neat film. Similarly, the binding energy of the Au electrons in the $4f_{5/2}$ state is found to decrease in the composite. Finally, microscopic probing of the NPs using conducting tip atomic force microscopy shows a switching behavior similar to that observed by the film as a whole (Tseng et al., 2005). Taken together, these data provide a compelling case for charge transfer switching, consistent with phenomena observed for the other resistive organic memories described above.

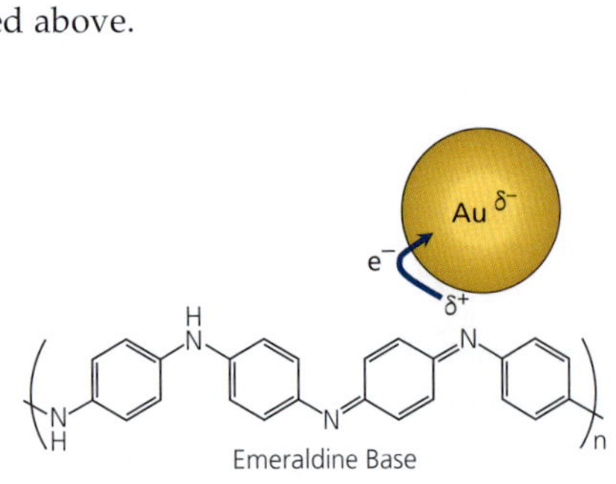

Scheme 1 Charge transfer between emeraldine base PANI and a Au NP. $\delta =$ fractional charge transfer (Tseng et al., 2005).

9.4.4 Ferroelectric capacitors and diodes

The last two-terminal structure to be considered is based on the *ferroelectric effect*. Analogous to a ferromagnetic material, a ferroelectric is a dipolar material that has a spontaneous polarization below its *Curie temperature*. When an external electric field of sufficient magnitude is applied opposite to the direction of the dipoles, the dipole moment flips to find a new equilibrium. The field required to rotate the dipole moment is the *coercive field*, or F_C. Once the field is removed, the polarization relaxes to its *remnant polarization*, P_r. Since ferroelectric polarization requires alignment of molecular dipoles, the materials must be ordered, and preferably crystalline. Also, polarization switching entails the physical rotation of the molecules, and hence it is slow compared to charge transfer in resistive memories.

The polarization is related to the electric displacement vector, **D**, via

$$\mathbf{D} = \overleftrightarrow{\varepsilon}_r \mathbf{F} + \mathbf{P}, \tag{9.33}$$

where $\overleftrightarrow{\varepsilon}_r$ is the relative permittivity tensor, **F** is the field, and **P** is the polarization vector. The characteristic D vs. F behavior of an archetype ferroelectric is shown in Fig. 9.54a. At electric fields that are insufficient to switch the polarization state, or for a completely depolarized sample, the relationship between D and F is approximately linear and reversible, as shown by the dotted line near the origin. In a spontaneously polarized sample (point A in Fig. 9.54a) or after applying a high electric field, most but not all of the molecules are aligned (inset A, equivalent to logical **0**), resulting in a net polarization $-P_r$. Applying an electric field in the direction opposite to $-P_r$ results in a monotonic reduction of the displacement, until $F = F_C$. At this point (B), the material is completely depolarized. Further increases in F result in preferred alignment along the field direction until the displacement saturates at voltage $V_{CC} = F_{CC} \cdot t$, where t is the film thickness. The polarization then relaxes to a new remnant value, P_r, with the dipoles pointing opposite to that of their original orientation (inset D, logical **1**). The hysteresis is complete by reversing the field to $-F_C$ (point E), and then to $-V_{CC}$ at point F, and back to $F = 0$. The open loop created by the variation in D with F provides a memory window of width $2F_Ct$. The read–write memory function is achieved, therefore, by sweeping the external field between $\pm F_C$. A larger F_C results in a lower error rate in distinguishing between the **1** and **0** states. Note that at temperatures larger than the Curie temperature, the memory window vanishes, although the D vs. F characteristic retains its S-shape. At these temperatures, the material is *paraelectric*.

The field applied to a ferroelectric changes its polarization, and hence the capacitance of the element. The memory state is determined by measuring its capacitance, C_T, employing the Sawyer–Tower circuit in Fig. 9.54b (Sawyer and Tower, 1930). This is a capacitive voltage divider network comprising the unknown ferroelectric capacitor, and a known reference, C_{ref}. The state of the ferroelectric capacitor is set by closing the switch, applying a d.c. voltage, V_x, and an a.c. test signal supplies the series capacitance network. Then, C_T is determined by measuring the voltages, V_x and V_T. Since a small voltage is applied during the read cycle, the polarization state of the capacitor is changed, degrading its contents. After a number of read cycles, the memory is refreshed to restore it to its original logic state.

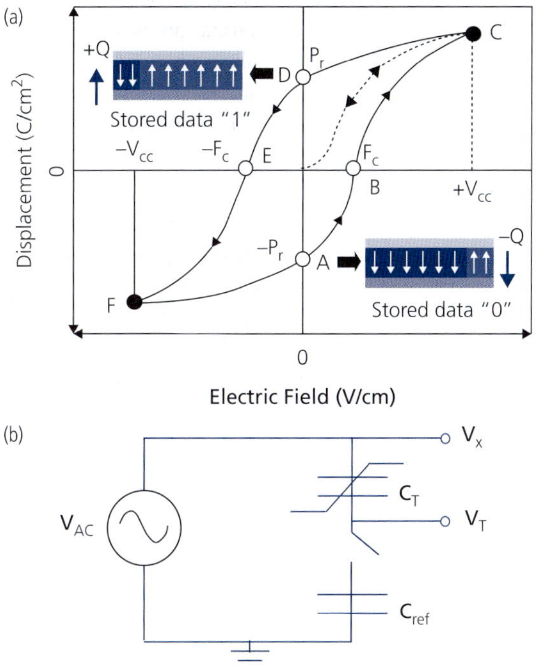

Figure 9.54 (a) Hysteresis in the electrical coercivity of a ferroelectric. The remnant polarization, P_r, and the coercive field F_c are shown. The dotted line from the origin to point C shows the displacement, D, vs. electric field in the initially unpolarized state. The orientations of the electric dipoles in the film are illustrated in the insets. (b) Sawyer–Tower circuit used for measuring the test capacitance, C_T, of the ferroelectric using a second reference capacitor that can be switched out during ferroelectric polarization. Applying a voltage $V_x = \pm V_{cc}$ polarizes C_T.

The ferroelectric effect in organics is obtained using dipolar molecules that undergo a conformational reorientation when placed in an electric field. Ferroelectricity was noted in polyvinylidene fluoride by Kawai (1969) and later in various nylons. Today, the most commonly employed ferroelectric in both two and three terminal devices is a copolymer of VDF and TrFE (P(VDF-TrFE)) in a molar ratio of 2:8. This material has a large remnant polarization of approximately 10 $\mu C/cm^2$ and is stable in air. Small molecules have also shown ferroelectric effects. However, given their steric bulk, it is difficult to reorient the molecules in the solid state, whereas polymer chains only need to be twisted about the backbone axis. Liquid crystalline small molecules have the ability to undergo rotations required to exhibit ferroelectric dipole reorientation.

A segment of the copolymer P(VDF-TrFE) in Fig. 9.55a illustrates the atomic orientation along the backbone. Its ferroelectric properties are inferred from

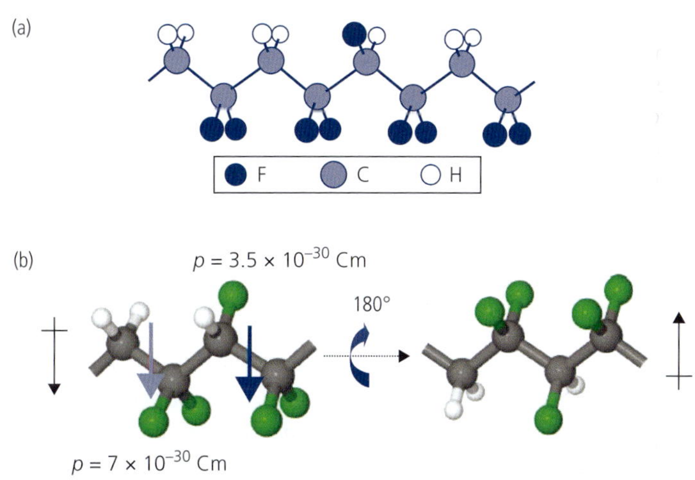

Figure 9.55 (a) Structure of ferroelectric polymer P(VDF-TrFE). (b) The molecule is twisted about its molecular backbone by 180° under the influence of an external electric field. The dipole moments due to the regular orientation of the electron withdrawing F atoms (green) and the donating C atoms (gray) along the backbone segment are large, giving this molecule its ferroelectric properties. Hydrogen atoms are shown in white. The total molecular dipole moment orientation is shown by the arrows to the left and right of the molecules. Adapted from Heremans et al. (2011).

Fig. 9.55b. The *trans* conformation of both the VDF and TrFE segments results in parallel dipole moments pointing between the electron rich F atoms, and electron deficient C atoms. The dipoles of the individual segments add to create a large, spontaneous polarization when the film is deposited and thermally annealed at 140–150°C, or by stretching to elongate the polymer chains (Naber et al., 2004, Nguyen et al., 2008). The higher dipole moment of 7×10^{-30} C m is contributed by VDF, and a similarly oriented TrFE has a dipole moment of 3.5×10^{-30} C m when oriented in the all-*trans* configuration. A coercive field of 500–700 kV/cm directed opposite to the dipole can twist the chain to lower the dipole energy. The information state is changed by the physical rotation of the polymer chain to reverse its polarization state. For P(VDF-TrFE), this takes approximately 160 μs (i.e. ~6 kHz) (Kim et al., 2009). The large coercive fields require high switching voltages, or the use of very thin films. The magnitude of the ferroelectric polarization is reduced in the thinnest films, perhaps due to surface roughness and pinning of the dipole near the polymer/electrode interface.

The switching behavior of a Au/P(VDF-TrFE)/PEDOT:PSS/ITO capacitor is shown for two different ferroelectric layer thicknesses in Fig. 9.56a. The low switching voltages, even for films as thin as 65 nm, is attributed to the smooth surface presented by PEDOT:PSS along with the Au electrode that do not react with the P(VDF-TrFE) to create a non-ferroelectric dead zone near its surface, as is observed using an Al electrode. The switching time ranges between 100 μs to 16 μs as the coercive field is increased from 800 to 1400 kV/cm. Over this range, no significant penalty in the memory window width is observed (Naber et al., 2004).

Figure 9.56b shows the memory retention (or *fatigue*) of a substrate/Ti/Ppy:PSS (9 nm)/P(VDF-TrFE) (50 nm)/Ppy:PSS (15 nm)/Ti capacitor at room temperature, 45°C, and 60°C. The Ppy:PSS film prevents reaction with the electrode metals, or their diffusion into the ferroelectric film. This permits the use of very thin P(VDF-TrFE) layers, leading to switching at only 2.6 V and $2F_Ct = 0.85$ V. The thin film device also features a switching time of 30 μs (30 kHz). At room temperature, the device suffers a 20% loss in remnant polarization after 10^7 cycles, with an increase in coercive voltage to 4.5 V and a concomitant closing of the memory window to $2F_Ct = 0.32$ V. The maximum P_r and the retention time decrease as the Curie temperature is approached (Xu et al., 2007).

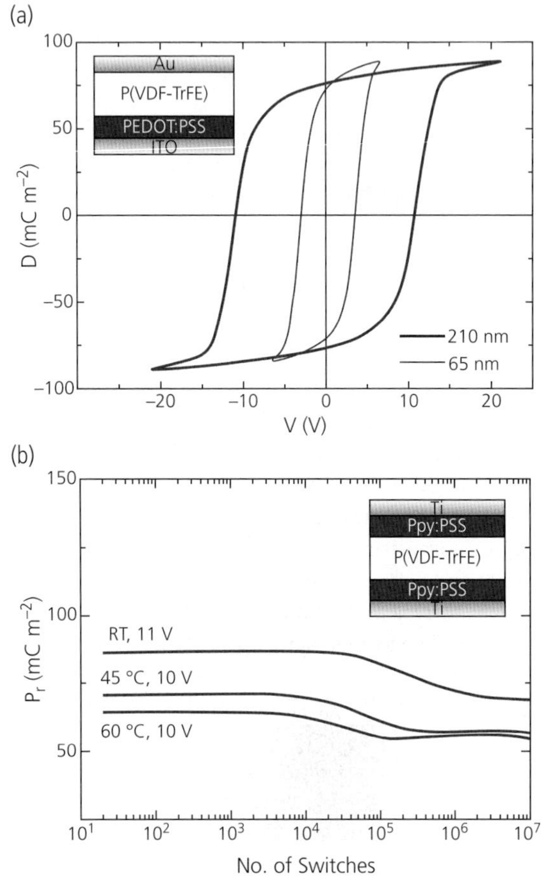

Figure 9.56 (a) Hysteresis loops of the ferroelectric capacitor shown in inset for two different P(VDF-TrFE) film thicknesses (Naber et al., 2010). (b) Remnant polarization vs number of switching cycles and temperature of the device in the inset (Xu et al., 2007).

A shortcoming of the capacitive memory is the destructive nature of the read-out cycle: a voltage pulse required to read the memory state leads to degradation or loss of that state. A diode composed of a ferroelectric material, however, serves as an electrically switchable resistive memory element whose logic state can, in principle, be non-destructively queried. Rectification also is useful for reducing sneak currents in memory arrays (see Fig. 9.46).

Rectification can be achieved by blending a second component with the ferroelectric, thereby forming a junction, or the ferroelectric material itself can form a rectifying Schottky barrier with its contacts. The diode conductivity is then switched as follows. Under forward bias, polarization oriented parallel to the electric field and perpendicular to the junction, decreases the local depletion region width. This permits tunneling across the junction, leading to a large forward current. When the polarization is reversed, the depletion region becomes large, consequently increasing the resistance and hence decreasing the conductivity. Since the diode forward current is exponentially dependent on voltage, the difference in current between logic states can be several orders of magnitude. However, ferroelectrics tend to have very low conductivities, which reduce the contrast between logic states. To increase the conductivity, the material needs to be doped, although mobile charges from the dopant screen the dipole field, thus reducing P_r.

The tradeoff between conductivity and polarization can be substantially reduced by using blends of conducting and ferroelectric polymers. The phase separation of the components results in a bicontinuous network that forms a bulk heterojunction between the electrodes. An example BHJ switching diode structure is illustrated in Fig. 9.57a. The conductive region comprising P3HT is contacted by blocking anode and cathode (e.g. Ag) contacts, or a Ag contact on one surface and an ohmic (e.g. LiF/Al) contact on the other. The former, symmetrically contacted device is resistive, whereas the latter is a diode useful in cross-point memory arrays.

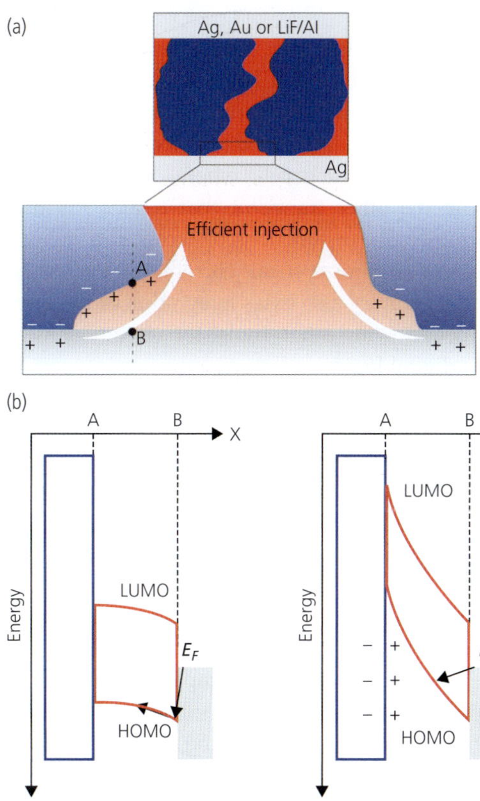

Figure 9.57 (a) Illustration of the nanostructure domains of a phase separated P3HT (red regions)/P(VDF-TrFE) (blue regions) diode memory element. (b) Unpolarized (*left*) and polarized (*right*) energy level diagrams of the memory element. The diagrams correspond to the region marked A-B in (a), with the ferroelectric to the left of point A, the polymer between A and B, and the metal to the right of B (Asadi et al., 2008).

The memory diode operation is understood from the energy level diagrams in Fig. 9.57b. In the unpolarized state (Fig. 9.57b, left), the metal/polymer barrier prevents charge injection, and the diode is nonconducting. The ferroelectric is polarized by applying a voltage pulse to result in an internal field with $F > F_C$ ~ 500 kV/cm. The polarization field tips the bands of the conducting polymer, resulting in a large tunneling injection current into the HOMO, as shown in Fig. 9.57b, right. Note that a polarization pulse of the opposite polarity can also induce charge injection, but into the LUMO instead (with a consequent reversal of the current direction). The magnitude of the tunneling current is exponentially dependent on the barrier width. Hence, the nanostructure of the interpenetrating polymers must be thin (typically $\leq$ 20 nm). The structure successfully separates the conductive and ferroelectric regions, with the high energy barriers between P3HT and P(VDF-TrFE) preventing neutralization of the dipoles by free charges that reside within the separate thin film domains.

The current–voltage characteristics of an asymmetric Ag/P3HT:P(VDF-TrFE) (1:10, 14–25 nm)/LiF/Al diode are shown both before and after poling at ± 20 V in Fig. 9.58a. The current ratio is I_{on}/I_{off} ~ 50 at +3 V, with a reverse leakage current of approximately 300 nA/cm^2. The switching time between the states is 500 μs. The memory state of the diode extends to more than 1000, −20 V write/3 V read/+20 V write cycles, as shown in Fig. 9.58b. The I_{on}/I_{off} shows no change over this large number of pulses, indicating that the cycle scheme is nondestructive of the contents of the memory. Furthermore, the data retention time is >11 days, comparable to that of P(VDF-TrFE) capacitive memories. This device shows that separating the conductive and ferroelectric memory regions is possible using organic thin films. However, the switching times and on/off ratios still fall short of capacitive memories using this same ferroelectric materials system. This is compensated, to some degree, by the simplicity of integration of the resistive memory element with a diode in a single device.

9.4.5 Transistor memories

Organic thin film transistors present an alternative architecture for achieving a resistive memory. In a transistor memory, charge is introduced and retained in the gate insulator, resulting in a hysteretic shift in the threshold voltage (V_T). Equivalently, the dielectric

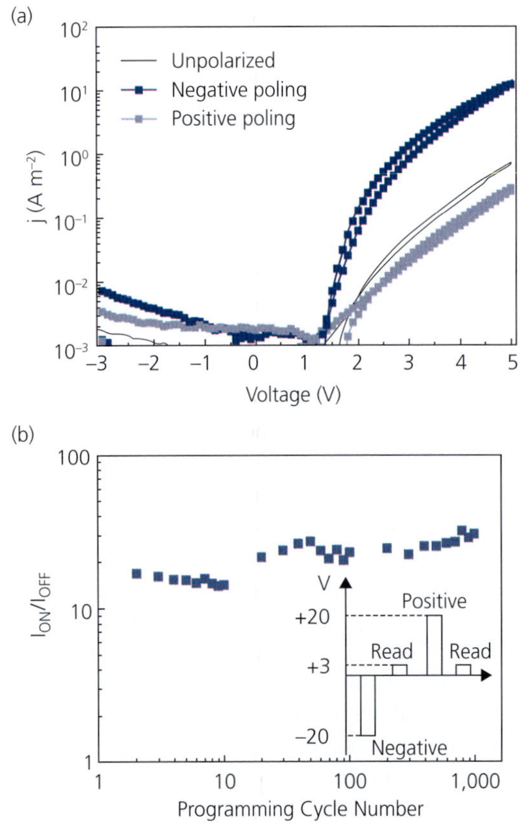

Figure 9.58 (a) Current–voltage characteristics of a Ag/P3HT:P(VDF-TrFE) (1:10)/LiF/Al diode in the unpolarized, negatively and positively polarized states. (b) On/off current ratio vs. number of programming cycles. Inset: Programming cycle pulse sequence (Asadi et al., 2008).

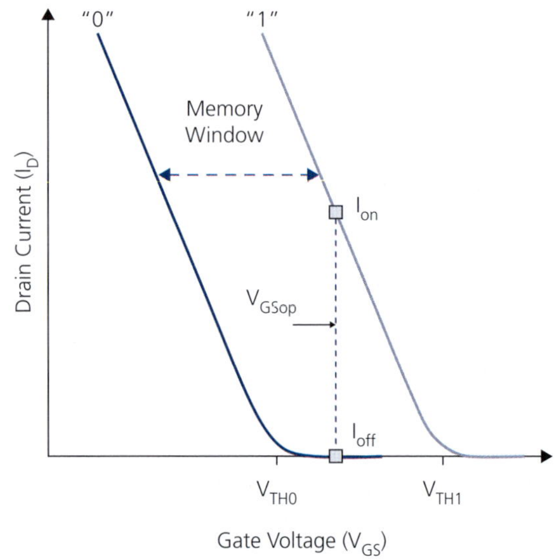

Figure 9.59 Drain–gate transfer characteristics of a memory transistor whose threshold voltage is increased from V_{TH0} to V_{TH1} when switched from the **0** to the **1** logic state, respectively. The corresponding currents at the operating voltage, V_{GSop}, are I_{on} and I_{off}.

constant of the insulator can be changed by the excess charge, or by changing its dielectric properties. The change in dielectric properties changes the potential at the channel/insulator interface. This in turn results in switching the transistor channel between a conducting (logical **1** state) and a non-conducting (logical **0**) state at the same gate voltage. The drain current (I_D) vs. gate voltage (V_{GS}) characteristics of a transistor memory element are shown in Fig. 9.59. By shifting the threshold voltage by $\Delta V_T = |V_{T1} - V_{T0}|$, the current at the operating voltage of V_{GSop} can differ by several orders of magnitude between I_{on} and I_{off}. The memory window of the transistor is ΔV_T, and serves the same purpose of memory retention as do the hysteresis loops that characterize WORM, filamentary and NP resistive memories.

An advantage of using transistor memory elements is that they combine the variable resistive medium (i.e. the channel) with a nonlinear switch (activated by the gate potential) used to access the element. By using three interconnections for each memory cell as illustrated in Fig. 9.47, the gate potential can switch the channel from I_{on} to I_{off}. To be effective, the change in gate voltage must not simultaneously change its memory state. This is avoided by integrating a second transistor within each cell, where one is used for accessing the other transistor that retains the data. The combination of functions of access and memory storage can minimize sneak currents in arrays. The transistor memory also eases integration of the memory function with other transistor circuit elements such as digital logic.

Switching the charge trapped within the gate insulator changes the potential at the insulator/semiconductor interface. The charge modulates the voltage required to achieve a desired drain current. In Eq. 8.75, we showed that the change in threshold voltage is given by

$$\Delta V_T = \frac{q\Delta N_{tr}}{C_i}, \tag{9.34}$$

where ΔN_{tr} is the change in trapped charge in the insulator and C_i is the insulator capacitance. A memory transistor relies on the reversible and intentional injection of charges into deep traps that are controllably introduced into the dielectric during fabrication to achieve reproducible retention times and a large memory window.

A dielectric whose charge density is controlled, and that can retain the charge or polarization state (and hence its value of $C_i \propto \varepsilon_r/t$, where t is the insulator thickness), is called an *electret*. An electret can take on many different forms, although the most common realization in organic memory transistors are based on (i) charge trapping at interfaces within the dielectric layer, (ii) floating gates, (iii) NP dispersions, and (iv) ferroelectric gate dielectrics. The transistor memory is erased when the charge stored in the electret is neutralized, for example by reversing the gate "writing" voltage. Transistor memories, like resistive memories, can therefore be both non-volatile and programmable.

9.4.5.1 Interface charge trapping

An electret can be formed by using a gate insulator comprising two different regions. One is an ultrathin dielectric that contacts the semiconducting channel, and a second is a thick dielectric of a different composition that forms the bulk of the gate insulator. The interface between the two dielectric layers provides a site for charge trapping, and hence long-term storage. The traps are charged by applying a sufficiently large gate potential to cause tunneling from the channel into the traps through the thin dielectric. To erase the memory, the gate potential is reversed to attract the opposite charges, thereby neutralizing the traps.

An example of a charge trap OTFT is illustrated in the inset of Fig. 9.60. The bottom gate, bottom contact (BG/BC) transistor consists of a n^{++}Si gate contact coated with a 20 nm thick, thermally grown SiO_2, first insulator. The Au source (S) and drain (D)

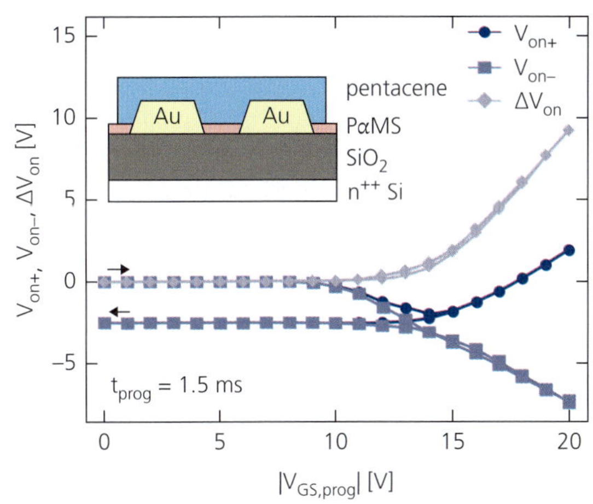

Figure 9.60 Negative and positive turn on voltages (V_{on-} and V_{on+}, respectively), and their difference (ΔV_{on}) vs. gate-source programming voltage ($|V_{GS,prog}|$) for a memory transistor employing a charge trap layer at the interface between two different gate insulators. Inset: Illustration of the OTFT with a bottom Si gate, and bottom Au S and D contacts. The traps exist at the interface between the PαMS and SiO_2 gate dielectrics. Adapted from Debucquoy et al. (2009).

contacts are deposited, followed by dipping the structure into a solution containing a fluorinated thiol SAM that selectively attaches only to the Au surfaces. This monolayer serves as a "resist" that prevents attachment of the 4 nm thick PαMS second insulating layer that is deposited from a 0.1 wt% toluene solution. The pentacene channel coats the entire substrate to form OTFTs with gate lengths of $L = 10$ μm and widths of $W = 1$ mm. The PαMS is sufficiently thin to permit tunneling of electrons and holes from the pentacene to the trap layer that resides at the PαMS/SiO_2 interface.

The switching characteristics of the memory OTFT are shown in Fig. 9.60. Two onset gate voltages, V_{on-} and V_{on+} are defined at the point where I_D reaches 100 pA. The memory window is $\Delta V_{on} = |V_{on+} - V_{on-}|$, which is approximately equal to ΔV_T. Here, the onset voltages are obtained by applying a programming gate potential, $V_{GS,prog}$, over a time window of length, t_{prog}. V_{on-} is the onset voltage after the negative write pulse of $-V_{GS,prog}$ that attracts holes from the channel into the dielectric traps. By reversing the voltage to $+V_{GS,prog}$ for the same duration, electrons are attracted to the traps. This opens a memory window by shifting the onset voltage to V_{on+}. The onset voltages change at $V_{GS,prog} = 10$–14 V. At this point, there is hysteresis in the onset voltages but there is no memory retention since $\Delta V_{on} = 0$. However, above 14 V, ΔV_{on} increases to 10 V when $V_{GS,prog} = 20$ V, indicating that charge is stored via tunneling, and moreover, it can be removed when the programming voltage is reversed. The switching time of the memory state is $t_{prog} =$ 1.5 ms. The mobility in the channel is also a function of the memory state. That is, the hole mobility decreases from its peak of 0.28 cm^2/V s to 0.16 cm^2/V s when the traps are negatively charged, presumably due to increased scattering at the insulator/pentacene channel interface. However, there is no such effect on the mobility for positively charged, or neutral traps (i.e. when the memory state is erased) (Debucquoy et al., 2009).

Two important characteristics of a memory element are the fatigue experienced by the device after repeated read/write cycles, and the retention time after the memory state has been established. Repetitious cycling of a memory can induce irreversible damage to the interface, resulting in a decrease in the memory window. The current onset voltage and the memory window up to 800 read/erase cycles of the device in Fig. 9.60 are shown in Fig. 9.61a. The voltages are stable over the first 100 cycles, at which point the contrast between them begins to fade. However, even after 800 cycles, the memory window is only reduced from 1.8 V to 1.4 V. The drain current ratio

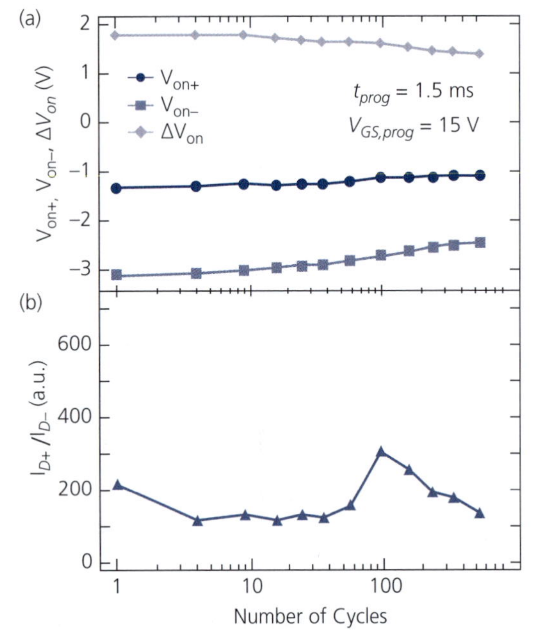

Figure 9.61 (a) Onset and memory window voltages as functions of the number of read/erase cycles of the device in Fig. 9.60. (b) The ratio in drain currents obtained at a transistor operating point of $V_{GS} = -1.5$ V and $V_{DS} = -0.02$ V. The write and erase states are obtained using programming voltages of $V_{GS,prog} = \pm 20$ V (Debucquoy et al., 2009).

between the two states measured at an operating point of $V_{GS} = -1.5$V and $V_{DS} = -0.02$V remains stable at approximately $I_{D+}/I_{D-} = 100$. By comparison, Si flash memories tolerate 10^5 program cycles without significant closure of the memory windows (Bez et al., 2003).

Data retention provides a measurement of the trap discharge time. The retention of both the erased and written states of the transistor in Fig. 9.60 after storage in the dark in N_2 is 3 mo. before all of the trapped electrons are lost, giving $V_{on+} \rightarrow 0$ V. These data indicate that the transistors employing trapped charge layers are both programmable and non-volatile for read/write voltages ±15–20 V.

An interesting variant of the charge trap memory transistor is one that is optically programmable, shown in Fig. 9.62a. The Si substrate forms the gate electrode. A 5.1 nm thick HfO_3 layer is deposited for the first insulator, followed by deposition of a diarylethene phosphonic acid SAM (DAE) that attaches to the HfO_3 to provide the second, 2.6 nm thick dielectric. The pentacene channel is subsequently deposited by vacuum thermal evaporation, and Au stripes formed the bottom gate/top gate contact (BG/TC) transistor. The channel has a mobility of 0.7 cm^2/V

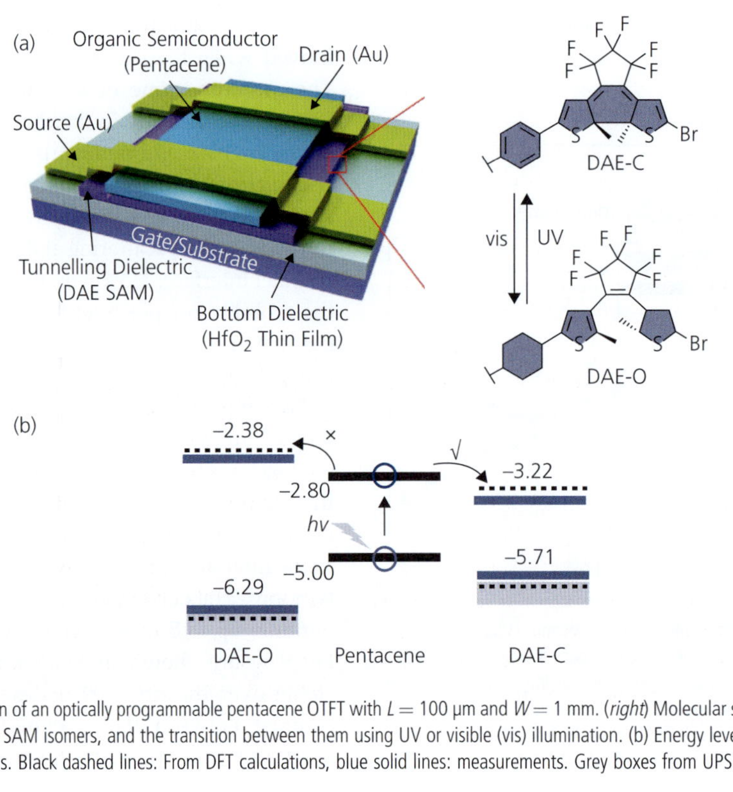

Figure 9.62 (a) Illustration of an optically programmable pentacene OTFT with $L = 100$ μm and $W = 1$ mm. (*right*) Molecular structures show the closed (DAE-C) and open (DAE-O) SAM isomers, and the transition between them using UV or visible (vis) illumination. (b) Energy level diagrams of the DAE-O, pentacene, and DAE-C films. Black dashed lines: From DFT calculations, blue solid lines: measurements. Grey boxes from UPS and PES measurements. From Chen et al. (2015a).

s, typical of conventional pentacene OTFTs. The DAE is optically converted by UV illumination between the open C–C bond (DAE-O) and closed-bond (DAE-C) isomers. The increased π-electron delocalization of DAE-C leads to a reduced energy gap of 1.49 eV, compared to 3.91 eV for DAE-O (see Fig. 9.62b). The bond is reopened to convert DAE-C back to DAE-O using lower energy, visible radiation that is preferentially absorbed in the former isomer. The increase and decrease of the energy gap results in an optically controlled barrier to electron transport from the pentacene channel to the trap interface formed between the SAMs and the HfO_3. By this means, a persistent charge is maintained on the traps that can be written or erased based on the wavelength of the programming illumination.

The energetics of the memory write/erase process are shown in Fig. 9.63. In step **a**, photocurrent is generated by illuminating the pentacene channel with 633 nm wavelength light. The high energy barrier presented by the initially preset DAE-O layer prevents charge from reaching the trap levels between the DAE and the HfO_3 dielectrics. The DAE-O is converted to the low energy gap DAE-C using UV illumination at 365 nm for several hundred seconds. This allows charge transport via tunneling from the SAM into the trap states, thereby writing the memory state (step **b**). The negative charge on the gate insulator is summed with the photocurrent to increase I_D. The channel current is progressively increased by generating further charge in the channel by continued illumination at 633 nm (step **c**). Applying $V_{GS} < 0$ in step **d** discharges the traps to erase the memory contents. Finally, visible light at > 420 nm is used to recover the DAE-O isomer, restoring the large energy barrier and thus the memory to its initial, preset state, **a**.

A shortcoming of this structure is that it takes several hundred seconds to program and erase, even at large optical powers (e.g. >1 mW/cm^2 at 633 nm). Furthermore, the state of the entire memory is changed by blanket illumination over the array. As we will show below, this feature is useful for recording grayscale analog images, but cannot enable customized data storage unless each memory element is separately illuminated.

The drain current characteristics of the optical memory OTFT are provided in Fig. 9.64a. The memory window is approximately $\Delta V_T = 4$ V. The very low voltage operation is enabled by the thin, high dielectric constant gate insulator. Applying $V_{GS} = -3$ V erases the information before the DAE is converted back to its open-bond isomer. The initial and erased states show nearly identical I_D vs.

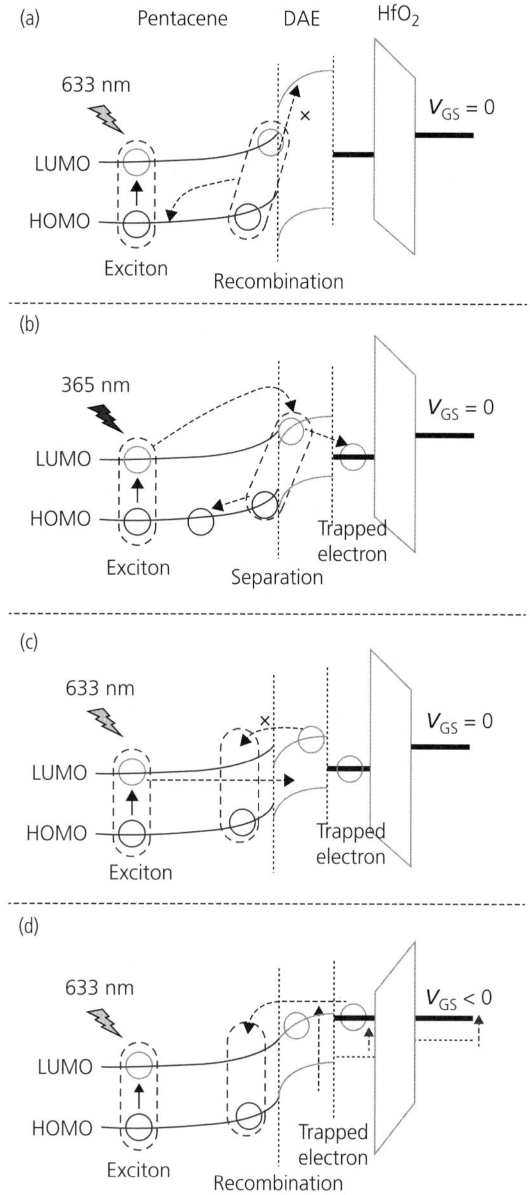

Figure 9.63 Energy levels of the optically activated memory transistor in Fig. 9.62. The processes of charge storage (memory writing) and charge loss from the traps at the DAE/HfO_3 interface are described in the text (Chen et al., 2015a).

V_{GS} characteristics, indicating a lack of hysteresis between the programming and erase cycles. The drain current contrast between the programmed (I_1) and erased states (I_0) is $I_1/I_0 = 4400$ under intense, 9.27 mW/cm^2 illumination at 633 nm (see Fig. 9.64b). Furthermore, there is only a small reduction in ΔV_T after 10^4 cycles, with a retention time of at least 7 days (Chen et al., 2015a). This shows the ability of ultrathin SAM and dielectric layers to retain charge with little or no leakage over long periods, suggesting a notable lack of pinholes or other defects between the traps and the channel.

This optical memory has been integrated into a (5 × 5) cm^2, 30 × 30 transistor array on a flexible PET substrate. The ITO control gate is coated on the back substrate surface, followed the deposition of the dielectrics and a 40 nm thick pentacene channel. Deposition of the Au S and D contacts completed the L/W = 120 μm/800 μm array. A patterned mask (Fig. 9.64b, upper inset) was placed onto the memory surface, which was then exposed to 365 nm wavelength illumination at 100 μW/cm^2 for 200 s to create a latent image replica on the underlying transistors. The shape was retained for 24 h, and was repeatedly cycled between a programmed image and a blank, erased memory array by applying a $V_{GS} = -3$ V, 1 s erase pulse. The array current output shows the stored image of the mask in the lower inset, Fig. 9.64b.

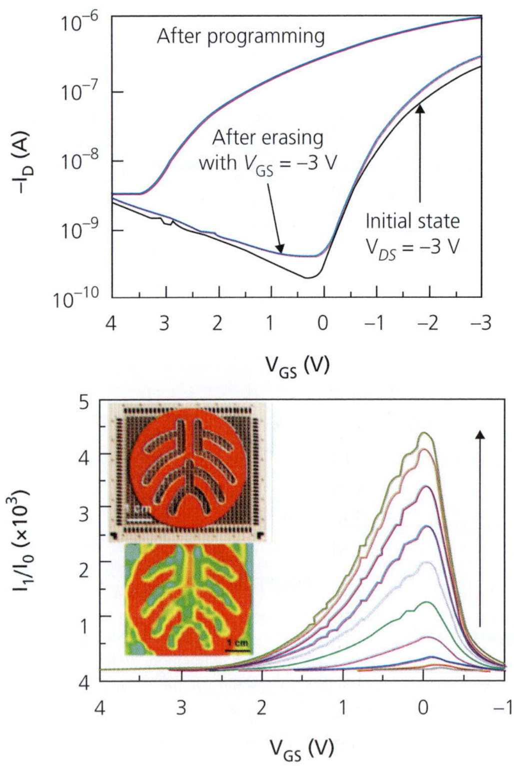

Figure 9.64 Drain current characteristics of an optical memory OTFT at $V_{DS} = -3$ V in its initial state, after programming by illumination from UV and 633 nm sources, and after memory erasure for 1 s at $V_{GS} = -3$ V. (b) The ratio of drain currents in the written and erased states (I_1 and I_0, respectively) vs. 633 nm wavelength illumination power from 0 to 9.27 mW/cm^2 (increasing in the direction of the arrow). Illumination time is 200 s at $V_{DS} = -3$ V. Inset: (*top*) Image of a 5 × 5 cm^2, 30 × 30 transistor array with a patterned mask laid onto its surface. (*bottom*) Latent image written on the memory. The color scale varies from $I_1 = 0$ (red) to 600 pA (green) (Chen et al., 2015a).

9.4.5.2 Floating gate and metal NP charge storage

Charge trapping on an ultrathin metal sheet known as a *floating gate*, is a common means for data storage in semiconductor memories. The sheet is embedded in the gate dielectric within a tunneling distance of charge from the channel. Similar to dielectric interfacial traps, the charge on the metal layer is neutralized by applying a gate potential opposite to that used for writing, thus providing programming capability. Metal NP layers can substitute for a continuous metal sheet deposited from solution between the first and second layers. Both floating gate and metal NP memory transistors operate at very low voltages when sandwiched between thin dielectric layers.

An example of a floating gate memory OTFT is shown in Fig. 9.65a, with a cross-sectional image of the gate region in Fig. 9.65b. The 16 nm thick floating Al gate is set back from the pentacene channel by a compound insulator comprising oxygen plasma-deposited AlO_x (4 nm), and a solution deposited phosphonic acid SAM (2 nm) in contact with the channel. The floating gate comprises a 16 nm thick Al layer separated from the Al control gate (also 16 nm thick) by a second AlO_x (4 nm)/SAM (2 nm) dielectric. The BG/TC transistor is completed by depositing the Au S and D onto the pentacene channel.

The drain current characteristics in Fig. 9.65c show a memory window of 2 V. The transistor is written by hole tunneling into the floating gate at $V_{GS} = -6$ V and $V_{DS} = -1$ V. The memory state is erased by reversing the gate voltage to $V_{GS} = +3$ V. The $I_{on}/I_{off} = 10^5$ provides a suitable contrast for large memory array applications. Also, the I_{off} is apparently almost entirely due to ~10 pA gate leakage that results in a slow discharge of the memory contents during each read cycle, requiring periodic refreshing of its contents. A notable feature of this architecture is the time (~1 s) required for programming, shown in Fig. 9.65d. Finally, fatigue in the threshold voltage difference between the written and erased modes is apparent after 1000 cycles due to partially irreversible charge trapping along with damage to the floating gate after many programming cycles. Annealing the devices at 140°C in nitrogen recovers the initial value of ΔV_T. Memory retention times of 10^3–10^4 s are observed (Sekitani et al., 2009a).

Flexible memories employing the transistors in Fig. 9.65 consisting of 26 × 26 cells on a 5 × 5 cm^2 PEN substrate have been demonstrated. Each cell comprises a pair of floating gate transistors sharing a

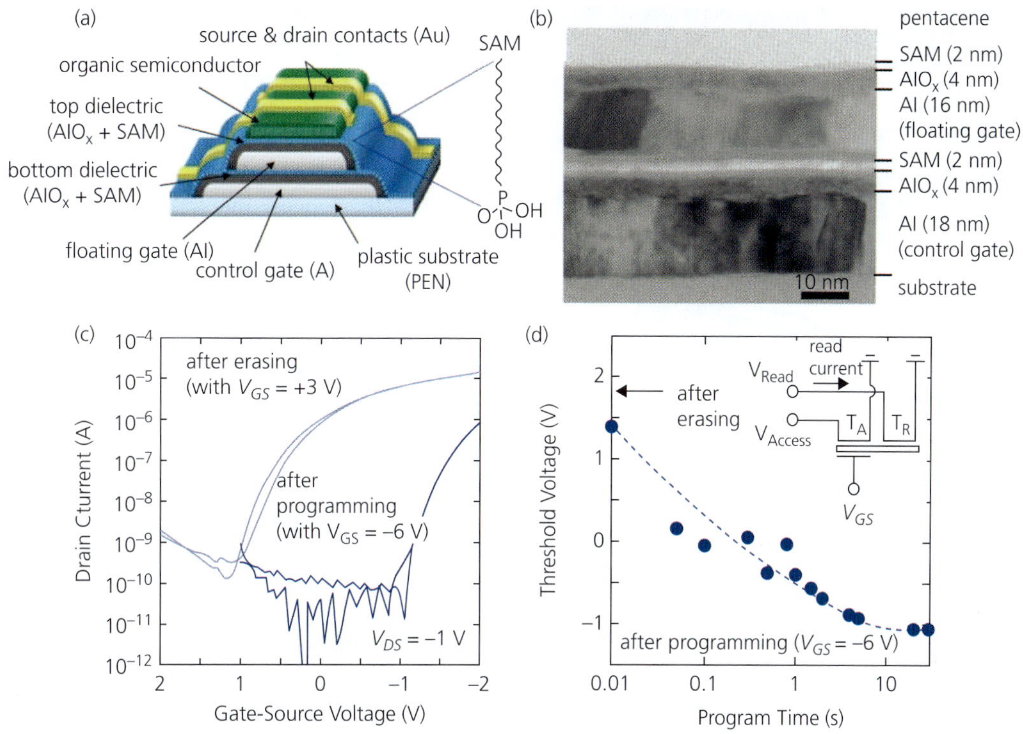

Figure 9.65 (a) Schematic illustration of a floating gate, pentacene OTFT with $L = 50$ μm and $W = 500$ μm on a PEN substrate. The phosphonic acid SAM molecular structure is shown at the right. (b) Cross-sectional TEM image of the transistor. (c) Drain current characteristics at $V_{DS} = -1$ V after programming and erasing of the transistor memory using the gate voltages indicated. (d) Threshold voltage after programming and erasing. The erase voltage is +3 V. Inset: Memory cell circuit schematic used in a flexible, 26 × 26 cell, 50 × 50 mm² pressure sensitive array memory. After Sekitani et al. (2009a).

common channel, as shown in Fig. 9.65d, inset. The transistor, T_A, provides access to the readout transistor (T_R) where the information is stored. At two transistors per bit, a total of 1352 transistors are used in the array. Each array cross-point is combined with a pressure-sensitive resistive element. Hence the circuit stores tactile information patterns at each memory site. The tactile pattern was stored for 12 h although there was considerable "image" distortion due to device leakage over that time period.

The dependence of the memory properties of OTFTs on metal NP composition and the gate bilayer dielectric structure has been widely investigated (Kim and Lee, 2010, Kang et al., 2013). We have seen in Section 7.5.2 that very thin layers of metal do not fully wet the organic surface. Hence, Au NP clusters form as shown in Fig. 9.66a and b for two different deposition rates on a PS dielectric.

These NP layers were used in TG/BC OTFTs with a P(NDI2OD-T2) n-channel semiconductor. The dielectric bilayers comprise a 40 nm thick PS layer, a 1 nm thick layer of vacuum deposited metal, followed by a 260 nm thick PMMA secondary dielectric. Write and erase require $V_{GS} = \pm$ 80 V due to the thick PS/PMMA gate dielectric. Indeed, even the 40 nm PS first dielectric may be too thick to charge the NPs. Micrographs in Fig. 7.180 show that Ag NPs can diffuse deep into the organic layer, which may significantly reduce the tunneling distance. Figure 9.66c shows that the drain current characteristics are sensitive to deposition rate: the higher deposition rate results in a lower density of larger NPs than at lower rates. As a result, the memory window and I_{on}/I_{off} are reduced due to the lower density of trapped charge in the films with lower NP densities.

The dependence of memory transistor properties on metal (Au, Ag, Cu, and Al) composition are summarized in Table 9.4. The Au and Ag films coalesce into nanometer clusters that are uniformly distributed across the polymer surface. Copper, on the other hand, forms larger clusters at a lower density, whereas Al wets the polymer surface and thus no particles are observed. Higher densities of NPs results in a larger memory window, with the largest $\Delta V_T = 51.7$ V for the Au NP device with a write-erase gate voltage of ±80V. In contrast, no memory window is

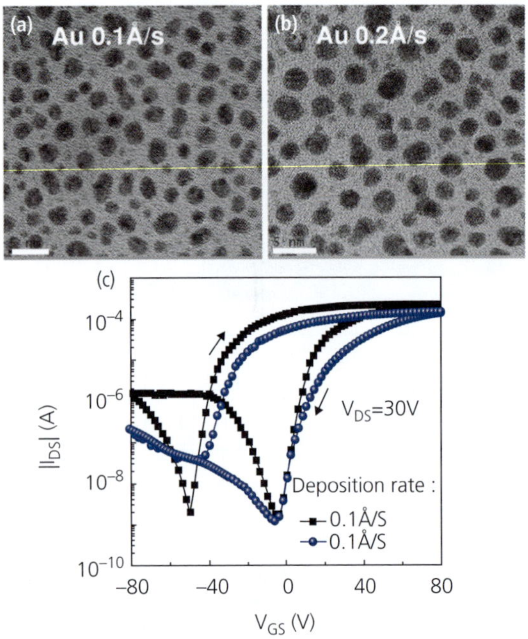

Figure 9.66 Transmission electron micrographs of Au NPs formed by vacuum thermal deposition of 1 nm thick films on PS at (a) 0.1 Å/s and (b) 0.2 Å/s. The scale bars are 5 nm. (c) Drain current characteristics of PS/Au NP/PMMA transistors with $L = 10$ μm and $W = 1$ mm at two different NP film growth rates (Kang et al., 2013).

Table 9.4 Characteristics of n-channel P(NDI2OD-T2) memory OTFTs employing different metal films at the PS/PMMA gate dielectric interface (Kang et al., 2013)

Parameter	Au	Ag	Cu	Al
NP size[a] (ave.) (nm)	4.6	6.8	13	~200
NP surface density $\times 10^{-12}$ (cm^{-2})	3.0	1.5	1.0	NA
Memory window (V), $V_{GS} = \pm 50$ V	0.9	0	0	0
Memory window (V), $V_{GS} = \pm 80$ V	51.7	29.1	22.4	2
Charge/NP at $V_{GS} = \pm 80$ V	1.3	4.5	13.8	NA
Retention time (s)	10^7	$>10^7$	$>10^8$	NA

[a] Nanoparticle sizes observed after deposition of 1 nm thick metal films.

developed for Al, presumably due to the formation of a continuous film at the PS/PMMA interface. However, if the layer were perfectly flat, it would form a floating gate, in which case some programmability would remain. Since this is not observed, penetration of Al through the PS first dielectric may cause leakage that prevents charge storage. The memory window is progressively smaller for Ag and Cu due to the smaller NP surface density, which reduces the amount of charge that can be stored. Also, even $V_{GS} = \pm 50$ V is insufficient to change the memory state of the transistors, again due to the use of thick dielectrics (Kang et al., 2013).

The retention time is lowest for the Ag NP device, possibly due to tunneling between the high density of particles at the interface that leads to their relatively rapid discharge. The lowest density Cu interlayer has the longest memory retention since the clusters are too far apart to allow for significant tunneling between them. Apparently, the NP size, shape and particularly distribution within the film bulk determine how rapidly charge can dissipate, and how large a memory window can be achieved. Indeed, the particle spatial distribution may be the most important attribute of NP-based organic memories. Diffusion into the dielectric bulk can be avoided using solution-deposited, non-metallic NPs.

Precisely controlled ferritin NP multilayer films on a SiO_2 or Al_2O_3 gate dielectric have been used to store charge supplied from a pentacene channel (see Fig. 9.67). Ferritin is an iron-containing protein that forms 12 nm diameter spherical particles with a shell thickness of 2 nm. The particles are deposited one layer at a time from solution to form a multilayer stack whose depth is precisely controlled to achieve the desired charge density. Charges are stored in the volume comprising alternating layers of anionic PAH and cationic ferritin. The switching times of devices comprising 40 bilayers in pentacene OTFTs are > 10 μs which is considerably less than for the floating gate transistors discussed above. The memory windows are approximately 20 V, $I_{on}/I_{off} = 10^4$, and retention and fatigue characteristics are similar to those obtained for thin film floating gate devices (Kim et al., 2013a). These non-metallic NPs, therefore, show promise for charge storage applications, although the long term stability of the transistors has not been explored.

Ambipolar conduction of the channel is an important for to rapid charging and discharging of the floating gate or NP layer. We have seen that these two processes entail the attraction of charge of one sign from the channel during the program cycle, followed by the opposite sign charge during erase. Both charge types, therefore, must be transported through the channel, which can cause significant cycle time delays if the mobilities of electrons and holes are markedly different. Pentacene is an example material that has a higher hole than electron mobility, although the asymmetry is small enough that it is useful in memory transistors. Likewise, PDPP-TBT is an ambipolar polymer with nearly equal electron and hole mobilities of 0.051 cm^2/V s and 0.037 cm^2/V s, respectively, that has been

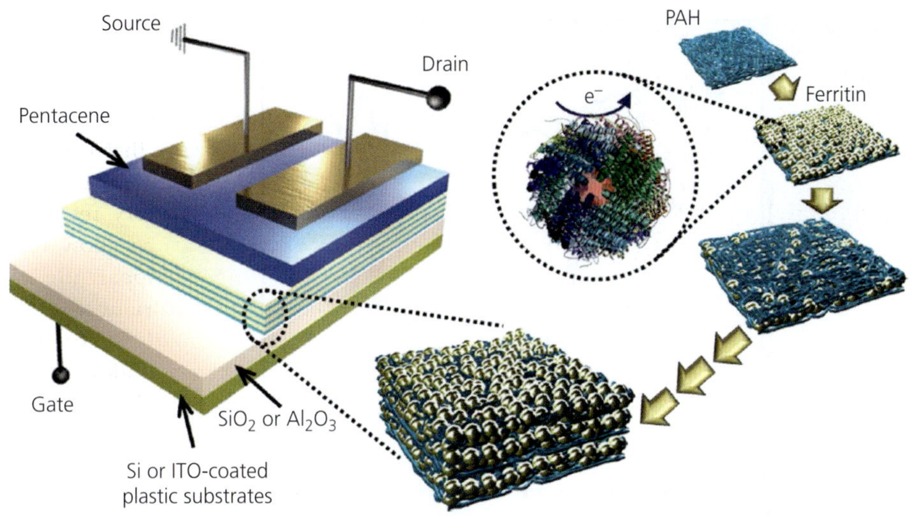

Figure 9.67 Schematic illustration of the fabrication of PAH/ferritin dielectrics by layer-by-layer solution processing. The multilayer stack is deposited onto a SiO_2 (200 nm) or Al_2O_3 (50 nm) dielectric, and coated by the pentacene channel. Gold source and drain electrodes are deposited to complete the BG/TC memory transistor with $L = 100$ μm and $W = 800$ μm (Kim et al., 2013a).

used with Au NP charge trapping layers (Zhou et al., 2013).

9.4.5.3 Ferroelectric memory transistors

We showed in Section 9.4.4 that the dipole moment of a ferroelectric can be reversibly switched by alternately applying a positive or negative voltage that exceeds the coercive field strength (~50 MV/m). Higher fields are required for the thinnest films where the dipolar molecules become pinned at the film surfaces. The *ferroelectric memory* OTFT, or OFeFET, operates by applying a gate voltage that can switch the dipole direction within a ferroelectric gate insulator. This screens the gate potential at the channel/insulator interface, which results in an increase or decrease in charge density (and hence a shift in V_T and I_D), depending on the dipole direction. Since the dipole orientation is bistable, the OFeFET can assume one of two different threshold voltages corresponding to a logical **1** or **0**. The device memory state is read by applying a gate voltage near V_T, and measuring the drain current.

We noted previously that ferroelectrics spontaneously polarize when heated above their Curie temperature, T_C, and then cooled. This process aligns the molecules into a crystalline film, or in a polycrystalline structure with low angle grain boundaries between domains. If unpolarized (i.e. at $T > T_C$, or before annealing), the materials are paraelectric, and the polarization depends linearly on the field strength. Alternatively, the crystalline domains can be aligned using mechanical strain. This is illustrated for P(VDF-TrFE) deposited on PET films laminated with an Al contact layer in Fig. 9.68. The PET is a thermoformable plastic that can apply strains to the ferroelectric layer that approach 100% when stretched at elevated temperatures. The Fourier transform infrared (FTIR) spectra of several 900 nm thick P(VDF-TrFE) films subjected to different thermal treatments are shown in Fig. 9.68a. The 1400 cm^{-1} band intensity is most sensitive to ordering. The lower energy bands at 850 and 883 cm^{-1} are associated with the *trans-gauche* and the highly ferroelectric all-*trans* configurations, respectively. The peak at 1286 cm^{-1} corresponds to an extended *trans* sequence of more than four monomers. This latter feature indicates the state of polarization switching of the film. The conformational changes into the ferroelectric *trans* configuration require readjustment of the polymer chain organization. This is a thermally activated process that can be accelerated by applying physical strain. Spectrum (i) corresponds to the unannealed sample. Its low intensity is a consequence of a disordered structure. Likewise, the film stretched at room temperature (iii) shows relatively weak spectral intensity compared to an unstrained film that is ordered by heating in N_2 for 2 h at $T = 140°C > T_C \sim 100°C$ (spectrum (ii)). The film that is tensile strained by 50% at 70°C (spectrum (iv)) shows the greatest degree of crystallinity in the all-*trans* phase, even though the temperature is below T_C.

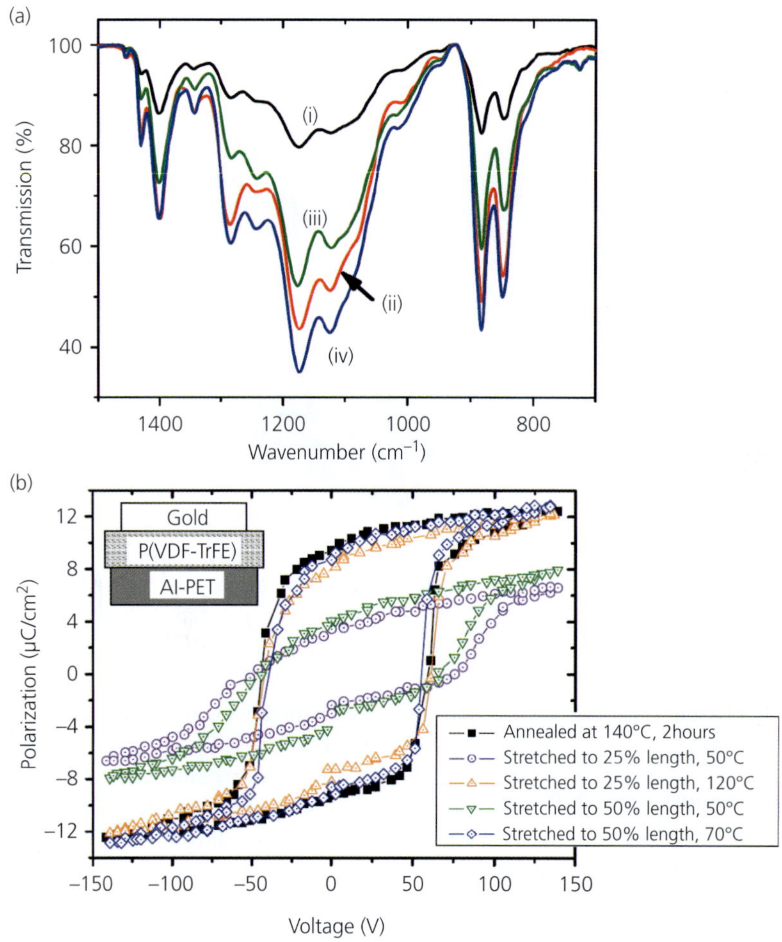

Figure 9.68 (a) Fourier transform infrared spectra of P(VDF-TrFE) on Al-coated PET films. (i) As-deposited, (ii) annealed at 140°C, (iii) stretched by 50% at 25°C, and (iv) stretched by 50% at 70°C. (b) Hysteresis loops for ferroelectric capacitors whose films are treated as indicated in legend. Inset: Schematic diagram of the capacitor (Nguyen et al., 2008).

The ferroelectric hysteresis loops of capacitors using films strained to 25% and 50% at different temperatures are shown in Fig. 9.68b. The capacitive memories comprise a P(VDF-TrFE) film sandwiched between an Al back contact and a Au top contact. The remnant polarization of the high temperature annealed sample, as well as those strained at 50% at 70°C and at 25% at 120°C are roughly equal, at 10 μC/cm^2. However, for 25% or 50% strain at 50°C, $P_r = 4$ μC/cm^2 with a correspondingly higher coercive field than those exposed to higher temperatures. The results are consistent with the FTIR data: both strain and elevated temperature result in a higher degree of polarization. In addition, the strain and temperature effects reinforce each other. A higher strain requires a lower temperature than a lower strain (Nguyen et al., 2008).

The operating characteristics of pentacene channel BG/TC OFeFETs with stretched and annealed P(VDF-TrFE) are shown in Fig. 9.69a. They show a hole mobility of 0.072–0.12 cm^2/V s after stretching. The device with 50% strain and annealed at 70°C has the largest memory window of $\Delta V_T = 60$ V and $I_{on}/I_{off} = 10^4$, which is even larger than obtained for the unstretched transistor annealed at 140°C. Furthermore, the 50% strain device showed an extrapolated data retention time of 10^4 s, at which point $I_{on}/I_{off} = 10^2$. Interestingly, the memory window is independent of the orientation of the channel relative to the stretching direction. Figure 9.69b shows the drain current characteristics of two OFeFETs whose channels are oriented perpendicular (circles) and parallel (triangles) to the direction that the substrate is distorted. In both cases, the performance is nearly identical, indicating that the polymer chains are equally poled along the ferroelectric *trans* direction. This directional independence makes the process adaptable to fabrication of large-scale memories without being

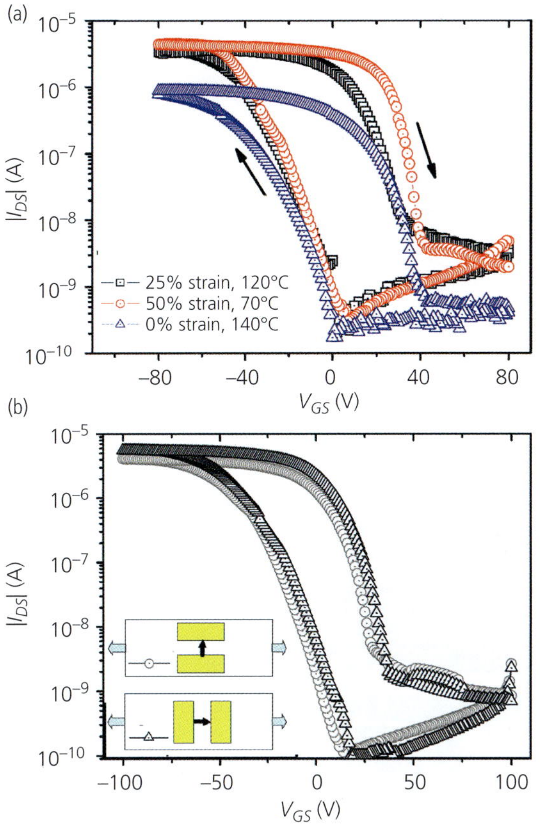

Figure 9.69 (a) Drain current characteristics of pentacene/P(VDF-TrFE) OFeFETs with $L = 7$ μm and $W = 4$ mm on PET substrates strained at different temperatures. (b) Dependence of the drain current characteristics on orientation of the channel relative to the stretch direction of a 50% strained OFeFET at 70°C. Inset: Top illustration shows the channel (black arrows) normal to the stretching direction (colored arrows, circle symbols in plot); bottom illustration shows the channel oriented parallel to the stretching direction (triangle symbols) (Nguyen et al., 2008).

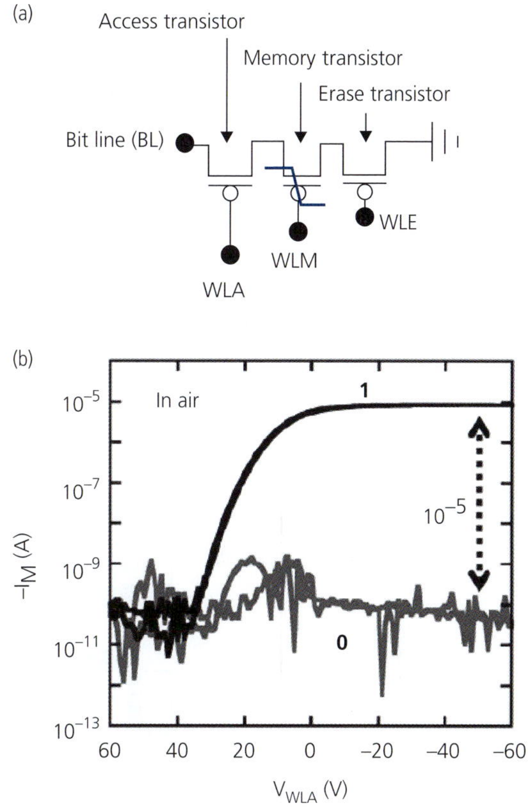

Figure 9.70 (a) Three transistor (3T) memory circuit comprising an access, memory, and erase transistors. (b) Drain current along the bit line vs. the access transistor gate voltage (V_{WLA}) for the memory transistor in the **1** and **0** states. $I_{on}/I_{off} = 10^5$ (Sekitani et al., 2009b).

constrained by the orientation of one device block relative to others on the same substrate.

A P(VDF-TrFE) OFeFET memory array on flexible polyimide was implemented using the three transistor (3T) circuit for each memory cell shown in Fig. 9.70a. The 3T architecture is immune from destructive memory readout since it separates data storage, cell addressing (or access), and the erase functions between independent devices. This is a more robust design than either a 1T memory element where all functions are combined in a single device, or even a 1T1C (i.e. 1 transistor, 1 capacitor) network.

An interesting application of the OFeFET memory is the multilayer "communication system" in Fig. 9.71a, where a radiofrequency (r.f.) wireless transmitter signal is received by an integrated coil antenna. An r.f. receiver is positioned over a second coil on the array. This activates two mechanical switches to form an *ad hoc*, hard-wired network between the transmit and receive locations whose positions are retained by the corresponding memory OFeFETs.

A conceptual diagram of the circuit architecture and one multilayer array cell is shown in Figs. 9.71a and b, respectively (Sekitani et al., 2009b). The circuit employs a 8 x 9 array with dimensions of (25 × 25) cm². The cell pitch is 2.5 cm along both the rows and columns. The Ag BG/Ag TC transistors comprised a 50 nm thick vacuum-deposited pentacene channel on a 75 μm thick polyimide substrate. The contacts and the array word lines used Ag NPs that are sintered at 180°C. The 400 nm thick P(VDF-TrFE) film used for the memory transistor is cured at 90°C and then annealed above T_C at 135°C for 2 h in air. The channel lengths of all the transistors were $L = 30$ μm, with $W = 16.5$ mm, 45 mm, and 300 μm for the access, memory and erase transistors, respectively. Finally, the transistors were encapsulated by depositing a

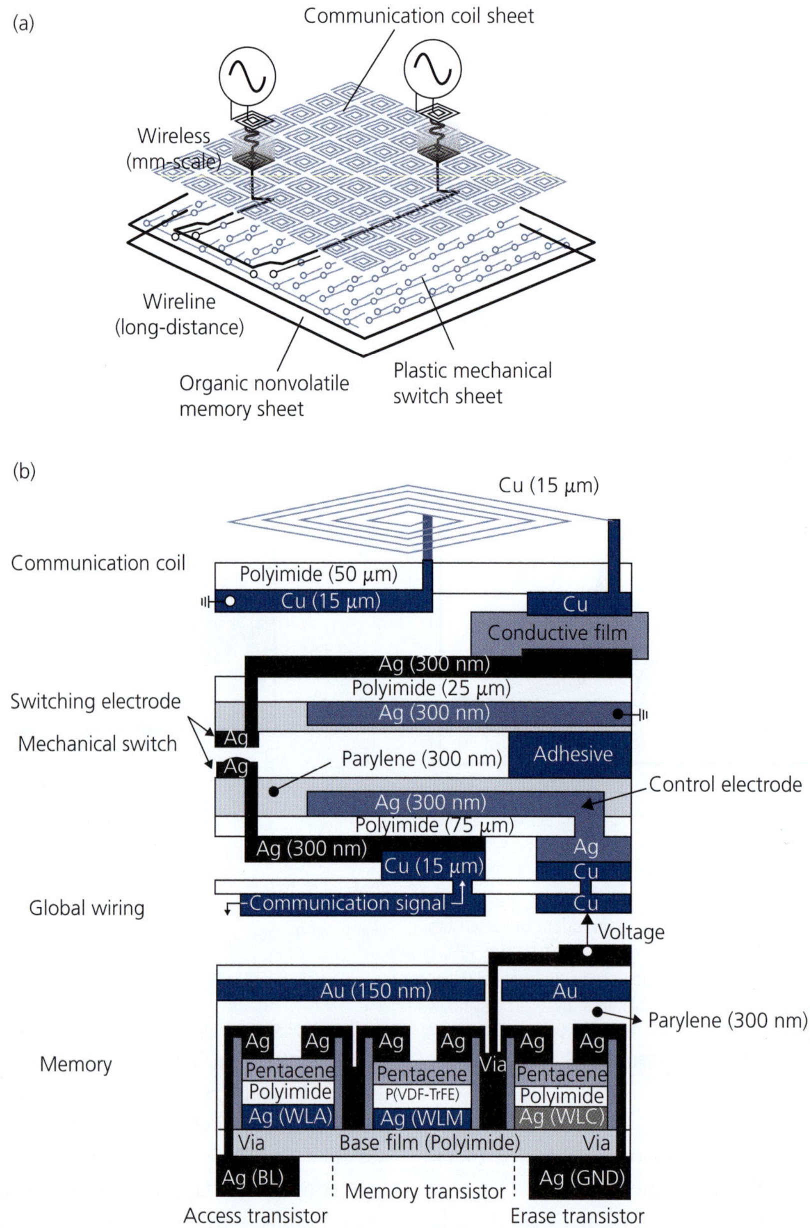

Figure 9.71 (a) Communication system architecture illustrating the coupling of a wireless transmitter and receiver positioned over the communication coil sheet to establish the *ad hoc* hard-wired connection between their positions. The connection is retained by the non-volatile OFeFET 3T memory. (b) The multilayer array element comprising a memory coil, an electrically activated mechanical switch, a global wiring layer, and the 3T memory cell (Sekitani et al., 2009b).

parylene/Au bilayer to prevent exposure of their active regions to the ambient.

All layers use printing techniques for patterning: screen printing for the Cu coil, inkjet printing for the Ag NP conducting lines, and a combination of inkjet and microdroplet printing for the 3T memory cell. The function of this circuit is as follows: The 2 MHz r.f. signals are inductively coupled to the adjacent Cu coil, inducing a voltage that appears on one electrode of a mechanical relay. The opposite terminal of the relay is connected to the drain of the memory transistor. When the transistor is in the **1** state, 30 V from the bit line (BL) is supplied to the relay, closing the switch and connecting the cell to the "global wiring sheet" (Sekitani et al., 2007). A similar sequence occurs at the receive end, thereby completing a hard-wired interconnection through the wiring sheet between these two locations. The interconnection is maintained until an erase signal is sent to either the transmit or receive array cross-points. This causes the memory voltage to return to the **0** state, which in turn disconnects the BL from the switch, opening the relay and dropping the connection to the other array location. The mechanical relays, comprising electrostatically deflected perylene arms, presents a resistance of ~1 Ω.

The memory state is switched by applying V_{WLM} to the word line connected to the gate of the memory transistor, and its state is queried according to the voltages applied to word lines at the gates of the access and erase transistors (V_{WLA} and V_{WLE}, respectively). The bias conditions leading to read, write and erase are summarized in Table 9.5. The currents in the **1** and **0** states are shown in Fig. 9.70b as a function of the access word line voltage, V_{WLA}. The current contrast ratio is $I_{on}/I_{off} = 10^5$. The write time is 40 ms and the erase time is 10 ms, both limited by parasitic capacitances of the 3T circuit.

Hard wired interconnections in Fig. 9.71 are dynamic, secure, and non-volatile. That is, the interconnect is maintained by the memory state of the OFeFETs, and hence is sustained until an erase signal is sent. The interconnections do not require continual, free space communication, releasing the address electronics to perform other functions, such as establishing other routing paths within the same array. This circuit shows the degree of sophistication that can be achieved using flexible electronics, and the ability of integrated memories to enable new applications.

In concluding this discussion, we have seen that organic memories based on charge trapping, filamentary conduction and ferroelectric effects can have long data retention times, high stability, and fast write/read/erase cycle performance. The memory element itself can be a resistive or capacitive switch, a diode or a transistor. Both volatile and non-volatile memories are possible leading to WORM and DRAM architectures. A nonlinear element such as a diode or access transistor must be used in conjunction with the resistive or reactive memory element to permit scaling to large, high capacity memory arrays. While their performance and bit density falls short of that attained with conventional Si technology, organic memories can be flexible and can extend over large areas. We have shown that both of these attributes enable special purpose applications that are not served by high-density Si-based memories. Organic memories have, indeed, made substantial progress during the last decade, yet further developments are needed to improve their speed, data retention time, and to reduce memory fatigue.

Table 9.5 Voltages used to write, erase, and read the 3T memory element in Fig. 9.70 (Sekitani et al., 2009b)

Action	State	V_{BL} (V)	V_{WLA} (V)	V_{WLM} (V)	V_{WLE} (V)	I_M (A)
Write	**0→1**	40	−40	−60	−40	10^{-5}
Erase	**1→0**	−40	−40	+60	−40	10^{-10}
Read		−10	−60 to 60	0	−40	

9.5 Organic/two-dimensional semiconductor heterojunctions

While previous discussions have dealt with 3D electronic systems, that is, those systems whose characteristics are controlled by bulk properties of the semiconductor, limited dimensional systems are playing an increasingly important role in modern optics and electronics. In particular, 2D and one-dimensional transport of charge and excited states may offer ultrahigh 2D mobilities, or 1D quantum interference effects. Much of the interest in limited dimensional systems stems from the fact that Moore's law of transistor scaling is nearing an end due to practical limitations on shrinking gate lengths to the atomic scale. The powerful economic drivers underlying Moore's law are, nevertheless, compelling scientists and engineers to seek methods to circumvent this limit by using materials and systems with superior electronic properties than Si. Two-dimensional materials, that is, those that comprise only a single or a very few number of monolayers, are such a class of solids. The archetype example of a 2D material is graphene—a single layer of fused, interconnected carbon rings. Graphene has exhibited an exceptionally high charge mobility and charge density (Bolotin et al., 2008). In its pristine form, it is a semimetal, although doping, strain, or externally applied fields can induce a band gap. Another example is hexagonal boron nitride (*h*-BN). This graphene isomorph comprises alternating B and N atoms that occupy inequivalent lattice positions. Unlike graphene, *h*-BN is an insulator.

Perhaps the most interesting class of materials for use in active organic electronic devices comprises 2D monolayers of transition metal dichalcogenides (TMDCs). Their molecular formula is MX_2, where M is a group IV, V, or VI transition metal and X is a chalcogen (S, Se, or Te). These materials are characterized by a direct or indirect bandgap whose magnitude depends on the number of monolayers comprising the film. Like graphene, they have high mobilities and mechanical flexibility, and can be intensely luminescent.

Two-dimensional W-M excitons are supported by TMDC monolayers. The layers feature a large spin–orbit coupling that allows for rapid transitions between the singlet and triplet excited states. In this regard, HJs between 2D TMDCs and organics have the potential to create structures that combine the advantages of both systems while reducing their individual disadvantages. The Frenkel-like excitonic nature of organics gives them a high oscillator strength, thus providing strong optical absorption. Illumination generates excitons that dissociate at a TMDC/organic interface, and can be transported in the very high mobility 2D film. The large spin–orbit coupling of TMDCs may lead to efficient light emission from both singlet and triplet states generated within the organic region of the HJ. Both materials are flexible, and layering one on the other is possible using techniques such as vapor deposition of organics, and exfoliation of monolayers from TMDCs.

Monolayer TMDCs have a small absorbance even though they might have an exceptionally high absorption coefficient. Furthermore, it has not yet been possible to grow defect-free films over even modest areas. Films grown by CVD are polycrystalline, and consist of one or more atomic layers in island-like domains of unpredictably varying sizes. Each domain may possess significantly different electronic characteristics, depending on the number of layers. Moreover, a high density of chalcogen vacancies and defects at domain edges and within the monolayers themselves reduces the charge mobility well below the theoretical limit predicted for perfect films, while also reducing the exciton lifetime. However, organics have been shown to effectively passivate the surface defects of semiconductors (Lunt et al., 1991, Li et al., 2011, Yong et al., 2012, Lee et al., 2018). Hence, a judicious choice of organic may reduce the effects of chalcogen vacancies in the TMDC films.

9.5.1 Electronic characteristics of transition metal dichalcogenides

Covalent bonds between the M and X atoms provide strong intralayer adhesive forces. However, far weaker van der Waals bonds between the layers permit them to be separated into monolayers via exfoliation with properties unique from those of the bulk. The layers themselves can assume either a hexagonal (H) or tetragonal (T) symmetry, where the metal atoms have either trigonal (1H/2H) or

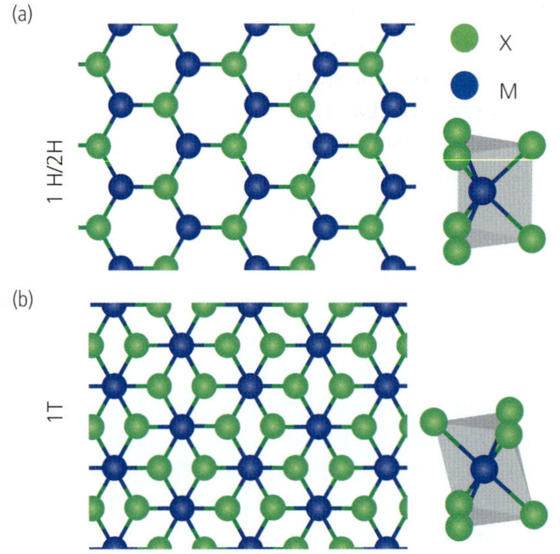

Figure 9.72 Two crystal structures possible for TMDCs. (a) Hexagonal in one or two repeat units along the vertical, (0001) direction (1H and 2H, respectively), and (b) tetragonal (1T).

octahedral (1T) coordination with the X atoms in the basal plane (see Fig. 9.72). The hexagonal lattice can comprise either one or two layers for each unit along the (0001) **c**-direction, corresponding to the 1H or 2H structures, respectively. For many TMDCs, the 1T structure is unstable at room temperature. A 3D view of a bilayer of an MX_2 compound is shown in Fig. 9.73a, showing the relative positions of the metal and chalcogen layers within and between the atomic planes. The lattice parameters for WS_2 are $a = 3.154$ Å and $c = 12.362$ Å, and for MoS_2, $a = 3.160$ Å and $c = 12.295$ Å, reflecting the large asymmetries of these 2D lattices (Wilson and Yoffe, 1969, Mattheiss, 1973).

An important feature of TMDCs is the tunability of their band gaps, depending on the number of layers. For example, WS_2 can be tuned from an indirect bandgap, bulk semiconductor with $E_G = 1.3$ eV to a direct bandgap monolayer with $E_G = 2.1$ eV. The energetics of bilayers, trilayers, and so on, monotonically transition from their monolayer to bulk values. The indirect band gap increases with the number of layers, while the direct gap stays almost unchanged. Hence, the material transitions from indirect to direct gap between a bilayer and a monolayer. The band structure calculated for WS_2 is shown in Fig. 9.73b, where the vertical axis is in Hartrees (Ha). A Hartree is given by $E_h = m_0\left(q^2/4\pi\varepsilon_0\hbar\right)^2 = 2R_\infty hc = 27.2$ eV, where m_0 is the electron rest mass, R_∞ is the Rydberg constant, and c is the speed of light. Interestingly, the direct band gap is not at $k = 0$ (Γ), but rather at the K symmetry point. Similar band structures and behavior with decreasing numbers of layers are found for MoS_2 (Kuc et al., 2011).

A simplified energy band diagram of a TMDC monolayer at the band extrema is shown in Fig. 9.74a. The exciton level is shown by the dashed line. Several features in the absorption spectrum of $MoSe_2$ in Fig. 9.74b are associated with transitions shown in the band diagram. Notably, the A and B excitons resulting from valence band splitting via a large spin–orbit coupling energy, Δ_{SO}, appear at 790 nm and 700 nm,

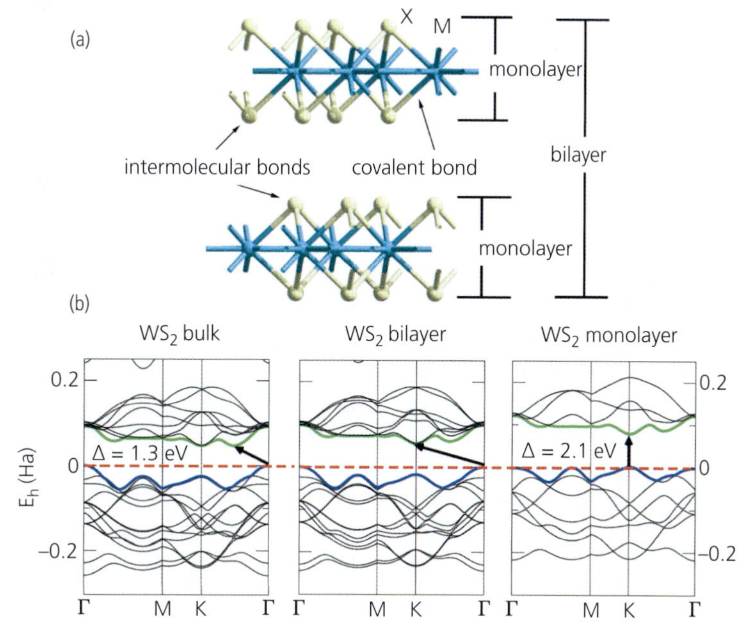

Figure 9.73 (a) Illustration of two monolayers of a TMDC. The individual layers are bonded via van der Waals forces between neighboring chalcogens (X), whereas the in-plane atoms are covalently bonded. M is the transition metal atom. (b) Calculated band structure of the WS_2 bulk, and a bilayer, and a monolayer. The conduction band minimum is shown by the green line, the Fermi energy by the dashed line, and the valence band maximum by the blue line. The energy gaps are indicated by arrows (vertical arrow for direct band gap and diagonal arrows for indirect band gaps) (Kuc et al., 2011).

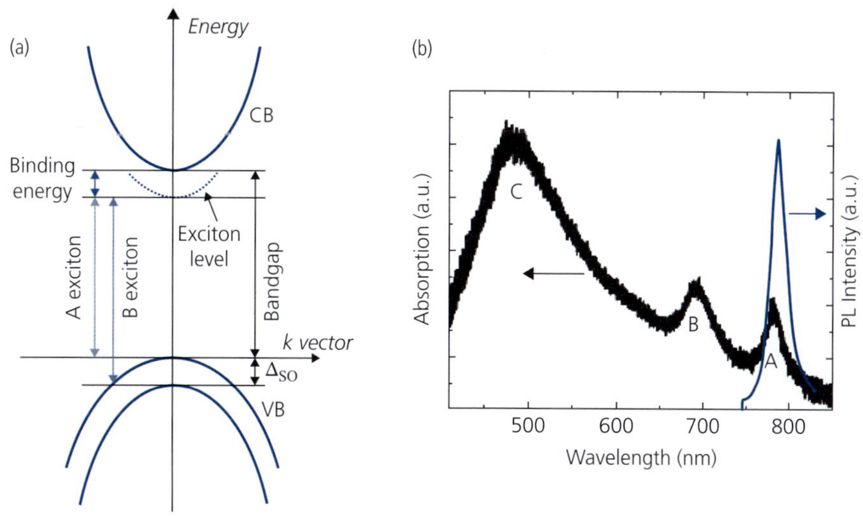

Figure 9.74 (a) Simplified energy band diagram of a monolayer TMDC at the band extrema identifying the A and B excitons, the conduction and valence bands (CB an VB, respectively) and the spin–orbit energy splitting, Δ_{SO}. (b) Absorption and PL spectra of $MoSe_2$ indicating the A, B, and C exciton features (Liao, 2018).

respectively. Near the band extrema, the valence and conduction bands are parallel and symmetric about their maxima. Hence, excitations generate electrons and holes that propagate at the same velocities but in opposite directions. This larger energy transition commonly observed in many TMDCs gives rise to the broad C feature in Fig. 9.74b. The PL spectrum is due to recombination of the lowest energy A exciton.

The mono- and multilayer samples contain a high density of surface and edge chalcogen vacancies that adversely affect the charge mobilities (Qiu et al., 2013) and exciton lifetimes (Amani et al., 2015), thus reducing conductivity and PL efficiency. The vacancies can be filled via passivation by thiolation. Indeed, attaching thiols or electron donating or withdrawing species onto TMDC monolayers can also result in n- or p-type doping that increases the layer conductivity.

A potential mechanism for defect passivation of a MoS_2 monolayer using thiol-based molecules is illustrated in Fig. 9.75. The S binds to Mo at the vacancy site, leaving the functionalizing R group exposed. Thiolation has been demonstrated using grinding-assisted ultrasonic exfoliation of MoS_2 in solution to generate a distribution of flakes ranging in size from 100–300 nm, and with thicknesses up to 5 nm. A monolayer (ML) of the TMDC is 0.7 nm thick. This suggests that the process generates a distribution of flake thicknesses from 1 to 6 ML. Exfoliation results in a high S vacancy that is repaired by introducing thiols into the suspension containing the flakes. Such thiols include *p*-mercaptophenol (**S1**), thiophenol (**S2**), 1-propanethiol (**S3**), 1-nonanethiol (**S4**), and 1-dodecanethiol (**S5**, see Fig. 9.76). Using a combination of infrared and X-ray photoelectron spectroscopy (XPS), the absorption at 2563 cm^{-1} due to the S–H bond is absent in the functionalized monolayers, indicative of binding of the thiols to the MoS_2 sheets.

The energetic shifts engendered by the attachment of the several thiolating compounds have been investigated by photoelectron and optical spectroscopy, with results shown in Fig. 9.76. The valence band energies are slightly destabilized by the thiol attachments due to their mild Lewis base nature. Figure 9.76a shows that the valence band energy is decreased relative to vacuum as the electron withdrawing strength of the functionalizing group decreases from the alcoholic substituted **S1** to the less electronegative aromatic thiol, **S2**. Similarly, the shorter alkane chain of **S3** is a more effective electron withdrawing moiety than the more extended chains of **S4** and **S5**. The energy gap is correspondingly affected by thiolation. Thiol attachment results in longer wavelength PL emission, which is a consequence of electron-withdrawing functionalizing groups attached to the S-atom in the thin film. These effects are confirmed by the Fermi-energies in Fig. 9.76b. Thus, the aromatic **S2** and longer alkane chain compounds **S4** and **S5** lead to pronounced n-type doping of the TMDC, shifting the Fermi energy toward the conduction band minimum, whereas the as-prepared film shows intrinsic conductive behavior with the Fermi energy at mid-gap (Nguyen et al., 2015).

Attachment of organic molecules with greater or lesser electron withdrawing character can be used to dope the 2D materials, thereby modifying their conductivity (and conductivity type) over a wide range. Density functional theory (DFT) calculations of monolayers of organic compounds on pristine (i.e. defect-free) MoS_2, for example, have shown that they can serve as either electron donors or acceptors, depending on the positions of the HOMO and LUMO energies within the band gap of the monolayer (cf. Section 4.4.1). Examples of electron donating DMPD and accepting TCNE are shown in Fig. 9.77. The relative positions of the HOMO and LUMO energies within the bandgap of the monolayer film show that DMPD is an *n*-type dopant, contributing electrons from its filled HOMO into the conduction band of MoS_2. Thus, the

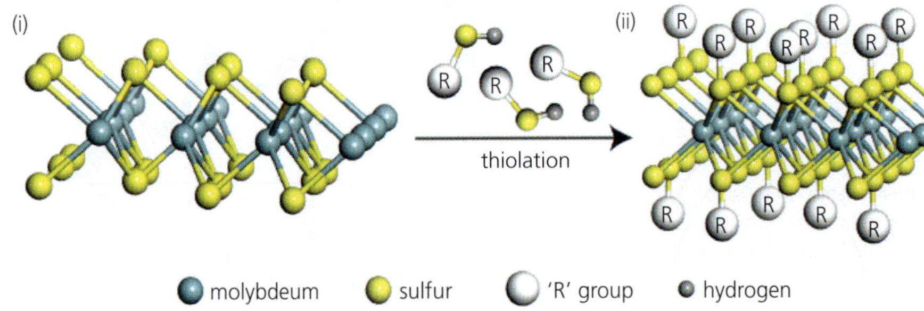

Figure 9.75 (i) A monolayer of MoS_2 showing numerous S vacancies. (ii) Immersing the TMDC in a solution containing functionalized thiol moieties results in attachment of the S-atom to the TMDC, exposing the R-group (Nguyen et al., 2015).

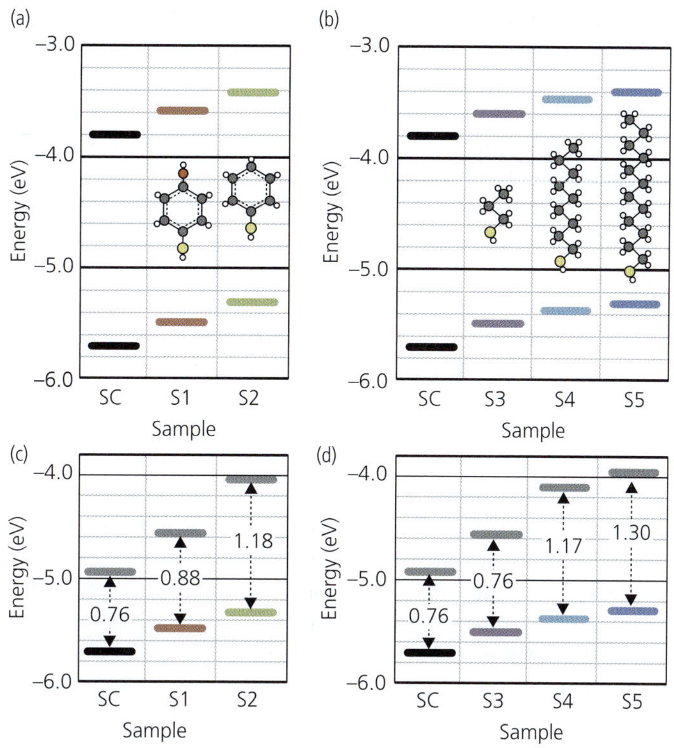

Figure 9.76 Valence band maxima and conduction band minima for the (a) as-prepared MoO_2 (**SC**), and compounds **S1** and **S2**, and (b) compounds **S3–S5**. The thiols are shown in each figure, containing oxygen (red), carbon (purple), sulfur (yellow), and hydrogen (white). (c, d) As in (a, b), showing the valence band maxima and Fermi energies (purple lines) (Nguyen et al., 2015).

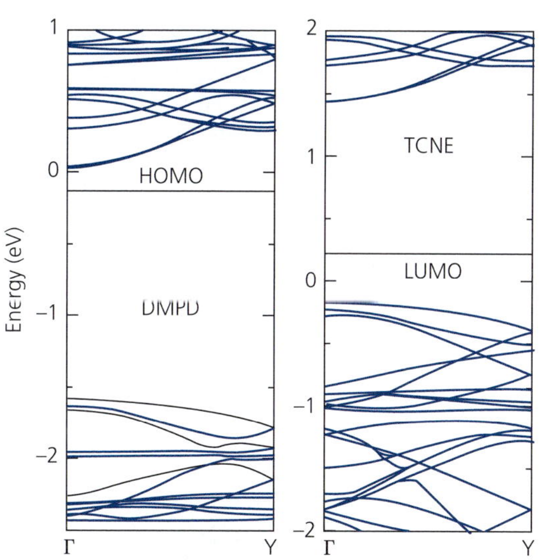

Figure 9.77 Band structure relative to the Fermi energy at 0 eV for a monolayer of MoS_2 with an adsorbed monolayer of DMPD and TCNE on its surface. The HOMO and LUMO energies of the respective molecules are indicated (Cai et al., 2016).

Fermi energy is positioned approximately at the conduction band minimum. Likewise, the LUMO energy of TCNE is within 0.5 eV of the MoS_2 valence band maximum, at the high symmetry Γ-point, behaving as an acceptor.

The amount of charge transfer from an adsorbed organic layer is calculated by assuming a multistep process whereby charge transfer across the interface is followed by electronic polarization. That is, the total average in-plane charge density transferred at position z normal to the film plane is

$$\Delta\rho(z) = \rho_{org/TMD}(z) - \rho_{org}(z) - \rho_{TMD}(z), \qquad (9.35)$$

where $\rho_{org/TMD}(z)$ is the charge density of the combined organic/TMDC system and $\rho_{org}(z)$ and $\rho_{TMD}(z)$ are the charge densities of the respective, non-interacting systems. The total charge transferred is then

$$\Delta n(z_0) = \int_{-\infty}^{z_0} \Delta\rho(z)dz, \qquad (9.36)$$

where z_0 is the equilibrium distance of the organic molecule to the film plane (Cai et al., 2016, Shen and Tao, 2017). The amount of charge transfer depends sensitively on the fraction of the TMDC surface that is covered, the sites occupied by the organic molecules, and whether or not the dopant is positioned above an atomic vacancy. For DMPD and TCNE, for example, the total charge density transferred to pristine MoS_2 at low molecular coverage is 5×10^{12} cm^{-2} electrons and ~1×10^{12} cm^{-2} holes, respectively, whereas the density of electrons is approximately doubled if the dopants are positioned above a S vacancy (Cai et al., 2016).

As noted, surface passivation can have dramatic effects on the exciton lifetime and charge concentration within the monolayer. In Fig. 9.78a we show that exfoliated MoS_2 films treated with the organic superacid, TFSI, show a 190-fold increase in PL efficiency compared to the as-prepared sample. Furthermore, neither PL peak shifts nor line shape changes are observed due to coverage by TFSI. The PL quantum yields (PLQYs) of the two samples are plotted vs. charge generation rate, G (proportional to the pump intensity) are shown in Fig. 9.78b. The PLQY is possibly >95% for the TFSI-treated 2D film at low fluence, compared to < 1% for the as-prepared sample. Caution, however, is warranted in claiming such high efficiencies in samples with unusual geometries as in 2D films. We can conclude, however, that the relative efficiency of the treated film is much larger than for the untreated film.

At high fluence, the number of photogenerated charges exceeds that of the background carrier concentration in MoS_2. In this case, the number of electrons (n) and holes (p) are equal, and the recombination rate is proportional to $np = n^2 \propto G^2$. As shown in Section 3.10, at high fluence we expect that the exciton annihilation rate is proportional to the square of the number of excitons, N_{ex}. Hence, the PL quantum efficiency is equal to the ratio of the radiative recombination rate to the total recombination rate, viz.

$$\eta_{PL} = \frac{k_r N_{ex}}{k_r N_{ex} + k_{nr} n^2 + k_{ex} N_{ex}^2}, \qquad (9.37)$$

where k_r is the fluorescence recombination rate, k_{nr} is the charge recombination rate at defects, and k_{ex} is the biexcitonic recombination rate. For the treated sample, it was assumed that $k_{nr} \to 0$. Fits to the data in Fig. 9.78b using Eq. 9.37, as shown by the dashed lines yield $k_{ex} = 2.8$ cm^2/s and $k_{nr} = 1.5 \times 10^6$ cm^2/s. We note that the exciton lifetimes are similarly enhanced for the surface passivated samples. Transient luminescence measurements yield $k_r =$ 0.3 ns for the untreated, and 10.8 ns for the treated MoS_2 film, which is consistent with the PLQY results (Amani et al., 2015).

An additional consequence of surface passivation by TFSI of transistors employing a MoS_2 channel is a threshold voltage shift of 2.5 V in Fig. 9.79. This suggests that passivation dedopes the thin film, reducing the electron concentration in the channel compared to the as-prepared sample.

Questions remain as to what role is played by TFSI in passivating the TMDC surface. Apparently, TFSI does not directly affect the surface vacancies as in the case of the thiolated samples. However, this strong acid can reduce adsorbed –OH and water, agents that also introduce midgap states. Yet, to achieve such high

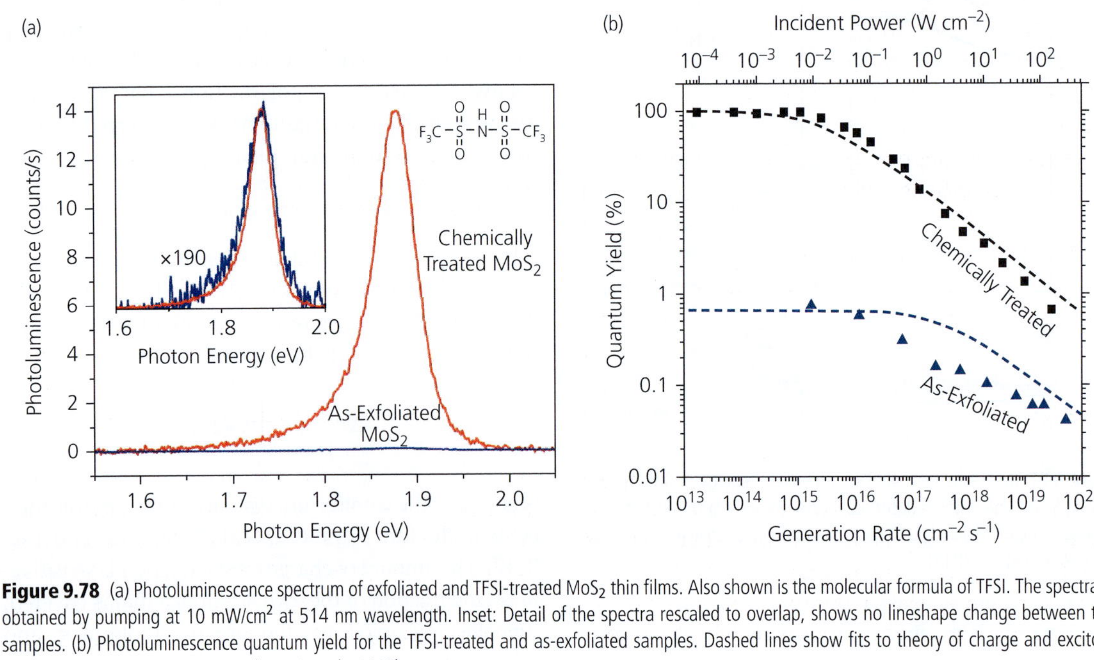

Figure 9.78 (a) Photoluminescence spectrum of exfoliated and TFSI-treated MoS_2 thin films. Also shown is the molecular formula of TFSI. The spectra are obtained by pumping at 10 mW/cm^2 at 514 nm wavelength. Inset: Detail of the spectra rescaled to overlap, shows no lineshape change between the samples. (b) Photoluminescence quantum yield for the TFSI-treated and as-exfoliated samples. Dashed lines show fits to theory of charge and exciton recombination vs. pump intensity (Amani et al., 2015).

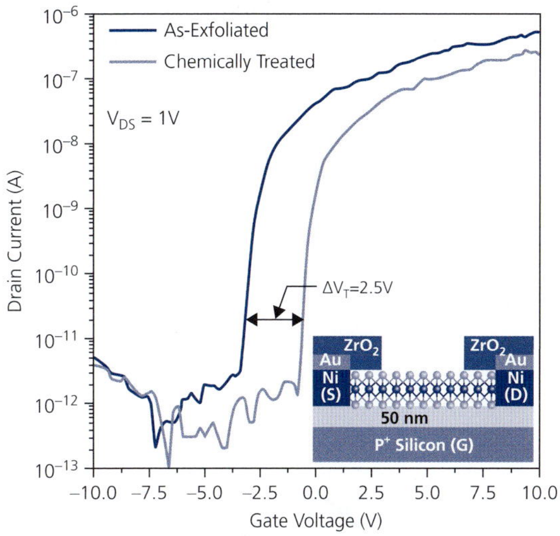

Figure 9.79 Drain current characteristics of MoS_2 BG/TC transistors employing exfoliated and TFSI-treated thin films. The ZrO_2 coating protects the Au/Ni contacts from attack by the TFSI superacid (Amani et al., 2015).

quantum yields, TFSI must have some effect in reducing the S-vacancy density since these also introduce midgap recombination centers. One possibility is that the elimination of the atmospheric surface contaminants by protonation by TFSI makes the dangling bonds available for surface reconstruction. This may result in clustering of S atoms, thus reducing the opportunities for exciton migration to randomly distributed defects at the MoS_2 surface (Amani et al., 2015).

Band gap narrowing, energy level alignment, exciton binding, and other interactions between adsorbed organic layers and TMDCs are strongly influenced by electronic polarization. Indeed, even single monolayer TMDCs can effectively screen electrostatic interactions between adsorbed organic layers from the underlying substrate. This is illustrated by dramatic changes in the energy gap of PTCDA when adsorbed onto the surface of WSe_2 on graphite, vs. when it is in direct contact with a substrate. Figure 9.80a shows *scanning tunneling spectroscopy* data for monolayers of PTCDA physisorbed directly onto a metallic Au(111) surface, a semimetallic graphite surface, and on a semiconducting WSe_2 monolayer on graphite. Scanning tunneling spectroscopy provides a relative measure of the electronic DOS at the film surface. Applying a time varying voltage between the microscope tip and the sample, and then measuring the magnitude of the tunneling current provides spatial information on the DOS. The d.c. tip bias offset corresponds to its energy relative to E_F, whereas the current is proportional

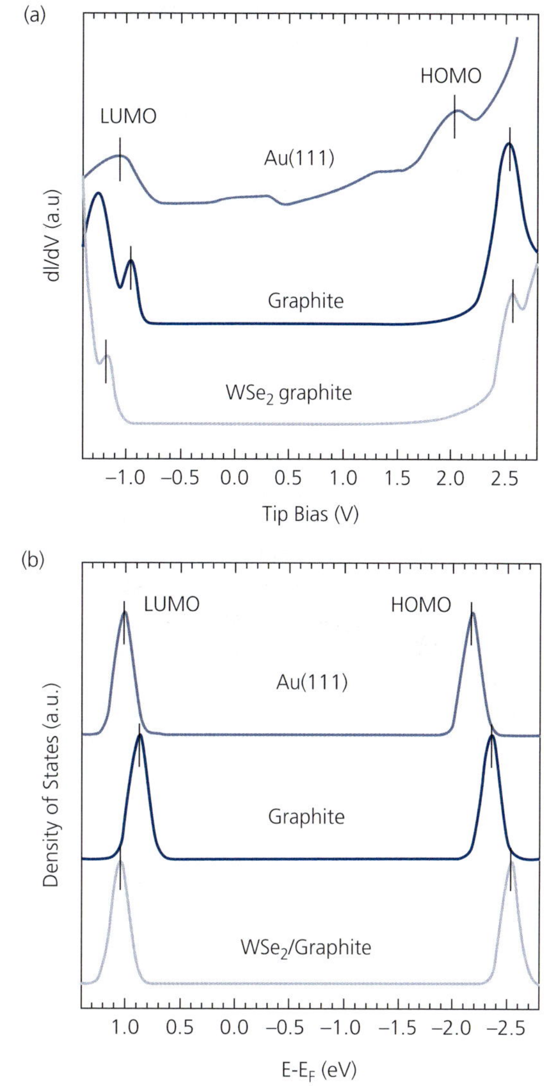

Figure 9.80 (a) Scanning tunneling spectroscopy data for a monolayer of PTCDA on Au(111), graphite, and WSe_2/graphite. (b) DFT calculations of the LUMO and HOMO densities of states for a flat-lying PTCDA molecule on these same substrates (Zheng et al., 2016).

to the magnitude of the local DOS (Zandvliet and van Houselt, 2009). For example, within the energy gap, the current is zero, whereas the current is large at the peak HOMO and LUMO densities of states.

Spectra for PTCDA molecules absorbed onto various substrates show significant energy gap narrowing, from 3.73 eV for the WSe_2/graphite sample, to 3.49 eV for PTCDA directly on graphite, and 3.10 eV on the Au(111) substrate. This is compared with 4.7 eV for isolated PTCDA molecules in the gas phase (Dori et al., 2006). The calculated HOMO-LUMO gaps using DFT that include polarization effects shown in Fig. 9.80b yield 2.93 eV (Au(111)),

2.99 eV (graphite), and 3.40 eV (WSe_2/graphite) (Zheng et al., 2016). These calculations are for isolated molecules, and hence ignore mutual polarization between molecules, thus underestimating the energy gap by approximately 0.3 eV. Nevertheless, in both the measured and calculated data, the WSe_2 monolayer screens image charges from metallic and semimetallic substrates, considerably reducing energy gap narrowing. This example is evidence that band alignments and energy gaps of adsorbed organic molecules on the surfaces of 2D semiconductors are influenced by charge exchange and associated electronic interactions at the HJ.

9.5.2 Charge transfer at organic/2D semiconductor heterojunctions

Thus far we have considered the properties of monolayer MX_3 thin films and their modification by adsorption or chemical reaction with organics. We now consider the charge transport characteristics across organic/2D semiconductor HJs, and how these junctions can be exploited in practical devices. The reader is referred to Section 4.7.3 where the topic of hybrid organic/inorganic HJs is developed. Many of the phenomena and analyses discussed in the context of bulk structures are also applied to organic/2D HJs.

A disadvantage of using TMDCs in active photonic applications is the low optical absorption of single monolayers, even though they have exceptionally large oscillator strengths. Nevertheless, 2D materials have the advantage of large in-plane charge mobilities. The shortcoming of a low absorbance can be mitigated by using an efficient light absorbing organic layer that transfers energy to the TMDCs. Such hybrid systems can therefore benefit from broad absorption combined with efficient energy harvesting. Energy transfer from an organic to a TMDC can occur either via charge exchange across a type II HJ (Section 4.7.3), or if the emission spectrum of the donor (e.g. the organic) overlaps the absorption spectrum of the acceptor (e.g. the TMDC), Förster energy transfer may be active.

Efficient Förster energy transfer (FRET) has been observed for the archetype monolayer $MoSe_2$/2 nm thick PTCDA HJ system. The excited state dipole moment of PTCDA grown by VTE lies in-plane, thus maximizing coupling to the W-M exciton of the TMDC (Gu et al., 2017). Previously, we have seen that PTCDA molecules prefer a flat-lying morphology on most substrates. Since the excited state dipole moment of PTCDA lies within the molecular plane, it stands to reason that the dipole will lie parallel to the 2D inorganic exciton dipole moment as well.

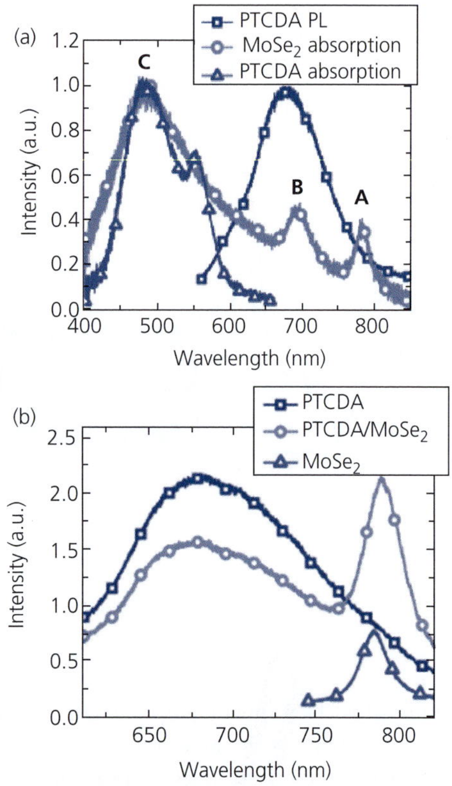

Figure 9.81 (a) Normalized differential reflectivity spectra of PTCDA (triangles), and $MoSe_2$ (circles), and photoluminescence spectrum of PTCDA (squares). The A, B, and C excitons of $MoSe_2$ are indicated. (b) Photoluminescence spectra of PTCDA (squares), $MoSe_2$ (triangles), and $MoSe_2$/PTCDA (circles). The measurements use a 510 nm laser source with a power of 1.28μW/mm^2 (Gu et al., 2017).

The absorption spectra of the organic and inorganic films are shown in Fig. 9.81a, along with the emission spectrum of PTCDA. The PTCDA PL overlaps the A and B exciton absorption features of $MoSe_2$, due to direct bandgap transitions at its high symmetry K-point, and the C feature is due to band nesting (cf. Fig. 9.74). The overlap between the PTCDA emission and TMDC absorption indicates that efficient energy transfer occurs, as observed in the PL spectra of the individual layers and for the HJ shown in Fig. 9.81b. The PL intensity of PTCDA is quenched in the hybrid sample, while that of $MoSe_2$ increases. The FRET efficiency, $\eta = 1 - I_{DA}/I_D = 30\%$, where I_{DA} and I_D are the spectrally integrated donor emission intensities with and without the acceptor layer on the PTCDA surface, respectively.

Contrary to PTCDA, vacuum deposited tetracene molecules stand up on WS_2, as shown in Fig. 9.82a. These data also indicate efficient and rapid energy and charge transfer in this system. The steady state

PL spectra for the constituents of the HJ and of a 20 nm thick Tc/1 ML WS_2 are shown in Fig. 9.82b. Due to the band alignment at the type II HJ, excitons generated by 2.8 eV photons in the Tc migrate to the HJ, and then dissociate into more weakly bound HCTEs that are precursors to free electrons in the TMDC and holes in the Tc. The spectra explicitly show that the 2.0 eV, A exciton of WS_2 dominates the WS_2/Tc HJ spectrum along with suppression of higher energy vibronics of Tc. Both spectral changes are expected when energy transfer is efficient. A CT feature appears at 1.7 eV, from which its binding energy of 0.3 eV is inferred. Transient absorption spectra of selectively pumped Tc and WS_2 layers shows that hole transfer occurs within 3.2 ps of the CT state formation, whereas electron transfer in the opposite direction may take the marginally longer time of 2.2 ps.

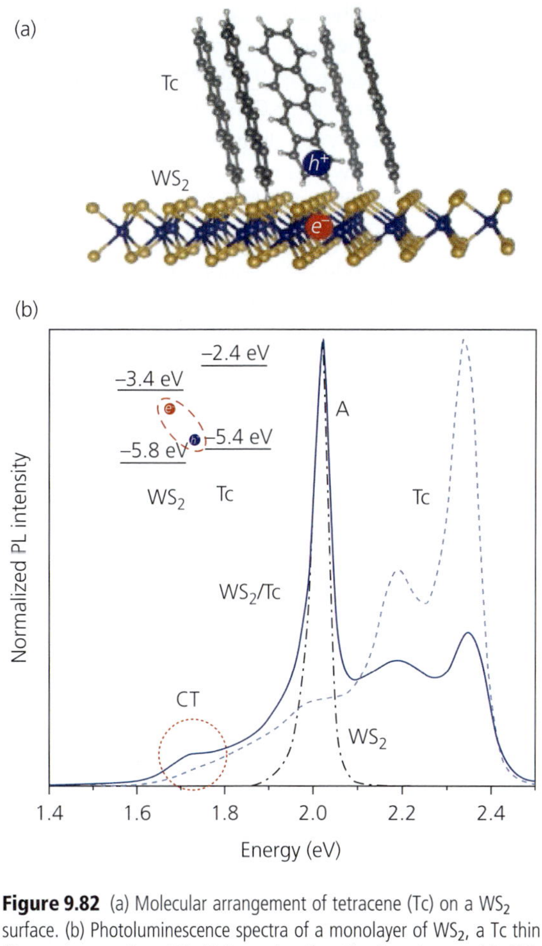

Figure 9.82 (a) Molecular arrangement of tetracene (Tc) on a WS_2 surface. (b) Photoluminescence spectra of a monolayer of WS_2, a Tc thin film, and a monolayer WS_2/Tc heterojunction. The A exciton peak in WS_2 and the CT feature in the HJ spectra are indicated. Inset: Proposed energy level diagram of the WS_2/Tc heterojunction illustrating the HCTE (dashed oval). After Zhu et al. (2018).

The diffusion of excitons at WS_2/Tc HJs have been spatially and temporally mapped using a pump source that locally generates excitons, and then positioning a probe beam a prescribed distance from the pump to observe the reflectance as a function of time. The pump energy at 3.1 eV excites both the TMDC and Tc, whereas the probe at 2.0 eV is sensitive to charges in the WS_2. The evolution of the exciton density is modeled using solutions to the time dependent diffusion equation (Section 7.1.4). The change in reflection (i.e. the ratio of the difference in reflectance with and without the pump beam, to the reflectance without the pump) obtained vs. position and time after the trailing edge of the pump pulse is shown in Fig. 9.83. Since the differential resistance is a function of exciton density, the profiles can be fit to solutions of the diffusion equation. Assuming two different Gaussian widths, $\sigma_{1,t}$ and $\sigma_{2,t}$, the exciton density is

$$N(x,t) = N_1 \exp\left(-\frac{(x-x_0)^2}{2\sigma_{1,t}}\right) + N_2 \exp\left(-\frac{(x-x_0)^2}{2\sigma_{2,t}}\right), \tag{9.38}$$

where N_1 and N_2 are the exciton populations at $x = x_0$, and x_0 is the position of the pump. The implication of this fit is that two different exciton populations coexist, each with a different diffusion constant given by

$$D_{1,2} = \frac{\sigma_{1,2t}^2 - \sigma_0^2}{2t}, \tag{9.39}$$

where σ_0 is the time-independent spatial variance of the distribution. The fits yield $N_1/N_2 = 2/3$, with corresponding diffusion constants of $D_1 = 1\ \text{cm}^2/\text{s}$ observed for the fast-moving population at $t > 100$ ps, and $D_2 = 0.04\ \text{cm}^2/\text{s}$ for the more slowly moving excitons. Indeed, using an exciton lifetime of $\tau = 2$ ns

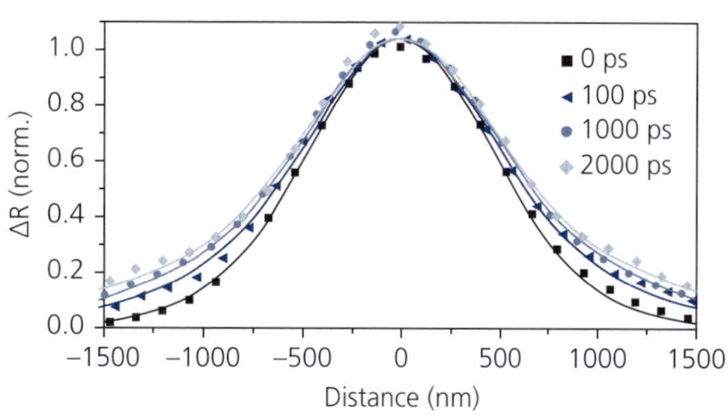

Figure 9.83 Reflectance change of a monolayer WS_2/Tc HJ as a function of distance and time relative to the position and trailing edge of the pump beam, respectively. The lines are fits to a double Gaussian function (Zhu et al., 2018).

taken from transient data, the diffusion length for the N_1 population is $L_{D1} = \sqrt{D_1 \tau} = 450$ nm. This large diffusion length is not unexpected for 2D thin films, although it is undoubtedly less than can be achieved in defect-free films where diffusion constants are expected to be at least an order of magnitude higher. Using a similar analysis, $L_D = 90$ nm, assuming that both populations have the same lifetime.

It has been proposed that the N_1 population with its long diffusion length is due to free excitons, while the N_2 population is due to excitons in energy sites extending deep into the WS_2 energy gap (Zhu et al., 2018). An alternative explanation is that this slower moving population is due to states trapped at surface S-vacancies. Trapped CT excitons have been predicted and observed at organic/GaN and organic/ZnO HJs (Panda et al., 2016, Panda and Forrest, 2017), and hence it is possible that similar effects are active at WS_2/Tc interfaces.

Spectroscopic evidence for HCTEs and efficient charge transfer at pentacene/MoS_2 (Homan et al., 2017) and WS_2/PTCDA HJs has also been reported (Liu et al., 2017). Excitons bound at the HJ can be modeled by assuming that the TMDC has only in-plane polarizability, and that holes in the TMDC bound to electrons on the organic side of the HJ are delocalized in the 2D film plane. Due to the asymmetry of the dielectric constants and effective masses of charges in the two materials, the HCTE wavefunction on the inorganic side has 2D W-M character, whereas on the organic side it has local Frenkel character (Renshaw and Forrest, 2014). Thus, the hole is Coulombically bound to an electron fixed on a molecule across the HJ. This assumption is justified by the lower dielectric constant and larger effective mass of charges in organics as compared to holes in TMDCs.

The potential energy of the hole is $V(r) = V_{xy}(x, y) + V_z(z)$, where $V_{xy}(x, y)$ is the Coulomb potential between the electron and the hole which primarily depends on the 2D distribution of the hole in x–y TMDC plane, and $V_z(z)$ includes other potential terms such as the image charge potential and energy level offsets, which are constant in the x-y plane. To simplify the analysis, the hole wavefunction is written as $\psi(r) = \psi_{xy}(x, y)\, \psi_z(z)$, where $r = \sqrt{x^2 + y^2 + z^2}$. Schrödinger's equation is then

$$\left(-\frac{\hbar^2}{2m^*}\nabla^2_{xy} + V_{xy}(x,y)\right)\psi_{xy}(x,y) = E'\psi_{xy}(x,y), \tag{9.40}$$

where m^* is the reduced effective mass of the HCTE and $E' = E + \frac{h}{2m^*}\frac{\psi_z''(z)}{\psi_z(z)} - V_z(z) = E - E_{free}$, which is approximately the HCTE relative to the free hole energy. Also, $\psi_z''(z)$ is the second spatial derivative of the wavefunction with respect to z. The Coulomb potential is $V_{xy}(x,y) = -\sum_{i=1}^{n}\frac{q^2/n}{4\pi\varepsilon_0\varepsilon_r|\mathbf{r}_h - \mathbf{r}_i|}$, where ε_r is the dielectric constant of WS_2, and $r_h = \sqrt{x^2 + y^2}$ is the position of the hole in the 2D film plane.

This analysis has been applied to electron and hole bound pairs at PTCDA/WS_2 type II HJs. The charge distribution of the anionic PTCDA is found via DFT, and is simplified by using a summation over n discrete, fractional point charges at positions $\mathbf{r}_i$ on the molecule. Due to the mirror symmetry of PTCDA, it is adequate to use $n = 2$ separated by 0.56 nm on the two halves of the perylene core.

The energy eigenvalues of Eq. 9.40 yields the energy levels and the 2D eigenfunctions shown in Fig. 9.84a, using the energy barriers at the PTCDA/WS_2 heterointerface in Fig. 9.84b. The lowest CT state has a binding energy of 38.2 meV, making it stable at room temperature. Also shown are the hole wavefunctions of the lowest four states corresponding to the 1s, 1p, 2s, and 1d orbitals, with average electron–hole separations of r_0 = 2.2, 7.5, 11.0, and 18.3 nm, respectively. The relative sizes of these several states are shown (Liu et al., 2017). Note, that in 2D systems, the angular momentum quantum number is not necessarily smaller than the principal quantum number, and that the eigenstates are spread over large distances along the 2D monolayer plane. This is characteristic of a large radius W-M state that is quantum-confined to a limited dimensional space, similar to the situation that obtains in inorganic quantum well structures.

9.5.3 Two-dimensional organic/inorganic heterojunction devices

The efficient charge transfer across the PTCDA/WS_2 heterointerface has been used for charge generation in a hybrid organic/inorganic photodiode, schematically illustrated in the inset, Fig. 9.85. The photodiode structure is ITO/TMDC/10 nm PTCDA/8 nm BPhen/100 nm Al. As in OPVs, the BPhen layer is an exciton blocking layer that also provides protection of the active layers during Al cathode deposition. The top electrode is patterned by a shadow mask with 38 × 38 μm^2 square holes aligned to the exfoliated

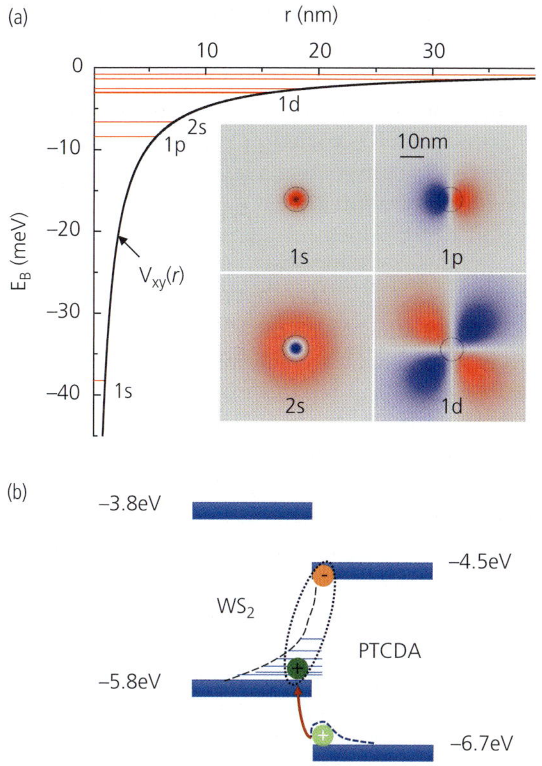

Figure 9.84 (a) Energy eigenvalues of HCTE (E_B) vs. Coulomb potential, $V_{xy}(r)$, for an electron on PTCDA. Inset: Hole wavefunction of lowest four eigenstates within the 2D WS_2 (x–y) plane. The black circle in the middle of each graph is a 10 nm-diameter scale. (b) Energy level diagram of the type II PTCDA/WS_2 heterojunction showing the HCTE state and the process of hole capture (Liu et al., 2017).

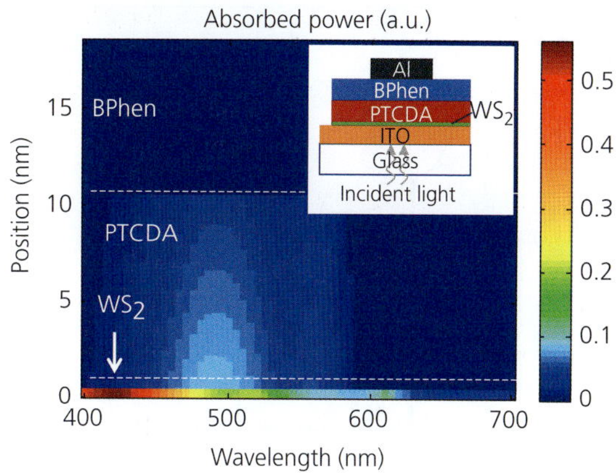

Figure 9.85 Absorbed power distribution in a PTCDA/WS_2 photodiode. The dashed lines demark the boundaries between the three films comprising the device. The color bar at right shows the relative intensity of the absorbed optical power. Inset: Schematic cross-section of the photodiode. Adapted from Liu et al. (2017).

WS_2 islands. Figure 9.85 shows the absorbed power for light incident normal to the substrate via the ITO anode vs. position along the device cross-section, calculated using the transfer matrix model in Section 7.1.4. Greater than 60% of the exciton population is generated in the WS_2 layer when the junction is excited at 450 nm $> \lambda >$ 600 nm, whereas > 85% of excitons are generated in the PTCDA at the intermediate wavelength range of $450 < \lambda < 600$ nm. It is indeed remarkable that only a monolayer of the TMDC absorbs such a large fraction of the light—a result of its exceptionally high oscillator strength.

The peak external quantum efficiency is $\eta_{ext} = 1.8\%$ at $\lambda = 430$ nm, as shown in Fig. 9.86a. The photoresponse maxima at $\lambda = 430$, 520, and 625 nm correspond to the absorption maxima of WS_2, while others match those of PTCDA. Figure 9.86b shows the internal quantum efficiency calculated using $\eta_{int} = \eta_{ext}/\eta_A$, where η_A is the absorption efficiency calculated from ellipsometric measurements of the optical constants of the various layers. The η_{int} is approximately 2% between $\lambda = 450$ nm and 600 nm, and increases to 11% at $\lambda < 450$ nm and $\lambda > 600$ nm.

The internal efficiency has been analyzed using an empirical fit to the diffusion efficiency of excitons in the TMDC and the organic layer, the dissociation efficiency at the HJ, and the charge collection efficiency using a procedure similar to that adopted for the analysis of OPDs in Section 7.1. The procedure yields an accurate fit to the data shown by the dashed line in Fig. 9.86b. From these results we come to two conclusions. The first is that, despite the limited dimensionality of the organic/TMDC HJ, it behaves in many ways like bulk (3D) organic/inorganic junctions in that excitons are photogenerated in both materials. The excitons drift to the HJ where a HCTE forms prior to dissociation into charges that are detected in the external circuit. The second is that the quantum efficiencies of organic/TMDC are remarkably high considering that only a monolayer of material is needed to achieve an external efficiency of 2%, and an internal efficiency of > 10%.

While these results illustrate the benefits of combining absorptive organic thin films with 2D materials, the structures are not a practical solution for photodetections. The TMDCs are polycrystalline, and the exfoliated films result in very small flakes that are difficult to work with. The flakes come in a variety of shapes and sizes (most less than a few tens of microns across) that cannot be precisely positioned on a substrate, which is an essential first step in batch fabrication processes. Before we can begin to consider TMDC-based devices as useful for electronic

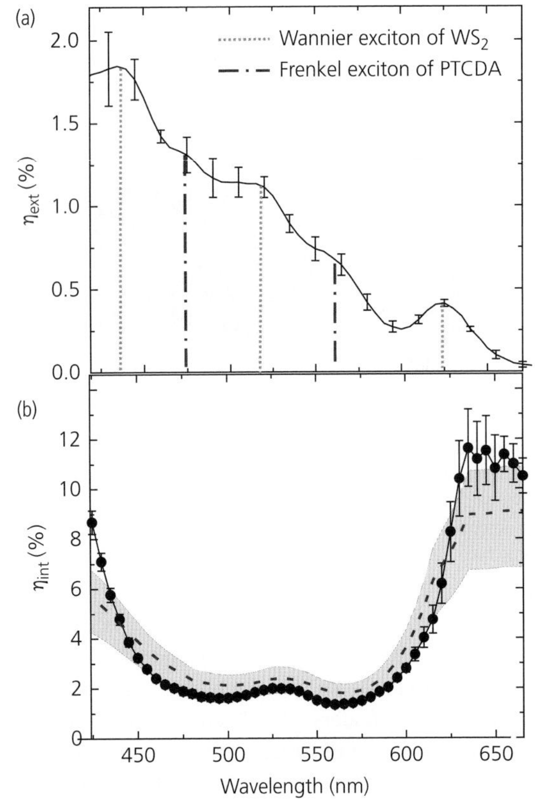

Figure 9.86 (a) External quantum efficiency spectrum of the photodiode in Fig. 9.85, identifying the spectral features in the stand-alone layers. (b) Internal quantum efficiency spectrum. The dashed line is the fit to the data, and the 95% confidence interval is delineated by the shaded region (Liu et al., 2017).

applications, large monolayer films, preferably single crystals, will need to be grown on substrates that provide stability as well as flexibility. Furthermore, a reproducible route to passivating chalcogen vacancies is required to achieve the high electronic performance that is projected for 2D materials. CVD appears to be the most promising growth method that may ultimately reach this goal, although defect-free, large area films have yet to be demonstrated even by this process.

In spite of the challenges, CVD grown monolayers of MoS_2 have been used in detectors employing a 16 nm thick PTB7 polymer. The CVD process uses hot (~500°C) MoO_3 and sulfur powder sources vaporized into an inert (Ar) carrier gas in a quartz reactor. As in organic vapor phase deposition (Section 5.4.2), the elemental source vapors are carried to the substrate where they react to form the MoS_2. Raman spectroscopy indicates that ~80% of the substrate is covered by monolayer films, with the other 20% covered by bilayer and higher order multilayers. The CVD films are typically polycrystalline, with domain sizes approximating those obtained by exfoliation, that is, ~10–50 μm.

The process sequence used to fabricate the photodiodes is illustrated in Fig. 9.87. The CVD films are (i) grown on SiO_2 coated Si substrates, and (ii) are protected by spin coating a layer of PMMA onto the growth surface. (iii) The SiO_2 is removed by immersion in KOH, which results in the delamination of the monolayer film. Next, (iv) the free-standing MoS_2 is transferred onto an ITO anode-coated glass surface, (v) after which the PMMA protective coating is removed in a chloroform solution. (vi) The TMDC is probed for its properties before (vii) a 16 nm thick layer of PTB7 is spin cast onto its surface. (viii) The 1.6 mm^2 device is completed by vapor deposition of a MoO_3 buffer and a Ag cathode (Shastry et al., 2016). Although the films cover the surface of the parent substrate, the devices are kept small to avoid excess current leakage from high densities of defects at the grain boundaries, as well as along the film surface.

The proposed energy level diagram for the device is shown in Fig. 9.88a. The energy level offsets show that PTB7 functions as the donor, and MoS_2 as the acceptor. The PL of the HJ shows complete quenching of emission from the polymer due to efficient energy transfer to MoS_2. Indeed, there is almost no emission from the HJ suggesting that the CT state itself does not radiatively combine, in contrast to the behavior of the Tc/WS_2 junction discussed previously.

The detector itself shows both a high absorbance from both the PTB7 between 500 nm and 750 nm, and the MoS_2 between 350 nm and 600 nm, with reduced absorption at longer wavelengths (see Fig. 9.88b). The external quantum efficiency features a peak of 20% at 450 nm, almost entirely due to absorption in the monolayer of MoS_2. This corresponds to a calculated internal efficiency of 40%. Due to its broad spectral coverage, the same device has been tested under simulated solar illumination. The resulting solar cell characteristics are an open circuit voltage of V_{OC} = 0.21 V, a short circuit current density of j_{SC} = 1.98 mA/cm^2 and a fill factor of $FF = 0.24$, yielding a power conversion efficiency of $PCE = 0.1\%$.

These performance characteristics point to several challenges confronting TMDC-based HJ devices. The first is the low V_{OC} compared to the HOMO-LUMO HJ offset potential of $\Delta E_{HL}/q = 0.8$ V. This is a direct consequence of significant non-radiative recombination at the organic/TMDC HJ. Non-radiative processes are likely due to sulfur surface and edge vacancies discussed above. Secondly, the very low

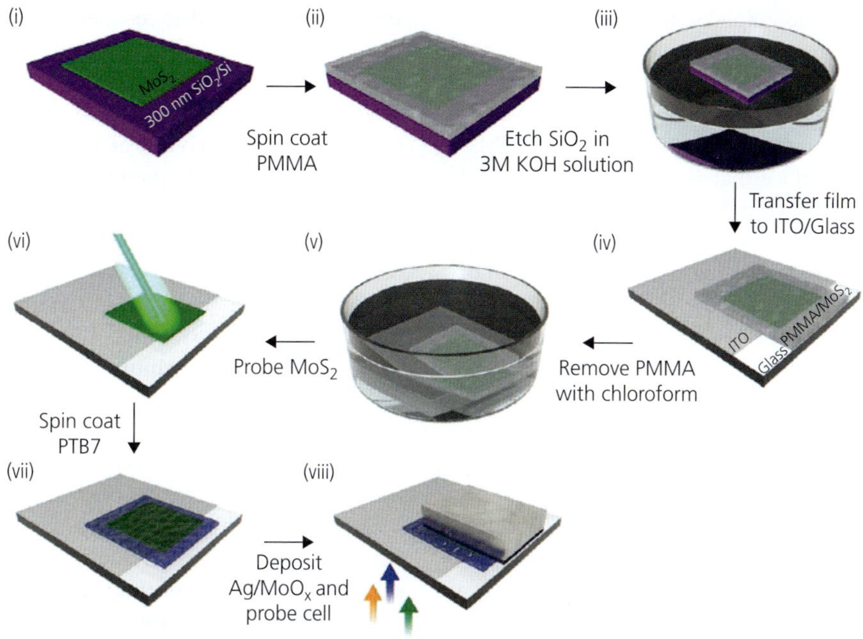

Figure 9.87 Process sequence for fabricating a PTB7/MoS_2 photodetector (Shastry et al., 2016).

FF results from poor rectifying characteristics, possibly due to a large shunt leakage current through grain boundary defects. This is apparent from the nearly symmetrical *j–V* characteristics in the reverse and forward biased directions (Shastry et al., 2016). Note that a similarly low fill factor has been reported for C_8-BTBT/MoS_2 HJ photovoltaics. In that case, the energy level offset between the TMDC and the organic presents a large barrier that impedes electron transport, requiring tunneling to extract the charges (He et al., 2015). A large series resistance results in a very large forward biased ideality factor of $n \approx 21$, which may also manifest as a large *FF*. However, significant contact resistances are not readily apparent in the device in Fig. 9.88a, although injection limited conduction cannot be ruled out as being responsible for the low *FF* and large series resistance. Finally, the low j_{SC} is due to the limited quantum efficiency from the ultrathin layers.

MoS_2 has been used as a source injection contact in a pentacene channel in a BG/BC thin film transistor with a SiO_2 dielectric. As above, the MoS_2 forms a rectifying, type II HJ with pentacene. The ambipolar OTFT showed a low on-off drain current ratio of ~100 to 200. This junction exhibits a source-drain photocurrent when illuminated due to charge dissociation at the HJ (Jariwala et al., 2015).

Significantly improved characteristics are achieved using graphene monolayers as the S and D electrodes

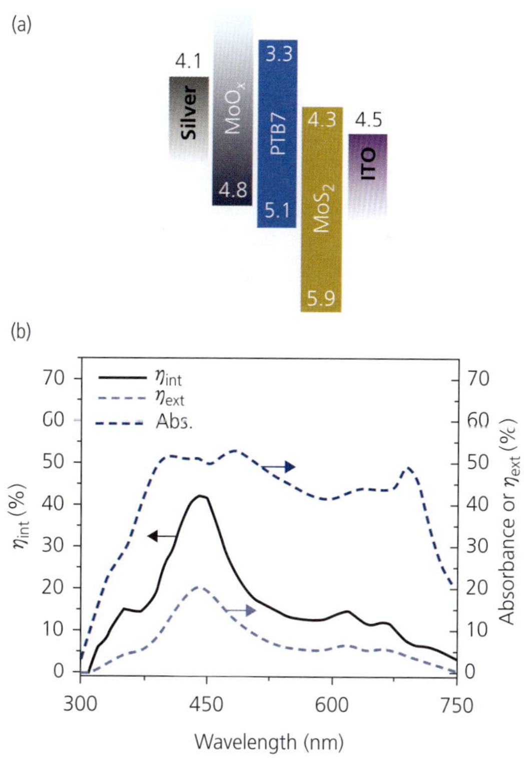

Figure 9.88 (a) Proposed energy level diagram of the PTB7/MoS_2 photodiode showing frontier orbital energies in eV. (b) Internal (black line) and external (light blue line) quantum efficiencies, and the absorbance spectra (dark blue line) of the active region of the device in (a) (Shastry et al., 2016).

(Lee et al., 2011). The graphene is initially grown via CVD on a Cu substrate, and then detached using a PMMA "handle." The film is transferred to a 300 nm thick SiO_2 gate insulator, after which the PMMA is removed with acetone. However, a residue of PMMA remains on the surface that is removed by thermal desorption. The device is then completed by S and D patterning to form a $L = 100$ μm and $W = 1$ mm gap between the electrodes. The last step is deposition of the pentacene channel.

A fused polyaromatic hydrocarbon such as pentacene has a strong tendency to form π-stacks with itself, as well as with substrates that present the same fused phenyl group configuration. Hence, pentacene deposited on graphene forms elongated crystals with the molecules lying in the substrate plane. This stacking habit results in the highest hole mobility perpendicular to the substrate, which is disadvantageous for transistors whose channels are oriented along the plane. The flat lying configuration is frustrated, however, by the PMMA contaminated surface. Molecules that deposit onto PMMA islands that are left from the transfer process stand on end, as shown in Fig. 9.89a. Molecules that land in the gaps between PMMA islands are also held in the vertical orientation due to interactions with neighboring, vertical standing pentacene molecules that are on the islands. Hence, the partial coverage of graphene by the residual PMMA has the combined effects of creating a vertical pentacene orientation which leads to a high channel mobility, as well as providing sufficiently large graphene regions between the islands to make low resistance contact to the channel.

The output characteristics of the graphene/pentacene OTFT are shown in Fig. 9.89b. The output conductance is low, as required for high gain circuit applications. More importantly, the channel mobility is 1.2 cm^2/V s, which is among the highest values achieved for pentacene OTFTs, primarily due to the order imposed by the electrodes. Furthermore, $I_{on}/I_{off} = 10^8$, which is competitive with the best pentacene-based OTFTs. Finally, the threshold voltage is $V_T = 18$ V. Interestingly, the characteristics of pentacene deposited onto thermally treated (and hence uncontaminated) graphene has a field effect mobility of only 0.4 cm^2/V-s, although I_{on}/I_{off} is comparable to that obtained for the OTFT with residual PMMA. The lower mobility is consistent with channels where the flat-lying pentacene molecules have a

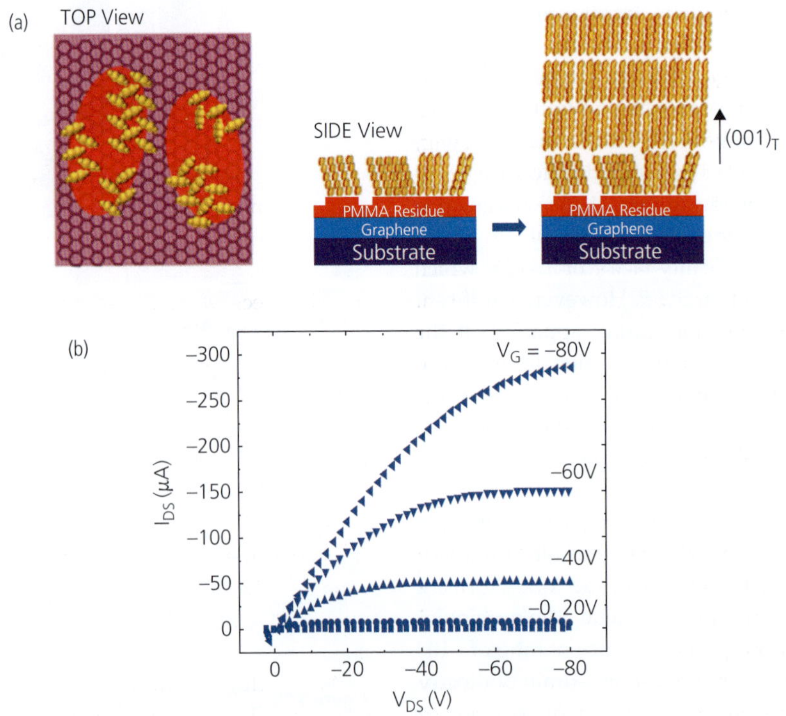

Figure 9.89 (a) Illustrations of the orientation of pentacene (gold structures) on graphene with a PMMA residue (red regions). The diagrams to the left are for a single pentacene layer, and on the right for a multilayer. (b) Output characteristics of the BG/BC pentacene OTFT with graphene S and D electrodes (Lee et al., 2011).

higher in-plane resistance than those where the molecules are up-standing (Lee et al., 2011).

The principal advantage of using graphene in thin film transistors is, therefore, its ability to template the growth of organic layers with large crystalline domains whose structure is guided by the 6-fold symmetry of the monolayer sheet. Van der Waals bonding between layers leads to nearly perfect interfaces with graphene contacts while also resulting in an organic crystal whose highest mobility direction is along the channel.

The graphene conductivity limits its use to a contact material rather than as the channel semiconductor itself (Lee et al., 2011, He et al., 2014). Electrically insulating *h*-BN has a similar symmetry as graphene, and thus also can serve as a growth template for the channel semiconductor. Thus, organics that are π-bonded to an *h*-BN monolayer can benefit from a high quality, low-defect interface. An example OTFT with a <10 nm thick *h*-BN gate insulator consists of a single monolayer of C_8-BTBT grown by sublimation in a vacuum tube furnace on exfoliated *h*-BN, previously transferred onto 285 nm SiO_2 on a Si bottom gate electrode. The top Au contacts are deposited onto the organic surface to form an OTFT with $W \simeq L \simeq 15$ μm.

The organic molecules can assume several different registries on the 2D template, one of which is shown in the inset, Fig. 9.90a. The molecules nucleate at random locations, but in general lie nearly flat to maximize their π-system contact with the *h*-BN lattice. This results in a polycrystalline film grown layer-by-layer, with single and double layer domain dimensions of several microns. The high quality interface between the organic monolayer and *h*-BN results in a field effect hole mobility of 10 cm^2/V s extracted from the drain current-gate voltage characteristics in Fig. 9.90a. This is considerably higher than expected for similar organic layers deposited directly onto the SiO_2 gate insulator. Also, the mobility decreases to approximately half this value as the organic thickness is increased to create both monolayer and bilayer domains. The ratio, $I_{on}/I_{off} = 10^6$, is indicative of a high quality interface that prevents parasitic channel leakage. The transfer characteristics in Fig. 9.90b show very low output conductance in the drain current saturation regime, with a threshold voltage of -15 V and a turn-on voltage of 8 V.

The transistor characteristics indicate a low density of grain boundaries in the channel, which is in part due to templated growth on the *h*-BN, as well as the relatively short channel length (He et al., 2014). While the first monolayer stacking habit is strongly

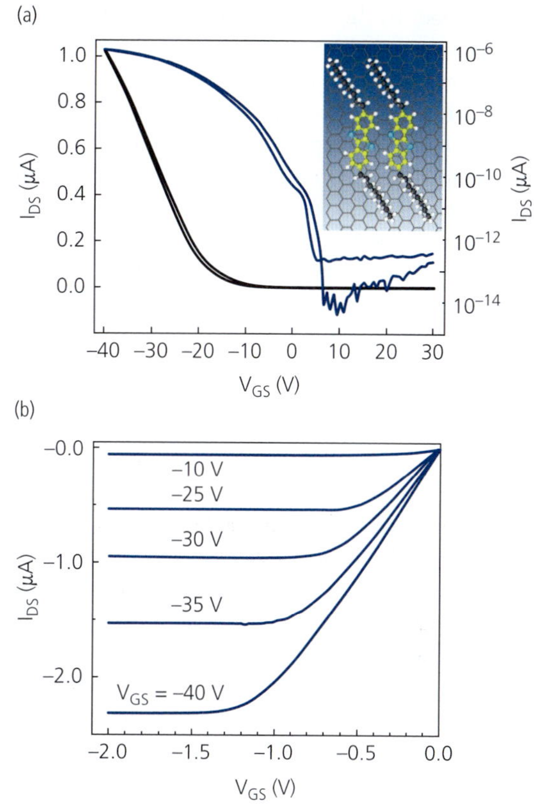

Figure 9.90 (a) Drain current–gate voltage characteristics of an ultrathin *h*-BN/C_8-BTBT OTFT. Inset: One of several possible substrate registrations of C_8-BTBT on graphene, which is similar to that found on *h*-BN. (b) Drain current transfer characteristics of the device in (a) (He et al., 2014).

influenced by the substrate, the crystal relaxes to its equilibrium structure as the number of layers increases. Similar thickness dependences are observed in *h*-BN/pentacene OTFTs, with the pentacene hole mobility decreasing abruptly as the number of layers increases from one to two at a thickness of only 3 nm (Zhang et al., 2016). The changes are correlated with the mode of hole transport: it is primarily via hopping in the monolayer film, becoming more band-like in the multilayers. Also, recall that the monolayer films can significantly screen the electric field from the substrate. Field-screening, particularly from charges trapped at the SiO_2 surface, cannot be ruled out as a source of the exceptionally high mobilities observed in channels employing a compound SiO_2/*h*-BN gate insulator.

We conclude from the foregoing discussion that organic/2D materials offer possibilities for understanding and exploiting the unique and significantly divergent properties of both materials systems. In Chapter 4, we saw that organic/inorganic HJs are characterized by a combination of Frenkel- and band-like behaviors that define these two classes of semiconductors. In addition, the ultrathin inorganic layers have remarkable properties such as high oscillator strengths, charge mobilities, and room-temperature-stable quantum-confined excited states. Their thinness, combined with the compliant mechanical properties of organics suggests that organic/2D HJs may find uses in ultrathin, highly flexible electronic devices. Yet 2D materials also face significant challenges. The semiconducting transition metal dichalcogenides are plagued by a high density of atomic vacancies, both within the monolayer as well as at its edges. And it remains a challenge to grow large, crystalline sheets of any of these materials. Very often the crystals come in small flakes or domains only a few microns in diameter. While organic molecules can passivate surface defects, the small crystal sizes make it difficult to fabricate photonic or electronic devices that provide reproducibly high performance. Hence, while the science is rich with much yet to be discovered in the nascent field of limited-dimensional organic electronics, the exploitation of the organic/inorganic HJ in practical detectors, emitters and transistors remains only a promise.

9.6 Single molecule electronics

In Section 9.5, we introduced many interesting and emergent physical phenomena found in 2D systems such as organic HJs with transition metal dichalcogenides, graphene and hexagonal boron nitride. The final step in this progression to limited dimensionality is in 1D (i.e. quantum wire) systems, which is the domain of single molecule electronics, or sometimes, *molecular electronics*. We note that zero-dimensional (0D) systems have been described in Chapters 6 and 7 in the context of quantum dot-based devices and associated phenomena used to describe single molecule luminescent centers in OLEDs, organic semiconductor lasers, and detectors. The 0D phenomena in those cases are observed by embedding the NPs or molecules in a bulk, host matrix. They fundamentally differ from the behavior of limited dimensional systems treated here in that 0D systems do not support charge transport.

The scale of 1D molecular wires is such that only quantum mechanical descriptions are adequate for their understanding. Single molecules bridging metal electrodes, or forming the channels in single molecule transistors, have exhibited quantum interference (Aradhya and Venkataraman, 2013), the Kondo effect (Scott and Natelson, 2010), tunneling between molecular states (Nichols and Higgins, 2015, Perrin et al., 2015), and so on. In spite of many difficulties encountered in their handling, single molecule transistors and electronic circuits have been demonstrated in many laboratories over several decades (Joachim et al., 2000). While these devices may eventually lead to scaling of gates with lengths <1 nm, more immediate applications may emerge in chemical sensing. That is, the presence or absence of a single molecule between the nanometer scale gap between electrodes provides unprecedented sensitivity for detection of specific molecular species that selectively attach to functionalized contact surfaces (Nichols and Higgins, 2015).

The single molecule "p-n" junction device was originally envisioned to consist of a hemiquinone comprising an electron deficient quinone acceptor and an electron rich methoxy donor moiety bridged by a non-conjugated methylene bridge (see **Scheme 2**) (Aviram and Ratner, 1974). While this particular configuration is not the focus of current research on molecular devices, it is instructive to understand its operational concept. The equilibrium energy level diagram for the device is shown in Fig. 9.91a. The molecular contact with the cathode introduces a narrow energy barrier between the metal with work function ϕ_M, and the π-electron system of the acceptor group. The $-CH_2-$ σ-bridge presents a tunneling barrier between the acceptor and the donor groups. A second barrier exists at the donor/anode interface. The LUMO energy of the acceptor and the HOMO energy of the donor lie near to the Fermi energy. By applying a negative voltage to the cathode (i.e. the molecule is forward biased), the LUMO of the acceptor aligns with ϕ_M, allowing electrons to resonantly tunnel into the acceptor group (process A, Fig. 9.91b). Similarly, the filled HOMO of the donor is brought into alignment with the anode, allowing for electron tunneling (process C). The circuit is completed across the molecule by tunneling across the bridge (process B) from acceptor to donor. In contrast,

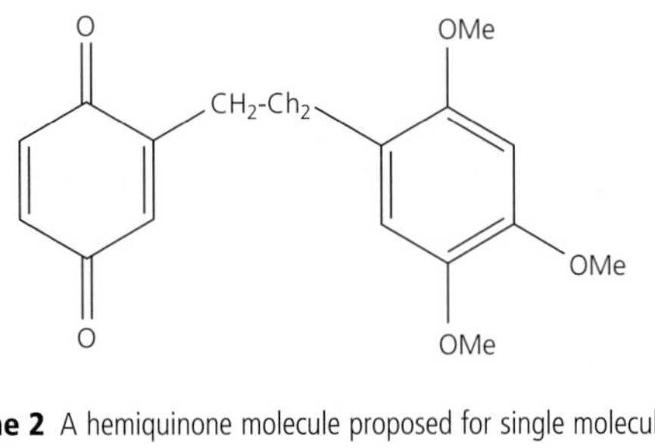

Scheme 2 A hemiquinone molecule proposed for single molecule rectifiers.

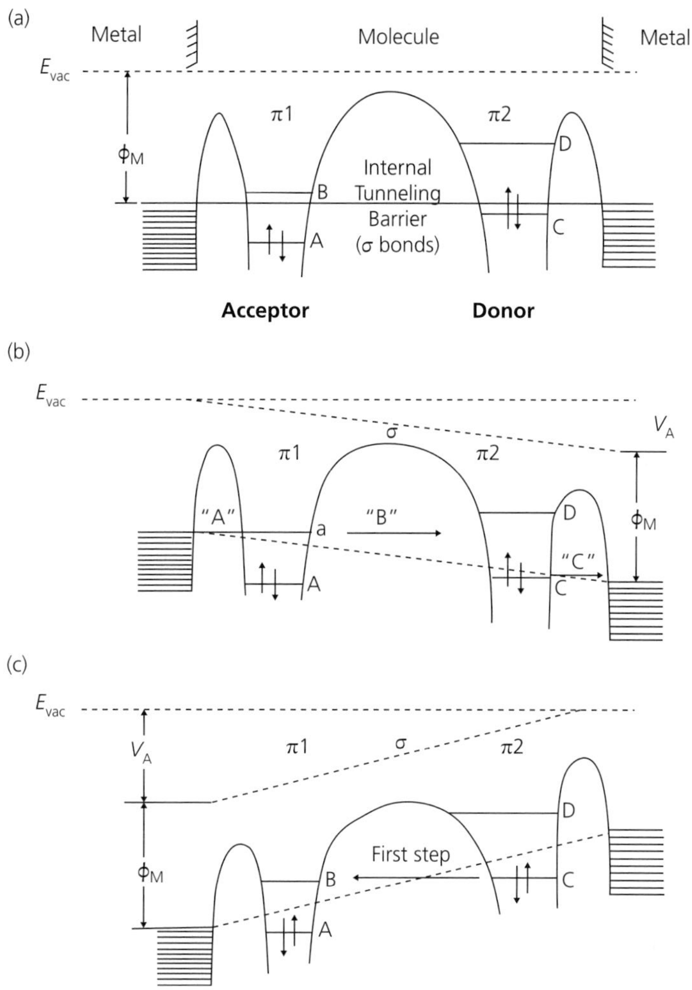

Figure 9.91 Conceptual energy level diagram of a molecular rectifier (a) at equilibrium and (b) in forward and (c) reverse bias. Filled energy levels have electrons with spins indicated by arrows. The π- and σ-bonds of the several moieties comprising the molecules are shown (Aviram and Ratner, 1974).

reverse bias in Fig. 9.91c results in misalignment between the metal work function of the anode and the donor, or between the energy levels of the donor and acceptor molecular units, thus inhibiting charge transport.

The architecture in Fig. 9.91 differs from modern molecular electronic devices that focus on molecules that lack separate donor and acceptor units. Nevertheless, it illustrates several features of all such devices. The contact barriers between electrodes influence the transport properties of molecular wires. Similarly, internal energy level alignments, intramolecular charge transfer, the width of molecular energy levels (and hence their overlap with the metal Fermi energy), and resonant tunneling are all important to device operation.

The 1D molecular wire can either contain an excess electron or hole via reduction or oxidation. The electron energy is estimated by assuming that it is contained along a wire of molecular length, L. The energy difference between quantum states assuming a particle in a 1D box, is $\Delta E = \frac{h^2}{2m^*}\left(\frac{\pi}{L}\right)^2$. For $L = 1$ nm, which is typical of many small molecules, spatial confinement leads to $\Delta E \approx 0.4$eV. A second important energy scale is the molecular charging energy. The large Coulomb repulsion between closely spaced charges makes it unlikely that more than a single excess charge can reside on the molecule at one time. The charging energy of the molecule is given by $E_C = \frac{1}{2}CV^2 = \frac{1}{2}\frac{q^2}{C}$, where C is the molecular capacitance. The charging energy is typically several hundreds of meV. The final consideration is the thermal energy, which determines molecular level broadening as well as occupation of the vibronics within a given electron orbital. The room temperature thermal energy is $k_BT = 25.9$ meV, but can be reduced when the sample is cooled.

A simple model that defines the energies of a single molecule device is shown in Fig. 9.92. The molecule is assumed to bridge the gap between two identical metal S and D electrodes. A barrier between the molecule and the contacts has coupling coefficients, Γ_S and Γ_D, respectively. Since the molecule, or its contact anchoring points may not be symmetric, we assume that Γ_S also differs from Γ_D. Furthermore, the neutral molecule has N electrons and a chemical potential of μ_N. The molecule can attach to the contacts either via reaction with the electrode surfaces, or via electrostatic bonds. Adding an electron via tunneling or hopping from the S electrode to the molecule thereby shifts its chemical potential to $\mu_{N\rightarrow N+1}$. The imperfect energy alignment between the contacts (with chemical potentials μ_S and μ_D) produces contact energy barriers, as illustrated for a device at equilibrium in Fig. 9.92b. In forward bias, the molecular energy level aligns with the S contact as shown in Fig. 9.92c, thereby promoting electron injection. In the diagram, it is assumed that the S and D electrode compositions and contact barriers are similar, and thus the voltage is divided equally between them.

We define three regions for injection and transport. In the *weak coupling* regime, the coupling coefficient is $\Gamma_{S,D} \ll \Delta E, E_C, k_BT$. Here, the molecule is only weakly coupled to the contacts, possibly due to physisorption via van der Waals bonds. The molecule thus retains its discretely defined energy levels that it has when isolated, and consequently the level broadening is small. Furthermore, the molecule can support only integer charges, and indeed only a singly ionized state is likely to exist due to electrostatic repulsion from the charge occupying the molecule. This creates a *Coulomb blockade* that arrests further charge injection. Charge transport between the contacts is due to incoherent hopping.

Level broadening is increased in the *intermediate coupling* regime. Molecules covalently bound to the electrodes exhibit $\Gamma_{S,D}, k_BT \leq \Delta E, E_C$. In this regime, non-integer charges can reside on the molecule by sharing orbitals with the contacts. The shared valences result in coherent transport across the molecule.

Small molecules are commonly used for metal–organic–metal devices. These molecules have widely separated energy levels. Thus, only a single energy level is in resonance with the chemical potentials of the electrodes at voltages too low to allow for direct tunneling between S and D. Then, the conductance of the device is given by

$$G = G_0 T_S T_{mol} T_D. \tag{9.41}$$

Here, T_S and T_D are the electronic transmission coefficients of the molecular linkages to the source and drain, respectively, and T_{mol} is the transmission coefficient of the molecule itself. Also, G_0 is the *conductance quantum*, given by

$$G_0 = \frac{2q^2}{h} = 77.5\ \mu\text{S}. \tag{9.42}$$

Now G_0 is obtained from the 1D DOS given by $g(E) = dn/dE = 2/hv$, where v is the electron velocity, n is the charge density, and the factor of two results from the spin degeneracy of the state. Since the difference between energy levels is $dE = -qdV$, then from Ohm's law the conductance quantum follows from $di = -qdE\left(\frac{dn}{dE}\right)v = G_0dV$, where di is the differential current.

The transmission $T(E, V_A)$ through a molecular wire can be described in the intermediate coupling limit using the non-equilibrium Green's function approach (Haug and Jauho, 2008). The resulting Breit–Wigner probability distribution gives a DOS that has a Lorentzian line shape, from which we

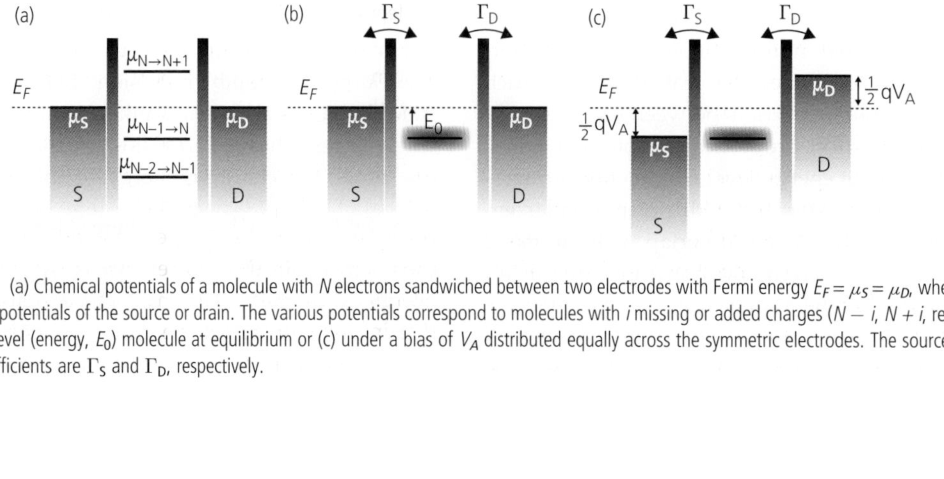

Figure 9.92 (a) Chemical potentials of a molecule with N electrons sandwiched between two electrodes with Fermi energy $E_F = \mu_S = \mu_D$, where $\mu_{S,D}$ are the chemical potentials of the source or drain. The various potentials correspond to molecules with i missing or added charges ($N - i$, $N + i$, respectively). (b) A single level (energy, E_0) molecule at equilibrium or (c) under a bias of V_A distributed equally across the symmetric electrodes. The source and drain coupling coefficients are Γ_S and Γ_D, respectively.

obtain the total electronic transmission function, $T(E, V_A) = T_S T_{mol} T_D$ (Breit and Wigner, 1936):

$$T(E, V_A) = \frac{4\Gamma_S\Gamma_D}{(E - E_0(V_A))^2 + (\Gamma_S + \Gamma_D)^2}. \quad (9.43)$$

Here, $T(E, V_A)$ represents the probability that an electron at energy, E, and applied voltage, V_A, will tunnel across the molecular junction. The alignment of the Fermi energy with the molecular level is given by the first term in the denominator, and the second term accounts for energy level broadening. Integrating this function over all energies gives the current–voltage characteristics (Haug and Jauho, 2008):

$$I(V_A) = \frac{q}{h}\int T(E, V_A)\left[f_S(E) - f_D(E)\right]dE, \quad (9.44)$$

where $f_S(E)$ and $f_D(E)$ are the Fermi–Dirac functions at the source and drain, respectively. Then, the current can be solved, to obtain

$$I = \frac{4G_0}{q}\frac{\Gamma_S\Gamma_D}{\Gamma_S + \Gamma_D} \times \left(\tan^{-1}\left(\frac{E_0 + \frac{1}{2}qV_A}{\Gamma_S + \Gamma_D}\right) - \tan^{-1}\left(\frac{E_0 - \frac{1}{2}qV_A}{\Gamma_S + \Gamma_D}\right)\right). \quad (9.45)$$

The current–voltage characteristics following Eq. 9.45 are plotted for several different molecular energies, E_0, and for a fixed $\Gamma_S = \Gamma_D = 10$ meV in Fig. 9.93a. As $E_0 \to 0$, the voltage of the current step decreases. That is, the molecular energy comes into resonance with the metal Fermi energy at lower voltages (assumed to be the same for S and D for symmetric junctions), thereby lifting the Coulomb blockade. The energy dependence of the blockade is found in the plot of the transmission function, $T(E,\ V_A)$, Eq. 9.43, in the inset of Fig. 9.93a. The peak transmission occurs at E_0, and has a Lorentzian shape. The width of the transmission function increases with $\Gamma_{S,D}$ as shown in the inset in Fig. 9.93b. Indeed, as the energy levels are broadened by approximately $\Gamma_S + \Gamma_D$, the current step is smeared out over a larger range of voltage, changing from an abrupt, step-like function for $\Gamma_{S,D} = 1$ meV, to a smooth "S" shaped curve for $\Gamma_{S,D} = 20$ meV at $E_0 = -0.2$ eV.

Finally, the *strong coupling* regime is accessed when $\Gamma, k_BT > E_C, \Delta E$. In this limit, the energy levels of the organic are strongly influenced by those of the metal. This results from a pronounced hybridization of the energy levels due, for example, to covalent bonding that significantly broadens the molecular energy levels. The resulting transport is coherent, and the blockade is eliminated since the molecular conductor loses its distinctive character that is now fully shared with the contacts.

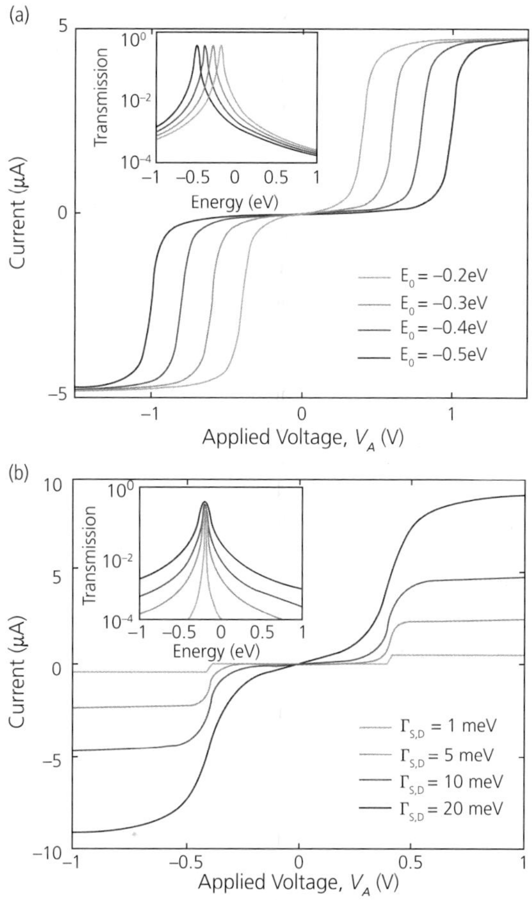

Figure 9.93 Current–voltage characteristics calculated for a single molecule spanning a junction between symmetric S and D electrodes as a function of (a) molecular energy level, E_0 for $\Gamma_{S,D} = 10$ meV, and (b) electrode coupling constant, $\Gamma_{S,D}$ for $E_0 = -0.2$ eV. Insets: Transmission vs. E_0 for the corresponding I–V characteristics (Perrin et al., 2015).

Thus far we have been concerned with two terminal devices. A three terminal molecular transistor can also be constructed, whereby the potential of the molecule bridging the gap between S and D is independently varied by an electrostatically coupled gate. The molecule may be suspended over the gate contact, the gate may be positioned laterally relative to the molecule, or the molecule can lie on the surface of the dielectric gate insulator with the gate electrode located on its opposite surface. In any case, the gate potential is used to change the energy levels of the molecule within the gap relative to the metal Fermi energy. The channel current thus provides quantitative information about the molecular energy levels. A negative gate potential repels the energy levels, leading to oxidation. Conversely, a positive potential lowers the energy, with the effect of reducing the molecule. The molecular transistor, therefore, is a spectroscopic tool for revealing the density and energetic distribution of states of the molecule.

The interdependence of the source-drain voltage, and gate voltage leads to the diamond-shaped *stability diagram* in Fig. 9.94a. The high current regions (white in the diagram) correspond to bias conditions leading to resonant *single electron tunneling* (SET). Current in this region is supported under bias conditions that lead to alignment between the chemical potential of a molecular level and the electrodes. In the absence of alignment, the molecule forms a conductance blockade. The current is significantly decreased, forming a pattern of *Coulomb diamonds* (black regions). These regions are characterized by a suppression of tunneling, leading to a stable charge state of N electrons. The electron addition energy, ΔE_{add}, is required to overcome the charge blockade. Additional states corresponding to $N \pm i$ electrons contact each other at the charge degeneracy points at $V_{GS} = \pm 1$ V. Thus, the number of Coulomb diamonds is equal to the number of charge states accessible to the molecule.

Additional features become apparent when the differential conductance, $G = dI/dV$, is mapped as in Fig. 9.94b. The edges of the regions are readily visible as high conductance boundaries to the diamonds. Lines running parallel to the diamonds are due to tunneling through additional excited states in the molecule. Such paths are enabled by higher electronic states, or molecular vibronics. Indeed, the existence of such features provide a spectroscopic tool for revealing the detailed energy landscape of the molecule forming the transistor channel.

The horizontal line of intermediate conductance connecting the diamonds is due to second order isoenergetic *co-tunneling*, illustrated in Fig. 9.95. The initial, charged molecular state is shown in Fig. 9.95a. In the elastic process, the misalignment of the initial state with the chemical potential of the drain electrode (μ_D) indicates that tunneling cannot occur between these states. However, it can occupy a virtual state under the condition that it does not violate the Heisenberg uncertainty principle, where its lifetime is shorter than $\Delta_\mathcal{T} = \Delta E/\hbar$. Here, ΔE is the energy between the initial and the virtual state (dashed line, Fig. 9.95b). In elastic co-tunneling, the charge transfers back onto the molecule at $t < \Delta_\mathcal{T}$, returning it to its original state. By this sequence, an electron is transported from the S to D, as shown by the final state in Fig. 9.95c. Inelastic co-tunneling is identical except that the molecule does not return to its initial

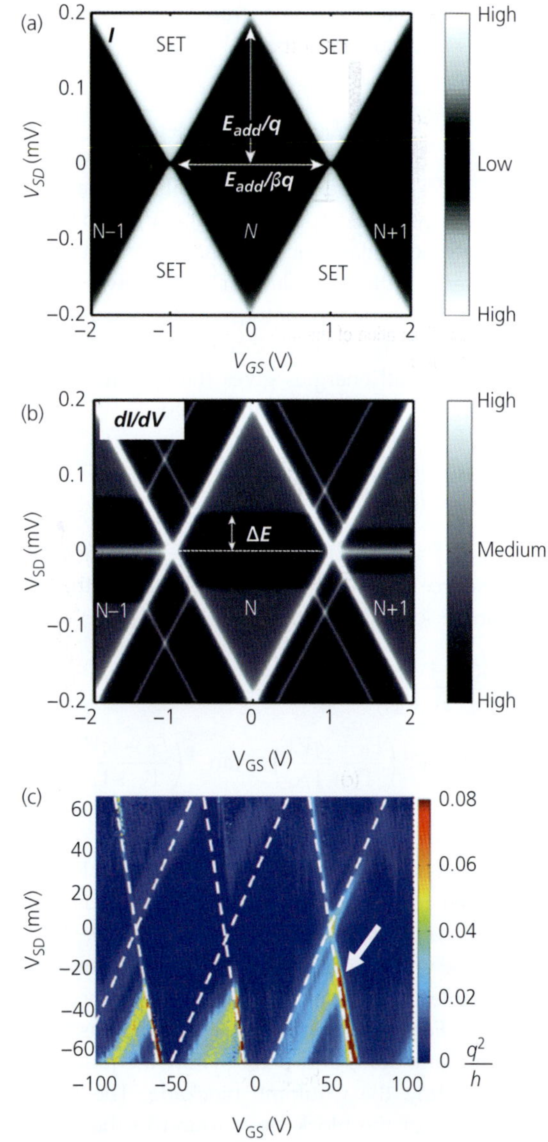

Figure 9.94 Example stability diagrams of V_{DS} vs. V_{GS} for single molecule transistors. (a) Drain current map indicating the energy required to add an electron (E_{add}) to the molecule. The high current single electron tunneling (SET) regions are separated from the low current non-resonant regions (black Coulomb diamonds) at the charge degeneracy points. Also, β is the gate efficiency parameter, which is the ratio of the energy change at the molecule to changes in gate potential. (b) As in (a) except the differential conductance is plotted to highlight additional features (Perrin et al., 2015). (c) Measured stability diagram of a single C_{60} molecule transistor. The white arrow at $V_{GS} = 50$ V indicates the 33 meV intracage vibrational mode of the fullerene (Scott and Natelson, 2010).

state, but rather it occupies a higher vibrational or electronic manifold. Since the inelastic final state is higher than the initial state, it requires an applied voltage $V_{GS} > \Delta E/q$. This gate voltage dependence results in the horizontal lines in the differential stability diagram in Fig. 9.94b.

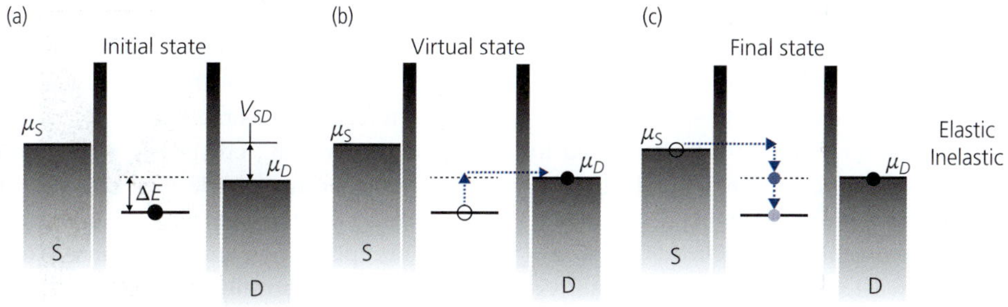

Figure 9.95 Illustration of the molecule in its (a) initial, (b) intermediate or virtual, and (c) final states for elastic and inelastic co-tunneling in a single molecule transistor.

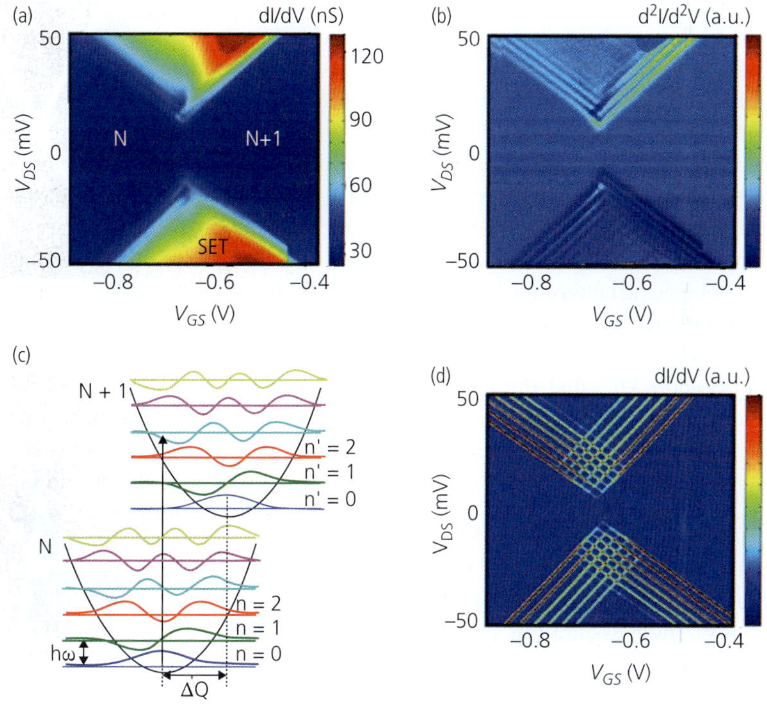

Figure 9.96 (a) Low temperature (1.8 K) differential conductance stability diagram of a Fe_4SMM molecular transistor. (b) Second derivative of the data in (a) showing by the parallel lines along the edges of the Coulomb diamond corresponding to multiple excited state transitions. (c) Configuration space diagram of the molecule in the *N* and *N* + 1 states showing the large Frank–Condon shift due to addition of an electron. ΔQ is the displacement of nuclear coordinates. (d) Calculated stability diagram that includes multiple electron–vibronic states for an electron–vibron coupling of $\lambda' = 2.21$ meV, and a vibronic energy quantum of $\Delta\varepsilon = 2.6$ meV (Burzurí et al., 2014).

An experimental stability diagram for a C_{60} device is shown in Fig. 9.94c. This diagram exhibits some of the features of the calculated plots, in that there are three charge states (three diamonds) accessed at different V_{GS}. Also, the secondary parallel diagonal line at $V_{GS} = 50$ V shows current via the higher 33 meV intracage vibronic of C_{60} (Park et al., 2000).

The high spectral resolution afforded by single molecule transistors is amply illustrated by the stability diagrams for the small molecule magnet, Fe_4SMM, in Fig. 9.96. This molecule has an approximate length of 1.9 nm, with the structure of $Fe_4L_2(dpm)_6$ (L is a tripodal ligand, $-C(CH_2OH)_3$, and dpm = dipivaloylmethane). The total spin of the molecule in its ground state is $S = 5$ (Accorsi et al., 2006). The device comprises a single molecule that bridges Au S and D electrodes, and is suspended over a third, gate electrode in a planar configuration. A notable feature of the stability diagram in Fig. 9.96a is that the SET regions do not share a charge degeneracy point. This is made more obvious by the second derivative of the current in Fig. 9.96b. Several equally spaced features parallel to the Coulomb diamonds are attributed to vibronic substates with resonant tunneling levels that are accessed as the drain and gate voltages are swept through 0 V. Low temperature (1.8 K) measurement is required to avoid washing out of the individual vibronic features (Burzurí et al., 2014).

The charge degeneracy point cannot be restored at any value of gate or drain potential. It is proposed that this results from a "Franck–Condon blockade," understood from the configuration diagram in Fig. 9.96c. The bulky Fe_4SMM undergoes a significant Franck–Condon shift from the N to the $N+1$ state due to strong electron–vibron coupling of λ'. The wavefunction overlap between the two states renders the lowest vibronics ($n = 0$, 1, etc.) inaccessible to direct transitions, thus forbidding resonant tunneling. This energy region is "vibronically blocked," thereby eliminating the degeneracy point and shifting the threshold gate voltage to ± 7.4 meV. The energy level spacing between the parallel current lines gives a vibronic level separation of $\Delta\varepsilon = \hbar\omega = 2.6$ meV.

We have shown in Section 3.6.1 that the Franck–Condon factor responsible for the configurational shift in Fig. 9.96c is:

$$FC_{n,0} = \frac{\lambda'^{2n}}{n!}\exp\left(-\lambda'^2\right). \tag{9.46}$$

The stability diagram in Fig. 9.89d is calculated based on DFT using the experimental values for $\Delta\varepsilon$, and assuming $\Gamma_{S,D} = 1$ meV. These assumptions yield a best fit to the data in Fig. 9.96b of $\lambda' = 2.21$ meV.

9.6.1 Fabricating and characterizing single molecule devices

Numerous methods have been devised to create suitably small gaps, and then to fill the gaps with only a single molecule. This is a challenging process; one that must clearly distinguish the conduction current from direct tunneling between electrodes (avoiding the molecule completely), or via residues or other impurities between the closely spaced S and D contacts. A practical solution must avail itself to statistical analysis of a large number of junctions since the gap may be empty (but still conductive, as mentioned), or it may be filled with more than a single molecule.

The most common electrode material is Au due to its lack of reactivity and its ductility. One preferred method of forming a gap is by slowly withdrawing a Au-coated scanning tunneling microscope (STM) tip from a Au-metallized surface in a solution containing a suspension of the target molecules. As the tip is withdrawn, at some point the continuous metal contact is broken, creating a free-standing gap. A molecule, or many molecules, can then attach from the solution to bridge the gap. Its presence is detected by an abrupt increase in conductance. The process is illustrated schematically in Fig. 9.97a. The target molecule (**X**) is functionalized to anchor itself to the opposing Au surfaces. Conductance data obtained by separating the Au tip in vacuum are shown in Fig. 9.97b. As the distance of the tip is increased to ~1 nm, the gap opens and the conductance vanishes. Repeated separations of the tip from the substrate results in the same conductance decrease, providing statistical assurance that the tip can be reproducibly opened by a comparable increase in tip distance. If the same procedure is performed in a 1 mM solution of 1,4-diaminobenzene molecules in a 1,2,4-trichlorobenzene solution, the conductance falls to a plateau at $10^{-2}G_0$ for a tip separation of approximately one molecular length. As the tip is separated further, there is no change in conductance; this is known as the *molecular lock-in state*, where the molecule spans the gap even though there are small increases in gap length. Finally, the gap is made so large that the molecule breaks its connection to one or both electrodes, and the conductance falls to that of an empty gap. The fact that the conductance plateau reached on each separation event is approximately the same length and at the same conductance indicates that only a single molecule fills the gap.

An interesting application of this point-contact gap formation technique is to measure the conductance of conjugated molecules with varying twist angles between two anchoring phenyl groups. The twist angle is varied by the choice of substituents on the neighboring phenyls: the larger twist angles result in interrupting the conjugation of the electronic systems across the molecule. Several representative molecules used are shown in Fig. 9.98a. Molecule **2** connects the phenyls with a fluorene moiety, thereby planarizing the structure to create a continuous conjugated path cross the molecular backbone. Attaching various groups induces a twist along the molecule via their steric hindrance. For example, methyl groups in compound **8** induce a $\theta = 90°$ twist. Each of the molecules is terminated with an amine that anchors to the Au electrodes.

Each of the molecular species is suspended in 1,2,4-trichlorobenzene solution during the formation of the gap. Conductance histograms taken at $V_{SD} = 20$ mV

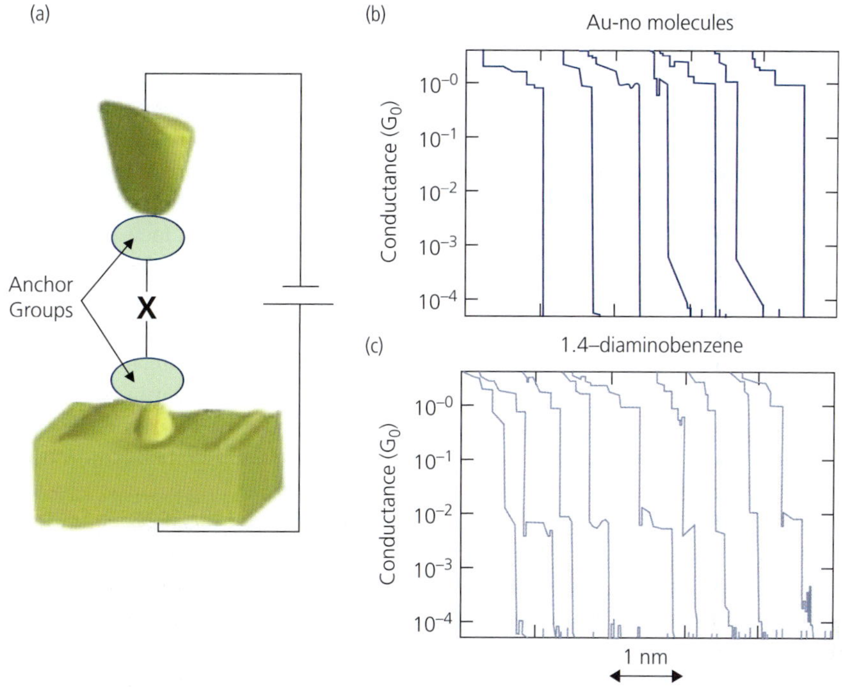

Figure 9.97 (a) Illustration of a STM-tip junction bridged by a molecule (**X**) bonded to both Au surfaces via functionalized anchor groups. (b) Data from a series of gaps formed by repetitively separating the tip from the Au film surface in the absence of a target molecule, **X**. Each line in the series shows the conductance (in units of G_0) as an approximately 1 nm wide gap is opened. (c) As in (b) with the tip immersed in a solution containing 1,4-diaminobenzene. The step at $10^{-2}G_0$ is due to filling of the gap with a single organic molecule (Venkataraman et al., 2006).

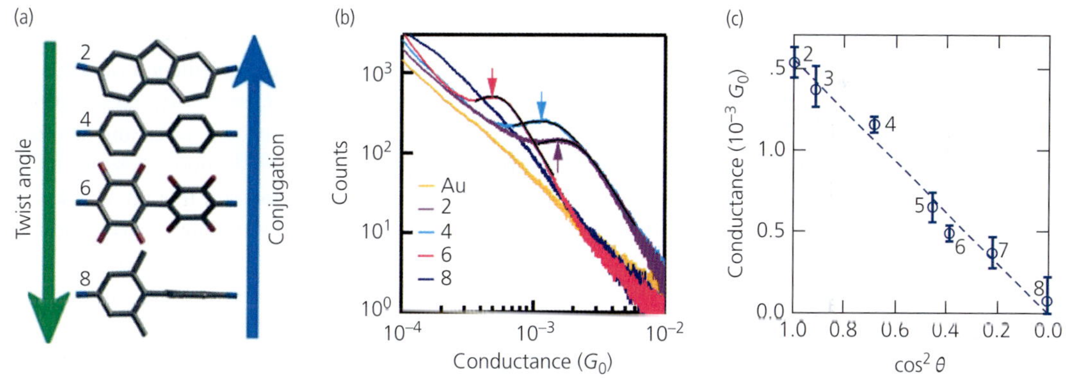

Figure 9.98 (a) Representative structures of molecules with increasing twist angles from compound **2** (angle $\theta = 0°$) to **8** ($\theta = 90°$). (b) Conductance histograms for four different compounds in (a), including that of the Au contacts alone. Arrows point to the conductance plateaus characteristic of the molecule spanning the gap between the STM tip and the substrate. (c) Conductance vs. $\cos^2\theta$ for compounds with various twist angles (Venkataraman et al., 2006).

comprising >10,000 contact breakages followed by loading the gap with the molecules are shown in Fig. 9.98b. The arrows locate the peak conductances that are obtained by fitting to a Lorentzian line shape according to Eq. 9.43. The peak widths are broadened by the distribution in conductances measured for the several devices, as well as to broadening by orbital hybridization with the contacts. A plot of conductance vs. twist angle is shown in Fig. 9.98c for seven of the nine molecular species investigated. The straight line fit follows the theoretical expectation that the conductance is a function of orbital overlap across the molecule, which attenuates exponentially with the length of the molecule (Mujica et al., 1999), and follows an approximately cos $^2\theta$ dependence on twist angle between conjugating groups (Venkataraman et al., 2006).

Another common method for gap formation is the *break junction*. This comprises a thin metal wire that is broken by applying mechanical stress (Tsutsui et al., 2008, Perrin et al., 2013, Xiang et al., 2013), or by a current that induces Joule heating (and hence melting or burning) (Prins et al., 2011) or electromigration (Strachan et al., 2005, O'Neill et al., 2007, Prins et al., 2009). *Electromigration* is the process whereby current plus the applied electric field displaces metal ions in the solid via collisions by the electrons carrying the current. Generally, electromigration is a parasitic effect in semiconductor devices, whereby a high current causes growth of metal filaments that eventually short circuit the junction. In single molecule devices, it is used to the opposite effect. That is, a constricted metal region along a wire spanning the gap between S and D is thinned by current, and eventually breaks.

An example process for forming a molecular transistor via electromigration is shown in Fig. 9.99a. Step 1: Large contact pads are deposited onto the surface of the SiO_2 gate dielectric grown on a Si substrate. Step 2: A thin film of Al is deposited in a strip, and its surface is oxidized. The Al is the gate contact coated by the thin Al_2O_3 gate dielectric. Step 3: The thin (~10 nm) constricted Au bridge is deposited, followed by nanopatterning using e-beam lithography or other suitably high resolution method. Step 4: The transistor is completed by depositing and patterning Au connecting pads to the bridge. A micrograph of the completed device (before gap formation) is shown in Fig. 9.99b. The planar structure puts the molecule in direct contact with the gate oxide. Alternatively, the metal bridge can be suspended over an open gap, using air or vacuum as the dielectric, or the gate potential can be applied from the side using a separate, narrow gate electrode positioned close to the gap (Xiang et al., 2013). For broken gap structures, the molecules are introduced by immersing the contacts for several hours in a solvent containing the target molecular species, and then rinsing the sample to remove excess material once the gap is filled.

The controlled nanogap formation process proceeds by the methodical, stepwise application of a voltage ramp. Similar procedures are used in forming gaps by electromigration, Joule heating, or electroburning. An example current-voltage characteristic of a gap formed in graphene is shown in Fig. 9.100. The device consists of a few-layer mechanically exfoliated graphene flake placed on a degenerately doped Si gate/280 nm thick SiO_2 gate insulator. Contact to the graphene is made by depositing and patterning Cr/Au electrodes. To form the nanogap, the voltage between the electrodes is ramped at 1 V/s while measuring the drain current (red line). The sweep is halted once the drain current drops to 10% of its peak value, after which V_{DS} is returned to zero. The voltage is then swept repeatedly until the drain current no longer increases with V_{DS} (black lines). The decrease in I_D with each successive sweep is due to electroburning, that is, the graphene undergoes Joule heating that causes it to react with oxygen. On each sweep, the graphene is further eroded and can support less current, leading to additional loss of material. Since only a small amount of material is lost on each cycle, the process can be immediately halted following

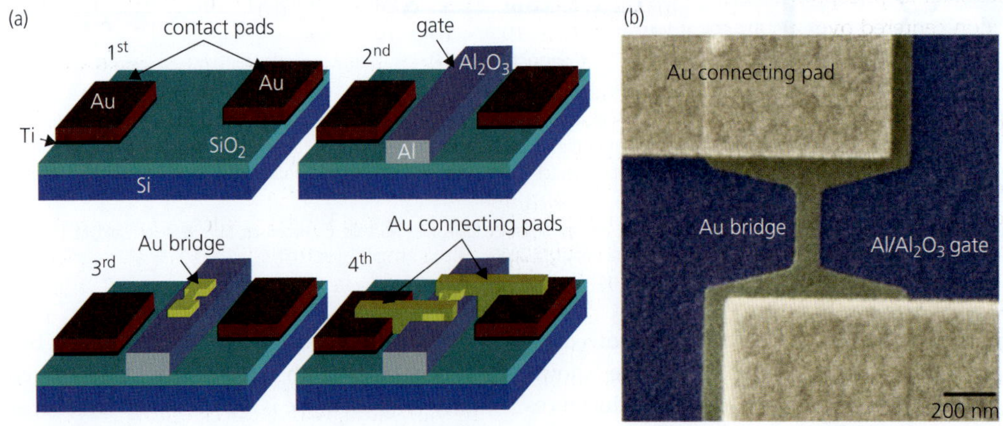

Figure 9.99 (a) Process sequence for fabricating a single molecule transistor. (b) Helium ion microscope image of the transistor prior to forming the gap by electromigration (Perrin et al., 2015).

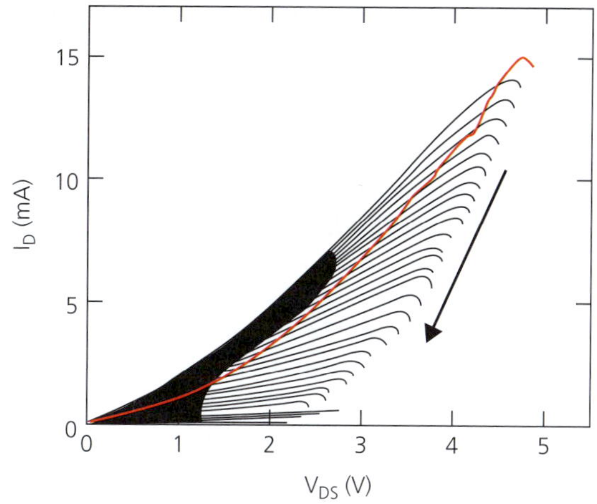

Figure 9.100 Electroburning formation of a few-layer graphene S and D contact. The first sweep is shown in red. Successive voltage sweeps follow the direction of the arrow until the gap is completely opened, and $I_D \rightarrow 0$ (Prins et al., 2011).

the opening of only the smallest gap, thereby controllably forming gaps of the required size. Note that the initial sweep current (red line) has a different, nonlinear dependence on voltage than subsequent sweeps. This is attributed to volatilization of surface impurities during the first application of thermal stress to the graphene. Analogous procedures have been employed in forming Au junctions by electromigration (Perrin et al., 2015) and in single walled nanotube electrodes (Marquardt et al., 2010).

The mechanically controlled break junction (MCBJ) physically fractures a constricted region of the metal bridge, with the dimensions of the resulting gap determined by the stress applied. The MCBJ formation process is shown in Fig. 9.101a. The device (a simple junction or transistor) is fabricated on a flexible metal substrate such as phosphor-bronze, with the contact constriction centered over a stationary support. Pressure is applied via piezoelectric actuators to translate the edges downward (or in some configurations, in the opposite, upward direction via translation of the center support), creating a maximum stress at the constriction. Pressure is increased until the substrate is sufficiently flexed to break the contact. The magnitude of the translation of the center to the edges of the device determines the gap opening. This method can produce sub-Ångstrom control of the gap length (Tsutsui et al., 2008). A profile view of a Au MCBJ suspended over a preformed cavity on a polyimide substrate is shown in Fig. 9.101b.

Data for a MCBJ with and without a ZnTPPdT porphyrin bridging two Au electrodes are shown in Fig. 9.102a and b, respectively. The ZnTPPdT is

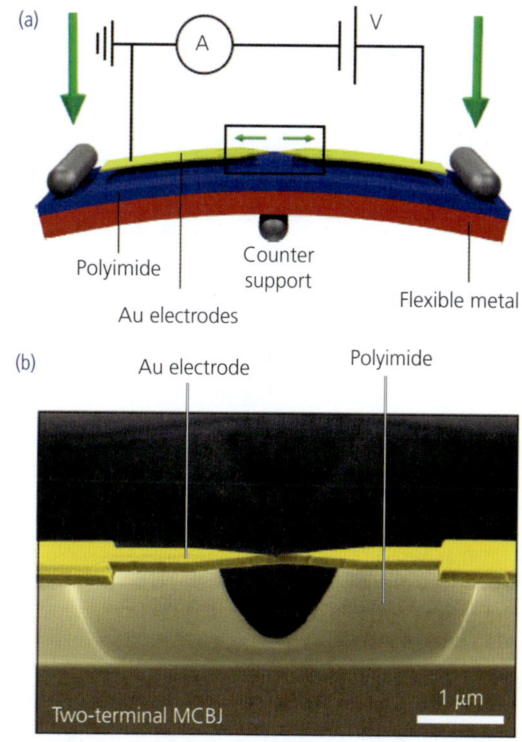

Figure 9.101 (a) Experimental configuration for forming a mechanically controlled break junction (MCBJ). (b) False color SEM image of a MCBJ (Perrin et al., 2013).

anchored to the electrodes via thiolphenyl terminating groups. The empty device shows that at the widest, initial displacement, d_0, the current is nearly zero, independent of V_{DS}, and it increases with decreasing gap width, $d_0 - \delta$. The small changes of δ, from 0.02 to 0.11 nm, indicate the exquisite precision with which the nanogap can be controlled. As the gap is narrowed, the current at high voltage increases exponentially, characteristic of tunneling between electrodes.

Filling the gap with ZnTPPdT results in the appearance of steps in the I–V characteristics in Fig. 9.102b. At the largest separation ($\delta = 0$), the current around $V_{DS} = 0$ is very small, indicative of weak coupling between contacts and molecule. In this regime, current is limited by the Coulomb blockade that allows only a single electron to reside on the molecule at one time. At $V_{DS} = \pm 0.6$ V, the blockade is lifted and the current is increased by sequential electron tunneling. Interestingly, the voltage corresponding to the onset of conduction monotonically decreases with decreasing gap width, with a "mechanical gate efficiency" of ~1 V/nm. As the current onset voltage decreases, the device shifts from the weak, to the intermediate coupled regime.

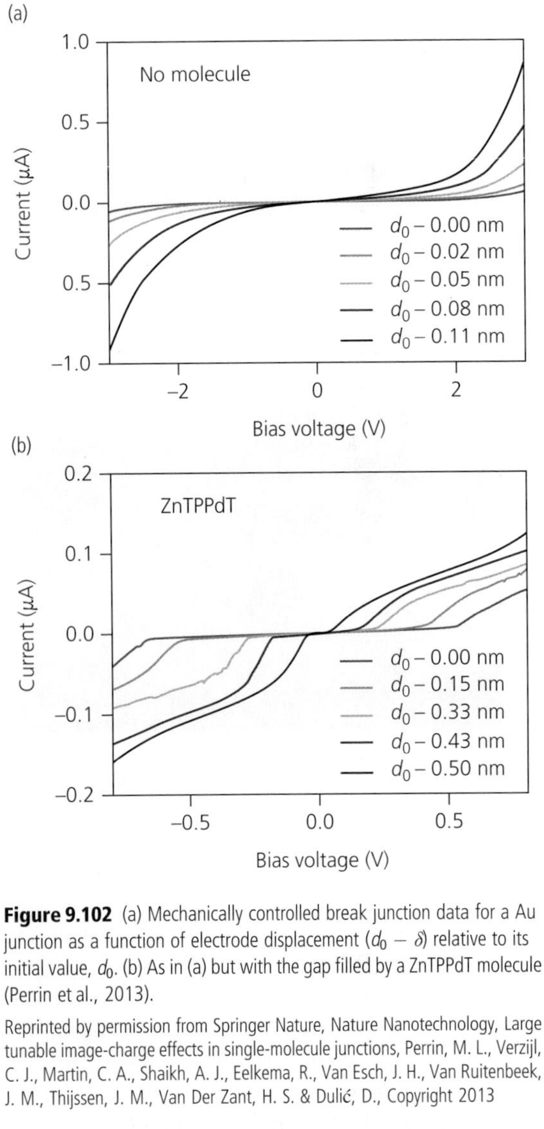

Figure 9.102 (a) Mechanically controlled break junction data for a Au junction as a function of electrode displacement ($d_0 - \delta$) relative to its initial value, d_0. (b) As in (a) but with the gap filled by a ZnTPPdT molecule (Perrin et al., 2013).

The dependence of the current onset vs. gap width is due to image charge interactions between molecular orbitals and those on the metal electrodes. The Zn porphyrin structure attached to two Au atoms on either side due to thiolation reactions is depicted in Fig. 9.103a. The calculated electron distribution for the HOMO is given in Fig. 9.103b, for the molecule within the gap in Fig. 9.103c, and for the LUMO, the distribution is shown in Fig. 9.103d. Notable features are that electrons on the HOMO are distributed across the molecule. In contrast, the LUMO distribution is to some degree, protected from exposure to the Au atoms by the thiol linkages since the electron distribution is localized on the porphyrin core. Within the gap, however, the electrons are attracted toward the anchoring groups due to image forces at the large Au electrode metal surface. For the calculation, it is assumed that the electrodes are infinite metal planes positioned approximately 1 Å from the first atomic layer of the molecule. As the distance between the molecule and the electrodes is reduced (i.e. increasing δ), the image charge attraction increases, further hybridizing the Au and molecular orbitals along the anchoring thiol linkages. The net effect is to increase the HOMO–LUMO gap, as shown in Fig. 9.103e. The HOMO level shift is considerably larger than the LUMO due to the aforementioned greater exposure of the electron distribution in the HOMO orbital. Indeed, the LUMO states are positioned well above the Fermi energy, whereas the gap and HOMO levels are close to E_F, indicating greater participation in charge conduction between the contacts. The calculated shifts are approximately double that observed in the experiment. This is due to the assumption of an infinite metal plane in the simulations, which is expected to have a larger image force than the actual, sharply pointed electrodes formed at the break junction. Note that there are additional energetic consequences arising from molecular deformation as it is squeezed, or elongated within the electrode gap. However, DFT calculations show that these effects are far smaller (50–60 meV) than those due to image charges, and in fact only results in a rigid shift of the HOMO and LUMO energies without affecting the energy gap between them (Perrin et al., 2013).

From the foregoing, we see that anchoring of the molecule to the electrodes determines the degree to which the molecular and metal levels are hybridized, and hence whether the device itself is operating in the weak, intermediate or strong coupling regimes (Chen et al., 2006, Martín et al., 2010, Briechle et al., 2012). This assumption is tested using asymmetrically functionalized triphenylenethynylene (TPE) in **Scheme 3**. This molecule has two different anchoring moieties to a Au junction. The S-termination forms a covalent bond, whereas the phenyl moiety forms a weak, facial π-bond with the opposing electrode. Consequently, the coupling strength to the electrodes is highly asymmetric. The behavior of this compound (compound (i), labeled Au|S-TPE|

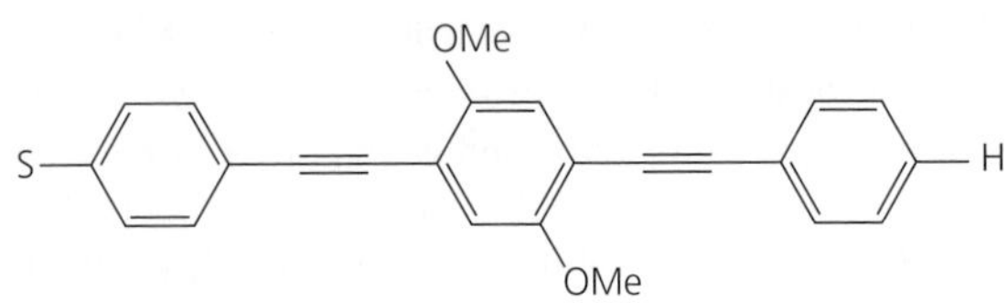

Scheme 3 Asymmetrically terminated oligophenylenethynylene.

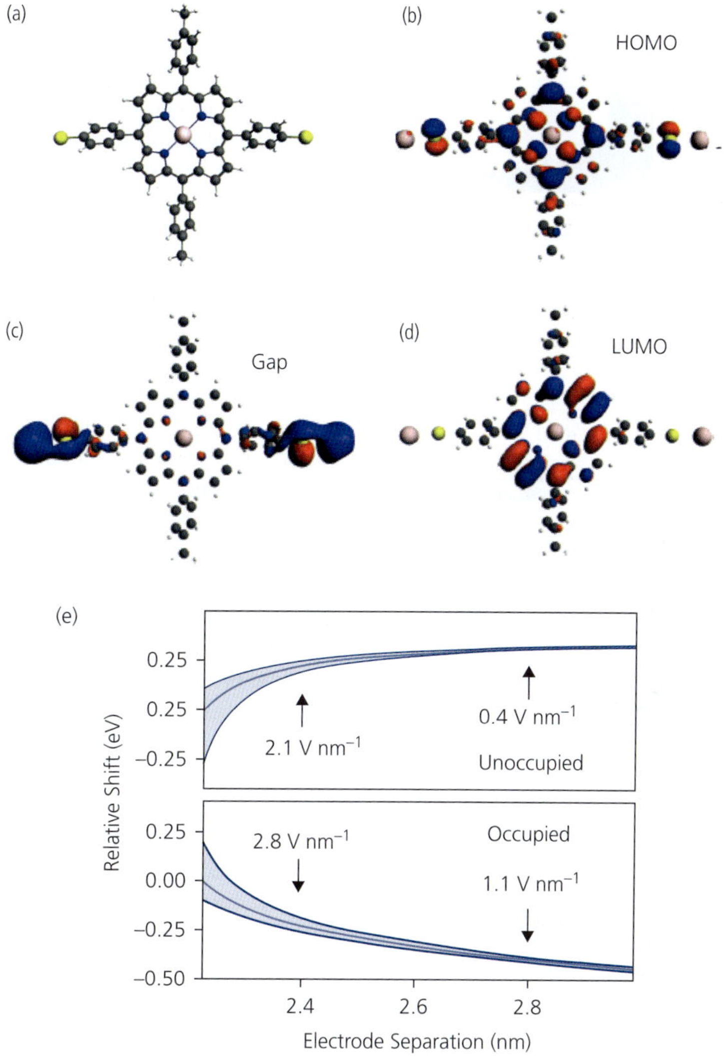

Figure 9.103 (a) Molecular structure of ZnTPPdT showing anchoring to Au atoms (yellow) via thiolation with the molecular end groups. Calculated (b) HOMO, (c) in-gap, and (d) LUMO electron distributions. Red and blue regions correspond to different electron wavefunction phases. (e) Calculated energy level shifts of the occupied and empty orbitals of ZnTPPdT vs. electrode separation relative to the initial separation. Uncertainties are shown by the shaded bands. Adapted from Perrin et al. (2013).

Au) is compared in Fig. 9.104 to that of the symmetric Au|S-TPE-S|Au molecule (compound (ii)) in a gap formed between an STM tip and a flat Au surface. Conductance histograms of the two compounds obtained by measuring *I* vs. separation distance, *s*, between the tip and planar contacts of approximately 450 junctions are shown in Fig. 9.104a. Typically, it takes ~1500 measurement attempts to achieve a data set for 400–500 devices. The measurement failures are due to obtaining the conductance from direct tunneling between contacts across an empty gap, multiple molecules filling the gap, and noisy data from poor contacts to a molecule, or to mechanical vibrations.

Several features appear for the TPE-based devices, depending on whether the current set-point of the tip is low (5 nA, left hand histograms) or high (60 nA). For example, at a low tip current, both molecules exhibit a low conductance ($5 \times 10^{-6}G_0$) peak **π-1**, whereas only (i) shows peak **π-2**. Also, a third peak, **A**, is apparent at a higher conductance ($2 \times 10^{-5}G_0$) for only compound (ii). At higher tip currents, the measured conductance is ten times that at lower currents, possibly indicating multiple electron

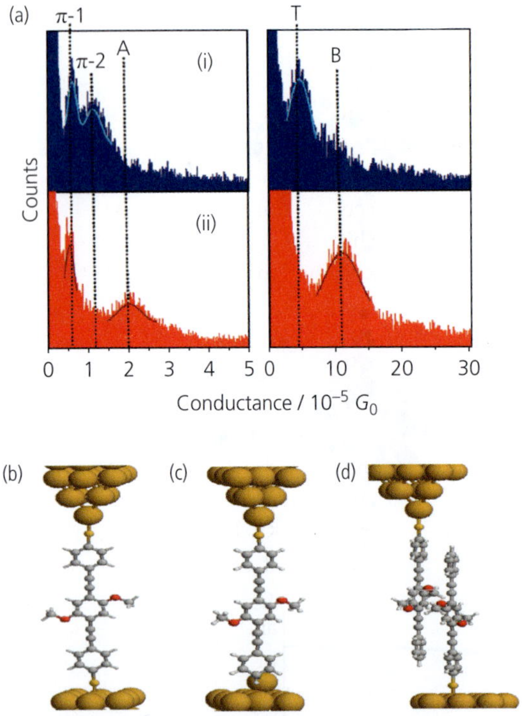

Figure 9.104 (a) Conductance histograms for compound (i) Au|S-TPE-S|Au. and (ii) Au|S-TPE|Au between Au contacts. Panels on the left correspond to low STM tip currents of 5 nA, and on the right for 60 nA, with a tip voltage of 0.6 V in both cases. Configuration between Au contacts of (b) Au|S-TPE-S|Au, (c) Au|S-TPE|Au, and (d) Au|S-TPE-π-TPE-S|Au (Martín et al., 2010).

Reprinted with permission from Martín, S., Grace, I., Bryce, M. R., Wang, C., Jitchati, R., Batsanov, A. S., Higgins, S. J., Lambert, C. J. & Nichols, R. J. 2010. Identifying diversity in nanoscale electrical break junctions. Journal of the American Chemical Society, 132, 9157-9164. Copyright 2010 American Chemical Society

Table 9.6 Characteristics and measured properties of compounds (i) and (ii) used in single molecule devices (Martín et al., 2010)

Molecule	Peak	G ($\times 10^{-5}G_0$)	s_{BO} (nm)	Mol. length (Au...Au) (nm)
i	π-1	0.61	2.92	2.84
	π-2	1.19	2.65	-
	T	4.86	2.10	2.19
ii	π-1	0.57	2.82	2.82
	A	2.02	2.26	2.51 .
	B	10.7	2.02	2.27

conduction. There, peak **T** is found on the asymmetric compound (i), whereas compound (ii) exhibits peak **B** at double the conductance of peak **T**. A summary of these results, along with a measurement of their junction breakoff distances, s_{BO}, that corresponds to the tip separation where the conductance, $G \to 0$, is provided in Table 9.6.

The data in Fig. 9.104a and in the table reveal that the peaks arise from very different bonding properties of the covalently attached thiol groups of compound (ii), the asymmetric attachment including a facial π-bonding of compound (i), and a third species of two laterally π-stacked TPE molecules in a configuration labeled Au | S-TPE-π-TPE-S | Au, shown in Figs. 9.104b-d. A clear picture begins to emerge from the relationship between the breakoff distance and the conductance of the various peaks. The calculated length of (i) is 2.19 nm, which is equal to s_{BO} for peak **T**. Similarly, the molecular length of (ii) is 2.5–2.3 nm, which is comparable to the **A** and **B** breakoff distances of 2.2.6 and 2.02 nm, respectively (Table 9.6). Hence, features **A**, **B**, and **T** are associated with a single molecule of either type spanning the gap between the STM tip and the Au base. Note, however, that the higher conductance of **B** compared to **T** at a high tip current suggests that the stronger anchoring of the two thiol groups of (ii) provide improved coupling compared to that afforded by the weak electrostatic facial coupling of the H-terminated phenyl group of (i). The conductances of these features along with **A** are all considerably higher than for **π-1** and **π-2** for either compound. (It is speculated that **π-2** may also be present in (ii) but is obscured in the shoulder of **π-1**.) Further, the breakoff distances of the **π-1** and **π-2** peaks are 2.65–2.92 nm, which are considerably longer than the corresponding molecular lengths. This suggests that the gap contains a pair of π-stacking molecules that are anchored at their ends by thiol attachment to the Au contacts. This configuration is illustrated in Fig. 9.104d. Conductivity is considerably reduced from that achieved by direct Au | S-TPE-S | Au coupling. The inference of a π-stacked bridging molecular pair is supported by the absence of similar conductance features and breakoff distances for molecules that replace the –OMe functionalizing group with *tert*-butyl; a bulky group that sterically inhibits π-stacking. The conclusions are further supported by DFT calculations of the electron transmission coefficients for the various configurations and anchoring groups (Martín et al., 2010).

Different bonding configurations have also been explored using *N*-alkane chains between Au electrodes symmetrically terminated with carboxylic acid (–COOH), amine (–NH_2), and thiol (–SH) groups. As noted above, thiols form strong covalent bonds with Au, whereas amine bonds are weaker. It is probable that carboxylic acid is bonded by a combination of ionic and coordinating interactions with Au, which forms the weakest coupling of the three terminations. Thus, varying the anchor groups can shift the coupling from strong to weak. This is directly observed by comparing the breakoff distances and conductivities for the same molecular core. Indeed, it is found that the conductance of N-alkane molecules follows a relationship characteristic of tunneling, viz. $G_N = A_N \exp(-\beta_N N)$. Tunneling transport is via the alkane linkage comprising N $-CH_2$ groups spanning the contact gap. The prefactor, A_N, depends on the strength of the bond between Au and the anchoring units. Thus, its magnitude follows the progression, A_N(Au-S) > A_N(Au-NH$_2$) > A_N(Au-COOH) (Chen et al., 2006).

9.6.2 Unique properties of single molecule devices

The conduction of single molecule devices is determined by the tunneling probability between electrodes via the molecular electronic states. Alignment of these electronic states with the contacts results in resonant tunneling that substantially increases conductivity. Quantum interference along multiple tunneling paths between electrodes will be present if the paths are electronically coupled. Consider N molecules, with the conductance of the ith molecule equal to G_i. Then, in a classical, non-interacting system, the total electronic transmission of the system is

$$T_N(E) = \sum_{i=1}^{N} T_i(E). \tag{9.47}$$

If all paths are equivalent, the transmission becomes, trivially, $T_N(E) = NT(E)$, where $T(E)$ is the transmission coefficient of the i_{th} molecule. Then, following Eq. 9.41, near $V = 0$:

$$G(E) = NT(E)G_0. \tag{9.48}$$

The case of quantum-coupled paths is considerably different, since the electronic wavefunction can experience constructive or destructive interference at the junctions, or nodes. These interference terms must be included in the total tunneling transparency, yielding (Magoga and Joachim, 1999)

$$T(E_F) = \sum_{i=1}^{N} T_i(E) + 2\sum_{\substack{i>1 \\ i\neq j}}^{N} e^{j\Delta\phi_{ij}} \sqrt{T_i(E_F)T_j(E_F)}. \tag{9.49}$$

Here, $\Delta\phi_{ij}$ is the relative phase of the wavefunctions at the nodes, which depends on energy as well as the interaction strengths between the several electronic levels of the molecules. In the ballistic transport regime, the output of such a "molecular interferometer" cannot be larger than the sum of the inputs, yet in the tunneling regime (corresponding to moderate to strong molecule/electrode hybridization), the constructive interference term can result in an output that is significantly greater than the sum of the inputs. Interference is expressed by the term on the far right of Eq. 9.49.

If the contributions from all levels are equal, and the levels are centered about the Fermi energy, then the phase shift is $\Delta\phi_{ij} = 0$. This leads to constructive interference, in which case the conductance is simplified to

$$G = \sum_{i=1}^{N} G_i + 2\sum_{\substack{i>1 \\ i\neq j}}^{N} \sqrt{G_i G_j}. \tag{9.50}$$

Thus, the total conductance is higher than the sum of the conductances of the individual molecular "arms." This violates the classical Kirchhoff's circuit law that states that the total current in a parallel network is equal to the sum of the currents in the branches. Observation of a conductance larger than the sum immediately to the right in Eq. 9.50 establishes the presence of coherent tunneling along multiple, parallel paths.

Closely packed self-assembled monolayers attached to a Au substrate on one end and a Au-coated AFM tip on the other provide a model system with which to observe electron coherence between parallel molecules. A set of molecules used for these studies is shown in **Scheme 4**. The anthraquinone structures of AQ-DT and AQ-MT result in cross-conjugation that is absent in AC-DT. Here, *cross-conjugation* exists

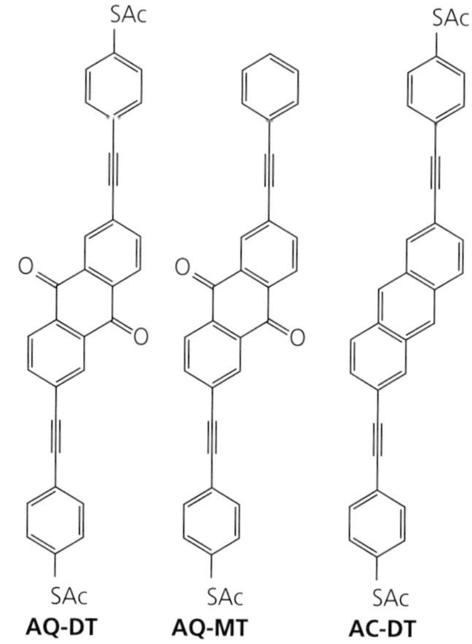

Scheme 4

in molecules where only two of three conjugated π-bonds interact. In contrast to linearly conjugated systems such as AC-DT, the π-bonds in AQ-DT and AQ-MT branch off, reducing the degree of conjugation. Thus, a comparison of the behavior of anthraquinone SAMs versus that of the fully conjugated AC can be used to quantify the degree of quantum interference. Note that both the dithiolated molecules AC-DT and AQ-DT are symmetrically and covalently bonded to the Au electrodes, whereas AQ-MT is monothiolated, with the opposing side of the molecule forming a weak electrostatic bond to Au by a non-hybridizing phenyl group.

Calculated transmission spectra of the three molecules are shown in Fig. 9.105a. The linearly conjugated AC-DT shows no particularly notable features. Indeed, its transmission smoothly varies across the energy gap, with peaks near the molecular orbitals due to resonant tunneling. This differs markedly from the cross-conjugated AQ compounds that have a pronounced dip at $E - E_F = 0.5$–0.6 eV resulting from destructive interference between the two possible electronic paths taken along the anthraquinone core. The calculated orbitals for AC-DT and AQ-DT are shown in Fig. 9.105b and d, respectively. In linearly conjugated AC-DT, the orbitals are symmetrically localized near their anchor groups, leading to a single transmission path illustrated by the energy level scheme in Fig. 9.105b. However, the cross-conjugated molecule has two different energy states symmetrically disposed about E_F. The high-energy path is via the core, and the lower energy route via direct tunneling between the anchors. These two paths undergo a π phase difference, and hence destructively interfere. The transmission calculated for these alternative paths is shown in Fig. 9.105c. Destructive and constructive interference results in dips and peaks in the transmission of the three-site process that are absent via tunneling through a single route comprising either the lower, direct tunneling process, or via only the upper level.

Since the conductance is $G = G_0 T$, the existence of interference should be reflected in the conductance spectra near $V = 0$ (Guédon et al., 2012). Figure 9.106a shows a group of conductance spectra vs. voltage taken for AC-DT using a Au-coated AFM tip break junction.

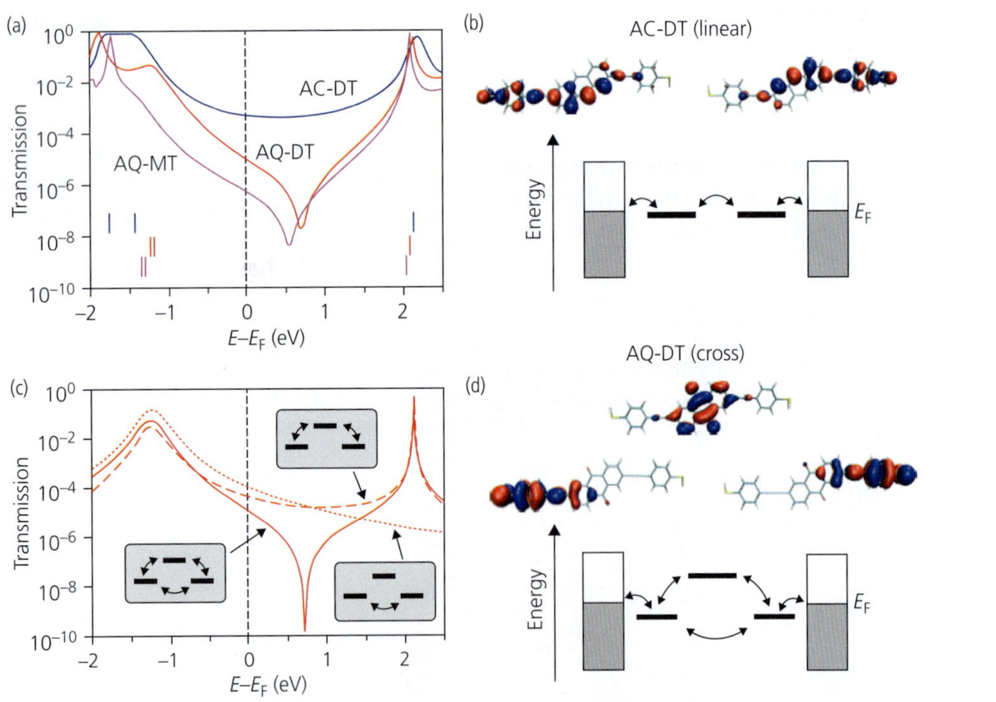

Figure 9.105 (a) Calculated transmission of the molecules in **Scheme 4**. The frontier HOMO and HOMO-1 orbital energies at $E < E_F$, and the LUMO energies at $E > E_F$ are shown by lines in the lower quadrants of the plot. (b) Calculated electronic orbitals in AC-DT and the proposed energy level scheme of the molecule anchored via its thiolating groups to the Au electrodes. (c) Electronic transmission via upper and lower electronic pathways for the cross-conjugated AQ-DT shown in (d). Interference between the upper and lower energy branches results in pronounced peaks in $T(E)$ (Guédon et al., 2012).

The conductance traces have symmetric bowing that is consistent with the smooth transmission vs. energy characteristic in Fig. 9.105a. Conductance histograms at $V = 0$V are compared in Fig. 9.106b. The conductance of the cross-conjugated AQ-DT is two orders of magnitude lower than for AC-DT, suggesting destructive interference between the upper and lower energy branches of the former molecule. The $G(V)$ data of AQ-DT are identical in shape to Fig. 9.106a, which is surprising given the strong anti-resonance in transmission predicted by the calculations. In contrast, the anti-resonance is clearly apparent by the pronounced dip in the conductance data in Fig. 9.106c for the asymmetrically anchored AQ-MT. Indeed, contrary to the constant curvature of the conductance of AQ-DT and AC-DT, the curvature is negative across the entire voltage domain $V = \pm 1$ V for AQ-MT. This is comparable to the width of the antiresonance predicted in Fig. 9.105a. The symmetry of the data, even for an antisymmetrically contacted molecule, indicates that the transmission coefficient is nevertheless symmetric about E_F, which itself lies near the interference minimum.

While the markedly lower conductance of the symmetric AQ-DT compared to AC-DT suggests that it, too, undergoes destructive interference, the question remains as to why there is no dip in conductance similar to that found for AQ-MT. It is speculated that this is due to a shift in the transmission dip to above E_F in the symmetric molecule, making it beyond the reach of the experimental voltage range of ± 1 V (Guédon et al., 2012).

A more direct test of quantum interference is achieved by comparing the conductance of otherwise similar, single and doubly connected moieties symmetrically contacted between an STM tip and a Au substrate. Such a pair of molecules is shown in **Scheme 5**. Both molecules are anchored between Au contacts via their thiol groups. Whereas MTMB offers only a single conductive path via its phenyl

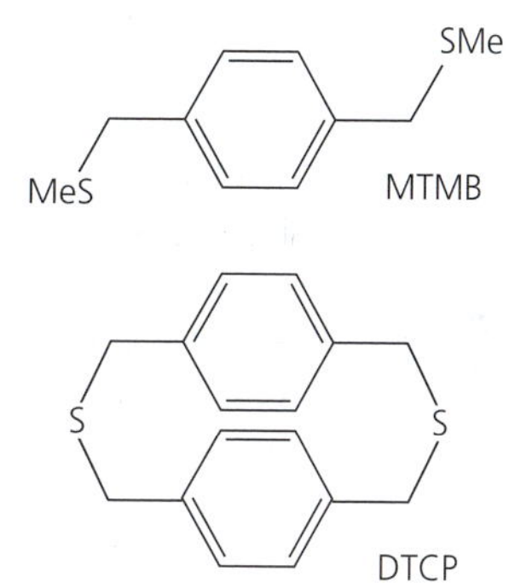

Scheme 5

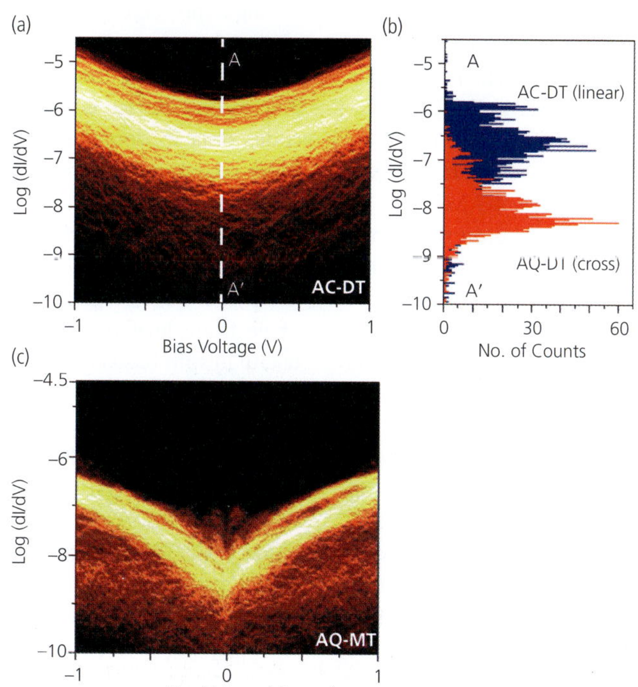

Figure 9.106 (a) Conductance vs. bias voltage for AC-DT. (b) Conductance histograms at $V = 0$ V for AC-DT (along the vertical dashed line in (a)) and AQ-DT. (c) Conductance vs. bias voltage for AQ-MT (Guédon et al., 2012).

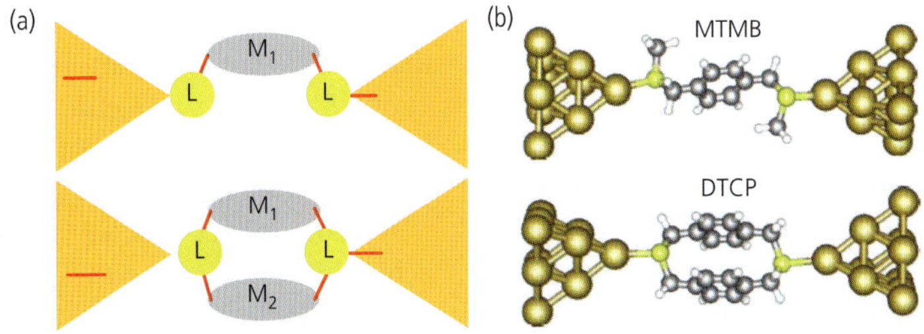

Figure 9.107 (a) Illustration of the single and dual transport paths with similar linking moieties (L) between electrodes. (b) A molecular structural diagram of the two molecules in **Scheme 5** spanning the gap between Au contacts (Vazquez et al., 2012).

group, DTCP links a pair of phenyls that split the orbitals, to result in a quantum analog of a double-branched, optical *Mach–Zehnder interferometer* (see Fig. 9.107a). Equation 9.50 suggests that if the conductances of the two branches are similar, that is, $G_M = G_{M1} = G_{M2}$, then constructive interference between them should lead to a total resonant conductance of $G = G_{M1} + G_{M2} + 2\sqrt{G_{M1}G_{M2}} = 4G_M$. This is double the conductance possible assuming Kirchhoff's law of parallel circuit branches. This idealized situation assumes that the linkers and anchor branches are equivalent, E_F is at midgap, the electrode couplings, Γ, are independent of energy, and the coupling in the single molecular backbone species to the contact gateway states is small. These conditions are not completely met by MTMB and DTCP in the configurations shown in Fig. 9.107b. For example, the thiolating group in the singly branched MTMB is terminated by a methyl (Me) group, which is similar but not identical to the methyl linker in DTCP.

Transmission spectra calculated for the two molecular species are shown in Fig. 9.108a. These idealized results show that $G(\text{DTCP}) = 4G(\text{MTMB})$, as predicted. Furthermore, the transmission resonance corresponding to the LUMO is twice as broad as for the single branched MTMB. The splitting between the bonding and antibonding orbitals of the cyclic DTCP is Δ, although only the bonding orbital is coupled to the gateway states, and thus it alone contributes to the junction conductance.

The constructive superposition of electronic states is inferred from the conductance histograms in Fig. 9.108b. The histograms each comprise >6000 measurements. They give the conductance as a function of tip/substrate separation, d, starting at a high conductance at $d = 0$ where the connection between contacts is broken. The conductance decreases until the molecules are fully extended at $d = 0.5$–0.55 nm, corresponding to molecular breakoff. The residual conductance beyond this separation is due to direct tunneling between the STM tip and the Au film substrate.

Histograms for the lowest conductance of both molecules near breakoff are replotted at the right in Fig. 9.108b. It is found that the minimum conductance for MTMB is $2.8 \times 10^{-4}G_0$, and for DTCP it is $9.7 \times 10^{-4}G_0$. Thus, the ratio of the conductance of the double branched DTCP to that of the single-path molecule, MTMB, is $G(\text{MTMB})/G(\text{DTCP}) = 2.8$. While this falls short of the theoretical ratio of 4, it is consistent with DFT calculations for this molecular system that yield a conductance ratio of 3.3. Differences may arise from non-idealities in molecular coupling to the contacts. This value is convincingly larger than predicted by Kirchhoff's law, providing evidence for quantum mechanical interference arising from tunneling across the closely coupled orbital systems of the double-backbone DTCP molecule (Vazquez et al., 2012).

Single molecule *spintronic* devices are those that exploit the quantum mechanical spin properties of the materials of which they are comprised. An example of a spintronic device is the magnetic memory that encodes data in domains in a ferromagnetic medium. More recently, coherent spin transport and detection has formed the foundation of quantum information processing, whereby data is encoded into the spin states of a population of atoms or molecules. In Chapter 3 we showed that the non-degenerate nature of electronic levels in organics make them particularly useful in spin-based devices. An obvious example of a spin-controlled organic device is the electrophosphorescent OLED that can achieve 100% internal quantum efficiency via spin–orbit coupling to the triplet state. In this context, a property that can be exploited in single molecule memory devices is *giant magnetoresistance* (GMR). GMR is exhibited when two ferromagnetic layers are separated by a non-magnetic, conducting spacer. If the spacer is an insulator, the effect is termed *tunneling magnetoresistance*, or TMR.

The resistance of the junction is determined by the spin polarized DOS that increases or decreases if the spin polarization of the two contacts is parallel, or antiparallel, respectively. This can be understood

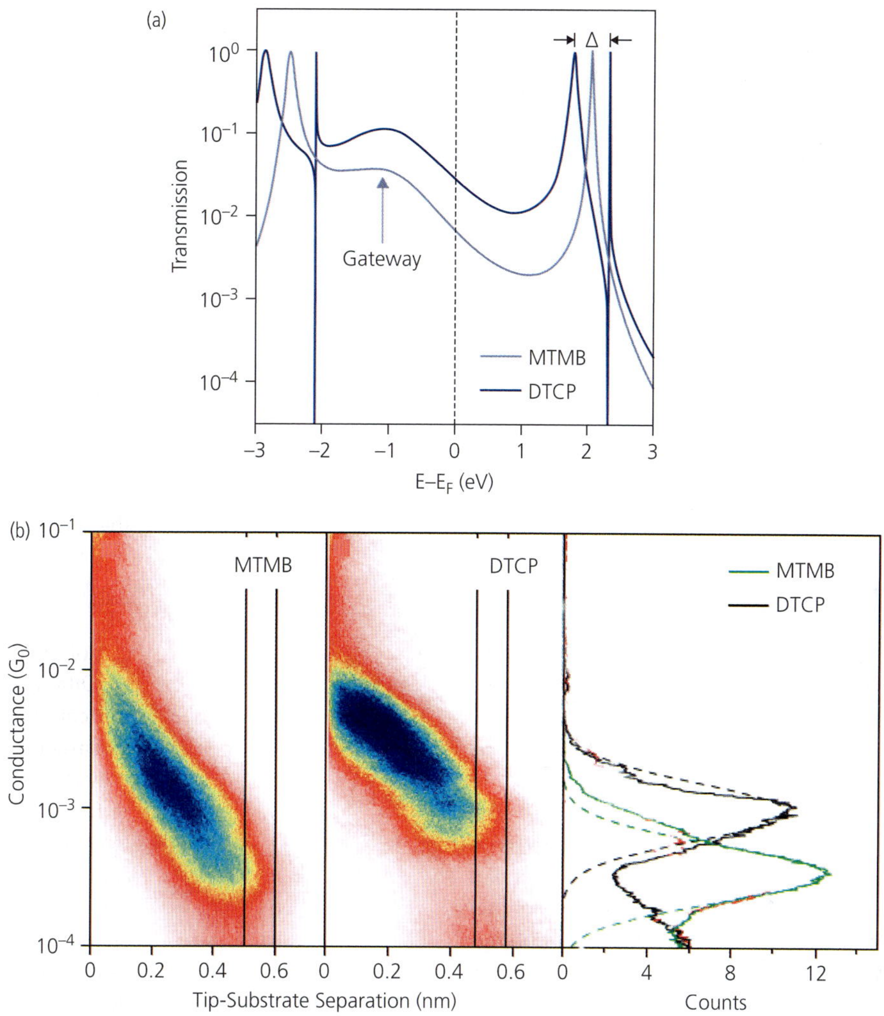

Figure 9.108 (a) Calculated transmission of MTMB and DTCP. The gateway energy of the anchor groups is indicated. Δ is the bonding/antibonding orbital splitting of DTCP. (b) Conductance histograms of >6000 measurements for the MTMB and DTCP from the point of the Au contact breakage to molecular breakoff at 0.50–0.55 nm (left and center, respectively). Histograms taken for the lowest conductance near breakoff (as defined by the vertical lines) are shown at right along with Gaussian fits (dashed lines). Adapted from Vazquez et al. (2012).

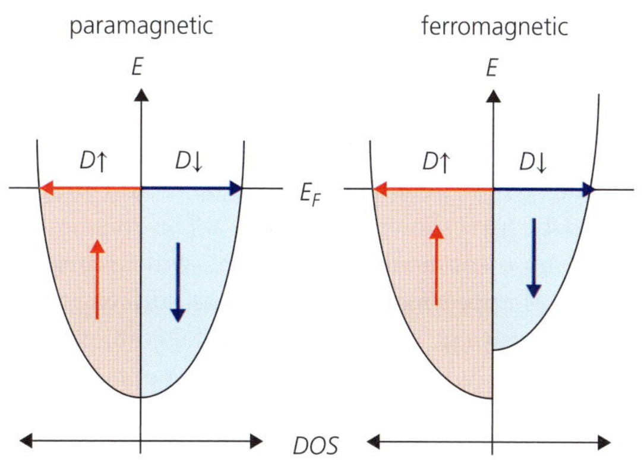

Figure 9.109 Spin polarized densities of states (DOS) for paramagnetic and ferromagnetic materials. Here, $D\uparrow$ indicates spins in the "up" direction, and $D\downarrow$ indicates spins in the "down" direction.

from the densities of states shown for paramagnets and ferromagnets in Fig. 9.109. The net electron *spin polarization*, P, of a magnetic solid is given by the difference between spins in the "up" vs. the "down" direction, $D\uparrow$ and $D\downarrow$, respectively. That is,

$$P = \frac{D\uparrow(E_F) - D\downarrow(E_F)}{D\uparrow(E_F) + D\downarrow(E_F)}. \tag{9.51}$$

Paramagnetic materials have zero net polarization, whereas $P < 1$ for ferromagnets due to an imbalance of the majority ($D\uparrow$) and minority ($D\downarrow$) densities of states.

A single molecule device can exhibit GMR if a semiconducting organic molecule bridges the gap between spin-polarized electrodes. This has been demonstrated in ultrahigh vacuum at 4 K using a spin-polarized

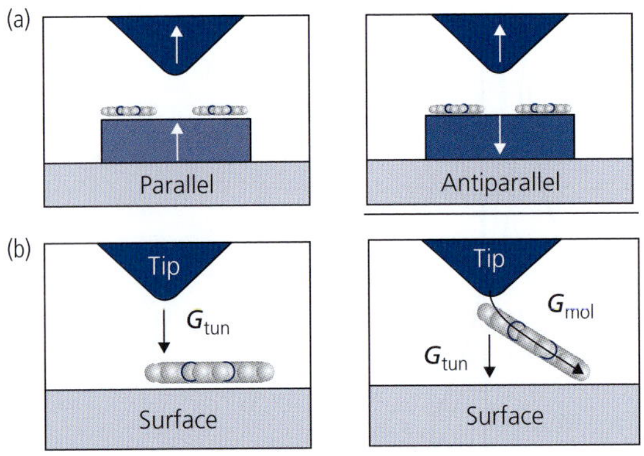

Figure 9.110 (a) Parallel and antiparallel orientations of a spin-polarized STM (Sp-STM) tip relative to a ferromagnetic domain. (b) Configuration of a molecule on a substrate before (left) and after (right) it jumps to contact the tip as the gap with the substrate is decreased. Before pick-up, the conductance is due only to tunneling (G_{tun}). After pick up the conductance is $G_{tun} + G_{mol}$, where G_{mol} is the molecular conductance (Schmaus et al., 2011).

cobalt-coated tungsten STM tip as one electrode, and a single crystalline Cu(111) substrate coated with single domain ferromagnetic Co nanoislands. The tips are polarized out of plane, whereas the dipole orientation of the nanoislands is either normal, or antinormal to the substrate plane, as illustrated in Fig. 9.110a. The islands were coated with H_2Pc at submonolayer coverage. The larger DOS that results from parallel dipoles leads to a higher tunneling current, and hence a higher tunneling conductance, G_{tun}, compared to the antiparallel configuration (Nguyen-Manh et al., 1998). When the tip is distant from the substrate surface, the current is due to direct tunneling from the tip. This current is due solely to TMR. However, if a H_2Pc molecule lies within the gap, the molecule will "jump-to-contact" the tip as it is brought near to the substrate, as shown in Fig. 9.110b. This leads to a higher total conductance equal to $G_{tot} = G_{tun} + G_{mol}$, where G_{mol} is the conductance of the molecule alone. The molecular conductance is then simply the difference between the non-contacted and the contacted conductances, and corresponds to GMR.

Prior to the molecular jump, the conductance increases exponentially as the tip distance is decreased, as is characteristic of tunneling. However, after the molecule jumps, the total conductance increases very slowly with decreasing distance since conduction now occurs via the molecular π-system. Note that the jump-to-contact method is opposite to the previously described break junction procedure, since in the former the starting configuration is with the tip and substrate in a non-contact configuration, whereas for the break junction, the tip and substrate initially form a welded contact.

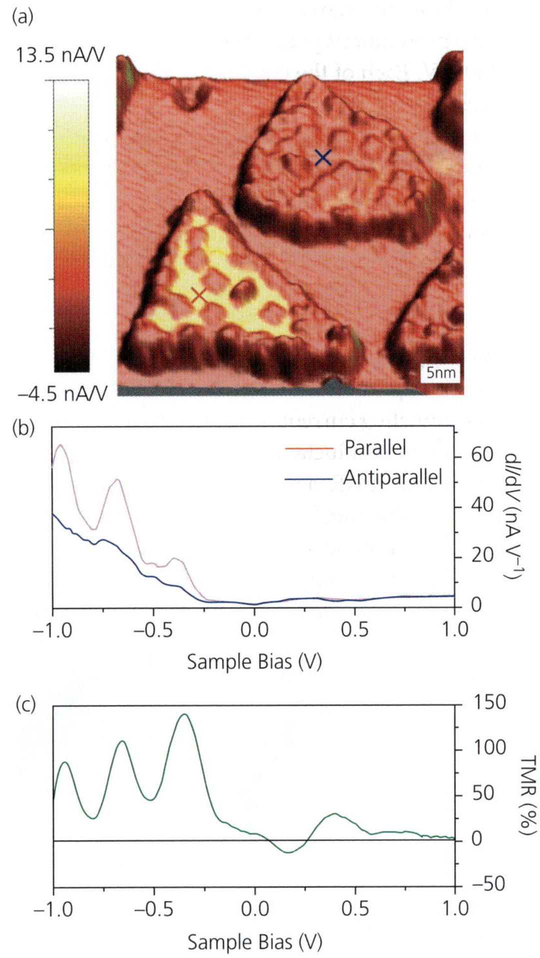

Figure 9.111 (a) Co nanoislands coated with a submonolayer of vacuum-deposited H_2Pc with the magnetic domain parallel (bottom island) and antiparallel (top island) to the Sp-STM tip polarization direction. High current (yellow regions) due to GMR is apparent in the bottom island. (b) Differential conductance vs. bias for the parallel and antiparallel domains in (a). The measurement points in both are shown by "X." (c) Difference between the conductance of the antiparallel and parallel domains from the data in (b). The most optimistic estimate of TMR is due to the antiparallel orientation, and is plotted on the right. The tip bias is 10 mV (Schmaus et al., 2011).

A spin-polarized STM (Sp-STM) image of two Co domains whose polarizations are parallel and antiparallel to the tip is shown in Fig. 9.111a. The domains are coated by a submonolayer of H_2Pc, with individual molecules appearing in the image as surface roughness. The current from the antiparallel configuration is lower than when the tip and domain polarizations are parallel. The conductance of the two islands after the contacts are bridged by an H_2Pc molecule differ considerably, as expected due to the lower spin-polarized DOS of the antiparallel island.

The tip current for both tip/domain dipole orientations is shown in Fig. 9.111b, where the data are taken from the locations marked by "X" in Fig. 9.111a. There are three pronounced peaks, with the largest at a bias of −350 mV. Each of the conductance peaks is due to transmission maxima resulting from the molecular orbitals and molecule/contact couplings coming into resonance. The GMR is given by the ratio of the difference between the antiparallel (G_{AP}) and parallel (G_P) conductances:

$$GMR = \frac{G_P - G_{AP}}{G_{AP}}. \tag{9.52}$$

For the sample in Fig. 9.111b, $G_{AP} = 0.16G_0$ and $G_P = 0.25G_0$, giving $GMR = 61\%$. This is 10 times higher than the tunneling current. Hence, it is concluded that the change in conductance is indeed due to GMR as opposed to TMR across an insulating gap. For comparison, the most optimistic TMR response is plotted on the right of Fig. 9.111c.

The origin of the magnetoresistance can be understood using DFT calculations of the orbital structure of H_2Pc. This analysis reveals that there is a doubly degenerate LUMO level on pairs of aromatic rings forming the Pc core structure. The N on the phthalocyanine coordinates with Co, resulting in a broad transmission peak within 100 meV of the Fermi energy. This hybridization, in turn, broadens the couplings, Γ, to both substrate and tip. Equation 9.43 implies that broadening increases T, and therefore the conductance, as observed. The modeling results indicate that hybridization of the anchoring bonds makes GMR relatively insensitive to the difference between E_F and the resonance energy, as long as these two energies are less than Γ. Furthermore, GMR depends primarily on the ratio of the majority to minority orbital broadening energies (Schmaus et al., 2011).

Quantum interference, energy level spectroscopy, and GMR are only three of the many unique properties that also include thermoelectricity (Hicks and Dresselhaus, 1993), optical emission and absorption, and a plethora of other phenomena exhibited by single molecule devices (Aradhya and Venkataraman, 2013). The 1D structures are governed primarily by quantum effects, and hence understanding their behavior requires a comprehensive knowledge of the energetics of the molecules and the systems. While it is challenging to control the gaps containing only a single molecule, progress in the technology of break junctions, e-beam nanopatterning, and single molecule transistor designs have brought this field to an impressive stage of maturity. While it remains unclear if there will be practical applications beyond molecular characterization, spin transport for quantum computing, and single molecule detection or "fingerprinting," these limited dimensional devices provide ample opportunities for discovering new physical properties and phenomena of molecular electronic materials (Joachim et al., 2000, Cuniberti et al., 2005).

9.7 Summing up

This chapter has focused on applications and phenomena that do not fit neatly into the main themes of the previous chapters. This does not by any means suggest that the selected topics are less important. They are simply included here to give a taste of the incredible breadth of the field known as organic electronics. Furthermore, the subjects that are considered cannot be viewed as representative of the vast array of interesting material that could have been included.

It is worthwhile to reiterate at this point how the topics were chosen for inclusion in this chapter. The first criterion is that the subject is clearly an organic electronic device, that is to say it must be current driven, and it should also exhibit an interesting excitonic property (i.e. light emission, light field interactions, etc.). Of course, there are numerous examples that could rightfully fall into the category of organic electronics that do not feature a conductive and/or excitonic material property. Major examples that are primarily organic electronic in nature without such characteristics are electro-optical devices exploiting nonlinear optical effects, electrochromic devices and phenomena, organic magnetic materials, and so on. Even among current-based devices, many important topics have been omitted, such as polymer batteries and spintronic devices (although this was briefly touched on in Section 9.6). Clearly, the list of topics is vast and growing every day. And many could easily fill a volume at least as long as this one!

Yet, it is hoped that acquainting the reader with ionic conductivity in LECs, strongly coupled states in optical microcavities, organic thermoelectricity and memories, and organics in limited dimensional (i.e. 2D and 1D) systems will encourage further exploration into the endless possibilities offered by organic electronics. More information on these and other topics can be found in the diversity of sources available in the ever-growing body of scientific literature. Several such references are provided in the following section on *Further reading*.

Further reading

1. V. M. Agranovich, 2009. *Excitations in Organic Solids*. Oxford: Oxford University Press.
2. J. C. Cuevas and E. Scheer, 2017. *Molecular Electronics: An Introduction to Theory and Experiment*, 2nd ed., World

Scientific Series in Nanoscience and Nanotechnology, Vol. 15. Danvers, MA: World Scientific Publishing.
3. G. Cuniberti, G. Fagas, and K. Richter (Eds.), 2005. *Introducing Molecular Electronics*. Heidelberg: Springer.
4. D. M. Guldi, H. Nishihara, and L. Venkataraman (Eds.), 2015. Molecular wires: themed issue. *Chem. Soc. Rev.*, 44.
5. P. Heremans, G. H. Gelinck, R. Müller, K.-J. Baeg, D.-Y. Kim, and Y.-Y. Noh, 2011. Polymer and organic nonvolatile memory devices. *Chem. Mater.*, 23, 341.
6. A. V. Kavokin, J. J. Baumberg, G. Malpuech, and F. P. Laussy, 2017. *Microcavities*. Oxford: Oxford University Press.
7. D. K. C. MacDonald, 2006. *Thermoelectricity: An Introduction to the Principles* Mineola, NY: Dover Publications.
8. J. S. Moodera, B. Koopmans, and P. M. Oppeneer (Eds.), 2014. Organic spintronics. *MRS Bull*, 39, 578.
9. N. F. Mott and E. A. Davis, 1979. *Electronic Processes in Non-Crystalline Materials*. Oxford: Clarendon Press.
10. S. Paul (Ed.), 2009. Making nano-bits remember: a recent development in organic electronic memory devices. *Phil. Trans. R. Soc. A*, 367.
11. P. N. Prasad and D. J. Williams, 1991. *Introduction to Nonlinear Optical Effects in Molecules and Polymers*. New York: Wiley.
12. D. W. Snoke and J. Keeling, 2017. The new era of polariton condensates. *Phys. Today*, 70, 54.

Problems

1. A light emitting electrochemical cell consists of an anion–cation pair and a luminescent molecule. We want to consider the performance of such a device of length, $L =$ 500 μm, and that has an anion/cation mobility ratio of $\mu_A/\mu_C = 2$, with $\mu_A = 10^{-7}$ cm^2/V s. Assuming that the density of salts is $n_S = 10^{18}$ cm^{-3}, and the applied voltage is $V_a = 5$ V between Ohmic contacts,
   (a) Write an expression for the potential and electric field distributions vs. time, that is, $V(x, t)$ and $F(x, t)$ across the active layer. Assume that the charge mobilities are 100 times higher than the ionic mobilities.
   (b) Write an expression for the width of the emission zone vs. time. Assume the zone width is determined by the point at which the electron and hole concentrations drop to 30% of their peak values.
   (c) Assuming that the emission intensity is proportional to the width of the emission zone, plot the time evolution of the intensity.
   (d) What is the time needed to reach 50% of the peak intensity of this LEC?
2. In an LEC, we find that the emission zone does not form into a straight line, but rather something like a zigzag.
   (a) Give three potential sources of this spatial variation between parallel contacts.
   (b) If the ratio of the radiative to the non-radiative rate in an LEC is $k_r/k_{nr} = 10$, and the activation energy of the anion and cation mobilities is 1.0 eV, plot the evolution of the peak luminescence efficiency vs. temperature for the device in Prob. 1.
3. A Bloch surface wave (BSW) can be excited with light incident on the glass substrate side of a DBR at angle θ.
   (a) Calculate the optical intensity profile shown for the DBR whose index profile is provided in in Fig. P9.3. The layer thicknesses are: 6 pairs of 100 nm SiN_x and 135 nm SiO_2 (index = 2.0 and 1.5, respectively) as well as a thin top-most pair of 11.5 nm SiN_x and 15 nm SiO_2.
   (b) An organic thin film with absorption coefficient of $\alpha = 2.5 \times 10^5$ cm^{-1} for an exciton with energy 2.2 eV, and index of refraction, $n = 1.6$ is deposited onto the surface of the DBR. Calculate the total fraction of light absorbed vs. film thickness.

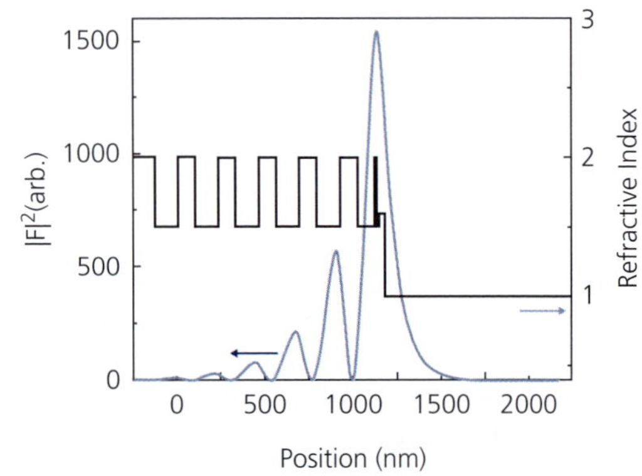

Figure P9.3 Index profile and optical intensity profile of a surface Bloch wave.

   (c) Assuming that the BSW is excited by a laser at a wavelength of 544 nm, what is the angle at which the BSW can be excited using the Kretschmann configuration with a prism with index $n = 1.75$. The excitation field is TE polarized.
   (d) For a Rabi splitting of 250 meV, calculate and plot the reflectivities of the lower and upper polariton branches vs. angle for the material in b. in a SBW cavity.
4. Show that:
   (a) The maximum efficiency of a thermoelectric refrigerator is given by

$$\eta_{COP} = \frac{T_C}{\eta_C T_H} \frac{\sqrt{1+Z\bar{T}} - T_H/T_C}{\sqrt{1+Z\bar{T}}+1},$$

   (b) The refrigerator can support a maximum temperature difference of $\Delta T_{\max} = \frac{1}{2} Z T_C^2$.
   (c) Plot η_{COP} vs. Z from to 1 to 5×10^{-3} K^{-1}.
5. Assume that a memory consists of rectifiers in series with resistive elements in an $N \times N$ array. If the reverse saturation current of the diode accompanying each element is I_0:
   (a) How does the size of the array (N) scale with I_0, assuming that the memory current must be S times larger than the background current?
   (b) For a resistance R_{min} of each resistive memory element in the **1** logic state, and $R_{max} = 10^3 R_{min}$ in the **0** logic state, how does the readout time scale with N? Assume a junction area of A and a depletion region width of W for an ideal organic diode with a series resistance, R_{ser}.

(c) Assuming $I_0 = 1$ nA for a 0.01 mm^2 diode with an active region thickness of 300 nm, calculate the largest array size, N_{max}, for $R_{min} = 100\ \Omega$.

(d) Calculate the memory bit readout time for the device in (c).

6. Figure P9.6 shows the absorption and PL spectra of the molecule, PTCDA, and a monolayer of the TMDC, $MoSe_2$. The transition dipole moment lies within the plane of PTCDA, that is, along its perylene core. Furthermore, the transition dipole of the TMDC lies along the film plane. Assume a TMDC peak absorption coefficient of $\alpha = 7 \times 10^5\ cm^{-1}$, and for PTCDA, $\alpha = 2.5 \times 10^5\ cm^{-1}$. The in-plane dielectric constants are $\varepsilon = 4$ for $MoSe_2$, and for PTCDA, the in-plane dielectric constant is $\varepsilon_{\|} = 4.5$ and perpendicular to the plane $\varepsilon_{\perp} = 1.9$. The natural decay lifetimes of PTCDA and the TMDC are 5 ns and 100 ps, respectively.

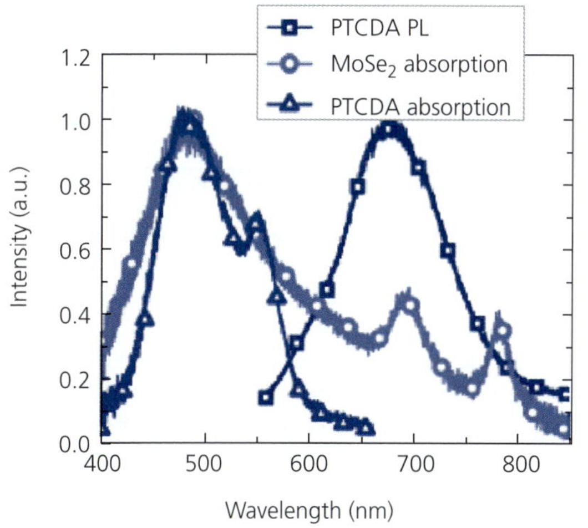

Figure P9.6 Absorption and PL spectra of materials used in Problem 6.

(a) Calculate the Förster radius and FRET transfer rate for a Frenkel exciton from PTCDA to $MoSe_2$. Assume that the PTCDA molecule lies flat on the TMDC surface.

(b) Do the calculation in (a) but assume that the PTCDA molecular plane is parallel to, and is tilted at 30° from the plane of the 2D film.

(c) An ultrathin film of Al_2O_3 ($\varepsilon = 9.1$) is interposed between the organic and TMDC thin films. Calculate and plot the Förster radius and FRET transfer rates vs. Al_2O_3 thickness, δ. At what Al_2O_3 thickness does the transfer rate decrease to 1% of its value at $\delta = 0$?

7. Assume that two conjugated molecules are separated by a σ-bridge linkage. This type of compound may be approximated as two quantum wells separated by a barrier, as shown in Fig. P9.7. Assume that the free carrier density inside the wells and barriers are negligible.

(a) At what applied voltage(s) will resonant tunneling be observed?

(b) Write an expression for the electronic transmission coefficient for this molecular system. Assume that the quantized energy levels are thermally broadened, and can be approximated by a Gaussian distribution. The effective mass is equal to the rest mass of the electron.

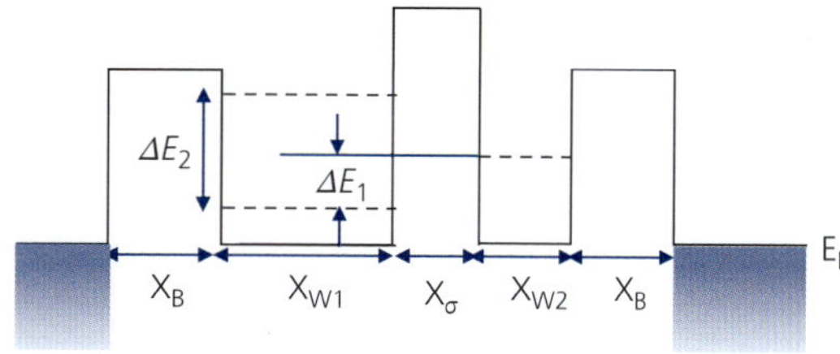

Figure P9.7 Energy level diagram for a molecule between two contacts. The molecule has two conjugated segments of widths X_{W1} and X_{W2} bridged by a non-conjugated segment of width X_σ. The barriers to the metal contacts (at far left and right) are of equal height and width. Quantum states are shown by dashed lines.

8. Assume that the energy levels of a material can be described by a simple model of a particle in a cubic box of infinite potential depth, and of length L.

(a) Calculate the density of states and the energy levels for the bulk material. Plot your results.

(b) Repeat (a) for a 2D solid (e.g. a TMDC).

(c) Repeat (a) for a 1D solid (e.g. a molecular wire)

(d) Repeat (a) for a 0D solid (e.g. a quantum dot within a polymer matrix).

9. Given the transmission coefficient (Eq. 9.43) for a single molecule device:

(a) Derive the current–voltage function:

$$I = \frac{4G_0}{q}\frac{\Gamma_S\Gamma_D}{\Gamma_S+\Gamma_D}\left(\tan^{-1}\left(\frac{E_0+\frac{1}{2}qV_A}{\Gamma_S+\Gamma_D}\right)\right.$$
$$\left.-\tan^{-1}\left(\frac{E_0-\frac{1}{2}qV_A}{\Gamma_S+\Gamma_D}\right)\right).$$

(b) Calculate the conductance, G, of the device at applied voltage, V_A.

(c) Plot G *vs.* V_A for $E_0 = 40$ meV, and $\Gamma_S = \Gamma_D = 1, 5, 10,$ and 20 meV.

10. In Fig. 9.98, it was suggested that the conductance, $G \propto \cos^2\theta$, where θ is the twist angle between conjugated groups forming the molecule.

(a) Give a plausible physical explanation why this relationship should be true.

(b) Assuming the molecule is the biphenyl moiety, that is, compound **4** in Fig. 9.98, derive a simple expression for $I(V)$ and G at $V \to 0$ that reproduces the $\cos^2\theta$ behavior observed.

(c) Plot $I(V)$ and G vs. $\cos^2\theta$ for a triphenyl analog of compound **4**. For this, two angles, θ and ϕ, are required.

References

Accorsi, S., Barra, A.-L., Caneschi, A., Chastanet, G., Cornia, A., Fabretti, A. C., Gatteschi, D., Mortalò, C., Olivieri, E., Parenti, F., Rosa, P., Sessoli, R., Sorace, L., Wernsdorfer, W. & Zobbi, L. 2006. Tuning anisotropy barriers in a family of tetrairon(III) single-molecule magnets with an $S = 5$ ground state. *J. Am. Chem. Soc.*, 128, 4742.

AlTal, F. & Gao, J. 2016. High resolution scanning optical imaging of a frozen polymer p-n junction. *J. Appl. Phys.*, 120, 115501.

Amani, M., Lien, D.-H., Kiriya, D., Xiao, J., Azcatl, A., Noh, J., Madhvapathy, S. R., Addou, R., Santosh, K. & Dubey, M. 2015. Near-unity photoluminescence quantum yield in MoS_2. *Science*, 350, 1065.

Aradhya, S. V. & Venkataraman, L. 2013. Single-molecule junctions beyond electronic transport. *Nat. Nanotechnol.*, 8, 399.

Asadi, K., De Leeuw, D. M., De Boer, B. & Blom, P. W. 2008. Organic non-volatile memories from ferroelectric phase-separated blends. *Nat. Mater.*, 7, 547.

Aviram, A. & Ratner, M. A. 1974. Molecular rectifiers. *Chem. Phys. Lett.*, 29, 277.

Baranov, D. G., Wersall, M., Cuadra, J., Antosiewicz, T. J. & Shegai, T. 2017. Novel nanostructures and materials for strong light–matter interactions. *ACS Photonics*, 5, 24.

Bez, R., Camerlenghi, E., Modelli, A. & Visconti, A. 2003. Introduction to flash memory. *Proc. IEEE*, 91, 489.

Billen, J., Steudel, S., Müller, R., Genoe, J. & Heremans, P. 2007. A comprehensive model for bipolar electrical switching of CuTCNQ memories. *Appl. Phys. Lett.*, 91, 263507.

Bolotin, K. I., Sikes, K. J., Jiang, Z., Klima, M., Fudenberg, G., Hone, J., Kim, P. & Stormer, H. L. 2008. Ultrahigh electron mobility in suspended graphene. *Solid State Commun.*, 146, 351.

Bozano, L. D., Kean, B. W., Beinhoff, M., Carter, K. R., Rice, P. M. & Scott, J. C. 2005. Organic materials and thin-film structures for cross-point memory cells based on trapping in metallic nanoparticles. *Adv. Funct. Mater.*, 15, 1933.

Breit, G. & Wigner, E. 1936. Capture of slow neutrons. *Phys. Rev.*, 49, 519.

Briechle, B. M., Kim, Y., Ehrenreich, P., Erbe, A., Sysoiev, D., Huhn, T., Groth, U. & Scheer, E. 2012. Current–voltage characteristics of single-molecule diarylethene junctions measured with adjustable gold electrodes in solution. *Beilstein J. Nanotechno.*, 3, 798.

Bubnova, O., Khan, Z. U., Malti, A., Braun, S., Fahlman, M., Berggren, M. & Crispin, X. 2011. Optimization of the thermoelectric figure of merit in the conducting polymer poly(3,4-ethylenedioxythiophene). *Nat. Mater.*, 10, 429.

Bubnova, O., Khan, Z. U., Wang, H., Braun, S., Evans, D. R., Fabretto, M., Hojati-Talemi, P., Dagnelund, D., Arlin, J.-B., Geerts, Y. H., Desbief, S., Breiby, D. W., Andreasen, J. W., Lazzaroni, R., Chen, W. M., Zozoulenko, I., Fahlman, M., Murphy, P. J., Berggren, M. & Crispin, X. 2013. Semi-metallic polymers. *Nat. Mater.*, 13, 190.

Burzurí, E., Yamamoto, Y., Warnock, M., Zhong, X., Park, K., Cornia, A. & van der Zant, H. S. J. 2014. Franck–Condon blockade in a single-molecule transistor. *Nano Lett.*, 14, 3191.

Cai, Y., Zhou, H., Zhang, G. & Zhang, Y.-W. 2016. Modulating carrier density and transport properties of MoS_2 by organic molecular doping and defect engineering. *Chem. Mater.*, 28, 8611.

Chen, F., Li, X., Hihath, J., Huang, Z. & Tao, N. 2006. Effect of anchoring groups on single-molecule conductance: comparative study of thiol-, amine-, and carboxylic acid-terminated molecules. *J. Am. Chem. Soc.*, 128, 15874.

Chen, H., Cheng, N., Ma, W., Li, M., Hu, S., Gu, L., Meng, S. & Guo, X. 2015a. Design of a photoactive hybrid bilayer dielectric for flexible nonvolatile organic memory transistors. *ACS Nano*, 10, 436.

Chen, Y., Zhao, Y. & Liang, Z. 2015b. Solution processed organic thermoelectrics: towards flexible thermoelectric modules. *Energy Environm. Sci.*, 8, 401.

Cheng, C.-Y., Dhanker, R., Gray, C. L., Mukhopadhyay, S., Kennehan, E. R., Asbury, J. B., Sokolov, A. & Giebink, N. C. 2018. Charged polaron polaritons in an organic semiconductor microcavity. *Phys. Rev. Lett.*, 120, 017402.

Ciuti, C., Bastard, G. & Carusotto, I. 2005. Quantum vacuum properties of the intersubband cavity polariton field. *Phys Rev B*, 72, 115303.

Cuniberti, G., Fagas, G. & Richter, K. (Eds.) 2005. *Introducing Molecular Electronics*. Heidelberg: Springer.

Cutler, M., Leavy, J. F. & Fitzpatrick, R. L. 1964. Electronic Transport in Semimetallic Cerium Sulfide. *Phys. Rev.*, 133, A1143.

Daskalakis, K., Maier, S., Murray, R. & Kéna-Cohen, S. 2014. Nonlinear interactions in an organic polariton condensate. *Nat. Mater.*, 13, 271.

Debucquoy, M., Bode, D., Genoe, J., Gelinck, G. H. & Heremans, P. 2009. An organic charge trapping memory transistor with bottom source and drain contacts. *Appl. Phys. Lett.*, 95, 239.

Deng, H., Haug, H. & Yamamoto, Y. 2010. Exciton-polariton Bose–Einstein condensation. *Rev. Mod. Phys.*, 82, 1489.

Deng, H., Weihs, G., Snoke, D., Bloch, J. & Yamamoto, Y. 2003. Polariton lasing vs. photon lasing in a semiconductor microcavity. *Proc. Natl. Acad. Sci.*, 100, 15318.

Dori, N., Menon, M., Kilian, L., Sokolowski, M., Kronik, L. & Umbach, E. 2006. Valence electronic structure of gas-phase 3,4,9,10-perylenetetracarboxylic acid dianhydride: Experiment and theory. *Phys. Rev. B*, 73, 195208.

Eizner, E., Brodeur, J., Barachati, F., Sridharan, A. & Kéna-Cohen, S. 2018. Organic photodiodes with an extended responsivity using ultrastrong light–matter coupling. *ACS Photonics*, 5, 2921.

Fritzsche, H. 1971. A general expression for the thermoelectric power. *Solid State Commun.*, 9, 1813.

Genco, A., Ridolfo, A., Savasta, S., Patanè, S., Gigli, G. & Mazzeo, M. 2018. Bright polariton coumarin-based OLEDs operating in the ultrastrong coupling regime. *Adv. Opt. Mater.*, 6, 1800364.

Giebink, N. C., Wiederrecht, G. P., Wasielewski, M. R. & Forrest, S. R. 2010. The ideal diode equation for organic heterojunctions. I. Derivation and application. *Phys. Rev. B*, 82, 155305.

Glaudell, A. M., Cochran, J. E., Patel, S. N. & Chabinyc, M. L. 2015. Impact of the doping method on conductivity and thermopower in semiconducting polythiophenes. *Adv. Energy Mater.*, 5, 1401072.

Greczynski, G., Kugler, T., Keil, M., Osikowicz, W., Fahlman, M. & Salaneck, W. R. 2001. Photoelectron spectroscopy of thin films of PEDOT-PSS conjugated polymer blend: a mini-review and some new results. *J. Electron Spectrosc. Rel. Phenom.*, 121, 1.

Gu, J., Liu, X., Lin, E.-c., Lee, Y.-H., Forrest, S. R. & Menon, V. M. 2017. Dipole-aligned energy transfer between excitons in two-dimensional transition metal dichalcogenide and organic semiconductor. *ACS Photonics*, 5, 100.

Gubbin, C. R., Maier, S. A. & Kéna-Cohen, S. 2014. Low-voltage polariton electroluminescence from an ultrastrongly coupled organic light-emitting diode. *Appl. Phys. Lett.*, 104, 233302.

Guédon, C. M., Valkenier, H., Markussen, T., Thygesen, K. S., Hummelen, J. C. & Van Der Molen, S. J. 2012. Observation of quantum interference in molecular charge transport. *Nat. Nanotechnol.*, 7, 305.

Guillet, T. & Brimont, C. 2016. Polariton condensates at room temperature. *C. R. Phys.*, 17, 946.

Hack, M. & Street, R. A. 1992. Analysis of double injection in amorphous silicon p-i-n diodes. *J. Appl. Phys.*, 72, 2331.

Haug, H. J. W. & Jauho, A.-P. 2008. *Quantum Kinetics in Transport and Optics of Semiconductors*. Heidelberg: Springer Science, Ch. 12.

He, D., Pan, Y., Nan, H., Gu, S., Yang, Z., Wu, B., Luo, X., Xu, B., Zhang, Y. & Li, Y. 2015. A van der Waals pn heterojunction with organic/inorganic semiconductors. *Appl. Phys. Lett.*, 107, 183103.

He, D., Zhang, Y., Wu, Q., Xu, R., Nan, H., Liu, J., Yao, J., Wang, Z., Yuan, S. & Li, Y. 2014. Two-dimensional quasi-freestanding molecular crystals for high-performance organic field-effect transistors. *Nat. Commun.*, 5, 5162.

Heremans, J. P., Dresselhaus, M. S., Bell, L. E. & Morelli, D. T. 2013. When thermoelectrics reached the nanoscale. *Nat. Nanotechnol.*, 8, 471.

Heremans, P., Gelinck, G. H., Müller, R., Baeg, K.-J., Kim, D.-Y. & Noh, Y.-Y. 2011. Polymer and organic nonvolatile memory devices. *Chem. Mater.*, 23, 341.

Heuer, H. W., Wehrmann, R. & Kirchmeyer, S. 2002. Electrochromic window based on conducting poly(3,4- ethylenedioxythiophene)poly(styrene sulfonate). *Adv. Funct. Mater.*, 12, 89.

Hicks, L. D. & Dresselhaus, M. S. 1993. Thermoelectric figure of merit of a one-dimensional conductor. *Phys. Rev. B*, 47, 16631.

Hobson, P. A., Barnes, W. L., Lidzey, D., Gehring, G., Whittaker, D., Skolnick, M. & Walker, S. 2002. Strong exciton–photon coupling in a low-Q all-metal mirror microcavity. *Appl. Phys. Lett.*, 81, 3519.

Holmes, R. & Forrest, S. 2004. Strong exciton–photon coupling and exciton hybridization in a thermally evaporated polycrystalline film of an organic small molecule. *Phys. Rev. Lett.*, 93, 186404.

Holmes, R. J., Kéna-Cohen, S., Menon, V. M. & Forrest, S. R. 2006. Strong coupling and hybridization of Frenkel and Wannier–Mott excitons in an organic–inorganic optical microcavity. *Phys. Rev. B*, 74, 235211.

Homan, S. B., Sangwan, V. K., Balla, I., Bergeron, H., Weiss, E. A. & Hersam, M. C. 2017. Ultrafast exciton dissociation and long-lived charge separation in a photovoltaic pentacene–MoS_2 van der Waals heterojunction. *Nano Lett.*, 17, 164.

Hou, S., Qu, Y., Liu, X. & Forrest, S. R. 2019. Weak to ultrastrong coupling of vibrationally dressed Frankel excitons and Bloch surface waves in one-sided dielectric optical cavities. *Phys. Rev. B*, 100, 045410.

Hutchison, J. A., Schwartz, T., Genet, C., Devaux, E. & Ebbesen, T. W. 2012. Modifying chemical landscapes by coupling to vacuum fields. *Angew. Chem. Int. Ed.*, 51, 1592.

Jariwala, D., Howell, S. L., Chen, K.-S., Kang, J., Sangwan, V. K., Filippone, S. A., Turrisi, R., Marks, T. J., Lauhon, L. J. & Hersam, M. C. 2015. Hybrid, gate-tunable, van der Waals p-n heterojunctions from pentacene and MoS_2. *Nano Lett.*, 16, 497.

Joachim, C., Gimzewski, J. K. & Aviram, A. 2000. Electronics using hybrid-molecular and mono-molecular devices. *Nature*, 408, 541.

Johansson, T., Pettersson, L. A. A. & Inganas, O. 2002. Conductivity of de-doped poly(3,4-ethylenedioxythiophene). *Synth. Met.*, 129, 269.

Kaiser, A. B. 1989. Thermoelectric power and conductivity of heterogeneous conducting polymers. *Phys. Rev. B*, 40, 2806.

Kang, M., Baeg, K. J., Khim, D., Noh, Y. Y. & Kim, D. Y. 2013. Printed, flexible, organic nano-floating-gate memory: effects of metal nanoparticles and blocking dielectrics on memory characteristics. *Adv. Funct. Mater.*, 23, 3503.

Kawai, H. 1969. The piezoelectricity of poly(vinylidene fluoride). *Jpn. J. Appl. Phys.*, 8, 975.

Kéna-Cohen, S. & Forrest, S. 2008. Giant Davydov splitting of the lower polariton branch in a polycrystalline tetracene microcavity. *Phys. Rev. B*, 77, 073205.

Kéna-Cohen, S. & Forrest, S. R. 2010. Room-temperature polariton lasing in an organic single crystal microcavity. *Nat. Photonics*, 4, 371.

Kéna-Cohen, S., Davanço, M. & Forrest, S. R. 2008. Strong exciton–photon coupling in an organic single crystal microcavity. *Phys. Rev. Lett.*, 101, 116401.

Kéna-Cohen, S., Maier, S. A. & Bradley, D. D. 2013. Ultrastrongly coupled exciton–polaritons in metal-clad organic semiconductor microcavities. *Adv. Opt. Mater.*, 1, 827.

Kever, T., Nauenheim, C., Böttger, U. & Waser, R. 2006. Preparation and characterisation of amorphous Cu:7,7,8,8-tetracyanoquinodimethane thin films with low surface roughness via thermal co-deposition. *Thin Solid Films*, 515, 1893.

Kim, B. J., Ko, Y., Cho, J. H. & Cho, J. 2013a. Organic field-effect transistor memory devices using discrete ferritin nanoparticle-based gate dielectrics. *Small*, 9, 3784.

Kim, G.-H., Shao, L., Zhang, K. & Pipe, K. P. 2013b. Engineered doping of organic semiconductors for enhanced thermoelectric efficiency. *Nat. Mater.*, 12, 719.

Kim, S.-J. & Lee, J.-S. 2010. Flexible organic transistor memory devices. *Nano Lett.*, 10, 2884.

Kim, W. Y., Cho, B., Kim, S. Y., Lee, Y. S. & Lee, H. C. 2009. Retention performance of ferroelectric polymer film for nonvolatile memory devices. *IEEE Electron Device Lett.*, 30, 822.

Kuc, A., Zibouche, N. & Heine, T. 2011. Influence of quantum confinement on the electronic structure of the transition metal sulfide TS_2. *Phys. Rev. B*, 83, 245213.

Lee, B., Liu, X., Lee, K., Fan, D., Song, B., Jung, B. J. & S. R. Forrest 2018. Surface passivation of InP using an organic thin film. *J. Cryst. Growth*, 503, 9.

Lee, J.-K., Yoo, D. S., Handy, E. S. & Rubner, M. F. 1996. Thin film light emitting devices from an electroluminescent ruthenium complex. *Appl. Phys. Lett.*, 69, 1686.

Lee, W. H., Park, J., Sim, S. H., Lim, S., Kim, K. S., Hong, B. H. & Cho, K. 2011. Surface-directed molecular assembly of pentacene on monolayer graphene for high-performance organic transistors. *J. Am. Chem. Soc.*, 133, 4447.

Lerario, G., Ballarini, D., Fieramosca, A., Cannavale, A., Genco, A., Mangione, F., Gambino, S., Dominici, L., De Giorgi, M. & Gigli, G. 2017. High-speed flow of interacting organic polaritons. *Light: Sci. Appl.*, 6, e16212.

Li, N., Lee, K., Renshaw, C. K., Xiao, X. & Forrest, S. R. 2011. Improved power conversion efficiency of InP solar cells using organic window layers. *Appl. Phys. Lett.*, 98, 053504.

Liao, C.-T., Chen, H.-F., Su, H.-C. & Wong, K.-T. 2011. Tailoring balance of carrier mobilities in solid-state light-emitting electrochemical cells by doping a carrier trapper to enhance device efficiencies. *J. Mater. Chem.*, 21, 17855.

Liao, X. 2018. *Energy transfer in heterogeneous organic material systems*. PhD Thesis, University of Michigan.

Lidzey, D., Bradley, D., Virgili, T., Armitage, A., Skolnick, M. & Walker, S. 1999. Room temperature polariton emission from strongly coupled organic semiconductor microcavities. *Phys. Rev. Lett.*, 82, 3316.

Lidzey, D. G., Bradley, D., Skolnick, M., Virgili, T., Walker, S. & Whittaker, D. 1998. Strong exciton–photon coupling in an organic semiconductor microcavity. *Nature*, 395, 53.

Lin, H., Pei, Z., Chen, J., Hwang, G., Fan, J. & Chan, Y. 2007a. A new nonvolatile bistable polymer-nanoparticle memory device. *IEEE Electron Device Lett.*, 28, 951.

Lin, H.-T., Pei, Z. & Chan, Y.-J. 2007b. Carrier transport mechanism in a nanoparticle-incorporated organic bistable memory device. *IEEE Electron Device Lett.*, 28, 569.

Liu, X., Gu, J., Ding, K., Fan, D., Hu, X., Tseng, Y.-W., Lee, Y.-H., Menon, V. & Forrest, S. R. 2017. Photoresponse of an organic semiconductor/two-dimensional transition metal dichalcogenide heterojunction. *Nano Lett.*, 17, 3176.

Lunt, S. R., Ryba, G. N., Santangelo, P. G. & Lewis, N. S. 1991. Chemical studies of the passivation of GaAs surface recombination using sulfides and thiols. *J. Appl. Phys.*, 70, 7449.

Ma, L., Liu, J. & Yang, Y. 2002. Organic electrical bistable devices and rewritable memory cells. *Appl. Phys. Lett.*, 80, 2997.

Magoga, M. & Joachim, C. 1999. Conductance of molecular wires connected or bonded in parallel. *Phys. Rev. B*, 59, 16011.

Marquardt, C. W., Grunder, S., Błaszczyk, A., Dehm, S., Hennrich, F., Löhneysen, H. v., Mayor, M. & Krupke, R. 2010. Electroluminescence from a single nanotube–molecule–nanotube junction. *Nat. Nanotechnol.*, 5, 863.

Martién, S., Grace, I., Bryce, M. R., Wang, C., Jitchati, R., Batsanov, A. S., Higgins, S. J., Lambert, C. J. & Nichols, R. J. 2010. Identifying diversity in nanoscale electrical break junctions. *J. Am. Chem. Soc.*, 132, 9157.

Mattheiss, L. F. 1973. Band structures of transition-metal-dichalcogenide layer compounds. *Phys. Rev. B*, 8, 3719.

Meier, S. B., Tordera, D., Pertegas, A., Roldan-Carmona, C., Ortí, E. & Bolink, H. J. 2014. Light-emitting electrochemical cells: recent progress and future prospects. *Mater. Today*, 17, 217.

Mills, T. J. & Lonergan, M. C. 2012. Charge injection and transport in low-mobility mixed ionic/electronic conducting systems: regimes of behavior and limiting cases. *Phys. Rev. B*, 85, 035203.

Mindemark, J., Tang, S., Wang, J., Kaihovirta, N., Brandell, D. & Edman, L. 2016. High-performance light-emitting electrochemical cells by electrolyte design. *Chem. Mater.*, 28, 2618.

Möller, S., Perlov, C., Jackson, W., Taussig, C. & Forrest, S. R. 2003. A polymer/semiconductor write-once read-many-times memory. *Nature*, 426, 166.

Mott, N. F. & Davis, E. A. 1979. *Electronic Processes in Non-Crystalline Materials*. Oxford: Clarendon Press.

Muallem, M., Palatnik, A., Nessim, G. D. & Tischler, Y. R. 2016. Strong light-matter coupling and hybridization of molecular vibrations in a low-loss infrared microcavity. *J. Phys. Chem. Lett.*, 7, 2002.

Mujica, V., Nitzan, A., Mao, Y., Davis, W., Kemp, M., Roitberg, A. & Ratner, M. 1999. Electron transfer in molecules and molecular wires: geometry dependence, coherent transfer, and control. *Adv. Chem. Phys.*, 107, 403.

Muller, R., Krebs, C., Goux, L., Wouters, D. J., Genoe, J., Heremans, P., Spiga, S. & Fanciulli, M. 2009. Bipolar resistive electrical switching of CuTCNQ memories incorporating a dedicated switching layer. *IEEE Electron Device Lett.*, 30, 620.

Naber, R., Blom, P., Marsman, A. & De Leeuw, D. 2004. Low voltage switching of a spin cast ferroelectric polymer. *Appl. Phys. Lett.*, 85, 2032.

Naber, R. C., Asadi, K., Blom, P. W., de Leeuw, D. M. & de Boer, B. 2010. Organic nonvolatile memory devices based on ferroelectricity. *Adv. Mater.*, 22, 933.

Nguyen, C. A., Mhaisalkar, S. G., Ma, J. & Lee, P. S. 2008. Enhanced organic ferroelectric field effect transistor characteristics with strained poly(vinylidene fluoride-trifluoroethylene) dielectric. *Org. Electron.*, 9, 1087.

Nguyen, E. P., Carey, B. J., Ou, J. Z., van Embden, J., Gaspera, E. D., Chrimes, A. F., Spencer, M. J., Zhuiykov, S., Kalantar-zadeh, K. & Daeneke, T. 2015. Electronic tuning of 2D MoS_2 through surface functionalization. *Adv. Mater.*, 27, 6225.

Nguyen—Manh, D., Tsymbal, E. Y., Pettifor, D. G., Arcangeli, C., Tank, R., Andersen, O. K. & Pasturel, A. 1998. Spin-polarized density of states and electron tunneling from the Co/Al_2O_3 interface. *Mater. Res. Soc. Proc.*, 492, 319.

Nichols, R. J. & Higgins, S. J. 2015. Single-molecule electronics: chemical and analytical perspectives. *Annu. Rev. Anal. Chem.*, 8, 389.

O'Neill, K., Osorio, E. A. & Zant, H. S. J. v. d. 2007. Self-breaking in planar few-atom Au constrictions for nanometer-spaced electrodes. *Appl. Phys. Lett.*, 90, 133109.

Panda, A., Ding, K. & Forrest, S. R. 2016. Free and trapped charge transfer excitons at a ZnO/small molecule heterojunction. *Phys. Rev. B*, 94, 125429.

Panda, A. & Forrest, S. R. 2017. Quantum confinement of hybrid charge transfer excitons in GaN/InGaN/organic semiconductor quantum wells. *Nano Lett.*, 17, 7853.

Park, H., Park, J., Lim, A. K. L., Anderson, E. H., Alivisatos, A. P. & McEuen, P. L. 2000. Nanomechanical oscillations in a single-C_{60} transistor. *Nature*, 407, 57.

Pei, Q., Yu, G., Zhang, C., Yang, Y. & Heeger, A. J. 1995. Polymer light-emitting electrochemical cells. *Science*, 269, 1086.

Pei, Q. B., Zuccarello, G., Ahlskog, M. & Inganas, O. 1994. Electrochromic and highly stable poly(3,4- ethylenedioxythiophene) switches between opaque blue-black and transparent sky blue. *Polymer*, 35, 1347.

Perrin, M. L., Burzurí, E. & van der Zant, H. S. J. 2015. Single-molecule transistors. *Chem. Soc. Rev.*, 44, 902.

Perrin, M. L., Verzijl, C. J., Martin, C. A., Shaikh, A. J., Eelkema, R., Van Esch, J. H., Van Ruitenbeek, J. M., Thijssen, J. M., Van Der Zant, H. S. & Dulić, D. 2013. Large tunable image-charge effects in single-molecule junctions. *Nat. Nanotechnol.*, 8, 282.

Plumhof, J. D., Stöferle, T., Mai, L., Scherf, U. & Mahrt, R. F. 2014. Room-temperature Bose–Einstein condensation of cavity exciton–polaritons in a polymer. *Nat. Mater.*, 13, 247.

Prime, D. & Paul, S. 2009. Overview of organic memory devices. *Philos. Trans. R.* .367, 4141.

Prins, F., Barreiro, A., Ruitenberg, J. W., Seldenthuis, J. S., Aliaga-Alcalde, N., Vandersypen, L. M. K. & van der Zant, H. S. J. 2011. Room-temperature gating of molecular junctions using few-layer graphene nanogap electrodes. *Nano Lett.*, 11, 4607.

Prins, F., Hayashi, T., Steenwijk, B. J. A. d. V. v., Gao, B., Osorio, E. A., Muraki, K. & Zant, H. S. J. v. d. 2009. Room-temperature stability of Pt nanogaps formed by self-breaking. *Appl. Phys. Lett.*, 94, 123108.

Qiu, H., Xu, T., Wang, Z., Ren, W., Nan, H., Ni, Z., Chen, Q., Yuan, S., Miao, F., Song, F., Long, G., Shi, Y., Sun, L., Wang, J. & Wang, X. 2013. Hopping transport through defect-induced localized states in molybdenum disulphide. *Nat. Commun.*, 4, 2642.

Renshaw, C. K. & Forrest, S. R. 2014. Excited state and charge dynamics of hybrid organic/inorganic heterojunctions. I. Theory. *Phys. Rev. B*, 90, 045302.

Rowe, D. M. 2005. General principles and theoretical considerations. In: Rowe, D. M. (ed.) *Thermoelectrics Handbook: Macro to Nano*. 1st ed. Boca Raton, FL: CRC Press.

Rozenman, G. G., Akulov, K., Golombek, A. & Schwartz, T. 2017. Long-range transport of organic exciton-polaritons revealed by ultrafast microscopy. *ACS Photonics*, 5, 105.

Russ, B., Glaudell, A., Urban, J. J., Chabinyc, M. L. & Segalman, R. A. 2016. Organic thermoelectric materials for energy harvesting and temperature control. *Nat. Rev. Mater.*, 1, 16050.

Sanvitto, D. & Kéna-Cohen, S. 2016. The road towards polaritonic devices. *Nat. Mater.*, 15, 1061.

Sawyer, C. B. & Tower, C. 1930. Rochelle salt as a dielectric. *Phys. Rev.*, 35, 269.

Schmaus, S., Bagrets, A., Nahas, Y., Yamada, T. K., Bork, A., Bowen, M., Beaurepaire, E., Evers, F. & Wulfhekel, W. 2011. Giant magnetoresistance through a single molecule. *Nat. Nanotechnol.*, 6, 185.

Scott, G. D. & Natelson, D. 2010. Kondo Resonances in Molecular Devices. *ACS Nano*, 4, 3560.

Sekitani, T., Takamiya, M., Noguchi, Y., Nakano, S., Kato, Y., Sakurai, T. & Someya, T. 2007. A large-area wireless power-transmission sheet using printed organic transistors and plastic MEMS switches. *Nat. Mater.*, 6, 413.

Sekitani, T., Yokota, T., Zschieschang, U., Klauk, H., Bauer, S., Takeuchi, K., Takamiya, M., Sakurai, T. & Someya, T. 2009a. Organic nonvolatile memory transistors for flexible sensor arrays. *Science*, 326, 1516.

Sekitani, T., Zaitsu, K., Noguchi, Y., Ishibe, K., Takamiya, M., Sakurai, T. & Someya, T. 2009b. Printed nonvolatile memory for a sheet-type communication system. *IEEE Trans. Electron Devices*, 56, 1027.

Shastry, T. A., Balla, I., Bergeron, H., Amsterdam, S. H., Marks, T. J. & Hersam, M. C. 2016. Mutual photoluminescence quenching and photovoltaic effect in large-area single-layer MoS_2–polymer heterojunctions. *ACS Nano*, 10, 10573.

Shen, N. & Tao, G. 2017. Charge transfer and interface engineering of the pentacene and MoS_2 monolayer complex. *Adv. Mater. Interfaces*, 4, 1601083.

Skolnick, M. S., Fisher, T. A. & Whittaker, D. M. 1998. Strong coupling phenomena in quantum microcavity structures. *Semiconductor Sci. Technol.*, 13, 645.

Slootsky, M., Liu, X., Menon, V. & Forrest, S. R. 2014. Room temperature Frenkel–Wannier–Mott hybridization of degenerate excitons in a strongly coupled microcavity. *Phys. Rev. Lett.*, 112, 076401.

Slootsky, M., Zhang, Y. & Forrest, S. R. 2012. Temperature dependence of polariton lasing in a crystalline anthracene microcavity. *Phys. Rev. B*, 86, 045312.

Snoke, D. W. & Keeling, J. 2017. The new era of polariton condensates. *Phys. Today*, 70, 54.

Snyder, G. J. & Toberer, E. S. 2011. Complex thermoelectric materials. In: *Materials For Sustainable Energy: A Collection of Peer-Reviewed Research and Review Articles from Nature Publishing Group*. Singapore: World Scientific.

Song, S., Cho, B., Kim, T.-W., Ji, Y., Jo, M., Wang, G., Choe, M., Kahng, Y. H., Hwang, H. & Lee, T. 2010. Three-dimensional integration of organic resistive memory devices. *Adv. Mater.*, 22, 5048.

Strachan, D., Smith, D., Johnston, D., Park, T.-H., Therien, M. J., Bonnell, D. & Johnson, A. 2005. Controlled fabrication of nanogaps in ambient environment for molecular electronics. *Appl. Phys. Lett.*, 86, 043109.

Sun, T., Peavey, J. L., David Shelby, M., Ferguson, S. & O'Connor, B. T. 2015. Heat shrink formation of a corrugated thin film thermoelectric generator. *Energy Convers. Manage.*, 103, 674.

Tassone, F. & Yamamoto, Y. 1999. Exciton–exciton scattering dynamics in a semiconductor microcavity and stimulated scattering into polaritons. *Phys. Rev. B*, 59, 10830.

Tischler, J. R., Bradley, M. S., Bulović, V., Song, J. H. & Nurmikko, A. 2005. Strong coupling in a microcavity LED. *Phys. Rev. Lett.*, 95, 036401.

Tischler, J. R., Scott Bradley, M., Zhang, Q., Atay, T., Nurmikko, A. & Bulović, V. 2007. Solid state cavity QED: Strong coupling in organic thin films. *Org. Electron.*, 8, 94.

Toshima, N. 2017. Recent progress of organic and hybrid thermoelectric materials. *Synth. Met.*, 225, 3.

Toshima, N. & Ichikawa, S. 2015. Conducting polymers and their hybrids as organic thermoelectric materials. *J. Electron. Mater.*, 44, 384.

Toshima, N., Oshima, K., Anno, H., Nishinaka, T., Ichikawa, S., Iwata, A. & Shiraishi, Y. 2015. Novel hybrid organic thermoelectric materials: three-component hybrid films consisting of a nanoparticle polymer complex, carbon nanotubes, and vinyl polymer. *Adv. Mater.*, 27, 2246.

Tseng, R. J., Huang, J., Ouyang, J., Kaner, R. B. & Yang, Y. 2005. Polyaniline nanofiber/gold nanoparticle nonvolatile memory. *Nano Lett.*, 5, 1077.

Tsutsui, M., Shoji, K., Taniguchi, M. & Kawai, T. 2008. Formation and self-breaking mechanism of stable atom-sized junctions. *Nano Lett.*, 8, 345.

van Reenen, S., Akatsuka, T., Tordera, D., Kemerink, M. & Bolink, H. J. 2013. Universal transients in polymer and ionic transition metal complex light-emitting electrochemical cells. *J. Am. Chem. Soc.*, 135, 886.

van Reenen, S., Matyba, P., Dzwilewski, A., Janssen, R. A., Edman, L. & Kemerink, M. 2010. A unifying model for the operation of light-emitting electrochemical cells. *J. Am. Chem. Soc.*, 132, 13776.

Vazquez, H., Skouta, R., Schneebeli, S., Kamenetska, M., Breslow, R., Venkataraman, L. & Hybertsen, M. 2012. Probing the conductance superposition law in single-molecule circuits with parallel paths. *Nat. Nanotechnol.*, 7, 663.

Venkataraman, L., Klare, J. E., Nuckolls, C., Hybertsen, M. S. & Steigerwald, M. L. 2006. Dependence of single-molecule junction conductance on molecular conformation. *Nature*, 442, 904.

Wilson, J. A. & Yoffe, A. D. 1969. The transition metal dichalcogenides discussion and interpretation of the observed optical, electrical and structural properties. *Adv. Phys.*, 18, 193.

Xiang, D., Jeong, H., Kim, D., Lee, T., Cheng, Y., Wang, Q. & Mayer, D. 2013. Three-terminal single-molecule junctions formed by mechanically controllable break junctions with side gating. *Nano Lett.*, 13, 2809.

Xing, K. Z., Fahlman, M., Chen, X. W., Inganas, O. & Salaneck, W. R. 1997. The electronic structure of poly (3,4-ethylene-dioxythiophene): studied by XPS and UPS. *Synth. Met.*, 89, 161.

Xu, H., Zhong, J., Liu, X., Chen, J. & Shen, D. 2007. Ferroelectric and switching behavior of poly(vinylidene fluoride-trifluoroethylene) copolymer ultrathin films with polypyrrole interface. *Appl. Phys. Lett.*, 90, 092903.

Xu, X., Register, R. A. & Forrest, S. R. 2006. Mechanisms for current-induced conductivity changes in a conducting polymer. *Appl. Phys. Lett.*, 89, 142109.

Yang, Y., Ouyang, J., Ma, L., Tseng, R. H. & Chu, C. W. 2006. Electrical switching and bistability in organic/polymeric thin films and memory devices. *Adv. Funct. Mater.*, 16, 1001.

Yong, C. K., Noori, K., Gao, Q., Joyce, H. J., Tan, H. H., Jagadish, C., Giustino, F., Johnston, M. B. & Herz, L. M. 2012. Strong carrier lifetime enhancement in GaAs nanowires coated with semiconducting polymer. *Nano Lett.*, 12, 6293.

Yoshida, A. & Toshima, N. 2014. Gold nanoparticle and gold nanorod embedded PEDOT: PSS thin films as organic thermoelectric materials. *J. Electron. Mater.*, 43, 1492.

Yu, C., Kim, Y. S., Kim, D. & Grunlan, J. C. 2008. Thermoelectric behavior of segregated-network polymer nanocomposites. *Nano Lett.*, 8, 4428.

Zandvliet, H. J. W. & van Houselt, A. 2009. Scanning tunneling spectroscopy. *Annu. Rev. Anal. Chem.*, 2, 37.

Zhang, Y., Qiao, J., Gao, S., Hu, F., He, D., Wu, B., Yang, Z., Xu, B., Li, Y. & Shi, Y. 2016. Probing carrier transport and structure-property relationship of highly ordered organic semiconductors at the two-dimensional limit. *Phys. Rev. Lett.*, 116, 016602.

Zheng, Y. J., Huang, Y. L., Chen, Y., Zhao, W., Eda, G., Spataru, C. D., Zhang, W., Chang, Y.-H., Li, L.-J. & Chi, D. 2016. Heterointerface screening effects between organic monolayers and monolayer transition metal dichalcogenides. *ACS Nano*, 10, 2476.

Zhou, Y., Han, S.-T., Sonar, P. & Roy, V. A. L. 2013. Nonvolatile multilevel data storage memory device from controlled ambipolar charge trapping mechanism. *Sci. Rep.*, 3, 2319.

Zhu, T., Yuan, L., Zhao, Y., Zhou, M., Wan, Y., Mei, J. & Huang, L. 2018. Highly mobile charge-transfer excitons in two-dimensional WS_2/tetracene heterostructures. *Sci. Adv.*, 4, eaao3104.

Zhu, Y., Gauthier, D. J., Morin, S., Wu, Q., Carmichael, H. & Mossberg, T. 1990. Vacuum Rabi splitting as a feature of linear-dispersion theory: analysis and experimental observations. *Phys. Rev. Lett.*, 64, 2499.

APPENDIXES

APPENDIX A

Glossary of frequently used and common abbreviations

Acronym	Definition
A	Acceptor
AFM	Atomic force microscope
ALD	Atomic layer deposition
AM1.5G	Air mass 1.5 Global.
AMOLED	Active matrix OLED (display)
ANSI	American National Standards Institute
ARC	Antireflection coating
ASE	Amplified spontaneous emission
ASTM	American Section of the International Association of Testing Materials
BAPV	Building applied photovoltaic
BC, TC	Bottom contact, top contact
BG, TG	Bottom gate, top gate
BHJ	Bulk heterojunction
BIOS	Basic input–output system
BIPV	Building integrated photovoltaic
BP	Boiling point
BSR	Back surface reflector
BSW	Bloch surface wave
BW	Energy bandwidth
CCM	Color changing media
CCT	Correlated color temperature
CELIV	Charge extraction by linearly increasing voltage
CF	Color filter
CGL	Charge generation layer
CIE	Commission Internationale d'Eclairage
CMOS	Complementary metal oxide semiconductor
CNT	Carbon nanotube
CPS	Chance, Prock, and Sibley
CRI	Color rendering index
CRZ	Charge recombination zone
CT	Charge transfer
CVD	Chemical vapor deposition
D	Drain, donor

Continued

Acronym	Definition
D-A HJ	Donor–acceptor heterojunction
DBR	Distributed Bragg reflector
DFB	Distributed feedback
DFT	Density functional theory
DOS	Density of states
DRAM	Dynamic random access memory
DSMC	Direct simulation Monte Carlo
DUT	Device under test
EA	Electron affinity
EIL	Electron injection layer
EL	Electroluminescence
EML	Emission layer
EQE	External quantum efficiency
ETL	Electron transport layer
FC	Franck–Condon integral
FeFET	Ferroelectric field effect transistor
FET	Field effect transistor
FIB	Focused ion beam
FOLED	Flexible OLED
FPA	Focal plane array
FRET	Fluorescence (or Förster) resonant energy transfer
FTIR	Fourier transform infrared spectroscopy
FWHM	Full width at half maximum
G	Gate
GDM	Gaussian disorder model
GISAXS	Glancing incidence small angle X-ray scattering
GIWAXS	Grazing incidence wide angle X-ray scattering
GIXD	Grazing incidence X-ray diffraction
GMR	Giant magnetoresistance
HCTE	Hybrid charge transfer exciton
HFPA	Hemispherical focal plane array
HIL	Hole injection layer
HJ	Heterojunction
HOMO	Highest occupied molecular orbital
HTL	Hole transport layer
ICT	Intramolecular charge transfer
IJP	Ink jet printing
IMF	Index matching fluid
IP	Ionization potential
IPES	Inverse photoelectron spectroscopy
IQE	Internal quantum efficiency
IR	Infrared
ISC	Intersystem crossing

(Continued)

Continued

Acronym	Definition
j–V	Current density vs. voltage
L-B	Langmuir–Blodgett
LC	Ligand centered
LCAO	Linear combination of atomic orbitals
LCD	Liquid crystal display
LDI-TOF-MS	Laser desorption ionization-time of flight-mass spectroscopy
LED	Light emitting diode
LEP	Light emitting polymer
LIG	Low index grid
LP or LPB	Lower polariton branch
LTPS	Low temperature polycrystalline silicon
LUMO	Lowest unoccupied molecular orbital
MC	Metal centered
MCBJ	Mechanically controlled break junction
MD	Molecular dynamics
MIS	Metal–insulator–semiconductor
MISFET	Metal–insulator–semiconductor FET
ML	Monolayer
MLA	Microlens array
MLCT	Metal ligand charge transfer
MMLCT	Metal–metal ligand charge transfer
MNR	Meyer–Neldel rule
MO	Molecular orbital
MOSFET	Metal–oxide–semiconductor FET
MPP	Maximum power point
MWIR	Mid wavelength infrared
MWNT	Multiwalled nanotube
NFA	Non-fullerene acceptor
NIR	Near infrared
NP	Nanoparticle
NTSC	National Television Standards Committee
O-B	Onsager–Braun
OFeFET	Organic ferroelectric field effect transistor
OI-HJ	Organic/inorganic heterojunction
OLED	Organic light emitting device (or diode)
OMBD	Organic molecular beam deposition
OPC	Organic photoconductor
OPD	Organic photodetector
OPT	Organic phototransistor
OPV	Organic photovoltaic cell
OSC	Organic semiconductor
OSL	Organic semiconductor laser
OTFT	Organic thin film transistor

Continued

Acronym	Definition
OTR	Oxygen transmission rate
OVJP	Organic vapor jet printing
OVPD	Organic vapor phase deposition
PAH	Polyaromatic hydrocarbon
PES	Photoelectron spectroscopy
PFEO	Perimeter free electron orbital
PHOLED	Phosphorescent organic light emitting device (or diode)
PL	Photoluminescence
PLQY	Photoluminescence quantum yield
PM-HJ	Planar-mixed heterojunction
PP	Polaron pair
PR	Photoresist
PVD	Plasma vapor deposition
QD	Quantum dot
QVGA	Quarter video graphics array
R2R	Roll-to-roll
RAM	Random access memory
RGA	Residual gas analysis (or analyzer)
RGB	Red, green, blue
RHEED	Reflection high energy electron diffraction
ROIC	Readout integrated circuit
ROM	Read only memory
S	Source
S/N	Signal-to-noise ratio
SAED	Selected area electron diffraction
SAM	Self-assembled monolayer
SCCM	Standard cubic centimeters per minute
SCL	Space charge limit
SEM	Scanning electron microscope
SF	Singlet fission
SOC	Spin–orbit coupling
SOLED	Stacked OLED
SOMO	Singly occupied molecular orbital
SPA	Singlet-polaron annihilation
SPP	Surface plasmon polariton
SQ	Shockley–Queisser (limit)
SRAM	Static random access memory
SRC	Standard reporting conditions
sRGB	Super RGB color gamut
SRH	Shockley–Read–Hall (recombination)
SSA	Singlet-singlet annihilation
SSSE	Solid state solvation effect
STA	Singlet–triplet annihilation

(Continued)

Continued

Acronym	Definition
STM	Scanning tunneling microscope
SVA	Solvent vapor annealing
SWIR	Short wavelength infrared
SWNT	Single walled nanotube
TADF	Thermally assisted delayed fluorescence
TD-DFT	Time dependent density functional theory
TDM	Transition dipole moment
TEC	Thermoelectric cooler
TEG	Thermoelectric generator
TEM	Transmission electron microscope
TFL	Trap filled limit
TFT	Thin film transistor
TIR	Total internal reflection
TMDC	Transition metal dichalcogenide
TMR	Tunneling magnetoresistance
TOF	Time of flight
TOF-SIMS	Time of flight secondary ion mass spectroscopy
TOLED	Transparent OLED
TPA	Triplet–polaron annihilation
TTA	Triplet–triplet annihilation
UHV	Ultrahigh vacuum
UP or UPB	Upper polariton branch
UPS	Ultraviolet photoelectron spectroscopy
UV	Ultraviolet
Vis	Visible
VTE	Vacuum thermal evaporation
W-M	Wannier–Mott
WOLED	White emitting OLED
WORM	Write once read many times
WVTR	Water vapor transmission rate

APPENDIX B

Identification of chemical names

Chemical name	IUPAC nomenclature
1-NPSQ	2,4-bis[4-(*N*-phenyl-1-naphthylamino)-2,6-dihydroxyphenyl]squaraine
26nCPy	2,6-bis(*N*-carbazolyl)pyridine
2CzPN	1,2-bis(carbazol-9-yl)-4,5-dicyanobenzene
2T-NATA	4,4′,4″-tris(*N*-2-naphthyl-*N*-phenylamino)triphenylamine
44TCzPN	6,6′-(9H,9′H-[4,40-bicarbazole]-9,9′-diyl)bis(4-(9H-carbazol-9-yl) isophthalonitrile)
4CzCNPy	2,3,5,6-tetracarbazole-4-cyano-pyridine
4CzIPN	1,2,3,5-tetrakis(carbazol-9-yl)-4,6-dicyanobenzene
4CzPNPh	3,4,5,6-tetrakis(3,6-diphenylcarbazol-9-yl)-1,2-dicyanobenzene
4CzTPN-Ph	2,3,5,6-tetrakis(3,6-diphenylcarbazol-9-yl)-1,4-dicyanobenzene
4P-NPD	*N,N*′-di-1-naphthalenyl-*N,N*′-diphenyl-[1,1′:4′,1″:4″,1‴-quaterphenyl]-4,4‴-diamine
4TBPPZn	tetra-(2,6-*tert*-butyl)phenolporphyrin zinc
AC-DT	2,6-bis[(4-acetylthiophenyl)ethynyl]anthracene
ACRSA	10-phenyl-10H,10′H-spiro[acridine-9,9′-anthracen]-100-one
ACRXTN	3-(9,9-dimethylacridin-10(9H)-yl)-9H-xanthen-9-one
ADN	9,10-di(naphtha-2-yl)anthracene
AIDCN	2-amino-4,5-imidazoledicarbonitrile
Alq_3	tris(8-hyroxyquinoline)aluminum
AMS	poly(α-methylstyrene)
APFO3	poly(2,7-(9,9-dioctyl-fluorene)-*alt*-5,5-(4′,7′-di-2-thienyl-2′,1′,3′-benzothiadiazole))
AQ-DT	2,6-bis[(4-acetylthiophenyl)ethynyl]-9,10-anthraquinone
AQ-MT	2-[(4-acetylthiophenyl)ethynyl]-6-(phenylethynyl)-9,10-anthraquinone
ASSQ	2-[4-(*N,N*-diisobutylamino)-2,6-dihydroxyphenyl]-4-diphenylaminosquaraine
B3PYMPM	4,6-bis(3,5-di-3-pyridylphenyl)-2-methylpyrimidine
BAlq	bis(2-methyl-8-quinolinato)4-pheylphenolato aluminum(III)
BCP	(2,9-dimethyl-4,7-diphenyl-1,10-phenanthroline), or bathocuproine
BCzVBi	4,4′-bis(9-ethyl-3-carbazovinylene)-1,1′-biphenyl
BD-6MDPA	*N,N*′-diphenyl-*N,N*′-di-*m*-tolylspiro[fluorene-7,9′-benzofluorene]-5,9-diamine
BDCP-IC	8,15-dihexyloxyltetrakis(4-hexylphenyl)-(5,5,12,12-tetrakis(4-hexylphenyl)-dithienyl [1,2-b:4,5-b′]benzodithieno[3,2-b]thiophene-2,7-diyl)bis(2-(3-oxo-2,3-dihydroinden-1-ylidene)malononitrile)
BDN	bis-(4-dimethylaminodithiolbenzil)-nickel(II)
BDT-IC	(5,5,12,12-tetrakis(4-hexylphenyl)-dithienyl[1,2-b:4,5-b′]benzodithiophene-2,7-diyl)bis (2-(3-oxo-2,3-dihydroinden-1-ylidene)malononitrile)
BeEH-PPV	poly(2-butyl,5-(2′-ethylhexyl)-1,4-phenylenevinylene)

(*Continued*)

Continued

Chemical name	IUPAC nomenclature
BF-DPB	*N,N′*-(diphenyl-*N,N′*-bis-9,9-dimethyl-fluoren-2-yl)-benzidine
bis-PXD-OXD	2,5-bis(4-(10-phenoxazyl)phenyl)-1,3,4-oxadiazole
BN1	4″,4‴-*N,N*-diphenylamine-4,4′-diphenyl-1,1′-binaphthyl
BN2	4″,4‴-(9*H*-9-carbazole)-4,4′-diphenyl-1,1′-binaphthyl
BNP-IC	3,9-bis(2-methylene-(3-(1,1-dicyanomethylene)-indanone))-tetrakis(4-hexylphenyl)-di-5-(7,7,15,15-tetrakis(4-hexylphenyl)-thieno[3,2-b:2′,3′-d]pyran-1,9-dimethoxynaphthalene)
BP1T	2,5-bis(4-biphenylyl)thiophene
BP2T	2,5-bis(4-biphenylyl)bithiophene
BP3T	5,5″-bis(4-biphenylyl)-2,2′:5′,2″-terthiophene
BP4mPy	3,5,3′,5′-tetra(m-pyrid-3-yl)phenyl[1,1′]biphenyl
BPhen	bathophenanthroline
BPTT	2,5-bis-biphenyl-4-yl-thieno[3,2,b]thiophene
BPyTP2	2,7-bis(2,20-bipyridine-5-yl)triphenylene
Br_2-DBP	dibenzo([f,f′]-4,7′-di[4-bromophenyl]-4′,7-diphenyl)diindeno[1,2,3-cd:1′,2′,3′-lm] perylene
BT-CIC	4,4,10,10-tetrakis(4-hexylphenyl)-5,11-(2-ethylhexyloxy)-4,10-dihydro-dithienyl[1,2-b:4,5b]benzodithiophene-2,8-diyl)bis(2-(3-oxo-2,3-dihydroinden-5,6-dichloro-1-ylidene) malononitrile
BT-IC	8,15-dihexyloxyl-tetrakis(4-hexylphenyl)-(5,5,12,12-tetrakis(4-hexylphenyl)-dithienyl[1,2-b:4,5-b′]benzodithiophene-2,7-diyl)bis(2-(3-oxo-2,3-dihydroinden-1-ylidene) malononitrile)
Bt_2Ir(acac)	bis(2-phenylbenzothiozolato-N,$C^{2'}$)(acetylacetonate)
BTDC (anti & syn)	2-((7-(*N*-(2-ethylhexyl)-benzothieno[3,2-*b*]thieno[2,3-d]pyrrol-2-yl)benzo[c][1,2,5] thiadiazol-4-yl)methylene)malononitrile
Btp_2Ir(acac)	bis[2-(2′-benzothienyl)-pyridinato-N,$C^{2'}$](acetylacetonate)
BTQBT	bis(1,2,5-thaidiazolo)-*p*-quinobis(1,3-dithiole)
C_{10}-DNTT	2,9-didecyl-dinapthol[2,3-b:2′,3′-f]thieno[3,2-b]thiophene
C_8-BTBT	dioctylbenzothienobenzothiophene
CB	chlorobenzene
CBP	4,4′-bis(9-carbazolyl)-1,1′-biphenyl
CF	chloroform
CH4T	cyclohexylquaterthiophene
ClAlPc	chloroaluminum phthalocyanine
CN-MBE	1-cyano-*trans*-1,2-bis(4′-methylbiphenyl)ethylene
CN-TFMBE	1-cyano-*trans*-1,2-bis(3′,5′-bis-trifluoromethylbiphenyl)ethylene
CN-TFPA	3,3′-(1,4-phenylene)bis(2-(3,5-bis-trifluoromethyl)phenyl)acrylonitrile
Coumarin 540	3-(2-benzothiazolyl)-*N,N*-diethylumbelliferylamine, 3-(2-benzothiazolyl)-7-(diethylamino) coumarin
Coumarin 545P	10-(2-benzothiazolyl)-1,3,3,7,7-pentamethyl-2,3,6,7-tetrahydro-1*H*,5*H*,11*H*-benzo[*l*]-pyrano[6,7,8-*ij*]-quinolizin-11-one
Coumarin 545T	10-(2-benzothiazolyl)-1,1,7,7-tetramethyl-2,3,6,7-tetrahydro-1*H*,5*H*,11*H*-benzo[*l*]-pyrano[6,7,8-*ij*]-quinolizin-11-one
CPD	*N,N′*-di(phenylcarbazole)-*N,N′*-bis-phenyl-(1,1′-biphenyl)-4,4′-diamine
CuPc	copper phthalocyanine
CzP-BZP	4-(4-9H-carbazol-9-yl)phenyl-(7-phenylbenzo[c][1,2,5]thiadiazole)

Continued

Chemical name	IUPAC nomenclature
CzSi	9-(4-tert-butylphenyl)-3,6-bis(triphenylsilyl)-9H-carbazole
D_{3A}	9,10-bis[(10-decylanthracen-9-yl)ethynyl]anthracene
DAE	diarylethene phosphonic acid
DAST	4-dimethylamino-*N*-methyl-4-stilbazolium tosylate
DB-TCNQ	dibenzotetrathiafulvalene-tetracyanoquinodimethane
dbo-P2TP	5,5′-bis(4-butoxyphenyl)-2,2′-bithiophene
DBP	tetraphenyldibenzoperiflanthene
DBT	2-((7-(*N*-(2-ethylhexyl)-dithieno[3,2-b:2′,3′-d]pyrrol-2-yl)benzo[c][1,2,5]thiadiazol-4-yl) methylene)malononitrile
DCB	1,4-dichlorobenzene
DCJTB	4-(dicyanomethylene)-2-*tert*-butyl-6-(1,1,7,7-tetramethyljulolidyl-9-enyl)-4*H*-pyran
DCM	dichloromethane
DCM1	[2-[2-[4-(dimethylamino)phenyl]ethenyl]-6-methyl-4H-pyran-4-ylidene]-propanedinitrile
DCM2	4-(dicyanomethylene)-2-methyl-6-julolidyl-9-enyl-4H-pyran
DCMT	3′,4′-dibutyl-5,5″-bis(dicyanomethylene)-5,5″-dihydro-2,2′:5′,2″-terthiophene
DCzIPN	4,6-di(9H-carbazol-9-yl)isophthalonitrile
DEH	diethylaminobenzaldehyde diphenylhydrazone
DH7T	dihexyl-7-thiophene
diF-TES-ADT	difluorinated 5,11-bis(triehtylsilylethynyl)anthradithiophene
DIO	1,8-diiodooctane
DIP	diindenoperylene
DM-2,9-DMQA	*N,N′*-dimethyl-2,9-dimethylquinacridone
DMAC-BP	bis[4-(9,9-dimethyl-9,10-dihydroacridine)phenyl]methanone
DMAC-DPS	bis[4-(9,9-dimethyl-9,10-dihydroacridine)phenyl]sulfone
DMACDPS	bis[4-(9,9-dimethyl-9,10-dihydroacridine)phenyl]sulfone
DMC	decamethylcobaltocene
DMF	dimethylformamide
DMPD	dimethyl-*p*-phenylenediamine
DMPPP	1-(2,5-dimethyl-4-(1-pyrenyl)phenyl)pyrene
DMQA	*N,N′*-dimethylquinacridone
DMSO	dimethyl sulfoxide
DNB	2,4-diaminonitrobenzene
DNTT	dinaphtho-[2,3-b:20,30-f]thieno[3,2-b]thiophene
DOTCI	3,3′-diethyl-2,2′-oxathiacarbocyanine iodide
DPA	5,12-diphenylanthracene
DPASQ	2-[4-(*N,N*-diphenylamino)-2,6-dihydroxyphenyl]-4-(4-diphenyliminio)-2,5-dien-1-ylidene}-3-oxocyclobut-1-en-1-olate
DPEPO	bis-(2-(diphenylphosphino)phenyl)ether oxide
DPFS	9-[4-(2-diphenylamino-9,9-diethylfluoren-7-ylethenyl)phenyl]-9-phenylfluorene
DPh-BTBT	2,7-diphenyl[1]benzothieno[3,2-b]benzothiophene
DPM	2-((7-(*N*-(2-ethylhexyl)-6-p-tolyl-dithieno[3,2-b:2′,3′-d]pyrrol-2-yl)pyrimidin-5-yl) methylene)malononitrile
DPSQ	2-[4-(*N,N*-diphenylamino)-2,6-dihydroxyphenyl]-4-[(4-(*N,N*-diphenyliminio)-2,6-dihydroxyphenyl)-2,5-dien-1-ylidene]-3-oxocyclobut-1-en-1-olate

(Continued)

Continued

Chemical name	IUPAC nomenclature
Dpt	5,6-diphenyltetracene
DT-IC	5,5,10,10-tetrakis(4-hexylphenyl)-thiophene-thieno[3,2-b]thiophene-thiophene)bis(2-(3-oxo-2,3-dihydroinden-1-ylidene)malononitrile
DT-TTF	Dithiotetrathiafulvalene
DTCP	2,11-dithia(3,3)paracyclophane
DTDCPB	2-([7-(4-*N,N*-ditolylaminophenylen-1-yl)-2,1,3-benzothiadiazol-4-yl]methylene) malononitrile
DTDCTB	2-((7-(5-(dip-tolylamino)thiophen-2-yl)benzo[c][1,2,5]thiadiazol-4yl)methylene) malononitrile
DTffBT	poly(benzodithiophene-dithienyldifluorobenzothiadiazole)
DTPyT	poly(benzodithiophene-dithienylthiadiazolopyridine)
DTS	decyltrichlorosilane
EG	ethylene glycol
Erq_3	erbium(III) tris(8-hydroxyquinoline)
$F_{16}CuPc$	fluorinated copper phthalocyanine
F_4-TCNQ	2,3,5,6-tetrafluoro-7,7,8,8-tetracyanoquinodimethane
F_6TNAP	1,3,4,5,7,8-hexafluorotetracyanonaphthaquinodimethane
F8BT	poly[(9,9-di-*n*-octylfluorenyl-2,7-diyl)-*alt*-(benzo[2,1,3]thiadiazol-4,8-diyl)]
F8T2	poly(9,9-diocytlfluorene-co-bithiophene)
F8TBT	poly[(9,9-dioctylfluorene)-2,7-diyl-alt-(4,7-bis(3-hexylthien-5-yl)-2,1,3-benzothiadiazole)-2′,2″-diyl]
FIPAC	9-(9,9-dimethyl-9H-fluoren-3yl)-14-phenyl-9,14-dihydrodibenzo[a,c]phenazine
FIr6	bis-(4′,6′-difluorophenylpyridinato)tetrakis(1-pyrazolyl)borate iridium(III)
FIrpic	bis[2-(4,6-difluorophenyl)pyridinato-C^2,N](picolinato)iridium(III)
FIzIr	tris(1-(9′,9′-dimethyl-2′-fluorenyl)pyrazolyl-N,$C^{2'}$)iridium(III)
FPt1	platinum (II) (2-(4′,6′-difluorophenyl) pyridinato-N,$C^{2'}$)(2,4-pentanedionato)
H_2Pc	metal-free phthalocyanine
HATCN	hexaazatriphenylene hexacarbonitrile
HMDS	hexamethyldisilazane
HMTPD	4,4′-bis[*N,N*′-(3-tolyl)amino]-3,3′-dimethylbiphenyl
HMTTF	hexamethylenetetrathiafulvalene
iBuBTDC	2-((7-(*N*-(isobutyl)-benzothieno[3,2-*b*]thieno[2,3-*d*]-pyrrol-2-yl)benzo[*c*][1,2,5]thiadiazol-4-yl)methylene)malononitrile
ICBA	indene-C_{60}-bisadduct
IDT-IC	(4,4,9,9-tetrakis(4-hexylphenyl)-4,9-dihydro-s-indaceno[1,2-b:5,6-b′]dithiophene-2,7-diyl)bis-2-(3-oxo-2,3-dihydroinden-1-ylidene)malononitrile
IDTIDT-IC	3,8,11,16-tetrakis(4-hexylphenyl)- thieno[2,3-d:5,6-d′]-di-s-indaceno[1,2-b:5,6-b′] dithiophene)bis(2-(3-oxo-2,3-dihydroinden-1-ylidene)malononitrile
IEICO	2,2′-((2Z,2′Z)-((5,5′-bis(4,4,9,9-tetrakis(4-hexylphenyl)-4,9-dihydro-*s*-indaceno[1,2-b:5,6-*b*′]dithiophene-2,7-diyl) bis (4-((2-ethylhexyl)oxy)thiophene-5,2-diyl))bis (methanylylidene))bis-(3-oxo-2,3-dihydro-1*H*-indene-2,1-diylidene))dimalononitrile
$Ir(5\text{-ph-ppy})_3$	tris[2-(5′-phenyl)phenylpyridine]iridium(III)
$Ir(dbi)_3$	tris[1-(2,4-diisopropyldibenzo[b,d]furan-3-yl)-2-phenylimidazole]iridium(III)
$Ir(dfppz)_2$(dtb-bpy)	(bis(1-(2,4-difluorophenyl)pyrazole))-4,4′-di(*tert*-butyl)-2,20-bipyridineiridium(III)
$Ir(dhfpy)_2$(acac)	bis(2-(9,9-dihexylfluorenyl)-1-pyridine)(acetylacetonate)iridium(III)

Continued

Chemical name	IUPAC nomenclature
$Ir(dmp)_3$	tris[3-(2,6-dimethylphenyl)-7-methylimidazo[1,2-f] phenanthridine]iridium(III)
$Ir(F_2ppy)_3$	tris[2-(4,6-difluorophenyl)pyridyl]iridium(III)
$Ir(ipripmi)_3$	tris[1-(2,6-diisopropylphenyl)-2-phenyl-1H-imidazole]iridium(III)
$Ir(iprpmi)_3$	*fac*-tris[(2,6-diisopropylphenyl)-2-phenyl-1H-imidazole]iridium(III)
$Ir(MDQ)_2(acac)$	bis(2-methyldibenzo[f,h]quinoxaline)(acetylacetonate)iridium(III)
$Ir(pmb)_3$	tris-(*N*-phenyl-*N*-methylbenzimidazol-2-yl)iridium(III)
$Ir(pmp)_3$	tris-(*N*-phenyl-N-methylpyridoimidazol-2-yl)iridium(III)
$Ir(PPQ)_2$ (acac)	bis-(2,4-diphenylquinolyl-N,$C^{2'}$)(acetylacetonate)iridium(III)
$Ir(ppy)_2acac$	bis[2-(2-pyridinyl-N)phenyl-C](acetylacetonato)iridium(III)
$Ir(ppy)_2bpy$	(bis(2-phenylpyridine)-2,2-bipyridine)iridium(III)
$Ir(ppy)_3$	tris[2-phenylpyridinato-C^2,N]iridium(III)
$Ir(ppz)_3$	tris(phenylpyrazole)iridium(III)
$Ir(tBuCN-F)_2(acac)$	bis(3-[4-(tert-butyl)pyridin-2-yl]-2,6-difluorobenzonitrile)(acetylacetonate)iridium(III)
IT-IC	4,4,9,9-tetrakis(4-hexylphenyl)-4,9-dihydro-s-indaceno[1,2-b:5,6-b′]dithieno[3,2-b] thiophene-2,7-diyl) bis(2-(3-oxo-2,3-dihydroinden-1-ylidene)malononitrile
IT-IC-4F	(2-(3-oxo-2,3-dihydroinden-5,6-difluoro-1-ylidene)malononitrile)-5,5,11,11-tetrakis(4-hexylphenyl)-dithieno[2,3-d:2′,3′-d′]-s-indaceno[1,2-b:5,6-b′]dithiophene
IT-M	3,9-bis(2-methylene- ((3-(1,1-dicyanomethylene)-6/7-methyl)-indanone))-5,5,11,11-tetrakis(4-hexylphenyl)-dithieno [2,3-d:2′,3′-d′]-s-indaceno[1,2-b:5,6-b′]dithiophene
ITCC-M	3,9-bis((Z)-1-(6-(dicyanomethylene)-2-methyl-5,6-dihydro-6H-cyclopenta[*b*]thiophen-6-one-5-yl) ethylene)-5,5,11,11-tetrakis(4-hexylphenyl)dithieno[2,3-*d*:2′,3′-*d*′]-*s*-indaceno [1,2-b:5,6-b′]dithiophene
ITO	indium tin oxide
Liq	lithium quinolate
m-MTDATA	4,4′,4″-tris(3-methylphenylphenylamino)-triphenylamine
MADN	2-methyl-9,10-di(2-naphthyl)anthracene
mCP	bis(9-carbazolyl)benzene
MDMO-PPV	poly[2-methoxy-5-(3′,7′-dimethyloctyloxy)-1,4-phenylenevinylene]
Me-PTC	dimethyl-3,4,9,10 perylenetetracarboxylic diimide
MEH-PPV	poly(2-methoxy-5-(2′-ethyl)-hexyloxy-*p*-phenylenevinylene)
MeLPPP	methyl-substituted ladder type poly(*p*-phenylene)
MeO-TPD	*N,N,N′,N′*-tetrakis(4-methoxyphenyl)-benzidine
MeTHF	2-methyltetrahydrofuran
$Mo(tfd)_3$	molybdenum tris-[1,2-bis(trifluormethyl)ethane-1,2-dithiolene]
MTDATA	Same as m-MTDATA
MTMB	1,4-bis(methyl(thio)methyl)–benzene
N1100	*N,N′*-bis(2,2,3,3,4,4,4-heptfluorobutyl-1,7-dicyanoperylene-3,4,9,10-tetracarboxylic) diimide
$NOPF_6$	nitrosonium hexafluorophosphate
NPD (also NPB)	*N,N′*-di(1-naphthyl)-*N,N′*-diphenyl-(1,1′-biphenyl)-4,4′-diamine
NRS-PPV	poly[4′-(3,7-dimethyloctyloxy)-1,1′-biphenylene-2,5-vinylene]
NSM	4-((1,3-dioxo-1,3-dihydro-2H-inden-2-ylidene)methyl)benzoic acid
NSN	bis(4-(4,6-diphenyl-1,3,5-triazine-2-yl)phenyl)diphenylsilane
NTCDA	3,4,7,8-naphthalenetetracarboxylic dianhydride

(*Continued*)

Continued

Chemical name	IUPAC nomenclature
OC_1C_{10}-PPV	poly[2-methoxy-5-(3′,7′-dimethyloctyloxy)-*p*-phenylenevinylene]
ODPA	octadecylphosphonic acid
ODTS	octadecyltrichlorosilane
OJ2	4-(dicyanomethylidene)-2-*tert*-butyl-6-(2-(1-(4-*tert*-butylphenyl)-2,3-dihydro-2-methylindolin-5-yl)vinyl)-4*H*-pyran
OTMS	octadecyltrimethoxysilane
OTS	octyltrichlorosilane
OXD	oxadiazole
OXD-7	1,3-bis[(4-tert-butylphenyl)-1,3,4-oxadiazolyl]phenylene
P(NDI2OD-T2)	poly[[*N,N′*-bis(2-octyldodecyl)-naphthalene-1,4,5,8-bis(dicarboximide)-2,6-diyl]-*alt*-5,5′-(2,2-bithiophene)]
P(VDF-TrFE)	poly(vinylidene fluoride-trifluoroethylene)
P3HT	poly(3-hexylthiophene-2,5-diyl)
P3HT75-co-EHT25	poly(3-hexylthiophene-co-3-(2-ethylhexyl)-thiophene)
P3HTTDPP	poly(3-hexylthiophene-thiophene-diketopyrrolopyrrole)
P3OT	poly(3-octylthiophene)
P3TI	poly[*N,N′*-bis(2-hexyldecyl)isoindigo-6,6′-diyl-*alt*-thiophene-2,5-diyl]
P6D	1-phosphonohexadecane
p6P	para-sexiphenyl
PAH	poly(allylamine hydrochloride)
PANI	polyaniline
PB16TTT	poly(2,5-bis(3-hexadecylthiophene-2-yl)thieno[3,2-b]thiophene)
pBCb2Cz	9-(4-(9H-pyrido[2,3-b]indol-9-yl)phenyl)-9H-3,9′-bicarbazole
PBD	2-biphenyl-4-yl-5-(4-tert-butyl-phenyl)-[1,3,4]oxadiazole
PBDB-T	poly[(2,6-(4,8-bis(5-(2-ethylhexyl)thiophen-2-yl)benzo[1,2-b:4,5-b′] dithiophene)-*co*-(1,3-di(5-thiophene-2-yl)-5,7-bis(2-ethylhexyl)benzo[1,2-c:4,5-c′]dithiophene-4,8-dione)]
PBDTT-DPP	poly(2,60-4,8-bis(5-ethylhexylthienyl)benzo-[1,2-b;3,4-b]dithiophene-alt-5-dibutyloctyl-3,6-bis(5-bromothiophen-2-yl)pyrrolo[3,4-c]pyrrole-1,4-dione)
PBDTT-DPP	poly(2,6′-4,8-di(5-ethylhexylthienyl)benzo[1,2-b;3,4-b]dithiophene-alt-5-dibutyloctyl-3,6-bis(5-bromothiophen-2-yl)pyrrolo[3,4-c]pyrrole-1,4-dione)
PBnDT-DTffBT	poly(benzo[1,2-*b*:4,5-*b′*]dithiophene-*alt*-DTffBT)
PBTDTTT-S-T	poly[(((2-hexyldecyl)sulfonyl)-4,6-di(thiophen-2-yl)thieno[3,4-b]thiophene-2,6-diyl)-alt-(4,8 bis((2-ethylhexyl)oxy)benzo[1,2-b:4,5-b′]dithiophene-2,6-diyl)]
PBTTT	poly(2,5-bis(3-alkylthiophen-2-yl)thieno[3,2-b]thiophene)
pBTTT	poly[2,5-bis(3-alkylthiophen-2-yl)thieno(3,2-b)thiophene]
PC	polycarbonate
$PC_{61}BM$	[6,6]-phenyl-C_{61} butyric acid methyl ester
$PC_{71}BM$	[6,6]-phenyl-C_{71} butyric acid methyl ester
PCBA	phenyl-C_{61} butyric acid
PCE-10	poly[4,8-bis(5-(2-ethylhexyl)thiophen-2-yl)benzo[1,2-b:4,5-b′]dithiophene-co-3-fluorothieno[3,4-b]thio-phene-2-carboxylate]
PCPDTBT	poly[2,6-(4,4-bis(2-ethylhexyl)-4H-cyclopenta[2,1-b;3,4-b]-dithiophene)-alt-4,7- (2,1,3-benzothiadiazole)]
PDA	2,7-bis(4′-*n*-hexylphenyl)-4,9-diphenylanthrazoline
PDAC	poly(diallyldimethylammonium chloride)

Continued

Chemical name	IUPAC nomenclature
PDDTT	poly(5,7-bis(4-decanyl-2-thienyl)-thieno (3,4-b)diathiazole-thiophene-2,5)
PDI-2DTT	poly([*N,N′*-bis(2-decyl-tetradecyl)-3,4,9,10-perylene diimide-1,7-diyl]-alt-[[2,2′]-bi(dithieno[3,2-b:2′,3′-d]thiophenyl)-6,6′-diyl])
PDMS	polydimethylsiloxane
PDOFA	poly(dioctylfluorene-*co*-anthracene)
PDOFN	poly(dioctylfluorene-*co*-naphthacene)
PDOFP	poly(dioctylfluorene-*co*-pentacene)
PDPP-TBT	poly(diketopyrrolopyrrole-thiophenebenzothiadiazolethiophene)
PDPP3T	poly(diketopyrrolopyrrole-terthiophene)
PDPP3T	poly-(diketopyrrolopyrrole-terthiophene)
PDTG-TPD	poly(dithieno[3,2-b:2,3-d]germole thieno[3,4-c]pyrrole-4,6-dione)
PEDOT:PSS	poly(3,4-ethylenedioxythiophene):poly(styrenesulfonate)
PEIE	polyethylenimine 80% ethoxylated aqueous solution
PEN	polyethylene naphthalate
PEO	poly(ethylene oxide)
PES	polyether sulfone
PET	polyethylene terephthalate
PETT	poly(nickel 1,1,2,2-ethenetetrathiolate)
PF12TBT	poly(2,7-(9,9-didodecylfluorene)-alt-5,5-[4′,7′-bis-(2-thienyl)-2′,1′,3′-benzothiadiazole])
PF_6	hexfluorophosphate
PFBDB-T	poly[(2,6-(4,8-bis(5-(2-ethylhexyl)thiophen-2-yl)-benzo[1,2-b:4,5-b′]dithiophene))-alt-(5,5-(1′,3′-di-2-thienyl)-5′,7′-bis(2-ethylhexyl)benzo[1′,2′-c:4′,5′-c′]dithiophene-4,8-dione)]
PFBT5	poly[2,7-(9,9′-diheylfluorene)-co-4,7-(2,1,3-benzothiadiazole)]
PffBT4T (PCE11)	poly[(5,6-difluoro-2,1,3-benzothiadiazol-4,7-diyl)-*alt*-(3,3‴-di(2-octyldodecyl)-2,2′;5′,2″;5″,2‴-quaterthiophen-5,5‴-diyl)]
PFN	poly[(9,9-bis(30-(*N,N*-dimethylamino)-propyl)-2,7-fluorene)-*alt*-2,7-(9,9-dioctylfluorene)]
PFN-Br	poly[9,9-bis(6′-(*N,N*-diethylamino)propyl)fluorene]-Br
PFO	poly(3,4-ethylenedioxythiophene)
PFO-DHBT10	polyflourene-*co*-4,7-di(4-hexylthien-2-yl)-2,1,3-benzothiadiazole
PFPA-1	poly(3,3′-([(9′,9′-dioctyl-9H,9′H-[2,2′-bifluorene]-9,9-diyl)bis(4,1-phenylene)]bis(oxy))bis(N,N-dimethylpropan1-amine))
PI	polyimide
PIBM	poly(isobutyl methacrylate)
PIDT-phanQ	poly(indacenodithiophene-co-phananthrene-quinoxaline)
PMDA	pyromellitic dyanhydride
PMDPP3T	poly[[2,5-bis(2-hexyldecyl)-2,3,5,6-tetrahydro-3,6-dioxopyrrolo[3,4-c]pyrrole-1,4-diyl]-*alt*-[3′,3″-dimethyl-2,2′:5′,2″-terthiophene]-5,5″-diyl]
PMMA	poly(methyl methacrylate)
PMPSi	poly(methylphenylsiline)
Pn	Pentacene
PO-01	bis(4-phenylthieno[3,2-c]pyridinato-N,$C^{2'}$)acetylacetonateiridium(III)
PO15	2,8-bis(diphenylphosphoryl)-dibenzothiophene
Poly-TPD	poly(*N,N′*-bis(4-butylphenyl-N,N′-bis(phenyl)benzidine))

(*Continued*)

Continued

Chemical name	IUPAC nomenclature
PPE	poly(*p*-phenyleneethynylene)
PPI	1,2,-diphenyl-1H-phenanthro[9,10-d]imidazole
PPIP	4,4′-bis(1-phenyl-1*H*-phenanthro[9,10-*d*]imidazole-2-yl)biphenyl
PPP	poly-*p*-phenylene
PPT	2,8-bis(diphenylphosphoryl) dibenzo[*b,d*]thiophene
PPV	poly(phenylene vinylene)
Ppy:PSS	polypyrrole:poly(styrenesulfonic acid)
PQIr	bis(2-phenylquinolyl-N,$C^{2'}$)acetylacetonateiridium(III)
PQT-12	poly(3,3″-didodecylquaterthiophene)
PS	polystyrene
PSF	poly(9,9-dialkocyphenyl-2,7-silafluorene)
PSQ	2,4-bis[4-(*N,N*-dipropylamino)-2,6-dihydroxyphenyl] squaraine
Pt-17	platinum(II) bis(*N*-methylimidazolyl)-toluene chloride
Pt(TPBP)	platinum tetraphenyltetrabenzoporphyrin
Pt707	platinum(II) 1,1′-(3,3′-oxybis(3,1-phenylene))bis(3-methyl-1 H-imidazol-3-ium) hexafluorophosphate(V)
PTAA	poly(triarylamine)
PTB7	poly[[4,8-bis[(2-ethylhexyl)oxy]benzo[1,2-b:4,5-b′]dithiophene-2,6-diyl][3-fluoro-2-[(2-ethylhexyl)carbonyl]thieno[3,4-b]thiophenediyl]]
PTB7-Th	poly[4,8-bis(5-(2-ethylhexyl)thiophen-2-yl)benzo[1,2-b;4,5-b′]dithiophene-2,6-diyl-alt-(4-(2-ethylhexyl)-3-fluorothieno[3,4-*b*]thiophene)-2-carboxylate-2-6-diyl]
PTCBI	3,4,9,10-perylenetetracarboxylic bisbenzimidazole
PTCDA	3,4,9,10-perylenetetracarboxylic dianhydride
PtD	(3-(trifluoromethyl))(5-(pyridyl)-pyrazolate 5-pyridyl-tetrazolate)platinum(II)
PTES	phenyltriethoxysilane
PTFE	polytetrafluoroethylene
PtNON	platinum(II) 9-(pyridin-2-yl)-2-(9-(pyridin-2-yl)-9H-carbazol-2-yloxy)-9H-carbazole
PtOEP	platinum(II) 2,3,7,8,12,13,17,18-octaethyl-21*H*,23*H*-porphyrin
PTT	polythieno[3,4-b]thiophene
PTV	polythienylene vinylene
PVA	polyvinyl alcohol
PVC	poly(vinyl chloride)
PVK	poly(*N*-vinylcarbazole)
PVP	poly(4-vinylphenol)
PVP-co-HEM	PVP-co-2-hydroxyethyl methacrylate
PXZ-DPS	bis[4-(phenoxazine)phenyl]sulfone
PXZ-TRX	2-phenoxazine-4,6-diphenyl-1,3,5-triazine
Py-PTC	2,9-di(pyrid-2-yl)-anthra[2,1,9-def:6,5,10-d′e′f′]diisoquinoline-1,3,8,10-tetrone
PyCN-ACR	PyCN: dicyanopyrizine; ACR: 9,10-dihyroacridine
PYDC	2-((2-(4-(2-ethylhexyl)-4H-benzothieno[3,2-b]thieno[2,3-d]pyrrol-2-yl)pyrimidin-5-yl) methylene)malononitrile
PαMS	poly(α-methylstyrene)
Rbn	rubrene
Red 2	adamantyl1-6-(3,3,7,7-tetrametyljulolidyl)vinyl-4-(2-methylene-1,3-indanedione)-4*H*-pyran

Continued

Chemical name	IUPAC nomenclature
$Ru(acac)_3$	tris(acetylacetonato)ruthenium(III)
$Ru(terpy)_2$	bis(terpyridine)ruthenium(II)
RuCp*(mes)	ruthenium(pentamethylcyclopentadienyl)(1,3,5-trimethylbenzene)
SF-PDI_2	9,9′-spirobi[fluorene]-2,7-di-1-N,N-Bis(1-hexylheptyl)-perylene-3,4:9,10-bis-(dicarboximide)
SFXSPO	4′-diphenylphosphinoylspiro(fluorine-9,9′-xanthene)
SIMEF	silylmethylfullerene
SnPc	tin phthalocyanine
Spiro-BTBi	2,7-bis(biphenyl-4-yl)-2′,7′-di-*tert*-butyl-9,9′-spirobifluorene
Spiro-DPVBi	2,2′,7,7′-tetrakis(2,2-diphenylvinyl)-9,9′-spirobifluorene
Spiro-TAD	2,2′,7,7′-tetrakis(*N,N*-diphenylamino)-9,9′-spirobifluorene
Spiro2-CBP	2,7-bis(carbazol-9-yl)-9,9′-spirobifluorene
SQ	2,4-bis[4-(*N,N*-diisobutylamino)-2,6-dihydroxylphenyl]squaraine
SubNc	boron subnaphthalocyanine chloride
SubPc	boron subphthalocyanine chloride
T2T	2,4,6-tris(biphenyl-3-yl)-1,3,5-triazine
TAPC	4,4′-cyclohexylidenebis[*N,N*-bis(4-methylphenyl)benzenamine]
TAZ	(1′-naphthyl)-5-phenyl-1,2,3-triazole
TBDNPA	2-*tert*-butyl-9,10-bis[4-(2-naphthyl)phenyl]anthracene
TBPe	2,5,8,11-tetra-*tert*-butylperylene
TBRb	2,8-di-*tert*-butyl-5,11-bis(4-*tert*-butylphenyl)-6,12-diphenyltetracene
TCNE	tetracyanoethylene
TCNQ	tetracyanoquinodimethane
TCTA	4,4′,4″-tri(*N*-carbazolyl)triphenylamine
TDAF	2,7-bis[9,9-di(4-methylphenyl)-fluoren-2-yl]-9,9-di(4-methylphenyl)fluorene
TDBC	(5,6-dichloro-2-[3-[5,6-dichloro-1-ethyl-3-(3-sulfopropyl)-2(3H)-benzimidazolidene]-1-propenyl]-1-ethyl-3-(3-sulfopropyl) benzimidazolium hydroxide, inner salt, sodium salt)
TDBT	2-((7-(*N*-(2-ethylhexyl)-6-*p*-tolyldithieno[3,2-b:2′,3′-d]pyrrol-2-yl)benzo[c][1,2,5] thiadiazol-4-yl)methylene)malononitrile
TDPM	2-((2-(*N*-(2-ethylhexyl)-6-*p*-tolyldithieno[3,2-b:2′,3′-d]pyrrol-2-yl)pyrimidin-5-yl) methylene)malononitrile
TFP	poly[(9,9- dioctylfluorenyl-2,7-diyl)-co-(4,4′-bis(*N*-(4-*sec*-butyl)-phenyl)-diphenylamine)]
TFSI	bis(trifluoromethane)sulfonimide
THAP	tris[2,5-bis(3,5-bis(trifluoromethyl)-phenyl)-thieno](3,4-*b,h,n*)-1,4-5,8,9,12-hexaazatriphenylene
THF	tetrahydrofuran
TiOPc	titanium oxide phthalocyanine
TIPS-Pn	6,13-bis-(triisopropylsilylehthynyl)pentacence
TMPE-OH	triethylolpropane ethoxylate
TmPyPB	1,3,5-tri((3-pyridyl)-phen-3-yl)benzene
TMS	bis(trimethylsilyl) sulfide
TPA-BZP	*N,N*-diphenyl-4-(7-phenylbenzo[c][1,2,5]thiadiazol-4-yl)aniline
TPA-DCPP	TPA: triphenylamine; DCPP: 2,3-dicyanopyrazinophenanthrene
TPA-PPI	*N,N*-diphenyl-4′-(1-phenyl-1H-phenanthro[9,10-d]imidazol-2-yl)biphenyl-4-amine

(*Continued*)

Continued

Chemical name	IUPAC nomenclature
TPAXAN	2′,5′-dimethyl-4′-(10-(naphthalen-2-yl)anthracen-9-yl)-*N,N*-diphenyl-[1,1′-biphenyl]-4-amine
TPBA	9,9′,10,10′-tetraphenyl-2,2′-bianthracene
TPBi	2,2′,2″-(1,3,5-benzinetriyl)-tris(1-phenyl-1-*H*-benzimidazole)
TPD	*N,N*′-diphenyl-*N,N*′-bis(3-methylphenyl)-1,1′-biphenyl-4,4′-diamine
TPP	tetraphenylporphyrin
TPT-IC	3,9-bis(2-methylene-(3-(1,1-dicyanomethylene)-indanone))-3′,8′-5,5,10,10-tetrakis(4-hexylphenyl)dihydropyran[5,6]thieno[3,2-b]pyran[d]thiophene
TPTPA	tris[4-(5-phenylthiophen-2-yl)phenyl]amine
TQ1	poly[2,3-bis-(3-octyloxyphenyl)quinoxaline-5,8-diyl-*alt* thiophene-2,5-diyl]
TSPO1	diphenyl-4-triphenylsilylphenylphosphine oxide
TT-FIC	(4,4,10,10-tetrakis(4-hexylphenyl)-4,10-dihydrothieno[2″,3″:4′,5′]thieno [3′,2′:4,5] cyclopenta[1,2-b]thieno[2,3-d]thiophene-2,8-diyl)bis(2-(3-oxo-2,3-dihydro)inden-5,6-difluoro-1-ylidene)malononitrile
TTB	*N,N*′,*N*′,*N*′-tetrakis(4-methylphenyl)-(l,l′-biphenyl)-4,4′-diamine
TTF	tetrathiafulvalene
TTF-4SC18	tetrakis-(octadecylthio)-tetrathiofulvalene
TTF-TCNQ	tetrathiofulvalinium 7,7′,8,8′-tetracyanoquinodimethane
Ttn	tetracene
TTN	tetrathianaphthacene
TTPA	9,10-bis[*N,N*-di-(*p*-tolyl)-amino]anthracene
UGH2	*p*-bis(triphenylsilyly)benzene
$W_2(hpp)_4$	ditungsten tetra-hexa-hydropyrimidopyrimidine
Y6	2,2′-(2Z,2′Z)-((12,13-bis(2-ethylhexyl)-3,9-diundecyl-12,13-dihydro-[1,2,5]thiadiazolo [3,4-e]thieno[2″,30′:4′,5′]thieno[2′,3′:4,5]pyrrolo[3,2-g]thieno[2′,3′:4,5]thieno[3,2-b] indole-2,10-diyl)bis(methanylylidene))bis(5,6-difluoro-3-oxo-2,3-dihydro-1H-indene-2,1-diylidene))dimalononitrile
ZnPc	zinc phthalocyanine
ZnTPPdT	zinc (5,15-di(*p*-thiolphenyl)-10,20-di(*p*-tolyl)porphyrin)
α-NPD	2,2′-dimethyl-*N,N*′-di-[(1-naphthyl)-*N,N*′-diphenyl]-1,1′-biphenyl-4,4′-diamine
$\alpha-6\tau$	α-sexithiophene

APPENDIX C

Measuring ionization potentials and electron affinities

Measuring molecular frontier orbital energies is key to understanding the fundamental optical and electronic properties of a material. Once the transport energy levels are defined, the appropriate materials can be chosen for the design of high performance devices. There are numerous ways for measuring molecular energetics, ranging from ultraviolet and x-ray photoelectron spectroscopies, to scanning Kelvin probe microscopy and cyclic voltammetry. Here we discuss the two most common methods: ultraviolet photoelectron spectroscopy (UPS or PES), and cyclic voltammetry (CV), and their relationships to each other.

Photoelectron spectroscopy is a convenient method for determining the ionization (i.e. HOMO) energies of organic compounds. The complementary process of inverse photoelectron spectroscopy (IPES) is used for measuring the LUMO energies. The energetics of both processes are illustrated in Fig. C.1a for an organic semiconductor of energy gap, E_G, ionization energy, $IE = q(IP)$, where IP is the ionization potential, electron affinity, EA, and work function, ϕ. Photoelectron spectroscopy is based on the photoelectric effect: a photon of kinetic energy, $h\nu$, is incident on the semiconductor, exciting a bound electron at initial energy E_i into the vacuum at final energy E_f. The initial state is that of an N-electron system. After the electron is ejected, the final state corresponds to a singly-ionized, $N-1$ state. It leaves behind one electron in the singly occupied molecular orbital, or SOMO. The photoelectron has kinetic energy E_{kin} given by

$$E_{kin} = h\nu - \left[E_f(N-1) - E_i(N)\right]. \tag{C.1}$$

The binding energy, E_{bin}, of the electron in the ith orbital at energy, E_i, referenced to the Fermi energy, E_F, is thus given by the term in brackets in Eq. C.1. By inspection of Fig. C.1a, this yields

$$E_{kin} = h\nu - E_{bin} - \phi. \tag{C.2}$$

Furthermore, we obtain the ionization energy via

$$IE = h\nu - E_{kin} = h\nu - \left[E_{cutoff} - E_{HOMO}\right], \tag{C.3}$$

where the cutoff energy, E_{cutoff}, is energy of the photon that results in the highest kinetic energy electron, and E_{HOMO} corresponds to the electron emitted with $E_{kin} = 0$. Thus, the ionization energy is determined from the difference between the onset and the cutoff energies in the PES spectrum in Fig. C.1b. The energies are typically referenced to E_F, which is determined by measuring the work function of a metal film on the substrate prior to deposition of the organic.

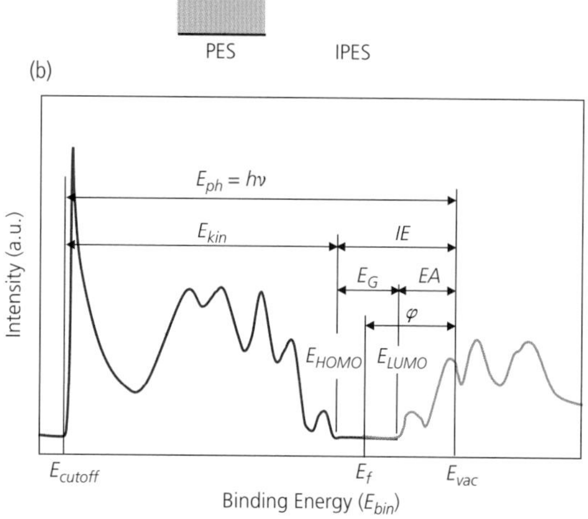

Fig. C.1 (a) Energetics of photoelectron (PES) and inverse photoelectron spectroscopy (IPES) in relation to an organic semiconductor. (b) Example PES and IPES spectra, and the associated energies.

The binding energies of the bound-state orbitals can be obtained from the spectral features at $E_{bin} > E_{HOMO}$. Similarly, surface states, or defect states within the energy gap may also be apparent, although the sensitivity of the technique is limited only to states whose density is typically > 1% of the HOMO density of states. The measurement resolution is typically 50–100 meV. Finally, since the electron penetration depths within the solid are only several Ångstroms, PES is inherently a surface-sensitive probe. Hence, the measurements should be conducted in ultrahigh vacuum, where surface contamination and surface charging are minimized.

The electron energies are typically measured by an energy analyzer such as that shown in Fig. C.2. The UV light source used is frequently the 21.22 eV emission from the HeIα transition. It impinges onto the surface of a thin film sample, ejecting photoelectrons into an electron lens. The kinetic energies of the electrons, all travelling at different velocities, are dispersed in the hemispherical "clam shell" analyzer that has a field between the inner and outer hemispheres. The higher energy electrons travel near the outer hemisphere, with the lower energy electrons describing tighter radii near the inner hemisphere. The dispersed electrons are then detected by a multichannel detector to resolve the PES spectrum, yielding the integrated density of states. Rotating the sample relative to the input aperture, resolves the electron momenta parallel and perpendicular to the sample ($k_{\|}$ and $k_{\perp}$, respectively). This variation on PES is known as angle resolved ultraviolet electron spectroscopy, or ARUPS.

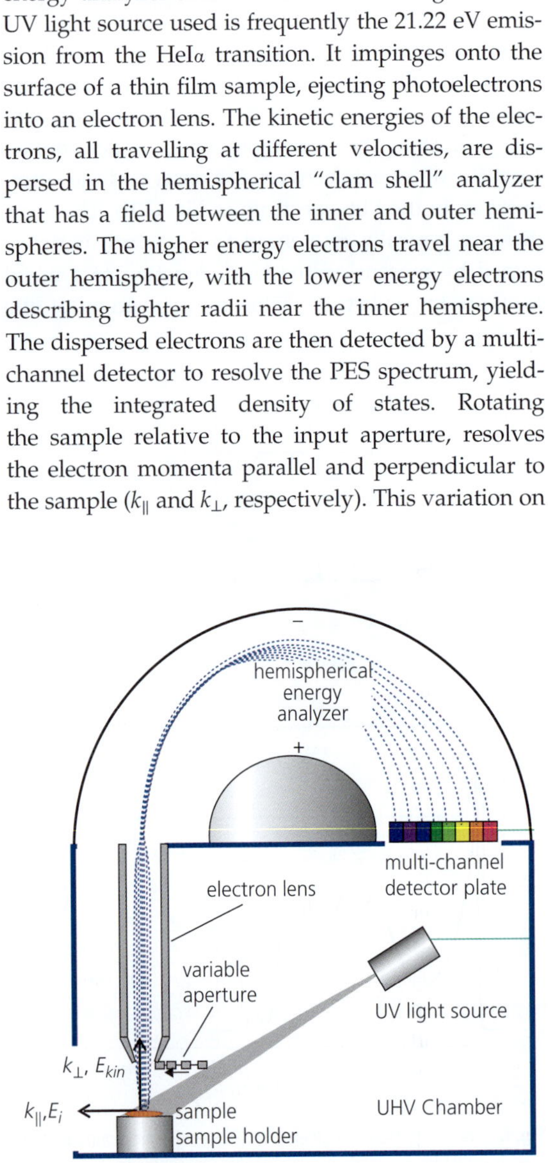

Fig. C.2 Experimental configuration used in PES.

The EA, and hence E_{LUMO}, are measured using the complementary technique of inverse photoelectron spectroscopy. In IPES, an electron beam is incident on the sample, populating high energy initial states, shown on right hand side of Fig. C.1a. As the electrons relax, they emit light of energy $h\nu$ whose spectrum is measured using a monochromator. The lowest energy photons correspond to the electron affinity, and the various high energy LUMO level densities of states can be inferred from the spectral features at energies higher than E_{LUMO}. Since the density of electrons is limited, the signal from the IPES spectrum is considerably smaller than from PES, and hence the accuracy of the method is correspondingly limited to 100–150 meV. The measured difference between E_{HOMO} and E_{LUMO} using a combination of PES and IPES, respectively, gives the energy gap.

Cyclic voltammetry is unlike PES, in that it is carried out in solution on individual molecules within a room ambient, eliminating the need for ultraclean environments and the associated costly equipment. Cyclic voltammetry is an electrochemical process whereby a molecular species dissolved in solution is reduced or oxidized, depending on its energy relative to a second molecular species or an electrode. A singly ionized molecule corresponds to its +1 oxidation state, or its −1 reduced state. The energies required for oxidation and reduction therefore have a one-to-one correspondence with the IE and the EA, respectively. The reduction of the singly ionized molecule, M_1^+, due to charge transfer from neutral M_2^0 is shown in inset (i) of Fig. C.3. This so-called homogeneous electron transfer occurs when the HOMO of M_2 is higher than that of M_1. Heterogeneous charge transfer from an electrode that occurs in CV is shown in inset (ii). At a particular bias applied between the so-called working electrode and the solute (or analyte), the metal work function is higher than the HOMO of the analyte, promoting charge transfer that results in a reduction reaction. The potential that leads to oxidation or reduction thus provides a relative measurement of the ionization and electron affinity energies, respectively.

One-electron transfer results in an energy in the electrochemical cell, E, governed by the Nernst equation, viz. (Bard and Faulkner, 1980)

$$E = E^{0'} + \frac{RT}{F}\ln\frac{[M^+]}{[M^0]}. \tag{C.4}$$

Here, $E^{0'}$ is the formal potential determined by the details of the electrochemical cell, R is the universal

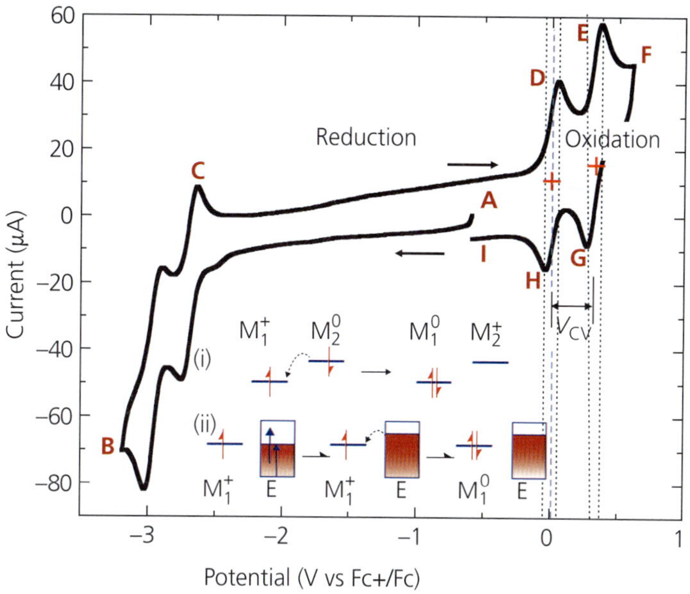

Fig. C.3 Cyclic voltammogram of a subject molecular solute (Ir(ppy)3) in a solution containing the reference, ferrocene/ferrocenium (Fc/Fc$^+$). $V_{CV} = 0.32$ V is the voltage difference between the oxidation potentials of the reference and the solute. The blue arrow is the direction of the voltage sweep. Data from D'Andrade et al. (2005). Inset: (i) Homogeneous electron transfer from the HOMO of molecule M_2 to M_1; (ii) heterogeneous electron transfer from an electrode, E, to M_1. The horizontal line in E is its work function.

gas constant, and $F = qN_A$ is Faraday's constant where N_A is Avogadro's number. Also, $[M^+]$ and $[M^0]$ are concentrations of the oxidized and neutral analytes, respectively. Thus, $E = E^{0'}$ when the concentrations of the oxidized and neutral species at the electrode are equal.

A representative voltammogram for micromolar concentrations of Ir(ppy)3 with a ferrocene (Fc) reference in a DMF solution with a tetra(*n*-butyl)ammonium hexafluorophosphate salt to increase solution conductivity, is shown in Fig. C.3. The voltage is swept in the direction of the blue arrow at 100 mV/s starting at point **A**, reversing sweep direction at **B**, sweeping in the positive direction until point **F**, once more reversing direction, and completing the cycle at **I**. The sweep is reversible, indicating that the molecule is stable under oxidation and reduction at the working electrode.

The oxidation potential of the analyte relative to the Fc reference is obtained from the several peaks **D**, **E**, **G**, and **H**. The anodic scan in the positive direction generates a peak anodic current at **D** corresponding to the point where the transport of Fc to the electrode is maximum. At higher potentials, Fc becomes depleted from solution near the electrode, and the current drops until it once more begins to increase to a second peak at **E**. This second peak is due to oxidation of the analyte, Ir (ppy)3. In the reverse, cathodic scan, the opposite reactions occur in reverse order, where the analyte is now reduced. The midpoints between the corresponding peaks, indicated by "+" are $E_{1/2} = E^{0'}$, the formal potential. From these data, the relative oxidation potentials of the analyte and Fc is $V_{CV} = 0.32$ V. Similar reduction peaks are found by the structure in the scan near $V = -3$ V. The first reduction peak of $\mathrm{Ir(ppy)}_3^+ \to \mathrm{Ir(ppy)}_3$ is at point **C**.

The peak positions are considerably easier to resolve using differential pulse voltammetry (DPV). The DPV data are a differential of the conventional voltammogram that are obtained by plotting the current immediately prior to increasing the baseline voltage using a linear voltage ramp or a pulse staircase. A differential voltammogram of the sample in Fig. C.3 is shown in Fig. C.4. The resolution of this technique is approximately 50 meV, or at least twice that achieved using PES. From the oxidation and reduction peaks of the analyte, we obtain a redox potential difference of $\Delta E_{redox} = 3.01$ eV.

The experimental setup for CV measurements is illustrated in Fig. C.5. It consists of a working electrode that is resistant to corrosion in the electrolytic solution (typically Pt or glassy carbon), and a counter electrode often of the same composition. The reference electrode should have a stable equilibrium potential. The voltammograms in Figs. C.3 and C.4 use a Fc$^+$/Fc couple that is commonly used for analyzing organic electronic materials. The electrolyte itself comprises a solvent such as DMF, DCM, acetonitrile,

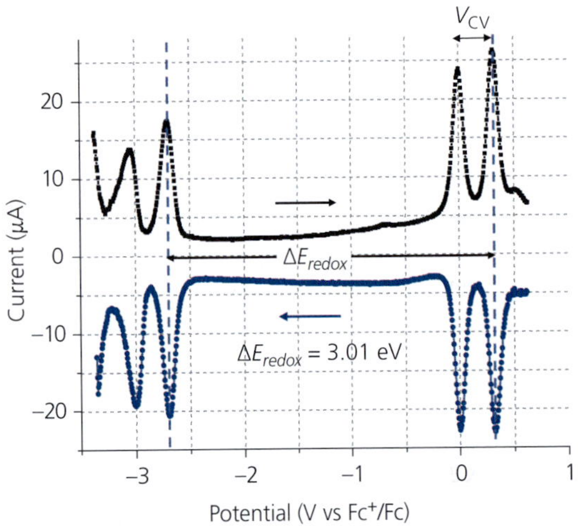

Fig. C.4 Differential pulse voltammograms for the anodic (top) and cathodic (bottom) sweeps of the sample in Fig. C.3. The sweep directions are shown by arrows. Data courtesy of P. Djurovich (private communication, 2019).

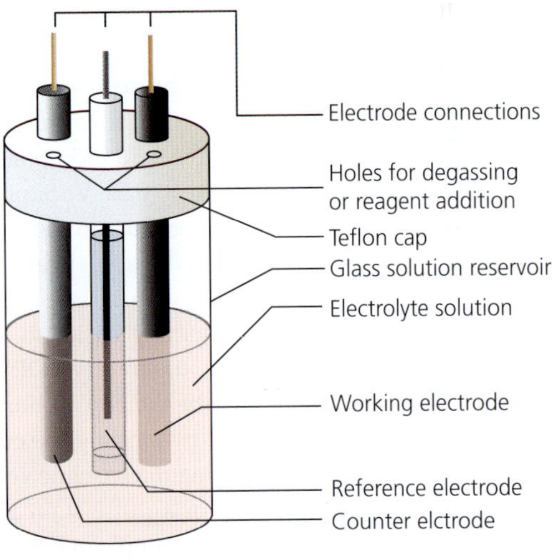

Fig. C.5 Electrochemical test cell used in cyclic voltammetry (Elgrishi et al., 2017).

etc. The solvent also contains a high concentration of an electrolyte to minimize resistance that might distort the data by increasing the voltages of the peak anodic and cathodic currents. Ammonium salts are preferred for this purpose. Similarly, the voltage sweep must be sufficiently slow to avoid overtaking the rate of ionic diffusion. This can flatten the peaks, reducing the measurement resolution (Elgrishi et al., 2017). Finally, accurate data are obtained only for molecules that can be reversibly cycled. Irreversible oxidation or reduction implies that the molecule

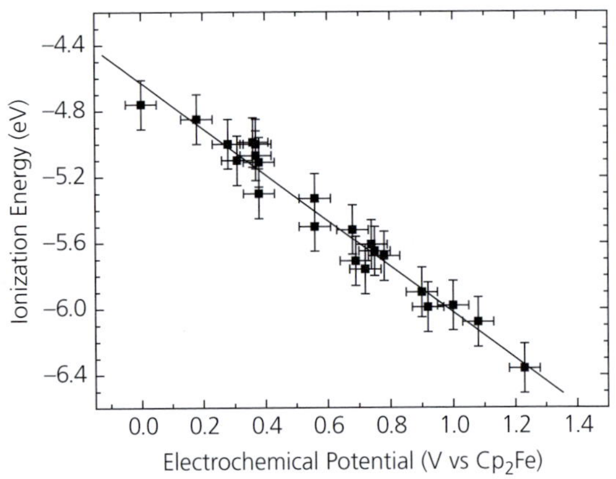

Fig. C.6 Ionization energy measured by PES vs. the electrochemical potential measured in DMF referenced to biphenyl-substituted ferrocene for a population of common organic electronic molecules (data points). The line is a fit to Eq. C.6 (D'Andrade et al., 2005).

decomposes during measurement. Irreversibility results in an open voltammogram.

A correspondence between V_{CV}, the ionization energy measured by PES, and E_{HOMO} is obtained by compensating for image charge effects between the analyte and the working electrode, and solvation effects that depend on the relative dielectric constants of the solution (ε_{CV}) used in CV, and the thin film (ε_{film}) analyzed by photoelectron spectroscopy. It has been shown that CV and PES potentials are related by (D'Andrade et al., 2005)

$$V_{CV} = -\frac{1}{qf(r,d)}\frac{\varepsilon_{film}}{\varepsilon_{CV}}E_{HOMO} - V_{REF}, \qquad \text{(C.5)}$$

where $f(r, d)$ is a function due to Coulomb attraction between a "spherical" molecule of effective radius r forming an adsorbed film on the electrode of approximate thickness, d, and its image charge in the electrode. Also, V_{REF} is the oxidation potential of the reference molecule. Figure C.6 shows a plot of the ionization potential vs. V_{CV} referenced to Fc^+/Fc of a population of common organic electronic molecules. The reference ionization potential is −4.76 eV for a biphenyl-substituted ferrocene (Cp_2Fe). This reference is indicated by the point at a $V_{CV} = 0$ V. The straight line is a best fit to the data using:

$$E_{HOMO} = -(1.4\pm0.1)\times(qV_{CV}) - (4.60\pm0.08)\ \text{eV}. \qquad \text{(C.6)}$$

This functional form is within measurement error of Eq. C.5, where $qV_{REF} = -4.76$ eV. The fits are largely independent of the solvent used, suggesting that image charge effects have a more significant contribution to the slope than solvation effects.

Empirical fits between the LUMO energies measured by IPES and the reduction potentials (again relative to Fc^+/Fc) in a DMF solution using a tetra(n-butyl)ammonium hexafluorophosphate electrolyte of a similarly large population of organic electronic small molecules results in the following linear relationship (Djurovich et al., 2009):

$$E_{LUMO}(\text{IPES}) = -(1.19\pm0.08)\times qV_{CV} - (4.78\pm0.17). \tag{C.7}$$

Note that reduction potentials are measured at $V_{CV} < 0$.

Finally, it is reasonable to expect that the redox potential, which is equal to the difference between adding and subtracting an electron from the molecule, should be related to $E_{LUMO} - E_{HOMO} = E_G$, the transport energy gap. Indeed, a statistical analysis of organic electronic compounds results in a correlation that is accurately fit by (Sworakowski, 2018)

$$E_G = 1.16\Delta E_{redox}. \tag{C.8}$$

The 16% difference in these values is due to dielectric polarization. In the thin film, the molecules must collectively respond to the presence or absence of charge in the solid. However, in CV, the molecular species are diluted (often at micromolar concentrations), and thus the redox energies correspond to those of isolated molecules within the polar dielectric medium of the solvent. The solvent dielectric constant is generally considerably higher than that of the solid ($\varepsilon_{CV} \approx 30$ vs. $\varepsilon_{film} \approx 3$–4). This substantially screens the electric field in the solution, reducing the energy required to remove, or add a charge. Quantitative analysis shows that this leads to a systematically smaller ΔE_{redox} compared to E_G whose magnitude is consistent with the empirical fit in Eq. C.8.

References

Bard, A. J. & Faulkner, L. R. 1980. *Electrochemical Methods: Fundamentals and Applications*. New York: John Wiley and Sons.

D'Andrade, B. W., Datta, S., Forrest, S. R., Djurovich, P., Polikarpov, E. & Thompson, M. E. 2005. Relationship between the ionization and oxidation potentials of molecular organic semiconductors. *Org. Electron.*, 6, 11.

Djurovich, P. I., Mayo, E. I., Forrest, S. R. & Thompson, M. E. 2009. Measurement of the lowest unoccupied molecular orbital energies of molecular organic semiconductors. *Org. Electron.*, 10, 515.

Elgrishi, N., Rountree, K. J., McCarthy, B. D., Rountree, E. S., Eisenhart, T. T. & Dempsey, J. L. 2017. A practical beginner's guide to cyclic voltammetry. *J. Chem. Educ.*, 95, 197.

Sworakowski, J. 2018. How accurate are energies of HOMO and LUMO levels in small-molecule organic semiconductors determined from cyclic voltammetry or optical spectroscopy? *Synth. Met.*, 235, 125.

General index

f after a page number indicates a figure; t after a page number indicates a table.

D

E

F

G

H

T

Chemical name index

H

I

M

N

O

P